Comprehensive Organometallic Chemistry II

A Review of the Literature 1982–1994

Comprehensive Organometallic Chemistry II

A Review of the Literature 1982–1994

Editors-in-Chief

Edward W. Abel
University of Exeter, UK

F. Gordon A. Stone
Baylor University, Waco, TX, USA

Geoffrey Wilkinson
Imperial College of Science, Technology and Medicine, London, UK

Volume 5

VANADIUM AND CHROMIUM GROUPS

Volume Editors

Jay A. Labinger
California Institute of Technology, Pasadena, CA, USA

Mark J. Winter
University of Sheffield, UK

PERGAMON

UK Elsevier Science Ltd., The Boulevard, Langford Lane, Kidlington, Oxford OX5 1GB, UK

USA Elsevier Science Inc., 660 White Plains Road, Tarrytown, New York 10591-5153, USA

JAPAN Elsevier Science Japan, Tsunashima Building Annex, 3-20-12 Yushima, Bunkyo-ku, Tokyo 113, Japan

---

First edition 1995

Library of Congress Cataloging in Publication Data
Comprehensive organometallic chemistry II : a review of the literature 1982–1994 / editors-in-chief, Edward W. Abel, F. Gordon A. Stone, Geoffrey Wilkinson
p. cm.
Includes indexes.
1. Organometallic chemistry. I. Abel, Edward W. II. Stone, F. Gordon A. III. Wilkinson, Geoffrey.
QD411.C652 1995
547'.05—dc20 95–7030

British Library Cataloguing in Publication Data
A catalogue record for this book is available from the British Library.

ISBN 0–08–040608–4 (set : alk. paper)
ISBN 0–08–042312–4 (Volume 5)

Important note
For safety reasons, readers should always consult the list of abbreviations on p. xi before making use of the experimental details provided.

∞™ The paper used in this publication meets the minimum requirements of the American National Standard for Information Sciences—Permanence of Paper for Printed Library Materials, ANSI Z39.48–1984.

Chemical structures drawn by Synopsys Scientific Systems Ltd., Leeds, UK.

Printed and bound in Great Britain by BPC Wheatons Ltd., Exeter, UK.

Contents

Preface

'Comprehensive Organometallic Chemistry', published in 1982, was well received and remains very highly cited in the primary journal literature. Since its publication, studies on the chemistry of molecules with carbon–metal bonds have continued to expand rapidly. This is due to many factors, ranging from the sheer intellectual challenge and excitement provided by the continuing production of novel results, which demand new ideas, through to the successful application of organometallic species in organic syntheses, the generation of living catalysts for polymerization, and the synthesis of precursors for materials employed in the electronic and ceramic industries. For many reasons, therefore, we judged it timely to update 'Comprehensive Organometallic Chemistry' with a new work.

Due to the scope and depth of this area of chemistry, to have merely revised each of the original nine volumes did not seem the most user-friendly or cost-effective procedure to follow. As a consequence of the sheer bulk of the literature of the subject, a revised edition would necessarily require either the elimination of much chemistry of archival value but which is still important, or the production of a set of volumes significantly larger in number than the original nine. Accordingly, we decided it would be best to use the original work as a basis for new volumes focusing on organometallic chemistry reported since 1982, with reference back to the original work when necessary. For ease of use the new volumes maintain the same general structure as employed previously but reflect the changes in substance and direction the field has undergone in the last ten years. Thus it is not surprising that the largest volume in the new work concerns the role of the transition elements in metal-mediated organic syntheses.

The expansion of organometallic chemistry since the early 1980s also led us to decide that an updating of 'Comprehensive Organometallic Chemistry' would be more effectively accomplished if each volume had one or two editors who would be responsible both for recruiting experts for the Herculean task of writing the many chapters of each volume and for overseeing the content. We are deeply indebted to the volume editors and their authors for the time and effort they have given to the project.

As with the original 'Comprehensive Organometallic Chemistry', published some thirteen years ago, we hope this new version will serve as a pivotal reference point for new work and will function to generate new ideas and perceptions for the continued advance of what will surely continue as a vibrant area of chemistry.

Edward W. Abel
Exeter, UK

F. Gordon A. Stone
Waco, Texas, USA

Geoffrey Wilkinson
London, UK

Preface to 'Comprehensive Organometallic Chemistry'

Although the discovery of the platinum complex that we now know to be the first π-alkene complex, $K[PtCl_3(C_2H_4)]$, by Zeise in 1827 preceded Frankland's discovery (1849) of diethylzinc, it was the latter that initiated the rapidly developing interest during the latter half of the nineteenth century in compounds with organic groups bound to the elements. This era may be considered to have reached its apex in the discovery by Grignard of the magnesium reagents which occupy a special place because of their ease of synthesis and reactivity. With the exception of trimethylplatinum chloride discovered by Pope, Peachy and Gibson in 1907 by use of the Grignard reagent, attempts to make stable transition metal alkyls and aryls corresponding to those of main group elements met with little success, although it is worth recalling that even in 1919 Hein and his co-workers were describing the 'polyphenyl-chromium' compounds now known to be arene complexes.

The other major area of organometallic compounds, namely metal compounds of carbon monoxide, originated in the work starting in 1868 of Schützenberger and later of Mond and his co-workers and was subsequently developed especially by Hieber and his students. During the first half of this century, aided by the use of magnesium and, later, lithium reagents the development of main group organo chemistry was quite rapid, while from about 1920 metal carbonyl chemistry and catalytic reactions of carbon monoxide began to assume importance.

In 1937 Krause and von Grosse published their classic book 'Die Chemie der Metallorganischen Verbindungen'. Almost 1000 pages in length, it listed scores of compounds, mostly involving metals of the main groups of the periodic table. Compounds of the transition elements could be dismissed in 40 pages. Indeed, even in 1956 the stimulating 197-page monograph 'Organometallic Compounds' by Coates adequately reviewed organo transition metal complexes within 27 pages.

Although exceedingly important industrial processes in which transition metals were used for catalysis of organic reactions were developed in the 1930s, mainly in Germany by Reppe, Koch, Roelen, Fischer and Tropsch and others, the most dramatic growth in our knowledge of organometallic chemistry, particularly of transition metals, has stemmed from discoveries made in the middle years of this century. The introduction in the same period of physical methods of structure determination (infrared, nuclear magnetic resonance, and especially single-crystal X-ray diffraction) as routine techniques to be used by preparative chemists allowed increasingly sophisticated exploitation of discoveries. Following the recognition of the structure of ferrocene, other major advances quickly followed, including the isolation of a host of related π-complexes, the synthesis of a plethora of organometallic compounds containing metal–metal bonds, the characterization of low-valent metal species in which hydrocarbons are the only ligands, and the recognition from dynamic NMR spectra that ligand site exchange and tautomerism were common features in organometallic and metal carbonyl chemistry. The discovery of alkene polymerization using aluminium alkyl–titanium chloride systems by Ziegler and Natta and of the Wacker palladium–copper catalysed ethylene oxidation led to enormous developments in these areas.

In the last two decades, organometallic chemistry has grown more rapidly in scope than have the classical divisions of chemistry, leading to publications in journals of all national chemical societies, the appearance of primary journals specifically concerned with the topic, and the growth of annual review volumes designed to assist researchers to keep abreast of accelerating developments.

Organometallic chemistry has become a mature area of science which will obviously continue to grow. We believe that this is an appropriate time to produce a comprehensive review of the subject, treating organo derivatives in the widest sense of both main group and transition elements. Although advances in transition metal chemistry have appeared to dominate progress in recent years, spectacular progress has, nevertheless, also been made in our knowledge of organo compounds of main group elements such as aluminium, boron, lithium and silicon.

In these Volumes we have assembled a compendium of knowledge covering contemporary organometallic and carbon monoxide chemistry. In addition to reviewing the chemistry of the elements individually, two Volumes survey the use of organometallic species in organic synthesis and in catalysis, especially of industrial utility. Within the other Volumes are sections devoted to such diverse topics as the nature of carbon–metal bonds, the dynamic behaviour of organometallic compounds in solution, heteronuclear metal–metal bonded compounds, and the impact of organometallic compounds on the environment. The Volumes provide a unique record, especially of the intensive studies conducted during the past 25 years. The last Volume of indexes of various kinds will assist readers seeking information on the properties and synthesis of compounds and on earlier reviews.

As Editors, we are deeply indebted to all those who have given their time and effort to this project. Our Contributors are among the most active research workers in those areas of the subject that they have reviewed and they have well justified international reputations for their scholarship. We thank them sincerely for their cooperation.

Finally, we believe that 'Comprehensive Organometallic Chemistry', as well as providing a lasting source of information, will provide the stimulus for many new discoveries since we do not believe it possible to read any of the articles without generating ideas for further research.

E. W. ABEL
Exeter

F. G. A. STONE
Bristol

G. WILKINSON
London

Contributors to Volume 5

Dr. P. Berno
Department of Chemistry, University of Ottawa, Ottawa, Ontario K1N 6N5, Canada

Professor S. Gambarotta
Department of Chemistry, University of Ottawa, Ottawa, Ontario K1N 6N5, Canada

Dr. S. D. Gray
Department of Chemistry, University of Arizona, Tucson, AZ 85721, USA

Dr. M. J. Morris
Department of Chemistry, The University of Sheffield, Sheffield, S3 7HF, UK

Dr. D. Richeson
Department of Chemistry, University of Ottawa, Ottawa, Ontario K1N 6N5, Canada

Dr. M. W. Whiteley
Department of Chemistry, Faculty of Science, The University of Manchester, Oxford Road, Manchester, M13 9PL, UK

Professor D. E. Wigley
Department of Chemistry, University of Arizona, Tucson, AZ 85721, USA

Dr. M. J. Winter
Department of Chemistry, The University of Sheffield, Sheffield, S3 7HF, UK

Dr. S. Woodward
School of Chemistry, University of Hull, Cottingham Road, Kingston upon Hull, HU6 7RX, UK

Abbreviations

The abbreviations used throughout 'Comprehensive Organometallic Chemistry II' are consistent with those used in 'Comprehensive Organometallic Chemistry' and with other standard texts in this area. The abbreviations in some instances may differ from those commonly used in other branches of chemistry.

Ac	acetyl
acac	acetylacetonate
AIBN	2,2'-azobisisobutyronitrile
Ar	aryl
arphos	1-(diphenylphosphino)-2-(diphenylarsino)ethane
Azb	azobenzene
9-BBN	9-borabicyclo[3.3.1]nonyl
9-BBN-H	9-borabicyclo[3.3.1]nonane
BHT	2,6-di-*t*-butyl-4-methylphenol (butylated hydroxytoluene)
bipy	2,2'-bipyridyl
t-BOC	*t*-butoxycarbonyl
bsa	*N*,*O*-bis(trimethylsilyl)acetamide
bstfa	*N*,*O*-bis(trimethylsilyl)trifluoroacetamide
btaf	benzyltrimethylammonium fluoride
Bz	benzyl
can	ceric ammonium nitrate
cbd	cyclobutadiene
1,5,9-cdt	cyclododeca-1,5,9-triene
chd	cyclohexadiene
chpt	cycloheptatriene
[Co]	cobalamin
(Co)	cobaloxime $[Co(DMG)_2]$ derivative
cod	1,5-cyclooctadiene
cot	cyclooctatetraene
Cp	η^5-cyclopentadienyl
Cp*	pentamethylcyclopentadienyl
18-crown-6	1,4,7,10,13,16-hexaoxacyclooctadecane
CSA	camphorsulfonic acid
csi	chlorosulfonyl isocyanate
Cy	cyclohexyl
dabco	1,4-diazabicyclo[2.2.2]octane
dba	dibenzylideneacetone
dbn	1,5-diazabicyclo[4.3.0]non-5-ene
dbu	1,8-diazabicyclo[5.4.0]undec-7-ene
dcc	dicyclohexylcarbodiimide
dcpe	1,2-bis(dicyclohexylphosphino)ethane
ddq	2,3-dichloro-5,6-dicyano-1,4-benzoquinone
deac	diethylaluminum chloride
dead	diethyl azodicarboxylate
depe	1,2-bis(diethylphosphino)ethane
depm	1,2-bis(diethylphosphino)methane
det	diethyl tartrate (+ or –)

DHP	dihydropyran
diars	1,2-bis(dimethylarsino)benzene
dibal-H	diisobutylaluminum hydride
dien	diethylenetriamine
DIGLYME	bis(2-methoxyethyl)ether
diop	2,3-*O*-isopropylidene-2,3-dihydroxy-1,4-bis(diphenylphosphino)butane
dipt	diisopropyl tartrate (+ or –)
dma	dimethylacetamide
dmac	dimethylaluminum chloride
DMAD	dimethyl acetylenedicarboxylate
dmap	4-dimethylaminopyridine
DME	dimethoxyethane
DMF	*N,N'*-dimethylformamide
DMG	dimethylglyoximate
DMI	*N,N'*-dimethylimidazalone
dmpe	1,2-bis(dimethylphosphino)ethane
dmpm	bis(dimethylphosphino)methane
DMSO	dimethyl sulfoxide
dmtsf	dimethyl(methylthio)sulfonium fluoroborate
dpam	bis(diphenylarsino)methane
dppb	1,4-bis(diphenylphosphino)butane
dppe	1,2-bis(diphenylphosphino)ethane
dppf	1,1'-bis(diphenylphosphino)ferrocene
dpph	1,6-bis(diphenylphosphino)hexane
dppm	bis(diphenylphosphino)methane
dppp	1,3-bis(diphenylphosphino)propane
eadc	ethylaluminum dichloride
edta	ethylenediaminetetraacetate
eedq	*N*-ethoxycarbonyl-2-ethoxy-1,2-dihydroquinoline
en	ethylene-1,2-diamine
Et_2O	diethyl ether
F_6 acac	hexafluoroacetylacetonate
Fc	ferrocenyl
Fp	$Fe(CO)_2Cp$
HFA	hexafluoroacetone
hfacac	hexafluoroacetylacetonate
hfb	hexafluorobut-2-yne
HMPA	hexamethylphosphoramide
hobt	hydroxybenzotriazole
$IpcBH_2$	isopinocampheylborane
Ipc_2BH	diisopinocampheylborane
kapa	potassium 3-aminopropylamide
K-selectride	potassium tri-*s*-butylborohydride
LAH	lithium aluminum hydride
LDA	lithium diisopropylamide
LICA	lithium isopropylcyclohexylamide
LITMP	lithium tetramethylpiperidide
L-selectride	lithium tri-*s*-butylborohydride
LTA	lead tetraacetate
mcpba	*m*-chloroperbenzoic acid
MeCN	acetonitrile
MEM	methoxyethoxymethyl
MEM-Cl	β-methoxyethoxymethyl chloride

Mes	mesityl
mma	methyl methacrylate
mmc	methylmagnesium carbonate
MOM	methoxymethyl
Ms	methanesulfonyl
MSA	methanesulfonic acid
MsCl	methanesulfonyl chloride
nap	1-naphthyl
nbd	norbornadiene
NBS	*N*-bromosuccinimide
NCS	*N*-chlorosuccinimide
nmo	*N*-methylmorpholine *N*-oxide
NMP	*N*-methyl-2-pyrrolidone
Nu^-	nucleophile
ox	oxalate
pcc	pyridinium chlorochromate
pdc	pyridinium dichromate
phen	1,10-phenanthroline
phth	phthaloyl
ppa	polyphosphoric acid
ppe	polyphosphate ester
$[PPN]^+$	$[(Ph_3P)_2N]^+$
ppts	pyridinium *p*-toluenesulfonate
py	pyridine
pz	pyrazolyl
Red-Al	sodium bis(2-methoxyethoxy)aluminum dihydride
sal	salicylaldehyde
salen	*N,N'*-bis(salicylaldehydo)ethylenediamine
SEM	β-trimethylsilylethoxymethyl
tas	tris(diethylamino)sulfonium
tasf	tris(diethylamino)sulfonium difluorotrimethylsilicate
tbaf	tetra-*n*-butylammonium fluoride
TBDMS	*t*-butyldimethylsilyl
TBDMS-Cl	*t*-butyldimethylsilyl chloride
TBDPS	*t*-butyldiphenylsilyl
tbhp	*t*-butyl hydroperoxide
TCE	2,2,2-trichloroethanol
TCNE	tetracyanoethene
TCNQ	7,7,8,8-tetracyanoquinodimethane
terpy	2,2':6',2"-terpyridyl
tes	triethylsilyl
Tf	triflyl (trifluoromethanesulfonyl)
TFA	trifluoracetic acid
TFAA	trifluoroacetic anhydride
tfacac	trifluoroacetylacetonate
THF	tetrahydrofuran
THP	tetrahydropyranyl
tipbs-Cl	2,4,6-triisopropylbenzenesulfonyl chloride
tips-Cl	1,3-dichloro-1,1,3,3-tetraisopropyldisiloxane
TMEDA	tetramethylethylenediamine [1,2-bis(dimethylamino)ethane]
TMS	trimethylsilyl
TMS-Cl	trimethylsilyl chloride
TMS-CN	trimethylsilyl cyanide
Tol	tolyl
tpp	*meso*-tetraphenylporphyrin

Tr	trityl (triphenylmethyl)
tren	2,2',2''-triaminotriethylamine
trien	triethylenetetraamine
triphos	1,1,1-tris(diphenylphosphinomethyl)ethane
Ts	tosyl
TsMIC	tosylmethyl isocyanide
ttfa	thallium trifluoroacetate

Contents of All Volumes

Volume 13 Structure Index

Volume 14 Cumulative Indexes

1
Vanadium

P. BERNO, S. GAMBAROTTA, and D. RICHESON
University of Ottawa, ON, Canada

1.1 INTRODUCTION

The aim of the present review of the organometallic chemistry of vanadium is to extensively cover the literature period between 1982 and 1992. For convenience, a few relevant references prior to 1982 have been included. For a more in-depth view of pre-1982 material, the reader is advised to consult the first edition of *Comprehensive Organometallic Chemistry.*

1.2 CARBONYL AND RELATED COMPOUNDS

1.2.1 $V(CO)_6$ and Derivatives

1.2.1.1 Synthesis, structure, and bonding

In spite of continuous interest in $V(CO)_6$ and its derivatives, an improved synthesis has yet to be discovered. The current procedure requires the initial large-scale preparation of $NaV(CO)_6$ under high CO pressure and elevated temperatures to achieve the reductive carbonylation of VCl_3 (Scheme 1).[1,2] The reduction of VCl_3 with sodium sand may be performed at room temperature in THF under CO at atmospheric pressure, in the presence of a fairly large amount of cot (cot = cyclooctatetraene) which acts as a catalyst (V/cot molar ratio 0.5).[3] Interestingly, no appreciable carbonylation is obtained either when $cotNa_2$ is employed in place of cot or when other radical anion species, such as (anthracene)Na are used. This suggests that intermediate (cot)V species undergo carbonylation at an early stage of the reaction. The major drawback of this synthetic procedure, which is otherwise very effective, lies in the considerable price of the cot catalyst which makes large-scale preparation unrealistic. The yield of $[V(CO)_6]^-$ obtained in the absence of cot catalyst is as low as 2%.[4]

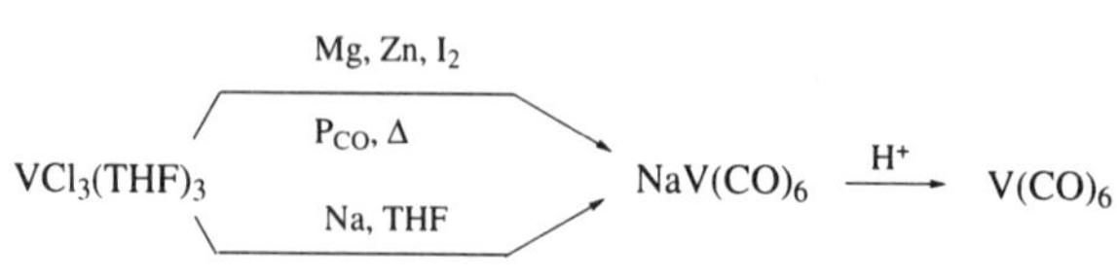

Scheme 1

Transformation of $Na[V(CO)_6]$ into $V(CO)_6$ is obtained via acidification of the corresponding anion with a solution of HCl (4 M) in diethyl ether[5] or, more conveniently, via direct sublimation of the $[R_4N][V(CO)_6]$ salts carried out in the presence of 100% H_3PO_4.[6]

Highly reduced vanadium carbonyl $Na_3[V(CO)_5]$, which formally contains vanadium in its lowest known oxidation state (−3), can be obtained by further reduction of $[Na(diglyme)_2][V(CO)_6]$ in liquid ammonia at −78 °C.[7] Addition of alkali metal iodide allows the preparation and isolation of the corresponding $M_3[V(CO)_5]$ (M = K, Rb, Cs) (Scheme 2).[8]

Treatment in liquid ammonia of $Na_3[V(CO)_5]$ prepared *in situ* with one equivalent of EtOH at −50 °C affords $Na_2[HV(CO)_5]$ which can be identified on the basis of the ^1H, ^{13}C, and ^{51}V-NMR spectra. A gold–vanadium cluster, $[[(Ph_3P)Au]_3[V(CO)_5]]$, is obtained upon reaction with Ph_3PAuCl.[9] Photolysis of $Ph_3PAuV(CO)_6$ carried out in THF solution and in the presence of $Ph_3PAuNCO$ and of $(Ph_3P)AuN_3$ followed by chromatography on Al_2O_3 also yields $[(Ph_3P)_3Au]_3V(CO)_5$, $[[(Ph_3P)Au]_6V(CO)_4]NCO$ and homometallic gold clusters in about equal proportions. The crystal structure of the $[[(Ph_3P)Au]_6V(CO)_4]^+$ cation obtained upon anion exchange with PF_6, has been determined by x-ray diffraction analysis.

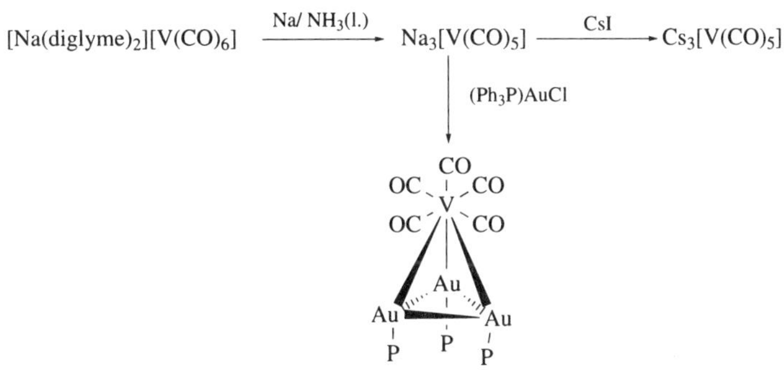

Scheme 2

New organometallic derivatives $[(R_3E)V(CO)_5]^{2-}$ (R = alkyl or aryl; E = Sn, Pb) and $(R_3E)_2[V(CO)_5]^-$ have been prepared from R_3EX and $[V(CO)_5]^{3-}$ in moderate yield.[10] A crystalline sample of $[Et_4N][(Ph_3Sn)_2V(CO)_5]$ has been prepared from photolysis of $[Et_4N][V(CO)_6]$ carried out in the presence of Ph_6Sn_2. The crystal structure shows heptacoordinated vanadium (V–Sn = 0.2757 and 0.2785 nm).

1.2.1.2 Disproportionation reactions

The 17-electron configuration of the d^5 vanadium(0) of $V(CO)_6$ makes this long-lived radical particularly inclined towards one-electron reduction to form the 18-electron closed-shell anion $[V(CO)_6]^-$. For example, simple treatment with 1,2,4,5-tetramethylbenzene[11] in heptane, leads to the formation of $[(\eta^6\text{-arene})V(CO)_4][V(CO_6)]$ in 58% yield (Scheme 3). The crystal structure, consisting of a symmetrical octahedral $V(CO)_6$ anion and a D_{4v} piano stool $[(\eta^6\text{-arene})V(CO)_4]$ cation, is in agreement with infrared data in solution. In contrast to the anionic $[V(CO)_6]^-$ unit which is remarkably stable, the 18-electron $[(\eta^6\text{-arene})V(CO)_4]$ cation is readily transformed into $[V(THF)_4]^+$ under mild conditions.

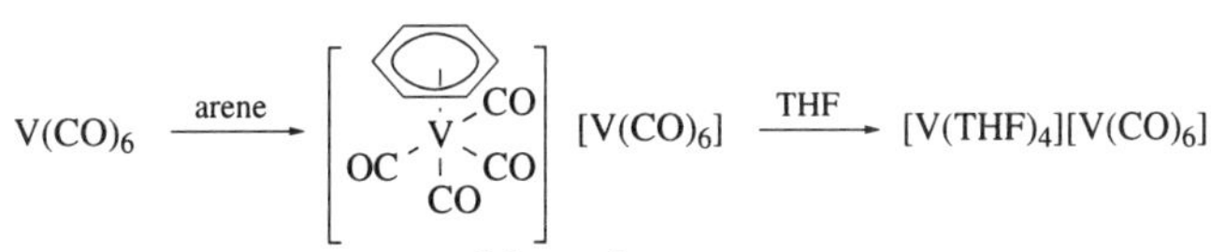

Scheme 3

Oxygen and nitrogen Lewis bases promote the disproportionation of $V(CO)_6$ to $[VB_n][V(CO)_6]$ (Scheme 4).[12] An isocarbonyl species was obtained from the reaction of $V(CO)_6$ with decamethylvanadocene. Here, the ability of $V(CO)_6$ to function as a one-electron oxidant, while decamethylvanadocene acts as one-electron reductant, results in the formation of an interesting $Cp_2{}^*V(\mu\text{-}CO)V(CO)_5$ complex.[13]

The disproportionation reactions of $V(CO)_{6-n}P_n$ phosphine derivatives are considerably slower. For example, the disproportionation of $V(CO)_5[P(Bu^n{}_3)]$ induced by CH_3CN is five orders of magnitude slower than that of $V(CO)_6$.[14]

$$2\,V(CO)_6 + nB \longrightarrow (CO)_5V\text{-}(CO)\text{-}VB_n \longrightarrow [B_nV][V(CO)_6]$$

Scheme 4

Disproportionation reactions have been used for the preparation of cationic vanadium(II) complexes. The reaction between $[Et_4N][V(CO)_6]$, $(Bu^t)NC$ and iodobenzene dichloride in EtOH affords $[V[CN(Bu^t)]_6][V(CO)_6]_2$ in crystalline form.[15]

Although vanadium isocyanide binary complexes are rare, cyclic voltammetry experiments on $[V[CN(Bu^t)]_6]^{2+}$ have shown that the 0, 1, 2, and 3 oxidation states may be readily accessible.[16]

1.2.1.3 CO-Displacement reactions

A variable number of CO molecules can be replaced in both $[V(CO)_6]^-$ and $H[V(CO)_6]$ upon photolysis in the presence of mono- and polydentate phosphines,[17] forming the corresponding $[V(CO)_{6-n}(\text{phosphine})_n]^-$ derivatives (Table 1). The substitution of carbon monoxide in the 17-electron $V(CO)_6$ proceeds at or below room temperature to form monosubstitution products $V(CO)_5L$ (L = phosphine, phosphite).[14,18] A rather general alternative synthesis for the preparation of carbonyl diphosphine complexes of vanadium(0) involves oxidation of $[V(CO)_4L]^-$ ($L = Ph_2P(CH_2)_nPPh_2$ (n = 1–4)) by $[C_7H_7][BF_4]$.[19]

Table 1 Spectroscopic data for $[V(CO)_{6-n}L_n]$.

Complex	ν_{CO} (cm^{-1})	$\delta(^{51}V)$ (ppm)	*Ref.*
$[V(CO)_5(PPh_2Et)]$	1963, 1859, 1818	−1849	17
$[V(CO)_5(PPh_2Me)]$	1960, 1860, 1825, 1805	−1839	17
$[V(CO)_5(PCy_3)]$	1955, 1855, 1815	−1864	22
$[V(CO)_5[P(OMe)_3]]$	1974, 1856, 1829		14
$[V(CO)_5[(Ph_2PCH_2CH_2)_2PCH_2]_2]$	1960, 1854, 1815	−1866, −1857	17
$[V(CO)_4(PPh_2Et)_2]$	1898, 1792, 1727, 1734	−1685	17
$[V(CO)_4[(Ph_2PCH_2CH_2)_2PCH_2]_2]$	1889, 1780, 1755, 1735	−1831	17
$[V(CO)_4(Ph_2PCH_2CH_2PPh_2)]$	1903, 1799, 1780, 1747	−1736	17
cis-$[V(CO)_4[P(OMe)_3]_2]$	2007, 1905, 1885, 1865		14
$[V(CO)_3[(Ph_2PCH_2CH_2)_2PCH_2]_2]$	1890, 1798, 1700	−1778, −1734	17

Further substitution proceeds by an associative mechanism but with a rate approximately three orders of magnitude slower than for $V(CO)_6$. Neutral vanadium carbonyl complexes of group 14 triorganyls $[V(CO)_5(ER_3)]$ (E = Sn, Pb) can be obtained by reaction of $NaV(CO)_6$ with R_3ECl (E = Sn, Pb).[20]

While good to moderate acceptor ligands (AsR_3, SbR_3) can be introduced into $[V(CO)_6]^-$ photochemically,[21] complexes of weak π-acceptor ligands become more labile under ultraviolet irradiation conditions. This obstacle has been avoided with the synthesis of labile vanadium carbonyl complexes, enabling the preparation of other vanadium carbonyl derivatives via mild-condition ligand exchange reactions (Table 2).[22–7]

A useful alternative to the photochemical preparation of neutral zerovalent vanadium dicarbonyl phosphine derivatives is the reduction of low-valent vanadium phosphine complexes. $VCl_2(dmpe)_2$ reacts in THF with Na/Hg under a CO atmosphere, forming crystalline *trans*-$(CO)_2V(dmpe)_2$ in good yield (Scheme 5).[28] Further reaction with Ag^+ in acetonitrile gives the corresponding cationic complex $[cis\text{-}(CO)_2V(dmpe)_2(MeCN)]^+$ which has been isolated as the tetraphenylborate salt. Interaction with acids gives neutral complexes $V(CO)_2(dmpe)_2X$ (X = Cl, $MeCO_2$, $EtCO_2$, CF_3CO_2, $PhPO_2H$, NH_2SO_3). The chloride in $V(CO)_2(dmpe)_2X$ can be exchanged in methanol for other nucleophiles (Table 3).

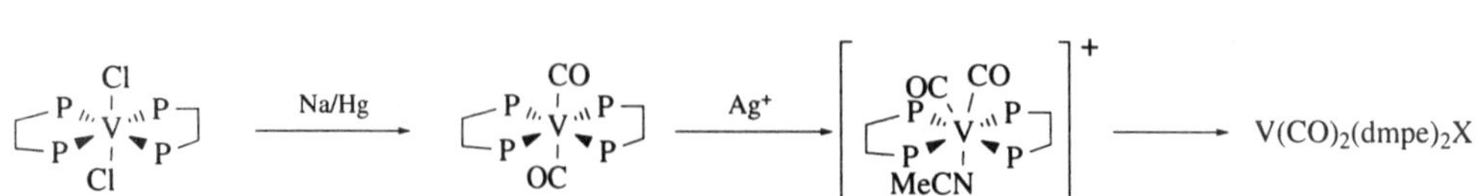

X = Cl, RCO_2, $PhPO_2H$, NH_2SO_3

Scheme 5

Table 2 Spectroscopic data for $[V(CO)_{6-n}L_n]^-$.

L	*Coordination* via	ν_{CO} (cm^{-1})	$\delta(^{51}V)$ (ppm)
O_2NPh	O	2005, 1880	–534
2-C(Ph)(O)py	O	1950, 2855	–781
THF	O	1948, 1803, 1750	–1367
$S(O)Me_2$	O	1977, 1870, 1827	–1390
$S(O)Ph_2$	O	1978, 1835	–1657
SO_2	O	2014, 1832, 1940	–1496
OPb_2	O	1800, 1735	–1339
Acetone	O	1955, 1795	–1423
py	N	1960, 1800, 1700	–1473
$2NH_2$-py	N(py)	1955, 1800, 1750	–1493
4NC-py	N(py)	1350, 1810	–1440
MeCN	N	1965, 1805, 1785	–1615
PhCN	N	1957, 1813, 1760	–1606
NEt_3	N	1958, 1800, 1750	–1482
$NHEt_2$	N	1953, 1792, 1738	–1522
NH_2Ph	N	1950, 1802, 1755	–1506
NEt_2Ph	N	1953, 1792, 1738	–1505
N_2	N	1972, 1843, 1805	–1091
NH_3	N	1965, 1780, 1750	–1587
CN	C	1950, 1790, 1745	–1864
CNCy	C	1953, 1823	–1400
CN(*t*-Bu)	C	1942, 1823	–1898
CN(*i*-Pr)	C	1949, 1825	–1901
$CNCH_2CO_2Et$	C	1948, 1867, 1830	–1906
η^2-CS_2		2022, 1932, 1890	–1390
η^2-MeC≡CEt			–1719
η^2-EtC≡CEt			–1740
η^2-H_2C=CHPr			–1808
SMePh	S	1963, 1807, 1770	–1628
SPh_2	S	1968, 1820, 1773	–1618
$SePh_2$	Se	1963, 1818, 1780	–1658
$TePh_2$	Te	1958, 1860, 1820, 1790	–1753
$BiEt_3$	Bi		–1740
$SbPh_3$	Sb		–1881
$AsPh_2CH_2R$	As		–1824

Table 3 Spectroscopic data for $(CO)_2$ $V(dmpe)_2$ derivatives.[28]

Complex	ν_{CO} (cm^{-1})	$\delta(^{31}P)$	$\delta(^{51}V)$
[*trans*-$V(CO)_2dmpe_2MeCN]BPh_4$	1836, 1774		
trans-$V(CO)_2dmpe_2$	1820, 1700		
cis-$V(CO)_2dmpe_2Cl$	1811, 1748	53.7	–1134
cis-$V(CO)_2dmpe_2(MeCO_2)$	1809, 1741	52.9	–1074
cis-$V(CO)_2dmpe_2(EtCO_2)$	1808, 1750	56.7	–1083
cis-$V(CO)_2dmpe_2(CF_3CO_2)$	1816, 1752	54.0	–2009
cis-$V(CO)_2dmpe_2(N_3)$	1800, 1745	53.6	–1161
cis-$V(CO)_2dmpe_2(CN)$	1815, 1765	57.4	–1248
cis-$V(CO)_2dmpe_2(PhHPO_2)$	1808, 1743	53.9	–975
cis-$V(CO)_2dmpe_2(NH_2SO_3)$	1816, 1750	52.6	–906

1.2.1.4 Redox reactions

$V(CO)_6$ has been employed for the selective oxidation of M(–1) carbonylmetalates (M = Co, Mn, Cr, Mo, W) (Scheme 6).[29] For $V(CO)_6/[Mn(CO)_5]^-$ electron transfer proceeds via outer sphere oxidation: with labeled ^{13}CO, no carbonyl exchange takes place between the two different transition metals.[30] A similar reactivity pattern was observed with anionic cyclopentadienyl derivatives of the group 6 metals $[MCp(CO)_3]^-$ (M = Cr, Mo, W), which formed $V(CO)_6^-$ and $[CpM(CO)_2]_3$.

Electron transfer from $V(CO)_6$ to the substrate is also evident in the reaction with cyclopentadienyl cobalt derivatives of η^4-alkylcyclopentadienes.[31] The reaction results in the removal of the alkyl group

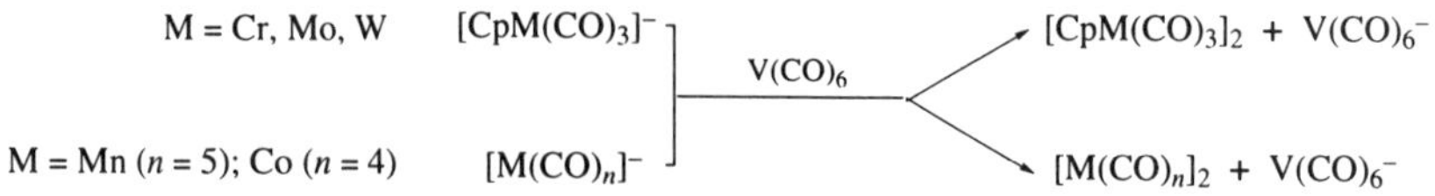

Scheme 6

from the ring, probably via a radical mechanism, as indicated by the formation of the coupling products R–R. In the case of ECpH (E = Et_3Sn, $Bu^n{}_3Sn$, Ph_3Pb), it is possible to isolate the corresponding E–$V(CO)_6$ derivatives.[32] The x-ray crystal structure of $Ph_3PbV(CO)_6$[31] has shown a heptacoordinated vanadium atom where the six CO molecules and the lead donor atom define a distorted capped octahedral geometry, the lead atom occupying the capping position. $(Ph_3Sn)V(CO)_6$ can be obtained from the reaction of Ph_3SnH with $V(CO)_6$.

The complexes $(R_3Sn)V(CO)_5PPh_3$ and $(R_3Sn)V(CO)_6$ (R = Ph, Et, Bu^n, Bz) react with pentafulvene $C_5H_4CR_2$ to give ring-substituted cyclopentadienyl or fulvene complexes.[33] Whereas less polar fulvenes (R = Ph, Me, $\frac{1}{2}(CH_2)_4$) form alkyl–Cp complexes with elimination of the stannyl ligand, the reaction with polar fulvenes ($R_2 = S(CH_2)_2S$, NMe_2) gives the fulvene derivatives where the stannyl ligand is still coordinated to the vanadium.

The versatility of $V(CO)_6$ as a reducing agent is also shown by the reductive coupling (Würtz-like) of alkyl chlorides RX (R = Ph, Bz, Ph_3C, Ph_2P) with the formation of R–R and VCl_3.[31]

Irradiation of $V(CO)_4L_2$ (L = dppe, dppp, dppm, diars, arphos, dpase) in the presence of various substituted benzoyl chlorides and cyclopropanoyl chloride affords the corresponding η^2-bonded acyls $(\eta^2\text{-RCO})V(CO)_3L_2$.[34]

An intriguing dioxygen complex $(O_2)V(CO)_4$ has been detected, during EPR experiments carried out at 100 K, in crystals of $Mo(CO)_6$ doped with $V(CO)_6$ (^{13}C-enriched) and γ-irradiated for several hours.[35]

A formal transfer of two electrons in photolysis of $NaV(CO)_6$ with the conjugated alkyl chloride 1-chloro-2,4-pentadiene in THF at room temperature (Equation (1)),[36] yields an unusual dienyl $(\eta^5\text{-}C_5H_7)V(CO)_4$ complex which can be isolated and purified by sublimation in vacuum to give a dark red crystalline solid in 37% yield.

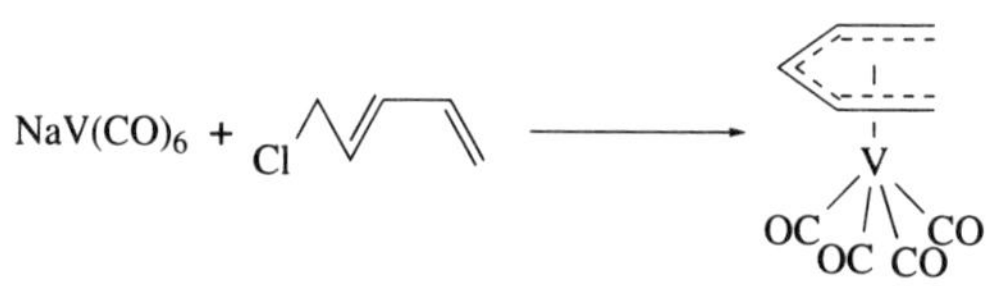

(1)

The formation of a heterobimetallic triangular cluster, from the reaction of $[Na(diglyme)_2][V(CO)_6]$ with the halide-bridged dinuclear complex $(\mu\text{-Cp})(\mu\text{-Cl})Pd_2(PR_3)_2$, involves formal transfer of one electron from $[V(CO)_6]^-$ to the dipalladium unit (Equation (2)).[37] The structure is based on infrared and NMR data, and the analogy with $[(\mu\text{-Cp})Pd_2(PR_3)_2](\mu_3\text{-CO})[CpMo(\mu_2\text{-CO})_2]$ for which an x-ray crystal structure has been reported.[37]

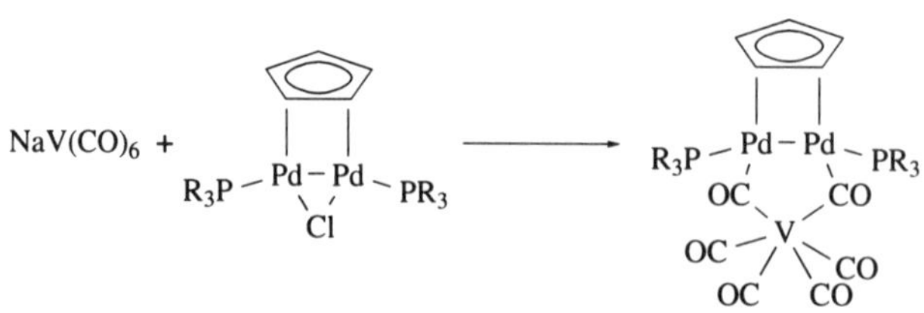

(2)

Oxidation of $NaV(CO)_4(dppe)_2$ with either H_2S at pH 5 or with $Na_2E_2O_3$ (E = S, Se, Te) at pH 5 and 10 affords a series of isostructural μ-bridged chalcogenide divanadium derivatives $(\mu\text{-E})\text{-}[V(CO)_3(dppe)_2]_2$ (E = S, Se, Te).[38,39]

One of the most important and promising achievements obtained in the chemistry of electron-rich vanadium carbonyls is their ability to promote carbon monoxide reductive coupling. Reaction of $NaV(CO)_2(dmpe)_2$ with trimethylsilyl reagents, such as TMS-OTf leads to the reductive coupling of two molecules of CO with formation of a paramagnetic bis(trimethylsiloxy)acetylene complex (Scheme 7).[40] Use of TMS-Br forms the diamagnetic $V(TMS\text{-}OC{\equiv}CO\text{-}TMS)(dmpe)_2Br$. Both complexes react with H_2 at room temperature and mild pressure to form exclusively *cis*-bis(trimethylsiloxy)ethylene in good yield.

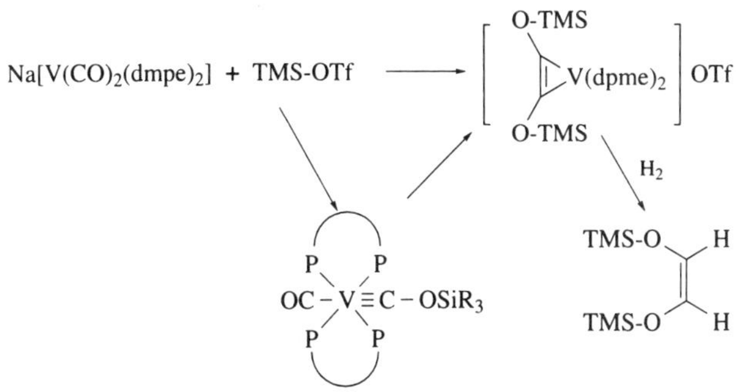

Scheme 7

Examples of metal–metal bonds formed by low-valent vanadium carbonyls are rare. Vahrenkamp has reported the preparation and crystal structure of dimeric $V_2(CO)_8(\mu\text{-}PPh_2)_2$ containing a short V–V contact.[41,42]

1.2.2 Carbonyl Hydrides and Derivatives

Protonation of $NaV(CO)_6$ with stoichiometric amounts of HCl in hydrocarbon solvents at –30 °C gives $HV(CO)_6$. Although the analogous niobium and tantalum congeners may be isolated and characterized, $HV(CO)_6$ is rather labile readily decomposing to $V(CO)_6$ and H_2. Stoichiometric amounts of water promote its fast disproportionation to $[H_3O]^+[V(CO)_6]^-$.[43] Unfortunately, conformational disorder in the crystal structure of $[pyH][V(CO)_6]$[44] makes it impossible to draw definitive conclusions about the position of the hydrogen atoms; however, it is reasonable to assume that hydrogen bridges are formed between the carbonyl oxygen atoms and the nitrogen of pyridine. While the 2-bromopyridinium compound decomposes spontaneously to H_2 and $V(CO)_6$, the acridinium and quinolinium derivatives decompose at 0 °C in THF or ether (Scheme 8). The successful isolation of the unstable (acridineH)$[V(CO)_6]$ has been possible only in water due to its extremely low solubility, which allows prompt separation from the other reaction products. Otherwise, fast inter-ionic electron transfer takes place giving C–C coupling and formation of 9,9'-biacridan in high yield (90%).

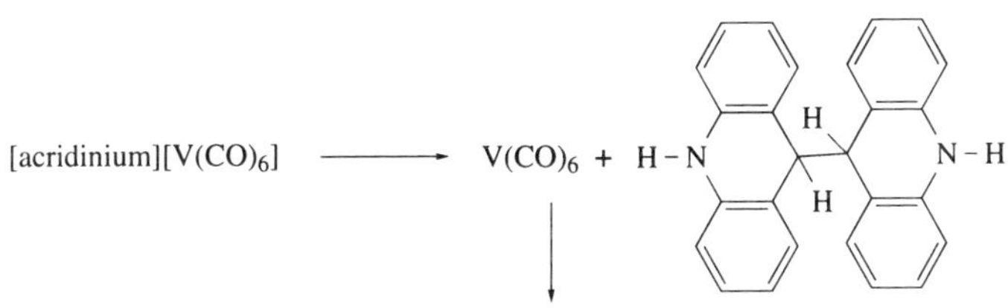

Scheme 8

Hydride complexes stabilized by phosphines $HV(CO)_{6-n}(L)_n$ (Table 4) have been obtained by treatment of the anionic derivatives $[V(CO)_{6-n}L_n]^-$ with Bu^tCl/H_2O mixtures or, more conveniently, by ion exchange on silica gel and subsequent ultraviolet irradiation.[17,45,46] In the case of *facial* tricarbonyl vanadium hydride polydentate phosphine complexes, ^{1}H-NMR data suggest face-capped octahedral structures with restricted dynamic behavior.[17,46] According to this model, the hydride caps a $V(CO)_nP_m$ octahedron, migrating between all faces not edge-spanned by the phosphine. By contrast, infrared studies have shown that in the monophosphine anions $[(R_3P)V(CO)_5]Na$ (R = Ph, PhO, Bu^n) the Na^+ cation interacts specifically with the CO group *trans* to L.[47,48]

Thermally stable vanadium hydride derivatives of dppe and of the tripodal trimpsi (trimpsi = $Bu^tSi(CH_2PMe_2)_3$), $HV(CO)_4$(dppe), and (trimpsi)$V(CO)_3H$, have been isolated and crystallographically characterized (Equation (3)).[49–51] (trimpsi)$V(CO)_3H$ has been prepared by reductive carbonylation of (trimpsi)VCl_3(THF) in THF with sodium naphthalenide. The solvent, or traces of water are the most likely sources of the hydride ligand. Although the hydrogen atom position could not conclusively be located by the x-ray crystal structure, the hydride resonance can be clearly observed in the ^{1}H-NMR spectrum as a broad singlet at δ –5.59 that resolves into a quartet ($J_{P\text{-}H}$ = 26 Hz) upon ^{51}V decoupling.

Table 4 Spectroscopic data for $HV(CO)_{6-n}L_n$.

Complex	ν_{CO} (cm^{-1})	$\delta(^{51}V)$ (ppm)	$\delta(^{1}H)$ (ppm)	*Ref.*
$HV(CO)_4(P_1)_2$	1984s, 1855 vs,br	−1685		17
$HV(CO)_4P_2P_4$	1975m, 1855s	−1850		17
$HV(CO)_3P_2P_4$	1901m,br 1798vs	−1717		17
$HV(CO)_2P_2P_4$	1745	−1485		17
$HV(CO)p_2P_4$	1720			
HV(CO)(dppe)	1990s,1878vs,br	−1680		49
$HV(CO)_3$(trimpsi)	1880, 1785, 1740	−1583		51
$HV(CO)_nP_m$				52
$n = 4$ P_m = dppe	1990, 1878	−1690	−4.91(t)	
= dcype	1975, 1880, 1857, 1811	−1745	−5.52	
= bdmpm	1979, 1865	−1716		
= tdep	1979, 1862			
$n = 3$ P_m = bdmpm	1931, 1835, 1810	−1582	−5.61	
= bdepm	1931, 1834, 1800	−1606	−6.14	
= bdepe	1930, 1835, 1790	−1581	−5.70	
$n = 2$ P_m = tdep	1791, 1764	−1393		

$P_1 = Ph_2PEt$. $P_2(n) = Ph_2P(CH_2)_nPPh_2$. $P_2P_4 = (Ph_2PCH_2CH_2)_2PCH_2P(CH_2CH_2PPh_2)_2$.

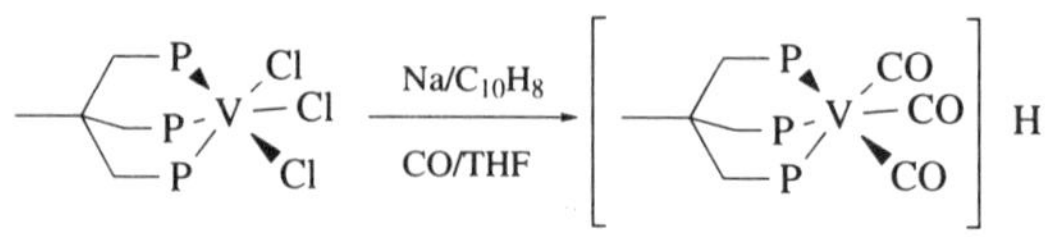

(3)

1.2.3 Carbonyl Alkyls

Two synthetic routes to vanadium(I) alkyl carbonyl derivatives have been described. Various alkyls of general formula $RV(CO)_nP_m$ (P_m = polydentate phosphine) have been prepared by photoreactions of $HV(CO)_nP_m$ and alkenes.[52] Reaction of $[Et_4N][V(CO)_4(Ph_2PCH_2CH_2PPh_2)]$ with RI (R = Me, Pr^i, Bu^t, TMS) leads to the corresponding $RV(CO)_4$(dppe) complexes.[46] Several alkyl $RV(CO)_4$(dppe) compounds have been isolated and thoroughly chracterized via infrared and ^{51}V NMR spectroscopy. Relevant data are reported in Table 5.

Table 5 Spectroscopic data for $RV(CO)_4$dppe.

R	ν_{CO} (cm^{-1})	$\delta(^{51}V)$	*R*	ν_{CO} (cm^{-1})	$\delta(^{51}V)$
H		−1695	Me	1934, 1850, 1836	−1032
Ph_3Pb		−1665	η^3-C_3H_5		−1420
Ph_3Sn		−1651	Neohexyl	1930, 1825, 1840	−1044
TMS	1920, 1820	−1235	2-Me-2-butyl	1930, 1828, 1845	−1062
Bu^t	1920, 1845	−1096	1-octyl	1925, 1830, 1845	−1074
Pr^i	1933, 1850, 1836	−1048			

1.2.4 Spectroscopy

Precautions should be taken in the interpretation of the ^{51}V-NMR data of $[LV(CO)_5]^-$ complexes. The large range of shifts observed for the series of structurally equivalent species reported in Table 2, demonstrates that the commonly applied concept of the participation of variable local diamagnetic contribution to the local shielding is misleading.[53,54] The metal shielding is influenced by a combination of factors including polarizability, covalency, electronegativity, and ligand strength. A useful feature is that the shift ranges are very distinct for the various ligand types [CO > CNR > PR_3 > AsR_3 > BiR_3 > DMSO > NCR > py]. An important consequence is that for weak to medium π-interaction, there is a decrease in shielding with decreasing π-acceptor power of the ligand, increasing ligand bulkiness, increasing ring strain in the chelate structure and increasing degree of substitution.[55] These factors may be accurately evaluated via ^{51}V-NMR spectroscopy, thus allowing rather precise structural characterizations. For example, a ^{51}V-NMR spectroscopic study (Table 6),[56] has pointed out that shielding of the vanadium nucleus decreases in the order Sn > Ge > Si > Pb, as a result of the counteracting effects upon the paramagnetic deshielding term. In agreement with these considerations which predict more effective covalent interactions of $V(CO)_6$ with elements with higher metallic character, such as lead and gold, the crystal structure of $[V(CO)_6][AuPPh_3]$ has shown 7-coordinated gold-capped octahedral vanadium (Scheme 1).[57]

Table 6 ^{51}V NMR and IR data for $(ER_m)V(CO)_6$.[56]

Complex	*ν(CO)* (cm^{-1})	*δ(^{51}V)* (ppm)
$[V(CO)_6]^-$		−1954
$[V(ER_m)(CO)_6]$		
$ER_m = SnPh_3$		−1953
$ER_m = SnEt_3$	2060vs 1980m 1910s 1830m (CH_2Cl_2)	−1954
$ER_m = PbPh_3$	2065vs 2020m 1985s 1955vs 1940sh 1903m (KBr)	−1750
$ER_m = AuPPh_3$		−1836
$[V(CO)_5PPh_3]^-$		−1813
$[V(EPh_3)(CO)_5PPh_3]$		
E = Sn		−1651
E = Pb	2013s 1975m 1910vs 1890sh 1855s (KBr)	−1565
$[V(EPh_3)(CO)_5]^{2-}$		
E = Sn		−1994
E = Pb		−1967
$[V(Cp)H(CO)_3]^-$		−1730
$[V(Cp)EPh_3(CO)_3]^-$		
E = Si	1945s 1840s 1810sh (CH_2Cl_2)	−1860
E = Ge	1947s 1875m 1830vs 1785s (CH_2Cl_2)	−2026
E = Sn	1969w 1880s 1790sh 1775vs (CH_2Cl_2/CH_3CN)	−2054
E = Pb	1968, 1892s 1868vs 1800vs (CH_2Cl_2/CH_3CN)	−1789

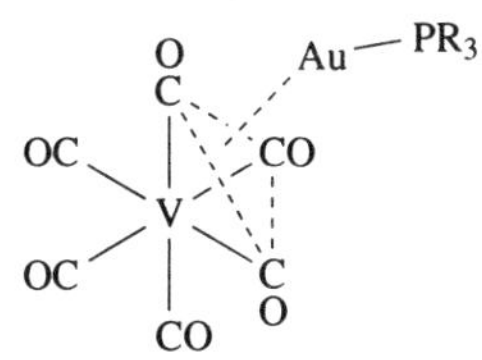

(1)

1.2.5 Binary Nitrosyl Carbonyl Derivatives

Reaction of equimolar $[Et_4N][V(CO)_6]$ and $[NO][BF_4]$ in methylene chloride at −40 °C, affords the thermally unstable $V(CO)_5NO$, which can be isolated in high yield as deep red-violet crystals.[58] This volatile material undergoes CO replacement with a large variety of neutral and anionic bases at temperatures ranging from −30 °C to 0 °C, to form a series of substitution products (Table 7) in good yield (Scheme 9).

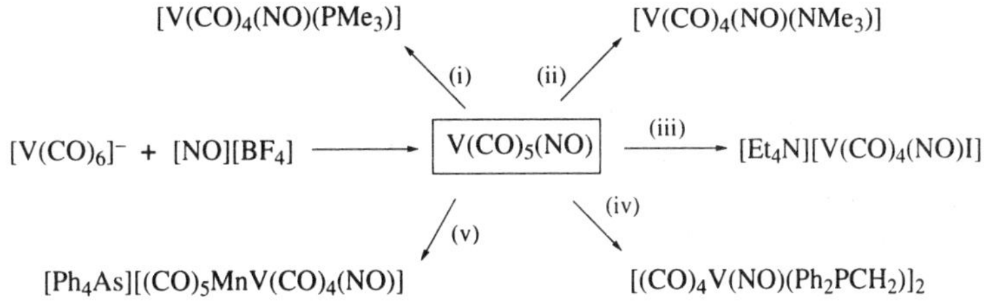

i, PMe_3; ii, NMe_3; iii, $[Et_4N]I$; iv, dppe; v, $[Ph_4As][Mn(CO)_5]$

Scheme 9

The definitive confirmation of a strong *trans* effect in these complexes[59] is provided by the x-ray crystal structure of $[V(CO)_4(NO)(PMe_3)]$.[60] The phosphine ligand, *trans* to the NO group, shows a very long V–P bond (V–P = 0.2543 nm).

Table 7 Spectroscopic data for V(CO)(NO)L.

Complex	*Phase*	ν_{CO} (cm^{-1})	ν_{NO} (cm^{-1})	$\delta(^{51}V)$ (ppm)
$[V(CO)_4(NO)L]$				
$L = PPh_3$	CCl_4	1935vs	1630s	
dppm	CH_2Cl_2	1955vs	1620s	
PMe_2H	cyclohexane	1976m, sh 1959vs	1661s	
$P(OMe)_3$	cyclohexane	1988m, 1961vs	1668s	
$[V(CO)_3(NO)L_2]$				
$L_2 = 2PMe_3$	CH_2Cl_2	2009m, 1914vs	1582s	−1337
$2PMe_2H$	cyclohexane	2000m, 1910vs	1575s	
$2P(OMe)_3$	KBr	2010m, 1925vs	1610s	
dppm	film	2000m, 1915vs 1870s, sh	1580s	
dppe	THF	2004m, 1919vs	1599s	−1379
dppp	THF	1994m, 1937vs	1605s	−1340
dmpe	THF	2000m, 1937vs	1601s	−1480
pab	Nujol	2009m, 1928vs 1918sh 1883vs	1595s 1575sh	−160
dpae	THF	1993m, 1924vs	1590s	−140
diars	THF	202m, 1943vs	1597s	−187
$[V(CO)_2(NO)L_3]$				
$L_3 = cp_3$	THF	1977s 1954s 1930sh	1573s	−1242
1.5 diars	THF	1989vs	1560s	−269
$[V(CO)_4(NO)PMe_3]$	hexane	1960, 1960	1663	
$[V(CO)_4(NO)dppe]$	hexane	1980	1660	
$[V(CO)_4(NO)NMe_3]$	hexane	1990, 1960	1650	

Another convenient synthetic route to $V(CO)_{5-n}(NO)L_n$ is the nitrosylation of $V(CO)_{6-n}L_n$ and of $VH(CO)_{6-n}L_n$ ($L = \frac{1}{n}$ di- and tritertiary phosphine) with NOX ($X = Cl, BF_4$) or $[Co(NO)_2X]_2$ (X = Cl, Br, I).[61–4] Relevant infrared and ^{51}V-NMR data are reported in Table 7. Reactions in THF yield a mixture of $[V(CO)_{5-n}L_n(NO)]$ and $[V(NO)_2(THF)_4]X$ as the main products. The replacement of THF is easily achieved in $[V(NO)_2(THF)_4]X$ by a number of ligands with a variety of donor atoms, enabling the preparation of the corresponding substitution compounds (Tables 8 and 9).

Table 8 Spectroscopic data for $[VL_{4-n}(L')_n(NO)_2]X$ (L = THF).

L'	ν_{NO} (cm^{-1})	$[VL(L')_3(NO)_2]X\delta(^{51}V)$ (ppm)	$[VL_2(L')_2(NO)_2]X\delta(^{51}V)$ (ppm)
PMe_3	1688, 1570	−759	
PEt_2Ph	1691, 1564	−760	−544
$PEtPh_2$		−726	−483
	1697, 1571	−730	−479
		−754	−476
$P(OMe)_3$	1697, 1582	−577	
$SbEt_3$	1701, 1567	−735	
$AsEt_3$	1698, 1568	−723	−397
$\frac{1}{2}$ dmpe	1686, 1590	−894	
$\frac{1}{2}$ phpp	1699, 1582	−849	
$\frac{1}{2}$ dppm	1709, 1568	−726	−483
$\frac{1}{2}$ dppe	1697, 1568	−731	−544
$\frac{1}{2}$ dppp	1701, 1569	−734	
$\frac{1}{2}$ dppb	1698, 1569	−737	−486
$\frac{1}{2}$ ppb	1702, 1563	−707	−511
$\frac{1}{2}$ arphos	1702, 1572		
$\frac{1}{2}$ mab	1701, 1580	−790	

The related reaction of $V(CO)_6$ with 2,2,6,6-tetramethylpiperidine N-oxide (TMPO) gives $(TMPO)V(CO)_3$ in almost quantitative yield.[65,66]

Table 9 Spectroscopic data for $[VL_4(NO)_2]X$.

L	*X*	$\delta(^{51}V)$ (ppm)	*Solvent*	ν_{NO} (cm^{-1})	*Solvent*
$\frac{1}{2}$ $H_2N(CH_2)_2NH_2$	Br	−446	MeOH	1687, 1538	KBr
$\frac{1}{2}$ *o*-Phenanthroline	Br	−394	NCMe	1695, 1565	MeOH
$\frac{1}{2}$ *a,a'*-Bipyridyl	Br	−395	NCMe	1700, 1582	NCMe
				1693, 1562	KBr
$\frac{1}{2}$ TolN=CH–CH=NTol	Br	−332	THF	1707, 1578	THF
$\frac{1}{2}$ Pr^iN=CH–CH=NiPr	Br	−328	THF	1706, 1583	THF
NCMe	Br	−275	NCMe	1736, 1608	NCMe
NC_5H_5	Br	−174	pyridine	1697, 1550	pyridine
NH_2Ph	Br	−130	NH_2Ph	1695	NH_2Ph
$\frac{1}{4}$ n_4	Br	−176	THF	1705, 1589	THF
SPh_2	Br	−23	THF		
Thiophene	Br	−20	thiophene		THF
$OPEt_3$	Br	+71	THF	1700, 1543	Nujol
$OSMe_2$	Br	+74	DMSO	1687, 1540	acetone
$OCMe_2$	Br	+142	acetone	1567	$OC(Pr^i)_2$
$OC(Pr^i)_2$	Br	+210	$OC(Pr^i)_2$	1568	THF
THF	Br	+296	THF	1708, 1567	
Bu^tNC	PF_6	−1040		1729, 1619	
	Cl	−1038			
CyNC	I	−1042		1714, 1600	
MeCN	Br	−272		1736, 1608	
THF	Br	+296		1708, 1565	

1.3 σ-BONDED ALKYL AND ARYL COMPLEXES

1.3.1 Vanadium(I) Complexes

The few examples of vanadium(I) alkyls available in the literature are provided by complexes in which the metal is stabilized by both phosphine and CO ligands. The synthesis, reactivity, and spectroscopy of these species have been already discussed in the section dedicated to the vanadium carbonyls.

1.3.2 Vanadium(II) Complexes: Synthesis, Structure, and Bonding

Unless supported by cyclopentadienyl ligands, vanadium(II) and vanadium(III) alkyls and aryls remain rare.

Crystallization of $Ph_6VLi_4(Et_2O)_4$,[67] previously characterized by EPR,[68] allowed the structural elucidation of the sole example of a homoleptic vanadium(II) aryl (Equation (4)). The crystal structure showed that the lithium cations are bound in a tetrahedral arrangement which interlocks with the VPh_6 octahedron.

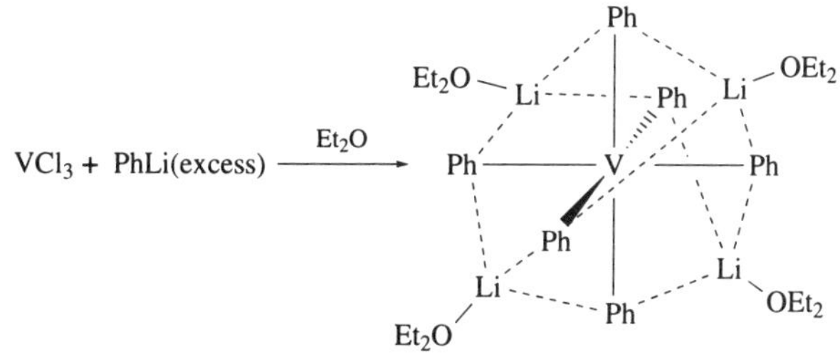

(4)

A neutral vanadium(II) alkyl has been prepared by reaction of $VCl_2(dmpe)_2$ with MeLi in ether and characterized on the basis of the analytical, spectroscopic, and magnetic data.[69] Although no crystal structure has been reported for this compound, it is expected to be isostructural with *trans*-$Me_2M(dmpe)_2$ (M = Ti, Cr, Mn). Dinitrogen fixation has been achieved by reaction of $VCl_2(TMEDA)_2$[70] with two equivalents of $[o\text{-}C_6H_4CH_2N(CH_3)_2]Li$ and subsequent addition of one equivalent of pyridine

(Scheme 10). The reaction initially forms an intermediate complex which, upon treatment with one equivalent of pyridine, reacts with N_2 to form the final dinitrogen complex (μ-N_2)[[(*o*-Me_2NCH_2)C_6H_4]$_2$V(py)]$_2$(THF)$_2$.[71,72] The magnetic moment ($\mu_{eff} = 3.47\ \mu_B$) is significantly lower than expected for a d^3 electronic configuration, confirming that no significant reduction of dinitrogen occurred. The coordination of dinitrogen appears to be rather robust since the N_2 unit cannot be removed even *in vacuo*. However, the N_2 moiety may be replaced only by Lewis bases such as CO, alkynes, and pyridine. In particular, the reaction with excess pyridine forms the monomeric vanadium(II) complex [(*o*-Me_2NCH_2)C_6H_4]$_2$V(py)$_2$.[70]

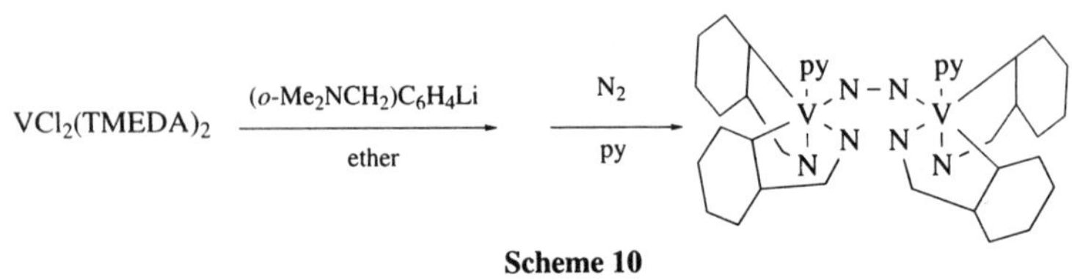

Scheme 10

The V≡V triple bonds observed in a few divalent vanadium complexes, may be ascribed to the d^3 electronic configuration. The presence of this functionality has been postulated in the diamagnetic [2,6-$(MeO)_2C_6H_3$]$_4$$V_2$. According to the original preparation,[73] this compound was synthesized by the reaction of VCl_3(THF)$_3$ with 2,6-$(MeO)_2C_6H_3Li$ at low temperature in THF (Equation (5)). As is the case for $Ph_6VLi_3(Et_2O)_3$,[67,68] the reaction is accompanied by reduction of the metal center forming the final diamagnetic vanadium(II) complex formulated as [2,6-$(MeO)_2C_6H_3$]$_4$$V_2$. The crystal structure of this and other closely related compounds[74] have shown very short V≡V triple bonds (V–V = 0.2200 nm) and the typical lantern-type structure of the closely related quadruply bonded dichromium complexes. Fenske–Hall and Hartree–Fock calculations carried out on model systems have confirmed the possibility of a V≡V triple bond. However, the single-bond configuration has been calculated to be significantly more stable than the triply bonded one.[75]

Related to the three-center chelating ligands, which are widely employed for the formation of short and supershort metal–metal contacts, the organolithium [$Ph_2PCHPPh_2$]Li reacts with VCl_2(TMEDA)$_2$ to give a monomeric anionic species [[$Ph_2PCHPPh_2$]$_3$V][Li(THF)$_4$] (Equation (6)). The vanadium anion is octahedrally coordinated, with the shortest ligand–metal interatomic contacts being those formed with the phosphorus atoms. In spite of the fact that the negative charge is formally born by the carbon atom, the V–C distances (V–C = 3.281 nm) appear to be out of the bonding range.[76]

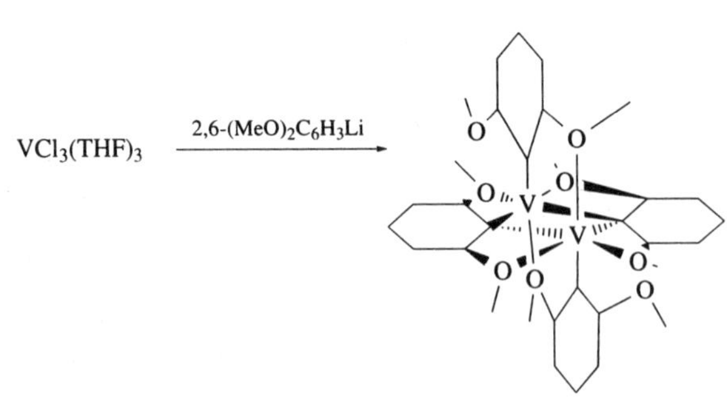

(5)

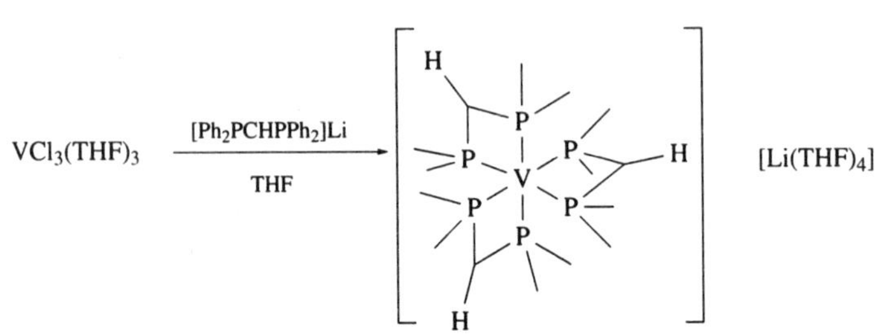

(6)

1.3.3 Homoleptic V(III) Alkyl Complexes

1.3.3.1 Synthesis and structure

Reactions of $VCl_3(THF)_3$ with RM (R = Bz, CH_2TMS, C_6F_5; M = MgX, Li), carried out in either diethyl ether or THF, in the temperature range −78 to −10 °C and with a stoichiometric ratio of 1:3, form a series of complexes formulated as $R_3V(THF)$ (Scheme 11).[77] Different stoichiometric ratios lead to the formation of covalent compounds of the type R_nVCl_{3-n} (n = 1, 2, 3; R = Bz, CH_2TMS, C_6F_5). These species spontaneously disproportionate, via formation of an intermediate dinuclear complex, to equimolar amount of VR_4 and "VCl_2". The presence of "VCl_2" was indicated by the formation of the $VCl_2(py)_4$ complex, upon treatment with pyridine. The complex Bz_3V can be prepared in THF at −78 °C, but appears to be thermally unstable at 0 °C where it decomposes forming black precipitates, probably containing vanadium in its lower oxidation states. Higher thermal stability has been observed for the TMS-CH_2- and C_6F_5-crystalline derivatives. Reaction of these $R_3V(THF)$ species with one equivalent of Bu^tOH[78] results in the formation of an unstable $R_2V(OBu^t)$. THF species which readily disproportionates to vanadium(IV) and vanadium(II) compounds. The anionic vanadate species $[(TMS\text{-}CH_2)V(OBu^t)_3]Li$, obtained by treatment of $(Bu^tO)_3V$ with TMS-CH_2Li, also decomposes to a mixture of $(TMS\text{-}CH_2)_2V(OBu^t)_2$ and $[V(OBu^t)_3]Li$.

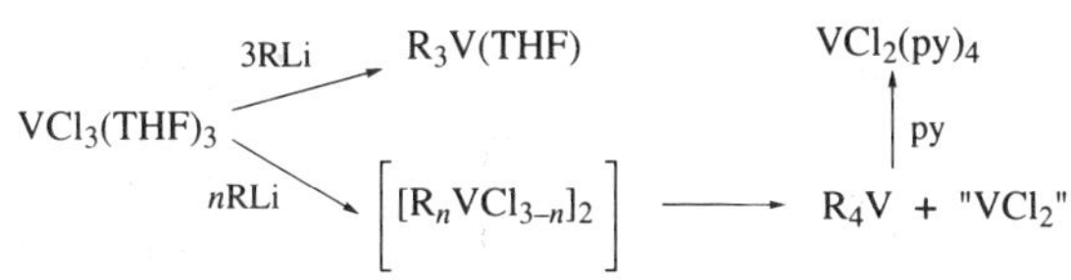

Scheme 11

Fixation of dinitrogen has been achieved from the low-temperature reaction of $VCl_3(THF)_3$ with 3 equivalents of NpLi (Np = neopentyl) (Equation (7)).[79] The reaction carried out in Et_2O forms a reddish-brown dinuclear complex where the two tetrahedral vanadium atoms are linked by one linear bridging molecule of dinitrogen. As with $(\mu\text{-}N_2)[[(o\text{-}Me_2NCH_2)C_6H_4]_2V(py)]_2(THF)_2$,[71] dinitrogen is easily lost upon treatment with a Lewis base (Me_3P, py or Bu^tCN).

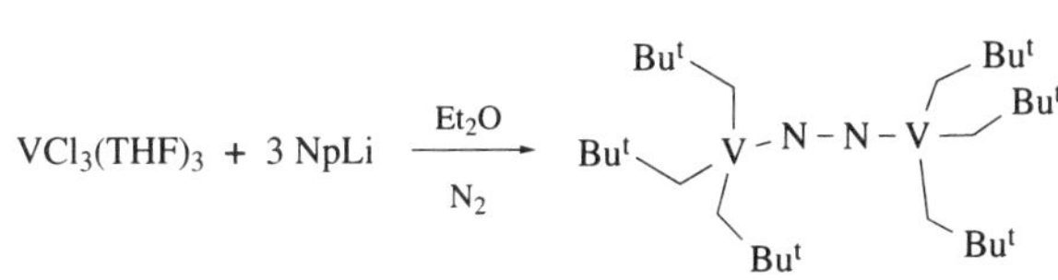

(7)

The relatively facile reduction of vanadium(III) to vanadium(II) is the key to understanding the extreme paucity of trialkyl and triaryl vanadium compounds. The only compound for which a crystal structure has been reported and the chemical reactivity widely investigated is $Mes_3V(THF)$ (Mes = 2,4,6-$Me_3C_6H_2$). The rather straightforward preparation is carried out by reacting $VCl_3(THF)_3$ with three equivalents of MesMgBr in THF.[80] $Mes_3V(THF)$[81] is obtained, after elimination of the magnesium salts via addition of dioxane, as deep-blue prisms from toluene. This compound is thermally stable and can be heated to 100 °C in toluene for short periods.

1.3.3.2 Reactivity

Several reactions discovered for $Mes_3V(THF)$ may be grouped into three main classes of reactions: (i) proton-transfer, (ii) oxidative addition, and (iii) insertion reactions (Scheme 12).

Proton transfer reactions have been carried out with a number of reagents containing "acidic" hydrogens. Acidolysis of $Mes_3V(THF)$ with three equivalents of HCl or carboxylic acids affords the corresponding vanadium(III) chloride or carboxylate.[82] Treatment of $Mes_3V(THF)$ with one equivalent of acid forms an equimolar mixture of R_4V and $[RC(O)O]_2V$. As is the case for the R_3V complexes discussed above, the reaction is assumed to proceed via formation of an unstable $[R_2VX]_2$ intermediate. Reaction of $Mes_3V(THF)$ with acetylacetone forms $Mes_{3-n}V(acac)_n(solv)$.[82]

Similarly, reaction with TBDMS-OH allows the isolation of the unique homoleptic and neutral vanadium(III) alkoxide $V(O\text{-}TBDMS)_3$.[83] While the formulation is supported by analytical and spectroscopic data, a dimeric structure has been proposed on the basis of the low magnetic moment

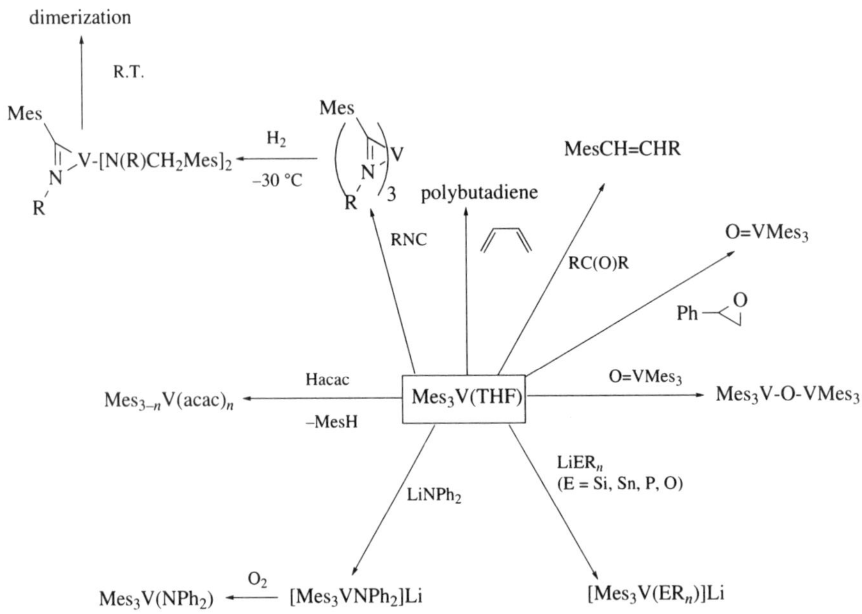

Scheme 12

($\mu_{eff} = 2.2\ \mu_B$). Reaction with coordinating molecules such as THF, pyridine, and bipyridine forms monomeric species with a magnetic moment as expected for a high-spin d^2 electronic configuration of a monomeric vanadium(III) center. Further reaction with one equivalent of lithium silanoate gives the homoleptic anionic metallate LiV(O-TBDMS)$_4$ which probably possesses a tetrahedral coordination geometry.

Diamagnetic (Mes)$_3$V(O) is formed during a smooth reaction of Mes$_3$V(THF) with styrene oxide.[84] The same product is obtained upon treatment with O_2. The tetrahedral coordination geometry of the vanadium atom has been demonstrated by an x-ray crystal structure (V–O = 0.1578 nm, V–C = 0.2022–0.2079 nm). This complex[85a] reacts with an equimolar amount of Mes$_3$V to form paramagnetic (Mes)$_3$V-O-V(Mes)$_3$ which was isolated as a red crystalline solid.[85b]

Anionic vanadium(III) alkyl metallates are obtained upon treatment of Mes$_3$V(THF) with LiER$_n$ (ER$_n$ = SiPh$_3$, SnPh$_3$, NPh$_2$, PPh$_2$, O(Bu^t)) to form complexes of the type LiVMes$_3$(ER$_n$). The characterization of these paramagnetic compounds has been carried out on the basis of the infrared spectral properties and analytical data.[86]

Isocyanides, such as Bu^{t}NC, insert into the three V–C bonds of V(Mes)$_3$(THF) to form the tris(η^2-iminoacyl) complex V[η^2-(Bu^t)NC-Mes]$_3$ the structure of which has been demonstrated by both chemical degradation and x-ray analysis.[87] Further insertion of CO_2 takes place into the V–C bond to form η^2-[Bu^t)NC-Mes]V[η^2-(Bu^t)NC(Mes)C(O)O]$_2$. This complex undergoes hydrolysis to form (Bu^t)-N=CH-Mes, or oxidation followed by hydrolysis to produce (Bu^t)NH-C(O)-Mes. Complex V[η^2-(Bu^t)NC-Mes]$_3$ also undergoes rapid hydrogenolysis at –30 °C, to form the paramagnetic [η^2-(Bu^t)NC-Mes]V[MesCH$_2$N(Bu^t)]$_2$ ($\mu_{eff} = 2.77\ \mu_B$) which, due to its thermal instability, can only be isolated at 253 K as a red crystalline solid.[88] A complicated thermal rearrangement takes place at room temperature to yield the dinuclear diamagnetic [[[CH$_2$(2,4-Me)$_2$C$_6$H$_2$(6,μ-CH$_2$)]N(Bu^t)]V[μ-N(Bu^t)]]$_2$. The presence of a V–V single bond in this complex has been strongly suggested by both the diamagnetism and the short V–V distance (0.2460(1) nm).

1.3.4 Heteroleptic Vanadium(III) Alkyl Complexes

Tetradentate Schiff bases are able to stabilize both vanadium(III) and vanadium(IV) alkyl species. Reaction of (Schiff)VCl(L) (Schiff = ethylenebis(acetylacetone iminato) dianion, L = THF) with Grignard reagents RMgX (R = Me, Bz, Ph, Mes) occur exclusively on the transition metal with concomitant loss of the axially coordinated molecule of solvent and formation of the corresponding (Schiff)VR complex (Equation (8)).[89,90] The magnetic moments of the methyl and benzyl derivatives range between 2.16 and 2.40 BM, thus being in agreement with the presence of weak V–V coupling.

(Schiff)VCl + RLi ⟶ 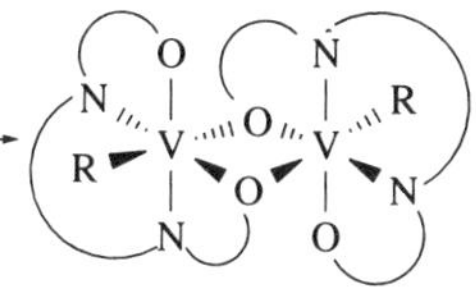 (8)

Pentacoordinated structures have been obtained in the case of macrocyclic ligands such as dibenzotetramethyltetraaza[14]annulene (TAA).[91,92] Reaction of $VCl_3(THF)_3$ with $TAALi_2$ gives the corresponding TAAVCl which can be converted into the corresponding monomeric and paramagnetic alkyl (TAA)VR (R = Me, Bz, Ph, Mes, Cp) upon reaction with the corresponding Grignard or alkyl lithium derivative (Equation (9)). Similar results may be more conveniently obtained by using (TAA)VO starting materials.

(TAA)VCl + RLi ⟶ 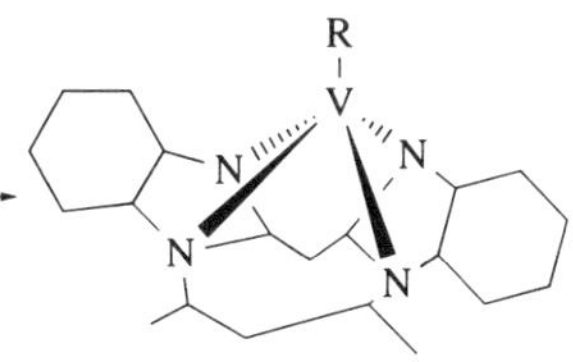 (9)

Terdentate chelating diaminoamido or diphosphinoamido ligands $[N(CH_2CH_2ER_2)_2]^-$ (E = N, P; R = Me, Et) have enabled the preparation of the corresponding $LVCl_2$ complexes via a simple metathesis reaction of $VCl_3(THF)_3$. Subsequent treatment with RLi (R = Me, Ph) affords the corresponding dialkyl derivatives LVR_2 (Scheme 13).[93–5] The monomeric nature of these species has been confirmed by cryoscopy, mass spectroscopy data, and by the crystal structure obtained for the diphenyl derivative $LVPh_2$ (V–C = 0.2140–0.2162 nm).[95]

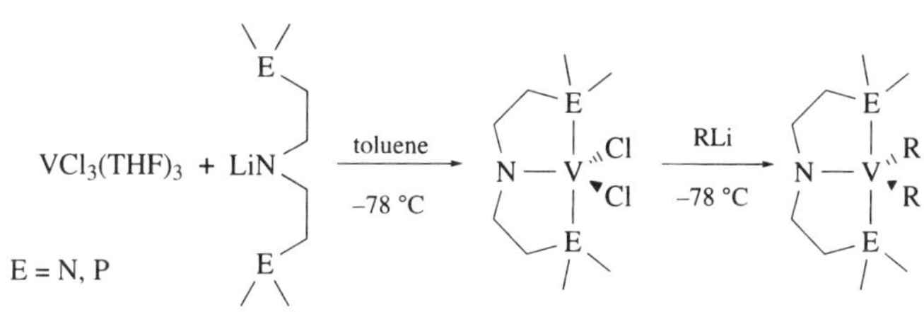

Scheme 13

The involvement of trivalent vanadium as an active center for alkene polymerization reactions has been demonstrated by kinetic and spectroscopic studies on the reaction of (2-methyl-1,3-butanedionato)$_3$V with $AlEt_2Cl$.[96] Although the structure of the active species is unknown, this toluene-soluble reagent acts as an efficient catalyst for the living polymerization of propylene,[97] to give syndiotactic polypropylene of narrow molecular-weight distribution. Secondary vanadium(III) alkyl complexes, derived from propene insertion into $V(acac)_3/AlEt_2Cl$ system, have been postulated as the active species.[98]

1.3.5 Vanadium(IV) and Vanadium(V) Alkyl and Aryl Complexes

A thermally robust and hydrolysis-resistant vanadium(IV) dialkyl complex has been obtained by using a tetradentate Schiff base as a supporting ligand. These monomeric and paramagnetic complexes have been prepared by treatment of the corresponding (acacen)VCl_2 with PhLi.[99]

Tetra- and pentavalent vanadium alkyl derivatives $Li[(Bu^n)VO(Bu^tO)_3]$, TMS-$CH_2V(Bu^tO)_3$, and $(CH_3)_2V(O\text{-}TMS)_2$ are obtained in the reaction of $OV(OR)_3$ (R = Bu^tO, TMS) with the corresponding RLi reagents (R = Bu^n, Me, CH_2TMS). These compounds are obtained in low yield (15–20%) as colored liquids after distillation (Scheme 14).[100]

Studies on the alkylation reactions of $VOCl_3$ with alkyl lithium reagents have established that only partially alkylated compounds $R_{3-n}VOCl_n$ are formed. In contrast, the reaction of $VOCl_2(THF)_2$ with an excess of RLi (R = Mes, Ph) forms paramagnetic $Mes_3VOLi(THF)_{3.25}$ and $Ph_4VOLi_2(THF)_{4.25}$ (μ_{eff} = 1.82 and 1.75 μ_B, respectively).[101] Oxidation to the corresponding pentavalent vanadium species Mes_3VO can be achieved upon treatment of Mes_3VOLi with chloranil.[85a] Comproportionation reaction between

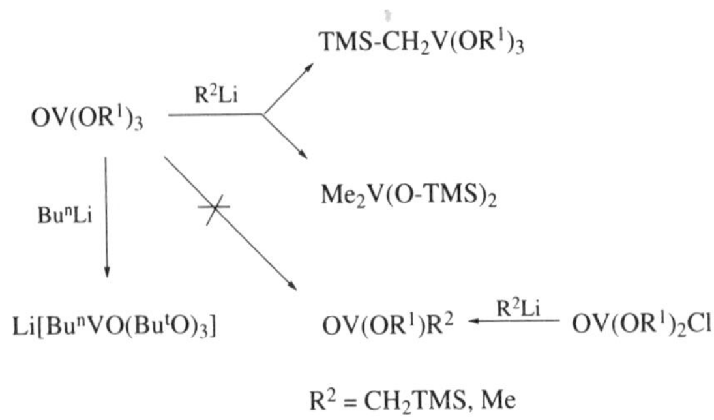

Scheme 14

Mes_3VO and $Mes_3V(THF)$ in ether at room temperature affords $Mes_3V(\mu\text{-}O)VMes_3$ in 93% yield. Mes_3VO reacts also with $HgCl_2$ or I_2 to form $VOCl_3$ and VOI_2, respectively.[85a] Treatment of $VOCl_3$ with dimesitylzinc in hexane at −20 °C affords the thermally unstable $MesVOCl_2$ (Scheme 15). This compound is very reactive and decomposes above room temperature with elimination of mesityl chloride.[102]

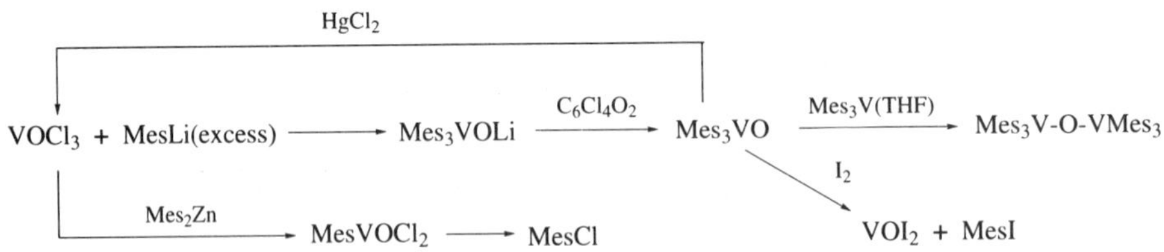

Scheme 15

One of the stimuli for developing the chemistry of vanadyl alkoxide derivatives $(O)V(OR)_3$, is their ability to form acid/base aggregates with AlR_3 which act as efficient catalysts for alkene polymerization. Reaction of $(Pr^i)_3VO$ with silsesquioxane $[(C_6H_{11})_7Si_7O_9(OH)_3]$[103] affords $[(C_6H_{11})_7Si_7O_{12}V(O)]$ which is in equilibrium with the corresponding dimer.[97] Reaction of $[(C_6H_{11})_7Si_7O_{12}V(O)]$ with AlR_3 gives an adduct through coordination of aluminum to the VO group which acts as a very efficient catalyst for the polymerization of ethylene and propylene at low pressure and room temperature.[104] The coordination of AlR_3 to the oxo ligands facilitates the alkene insertion into the V–R bond formed by the transfer of one alkyl group from aluminum to the vanadium atom.[105]

Analogous imidoalkyl complexes, such as $(Bu^tN)VR(Bu^tO)_2$ (R = Me, Bu^n, CH_2TMS, Mes),[106] have been prepared by treatment of $(Bu^tN)VCl(Bu^tO)_2$ with the corresponding alkyl lithium reagent (Scheme 16). Imido dialkyl species $(Bu^tN)VR_2(Bu^tO)$ (R = alkyl, aryl) can be obtained from the corresponding dichloride species $(Bu^tN)VCl_2(Bu^tO)$. Tris-alkyl derivatives $(Bu^t)NVR_3$ can be obtained in good yield from $(Bu^t)NVCl_3$ starting materials. These products have been characterized by elemental analysis, ^1H, ^{13}C, and ^{51}V-NMR spectroscopy.[106]

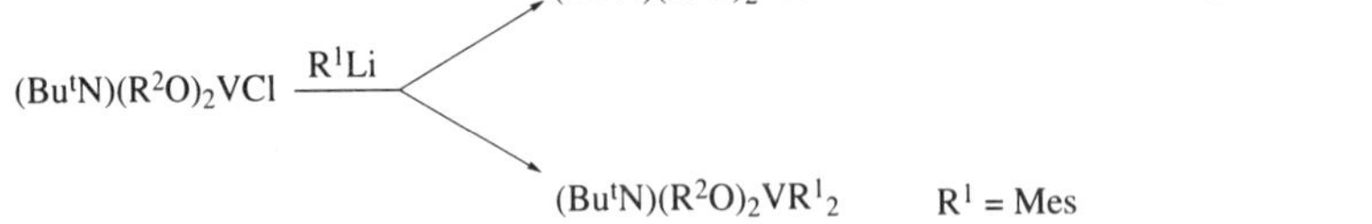

Scheme 16

Sterically crowded imino-vanadium precursors are prepared by treatment of $VOCl_3$ with equimolar amounts of $(Bu^t)_3SiNHLi$. Alkylation of the resulting $[(Bu^t)_3SiN][(Bu^t)_3SiNH]VCl_2$ with Me_2Mg(dioxane), Np_2Mg (Np = neopentyl) or $(TMS\text{-}CH_2)_2Mg$ yields the corresponding thermally stable products $[(Bu^t)_3SiN][(Bu^t)_3SiNH]VRCl$. When MeLi has been used as the alkylating agent (ether at −78 °C) the unprecedented complex $[[(Bu^t)_3SiN]_2VMe_2]Li$ containing a ligand-shielded solvent-free lithium ion has been obtained (Scheme 17).[107]

The V–C bond of an imido–amido vanadium alkyl complex, $R^1N{=}V(NHR^1)_2R^2$, has been shown to perform C–H σ-bond metathesis with a number of hydrocarbons including CH_4, C_6H_{12}, C_6H_6, mesitylene, and alkenes.[108] The metathesis and the insertion of alkynes into the V–C bond proceeds via cycloaddition of the unsaturated substrate to the V=NR double bond (Scheme 17).[109]

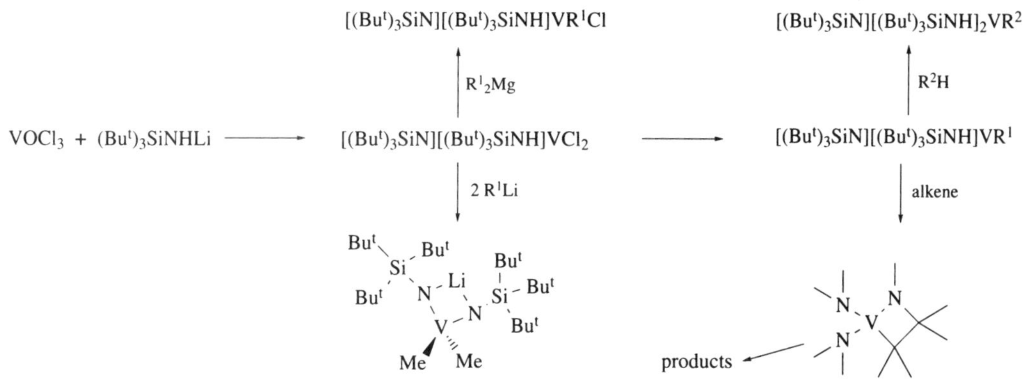

Scheme 17

Reaction of $VOCl_3$ with (*p*-Tol)NCO forms (*p*-Tol)$NVCl_3$ evolving CO_2. Further alkylation with (TMS-CH_2)$_2$Mg gives the corresponding [(*p*-Tol)NV(CH_2TMS)$_x$$Cl_{3-x}$] ($x$ = 1–3) depending on the stoichiometric ratio. Reduction to vanadium(IV) species has been observed when organolithium derivatives have been employed.[110] The relevant NMR data are reported in Tables 10 and 11.

Table 10 NMR data for $RVO(OR)_2$.

Complex	*^1H*	*^{13}C*	*^{51}V*
TMS-CH_2VO(Bu^{t}O)$_2$	0.23, 1.38, 2.0	1.3, 31.7, 84.4	−210
MeVO(Bu^{t}O)$_2$	1.33	31.7, 42.7, 84.5	−236
MeVO(Pr^{i}O)$_2$	1.28, 1.5, 4.87	26.0, 44.5, 81.2	−212

Table 11 NMR data for imido alkyl vanadium complexes.

R	*^1H*	*^{13}C*	*^{51}V*
(Bu^t)NVR(Bu^{t}O)$_2$			
Me	1.55	75.4, 80.1, 35.9	−377
Bu^n	2.32	75.2, 79.8, 64.4	−357
CH_2TMS	1.6	77.8, 80.3, 50.7	+242
Mes		76.4, 81.5, 188.0	−451
[(Bu^t)$_3$SiN][(Bu^t)$_3$SiNH]VRCl			
Me	9.2, 1.39, 1.34	50.5, 29.2, 23.7, 22.5	570
Np	11.6, 3.5, 2.5, 1.41, 1.30, 1.29	101–90, 38.1, 34.1, 31.1, 30.9, 25.5, 24.4	386
CH_2TMS	11.7, 3.8, 2.8, 1.39, 1.28, 0.34	83–73, 31.2, 25.8, 24.4, 2.8	279
[[(Bu^t)$_3$SiN]$_2$VMe$_2$]Li	1.27, 0.93	41, 29.7, 23.6	417
(*p*-Tol)NV(CH_2TMS)$_3$			1048

1.4 VANADIUM(I) MONO(η^5-CYCLOPENTADIENYL) COMPLEXES

1.4.1 CpV(CO)$_4$ and its Derivatives

1.4.1.1 Synthesis, structure, and bonding

Several synthetic routes have been developed for the preparation of CpV(CO)$_4$ and its derivatives.[111] The reduction with potassium is typically carried out at −80 °C,[112] initially leading to the quantitative formation of [Cp_2V]K (Scheme 18). This new species is stable at low temperature but it reacts rapidly with CO on warming to room temperature, to afford CpV(CO)$_4$ and CpK. CpV(CO)$_4$ can be isolated in good yield by sublimation *in vacuo*.[112]

The alkylcyclopentadienyl derivatives Cp'V(CO)$_4$ cannot be prepared using the above procedures. Instead, the preparation requires direct reaction of V(CO)$_6$ with the corresponding alkylcyclopentadiene (Equation (10)). The permethyl derivative[113] is also accessible in 45% yield by following this synthetic

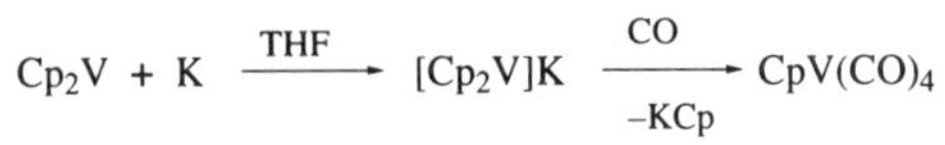

Scheme 18

procedure after purification via column chromatography on silica gel. Other substituted derivatives may be prepared in yields ranging from 55% to 85% (Table 12).[114] However, small and variable amounts of $[Cp'_2V(CO)_2][V(CO)_6]$, which is favored by lower reaction temperatures coupled with longer reaction times, are also formed in a side reaction. All of these derivatives have in common a marked sensitivity to light.[114]

$$V(CO)_6 + Cp'H \longrightarrow Cp'V(CO)_4 + 2\,CO + 0.5\,H_2 \quad (10)$$

An interesting variation of this procedure consists of the reaction of $V(CO)_6$ with acenaphthylene. The reaction forms a dinuclear compound via the reductive coupling of two molecules of acenaphthylene, where the two $V(CO)_4$ moieties bind the five-membered rings from the opposite sides of the ligand plane (Equation (11)).[115] The yield of annelated Cp'' derivatives (Cp'' = indenyl, fluorenyl) may be significantly improved by reacting $V(CO)_6$ with the bromide derivatives Cp''Br.[116] This method is rather versatile and it may be applied even to the reaction of $V(CO)_6$ with perhalogenated Cp, to yield $C_5X_5V(CO)_4$ (X = Cl, Br), and with halogenated fulvenes,[117] such as 1,2,3,4-tetrachloro-6-phenylpentafulvene, to form $Cl_4CBzV(CO)_4$. The indenyl (Ind)$V(CO)_4$ derivative is more conveniently prepared from $[V(CO)_6]^-$ and (indenyl)HgCl.[118]

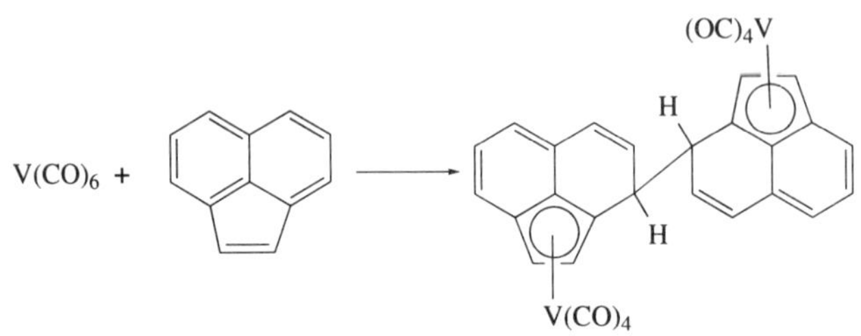

(11)

Table 12 Spectroscopic data for $CpV(CO)_4$ derivatives.

Cp'	*Color*	ν_{CO} (cm^{-1})	$\delta(^{51}V)$
Cp*	orange	2023, 1913, 1885	+44
Indenyl	orange	2037, 1950, 1929	+159
(trityl)C_5H_4	yellow	2030, 1945, 1925, 1888	+50
(Et)C_5H_4	orange	2024, 1914, 1885	+31
(cetyl)MeC_5H_3	yellow	2034, 1942, 1921, 1893	+19.1, +16.0
(Cy)MeC_5H_3	orange	2022, 1920, 1895	+18.9, +16.6
MeC_5H_4	orange	2038, 1927, 1898	+9.1
Fluorenyl	red	2024, 1954, 1932, 1904	−1118
(C_5Cl_5)	orange-red	2046, 1982, 1963, 2039, 1932	−1010
(C_5Br_5)	red	2043, 1978, 1963, 2036, 1931	−1029

A convenient alternative route to a series of monocyclopentadienyl carbonyl–phosphine (η^5-$C_5H_4CHR_2$)$V(CO)_2P_n$ derivatives (Table 13) consists of the photoreaction of $HV(CO)_{6-n}P_n$ (P_n = $Ph_2PCH_2PPh_2$, $Ph_2P(CH_2)_2PPh_2$, $MeC(CH_2PPh_2)_3$) complexes with fulvene $C_5H_4CR_2$ (R = Me, Ph, p-C_6H_4OMe, p-$C_6H_4NMe_2$).[119] These complexes have been obtained in yields ranging between 15 and 50% as red-brown air-sensitive, diamagnetic solids and have been spectroscopically characterized.

1.4.1.2 Reactivity at the metal center

Among the reactions observed for the $Cp'V(CO)_4$ derivatives, the replacement of CO by various nucleophiles (Equation (12)) is certainly the most widely investigated. Kinetic studies have shown that these reactions proceed through a purely dissociative mechanism as indicated by the first-order rate law,[118,120] and by the activation parameters ($\Delta H^{\ddagger}$ = 35.3 kcal mol^{-1} and $\Delta S^{\ddagger}$ = 14.6 esu for the Cp

Table 13 Spectroscopic data for $Cp'V(CO)_2P_n$ derivatives.

Cp'	ν_{CO} (cm^{-1})	$\delta(^{51}V)$
$C_5H_4CHPh_2$		
dppm	1862, 1800	−812
dppe	1863, 1791	−1083
triphos	1862, 1775	
$C_5H_4CHMe_2$		
dppe	1859, 1787	−1058
$C_5H_4CH(p\text{-}C_6H_4OMe)_2$		
dppe	1860, 1788	
$C_5H_4CH(p\text{-}C_6H_4NMe_2)_2$		
dppe	1861, 1790	−1082

derivative, and $\Delta H^{\ddagger} = 31.9$ kcal mol^{-1} and $\Delta S^{\ddagger} = 10.5$ esu for the indenyl complex). Spectroscopic investigations using ^{13}CO in frozen gas matrices (12 K),[121] have shown that a second photochemical reaction, involving a change in the ring hapticity and formation of a $(\eta^3\text{-Cp})V(CO)_n$ (n = 4–5) may play some role, thus suggesting that the contribution of S_N2 mechanisms cannot be ignored.

$$Cp'V(CO)_4 + Nu \xrightarrow{h\nu} Cp'V(CO)_3Nu + CO \qquad (12)$$

Cp' = Cp, Ind

Nu = Lewis base

The replacement of one or more molecules of CO can be achieved with a number of nucleophiles via low-temperature photolysis of $CpV(CO)_4$ (Table 14). For example, low-temperature photoreaction of $Cp'V(CO)_4$ with alkynes affords the corresponding *cis*-$Cp'V(CO)_2(RCCR)$ (R = C_6F_5, H, Me, Ph, TMS; Cp' = Cp, CpR, Cp*, Ind) complexes.[122–4] One of the two residual molecules of CO may readily and quantitatively be displaced by PMe_3 or Bu^{t}NC, to form the corresponding CpV(CO)(L)(RCCR) complexes (Table 15). In the case of the pentafluorophenylalkyne, a second molecule of acetylene may be introduced by photolysis in the presence of two equivalents of acetylene, forming the orange crystalline and air-stable $CpV(CO)(F_5C_6CCC_6F_5)_2$.[122] The reaction can be reversed by photolysis in the presence of CO gas.

Table 14 Spectroscopic data for $CpV(CO)_3L$ complexes.

Complex, L	$\delta(^{51}V)$	ν_{CO}	*Complex, L*	$\delta(^{51}V)$	ν_{CO}
H	−1730	1889, 1775	dppm	−1340	1950, 1872
CO	−1534	2027, 1917	dpsm	−1368	
$P(OMe)_3$	−1496	1963, 1882, 1863	THF	−217	1950, 1835
CN	−1394		GeH_3^-		1950, 1878, 1782
I	−742	1935, 1835, 1805	NCMe	−719	
Br	−578	1943, 1840, 1800	N_2H_4	−679	
Cl	−515	1948, 1840, 1805	Imidazole	−659	
F	+417	1967, 1862, 1815	Pyridine	−519	
PPh_2Me	−1368	1950, 1863, 1850	SPh_2	−843	
$AsPh_3$	−1258.8	1954, 1856,	$SePh_2$	−906	
As_2Ph_4	−1145	1955, 1870, 1859	$TePh_2$	−1145	
$SbPh_3$	−1430	1949, 1871, 1851			

Photolysis of $CpV(CO)_4$ in liquid ammonia produces $CpV(CO)_3(NH_3)$. This nonvolatile material loses NH_3 upon sublimation at 200 °C. Pyrolysis at 700 °C produced V_4C_3 as the main component.[125] Gas-phase pyrolysis of $CpV(CO)_4$ at 700 °C also produces V_4C_3.[125]

Photoreaction of $CpV(CO)_4$ in the presence of $Ph_2ECH_2EPh_2$ yields mononuclear (E = P) or dinuclear ligand-bridged complexes (E = As, Sb) depending on the nature of the donor atom.[126,127]

Labilization of CO in $CpV(CO)_4$ under photochemical conditions has been demonstrated by spectroscopic studies carried out with labeled ^{13}CO. A mixture of $CpV(^{12}CO)_{4-n}(^{13}CO)_n$ has been obtained upon exposure of the *in situ* generated $CpV(CO)_3(THF)$ to ^{13}CO. The ^{51}V NMR spectrum shows the expected binominal multiplets for the various isotopomers[128,129] while the ^{13}C NMR spectrum shows an eight-line pattern due to the coupling of ^{13}C to the 7/2 spin nucleus of ^{51}V. The necessity of

Table 15 Spectroscopic data for alkyne complexes.

Complex	$\delta(^{51}V)$	$\delta(^{13}C\text{-}Cp)$	$\delta(^{1}H\text{-}Cp)$	ν_{CO}
$CpV(CO)_2LL$				
LL = $(CO)_2$	−1560	91.7	5.27	2020, 1922
C_2H_2	−598	97.9	5.66	1995, 1915
C_2Me_2	−658	97.8	5.57	1978, 1898
HC_2Ph	−633	98.0	5.74	1992, 1918
C_2Ph_2	−626	98.4	5.75	1985, 1918
$C_2(TMS)_2$	−265	98.3	5.74	1966, 1892
$CpV(CO)(PMe_3)LL$				
LL = C_2H_2	−470	96.0	5.26	1855
C_2Me_2	−553	95.7	5.21	1840
HC_2Ph	−490	96.2	5.33	1863
C_2Ph_2	−498	96.1	5.32	1870
$C_2(TMS)_2$	−260	95.5	5.22	1848
$CpV(CO)(Bu^tCN)LL$				
LL = C_2H_2	−505	96.2	5.32	1880
C_2Me_2	−586	96.1	5.24	1868
HC_2Ph	−513	96.2	5.39	1886
C_2Ph_2	−487	96.7	5.46	1892
$C_2(TMS)_2$	−216	95.9	5.33	1885
$Cp^*V(CO)_2LL$				
LL = $(CO)_2$	−1499	104.7, 10.9	1.93	2016, 1917
C_2H_2	−560	109.1, 11.1	1.87	1985, 1907
C_2Me_2	−648	109.1, 11.3	1.83	1971, 1889
HC_2Ph	−557	110.1, 11.5	1.87	1965, 1891
$Cp^*V(CO)(PMe_3)(C_2H_2)$	−296	108.1, 11.8	1.78	1850
$Cp^*V(CO)(PMe_3)(C_2HPh)$	−333	108.0, 11.9	1.79	1856
$IndenylV(CO)_2LL$				
LL = $(CO)_2$	−1398			2026, 1934
C_2H_2	−455			1992, 1917
C_2Me_2	−560			1979, 1907, 1901
HC_2Ph	−490			1985, 1915

photochemical methods has certainly been the factor limiting the development of this chemistry since incoming ligands as well as the final complexes may be unstable under photochemical conditions. This problem has been overcome with the successful preparation of labile $CpV(CO)_3L$ adducts. The most widely employed compound is the thermolabile $CpV(CO)_3(THF)$, which is produced by low-temperature photolysis of $CpV(CO)_4$ in THF.[130,131] An alternative preparation, the irradiation of $[CpV(H)(CO)_3][Ph_3C]$, has a significant drawback in the ability of the final product to react with the starting material to form $[CpV(CO)_3]_2(\mu\text{-H})$.[130] The thermal lability of $CpV(CO)_3(THF)$ is ascribed to the easy dissociation of the coordinated THF to form $CpV(CO)_3$ which rapidly rearranges into a mixture of $CpV(CO)_4$ and the green $Cp_2V_2(CO)_5$. Replacement of the cyclopentadienyl ligand with RCp (RCp = Cp*, Ind, C_5H_4TMS, $C_5H_3(Me)(cetyl)$) increases the thermal stability of the corresponding $Cp'V(CO)_3(THF)$ complexes, while the employment of substituted tetrahydrofuran [THF" = 2-MeTHF, 2,2'-Me_2THF] has the opposite effect.[131] All these complexes decompose at temperatures below ambient and a full characterization has not yet been obtained.

Under ordinary reaction conditions, a large number of Lewis base adducts $CpV(CO)_3L$ and *cis*-$CpV(CO)_2LL$ (LL = bipy, *o*-phen, tolane, Sn, N, O, S, Se, or Te functionalities) have been prepared from $CpV(CO)_3(THF)$. A series of crystallographically characterized derivatives, has been obtained from the reaction with diazabutadienes,[132] azines,[132] and hydrazines (Table 16).[131] The low-temperature reaction yields the complexes $CpV(CO)_2L$ (L = $R^1N{=}CHCH{=}NR^1$, $R^2HC{=}NN{=}CHR^2$; R^1 = *n*-C_5H_{11}, Pr^i, Bu^t, Ph, 4-$MeOC_6H_4$; R^2 = *p*-Tol, 4-HOC_6H_4, 4-$MeOC_6H_4$).[132] Analogous complexes may be prepared from direct irradiation of $Cp^*V(CO)_4$ in the presence of L. The substituents of the diazabutadiene and of the azine ligands apparently have an important role in determining the structure of the final compound (mono- versus disubstituted). In the case of MeHC=NN=CHMe the monosubstituted complex $CpV(CO)_3L$ has been obtained and characterized.[132]

Reaction of $CpV(CO)_3(THF)$ with chiral bidentate amines gives rise to a mixture of diastereomeric complexes, $CpV(CO)_2(NN^*)$ (NN* = chiral bidentate amine), composed of two pairs of enantiomers with two chiral centers situated, respectively, on vanadium and the carbon backbone of the bidentate nitrogen ligand. Distinct chemical shifts for the two couples of enantiomers may be observed in the ^{51}V-NMR spectrum.[133]

Table 16 Spectroscopic data for Cp'V(CO)$_3$(hydrazine) complexes.

Cp'	$\delta(^{51}V)$	ν_{CO}	*Cp'*	$\delta(^{51}V)$	ν_{CO}
Cp*			MeC_5H_3(cetyl)		
N_2H_4	−680	1933, 1825	N_2H_4	−676, −679	1945, 1835
N_2H_3Me	−687	1934, 1825	N_2H_3Me	−664	
NH_2Me_2	−623	1935, 1855	N_2H_3Ph	−620	
NHMeNHMe	−570	1933, 1827	$C_5(TMS)H_4$		
N_2H_3Ph	−649	1936, 1833	N_2H_4	−661	1950, 1840
Indenyl			N_2H_3Me	−664	
N_2H_4	−661	1953, 1857	N_2H_3Ph	−620	

Replacement of THF with tetrahydrothiophene allows the large-scale preparation in crystalline form of the intense red CpV(CO)$_3$THT.[134] In spite of its higher stability, which has allowed structural characterization, THT-substitution reactions are very fast with several ligands, including phosphine, isocyanide, bipyridine, and amines, thus enabling the preparation of the corresponding $CpV(CO)_{4-n}L_n$ derivatives (L = PR_3, NR_3H, RNC, α,α-bipyridine). Higher lability has been observed for the isostructural Cp*V(CO)$_3$(SMe$_2$) complex.[135] A solution of this compound reacts with elemental tellurium in the presence of sodium polytelluride to give the binuclear μ-oxo species $Cp^*_2V_2Te_3O$.[136] Further reaction with chalcogen sources (S_8, Se_n), chalcogen hydrides (H_2S, H_2Se), chalcogenides (Na_2S) and polychalcogenides (Na_2S_{22}, Na_2Se_5), leads to mixed tetrachalcogenides complexes of the type $Cp^*_2V_2E_4$ and $Cp^*_2V_2E_3O$.

Isocyanides may be coordinated to the "CpV(CO)$_3$" unit via thermal reaction of CpV(CO)$_4$ in boiling toluene in the presence of Bu^{t}NC. Monitoring of the reaction progress via infrared spectroscopy indicates that the CpV(CO)$_3$(CNR) is only formed in 20% yield. The reaction yield may be improved by the employment of PdO as catalyst.[137] Di- and trisubstituted products have been obtained under these conditions (Table 17).

Table 17 Spectroscopic data for isonitrile complexes.

Complex	$\delta(^{51}V)$	$\delta(^{1}H\text{-}Cp, other)$	ν_{CO}	ν_{NC}
CpV(CO)$_3$(CNR)				
R = Me		4.57, 2.10	1974, 1900, 1894	2130
Pr^i	−1403	4.58, 3.01, 0.71	1972, 1900, 1889	2100
Bu^t	−1404	4.58, 0.90	1973, 1898, 1890	2094
C_6H_{11}	−1401	4.62, 1.0 − 1.2	1971, 1896, 1885	2100
$CH_2C_6H_5$		4.55, 3.90, 6.8 − 7.1	1977, 1903, 1894	2114
2,6-Me$_2$-C$_6$H$_3$	−1430	4.59, 2.10, 6.3 − 6.7	1968, 1906, 1896	2075
CpV(CO)$_2$(CNR)$_2$				
R = Bu^t	−1268, −1301	4.84, 1.08	1880, 1831	2090, 2010
2,6-Me$_2$-C$_6$H$_3$	−1327, −1342	4.88, 2.28, 6.8	1903, 1868	2060, 2000, 1990
CpV(CO)(CNR)$_3$				
2,6-Me$_2$-C$_6$H$_3$	−1235	5.07, 2.31, 6.8	1840	1974

Spectroscopic evidence has been reported for the formation of a nonclassical dihydrogen adduct CpV(CO)$_3$(H$_2$) by UV photolysis of CpV(CO)$_4$ in the presence of H_2. This species has been detected in both *n*-heptane solution at 25 °C and in liquid xenon at −78 °C.[138,139] Thermal displacement of H_2 may be achieved with N_2 in liquid xenon resulting in the formation of the corresponding CpV(CO)$_3$N$_2$ ($\nu_{N\text{-}N}$ = 2199).[138] The dinitrogen complex can also be generated by direct irradiation of CpV(CO)$_4$ in a N_2 matrix.[121]

The replacement of one CO molecule in CpV(CO)$_4$ may also be achieved with anionic ligands including halides.[140] Reaction of CpV(CO)$_4$ with $KGeH_3$ affords the thermally unstable $[CpV(CO)_3(GeH_3)]^-K^+$.[141] The alkali cation may be exchanged in aqueous solution with PPh_4Br giving a more stable and crystalline compound, allowing x-ray crystallographic characterization.

The 18-electron configuration of CpV(CO)$_4$ is certainly responsible for its relative inertness and consequent rather poor chemical reactivity. A rare example of insertion of CO into organic substrates promoted by CpV(CO)$_4$ is provided by the photoreaction of CpV(CO)$_4$ in the presence of triphenylcyclopropene affording the corresponding vinylketene complex in 76% yield as a yellow-brown solid (Scheme 19).[142]

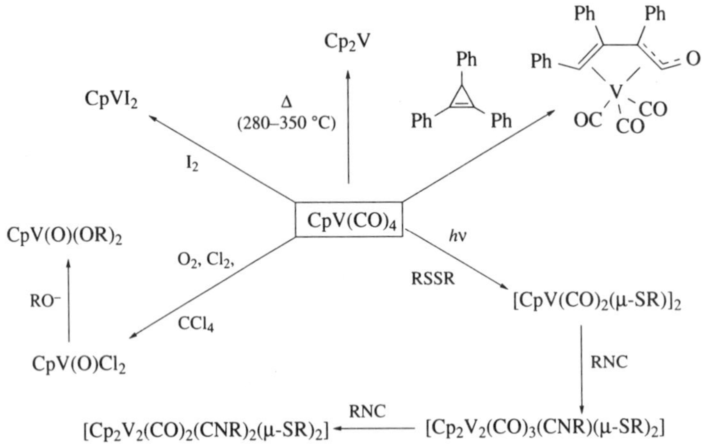

Scheme 19

Treatment of $CpV(CO)_4$ with I_2 in CH_2Cl_2 at room temperature gives a CO-free $CpVI_2$ complex in good yield, and there is no evidence that any vanadium carbonyl iodide might be formed as an intermediate.[143] The oxychlorination reaction of $CpV(CO)_4$ carried out with Cl_2/O_2 mixtures in CCl_4 provides a general synthetic route to vanadium compounds of composition $Cp'V(O)Cl_2$ (Cp' = Cp, Cp*, C_5H_4Me, C_5Me_4Et).[144] Further reaction with phenolate or thiophenolate gives the corresponding substitution products $CpV(O)(X)_2$ (X = OPh, SPh). The structure of the thiophenolate compound has been demonstrated by x-ray analysis.[145] A dinuclear species $[(\eta^5\text{-}C_5R_5)V(O)]_2(\mu\text{-}O)(\mu,\eta^2\text{-}OCH_2CH_2O)$ is obtained upon treatment of $CpV(O)Cl_2$ with glycol in the presence of triethylamine.[144]

Photoinduced decarbonylation of $CpV(CO)_4$ in the presence of organodisulfide RSSR (R = Ph) or a thiol (R = Pr, Bu^t, Et) at room temperature and in THF gives the binuclear diamagnetic compound $[Cp_2V_2(CO)_4(\mu\text{-}SR)_2]$.[146] Subsequent substitution of CO with RNC may be achieved thermally forming isomeric substitution products $[Cp_2V_2(CO)_3(RNC)(\mu\text{-}SR)_2]$.

Measurements of spin–lattice relaxation time and atom–atom potential calculations for monocyclopentadienyl vanadium carbonyl derivatives have shown that the barrier to cyclopentadienyl rotation is about 7.07 kJ mol^{-1} in the solid phase, thus indicating fast rotation at room temperature.[147]

1.4.2 Nitrosyl Carbonyl Derivatives

Eighteen-electron, diamagnetic monocyclopentadienyl vanadium nitrosyl complexes of the type $CpV(NO)_2L$ (L = two-electron Lewis base) are well represented in the family of monocyclopentadienyl vanadium systems.[148] Among these species, $CpV(NO)_2(CO)$ is very labile and affords replacement reaction with a series of Lewis bases (Table 18) in the dark and at room temperature. The reaction may be monitored via either NMR or IR.[149] The $\delta(^{51}V)$ of these species fall in the range −1300 to −500 ppm (referred to $VOCl_3$) whereas the ^{13}C-chemical shift changes only in the range 98–102 ppm depending on the electronegativity of the atom bound to the metal.

1.4.3 Polynuclear Carbonyls

A crystallographic investigation has revealed the dinuclear vanadium(I) carbonyl complex $Cp_2V_2(CO)_5$[150] to be formed during the initial step of the photolysis of $CpV(CO)_4$ when the transient $CpV(CO)_3$ reacts with remaining starting material (Scheme 20). The complex possesses a V=V double bond supported by two semibridging CO ligands. The complex may be cleaved by strong nucleophiles to form monomeric species. Further photolysis in the presence of a Lewis base allows substitution of one terminal CO. Large electron-deficient alkynes react with $Cp_2V_2(CO)_5$ to give cyclobutadiene complexes $CpV(CO)_2(R_4C_4)$ which cannot be formed upon photolysis of $CpV(CO)_4$ in the presence of dienes. Conversely, smaller and more basic alkynes give simple monomeric adducts $CpV(CO)_2(RCCR)$.[150]

Heterobimetallic complexes have been prepared by carrying out the photolysis of $CpV(CO)_4$ in the presence of different organometallic complexes. For example, photolysis with Cp_2NbH_3 gives the μ-H

Table 18 Spectroscopic data for $CpV(NO)_2L$.

L	$\delta(^{51}V)$	$\delta(^{13}C)$	$\delta(^{1}H)$	ν_{NO}
CO	−1292	99.3	5.71	1725, 1637
$P(OMe)_3$	−1265	98.9	5.47	1694, 1599
$P(OPh)_3$	−1213	98.2	4.90	1709, 1611
$P(Et)_3$	−1170	98.2	5.45	1681, 1585
$P(Me)_3$	−1170	99.0	5.43	1685, 1586
$P(n\text{-}C_3H_7)_3$	−1162	98.3	5.40	16.84, 1583
$C_6H_{11}NC$	−1157	98.7	5.50	1697, 1599
Bu^tNC	−1152	98.7	5.48	1694, 1597
$P(Bu^i)_3$	−1126	98.6	5.41	1682, 1584
$P(Ph)_3$	−1116	97.8	5.34	1689, 1593
$S(Et)_2$	−877	100.1	5.48	1686, 1587
$S(Me)_2$	−870	100.1	5.53	1688, 1587
$S{=}C(NH_2)_2$	−854	100.6	5.39	1674, 1567
$S(CH_2)_4$	−851	100.1	5.53	1686, 1575
PhCN	−822	100.9	5.36	1682, 1578
CD_3CN	−802	100.6	5.49	1693, 1588
$(Et)_3N$	−797	101.1	5.54	1678, 1570
$(Et)_2NH$	−745	101.3	5.46	1679, 1571
C_5H_5N	−744	101.3	5.53	1681, 1578
$(Ph)NH_2$	−727	101.0	5.36	1677, 1580
$O{=}C(NH_2)_2$	−618	101.8	5.43	1683, 1563
ONC_5H_5	−595	102.1	5.60	1689, 1571
$O{=}S(Me)_2$	−583	101.9	5.48	1674, 1564
$O{=}C(Me)_2$	−577	101.8	5.50	
$O{=}P(Ph)_3$	−539	101.7	5.33	1675, 1564
THF	−511	101.2	5.54	1686, 1578

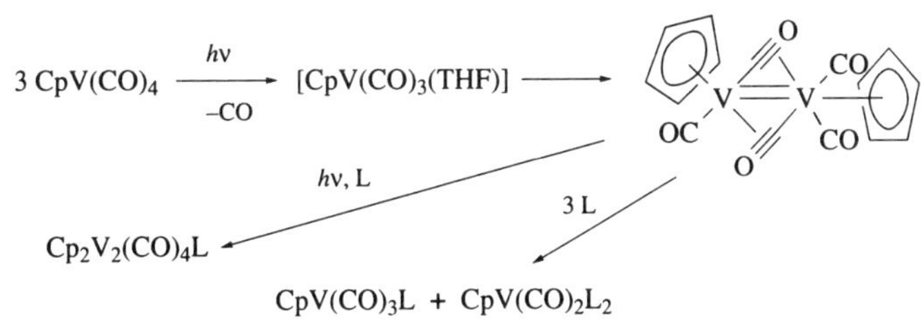

Scheme 20

complex $CpV(CO)_3$-(μ-H)-$NbCp_2(CO)$ (Scheme 21).[151] The closely related complex $[(Ph_2P)_2N]$-$[CpNb(CO)_3$-(μ-H)-$V(CO)_3Cp]$ may be obtained from reaction of $CpV(CO)_3(THF)$ with $[(Ph_2P)_2N][C$-$pNb(CO)_3H]$ (δ(μ-H) = −19.7 ppm).[152]

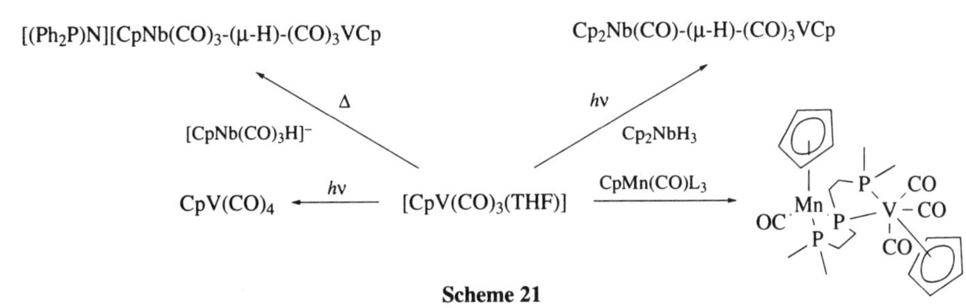

Scheme 21

Monometallic and monoligated complexes $CpMn(CO)_2P_2$ and $CpMn(CO)P_3$ are easily coupled with other transition metal carbonyls such as $CpV(CO)_3(THF)$ to afford $CpMn(CO)P_3(CO)_3VCp$. The complex can be isolated as a red-brown powder in 74% yield. Dinuclear complexes containing $CpV(CO)_2$ moieties may be conveniently prepared by reaction of $CpV(CO)_2P_3$ ($P_3 = Ph_2P$-$(CH_2)_nP(Ph)(CH_2)_nPPh_2$ $n = 2$, 3) with $M(CO)_5(THF)$ (M = Cr, Mo, W).[153]

Nucleophilic cyclopentadienyl carbonyl anions such as $[CpM(CO)_3]^-$ may be used as nucleophiles in the reaction with VCl_3 to form "open cluster" complexes $[CpM(CO)_3]_3(\mu_3\text{-}V)$ (M = Mo, W).[154]

The possibility of bonding two transition metals of very different nature to the two rings of the fulvalene ligand, has been achieved by using a cyclopentadienyl-substituted cyclopentenone as a precursor (Scheme 22).[155] After complexing the cyclopentadienyl moiety to a $(CO)_3WMe$ unit, the residual cyclopentenone ring may be deoxygenated and aromatized in a series of steps and then used to bind different metal carbonyls including $V(CO)_6$. The resulting diamagnetic *anti*-(η^5,η^5-$C_{10}H_8$)[W(CO)$_3$Me][V(CO)$_4$] can be isolated in 30% yield and spectroscopically characterized.

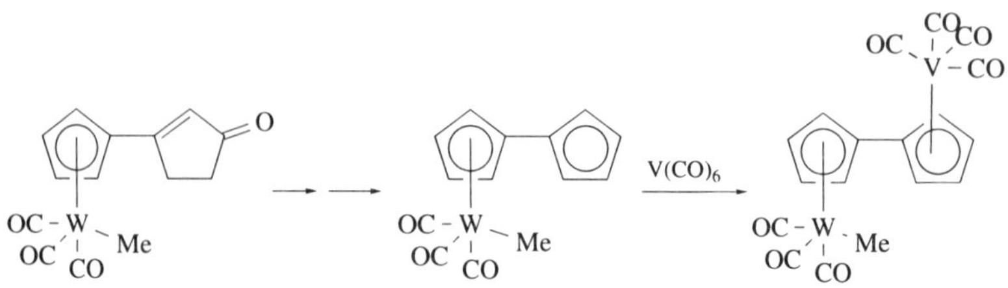

Scheme 22

A rare example of activation of CO coordinated to the CpV moiety is provided by the reaction of $CpV(CO)_4$ with Cp_2Zr(*trans*-butadiene). The reaction yields an unprecedented [(π-allyl)zirconoxycarbene]vanadium species (Scheme 23).[156,157]

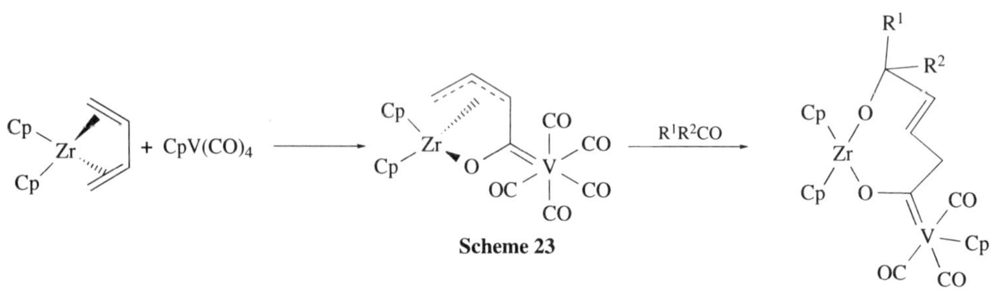

Scheme 23

Reduction of $CpV(CO)_4$ in THF by alkali metals generates the dianionic species $[CpV(CO)_3]Na_2$, which reacts with H_2O to give the monohydride $[CpV(CO)_3H]Na$.[158] The same species may be prepared under milder phase transfer catalysis conditions by treating $CpV(CO)_4$ in benzene with 5 N aqueous solution of NaOH in the presence of $[Bu_4N][HSO_4]$ as a phase-transfer agent.[159]

1.4.4 Monocyclopentadienyl Derivatives Supported by Ligands other than CO

Noncarbonyl monocyclopentadienyl vanadium alkene and alkyne complexes may be prepared via reduction of $CpVCl_2(PMe_3)_2$ in the presence of the appropriate substrate (Scheme 24). The nature of the alkyne is central to the determination of the final product. For example, while reduction with magnesium in the presence of tolane or diethylacetylene leads to the formation of the corresponding $CpV(RCCR)(PMe_3)$ complex, the reaction between magnesium and MeCCMe gives $CpV(\eta^6$-$C_6Me_6)$.[160,161] A diene complex $CpV(diene)(PMe_3)$ is obtained upon treatment of $CpVCl_2(PMe_3)_2$ with magnesium in THF and in the presence of diene. The same complex was previously obtained by reaction of $K[Cp_2V]$ with butadiene.[112] The vanadium(II) species $CpVCl(PMe_3)_2$, prepared via Na/Hg reduction of $CpVCl_2(PMe_3)_2$, reacts with $(BrMg)_2(CH_2)_4$ to yield a unique example of a vanadium ethylene complex $CpV(ethylene)(PMe_3)_2$ (Scheme 25).[160] The corresponding alkyne complex $CpV(PhCCPh)(PMe_3)_2$ reacts with unsaturated substrates resulting in C–C bond formation reactions (Scheme 26). Reaction with 1,3-butadiene affords the hexadienyl complex $CpV(\eta^1,\eta^3$-$C_2Ph_2C_4H_6)(PMe_3)$, while a diphenylhexadiene $CpV[\eta^4$-CH_2=CHC(Ph)=C(Ph)Et] (PMe_3) complex is obtained from the cotrimerization with ethene in the ratio 2:1.[162] Metallacyclic species $CpV(C_4Me_2Ph_2)PMe_3$ are formed upon reaction with 2-butyne.

Reduction of Cp_2V with potassium generates $[Cp_2V]K$ which reacts with coordinating substrates such as butadiene or Ph_3P, giving $[CpV(C_8H_{12})]K$[112] and $CpV(\eta$-Ph)PPh_2,[112] respectively. Reaction of $[Cp_2V]K$ with cyclohexadiene affords a unique vanadium hydride $(CpVH)_2(\mu$-$C_6H_6)$ via oxidative addition into the cyclohexadiene C–H bonds and consequent aromatization to a benzene system (Scheme 27).[163] This complex reacts with CO to form $CpV(CO)_4$.

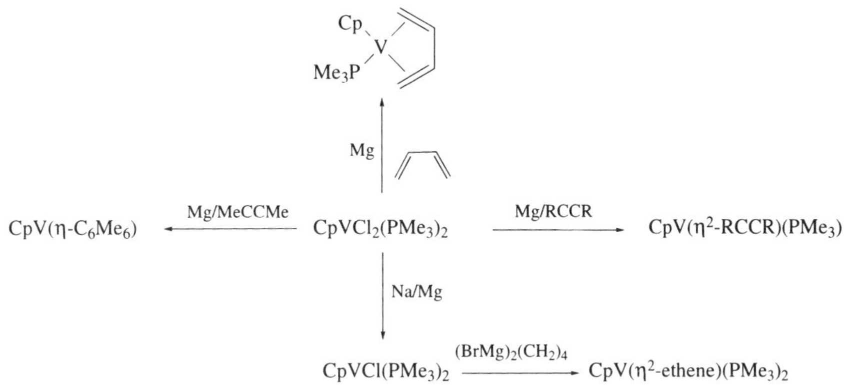

Scheme 24

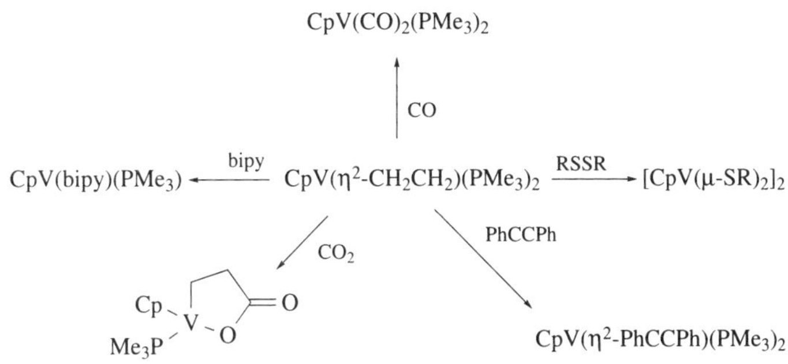

Scheme 25

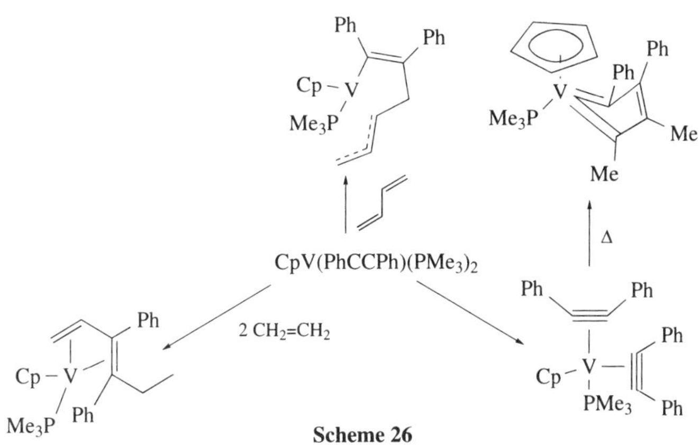

Scheme 26

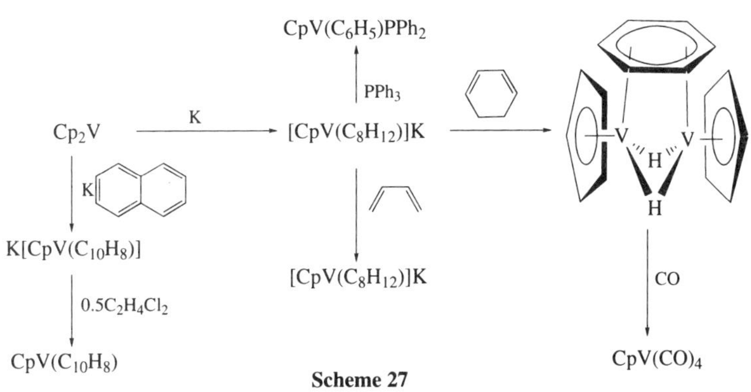

Scheme 27

Reduction of Cp_2V with potassium naphthalenide produces an intermediate formulated as $K[CpV(C_{10}H_8)]$ which reacts with dichloroethane to form $CpV(C_{10}H_8)$.[164]

1.4.5 Multidecker Compounds

Arene-bridged binuclear vanadium compounds with triple-decker sandwich structures, formally unsaturated 26-electron species, may be conveniently prepared in about 50% yield by reaction of $CpV(C_3H_5)_2$ with cyclohexadiene (heptane at 100 °C) to form green–blue crystals of $(CpV)_2[\mu\text{-}(\eta\text{-}C_6H_6)]$ (Scheme 28).[165] The coordinated benzene ring is rather labile and can be easily replaced by toluene or mesitylene upon heating at 110 °C in the appropriate aromatic hydrocarbon. Oxidative addition reactions are observed upon treatment of $(CpV)_2[\mu\text{-}(\eta\text{-}C_6H_6)]$ with C_3H_5Cl, $ClCH_2CH_2Cl$, or I_2 forming $CpV(C_3H_5)Cl(THF)$, CpVCl(THF),[164] and CpVI(THF), respectively. Reaction with CpLi in THF at 60 °C affords a mixture of Cp_2V and $CpV(C_6H_6)Li(THF)_2$. The second component may be crystallized after treatment with TMEDA and transformation into $CpV(C_6H_6)Li(TMEDA)$.[166]

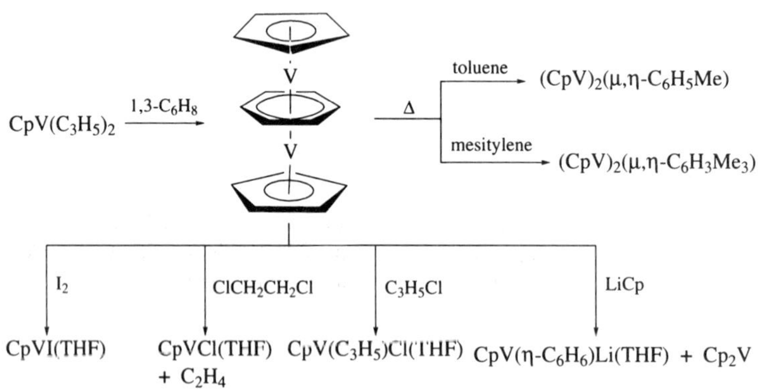

Scheme 28

The reaction of Cp_2V with $C_{10}H_8Yb(THF)_3$ affords the triple-decker compound $CpV(C_{10}H_8)VCp$ which has been characterized by x-ray diffraction methods.[167] The two vanadium atoms each coordinate to both rings of the naphthalene molecule in a *trans* configuration.

A curious reaction has been reported for the $Cp^*V(C_{10}H_8)$ complex[164] with *t*-butylphosphacetylene $Bu^tC\equiv P$. The reaction leads to the formation of a complicated structure with the formula $Cp^*V(Bu^tCP)_3$ in which a CpV moiety is coordinated to a $(Bu^tCP)_3$ unit (Equation (13)).[168]

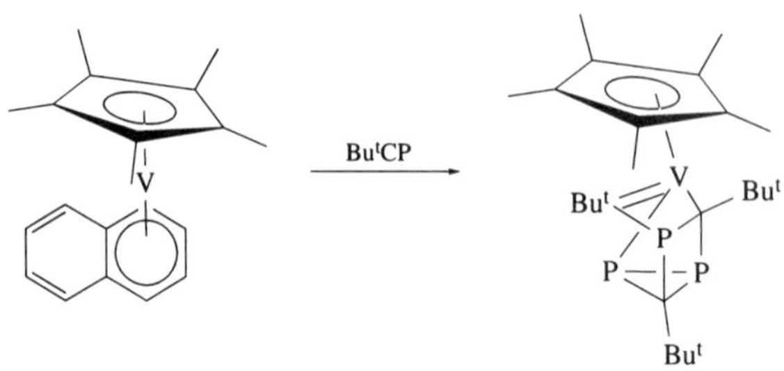

(13)

The cothermolysis of white phosphorus (P_4) with $(Me_4C_5Et)V(CO)_4$ affords the triple-decker compounds $[(Me_4C_5Et)V]_2(\mu,\eta^6\text{-}P_6)$ (Equation (14)), the structure of which has been elucidated by x-ray analysis.[169]

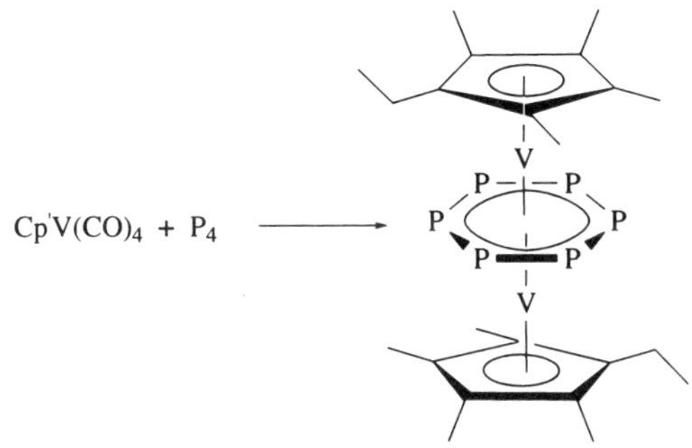

(14)

1.5 OTHER MONO(η^5-CYCLOPENTADIENYL) COMPLEXES

1.5.1 Vanadium(II) and Vanadium(III) Complexes

1.5.1.1 Halide derivatives

Monocyclopentadienyl vanadium halides of general formula $Cp'VX_2(PR_3)_2$ (Cp' = Cp, C_5H_4Me; R = Me, Et; X = Cl, Br) have been prepared by reaction of $VX_3(PR_3)_2$ with Cp'M (M = Na, Tl, $SnBu_3$, $\frac{1}{2}$Mg) (Scheme 29).[170] The formulation of these compounds has been inferred from the analytical and spectroscopic data and demonstrated by the x-ray crystal structure obtained in the case of the trimethylphosphine adduct $CpVCl_2(PMe_3)_2$.[170] In the absence of phosphine, a ready disproportionation reaction takes place in THF forming an equimolar mixture of $VX_3(THF)_3$ and Cp_2VX. The bromo-derivative may be stabilized when incorporated into an heterodimetallic frame, as in the case of $CpVBr_2{\cdot}ZnBr_2{\cdot}3THF$.[171] Direct iodination of $CpV(CO)_4$ with I_2 does not yield $CpVI_2$ as originally reported.[172] A subsequent crystallographic investigation demonstrated that the correct formulation of the complex is $CpVI_3$.[173] A THF adduct $CpVI_2(THF)$ may be prepared and isolated in crystalline form by iodination of $(CpV)_2(C_6H_5Pr)$.[163]

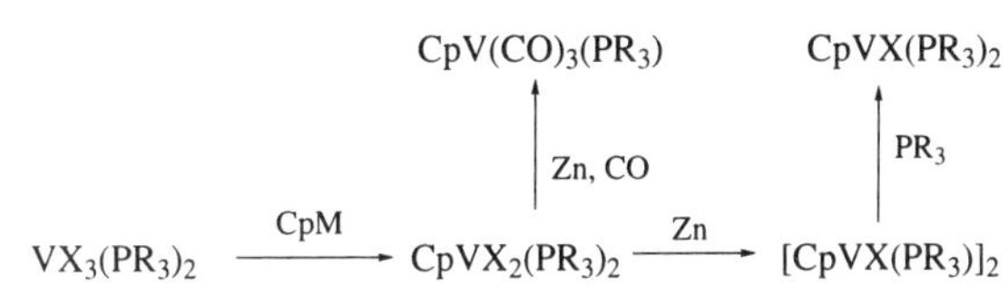

Scheme 29

Pentamethylcyclopentadienyl derivatives appear to have a higher stability than their Cp analogues. Both the bromide and chloride species $[Cp^*VBr_2]_2$ and $[Cp^*VCl_2]_3$ have been prepared by sodium amalgam reduction of the corresponding trihalide or by reaction of $VX_3(THF)_3$ with Cp^*SnBu_3.[174]

While reduction of $CpVCl_2(PEt_3)_2$ with zinc or aluminum in THF under CO offers a simple and effective method for the preparation of $CpV(CO)_3(PEt_3)$, dimeric and paramagnetic $[CpVX(PR_3)]_2$ are formed when the reduction reactions are carried out under nitrogen atmosphere (Scheme 29).[175] The dinuclear structure of $[CpVCl(PMe_3)]_2$ has been clarified by x-ray analysis. The rather short V–V distance (V–V = 0.3245 nm) is probably at the origin of the antiferromagnetism observed for this dimeric species, as indicated by low-temperature ESR and magnetic susceptibility studies.[176] The dimer may be cleaved upon treatment with excess PR_3 or by reaction with bidentate phosphine such as DMPE or DPPE, forming the corresponding $CpVX(PR_3)_2$ and CpVX(dmpe).[175] The MeCp analogues also are accessible by this synthetic procedure. A THF adduct CpVCl(THF) may be isolated as mauve crystals from the reaction of $(CpV)_2(\mu,\eta^6{:}\eta^6\text{-}C_6H_6)$ with 1,2-dichloroethane.[163] The formulation of this complex is also indicated by the reaction of CpVCl(THF) with Cp*Li which affords the corresponding vanadocene CpCp*V in high yield.

1.5.1.2 Carboxylato derivatives

Reaction of vanadocene and decamethylvanadocene with carboxylic acids affords the corresponding dinuclear $Cp_2V_2(RCO_2)_4$.[112,177] In a similar fashion, reaction of Cp_2V with titanium *N,N*-dialkylcarbamates affords the dinuclear vanadium(III) derivative $Cp_2V_2(O_2CNEt_2)_4$, which is basically isostructural with the carboxylates.[178] The pentamethylcyclopentadienyl derivative $[Cp^*V(O_2CNEt_2)]_2$ may be prepared in similar manner from the reaction of $(Me_5C_5)_2V$ with $[Ti(O_2CNEt_2)_4]$.[179]

1.5.1.3 Thiolato and sulfido derivatives

Dinuclear diamagnetic $Cp'_2V_2(CO)_4(\mu\text{-}ER)_2$ (Cp' = Cp, Cp*) can be prepared via room temperature photolysis of either $CpV(CO)_4$ or $Cp^*V(CO)_4$ in the presence of the corresponding diorganochalcogenide E_2R_2 (E = S, Se; R = Me, Ph).[180] The x-ray crystal structures of $Cp^*_2V_2(CO)_4(\mu\text{-}SR)_2$ (R = Me, Ph) have demonstrated the nuclearity of these compounds and revealed the presence of bent $V_2(SR)_2$ cores. Reacting the thermally labile derivatives $Cp'V(CO)_3(Me_2S)$ with E_2R_2 under ordinary reaction conditions provides an alternative preparation (Scheme 30). In the presence of an excess of E_2R_2, further decarbonylation takes place to form the paramagnetic $[Cp'V(ER)_2]_2$. Reaction of either $Cp'V(CO)_4$ or $Cp'V(CO)_3(Me_2S)$ with H_2S forms $Cp_2'V_2(CO)_4(\mu\text{-}S)_2$ which shows a tendency to lose CO. This species cannot be formed by reacting $Cp'V(CO)_4$ or $Cp'V(CO)_3L$ (L = THF, MeCN) with excess sulfur or selenium, since this reaction forms the $Cp_2V_2E_4$ clusters.[136a] Relevant spectroscopic data are reported in Table 19. Thermal or photochemical reaction of $CpV(CO)_4$ with the dithiols $C_2H_4(SH)_2$ or $MeC_2H_3(SH)_2$ gives the dinuclear complex $Cp_2V_2(RC_2H_3S_2)_2$.

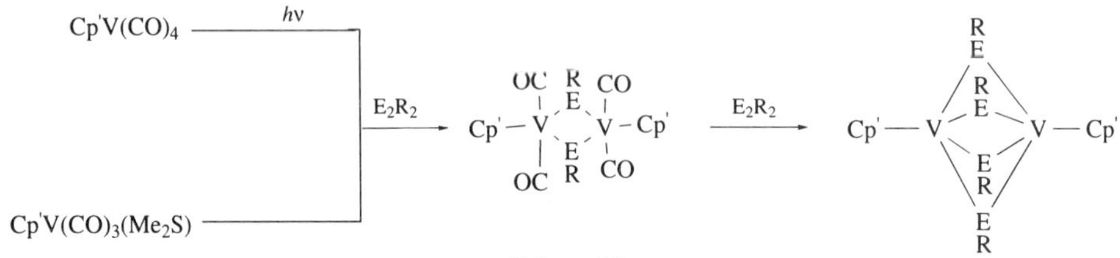

Scheme 30

Table 19 Spectroscopic data for $Cp'_2V_2(CO)_4(ER)_2$.

Complex	ν_{CO} (cm^{-1})	$\delta(^{51}V)$	$\delta(^{13}C)$	$\delta(^{1}H)$
$Cp_2V_2(CO)_4(\mu\text{-}SMe)_2$	1997, 1990, 1933	−788	94.03, 23.48	5.03, 2.1
$(Cp^*)_2V_2(CO)_4(\mu\text{-}SMe)_2$	1971, 1910	−777	11.2, 103.70, 22.5	1.75, 1.9
$Cp_2V_2(CO)_4(\mu SeMe)_2$	1992, 1988, 1935	−815	92.38, 92.47, 6.78	5.05, 2.0
$(Cp^*)_2V_2(CO)_4(\mu\text{-}SeMe)_2$	1969, 1912	−772	11.9, 103.08, 4.66	1.81, 1.9
$Cp_2V_2(CO)_4(\mu\text{-}SPh)_2$	1995, 1955, 1916	−686	95.33, 143.12, 131.51	5.28, 5.7
$(Cp^*)_2V_2(CO)_4(\mu\text{-}SPh)_2$	1971, 1933, 1886	−712	10.94, 102.75, 141.66, 133.44	
$Cp_2V_2(CO)_4(\mu\text{-}SePh)_2$	1993, 1948, 1917	−727	93.31, 133.43	5.23, 7.3
$(Cp^*)_2V_2(CO)_4(\mu\text{-}SePh)_2$	1972, 1933, 1887	−723	11.67, 103.94, 130.90, 134.48	
$(Cp^*)_2V_2(CO)_4(\mu\text{-}S)_2$	1920, 1983	−721	11.82, 103.65	1.60

1.5.1.4 Hydride, alkyl, and alkylidene derivatives

Both $CpVX_2(PR_3)_2$ and $CpVX(PR_3)_2$ are used as starting materials for the preparation of paramagnetic high-spin vanadium(II) and vanadium(III) alkyl derivatives $CpVR^1_2(PR^2_3)_2$ (R^1 = Me, Ph, Bz, allyl; R^2 = Me, Et) and CpVR(DMPE) (R = Me, Pr, Ph) via a halide replacement reaction carried out with the appropriate alkyl lithium or Grignard reagent (Scheme 31).[181,182] Both vanadium(II) and vanadium(III) alkyl complexes act as catalysts for alkene hydrogenation and isomerization.[181] The diamagnetic alkylidene derivative $CpV(CHBu^t)$(dmpe) is prepared via thermal decomposition of the thermolabile $CpV(CH_2Bu^t)_2(PMe_3)$ at room temperature and in the presence of chelating phosphine.[183] Comproportionation of $CpVMe_2(PR_3)_2$ with $CpVX_2(PR_3)_2$ gives selective conversion to $CpVMeCl(PMe_3)_2$[184] which may be subsequently reacted with (allyl)MgBr to form the mixed hydrocarbyl species CpV(allyl)Me(PMe_3)$_2$. The vanadium(III) alkyl species $CpVR_2(PR_3)_2$ react readily with H_2 in benzene or in the presence of 2,3-dimethyl-1,3-butadiene to form $(CpV)_2(C_6H_6)$ and $CpV(C_6H_{10})PMe_3$, respectively. Attempts to prepare vanadium(III) hydride species via reaction of $CpVCl_2$(DMPE) with $LiBH_4$ led only to the formation of the low-spin monomeric CpV(DMPE)(η^2-BH_4).[181] However, two structurally characterized vanadium(II) hydride species, diamagnetic [CpV(μ-H)(dmpe)]$_2$ and paramagnetic [CpV(H)(dmpe)]$_2$(μ-dmpe), have been obtained by hydrogenolysis of CpV(Pr^n)(dmpe).[185] Although the hydride could not be located in the x-ray crystal structure of CpV(H)(dmpe)]$_2$(μ-dmpe), its presence is inferred on the basis of ESR data. The preparation and characterization of the dinuclear hydride $(Cp_2V)_2(\mu\text{-}H)_2(\mu\text{-}C_6H_6)$[163] has been described above.

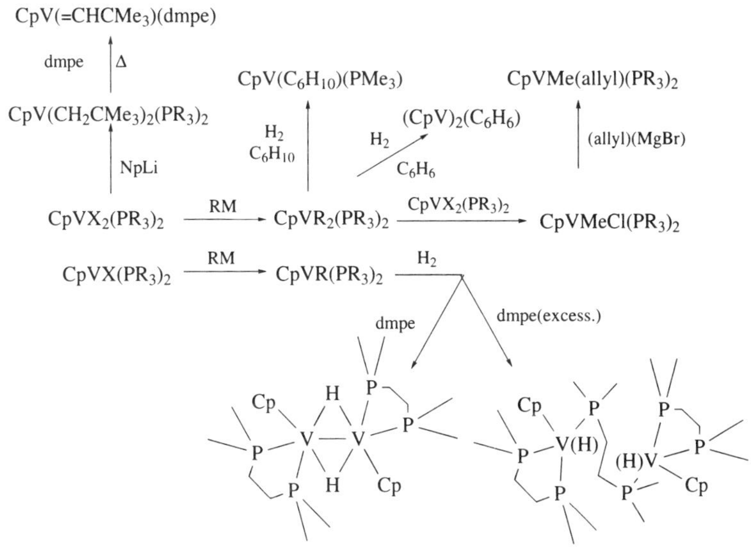

Scheme 31

A dinuclear vanadium(III) bisbutadienyl [CpV[μ,μ-$(CH_2)_4$]]$_2$ complex has been obtained via ethylene reductive coupling performed by $CpV(C_{10}H_8)$ ($C_{10}H_8$ = naphthalene) (Scheme 32).[186]

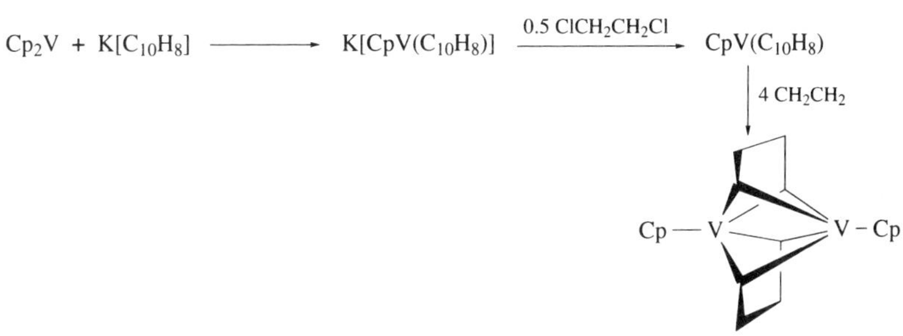

Scheme 32

1.5.1.5 Cyclooctatetraenyl derivatives

The preparation of the dinuclear $(CpV)_2(\mu,\eta^5{:}\eta^5\text{-cot})$ complex has been carried out by a one-pot synthesis by using a mixture of "$VCl_2(THF)_2$," NaCp, and $K_2C_8H_8$ in THF.[187] The crystal structure of the resulting reddish-brown product revealed a dinuclear structure where the two CpV units are bridged by a folded $\eta^5{:}\eta^5$-cot ligand (Scheme 33). The very short V–V distance (V–V = 0.2439 nm) may be explained either in terms of a long V≡V triple bond or in terms of a short V–V single bond.[188–92] There is agreement that the V–V interaction is probably rather weak as indicated by the small HOMO (δ) LUMO (δ*) gap determined by NMR spectroscopic studies.[193]

When $VCl_3(THF)_3$ is substituted for "$VCl_2(THF)_2$" in the above reaction, a red crystalline complex formulated as [CpV(cot)]$_2$ was isolated (Scheme 33).[194] The crystal structure indicates a dimeric complex where two CpV units are placed in a *trans* orientation with respect to the bicyclic ligand (bis-cot) formed by the oxidative coupling of two molecules of cot. A green crystalline isomeric complex

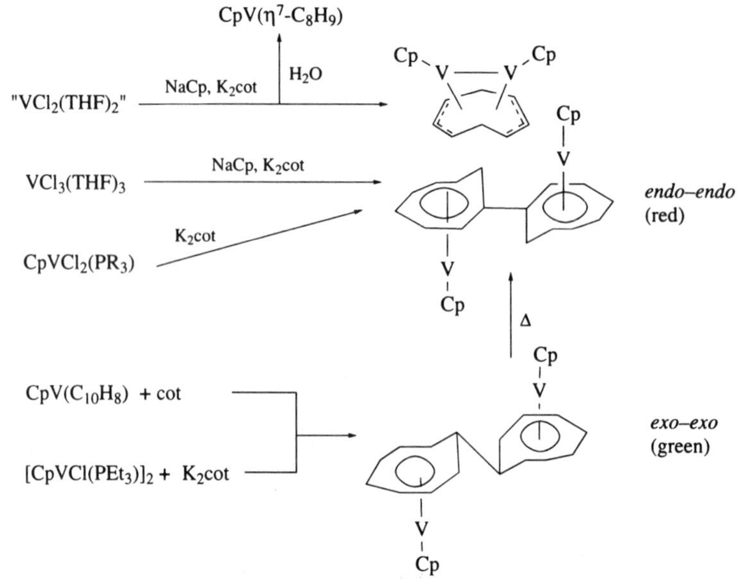

Scheme 33

may also be isolated from either the reaction of $[CpVCl(PEt_3)]_2$ with K_2cot or the reaction of $CpV(C_{10}H_8)$ ($C_{10}H_8$ = naphthalene) with cot. The crystal structure revealed the same basic structure of the red isomer, where the only significant difference is provided by the position of the out-of-plane carbon atoms which are directly engaged in the formation of the C–C *endo-endo* linkage between the two rings.

The monomeric, neutral, paramagnetic cyclooctatrienyl complex $CpV(\eta^7-C_8H_9)$ was obtained from a one-pot reaction starting from VCl_2, NaCp, K_2cot followed by controlled hydrolysis.[195] Both ^{1}H and ^{2}H-NMR spectroscopy provide strong evidence that a stereoselective exo addition of H^+ to the $[CpV(cot)]^-$ anion with subsequent metal-assisted H-migration takes place.

Synfacial binuclear μ-cot complexes of the general formula $X(Cp'V)_2(\mu\text{-cot})$ with Cp'-X-Cp' as an additional bridging ligand (X = CH_2, $SiMe_2$, $GeMe_2$) have been prepared from "$VCl_2(THF)_2$," K_2cot, and $Li_2[(C_5H_4)_2X]$ in a one-pot reaction (Equation (15)). The presence in the reaction mixture of vanadocenophane $\{[(C_5H_4)_2X]V\}_2$ (X = $SiMe_2$, $GeMe_2$) as a by-product was demonstrated.[196]

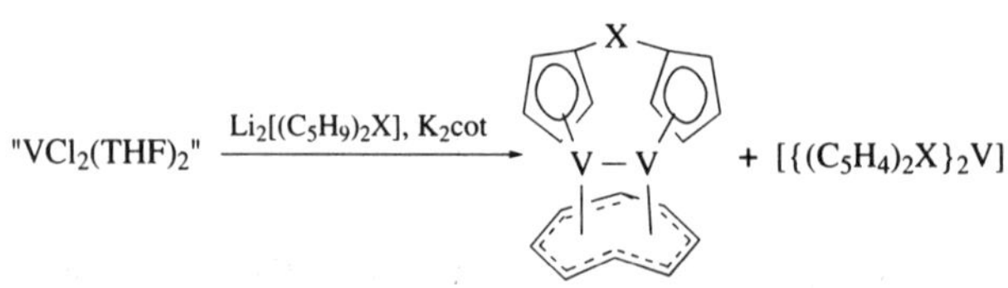

(15)

1.5.2 Vanadium(IV) and Vanadium(V) Complexes

1.5.2.1 Halide complexes

Monomeric vanadium(IV) trihalide derivatives $Cp'VX_3$ (Cp' = Cp, Cp*, MeCp, Me_4EtC_5) can be prepared in high yield by oxidation of $CpV(CO)_4$ with halogens[174,197,198] or by treatment of Cp_2VX_2 with $SOCl_2$ (Scheme 34).[171,199,200] The formulation of these compounds, initially believed to be the vanadium(III) complexes $CpVX_2$, has been demonstrated by the x-ray analysis obtained in the case of the iodo-derivative.[174] The reversible reduction wave of $MeCpVCl_3$ has been observed at +425 mV (vs. Ag/AgCl), which is higher than the strongly oxidizing vanadium(V) $MeCpVOCl_2$ (+270 mV). Among the $CpVX_3$ series (X = Cl, Br, I), $CpVBr_3$ is the most oxidizing species (+510 mV) (Ag/AgCl) being able to oxidize ferrocene and cobaltocene. The resulting paramagnetic salt $[Cp_2Fe][CpVBr_3]$ has been characterized by infrared, Mössbauer, and NMR spectroscopy.[174] Similar results were obtained in the reaction of $CpVCl_3$ with Cp_2Co to form $[Cp_2Co][CpVCl_3]$ which is probably isostructural with the ferrocenium salt.[201] The $CpVI_3$ species, which exists in solution as two distinct species, appears to possess a lower oxidizing power (+425 mV).

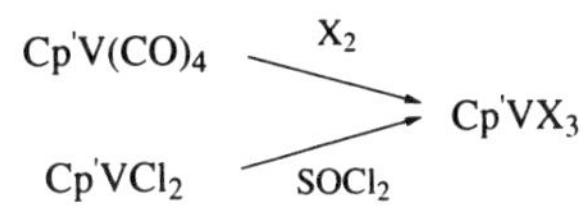

Scheme 34

The large positive oxidation potential is the key to interpreting the chemical reactivity of these species. An unusual charge transfer salt [TmTTF][(MeCp)VCl_3] was obtained in the reaction of (MeCp)VCl_3 with tetramethyltetrathiafulvalene (TmTTF).[202]

The monocyclopentadienyl pentavalent vanadium Cp'V(O)Cl_2 (Cp' = Cp, MeCp) complex has been obtained as the major component of the reaction mixture obtained from the oxidation of Cp'_2VCl_2 with NO.[200] The presence in the reaction mixture of small amount of tetrameric [Cp'VCl(μ-O)]$_4$ with an almost planar [V(μ-O)]$_4$ core has also been ascertained. Reaction of Cp*VI_2 with O_2 in 2:1 stoichiometric ratio gives the green [Cp*VI(O)]$_2$(μ-O) while [Cp*VI_2]$_2$(μ-O) has been obtained in the case of 4:1 stoichiometric ratio. Mononuclear Cp*V(O)X_2 complexes are obtained in a fast reaction upon exposure of Cp*VX_3 to air, or in a one-step synthesis upon reaction of CpV(CO)$_4$ with O_2 and treatment of the resulting black, diamagnetic $Cp_3V_4O_9$ with X_2 (Scheme 35).[197,203] Replacement of the chlorine atom of CpV(O)Cl_2 may be achieved with phenoxides or thiophenoxides with formation of CpV(O)(OR)$_2$ and CpV(O)(SR)$_2$, respectively.[203] Reaction with glycol in the presence of Et_3N forms the dinuclear [CpV(O)]$_2$(μ-O)(μ,η^2-OCH_2CH_2O) with a seven-membered $V_2O_3C_2$ core.[203] A dinuclear μ-oxo complex [Cp*VOBr]$_2$(μ-O) has been obtained when Cp*VBr_3 is reacted with Cp*$_3V_4O_9$ with a stoichiometric ratio of 2:7. The formation of the μ-oxo species can be reversed in the case of the fluoride derivative by treatment with HF. The cyclic trimeric oxo-species [(Cp*V(O)(μ-O)]$_3$ may be prepared by treatment of Cp*VCl_2O with Ag_2CO_3.[204] The nuclearity of this cluster may be increased via partial deoxygenation reaction, achieved by treatment with PMe_2Ph. The reaction forms the paramagnetic tetrameric [(Cp*V]$_4$(μ-O)$_6$ which has been isolated as dark-green crystals.[205]

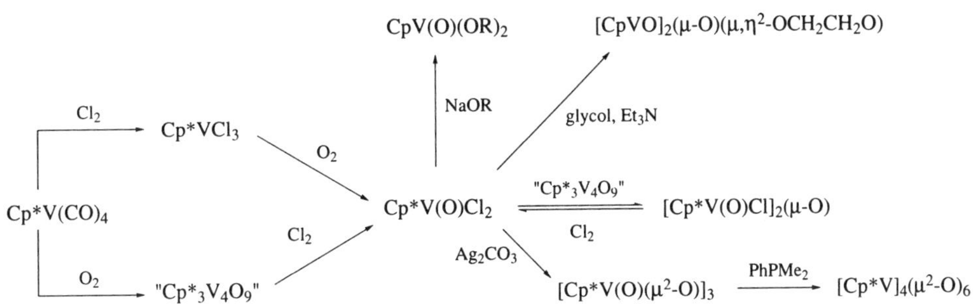

Scheme 35

Table 20 Spectroscopic data for CpV(O)X_2 complexes.

Complex	$\delta(^{51}V)$	$\delta(^{13}C)$	$\delta(^{1}H)$	$\nu_{(V-O)}$	$\nu_{(V-O-V)}$
Cp*VOF_2	−404		2.25	965	
Cp*$VOCl_2$	−25	133.7, 13.3	2.32	967	
Cp*$VOBr_2$	+119	133.5, 13.3	2.29	966	
[Cp*VOF]$_2$(μ-O)	−509	128.7, 11.3	2.16	962	770
[Cp*VOCl]$_2$(μ-O)	−360	129.7, 12.9	2.21	963	730
[Cp*VOBr]$_2$(μ-O)	−309	129.7, 12.9	2.22	966	727
[Cp*VOI]$_2$(μ-O)	−238	129.1, 14.0	2.23	967	720
"Cp*$_3V_4O_9$"	−573	129.7, 11.9	2.11	975, 957	

Pentavalent vanadium imido derivatives have been obtained upon treatment of Bu^{t}N=VCl_3 with CpLi.[206] The resulting (Bu^{t}N)VCpCl_2 complex undergoes further ligand replacement reaction with Bu^{t}OLi or LiNHR^2 to form (Bu^{t}N)VCpCl(Bu^{t}O) and (Bu^{t}N)VCpCl(NHR^2),[207] respectively. Alkylation reactions carried out on both complexes with several alkyl lithium R^{1}Li reagents (R^1 = Me, Bu^n, CH_2TMS, Mes, Cp; R^2 = Me, Bu^n, Bu^s, Bu^t, CH_2TMS) gives the corresponding CpV=N-(Bu^t)(Bu^{t}O)(alkyl) derivatives. Relevant spectroscopic data for CpV^V complexes are reported in Tables 20–22.

Table 21 Spectroscopic data for $(Bu^tN)VCpX_2$ complexes.

Complex	*δ(⁵¹V)*	*δ(¹H)*
$(Bu^tN)VCpCl_2$	−457	5.95, 1.02
$(Bu^tN)VCp(BuO^t)Cl$	−763	6.01, 1.42, 1.05
$(Bu^tN)VCp(BuO^t)Me$	−706	5.72, 1.37, 1.29, 1.12
$(Bu^tN)VCp(BuO^t)(Bu^n)$	−622	5.71, 1.35, 2.32, 1.85, 1.40, 1.15, 1.01
$(Bu^tN)VCp(BuO^t)(CH_2TMS)$	−682	5.78, 1.85, 1.34, 1.11, 0.20
$(Bu^tN)VCp(BuO^t)Mes$	−758	

Table 22 Spectroscopic data for $(Bu^tN)VCp(Bu^tNH)X$ complexes.

Complex	*δ(⁵¹V)*	*δ(¹H)*
$(Bu^tN)VCp(Bu^tNH)Cl$	−785	11.3, 5.75, 1.34, 1.17
$(Bu^tN)VCp(Bu^tNH)_2$	−894	
$(Bu^tN)VCp(Bu^tNH)Me$	−754	9.18, 5.52, 1.30, 1.21, 0.65
$(Bu^tN)VCp(Bu^tNH)(Bu^n)$	−688	8.93, 5.51, 1.88, 1.42, 1.30, 1.22, 1.03
$(Bu^tN)VCp(Bu^tNH)(Bu^s)$	−617	8.64, 5.52, 2.81, 2.05/1.75, 1.55/1.47, 1.27, 1.24, 1.05/0.95
$(Bu^tN)VCp(Bu^tNH)(Bu^t)$	−566	8.33, 5.64, 1.65, 1.26, 1.21
$(Bu^tN)VCp(Bu^tNH)(CH_2TMS)$	−750	9.33, 5.59, 1.34, 1.18, 0.23, 0.10

1.5.2.2 Thiolato and sulfido complexes

Both thermal and photoinduced decarbonylation reactions of $Cp^*V(CO)_4$ in the presence of excess sulfur or selenium affords binuclear chalcogeno-bridged complexes such as $Cp^*_2V_2E_5$ (E = S, Se) and $Cp^*_2V_2E_4$. Although oxo-bridged products such as $Cp^*_2V_2E_4O$ and $Cp^*_2V_2E_3O$ have occasionally been isolated from the reaction mixtures and fully characterized, the origin of the bridging oxygen atom is unclear.[208] Elimination of chalcogen atoms from $Cp^*_2V_2E_5$ by Bu^n_3P results in the formation of $Cp^*_2V_2E_3$ and $Cp^*_2V_2(CO)_4S_2$ as the major products. Further reaction of these complexes with polysulfide or polyselenide (Na_2S_2 or Na_2Se_5) form mixed binuclear complexes which may contain both sulfur and selenium in the chalcogenovanadium framework.[209]

Thermal rearrangement of $(MeCp)_2VS_5$[210] yields the dinuclear pentasulfide $(MeCp)_2V_2S_5$ in about 85% yield.[211] This dimer reacts with alkynes to give the piano-stool dithiolene $(MeCp)V[SC(R)C(R)S]_2$ complexes,[212] and with Bu_3P to give $(MeCp)_2V_2S_4$ through a reversible desulfurization reaction (Scheme 36).[213,214] Reactions of $(MeCp)_2V_2S_4$[214] with $Fe(CO)_5$, $Fe(CO)_3(NO)_2$, and $IrCl(PPh_3)(N_2)$ yield the mixed-metal clusters $(MeCp)_2V_2S_4Fe(CO)_3$, $(MeCp)_2V_2S_4Fe(NO)_2$, and $(MeCp)_2V_2S_4IrCl(PPh_3)$, respectively.[215,216] Further reaction of $(MeCp)_2V_2S_4Fe(NO)_2$ with $Hg[Fe(CO)_3(NO)]_2$ forms the 58-electron $(MeCp)_2V_2S_4[Fe(NO)]_2$ complex with a $V_2Fe_2S_4$ cubane-like structure (Scheme 37).[216] The replacement of iron with cobalt in $(MeCp)_2V_2S_4Fe(NO)_2$ and formation of a 60-electron cluster $(MeCp)_2V_2S_4[Co(NO)]_2$ can be carried out with $Co(NO)(CO)_3$ in the presence of Me_3NO. Thiocubane structures of the type $(MeCp)_2V_2S_4M_2L_2$ (M = Co, Ni, Cu; L = NO, PR_3) have been conveniently prepared when the reactions of $(MeCp)_2V_2S_4$ with $CoI_2(NO)_2$, $[NiI(NO)]_x$, or $CuCl(PPh_3)$ have been carried out in the presence of zinc as a halide receptor.[217] Treatment of $(MeCp)_2V_2S_4$ with Bu_3P, gives the electron-deficient cubane $(MeCp)_4V_4S_4$ and the polyhedral $(MeCp)_5V_5S_6$.[218] The isostructural cyclopentadienyl derivative $Cp_4V_4S_4$ was prepared by reaction of vanadocene with either dithioacetic acid[219] or Bu^tSH in boiling heptane.[220] In this second case, variable amounts of the diamagnetic pentanuclear cluster $Cp_5V_5(\mu^3\text{-}S)_6$ are also formed. The structure of this cluster[221] consists of a trigonal bipyramid of five vanadium atoms, each connected to one Cp ring, with the six face-capping sulfur atoms. The cationic species $[(MeCp)_4V_4S_4]^+$ can be prepared by treating the neutral $(MeCp)_4V_4S_4$ with Ph_3CBF_4.[222] Reaction of $(MeCp)_5V_5S_6$ with TCNQ (TCNQ = tetracyanoquinonedimethane) forms the semiconducting charge-transfer salt $[(MeCp)_5V_5S_6](TCNQ)_2$.[218]

Reaction of Cp^*VOCl_2 with THF solutions of K_2S_x leads to the pentasulfidovanadium complex Cp^*VOS_5, which contains a six-membered metallacycle ring in chair conformation.[223]

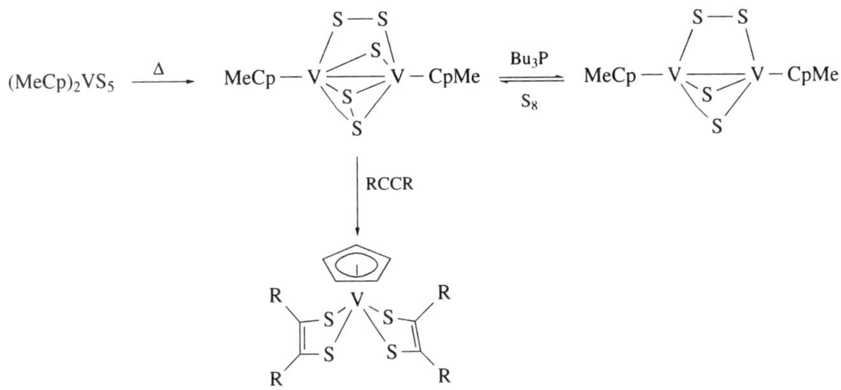

Scheme 36

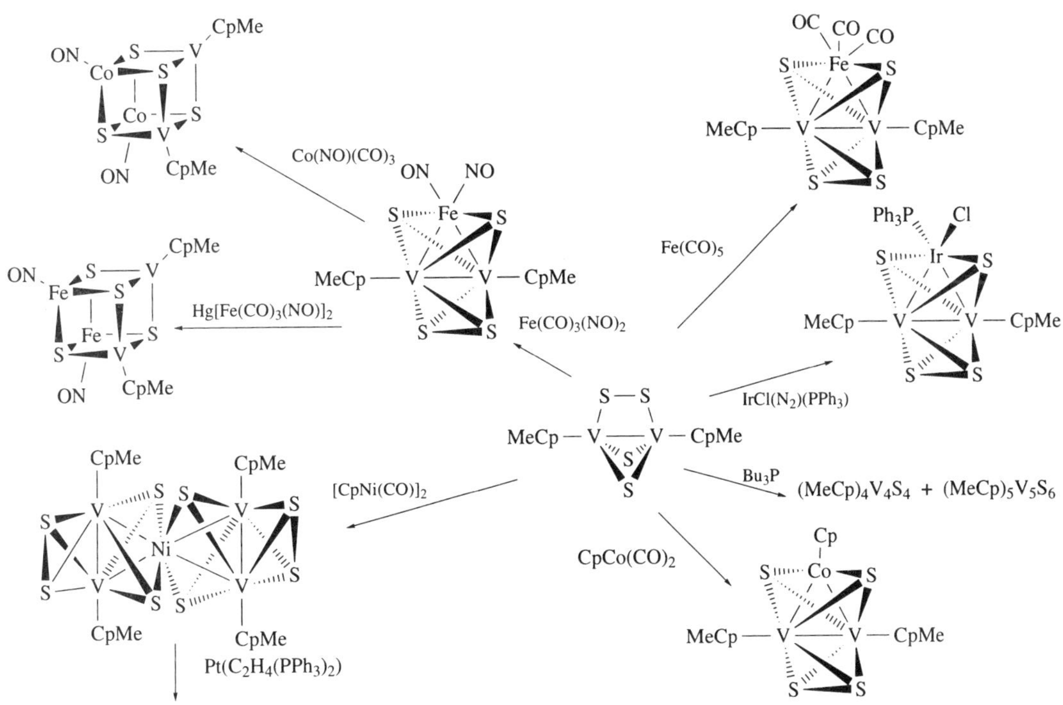

Scheme 37

1.6 BIS(η^5-CYCLOPENTADIENYL) COMPLEXES AND DERIVATIVES

1.6.1 Synthesis and Properties of Vanadium(II) Complexes

The general synthetic methods to ring-substituted vanadocenes involve the *in situ* reduction of vanadium(III) chloride and reaction with the appropriate lithium salt. The most general method employed is based on the reduction of VCl_3 in THF with zinc dust.[224–8] Subsequent reaction with the appropriate salt of the ring-substituted cyclopentadienyl anion yields the substituted species. In the case of decamethylvanadocene, Cp^*_2V, a synthetic procedure involving the use of excess LiCp* and VCl_3 gives a 70% yield of product.[229] A series of 1,1' ring-substituted vanadocenes with the general formula $[C_5H_4EMe_3]_2V$ (E = C, Si, Ge, Sn) have been synthesized from $[C_5H_4EMe_3]Li$ and $V_2Cl_3THF_6[Zn_2Cl_6]$ in the yield range 30–50%.[226] With the exception of the di-*t*-butyl derivatives, these compounds were all obtained as oils.

Decamethylvanadocene is a dark red, air-sensitive material that has been prepared by the two methods outlined above in 65–70% yield.[224,229] The hexane soluble solid can be isolated by sublimation or crystallization and is thermally stable with a melting point of 300 °C. The molecular structure of Cp^*_2V exhibits a sandwich structure with planar Cp* rings (average $V–C_{ring}$ distance of 0.191(5) nm).[229]

Variable and room temperature magnetic susceptibility data show that the complex obeys the Curie–Weiss law with a magnetic moment indicative of three unpaired electrons. Magnetic measurements and EPR data are consistent with a ground-state electronic configuration of $^4A[e_g{}^2a_{1g}{}^1]$ for the 15-electron complex.

From the ultraviolet–visible spectrum, the ligand field parameters Δ_1,Δ_2, and the Racah electron repulsion parameter, B, were derived. The large value of B was interpreted as an indication of increased electron density on the metal for the methylated derivative.[224]

Paramagnetic ^1H and ^{13}C NMR spectra of $[C_5Me_4Et]_2V$ and $[C_5H_4EMe_3]_2V$ (E = C, Si, Ge, Sn) have been analyzed and assigned.[225,226] Based on these assignments, the Bu^t group was shown to be electron releasing while the others were electron withdrawing in the sequence $Me_3Si > Me_3Ge \approx Me_3Sn$.

Syntheses of the deep-green complex, bis(indenyl)vanadium (indenyl = η^5-C_9H_7) have been reported with yields varying from 20 to 85%.[227,228] In general, the synthetic methods are based on the reaction of (indenyl)Li and VCl_3/Zn suspensions in THF. The complex can be crystallized from toluene or sublimed. Bis(indenyl)vanadium has a room temperature magnetic moment of 3.97 μ_B (294 K) consistent with three unpaired electrons. An attempt at structural elucidation by x-ray diffraction was unsuccessful due to the presence of two separate orientations in the crystal lattice. However, the data did show that this compound exhibited a sandwich-type structure.[228]

1.6.2 Coordination Compounds

The coordination compounds of vanadocene and ring-substituted derivatives have been quite extensively explored and are the subject of this section. This section also covers the oxidation chemistry of these compounds and the coordination chemistry of the oxidized products. Oxidative addition reactions as routes to higher valent vanadium complexes will be discussed below.

Unstable complexes are formed between N_2, O_2, H_2, and CH_4 and vanadocene and decamethylvanadocene.[230] The binding of these small molecules to the metallocene units has been investigated at low temperature by infrared spectroscopy and isotopic substitution.

1.6.2.1 Carbonyl complexes

Decamethylvanadocene reacts directly with carbon monoxide leading to $Cp^*_2V(CO)$.[229] This behavior parallels that of vanadocene which reacts rapidly with CO at room temperature.[231] This 17-electron complex is paramagnetic with one unpaired electron ($\mu_{eff} = 1.75\ \mu_B$) and has a C–O stretching frequency of 1845 cm^{-1} (for $Cp_2V(CO)$, $\nu_{CO} = 1881\ cm^{-1}$). An x-ray structural study revealed a bent metallocene arrangement.

The interaction of silica-supported vanadocene with CO was investigated by infrared spectroscopy.[232] When these solids were exposed to CO, infrared spectra indicated the formation of $Cp_2V(CO)$ and/or $[Cp_2V(CO)_2]^+$.

Small hyperfine coupling in the EPR spectra of $Cp_2V^{13}CO$ ($A^C_{iso} = 9.3 \times 10^{-4}\ cm^{-1}$) and $Cp^*_2V(^{13}CO)$ ($A^C_{iso} = 10.7 \times 10^{-4}\ cm^{-1}$) arises from delocalization of the unpaired electron onto the ^{13}CO ligand.[233]

The electronic structure of $Cp_2V(CO)$ has been modeled with a spin-restricted and spin-polarized SCF-Xα-DV calculations.[233] These calculations yield a $^2A_1$ ground state with an a_1 type HOMO. Orbital contour plots show that the odd electron occupies a d-orbital hybrid localized between the Cp rings.

The kinetics and mechanism of CO substitution for Cp^*_2VCO were investigated by monitoring the substitution reactions with dimethyl acetylenedicarboxylate and with ^{13}CO.[233,234] The reactions proceed via an associative (S_N2) mechanism.

In order to elucidate possible intermediate species in the associative reaction of Cp^*_2V with CO, the reaction of CO and (η^5-indenyl)$_2$V was investigated.[235] Under one atmosphere of CO, (η^5-indenyl)$_2$V was converted to the dicarbonyl adduct. To avoid the formally 19-electron configuration, one of the indenyl rings slips to an η^3-mode, a feature confirmed by the molecular structure which is indicative of (η^3-indenyl)(η^5-indenyl)$V(CO)_2$ (**2**).

The carbon monoxide adduct of vanadocene, $Cp_2V(CO)$, reacts with N_2O, slowly but quantitatively at room temperature, to yield $[[Cp_2V]_2(CO_3)]_n$ where $n = 1$ or 2.[236] The steel-blue, extremely air-sensitive complex is formulated mainly as a monomer ($n = 1$), based on the measured magnetic moment of 2.62 μ_B. However, the infrared spectrum exhibited four absorption bands which would indicate the presence of some dimeric, $[[Cp_2V]_2(CO_3)]_2$, species. A similar but nonquantitative reaction was also observed with O_2.

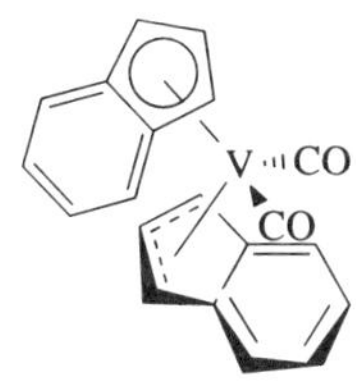

(2)

The oxidation of $Cp^*_2V(CO)$ with N_2O proceeds along a completely different course than that of the cyclopentadienyl analogue.[237] The major product of this reaction was a green THF-soluble species, $Cp^*_2V(CO_3)$, which could be isolated in a 70% yield. The infrared spectrum ($\nu_{CO_3} = 1660\ cm^{-1}$) indicates a complex with CO_3^{2-} acting as a bidentate ligand (**3**).

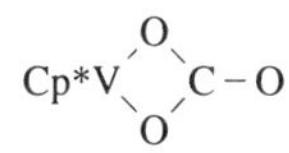

(3)

Consistent with this proposal, the EPR spectrum exhibited a low value for the isotropic coupling with ^{51}V ($A_{iso} = 32.05$ G) indicating that the single unpaired electron from the vanadium(IV) center is delocalized into the π-system of the carbonate ligand. The other product of the reaction (17%) was a purple polymeric material proposed to be $[\{[Cp^*V(\mu\text{-}O)]_4(OH)_2\}(\mu\text{-}O)]_n$. Reaction of O_2 with $Cp^*_2V(CO)$ also gave a mixture of these two compounds.

1.6.2.2 Reactions of vanadocenes with unsaturated molecules

Complexes of vanadocene with activated alkenes (acrylonitrile, acrolein, crotonaldehyde, and methyl vinyl ketone) have been investigated.[238] In all cases a brown air-sensitive material precipitates from an ether solution of Cp_2V upon exposure to the alkene. The infrared spectra of these products does not exhibit a C=C stretch which was interpreted as evidence for formation of a metallacyclopropane derivative. EPR spectra agree with this assignment by showing a decrease from three to one unpaired electrons upon coordination.

Reaction of Cp^*_2V with isocyanides, RNC (R = C_6H_{11}, Bu^t), resulted in the cleavage of the R–N bond and formation of $Cp^*_2V(CN)(CNR)$ ($\nu_{C\text{-}N}$: R = C_6H_{11}, 2050, 2070 cm^{-1}; R = Bu^t, 2050, 2070 cm^{-1}).[229] The structure of the cyclohexyl complex was determined.

The reaction between vanadocene and nitrilium salts ($RC{\equiv}NMe^+$; R = C_6H_5, Me) produces spectroscopic features characteristic of an iminoacyl ligand.[239] From the reaction of $C_6H_5C{\equiv}NMe^+BF_4^-$ a maroon complex was isolated and shown to be $[Cp_2V(\eta^2\text{-}C(Ph)NCH_3]BF_4$ by x-ray analysis. The bent metallocene structure is depicted in Structure (**4**) ($V\text{–}C_1 = 0.2054(4)$ nm, V–N = 0.2030(4) nm, C–N = 0.1245(7) nm, $C_1\text{–}V\text{–}N = 35.2(2)°$).

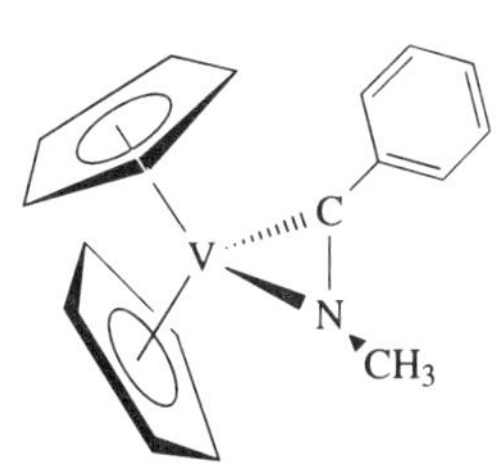

(4)

The direct reaction of Cp_2V with $C_6H_5N{=}NC_6H_5$ yields the azobenzene adduct of vanadocene as a black solid in 85% yield.[240] Hydrolysis with HCl or H_2O yields Cp_2VCl_2 and hydrazobenzene.

Vanadocene reacts with NO in a 1:1 stoichiometry to yield a mixture of products from disproportionation of $Cp_2V(NO)$.[241] The reactions between $Cp_2V(CO)$ and NO or NOCl yield complexes that contain the NCO^- ligand.[241] The reaction with NO proceeds with a 2:1 stoichiometry to give an inseparable mixture of polymeric oxide and a complex containing the NCO ligand. In contrast, the reaction with NOCl gave a mixture of products with at least one containing the cyanate ligand. $Cp_2V(CO)_2^+$ does not react with NO.

Paraformaldehyde reacts readily with vanadocene to yield a deep green complex containing an η^2 coordinated formaldahyde, $Cp_2V(CH_2O)$, and methyl formate.[242,243] The crystal structure of this compound was determined by x-ray diffraction. Examination of the structure reveals a lengthened C–O bond (C–O = 0.135 nm versus 0.1225 nm in free formaldehyde) and a V–C bond distance (0.209 nm) indicative of some double-bond character. The reactivity of this complex is summarized in Scheme 38. Lewis acids (BF_3, $TiCl_4$) and alkylating reagents (Et_3OBF_4, $MeSO_3F$) react with the formaldehyde complex to yield unstable oxymethylene ligands. Stable, isolable oxymethylene complexes, $Cp_2V(Cl)(CH_2OC(O)R)$, were synthesized from acetyl and benzoyl chlorides.

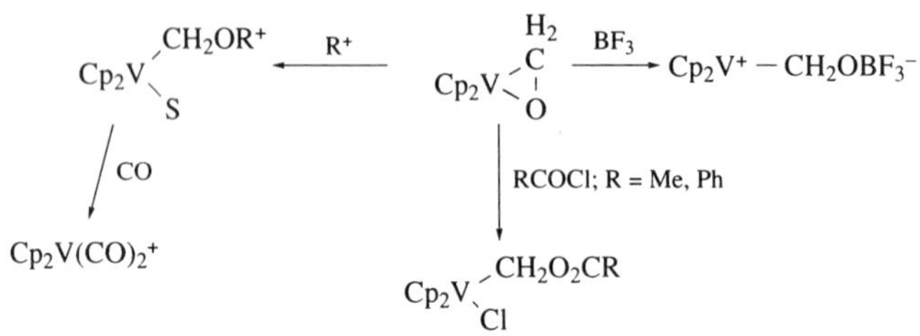

Scheme 38

Thiobenzophenone reacts rapidly with vanadocene to yield a red, air-sensitive, paramagnetic (μ_{eff}(292 K) = 1.74 μ_B) complex.[244] Characterized by x-ray diffraction as η^2-coordinated thiobenzophenone (Equation (16)) (V–C = 0.2306(4) nm, V–S = 0.2374(2) nm, C–S = 0.1762(4) nm, C_1–V–N = 35.2(2)°), this complex decomposes at 100 °C to yield diphenylfulvene and $[CpVS]_n$.

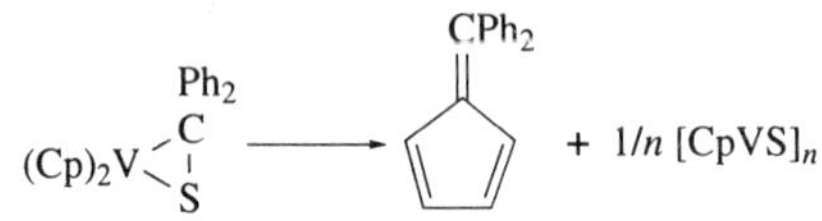

(16)

1.6.2.3 Heteroallenes

The reactivity of vanadocene and ring-substituted derivatives with a variety of heteroallenes exemplifies the "carbenoid" nature of the Cp_2V unit.

When vanadocene is reacted with thioketenes $R_2C{=}C{=}S$ (R = TMS, CF_3, Bu^t, $C(Me)_2CH_2CH_2CH_2C(Me)_2$) a series of complexes with η^2-C,S bonded ketenes are formed.[245,246] These compounds have been characterized by single crystal x-ray diffraction (**5**). The α angles fall in the range of 132.5–134.3°; the S–C_1, V–C_1, and V–S distances fall in the ranges of 0.168–0.173 nm, 0.208–0.217 nm, and 0.237–0.241 nm, respectively.

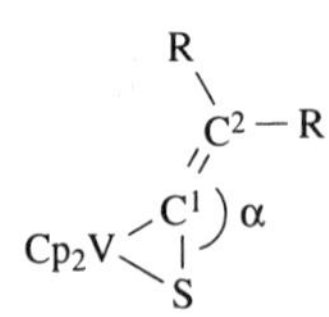

(5)

Similarly, reaction of vanadocene with ketene imines $R^1_2C{=}C{=}NR^2$ (R^1 = Me, Ph; R^2 = Ph) leads to adducts containing an η^2-C,N-bonded ketene imine.[247] These compounds can be isolated as black crystals in 87–95% yield. The structures of these complexes have been confirmed by infrared ($\nu_{C=C=N}$ = 1670, 1680, 1685 cm^{-1}) and x-ray diffraction.

Complexes of vanadocene containing an η^2-*N,N*-diazo unit bound to the Cp_2V moiety can be synthesized by direct reaction of the metallocene with diethyl diazoketomalonate (dedm) and diphenyldiazomethane (dpdm).[248] Confirmation of the presence of the intact diazo moiety is indicated by the reaction of these adducts with O_2 which releases free dedm and dpdm.

Cp^*_2V reacts with phenyl isothiocyanate to give a bright green η^2-C,S bound isothiocyanate complex, $Cp^*_2V(SCNPh)$.[249] The nature of this complex is confirmed through infrared spectroscopy ($\nu_{C=N}$ = 1610 cm^{-1}), magnetic susceptibility (1.77 μ_B) and single-crystal x-ray diffraction. This complex is transformed in refluxing *n*-heptane through the migration of the metal bonded carbon to one of the Cp* rings to give $Cp^*V[PhN\text{-}C(S)\text{-}Cp^*]_2$ (Equation (17)).

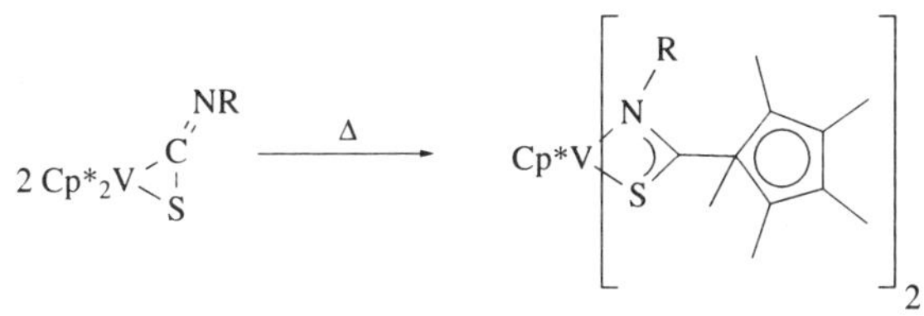

(17)

Reaction of decamethylvanadocene with carbonyl sulfide proceeds through a labile crystalline complex $Cp^*_2V(COS)$ ($\nu_{CO} = 1625\ cm^{-1}$) to give $Cp^*_2V(S_2)$ and $Cp^*_2V(CO)$, the products of reductive disproportionation.[249] This contrasts with the behavior of vanadocene which reacts with COS according to Equation (18), to form a mixture of dinuclear dithiocarbonato complex and the carbonyl adduct of vanadocene.[250]

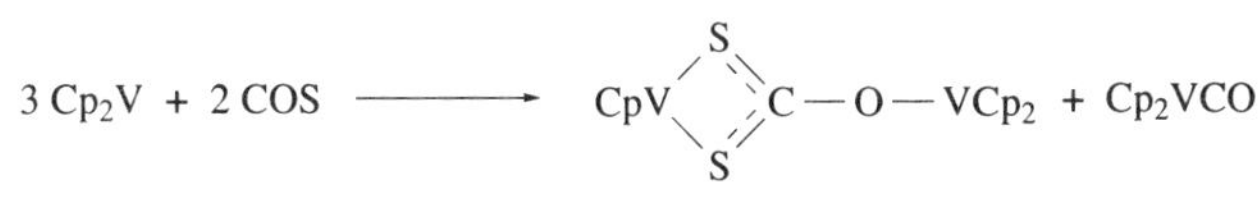

(18)

1.6.3 Bis(η^5-cyclopentadienyl) Vanadium (III), (IV), and (V)

1.6.3.1 Vanadocene halides and ring-substituted derivatives

The two series of complexes Cp_2VX and Cp_2VX_2 (X = halide) are important precursors for a variety of other species. These complexes, like their $Cp_2Ti^{IV}X_2$ analogues, have also been investigated as potential Ziegler–Natta alkene polymerization catalysts. In addition, several recent investigations indicate that vanadocene dihalides may be antitumor agents.

As described in *COMC-I* (Vol. 3, Chapter 24, Section 6.8), the general synthetic route to Cp_2VX (X = Cl, Br, I) is the oxidation of Cp_2V with RX (R = alkyl; X = Cl, Br, I), CH_2X_2 (X = Cl, Br), or with I_2. Similarly, one-electron oxidation of decamethylvanadocene with X_2 (X = Cl and I)[229] or HCl[251] yields the corresponding Cp^*_2VX complexes. The complex $(\eta^5\text{-}C_9H_7)_2VI$ has been prepared by oxidation of $(\eta^5\text{-}C_9H_7)_2V$ with I_2. The ring-substituted vanadocene, $[\eta^5\text{-}C_5Me_4Et]_2VBr$, was synthesized by the reaction of $VBr_3(THF)_3$ with $Li[C_5Me_4(Et)]$.[225]

For the formation of Cp_2VCl and Cp^*_2VCl, the reaction between the appropriate metallocene and $PbCl_2$ has been found to be a convenient alternative to the above procedures.[252] In this case, the by-product of the reaction is elemental lead. It is not possible to generate vanadium(IV) species by this method.

Reaction of VCl_4 with an appropriate cyclopentadienyl anion source provides a general synthetic route to Cp_2VX_2 (X = Cl, Br). A series of 1,1'-ring-substituted vanadocene dihalides, $(R_3EC_5H_4)_2VX_2$ (E = C, Si, Ge; X = Cl, Br, I), have been synthesized using analogous procedures.[253] Ring-bridged vanadocene dichlorides $[Me_2E(\eta^5\text{-}C_5H_4)_2]VCl_2$ (E = Si, Ge) were isolated as green, air-stable solids from the reaction of the dilithium salt of $Me_2E(C_5H_4)_2$ and VCl_4.[254,255] Synthesis of Cp_2VBr_2 has been reported and involves the reaction of Cp_2VCl_2 with BBr_3 or Br_2.[256] Solution EPR spectra for Cp_2VCl_2, Cp_2VClBr, and Cp_2VBr_2 show that A_{iso} (^{51}V) decreases in the order $Cp_2VCl_2 > Cp_2VClBr > Cp_2VBr_2$ while the average g values increase in the opposite order. These observations have been rationalized in terms of greater spin delocalization onto the halide ligand possibly through a d_π–p_π interaction.

Photoelectron spectra have been measured for a series of Cp_2VX species (X = Cl, Br, I, CH_2TMS, CH_2Bu^t, C_6F_5, Ph, $o\text{-}MeC_6H_4$, $m\text{-}MeC_6H_4$, $2,6\text{-}Me_2C_6H_3$).[257] π-Bonding interactions were found to decrease in the order halide > alkyl > aryl.

The electrochemical behavior of Cp_2VCl, Cp_2VCl_2, $Cp^*_2VCl_2$ has been examined.[258] Reduction of Cp_2VCl (−1.63 V) and Cp_2VCl_2 (−0.56 V) is followed by rapid loss of Cl^- and conversion to Cp_2V and Cp_2VCl, respectively. Oxidation of Cp_2VCl (−0.08 V) gives the corresponding cation which rapidly converts to Cp_2VCl_2 and unstable Cp_2V^{2+}. Slurries of aluminum, prepared by condensation of metal vapor with organic solvent at −196 °C, have been used as reducing agents for the conversion of Cp_2VX (X = Cl, Me) to vanadocene.[259]

A series of high-valent vanadocene complexes with fluorinated group 15 ligands have been synthesized and characterized (Scheme 39 and Table 23).[260–3] The reaction of AsF_5 with Cp_2VCl_2 in liquid SO_2 leads to $[Cp_2VCl_2]AsF_6$ and $[Cp_2V(AsF_6)_2]AsF_6$.[260] These black microcrystalline solids can be crystallized from SO_2 in high yield (>75%). A similar procedure employing $AgSbF_6$ yields $Cp_2V(SbF_6)_2$ which can be recrystallized from SO_2/SO_2ClF. The structures of both $Cp_2V(SbF_6)_2$[262] and $[Cp_2VCl_2]AsF_6$[263] have been determined by x-ray crystallography.

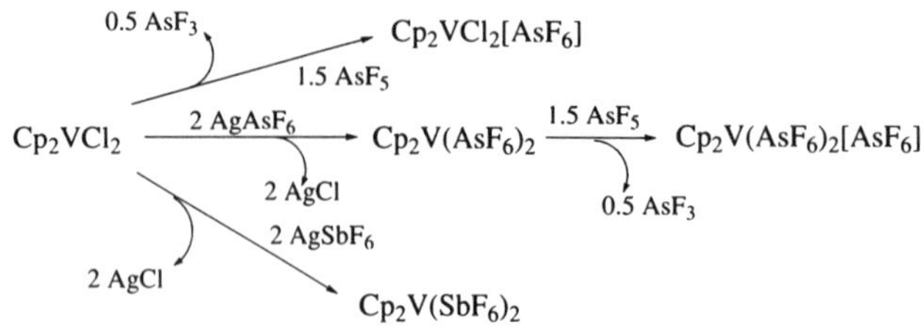

Scheme 39

Table 23 Spectroscopic data for $(\eta^5\text{-}C_5H_5)_2V^{IV}$ and $(\eta^5\text{-}C_5H_5)_2V^{V}$ complexes.

Complex	*Infrared* (cm^{-1})	*^{19}F NMR (SO_2, relative $CFCl_3$)* (ppm)
$[Cp_2VCl_2]AsF_6$	$\nu_{AsF_6^-} = 700, 398$	−42.15, −53.25, −64.25, −75.35, J = 937 H
	$\nu_{VCl} = 320, 268, 225$	
$[Cp_2V(AsF_6)_2]AsF_6$	$\nu_{AsF_6^-} = 700, 395$	−47.55 broad singlet
$Cp_2V(SbF_6)_2$	$\nu_{SbF_6^-} = 665, 660, 640$	

1.6.3.2 Coordination chemistry

In contrast to vanadocene, the cyclic voltammogram of Cp^*_2V in acetonitrile exhibits no reversible one-electron waves.[224] In acetone, ether, and acetonitrile, both vanadocene and decamethyl vanadocene are readily oxidized by $[Cp_2Fe]PF_6$. An air-sensitive, green crystalline complex, $[Cp^*_2V(NCCH_3)]PF_6$, was isolated from the reaction in acetonitrile.[224] Reaction of this complex with carbon monoxide resulted in the formation of a yellow, 18-electron, diamagnetic complex, $[Cp^*_2V(CO)_2]PF_6$. The analogous complex, $[Cp_2V(CO)_2]PF_6$, is isolated when the oxidation is performed in the presence of CO.[264]

Bis(indenyl)vanadium(II) can be oxidized with I_2 to yield crystalline $(\eta^5\text{-}C_9H_7)_2VI$, which has two unpaired electrons.[228] The oxidation can also be effected using $[Cp_2Fe]PF_6$.[264] Performing the oxidation in the presence of CO and a suitable anion (BPh_4^-, PF_6^-) yields the dicarbonyl cation, $(\eta^5\text{-}C_9H_7)_2V(CO)_2^+$. X-ray crystallography reveals that for both cases there is slippage of the η^5-indenyl toward an η^3-coordination mode.[228] The cyclic voltammogram of $[(\eta^5\text{-}C_9H_7)_2V(CO)_2]PF_6$ shows two reversible one-electron reduction waves.[264]

In nonpolar solvents, Cp^*_2VI reacts readily with CO to yield the yellow, diamagnetic complex, $Cp^*_2V(CO)I$ ($\nu_{CO} = 1890$ cm^{-1}).[229] Gentle heating of solutions of this complex *in vacuo* regenerates the iodide complex in good yield. If the reaction with CO is carried out in a polar solvent, $Cp^*_2V(CO)_2^+$ is obtained. The analogous reaction of Cp^*_2VX (X = Cl or I) with *t*-butyl isocyanide results in the formation of $[Cp^*_2V(CNBu^t)_2]^+$.[229]

Table 24 CO stretching frequencies for various carbonyl complexes (see Section 1.6.3.2).

Complex	ν_{CO} (cm^{-1})	*Ref.*
$(Cp^*)_2VCO$	1845	229
$(\eta^5\text{-}C_9H_7)_2V(CO)_2$	1978, 1926	235
$(Cp^*)_2V(CO)I$	1890	229
$[Cp_2V(CO)_2]PF_6$	1990, 1954	229
$[(\eta^5\text{-}C_9H_7)_2V(CO)_2]PF_6$	2037, 1983	264
$[(\eta^5\text{-}C_9H_7)_2V(CO)_2]BPh_4$	2041, 2001	228

The reactions of vanadocene halides, Cp_2VX and Cp_2VX_2, and NO have been investigated. The reaction between Cp_2VI and NO produced two products formulated as $Cp_2VI(NO)$.[241] One of these complexes was a brown insoluble polymer which contains ionic iodide. The other product is green monomeric $Cp_2VI(NO)$. Solutions of $Cp_2VI(NO)$ (THF, room temperature) slowly convert to diamagnetic $[[CpVI]_2[CpV(NO)]_2(\mu\text{-}O)_4]$. The structure of this compound as determined by x-ray crystallography, shows an eight-membered alternating ring of vanadium and oxygen atoms.[241]

Reaction between Cp_2VBr and NO gave a complicated mixture of insoluble products with one soluble green monomeric species $Cp_2VBr(NO)$.[241] In addition to insoluble nitrosyls, the products of the reaction of Cp_2VCl and NO were N_2O, and $Cp_3V_2Cl_2O_2$ which could also be obtained from Cp_2VCl and O_2 (see Section 1.1.6.3.6).[236,241]

The reactions of NO with decamethylcyclopentadienyl analogues were investigated and found to yield a series of thermally unstable (dec. > −30 °C) nitrosyl complexes.[265] The reaction of Cp^*_2VI with NO gave the 19-electron complex $Cp^*_2VI(NO)$ (ν_{NO} 1710 cm^{-1}). In contrast, Cp^*_2VBr reacts with NO to give the ionic complex $[Cp^*_2VBr(NO)][Cp^*_2VBr_2(NO)]$. The cation is diamagnetic with ν_{NO} at 1755 cm^{-1} while the anion is paramagnetic with ν_{NO} at 1715 cm^{-1} and a magnetic moment of 2.55 μ_B. This complex was also characterized by x-ray diffraction.

1.6.3.3 Synthesis and properties of pseudohalide derivatives

Reaction of Cp_2V with HN_3 and NCN_3 results in oxidation to vanadium(IV) derivatives, $Cp_2V(N_3)_2$ and $Cp_2V(CN)(N_3)$, respectively.[266] These brown, insoluble, explosive species have been identified by infrared spectroscopy. Similarly, reaction of Cp_2V with dicyanogen, $(CN)_2$, and cyanogen halides, XCN (X = Cl, Br, or I), leads to the formation of $Cp_2V(X)CN$ (X = CN, Cl, Br, or I).[267]

Vanadocene reacts with isocyanates and thioisocyanates of the formulae R_3ENCX and $R_2EN(CX)_2$ (E = Sn, Si; X = O, S).[268] The reactions proceed by oxidative addition of the pseudohalide ligand to the vanadium center to yield violet Cp_2VNCX complexes and $(R_3E)_2$ (Scheme 40). The same complexes can be prepared in high yield (80%) by direct reaction of Cp_2VCl with R_3ENCX (Scheme 40). Vanadium (IV) isocyanates and isothiocyanates can be prepared in an analogous fashion (Equation (19)).[268]

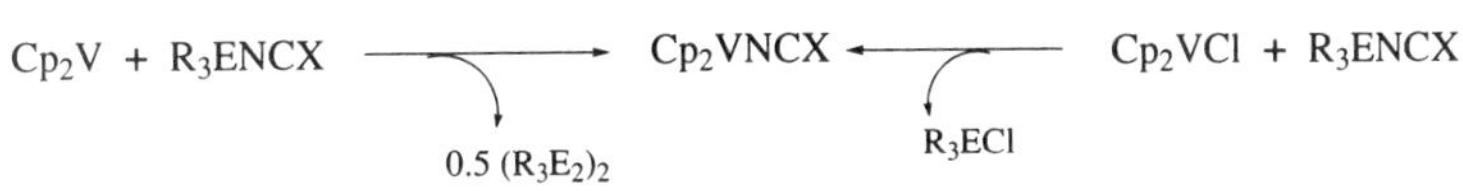

Scheme 40

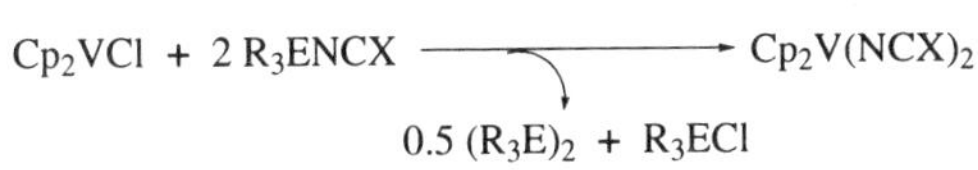

(19a)

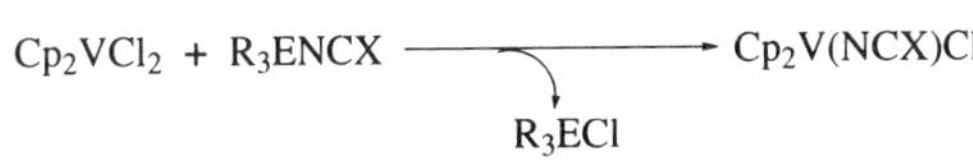

(19b)

1.6.3.4 Antitumor activity of vanadocene halides and pseudohalides

As part of a general investigation of the antitumor activity of metallocene dihalides and bis(pseudohalides), there has been quite an extensive exploration of the *in vitro* and *in vivo* effects of vanadocene dichloride as well as the aqueous coordination chemistry of this complex. The cellular distribution of vanadium and its physiological effects have been investigated after treatment with Cp_2VCl_2.[269–73] The accumulation of vanadium in the nucleus of the cell has been taken as an indication of an interaction between the metal-containing species and nucleic acids, especially DNA.[272,273] The biodistribution and toxicological effects of vanadium on mice, after injection with Cp_2VCl_2, have been investigated.[274–6] The *in vitro* antiviral activity of Cp_2VCl_2 was found to be marginal for most DNA and RNA viruses; however, the complex was found to be weakly inhibitory towards HIV reverse transcriptase.[277] The coordination chemistry of adriamycin, an antibiotic agent, with vanadocene dichloride has been investigated in order to synthesize bifunctional antitumor compounds.[278]

In order to probe the origin of the carcinostatic activity of Cp_2MCl_2 (M = Ti, V, Nb, Mo) agents, the aqueous coordination chemistry of these species was investigated.[274] In aqueous solution, the chloride ions of Cp_2VCl_2 are rapidly substituted by water molecules. The V–Cp bond was found to be stable at physiological pH.[279] Titrimetric studies were used to measure the acidity of the CpM^{2+}-bound water molecules (pK_a = 4.73(3) and 5.15(13)). In strong acid solution, $[Cp_2V(H_2O)_2]^{2+}$ is the only vanadium species present.

Significant reduction of the ^{31}P nuclear relaxation times for nucleotides in the presence of vanadocene dichloride indicates that the paramagnetic complex interacts selectively with the phosphate of the nucleotide in aqueous solution.[280] An x-ray structural analysis of a model complex, $[Cp_2V(OH_2)_2][O_2P(OPh)_2]$, exhibits a pseudotetrahedral dication, $Cp_2V(OH_2)_2^{2+}$, connected to the diphenyl phosphate dianion by hydrogen bonds.[280] Inductively coupled plasma spectroscopy has been used to measure the binding ability of several metallocene dichlorides, Cp_2MCl_2 (M = Ti, V, Nb, Mo) with DNA and vanadium was found to be the only metal that did not make a stable DNA–metal adduct.[281]

1.6.3.5 Carboxylato and N,N-dialkyl carbamato complexes

Reaction of Cp_2V with $Ti(O_2CNEt_2)_4$ yields a dimeric vanadium(III) *N,N*-diethyl carbamato complex, $Cp_2V_2(O_2CNEt_2)_4$, by simultaneous electron- and ligand-transfer reaction.[282] The structure of this complex has been determined by x-ray crystallography. In contrast, Cp^*_2V reacts with $Ti(O_2CNEt_2)_4$ to yield a monomeric (cryoscopic measurement) species $Cp^*_2V(O_2CNEt_2)$ as a blue, oxygen-sensitive solid.[283]

1.6.3.6 Oxo, alkoxo, and aryloxo complexes

Reaction of molecular oxygen with Cp_2VCl yielded a green solid which was characterized as $Cp_3V_2Cl_2O_2$ based on mass spectral and microanalytical results.[236] Structure (**6**) is suggested for this species. In contrast, reaction of O_2 and Cp_2VI leads to the formation of an intractable polymeric material.

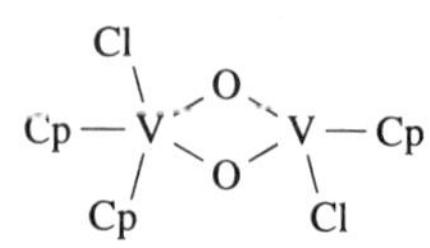

(**6**)

Oxidation of Cp^*_2V with molecular oxygen in a 1:2 ratio leads to the unusual complex $[\mu\text{-}\eta^3\text{-}Cp^*O_3)V(O)]_2$.[284] The nature of this hexane soluble, ruby-red, air-stable product was established by spectroscopy ($\nu_{V=O}$ = 935 cm^{-1}, ^{51}V NMR (C_6H_{14}) = −672.5 ppm) and x-ray diffraction (**7**).

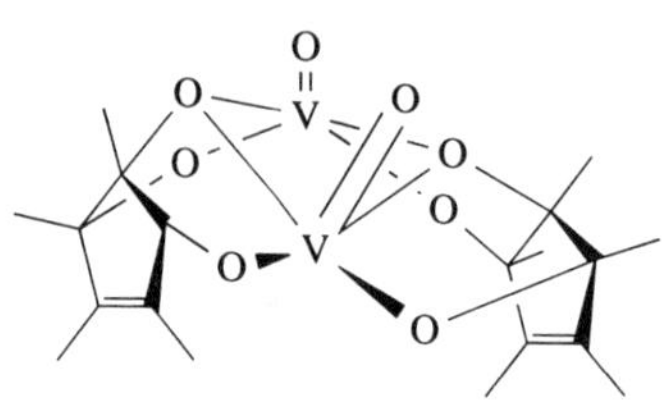

(**7**)

A series of organovanadium oxo clusters can be synthesized from vanadocene and N_2O or amine *N*-oxides (Scheme 41 and Structure (**8**)).[285–7] X-ray crystallography gave definitive assignment of the structure for these clusters. $[CpV]_5(\mu_3\text{-}O)_6$, the product of vanadocene and N_2O, is a trigonal bipyramid of octahedrally coordinated vanadium atoms with each triangular face capped by an oxygen (**8**).

Substituting $O\text{-}N(CH_3)_3$ in the reaction causes deposition a black solid which by mass spectrum, elemental analysis, and the single ^1H NMR peak (−171.4 ppm) was formulated as $[CpV]_6(\mu_3\text{-}O)_8$. The remaining brown-black solution yields crystals of $[(Cp)_5V_6(\mu_3\text{-}O)_8]_2[(\mu\text{-}O)_2VCp(NMe_3)_2]$ upon cooling. When the oxidation is carried out in THF the product is a black diamagnetic crystalline solid, $[\{Cp_5V_6(\mu_3\text{-}O)_8\}_2\{\mu\text{-}[(\mu_2\text{-}O)_8V_4Cp_4\}]$. Longer reaction times (72 h) lead to the formation of $Cp_5(O)V_6(\mu_3\text{-}O)_8$ the structure of which is related to that of $[Cp_5V_6(\mu_3\text{-}O)_8]_2(\mu\text{-}O)$.[287]

Oxidation of Cp_2V with $(C_2H_5)_3N\text{-}O$ in a toluene solution yields a mixture of products from which $Cp_4V_4(\mu_3\text{-}O)_4$ can be isolated by vacuum sublimation at 275 °C. This species was identified by mass spectrum, EPR, and magnetic susceptibility. Crystals of $\{Cp_5V_6(\mu_3\text{-}O)_8\}_2(\mu\text{-}O)$ can be obtained from the toluene solution.

Green $Cp_3V_2Cl_2O_2$ was the isolated product from the reaction of Cp_2VCl and NO.[241] The same product had been previously isolated from the reaction of Cp_2VCl and O_2.[236]

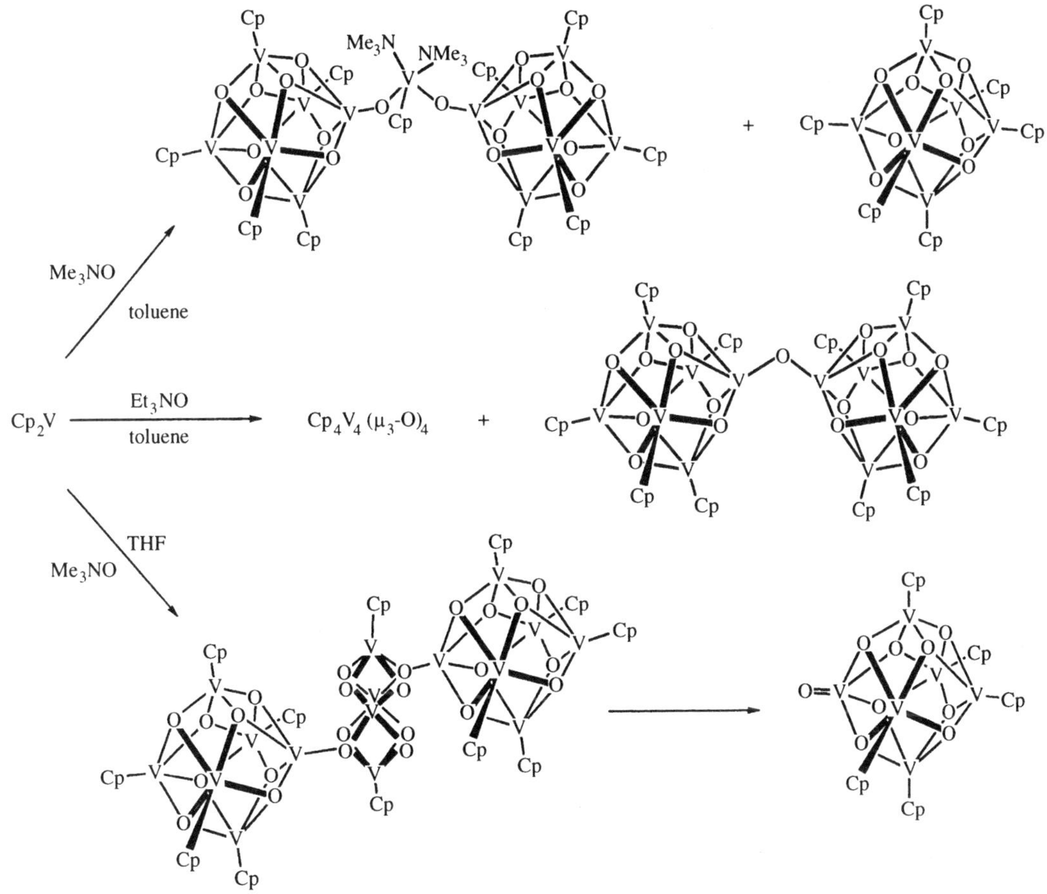

Scheme 41

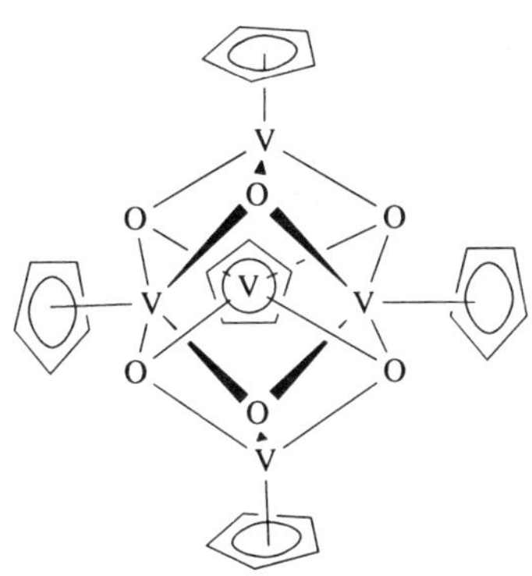

(8)

Using Ag_2O or Ag_2CO_3 as the oxidant for an equimolar mixture of Cp_2VCl_2 and $CpVCl_3$ the unsymmetrical dimer $[Cp_2VCl](\mu\text{-}O)[CpVCl_2]$ can be obtained in 27% yield.[288] The product consists of a linear bridged V–O–V (179.8°) with the $Cp_2(Cl)V\text{–}O$ distance of 0.1993(5) nm and a $Cp(Cl_2)V\text{–}O$ distance of 0.1653(5) nm.

Alkoxo derivatives of vanadium(IV) are poorly documented and conventional methods of obtaining the alkoxides from Cp_2VX_2 (X = halide) have been unsuccessful. Reaction of Cp_2VCl_2 with oxiranes yield $Cp_2V(OCH_2CH_2Cl)_2$ and $Cp_2V(OCH(Me)CH_2Cl)_2$.[289] EPR spectra for these compounds were obtained in dichloromethane or in pentane and show eight lines due to ^{51}V hyperfine coupling as expected for a monomeric species. Examination of the regiochemistry for the ring-opening reaction of Cp_2VCl_2 with propylene oxide reveals the exclusive formation of $Cp_2V(OCH(Me)CH_2Cl)_2$.[290]

1.6.3.7 Thiolato and sulfide complexes

Polysulfide complexes have been synthesized from metallocene and metallocene halides with elemental sulfur[291,292] and with sulfide anions.[293] Cp^*_2V and Cp^*_2VCl react with elemental sulfur, in THF to form $Cp^*_2V(\eta^2\text{-}S_2)$ (S–S = 0.2028 nm).[291,292] Alkylation of the disulfido with $MeSO_3F$ gave $[Cp^*_2V(S_2Me)][SO_3F]$. If the reaction of Cp^*_2VCl with sulfur is carried out in refluxing benzene a diamagnetic dimeric vanadium(IV) complex is formed, $[Cp^*VCl(\mu\text{-}S)]_2$ ($\mu_{eff} = 1.89\ \mu_B$; V–S 0.216(2) nm, 0.223(1) nm; V···V 0.2829(3) nm).[292]

The disulfido complex, $Cp^*_2V(\eta^2\text{-}S_2)$, can also be prepared from the reaction of $Cp^*_2VCl_2$ and NH_4S_5 in a water–acetone mixture in 50% yield ($\nu_{S-S} = 552\ cm^{-1}$, S–S 0.2053(3) nm).[293] In contrast to this result, the bridged vanadocene complex $[Me_2E(\eta^5\text{-}C_5H_4)_2]VCl_2$ (E = Si, Ge) reacts with NH_4S_5 to yield a pentasulfide chelate complex, $[Me_2E(\eta^5\text{-}C_5H_4)_2]V(S_5)$.[255]

Thiolato complexes, $Cp_2V(SPh)_2$[294] and $Cp_2V(SMe)_2$[295], can be prepared from Cp_2VCl_2 and the appropriate thiolate salt, NaSR (R = Ph, Me). The latter complex can be isolated in 61% as dark blue-black crystals. Alternatively, one-electron oxidation of decamethylvanadocene with *p*-Tol disulfide gave $(C_5Me_5)_2V(SC_6H_4Me)$ in 72% yield.[229] This complex reacts with CO to produce $Cp^*_2V(SC_6H_4Me)(CO)$ ($\nu_{CO} = 1890\ cm^{-1}$). Reaction with Bu^tNC promotes ionization of this complex to $[Cp^*_2V(CNBu^t)_2](SC_6H_4Me)$. Mass spectral characterization of several metallocene dithiolene complexes has been reported.[296]

Vanadocene reacts with $[Cp_2Co][Hg[SGe(C_6F_5)_3][Ge(C_6F_5)_3]_2]$ to yield a vanadium(III) thiolate complex, $Cp_2V(SGe(C_6F_5)_3)$.[297] The thermal decomposition of the selenium analogue $Cp_2V(SeGeEt_3)$ proceeds to yield $HGeEt_3$, Cp_2V, and VSe_2.[298]

Spectral studies and preparation of several chelating dithiocarbamate, dithiophosphonate, and dithioarsenate complexes of the general formula $[Cp_2VL_2]^+$ (L_2 = one bidentate ligand) have been described (Scheme 42).[299,300] Trends in the EPR spectral parameters correlate increases in the L–V–L angle and the π-acceptor character of L with greater spin delocalization onto the ligand while a relatively more π-donating ligand promotes localization of the unpaired spin on the vanadium center.

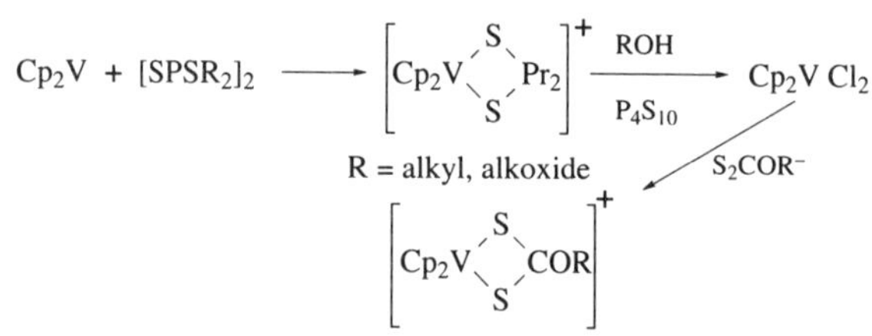

Scheme 42

1.6.3.8 Amido and imido complexes

The syntheses of $Cp^*_2V{=}NR$ (R = C_6H_5, 2,6-$(Me)_2C_6H_3$, 2-$(Ph)C_6H_4$) have been achieved via reaction of Cp^*_2V with organic azides.[301,302] X-ray structures of the phenyl and 2,6-dimethylphenyl complexes reveal a linear V–N–C arrangement with a short V–N distance (0.1707(6) nm, 0.1730(5) nm).[301,302] An assignment of the V=N stretching frequency was made by preparation of the labeled complex $Cp^*_2V{=}^{15}NPh$.[303] These formally 19-electron complexes exhibit μ_{eff} values of 1.56–1.79 μ_B, consistent with the presence of one unpaired electron.[302] Room temperature, solution EPR spectra display $g = 1.986–1.988$, with $A_V = (4.6–9.8) \times 10^{-4}\ cm^{-1}$ and $A_N = (4.6–4.9) \times 10^{-4}\ cm^{-1}$.[302] Cyclic voltammograms of these compounds in CH_2Cl_2 show quasireversible oxidations at potentials approximately 750 mV more negative than ferrocene.[304]

1.6.3.9 σ-Bonded alkyl, dialkyl, allyl, acetylides, hydrides, alkylidenes, alkylidynes

Consistent with the previous report in *COMC-I* (Volume 3, Chapter 24, Section 6.4.1) most of the compounds of this class are trivalent complexes of Cp_2VR or ring-substituted derivatives. In addition, the synthetic methods employed in the synthesis of these complexes are analogous to the previous report.

In general, the reactions of organic halides with vanadocene proceed according to Scheme 43. The reaction of Cp_2V with benzyl bromide results in the formation of bibenzyl and Cp_2VBr.[305] Primary, secondary, and tertiary aliphatic halides react with Cp_2V by oxidative addition to give $Cp_2V(R)X$.[306] The primary product is stable if the alkyl group is methyl but in the cases where the alkyl group has a β proton, alkene and alkane are produced together with Cp_2VX. In some cases significant amounts of dihydrogen are also produced.

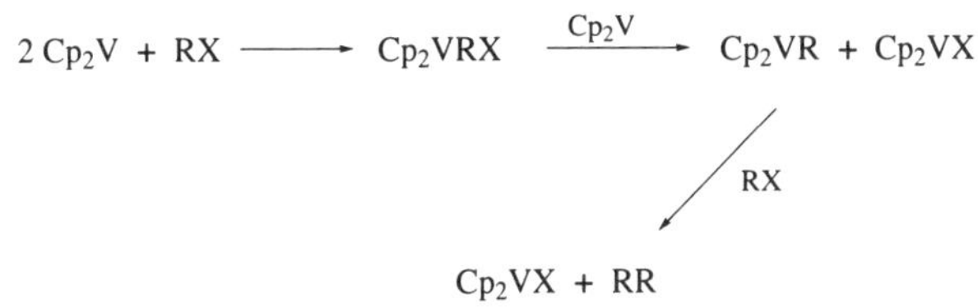

Scheme 43

An alkynide complex, $Cp_2VC{\equiv}CBu^t$(ν_{CC} = 2075 cm^{-1}), was isolated as a black paramagnetic complex (μ_{eff} = 2.89 μ_B) from the reaction of Cp_2VCl with $LiC{\equiv}CBu^t$.[307] Crystallographic characterization of this complex confirms the monomeric nature of the product with a bent-metallocene configuration and a $V{-}C_{alkynide}$ distance of 0.2075(5) nm.

Vanadocene allyl, $Cp_2V(\eta^1\text{-}C_3H_5)$, reacts with strong π-acceptor ligands such as CO, CO_2 and isonitriles (Scheme 44).[308] Reaction with CO yields $Cp[\eta^4\text{-}Cp(C_3H_5)]V(CO)_2$, the product of allyl migration to a Cp ring. A carboxylato complex, $Cp_2V(O_2CC_3H_5)$, was formed when the allyl complex was reacted with CO_2. Insertion of an isocyanide moiety to yield stable complexes, $Cp_2V[\eta^1\text{-}C(R){=}N\text{-}2,6\text{-}(Me)_2C_6H_3]$, was observed when Cp_2VR (R = Me, Ph, $\eta^1\text{-}C_3H_5$) were reacted with 2,6-xylylisocyanide.[308]

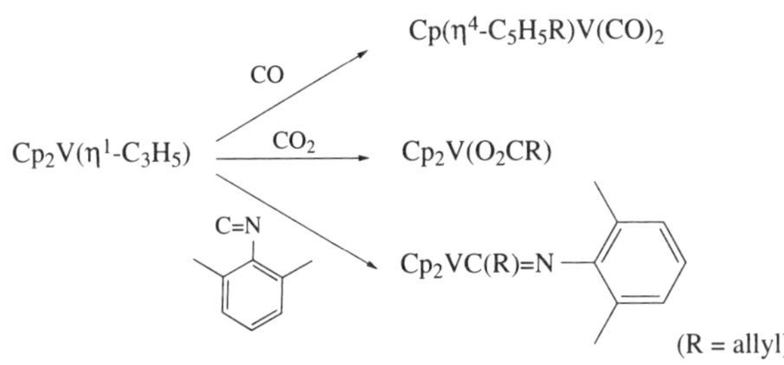

Scheme 44

The alkylation of Cp^*_2VCl with methyllithium yields the paramagnetic complex $Cp^*_2VCH_3$.[251] The reactivity of this complex is summarized in Scheme 45. Treatment with excess CO at atmospheric pressure afforded the diamagnetic acetyl carbonyl complex $Cp^*_2V(CO)(C(O)Me)$. Low-temperature spectroscopic investigation of the reaction suggests the presence of a diamagnetic compound $Cp^*_2V(CO)CH_3$ (^1H NMR δ = 0.26 ppm (V-Me) and a paramagnetic acetyl complex $Cp^*_2VC(O)CH_3$(IR). Under hydrogen pressures of 17–25 atm, the methyl complex can be hydrogenolyzed to a paramagnetic monohydride complex Cp^*_2VH ($\nu_{V\text{-}H}$ = 1625 cm^{-1}).[251] This complex will undergo exchange with a deuterium atmosphere to yield the monodeuteride ($\nu_{V\text{-}D}$ = 1177 cm^{-1}). The hydride complex does not react with ethylene up to 17 atm.

Photolytic decomposition or oxidation of Cp_2VMe_2 results in the formation of methane and ethane.[309] Reduction of Cp_2VMe_2 with sodium results in decomposition and evolution of methane. The thermal decomposition of $Cp_2V(CH_2C(Me)_2Et)$ yields CH_4, C_2H_4, $CH_2{=}CMe_2$, and unspecified C_3 and C_4 species.[310]

The formation of metallocyclic complexes with the Cp_2V^{IV} moiety was attempted by reacting vanadocene dihalides with di-Grignards.[311,312] Reaction of vanadocene dichloride with the di-Grignard reagent $[o\text{-}[(THF)_nClMgCH_2]C_6H_4]_2$ in THF did not yield a metallacyclic complex, as was found for the niobium and tantalum analogues; instead the product was a bimetallic complex $[Cp_2V(Cl)]_2\text{-}[[CH_2C_6H_4]_2]$.[311]

A dimethylmetallacyclobutane complex of vanadium(IV), $Cp_2V(CH_2C(Me)_2CH_2)$ was synthesized by the reaction of 1,3-bis(bromomagnesio)-2,2-dimethylpropane with Cp_2VCl_2.[312] At 25 °C the metallocyclic complex decomposes with a half-life of 4 h. Based on reactivity studies, it was concluded that this compound does not appear to fragment to a metallacarbene.

Reaction of vanadocene mono- and dihalides with ferrocenyl lithium reagents have been investigated. The reaction of ferrocenyl dilithium with $(\eta^5\text{-}RC_5H_4)_2VX$ (X = Cl, Br; R = H, Me) yields complexes in

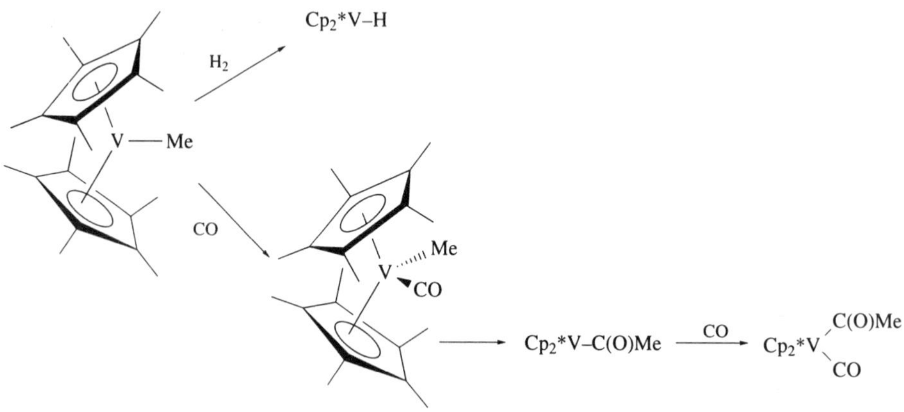

Scheme 45

which the paramagnetic vanadocene units are bridged by ferrocene.[313] In contrast, the reaction of Cp_2VX_2 with ferrocenyl lithium produced a disubstituted product, $Cp_2V(C_{10}H_9Fe)_2$.[314] The reaction between Cp_2VCl and 2-dimethylaminomethylferrocenyl lithium produces a black paramagnetic solid ($\mu_{eff} = 2.83\ \mu_B$) with the formula $Cp_2V(C_5H_3(CH_2NMe_2)FeCp)$ (microanalysis, ^1H NMR) in 65% yield (Equation (20)).[304]

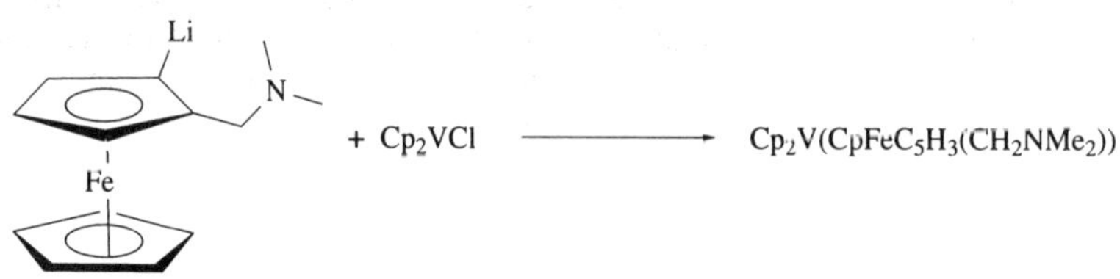

(20)

1.6.3.10 *Other group IV elements*

Cp_2V reacts with GeH_4 to give a germylene bridged species $[Cp_2V]_2GeH_2$.[315] Thermal decomposition of a series of compounds of general formula $Cp_2V(ER_3)$ ($ER_3 = GeEt_3$, $SnEt_3$) has been investigated in both solids and solutions.[298] For $Cp_2V(SnEt_3)$, the main decomposition products are vanadocene and hexaethyldistannane. In contrast, Et_3GeH, Et_3GeCp, Cp_2V, and $CpV(C_5H_4GeEt_3)$ are the major products of $Cp_2V(GeEt_3)$ decomposition. Vanadocene can be oxidized by $[Cp_2Co]^+[Hg[Ge(C_6F_5)]_3]^-$ or $Hg[Ge(C_6F_5)]_2$ to yield a vanadium(III) germyl complex, $Cp_2VGe(C_6F_5)_3$.[297,316]

1.6.4 Polynuclear and Heterobimetallic Compounds

Paramagnetic decamethylbivanadocenes have been synthesized according to Equation (21).[317] The complex was isolated from hexane as brown, analytically pure microcrystals in 25% yield.

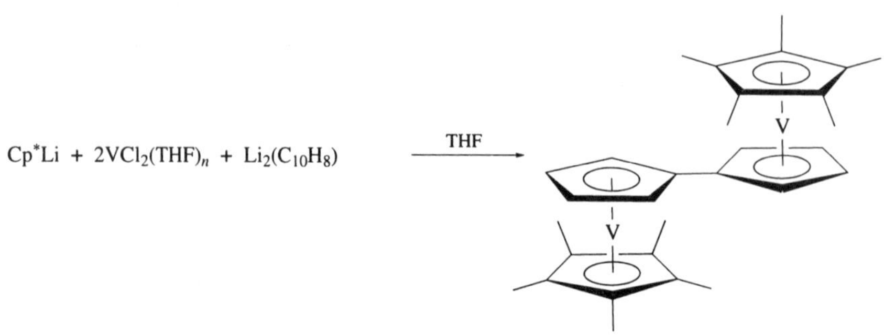

(21)

Vanadocenes and their coordination complexes have been used as starting materials for the synthesis of heterobimetallic complexes. The μ-isocarbonyl complex, $Cp^*_2V(\mu\text{-}OC)V(CO)_5$, resulting from the reaction of decamethylvanadocene and $V(CO)_6$ has been discussed in Section 1.1.2.1.2.[13]

The compounds [W(≡CR)(CO)$_2$Cp] (R = Me, *p*-Tol) react slowly with Cp$_2$V to give black crystals of a heterobimetallic complex, Cp$_2$V(μ-CR)W(CO)$_2$Cp).[318] The molecular structure of this unsymmetrical complex was established by x-ray crystallography (**9**).

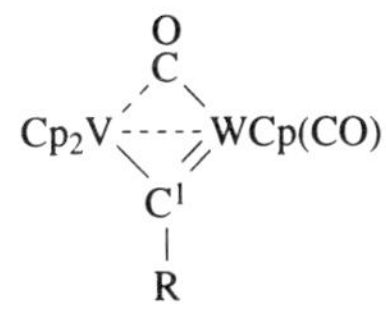

(9)

Reaction of Cp$_2$V(η^2-CS$_2$)[319] with several Lewis acid metal complexes ML$_n$ (ML$_n$ = [Co(CO$_4$]$^+$, Cr(CO)$_5$, Mo(CO)$_5$, CpMn(CO)$_2$, and Fe(CO)$_5$) has been reported to yield bimetallic complexes with a bridging CS$_2$ ligand.[320]

Preparation of a bimetallic complex, Cp$_2$V[Co(CO)$_4$], was claimed based on an infrared spectroscopic investigation of the reaction of Cp$_2$VCl with [Co(CO)$_4$]$^-$ and by the reaction of Cp$_2$V with Co$_2$(CO)$_8$.[321] Subsequent reaction of this complex with Co$_2$(CO)$_8$ leads to Cp$_2$V[OCCo$_3$(CO)$_9$].

1.6.5 Catalytic Applications

Vanadocene dichloride has been found to be an effective catalyst for the methylation of 1,4-bis(trimethylsilyl)butadiyne with AlMe$_3$ (Equation (22)).[322] This is in contrast to the reaction of 1,4-bis(trimethylsilyl)butadiyne with AlMe$_3$ in the presence of zirconocene and titanocene dichloride catalysts which yields selectively (*E*)-2-methyl-1,4-bis(trimethylsilyl)-1-buten-3-yne.

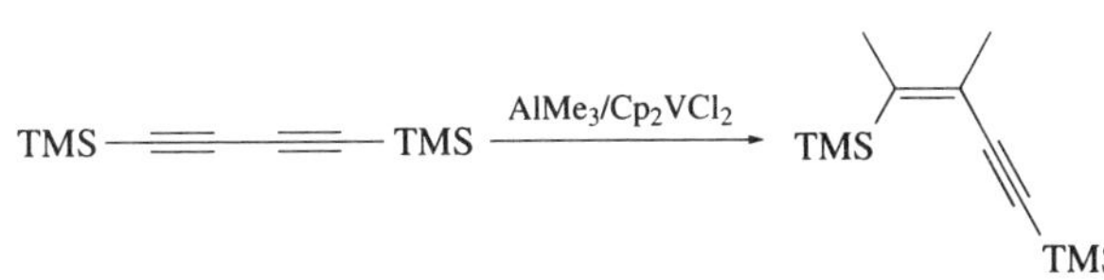

(22)

Polymerization of propylene was conducted at 40 °C and 1 atm in the presence of a MgCl$_2$–VCl$_3$–Cp$_2$V(Me)$_2$ catalyst system.[323]

Both Cp$_2$V and Cp$_2$VMe$_2$ have been investigated for their ability to catalyze the dehydrocoupling of phenylsilane.[324] Both complexes exhibit essentially the same reactivity; in refluxing toluene the predominant products are 1,2-diphenyldisilane and 1,2,3-triphenyltrisilane, respectively.

1.7 ARENE, DIENE, AND POLYENE COMPOUNDS

1.7.1 Bis(η^6-arene) Complexes

Two general synthetic routes to η^6-arene complexes are (i) reaction of metal atoms with arenes, and (ii) the reduction of VCl$_4$ with AlCl$_3$/Al in refluxing arene followed by hydrolysis (Friedel–Crafts method). Both of these procedures are discussed in *COMC-I* (Volume 3, Chapter 24, Section 1.7.1.1). The following sections review the progress made in expanding upon these synthetic methods as well as the chemical reactivity and physical properties of bis(arene) vanadium complexes.

1.7.1.1 Synthesis and properties

The interaction of vanadium metal vapors with naphthalene has been studied by matrix isolation ultraviolet-visible spectroscopy.[325] The development of a new rotary metal atom reactor that allows electron-beam vaporization of the metal has been successfully employed in the synthesis of several bis(η^6-arene) complexes of vanadium.[326]

Interaction of vanadium atom vapors with the naphthalene groups of both cyclic tetrameric naphthoxymethylsiloxanes and polymers which possess this moiety produced poly- and cyclosiloxane bound η^6-naphthalene complexes.[327] Aggregates with the general formula (arene)$_n$M$_m$ have been invoked.[328]

Treatment of the red-brown oil obtained from the reduction of VCl_3 using $Al/AlCl_3/1,3,5\text{-}Me_3C_6H_3$ with THF affords a brown solid of the cationic 16-electron complex $[(1,3,5\text{-}Me_3C_6H_3)_2V]AlCl_4$.[329] The structure of this complex was determined by x-ray diffraction and consists of discrete $V(1,3,5\text{-}Me_3C_6H_3)_2^+$ and $AlCl_4^-$ ions. The iodide complex, $VI(1,3,5\text{-}Me_3C_6H_3)_2$[330] can be reduced by a variety of reducing agents (Cp_2Co, MeLi, Na, or $Na[AlH_2(OCH_2CH_2OMe)_2]$) or reacted with compounds containing donor atoms (THF, $P(OMe)_3$, MeCN, or py) to give the neutral bis(arene) complex $V(1,3,5\text{-}Me_3C_6H_3)_2$ (Scheme 46).[331] The latter reaction appears to be the result of a disproportionation reaction.

$$(1,3,5\text{-}Me_3C_6H_3)_2V \xleftarrow{\text{reducing agents}} (1,3,5\text{-}Me_3C_6H_3)_2VI \xrightarrow[\text{}]{\text{THF, P(OMe)}_3\text{, MeCN, py}} (1,3,5\text{-}Me_3C_6H_3)_2V \ (40\text{–}45\%)$$

Scheme 46

An alternative preparation of bis(arene)vanadium complex consists of dissolution of vanadium(III) chloride in a mixture of THF and arene followed by cooling to −110 °C and condensation of excess potassium atoms forming what is believed to be anionic bis(arene)vanadium complexes.[332] Subsequent oxidation with chromatographic alumina yields $(toluene)_2V$ and $[1\text{-}(Me)naphthalene]_2V$ in 15 and 25% yields, respectively.

Ring-bridged and bis(triphenylsilyl) substituted arene complexes have been synthesized by lithiation of the parent bis(benzene) vanadium complex and subsequent reaction with diphenyldichlorosilane and triphenylchlorosilane, respectively.[333] A decrease in $A(^{51}V)$ for these complexes suggests an increased metal-to-ligand electron delocalization for the ring-bridged species relative to the parent $(C_6H_6)_2V$.

1.7.1.2 Reactivity

The potassium salts of the anionic complexes $[(C_6H_6)_2V]^-$ and $[(mesitylene)_2V]^-$ can be isolated as black solids from the reduction of $(arene)_2V$ with potassium metal in THF.[334–6] A single-crystal x-ray diffraction study of the solvated $K[(mesitylene)_2V]$ shows the anion possesses a sandwich structure.[334,335] THF solutions of this species are readily oxidized by both CO and CO_2 to give $(benzene)_2V$.[336] An examination of the EPR spectrum of the 18-electron anions in the presence of HMPA revealed an unexplained strong, finely structured signal with $g = 2.023$.[336]

Oxidation of $(1,3,5\text{-}Me_3C_6H_3)_2V$ has been carried out with allyl chloride to yield vanadium(II) chloride (75%) and mesitylene.[329] Bis(ethylbenzene)vanadium is oxidized by $[(C_6F_5)_3Ge]_2Hg$ to give the ionic complex $[[(C_6F_5)_3Ge]_3Hg]^-[V(ethylbenzene)_2]^+$ (87% yield).[337]

The reaction of $(C_6H_6)_2V$ and dithioacetic acid, MeCSSH, affords a novel dimeric species, $[(MeCS_2)_2V]_2(\mu,\eta^2\text{-}S_2)$, in approximately 60% yield.[338] The structure of this brown diamagnetic product has been characterized by x-ray crystallography (V–V = 0.24 nm).

Reaction of $(mesitylene)_2V$ with $Ti(O_2CNEt_2)_4$ yields an unseparated mixture of Ti^{III} and V^{III} *N,N*-diethylcarbamato complexes derived from a simultaneous electron- and ligand transfer reaction.[339] The infrared spectrum of this mixture appears to be the sum of the spectra for $[Ti(O_2CNEt_2)_3]_n$ and $[V(O_2CNEt_2)_3]_n$.

1.7.1.3 Physical properties

The infrared absorption intensities of $(benzene)_2V$ have been used to calculate the effective charge on the vanadium atom (absolute value = 0.5–0.9 electrons).[340] The responses of $(benzene)_2V$, $(toluene)_2V$, and $(mesitylene)_2V$ to laser irradiation while isolated in nitrogen matrices have been investigated.[341,342] Vapor-phase electronic absorption spectra for a series of $(arene)_2V$ complexes yield ionization limits for these complexes.[343] These limits correspond to the ionization from the highest occupied orbital and are in good agreement with those known from photoelectron spectroscopy.

A cyclovoltammetric study of the redox behavior of bis(benzene)vanadium revealed both a reversible one-electron oxidation (−0.35 V vs. SCE) and a reversible one-electron reduction (−2.71 V vs. SCE).[344] In addition to the reversible wave for the $(C_6H_6)_2V/(C_6H_6)_2V^+$ couple, an irreversible oxidation to the dication was observed at 0.35 V vs. SCE.

Low-temperature and variable-temperature proton ENDOR spectra on bis(benzene) vanadium isolated in an argon matrix or diluted in 2,2-paracyclophane or ruthenocene single crystals yield information on site symmetry (D_{6h}) of the matrix crystals and an estimation of the activation parameters for the rotational reorientation of the benzene rings with respect to the matrix crystal.[345,347]

The electronic structure of group 5 bis(arene)metal complexes has been examined by EPR and photoelectron spectroscopy (PES).[348,349] A substantial increase in the proton coupling constant was observed down the series $(C_6H_6)_2V$, $(C_6H_6)_2Nb$, $(C_6H_6)_2Ta$. From PES the ion ground state for $(C_6H_6)_2V$ and $(C_6H_5Me)_2V$ were assigned as $^3E_2$ while for $(C_6H_6)_2Nb$, $(C_6H_5Me)_2Nb$ and $(C_6H_3Me_3)_2Nb$ a $^1A_1$ ground state was assigned. Magnetic circular dichroism (MCD) has been used to determine the magnetization curve and isotropic g values for matrix isolated (η^6-toluene)$_2$V, and the value (1.94 ± 0.03) is in excellent agreement with that obtained by EPR.[350]

Extended Hückel calculations on the conformation of bis(1,4-$F_2C_6H_4$)vanadium have been used to rationalize the observed dihedral angle (53°) between the fluorine atoms.[351]

1.7.2 Mono(η^6-arene) and "Half-sandwich Complexes"

An unusual bimetallic arene complex, (η^6,η^5-C_9H_7)$_2V_2$, has been synthesized in crystalline form by the reduction of (η^5-C_9H_7)$_2$V with potassium in THF and structurally characterized (**10**).[352] This molecule is a unique example of a bimetallic sandwich complex where the two metal atoms are bridged through binding of one atom to the cyclopentadienyl moiety of the indenyl and the other to the six-membered ring. The complex possesses a center of inversion with a V–V distance of 0.2351 nm. The six-membered ring is nonplanar and appears to be distorted in a dienelike configuration with a planar four-carbon unit (2,3,4,5) forming a 13.5° angle with the two bridging carbon atoms (1,6).

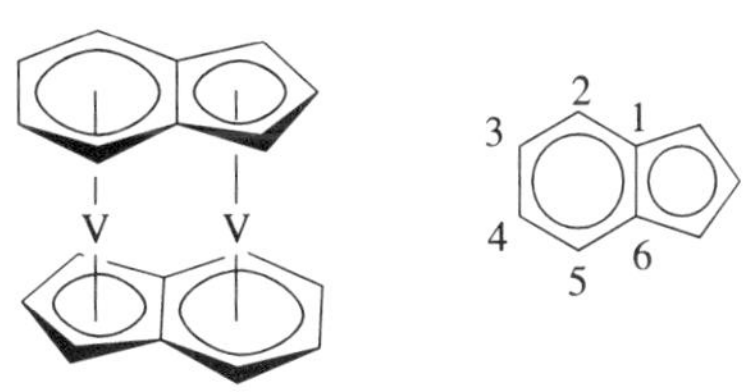

(10)

Improved syntheses for a series of tetracarbonylvanadium arene cation complexes, $[(\eta^6\text{-}C_6H_{6-n}Me_n)V(CO)_4]^+[V(CO)_6^-]$ (n = 1,3,4), have been reported.[353] The x-ray crystal structure of the 1,2,4,5-tetramethylbenzene complex confirms the half-sandwich nature of the cation.

The half-sandwich zero-valent complex $(C_6H_6)V$ has been synthesized by the cocondensation of vanadium atoms with C_6H_6/Ar mixtures at 12 K yielding the matrix-isolated species which was first identified by a characteristic absorption in the ultraviolet–visible spectrum at 550 nm.[354–6] EPR and ultraviolet–visible spectroscopies confirm that the ground-state electronic configuration, structure, and bonding properties for $(C_6H_6)V$ are not greatly perturbed from those of $(C_6H_6)_2V$.[354,355] This proposal receives support from the results of theoretical (SW-SCF-Xα and LSD-LCAO) calculations for a C_{6v} geometry.[355,356] The infrared resonances of $(C_6H_6)V$ have been thoroughly assigned.[355] The molecule $(C_6F_6)V$ was also synthesized and identified on the basis of EPR spectroscopy.[355]

In addition to $(C_6H_6)V$, the complexes $(C_6H_6)V_2$ and $(C_6H_6)_2V_2$ were also isolated in argon matrices.[354,355] In order to provide some insight into the possible modes of bonding in $(C_6H_6)V_2$, geometry optimization and electronic structure of this unusual complex have been investigated by the LSD-LCAO[357] and SW-SCF-Xα methods.[355,358]

1.7.3 Heteroatom-substituted Arenes

Reaction of alkali metal borinates, $[C_5H_5BR]M$ (R = Me, Ph), with VCl_3 yields paramagnetic sandwich compounds $V(C_5H_5BR)_2$ (Scheme 47).[359] These complexes can be converted to the tetracarbonyl half-sandwich complexes by reaction with $HgCl_2$ and $[V(CO)_6]^-$. Heating $(C_5H_5BR)V(CO)_4$ with cycloheptatriene converts the tetracarbonyl to $(C_5H_5BR)V(C_7H_7)$.

The interaction of vanadium metal vapors and 2,6-dimethylpyridine results in the isolation of (2,6-dimethylpyridine)$_2$V.[360] The spectral properties (IR, EPR) of this compound are quite similar to bis(benzene)vanadium indicating structural and electronic similarities between these complexes. The complex displays the expected sandwich-type structure, although the metal–ring π-coordination is unsymmetric with the vanadium atom displaced slightly toward the nitrogen atoms.[361] Similarly, reaction of vanadium metal vapors and phosphabenzene (C_5H_5P) yields a red-brown, sublimable solid $(C_5H_5P)_2V$.[362] This complex exhibits a melting point of 210 °C and apparently air-stable solutions. The

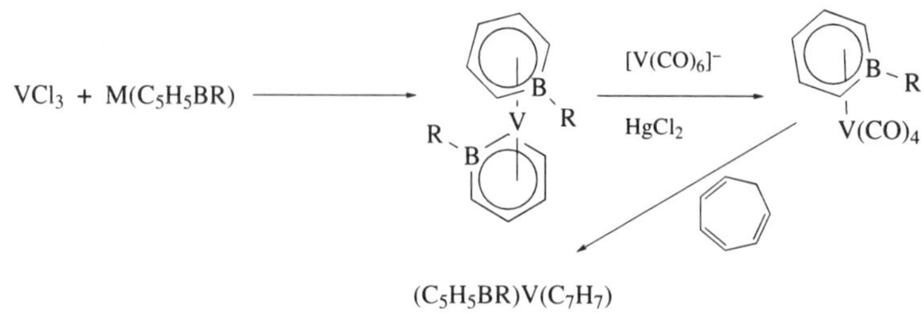

Scheme 47

x-ray structure confirms an eclipsed sandwich configuration with some disorder in the P/C positions of the ring. Electrochemical investigations reveal both a reversible one-electron oxidation and reduction at 0.18 V and space −1.99 V vs. SCE, respectively.

1.7.4 Dienyl ("Open-metallocene") Complexes

1.7.4.1 Pentadienyl

The pentadienyl moiety has been employed as a π-bonded hydrocarbon ligand in the preparation of a new family of vanadium complexes. These ligands have the potential to bind in an η^1-, η^3-, or η^5-mode, thus relating them to both the cyclopentadienyl and allyl groups.

Reaction of 2 equivalents of K(2,4-dimethylpentadienyl) with "VCl_2" leads to the formation of green bis(2,4-C_7H_{11})V.[363] In contrast to Cp_2V, (2,4-C_7H_{11})$_2$V has a low spin configuration. An x-ray crystallographic investigation revealed that the two pentadienyl ligands have an ideally staggered orientation with a very short average V–C distance of 0.2211(2) nm (**11**).[364]

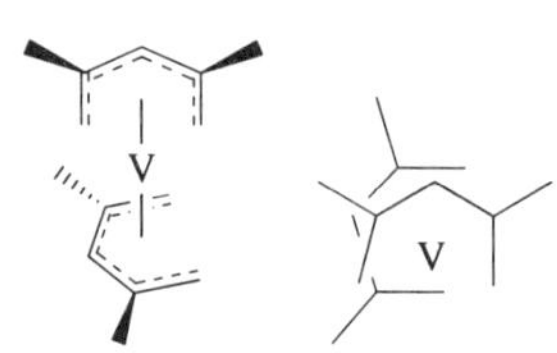

(11)

The only other isolated sandwich complex with a pentadienyl ligand is the half-open metallocene Cp(1,5-(TMS)$_2$$C_5H_5$)V. This complex was prepared via the reduction of $CpVI_2$ followed by reaction with K(1,5-(TMS)$_2$$C_5H_5$).[365] Like (2,4-$C_7H_{11}$)$_2$V, this species appears to have a low spin electronic configuration ($\mu_{eff} = 1.3\ \mu_B$); however, the possibility of a dimeric structure was not eliminated.

1.7.4.2 Cyclohexadienyl

The 6,6-dimethylcyclohexadienyl anion ($dmCh^-$), a cyclic dienyl ligand, has been employed in the synthesis of paramagnetic $(dmCh)_2V$.[366] This air-sensitive material reacts readily with CO and PMe_3 and PF_3 to yield the corresponding adducts (Tables 25 and 26).

1.7.4.3 Adducts and mixed ligand systems

While 2,4-dimethylpentadienyl proved to be quite versatile in the synthesis of metallocene analogues of vanadium, attempts to isolate complexes with pentadienyl ligands having fewer methyl groups have apparently been unsuccessful.[367] The addition of phosphine or carbonyl ligands increases the stability of these metallocene species allowing the isolation of several adducts of open-vanadocenes.[365,368] In addition, use of the pentadienyl group has been extended to mixed π-ligand systems such as Cp[1,5-(TMS)$_2$$C_5H_5$]VCO and Cp[1,5-(TMS)$_2$$C_5H_5$]$VPEt_3$ (Tables 25 and 26).[365]

The syntheses of a variety of 17-electron open-vanadocene adducts have been reported incorporating a variety of pentadienyl groups (C_5H_7, 1-, 2-, and 3-C_6H_9, 2,3-C_7H_{11}, 2,4-C_7H_{11}) and ligands (CO, PMe_3,

Table 25 EPR data for dienyl complexes and adducts.

Compound	*Color*	*A(^{51}V), A(^{31}P)* (10^{-4} T)
$Cp(1,5\text{-}(TMS)_2C_5H_5)V$	yellow–purple	63.4
$Cp(1,5\text{-}(TMS)_2C_5H_5)VCO$	orange–red	67.1
$Cp(1,5\text{-}(TMS)_2C_5H_5)VPEt_3$	dark brown	74.0, 28.2
$(dmCh)_2V$	green	54
$(dmCh)_2VCO$	yellow–green	71
$(dmCh)_2VPMe_3$[a]	red–brown	76
$(dmCh)_2VPF_3$[a]		79

[a] Stable only at low temperature.

Table 26 CO stretching frequencies for dienyl compounds.

Compound	ν_{CO} (cm^{-1})
$(C_5H_7)_2VCO$	1962
$(1\text{-}C_6H_9)_2VCO$	1944
$(2\text{-}C_6H_9)_2VCO$	1958
$(3\text{-}C_6H_9)_2VCO$	1952
$(2,3\text{-}C_7H_{11})_2VCO$	1946
$(2,4\text{-}C_7H_{11})_2VCO$	1948
$Cp(1,5\text{-}(TMS)_2C_5H_5)VCO$	1938
$(dmCh)_2VCO$	1917

PEt_3, $P(OMe)_3$, PF_3, dmpe).[367] Single-crystal structure determinations of several of the CO adducts are reported as well as the EPR and infrared spectra (Tables 26 and 27). The CO lability in the complexes $(C_5H_7)_2VCO$ and $(2,4\text{-}C_7H_{11})_2VCO$ as well as the mixed complexes $Cp(C_5H_7)VCO$ and $Cp(2,4\text{-}C_7H_{11})VCO$ has been investigated.[368,369] These compounds exchange with CO 10^3–10^4 times slower than the bis(cyclopentadienyl) analogues. The associative CO exchange rate for $(dmCh)_2VCO$ is about 10^{-4} times that of $Cp^*{}_2VCO$.[370]

Table 27 EPR data for various adducts of dienyl compounds.

L	*A(^{51}V), A(^{31}P)* (10^{-4} T) *C_5H_7*	*3-C_6H_9*	*2,4-C_7H_{11}*
CO	79.1	78.4	78.4
PMe_3	77.7, 29.7		77.6, 32.5
PEt_3	78.4, 26.4	77.3, 28.8	77.8, 32.7
$P(OMe)_3$	78.0, 38.7	77.5, 41.8	77.9, 41.2
PF_3	77.7, 47.6	76.7, 51.4	76.7, 53.6

The electronic structures of $(C_5H_7)_2V$ and $(C_5H_7)_2VCO$ have been modeled theoretically (SCF-Xα-DV) and the results compared with their Cp analogs.[371] In contrast to Cp_2VCO, orbital contour plots for $(C_5H_7)_2VCO$ show the unpaired electron density to be pointing toward the pentadienyl ligands. EPR spectra gave no evidence for hyperfine splitting due to delocalization of the unpaired electron onto the ^{13}CO ligands of either $(C_5H_7)_2V^{13}CO$, $(2,4\text{-}Me_2C_5H_5)_2V^{13}CO$, or $Cp[1,5\text{-}(TMS)_2C_5H_5)V^{13}CO$.

1.7.5 Cycloheptatrienyl and Cyclooctatetraenyl Complexes

A structurally characterized complex of vanadium and cyclooctatetraenide dianion, $(\eta^8\text{-}C_8H_8)V[Et_2C_2B_4H_4]$, has been prepared by the reaction of $K_2C_8H_8$ and the nido-carborane anion $[Et_2C_2B_4H_4]^-$ with VCl_3 (Scheme 48).[372,373] The complex is a dark-green paramagnetic solid which exhibits no air sensitivity even in solution. Magnetic susceptibility and EPR support the formulation of vanadium(IV) with 17 electrons in its valence shell. The x-ray structure consists of a *closo* pentagonal-bipyramidal VC_2B_4 metallacarborane cage with the metal coordinated to a planar $\eta^8\text{-}C_8H_8$ ligand.

$$VCl_3 + K_2C_8H_8 + [Et_2C_2B_4H_5]Na \xrightarrow{\text{i, THF, ii, } O_2} C_8H_8V[Et_2C_2B_4H_5]$$

Scheme 48

The reaction of cycloheptatriene and $(C_5H_5BR)V(CO)_4$ leads to formation of the paramagnetic sandwich complexes, $(C_5H_5BR)V(C_7H_7)$ (Scheme 47).[359]

ACKNOWLEDGEMENTS

Our grateful acknowledgement is due to Prof. D. Rehder (Hamburg), Dr. A. Horton (Shell, Amsterdam), Prof. J. Teuben (Groningen), Prof. F. Preuss (Kaiserlautern), Dr. J. Jubb (Shell, Amsterdam), Prof. T. Rauchfuss (Urbana), Prof. R. Ernst (Salt Lake), Prof. J. Heck (Hamburg), Prof. G. Kreisel (Jena), Prof. K. Jonas (Max-Planck), Dr. G. Fochi (Pisa), Prof. M. Rausch (Amherst), Prof. F. Bottomley (Fredericton), Prof. W. Trogler (San Diego), and Prof. G. Ozin (Toronto) for their careful reading and constructive criticism.

1.8 REFERENCES

1. R. M. P. Werner and H. E. Podall, *Chem. Ind. (London)*, 1961, **80**, 144.
2. R. Ercoli, F. Calderazzo and A. Alberola, *J. Am. Chem. Soc.*, 1960, **82**, 2966.
3. F. Calderazzo and G. Pampaloni, *J. Organomet. Chem.*, 1983, **250**, C33.
4. F. Calderazzo, U. Englert, G. Pampaloni, G. Pellizi and R. Zamboni, *Inorg. Chem.*, 1983, **22**, 1865.
5. R. B. King, 'Organometallic Synthesis', Academic Press, New York, 1965, vol. 1, p. 82.
6. (a) J. E. Ellis, R. A. Faltyneck, G. L. Rochfort, R. E. Stevens and G. A. Zank, *Inorg. Chem.*, 1980, **19**, 1082; (b) G. Silvestri, S. Gambino, M. Guainazzi and R. Ercoli, *J. Chem. Soc. Dalton Trans.* 1972, 2558.
7. G. F. P. Warnock, S. B. Philson and J. E. Ellis, *J. Chem. Soc. Chem. Commun.*, 1984, 893.
8. J. E. Ellis, K. L. Fjare and T. G. Hayes, *J. Am. Chem. Soc.*, 1981, **103**, 6100.
9. J. E. Ellis, *J. Am. Chem. Soc.*, 1981, **103**, 6106.
10. J. E. Ellis, T. G. Hayes and R. E. Stevens, *J. Organomet. Chem.*, 1981, **216**, 191.
11. F. Calderazzo, G. Pampaloni, D. Vitali and P. F. Zanazzi, *J. Chem. Soc., Dalton Trans.*, 1982, 1993.
12. T. G. Richmond, Q. Z. Shi, W. C. Trogler and F. Basolo, *J. Am. Chem. Soc.*, 1984, **106**, 76.
13. J. H. Osborne, A. L. Rheingold and W. C. Trogler, *J. Am. Chem. Soc.*, 1985, **107**, 6292.
14. Q. Z. Shi, T. G. Richmond, W. C. Trogler and F. Basolo, *J. Am. Chem. Soc.*, 1984, **106**, 71.
15. L. D. Silverman, P. W. K. Corfield and S. J. Lippard, *Inorg. Chem.*, 1981, **20**, 3106.
16. S. S. Shah and A. W. Maverick, *J. Chem. Soc. Dalton Trans.*, 1987, 2881.
17. D. Rehder and U. Puttfarcken, *Z. Naturforsch.* 1982, **37b**, 348.
18. Q. Z. Shi, T. G. Richmond, W. C. Trogler and F. Basolo, *J. Am. Chem. Soc.*, 1982, **104**, 4032.
19. K. Ihmels and D. Rehder, *J. Organomet. Chem.*, 1981, **218**, C54.
20. R. Talay and D. Rehder, *J. Organomet. Chem.*, 1984, **262**, 25.
21. R. Talay and D. Rehder, *Z. Naturforschung*, 1981, **36b**, 451.
22. K. Ihmels and D. Rehder, *J. Organomet. Chem.*, 1983, **232**, 151.
23. J. E. Ellis and K. L. Fjare, *Organometallics*, 1982, **1**, 898.
24. J. E. Ellis and K. L. Fjare, *J. Organomet. Chem.*, 1981, **214**, C33.
25. K. Ihmels and D. Rehder, *Chem. Ber.*, 1985, **118**, 895.
26. K. Ihmels and D. Rehder and V. Pank, *Inorg. Chim. Acta*, 1985, **96**, L69.
27. K. Ihmels and D. Rehder, *Organometallics*, 1985, **4**, 1340.
28. F. J. Wells, G. Wilkinson M. Motevalli and M. B. Hursthouse, *Polyhedron*, 1987, **6**, 1351.
29. F. Calderazzo, G. Pampaloni and P. F. Zanazzi, *Chem. Ber.*, 1986, **119**, 2796.
30. F. Calderazzo and G. Pampaloni, *J. Chem. Soc. Chem. Commun.*, 1984, 1249.
31. F. Calderazzo and G. Pampaloni, G. Pelizzi and F. Vitali, *Polyhedron*, 1988, **7**, 2039.
32. A. Duch, M. Hoch and D. Rehder, *Chimia*, 1988, **42**, 179.
33. D. Rehder and D. Wenke, *J. Organomet. Chem.*, 1988, **348**, 205.
34. J. Schiemann and E. Weiss, *J. Organomet. Chem.*, 1983, **255**, 179.
35. J. R. Morton, K. F. Preston and Y. P. Vaills, *Can. J. Spectrosc.*, 1988, **33**, 43.
36. W. J. Lin, G. H. Lee, S. M. Peng and R. S. Liu, *Organometallics*, 1991, **10**, 2310.
37. H. Werner, P. Thometzek, C. Krüger and H. J. Kraus, *Chem. Ber.*, 1986, **119**, 2777.
38. N. Albrecht, P. Hubener, U. Behrens and E. Weiss, *Chem. Ber.*, 1985, **118**, 4059.
39. J. Schiemann, P. Hubener and E. Weiss, *Angew. Chem.*, 1983, **95**, 1021.
40. J. D. Protasiewicz and S. J. Lippard, *J. Am. Chem. Soc.*, 1991, **113**, 6564.
41. H. Vahrenkamp, *Chem. Ber.*, 1978, **111**, 3472.
42. T. Kawamura, S. Enoki, S. Hayashida and T. Yonezawa, *Bull. Chem. Soc. Jap.*, 1982, **55**, 3417.
43. (a) F. Calderazzo, M. Castellani and G. Pampaloni, *J. Chem. Soc., Dalton Trans.* 1985, 1995; (b) F. Calderazzo, *Ann. N.Y. Acad. Sci.*, 1983, **415**, 37.
44. (a) F. Calderazzo, G. Pampaloni, M. Lanfranchi, G. Pelizzi, *J. Organomet. Chem.*, 1985, **296**, 1; (b) F. Calderazzo, G. Pampaloni, G. Pelizzi and F. Vitali, *Organometallics*, 1988, **7**, 1083.
45. D. Rehder, P. Oltmanns, M. Hoch, C. Weidemann and W. Priebsch, *J. Organomet. Chem.*, 1986, **308**, 19.

46. F. Süssmilch, W. Glöckner and D. Rehder, *J. Organomet. Chem.*, 1990, **388**, 95.
47. M. Darensbourg and J. M. Hanckel, *Organometallics*, 1982, **1**, 82.
48. M. Y. Darensbourg and J. M. Hanckel, *J. Organomet. Chem.*, 1981, **217**, C9.
49. U. Puttfarcken and D. Rehder, *J. Organomet. Chem.*, 1980, **185**, 219.
50. T. Greiser, U. Puttfarcken and D. Rehder, *Trans. Met. Chem.* 1979, **4**, 168.
51. T. G. Gardner and G. S. Girolami, *Organometallics*, 1987, **6**, 2551.
52. D. Rehder, F. Süssmilch, W. Priebsch and M. Fornalczyk, *J. Organomet. Chem.*, 1991, **411**, 357.
53. D. Rehder and K. Ihmels, *Inorg. Chim. Acta*, 1983, **76**, L313.
54. C. J. Jameson, D. Rehder and M. Hoch, *J. Am. Chem. Soc.*, 1987, **109**, 2589.
55. D. Rehder, H. C. Bechthold, A. Kececi, H. Schmidt and M. Siewing, *Z. Naturforschung*, 1982, **37b**, 631.
56. R. Talay, D. Rehder, *J. Organomet. Chem.*, 1984, **262**, 25.
57. M. G. B. Drew, *Acta Cryst.*, 1982, **B38**, 254.
58. K. L. Fjare and J. E. Ellis, *J. Am. Chem. Soc.*, 1983, **105**, 2303.
59. Q. Z. Shi, T. G. Richmond, W. C. Trogler and F. Basolo, *Inorg. Chem.*, 1984, **23**, 957.
60. U. Behrens and S. Tofke, *Acta Cryst.* 1986, **C42**, 161.
61. J. Schiemann, E. Weiss, F. Näumann and D. Rehder, *J. Organomet. Chem.*, 1982, **232**, 219.
62. F. Näumann, D. Rehder and V. Pank, *Inorg. Chim. Acta*, 1984, **84**, 117.
63. F. Näumann and D. Rehder, *Z. Naturforschung*, 1984, **39b**, 1647.
64. F. Näumann and D. Rehder, *Z. Naturforschung*, 1984, **39b**, 1654.
65. P. Jaitner and W. Huber, *Inorg. Chim. Acta*, 1987, **129**, L45.
66. D. Rehder and P. Jaitner, *J. Organomet. Chem.*, 1987, **329**, 337.
67. M. M. Olmstead, P. P. Power and S. C. Shoner, *Organometallics*, 1988, **7**, 1380.
68. R. Kirmse, J. Stach and G. Kreisel, *Z. Chemistry*, 1980, **20**, 420.
69. G. S. Girolami, G. Wilkinson, A. M. R. Galas, M. Thornton-Pett and M. B. Hursthouse, *J. Chem. Soc., Dalton Trans.*, 1985, 1339.
70. J. J. H. Edema, W. Stauthamer, F. van Bolhuis, S. Gambarotta, W. J. J. Smeets and A. L. Spek, *Inorg. Chem.*, 1990, **29**, 1302.
71. J. J. H. Edema, A. Meetsma and S. Gambarotta, *J. Am. Chem. Soc.*, 1989, **111**, 6878.
72. G. J. Leigh, R. Prieto-Alcon and J. R. Sanders, *J. Chem. Soc. Chem. Commun.*, 1991, 921.
73. W. Seidel, G. Kreisel and H. Z. Mennega, *Z. Chemistry* 1976, **16**, 492.
74. (a) F. A. Cotton and M. Millar, *J. Am. Chem. Soc.*, 1977, **99**, 7886; (b) F. A. Cotton, G. E. Lewis and G. N. Mott, *Inorg. Chem.*, 1983, **22**, 560.
75. F. A. Cotton and M. P. Diebold, I. Shim, *Inorg. Chem.*, 1985, **24**, 1510.
76. J. J. H. Edema, A. Meetsma, F. vanBolhuis and S. Gambarotta, *Inorg. Chem.*, 1991, **30**, 2056.
77. (a) G. A. Razuvaev, V. N. Latyaeva, L. I. Vyshinskaya and V. V. Drobotenko, *J. Organomet. Chem.*, 1981, **208**, 169; (b) W. Seidel and G. Kreisel, *Z. Anorg. Allg. Chem.*, 1977, **435**, 153.
78. G. A. Razuvaev and L. I. Vyshinskaya, V. V. Drobotenko, G. Y. Malkova and N. N. Vyshinsky, *J. Organomet. Chem.*, 1982, **239**, 335.
79. J. K. Buijink, A. Meetsma and J. H. Teuben, *Organometallics*, 1993, **12**, 2004.
80. W. Seidel and G. Kreisel, *Z. Anorg. Allg. Chem.*, 1977, **435**, 146.
81. S. Gambarotta, C. Floriani, A. Chiesi-Villa and C. Guastini, *J. Chem. Soc. Chem. Commun.*, 1984, 886.
82. G. Kreisel and W. Seidel, *Z. Chemistry*, 1986, **26**, 260.
83. W. Seidel and U. Weinzierl, *Z. Chemistry* 1989, **29**, 29.
84. J. Ruiz, M. Vivanco, C. Floriani, A. Chiesi-Villa and C. Guastini, *J. Chem. Soc. Chem. Commun.*, 1991, 762.
85. (a) W. Seidel and G. Kreisel, *Z. Chemistry*, 1982, **22**, 113; (b) W. Seidel and G. Kreisel, *Z. Anorg. Allg. Chem.*, 1988, **559**, 118.
86. G. Kreisel and W. Seidel, *Z. Anorg. Allg. Chem.*, 1981, **478**, 106.
87. M. Vivanco, J. Ruiz, C. Floriani, A. Chiesi-Villa and C. Guastini, *Organometallics*, 1990, **9**, 2185.
88. J. Ruiz, M. Vivanco, C. Floriani, A. Chiesi-Villa and C. Guastini, *J. Chem. Soc., Chem. Commun.*, 1991, 214.
89. S. Gambarotta, M. Mazzanti, C. Floriani, A. Chiesi-Villa, C. Guastini, *J. Chem. Soc., Chem. Commun.*, 1985, 829.
90. J. M. Rosset, C. Floriani, M. Mazzanti, A. Chiesi-Villa and C. Guastini, *Inorg. Chem.*, 1990, **29**, 3991.
91. E. Solari, S. De Angelis, C. Floriani, A. Chiesi-Villa and C. Rizzoli, *Inorg. Chem.*, 1992, **31**, 96.
92. S. Ciurli, C. Floriani, A. Chiesi-Villa and C. Guastini, *J. Chem. Soc., Chem. Commun.*, 1986, 1401.
93. A. R. Wills and P. G. Edwards, *J. Chem. Soc., Dalton Trans.*, 1989, 1253.
94. A. A. Danopoulos and P. G. Edwards, *Polyhedron*, 1989, **8**, 1339.
95. A. R. Wills, P. G. Edwards, M. Harman and M. B. Hursthouse, *Polyhedron*, 1989, **8**, 1457.
96. Y. Doi, S. Suzuki and K. Soga, *Macromolecules*, 1986, **19**, 2896.
97. F. J. Feher, J. F. Walzer and R. L. Blanski, *J. Am. Chem. Soc.*, 1991, **113**, 3618.
98. Y. Doi, T. Keii, *Adv. Polym. Sci.*, 1986, **73/74**, 201.
99. R. Seangprasertkij and T. L. Riechel, *Inorg. Chem.*, 1986, **25**, 3121.
100. F. Preuss and L. Ogger, *Z. Naturforschung*, 1982, **37b**, 957.
101. W. Seidel and G. Kreisel, *Z. Chem.*, 1981, **21**, 295.
102. W. Seidel and G. Kreisel, *Z. Anorg. Allg. Chem.*, 1988, **559**, 118.
103. F. J. Feher, D. A. Newman and J. F. Walzer, *J. Am. Chem. Soc.*, 1989, **111**, 1741.
104. F. J. Feher and R. L. Blanski, *J. Am. Chem. Soc.*, 1992, **114**, 5886.
105. F. J. Feher and R. L. Blanski, *Organometallics*, 1993, **12**, 958.
106. F. Preuss and H. Becker, *Z. Naturforschung*, 1986, **41b**, 185.
107. J. de With, A. D. Horton and A. G. Orpen, *Organometallics*, 1990, **9**, 2207.
108. J. de With and A. D. Horton, *Angew. Chem., Int. Ed. Engl.*, 1993, **32**, 903.
109. J. de With, A. D. Horton and A. G. Orpen, *Organometallics*, 1993, **12**, 1493.
110. D. D. Devore, J. D. Lichtenhan, F. Takusagawa and E. A. Maatta, *J. Am. Chem. Soc.*, 1987, **109**, 7408.
111. R. B. King, 'Organometallic Synthesis', Academic Press, New York, 1965, Vol. 1, 105.
112. K. Jonas and V. Wiskamp, *Z. Naturforschung*, 1983, **38b**, 1113.

113. W. A. Herrmann and W. Kalcher, *Chem. Ber.*, 1982, **115**, 3886.
114. M. Hoch, A. Duch and D. Rehder, *Inorg. Chem.*, 1986, **25**, 2907.
115. D. Rehder, M. Hoch and M. Link, *Organometallics*, 1988, **7**, 233.
116. M. Hoch and D. Rehder, *Chem. Ber.*, 1988, **121**, 1541.
117. W. Priebsch, M. Hoch and D. Rehder, *Chem. Ber.*, 1988, **121**, 1971.
118. R. M. Kowaleski, D. O. Kipp, K. J. Stauffer, P. N. Swepston and F. Basolo, *Inorg. Chem.*, 1985, **24**, 3750.
119. D. Wenke and D. Rehder, *J. Organomet. Chem.*, 1984, **273**, C43.
120. J. W. Freeman and F. Basolo, *Organometallics*, 1991, **10**, 256.
121. R. B. Hitam and A. J. Rest, *Organometallics*, 1989, **8**, 1598.
122. D. F. Foust, M. D. Rausch, *J. Organomet. Chem.*, 1982, **239**, 321.
123. H. G. Alt and H. E. Engelhardt, *Z. Naturforsch.*, 1985, **40b**, 1134.
124. H. G. Alt and H. E. Engelhardt, A. Razavi, M. D. Rausch, R. D. Rogers, *Z. Naturforsch.*, 1988, **43b**, 438.
125. G. M. Brown and L. Maya, *Inorg. Chem.*, 1989, **28**, 2007.
126. R. Borowski, D. Rehder and K. Von Deuten, *J. Organomet. Chem.*, 1981, **220**, 45.
127. R. A. Bartlett, H. V. R. Dias and P. P. Power, *J. Organomet. Chem.*, 1989, **362**, 87.
128. M. Hoch and D. Rehder, *Inorg. Chim. Acta*, 1986, **111**, L13.
129. D. Rehder, M. Hoch and C. J. Jameson, *Magn. Res. Chem.*, 1990, **28**, 138.
130. M. Hoch, D. Rehder, *J. Organometal. Chem.*, 1985, **288**, C25.
131. C. Woitha and D. Rehder, *J. Organomet. Chem.*, 1988, **353**, 315.
132. C. Woitha and U. Behrens, V. Vergopoulos and D. Rehder, *J. Organomet. Chem.*, 1990, **393**, 97.
133. M. Hoch and D. Rehder, *Inorg. Chim. Acta*, 1986, **115**, L23.
134. S. Gambarotta, A. Chiesi-Villa and C. Guastini, *Inorg. Chem.*, 1988, **27**, 99.
135. F. Calderazzo, A. Belforte and P. F. Zanazzi, *Gazz. Chim. Ital.*, 1985, **115**, 71.
136. (a) M. Herberhold and M. Schrepfdermann, *J. Organomet. Chem.*, 1991, **419**, 85; (b) O. A. Rajan, M. McKenna, J. Noordik, R. C. Haltiwanger and M. Ratowski Dubois, *Organometallics*, 1984, **3**, 831.
137. N. J. Coville, G. W. Harris and D. Rehder, *J. Organomet. Chem.*, 1985, **293**, 365.
138. M. T. Haward, M. W. George, S. M. Howdle and M. Poliakoff, *J. Chem. Soc., Chem. Commun.*, 1990, 913.
139. M. T. Haward, M. W. George, P. Hamley and M. Poliakoff, *J. Chem. Soc., Chem. Commun.*, 1991, 1101.
140. R. Talay and D. Rehder, *Inorg. Chim. Acta*, 1983, **77**, L175.
141. H. Pohlmann and E. Weiss, *Chem. Ber.*, 1988, **121**, 1427.
142. K. J. Jens and E. Weiss, *Chem. Ber.*, 1984, **117**, 2469.
143. R. B. King and C. D. Hoff, *J. Organomet. Chem.*, 1982, **225**, 245.
144. W. A. Herrmann, E. Herdtweck and G. Weichselbaumer, *J. Organomet. Chem.*, 1989, **362**, 321.
145. J. Kulpe, E. Herdtweck, G. Weichselbaumer and W. A. Herrmann, *J. Organomet. Chem.*, 1988, **348**, 369.
146. F. Y. Petillon, P. Schollhammer and J. Talarmin, *J. Organomet. Chem.*, 1991, **411**, 159.
147. D. F. R. Gilson, G. Gomez, I. S. Butler and P. J. Fitzpatrick, *Can. J. Chem.*, 1983, **61**, 737.
148. F. J. Regina and A. Wojcicki, *Inorg. Chem.*, 1980, **19**, 3803.
149. M. Herberhold and H. Trampisch, *Z. Naturforschung*, 1982, **37b**, 614.
150. L. N. Lewis and K. G. Caulton, *J. Organomet. Chem.*, 1983, **252**, 57.
151. W. A. Herrmann, H. Biersack, B. Balbach, P. Wulknitz and M. L. Ziegler, *Chem. Ber.*, 1984, **117**, 79.
152. P. Oltmanns and D. Rehder, *J. Organomet. Chem.*, 1988, **345**, 87.
153. A. Kececi, D. Rehder, W. Roose and R. Talay, *Chem. Ber.*, 1982, **115**, 3257.
154. V. Grossrucchhaus and D. Rehder, *Inorg. Chim. Acta*, 1990, **172**, 141.
155. A. P. Kahn, D. A. Newman and K. P. C. Vollhardt, *Synth. Lett.*, 1990, **3**, 141.
156. G. Erker, R. Pfaff, C. Krüger and S. Werner, *Organometallics*, 1991, **10**, 3559.
157. G. Erker, R. Lecht, R. Schlund, K. Angermund and C. Krüger, *Angew. Chem., Int. Ed., Engl.*, 1987, **26**, 666.
158. R. J. Kinney, W. D. Jones and R. G. Bergmann, *J. Am. Chem. Soc.*, 1978, **100**, 7902.
159. S. Falicki and H. Alper, *Organometallics*, 1988, **7**, 2548.
160. B. Hessen, A. Meetsma and J. H. Teuben, *J. Am. Chem. Soc.*, 1988, **110**, 4860.
161. B. Hessen and J. H. Teuben, *J. Organomet. Chem.*, 1989, **367**, C18.
162. B. Hessen, A. Meetsma, F. van Bolhuis, J. H. Teuben, G. Helgenson and S. Jagner, *Organometallics*, 1990, **9**, 1925.
163. K. Jonas, V. Wiskamp, Y. H. Tsay and C. Krüger, *J. Am. Chem. Soc.*, 1983, **105**, 5480.
164. K. Jonas, W. Rüsseler, C. Krüger and E. Raabe, *Angew. Chem., Int. Ed. Engl.*, 1986, **25**, 925.
165. A. W. Duff, K. Jonas, R. Goddard, H. J. Kraus and C. Krüger, *J. Am. Chem. Soc.*, 1983, **105**, 5479.
166. K. Jonas, W. Rüsseler, K. Angermund and C. Krüger, *Angew. Chem., Int. Ed. Engl.*, 1986, **25**, 927.
167. M. N. Bochkorev, I. L. Fedushkin, H. Schumann and J. Loebel, *J. Organomet. Chem.*, 1991, **410**, 321.
168. R. Milczarek, W. Rüsseler, P. Binger, K. Jonas, K. Angermund, C. Kruger and M. Regitz, *Angew. Chem.*, 1987, **99**, 957.
169. O. J. Scherer, J. Schwalb, H. Swarowsky, G. Wolmershauser, W. Kaim and R. Gross, *Chem. Ber.*, 1988, **121**, 443.
170. J. Nieman, J. H. Teuben, J. C. Huffman and K. G. Caulton, *J. Organomet. Chem.*, 1983, **255**, 193.
171. K. H. Thiele and L. Oswald, *Z. Anorg. Allg. Chem.*, 1976, **423**, 231.
172. R. B. King and C. D. Hoff, *J. Organomet. Chem.*, 1982, **225**, 245.
173. D. B. Morse, T. B. Rauchfuss and S. R. Wilson, *Inorg. Chem.*, 1991, **30**, 775.
174. R. Poli, *Chem. Rev.*, 1991, **91**, 509.
175. J. Nieman and J. H. Teuben, *Organometallics*, 1986, **5**, 1149.
176. J. Nieman, J. H. Teuben, F. B. Hulsbergen, R. A. G. de Graaff and J. Reedijk, *Inorg. Chem.*, 1987, **26**, 2376.
177. F. A. Cotton, S. A. Duraj and W. J. Roth, *Organometallics*, 1985, **4**, 1174.
178. F. Calderazzo, G. Pampaloni, M. Sperrle and U. Englert, *Z. Naturforschung*, 1992, **47b**, 389.
179. F. Calderazzo, S. Ianelli, G. Pampaloni, G. Pelizzi and M. Sperrle, *J. Chem. Soc., Dalton. Trans.*, 1991, 693.
180. M. Herberhold, M. Kuhnlein, W. Kremnitz and A. L. Rheingold, *J. Organomet. Chem.*, 1990, **383**, 71.
181. B. Hessen *et al.*, *Organometallics*, 1987, **6**, 2354.
182. J. Nieman, J. W. Pattiasina and J. H. Teuben, *J. Organomet. Chem.*, 1984, **262**, 157.
183. B. Hessen, A. Meetsma and J. H. Teuben, *J. Am. Chem. Soc.*, 1989, **111**, 5977.
184. B. Hessen, J. H. Teuben, T. H. Lemmen, J. C. Huffman and K. G. Caulton, *Organometallics*, 1985, **4**, 946.

185. B. Hessen, F. van Bolhuis and J. H. Teuben, *J. Am. Chem. Soc.*, 1988, **110**, 295.
186. K. Jonas, W. Rüsseler, C. Krüger and E. Raabe, *Angew. Chem., Int. Ed. Engl.*, 1986, **25**, 925.
187. Ch. Elschenbroich, J. Heck, W. Massa, E. Nun and R. Schmidt, *J. Am. Chem. Soc.*, 1983, **105**, 2905.
188. H. P. Luthi and C. W. Bauschlicher, Jr., *J. Am. Chem. Soc.*, 1987, **109**, 2046.
189. J. Weber, H. Chermette and J. Heck, *Organometallics*, 1989, **8**, 2544.
190. P. Mougenot, J. Demuynck, M. Benard and C. W. Bauschlicher, Jr., *J. Am. Chem. Soc.*, 1988, **110**, 4503.
191. C. Poumbga, C. Daniel and M. Benard, *Inorg. Chem.*, 1990, **29**, 2387.
192. C. Poumbga, C. Daniel and M. Benard, *J. Am. Chem. Soc.*, 1991, **113**, 1090.
193. B. Bachmann, F. Hahn, J. Heck and M. Wünsch, *Organometallics*, 1989, **8**, 2523.
194. B. Bachmann, J. Heck, G. Meyer, J. Pebler and T. Schleid, *Inorg. Chem.*, 1992, **31**, 607.
195. B. Bachmann and J. Heck, *Organometallics*, 1991, **10**, 1373.
196. B. Bachmann, G. Baum, J. Heck, W. Massa and M. L. Ziegler, *Z. Naturforsch.*, 1990, **45b**, 221.
197. M. Herberhold, W. Kremnitz, M. Kuhnlein, M. L. Ziegler and K. Brunn, *Z. Naturforschung*, 1987, **42b**, 1520.
198. M. S. Hammer and L. Messerle, *Inorg. Chem.*, 1990, **29**, 1780.
199. D. B. Morse, D. N. Hendrickson, T. B. Rauchfuss and S. R. Wilson, *Organometallics*, 1988, **7**, 496.
200. F. Bottomley, J. Darkwa, L. Sutin and P. S. White, *Organometallics*, 1986, **5**, 2165.
201. D. L. Hughes, M. Jimenez-Tenorio and G. J. Leigh, *Polyhedron*, 1989, **8**, 1780.
202. D. B. Morse, T. B. Rauchfuss and S. R. Wilson, *J. Am. Chem. Soc.*, 1988, **110**, 2646.
203. W. A. Herrmann, E. Herdtweck and G. Weichselbaumer, *J. Organomet. Chem.*, 1989, **362**, 321.
204. F. Bottomley and L. Sutin, *J. Chem. Soc., Chem. Commun.*, 1987, 1112.
205. F. Bottomley, C. P. Magill and B. Zhao, *Organometallics*, 1990, **9**, 1700.
206. F. Preuss, H. Becker and H. J. Hausler, *Z. Naturforsch.*, 1987, **42b**, 881.
207. F. Preuss, H. Becker and T. Wieland, *Z. Naturforschung*, 1990, **45b**, 191.
208. M. Herberhold and M. Kuhnlein, *New. J. Chem.*, 1988, **12**, 357.
209. M. Herberhold, M. Kuhnlein, M. Schrepfermann, M. L. Ziegler and B. Nuber, *J. Organomet. Chem.*, 1990, **398**, 259.
210. K. G. Muller, J. L. Petersen and L. F. Dahl, *J. Organomet. Chem.*, 1976, **111**, 91.
211. C. M. Bolinger, T. B. Rauchfuss and A. L. Rheingold, *Organometallics*, 1982, **1**, 1551.
212. J. Darkwa, D. M. Giolando, C. Murphy and T. B. Rauchfuss, *Inorg. Synth.*, 1990, **27**, 51.
213. C. M. Bolinger, T. B. Rauchfuss and S. R. Wilson, *J. Am. Chem. Soc.*, 1982, **104**, 7313.
214. C. M. Bolinger, T. B. Rauchfuss and A. L. Rheingold, *J. Am. Chem. Soc.*, 1983, **105**, 6321.
215. C. M. Bolinger, T. D. Weatherill, T. B. Rauchfuss, A. L. Rheingold, C. S. Day and S. R. Wilson, *Inorg. Chem.*, 1986, **25**, 634.
216. T. B. Rauchfuss, T. D. Weatherill, S. R. Wilson and J. P. Zebrowski, *J. Am. Chem. Soc.*, 1983, **105**, 6508.
217. T. B. Rauchfuss, S. D. Gammon, T. D. Weatherill and S. R. Wilson, *New J. Chem.*, 1988, **12**, 373.
218. C. M. Bolinger *et al.*, *Organometallics*, 1986, **5**, 2386.
219. S. A. Duraj, M. T. Andras and B. Rihter, *Polyhedron*, 1989, **8**, 2763.
220. A. A. Pasynskii *et al.*, *J. Organomet. Chem.*, 1988, **344**, 195.
221. I. L. Eremenko, A. S. Katugin, A. A. Pasynskii, Y. T. Struchkov and V. E. Shklover, *J. Organomet. Chem.*, 1988, **345**, 79.
222. J. Darkwa, J. R. Lockemeyer, P. D. W. Boyd, T. B. Rauchfuss and A. L. Rheingold, *J. Am. Chem. Soc.*, 1988, **110**, 141.
223. M. Herberhold, M. Kuhnlein, M. Ziegler and B. Nuber, *J. Organomet. Chem.*, 1988, **349**, 131.
224. J. L. Robbins, N. Edelstein, B. Spencer and J. C. Smart, *J. Am. Chem. Soc.*, 1982, **104**, 1882.
225. F. H. Köhler, K. H. Doll and W. Prössdorf, *J. Organomet. Chem.*, 1982, **224**, 341.
226. F. H. Köhler and W. A. Geike, *J. Organomet. Chem.*, 1987, **328**, 35.
227. K. Jonas, W. Rüsseler, C. Krüger and E. Raabe, *Angew. Chem.*, 1986, **98**, 905.
228. J. R. Bocarsley, C. Floriani, A. Chiesi-Villa and C. Guastini, *Inorg. Chem.*, 1987, **26**, 1871.
229. S. Gambarotta, C. Floriani, A. Chiesi-Villa and C. Guastini, *Inorg. Chem.*, 1984, **23**, 1739.
230. B. V. Lokshin, I. I. Greenwald, *J. Mol. Struct.*, 1990, **222**, 11.
231. F. Calderazzo, G. Fachinetti and C. Floriani, *J. Am. Chem. Soc.*, 1974, **96**, 3695.
232. B. Jonson, B. Rebenstorf and R. Larsson, *Acta Chem. Scand.*, 1988, **A42**, 156.
233. R. M. Kowaleski, F. Basolo J. H. Osborne and W. Trogler, *Organometallics*, 1988, **7**, 1425.
234. R. M. Kowaleski, W. Trogler and F. Basolo, *Gazz. Chim. Ital.*, 1986, **116**, 105.
235. R. M. Kowaleski, A. L. Rheingold, W. Trogler and F. Basolo, *J. Am. Chem. Soc.*, 1986, **108**, 2460.
236. F. Bottomley and J. Darkwa, *J. Chem. Soc., Dalton Trans.*, 1983, 399.
237. F. Bottomley and J. Darkwa, *J. Chem. Soc., Dalton Trans.*, 1988, 2505.
238. M. Morán, J. J. Santos-Garcia, J. R. Masaguer and V. Fernández, *J. Organomet. Chem.*, 1985, **295**, 327.
239. A. M. Carrier, J. G. Davidson, E. K. Barefield and D. G. Van Derveer, *Organometallics*, 1987, **6**, 454.
240. G. Fochi, C. Floriani, J. C. J. Bart and G. Giunchi, *J. Chem. Soc., Dalton Trans.*, 1983, 1515.
241. (a) F. Bottomley, J. Darkwa and P. S. White, *J. Chem. Soc., Dalton Trans.*, 1985, 1435; (b) F. Bottomley, J. Darkwa and P. S. White, *J. Chem. Soc., Chem. Commun.*, 1982, 1039.
242. S. Gambarotta, C. Floriani, A. Chiesi-Villa and C. Guastini, *Organometallics*, 1986, **5**, 2425.
243. S. Gambarotta, C. Floriani, A. Chiesi-Villa and C. Guastini, *J. Am. Chem. Soc.*, 1982, **104**, 2019.
244. M. Pasquali, P. Leoni, C. Floriani, A. Chiesi-Villa and C. Guastini, *Inorg. Chem.*, 1983, **22**, 841.
245. R. Drews, D. Wormsbächer and U. Behrens, *J. Organomet. Chem.*, 1984, **272**, C40.
246. J. Benecke, R. Drews, U. Behrens, F. Edelmann, K. Keller and H. W. Roesky, *J. Organomet. Chem.*, 1987, **320**, C31.
247. T. Sielisch and U. Behrens, *J. Organomet. Chem.*, 1986, **310**, 179.
248. S. Gambarotta, C. Floriani, A. Chiesi-Villa and C. Guastini, *J. Am. Chem. Soc.*, 1982, **104**, 1918.
249. S. Gambarotta, M. L. Fiallo, C. Floriani, A. Chiesi-Villa and C. Guastini, *Inorg. Chem.*, 1984, **23**, 3532.
250. M. Pasquali, C. Floriani, A. Chiesi-Villa and C. Guastini, *Inorg. Chem.*, 1980, **19**, 3847.
251. C. J. Curtis, J. C. Smart and J. L. Robbins, *Organometallics*, 1985, **4**, 1283.
252. G. A. Luinstra and J. H. Teuben, *J. Chem. Soc., Chem. Commun.*, 1990 1470.
253. H. Köpf and N. Klouras, *Chem. Scripta*, 1982, **19**, 122.
254. H. Köpf and N. Klouras, *Bull. de la Soc. Chim., Beograd* 1981, **46**, 709.
255. N. Klouras, *Z. Naturforschung*, 1991, **46b**, 650.

256. S. P. Mishra, S. N. Singh, J. C. Evans and C. C. Rowlands, *J. Mol. Struct.*, 1984, **112**, 59.
257. J. C. Green, M. P. Payne and J. H. Teuben, *Organometallics*, 1983, **2**, 203.
258. Y. Mugnier, C. Moise and E. Laviron, *Nouv. J. Chim.*, 1982, **6**, 197.
259. W. E. Lindsell and R. A. Parr, *Polyhedron*, 1986, **5**, 1259.
260. P. Gowik, T. Klapötke and J. Pickardt, *Organometallics*, 1989, **8**, 2953.
261. P. Gowik and T. Klapötke, *Monat. Chem.*, 1989, **120**, 711.
262. P. Gowik and T. Klapötke, *J. Organomet. Chem.*, 1990, **385**, 345.
263. F. H. Görlitz, P. Gowik, T. Klapötke, D. Wang, R. Meier and J. von Welzen, *J. Organomet. Chem.*, 1991, **408**, 343.
264. G. A. Miller, M. J. Therien and W. C. Trogler, *J. Organomet. Chem.*, 1990, **383**, 271.
265. F. Bottomley, J. Darkwa and P. S. White, *Organometallics*, 1985, **4**, 961.
266. M. Morán and M. Gayoso, *J. Organomet. Chem.*, 1983, **243**, 423.
267. M. Morán and M. Gayoso, *Z. Naturforschung* 1983, **38b**, 177.
268. G. A. Razuvaev *et al.*, *J. Organomet. Chem.*, 1987, **329**, 327.
269. P. Köpf-Maier, W. Wagner and E. Liss, *J. Cancer Res. Clin. Oncol.*, 1981, **201**, 21.
270. P. Köpf-Maier, *J. Cancer Res. Clin. Oncol.*, 1982, **103**, 145.
271. P. Köpf-Maier, W. Wagner and E. Liss, *J. Cancer Res. Clin. Oncol.*, 1983, **106**, 44.
272. P. Köpf-Maier and D. Krahl, *Chem.-Biol. Interactions*, 1983, **44**, 317.
273. P. Köpf-Maier and G. Hermann, *Virchows Arch. [Cell Pathol.]*, 1984, **47**, 107.
274. J. H. Toney, M. S. Murthy and T. J. Marks, *Chem.-Biol. Interactions*, 1985, **56**, 45.
275. P. Köpf-Maier and P. Funke-Kaiser, *Toxicology*, 1986, **38**, 81.
276. M. S. Murthy, L. N. Rao, L. Y. Kuo, J. H. Toney and T. J. Marks, *Inorg. Chim. Acta*, 1988, **152**, 117.
277. S. G. Ward, R. C. Taylor, P. Köpf-Maier, H. Köpf, J. Balzarini and E. De Clercq, *Appl. Organomet. Chem.*, 1989, **3**, 491.
278. A. Moustatih, M. M. L. Fiallo, A. Garnier-Suillerot, *J. Med. Chem.*, 1989, **32**, 336.
279. J. H. Toney and T. J. Marks, *J. Am. Chem. Soc.*, 1985, **107**, 947.
280. J. H. Toney C. P. Brock and T. J. Marks, *J. Am. Chem. Soc.*, 1986, **108**, 7263.
281. M. L. McLaughlin, J. M. Cronan, Jr., T. R. Schaller and R. D. Snelling, *J. Am. Chem. Soc.*, 1990, **112**, 8949.
282. F. Calderazzo, G. Pampaloni, G. Pelizzi, M. Sperrle and U. Englert, *Z. Naturforschung*, 1992, **47b**, 389.
283. F. Calderazzo, S. Ianelli, G. Pampaloni G. Pelizzi and M. Sperrle, *J. Chem. Soc., Dalton Trans.*, 1991, 693.
284. F. Bottomley, C. P. Magill and P. S. White, *J. Am. Chem. Soc.*, 1989, **111**, 3070.
285. F. Bottomley, D. E. Paez and P. S. White, *J. Am. Chem. Soc.*, 1982, **104**, 5651.
286. F. Bottomley, D. E. Paez and P. S. White, *J. Am. Chem. Soc.*, 1985, **107**, 7226.
287. F. Bottomley, D. F. Drummond, D. E. Paez and P. S. White, *J. Chem. Soc. Chem. Commun.*, 1986, 1752.
288. F. Bottomley, L. Sutin and P. S. White, *J. Organomet. Chem.*, 1988, **338**, 205.
289. S. Sabo, R. Choukroun and D. Gervais, *J. Chem. Soc., Dalton Trans.*, 1983, 2695.
290. R. Choukroun, *Inorg. Chim. Acta*, 1981, **58**, 121.
291. S. Gambarotta, C. Floriani, A. Chiesi-Villa and C. Guastini, *J. Chem. Soc., Chem. Commun.*, 1983, 184.
292. C. Floriani, S. Gambarotta, A. Chiesi-Villa and C. Guastini, *J. Chem. Soc., Dalton Trans.*, 1987, 2099.
293. S. A. Koch and V. Chebolu, *Organometallics*, 1983, **2**, 350.
294. M. J. Bader and A. H. Al-Mowali, *Chem. Scripta*, 1983, **22**, 241.
295. T. A. Wark and D. W. Stephan, *Organometallics*, 1989, **8**, 2836.
296. T. Klapötke and H. Köpf, *Z. Anorg. Allg. Chem.*, 1988, **558**, 217.
297. L. V. Pankratov *et al.*, *J. Organomet. Chem.*, 1987, **335**, 313.
298. G. A. Razuvaev *et al.*, *J. Organomet. Chem.*, 1982, **225**, 233.
299. A. T. Casey and J. B. Raynor, *J. Chem. Soc., Dalton Trans.*, 1983, 2057.
300. M. Morán and I. Cuadrado, *J. Organomet. Chem.*, 1986, **311**, 333.
301. S. Gambarotta, A. Chiesi-Villa and C. Guastini, *J. Organomet. Chem.*, 1984, **270**, C49.
302. J. H. Osborne, A. L. Rheingold and W. C. Trogler, *J. Am. Chem. Soc.*, 1985, **107**, 7945.
303. J. H. Osborne and W. C. Trogler, *Inorg. Chem.*, 1985, **24**, 3098.
304. K. Jacob, W. Kretschmer, K.-H. Thiele, I. Pavlik, A. Lycka and J. Holecek, *Z. Anorg. Allg. Chem.*, 1991, **606**, 133.
305. R. Sustmann and G. Kopp, *J. Organomet. Chem.*, 1988, **347**, 313.
306. R. Sustmann and G. Kopp, *J. Organomet. Chem.*, 1988, **347**, 325.
307. W. J. Evans I. Bloom and R. J. Doedens, *J. Organomet. Chem.*, 1984, **265**, 249.
308. J. Nieman and J. H. Teuben, *J. Organomet. Chem.*, 1985, **287**, 207.
309. D. F. Foust and M. D. Rausch, *J. Organomet. Chem.*, 1985, **287**, 195.
310. B. H. Chang, H.-S. Tung and C. H. Brubaker, Jr., *Inorg. Chim. Acta*, 1981, **51**, 143.
311. S. I. Bailey, L. M. Engelhardt, W.-P. Leung, C. L. Raston, I. M. Ritchie and A. I. White, *J. Chem. Soc., Dalton Trans.*, 1985, 1747.
312. J. W. F. L. Seetz, B. J. J. Van de Heisteeg, G. Schat, O. S. Akkerman and F. J. Bickelhaupt, *Organomet. Chem.*, 1984, **275**, 173.
313. F. H. Köhler, W. A. Geike, P. Hofmann, U. Schubert and P. Stauffert, *Chem. Ber.*, 1984, **117**, 904.
314. M. Wedler, H. W. Roesky, F. T. Edelmann and U. Behrens, *Z. Naturforschung*, 1988, **43b**, 1461.
315. D. Melzer and E. Weiss, *J. Organomet. Chem.*, 1984, **263**, 67.
316. M. N. Bochkarev, L. V. Pankratov, V. K. Cherkasov, G. A. Razuvaev, V. N. Latyaeva, A. N. Lineva, *J. Organomet. Chem.*, 1984, **263**, 21.
317. P. Hudeczek and F. H. Köhler, *Organometallics*, 1992, **11**, 1773.
318. U. Behrens and F. G. A. Stone, *J. Chem. Soc., Dalton Trans.*, 1984, 1605.
319. G. Fachinetti and S. Del Nero and C. Floriani, *J. Chem. Soc., Dalton Trans.*, 1976, 1046.
320. C. Moise, *J. Organomet. Chem.*, 1983, **247**, 27.
321. J. Martin, M. Fauconet, C. Moise, *J. Organomet. Chem.*, 1989, **371**, 87.
322. T. Kusumoto, K. Nishide and T. Hiyama, *Bull. Chem. Soc. Jpn.*, 1990, **63**, 1947.
323. K. Soga, T. Uozumi and H. Yanagihara, *Makromol. Chem.*, 1989, **190**, 31.
324. C. Aitken, J.-P. Barry, F. Gauvin, J. F. Harrod, A. Malek and D. Rousseau, *Organometallics*, 1989, **8**, 1732.
325. P. D. Morand and C. G. Francis, *Inorg. Chem.*, 1985, **24**, 56.

326. G. A. Ozin, M. P. Andrews, C. G. Francis, H. X. Hüber and K. Molnar, *Inorg. Chem.*, 1990, **29**, 1068.
327. C. G. Francis, P. D. Morand and N. J. Spare, *Organometallics*, 1985, **4**, 1958.
328. G. A. Ozin and M. P. Andrews, *Angew. Chem., Int. Ed. Engl.*, 1982, **21**, 212.
329. F. Calderazzo, G. Pampaloni, L. Rocchi and J. Marchetti, *J. Organomet. Chem.* 1991, **417**, C16.
330. F. Calderazzo, *Inorg. Chem.*, 1964, **6**, 810.
331. T. Avilés and J. H. Teuben, *J. Organomet. Chem.*, 1983, **253**, 39.
332. P. N. Hawker and P. L. Timms, *J. Chem. Soc., Dalton Trans.*, 1983, 1123.
333. C. Elschenbroich, J. Hurley, B. Metz, W. Massa and G. Baum, *Organometallics*, 1990, **9**, 889.
334. G. Fochi, D. Braga and P. Sabatino, *Organometallics*, 1988, **7**, 565.
335. D. Braga and P. Sabatino, *Acta Cryst.*, 1990, **C46**, 2308.
336. G. Fochi, X. Runjan and A. Colligiani, *J. Chem. Soc., Dalton Trans.*, 1990, 2551.
337. M. N. Bochkarev, L. V. Pankratov, V. K. Cherkasov, G. A. Razuvaev, V. N. Latyaeva and A. N. Lineva, *J. Organomet. Chem.*, 1984, **263**, 21.
338. S. A. Duraj, M. T. Andras and P. A. Kibala, *Inorg. Chem.*, 1990, **29**, 1232.
339. F. Calderazzo, G. Pampaloni, M. Sperrle and U. Englert, *Z. Naturforschung*, 1992, **47b**, 389.
340. I. I. Greenwald and B. V. Lokshin, *J. Mol. Struct.* 1990, **218**, 255.
341. A. McCamley and R. N. Perutz, *J. Phys. Chem.*, 1991, **95**, 2738.
342. R. N. Perutz, J. N. Hill and A. McCamley, *Coord. Chem. Rev.*, 1991, **111**, 111.
343. S. Yu. Ketkov, G. A. Domrachev and G. A. Razuvaev, *J. Mol. Structure* 1989, **195**, 175.
344. C. Elschenbroich, E. Bilger and B. Metz, *Organometallics*, 1991, **10**, 2823.
345. G. Cirelli, A. Russu, R. Wolf, M. Rudin, A. Schweiger and H. H. Günthard, *Chem. Phys. Lett.*, 1982, **92**, 223.
346. R. Wolf, A. Schweiger and H. H. Günthard, *Mol. Physics*, 1984, **53**, 567.
347. R. Wolf, A. Schweiger and H. H. Günthard, *Mol. Physics*, 1984, **53**, 585.
348. F. G. N. Cloke, A. N. Dix, J. C. Green, R. N. Perutz and E. A. Seddon, *Organometallics*, 1983, **2**, 1150.
349. M. P. Andrews, S. M. Mattar and G. A. Ozin, *J. Phys. Chem.*, 1986, **90**, 1037.
350. R. G. Graham, R. Grinter and R. N. Perutz, *J. Am. Chem. Soc.*, 1988, **110**, 7036.
351. M.-D. Su and S.-Y. Chu, *J. Organomet. Chem.*, 1990, **384**, 71.
352. K. Jonas, W. Rüsseler, C. Krüger and E. Raabe, *Angew. Chem.*, 1986, **98**, 905.
353. F. Calderazzo, G. Pampaloni, D. Vitali and P. F. Zanazzi, *J. Chem. Soc., Dalton Trans.*, 1982, 1993.
354. M. P. Andrews, H. X. Hüber, S. M. Mattar, D. F. McIntosh and G. A. Ozin, *J. Am. Chem. Soc.*, 1983, **105**, 6170.
355. M. P. Andrews, S. M. Mattar and G. A. Ozin, *J. Phys. Chem.*, 1986, **90**, 744.
356. S. M. Mattar and W. Hamilton, *J. Phys. Chem.*, 1989, **93**, 2997.
357. S. M. Mattar and S. E. Brewer, *J. Phys. Chem.*, 1992, **96**, 1611.
358. M. P. Andrews and G. A. Ozin, *J. Phys. Chem.*, 1986, **90**, 1245.
359. G. E. Herberich, W. Boveleth, B. Hessner, W. Koch, E. Raabe and D. Schmitz, *J. Organomet. Chem.*, 1984, **265**, 225.
360. E. J. Wucherer and E. L. Muetterties, *Organometallics*, 1987, **6**, 1691.
361. E. J. Wucherer and E. L. Muetterties, *Organometallics*, 1987, **6**, 1696.
362. C. Elschenbroich, M. Nowotony, B. Metz, W. Massa and J. Graulich, *Angew. Chem.*, 1991, **103**, 601.
363. D. R. Wilson, J.-Z. Liu and R. D. Ernst, *J. Am. Chem. Soc.*, 1982, **104**, 1120.
364. C. F. Campana, R. D. Ernst, D. R. Wilson and J.-Z. Liu, *Inorg. Chem.*, 1984, **23**, 2732.
365. P. T. DiMauro and P. T. Wolczanski, *Organometallics*, 1987, **6**, 1947.
366. J. W. Freeman, N. C. Hallinan, A. M. Arif, R. W. Gedridge, R. D. Ernst and F. Basolo, *J. Am. Chem. Soc.*, 1991, **113**, 6509.
367. T. D. Newbound, A. L. Rheingold and R. D. Ernst, *Organometallics*, 1992, **11**, 1693.
368. R. M. Kowaleski, F. Basolo, W. Trogler, R. W. Gedridge, T. D. Newbound and R. D. Ernst, *J. Am. Chem. Soc.*, 1987, **109**, 4860.
369. R. M. Kowaleski, F. Basolo, W. Trogler and R. D. Ernst, *J. Am. Chem. Soc.*, 1986, **108**, 6046.
370. N. C. Hallinan, G. Morelli and F. Basolo, *J. Am. Chem. Soc.*, 1988, **110**, 6585.
371. R. M. Kowaleski, F. Basolo, J. H. Osborne and W. Trogler, *Organometallics*, 1988, **7**, 1425.
372. R. G. Swisher, E. Sinn, G. A. Brewer and R. N. Grimes, *J. Am. Chem. Soc.*, 1983, **105**, 2079.
373. R. G. Swisher, E. Sinn and R. N. Grimes, *Organometallics*, 1984, **3**, 599.

2
Niobium and Tantalum

DAVID E. WIGLEY and STEVEN D. GRAY
University of Arizona, Tucson, AZ, USA

2.1 INTRODUCTION AND ORGANIZATION

2.1.1 Overview

Although the rapid growth in niobium and tantalum organometallic chemistry is often considered more contemporary than that of their neighboring elements, these metals have been of singular importance in affording paradigms for many new ligands, structures, reactions, and complexes. Much of the chemistry of niobium and tantalum is dominated by the thermodynamic driving force to attain their highest oxidation state, in addition to their Lewis acidity in their high-valent complexes. When coupled with their facility for forming metal–ligand multiple bonds, these properties reinforce the viability of certain complexes as models for supported early metal catalysts, in addition to imparting considerable use to others for carbon–carbon bond formations in organic synthesis. At the same time, the propensity of these metals to form dinuclear or cluster complexes, including a sizable number of heterobimetallic compounds, in lower oxidation states has also afforded a range of heretofore-unknown structures and unusual reaction chemistry.

Complexes of niobium and tantalum have been essential to understanding the chemistry of a number of metal-catalyzed processes. For example, the development of high-oxidation-state niobium and tantalum alkylidene chemistry has been central to attaining a firm grasp of alkene metathesis. Certain tantalum complexes have provided exceptional models for Fischer–Tropsch catalysis, especially by affording examples of the elusive carbon–carbon bond formation steps. The evasive carbon–nitrogen bond scission step of hydrodenitrogenation catalysts has been successfully modeled with complexes of tantalum, while other tantalum species constitute highly selective catalysts for alkene dimerization.

The growth of the field since the first edition of this work, as simply judged from the numbers of papers published, has been remarkable. The niobium and tantalum chapter in *COMC-I* included coverage through 1980. Since that time, the total number of papers in this area has roughly tripled. Important general reference works for niobium and tantalum organometallic compounds include the outstanding chapter in *COMC-I* by Labinger.[1] The reader is referred to that chapter for a thorough introduction to this area, for an excellent historical context, for discussion of the unusual NMR properties of niobium complexes, and for the exhaustive collection of organometallic compounds reported through 1980. Reviews of the structural chemistry of niobium[2] and tantalum[3] are also of note. The *Dictionary of Organometallic Compounds*, first published in 1984, provides a valuable compilation of niobium and tantalum compounds[4] in a series supplemented with annual updates. Several relevant review articles have appeared over the years and these reviews and general reference works for specific ligand or compound types are noted in each section.

2.1.2 Organization and Coverage

The heart of this chapter consists of a series of compound tables, arranged as indicated in the table of contents, which afford an organizational framework on which to found discussions. Since heteroleptic compounds (containing more than one type of ligand) are usually described in only one section of this chapter, a set of "Classification Priorities" as reported in Table 1 has been adapted from Labinger's contribution in *COMC-I* and serves to categorize these compounds. Thus, complexes containing the highest priority alkylidene ligands are described in the alkylidene section and appear only in the alkylidene table. Note that compounds are not discussed in this priority, but are rather organized in a progression according to ligand size or complexity, as outlined in the table of contents. For clarity, additional comments which serve to clarify a specific ligand definition or compound type will be noted in each section.

Table 1 Classification priorities for heteroleptic complexes.

Priority	*Carbon ligand*	*Comments*
1	Alkylidenes and alkylidynes	M=C bonds in metallacyclic environments described as metallacycles. Includes μ-CHR and μ-CR ligands
2	π-Bonded ligands other than cyclopentadienyl	π-ligand must have at least one carbon directly bound to metal
3	η^1 Ligands	alkyls are higher in priority than hydrides, which are higher in priority than metallacycles. Only metallacycles with at least one direct M–C bond are included here
4	Carbonyls and isocyanides	CO or CNR must be bound to Nb or Ta in heterobimetallic complexes
5	Cyclopentadienyls	other η^5 ligands are included here, e.g., pentadienyl and $[\eta^5\text{-}C_2B_9H_{11}]^{2-}$ (dicarbollide)

In order to make each ligand type (and the corresponding discussion) more tractable, most are further subdivided. For example, larger tables may be subdivided according to the ancillary ligand set, formal oxidation state, or some other convenient classification. Each table includes a "Comments" section, which notes if a structural study has been undertaken, specified as either "x ray" or "neutron diffraction." Occasionally synthetic strategies (noted as "prep") are included and any additional information which may be useful is provided. Throughout this chapter, M refers to both niobium and tantalum.

Labinger's chapter on niobium and tantalum in *COMC-I* included coverage through 1980, with a few papers from 1981. To insure that no gap in the record exists, we have included compounds which appeared in 1980 in the tables, and therefore consider coverage from 1980 through 1993, but discussion of the 1980 chemistry is limited (except where especially relevant) and reference is made to Labinger's chapter in *COMC-I* where certain topics are covered in more detail.

2.2 CARBONYLS AND ISOCYANIDES

2.2.1 Carbonyls and Substituted Carbonyls

Reviews on advances in group 5 carbonyl chemistry appeared in 1988[5] and 1992.[6] Since 1980, significant progress in the carbonyl chemistry of these elements has been made, including the development of low-pressure syntheses of the hexacarbonyl anions and considerable work on the intermediate oxidation states of the monokis η^5-Cp carbonyl complexes.

2.2.1.1 Carbonyl complexes without cyclopentadienyl ligands

Carbonyl complexes that do not contain cyclopentadienyl ligands appear in Table 2, where these species are further sorted according to their ancillary ligand set.

In contrast to the stability of $V(CO)_6$, neutral binary carbonyls of niobium or tantalum have not been characterized (except in matrix isolation experiments[5,36]) and neutral mixed-ligand species of the type $M(CO)_n(PR_3)_{6-n}$ also remain elusive. However, recently isolated complexes $Ag_3M_3(CO)_{12}(dmpe)_3$ (**1**) do appear to contain M^0 [M(CO)$_4$(dmpe)] units which are loosely coordinated to a trinuclear metallic silver cluster,[18] and completely substituted $Ta(dmpe)_3$ has been described.[37] In spite of the lack of neutral carbonyls, salts of the 18-electron anions $[M(CO)_6]^-$ are readily accessible and constitute useful precursors in carbonyl chemistry. Their preferred method of preparation now involves the low-temperature, alkali metal naphthalenide reduction of the pentahalides, followed by the addition of atmospheric pressure CO, from which multigram quantities of $[M(CO)_6]^-$ salts can be generated in good yields (Equation (1))[10] This method, first described in 1983, supplants the low-yield, high-pressure carbonylation strategies previously employed.[1] Numerous mixed-ligand substitution derivatives of the type $[M(CO)_nL_{6-n}]^-$ (L = neutral, two-electron ligand) have also been reported (Table 2), and a useful theoretical study on hexacarbonyl substitution reactions has appeared.[38]

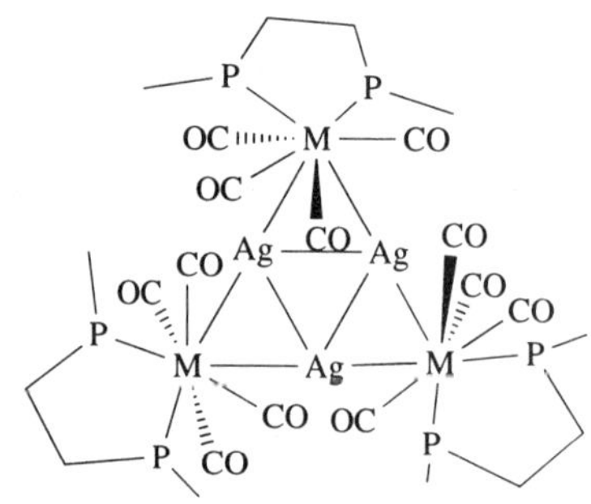

(**1**)

$$MCl_5 + Na[C_{10}H_8] + CO\ (1\ atm) \longrightarrow Na[M(CO)_6] \qquad (1)$$

Both oxidized and reduced derivatives of $[M(CO)_6]^-$ have been studied; for example, the "highly reduced" carbonyls $[M(CO)_5]^{3-}$ have been described by Ellis.[39] Both niobium and tantalum complexes $[Na(DIGLYME)_2][M(CO)_6]$ can be smoothly reduced with sodium in liquid NH_3 to provide unstable $Na_3M(CO)_5$, which is metathesized to afford shock-sensitive, explosive $Cs_3M(CO)_5$ in moderate yields.[11] These species are subject to electrophilic attack, for example, the *in situ* treatment of the trisodium salt with Ph_3SnCl provides $[Et_4N][M(CO)_5(SnPh_3)]$ after metathesis, while treatment with $[NH_4]Cl$ provides unstable $Na[M(CO)_5(NH_3)]$.

Upon oxidizing salts of $[M(CO)_6]^-$ with HCl or halogens, the M^I dinuclear complex anions $[M_2(\mu\text{-}X)_3(CO)_8]^-$ (**2**) are prepared (Scheme 1).[17] For example, $[Na(THF)_2][Nb_2Cl_3(CO)_8]$ is available from reacting $[Na(THF)_n][Nb(CO)_6]$ with $PhICl_2$, while $[H(THF)_2][M_2Cl_3(CO)_8]$ are the products of the reaction between $[Na(THF)_n][M(CO)_6]$ and HCl. Simple metathesis reactions afford the μ-acetate or μ-OMe derivatives $Na[Nb_2(\mu\text{-}X)_3(CO)_8]$ ($X^- = OMe^-$ or O_2CMe^-). The dinuclear complexes $[H(THF)_2][M_2Cl_3(CO)_8]$ are readily reduced with sodium in the presence of CO to re-form the corresponding $[M(CO)_6]^-$ complexes (Scheme 1).

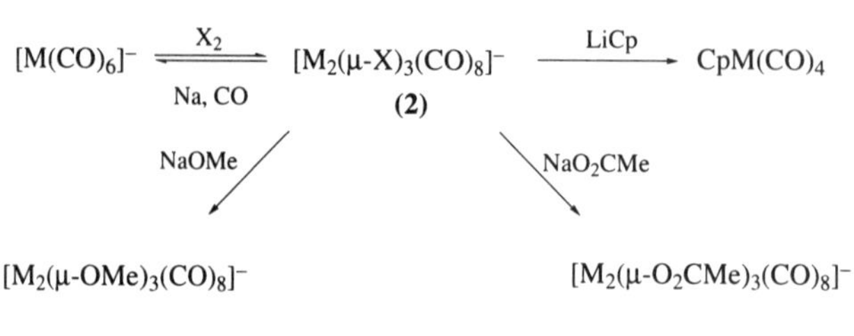

Scheme 1

Finally, the double salts $[M(CO)_3(PMe_3)_4]^+[M(CO)_5(PMe_3)]^-$ have been described for both niobium[26] and tantalum.[25] The niobium analogue has been obtained, albeit irreproducibly and in low yield, by

Table 2 Carbonyl complexes of niobium and tantalum without cyclopentadienyl ligands.

Compound	*Comments*	*Ref.*
Binary carbonyls		
$Na[M(CO)_6]\cdot n$THF	prep: MCl_5 + Na + CO + cot	7
	prep: MCl_5 + Mg/Zn/py + CO	8
	prep: $[H(THF)_2][M_2(CO)_8Cl_3]$ + Na + CO	9
$Na[Ta(CO)_6]$		10
$[Na(diglyme)_2][M(CO)_6]$		10,11
$[NR_4][M(CO)_6]$	R = Et, Bu^n; 1 atm CO preparation	10
$[PPN][Nb(CO)_6]$	x ray	12,8
$[NEt_4][Nb(CO)_6]$	^{93}Nb NMR study	13
$[PPh_4][Ta(CO)_6]$		10
$[Q][M(CO)_6]$	$[Q]^+ = [Me\text{-}py]^+$, $[Me\text{-}Nic]^+$, $[Me\text{-}QUIN]^+$, $[Me_2\text{-}4,4'\text{-}bipy]^+$	14
$[Me_2\text{-}bipy][M(CO)_6]_2$		14
$[Ni(phen)_3][Nb(CO)_6]_2$	prep: $[Na(THF)][Nb(CO)_6]$ + $[Ni(phen)_3]Cl_2$	12
$Na_3M(CO)_5$	unstable	11
$Cs_3M(CO)_5$	explosive	11
Isocyanide carbonyls		
$[AsPh_4][M(CO)_5(CNBu^t)]$		11
$Nb(CNMe)(CO)(dmpe)_2Cl$		15
$Na[M(CNMe)(CO)(dmpe)_2]$		16
Halide complexes		
$[H(THF)_2][M_2(\mu\text{-}Cl)_3(CO)_8]$	M = Nb (x ray), Ta; prep: $[Na(THF)][Nb(CO)_6]$ + HCl	9,17
$[PPN][M_2(\mu\text{-}Cl)_3(CO)_8]$		9,17
$[Na(THF)_2][Nb_2(\mu\text{-}Cl)_3(CO)_8]$		17
$Na[Nb_2(\mu\text{-}X)_3(CO)_8]$	X = Br, I	17
$Na[M_2(\mu\text{-}X)_3(CO)_8]$	X = Cl, I	14
Nitrogen-donor ligands		
$[M^1][M^2(CO)_5(NH_3)]$	$[M^1]^+ = Na^+$ (unstable), $[AsPh_4]^+$; M^2 = Nb, Ta	11
$Ta(CO)_4(L_2)I$	L_2 = TMEDA, bipy	18
Phosphorus-donor ligands		
$[AsPh_4][M(CO)_5(PR_3)]$	$PR_3 = PR_3$, $P(OR)_3$; R not specified	11
$[NEt_4][Nb(CO)_5(PR_3)]$	$PR_3 = PEt_3$, $P(NEt_3)_3$, $P(OMe)_3$	19
$[M^1][M^2(CO)_4(dppe)]$	$[M^1]^+ = [Cp^*{}_2Co]^+$, $[NEt_4]^+$; M^2 = Nb, Ta	18
$[NEt_4][Nb(CO)_4L_2]$	$L_2 = Ph_2P(CH_2)_nPPh_2$ for n = 1–4, $Ph_2As(CH_2)_2AsPh_2$, or *cis*-$Ph_2PCH{=}CHPPh_2$; ^{93}Nb NMR studies	19
	L_2 = dppe, $PhP(CH_2CH_2PPh_2)_2$, $P(CH_2CH_2PPh_2)_3$	20
$Nb(CO)_4(dppe)X$	X = Cl, Br, I	14
$Ta(CO)_4(dppe)X$	X = Cl, Br, I (x ray)	14
$Ta(CO)_4(dppm)I$		18
$[NEt_4][Ta(CO)_4(L_2)]$	$L_2 = \eta^2\text{-}P(CH_2CH_2PPh_2)_3$, dppe, $\eta^2\text{-}PhP(CH_2CH_2PPh_2)_2$, $\eta^2\text{-}(Ph_2PCH_2CH_2PPhCH_2)_2$	21
$Ta(CO)_3(PMe_3)_3X$	X = Cl (x ray), Br, I	22
	X = Cl, x ray	23
$Ta(CO)_3[MeP(CH_2CH_2CH_2PMe_2)_2]Cl$		24
$[Ta(CO)_3(PMe_3)_4][Ta(CO)_5(PMe_3)]$	x ray; prep: $TaH_2(\eta^2\text{-}BH_4)(PMe_3)_4$ + CO	25
$[Nb(CO)_3(PMe_3)_4][Nb(CO)_5(PMe_3)]$	prep: $(\eta\text{-}PhMe)Nb(PMe_3)_3H$ + CO	26
$[NEt_4][Nb(CO)_3(PF_3)_3]$	^{93}Nb NMR study	13
$Nb(CO)_2(dmpe)_2Cl$		27,28
$M(CO)_2(dmpe)_2Cl$	M = Nb (x ray), Ta; low-pressure preparation	29
$Nb(CO)_2(dmpe)_2X$	X = Br, N_3	28
$Nb(CO)_2(dmpe)_2I$	x ray	29
$Ta(CO)_2(depe)_2Cl$	x ray	29
$Ta(CO)_2(dppe)_2Cl$		24
$Ta(CO)_2(dmpe)_2X$	X = Cl, Br, I	27
$Ta(CO)_2(dmdepe)_2Cl$[a]		27
$Na[Ta(CO)_2(dmpe)_2]$		30,31
$[Hx^n{}_4N][Ta(CO)_2(dmpe)_2]$	x ray	30
$[NEt_4][Ta(CO)_4L_2]$	L_2 = 11 different bidentate phosphine ligands	24
$[NEt_4][Ta(CO)_5(\eta^1\text{-}L_2)]$	$L_2 = Ph_2PCH_2CH_2PEt_2$, $Cy_2PCH_2CH_2PCy_2$	24
Oxygen-donor ligands		
$Na[Nb_2(OAc)_3(CO)_8]$		9,17
$Na[Nb_2(OMe)_3(CO)_8]$		17
$M(CO)_4(acac)(THF)$	prep: $[M(CO)_6]^-$ + $Ni(acac)_2$ or $Fe(acac)_3$; x ray of $Ta(CO)_4(acac)(THF)$	32
$Ta(CO)(silox)_3$[b]	unstable	33
$Ta(CO)(silox)_3(THF)$		33
$Ta(CO)_2(silox)_3$		33
"$[Ta(CO)(silox)_3]_n$"	structure unknown, but probably not simple CO adduct; n = 2 likely	33

Table 2 (continued)

Compound	*Comments*	*Ref.*
Heterobimetallic carbonyls		
$[NEt_4][M(CO)_5(SnPh_3)]$		11
$Ag_3M_3(CO)_{12}(dmpe)_3$	M = Nb (x ray), Ta	18
	M = Nb, Ta (x ray)	34
$(Ph_3PAu)_3Ta(CO)_5$	prep: $[NEt_4]_2[Ta(H)(CO)_5] + 3Ph_3PAuCl$	35
$[Ta(CO)(dmpe)_2](\mu\text{-}CO)(Cp^*{}_2ZrCl)$	prep: $Na[Ta(CO)_2(dmpe)_2] + Cp^*{}_2ZrCl_2$; x ray	31

[a] dmdepe = $Me_2PCH_2CH_2PEt_2$. [b] silox = $[OSiBu^t{}_3]^-$.

displacement of the arene in $(\eta\text{-}C_6H_5Me)Nb(PMe_3)_3H$ with CO[26] while the tantalum complex has been prepared in high yield by the addition of CO to $Ta(\eta^2\text{-}BH_4)H_2(PMe_3)_4$.[25]

2.2.1.2 Cyclopentadienyl carbonyl complexes

Cyclopentadienyl carbonyl complexes are listed in Table 3 and are also sorted according to ancillary ligand.

The cyclopentadienyl derivatives $(\eta\text{-}C_5R_5)M(CO)_4$ remain useful precursors and reagents in niobium and tantalum carbonyl chemistry. These species are prepared by reductive carbonylation of $(\eta\text{-}C_5R_5)MCl_4$ (using Na or Zn) under high CO pressures, often in the presence of aluminum–copper as halogen acceptor,[40,41] although a low-pressure route has been reported. Thus, high yields of $CpM(CO)_4$ can also be obtained from reacting dinuclear $[M_2(\mu\text{-}X)_3(CO)_8]^-$ with LiCp.[17] $(\eta\text{-}C_5R_5)M(CO)_4$ compounds exhibit a rich reactivity; like $[M(CO)_6]^-$, $(\eta\text{-}C_5R_5)M(CO)_4$ species undergo both oxidation and reduction. Photolytic substitution of one or more of the CO ligands in $(\eta\text{-}C_5R_5)M(CO)_4$ complexes

Table 3 Carbonyl complexes of niobium and tantalum containing cyclopentadienyl ligands.

Compound	*Comments*	*Ref.*
Complexes with carbonyl and cyclopentadienyl ligands only		
$CpM(CO)_4$	M = Nb (x ray), Ta	40
$CpNb(CO)_4$	prep: $CpNbCl_4$ + Na + Cu/Al + CO 25.5 MPa	41
	prep: $NbCl_5$ + NaK + CO (100 atm) + CpHgCl	42
	^{93}Nb NMR study	13,43
	structural and MO analysis	44
$Cp'M(CO)_4$	prep: $Cp'MCl_4$ + Zn + CO	40
$Cp'Nb(CO)_4$	prep: $NbCl_5$ + NaK + CO (100 atm) + Cp'HgCl	42
$Cp^*M(CO)_4$		40
$Cp^*Ta(CO)_4$	prep: Cp^*TaCl_4 + NaHg + CO (1.5 atm) + PMe_3 (improved procedure)	45,46
$(\eta^5\text{-}C_9H_7)Nb(CO)_4$	C_9H_7 = indenyl	47
$(\eta\text{-}C_5H_3\text{-}1,3\text{-}Bu^t{}_2)Nb(CO)_4$		48
$[CpM(CO)_4]^-$	electrochemically generated	49
$Na_2[CpNb(CO)_3]$	IR only	50
$[M^1]_2[CpM^2(CO)_3]$	$[M^1]^+ = Na^+, Cs^+$; M^2 = Nb, Ta	51
$Cp_2M(CO)$	electrochemically generated transient	49
	photochemical generation in matices	52
$[Cp_2Nb(CO)]^-$	electrochemically generated	49
$[CpNb(CO)_2]_3(\mu_3\text{-}CO)$	prep: $CpNb(CO)_4 + h\nu$ NMR studies	53,42
$[Cp'Nb(CO)_2]_3(\mu_3\text{-}CO)$	prep: $Cp'Nb(CO)_4 + h\nu$	42
Halide complexes		
$CpNb(CO)_3I_2$	prep: $CpNb(CO)_4 + I_2$	54
$[CpMCl(\mu\text{-}Cl)(CO)_2]_2$	reformulation of "$CpM(CO)_3Cl_2$" reported previously	55,56,57
$[Cp'NbCl(\mu\text{-}Cl)(CO)_2]_2$	x ray	56,57
$[Cp'Nb(\mu\text{-}Cl)(CO)_2]_2$	x ray	56,58
$[Cp^*Ta(\mu\text{-}Cl)(CO)_2]_2$	x ray	58
$Cp_2Nb(CO)Cl$	prep: $Cp_2NbH(CO) + CHCl_3$	59
	x ray	60,61
	improved preparation and uses	62,63
$Cp_2M(CO)Cl$	prep: $Cp_2MH(CO)$ + RCOCl	64
	prep: $[Cp_2MCl]$ + CO	49
	other preparations	65

Table 3 (continued)

Compound	*Comments*	*Ref.*
$Cp_2Ta(CO)Cl$	prep: $[Cp_2TaCl]_n + CO$	66
$Cp_2Nb(CO)Br$		67
$Cp_2Ta(CO)Br$		68
$Cp'_2Nb(CO)Cl$	prep: $Cp'_2Nb[P(OMe)_3]Cl + CO$	69
$(\eta\text{-}C_5R_5)_2Nb(CO)Cl$	$C_5R_5 = C_5H_4\text{-TMS}, C_5H_3(TMS)_2$; prep: $(\eta\text{-}C_5R_5)_2NbCl_2 + NaHg + CO$	62
$Cp^*_2Nb(CO)Cl$	prep: $Cp^*_2Nb(CO)H + HCl$	70
Silyl complexes		
$Cp_2Nb(TMS)(CO)$	prep: $Cp_2Nb(TMS)Cl + NaHg + CO$	71
$Cp_2Ta(SiMe_2H)(CO)$		72
Nitrogen-donor ligands		
$CpM(CO)_3(N_2)$	prep: $CpM(CO)_4 + N_2 + h\nu$	73
$CpNb(CO)_2(N_2)_2$	transient, IR only	73
$CpNb(=NAr)(CO)(PMe_3)$[a]	prep: $CpNb(=NAr)(CH_2=CHR)(PMe_3) + CO$	74
$Cp^*Nb(=NAr)(CO)(PMe_3)$		75
$Cp_2Nb(CO)[N(Tol)CH=CPh_2]$		76
$Cp_2M(CO)[N(Tol)CH=NTol]$		77
Phosphorus- and arsenic-donor ligands		
$CpMCl_2(CO)(PMe_3)_2$		78
$CpNbCl_2(CO)(PR_3)_2$	$PR_3 = PMe_3, PMe_2Ph$, 1/2 dppe	79,80
	$PR_3 = PMe_3, PEt_3$	28
	$PR_3 = PMe_3$	57
$CpNbX_2(CO)(PR_3)_2$	X = Cl, Br, I; $PR_3 = PMe_3, PMe_2Ph$	81
$CpTaCl_2(CO)(PR_3)_2$	$PR_3 = PMe_3$	46
	$PR_3 = PMe_3, PEt_3, PPhMe_2$ (x ray), PPh_3	28
$CpNbCl_2(CO)(dmpe)$	prep: $CpNb(dmpe)Cl_3 + MgHg + CO$	82
$CpNbX_2(CO)(dmpe)$	X = Cl, Br, I	83
$CpNbI_2(CO)(dmpe)$		54
$Cp^*NbCl_2(CO)_2(PR_3)$	$PR_3 = PMe_3, PMe_2Ph$	84
	$PR_3 = PMe_3$, prep: $Cp^*NbCl_2(PMe_3)_2 + CO$	85
$Cp^*TaCl_2(CO)(PMe_3)_2$		78
$Cp^*TaCl_2(CO)_2(PMe_3)$	x ray	45,78,86,87,46,88
$Cp^*NbCl_2(CO)(dmpe)$		84
$Cp^*TaCl_2(CO)(dmpe)$	x ray; prep: $Cp^*TaCl_2(CO)_2(THF) + dmpe$	88
$(\eta\text{-}C_5H_4TMS)NbCl_2(CO)(PR_3)_2$	$PR_3 = PMe_2Ph$, 1/2 dppe	79
$CpNb(CO)_3(PR_3)$	^{31}P NMR study of 24 PR_3 ligands	89
$CpTa(CO)_3(PR_3)$	$PR_3 = PMe_2Ph, P(OPh)_3, PBu^n_3$	90
$(\eta\text{-}C_5H_4CHMe_2)Ta(CO)_3(\eta^1\text{-dppm})$		24
$Cp^*Ta(CO)_3(PMe_3)$		45,91
$(\eta^5\text{-}C_9H_7)Nb(CO)_3(PMe_3)$	C_9H_7 = indenyl	92
$CpNb(CO)_2(PR_3)_2$	*cis*-isomer; ^{31}P NMR study of 19 PR_3 ligands	89
$CpNb(CO)_2L_2$	$L_2 = Ph_2P(CH_2)_nPPh_2$ for n = 1–4, $Ph_2PCH_2CH_2AsPh_2$; ^{93}Nb NMR study	13
	$L_2 = (PEt_3)_2, Ph_2P(CH_2)_nPPh_2$ for n = 1–5, $Ph_2PCH=CHPPh_2$, $Ph_2PCH_2CH_2AsPh_2$, $o\text{-}C_6H_4(AsPh_2)(PPh_2)$; ^{93}Nb NMR study	93
$Cp^*Ta(CO)_2(PR_3)_2$	$PR_3 = PMe_3$, 1/2 dmpe	94
	$PR_3 = PMe_3$, prep: $Cp^*Ta(\eta^2\text{-}CHPMe_2)(H)_2(PMe_3) + CO$	46,91,95,96
	$PR_3 = PMe_3$, prep: $[\eta^7\text{-}C_5Me_3(CH_2)_2]Ta(H)_2(PMe_3)_2 + CO$	97
$CpNb(CO)_2(dmpe)$		82,83,98
$CpNb(CO)_2(dppe)$		99
$CpTa(CO)_2(dmpe)$		83
$CpNb(CO)_2[\eta^2\text{-}Ph_2P(CH_2)_2PR\text{-}(CH_2)_2PPh_2]$	R = Ph or Cy; ^{93}Nb NMR studies	100
$CpNb(CO)_2[\mu\text{:}\eta^2,\eta^1\text{-}Ph_2P(CH_2)_2PPh\text{-}(CH_2)_2PPh_2]CpNb(CO)_3$	^{93}Nb NMR studies	100
$(\eta\text{-}C_5H_4CHR_2)Ta(CO)_2[\eta^2\text{-}PhP\text{-}(CH_2CH_2PPh_2)_2]$	$R_2 = Me_2, Ph_2, (NMe_2)(H)$	24
$(\eta\text{-}C_5H_4CHMe_2)Ta(CO)_2(dppe)$		24
$[CpNb(CO)_2(dppe)]PF_6$		99
$CpNb(CO)(PR_3)_3$	$PR_3 = PMe_3, PMe_2Ph$	81
$Cp_2M(PR_2)(CO)$	R = Ph	101
	R = Me, Ph	102
$[Cp_2M(PPh_2H)CO]Cl$		101
$[Cp_2M(PPh_2Me)CO]I$	prep: $Cp_2M(PPh_2)(CO) + MeI$	101
$[Cp_2M(CO)(PPhClH)]Cl$	prep: $Cp_2M(CO)H + PPhCl_2$	103
$Cp_2M(CO)[P(O)PhH]$		103
$Cp_2M(CO)[P(X)R_2]$	X = S for R = Me, Ph; X = O for R = Ph	102

Table 3 (continued)

Compound	*Comments*	*Ref.*
$[Cp_2Nb(CO)(PPr^iPhH)]Cl$		104
$Cp_2Nb(CO)(PPr^iPh)$	x ray	104
$[Cp^*CpTa(CO)(PPhClH)]Cl$	two diastereomers	103
$Cp^*CpTa(CO)[P(O)PhH]$	prep: $[Cp^*CpTa(CO)(PPhClH)]Cl + OH^-$	103
$Cp^*Nb(\eta^4\text{-}P_4)(CO)_2$	x ray	105
$[Cp^*Nb(CO)(\mu,\eta^2\text{-}P_2)]_2$		105
$Cp^*Nb(\eta^4\text{-}As_4)(CO)_2$	x ray	106
$[Cp^*Nb(CO)_2]_2(As_2)_2$	prep: $Cp^*Nb(CO)_4 + As_4 + h\nu$	106
Oxygen-donor ligands		
$CpNb(CO)_3(THF)$	prep: $CpNb(CO)_4 + THF + h\nu$	41,107
$CpM(CO)_3(THF)$	substitution kinetics	90
$(\eta\text{-}C_5R_5)M(CO)_3(THF)$	$C_5R_5 = Cp', Cp^*$	40
$Cp^*Ta(CO)_2Cl_2(THF)$	prep: $Cp^*Ta(TMS)Cl_3 + CO + THF$	86,108
	x ray; prep: $Cp^*TaCl_4 + Al/HgCl_2 + CO + THF$	88
$Cp^*{}_2Ta(CO)(OMe)$		109
$Cp_2Nb(CO)(OCH{=}CPh_2)$		76
$Cp_2Nb(CO)(OCH{=}NTol)$	prep: $Cp_2NbH(CO) + O{=}C{=}NTol$	76
$Cp_2M(CO)(OCH{=}NPh)$		77
Sulfur-, selenium-, and tellurium-donor ligands		
$CpNb(CO)_3(SMe_2)$	x ray	110
$CpNb(CO)_3(L)$	L = THT, C_4H_8Se, C_4H_8Te; substitution kinetics	90
$CpTa(CO)_3(THT)$	substitution kinetics	90
$[CpNb(CO)_2]_2(\mu\text{-}S)_2$	x ray	111
$[CpNb(CO)_2]_2(\mu\text{-}S)_3$	x ray; prep: $CpNb(CO)_3(THF) + H_2S$	111
$[CpNb(CO)_2]_2(\mu\text{-}SCH_3)_2$	x ray	111
$Cp_2Nb(CO)(SCH{=}NTol)$	prep: $Cp_2NbH(CO) + S{=}C{=}NTol$	76
$Cp_2M(CO)(SCH{=}NPh)$		77
$Cp_2M(CO)(\eta^1\text{-}SCH{=}S)$		77
Heterobimetallic carbonyls		
$Cp_2(CO)Nb(\mu\text{-}CO)Co(CO)_3$		112
$Cp_2(CO)Nb(\mu\text{-}CO)Mn(CO)_4$	x ray	113,114
$(\eta\text{-}C_5H_4Bu^t)_2(CO)MCo(CO)_4$	M = Nb, Ta	115
$(\eta\text{-}C_5H_4Bu^t)_2(CO)M(OC)Co_3(CO)_9$	M = Nb, Ta	115
$Cp(CO)_2Nb(\mu_3\text{-}S)_2Fe_2(CO)_6$	x ray	116
$(\eta\text{-}C_5H_4CHMePh)CpNb(CO)\{ML_n\}$	$ML_n = Co(CO)_4$, $HFe(CO)_4$, $HFe_3(CO)_{11}$	117
$Cp_2M^2M^1Cp(CO)_3$	M^1 = Mo, W; M^2 = Nb, Ta	118
$Cp_2Nb(CO)(\mu\text{-}PPh_2)Cr(CO)_5$		101
$Cp_2M^2(CO)[\mu\text{-}\eta^1{:}\eta^1\text{-}P(S)R_2]ML_n$	all combinations of R = Me, Ph; $ML_n = Cr(CO)_5$, $CpMn(CO)_2$, $Fe(CO)_4$; M^2 = Nb, Ta	102
$M(CO)_4(\mu\text{-}\eta^5\text{-}C_4H_4BPh)CpFe$		119
$[NEt_4][CpNb(CO)_3(AuPPh_3)]$		120
$Cp_2Nb(CO)(SnCl_3)$		59
$Cp'_2Nb(CO)(SnCl_3)$	prep: $Cp'_2Nb(CO)H + SnCl_4 + NEt_3$; x ray	59,121
$Cp_2Nb(CO)(SnEt_2Cl)$		59
$Cp_2Nb(CO)(SnPh_nCl_{3-n})$	$n = 1, 2$	59
$Cp_2Nb(CO)(SnMe_3)$	prep: $Cp_2Nb(CH_2{=}CHPh)(SnMe_3) + CO + h\nu$	122
$Cp_2Nb(CO)(SnPh_3)$	x ray	59,121
$[NEt_4][CpNb(CO)_3(SnPh_3)]$	x ray	123
$[NEt_4][CpNb(CO)_3(SnR_3)]$	R = Et, CH_2Ph, Ph	120
$[NEt_4][CpM(CO)_3(SnPh_3)]$		51
$[(\eta\text{-}C_5H_4R)_2Nb(CO)]_2(\mu\text{-}SnCl_2)$	R = H, Me (x ray)	124
$[NEt_4][CpNb(CO)_3)(GePh_3)]$		120
$[NEt_4][CpNb(CO)_3(PbR_3)]$	R = Et, Ph	120

[a] $Ar = 2,6\text{-}C_6H_3Pr^i{}_2$.

also occurs readily. $CpNb(CO)_4$ can be photolyzed in the presence of cyclooctatetraene to provide $CpNb(CO)_2(\eta^4\text{-}C_8H_8)$; $CpNb(\eta^4\text{-}C_7H_8)(CO)_2$ is prepared similarly from $CpNb(CO)_4$ and excess cycloheptatriene.[54] Accordingly, a large number of $(\eta\text{-}C_5R_5)M(CO)_nL_{4-n}$ complexes are accessible by the photolysis route (Table 3) and the ligand substitution pathways of these reactions have been examined.[90]

The M^{III} carbonyls "$CpM(CO)_3Cl_2$" have been reported from $Al/HgCl_2$ reduction of $CpMCl_4$ under an atmosphere of CO,[55] but subsequent work has resulted in the reformulation of these complexes as dicarbonyl dimers. Thus, $[CpMCl(\mu\text{-}Cl)(CO)_2]_2$ and $[Cp'NbCl(\mu\text{-}Cl)(CO)_2]_2$ ($Cp' = \eta\text{-}C_5H_4Me$) have been prepared in this manner and fully characterized by Curtis *et al.*[57] These compounds appear to exist as dimers in noncoordinating solvents (in which they are sparingly soluble), but exhibit a reversible coordination equilibrium in THF and acetonitrile, forming $CpMCl_2(CO)_2L$ and $Cp'NbCl_2(CO)_2L$ (for L = THF or NCMe). However, when the $[\eta\text{-}C_5Me_5]^-$ complex Cp^*TaCl_4 is reduced under similar conditions (1 atm CO, THF), monomeric $Cp^*TaCl_2(CO)_2(THF)$ is isolated.[88] The M^{II} oxidation state is especially rare for organometallic compounds of niobium and tantalum, but M^{II} organometallics are accessible by reducing M^{III} carbonyl precursors.[58] Thus, $[CpNbCl(\mu\text{-}Cl)(CO)_2]_2$ affords $[CpNb(\mu\text{-}Cl)(CO)_2]_2$ upon reduction with NaHg, whereas reducing $Cp^*TaCl_2(CO)_2(THF)$ provides the Ta^{II} analogue $[Cp^*Ta(\mu\text{-}Cl)(CO)_2]_2$.[58]

Bis(cyclopentadienyl) carbonyl complexes continue to be reported and used, with M^{III} species $(\eta\text{-}C_5R_5)_2M(CO)X$ constituting the most often encountered stoichiometry. Improved preparations of the halides $Cp_2M(CO)X$ (X = Cl or Br) via reductive carbonylation of Cp_2MX_2 have been described.[62] The complexes $(\eta\text{-}C_5R_5)_2Nb(CO)Cl$ (R = H or Me) are also readily prepared by the low-temperature reaction of $(\eta\text{-}C_5R_5)_2Nb(CO)H$ with HCl.[70] Several heterobimetallic derivatives containing the $Cp_2Nb(CO)$ unit are also well established. For example, Cp_2NbH_3 reacts with $Mn_2(CO)_{10}$ in the presence of NEt_3 to provide $Cp_2Nb(CO)(\mu\text{-}CO)Mn(CO)_4$.[113,114] Similarly, $(\eta\text{-}C_5H_4Bu^t)_2MH_3$ or $(\eta\text{-}C_5H_4Bu^t)_2MH(CO)$ react in heptane solution with $Co_2(CO)_8$ to give the heterobimetallic complexes $(\eta\text{-}C_5H_4Bu^t)_2(CO)MCo(CO)_4$.[115] Further reaction with excess $Co_2(CO)_8$ affords species formulated as the oxymethylidyne complex $(\eta\text{-}C_5H_4Bu^t)_2(CO)M(OC)Co_3(CO)_9$.

2.2.2 CO Reductive Coupling

While most homogeneous studies of this chemistry have centered on H_2 or metal hydride reductions of coordinated CO ligands, the often elusive carbon–carbon bond-forming steps have been firmly established using complexes of niobium and tantalum. Section 2.8.1 provides a further discussion of Fischer–Tropsch (F–T) model chemistry using alkoxide-supported complexes, whereas this section describes one means of forming new carbon–carbon bonds, by the reductive coupling of CO ligands.

A molecular orbital study identified seven-coordinate, electron-rich M^I complexes such as $M(CO)_2(dmpe)_2Cl$ as likely candidates to couple reductively their carbonyl ligands.[125] Accordingly, Lippard and co-workers[126] reported that upon reduction of $Ta(CO)_2(dmpe)_2Cl$ with NaHg, an intermediate species formulated as $[Ta(CO)_2(dmpe)_2]^-$ is generated, which reacts with TMS-Cl to form $Ta(TMS\text{-}OC{\equiv}CO\text{-}TMS)(dmpe)_2Cl$ (Scheme 2). The bis(trimethylsiloxy) groups in $Ta(TMS\text{-}OC{\equiv}CO\text{-}TMS)(dmpe)_2Cl$ undergo rapid acidolysis with dilute HCl to form the hydride dihydroxyalkyne complex, $[Ta(H)(HOC{\equiv}COH)(dmpe)_2Cl]Cl$.[127] The mechanism of carbon–carbon bond formation in this reaction, particularly any identifiable intermediate species, is of obvious interest with respect to Fischer–Tropsch (F–T) chemistry. Thus, when the more sterically bulky electrophile Pr^i_3SiCl is added to a solution of $Na[Ta(CO)_2(dmpe)_2]$, the plum-red carbyne complex $Ta({\equiv}COSiPr^i_3)(dmpe)_2(CO)$ (**3**) can be isolated in high yield.[128] Upon reaction of (**3**) with TMS-Cl, coupling of the carbyne and carbonyl ligands ensues to provide $Ta(Pr^i_3SiOC{\equiv}CO\text{-}TMS)(dmpe)_2Cl$ (**4**), thereby supporting the intermediacy of (**3**) in the formation of the alkyne complexes. The niobium analogue $Nb({\equiv}COSiPr^i_3)(dmpe)_2(CO)$ has been prepared and undergoes the CO coupling reaction similarly.[128]

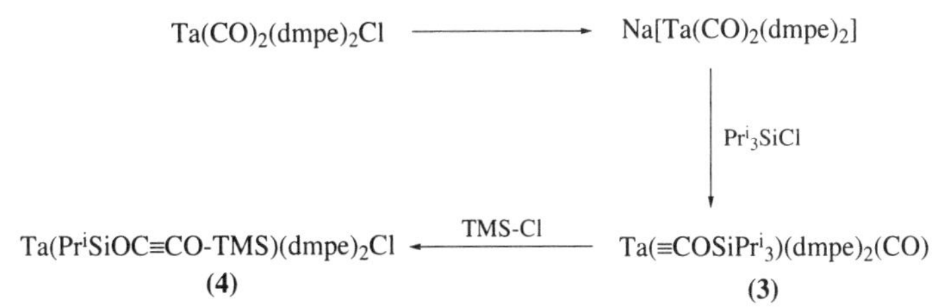

Scheme 2

2.2.3 Isocyanides and CNR Reductive Coupling

As can be seen from Table 4, the isocyanide chemistry of niobium and tantalum is limited. Early reactions of niobium and tantalum complexes with CNR ligands yielded incompletely characterized species formulated as iminoacyl derivatives arising from insertion of the isocyanide into metal–halide or metal–alkyl bonds,[1] although later evidence, especially in vanadium isocyanide chemistry,[129] makes the haloacyl formulations suspect. Iminoacyl complexes of niobium and tantalum are described in Section 2.5.3.

Table 4 Isocyanide complexes of niobium and tantalum.

Compound	*Comments*	*Ref.*
$Cl_2(RNC)_4Nb(\mu\text{-}\eta^2{:}\eta^2\text{-}RNCCNR)NbCl_4$	$R = Bu^t$; x ray	130,131
	$R = Pr^i$, Cy; prep: $M_2Cl_6(SMe_2)_3 + CNR$	131
$Nb_3Cl_8(CNBu^t)_5$	x ray; contains one $\mu_3\text{-}\eta^3(C){:}\eta^2(N)$-CNR ligand	130,131
$[Ph_4As][M(CO)_5(CNBu^t)]$		11
$Nb(CNMe)(CO)(dmpe)_2Cl$		15
$Na[M(CNMe)(CO)(dmpe)_2]$		16
$[CpNb(CNBu^t)_4Cl][NbOCl_4]$	x ray	132
$Cp_2Nb(CNR)Cl$	R = Ph, Cy	133
	$R = Bu^t$, Ar'^a	134
$(\eta\text{-}C_5H_4TMS)_2Nb(CNR)Cl$	$R = Bu^n$	134
	$R = Cy, Bu^t$, Ph, Ar'	135
	R = Cy	136
$[\eta\text{-}C_5H_3(TMS)_2]_2Nb(CNR)Cl$	$R = Bu^n$, Ph, Ar'	134
	$R = Cy, Bu^t$	135
$[(\eta\text{-}C_5H_4TMS)_2Nb(CNR)Cl]BF_4$	R = Ph, Ar'	134
$\{[\eta\text{-}C_5H_3(TMS)_2]_2Nb(CNR)Cl\}BF_4$	R = Ph, Ar'	134
$\{[\eta\text{-}C_5H_4(TMS)]_2Nb(CNCy)Cl\}^+$	electrochemically generated	136
$[Cp_2Nb(CNMe)_2][BF_4]_2$	x ray; prep: $Cp_2NbCl_2 + CNMe + Tl[BF_4]$	137
$Cp^*_2Ta(CNMe)(OMe)$		109
$Cp_2Nb(CNMe)(NMe_2)$	prep: $Cp_2NbH_3 + CNMe$	138
$Cp_2Nb(CNCMe_2CH_2Bu^t)(CN)$	prep: $Cp_2NbH_3 + CNCMe_2CH_2Bu^t$	139
$Cp_2Nb(CNAr')(SnMe_3)$		122
$Cp^*_2Ta(CNMe)(SCH_2CH_2Ph)$		140

[a] $Ar' = 2,6\text{-}C_6H_3Me_2$.

Although no congeners of $V(CNBu^t)_3Cl_3$ or $[V(CNBu^t)_6]^{2+}$ [129] have been characterized for niobium or tantalum, several of their isocyanide complexes have been described. The alkyl derivatives $(\eta\text{-}C_5R^1{}_5)_2M(CH_2CH_2R^1)(CNR^2)$ are formed upon treating $(\eta\text{-}C_5R^1{}_5)_2M(CH_2{=}CHR^1)H$ with alkyl or aryl isocyanides and such reactions have played a central role in investigating the kinetics and thermodynamics of the migratory insertion ⇌ β-hydride elimination pathway.[138] Back-bonding from M^{III} to the $C{\equiv}NR^2$ π^* level is substantial in these complexes such that carbene-like character and considerable electron density on the CNR^2 nitrogen are observed. Therefore, Lewis acids such as $AlEt_3$ form 1:1 adducts with some compounds (consider $Cp_2Ta(CH_2CH_2R^1)(CNR^2{\cdot}AlEt_3)$), via the isocyanide nitrogen.[141] Related halide derivatives $Cp_2Nb(CNR)Cl$ (R = Ph or Cy)[133] and $[CpNb(CNBu^t)_4Cl]^+$ [132] are available upon reducing Cp_2NbCl_2 or $CpNbCl_4$ in the presence of the appropriate isocyanide. The reductive coupling of CNR ligands has also been observed. When $CNBu^t$ reacts with $Nb_2Cl_6(SMe_2)_3$ (Nb=Nb), two cluster species can be isolated, trinuclear $Nb_3Cl_8(CNBu^t)_5$ (**5**) and dinuclear $Nb_2Cl_6(CNBu^t)_6$ (**6**), in a ratio that is highly condition dependent.[130,131] Red-brown (**5**) contains a $\mu_3\text{-}\eta^3(C){:}\eta^2(N)$-CNR ligand reminiscent of the unique CO ligand in $Cp_3Nb_3(CO)_7$. Purple (**6**) can be described as $Cl_2(Bu^tNC)_4Nb(\mu\text{-}\eta^2{:}\eta^2\text{-}Bu^tNCCNBu^t)NbCl_4$ as it contains an unusual bridging ligand arising from the reductive coupling of two $CNBu^t$ ligands. One simple valence bond description of this ligand is presented in metallacycle (**6**), but a highly delocalized view of the bonding within this ligand is clearly necessary. Carnahan and Lippard[16] reported that the reduction of $Ta(CNMe)_2(dppe)_2Cl$ with NaHg, followed by the addition of TMS-Cl, provides $[TMS(Me)NC{\equiv}CN(Me)TMS]Ta(dmpe)_2Cl$, presumably by a reaction sequence analogous to that discovered in the carbonyl complexes. Reductive cross-coupling of CO and CNR ligands at the $[Ta(dmpe)_2Cl]$ template has also been demonstrated.[16]

2.3 η^1 COMPLEXES

As indicated in Table 1, alkyl, aryl, and related σ-bonded carbon ligands take priority over hydride ligands in arranging η^1 complexes. Both alkyls and hydrides are reported in this section, whereas metallacycles with at least one metal–carbon bond are described in Section 2.6.

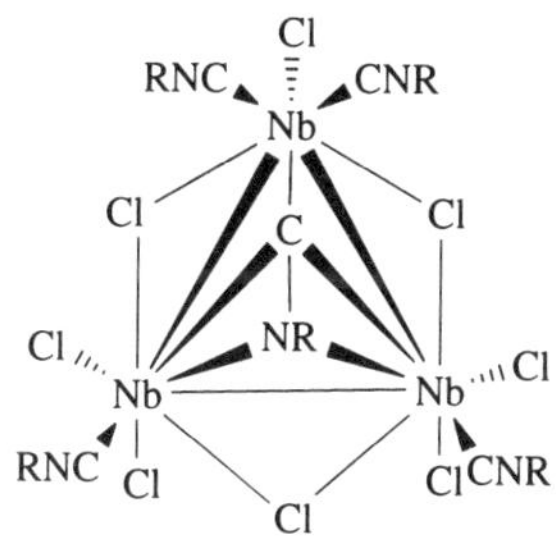

(5)

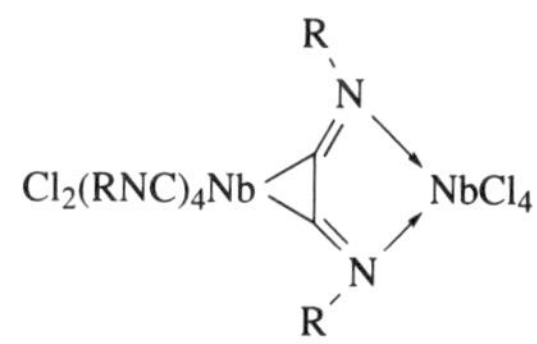

(6)

2.3.1 Alkyls, Aryls, and Related σ Complexes

Typical preparative methods for introducing alkyl ligands and fundamental reactions of alkyl complexes were detailed by Labinger.[1] This section highlights the more recent descriptive chemistry of alkyl complexes.

2.3.1.1 Alkyl and aryl complexes without cyclopentadienyl ligands

Table 5 summarizes the alkyl, aryl, and related σ complexes of niobium and tantalum without cyclopentadienyl ligands.

Table 5 Alkyl, aryl, and related complexes of niobium and tantalum without cyclopentadienyl ligands.

Compound	*Comments*	*Ref.*
$TaMe_5$	*ab initio* MO calculations; electron diffraction studies	142–4
$Ta(CH_2\text{-}4\text{-}C_6H_4Me)_5$	x ray	145
$Ta(CH_2Bu^t)_2Et_3$		146
Halide complexes		
$Ta(CH_2Ph)_nCl_{5-n}$	$n = 1, 2, 4$	147
$Ta(CH_2D)_nCl_{5-n}$	$n = 1$–3; NMR study	148
$M(CH_2C_6H_5\text{-polymer})Cl_4$	poly(styrene–divinylbenzene) attached $M(CH_2Ph)Cl_4$	149
$Ta[2\text{-}C_6H_4(CH_2NMe_2)]Cl_4$	pendant amine ligand	150
$Ta[2\text{-}C_6H_4(CHMeNMe_2)]Cl_4$		150
$Ta[2\text{-}C_6H_3(CH_2NMe_2)\text{-}6\text{-}Me]Cl_4$		150
$Ta(CH_2Ph)_2Br_3$		151
$TaMe_3Cl_2$	SCF Xα calculations and He^{I} and He^{II} PES	152
$Ta(CH_2\text{-}4\text{-}C_6H_4Me)_3Cl_2$	proposed	145
$Ta(CH_2\text{-adamantyl})_3Cl_2$		153
$Ta(CH_2Bu^t)_2(Et)Cl_2$	prep: $Ta(CH_2Bu^t)_2Cl_3 + ZnEt_2$	154
$Ta[2\text{-}C_6H_4(CH_2NMe_2)](CH_2Ph)Cl_3$	pendant amine ligand	150
$Ta[2\text{-}C_6H_3(CH_2NMe_2)\text{-}6\text{-}Me](CH_2Ph)Cl_3$		150
$Ta(CH_2TMS)_3Cl_2$	UV–vis study	155
$Ta[2\text{-}C_6H_4(CH_2NMe_2)](CH_2Ph)_2Cl_2$	x ray	150
$Ta[2\text{-}C_6H_4(CHMeNMe_2)](CH_2Ph)_2Cl_2$		150
$Ta[2\text{-}CH_2C_6H_4(NMe_2)](CH_2Ph)_2Cl_2$		150
$Ta[1\text{-}C_{10}H_6\text{-}8\text{-}NMe_2](CH_2Ph)_2Cl_2$		150
Phosphorus-donor ligands		
$TaMe(CO)_2(dmpe)_2$	x ray	27,29
$Ta(CHMeCHMe_2)(CO)_4(Ph_2PCH_2CH_2PEt_2)$		24
$Ta(CH_2R)_2Cl_3(PMe_3)$	R = Ph, Bu^t	151
$MMe_5(PMe_3)_2$	M = Nb, Ta	156
Oxygen-donor ligands		
$TaMe(OAr')_4$	x ray	157
	UV–vis study	155
$TaR(O_2CNMe_2)_4$	R = Bu^t, Pr^i, Et, CH_2TMS (x ray)	158
$TaEt(H)(silox)_3$		159,160
$Ta(C_5H_2Me_2N)(H)(silox)_3$		161
$Ta(CPh{=}CPhCR{=}CRI)(OAr)_3I$	R = Me, Et	162

Although complexes containing only alkyl ligands have long been known, few have appeared since 1980. These include $Ta(CH_2Bu^t)_2Et_3$[146] and structurally characterized $Ta(CH_2\text{-}4\text{-}C_6H_4Me)_5$.[145] While the

Table 5 (continued)

Compound	*Comments*	*Ref.*
Ta(TMS-C=CMeCPh=CHI)(OAr)$_3$I	butadienyl complex	162
TaPh(OR)$_2$Cl$_2$	R = Ar''^{a}, 2,6-C$_6$H$_2$Bu$^t_2$-4-Me, 2,6-C$_6$H$_2$Bu$^t_2$-4-OMe; implicated, not observed	163
Ta(CH$_2$Ph)(silox)$_2$Cl$_2$		160,164
[Ta(H)(silox)$_3$]$_2$[μ:η^1,η^1-CH$_2$(CH$_2$)$_3$O]	prep: Ta(H)I(silox)$_3$ + NaHg + THF	160
[Ta(Et)(silox)$_2$]$_2$(μ-O)$_2$	prep: [TaH(silox)$_2$]$_2$(μ-O)$_2$ + CH$_2$=CH$_2$	160
[TaH(silox)$_2$](μ-O)$_2$[TaMe(silox)$_2$]		160
Nb$_2$Cl$_4$Me$_2$(NPh)(μ-O)(PMe$_2$Ph)$_2$		165
Ta(CH$_2$Bu^t)$_2$(O-2,6-C$_6$H$_2$Bu$^t_2$-4-Me)Cl$_2$	prep: Ta(CH$_2$Bu^t)$_2$Cl$_3$ + Li(O-2,6-C$_6$H$_2$Bu$^t_2$-4-Me) at 0 °C	166
Nb(CH$_2$adamantyl)$_2$(OBu^t)$_3$	prep: Nb(OBu^t)$_4$ + Li[CH$_2$-adamantyl]	153
TaMe$_2$(OAr')$_3$		157
Ta(CH$_2$-4-C$_6$H$_4$Me)$_2$(OAr)$_3$		167
MR$_2$(O-2,6-C$_6$H$_3$Ph$_2$)$_3$	R = Me, CH$_2$-4-C$_6$H$_4$Me	168
	M = Ta; R = Me, CH$_2$-4-C$_6$H$_4$Me	169
	R = Me	170
TaMe$_2$(O-2,6-C$_6$H$_3$Ph$_2$)$_3$	prep: Ta(O-2,6-C$_6$H$_3$Ph$_2$)$_3$Cl$_2$ + MeLi	171
Ta(CD$_3$)$_2$(O-2,6-C$_6$H$_3$Ph$_2$)$_3$		168
Ta(CH$_2$-4-C$_6$H$_4$Me)$_2$(O-2,6-C$_6$H$_3$Ph$_2$)$_3$		170
Ta(CH$_2$Bu^t)$_2$(OBu^t)Cl$_2$		172
Ta(CH$_2$Bu^t)$_2$(silox)$_2$Cl		164
Ta(CH$_2$TMS)$_2$(OAr)$_2$Cl		173
TaMe$_2$[OSi(2-C$_6$H$_4$Me)$_3$]$_2$Cl		174
NbMe$_3$(OAr')$_2$		157
TaR$_3$(OAr')$_2$	R = Me, CH$_2$TMS, CH$_2$Ph (x ray)	157
	R = Me; UV–vis study	155
TaR$_3$(OAr)$_2$	R = CH$_2$TMS	173,175
	R = Me, CH$_2$TMS, CH$_2$Ph	155,157
	R = Me	176
TaMe$_3$(OAr'')$_2$	prep: Ta(OAr'')$_2$Cl$_3$ + LiMe; x ray	155,175 177–9
TaMe$_3$(O-2,6-C$_6$H$_3$Ph$_2$)$_2$		169,171
TaMe$_3$(O-2,6-C$_6$H$_2$Bu$^t_2$-4-OMe)$_2$		155
TaMe$_3$(OPr^i)$_2$		177
M(CH$_2$TMS)$_3$(OPr^i)$_2$		177
Ta(CH$_2$TMS)$_3$(OMe)$_2$		177
TaR$_3$(OAr')$_2$	R = CH$_2$TMS, CH$_2$Bu^t; prep: Ta(OAr')$_2$Cl$_3$ + LiR	177
{TaMe$_3$[OSi(2-C$_6$H$_4$Me)$_3$]}$_2$	prep: thermolysis of TaMe$_4$[OSi(2-C$_6$H$_4$Me)$_3$]	174,180
TaMe$_3$[OSi(2-C$_6$H$_4$Me)$_3$]X	X = Cl, I; x ray of cocrystallized TaMe$_3$[OSi(2-C$_6$H$_4$Me)$_3$]X for X = 2/3 Cl, 1/3 I	174
TaMe$_3$(OCMe$_2$CH$_2$PPh$_2$)$_2$	prep: TaMe$_3$Cl$_2$ + LiOCMe$_2$CH$_2$PPh$_2$	181
M(CH$_2$-4-C$_6$H$_4$Me)$_3$(O-2,6-C$_6$H$_3$Ph$_2$)$_2$	M = Nb, Ta; x ray of both	182
	M = Ta	169
	M = Nb; procatalyst for aromatic hydrocarbon hydrogenation	183,184
Ta(CH$_2$-4-C$_6$H$_4$Me)$_3$(O-2,4,6-C$_6$H$_2$Ph$_3$)$_2$		182
Ta(CH$_2$-4-C$_6$H$_4$Me)$_3$[O-2,4-C$_6$H$_2$Ph$_2$-6-(C$_6$H$_4$-4-X)]$_2$	X = Me, OMe, Cl	182
TaR$_3$(O-2,6-C$_6$H$_3$Ph$_2$)$_2$	R = CH$_2$-4-C$_6$H$_4$Me, CH$_2$TMS	173
M(CH$_2$TMS)$_3$(OPr^i)$_2$	prep: [M(OPr^i)$_5$]$_2$ + LiCH$_2$TMS; UV–vis study	155
MR$_3$(O-2,6-C$_6$H$_3$Ph$_2$)$_2$	R = Me, CH$_2$-4-C$_6$H$_4$Me	168
Ta(CH$_2$TMS)$_3$(OTp)$_2$	OTp = 9-oxotriptycene	164
TaR$_3$(silox)$_2$	R = Me, CH$_2$Ph	164
Ta(CH$_2$Bu^t)$_3$(silox)Cl		164
Ta(CH$_2$TMS)$_4$(OMe)	prep: Ta(OMe)$_5$ + LiCH$_2$TMS	177
TaMe$_4$[OSi(2-C$_6$H$_4$Me)$_3$]	thermally unstable	174,180
{Ta[2,6-C$_6$H$_3$(CH$_2$NMe$_2$)$_2$]Cl$_2$(μ-O)}$_2$	x ray	185,186
Nb[ON(Me)NO]$_3$Me$_2$	prep: NbMe$_5$ + NO(g.)	187
Nb[ON(Me)NO]$_2$MeCl$_2$	prep: NbMe$_3$Cl$_2$ + NO(g.)	187
{Ta[ON(Me)NO]$_2$Me$_3$}$_2$	prep: TaMe$_5$ + NO(g.)	187
Nitrogen-donor ligands		
TaR(NMe$_2$)$_4$	R = Bu^t (x ray), Pr^i, Et, CH$_2$TMS; prep: Ta(NMe$_2$)$_4$Cl + LiR	158
Ta(4-C$_6$H$_4$Me)(NMe$_2$)$_3$Br	x ray	158
Ta(CH$_2$Bu^t)$_2$(NMe$_2$)$_3$		188
[Ta(CH$_2$Bu^t)$_2$N]$_5$	x ray; prep: Ta(=CHBu^t)(CH$_2$Bu^t)$_3$ + NH$_3$	188
Ta[2-C$_6$H$_4$(CHRNMe$_2$)](CH$_2$Ph)$_2$Cl$_2$	R = H (x ray), Me	189
LTaMe$_3$Cl	L = Tp, Tp' (x ray)	190
[η^3-B(pz)$_4$]TaMe$_3$Cl		190
[H$_2$B(3,5-Me$_2$pz)$_2$]TaMe$_3$Cl	TaMe$_3$Cl$_2$ + K[H$_2$B(3,5-Me$_2$pz)$_2$]; x ray	191

Table 5 (continued)

Compound	*Comments*	*Ref.*
$Ta(=NPh)(CH_2Bu^t)_3(THF)$	prep: $Ta(=NPh)Cl_3(THF)_2 + Mg(CH_2Bu^t)_2 \cdot (diox)$	192
$[Ta(CH_2Bu^t)_3(THF)]_2(\mu-N_2)$		192
$Ta(=NSiBu^t_3)(NHSiBu^t_3)Me_2(py)_n$	$n = 0, 1$	193
$Ta(=NSiBu^t_3)(NHSiBu^t_3)_2R$	R = Me, Ph, CH_2Ph, or CH_2Bu^t	193
$[Ta(=NSiBu^t_3)(NHSiBu^t_3)_2]_2(\mu:\eta^1,\eta^1-C_6H_4)$		193
$Ta(=NSiBu^t_3)(NHSiBu^t_3)_2(C_6H_4Me)$		193
$Ta(=NSiBu^t_3)_2MeL_2$	L = py (x ray), THF	193
$Ta[2,6-C_6H_3(CH_2NMe_2)_2](=NR)Cl_2$	R = Me, TMS	185,186
$Nb_2Cl_4Me_2(NPh)(\mu-O)(PMe_2Ph)_2$		165
Alkyl complexes containing cyclometallated ligands		
$Ta[O-2,6-C_6H_3Bu^t(CMe_2CH_2)](OAr'')R_2$	R = Me; cyclometallated OAr" ligand; prep: thermolysis of $TaMe_3(OAr'')_2$	175,178,179
	R = Ph; x ray	163,178
	R = *o*-, *m*-, *p*-C_6H_4Me	163,194
$Ta[O-2,6-C_6H_2Bu^t(CMe_2CH_2)-4-Me](O-2,6-C_6H_2Bu^t_2-4-Me)R_2$	R = Ph, C_6D_5, 2-C_6H_4Me, 3-C_6H_4Me, 4-C_6H_4Me, 3-C_6H_4F, 4-C_6H_4F	163
$Ta[O-2,6-C_6H_2Bu^t(CMe_2CH_2)-4-OMe](O-2,6-C_6H_2Bu^t_2-4-OMe)Ph_2$		163
$Ta[O-2,6-C_6H_3Bu^t(CMe_2CH_2)]_2R$	R = Me	178,179
	R = Ph	178
	R = CH_2TMS	157,195,196
	R = *m*-, *p*-C_6H_4Me	163,194
$Ta[O-2,6-C_6H_3Bu^t(CMe_2CH_2)](OAr'')PhCl$	prep: $Ta(OAr'')_2Cl_3$ + 2 PhLi	163
$Ta[O-2,6-C_6H_2Bu^t(CMe_2CH_2)-4-X](O-2,6-C_6H_2Bu^t_2-4-X)PhCl$	X = Me, OMe	163
$Ta[O-2,6-C_6H_2Bu^t(CMe_2CH_2)-4-Me]-(CH_2Bu^t)Cl_2$	prep: $Ta(CH_2Bu^t)_2Cl_3 + Li(O-2,6-C_6H_2Bu^t_2-4-Me)$; x ray	166
$Ta[O-2,6-C_6H_3Bu^t(CMe_2CH_2)]_2Ph$		163
$Ta[O-2,6-C_6H_2Bu^t(CMe_2CH_2)-4-Me]_2R$	R = Ph, C_6D_5, C_6D_4H, 3-C_6H_4Me, 4-C_6H_4Me, 3-C_6H_4F, 4-C_6H_4F	163
$Ta[O-2,6-C_6H_2Bu^t(CMe_2CH_2)-4-OMe]_2Ph$		163
$M[O-2,6-C_6H_3Ph(C_6H_4)](O-2,6-C_6H_3Ph_2)_2R$	R = Me, CH_2-4-C_6H_4Me	168
	M = Ta; R = Me (x ray), CH_2-4-C_6H_4Me	169
	R = Me; x ray of M = Ta compound	170
	R = CH_2-4-C_6H_4Me	170
$Ta[O-2,6-C_6H_3Ph(C_6H_4)](O-2,6-C_6H_3Ph_2)_2Bu^n$	x ray	170
$Ta[2-C_6H_4(CHMeNMeCH_2)(CH_2Ph)Cl_2(THF)$	x ray	150
$Ta[1-C_{10}H_6-8-NMeCH_2](CH_2Ph)Cl_2(THF)$		150
$M[O-2,6-C_6H_3Ph(C_6H_4)](O-2,6-C_6H_3Ph_2)R_2$	R = Me, CH_2-4-C_6H_4Me	168
	M = Ta; R = Me, CH_2-4-C_6H_4Me	169
	M = Ta; R = CH_2-4-C_6H_4Me	182
$Ta[O-2,4,6-C_6H_2Ph_2(C_6H_4)](O-2,4,6-C_6H_2Ph_3)(CH_2-4-C_6H_4Me)_2$		182
$Ta[O-2,4,6-C_6H_2Ph_2(C_6H_3X)][O-2,4-C_6H_2Ph_2-6-(C_6H_4-4-X)](CH_2-4-C_6H_4Me)_2$	X = Me, OMe, Cl	182
$Ta[O-2,4,6-C_6H_2Ph(C_6H_4X)(C_6H_4)][O-2,4-C_6H_2Ph_2-6-(C_6H_4-4-X)](CH_2-4-C_6H_4Me)_2$	X = Me, OMe, Cl	182
Miscellaneous		
$(\eta^5-C_2B_9H_{11})TaMe_3$	prep: $(\eta^5-C_2B_9H_{11})TaCl_3$ + MeMgBr or $TaMe_3Cl_2 + Li[C_2B_9H_{11}]$	197

[a] $Ar'' = 2,6-C_6H_3Bu^t_2$.

simplest bonding arguments might suggest a trigonal bipyramidal structure for these d^0, M(σ-donor)$_5$ complexes, in the solid state the latter complex adopts a distorted square-pyramidal geometry.[145] Gas-phase electron diffraction studies on $TaMe_5$, a complex long believed to be a trigonal bipyramid, indicate that it too is best described as a distorted square pyramid.[144] *Ab initio* calculations for $TaMe_5$ are consistent with this observed ground-state structure for $TaMe_5$.[142,143]

Far more common than the binary alkyls are the halide complexes of the type MR_nX_{5-n} (X = Cl or Br) and their base adducts. The conventional synthetic approach to these complexes involves simple metathesis reactions of their halides (and sometimes their alkoxides[153,177]) with main-group alkyls. Mild alkylating agents such as organozinc, -mercury, and -tin compounds are often effective for the selective preparation of these mixed ligand MR_nX_{5-n} derivatives, while Grignard reagents or alkyllithiums are typically required to drive the substitution to completeness. This point is best illustrated by the complete series of substituted benzyl complexes $Ta(CH_2Ph)_nCl_{5-n}$ (n = 1–5), prepared from $TaCl_5$ and $Zn(CH_2Ph)_2$ (for n = 1, 2, and 3) followed by $Mg(CH_2Ph)_2(THF)_2$ (for n = 5).[147] The tetraalkyl $Ta(CH_2Ph)_4Cl$ is available from the reaction of $Ta(CH_2Ph)_5$ with HCl.[147]

Alkyl and aryl complexes containing hard, π-donors such as alkoxide, amide, and imide ligands, have come to play a central role in C–H bond activation studies (Section 2.8.2). It has been proposed that the inclusion of multiple π donors in the ligand set precludes the formation of agostic M ⇌ H–C interactions which may presage alkyl degradations.[158] For example, Chisholm *et al.*[158] have prepared the complexes $TaR(NMe_2)_4$ (for R = Et, Pr^i, Bu^t, or CH_2TMS), which do not undergo a facile β-hydrogen elimination. Other representative complexes of this class include $Ta(C_6H_4\text{-}4\text{-}Me)(NMe_2)_3Br$,[158] $TaR_3(OAr)_2$ (for R = Me, CH_2TMS, or CH_2Ph and Ar = 2,6-$C_6H_3Pr^i_2$),[157] and $Ta(=NSiBu^t_3)(NHSiBu^t_3)_2R$ (R = Me, Ph, CH_2Ph, or CH_2Bu^t).[193]

2.3.1.2 Cyclopentadienyl alkyl and aryl complexes

Alkyl, aryl, and related σ complexes of niobium and tantalum which also contain cyclopentadienyl ligands are outlined in Table 6.

Many new half-sandwich complexes of the type $(\eta\text{-}C_5R_5)MR_nCl_{4-n}$ have been prepared, either by the addition of main-group alkyls to $(\eta\text{-}C_5R_5)MCl_4$ species or by the reaction of MR_nCl_{5-n} with cyclopentadienyl transfer reagents. For example, the stoichiometric addition of MeLi to Cp^*TaCl_4 provides Cp^*TaMe_4,[198] while adding Cp*Li to MMe_2Cl_3 affords $Cp^*MMe_2Cl_2$.[209,210] Mixed-ligand complexes containing alkoxide or amide ligands may also be obtained by straightforward metathetical procedures. For example, Cp^*TaMe_3Cl reacts with LiOR (R = Me, Pr^i, or Bu^t) and with $LiNMe_2$ to provide $Cp^*TaMe_3(OR)$ and $Cp^*TaMe_3(NMe_2)$, respectively.[199,211] A number of imido complexes of the general formula $(\eta\text{-}C_5R_5)M(=NR)R_2$ have been prepared (Table 6).[215,257]

Bis(cyclopentadienyl) alkyl and aryl complexes are especially common. Unlike the typical d^0 monokis η^5 complexes described above, many d^1 and d^2 $(\eta\text{-}C_5R_5)_2M$ alkyl species are known. Neutral d^0 complexes of the type $(\eta\text{-}C_5R_5)_2MR_nX_{3-n}$ (n = 1–3) are still fairly rare (cf., $Cp_2NbMe_2(\eta^1\text{-}S_2CNEt_2)$[217]),[1] although cationic, 16-electron complexes of the type $[(\eta\text{-}C_5R_5)_2MR_nX_{2-n}]^+$ (n = 1 or 2) are well established. The d^0 complexes $[Cp_2TaMe_2]X$ (X^- = $[BF_4]^-$, $[PF_6]^-$, $[BPh_4]^-$) have been made by oxidizing Cp_2TaMe_2 with ferrocenium salts $[Cp_2Fe]X$.[219] Many 17-electron d^1 complexes such as $Cp_2Nb(CH_2TMS)Cl$ and Cp_2NbMe_2 are accessible by the selective alkylation of the appropriate $(\eta\text{-}C_5R_5)_2MCl_2$ compound.[231] Finally, 18-electron d^2 complexes of the general form $(\eta\text{-}C_5R_5)_2M(R)L$ (e.g., L = CO, CNR, PR_3) may be obtained by the addition of L to $(\eta\text{-}C_5R_5)_2M(H)(R_2C{=}CR_2)$ (see Section 2.5.1.3) or by the reduction of $Cp_2M(R)Cl$ complexes in the presence of donor ligands.

Although originally considered exceptional, terminal oxo complexes of the form $(\eta\text{-}C_5R_5)_2M(=O)R$ are now frequently encountered. $Cp^*_2Ta(=O)Me$ is prepared either by thermolysis of the formaldehyde complex $Cp^*_2Ta(\eta^2\text{-}O{=}CH_2)H$[226] or by the Wittig-like reaction of $Cp^*_2Ta(=CH_2)Me$ with organic carbonyls.[228] The former reaction is especially interesting since it involves a transient d^2 complex which undergoes a formal α-alkyl elimination. Thus, the $[Cp^*_2Ta(=CH_2)H \rightleftharpoons Cp^*_2TaMe]$ equilibrium mixture has been shown to react with MeOH to afford transient $[Cp^*_2TaOMe]$, which undergoes β-H elimination to provide the kinetic product $Cp^*_2Ta(\eta^2\text{-}O{=}CH_2)H$.[109,226] Upon heating $Cp^*_2Ta(\eta^2\text{-}O{=}CH_2)H$, an equilibrium concentration of $[Cp^*_2TaOMe]$ is regenerated, which undergoes a rate-limiting α-alkyl elimination to generate the thermodynamic product $Cp^*_2Ta(=O)Me$.[109]

2.3.1.3 Reactivity of alkyl and aryl complexes

Many of the reactivity studies of alkyl complexes, particularly of their α- and β-hydrogen elimination processes which are often evoked in catalytic cycles (Section 2.8), were described in detail in *COMC-I*.[1] In addition, α-processes leading to new alkylidene complexes will be covered in Section 2.4.1, and β-processes leading to alkene and benzyne complexes will be described in Sections 2.5.1 and 2.5.2, respectively. Alkyl and aryl degradations in σ-bond metathesis reactions resulting in C–H bond activations are outlined in Section 2.8.2.

The polarity of the niobium– and tantalum–carbon bond in their complexes renders the alkyl ligand in L_nMR susceptible to attack by electrophiles, especially protonolysis reactions by HX (including H_2O) to afford alkane and L_nMX. Hydrogenolysis also occurs to provide alkane and metal hydride. For example, $Ta(CH_2\text{-}4\text{-}C_6H_4Me)_2(OAr)_3$ undergoes clean hydrogenolysis in the presence of phosphines to provide the dihydride complexes $TaH_2(OAr)_3(PR_3)$.[167] Similarly, Cp^*TaMe_4 reacts with H_2 and phosphines to afford $Cp^*TaH_4(PR_3)_2$.[94]

Table 6 Alkyl, aryl, and related complexes of niobium and tantalum containing cyclopentadienyl ligands.

Compound	*Comments*	*Ref.*
Monokis(cyclopentadienyl) M^V *complexes*		
Cp*TaMe$_4$	prep: Cp*TaCl$_4$ + MeLi	198
Cp*Ta(R)Me$_3$	R = CH$_2$Bu^t, CH$_2$Ph	199
CpTa(=CRCR=CMeNBu^t)Me	R = Ph (x ray), Tol; metallacyclopentatriene complex	200
Cp*Ta(CH$_2$PPh$_2$CH$_2$)Me$_n$Cl$_{3-n}$	n = 1, 3	201
Cp*Ta(=CHPPh$_2$CH$_2$)Me$_2$	x ray	201
Cp*TaCl$_4$(CH$_2$=PMe$_{3-n}$Ph$_n$)	n = 0, 1, 2 (x ray), 3; phosphorus ylide complex	202
Cp*Ta(CH$_2$Ph)Cl$_3$		147
(η-C$_5$Me$_4$R)Ta(CH$_2$C$_6$H$_4$-4-Me)Cl$_3$	R = Me, Et	203
(η-C$_5$Me$_4$R)Ta(CH$_2$Bu^t)Cl$_3$	R = Me, Et	203
	R = Me	204,205
Cp*Ta(Pr^n)Cl$_3$		204
[η-C$_5$Me$_3$(TMS)$_2$]Ta(CH$_2$-TMS)Cl$_3$		206
Cp*TaCl$_3$[CH$_2$P(=X)Ph$_2$]	X = O, S	207
(η-C$_5$Me$_4$R)Ta(CH$_2$Ph)$_2$Cl$_2$	R = Me, Et	147
Cp*Ta(CH$_2$Bu^t)$_2$Cl$_2$		208
Cp*NbMe$_2$Cl$_2$	prep: NbMe$_2$Cl$_3$ + LiCp*	209
Cp*Ta(CH$_2$Bu^t)(H)Cl$_2$		204
Cp*TaMe$_3$Cl	prep: TaMe$_3$Cl$_2$ + LiCp*	210
(η-C$_5$H$_4$R)Ta(CH$_2$Ph)$_3$Cl	R = H, Me	147
Cp*TaMe$_3$(NMe$_2$)		199
Cp*TaMe$_3$(OR)	R = Bu^t, Pr^i, Me	199
	R = Bu^t	211
Cp*TaMe$_3$(TMS)		212
CpNb(=NMe)Me$_2$		213
Cp*Nb(=NAr)(CH$_2$Ph)$_2$		214
CpNb(=NAr)Ph$_2$(PMe$_3$)		214
Cp*Ta(=NR)Me$_2$	R = Bu^t, CH$_2$Bu^t, Me	199
CpNb(=NBu^t)(CH*$_2$Bu^t)$_2$	H*$_2$ = H$_2$, HD	215
CpNb(=NAr)(CH*$_2$Bu^t)$_2$	H*$_2$ = H$_2$ (x ray), HD	215
CpNb(=NAr)(CH$_2$Bu^t)X	X = Cl, Br	215
[Cp*TaMeN]$_3$	prep: Cp*TaMe$_4$ + NH$_3$ + Δ; x ray	216
Cp′(η^5-C$_2$B$_9$H$_{11}$)TaR$_2$	R = Me, CH$_2$-TMS	197
Monokis(cyclopentadienyl) M^V/M^{IV} *complexes*		
[K(OEt$_2$)$_n$][Cp*TaMeN]$_3$	prep: [Cp*TaMeN]$_3$ + NaK + Et$_2$O	216
Monokis(cyclopentadienyl) M^{III} *complexes*		
Cp*TaMe$_2$(bipy)	prep: Cp*TaMe$_4$ + H$_2$ + bipy	94
CpNbMe(I)(CO)(PMe$_2$Ph)$_2$		81
[CpNbMe(CO)(PMe$_3$)$_3$]I		81
[CpNbMe$_2$(NO)]	proposed intermediate from Cp$_2$NbMe$_2$ + NO	187
Bis(cyclopentadienyl) M^V *complexes*		
Cp$_2$NbMe$_2$[η^1-S$_2$CNEt$_2$]		217
Cp*$_2$TaMe(X)H	X = SiH$_3$, PH$_2$, OMe, SMe, NMe$_2$	109
Cp*$_2$Ta(CH=CH$_2$)(SBn)H		140
[Cp$_2$Nb(CH$_2$PPh$_2$)$_2$]BF$_4$		218
[Cp$_2$TaMe$_2$]X	X$^-$ = [BF$_4$]$^-$, [PF$_6$]$^-$, [BPh$_4$]$^-$	219
[Cp*$_2$TaMe$_2$]BF$_4$		109
[Cp$_2$Nb(CH$_2$TMS)Cl]PF$_6$		71
[Cp$_2$NbMeI]I		220
[Cp′$_2$Nb(CH$_2$TMS)F]X	X$^-$ = [PF$_6$]$^-$, [BF$_4$]$^-$	221
{Cp$_2$NbMe[η^2-S$_2$CNMe$_2$]}[S$_2$CNMe$_2$]		217
Cp$_2$Nb[η^1-CH(SH)(SMe)](Bu^n)(η^1-BH$_4$)		222
Cp$_2$Nb(=O)Me	prep: thermolysis of Cp$_2$NbMe(η^2-ONMe)	187
Cp$_2$Ta(=O)Me		223
Cp$_2$Nb(=O)[C$_7$H$_5$(CF$_3$)$_2$]	x ray; prep: Cp$_2$Nb(η^1-Cp)$_2$ + CF$_3$C≡CCF$_3$ + O$_2$	224
Cp′$_2$Nb(=O)(CH$_2$TMS)	x ray	221,225
(η-C$_5$H$_4$TMS)$_2$Nb(=O)R	R = Me (x ray), Bu^n, CH$_2$Bu^t, CH$_2$TMS	62
Cp*$_2$Ta(=O)Me		109,226–8
{[Cp$_2$NbMe]$_2$(μ-O)}X$_2$	X$^-$ = I$^-$, [BF$_4$]$^-$	220
[Cp*TaMe$_3$]$_2$(μ-O)	x ray	229,230
[Cp′$_2$Nb(CH$_2$TMS)(μ-O)ZnCl$_2$]$_2$	x ray	221
[Cp$_2$Nb(S)Me]$_n$		220
Cp*$_2$Ta(=S)Me		109,140
Cp*$_2$Ta(=S)R	R = Ph, Bn, CH$_2$Bu^t, CH$_2$CH$_2$Ph, CH$_2$CH$_2$Bu^t	140
Cp*$_2$Ta(=NMe)Me		109
(η-C$_5$H$_4$TMS)$_2$Nb(η^2-O$_2$)Me		62
Cp$_2$NbMe(η^2-ONMe)	prep: Cp$_2$NbMe$_2$ + NO; NMR only	187
Cp*$_2$Ta(η^2-O$_2$)R	R = Me, Et, Pr^n, Ph, CH$_2$Ph (x ray)	227

Table 6 (continued)

Compound	*Comments*	*Ref.*
$Cp_2TaMe(\mu:\eta^2,\eta^3\text{-}C_3H_2O_2)Co_2(CO)_6$	x ray	223
$Cp_2TaMe(\mu:\eta^2,\eta^3\text{-}C_3H_2O_2)\text{-}Fe_2(CO)_7$	x ray	223
$Cp_2Nb(\eta^2\text{-}S_2)R$	R = Me, η^1-Cp, Bu^n	220
$Cp_2Ta(\eta^2\text{-}S_2)Me$		220
$Cp_2Nb(\eta^2\text{-}Se_2)Me$		220
Bis(cyclopentadienyl) M^{IV} complexes		
$(\eta\text{-}C_5H_4R)_2Nb(CH_2TMS)_2$	R = H, Me, TMS	231
$(\eta\text{-}C_5H_4R)_2NbMe_2$	R = H, TMS	231
$(\eta\text{-}C_5H_4R)_2Nb(CH_2Ph)_2$	R = H (x ray), TMS	231
$Cp_2Nb(CH_2PPh_2)_2$		218
$Cp_2Nb(CH_2Ph)_2$		232
$Cp_2Nb(\eta^1\text{-}Cp)_2$		233
$Cp_2Ta(\eta^1\text{-}Cp)_2$		233,234
$Cp_2Nb(Ph)Cl$		232
$Cp_2Ta(CH_2Ph)Cl$		99
$(\eta\text{-}C_5H_4Bu^t)CpNb(\eta^1\text{-}C_5H_4R)Cl$	R = H, Bu^t	235
$Cp_2Nb(\eta^1\text{-}C_5H_4Bu^t)Cl$		235
$(\eta\text{-}C_5H_4R)_2Nb(CH_2TMS)Cl$	R = H, Me, TMS	231
$(\eta\text{-}C_5H_4R)_2Nb(CH_2Bu^t)Cl$	R = Me, TMS	231
$Cp_2Nb(CCF_3{=}CHCF_3)F$		236
$Cp_2Nb(2\text{-}SC_4Cl_3)Cl$		237
$Cp_2Nb(CH_2PPh_2)Cl$		218
$Cp'_2Nb(CO_3)(CH_2TMS)$	prep: disproportionation of $Cp'_2Nb(\eta^2\text{-}CO_2)(CH_2TMS)$	228
$[(\eta^5:\eta^1\text{-}C_5H_4)CpNbH]_2$	niobocene dimer	239
$\{[Cp_2NbCl]_2(\mu\text{-}CH_2{=}CHCN)\}[BF_4]_2$	prep: $Cp_2NbCl_2 + AgBF_4 + CH_2{=}CHCN$	240
$Cp^*_2Ta_2Cl_3(R)(H)_2$	R = CH_2Bu^t, CH_2TMS, Me	204
$(\eta\text{-}C_5Me_4Et)_2Ta_2Cl_3(R)(H)_2$	R = CH_2Bu^t, CH_2TMS, Et, Me (x ray)	204
$Cp_2(CO)M^1CH(Me)OM^2(Me)Cp_2$	M^1 = Nb, Ta; M^2 = Zr, Hf	65
$Cp_2Nb(H)(\eta^1:\eta^5\text{-}C_5H_4)Fe(CO)_2$	x ray	241
$Cp_2Nb(\mu\text{-}CH_2PPh_2)_2Mo(CO)_4$	x ray	218
Bis(cyclopentadienyl) M^{III} complexes		
$(\eta\text{-}C_5H_4Pr^i)_2TaMe(PMe_3)$		242
$(\eta\text{-}C_5H_4TMS)_2NbMe(PR_3)$	$PR_3 = PMe_3$, PMe_2Ph	243
$Cp_2Ta(CH_2SiHBu^t_2)(PMe_3)$		244
$Cp_2NbR(CO)$	R = Me, CH_2Ph, CD_2Ph, $CH_2\text{-}4\text{-}C_6H_4OMe$, Ph, $CH_2OZr(H)Cp^*_2$	232
	R = Me, CH_2Bu^t, CH_2TMS, CH_2Ph, C_6F_5	62
	R = Pr^n, Pr^i, Bu^n, Bu^s, or Pn^n	245
	R = Me; UV PES study	246
$Cp'_2NbR(CO)$	R = CH_2Ph, CH_2Bu^t, CH_2TMS	247
	R = CH_2TMS	238,248
$Cp^*_2Nb(CH_2CH_2R)(CO)$	R = H, Me, Ph, $4\text{-}C_6H_4NMe_2$, $4\text{-}C_6H_4OMe$, Tol, $4\text{-}C_6H_4CF_3$	249
	R = $3\text{-}C_6H_4CF_3$, Ph, $3\text{-}C_6H_4Me$, $3\text{-}C_6H_4NMe_2$	138
$(\eta\text{-}C_5H_4TMS)_2NbR(CO)$	R = Me, CH_2Bu^t, CH_2TMS, CH_2Ph, C_6F_5 (x ray)	62
	R = CHEtPh, $CHPh_2$, $CHMe_2$	250
$Cp_2M(CH_2CH_2Me)(CO)$	UV PES studies	251
$Cp'_2Ta(CH_2CH_2Me)(CO)$	UV PES studies	251
$Cp_2Ta(CH_2Me)(CO)$		252
$Cp_2Ta(CH_2SiHBu^t_2)(CO)$	x ray	244
$Cp^*_2NbMe(CO)$		70
$Cp_2Nb(Bu)(CNR)$	R = Me, Bu^t, Cy, Ar'	245
$(\eta\text{-}C_5H_4\text{-}TMS)_2NbMe(CNBu^t)$		135
$Cp^*_2Nb(CH_2CH_2R)(CNMe)$	R = H, Me, Ph, $4\text{-}C_6H_4NMe_2$, $4\text{-}C_6H_4OMe$, Tol, $4\text{-}C_6H_4CF_3$	249
	R = $3\text{-}C_6H_4CF_3$, Ph, $3\text{-}C_6H_4Me$, $3\text{-}C_6H_4NMe_2$	138
$Cp_2Ta(CH_2CH_2R^1)(CNR^2)$	R^1 = Me, Et; R^2 = Me, Bu^t, Cy, Ar'	141
$Cp_2Ta(CH_2CH_2R^1)(CNR^2{\cdot}AlEt_3)$	R^1 = Me, Et; R^2 = Me, Bu^t, Cy, Ar'	141
$Cp^*_2Nb(CO{\cdot}AlEt_3)R$	R = Me, Et, CH_2CH_2Ph, CMe=CHMe	253
$Cp^*_2Nb(CO{\cdot}AlMe_3)R$	R = Me, CH_2CH_2Ph	253
$Cp^*_2Nb[(CO{\cdot}AlMe_2(CH_2CH_2Ph)]Me$		253
$Cp_2Nb[C(CO_2Me){=}CH(CO_2Me)](CO)$	x ray	254
$Cp_2Nb(CR{=}CH_2)(CO)$	R = CO_2Me, C(O)Me	254
$Cp_2M(CH{=}CH_2)(CO)$		65
$Cp_2M(CR{=}CH_2)(CO)$	R = CN, CF_3	255
$Cp_2M(CR{=}CHR)(CO)$	R = CN, CF_3	255
$Cp^*_2Nb(CMe{=}CHMe)(CO)$		253
$[(\eta\text{-}C_5H_4TMS)_2Nb((\eta^1\text{-}C)\text{-}C(O){=}CR_2)Cl]^-$	$R_2 = Ph_2$, Me_2, EtPh; electrochemically generated	250
$[Na(18\text{-crown-}6)(THF)_2][Cp_2Nb(CH_2PPh_2)_2]$	x ray	218
$Na[Cp'_2Nb(CO_3)(CH_2TMS)]$		238
$\{[(\eta^5:\eta^1\text{-}C_5H_4)CpNbH]_2\}^{2-}$	electrochemically generated	239
$Cp_2NbCH_2OZr(H)Cp^*$	tautomer of $Cp_2(H)Nb{=}CHOZr(H)Cp^*_2$	232
$Cp_2(CO)NbCH(Me)OZrCp_2$		256

Bercaw and co-workers[199] employed the [Cp^*TaMe_3] template to study the propensity of methyl groups to abstract α- or β-hydrogens from pendent alkyl, allyl, amide, and alkoxide ligands (Scheme 3). Thus, Cp^*TaMe_3Cl reacts with LiNHR to provide imido complexes of the type $Cp^*Ta(=NR)Me_2$,[257] along with one equivalent of CH_4. This formal α-hydrogen abstraction does not occur when Cp^*TaMe_3Cl is treated with lithium alkyls; reaction with $LiCH_2Bu^t$ provides $Cp^*Ta(CH_2Bu^t)Me_3$ and not the alkylidene $Cp^*Ta(=CHBu^t)Me_2$. Upon reacting Cp^*TaMe_3Cl with $CH_2{=}CMeCH_2MgBr$, the unstable allyl complex $Cp^*Ta(\eta^3\text{-}CH_2CMeCH_2)Me_3$ is formed, which slowly undergoes a β-hydrogen transfer to release CH_4 and provide the trimethylene methane complex $Cp^*Ta[\eta^4\text{-}C(CH_2)_3]Me_2$.[199] Similarly, the reaction of Cp^*TaMe_3Cl with $LiNMe_2$ provides $Cp^*TaMe_3(NMe_2)$, which decomposes to the η^2-imine complex $Cp^*Ta(\eta^2\text{-}MeN{=}CH_2)Me_2$. Interestingly, $Cp^*TaMe_3(OPr^i)$ does not decompose to provide the known acetone complex $Cp^*Ta(\eta^2\text{-}O{=}CMe_2)Me_2$. Bercaw and co-workers[199] found that the rates of hydrogen transfer correlate with the particular α-atom bound to the metal center in the ligand which transfers its hydrogen. Thus, for reactions involving α-hydrogen transfer, the observed rates of transfer are N-ligand > C-ligand, which correlates directly with the strength of the metal–ligand multiple bond formed in the α-hydrogen transfer, that is, Ta=NR > $Ta{=}CR_2$. Conversely, for reactions involving a β-hydrogen transfer, the observed order is reversed: C-ligand > N-ligand > O-ligand.[199]

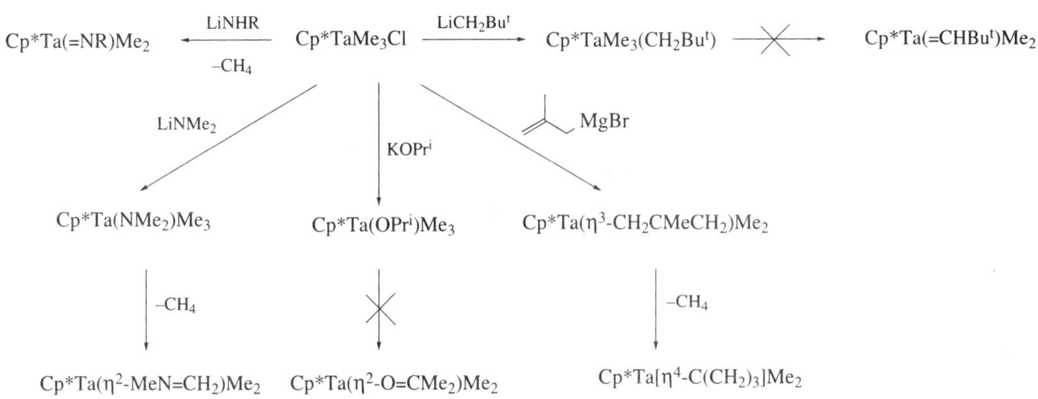

Scheme 3

2.3.2 Hydride Complexes

The chemistry of niobium and tantalum hydrides has developed especially rapidly since the early 1980s. This section emphasizes the descriptive chemistry of hydride complexes, while their catalytic applications are covered in Section 2.8. According to our ligand priority scheme, hydrides follow alkyls in precedence.

2.3.2.1 Hydride complexes without cyclopentadienyl ligands

Table 7 outlines the hydride complexes without cyclopentadienyl ligands. The phosphine hydrides and the alkoxide hydrides are seen to constitute the largest groups of hydride complexes without η^5 ligands.

(i) Phosphine hydrides

Considerable advances have been made in the phosphine hydrides of niobium and tantalum with a range of oxidation states in mono- and dinuclear compounds being developed. The d^3 complexes $TaCl_2(PMe_3)_4$ and $MCl_2(dmpe)_2$ react with H_2, forming eight-coordinate M^{IV} hydrides.[259–61] Thus, $TaH^*_2Cl_2(PMe_3)_4$, $NbH^*_2Cl_2(dmpe)_2$, and $TaH^*_2Cl_2(dmpe)_2$ (H^* = H or D) are all prepared from the appropriate precursor and H_2 (or D_2) at ~275 kPa in ether or alkane solvents (Scheme 4). These well-characterized, paramagnetic M^{IV} hydrides provide beautifully informative ESR spectra in solution, featuring hyperfine coupling to the metal [$I(^{181}Ta) = 7/2$, 100%] and also coupling to the ligand ^{31}P nuclei.[260] The structures of both $TaH_2Cl_2L_4$ (L = PMe_3, 1/2 dmpe) complexes have been reported; for example $TaH_2Cl_2(dmpe)_2$ was shown to possess a distorted square antiprismatic structure.[259]

Table 7 Hydride complexes of niobium and tantalum without cyclopentadienyl ligands.

Compound	*Comments*	*Ref.*
TaH_5	hypothetical; MO calculations	142,143
Phosphine hydrides (with or without halide ligands)		
$NbH_5(PMe_3)_4$	prep: $NbMe_5(PMe_3)_2 + H_2$ + excess PMe_3	156
$TaH_5(PMe_3)_4$	prep: $TaH_2(\eta^2\text{-}BH_4)(PMe_3)_4 + H_2 + PMe_3$	25
$TaH(PPh_2)_2(dmpe)_2$	x ray	258
$NbH^*_2Cl_2(dmpe)_2$	H* = H, D; prep: $NbCl_2(dmpe)_2 + H^*_2$	259,260
$TaH^*_2Cl_2(dmpe)_2$	H* = H (x ray), D	259,260
$TaH^*_2Cl_2(PMe_3)_4$	H* = H (x ray), D	23,260,261
$TaH_2Cl(PMe_3)_4$	x ray; neutron diffraction structure at 50 K	22,23,261
$TaH_2Cl(dmpe)_2$		23
$[TaH_2Cl(PMe_3)_3]_2(\mu\text{-}N_2)$		23
$[TaCl_2(PMe_3)_2]_2(\mu\text{-}H)_2(\mu\text{-}Cl)_2$	x ray; prep: $Ta_2Cl_6(PMe_3)_4 + H_2$	262–5
	prep: $TaCl_3(PMe_3)_3 + H_2$	266
$[TaCl_2(PMe_3)_2]_2(\mu\text{-}D)_2(\mu\text{-}Cl)_2$		265
$[TaCl_2(PMe_3)_2]_2(\mu\text{-}H)_2$	x ray	261,264,265
$[NbCl_2(PMe_3)_2]_2(\mu\text{-}H)_4$		260
$[TaCl_2(PMe_3)_2]_2(\mu\text{-}H)_4$	x ray	264,265
$[TaCl_2(PMe_3)_2]_2(\mu\text{-}H)_3(\mu\text{-}Cl)$	prep: $[TaCl_2(PMe_3)_2]_2(\mu\text{-}H)_2$ + HCl	264,265
$Ta_2H_2Cl_6(PMe_2Ph)_4$	possibly isostructural with $[TaCl_2(PMe_3)_2]_2(\mu\text{-}Cl)_2(\mu\text{-}H)_2$	267
Carbonyl hydrides (with or without phosphine ligands)		
$TaH(CO)_2(dmpe)_2$		27
$TaH(CO)_2(dppe)_2$		21
$TaH(CO)_2L_4$	$L_4 = (dppe)_2$ (x ray), $(Ph_2PCH_2CH_2PEt_2)_2$, $(Ph_2PCH_2CH_2PPhCH_2)_2$	24
$NbH(CO)_2L_4$	$L_4 = (Ph_2PCH_2CH_2PPhCH_2)_2$, $P(CH_2CH_2PPh_2)_3$	268
$TaH(CO)_3L_3$	$L_3 = P(CH_2CH_2PPh_2)_3$, $PhP(CH_2CH_2PPh_2)_2$, $(Ph_2PCH_2CH_2PPhCH_2)_2$	21
	L_3 = six different tridentate phosphine ligands	24
$TaH(CO)_3[PhP(CH_2CH_2PPh_2)_2]$		268
$NbH(CO)_3L_3$	$L_3 = PhP(CH_2CH_2PPh_2)_2$, $P(CH_2CH_2PPh_2)_3$	20
	$L_3 = P(CH_2CH_2PPh_2)_3$, $PhP(CH_2CH_2PPh_2)_2$, $(Ph_2PCH_2CH_2PPhCH_2)_2$	268
$TaH(CO)_4(dppe)$		21,14
$TaH(CO)_4L_2$	L_2 = eight different bidentate phosphine ligands	24
$NbH(CO)_4(dppe)$		20
$[NEt_4]_2[MH(CO)_5]$		35
$Na_2[MH(CO)_5]$	thermally unstable	35
Alkoxide hydrides		
$[NbH_2(silox)_2]_2$		159
$[TaH_2(silox)_2]_2$	terminal hydrides; x ray	33,159, 160,269
$[TaH^*_2(silox)_2]_2$	H* = H or D; isotopomers $Ta_2H_nD_{4-n}(silox)_4$ observed	160
$[Ta(silox)_2Cl]_2(\mu\text{-}H)_2$	prep: $[TaH_2(silox)_2]_2$ + HCl	159,160, 269
$[TaH(silox)_2]_2(\mu\text{-}O)_2$		160
$[TaH(silox)_2]_2(\mu\text{-}H)_2(\mu\text{-}O)$		160,269
$TaH_2(silox)_3$		159,160, 270
$TaH^*_2(silox)_3$	H^*_2 = HD or D_2	160
$TaHX(silox)_3$	X = Cl, I	160
$(silox)_2(H)\overline{Ta(OSiBu^t_2CMe_2CH_2)}$	cyclometallated ligand	160
$NbH(O\text{-}2,6\text{-}C_6H_3Cy_2)_4$	prep: $NbH_2(O\text{-}2,6\text{-}C_6H_3Ph_2)_3 + H_2$; x ray	168
$NbH_2(O\text{-}2,6\text{-}C_6H_3Ph_2)_3$		168
$\overline{NbH[O\text{-}2,6\text{-}C_6H_3Ph(C_6H_4)]}(O\text{-}2,6\text{-}C_6H_3Ph_2)_2$	cyclometallated ligand	271
$TaH_2Cl(OAr)_2(PR_3)_2$	$PR_3 = PMe_3$, PMe_2Ph (x ray), $PMePh_2$	173
$TaH_3(OAr)_2(PMe_2Ph)_2$	prep: $Ta(CH_2\text{-}TMS)_3(OAr)_2 + H_2 + PMe_2Ph$	173
$TaH_3(O\text{-}2,6\text{-}C_6H_3Cy_2)_2(PR_3)_2$	$PR_3 = PMe_2Ph$ (x ray), $PMePh_2$	173
	$PR_3 = PMe_2Ph$; procatalyst for aromatic hydrocarbon hydrogenation	183
$TaH_2(OAr)_3(PR_3)$	$PR_3 = PMe_3$, PBu_3, $PPhMe_2$ (x ray), PMe_2Cy, PPh_2Me, PPhMeCy, $PMeCy_2$	167
$TaH(OAr)_4$		167
Miscellaneous		
$[TaH(\mu\text{-}H)_2(NC_{12}H_8)(PMe_2Ph)_2]_2$	x ray; $[NC_{12}H_8]^-$ = 9*H*-carbazole amide	272

An elegant study of dinuclear hydrides has been reported by Sattelberger and co-workers.[265] Dinuclear $[TaCl_2(PMe_3)_2]_2(\mu\text{-}Cl)_2$ (**7**) (Ta=Ta) has been prepared from the reduction of $TaCl_5$ with two equivalents of NaHg in the presence of PMe_3.[273] The Ta=Ta bond of this complex is readily

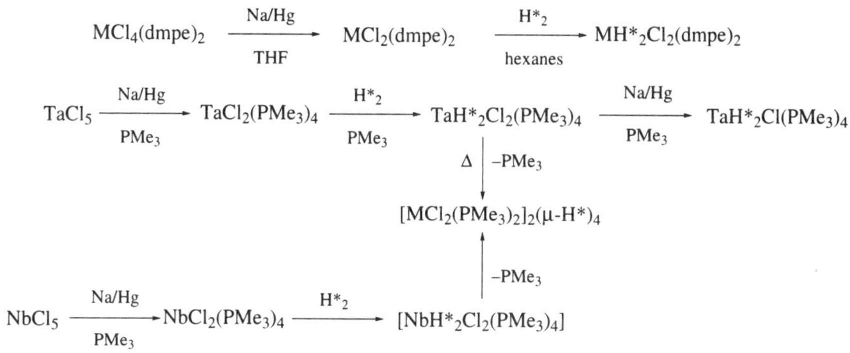

Scheme 4

hydrogenated to afford the Ta^{IV}–Ta^{IV} dimer $[TaCl_2(PMe_3)_2]_2(\mu\text{-}H)_2(\mu\text{-}Cl)_2$ (**8**).[262,263,265] Upon reducing this complex with NaHg, the dark green Ta^{III}–Ta^{III} dimer $[TaCl_2(PMe_3)_2]_2(\mu\text{-}H)_2$ (**9**) is formed,[264] which reacts cleanly with Cl_2, HCl, or H_2 to form the series of quadruply bridged dimers $[TaCl_2(PMe_3)_2]_2(\mu\text{-}H)_2(\mu\text{-}Cl)_2$ (**8**), $[TaCl_2(PMe_3)_2]_2(\mu\text{-}H)_3(\mu\text{-}Cl)$ (**10**), and $[TaCl_2(PMe_3)_2]_2(\mu\text{-}H)_4$ (**11**) respectively (Scheme 5).[265]

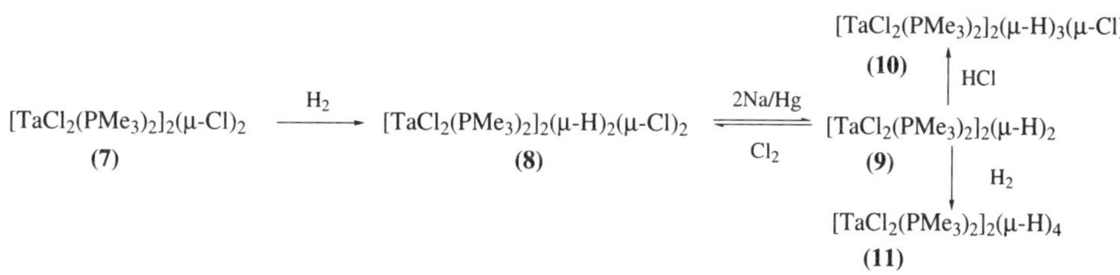

Scheme 5

Finally, the polyhydride complexes $NbH_5(PMe_3)_4$[156] and $TaH_5(PMe_3)_4$[25] have been described. The alkyl complex $NbMe_5(PMe_3)_2$ can be hydrogenated under 3 atm of H_2, in the presence of excess PMe_3, to afford the pentahydride $NbH_5(PMe_3)_4$.[156] $TaH_2Cl_2(PMe_3)_4$ reacts with $LiBH_4$ in the presence of PMe_3 to yield $TaH_2(\eta^2\text{-}BH_4)(PMe_3)_4$ and reaction of this complex with H_2 (276 kPa) in the presence of PMe_3 produces $TaH_5(PMe_3)_4$.[25]

(ii) Alkoxide hydrides

As described in Section 2.8, alkoxide hydrides have become important in catalytic arene hydrogenation studies.

The three-electron reduction of $M(silox)_2Cl_3$ (silox = $[OSiBu^t{}_3]^-$) in the presence of H_2 leads to the formation of M^{IV} dimers of the type $[MH_2(silox)_2]_2$ (**12**) which are proposed to contain terminal M–H bonds (IR) and an unsupported metal–metal bond leading to overall D_{2d} symmetry.[159] The tantalum complex reacts with four equivalents of HCl with loss of 2 mol of H_2 to produce $[Ta(silox)_2Cl_2]_2(\mu\text{-}H)_2$ and with six equivalents of HCl to re-form $Ta(silox)_2Cl_3$. More sterically encumbered $Ta(silox)_3Cl_2$ is reduced under an H_2 atmosphere to form monomeric $TaH_2(silox)_3$.[159] $Ta(OAr)_2Cl_{3-n}(CH_2\text{-TMS})_n$ (n = 2 or 3) can be hydrogenated in the presence of phosphines to form the 7-coordinate complexes $Ta(OAr)_2H_2Cl(PR_3)_2$ (PR_3 = PMe_3, PMe_2Ph, or $PMePh_2$) and $Ta(OAr)_2H_3(PMe_2Ph)_2$.[173]

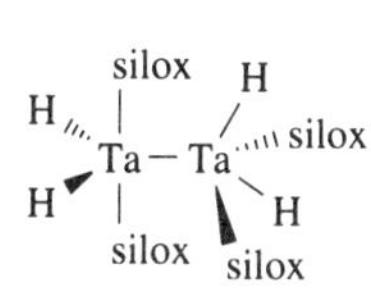

(12)

2.3.2.2 Cyclopentadienyl hydrides

Table 8 summarizes the cyclopentadienyl hydrides which for convenience are sorted according to ancillary ligand set beyond η^5 and hydride ligands.

(i) Compounds containing only cyclopentadienyl and hydride ligands

The hydrides $(\eta\text{-}C_5R_5)_2NbH_3$ and $(\eta\text{-}C_5R_5)_2TaH_3$ continue to be extremely useful starting complexes for a wealth of chemistry; Table 8 includes references to various preparations of these compounds and selected references for their use in synthesis. A number of new substituted η^5 ligands have appeared in compounds of the $(\eta\text{-}C_5R_5)_2MH_3$ stoichiometry, and several M^{III} species of stoichiometry $[(\eta\text{-}C_5R_5)_2MH_2]^-$ with a range of counter ions have been prepared from deprotonation of $(\eta\text{-}C_5R_5)_2MH_3$.[294]
Perhaps the most important development in the hydride chemistry of these metals has been the interpretation of the unusual NMR properties in certain niobium hydrides as due to bonding interactions between the hydride ligands.[278,281,305] Both nonclassical dihydrogen ($\eta\text{-}H_2$) complexes and trihydride (coordinated $[\eta\text{-}H_3]^-$) structures have been suggested in complexes of niobium. Labinger[1] first outlined some of the unusual NMR properties of Cp_2NbH_3, including an unusually large $^1J_{H\text{-}H}$ coupling constant in its $AlEt_3$ adduct and Curtis *et al.*[308] first reported a temperature dependence to the $^1J_{H\text{-}H}$ coupling constant in Cp_2NbH_3. The niobium hydrides $(\eta\text{-}C_5R_5)_2NbH_3$ have also been suggested by theoretical calculations and NMR experiments to possess a thermally accessible dihydrogen state slightly higher in energy from the hydride (Equation (2))[278] More recently, further experiments and calculations support the occurrence of exchange couplings observed in these complexes because of rotational tunneling of H_2 in the dihydrogen state.[325–7]

$$(\eta\text{-}C_5H_4TMS)_2NbH_3 \rightleftharpoons (\eta\text{-}C_5H_4TMS)_2Nb(\eta\text{-}H_2)H \qquad (2)$$

Heinekey[274] has carefully examined the ^1H NMR spectra of Cp_2MH_3 (M = Nb, Ta), which are shown to display an AB_2 spin system at ambient temperature. A temperature-dependent $J_{A\text{-}B}$ is observed for niobium (13.6 Hz at 293 K) and a temperature-independent $J_{A\text{-}B}$ (9.5 Hz) for Cp_2TaH_3.[274] Upon lowering the temperature, the $J_{A\text{-}B}$ coupling for Cp_2NbH_3 decreases until it becomes unresolved at 243 K; further lowering of the temperature allows coupling to reappear. These observations are attributed to both a magnetic component (conventional Fermi contact mechanism) and a quantum-mechanical exchange component (temperature dependent) of the coupling, which have opposite signs. Thus, the observed coupling is given by $J_{obs} = -2J_{exch} + J_{magn}$ and for Cp_2NbH_3 the exchange coupling is of the same order of magnitude as the magnetic coupling and the two-bond magnetic coupling is negative. As temperature is lowered, J_{exch} goes to 0 at ~173 K where coupling, now due entirely to J_{magn}, reappears.[274]
Relevant to this discussion is a new series of heterobimetallic adducts of niobocene trihydrides with Cu^+ and Ag^+ in which exchange coupling is also observed and an accessible dihydrogen state is proposed. These compounds are proposed to have a centrally located Cu^+ or Ag^+ ion with two μ-H ligands bonded from each $(\eta\text{-}C_5R_5)_2NbH_3$ moiety, that is, structurally, $\{[(\eta\text{-}C_5R_5)_2NbH(\mu\text{-}H)_2]_2Cu\}PF_6$ and $\{[(\eta\text{-}C_5R_5)_2NbH(\mu\text{-}H)_2]_2Ag\}BF_4$. The silyl-substituted complexes $\{[(\eta\text{-}C_5H_4\text{-}TMS)_2NbH(\mu\text{-}H)_2]_2Cu\}PF_6$ (**13**) and $\{[(\eta\text{-}C_5H_3(TMS)_2)_2NbH(\mu\text{-}H)_2]_2Cu\}PF_6$ (**14**) show AB_2 spectra at room temperature and ABC spectra at low temperature, with no H···H coupling observed despite narrow linewidths. Of the compounds $\{[(\eta\text{-}C_5R_5)_2NbH(\mu\text{-}H)_2]_2Cu\}PF_6$, only $\{[Cp_2NbH(\mu\text{-}H)_2]_2Cu\}PF_6$ shows a typical AB_2 pattern ($J_{A\text{-}B}$ = 15.8 Hz) with no temperature dependence in the spectrum, demonstrating the suppression in the exchange coupling which allows the observation of the magnetic coupling.[278] This phenomenon is not observable in the Ag^+ complexes $\{[(\eta\text{-}C_5R_5)_2NbH(\mu\text{-}H)_2]_2Ag\}BF_4$, since the chemical shift difference between the hydrides, $\Delta\nu$, is too small. These observations support Heinekey's suggestion that the exchange and magnetic coupling constants have opposite signs.[274] To account for these observations, the authors propose an equilibrium between structures in which a dihydrogen complex (which allows a direct interaction between the niobium lone pair and Lewis acid) and the hydride are exchanging (Equation (3)).[278]

$$Nb(H)(\mu\text{-}H)_2M \rightleftharpoons Nb(H_2)(\mu\text{-}H)M \qquad (3)$$

Nonclassical trihydride $[\eta\text{-}H_3]^-$ structures have also been suggested in silyl-substituted cyclopentadienyl complexes of niobium, $(\eta\text{-}C_5H_4TMS)_2NbH_3$ and $[\eta\text{-}C_5H_3(TMS)_2]_2NbH_3$.[281] In contrast, the tantalum hydrides $(\eta\text{-}C_5H_4TMS)_2TaH_3$ and $[\eta\text{-}C_5H_3(TMS)_2]_2TaH_3$ are found to have NMR properties consistent with true hydride behavior, that is, typical AB_2 coupling ($^2J_{A\text{-}B}$ ~ 10 Hz) which is temperature

Table 8 Hydride complexes of niobium and tantalum containing cyclopentadienyl ligands.

Compound	*Comments*	*Ref.*
Cp_2MH_3	He^I and He^{II} PES study	246
	NMR properties	274
Cp_2NbH_3	use in CO reduction studies	112,275
	various preparations and uses of	232,256,276,277
	NMR properties; MO calculations	278
Cp_2TaH_3	various preparations and uses of	68,72,99,276,279,280
Cp_2TaD_3		68
Cp'_2NbH_3		281
$Cp^*_2NbH_3$	prep: $Cp^*_2Nb(\eta^2\text{-}BH_4) + H_2 + py$	70
$Cp^*_2TaH_3$		109,282
$(\eta\text{-}C_5H_4Pr^i)_2TaH_3$		242
$(\eta\text{-}C_5H_4Bu^t)_2NbH_3$		115,283–5
$(\eta\text{-}C_5H_4Bu^t)_2TaH_3$		115,284
$(\eta\text{-}C_5H_4TMS)_2NbH_3$	NMR properties	278,281,286
	electrochemistry	285
	reactions of	287
$(\eta\text{-}C_5H_4R)_2TaH_3$	$R = Bu^t$, TMS, CMe_2Ph, CHMePh	288
$(\eta\text{-}C_5H_4TMS)_2TaH_3$	NMR properties	281
$[\eta\text{-}C_5H_3(TMS)_2]_2NbH_3$	NMR properties	278,281,286
	reactions of	287
$[\eta\text{-}C_5H_3(TMS)_2]_2TaH_3$	NMR properties	281
$(\eta\text{-}C_5H_4CHMePh)CpNbH_3$		117
Cp^*CpTaH_3	prep: $Cp^*TaCl_3(PMe_3) + NaCp$	282
$[(\eta^5{:}\eta^1\text{-}C_5H_4)CpNbH]_2$	niobocene dimer	276,277,289–91
$Na_2[(\eta^5{:}\eta^1\text{-}C_5H_4)_2Cp_2Nb_2H_2]\cdot THF$		289
$(\eta\text{-}C_5H_4SiMe_2OSiMe_2\text{-}\eta^5{:}\eta^1\text{-}C_5H_3)_2Nb_2H_2$	prep: $(\eta^5{:}\eta^5\text{-}C_5H_4SiMe_2OSiMe_2C_5H_4)NbCl_2 + Na$	289
$[Na(OEt_2)_2]_2[(\eta\text{-}C_5H_4SiMe_2OSiMe_2OSi\text{-}Me_2\text{-}\eta^5{:}\eta^1\text{-}C_5H_3)_2Nb_2H_2]$	x ray	289
$(\eta^5{:}\eta^5\text{-}C_{10}H_8)Cp_2Nb_2(\mu\text{-}H)_2$	$\eta^5{:}\eta^5\text{-}C_{10}H_8$ = fulvalene	290–3
$[(\eta\text{-}C_5H_4R)_2NbH(\mu\text{-}H)]_2$	R = TMS, Bu^t; electrochemically generated	285
$[Na(benzo\text{-}15\text{-}crown\text{-}5)][Cp_2NbH_2]$	x ray	291
$[M][Cp_2NbH_2]$	$M^+ = Li^+$, Na^+, $[Li(12\text{-}crown\text{-}4)]^+$, $[Na(15\text{-}crown\text{-}5)]^+$, $[K(18\text{-}crown\text{-}6)]^+$	294
$Li[Cp_2NbH_2]$		295
$Na[Cp_2NbH_2]$	prep: $Cp_2NbH_3 + NaH$	291
$K[Cp_2NbH_2]$		296
$[Cp_2NbH_2]_2Yb(diglyme)$	x ray	296
$(Cp_2NbH_2)_3Ce\cdot 6THF$		295
Cp_2MH	photochemical generation in matices	52
$(\eta\text{-}C_5H_4TMS)_2NbH$		285
Halide complexes		
$[CpTaCl_2H]_n$		204
$[Cp^*TaCl_2H^*]_2$	$H^* = H$, D (HD compound also prepared); prep: $Cp^*Ta(CH_2{=}CHR)Cl_2 + H_2$	204,206,208
$[(\eta\text{-}C_5Me_4Et)TaCl_2H^*]_2$	$H^* = H$, D	204,206,208
$[(\eta\text{-}C_5Me_4R)Ta(\mu\text{-}H)Cl_2]_2$	R = Me, Et; prep: $[(\eta\text{-}C_5Me_4R)Ta(\mu\text{-}Cl)]_2 + H_2$	297
$[(\eta\text{-}C_5Me_4Et)TaBr_2H]_2$		204
$(\eta\text{-}C_5H_4Bu^t)_2TaCl_2H$		298
$\{[\eta\text{-}C_5Me_3(TMS)_2]TaCl_2H\}_2$	prep: $[\eta\text{-}C_5Me_3(TMS)_2]Ta(CH_2{=}CHMe)Cl_2 + H_2$	206
$Cp^*(\eta\text{-}C_5Me_4Et)Ta_2Cl_4H_2$		204
$(\eta\text{-}C_5Me_4Et)[\eta\text{-}C_5Me_3(TMS)_2]Ta_2Cl_4H_2$		206
$Cp^*_2Ta_2Cl_5H$		208
$(\eta\text{-}C_5Me_4Et)_2Ta_2Cl_5H$		204,208
$[(\eta\text{-}C_5Me_4Et)TaCl_2H]_2(py)$	R = Me, Et	206
$(\eta^5{:}\eta^5\text{-}C_{10}H_8)Cp_2Nb_2(\mu\text{-}H)(\mu\text{-}X)$	X = Cl, Br, I	290
	X = Cl	293,299
Carbonyl complexes		
$Cp_2NbH(CO)$	prep: $Cp_2Nb(CH_2{=}CHR)H + CO$	245
	prep: $Cp_2NbH_3 + Cr(CO)_6$	112
	x ray	60
	prep: $Cp_2NbCl_2 + Na[AlH_2(OCH_2CH_2OMe)_2]$	276
	uses of	59,76,256
	electrochemical study	49
$Cp_2TaH(CO)$	x ray	276,300
	preparations and uses of	66,68,252,301,302
	UV PES study	251
$Cp_2MH(CO)$	photochemical generation in matices	52
	He^I and He^{II} PES study	246
	uses of	64,65,303

Table 8 (continued)

Compound	*Comments*	*Ref.*
$Cp'_2MH(CO)$		303
$(\eta\text{-}C_5H_4CHMePh)CpNb(CO)H$		117
$(\eta\text{-}C_5H_4Bu^t)_2TaH(CO)$		302
$(\eta\text{-}C_5H_4TMS)_2NbH(CO)$		304
$[\eta\text{-}C_5H_3(TMS)_2]_2NbH(CO)$		304
$(\eta\text{-}C_5H_4R)_2TaH(CO)$	$R = Bu^t$, TMS, CMe_2Ph, CHMePh	288
$Cp^*_2Nb(CO)H$		70
$Cp^*_2Ta(CO)H$		282
$Cp^*CpTa(CO)H$		282
$[Et_4N][CpMH(CO)_3]$		51
$[NEt_4][CpNb(H^*)(CO)_3]$	$H^* = H$, D	50
$[PPN][CpNbH(CO)_3]$[a]	prep: $CpNb(CO)_4$ + NaHg + THF + [PPN]Cl	107
$Na[CpNbH(CO)_3]$	prep: $CpNb(CO)_4$ + NaHg + THF	120
$[MR_n]^+[CpNbH(CO)_3]^-$	ion-paired; $[MR_n]^+ = (AuPPh_3)^+$, $(SnR_3)^+$, $(GePh_3)^+$, $(PbR_3)^+$	120
$CpTaH_2(CO)_3$		73,305
$CpNbH_2(CO)_3$	in equilibrium with $CpNb(\eta\text{-}H_2)(CO)_3$	305
$CpNb(\eta\text{-}H_2)(CO)_3$		305
$(\eta\text{-}C_5H_4Pr^i)_2TaH(CO)$		242
$(\eta\text{-}C_5H_4Bu^t)_2MH(CO)$		115
$[(\eta\text{-}C_5Me_4Et)TaCl_2H]_2(CO)$	^{13}CO also prepared	206
$Cp^*_2NbH(CO\cdot AlREt_2)$	R = Me, Et, CH_2CH_2Ph	253
$Cp^*_2NbH(CO\cdot AlR_3)$	R = Me, Et	253
$Cp^*TaH(CO\cdot AlEt_3)$		253
$Cp_2Nb(\mu\text{-}H)MCp(CO)_3$	M = Nb (x ray), Ta; prep: Cp_2MH_3 + $CpM(CO)_4$ + $h\nu$	306
$Cp_2Nb(CO)(\mu\text{-}H)NbCp(CO)_3$	x ray	307
$[PPN]\{[CpNb(CO)_3]_2(\mu\text{-}H)\}$		107
Silyl complexes		
$Cp_2MH_2(SiMe_2Ph)$	M = Nb, Ta (x ray)	308
$Cp_2NbH_2(SiMe_2O\text{-}TMS)$		308
$Cp_2TaH(SiR_3)_2$	$SiR_3 = SiHMe_2$, TMS	244
Nitrogen-donor ligands		
$Cp^*Nb(=NAr)H_2(PMe_3)$	prep: $Cp^*Nb(=NAr)Cl_2$ + Mg + PMe_3	75
$Cp^*Ta(=NBu^t)H_2(PR_3)$	$PR_3 = PMe_3$, PMe_2Ph	199
$Cp^*Ta(=NCH_2Bu^t)H_2(PR_3)$	$PR_3 = PMe_3$, PMe_2Ph	199
$Cp^*_2Ta(=NR)H$	R = Ph; x ray	309–11
	R = Me, Ph; MO calculations	312
	$R = Bu^t$	310
	R = H	309
$Cp_2M(=NN=CPh_2)H$		313
$Cp_2Nb[=NN=CPh(2\text{-}C_2H_4Br)]H$		313
$[(\eta^5{:}\eta^5\text{-}C_{10}H_8)Cp_2Nb_2(NHR)_2(\mu\text{-}H)][X]_3$	R = Ph, Tol; $X^- = Cl^-$, $O_2CCF_3^-$, 1/2 SO_4^{2-}	314
$Cp_2MH[\eta^2(N,N)\text{-}N_2CPh_2]$	prep: Cp_2MH_3 + N_2CPh_2	139
$(\eta^5\text{-}C_9H_7)_2NbH(py)$		47
Phosphorus-donor ligands		
$Cp_2MH(PEt_3)$	prep: $Cp_2MH(CO)$ + PEt_3 + $h\nu$	276
$Cp_2NbH(PPh_2H)$		315
$Cp_2TaH(PMe_3)$	prep: $Cp_2Ta(CH_2{=}CH_2)H$ + PMe_3 + $h\nu$	252
	uses of	72
	prep: $CpTaCl_4$ + Na + neat PMe_3	316
$Cp_2TaH(PR_3)$	$PR_3 = PMe_2Ph$, $P(OMe)_3$	317
	$PR_3 = PEt_3$, PPh_3	66
$(\eta\text{-}C_5H_4TMS)_2NbH[P(OMe)_3]$		304,318
$\{(\eta\text{-}C_5H_4TMS)_2NbH[P(OMe)_3]\}^+$	electrochemically generated	318
$(\eta^5\text{-}C_9H_7)_2NbH(PMe_2Ph)$	C_9H_7 = indenyl	47
$[Cp_2TaH_2(PR_3)]PF_6$	$PR_3 = PMe_2Ph$, $P(OMe)_3$; both *endo*- and *exo*-isomers	317
	$PR_3 = PMe_2Ph$, $P(OMe)_3$	319
$Cp^*NbH_4(PR_3)_2$	$PR_3 = PMe_3$, 1/2 dmpe	211
$Cp^*TaH_4(PR_3)_2$	$PR_3 = PMe_3$, PMe_2Ph, $P(OMe)_3$, 1/2 dmpe	94
	$PR_3 = PMe_3$ (x ray), $P(OMe)_3$, 1/2 dmpe	211
	$PR_3 = PMe_3$	91,95
$Cp^*TaH_3Cl(PMe_3)_2$	prep: Cp^*TaMe_3Cl + H_2 + PMe_3	94,211
"$CpNbH_2(dmpe)$"		98
$\{[(\eta\text{-}C_5H_4TMS)_2NbH_2[P(OMe)_3]\}X$	$X^- = [PF_6]^-$, $[BF_4]^-$; electrochemically generated	285
$Cp^*Ta(PPh_2)_3H$		210
Oxygen- and sulfur-donor ligands		
$Cp^*(Bu^tO)_2(H)Ta(\mu\text{-}H)_2TaH_3Cp^*$	prep: $Cp^*Ta(OBu^t)Me_3$ + H_2 + PMe_3	211
$(\eta^5{:}\eta^5\text{-}C_{10}H_8)Cp_2Nb_2(\mu\text{-}H)(\mu\text{-}OR)$	R = Me, Et (x ray)	290,293,299

Table 8 (continued)

Compound	*Comments*	*Ref.*
$Cp^*TaH_2(OR)_2$	R = Ar', Mes, Ar	320
$Cp^*_2Ta(=O)H$	x ray; ^{16}O and ^{18}O isotopomers	226,311,309
	^1H and ^2H isotopomers	109
$(\eta\text{-}C_5H_4Bu^t)_2Ta(=O)H$		298
$[(\eta\text{-}C_5H_4TMS)_2NbH_2(THF)]PF_6$	electrochemically generated	285
$[(\eta\text{-}C_5H_4Bu^t)_2NbH_2(THF)]PF_6$	electrochemically generated	285
$(\eta\text{-}C_5Me_4Et)_2Nb(\eta\text{-}S_2)H$		321
$(\eta\text{-}C_5H_4Bu^t)_2Ta(\eta\text{-}S_2)H$	x ray	322
$Cp^*_2Ta(=S)H$		140
$(\eta\text{-}C_5H_4Bu^t)_2Ta(=S)H$	prep: $(\eta\text{-}C_5H_4Bu^t)_2Ta(\eta\text{-}S_2)H + PR_3$	298,322
Mixed-ligand complexes		
$[CpNb(CO)_2(dmpe)H]PF_6$		82,83
$[CpNb(dmpe)H(\mu\text{-}H)]_2$	prep: $CpNbCl_2(CO)(dmpe) + NaAlH_2(OCH_2CH_2OMe)_2$	83
$CpNb(dmpe)H(\eta^2\text{-}BH_4)$	prep: $CpNbCl_2(CO)(dmpe) + NaBH_4$	83
$Cp^*Ta(CO)(PR_3)_2H_2$	$PR_3 = PMe_3$, 1/2 dmpe	94
$Cp^*Ta(CO)_n(PMe_3)_{3-n}(H)Cl$	$n = 1, 2$	94
$CpNb(H)Cl(CO)(PR_3)_2$	$PR_3 = PMe_3$ (x ray), PMe_2Ph	81
$[CpNbH(CO)(PR_3)_3]BF_4$	$PR_3 = PMe_3$, PMe_2Ph	81
$[(\eta\text{-}C_5Me_4Et)TaCl_2(H)]_2(py)$	R = Me, Et	206
Miscellaneous		
$[\eta^5{:}\eta^1\text{-}C_5Me_4(CH_2)_2CH{=}]Cp^*TaH$		109
$Cp_2NbH(\eta^2\text{-}Bu^tB{\equiv}NBu^t)$	x ray; prep: $Cp_2NbH_3 + Bu^tB{\equiv}NBu^t$	323
$Cp_2NbH[\eta^2\text{-}(TMS)Bu^tNB{\equiv}NBu^t]$		323
$(\eta\text{-}C_5H_4Bu^t)_2Ta[\eta^2(N,S)\text{-}NPhC(=S)S]H$	prep: $(\eta\text{-}C_5H_4Bu^t)_2Ta(=S)H + S{=}N{=}CPh$	298
$\{[Cp_2(CO)Nb]_2(\mu\text{-}H)\}^+[Mn(CO)_5]^-$	x ray; prep: $Cp_2NbH_3 + Mn_2(CO)_{10} + h\nu$	324

[a] $[PPN] = [(Ph_3P)_2N]$.

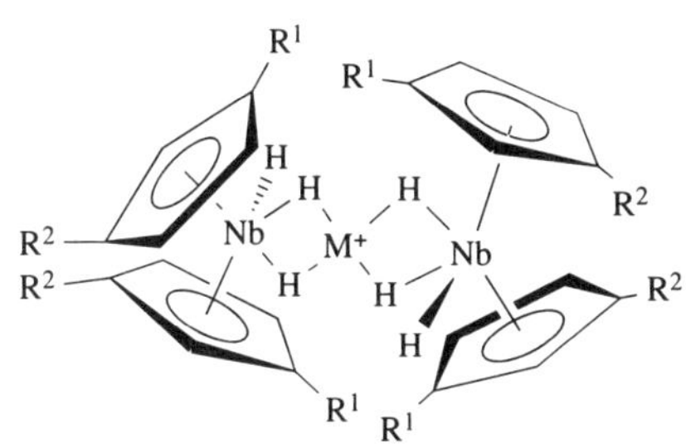

(13) $R^1 = H$, $R^2 = TMS$, $M^+ = Cu^+$
(14) $R^1 = R^2 = TMS$, $M^+ = Cu^+$

invariant.[281] However, the niobium complexes $(\eta\text{-}C_5H_4TMS)_2NbH_3$ and $[\eta\text{-}C_5H_3(TMS)_2]_2NbH_3$ have unusual NMR properties, notably (a) short T_1 values, (b) unusually large $^1J_{H\text{-}H}$ coupling constants (36.5 Hz and 70.0 Hz for $(\eta\text{-}C_5H_4TMS)_2NbH_3$ and $[\eta\text{-}C_5H_3(TMS)_2]_2NbH_3$, respectively at room temperature), and (c) extremely temperature-dependent $^1J_{H\text{-}H}$ coupling constants (measured from 3 to ~100 Hz for $[\eta\text{-}C_5H_3(TMS)_2]_2NbH_3$). Collectively, these data are interpreted as arising from H···H bonding interactions among the hydrides induced by the electron-withdrawing effects of the TMS groups. Antiñolo *et al.*[281] therefore suggested the formation of "nonclassical" trihydride $[H_3]^-$ structures in niobium to explain these data, that is, $(\eta\text{-}C_5H_4TMS)_2Nb(\eta\text{-}H_3)$ and $[\eta\text{-}C_5H_3(TMS)_2]_2Nb(\eta\text{-}H_3)$, while the classical hydride structures for tantalum seem apparent.

An alternative explanation of these unusual NMR properties in $(\eta\text{-}C_5H_4TMS)_2NbH_3$ and $[\eta\text{-}C_5H_3(TMS)_2]_2NbH_3$ has been proposed by Zilm *et al.*[286] They presented experimental and theoretical evidence which suggests these couplings in the niobium complexes are not magnetic in origin. The fact that $J(^1H\text{-}^3T)$ couplings are not observed in the partially tritiated isotopomers supports their proposal that the observed couplings in $(\eta\text{-}C_5H_4TMS)_2NbH_3$ and $[\eta\text{-}C_5H_3(TMS)_2]_2NbH_3$ are in fact quantum mechanical exchange couplings analogous to exchange couplings between electrons in radical pairs observed in ESR experiments. Thus, Zilm *et al.*[286] proposed that the observed couplings are a simple manifestation of a quantum mechanical motion associated with the hydrogen nuclei in the trihydride structure.

(ii) Cyclopentadienyl halide hydrides

The cyclopentadienyl hydride compounds which also contain chloride or bromide ligands commonly exist as dinuclear species but are usually synthesized from mononuclear precursors. For example, the alkene complexes $Cp^*Ta(CH_2{=}CHMe)Cl_2$ and $Cp^*Ta(CH_2{=}CHPh)Cl_2$ react with H_2 (pentane solution) to afford green, insoluble $[Cp^*TaCl_2H]_2$, for which a bridging hydride structure is proposed.[204] More soluble $[(\eta\text{-}C_5Me_4Et)TaCl_2H]_2$ is available from hydrogenation of $(\eta\text{-}C_5Me_4Et)Ta(alkene)Cl_2$ or from reacting $(\eta\text{-}C_5Me_4Et)Ta(CH_2Bu^t)_2Cl_2$ with H_2. In large-scale preparations, a small yield of side-product $(\eta\text{-}C_5Me_4Et)_2Ta_2Cl_5H$ is also isolable from the $(\eta\text{-}C_5Me_4Et)Ta(CH_2Bu^t)_2Cl_2 + H_2$ synthesis. These compounds are especially of interest for the production of stabilized formyls, for example, $[(\eta\text{-}C_5Me_4Et)TaCl_2(H)]_2$ reacts rapidly with CO to form $[(\eta\text{-}C_5Me_4Et)TaCl_2]_2(\mu\text{-}H)(\mu\text{-}CHO)$.[206]

(iii) Cyclopentadienyl carbonyl hydrides

Reports of the hydrides $(\eta\text{-}C_5R_5)_2MH(CO)$ have continued with detailed physical characterization by electrochemistry, photoelectron spectroscopy, x-ray structural studies, etc., constituting highly significant contributions. Table 8 includes references to various preparations of these compounds, physical studies, and selected references for their use in synthesis. As described above for the $(\eta\text{-}C_5R_5)_2MH_3$ compounds, nonclassical hydride structures have been observed in $CpNbH_2(CO)_3$ in matrix isolation experiments. Thus, the photolysis of $CpM(CO)_4$ and H_2 have been examined in solution at room temperature (in heptane) and at low temperatures (in liquid xenon).[305] For tantalum, the classical dihydride complex $CpTa(H)_2(CO)_3$ is formed, whereas for niobium, a temperature-dependent equilibrium between a classical dihydride and the dihydrogen complex is observed (Equation (4)). Note that the dihydrogen complex of vanadium, $CpV(\eta\text{-}H_2)(CO)_3$, is formed under these conditions using $CpV(CO)_4$.[305]

$$CpNb(H)_2(CO)_3 \rightleftharpoons CpNb(\eta\text{-}H_2)(CO)_3 \qquad (4)$$

The carbonyl hydride compounds $Cp^*_2M(CO)H$ (M = Nb, Ta) have been observed to bind reversibly aluminum alkyls AlR_3 at the carbonyl carbon to afford compounds of general formula $Cp^*_2M(CO{\cdot}AlR_3)(H)$.[253] An interesting preparation of such a hydride is observed from the alkyl complexes $Cp^*_2Nb(CO{\cdot}AlEt_3)R$ (R = Me, Et, CH_2CH_2Ph or CMe=CHMe) which form from $Cp^*_2Nb(CO)R$ and $AlEt_3$.[253] The room-temperature thermolysis of $Cp^*_2Nb(CO{\cdot}AlEt_3)R$ (R = Me, Et, or CH_2CH_2Ph) induces an exchange between niobium-bound and aluminum-bound alkyls to form $Cp^*_2Nb(CO{\cdot}AlREt_2)(Et)$, which then β-H eliminates to afford $Cp^*_2Nb(CO{\cdot}AlREt_2)H$ (Equation (5)).[253]

$$Cp^*_2Nb(CO)R + AlEt_3 \longrightarrow Cp^*_2Nb(CO{\bullet}AlEt_3)R \longrightarrow Cp^*_2Nb(CO{\bullet}AlREt_2)Et \longrightarrow Cp^*_2Nb(CO{\bullet}AlREt_2)H \qquad (5)$$

R = Me, Et, or CH_2CH_2Ph

(iv) Other cyclopentadienyl hydrides

Table 8 has further divided the cyclopentadienyl hydrides by Si-, N-, P-, O-, or S-donor ligands and also combinations of these ligand sets. Although only a handful of silyl- and sulfur-donor complexes are known (compared to η^5 complexes without hydrides, Section 2.7.3.2), an array of compounds with N-, P-, and O-donors have been examined. Selected examples are described below.

Imido hydride complexes of tantalum have been of considerable interest. The protonation of $Cp^*_2Ta(=CH_2)H$ with aniline affords the imido hydride $Cp^*_2Ta(=NPh)H$ while the unsubstituted imide $Cp^*_2Ta(=NH)H$ is provided in low yield from $Cp^*_2Ta(=CH_2)H$ and ammonia (or from $Cp^*_2TaCl_2$ and $NaNH_2$).[309] The protonation reaction involves an initial tautomerization of $Cp^*_2Ta(=CH_2)H$ to d^2 $[Cp^*_2Ta(Me)]$, followed by reaction with H_2NPh to generate $Cp^*_2Ta(=NPh)H$ (Scheme 6). Molecular orbital calculations on the model compound $Cp_2Ta(=NPh)H$ have been described by Jørgensen.[312]

While reports and the use of the phosphine hydrides $(\eta\text{-}C_5R_5)_2MH(PR_3)$ have continued, several monokis(cyclopentadienyl) derivatives have been of particular interest. The high-pressure (100 atm) hydrogenation of Cp^*TaMe_4 or Cp^*TaMe_3Cl in the presence of PR_3 ligands affords the hydrido complexes $Cp^*TaH_4(PR_3)_2$ ($PR_3 = PMe_3$, PMe_2Ph, $P(OMe)_3$, 1/2 dmpe) and $Cp^*TaH_3Cl(PMe_3)_2$, respectively.[94] Low-temperature, high-field NMR spectra indicate structures with C_s symmetry and equivalent phosphines for these monodentate PR_3 complexes, consistent with the x-ray structure of

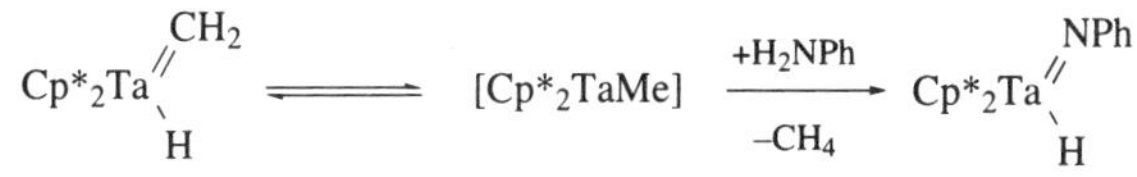

Scheme 6

$Cp^*TaH_4(PMe_3)_2$ (**15**) which indicates a tricapped trigonal prism.[211] The niobium analogues $Cp^*NbH_4(PR_3)_2$ ($PR_3 = PMe_3$, 1/2 dmpe) are formed similarly from the hydrogenation of Cp^*NbMe_4.[211] The hydride ligands in $Cp^*TaH_4(PR_3)_2$ are found to be more hydridic than acidic.

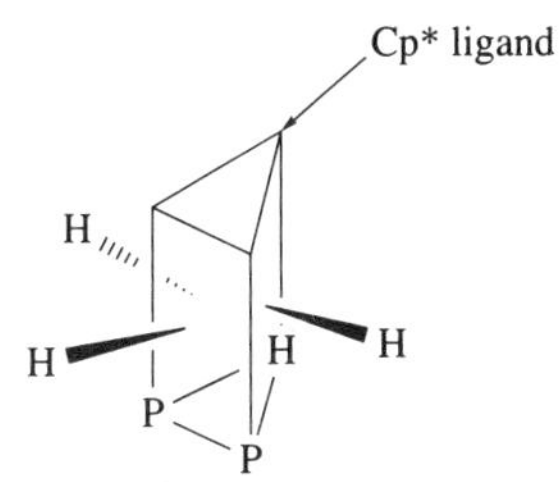

(15)

2.3.2.3 Heterobimetallic hydride complexes

Table 9 shows the heterobimetallic hydride complexes. This category of hydride complex includes not only μ-hydrido heterobimetallic compounds (e.g., $Cp_2(CO)Nb(\mu\text{-}H)Mo(CO)_5$), but also metal–metal-bonded species which contain terminal Nb–H or Ta–H ligands (e.g. $Cp'_2TaH(ZnCp)_2$). Table 9 separates these complexes by the IUPAC group of the heteroatom, which allows us a niche for compounds containing coordinated $[BH_4]^-$ ligands, and emphasizes the range of complexes prepared. A small number of these species also appear in Table 8, since structural data are often lacking and the distinction between anionic hydrides, ion pairs in solution, and coordinated counterions may be indeterminate.

Irradiating Cp_2NbH_3 and $M(CO)_6$ (M = Cr, Mo, W) in THF affords the complexes $Cp_2(CO)Nb(\mu\text{-}H)M(CO)_5$.[330] These structures constitute examples of the general category $Cp_2(CO)Nb(\mu\text{-}H)ML_n$, where the 18-electron moiety $Cp_2NbH(CO)$ acts as a two-electron donor ligand to a formal 16-electron ML_n fragment.[1,112] Photolysis of Cp_2NbH_3 and $Mn_2(CO)_{10}$ under similar conditions (toluene, near-UV irradiation, −15 °C) affords an unstable complex which cannot be isolated, but which upon standing forms the salt $\{[Cp_2(CO)Nb]_2(\mu\text{-}H)\}^+[Mn(CO)_5]^-$.[324,330] A detailed look at the symmetry of $\{[Cp_2(CO)Nb]_2(\mu\text{-}H)\}^+$ also suggests a complex of the general category $Cp_2(CO)Nb(\mu\text{-}H)ML_n$, where 18-electron $Cp_2NbH(CO)$ acts as a two-electron donor to 16-electron $Cp_2Nb(CO)$.

While transition metal heterobimetallics typically involve hydride ligands in a bridging environment, there are many main-group element heterobimetallics with direct Nb–M or Ta–M bonds which contain only terminal hydrides bound to niobium or tantalum. For example, Cp_2NbH_3 reacts with Cp_2Zn to form compounds containing no μ-H ligands. Excess Cp_2NbH_3 reacts with Cp_2Zn (benzene) to afford $[Cp_2NbH_2]_2(\mu\text{-}Zn)$, but when Cp_2Zn is in excess, *exo*-$Cp_2NbH(ZnCp)_2$ is isolated.[336] In THF solution, *endo*-$Cp_2NbH_2(ZnCp)$ is the only product isolated, regardless of the ratio of reactants.[336] $[Cp'_2TaH_2]_2(\mu\text{-}Zn)$, *endo*-$Cp'_2TaH_2(ZnCp)$, and *endo*- and *exo*-$Cp'_2TaH(ZnCp)_2$ are prepared similarly and also contain no bridging hydride ligands.[337]

Table 9 provides a means to situate logically $[BH_4]^-$ ligands, which have been characterized in the η^1 and η^2 binding modes for niobium and tantalum. For example, $TaH_2Cl_2(PMe_3)_4$ reacts with $LiBH_4$ in the presence of PMe_3 to yield $TaH_2(\eta^2\text{-}BH_4)(PMe_3)_4$.[25] Reaction of this complex with CO (276 kPa) produces initially $Ta(\eta^1\text{-}BH_4)(CO)_3(PMe_3)_3$ which, if not isolated from the reaction mixture, converts to $[Ta(CO)_3(PMe_3)_4][Ta(CO)_5PMe_3]$. A provocative NMR study of $Cp^*CpTa(\eta^2\text{-}BH_4)$, including quantitative two-dimensional exchange studies, showed that the scrambling of the terminal and bridging hydrides occurs by a mechanism in which there is no direct exchange between the two terminal hydrogens.[344] These results are interpreted as arising from an intermediate η^3-BH_4 complex with a "slipped" cyclopentadienyl ring, that is, intermediate $(\eta^3\text{-}Cp)Cp^*Ta(\eta^3\text{-}BH_4)$ or $Cp(\eta^3\text{-}Cp^*)Ta(\eta^3\text{-}BH_4)$ which can scramble hydrogens via rotation about the Ta···B axis.[343,344]

Table 9 Heterobimetallic hydride complexes of niobium and tantalum.

Compound	*Comments*	*Ref.*
Group 2		
$Cp_2TaH_3{\cdot}MgI_2$		280
Group 4		
$(\eta^5{:}\eta^5\text{-}C_{10}H_8)(CpTi)(CpNb)(\mu\text{-}H)_2$	prep: $Cp_2NbH_2Li + Cp_2TiCl_n$ $(n = 1, 2)$	295,328
$(\eta^5{:}\eta^5\text{-}C_{10}H_8)Cp_2TiNb(\mu\text{-}H)(\mu\text{-}X)$	X = Cl, Br	295
Group 5		
$Cp_2Nb(\mu\text{-}H)VCp(CO)_3$	x ray	306
$[PPN][Cp(CO)_3Nb(\mu\text{-}H)V(CO)_3Cp]$		107
Group 6		
$[Cp_2(CO)Nb(\mu\text{-}H)]_2Mo(CO)_4$	x ray	329
$[Cp_2(CO)Nb(\mu\text{-}H)_2]_2M(CO)_4$	M = Mo (x ray), W	114
$Cp_2(CO)Nb(\mu\text{-}H)M(CO)_5$	M = Cr (x ray), Mo, W	330
$Cp_2(CO)Ta(\mu\text{-}H)M(CO)_5$	M = Cr, Mo, W	331
$(\eta^5\text{-}C_5H_4R)_2(CO)Ta(\mu\text{-}H)M(CO)_5$	R = H, Bu^t; M = Cr, Mo, W	302
$[PPN][Cp(CO)_3Nb(\mu\text{-}H)M(CO)_5]$	M = Cr, Mo, W	107
Group 7		
$\{[Cp_2(CO)Nb]_2(\mu\text{-}H)\}^+[Mn(CO)_5]^-$	x ray	324,330
$Cp_2(CO)Ta(\mu\text{-}H)Mn_2(CO)_{10}$	x ray; prep: $Cp_2TaH_3 + Mn_2(CO)_{10} + h\nu$	330,332
$Cp_2(CO)M(\mu\text{-}H)Mn(CO)_2Cp$	M = Nb (x ray), Ta	333
$Cp_2(CO)Ta(\mu\text{-}H)Mn(CO)_2Cp$		331
$(\eta\text{-}C_5H_4R)_2(CO)Ta(\mu\text{-}H)Mn(CO)_2Cp$	R = H, Bu^t	302
$[PPN][Cp(CO)_3Nb(\mu\text{-}H)Mn(CO)_2(\eta\text{-}C_5H_4R)]$	R = H, Me	107
Group 8		
$(Cp_2NbH)(\mu\text{-}C_5H_4)Fe(CO)_2$	prep: $Cp_2Nb(\eta^2\text{-}BH_4) + NEt_3 + CpFe(CO)_2Me$	241
$Cp_2(CO)Nb(\mu\text{-}H)Fe(CO)_4$	prep: $Cp_2NbH_3 + Fe(CO)_5$	112
$(\eta\text{-}C_5H_4R)_2(CO)M(\mu\text{-}H)Fe(CO)_4$	R = H, Me	303
$(\eta\text{-}C_5H_4R)_2(CO)M[(\mu\text{-}CO)(\mu\text{-}H)Fe_3(CO)_{10}]$	R = H, Me	303
Group 9		
$Cp_2Ta(\mu\text{-}H)(\mu\text{-}Cl)Rh(cod)$	prep: $Cp_2TaH_3 + [RhCl(cod)]_2$	301
$Cp_2Ta(\mu\text{-}H)(\mu\text{-}Cl)Rh(dppe)$		301
Group 10		
$Cp_2(CO)Nb(\mu\text{-}H)Ni(CO)_3$	x ray	114,334
Group 11		
$\{[(\eta\text{-}C_5R_5)_2NbH(\mu\text{-}H)_2]_2Cu\}PF_6$	C_5R_5 = Cp, C_5H_4TMS, $C_5H_3(TMS)_2$	278
$Cp_2(H)Ta[(\mu\text{-}H)Cu(\mu\text{-}I)_2Cu(\mu\text{-}H)]_2Ta(H)Cp_2$	x ray	284
$(\eta\text{-}C_5H_4Bu^t)_2(H)Ta(\mu\text{-}H)_2Cu(\mu\text{-}I)_2Cu(\mu\text{-}H)_2Ta(H)(\eta^5\text{-}C_5H_4Bu^t)_2$	x ray	284
$Cp_2TaH_3{\cdot}2CuCl$		284
$\{[(\eta\text{-}C_5R_5)_2NbH(\mu\text{-}H)_2]_2Ag\}BF_4$	C_5R_5 = Cp, C_5H_4TMS, $C_5H_3(TMS)_2$	278
$[(\eta\text{-}C_5H_4TMS)_2NbAuH_2]_3$	x ray	335
Group 12		
$[Cp_2NbH_2]_2(\mu\text{-}Zn)$	prep: excess $Cp_2NbH_3 + Cp_2Zn$	336
$Cp_2NbH(ZnCp)_2$		336
Cp_2NbH_2ZnCp	*endo*-isomer; x ray	336
$[Cp'_2TaH_2]_2(\mu\text{-}Zn)$		337
Cp'_2TaH_2ZnCp	*endo*-isomer	337
$Cp'_2TaH(ZnCp)_2$	*exo*-isomer (x ray), *endo*-isomer	337
$Cp_2(H)Ta(\mu\text{-}H)_2ZnCl_2(THF)$		338
$Cp_2TaH_3{\cdot}ZnX_2(OEt_2)$	X = Cl, I	280
$Cp_2TaH_3{\cdot}ZnX_2(THF)$	X = Cl (x ray), I	280
$[(\eta^5\text{-}C_5H_4Bu^t)_2Ta(\mu\text{-}H)_2Zn(\mu\text{-}Cl)]_2$	x ray	339
$Cp_2TaH_3{\cdot}CdI_2$		280
$Cp_2NbH(HgCl_2)_2{\cdot}0.5HgCl_2$		340
$Cp_2NbH(HgBr_2)_2{\cdot}HgBr_2$		340
$Cp_2NbH(HgI_2)_2{\cdot}0.66HgI_2$		340
$Cp_2NbH[Hg(SR)]$	R = Et, Bu^t	340
Group 13		
$Cp_2Nb(\eta^2\text{-}BH_4)$	use of	341
$Cp^*_2Nb(\eta^2\text{-}BH_4)$	prep: $NbCl_5 + LiCp^* + NaBH_4$	70
$(\eta\text{-}C_5Me_4Et)_2Nb(\eta^2\text{-}BH_4)$		342
$(\eta\text{-}C_5H_4Bu^t)_2Nb(\eta^2\text{-}BH_4)$		283
$(\eta^5\text{-}C_9H_7)_2Nb(\eta^2\text{-}BH_4)$	$[C_9H_7]^-$ = indenyl	47
$(\eta\text{-}C_5H_4\text{-}CH_2C_6H_5\text{-polymer})_2Nb(BH_4)$	poly(styrene–divinylbenzene) attached $Cp_2Nb(BH_4)$	149
$Cp^*CpTa(\eta^2\text{-}BH_4)$	NMR study	343,344
$CpNb(dmpe)H(\eta^2\text{-}BH_4)$		83
$Ta(\eta^1\text{-}BH_4)(CO)_3(PMe_3)_3$		25
$TaH_2(\eta^2\text{-}BH_4)(PMe_3)_4$		25
$Cp^*_2Nb(H{\cdot}AlEt_3)H_2$		253

Table 9 (continued)

Compound	*Comments*	*Ref.*
$Cp^*CpTa(CO)(H{\cdot}AlEt_3)$	prep: $Cp^*CpTa(CO)H + AlEt_3$	253
$(\eta\text{-}C_5H_4TMS)_2Nb(\eta^2\text{-}AlH_4)$		345
$(\eta^5{:}\eta^5\text{-}C_{10}H_8)Cp_2Nb_2(\mu\text{-}AlH_4)_2$		290
$[Cp'_2NbH]_3(InCl)_2$	x ray	61
Group 14		
$Cp_2Nb(SnMe_3)H_2$	HD complex also prepared	122,294
$[M][Cp_2Nb(SnMe_3)H]$	$M^+ = [Li(12\text{-crown-}4)]^+$, $[Na(15\text{-crown-}5)]^+$, $[K(18\text{-crown-}6)]^+$	294
$Cp_2Nb(SnMe_3)_2H$		122,294
$Cp_2TaH_2(SnCl_3)$	prep: $Cp_2TaH_3 + SnCl_4$	346,347
$Cp_2TaH_2(SnMeCl_2)$	x ray	346,347
$Cp_2TaH_2(SnMe_2Cl)$		346,347
$[Cp_2NbH]_2(\mu\text{-}SnEtCl)$	x ray	124
Lanthanides		
$[Cp_2NbH_2]_2Yb(diglyme)$		296

2.4 ALKYLIDENES AND ALKYLIDYNES

In the mid-1970s to early-1980s, niobium and tantalum alkylidenes acquired an extraordinary significance as prototypes of metal–carbon double bonds in high-oxidation-state complexes that were thought to model the active sites in alkene metathesis and F–T catalysis. In *COMC-I*[1] and in other reports,[348,349] this earlier alkylidene and alkylidyne chemistry has been reviewed in great detail, including their applications in alkene dimerization (Section 2.8.4) and metathesis (Section 2.8.5). More recently, alkylidene complexes of these metals have served as important elements in C–H bond activation systems (Section 2.8.2), have gained a renewed significance in F–T chemistry (Section 2.8.1), and have been used in alkyne polymerizations (Section 2.8.3) and ring-opening metathesis polymerization (ROMP) of cyclic alkenes (Section 2.8.5). Applications of alkylidenes in these processes are described in their respective sections, while this section emphasizes their synthesis and reactivity. Alkylidene compounds are summarized in Table 10 and alkylidynes in Table 11.

2.4.1 Synthesis

The most commonly employed strategy for the preparation of alkylidene complexes is the abstraction of an α-hydrogen from an alkyl ligand. α-H abstraction may take place by either (i) an intermolecular deprotonation of an alkyl ligand using an external base or (ii) an intramolecular α-H transfer in a sterically congested bis(alkyl) complex to afford an alkylidene ligand and free alkane. For example, $Cp'Ta(CH_2Ph)_3Cl$ can be dehydrohalogenated to provide $Cp'Ta({=}CHPh)(CH_2Ph)_2$ as indicated in Equation (6);[147] conversely, intramolecular α-H abstraction occurs upon thermolysis or photolysis of $Cp^*Ta(CH_2Ph)_2Cl_2$ to provide $Cp^*Ta({=}CHPh)Cl_2$ (Equation (7)).[147] A distinction is made between the complete removal of an α-hydrogen from the complex, that is, α-H abstraction, and the transfer of the α-H to the metal center with the formation of a discrete hydride complex, that is, α-H elimination (by analogy with β-H elimination). Thus, α-H eliminations may occur in $d^{n\geq 2}$ alkyl complexes, since this reaction represents a formal two-electron oxidation of the metal center. For example, the Ta^{III} complex $Cp^*_2Ta(CO)Me$ can be photolyzed under N_2 to provide the methylidene hydride $Cp^*_2Ta({=}CH_2)H$ (Equation (8)).[226] Likewise, $Ta({=}CHBu^t)(H)Cl_2(PMe_3)_3$ is obtained on two-electron reduction of $Ta(CHBu^t)Cl_4$ in the presence of PMe_3.[358]

$$Cp'TaBz_3Cl + Ph_3P{=}CH_2 \longrightarrow Cp'Ta({=}CHPh)Bz_2 + [Ph_3PMe]Cl \quad (6)$$

$$Cp^*TaBz_2Cl_2 + \Delta \text{ or } h\nu \longrightarrow Cp^*Ta({=}CHPh)Cl_2 + PhMe \quad (7)$$

$$Cp^*_2Ta(CO)Me + h\nu \longrightarrow [Cp^*_2TaMe] + CO \longrightarrow Cp^*_2Ta({=}CH_2)H \quad (8)$$

Other methods by which alkylidene ligands may be introduced to the metal center include the following. (i) The two-electron oxidation by an alkylidene-transfer reagent such as a phosphorus ylide has been employed in the production of alkylidenes. For example, photolysis of $Cp_2TaPh(PMe_3)$ in the presence of $Me_3P{=}CH_2$ provides $Cp_2Ta({=}CH_2)Ph$[244] and adding $Me_3P{=}CH_2$ to $Cp^*_2Ta({=}CD_2)D$ results in

Table 10 Alkylidene and related complexes of niobium and tantalum.

Compound	*Comments*	*Ref.*
Halide complexes and their adducts		
$Ta(=CHBu^t)Cl_3(THF)_2$	*cis,mer*-isomer	151,350,351
	metallacyclization reactions	352,353
$Ta(=CHBu^t)Cl_3(DME)$	prep: $Ta(CH_2Bu^t)_2Cl_3$ + DME + THF	354
$Ta(=CHBu^t)Cl_3L_2$	L = PMe_3, PMe_2Ph, $PMePh_2$, py, 1/2 dmpe, 1/2 TMEDA, 1/2 bipy, 1/2 *o*-$C_6H_4(AsMe_2)_2$	151
	L = PMe_3	351
$Ta(=CHBu^t)Cl_3(PEt_3)_2$		355
$Ta(=CHBu^t)Cl_3(dmpe)$		356
$Ta(=CHR)Cl_3(PMe_3)_2$	R = TMS, Ph	151
	R = Ph; reactions	352
$Ta(=CHPh)Br_3(PMe_3)_2$		151
$Ta(=CHBu^t)Br_3L_2$	L = PMe_3, THF, 1/2 dmpe, 1/2 dppe	151
$Nb(=CHBu^t)Cl_3L_2$	L = PMe_3, THF, PMe_2Ph, 1/2 dmpe	151
	L = PMe_3	172
$[Ta(=CHBu^t)Cl_3L]_2$	L = PMe_3, PMe_2Ph, $PMePh_2$	151
	L = PMe_3; neutron diffraction structure	357
$[Ta(=CHBu^t)Br_3(PMe_3)]_2$		151
$[Nb(=CHBu^t)Cl_3L]_2$	L = PMe_3, PMe_2Ph	151
$Ta(=CHBu^t)Cl(PMe_3)_4$	prep: $Ta(=CHBu^t)Cl_3(PMe_3)_2$ + NaHg + PMe_3 (under Ar)	358,359
$Ta(=CHBu^t)X(dmpe)_2$	X = Cl, I	356
$Ta(=CHBu^t)_2Cl(PMe_3)_2$	prep: $Ta(CH_2Bu^t)_4Cl$ + PMe_3	146
$(\eta-C_5R_5)Nb(=CHBu^t)Cl_2$	R = H, Me	151
	R = Me; alkene dimerization catalyst precursor	360
$CpTa(=CHBu^t)Cl_2$	metallacyclization reactions	352,353
	alkene dimerization catalyst precursor	360
$CpTa(=CHBu^t)Br_2$	alkene dimerization catalyst precursor	360
$(\eta-C_5R_5)Ta(=CHCMe_2Ph)Cl_2$	R = H, Me	361
$(\eta-C_5Me_4R)Ta(=CHPh)Cl_2$	R = Me, Et	147
Alkene complexes		
$Ta(=CHBu^t)(CH_2=CH_2)R(PMe_3)_2$	R = CH_2Bu^t, Et	146
$Cp^*Ta(=CHBu^t)(CH_2=CH_2)(PMe_3)$	neutron diffraction structure	357
Alkyl, aryl, and related complexes		
$Ta(=CHBu^t)(Et)Cl_2(PMe_3)_2$	prep: $Ta(CH_2Bu^t)_2(Et)Cl_2$ + PMe_3	154
$Ta(=CHBu^t)(CH_2Bu^t)_3$	metallacyclization reactions	352
$Ta(=CH\text{-}TMS)(CH_2TMS)(OAr'')_2$	prep: $Ta(OAr'')_2Cl_3$ + $LiCH_2TMS$; x ray	157,177,195
$Ta(=CH\text{-}TMS)(CH_2TMS)(OAr)_2$		157,175
$Ta(=CHPh)(CH_2Ph)(OAr'')_2$		157
$Ta(=CH_2)(Me)(OAr'')_2$	prep: photolysis of $TaMe_3(OAr'')_2$	175,179
$Ta(=CH_2)(Me)(O\text{-}2,6\text{-}C_6H_2Bu^t_2\text{-}4\text{-}OMe)_2$		179
$Ta(=CHPh)(CH_2Ph)(silox)_2$		164
$Ta(=CHBu^t)(CH_2Bu^t)(silox)_2$		164
$Ta(=CHBu^t)(CH_2Bu^t)_2(silox)$		164
$Ta(=CHBu^t)[2,6\text{-}C_6H_3(CH_2NMe_2)_2]Cl_2$	x ray	185,186
$Ta[CH_2SiMe_2N(Bu^t)B(CH_2Bu^t)=NBu^t](=CHBu^t)(CH_2Bu^t)$	metallacycle; x ray	352
$Ta(=CHBu^t)(CH_2=CH_2)R(PMe_3)_2$	R = CH_2Bu^t, Et	146
$[Ta(=CHBu^t)R(PMe_3)_2]_2(\mu-N_2)$	R = Me, CH_2Bu^t (x ray)	359,266
	R = CH_2Bu^t, x ray	362
$Ta(=CHBu^t)_2R(PMe_3)_2$	R = Me, Et, Bu, Mes, CH_2Bu^t	146
$(\eta^5-C_5Me_4R)Ta(=CHPh)(CH_2Ph)Cl$	R = Me, Et	147
$Cp'Ta(=CHPh)(CH_2Ph)_2$	prep: $Cp'Ta(CH_2Ph)_3Cl$ + $Ph_3P=CH_2$	147
$Cp^*Ta(=CHPh)(CH_2Ph)_2$	x ray	147
	prep: Cp^*TaCl_4 + $PhCH_2MgCl$	198
$Cp^*M(=CH\text{-}TMS)(CH_2TMS)_2$	x ray for M = Nb	363
$(\eta-C_5H_4TMS)Ta(=CH\text{-}TMS)(CH_2TMS)_2$		363
$[\eta-C_5H_3(TMS)_2]Ta(=CH\text{-}TMS)(CH_2TMS)_2$		363
$Cp^*Ta(=CHPh)(CH=PPh_3)Cl$		147
$Cp_2Ta(=CH_2)(Me)$		223,244,364
$Cp^*_2Ta(=CH_2)(Me)$	prep: $[Cp_2^*TaMe_2]BF_4$ + base or $Cp^*_2Ta(=CH_2)H$ + $Me_3P=CH_2$	109,226
$Cp_2Ta(=CH_2)Ph$		244
$(\eta^5-C_9H_7)CpTa(=CH_2)Me$	C_9H_7 = indenyl	365
$Cp_2(R)Nb=CHOZr(H)Cp^*_2$	R = H, Me, CH_2Ph, CD_2Ph, $CH_2\text{-}4\text{-}C_6H_4OMe$, Ph, $CH_2OZr(H)Cp^*_2$	232
Hydride complexes		
$Ta(=CHBu^t)(H)Cl_2(PMe_3)_3$	$Ta(=CHBu^t)(H)Cl_2(PMe_3)_2$	358,205
$Ta(=CHBu^t)(H)I_2(PMe_3)_3$		205,366

Table 10 (continued)

Compound	*Comments*	*Ref.*
$Ta(=CHBu^t)(H)(OTf)_2(PMe_3)_3$[a]	prep: $Ta(=CHBu^t)(H)Cl_2(PMe_3)_3$ + TMS-OTf	205
$[NEt_4][Ta(=CHBu^t)(H)Cl_3(PMe_3)_2]$		205
$[Ta(=CHBu^t)(H)Cl(dmpe)_2]X$	X = Cl^-, $[O_2CCF_3]^-$	205
$Cp^*Ta(=CHBu^t)(H)Cl(PMe_3)$		358,205
$Cp^*_2Ta(=CH_2)(H)$	prep: $Cp^*_2TaCl_2$ + 2 MeLi	109,226,227
$Cp^*(\eta\text{-}C_5Me_4Ph)Ta(=CH^*_2)H^*$	H* = H, D	109
$[\eta^5\text{:}\eta^1\text{-}C_5Me_4(CH_2)_2CH=]Cp^*TaH$		109
$Cp^*_2Ta(=CHPh)(H)$		227
$Cp_2Ta(=CH\text{-}TMS)(H)$		244
$Cp^*_2Ta(=C=CH_2)(H)$		109,226
$(\eta\text{-}C_5H_4TMS)_2Nb(=C=CPh_2)(H)$		367
$Cp_2(H)Nb=CHOZrClCp^*_2$		232
$Cp_2(H)Nb=CDOZr(D)Cp^*_2$	prep: $Cp_2Nb(CO)R$ + $Cp^*_2ZrD_2$	232
Oxygen- and sulfur-donor ligands (without alkyl ligands)		
$Ta(=CHBu^t)(OBu^t)Cl_2(PMe_3)$		172,351
$M(=CHBu^t)(OBu^t)_2Cl(PMe_3)$		172,351
$M(=CHBu^t)(OBu^t)_2Cl(PMe_3)_2$	observed by NMR only	172
$Ta(=CHPh)(OBu^t)_2Cl(PMe_3)_2$		172,351
$[Ta(=CHPh)(OBu^t)_2Cl(PMe_3)]_n$	observed by NMR only	172
$Ta(=CHBu^t)(OBu^t)_3$		172
$Ta(=CHR)(OAr)_3(THF)$	R = Bu^t, Ph	368
$Ta(=CHR)(OAr)_3(THF)$	R = Ph, CH-TMS	369
$Ta(=CHBu^t)(OAr)_3L$	L = THF or py	369
	L = THF; use as ROMP catalyst	370
	L = THF; reactivity	371
$Ta(=CHRCR=CHBu^t)(OAr)_3(py)$	R = Me, (x ray), Ph	372
$Ta(=CHBu^t)(OAr')_3L$	L = THF or py	369
	L = THF; use as ROMP catalyst	370
$(silox)_3Ta=C=C=Ta(silox)_3$	prep: $Ta(silox)_3$ + CO; x ray	33,270
$Ta(=CHBu^t)(S\text{-}2,4,6\text{-}C_6H_2Pr^i_3)_3L$	L = THF, py	369
$Ta(=CH^tBu)(S\text{-}2,4,6\text{-}C_6H_2Pr^i_3)_3(SEt_2)$	x ray	373
Nitrogen-donor ligands		
$Ta(=CHBu^t)(=N\text{-}TMS)Cl(PMe_3)_2$	prep: $Ta(=CHBu^t)Cl(PMe_3)_4$ + N_3TMS	350
$[Ta(=CHBu^t)Cl(PMe_3)_2]_2(\mu\text{-}N_2)$	both ^{14}N and ^{15}N isotopomers prepared	266,359
$[Ta(=CHBu^t)R(PMe_3)_2]_2(\mu\text{-}N_2)$	R = Me, CH_2Bu^t (x ray)	266,359
$Ta(=CHR)Cl_2[N(SiMe_2CH_2PPr^i_2)_2]$	R = Bu^t, Ph	374
$\{Ta(=CHR)[N(SiMe_2CH_2PPr^i_2)_2]\}_2(\mu\text{-}N_2)$	R = Bu^t, Ph	374
$Cp^*Nb(=NAr)(=CHPh)(PMe_3)$	x ray	214
Bridging alkylidene complexes		
$[CpTa(\mu_2\text{-}\eta^1\text{:}\eta^5\text{-}C_5H_4)]_2(\mu\text{-}CHSiHBu^t_2)$	x ray	244
$[Cp^*TaMe_2]_2(\mu\text{-}CH_2)(\mu\text{-}H)_2$		199
$[TaH(silox)_2]_2(\mu\text{-}CH_2)(\mu\text{-}O)$		160
$[Ta(silox)_2]_2(\mu\text{-}O)_2(\mu\text{-}CHMe)$	$\mu\text{-}^{13}CHMe$ analogue also prepared	160,269
$Me_2Ta(\mu\text{-}CH_2)(\mu\text{-}OCMe_2CH_2PPh_2)_2PtMe$	*trans*-phosphines about platinum; prep: $TaMe_3$-$(OCMe_2CH_2PPh_2)_2$ + $(cod)PtMe_2$ + Δ	181
$(TMEDA)Ta(\mu\text{-}CH_2)(\mu\text{-}Me)(\mu\text{-}OCMe_2CH_2PPh_2)_2Ni$	prep: $TaMe_3(OCMe_2CH_2PPh_2)_2$ + $(TMEDA)NiMe_2$	181
$Cp_2Ta(\mu\text{-}CH_2)_2Ir(CO)_2$		365
$(\eta^5\text{-}C_9H_7)CpTa(\mu\text{-}CH_2)_2Ir(CO)_2$	C_9H_7 = indenyl	365
$Cp_2MeTa(\mu\text{-}CH_2)Pt(PMe_3)_2$	x ray	364
$Cp_2Ta(\mu\text{-}CH_2)_2Pt(H)(PMe_3)_n$	*n* = 1 (x ray), 2	364
$Cp_2Ta(\mu\text{-}CH_2)_2Pt(H)(D)_2(PMe_3)$		364
$Cp_2MeTa(\mu\text{-}CH_2)Pt(PMe_3)$		364
$Cp_2Ta(\mu\text{-}CH_2)_2PdCp$	prep: $Cp_2Ta(=CH_2)(Me)$ + $CpPd(\eta\text{-}C_3H_5)$	375
$Cp_2Ta(\mu\text{-}CH_2)_2Pd(L)Cl$	L = PMe_3, $P(OMe)_3$	375
$[Cp_2Ta(\mu\text{-}CH_2)_2PdL_2]Cl$	L = PMe_3, $P(OMe)_3$, 1/2 dmpe (x ray)	375
$[Cp_2Ta(\mu\text{-}CH_2)_2Pd(CH_2CN)[P(OMe)_3]$		375
$[Cp_2Ta(\mu\text{-}CH_2)_2PdL_2][C_5H_5]$	x ray	375
Miscellaneous		
$(silox)_3Ta=C=C=O$	ketenylidene complex	33
$(silox)_3Ta[\eta^2(O,C)\text{-}OC=C=Ta(silox)_3]$	prep: $(silox)_3Ta=C=C=O$ + $Ta(silox)_3$	33
$(silox)_3Ta=C=C\text{:}$	transient vinylidene, not observed	33

[a] OTf = OSO_2CF_3.

the formation of $Cp^*_2Ta(=CH_2)CD_3$ by the equilibrium shown in Equation (9).[226] (ii) Alkylidene complexes may also arise from metathesis methods involving the decomposition of metallacyclic species, as detailed in Section 2.6. For example, excess THF disrupts the metallacyclobutane

Table 11 Alkylidyne complexes of niobium and tantalum.

Compound	*Comments*	*Ref.*
Terminal alkylidynes		
$Cp^*Ta(\equiv CBu^t)(H)(PMe_3)_2$	prep: $Cp^*Ta(=CHBu^t)Br_2 + NaHg + PMe_3$ or $Cp^*Ta(\equiv CBu^t)Cl(PMe_3)_2 + LiEt$	358
$Cp^*Ta(\equiv CBu^t)Cl(PMe_3)_2$		358
$Cp^*Ta(\equiv CBu^t)(H)(dmpe)$		358
$Ta(\equiv CBu^t)(H)(dmpe)_2(ClAlMe_3)$	x ray; prep: $Ta(=CHBu^t)Cl(dmpe)_2 + AlMe_3$	356
$Ta(\equiv CBu^t)(H)I(dmpe)_2$	in equilibrium with $Ta(=CHBu^t)I(dmpe)_2$	356
$Ta(\equiv CBu^t)(H)(OTf)(dmpe)_2$	prep: $Ta(=CHBu^t)Cl(dmpe)_2$ + TMS-OTf	356
$M(\equiv COSiPr^i_3)(CO)(dmpe)_2$	M = Nb, Ta (x ray)	30,128
	M = Ta	376
$Ta(\equiv CO\text{-}TMS)(CO)(dmpe)_2$		31
$Ta(\equiv COSiBu^tPh_2)(CO)(dmpe)_2$	x ray	376
$Ta(\equiv COSiBu^tPh_2)(CO{\cdot}AlEt_3)(dmpe)_2$	x ray	376
$Nb[\equiv CN(SiPh_3)Me](CO)(dmpe)_2$	x ray	16
$Nb[\equiv CN(SiMe_2Bu^t)Me](CO)(dmpe)_2$		16
Bridging alkylidynes		
$[M(\mu\text{-C-TMS})(NC_{12}H_8)_2]_2$	M = Nb, Ta; $[NC_{12}H_8]^-$ = carbazole amide	377
	M = Ta; prep: $[Ta(\mu\text{-C-TMS})(CH_2TMS)_2]_2$ + 9H-carbazole; x ray; $[NC_{12}H_8]^-$ = carbazole amide	272,377
$[M(\mu\text{-C-TMS})(CH_2TMS)_2]_2$	supported on silica for hydrogenation catalyst	378
$Ta_2(\mu\text{-C-TMS})_2(CH_2TMS)_3(O\text{-}2,6\text{-}C_6H_3Ph_2)$	x ray	196
$Ta_2(\mu\text{-C-TMS})_2(CH_2TMS)_2(OR)_2$	R = 2,6-$C_6H_3Ph_2$ (isomeric mixture), Ar''	196
$Ta_2(\mu\text{-C-TMS})_2(O\text{-}2,6\text{-}C_6H_3Ph_2)_4$	x ray	196
$[Ta(\mu\text{-}CBu^t)Cl_2(DME)]_2[Zn(\mu\text{-}Cl)_2]$	x ray	354,379
$[Ta(\mu\text{-}CBu^t)Cl_2(DME)]_2ZnClBu^t$		354
$[Ta(\mu\text{-}CBu^t)(CH_2Bu^t)_3][Li(DME)]$	prep: $Ta(=CHBu^t)Cl_3(DME) + Li(CH_2Bu^t)$	354
$[(\eta\text{-}C_5Me_4Et)TaCl_2]_2(\mu\text{-H})(\mu\text{-}CHPMe_3)(\mu\text{-O})$	μ-D and μ-$CDPMe_3$ compounds also prepared	206,208
	x ray	380
$\{[\eta\text{-}C_5Me_3(TMS)_2]TaCl_2\}_2(\mu\text{-H})(\mu\text{-}^{13}CHPMe_3)(\mu\text{-O})$		206
$(\eta\text{-}C_5Me_4Et)[\eta\text{-}C_5Me_3(TMS)_2]Ta_2Cl_4(\mu\text{-H})(\mu\text{-}^{13}CHPMe_3)(\mu\text{-O})$		206

$\overline{Ta(CHPhCHBu^tCH_2})(OAr)_3$ and sets up the equilibrium shown in Equation (10).[368,369] (iii) Finally, 1,2-addition across the $L_nM{\equiv}CR$ triple bond of an alkylidyne can afford alkylidenes.[349] For example, $Cp^*Ta(=CHBu^t)(CH_2{=}CH_2)(PMe_3)$ is formed upon reacting $Cp^*Ta(\equiv CBu^t)(PMe_3)_2Cl$ with $MgEt_2(diox)$ (diox = dioxane) as suggested in Equation (11).[357]

$$[Cp_2^*Ta(=CD_2)D \rightleftharpoons Cp_2^*Ta(CD_3)] + Me_3P{=}CH_2 \longrightarrow Cp_2^*Ta(=CH_2)CD_3 + PMe_3 \tag{9}$$

$$\overline{Ta(CHPhCHBu^tCH_2})(OAr)_3 + THF \rightleftharpoons Ta(=CHPh)(OAr)_3(THF) + CH_2{=}CHBu^t \tag{10}$$

$$Cp^*Ta(\equiv CBu^t)(PMe_3)_2Cl + MgEt_2(diox) \longrightarrow [Cp^*Ta(\equiv CBu^t)(CH_2Me)(PMe_3)]_n \longrightarrow$$
$$Cp^*Ta(=CHBu^t)(CH_2{=}CH_2)(PMe_3) + PMe_3 \tag{11}$$

Alkylidyne complexes may often be prepared by analogous α-hydrogen processes starting from alkylidenes. For example, abstracting a chloride ion from $Ta(=CHBu^t)Cl(dmpe)_2$ with $AlMe_3$ facilitates α-H elimination from nascent d^2 $[Ta(=CHBu^t)(dmpe)_2]^+$ to afford the alkylidyne hydride $[Ta(\equiv CBu^t)(H)(dmpe)_2][AlMe_3Cl]$.[356] As outlined in Section 2.2.2, carbyne complexes of niobium and tantalum have been prepared by electrophilic attack on coordinated CO ligands. Thus, the anionic complexes $Na[M(CO)_2(dmpe)_2]$ react with Pr^i_3SiCl to provide $M(\equiv COSiPr^i_3)(CO)(dmpe)_2$.[30,128]

A study of α-H vs. β-H elimination and migratory insertion pathways has been reported by Bercaw and co-workers[109] for both d^0 Cp^*_2ScR and transient d^2 $[Cp^*_2TaR]$ systems. The 16-electron $[Cp^*_2TaR]$ are unstable to elimination but, perhaps unexpectedly, α-H elimination is found to be more kinetically favored than β-H elimination. Both processes could be studied from a common alkyl intermediate which arises from the thermal decomposition of $Cp^*_2Ta(=C{=}CH_2)H$ in the absence of a trapping agent (Scheme 7). Transient $[Cp^*_2TaCH{=}CH_2]$ gives a kinetic α-H elimination product $[\eta^5{:}\eta^1\text{-}C_5Me_4(CH_2)_2CH{=}]Cp^*TaH$, which rearranges to the β-H elimination product $[\eta^5{:}\eta^2\text{-}C_5Me_4CH_2CH{=}CH_2]Cp^*TaH$ as shown.[109]

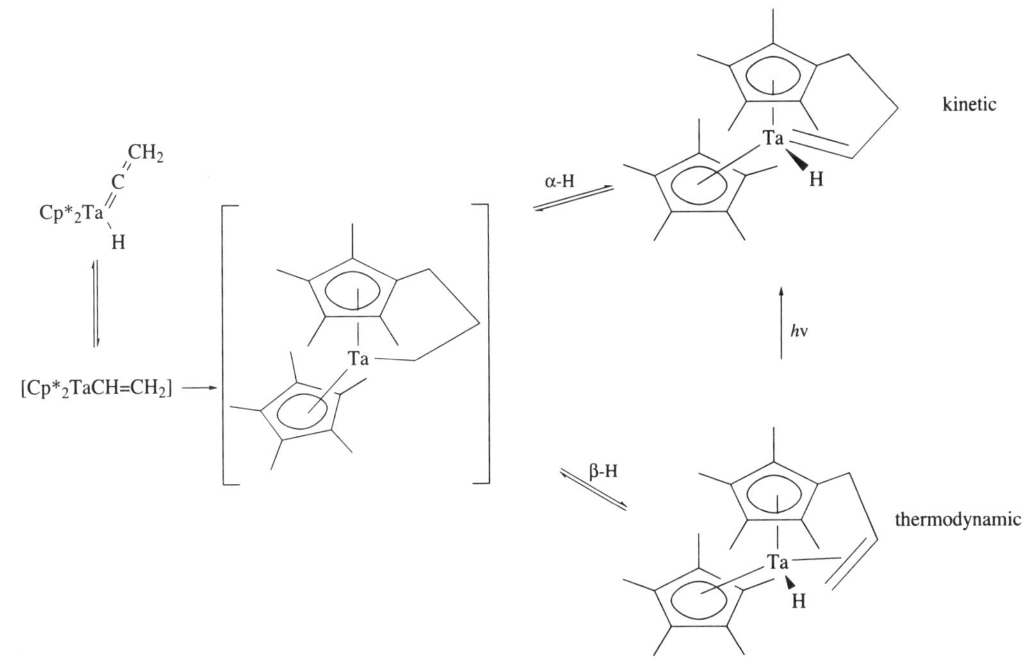

Scheme 7

2.4.2 Structural and Spectroscopic Properties

The diagnostic structural and spectroscopic properties of alkylidenes and alkylidynes are briefly outlined here; a detailed analysis was given in *COMC-I*.[1] Diagnostic structural features of alkylidenes and alkylidynes include M=CHR and M≡CR distances indicative of multiple bonding and, for some complexes, a geometry about C-α appropriate for sp^2 or *sp* hybidization. In "electron-rich" (18-electron) alkylidenes, the M–C-α–R bond angle may increase from that expected as R increases in size. However, in "electron-deficient" (<18-electron) species, substantial increases in the M–C-α–R bond angles over predicted values are observed, accompanied by a decrease in the M–C-α–H angle.[357] This distortion towards a semibridging CH-α seems to arise from the electrophilic metal center accommodating some of its electron deficiency through a M ← H-α agostic interaction.

Perhaps the most diagnostic spectroscopic feature of niobium and tantalum alkylidene and alkylidyne complexes is the large downfield shift of the C-α resonance in their ^{13}C NMR spectra. Alkylidene ^{13}C-α resonances typically appear in the range 220–280 ppm downfield of $SiMe_4$, whereas those for alkylidynes generally occur at >300 ppm. The alkylidene $J_{C\text{-}H\text{-}\alpha}$ coupling constant correlates well with the degree of distortion in an agostic alkylidene. Small values of $J_{C\text{-}H\text{-}\alpha}$ (~75–100 Hz) are indicative of distortions towards an agostic CH-α with long, relatively weak C–H bonds and higher values of $J_{C\text{-}H\text{-}\alpha}$(≥120 Hz) are typical of the undistorted alkylidenes. Less diagnostic is the ^1H NMR chemical shift of the alkylidene CH-α. However, values for the undistorted alkylidene CH-α may occur from 10 to 12 ppm, while significant upfield shifts occur for the distorted, more "hydride-like" agostic alkylidene. The CH-α stretching mode in the IR spectrum also reflects these distortions when H-α is pulled into a semibridging position (and the C–H bond is weakened) by a highly electrophilic metal center.

Finally, a few theoretical studies of alkylidenes have been reported. In particular, calculations on unusual M–C–H bond angles and α-H abstraction in alkylidenes,[381] *ab initio* calculations on ligand and substituent effects in alkylidene complexes,[382] and calculations on model bis(alkylidene) $L_nM(=CHR)_2$ compounds[383] have all been described.

2.4.3 Complexes

The years following the 1974 discovery of alkylidenes were accompanied by the development of numerous d^0 alkylidenes of the type $M(=CHR)X_3L_2$, d^2 alkylidenes $Ta(=CHR)L_4X$, and bis(alkylidene) complexes of the type $Ta(=CHR)_2XL_2$, prepared by the strategies outlined in Section 2.4.1.[1] Many of these complexes still serve as valuable starting materials in the preparation of new alkylidene complexes. More recently, alkylidene complexes containing π-donor ligands, particularly alkoxides, have been developed and used in several catalytic applications. For example, $Ta(=CHBu^t)Cl_3(THF)_2$ is alkoxylated with LiOAr to afford $Ta(=CHBu^t)(OAr)_3(THF)$, which acts as a ROMP catalyst for norbornenes.[368,369] Other alkoxide derivatives arise from alkoxide starting materials; for example photolyzing $Ta(CH_2TMS)_3(OAr)_2$ provides the alkylidene $Ta(=CH\text{-}TMS)(CH_2TMS)(OAr)_2$.[175] Imido derivatives have also been examined, for example, oxidizing $Ta(=CHBu^t)Cl(PMe_3)_3$ with N_3TMS provides $Ta(=CHBu^t)(=N\text{-}TMS)Cl(PMe_3)_2$,[350] and treatment of $Ta(=CHBu^t)Cl(PMe_3)_4$ with dinitrogen provides $[Ta(=CHBu^t)Cl(PMe_3)_2]_2(\mu\text{-}N_2)$.[359]

Monokis(cyclopentadienyl) complexes of the general type $(\eta\text{-}C_5R_5)M(=CHR)(CH_2R)_nX_{2-n}$ (n = 0–2) have been prepared in reactions which emphasize the role of steric congestion in inducing intramolecular α-H abstraction. For example, the addition of LiCp* to solutions of $Ta(CH_2Ph)_3Cl_2$ affords $Cp^*Ta(=CHPh)(CH_2Ph)Cl$[147] while the halides Cp^*MCl_4 react with four equivalents of $Li(CH_2\text{-}TMS)$ to provide $Cp^*M(=CH\text{-}TMS)(CH_2TMS)_2$.[363] Several new bis(cyclopentadienyl) complexes of the type $(\eta\text{-}C_5R'_5)_2M(=CHR^3)R^2$ (for R^2 = alkyl or hydride) have appeared since 1980. For example, treatment of $Cp^*_2TaCl_2$ with MeLi or CH_2=CHMgBr results in the formation of the hydrido alkylidene complexes $Cp^*_2Ta(=CH_2)H$ and $Cp^*_2Ta(=C=CH_2)H$, respectively.[226] These compounds exist in a 1,2-hydrogen shift (i.e., retro α-H elimination) equilibrium with their corresponding alkyls. Thus, the d^2 tautomer can be trapped upon adding CO to afford $Cp^*_2Ta(CO)R$ (R = Me or $CH=CH_2$) (Equation (12)[226]). The reaction of $Cp_2Ta(=CH_2)Me$ and $Pt(CH_2=CH_2)(PMe_3)_2$ affords $Cp_2TaMe(\mu\text{-}CH_2)Pt(PMe_3)_2$ which itself reacts with excess PMe_3 to regenerate $Cp_2Ta(=CH_2)Me$ along with $Pt(PMe_3)_4$.[364] This latter reaction represents a highly unusual bridging → terminal alkylidene transformation upon fragmentation of a dinuclear μ-alkylidene species.

$$[Cp^*_2Ta(=C=CH_2)H \rightleftharpoons Cp^*_2Ta(CH=CH_2)] + CO \longrightarrow Cp^*Ta(CO)(CH=CH_2) \quad (12)$$

Alkylidyne complexes of niobium and tantalum still remain relatively rare. Complexes of the type $Ta(\equiv CBu^t)(H)(dmpe)_2X$ (X^- = $ClAlMe_3^-$, I^-, or OTf^-) have been reported and exist in equilibrium with their insertion or retro α-hydrogen elimination tautomers $Ta(=CHBu^t)(dmpe)_2X$ (Equations (13) and (14)[356]). Similarly, $Cp^*Ta(\equiv CBu^t)(H)(PMe_3)_2$ may be prepared by reducing either $Cp^*Ta(=CHBu^t)Br_2$ or $Cp^*Ta(CH_2Bu^t)_2Cl_2$ in the presence of PMe_3.[358] Many carbyne complexes of the type $M(\equiv COSiR_3)(CO)(dmpe)_2$ have also appeared in studies related to the reductive coupling of CO, as described in Section 2.2.2.[30,31,128,376]

$$Ta(\equiv CBu^t)(H)(dmpe)_2I \rightleftharpoons Ta(=CHBu^t)(dmpe)_2I \quad (13)$$

$$Ta(=CHBu^t)(dmpe)_2Cl + AlMe_3 \rightleftharpoons Ta(\equiv CBu^t)(H)(dmpe)_2(ClAlMe_3) \quad (14)$$

2.4.4 Reactivity

The alkylidene metal–carbon double bond represents a site of reactivity which may undergo, *inter alia*, the following reactions: (i) 1,2-additions across the L_nM=CHR double bond; or (ii) formal [2 + 2] cycloadditions or "metallacyclization" reactions with unsaturated molecules. These latter "metallacyclization" reactions may involve the formation of complexes such as the metallacyclobutenes $Ta(CR=CRCHBu^t)(OAr)_3$ prepared from $Ta(=CHBu^t)(OAr)_3(THF)$ and RC≡CR,[372] or heteroatom-substituted metallacycles such as metallaoxetanes $Cp^*_2Ta[OCH(C_6H_4\text{-}4\text{-}X)CH_2]Me$, formed from $Cp^*_2Ta(=CH_2)Me$ with substituted benzaldehydes $O=CH(C_6H_4\text{-}4\text{-}X)$.[228] Cycloaddition reactions such as these will be detailed in Section 2.6.

Like alkyl ligands, the α-carbon of d^0 alkylidene complexes is susceptible to electrophilic attack and, accordingly, HX reagents (HCl, HOR, etc.) often add across highly polar $L_nM^{\delta+}=C^{\delta-}HR$ bonds to afford $L_n(X)M(CH_2R)$. Alkylidyne complexes may also react with HX in this fashion to afford $L_n(X)M(=CHR)$ species, which may then react further with HX. Note, however, that d^2 alkylidenes such as $Ta(=CHBu^t)Cl(dmpe)_2$ are typically protonated by HX at the metal to afford hydrides such as $[Ta(=CHBu^t)(H)Cl(dmpe)_2]X$.[205] The C–H bonds of appended ligands may engage in 1,2-additions across L_nM=CHR bonds to convert the alkylidene into an alkyl and generate cyclometallated ligands (see Section 2.8.2). For example, intermediate $Ta(OAr'')_2(=CH_2)Me$ (Ar'' = 2,6-$C_6H_3Bu^t_2$) can be produced photolytically and rapidly adds a *t*-butyl C–H bond (from O-Ar'') across $Ta=CH_2$ to provide $Ta[OC_6H_3Bu^t(CMe_2CH_2)](OAr'')Me_2$.[179]

The reactivity of alkylidyne complexes is less well developed. As described in Section 2.2.2, carbyne carbonyls react with electrophiles to afford alkyne compounds.[31] For example, $Ta(\equiv COSiPr^i_3)(CO)(dmpe)_2$ reacts with TMS-Cl to afford the asymmetric disiloxyethyne complex $Ta(Pr^i_3SiOC\equiv CO\text{-}TMS)(dmpe)_2Cl$.[30] The role of Lewis acids in this process is discussed.[31] The μ-alkylidyne ligand in $[Cl_2(DME)Ta(\mu\text{-}CBu^t)]_2[Zn(\mu\text{-}Cl)_2]$[354] is reported to undergo cycloaddition with two equivalents of alkyne to form cyclopentadienyl species.

2.5 η^2 COMPLEXES

2.5.1 Alkene Complexes

Complexes of alkenes are implicated in their catalytic polymerization, metathesis, and dimerization reactions, so the chemistry of isolable alkene complexes is of considerable interest. As can be seen from Table 12, the number of alkene complexes of niobium and tantalum reported from 1980 is relatively small, even compared with the alkyne derivatives of these metals. The principal pathways by which niobium and tantalum alkene complexes arise include (i) the coordination of the alkene to a $d^{n\geq 2}$ metal center (which ensures that the resulting adduct is stabilized by π back-bonding), including simple ligand exchange reactions, (ii) β-hydride "elimination" in an unsaturated, $d^{n\geq 2}$ alkyl complex, (iii) β-hydride "abstraction" in a d^0 dialkyl complex, and (iv) fragmentation of a d^0 metallacyclopentane (see Section 2.6.1).

2.5.1.1 Alkene complexes without cyclopentadienyl ligands

Several examples of alkene complexes of the type $Ta(\eta^2\text{-alkene})X_3L_2$ arise from either the interaction of d^2 compounds with alkene or from the reduction of d^0 complexes in the presence of alkene. For example, the alkylidene complexes $M(=CHR)X_3(PMe_3)_2$ (X = Cl or Br) react with terminal alkenes to form the organic products arising from rearrangement of an intermediate metallacyclobutane (see Section 2.6.1), with the alkene adduct constituting the organometallic product.[172] Therefore, for tantalum, reacting $Ta(=CHBu^t)Cl_3(PMe_3)_2$ with ethene, propene, or styrene affords the alkene adducts $Ta(CH_2{=}CHR)Cl_3(PMe_3)_2$ (R = H, Me, or Ph).[172] The alkoxide complexes $M(=CHBu^t)(OBu^t)_2Cl(PMe_3)$ react with ethene in the presence of excess PMe_3 to form $M(CH_2{=}CH_2)(OBu^t)_2Cl(PMe_3)_2$,[172] while related compounds are prepared upon reaction of $Ta(silox)_3$ with alkenes ($CH_2{=}CH_2$, $CH_2{=}CHMe$, $CH_2{=}CHEt$, or *cis*-MeCH=CHMe) to give $Ta(\eta^2\text{-alkene})(silox)_3$.[161]

The d^4 complexes of the form $Ta(\eta^2\text{-alkene})L_4X$ have also been studied. Thus, reduction of $Ta(CH_2{=}CH_2)(PMe_3)_2X_3$ (X = Cl or Br) under N_2 in the presence of excess PMe_3 gives orange crystals of $[Ta(CH_2{=}CH_2)X(PMe_3)_3]_2(\mu\text{-}N_2)$, which also form upon exposing $Ta(CH_2{=}CH_2)(PMe_3)_4X$ to N_2.[266,359] The latter method has proven useful in generating the $^{15}N_2$ complexes for IR and NMR studies.

2.5.1.2 Monokis(cyclopentadienyl) alkene complexes

Early in the 1980s, a series of compounds of the form $Cp^*Ta(\eta^2\text{-alkene})Cl_2$ were reported and studied in great detail by Schrock and co-workers,[360,384,385] especially with regard to their metallacyclization chemistry. Several of these compounds served as catalyst precursors for the dimerization of alkenes as outlined in Section 2.8.4.[360] The chemistry of most of these compounds has been described in detail by Labinger[1] and the bulk of our coverage will be presented in Section 2.6.1.

A number of other monokis(cyclopentadienyl) alkene complexes of niobium and tantalum have been reported with other ancillary ligands, for example, the d^2 complex $Cp^*Nb(=NAr)(PMe_3)_2$ reacts with $CH_2{=}CH_2$ to displace one PMe_3 ligand to provide the formal d^2 compound $Cp^*Nb(=NAr)(PMe_3)(CH_2{=}CH_2)$.[75] Reacting $CpNb(=NAr)Cl_2$[390] and two equivalents of the Grignard reagent EtMgCl or Pr^nMgCl (in the presence of PMe_3) forms similar d^2 complexes $CpNb(=NAr)(CH_2{=}CHR)(PMe_3)$ (for R = H or Me, respectively).[74] A dialkyl intermediate which undergoes β-hydride abstraction is suggested by the detection of propane from the reaction with Pr^nMgCl. All four possible isomers of the propene complex are observed (**16**)–(**19**) and one of these (**16**), with the methyl group oriented towards the NAr ligand, away from the Cp and PMe_3 ligands) can be crystallized selectively from concentrated pentane solutions (–78 °C). The structure of (**16**) supports the analogy between the CpNb(=NR) fragment and the bent metallocenes.[74,390]

Table 12 Alkene complexes of niobium and tantalum.

Compound	*Comments*	*Ref.*
Alkene complexes without cyclopentadienyl ligands		
$Ta(CH_2{=}CH_2)X_3(PMe_3)_2$	X = Cl; prep: $Ta_2Cl_6(PMe_3)_4 + CH_2{=}CH_2$ or $Ta({=}CHBu^t)Cl_3(PMe_3)_2 + CH_2{=}CH_2$	172,262,359
	X = Cl, Br	266
$Ta(CH_2{=}CHR)Cl_3(PMe_3)_2$	R = Me, Ph	172
$[Ta(CH_2{=}CH_2)Cl_3(PMe_3)]_2$		172
$Ta(CH_2{=}CH_2)X(PMe_3)_4$	X = Cl; prep: $TaH_2Cl(PMe_3)_4 + CH_2{=}CH_2$	23
	X = Cl, Br	266
$Ta(CH_2{=}CH_2)Cl(dmpe)_2$		266
$Ta(CH_2{=}CH_2)RCl_2(PMe_3)_2$	R = Et, CH_2Bu^t	154
$Ta(CH_2{=}CH_2)(Et)I_2(PMe_3)_2$		205
$Ta(CH_2{=}CHR)[2,6\text{-}C_6H_3(CH_2NMe_2)_2]Cl_2$	R = Bu^t (x ray), Me	185
$Ta(CH_2{=}CH_2)_2Et(PMe_3)_2$		146
$Ta(CH_2{=}CH_2)(OBu^t)_nCl_{3-n}(PMe_3)_2$	$n = 1, 2$	172,351
$Nb(CH_2{=}CH_2)(OBu^t)_2Cl(PMe_3)_2$		172
$Ta(alkene)(silox)_3$	alkene = $CH_2{=}CH_2$, $CH_2{=}CHMe$	270
	alkene = $CH_2{=}CH_2$, $CH_2{=}CHMe$, $CH_2{=}CHEt$, *cis*-$MeCH{=}CHMe$	161
$[Ta(CH_2{=}CH_2)X(PMe_3)_3]_2(\mu\text{-}N_2)$	X = Cl; prep: $Ta(CH_2{=}CH_2)(PMe_3)_2Cl_3 + NaHg + PMe_3 + N_2$	359
	X = Cl, Br	266
$Ta(CH_2{=}CH_2)(\eta\text{-}C_4H_6)Cl(PMe_3)_2$	X = Cl, Et	146
$Ta({=}NPh)(CH_2{=}CHR)(PMe_3)_3Cl$	R = H, Ph; prep: $Ta({=}NPh)(PMe_3)_4Cl + CH_2{=}CHR$	192,350
Monokis(cyclopentadienyl) alkene complexes		
$Cp^*Ta(CH_2{=}CHMe)Cl_2$		360
$Cp^*Ta(CH_2{=}CH_2)H_2(dmpe)$		94
$Cp^*Ta(cyclooctene)X_2$	$X_2 = Cl_2$, Br_2, (OMe)Cl; catalyst precursors for alkene dimerization	360
$Cp^*Ta(alkene)Cl_2$	alkene = $CH_2{=}CH_2$, $CH_2{=}CHMe$, $CH_2{=}CHPr^n$, $CH_2{=}CHBu^i$ $CH_2{=}CHPh$, $CH_2{=}CHBu^t$, *cis*-$MeCH{=}CHEt$, cyclooctene	384
	alkene = $CH_2{=}CHPh$; catalyst precursor for α,ω-diene cyclization	385
$Cp^*Ta(CH_2{=}CH_2)Me_2$	prep: $Cp^*TaMe_3Cl + LiEt$	210
$Cp^*Ta({=}CHBu^t)(CH_2{=}CH_2)(PMe_3)$	x ray	357
$CpNb({=}NAr)(alkene)(PMe_3)$	alkene = $CH_2{=}CH_2$, $CD_2{=}CD_2$, $CH_2{=}CHMe$ (four isomers, x ray)	74
$Cp^*Nb({=}NAr)(CH_2{=}CH_2)(PMe_3)$	prep: $Cp^*Nb({=}NAr)(PMe_3)_2 + CH_2{=}CH_2$	75
$[\eta^5\text{-}C_5Me_3(TMS)_2]Ta(CH_2{=}CHMe)Cl_2$		206
$Cp^*X_2Ta(\mu\text{-}\eta^1{:}\eta^2\text{-}CH{=}CH_2)(\mu\text{-}H)\text{-}(\mu\text{-}X)TaXCp^*$	X = Cl, Br; prep: $[Cp^*Ta(\mu\text{-}X)_2]_2 + CH_2{=}CH_2$	386
$(\eta^5\text{-}C_5Me_4Et)X_2Ta(\mu\text{-}\eta^1{:}\eta^2\text{-}CH{=}CH_2)(\mu\text{-}H)(\mu\text{-}X)TaX(\eta\text{-}C_5Me_4Et)$	X = Cl, Br	386
$Cp^*TaMe(PPh_2)(CH_2{=}CH_2)$	prep: $Cp^*TaMe_2(CH_2{=}CH_2) + HPPh_2$	210
$CpTa(\eta^2\text{-}C_4H_6)Cl_2(PMe_3)_2$		46
Bis(cyclopentadienyl) alkene complexes		
$Cp_2Ta(CH_2{=}CH_2)X$	X = Br, I	68
$Cp_2Ta(CH_2{=}CH_2)Me$		244
$Cp_2Ta(CH_2{=}CH_2)Et$		138,246
$Cp_2Ta(RO_2CCH{=}CHCO_2R)H$	R = Me, TMS; *trans*-isomers	387
$Cp_2Ta(PhCH{=}CHPh)(H)$	*trans*-isomer	387
$Cp_2Ta(CH_2{=}CHR)H$	R = H; prep: Cp_2TaCl_2 + 2 EtMgBr	252
	R = H, Me	138
	R = Me; *exo*-isomer	388,389
	R = Me, Et	141,251
	R = H; UV PES study	246
$Cp'_2Ta(CH_2{=}CHEt)H$		251
$(\eta^5\text{-}C_9H_7)_2Nb(CH_2{=}CHPh)H$	C_9H_7 = indenyl	47
$Cp_2Nb(CH_2{=}CHR)H$	R = Me, Et, Pr^n	245
	R = H, Me, 4-$C_6H_4CF_3$, Ph, 4-C_6H_4OMe, 4-$C_6H_4NMe_2$	138
	R = Me; *exo*-isomer	388,389
	R = H; UV PES study	246
$Cp^*_2Ta(CH_2{=}CHR)H$	R = H, Me, Ph	282
	R = H, Me	138
	R = H	109
$Cp^*_2Nb(CH_2{=}CHR)H$	R = H, Me, Ph, 4-$C_6H_4NMe_2$, 4-C_6H_4OMe, Tol, 4-$C_6H_4CF_3$	249
	R = H; prep: $Cp^*_2NbH_3 + CH_2{=}CH_2$	70
	R = 3-$C_6H_4CF_3$, Ph (x ray), 3-C_6H_4Me, 3-$C_6H_4NMe_2$	138
$[\eta^5{:}\eta^2\text{-}C_5Me_4CH_2CH{=}CH_2]Cp^*TaH$		109
$[\eta^5,\eta^1\text{-}C_5Me_4CH_2]Cp^*Ta(CH_2{=}CH_2)$		109
$Cp^*_2Ta(C_2H_4{\cdot}AlEt_3)(H)$	prep: $Cp^*_2Ta(CH_2{=}CH_2)H + AlEt_3$; x ray	253
$Cp^*_2Nb(C_2H_4{\cdot}AlR_3)(H)$	R = Me, Et	253
$Cp^*(\eta\text{-}C_5Me_4Ph)Ta(CH_2{=}CH_2)H$		109
$Cp_2Nb(CH_2{=}CH_2)(TMS)$	x ray	71
$Cp_2Nb(CH_2{=}CHPh)(SnMe_3)$	two isomers	122
$Cp_2Nb(\eta^2\text{-}CH_2{=}CHCMe{=}CH_2)(SnMe_3)$		122

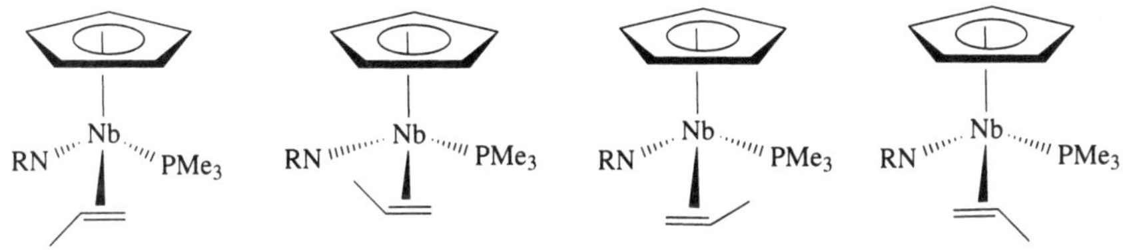
(16) (17) (18) (19)

2.5.1.3 Bis(cyclopentadienyl) alkene complexes

Bis(cyclopentadienyl) alkene complexes are some of the earliest reported organometallic compounds of these metals. The reactions of Cp_2NbCl_2 with the Grignards RMgCl (for R = Pr^n, Pr^i, Bu^n, Bu^s, or Pn^n) are reported to afford the complexes Cp_2Nb(alkene)H.[245] These reactions presumably proceed via the Nb^{III} intermediate "Cp_2NbR" which β-H eliminates stereospecifically to form the alkene hydrides. Thus, the *n*-propyl complex decomposes initially only to the *endo*-alkene hydride complexes (with the substituted end of the alkene towards the hydride ligand), but slowly rearranges to a mixture of *endo*- and *exo*-isomers. However, the isopropyl complex affords predominantly the *exo*-alkene hydride complex as the kinetic product (with the substituted end of the alkene away from the hydride ligand), which slowly equilibrates to the same mixture of *endo*- and *exo*-isomers (Scheme 8).

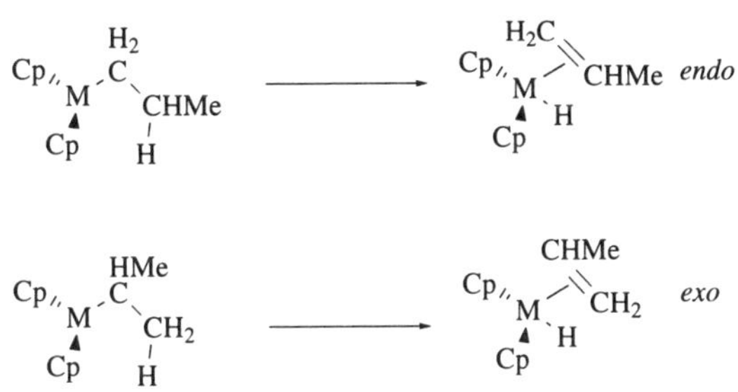

Scheme 8

Doherty and Bercaw[249] have reported extensive studies of the alkene hydride compounds $Cp^*{}_2Nb(CH_2{=}CHR)H$, which are prepared from reacting $Cp^*{}_2NbH_3$ with the corresponding alkene (for R = H, Ph, 4-$C_6H_4NMe_2$, 4-C_6H_4OMe, Tol, 4-$C_6H_4CF_3$) or from the reaction of $Cp^*{}_2NbCl_2$ with RCH_2CH_2MgCl (for R = H, Me). The latter reaction is related to those reported by Klazinga and Teuben[245] (see above) and are also presumed to proceed via "$Cp^*{}_2NbR$" intermediates which β-H eliminate to form the alkene hydrides. A single isomer is obtained and models indicate that the *endo*-isomer alone is isolated.[249]

To attain a firm understanding of the β-hydrogen elimination reaction, Doherty and Bercaw[249] carefully examined its microscopic reverse, the migratory insertion which occurs upon reaction of $Cp^*{}_2Nb(CH_2{=}CHR)H$ with L = CO or CNMe to afford $Cp^*{}_2Nb(CH_2CH_2R)L$. Magnetization transfer experiments demonstrate a reversible migratory insertion ⇌ β-H elimination equilibrium shown in Equation (15), although the β-elimination rate constant is orders of magnitude greater than the insertion rate constant.[249] The migratory insertion reactions of the series of *para*-substituted styrene complexes $Cp^*{}_2Nb(CH_2{=}CHR)H$ (R = Ph, 4-$C_6H_4NMe_2$, 4-C_6H_4OMe, Tol, 4-$C_6H_4CF_3$) in the presence of L were examined in a detailed kinetic and mechanistic study, and found to be consistent with a four-center transition state in which bond making and breaking are concerted (Scheme 9). A small degree of positive charge builds up on the alkene C-β and negative charge on the migrating hydrogen, more consistent with its migration as a hydride H^-.

$$Cp^*{}_2Nb(CH_2{=}CH_2)H \rightleftharpoons Cp^*{}_2Nb(Et) \qquad (15)$$

A closer examination of $Cp_2M(exo\text{-}CH_2{=}CHR)H$ using spin-saturation transfer NMR experiments has provided compelling evidence for intermediates with agostic M ← H–C bonding in the migratory insertion ⇌ β-H elimination reactions.[388] Initially, a mechanism for the exchange of NbH and the CH_2 hydrogens in *exo*-$Cp_2Nb(CH_2{=}CHMe)H$ was proposed based upon its rate being two orders of magnitude faster than the exchange of inequivalent Cp rings (Scheme 10).[138] This mechanism involves a simple insertion followed by inversion about niobium, and reinsertion to another *exo*-isomer.

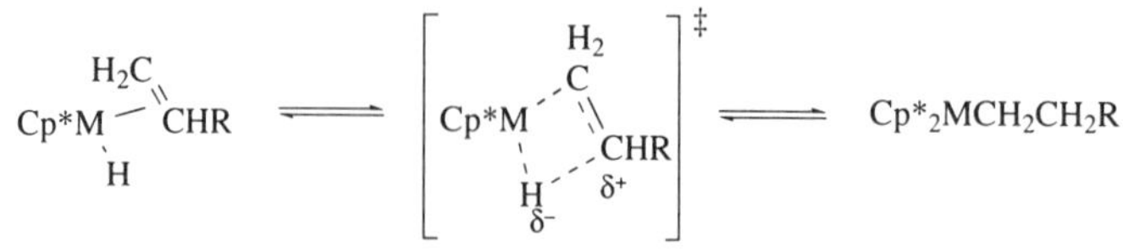

Scheme 9

However, in a later report,[388] a new mechanism was proposed which involves an agostic alkyl ligand in intermediate $Cp_2Nb(CHMe_2)$ that is formed upon insertion (Scheme 11). Thus, the exchange of NbH and the CH_2 hydrogens occurs by in-place rotation of this agostic alkyl intermediate. To exchange Cp ligands, the niobium center becomes planar as the alkyl group "slides" across the face of the metal to form a complex (**20**) with a molecular plane of symmetry in which now two agostic interactions (one from each Me group) are present.

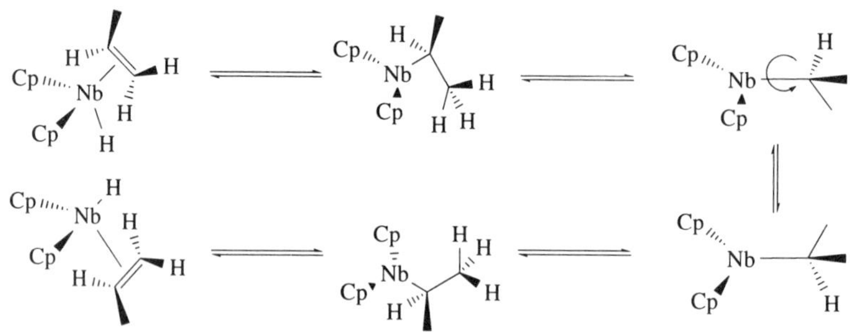

Scheme 10

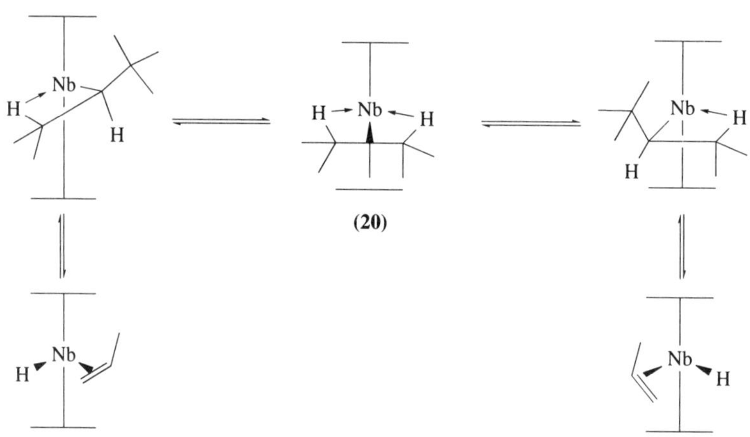

Scheme 11

Perhaps the most intriguing series of alkene compounds are those formed from the ethene hydrides $Cp^*_2M(CH_2{=}CH_2)H$, which are reported to bind aluminum alkyls reversibly to afford species of general formula $Cp^*_2M(C_2H_4{\cdot}AlR_3)(H)$ (**21**), in which AlR_3 coordinates to the opposite side of the ethene π electrons.[253] NMR data and the structure of $Cp^*_2Ta(C_2H_4{\cdot}AlEt_3)(H)$ are consistent with the closer association of aluminum with the *endo*-carbon of the ethene. The mixed-ring compound Cp*CpTa(CO)H, however, coordinates $AlEt_3$ at the hydride site to afford $Cp^*CpTa(CO)(H{\cdot}AlEt_3)$.

2.5.2 Alkyne and Benzyne Complexes

Alkyne complexes of niobium and tantalum have attracted considerable attention because of their use in organic synthesis and alkyne cyclooligomerization chemistry. The alkyne chemistry of niobium and tantalum is now better developed than their alkene chemistry; Table 13 summarizes alkyne and benzyne complexes of these metals.

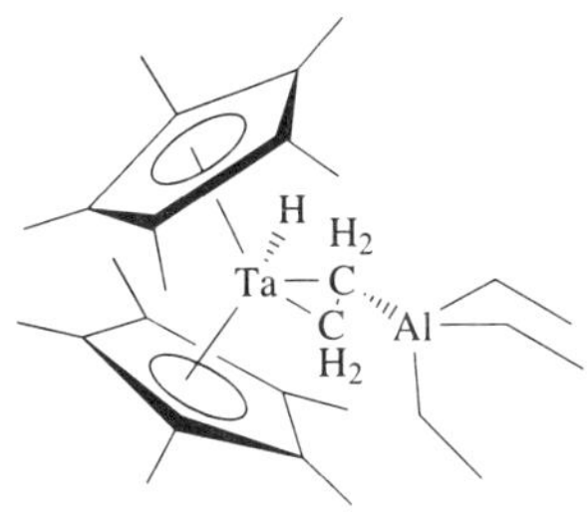

(21)

2.5.2.1 Alkyne complexes without cyclopentadienyl ligands

Most of the niobium and tantalum alkyne complexes without cyclopentadienyl ligands may be categorized according to the following general forms: (i) six-coordinate complexes $M(RC{\equiv}CR)X_3L_2$, which includes anionic $[M(RC{\equiv}CR)X_4L]^-$ and dimeric $[M(RC{\equiv}CR)X_2(\mu\text{-}X)L]_2$; (ii) formal d^4 compounds $M(RC{\equiv}CR)L_4X$; (iii) alkoxide derivatives $M(RC{\equiv}CR)(OR)_3L_n$ ($n = 0$ or 1); and (iv) dinuclear species with μ-RC≡CR ligands of the type $M_2(\mu\text{-}RC{\equiv}CR)X_6L_2$.

Some of the most synthetically useful alkyne compounds are formed from the reaction of d^2 $NbCl_3(DME)$ with alkynes in CH_2Cl_2 to afford the adducts $Nb(RC{\equiv}CR)Cl_3(DME)$.[392] A number of structures of the related THF adducts $Nb(RC{\equiv}CR)Cl_3(THF)_2$ and $[Nb(RC{\equiv}CR)Cl_4(THF)]^-$ have been reported.[391] In most cases, structural,[391] theoretical,[88] and reactivity[407] studies of these formal d^2 derivatives suggest that they are best described as d^0 metallacyclopropenes.

An interesting structural and reactivity study has been made of dinuclear metal–metal-bonded compounds with alkynes. $Nb_2Cl_6(THT)_3$ (Nb=Nb) reacts with PhC≡CPh to afford $[Nb(PhC{\equiv}CPh)Cl_2(\mu\text{-}Cl)(THT)]_2$[394] and $Ta_2Cl_6(THT)_3$ reacts with $Bu^tC{\equiv}CMe$ to form $[Ta(Bu^tC{\equiv}CMe)Cl_2(\mu\text{-}Cl)(THT)]_2$ in which the alkynes are bound in a terminal rather than bridging fashion.[399] However when $Ta_2Cl_6(THT)_3$ and $Bu^tC{\equiv}CBu^t$ react, the alkyne-bridged species $Ta_2(\mu\text{-}Bu^tC{\equiv}CBu^t)Cl_6(THF)_2$ is isolated.[396] An extended Hückel molecular orbital study of a large number of perpendicular and parallel ethyne complexes has been reported where bonding mode vs. d^n configuration, etc., was examined.[397] Specifically, Cotton and Hall's $Ta_2(\mu\text{-}Bu^tC{\equiv}CBu^t)Cl_6(THF)_2$[396] and examples of $Cp_2Nb_2(\mu\text{-}RC{\equiv}CR)_2(CO)_2$ from the pre-1970 Russian literature[1] were examined. Several alkyne complexes of Ta^I have also been reported. Thus, $Ta(CO)_3(PMe_3)_3I$ reacts with RC≡CR to form the d^4 complexes $Ta(RC{\equiv}CR)(CO)_2(PMe_3)_2I$.[404]

Finally, the alkyne compounds $[(\eta\text{-}C_6H_3Me_3)_2Nb(RC{\equiv}CR)]BPh_4$ (R = Me or Ph) are available as red or red-brown microcrystals from reacting $[(\eta\text{-}C_6H_3Me_3)_2Nb]BPh_4$ with one equivalent of RC≡CR.[412] When toluene solutions of $[(\eta\text{-}C_6H_3Me_3)_2Nb(RC{\equiv}CR)]BPh_4$ are heated to 100 °C, red to orange solutions are obtained from which $[Ph_2B(\eta\text{-}C_6H_5)_2]Nb(RC{\equiv}CR)$ (**22**) can be crystallized.

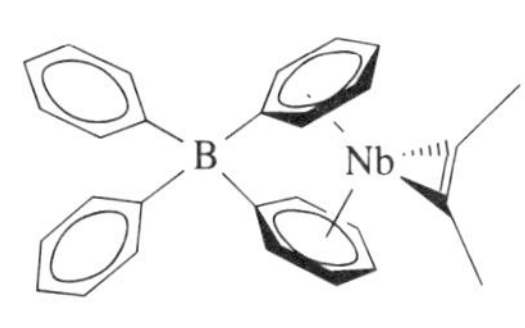

(22)

2.5.2.2 Cyclopentadienyl alkyne complexes

Early reports of niobium and tantalum alkynes in the Russian literature in the late 1960s describe monokis(cyclopentadienyl) carbonyl complexes with stoichiometries such as $CpM(RC{\equiv}CR)(CO)_2$, $CpM(RC{\equiv}CR)_2(CO)$, and $[CpM(\mu\text{-}RC{\equiv}CR)(CO)]_2$, typically prepared by photolysis of $CpM(CO)_4$ in the presence of an alkyne. This chemistry was described in detail by Labinger.[1] By far the majority of cyclopentadienyl alkyne complexes of niobium and tantalum that have been prepared may be categorized according to one of the general forms $(\eta\text{-}C_5R_5)_2M(RC{\equiv}CR)X$ or $(\eta\text{-}C_5R_5)M(RC{\equiv}CR)X_2$. The anionic ligand X^- in these very stable d^2 complexes may be halide, alkyl, vinyl, hydride, alkoxide, thiolate, etc., as outlined in Table 13.

Table 13 Alkyne and benzyne complexes of niobium and tantalum.

Compound	*Comments*	*Ref.*
Alkyne complexes without cyclopentadienyl ligands		
Nb(RC≡CR)Cl$_3$(THF)$_2$	RC≡CR = PhC≡CPh, PhC≡CMe (x ray of both)	391
	RC≡CR = MeC≡CMe, PhC≡CEt, Bu^{t}C≡CMe, PhC≡CSiMe$_2$Bu^t (x ray)	392
[Mg$_2$Cl$_3$(THF)$_6$][Nb(RC≡CR)Cl$_4$(THF)]	x ray	391
Nb(RC≡CR)Cl$_3$(DME)	RC≡CR = PhC≡CPh, PhC≡CEt, EtC≡CEt, MeC≡CMe, TMS-C≡CH, TMS-C≡CPr^n	392
Nb(RC≡CR)Br$_3$(DME)	RC≡CR = PhC≡CSiMe$_2$Bu^t, EtC≡CEt	392
[Ta(PhC≡CPh)Cl$_2$(μ-Cl)(THT)]$_2$[a]	originally formulated as Ta(PhC≡CPh)Cl$_3$(THT)$_3$	393,394
[Hpy][Ta(PhC≡CPh)Cl$_4$(py)]	x ray	393
	bonding study	395
[Hpy][Nb(PhC≡CPh)Cl$_4$(py)]		394
Ta$_2$(μ-Bu^{t}C≡CBu^t)Cl$_6$(THF)$_2$	prep: Ta$_2$Cl$_6$(THT)$_3$ + Bu^{t}C≡CBu^t; x ray	396
	MO studies	397,398
[Ta(Bu^{t}C≡CMe)Cl$_2$(μ-Cl)(THT)]$_2$	x ray	399
[Nb(PhC≡CPh)Cl$_2$(μ-Cl)(THT)]$_2$	x ray	394
[(PhC≡CPh)NbCl$_3$]$_4$	x ray	400,401
M(TMS-OC≡CO-TMS)(dmpe)$_2$X	M = Ta, X = Cl; x ray	31,126,128, 402,403
	M = Ta, X = TMS	402
	M = Nb, X = Cl; x ray	126,128,403
	X = OTf; M = Nb, Ta (x ray)	126
	M = Ta; X = Cl$_{0.8}$I$_{0.2}$; x ray	126
[M(HOC≡COH)(H)(dmpe)$_2$Cl]Cl	M = Ta; x ray	127,403
	M = Nb	403
M(Pr$^i_3$SiOC≡COSiPr$^i_3$)(dmpe)$_2$(OTf)		128
Ta(Pr$^i_3$SiOC≡CO-TMS)(dmpe)$_2$Cl		30,376
Nb[(TMS)MeNC≡CO-TMS](dmpe)$_2$Cl	x ray	16,251
Ta[(TMS)MeNC≡CO-TMS](dmpe)$_2$Cl		16
[Ta(HOC≡CNHMe)(H)(dmpe)$_2$Cl]Cl	x ray	16
Ta[(TMS)MeNC≡CN(TMS)Me](dmpe)$_2$Cl		16
Ta(TMS-OC≡COBF$_3$)(H)(dmpe)$_2$Cl	x ray; prep: Nb(CNMe)(CO)(dmpe)$_2$Cl + NaHg + TMS-Cl	403
Ta(Bu^{t}Me$_2$SiOC≡COSiMe$_2$Bu^t)(dmpe)$_2$Cl	prep: Ta(CO)$_2$(dmpe)$_2$Cl + NaHg + Bu^{t}Me$_2$SiCl	403
Nb[(Bu^{t}Me$_2$Si)MeNC≡CO-TMS](dmpe)$_2$Cl	x ray	16
[Ta(CO)(dmpe)$_2$](μ-CO)[Ta(ROC≡COR)(dmpe)$_2$]	R = TMS, SiBu^{t}Me$_2$ (x ray)	31
Ta(RC≡CR)(CO)$_2$(PMe$_3$)$_2$I	RC≡CR = PhC≡CPh (x ray), PhC≡CH, HC≡CH, Ph$_2$PC≡CPPh$_2$	404
Ta(PhC≡CH)(^{13}CO)$_2$(PMe$_3$)$_2$I		404
Ta(PhC≡CPh)(CO)$_2$[P(OMe)$_3$]$_2$I		404
Nb(PhC≡CPh)(CO)$_2$(PEt$_3$)$_2$X	R = Ph; X = Cl, Br, I	405
	R = Et; X = Cl, Br, I	405
	R = H, Me; X = I	405
Nb(PhC≡CPh)(CO)$_2$(dppe)X	X = Br, I	405
[Nb(RC≡CR)$_2$(CO)$_2$(μ-X)]$_2$	R = Ph; X = Br	406
	R = Ph, Et, Me; X = I	406
	RC≡CR = PhC≡CH; X = I	406
Nb(RC≡CR)(CO)$_2$(PEt$_3$)$_2$I	R = Ph, Et, Me	406
Ta(RC≡CR)(OAr)$_3$	RC≡CR = PhC≡CPh (x ray), TMS-C≡CMe	162,407
	RC≡CR = HC≡CBu^t, implicated	408
Ta(PhC≡CPh)(OAr)$_3$(N≡CR)	R = Me, CD$_3$, Et, Bu^t, Ph	409
Ta(RC≡CR)(silox)$_3$	R = Me	270
	R = H, Me, CF$_3$	161
Tp'Nb(PhC≡CMe)Cl$_2$	x ray; prep: Nb(PhC≡CMe)Cl$_3$(DME) + KTp'	410
Tp'Nb(PhC≡CMe)Me$_2$		410
Ta(=NAr)(RC≡CR)Cl(py)$_2$	RC≡CR = EtC≡CEt, PhC≡CPh, TMS-C≡CMe	411
Ta(=NAr)(RC≡CR)(OAr)(py)$_2$	R = Et, Pr^n	411
[(η-C$_6$H$_3$Me$_3$)$_2$Nb(RC≡CR)]BPh$_4$	R = Me, Ph	412
[Ph$_2$B(η-C$_6$H$_5$)$_2$]Nb(RC≡CR)	R = Me (x ray), Ph	412
Monokis(cyclopentadienyl) alkyne complexes		
CpNb(RC≡CR)$_2$(CO)	RC≡CR = HC≡CH, MeC≡CMe, HC≡CMe, HC≡CPh or HC≡CHx^n	413
CpTa(HC≡CH)$_2$(CO)	prep: CpTa(CO)$_4$ + HC≡CH + *hν*	413
Cp*Nb(EtC≡CEt)Cl$_2$	prep: [Cp*NbCl$_2$]$_2$ + EtC≡CEt	209
Cp'Nb(RC≡CR)Cl$_2$	R = Tol	56
	R = Ph, Tol (x ray)	57
[Cp'Nb(RC≡CR)(μ-Cl)]$_2$	R = Tol; x ray	56,58
	R = Ph	58
CpNb(RC≡CR)Cl$_2$	R = Ph, Tol, Et	57
	R = Ph	414

Table 13 (continued)

Compound	*Comments*	*Ref.*
$CpNb(RC{\equiv}CH)Cl_2$	R = Ph, Bu^n (isolated); R = $(CH_2)_3OSiBu^tMe_2$, $(CH_2)_3OCOBu^i$, TMS (implicated)	415
$CpNb(RC{\equiv}CR)Me_2$	R = Ph, Tol	416
$CpNb(PhC{\equiv}CPh)(SPr^i)_2$	prep: $CpNb(PhC{\equiv}CPh)Cl_2$ + $LiSPr^i$	414
$CpNb(SPr^i)_2(PhC{\equiv}CPh)Co_2(CO)_4$		414
$CpNb(\eta\text{-}C_4H_6)(RC{\equiv}CR)(PMe_3)$	RC≡CR = PhC≡CPh, PhC≡CMe (x ray), TMS-C≡CPh	417
$CpTa(RC{\equiv}CR)Cl_2$	R = Ph, Tol; prep: $[CpTaCl(\mu\text{-}Cl)(CO)_2]_2$ + RC≡CR	57
$CpTa(RC{\equiv}CR)Me_2$	R = Ph, Tol	200,416
$CpTa(RC{\equiv}CR)[\eta^2\text{-}C(=NBu^t)Me]Me$	R = Ph, Tol (x ray); prep: $CpTa(RC{\equiv}CR)Me_2$ + $C{\equiv}NBu^t$	200,416
$Cp^*Ta(RC{\equiv}CR)Cl_2$	RC≡CR = PhC≡CPh (x ray), PhC≡CH, MeC≡CMe, EtC≡CEt, HC≡CH	418
	RC≡CR = PhC≡CPh, HC≡CPh, MeC≡CMe, TMS-C≡C-TMS	212
	RC≡CR = PhC≡CPh, HC≡CPh	46
	R = Ph, Tol; prep: $Cp^*TaCl_2(CO)_2(THF)$ + RC≡CR	88
$(\eta\text{-}C_5Me_4Bu^t)Ta(MeC{\equiv}CMe)Cl_2$		379
$(\eta\text{-}C_5H_2Bu^t{}_3)Ta(Bu^tC{\equiv}CH)Cl_2$		379
$Cp^*Ta(RC{\equiv}CR)Br_2$	R = Et	418
	R = Me	297
$Cp^*Ta(PhC{\equiv}CPh)(OMe)_nCl_{2-n}$	n = 1, 2	418
$CpNb(=NAr)(RC{\equiv}CR)(PMe_3)$	R = Me, Ph	74
$Cp^*Nb(=NAr)(PhC{\equiv}CPh)(PMe_3)$	prep: $Cp^*Nb(=NAr)(PMe_3)_2$ + PhC≡CPh	75
$[CpNb(CO)]_2(\mu\text{-}PhC{\equiv}CPh)_2$	MO study	419
$Cp(PhC{\equiv}CPh)Nb(\mu\text{-}SPr^i)_2Mo(CO)_4$	x ray	414
Bis(cyclopentadienyl) alkyne complexes		
$Cp_2Nb(RC{\equiv}CR)Cl$	RC≡CR = PhC≡CPh, PhC≡CH, HC≡CH	133,420
$Cp_2Nb(RC{\equiv}CPh)Br$	R = H, Ph	67
$Cp_2Nb(RC{\equiv}CR)H$	R = Ph	232
	R = Me, Et, Ph, TMS	421
	RC≡CR = $Bu^tC{\equiv}CH$, TMS-C≡CH (x ray of both), PhC≡CPh, PhC≡CMe	422
$Cp_2Nb(TMS\text{-}C{\equiv}C\text{-}TMS)[CR{=}CH_2]$	R = CO_2Me, COMe	421
$Cp_2Nb(TMS\text{-}C{\equiv}C\text{-}TMS)[C(CO_2Me){=}CH(CO_2Me)]$		421
$Cp_2Nb(CF_3C{\equiv}CCF_3)H$	prep: Cp_2NbH_3 + $CF_3C{\equiv}CCF_3$	236
$Cp_2Nb(RC{\equiv}CCF_3)F$	R = H, CF_3	236
$Cp_2Nb(MeC{\equiv}CMe)(SnMe_3)$	prep: $Cp_2Nb(SnMe_3)H_2$ + MeC≡CMe	122
$Cp_2Nb(RC{\equiv}CH)(Et)$	R = Me, Et, Ph	422
$Cp_2Nb(MeC{\equiv}CMe)(CH_2CH_2R)$	R = H (x ray), Me, Ph	422
$(\eta\text{-}C_5H_4TMS)_2Nb(RC{\equiv}CR)Cl$	RC≡CR = PhC≡CPh (x ray), HC≡CPh, or HC≡CH	243
$(\eta\text{-}C_5H_4TMS)_2Nb(PhC{\equiv}CPh)X$	X = Br, I, H	304
$(\eta\text{-}C_5H_4TMS)_2Nb(PhC{\equiv}CPh)$	electrochemically generated	423
$(\eta\text{-}C_5H_4Pr^i)_2Ta(MeC{\equiv}CMe)Cl$		242
$Cp_2Ta(PhC{\equiv}CPh)Cl$	prep: $[Cp_2TaCl]_n$ + PhC≡CPh	66
$Cp_2Ta(Pr^nC{\equiv}CPr^n)X$	X = Br, I	68
$Cp_2Ta(RC{\equiv}CR)H$	R = Me, Ph, TMS, Bu^t	424
	R = Me, TMS, Ph	387
	R = Pr^n	68
$Cp_2Ta(RC{\equiv}CR)[C(CO_2Me){=}CH_2]$	R = Me, Ph, TMS, Bu^t	424
$Cp_2Ta(RC{\equiv}CR)[C(CN){=}CH_2]$	R = Me (x ray), Ph, TMS, Bu^t	424
$Cp_2Ta(RC{\equiv}CR)[E\text{-}C(CO_2Me){=}CH(CO_2Me)]$	R = Ph, Bu^n	424
$Cp_2Ta(RC{\equiv}CR)(Et)$	RC≡CR = MeC≡CH, MeC≡CMe	422
$Cp^*{}_2Nb(MeC{\equiv}CMe)H$		253
$Cp^*{}_2Ta(MeC{\equiv}CMe)H$		282
$Cp^*CpTa(MeC{\equiv}CMe)H$		282
$Cp_2Nb(RC{\equiv}CR)(SMe)$	R = CF_3, CO_2Me	425
$Cp_2Ta(\mu\text{-}C_3H_3)Re_2(CO)_9$	x ray; $\eta^2(C,C)$ bound to tantalum	223
Benzyne complexes		
$[Nb(\eta^2\text{-}C_6H_4)_2Ph_3(LiPh{\cdot}THF)(Li{\cdot}THF)_4]$	prep: $NbCl_5$ + excess PhLi + THF; x ray	426,427
$[Li_4Cl_2(THF)_{10}][Ta(\eta^2\text{-}C_6H_4)_2Ph_4(Li{\cdot}THF)_2]_2$	x ray	426,427
$Ta[O\text{-}2,6\text{-}C_6H_3Bu^t(CMe_2CH_2)](\eta^2\text{-}C_6H_3Me)(OAr'')$	two isomers; not observed directly	163,194
$Ta[O\text{-}2,6\text{-}C_6H_3Bu^t(CMe_2CH_2)](\eta^2\text{-}C_6H_4)(OAr'')$	implicated, not observed directly	163
$Ta[O\text{-}2,6\text{-}C_6H_2Bu^t(CMe_2CH_2)\text{-}4\text{-}X](\eta^2\text{-}C_6H_4)(O\text{-}2,6\text{-}C_6H_2Bu^t{}_2\text{-}4\text{-}X)$	x = Me, OMe; implicated, not observed directly	163
$M(\eta^2\text{-}2,5\text{-}C_6H_2Me_2)(2,5\text{-}C_6H_3Me_2)_2(PMe_2Ph)_2$	x ray of both M = Nb, Ta	428
$M(\eta^2\text{-}2\text{-}C_6H_3Me)(2\text{-}C_6H_4Me)_2(PMe_2Ph)_2$		428
$Ta(\eta^2\text{-}C_6H_4)Ph_2(PMe_3)_2$	prep: $TaCl_5$ + LiPh + PMe_3	428
$CpNb(=NAr)(\eta^2\text{-}C_6H_4)(PMe_3)$	prep: thermolysis of $CpNb(=NAr)Ph_2(PMe_3)$; x ray	214
$Cp^*{}_2Ta(\eta^2\text{-}C_6H_4)H$		109,227

[a] THT = tetrahydrothiophene.

(i) Monokis(cyclopentadienyl) alkyne complexes

Alkyne complexes of the form (η-C_5R_5)M(RC≡CR)X_2 have been prepared and studied extensively. Alkenes (typically styrene or cyclooctene) are readily displaced from Cp*Ta(η^2-alkene)X_2 (X = Cl or Br) to afford Cp*Ta(RC≡CR)Cl_2 (RC≡CR = PhC≡CPh, PhC≡CH, MeC≡CMe, EtC≡CEt, or HC≡CH).[418] It is interesting that none of these complexes reacts with excess alkyne to form metallacyclopentadienes. The alkoxide derivatives Cp*Ta(PhC≡CPh)(OMe)Cl and Cp*Ta(PhC≡CPh)$(OMe)_2$ are available from Cp*Ta(PhC≡CPh)Cl_2 and NaOMe.[418] When Cp*Ta(TMS)Cl_3 is refluxed in benzene in the presence of alkyne, reductive elimination of TMS-Cl occurs and the d^2 adducts Cp*Ta(RC≡CR)Cl_2 (RC≡CR = PhC≡CPh, HC≡CPh, MeC≡CMe, or TMS-C≡C-TMS) are available.[212]

Perhaps the most intriguing reaction to give rise to cyclopentadienyl alkyne complexes is the cycloaddition sequence involving the tantalum alkylidyne complex $[Cl_2(DME)Ta(\mu\text{-}CBu^t)]_2[Zn(\mu\text{-}Cl)_2]$ (**23**).[354,379] Thus, the alkylidyne $[Cl_2(DME)Ta(\mu\text{-}CBu^t)]_2[Zn(\mu\text{-}Cl)_2]$ and MeC≡CMe (neat) or Bu^{t}C≡CH (in toluene) afford (η-$C_5Me_4Bu^t$)Ta(MeC≡CMe)Cl_2 and (η-$C_5H_2Bu^t_3$)Ta(Bu^{t}C≡CH)Cl_2, respectively.[379] (See also Section 2.7.3.1.)

The reaction of dinuclear complexes with alkynes can also lead to (η-C_5R_5)M(RC≡CR)X_2 derivatives. Cp*Ta(MeC≡CMe)Br_2 is formed from dinuclear $[Cp^*Ta(\mu\text{-}Br)_2]_2$ and MeC≡CMe[297] while $[Cp^*NbCl_2]_2$ (also presumably chloride bridged) reacts with EtC≡CEt to form Cp*Nb(EtC≡CEt)Cl_2.[209] Dinuclear carbonyl compounds may also serve as very useful precursors to alkyne derivatives. Thus, compounds such as Cp'Nb(TolC≡CTol)Cl_2 are available directly from Cp'$NbCl_4$, aluminum, and TolC≡CTol; however the preferred route to such adducts is through displacement of CO ligands from $[CpMCl(\mu\text{-}Cl)(CO)_2]_2$ (M = Nb, Ta) and $[Cp'NbCl(\mu\text{-}Cl)(CO)_2]_2$.[57] The alkyne adducts Cp'Nb(RC≡CR)Cl_2 (for R = Ph, Tol) can be reduced with NaHg in toluene, forming dark green-brown, diamagnetic $[Cp'Nb(RC{\equiv}CR)(\mu\text{-}Cl)]_2$ (Nb–Nb single bond), which was structurally characterized for the R = Tol complex.[58] Structural data for these complexes are consistent with the alkyne ligand being highly reduced, and therefore best described as a d^0 metallacyclopropene (**24**), rather that a d^2 alkyne ligand (**25**), as a result of extensive π back-bonding from the formally $d^{n \geq 2}$ metal center to the alkyne $\pi_{\parallel}$* orbital. In addition, MO calculations suggest additional π donation from the $\pi_{\perp}$ orbital of the alkyne to the metal occurs as depicted in (**26**), consistent with their description as a formal "four-electron-donor" alkyne ligand.

(24)

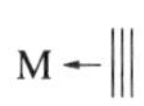

(25)

(26)

(ii) Bis(cyclopentadienyl) alkyne complexes

A study of bis(cyclopentadienyl) alkyne complexes underscored two principal methods of preparing such compounds from precursors containing an unsaturated ligand, in this case $Cp_2M(\eta^2$-alkene)H:[422] (a) the reaction of the alkene hydrides $Cp_2M(\eta^2$-alkene)H[429] with alkynes can induce hydride migration to the alkene to generate Cp_2M(RC≡CR)(alkyl); or (b) in the case of bulkier alkynes, simple substitution of the alkene may occur to form Cp_2M(RC≡CR)H.[422] For example, $Cp_2Nb(CH_2{=}CH_2)H$ reacts with RC≡CH (for R = Me, Et, or Ph; sealed tube, benzene solution, 60 °C) to form Cp_2Nb(RC≡CH)(Et); Cp_2Ta(MeC≡CH)(Et) is prepared similarly. All these products consist of a mixture of *endo*- and *exo*-isomers differing by the rotational orientation of the RC≡CH ligand within the molecular plane of symmetry. The alkene complexes $Cp_2Nb(CH_2{=}CHR)H$ (R = H, Me, or Ph) react with MeC≡CMe to form the corresponding complexes $Cp_2Nb(MeC{\equiv}CMe)(CH_2CH_2R)$ (Scheme 12). Both NMR and structural studies of Cp_2Nb(MeC≡CMe)(Et) (**27**) reveal an unusual α-H agostic interaction between the ethyl CH-α and the metal center. However, with the bulkier alkynes Bu^{t}C≡CH or TMS-C≡CH, the reactions with $Cp_2Nb(CH_2{=}CHMe)H$ afford the substitution products Cp_2Nb(RC≡CH)(H) (for R = Bu^t or TMS), both isolated as exclusively the *exo* products.[422]

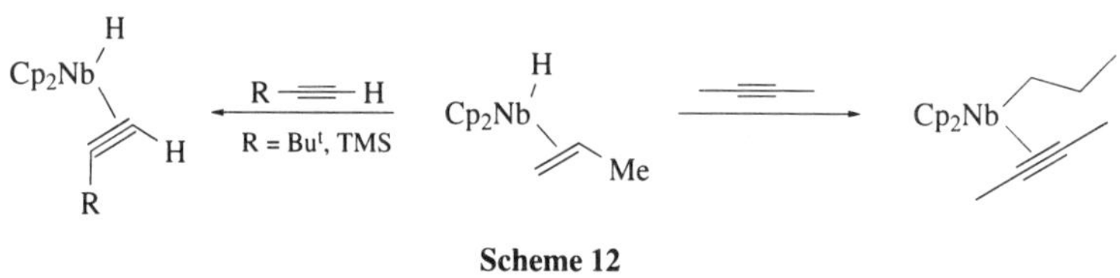

Scheme 12

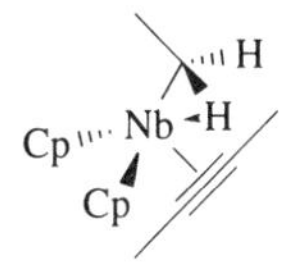

(27)

The thermal reaction between Cp_2NbH_3 and RC≡CR affords $Cp_2Nb(RC{\equiv}CR)H$ (for R = Et, Ph, or TMS) while $Cp_2Nb(MeC{\equiv}CMe)H$ is obtained from photolysis of $Cp_2Nb(TMS\text{-}C{\equiv}C\text{-}TMS)H$ in the presence of MeC≡CMe.[421] $Cp_2Nb(TMS\text{-}C{\equiv}C\text{-}TMS)H$ also inserts alkynes (e.g., HC≡CCOMe) to afford the vinyl complexes (e.g., $Cp_2Nb(TMS\text{-}C{\equiv}C\text{-}TMS)[C(COMe){=}CH_2]$), but the authors speculate that this proceeds by a nonmigratory insertion pathway).[421]

2.5.2.3 Benzyne complexes

A number of provocative benzyne complexes have been reported (Table 13). The 1981 report of the preparation of niobium and tantalum compounds containing two coordinated benzyne ligands[426] was followed in 1988 by the structural determination of these species.[427] Thus, when $NbBr_5$ or $TaBr_5$ is reacted with excess PhLi in THF, species originally formulated as the anionic bis(benzyne) complexes $[Li(THF)]_3[M(\eta^2\text{-}C_6H_4)_2Ph_4]$ are formed.[426] Some years later, Power and co-workers[427] examined these complexes (in this case prepared from MCl_5 and excess PhLi in THF) by x-ray structural methods. These compounds were found to be the anionic bis(benzyne) complexes $[Nb(\eta^2\text{-}C_6H_4)_2Ph_3(LiPh{\cdot}THF)(Li{\cdot}THF)_4]$ and $[Ta(\eta^2\text{-}C_6H_4)_2Ph_4(Li{\cdot}THF)_2]^-$ (isolated as its $[Li_4Cl_2(THF)_{10}]^{2+}$ salt) which contain PhLi and $[Li(THF)_n]^+$ ions as integral, bridging components within the structure.[427]

Wilkinson and co-workers[428] prepared the tetraaryls $M(2,5\text{-}C_6H_3Me_2)_4$ and $M(2\text{-}C_6H_4Me)_4$ from $NbCl_4(THF)_2$ or $TaCl_4$ and LiR or RMgBr; they react with PMe_2Ph to afford the η^2-benzyne (or η^2-aryne) complexes $M(\eta^2\text{-}2,5\text{-}C_6H_2Me_2)(2,5\text{-}C_6H_3Me_2)_2(PMe_2Ph)_2$ and $M(\eta^2\text{-}2\text{-}C_6H_3Me)(2\text{-}C_6H_4Me)_2(PMe_2Ph)_2$, upon elimination of the free arene. $Ta(\eta^2\text{-}C_6H_4)Ph_2(PMe_3)_2$ is prepared similarly from $TaCl_5$, LiPh, and PMe_3 in Et_2O.[428] Gibson and co-workers[214] prepared the diphenyl complex $CpNb(=NAr)Ph_2(PMe_3)$, which upon gentle heating (60 °C) eliminates benzene to form the η^2-benzyne compound $CpNb(=NAr)(\eta^2\text{-}C_6H_4)(PMe_3)$ (**28**). The delocalized benzyne aromatic ring is consistent with a two-electron donor alkyne, that is, donation from $\pi_{\|}$ with back-donation into the benzyne π^* orbital.

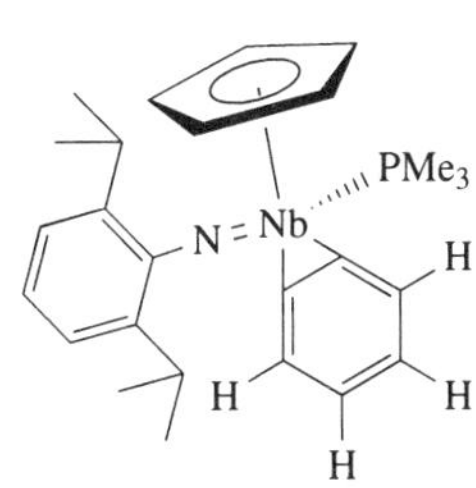

(28)

2.5.3 η^2-Acyls, η^2-Iminoacyls, and Related η^2 Complexes

The interest in acyl [MC(=O)R] and iminoacyl [MC(=NR)R] ligands generally derives from their potential relevance to catalytic CO reduction, especially the C–C bond-forming reactions of F–T chemistry.[430] In high-valent niobium and tantalum complexes, these ligands coordinate η^2 through both the carbon and the heteroatom (O or N), reflecting the Lewis acidity of the metal. There now exists a broad range of η^2 compounds as outlined in Table 14. We include in this section compounds containing η^2 ligands bound to the metal by at least one carbon atom. Therefore, we are also concerned with isocyanate, imine, ketene, etc., complexes, and also unusual η^2 compounds containing η^2-$CHPMe_2$ or η^2-CH_2PMe_2 ligands. Bridging η^2 ligands ($\mu{:}\eta^2,\eta^x$ for x = 1 or 2) are also included.

2.5.3.1 Formyl and acyl complexes

Insertions into metal–hydride bonds to generate formyl complexes [MC(=O)H] are thermodynamically unfavorable under most conditions; however, a "stabilized formyl" can be generated upon reacting dimeric $[(\eta\text{-}C_5Me_4Et)TaCl_2(H)]_2$ (presumably hydride bridged) with one equivalent of CO.[206,208] The resulting complex $[(\eta\text{-}C_5Me_4Et)TaCl_2]_2(\mu\text{-}H)(\mu\text{-}CHO)$ contains a $\mu{:}\eta^2,\eta^2$-CHO ligand which bridges the two tantalum atoms and has an unusually long C–O bond.[432] More recently, when $[Ta(silox)_2Cl]_2(\mu\text{-}H)_2$ was treated with excess CO, the μ-formyl complex $[Ta(silox)_2Cl]_2(\mu\text{-}H)(\mu{:}\eta^2,\eta^2\text{-}CHO)$ was isolated; it releases MeOH upon treatment with H_2O.[160] In contrast, 1 atm of CO reacts with the dihydride dimer $[TaH_2(silox)_2]_2$ to form the CO bond scission product $[TaH(silox)_2]_2(\mu\text{-}CH_2)(\mu\text{-}O)$.

Acyl and silaacyl complexes of tantalum and acyls of niobium are also well established. When $Cp^*Ta(CH_2Bu^t)Cl_3$ reacts with CO, the acyl $Cp^*Ta[\eta^2\text{-}C(O)CH_2Bu^t]Cl_3$ is isolated, which has been structurally characterized as containing a very short Ta–O bond.[203] One of the most interesting reaction features of $Cp^*Ta[\eta^2\text{-}C(O)CH_2Bu^t]Cl_3$ is its magnesium reduction, which results in acyl C–O bond cleavage and intramolecular "deoxygenative coupling" with a C–C bond of the Cp* ligand to form the substituted benzene $C_6Me_5(CH_2Bu^t)$ and unidentified metal products (Scheme 13).[433] The unstable η^2-silaacyl complex $Cp^*Ta[\eta^2\text{-}C(O)TMS]Cl_3$ is readily formed upon reaction of limited quantities of CO with $Cp^*Ta(TMS)Cl_3$.[86,108,434] This species displays electrophilic behavior at its acyl carbon, as evidenced by its reaction with pyridine to form the adduct $Cp^*Ta[\eta^2\text{-}C(O)TMS(NC_5H_5)]Cl_3$, where the structural parameters about the carbonyl group indicate a saturated carbon.[434]

An interesting route to niobium acyl complexes of the form $[(\eta\text{-}C_5H_4TMS)_2Nb[\eta^2\text{-}C(O)CHR_2)Br]BF_4$ has been uncovered.[436] Thus, protonation of the $\eta^2(O,C)$-ketene species of the form $(\eta\text{-}C_5H_4TMS)_2Nb[\eta^2(O,C)\text{-}O{=}C{=}CR_2]Br$ with $HBF_4{\cdot}OEt_2$ affords the BF_4^- salts of the acyl cations. The cationic acyl complexes $\{(\eta\text{-}C_5H_4TMS)_2Nb[\eta^2\text{-}C(O)CHR_2]Cl\}BF_4$ (R_2 = MePh, EtPh, Me_2, Ph_2) and $\{Cp_2Nb[\eta^2\text{-}C(O)CHEtPh]Cl\}BF_4$ are prepared similarly.[437]

Table 14 η^2-Acyls, η^2-iminoacyls, and related η^2 complexes of niobium and tantalum.

Compound	*Comments*	*Ref.*
Formyl and related complexes		
$[(\eta\text{-}C_5Me_4Et)TaCl_2]_2(\mu\text{-}H)(\mu\text{-}\eta^2\text{-}CHO)$	prep: $[(\eta\text{-}C_5Me_4Et)TaCl_2(H)]_2$ + 1 equiv. CO; x ray	206,208, 431,432
	(μ-^{13}CHO) and (μ-D)(μ-CDO) analogue	206
$[Cp^*TaCl_2]_2(\mu\text{-}H)(\mu\text{-}\eta^2\text{-}CHO)$		206
$\{[\eta\text{-}C_5Me_3(TMS)_2]TaCl_2\}_2(\mu\text{-}H)(\mu\text{-}\eta^2\text{-}^{13}CHO)$	prep: $\{[\eta\text{-}C_5Me_3(TMS)_2]TaCl_2(H)]\}_2$ + ^{13}CO	206
$(\eta\text{-}C_5Me_4Et)[\eta\text{-}C_5Me_3(TMS)_2]Ta_2Cl_4(\mu\text{-}H)(\mu\text{-}\eta^2\text{-}^{13}CHO)$		206
$[(\eta\text{-}C_5Me_4Et)TaCl_2]_2(\mu\text{-}Cl)(\mu\text{-}\eta^2\text{-}CHO)$	prep: $(\eta\text{-}C_5Me_4Et)_2Ta_2Cl_5H$ + CO	208
$(\eta\text{-}C_5Me_4Et)_2Ta_2Cl_3(H)(CHO)(CH_2TMS)$		206
$[Ta(silox)_2Cl]_2(\mu\text{-}H)(\mu{:}\eta^2,\eta^2\text{-}CHO)$		160,269
Acyl complexes		
$Cp^*Ta[\eta^2\text{-}C(O)TMS]Cl_3$	^{13}C(O)R analogue also reported	86,108
$Cp^*Ta[\eta^2\text{-}C(O)CH_2Bu^t]Cl_3$	^{13}C(O)R analogue also reported; x ray	86,203,433
$(\eta\text{-}C_5Me_4Et)Ta[\eta^2\text{-}C(O)CH_2Bu^t]Cl_3$		203,433
$(\eta\text{-}C_5R_5)Ta[\eta^2\text{-}C(O)CH_2\text{-}4\text{-}C_6H_4Me]Cl_3$	η-C_5R_5 = Cp*, C_5Me_4Et; implicated, not observed	203
$Cp^*Ta[\eta^2\text{-}^{13}C(O)CH_2Bu^t]Cl_3(py)$		86
$Cp^*Ta[\eta^2\text{-}C(O)TMS(py)]Cl_3$	prep: $Cp^*Ta[\eta^2\text{-}C(O)TMS]Cl_3$ + py; x ray	86,434
$Cp^*Ta[\eta^2\text{-}C(O)TMS(L)]Cl_3$	L = py-2,6-Me_2, py-4-Me	86
	L = PMe_3, PEt_3 (x ray), $P(OMe)_3$	86,434
	L = PEt_3; redetermination of space group	435
$Cp^*Ta[\eta^2\text{-}^{13}C(O)TMS(L)]Cl_3$	L = py, py-2,6-Me_2, PMe_3, PEt_3	86
$\{(\eta\text{-}C_5H_4TMS)_2Nb[\eta^2\text{-}C(O)CHR_2]Br\}BF_4$	R_2 = Ph_2 Me_2, MePh, EtPh	436
$\{(\eta\text{-}C_5H_4TMS)_2Nb[\eta^2\text{-}C(O)CHR_2]Cl\}BF_4$	R_2 = Me_2, EtPh, Ph_2	437,438
	R_2 = EtPh; x ray	439
	R_2 = MePh	437
	R_2 = $Me(CH_2Bu^t)$	438
$\{Cp_2Nb[\eta^2\text{-}C(O)CHEtPh]Cl\}BF_4$		437
$\{(\eta\text{-}C_5H_4TMS)_2Nb[\eta^2\text{-}C(O)CHR_2]H\}BF_4$	R_2 = EtPh	439
	R_2 = MePh, EtPh, Me_2, Ph_2	437
	R_2 = $Me(CH_2Bu^t)$, EtPh	438

Table 14 (continued)

Compound	*Comments*	*Ref.*
Iminoacyl complexes		
$Ta[\eta^2\text{-}C(=NBu^t)Me]_3Cl_2$	prep: $TaMe_3Cl_2 + Bu^tNC$	440
$Ta[\eta^2\text{-}C(=NAr')R]_2(OAr')_2R$	R = Me (x ray), CH_2Ph	441–3
$Ta[\eta^2\text{-}C(=NAr)R]_2(OAr')_2R$	R = Me, CH_2Ph	441–3
$Ta[\eta^2\text{-}C(=NAr')CH_2Ph]_2(OAr')_3$		441,442
$Ta[\eta^2\text{-}C(=NBu^t)Ph](\eta^2\text{-}Bu^tN{=}CPh_2)(OAr')_2$	prep: $Ta(OAr')_2Ph_3$ + excess $CNBu^t$	443
$Ta[\eta^2\text{-}C(=NBu^t)Ph]_2[\eta^2\text{-}Bu^tN{=}CPh_2](CNBu^t)$	x ray	428
$CpTa(RC{\equiv}CR)[\eta^2\text{-}C(=NBu^t)Me]Me$	R = Ph, Tol (x ray); prep: $CpTa(RC{\equiv}CR)Me_2 + CNBu^t$	200,416
$(\eta\text{-}C_5H_4TMS)_2Nb[\eta^2\text{-}C(=NPh)CHPh_2]$		444
$(\eta\text{-}C_5H_4TMS)_2Nb[\eta^2\text{-}C(=NR)Me]$	R = Ph, Cy	135
$(\eta\text{-}C_5H_4TMS)_2Nb[\eta^2\text{-}C(=NCy)CH_2TMS]$	prep: $(\eta\text{-}C_5H_4TMS)_2Nb(CNCy)Cl + LiCH_2TMS$	135
$\{(\eta\text{-}C_5H_4TMS)_2Nb[\eta^2\text{-}C(=NPh)CHPh_2]X\}BF_4$	X = Cl, Br; prep: $(\eta\text{-}C_5H_4TMS)_2Nb[\eta^2(N,C)\text{-}PhN{=}C{=}CPh_2]X + HBF_4{\cdot}OEt_2$	445,446
	X = Cl	438
$\{(\eta\text{-}C_5H_4TMS)_2Nb[\eta^2\text{-}C(=NEt)CHPh_2]Cl\}BF_4$		438
$\{(\eta\text{-}C_5H_4TMS)_2Nb[\eta^2\text{-}C(=NPh)CHPh_2]F\}BF_4$	x ray	446
$\{(\eta\text{-}C_5H_4TMS)_2Nb[\eta^2\text{-}C(=NPh)CHEtPh]X\}BF_4$	X = Cl (x ray), Br	445,446
$\{(\eta\text{-}C_5H_4TMS)_2Nb[\eta^2\text{-}C(=NPh)CHMePh_2]X\}BF_4$	X = Cl, Br	446
$\{(\eta\text{-}C_5H_4TMS)_2Nb[\eta^2\text{-}C(=N\text{-}4\text{-}C_6H_4Br)\text{-}CHEtPh]X\}BF_4$	X = Cl (x ray), Br	446
η^2-Imine and η^2-nitrile complexes		
$Cp^*Ta(\eta^2\text{-}RN{=}CH_2)Me_2$	R = Me, Pr^n	199
$Cp^*Ta(\eta^2\text{-}MeN{=}CH_2)(Me)(H)(PMe_3)$	prep: $Cp^*Ta(\eta^2\text{-}MeN{=}CH_2)Me_2 + PMe_3 + H_2$	199
$Cp^*Ta(\eta^2\text{-}MeN{=}CH_2)(Me)(OPr^i)$	prep: $Cp^*Ta(\eta^2\text{-}MeN{=}CH_2)(Me)(H)(PMe_3) + O{=}CMe_2$	199
$Cp_2Ta[\eta^2\text{-}Bu^tCH_2CMe_2N{=}CH(CN)]H$		139
$Cp_2Nb(\eta^2\text{-}NCMe)(SnMe_3)$	two isomers	122
$Cp^*{}_2Ta(\eta^2\text{-}MeN{=}CH_2)H$	kinetic decomposition product of $[Cp^*{}_2Ta(NMe_2)]$	109
$Cp^*{}_2Ta(\eta^2\text{-}BzN{=}CH_2)H$	prep: $Cp^*{}_2Ta(=CH_2)Me + HNMeBz$	140
$Ta[\eta^2\text{-}C(=NBu^t)Ph](\eta^2\text{-}Bu^tN{=}CPh_2)(OAr')_2$		443
CO_2, CS_2, isocyanate, and related complexes		
$Cp'_2Nb(\eta^2\text{-}CO_2)R$	R = CH_2TMS; x ray	221,225,238
	R = R = CH_2Ph (x ray), CH_2Bu^t, CH_2TMS	247
$Cp_2Nb(\eta^2\text{-}CO_2)Cl$		63
$Cp'_2Nb(CH_2TMS)(\mu,\eta^2\text{-}CO_2)ZnCl_2$		221
$Cp'_2Nb(\eta^2(S,C)\text{-}SCO)(CH_2TMS)$	prep: $Cp'_2Nb(CO)(CH_2TMS) + S_8$	248
$Cp_2Nb(\eta^2\text{-}CS_2)R$	R = Me, Bu^n	447
$(\eta\text{-}C_5H_4TMS)_2Nb(\eta^2\text{-}CS_2)Br$	x ray	304
$(\eta\text{-}C_5H_4TMS)_2Nb(OCNPh)Cl$		448
$(\eta\text{-}C_5H_4TMS)_2Nb(\eta^2\text{-}OCNPh)Br$		304
$(\eta\text{-}C_5H_4TMS)_2Nb(\eta^2\text{-}SCNPh)Br$		304
$\{Cp_2Nb[\eta^2\text{-}C(S)SMe]R\}I$	R = Me; prep: $Cp_2Nb(\eta^2\text{-}CS_2)Me + MeI$	447
	R = Bu^n	222,447
$\{Cp_2Nb[\eta^2\text{-}C(S)SR]Bu^n\}I$	R = Et, Pr^i, CH_2CO_2Et, 1/2 CH_2	222
$\{Cp_2Nb[\eta^2\text{-}C(S)SR]Bu^n\}Br$	R = CH_2Ph, $CH_2CH_2CH_2Cl$, $CH_2CH{=}CH_2$; prep: $Cp_2Nb(\eta^2\text{-}CS_2)(Bu^n) + RBr$	222
$\{Cp_2Nb[\eta^2\text{-}C(S)SCH_2CH{=}CH_2]Me\}Br$		447
$Cp_2Nb(Bu^n)(\eta^2\text{-}CS_2)ML_n$	ML_n = $Cr(CO)_5$, $Mo(CO)_5$, $W(CO)_5$ (x ray), $Fe(CO)_4$, $Mn(CO)_2Cp'$, $Nb(Bu^n)Cp_2$	447
Ketene and ketenimine complexes		
$Cp_2Nb[\eta^2(O,C)\text{-}O{=}C{=}CR_2]Cl$	R_2 = EtPh, Me_2, MeEt	437
$(\eta\text{-}C_5H_4TMS)_2Nb[\eta^2(O,C)\text{-}O{=}C{=}CR_2]Cl$	R_2 = Me_2, Ph_2, MePh, EtPh	345,437
	R_2 = EtPh	439
	R_2 = $Me(CH_2Bu^t)$, MeEt	437
$(\eta\text{-}C_5H_4TMS)_2Nb[\eta^2(O,C)\text{-}O{=}C{=}CR_2]Br$	R_2 = Ph_2 (x ray), Me_2, MePh, EtPh	436
$(\eta\text{-}C_5H_4TMS)_2Nb[\eta^2(O,C)\text{-}O{=}C{=}CR_2)]\text{-}(CCO_2Me{=}CHCO_2Me)$	R_2 = MePh, EtPh, Ph_2	437
$(\eta\text{-}C_5H_4TMS)_2Nb[\eta^2(O,C)\text{-}O{=}C{=}CR_2](H)$	R_2 = MePh	345,437
	R_2 = EtPh	439
	R_2 = EtPh, Ph_2, Me_2	250,437
	R_2 = $Me(CH_2Bu^t)$	437
$\{(\eta\text{-}C_5H_4TMS)_2Nb[\eta^2(O,C)\text{-}O{=}C{=}CPh_2]\}^-$	electrochemically generated	436
$(\eta\text{-}C_5H_4TMS)_2Nb[\eta^2(N,C)\text{-}PhN{=}C{=}CPh_2]X$	X = Cl (x ray), Br	444–6
$(\eta\text{-}C_5H_4TMS)_2Nb[\eta^2(N,C)\text{-}PhN{=}C{=}CPh_2]F$		446
$(\eta\text{-}C_5H_4TMS)_2Nb[\eta^2(N,C)\text{-}TolN{=}C{=}CPh_2]X$	X = Cl, Br	444
$(\eta\text{-}C_5H_4TMS)_2Nb[\eta^2(N,C)\text{-}PhN{=}C{=}CMePh]X$	X = Cl, Br; prep: $(\eta\text{-}C_5H_4TMS)_2NbX + PhN{=}C{=}CMePh$	444,446
$(\eta\text{-}C_5H_4TMS)_2Nb[\eta^2(N,C)\text{-}PhN{=}C{=}CEtPh]X$	X = Cl, Br	445,446
$(\eta\text{-}C_5H_4TMS)_2Nb[\eta^2(N,C)\text{-}(4\text{-}C_6H_4Br)N{=}C{=}CPh_2]X$	X = Cl, Br	446
$(\eta\text{-}C_5H_4TMS)_2Nb[\eta^2(N,C)\text{-}PhN{=}C{=}CPh_2]$		444

Table 14 (continued)

Compound	*Comments*	*Ref.*
$Ta[\eta^2(O,C)\text{-}O{=}C{=}CPh_2](silox)_3$	prep: $Ta(silox)_3 + O{=}C{=}CPh_2$	33
$(silox)_3Ta[\eta^2(O,C)\text{-}OC{=}C{=}Ta(silox)_3]$	prep: $(silox)_3Ta{=}C{=}C{=}O + Ta(silox)_3$	33
Ketone, aldehyde, and related complexes		
$(\eta\text{-}C_5H_4TMS)_2Nb(\eta^2\text{-}O{=}CHCHPh_2)$	electrochemically generated	449
$Cp^*Ta(\eta^2\text{-}O{=}CMe_2)(CO)(bipy)$		94
$Cp^*_2Ta(\eta^2\text{-}O{=}CH_2)H$		109,226
$Cp^*_2Ta(\eta^2\text{-}O{=}CHR)Me$	R = H, Ph	228
$Ta(\eta^2\text{-}O{=}CH_2)(silox)_3$		159,160,270
$Ta(\eta^2\text{-}O{=}CHEt)(silox)_3$		160
$[Ta(silox)_2]_2(\mu{:}\eta^1,\eta^1\text{-}CH{=}CHO)(\mu{:}\eta^1,\eta^2\text{-}CH_2O)(\mu\text{-}O)$		160
$[TaH(silox)_2](\mu{:}\eta^2,\eta^2\text{-}CHO)(\mu{:}\eta^1,\eta^2\text{-}CH_2O)\text{-}[Ta(silox)_2]$	$\mu{:}\eta^2,\eta^2\text{-}^{13}CHO$ analogue also prepared	160,269
$[Ta(silox)_2]_2(\mu\text{-}O)_2(\mu\text{-}RCHO)$	R = H, Me	160
$Cp^*_2Ta(\eta^2\text{-}S{=}CH_2)H$		109
$Cp^*_2Ta(\eta^2\text{-}S{=}CHR)H$	R = H, Ph, CH_2Bu^t	140
$Cp^*_2Ta[\eta^2\text{-}S{=}CHCH_2Ph]H$	x ray	140,450
$Cp_2Nb[\eta^2\text{-}S{=}C(C_6H_6Me_4)][SCH{=}C(C_6H_6Me_4)]$	prep: $Cp_2NbH_3 + (C_6H_6Me_4)C{=}S$	451
$Cp_2Nb[\eta^2\text{-}S{=}C{=}CBu^t_2](SCH{=}CBu^t_2)$	x ray	451
$\{Cp_2Nb[\eta^2\text{-}S{=}C(C_6H_6Me_4)]\}I$		451
$Cp_2Nb[\eta^2\text{-}S{=}CH(SR)](Bu^n)$	R = Me, Et, Pr^i, CH_2CO_2Et, CH_2Ph, $CH_2CH_2CH_2Cl$, 1/2 CH_2	222
Miscellaneous		
$[(\eta\text{-}C_5Me_4Et)TaCl_2][\mu\text{-}\eta^1(N),\eta^2(N,C)\text{-}NCHMe](\mu\text{-}Cl)(\mu\text{-}H)[(\eta\text{-}C_5Me_4Et)TaCl]$	nominally $(\eta\text{-}C_5Me_4Et)_2Ta_2Cl_4H(NCHMe)$; x ray; $(\eta\text{-}C_5Me_4Et)_2Ta_2Cl_4D(NCDMe)$ also prepared	452,453
$Cp^*_2Ta_2Cl_4H(NCHMe)$		452
Complexes containing cyclometallated PMe_3		
$Ta(\eta^2\text{-}CHPMe_2)Cl(PMe_3)_4$	prep: $TaCl_5$ + neat PMe_3 + Na	454
$Ta(\eta^2\text{-}CHPMe_2)(\eta^4\text{-}C_4H_6)Cl(PMe_3)_2$	prep: $TaCl_5$ + neat PMe_3 + Na	454
$Ta(\eta^2\text{-}CHPMe_2)(\eta^2\text{-}CH_2PMe_2)(PMe_3)_3$	prep: $TaCl_5$ + neat PMe_3 + Na; x ray	454,455
$Ta(\eta^2\text{-}CHPMe_2)(\eta^2\text{-}CH_2PMe_2)H_2(PMe_3)_2$	prep: $TaCl_5$ + neat PMe_3 + Na	454
$CpTa(\eta^2\text{-}CH_2PMe_2)(H)(PMe_3)_2$	prep: $CpTaCl_4$ + Na + neat PMe_3	316
$Cp^*Ta(\eta^2\text{-}CH_2PMe_2)(H)(dmpe)$		91
$Cp^*Ta(\eta^2\text{-}CHPMe_2)(H)_2(PMe_3)$	nominally "$Cp^*Ta(PMe_3)_2$"; x ray	95–7,316, 320,456,457
$Cp^*Ta(\eta^2\text{-}CHPMe_2)(H)_2(PR_3)$	$PR_3 = PMe_2Ph$, $P(OMe)_3$, PEt_3, $PMePh_2$, $P(OPh)_3$, η^1-dmpe	91
$Cp^*Ta(\eta^2\text{-}CHPMe_2)(H)(X)(PMe_3)$	X = Cl, Br, I	91,95
$Cp^*Ta(\eta^2\text{-}CHPMe_2)(H)(Br)(PMe_3)$	prep: $Cp^*Ta(\eta^2\text{-}CHPMe_2)(H)_2(PMe_3)$ + MeBr; alkene dimerization catalyst	457
$Cp^*Ta(\eta^2\text{-}CHPMe_2)X_2(PMe_3)$	X = Cl, Br, I	91
$Cp^*Ta(\eta^2\text{-}CHPMe_2)(CH_2CH_2Bu^t)(H)$	prep: $Cp^*Ta(\eta^2\text{-}CHPMe_2)(H)_2(PMe_3) + CH_2{=}CHBu^t$	456
$Cp^*Ta(\eta^2\text{-}CHPMe_2)(CH_2CH_2Bu^t)_2$	x ray	456
$Cp^*Ta(\eta^2\text{-}CHPMe_2)(CH_2CH_2R)Br$	R = H, Me, Bu^t, Ph	457

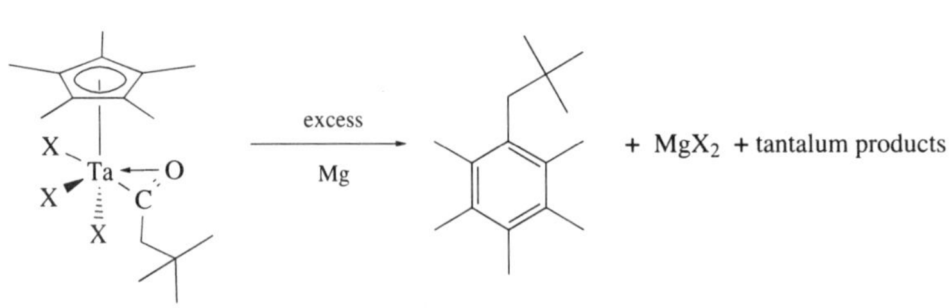

Scheme 13

2.5.3.2 *Iminoacyl, η²-imine, and η²-nitrile complexes*

Iminoacyl derivatives of niobium and tantalum have also been reported; for example, when toluene solutions of $CpTa(RC{\equiv}CR)Me_2$ are reacted with the isocyanide $CNBu^t$, the iminoacyl complexes $CpTa(RC{\equiv}CR)[\eta^2\text{-}C({=}NBu^t)Me]Me$ (R = Ph, Tol) are formed.[200,416] Thermolysis of these compounds in toluene results in coupling of the iminoacyl and alkyne ligands to form the metallacyclopentatriene-type derivatives $CpTa\overline{({=}CRCR{=}CMeN}Bu^t)Me$ (R = Ph, Tol).

An interesting sequence of reactions has been uncovered by Rothwell and co-workers in which both η^2-iminoacyl and then η^2-imine ligands are synthesized by successive alkyl ligand migratory insertions. For example, the bis(iminoacyl) complex $Ta[\eta^2\text{-}C(=NAr')CH_2Ph]_2(OAr')_3$ (Ar' = 2,6-$C_6H_3Me_2$) is formed upon reacting CNAr' with $Ta(CH_2Ph)_2(OAr')_3$.[441,442] $Ta[\eta^2\text{-}C(=NR)Me]_2(OAr')_2Me$ and $Ta[\eta^2\text{-}C(=NR)CH_2Ph]_2(OAr')_2(CH_2Ph)$ are prepared by the analogous reactions of $TaMe_3(OAr')_2$ or $Ta(CH_2Ph)_3(OAr')_2$, respectively, with CNR (R = Ar or Ar').[443] However, excess $CNBu^t$ reacts with $Ta(OAr')_2Ph_3$ to give a mixed η^2-iminoacyl–η^2-imine derivative, $Ta[\eta^2\text{-}C(=NBu^t)Ph](\eta^2\text{-}Bu^tN{=}CPh_2)(OAr')_2$, resulting from a "double migration" of alkyl to isocyanide. Thermolyzing $Ta[\eta^2\text{-}C(=NAr')Me]_2(OAr')_2Me$ or $Ta[\eta^2\text{-}C(=NBu^t)Ph](\eta^2\text{-}Ph_2C{=}NBu^t)(OAr')_2$ results in isolation of the imido–amido complexes $Ta(=NAr')[N(Ar')CMe{=}CMe_2](OAr')_2$ and $Ta(=NBu^t)[N(Bu^t)CPh{=}CPh_2](OAr')_2$, presumably via iminoacyl–imine coupling followed by an electrocyclic rearrangement. Thermolyzing the benzyl derivatives results in another type of rearrangement to give $Ta(=NR)[N(R)CH_2CH_2Ph](OAr')_2$ (R = Ar or Ar'). These reactions are also thought to proceed via η^2-imine η^2-iminoacyl complexes, followed by sequential transfer of benzyl α-hydrogens of the η^2-imine to the η^2-iminoacyl (Scheme 14).[443] Table 14 includes the η^2-nitrile complex $Cp_2Nb(\eta^2\text{-}NCMe)(SnMe_3)$ (two isomers) formed from photolysis of $Cp_2Nb(\eta^2\text{-}CH_2{=}CHPh)(SnMe_3)$ in the presence of N≡CMe.[122]

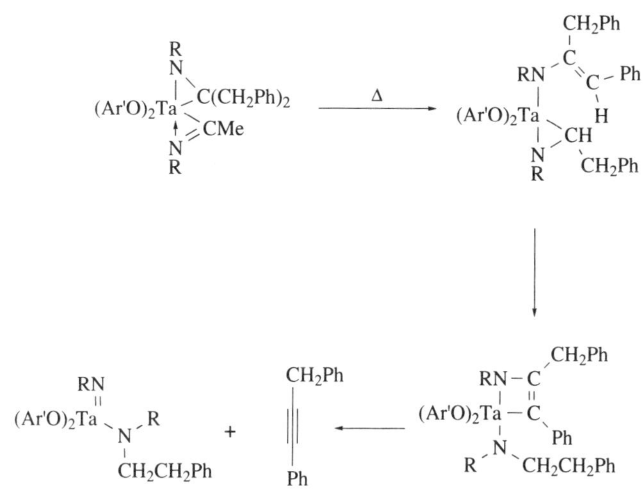

Scheme 14

2.5.3.3 CO_2, isocyanate, and related complexes

Several η^2-isocyanate compounds have been reported in which the ligand is characteristically bound η^2(*O,C*)-O=C=NR to the d^2 niobium fragment $(\eta\text{-}C_5R_5)_2NbX$. For example, reaction of the reduced species $(\eta\text{-}C_5H_4TMS)_2NbCl$ with PhNCO affords the adduct $(\eta\text{-}C_5H_4TMS)_2NbCl(\eta^2\text{-}OCNPh)$, which, upon thermolysis, forms the imido complex $(\eta\text{-}C_5H_4TMS)_2Nb(=NPh)Cl$.[448] Several examples of CO_2 and CS_2 complexes have been uncovered, with perhaps the most intriguing preparation of such compounds being the clean reaction between $Cp'_2Nb(CO)R$ (R = CH_2Ph, CH_2Bu^t, CH_2TMS) and O_2, which provides $Cp'_2Nb(\eta^2\text{-}CO_2)R$ in good yield.[247] Isotopic labeling studies indicate that the new oxygen atom is derived from O_2 and that this atom is equally distributed between coordinated and uncoordinated oxygen sites of the η^2(*C,O*)-CO_2 group. Moreover, incorporation occurs during ligand formation, since no assimilation of ^{16}O is noted in the presence of $C^{16}O_2$.[247]

2.5.3.4 Ketene and ketenimine complexes

The η^2-ketene and -ketenimine compounds which are known typically have the η^2 ligand π bound to a d^2 $(\eta\text{-}C_5H_4TMS)_2NbX$ (X = Cl, Br) moiety. The d^2 species $(\eta\text{-}C_5H_4TMS)_2NbX$ (X = Cl or Br) can be generated by reduction of the d^1 dihalide compounds, and coordination of $O{=}C{=}CR_2$ or $RN{=}C{=}CR_2$ occurs exclusively in the η^2(*O,C*) and η^2(*N,C*) mode, reflecting the "oxophilicity" of the metal. For example, $(\eta\text{-}C_5H_4TMS)_2NbCl$ coordinates PhN=C=CMePh to form only $(\eta\text{-}C_5H_4TMS)_2Nb[\eta^2(N,C)\text{-}PhN{=}C{=}CMePh]Cl$.[444,446]

2.5.3.5 Ketone, aldehyde, and related complexes

A handful of η^2-ketone and -aldehyde compounds are known; these can be prepared by two successive migratory insertions of CO into M–R or M–H bonds, by coordination of the free ligand to a d^2 fragment, or by β-elimination from an alkoxide complex. For example, carbonylation of $Cp^*TaMe_2(bipy)$ affords $Cp^*Ta(\eta^2\text{-}O{=}CMe_2)(CO)(bipy)$[94] while nascent $[Cp^*_2TaMe]$ (in equilibrium with $Cp^*_2Ta({=}CH_2)H$) reacts with MeOH to afford the hydrido formaldehyde complex $Cp_2^*Ta(\eta^2\text{-}O{=}CH_2)H$.[109,226] Similarly, treatment of the $[Cp^*_2Ta({=}C{=}CH_2)H \rightleftharpoons Cp^*_2TaCH{=}CH_2]$ equilibrium mixture with thiols $HSCH_2R$ affords the η^2-thioaldehyde compounds $Cp^*_2Ta(\eta^2\text{-}S{=}CHR)H$ (for R = H, Ph, Bz, or CH_2Bu^t) which constitute the kinetic product of this reaction.[140] Upon heating these compounds, equilibrium concentrations of $[Cp^*_2TaSCH_2R]$ are generated, which undergo a rate-determining α-alkyl elimination to afford $Cp^*_2Ta({=}S)CH_2R$. The complexes $Ta(\eta^2\text{-}O{=}CH_2)(silox)_3$ and $Ta(\eta^2\text{-}O{=}CHEt)$-$(silox)_3$ are available either from carbonylation of $TaH_2(silox)_3$ or $Ta(H)(Et)(silox)_3$ or from reacting $Ta(silox)_3$ with $[H_2CO]_n$ or EtCHO, respectively.[160]

2.5.3.6 Complexes containing cyclometallated PMe_3

Green, Gibson, and co-workers have prepared and examined a series of complexes in which the η^2 ligand constitutes a cyclometallated PMe_3 ligand, either $\eta^2\text{-}CH_2PMe_2$ (formally alkyl carbon) or η^2-$CHPMe_2$ (formally alkylidene carbon). For example, Green *et al.*[454] reported a series of lower-valent tantalum compounds from reducing $TaCl_5$ in liquid PMe_3 using sodium sand. This reduction produces the following compounds in sequence: $TaCl_5(PMe_3)_2$, $TaCl_4(PMe_3)_3$, $TaCl_3(PMe_3)_3$, $TaCl_2(PMe_3)_4$, $Ta(\eta^2\text{-}CHPMe_2)Cl(PMe_3)_4$, and finally $Ta(\eta^2\text{-}CHPMe_2)(\eta^2\text{-}CH_2PMe_2)(PMe_3)_3$. Gibson *et al.*[316] performed similar reductions of $CpTaCl_4$ and Cp^*TaCl_4. Thus, reducing $CpTaCl_4$ with sodium sand in neat PMe_3 affords a mixture from which $Cp_2Ta(H)PMe_3$ is isolated and $CpTa(\eta^2\text{-}CH_2PMe_2)(H)(PMe_3)_2$ is identified. Reducing Cp^*TaCl_4 under similar conditions gives $Cp^*Ta(\eta^2\text{-}CHPMe_2)(H)_2(PMe_3)$. The nominal bis(phosphine) complex "$Cp^*Ta(PMe_3)_2$" has been shown to exist in the bis(cyclometallated) form $Cp^*Ta(\eta^2\text{-}CHPMe_2)(H)_2(PMe_3)$ in which the metal is d^0.[96,316,320]

2.5.4 η^2-Arene, η^2-Pyridine, and Related Complexes

The discovery that certain d^2 tantalum species can bind aromatic organic molecules in an η^2 fashion, and thereby interrupt their aromaticity, constitutes one of the most significant advances in group 5 chemistry. Table 15 outlines the η^2-arene, η^2-pyridine, and related complexes of tantalum.

Table 15 η^2-Arene, η^2-pyridine, and related complexes of niobium and tantalum.

Compound	*Comments*	*Ref.*
$\{\mu\text{-}\eta^2(1,2):\eta^2(4,5)\text{-}C_6H_6\}[Ta(silox)_3]_2$	x ray	458
$[\eta^2(N,C)\text{-}NC_5H_5]Ta(silox)_3$	prep: $Ta(silox)_3$ + py; x ray	161,458
$[\eta^2(N,C)$-heterocycle$]Ta(silox)_3$	heterocycle = NC_5D_5, 6-NC_5H_4Me, 2,6-$NC_5H_3Me_2$, 1,3-$N_2C_4H_4$	161
$[\eta^2(N,N)$-heterocycle$]Ta(silox)_3$	heterocycle = 1,2-$N_2C_4H_4$	161
$[\eta^2(N,C)\text{-}2,4,6\text{-}NC_5H_2Bu^t_3]Ta(OAr)_2Cl$	prep: $Ta(CBu^t{=}CHCBu^t{=}CH)(OAr)_2Cl + N{\equiv}CBu^t$; x ray	408,459
$[\eta^2(N,C)$-heterocycle$]Ta(OAr)_3$	heterocycle = NC_9H_7 (quinoline), $NC_{10}H_9$ (6-methylquinoline); prep: $[\eta^1(N)$-heterocycle$]Ta(OAr)_3Cl_2$ + NaHg	460
$[\eta^2(N,C)\text{-}NC_{10}H_9]Ta(OAr)_3(PMe_3)$	$NC_{10}H_9$ = 6-methylquinoline; x ray	460
$[\eta^2(N,C)$-heterocycle$]Ta(OAr)_2Cl(OEt_2)$	heterocycle = NC_9H_7 (quinoline), $NC_{10}H_9$ (6-methylquinoline; x ray)	461
$[\eta^2(N,C)\text{-}2,4,6\text{-}NC_5H_2Bu^t_3]Ta(OAr)_2R$	R = Me, Et, Pr^i, Ph	461

The first compounds of this type were discovered by Wolczanski and co-workers[161,458] upon reacting d^2 $Ta(silox)_3$ with pyridine, which afforded the η^2-pyridine complex $[\eta^2(N,C)\text{-}NC_5H_5]Ta(silox)_3$ (**29**). Diagnostic of the η^2-binding mode is the upfield shift of the ^1H and ^{13}C NMR resonances for the ring CH bound directly to the metal. The structural and spectroscopic features of the $\eta^2(N,C)\text{-}NC_5H_5$ ligand are more indicative of a formal Ta^V metallacycle rather that a Ta^{III} π complex, which results from the substantial π back-bonding into a portion of the aromatic ring. Additionally, when $Ta(silox)_3$ is allowed to stand in benzene for extended periods, brown crystals of $\{\mu\text{-}\eta^2(1,2):\eta^2(4,5)\text{-}C_6H_6\}[Ta(silox)_3]_2$ (**30**) precipitate in low yield, with competing cyclometallation of a silox ligand to form $(silox)_2(H)Ta[OSiBu^t_2CMe_2CH_2]$.

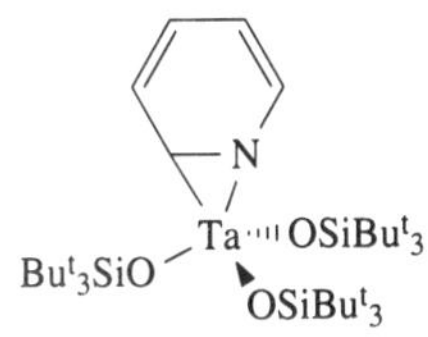

(29)

(30)

Treating solutions of $Ta(silox)_3$ with a range of nitrogen heterocycles is shown to afford $\eta^2(N,C)$- or $\eta^2(N,N)$- but not $\eta^2(C,C)$-heterocycles.[161] Thus, $Ta(silox)_3$ reacts with pyridine, 2-picoline (6-NC_5H_4Me), 2,6-lutidine (2,6-$NC_5H_3Me_2$), pyridazine (1,2-$N_2C_4H_4$), and pyrimidine (1,3-$N_2C_4H_4$) to form $[\eta^2(N,C)$-$NC_5H_5]Ta(silox)_3$, $[\eta^2(N,C)$-6-$NC_5H_4Me]Ta(silox)_3$, $[\eta^2(N,C)$-2,6-$NC_5H_3Me_2]Ta(silox)_3$, $[\eta^2(N,N)$-1,2-$N_2C_4H_4]Ta(silox)_3$, and $[\eta^2(N,C)$-1,3-$N_2C_4H_4]Ta(silox)_3$, respectively. Prior to the formation of the η^2-2,6-lutidine complex $[\eta^2(N,C)$-2,6-$NC_5H_3Me_2]Ta(silox)_3$, the pyridyl hydride complex $Ta(C_5H_2Me_2N)(H)(silox)_3$ forms and equilibrates with the η^2 compound. This observation, along with MO studies, is consistent with the η^2-heterocycle adducts forming by nucleophilic attack of $Ta(silox)_3$ (using its filled d_{z^2}) on the heterocycle π^*, which is predominantly C=N π^* in nature.

In studies related to the hydrodenitrogenation catalysis (see Section 2.8.6), Wigley and co-workers have demonstrated an $\eta^1(N)$ to $\eta^2(N,C)$ bond mode rearrangement upon reducing d^0 quinoline adducts to the d^2 oxidation state. Thus, the reaction of $Ta(OAr)_3Cl_2(OEt_2)$ with quinoline (NC_9H_7) and 6-methylquinoline ($NC_{10}H_9$) in pentane affords the adducts $[\eta^1(N)$-$NC_9H_7]Ta(OAr)_3Cl_2$ and $[\eta^1(N)$-$NC_{10}H_9]Ta(OAr)_3Cl_2$.[460] Upon reduction of these compounds by two electrons, an $\eta^1(N) \rightarrow \eta^2(N,C)$ bonding mode rearrangement is effected and $[\eta^2(N,C)$-$NC_9H_7]Ta(OAr)_3$ and $[\eta^2(N,C)$-$NC_{10}H_9]Ta(OAr)_3$ are isolated.[460] The related η^2-pyridine complex $[\eta^2(N,C)$-2,4,6-$NC_5H_2Bu^t_3]Ta(OAr)_2Cl$ has been prepared from cycloaddition reactions.[408,459] Thus, the α,β'-disubstituted metallacyclopentadiene complex $Ta(CBu^t{=}CHCBu^t{=}CH)(OAr)_2Cl$ forms an isolable adduct with $N{\equiv}CBu^t$, $Ta(CBu^t{=}CHCBu^t{=}CH)(OAr)_2Cl(N{\equiv}CBu^t)$, that undergoes a cycloaddition reaction upon warming to room temperature to afford $[\eta^2(N,C)$-2,4,6-$NC_5H_2Bu^t_3]Ta(OAr)_2Cl$.

2.6 METALLACYCLIC COMPLEXES

Metallacyclic complexes of niobium and tantalum have come to occupy a central role in understanding several catalytic and stoichiometric transformations of organic molecules. Their intermediacy has been demonstrated, for example, in the dimerization and metathesis of alkenes and in their ring-opening metathesis polymerization (ROMP). Selective carbon–carbon bond-forming reactions have been developed using these metals in reactions which proceed via metallacyclic intermediates. These processes are discussed in Section 2.8 while the present section considers their formation and reactions. To be included in this section, the metallacyclic ligand must have at least one carbon atom bound to the metal (Table 16).

2.6.1 Metallacyclobutanes

The first isolable metallacyclobutanes of tantalum or niobium were reported only in 1986, despite the overwhelming evidence for their formation from alkylidene complexes and alkenes.[1] Thus, although metallacyclobutanes are of considerable importance in alkene metathesis, most tantalum and niobium alkylidene complexes do not afford metathesis products upon reaction with alkenes. Just as their instability is noted (Section 2.6.4) in the "ring contraction" decomposition of tantallacyclopentanes, when tantallacyclobutanes are generated from an alkylidene and an alkene, they are typically unstable with respect to the β-hydrogen shift shown in Scheme 15. Thus, the first-formed products from the reaction of $CpTa(=CHBu^t)Cl_2$ and terminal alkenes $CH_2{=}CHR$ are the metallacycles $CpCl_2$-$Ta(CHBu^tCH_2CHR)$ and its regioisomer $CpCl_2Ta(CHBu^tCHRCH_2)$, which are unstable with respect to β-hydrogen shift to generate the alkenes shown.

When the ring cleavage step in Scheme 15 is slowed relative to metathesis, discrete metallacyclobutanes can been obtained. Thus, metallacyclobutanes are isolated from the metallacyclization reactions of the alkylidene complex $Ta(=CHBu^t)(OAr)_3(THF)$ and alkenes.[368,369] $Ta(=CHBu^t)(OAr)_3(THF)$ and ethene react to form colorless crystals of the stable metallacyclobutane $Ta(CH_2CH_2CH_2)(OAr)_3$, along with by-product $CH_2{=}CHBu^t$. $Ta(=CHBu^t)(OAr)_3(THF)$ reacts with one equivalent of styrene to provide orange crystals of $Ta(CHPhCHBu^tCH_2)(OAr)_3$. This latter complex is

Table 16 Metallacyclic complexes of niobium and tantalum.

Compound	*Comments*	*Ref.*
Metallacyclobutanes		
Ta($CH_2CH_2CH_2$)$(OAr)_3$	prep: Ta(=CHBu^t)$(OAr)_3$(THF) + CH_2=CH_2	368,369
Ta($CH_2CH_2CH_2$)$(OAr)_3$(py)		368,369
Ta(CHPhCHBu^{t}CH$_2$)$(OAr)_3$	x ray	368,369
Ta(CH-TMS-CH-TMS-CH$_2$)$(OAr)_3$		369
Ta[CH(C_5H_8)CHCHBu^t]$(OAr)_3$	x ray	369,370
Ta[CH(C_5H_8)CHCH(C_5H_8)CH=CHBu^t]$(OAr)_3$		369
Ta(CHPhCHBu^{t}CH$_2$)(OAr')$_3$		369
Metallacyclobutenes		
Ta(CR=CRCHBu^t)$(OAr)_3$	R = Me, Ph, TMS	372
Ta[C(CH=CHBu^t)=CMeCH$_2$]$(OAr)_3$(py)$_n$	n = 0, 1	372
Ta(C-TMS=C-TMS-CH$_2$)$(OAr)_3$(py)		372
Metallacyclobutadienes		
Ta[=CBu^{t}C-TMS=C-TMS](TMS-C≡C-TMS)Cl_2	possible formulation; prep: [Cl_2(DME)Ta(μ-CBu^t)]$_2$[Zn(μ-Cl)$_2$] + TMS-C≡C-TMS; see text	379
Metallacyclopentanes		
CpTa($CH_2CH_2CH_2CH_2$)Cl_2	use as catalyst precursor	360
CpTa[CH$_2$(C_6H_{10})CH$_2$]Cl_2	from 1,7-octadiene; *cis–trans* mixture	385
(η-C_5Me_4R)Ta($CH_2CH_2CH_2CH_2$)Cl_2	R = Me; x ray	462
	R = Me, Et	452
	R = Me; structural and MO analysis	44
Cp*Ta[CH$_2$(C_5H_8)CH$_2$]Cl_2	*cis*-isomer	385,462
	structural and MO analysis	44
Cp*Ta[CH$_2$(C_6H_{10})CH$_2$]Cl_2	from 1,7-octadiene; *cis–trans* mixture	385
Cp*Ta[CH$_2$(C_7H_{12})CH$_2$]Cl_2	*trans*-isomer	385
Cp*Ta(CH$_2$CHMeCHMeCH$_2$)Cl_2	β,β'-isomer	204,360
Cp*Ta(CHMeCH$_2$CHMeCH$_2$)Cl_2	α,β'-isomer	360
Cp*Ta(CH$_2$CHRCHRCH$_2$)Cl_2	R = H, Me, Pr^n, Bu^i, CH$_2$Bu^t	384
Cp*Ta(CH$_2$CHMeCHMeCH$_2$)(OMe)Cl	use as catalyst precursor	360
Cp*Ta($CH_2CH_2CH_2CH_2$)(η-C_4H_6)		94
Ta($CH_2CH_2CH_2CH_2$)$(OAr)_2$Cl		463
Metallacyclopentenes		
Cp*Ta(CH=CHCH$_2$CH$_2$)Cl_2	prep: Cp*Ta(HC≡CH)Cl_2 + CH_2=CH_2	418
Ta[CH$_2$C(=CMeCH$_2$Bu^t)CMe=CMe]$(OAr)_3$	x ray	372
Metallacyclopentadienes		
Ta(CPh=CPhCR=CR)$(OAr)_3$	R = Me, Et	162,407
Ta(CPh=CPhCH=CR)$(OAr)_3$	R = Bu^t, TMS, Ph	162,407
Ta(CEt=CEtCEt=CEt)$(OAr)_3$	x ray	162,407
Ta(CR=CHCR=CH)$(OAr)_3$	R = TMS, Bu^t	162
Ta(CMe=C-TMS-CMe=CMe)$(OAr)_3$		162
Ta(C-TMS=CMeCPh=CH)$(OAr)_3$		162
Ta(CBu^t=CHC-TMS=CMe)$(OAr)_3$		162
(ArO)$_3$Ta(CPh=CPhCHC)μ-(CH$_4$)$_4$-(C=CHCPh=CPh)Ta$(OAr)_3$	prep: Ta(PhC≡CPh)$(OAr)_3$ + 1,7-octadiyne	162
Ta(CBu^t=CHCH=CBu^t)$(OAr)_2$Cl	α,α,'-isomer; Ta(CBu^t=CDCD=CBu^t) analogue prepared	408,459
Ta(CBu^t=CHCBu^t=CH)$(OAr)_2$Cl	α,β'-isomer; Ta(CBu^t=CDCBu^t=CD) analogue prepared	408,459
Ta(CBu^t=CHCBu^t=CH)$(OAr)_2$Cl(N≡CBu^t)		408,459
Ta(CBu^t=CHCBu^t=CH)$(OAr)_2$Cl(THF)		408
Ta(CBu^t=CHCH=CBu^t)(=NAr)Cl(py)$_2$		411
Cp$_2$Nb(CCF$_3$=CCF$_3$CCF$_3$=CCF$_3$)	x ray	236
Cp$_2$Nb[CEtCEtC(CO$_2$Me)CH(CO$_2$Me)]	alkylidene–alkyl metallacyde	421
Cyclometallated ligands		
Ta[O-2,6-C_6H_3Bu^t(CMe$_2$CH$_2$)](OAr'')Cl_2	prep: Ta(OAr'')$_2$Cl$_3$ + 1 PhLi	163
Ta[O-2,6-C_6H_2Bu^t(CMe$_2$CH$_2$)-4-X](O-2,6-C_6H_2Bu$^t_2$-4-X)Cl_2	X = Me, OMe	163
Ta[O-2,6-C_6H_3Bu^t(CMe$_2$CH$_2$)]$_2$Cl		163
Ta[O-2,6-C_6H_2Bu^t(CMe$_2$CH$_2$)-4-Me]$_2$Cl	X = Me, OMe	163
Nb[O-2,6-C_6H_3Ph(C_6H_4)](O-2,6-C_6H_3Ph$_2$)$_2$Cl	x ray	170
M[O-2,6-C_6H_3Ph(C_6H_4)]$_2$(O-2,6-C_6H_3Ph$_2$)	M = Nb, Ta	168
	M = Ta	169

Table 16 (continued)

Compound	*Comments*	*Ref.*
$Nb[O\text{-}2,6\text{-}C_6H_3Ph(C_6H_4)](H)(O\text{-}2,6\text{-}C_6H_3Ph_2)_2$		271
$Ta(OSiBu^t{}_2CMe_2CH_2)(H)(silox)_2$		160,161, 270,458
$[(\eta^7\text{-}C_5(Bu^t)(CH_2Bu^t)_2(CH_2CMe_2CH_2)_2]TaCl_2$	x ray	379
$[(\eta^6\text{-}C_5(Bu^t)(CH_2Bu^t)_3(CH_2CMe_2CH_2)]TaCl_3$		379
Heterocyclic metallacycles		
$CpTa(=CRCR=CMeNBu^t)Me$	R = Ph (x ray), Tol; prep: thermolysis of $CpTa(RC{\equiv}CR)(\eta^2\text{-}MeC{=}NBu^t)Me$	200,416
$CpTa(=CRCR=CMeO)Me$	R = Ph, Tol; prep: $CpTa(RC{\equiv}CR)Me_2$ + CO	416
$Ta(OCHPhCH_2CH_2CH_2)(OAr)_3(py)$		368,369
$Ta(OCMe_2CH_2CH_2CH_2)(OAr)_3$	prep: $Ta(CH_2CH_2CH_2)(OAr)_3 + O{=}CMe_2$	368,369
$Ta(OCHBu^tCH_2CH_2CH_2)(OAr)_3$		369
$Ta[CPh=CPhC(=CH^*{}_2)NH^*](OAr)_3$	H* = H, D	409
$Ta[CPh=CPhC(=CHMe)NH](OAr)_3$	prep: $Ta(PhC{\equiv}CPh)(OAr)_3 + N{\equiv}CEt$	409
$Cp^*{}_2Ta(OCH_2CH_2)Me$	metallaoxetane	228
$Cp^*{}_2Ta[OCH(4\text{-}C_6H_4X)CH_2)Me$	X = H (x ray), Cl, CF_3, CN, NO_2	228
$CpTa(OCPr^i{}_2CH_2CH=CHCH_2)(\eta^4\text{-}2,3\text{-}Me_2C_4H_4)$		464
$Cp(OCPr^i{}_2CH_2CH=CHCH_2)Ta(OCPr^i{}_2CH_2\text{-}CMe=CMeCH_2)$	two metallacyclic ligands	464
$CpTa[CHBu^tBRNBu^t]Cl_2$	R = Et, Pr, Bu; prep: $CpTa(=CHBu^t)Cl_2 + RB{=}NBu^t$	352
$CpTa[CHBu^tBRNR]Cl_2$	R = Ph, Pr^i	353
$CpTa[=CBu^tBRNBu^t]Cl$	R = Et, Pr, Bu; prep: thermolysis of $CpTa[CHBu^tB(R)NBu^t]Cl$	352
$CpTa[=CBu^tBRNR]Cl$	R = Ph (x ray), Pr^i (x ray)	353
$Ta[NBu^t{=}BRCHBu^tBR{=}NBu^t]Cl_3$	R = Et, Pr, Bu; prep: $Ta(=CHBu^t)Cl_3(THF)_2$ + 2 $RB{=}NBu^t$ R = Ph, Pr^i	352 353
$CpTa[CHBu^tBNBu^t(TMS)NBu^t]Cl_2$		352
$Ta(=NCBu^t{=}CHCBu^t{=}CHCHBu^t)(OAr)_2$	metallacyclic imide	460
$Cp^*Ta(CH_2PPh_2CH_2)Cl_3$	x ray; zwitterionic metallacycle	201,207
$Cp^*Ta(CH_2PMe_2CH_2)Cl_3$	in equilibrium with dimer $[Cp^*TaCl_3(\mu\text{-}CH_2PMe_2CH_2)]_2$	207
$Cp^*Ta(CH_2PMePhCH_2)Cl_3$	in equilibrium with dimer $[Cp^*TaCl_3(\mu\text{-}CH_2PMePhCH_2)]_2$	207
$Cp_2Nb[OC(=CPh_2)OC(=CPh_2)]Cl$		437
Miscellaneous		
$(C_{12}H_8N)_2Ta(C\text{-}TMS)Ta(C\text{-}TMS\text{-}CEtCEt)(NC_{12}H_8)_2$	x ray; $[NC_{12}H_8]^-$ = carbazole amide	377
$(C_{12}H_8N)_2Nb(C\text{-}TMS)Nb(C\text{-}TMS\text{-}CEtCEt)(NC_{12}H_8)_2$	prep: $[Nb(\mu\text{-}C\text{-}TMS)(NC_{12}H_8)_2]_2 + EtC{\equiv}CEt$	377
$(C_{12}H_8N)_2M(C\text{-}TMS)M(C\text{-}TMS\text{-}CHC\text{-}TMS)(NC_{12}H_8)_2$		377

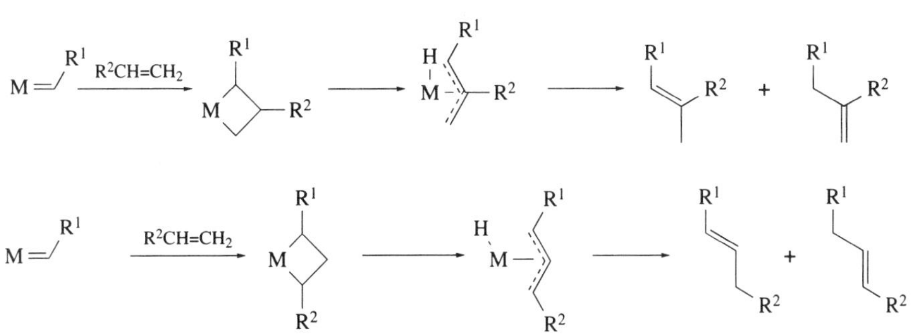

Scheme 15

presumed to arise when one of the two possible initial metallacycles breaks apart and the resulting alkene readds (Scheme 16).

2.6.2 Metallacyclobutenes

Isolable metallacyclobutenes of niobium and tantalum are rare, but can be formed from cyclization of an appropriate alkylidene complex and an alkyne. Thus, $Ta(=CHBu^t)(OAr)_3(THF)$ reacts with $RC{\equiv}CR$ (R = Me, Ph, or TMS) to yield the corresponding, THF-free, metallacyclobutene complexes

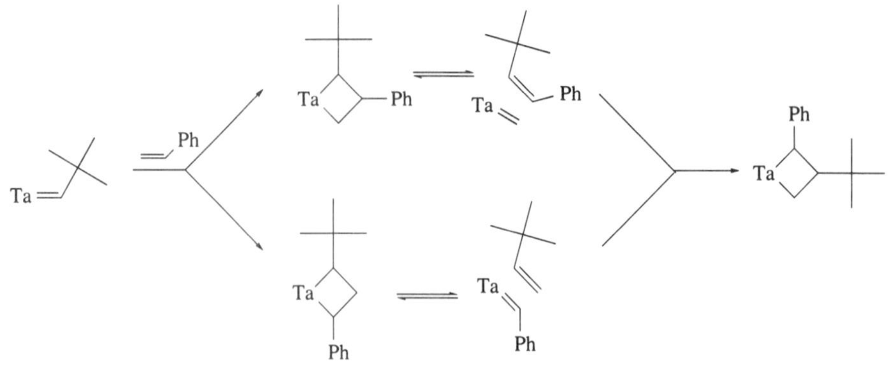

Scheme 16

Ta(CR=CRCHBu^t)(OAr)$_3$.[372] When 2-methyl-1-buten-3-yne (HC≡CC(=CH$_2$)Me) is reacted with Ta(=CHBu^t)(OAr)$_3$(THF), the first-formed metallacycle rearranges and a different metallacyclobutene, Ta[CH$_2$CMe=C(CH=CHBu^t)](OAr)$_3$, is isolated. Pyridine forms a simple adduct with Ta(C-TMS=C-TMS-CHBu^t)(OAr)$_3$ and with Ta[CH$_2$CMe=C(CH=CHBu^t)](OAr)$_3$; however, methyl- or phenyl-substituted metallacyclobutenes Ta(CR=CRCHBu^t)(OAr)$_3$ (R = Me or Ph) are decomposed in the presence of pyridine by an electrocyclic process to provide the vinylalkylidene complexes Ta(=CRCR=CHBu^t)(OAr)$_3$(py).[372] Finally, we note a computational study on metallacyclobutene vs. η^3-vinylalkylidene structures, including an examination of models of these group 5 compounds.[465]

2.6.3 Metallacyclobutadienes

Although many metallacyclobutadienes are known for molybdenum and tungsten, these metallacycles are rare for the group 5 metals. One metallacyclobutadiene has probably been isolated in an unusual cycloaddition assembly of an η-C$_5$R$_5$ ligand from a tantalum alkylidyne complex and two molecules of an alkyne; this process is also described briefly in Section 2.7.3.1.[379] Thus, [Cl$_2$(DME)Ta(μ-CBu^t)]$_2$[Zn(μ-Cl)$_2$][354] reacts with alkynes RC≡CR in inert solvents to give the alkyne adducts (η-C$_5$R$_4$Bu^t)Ta(RC≡CR)Cl$_2$ and in chlorinated solvents to afford the tetrachlorides (η-C$_5$R$_4$Bu^t)TaCl$_4$.[379] Evidence that these reactions proceed via the metallacyclobutadiene is suggested by the reaction of the bulky alkyne TMS-C≡C-TMS with [Cl$_2$(DME)Ta(μ-CBu^t)]$_2$[Zn(μ-Cl)$_2$] which forms an impure product proposed to be the metallacyclic compound Ta[CBu^{t}C-TMS-C-TMS](TMS-C≡C-TMS)Cl$_2$.[379]

2.6.4 Metallacyclopentanes

Compared with the other metallacycles outlined in this section, very few new metallacyclopentanes of niobium and tantalum have been described since *COMC-I* was published; many of the papers on these species were published around 1980.[172,384,466] Metallacyclopentanes are, however, of considerable importance as intermediates in catalytic alkene dimerization (Section 2.8.4). Perhaps the most detailed analysis of such processes came from examining a series of metallacyclopentanes of the type Cp*Cl$_2$Ta(CH$_2$CHRCHRCH$_2$). The *trans*-β,β'-disubstituted metallacyclopentanes Cp*Cl$_2$-Ta(CH$_2$CHRCHRCH$_2$) (R = H, Me, Pr^n, or CH$_2$Bu^t) are prepared from the reactions of the alkene complexes Cp*Ta(CH$_2$=CHR)Cl$_2$ with another equivalent of alkene; these metallacycles are shown to be in equilibrium with the alkene complex Cp*Ta(CH$_2$=CHR)Cl$_2$ and free alkene CH$_2$=CHR (Equation (16)).[360,384] This equilibrium lies far to the right (favoring the metallacycle) for the unsubstituted (ethene) metallacycle and is characterized by $K_{eq} \approx 1.5$ for the R = CH$_2$Bu^t metallacycle.[466] In addition, alkene substitution reactions are found to proceed largely via intermediate metallacyclopentanes since interconversions between alkene complex and metallacycles are facile at low temperatures (Equation (17)).

Another pathway exists by which tantallacyclopentanes decompose to an alkene complex (distinctively different from the retrocyclization described above) which is central to the mechanism of alkene dimerization. At higher temperatures, metallacycles Cp*Cl$_2$Ta(CH$_2$CHRCHRCH$_2$) decompose thermally to afford 0.5 equivalents of the alkene dimer along with alkene complex. Both the α,β'-

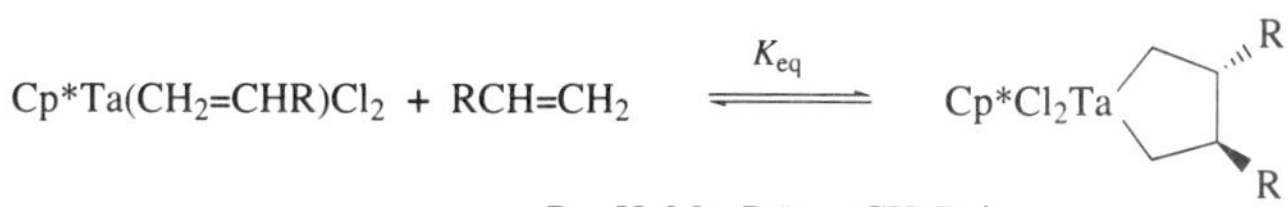

(16)

R = H, Me, Pr^n, or CH_2Bu^t

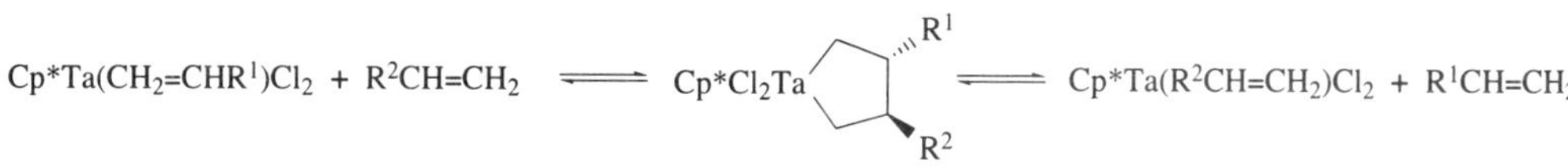

(17)

disubstituted isomer $Cp^*Ta(CHMeCH_2CHMeCH_2)Cl_2$ and the β,β'-disubstituted isomer $Cp^*Ta(CH_2CHMeCHMeCH_2)Cl_2$ are effective for the catalytic dimerization of terminal alkenes $CH_2{=}CHR$ (for R = H, Me, Pr^n, Hx^n, CH_2Pr^i, CH_2Bu^t). As the size of R in the monomer $CH_2{=}CHR$ increases, the proportion of the head-to-tail dimer ($CH_2{=}CRCH_2CH_2R$) increases over that observed for propene (R = Me) dimerization, affording indirect evidence that the precursors to both dimers, viz., $Cp^*Cl_2Ta(CH_2CHRCHRCH_2)$ and $Cp^*Cl_2Ta(CHRCH_2CHRCH_2)$, are in equilibrium.[360]

Deuterium-labeling studies using 2-deuterio-1-pentene do not support a simple reductive coupling → β-H elimination → reductive elimination pathway (Scheme 17). Instead, the unusual "ring contraction" of a metallacyclopentane into a substituted metallacyclobutane is consistent with the available data. The new metallacyclobutane can then undergo a metal-mediated β-hydrogen shift, that is, β-elimination followed by reductive elimination to generate the alkene dimers shown in Scheme 18.[360] The catalytic cycle is completed by reductive coupling of two more alkenes to regenerate the metallacyclopentane.

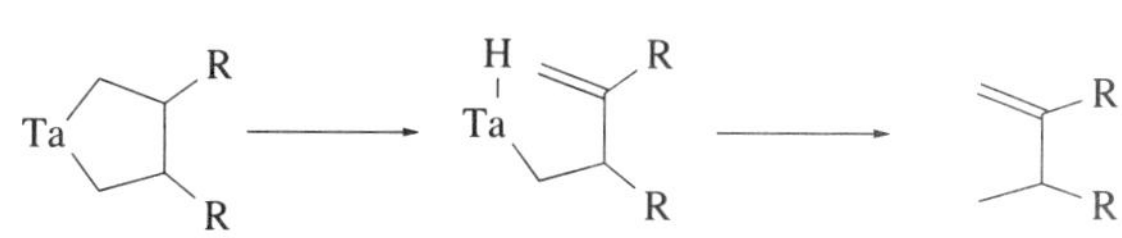

Scheme 17

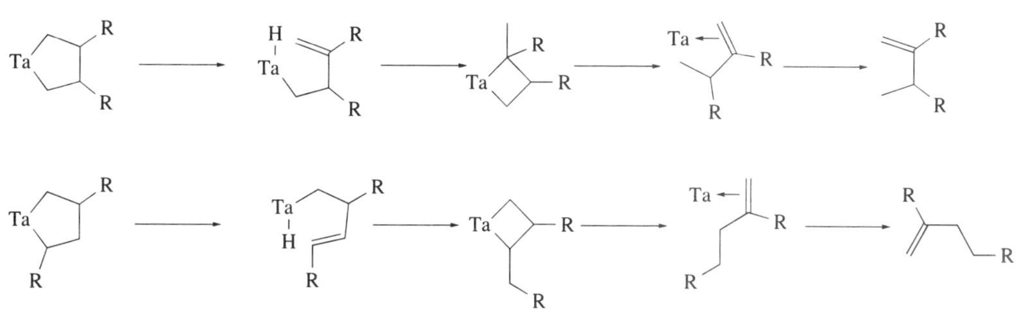

Scheme 18

2.6.5 Metallacyclopentenes

These metallacycles are rare. Two methods have been described for their isolation. First, the reductive coupling of the appropriate unsaturated ligands can lead to metallacyclopentene formation. Thus, Schrock and co-workers[418] reported the slow formation of $Cp^*Ta(CH{=}CHCH_2CH_2)Cl_2$ from the reaction of $Cp^*Ta(HC{\equiv}CH)Cl_2$ and $CH_2{=}CH_2$. It is interesting that $Cp^*Ta(HC{\equiv}CH)Cl_2$ does not metallacyclize with alkynes to afford metallacyclopentadienes. Second, the electrocyclic rearrangement of an unstable metallacyclobutene is reported to give rise to a metallacyclopentene. Thus, upon adding one equivalent of 2-butyne to the vinyl alkylidene $Ta({=}CMeCMe{=}CHBu^t)(OAr)_3(py)$, the metallacyclopentene complex $Ta[CH_2C({=}CMeCHBu^t)CMe{=}CMe](OAr)_3$ can be isolated.[372]

2.6.6 Metallacyclopentadienes

The first metallacyclopentadienes of tantalum were reported only in 1988[407] and the only niobacyclopentadienes known prior to 1981 were $Cp_2\overline{Nb(CR{=}CRCR{=}C}R)$ (for R = Ph or CF_3); only the trifluoromethyl complex $Cp_2\overline{Nb(CCF_3{=}CCF_3CCF_3{=}C}CF_3)$ is prepared by cyclizing the alkyne.[1] Since then, many metallacyclopentadienes of tantalum have been reported, all prepared from the reductive coupling of two alkynes, and in many cases "cross-coupled" products are accessible from the reactions of discrete alkyne adducts with a different alkyne.

Reducing Et_2O solutions of $Ta(OAr)_3Cl_2(OEt_2)$ with NaHg in the presence of PhC≡CPh or TMS-C≡CMe affords $Ta(PhC{\equiv}CPh)(OAr)_3$ and $Ta(TMS\text{-}C{\equiv}CMe)(OAr)_3$ in good yield.[162,407] Both compounds readily undergo metallacyclization reactions with other alkynes to form the metallacyclopentadienes. Thus, $\overline{Ta(CPh{=}CPhCR{=}C}R)(OAr)_3$ (R = Me, Et) and $\overline{Ta(CPh{=}CPhCH{=}C}R)(OAr)_3$ (R = Bu^t, TMS, Ph) are isolated upon addition of the appropriate alkyne to $Ta(PhC{\equiv}CPh)(OAr)_3$. When $Ta(OAr)_3Cl_2(OEt_2)$ is reduced with NaHg in the presence of EtC≡CEt, $Bu^tC{\equiv}CH$, or TMS-C≡CH, no alkyne adduct is detected but the metallacycles $\overline{Ta(CEt{=}CEtCEt{=}C}Et)(OAr)_3$ and $\overline{Ta(CR{=}CHCR{=}C}H)(OAr)_3$ (for R = TMS, Bu^t) are isolated directly. These compounds can all be quenched with water to release the corresponding butadiene.[162]

Perhaps the best studied metallacyclopentadiene complexes of these metals have been examined with regard to the metallacycle formation and fragmentation reactions occurring during isomerization.[408] The kinetic product from reducing $Ta(OAr)_2Cl_3(OEt_2)$ with two equivalents of NaHg in the presence of $HC{\equiv}CBu^t$ is the α,α'-metallacyclopentadiene complex $\overline{Ta(CBu^t{=}CHCH{=}C}Bu^t)(OAr)_2Cl$ (**31**).[408,459] Thermolysis of this α,α'-isomer (**31**) affords the α,β'-isomer $\overline{Ta(CBu^t{=}CHCBu^t{=}C}H)(OAr)_2Cl$ (**32**) cleanly and quantitatively. Some of the intermediates which might be invoked in the α,α' (**31**) → α,β' (**32**) isomerization are presented in Scheme 19. Either pathway via (**33**) or (**34**) requires the disruption of the metallacycle and the formation of two alkyne molecules, either both coordinated (**33**) or with only one coordinated (**34**). Specifically, pathway (**31**) → (**34**) → (**32**) does not require a bis(alkyne) intermediate; however, the (**33**) ⇌ (**34**) equilibrium is possibly operative as indicated.

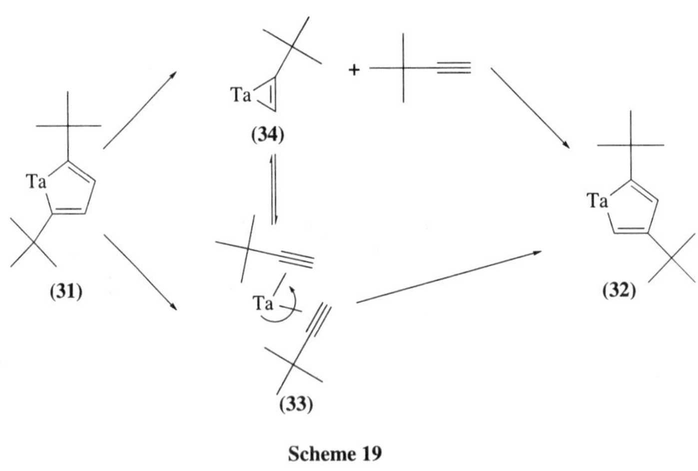

Scheme 19

The crossover experiment presented in Scheme 20 argues strongly against intermediate (**33**) only, that is, against (**31**) → (**33**) → (**32**). Perhaps most significant is the statistical crossover presented in Scheme 20. Thus, when metallacycle (**31**) disrupts into "$Ta(HC{\equiv}CBu^t)(ArO)_2Cl$" (**34**) plus $HC{\equiv}CBu^t$, the formation of (**32**) by recoupling that same alkyne is no more favored than coupling a different alkyne which originated from a neighboring metallacycle's disruption. These experiments are most consistent with the α,α' (**31**) → α,β' (**32**) rearrangement proceeding via (**31**) → (**34**) → (**32**). If intermediate (**33**) (viz., a complex with equivalent, coordinated alkynes) is involved at all, that is, if the rearrangement occurs via (**31**) → (**33**) → (**34**) → (**32**), then the (**33**) → (**34**) step must be significantly faster than (**33**) → (**32**), otherwise metallacycle (**32**) would retain some "memory" of (**33**) and nonstatistical crossover would have resulted.[408]

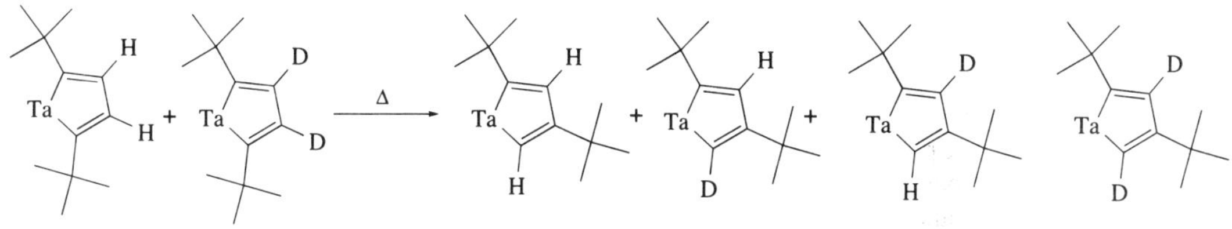

(31) (31-d2) (32) (32-d1(α)) (32-d1(β)) (32-d2)

Scheme 20

2.6.7 Other Metallacycles, Including Cyclometallated Ligands

Several other types of metallacycles of niobium and tantalum have been reported; these are outlined in Table 16 under various headings. For example, thermolysis of the iminoacyl complexes $CpTa(RC{\equiv}CR)[\eta^2\text{-}C(=NBu^t)Me]Me$ (R = Ph, Tol) in toluene results in coupling of the iminoacyl and alkyne ligands to form the metallacyclopentatriene-type derivatives $CpTa(=CRCR=CMeNBu^t)Me$ (**35**).[200,416] "Heterocyclic" metallacycles are formed from reactions of the alkylidene complex $CpTa(=CHBu^t)Cl_2$ with iminoboranes $RB{=}NBu^t$, which forms "azaborametallacyclobutenes" of the type $CpTa[CHBu^tB(R)NBu^t]Cl_2$.[352] Unusual "dimetallabenzene" complexes have been prepared and the tantalum species structurally characterized.[377]

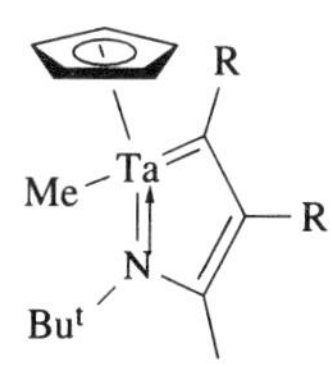

(35)

Niobium and tantalum alkoxides have also served as templates for cyclometallation reactions. For example, thermolysis of the trialkyls $M(O\text{-}2,6\text{-}C_6H_3Ph_2)_2R_3$ is reported to afford metallacyclic $M[OC_6H_3Ph(C_6H_4)](O\text{-}2,6\text{-}C_6H_3Ph_2)R_2$.[168] However, thermolysis of the dialkyls $M(O\text{-}2,6\text{-}C_6H_3Ph_2)_3R_2$ gives, sequentially, mono(cyclometallated) and bis(cyclometallated) products $M[OC_6H_3Ph(C_6H_4)](O\text{-}2,6\text{-}C_6H_3Ph_2)_2R$ (R = Me, $CH_2\text{-}4\text{-}C_6H_4Me$) and $M[OC_6H_3Ph(C_6H_4)]_2(O\text{-}2,6\text{-}C_6H_3Ph_2)$. The bis-(cyclometallated) products $M[OC_6H_3Ph(C_6H_4)]_2(O\text{-}2,6\text{-}C_6H_3Ph_2)$ are also available from reduction of $M(O\text{-}2,6\text{-}C_6H_3Ph_2)_3Cl_2$ with two equivalents of NaHg. Similar cyclometallations involving $Ta(O\text{-}2,6\text{-}C_6H_3Bu^t_2)_2Me_3$ have been reported and are outlined in Section 2.8.2.[178,179]

Metallacyclization occurs when $Ta(CH_2CH_2CH_2)(OAr)_3(py)$ reacts with O=CHPh, to give oxymetallacycle $Ta(OCHPhCH_2CH_2CH_2)(OAr)_3(py)$.[368] Similarly, acetone and $Ta(CH_2CH_2CH_2)(OAr)_3$ form $Ta(OCMe_2CH_2CH_2CH_2)(OAr)_3$. Bercaw and co-workers[228] have studied the Wittig-like reactions of alkylidene complexes with organic carbonyls employing $Cp^*_2Ta(=CH_2)Me$. Thus, treatment of $Cp^*_2Ta(=CH_2)Me$ with substituted benzaldehydes $O{=}CH(C_6H_4\text{-}4\text{-}X)$ (X = H, Cl, CF_3, CN or NO_2) provides the isolable metallaoxetane complexes, $Cp^*_2Ta[OCH(C_6H_4\text{-}4\text{-}X)CH_2]Me$. Upon mild thermolysis these metallacycles cleanly and irreversibly undergo a first-order, retro-[2 + 2] cycloaddition to provide $Cp^*_2Ta(=O)Me$ and the substituted styrenes, $CH_2{=}CH(C_6H_4\text{-}4\text{-}X)$. The relative rates of these decompositions depend only slightly upon the nature of the *para* substituent of the metallaoxetane phenyl group and little on the solvent polarity, indicating only minor charge buildup in the transition state.[228]

Finally, a number of other complexes containing cyclometallated ligands are described in other sections. Thus, the cyclometallated PMe_3 ligands $\eta^2\text{-}CH_2PMe_2$ and $\eta^2\text{-}CHPMe_2$ are outlined in Section 2.5.3.6 and cyclometallated C_5Me_5 ligands, for example, $\eta^6\text{-}C_5Me_4CH_2$ and $\eta^7\text{-}C_5Me_3(CH_2)_2$, are described elsewhere.

2.7 π COMPLEXES OF CARBOCYCLIC, -ENE, AND -ENYL LIGANDS

2.7.1 η^3 Complexes

Table 17 outlines the relatively small number of η^3 complexes of niobium and tantalum. Some of the earliest examples of η^3 complexes were unusual cyclooctatetraene species, for instance $[Nb(C_8H_8)_3]^-$, which is highly fluxional in solution, but is best described as $[(\eta^4\text{-}C_8H_8)Nb(\eta^3\text{-}C_8H_8)_2]^-$ in the solid state.[467] More recently, metal vapor synthesis methods and reactions of hydride complexes with cumulenes or butadienes have afforded entries into η^3 species, but the chemistry of η^3 compounds remains comparatively underdeveloped.

Table 17 η^3 Complexes of niobium and tantalum.

Compound	*Comments*	*Ref.*
$M(\eta^3\text{-}CH_2CHCHMe)(\eta\text{-}C_4H_6)_2$	metal vapor synthesis preparation	468,469
$(\eta\text{-}C_5R_5)_2M(\eta^3\text{-}CH_2CHCH_2)$	C_5R_5 = Cp, Cp'; PES study	251
$Cp_2Nb(\eta^3\text{-}CH_2CHCHMe)$	prep: $Cp_2Nb(CH_2{=}CH_2)H + CH_2{=}CHCH{=}CH_2$	429
$Cp_2Ta(\eta^3\text{-}CH_2CHCHMe)$	prep: Cp_2TaH_3 + butadiene	99
$Cp_2Nb(\eta^3\text{-}MeCHCHCHEt)$		429
$Cp_2Nb(\eta^3\text{-}MeCHCRCHMe)$	R = H, Me	429
$Cp_2Nb(\eta^3\text{-}CH_2CMeCHMe)$		429
$Cp_2Nb(\eta^3\text{-}CH_2CHCHEt)$		429
$Cp_2Nb(\eta^3\text{-}CH_2CHCR_2)$	$R_2 = Me_2$, MeEt	429
$Cp_2Nb(\eta^3\text{-}CH_2CRCH_2)$	R = H, Me	429
$Cp_2Ta(\eta^3\text{-}CH_2CHCHMe)$	prep: $Cp_2TaCl_2 + Mg(C_4H_6)$	429
$Cp^*Ta(\eta^3\text{-}CH_2CMeCH_2)Me_3$	unstable with respect to CH_4 elimination	199
$Nb(\eta^3\text{-}CH_2CHCHMe)(\eta\text{-}C_4H_6)(dmpe)$	prep: $Nb(\eta^3\text{-}CH_2CHCHMe)(\eta\text{-}C_4H_6)_2$ + dmpe	469
$Cp_2Ta(\eta^3\text{-}2,3\text{-}C_7H_{11})$	$2,3\text{-}C_7H_{11}$ = 2,3-dimethylpentadienyl; localized alkylalkene bonding to Ta	470
$Cp_2Ta(\eta^3\text{-}2,4\text{-}C_7H_{11})$	$2,4\text{-}C_7H_{11}$ = 2,4-dimethylpentadienyl; presumed to be η^3-bound	470
$(\eta^5\text{-}C_9H_7)_2Nb(\eta^3\text{-}CH_2CHCH_2)$	C_9H_7 = indenyl; prep: $(\eta^5\text{-}C_9H_7)_2Nb(\eta^2\text{-}BH_4) + NMe_3 + CH_2{=}C{=}CH_2$	47
$(\eta^5\text{-}C_9H_7)_2Nb(\eta^3\text{-}MeCHCHCHEt)$	C_9H_7 = indenyl	47
$Cp_2M(\eta^3\text{-}OCHNPh)$	prep: $Cp_2M(\eta^1\text{-}OCH{=}NPh)(CO) + h\nu$	77
$Cp_2Nb(\eta^3\text{-}SCHNPh)$		77
$Cp_2M(\eta^3\text{-}TolNCHNTol)$	prep: $Cp_2M[\eta^1\text{-}N(Tol)CH{=}NTol](CO) + h\nu$	77

Sixteen-electron allyl–butadiene complexes, purple $Nb(\eta^3\text{-}CH_2CHCHMe)(\eta^4\text{-}C_4H_6)_2$ and orange $Ta(\eta^3\text{-}CH_2CHCHMe)(\eta^4\text{-}C_4H_6)_2$, have been prepared from the cocondensation of butadiene (in THF) with Nb or Ta atoms in metal vapor synthesis preparations.[468,469] Pentane solutions of $Nb(\eta^3\text{-}CH_2CHCHMe)(\eta^4\text{-}C_4H_6)_2$ react with dmpe to give an orange oil, shown to be $Nb(\eta^3\text{-}CH_2CHCHMe)(\eta^4\text{-}C_4H_6)(dmpe)$.[469]

Reactions of $(\eta^5\text{-}C_9H_7)_2Nb(\eta^2\text{-}BH_4)$ (where $[C_9H_7]^-$ = indenyl) with NMe_3 are used to produce solutions of nascent $[(\eta^5\text{-}C_9H_7)_2NbH]$ which undergoes further reaction with unsaturated substrates.[47] Thus, $(\eta^5\text{-}C_9H_7)_2Nb(\eta^2\text{-}BH_4)$ reacts with $CH_2{=}C{=}CH_2$ in the presence of NMe_3 to afford $(\eta^5\text{-}C_9H_7)_2Nb(\eta^3\text{-}CH_2CHCH_2)$, whereas reaction with *trans,trans*-2,4-hexadiene forms $(\eta^5\text{-}C_9H_7)_2Nb(\eta^3\text{-}MeCHCHCHEt)$.[47] $Cp_2Ta(\eta^3\text{-}CH_2CHCHMe)$ is prepared similarly from Cp_2TaH_3 and butadiene.[99] The ethene complex $Cp_2Nb(CH_2{=}CH_2)H$ may also serve as a source of $[Cp_2NbH]$ upon extrusion of ethene; thus, $Cp_2Nb(CH_2{=}CH_2)H$ reacts by migratory insertion of a range of conjugated dienes into the Nb–H bond to afford η^3 allyl species.[429] For example, $Cp_2Nb(CH_2{=}CH_2)H$ reacts with butadiene to afford $Cp_2Nb(\eta^3\text{-}CH_2CHCHMe)$ and with 1-methylbutadiene to form $Cp_2Nb(\eta^3\text{-}MeCHCHCHMe)$.[429]

Heteronuclear η^3 complexes have also been described. Thus, insertion reactions of heterocumulenes O=C=NPh, S=C=NPh, and TolN=C=NTol into Nb–H and Ta–H bonds in $Cp_2MH(CO)$ give the $\eta^1(O)$, $\eta^1(S)$, and $\eta^1(N)$ insertion products $Cp_2M(\eta^1\text{-}OCH{=}NPh)(CO)$, $Cp_2Nb(\eta^1\text{-}SCH{=}NPh)(CO)$, and $Cp_2M(\eta^1\text{-}N(Tol)CH{=}NTol)(CO)$, respectively.[77] Photolysis of toluene solutions of these η^1 complexes results in the loss of CO and an $\eta^1 \rightarrow \eta^3$ conversion to afford the complexes $Cp_2M(\eta^3\text{-}OCHNPh)$, $Cp_2Nb(\eta^3\text{-}SCHNPh)$, and $Cp_2M(\eta^3\text{-}TolNCHNTol)$.[77]

2.7.2 η^4 Complexes

Although η^4 ligands are not especially common in niobium and tantalum complexes, their chemistry has grown considerably as compared with the η^3 ligands, with advances in the η^4-butadiene complexes being of particular note (Table 18).

Table 18 η^4 Complexes of niobium and tantalum.

Compound	*Comments*	*Ref.*
$(\eta^4\text{-}C_{10}H_8)Ta(dmpe)_2Cl$	$C_{10}H_8$ = naphthalene; electrochemical studies	27
$(\eta^4\text{-}C_{14}H_{10})Ta(dmpe)_2Cl$	$C_{14}H_{10}$ = anthracene; x ray	29
$(\eta\text{-}C_4H_6)Nb(dmpe)_2Cl$	prep: $NbCl_4(dmpe)_2 + Mg(C_4H_6)\cdot 2THF$	471
$M(\eta^3\text{-}1\text{-}MeC_3H_4)(\eta\text{-}C_4H_6)_2$	metal vapor synthesis preparation	468,469
$Nb(\eta^3\text{-}1\text{-}MeC_3H_4)(\eta\text{-}C_4H_6)(dmpe)$	prep: $Nb(\eta^3\text{-}1\text{-}MeC_3H_4)(\eta\text{-}C_4H_6)_2$ + dmpe	469
$Nb(\eta\text{-}C_4H_6)(H^*)(dmpe)_2$	H* = H (x ray), D	469
$Ta(\eta\text{-}C_4H_6)(CH_2{=}CH_2)X(PMe_3)_2$	X = Cl, Et	146
$CpNb(\eta^4\text{-}CHMe{=}CMeCMe{=}CHMe)Cl_2$	x ray	82,83
$[(\eta\text{-}C_5H_5)Nb(\eta^4\text{-}CHMe{=}CMeCMe{=}CHMe)(\mu\text{-}Cl)]_2$	prep: $CpNb(\eta^4\text{-}CHMe{=}CMeCMe{=}CHMe)Cl_2$ + NaHg	83
$CpTa[\eta^4\text{-}CMe_2{=}CHCMe{=}O]Cl_2$	x ray	472
$CpNb(\eta^4\text{-}C_7H_8)(CO)_2$		54
$CpNb(\eta^4\text{-}C_8H_8)(CO)_2$	prep: $CpNb(CO)_4 + C_8H_8 + h\nu$	54
$Cp^*Ta(CH_2CH_2CH_2CH_2)(\eta\text{-}C_4H_6)$	prep: $Cp^*TaH_4(PMe_3)_2 + CH_2{=}CH_2$	94
$(\eta\text{-}C_7H_7)Nb(\eta^4\text{-}C_7H_8)(PMe_3)$	prep: $NbCl_4(THF)_2 + Mg + C_7H_8 + PMe_3$	26,92,473
$(\eta\text{-}C_7H_7)Nb(\eta\text{-}C_4H_6)(PMe_3)$	prep: $(\eta^6\text{-}C_7H_8)Nb(PMe_3)_2Cl_2 + Mg(C_4H_6)\cdot 2THF$	473
$Cp^*Ta[\eta^4\text{-}C(CH_2)_3]Me_2$	trimethylenemethane complex	199
$CpTa(\eta\text{-butadiene})Cl_2$	butadiene = C_4H_6 (x ray), $2\text{-}MeC_4H_5$, $2,3\text{-}Me_2C_4H_4$	474
	butadiene = $2,3\text{-}Me_2C_4H_4$	475
$Cp^*Ta(\eta\text{-butadiene})Cl_2$	butadiene = C_4H_6, $2\text{-}MeC_4H_5$, $2,3\text{-}Me_2C_4H_4$; prep: Cp^*TaCl_4 + Mg(butadiene)	474
	butadiene = C_4H_6, $2\text{-}MeC_4H_5$, $1\text{-}MeC_4H_5$; prep: Cp^*TaCl_4 + 2 equiv. butenyl Grignard	475
	butadiene = C_4H_6; prep: $Cp^*TaCl_2(PMe_3)_2 + C_4H_6$	46
	butadiene = C_4H_6; prep: $Cp^*TaCl_2(CO)_2(THF) + C_4H_6$	57,88
$CpTa(\eta\text{-butadiene})_2$	butadiene = C_4H_6 (x ray), $2,3\text{-}Me_2C_4H_4$	474
$Cp^*Ta(\eta\text{-butadiene})_2$	butadiene = C_4H_6, $2\text{-}MeC_4H_5$, $2,3\text{-}Me_2C_4H_4$	474
$Cp^*Ta(\eta\text{-}C_4H_6)(\eta\text{-}2,3\text{-}Me_2C_4H_4)$	x ray	464,474,476
$Cp^*Ta(\eta\text{-butadiene})(\eta\text{-butadiene}')$	(butadiene)(butadiene') = $(C_4H_6)(2\text{-}MeC_4H_5)$; $(2\text{-}MeC_4H_5)(2,3\text{-}Me_2C_4H_4)$	464
$CpTa(\eta\text{-butadiene})(\eta\text{-butadiene}')$	(butadiene)(butadiene') = $(C_4H_6)(2,3\text{-}Me_2C_4H_4)$; $(C_4H_6)(2\text{-}MeC_4H_5)$; $(2\text{-}MeC_4H_5)(2,3\text{-}Me_2C_4H_4)$; $(2,3\text{-}Me_2C_4H_4)(2,3\text{-}Me_2C_4D_4)$	464
$CpTa(\eta^2\text{-}OCPr^i{}_2CH_2CH{=}CHCH_2)(\eta\text{-}2,3\text{-}Me_2C_4H_4)$		464
$CpNb[\eta^4\text{-}CH_2{=}CHC({=}O)OR]Cl_2$	R = Me, Et, Bu	57
$Cp^*Ta[\eta^4\text{-}CH_2{=}CHC(OMe){=}O]Cl_2$		88
$Ta(\eta\text{-}C_4H_6)(\eta^2\text{-}CHPMe_2)Cl(PMe_3)_2$	x ray	454
$CpNb(\eta^8\text{-}C_{14}H_{22})$	contains two η^4 interactions	470
$CpNb(\eta\text{-butadiene})Cl_2$	butadiene = C_4H_6, $2\text{-}MeC_4H_5$, $2,3\text{-}Me_2C_4H_4$	477
$Cp^*Nb(\eta\text{-butadiene})Cl_2$	butadiene = C_4H_6, $2\text{-}MeC_4H_5$, $2,3\text{-}Me_2C_4H_4$	477
$[CpNb(\eta\text{-butadiene})(\mu\text{-}Cl)]_2$	butadiene = C_4H_6, $2\text{-}MeC_4H_5$	477
$CpNb(\eta\text{-}C_4H_6)_2$	two isomers: (i) both *s-cis* and (ii) one *s-cis*, one *s-trans* $\eta\text{-}C_4H_6$	477
	x ray	417
$(\eta\text{-}C_5R_5)Nb(\eta\text{-}C_4H_6)_2$	C_5R_5 = Cp', C_5H_4TMS	417
$CpNb(\eta\text{-butadiene})_2$	butadiene = $2\text{-}MeC_4H_5$, $2,3\text{-}Me_2C_4H_4$	477
$Cp^*Nb(\eta\text{-butadiene})_2$	butadiene = C_4H_6, $2\text{-}MeC_4H_5$, $2,3\text{-}Me_2C_4H_4$	477
$CpNb(\eta\text{-}C_4H_6)(\eta\text{-}2,3\text{-}Me_2C_4H_4)$	*supine–prone* (*exo–endo*) geometry	477
$CpNb(\eta\text{-}2,3\text{-}Me_2C_4H_4)(\eta\text{-}2,3\text{-}Me_2C_4D_4)$		477
$CpNb(\eta\text{-}C_4H_6)R_2$	R = Me, Bu	477
$CpNb(\eta\text{-}C_4H_6)L_2$	L = CO, $CNBu^t$, $P(OMe)_3$, PMe_3	417
$Cp'Nb(\eta\text{-}C_4H_6)(PMe_3)_2$		417
$(\eta\text{-}C_7H_7)(\eta^4\text{-}C_7H_8)Nb(PMe_3)$		92
$CpNb(\eta\text{-}C_4H_6)(alkyne)(PMe_3)$	alkyne = $PhC{\equiv}CPh$, $PhC{\equiv}CMe$ (x ray), $TMS\text{-}C{\equiv}CPh$	417
$Nb(OC_6H_3Ph\text{-}\eta^4\text{-}C_6H_7)(O\text{-}2,6\text{-}C_6H_3Ph_2)_2$	coordinated cyclohexadiene ring; x ray	168,271
$Nb(OC_6H_2Ph_2\text{-}\eta^4\text{-}C_6H_7)(O\text{-}2,4,6\text{-}C_6H_3Ph_3)_2$	prep: $Nb(O\text{-}2,4,6\text{-}C_6H_3Ph_3)_3Cl_2 + NaHg + H_2$	478
$(Bu^n{}_3Sn)[Nb_6(\eta\text{-}C_4Ph_4)_3Cl_{19}]$	x ray; cyclobutadiene complex	479
$[Nb_2(\mu\text{-}C_4Ph_4)Cl_2(\mu\text{-}Cl)(THF)_4]_2[Nb_4(\mu\text{-}C_4Ph_4)_2(\mu\text{-}O)Cl_4(\mu\text{-}Cl)_4]$	x ray; $\mu,\eta^4{:}\eta^4\text{-}C_4Ph_4$ ligands; monocation, dianion	480
$Nb_4(\mu\text{-}C_4Ph_4)_2O_2Cl_4(THF)_4$	$\mu,\eta^4{:}\eta^4\text{-}C_4Ph_4$ ligands; x ray	481
$[Mg_2Cl_3(THF)_6][Nb_4(\mu\text{-}C_4Ph_4)_2(\mu\text{-}O)_2Cl_7\{Mg(THF)_3\}]$	$\mu,\eta^4{:}\eta^4\text{-}C_4Ph_4$ ligands; prep: $NbCl_4(THF)_2 + PhC{\equiv}CPh$ + Mg; x ray	481
$[M]^+{}_2[Nb_4OCl_8(C_4Ph_4)_2]^{2-}$	$[M]^+ = [Mg_2Cl_3(THF)_6]^+$, Na^+, $[HPEt_3]^+$; x ray of each	482
$[Nb_4OCl_8(C_4Ph_4)_2]^{2-}$	Fenske–Hall MO calculations	483

As noted by Labinger,[1] many of the early η^4 complexes were obtained unexpectedly in nonrational syntheses; examples from the 1980s include the following: (i) the attempted reduction of $TaCl_4(dmpe)_2$ by sodium naphthalenide affords the η^4-naphthalene complex $(\eta^4\text{-}C_{10}H_8)TaCl(dmpe)_2$;[27] (ii) $Cp^*TaH_4(PMe_3)_2$ reacts with ethene to afford, *inter alia*, $Cp^*\overline{Ta(CH_2CH_2CH_2CH_2)}(\eta^4\text{-}C_4H_6)$,[94] in which the butadiene ligand possibly arises by a double β-hydrogen elimination from an intermediate metallacyclopentane; and (iii) the niobium derivative $CpNb(\eta\text{-}C_4Me_4H_2)Cl_2$ is obtained upon treatment of $CpNb(dmpe)Cl_2(CO)$ with NaHg and MeC≡CMe, followed by HCl (aq.) workup.[82,83] In the past decade, some general classes of η^4 compounds have emerged, including $CpM(\eta\text{-butadiene})Cl_2$, $CpM(\eta\text{-butadiene})_2$, and complexes containing unusual $\mu,\eta^4{:}\eta^4\text{-}C_4Ph_4$ ligands such as $[Nb_4OCl_8(C_4Ph_4)_2]^{2-}$.

2.7.2.1 *η⁴-Butadiene complexes*

Perhaps the best characterized η^4 complexes of group 5 are the η-butadiene species, which are usually prepared by rational methods. For example, both niobium[477] and tantalum[474] Cp and Cp* complexes $(\eta\text{-}C_5R_5)M(\eta\text{-}C_4H_6)Cl_2$ and $(\eta\text{-}C_5R_5)M(\eta\text{-}C_4H_6)_2$ are prepared by the reaction of $(\eta\text{-}C_5R_5)MCl_4$ with one and two equivalents of magnesium butadiene (2-butene-1,4-diylmagnesium), respectively. The monokis(butadiene) complexes are characterized by ligands which assume the *supine* (*exo*) conformation (**36**), while the $\eta\text{-}C_4R_6$ ligands of the bis(butadiene) derivatives prefer the *supine–prone* (*exo–endo*) geometry (**37**). In structurally characterized complexes, the niobium or tantalum center binds the butadiene ligand in such a fashion as to favor the metallacyclopent-3-ene structure (**39**) over the simple η^4-butadiene adduct (**38**). This structure reflects the metal center attaining a higher oxidation state through back-bonding into empty ligand π* orbitals, as supported by molecular orbital studies.[474]

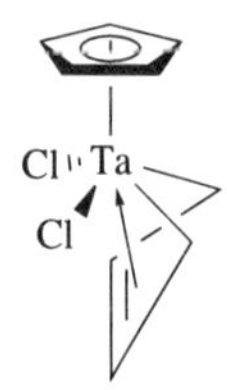

(36)

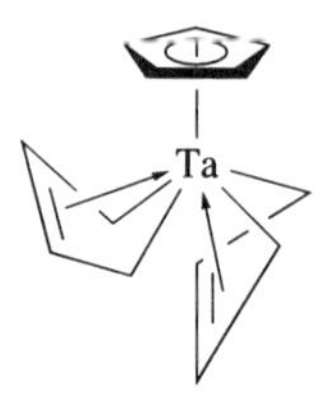

(37)

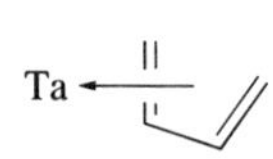

(38)

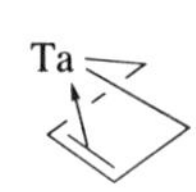

(39)

Under some conditions, reduction of Nb^V to Nb^{IV} occurs in the reaction of $CpNbCl_4$ with $Mg(C_4R_6)$ reagents to form the diamagnetic, binuclear complexes $[CpNb(\eta^4\text{-}C_4R_6)(\mu\text{-}Cl)]_2$ (**40**) in which the butadienes are situated mutually *cis*.[477] Although all of the aforementioned compounds contain *s-cis*-butadiene ligands, in the preparation of $CpNb(\eta\text{-}C_4H_6)_2$, another isomer characterized as $CpNb(\eta\text{-}s\text{-}cis\text{-}C_4H_6)(\eta\text{-}s\text{-}trans\text{-}C_4H_6)$ (**41**) is also obtained.[477] Interestingly, $CpNb(\eta\text{-}C_4H_6)_2$ has been shown by NMR to exist as two isomers in solution: bis(*cis*-butadiene) (which is structurally characterized) and the *cis*-butadiene,*trans*-butadiene isomer.[417] One of the C_4H_6 ligands in these complexes can be substituted by CO, $CNBu^t$, $P(OMe)_3$, or PMe_3 to afford the corresponding adducts $CpNb(\eta\text{-}C_4H_6)L_2$.

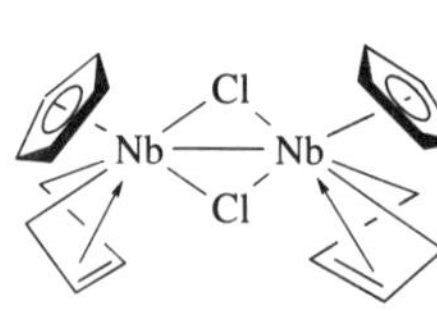

(40)

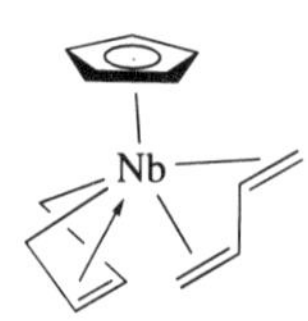

(41)

A number of mixed-ligand butadiene complexes, for example, $Cp^*Ta(\eta\text{-}C_4H_6)(\eta\text{-}2,3\text{-}Me_2C_4H_4)$ and $CpTa(\eta\text{-}2\text{-}MeC_4H_5)(\eta\text{-}2,3\text{-}Me_2C_4H_4)$, are available from $(\eta\text{-}C_5R_5)Ta(\eta\text{-butadiene})Cl_2$ and one equivalent of a different Mg(butadiene), for butadiene = C_4H_6, $2\text{-}MeC_4H_5$, or $2,3\text{-}Me_2C_4H_4$.[464] All structures examined show an $(\eta\text{-}C_5R_5)Ta$(*supine*-butadiene)(*prone*-butadiene') geometry in which the larger of the two dienes appears to prefer the *supine* (*exo*) position. Since the mono(butadiene) complexes contain a supine butadiene, when $(\eta\text{-}C_5R_5)Ta(\eta\text{-}C_4H_6)(\eta\text{-}2,3\text{-}Me_2C_4H_4)$ is prepared from $(\eta\text{-}C_5R_5)Ta(\eta\text{-}supine\text{-}C_4H_6)Cl_2$ and $Mg(2,3\text{-}Me_2C_4H_4)$, the initial geometry of the coordinated $\eta\text{-}supine\text{-}C_4H_6$ rearranges to *prone* in the final product. However, when $(\eta\text{-}C_5R_5)Ta(\eta\text{-}C_4H_6)(\eta\text{-}2,3\text{-}Me_2C_4H_4)$ is prepared from $(\eta\text{-}C_5R_5)Ta(\eta\text{-}supine\text{-}2,3\text{-}Me_2C_4H_4)Cl_2$ and $Mg(C_4H_6)$, no such transformation takes place.[464]

Another synthetic method for preparing compounds of the (η-C_5R_5)Ta(η-butadiene)Cl_2 class has been developed from the reaction of (η-C_5R_5)$TaCl_4$ with two equivalents of methylated allyl Grignards (Scheme 21).[475] For example, Cp*$TaCl_4$ reacts with two equivalents of 2-butenyl Grignard [ClMgCH(Me)CH=CH_2] to afford a presumed bis(η^1-allyl) complex which undergoes a β-H abstraction by one ligand to afford Cp*Ta(η^4-C_4H_6)Cl_2. Cp*Ta(η^4-2-MeC_4H_5)Cl_2, CpTa(η-2,3-$Me_2C_4H_4$)Cl_2, and Cp*Ta(η-1-MeC_4H_5)Cl_2 are also prepared in this fashion.[475] Curtis and co-workers[57,88] have reported that Cp*Ta(η-C_4H_6)Cl_2 is accessible upon reaction of butadiene with Cp*$TaCl_2(CO)_2$(THF). Similarly, Cp*$TaCl_2(PMe_3)_2$ reacts with butadiene to form Cp*Ta(η^4-C_4H_6)Cl_2.[46]

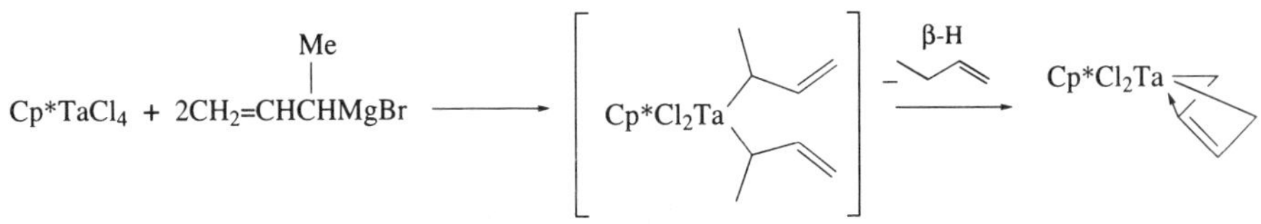

Scheme 21

2.7.2.2 η^4-Cyclobutadiene complexes

Only a few cyclobutadiene derivatives have been reported and are typically prepared from the cycloaddition of alkynes. An early report of the reaction of CpNb(PhC≡CPh)$_2$(CO) and PhC≡CPh (in refluxing benzene) to form CpNb(η-C_4Ph_4)(PhC≡CPh)(CO) was described, along with the thermal decomposition of this compound to provide C_6Ph_6.[484] Cotton and Shang[479] have prepared and structurally characterized the cluster complex (Bu^n_3Sn)[Nb_6(η-C_4Ph_4)$_3Cl_{19}$] formed from the reaction of $NbCl_5$, Bu^n_3SnH, and PhC≡CPh in toluene.

2.7.2.3 Other η^4 ligands

Perhaps the most unusual η^4 complexes are those formed by coupling two alkyne molecules in a "C"-shaped C_4R_4 ligand which bridges two metal centers along the edge of a cluster.[482] The most common structure of this type is the anionic niobium species [$Nb_4OCl_8(C_4Ph_4)_2$]$^{2-}$ (**42**), of which several salts have been isolated.[482] Its electronic structure has been examined using Fenske–Hall calculations and is consistent with the formal Nb^{III} oxidation state and a formal double bond between close pairs of metal atoms in the tetranuclear cluster.[483] Thus, reducing $NbCl_4(THF)_2$–PhC≡CPh mixtures or reducing Nb(PhC≡CPh)$Cl_3(THF)_2$ affords salts of [$Nb_4OCl_8(C_4Ph_4)_2$]$^{2-}$ (**42**) containing these unusual μ,η^4:η^4-C_4Ph_4 ligands, that is, bonding η^4 to two metals. Several of these complexes are now known, including the [$Nb_2(\mu$-$C_4Ph_4)Cl_2(\mu$-Cl)(THF)$_4$]$^+$ cation, which contains a μ,η^4:η^4-C_4Ph_4 ligand in a binuclear cluster,[480] and neutral $Nb_4(\mu$-$C_4Ph_4)_2O_2Cl_4(THF)_4$, which is isolated in 35% yield from the reaction of Nb(PhC≡CPh)$Cl_3(THF)_2$, NaHg, and PMe_2Ph in THF–benzene solutions.[481]

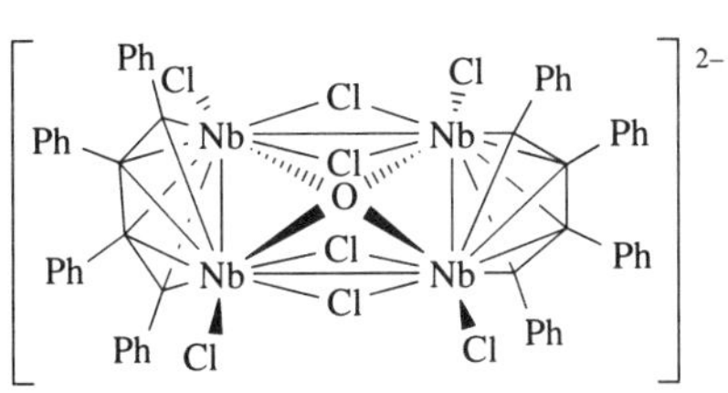

(42)

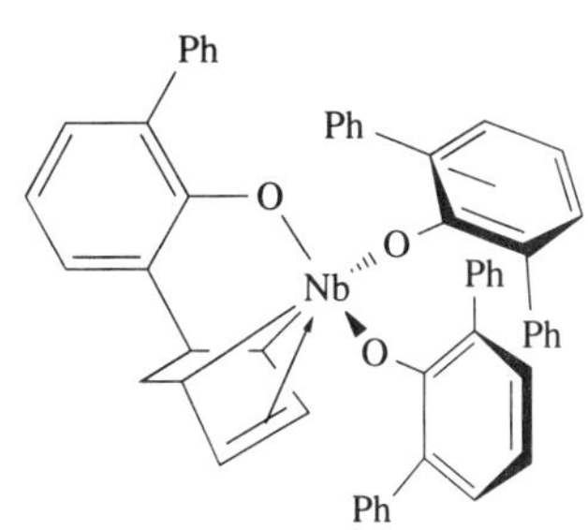

(43)

A complex containing a coordinated cyclohexadiene ring, viz., Nb(OC_6H_3Ph-η^4-C_6H_7)(O-2,6-$C_6H_3Ph_2$)$_2$ (**43**), can be isolated upon reducing Nb(O-2,6-$C_6H_3Ph_2$)$_3Cl_2$ with NaHg under an H_2 atmosphere.[271] This species can be considered an intermediate in the complete hydrogenation reaction which converts the O-2,6-$C_6H_3Ph_2$ ligands into O-2,6-$C_6H_3Cy_2$ ligands.[271] Upon reducing Nb(O-2,4,6-$C_6H_2Ph_3$)$_3Cl_2$ with NaHg under an H_2 atmosphere, an analogous cyclohexadiene complex, Nb($OC_6H_2Ph_2$-η^4-C_6H_7)(O-2,4,6-$C_6H_2Ph_3$)$_2$, is formed.[478]

2.7.3 η[5] Complexes

Many niobium and tantalum compounds containing cyclopentadienyl ligands have been described in previous sections, and in accordance with our ligand priority scheme (Section 2.1.2), the η-C_5R_5 complexes without other carbon-bound ligands are outlined here. We have divided these derivatives between the monokis- and bis(cyclopentadienyl) complexes. Labinger[1] has described the chemistry derived from attempts to prepare simple metallocenes, which includes the formation of higher oxidation state species such as $Cp_2M(\eta^1\text{-}C_5H_5)_2$ or Cp_2MH_3 (in the presence of H_2 and/or $NaBH_4$). The reader is referred to *COMC-I* for a complete discussion of these compounds.[1]

2.7.3.1 Monokis(cyclopentadienyl) complexes

Table 19 summarizes the monokis η[5] compounds of niobium and tantalum, sorted according to ancillary ligands.

(i) Halide complexes

Perhaps the most significant advances in the η[5] chemistry of these metals are centered around the monokis(cyclopentadienyl) complexes, made possible by the ease of preparation and availability of (η-C_5R_5)MX_4 derivatives. A review of monokis η[5] halide complexes of the transition metals has appeared.[530] In addition, a review covering organometallic dinuclear complexes, including metal–metal-bonded species, encompasses several monokis(cyclopentadienyl) niobium and tantalum species.[531]

The chemistry of monokis(cyclopentadienyl) niobium and tantalum complexes developed after that of their bis(cyclopentadienyl) analogues since a convenient entry into (η-C_5R_5)MX_4 directly from the pentahalides was uncovered later. Typically, tin–[99,198,488] or silicon–cyclopentadienyl reagents are required for the selective substitution of one cyclopentadienyl ligand at MX_5 to form (η-C_5R_5)MX_4. These reactions can be solvent and stoichiometry dependent. For example, treating MCl_5 with one equivalent of $Bu^n{}_3Sn(C_5R_5)$ reagents affords high yields of the monokis η[5] derivatives (η-C_5R_5)MCl_4,[99] while reacting MCl_5 with excess $Bu^n{}_3Sn(C_5R_5)$ induces clean reduction and now constitutes the preferred route to the d^1 compounds (η-C_5R_5)$_2MCl_2$.[242,308] Few structures of (η-C_5R_5)MX_4 derivatives are known: $[Cp^*MF_4]_2$ is dimeric[487] and Cp^*TaBr_4 is monomeric[297] while the unsubstituted η-C_5H_5 complexes $CpMX_4$ are presumed to be dimeric.

Lower oxidation state (η-C_5R_5)MX_n complexes often form interesting dinuclear halide-bridged and/or metal–metal-bonded species.[531] For example, Messerle and co-workers[297] reported that reducing mononuclear Cp^*TaX_4 or (η-C_5Me_4Et)TaX_4 (X = Cl or Br) with two equivalents of sodium affords the d^2 dinuclear complexes $[Cp^*Ta(\mu\text{-}X)_2]_2$ and $[(\eta\text{-}C_5Me_4Et)Ta(\mu\text{-}X)_2]_2$, respectively. These complexes are characterized by a metal–metal double bond and constitute rare examples of complexes with four μ-halide ligands. $[Cp^*NbCl_2]_2$ is likely to have a similar structure.[84,209] Reducing (η-C_5Me_4R)TaX_4 (R = Me or Et; X = Cl or Br) with one equivalent of NaHg affords a 70% yield of the paramagnetic Ta^{IV} dimers $[Cp^*Ta(\mu\text{-}X)X_2]_2$ and $[(\eta\text{-}C_5Me_4Et)Ta(\mu\text{-}X)X_2]_2$, which are characterized by no metal–metal interaction.[493] Structurally characterized complexes of this series include d^0 Cp^*TaBr_4 (**44**), d^1 $[(\eta\text{-}C_5Me_4Et)Ta(\mu\text{-}Br)Br_2]_2$ (**45**), and d^2 $[Cp^*Ta(\mu\text{-}Br)_2]_2$ (**46**).

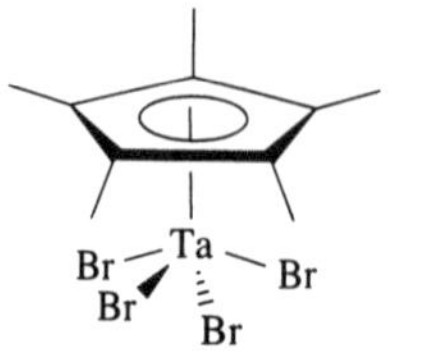

(44)

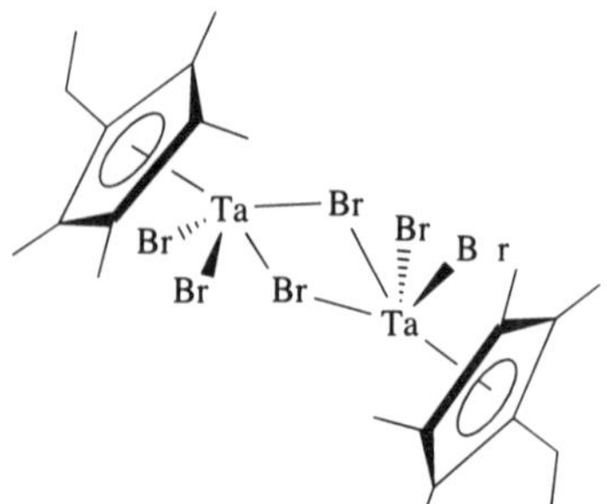

(45)

Perhaps the most unusual preparation of a cyclopentadienyl compound is seen in the cycloaddition assembly of an η-C_5R_5 ligand from the tantalum alkylidyne complex $[Cl_2(DME)Ta(\mu\text{-}CBu^t)]_2[Zn(\mu\text{-}$

Table 19 Monokis η^5 complexes of niobium and tantalum.

Compound	*Comments*	*Ref.*
Halide complexes		
$(\eta\text{-}C_5R_5)NbF_4$	R = H, Me	485
Cp^*TaF_4		486,487
$[Cp^*NbF_4\cdot 2AsF_3]_2[Cp^*NbF_4(HF)\cdot AsF_3]_2$	x ray	485
$[Cp^*TaF_3(\mu\text{-}F)]_2\cdot 2AsF_3$	prep: $Cp^*TaCl_4 + AsF_3$; x ray	487
$CpMCl_4$	prep: $MCl_5 + Bu^n{}_3SnCp$	99
	prep: MCl_5 + TMS-Cp	488,489
$Cp'MCl_4$	prep: $MCl_5 + Bu^n{}_3SnCp'$	99
Cp^*NbCl_4	prep: $NbCl_5$ + TMS-Cp*	490,491
Cp^*TaCl_4	prep: $TaCl_5 + Bu^n{}_3SnCp^*$	198
	photo- and electrochemical studies	492
$[Cp^*TaCl_4]^-$	electrochemically generated	492
$(\eta\text{-}C_5Me_4Et)NbCl_4$		491
$(\eta\text{-}C_5Me_4R)TaCl_4$	R = Me, Et	297
$(\eta\text{-}C_5H_4TMS)NbCl_4$		79,80
$(\eta\text{-}C_5H_3\text{-}1,3\text{-}Bu^t{}_2)NbCl_4$	prep: $NbCl_5 + Bu^n{}_3Sn(C_5H_3\text{-}1,3\text{-}Bu^t{}_2)$	48
$(\eta\text{-}C_5R_4Bu^t)TaCl_4$	R = Me, Et, Pr^n; prep: $[Cl_2(DME)Ta(\mu\text{-}CBu^t)]_2[Zn(\mu\text{-}Cl)_2]$ + 2 RC≡CR	379
$(\eta\text{-}C_5H_2R_3)TaCl_4$	R = Bu^t, Bu^i	379
$(\eta\text{-}C_5H_4CH_2C_6H_5\text{-polymer})MCl_4$	poly(styrene–divinylbenzene)-attached $CpMCl_4$	149
$(\eta\text{-}C_5H_4CH_2C_6H_5\text{-polymer})NbCl_3$	poly(styrene–divinylbenzene)-attached $CpNbCl_3$	149
$[(\eta\text{-}C_5Me_4R)Ta(\mu\text{-}Cl)Cl_2]_2$	R = Me, Et	493
$[Cp^*NbCl_2]_2$	prep: $Cp^*NbMe_2Cl_2 + H_2$	209
	prep: Cp^*NbCl_4 + 2 NaHg	84
$[(\eta\text{-}C_5Me_4R)Ta(\mu\text{-}Cl)_2]_2$	R = Me, Et; prep: $(\eta\text{-}C_5Me_4R)TaCl_4$ + 1 NaHg	297,386
$(\eta\text{-}C_5Me_4R)TaBr_4$	R = Me (x ray), Et	297
$[(\eta\text{-}C_5Me_4R)Ta(\mu\text{-}Br)Br_2]_2$	R = Me, Et (x ray)	493
$[(\eta\text{-}C_5Me_4R)Ta(\mu\text{-}Br)_2]_2$	R = Me, Et (x ray)	297,386
$(\eta^5\text{-}C_2B_9H_{11})TaCl_3$	dicarbollide complex; x ray	197
Silyl complexes		
$Cp^*Ta(TMS)Cl_3$	x ray; prep: $Cp^*TaCl_4 + Al(TMS)_3\cdot OEt_2$	108,212
	MO calculations and structural analysis	494,495
$Cp^*Ta(TMS)(OR)Cl_2$	R = TMS, Bu^t	212
$Cp^*Ta(TMS)Cl_2(PR_3)$	$PR_3 = PMe_3$ (x ray), PEt_3	212
	MO calculations and structural analysis	494,495
Nitrogen donors		
$Cp'NbCl_4L$	L = NCMe, NCPh	99
$[CpTaCl_4]_2(bipy)$		28
$[(\eta\text{-}C_5H_4R)NbCl_3(bipy)]_2(\mu\text{-}O)$	R = H, Me, TMS	495
$[(\eta\text{-}C_5H_4R)NbCl_3(phen)]_2(\mu\text{-}O)$		495
$CpNb(NMeNMe_2)Cl_3$	prep: $CpNbCl_4 + Li(NMeNMe_2)$	98
$CpNb(NPhNH_2)Cl_3$		98
$(\eta\text{-}C_5Me_4R)Nb(N{=}PPh_3)Cl_3$	R = Me, Et (x ray)	491
$Cp^*Nb(N{=}S(O)Me_2)Cl_3$		491
$[(\eta\text{-}C_5Me_4Et)NbCl_3(N{=}PPh_2)]_2(C_2H_2)$	x ray	491
$Cp^*TaF_4(HN{=}PPh_3)$	x ray	496
$CpTa[\eta^2(N,N)\text{-}N(TMS)NCPh_2]Cl_3$	prep: $Cp^*Ta(TMS)Cl_3 + N_2CPh_2$	497
$Cp^*Ta[(TMS\text{-}N)_2CR]F_3$	R = Ph, 4-$C_6H_4CF_3$	486
$Cp^*Ta[(TMS\text{-}N)_2C\text{-}4\text{-}C_6H_4X]F_3$	X = H, CF_3, NMe_2, Me, CN, OMe (x ray)	498
$Cp^*Ta[OC(TMS)NN(CPh_2)]Cl_3$	$O^{13}C(TMS)$ analogue also reported	86
$Cp^*Ta[OCH(TMS)\text{-}o\text{-}C_5H_4N]Cl_3$	$O^{13}CH(TMS)$ analogue also reported	86
Imido complexes		
$CpNb({=}NR)Cl_2$	R = Me; x ray; prep: $CpNbCl_4 + (TMS)_2NMe$	213,390
	R = Bu^t, Me, Ar	361
	R = Bu^t, Ar; x ray of both	390
$Cp^*Nb({=}NR)Cl_2$	R = Me, Bu^t	361
	R = Me, Ar	390
	R = Ar	75
$CpNb({=}NR)Cl_2(PMe_3)$	R = Me; x ray	213,390
	R = Bu^t, Ar	390
$CpNb({=}NMe)Cl_2(PPh_3)$		390
$Cp^*Nb({=}NMe)Cl_2(PMe_3)$		390
$Cp^*Nb({=}NAr)({=}E)(PMe_3)$	E = S, Se, Te	499
$CpNb({=}NMe)(OAr')_2$		213,390
$CpNb({=}NMe)(OBu^t)_2$		390
$CpNb({=}NR)(OBu^t)_2$	R = Bu^t, Ar	361
$CpNb({=}NBu^t)(OR)_2$	R = Ar, 2,6-$C_6H_3Ph_2$	390
$CpNb({=}NAr)_2(py)$	prep: $Nb({=}NAr)_2Cl(py)_2$ + NaCp	500
$CpTa({=}NAr)Cl_2$		361

Table 19 (continued)

Compound	*Comments*	*Ref.*
$Cp^*Ta(=NR)Cl_2$	R = Me, Ar	361
	R = Ar; x ray	390,501
$Cp^*Ta(=NR)(OPr^i)_2$	$R = Bu^t, CH_2Bu^t$	199
$(\eta-C_5Me_4R)Ta(=NEt)Cl_2$	R = Me, Et	452
$Cp^*Ta(=NAr)(NEt_2)Cl$		501
$Cp^*_2Ta(=NR)Cl$	$R = Ph, CH_2Ph$	310
$[Li(L)][Cp^*Ta(=NAr)_2Cl]$	$L = Et_2O$ (x ray), THF	501
$Cp^*Nb(=NAr)(PMe_3)_2$		75
$[\eta-C_5H_4(CH_2)_3N=]NbCl_2$	prep: $NbCl_5 + (TMS)_2N(CH_2)_3C_5H_4(TMS)$	502
$[\eta-C_5H_4(CH_2)_3N=]NbCl_2(PMe_3)$		502
Phosphorus donors		
$Cp'NbCl_4L$	$L = P(OMe)_3$, dppe	99
$CpTaCl_4L$	$L = PMe_3$	46
	$L = PMe_3, PCy_3$	28
Cp^*TaCl_4L	$L = PMe_3, P(OMe)_3, \eta^1$-dppe	198
	$L = PMe_3$	212,46
$[(\eta-C_5H_4TMS)NbCl_3]_2(\mu\text{-dppe})(\mu-O)$		495
$[(\eta-C_5H_4TMS)NbCl_3]_2(\mu\text{-dppm})(\mu-O)$		495
$CpNbCl_3L_2$	$L = PMe_3, PMe_2Ph$, 1/2 dppe	79
$CpNbCl_3L$	$L = PMe_2Ph, PMePh_2$	28
$(\eta-C_5H_4TMS)NbCl_3L_2$	$L = PMe_2Ph$, 1/2 dppe	79
$CpTaCl_3(PMe_3)$		46
$Cp^*TaCl_3(PMe_3)$	prep: $Cp^*TaCl_4 + Mg + PMe_3$; x ray	46,282
$(\eta-C_5Me_4R)TaCl_3(PMe_3)$	R = Me, Et; prep: $[(\eta-C_5Me_4R)Ta(\mu-X)_2]_2 + PMe_3$	297
$CpNbCl_3(dmpe)$	prep: $CpNbCl_4 + dmpe + EtAlCl_2$	82,83
$Cp^*NbCl_3(dmpe)$		84
$Cp^*TaCl_3(dmpe)$		46
$CpNbX_3(dppe)$	X = Cl, Br	99
$(\eta-C_5Me_4R)TaBr_3(PMe_3)$	R = Me, Et	297
$CpNbCl_2(PMe_3)_3$	x ray	78,503
$Cp^*NbCl_2(PMe_3)_2$	prep: $Cp^*NbCl_4 + Mg + PMe_3$	85
$Cp^*NbCl_2(PR_3)_2$	$PR_3 = PMe_3, PMe_2Ph$, 1/2 dmpe	84
$Cp_2Ta_2Cl_4(PMe_2Ph)_2$		267
$CpTaCl_2(PMe_3)_3$		46,57,78
$Cp^*TaCl_2(PMe_3)_2$		45,46,87
$Cp^*TaCl_2(PMe_3)_3$		78
$[CpNb(dmpe)Cl]Cl$		83
$CpNb(PF_3)_4$	^{93}Nb NMR study	13
$CpNb(dmpe)_2$		83
$CpNb(dppe)_2$		98
Oxygen donors		
$CpNbCl_4(THF)$		28
$CpTaCl_4(DME)$		28
$Cp^*NbCl_3(OH)$	prep: $Cp^*NbCl_4 + HNEt_2 + H_2O$	490
$(\eta-C_5H_4R)Nb(OPh)Cl_3$	R = H, TMS	504
$Cp^*Ta(OR)Cl_3$	R = Me, Et	87
$Cp^*Ta(OR)_3Cl$	$R = Pr^i$, Me	94
$Cp^*Ta(OR)_4$	$R = Me, Pr^i$	94,96
	$R = Me, Pr^i, Ph$	320
$CpNb(O_2C_6H_4)Cl_2$	prep: $CpNbCl_4 + 1,2-(HO)_2C_6H_4$	504
$CpNbCl_2(O_2C_2Ph_2)L$	$L = PMe_3, PMe_2Ph$	505
$CpNbCl_2(O_2C_6H_2Bu^t_2)L$	$L = PMe_3, PMe_2Ph$	505
$Cp^*Nb[O=CMeCH=CMeO]F_3$		506
$Cp^*Nb[O=CPhCH=CPhO]F_3$	x ray	506
$Cp^*Ta(OCH=CHBu^t)Cl_3(py)$	*cis*-isomer; $Ta(O^{13}CH=CHBu^t)$ analogue prepared	86
$Cp^*Ta[O=CPhCH=CPhO]F_3$	x ray	486
$[Cp^*Ta(O_2C_2H_4)_2]_n$		320
$[Cp^*Ta(1,2-O_2C_6H_4)_2(PMe_3)]_n$	prep: $Cp^*Ta(\eta^2-CHPMe_2)(H)_2(PMe_3) + 1,2-(HO)_2C_6H_4$	320
$Cp^*Ta(OCH=CHC_6H_4\text{-}4\text{-}Me)Cl_3$		203
$(\eta-C_5Me_4R)Ta(OCH=CHC_6H_4\text{-}4\text{-}Me)Cl_3(THF)$	R = Me (x ray), Et	203
$Cp^*TaCl_3(O_2CR)$	$R = Ph, C_6H_4\text{-}4\text{-}Bu^t, Me, CD_3$; photochemical studies	492
$Cp^*TaCl_3(O_3SR)$	$R = Me, CF_3$; photochemical studies	492
$[Cp^*Ta(O)Cl_2]_n$		86
$Cp^*Ta(=O)Cl_2(PMe_3)$	prep: $Cp^*TaCl_2(PMe_3)_3 + CO_2$	78
$Cp^*Ta(OCR^1R^2TMS)Cl_3$	R^1,R^2 = H, H; H, Me; Me, Me; Me, Ph; $-(CH_2)_5-$, H, $CH=CH_2$; Me, $CH=CH_2$	497
$Cp^*Ta[OC(TMS)NN(CPh_2)]Cl_3$	x ray	507
$Cp^*Ta[OCH(TMS)C(O)OEt]Cl_3$	$O^{13}CH(TMS)^{13}C(O)OEt$ analogue also reported	86,108

Table 19 (continued)

Compound	*Comments*	*Ref.*
$Cp^*Ta[OCH_2C(O)OEt]Cl_3$		86,108
$Cp^*Ta[OCH(TMS)C(O)O(CH_2)_3CH{=}CH_2]Cl_3$		86,108
$Cp^*Ta[OC(TMS)C(py\text{-}2,6\text{-}Me_2)O]Cl_3$	$O^{13}C(TMS)^{13}C(py\text{-}2,6\text{-}Me_2)O$ analogue also reported	86
$Cp^*Ta[OC(TMS)C(PCy_3)O]Cl_3$	x ray; $O^{13}C(TMS)^{13}C(PCy_3)O$ analogue reported	86
$Cp^*Ta[OC(TMS)NN(CPh_2)]Cl_3$	$O^{13}C(TMS)$ analogue also reported	86
$Cp^*Ta[OCH(TMS)\text{-}o\text{-}C_5H_4N]Cl_3$	$O^{13}CH(TMS)$ analogue also reported	86
$[Cp'NbCl_3(OH_2)]_2(\mu\text{-}O)$		99
$[(\eta\text{-}C_5H_4R)NbCl_3(OH_2)]_2(\mu\text{-}O)$	R = H, TMS	79
$[CpNbCl_3(OH_2)]_2(\mu\text{-}O)\cdot 2THF$	x ray	508
$[(\eta\text{-}C_5H_4R)NbCl_2(\mu\text{-}Cl)]_2(\mu\text{-}O)$	R = H, TMS (x ray)	79
	R = TMS	504
	R = H, Me, TMS	495
$[Cp^*NbCl_3]_2(\mu\text{-}O)$		490
$[Cp^*NbCl_2(OH)]_2(\mu\text{-}O)$		490
$[Cp^*NbCl_2]_2(\mu\text{-}Cl)(\mu\text{-}OH)(\mu\text{-}O)$	x ray	509,510
$[(\eta\text{-}C_5H_4TMS)NbCl]_2(\mu\text{-}O)(\mu\text{-}O_2C_6H_4)_2$	x ray	504
$[Cp^*TaCl_3]_2(\mu\text{-}O)$	prep: $Cp^*TaCl_4 + H_2O$	229,230
	prep: $Cp^*TaCl_4 + (TMS)_2O$	87
$[Cp^*Ta(\mu\text{-}O)Cl_2]_2$		46,87
$[Cp^*TaCl_2(\mu\text{-}OH)]_2(\mu\text{-}O)$	prep: $[Cp^*TaCl_3]_2(\mu\text{-}O) + H_2O$	229,230
	prep: $Cp^*Ta(CO)_2(THF)$ + air; x ray	511
Sulfur and selenium donors		
$CpNb[\eta^2\text{-}S_2P(OPr^i)_2]Cl_3$	x ray; prep: $CpNbCl_4 + (Pr^iO)_2P(=S)(SH) + NEt_3$	512
$CpNbCl(\mu\text{-}Cl)_2Nb[\eta^2\text{-}S_2P(OPr^i)_2]Cp$	x ray	512
$[Li_2(THF)_2Cp^*NbS_3]_2$		513
$(\eta\text{-}C_5Me_4R)Ta(SCH{=}CHS)_2$	R = Me, Et	514
$Cp^*Ta(SCH_2CH_2S)_2$	prep: $Cp^*TaCl_4 + Na_2S_2C_2H_4$	514
$Cp'Ta(SPh)_4$	prep: $Cp'TaCl_4$ + NaSPh	515
$Cp^*Ta(S_2C_7H_{10})_2$	$S_2C_7H_{10}$ = norbornane-*exo*-2,3-dithiolate	516
$[Li_2(THF)_2Cp^*TaS_3]_2$	x ray	513,517
$Li_2(TMEDA)_2Cp^*TaS_3$		513,517
$[Li_2(THF)_3][Cp^*Ta(S_2)(\mu\text{-}S_2)]_2(\mu\text{-}O)$	x ray	513
$Li_2(DME)Cp^*TaS_3$		513
$[Li_3(TMEDA)_2Cp^*TaS_3Cl]_2(\mu\text{-}TMEDA)$	x ray	513
$Cp^*Ta(Se)_3Li_3Cl(THF)$	prep: $Cp^*TaCl_4 + Li_2Se_2$ + THF	518
$Cp^*Ta(Se)_3Li_2(TMEDA)_2$		518
Cluster and heterobimetallic compounds		
$[Cp_3Cr_3(\mu_3\text{-}S)_4]NbCp$	prep: $[CpCr(\mu\text{-}SBu^t)]_2(\mu\text{-}S) + CpNb(CO)_4 + h\nu$	519
$CpNbCl_2(\mu\text{-}SEt)_2NiCp$	x ray	414
$[Cp'NbCl_2]_3(\mu_3\text{-}O)$		520
$[CpNbCl(\mu\text{-}Cl)]_3(\mu_3\text{-}O)(\mu_3\text{-}OH)$	x ray	508
$[Cp'NbCl(\mu_2\text{-}Cl)]_3(\mu_3\text{-}O)(\mu_3\text{-}OH)$	x ray	508,520
$[Cp'NbCl(\mu_2\text{-}Cl)]_3(\mu_3\text{-}O)(\mu_3\text{-}Cl)$	x ray	520
$[CpNbCl(\mu\text{-}OH)]_3(\mu_3\text{-}OH)(\mu_3\text{-}O)$		508
$[Cp^*NbCl]_3(\mu_2\text{-}O)_3(\mu_3\text{-}O)(\mu_2\text{-}Cl)$	x ray	490,521
$\{[Cp^*Nb]_3(OH)_2(\mu_2\text{-}OH)(\mu_3\text{-}OH)(\mu_2\text{-}O)_2(\mu_3\text{-}O)Cl\}Cl$	x ray	521
$\{[Cp^*Nb(\mu\text{-}Cl)(\mu\text{-}O)]_3\}\{[Cp^*Nb]_3(\mu\text{-}Cl)_2(\mu\text{-}OH)(\mu\text{-}O)_3\}[Zn_4Cl_{10}]$	x ray	509
$\{[Cp^*Nb(\mu\text{-}Cl)(\mu\text{-}O)]_3\}_2[A]$	$[A]^{2-} = [Zn_2Cl_6]^{2-}$, $[Zn_4Cl_{10}]^{2-}$, $[Zn_6Cl_{14}]^{2-}$; x ray of each	510
$\{[Cp^*NbCl]_3(\mu\text{-}Cl)(\mu\text{-}O)_2(\mu_3\text{-}OH)(\mu_3\text{-}O)\}_2[Zn_2Cl_6]$	x ray	510
$(\eta\text{-}C_5H_4R)_4Nb_5Cl_5(OH)_9O(THF)_5$	R = H, Me	508
$(\eta\text{-}C_5Me_4Et)_3Nb_3S_7$		321
$(\eta\text{-}C_5H_4Bu^t)_3Nb_3S_{10}O$	x ray; two isomers	322,522
$(\eta\text{-}C_5H_4Bu^t)_3Nb_3S_{12}$	x ray	522
$(\eta\text{-}C_5Me_4Et)_3Nb_3S_{12}$		321
$(\eta\text{-}C_5H_4Bu^t)_4Nb_4S_{13}$	prep: thermolysis of $(\eta\text{-}C_5H_4Bu^t)_4Nb_2S_n$ (n = 8, 9)	322,522
$[(\eta\text{-}C_5Me_4Et)Nb]_3(\mu_2\text{-}S)_3(\mu\text{-}S_3BSH)$		523
$Cp_3Nb_3Se_5Cl_2$	x ray; prep: $CpNbCl_4 + Se(TMS)_2$	524
$[(\eta\text{-}C_5Me_4R)Nb]_2(\mu\text{-}\eta^6{:}\eta^6\text{-}P_6)$	R = Me, Et (x ray); prep: $(\eta\text{-}C_5Me_4R)Nb(CO)_4 + P_4$	105
$[(\eta\text{-}C_5H_3\text{-}1,3\text{-}Bu^t_2)Nb]_2(\mu\text{-}\eta^6{:}\eta^6\text{-}P_6)$	x ray	48
$[Cp_8Ta_6S_{10}]^{2+}[TaSCl_5]^{2-}$	x ray; prep: $CpTaCl_4 + S(TMS)_2$	524
$Cp_3Ta_3S_7Cl_2$	x ray	524
$Cp_4Ta_4S_{13}$	x ray	524
$Li_2(THF)_2Cp^*{}_3Ta_3S_6$		513
$[Cp^*TaCl_2(\mu\text{-}Cl)_3]_2[\{TaCl_5(\mu\text{-}O)TaCl_3\}_2(\mu\text{-}Cl)_2]$	x ray	525,526

Table 19 (continued)

Compound	*Comments*	*Ref.*
$[Cp^*TaCl(\mu_2\text{-}O)]_3(\mu_3\text{-}O)(\mu\text{-}Cl)$	x ray	229,230
$\{[Cp^*Ta(\mu_2\text{-}O)]_3(\mu_3\text{-}O)_2Cl(H_2O)_2\}Cl$	x ray	229,230
$[Cp^*Ta]_4(\mu_2\text{-}O)_4(\mu_3\text{-}O)_2(\mu_4\text{-}O)(OH)_2$	x ray; prep: $Cp^*Ta(\eta^2\text{-}CHPMe_2)(H)_2(PMe_3) + H_2O$	96
$[Cp^*ClTa(N)]_3$	prep: $Cp^*TaCl_4 + N(SnMe_3)_3$; x ray	527
Boron hydride complexes		
$(\eta\text{-}C_5Me_4Et)_2Nb_2Cl_2(B_2H_6)$	prep: $NbCl_5 + Li(C_5Me_4Et) + NaBH_4$	342
$(\eta\text{-}C_5Me_4R)_2Nb_2(B_2H_6)_2$	R = Me, Et (x ray), Bu^t	342
$(\eta\text{-}C_5Me_4R)_2Ta_2(\mu\text{-}X)_2(B_2H_6)$	R = Me, Et; X = Cl, Br; two isomers	528
$(\eta\text{-}C_5Me_4R)_2Ta_2(B_2H_6)$	R = Me, Et	528
Miscellaneous		
$Cp^*TaCl_4(CH_2{=}PMe_{3-n}Ph_n)$	$n = 0, 1, 2$ (x ray), 3; phosphorus ylide complex	202
$(\eta\text{-}C_7H_7)NbCp$	prep: $(\eta^6\text{-}C_7H_8)Nb(PMe_3)_2Cl_2 + NaCp$	473,529
$(\eta\text{-}C_7H_7)NbCp'$		529
$(\eta\text{-}C_7H_7)NbCpBr$	prep: $(\eta\text{-}C_7H_7)NbCp + CH_2{=}CHCH_2Br$	473

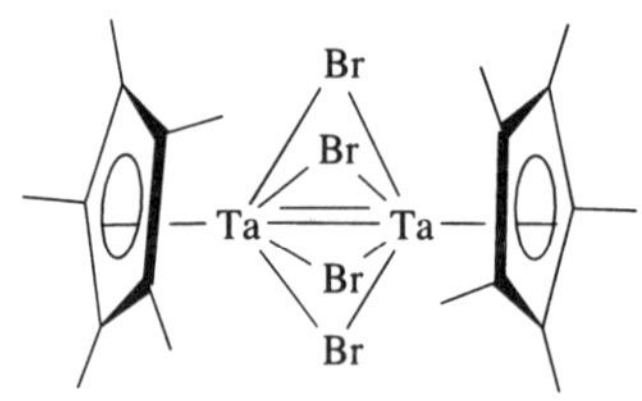

(46)

$Cl)_2]$[354] and two molecules of an alkyne in chlorinated solvents.[379] For example, $[Cl_2(DME)Ta(\mu\text{-}CBu^t)]_2[Zn(\mu\text{-}Cl)_2]$ reacts with RC≡CR (in CH_2Cl_2 solution) to afford $(\eta\text{-}C_5R_4Bu^t)TaCl_4$ (for R = Me, Et, or Pr^n) and with RC≡CH to form $(\eta\text{-}C_5H_2R_2Bu^t)TaCl_4$ (for R = Bu^t or Bu^i) (Scheme 22). For $(\eta\text{-}C_5H_2Bu^t_3)TaCl_4$, the 1,2,4-isomer is formed while two isomers are formed for $(\eta\text{-}C_5H_2Bu^i_2Bu^t)TaCl_4$ (see also Section 2.5.2.2).

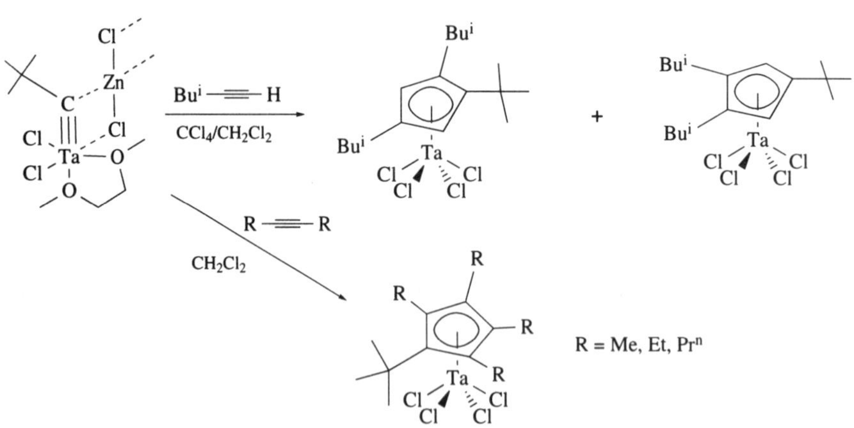

Scheme 22

(ii) Silyl complexes

Only a handful of silyl complexes have been reported, although $Cp^*Ta(TMS)Cl_3$ has served as an important starting material for metallacyclic and η^2 derivatives. Thus, Cp^*TaCl_4 and $Al(TMS)_3{\cdot}OEt_2$ are found to react (in pentane) to form the silyl complex $Cp^*Ta(TMS)Cl_3$.[108,212] KO-TMS or $LiOBu^t$ reacts with $Cp^*Ta(TMS)Cl_3$ forming $Cp^*Ta(TMS)(OR)Cl_2$ (R = TMS, OBu^t),[212] while a one-electron reduction of $Cp^*Ta(TMS)Cl_3$ occurs upon stirring with PMe_3 to afford $Cp^*Ta(TMS)(PMe_3)Cl_2$.[212]

(iii) Nitrogen donors

The monokis(cyclopentadienyl) complexes containing nitrogen-donor ligands include only a few species containing neutral N-donor ligands (e.g., the $\eta^1(N)$-phosphine imine adduct $Cp^*TaF_4(HN{=}PPh_3)$[496]), and most compounds involve formally anionic N-donor ligands (Table 19). For example, hydrazido(−1) complexes $CpNb(NR^1NR^2_2)Cl_3$ are available from reacting $CpNbCl_4$ with $Li(NR^1NR^2_2)$ reagents in THF.[98] Upon treatment of Cp^*NbCl_4 with TMS-N=S(O)Me_2 or TMS-N=PPh_3, the metathesis products $Cp^*Nb(N{=}S(O)Me_2)Cl_3$ and $Cp^*Nb(N{=}PPh_3)Cl_3$, respectively, are isolated.[491]

(iv) Imido complexes

Cyclopentadienyl complexes containing organoimido ligands comprise a substantial group. The tetrachlorides $(\eta\text{-}C_5R_5)MCl_4$ are the principal starting materials for d^0 CpM(=NR) complexes and reactions in which cleavage of an amide N–α-substituent bond results in Nb=NR ligand formation are typical. For example, $(\eta\text{-}C_5R_5)NbCl_4$ (R = H, Me) reacts with $(TMS)_2NMe$ to afford $(\eta\text{-}C_5R_5)Nb({=}NMe)Cl_2$, while TMS-NHAr provides convenient routes to $(\eta\text{-}C_5R_5)Nb({=}NAr)Cl_2$.[213,390] Alkoxide and alkyl derivatives of these chlorides are readily available via metathesis reactions. Fenske–Hall MO calculations on $CpNb({=}NMe)Cl_2$[390] allowed Gibson and co-workers to identify a correspondence between the frontier orbitals of the group 4 "bent metallocene" moieties Cp_2M[532] and the group 5 Cp–imido complexes CpM(=NR),[390] based upon the formal $1\sigma,2\pi$ orbital analogy between $[\eta\text{-}Cp]^-$ and $[NR]^{2-}$ ligands (see also Section 2.5.2.3). The chlorides $(\eta\text{-}C_5R_5)Ta({=}NR)Cl_2$, have been reported from reactions of $(\eta\text{-}C_5R_5)TaCl_4$ with TMS-NHR[390] or LiNHR[501] (Equation (18)) or observed via intermetal transfer reactions of multiply bonded ligands (Equation (19)).[361]

$$Cp^*TaCl_4 + 2LiNHR \longrightarrow Cp^*Ta({=}NR)Cl_2 + 2LiCl + H_2NR \qquad (18)$$

$$CpNb({=}NMe)Cl_2 + Cp^*Ta({=}NAr)Cl_2 \rightleftharpoons CpNb({=}NAr)Cl_2 + Cp^*Ta({=}NMe)Cl_2 \qquad (19)$$

(v) Phosphorus donors

Phosphine adducts of the monokis(cyclopentadienyl) halide complexes have been described for a range of oxidation states and are usually prepared from reducing $(\eta\text{-}C_5R_5)MX_4$ compounds in the presence of PR_3 ligands. Complexes of the form $(\eta\text{-}C_5R_5)MX_4(PR_3)$, $(\eta\text{-}C_5R_5)MX_3(PR_3)_n$ (n = 1 or 2), $(\eta\text{-}C_5R_5)MX_2(PR_3)_n$ (n = 2 or 3), and $(\eta\text{-}C_5R_5)M(PR_3)_4$ have been reported (Table 19). For example, $CpTaCl_3(PMe_3)$, $CpTaCl_2(PMe_3)_3$, $Cp^*TaCl_2(PMe_3)_2$, and $Cp^*TaCl_3(dmpe)$ are all available from the Mg reduction of the appropriate $(\eta\text{-}C_5R_5)TaCl_4$ precursor in the presence of a trialkyl phosphine.[46] $CpNbCl_4$ can be reduced with two equivalents of NaHg in the presence of excess PMe_3 to afford $CpNbCl_2(PMe_3)_3$, which is structurally characterized as a six-coordinate, *cis*-Cl, *mer*-PMe_3 adduct.[503]

(vi) Oxygen donors

Perhaps the most extensive array of monokis(cyclopentadienyl) derivatives of niobium and tantalum are those compounds containing oxygen-donor ligands, including alkoxide, hydroxide, oxo, enolate, carboxylate, and metallacyclic compounds with coordinated oxygen atoms, and also simple ether adducts (Table 19). Many of these compounds derive from the appropriate $(\eta\text{-}C_5R_5)MX_4$ halide precursor (e.g., $Cp^*Ta(OEt)Cl_3$ is formed from Cp^*TaCl_4 and TMS-OEt[87]), but less rational syntheses have also been uncovered. For example, protonation of $Cp^*Ta(\eta^2\text{-}CHPMe_2)(H)_2(PMe_3)$ with HOR (R = Me, Pr^i, or Ph) in benzene affords $Cp^*Ta(OR)_4$.[96,320]

An area of increasing interest is the oxo chemistry of these metals and many monokis(cyclopentadienyl) oxo compounds have been accessed by hydrolysis of the monokis η^5 halides. For example, upon adding one equivalent each of $HNEt_2$ and H_2O successively to toluene solutions of Cp^*NbCl_4, $Cp^*NbCl_3(OH)$ can be isolated as green microcrystals.[490] When 0.5 equivalents each of $HNEt_2$ and H_2O are used, the μ-O complex $[Cp^*NbCl_3]_2(\mu\text{-}O)$ is formed. A 90% yield of $[Cp^*NbCl_2(OH)]_2(\mu\text{-}O)$ is available upon stirring toluene solutions of either $Cp^*NbCl_3(OH)$ or $[Cp^*NbCl_3]_2(\mu\text{-}O)$ in air for 15 min (Scheme 23).[490] A similar series has also been reported for tantalum; thus, reacting Cp^*TaCl_4 with H_2O in the solid state (or in solution) affords the μ-oxo complexes $[Cp^*TaCl_3]_2(\mu\text{-}O)$ and $[Cp^*TaCl_2(OH)]_2(\mu\text{-}O)$.[229,230]

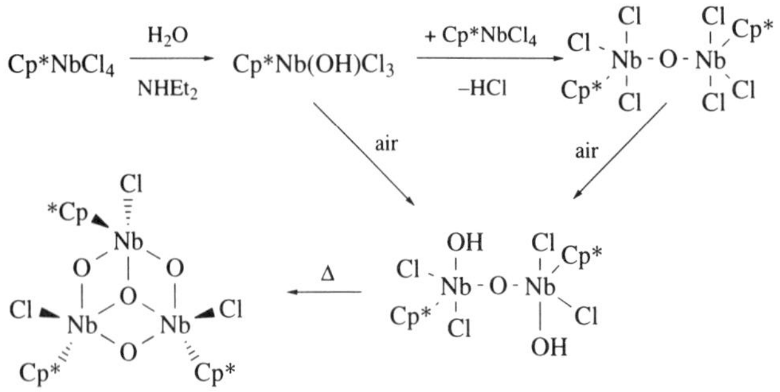

Scheme 23

An assortment of metallacyclic complexes with oxygen donors are available from reactions of the η^2-silaacyl complex Cp*Ta[η^2-C(O)TMS]Cl_3 with unsaturated molecules.[86,108] For example, Cp*Ta[η^2-C(O)TMS]Cl_3 is found to react rapidly with N_2CPh_2 to form metallacyclic Cp*Ta[OC(TMS)NN(CPh_2)]Cl_3, arising from a 1,2-insertion of the N–N bond into the Ta–C bond. Pyridine reacts with Cp*Ta[η^2-C(O)TMS(PR_3)]Cl_3 (PR_3 = PMe_3 or PEt_3) to give a product resulting from the silaacyl carbon inserting into the *o*-C–H bond of the pyridine, viz., Cp*Ta[OCH(TMS)-*o*-C_5H_4N]Cl_3.[86]

When the η^2-silaacyl complex Cp*Ta[η^2-C(O)TMS]Cl_3 is generated in the presence of excess CO, a C–C bond-forming reaction occurs to generate a ketene-like intermediate which cleaves ethers with β-hydrogens. Thus, subjecting Cp*Ta(TMS)Cl_3 to ≥ 69 kPa CO in Et_2O, orange-yellow Cp*Ta[OCH(TMS)C(O)OEt]Cl_3 forms in 35–50% yield.

(vii) Sulfur and selenium donors

Monokis(cyclopentadienyl) derivatives of niobium and tantalum containing sulfur or selenium donors, including thiolate, phosphorodithiolate, sulfide, and selenide complexes have been reported (Table 19). For example, the dithiolate compounds of Ta^V Cp*Ta(SCH=CHS)$_2$, (η-C_5Me_4Et)-Ta(SCH=CHS)$_2$, and Cp*Ta(SCH_2CH_2S)$_2$ are prepared by reacting Cp*$TaCl_4$ or (η-C_5Me_4Et)$TaCl_4$ with $Na_2S_2C_2H_2$ (THF, −10 °C), and from Cp*$TaCl_4$ with $Na_2S_2C_2H_4$ (CH_2Cl_2, room temperature).[514] Cp*Ta(SCH=CHS)$_2$ shows the existence of two conformers in solution.

A fascinating series of sulfide and selenide di- and trinuclear compounds have been described. The S_2^{2-} (disulfide) complex [$Li_2(THF)_3$][Cp*Ta(S_2)(μ-S_2)]$_2$(μ-O) is formed from the reaction of [Cp*$TaCl_3$]$_2$(μ-O) with four equivalents of Li_2S_x (x = 2–5).[513] However, Cp*MCl_4 react with Li_2S_2 to form sulfido complexes [$Li_2(THF)_2$Cp*MS_3]$_2$. The tantalum complex (**47**) has been structurally characterized as containing two [Cp*TaS_3]$^{2-}$ units with [Li(THF)]$^+$ bridges in a hexagonal prismatic $Li_4Ta_2S_6$ core.[513,517] The THF molecules in [$Li_2(THF)_2$Cp*TaS_3]$_2$ are displaced by DME or TMEDA to form Li_2(DME)Cp*TaS_3 and $Li_2(TMEDA)_2$Cp*TaS_3, respectively; however, when Cp*$TaCl_4$ reacts with four to five equivalents of Li_2S_2 in the presence of TMEDA, LiCl is retained and [$Li_3(TMEDA)_2$Cp*TaS_3Cl]$_2$(μ-TMEDA) is isolated.[513] Similarly, reacting Cp*$TaCl_4$ with four to five equivalents of Li_2Se_2 in THF affords yellow crystals of Cp*$TaSe_3Li_3Cl(THF)_3$ (**48**).[518] When Cp*$TaCl_4$ reacts with Li_2Se in THF, followed by treatment with TMEDA, the final "LiCl" unit was removed and yellow crystals of Cp*$TaSe_3Li_2(TMEDA)_2$ are isolated in high yield. These complexes share the common [Cp*$TaSe_3$]$^{2-}$ unit and both have been structurally characterized.[518]

(viii) Cluster and heterobimetallic compounds

The μ-O dinuclear compounds described above may be hydrolyzed further to afford tri- or tetranuclear cluster species. For example, the cluster compound [Cp*TaCl(μ_2-O)]$_3$(μ_3-O)(μ-Cl) is formed when [Cp*$TaCl_2$(OH)]$_2$(μ-O) loses H_2O and HCl upon heating.[229,230] Similarly, upon heating [Cp*$NbCl_2$(OH)]$_2$(μ-O) to 170 °C *in vacuo*, yellow-orange [Cp*NbCl(μ_2-O)]$_3$(μ_3-O)(μ_2-Cl) is formed.[490] {[Cp*Ta(μ_2-O)]$_3$(μ_3-O)$_2$Cl(H_2O)$_2$}Cl is produced when [Cp*TaCl(μ_2-O)]$_3$(μ_3-O)(μ-Cl) is hydrolyzed.[229,230] Controlled hydrolysis of Cp*Ta(η^2-$CHPMe_2$)(H)$_2$(PMe_3) in deoxygenated H_2O (excess) gives the unusual oxo butterfly cluster [Cp*Ta]$_4$(μ_2-O)$_4$(μ_3-O)$_2$(μ_4-O)(OH)$_2$ in which doubly, triply, and quadruply bridged oxo ligands are present.[96]

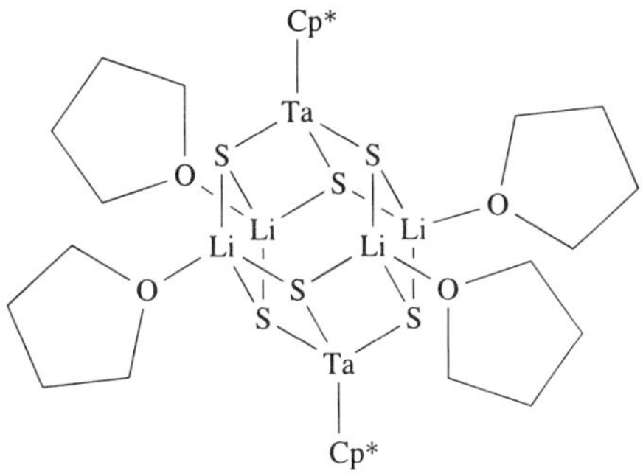

(47)

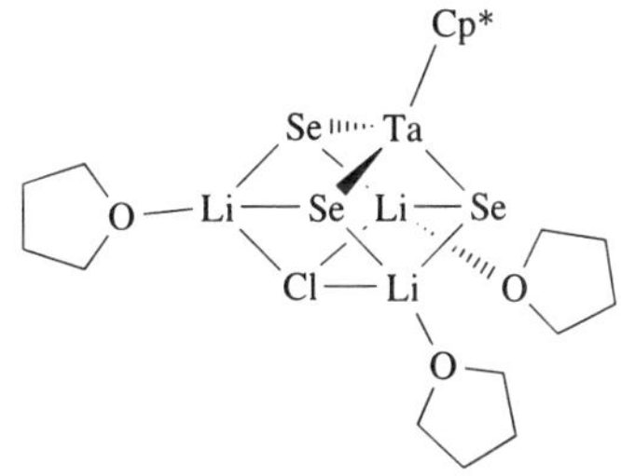

(48)

Organometallic oxide clusters of niobium also form from O_2 oxidation of $Cp^*{}_2NbCl_2$.[509,510] Table 19 also includes trinuclear sulfur and selenium cluster complexes, for example, $(\eta\text{-}C_5Me_4Et)_3Nb_3S_7$ and $(\eta\text{-}C_5Me_4Et)_3Nb_3S_{12}$, which are formed on reacting $(\eta\text{-}C_5Me_4Et)_2Nb(\eta^2\text{-}S_2)H$ with S_8.[321]

Finally, beautiful triple-decker sandwich complexes containing monokis η^5 moieties have been characterized. Under photolytic conditions, hexane solutions of $(\eta\text{-}C_5Me_4Et)Nb(CO)_4$ react with white phosphorus to afford $[(\eta\text{-}C_5Me_4Et)Nb]_2(\mu\text{-}\eta^6{:}\eta^6\text{-}P_6)$ (**49**).[105] Refluxing $(\eta\text{-}C_5H_3\text{-}1,3\text{-}Bu^t{}_2)Nb(CO)_4$ in decalin in the presence of P_4 affords a low yield of orange, crystalline $[(\eta\text{-}C_5H_3\text{-}1,3\text{-}Bu^t{}_2)Nb]_2(\mu\text{-}\eta^6{:}\eta^6\text{-}P_6)$. Molecular orbital calculations on these sandwich compounds were reported comparing planar (as is observed) vs. puckered P_6 ring structures.[48]

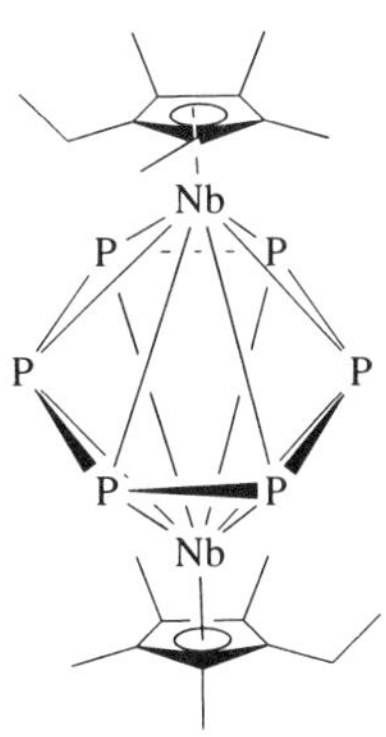

(49)

(ix) Boron hydride complexes

When two equivalents of $LiBH_4$ react with the metal–metal-bonded complexes $[(\eta\text{-}C_5Me_5)Ta(\mu\text{-}X)_2]_2$ and $[(\eta\text{-}C_5Me_4Et)Ta(\mu\text{-}X)_2]_2$ (X = Cl or Br), the hexahydridodiborate (formally $[B_2H_6]^{2-}$) adducts $(\eta\text{-}C_5Me_4R)_2Ta_2(\mu\text{-}X)_2(B_2H_6)$ (two isomers) are formed.[528] Further reaction of $(\eta\text{-}C_5Me_4R)_2Ta_2(\mu\text{-}X)_2(B_2H_6)$ with $LiBH_4$ affords $(\eta\text{-}C_5Me_4R)_2Ta_2(B_2H_6)_2$ (R = Me, Et) with two B_2H_6 ligands, each containing B–B bonds. The classical preparation of $(\eta\text{-}C_5Me_4Et)_2NbCl_2$ using $NbCl_5$, $Li[C_5Me_4Et]$, and $NaBH_4$ is found to be very sensitive to conditions and in some cases also provides boron hydride complexes.[342] Thus, by varying the stoichiometries and temperature, the isolation of dinuclear $(\eta\text{-}C_5Me_4Et)_2Nb_2Cl_2(B_2H_6)$ and $(\eta\text{-}C_5Me_4Et)_2Nb_2(B_2H_6)_2$, containing the formal $[B_2H_6]^{2-}$ dianion (which is not stable in the uncomplexed state), can be effected.[342]

2.7.3.2 Bis(cyclopentadienyl) complexes

Table 20 summarizes the bis η^5 halide (only) complexes of niobium and tantalum; all other bis η^5 complexes are included in Table 21.

Table 20 Bis η^5 halide complexes of niobium and tantalum.

Compound	*Comments*	*Ref.*
M^V complexes		
Cp_2MCl_3		533
Cp_2NbCl_3	electrochemically generated	534
$Cp_2MCl_nBr_{3-n}$	$n = 1, 2$	533
$Cp_2MCl_nI_{3-n}$	$n = 1, 2$	533
Cp_2NbClX_2	X = Br, I	535
Cp_2NbBr_3	prep: $Cp_2Nb(CH_2Ph)_2 + Br_2$	240
	prep: $Cp_2Nb(OC_6Cl_5)_2 + Br_2$	535
Cp_2NbI_3		535–7
$[Cp_2MCl_2][I_3]$		533
$[Cp_2NbCl_2]X$	$X^- = [PF_6]^-$	71
	$X^- = [SbF_6]^-$; x ray	538
	$X^- = [SbF_6]^-, [AlCl_4]^-, [WF_6]^-$	539
	$X^- = [BF_4]^-, [PF_6]^-, [BPh_4]^-$	219
$[Cp_2NbCl_2]X{\cdot}NCMe$	$X^- = [BF_4]^-$ (x ray), $[PF_6]^-$	219
$[(\eta^5\text{-}C_5H_4TMS)_2NbX_2][PF_6]$	X = Cl, Br, I	540,541
$[Cp_2TaCl_2]X$	$X^- = [BF_4]^-, [PF_6]^-, [BPh_4]^-$	219
$[Cp_2TaCl_2]X{\cdot}NCMe$	$X^- = [BF_4]^-, [PF_6]^-$	219
$[Cp_2TaCl_2][Br_3]$		533
$[Cp_2MClX][X_3]$	X = Br, I	533
$[Cp_2NbX_2][X_3]$	X = Br, I	240
$[Cp_2NbCl_2]^+$	electrochemically generated	534
$[Cp_2MX_2]^+$	X = Cl, Br, I; electrochemically generated	542
$(\eta\text{-}C_5R_5)(\eta^5\text{-}C_2B_9H_{11})TaCl_2$	C_5R_5 = Cp, Cp' (x ray); dicarbollide(2–) complex	197
$(\eta^5\text{-}C_2B_9H_{11})_2TaCl$	dicarbollide(2–) complex	197
M^{IV} complexes		
Cp_2NbF_2		236
Cp_2NbCl_2	prep: $CpNbCl_4 + Bu^n{}_3SnCp$	242,308
	prep: $NbCl_4(THF)_2 + LiCp$	231
	other preparations	71,235,542,543
	ESR studies	233,544
	PES studies	251
	electrochemical studies	534,545
	antitumor activity	546
	use in CO reduction	547
Cp_2TaCl_2	prep: $TaCl_5 + Bu^n{}_3SnCp$	99,308
	other preparations	252,280,542
	ESR studies	233,544
	PES studies	251,548
Cp'_2NbCl_2	prep: $NbCl_5 + Bu^n{}_3SnCp'$	99
	prep: $NbCl_4(THF)_2 + LiCp'$	231
	electrochemical studies	69
	PES studies	251,548
	ESR study	549
Cp'_2TaCl_2	PES studies	251
	x ray	550
$Cp^*{}_2NbCl_2$	prep: $Cp^*{}_2Nb(\eta^2\text{-}BH_4)$ + dil. HCl	70
$Cp^*{}_2TaCl_2$	prep: $Cp^*TaCl_3(PMe_3) + KCp^*$	282
$(\eta\text{-}C_5Me_4Et)_2NbCl_2$		342
$(\eta\text{-}C_5H_4Bu^t)_2NbCl_2$		235
$(\eta\text{-}C_5H_4Bu^t)_2TaCl_2$		298
$(\eta\text{-}C_5H_4TMS)_2NbCl_2$	prep: $NbCl_4(THF)_2 + Li(C_5H_4TMS)$	231
	prep: $(\eta\text{-}C_5H_4TMS)_2NbCl + PhCH_2Cl$	287
	x ray	62,551
	electrochemical studies	552
$[\eta\text{-}C_5H_3(TMS)_2]_2NbCl_2$	1,3-isomer	62
$(\eta\text{-}C_5H_4CH_2C_6H_5\text{-polymer})_2NbCl_2$	poly(styrene–divinylbenzene)-attached Cp_2NbCl_2	149
$(\eta{:}\eta\text{-}C_5H_4SiMe_2OSiMe_2C_5H_4)NbCl_2$		289
$(\eta^5\text{-}C_9H_7)_2NbCl_2$	C_9H_7 = indenyl	47
$(\eta\text{-}C_5H_4R)CpNbCl_2$	R = Bu^t	235
	R = Me, Bu^t, CHMePh	553
$(\eta\text{-}C_5H_4R)Cp'NbCl_2$	R = Bu^t, CHMePh	553
$(\eta\text{-}C_5H_4Bu^t)[\eta\text{-}C_5H_4(CHMePh)]NbCl_2$		553
$(\eta\text{-}C_5H_4R^1)(\eta\text{-}C_5H_4R^2)NbCl_2$	$R^1 = Bu^t$ and R^2 = H, Me, CHMePh; R^1 = Me and R^2 = H	554
$(\eta\text{-}C_5H_4CHMePh)CpNbCl_2$		117
$Cp^*CpTaCl_2$	prep: $Cp^*TaCl_3(PMe_3) + NaCp$	282
$(\eta\text{-}C_5H_4Pr^i)_2TaCl_2$	prep: $TaCl_5 + Bu^n{}_3Sn(C_5H_4Pr^i)$	242
$(\eta\text{-}C_5H_4R)_2TaCl_2$	R = Bu^t, TMS, CMe_2Ph, CHMePh	288

Table 20 (continued)

Compound	*Comments*	*Ref.*
Cp_2NbBr_2	prep: $NbBr_5$ + excess NaCp + HBr	67
	other preparations	542
Cp_2TaBr_2	prep: $TaBr_5 + Bu^n{}_3SnCp$	99
	PES studies	548
	other preparations	542
$(\eta\text{-}C_5H_4TMS)_2NbX_2$	X = Br, I	540
	X = Br, I; prep: $(\eta\text{-}C_5H_4TMS)_2NbX$ + RX	287
$[\eta\text{-}C_5H_3(TMS)_2]_2NbX_2$	X = Br, I	287,540
Cp_2NbI_2	prep: $Cp_2Nb(PPh_2)_2 + I_2$	536
	prep: $Cp_2Nb(CH_2Ph)_2 + I_2$	240
	other preparations	542
Cp_2TaI_2	prep: Cp_2TaCl_2 + KI	542
M^{III}/M^{IV} complexes		
$Cp_4Nb_2Cl_3$	$[Cp_2NbCl]_2(\mu\text{-}Cl)$ proposed structure	49,555
	electrochemical studies	545
$Cp_4Ta_2Cl_3$	$[Cp_2TaCl]_2(\mu\text{-}Cl)$ proposed structure	49,66
$Cp_4Nb_2Br_3$		67
$[Cp_4Nb_2Cl_2]ClO_4$		555
$[Cp_4Ta_2Cl_2]ClO_4$		66
M^{III} complexes		
Cp_2NbCl		49,437
$[Cp_2NbCl]_2$		239
$[Cp_2TaCl]_n$	structure unknown	49,66
$(\eta\text{-}C_5H_4TMS)_2NbCl$	formulated as dimer in one case	436,448,552
	improved preparation	437
$[\eta\text{-}C_5H_3(TMS)_2]_2NbCl$	1,3-isomer; prep: $[\eta\text{-}C_5H_3(TMS)_2]_2NbCl_2$ + NaHg	556
$(\eta\text{-}C_5H_4Bu^t)_2TaCl$		298
$[Cp_2NbBr]_2$		67
$(\eta\text{-}C_5H_4TMS)_2NbBr$		304,436
$(\eta\text{-}C_5H_4TMS)_2NbI$		304
$[Cp_2TaX]$	X = Br, I	68
$[Cp_2NbCl_2]^-$	electrochemically generated	49,545
$[Cp_2TaCl_2]^-$	electrochemically generated	49
$[(\eta\text{-}C_5H_4TMS)_2NbCl_2]^-$	electrochemically generated	552
$Cp_4Nb_2Cl_2(DMF)$	$Cp_2ClNb(\mu\text{-}Cl)Nb(DMF)Cp_2$ proposed structure	557
$[Cp_4Nb_2Cl_3]^-$	electrochemically generated	239,545
M^{III}/M^{II} complexes		
$(\eta^5{:}\eta^5\text{-}C_{10}H_8)Cp_2Nb_2(\mu\text{-}Cl)$	x ray	558,559
$(\eta^5{:}\eta^5\text{-}C_{10}H_8)Cp_2Nb_2(\mu\text{-}X)$	X = Br, SCN	559
M^{II} complexes		
$[Cp_2NbCl_2]^{2-}$	electrochemically generated	49
$[Cp_2TaCl_2]^{2-}$	electrochemically generated	49
Heterobimetallic complexes		
$Cp_2Nb(\mu\text{-}Cl)_2FeCl_2$	prep: $Cp_4Nb_2Cl_3 + FeCl_3$	555
$Cp_2Nb(\mu\text{-}Cl)(\mu\text{-}Br)Tl(C_6F_5)_2$		555
$Cp_2ClNb(\mu\text{-}Cl)MX_n$	$MX_n = AlCl_3, FeCl_3, TlCl_3, TlBr(C_6F_5)_2$	555
$Cp_2Ta(\mu\text{-}Cl)_2MCl_2$	M = Fe, Al, Tl	66
$Cp_2ClTa(\mu\text{-}Cl)MCl_3$	M = Fe, Al, Tl	66
$[Cp_2Ta(\mu\text{-}Cl)_2MCl_2][MCl_4]$	M = Fe, Al	66
$Cp_2Ta(\mu\text{-}Cl)(\mu\text{-}Br)Tl(C_6F_5)_2$		66
$Cp_2ClTa(\mu\text{-}Cl)Tl(C_6F_5)_2Br$		66

Table 21 Other bis η^5 complexes of niobium and tantalum.

Compound	*Comments*	*Ref.*
Silyl complexes		
$Cp_2Nb(SiPh_3)_2$	prep: $Cp_2NbCl_2 + LiSiPh_3$	536
$[Cp_2Nb(SiPh_3)_2]BF_4$		536
$[Cp_2Nb(SiPh_3)_2]_2(\mu\text{-}O)$	prep: $Cp_2Nb(SiPh_3)_2 + O_2$	536
$Cp_2Ta(SiMe_2H)(PMe_3)$		72
Nitrogen-donor ligands, including imido complexes		
$[Cp_2NbL_2][PF_6]_2$	L = 1/2 bipy (x ray), 1/2 phen	560

Table 21 (continued)

Compound	*Comments*	*Ref.*
$[Cp_2Nb(NCMe)_2][BF_4]_2$	x ray	137
$[Cp_2NbCl(NCMe)]BF_4$		240
$Cp_2Nb(\eta^1\text{-}NN{=}CPh\text{-}2\text{-}C_6H_4)_2$	x ray	313
$[Cp_2NbL_2]^+$	L = 1/2 bipy, 1/2 phen; electrochemically generated	560
$Cp_2Nb(8\text{-}NC_9H_6O)_2Cl$	$8\text{-}NC_9H_6OH$ = 8-quinolinol	561
$Cp_2Nb(8\text{-}NC_9H_6O)_2(OEt)$	$8\text{-}NC_9H_6OH$ = 8-quinolinol	561
$Cp_2Nb(NCS)_2$	ESR studies	233
$Cp_2Nb(bipy)$		562
$(\eta^5{:}\eta^5\text{-}C_{10}H_8)Cp_2Nb_2(\mu\text{-}NR)_2$	R = Ph, $4\text{-}C_6H_4OMe$ (x ray)	563,564
	R = Ph, Tol	314,565
$[(\eta^5{:}\eta^5\text{-}C_{10}H_8)Cp_2Nb_2(\mu\text{-}NPh)_2][BF_4]$	R = Ph, $4\text{-}C_6H_4OMe$ (x ray)	564
$[(\eta^5{:}\eta^5\text{-}C_{10}H_8)Cp_2Nb_2(NHR)_2]X_2$	R = Ph, Tol; $X^- = Cl^-$, $O_2CCF_3^-$, 1/2 SO_4^{2-}	314
$\{(\eta^5{:}\eta^5\text{-}C_{10}H_8)Cp_2Nb_2[N(Et)Tol]_2\}[BF_4]_2$		314
$Cp_2M({=}NN{=}CPh_2)Cl$		313
$Cp_2Nb[{=}NN{=}CPh(2\text{-}C_6H_4Br)]Cl$		313
$Cp_2Nb_2[\eta\text{-}C_5H_4C(CO_2Et){=}N\text{-}\mu\text{-}N]_2$	μ-hydrazido complex	313
$(\eta\text{-}C_5H_4TMS)_2Nb({=}NPh)Cl$	prep: thermolysis of $(\eta\text{-}C_5H_4TMS)_2NbCl(PhNCO)$	448
$Cp^*_2Ta({=}NPh)Cl$	x ray	310
$Cp^*_2Ta({=}NCH_2Ph)Cl$	thermally unstable	310
Phosphorus-donor ligands		
$Cp_2NbCl(PR_3)$	$PR_3 = PMe_2Ph$, $P(OMe)_3$	133
	$PR_3 = P(OMe)_3$, PMe_3	239
$Cp_2NbX(PMe_3)$	X = Cl, Br	240
$[Cp_2NbCl(PMe_3)]BF_4$		240
$Cp'_2NbCl[P(OMe)_3]$	prep: Cp'_2NbCl_2 + NaHg + $P(OMe)_3$	69
$Cp_2TaCl(PR_3)$	$PR_3 = PMe_2Ph$, PPh_3	66
$PR_3 = PMe_3$	454	
$Cp'_2TaCl(PMe_3)$		240
$(\eta\text{-}C_5H_4Pr^i)_2TaCl(PMe_3)$		242
$(\eta\text{-}C_5H_4TMS)_2NbCl(PR_3)$	$PR_3 = PMe_3$, PMe_2Ph, $P(OMe)_3$, $P(OEt)_3$	243
$(\eta\text{-}C_5H_4TMS)_2NbCl[P(OMe)_3]$		136
$Cp_2NbBr(PR_3)$	$PR_3 = PMe_2Ph$, $P(OMe)_3$	67
$(\eta\text{-}C_5H_4TMS)_2NbX[P(OMe)_3]$	X = Br, I	304
$\{(\eta\text{-}C_5H_4TMS)_2Nb[P(OMe)_3]\}^{2+}$	electrochemically generated	318
$[(\eta\text{-}C_5H_4TMS)_2NbX(DMF)]^{2+}$	X = Cl, Br, I; electrochemically generated	541
$[Cp'_2Ta(dmpe)]BF_4$		27
$[Cp_2Ta(dmpe)]PF_6$		27
$[Cp'_2Ta(dmpe)][BF_4]_2$		27
$[Cp_2M(PPh_2H)(PR_3)]Cl$	$PR_3 = PMe_2Ph$, $P(OEt)_3$	101
$Cp_2M(PPh_2)(PR_3)$	$PR_3 = PMe_2Ph$, $P(OEt)_3$	101
$Cp_2Nb(PPh_2)_2$	prep: $Cp_2NbCl_2 + LiPPh_2$	536
$[Cp_2Nb(PPh_2)_2]BF_4$	prep: $Cp_2Nb(PPh_2)_2 + [Cp_2Fe]BF_4$	536
$[Cp_2Nb(PPh_2)_2]_2(\mu\text{-}O)$		536
$\{(\eta\text{-}C_5H_4TMS)_2NbCl[P(OMe)_3]\}PF_6$	chemically and electrochemically generated	136
$Cp_2Ta(SiMe_2H)(PMe_3)$		72
$[Cp_2Nb(OPPh_3)(PMe_3)]X_2$	$X_2 = (Cl)(BF_4)$	240
$Cp_2Ta(\eta^2\text{-}PhP{=}PPh)H$		315
$(\eta^5\text{-}2,4\text{-}C_7H_{11})_2Nb(PEt_3)$	x ray	566,567
Oxygen-donor ligands		
$Cp_2Nb(\eta^2\text{-}O_2)Cl$	x ray	568
$(\eta\text{-}C_5H_4TMS)_2Nb(\eta\text{-}O_2)Cl$		62,134
$[\eta\text{-}C_5H_3(TMS)_2]_2Nb(\eta\text{-}O_2)Cl$		134
$[Cp_2NbCl_2]_2(\mu\text{-}O)$	prep: $Cp_2NbCl_2 + O_2$	546
$[Cp'_2NbCl]_2(\mu\text{-}O)$	x ray	61
$[Cp_2NbBr_2]_2(\mu\text{-}O)$	prep: $Cp_2NbBr_2 + O_2$	67
$[Cp_2Nb(PPh_2)_2]_2(\mu\text{-}O)$		536
$(\eta^5{:}\eta^5\text{-}C_{10}H_8)Cp_2Nb_2Cl_2(\mu\text{-}O)$	prep: $(\eta^5{:}\eta^5\text{-}C_{10}H_8)Cp_2Nb_2(\mu\text{-}H)(\mu\text{-}Cl) + H_2O$	290
$Cp_2Nb({=}O)Cl$	x ray	62,555,569
$(\eta\text{-}C_5H_4R^1)(\eta^5\text{-}C_5H_4R^2)Nb({=}O)Cl$	$R^1 = Bu^t$ and R^2 = H, Me, CHMePh; R^1 = Me and R^2 = H	554
$[\eta\text{-}C_5H_{5-n}(TMS)_n]_2Nb({=}O)Cl$	$n = 1, 2$	62,134
$Cp_2Nb({=}O)(OC_6Cl_5)$		535
$Cp^*_2Ta({=}O)X$	X = OMe, OSO_2Me, OSO_2Tol	227
$Cp_2Nb[ON(Me)NO]({=}O)$	prep: $Cp_2Nb({=}O)Me$ + NO(g.)	187
$[(\eta\text{-}C_5H_4TMS)_2NbCl(OR)][PF_6]$	R = Me, Et, Pr^i	540,541
$Cp_2Nb(OC_6Cl_5)_nCl_{2-n}$	$n = 1, 2$	535
$[Cp_2Nb(OC_6Cl_5)_2](\mu\text{-}O)$		535
$[Cp^*_2TaCl(OH)][OTf]$	x ray	570

Table 21 (continued)

Compound	Comments	Ref.
$[Cp^*_2Ta(OH)_2][OTf]$	x ray	570
$Cp_2Nb(O_2CR)_2$	R = H, Me, Et, Ph	571
$(\eta^5\text{-}C_9H_7)_2Nb(O_2CR)_2$	R = H, Me, Et, Ph	571
$(\eta\text{-}C_5R_5)_2NbX_2Cl$	C_5R_5 = Cp, C_9H_7; X = ($O_2C_4H_7N_2$, $O_2C_{14}H_{11}N_2$	572
$(\eta\text{-}C_5R_5)_2NbX_2$	C_5R_5 = Cp, C_9H_7; X_2 = ($O_2C_4H_7N_2$)($O_2C_4H_6N_2$), ($O_2C_{14}H_{11}N_2$)($O_2C_{14}H_{10}N_2$), ($O_2C_{14}H_{10}N_2$)Cl	572
	C_5R_5 = Cp, C_9H_7; X_2 = O_2CCO_2, $O_2CCH_2CO_2$, $O_2C(CH_2)_2CO_2$, $O_2C(CHOH)_2CO_2$, $O_2C(o\text{-}C_6H_4)CO_2$	573
$Cp_2Nb(8\text{-}NC_9H_6O)_2Cl$	$8\text{-}NC_9H_6OH$ = 8-quinolinol	561
$Cp_2Nb(8\text{-}NC_9H_6O)_2(OEt)$	$8\text{-}NC_9H_6OH$ = 8-quinolinol	561
$[Cp_2NbCl(OPPh_3)]BF_4$		240
Sulfur-donor ligands		
$Cp_2Nb(SPh)_2$	x ray	574
	MO study	575
$[Cp_2Nb(SPh)_2]PF_6$	x ray	574
$Cp_2Nb(S\text{-}4\text{-}C_6H_4X)_2$	X = NO_2, Cl (x ray), H, Me (x ray), OMe (x ray), NH_2	576
	X = Cl; MO study	575
$[Cp_2Nb(SC_6H_4\text{-}4\text{-}X)_2]BF_4$	X = H, Me, Cl (x ray)	576
$Cp_2Nb(2\text{-}SC_4H_3)_nCl_{2-n}$	n = 1, 2	537
$Cp_2Nb(2\text{-}SC_4H_3)(SPh)$		537
$Cp^*_2Nb(SR)_2$	R = Me, Ph; ESR studies	577
$(\eta\text{-}C_5R^1_5)_2Nb(SR^2)_2$	$C_5R^1_5$ = Cp, C_9H_7; R^2 = Me, Et, Pr^n, Pr^i, Bu, Bu^i, Pn^i	578
$[Cp_2Nb(SPh)]_n$	polymeric structure proposed	341
$[Cp_2Nb(SMe)_2]^{0,1+}$	MO study	575
$[Cp_2Nb(1,4\text{-}SC_6H_4S)]_n$	polymer	579
$Cp_2Ta(SR)_2$	R = Me	99
	R = Me, Ph; ESR spectra	580
$[Cp_2Nb(S_2C_2R_2)]PF_6$	R = Me, Ph	581
$Cp_2Nb(S_2C_2R_2)$	R = Me, Ph; electrochemically generated	581
$Cp_2Nb(2\text{-}SC_4H_3)_nCl_{2-n}$	n = 1, 2	535
$(\eta\text{-}C_5Me_4Et)_2Nb(=S)(SH)$	prep: $(\eta\text{-}C_5Me_4Et)_2Nb(BH_4)$ + S_8	321
$Cp_2Nb(\eta\text{-}S_2)Cl$	x ray	582
$(\eta\text{-}C_5Me_4Et)_2Nb(\eta\text{-}S_2)(SH)$		321
$Cp^*_2Nb(\eta\text{-}S_2)(SH)$	x ray	283,322
$[Cp_2Nb(\mu\text{-}S)]_2$	x ray	341
$(\eta\text{-}C_5H_4Bu^t)_4Nb_2S_n$	n = 8, 9	522
$[(\eta\text{-}C_5H_4Bu^t)_2Nb(\eta\text{-}S_2)]_2(\mu\text{-}S_5)$	x ray	283
$Cp^*_4Nb_2S_5$	possibly $[Cp^*_2Nb(\eta\text{-}S_2)]_2(\mu\text{-}S)$	283
$(\eta\text{-}C_5H_4Bu^t)_2Ta(=S)I$		322
$Cp_2Nb[SCH{=}N(Tol)]$		76
$Cp_2Nb(\eta^2\text{-}SSCH_2CCF_3{=}CHCF_3)$	prep: $Cp_2Nb(\eta\text{-}S_2)Me$ + $CF_3C{\equiv}CCF_3$	220
$Cp_2M[\eta^1(S)\text{-}SC(O)Ph]_2$	prep: Cp_2MCl_2 + Tl[SC(O)Ph]	542
$\{Cp_2M[\eta^1(S)\text{-}SC(O)Ph]_2\}^+$	electrochemically generated	542
$[Cp_8Ta_6S_{10}]^{2+}[TaSCl_5]^{2-}$	x ray	524
$Cp_2Nb[S_2P(OR)_2]$	R = Me, Et, Bu^i; ESR studies	233
$\{Cp_2Nb[\eta\text{-}S_2P(OR)_2]\}PF_6$	R = Me, Et, Pr^i	583,584
$Cp_2Nb(\eta\text{-}S_2)[\eta^1\text{-}S_2P(OR)_2]$	R = Me, Et, Pr^i, Bu^t	583
	R = Et, Pr^i	217
$Cp_2Nb[\eta^1\text{-}S_2P(OR)_2]_2$	R = Et, Pr^i	217
$Cp_2Nb(S_2CNR_2)$	R = Me, Et; ESR studies	233
$Cp_2Nb(S_2COR)$	R = Me, Et; ESR studies	233
$(\eta^5{:}\eta^5\text{-}C_{10}H_8)Cp_2Nb_2(\mu\text{-}SCN)$		559
$[(\eta^5{:}\eta^5\text{-}C_{10}H_8)Cp_2Nb_2(\mu\text{-}S)_2]Cl$		559
$Pt[Cp_2Ta(SMe)_2]_2$		99
$[Cp_2Nb(\mu\text{-}SPh)_2Mo(CO)_4]$		574
$[Cp_2Nb(\mu\text{-}SPh)_2Mo(CO)_4]PF_6$	x ray	574
Metal–metal-bonded complexes		
$Cp_2Nb(SnMe_3)(PMe_3)$		122
$[Cp_2Nb(SnMe_3)]_2(\mu\text{-}O)$	x ray	122
$[M][Cp_2Nb(SnMe_3)_2]$	M^+ = [Li(12-crown-4)]$^+$, [Na(15-crown-5)]$^+$, [K(18-crown-6)]$^+$	294
$Cp'_2Ta(ZnCp)_3$		337
$Cp_2Nb[Hg(S_2CNEt_2)]_3$	x ray	340
$\{Cp_2Nb[Au(PPh_3)]_2\}X$	X^- = $[ClO_4]^-$, $[PF_6]^-$; prep: Cp_2NbH_3 + $[Au(PPh_3)]X$	585
$\{(\eta\text{-}C_5H_4TMS)_2Nb[Au(PPh_3)]_2\}X$	X^- = $[ClO_4]^-$, $[PF_6]^-$ (x ray)	585
Miscellaneous		
$Cp_2Nb(CN)_2$		139,233

(i) Bis(cyclopentadienyl) halide complexes

Table 20 lists these complexes by oxidation state and includes assorted mixed-valence dinuclear species and heterobimetallic complexes which contain bridging halide ligands. Although the d^0 complexes Cp_2MBr_3 were first described in 1954, new d^0 bis(cyclopentadienyl) halides and new preparations of known compounds continue to be reported. For example, the cationic complexes $[Cp_2MCl_2]X$ ($X^- = [BF_4]^-$, $[PF_6]^-$, etc.) are prepared from oxidizing the metallocene dichlorides Cp_2MCl_2 with $[Cp_2Fe]X$ in CH_2Cl_2 or THF.[71,219] The solvates $[Cp_2MCl_2]X{\cdot}NCMe$ are obtained when these oxidations are effected in NCMe and $[Cp_2NbCl_2]BF_4{\cdot}NCMe$ has been structurally characterized.[219] The electrochemical generation of both neutral[534] and cationic[534,542] M^V species has also been reported.

The d^1 M^{IV} $(\eta\text{-}C_5R_5)_2MX_2$ species are the largest group of bis η^5 halides, which mirrors the redox stability of the d^1 oxidation state, especially for niobium. Thus, reactions using a d^0 precursor often lead to reduction of the metal center and isolation of a d^1 Cp_2MX_2 compound. For example, $Bu^n{}_3Sn(C_5H_4Pr^i)$ reacts with excess $TaCl_5$ to form $(\eta\text{-}C_5H_4Pr^i)_2TaCl_2$,[242] Cp_2TaCl_2, Cp_2TaBr_2, and Cp'_2NbCl_2 are all available from the reactions of the appropriate MX_5 halide with $Bu^n{}_3Sn(C_5H_4R)$ (for R = H or Me).[99] Simple metathesis reactions of d^1 precursors and oxidation reactions of d^2 precursors also constitute viable routes to d^1 Cp_2MX_2 compounds, as indicated in Equations (20),[231] (21),[287] (22)[240] and (23).[287]

$$NbCl_4(THF)_2 + 2Li(C_5H_4R) \longrightarrow (\eta\text{-}C_5H_4R)_2NbCl_2 \quad (20)$$

R = H, Me, or TMS

$$(\eta\text{-}C_5H_4TMS)_2NbCl + PhCH_2Cl \longrightarrow (\eta\text{-}C_5H_4TMS)_2NbCl_2 \quad (21)$$

$$Cp_2Nb(CH_2Ph)_2 + I_2 \longrightarrow Cp_2NbI_2 + \text{excess } I_2 \longrightarrow [Cp_2NbX_2][X_3] \quad (22)$$

$$(\eta\text{-}C_5H_4TMS)_2NbH_3 + EtBr \longrightarrow (\eta\text{-}C_5H_4TMS)_2NbBr_2 \quad (23)$$

The pentamethylcyclopentadienyl derivative $Cp^*{}_2NbCl_2$ is available by reacting $Cp^*{}_2Nb(\eta^2\text{-}BH_4)$ with HCl (aq.)[70] while the tantalum analogue $Cp^*{}_2TaCl_2$ must be prepared from $Cp^*TaCl_3(PMe_3)$ and KCp^*.[282] This latter reaction has allowed the facile preparation of mixed-ligand compounds as indicated in Equation (24).[282] Other d^1 mixed Cp complexes (e.g., $(\eta\text{-}C_5H_4R)Cp'NbCl_2$ for R = Bu^t or CHMePh) have been prepared by adding one $\eta\text{-}C_5H_4R$ ligand to $NbCl_5$ (in heptane) using $Bu^n{}_3Sn(C_5H_4R)$, followed by *in situ* reduction to Nb^{IV} and metathesis with $Li(C_5H_4R)$.[553]

$$Cp^*TaCl_3(PMe_3) + NaCp \longrightarrow Cp^*CpTaCl_2 + PMe_3 + NaCl \quad (24)$$

An interesting series of formal mixed-valence M^{III}–M^{IV} compounds $Cp_4M_2Cl_3$ have been reported. The NaHg reduction of Cp_2NbCl_2 (a THF suspension) for short periods of time (~30 min) affords a purple compound upon evaporating the solution to dryness.[555] This complex is formulated as the μ-Cl species $[Cp_2NbCl]_2(\mu\text{-}Cl)$ (or "$Cp_4Nb_2Cl_3$"), which is proposed to contain one $Cp_2Nb^{III}Cl$ moiety and one $Cp_2Nb^{IV}Cl_2$ moiety, although its exact nature is not described. This compound oxidizes in air to $Cp_2Nb(=O)Cl$ and reacts with $AgClO_4$ to form $[Cp_4Nb_2Cl_2]ClO_4$. Moreover, Cp_2NbCl_2 can be "displaced" from $[Cp_2NbCl]_2(\mu\text{-}Cl)$ upon reaction with $FeCl_3$ and $Tl(C_6F_5)_2Br$, forming the Cp_2NbCl adducts $Cp_2Nb(\mu\text{-}Cl)_2FeCl_2$ and $Cp_2Nb(\mu\text{-}Cl)(\mu\text{-}Br)Tl(C_6F_5)_2$, respectively.[555] Similarly, reducing Cp_2NbBr_2 with two equivalents of NaHg affords dinuclear $Cp_4Nb_2Br_3$, which is proposed to take on a $Cp_2BrNb(\mu\text{-}Br)NbBrCp_2$ structure with one formally Nb^{III} and one Nb^{IV} metal center.[67]

An equally intriguing series of $(\eta\text{-}C_5R_5)_2M^{III}X$ complexes has been prepared from the chemical or electrochemical reduction of $(\eta\text{-}C_5R_5)_2M^{IV}X_2$. The structures and molecularities of these compounds are often unknown; indeed, species of identical composition prepared similarly have been reported as monomeric in one report and dimeric in another. In the case of bulkier cyclopentadienyl ligands, such as $\eta\text{-}C_5H_4TMS$, the indications are that a very sensitive monomeric complex is indeed isolable and for less bulky groups, monomer ⇌ dimer equilibria may be set up in solution in which the slow formation of a dimer or oligomer may be observed. Since details which might permit such a determination are often lacking, compounds reported in Table 20 are presented as they are in the original literature.

For example, the NaHg reduction of the complexes $(\eta\text{-}C_5H_4TMS)_2NbX_2$ (X = Br or I) affords the extremely sensitive, 16-electron Nb^{III} compounds $(\eta\text{-}C_5H_4TMS)_2NbBr$ and $(\eta\text{-}C_5H_4TMS)_2NbI$.[304] These 16-electron species are readily complexed by CO, CNR, PhC≡CPh, etc., to form the corresponding $(\eta\text{-}C_5H_4TMS)_2NbX(L)$ compounds,[304] by PhNCO to provide $(\eta\text{-}C_5H_4TMS)_2Nb[\eta^2(O,C)\text{-}OCNPh]Cl$,[448] by various ketenes to afford $(\eta\text{-}C_5H_4TMS)_2Nb[\eta^2(O,C)\text{-}O{=}C{=}CR_2]X$,[345,436] and by ketenimines such as $PhN{=}C{=}CPh_2$ to give $(\eta\text{-}C_5H_4TMS)_2Nb[\eta^2(N,C)\text{-}PhN{=}C{=}CPh_2]X$.[444]

An improved preparation of the Nb^{III} complex $(\eta\text{-}C_5H_4TMS)_2NbCl$ has been described from Pedersen's complex $NbCl_3(DME)$.[437] Thus, $NbCl_3(DME)$ and two equivalents of $Li[C_5H_4TMS]$ react to form $(\eta\text{-}C_5H_4TMS)_2NbCl$; Cp_2NbCl is prepared similarly from $NbCl_3(DME)$ and NaCp, but must be used *in situ* since oligomerization reactions occur upon attempted isolation.[437]

Finally, a series of halide-bridged, heterobimetallic compounds have been prepared from M^{III} and M^{IV} halides and other transition and main-group elements (Table 20). The adducts $Cp_2Nb(\mu\text{-}Cl)_2FeCl_2$ and $Cp_2Nb(\mu\text{-}Cl)(\mu\text{-}Br)Tl(C_6F_5)_2$ can be prepared as described above.[555] Cp_2NbCl_2 also reacts with metal halides MX_n (for $MX_n = AlCl_3$, $FeCl_3$, $TlCl_3$, or $Tl(C_6F_5)_2Cl$), forming $Cp_2ClNb(\mu\text{-}Cl)MX_n$.[555]

(ii) Other bis(cyclopentadienyl) complexes

Table 21 lists all other bis η^5 complexes, including those containing Si-, N-, P-, O-, S-, and so on, donors either alone or in combination with halide ligands.

(a) Silyl complexes. Only a handful of bis η^5 silyl compounds are known; for example, $Cp_2Nb(SiPh_3)_2$ is formed from Cp_2NbCl_2 and $LiSiPh_3$ and can be oxidized with $[Cp_2Fe]BF_4$ to afford cationic $[Cp_2Nb(SiPh_3)_2]BF_4$.[536] Treating $Cp_2Nb(SiPh_3)_2$ with O_2 is reported to form $[Cp_2Nb(SiPh_3)_2]_2(\mu\text{-}O)$.

(b) Nitrogen donors. The bis η^5 complexes with nitrogen donor ligands include both neutral and cationic nitrile, bipy, and phen adducts. For example, Cp_2NbCl_2 reacts with L_2 = bipy or phen in the presence of $TlPF_6$ to replace chlorides for nitrogen donors and form $[Cp_2NbL_2][PF_6]_2$.[560] Cyclic voltammetric studies of both complexes revealed two reduction processes, the first corresponding to the reversible metal-based reduction and generation of $[Cp_2NbL_2]^+$, while the more negative (and less reversible) second reduction is assigned a ligand-based reduction. The absence of the usual $Nb^{IV} \rightarrow Nb^V$ oxidation process is attributed to the dicationic character of $[Cp_2NbL_2]^{2+}$.[560]

Both terminal and bridging imido-like complexes of bis η^5 moieties are also reported, with the μ-NNR or μ-NR ligands occurring in fulvalene-bridged niobium centers. There are considerably fewer imido complexes coordinated to bis η^5 fragments than to the monokis η^5 compounds. Upon reaction of THF solutions of $Cp^*_2TaCl(THF)$ with N_3Ph, an orange solid, proposed to be $Cp^*_2Ta(N_3Ph)Cl$, is isolated.[310] Thermolysis of this complex in toluene (90 °C, 10 min) affords the oxidation product $Cp^*_2Ta(=NPh)Cl$. An x-ray structure of $Cp^*_2Ta(=NPh)Cl$ reveals the expected bent metallocene structure with a $Ta\text{–}N\text{–}C_{ipso}$ angle of 176.4(4)°. Displacement of chloride ligand from $Cp^*_2TaCl(THF)$ by LiNHR (R = Ph, Bu^t) provides a high-yield route to $Cp^*_2Ta(=NR)H$, presumably via intermediate $[Cp^*_2Ta(NHR)]$.[310] An x-ray structural determination of $Cp^*_2Ta(=NPh)H$ reveals a long Ta–N bond of 0.1831 (10) nm but a nearly linear Ta–N–C angle of 177.8 (9)°. While an *sp*-hybridized nitrogen is implied by the Ta–N–C linearity, a Ta≡N triple bond is not required and the suggestion has been made that the second π bond from nitrogen to tantalum is extremely weak, as the Ta–N distance suggests.[309] Molecular orbital calculations on the model compound $Cp_2Ta(=NPh)H$ have been described by Jørgensen.[312]

(c) Phosphorus donors. Phosphine adducts of the bis η^5 complexes have been described with complexes of the form $(\eta\text{-}C_5R_5)_2MX(PR_3)$ constituting the largest group. Examples include $Cp_2NbBr(PMe_3)$,[240] $(\eta\text{-}C_5H_4Pr^i)_2TaCl(PMe_3)$,[242] and $(\eta\text{-}C_5H_4TMS)_2NbCl[P(OMe)_3]$,[243] which are all prepared by reducing the appropriate Cp_2MCl_2 complex in the presence of PR_3. Less common are the cationic M^{IV} $[(\eta\text{-}C_5R_5)_2MX(PR_3)]^+$ species (e.g., $\{(\eta\text{-}C_5H_4TMS)_2NbCl[P(OMe)_3]\}PF_6$[136]) and the M^{III} or M^{IV} $[(\eta\text{-}C_5R_5)_2M(PR_3)_2]^{n+}$ derivatives (e.g., $[Cp'_2Ta(dmpe)][BF_4]_n$ for n = 1 or 2[27]). Pentadienyl or "open cyclopentadienyl" complexes have also been prepared, for example, $NbCl_4(PEt_3)_2$ reduces upon reaction with four equivalents of $K[2,4\text{-}C_7H_{11}]$ to afford $(2,4\text{-dimethylpentadienyl})_2Nb(PEt_3)$.[586] Phosphido $[PR_2]^-$ adducts of bis η^5 compounds have been described, for example, $Cp_2Nb(PPh_2)_2$ and its oxidation products $[Cp_2Nb(PPh_2)_2]BF_4$ and $[Cp_2Nb(PPh_2)_2]_2(\mu\text{-}O)$ have been reported.[536]

(d) Oxygen donors. A diverse group of bis η^5 derivatives of niobium and tantalum are known which contain oxygen-donor ligands, including $\eta^2\text{-}O_2$, μ-O, terminal oxo, alkoxide, hydroxide, carboxylate, and metallacyclic compounds with coordinated oxygen atoms (Table 21). Terminal oxo compounds have been prepared upon O_2 oxidation of d^2 species. Thus, reacting $(\eta\text{-}C_5R_5)_2Nb(CO)Cl$ species with dry O_2 is reported to form oxo derivatives $Cp_2Nb(=O)Cl$, $(\eta\text{-}C_5H_4TMS)_2Nb(=O)Cl$, or $[\eta\text{-}C_5H_3(TMS)_2]_2Nb(=O)Cl$.[62] Note that the μ-oxo species $[Cp_2NbBr_2]_2(\mu\text{-}O)$ forms when the d^1 complex Cp_2NbBr_2 is exposed to O_2.[67] $(\eta\text{-}C_5H_4TMS)_2Nb(\eta\text{-}O_2)Cl$ is formed when H_2O_2 is used to oxidize d^1 $(\eta\text{-}C_5H_4TMS)_2NbCl_2$.[62] A terminal oxo complex, $Cp_2Nb(=O)Cl$ (prepared from air oxidation of $Cp_2Nb(H)(TolAs)_2$) has been structurally characterized with Nb–O = 0.1737 (6) nm.[569]

In contrast to the oxidation of the carbonyl compounds, O_2 oxidation of certain isocyanide complexes $(\eta\text{-}C_5R_5)_2NbCl(CNR)$ affords peroxo species, for example, $(\eta\text{-}C_5H_4TMS)_2NbCl(\eta\text{-}O_2)$ and [η-

$C_5H_3(TMS)_2]_2NbCl(\eta\text{-}O_2)$.[134] Both peroxo complexes react with PPh_3 to abstract an oxygen (forming $Ph_3P{=}O$) to generate $(\eta\text{-}C_5H_4TMS)_2Nb({=}O)Cl$ and $[\eta\text{-}C_5H_3(TMS)_2]_2Nb({=}O)Cl$ (Equations (25) and (26)). *t*-Butyl peroxide (Bu^tOOH) also reacts with $[\eta\text{-}C_5H_3(TMS)_2]_2NbCl(CNR)$ (for R = Ph or 2,6-$C_6H_3Me_2$) to form the same oxo complex $[\eta\text{-}C_5H_3(TMS)_2]_2Nb({=}O)Cl$.[134]

$$(\eta\text{-}C_5H_4TMS)_2NbCl(CNR) + O_2 \longrightarrow (\eta\text{-}C_5H_4TMS)_2NbCl(\eta\text{-}O_2) + CNR \qquad (25)$$

$$(\eta\text{-}C_5H_4TMS)_2NbCl(\eta\text{-}O_2) + PPh_3 \longrightarrow (\eta\text{-}C_5H_4TMS)_2Nb({=}O)Cl + Ph_3P{=}O \qquad (26)$$

(e) Sulfur donors. Sulfur-donor ligands are encountered more frequently in niobium and tantalum as bis η^5 complexes rather than monokis η^5 species. The most common stoichiometry is the bis(thiolate) $[(\eta\text{-}C_5R^1{}_5)_2Nb(SR^2)_2]^{0/1+}$ class, although $\eta\text{-}S_2$, bridging and terminal sulfido, and various sulfur chelates are also observed.

$Cp^*{}_2Nb(\eta^2\text{-}BH_4)$ reacts with S_8 to form, *inter alia*, a complex which analyzes as $Cp^*{}_2NbS_3H$.[283] Although the question of whether this product should be formulated as $Cp^*{}_2Nb(H)(S_3)$ or $Cp^*{}_2Nb(SH)(\eta\text{-}S_2)$ was originally unanswered, a later crystallographic study revealed this species to be $Cp^*{}_2Nb(SH)(\eta\text{-}S_2)$.[322]

Neutral $Cp_2Nb(SPh)_2$ and $Cp_2Nb(\mu\text{-}SPh)_2Mo(CO)_4$ (prepared by the procedure of Douglas and Green[587]) can be oxidized with $NOPF_6$ to form the cationic derivatives $[Cp_2Nb(SPh)_2]PF_6$ and $[Cp_2Nb(\mu\text{-}SPh)_2Mo(CO)_4]PF_6$.[574] Cyclic voltammetric and IR data for diamagnetic $[Cp_2Nb(\mu\text{-}SPh)_2Mo(CO)_4]PF_6$ support the positive charge of this complex being delocalized over both metal complex portions.[574] A related series of d^1 vs. d^0 complexes, $Cp_2Nb(SC_6H_4\text{-}4\text{-}X)_2$ (X = NO_2, Cl, H, Me, OMe, NH_2), are available by reacting Cp_2NbCl_2 with HSC_6H_4-4-X in the presence of NEt_3 (in toluene solution).[576] Cyclic voltammetric measurements on these species revealed that both $Nb^{III} \rightleftharpoons Nb^{IV}$ and $Nb^{IV} \rightleftharpoons Nb^{V}$ couples are observed and are reversible.[576] Structurally, the d^0 complexes and $Cp_2Nb(SC_6H_4\text{-}4\text{-}Cl)_2$ all have *endo* arrangements of their SR ligands (relative to the Cp rings) and large S–Nb–S angles (**50**). However, all other d^1 compounds are characterized by an *exo* configuration with phenyl rings bent back into the cleft formed by the Cp ligands (**51**). The molecular orbital origins of this bending have been investigated.[575,576]

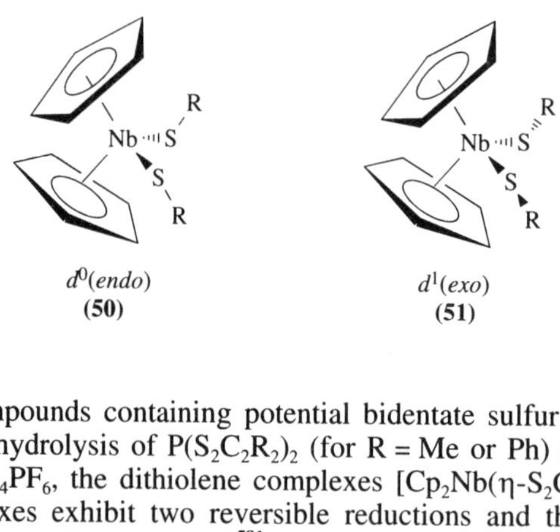

d^0(*endo*) **(50)** d^1(*exo*) **(51)**

Finally, a number of compounds containing potential bidentate sulfur ligands have been reported, bound either η^1 or η^2. The hydrolysis of $P(S_2C_2R_2)_2$ (for R = Me or Ph) in the presence of Cp_2NbCl_2 forms, upon addition of NH_4PF_6, the dithiolene complexes $[Cp_2Nb(\eta\text{-}S_2C_2R_2)]PF_6$.[581] Electrochemical studies reveal these complexes exhibit two reversible reductions and the neutral species $Cp_2Nb(\eta\text{-}S_2C_2R_2)$ have been electrochemically generated.[581]

(f) Metal-metal bonded complexes. A few heterobimetallic metal–metal-bonded complexes containing tin, zinc, mercury, or gold directly bound to Cp_2M moieties have been reported. Photolysis of $Cp_2Nb(\eta^2\text{-}CH_2{=}CHPh)(SnMe_3)$ in the presence of CO_2 forms a mixture of $Cp_2Nb(CO)(SnMe_3)$ and $[Cp_2Nb(SnMe_3)]_2(\mu\text{-}O)$, formed by CO_2 cleavage.[122] The hydrides $Cp_2Nb(SnMe_3)H_2$ and $Cp_2Nb(SnMe_3)_2H$ are deprotonated by BuLi-12-crown-4, $NaNH_2$-15-crown-5 and KH-18-crown-6 to form $[Cp_2Nb(SnMe_3)H][M]$ and $[Cp_2Nb(SnMe_3)_2][M]$ [for M^+ = $[Li(12\text{-crown-}4)]^+$, $[Na(15\text{-crown-}5)]^+$, $[K(18\text{-crown-}6)]^+$], respectively.[294] The structure of $[Cp_2Nb(SnMe_3)_2][K(18\text{-crown-}6)]$ shows well separated ions. In the appropriate ratios, Cp'_2TaH_3 and Cp_2Zn react to afford $Cp'_2Ta(ZnCp)_3$ in refluxing benzene; the same product forms from $Cp'_2Ta(CH_2{=}CH_2)H$ and Cp_2Zn.[337] Similar reactions between Cp_2NbH_3 or $(\eta\text{-}C_5H_4TMS)_2NbH_3$ and solutions of $[Au(PPh_3)]X$ (X^- = $[ClO_4]^-$ or $[PF_6]^-$) result in the formation of the $NbAu_2$ clusters $\{Cp_2Nb[Au(PPh_3)]_2\}X$ and $\{(\eta\text{-}C_5H_4TMS)_2Nb[Au(PPh_3)]_2\}X$, which have been characterized with triangular $NbAu_2$ frameworks.[585]

2.7.4 η⁶ Complexes

There have been several successful forays into niobium and tantalum arene chemistry as a result of improved methods for the initial coordination of an arene ligand. The primary methods of introducing the arene ligand to these metals include (i) by Al/AlX_3 reduction of a metal halide in the presence of the arene (Fischer–Hafner conditions); (ii) by metal vapor synthesis reactions, and (iii) by alkyne cyclotrimerization chemistry. Many of these complexes exhibit redox behavior that has allowed access to a range of oxidation states (from M^{-I} to M^{IV}). Table 22 summarizes η^6 complexes of niobium and tantalum. Some of the low-valent compounds have appeared in a short review.[6]

2.7.4.1 *Fischer–Hafner preparations*

The earliest report of isolable niobium and tantalum arene complexes by Fischer and Röhrscheid[599] included the trinuclear clusters $[(\eta\text{-}C_6Me_6)_3M_3Cl_6]^+$ isolated upon quenching the primary reaction product obtained from MCl_5 and $Al/AlCl_3$ in a C_6Me_6 melt. Another complex, proposed to be dinuclear $(\eta\text{-}C_6Me_6)_2M_2Cl_4$, was also isolated from these reactions but incompletely characterized. The initial reaction product from MCl_5, $Al/AlCl_3$, C_6Me_6 mixtures was later identified by Calderazzo *et al.*[590,591] as $(\eta\text{-}C_6Me_6)Nb(\eta^2\text{-}AlX_4)_2$ (X = Cl, Br) and was shown (for the bromide complex) to convert into $[(\eta\text{-}C_6Me_6)_3Nb_3Br_6]Br$ upon hydrolysis. The centrosymmetric dimer $[(\eta\text{-}C_6Me_6)Nb(\mu\text{-}Br)_2]_2$ has been shown to arise from THF treatment of $(\eta\text{-}C_6Me_6)Nb(\eta^2\text{-}AlBr_4)_2$ (Scheme 24) and is also confirmed to react with H_2O to form trinuclear $[(\eta\text{-}C_6Me_6)_3Nb_3Br_6]Br$ in 50% yield, for which a possible reaction stoichiometry has been suggested (Equation (27)).[591]

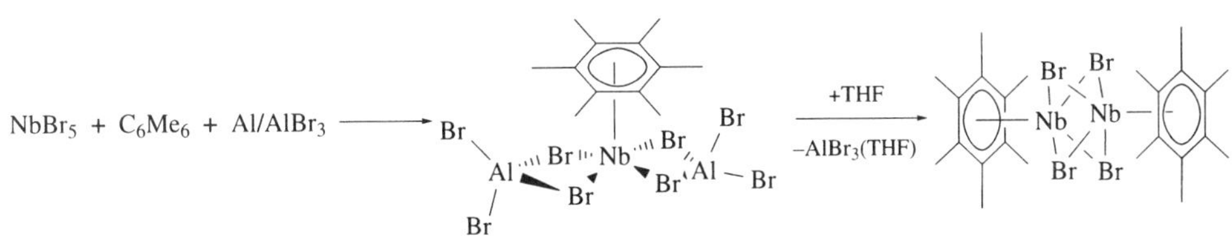

Scheme 24

$$3(\eta\text{-}C_6Me_6)Nb(\eta^2\text{-}AlBr_4)_2 + H_2O \longrightarrow 1/2H_2 + [(\eta\text{-}C_6Me_6)_3Nb_3Br_6]Br + 5AlBr_3 + AlBr_2(OH) \quad (27)$$

The trinuclear cations $[(\eta\text{-}C_6Me_6)_3M_3Cl_6]^+$ can be oxidized to afford salts of $[(\eta\text{-}C_6Me_6)_3M_3Cl_6]^{2+}$, some of which have been structurally characterized. For example, Stollmaier and Thewalt[593] described a high-quality crystal structure of the monocation $[(\eta\text{-}C_6Me_6)_3Nb_3Cl_6]Cl$ (originally prepared by Fischer and Röhrscheid[599]), for which an imprecise structure was communicated some years earlier.[600] The arene ligand in $[(\eta\text{-}C_6Me_6)_3Nb_3Cl_6]^+$ is characterized by the folding and structural distortions described below for complexes such as $(\eta\text{-}C_6R_6)Ta(OAr)_2Cl$.[463,594]

2.7.4.2 *Metal vapor synthesis preparations*

Zerovalent bis(arene) complexes can be prepared from cocondensation of niobium or tantalum atoms with arene; thus $(\eta\text{-arene})_2Nb$ (arene = C_6H_6, C_6H_5Me, 1,3,5-$C_6H_3Me_3$, 1,3,5-$C_6H_3Bu^t_3$)[588,589] and $(\eta\text{-}C_6H_6)_2Ta$[588] can be isolated in modest yields. It was later found that a 30–50% yield of $(\eta\text{-}C_6H_3Me_3)_2Nb$ can be isolated from the reaction of $NbCl_5$, Al, AlX_3, and mesitylene in DME at −15 °C, which is a rare example of a bis(arene) group 5 complex being accessible via solution methods rather than metal vapor synthesis techniques.[590,591]

The bis(toluene) complex $(\eta\text{-}C_6H_5Me)_2Nb$, and its analogues originating from metal vapor syntheses, have provided a rich reaction chemistry. Thus, $(\eta\text{-}C_6H_5Me)_2Nb$ serves as a useful precursor for a range of other arene complexes such as $[(\eta\text{-}C_6H_5Me)_2Nb(PR_3)]BF_4$, $(\eta\text{-}C_6H_5Me)Nb(PMe_3)_3H$, and $(\eta\text{-}C_6H_5Me)_2NbMe$, as outlined in Scheme 25.[26,92]

Table 22 η^6 Complexes of niobium and tantalum.

Compound	*Comments*	*Ref.*
$(\eta\text{-arene})_2Nb$	arene = C_6H_6, C_6H_5Me, 1,3,5-$C_6H_3Me_3$	588
	arene = 1,3,5-$C_6H_3Bu^t_3$	589
	arene = 1,3,5-$C_6H_3Me_3$	590,591
$(\eta\text{-}C_6H_6)_2Ta$	prep: Ta(atoms) + C_6H_6; metal vapor synthesis	588
$[(\eta\text{-}1,3,5\text{-}C_6H_3Bu^t_3)_2Nb]BF_4$		589
$(\eta\text{-}C_6H_5Me)_2NbR$	R = Me, Ph	92
	R = Me, Ph, CH_2Ph	26
$(\eta\text{-}C_6H_5Me)_2NbX$	X = Br, I	26,92
$K[(\eta\text{-}C_6H_5Me)_2Nb]$	reactive intermediate	92
$(\eta\text{-}C_6H_5Me)Nb(PMe_3)_3H$		26,92
$(\eta\text{-}C_6H_5Me)Nb(PMe_3)_3(TMS)$		26,92
$K[(\eta\text{-}C_6H_5Me)Nb(PMe_3)_3]$	reactive intermediate	26,92
$[(\eta\text{-}C_6H_5Me)_2Nb(PR_3)]BF_4$	PR_3 = PMe_3 (x ray), PMe_2Ph	592
$[K(18\text{-crown-}6)][(\eta\text{-}1,3,5\text{-}C_6H_3Me_3)_2Nb]$		592
$(\eta\text{-}C_6H_5Me)Nb[\eta^3\text{-}(Me_2PCH_2)_3SiMe]$	x ray; prep: $(\eta\text{-}C_6H_5Me)_2Nb + (Me_2PCH_2)_3SiMe$	26,92
$K[(\eta\text{-}C_6H_5Me)Nb\{\eta^3\text{-}(Me_2PCH_2)_3SiMe\}]$	*in situ* generation	26
$(\eta\text{-}C_6H_5Me)Nb[\eta^3\text{-}(Me_2PCH_2)_3SiMe]H$		26,92
$(\eta\text{-}C_6H_5Me)CpNb(PMe_3)$	prep: $(\eta\text{-}C_6H_5Me)Nb(PMe_3)_3H + C_5H_6$	26,92
$(\eta\text{-}C_6H_5Me)(\eta^5\text{-}C_9H_7)Nb(PMe_3)$	$\eta^5\text{-}C_9H_7 = \eta^5$-indenyl	26,92
$(\eta\text{-}C_6H_5Me)(\eta^5\text{-}C_7H_9)Nb(PMe_3)$	x ray; prep: $(\eta\text{-}C_6H_5Me)Nb(PMe_3)_3H + C_7H_8$	26,92
$(\eta\text{-}C_6H_5Me)_2Nb_2(\mu\text{-}SR)_4$	R = CH_2Ph, Me, Bu^n; prep: $(\eta\text{-}C_6H_5Me)_2Nb$ + RSSR	26
$[(\eta\text{-}C_6Me_6)_3Nb_3Cl_6]Cl$	x ray	593
$[(\eta\text{-}C_6Me_6)_3Nb_3Br_6]Br$	prep: $(\eta\text{-}C_6Me_6)_2Nb_2(\mu\text{-}Br)_4 + H_2O$	590,591
$[(\eta\text{-arene})Nb(CO)_4][Al_2Br_6Cl]$	arene = 1,4-$C_6H_4Me_2$, 1,3,5-$C_6H_3Me_3$, 1,2,4,5-$C_6H_2Me_4$, C_6Me_6	17
$[(\eta\text{-arene})Ta(CO)_4][Al_2Br_6Cl]$	arene = C_6H_5Me, 1,4-$C_6H_4Me_2$, 1,3,5-$C_6H_3Me_3$, 1,2,4,5-$C_6H_2Me_4$, C_6Me_6	17
$[(\eta\text{-}C_6H_3Me_3)_2Nb(CO)]X$	$X^- = [Nb(CO)_6]^-$; prep: $(\eta\text{-}C_6H_3Me_3)_2Nb$ + CO	590,591
	$X^- = [V(CO)_6]^-$	591
	$X^- = [BPh_4]^-$; x ray	591
$[(\eta\text{-}C_6H_3Me_3)_2Nb(THF)]BPh_4$	*in situ* preparation	591
$(\eta\text{-arene})Nb(\eta^2\text{-}AlCl_4)_2$	arene = C_6H_6, C_6Me_6	590
	arene = C_6H_6, C_6Me_6, $C_6H_3Me_3$	591
$(\eta\text{-}C_6Me_6)Nb(\eta^2\text{-}AlBr_4)_2$		590
$(\eta\text{-}C_6Me_6)_2Nb_2(\mu\text{-}Br)_4$	prep: $(\eta\text{-}C_6Me_6)Nb(\eta^2\text{-}AlBr_4)_2$ + THF; x ray	590,591
$[(\eta\text{-}C_6H_3Me_3)_2Nb(RC{\equiv}CR)]BPh_4$	R = Me, Ph	412
$[Ph_2B(\eta\text{-}C_6H_5)_2]Nb(RC{\equiv}CR)$	R = Me (x ray), Ph	412
$(\eta\text{-}C_6Me_6)Ta(OAr)_2Cl$	x ray	463,594
$(\eta\text{-}C_6Et_6)Ta(OAr)_2Cl$	prep: $Ta(OAr)_2Cl_3(OEt_2) + 2\ NaHg + 3\ EtC{\equiv}CEt$	594
$[\eta\text{-}C_6(CD_3)_6]Ta(OAr)_2Cl$		463
$(\eta\text{-}C_6Et_6)Ta(OAr')_2Cl$		595
$(\eta\text{-arene})Ta(OAr)Cl_2$	arene = C_6Me_6, C_6Et_6	595
$(\eta\text{-}C_6Et_6)Ta(OAr)_2X$	X = Cl, Br, I	596,597
$(\eta\text{-}C_6Et_6)Ta(OAr')Cl_2$	prep: $(\eta\text{-}C_6Et_6)Ta(OAr')_2Cl + Ta(OAr')_2Cl_3(OEt_2)$	595
$(\eta\text{-}C_6Et_6)Ta(OAr)Br_2$		597
$(\eta\text{-}C_6Me_6)Ta(OAr)R_2$	R = Me, Ph	598
$(\eta\text{-}C_6Me_6)Ta(OAr)Et_2$	x ray	598
$(\eta\text{-}C_6Me_6)Ta(OAr)(Et)Br$		598
$(\eta\text{-}C_6Me_6)Ta(OAr)(R)H$	R = Et, CH_2TMS	598
$(\eta\text{-}C_6Me_6)Ta(OAr)(CH_2TMS)Cl$		598
$(\eta\text{-arene})Ta(OAr)_2$	arene = C_6Me_6, C_6Et_6 (x ray)	596,597
$(\eta\text{-}1,3,5\text{-}C_6H_3Bu^t_3)Ta(OAr)_2Cl$	prep: $(ArO)_2ClTa(CBu^t{=}CHCBu^t{=}CH) + HC{\equiv}CBu^t$	408,459
$(\eta\text{-}1,3,5\text{-}C_6H_3Bu^t_3)Ta(OAr)_2Me$	x ray	408
$[\eta^6\text{-}C_5(Bu^t)(CH_2Bu^t)_3(CH_2CMe_2CH_2)]TaCl_3$		379
$[\eta^6\text{-}C_5Me_4CH_2]_2TaH$	thermodynamic product from thermolysis of $Cp^*_2Ta({=}CH_2)H$	109
$[\eta^6\text{-}C_5Me_4CH_2]Cp^*Ta(CH{=}CH_2)H$		109
$[\eta^6\text{-}C_5Me_4CH_2]Cp^*Ta(CH_2{=}CH_2)$	unstable; rearranges to $[\eta^5{:}\eta^1\text{-}C_5Me_4(CH_2)_3]Cp^*Ta$	109
$(\eta^6\text{-}C_7H_8)Nb(PMe_3)_2Cl_2$	cycloheptatriene complex	473

2.7.4.3 Cycloaddition preparations

Another incursion into niobium and tantalum arene chemistry which has been especially successful came with the discovery that Ta^{III} arene complexes could be isolated from alkyne cyclotrimerization reactions. Thus, upon reducing $Ta(OAr)_2Cl_3(OEt_2)$ by two electrons in the presence of $RC{\equiv}CR$ (R = Me, Et), blue crystals of $(\eta\text{-}C_6R_6)Ta(OAr)_2Cl$ (**52**) are isolated.[463,594] One of the most significant structural

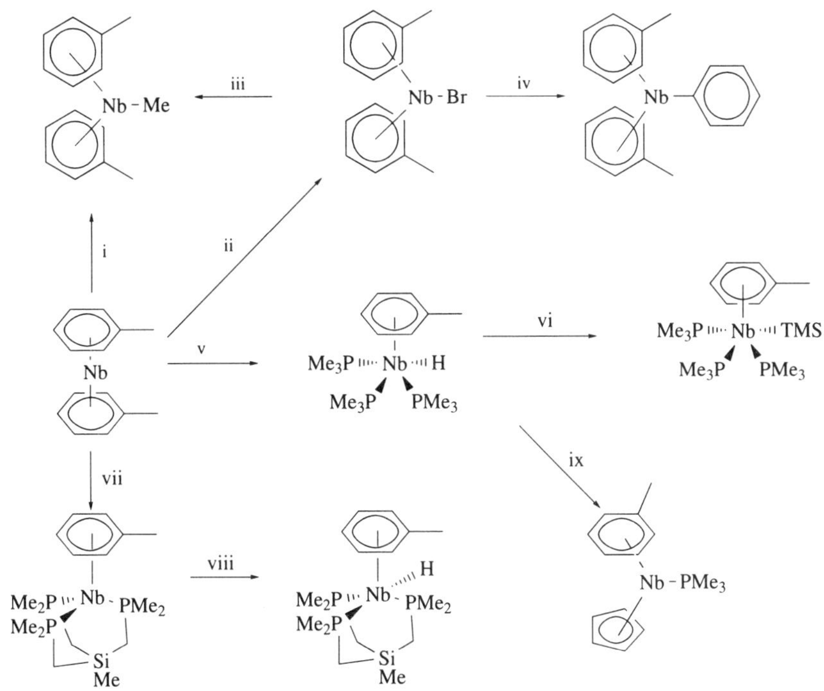

i, K film, then MeI; ii, CH_2=$CHCH_2Br$; iii,, MeLi; iv, PhLi; v, PMe_3; vi, K film, then TMS-Cl
vii, $MeSi(CH_2CH_2PMe_2)_3$; viii, K film, then H_2O; ix, C_5H_6

Scheme 25

features of these compounds is a coordinated C_6R_6 ligand which shows (i) substantial folding along the C-1–C-4 axis (the dihedral angle can be as large as 34°); (ii) an interruption of aromaticity within the C_6R_6 ring which can be described as a 1,4-diene-type π localization; and (iii) the close approaches of C-1 and C-4 to the metal which compare to the values of Ta–C(sp^3) single bond lengths, as indicated in (**53**). These features may be attributed to a back-bonding interaction between filled metal δ functions and the arene LUMO as suggested in (**54**).[595]

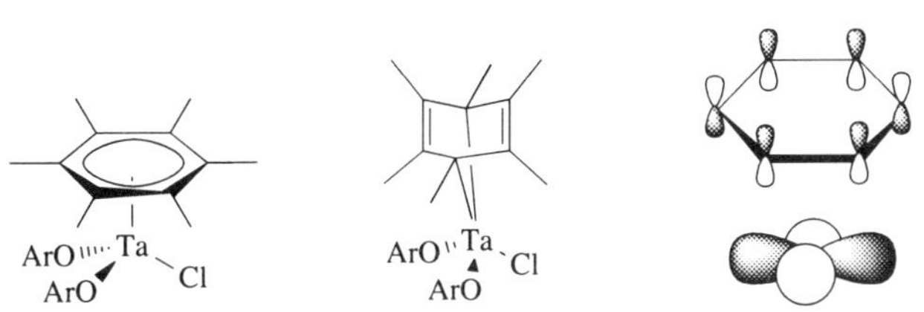

(52) **(53)** **(54)**

Both Ta^{II} and Ta^{IV} arenes are available from redox reactions of these complexes. Thus, stable Ta^{II} complexes (η-C_6R_6)$Ta(OAr)_2$ (R = Me, Et) are prepared from the one-electron reduction of (η-C_6R_6)$Ta(OAr)_2Cl$[596,597] and electrochemical evidence has been presented for the existence of unstable [(η-C_6Me_6)Ta(OAr)R_2]$^+$ complexes (R = Me or Ph).[598]

2.7.4.4 Miscellaneous preparations

Finally, a number of preparations of arene compounds have been effected in recent years by means other than those outlined above. The tetracarbonyl cations [(η-arene)$M(CO)_4$][Al_2Br_6Cl] have been prepared by Calderazzo *et al.*[17] by reacting [$M_2(\mu\text{-}Cl)_3(CO)_8$]$^-$ with arenes in the presence of a halide acceptor such as $AlBr_3$. The [(η-arene)$M(CO)_4$]$^-$ anions can be compared with their isoelectronic, neutral cyclopentadienyl analogues. Perhaps the most unusual arene compounds are available from the arene alkyne derivatives [(η-$C_6H_3Me_3$)$_2$Nb(RC≡CR)]BPh_4 (R = Me or Ph), themselves formed from reacting [(η-$C_6H_3Me_3$)$_2$Nb]BPh_4 with one equivalent of RC≡CR.[412] (These species are mentioned briefly in Section 2.5.2.1.) When toluene solutions of [(η-$C_6H_3Me_3$)$_2$Nb(RC≡CR)]BPh_4 are heated to 100 °C, red to orange solutions are obtained from which $[Ph_2B(\eta\text{-}C_6H_5)_2]Nb(RC{\equiv}CR)$ can be isolated. These compounds are characterized by two of the four phenyl rings of the $[BPh_4]^-$ ion having coordinated the niobium atom; the adduct $[Ph_2B(\eta\text{-}C_6H_5)_2]Nb(MeC{\equiv}CMe)$ has been structurally characterized.[412]

2.7.5 η^7 Complexes

Although $(\eta\text{-}C_7H_7)NbCp$ was reported some years ago, the chemistry of η^7 complexes of niobium and tantalum at first remained undeveloped. Table 23 summarizes the η^7 complexes of niobium and tantalum that have been reported.

Table 23 η^7 Complexes of niobium and tantalum.

Compound	*Comments*	*Ref.*
$(\eta\text{-}C_7H_7)Nb(CO)_2(PMe_3)$	prep: $(\eta\text{-}C_7H_7)(\eta^4\text{-}C_7H_8)Nb(PMe_3)$ + CO	26,92
$(\eta\text{-}C_7H_7)Nb(\eta^4\text{-}C_7H_8)(PMe_3)$	prep: $(\eta\text{-}C_6H_5Me)Nb(PMe_3)_3H + C_7H_8$	26,92
	prep: $NbCl_4(THF)_2 + Mg + C_7H_8 + PMe_3$	473
$[(\eta\text{-}C_7H_7)Nb(\eta^5\text{-}C_7H_9)Nb(PMe_3)]PF_6$	prep: $(\eta\text{-}C_7H_7)Nb(\eta^4\text{-}C_7H_8)Nb(PMe_3) + NH_4PF_6$	473
$(\eta\text{-}C_7H_7)Nb(\eta\text{-}C_4H_6)(PMe_3)$	prep: $(\eta^6\text{-}C_7H_8)Nb(PMe_3)_2Cl_2 + Mg(C_4H_6){\cdot}2THF$	473
$(\eta\text{-}C_7H_7)Nb(dmpe)(CO)$	x ray	473
$(\eta\text{-}C_7H_7)Nb(PMe_3)_2I$	x ray; prep: $(\eta\text{-}C_7H_7)Nb(\eta^4\text{-}C_7H_8)Nb(PMe_3)$ + MeI	473
$(\eta\text{-}C_7H_7)NbCp$	prep: $(\eta^6\text{-}C_7H_8)Nb(PMe_3)_2Cl_2$ + NaCp	473,529
$(\eta\text{-}C_7H_7)NbCp'$		529
$[(\eta\text{-}C_7H_7)Nb(\eta\text{-}C_5R_5)(THF)]^+$	C_5R_5 = Cp, Cp'; electrochemically generated	529
$[(\eta\text{-}C_7H_7)NbCp'(THF)]PF_6$	prep: $(\eta\text{-}C_7H_7)NbCp' + [Cp_2Fe]PF_6$; x ray	529
$(\eta\text{-}C_7H_7)NbCpBr$	prep: $(\eta\text{-}C_7H_7)NbCp + CH_2{=}CHCH_2Br$	473
$[K(18\text{-crown-}6)][(\eta\text{-}C_7H_7)NbCp]$	x ray; prep: $(\eta\text{-}C_7H_7)NbCp$ + K film + 18-crown-6	529
$[\eta^7\text{-}C_5Me_3(CH_2)_2]Ta(H)_2(PMe_3)_2$	x ray; doubly cyclometallated Cp* ligand	97,316
$[\eta^7\text{-}C_5Me_3(CH_2)_2]Cp^*TaH$	kinetic product from thermolyzing $Cp^*_2Ta({=}CH_2)H$	109
$[\eta^7\text{-}C_5(Bu^t)(CH_2Bu^t)_2(CH_2CMe_2CH_2)_2]TaCl_2$	prep: $[Cl_2(DME)Ta(\mu\text{-}CBu^t)]_2[Zn(\mu\text{-}Cl)_2] + Bu^tCH_2C{\equiv}CCH_2Bu^t$	379

The most significant results for η^7-cycloheptatrienyl complexes appeared in a series of papers from Green's group at Oxford, a portion of which is summarized in Scheme 26. Green and co-workers reacted $(\eta\text{-}C_6H_5Me)Nb(PMe_3)_3H$ with cycloheptatriene to form, *inter alia*, $(\eta\text{-}C_7H_7)Nb(\eta^4\text{-}C_7H_8)PMe_3$; this complex reacts further with CO to form $(\eta\text{-}C_7H_7)Nb(CO)_2(PMe_3)$.[92] However, since $(\eta\text{-}C_6H_5Me)Nb(PMe_3)_3H$ originates from the metal vapor synthesis preparation of $(\eta\text{-}C_6H_5Me)_2Nb$, a more convenient route to η^7 compounds was needed. Green crystals of paramagnetic $(\eta^6\text{-}C_7H_8)Nb(PMe_3)_2Cl_2$ are available from reducing $NbCl_4(THF)_2$ with two equivalents of NaHg in the presence of cycloheptatriene (C_7H_8) and PMe_3.[473] When this reaction is carried out using an excess of NaHg, $(\eta\text{-}C_7H_7)Nb(\eta^4\text{-}C_7H_8)(PMe_3)$ is isolated in low yield. $(\eta^6\text{-}C_7H_8)Nb(PMe_3)_2Cl_2$ appears to be an intermediate in the formation of $(\eta\text{-}C_7H_7)Nb(\eta^4\text{-}C_7H_8)(PMe_3)$ since it can be reduced with NaHg in the presence of C_7H_8 to form the latter compound. However, $(\eta\text{-}C_7H_7)Nb(\eta^4\text{-}C_7H_8)(PMe_3)$ is prepared in high yield from the magnesium reduction of $NbCl_4(THF)_2$ (in THF) in the presence of cycloheptatriene and PMe_3.

When $(\eta\text{-}C_7H_7)Nb(\eta^4\text{-}C_7H_8)(PMe_3)$ is treated with NH_4PF_6, orange crystals of $[(\eta\text{-}C_7H_7)Nb(\eta^5\text{-}C_7H_9)(PMe_3)]PF_6$ are isolated.[473] $(\eta\text{-}C_7H_7)Nb(\eta^4\text{-}C_7H_8)(PMe_3)$ reacts with CO and dmpe forming $(\eta\text{-}C_7H_7)Nb(dmpe)(CO)$ ($\nu(CO) = 1785\ cm^{-1}$) and with methyl iodide to form $(\eta\text{-}C_7H_7)Nb(PMe_3)_2I$, both of which were structurally characterized. Upon treating $(\eta^6\text{-}C_7H_8)Nb(PMe_3)_2Cl_2$ with $Mg(C_4H_6){\cdot}2THF$, lime-green crystals of $(\eta\text{-}C_7H_7)Nb(\eta^4\text{-}C_4H_6)(PMe_3)$ can be isolated. Excess NaCp and $(\eta^6\text{-}C_7H_8)Nb(PMe_3)_2Cl_2$ react to afford the long-known $(\eta\text{-}C_7H_7)NbCp$, which itself reacts with allyl bromide forming red crystals of 18-electron $(\eta\text{-}C_7H_7)NbCpBr$.

The redox chemistry of $(\eta\text{-}C_7H_7)NbCp$ and its C_5H_4Me analogue, $(\eta\text{-}C_7H_7)NbCp'$, have also been examined.[529] These compounds are reported to undergo reversible, one-electron oxidation (–0.82 V vs. SCE in THF) and reversible, one-electron reduction (–2.40 V vs. SCE in THF) in cyclic voltammetric experiments.

Another form of an η^7 ligand is seen in complexes containing a doubly "tucked-in" or doubly cyclometallated Cp* ligand. In THF solution, the NaHg reduction of Cp^*TaCl_4 in the presence of PMe_3 provides, *inter alia*, the complex $[\eta^7\text{-}C_5Me_3(CH_2)_2]Ta(H)_2(PMe_3)_2$ (**55**) containing such a ligand.[97,316] $[\eta^7\text{-}C_5Me_3(CH_2)_2]Ta(H)_2(PMe_3)_2$ was structurally characterized and the authors suggested that the adjacent ring methylenes are best described as sp^2 alkene-type groups. Upon reacting $[\eta^7\text{-}C_5Me_3(CH_2)_2]Ta(H)_2(PMe_3)_2$ with CO, migration of the hydrides back to the ring is observed and the carbonyl complex $Cp^*Ta(CO)_2(PMe_3)_2$ is isolated. Although one might expect $Cp^*Ta(\eta^2\text{-}$

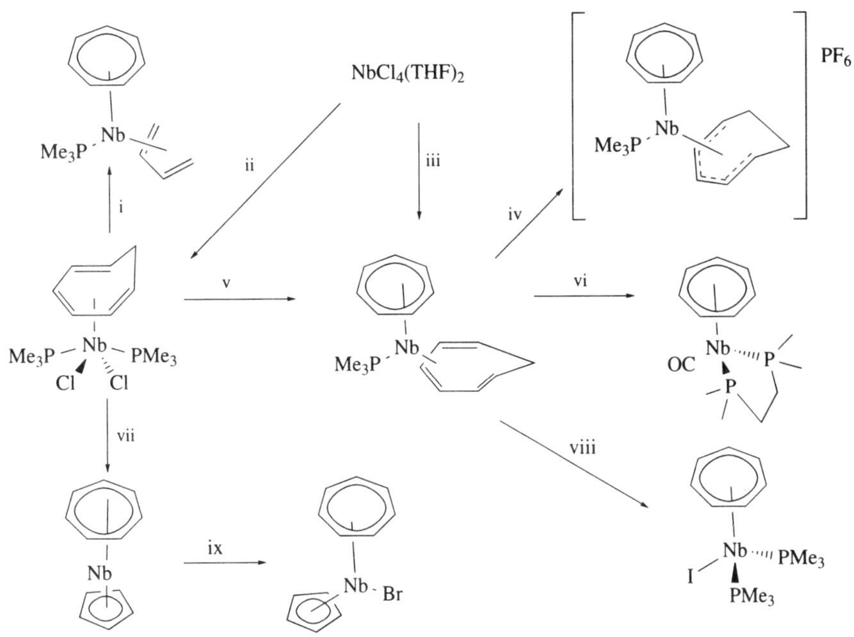

i, $Mg(C_4H_6)_2$•THF; ii, Na/Hg, PMe_3, C_7H_8; iii, Mg, PMe_3, C_7H_8; iv, NH_4PF_6; v, Na/Hg, C_7H_8
vi, dmpe, CO; vii, NaCp; viii, MeI; ix, CH_2=$CHCH_2Br$

Scheme 26

$CHPMe_2)(H)_2(PMe_3)$ and $[\eta^7\text{-}C_5Me_3(CH_2)_2]Ta(H)_2(PMe_3)_2$ to undergo conversion in solution, no evidence for such a process has been obtained.[316]

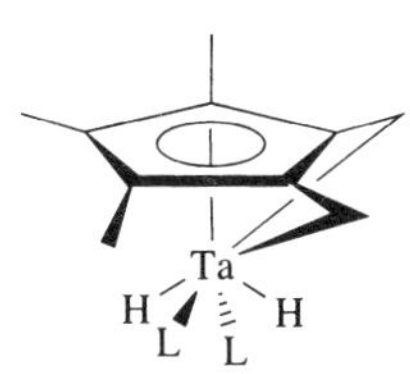

L = PMe_3
(55)

2.8 NIOBIUM AND TANTALUM COMPLEXES IN CATALYSIS AND ORGANIC SYNTHESIS

Although only occasionally acting as catalysts themselves, niobium and tantalum complexes have proved exceedingly beneficial in modeling and understanding the chemistry of a number of catalytic processes. A detailed discussion of each process is beyond the scope of this chapter; however, several cases in which niobium and tantalum have provided a more complete understanding of a catalytic process are outlined here. In addition, a few applications in which these metals are effective catalysts are related. It is not our intent to be comprehensive.

2.8.1 Fischer–Tropsch Chemistry

The catalytic conversion of CO and H_2 (synthesis gas) to hydrocarbons and oxygenates, that is, F–T (Fischer–Tropsch) chemistry, can become economically feasible for the production of petrochemicals when petroleum feedstocks are especially costly.[601] Although high metal–oxygen bond energies typically preclude catalytic CO reduction by niobium or tantalum, their complexes have proven extremely useful in understanding F–T chemistry, providing the most inclusive F–T model systems to be developed. Much of the early work on CO reduction in niobium and tantalum modeled individual steps in the F–T process. For example, H_2 reduces $Cp_2Nb(^{13}CO)H$ to form $^{13}CH_4$ and $Cp^*_2ZrH_2$ reduces $Cp_2Nb(CO)H$ to form the oxycarbene $Cp_2(H)Nb$=$CHOZr(H)Cp^*_2$. This oxycarbene itself forms $Cp_2(CO)$-

$NbCH_2OZr(H)Cp*_2$ upon reaction with CO and affords Cp_2NbH_3 and $Cp_2Zr(OMe)(H)$ upon reaction with H_2.[275]

Perhaps the most comprehensive F–T model which affords a paradigm for all of the F–T steps of C≡O bond scission, H transfer to C, and C–C bond formation, has been developed by Wolczanski and co-workers, using alkoxide-supported tantalum complexes.[33,159,160,269,270]

When $TaH_2(silox)_3$ or $Ta(H)(Et)(silox)_3$ is carbonylated, the CO insertion products, viz., η^2-aldehyde complexes $Ta(\eta^2\text{-}O{=}CH_2)(silox)_3$ and $Ta(\eta^2\text{-}O{=}CHEt)(silox)_3$, are formed. Intramolecular reduction of CO is evident from the carbonylation of a 1:1 mixture of $TaH_2(silox)_3$ and $TaD_2(silox)_3$, which affords $Ta(\eta^2\text{-}O{=}CH_2)(silox)_3$ and $Ta(\eta^2\text{-}O{=}CD_2)(silox)_3$ as the only products.

The dihydride dimer $[TaH_2(silox)_2]_2$ is especially instrumental in the development of F–T reactions (Scheme 27).[160,269] Thus, when $[Ta(silox)_2Cl]_2(\mu\text{-}H)_2$ is treated with excess CO, the μ-formyl complex $[Ta(silox)_2Cl]_2(\mu\text{-}H)(\mu{:}\eta^2,\eta^2\text{-}CHO)$ is isolated (which releases MeOH upon treatment with H_2O); however, 1 atm of CO reacts with the dihydride dimer $[TaH_2(silox)_2]_2$, forming the CO bond scission product $[TaH(silox)_2]_2(\mu\text{-}CH_2)(\mu\text{-}O)$ (**56**) (which releases CH_2D_2 upon D_2O hydrolysis). Re-formation of the CO bond in $[TaH(silox)_2]_2(\mu\text{-}CH_2)(\mu\text{-}O)$ occurs upon its reaction with a slight excess of CO, to afford the formyl formaldehyde complex $[TaH(silox)_2](\mu{:}\eta^2,\eta^2\text{-}CHO)(\mu{:}\eta^1,\eta^2\text{-}CH_2O)[Ta(silox)_2]$ (**57**) in 55% yield. Mild thermolysis of (**57**) forms the μ-ethylidene species $[Ta(silox)_2]_2(\mu\text{-}O)_2(\mu\text{-}CHMe)$ (**58**). The reaction sequence (**56**) + CO → (**57**) → (**58**) is highly significant in that it demonstrates the C–O bond scission and C–H and C–C bond-making reactions of the F–T reaction (Scheme 27).

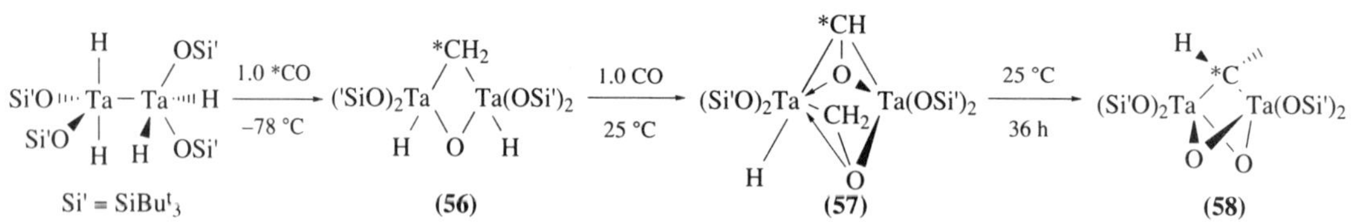

Scheme 27

Another model system has been described in the remarkable formation of the μ-dicarbide complex $[Ta(silox)_3]_2(\mu\text{-}C_2)$ from the d^2 complex $Ta(silox)_3$ and CO in a reaction which encompases both F–T steps of C–O bond scission and C–C bond formation.[33,270] Upon reducing THF solutions of $Ta(silox)_3Cl_2$ with NaHg, the d^2 complex $Ta(silox)_3$ is formed.[33,270] At room temperature, carbonylating $Ta(silox)_3$ under 0.1–1.0 atm of CO affords $[Ta(silox)_3]_2(\mu\text{-}C_2)$ according to Equation (28), consistent with the experimentally observed uptake of 0.47 mol of CO per mole of $Ta(silox)_3$.

$$Ta(silox)_3 + CO\ (<1\ atm) \longrightarrow 1/4[Ta(silox)_3]_2(\mu\text{-}C_2) + 1/2(silox)_3Ta{=}O \qquad (28)$$

When $Ta(silox)_3$ is treated with 1 atm of CO at low temperature (–78 °C), followed by warming to room temperature, 2 mol of CO per mole of $Ta(silox)_3$ are taken up and a mixture of $(silox)_3Ta{=}O$ and the ketenylidene $(silox)_3Ta{=}C{=}C{=}O$ is formed. A detailed mechanistic investigation of the reactions of $Ta(silox)_3$ with CO has revealed the probable sequence for the genesis of the alkylidene $[Ta(silox)_3]_2(\mu\text{-}C_2)$ and the ketenylidene $(silox)_3Ta{=}C{=}C{=}O$.[33] $[Ta(CO)(silox)_3]_n$ is the key intermediate through which the formation of $(silox)_3Ta{=}C{=}C{=}O$ and $[Ta(silox)_3]_2(\mu\text{-}C_2)$ is propagated, although the nature of the CO ligand is unknown. Scheme 28 outlines the stages of CO scission from $[Ta(CO)(silox)_3]_n$, based upon the most likely structure of this complex, through the ketenylidene $(silox)_3Ta{=}C{=}C{=}O$ to the formation of $[Ta(silox)_3]_2(\mu\text{-}C_2)$.[33]

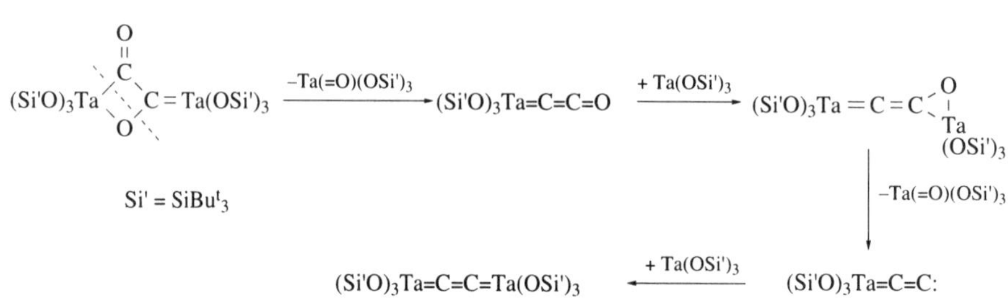

Scheme 28

2.8.2 C–H Bond Activation

Niobium and tantalum have also played an important role in understanding C–H bond activation. Of particular note is a series of C–H activation studies by Rothwell and co-workers using d^0, aryl oxide-supported tantalum complexes.[163,175,178,179,194,195] In addition, a brief overview of C–H bond activation by the early transition metals has been published.[602]

Rothwell and co-workers have found that upon thermolysis, $TaMe_3(OAr'')_2$ undergoes two sequential C–H bond activation steps that form two equivalents of CH_4 and mono- and bis(cyclometallated) species, $\overline{Ta[OC_6H_3Bu^t(CMe_2CH_2)]}(OAr'')Me_2$ and $\overline{Ta[OC_6H_3Bu^t(CMe_2CH_2)]}_2Me$, respectively (Scheme 29).[175,178,179] The fact that $Ta(CD_3)_3(OAr'')_2$ thermolyzes to form two equivalents of CD_3H, in addition to other mechanistic data, supports the [2 + 2], σ-bond metathesis nature of these C–H activations. However, upon photolysis of $Ta(CD_3)_3(OAr'')_2$, CD_4 is produced as the primary organic product, suggesting that an α-hydrogen abstraction produces intermediate $Ta(OAr'')_2(=CD_2)(CD_3)$ which undergoes a C–H addition across its Ta=C double bond.[179] The addition of a C–H bond across a Ta=C bond must also be operative in the formation of the bis(cyclometallated) species $\overline{Ta[OC_6H_3Bu^t(CMe_2CH_2)]}_2(CH_2TMS)$ upon thermolyzing $Ta(=CH\text{-}TMS)(CH_2TMS)(OAr'')_2$.[175,195]

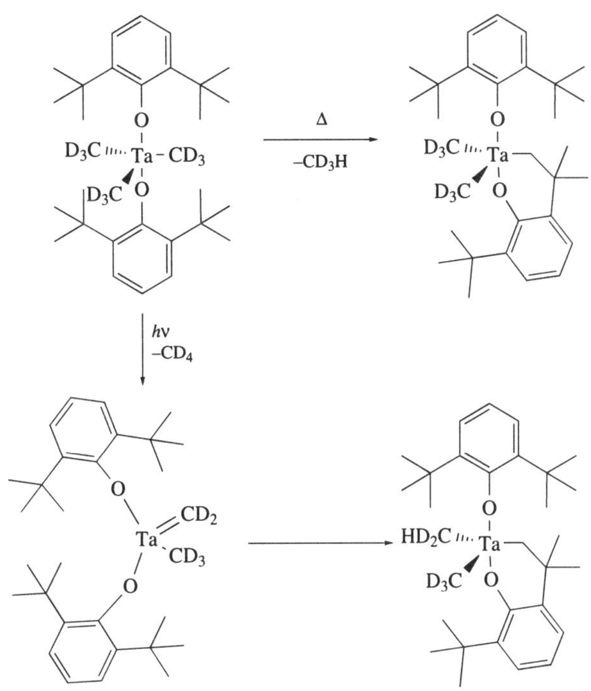

Scheme 29

The intermediacy of benzyne complexes has been demonstrated in related intramolecular C–H activation.[163,194] Thus, $Ta(OAr'')_2Cl_3$ can be cleanly arylated with three equivalents of LiR (R = *o*-, *m*-, or *p*-C_6H_4Me) to yield the mono(cyclometallated) $\overline{Ta[OC_6H_3Bu^t(CMe_2CH_2)]}(OAr'')R_2$.[194] Upon extended thermolysis, one equivalent of toluene is lost and the bis(cyclometallated) complexes $\overline{Ta[OC_6H_3Bu^t(CMe_2CH_2)]}_2R$ are formed, in which the tolyl ligand has isomerized to the thermodynamic mixture of *m*- or *p*-tolyl (but not *o*-tolyl) complexes. This observation, along with labeling studies, indicates that the bis(cyclometallated) complex $\overline{Ta[OC_6H_3Bu^t(CMe_2CH_2)]}_2R$ is in equilibrium with an intermediate mono(cyclometallated) benzyne complex, $\overline{Ta[OC_6H_3Bu^t(CMe_2CH_2)]}(\eta^2\text{-}C_6H_3Me)(OAr'')$, through which this isomerization can take place (Scheme 30).[194]

Schaller and Wolczanski[193] reported a careful C–H activation study using the imido complexes $Ta(=NSiBu^t_3)(NHSiBu^t_3)_2R$ (for R = Me, Ph, CH_2Ph, and CH_2Bu^t). Thermolysis of these alkyls results in a 1,2-elimination of RH to form transient $[Ta(=NSiBu^t_3)_2(NHSiBu^t_3)]$ which is able to add another C–H bond across a Ta=N bond. Despite the slow eliminations (moderate rates are attained only above 180 °C), mechanistic studies have been reported; for example, ground-state effects have been probed by observing the approach to equilibrium of $Ta(=NSiBu^t_3)(NHSiBu^t_3)_2R$ and CH_4. The authors reported that transition-state energy differences account for the considerable rate differences, arguing against a late transition state, despite the approximate correlation between rates and C–H bond energies of the eliminated R–H.[193]

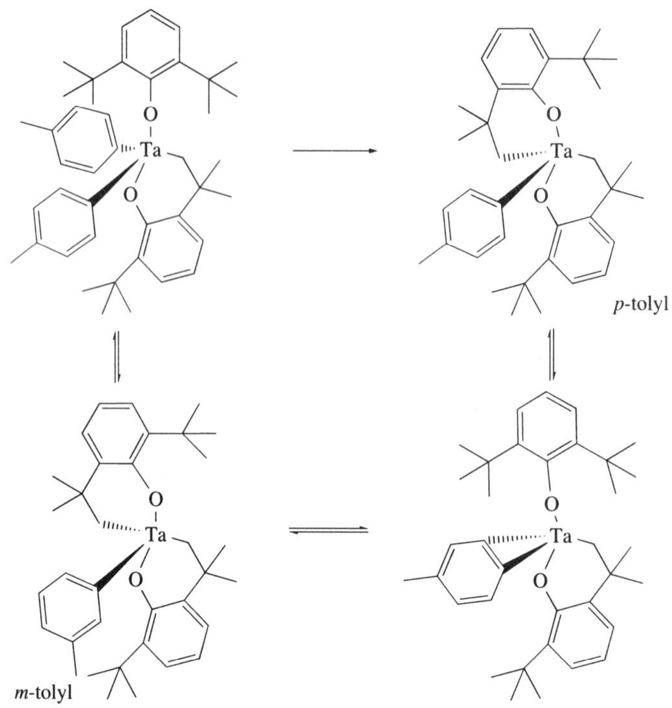

Scheme 30

An interesting C–H bond activation, which appears to proceed via an oxidative addition to a d^2 metal, has also been described. The hexamethylbenzene complex $(\eta\text{-}C_6Me_6)Ta(OAr)_2Cl$ is proposed to engage in a "tuck-in," or intramolecular C–H activation reaction to afford unstable $(\eta^6,\eta^1\text{-}C_6Me_5CH_2)Ta(H)(OAr)_2Cl$.[463] This purported hydride complex is not observed, but can be trapped with $HC{\equiv}CBu^t$ or $CH_2{=}CHCH_2Cl$ to form the η^1 compounds $(\eta^1\text{-}C_6Me_5CH_2)Ta(CH{=}CHBu^t)(OAr)_2Cl$ and $(\eta^1\text{-}C_6Me_5CH_2)Ta(CH_2CH_2CH_2Cl)(OAr)_2Cl$, respectively. Finally, both C–C and C–H bond activation of substituted arenes has been observed using $TaCl_5$ (in CH_2Cl_2 solvent) in an attempt to form cationic d^0 arene compounds.[603]

2.8.3 Alkene and Alkyne Polymerization

Most Ziegler–Natta alkene polymerizations seem to propagate by the direct insertion of alkene into a metal–carbon bond of a highly electrophilic and possibly cationic metal center (the classical Cossee-type mechanism). A well-defined tantalum catalyst has been reported[205,366] which may effect alkene polymerization by an alternative mechanism, first suggested by Rooney and co-workers.[604] This mechanism was originally proposed on the basis of the considerable precedent which is available for each proposed step, a feature which has eluded the most important insertion step of the classical Cossee-type description.

The two-electron reduction of $Ta(CH_2Bu^t)Cl_4$ in the presence of PMe_3 results in the formation of the alkylidene hydride complex $Ta({=}CHBu^t)(H)Cl_2(PMe_3)_3$.[358] This compound undergoes metathesis with TMS-I to afford its iodo congener $Ta({=}CHBu^t)(H)I_2(PMe_3)_3$.[366] Upon reacting $Ta({=}CHBu^t)(H)I_2(PMe_3)_3$ with n equivalents of $CH_2{=}CH_2$, a product is isolated that still contains alkylidene and hydride functionalities and is identified as a "living" polymer of ethene, viz., $Ta[{=}CH(CH_2CH_2)_nBu^t](H)I_2(PMe_3)_3$.[366] No products of metathesis or metallacyclobutane rearrangement (see Section 2.6.2) are observed in the reaction. Since niobium and tantalum alkylidenes form metallacycles rapidly upon their reaction with alkenes,[349,605] and since d^2 tantalum alkyls are known to α-hydrogen eliminate to afford alkylidene hydrides,[205,358] the Rooney–Green type of mechanism in Scheme 31 has been suggested for this polymerization.[604] Note, however, that a rapid $Ta^V({=}CHR)(H) \rightleftharpoons Ta^{III}(CH_2R)$ equilibrium which permits ethene insertion into a Ta–C single bond cannot unequivocally be ruled out.

The living polymerization of 2-butyne catalyzed by vinylalkylidene complexes such as $Ta({=}CMeCMe{=}CHBu^t)(OAr)_3(py)$ has also been described.[372] These catalysts are formed from the metallacyclization of an alkylidene complex with an alkyne, followed by the rearrangement of the

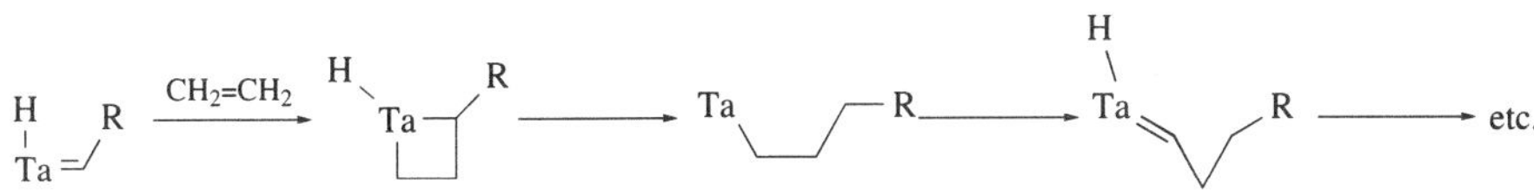

Scheme 31

resulting metallacyclobutene. Thus, Ta(=CHBu^t)(OAr)$_3$(THF) reacts with RC≡CR (R = Me, Ph, or TMS) to yield the corresponding, THF-free, metallacyclobutene complexes Ta(CR=CRC-HBu^t)(OAr)$_3$.[372] The substituted metallacyclobutenes Ta(CR=CRCHBu^t)(OAr)$_3$ (R = Me or Ph) are decomposed in the presence of pyridine by a four-electron, electrocyclic process to provide the vinylalkylidene complexes.[372] These compounds are in equilibrium with their corresponding metallacyclobutenes as indicated in Scheme 32. Ta(=CMeCMe=CHBu^t)(OAr)$_3$(py) acts as a living catalyst for the polymerization of 2-butyne (polydispersity index < 1.05) in reactions involving metallacyclobutene formation and rearrangement which re-forms the alkylidene active site (Scheme 33). Thus, Ta(=CMeCMe=CHBu^t)(OAr)$_3$(py) reacts with up to 200 equivalents of MeC≡CMe to afford living polymers of the form Ta[(=CMeCMe)$_n$=CHBu^t](OAr)$_3$(py); the polymer is readily cleaved from the catalyst by treatment with benzaldehyde.

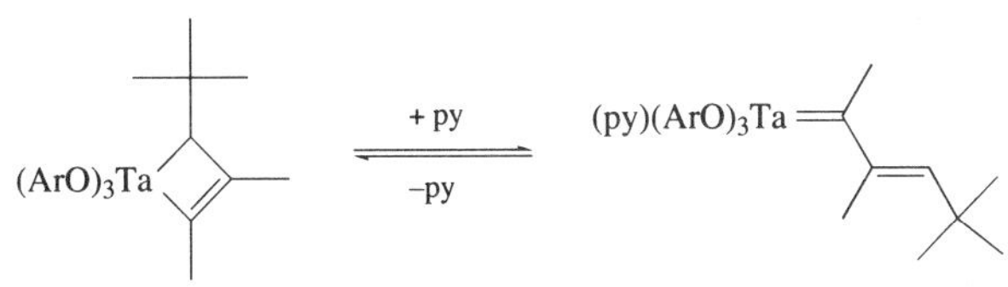

Scheme 32

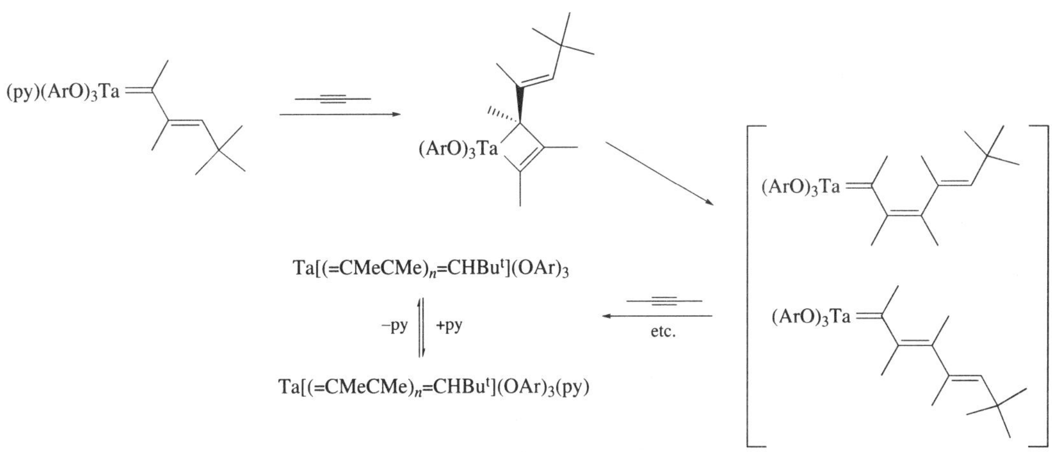

Scheme 33

2.8.4 Alkene Dimerization

One of the few processes for which niobium or tantalum complexes can serve as catalysts is the selective dimerization of alkenes. One mechanism by which terminal alkenes are dimerized, described in Section 2.6.4, involves the unusual "ring contraction" of a metallacyclopentane into a substituted metallacyclobutane. This new metallacycle can then undergo a metal-mediated β-hydrogen shift, that is, β-elimination followed by reductive elimination to generate the alkene dimer as shown in Scheme 18 in Section 2.6.4.[360]

In addition to the alkene complexes and metallacyclopentanes outlined in Section 2.6.4, there are several other catalyst precursors which effect alkene dimerization by generating the transient, d^2 active catalyst. For example, Cp*Ta(cyclooctene)X$_2$ (X = Cl or Br) and Cp*Ta(cyclooctene)(OMe)Cl, and also their metallacyclic products such as Cp*Ta(CH$_2$CHMeCHMeCH$_2$)(OMe)Cl, have been used in this regard.[360]

Another mechanism by which tantalum dimerizes ethene has been described as the "metallacyclopentane mechanism" (vis-á-vis the "ring contraction" mechanism) in which

$Ta(CH_2{=}CH_2)_2(Et)(PMe_3)_2$, prepared from $Ta(CH_2{=}CH_2)Cl_3(PMe_3)_2$ and 1.5 Et_2Mg(diox), serves as a precursor to the active catalyst.[146] Thus, Schrock and co-workers[146] reported that upon reaction of $Ta(CH_2{=}CH_2)_2(Et)(PMe_3)_2$ with ethene, the *n*-butyl complex $Ta(Bu^n)(CH_2{=}CH_2)_2(PMe_3)_2$ (**59**) is formed as the major organometallic product. Scheme 34 represents the proposed catalytic cycle for ethene dimerization through the intermediacy of $Ta(Bu^n)(CH_2{=}CH_2)_2(PMe_3)_2$ (**59**). The key step of this cycle involves a β-hydrogen "abstraction" or transfer from the *n*-butyl group to the C-α of the metallacyclopentane ligand of (**60**) to afford the dimer, coordinated 1-butene. The observed selectivity depends upon the rapid displacement of 1-butene from (**61**) by ethene to re-form (**59**) in Scheme 34.

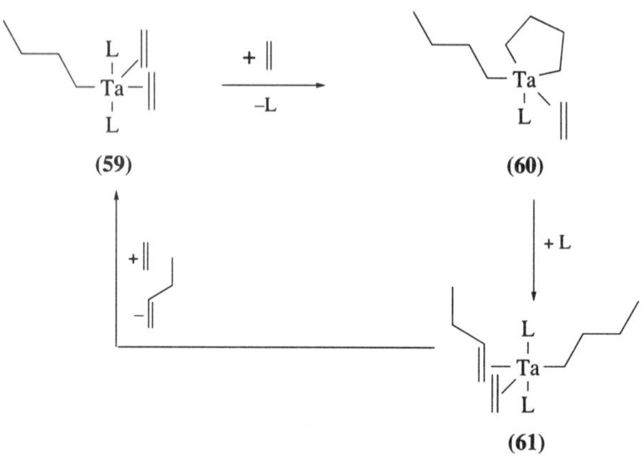
(**59**) (**60**) (**61**)

Scheme 34

A selective (although slow) dimerization catalyst has been reported using a complex containing the cyclometallated PMe_3 ligand $Cp^*Ta(\eta^2\text{-}CHPMe_2)(H)(Br)(PMe_3)$ (**62**). This species is observed to react with ethene or terminal alkenes $CH_2{=}CHR$ to afford the insertion products $Cp^*Ta(\eta^2\text{-}CHPMe_2)(CH_2CH_2R)Br$ (**63**) (for R = H, Me, Bu^t, or Ph) plus PMe_3.[457] The reverse, β-H elimination reaction results in these complexes being unstable; the equilibrium shown in Equation (29) shifts more towards the alkyl (**63**) side as the size of R decreases. However, warming the mixture to 70 °C in the presence of excess ethene results in the slow, although catalytic and selective, dimerization of $CH_2{=}CH_2$ to $CH_2{=}CHEt$. No intermediates are observed in this reaction; however, the fact that dimerizing $CD_2{=}CD_2$ results in the incorporation of deuterium into the H* (metallacycle methine) site of $Cp^*Ta(\eta^2\text{-}CH^*PMe_2)(H)(Br)(PMe_3)$, suggests that the $Ta(\eta^2\text{-}CHPMe_2)(H^*)(Br)(PMe_3) \rightleftharpoons Ta(\eta^2\text{-}CHH^*PMe_2)(Br)(PMe_3)$ equilibrium may be in effect.[457]

$$\underset{(\mathbf{62})}{Cp^*Ta(\eta^2\text{-}CHPMe_2)(H)(Br)(PMe_3)} + CH_2{=}CHR \rightleftharpoons \underset{(\mathbf{63})}{Cp^*Ta(\eta^2\text{-}CHPMe_2)(CH_2CH_2R)Br} + PMe_3 \quad (29)$$

2.8.5 Alkene Metathesis and Metathesis Polymerizations

Few catalytic reactions have been the subject of such intense scrutiny as alkene metathesis and the alkylidenes of niobium and tantalum have been central to this inquiry.[172,348,351,368,369,605] Again, the reader is referred to *COMC-I* where this metathesis chemistry, in particular the study of metallacyclobutane formation and fragmentation in tantalum complexes, has been described in some detail.[1] Although alkylidene complexes of the type $M(=CHBu^t)X_3(PMe_3)_2$ (X = Cl, Br) form metallacyclobutanes upon their reaction with terminal alkenes, they do not metathesize alkenes because the metallacycle rapidly rearranges by β-hydrogen transfer as described in Section 2.6.1.[172] However, by replacing PMe_3 with THF, metathesis products are observed; for example, $Ta(=CHBu^t)Cl_3(THF)_2$ is a short-lived metathesis catalyst for *cis*-MeCH=CHEt. Replacing chloride ligands with *t*-butyl alkoxide, for example, in complexes such as $Ta(=CHBu^t)(OBu^t)_2Cl(PMe_3)$ and $Nb(=CHBu^t)(OBu^t)_2Cl(PMe_3)$, results in the formation of a longer-lived catalyst for metathesis of internal alkenes.[172] Inquiries along the line of modifying niobium and tantalum complexes to speed up the metathesis over rearrangement have continued.[172,348,368,369]

More recently, the reactions of alkenes with tantalum alkylidene complexes containing hindered phenoxide ancillary ligands (e.g., $Ta(=CHBu^t)(OAr)_3(THF)$) have been used to examine metallacycle formation and fragmentation.[368,369] $Ta(=CHBu^t)(OAr)_3(THF)$ is prepared straightforwardly from the reaction of $Ta(=CHBu^t)Cl_3(THF)_2$ with $LiOAr(OEt_2)$. Metallacyclization reactions occur upon reaction of this alkylidene with alkenes. For example, $Ta(=CHBu^t)(OAr)_3(THF)$ gives the metallacyclobutane $\overline{Ta(CH_2CH_2CH_2)}(OAr)_3$ along with by-product $CH_2{=}CHBu^t$ with ethene, and is capable of metathesizing *cis*-MeCH=CHEt.[369] $Ta(=CHBu^t)(OAr)_3(THF)$ reacts with one equivalent of styrene to provide orange crystals of $\overline{Ta(CHPhCHBu^tCH_2)}(OAr)_3$, which is presumed to arise when one of the two possible initial metallacycles breaks apart and the resulting alkene readds (Scheme 16). A large excess of THF breaks apart metallacyclic $\overline{Ta(CHPhCHBu^tCH_2)}(OAr)_3$ and sets up the equilibrium shown in Equation (10).

Tantalum alkylidenes have also been used in the ring-opening metathesis–polymerization (ROMP) of cyclic alkenes via intermediate metallacyclobutane complexes.[369,370] Thus, $Ta(=CHBu^t)(OAr)_3(THF)$ reacts with norbornene to form the expected metallacyclobutane $\overline{Ta[CH(C_5H_8)CHCHBu^t]}(OAr)_3$, which has been structurally characterized. $\overline{Ta[CH(C_5H_8)CHCHBu^t]}(OAr)_3$ reacts with another equivalent of norbornene to form metallacyclic $\overline{Ta[CH(C_5H_8)CHCH(C_5H_8)CH{=}CHBu^t]}(OAr)_3$; this product can be considered as the "second metallacycle" formed between the first insertion product, $Ta[=CH(C_5H_8)CH{=}CHBu^t](OAr)_3$, and the second equivalent of norbornene. When $Ta(=CHBu^t)(OR)_3(THF)$ (R = Ar or Ar') are treated with excess norbornene and heated to 50 °C, polynorbornene is produced as demonstrated in Scheme 35.[370] The thiolate-supported alkylidene $Ta(=CHBu^t)(S\text{-}2,4,6\text{-}C_6H_2Pr^i_3)_3(py)$ also effects the ROMP of norbornene.[369]

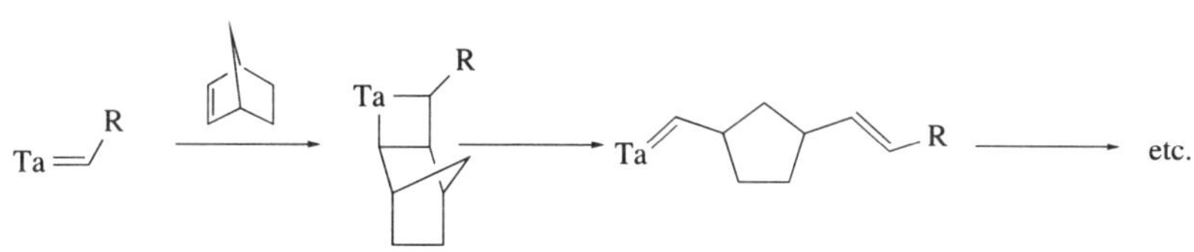

Scheme 35

2.8.6 Alkyne Cyclotrimerization

The net [2 + 2 + 2] cyclotrimerization of alkynes to arenes constitutes a process which has found some use in organic synthesis. Metallacyclopentadienes, generated from the reductive coupling of two alkyne molecules, constitute viable intermediates in metal-catalyzed alkyne cyclotrimerization catalysis. While their participation in the catalytic chemistry is apparent, complexes of tantalum have afforded a rare observation of the direct conversion of metallacyclopentadienes into a higher, complexed alkyne cyclooligomer. Scheme 36 summarizes the formation of η-arene complexes of tantalum from a metallacycle.[408,459] A significant feature of this chemistry is the isomerization of the α,α'-substituted metallacyclopentadiene (**31**) to its α,β'-isomer (**32**) upon thermolysis (see Section 2.6.3), a process which is required before the third alkyne can approach the metal and react. Thus, if this reaction is performed at temperatures high enough to induce this isomerization, the cyclotrimerization becomes catalytic in tantalum.[408]

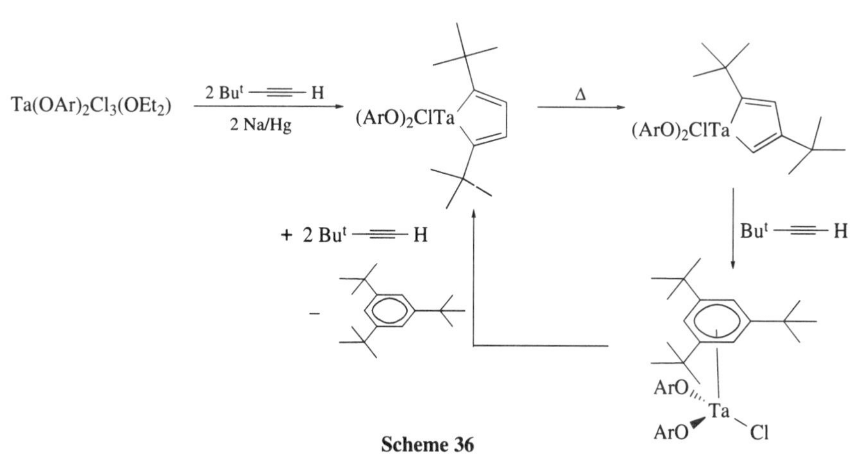

Scheme 36

The structural features noted in Section 2.7.4 for Ta^{III} arene complexes (which can be attributed to an interaction between metal δ functions and the arene LUMO) strongly resemble those of the

Diels–Alder adduct formed in [4 + 2] cycloaddition between a metallacyclopentadiene and an ethyne. These features have resulted in such structures being termed "7-metallanorbornadienes" (**53**) and are represented in Section 2.7.4.3.

Several other complexes are known to produce alkylbenzenes catalytically upon treatment with alkynes, most notably $M_2Cl_6(THT)_3$[606] and $NbCl_3(DME)$.[392] Catalytic reactions using $M_2Cl_6(THT)_3$ produce approximately equal ratios of the 1,2,4- and 1,3,5-$C_6H_3R_3$ isomers upon cyclizing terminal alkynes RC≡CH (for R = Ph, Et, or Pr^i); both isomers were also reported using $NbCl_3(DME)$, although no arene complexes were isolated from either system. $[CpNbCl(\mu\text{-}Cl)(CO)_2]_2$[57] is also a precursor to an alkyne cyclotrimerization catalyst which cyclizes HC≡CPh to form primarily 1,3,5-$C_6H_3Ph_3$.[57] While treatment of $[NEt_4][Nb_2(\mu\text{-}Br)_3(CO)_8]$ with internal and terminal alkynes afford alkyne adducts, the reaction of $[NEt_4][Nb_2(\mu\text{-}Br)_3(CO)_8]$ with HC≡CH itself forms only C_6H_6 and polyethyne.[406] Catalytic amounts of the niobium butadiene complexes of the type $CpNb(butadiene)Cl_2$, $CpNb(butadiene)_2$, and $[CpNb(butadiene)(\mu\text{-}Cl)]_2$ are reported to afford both 1,3,5- and 1,2,4-$C_6Et_3H_3$ upon cyclizing EtC≡CH.[477] Upon treatment of $NbCl_5$ with alkynes, a cyclotrimerization catalyst is generated (in CCl_4) after a long induction period which the authors proposed to be an Nb^{III} alkyne complex.[607] (Similar results have been reported in the cyclotrimerization of 1-hexyne with $NbCl_5$ and $TaCl_5$.[608]) Dehnicke and co-workers[401] described the reduction of $NbCl_5$ by PhC≡CPh to form the cluster $[(PhC{\equiv}CPh)NbCl_3]_4$, an observation which lends precedent to the suggestion that d^2 complexes are active catalysts in this system, even in the absence of a reducing agent other than alkyne itself.[607] The M^{III} oxidation state is apparently required for cyclization, since the $d^0 \rightleftharpoons d^2$ redox shuttle must be executed for the completion of a catalytic cycle (see Scheme 36).

A range of terminal alkynes RC≡CH (for R = Ph, Bu^n, $(CH_2)_3OSiBu^tMe_2$, $(CH_2)_3OCOBu^i$, or TMS) are catalytically cyclotrimerized by a species formed upon reducing $CpNbCl_4$ (with magnesium) in THF.[415] Prereduced $CpNbCl_4$ (from the $CpNbCl_4$–Mg system) is shown to effect benzocycle annulation reactions upon being treated with diacetylenes and RC≡CR mixtures, as described in Section 2.8.9.4.[415]

2.8.7 Hydrodenitrogenation

Catalytic hydrodenitrogenation (HDN) is the process by which organic nitrogen is removed as NH_3 from petroleum and coal-derived liquids, under hydrogenation conditions, to provide more processable and environmentally sound liquid fuel feedstocks.[609,610] By far the most difficult contaminants to process are the heterocyclic compounds, principally those containing six-membered pyridine and five-membered pyrrole rings.[611,612] The HDN process may be delineated into two primary steps: (i) aromatic ring hydrogenation and (ii) C–N bond cleavage by hydrogenolysis. Although it is difficult to make generalizations from one set of substrate/catalyst/conditions to another, the collective evidence points to the quinoline HDN reaction scheme shown in Scheme 37.[611–14] The most efficient and selective method of quinoline HDN involves the (**64**) → (**65**) → (**66**) → (**67**) pathway of Scheme 37 in which the nonheteroatom ring is not hydrogenated. This path represents a considerable saving in hydrogen and therefore provides a cheaper and higher quality (higher octane) product. However, most of the quinoline which undergoes HDN under industrial conditions (e.g., 13.8 MPa, H_2, 350 °C) is hydrogenated along the (**64**) → (**65**) → (**68**) → (**69**) → (**70**) pathway where the nonheteroatom ring is also hydrogenated before C–N bonds are cleaved.[615–17] Studies on tantalum chemistry have provided outstanding models of the HDN process at metal centers resembling those employed in hydrotreating.[460,461]

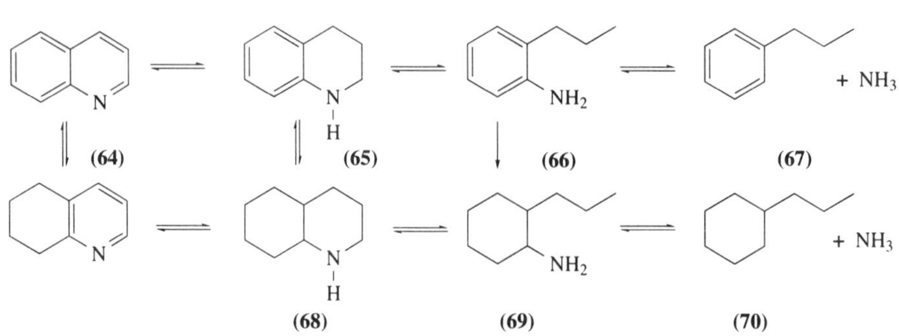

Scheme 37

When $Ta(OAr)_3Cl_2(OEt_2)$ reacts with quinoline (QUIN, NC_9H_7), Et_2O is readily displaced and pale-yellow $[\eta^1(N)\text{-QUIN}]Ta(OAr)_3Cl_2$ is isolated in high yield (Scheme 38).[460] The analogous reaction is effected using 6-methylquinoline (6MQ, $NC_{10}H_9$) to afford $[\eta^1(N)\text{-6MQ}]Ta(OAr)_3Cl_2$. Upon reduction of $[\eta^1(N)\text{-QUIN}]Ta(OAr)_3Cl_2$ with two equivalents of NaHg, and $\eta^1(N) \rightarrow \eta^2(N,C)$ bonding rearrangement is effected and $[\eta^2(N,C)\text{-QUIN}]Ta(OAr)_3$ is obtained. The C-α*H* resonance at δ 9.62 in the ^1H NMR spectrum of d^0 $[\eta^1(N)\text{-QUIN}]Ta(OAr)_3Cl_2$ shifts to δ 4.08 in d^2 $[\eta^2(N,C)\text{-QUIN}]Ta(OAr)_3$ (both in C_6D_6), very diagnostic of this $\eta^2(N,C)$ structure. The 6-methylquinoline complex $[\eta^1(N)\text{-6MQ}]Ta(OAr)_3Cl_2$ is similarly reduced to $[\eta^2(N,C)\text{-6MQ}]Ta(OAr)_3$, but is most easily isolated in crystalline form as its PMe_3 adduct, $[\eta^2(N,C)\text{-6MQ}]Ta(OAr)_3(PMe_3)$. An x-ray structural determination of $[\eta^2(N,C)\text{-6MQ}]Ta(OAr)_3(PMe_3)$ provides evidence for the disruption of aromaticity in the heterocyclic ring, where π electron localization is evident in the heterocycle, as the VB representations shown in Scheme 38 suggest. Hydrogenation studies on the η^2-(*N,C*) quinoline complex indicate that under mild conditions (13.8×10^5 Pa H_2, room temperature), selective hydrogenation of the nitrogen-containing ring only occurs, while no products arising from hydrogenating the C_6 ring are observed.

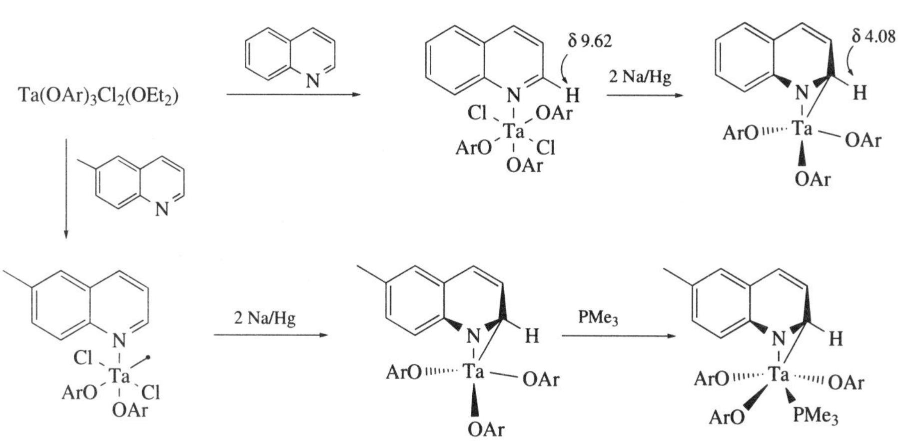

Scheme 38

This tantalum system has also led to the discovery that the $\eta^2(N,C)$-coordination mode of a pyridine ligand renders the C–N bond of the heterocycle subject to cleavage, thereby addressing one of the lingering questions in HDN catalysis.[609] Upon reaction of $[\eta^2(N,C)\text{-2,4,6-NC}_5H_2Bu^t_3]Ta(OAr)_2Cl$ with $Li[BEt_3H]$, orange crystals of "$(NC_5H_3Bu^t_3)Ta(OAr)_2$" are obtained in high yield, along with LiCl.[460] The NMR spectra of "$(NC_5H_3Bu^t_3)Ta(OAr)_2$" and "$(NC_5H_2DBu^t_3)Ta(OAr)_2$," which is prepared analogously using $LiBEt_3D$, are consistent with hydride attack at the bound carbon of the $\eta^2(N,C)$ pyridine ligand. X-ray crystallography, however, provides the dramatic evidence that the nitrogen–carbon bond in the pyridine has been cleaved and the resulting product is the metallacycle $\overline{Ta(=NCBu^t=CHCBu^t=CH\dot{C}HBu^t)}(OAr)_2$ (Scheme 39).

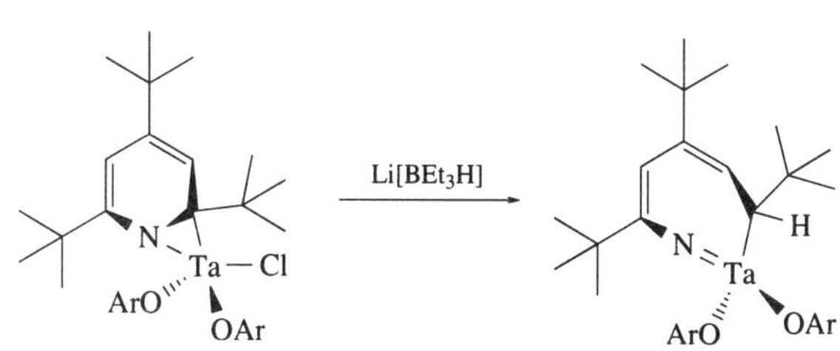

Scheme 39

2.8.8 Arene Hydrogenation

Another reaction which is effected in petroleum hydrotreating or coal liquefaction is the hydrogenation of aromatics, which is carried out simultaneously with HDN and HDS under high temperatures and H_2 pressures. Compounds of niobium and tantalum have served as excellent models and indeed catalysts for arene hydrogenation.[168,173,183,184,271,478] Initially, pendant phenyl rings of the 2,6-

diphenylphenoxide ligand or an arylphosphine ligand have been examined, which are hydrogenated intramolecularly to cyclohexyl groups.

Hydrogenation (8.3 MPa, 80 °C, 24 h) of the alkoxide alkyls $Ta(OAr)_2(CH_2TMS)_2Cl$ and $Ta(OAr)_2(CH_2TMS)_3$, in the presence of trialkylphosphines, results in the formation of the hydride compounds $Ta(OAr)_2H_2Cl(PR_3)_2$ (for $PR_3 = PMe_3$, PMe_2Ph, or $PMePh_2$) and $Ta(OAr)_2H_3(PMe_2Ph)_2$.[173] Under similar conditions (and in the presence of PR_3 ligands), the 2,6-diphenylphenoxide complexes $Ta(O\text{-}2,6\text{-}C_6H_3Ph_2)_2(CH_2\text{-}4\text{-}C_6H_4Me)_3$ and $Ta(O\text{-}2,6\text{-}C_6H_3Ph_2)_2(CH_2TMS)_3$ are hydrogenated to form seven-coordinate hydrides, $Ta(O\text{-}2,6\text{-}C_6H_3Cy_2)_2H_3(PR_3)_2$ ($PR_3 = PMe_2Ph$, $PMePh_2$) in which the 2,6-substituted phenyl rings are themselves hydrogenated to cyclohexyl groups. Solutions of $Ta(O\text{-}2,6\text{-}C_6H_3Cy_2)_2H_3(PMe_2Ph)_2$ catalytically hydrogenate naphthalene to tetralin in a reaction which is inhibited by excess PMe_2Ph.[173]

A complex containing a coordinated cyclohexadiene ring is isolated upon reducing $Nb(O\text{-}2,6\text{-}C_6H_3Ph_2)_3Cl_2$ with NaHg under an H_2 atmosphere, viz., $Nb(OC_6H_3Ph\text{-}\eta^4\text{-}C_6H_7)(O\text{-}2,6\text{-}C_6H_3Ph_2)_2$ (see Section 2.7.2.3).[168,271] This complex can be considered as an intermediate in the complete hydrogenation reactions, since one of the 2,6-disubstituted phenyl rings is partially hydrogenated and binds η^4 as a cyclohexadiene. The stereochemistry of hydrogenation using D_2 has been examined by NMR.[478] More recently, intermolecular hydrogenation of polyaromatic compounds was effected catalytically using $Nb(CH_2\text{-}4\text{-}C_6H_4Me)_3(O\text{-}2,6\text{-}C_6H_3Ph_2)_2$ in a reaction that can be extended to the preparation of cyclohexylphosphine ligands starting from the aryl phosphine.[184]

2.8.9 Niobium and Tantalum Complexes in Organic Synthesis

Only a handful of applications of niobium and tantalum complexes to new carbon–carbon bond forming reactions have been developed and a brief overview of some of these is presented here. The most useful of these are a series of cross-coupling reactions which exploit the redox potential of lower-valent niobium compounds for reductively coupling unsaturated organic substrates. Many of these applications employ alkynes as one of the unsaturated moieties and the use of alkyne complexes (i.e., metallacyclopropenes) as dianion equivalents has been reviewed.[618]

2.8.9.1 Cross-coupling of unsaturated organic substrates

The most significant advance in applying niobium or tantalum complexes in organic synthesis came with the report by Roskamp and Pedersen[619] of the use of d^2 $NbCl_3(DME)$, a reagent which is available on a large scale by reducing $NbCl_5$ with $Bu^n{}_3SnH$ in DME. Although the structure of $NbCl_3(DME)$ is not known (in part owing to its insolubility in solvents with which it does not react), its formulation as a d^2 species rather than an Nb^{IV} hydride is evident.[619] Roskamp and Pedersen's reagent has afforded, *inter alia*, an extremely convenient route to 2-amino alcohols. Thus, upon reaction of $NbCl_3(DME)$ with imines, the d^2 reagent is formally oxidized to a "metallaziridine," which couples carbonyl groups to form a new carbon–carbon bond (Scheme 40). The reaction no doubt proceeds via the metallacycle shown in Scheme 40, as base workup of the reaction affords the 2-amino alcohols in good yield. Cross-coupling reactions have also been described between alkynes and aldehydes. Thus, d^2 tantalum alkyne compounds can be prepared *in situ* by reducing $TaCl_5$ with Zn (in DME–benzene solvent) in the presence of alkyne.[620,621] Reacting solutions of these complexes with aldehydes, followed by an aqueous base quench, affords the corresponding allylic alcohols.

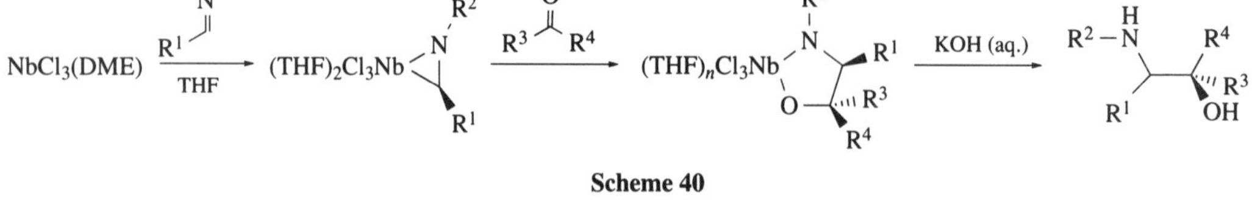

Scheme 40

Pedersen's reagent, $NbCl_3(DME)$, has also been useful in the regioselective synthesis of 2,3-disubstituted-1-naphthols *via* the cross-coupling of alkynes and 1,2-aryldialdehydes.[622] Takai and co-workers[623] have developed a series of similar cross-couplings, for example, the preparation of 1-naphthols from alkynes and phthalic dicarboxaldehyde, using alkyne complexes prepared *in situ* by reducing $TaCl_5$ or $NbCl_5$ with zinc in the presence of an alkyne.[623]

A very different type of coupling reaction has been developed by Roskamp and Pedersen[624] which effects the coupling of *N*-(trimethylsilyl)imines to give vicinal diamines. Using the d^1 complex $NbCl_4(THF)_2$, *N*-(trimethylsilyl)imines react to provide a d^1 niobium–*N*-imine complex, which dimerizes at the α-carbon owing to the localization of the unpaired electron at that carbon (Scheme 41). Hydrolysis of the resulting metallacycle affords the diamines. Since many of the synthetically desirable imines are not readily accessible, Roskamp and Pedersen[624] also developed an *in situ* generation of a niobium hydride which inserts nitriles to generate the niobium imines. Thus, $NbCl_4(THF)_2$ was shown to react with $Bu^n{}_3SnH$ to afford an intermediate hydride complex which inserts N≡CR into the Nb–H bond to form the imine; this process thereby ensures the generality of this route to vicinal diamines. In the case of R = Ph, the intermolecularly coupled product, that is, the bridging diimido complex $[NbCl_3(DME)]_2$(μ-NCHPhCHPhN), was isolated and fully characterized.[624]

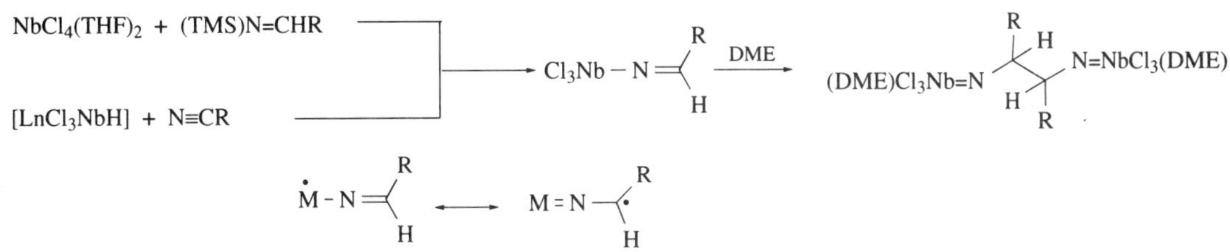

Scheme 41

2.8.9.2 Cyclization of α,ω-dienes

The same d^2 catalysts (typically transient, d^2 $[Cp^*TaCl_2]$) that are used to dimerize alkenes (Section 2.8.4) have also been used to cyclize α,ω-dienes.[385] Thus, the tantallacyclopentane $CpTa[CH_2(C_6H_{10})CH_2]Cl_2$ is prepared from either $CpTa(=CHBu^t)Cl_2$ or $CpTa(CH_2CH_2CH_2CH_2)Cl_2$ (in lower yield) and 1,7-octadiene. $Cp^*Ta[CH_2(C_5H_8)CH_2]Cl_2$ is available from 1,6-heptadiene and $Cp^*Ta(CH_2{=}CHMe)Cl_2$, and *trans*-$Cp^*Ta[CH_2(C_7H_{12})CH_2]Cl_2$ is similarly accessible from 1,8-nonadiene and $Cp^*Ta(CH_2{=}CHPh)Cl_2$. Under catalytic conditions, these complexes are used (either isolated or generated *in situ* from $Cp^*Ta(CH_2{=}CHPh)Cl_2$) as catalyst precursors for the cyclization of α,ω-dienes, in which elimination of the metallacyclopentane to the exocyclic methylene compound can be effected (Scheme 42).[385]

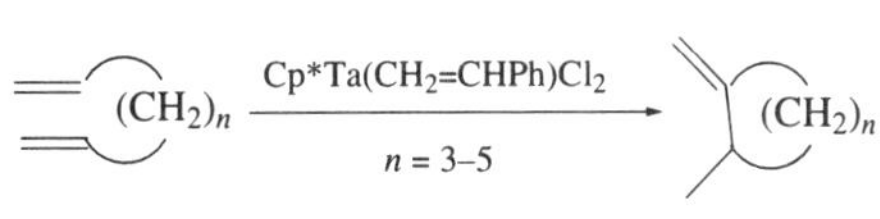

Scheme 42

2.8.9.3 Alkene synthesis from alkylidene complexes

Although not constituting a synthetic method of great use, alkenes are accessible upon reacting alkylidenes with aldehydes, ketones, or imines.[349] This reaction presumably proceeds via an intermediate metallacycle in a [2 + 2] cycloaddition reaction where the driving force for this reaction is the formation of a strong Ta=O or Ta=NR multiple bond in the by-product. Thus, $Ta(=CHCMe_3)(CH_2CMe_3)_3$ has been shown to function as "transition metal ylide" in a Wittig-like reaction; for example, upon reacting O=CHR and $Ta(=CHCMe_3)(CH_2CMe_3)_3$, high yields of $RCH{=}CHCMe_3$ and $[O{=}Ta(CH_2CMe_3)_3]_n$ are produced. $Ta(=CHBu^t)X_3L$ (X = OAr, OAr', or S-2,4,6-$C_6H_2Pr^i{}_3$; L = THF or py) are also obtained by methathetical procedures from $Ta(=CH^tBu)Cl_3(THF)_2$ and react readily with organic carbonyls to provide the expected "Wittig-like" products.[369]

2.8.9.4 Benzocycle annulation reactions

A series of benzocycle annulation reactions have been described from cocyclizing diynes and monoalkynes using a $CpNbCl_4$–Mg catalyst system.[415] Note that this system also effects the catalytic cyclotrimerization of terminal alkynes RC≡CH, as described in Section 2.8.6. Thus, upon reduction of

$CpNbCl_4$ with magnesium (in THF) in the presence of an alkyne, the adduct $CpNb(RC{\equiv}CR)Cl_2$ is presumed to form but is typically fairly labile. When a diyne is added to a prereduced $CpNbCl_4$–$RC{\equiv}CR$ solution, benzocycle annulation is effected and products such as that shown in Scheme 43 are readily isolated. The mechanism of such a reaction is no doubt related to that described for alkyne cyclotrimerization in Section 2.8.6, suggesting the intermediacy of alkyne adducts and metallacyclopentadienes. Other examples of this reaction indicate compatibility of this system with ester and urethane functionalities.[415]

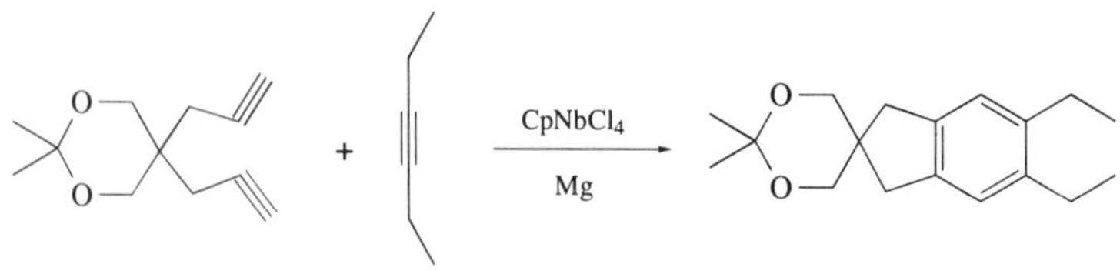

Scheme 43

2.9 REFERENCES

1. J. A. Labinger, in 'COMC-I' vol. 3, p. 705.
2. C. E. Holloway and M. Melnik, *J. Organomet. Chem.*, 1986, **303**, 1.
3. C. E. Holloway and M. Melnik, *J. Organomet. Chem.*, 1986, **303**, 39.
4. J. Buckingham (ed.), 'Dictionary of Organometallic Compounds', Chapman and Hall, London, 1984, vol. 2.
5. S. C. Srivastava and A. K. Shrimal, *Polyhedron*, 1988, **7**, 1639.
6. F. Calderazzo and G. Pampaloni, *J. Organomet. Chem.*, 1992, **423**, 307.
7. F. Calderazzo and G. Pampaloni, *J. Organomet. Chem.*, 1983, **250**, C33.
8. (a) F. Calderazzo, U. Englert, G. Pampaloni, G. Pelizzi and R. Zamboni, *Inorg. Chem.*, 1983, **22**, 1865; (b) G. Pampaloni and F. Calderazzo, *Inorg. Synth.*, 1990, **28**, 192.
9. F. Calderazzo, G. Pampaloni and P. F. Zanazzi, *J. Chem. Soc., Chem. Commun.*, 1982, 1304.
10. C. G. Dewey, J. E. Ellis, K. L. Fjare, K. M. Pfahl and G. F. P. Warnock, *Organometallics*, 1983, **2**, 388.
11. G. F. P. Warnock, J. Sprague, K. L. Fjare and J. E. Ellis, *J. Am. Chem. Soc.*, 1983, **105**, 672.
12. F. Calderazzo, G. Pampaloni and G. Pelizzi, *J. Organomet. Chem.*, 1982, **233**, C41.
13. D. Rehder, H.-C. Bechthold, A. Kececi, H. Schmidt and M. Siewing, *Z. Naturforsch., Teil B*, 1982, **37**, 631.
14. F. Calderazzo, G. Pampaloni, G. Pelizzi and F. Vitali, *Organometallics*, 1988, **7**, 1083.
15. E. M. Carnahan and S. J. Lippard, *J. Am. Chem. Soc.*, 1990, **112**, 3230.
16. E. M. Carnahan and S. J. Lippard, *J. Am. Chem. Soc.*, 1992, **114**, 4166.
17. F. Calderazzo, M. Castellani, G. Pampaloni and P. F. Zanazzi, *J. Chem. Soc., Dalton Trans.*, 1985, 1989.
18. F. Calderazzo, G. Pampaloni, U. Englert and J. Strähle, *J. Organomet. Chem.*, 1990, **383**, 45.
19. H.-C. Bechthold and D. Rehder, *J. Organomet. Chem.*, 1982, **233**, 215.
20. K. Bachmann and D. Rehder, *J. Organomet. Chem.*, 1984, **276**, 177.
21. D. Rehder, M. Fornalczy and P. Oltmanns, *J. Organomet. Chem.*, 1987, **331**, 207.
22. M. L. Luetkens, Jr., D. J. Santure, J. C. Huffman and A. P. Sattelberger, *J. Chem. Soc., Chem. Commun.*, 1985, 552.
23. M. L. Luetkens, Jr. *et al.*, *Inorg. Chem.*, 1987, **26**, 2430.
24. M. Fornalczyk, F. Suessmilch, W. Priebsch and D. Rehder, *J. Organomet. Chem.*, 1992, **426**, 159.
25. M. L. Luetkens, Jr., J. C. Huffman and A. P. Sattelberger, *J. Am. Chem. Soc.*, 1985, **107**, 3361.
26. M. L. H. Green, D. O'Hare, P. Mountford and J. G. Watkin, *J. Chem. Soc., Dalton Trans.*, 1991, 1705.
27. A. M. Bond, J. W. Bixler, E. Mocellin, S. Datta, E. J. James and S. S. Wreford, *Inorg. Chem.*, 1980, **19**, 1760.
28. R. J. Burt, G. J. Leigh and D. L. Hughes, *J. Chem. Soc., Dalton Trans.*, 1981, 793.
29. J. D. Protasiewicz, P. A. Bianconi, I. D. Williams, S. Liu, C. P. Rao and S. J. Lippard, *Inorg. Chem.*, 1992, **31**, 4134.
30. R. N. Vrtis, S. Liu, C. P. Rao, S. G. Bott and S. J. Lippard, *Organometallics*, 1991, **10**, 275.
31. R. N. Vrtis, S. G. Bott and S. J. Lippard, *Organometallics*, 1992, **11**, 270.
32. F. Calderazzo, G. Pampaloni and P. F. Zanazzi, *Chem. Ber.*, 1986, **119**, 2796.
33. D. R. Neithamer, R. E. LaPointe, R. A. Wheeler, D. S. Richeson, G. D. Van Duyne and P. T. Wolczanski, *J. Am. Chem. Soc.*, 1989, **111**, 9056.
34. F. Calderazzo, G. Pampaloni, U. Englert and J. Strähle, *Angew. Chem., Int. Ed. Engl.*, 1989, **28**, 471.
35. G. F. P. Warnock and J. E. Ellis, *J. Am. Chem. Soc.*, 1984, **106**, 5016.
36. R. L. DeKock, *Inorg. Chem.*, 1971, **10**, 1205.
37. F. G. N. Cloke *et al.*, *J. Organomet. Chem.*, 1984, **277**, 61.
38. Z. Lin and M. B. Hall, *Inorg. Chem.*, 1992, **31**, 2791.
39. J. E. Ellis, *Polyhedron*, 1989, **8**, 1611.
40. W. A. Herrmann, W. Kalcher, H. Biersack, I. Bernal and M. Creswick, *Chem. Ber.*, 1981, **114**, 3558.
41. W. A. Herrmann and H. Biersack, *J. Organomet. Chem.*, 1980, **191**, 397.
42. L. N. Lewis and K. G. Caulton, *Inorg. Chem.*, 1980, **19**, 3201.
43. D. Rehder, K. Paulsen and W. Basler, *J. Magn. Reson.*, 1985, **53**, 500.
44. R. Poli, *Organometallics*, 1990, **9**, 1892.
45. V. C. Gibson, T. P. Kee and W. Clegg, *J. Organomet. Chem.*, 1988, **353**, C23.
46. V. C. Gibson, T. P. Kee and W. Clegg, *J. Chem. Soc., Dalton Trans.*, 1990, 3199.
47. M. L. H. Green and A. K. Hughes, *J. Chem. Soc., Dalton Trans.*, 1992, 527.
48. A. Reddy *et al.*, *Organometallics*, 1992, **11**, 3894.

49. S. V. Kukharenko, G. L. Soloveichik and V. V. Strelets, *Metalloorg. Khim.*, 1990, **3**, 88 (*Chem. Abstr.*, 1990, **113**, 6512s).
50. F. Näumann, D. Rehder and V. Pank, *J. Organomet. Chem.*, 1982, **240**, 363.
51. K. M. Pfahl and J. E. Ellis, *Organometallics*, 1984, **3**, 230.
52. R. F. G. Baynham, J. Chetwynd-Talbot, P. Grebenik, R. N. Perutz and M. H. A. Powell, *J. Organomet. Chem.*, 1985, **284**, 229.
53. W. A. Herrmann, H. Biersack, M. L. Ziegler, K. Weidenhammer, R. Siegel and D. Rehder, *J. Am. Chem. Soc.*, 1981, **103**, 1692.
54. R. B. King and C. D. Hoff, *J. Organomet. Chem.*, 1982, **225**, 245.
55. A. M. Cardoso, R. J. H. Clark and S. Moorhouse, *J. Organomet. Chem.*, 1980, **186**, 237.
56. M. D. Curtis and J. Real, *Organometallics*, 1985, **4**, 940.
57. M. D. Curtis, J. Real and D. Kwon, *Organometallics*, 1989, **8**, 1644.
58. D. Kwon, J. Real, M. D. Curtis, A. L. Rheingold and B. S. Haggerty, *Organometallics*, 1991, **10**, 143.
59. Y. V. Skripkin *et al.*, *J. Organomet. Chem.*, 1984, **263**, 345.
60. A. S. Antsyshkina *et al.*, *Sov. J. Coord. Chem.* (*Engl. Transl.*), 1984, **10**, 872.
61. Y. V. Skripkin *et al.*, *Koord. Khim.*, 1985, **11**, 995.
62. A. Antiñolo, J. Martínez de Ilarduja, A. Otero, P. Royo, A. M. Manotti Lanfredi and A. Tiripicchio, *J. Chem. Soc., Dalton Trans.*, 1988, 2685.
63. A. Fakhr, Y. Mugnier, L. Roullier, R. Broussier, B. Gautheron and E. Laviron, *New. J. Chem.*, 1988, **12**, 213.
64. J. C. Leblanc, C. Moise and J. Tirouflet, *J. Organomet. Chem.*, 1985, **292**, 225.
65. A. Dormond and C. Moise, *Polyhedron*, 1989, **8**, 2189.
66. A. Antiñolo, M. Fajardo, A. Otero and P. Royo, *J. Organomet. Chem.*, 1983, **246**, 269.
67. L. Acedo, A. Otero and P. Royo, *J. Organomet. Chem.*, 1983, **258**, 181.
68. P. P. Deutsch, J. A. Maguire, W. D. Jones and R. Eisenberg, *Inorg. Chem.*, 1990, **29**, 686.
69. A. Fakhr, Y. Mugnier, R. Broussier, B. Gautheron and E. Laviron, *J. Organomet. Chem.*, 1983, **255**, C8.
70. R. A. Bell, S. A. Cohen, N. M. Doherty, R. S. Threlkel and J. E. Bercaw, *Organometallics*, 1986, **5**, 972.
71. J. Arnold, T. D. Tilley, A. L. Rheingold and S. J. Geib, *Organometallics*, 1987, **6**, 473.
72. D. H. Berry and Q. Jiang, *J. Am. Chem. Soc.*, 1987, **109**, 6210.
73. M. W. George *et al.*, *J. Am. Chem. Soc.*, 1993, **115**, 2286.
74. A. D. Poole, V. C. Gibson and W. Clegg, *J. Chem. Soc., Chem. Commun.*, 1992, 237.
75. U. Siemeling and V. C. Gibson, *J. Organomet. Chem.*, 1992, **426**, C25.
76. G. E. Herberich and H. Mayer, *J. Organomet. Chem.*, 1987, **322**, C29.
77. J. F. Leboeuf, J. C. Leblanc and C. Moise, *J. Organomet. Chem.*, 1987, **335**, 331.
78. H. G. Alt and H. E. Engelhardt, *Z. Naturforsch., Teil B*, 1989, **44**, 367.
79. A. M. Andreu, F. A. Jalón, A. Otero, P. Royo, A. M. M. Lanfredi and A. Tiripicchio, *J. Chem. Soc., Dalton Trans.*, 1987, 953.
80. D. E. Wigley, in 'Encyclopedia of Inorganic Chemistry', ed. R. B. King, Wiley, New York, 1994, vol. 5, p. 2462.
81. F. A. Jalón, A. Otero, P. Royo, J. L. Balcazar, F. Florencio and S. Garcia-Blanco, *J. Chem. Soc., Dalton Trans.*, 1989, 79.
82. M. J. Bunker, M. L. H. Green, C. Couldwell and K. Prout, *J. Organomet. Chem.*, 1980, **192**, C6.
83. M. J. Bunker and M. L. H. Green, *J. Chem. Soc., Dalton Trans.*, 1981, 85.
84. J. De la Mata, M. V. Galakhov, M. Gomez and P. Royo, *Organometallics*, 1993, **12**, 1189.
85. U. Siemeling and V. C. Gibson, *J. Organomet. Chem.*, 1992, **424**, 159.
86. J. Arnold, T. D. Tilley, A. L. Rheingold, S. J. Geib and A. M. Arif, *J. Am. Chem. Soc.*, 1989, **111**, 149.
87. V. C. Gibson and T. P. Kee, *J. Chem. Soc., Chem. Commun.*, 1989, 656.
88. D. Kwon and M. D. Curtis, *Organometallics*, 1990, **9**, 1.
89. D. Rehder and H. C. Bechthold, *Z. Naturforsch., Teil B*, 1984, **39**, 323.
90. J. W. Freeman and F. Basolo, *Organometallics*, 1991, **10**, 256.
91. H. M. Anstice, H. H. Fielding, V. C. Gibson, C. E. Housecroft and T. P. Kee, *Organometallics*, 1991, **10**, 2183.
92. M. L. H. Green, D. O'Hare and J. G. Watkin, *J. Chem. Soc., Chem. Commun.*, 1989, 698.
93. H. C. Bechthold and D. Rehder, *J. Organomet. Chem.*, 1981, **206**, 305.
94. J. M. Mayer and J. E. Bercaw, *J. Am. Chem. Soc.*, 1982, **104**, 2157.
95. T. P. Kee, V. C. Gibson and W. Clegg, *J. Organomet. Chem.*, 1987, **325**, C14.
96. V. C. Gibson, T. P. Kee and W. Clegg, *J. Chem. Soc., Chem. Commun.*, 1990, 29.
97. S. T. Carter, W. Clegg, V. C. Gibson, T. P. Kee and R. D. Sanner, *Organometallics*, 1989, **8**, 253.
98. M. Jimenez-Tenorio and G. J. Leigh, *Polyhedron*, 1989, **8**, 1784.
99. M. J. Bunker, A. De Cian, M. L. H. Green, J. J. E. Moreau and N. Siganporia, *J. Chem. Soc., Dalton Trans.*, 1980, 2155.
100. M. Hoch and D. Rehder, *Z. Naturforsch., Teil B*, 1983, **38**, 446.
101. G. Bonnet, O. Lavastre, J. C. Leblanc and C. Moise, *New. J. Chem.*, 1988, **12**, 551.
102. S. Challet, J. C. Leblanc and C. Moise, *New. J. Chem.*, 1993, **17**, 251.
103. G. Bonnet, O. Lavastre, J. C. Leblanc, C. Moise and G. Vitulli, *J. Organomet. Chem.*, 1988, **347**, C21.
104. G. Bonnet, M. M. Kubicki, C. Moise, R. Lazzaroni, P. Salvadori and G. Vitulli, *Organometallics*, 1992, **11**, 964.
105. O. J. Scherer, J. Vondung and G. Wolmershäuser, *Angew. Chem., Int. Ed. Engl.*, 1989, **28**, 1355.
106. O. J. Scherer, J. Vondung and G. Wolmershäuser, *J. Organomet. Chem.*, 1989, **376**, C35.
107. P. Oltmanns and D. Rehder, *J. Organomet. Chem.*, 1988, **345**, 87.
108. J. Arnold and T. D. Tilley, *J. Am. Chem. Soc.*, 1985, **107**, 6409.
109. G. Parkin, E. Bunel, B. J. Burger, M. S. Trimmer, A. van Asselt and J. E. Bercaw, *J. Mol. Catal.*, 1987, **41**, 21.
110. A. Belforte, F. Calderazzo and P. F. Zanazzi, *Gazz. Chim. Ital.*, 1985, **115**, 71.
111. W. A. Herrmann, H. Biersack, M. L. Ziegler and B. Balbach, *J. Organomet. Chem.*, 1981, **206**, C33.
112. K. S. Wong and J. A. Labinger, *J. Am. Chem. Soc.*, 1980, **102**, 3652.
113. A. A. Pasynskii *et al.*, *Zh. Neorg. Khim.*, 1981, **26**, 2435 (*Chem. Abstr.*, 1982, **96**, 122 937p).
114. Y. V. Skripkin *et al.*, *J. Organomet. Chem.*, 1982, **231**, 205.
115. C. Moise, J. F. Reynoud, J. C. Leblanc and R. Broussier, *J. Organomet. Chem.*, 1982, **240**, C15.
116. A. A. Pasynskii, B. I. Kolobkov, I. L. Eremenko, S. E. Nefedov, S. B. Katser and M. A. Porai-Koshits, *Zh. Neorg. Khim.*, 1992, **37**, 563.

117. J. F. Reynoud, J. C. Leblanc and C. Moise, *Transition Met. Chem. (Weinheim, Ger.)*, 1985, **10**, 291.
118. D. Perrey, J. C. Leblanc, C. Moise and J. Martin-Gil, *J. Organomet. Chem.*, 1991, **412**, 363.
119. G. E. Herberich, I. Hausmann and N. Klaff, *Angew. Chem., Int. Ed. Engl.*, 1989, **28**, 319.
120. I. Pforr, F. Naeumann and D. Rehder, *J. Organomet. Chem.*, 1983, **258**, 189.
121. A. A. Pasynskii *et al.*, *Izv. Akad. Nauk SSSR, Ser. Khim.*, 1983, **5**, 1207.
122. M. L. H. Green, A. K. Hughes and P. Mountford, *J. Chem. Soc., Dalton Trans.*, 1991, 1407.
123. F. Näumann, J. Kopf and D. Rehder, *J. Organomet. Chem.*, 1984, **267**, 249.
124. Y. V. Skripkin *et al.*, *Koord. Khim.*, 1985, **11**, 1136.
125. R. Hoffmann, C. N. Wilker, S. J. Lippard, J. L. Templeton and D. C. Brower, *J. Am. Chem. Soc.*, 1983, **105**, 146.
126. P. A. Bianconi, R. N. Vrtis, Ch. P. Rao, I. D. Williams, M. P. Engeler and S. J. Lippard, *Organometallics*, 1987, **6**, 1968.
127. R. N. Vrtis, Ch. P. Rao, S. G. Bott and S. J. Lippard, *J. Am. Chem. Soc.*, 1988, **110**, 7564.
128. R. N. Vrtis, Ch. P. Rao, S. Warner and S. J. Lippard, *J. Am. Chem. Soc.*, 1988, **110**, 2669.
129. L. D. Silverman, J. C. Dewan, C. M. Giandomenico and S. J. Lippard, *Inorg. Chem.*, 1980, **19**, 3379.
130. F. A. Cotton and W. Roth, *J. Am. Chem. Soc.*, 1983, **105**, 3734.
131. F. A. Cotton, S. A. Duraj and W. J. Roth, *J. Am. Chem. Soc.*, 1984, **106**, 6987.
132. H. C. Aspinall, M. M. Roberts and S. J. Lippard, *Inorg. Chem.*, 1984, **23**, 1782.
133. R. Serrano and P. Royo, *J. Organomet. Chem.*, 1983, **247**, 33.
134. M. Gomez, J. M. Martínez de Ilarduya and P. Royo, *J. Organomet. Chem.*, 1989, **369**, 197.
135. J. M. Martínez de Ilarduya, A. Otero and P. Royo, *J. Organomet. Chem.*, 1988, **340**, 187.
136. D. Lucas, Y. Mugnier, A. Antiñolo, A. Otero and M. Fajardo, *J. Organomet. Chem.*, 1992, **435**, C3.
137. M. A. A. F. de C. T. Carrondo, J. Morais, C. C. Romao, M. J. Romao and L. F. Veiros, *Polyhedron*, 1993, **12**, 765.
138. B. J. Burger, B. D. Santarsiero, M. S. Trimmer and J. E. Bercaw, *J. Am. Chem. Soc.*, 1988, **110**, 3134.
139. J. F. Leboeuf, J. C. Leblanc and C. Moise, *J. Organomet. Chem.*, 1989, **364**, C22.
140. J. E. Nelson, G. Parkin and J. E. Bercaw, *Organometallics*, 1992, **11**, 2181.
141. A. H. Klazinga and J. H. Teuben, *J. Organomet. Chem.*, 1980, **192**, 75.
142. T. Albright and H. Tang, *Angew. Chem., Int. Ed. Engl.*, 1992, **31**, 1462.
143. S. K. Kang, H. Tang and T. A. Albright, *J. Am. Chem. Soc.*, 1993, **115**, 1971.
144. C. Pulham, A. Haaland, A. Hammel, K. Rypdal, H. P. Verne and H. V. Volden, *Angew. Chem., Int. Ed. Engl.*, 1992, **31**, 1464.
145. C. J. Piersol, R. D. Profilet, P. E. Fanwick and I. P. Rothwell, *Polyhedron*, 1993, **12**, 1779.
146. J. D. Fellmann, R. R. Schrock and G. A. Rupprecht, *J. Am. Chem. Soc.*, 1981, **103**, 5752.
147. L. W. Messerle, P. Jennische, R. R. Schrock and G. J. Stucky, *J. Am. Chem. Soc.*, 1980, **102**, 6744.
148. M. L. H. Green, A. K. Hughes, N. A. Popham, A. H. H. Stephens and L. L. Wong, *J. Chem. Soc., Dalton Trans.*, 1992, 3077.
149. B.-H. Chang, C.-P. Lau, R. H. Grubbs and C. H. Brubaker, Jr., *J. Organomet. Chem.*, 1985, **281**, 213.
150. H. C. L. Abbenhuis *et al.*, *Organometallics*, 1993, **12**, 210.
151. G. A. Rupprecht, L. W. Messerle, J. D. Fellmann and R. R. Schrock, *J. Am. Chem. Soc.*, 1980, **102**, 6236.
152. S. Elbel, M. Grodzicki, L. Pille and G. Ruenger, *J. Mol. Struct.*, 1988, **175**, 441.
153. M. Bochman, G. Wilkinson and G. B. Young, *J. Chem. Soc., Dalton Trans.*, 1980, 1879.
154. J. D. Fellmann, R. R. Schrock and D. D. Traficante, *Organometallics*, 1982, **1**, 481.
155. L. R. Chamberlain and I. P. Rothwell, *J. Chem. Soc. Dalton Trans.*, 1987, 163.
156. K. W. Chiu, R. A. Jones, G. Wilkinson, A. M. R. Galas and M. B. Hursthouse, *J. Chem. Soc., Dalton Trans.*, 1981, 1892.
157. L. R. Chamberlain, I. P. Rothwell, K. Folting and J. C. Huffman, *J. Chem. Soc., Dalton Trans.*, 1987, 155.
158. M. H. Chisholm, L. S. Tan and J. C. Huffman, *J. Am. Chem. Soc.*, 1982, **104**, 4879.
159. R. E. LaPointe and P. T. Wolczanski, *J. Am. Chem. Soc.*, 1986, **108**, 3535.
160. R. L. Miller, R. Toreki, R. E. LaPointe, P. T. Wolczanski, G. D. Van Duyne and D. C. Roe, *J. Am. Chem. Soc.*, 1993, **115**, 5570.
161. K. J. Covert, D. R. Neithamer, M. C. Zonnevylle, R. E. LaPointe, C. P. Schaller and P. T. Wolczanski, *Inorg. Chem.*, 1991, **30**, 2494.
162. J. R. Strickler, P. A. Wexler and D. E. Wigley, *Organometallics*, 1991, **10**, 118.
163. L. R. Chamberlain, J. Kerschner, A. P. Rothwell, I. P. Rothwell and J. C. Huffman, *J. Am. Chem. Soc.*, 1987, **109**, 6471.
164. R. E. LaPointe, P. T. Wolczanski and G. D. Van Duyne, *Organometallics*, 1985, **4**, 1810.
165. L. G. Hubert-Pfalzgraf and G. Aharonian, *Inorg. Chim. Acta*, 1985, **100**, L21.
166. A. S. Baley, Y. Chauvin, D. Commereuc and P. B. Hitchcock, *New J. Chem.*, 1991, **15**, 609.
167. V. M. Visciglio, P. E. Fanwick and I. P. Rothwell, *J. Chem. Soc., Chem. Commun.*, 1992, 1505.
168. R. W. Chesnut, B. D. Steffey and I. P. Rothwell, *Polyhedron*, 1989, **8**, 1607.
169. R. W. Chesnut, B. D. Steffey, I. P. Rothwell and J. C. Huffman, *Polyhedron*, 1988, **7**, 753.
170. R. W. Chesnut, J. S. Yu, P. E. Fanwick, I. P. Rothwell and J. C. Huffman, *Polyhedron*, 1990, **9**, 1051.
171. R. W. Chesnut, L. D. Durfee, P. E. Fanwick, I. P. Rothwell, K. Folting and J. C. Huffman, *Polyhedron*, 1987, **6**, 2019.
172. S. M. Rocklage, J. D. Fellmann, G. A. Rupprecht, L. W. Messerle and R. R. Schrock, *J. Am. Chem. Soc.*, 1981, **103**, 1440.
173. B. C. Ankianiec, P. E. Fanwick and I. P. Rothwell, *J. Am. Chem. Soc.*, 1991, **113**, 4710.
174. S. G. Bott, M. A. Mazid, M. B. Hursthouse and A. C. Sullivan, *J. Chem. Soc., Dalton Trans.*, 1991, 355.
175. L. R. Chamberlain, A. P. Rothwell and I. P. Rothwell, *J. Am. Chem. Soc.*, 1984, **106**, 1847.
176. R. Wang, K. Folting, J. C. Huffman, L. R. Chamberlain and I. P. Rothwell, *Inorg. Chim. Acta*, 1986, **120**, 81.
177. L. Chamberlain, J. Keddington and I. P. Rothwell, *Organometallics*, 1982, **1**, 1098.
178. L. R. Chamberlain, J. Keddington, I. P. Rothwell and J. C. Huffman, *Organometallics*, 1982, **1**, 1538.
179. L. R. Chamberlain, I. P. Rothwell and J. C. Huffman, *J. Am. Chem. Soc.*, 1986, **108**, 1502.
180. S. G. Bott and A. C. Sullivan, *J. Chem. Soc., Chem. Commun.*, 1988, 1577.
181. S. M. Baxter and P. T. Wolczanski, *Organometallics*, 1990, **9**, 2498.
182. R. W. Chesnut, G. G. Jacob, J. S. Yu, P. E. Fanwick and I. P. Rothwell, *Organometallics*, 1991, **10**, 321.
183. J. S. Yu, B. C. Ankianiec, M. T. Nguyen and I. P. Rothwell, *J. Am. Chem. Soc.*, 1992, **114**, 1927.
184. J. S. Yu and I. P. Rothwell, *J. Chem. Soc., Chem. Commun.*, 1992, 632.
185. H. C. L. Abbenhuis *et al.*, *J. Am. Chem. Soc.*, 1992, **114**, 9773.

186. H. C. L. Abbenhuis, D. M. Grove, P. Van der Sluis, A. L. Spek and G. Van Koten, *Recl. Trav. Chim. Pays-Bas*, 1990, **109**, 446.
187. A. R. Middleton and G. Wilkinson, *J. Chem. Soc., Dalton Trans.*, 1980, 1888.
188. M. M. Banaszak Holl, P. T. Wolczanski and G. D. Van Duyne, *J. Am. Chem. Soc.*, 1990, **112**, 7989.
189. H. C. L. Abbenhuis, D. M. Grove, G. P. M. van Mier, A. L. Spek and G. van Koten, *J. Chem. Soc., Chem. Commun.*, 1989, 1581.
190. D. L. Reger, C. A. Swift and L. Lebioda, *Inorg. Chem.*, 1984, **23**, 349.
191. D. L. Reger, C. A. Swift and L. Lebioda, *J. Am. Chem. Soc.*, 1983, **105**, 5343.
192. S. M. Rocklage and R. R. Schrock, *J. Am. Chem. Soc.*, 1982, **104**, 3077.
193. C. P. Schaller and P. T. Wolczanski, *Inorg. Chem.*, 1993, **32**, 131.
194. L. R. Chamberlain and I. P. Rothwell, *J. Am. Chem. Soc.*, 1983, **105**, 1665.
195. L. R. Chamberlain, I. P. Rothwell and J. C. Huffman, *J. Am. Chem. Soc.*, 1982, **104**, 7338.
196. P. E. Fanwick, A. E. Ogilvy and I. P. Rothwell, *Organometallics*, 1987, **6**, 73.
197. R. Uhrhammer, D. J. Crowther, J. D. Olson, D. C. Swenson and R. F. Jordan, *Organometallics*, 1992, **11**, 3098.
198. R. D. Sanner, S. T. Carter, W. J. Bruton, Jr., *J. Organomet. Chem.*, 1982, **240**, 157.
199. J. M. Mayer, C. J. Curtis and J. E. Bercaw, *J. Am. Chem. Soc.*, 1983, **105**, 2651.
200. M. D. Curtis and J. Real, *J. Am. Chem. Soc.*, 1986, **108**, 4668.
201. M. Gomez, G. Jimenez, P. Royo, M. A. Pellinghelli and A. Tiripicchio, *J. Organomet. Chem.*, 1992, **439**, 309.
202. R. Fandos, M. Gomez, P. Royo, S. Garcia-Blanco, S. Martinez-Carrera and J. Sanz-Aparicio, *Organometallics*, 1987, **6**, 1581.
203. T. Y. Meyer, L. R. Garner, N. C. Baenziger and L. Messerle, *Inorg. Chem.*, 1990, **29**, 4045.
204. P. A. Belmonte, R. R. Schrock and C. S. Day, *J. Am. Chem. Soc.*, 1982, **104**, 3082.
205. H. W. Turner, R. R. Schrock, J. D. Fellmann and S. J. Holmes, *J. Am. Chem. Soc.*, 1983, **105**, 4942.
206. P. A. Belmonte, F. G. N. Cloke and R. R. Schrock, *J. Am. Chem. Soc.*, 1983, **105**, 2643.
207. R. Fandos, M. Gomez and P. Royo, *Organometallics*, 1989, **8**, 1604.
208. P. Belmonte, R. R. Schrock, M. R. Churchill and W. J. Youngs, *J. Am. Chem. Soc.*, 1980, **102**, 2858.
209. P. A. Belmonte, F. G. N. Cloke, K. H. Theopold and R. R. Schrock, *Inorg. Chem.*, 1984, **23**, 2365.
210. R. T. Baker, J. C. Calabrese, R. L. Harlow and I. D. Williams, *Organometallics*, 1993, **12**, 830.
211. J. M. Mayer, P. T. Wolczanski, B. D. Santarsiero, W. A. Olson and J. E. Bercaw, *Inorg. Chem.*, 1983, **22**, 1149.
212. J. Arnold, D. N. Shina, T. D. Tilley and A. M. Arif, *Organometallics*, 1986, **5**, 2037.
213. V. C. Gibson, D. N. Williams, W. Clegg and D. C. R. Hockless, *Polyhedron*, 1989, **8**, 1819.
214. J. K. Cockcroft, V. C. Gibson, J. A. K. Howard, A. D. Poole, U. Siemeling and C. Wilson, *J. Chem. Soc., Chem. Commun.*, 1992, 1668.
215. A. D. Poole *et al.*, *Organometallics*, 1993, **12**, 2549.
216. M. M. Banaszak Holl, M. Kersting, B. D. Pendley and P. T. Wolczanski, *Inorg. Chem.*, 1990, **29**, 1518.
217. J. Sala-Pala, J. L. Migot, J. E. Guerchais, L. Le Gall and F. Grosjean, *J. Organomet. Chem.*, 1983, **248**, 299.
218. D. R. Tueting, M. M. Olmstead and N. E. Schore, *Organometallics*, 1992, **11**, 2235.
219. K. H. Thiele, W. Kubak, J. Sieler, H. Borrmann, A. Simon, *Z. Anorg. Allg. Chem.*, 1990, **587**, 80.
220. J. L. Migot, J. Sala-Pala and J. E. Guerchais, *J. Organomet. Chem.*, 1983, **243**, 427.
221. P.-F. Fu, M. A. Khan and K. M. Nicholas, *Organometallics*, 1992, **11**, 2607.
222. J. Amaudrut, A. Kadmiri, J. Sala-Pala and J. E. Guerchais, *J. Organomet. Chem.*, 1984, **266**, 53.
223. G. Proulx and R. G. Bergman, *Science*, 1993, **259**, 661.
224. R. Mercier, J. Douglade, J. Amaudrut, J. Sala-Pala and J. E. Guerchais, *J. Organomet. Chem.*, 1983, **244**, 145.
225. P.-F. Fu, M. A. Khan and K. M. Nicholas, *Organometallics*, 1991, **10**, 382.
226. A. van Asselt, B. J. Burger, V. C. Gibson and J. E. Bercaw, *J. Am. Chem. Soc.*, 1986, **108**, 5347.
227. A. van Asselt, M. S. Trimmer, L. M. Henling and J. E. Bercaw, *J. Am. Chem. Soc.*, 1988, **110**, 8254.
228. L. L. Whinnery, Jr., L. M. Henling and J. E. Bercaw, *J. Am. Chem. Soc.*, 1991, **113**, 7575.
229. P. Jernakoff, C. de M. de Bellefon, G. L. Geoffroy, A. L. Rheingold and S. J. Geib, *Organometallics*, 1987, **6**, 1362.
230. P. Jernakoff, C. De Meric de Bellefon, G. L. Geoffroy, A. L. Rheingold and S. J. Geib, *New J. Chem.*, 1988, **12**, 329.
231. P. B. Hitchcock, M. F. Lappert and C. R. C. Milne, *J. Chem. Soc., Dalton Trans.*, 1981, 180.
232. R. S. Threlkel and J. E. Bercaw, *J. Am. Chem. Soc.*, 1981, **103**, 2650.
233. A. T. Casey and J. B. Raynor, *J. Chem. Soc., Dalton Trans.*, 1983, 2057.
234. A. H. Al-Mowali, *J. Chem. Soc., Dalton Trans.*, 1980, 426.
235. R. Broussier, J. D. Olivier and B. Gautheron, *J. Organomet. Chem.*, 1981, **215**, 59.
236. J. Sala-Pala, J. Amaudrut, J. E. Guerchais, R. Mercier, J. Douglade and J. G. Theobald, *J. Organomet. Chem.*, 1981, **204**, 347.
237. V. A. Knizhnikov, V. I. Potkin, R. V. Kaberdin and Y. A. Ol'dekop, *Zh. Obshch. Khim.*, 1991, **61**, 1365 (*Chem. Abstr.*, 1992, **116**, 21 187v).
238. G. S. Bristow, P. B. Hitchcock and M. F. Lappert, *J. Chem. Soc., Chem. Commun.*, 1982, 1145.
239. A. Fakhr, Y. Mugnier, R. Broussier, B. Gautheron and E. Laviron, *J. Organomet. Chem.*, 1987, **317**, 201.
240. F. A. Urbanos, M. Mena, P. Royo and A. Antiñolo, *J. Organomet. Chem.*, 1984, **276**, 185.
241. A. A. Pasynskii *et al.*, *J. Organomet. Chem.*, 1980, **201**, 269.
242. M. L. H. Green and B. Josseaume, *J. Organomet. Chem.*, 1980, **193**, 339.
243. A. Antiñolo *et al.*, *J. Chem. Soc., Dalton Trans.*, 1987, 975.
244. D. H. Berry, T. S. Koloski and P. J. Carroll, *Organometallics*, 1990, **9**, 2952.
245. A. H. Klazinga and J. H. Teuben, *J. Organomet. Chem.*, 1980, **194**, 309.
246. D. L. Lichtenberger, G. P. Darsey, G. E. Kellogg, R. D. Sanner, V. G. Young, Jr. and J. R. Clark, *J. Am. Chem. Soc.*, 1989, **111**, 5019.
247. P. F. Fu, M. A. Khan and K. M. Nicholas, *J. Am. Chem. Soc.*, 1992, **114**, 6579.
248. P. F. Fu, M. A. Khan and K. M. Nicholas, *Organometallics*, 1993, **12**, 3790.
249. N. M. Doherty and J. E. Bercaw, *J. Am. Chem. Soc.*, 1985, **107**, 2670.
250. A. Savaranamuthu, A. E. Bruce, M. R. M. Bruce, M. C. Fermin, A. S. Hneihen and J. W. Bruno, *Organometallics*, 1992, **11**, 2190.

251. H. Van Dam, A. Terpstra, A. Oskam and J. H. Teuben, *Z. Naturforsch., Teil B*, 1981, **36**, 420.
252. M. E. Eichner, H. G. Alt and M. D. Rausch, *J. Organomet. Chem.*, 1984, **264**, 309.
253. C. McDade, V. C. Gibson, B. D. Santarsiero and J. E. Bercaw, *Organometallics*, 1988, **7**, 1.
254. G. E. Herberich and H. Mayer, *J. Organomet. Chem.*, 1988, **347**, 93.
255. J. Amaudrut, J.-C. LeBlanc, C. Moise and J. Sala-Pala, *J. Organomet. Chem.*, 1985, **295**, 167.
256. J. A. Marsella, J. C. Huffman, K. Folting and K. G. Caulton, *Inorg. Chim. Acta*, 1985, **96**, 161.
257. D. E. Wigley, *Prog. Inorg. Chem.*, 1994, **42**, 239.
258. P. J. Domaille, B. M. Foxman, T. J. McNeese and S. S. Wreford, *J. Am. Chem. Soc.*, 1980, **102**, 4114.
259. M. L. Luetkens, Jr., W. L. Elcesser, J. C. Huffman and A. P. Sattelberger, *J. Chem. Soc., Chem. Commun.*, 1983, 1072.
260. M. L. Luetkens, Jr., W. L. Elcesser, J. C. Huffman and A. P. Sattelberger, *Inorg. Chem.*, 1984, **23**, 1718.
261. M. L. Luetkens, Jr., J. C. Huffman and A. P. Sattelberger, *J. Am. Chem. Soc.*, 1983, **105**, 4474.
262. A. P. Sattelberger, R. B. Wilson, Jr. and J. C. Huffman, *J. Am. Chem. Soc.*, 1980, **102**, 7111.
263. A. P. Sattelberger, R. B. Wilson, Jr. and J. C. Huffman, *Inorg. Chem.*, 1982, **21**, 4179.
264. R. B. Wilson, Jr., A. P. Sattelberger and J. C. Huffman, *J. Am. Chem. Soc.*, 1982, **104**, 858.
265. A. J. Scioly, M. L. Luetkens, Jr., R. B. Wilson, Jr., J. C. Huffman and A. P. Sattelberger, *Polyhedron*, 1987, **6**, 741.
266. S. M. Rocklage, H. W. Turner, J. D. Fellmann and R. R. Schrock, *Organometallics*, 1982, **1**, 703.
267. L. G. Hubert-Pfalzgraf and J. G. Riess, *Inorg. Chim. Acta*, 1981, **52**, 231.
268. D. Rehder, P. Oltmanns, M. Hoch, C. Weidemann and W. Priebsch, *J. Organomet. Chem.*, 1986, **308**, 19.
269. R. Toreki, R. E. LaPointe and P. T. Wolczanski, *J. Am. Chem. Soc.*, 1987, **109**, 7558.
270. R. E. LaPointe, P. T. Wolczanski and J. F. Mitchell, *J. Am. Chem. Soc.*, 1986, **108**, 6382.
271. B. D. Steffey, R. W. Chesnut, J. L. Kerschner, P. J. Pellechia, P. E. Fanwick and I. P. Rothwell, *J. Am. Chem. Soc.*, 1989, **111**, 378.
272. R. D. Profilet, P. E. Fanwick and I. P. Rothwell, *Polyhedron*, 1992, **11**, 1559.
273. A. P. Sattelberger, R. B. Wilson, Jr. and J. C. Huffman, *Inorg. Chem.*, 1982, **21**, 2392.
274. D. M. Heinekey, *J. Am. Chem. Soc.*, 1991, **113**, 6074.
275. J. A. Labinger and K. S. Wong, *ACS Symp. Ser.*, 1981, **152**, 253.
276. D. F. Foust, R. D. Rogers, M. D. Rausch and J. L. Atwood, *J. Am. Chem. Soc.*, 1982, **104**, 5646.
277. J. F. Leboeuf, J. C. Leblanc and C. Moise, *C. R. Acad. Sci., Ser. 2*, 1988, **307**, 1757.
278. A. Antiñolo, F. Carrillo, J. Fernandez-Baeza, A. Otero, M. Fajardo and B. Chaudret, *Inorg. Chem.*, 1992, **31**, 5156.
279. T. M. Zvukova, A. I. Sizov, Y. K. Gun'ko, B. M. Bulychev and G. L. Soloveichik, *Metalloorg. Khim.*, 1988, **1**, 1179 (*Chem. Abstr.*, 1989, **111**, 194 955u).
280. T. M. Arkhireeva, B. M. Bulychev, T. A. Sokolova, G. L. Soloveichik, V. K. Belsky and G. N. Boiko, *Inorg. Chim. Acta*, 1988, **141**, 221.
281. A. Antiñolo *et al.*, *J. Chem. Soc., Chem. Commun.*, 1988, 1210.
282. V. C. Gibson, J. E. Bercaw, W. J. Bruton, Jr. and R. D. Sanner, *Organometallics*, 1986, **5**, 976.
283. H. Brunner, U. Klement, J. Wachter, M. Tsunoda, J. C. Leblanc and C. Moise, *Inorg. Chem.*, 1990, **29**, 584.
284. T. M. Arkhireeva, B. M. Bulychev, A. I. Sizov, T. A. Sokolova, V. K. Bel'skii and G. L. Soloveichik, *Inorg. Chim. Acta*, 1990, **169**, 109.
285. L. Roullier, D. Lucas, Y. Mugnier, A. Antiñolo, M. Fajardo and A. Otero, *J. Organomet. Chem.*, 1991, **412**, 353.
286. K. W. Zilm *et al.*, *J. Am. Chem. Soc.*, 1990, **112**, 920.
287. A. Antiñolo, M. Fajardo, A. Otero, M. C. Puerta and Y. Mugnier, *J. Organomet. Chem.*, 1990, **382**, 389.
288. J. C. Leblanc, J. F. Reynoud and C. Moise, *C. R. Acad. Sci., Ser. 2*, 1982, **295**, 755.
289. D. A. Lemenovskii, V. P. Fedin, A. V. Aleksandrov, Yu. L. Slovokhotov and Yu. T. Struchkov, *J. Organomet. Chem.*, 1980, **201**, 257.
290. D. A. Lemenovskii, I. F. Urazovskii, I. E. Nifant'ev and E. G. Perevalova, *J. Organomet. Chem.*, 1985, **292**, 217.
291. D. A. Lemenovskii *et al.*, *J. Organomet. Chem.*, 1988, **342**, 31.
292. D. A. Lemenovskii, I. F. Urazowski, Y. K. Grishin and V. A. Roznyatovskii, *J. Organomet. Chem.*, 1985, **290**, 301.
293. D. A. Lemenovskii, I. F. Urazovskii and E. G. Perevalova, *Koord. Khim.*, 1985, **11**, 1079.
294. M. L. H. Green, A. K. Hughes and P. Mountford, *J. Chem. Soc., Dalton Trans.*, 1991, 1699.
295. D. A. Lemenovskii, I. E. Nifant'ev, E. G. Perevalova, S. N. Tsiomo, N. K. Efimov and V. I. Zubanova, *Koord. Khim.*, 1987, **13**, 1623.
296. M. L. H. Green, A. K. Hughes, D. M. Michaelidou and P. Mountford, *J. Chem. Soc., Chem. Commun.*, 1993, 591.
297. C. Ting, N. C. Baenziger and L. Messerle, *J. Chem. Soc., Chem. Commun.*, 1988, 1133.
298. H. Brunner, M. M. Kubicki, J. C. Leblanc, C. Moise, F. Volpato and J. Wachter, *J. Chem. Soc., Chem. Commun.*, 1993, 851.
299. E. G. Perevalova, I. F. Urazowski, D. A. Lemenovskii, Y. L. Slovokhotov and Y. T. Struchkov, *J. Organomet. Chem.*, 1985, **289**, 319.
300. J. A. Gagliardi, R. G. Teller, P. A. Vella and J. M. Williams, *Cryst. Struct. Commun.*, 1982, **11**, 861.
301. C. P. Casey and G. T. Whiteker, *Inorg. Chem.*, 1990, **29**, 876.
302. J. F. Reynoud, J.-C. Leblanc and C. Moise, *J. Organomet. Chem.*, 1985, **296**, 377.
303. J. F. Reynoud, J.-C. Leblanc and C. Moise, *Organometallics*, 1985, **4**, 1059.
304. A. Antiñolo, M. Fajardo, F. A. Jalón, C. Lopez Mardomingo, A. Otero and C. Sanz-Bernabe, *J. Organomet. Chem.*, 1989, **369**, 187.
305. M. T. Haward, M. W. George, P. A. Hamley and M. J. Poliakoff, *J. Chem. Soc., Chem. Commun.*, 1991, 1101.
306. W. A. Herrmann, H. Biersack, B. Balbach, P. Wülknitz and M. L. Ziegler, *Chem. Ber.*, 1984, **117**, 79.
307. W. A. Herrmann, H. Biersack, M. L. Ziegler and P. Wülknitz, *Angew. Chem., Int. Ed. Engl.*, 1981, **20**, 388.
308. M. D. Curtis, L. G. Bell and W. M. Butler, *Organometallics*, 1985, **4**, 701.
309. G. Parkin *et al.*, *Inorg. Chem.*, 1992, **31**, 82.
310. D. M. Antonelli, W. P. Schaefer, G. Parkin and J. E. Bercaw, *J. Organomet. Chem.*, 1993, **462** 213.
311. G. Parkin and J. E. Bercaw, *Polyhedron*, 1988, **7**, 2053.
312. K. A. Jørgensen, *Inorg. Chem.*, 1993, **32**, 1521.
313. D. A. Lemenovskii, M. Putala, G. I. Nikonov, N. B. Kazennova, D. S. Yufit and Y. T. Struchkov, *J. Organomet. Chem.*, 1993, **454**, 123.

314. D. A. Lemenovskii, S. A. Konde and E. G. Perevalova, *Koord. Khim.*, 1986, **12**, 703.
315. J. C. Leblanc and C. Moise, *J. Organomet. Chem.*, 1989, **364**, C3.
316. V. C. Gibson, T. P. Kee, S. T. Carter, R. D. Sanner and W. Clegg, *J. Organomet. Chem.*, 1991, **418**, 197.
317. J.-F. Leboeuf, O. Lavastre, J.-C. Leblanc and C. Moise, *J. Organomet. Chem.*, 1991, **418**, 359.
318. L. Roullier, D. Lucas, Y. Mugnier, A. Antinolo, M. Fajardo and A. Otero, *J. Organomet. Chem.*, 1990, **396**, C12.
319. B. Chaudret, H. H. Limbach and C. Moise, *C. R. Acad. Sci., Ser. 2*, 1992, **315**, 533.
320. V. C. Gibson and T. P. Kee, *J. Organomet. Chem.*, 1993, **444**, 91.
321. H. Brunner, G. Gehart, W. Meier, J. Wachter and B. Nuber, *J. Organomet. Chem.*, 1993, **454**, 117.
322. H.-J. Bach *et al.*, *Organometallics*, 1992, **11**, 1403.
323. E. Bulak, G. E. Herberich, I. Manners, H. Mayer and P. Paetzold, *Angew. Chem., Int. Ed. Engl.*, 1988, **27**, 958.
324. S. Baral, J. A. Labinger, W. R. Scheidt and F. J. Timmers, *J. Organomet. Chem.*, 1981, **215**, C53.
325. D. H. Jones, J. A. Labinger and D. P. Weitekamp, *J. Am. Chem. Soc.*, 1989, **111**, 3087.
326. K. W. Zilm, D. M. Heinekey, J. M. Millar, N. G. Payne and P. Demou, *J. Am. Chem. Soc.*, 1989, **111**, 3088.
327. D. M. Heinekey, J. M. Millar, T. F. Koetzle, N. G. Payne and K. W. Zilm, *J. Am. Chem. Soc.*, 1990, **112**, 909.
328. D. A. Lemenovskii, I. E. Nifant'ev and D. V. Zagorevskii, *Izv. Akad. Nauk SSSR, Ser. Khim.*, 1985, **8**, 1933.
329. A. S. Antsyshkina *et al.*, *Koord. Khim.*, 1982, **8**, 1423.
330. B. Balbach *et al.*, *Organometallics*, 1988, **7**, 325.
331. J. C. Leblanc, J. F. Reynoud and C. Moise, *J. Organomet. Chem.*, 1983, **244**, C24.
332. F. J. Timmers, W. R. Scheidt, J. A. Labinger and S. Baral, *J. Organomet. Chem.*, 1982, **240**, 153.
333. W. A. Herrmann, H. Biersack, B. Balbach and M. L. Ziegler, *Chem. Ber.*, 1984, **117**, 95.
334. A. S. Antsyshkina, M. A. Porai-Koshits, V. N. Ostrikova, A. A. Pasynskii, Yu. V. Skripkin and V. T. Kalinnikov, *Koord. Khim.*, 1982, **8**, 1552.
335. A. Antiñolo *et al.*, *J. Chem. Soc., Chem. Commun.*, 1990, 17.
336. P. H. M. Budzelaar, K. H. den Haan, J. Boersma, G. J. M. van der Kerk and A. L. Spek, *Organometallics*, 1984, **3**, 156.
337. P. H. M. Budzelaar, A. A. H. van der Zeijden, J. Boersma, G. J. M. van der Kerk, A. L. Spek and A. J. M. Duisenberg, *Organometallics*, 1984, **3**, 159.
338. T. M. Arkhireeva, V. K. Bel'skii, B. M. Bulychev and G. L. Soloveichik, *Izv. Akad. Nauk SSSR, Ser. Khim.*, 1986, **12**, 2819.
339. Y. K. Gun'ko, V. K. Bel'skii, A. I. Sizov, G. L. Soloveichik and B. M. Bulychev, *Metalloorg. Khim.*, 1989, **2**, 1121 (*Chem. Abstr.*, 1990, **112**, 217 133m).
340. R. Kergoat, M. M. Kubicki, J. E. Guerchais, N. C. Norman and A. G. Orpen, *J. Chem. Soc., Dalton Trans.*, 1982, 633.
341. Y. V. Skripkin, I. L. Eremenko, A. A. Pasynskii, Y. T. Struchkov and V. E. Shklover, *J. Organomet. Chem.*, 1984, **267**, 285.
342. H. Brunner *et al.*, *J. Organomet. Chem.*, 1992, **436**, 313.
343. M. L. H. Green and L.-L. Wong, *J. Chem. Soc., Dalton Trans.*, 1989, 2133.
344. M. L. H. Green and L.-L. Wong, *J. Chem. Soc., Chem. Commun.*, 1989, 571.
345. S. E. Halfon, M. C. Fermin and J. W. Bruno, *J. Am. Chem. Soc.*, 1989, **111**, 5490.
346. T. M. Arkhireeva, B. M. Bulychev, A. N. Protsky, G. L. Soloveichik and V. K. Bel'sky, *J. Organomet. Chem.*, 1986, **317**, 33.
347. T. M. Arkhireeva, B. M. Bulychev, A. N. Protskii, G. L. Soloveichik and V. K. Bel'skii, *Koord. Khim.*, 1987, **13**, 986.
348. R. R. Schrock, *ACS Symp. Ser.*, 1983, **211**, 369.
349. R. R. Schrock, in 'Reactions of Coordinated Ligands', ed. P. S. Braterman, Plenum, New York, 1986, p. 221.
350. S. M. Rocklage and R. R. Schrock, *J. Am. Chem. Soc.*, 1980, **102**, 7808.
351. R. Schrock, S. Rocklage, J. Wengrovius, G. Rupprecht and J. Fellmann, *J. Mol. Catal.*, 1980, **8**, 73.
352. H. Braunschweig, P. Paetzold and T. P. Spaniol, *Chem. Ber.*, 1993, **126**, 1565.
353. H. Braunschweig, P. Paetzold and R. Boese, *Chem. Ber.*, 1993, **126**, 1571.
354. A. W. Gal and H. Van der Heijden, *J. Chem. Soc., Chem. Commun.*, 1983, 420.
355. J. Wengrovius, R. Schrock, M. R. Churchill, J. R. Missert and W. J. Youngs, *J. Am. Chem. Soc.*, 1980, **102**, 4515.
356. M. R. Churchill, H. J. Wasserman, H. W. Turner and R. R. Schrock, *J. Am. Chem. Soc.*, 1982, **104**, 1710.
357. A. J. Schultz, R. K. Brown, J. M. Williams and R. R. Schrock, *J. Am. Chem. Soc.*, 1981, **103**, 169.
358. J. D. Fellmann, H. W. Turner and R. R. Schrock, *J. Am. Chem. Soc.*, 1980, **102**, 6608.
359. H. W. Turner, J. D. Fellmann, S. M. Rocklage, R. R. Schrock, M. R. Churchill and H. J. Wasserman, *J. Am. Chem. Soc.*, 1980, **102**, 7809.
360. S. J. McLain, J. Sancho and R. R. Schrock, *J. Am. Chem. Soc.*, 1980, **102**, 5610.
361. M. Jolly, J. P. Mitchell and V. C. Gibson, *J. Chem. Soc., Dalton Trans.*, 1992, 1331.
362. M. R. Churchill and H. J. Wasserman, *Inorg. Chem.*, 1981, **20**, 2899.
363. I. De Castro, J. de la Mata, P. Gomez-Sal, P. Royo and J. M. Selas, *Polyhedron*, 1992, **11**, 1023.
364. E. N. Jacobsen, K. I. Goldberg and R. G. Bergman, *J. Am. Chem. Soc.*, 1988, **110**, 3706.
365. M. J. Hostetler and R. G. Bergman, *J. Am. Chem. Soc.*, 1990, **112**, 8621.
366. H. W. Turner and R. R. Schrock, *J. Am. Chem. Soc.*, 1982, **104**, 2331.
367. M. C. Fermin and J. W. Bruno, *J. Am. Chem. Soc.*, 1993, **115**, 7511.
368. K. C. Wallace, J. C. Dewan and R. R. Schrock, *Organometallics*, 1986, **5**, 2162.
369. K. C. Wallace, A. H. Liu, J. C. Dewan and R. R. Schrock, *J. Am. Chem. Soc.*, 1988, **110**, 4964.
370. K. C. Wallace and R. R. Schrock, *Macromolecules*, 1987, **20**, 448.
371. R. R. Schrock, M. Wesolek, A. H. Liu, K. C. Wallace and J. C. Dewan, *Inorg. Chem.*, 1988, **27**, 2050.
372. K. C. Wallace, A. H. Liu, W. M. Davis and R. R. Schrock, *Organometallics*, 1989, **8**, 644.
373. K. C. Wallace, W. M. Davis and R. R. Schrock, *Inorg. Chem.*, 1990, **29**, 1104.
374. M. D. Fryzuk, T. S. Haddad, M. Mylvaganam, D. H. McConville and S. J. Rettig, *J. Am. Chem. Soc.*, 1993, **115**, 2782.
375. M. D. Butts and R. G. Bergman, *Organometallics*, 1993, **12**, 4269.
376. J. D. Protasiewicz, A. Masschelein and S. J. Lippard, *J. Am. Chem. Soc.*, 1993, **115**, 808.
377. R. D. Profilet, P. E. Fanwick and I. P. Rothwell, *Angew. Chem., Int. Ed. Engl.*, 1992, **31**, 1261.
378. R. D. Profilet, A. P. Rothwell and I. P. Rothwell, *J. Chem. Soc., Chem. Commun.*, 1993, 42.
379. H. van der Heijden, A. W. Gal, P. Pasman and A. G. Orpen, *Organometallics*, 1985, **4**, 1847.
380. M. R. Churchill and W. J. Youngs, *Inorg. Chem.*, 1981, **20**, 382.
381. R. J. Goddard, R. Hoffmann and E. D. Jemmis, *J. Am. Chem. Soc.*, 1980, **102**, 7667.

382. T. R. Cundari and M. S. Gordon, *J. Am. Chem. Soc.*, 1992, **114**, 539.
383. D. Cauchy, Y. Jean, O. Eisenstein and F. Volatron, *Organometallics*, 1988, **7**, 829.
384. R. Schrock, S. McLain and J. Sancho, *Pure Appl. Chem.*, 1980, **52**, 729.
385. G. Smith, S. J. McLain and R. R. Schrock, *J. Organomet. Chem.*, 1980, **202**, 269.
386. C. Ting and L. Messerle, *J. Am. Chem. Soc.*, 1987, **109**, 6506.
387. G. E. Herberich and I. Savvopoulos, *J. Organomet. Chem.*, 1989, **362**, 345.
388. J. E. Bercaw *et al.*, *J. Chem. Soc., Chem. Commun.*, 1989, 734.
389. M. L. H. Green, A. Sella and L.-L. Wong, *Organometallics*, 1992, **11**, 2650.
390. D. N. Williams *et al.*, *J. Chem. Soc., Dalton Trans.*, 1992, 739.
391. F. A. Cotton and M. Shang, *Inorg. Chem.*, 1990, **29**, 508.
392. J. B. Hartung, Jr. and S. F. Pedersen, *Organometallics*, 1990, **9**, 1414.
393. F. A. Cotton and W. T. Hall, *Inorg. Chem.*, 1980, **19**, 2352.
394. F. A. Cotton and W. Roth, *Inorg. Chim. Acta*, 1984, **85**, 17.
395. E. A. Robinson, *J. Chem. Soc., Dalton Trans.*, 1981, 2373.
396. F. A. Cotton and W. T. Hall, *Inorg. Chem.*, 1980, **19**, 2354.
397. D. M. Hoffman, R. Hoffmann and C. R. Fisel, *J. Am. Chem. Soc.*, 1982, **104**, 3858.
398. F. A. Cotton and X. Feng, *Inorg. Chem.*, 1990, **29**, 3187.
399. F. A. Cotton and W. T. Hall, *Inorg. Chem.*, 1981, **20**, 1285.
400. E. Hey, F. Weller and K. Dehnicke, *Naturwissenschaften*, 1983, **70**, 41.
401. E. Hey, F. Weller and K. Dehnicke, *Z. Anorg. Chem. Allg. Chem.*, 1984, **514**, 25.
402. P. A. Bianconi, I. D. Williams, M. P. Engeler and S. J. Lippard, *J. Am. Chem. Soc.*, 1986, **108**, 311.
403. R. N. Vrtis, S. G. Bott, R. L. Rardin and S. J. Lippard, *Organometallics*, 1991, **10**, 1364.
404. M. J. McGeary, A. S. Gamble and J. L. Templeton, *Organometallics*, 1988, **7**, 271.
405. F. Calderazzo, C. Felten, G. Pampaloni and D. Rehder, *J. Chem. Soc., Dalton Trans.*, 1992, 2003.
406. C. Felten, D. Rehder, G. Pampaloni and F. Calderazzo, *Inorg. Chim. Acta*, 1992, **202**, 121.
407. J. R. Strickler, P. A. Wexler and D. E. Wigley, *Organometallics*, 1988, **7**, 2067.
408. D. P. Smith, J. R. Strickler, S. D. Gray, M. A. Bruck, R. S. Holmes and D. E. Wigley, *Organometallics*, 1992, **11**, 1275.
409. J. R. Strickler and D. E. Wigley, *Organometallics*, 1990, **9**, 1665.
410. M. Etienne, P. S. White and J. L. Templeton, *Organometallics*, 1991, **10**, 3801.
411. Y.-W. Chao, P. A. Wexler and D. E. Wigley, *Inorg. Chem.*, 1989, **28**, 3860.
412. F. Calderazzo, U. Englert, G. Pampaloni and L. Rocchi, *Angew. Chem., Int. Ed. Engl.*, 1992, **31**, 1235.
413. H. Alt and H. E. Engelhardt, *Z. Naturforsch., Teil B*, 1987, **42**, 711.
414. S. Woodward, M. D. Curtis, A. L. Rheingold and B. S. Haggerty, *Organometallics*, 1992, **11**, 2140.
415. A. C. Williams, P. Sheffels, D. Sheehan and T. Livinghouse, *Organometallics*, 1989, **8**, 1566.
416. M. D. Curtis, J. Real, W. Hirpo and W. M. Butler, *Organometallics*, 1990, **9**, 66.
417. G. E. Herberich, U. Englert, K. Linn, P. Roos and J. Runsink, *Chem. Ber.*, 1991, **124**, 975.
418. G. Smith, R. R. Schrock, M. R. Churchill and W. J. Youngs, *Inorg. Chem.*, 1981, **20**, 387.
419. M. C. Milletti and M. D. Curtis, *Polyhedron*, 1990, **9**, 1855.
420. P. Royo, T. Cuenca, E. Hernandez and R. Serrano, *Rev. R. Acad. Cienc. Exactas, Fis. Nat. Madrid*, 1984, **78**, 391 (*Chem. Abstr.*, 1986, **105**, 115 156j).
421. G. E. Herberich and H. Mayer, *Organometallics*, 1990, **9**, 2655.
422. H. Yasuda *et al.*, *Organometallics*, 1991, **10**, 4058.
423. H. Chollet, D. Lucas, Y. Mugnier, A. Antiñolo, A. Otero and M. Fajardo, *J. Organomet. Chem.*, 1992, **441**, 45.
424. G. E. Herberich, U. Englert, M. Hoeveler and I. Savvopoulos, *J. Organomet. Chem.*, 1990, **399**, 35.
425. J. L. Le Quere, F. Y. Petillon, J. E. Guerchais and J. Sala-Pala, *Inorg. Chim. Acta*, 1980, **43**, 5.
426. B. Sarry and R. Schaeffernicht, *Z. Naturforsch., Teil B*, 1981, **36**, 1238.
427. R. A. Bartlett, P. P. Power and S. C. Shoner, *J. Am. Chem. Soc.*, 1988, **110**, 1966.
428. S. U. Koschmieder, B. Hussain-Bates, M. B. Hursthouse and G. Wilkinson, *J. Chem. Soc., Dalton Trans.*, 1991, 2785.
429. H. Yasuda, T. Arai, T. Okamoto and A. Nakamura, *J. Organomet. Chem.*, 1989, **361**, 161.
430. L. D. Durfee and I. P. Rothwell, *Chem. Rev.*, 1988, **88**, 1059.
431. M. R. Churchill and H. J. Wasserman, *J. Chem. Soc., Chem. Commun.*, 1981, 274.
432. M. R. Churchill and H. J. Wasserman, *Inorg. Chem.*, 1982, **21**, 226.
433. T. Y. Meyer and L. Messerle, *J. Am. Chem. Soc.*, 1990, **112**, 4564.
434. J. Arnold, T. D. Tilley and A. L. Rheingold, *J. Am. Chem. Soc.*, 1986, **108**, 5355.
435. A. L. Rheingold, *Acta Crystallogr., Sect. C*, 1990, **46**, 2374.
436. A. Antiñolo *et al.*, *J. Organomet. Chem.*, 1992, **435**, 55.
437. M. C. Fermin, A. S. Hneihen, J. J. Maas and J. W. Bruno, *Organometallics*, 1993, **12**, 1845.
438. M. C. Fermin, B. Thiyagarajan and J. W. Bruno, *J. Am. Chem. Soc.*, 1993, **115**, 974.
439. J. W. Bruno, M. C. Fermin, S. E. Halfon and G. K. Schulte, *J. Am. Chem. Soc.*, 1989, **111**, 8738.
440. K. W. Chiu, R. A. Jones, G. Wilkinson, A. M. R. Galas and M. B. Hursthouse, *J. Chem. Soc., Dalton Trans.*, 1981, 2088.
441. L. R. Chamberlain, I. P. Rothwell and J. C. Huffman, *J. Chem. Soc., Chem. Commun.*, 1986, 1203.
442. L. R. Chamberlain *et al.*, *J. Am. Chem. Soc.*, 1987, **109**, 390.
443. L. R. Chamberlain, B. D. Steffey, I. P. Rothwell and J. C. Huffman, *Polyhedron*, 1989, **8**, 341.
444. A. Antiñolo, *et al.*, *Organometallics*, 1990, **9**, 2919.
445. A. Antiñolo *et al.*, *Organometallics*, 1991, **10**, 3435.
446. A. Antiñolo *et al.*, *Organometallics*, 1993, **12**, 381.
447. J. Amaudrut, J. Sala-Pala, J. E. Guerchais, R. Mercier and J. Douglade, *J. Organomet. Chem.*, 1982, **235**, 301.
448. A. Antiñolo, S. Garcia-Ledó, J. Martínez de Ilarduya and A. Otero, *J. Organomet. Chem.*, 1987, **335**, 85.
449. D. Lucas, H. Chollet, Y. Mugnier, A. Antiñolo, A. Otero and M. Fajardo, *J. Organomet. Chem.*, 1992, **426**, C4.
450. J. E. Nelson, J. E. Bercaw, R. E. Marsh and L. M. Henling, *Acta Crystallogr., Sect. C.*, 1992, **48**, 1023.
451. K. Seitz and U. Behrens, *J. Organomet. Chem.*, 1985, **294**, C9.
452. M. R. Churchill, H. J. Wasserman, P. A. Belmonte and R. R. Schrock, *Organometallics*, 1982, **1**, 559.
453. M. R. Churchill and H. J. Wasserman, *J. Organomet. Chem.*, 1983, **247**, 175.

454. M. L. H. Green, P. M. Hare and J. A. Bandy, *J. Organomet. Chem.*, 1987, **330**, 61.
455. V. C. Gibson, P. Grebenik and M. L. H. Green, *J. Chem. Soc., Chem. Commun.*, 1983, 1101.
456. V. C. Gibson, T. P. Kee and W. Clegg, *J. Chem. Soc., Chem. Commun.*, 1990, 313.
457. V. C. Gibson, T. P. Kee and A. D. Poole, *J. Chem. Soc., Chem. Commun.*, 1990, 1720.
458. D. R. Neithamer, L. Párkányi, J. F. Mitchell and P. T. Wolczanski, *J. Am. Chem. Soc.*, 1988, **110**, 4421.
459. J. R. Strickler, M. A. Bruck and D. E. Wigley, *J. Am. Chem. Soc.*, 1990, **112**, 2814.
460. S. D. Gray, D. P. Smith, M. A. Bruck and D. E. Wigley, *J. Am. Chem. Soc.*, 1992, **114**, 5462.
461. K. D. Allen *et al.*, *Polyhedron*, 1995, **14**, in press.
462. M. R. Churchill and W. J. Youngs, *Inorg. Chem.*, 1980, **19**, 3106.
463. K. R. Ballard, I. M. Gardiner and D. E. Wigley, *J. Am. Chem. Soc.*, 1989, **111**, 2159.
464. T. Okamoto, H. Yasuda, A. Nakamura, Y. Kai, N. Kanehisa and N. Kasai, *Organometallics*, 1988, **7**, 2266.
465. P. Hofmann, M. Hammerle and G. Unfried, *New J. Chem.*, 1991, **15**, 769.
466. S. J. McLain, C. D. Wood and R. R. Schrock, *J. Am. Chem. Soc.*, 1979, **101**, 4558.
467. L. J. Guggenberger and R. R. Schrock, *J. Am. Chem. Soc.*, 1975, **97**, 6693.
468. P. R. Brown, F. G. N. Cloke and M. L. H. Green, *J. Chem. Soc., Chem. Commun.*, 1980, 1126.
469. P. R. Brown, M. L. H. Green, P. M. Hare and J. A. Bandy, *Polyhedron*, 1988, **7**, 1819.
470. E. Melendez, A. M. Arif, A. L. Rheingold and R. D. Ernst, *J. Am. Chem. Soc.*, 1988, **110**, 8703.
471. S. S. Wreford and J. F. Whitney, *Inorg. Chem.*, 1981, **20**, 3918.
472. E. Guggolz, M. L. Ziegler, H. Biersack and W. A. Herrmann, *J. Organomet. Chem.*, 1980, **194**, 317.
473. M. L. H. Green, P. Mountford, P. Scott and V. S. B. Mtetwa, *Polyhedron*, 1991, **10**, 389.
474. H. Yasuda *et al.*, *J. Am. Chem. Soc.*, 1985, **107**, 2410.
475. K. Mashima, Y. Yamanaka, S. Fujikawa, H. Yasuda and A. Nakamura, *J. Organomet. Chem.*, 1992, **428**, C5.
476. H. Yasuda and A. Nakamura, *Angew. Chem., Int. Ed. Engl.*, 1987, **26**, 723.
477. T. Okamoto, H. Yasuda, A. Nakamura, Y. Kai, N. Kanehisa and N. Kasai, *J. Am. Chem. Soc.*, 1988, **110**, 5008.
478. B. D. Steffey and I. P. Rothwell, *J. Chem. Soc., Chem. Commun.*, 1990, 213.
479. F. A. Cotton and M. Shang, *Inorg. Chem.*, 1990, **29**, 2614.
480. F. A. Cotton and M. Shang, *Inorg. Chem.*, 1990, **29**, 2619.
481. F. A. Cotton and M. Shang, *Organometallics*, 1990, **9**, 2131.
482. F. A. Cotton and M. Shang, *J. Am. Chem. Soc.*, 1990, **112**, 1584.
483. F. A. Cotton and X. Feng, *Inorg. Chem.*, 1990, **29**, 3697.
484. A. A. Nesmeyanov, A. I. Gusev, A. A. Pasynskii, K. N. Anisimov, N. E. Kolobova and Yu. T. Struchkov, *J. Chem. Soc., Chem. Commun.*, 1969, 739.
485. H. W. Roesky, M. Sotoodeh, Y. M. Xu, F. Schrumpf and M. Noltemeyer, *Z. Anorg. Allg. Chem.*, 1989, **580**, 131.
486. H. W. Roesky, F. Schrumpf and M. Noltemeyer, *Z. Naturforsch., Teil B*, 1990, **44**, 1369.
487. H. W. Roesky, F. Schrumpf and M. Noltemeyer, *J. Chem. Soc., Dalton Trans.*, 1990, 713.
488. A. M. Carduso, R. J. H. Clark and S. Moorhouse, *J. Chem. Soc., Dalton Trans.*, 1980, 1156.
489. H. Yasuda, T. Okamoto and A. Nakamura, in 'Organometallic Syntheses', eds. R. B. King and J. J. Eisch, Elsevier, Amsterdam, 1988, vol. 4, p. 20.
490. J. De la Mata, R. Fandos, M. Gomez, P. Gomez-Sal, S. Martinez-Carrera and P. Royo, *Organometallics*, 1990, **9**, 2846.
491. I. Leichtweis, H. W. Roesky, M. Noltemeyer and H. G. Schmidt, *Z. Naturforsch., Teil B*, 1991, **46**, 425.
492. S. Paulson, B. P. Sullivan and J. V. Caspar, *J. Am. Chem. Soc.*, 1992, **114**, 6905.
493. C. Ting and L. Messerle, *Inorg. Chem.*, 1989, **28**, 171.
494. J. S. Lin and J. V. Ortiz, *Organometallics*, 1988, **7**, 1229.
495. A. M. Andreu, F. A. Jalón and A. Otero, *J. Organomet. Chem.*, 1988, **353**, 185.
496. F. Schrumpf, H. W. Roesky and M. Noltemeyer, *Z. Naturforsch., Teil B*, 1990, **45**, 1600.
497. J. Arnold and T. D. Tilley, *J. Am. Chem. Soc.*, 1987, **109**, 3318.
498. F. Schrumpf, H. W. Roesky, T. Subrahmanyan and M. Noltemeyer, *Z. Anorg. Allg. Chem.*, 1990, **583**, 124.
499. U. Siemeling and V. C. Gibson, *J. Chem. Soc., Chem. Commun.*, 1992, 1670.
500. Y.-W. Chao, P. A. Wexler and D. E. Wigley, *Inorg. Chem.*, 1990, **29**, 4592.
501. T. C. Baldwin, S. R. Huber, M. A. Bruck and D. E. Wigley, *Inorg. Chem.*, 1993, **32**, 5682.
502. D. M. Antonelli, M. L. H. Green and P. Mountford, *J. Organomet. Chem.*, 1992, **438**, C4.
503. A. Zaki, L. G. Hubert-Pfalzgraf and L. Toupet, *Acta Crystallogr., Sect. C*, 1991, **47**, 533.
504. F. A. Jalón, A. Otero, P. Royo, J. M. Fernandez, M. J. Rosales and R. A. Toscano, *J. Organomet. Chem.*, 1987, **331**, C1.
505. F. A. Jalón, P. Gomez-Sal, A. Otero, P. Royo, S. Garcia-Blanco and S. Martinez-Carrera, *J. Organomet. Chem.*, 1987, **332**, 289.
506. Y. Xu, H. W. Roesky, F. Schrumpf and M. Noltemeyer, *Z. Naturforsch., Teil B*, 1990, **45**, 423.
507. J. Arnold, T. D. Tilley, A. L. Rheingold and S. J. Geib, *J. Chem. Soc., Chem. Commun.*, 1987, 793.
508. F. Bottomley, P. N. Keizer, P. S. White and K. F. Preston, *Organometallics*, 1990, **9**, 1916.
509. F. Bottomley and S. Karslioglu, *J. Chem. Soc., Chem. Commun.*, 1991, 222.
510. F. Bottomley and S. Karslioglu, *Organometallics*, 1992, **11**, 326.
511. D. Kwon, M. D. Curtis, A. L. Rheingold and B. S. Haggerty, *Inorg. Chem.*, 1992, **31**, 3489.
512. S. Woodward, U. Riaz, M. D. Curtis, B. S. Haggerty and A. L. Rheingold, *Organometallics*, 1990, **9**, 2703.
513. K. Tatsumi *et al.*, *Organometallics*, 1993, **12**, 352.
514. K. Tatsumi, J. Takeda, Y. Sekiguchi, M. Kohsaka and A. Nakamura, *Angew. Chem., Int. Ed. Engl.*, 1985, **24**, 332.
515. O. J. Curnow, M. D. Curtis, A. L. Rheingold and B. S. Haggerty, *Inorg. Chem.*, 1991, **30**, 4043.
516. K. Tatsumi, Y. Inoue and A. Nakamura, *J. Organomet. Chem.*, 1993, **444**, C25.
517. K. Tatsumi, Y. Inoue, A. Nakamura, R. E. Cramer, W. VanDoorne and J. W. Gilje, *J. Am. Chem. Soc.*, 1989, **111**, 782.
518. K. Tatsumi, H. Kawaguchi and K. Tani, *Angew. Chem., Int. Ed. Engl.*, 1993, **32**, 591.
519. A. A. Pasynskii, I. L. Eremenko, B. Orazsakhatov, V. T. Kalinnikov, G. G. Aleksandrov and Y. T. Struchkov, *J. Organomet. Chem.*, 1981, **216**, 211.
520. M. D. Curtis and J. Real, *Inorg. Chem.*, 1988, **27**, 3176.
521. I. Leichtweis, H. W. Roesky, M. Noltemeyer and H. G. Schmidt, *Chem. Ber.*, 1991, **124**, 253.
522. H. Brunner, W. Meier, J. Wachter, B. Nuber and M. L. Ziegler, *J. Organomet. Chem.*, 1990, **381**, C7.

523. H. Brunner, G. Gehart, B. Nuber, J. Wachter and M. L. Ziegler, *Angew. Chem., Int. Ed. Engl.*, 1992, **31**, 1021.
524. D. Fenske and P. G. Maue, *Z. Naturforsch., Teil B*, 1989, **44**, 531.
525. J. C. Green, C. P. Overton, K. Prout and J. M. Marin, *J. Organomet. Chem.*, 1983, **241**, C21.
526. J. M. Marin, V. S. B. Mtetwa and K. Prout, *Acta Crystallogr., Sect. C*, 1985, **41**, 55.
527. H. Plenio, H. W. Roesky, M. Noltemeyer and G. M. Sheldrick, *Angew. Chem., Int. Ed. Engl.*, 1988, **27**, 1330.
528. C. Ting and L. Messerle, *J. Am. Chem. Soc.*, 1989, **111**, 3449.
529. M. L. H. Green, P. Mountford, V. S. B. Mtetwa, P. Scott and S. J. Simpson, *J. Chem. Soc., Chem. Commun.*, 1992, 314.
530. R. Poli, *Chem. Rev.*, 1991, **91**, 509.
531. L. Messerle, *Chem. Rev.*, 1988, **88**, 1229.
532. J. W. Lauher and R. Hoffmann, *J. Am. Chem. Soc.*, 1976, **98**, 1729.
533. A. Antiñolo, M. Fajardo, A. Otero and P. Royo, *J. Organomet. Chem.*, 1984, **265**, 35.
534. A. Fakhr, Y. Mugnier, R. Broussier and B. Gautheron, *J. Organomet. Chem.*, 1985, **279**, C15.
535. V. A. Knizhnikov, S. A. Makhnach and Y. A. Ol'dekop, *Zh. Obshch. Khim.*, 1988, **58**, 2285.
536. Yu. A. Ol'dekop and V. A. Knizhnikov, *Zh. Obshch. Khim.*, 1981, **51**, 1723 (*Chem. Abstr.*, 1981, **95**, 220 117v).
537. V. A. Knizhnikov, M. G. Novikova, S. A. Makhnach and Y. A. Ol'dekop, *Zh. Obshch. Khim.*, 1990, **60**, 1339 (*Chem. Abstr.*, 1990, **113**, 231 580g).
538. P. Gowik, T. Klapötke and J. Pickardt, *Organometallics*, 1989, **8**, 2953.
539. P. Gowik and T. Klapötke, *J. Fluorine Chem.*, 1990, **47**, 273.
540. A. Antiñolo, M. Fajardo, F. A. Jalón, A. Otero, M. C. Puerta and Y. Mugnier, *Polyhedron*, 1989, **8**, 1848.
541. A. Antiñolo, M. Fajardo, A. Otero, Y. Mugnier, H. Nabaoui and H. Mourad, *J. Organomet. Chem.*, 1991, **414**, 155.
542. J. A. Hunter, W. E. Lindsell, K. J. McCullough, R. A. Parr and M. L. Scholes, *J. Chem. Soc., Dalton Trans.*, 1990, **7**, 2145.
543. C. R. Lucas, *Inorg. Synth.*, 1990, **28**, 267.
544. A. H. Al-Mowali and W. A. A. Kuder, *J. Organomet. Chem.*, 1980, **194**, 61.
545. A. Fakhr, Y. Mugnier, R. Broussier and B. Gautheron, *J. Organomet. Chem.*, 1984, **269**, 53.
546. P. Koepf-Maier, M. Leitner and H. Koepf, *J. Inorg. Nucl. Chem.*, 1980, **42**, 1789.
547. V. V. Strelets, S. V. Kukharenko and V. N. Tsarev, *Izv. Akad. Nauk SSSR, Ser. Khim.*, 1987, **2**, 258.
548. C. Cauletti *et al.*, *J. Electron Spectrosc. Relat. Phenom.*, 1980, **18**, 61.
549. J. L. Petersen and J. W. Egan, Jr., *Inorg. Chem.*, 1983, **22**, 3571.
550. F. M. Miao and K. Prout, *Cryst. Struct. Commun.*, 1982, **11**, 269.
551. A. M. Manotti Lanfredi, A. Tiripicchio, M. Kapon and G. M. Reisner, *J. Chem. Soc., Dalton Trans.*, 1990, 375.
552. H. Nabaoui *et al.*, *J. Organomet. Chem.*, 1988, **338**, C17.
553. R. Broussier, J. D. Olivier and B. Gautheron, *J. Organomet. Chem.*, 1981, **210**, 169.
554. R. Broussier, J. D. Olivier and B. Gautheron, *J. Organomet. Chem.*, 1983, **251**, 307.
555. A. Antiñolo, M. Fajardo, A. Otero and P. Royo, *J. Organomet. Chem.*, 1982, **234**, 309.
556. H. Nabaoui *et al.*, *J. Organomet. Chem.*, 1989, **375**, 67.
557. A. Fakhr *et al.*, *J. Organomet. Chem.*, 1989, **364**, 149.
558. D. A. Lemenovskii, V. P. Fedin, Y. L. Slovokhotov, A. I. Yanovskii and Y. T. Struchkov, *Koord. Khim.*, 1980, **6**, 882.
559. D. A. Lemenovskii, M. V. Tsikalova, S. A. Konde and E. G. Perevalova, *Koord. Khim.*, 1983, **9**, 1060.
560. M. J. Calhorda *et al.*, *J. Organomet. Chem.*, 1992, **426**, 195.
561. H. S. Malhotra, I. Rani and R. K. Multani, *J. Inst. Chem. (India)*, 1985, **57**, 105 (*Chem. Abstr.*, 1986, **105**, 226 858j).
562. N. B. Patrikeeva, O. N. Suvorova and D. A. Domrachev, *Metalloorg. Khim.*, 1991, **4**, 684 (*Chem. Abstr.*, 1991, **115**, 92 479g).
563. A. N. Nesmeyanov, E. G. Perevalova, V. P. Fedin, Yu. L. Slovokhotov, Yu. T. Struchkov and D. A. Lemenovskii, *Dokl. Akad. Nauk. SSSR*, 1980, **252**, 1141.
564. D. A. Lemenovskii, V. P. Fedin, Yu. L. Slovokhotov and Yu. T. Struchkov, *J. Organomet. Chem.*, 1982, **228**, 153.
565. D. A. Lemenovskii, S. A. Konde and E. G. Perevalova, *J. Organomet. Chem.*, 1982, **226**, 223.
566. L. Stahl, J. P. Hutchinson, D. R. Wilson, R. D. Ernst, *J. Am. Chem. Soc.*, 1985, **107**, 5016.
567. T. E. Waldman, L. Stahl, D. R. Wilson, A. M. Arif, J. P. Hutchinson and R. D. Ernst, *Organometallics*, 1993, **12**, 1543.
568. I. Bkouche-Waksman, C. Bois, J. Sala-Pala and J. E. Guerchais, *J. Organomet. Chem.*, 1980, **195**, 307.
569. A. L. Rheingold, J. B. Strong, *Acta Crystallogr., Sect. C*, 1991, **47**, 1963.
570. R. W. Quan, J. E. Bercaw and W. P. Schaefer, *Acta Crystallogr., Sect. C*, 1991, **47**, 2057.
571. R. S. Arora, S. C. Hari and R. K. Multani, *J. Inst. Chem. (India)*, 1981, **53**, 252 (*Chem. Abstr.*, 1982, **96**, 122 942m).
572. R. C. Sharma, A. C. Gupta and M. K. Rastogi, *Indian J. Chem., Sect. A*, 1982, **21**, 190.
573. H. S. Malhotra, R. S. Arora and R. K. Multani, *J. Inst. Chem. (India)*, 1983, **55**, 33 (*Chem. Abstr.*, 1983, **99**, 38 588e).
574. M. Y. Darensbourg, R. Silva, J. Reibenspies and C. K. Prout, *Organometallics*, 1989, **8**, 1315.
575. C. Q. Simpson, II and M. B. Hall, *New J. Chem.*, 1991, **15**, 763.
576. M. Y. Darensbourg, C. J. Bischoff, S. A. Houliston, M. Pala and J. Reibenspies, *J. Am. Chem. Soc.*, 1990, **112**, 6905.
577. M. J. Bader and A. H. Al-Mowali, *Chem. Scr.*, 1983, **22**, 241 (*Chem. Abstr.*, 1984, **100**, 156 739s).
578. H. S. Malhotra and R. K. Multani, *J. Inst. Chem. (India)*, 1984, **56**, 135 (*Chem. Abstr.*, 1984, **101**, 211 349n).
579. W. Douglas, *J. Organomet. Chem.*, 1985, **287**, C39.
580. A. H. Al-Mowali, *Inorg. Chim. Acta*, 1983, **77**, L51.
581. B. Viard, J. Amaudrut, J. Sala-Pala, A. Fakhr, Y. Mugnier and C. Moise, *J. Organomet. Chem.*, 1985, **292**, 403.
582. R. Mattes and M. Bradtke, *Z. Kristallogr.*, 1981, **157**, 95.
583. B. Viard, J. Sala-Pala, J. Amaudrut, J. E. Guerchais, C. Sanchez and J. Livage, *Inorg. Chim. Acta*, 1980, **39**, 99.
584. C. Sanchez, D. Vivien, J. Livage, J. Sala-Pala, B. Viard and J. E. Guerchais, *J. Chem. Soc., Dalton Trans.*, 1981, 64.
585. M. Fajardo, M. P. Gomez-Sal, P. Royo, S. Martinez Carrera and S. Garcia Blanco, *J. Organomet. Chem.*, 1986, **312**, C44.
586. R. D. Ernst, *Chem. Rev.*, 1988, **88**, 1255.
587. W. E. Douglas and M. L. H. Green, *J. Chem. Soc., Dalton Trans.*, 1972, 1796.
588. F. G. N. Cloke and M. L. H. Green, *J. Chem. Soc., Dalton Trans.*, 1981, 1938.
589. F. G. N. Cloke, K. A. E. Courtney, A. A. Sameh and A. C. Swain, *Polyhedron*, 1989, **8**, 1641.
590. F. Calderazzo, G. Pampaloni, L. Rocchi, J. Strähle and K. Wurst, *Angew. Chem., Int. Ed. Engl.*, 1991, **30**, 102.
591. F. Calderazzo, G. Pampaloni, L. Rocchi, J. Strähle and K. Wurst, *J. Organomet. Chem.*, 1991, **413**, 91.
592. J. A. Bandy, K. Prout, F. G. N. Cloke, H. C. de Lemos and J. M. Wallis, *J. Chem. Soc., Dalton Trans.*, 1988, 1475.

593. F. Stollmaier and U. Thewalt, *J. Organomet. Chem.*, 1981, **222**, 227.
594. M. A. Bruck, A. S. Copenhaver and D. E. Wigley, *J. Am. Chem. Soc.*, 1987, **109**, 6525.
595. D. J. Arney, P. A. Wexler and D. E. Wigley, *Organometallics*, 1990, **9**, 1282.
596. P. A. Wexler and D. E. Wigley, *J. Chem. Soc., Chem. Commun.*, 1989, 664.
597. P. A. Wexler, D. E. Wigley, J. B. Koerner and T. A. Albright, *Organometallics*, 1991, **10**, 2319.
598. D. J. Arney, M. A. Bruck and D. E. Wigley, *Organometallics*, 1991, **10**, 3947.
599. E. O. Fischer and F. Röhrscheid, *J. Organomet. Chem.*, 1966, **6**, 53.
600. M. R. Churchill and S. W.-Y. Chang, *J. Chem. Soc., Chem. Commun.*, 1974, 248.
601. M. Röper, in 'Catalysis in C_1 Chemistry', ed. W. Keim, Reidel, Dordrecht, 1983, p. 41.
602. I. P. Rothwell, *Polyhedron*, 1985, **4**, 177.
603. E. Solari, C. Floriani, A. Chiesi-Villa and C. Rizzoli, *J. Chem. Soc., Chem. Commun.*, 1991, 841.
604. K. J. Ivin, J. J. Rooney, C. D. Stewart, M. L. H. Green, R. Mahtab, *J. Chem. Soc., Chem. Commun.*, 1978, 604.
605. R. R. Schrock, *Science*, 1983, **219**, 13.
606. F. A. Cotton, W. T. Hall, K. J. Cann and F. J. Karol, *Macromolecules*, 1981, **14**, 233.
607. G. Lachmann, J. A. K. DuPlessis and C. J. DuToit, *J. Mol. Catal.*, 1987, **42**, 151.
608. T. Masuda, Y. Deng and T. Higashimura, *Bull. Chem. Soc. Jpn.*, 1983, **56**, 2798.
609. T. C. Ho, *Catal. Rev.-Sci. Eng.*, 1988, **30**, 117.
610. J. R. Katzer and R. Sivasubramanian, *Catal. Rev.-Sci. Eng.*, 1979, **20**, 155.
611. R. M. Laine, *Ann. N.Y. Acad. Sci.*, 1983, **415**, 271.
612. R. H. Fish, *Ann. N.Y. Acad. Sci*, 1983, **415**, 292.
613. R. H. Fish, A. D. Thormodsen, A. D. Moore, D. L. Perry and H. Heinemann, *J. Catal.*, 1986, **102**, 270.
614. C. N. Satterfield, M. Modell and J. A. Wilkers, *Ind. Eng. Chem. Process Des. Dev.*, 1980, **19**, 154.
615. C. N. Satterfield and S. H. Yang, *Ind. Eng. Chem. Process Des. Dev.*, 1984, **23**, 11.
616. J. T. Miller and M. F. Hineman, *J. Catal.*, 1984, **85**, 117.
617. C. N. Satterfield and J. F. Cocchetto, *Ind. Eng. Chem. Process Des. Dev.*, 1981, **20**, 53.
618. A. Maercker and M. Theis, *Top. Curr. Chem.*, 1987, **138**, 1.
619. E. J. Roskamp and S. F. Pedersen, *J. Am. Chem. Soc.*, 1987, **109**, 6551.
620. K. Takai, Y. Kataoka and K. Utimoto, *J. Org. Chem.*, 1990, **55**, 1707.
621. K. Takai, J. Miyai, Y. Kataoka and K. Utimoto, *Organometallics*, 1990, **9**, 3030.
622. J. B. Hartung, Jr. and S. F. Pedersen, *J. Am. Chem. Soc.*, 1989, **111**, 5468.
623. Y. Kataoka, J. Miyai, M. Tezuka, K. Takai, K. Oshima and K. Utimoto, *Tetrahedron Lett.*, 1990, 369.
624. E. J. Roskamp and S. F. Pedersen, *J. Am. Chem. Soc.*, 1987, **109**, 3152.

3
Hexacarbonyls and Carbonyl Complexes of Carbon σ-Bonded Ligands of Chromium, Molybdenum and Tungsten

MARK J. WINTER
University of Sheffield, UK

3.1 GENERAL REMARKS

The organization of the coverage of the group 6 metals is different from that in *COMC-I* in that all three metals are covered together in common chapters (based on ligand type). Since 1981 an immense volume of material on group 6 metals has appeared and as a result some selectivity of coverage has been applied.

3.2 THE METAL HEXACARBONYLS

The continued growth in interest of the chemistry of group 6 metal carbonyls is a consequence of many factors, including the ready commercial availability of $[M(CO)_6]$ (M = Cr, Mo or W), but in particular because of the remarkable variety of chemistry displayed. Despite the ready availability of the hexacarbonyls there is still considerable interest in new syntheses of hexacarbonyls.[1,2] Useful material is to be found in reviews on carbon- and oxygen-bonded metal carbonyls,[3] and *ab initio* calculations on organometallics.[4]

3.2.1 Calculations on and Spectroscopic Studies of $[M(CO)_6]$

The complexes $[M(CO)_6]$ are the subject of considerable theoretical and spectroscopic interest.[5–18] Thus, extended-basis-set Hartree–Fock–Roothan calculations suggest that the theoretical dissociation energy for the dissociation of the first carbonyl from $[Cr(CO)_6]$ (208.5 kJ mol^{-1}) is larger than the average dissociation energy (123.5 kJ mol^{-1}).[9] These values correspond closely to values obtained in other theoretical studies.[10] The σ-bond primarily involves a rehybridization of the carbonyl with little net transfer of electron density to chromium. However, there is a net transfer of electron density to the carbonyls through π-effects, meaning that the chromium atom possesses a net positive charge.[9] Other calculations reinforce this result.[11] These and other calculations point to a degree of double-bond character between the metal and the carbonyl.[12,13]

An *ab initio* analysis of the electronic spectra of the carbonyls $[Mn(CO)_6]^+$, $[Cr(CO)_6]$ and $[V(CO)_6]^-$ confirms the classical assignment of the two strongest bands as metal–ligand charge transfer (MLCT) bands. However, the calculations also suggest alternative assignments to those that are often made for the weaker bands,[19] including at least one band identified as a metal–ligand spin forbidden charge transfer band.[20] Other calculations suggest that the near-UV spectra of $[Cr(CO)_6]$ are determined exclusively by metal-to-ligand charge-transfer transitions and that $d \rightarrow p$ and $d \rightarrow s$ transitions do not contribute above 200 nm.[21]

Metal NMR spectroscopy is a valuable tool and has been reviewed.[22] Molybdenum-95, but to a much lesser extent ^{183}W NMR spectroscopies are becoming invaluable methods[23,24] for the analysis of molybdenum carbonyl[25–61] and tungsten carbonyl[62–73] complexes of many different types. It seems that the ^{95}Mo chemical shift cannot be used to distinguish between mononuclear Mo^0, Mo^{II} and Mo^{IV} species, but the shift is sensitive to small structural and electronic variations. For instance, within the series $[Mo(CO)_6]$, $[Mo(CO)_5L]$, $[Mo(CO)_4L_2]$ and $[Mo(CO)_3L_3]$ (L = PR_3 or $P(OR)_3$), the chemical shift range spans δ −1090 to −1870.[54]

The ^{183}W NMR spectrum can be measured directly in cases where the complex is very soluble and stable in the presence of a relaxation agent. Under appropriate conditions a satisfactory ^{183}W NMR signal for $[W(CO)_6]$ is seen in ~20 min. The spectra for phosphine carbonyl complexes emerge even more quickly, as a consequence of shorter relaxation times owing to dipole–dipole interactions involving ^{31}P and ^{183}W nuclei and superior solubility.[64] Values for these ^{183}W shifts are quoted conveniently relative to $[W(CO)_6]$.[65] The chemical shifts of the two nuclei ^{95}Mo and ^{183}W are parallel, with the ^{183}W shifts being ~1.7 times as sensitive as the ^{95}Mo shift to changes in chemical environment.[65]

An analysis of the ^{13}C NMR spectra and CO stretching force constants k(CO) of a series of substituted carbonyl complexes, $[W(CO)_5L]$, shows good correlation between δ_C and k(CO) for a broad range of ligand donor–acceptor ligand properties. Strong single-faced π-acceptor ligands give rise to an additional downfield shift of the ^{13}C resonance for the CO *trans* to the ligand. This is attributed to a breakdown of the cylindrical symmetry within the L–W–C–O fragment. Further analysis gives the following values for the *trans* influence of the ligand towards W^0: $Ph_2C > CO >$ alkenes $> CN^- > P(OR)_3 > PR_3 > H^-$, Me^-, NCS^-, $OC(O)R^-$, py $> RNH_2 > SC(S)R^-$, $AsR_3 > SbR_3$, $SH^- > Cl^- > Br^- > I^-$.[74] The hexacarbonyls have also been studied by ^{17}O NMR spectroscopy, allowing determinations of ^{17}O quadrupole coupling constants.[75,76] Single crystal IR–reflectance spectroscopy in polarized light allows the assignment of the spectrum of solid $[Mo(CO)_6]$.[77] The crystal structures of $[Mo(CO)_6]$[78] and $[W(CO)_6]$[79] have been examined.

3.2.2 Substitution Reactions

Substitution of the complexes $[M(CO)_6]$ (M = Cr, Mo or W) is one of the most intensively studied substitution processes.[80,81] Substitution is achieved through thermal activation, photochemical activation (see Section 3.2.3) or by chemical activation. The reader is referred to the appropriate section for specific information on the various reaction product types, which are categorized in this volume according to the nature of the ligand. Some platinum complexes, such as $[Pt(PPh_3)_4]$, are catalysts for the substitution of carbonyls in $[W(CO)_6]$.[82]

A good way of activating $[M(CO)_6]$ complexes chemically is through the addition of Me_3NO, which removes a carbonyl as CO_2, and in the absence of other ligands the result is the formation of $[M(CO)_5(NMe_3)]$ containing a relatively labile NMe_3 ligand.[83,84] A kinetic study shows that the reaction between $[M(CO)_6]$ and Me_3NO is first order in both species, but zero order in $[PPh_3]$ in the presence of PPh_3. The rate of reaction is $W > Mo \geq Cr$. Qualitatively, Me_3NO is of comparable nucleophilicity to MeLi in its reaction with $[M(CO)_6]$. Similar results are seen for the reaction of various substituted derivatives $[ML(CO)_5]$ with Me_3NO.[85,86] Various aryl amine *N*-oxides are suitable for these reactions. The rate of reaction increases with basicity of the *N*-oxide reagent.[87] The solid-state reaction between $[W(CO)_6]$ and Me_3NO has been investigated by thermogravimetric and derivative thermogravimetric methods. Under these conditions the optimum reactant ratio is ~1:4 $[W(CO)_6]$:Me_3NO.[88]

3.2.3 Photolysis of $[M(CO)_6]$

The reader is referred to a review on the photochemistry of organometallic compounds.[89] The hexacarbonyls $[M(CO)_6]$ have been subjected to all extremes of photolysis, both in the absence and presence of ligands, some of which are not renowned for their ligand characteristics. Thus, photolysis of $[Cr(CO)_6]$ in liquid xenon, or in liquid krypton doped with xenon, affords a relatively long-lived species ($t_{1/2} \approx 2$ s at −98 °C). This is the solvated pentacarbonyl $[Cr(CO)_5(Xe)]$ whose IR spectrum (ν_{CO}: 2086, 1960, 1934 cm^{-1}) is very similar to that of a pentacarbonyl generated in a xenon matrix.[90] If the xenon is doped with nitrogen, the result is the formation of a range of dinitrogen species $[Cr(CO)_{6-x}(N_2)_x]$ (x = 1–5). The pentacarbonyl is relatively stable (thermally stable at −35 °C in liquid xenon), while the monocarbonyl is thermally unstable at −90 °C.[91] A number of dihydrogen complexes are formed during photolysis in hydrogen doped solutions in xenon. These include the pentacarbonyls $[M(CO)_5(H_2)]$ (M = Cr, Mo or W) and the bis(dihydrogen) species $[Cr(CO)_4(H_2)_2]$.[92,93] With ethene doped xenon, the product is $[Cr(CO)_5(C_2H_4)]$ and *cis*- and *trans*-$[Cr(CO)_4(C_2H_4)_2]$.[94]

The hexacarbonyls have been subjected to photolysis in argon and methane matrices in the presence of oxygen,[95–8] carbon dioxide,[99] nitrous oxide[99] and CCl_4.[100] Again, these are largely IR studies, but reinforced by UV–visible and Raman spectroscopies. In the presence of oxygen, the eventual result is MO_2, and studies using ^{18}O enriched materials show that this is formed via four intermediates: $M(\eta^2\text{-}O_2)(CO)_4$, $M(O)_2(CO)_4$, $M(O)_2(CO)_3$ and $M(O)_2(CO)_2$.

One alternative to the use of frozen gas matrices is to photolyse [M(CO)$_6$] in a polyvinyl chloride matrix or in a nujol mull. This is suitable over the temperature range 12–298 K. One advantage of this approach is that the resulting species may be studied over a wide temperature range.[101,102]

Pulsed laser pyrolysis studies on the gas-phase thermal decomposition of [M(CO)$_6$] gives the first metal–carbonyl bond energies for the hexacarbonyls as 154.8, 167.5 and 192.5 kJ mol^{-1} (±8.4 in each case) for Cr, Mo and W, respectively. These values lead to a number of other derived bond energies.[103] The transient IR spectra obtained in excimer laser photolysis studies of gas-phase [Cr(CO)$_6$] show that the resultant [Cr(CO)$_5$] is C_{4v}, [Cr(CO)$_4$] is C_{2v} and [Cr(CO)$_3$] is C_{3v}, the same geometries displayed when observed in condensed-phase experiments.[104,105] Other studies show that the [Cr(CO)$_5$] produced by pulsed laser photoexcitation is formed with sufficient internal energy to undergo dissociation to [Cr(CO)$_4$] and CO.[106,107] The tetracarbonyl [Cr(CO)$_4$] is able to react with ligands such as NH_3, CO and ethene at rates that seem to be largely independent of the π-acceptor ability of the ligand and to steric effects. These studies give the M–L bond dissociation energies of $DH°$[(OC)$_5$Cr–CO] = 154.8 ± 21 kJ mol^{-1} and $DH°$[(OC)$_4$Cr–CO] = 104.6 ± 21 kJ mol^{-1}.[108,109] The first bond energies, determined by laser powered homogeneous pyrolysis, are very similar.[110] Photoproducts such as [Cr(CO)$_3$] or [Cr(CO)$_2$] are probably responsible for the isomerization of but-1-ene to but-2-ene.[111]

Photolysis of the hexacarbonyls [Mo(CO)$_6$] and [W(CO)$_6$] in the gas phase has also been examined.[8,112–15] Again, the pentacarbonyls recombine with ligands such as CO,[116] and react with H_2 to form [W(CO)$_5$(H$_2$)].[117] The pentacarbonyl [W(CO)$_5$] forms complexes reversibly with alkanes except CH_4 in the gas phase. The interaction of these alkanes with [W(CO)$_5$] seems to be consistent with a two-centre, three-electron C–H–M bond formed through σ → M electron donation. Fluorinated alkanes show related behaviour, but in contrast with CH_4, MeF does form an observable complex.[118,119]

The photolysis of [M(CO)$_6$] in alkane solvents has been studied; cyclohexane and methylcyclohexane seem to be the most commonly used solvents. In these cases, the result is photodissociation of a carbonyl ligand to form [M(CO)$_5$(cyclohexane)].[120–2] Labelling studies using [Cr(CO)$_5$(^{13}CO)] show that the pentacarbonyl is C_{4v} at room temperature and that the structure in solution is similar to that in low-temperature matrices. In the presence of water, a species [Cr(CO)$_5$(OH$_2$)] is formed, and [Cr(CO)$_5$(C$_6$H$_{12}$)] is considerably more reactive with H_2O than with CO.[122] In the presence of donor ligands such as 4-acetylpyridine, the solvent molecule is displaced to form [M(CO)$_5$L]. This reaction proceeds by the reversible dissociation of solvent to form [M(CO)$_5$] followed by "scavenging" of the [M(CO)$_5$] species to form [M(CO)$_5$L].[121] Similar results are found when using other solvents such as heptane and aromatic solvents in the presence of various donor ligands,[123–41] including N_2[142] and H_2.[143] Photoelectron calorimetric studies show that the enthalpies of CO dissociation from [Cr(CO)$_6$] in heptane and cyclohexane are endothermic, 113 and 100.5 kJ mol^{-1}, respectively, a result attributable to a stronger agostic interaction for the cyclohexane case. The dissociation of CO for [M(CO)$_6$] (M = Mo or W) is more endothermic (134 and 138 kJ mol^{-1}, respectively).[132] The mechanism of substitution depends on the solvent. Thus, reactions of fluorobenzene solvated [Cr(CO)$_5$] proceed by a dissociative process, while in the case of chlorobenzene[129] or heptane[144] dissociative and interchange paths are followed.

3.2.4 Studies on Supported [M(CO)$_6$]

There have been a number of studies of hexacarbonyls adsorbed[145] onto, in particular, alumina[146–63] and zeolites.[40,145,149,150,153–5,164–88] In other studies, [W(CO)$_6$] has been shown to physisorb onto porous Vycor glass to give a pentacarbonyl species whose electronic and EPR spectra suggest a square-pyramidal C_{4v} structure with either silanol or chemisorbed water occupying the sixth site. Activation parameters with CO suggest that the W(CO)$_5$–PVG interaction energy is ≤29.3 kJ mol^{-1}.[189] There have been a number of studies addressing the decomposition of [W(CO)$_6$] and [Mo(CO)$_6$] on surfaces such as Si(111)-7 × 7[190–2] and W(110).[193]

Alumina adsorbed species have been studied extensively. The material Mo(CO)$_3$(ads) made by adsorbing [Mo(CO)$_3$(η^6-C$_6$H$_6$)] onto alumina is largely identical to that made by the reaction of [Mo(CO)$_6$] with alumina, but the material made from the reaction of alumina with [Mo(CO)$_3$(NCMe)$_3$] is different. For instance, it is inactive as a catalyst for alkene metathesis.[147] Group 6 carbonyls when supported on dehydroxylated alumina are extremely active alkene hydrogenation and Fischer Tropsch catalysts.[156,161–3]

Cross-polarization–magic angle spinning ^{13}C NMR spectroscopy has been used to study the nature of the γ-alumina supported molybdenum subcarbonyl interaction. Enrichment of samples by ^{13}CO is required. The spectra show a Mo(CO)$_5$(ads) species, probably fluxional on activation at 30 °C and

$Mo(CO)_3$(ads) on activation at higher temperatures.[146,149] An analysis of high-resolution solid-state ^{13}C NMR magic angle spectra shows that there is very facile rotation of a $Mo(CO)_5$(ads) species about the surface–Mo bond.[160] The interaction of $[Mo(CO)_6]$ with hydroxylated alumina has been investigated by inelastic electron tunnelling spectroscopy. It chemisorbs to form a molybdenum carboxylic acid which desorbs at room temperature. Heating during vapour phase exposure to the hexacarbonyl gives multilayers of metallic molybdenum at 140 °C.[194]

In a number of cases, group 6 metal carbonyl–zeolite species have catalytic activity.[154,155,166–9,174,178,181–7] A number of different techniques have been used to study these materials, including NMR,[40,149,150,173,175,195] IR,[153,170,175,176,179,181,196] diffuse reflectance IR,[170,179] Raman,[180] EXAFS[172,175] and visible[149] spectroscopies.

In the case of Na–X and Na–Y zeolites, adsorption of $[M(CO)_6]$ leads to a rose-coloured surface subcarbonyl with $M(CO)_3$ stoichiometry.[149–51] The ^{13}C NMR shift of δ 247 suggests an anionic nature. Temperature programmed decomposition of $[Cr(CO)_6]$ on Na–X zeolite shows two peaks at 395 and 550 K as carbon monoxide is evolved.[149] Xenon-129 NMR spectroscopy suggests that molybdenum is able to migrate into zeolite cages.[173]

3.3 CHARGED CARBONYL COMPLEXES

By far the most important carbonyl complexes are anionic, and these have been reviewed.[197,198] A few cationic complexes $[M(CO)_n]^{m+}$ are seen as transient species in the gas phase and are discussed later in this section.

3.3.1 Syntheses and Structures

From a synthetic point of view, the most useful metal carbonylates are $[M(CO)_5]^{2-}$ and $[M_2(CO)_{10}]^{2-}$. These were first made well before 1981, but recent syntheses offer the advantages of convenience and purity over earlier methods. The geometry of $[Cr_2(CO)_{10}]^{2-}$ is closely related to that of $[Mn_2(CO)_{10}]$, $[Cr_2(CO)_9(NO)]^-$ and $[Cr(CO)_8(NO)]$.[199]

Reduction of the amine complexes $[M(CO)_5(NMe_3)]$ (M = Cr, Mo or W) with $NaC_{10}H_8$/THF affords the dianions $[M(CO)_5]^{2-}$. The reaction is more or less instantaneous at −78 °C and naphthalene may be removed by washing with ether at 0 °C. Clean reduction of $[W(CO)_6]$ to $[W(CO)_5]^{2-}$ is achieved under similar conditions, but since the hexacarbonyl is only slightly soluble in THF at −78 °C, an inconvenient slow addition of $NaC_{10}H_8$ is required. The molybdenum and tungsten dianions are isolable as $[Na(K\text{-}2.2.1)]_2[M(CO)_5]$ on addition of Kryptofix 2.2.1 to solutions of $Na_2[M(CO)_5]$. The only impurities are traces of the dianions $[M_2(CO)_{10}]^{2-}$. The crystal structures of these salts show that the metal lies in a slightly distorted trigonal bipyramidal environment. While the IR spectra of the $[Na(K\text{-}2.2.1)]^+$ salts show no ion pairing, there is considerable ion pairing evident in the spectra of simple alkali-metal salts, both as isocarbonyl interactions and as direct M–alkali-metal interactions.[200,201]

The highly reduced M^{IV} organometallates $[M(CO)_4]^{4-}$ (M = Cr, Mo or W) are also known. These are formed in the alkali metal reduction of $[M(CO)_4(TMEDA)]$ in liquid ammonia. These are shock-sensitive solids with very low ν_{CO} bands (e.g., $Na_4[Cr(CO)_4]$, ν_{CO} (silicone fluid mull) 1657w, 1462s cm^{-1}). It seems likely that the sodium ions interact strongly with the carbonyl oxygen atoms and that the salts could be formulated very reasonably as $[M(CONa)_4]_x$.[202] The even more reduced anions $[M(CO)_3]^{6-}$ are not known. However, reduction of $[W(CO)_3(pmta)]$ (pmta, 1,1,4,7,7-pentamethyldiethylenetriamine) by six equivalents of potassium in NH_3 does give a poorly characterized species which reacts with various electrophiles to form tricarbonyl derivatives. These include $[WH(SnPh_3)_2(CO)_3]^{3-}$.[203]

Solutions containing the dimetallic anions $(NEt_4)_2[M_2(CO)_{10}]$ are prepared conveniently using an ion-pair extraction method in which the hydrides $(NEt_4)[M_2(H)(CO)_{10}]$ are treated with $Et_4NHSO_4/KOH/NCMe/H_2O$.[204] An x-ray crystal structure of the complex $PPN[Mo_2(CO)_{10}]$ reveals a staggered conformation (D_{4d}) structure reminiscent of $[Mn_2(CO)_{10}]$ with a Mo–Mo separation of 315.3(2) pm.[205]

Other carbonylate anions are generally prepared as transient species in the gas phase. Thus, in a flowing afterglow apparatus, $[Cr(CO)_6]$ reacts with electrons to form the 17-electron radical anion $[Cr(CO)_5]^-$.[206] This method also affords the tricarbonyls $[Cr(CO)_3]^-$.[207] Other variations on this low-energy electron attachment and carbonyl dissociation involving $[Cr(CO)_6]$ are also known.[208–12] These processes involve electron attachment into an appropriate orbital and this has been examined through calculations on $[Cr(CO)_6]^-$.[213] γ-Irradiation of single crystals of $PPN[MH(CO)_5]$ (M = Cr, Mo or W) affords persistent signals in the anisotropic EPR spectra assigned to $[M(CO)_5]^-$ and an analysis of these spectra indicate square-pyramidal structures with unpaired spin densities largely localized in the d_{z^2} orbital.[214] Irradiation of methanol or ethanol solutions of $[Mo(CO)_6]$ affords $[Mo(CO)_6]^-$.[215,216]

There is far less information available for cationic metal carbonyl complexes. The cation $[Cr(CO)_6]^+$ is formed through the decomposition of $[CrF(CO)_5]$ (observed in electrochemical studies of [dibenzo-18-crown-6-K]$[CrF(CO)_5]$). The same ion is a product in the controlled potential electrolysis of $[Cr(CO)_6]$ in CH_2Cl_2.[217] The 17-electron ion $[W(CO)_6]^+$ is formed during the collision-induced dissociation of $[W(CO)_6]$ in a mass spectrometer using helium or argon as target gases.[218] This ion is the target of several other mass spectrometric studies.[5,219–24] The unsaturated dimetal ions $[Cr_2(CO)_n]^+$ (n = 0–6) are observed in electron-impact ionization and other mass spectrometric studies on $[Cr(CO)_6]$.[225,226]

3.3.2 Reactions

The dianions $[M(CO)_5]^{2-}$ show reactions characteristic of a highly nucleophilic species. Reaction of $Li_2[W(CO)_5]$ with CO_2 affords carbonate and $[W(CO)_6]$ in a reaction that proceeds via the η^1-CO_2 complex $[W(CO)_5(CO_2)]^{2-}$.[227] In related reactions, carbon (monooxide)(monosulfide) (COS) reacts with $[W(CO)_5]^{2-}$ to give $[W(CO)_6]$. However, thiocarbonyl species are not formed in reactions involving CS_2. The reaction between $[W(CO)_5]^{2-}$ and PhNCO gives 84% and 5% yields of $[W(CO)_6]$ and $[W(CO)_5(CNPh)]$, respectively.[228]

The dianion $[M(CO)_5]^{2-}$ is a useful carbene precursor (Equations (1)[229] and (2)[230]). The methodology of Equation (1) gives yields comparable with standard routes (Section 3.10), and also allows the formation of otherwise inaccessible species such as $[Cr(CO)_5\{=C(OMe)C_6H_3(OAc)_2\text{-}2,5\}]$. The reaction of $[Cr(CO)_5]^{2-}$ with $InCl_3$ followed by Br^- affords the dianion $[InBr\{Cr(CO)_5\}_2]^{2-}$ (Equation (3)). The corresponding reaction with $TlCl_3$ gives similar products.[231] The reaction of equimolar amounts of $Na_2[W(CO)_5]$ and $Na_2[W_2(CO)_{10}]$ with $CpLuCl_2$ (Equation (4)) is remarkable for the formation of the dianion $[W_3(CO)_{14}]^{2-}$. Its x-ray crystal structure shows a linear array of three tungsten atoms, each of which is octahedrally coordinated. The three $W(CO)_4$ fragments are eclipsed and the W–W lengths are ~279 pm.[232]

$$[M(CO)_5]^{2-} \xrightarrow{R^1C(=O)Cl} [M(COR^1)(CO)_5]^- \xrightarrow{[R^2]^+} [M\{=C(OR^1)R^2\}(CO)_5] \qquad (1)$$

$$Na_2[Cr(CO)_5] + R^1C(=O)NR^2_2 \xrightarrow[\text{ii, } Al_2O_3]{\text{i, TMS-Cl}} [Cr(CO)_5\{=C(R^1)NR^2_2\}] \qquad (2)$$

$$[Cr(CO)_5]^{2-} \xrightarrow[\text{ii, } Br^-]{\text{i, } InCl_3} [InBr\{Cr(CO)_5\}_2]^{2-} \qquad (3)$$

$$Na_2[W(CO)_5] + Na_2[W_2(CO)_{10}] + CpLuCl_2 \xrightarrow{\text{DME}}$$ 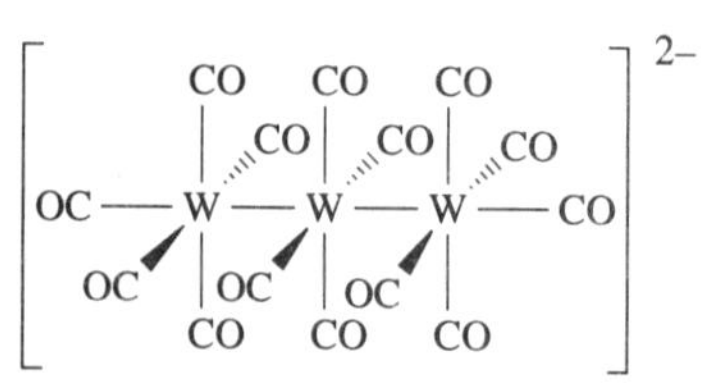 (4)

The chemistry of the tetraanions $[M(CO)_4]^{4-}$ is less well defined. However, the reaction of $[M(CO)_4]^{4-}$ (M = Cr or W) with $SnClPh_3$ gives respectable yields of $[M(SnPh_3)_2(CO)_4]^{2-}$, and these anions react further to form the monanionic seven-coordinate $[M(SnPh_3)_3(CO)_4]^-$.[233] The dihydride $[WH_2(CO)_4]^{2-}$ represents formally doubly protonated $[W(CO)_4]^{4-}$, but it is obtained by the reaction of $[W(CO)_4(TMEDA)]$ with $K[BHBu^s_3]$.[202] Small amounts of the hydride $(NEt_4)_4[MoH(CO)_3]_4$ are formed in the reaction of $Na_4[Mo(CO)_4]$ with excess NCMe followed by cation exchange.[234]

The reaction of $[Fe_2(CO)_8]^{2-}$ with CH_2I_2 is known to give the μ-methylene complex $[Fe_2(CO)_8(\mu\text{-}CH_2)]$, but the corresponding reaction of the dimetal anions $[M_2(CO)_{10}]^{2-}$ with CH_2X_2 (X = Br or I) gives only $[MX(CO)_5]^-$.[204] These dianions are, however, a source of metal cluster compounds in their reactions with metal halide complexes $MClL_n$ (see Chapter 4, Volume 10).

The dianion $[W_2(CO)_{10}]^{2-}$ possesses a photochemistry related to other M–M bonded complexes. Thus, irradiation into the low-energy tail of the $d\pi$–σ^* transition leads to W–W homolysis. The resulting

17-electron anion $[W(CO)_5]^-$ reacts with ligands such as 4-cyanopyridine to form 19-electron adducts.[235] Treatment of $Na_2[Cr_2(CO)_{10}]$ with $AsClBu^t_2$ affords the arsinidene complex $[AsBu^t\{Cr(CO)_5\}_2]$. This complex contains no Cr–Cr bond.[236]

Studies on the reactions of the radical anions $[Cr(CO)_5]^-$ with O_2 in the gas phase show that the eventual products are CrO_n ($n = 2$–5) and $[Cr(CO)_3(O_2)]^-$ via the primary products $[Cr(CO)_3(O)_m]^-$ ($m = 1$, 2).[210–12] Reactions with various organic electrophiles in the gas phase seem to proceed by an initial electron transfer process followed by carbonyl substitution processes.[206,209]

3.4 CARBONYL DERIVATIVES WITH CARBON σ-DONOR LIGANDS

The most numerous compounds within this class are of the type $[MR(CO)_2LCp]$ (R = alkyl, aryl, formyl, acyl, etc.; L = CO, PR_3, etc.) and are discussed in Section 3.7.2. The chemistry of halogenoalkyl complexes[237] and bridged hydrocarbyl binuclear transition metal complexes has been reviewed,[238] as has the photochemistry of alkyl, alkylidene and alkylidyne complexes.[239]

Acyl derivatives $[M(COR)(CO)_5]^-$ are used principally for carbene synthesis and examples are given in Section 3.10, while η^2-acyl and η^2-iminoacyl complexes have been reviewed.[240] Formyl complexes such as $[Cr(CHO)(CO)_5]^-$ are uncommon, but a new way to make this species has been reported; this involves the electroreduction of $[Cr(CO)_6]$ to form the 19-electron anion $[Cr(CO)_6]^-$ and subsequent hydrogen donation to that species from $SnHBu_3$.[241] Formyl complexes have been reviewed.[242] Other reports of formyl complexes include that of the fluxional [M(CHO)(CO)(NPh)Tp'], formed in the reaction of the cation $[M(CO)_2(NPh)Tp']PF_6$ with excess $LiBH_4$.[241]

Of the remainder, complexes of the type $[WR(CO)_5]^-$ are perhaps most important. The methyl derivative $[WMe(CO)_5]^-$ is available, for instance, by the reaction of $[WCl(CO)_5]^-$ with MeLi.[243] The x-ray crystal structures of $[Na(18\text{-crown-}6)][WMe(CO)_5]$[244] and the PMe_3 derivative $[PPN][WMe(CO)_4(PMe_3)]$[243] are both known. In both cases the anions are substantially conventional octahedra similar to $[ReMe(CO)_5]$. The W–Me bond is somewhat shorter in the PMe_3 derivative (218(3) pm) than in the pentacarbonyl (231.3(17) pm). The IR spectrum of $Na[WMe(CO)_4(PMe_3)]$ shows that there is ion pairing between the sodium and the CO site *trans* to the PMe_3. The electron-donating group PMe_3 promotes insertion of CO_2 into the W–Me bond by a factor of some 243.[243] The corresponding derivative $[PPN][W(CH_2CN)(CO)_5]$ does not undergo a related insertion reaction with CO_2.[245] The reaction of $[WMe(CO)_5]^-$ with CS_2 affords the insertion product $[PPN][W(CO)_4(\eta^2\text{-}S_2CMe)]$ after photolysis of the primary product $[PPN][W(CO)_5(S_2CMe)]$.[246]

3.5 CARBONYL DERIVATIVES WITH CYANIDE LIGANDS

Group 6 metal carbonyl complexes containing cyanide ligands are not particularly common. The pentacarbonyls $[M(CO)_5(CN)]^-$ are good precursors for isocyanide complexes (see Section 3.6). Oxidation of $[Cr(CN)(CO)_5]^-$ in nonaqueous media with O_2 affords the paramagnetic species $[Cr(CN)(CO)_5]$.[247] Nitrosation of $Na[W(CN)(CO)_5]$ gives a trinuclear complex (**1**) (Equation (5)).[248]

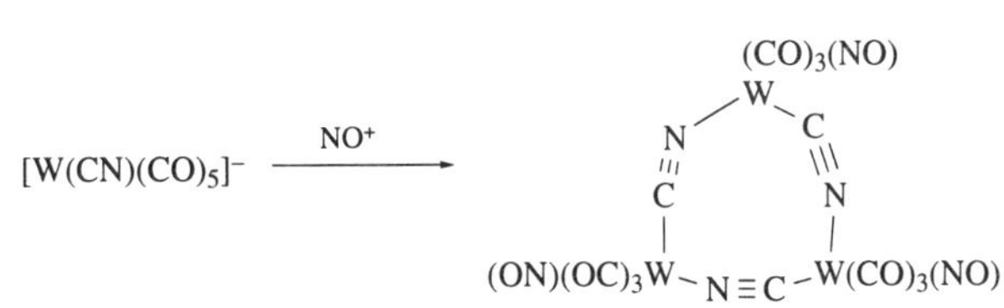

(5)

(**1**)

Related ligands include CNO^-. The cyanate complex $NEt_4[W(CNO)(CO)_5]$ is available from the reaction of $NEt_4[WBr(CO)_5]$ and AgCNO. It isomerizes to the corresponding isocyanate species.[249]

3.6 ISOCYANIDE COMPLEXES

The chemistry of isocyanide complexes has been reviewed by Singleton and Oosterhuizen.[250]

3.6.1 Syntheses

A number of examples of the syntheses of isocyanide complexes are given in Table 1. When the free isocyanide is available, isonitrile derivatives of $[M(CO)_6]$ are made typically by substitution of the parent hexacarbonyl. Simple thermal substitution tends to give a mixture of products and requires forcing conditions. The use of a catalyst such as PdO affords clean conversion to the required product $[M(CO)_{6-n}(CNR)_n]$ (n = 1–6), and the method is judged sufficiently reproducible to be used as an undergraduate laboratory experiment.[72,251,252]

Two other notable methods (Table 1) are the nucleophilic attack by the complex $[Cr(CN)(CO)_5]^-$ on suitable species RX (ClCN, BXR_2, etc.), and the replacement of the tridentate ligand η^6-C_7H_8 in $[M(CO)_3(\eta^6\text{-}C_7H_8)]$ by various tridentate isocyanide ligands.

3.6.2 Properties

Molybdenum complexes of the type $[Mo(CO)_n(CNR)_{6-n}]$ are particularly amenable to analysis by ^{95}Mo NMR spectroscopy. The linewidths are narrow. The linewidths of isocyanide complexes are a function of scalar coupling between the ^{95}Mo and ^{14}N nuclei, which is inversely proportional to the number of isocyanide substituents.[55] The chemical shift is not particularly dependent on R. However, the values of δ_{Mo} show that the metal centres are progressively deshielded as successive CNR substituents replace carbonyl ligands. Therefore, ^{95}Mo NMR spectroscopy is a convenient way of monitoring these species. The ^{13}C NMR signals of the isocyanides also become more deshielded with increasing number of isocyanide ligands in the molecule.[55,72] The use of molybdenum and tungsten NMR methods in this and other contexts has been reviewed.[23] Other studies using ^{13}C, ^{17}O and ^{14}N NMR show that alkyl substituted isocyanides are better net donors than aryl isocyanides, and show increased electron density at the metal on successive substitutions by isocyanides.[253]

There are many results of x-ray crystal structure determinations in the literature, but comparatively few gas-phase studies. An analysis of $[Cr(CO)_5(C{\equiv}NCF_3)]$ by electron-diffraction methods suggests that the effective C–N–C angle is about 140°. This is an indication of strong backbonding and a significant contribution from a Cr=C=N–R resonance form.[254] These results are reinforced by the results of an x-ray photoelectron spectroscopy study.[255]

Vibrational spectroscopic data on the complexes $[Cr(CO)_5(C{\equiv}NCFH_2)]$ and $[Cr(CO)_5(C{\equiv}NCF_2H)]$ suggest that for both ligands the π-acceptor/σ-donor ratio is lower than that for $C{\equiv}NCF_3$.[260] Other IR and Raman data interpreted in terms of a valence force field show that the C≡NC(O)Ph ligand is rather similar to carbonyl for chromium.[256] The notable feature of the IR spectra of the isocyanogen complex $[Cr(CO)_5(CNCN)]$ is that the coordinating CN group possesses almost identical coordinating and vibrational properties to those of the carbonyl ligands in that complex.[257] Electrochemical oxidation data for the compounds $[Cr(CO)_n(CNR)_{6-n}]$ (R = Me or Ph) correlate well with an additive relationship addressing the energies of the three metal d_π orbitals in a d^6 system. They show that C≡NPh is a worse σ-donor and a worse π-acceptor than C≡NMe.[258]

One useful feature of isocyanide ligands is their potential for bidenticity, which leads to the question as to whether there is communication between the metals in complexes of diisocyanides such as L_nM–(CNC_6H_4NC)–ML_n. A spectroscopic examination of $[(OC)_5W(CNC_6H_4NC)]$ and $[(OC)_5W(CNC_6H_4NC)W(CO)_5]$ shows a pronounced difference in the W → π*(CNR) metal–ligand charge transfer bands between the two complexes (a red shift: 28 248–27 027 cm^{-1}). There is no such difference for the saturated ring analogues $[(OC)_5W(CNC_6H_{10}NC)]$ and $[(OC)_5W(CNC_6H_{10}NC)W(CO)_5]$. These results suggest that there is communication across the π-system of the ring. However, the environment of one metal in $[(OC)_5W(CNC_6H_4NC)WBr_2(PEt_3)_2]$ is insensitive to changes in the oxidation state of the second metal. The results of W(4*f*) x-ray photoelectron spectroscopic studies suggest that this is because the two metals, despite their differences in formal oxidation state, possess very similar charge densities.[259]

Some complexes of the type $[Mo(CO)_2(CNCy)_4]$ and $[Mo(CO)_2(CNR)_2(PR_3)_2]$ undergo electrochemically induced facile isomerizations to the corresponding all-*trans* complexes. In each case the critical step is oxidation which affords a 17-electron species, some of which have been isolated. The isomerizations are intramolecular and can be interpreted as series of trigonal twists.[291,292]

The NMR spectra of the cyclopentadienyl complexes $[Mo(CO)(CNR)_3Cp]^+$ (R = Me, Et or Bu^t) are consistent with conventional four-legged piano-stool geometries. On warming there is an exchange of *cis* and *trans* environments in the cases where R = Me or Et.[293]

Table 1 Examples of isocyanide complexes containing carbonyl ligands.

Synthesis	*Complex*	*Comments*	*Ref.*
$[Cr(C{\equiv}NCCl_3)(CO)_5] + RNH_2$	$[Cr(C{\equiv}NR)(CO)_5]$	R = many different alkyl, aryl	260
$[M(CO)_6] + CNR$	$[M(CO)_n(CNR)_{6-n}]$	M = Mo, W; R = $C_6H_3Me_2$-2,6, Bu^t, Cy, Pr^i; PdO catalyst	72,251
$[M(CO)_6] + CNBu^t$	$[M(CO)_n(CNBu^t)_{6-n}]$	M = Cr, Mo, W; n = 3–5; $[Ru_3(CO)_{12}]$, $[Co_2(CO)_8]$, $[M(CO)_2Cp]_2$ (M = Fe, Ru), $[M_4(CO)_{12}]$ (Co, Ir), $CoCl_2{\cdot}2H_2O$ are all catalysts	83,251, 261–3
$[Cr(CO)_5(C{\equiv}NCCl{=}CH_2)] + MeNH_2$	$[Cr(CO)_5(C{\equiv}NMe)]$		264
$[MI(CO)_5]^- + CNC_6H_4CH_2PR_3/AgBF_4$	$[M(CNC_6H_4CH_2PR_3)(CO)_5]$	M = Cr, Mo, W; $R_3 = Me_3, Ph_3, Ph_2Bz$	265
$[Cr(C{\equiv}NCClFCF_2Cl)(CO)_5] + Zn/Et_2O/MeCO_2H$	$[Cr(C{\equiv}NCF{=}CF_2)(CO)_5]$	x-ray	266
$[Cr(CO)_5(C{\equiv}NCCl_2F)] + SnHBu_3$	$[Cr(CO)_5(C{\equiv}NCH_2F)]$		267
$[Cr(CO)_5(C{\equiv}NCCl_2H)] + HgF_2$	$[Cr(CO)_5(C{\equiv}NCHF_2)]$		267
$[Cr(CN)(CO)_5]^- + PhN_2BF_4/HClFC{-}CF_2Cl$	$[Cr(C{\equiv}NCClFCF_2Cl)(CO)_5]$		266
$[Cr(CN)(CO)_5]^- + ArN_2BF_4/CHCl_2R$	$[Cr(C{\equiv}NCCCl_2R)(CO)_5]$	R = H, Cl	268,269
$[Cr(CN)(CO)_5]^- + RSO_2Cl$	$[Cr(C{\equiv}NCSO_2R)(CO)_5]$	R = Me, Et, Bz, Ph, Tol, C_6H_4F	270
$[Cr(CN)(CO)_5]^- + BXR_2$	$[Cr(C{\equiv}NBR_2)(CO)_5]$	R = Et, OMe, NMe_2	271
$[Cr(CN)(CO)_5]^- + ClCN$	$[Cr(CNCN)(CO)_5]$	x-ray (∠CNR = 168.5(4)°)	272
$[Cr(CN)(CO)_5]^-$ + cyanuric chloride	$[\{(OC)_5Cr\}_3\{(CN)_3C_3N_3\}]$		272
$[W(CN)(CO)_5]^- + HBF_4/CyNO$	$[W\{C{\equiv}N{-}C(O)NHCy\}(CO)_5]$		273
$[Cr(CN)(CO)_5]^- + [Cr(CNH)(CO)_5]$	$[(OC)_5Cr(C{\equiv}NHN{\equiv}C)Cr(CO)_5]^-$	x-ray	274
$[Cr(CO)_5(2\text{-}Me\text{-}THF)] + CNCN$ (–20 °C)	$[Cr(CNCN)(CO)_5]$		257
$[M(CO)_6] + MeC(CH_2N{=}C)_3$	$[M(CO)_3\{(C{\equiv}NCH_2)_3CMe\}]$	M = Cr, W, x-ray	275
$[M(CO)_3(\eta\text{-}C_7H_8)] + HC\{(CH_2)_3N{\equiv}C\}_3$	$[M(CO)_3(\{C{\equiv}N(CH_2)_3\}_3CH)]$	M = Cr, W	276
$[M(CO)_3(\eta\text{-}C_7H_8)] + N\{(CH_2)_3N{\equiv}C\}_3$	$[M(CO)_3(\{C{\equiv}N(CH_2)_3\}_3N)]$	M = Cr, Mo, W	276,277
$[M(CO)_3(\eta\text{-}C_7H_8)] + N\{(CH_2)_3OC_6H_4N{\equiv}C\}_3$	$[M(CO)_3(\{C{\equiv}NC_6H_4O\text{-}(CH_2)_3\}_3N)]$	M = Cr, Mo, W. *ortho* and *para* substituted	277
$[M(CO)_3(\eta\text{-}C_7H_8)] + C{\equiv}NCF_3$	*mer*-$[M(C{\equiv}NCF_3)_3(CO)_3]$	M = Mo, W	278
$[M(CO)_2(NCMe)(\eta\text{-}C_7H_7)]^+ + CNBu^t$	$[M(CO)_2(CNBu^t)(\eta\text{-}C_7H_7)]^+ + [M(CO)_2(CNBu^t)_3(\eta^3\text{-}C_7H_7)]^+$	M = Mo, W. The ring-slipped product losses one CO and one $CNBu^t$ to form $[M(CO)(CNBu^t)_2(\eta\text{-}C_7H_7)]^+$	279
$[Cr(C{\equiv}NCCCl_2CHCl_2)(CO)_5] + Zn/Et_2O/MeCO_2H$	$[Cr(C{\equiv}NCCl{=}CH_2)(CO)_5]$	*E* + *Z*	264
mer-$[M(C{\equiv}NCF_3)_3(CO)_3] + C{\equiv}NCF_3/h\nu$	$[M(C{\equiv}NCF_3)_{6-n}(CO)_n]$	M = Mo, W	278
$[Cr(CO)_5(CNCF_3)] + CNR/h\nu$	$[Cr(CO)_4(CNCF_3)(CNR)]$	R = Me, Ph	280
$[M(CO)_3(NCMe)_3] + CNR^1/R^2N{=}CH{-}CH{=}NR^2$	*fac*-$[M(CO)_3(CNR^1)(R^2N{=}CH{-}CH{=}NR^2)]$	M = Cr, Mo, W; R^1 = Me, Pr^i, Bu^t, Xy	281–4
cis-$[M(CO)_2(bipy)_2] + CNEt$	*cis,cis*-$[M(CO)_2(CNEt)_2bipy]$	M = Mo, W	285
$[W({=}CPh_2)(CO)_5] + Me_2(Et_2N)CC{\equiv}CN$	$[W(C{\equiv}NCHPh_2)(CO)_5]$		286
$[WI(CO)_2py_2({\equiv}CPh)] + CNBu^t$	$[WI(CO)_2(CNBu^t)_2({\equiv}CPh)]$		287
$[MCl_2(CO)_2(PMe_3)_3] + CNR/PMe_3$	$[MCl_2(CO)(CNR)(PMe_3)_3]$	M = Mo, W; R = Bu^t, Bz, Cy	288
trans-$[M(CO_2)_2(PMe_3)_4] + CNR$	*trans,mer*-$[M(CO_2)_2(CNR)(PMe_3)_3]$	R = Me, Pr^i, Bu^t	289
$[W(OEtH)(PR_3)(CO)_4] + C{\equiv}NC_6H_4N{\equiv}C$	$[(R_3P)(OC)_4W(C{\equiv}NC_6H_4N{\equiv}C)\text{-}W(CO)_4(PR_3)]$	R = Et, Ph	290

3.6.3 Reactions

Chromium isocyanide complexes can provide routes to new and interesting isocyanides. For instance, thermal degradation of [Cr{C≡N–C(Cl)=CH(Cl)}$(CO)_5$] (quartz tube, 10 Pa, 510 K) affords the free isocyanide HC≡C–N≡C, which is perhaps an interstellar molecule.[294] The fluorinated vinyl isocyanide CF_2=CF(N≡C) is formed on flash vacuum pyrolysis of [Cr(C≡N–CF=CF_2)$(CO)_5$] (Equation (6)).[266]

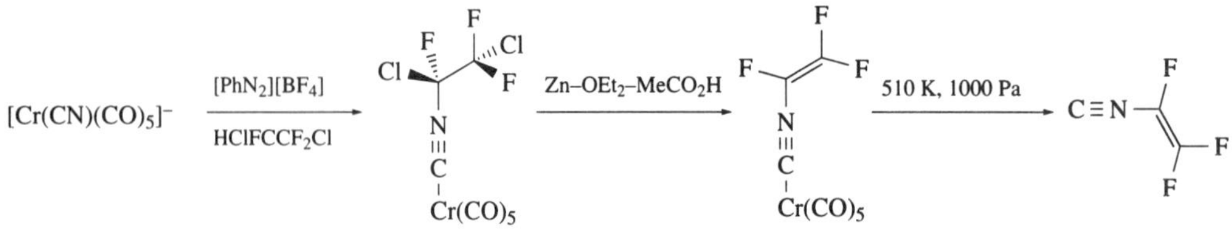

(6)

The complex [Cr(C≡$NCCl_3$)$(CO)_5$] is versatile. It reacts with many different primary amines to form isocyanide complexes [Cr(C≡NR)$(CO)_5$] in a complex process, but in high yield.[260] The carbyne capped cluster (**2**) is formed in its reaction with [$Co_2(CO)_8$] (Equation (7)). The IR spectrum of this product consists of essentially independent $Cr(CO)_5$ and $Co_3(CO)_9$ fragments.[295] The reaction of [Cr(C≡$NCCl_3$)$(CO)_5$] with 1,2-dithiols in the presence of NEt_3 gives carbene complexes (Equation (8)).[296]

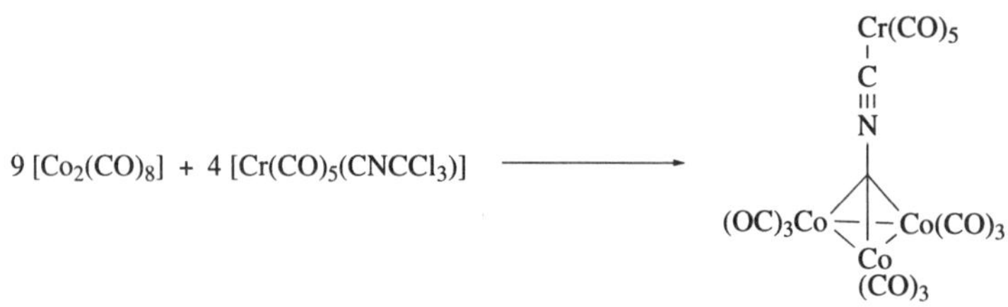

(7)

(**2**)

[Cr(C≡$NCCl_3$)$(CO)_5$] + 2 $HSCH_2CHRSH$ + 3NEt_3 ⟶ [$Cr(CO)_5${=$CSCH_2CH(R)S$}] (8)

Carbene complexes are also formed though deprotonation of [M($CNC_6H_4CH_2PR_3$)$(CO)_5$] (M = Cr, Mo or W) with Na[N$(TMS)_2$]; the product [M{$CNHC_6H_4C(PR_3)$}$(CO)_5$] is formed in addition to indole and protonated indole type complexes.[265] Addition of $H_2NCH_2CH_2NH_2$ to [$Cr(CO)_5$(C≡NCCl=CH_2)] results in a cyclic carbene complex (Equation (9)).[264] A complex rearrangement of various thioether complexes of the type *cis*-[M(CNR)$(CO)_4${S(R^1)CH_2R^2}] after treatment with BuLi, PEt_3 or Et_3O^+ affords trisubstituted aminocarbene thioether chelates (Equation (10)).[297] Addition of Ph_2C=C=NPh to the metallonitrile ylide [$M(CO)_5$(C≡$NCHPPh_3$)] affords the carbenoid species (**3**) (Equation (11)). The PPh_3 group is lost on treatment with water as O=PPh_3.[298]

[$Cr(CO)_5$(C≡NCCl=CH_2)] + $H_2NCH_2CH_2NH_2$ ⟶ [$Cr(CO)_5${=$CNHCH_2CH_2NH$)] (9)

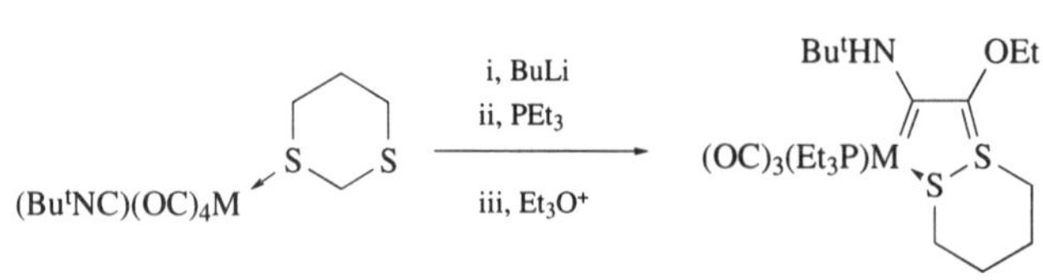

(10)

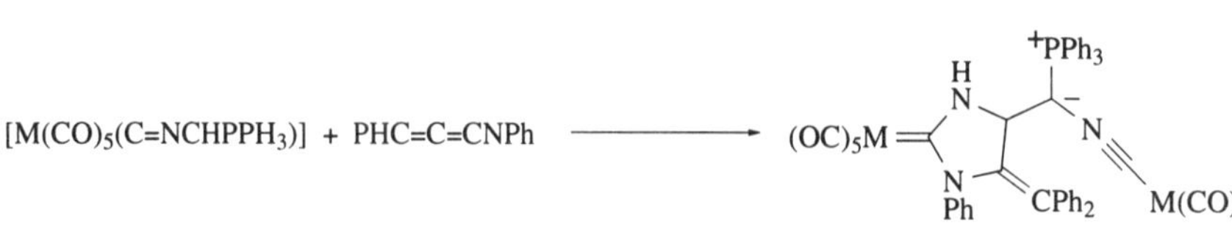

(11)

(**3**)

There is also considerable interest in the chemistry of coordinated hydrogen isocyanide as $[M(CO)_5(CNH)]$. The complex $[W(CO)_5(CNH)]$ adds regiospecifically by a 1,2-addition process across various internal alkenes $R^1R^2C{=}CR^3R^4$ to form the new isonitrile complexes $[W(CO)_5(CNCR^1R^2CHR^3R^4)]$.[299] Related reactions with CyNCNCy afford the isocyano carboxamides $[M(CO)_5\{CNC({=}O)NHCy\}]$ (M = Cr or W),[266] while ring opening of epoxides by $[M(CO)_5(CNH)]$ (M = Cr or W) affords the 2-hydroxy alkyl of oxazolin-2-ylidene complexes.[300] Corresponding reactions with aziridines affords cyclic diaminocarbene complexes.[301]

Oxidation of $[W(CNBu^t)(CO)_5]$ with I_2 affords the seven-coordinate W^{II} complex $[WI_2(CO)_4(CNBu^t)]$. Further treatment with $CNBu^t$ eventually leads to $[WI(CNBu^t)_6]I$.[302]

Protonation of the isonitrile alkylidyne species (**4**) (Equation (12)) results in protonation at the alkylidyne carbon and a coupling process to form complexes (**5**) and (**6**). Further experiments indicate that the first protonation stage affords an alkylidene species in which the hydrogen adopts an agostic posture (δ_H – 3.0).[303]

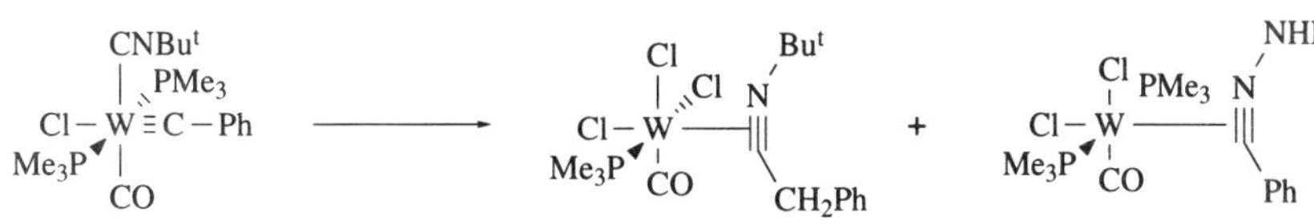

(12)

(**4**) (**5**) (**6**)

3.7 ALKYNIDE COMPLEXES

Most of the alkynide chemistry of the group 6 elements centres on the chemistry of derivatives of the cyclopentadienyl complex $[W(C{\equiv}CPh)(CO)_3Cp]$.[304]

3.7.1 Syntheses

The alkynide complex $[W(C{\equiv}CPh)(CO)_3Cp]$ is available by the reaction of $[WCl(CO)_3Cp]$ with $PhC{\equiv}CH$ in the presence of CuI/$NHEt_2$.[304] Furthermore, the reaction of $[MI(CO)_3Cp]$ (M = Mo or W) with $PhC{\equiv}CH$ gives good yields of $[M(C{\equiv}CPh)(CO)_3Cp]$ and is catalysed by CuI/NEt_3. This reaction appears to proceed via copper alkynide species.[305] Deprotonation of the alkynide $[Fe(C{\equiv}C–C{\equiv}H)(CO)LCp]$ (L = CO or PPh_3) with $LiBu^s$ affords an anion which reacts with $[M(CO)_3Cp]^-$ to form the alkynide complexes $[Cp(OC)LFeC{\equiv}C–C{\equiv}CM(CO)_3Cp]$.[306] Deprotonation of terminal alkyne complexes may result in alkynide formation. Thus, the reaction of the cationic bis(alkyne) derivative $[W(CO)(\eta^2\text{-}Bu^tC{\equiv}CH)_2Cp^*]^+$ with KH results in the reactive complex $[W(C{\equiv}CBu^t)(CO)(\eta^2\text{-}Bu^tC{\equiv}CH)Cp^*]$.[307] The related reaction between $[Mo(PR_3)_2(\eta^2\text{-}Bu^tC{\equiv}CH)Cp]^+$ ($PR_3 = P(OMe)_3$ or PMe_2Ph) and $NaN(TMS)_2$ in the presence CO affords *trans*-$[Mo(C{\equiv}CBu^t)(CO)(PR_3)_2Cp]$.[308] The reaction between $[W(CO)_3Cp]^-$ and CH_2I_2 leads to a ketene species $[\{CpW(CO)_2\}_2(\mu\text{-}CH_2CO)(\mu\text{-}CO)]$ which converts into the crystallographically characterized dinuclear species $[Cp(OC)_3WC{\equiv}CW(CO)_3Cp]$ at temperatures above –5 °C in MeCN.[309]

In a few instances the alkynide group bridges two metals. This is the case for $[(RC{\equiv}C)Pt(\mu\text{-}C{\equiv}CR)(\mu\text{-}dppm)_2M(CO)_3]$ (M = Mo or W; R = Me, Ph or Tol) compounds formed in the reaction of $[(RC{\equiv}C)_2Pt(\mu\text{-}dppm)_2AgX]$ with either $[Mo(CO)_3(\eta^6\text{-}C_7H_8)]$ or $[W(CO)_3(NCMe)_3]$. The chromium analogue was made but not isolated.[310] The M–M bond is preserved in complexes of the type $Li[Mo_2(\mu\text{-}\sigma\text{-}\eta^2\text{-}C{\equiv}CR)(CO)_4Cp_2]$, formed by the reaction of $LiC{\equiv}CR$ with $[Mo_2(CO)_4Cp_2]$ (see Section 3.9.2).[311] Deprotonation of the μ-alkyne $[Mo_2(CO)_4(\mu\text{-}HC{\equiv}CCH_2OMe)Cp_2]$ ($NaNH_2$) affords an anionic bridged alkynide complex structurally characterized as $[Na(15\text{-crown-}5)][Mo_2(CO)_4(\mu\text{-}\eta^1{:}\eta^2\text{-}C{\equiv}CCH_2OMe)Cp_2]$.[312]

3.7.2 Properties

The x-ray crystal structure of $[Cp(OC)_3WC{\equiv}CW(CO)_3Cp]$ shows an essentially linear WC≡CW unit with a short C≡C length (C≡C 118(3) pm). The metal fragments are twisted such that the dihedral angle between the planes of the Cp groups is 55.6(6)°.[309] The alkynyl group is also linear in $W(C{\equiv}CC_3H_5)(CO)_2(PMe_3)Cp$.[313] Spectroscopic studies suggest that the two metal centres in $[Cp(OC)LFeC{\equiv}C–C{\equiv}CM(CO)_3Cp]$ communicate through the butadiynediyl bridge.[306]

The x-ray structures of the pair of complexes [Mo(C≡CPh)(CO)$_2$(η-C$_7$H$_7$)] and the one-electron oxidation product [Mo(C≡CPh)(CO)$_2$(η-C$_7$H$_7$)][BF$_4$] show that there is an increase in M–C$_{alkynyl}$ bond length of 6.1(2) pm on oxidation, while the C≡C distance remains constant. This and other spectroscopic data suggest that the alkynyl ligand is largely only a σ-donor, with very little contribution from $d\pi \rightarrow \pi^*$ backbonding interactions.[314,315]

3.7.3 Reactions

Cycloaddition reactions of alkynide complexes have been reviewed.[316] One complex investigated in some depth is [W(C≡CPh)(CO)$_3$Cp],[304] which is a good precursor for cluster complexes (see Chapter 4, Volume 10). Other representative examples of the reactivity of [M(C≡CPh)(CO)$_3$Cp] (M = Mo or W) are shown in Scheme 1[304,308,317–22] and the reader is also referred to Section 3.7.2 for further discussion. Trimolybdenum 48-electron clusters [Mo$_3$(CO)$_5$(μ-CCR)Cp$_3$] (R = alkyl or aryl) are formed in the reaction of Li[Mo$_2$(μ-σ-η^2-C≡CR)(CO)$_4$Cp$_2$] with the cation [Mo(NCMe)$_2$(CO)$_2$Cp]$^+$.[312]

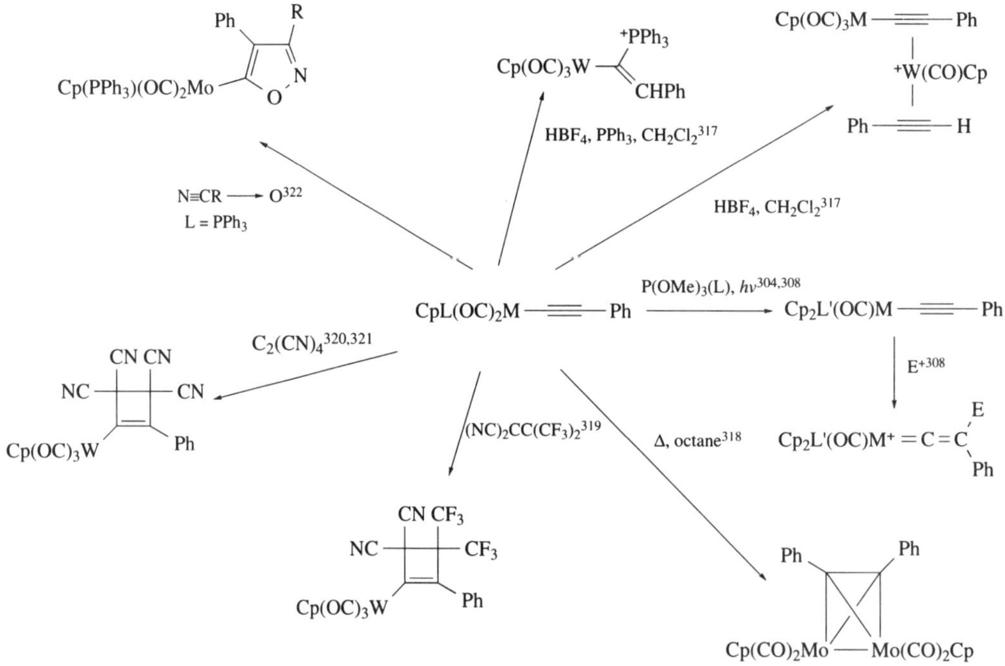

Scheme 1

Protonation or alkylation of alkynyl complexes proceeds at the β-carbon centre (Scheme 1). This is also reflected in other systems. Addition of CO$_2$ to the anionic alkynyls Li[W(C≡CR)(CO)$_3$(dppe)] (R = Me or Bu^n) results in formation of the vinylidenes Li[W{=C=CR(CO$_2$)}(CO)$_3$(dppe)]. The corresponding Ph$_3$PNPPh$_3$$^+$ salt does not react, suggesting the importance of ion pairing in this system. This anion alkylates (Me$_3$O$^+$) to form the neutral derivative [W{=C=CR(CO$_2$Me)}(CO)$_3$(dppe)].[323] Alkylation of the known anion [NEt$_4$][W(C≡CBu^t)(CO)$_5$] affords the vinylidenes [W(=C=CBu^{t}R)(CO)$_5$] (R = Me or Et).[324] When the alkynyl group in a molecule bridges, the resulting vinylidene might also bridge. This is the case for protonation of [(RC≡C)Pt(μ-C≡CR)(μ-dppm)$_2$M(CO)$_3$], which gives [(RC≡C)Pt(μ-C=CHR)(μ-dppm)$_2$M(CO)$_3$]$^+$.[310]

3.8 THIOCARBONYL AND SELENOCARBONYL COMPLEXES

Despite the potential richness of thiocarbonyl chemistry, there continues to be relatively little information available for group 6 thiocarbonyl complexes. The reason for this is largely because the only relatively accessible starting materials are [M(CO)$_5$(CS)] and [Cr(CO)$_2$)(CX)(arene)]. In turn, the lack of metal complex starting materials is due to the lack of sources of CS and CSe, neither of which is available as free diatomic molecules under normal conditions. Thiocarbonyl complexes have been well reviewed.[325]

3.8.1 Syntheses

The parent pentacarbonyls $[Cr(CO)_5(CX)]$ (X = S or Se) are available through arene replacement reactions of the arene complexes $[Cr(CO)_2(CX)(\eta\text{-}C_6H_6)]$ (Equation (13)). The thiocarbonyl is formed in 87% yield, and the selenocarbonyl in 45% yield. The arene complex $[Cr(CO)_2(CX)(\eta\text{-}C_6H_6)]$ precursor (Equation (14)) is formed through the reaction of a cyclooctene complex with CX_2 (X = S or Se). Typical yields are 30–50%.[326,327]

$$[Cr(CO)_2(CX)(\eta\text{-}C_6H_6)] + CO \xrightarrow[50\text{–}70\,°C]{20\text{ atm}} [Cr(CO)_5(CX)] \quad (13)$$

$$[Cr(CO)_3(\eta\text{-}C_6H_6)] \xrightarrow{C_8H_{14},\ C_6H_6,\ h\nu} [Cr(CO)_2(C_8H_{14})(\eta\text{-}C_6H_6)] \xrightarrow{CX_2} [Cr(CO)_2(CX)(\eta\text{-}C_6H_6)] \quad (14)$$

3.8.2 Properties

The IR spectra in the gas phase and solution phase suggest a C_{4v} structure for both $[Cr(CO)_5(CX)]$ (X = S or Se) species. In CS_2 the ν_{CX} stretches are at 1261.0 cm^{-1} (X = S) and 1077.4 cm^{-1} (X = Se).[326] A more detailed analysis of the absolute integrated IR intensities for the complexes $[Cr(CO)_5(CX)]$ and $[Cr(CO)_2(CX)(\eta\text{-}C_6H_6)]$ (X = O, S or Se) led to the conclusion that the π-acceptor abilities of the CX ligands decrease in the order CSe > CS > CO.[328] Assignments based on isotopic labelling studies are made for most of the fundamental vibrations of the pentacarbonyls $[M(CO)_5(CX)]$ (M = Cr, X = S or Se; M = W, X = S)[329] and for the arene complexes $[Cr(CO)_2(CX)(\eta\text{-}C_6H_6)]$ (X = O, S or Se).[330] High-pressure IR and micro-Raman studies on both $[Cr(CO)_5(CS)]$ and $[Cr(CO)_2(CS)(\eta\text{-}C_6H_6)]$ in the solid state show that there are no pressure-induced phase transitions for pressures up to 30 kbar. In both complexes, the values of $d\nu/dP$ suggest that Cr–CO backbonding is enhanced by increased pressure, as is the Cr–CS backbonding for $[Cr(CO)_5(CS)]$.[331] The ^{13}C chemical shift tensor of the CS ligand in $[Cr(CO)_2(CS)(\eta\text{-}C_6H_6)]$ has been measured,[332] and Fenske–Hall methods used to calculate its ^{13}C NMR shifts.[333]

3.8.3 Reactions

It has already been pointed out that arene displacement by CO in $[Cr(CO)_2(CS)(\eta\text{-}C_6H_6)]$ leads to $[Cr(CO)_5(CS)]$. A variation on this is the arene displacement by phosphite (Equation (15)). The product is octahedral with *mer* stereochemistry and mutually *trans* CO groups. The Cr–P bond *trans* to CS is somewhat longer than the other Cr–P bond. This is attributed to the stronger electron-withdrawing capacity of the CS ligand.[334]

$$[Cr(CO)_2(CS)(\eta\text{-}C_6H_6)] + 3\ P(OPh)_3 \longrightarrow [Cr(CO)_2(CS)\{P(OPh)_3\}_3] \quad (15)$$

The thiocarbonyl sulfur atom shows a definite tendency to function as a Lewis base. Thus, the anions $[Mo(CO)_2(CS)L]^-$ (L = Cp or Tp) react with $ClC_6H_3(NO_2)$-2,4, MeI or EtI to form the carbynes $[Mo(CO)_2(\equiv CSR)L]$ (R = $C_6H_3(NO_2)_2$, Me or Et). The Tp derivatives are more stable than the cyclopentadienyl derivatives.[335] The same tendency is shown (Equation (16)) when a $Cr(CO)_5$ group coordinates to a thiocarbonyl. The Cr–C–S link is linear, but the C–S–$Cr(CO)_5$ link is not. The S–Cr bond is long (0.248 6(2) pm), and this is reflected in a tendency for this bond to break in polar solvents.[336]

$$[Cr(CO)_2(CS)(\eta\text{-}C_6H_6)] + [Cr(CO)_5(THF)] \longrightarrow [Cr(CO)_2\{CSCr(CO)_5\}\{P(OPh)_3\}_3] \quad (16)$$

However, not all metal fragments bind at the sulfur atom. In the case of $[W(CO)_2(CS)Tp]^-$, reaction with $AuCl(PR_3)$ (R = Me or Ph)[337] or $[Mo(CO)_2(NCMe)_2(\eta^5\text{-}C_9H_7)]^+$ [338] results in the formation of heterobimetallic M–M' complexes containing semibridging thiocarbonyl groups, such as the example given in Equation (17).

Thiocarbonyl complexes can be precursor of carbenes. For instance, SMe^- reacts with $[W(CO)_5(CS)]$ to form $[W\{C(=S)SMe\}(CO)_5]^-$, which in turn reacts with MeI to form dithiocarbene derivatives $[W\{=C(SMe)_2\}(CO)_5]$.[339]

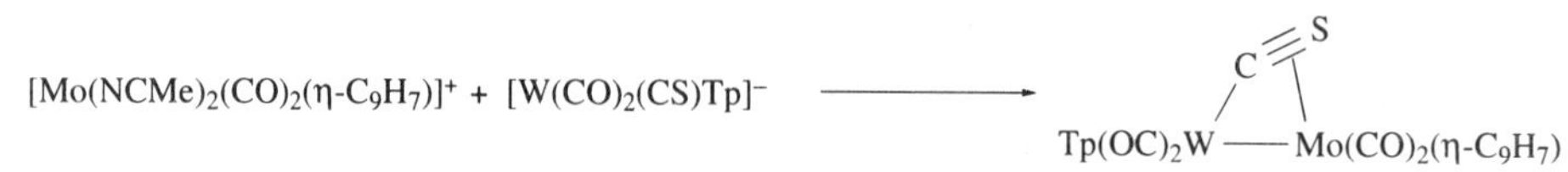

(17)

3.9 CARBENE COMPLEXES WITH NO HETEROATOM STABILIZING SUBSTITUENTS

Due to the amount of available information, the discussion of carbene complexes is divided into three sections: those in which there is no heteroatom stabilization, those with an oxygen heteroatom substituent (Section 3.10), and those with a nitrogen substituent (Section 3.11). Dihalocarbene complexes have been reviewed.[340] The formation of cyclopropanes from transition metal carbene complexes and alkenes,[341] and the catalytic transformations of carbene complexes have also been reviewed.[342]

Most carbonyl complexes of nonheteroatom stabilized carbenes are either derivatives of $[M(CO)_5\{=CR(Ar)\}]$ or those containing the $M(CO)_2\{=CR(Ar)\}Cp$ fragment. For convenience, they are described as alkylidenes in this chapter. Several reviews address the chemistry of alkylidene complexes.[239,343–5]

3.9.1 Synthesis and Properties

Complexes of the type $[Mo(CO)_3(=CH_2)Cp]^+$ enjoy only a brief existence and so must be generated and used within a few minutes at −78 °C. This is achieved conveniently through chloride abstraction from dichloromethane solutions of $[Mo(CH_2Cl)(CO)_3Cp]$ by $AgPF_6$. The consequence of leaving solutions too long is disproportionation to form $[Mo(CO)_3(C_2H_4)Cp]^+$ and solvated $[Mo(CO)_3Cp]^+$.[346] If a carbonyl group is substituted, this allows the synthesis of derivatives which may be characterized spectroscopically. Thus, hydride abstraction from $[W(CH_2R)(CO)_2LCp]$ (R = H, L = PPh_3 or PEt_3; R = Ph, L = PPh_3) at −50 °C affords the corresponding alkylidene cation in quantitative yield (Equation (18)). Under these conditions, they are sufficiently stable to be characterized spectroscopically. The molybdenum analogues are less stable and are best made by the reaction of $[Mo(CH_2OR)(CO)_2(PPh_3)Cp]$ with TMS-OSO_2CF_3 at −90 °C (Equation (19)).[347] An extension of this procedure to fluoride abstraction allows the isolation of the closely related $=CF_2$ carbenes $[Mo(=CF_2)(CO)_2LCp]^+$ (Equation (20)). This method also succeeds for the Cp* derivative.[348] Protonation of vinyl complexes can lead to cationic carbenes (Equations (21), (22)[349] and (23)[350]). The important feature of the second of these products (Equation (22)) is the agostic interaction between the β-hydrogen atoms and the metal. Rotation of the methyl group in the =CMePh derivative is fast.[349] Protonation by HBF_4 of the carbynes $[W(CO)_2(\equiv CTol)Cp]$ results in the formation of the transient carbenes $[W(CO)_2(=CHTol)Cp]^+$ (Equation (24)).[351]

$$[W(CH_2R)(CO)_2LCp] + [PH_3C]AsF_6 \longrightarrow [W(=CHR)(CO)_2LCp]^+ \quad (18)$$

$$[Mo(CH_2OR)(CO)_2(PPh_3)Cp] + Me_3SiOSO_2CF_3 \longrightarrow [Mo(=CH_2)(CO)_2LCp]^+ \quad (19)$$

$$[Mo(CF_3)(CO)_3Cp^*] + Me_3SiOSO_2CF_3 \longrightarrow [Mo(=CF_2)(CO)_3Cp^*]^+ \quad (20)$$

$$[W(\eta\text{-}CPh{=}CH_2)(CO)(PhCCH)Tp'] + HBF_4{\bullet}OMe_2 \longrightarrow [Mo(=CMePh)(CO)(PhCCH)Tp']^+ \quad (21)$$

$$[W(\eta^2\text{-}CPh{=}CHR)(CO)_2Tp'] + HBF_4{\bullet}OMe_2 \longrightarrow [Mo\{=C(CH_2R)Ph\}(CO)_2Tp'\}^+ \quad (22)$$

$$[Cr(CO)_2(\eta^2\text{-}HCCH)(\eta\text{-}C_6H_3Me_3)] \xrightarrow[\text{ii, }HBF_4]{\text{i, }PBu^n_3} [Cr(CO)_2\{=C(CH_2PBu^n_3)H\}(\eta\text{-}C_6H_3Me_3)]_3 \quad (23)$$

$$[W(CO)_2(\equiv CTol)Cp] + HBF_4 \longrightarrow [W(CO)_2(=CHTol)Cp]^+ \quad (24)$$

The neutral complex $[CrH(CO)_2(=CH_2)Cp]$ is formed through the photolysis of $[CrMe(CO)_3Cp]$ in gas matrices at 12 K. Gentle warming to 30 K results in reversal of this reaction.[352] Other examples of nonheteroatom stabilized carbenes $[MX(CO)_2(=CR^1R^2)Cp]$ are scarce. However, treatment of

$[W(SnPh_3)(CO)_2\{=C(OEt)Ph\}Cp]$ with R^- (R = H or Me) followed by acid leads to $[W(SnPh_3)(CO)_2\{=CRPh\}Cp]$ (Equation (25)). This reacts with I_2 to form the corresponding iodide.[353] The only example of anionic complexes $[M(CO)_2(=CR^1R^2)Cp]^-$ appears to be $[W(CO)_2(=CHPh)Cp]^-$, a species made by $NaC_{10}H_8$ reduction of the iodide (Equation (26)).[354] The mechanistically complex cyclization involving linkage of an alkyne, ammonia and a nitrile (Equation (27)) to form a metallabenzene derivative is also unusual.[355]

$$[W(SnPh_3)(CO)_2\{=C(OEt)Ph\}Cp] \xrightarrow[\text{ii, } H^+]{\text{i, } LiBEt_3H} [W(SnPh_3)(CO)_2\{=CHPh\}Cp] \quad (25)$$

$$[WI(CO)_2\{=CHPh\}Cp] + NaC_{10}H_8 \longrightarrow [W(CO)_2\{=CHPh\}Cp]^- \quad (26)$$

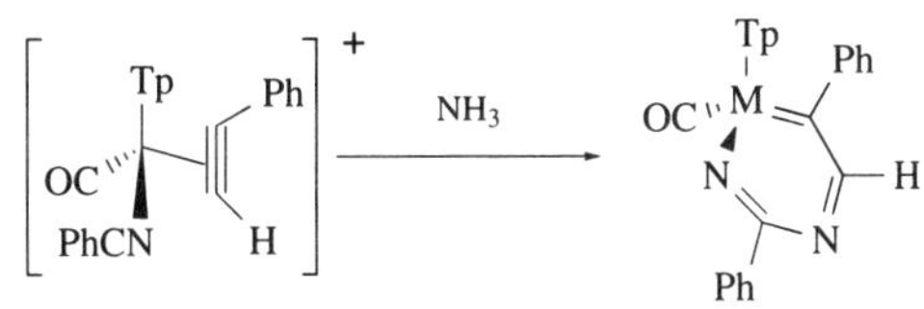

(27)

A common source of the $=CR^1R^2$ ligand is the corresponding diazo compound $CR^1R^2N_2$. Thus, the reaction of Ph_2CN_2 with [Cr(CO)(NO)(THF)Cp] leads to $[Cr(CO)(NO)(=CPh_2)Cp]$ (Equation (28)).[356] However, the reactions of diazoalkanes with $[Mo_2(CO)_4Cp_2]$ are complex, and unexpected products often arise,[357] but the reaction with 9-diazofluorene gives good yields of the corresponding μ-fluorenylidene species (Equation (29)).[358]

$$[Cr(CO)(NO)(THF)Cp] + PH_2CN_2 \longrightarrow \{Cr(CO)(NO)=CPh_2)Cp] \quad (28)$$

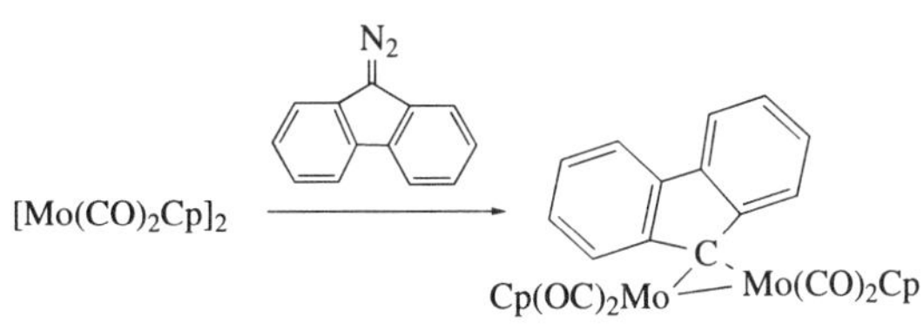

(29)

The complex $[W(CO)_5(=CPh_2)]$ has been known for some time. The molybdenum species $[Mo(CO)_5(=CPh_2)]$ has been prepared more recently (Equation (30)).[359] The benzylidene analogues are made conveniently by the consecutive reaction of readily available carbenes $[M(CO)_5\{=C(OMe)C_6H_4R\}]$ (M = W, R = H, Me or OMe; M = Cr, R = OMe) with hydride and acid (Equation (31)).[360,361] Other routes to carbene complexes of this type include metathesis reactions (Equation (32)). This photochemical process is suppressed by 1 atm of CO and it appears that photochemical excitation proceeds by loss of a *cis*-carbonyl ligand to form a $[W(CO)_4(=CPh_2)]$ fragment, which binds the alkene before undergoing metallacyclobutane formation and metathetical exchange.[362] Cyclopropylidene complexes (M = Cr, Mo or W) (Equation (23)) are formed in high yield through the use of LiC≡COEt as the organolithium source in a standard Fischer carbene type synthesis, which differs in that the expected product $[Cr(CO)_5\{=C(OEt)C{\equiv}COEt\}]$ is not formed.[363]

$$[Mo(CO)_5]^{2-} \xrightarrow[\text{ii, } 2\,H^+]{\text{i, } Ph_2C{=}NMe_2^+} [Mo(CO)_5(=CPh_2)] \quad (30)$$

$$[M(CO)_5\{=C(OMe)C_6H_4R\}] \xrightarrow[\text{ii, } HBF_4{\bullet}OEt_2]{\text{i, } [BH(OPr^i)_3]^-} [M(CO)_5\{=C(H)C_6H_4R\}] \quad (31)$$

$$\{W(CO)_5(=CPh_2)] + CH_2{=}CPh(C_6H_4OMe\text{-}4) \longrightarrow [W(CO)_5\{=CPh(C_6H_4OMe\text{-}4) + Ph_2C{=}CH_2 \quad (32)$$

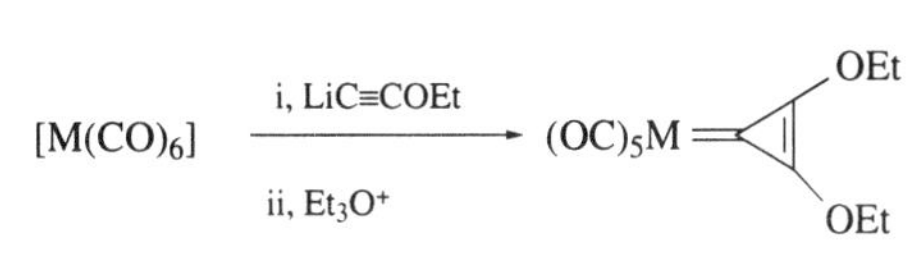

(33)

3.9.2 Reactions

Most reactions of $[M(CO)_5(=CR^1R^2)]$ reported involve $[W(CO)_5(=CPh_2)]$ or $[W(CO)_5(=CHPh)]$. Some representative examples of each are given in Schemes 2[364–73] and 3.[361,374–84]

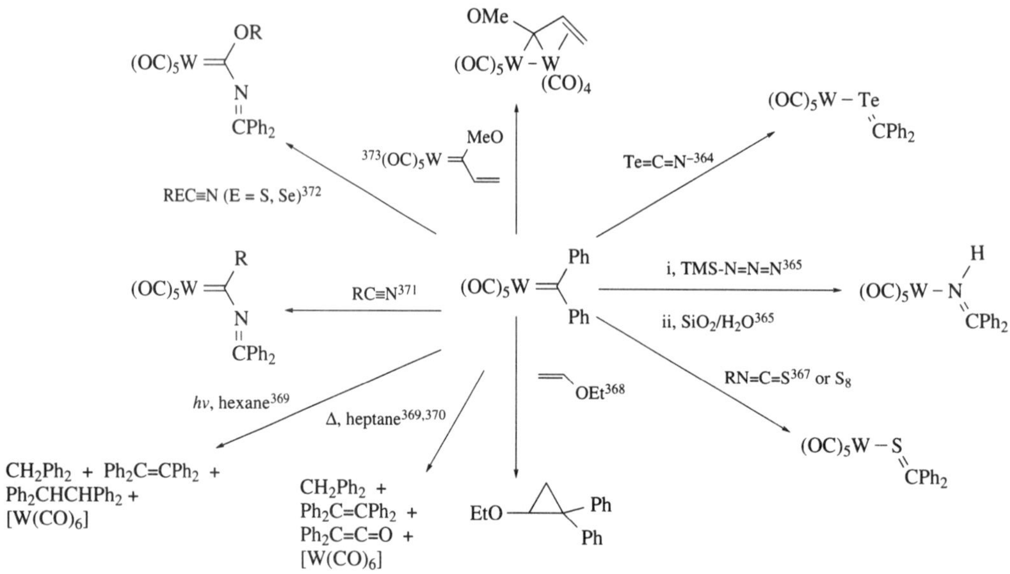

Scheme 2

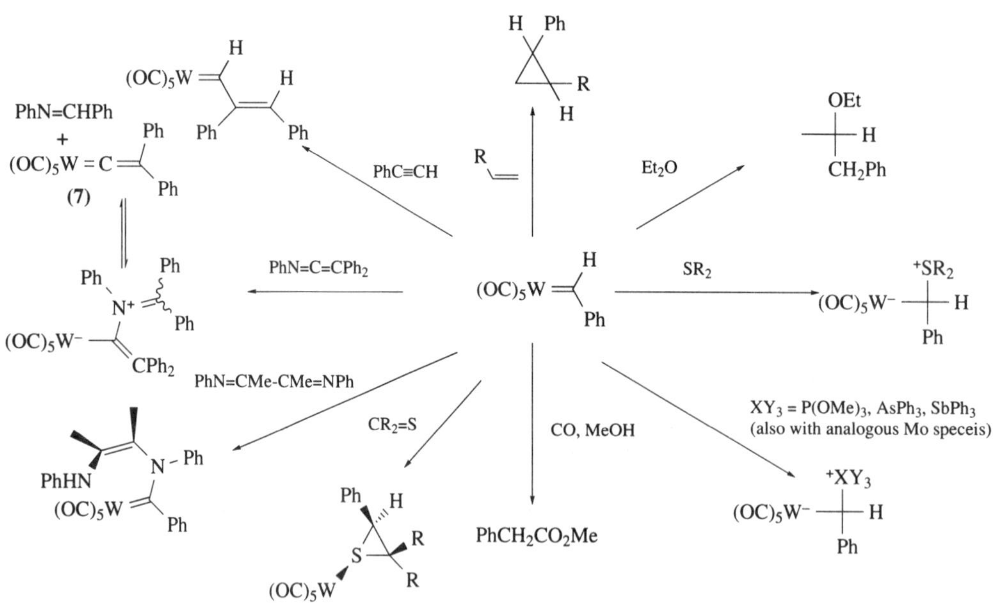

Scheme 3

Bridging alkylidenes have been reviewed.[344] There are relatively few examples of simple bridged M–M complexes for group 6 metals. However, some chemistry of the bridging alkylidene (**9**) is displayed in Scheme 4.[385–9] The reaction with alkynes is notable in that the structure of one of the products (**8**) contains the same bridging unit as the starting material and can therefore be regarded as a chain-carrier intermediate. Under more forcing conditions, polymers are formed.

The transiently produced cationic alkylidene $[W(=CHPh)(CO)_2Cp]^+$ readily incorporates alkynes. Thus, it reacts with PhC≡CPh by insertion into the W=C bond to form a vinyl carbene derivative (Equation (34)).[351] The anion $[W(=CHPh)(CO)_2Cp]^-$ reacts with acid to form the η^3-benzyl $[W(CO)_2(\eta^3\text{-}C_7H_8)Cp]$, probably via a hydride to alkylidene migration involving an intermediate hydride $[WH(CO)_2(=CHPh)Cp]$.[354] In a related process the stannyl complex $[Mo(SnPh_3)(CO)_2(=CHPh)Cp]$ exists in equilibrium with an η^3-benzyl complex (**10**) (Equation (35)).[353]

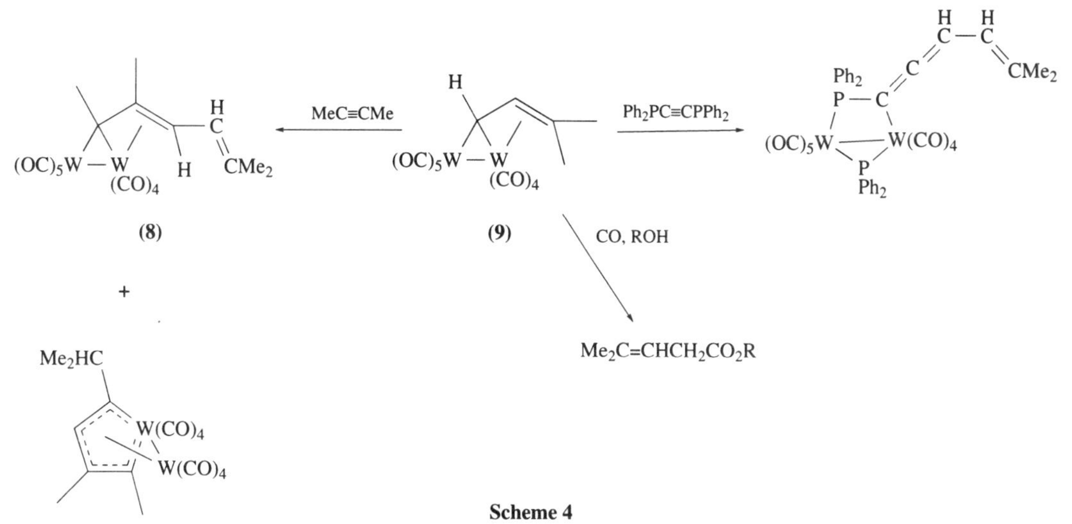

Scheme 4

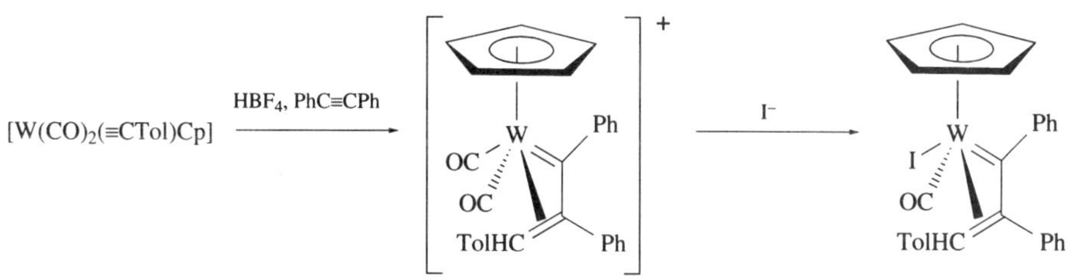 (34)

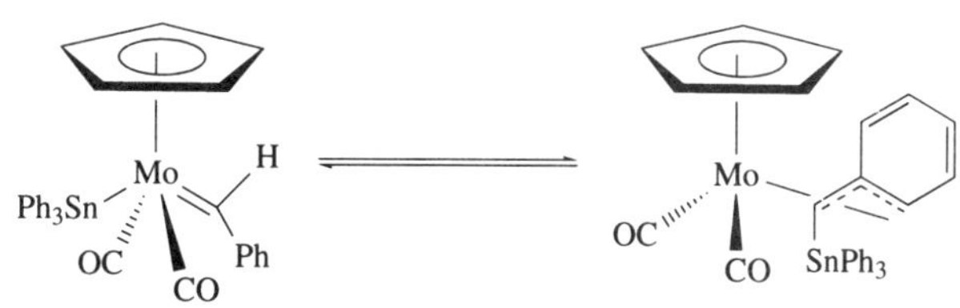 (35)

(10)

3.9.3 Vinylidenes and Related Species

Vinylidene complexes ($L_nM{=}C{=}CR^1R^2$) are comparatively unusual, but have been reviewed.[390–2] The pentacarbonyl (**7**) (Scheme 3) is formed in a metathesis reaction.[380] The dialkyl derivatives [W(CO)$_5$(=C=CBu^{t}R)] (R = Me or Et) are also known and are formed by β-additions to alkynide ligands.[324] Vinylidene complexes react with electron-rich alkynes to form cyclobutenylidene derivatives (Equation (36)). Oxidative cleavage of the cyclobutenylidene gives the corresponding cyclobutenones.[393] A variant of these pentacarbonyl structures is provided by the carbonyl-substituted complex *mer*-[W(CO)$_3$(=C=CHPh)(dppe)]. This species is formed by the water catalysed rearrangement of *fac*-[W(CO)$_3$(η^2-PhCCH)(dppe)].[394] The β-carbon atom is nucleophilic and so reacts with MeO$_2$CC≡CCO$_2$Me to form a vinylvinylidene (**11**) (Equation (37)).[395] Other routes to vinylidene complexes include the photochemical reaction between CR$_2$=CCl(TMS) and [Cr(CO)$_3$(η-C$_6$H$_6$)] (Equation (38)).[391,396]

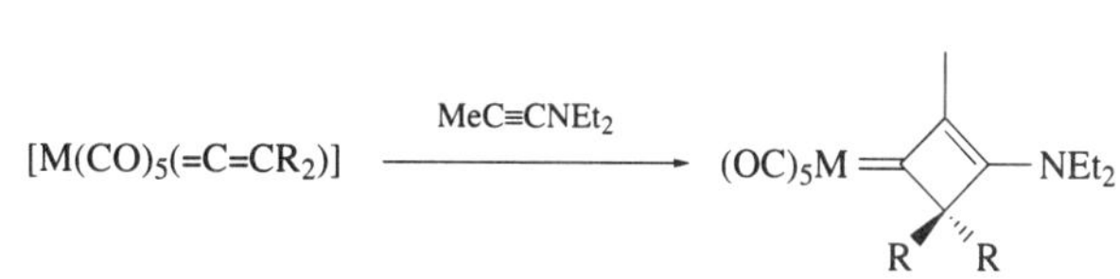 (36)

Allenylidene complexes are also known and formed initially from an alkynide species (Equation (39)). These interesting compounds add amines, imines and hydrazines. For instance, the addition of NEt$_2$H to [M(CO)$_5$(=C=CR$_2$)] (M = Cr or W) results in the formal addition of the amine across the

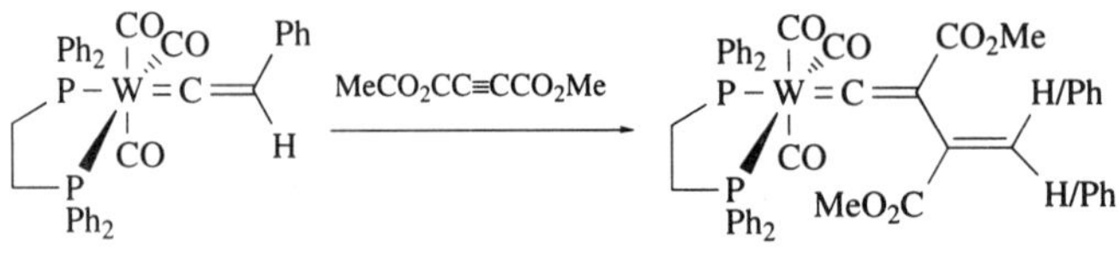

(11) (37)

$$[Cr(CO)_3(\eta^6\text{-}C_6H_6)] \xrightarrow{h\nu,\ CR_2=CCl\text{-}TMS} [Cr(CO)_2(=C=C=CR_2)(\eta^6\text{-}C_6H_6)] \quad (38)$$

α- and β-carbon atoms to form the carbenes $[M(CO)_5\{=C(NEt_2)\text{-}C=CR_2\}]$.[397] Heteroallenylidenes are rare. However, the azaallenylidene compounds $[M(CO)_5(=C=N=CR_2)]X$ are known (Scheme 5).[398–402]

$$[W(CO)_5(THF)] \xrightarrow[\text{ii, }COCl_2]{\text{i, }LiCCCR_2OLi} [W(CO)_5(=C=C=CR_2)] \quad (39)$$

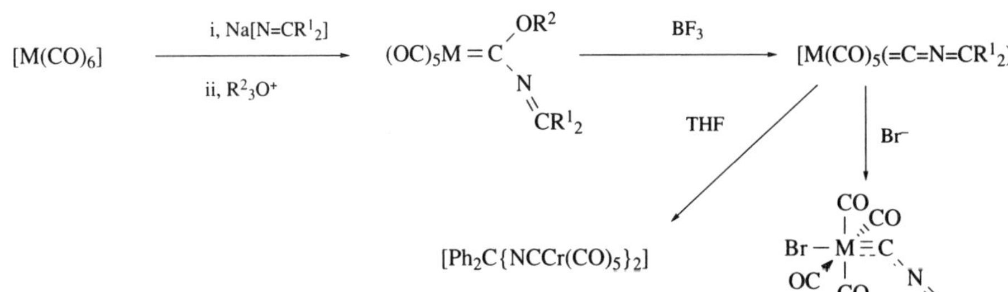

Scheme 5

The vinylidene ligand is capable of bridging the Mo–Mo bond and is formed, for instance, in the photochemical reaction of $[Mo(CO)_2Cp^*]_2$ with ethyne (Equation (40)).[403] A related species (**12**) is formed on alkylation of the species formed in the reaction of an alkynide with $[Mo(CO)_2Cp]_2$ (Equation (41)). Modification of this procedure to use $LiC{\equiv}C\text{-}CMe{=}CH_2$ followed by protonation (chromatography on Al_2O_3) results in the allenylidene complex $[Mo_2(CO)_4(\mu\text{-}\sigma,\eta^2\text{-}C=C=CMe_2)Cp_2]$. The tungsten derivative is also available.[404] Both of these compounds react with $HBF_4{\cdot}OEt_2$ by protonation at the β-carbon atom to form the cation $[M_2(CO)_4(\mu\text{-}HCCCMe_2)Cp_2]^+$. However, reaction of the molybdenum species with a hydride leads to attack at the γ-carbon atom position and consequent formation of the μ-alkynide anion $[Mo_2(CO)_4(\sigma,\mu\text{-}C{\equiv}CPr^i)Cp_2]^-$.[405]

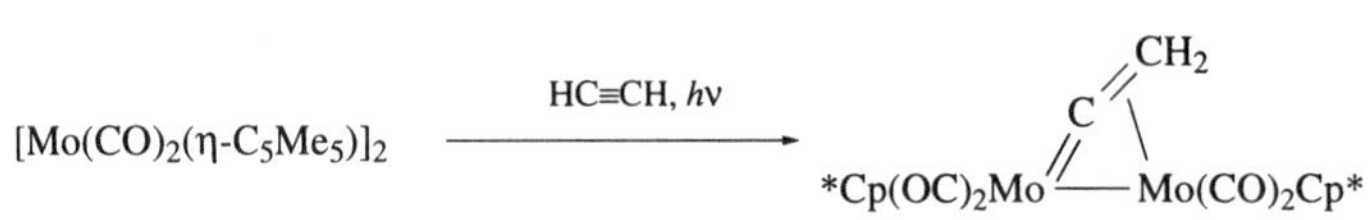

(40)

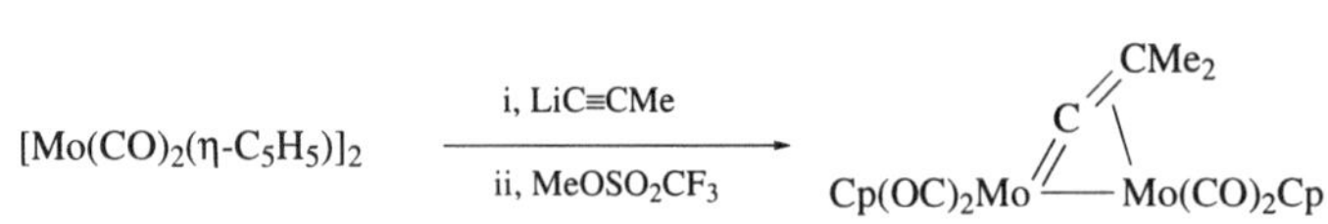

(12) (41)

3.10 OXACARBENE COMPLEXES

3.10.1 Syntheses

A number of new syntheses of oxacarbene complexes are given in Table 2, while new syntheses of thiocarbene complexes are given in Table 3. Syntheses of metal carbene complexes[406] and cluster complexes with heteroatom-stabilized carbene ligands have been reviewed.[407] There are many variations on the original synthesis of Fischer-type[408] carbene complexes involving the reaction of $[M(CO)_6]$ with an organolithium reagent followed by alkylation. The use of ^{13}C enriched metal carbonyls gives ^{13}C-labelled carbene complexes.[409] The use of alkyl halides to alkylate anionic acyls of the type $[M\{C(=O)R\}(CO)_5]^-$ (M = Cr, Mo or W) is generally unsuccessful. However, under phase transfer conditions in CH_2Cl_2/H_2O, or in just H_2O, in the presence of catalytic amounts of $Bu^n{}_4Br$, the reactions do work. This allows access to a variety of alkoxy carbenes.[410] The dianions $[M(CO)_5]^{2-}$ are useful carbene precursors (Section 3.3.2).

An interesting chemistry of metaloxy carbenes of the type $[M^1(CO)_5\{=CR^1OM^2L_n\}]$, examples of which are given in Table 2, is developing and has been well reviewed.[411,412] There are now many examples of carbene chelate complexes—complexes with heteroatoms or C=C bonds at the γ-position of the carbene substituents—and these have been reviewed. One reason for their importance is the tendency for the C=C bond to dissociate from the metal, leaving a vacant coordination site.[413] Their tendency to polymerize alkynes is enhanced by Lewis acids.[414]

3.10.2 Properties

The reader is referred to a review on the electronic structures of transition metal complexes.[406] Calculations on systems such as $[Cr(CO)_5\{=C(OMe)Me\}]$ or the model $[Cr(CO)_5\{=C(OH)H\}]$ suggest that Fischer-type carbenes are best regarded as singlet-type fragments with the π-cloud polarized towards the metal. Rotation about the M=C bond is essentially free. This is not a consequence of the amount of π-interaction but arises because there are two occupied *d* orbitals from which backbonding can occur with the carbene in any orientation. There is considerable multiple-bond character between the carbene atom and the oxygen atom. A localized MO approach aimed at a bonding description for $[Cr(CO)_5\{=C(OMe)Me\}]$ suggests that there is a highly localized C=O bond between the carbene and the methoxy group, and that the $Cr–C_{carbene}$ bond is essentially single.[491,492] Other calculations on $[M^1(CO)_5(=M^2H_2)]$ (M^2 = C, Si, Ge or Sn) suggest that here also rotation about the $M^1=M^2$ bond is essentially free, and also that the $M^1=M^2$ donor–acceptor interactions become less efficient for the heavier elements.[493] Calculations on the model system $[Mo(CO)_2(PH_3)(=CH_2)Cp]^+$ suggest that the preferred geometry is with the carbene lying so that it lies along the bisector of the $Mo(CO)_2$ plane.[494]

The use of ^{53}Cr NMR as a spectroscopic probe for carbene complexes is emerging. The ^{53}Cr NMR shift depends on the donor–acceptor properties of the carbene and there is also a dependence on the steric properties of the heterogroup substituents.[495]

The complexes $[M(CO)_5\{=C(OMe)Ph\}]$ may exist as *syn* or *anti* isomers related by the restricted rotation about the $C_{carbene}–O$ bond. Irradiation of $[M(CO)_5\{=C(OMe)Ph\}]$ (M = Cr or W) in inert gas matrices at 10 K results in complete conversion from the *anti* to the *syn* form. A similar rearrangement occurs on irradiation in the ligand field band at 350 nm, but here the reaction is also followed by the release of carbonyl as a secondary process. Notably, the resulting species is apparently still coordinatively saturated, apparently through an interaction of a $C_{methoxy}–H$ bond with the metal. In solvents such as MeCN the observed photochemical product is *cis*-$[M(CO)_4(NCMe)\{=C(OMe)Ph\}]$. Low-temperature photolysis in the presence of alkynes leads to labile complexes such as $[M(CO)_4\{=C(OMe)Ph\}(PhCCPh)]$.[496–9]

The carbenes $[M(CO)_5\{=C(OMe)Ph\}]$ (M = Cr, Mo or W), $[Cr(CO)_5\{=\overline{CCMe_2(CH_2)_2O}\}]$ and $[W(CO)_5(=CPh_2)]$ undergo exchange with ^{13}CO. All apart from the $=CPh_2$ complex undergo selective exchange into the *cis* position. Reaction rates are comparable with substitution reactions with phosphines. The five-coordinate intermediate is relatively rigid for M = Mo but not M = Cr.[500]

Phosphines are known to add to the carbene atom in the complexes $[M(CO)_5\{=C(OEt)R\}]$ (M = Cr, Mo or W; R = Me or Ph). The equilibrium constant and the reverse rate constant k_{-1} are affected markedly by steric influences. The forward rate constant measures intrinsic reactivity and is not so affected.[501]

3.10.3 Reactions

Oxacarbenes are extremely versatile. One of their principal uses is as reagents for organic chemical synthesis (see Chapters 5.1–5.5, Volume 12); here only reactions leading to new organometallic complexes are summarized. Benzannulation reactions of carbene complexes are particularly important. However, thermolysis of the biphenyl carbene (**13**) results in the loss of three CO groups and the formation of an arene complex. This contrasts with photolysis of the same material which affords benzannulation products (Equation (42)).[419]

The reaction of the carbene $[Cr(CO)_5\{=C(OMe)Me\}]$ with OH^- and piperidine results in reversible deprotonation of the methyl group to form the anion $[Cr(CO)_5\{=C(OMe)CH_2\}]^-$. The pK_a is 12.3 and the rate constants for the reaction are very low.[502] Other representative reactions of the anion are shown

Table 2 Examples of new oxacarbene complexes.

Synthesis	*Complex*	*Comments*	*Ref.*
$[M(CO)_6]$ + i, LiR^1; ii, $R^2{}_3O^+$ or R^2OTf	$[M(CO)_5\{=C(OR^2)R^1\}]$	M = Cr, Mo, W; R^2 = Me, Et; R^1 = alkyl, aryl, vinyl, alkynyl, cyclopropyl, pyrryl, furyl, thienyl, $SiHMes_2$, etc.	415–423
$[M(CO)_6]$ + $N{=}NC(OMe)Ph$, $h\nu$	$[M(CO)_5\{=C(OMe)Ph\}]$	M = Cr, W	424
$[W(CO)_6]$ + i, TMS-C≡CH, $h\nu$; ii, ROH, SiO_2	$[W(CO)_5\{=C(OR)Me\}]$	R = Me, allyl	425
$[W(CO)_6]$ + i, $H_2C{=}CH(CH_2)_2MgBr$; ii, R_3O^+	$[W(CO)_5\{=C(OR)CH_2CH_2CH{=}CH_2\}]$	R = Me, Et. Various substituents on the alkyl chain are possible	426–428
$[M(CO)_6]$ + i, $[Fe(\eta\text{-}C_5H_4Li)(\eta\text{-}C_5H_4PPh)]$; ii, Et_3O^+	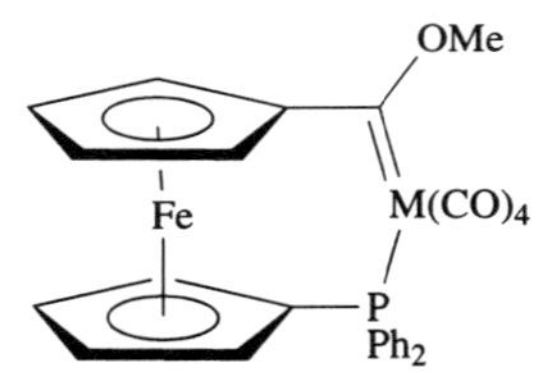	M = Cr, W	429
$[W(CO)_6]$ + i, $[Fe\{\eta\text{-}C_5H_3Li(CHMeNMe_2)\}Cp]$; ii, Et_3O^+	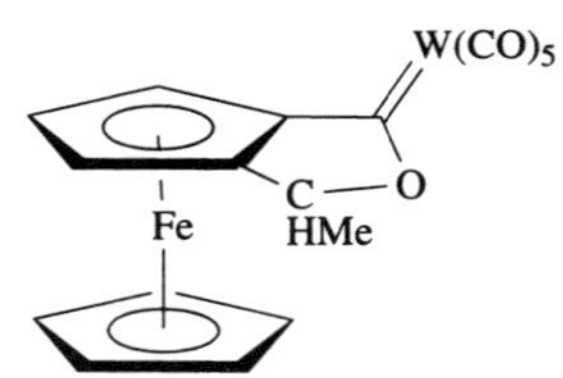	81:19 mixture of diastereomers	430
$[M(CO)_5L]$ + i, LiR^1; ii, $R^2{}_3O^+$	$[M(CO)_4L\{=C(OR^2)R^1\}]$	M = Cr, Mo, W; R^2 = Me, Et; R^1 = alkyl, aryl, etc; L = phosphirene, CNR	431, 432
$[Cr(COPh)(CO)_5]^-$ + NCMe, Et_4NClO_4, electrolysis, 0 °C, 0.40 V	$[Cr(CO)_5\{=C(OH)Ph\}]$	X-ray on NCMe solvate	433
$[M(CO)_6]$ + i, EtOC≡CLi; ii, Et_3O^+	$[M(CO)_5\{=CC(OEt){=}C(OEt)\}]$	M = Cr, Mo, W. X-ray for M = Cr	363
$[Mo(CO)_6]$ + i, $Li(CH_2)_nC{\equiv}CR$; ii, $MeOSO_2CF_3$	$[Mo(CO)_5\{=C(OMe)(CH_2)_nC{\equiv}CR\}]$	n = 2–4; R = Me, Pr	434
$[Cr(CO)_4(nbd)]$ + i, LiAr; ii, Et_3O^+	$[Cr(CO)_3\{=C(OEt)Ar\}(nbd)]$		435
$[M(CO)_3(\eta^6\text{-}C_7H_8)]$ + i, LiAr; ii, Et_3O^+	$[M(CO)_2\{=C(OEt)Ar\}(\eta^6\text{-}C_7H_8)]$	M = Cr, W	436
$[Cr(CO)_6]$ + 2 $PhC(=CH_2)O\cdots K(18\text{-crown-6})$	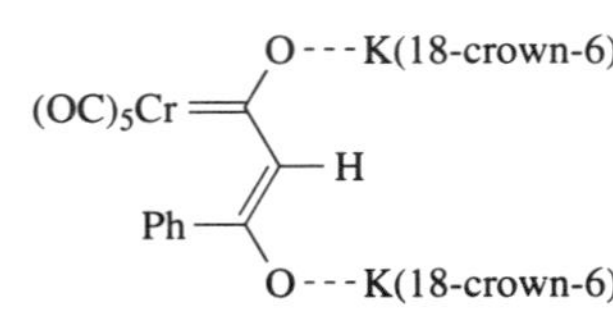	90% yield	437

$[M(CO)_6]$ + i, $C_6H_4(CH_2MgCl)_2$-2; ii, Et_3O^+	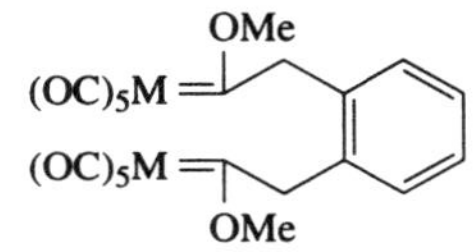	M = Cr, W	438
$[M(CO)_6]$ + i, $C_6H_4Li_2$-2; ii, Et_3O^+	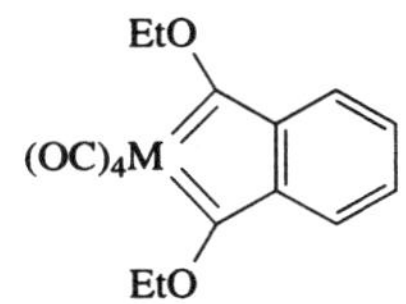	M = Cr, Mo, W	439–441
$[M(CO)_6]$ + i, PhLiNNLiPh; ii, Et_3O^+	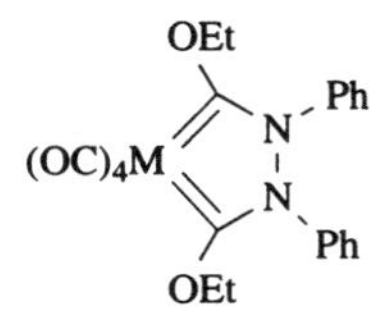	M = Cr, W	442
$[M(CO)_6]$ + i, $LiC_6H_4–C_6H_4Li$; ii, Et_3O^+	$[OC)_5W\{=C(OEt)C_6H_4–C_6H_4(EtO)C=\}W(CO)_5]$	M = Cr, W. Centrosymmetric molecules with virtually no delocalization between the aryl rings of the biphenyl group	443
$[M(CO)_6]$ + i, LiMe; ii, $[Re(FBF_3)(CO)_5]$	$[M(CO)_5\{=C[ORe(CO)_5]Me\}]$	M = Cr, W	444
$[M\{C(O)R\}(CO)_4L]^-$ + $[Re(FBF_3)(CO)_5]^+$	$[M(CO)_4L(=C\{ORe(CO)_5\}R)]$	M = Cr, Mo, W; R = Ph, Fc, Me	445
$[W(CO)_6]$ + $[U\{=CH(PMePhR)\}Cp_3]$	$[W(CO)_5\{=C(OUCp_3)CH(PMePhR)\}]$	Best regarded as an enolate	446
$[Zr(\eta^4\text{-}C_4H_6)Cp_2]$ + $[M(CO)_6]$	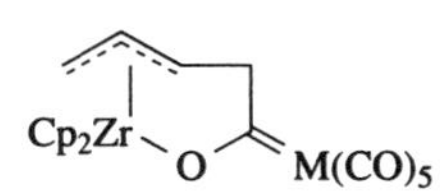	M = Cr, Mo, W. Also works with substituted butadiene complexes	447–9
$[Zr(\eta^4\text{-}C_4H_6)Cp_2]$ + i, $[W(CO)_6]$; ii, $R^1R^2C=O$	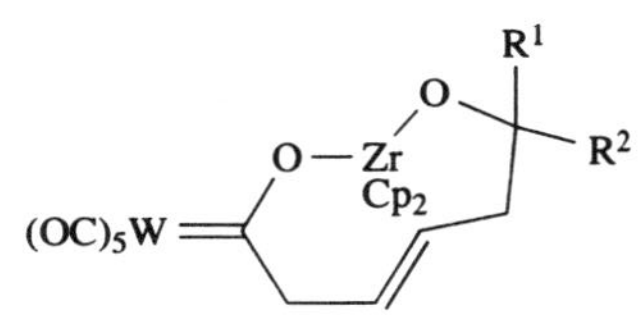	R^1 = Me; R^2 = Me, Bu^t	448, 450

Table 2 (continued)

Synthesis	*Complex*	*Comments*	*Ref.*
$[Ti(C_2H_4)Cp^*_2] + [M(CO)_6]$	$[M(CO)_5\{=\overline{COTi(Cp^*_2)CH_2CH_2}\}]$		451
$[ZrVi_2Cp^*_2] + [M(CO)_6]$	$[M(CO)_5\{=\overline{COZr(Cp^*_2)CHCH}\}]$	M = Cr, Mo, W	452
$ZrPh_2Cp_2 + [W(CO)_6]$, Δ	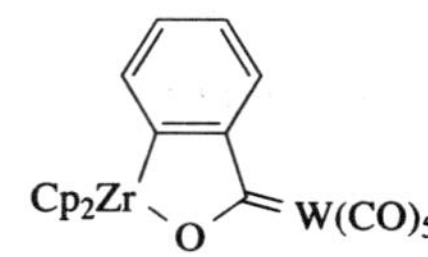	Via benzyne intermediate	453
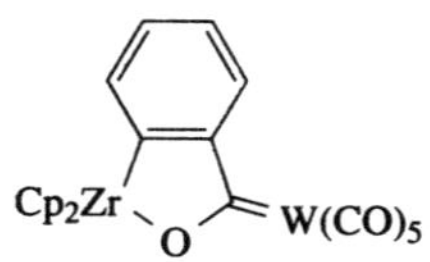 $+ H_2O$	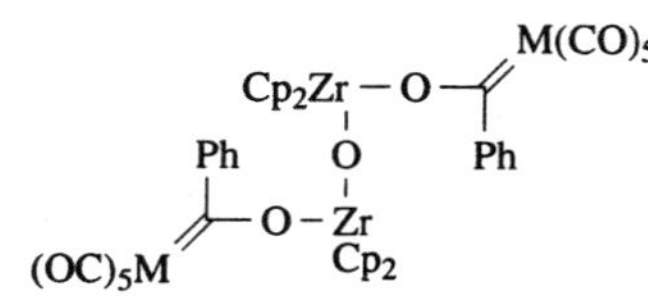	M = Mo, W	454
$Cp^*_2TiCl_2 + 2[Cr(COMe)(CO)_5]^-$	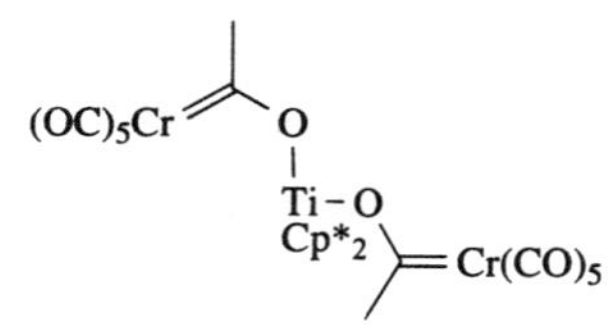	Methyl group wedged between Cp* groups	455
$[Cr(CO)_6]$ + i, LiR; ii, NMe_4^+; iii, [MX]	$[Cr(CO)_5\{=C(OM)R\}]$	MX = $[TiBr(OPr^i_3)]$; R = Bu^n, 2-furyl, $CMe{=}CH_2$. MX = $[TiClCp_2]$; R = $CMe{=}CH_2$	456
$[W(CO)_5\{=C(OEt)C{\equiv}CPh\}] + MeOH$	$W(CO)_5\{=C(OEt)CH{=}C(OMe)Ph\}]$	Also works with thiols	457
$[M(CO)_5\{=C(OR^2)C{\equiv}CR^3\}] + R^1OH$	$[M(CO)_5\{=C(OR^2)CH{=}CR^3(OR^1)\}]$	M = Cr, W	458
$[M(CO)_5\{=C(OMe)Ar\}] + CH_2{=}CH(CH_2)_nOH$	$[M(CO)_5\{=C[O(CH_2)_nCH{=}CH_2]Ar\}]$	M = Cr, W; n = 1, 2; Ar = Ph, Tol	459
$[W(CO)_5\{=C(OR)(CH_2)_nCH{=}CH_2\}]$ + heat	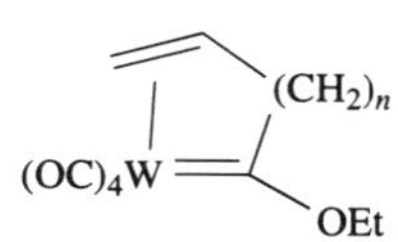	For $n = 3$, the x-ray structure shows the alkene and carbene are parallel; decomposition of this species affords the cyclopropane. For $n = 2$, the alkene and carbene are perpendicular; no cyclopropane is formed on decomposition or heating.	426–8, 460

$[W(CO)_5\{=C(NHCH_2CH{=}CH_2)Tol\}] + h\nu$	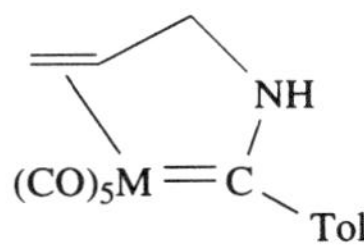		461
$[M(CO)_5\{=C(OMe)C_6H_4OMe\text{-}2\}]$ + heat	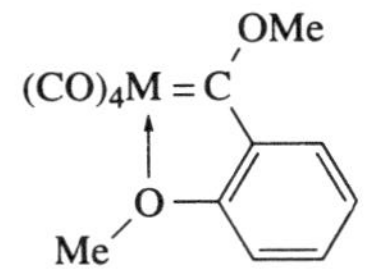	M = Cr, Mo, W	415, 462
$[W(CO)_5\{=C(OMe)C{\equiv}CPh\}] + 2CH_2N_2$	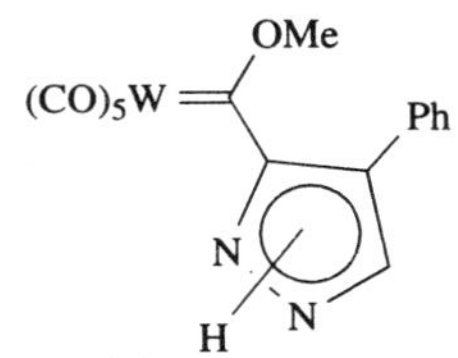		463
$[M(CO)_6]$ + i, $2Me_3P{=}CH_2$; ii, $(TMS)OSO_2CF_3$	$[M(CO)_5\{=C(O\text{-}TMS)CH{=}PMe_3\}]$	M = Cr, Mo, W; thermolabile. ^{13}C NMR and IR data suggest delocalization to give a metal vinyl	464
$[M\{=C(OR)C{\equiv}CPh\}(CO)_5] + \{Co_2(CO)_8\}$	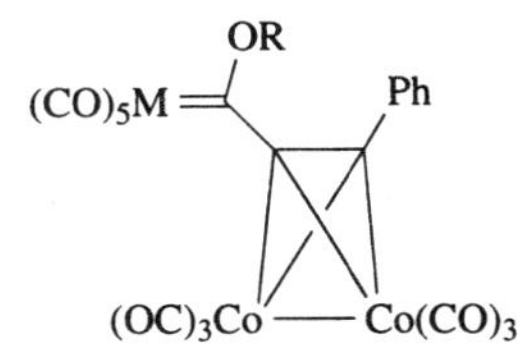	M = Cr, W; R = Me, Et	465
$[Cr(CO)_5\{=C(OR^2)C{\equiv}CR^1\}] + \overline{O(CH_2)_3CH{=}CH}$	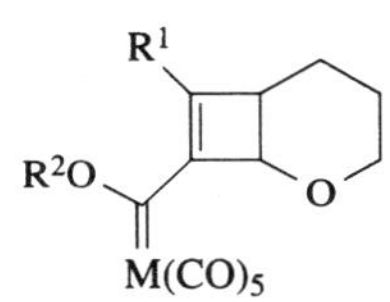	Cycloaddition is pressure sensitive	466

Table 2 (continued)

Synthesis	Complex	Comments	Ref.
$[Cr(CO)_5(\equiv CNEt_2)]^+ + [M(COR)(CO)_5]^-$	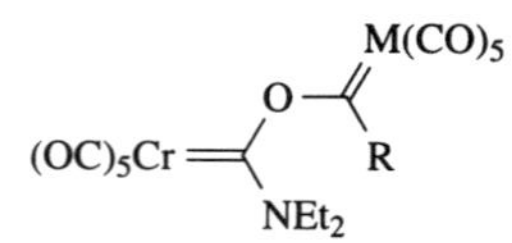	M = Cr; R = Tol. M = W; R = Me	467
$[W(COMe)(CO)_5]^- + BClBu^n{}_2$, Et_2O	$[W(CO)_5\{=C(OBBu^n{}_2)Me]\bullet OEt_2$		468
$[M(CO)_6] + HC\equiv C(CH_2)_nOH$, $h\nu$	$[M(CO)_5\{\overline{=C(CH_2)_{n+2}O}\}]$	M = Cr, W	469
$[W(CO)_6]$ + i, $LiC\equiv CCMe_2OLi$; ii, HCl	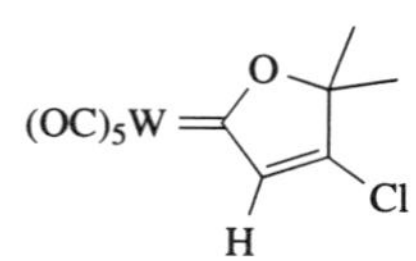		470
$[Cr(CO)_6]$ + i, $LiC\equiv CCMe_2OLi$; ii, HCl	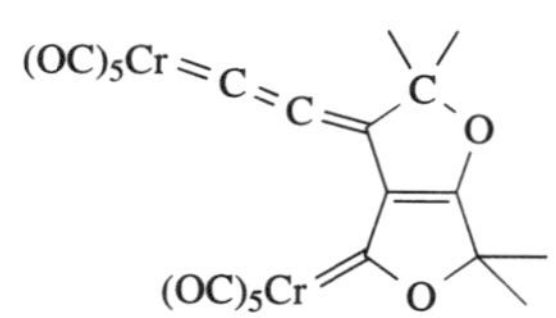		470

$[Cr(CO)_5\{=C(OMe)Me\}]$ + i, BuLi; ii, $BF_3 \cdot OEt_2$; iii, $\overline{CR^1R^2CR^3R^4O}$	$[Cr(CO)\{=\overline{CCH_2CR^1R^2CR^3R^4O}\}]$	Various (including optically pure) epoxides	471
$[M(CO)_3Cp]^-$ + $I(CH_2)_3I$	*trans*-$[MI(CO)_2\{=\overline{CO(CH_2)_2CH_2}\}Cp]$	M = Mo, W. Also Cp* systems	472, 473
$[W(CO)_3Cp]^-$ + $[Mo\{(CH_2)_3Br\}(CO)_3Cp]$	*trans*-$[W\{Mo(CO)_3Cp\}(CO)_2\{=\overline{CO(CH_2)_2CH_2}\}]$	Carbene shift from Mo to W	474
$[Mo_2(CO)_6(\eta^5:\eta^5\text{-fulvalene})]^{2-}$ + $I(CH_2)_3I$	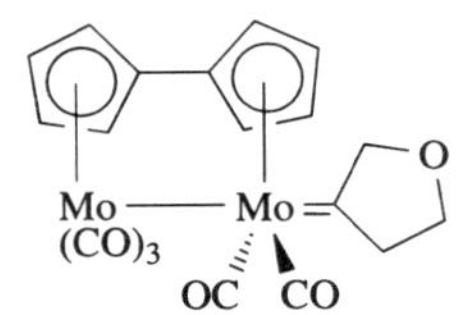		475
$[M(SnPh_3)(CO)_3L$ + i, LiPh; ii, Et_3O^+	*trans*-$[M(SnPh_3)(CO)_2\{=C(OEt)Ph\}L]$	L = Cp, Cp*, indenyl	476
$[Mo(CHO)(CO)_2LCp^*]$ + $MeSO_3F$	$[Mo(CO)_2\{=C(OMe)H\}LCp^*]^+$	L = PPh_3, PPh_2Me	477
$[Mo(CHO)(CO)_2(PPh_3)Cp^*]$ + H^+	$[Mo\{=C(OH)H\}(CO)_2(PPh_3)Cp^*]^+$	Similar reaction with $MeSO_3F$	477, 478
$[MR(CO)_3Cp]$ + i, PMe_3; ii, $MeOSO_2F$	*trans*-$[Mo(CO)_2(PMe_3)\{=C(OMe)Me\}Cp]^+$	M = Cr, Mo, W; R = Me, Et. Acyl intermediate	479
$[Mo(CF_3)(CO)_3Cp]$ + $h\nu$, matrix	*trans*-$[MoF(=CF_2)(CO)_2Cp]$	Not isolated	480

Table 3 Examples of new thiacarbene complexes.

Synthesis	*Complex*	*Comments*	*Ref.*
$[W(CO)_5(CS)]$ + i, SR^{1-}; ii, R^2I	$[W(CO)_5\{=C(SR^1)(SR^2)\}]$		481
$[M(CO)_6]$ + i, LiPh; ii, NEt_4^+; iii, MeCOCl; iv, RXH or i, LiAr; ii, TMEDA, MeCOCl; iii, RSH	$[M(CO)_5\{=C(XR)Ph\}]$	M = Cr, W; X = S, Se; R = alkyl, aryl	482, 483
$[W(CO)_5\{=C(OEt)Ph\}]$ + i, PhC≡CSLi; ii, Se_8, iii, SiO_2	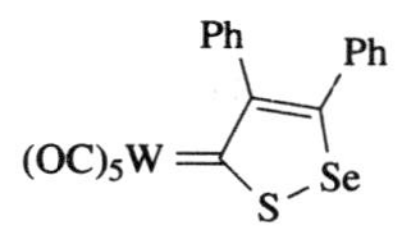		484
$[W(CO)_5\{=C(OEt)Ph\}]$ + i, PhC≡CSLi; ii, MeI	$[W(CO)_5\{=C(SMe)CPh{=}CPh(OEt)\}]$		485
$[W(CO)_5\{=C(OEt)Ph\}]$ + i, $LiCHS(CH_2)_3S$; ii, SiO_2	$[W(CO)_5\{=C(Ph)S(CH_2)_3SCH{=}C(OEt)Ph\}]$		486, 487
$[Cr(CO)_5(\equiv CNEt_2)]BF_4$ + LiSeAr	$[Cr(CO)_5\{(=NEt_2)SeAr\}]$		488
$[W(CO)_3(CS_2)(L_2)]$ + RC≡CR	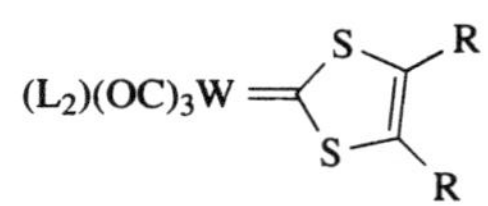	L_2 = chelating phosphine or $(PMe_3)_2$	489
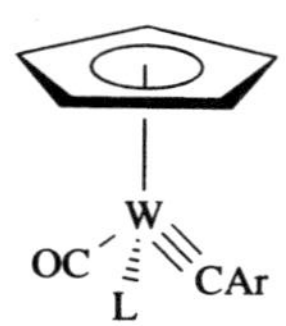 + SMe_3^+	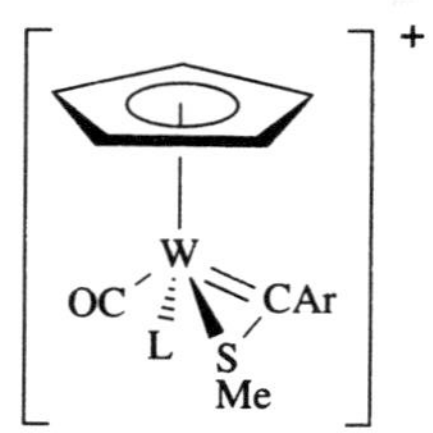	Ar = Ph, Tol; L = CO, PMe_3	490

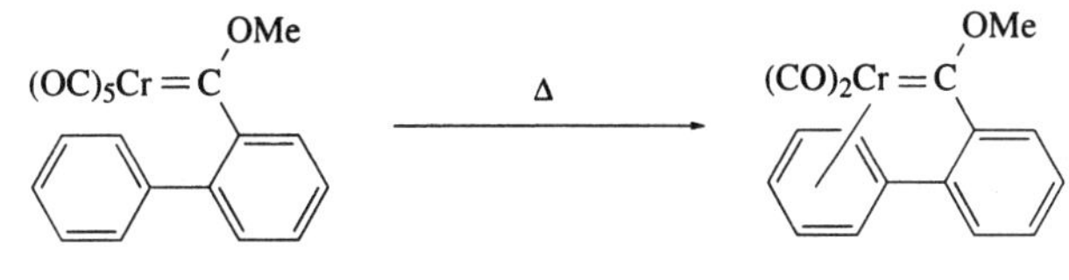

(42)

(13)

in Scheme 6.[503–6] An alternative way to modify the carbene functionality at the α position is shown in Equation (43). This reaction is quite tolerant to functionality (XR = OMe or SPh; R^1 = H) and involves the use of DMF/dimethyl acetal (DMA). When R^1 is an alkyl group, the products are $[Cr(CO)_5\{=C(OMe)(OR^1)\}]$.[507]

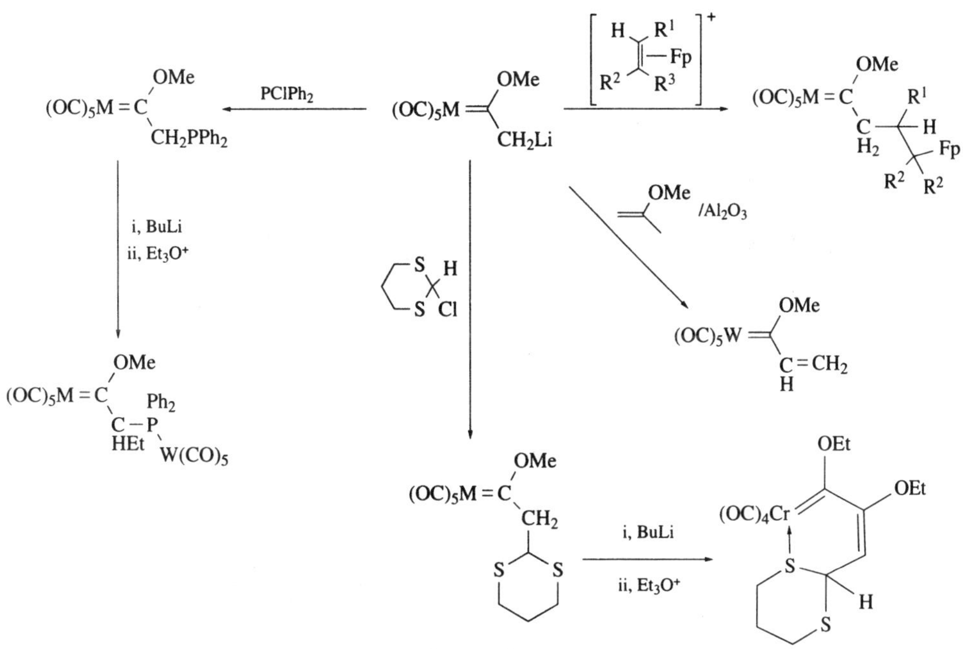

Scheme 6

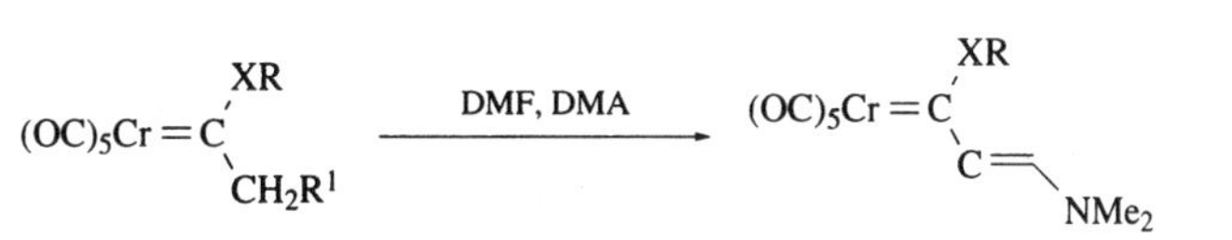

(43)

Carbene complexes $[M(CO)_5\{=C(OR)Ph\}]$ react with isocyanides by an insertion process to form ketenimine complexes (M = Cr or W; R = allyl, Cy or vinyl) (Equation (44)).[508–10] Various cyanamides and nitriles also undergo related insertion reactions, in these cases to form new carbene complexes (M = Cr or W; Equation (45)[511,512]) or nitrile complexes[511,513] (Equation (46)).[511] Insertion reactions with alkynes such as $MeC\equiv CNEt_2$ afford vinyl derivatives $[M(CO)_5\{=C(NEt_2)CMe=C(OMe)Ph\}]$. These reactions are accelerated by pressure.[514] The products react further with alkynes to form pyridines and pyridinium salts via the highly regioselective formation of pyridinylidene complexes.[515] A related reaction is seen in the reaction of the phosphinocarbene chelate complex with $MeC\equiv CNEt_2$ (Equation (47)).[516] Allenes are methylenated by carbene complexes in reactions that proceed initially by a cycloaddition to form a metallacyclobutane (M = Cr, Mo or W; Equation (48)). The resulting trimethylenemethane complex shown is accompanied by lesser amounts of a second isomer.[517]

$$[M(CO)_5\{=C(OR)Ph\}] + CNR \longrightarrow [M(CO)_5\{NR=C=C(OEt)Ph\}] \qquad (44)$$

$$[M(CO)_5\{=C(OR)Ar\}] + NCNMe_2 \longrightarrow [M(CO)_5\{=C(NMe_2)N=C(OR)Ar\}] \qquad (45)$$

$$[M(CO)_5\{=C(OMe)Ph\}] + [(Ph_3P)_2N][N=CBH_3] \longrightarrow [[(PH_3P)_2N][M(CO)_5\{NCC(OMe)(Ph)BH_3\}] \qquad (46)$$

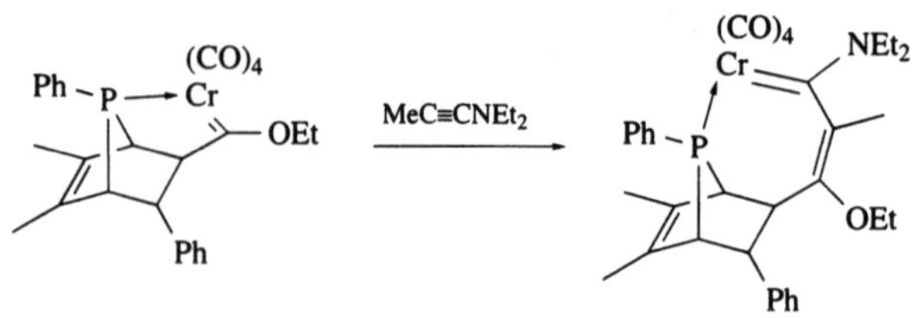

(47)

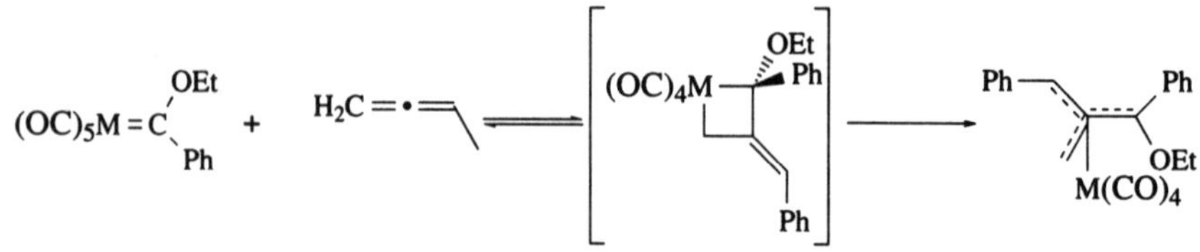

(48)

Some reactions of alkenyl carbenes are given in Scheme 7.[416,518–24] The sequence in Equation (49) is interesting. It involves conjugate addition of $EtMe_2CONa$ to the vinyl carbene, followed by a chain extension involving addition of further carbenes to the resulting anion. The products for $n = 2$ and 3 are isolable (for $n = 3$, in 15% yield).[525]

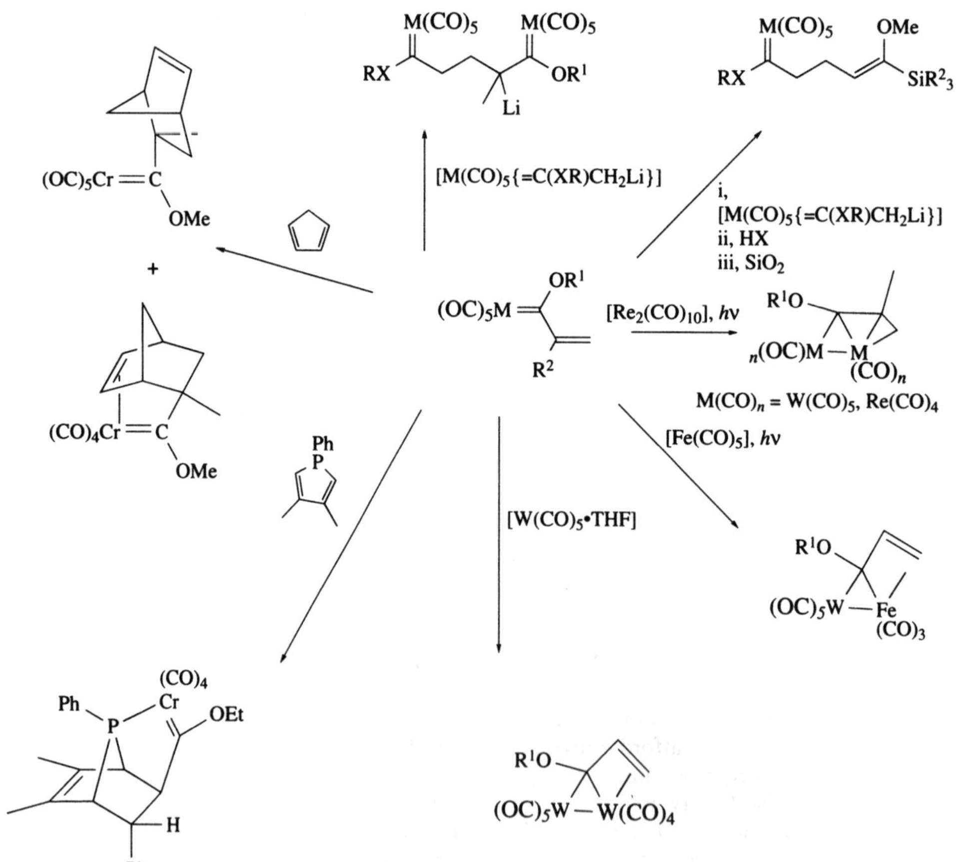

Scheme 7

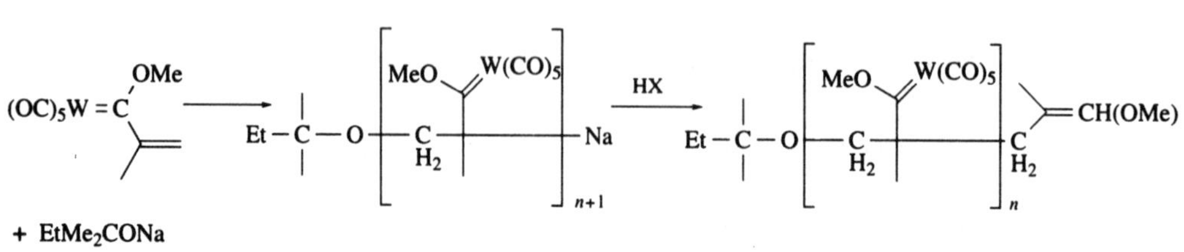

(49)

A few reactions of alkynyl carbenes $[M(CO)_5\{=C(OR)C{\equiv}CAr\}]$ are depicted in Scheme 8.[420,463,457,458,465,466,526–8] Reaction conditions are often considerably milder than if the metal carbene unit is absent. Typically, the alkynyl substituent adds reagents across the triple bond. In the reaction with alkynes, the net effect can be insertion of the alkyne into the metal–carbene bond. The reaction of $C(OEt)_2{=}CH(CO_2Et)$ with $[M(CO)_5\{=C(OR)C{\equiv}CPh\}]$ starts with a [2 + 2] cycloaddition in much the same way as the reaction of $Me_2C{=}C(OMe)(O\text{-}TMS)$, but eventually leads to pyranylidene complexes (Equation (50)).[529,530] Unusual products are derived from the reactions of carbohydrates (acting as alcohols ROH) with alkynyl complexes in reactions that parallel those with MeOH (Scheme 8).[531]

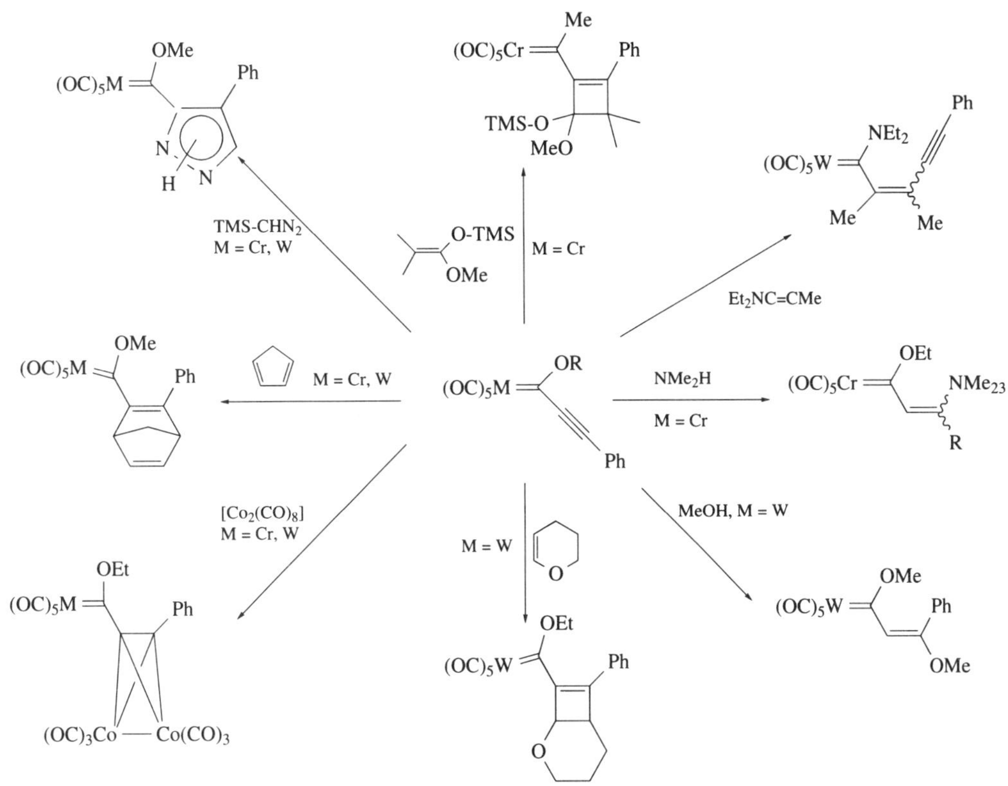

Scheme 8

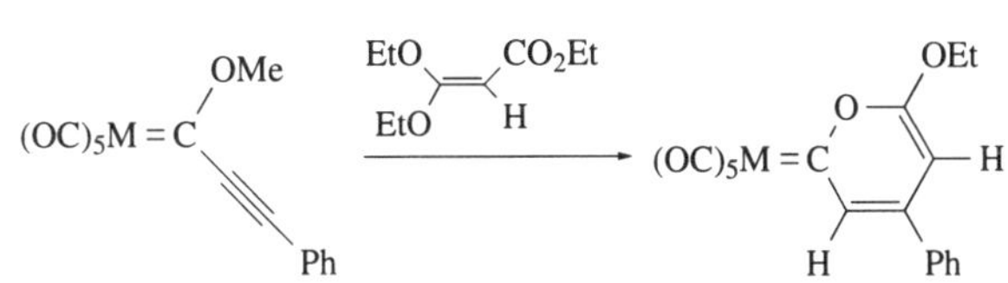

(50)

Bis(carbene) complexes are relatively unusual. In many ways their chemistry parallels that of monocarbenes. One of the carbene groups in complex (**14**) is susceptible to nucleophilic attack by PMe_3 (Scheme 9).[532] Activated alkynes insert into one of the metal–carbene bonds to form complex (**15**) (Scheme 9).[533] Thermolysis of complex (**14**) affords the diene (**16**), together with the its $Cr(CO)_3$ complex (**17**) and the corresponding complex containing two $Cr(CO)_3$ units (Scheme 9).[534]

Addition of $LiBEt_3H$ to the iodometal carbene complexes of the type $[MI(CO)_2\{=C(OR^1)R^2\}Cp]$ results in attack of hydride at the carbene, loss of I^- and oxygen coordination (Equation (51)).[535,536] However, reduction ($NaC_{10}H_8$) results in the formation of the anions $[M(CO)_2\{=C(OR^1)R^2\}Cp]^-$ (M = Mo or W).[537]

A few reactions of aryl thiocarbene complexes are depicted in Scheme 10.[483,538,539] Upon warming, the seleno derivative $[Cr(CO)_5\{=C(NEt_2)SeAr\}]$ decarbonylates and rearranges to a carbyne complex (Equation (52)).[488]

The dithiocarbene complex $[W(CO)_5\{=C(SMe)_2\}]$ reacts with amines RNH_2 to form isocyanide complexes $[W(CO)_5(C{\equiv}NR)]$ with MeSH as a by-product. Mechanistic studies show that the reactions proceed by a two-term rate law, with first- and second-order terms in amine concentration.[540] However,

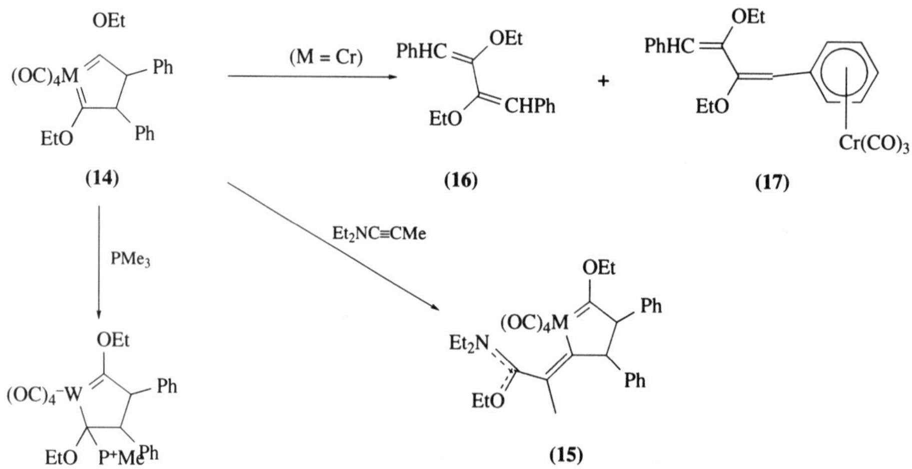

Scheme 9

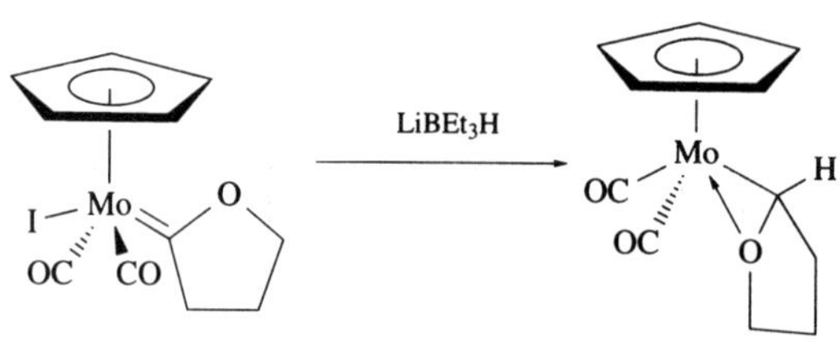 (51)

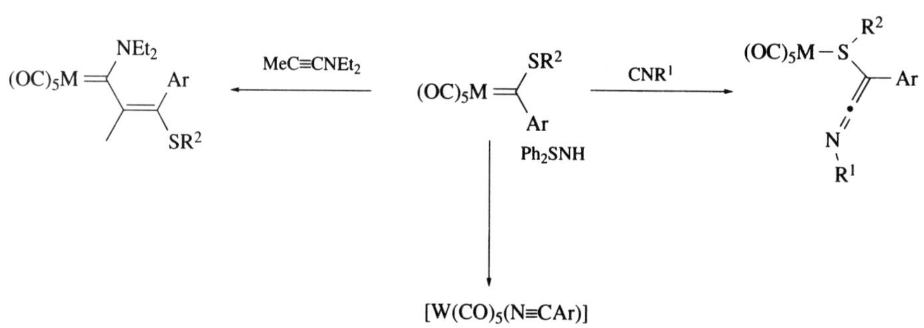

Scheme 10

$$[Cr(CO)_5\{=C(NEt_2)SeAr\}] \xrightarrow{\Delta} [Cr(SeAr)(CO)_4\{\equiv CNEt_2)\}] \quad (52)$$

secondary amines such as NMe_2H afford aminothiocarbene complexes (Equation (53)).[481] Related reactions with phosphines PR_3 also proceed by nucleophilic attack; in this case the resulting species $[W(CO)_5\{C(SMe)_2PR_3\}]$ rearranges to form a phosphorane species (Equation (54)).[339,541]

$$[W(CO)_5\{=C(SMe)_2\}] + Me_2NH \longrightarrow [W(CO)_5\{C(SMe)(NMe_2)\}] \quad (53)$$

$$[W(CO)_5\{=C(SMe)_2\}] + PR_3 \longrightarrow [W(CO)_5\{S(Me){=}PR_3\}] \quad (54)$$

Some thiocarbenes are known in which the sulfur is metal coordinated; one example is $[\overline{W\{\eta^2\text{-CHS}}Me\}(CO)_2Tp]$. This reacts with base to form the thiocarbyne $[W(CO)_2(\equiv CSMe)Tp]$, but with some nucleophiles (such as SMe^-) it reacts by addition at the carbene atom (Equation (55)).[542–4] Other cyclopentadienyl carbene complexes undergo substitution reactions on treatment with phosphines. Thus, addition of PMe_3 to $[Mo(CO)(NO)\{=C(OEt)Me\}Cp]$ results in the new carbene $[Mo(NO)(PMe_3)\{=C(OEt)Me\}Cp]$. Addition of $[MeSSMe_2]BF_4$ to this compound gives the cationic "molybdathiacyclopropane" complex $[Mo(NO)(PMe_3)\{\eta^2\text{-C(OEt)Me–SMe}\}Cp]^+$.[545] Sequential addition of NEt_3, PR_3 and HBF_4 to the η^2-thiocarbene $[\overline{W\{\eta^2\text{-CMeSMe}}\}(CO)_2Cp]$ results in the η^1-thiocarbene cation $[W(CO)_2(PMe_3)\{=C(SMe)Me\}Cp]^+$ (Equation (56)).[546] Addition of azide to the related η^2-thiocarbene $[\overline{W\{\eta^2\text{-CPhS}}Me\}(CO)_2Cp]$ results in the formation of a bicyclic system (Equation (57)).[547]

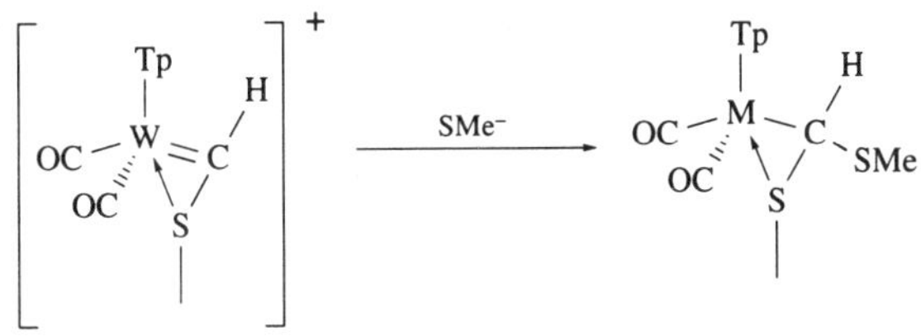

(55)

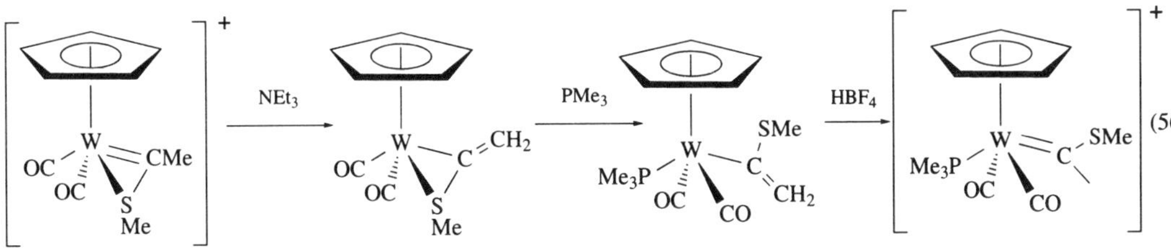

(56)

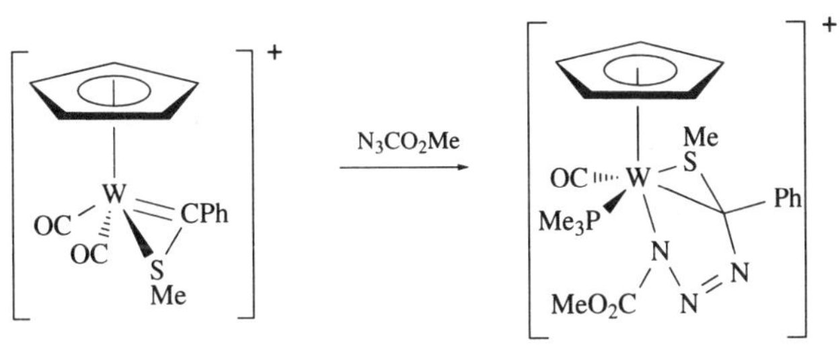

(57)

3.11 AZACARBENE COMPLEXES

The synthesis, physical properties, and spectroscopic properties of azacarbene complexes have been reviewed.[406,548,549] By far the most common method of synthesis of aminocarbenes is through aminolysis of an appropriate alkoxy carbene complex. Several examples are listed in Table 4. However, complexes of alkoxy carbenes $L_nM\{=C(OR)H\}$ are uncommon, and so do not provide good precursor molecules to complexes $L_nM\{=C(NR_2)H\}$. The complex $[Cr(CO)_5\{=C(NPh_2)H\}]$ is known[550] (Table 4) and is modifiable to other amino carbenes (Equation (58)).

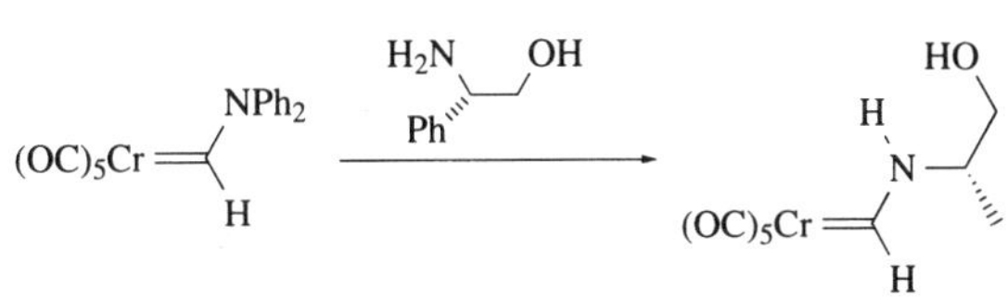

(58)

3.11.1 Properties

One interesting class of carbene is that in which a pendant group is coordinated *cis* to the carbene. The coordinating group is often an alkene and in such cases (for examples see Table 4) the alkene and the carbene are more or less perpendicular to each other.[413,417,428,461,551–4] These complexes are particularly useful for their reactivity towards alkynes (see Chapter 5.3, Volume 12).

The bis(carbene) complexes $[Mo(CO)_4\{=\overline{CN(R)(CH_2)_2N}R\}]$ (R = Me or Et) are derived from electron-rich alkenes (reviewed in Ref. 555) and exist as *cis* and *trans* isomers which interconvert so that the activation energy for the *trans* → *cis* reaction when R = Me is 70.4 ± 2 kJ mol^{-1} (slightly larger for R = Et). The Mo=C lengths are longer in the *cis* isomer.[556]

3.11.2 Reactions

Chromium aminocarbene complexes are prone to reaction with alkynes to form various chromium arene derivatives from which the arene may be liberated. The reader is referred to Volume 12 for further

Table 4 Examples of new azacarbene complexes.

Synthesis	*Complex*	*Comments*	*Ref.*
$K_2[Cr(CO)_5]$ + i, $R^1C(=O)NR^2_2$; ii, TMS-Cl	$[Cr(CO)_5\{=C(NR^2_2)R^1\}]$	$[Cr(CO)_5]^{2-}$ formed by $[Cr(CO)_6] + C_8K$	558, 559
$Na_2[Cr(CO)_5] + [R_2N{=}CHCl]Cl$	$[Cr(CO)_5\{=C(NR^2_2)H\}]$		495, 550, 560
$[Cr(CO)_6]$ + i, $LiNR^1R^2$; ii, SF_4	$[Cr(CO)_5\{=C(NR^1R^2)F\}]$	Carbenes with F substituents are rare. R^1, R^2 = alkyl	561
$[M(CO)_5\{=C(OMe)CH{=}CHMe\}] + MeNH_2$	$[M(CO)_5\{=(NHMe)CH{=}CHMe\}]$	M = Cr, W	562
$[Cr(CO)_5\{=C(OMe)Me\}] + H_2NCH_2CH{=}CH_2$	$[Cr(CO)_5\{=C(NHCH_2CH{=}CH_2)Me\}]$	99% yield	551
$[W(CO)_5\{=C(OMe)R\}] + H_2NCH_2CH{=}CH_2$	$[W(CO)_5\{=C(NHCH_2CH{=}CH_2)R\}]$	R = Me, Ph	552
$[W(CO)_5\{=C(OEt)Me\}] + H_2NCH_2CH{=}CH_2$	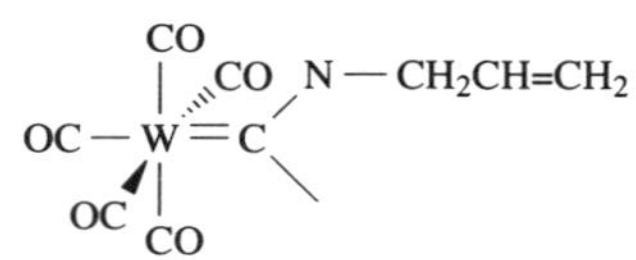	Two isomers ($Z + E$), both crystallographically characterized	553
$[Cr(CO)_5\{=C(OMe)C_6H_4OMe\text{-}2\}]$ + $HCl{\bullet}H_2N(CH_2)_2C{\equiv}CR$, Et_3N	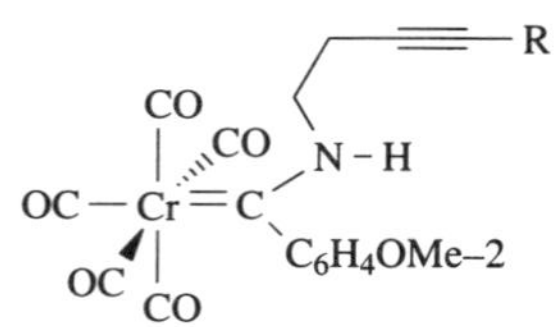		563
$[Cr(CO)_5\{=C(NMeCH_2CH{=}CH_2)Me\}] + \Delta$	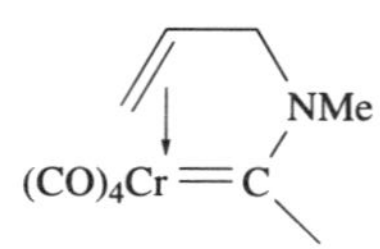		551
$[Mo(CO)_6]$ + All(H)N$(CH_2)_2$N(H)All + $CH(NMe_2)(OMe_2)$	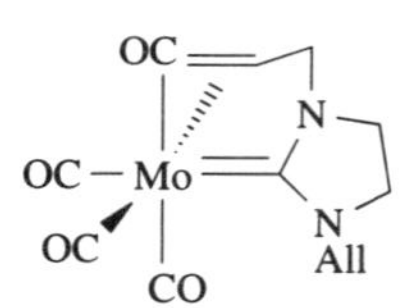	Under the reaction conditions, an electron-rich enetetraamine is formed	564

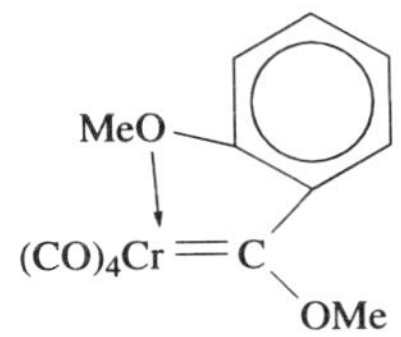 $+ NHMe_2$, –30 °C	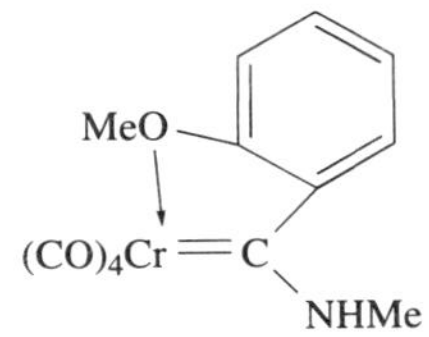	Amine complexes are formed as by-products	565
$[M(CO)_5\{=C(OEt)R\}]$ + 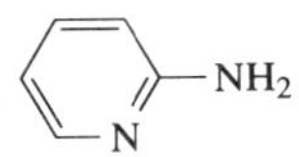	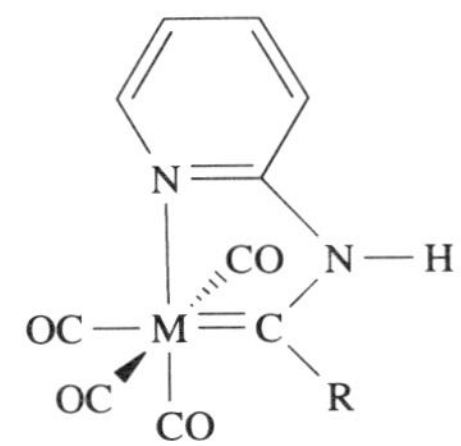	M = Cr, Mo; R = Me, Ph	566
$[M(CO)_5\{=C(OR)Tol\}]$ + R(H)N–CH=NMe	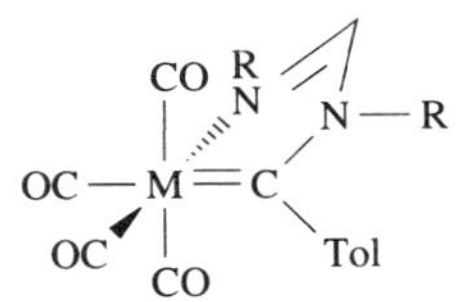	M = Cr, Mo; R = alkyl, Bz, allyl, propargyl	567
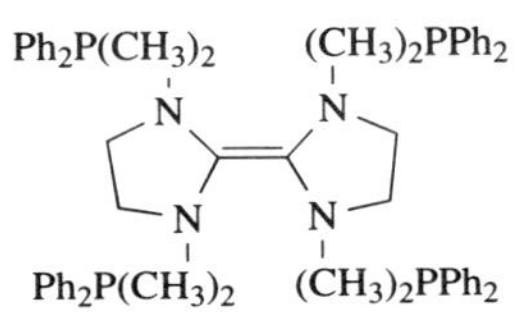 $+ [Mo(CO)_6]$	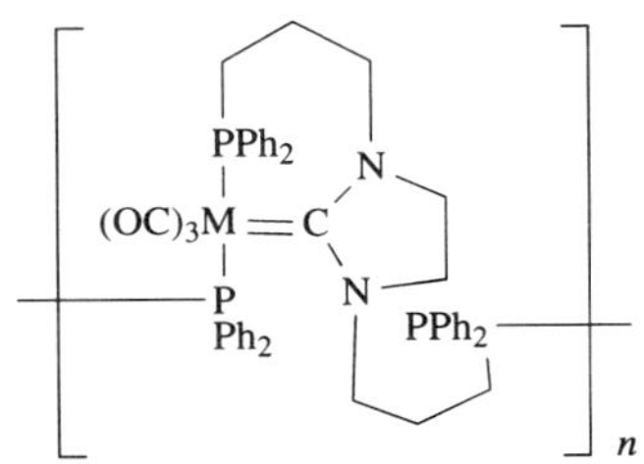		553

Table 4 (continued)

Synthesis	*Complex*	*Comments*	*Ref.*
$[Cr(CO)_5\{=C(OMe)R^1\}] + R^2CN$	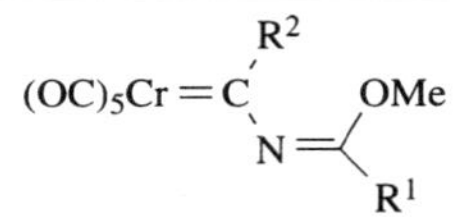	R^1 = Ph, Me; R^2 = alkyl, Ar	568
$[W(CO)_5\{=C(OR)C{\equiv}CPh\}] + Et_2NC{\equiv}CNEt_2$	$[W(CO)_5\{=C(NEt_2)[C(NEt_2{=}C(OR)C{\equiv}CPh]\}]$	R = Me, Et	529
$[M(CO)_5\{=C(OEt)SiR^1{}_3\}] + NHR^2{}_2$	$[M(CO)_5\{=C(NR^2{}_2)SiR^1{}_3\}]$	M = Cr, Mo, W. If $NHR^2{}_2$ is sterically demanding, the NHR^2 derivative is formed instead.	569
$[Cr(CO)_5\{=C(OR)\}] + \overline{HNCH_2CH}Me$	$[Cr(CO)_5\{=C(\overline{NCH_2CH}Me)R\}]$	R = Me, Ph, Tol	570
$[W(CO)_5\{=C(OEt)Ph\}] + RCH{=}CH{-}N{\equiv}C$	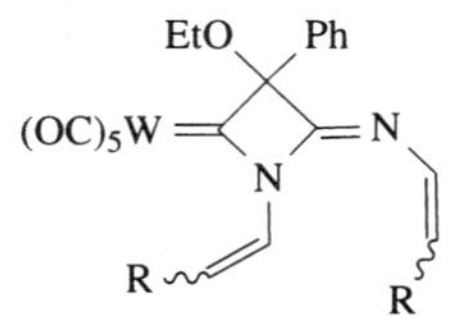	R = H, Me	508
$[W(THF)(CO)_5]$ + i, 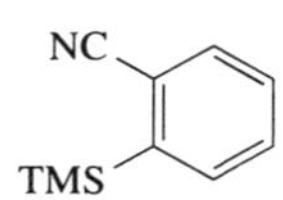 ; ii, MeOH, KF	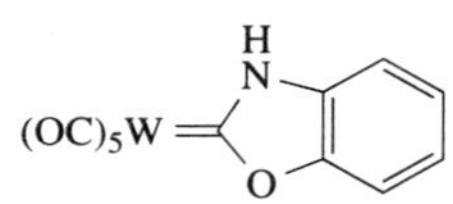	X-ray	571
$[M(CO)_6]$ + 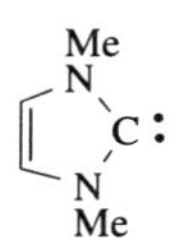	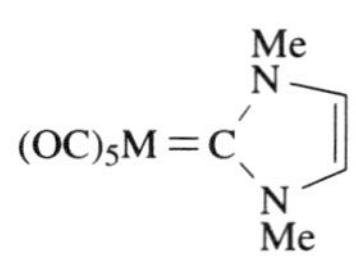	M = Cr, Mo, W. The free carbene resembles an electron-rich phosphine (e.g., PMe_3) in terms of ligand properties and metal complex synthesis	572

$[W(CNCH_2Tos)(CO)_5]$ + i, BuLi; ii, ArN_2^+	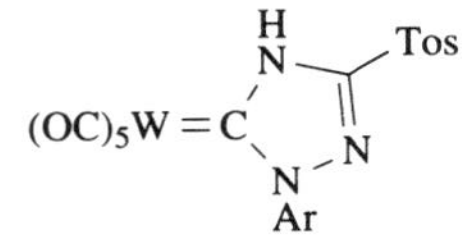		573
$[M(CO)_2(CNMe)Cp]^-$ + MeI, H^+	*cis*-$[MI(CO)_2\{=C(NHMe)Me\}(CO)_2Cp]$		574, 575
$[M(CO)_2(CNMe)Cp]^-$ + $I(CH_2)_nI$	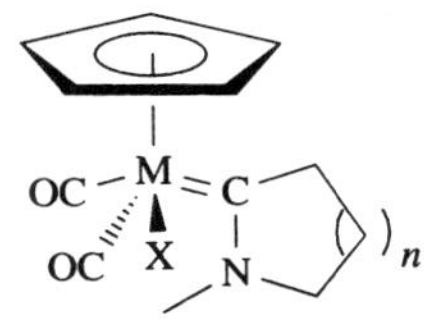	M = Mo, n = 1 or 2; M = W, n = 1	574, 575
$[Mo(CO)_3Cp]^-$ + PhC(Cl)=NMe	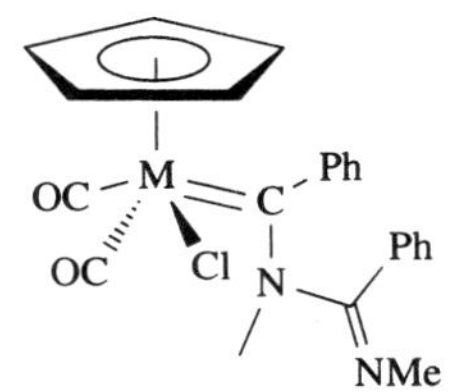	Heating this product in THF causes CO loss and coordination of pendant =NMe group	576

details on these reactions. The reader is again referred to Volume 12 for a discussion of chemistry which constitutes, in essence, a modification of the carbene substituents beyond the directly attached atoms, or which results in the liberation of organic molecules. The use of carbene complexes for the synthesis of ketenimine complexes has been reviewed.[557]

A characteristic reaction of carbenes of the type $[M\{=C(NR^1R^2)CHR^3R^4\}(CO)_5]$ is their tendency to deprotonate at the α-carbon position. The resulting anion is reactive towards electrophiles. Some examples are indicated in Scheme 11.[577–9] This type of behaviour is also displayed by cyclic carbenes such as $[Cr(CO)_5\{=C[\overline{N(CH_2)_3C}H_2]Me\}]$ which alkylates to form various derivatives $[Cr(CO)_5\{=C[\overline{N(CH_2)_3C}H_2]CH_2R\}]$.[580]

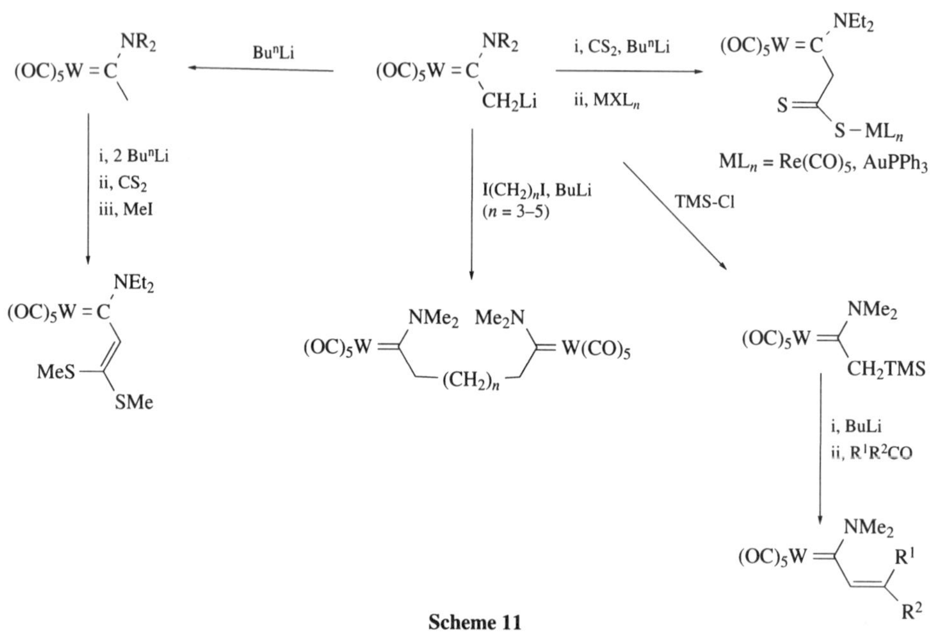

Scheme 11

When no acidic proton exists on the α-carbon atom, deprotonation, if it occurs, must take place elsewhere. In common with other azacarbenes, the resonance form (**18b**) (Equation (59)) is important, to the extent that deprotonation at the methyl group occurs only from the (*E*)-*N*-methyl group. Protonation gives only (**19**). Use of an alkylating agent such as BzBr also leads only to the *E* isomer.[581] Where the *E* site substituent is not a proton, then the *Z* site is used (Equation (60)).

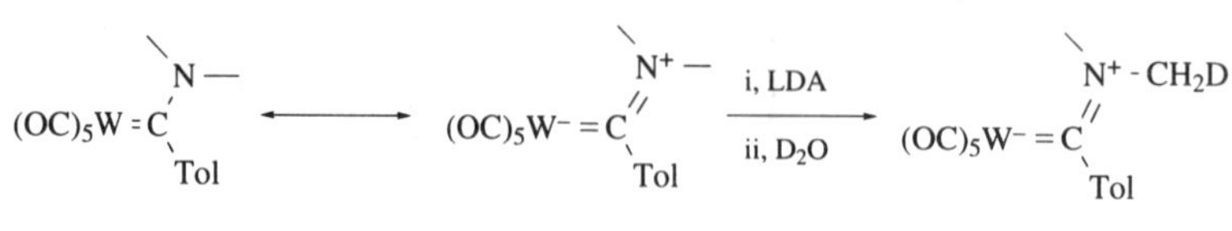

(59)

(18a) **(18b)** **(19)**

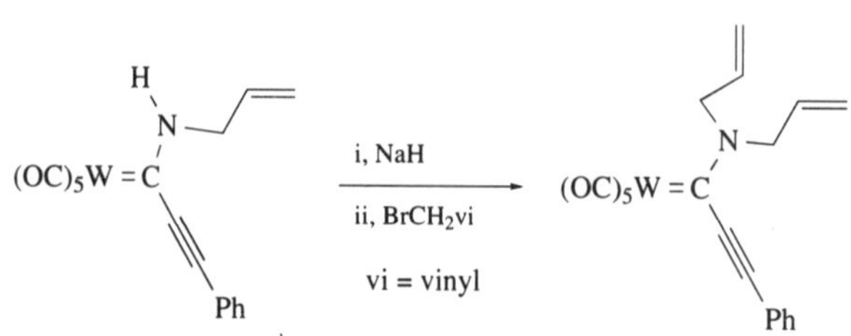

(60)

In other cases the negative charge is centred on the metal. Such anions are produced through reduction of a suitable carbene complex bearing halide ligands (Equation (61)).[537] Such anions then behave as metal-centred nucleophiles and the resulting products often rearrange further. In particular, groups such as hydride and acyl migrate to the carbene (Equation (61)).[582–4] Groups such as methyl and benzyl also migrate, but the primary products undergo further rearrangement to more complex products on warming to ambient temperature.[585,586] The products resulting from hydride migration revert back to the corresponding hydrido carbene complex by an α-elimination process on photochemical activation.[582,587]

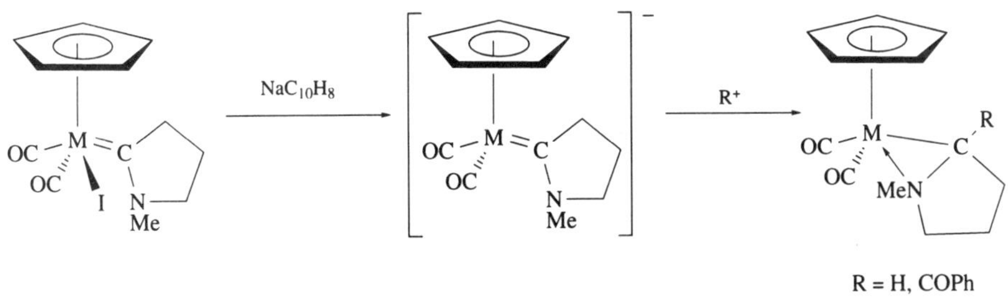

(61)

Under some conditions the amino group may be acylated. Thus reaction of various species $[Cr(CO)_5\{=C(NHR^1)R\}]$ with $\{R^2C(=O)\}_2O$ proceeds as in Equation (62). Thermolysis of the product results in loss of CO and intramolecular coordination of the acyl group.[588] Similar behaviour occurs when using $R^2C(=O)Cl$ as the acylating agent.[589]

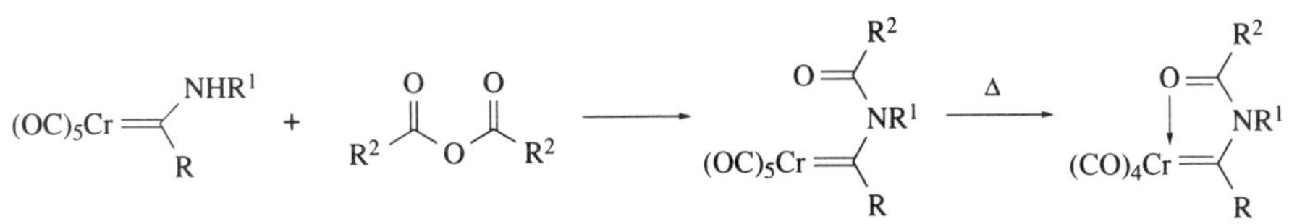

(62)

Sometimes group 6 metal carbene complexes transfer their carbene to another metal. Thus, the reaction between $[W(CO)_5\{=C(X)Ph\}]$ (X = OMe or NH_2) or $[W(CO)_5\{=C(NMe_2)Me\}]$ and $HAuCl_4$ results in the gold carbene complexes $[AuCl\{=C(X)Ph\}]$ or $[AuCl\{=C(NMe_2)Me\}]$.[590] Similar results are found for reactions of $[M(CO)_5\{=C(OR)NMe_2\}]$ (M = Cr, Mo or W) with $HAuCl_4$.[591,592]

The reaction of the alkynyl carbene (**20**) with urotropine in the presence of water is unusual (Equation (63)). Fragmentation of the urotropine skeleton occurs to give a binuclear complex.[593]

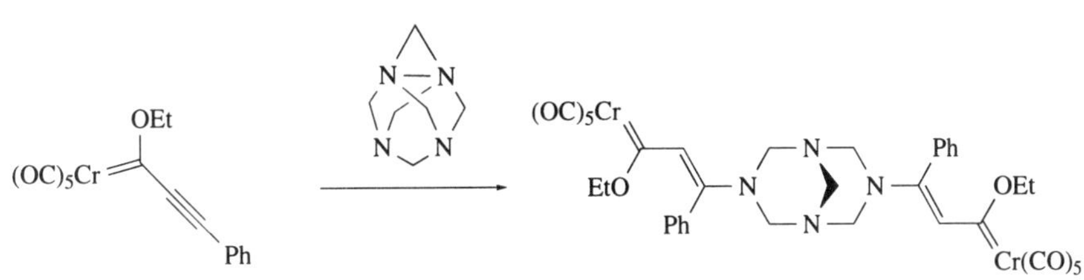

(63)

(**20**)

Silyl carbenes complexes are reviewed.[594] Thermolysis of the silyl carbenes $[M(CO)_5\{=C(NR^1_2)SiR^2_3\}]$ (M = Cr, Mo or W; $SiR^2_3 = SiPh_3$, $SiPh_2Me$ or $SiPhMe_2$; $NR^1_2 = NMe_2$, NMeEt, NC_4H_8 or NC_5H_{10}) results in loss of a *cis*-carbonyl and formation of stable, isolable, fluxional 16-electron complexes $M(CO)_4\{=C(NR^1_2)SiR^2_3\}]$. X-ray analysis shows that a phenyl group on the silyl group screens the vacant coordination site, and it appears that it is a relaxation of steric strain that is responsible for the facile decarbonylation.[565] Thermolysis of various monoalkylamino silyl carbene complexes $[M(CO)_5\{=C(NHR^1)SiR^2_3\}]$ results in loss of $SiHR^2_3$ and formation of $[M(CO)_6]$, and the isonitrile complexes $[M(CNR^1)(CO)_5]$ and $[M(CNR^1)_2(CO)_4]$. Labelling studies show that the silane proton arises from the NH group, probably by an intermolecular pathway.[595] Raman and IR studies on the photolysis of the silyl carbene $[W(CO)_5\{=C(NC_4H_8)SiPh_3\}]$ in noncoordinating solvents suggests that the product is the internally stabilized 16-electron complex $[W(CO)_4\{=C(NC_4H_8)SiPh_3\}]$, whereas in MeCN the product is the 18-electron $[W(CO)_4(NCMe)\{=C(NC_4H_8)SiPh_3\}]$.[596]

While there are many examples of azacarbenes $[L_nM\{=C(NR^1_2)R^2\}]$, there are very few complexes of the type $[L_nM\{=C(PR^1_2)R^2\}]$. Two such complexes are, however, formed as by-products in the reaction of the carbyne $[W(CO)_5(\equiv CNEt_2)]BF_4$ with KPMePh (Equation (64)).[597]

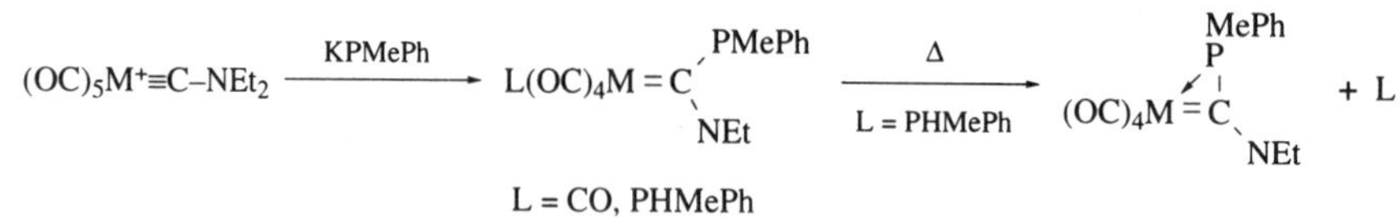

(64)

3.12 MX(CO)$_4$(≡CR) CARBYNE COMPLEXES

There is extensive literature on the chemistry of carbyne complexes. The reader is referred to a number of reviews on the subject.[239,598–603] The two most common complex classes are [MX(CO)$_4$(≡CR)] (X = halide or pseudohalide) and [M(CO)$_2$(≡CR)L] (L = cyclopentadienyl or a related ligand). The discussion here is therefore subdivided into a discussion of these two classes of complex. Carbyne complexes of the type [MX(CO)$_4$(≡CR)] are the subject of several reviews.[598,601]

3.12.1 Synthesis

Syntheses of carbyne complexes have been the subject of reviews.[601,604] Several syntheses are given in Table 5 together with some M≡C bond lengths and δ_C shifts.

The first syntheses of carbyne complexes of the type [MXL$_4$(≡CR)] involved the reaction of carbene complexes [M(CO)$_5${=C(OR^1)R^2}] with BX_3 (X = Cl, Br or I).[601] A number of new compounds have been made using this methodology (Table 5). An alternative approach involving the reaction of metal acyl complexes with Lewis acids such as $C_2O_2X_2$ (X = Cl or Br) offers considerable flexibility and scope (Equation (65)).[605] The products are transformed conveniently into more stable species by the addition of donor ligands L (Equation (66); L_2 = py$_2$, bipy or TMEDA). The use of $(CF_3CO)_2O$ in place of $C_2O_2X_2$ affords O_2CCF_3 derivatives (Equation (67)).[605]

$$[NMe_4][M\{C(O)R\}(CO)_5] + C_2O_2X_2 \longrightarrow [MX(CO)_4(\equiv CR)] \quad (65)$$

$$[MX(CO)_4(\equiv CR)] + 2\,L \longrightarrow [MX(CO)_2L_2(\equiv CR)] \quad (66)$$

$$[NMe_4][M\{C(O)R\}(CO)_5] + (CF_3CO)_2O \longrightarrow [M(O_2CCF_3)(CO)_4(\equiv CR)] \quad (67)$$

The low-temperature reaction of Br^-, I^- or SCN^- with the unstable [Mo(CO)$_5$(≡CNEt$_2$)]$^+$ affords [MoX(CO)$_4$(≡CNEt$_2$)]. On warming, these decarbonylate to form bridged complexes (Equation (68)). In other cases (X = SeC$_6$H$_4$CF$_3$-4 or OCN) the bridging species are produced directly.[606]

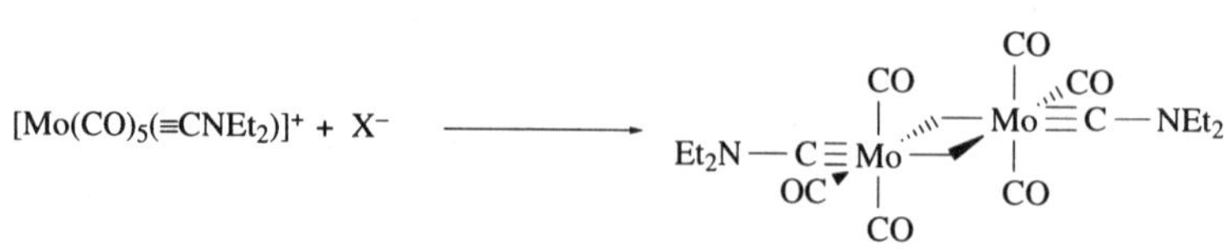

(68)

3.12.2 Structure

The nature of metal carbyne complexes is interesting and has attracted considerable attention. Many such complexes have been characterized by x-ray crystallography. The structure of [CrCl(CO)$_4$(≡CMe)] is notable.[607] The chromium–carbonyl distances of one pair of mutually *trans*-carbonyl groups (average 192.8(3) pm) are distinctly different from those of the other pair (average 198.4(3) pm). Raman and IR spectroscopic studies of the carbyne in the solid state [CrBr(CO)$_4$(≡CMe)] and its partially deuterated derivatives show that there are different C–H bond strengths. This could be a consequence of hyperconjugative effects or crystal packing forces.[608–10]

Several authors have analysed these compounds using computational methods.[611–15] For instance, calculations suggest that there is a negative net charge on the carbyne carbon atom in [CrCl(CO)$_4$(≡CMe)] and that the Cr≡C π bonds are quite covalent relative to the π back donation in the Cr(CO)$_4$ plane.[611] An investigation using an extension to the CNDO formalism suggests that the CMe group has electronic properties rather similar to those of the NO radical.[614] *Ab initio* full-gradient geometry optimization and CASSCF calculations suggest that the cause of two distinct pairs of Cr–C$_{CO}$

Table 5 Some carbyne complexes of the type $[MXL^1L^2L^3L^4(\equiv CR)]$ (X = halide or pseudohalide; L = two-electron ligand).

Synthesis	*Complex*	δ_C	*M≡C* (pm)	*Comments*	*Ref.*
	$[CrCl(CO)_4(\equiv CMe)]$		171.0		607, 611
$NMe_4[M(COR)(CO)_5] + C_2O_2X_2$	$[MX(CO)_4(\equiv CR)]$			M = Cr, Mo, W; X = Cl, Br; R = Ph (also Me for W)	605
Thermolysis of $[Cr(CO)_4\{=(CNEt_2)SeR\}]$	$[Cr(SeR)(CO)_4(\equiv CNEt_2)]$		175.0(12) R = Ph	R = aryl	616
$[Cr(CO)_4\{=(CNEt_2)SeR\}] + PPh_3$	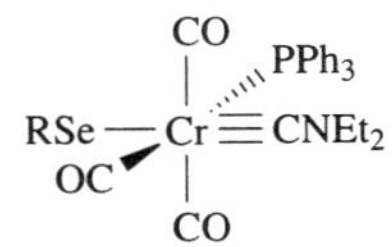				616
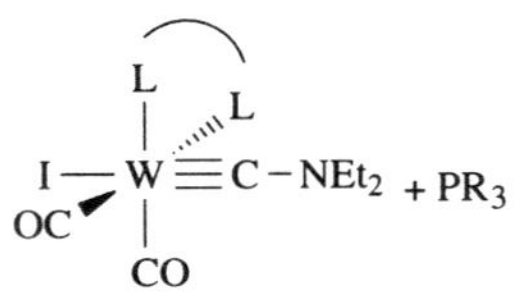	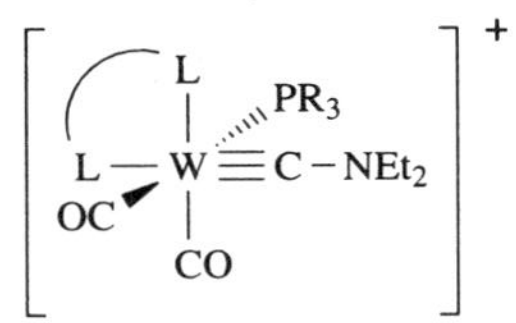			L = bipy, phen; R = Me, Et	617
$[MCl(CO)_4(\equiv CPh)] + P(OMe)_3$	$[MCl(CO)(POMe_3)_3(\equiv CPh)]$			M = Mo, W	618
$[WCl(CO)_2(PMe_3)_2(\equiv CR)] + PMe_3$	$[WCl(CO)(PMe_3)_3(\equiv CR)]$	261.9 ($CDCl_3$, R = Tol)		R = Ph, Tol	619
$[WI(CO)_2(PMe_3)_2(\equiv CNEt_2)] + MeSO_3CF_3$	$[W(OSO_2CF_3)(CO)_2(PMe_3)_2(\equiv CNEt_2)]$	247.5 (CD_2Cl_2)		Also CNR derivatives in place of PMe_3	620
$[Mo(CO)_5(\equiv NEt_2)]BF_4 + X^-$	*trans*-$[MoX(CO)_4(\equiv CNEt_2)]$	253.3 $(CD_3)_2C{=}O$		X = Br, I, SCN	606
$[Cr(CO)_5(\equiv NPr^i_2)]SbCl_6 + [N(PPh_3)_2]Cl$	*trans*-$[CrCl(CO)_4(\equiv CNPr^i_2)]$		174.7(5)		621
trans-$[Cr(CO)_4(PPh_3)(\equiv CNEt_2)]BF_4 + NEt_4NBr$	*mer*-$[CrBr(CO)_3(PPh_3)(\equiv CNEt_2)]$		175(1)		622–4
$[W(CO)_6]$ + i, $C_6H_4Li_2$-4; ii, Et_3O^+; iii, BBr_3	$[(OC)_4BrW(\equiv CC_6H_4C\equiv)WBr(CO)_4]$		189(3)		625
	$[WBr(CO)_2(\equiv CTol)(dppe)]$		179.8(5)		605, 626
$[WBr(CO)_2(\equiv CTol)(dmpe)] + [AsPh_4]Cl$	$[WCl(CO)_2(\equiv CTol)(dmpe)]$	268.0 ($CDCl_3$)			627
$[WBr(CO)_2(py)_2(\equiv CTol)]$ + dmpe	$[WBr(CO)_2(\equiv CTol)(dmpe)]$	267.1 ($CDCl_3$)			627
$[WBr(CO)_2(\equiv CTol)(dmpe)] + TlSCN$	$[W(NCS)(CO)_2(\equiv CTol)(dmpe)]$	277.3 ($CDCl_3$)	182.7(6)		627
$[WBr(CO)_4(\equiv CPh)] + [Mo_2(OPr^i)_6]$/py	$[WBr(CO)_2(py)_2(\equiv CPh)]$		184(2)		628
$[WI(CO)_2(py)_2(\equiv CNEt_2)] + P(OMe)_3$	$[WI(CO)_2\{P(OMe)_3\}_2(\equiv CNEt_2)]$	232.2 (*cis*-$CDCl_2$) 236.6 (*trans*-$CDCl_2$)			629
cis,cis-$[MoBr(CO)_2(PMe_3)_2(\equiv CPh)] + h\nu$	*trans,trans*-$[MoBr(CO)_2(PMe_3)_2(\equiv CPh)]$	280.9 (CD_2Cl_2)			630
$[WBr(CO)_2(py)_2(\equiv CNEt_2)] + KCN$	$[W(CN)(CO)_2(py)_2(\equiv CNEt_2)]$	257.9 (CD_2Cl_2)			631
$[WCl(CO)_4(\equiv CPh)] + Na[Co(CO)_4]$	$[W\{Co(CO)_4\}(CO)_4(\equiv CPh)]$	282.6 ($CDCl_2$)	181.1(12)		632
$[Cr(CO)_5\{=C(OEt)C$-*c*-$C_3H_5\}] + BBr_3$	$[Cr\{BrCr(CO)_5\}(CO)_4\{\equiv C$-*c*-$C_3H_5)]$		171.4(13)	μ-Br	421
$[MX(CO)_4(\equiv CPh)] + [Cr(CO)_5(CH_2Cl_2)]$	$[M\{XCr(CO)_5\}(CO)_4(\equiv CPh)]$			M = Cr; X = Cl, Br, I. M = W; X = Br	633
$[WBr(CO)_2(bipy)(\equiv CNEt_2)] + K[Cr(PPh_2)(CO)_5]$	$[W\{PPh_2[Cr(CO)_5]\}(CO)_2bipy(\equiv CNEt_2)]$		187.7(8)	Also $[M(EPh_2)(CO)_5]^-$. M = Cr, Mo, W; E = P, As, Sb	694

Table 5 (continued)

Synthesis	*Complex*	δ_C	*M≡C* (pm)	*Comments*	*Ref.*
Thermolysis of $[W(CO)_3(\equiv CBz)(dppe)][BF_4]$	$[W(CO)_2(\equiv CCH_2Ph)(dppe)][BF_4]$				394
$[W(CO)_2(\equiv CBz)(dppe)BF_4 + NEt_4F \cdot 2H_2O$	$[WF(CO)_2(\equiv CBz)(dppe)]$	279.0 ($CDCl_3$)		Also NEt_4X (X = Cl, Br, I)	394
$[WI(CO)_2(py)_2(\equiv CNEt_2)] + CNBu^t$	$[WI(CO)_2(CNBu^t)_2(\equiv CNEt_2)]$	236.0 (CD_2Cl_2)		Also CNMe	634
$[WCl(CO)_2(py)_2(\equiv CPh)]$ + maleic anhydride	$[WCl(CO)(C_4H_2O_3)(py)_2(\equiv CPh)]$	268.0 ($CDCl_3$)	180.1(6)		635, 636
$Mo(CO)_6$ + i, $Fe(\eta^5\text{-}C_5H_4Li)_2$; ii, $(CF_3CO)_2O$, TMEDA	$[Fe(\eta^5\text{-}C_5H_4\{C\equiv Mo(O_2CCF_3)(CO)_2(TMEDA)\}_2]$	286.9 ($CDCl_3$)			637
$M(CO)_6$ + i, $LiC\equiv CBu^t$; ii, $(CF_3CO)_2O$; iii, L_2	$M(O_2CCF_3)(CO)_2L_2(\equiv C\text{–}C\equiv CBu^t)$			M = Mo, W; $L_2 = py_2$, TMEDA	638
$[M(CO)_6]$ + i, $LiNPr^i_2$; ii, $(CF_3CO)_2O$; iii, PPh_3	$[M(OCOCF_3)(CO)_3(PPh_3)(\equiv NPr^i_2)]$	238.2 (M = Cr, $CDCl_3$)		M = Cr, Mo, W	639
$[M(CO)_5(\equiv CNEt_2)]^+ + X^-$	$[MX(CO)_3(\equiv CNEt_2)]_2$		179.5(18)	M = Mo; X = I	606
			181.6(16)	M = Mo; X = OCN	
			179.5(18)	M = W; X = N_3	
			183.9(25)	M = W; X = SPh	
			179.3(22)	M = W; X = SPh	
$[WCl_2(PMePh_2)_4] + C_3O_2$	$[WCl_2(CO)(PMePh_2)(\equiv CPMePh_2)]$	218.7 (CD_2Cl_2)	182.3(7)	Via ketene complex	640
$[WI(CO)_2bipy(\equiv CNEt_2)] + I_2$	$[WI_3(CO)bipy(\equiv CNEt_2)]$			Seven-coordinate complex	641
$[WI(CO)_2(CNBu^t)_2(\equiv CNEt_2)] + I_2$	$[WI_3(CO)(CNBu^t)_2(\equiv CNEt_2)$	257.4 (CD_2Cl_2)			634
$[WI_3(CO)bipy)\equiv CNEt_2] + PMe_3$	$[W(CO)(PMe_3)_2bipy(\equiv CNEt_2)]I$	254.1 (CD_2Cl_2)			641
$[WI(CO)_2(CNBu^t)_2(\equiv CNEt_2)] + CNBu^t$	*mer*-$[WI(CO)(CNBu^t)_3(\equiv CNEt_2)]$ + $[W(CO)(CNBu^t)_4(\equiv CNEt_2)]I$	236.7 (CD_2Cl_2) 250.1 (CD_2Cl_2)		Equal amounts of each formed	642
$[Cr(CO)_5\{=C(NPr^i_2)N=C(OMe)Ph\}]$ + i, BCl_3; ii, $SbCl_5$	$[Cr(CO)_5(\equiv CNPr^i_2)]SbCl_6$	285.1 (CD_2Cl_2)			643
$[M(CO)_5\{C(NEt_2)OEt\}] + BF_3$	$[M(CO)_5\{\equiv NEt_2)]BF_4$	256.9 (CD_2Cl_2) (M=W)	190(3), (180)3	M = Cr, Mo, W M = W; $SbCl_6^-$ counteranion	606, 621 622, 644
$[Cr(CO)_5(\equiv CNEt_2)]CF_4 + PPh_3$	*trans*-$[Cr(CO)_4(PPh_3)(\equiv CNEt_2)][BF_4]$		176(1)		622, 623
$[W(CO)_3(=C=CHPh)(dppe)] + HBF_4$	$[W(CO)_3(\equiv CBz)(dppe)][BF_4]$	317.4 ($CDCl_3$)			394
$[WBr(CO)_4(\equiv CNCy_2)] + NEt_4NBr$	$NEt_4[WBr_2(CO)_3(\equiv CNCy_2)]$	248.0 (CD_2Cl_2)		Similar result with PPNCl	645
$[WBr(CO)_2(bipy)(\equiv CNEt_2)]$ + i, *cis*-$[Mo(CO)_4(PPh_2K)]$; ii, NEt_4NBr	$NEt_4[(OC)_4Mo(\mu\text{-}PPh_2)_2W(CO)_2(\equiv CNEt_2)]$	290.8 (DMF-d_7)			646
$[W(STol)(CO)_4(\equiv CNEt_2)] + Et_4NSCN$	$(NEt_4)_2[W(NCS)_3(CO)_2(\equiv CNEt_2)]$	249.4 (CD_2Cl_2)		Dianion, very unusual	647
$[WI(CO)_2bipy(\equiv CNEt_2)]$ + i, $Na_2[S_2C_2(CN)_2]$; ii, NEt_4NBr	$NEt_4[W(CO)_2(\equiv CNEt_2)\{S_2C_2(CN)_2\}]$	298.2 (DMF-d_7)		Five-coordinate complex	648

distances in [CrCl(CO)$_4$(≡CMe)] is a consequence of a tilt of the methyl group so that its local C_{3v} axis is no longer collinear with the Cr≡C axis. The tilt is caused by close contact of the methyl group with the chlorine ligands of neighbouring molecules—a crystal packing effect.[611,615] Other calculations suggest that the two Cr–C(carbyne) π bonds are equivalent in the tetracarbonyl [Cr(CO)$_4$(PMe$_3$)(≡CMe)]$^+$, a little different in the corresponding CPh derivative, and very different in CNEt$_2$ complexes. The Cr≡C–R axis is normally slightly bent because of the low symmetry of the carbyne π clouds, crystal-packing effects, and nonbonding interactions.[613]

An analysis of the solid-state structures of the pair of compounds [WBr(≡CPh)(dmpe)$_2$]$^{n+}$ (n = 0 or 1) reveals that it is only the core bond distances that vary with n. These studies further suggest that it is the metal d_{xy} orbital that is the HOMO. This orbital is nonbonding for the halide and benzylidene groups. The LUMO seems to be W≡C antibonding.[649] Extended Hückel calculations suggest that the NEt$_2$ group in complexes of the type [Cr(CO)$_4$L(≡CNEt$_2$)]$^+$ (L = CO or PPh$_3$) and [CrX(CO)$_4$(≡CNEt$_2$)] (X = halide or pseudohalide) is in conjugation with the Cr≡C bond and the extent of conjugation is dependent on the *trans* ligand.

Alkylidyne complexes of the type [WX(CO)$_2$L$_2$(≡CR)] are luminescent. These luminescent properties are associated with the alkylidyne ligand, but are dependent on L and R.[650]

3.12.3 Reactions

Carbyne complexes of the type [MXL$_4$(≡CR)] display several characteristic reactions. Two particularly common processes are substitution of either X or L, or both. Coupling reactions of the carbyne ligand with other groups are not unusual.

In a number of cases, the *trans*-halide (or a pseudohalide) of *trans*-[WX(CO)$_2$L$_2$(≡CR)] (L$_2$ = 2CO or other donor ligands) is substituted by an appropriate donor ligand. The neutral complexes *trans*-[WX(CO)$_4$(≡CC$_6$H$_4$R-4)] (X = Cl, Br or I; R = H, Me or CF$_3$) undergo facile exchange of X$^-$ for a second halide Y$^-$ (Equation (69)). The substitution follows a second-order rate law, which is first order in the concentrations of each of the halide and the metal complex. Notable features of this reaction are that the influence of the leaving halide is not great, the influence of the substituent R is small, and the effect of the solvent is large. The rate-determining step seems to be a nucleophilic frontier-orbital controlled attack of halide at the complex.[651,652] In addition, substitution of a carbonyl group may occur, in which case the product is an anionic carbyne (Equation (70)).[645] The new *cis*-halo ligand is particularly prone to further substitution from ligands such as PPh$_3$.

$$\textit{trans-}[\mathrm{WX(CO)_4(\equiv CC_6H_4R\text{-}4)}] + \mathrm{Y^-} \rightleftharpoons \textit{trans-}[\mathrm{WY(CO)_4(\equiv CC_6H_4R\text{-}4)}] + \mathrm{X^-} \quad (69)$$

$$\mathrm{trans\text{-}[WX(CO)_4(\equiv CNR_2)]} + \mathrm{X^-} \longrightarrow \textit{mer-}[\mathrm{WX_2(CO)_3(\equiv CNR_2)}]^- + \mathrm{X^-} \quad (70)$$

The halide group also exchanges with cyanide, but in this case the carbyne functionality is also important. The product is the expected [W(CN)(CO)$_2$L$_2$(≡CNEt$_2$)] (CN now *cis* to the carbyne; L$_2$ = phen or bipy) in the case of the reaction of *trans*-[WBr(CO)$_4$(≡CNEt$_2$)],[631] but an anionic ketenyl in the case of *trans*-[WBr(CO)$_2$L$_2$(≡CR)] (Equation (71)); R = Me, Ph).[653,654] A related coupling is seen on the introduction of chelating Schiff base ligands (Equation (72)); H$_2$L$_2$ = HNCH=CH–CH=CH(CH=NMe).[655]

$$\textit{trans-}[\mathrm{WBr(CO)_2L_2(\equiv CR)}] + 2\,\mathrm{CN^-} \rightleftharpoons [\mathrm{W(CN)_2(CO)(\eta^2\text{-}OCCR)}]^{2-} + \mathrm{Br^-} \quad (71)$$

$$\textit{trans-}[\mathrm{WCl(CO)_2(py)_2(\equiv CR)}] + \mathrm{H_2L_2} \rightleftharpoons [\mathrm{W(CO)L_2(\eta^2\text{-}OCCR)}]^- \quad (72)$$

Other groups which exchange for halide include a number of metal complexes [M(EPh$_2$)(CO)$_5$]$^-$ (M = Cr, Mo or W; E = P, As or Sb).[656] The use of S$_2$CNEt$_2^-$ leads to a ketenyl [W(CO)(S$_2$CNEt$_2$)$_2$(PhCCO)]$^-$, probably via the anionic intermediate [W(CO)$_2$(η^1-S$_2$CNEt$_2$)(η^2-S$_2$CNEt$_2$)(PhCCO)]$^-$.[657] The behaviour of Na$_2$[S$_2$C$_2$(CN)$_2$] is more complex and leads to a five-coordinate anion (Equation (73)).[642] The pseudohalide STol$^-$ of *trans*-[W(STol)(CO)$_4$(≡CNEt$_2$)] and two carbonyl groups are replaced by NCS$^-$ in forming a dianionic compound [W(NCS)$_3$(CO)$_2$(≡CNEt$_2$)]$^{2-}$, isolated as the NEt$_4^+$ salt (Equation (74)).[641] The halide located *cis* to the carbyne in the anionic carbyne complexes [WX$_2$(CO)$_3$(≡CNR$_2$)]$^-$ is easily replaced by neutral ligands (Equation (75)).[639]

$$\textit{trans-}[\mathrm{WI(CO)_2L_2(\equiv CNEt_2)}] + \mathrm{Na[S_2C_2(CN)_2]} \longrightarrow [\mathrm{W(CO)_2(\equiv CNEt_2)\{S_2C_2(CN)_2\}}]^- \quad (73)$$

$$trans\text{-}[WI(CO)_2L_2(\equiv CNEt_2)] + 3NEt_4(SCN) \longrightarrow (NEt_4)_2[W(NCS)_3(CO)_2(\equiv CNEt_2)] \quad (74)$$

$$PPN[WCl_2(CO)_3(\equiv CNCy)] + PPh_3 \longrightarrow [WCl(CO)_3(PPh)_3(\equiv CNCy_2)] \quad (75)$$

Complexes derived from compounds with the general formula $[MX(CO)_4(\equiv CR)]$ generally react with cyclopentadienyl type salts by elimination of the halide *and* two carbonyl groups, resulting in the formation of carbynes of the general formula $M(CO)_2(\equiv CR)Cp$ (Section 3.13). There is an extensive carbonyl substitution chemistry of the complexes $[MX(CO)_4(\equiv CR)]$, $[M(CO)_5(\equiv CR)]^+$ and their derivatives. Thus, phosphines react with $[Cr(CO)_5(\equiv CNEt_2)]^+$ to form *trans* complexes (Equation (76)).[623,624] When the substituting group is Cl^-, the neutral *trans*-chloro species is produced, but in this case an intermediate is formed by addition of Cl^- to the carbyne which only collapses through CO loss to the final product on warming (Equation (77)).[621] In this case the rearrangement appears to be intramolecular, involving a concerted Cr–CO dissociation and C → Cr migration of the chloro group. In the case of SeR^- (R = aryl), a similar result is observed and the carbene species produced are isolable and fully characterized (Equation (78)).[488] The reaction of $[Cr(CO)_5(\equiv CNEt_2)]^+$ with $LiPbPh_3$ is most unusual in that the initial product is a plumbyl carbene complex. Here also, on warming (this time only to room temperature), decarbonylation and migration occurs to give a carbyne species (Equation (79)).[658]

$$[Cr(CO)_5(\equiv CNEt_2)]^+ + PPh_3 \longrightarrow trans\text{-}[Cr(CO)_4(PPh_3)(\equiv CNEt_2)]^+ \quad (76)$$

$$[Cr(CO)_5(\equiv CNEt_2)]^+ + Cl^- \longrightarrow [Cr(CO)_5\{=C(Cl)NEt_2\}] \longrightarrow trans\text{-}[CrCl(CO)_4(\equiv CNEt_2)] \quad (77)$$

$$[Cr(CO)_5(\equiv CNEt_2)]^+ + LiSeR \longrightarrow [Cr(CO)_5\{=C(SeR)NEt_2\}] \longrightarrow trans\text{-}[Cr(SeR)(CO)_4(\equiv CNEt_2)] \quad (78)$$

$$[Cr(CO)_5(\equiv CNEt_2)]^+ + LiPbPh_3 \longrightarrow [Cr(CO)_5\{=C(PbPh_3)NEt_2\}] \longrightarrow trans\text{-}[Cr(PbPh_3)(CO)_4(\equiv CNEt_2)] \quad (79)$$

Many complexes of the type *trans*-$[MX(CO)_4(\equiv CR)]$ (M = Cr or W; X = Cl, Br, I or SePh; R = Me, Ar or NEt_2) react with PPh_3 in CCl_3Me according to a rate law which is first order in the metal complex. The rate is largely independent of R but increases along the series I < Br < Cl < SePh, and is faster for chromium than tungsten by a factor of 30–50. The activation enthalpies lie within the range $\Delta H^{\ddagger} = 97$–$116\ kJ\ mol^{-1}$ and the entropies of activation lie within the range 34–72 $J\ mol^{-1}\ K^{-1}$.[659] A variation of this general theme is seen in the reaction of PMe_3 with $[M\{Re(CO)_5\}(CO)_4(\equiv CPh)]$ (M = Cr or W) which gives a μ-ylide complex (Equation (80)), a consequence of the noninnocence of the $Re(CO)_5$ substituent.[660]

$$[M\{Re(CO)_5\}(CO)_4(\equiv CPh)] + PMe_3 \longrightarrow [(OC)_4Re(\mu\text{-}CO)\{\mu\text{–}C(Ph)PMe_3\}M(CO)_4] \quad (80)$$

The reaction of *trans*-$[WI(CO)_4(\equiv CPh)]$ with picolin followed by $CNBu^t$ results in the carbonyl-free complex $[WI(CNBu^t)_4(\equiv CPh)]$ which is in equilibrium with the cation $[W(CNBu^t)_5(\equiv CPh)]I$. This reaction proceeds via the intermediates $[WI(CO)_2(pic)_2(\equiv CPh)]$ and $[WI(CNBu^t)_2(pic)_2(\equiv CPh)]$.[661] A common way to prepare substituted derivatives of $[MX(CO)_4(\equiv CR)]$ is to proceed via reasonably stable derivatives $[MX(CO)_2L_2(\equiv CR)]$ in which the ligands L are themselves very labile to substitution. Thus, the pyridine ligands of the complex $[WX(CO)_2(py)_2(\equiv CNEt_2)]$ are easily replaced by dppe,[605] dmpe,[627] bipy,[627] $2PMe_3$,[605] $2P(OMe)_3$[629] or 2CNR (R = Me or Bu^t).[634] In the last case, the remaining carbonyl groups are still liable to substitution. Treatment of the first product, $[WI(CO)(CNBu^t)_3(\equiv CNEt_2)]$ with $CNBu^t$ results in decarbonylation to form *trans*-$[WI(CNBu^t)_4(\equiv CNEt_2)]$. This reacts still further with $CNBu^t$ by loss of I^- to form $[W(CNBu^t)_5(\equiv CNEt_2)]I$.[642,662] Complexes of the type $[MX(CO)_2L_2(\equiv CR)]$ may exist with the two carbonyl ligands mutually *trans* and the two ligands L mutually *trans*, or both pairs mutually *cis*. Irradiation of various *cis*,*cis* complexes *cis*,*cis*-$[MX(CO)_2(PR_3)_2(\equiv CPh)]$ results in transformation to the respective *trans*, *trans* complex.[630]

In addition, a number of alkylidyne complexes such as $[WCl(CO)_2(py)_2(\equiv CPh)]$, $[WCl_2(CO)_2(CO)_2(py)(\equiv CPh)]^-$ and $[WCl(CO)_2(CO)_2(TMEDA)(\equiv CPh)]$ react with maleic anhydride or fumaronitrile by a net CO substitution to form the alkene alkylidyne complexes $[WCl(CO)(alkene)(py)_2(\equiv CPh)]$, $[WCl_2(CO)(alkene)(py)(\equiv CPh)]^-$ and $[WCl(CO)(alkene)$-$(TMEDA)(\equiv CPh)]$, respectively. In all cases there is a chlorine atom *trans* to the CPh group, while in the anions there is also a chlorine atom *trans* to the alkene. The fumaronitrile species are obtained as mixtures of diastereomers.[636]

Carbyne ligands are prone to coupling with other ligands such as carbonyl,[653,655] alkyne and carbyne. This is often induced by acid. Thus addition of HI to $[WI(CO)_2(CNBu^t)_2(\equiv CNEt_2)]$ results in proton induced coupling of one $CNBu^t$ with the $\equiv CNEt_2$ ligand to form a mixture of $[WI_2(CO)_2(CNBu^t)(\eta^2\text{-}Et_2NC\equiv CNHBu^t)]$ and $[WI_2(CO)(CNBu^t)_2(\eta^2\text{-}Et_2NC\equiv CNHBu^t)]$. The second of these products is available through the reaction of $[WI(CO)(CNBu^t)_3(\equiv CNEt_2)]$ with HI.[663] Proton induced coupling of Bu^tNC with a phenylcarbene ligand in the reactions of $WBr(CO)_2(CNBu^t)_2(\equiv CPh)$ results in $[W(CNBu^t)_3(Bu^tNHC\equiv CPh)]X_2$ (X = Br or I).[664] Protonation of the carbyne $[WCl(CO)_2(PMe_3)_2(\equiv CPh)]$ with HCl in the presence of visible light irradiation affords $[WCl_2(CO)_2(PMe_3)_2(\eta^2\text{-}PhC\equiv COH)]$,[665] while protonation of the related carbyne $[WCl(CO)(PMe_3)_2(CNBu^t)(\equiv CPh)]$ with HCl/MeOH affords the η^2-nitrilium complex $[WCl_3(CO)(PMe_3)(\eta^2\text{-}Bu^tN\equiv CBz)]$ as major product together with some $[WCl_2(CO)(PMe_3)_2(\eta^2\text{-}PhC\equiv CNHBu^t)]$. The first stage leading to these compounds is protonation at the carbyne to afford an alkylidene complex in which the alkylidene proton is agostically constrained to tungsten.[303]

The coupling of the carbyne in the anion $[(OC)_4Mo(\mu\text{-}PPh_2)_2W(CO)_2(\equiv CNEt_2)]^-$ with CO_2 to form $[(OC)_4Mo(\mu\text{-}PPh_2)_2\overline{W(CO)_2\{=C(NEt_2)C(O)O}\}]^-$ is unusual.[646] Carbyne–carbyne coupling is observed in a number of noncarbonyl complexes (Section 3.11). The reaction of the carbyne complex $[WBr(CO)_3(PPh_3)(\equiv CPh)]$ with (i) MeLi, (ii) $C_2O_2Br_2$ and (iii) PPh_3 affords the alkyne derivative $[WBr_2(CO)(PPh_3)(\eta^2\text{-}PhCCMe)]$, probably via coupling of two carbynes in an intermediate bis(carbyne),[666] as was suggested would happen by a theoretical analysis.[667] Coupling of a carbyne with both alkyne and carbonyl is observed in the reaction of $[WCl(CO)(PMe_3)_3(\equiv CPh)]$ with $PhC\equiv CH$ (Equation (81)).[668] Another kind of coupling occurs by reaction of the cation $[Cr(CO)_5(\equiv NEt_2)]^+$ with $LiAsMe_2$ (Equation (82)). The product of this reaction is effectively a bis(amino) carbene derivative.[669] Carbyne–alkene–carbyne coupling occurs in the reaction of $[WBr(CO)_4(\equiv CMe)]$ with the enzyne (**21**) (Equation (83)).[670] In other cases the reactions of $[WBr(CO)_4(\equiv CMe)]$ with various alkynes or cycloalkenes result in oligomerization. The carbyne seems to function as a reactive carbene in this regard and it may be that, under the reaction conditions, the carbyne functions as the 16-electron carbene $[W\{=C(Br)Me\}(CO)_4]$.[671] Carbynes also couple with metal–metal bonded species to form cluster compounds, many of which are heterometallic systems (see Chapter 4, Volume 10).

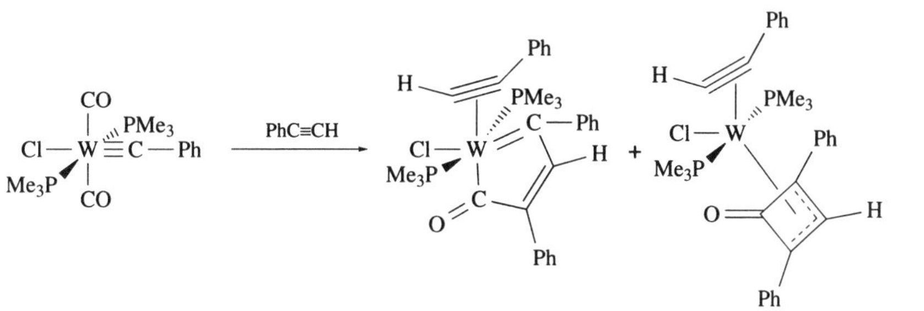

(81)

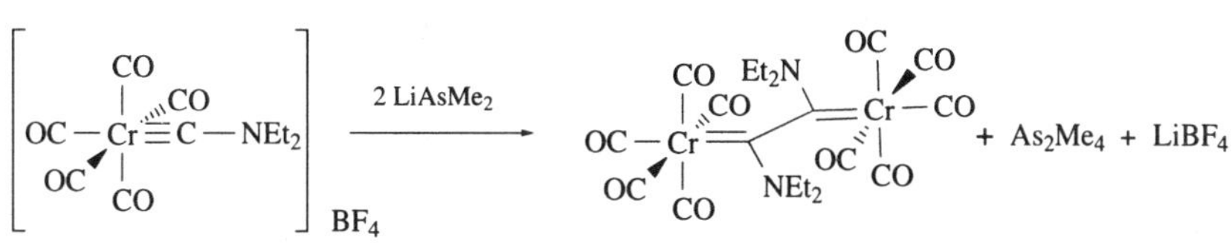

(82)

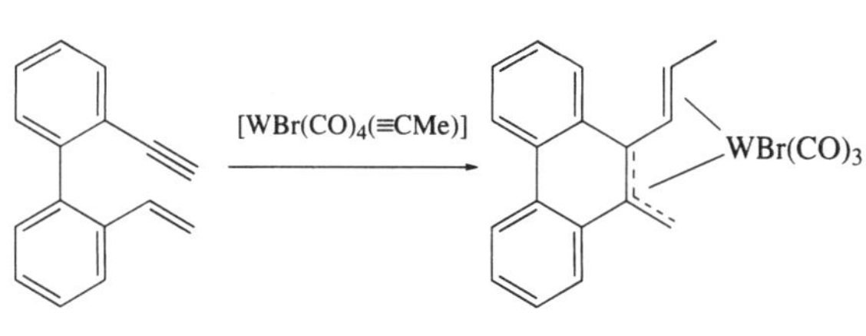

(83)

(21)

Carbyne complexes oxidize with reagents such as I_2. Thus, reaction of a mixture of *cis-* and *trans-* $[WI(CO)_2L_2(\equiv CNEt_2)]$ ($L_2 = P(OMe)_3$, bipy or phen) with I_2 results in $[WI_3(CO)L_2(\equiv CNEt_2)]$.[641,672] Notably, when L_2 = bipy or phen, these compounds react with PMe_3 by elimination of $[PMe_3I]I$ to form the cationic carbynes $[W(CO)(PMe_3)_2L_2(\equiv CNEt_2)]^+$.[641] The tetracarbonyls $[MBr(CO)_4(\equiv CR)]$ (M = Mo, R = Ph; M = W; R = Ph, Me or Pr^i) react with Br_2 in the presence of DME to form $[MBr_3(\equiv CR)(DME)]$ (Equation (84)). The reaction does not work well for M = Cr.[673]

$$[MBr(CO)_4(\equiv CR)] + Br_2/DME \longrightarrow [MBr_3(\equiv CR)(DME)] \qquad (84)$$

3.13 [M(CO)$_2$(≡CR)(η-C$_5$H$_5$)] AND RELATED COMPLEXES

3.13.1 Synthesis

There are two principal ways of synthesizing complexes of the type [M(CO)$_2$(≡CR)Cp]. Complexes based on [MX(CO)$_4$(≡CR)] react with cyclopentadienyl salts (or their analogues) by elimination of two carbonyl groups and the halide (Equation (85)). The other method involves the alkylation of a suitable ligand atom, generally the nitrogen atom of a coordinated isonitrile (Equation (86)), but occasionally the sulfur atom of a thiocarbonyl ligand. Examples of such syntheses are given in Table 6. There is considerable scope for variation of substituents. The choice of alkylating agent is often important. For instance, alkylation of [M(CO)$_2$(C≡NR)Cp*]$^-$ with MeI proceeds by alkylation at the metal to form [WR(CO)$_2$(CNEt)Cp*]. On warming, the latter undergoes a migratory insertion process to form [W(CO)$_2$(η^2-CMe=NEt)Cp*], which in turn converts into [W(CO)$_2$(η^3-CH$_2$CHNEt)Cp*].[674]

$$[MX(CO)_4(\equiv CR)] + NaCp \longrightarrow [M(CO)_2(\equiv CR)Cp] \qquad (85)$$

$$[M(CO)_2(C\equiv NR)Cp]^- + Et_3O^+ \longrightarrow [M(CO)_2(\equiv CNREt)Cp] \qquad (86)$$

Bis(carbyne) complexes are also known. These are of interest partly because of the tendency of two carbyne fragments to dimerize to an alkyne. In some cases alkylation of an existing isonitrile ligand can lead to bis(carbyne) species (Equation (87); L = Cp*, Tp').[675,676]

$$[W(CNEt)_2(\equiv CNEtR)Tp'] + Et_3OBF_4 \longrightarrow [W(CNEt)(\equiv CNEtR)(\equiv CNEt_2)Tp']^+ \qquad (87)$$

3.13.2 Structure

The structures of carbynes of the type [M(CO)$_2$(≡CR)L] (L = cyclopentadienyl type ligand) have been well reviewed.[601]

3.13.3 Reactions

The carbyne complex [M(CO)$_2$(≡CNEt$_2$)L] (M = Mo or W; L = Cp or Cp*) reacts with X$_2$ (X = Br or I) to form [MX$_2$(CO)(≡CNEt$_2$)L].[664,677–80] Some of these compounds, such as [WBr$_2$(CO)(≡CNEt$_2$)], undergo reduction by Na/Hg in the presence of a two-electron ligand L (L = phosphine or isonitrile) to form carbyne complexes [W(CO)(L)(≡CNEt$_2$)Cp] (Equation (88)).[680] The oxidation state of the metal in the carbyne species [M(CO)$_2$(≡CNEt$_2$)Cp] is increased on treatment with halogens (Equation (88); X = Br or I). These reactions are essentially quantitative and the halides are mutually *cis* in the products.[681] Carbyne complexes such as [WI$_2$(CO)(≡CNEt$_2$)Cp] react with CNBu^t in consecutive reactions to form [WI$_2$(CNBu^t)(≡CNEt$_2$)Cp] and then the cation [WI(CNBu^t)$_2$(≡CNEt$_2$)Cp]I.[682]

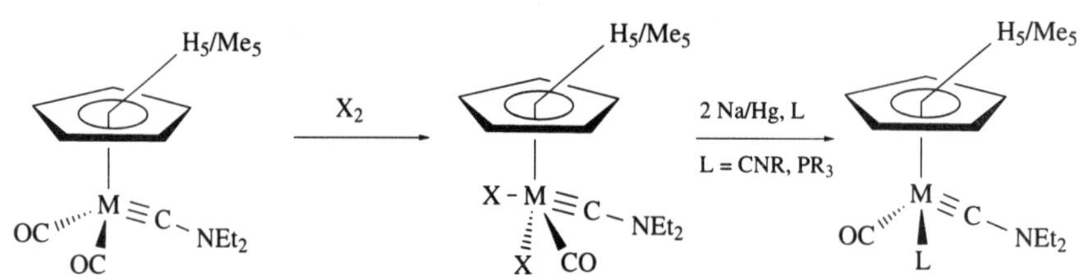

(88)

The result of treatment of complexes of the type [M(CO)$_2$(≡CR)Cp] with acid is dependent on both the functionality on the metal complex and the nature of the acid. For instance, protonation (HBF$_4$) of the carbynes [W(CO)$_2$(≡CTol)Cp] results in the formation of the transient carbenes [W(CO)$_2$(=CHTol)Cp]$^+$. These readily undergo insertion reactions with alkynes such as MeC≡CMe and PhC≡CPh to form naphthol derivatives. The use of terminal alkynes results in complete regioselectivity

Table 6 Syntheses of carbynes [M(CO)$_2$(≡CR)Cp].

Synthesis	*Complex*	*Comments*	*Ref.*
[MoCl(CO)$_4$(≡CTol) + NaCp	[Mo(CO)$_2$(≡CTol)Cp]		684
[WBr(CO)$_4$(≡CC$_6$H$_3$Me$_2$-2,6)] + NaCp·DME	[W(CO)$_2$(≡CC$_6$H$_3$Me$_2$-2,6)Cp]		685
[M(O$_2$CF$_3$)(CO)$_4$(≡CAr)] + NaCp·DME	[M(CO)$_2$(≡CAr)Cp]	M = Cr, Mo; R = C$_6$H$_3$Me$_2$-2,6. M = Mo; Ar = C$_6$H$_4$Me-2, C$_6$H$_4$OMe-2, C$_6$H$_4$NMe$_2$-2	685
[WI(CO)$_2$L$_2$(≡CNEt)] + KC$_5$R$_5$	[W(CO)$_2$(≡CNEt$_2$)(η-C$_5$R$_5$)]	R = H, Me; L = py, 4-Me-pyridine	677
[M(CO)$_2$(CNEt)L]$^-$ + [Et$_3$O]BF$_4$	[M(CO)$_2$(≡CNEt$_2$)L]	L = Cp, Cp*; M = Mo, W. η^2-iminoacyl [Mo(CO)$_2${η^2-C(Et)-=NEt}Cp*] by-product	664,674, 678,681, 686
[M(O$_2$CCF$_3$)(CO)$_2$(py)$_2$(≡C–C≡CBu^t)] + NaCp	[W(CO)$_2$(≡C–C≡CBu^t)Cp]		638
[M(O$_2$CCF$_3$)(CO)$_2$(pic)$_2$(≡C–C≡CTol)] + NaCp	[W(CO)$_2$(≡C–C≡CTol)Cp]		
[WI(CO)(CNBu^t)$_3$(≡CNEt$_2$)] + i, KCp; ii, I$_2$	[WI$_2$(CO)(≡CNEt$_2$)Cp]		682
trans-[MBr(CO)$_4$(≡CR^1)] + Na[Co{P(O)R$^2_2$}$_3$Cp]	[M(CO)$_2$(≡CR^1)-[Co{P(O)R$^2_2$}$_3$Cp]]	M = Mo, W; R^1 = Ph, Tol; R^2 = OMe, OPr^i	687,688
[W(CO)(CNEt)$_2$L]$^-$ + [Et$_3$O]BF$_4$	[W(CO)(CNEt)(≡CNEt$_2$)L]	L = Cp, Cp*	678,689
[Mo(CO){P(OMe)$_3$}(η^2-CPhCHPh)] + PhIO	[Mo(CO){P(OMe)$_3$}-(≡CCHPh$_2$)Cp]	Isomerization	690
[Cr(C≡CPh)(CO)$_3$Cp] + i, Na/Hg; ii, SiO$_2$	[Cp(OC)$_2$Cr≡CCPh=CPhC≡Cr(CO)$_2$Cp]		691
[CrBr(CO)$_2$(pic)$_2$(≡CNPr$^i_2$)] + C$_5$R$_5$$^-$	[Cr(CO)$_2$(≡CNPr$^i_2$)(η-C$_5$R$_5$)]	R = H, Me. Reaction also works with KTp'	692
[Cr(CO)(CNBu^t)$_4$(≡CNPr$^i_2$)]$^+$ + NaCp	[Cr(CO)(CNBu^t)(≡CNPr$^i_2$)Cp]	Reaction also works with KTp'	692
[WBr(CO)$_4$(≡CR)] + KTp'	[W(CO)$_2$(≡CR)Tp']	R = Tol, C$_6$H$_4$OMe-2, C$_6$H$_3$Me$_2$-2,6. The reaction works for several other pyrazolyl borate ligands	693,694
[Mo(CO)$_2$(CNEt)Tp']$^-$ + MeI	[Mo(CO)$_2$(≡CNREt)Tp']	M = Mo; R = Me, Ph, Bu^t. M = W; R = Me, Et	695,696
[Mo(CO)$_2$(η^2-COR)Tp'] + NaOEt	[Mo(CO)$_2$(≡CR)Tp']	R = Me, Et	697
[M(CO)$_3$Tp']$^-$ + [Ph$_2$I]$^+$/CH$_2$Cl$_2$	[M(CO)$_2$(≡CCl)Tp']	CH$_2$Cl$_2$ source of CHCl$_2$ radical	698
[M(CO)$_2$(≡CCl)Tp'] + ER$^-$	[M(CO)$_2$(≡CER)Tp']	E = S, Se; R = Me, Ph, C$_6$H$_4$NO$_2$-4	699
[M(CO)$_2$(≡CCl)Tp'] + OAr$^-$	[M(CO)$_2$(≡COAr)Tp']	Ar = Ph, Tol, C$_6$H$_4$OMe-4	700
[W(CO)$_2$(CS)Tp]$^-$ + MeI	[W(CO)$_2$(≡CSMe)Tp]		335
[MBr(CO)$_4$(≡CR)] + HC(pz)$_3$	[M(CO)$_2$(≡CR){HC(pz)$_3$}]$^+$	M = Mo, W; R = Me, Tol, C$_6$H$_3$Me$_2$-2,6 (W only)	701
[W(CO)$_2$(CS){HC(pz)$_3$}] + [Me$_3$O]BF$_4$	[W(CO)$_2$(≡CSMe){HC(pz)$_3$}]$^+$		702
[MBr(CO)$_4$(≡CR)] + TlBF$_4$/L	[M(CO)$_2$(≡CR)L]BF$_4$	M = Mo, W; L = HC(pz)$_3$, MeC(pz)$_3$, H$_2$C(pz)$_2$	701
[WBr(CO)$_4$(≡CR)] + K[H$_2$B(pz)$_2$]	[W(CO)$_3$(≡CR){H$_2$B(pz)$_2$}]	R = Me, Tol	703
[WCl(CO)$_2$(py)$_2$(≡CR)] + i, Na$_2$[C$_2$B$_{10}$H$_{10}$Me$_2$]; ii, NEt$_4$Cl	NEt$_4$[W(CO)$_2$(≡CR)(η^6-C$_2$B$_{10}$H$_{10}$Me$_2$)]	R = Tol, C$_6$H$_3$Me$_2$-2,6	704
[WBr(CO)$_4$(≡CR)] + Na$_2$[7,8-C$_2$B$_9$H$_9$Me$_2$]	[W(CO)$_2$(≡CR)(η-1,2-C$_2$B$_9$H$_9$Me$_2$)]$^-$	R = Me (using Tl$_2$[7,8-C$_2$B$_9$H$_9$Me$_2$]), Ph, Tol, C$_6$H$_4$Me-2, C$_6$H$_3$Me$_2$-2,6	705–8
[MoCl(CO){P(OMe)$_3$}$_3$(≡CTol)] + i, Na$_2$[7,8-C$_2$B$_9$H$_{10}$Me$_2$]; ii, NEt$_4$Cl	NEt$_4$[Mo(≡CTol)(CO)-{P(OMe)$_3$}(η^5-C$_2$B$_9$H$_9$Me$_2$)]		709
[Mo(CO)$_3$Cp]$^-$ + ClCOCF$_2$COCl	[Mo$_3$(CO)$_6$Cp$_3$(μ$_3$-CF)]	Decarbonylation, several products	710

in which the alkyne hydrogen locates adjacent to the naphthol hydroxy group. This is the opposite to the regiochemistry of conventional Dötz benzannulations.[351,683] Protonation (HBF$_4$) of [Mo(CO)$_2$(≡CNMeBu^t)Tp'] in the presence of the terminal alkyne PhC≡CH results in a complex process to form an η^2-vinyl [Mo(CO)$_2${η^2-C(Ph)=CHCH=NMeBu^t)}Tp']$^+$ formed by insertion of the alkyne into the metal–carbyne bond.[695] The transient $[W(CO)_2(=CHTol)Cp]^+$ reacts with iodide to form $[WI(CO)_2(=CHPh)Cp]$, a compound more conveniently prepared through the direct reaction of $[W(CO)_2(\equiv CTol)Cp]$ with aqueous HI. In the absence of iodide, a dimerization occurs (Equation (89)).[711,712] Although the cation $[W(CO)_2(=CHTol)Cp]^+$ possesses only transient stability, protonation of $[W(CO)_2(\equiv CC_6H_4OMe)Tp]$ with $HBF_4{\cdot}OEt_2$ affords the isolable complex $[W(CO)_2\{=CH(C_6H_4OMe)\}Tp]^+$. This 16-electron complex may be stabilized by intramolecular coordination of the methoxy oxygen atom.[694]

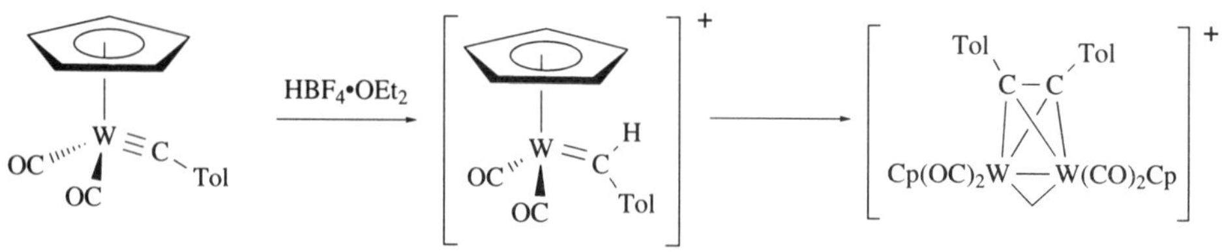

(89)

The reaction with HI is tolerant to the carbyne substituents. Thus, reaction of complex (**22**) (Equation (90)) with HI affords the alkylidene (**23**).[713] The reaction of $[W(CO)_2(\equiv CNEt_2)Cp]$ with HCl is related and results in formation of the carbene *trans*-$[WCl(CO)_2\{=CH(NEt_2)\}Cp]$.[714] This reacts further to form the η^2-acyl complex (**24**) (Equation (91)).[715] In many other cases, protonation of $[M(CO)_{2-n}(PMe_3)_n(\equiv CR)Cp]$ ($n = 0$, 1; M = Mo or W; R = Me, Ph, Tol or cyclopropyl) with HX (X = Cl, CF_3CO_2 or CCl_3CO_2) also gives η^2-acyl derivatives $[MX_2(CO)_{1-n}(PMe_3)_n(\eta^2\text{-}COCH_2R)Cp]$ in a process involving carbon–carbon bond formation.[715,716]

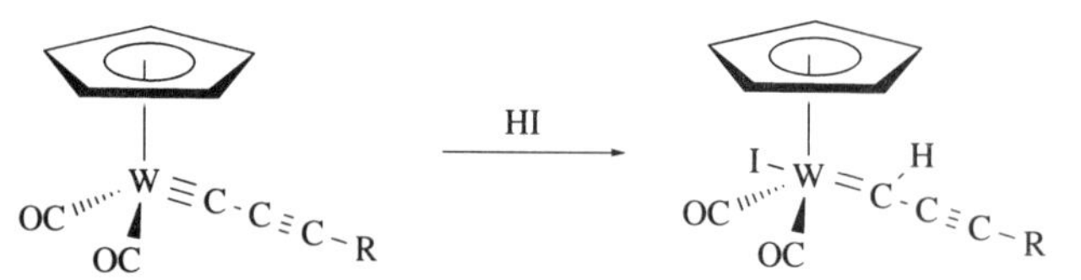

(90)

(**22**) R = Bu^t, Tol (**23**) R = Bu^t, Tol

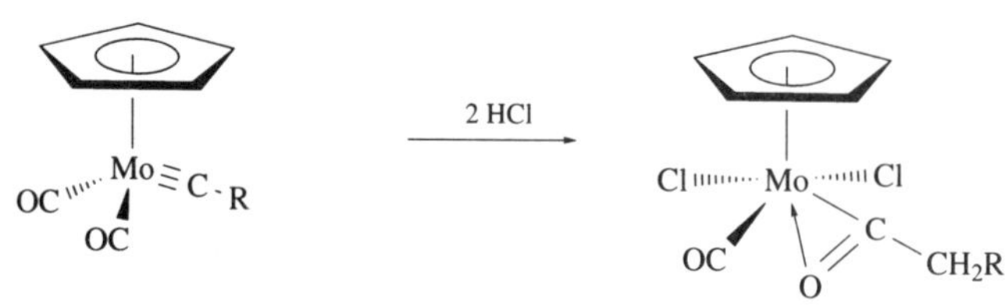

(91)

(**24**)

The carbyne cation $[W(CO)_2(\equiv CSMe)\{HC(pz)_3\}]^+$ protonates to form the dication $[W(CO)_2\{=CHSMe\}\{HC(pz)_3\}]^{2+}$, a species which reacts readily with nucleophiles.[702] A similar reaction occurs for the neutral species $[W(CO)_2(\equiv CSMe)Tp]$ which protonates to form the cationic η^2-thiocarbene $[W(CO)_2\{=CHSMe\}Tp]^+$.[542]

Protonation reactions involving carbaborane cyclopentadienyl analogues are complex and can result in migration of the carbyne to the carbaborane group, the formation of dimetallic species, protonation of the carbaborane, and carbaborane rearrangement.[717–19]

The chlorocarbyne complexes $[M(CO)_2(\equiv CCl)Tp']$ (M = Mo or W) react with a variety of nucleophiles by replacement of the chloride. Anionic nucleophiles result in the neutral complexes $[M(CO)_2(\equiv CNu)Tp']$ (Nu = OAr,[700] SR,[699] Fp,[720] or CHXY (X, Y = CN or CO_2Et)[721]). The iron compound $[Mo(CO)_2\{\equiv CFe(CO)_2Cp\}Tp']$ is interesting. Photolysis of this species in the presence of ligands PMe_3 or $CNBu^t$ results in substitution of both iron carbonyl groups, while CS_2 gives $[Mo(CO)_2\{\equiv CCS_2Fe(CO)Cp\}Tp']$. The reaction with HBF_4 gives an η^2-methyne derivative (**25**) (Equation (92)).[720]

The corresponding reactions with tertiary phosphines results in the cationic complexes $[W(CO)_2(\equiv CPR_3)Tp']^+$.[699] These cations may react further with phosphine. Thus, $[W(CO)_2(\equiv CPMe_3)Tp']^+$ reacts with excess PMe_3 to form the carbene $[W(CO)_2\{=C(PMe_3)_2\}Tp']^+$, a reaction reversed on addition of MeI which scavenges PMe_3 as $[PMe_4]I$.[722] The silylcarbynes $Tp'M(CO)_2(\equiv CSiMe_2Ph)$ (M = Mo or W) react with fluoride to form $[M(CO)_2(\equiv CH)Tp']$. These species undergoes slow dimerization to form $[Tp'(OC)_2M(\mu\text{-}C{=}CH_2)M(CO)_2Tp']$.[723]

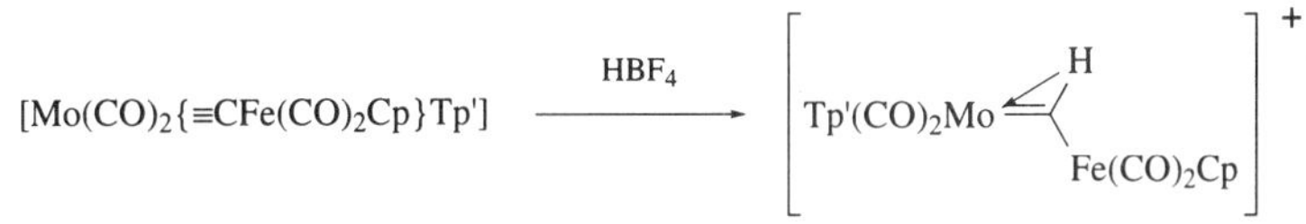

(92)

(25)

The products of reactions between complexes $[M(CO)_2(\equiv CR)Cp]$ for which nucleophilic replacement of R is not an option are varied. Under some conditions, carbonyl substitution can occur (M = Mo or W; Equation (93)).[684] Under other conditions, the result is the formation of ketenyl complexes (Equation (94)).[685,724–6] Similar ketenyl formation is observed by reaction of the anionic complex $NEt_4[Mo(CO)_2(\equiv CTol)(\eta^5\text{-}C_2B_9H_9Me_2)]$ with PMe_3.[727]

$$[M(CO)_2(\equiv CTol)Cp] + PMe_3 \longrightarrow [M(CO)(PMe_3)(\equiv CTol)Cp] \quad (93)$$

$$[M(CO)_2(\equiv CR^1)Cp] + PR^2_3 \longrightarrow [M(CO)(PR^2_3)(\eta^2\text{-}CR^1CO)Cp] \quad (94)$$

Halophosphines and arsines such as PMe_2Cl, PPh_2Cl and $AsMe_2I$ react with $[W(CO)_2(\equiv CR)Cp]$ (R = Me, Ph or Tol) to form η^3-coordinated phosphino or arsino ketene complexes,[728] whereas $PHPh_2$ adds across the W≡C bond of $[W(CO)_2(\equiv CTol)Cp]$ to form $[W(CO)_2\{\eta^2\text{-}CH(Tol)PPh_2\}Cp]$. This last compound is also available through nucleophilic attack by $LiPPh_2$ on $[W(CO)_2(\equiv CTol)Cp]$ followed by protonation (NH_4Br).[729] The reaction between $[Mo(CO)(PR_3)_2(\equiv CAr)\{H_2B(pz)_2\}]$ complexes and :P≡CBu^t results in the complexes $[Mo(CO)(PR_3)_2(\equiv CBu^t)\{H_2B(pz)_2\}]$, possibly via molybdacyclobutadiene intermediates.[730]

Various complexes of the type $[W(CO)_2(\equiv CR)Cp]$ react with one equivalent of $C_6H_4(PH_2)_2$-1,2 to form the metallaphosphine derivative $[W\{\sigma,\eta^3\text{-}C_6H_4(PH_2)(PC\{OH\}CHTol)\text{-}1,2\}(CO)Cp]$.[731] Treatment of $[W(CO)_2(\equiv CTol)Cp]$ with *rac*-$C_6H_4(PMePh)_2$-1,2 gives only one diastereomer of the ketenyl $[W\{\eta^1\text{-}C(Tol){=}C{=}O\}\{C_6H_4(PMePh)_2\}(CO)Cp]$.[731] Alternatively, an η^2-ketenyl derivative $[W(CO)(PEt_3)\{\eta^2\text{-}C(O)CSMe\}Tp']$ is formed in the reaction of $[W(CO)_2(\equiv CSMe)Tp']$ with PEt_3.[543]

The photochemistry of carbyne complexes in the presence of phosphines is complex. The UV photolysis of hexane solutions of $[W(CO)_2(\equiv CTol)Cp]$ in the presence of CO results in carbonylation of the carbyne ligand and coupling to form a ditungsten species (**26**) (Equation (95)). In the presence of PPh_3 rather than CO, the result is the η^2-ketenyl $[W(CO)(PPh_3)\{\eta^2\text{-}C(Tol)CO\}Cp]$.[732]

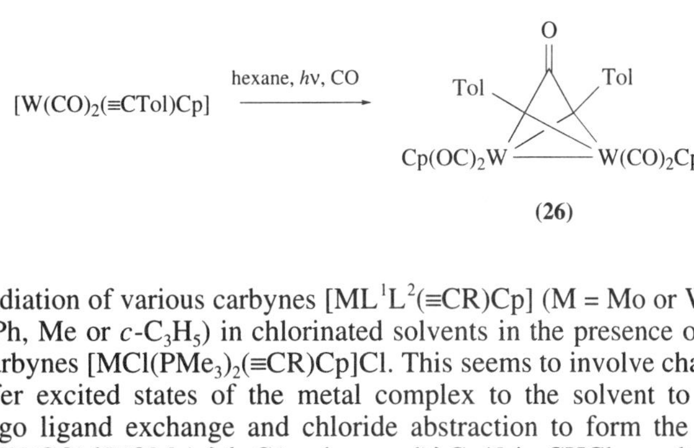

(95)

(26)

In related processes, irradiation of various carbynes $[ML^1L^2(\equiv CR)Cp]$ (M = Mo or W; $L^1 = P(OMe)_3$; L^2 = CO or $P(OMe)_3$; R = Ph, Me or *c*-C_3H_5) in chlorinated solvents in the presence of PMe_3 results in the formation of cationic carbynes $[MCl(PMe_3)_2(\equiv CR)Cp]Cl$. This seems to involve charge transfer from metal–ligand charge transfer excited states of the metal complex to the solvent to give 17-electron intermediates which undergo ligand exchange and chloride abstraction to form the final products.[733] Irradiation of the carbyne $[W(CO)\{P(OMe)_3\}\{\equiv C(cyclopropyl)\}Cp^*]$ in $CHCl_3$ results in the formation of $[W(Cl)_3\{P(OMe)_3\}Cp^*]$ together with cyclopentenone. The hydrogen of the cyclopentenone not derived from the cyclopropyl group is, in part, derived from the solvent.[734] On occasion, terminal phosphinidene complexes react with carbynes (Equation (96)).[735]

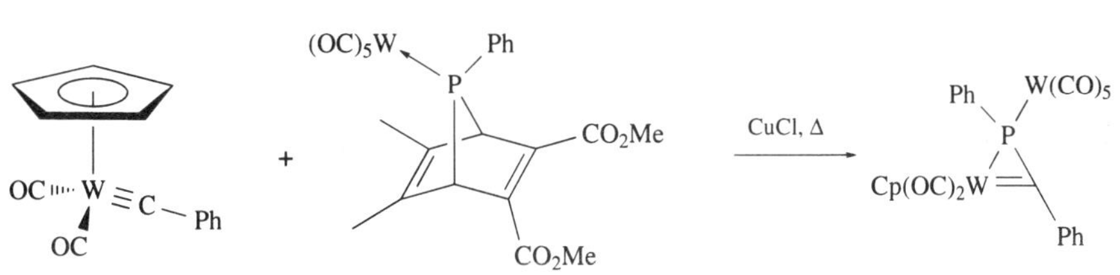

(96)

Treatment of [W(CO)$_2$(≡CR)Cp] (R = Tol or NEt$_2$) or [W(CO){P(OMe)$_3$}(≡CAr)Cp] (Ar = Ph or Tol) with [Me$_2$SSMe]$^+$ (a source of SMe$^+$) results in the η^2-thiocarbene cations [W(CO)$_2${=C(R)SMe}Cp]$^+$ or [W(CO){P(OMe)$_3$}{=C(Ar)SMe}Cp]$^+$.[490,736,737] These last compounds react with PMe$_3$ to form the ylide [M(CO)(PMe$_3$){η^2-C(Ar)(PMe$_3$)SMe}Cp]$^+$ (M = Mo or W).[490,738] The related Tp complexes behave in similar fashion. The addition of [Me$_2$SSMe]$^+$ to [W(CO)$_2$(≡CSMe)Tp] results in the η^2-dithiocarbene [W(CO)$_2${=C(SMe)SMe}Tp]$^+$. Thermolysis of this in ClCH$_2$CH$_2$Cl results in the carbyne complex [W(SMe)$_2$(≡CSMe)Tp] in a complex process involving CO loss, C–S cleavage and SMe migration. Alternatively, thermolysis of [W{η^2-C(SAr)(SMe)SMe}Tp] results in [W(CO)$_2$(≡CSMe)Tp], while its photolysis results in [W(SMe)(SAr)(≡CSMe)Tp].[739,740] The reaction between [W(CO)$_2$(≡CMe)Cp] and cyclohexene sulfide results in the dithioacetato-*S,S'* complex [W(CO)$_2$(S$_2$CMe)Cp].[741] Addition of cyclohexene sulfide or selenium to [W(CO)$_2${=C(SMe)SMe}Cp]$^+$ affords tungstabicyclo[1.1.0]butane species (Equation (97)).[742] Related behaviour is seen in reactions of complexes of this type with PClPh$_2$ (Equation (98)).[743]

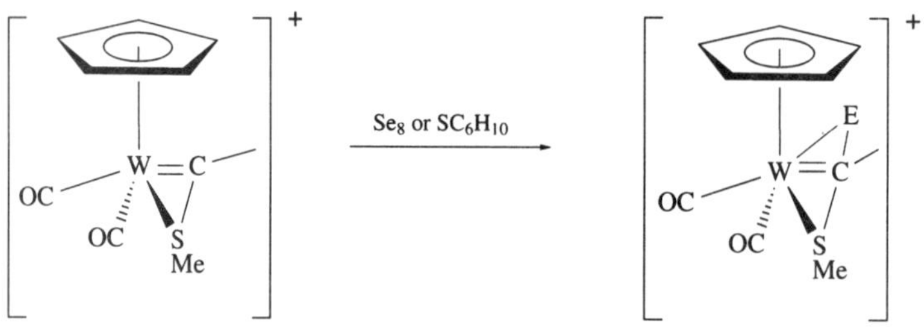

(97)

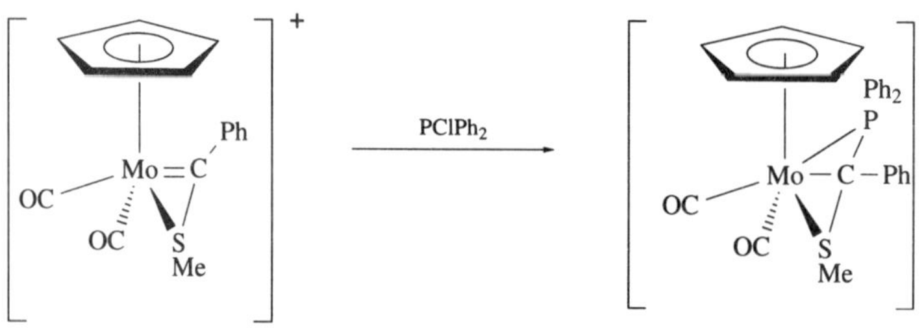

(98)

Carbyne complexes of the type [M(CO)$_2$(≡CR)Cp] and their derivatives are potent precursors for heteronuclear metal clusters (see Chapter 4, Volume 10). Mononuclear carbyne complexes are also good sources of bridging carbyne derivatives. Metal–metal bond formation is often accomplished by cation–anion coupling or elimination of labile ligands. Thus, the indenyl complex [Mo(CO)$_2$(NCMe)$_2$(Ind)]BF$_4$ reacts with the carbyne [(η^5-C$_2$B$_9$H$_9$Me$_2$)Mo(CO){P(OMe)$_3$}(≡CTol)]$^-$ by salt elimination and formation of dimetallic μ-carbyne complexes,[744] and the same anion reacts with Mo(CO)$_3$(NCMe)$_3$ to form trinuclear complexes [NEt$_4$][Mo$_3$(μ$_3$-CR)(μ-CO)(CO)$_8$(η^5-C$_2$B$_9$H$_9$Me$_2$)].[745] Treatment of mixtures of [NEt$_4$][W(≡Tol)(CO)$_2$(η^5-C$_2$B$_9$H$_9$Me$_2$)] and [W(≡CMe)(CO)$_2$Cp] with HBF$_4$ affords the complexes [W$_2$(μ-CR)(CO)$_3${η^5-C$_2$B$_9$H$_8$(CH$_2$Tol)Me$_2$}Cp].[746] Other representative examples of metal–metal bond formations are shown in Scheme 12.[747–52] The reaction with [WMe(CO)$_3$Cp] is an acyl–alkylidyne coupling process and is therefore rather unusual.[753]

In cases where the carbyne substituent contains a proton, then treatment with base can cause formation of vinylidene ligands. Thus, treatment of [Mo(CO){P(OMe)$_3$}(≡CR)Cp] (R = Me or cyclopropyl) with BuLi affords the corresponding anionic vinylidenes [Mo(CO){P(OMe)$_3$}(=C=CH$_2$)Cp]$^-$ and [Mo(CO){P(OMe)$_3$}(=C=CCH$_2$CH$_2$)Cp]$^-$, respectively. These vinylidenes tend to undergo electrophilic attack at the β-carbon atom. An example is shown in Equation (99).[754,755] In a similar fashion, the reaction of [Mo(CO)$_2$(≡CMe)Tp'] with Na[N(TMS)$_2$] forms the vinylidene anion [Mo(CO)$_2$(=C=CH$_2$)Tp']$^-$, which reacts with RX to form [Mo(CO)$_2$(≡CCH$_2$R)Tp'] (R = Me or Et).[697] In other cases, other parts of the molecule are more reactive, thus treatment of the cation [W(CO)$_2$(≡CR){HC(pz)$_3$}]BF$_4$ (R = Me or Tol) with BuLi followed by BF$_3$·OEt$_2$ results in neutral [W(CO)$_2$(≡CR){(F$_3$B)C(pz)$_3$}].[756]

The tendency for carbynes to couple was described in Section 3.12.3 for cases in which none of the other ligands are cyclopentadienyl or related ligands. Nucleophile induced carbyne–carbyne coupling to form complex (**27**) is displayed by the compound [W(CNEt)(≡CNEt)$_2$Cp*]BF$_4$ (**26**, Equation (100)), but not the corresponding Tp' derivative.[675,676] Calculations on the model system [W(CO)(CH)$_2$Cp]$^+$ predict spontaneous coupling to form the 16-electron system [CpW(CO)(HCCH)]$^+$, but that the introduction of amino substituents stabilizes the bis(carbyne) structure relative to the alkyne complex.[675]

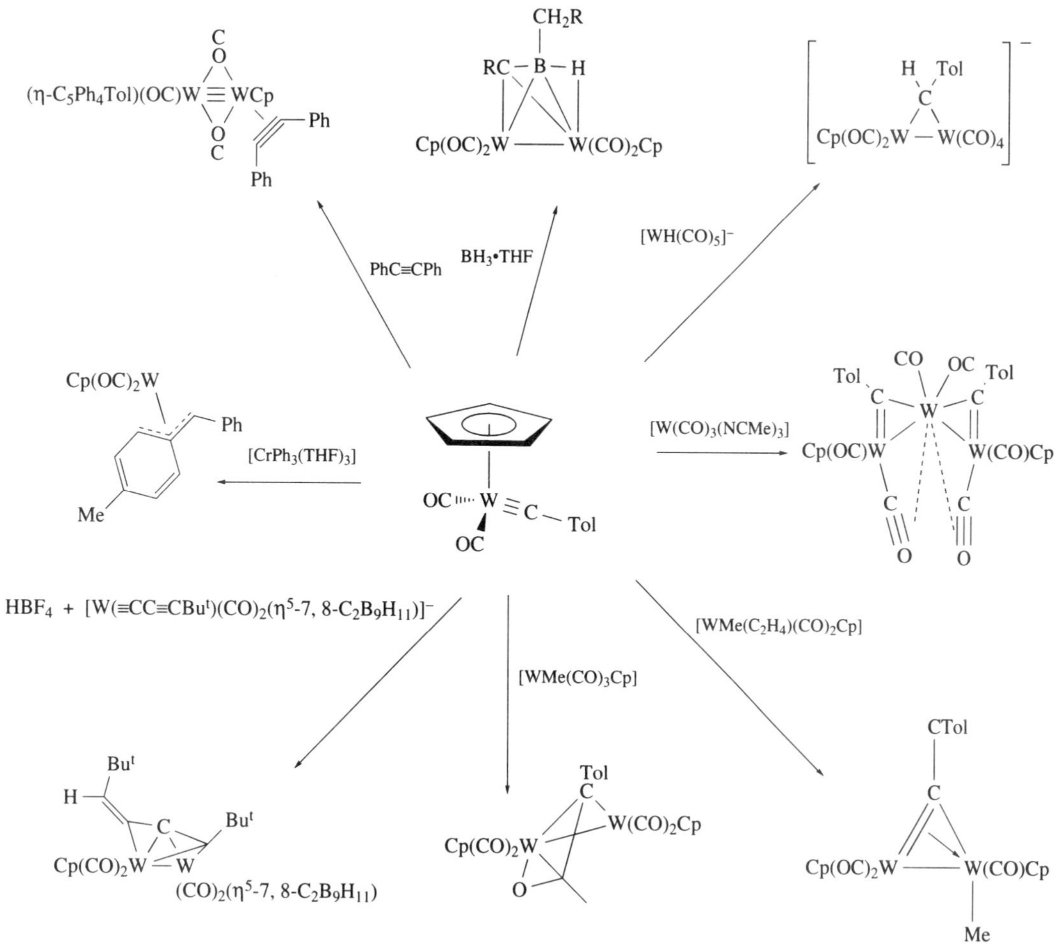

Scheme 12

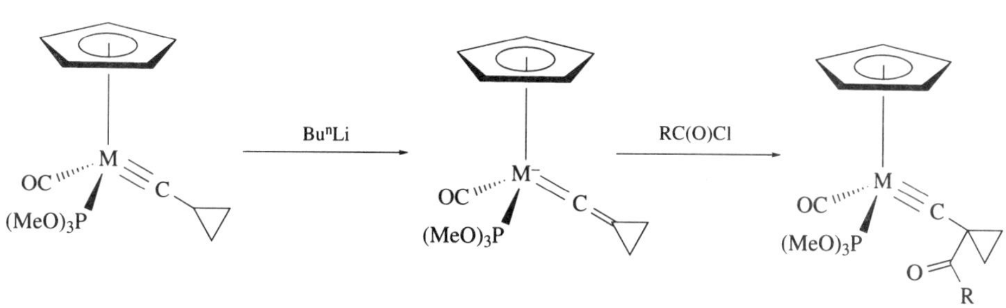

(99)

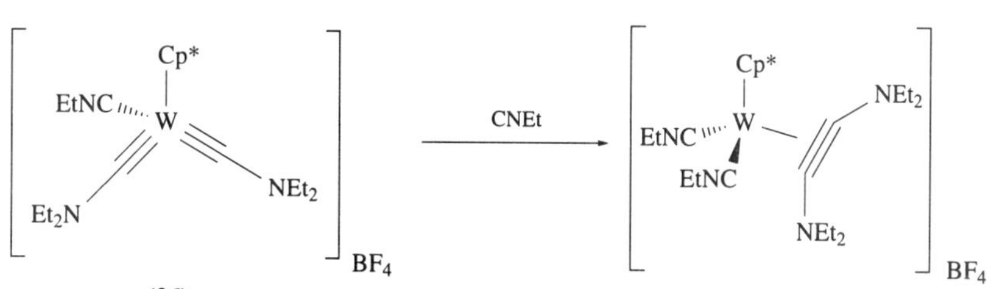

(26) (27)

(100)

3.14 REFERENCES

1. G. K.-I. Magomedov, *J. Organomet. Chem.*, 1990, **385**, 113.
2. G. K.-I. Magomedov, L. V. Morozova, S. A. Sigachev and E. A. Chernyshev, *Zh. Obshch. Khim.*, 1988, **58**, 497.
3. C. P. Horwitz and D. F. Shriver, *Adv. Organomet. Chem.*, 1984, **23**, 219.

4. A. Veillard, *Chem. Rev.*, 1991, **91**, 743.
5. S. R. Andrews and D. E. Parry, *Chem. Phys. Lett.*, 1993, **205**, 102.
6. L. A. Barnes, M. Rosi and C. W. Bauschlicher, *J. Chem. Phys.*, 1991, **94**, 2031.
7. M. J. Calhorda, C. F. Frazão and J. A. M. Simões, *J. Organomet. Chem.*, 1984, **262**, 305.
8. J. A. Ganske and R. N. Rosenfeld, *J. Phys. Chem.*, 1990, **94**, 4315.
9. D. E. Sherwood and M. B. Hall, *Inorg. Chem.*, 1983, **22**, 93.
10. T. Ziegler, V. Tschinke and C. Ursenbach, *J. Am. Chem. Soc.*, 1987, **109**, 4825.
11. D. Moncrieff, P. C. Ford, I. H. Hillier and V. R. Saunders, *J. Chem. Soc., Chem. Commun.*, 1983, 1108.
12. G. Al-Takhin, J. A. Connor, G. Pilcher and H. A. Skinner, *J. Organomet. Chem.*, 1984, **265**, 263.
13. D. J. Chornay, M. A. Coplan, J. A. Tossell, J. H. Moore, E. J. Baerends and A. Rozendaal, *Inorg. Chem.*, 1985, **24**, 877.
14. G. Cooper, K. H. Sze and C. E. Brion, *J. Am. Chem. Soc.*, 1990, **112**, 4121.
15. G. Cooper, J. C. Green, M. P. Payne, B. R. Dobson and I. H. Hillier, *J. Am. Chem. Soc.*, 1987, **109**, 3836.
16. M. Guerra, D. Jones, G. Distefano, A. Foffani and A. Modelli, *J. Am. Chem. Soc.*, 1988, **110**, 375.
17. J. A. Tossell, J. H. Moore and J. K. Olthoff, *J. Am. Chem. Soc.*, 1984, **106**, 823.
18. L. A. Barnes, B. Liu and R. Lindh, *J. Chem. Phys.*, 1993, **98**, 3978.
19. A. B. P. Lever, 'Inorganic Electronic Spectroscopy', 2nd edn., Elsevier, Amsterdam, 1984.
20. K. Pierloot, J. Verhulst, P. Verbeke and L. G. Vanquickenborne, *Inorg. Chem.*, 1989, **28**, 3059.
21. M. Kotzian, N. Rösch, H. Schröder and M. C. Zerner, *J. Am. Chem. Soc.*, 1989, **111**, 7687.
22. D. Rehder, *Coord. Chem. Rev.*, 1991, **110**, 161.
23. M. Minelli, J. H. Enemerk, R. T. C. Brownlee, M. J. O'Connor and A. G. Wedd, *Coord. Chem. Rev.*, 1985, **68**, 169.
24. J. Mason, *Polyhedron*, 1989, **8**, 1657.
25. E. W. Abel, D. E. Budgen, I. Moss, K. G. Orrell and V. Sik, *J. Organomet. Chem.*, 1989, **362**, 105.
26. R. B. Ali and J. Burgess, *Transition Met. Chem.*, 1993, **18**, 9.
27. E. C. Alyea and A. Somogyvari, *Inorg. Chim. Acta*, 1984, **83**, L49.
28. E. C. Alyea, J. Malito, S. D. Ernst, W. Kaim and S. J. Kohlmann, *Polyhedron*, 1989, **8**, 921.
29. E. C. Alyea, G. Ferguson, K. J. Fisher, R. A. Gossage and M. C. Jennings, *Polyhedron*, 1990, **9**, 2393.
30. E. C. Alyea, R. A. Gossage, J. Malito and Z. A. Munir, *Polyhedron*, 1990, **9**, 1059.
31. E. C. Alyea, K. J. Fisher, S. Foo and B. Philip, *Polyhedron*, 1993, **12**, 489.
32. F. Bonomi, S. Iametti and D. M. Kurtz, *Inorg. Chim. Acta*, 1992, **191**, 197.
33. R. T. C. Brownlee, A. F. Masters, M. J. O'Connor, A. G. Wedd, H. A. Kimlin and J. D. Cotton, *Org. Magn. Reson.*, 1982, **20**, 73.
34. R. T. C. Brownlee, M. J. O'Connor, B. P. Shehan and A. G. Wedd, *J. Magn. Res.*, 1985, **61**, 22.
35. R. T. C. Brownlee, M. J. O'Connor, B. P. Shehan and A. G. Wedd, *J. Magn. Res.*, 1985, **64**, 142.
36. R. T. C. Brownlee, M. J. O'Connor, B. P. Shehan and A. G. Wedd, *Aust. J. Chem.*, 1986, **39**, 931.
37. R. T. C. Brownlee, B. P. Shehan and A. G. Wedd, *Aust. J. Chem.*, 1988, **41**, 1457.
38. H. Brunner, R. Grassl, W. Meier, J. Wachter, B. Nuber and M. L. Ziegler, *J. Organomet. Chem.*, 1992, **434**, 63.
39. M. Cano, J. A. Campo, J. Y. Le Gall, R. Pichon, J. Y. Salaun and M. M. Kubicki, *Inorg. Chim. Acta*, 1992, **193**, 207.
40. P. A. Cybulski, D. J. Gillis and M. C. Baird, *Inorg. Chem.*, 1993, **32**, 460.
41. S. Dysart, I. Georgii and B. E. Mann, *J. Organomet. Chem.*, 1981, **213**, C10.
42. J. W. Faller and B. C. Whitmore, *Organometallics*, 1986, **5**, 752.
43. G. M. Gray and C. S. Kraihanzel, *Inorg. Chem.*, 1983, **22**, 2959.
44. G. M. Gray and R. J. Gray, *Organometallics*, 1983, **2**, 1026.
45. G. M. Gray, J. E. Whitten and J. W. Box, *Inorg. Chim. Acta*, 1986, **116**, 21.
46. G. M. Gray and W. Watt, *J. Organomet. Chem.*, 1992, **434**, 181.
47. L. Huis, P. J. W. Pouwels and C. Maclean, *Mol. Phys.*, 1990, **70**, 879.
48. P. Jaitner and W. Wohlgenannt, *Monatsh. Chem.*, 1982, **113**, 699.
49. M. M. Kubicki, R. Kergoat, J. Y. Le Gall, J. E. Guerchais, J. Douglade and R. Mercier, *Aust. J. Chem.*, 1982, **35**, 1543.
50. M. M. Kubicki, J. Y. Le Gall, R. Kergoat and L. C. G. De Lima, *Can. J. Chem.*, 1987, **65**, 1292.
51. M. M. Kubicki, J. Y. Le Gall, R. Pichon, J. Y. Salaun, M. Cano and J. A. Campo, *J. Organomet. Chem.*, 1988, **348**, 349.
52. M. M. Kubicki and P. Richard, *Acta Crystallogr., Sect. C*, 1991, **47**, 2362.
53. J. Malito, E. C. Alyea and R. A. Gossage, *Spectrochim. Acta, Part A*, 1992, **48**, 873.
54. A. F. Masters, G. E. Bossard, T. A. George, R. T. C. Brownlee, M. J. O'Connor and A. G. Wedd, *Inorg. Chem.*, 1983, **22**, 908.
55. C. A. Merlic and B. Adams, *J. Organomet. Chem.*, 1992, **431**, 313.
56. M. Minelli, J. L. Hubbard, K. A. Christensen and J. H. Enemark, *Inorg. Chem.*, 1983, **22**, 2652.
57. M. Minelli, J. H. Enemark, A. Bell and R. A. Walton, *J. Organomet. Chem.*, 1985, **284**, 25.
58. H. Brunner, W. Meier, J. Wachter, I. Bernal and E. Raabe, *J. Organomet. Chem.*, 1989, **362**, 95.
59. R. T. C. Brownlee, B. P. Shehan and A. G. Wedd, *Inorg. Chem.*, 1987, **26**, 2022.
60. M. Hughes, J. Mason, G. J. Leigh and R. L. Richards, *J. Organomet. Chem.*, 1988, **341**, 381.
61. J. Y. Le Gall, M. M. Kubicki and F. Y. Petillon, *J. Organomet. Chem.*, 1981, **221**, 287.
62. B. E. Mann, *Ann. Rep. NMR Spectrosc.*, 1991, **23**, 141.
63. N. M. Doherty and N. W. Hoffman, *Chem. Rev.*, 1991, **91**, 553.
64. R. L. Keiter and D. G. Van der Velde, *J. Organomet. Chem.*, 1983, **258**, C34.
65. G. T. Andrews, I. J. Colquhoun, W. McFarlane and S. O. Grim, *J. Chem. Soc., Dalton Trans.*, 1982, 2353.
66. R. Benn, H. Brenneke, J. Heck and A. Rufinska, *Inorg. Chem.*, 1987, **26**, 2826.
67. R. Benn, A. Rufinska, M. A. King, C. E. Osterberg and T. G. Richmond, *J. Organomet. Chem.*, 1989, **376**, 359.
68. D. J. Santure, J. C. Huffman and A. P. Sattelberger, *Inorg. Chem.*, 1985, **24**, 371.
69. C. G. Young and J. H. Enemark, *Aust. J. Chem.*, 1986, **39**, 997.
70. G. T. Andrews, I. J. Colquhoun and W. McFarlane, *Polyhedron*, 1983, **2**, 783.
71. I. J. Colquhoun, W. Mcfarlane and R. L. Keiter, *J. Chem. Soc., Dalton Trans.*, 1984, 455.
72. M. Minelli and W. J. Maley, *Inorg. Chem.*, 1989, **28**, 2954.
73. J. L. Bookham, W. McFarlane and I. J. Colquhoun, *J. Chem. Soc., Dalton Trans.*, 1988, 503.
74. W. Buchner and W. A. Schenk, *Inorg. Chem.*, 1984, **23**, 132.

75. E. Oldfield, M. A. Keniry, S. Shinoda, S. Schramm, T. L. Brown and H. S. Gutowsky, *J. Chem. Soc., Chem. Commun.*, 1985, 791.
76. S. P. Wang, P. Yuan and M. Schwartz, *Inorg. Chem.*, 1990, **29**, 484.
77. D. M. Adams, B. M. Peake and I. D. Taylor, *J. Chem. Soc., Chem. Commun.*, 1982, 240.
78. T. C. W. Mak, *Z. Kristallogr.*, 1984, **166**, 277.
79. F. Heinemann, H. Schmidt, K. Peters and D. Thiery, *Z. Kristallogr.*, 1992, **198**, 123.
80. K. G. Orell, *Mech. Inorg. Organomet. React.*, 1989, **6**, 249.
81. A. J. Poë, *Mech. Inorg. Organomet. React.*, 1991, **7**, 221.
82. S.-J. Wang and R. J. Angelici, *Inorg. Chem.*, 1987, **27**, 3233.
83. M. O. Albers and N. J. Coville, *Coord. Chem. Rev.*, 1984, **53**, 227.
84. T.-Y. Luh, *Coord. Chem. Rev.*, 1984, **60**, 255.
85. Y.-C. Gao, Q.-Z. Shi, D. L. Kershner and F. Basolo, *Inorg. Chem.*, 1988, **27**, 188.
86. Y.-L. Shi, Y.-C. Gao, Q.-Z. Shi, D. L. Kershner and F. Basolo, *Organometallics*, 1987, **6**, 1528.
87. J.-K. Shen, Y.-C. Gao, Q.-Z. Shi and F. Basolo, *J. Organomet. Chem.*, 1991, **401**, 295.
88. T. S. A. Hor, H. S. O. Chan, Y.-P. Leong and M.-M. Tan, *J. Organomet. Chem.*, 1989, **373**, 221.
89. C. Long, in 'The Chemistry of the Metal–Carbon Bond', ed. F. R. Hartley, Wiley, New York, 1989, vol. 5, p. 31.
90. M. B. Simpson, M. Poliakoff, J. J. Turner, W. B. Maier and J. G. McLaughlin, *J. Chem. Soc., Chem. Commun.*, 1983, 1355.
91. J. J. Turner, M. B. Simpson, M. Poliakoff, W. B. Maier and M. A. Graham, *Inorg. Chem.*, 1983, **22**, 911.
92. R. K. Upmacis, M. Poliakoff and J. J. Turner, *J. Am. Chem. Soc.*, 1986, **108**, 3645.
93. R. K. Upmacis *et al.*, *J. Chem. Soc., Chem. Commun.*, 1985, 27.
94. M. F. Gregory, S. A. Jackson, M. Poliakoff and J. J. Turner, *J. Chem. Soc., Chem. Commun.*, 1986, 1175.
95. J. A. Crayston, M. J. Almond, A. J. Downs, M. Poliakoff and J. J. Turner, *Inorg. Chem.*, 1984, **23**, 3051.
96. M. J. Almond, J. A. Crayston, A. J. Downs, M. Poliakoff and J. J. Turner, *Inorg. Chem.*, 1986, **25**, 19.
97. M. J. Almond and M. Hahne, *J. Chem. Soc., Dalton Trans.*, 1988, 2255.
98. M. J. Almond and A. J. Downs, *J. Chem. Soc., Dalton Trans.*, 1988, 809.
99. M. J. Almond, A. J. Downs and R. N. Perutz, *Inorg. Chem.*, 1985, **24**, 275.
100. T. Szymanska-Buzar, A. J. Downs, T. M. Greene and M. A. K. Wood, *J. Organomet. Chem.*, 1991, **419**, 151.
101. R. H. Hooker and A. J. Rest, *J. Organomet. Chem.*, 1983, **249**, 137.
102. J. Mascetti and A. J. Rest, *J. Chem. Soc., Chem. Commun.*, 1987, 221.
103. K. E. Lewis, D. M. Golden and G. P. Smith, *J. Am. Chem. Soc.*, 1984, **106**, 3905.
104. T. A. Seder, S. P. Church, A. J. Ouderkirk and E. Weitz, *J. Am. Chem. Soc.*, 1985, **107**, 1432.
105. T. A. Seder, S. P. Church and E. Weitz, *J. Am. Chem. Soc.*, 1986, **108**, 4721.
106. T. R. Fletcher and R. N. Rosenfeld, *J. Am. Chem. Soc.*, 1983, **105**, 6358.
107. T. R. Fletcher and R. N. Rosenfeld, *J. Am. Chem. Soc.*, 1985, **107**, 2203.
108. T. R. Fletcher and R. N. Rosenfeld, *J. Am. Chem. Soc.*, 1986, **108**, 1686.
109. T. R. Fletcher and R. N. Rosenfeld, *J. Am. Chem. Soc.*, 1988, **110**, 2097.
110. G. P. Smith, *Polyhedron*, 1988, **7**, 1605.
111. W. Tumas, B. Gitlin, A. M. Rosan and J. T. Yardley, *J. Am. Chem. Soc.*, 1982, **104**, 55.
112. J. P. Holland and R. N. Rosenfeld, *Chem. Phys. Lett.*, 1988, **145**, 481.
113. J. P. Holland and R. N. Rosenfeld, *J. Chem. Phys.*, 1988, **89**, 7217.
114. J. A. Ganske and R. N. Rosenfeld, *J. Phys. Chem.*, 1989, **93**, 1959.
115. Y. Ishikawa, P. A. Hackett and D. M. Rayner, *J. Phys. Chem.*, 1988, **92**, 3863.
116. Y. Ishikawa, P. A. Hackett and D. M. Rayner, *Chem. Phys. Lett.*, 1988, **145**, 429.
117. Y. Ishikawa, R. A. Weersink, P. A. Hackett and D. M. Rayner, *Chem. Phys. Lett.*, 1987, **142**, 271.
118. C. E. Brown, Y. Ishikawa, P. A. Hackett and D. M. Rayner, *J. Am. Chem. Soc.*, 1990, **112**, 2530.
119. Y. Ishikawa, C. E. Brown, P. A. Hackett and D. M. Rayner, *Chem. Phys. Lett.*, 1988, **150**, 506.
120. D. R. Tyler and D. P. Petrylak, *J. Organomet. Chem.*, 1981, **212**, 389.
121. A. J. Lees and A. W. Adamson, *Inorg. Chem.*, 1981, **20**, 4381.
122. S. P. Church, F.-W. Grevels, H. Hermann and K. Schaffner, *Inorg. Chem.*, 1985, **24**, 418.
123. S. Zhang, G. R. Dobson, V. Zang, H. C. Bajaj and R. van Eldik, *Inorg. Chem.*, 1990, **29**, 3477.
124. S. Zhang, V. Zang, H. C. Bajaj, G. R. Dobson and R. van Eldik, *J. Organomet. Chem.*, 1990, **397**, 279.
125. S. Zhang and G. R. Dobson, *Polyhedron*, 1990, **9**, 2511.
126. K. J. Asali, H. H. Awad, J. F. Kimbrough, B. C. Lang, J. M. Watts and G. R. Dobson, *Organometallics*, 1991, **10**, 1822.
127. G. R. Dobson, K. J. Asali, C. D. Cate and C. W. Cate, *Inorg. Chem.*, 1991, **30**, 4471.
128. S. Zhang and G. R. Dobson, *Inorg. Chim. Acta*, 1991, **181**, 103.
129. S. Zhang and G. R. Dobson, *Organometallics*, 1992, **11**, 2447.
130. U. Kunze and H. Jawad, *J. Organomet. Chem.*, 1984, **277**, C31.
131. S. K. Nayak and T. J. Burkey, *Organometallics*, 1991, **10**, 3745.
132. J. M. Morse, G. H. Parker and T. J. Burkey, *Organometallics*, 1989, **8**, 2471.
133. T. J. Burkey, *Polyhedron*, 1989, **8**, 2681.
134. D. E. Marx and A. J. Lees, *Organometallics*, 1987, **5**, 2072.
135. D. E. Marx and A. J. Lees, *Inorg. Chem.*, 1987, **26**, 2254.
136. M. Lee and C. B. Harris, *J. Am. Chem. Soc.*, 1989, **111**, 8963.
137. S. Oishi, *Organometallics*, 1988, **7**, 1237.
138. M. J. Schadt, N. J. Gresalfi and A. J. Lees, *Inorg. Chem.*, 1985, **24**, 2942.
139. M. J. Schadt, N. J. Gresalfi and A. J. Lees, *J. Chem. Soc., Chem. Commun.*, 1984, 506.
140. S.-C. Yu, X. Xu, R. Lingle and J. B. Hopkins, *J. Am. Chem. Soc.*, 1990, **112**, 3668.
141. F.-W. Grevels and V. Skibbe, *J. Chem. Soc., Chem. Commun.*, 1984, 681.
142. S. P. Church, F.-W. Grevels, H. Hermann and K. Schaffner, *Inorg. Chem.*, 1984, **23**, 3830.
143. S. P. Church, F.-W. Grevels, H. Hermann and K. Schaffner, *J. Chem. Soc., Chem. Commun.*, 1985, 30.
144. S. Zhang, H. C. Bajaj, V. Zang, G. R. Dobson and R. van Eldik, *Organometallics*, 1992, **11**, 3901.
145. D. C. Bailey and S. H. Langer, *Chem. Rev.*, 1981, **81**, 109.
146. G. W. Wagner and B. E. Hanson, *J. Am. Chem. Soc.*, 1989, **111**, 5098.

147. G. W. Wagner and B. E. Hanson, *Organometallics*, 1987, **6**, 2494.
148. S. Sivasanker and A. V. Ramaswamy, *Ind. J. Technol.*, 1983, **21**, 339.
149. W. M. Shirley, N. A.-M. and J. S. Frye, *Inorg. Chem.*, 1991, **27**, 3846.
150. W. M. Shirley, C. A. Powers and J. S. Frye, *Inorg. Chem.*, 1991, **30**, 4182.
151. W. M. Shirley, B. R. McGarvey, B. Maiti, A. Brenner and A. Cichowlas, *J. Mol. Catal.*, 1985, **29**, 259.
152. J. L. Portefaix, F. Figueras, M. Forissier and D. Foujols, *Bull. Soc. Chim. Fr.*, 1984, 33.
153. E. E. Platero, C. O. Areán, D. Scarano, G. Spoto and A. Zecchina, *Mater. Chem. Phys.*, 1991, **29**, 347.
154. Y. Okamoto, T. Imanaka, K. Asakura and Y. Iwasawa, *J. Phys. Chem.*, 1991, **95**, 3700.
155. Y. Okamoto, Y. Inui, H. Onimatsu and T. Imanaka, *J. Phys. Chem.*, 1991, **95**, 4596.
156. D. A. Hucul and A. Brenner, *J. Chem. Soc., Chem. Commun.*, 1982, 830.
157. J. Goldwasser, S. M. Fang, M. Houalla and W. K. Hall, *J. Catal.*, 1989, **115**, 34.
158. A. G. Anshits, S. N. Vereshchagin, N. G. Maksimov and W. K. Hall, *Kinet. Katal.*, 1990, **31**, 556.
159. A. G. Anshits, S. N. Vereshchagin, E. D. Korniets, G. D. Kholopova and W. K. Hall, *Kinet. Katal.*, 1988, **29**, 540.
160. T. H. Walter *et al.*, *J. Am. Chem. Soc.*, 1988, **110**, 1065.
161. S. Sivasanker, E. P. Yesodharan, C. Sudhakar and A. Brenner, *J. Catal.*, 1984, **87**, 514.
162. A. Cichowlas, E. P. Yesodharan and A. Brenner, *Appl. Catal.*, 1984, **11**, 353.
163. T. J. Thomas and A. Brenner, *J. Mol. Catal.*, 1983, **18**, 197.
164. J.-K. Wang, T. Komatsu, S. Namba, T. Yashima and T. Uematsu, *J. Mol. Catal.*, 1986, **37**, 327.
165. Y. I. Yermakov, B. N. Kuznetsove and V. A. Zakharov, 'Catalysis by Supported Complexes', Elsevier, New York, 1981.
166. A. Maezawa, H. Kane, Y. Okamoto and T. Imanaka, *Chem. Lett.*, 1988, 241.
167. R. F. Howe, in 'Tailored Metal Catalysts', ed. Y. Iwasawa, Reidel, Boston, MA, 1986, p. 141.
168. M. B. Logan, R. F. Howe and R. P. Cooney, *J. Mol. Catal.*, 1992, **74**, 285.
169. T. Chevreau, D. Cornet, M. Marzin and K. Roudias, *Bull. Soc. Chim. Fr.*, 1988, 483.
170. H. Boulet, C. Bremard, C. Depecker and P. Legrand, *J. Chem. Soc., Chem. Commun.*, 1991, 1411.
171. S. Abdo, J. Gosbee and R. F. Howe, *J. Chim. Phys. Phys.-Chim. Biol.*, 1981, **78**, 885.
172. K. Moller, T. Bein, S. Özkar and G. A. Ozin, *J. Phys. Chem.*, 1991, **95**, 5276.
173. C. L. Tway and T. M. Apple, *Inorg. Chem.*, 1992, **31**, 2885.
174. S. Namba, T. Komatsu and T. Yashima, *Chem. Lett.*, 1982, 115.
175. S. Özkar, G. A. Ozin and R. A. Prokopowicz, *Chem. Mater.*, 1992, **4**, 1380.
176. A. Zecchina, K. M. Rao, S. Coluccia, E. E. Platero and C. O. Areán, *J. Mol. Catal.*, 1989, **53**, 397.
177. A. Zecchina, S. Bordiga, E. E. Platero and C. O. Areán, *J. Catal.*, 1990, **125**, 568.
178. A. Kazusaka and R. F. Howe, *J. Catal.*, 1988, **111**, 50.
179. C. Bremard, C. Depecker, H. Desgrousilliers and P. Legrand, *Appl. Spectrosc.*, 1991, **45**, 1278.
180. C. Bremard and H. D. Grousilliers, *J. Raman Spectrosc.*, 1991, **22**, 125.
181. Y. Okamoto, H. Kane and T. Imanaka, *Chem. Lett.*, 1988, 2005.
182. Y. Okamoto, A. Maezawa, H. Kane, I. Mitsushima and T. Imanaka, *J. Chem. Soc., Faraday Trans. 1*, 1988, **84**, 851.
183. Y. Okamoto, A. Maezawa, H. Kane and T. Imanaka, *J. Chem. Soc., Chem. Commun.*, 1988, 380.
184. F. R. Hartley, 'Supported Metal Complexes', Reidel, Boston, MA, 1985.
185. Y. Okamoto, H. Onimatsu, M. Hori, Y. Inui and T. Imanaka, *Catal. Lett.*, 1992, **12**, 239.
186. Y. Okamoto, A. Maezawa, H. Kane and T. Imanaka, *J. Mol. Catal.*, 1989, **52**, 337.
187. Y. Okamoto, H. Kane and T. Imanaka, *Catal. Lett.*, 1989, **2**, 335.
188. Y. You-Sing and R. F. Howe, *J. Chem. Soc., Faraday Trans. 1*, 1986, **82**, 2887.
189. R. C. Simon, H. D. Gafney and D. L. Morse, *Inorg. Chem.*, 1985, **24**, 2565.
190. J. R. Swanson, F. A. Flitsch and C. M. Friend, *Surf. Sci.*, 1989, **215**, L293.
191. J. R. Swanson, F. A. Flitsch and C. M. Friend, *Surf. Sci.*, 1990, **226**, 147.
192. D. V. Chakarov, Z. C. Ying and W. Ho, *Surf. Sci.*, 1991, **255**, L550.
193. F. A. Flitsch, J. R. Swanson and C. M. Friend, *Surf. Sci.*, 1991, **245**, 85.
194. G. J. Gajda, R. H. Grubbs and W. H. Weinberg, *J. Am. Chem. Soc.*, 1987, **109**, 66.
195. R. Jelinek, S. Özkar, H. O. Pastore, A. Malek and G. A. Ozin, *J. Am. Chem. Soc.*, 1993, **115**, 563.
196. A. Zecchina, S. Bordiga, C. O. Areán and E. E. Platero, *J. Mol. Catal.*, 1991, **70**, 43.
197. J. E. Ellis, *Adv. Organomet. Chem.*, 1990, **31**, 1.
198. W. Beck, *Angew. Chem., Int. Ed. Engl.*, 1991, **30**, 168.
199. A. P. Masters, M. Parvez and T. S. Sorensen, *Can. J. Chem.*, 1992, **69**, 2136.
200. J. M. Maher, R. P. Beatty and N. J. Cooper, *Organometallics*, 1982, **1**, 215.
201. J. M. Maher, R. P. Beatty and N. J. Cooper, *Organometallics*, 1985, **4**, 1354.
202. J. T. Lin, G. P. Hagen and J. E. Ellis, *J. Am. Chem. Soc.*, 1983, **105**, 2296.
203. G. L. Rochfort and J. E. Ellis, *J. Organomet. Chem.*, 1983, **250**, 265.
204. D. H. Gibson, K. R. Phillips and F. U. Ahmed, *J. Organomet. Chem.*, 1984, **276**, 363.
205. R. E. Bachman and K. H. Whitmire, *Acta Crystallogr., Sect. C*, 1993, **49**, 327.
206. R. N. McDonald and P. Schell, *Organometallics*, 1988, **7**, 1820.
207. A. A. Bengali *et al.*, *J. Am. Chem. Soc.*, 1992, **114**, 5257.
208. P. M. George and J. L. Beauchamp, *J. Chem. Phys.*, 1982, 2959.
209. Y. H. Pan and D. P. Ridge, *J. Am. Chem. Soc.*, 1992, **114**, 2773.
210. C. E. C. A. Hop and T. B. Mcmahon, *J. Am. Chem. Soc.*, 1992, **114**, 1237.
211. D. L. Bricker and D. H. Russell, *J. Am. Chem. Soc.*, 1987, **109**, 3910.
212. K. Lane, L. Sallans and R. R. Squires, *J. Am. Chem. Soc.*, 1984, **106**, 2719.
213. L. G. Vanquickenborne and J. Verhulst, *J. Am. Chem. Soc.*, 1983, **105**, 1769.
214. R. C. Hynes, K. F. Preston, J. J. Springs and A. J. Williams, *Organometallics*, 1990, **9**, 2298.
215. J. O. Dziegielewski, *Polyhderon*, 1984, **3**, 1131.
216. J. O. Dziegielewski, R. Gil and P. Glowacki, *Radiochem. Radioanal. Lett.*, 1982, **50**, 251.
217. R. N. Bagchi, A. M. Bond, R. Colton, D. L. Luscombe and J. E. Moir, *J. Am. Chem. Soc.*, 1986, **108**, 3352.
218. S. R. Horning, M. Vincenti and R. G. Cooks, *J. Am. Chem. Soc.*, 1990, **112**, 119.
219. S. R. Horning, T. Kotiaho, L. E. Dejarme, J. M. Wood and R. G. Cooks, *Int. J. Mass Spectrom. Ion Proc.*, 1991, **110**, 1.

220. L. E. Dejarme, R. G. Cooks and T. Ast, *Org. Mass Spectrom.*, 1992, **27**, 667.
221. R. Susic, L. Lu, D. E. Riederer, D. Zigon, R. G. Cooks and T. Ast, *Org. Mass Spectrom.*, 1992, **27**, 769.
222. R. G. Cooks, T. Ast, B. Kralj, V. Kramer and D. Zigon, *J. Am. Soc. Mass Spectrom.*, 1990, **1**, 16.
223. F. M. Harris, P. J. Jackson and J. A. Rontree, *J. Am. Soc. Mass Spectrom.*, 1991, **2**, 108.
224. N. F. Zahran, S. K. Hindawi and A. I. Helal, *Ind. J. Chem., Sect. A*, 1989, **28**, 23.
225. L. Sellers-Hahn and D. H. Russell, *J. Am. Chem. Soc.*, 1990, **112**, 5953.
226. J. Allison, A. Mavridis and J. F. Harrison, *Polyhedron*, 1988, **7**, 1559.
227. J. M. Maher, G. R. Lee and N. J. Cooper, *J. Am. Chem. Soc.*, 1982, **104**, 6797.
228. G. R. Lee and N. J. Cooper, *Organometallics*, 1989, **8**, 1538.
229. M. F. Semmelhack and G. R. Lee, *Organometallics*, 1987, **6**, 1839.
230. R. Imwinkelried and L. S. Hegedus, *Organometallics*, 1988, **7**, 702.
231. O. J. Curnow, B. Schiemenz, G. Huttner and L. Zsolnai, *J. Organomet. Chem.*, 1993, **459**, 17.
232. I. P. Beletskaya, A. Z. Voskoboynikov, E. B. Chuklanova, A. I. Gusev and A. V. Kisin, *J. Organomet. Chem.*, 1993, **454**, 1.
233. J. T. Lin, G. P. Hagen and J. E. Ellis, *Organometallics*, 1983, **2**, 1145.
234. J. T. Lin and J. E. Ellis, *J. Am. Chem. Soc.*, 1983, **105**, 6252.
235. N. D. Silavwe, A. S. Goldman, D. R. Ritter and D. R. Tyler, *Inorg. Chem.*, 1989, **28**, 1231.
236. R. A. Jones and B. R. Whittlesey, *Organometallics*, 1984, **3**, 469.
237. H. B. Friedrich and J. R. Moss, *Adv. Organomet. Chem.*, 1991, **33**, 235.
238. J. Holton, M. F. Lappert, R. Pearce and P. I. W. Yarrow, *Chem. Rev.*, 1983, **83**, 135.
239. D. B. Pourreau and G. L. Geoffroy, *Adv. Organomet. Chem.*, 1985, **24**, 249.
240. L. D. Durfee and I. P. Rothwell, *Chem. Rev.*, 1988, **88**, 1059.
241. B. A. Narayanan, C. Amatore and J. K. Kochi, *Organometallics*, 1986, **5**, 926.
242. J. A. Gladysz, *Adv. Organomet. Chem.*, 1982, **20**, 1.
243. D. J. Darensbourg, R. K. Hanckel, C. G. Bauch, M. Pala, D. Simmons and J. N. White, *J. Am. Chem. Soc.*, 1985, **107**, 7463.
244. D. J. Darensbourg, C. G. Bauch and A. L. Rheingold, *Inorg. Chem.*, 1987, **26**, 977.
245. D. J. Darensbourg, J. A. Joyce and A. Rheingold, *Organometallics*, 1991, **10**, 3407.
246. D. J. Darensbourg, H. P. Wiegreffe and J. H. Reibenspies, *Organometallics*, 1991, **10**, 6.
247. W. P. Fehlhammer and C. Moinet, *J. Organomet. Chem.*, 1984, **260**, C55.
248. H. M. Dawes, M. B. Hursthouse, A. A. Del Paggio, E. L. Muetterties and A. W. Parkins, *Polyhedron*, 1985, **4**, 379.
249. F. Beck, W. P. Fehlhammer and W. Beck, *J. Organomet. Chem.*, 1985, **279**, C22.
250. E. Singleton and H. E. Oosterhuizen, *Adv. Organomet. Chem.*, 1982, **22**, 209.
251. M. O. Albers, E. Singleton and N. J. Coville, *J. Chem. Educ.*, 1986, **63**, 444.
252. N. J. Coville and M. O. Albers, *Inorg. Chim. Acta*, 1982, **65**, L7.
253. M. P. Guy, J. L. Coffer, J. S. Rommel and D. W. Bennett, *Inorg. Chem.*, 1988, **27**, 2942.
254. H. Oberhammer and D. Lentz, *Inorg. Chem.*, 1985, **24**, 1271.
255. D. B. Beach, W. L. Jolly and D. Lentz, *Inorg. Chem.*, 1985, **24**, 1892.
256. P. Caillet, J. Y. Le Marouille and P. Lemaux, *J. Organomet. Chem.*, 1985, **291**, 191.
257. G. C. Schoemaker, D. J. Stufkens, S. J. Goede, T. van der Does and F. Bickelhaupt, *J. Organomet. Chem.*, 1990, **390**, C1.
258. B. E. Bursten, *J. Am. Chem. Soc.*, 1982, **104**, 1299.
259. D. S. Grubisha, J. S. Rommel, T. M. Lane, W. T. Tysoe and D. W. Bennett, *Inorg. Chem.*, 1992, **31**, 5022.
260. W. P. Fehlhammer, S. Ahn and G. Beck, *J. Organomet. Chem.*, 1991, **411**, 181.
261. M. O. Albers, N. J. Coville and E. Singleton, *J. Organomet. Chem.*, 1987, **326**, 229.
262. M. O. Albers and N. J. Coville, *Inorg. Synth.*, 1990, **28**, 140.
263. M. O. Albers and N. J. Coville, *Reagents Transition Metal Complex Organomet. Synth.*, 1992, **28**, 140.
264. W. P. Fehlhammer and G. Beck, *J. Organomet. Chem.*, 1989, **379**, 97.
265. G. Facchin, M. Mozzon, R. A. Michelin, M. T. A. Ribeiro and A. J. L. Pombeiro, *J. Chem. Soc., Dalton Trans.*, 1992, 2827.
266. D. Lentz and D. Preugschat, *J. Chem. Soc., Chem. Commun.*, 1992, 1523.
267. D. Lentz and D. Preugschat, *J. Organomet. Chem.*, 1992, **436**, 185.
268. W. P. Fehlhammer, F. Degel and G. Beck, *Chem. Ber.*, 1987, **120**, 461.
269. G. Beck and W. P. Fehlhammer, *Angew. Chem., Int. Ed. Engl.*, 1988, **27**, 1344.
270. M. Höfler and H. Löwenich, *Chem. Ber.*, 1982, **115**, 2423.
271. M. Höfler and H. Löwenich, *J. Organomet. Chem.*, 1982, **226**, 229.
272. G. Christian, H. Stolzenberg and W. P. Fehlhammer, *J. Chem. Soc., Chem. Commun.*, 1982, 184.
273. E. Bär, W. P. Fehlhammer, W. Weigand and W. Beck, *J. Organomet. Chem.*, 1988, **347**, 101.
274. E. Bär, J. Fuchs, D. Rieger, F. Anguilar-Parrilla, H.-H. Limbach and W. P. Fehlhammer, *Angew. Chem., Int. Ed. Engl.*, 1991, **30**, 88.
275. F. E. Hahn and M. Tamm, *J. Organomet. Chem.*, 1991, **410**, C9.
276. F. E. Hahn and M. Tamm, *Angew. Chem., Int. Ed. Engl.*, 1992, **31**, 1212.
277. F. E. Hahn and M. Tamm, *Organometallics*, 1992, **11**, 84.
278. D. Lentz, *J. Organomet. Chem.*, 1990, **381**, 205.
279. J. R. Hinchliffe, A. Ricalton and M. W. Whiteley, *Polyhedron*, 1991, **10**, 267.
280. D. Lentz, B. Potter, R. Marschall, I. Brudgam and J. Fuchs, *Chem. Ber.*, 1990, **123**, 257.
281. L. D. Detter and R. A. Walton, *Polyhedron*, 1986, **5**, 1321.
282. A. Bell and R. A. Walton, *Polyhedron*, 1986, **5**, 845.
283. A. Bell and R. A. Walton, *Polyhedron*, 1986, **5**, 951.
284. A. Bell and R. A. Walton, *J. Organomet. Chem.*, 1985, **290**, 341.
285. J. A. Connor and C. Overton, *J. Chem. Soc., Dalton Trans.*, 1982, 2397.
286. H. Fischer and S. Zeuner, *J. Organomet. Chem.*, 1987, **321**, C6.
287. A. Mayr, S. M. Holmes and C. M. Bastos, *Organometallics*, 1992, **11**, 4358.
288. E. Carmona, L. Contreras, E. Gutiérrez-Puebla, A. Monge and L. Sánchez, *Inorg. Chem.*, 1990, **29**, 700.
289. R. Alvarez, E. Carmona, E. Gutiérrez-Puebla, J. M. Marín, A. Monge and M. L. Poveda, *J. Chem. Soc., Chem. Commun.*, 1984, 1326.

290. J. S. Rommel, J. B. Weinrach, D. S. Grubisha and D. W. Bennett, *Inorg. Chem.*, 1988, **27**, 2945.
291. K. A. Conner and R. A. Walton, *Inorg. Chem.*, 1986, **25**, 4422.
292. K. A. Conner and R. A. Walton, *Organometallics*, 1983, **2**, 169.
293. G. Lawless, G. McNally, A. R. Manning, D. Cunningham and P. McArdle, *Polyhedron*, 1986, **5**, 1741.
294. L. Zanathy, H. Bock, D. W. Lentz, D. Preugschat and P. Botschwina, *J. Chem. Soc., Chem. Commun.*, 1992, 403.
295. W. P. Fehlhammer, F. Degel and H. Stolzenberg, *Angew. Chem., Int. Ed. Engl.*, 1981, **20**, 214.
296. W. P. Fehlhammer and G. Beck, *J. Organomet. Chem.*, 1989, **369**, 105.
297. J. C. Viljoen, S. Lotz, L. Linford and H. G. Raubenheimer, *J. Chem. Soc., Dalton Trans.*, 1984, 1041.
298. W. P. Fehlhammer, G. Zinner, G. Beck and J. Fuchs, *J. Organomet. Chem.*, 1989, **379**, 277.
299. E. Bär and W. P. Fehlhammer, *J. Organomet. Chem.*, 1988, **353**, 197.
300. E. Bär, A. Völkl, F. Beck, W. P. Fehlhammer and A. Robert, *J. Chem. Soc., Dalton Trans.*, 1986, 863.
301. W. P. Fehlhammer, E. Bär and B. Boyadjiev, *Z. Naturforsch., Teil A*, 1986, **41**, 1023.
302. P. Umland and H. Vahrenkamp, *Chem. Ber.*, 1982, **115**, 3580.
303. A. Mayr and C. M. Bastos, *J. Am. Chem. Soc.*, 1990, **112**, 7797.
304. M. I. Bruce, M. G. Humphrey, J. G. Mattisons, S. K. Roy and A. G. Swincer, *Aust. J. Chem.*, 1984, **37**, 1955.
305. D. Villemin and E. Schigeko, *J. Organomet. Chem.*, 1988, **346**, C24.
306. A. Wong, P. C. W. Kang, C. D. Tagge and D. R. Leon, *Organometallics*, 1990, **9**, 1992.
307. A. K. McMullen, J. P. Selegue and J.-G. Wang, *Organometallics*, 1991, **10**, 3421.
308. P. N. Nickias, J. P. Selegue and B. A. Young, *Organometallics*, 1988, **7**, 2248.
309. M.-C. Chen *et al.*, *Organometallics*, 1991, **10**, 378.
310. A. Blagg, A. T. Hutton, P. G. Pringle and B. L. Shaw, *J. Chem. Soc., Dalton Trans.*, 1984, 1815.
311. S. F. T. Froom, M. Green, R. J. Mercer, K. R. Nagle, A. G. Orpen and R. A. Rodrigues, *J. Chem. Soc., Dalton Trans.*, 1991, 3171.
312. A. W. Al-Saadoon, M. Green, R. J. Mercer and A. G. Orpen, *J. Organomet. Chem.*, 1990, **384**, C12.
313. W. Sieber, M. Wolfgruber, D. Neugebauer, O. Orama and F. R. Kreissl, *Z. Naturforsch., Teil B*, 1983, **38**, 67.
314. R. L. Beddoes, C. Bitcon and M. W. Whiteley, *J. Organomet. Chem.*, 1991, **402**, 85.
315. J. S. Adams, C. Bitcon, J. R. Brown, D. Collison, M. Cunningham and M. W. Whiteley, *J. Chem. Soc., Dalton Trans.*, 1987, 3049.
316. M. E. Welker, *Chem. Rev.*, 1992, 97.
317. N. E. Kolobova, V. V. Skripkin, T. V. Rozantseva, Y. T. Struchkov, G. G. Aleksandrov and V. G. Andrianov, *J. Organomet. Chem.*, 1981, **218**, 351.
318. N. A. Ustynyuk, V. N. Vinogradova, V. N. Korneva, D. N. Kravtsov, V. G. Andrianov and Y. T. Struchkov, *J. Organomet. Chem.*, 1984, **277**, 285.
319. M. I. Bruce, M. J. Liddell, M. R. Snow and E. R. T. Tiekink, *Organometallics*, 1988, **7**, 343.
320. M. I. Bruce, T. W. Hambley, J. R. Rodgers, M. R. Snow and A. G. Swincer, *J. Organomet. Chem.*, 1982, **226**, C1.
321. M. I. Bruce, T. W. Hambley, M. R. Snow and A. G. Swincer, *Organometallics*, 1985, **4**, 494.
322. V. N. Kalinin, T. V. Rozantseva, P. V. Petrovskii, A. S. Batsanov and Y. T. Struchov, *J. Organomet. Chem.*, 1989, **372**, 287.
323. K. R. Birdwhistell and J. L. Templeton, *Organometallics*, 1985, **4**, 2062.
324. A. Mayr, K. C. Schaefer and E. Y. Huang, *J. Am. Chem. Soc.*, 1984, **106**, 1517.
325. P. V. Broadhurst, *Polyhedron*, 1985, **4**, 1801.
326. A. M. English *et al.*, *J. Organomet. Chem.*, 1981, **205**, 177.
327. I. S. Butler, A. M. English and K. R. Plowman, *Inorg. Synth.*, 1982, **21**, 1.
328. I. M. Baibich, A. M. English and I. S. Butler, *Organometallics*, 1984, **3**, 1786.
329. A. M. English, K. R. Plowman and I. S. Butler, *Inorg. Chem.*, 1981, **20**, 2553.
330. A. M. English, K. R. Plowman and I. S. Butler, *Inorg. Chem.*, 1982, **21**, 338.
331. Y. N. Huang, I. S. Butler and D. F. R. Gilson, *Inorg. Chem.*, 1992, **31**, 303.
332. Y. N. Huang, D. F. R. Gilson, I. S. Butler and F. G. Morin, *Inorg. Chem.*, 1992, **31**, 322.
333. P. T. Czech, X.-Q. Ye and R. F. Fenske, *Organometallics*, 1990, **9**, 2016.
334. P. H. Bird, A. A. Ismail and I. S. Butler, *Inorg. Chem.*, 1985, **24**, 2911.
335. W. W. Greaves and R. J. Angelici, *Inorg. Chem.*, 1981, **20**, 2983.
336. S. Lotz, R. R. Pille and P. H. Van Rooyen, *Inorg. Chem.*, 1986, **25**, 3053.
337. H. P. Kim, S. Kim, R. A. Jacobson and R. J. Angelici, *J. Am. Chem. Soc.*, 1986, **108**, 5154.
338. R. A. Doyle, L. M. Daniels, R. J. Angelici and F. G. A. Stone, *J. Am. Chem. Soc.*, 1989, **111**, 4995.
339. R. A. Pickering, R. A. Jacobson and R. J. Angelici, *J. Am. Chem. Soc.*, 1981, **103**, 817.
340. P. J. Brothers and W. R. Roper, *Chem. Rev.*, 1988, **88**, 1293.
341. M. Brookhart and W. B. Studebaker, *Chem. Rev.*, 1987, **87**, 411.
342. M. P. Doyle, *Chem. Rev.*, 1986, **86**, 919.
343. A. B. Antonova and A. A. Ioganson, *Russian Chem. Rev.*, 1989, **58**, 693.
344. W. A. Herrmann, *J. Organomet. Chem.*, 1983, **250**, 319.
345. J. E. Hahn, *Prog. Inorg. Chem.*, 1984, **31**, 205.
346. T. W. Bodnar and A. R. Cutler, *Organometallics*, 1985, **4**, 1558.
347. S. E. Kegley, M. Brookhart and G. R. Husk, *Organometallics*, 1982, **1**, 760.
348. J. D. Koola and D. M. Roddick, *Organometallics*, 1991, **10**, 591.
349. S. G. Feng, P. S. White and J. L. Templeton, *J. Am. Chem. Soc.*, 1992, **114**, 2951.
350. H. G. Alt, H. E. Engelhardt and A. C. Filippou, *J. Organomet. Chem.*, 1988, **355**, 139.
351. K. E. Garrett *et al.*, *J. Am. Chem. Soc.*, 1989, **111**, 8383.
352. K. A. Mahmoud, A. J. Rest and H. G. Alt, *J. Chem. Soc., Chem. Commun.*, 1983, 1011.
353. M. J. Winter and S. Woodward, *J. Chem. Soc., Chem. Commun.*, 1989, 457.
354. M. J. Winter and S. Woodward, *J. Organomet. Chem.*, 1989, **361**, C18.
355. S. G. Feng, P. S. White and J. L. Templeton, *Organometallics*, 1993, **12**, 1765.
356. W. A. Herrmann, J. L. Hubbard, I. Bernal, J. D. Korp, B. L. Haymore and G. L. Hillhouse, *Inorg. Chem.*, 1984, **23**, 2978.
357. M. J. Winter, *Adv. Organomet. Chem.*, 1989, **29**, 101.
358. J. J. D'Errico and M. D. Curtis, *J. Am. Chem. Soc.*, 1983, **105**, 4479.

359. R. P. Beatty, J. M. Maher and N. J. Cooper, *J. Am. Chem. Soc.*, 1981, **103**, 238.
360. H. Fischer, S. Zeuner and K. Ackermann, *J. Chem. Soc., Chem. Commun.*, 1984, 684.
361. H. Fischer and D. Reindl, *J. Organomet. Chem.*, 1990, **385**, 351.
362. L. K. Fong and N. J. Cooper, *J. Am. Chem. Soc.*, 1984, **106**, 2595.
363. K. N. Juneau, L. S. Hegedus and F. W. Roepke, *J. Am. Chem. Soc.*, 1989, **111**, 4762.
364. H. Fischer and S. Zeuner, *J. Organomet. Chem.*, 1983, **252**, C63.
365. H. Fischer and S. Zeuner, *J. Organomet. Chem.*, 1985, **286**, 201.
366. H. Fischer, *J. Organomet. Chem.*, 1981, **219**, C34.
367. H. Fischer, *J. Organomet. Chem.*, 1981, **222**, 241.
368. K. Weiss and K. Hoffmann, *J. Organomet. Chem.*, 1983, **255**, C24.
369. B. H. Edwards and M. D. Rausch, *J. Organomet. Chem.*, 1981, **210**, 91.
370. H. Fischer, *Angew. Chem., Int. Ed. Engl.*, 1983, **22**, 874.
371. H. Fischer and S. Zeuner, *J. Organomet. Chem.*, 1987, **327**, 63.
372. H. Fischer, S. Zeuner, K. Ackermann and U. Schubert, *J. Organomet. Chem.*, 1984, **263**, 201.
373. A. Parlier, F. Rose-Munch, M. Rudler and H. Rudler, *J. Organomet. Chem.*, 1982, **235**, C13.
374. H. Fischer, J. Schmid and R. Märkl, *J. Chem. Soc., Chem. Commun.*, 1985, 572.
375. H. Fischer, J. Schmid and S. Zeuner, *Chem. Ber.*, 1987, **120**, 583.
376. H. Fischer and J. Schmid, *J. Organomet. Chem.*, 1986, **306**, 203.
377. H. Fischer, H. Jungklaus and J. Schmid, *J. Organomet. Chem.*, 1989, **368**, 193.
378. H. Fischer, K. Treier and U. Gerbing, *J. Organomet. Chem.*, 1992, **433**, 127.
379. H. Fischer, K. Treier, E.-C. Plappert and J. Hofmann, *J. Organomet. Chem.*, 1990, **399**, 153.
380. H. Fischer, A. Schlageter, W. Bidell and A. Früh, *Organometallics*, 1991, **10**, 389.
381. H. Fischer, J. Hofmann and E. Mauz, *Angew. Chem., Int. Ed. Engl.*, 1991, **30**, 998.
382. H. Fischer, E. Mauz, M. Jaeger and R. Fischer, *J. Organomet. Chem.*, 1992, **427**, 63.
383. H. Fischer and M. Jaeger, *J. Organomet. Chem.*, 1992, **456**, 205.
384. M. P. Doyle, J. H. Griffin, V. Bagheri and R. L. Dorow, *Organometallics*, 1984, **3**, 53.
385. J. Levisalles, F. Rose-Munch, H. Rudler, J.-C. Daran and Y. Jeannin, *J. Organomet. Chem.*, 1985, **279**, 413.
386. D. Navarre, F. Rose-Munch and H. Rudler, *J. Organomet. Chem.*, 1985, **284**, C15.
387. J. Levisalles, F. Rose-Munch, H. Rudler, J.-C. Daran and Y. Jeannin, *J. Chem. Soc., Chem. Commun.*, 1981, 1057.
388. J. Levisalles *et al.*, *J. Chem. Soc., Chem. Commun.*, 1981, 1055.
389. J. Levisalles, F. Rose-Munch, H. Rudler, J.-C. Daran, Y. Dromzée and Y. Jeannin, *J. Chem. Soc., Chem. Commun.*, 1981, 152.
390. M. I. Bruce and A. G. Swincer, *Adv. Organomet. Chem.*, 1983, **22**, 59.
391. U. Schubert, U. Kirchgässner, J. Grönen and H. Piana, *Polyhedron*, 1989, **8**, 1589.
392. M. I. Bruce, *Chem. Rev.*, 1991, **91**, 197.
393. H. Fischer, O. Podschadly, A. Früh, C. Troll, R. Stumpf and A. Schlageter, *Chem. Ber.*, 1992, **125**, 2667.
394. K. R. Birdwhistell, T. L. Tonker and J. L. Templeton, *J. Am. Chem. Soc.*, 1985, **107**, 4474.
395. A. S. Gamble, K. R. Birdwhistell and J. L. Templeton, *Organometallics*, 1988, **7**, 1046.
396. U. Schubert and J. Grönen, *Chem. Ber.*, 1989, **122**, 1237.
397. H. Fischer, G. Roth, D. Reindl and C. Troll, *J. Organomet. Chem.*, 1993, **454**, 133.
398. F. Seitz and M. S. Wrighton, *Inorg. Chem.*, 1987, **26**, 64.
399. F. Seitz, H. Fischer, J. Riede and J. Vogel, *Organometallics*, 1986, **5**, 2187.
400. F. Seitz, H. Fischer and J. Riede, *J. Organomet. Chem.*, 1985, **287**, 87.
401. F. Seitz and H. Fischer, *J. Organomet. Chem.*, 1985, **290**, C31.
402. H. Fischer, F. Seitz and J. Riede, *J. Chem. Soc., Chem. Commun.*, 1985, 537.
403. N. M. Doherty, C. Eischenbroich, H.-J. Kneuper and S. A. R. Knox, *J. Chem. Soc., Chem. Commun.*, 1985, 170.
404. S. F. T. Froom, M. Green, R. J. Mercer, K. R. Nagle, A. G. Orpen and S. Schwiegk, *J. Chem. Soc., Chem. Commun.*, 1986, 1666.
405. S. F. T. Froom, M. Green, K. R. Nagle and D. J. Williams, *J. Chem. Soc., Chem. Commun.*, 1987, 1305.
406. K. H. Dötz, H. Fischer, P. Hofmann, F. R. Kreissl, U. Schubert and K. Weiss, in 'Transition Metal Carbene Complexes', ed. D. Seyferth, VCH, Weinheim, 1983.
407. R. D. Adams, *Chem. Rev.*, 1989, **89**, 1703.
408. H. Fischer, in 'The Chemistry of the Metal–Carbon Bonds', ed. S. Patai, Wiley, New York, 1982, vol. 1, p. 181.
409. E. Lastra and L. S. Hegedus, *J. Am. Chem. Soc.*, 1993, **115**, 87.
410. T. R. Hoye, K. Chen and J. R. Vyvyan, *Organometallics*, 1993, **12**, 2806.
411. G. Erker, *Angew. Chem., Int. Ed. Engl.*, 1989, **28**, 397.
412. G. Erker, *Polyhedron*, 1988, **7**, 2451.
413. K. H. Dötz, M. Popall and G. Müller, *J. Organomet. Chem.*, 1987, **334**, 57.
414. A. Parlier, H. Rudler, M. Fontanille and A. Soum, *J. Organomet. Chem.*, 1986, **312**, C11.
415. K. H. Dötz, W. Sturm, M. Popall and J. Riede, *J. Organomet. Chem.*, 1984, **277**, 267.
416. D. W. Macomber, M. Liang, P. Madhukar and A. G. Verma, *J. Organomet. Chem.*, 1989, **361**, 187.
417. C. A. Toledano, J. Levisalles, M. Rudler, H. Rudler, J.-C. Daran and Y. Jeannin, *J. Organomet. Chem.*, 1982, **228**, C7.
418. K. H. Dötz, M. Popall, G. Müller and K. Ackermann, *J. Organomet. Chem.*, 1990, **383**, 93.
419. C. A. Merlic, D. Xu and S. I. Khan, *Organometallics*, 1992, **11**, 412.
420. F. Stein, M. Duetsch, R. Lackmann, M. Noltemeyer and A. de Meijere, *Angew. Chem., Int. Ed. Engl.*, 1991, **30**, 1658.
421. E. O. Fischer, N. H. Tran-Huy and D. Neugebauer, *J. Organomet. Chem.*, 1982, **229**, 169.
422. K. H. Dötz and H. Larbig, *J. Organomet. Chem.*, 1992, **433**, 115.
423. U. Schubert, P. Wülfert and S. Mock, *J. Organomet. Chem.*, 1990, **459**, 55.
424. P. A. Chaloner, G. D. Glick and R. A. Moss, *J. Chem. Soc., Chem. Commun.*, 1983, 880.
425. A. Parlier and H. Rudler, *J. Chem. Soc., Chem. Commun.*, 1986, 514.
426. A. Parlier, H. Rudler, N. Platzer, M. Fontanille and A. Soum, *J. Chem. Soc., Dalton Trans.*, 1987, 1041.
427. C. Alvarez, A. Pacreau, A. Parlier, H. Rudler and J.-C. Daran, *Organometallics*, 1987, **6**, 1057.
428. C. Alvarez-Toledano *et al.*, *J. Organomet. Chem.*, 1987, **328**, 357.

429. I. Butler, W. R. Cullen, F. W. B. Einstein and A. C. Willis, *Organometallics*, 1985, **4**, 603.
430. R. Herman and I. Ugi, *Angew. Chem., Int. Ed. Engl.*, 1982, **21**, 788.
431. P. Le Floch, N. H. T. Huy and F. Mathey, *Organometallics*, 1988, **7**, 1293.
432. M. P. Guy, J. T. Guy and D. W. Bennett, *Organometallics*, 1985, **5**, 1696.
433. R. J. Klingler, J. C. Huffman and J. K. Kochi, *Inorg. Chem.*, 1981, **20**, 34.
434. D. F. Harvey and M. F. Brown, *J. Am. Chem. Soc.*, 1990, **112**, 7806.
435. J. B. Chen and B. Wang, *J. Organomet. Chem.*, 1992, **440**, 67.
436. C. Jiabi *et al.*, *J. Organomet. Chem.*, 1987, **329**, 69.
437. P. Veya, C. Floriani, A. Chiesi-Villa and C. Guastini, *Organometallics*, 1991, **10**, 1652.
438. D. M. Anderson, G. S. Bristow, P. B. Hitchcock, H. A. Jasim, M. F. Lappert and B. W. Skelton, *J. Chem. Soc., Dalton Trans.*, 1987, 2843.
439. E. O. Fischer, W. Röll, U. Schubert and K. Ackermann, *Angew. Chem., Int. Ed. Engl.*, 1981, **20**, 611.
440. U. Schubert, K. Ackermann, N. H. T. Huy and W. Röll, *J. Organomet. Chem.*, 1982, **232**, 155.
441. E. O. Fischer, W. Röll, N. H. T. Huy and K. Ackermann, *Chem. Ber.*, 1982, **115**, 2951.
442. N. H. T. Huy, C. Pascard, E. T. H. Dau and K. H. Dötz, *Organometallics*, 1988, **7**, 590.
443. N. H. T. Huy, P. Lefloch, F. Robert and Y. Jeannin, *J. Organomet. Chem.*, 1987, **327**, 211.
444. P. M. Fritz, M. Steimann and W. Beck, *Chem. Ber.*, 1987, **120**, 253.
445. P. M. Fritz, J. Breimair, B. Wagner and W. Beck, *J. Organomet. Chem.*, 1992, **426**, 343.
446. R. E. Cramer, J. H. Jeong and J. W. Gilje, *Organometallics*, 1986, **5**, 2555.
447. G. Erker, F. Sosna, R. Zwettler and C. Krüger, *Organometallics*, 1989, **8**, 450.
448. G. Erker, F. Sosna and U. Hoffmann, *J. Organomet. Chem.*, 1989, **372**, 41.
449. G. Erker *et al.*, *J. Organomet. Chem.*, 1990, **394**, 99.
450. G. Erker, F. Sosna, P. Betz, S. Werner and C. Krüger, *J. Am. Chem. Soc.*, 1991, **113**, 564.
451. K. Mashima, K. Jyodoi, A. Ohyoshi and H. Takaya, *J. Chem. Soc., Chem. Commun.*, 1986, 1145.
452. R. Beckhaus and K.-H. Thiele, *J. Organomet. Chem.*, 1989, **368**, 315.
453. G. Erker, U. Dorf, R. Mynott, Y.-H. Tsay and C. Krüger, *Angew. Chem., Int. Ed. Engl.*, 1985, **24**, 584.
454. G. Erker, U. Dorf, C. Krüger and Y.-H. Tsay, *Organometallics*, 1987, **6**, 680.
455. E. V. Anslyn, B. D. Santarsiero and R. H. Grubbs, *Organometallics*, 1988, **7**, 2137.
456. M. Sabat, M. F. Gross and M. G. Finn, *Organometallics*, 1992, **11**, 745.
457. A. Llebaria, J. M. Moretó, S. Ricart, J. Ros, J. M. Viñas and R. Yáñez, *J. Organomet. Chem.*, 1992, **440**, 79.
458. F. Camps *et al.*, *J. Organomet. Chem.*, 1991, **401**, C17.
459. C. P. Casey and A. J. Shusterman, *Organometallics*, 1985, **4**, 736.
460. C. A. Toledano, H. Rudler, J.-C. Daran and Y. Jeannin, *J. Chem. Soc., Chem. Commun.*, 1984, 574.
461. C. P. Casey, A. J. Shusterman, N. W. Vollendorf and K. J. Haller, *J. Am. Chem. Soc.*, 1982, **104**, 2417.
462. K. H. Dötz, H.-G. Erben, W. Staudacher, K. Harms, G. Müller and J. Riede, *J. Organomet. Chem.*, 1988, **355**, 177.
463. K. S. Chan and W. D. Wulff, *J. Am Chem. Soc.*, 1986, **108**, 5229.
464. S. Voran, H. Blau, W. Malisch and U. Schubert, *J. Organomet. Chem.*, 1982, **232**, C33.
465. A. Wienand, H.-U. Reissig, H. Fischer, D. Pflumm and C. Troll, *J. Organomet. Chem.*, 1992, **427**, C9.
466. R. Pipoh, R. van Eldik, S. L. B. Wang and W. D. Wulff, *Organometallics*, 1992, **11**, 490.
467. E. O. Fischer and W. Kleine, *J. Organomet. Chem.*, 1981, **208**, C27.
468. E. O. Fischer, S. G. Gibbins and W. Kellerer, *J. Organomet. Chem.*, 1981, **218**, C51.
469. K. H. Dötz, W. Sturm and H. G. Alt, *Organometallics*, 1987, **6**, 1424.
470. H. Berke, P. Härter, G. Huttner and J. von Seyerl, *J. Organomet. Chem.*, 1981, **219**, 317.
471. L. Lattuada, E. Licandro, S. Maiorana, H. Molinari and A. Papagni, *Organometallics*, 1991, **10**, 807.
472. N. A. Bailey *et al.*, *J. Chem. Soc., Dalton Trans.*, 1983, 2397.
473. N. A. Bailey, D. A. Dunn, C. N. Foxcroft, G. R. Harrison, M. J. Winter and S. Woodward, *J. Chem. Soc., Dalton Trans.*, 1988, 1449.
474. H. Adams, N. A. Bailey and M. J. Winter, *J. Chem. Soc., Dalton Trans.*, 1984, 273.
475. J. S. Drage and K. P. C. Vollhardt, *Organometallics*, 1985, **4**, 191.
476. H. Adams, N. A. Bailey, G. W. Bentley, G. Hough, M. J. Winter and S. Woodward, *J. Chem. Soc., Dalton Trans*, **1991**, 749.
477. A. Asdar, C. Lapinte and L. Toupet, *Organometallics*, 1989, **8**, 2708.
478. A. Asdar and C. Lapinte, *J. Organomet. Chem.*, 1987, **327**, C33.
479. G. Grötsch and W. Malisch, *J. Organomet. Chem.*, 1983, **258**, 297.
480. A. K. Campen, K. A. Mahmoud, A. J. Rest and P. A. Willis, *J. Chem. Soc., Dalton Trans.*, 1990, 2817.
481. R. A. Pickering and R. J. Angelici, *Inorg. Chem.*, 1981, **20**, 2977.
482. R. Aumann and J. Schröder, *J. Organomet. Chem.*, 1989, **378**, 57.
483. K. H. Dötz and V. Leue, *J. Organomet. Chem.*, 1991, **407**, 337.
484. H. G. Raubenheimer, G. J. Krüger, C. F. Marais, R. Otte and J. T. Z. Hattingh, *Organometallics*, 1988, **7**, 1853.
485. H. G. Raubenheimer *et al.*, *J. Chem. Soc., Dalton Trans.*, 1989, 1565.
486. H. G. Raubenheimer, E. O. Fischer, U. Schubert, C. Krüger and Y.-H. Tsay, *Angew. Chem., Int. Ed. Engl.*, 1981, **20**, 1055.
487. H. G. Raubenheimer, G. J. Krüger, A. van Lombard, L. Linford and J. C. Viljoen, *Organometallics*, 1985, **4**, 275.
488. E. O. Fischer, D. Himmelreich, R. Cai, H. Fischer, U. Schubert and B. Zimmer-Gasser, *Chem. Ber.*, 1981, **114**, 3209.
489. W. A. Schenk, D. Kuemmerle and C. Burschka, *J. Organomet. Chem.*, 1988, **349**, 183.
490. W. Schütt, N. Ullrich and F. R. Kreissl, *J. Organomet. Chem.*, 1991, **408**, C5.
491. D. S. Marynick and C. M. Kirkpatrick, *J. Am. Chem. Soc.*, 1985, **107**, 1993.
492. H. Nakatsuji, J. Ushio, S. Han and T. Yonezawa, *J. Am. Chem. Soc.*, 1983, **105**, 426.
493. A. Márquez and J. F. Sanz, *J. Am. Chem. Soc.*, 1992, **114**, 2903.
494. P. Kubácek, R. Hoffmann and Z. Havlas, *Organometallics*, 1982, **1**, 180.
495. A. Hafner, L. S. Hegedus, G. DeWeck, B. Hawkins and K. H. Dötz, *J. Am. Chem. Soc.*, 1988, **110**, 8413.
496. P. C. Servaas, D. J. Stufkens and A. Oskam, *J. Organomet. Chem.*, 1990, **390**, 61.
497. S. E. J. Bell, K. C. Gordon and J. J. McGarvey, *J. Am. Chem. Soc.*, 1988, **110**, 3107.
498. J. N. Bechara, S. E. J. Bell, J. J. McGarvey and J. J. Rooney, *J. Chem. Soc., Chem. Commun.*, 1986, 1785.

499. H. C. Foley, L. M. Strubinger, T. S. Targos and G. L. Geoffroy, *J. Am. Chem. Soc.*, 1983, **105**, 3064.
500. C. P. Casey and M. C. Cesa, *Organometallics*, 1982, **1**, 87.
501. H. S. Choi and D. A. Sweigart, *J. Organomet. Chem.*, 1982, **228**, 249.
502. J. R. Gandler and C. F. Bernasconi, *Organometallics*, 1989, **8**, 2282.
503. C. Kelley *et al.*, *Inorg. Chim. Acta*, 1992, **200**, 601.
504. M. Rudler-Chauvin and H. Rudler, *J. Organomet. Chem.*, 1981, **212**, 203.
505. H. G. Raubenheimer, G. J. Krüger and H. W. Viljoen, *J. Organomet. Chem.*, 1987, **319**, 361.
506. H. G. Raubenheimer, G. J. Krüger, H. W. Viljoen and S. Lotz, *J. Organomet. Chem.*, 1986, **314**, 281.
507. C. Baldoli, L. Lattuada, E. Licandro, S. Maiorana and A. Papagni, *Organometallics*, 1993, **12**, 2994.
508. R. Aumann, E. Kuckert, C. Krüger and K. Angermund, *Angew. Chem., Int. Ed. Engl.*, 1987, **26**, 563.
509. R. Aumann, E. Kuckert and H. Heinen, *Chem. Ber.*, 1987, **120**, 1293.
510. R. Aumann and E. Kuckert, *Chem. Ber.*, 1987, **120**, 1939.
511. H. Fischer, U. Schubert and R. Märkl, *Chem. Ber.*, 1981, **114**, 3412.
512. W. D. Wulff, V. Dragisich, J. C. Huffman, R. W. Kaesler and D. C. Yang, *Organometallics*, 1989, **8**, 2196.
513. H. Fischer, F. Seitz and G. Müller, *J. Organomet. Chem.*, 1987, **320**, 185.
514. K. J. Schneider, A. Neubrand, R. van Eldik and H. Fischer, *Organometallics*, 1992, **11**, 267.
515. R. Aumann and P. Hinterding, *Chem. Ber.*, 1992, **125**, 2765.
516. N. H. T. Huy, L. Ricard and F. Mathey, *Polyhedron*, 1990, **9**, 2193.
517. R. Aumann and J. Uphoff, *Angew. Chem., Int. Ed. Engl.*, 1987, **26**, 357.
518. D. W. Macomber, P. Madhukar and R. D. Rogers, *Organometallics*, 1991, **10**, 2121.
519. C. G. Kreiter, S. Schufft and G. Heckmann, *J. Organomet. Chem.*, 1992, **424**, 163.
520. C. Alvarez-Toledano *et al.*, *J. Organomet. Chem.*, 1987, **323**, 371.
521. D. W. Macomber, M. Liang and R. D. Rogers, *Organometallics*, 1988, **7**, 416.
522. N. H. T. Huy and F. Mathey, *Organometallics*, 1988, **7**, 2233.
523. K. H. Dötz, W. Kuhn, G. Müller, B. Huber and H. G. Alt, *Angew. Chem., Int. Ed. Engl.*, 1986, **25**, 812.
524. D. W. Macomber, M.-H. Hung, P. Madhukar, M. Liang and R. D. Rogers, *Organometallics*, 1991, **10**, 737.
525. D. W. Macomber and M.-H. Hung, *J. Organomet. Chem.*, 1989, **366**, 147.
526. H. Fischer, T. Meisner and J. Hofmann, *J. Organomet. Chem.*, 1990, **397**, 41.
527. K. H. Dötz and W. Kuhn, *J. Organomet. Chem.*, 1985, **286**, C23.
528. F. Camps, L. Jordi, J. M. Moretó, S. Ricart, A. M. Castaño and A. M. Echavarren, *J. Organomet. Chem.*, 1992, **436**, 189.
529. F. Camps, J. M. Moretó, S. Ricart, J. M. Viñas, E. Molins and C. Miravitlles, *J. Chem. Soc., Chem. Commun.*, 1989, 1560.
530. L. Jordi, J. M. Moretó, S. Ricart, J. M. Viñas, E. Molins and C. Miravitlles, *J. Organomet. Chem.*, 1993, **444**, C28.
531. R. Aumann, *Chem. Ber.*, 1992, **125**, 2773.
532. N. H. T. Huy, E. O. Fischer, H. G. Alt and K. H. Dötz, *J. Organomet. Chem.*, 1985, **284**, C9.
533. N. Huy, E. O. Fischer, J. Riede, U. Thewalt and K. H. Dötz, *J. Organomet. Chem.*, 1984, **273**, C29.
534. H. T. Huy, P. Lefloch, J. M. Louis and M. Fetizon, *J. Organomet. Chem.*, 1986, **311**, 79.
535. H. Adams, N. A. Bailey, P. Cahill, D. Rogers and M. J. Winter, *J. Chem. Soc., Dalton Trans.*, 1986, 2119.
536. H. Adams, N. A. Bailey, M. J. Winter and S. Woodward, *J. Organomet. Chem.*, 1991, **418**, C39.
537. V. A. Osborn and M. J. Winter, *J. Chem. Soc., Chem. Commun.*, 1985, 1744.
538. R. Aumann, J. Schröder, C. Krüger and R. Goddard, *J. Organomet. Chem.*, 1989, **378**, 185.
539. H. G. Raubenheimer, G. J. Krüger, F. Scott and R. Otte, *Organometallics*, 1987, **6**, 1789.
540. A. L. Steinmetz, S. A. Hershberger and R. J. Angelici, *Organometallics*, 1984, **3**, 461.
541. R. A. Pickering and R. J. Angelici, *J. Organomet. Chem.*, 1982, **225**, 253.
542. H. P. Kim, S. Kim, R. A. Jacobson and R. J. Angelici, *Organometallics*, 1984, **3**, 1124.
543. H. P. Kim, S. Kim, R. A. Jacobson and R. J. Angelici, *Organometallics*, 1986, **5**, 2481.
544. H. P. Kim and R. J. Angelici, *Organometallics*, 1986, **5**, 2489.
545. F. R. Kreissl and N. Ullrich, *J. Organomet. Chem.*, 1992, **440**, 335.
546. F. R. Kreissl, N. Ullrich, J. Ostermeier and W. Schütt, *J. Organomet. Chem.*, 1993, **459**, C6.
547. W. Schütt, E. Herdtweck and F. R. Kreissl, *J. Organomet. Chem.*, 1993, **456**, C15.
548. M. A. Schwindt, J. R. Miller and L. S. Hegedus, *J. Organomet. Chem.*, 1991, **413**, 143.
549. D. B. Grotjahn and K. H. Dötz, *Synlett.*, 1991, 381.
550. L. S. Hegedus, J. Montgomery, Y. Narukawa and D. C. Snustad, *J. Am. Chem. Soc.*, 1991, **113**, 5784.
551. C. Alvarez, A. Parlier, H. Rudler, R. Yefsah, J. C. Daran and C. Knobler, *Organometallics*, 1989, **8**, 2253.
552. A. Parlier, H. Rudler, J. C. Daran, C. Alvarez and F. Delgado Reyes, *J. Organomet. Chem.*, 1987, **327**, 339.
553. A. Parlier, H. Rudler, J. C. Daran and C. Alvarez, *J. Organomet. Chem.*, 1987, **333**, 245.
554. C. A. Toledano, A. Parlier, H. Rudler, J.-C. Daran and Y. Jeannin, *J. Chem. Soc., Chem. Commun.*, 1984, 576.
555. M. F. Lappert, *J. Organomet. Chem.*, 1988, **358**, 185.
556. M. F. Lappert, P. L. Pye, A. J. Rogers and G. M. McLaughlin, *J. Chem. Soc., Dalton Trans.*, 1981, 701.
557. R. Aumann, *Angew. Chem., Int. Ed. Engl.*, 1988, **27**, 1456.
558. M. A. Schwindt, T. Lejon and L. S. Hegedus, *Organometallics*, 1990, **9**, 2814.
559. L. S. Hegedus, M. A. Schwindt, S. DeLombaert and R. Imwinkelried, *J. Am. Chem. Soc.*, 1990, **112**, 2264.
560. C. Borel, L. S. Hegedus, J. Krebs and Y. Satoh, *J. Am. Chem. Soc.*, 1987, **109**, 1101.
561. J. T. Welch, E. M. Keskeny and R. A. Toscano, *J. Organomet. Chem.*, 1993, **461**, 95.
562. B. A. Anderson, W. D. Wulff, T. S. Powers, S. Tribbitt and A. L. Rheingold, *J. Am. Chem. Soc.*, 1992, **114**, 10 784.
563. K. H. Dötz, T. Schäfer and K. Harms, *Angew. Chem., Int. Ed. Engl.*, 1990, **29**, 176.
564. J. A. Chamizo, P. B. Hitchcock, H. A. Jasim and M. F. Lappert, *J. Organomet. Chem.*, 1993, **451**, 89.
565. K. H. Dötz and A. Rau, *J. Organomet. Chem.*, 1991, **418**, 219.
566. B. Denise *et al.*, *J. Organomet. Chem.*, 1991, **418**, 377.
567. K. H. Dötz, F. Kroll and K. Harms, *J. Organomet. Chem.*, 1993, **459**, 169.
568. D. C. Yang, V. Dragisich, W. D. Wulff and J. C. Huffman, *J. Am. Chem. Soc.*, 1988, **110**, 307.
569. W. Hepp and U. Schubert, *J. Organomet. Chem.*, 1990, **385**, 221.
570. B. Denise *et al.*, *J. Organomet. Chem.*, 1990, **386**, 51.
571. F. E. Hahn and M. Tamm, *J. Organomet. Chem.*, 1993, **456**, C11.

572. K. Öfele *et al.*, *J. Organomet. Chem.*, 1993, **459**, 177.
573. U. Eckhart, C. Robl and W. P. Fehlhammer, *Organometallics*, 1993, **12**, 3241.
574. H. Adams, N. A. Bailey, V. A. Osborn and M. J. Winter, *J. Chem. Soc., Dalton Trans.*, 1986, 2127.
575. V. A. Osborn and M. J. Winter, *Polyhedron*, 1986, **5**, 435.
576. H. Brunner, W. Meyer and J. Wachter, *J. Organomet. Chem.*, 1983, **243**, 437.
577. T. Weidmann, K. Sünkel and W. Beck, *J. Organomet. Chem.*, 1993, **459**, 219.
578. D. W. Macomber and P. Madhukar, *J. Organomet. Chem.*, 1992, **433**, 279.
579. D. W. Macomber, P. Madhukar and R. D. Rogers, *Organometallics*, 1989, **8**, 1275.
580. W. D. Wulff, B. A. Anderson and L. D. Isaacs, *Tet. Lett.*, 1989, **30**, 4061.
581. C. P. Casey, N. L. Hornung and N. W. Vollendorf, *J. Organomet. Chem.*, 1986, **303**, 375.
582. M. J. Winter, *Polyhedron*, 1989, **8**, 1583.
583. H. Adams, N. A. Bailey, M. J. Winter and S. Woodward, *J. Organomet. Chem.*, 1991, **410**, C21.
584. V. A. Osborn, C. A. Parker and M. J. Winter, *J. Chem. Soc., Chem. Commun.*, 1986, 1185.
585. H. Adams, N. A. Bailey, C. E. Tattershall and M. J. Winter, *J. Chem. Soc., Chem. Commun.*, 1991, 912.
586. H. Adams, N. A. Bailey, G. W. Bentley, C. E. Tattershall, B. F. Taylor and M. J. Winter, *J. Chem. Soc., Chem. Commun.*, 1992, 533.
587. C. E. Davey, R. Devonshire and M. J. Winter, *Polyhedron*, 1989, **8**, 1563.
588. K. H. Dötz, D. Grotjahn and K. Harms, *Angew. Chem., Int. Ed. Engl.*, 1989, **28**, 1384.
589. D. B. Grotjahn, F. E. K. Kroll, T. Schäfer, K. Harms and K. H. Dötz, *Organometallics*, 1992, **11**, 298.
590. R. Aumann and E. O. Fischer, *Chem. Ber.*, 1981, **114**, 1853.
591. E. O. Fischer and M. Böck, *Monatsh. Chem.*, 1984, **115**, 1159.
592. E. O. Fischer and M. Böck, *J. Organomet. Chem.*, 1985, **287**, 279.
593. R. Aumann, P. Hinterding, C. Krüger and R. Goddard, *J. Organomet. Chem.*, 1993, **459**, 145.
594. U. Schubert, *J. Organomet. Chem.*, 1988, **358**, 215.
595. W. Hepp and U. Schubert, *J. Organomet. Chem.*, 1987, **321**, 317.
596. A. D. Rooney, J. J. McGarvey, K. C. Gordon, R.-A. McNicholl, U. Schubert and W. Hepp, *Organometallics*, 1993, **12**, 1277.
597. E. O. Fischer, R. Reitmeier and K. Ackermann, *Angew. Chem., Int. Ed. Engl.*, 1983, **22**, 411.
598. A. Mayr and H. Hoffmeister, *Adv. Organomet. Chem.*, 1991, **32**, 227.
599. A. Z. Rybezhov, *Organomet. Chem. USSR*, 1991, **4**, 109.
600. S. A. Brew and F. G. A. Stone, *Adv. Organomet. Chem.*, 1993, **35**, 135.
601. H. Fischer, P. Hofmann, F. R. Kreissl, R. R. Schrock, U. Schubert and K. Weiss, 'Carbyne Complexes', VCH Verlagsgesellschaft, Weinheim, 1988.
602. M. Green, *J. Organomet. Chem.*, 1986, **300**, 93.
603. F. G. A. Stone, *Adv. Organomet. Chem.*, 1990, **31**, 53.
604. E. O. Fischer, D. Wittmann, A. Mayr and A. McDermott, *Inorg. Synth.*, 1989, **26**, 40.
605. G. A. McDermott, A. M. Dorries and A. Mayr, *Organometallics*, 1987, **6**, 925.
606. E. O. Fischer, D. Wittmann, D. Himmelreich, R. Cai, K. Ackermann and D. Neugebauer, *Chem. Ber.*, 1982, **115**, 3152.
607. C. Krüger, R. Goddard and K. H. Claus, *Z. Naturforsch., Teil B*, 1983, **38**, 1431.
608. G. P. Foulet-Fonseca, M. Jouan, N. Q. Dao, N. T. Huy and E. O. Fischer, *J. Chim. Phys. Phys.-Chim. Biol.*, 1990, **87**, 13.
609. G. P. Foulet-Fonseca, M. Jouan, N. Q. Dao, H. Fischer, J. Schmid and E. O. Fischer, *Spectrochim. Acta, Part A*, 1990, **46**, 339.
610. N. Q. Dao, M. Jouan, G. P. Fonseca, N. H. T. Huy and E. O. Fischer, *J. Organomet. Chem.*, 1985, **287**, 215.
611. A. Spasojevic-de Biré, N. Q. Dao, A. Strich, C. Thieffry and M. Bénard, *Inorg. Chem.*, 1990, **29**, 4908.
612. N. Q. Dao, H. Fevrier, M. Jouan and E. O. Fischer, *New. J. Chim.*, 1983, **7**, 719.
613. N. M. Kostic and R. F. Fenske, *Organometallics*, 1982, **1**, 489.
614. D. Saddei, H.-J. Freund and G. Hohlneicher, *J. Organomet. Chem.*, 1981, **216**, 235.
615. A. A. Low and M. B. Hall, *Organometallics*, 1990, **9**, 701.
616. H. Fischer, E. O. Fischer, D. Himmelreich, R. Cai, U. Schubert and K. Ackermann, *Chem. Ber.*, 1981, **114**, 3220.
617. A. C. Filippou, E. O. Fischer and H. G. Alt, *J. Organomet. Chem.*, 1988, **340**, 331.
618. A. Mayr, A. M. Dorries, G. A. McDermott and D. Van Engen, *Organometallics*, 1986, **5**, 1504.
619. A. Mayr, M. F. Asaro, M. A. Kjelsberg, K. S. Lee and D. Van Engen, *Organometallics*, 1987, **6**, 432.
620. A. C. Filippou and E. O. Fischer, *J. Organomet. Chem.*, 1988, **352**, 149.
621. H. Fischer, A. Motsch, R. Märkl and K. Ackermann, *Organometallics*, 1985, **4**, 726.
622. U. Schubert, D. Neugebauer, P. Hofmann, B. E. R. Schilling, H. Fischer and A. Motsch, *Chem. Ber.*, 1981, **114**, 3349.
623. H. Fischer, A. Motsch, U. Schubert and D. Neugebauer, *Angew. Chem., Int. Ed. Engl.*, 1981, **20**, 463.
624. H. Fischer and A. Motsch, *J. Organomet. Chem.*, 1981, **220**, 301.
625. N. Q. Dao, H. Fevrier, M. Jouan, E. O. Fischer and W. Röll, *J. Organomet. Chem.*, 1984, **275**, 191.
626. G. A. Carriedo, V. Riera, J. M. R. Gonzalez and M. G. Sanchez, *Acta. Crystallogr., Sect. C*, 1990, **46**, 581.
627. G. A. Carriedo, G. Sánchez, V. Riera, C. Bois, Y. Jeannin and D. Miguel, *J. Chem. Soc., Dalton Trans.*, 1990, 3355.
628. F. A. Cotton and W. Schwotzer, *Inorg. Chem.*, 1983, **22**, 387.
629. A. C. Filippou and E. O. Fischer, *J. Organomet. Chem.*, 1990, **382**, 143.
630. A. Mayr, M. A. Kjelsberg, K. S. Lee, M. F. Asaro and T.-C. Hsieh, *Organometallics*, 1987, **6**, 2610.
631. E. O. Fischer, A. C. Filippou and H. G. Alt, *J. Organomet. Chem.*, 1985, **296**, 69.
632. E. O. Fischer *et al.*, *J. Organomet. Chem.*, 1983, **247**, 239.
633. E. O. Fischer and J. K. R. Wanner, *J. Organomet. Chem.*, 1983, **252**, 175.
634. A. C. Filippou, E. O. Fischer and R. Paciello, *J. Organomet. Chem.*, 1988, **347**, 127.
635. A. Mayr, A. M. Dorries, G. A. McDermott, S. J. Geib and A. L. Rheingold, *J. Am. Chem. Soc.*, 1985, **107**, 7775.
636. A. Mayr, A. M. Dorries, A. L. Rheingold and S. J. Geib, *Organometallics*, 1990, **9**, 964.
637. S. J. Davies, A. F. Hill, M. U. Pilotti and F. G. A. Stone, *Polyhedron*, 1989, **8**, 2265.
638. I. J. Hart, A. F. Hill and F. G. A. Stone, *J. Chem. Soc., Dalton Trans.*, 1989, 2261.
639. S. Anderson and A. F. Hill, *J. Organomet. Chem.*, 1990, **394**, C24.
640. A. K. List, G. L. Hillhouse and A. L. Rheingold, *Organometallics*, 1989, **8**, 2010.

641. A. C. Filippou, E. O. Fischer and H. G. Alt, *J. Organomet. Chem.*, 1988, **344**, 215.
642. A. C. Filippou and E. O. Fischer, *J. Organomet. Chem.*, 1988, **352**, 141.
643. R. Märkl and H. Fischer, *J. Organomet. Chem.*, 1984, **267**, 277.
644. E. O. Fischer, D. Wittmann, D. Himmelreich, U. Schubert and K. Ackermann, *Chem. Ber.*, 1982, **115**, 3141.
645. A. C. Filippou, E. O. Fischer and H. G. Alt, *J. Organomet. Chem.*, 1986, **310**, 357.
646. E. O. Fischer, A. C. Filippou, H. G. Alt and U. Thewalt, *Angew. Chem., Int. Ed. Engl.*, 1985, **24**, 203.
647. E. O. Fischer and D. Wittmann, *J. Organomet. Chem.*, 1985, **292**, 245.
648. A. C. Filippou and E. O. Fischer, *J. Organomet. Chem.*, 1987, **330**, C1.
649. J. Manna, T. M. Gilbert, R. F. Dallinger, S. J. Geib and M. D. Hopkins, *J. Am. Chem. Soc.*, 1992, **114**, 5870.
650. A. B. Bocarsly, R. E. Cameron, H.-D. Rubin, G. A. McDermott, C. R. Wolff and A. Mayr, *Inorg. Chem.*, 1985, **24**, 3976.
651. H. Fischer and F. Seitz, *J. Organomet. Chem.*, 1984, **268**, 247.
652. H. Fischer and F. Seitz, *J. Organomet. Chem.*, 1984, **275**, 83.
653. E. O. Fischer, A. C. Filippou and H. G. Alt, *J. Organomet. Chem.*, 1984, **276**, 377.
654. A. C. Filippou, E. O. Fischer and J. Okuda, *J. Organomet. Chem.*, 1988, **339**, 309.
655. A. Mayr, G. A. McDermott, A. M. Dorries and D. Van Engen, *Organometallics*, 1987, **6**, 1503.
656. A. C. Filippou, E. O. Fischer, H. G. Alt and U. Thewalt, *J. Organomet. Chem.*, 1987, **326**, 59.
657. A. Mayr, G. A. McDermott, A. M. Dorries, A. K. Holder, W. C. Fultz and A. L. Rheingold, *J. Am. Chem. Soc.*, 1986, **108**, 310.
658. H. Fischer, E. O. Fischer and R. F. Cai, *Chem. Ber.*, 1982, **115**, 2707.
659. H. Fischer and B. Bühlmeyer, *J. Organomet. Chem.*, 1986, **317**, 187.
660. W. Uedelhoven, D. Neugebauer and F. R. Kreissl, *J. Organomet. Chem.*, 1981, **217**, 183.
661. A. C. Filippou and E. O. Fischer, *J. Organomet. Chem.*, 1990, **383**, 179.
662. A. C. Filippou and E. O. Fischer, *J. Organomet. Chem.*, 1989, **365**, 317.
663. A. C. Filippou, *Polyhedron*, 1990, **9**, 727.
664. A. C. Filippou and W. Grünleitner, *Z. Naturforsch., Teil A*, 1989, **44**, 1572.
665. A. Mayr, C. M. Bastos, R. T. Chang, J. X. Haberman, K. S. Robinson and D. A. Belle-Oudry, *Angew. Chem., Int. Ed. Engl.*, 1992, **31**, 747.
666. G. A. McDermott and A. Mayr, *J. Am. Chem. Soc.*, 1987, **109**, 580.
667. C. N. Wilker, R. Hoffmann and O. Eisenstein, *New J. Chem.*, 1983, **7**, 535.
668. A. Mayr, K. S. Lee and B. Kahr, *Angew. Chem., Int. Ed. Engl.*, 1988, **27**, 1730.
669. E. O. Fischer, D. Wittmann, D. Himmelreich and D. Neugebauer, *Angew. Chem., Int. Ed. Engl.*, 1982, **21**, 444.
670. T. M. Sivavec, T. J. Katz, M. Y. Chiang and G. X.-Q. Yang, *Organometallics*, 1989, **8**, 1620.
671. T. J. Katz, T. H. Ho, N.-Y. Shih, Y.-C. Ying and V. I. W. Stuart, *J. Am. Chem. Soc.*, 1984, **106**, 2659.
672. A. C. Filippou and E. O. Fischer, *J. Organomet. Chem.*, 1990, **382**, 143.
673. A. Mayr and G. A. McDermott, *J. Am. Chem. Soc.*, 1986, **108**, 548.
674. A. C. Filippou, W. Grünleitner and P. Kiprof, *J. Organomet. Chem.*, 1991, **410**, 175.
675. A. C. Filippou, P. Hofmann, P. Kiprof, H. R. Schmidt and C. Wagner, *J. Organomet. Chem.*, 1993, **459**, 233.
676. A. C. Filippou, W. Grünleitner, C. Völkl and P. Kiprof, *Angew. Chem., Int. Ed. Engl.*, 1991, **30**, 1167.
677. A. C. Filippou and E. O. Fischer, *J. Organomet. Chem.*, 1988, **349**, 367.
678. A. C. Filippou and W. Grünleitner, *J. Organomet. Chem.*, 1991, **407**, 61.
679. A. C. Filippou, W. Grünleitner and E. O. Fischer, *J. Organomet. Chem.*, 1991, **401**, C37.
680. A. C. Filippou and E. O. Fischer, *J. Organomet. Chem.*, 1988, **341**, C35.
681. A. C. Filippou and E. O. Fischer, *J. Organomet. Chem.*, 1990, **386**, 333.
682. A. C. Filippou, *Polyhedron*, 1989, **8**, 1285.
683. K. E. Garrett, W. C. Feng, H. Matsuzaka, G. L. Geoffroy and A. L. Rheingold, *J. Organomet. Chem.*, 1990, **394**, 251.
684. F. R. Kreissl, W. Uedelhoven and D. Neugebauer, *J. Organomet. Chem.*, 1988, **344**, C27.
685. S. J. Dossett, A. F. Hill, J. C. Jeffery, F. Marken, P. Sherwood and F. G. A. Stone, *J. Chem. Soc., Dalton Trans.*, 1988, 2453.
686. A. C. Filippou, W. Grünleitner, E. O. Fischer, W. Imhof and G. Huttner, *J. Organomet. Chem.*, 1991, **413**, 165.
687. W. Kläui and H. Hamers, *J. Organomet. Chem.*, 1988, **345**, 287.
688. W. Kläui, H. Hamers, M. Pfeffer, A. DeCian and J. Fischer, *J. Organomet. Chem.*, 1990, **394**, 213.
689. A. C. Filippou, W. Grünleitner and E. O. Fischer, *J. Organomet. Chem.*, 1991, **411**, C21.
690. W. Clegg, M. Green, C. A. Hall, D. Hockless, N. C. Norman and C. M. Woolhouse, *J. Chem. Soc., Chem. Commun.*, 1990, 1330.
691. N. A. Ustynyuk, V. N. Vinogradova, V. G. Andrianov and Y. T. Struchkov, *J. Organomet. Chem.*, 1984, **268**, 73.
692. A. C. Filippou, K. Wanninger and C. Mehnert, *J. Organomet. Chem.*, 1993, **461**, 99.
693. M. Green, J. A. K. Howard, A. P. James, A. N. D. M. Jelfs, C. M. Nunn and F. G. A. Stone, *J. Chem. Soc., Chem. Commun.*, 1984, 1623.
694. J. C. Jeffery, F. G. A. Stone and G. K. Williams, *Polyhedron*, 1991, **10**, 215.
695. A. S. Gamble, P. S. White and J. L. Templeton, *Organometallics*, 1991, **10**, 693.
696. A. C. Filippou, C. Wagner, E. O. Fischer and C. Völkl, *J. Organomet. Chem.*, 1992, **438**, C15.
697. D. C. Brower, M. Stoll and J. L. Templeton, *Organometallics*, 1989, **8**, 2786.
698. T. Desmond, F. J. Lalor, G. Ferguson and M. Parvez, *J. Chem. Soc., Chem. Commun.*, 1983, 457.
699. T. Desmond, F. J. Lalor, G. Ferguson and M. Parvez, *J. Chem. Soc., Chem. Commun.*, 1984, 75.
700. G. M. Jamison, P. S. White and J. L. Templeton, *Organometallics*, 1991, **10**, 1954.
701. P. K. Byers and F. G. A. Stone, *J. Chem. Soc., Dalton Trans.*, 1990, 3499.
702. R. A. Doyle and R. J. Angelici, *Organometallics*, 1989, **8**, 2207.
703. M. D. Bermúdez, F. P. E. Brown and F. G. A. Stone, *J. Chem. Soc., Dalton Trans.*, 1988, 1139.
704. D. D. Devore, S. J. B. Henderson, J. A. K. Howard and F. G. A. Stone, *J. Organomet. Chem.*, 1988, **358**, C6.
705. M. Green, J. A. K. Howard, A. P. James, C. M. Nunn and F. G. A. Stone, *J. Chem. Soc., Chem. Commun.*, 1984, 1113.
706. M. Green, J. A. K. Howard, A. P. James, C. M. Nunn and F. G. A. Stone, *J. Chem. Soc., Dalton Trans.*, 1987, 61.
707. F.-E. Baumann, J. A. K. Howard, R. J. Musgrove, P. Sherwood and F. G. A. Stone, *J. Chem. Soc., Dalton Trans.*, 1988, 1891.
708. F.-E. Baumann, J. A. K. Howard, O. Johnson and F. G. A. Stone, *J. Chem. Soc., Dalton Trans.*, 1987, 2661.

709. D. D. Devore, C. Emmerich, J. A. K. Howard and F. G. A. Stone, *J. Chem. Soc., Dalton Trans.*, 1989, 797.
710. W. Schulze, H. Hartl and K. Seppelt, *J. Organomet. Chem.*, 1987, **319**, 77.
711. J. A. K. Howard, J. C. Jeffery, J. C. V. Laurie, I. Moore, F. G. A. Stone and A. Stringer, *Inorg. Chim. Acta*, 1985, **100**, 23.
712. J. C. Jeffery, J. C. V. Laurie, I. Moore and F. G. A. Stone, *J. Organomet. Chem.*, 1983, **258**, C37.
713. G. C. Bruce and F. G. A. Stone, *Polyhedron*, 1992, **11**, 1607.
714. F. R. Kreissl, W. J. Sieber and M. Wolfgruber, *J. Organomet. Chem.*, 1984, **270**, C45.
715. F. R. Kreissl, H. Keller and W. Schütt, *J. Organomet. Chem.*, 1992, **441**, 75.
716. F. R. Kreissl, W. J. Sieber, H. Keller, J. Riede and M. Wolfgruber, *J. Organomet. Chem.*, 1987, **320**, 83.
717. A. P. James and F. G. A. Stone, *J. Organomet. Chem.*, 1986, **310**, 47.
718. J. C. Jeffery, S. Li and F. G. A. Stone, *Organometallics*, 1992, **11**, 1902.
719. S. A. Brew, N. Carr, J. C. Jeffery, M. U. Pilotti and F. G. A. Stone, *J. Am Chem. Soc.*, 1992, **114**, 2203.
720. M. Etienne, P. S. White and J. L. Templeton, *J. Am. Chem. Soc.*, 1991, **113**, 2324.
721. S. Chaona, F. J. Lalor, G. Ferguson and M. M. Hunt, *J. Chem. Soc., Chem. Commun.*, 1988, 1606.
722. A. E. Bruce, A. S. Gamble, T. L. Tonker and J. L. Templeton, *Organometallics*, 1987, **6**, 1350.
723. G. M. Jamison, A. E. Bruce, P. S. White and J. L. Templeton, *J. Am. Chem. Soc.*, 1991, **113**, 5057.
724. W. J. Sieber *et al.*, *J. Organomet. Chem.*, 1988, **340**, 341.
725. P. G. Byrne, M. E. Garcia, N. H. Tran-Huy, J. C. Jeffery and F. G. A. Stone, *J. Chem. Soc., Dalton Trans.*, 1987, 1243.
726. J. C. Jeffery, C. Sambale, M. F. Schmidt and F. G. A. Stone, *Organometallics*, 1982, **1**, 1597.
727. P. Dahlke, J. C. Jeffery, M. D. Mortimer and F. G. A. Stone, *Polyhedron*, 1992, **11**, 1587.
728. M. Wolfgruber, C. M. Stegmair and F. R. Kreissl, *J. Organomet. Chem.*, 1989, **376**, 45.
729. G. A. Carriedo, V. Riera, M. L. Rodriguez and J. C. Jeffery, *J. Organomet. Chem.*, 1986, **314**, 139.
730. A. F. Hill, J. A. K. Howard, T. P. Spaniol, F. G. A. Stone and J. Szameitat, *Angew. Chem., Int. Ed. Engl.*, 1989, **28**, 210.
731. J. C. Jeffery, D. W. I. Sams and K. D. V. Weerasuria, *J. Organomet. Chem.*, 1991, **418**, 395.
732. J. B. Sheridan, D. B. Pourreau, G. L. Geoffroy and A. L. Rheingold, *Organometallics*, 1988, **7**, 289.
733. J. D. Carter *et al.*, *J. Am. Chem. Soc.*, 1991, **113**, 2947.
734. K. B. Kingsbury, J. D. Carter and L. McElwee-White, *J. Chem. Soc., Chem. Commun.*, 1990, 624.
735. N. H. T. Huy, J. Fischer and F. Mathey, *Organometallics*, 1988, **7**, 240.
736. F. R. Kreissl, N. Ullrich and H. Keller, *J. Organomet. Chem.*, 1991, **403**, C29.
737. F. R. Kreissl and H. Keller, *Angew. Chem., Int. Ed. Engl.*, 1986, **25**, 904.
738. N. Ullrich, H. Keller, C. Stegmair and F. R. Kreissl, *J. Organomet. Chem.*, 1989, **378**, C19.
739. R. A. Doyle and R. J. Angelici, *J. Organomet. Chem.*, 1989, **375**, 73.
740. R. A. Doyle and R. J. Angelici, *J. Am. Chem. Soc.*, 1990, **112**, 194.
741. F. R. Kreissl and N. Ullrich, *J. Organomet. Chem.*, 1989, **361**, C30.
742. F. R. Kreissl and N. Ullrich, *J. Organomet. Chem.*, 1992, **426**, C8.
743. C. M. Stegmair and F. R. Kreissl, *J. Organomet. Chem.*, 1990, **397**, C6.
744. S. J. Dossett, I. J. Hart and F. G. A. Stone, *J. Chem. Soc., Dalton Trans.*, 1990, 3481.
745. S. J. Dossett, I. J. Hart, M. U. Pilotti and F. G. A. Stone, *J. Chem. Soc., Dalton Trans.*, 1990, 3489.
746. S. A. Brew, N. Carr, M. D. Mortimer and F. G. A. Stone, *J. Chem. Soc., Dalton Trans.*, 1991, 811.
747. G. A. Carriedo, J. A. K. Howard, D. B. Lewis, G. E. Lewis and F. G. A. Stone, *J. Chem. Soc., Dalton Trans.*, 1985, 905.
748. G. A. Carriedo *et. al.*, *J. Chem. Soc., Chem. Commun.*, 1982, 1006.
749. I. J. Hart, J. A. K. Howard, R. M. Lowry, T. P. Spaniol and F. G. A. Stone, *Polyhedron*, 1989, **8**, 2035.
750. G. C. Bruce, D. F. Mullica, E. L. Sappenfield and F. G. A. Stone, *J. Chem. Soc., Dalton Trans*, 1992, 2685.
751. J. C. Jeffery, A. L. Ratermann and F. G. A. Stone, *J. Organomet. Chem.*, 1985, **289**, 367.
752. D. Barratt, S. J. Davies, G. P. Elliot, J. A. K. Howard, D. B. Lewis and F. G. A. Stone, *J. Organomet. Chem.*, 1987, **325**, 185.
753. I. J. Hart, J. C. Jeffery, R. M. Lowry and F. G. A. Stone, *Angew. Chem., Int. Ed. Engl.*, 1988, **27**, 1703.
754. J. D. Carter, T. K. Schoch and L. McElwee-White, *Organometallics*, 1992, **11**, 3571.
755. M. D. Mortimer, J. D. Carter and L. McElwee-White, *Organometallics*, 1993, **12**, 4493.
756. P. K. Byers and F. G. A. Stone, *J. Chem. Soc., Dalton Trans.*, 1991, 93.

4

Carbonyl Complexes of Noncarbon σ-Bonded Ligands of Chromium, Molybdenum and Tungsten

SIMON WOODWARD
University of Hull, UK

4.1 GENERAL REMARKS

Although a complete listing of noncarbon ligand derivatives of $[M(CO)_6]$ (M = Cr, Mo, W) is beyond the scope of this chapter, attempts have been made to identify important papers published in the area from late 1981 to mid-1993. This chapter is designed to be used in conjunction with *COMC-I*;[1–3] repetition of discussions in the first edition is thus avoided.

4.1.1 Reaction Mechanisms

An overview of early investigations to 1982 on substitution reactions of octahedral carbonyl complexes, including group 6 species, is available in a comprehensive review by Howell and Burkinshaw.[4] Selected recent kinetic results leading to mechanistic conclusions are given in Table 1. Although dissociative behaviour dominates substitution processes in group 6 carbonyl complexes, a growing number of reactions for which outer-sphere encounter complexes are involved have been detected by kinetic studies. These reactions show the negative entropies and volumes of activation indicative of interchange, typically I_d, mechanisms. Such mechanisms are particularly prevalent where weak bonding interactions, such as hydrogen bonds, are possible in the encounter complex, as in the behaviour of $[W(CO)_5(NCCH_2CO_2H)]$ in THF.[5]

Table 1 Selected kinetic studies of the reactions of group 6 carbonyl complexes.

Reaction		$\Delta H^{\ddagger}$ (kJ mol^{-1})	$\Delta S^{\ddagger}$ (J K^{-1} mol^{-1})	*Ref.*
Displacement reactions				
$[Cr(CO)_4(C_5H_{10}NH)\{P(OPr^i)_3\}] + P(OPh)_3$	D	123(8)	86(25)	6
$[Cr(CO)_4(MeSCH_2CH_2SMe)] + P(OMe)_3$	D	96.6(50)	0(17)	7
$[Cr(CO)_4\{(Z)\text{-}Bu^tSCH{=}CHSBu^t\}] + P(OEt)_3$	D	92.5(58)	−8(5)	8
$[Mo(CO)_4(MeSCH_2CH_2SMe)] + P(OMe)_3$	I_d	72.4(63)	−59(21)	7
$[Mo(CO)_3(py)(dppe)] + P(OPr^i)_3$	D	106.59(8)	67.17(25)	9
$[W(CO)_4\{P(OPr^i)_3\}\{\eta^1\text{-}Bu^tSCH_2SBu^t\}] + P(OPr^i)_3$	D	118.0(25)	75.7(8)	10
$[W(CO)_5(NCMe)] + CO$	I_d	92.5 ± 5.0	−32 ± 15	5
$[W(CO)_5(CyH)] + 4\text{-}AcC_5H_4N$	I_d	14.2(8)	−57	11
$[W(CO)_4(ClPh)\{P(OPr^i)_3\}] + C_5H_{10}NH$	D	54.4(8)	23.0(46)	12
$[W(CO)_3(NCMe)(dppm)] + {}^{13}CO$	D	98 ± 6	−25 ± 18	13
Neucleophilic attack at M–CO				
$[Cr(CO)_6] + Ar_2TeO$	A	77 ± 3	−29 ± 2	14
$[Cr(CO)_6] + Me_3NO + PPh_3$	A	51 ± 2	−90 ± 7	15
$[Mo(CO)_6] + Me_3NO + PPh_3$	A	54.4 ± 3	−76 ± 12	15
$[Mo(CO)_5(PPh_3)] + Me_3NO + PPh_3$	A	64 ± 5	−69 ± 15	16
$[W(CO)_6] + Me_3NO + PPh_3$	A	70 ± 3	−19 ± 11	15
$[WI(CO)_4(NO)] + PBu_3$	A	68 ± 1	−85 ± 2	17
Chelate ring closures				
$[Mo(CO)_4(py)_2] + phen$	D	102 ± 2	56 ± 5	18
$[M(CO)_5(4\text{-}Mepy)] + PhN{=}C(Me)C(Me){=}NPh$	D	98 ± 5	0 ± 19	18
$[W(CO)_5\{\eta^1\text{-}2\text{-}pyCH({=}NPh)\}]$ -CO (irradiation)	I_d	45.6 ± 8	−140 ± 16	19
$[W(CO)_5\{\eta^1\text{-}2\text{-}pyCH({=}NBu^t)\}]$ -CO (irradiation)	I_d	82.0 ± 6	−57 ± 12	19
$[W(CO)_4(BrPh)(\eta^1\text{-}Bu^tSCH_2SBu^t)]$ -PhBr	I_d	26.4(13)	−46.8(33)	10

The transient pentacarbonyl intermediates $[M(CO)_5]$ are of pre-eminent importance in most dissociative substitutions of hexacarbonyls. Flash-photolysis studies demonstrate that the dissociative product $[M(CO)_5]$ is a high-energy entity indiscriminately reacting with whatever it collides. For example, the ratio of the rates of reaction with both solvent and added ligand molecules approach unity for competition reactions. For the substituted species $[M(CO)_4(L^1)(L^2)]$ ($L^2 = CO, PR_3$), the primary factor governing the rate of substitution is the ability of the unique ligand L^1 to stabilize the unsaturated $[M(CO)_4(L^1)]$ intermediates; efficient σ-donors are the most effective. Again, flash photolysis studies are invaluable in determining the stereochemistry of the unsaturated intermediate; a study of $[Cr(CO)_4(C_5H_{10}NH)(PR_3)]$ is typical.[6] The complexes (**1**) are also interesting probes for determining the stereochemical stability of related coordinatively unsaturated fragments. Reaction of complex (**1**) with ^{13}CO leads to the formation of the two diastereomers (**2**) and (**3**) in ratios of 80:20 (Mo) and 86:14 (W). These results are in accord with the mechanism of Scheme 1, where the rate of isomerization via five-coordinate (**4**) is comparable to its capture by ^{13}CO.[20]

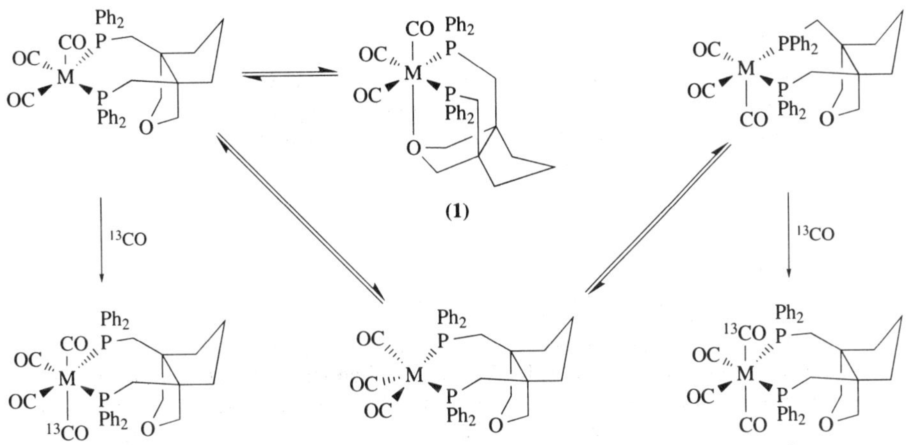

(1) (2) (4) (3)

M = Mo, W

Scheme 1

The bulk of current evidence suggests that associative processes can be of great importance in chelate ring closures of pendent *cis*-$[M(CO)_4(L)(\eta^1\text{-L–L})]$ ($L = CO, NR_3$, solvent; L–L = chelate ligand). Regardless of mechanism, these intramolecular reactions show first order kinetics. The reaction rates are highly dependent on the nature of the pendent entering group and show significantly negative entropies of activation, indicating extensive ordering in events up to and including the transition state.[21] When the leaving groups are particularly labile the reactions show dissociative tendencies (I_d or even D), as judged by $\Delta V^\ddagger$ and $\Delta S^\ddagger$ data.[21] The reactions of $[M(CO)_5(\eta^1\text{-N–N})]$ are the most studied. Complexes of bipy and diazabutadiene species ($R^1N{=}CR^2CR^2{=}NR^1$) exist in *s-cis* conformations slowing the reaction while the enforced *cis* geometry of phen ligands results in rapid ring closure. Under favourable conditions some degree of Cr···N interaction can apparently be detected in $[Cr(CO)_5(\eta^1\text{-phen})]$.[22]

4.2 CARBONYL DERIVATIVES WITH NONCARBON σ-DONOR LIGANDS

4.2.1 Carbonyl Halides

Complexes containing at least one M–X (X = Cl, Br, I) bond are the subject of this section; some additional information regarding $[MX(CO)_3\{HB(pz)_3\}]$ (pz = pyrazolyl) and related complexes is found in Section 4.3.1. In general, the inability of the smaller chromium species to support a seven-coordinate geometry leads to significant differences in reactivity between chromium and the other group 6 members.

4.2.1.1 Synthesis

(i) Carbonyl halides in oxidation state 0 and +1

Typical preparations of $[MX(CO)_5]^-$ involve carbonyl displacement from $[M(CO)_6]$ in the presence of $[NR_4]X$ or crown-ether complexes of KX. A typical preparative protocol is the *Inorganic Syntheses* preparation of $[NEt_4][CrCl(CO)_5]$.[23] Reactions conducted under two phase conditions are recommended as a convenient method of eliminating unreacted $[M(CO)_6]$ from the final product.[24] Two unusual preparations are the formation of the carbyne-containing complexes $[(CO)_4Cr(\mu\text{-}X)Cr(\equiv CR)(CO)_4]$ (X = Br, I; M = Cr, W; R = Me, Ph)[25] and the crown-ether complex *fac*-$[MoBr(CO)_3\{[15]P_2O_4Li\}]$.[26] Of greater interest is the development of a general route to chromium(I) complexes. Initial reaction of [K(dibenzo-18-crown-6)]X (X = F, Cl, Br, I) with $[Cr(CO)_6]$ fashions [K(dibenzo-18-crown-6][CrX(CO)$_5$] which is oxidized electrochemically to $[CrF(CO)_5]$ (for X = F). The other species (X = Cl, Br, I) are thermodynamically much less stable.[27]

(ii) Carbonyl halides in oxidation state +2

The smallest groups within this family are associated with chromium halocarbonyls and the carbonyl fluorides of the group 6 elements. Although $[Cr(CO)_6]$ does not undergo oxidative addition with elemental X_2 to give carbonyl products, *trans*-$[CrCl_2(CO)_4]$ is formed upon the interaction of $CrCl_2$ and CO under matrix isolation conditions, as judged by IR and EXAFS studies, but as expected the complex is not isolable.[28] A few well characterized metal(II) fluoro compounds have appeared recently, typical examples are $[MoF_2(CO)_3]_4$ prepared by XeF_2 oxidation of $[Mo(CO)_6]$,[29] $[MoF(CO)_2(dppe)_2]PF_6$ isolated from the reaction of $[Mo(CO)_2(dppe)_2]$ with $NOPF_6$,[30] and $[NEt_4][MoF(CO)_2(S_2CNEt_2)_2]$.[31] A comprehensive listing is available.[32]

Substituted halocarbonyls $[MX_2(CO)_n(L)_m]$ (M = Mo, W; X = Cl, Br, I) form the largest group of complexes in this section. One reason for this preponderance is the availability of several versatile starting materials. These include $[M_2(\mu\text{-}X)_2X_2(CO)_8]$ (M = Mo, W; X = Cl, Br, I) and $[MX_2(CO)_3(NCMe)_2]$ (M = Mo, W; X = Br, I) formed in high yield from $[M(CO)_3(NCMe)_3]$ (Equation (1)).[33] The complexes $[R_4N][MoX_3(CO)_4]$ (X = Br, I) are also effective '$MoX_2(CO)_n$' precursors.[34] For these and metal(0) starting materials, at least five general classes of reaction leading to carbonyl halides can be identified.

(a) Halogen addition to $[M(CO)_m(L)_n]$, demonstrated by the preparation of $[MoBr_2(CO)_3(bipy)]$ (Equation (2)).

(b) Carbonyl substitution in $[MX_2(CO)_m(L)_n]$ complexes promoted by ligand addition as in the preparation of (**5**) (Equation (3)).[35]

(c) Substitution reactions of $[M_2(\mu\text{-}X)_2X_2(CO)_8]$ and $[MX_3(CO)_4]^-$, although these involve ligand induced CO loss they are distinct from route (**b**) as they react initially by M–X bond cleavage, for example (Equation (4)).

(d) Decarbonylation of $[MX_2(CO)_n(L)_m]$ in the absence of ligands to provide mononuclear or oligomeric complexes as exemplified by Equation (5).

(e) Substitution of L in $[MX_2(CO)_n(L)_m]$ without CO loss, of which the formation of (**6**) is an unusual example in that one of the liberated PPh_3 ligands is incorporated into a new chelate (Equation (6)).

$$\underset{M = Mo, W}{[M(CO)_3(NCMe)_3]} \xrightarrow[-MeCN\ 100\%]{X_2\ 0\ ^\circ C} \underset{X = Br, I}{[MX_2(CO)_3(NCMe)_2]} \qquad (1)$$

$$[Mo(CO)_4(bipy)] \xrightarrow[-CO]{Br_2} [MoBr_2(CO)_3(bipy)] \qquad (2)$$

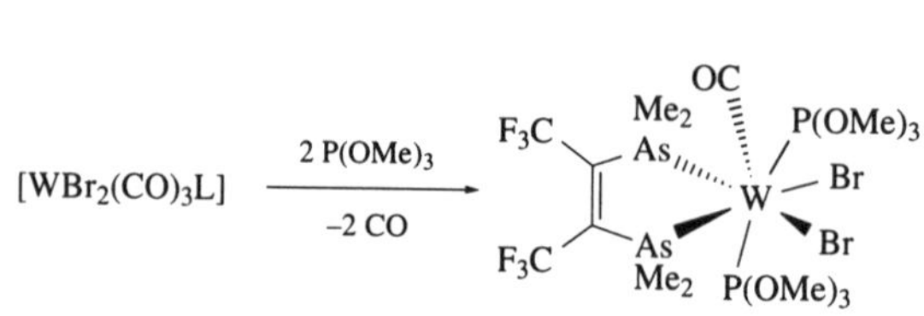

(3)

L = $Me_2As(CF_3)C{=}C(CF_3)AsMe_2$ (**5**)

$$[WBr_2(CO)_4]_2 \xrightarrow{PMe_3} [WBr_2(CO)_4(PMe_3)] \xrightarrow[-CO]{PMe_3} [WBr_2(CO)_3(PMe_3)_2] \qquad (4)$$

Table 2 Selected M^{II} halocarbonyl complexes.

Complex	*Synthesis*[a]	*Comments*	*M–X$_{av}$* (nm)	*Ref.*
N-ligated complexes				
$[MoBr(CO)_3(\mathbf{11})]X_3$	(a)	also I; macrocyclic ligand	0.2676	36
$[MoBr_2(CO)_2(CyN{=}CHCH{=}NCy)]$	(a)	also I and Mo^{III} species		37
$[MoBr_2(CO)_2\{H_2C_2(3,5\text{-}Me_2pz)_2\}]$	(a)	also I and related ligands	0.2538	38
$[MoBr_2(CO)(bipy)(dppm)]$	(b)		0.2676	39
$[WBr_2(CO)_3(PhN_3PPh_3)]$	(e)	also molybdenum	0.2620	40
$[WBr_2(CO)_3\{H_2C(3,4,5\text{-}Me_3pz)_2\}]$	(a)	also I	0.2658	38
$[WI_2(NCMe)_2(CO)_3]$	(a)	also Br	0.2822	33,41
P-, As- and Sb- ligated complexes				
$[MoBr_2(CO)_2(PEt_3)_2]$	(d)	also PPh_3; $16e^-$ species	0.2600	42,43
$[MoBr_2(CO)_2(PMePh_2)]$	(d)	also tricarbonyls		44
$[MoI_2(CO)_3\{P(OMe)_3\}_2]$	(e)	also dicarbonyl for tungsten		45
$[Mo_2I_4(CO)_6(PPh_3)_2(\mu\text{-}L_2)]$	(e)	L_2 = dppm, dppe, others		46
$[MoBr_2(CO)_2(PPh_3)(dppm)]$	(e)		0.2673	47
$[MoI_2(CO)_2(\eta^1\text{-}L_2)(L_2)]$	(c)	L_2 = dppe, dppm	0.2875	34
$[MoI_2(CO)(dmpm)_2]$	(c)		0.2925	48
$[MoI_2(CO)_3(Ph_2POPPh_2)]$	(a)	also Br derivative	0.2862	49
$[MoCl(CO)_2(dmpe)_2]PF_6$	(c)		0.2537	50
$[WX_2(CO)_4(EMe_3)]$	(c)	X = Cl, Br, I; E = P, As, Sb		51
$[WX_2(CO)_3(EMe_3)_2]$	(b) or (c)	X = Cl, Br, I; E = P, As, Sb		51
$[WBr_2(CO)_2(PMe_3)_3]$	(c)		0.2672	52
$[WBr_2(CO)_2(PPh_3)_2]$	(d)	16 e^- species	0.2568	53
$[WI_2(CO)(dppm)_2]$	(c) + (d)		0.2956	54
$[WI_2(CO)_3(\{P(OPh)_2\}_2NPh)]$	(e)	also molybdenum	0.2864	55

[a] For synthetic methods see text.

$$[MoBr_2(CO)_3(PMePh_2)_2] \xrightarrow[-CO]{\Delta\ CHCl_3} [MoBr_2(CO)_2(PMePh_2)_2]_n \qquad (5)$$

n = 1, 2, or oligomer

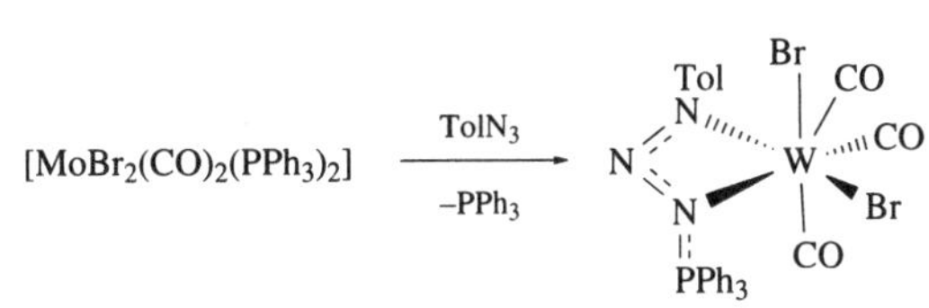

(6)

(6)

The synthetic approaches are further delineated by the examples in Table 2 which includes references for the examples above.

Although routes (a–e) cover the most commonly encountered preparations of M^{II} complexes many other routes are known. Carbonylation of high oxidation group 6 halo compounds is already well documented, a more unusual example of this chemistry is the preparation of $[WCl(CH_2PMe_3)(CO)_2(PMe_3)_3]OTf$.[56] Of wider applicability is the oxidative addition of aryl halides when held in close proximity to M^0 centres as in, for example, the reaction of $Me_2NCH_2CH_2N{=}CH(C_6F_5)$ with $[W(CO)_3(NCEt)_3]$.[57]

Coordination number and metal–carbonyl bond strength factors govern the synthesis of substituted carbonyl halides. Reaction coordinates leading to eight-coordinate intermediates are apparently not favoured; spontaneous expulsion of either CO[39] or halide[36,50] from the metal centre results. Alternatively, monodentate coordination of potential chelates occurs. Preparations involving the decarbonylation of $[MX_2(CO)_3(L)_2]$ species are sluggish for tungsten complexes due to high M–CO bond strengths. Spectator ligands that promote metal-to-CO backbonding also retard these reactions. For example, iodide complexes are often rather resistant to decarbonylation while the equivalent reactions of the bromide and chloride compounds are more facile. For reluctant systems, an inert gas purge directly into the reaction mixture often promotes CO expulsion. In these complexes the nature of the decarbonylation product depends on the cone angle (θ) of the phosphine. Mononuclear $[MX_2(CO)_2(PR_3)_2]$ species only result for ligands with θ > 136°; smaller phosphines result in dinuclear or oligomeric species associated through halide bridges.

4.2.1.2 Structure and properties

The crystal structures of numerous group 6 carbonyl halides have been analysed. Complexes of the type $[MX(CO)_5]^-$ show near octahedral geometries with the expected shortening of the M–C carbonyl distance *trans* to the halide caused by the poorer π-acceptor qualities of the halide ligand; a typical example is the crystal structure of [PPN][WCl(CO)$_5$], W–Cl 0.2560 nm, $\Delta(W{-}C_{cis}{-}W{-}C_{trans}) = 0.0055$ nm.[58]

From electron counting protocols it is predicted that seven-coordinate 18-electron M^{II} structures will predominate for group 6 complexes. In such instances, however, the situation is complicated by the absence of any uniquely favoured geometry.[59] Capped trigonal prismatic, pentagonal bipyramidal and capped octahedral stereochemistries are all known in the solid state and fluxional interconversions in solution are common. For stereochemically rigid capped octahedral stereochemistries a useful probe has been developed. In these complexes the presence of a unique CO capping ligand can be identified by its high-frequency ^{13}C NMR signal.[60] The MX_2 motif in seven-coordinate geometries is nearly always *cis*, but a transoid arrangement is demonstrated by $[MoBr_2(CO)_2\{CH_2(3,5\text{-}Me_2pz)_2\}]$ whose interhalogen angle is 175.5°. The near octahedral geometry of the complex leads to its observed paramagnetism contrasting with the diamagnetic behaviour of the Jahn–Teller distorted $[MX_2(CO)_2(PR_3)_2]$ family (M = Mo, W).[38]

The structures of the halide bridged complexes also deserve brief comment. As expected the M–X distances in the bridge are longer than for the equivalent terminally bonded interactions. These elongations can be quite dramatic if X–M–CO angles approaching 180° are realized in the dimer leading to strong *trans* influence. Structural characterizations have been reported for $[W(\mu\text{-}Br)_2Br_2(CO)_8]$, $[W(\mu\text{-}Br)_3Br(CO)_7]$,[61] $[Mo_2(\mu\text{-}I)_2I_2(CO)_8]$,[62] $[Mo_2(\mu\text{-}Br)(CO)_6(\mu_2\text{-}\eta^4\text{-}L)]$ (L = dianion of 2,6-$H_2NC_6H_3(PPh_2)_2$),[63] $[W_2(\mu\text{-}F)_3(CO)_4(PMePh_2)_4]BF_4$[64] and $[Mo_2(\mu\text{-}F)_3(CO)_4(PPh_3)_4]BF_4$.[65]

4.2.1.3 Reactions

The basic reactivity of $[MX(CO)_5]^-$ is dominated by halide and carbonyl displacement reactions. The *cis*-labilizing effect of the halide may be used to prepare *cis*-$[NEt_4][CrCl(CO)_4(L)]$ (L = PPh_3, $P(OPh)_3$, $AsPh_3$).[23] The versatility of halide displacement from $[MX(CO)_5]^-$ is best demonstrated by the preparation of a wide range of pseudohalide and alkene complexes prepared from $[NEt_4][WCl(CO)_5]$.[66]

The lability of the halide bridges and carbonyl ligands in $[M_2(\mu\text{-}X)_2X_2(CO)_8]$ leads to a wide range of reactivity; recent examples of which are summarized in Scheme 2. Related behaviour is demonstrated by the complexes $[MX_2(CO)_2(NCMe)_2]$ (M = Mo, W; X = Br, I) which contain highly labile acetonitrile ligands. Baker has exploited this reactivity to carry out numerous displacement reactions. For example, reaction with one equivalent of EPh_3 (E = P, As, Sb) leads to $[MI_2(CO)_3(NCMe)(EPh_3)]$. Subsequent reaction with two-electron donors leads to mixed systems $[MI_2(CO)_3(EPh_3)(L)]$. The nature of L may be varied widely: the reactions of PR_3, 1/2 $Ph_2P(CH_2)_4PPh_2$ and S_2CPCy_3 are typical.[46,67]

The deep blue–purple complexes $[MX_2(CO)_2(PR_3)_2]$ show an interesting range of reactivity. Being coordinatively unsaturated, these complexes readily coordinate two electron donors. However, additional complications in the displacement reactions of these species arise from the unexpected lability shown by the phosphine ligands in these complexes and their tendency to reduce to metal(0) species with some reagents. Recent results for these complexes are shown in abridged form in Scheme 3.

4.2.2 Carbonyl Hydrides

Hydride species that exist in equilibrium with dihydrogen (η^2-H_2) ligands are considered in Section 4.2.3. Anionic transition metal hydride complexes have been reviewed in general[81] and with special reference to CO_2 reduction.[82] The behaviour of $[M_4H_4(CO)_{12}]^{4-}$ (M = Mo, W) is included in a review on highly reduced organometallics.[83]

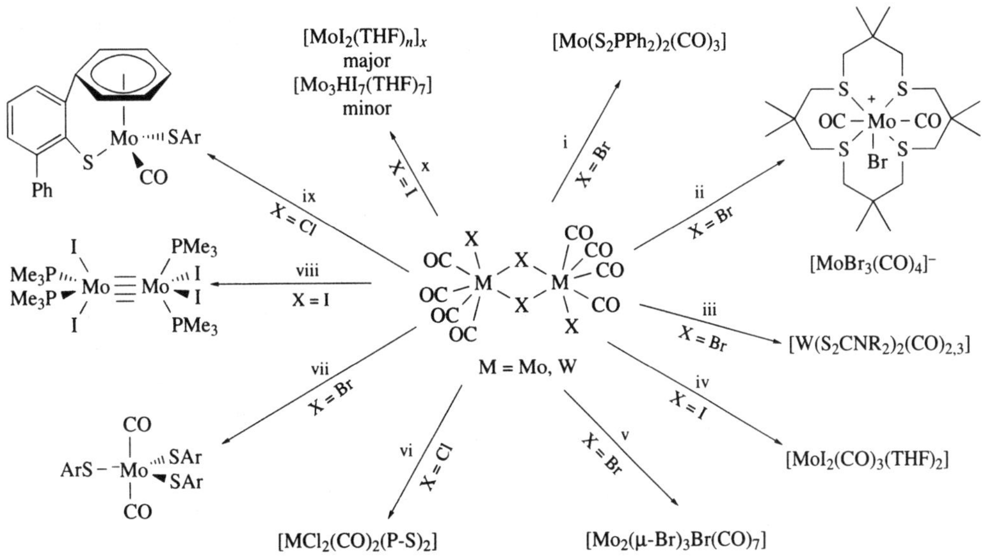

i, Ph_2PS_2H;[68] ii, S_4-Macrocycle;[69] iii, 2 NaS_2CNR_2;[70] iv, THF;[71] v, CO loss;[61] vi, $Ph_2PC(S)NHR$;[72] vii, Na[S-2,4,6-$C_6H_2Pr^i_3$];[73] viii, PMe_3 and heat;[74] ix, Na[S-2,6-$C_6H_3Ph_2$];[75] x, heat[76]

Scheme 2

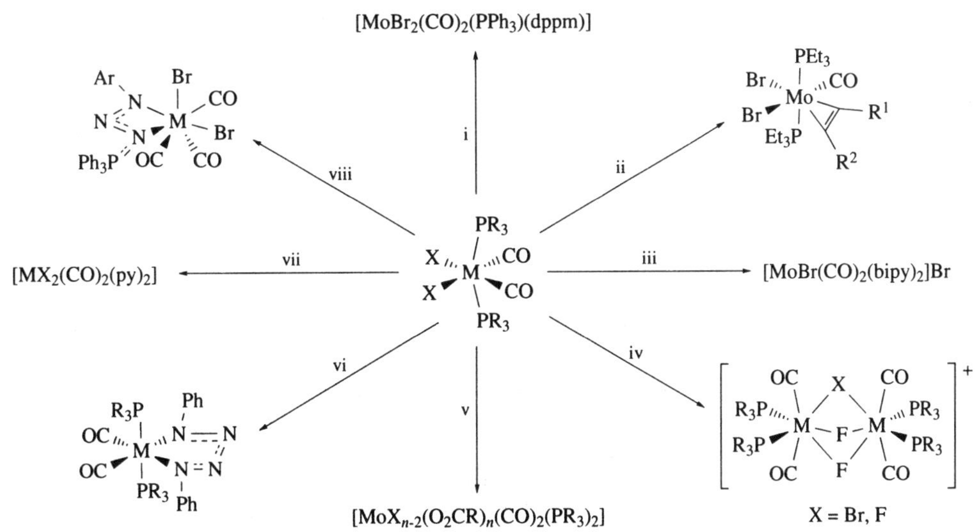

i, dppm;[47] ii, $R^1C{\equiv}CR^2$;[77] iii, 2 bipy;[47] iv, $AgBF_4$ or $NaBF_4$;[65,78] v, NaO_2CR;[79,80] vi, $Li_2[Ph_2N_4]$;[52] vii, py;[47] viii, ArN_3[40]

Scheme 3

4.2.2.1 Synthesis

Selected preparations for hydrido carbonyl complexes appear in Table 3 which includes recent reliable preparations for the often studied species $[MH(CO)_5]^-$ and $[M_2(\mu\text{-}H)(CO)_{10}]^-$. General synthetic routes to group 6 carbonyl hydrides are available and these are best illustrated with some recent examples; the relevant references are located in Table 3. These routes include:

(i) Protonation procedures. Both anionic and neutral complexes will accept H^+ if the metal is sufficiently electron rich. The electron density dictates the strength of acid required, thus the dianions $[M(CO)_5]^{2-}$ protonate from MeOH while neutral species require far stronger acids, typically HBF_4, CF_3SO_3H, or $HClO_4$. Protonation studies of $[W(CO)_3(P_3)]$ (P_3 = $3PR_3$ or tridentate ligand) substituted complexes reveal that one factor governing the ease of protonation in neutral complexes is relief of intraligand strain upon protonation. Unusual examples of this class are the syntheses of the complexes [WH(O_2CY)(CO)(dppe)$_2$] (Y = OMe, NR_2) prepared from MeOH or HNR_2 and CO_2. However, it is not clear whether this reaction proceeds via protonation by HY or the corresponding carbonic acids.

(ii) Attack of hydroxide on [M(CO)$_6$] to yield hydrides and CO_2 is well known; recent studies by Gibson have shown that such routes are improved by the use of two-phase reactions using [NEt$_4$]OH or equivalents (such as [K(cryptand-222)]OH) in place of KOH.

(iii) The reaction of metal carbonyl complexes with another transition metal or main group hydride form a third class of protocol. Two mechanistic pathways exist: either attack at coordinated CO (as in the preparation of [Mo$_2$(μ-H)(CO)$_{10}$]$^-$) or replacement of a metal-bound halide or pseudohalide.

(iv) Examples of oxidative additions to metal(0) centres are known for H_2, silanes and aldehydic C–H bonds. These reactions only proceed for coordinatively unsaturated species, such as those generated via photolysis of [M(CO)$_4$(dppe)] or by N_2 loss from [M(CO)(N_2)(dppe)$_2$].

Table 3 Selected group 6 carbonyl hydride complexes.

Complex	*Synthesis*[a]	*Comments*	*M–H$_{av}$*[b] (nm)	*Ref.*
Anionic complexes				
[PPN][CrH(CO)$_5$]	(i)	also molybdenum, tungsten	0.166	84,85
[Na(cryptand-222)][CrH(CO)$_5$]	(ii)	also molybdenum, tungsten		86
K[Cr$_2$(μ-H)(CO)$_{10}$]	(ii)	also tungsten	0.178	87,88
[NEt$_4$][Cr$_2$(μ-H)(CO)$_{10}$]	(ii)	two-phase reaction		89
[NEt$_4$][Mo$_2$(μ-H)(CO)$_{10}$]	(iii)			90
[PPN][WH(CO)$_4${P(OMe)$_3$}]	(i)		0.153	91
[PPh$_4$][W$_2$(μ-H)(CO)$_{10}$]	(iii)	neutron study	0.1897	92
[PPN]$_2$[W$_2$(μ-H)$_2$(CO)$_8$]	(iii)	neutron study	0.1927	93
[NPr$_4$]$_4$[Mo$_4$H$_4$(CO)$_{12}$]	(iii)	also tungsten	nd	94
Neutral complexes				
[CrH(SiCl$_3$)(CO)$_3$(dppe)]	(iv)	also molybdenum, tungsten		95
[MHCl(CO)$_3$(dppe)]	(i)	also other chelates	0.189	95
[WH(CO)$_2$(NO)(PR$_3$)$_2$]	(iii)	also chromium, molybdenum; R = alkyl, OR	nd	96–8
[WH(O$_2$CNR$_2$)(CO)(dppe)$_2$]	(i)	one η^1-dppe; R = various	nd	99
[WH(F-BF$_3$)(CO)$_3$(PCy$_3$)$_2$]	(i)	also HOTf product	nd	100
[WH(CO)$_3${HB(3,5-Me$_2$pz)$_3$}]	(i)		nd	107
[WH$_2$(CO)(dppe)$_2$]	(iv)			102,103
Cationic complexes				
[CrH(CO)$_2$(dmpe)$_2$]BPh$_4$	(i)		nd	104
[CrH(CO)$_2$(dppm)$_2$]OTf	(i)	also molybdenum, tungsten; other chelates		105
[WH(CO)$_3$(**11**)]$^+$	(i)	[WCl$_5$(=O)]$^{2-}$ salt	nd	106
[WH(CO)$_3$(PR$_3$)$_3$]OTf	(i)	several other complexes		107,108

[a] See text for synthetic method. [b] nd = hydride not detected in structure.

4.2.2.2 Structure and properties

Crystallographic studies of these complexes are complicated by the hydrides' poor diffraction characteristics. These problems may be partially overcome in low-temperature studies, and under fortunate circumstances accurate data may be obtained from neutron studies on perdeuterated complexes. Structurally characterized examples are gathered in Table 3, but in many cases the hydride ligand is not detected. Further selected data for species with detected hydride ligands are presented in Table 4. The complex [PPN][WH(CO)$_4${P(OMe)$_3$}] deserves special comment due to the presence of an apparent very short W–H distance (0.153 nm). Whether this an artefact of the crystal structure is unknown, further investigation would be helpful. Whatever the case, irradiation of the crystals with ^{60}Co gamma rays leads to the radical [W(CO)$_4${P(OMe)$_3$}]$^-$ trapped in the crystal matrix.[91] The parameter θ_{twist} is a valuable measure of the conformation of bridging hydride dimer complexes; this value is 0° for an eclipsed structure and around 45° for an ideal staggered conformation in [M$_2$(μ-H)(CO)$_{10}$] derived structures. All the dimer structures show bent M–H–M cores. While numerous calculations have been carried out into this bonding mode[109] a simple qualitative picture has been suggested by Crabtree[110] who argues that the principle driving force for M–H–M bending is maximization of backbonding from M($d\pi$)

to the M–H σ* orbital in much the same manner as in η^2-H_2 complexes. The solution structures of anionic carbonyl hydrides often show ion pairing effects. For example, Na[CrH(CO)$_5$] exists with an isocarbonyl (Na···OC) contact while the tungsten analogue and the related complexes Na[MH(CO)$_4$(PR$_3$)] (M = Cr, W) associate via Na–H–M bridges.[111]

Table 4 Structural studies of hydride-bridged complexes.

Complex	*M–H* (nm)		*M–M* (nm)	*M–H–M* (°)	θ_{twist} (°)	*Ref.*
K[Cr$_2$(μ-H)(CO)$_{10}$]	0.165	0.190	0.3260	132	36	88
[K(phen)$_3$][Cr$_2$(μ-H)(CO)$_{10}$]	0.171	0.174	0.3395	159	~0	112
[CrAu(μ-H)(CO)$_5$(PPh$_3$)]	0.164[a]	0.172[b]	0.2770	111		113
[Cr(μ-H)(CO)$_5${B$_2$H$_3$(PMe$_3$)$_2$}]	0.176[a]	0.128[c]	0.2876	141		114
[PPh$_4$][Mo$_2$(μ-H)(CO)$_8$(μ-dppm)]	0.185	0.189	0.3402 8	131	21.5	115
[NEt$_4$][Mo$_2$(μ-H)(CO)$_8$(μ-dppb)]	0.188	0.192	0.3499 5	133	41.7	115
[PPh$_4$][W$_2$(μ-H)(CO)$_{10}$][d]	0.1897		0.3340	123.4	~45	92
[NEt$_4$][W$_2$(μ-H)(CO)$_8$(μ-dppm)]	0.170	0.190	0.3428 6	140	20.6	116
[NEt$_4$][W$_2$(μ-H)(CO)$_8$(μ-dppp)]	0.190	0.200	0.3434	120	53.8	116
[W$_2$(μ-H)(CO)$_7$(NO)(μ-dppp)]	0.194		0.3315	117	45.5	116
[PPN]$_2$[W$_2$(μ-H)$_2$(CO)$_8$][d]	0.1920	0.193.3	0.3010	102.7		93

[a] Cr–H distance. [b] Au–H distance. [c] B–H distance. [d] Neutron study.

The carbonyl hydrides [MH(CO)$_5$]$^-$ are implicated as active catalysts in a number of organic transformations including photolytic water-gas-shift-reactions,[117] reductions of alkynes, aldehydes, ketones or α,β-unsaturated compounds using either hydrogen,[118] metal formates[119] or silanes.[120] Similarly, isolated [M$_2$(μ-H)(CO)$_{10}$]$^-$ complexes act as stoichiometric reducing agents for the equivalent transformations.[121] The mechanisms of these reactions are not known in detail but mechanistic studies of the reaction of [MH(CO)$_4$L]$^-$ (L = CO, PR$_3$) complexes with BuBr indicate that single electron transfer processes are likely to be important in these reductive reactions.[122]

4.2.2.3 Reactions

In addition to the organic insertion behaviour mentioned above, [CrH(CO)$_5$]$^-$ readily inserts COS leading to the crystallographically characterised thioformate [Cr{SC(O)H}(CO)$_5$]$^-$.[86] In a similar reaction [WH(CO)$_4${P(OMe)$_3$}]$^-$ inserts CO_2.[123] The complexes [MH(CO)$_4$L]$^-$ [L = CO, PR$_3$] also show reactivity with soft electrophiles, for example with AuCl(PPh$_3$) in the presence of TlPF$_6$ as a chloride abstractor to form the metal–metal bonded species [(Ph$_3$P)Au(μ-H)M(CO)$_4$L].[113] Although mononuclear anionic hydrides [MH(CO)$_5$]$^-$ do not undergo carbonyl substitution reactions, diphosphines react with [M$_2$(μ-H)(CO)$_{10}$]$^-$ complexes to bridge both metal centres.[115] Although this mode of reactivity is also shown by [W$_2$(μ-H)(CO)$_9$(NO)] species, additional products containing one phosphorus donor *cis* to the nitrosyl ligand can also be isolated.[124,125]

The complexes [MH(CO)$_2$(NO)(PR$_3$)$_2$] show a wide range of reactivity with unsaturated organic molecules. This is shown in abbreviated form in Scheme 4. The tungsten complexes are more reactive than their chromium analogues.

4.2.3 Dihydrogen Complexes

As the development of group 6 dihydrogen complexes is extremely well reviewed from the time of Kubas's paper[129] up to 1992, only an overview along with the most recent results are included here. Further information can be found in Jessop and Morris's extensive listing of η^2-H_2 complexes.[130]

4.2.3.1 Synthesis

All members of the series [M(η^2-H_2)(CO)$_3$(PCy$_3$)$_2$] (M = Cr, Mo, W) are isolable, but the chromium complex is highly labile.[131] For molybdenum and tungsten the PPr$^i_3$, P(*cyclo*-C$_5$H$_9$)$_3$ and PCy$_2$Pr^i complexes may also be accessed.[132] Kubas' original preparation ([M(CO)$_3$(chpt)] and PR$_3$ under H_2) is extremely sensitive to the phosphine bulk, those with cone angles (θ) less than ~160° lead to

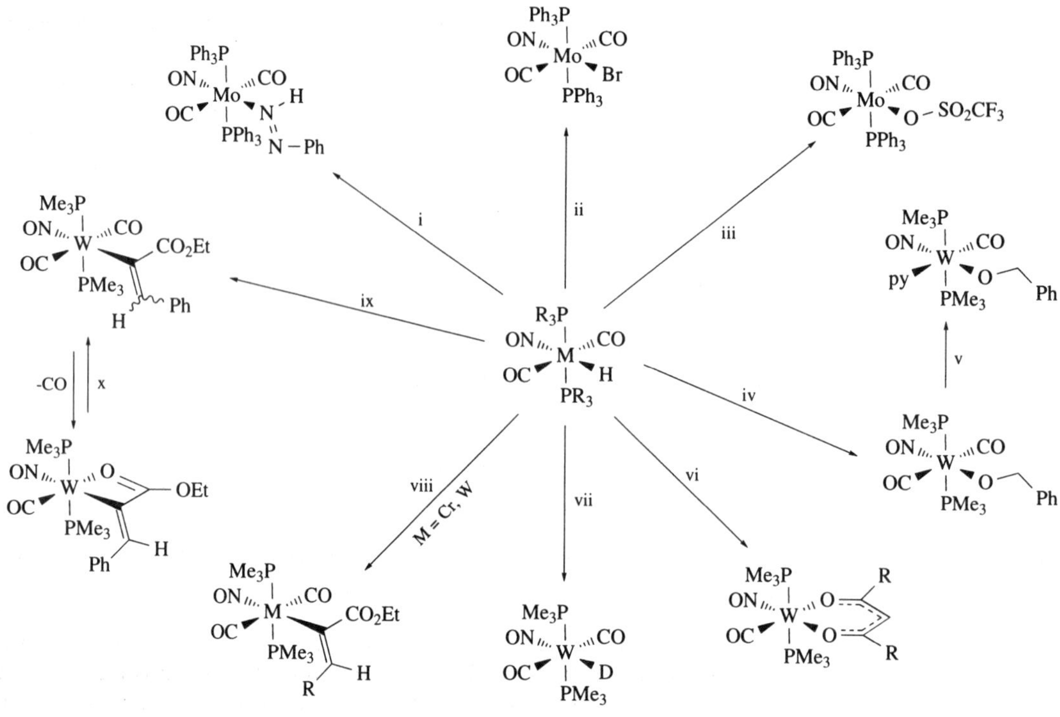

i, $[PhN_2]BF_4$;[97] ii, HBr;[97] iii, CF_3SO_3H;[97] iv, PhCHO;[126] v, py;[126] vi, $RC(O)CH_2C(O)R$;[127] vii, MeOD;[98] viii, $RC{\equiv}CCO_2Me$;[96,128] ix, $PhC{\equiv}CCO_2Et$;[128] x, CO[128]

Scheme 4

$[M(CO)_3(PR_3)_3]$ species whatever the stoichiometric ratio employed, while use of phosphines more sterically demanding than PPr^i_3 ($\theta = 163°$) or PCy_3 ($\theta = 170°$) yield uncharacterized products. All examples of group 6 dihydrogen complexes prepared to date have been prepared by direct interaction of H_2 gas with unsaturated or agostic precursors. The complex $[Mo(\eta^2\text{-}H_2)(CO)(dppe)_2]$ and its perdeutero analogue are the only other isolable complexes. A number of species including $[M(\eta^2\text{-}H_2)_n(CO)_{6-n}]$ (n = 1, M = Cr, Mo, W; n = 2, M = Cr),[133] $[Cr(\eta^2\text{-}H_2)(CO)_4(\textit{trans}\text{-cyclooctene})]$,[134] $[Cr(\eta^2\text{-}H_2)(CO)_2(C_6H_5Me)]$,[135] $[Mo(\eta^2\text{-}H_2)(CO)_3(nbd)]$,[136] may be observed in low-temperature photolysis experiments using liquified inert gas solvents or similar 'innocent' media.

4.2.3.2 Structure and properties

The bonding in dihydrogen complexes is described most simply as involving forward σ-donation from the H–H σ-orbital together with back donation into the dihydrogen σ* molecular orbital from a metal *d*-orbital. The degree of $M(d) \rightarrow H_2$ (σ*) donation dictates many of the bound dihydrogen properties. Complexes with strongly σ-donating ancillary ligands promote back donation and result in the formation of classical dihydrogen complexes via H_2 (σ*) orbital overpopulation. For example, while a bound dihydrogen is observed in $[Mo(\eta^2\text{-}H_2)(CO)(dppe)_2]$, related $[MoH_2(CO)(R_2PCH_2CH_2PR_2)_2]$ (R = Et, Pr^i) show classical dihydride structures.[137] In favourable cases, interconversions between the dihydrogen and dihydride tautomers may characterized. Notably for the complex $[W(\eta^2\text{-}H_2)(CO)_3(PPr^i_3)_2]$, full kinetic and thermodynamic parameters for the interconversion have been extracted.[138]

As the $M(\eta^2\text{-}H_2)$ bonding scheme is highly directional (colinear with a metal d_π orbital), rotation of the H_2 ligand is expected to be associated with an energy barrier. As well as transitions to the first rotational excited state, transitions within the ground state also arise originating from quantum mechanical effects. The ground-state wavefunctions for the H_2 molecule in rotation potential wells 180° apart overlap and this leads to rotational tunnelling transitions. The librational motion of the H_2 dumbbell can be interrogated by inelastic neutron scattering experiments. These experiments when linked to appropriate rotational models yield very low rotational barriers (typically ca. 2.5–7.0 kJ mol^{-1}) for group 6 complexes. A fuller discussion of these results is provided by Eckert and Kubas.[139]

The librational motion of dihydrogen in its bound complexes can cause complications in neutron structures of these species (the only reliable measure of the H–H distance). For example, in the complex $[Mo(\eta^2\text{-}H_2)(CO)(dppe)_2]$ a range of H–H distances (0.080–0.085 nm) is consistent with the data after correction; the librational motion (even at 12 K) results in artificial foreshortening of the H–H distance in the raw data. A similar value (0.082 nm) is found for $[W(H_2)(CO)_3(PPr^i_3)_2]$, the only other group 6 complex characterized by neutron diffraction. In the absence of neutron data, the best criterion for the assignment of coordinated dihydrogen remains the observation of a short ^{1}H NMR T_1 relaxation time supported by the observation of a large H–D coupling constant in the analogous HD complex. It is now recognized that care must be taken with T_1 measurements to work out all other relaxational processes apart from the H–H dipole relaxation. Further details may be found in a review by Jessop and Morris.[130]

4.2.3.3 *Reactions*

The principle reaction pathway shown by dihydrogen complexes is H_2 loss. Dihydrogen loss from $[M(\eta^2\text{-}H_2)(CO)_3(PCy_3)_2]$, (M = Cr, Mo), leads to the species $[M(CO)_3(PCy_3)_2]$. The tungsten complex is structurally characterized. The driving force for these reactions are the formation of strong agostic cyclohexyl–metal interactions. In the tungsten case, the hydrogen is bound some 0.230 nm from the metal with an enthalpy of ca. 40 kJ mol^{-1}.[140] The lability of the H_2 function occasionally shows some dependence on *trans* effects. Thus, while the D_2 ligand in *fac*-$[Mo(\eta^2\text{-}D_2)(CO)_3(nbd)]$ (D_2 *trans* to CO) is labile to N_2, that in the *mer*-isomer (D_2 *trans* to an alkene of the nbd ligand) is not.[136] Interestingly, the tungsten complexes $[W(\eta^2\text{-}H_2)(CO)_3(PR_3)_2]$ (R = Cy, Pr^i) promote H–D exchange with D_2/H_2 mixes both in solution and the solid state; these reactions are thought to be associative and involve formal 20-electron intermediates. Other patterns of reactivity can also be realized. Phosphine substitution occurs in the reaction of $[W(\eta^2\text{-}H_2)(CO)_3(PPr^i_3)_2]$ with PCy_3 to yield $[W(\eta^2\text{-}H_2)(CO)_3(PCy_3)_2]$.[141] Further examples of dihydrogen complex reactivity can be found in the review by Jessop and Morris.[130]

4.2.4 Carbonyl Complexes with Nitrogen σ-Donor Ligands

Complexes containing sp^3 nitrogen donor amines, imines, diazenes, nitriles and various inorganic nitrogen-based ligands are considered in this section. 1,4-Diimine complexes are considered in Section 4.3.1 of this chapter, for convenience all heterocyclic nitrogen ligands with extended π-systems are also considered as weak π-acceptors ligands and thus dealt with in Section 4.3.1. Reviews of carbonyl species containing Schiff-base[142] and diazene[143,144] ligands are available and these include information of relevance to their group 6 complexes.

4.2.4.1 *Synthesis*

Typical examples of recently prepared amine, imine, diazene and nitrile complexes appear in Table 5, which also contains the references for the examples cited below. A variety of synthetic approaches to nitrogen ligated complexes may be employed.

(i) Ubiquitous thermal, or less commonly photochemical, substitution of $[M(CO)_6]$.

(ii) Ligand exchange with labile carbonyl precursors such as $[M(CO)_5(THF)]$, $[M(CO)_3(chpt)]$, the nitrile complexes $[M(CO)_3(NCR)_3]$ (R = Me, Et), or more exotic molecules.

(iii) Modification of the carbene ligand in $[Cr(CO)_5\{=CR(OR)\}]$ complexes to $RN{=}C{=}CR_2$ and $RN{=}CR_2$ species. Insertion of isocyanides leads to reactive ketimine complexes which are not always isolable, Equation (7) is typical. Insertion reactions can also sometimes be carried out using aziridenes (Equation (8)) or sulfilimines (Equation (9)). These insertion reactions can result in products other than imines. For example, reaction of $[Cr(CO)_5\{=CAr(OR)\}]$ with 2-methylaziridine results in nitrile products.[145]

(iv) Imine template reactions at group 6 metals have been rarely employed but are known, for example, Equation (10).

(v) Diazene complex preparations are also amenable to nonsubstitution routes; oxidation of coordinated hydrazines with MnO_2 or $Pb(OAc)_4$ proving effective. For disubstituted products (RN=NR) normally an (*E*)/(*Z*) mixture is produced.

Nitrile complexes are normally prepared by substitution reactions; one alternative interesting variant is the electrochemical η^2–η^1 interconversion of $[W(CO)_5\{C_2(CN)_4\}]$. Addition of an electron to the

Table 5 Selected carbonyl complexes with sigma nitrogen donor ligands.

Complex	*Synthesis*[a]	*Comments*[b]	*M–N$_{av}$* (nm)	*Ref.*
Amines				
$[Mo(CO)_5(\mathbf{7})]$	(i)		0.2337	147
$[W(CO)_5\{HNCH(Me)(CH_2)_4\}]$	(i)		0.2340	148
$[Mo(CO)_3(\eta^3\text{-}(\mathbf{8})]$	(i)	free NH_2 reactive		149
$[Mo(CO)_3(\mathbf{9})]$	(ii)	long Mo–N bonds	0.2399	150
$[Mo(CO)_3(\mathbf{10})]$	(i)	N_2S–Mac	0.2313	151
$[M(CO)_3(\mathbf{11})]$	(i)	M = Cr, Mo, W; N_3–Mac		36,152–4
$[Mo(CO)_3(\eta^3\text{-}\mathbf{12})]$	(i)	N_4–Mac	0.2368	155
$[W(CO)_3\{MeN(CH_2CH_2NMe_2)_2\}]$	(i)	useful precursor		156
Imines				
$[Cr(CO)_2\{\eta\text{-}C_6H_5(CH_2)_2CPh(=NPh)\}]$	(ii)		0.2121	157
$[Cr(CO)_5\{RN=C=CMe(OMe)\}]$	(iii)	R = Me, Cy, Bu^t		158
$[W(CO)_5\{HN=CMe(OEt)\}]$	(iii)	also Cr	0.2249	145
$[W(CO)_5\{HN=CPh(OMe)\}]$	(iii)	also Cr, Mo	0.2249	159
$[Mo(CO)_4(\eta^2\text{-}(\mathbf{13})]$	(i)	N_2O_2–Mac	0.2319	160
$[Mo(CO)_3\{H_2NC_6H_4N=CHC_6H_4CHO\}]$	(iv)	template reaction	0.2239[c]	161
Diazenes				
$[M(CO)_5(\eta^1\text{-}RN=NR)]$	(v)	M = Cr, W; R = Me, Et		162
$[W_2(CO)_{10}(\mu\text{-}N=NCMe_2CMe_2)]$	(ii)		0.2207	163
$[W(CO)_2(NO)(\eta^1\text{-}HN=NMe)(PPh_3)_2]^+$	(v)	PF_6^- counter anion	0.2215	164
Nitriles				
$[Cr(CO)_5\{NCCH(=PPh_3)\}]$	(ii)	also Mo, W	0.2062	165
$[Cr(CO)_5\{NCN=C(NMe_2)_2\}]$	(ii)		0.2077	166
$[W(CO)_5(NCMe)(dppm)]$	(ii)		0.2190	13
$[W(CO)_5(NCCH_2CO_2H)]$	(ii)	also other derivatives	0.2178	5
fac-$[W(CO)_3(NCEt)_3]$	(i)	versatile precursor		167

[a] For synthetic methods see text. [b] Mac indicates macrocyclic ligand. [c] Mo–imine distance.

(7)

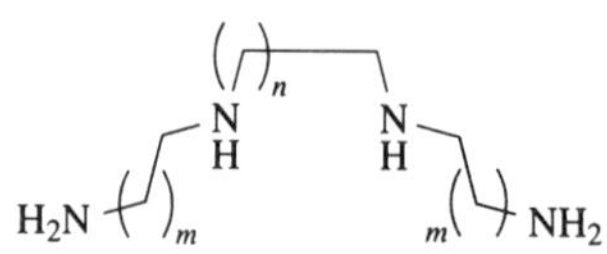

n 1 1 1
m 1 2 2

(8)

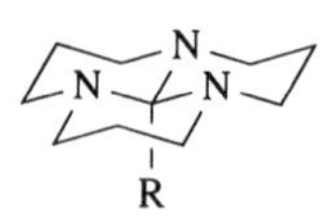

R = H, Me

(9)

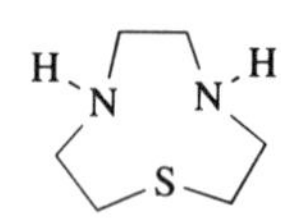

(10)

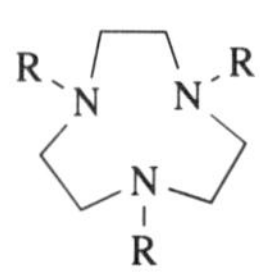

R = H, Me, CH$_2$Ph

(11)

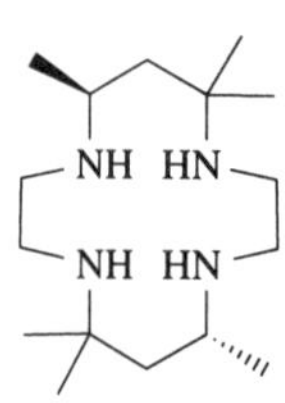

(12)

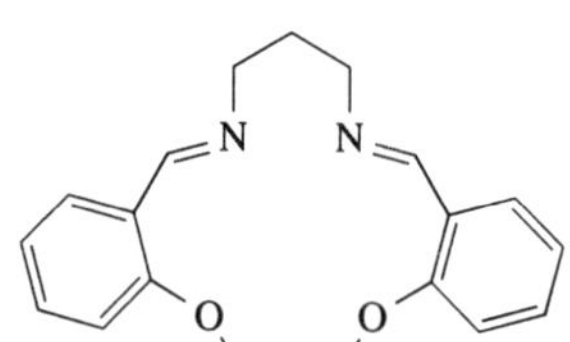

(13)

ligand π* orbital of the $(NC)_2C=C(CN)_2$ fragment results in diminished backbonding and leads to conversion to the η^1-nitrile complex.[146]

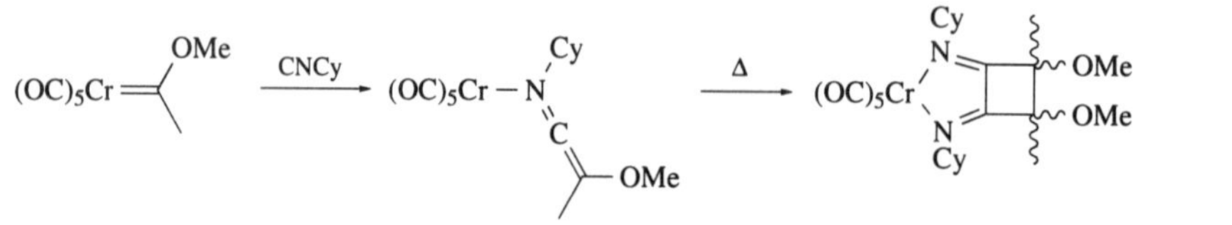

(7)

Some recent syntheses and structural characterizations of complexes of isocyanate (NCO^-), isothiocyanate (NCS^-) and azide (N_3^-) ligands are of note. Reactions of $[NEt_4]N_3$ with $[M(CO)_5(THF)]$ lead to the expected $[NEt_4][M(N_3)(CO)_5]$ complexes, the tungsten example having been structurally

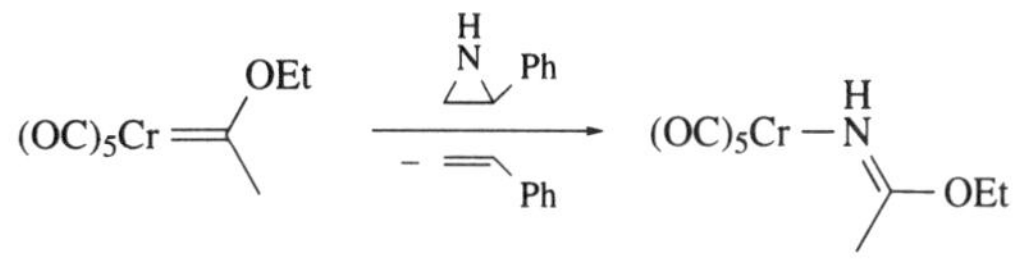

(8)

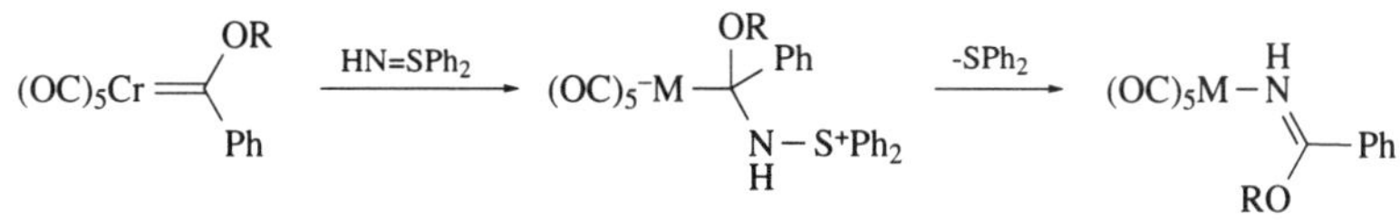

(9)

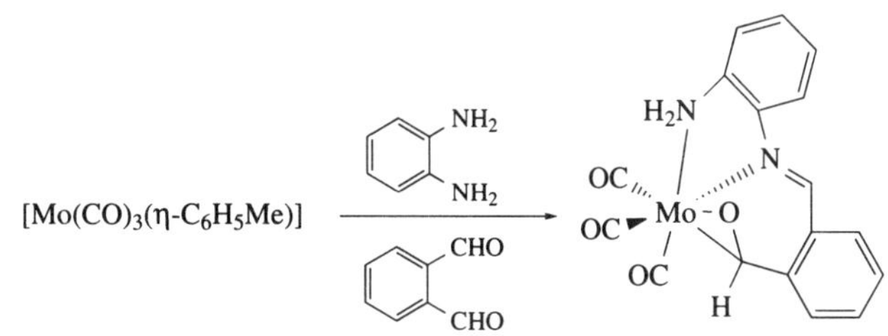

(10)

characterized (W–N 0.2236 nm; W–N–N 129.9°).[168] Alternatives to substitution reactions or nucleophilic additions at coordinated CO are oxidative procedures. For example, reaction of *mer*-[Mo(CO)$_3$(η^1-dppm)(dppm)] with Hg(SCN)$_2$ leads to [Mo(NCS)$_2$(CO)$_2$(η^1-dppm)(dppm)].[169] One of the isothiocyanate ligands is linear (Mo–N–C 175.8°) while the other is appreciably bent (Mo–N–C 159.7°). The preparation of [NEt$_4$][W(NCO)(CO)$_5$] by the reaction of [NEt$_4$][WBr(CO)$_5$] with AgCNO, interestingly, involves isomerization of the intermediate [NEt$_4$][W(CNO)(CO)$_5$] complex.[170]

4.2.4.2 Structure and properties

Representative M–N distances can be found in Table 5. Molecular modelling studies reveal a steady bond length increase with increasing amine bulk.[171] Inspection of the structural database also reveals that hydrogen bonding of coordinated primary and secondary amines is a common feature where electronegative hydrogen acceptors are available *cis* to the R$_2$NH donor. These effects can lead to such ligands being less labile than expected. In the presence of excess amine hydrogen bonding of the complexed and free amine may be detected by phonoacoustic spectroscopy; in the case of [Cr(CO)$_5$(Bu^{n}NH$_2$)] these interactions amount to ~20 kJ mol^{-1}.[172]

4.2.4.3 Reactions

Recent reactivity studies of amine complexes have concentrated on macrocyclic species. This chemistry is reviewed up to 1986;[173] selected examples are summarized in Scheme 5.

Imines often engender unusual reactivity by placing other ligand components in close proximity to the metal. Examples of such behaviour include the oxidative additions of aromatic chloro bonds, as in the reaction of (2,6-C$_6$H$_3$Cl$_2$)CH(=NAr) with [Mo(CO)$_3$(chpt)].[176] Even C–F bonds are reactive, as in the preparation of [WF(CO)$_3${2-C$_6$F$_4$CH(=NCH$_2$CH$_2$NMe$_2$)}].[57] One other unusual oxidation of note is the treatment of [Cr(CO)$_5$(NH$_2$Ph)] with Pb(OAc)$_4$ at low temperature which leads to the deeply coloured complex [Cr(CO)$_5$(1,4-HNC$_6$H$_4$NPh)]. The reaction is only applicable to chromium; molybdenum and tungsten starting materials just decompose.[177]

4.2.5 Carbonyl Derivatives with Oxygen Ligands

This section covers a wide range of both neutral and anionic ligands, including ethers, ketones, phosphine oxides (reviewed[178,179]), CO$_2$ (reviewed[82]), hydroxides, alkoxides (dimetal alkoxides reviewed[180,181]), carboxylates, carbonates and oxides. A general overview of the area is also available.[182] The coverage of ether complexes has been deliberately restricted, but many papers deal with [M(CO)$_5$(THF)] and related complexes.

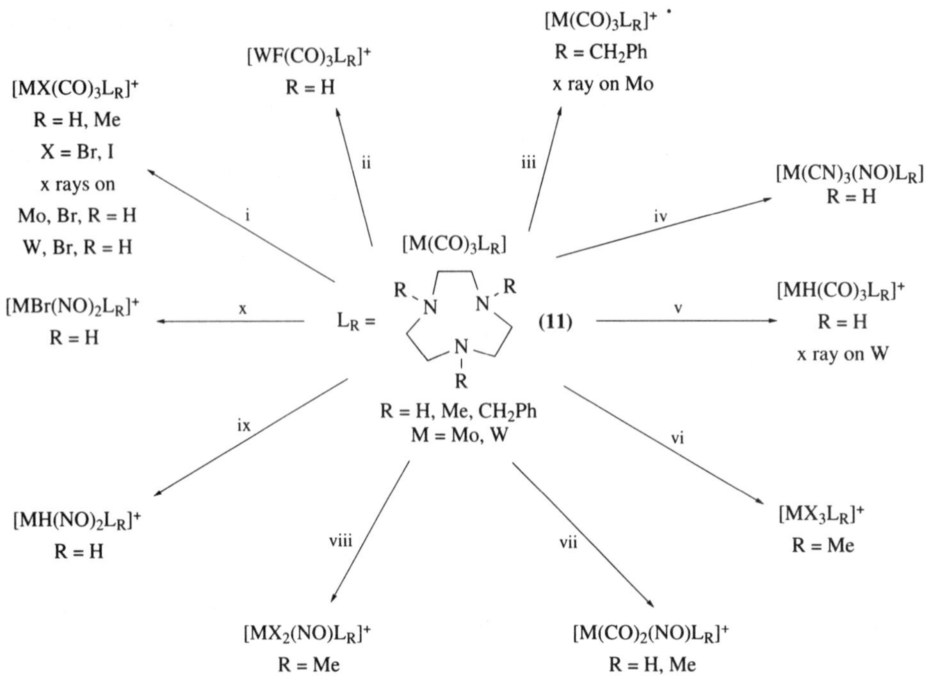

i, Br_2 or I_2;[36,106,152,174] ii, 20% HF;[106] iii, $[FeCp_2][PF_6]$;[154] iv, HNO_3–KCN;[36]
v, HCl[36,106] vi, excess Cl_2 or Br_2;[152,174] vii, 0.07 M HNO_3–$NaBF_4$ or $NO[BF_4]$;[36,153,175]
viii, HX–$NaNO_2$–$NaPF_6$;[152,153] ix, 0.4 M HNO_3–$NaClO_4$;[36] x, 0.4 M HNO_3–NaBr[36]

Scheme 5

4.2.5.1 Synthesis

Selected oxygen containing carbonyl complexes are provided in Table 6. Some of the ligands mentioned therein are shown in structures (**14**)–(**16**). The range of synthetic methods employed within this class is too wide to enable the specific categorization employed in other sections, selected routes are highlighted below. In addition to the ubiquitous substitution reactions of $[M(CO)_6]$ and $[M(CO)_{6-n}(L)_n]$, other routes encompass insertion and metathesis reactions, together with attack at coordinated ligands, and a number of nonstandard strategies. Several examples (Equations (11)–(16)) illustrate the general trends in the chemistry; these are referenced in Table 6.

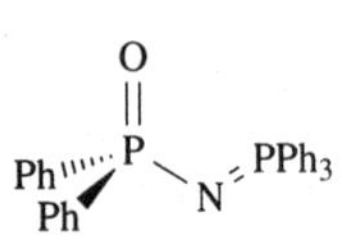

(14)

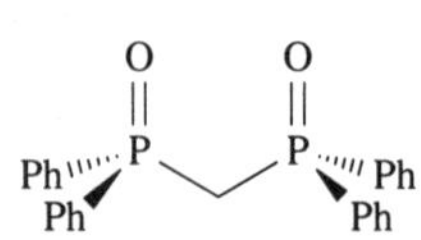

(15)

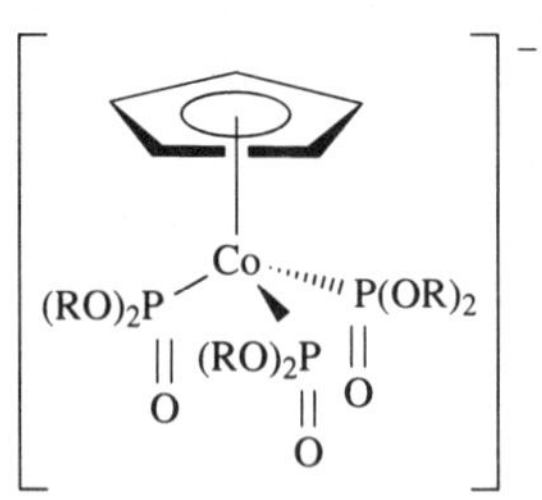

(16)

Carbon dioxide is fairly reactive towards electron rich group 6 complexes but well-characterized carbonyl complexes are only isolated rarely (Equation (11)). These reactions frequently lead to carbonate species (resulting from CO_2 disproportionation), and carboxylate species (from insertion reactions), the preparations of (**17**), (**18**) and (**19**) (Equations (12) and (13)) aretypical. Interest in CO_2 fixation, and in the mechanism of the water-gas-shift reaction, has also led to the preparation of a number of $[\{M(OR)(CO)_n\}_m]^{m-}$ (n = 5–3; m = 1–4) aggregates. These species are easily obtained by the reaction of labile '$M(CO)_n$' sources with $[NR_4]OR$. However, in the absence of crystallographic data, determining the degree of aggregation can be problematic. Molybdenum and tungsten diacetate and carboxylate species are best prepared by metathesis of carbonyl halides as, for example in Equation (14).

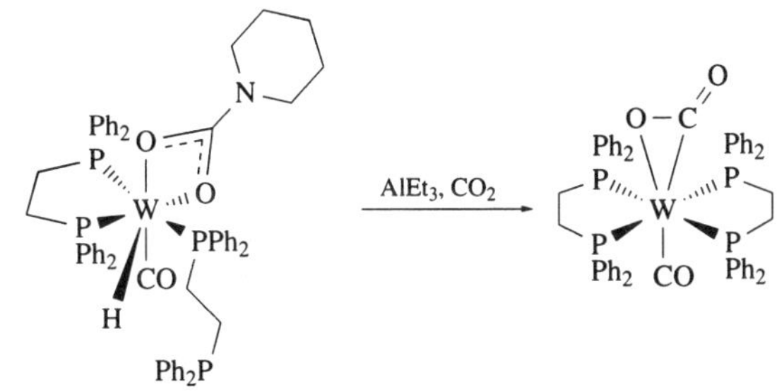

(11)

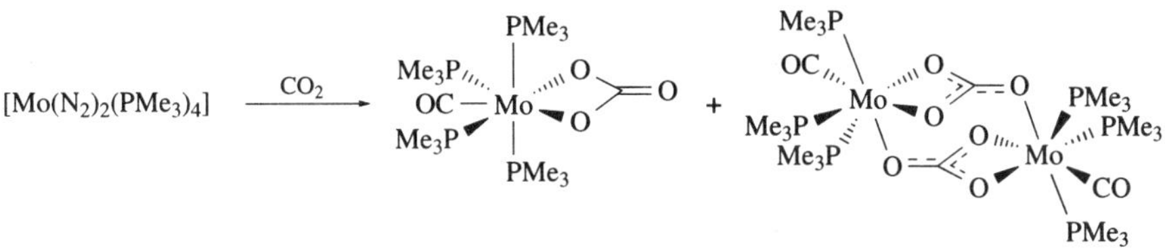

(12)

(17) **(18)**

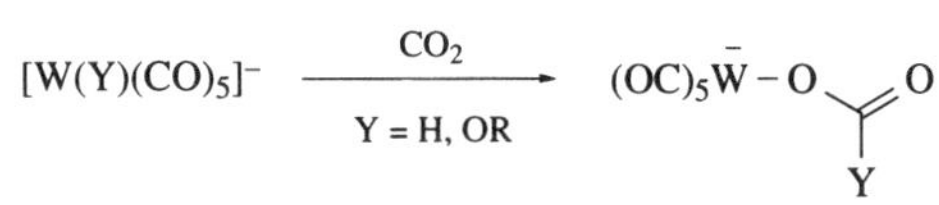

(13)

(19)

$$[MoBr_2(CO)_2(PR_3)_2] \xrightarrow{NaO_2CR} [Mo(O_2CR)_2(CO)_2(PR_3)_2] \qquad (14)$$

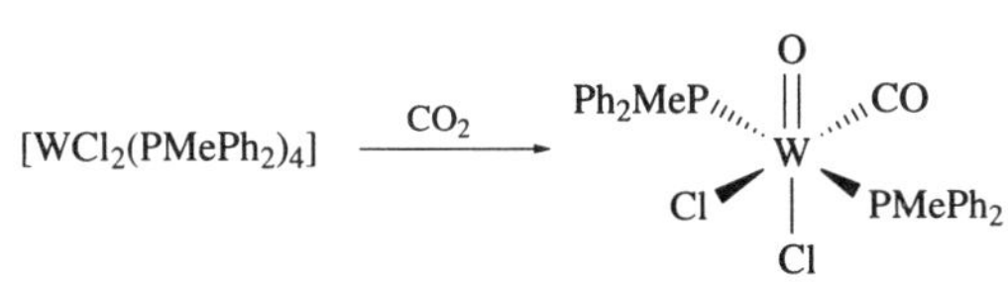

(15)

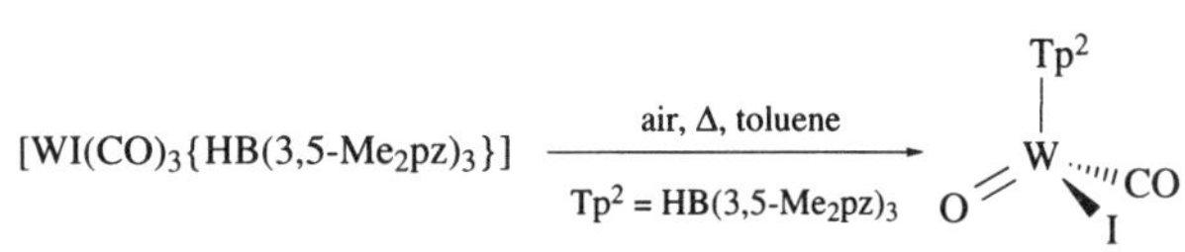

(16)

Alternative approaches involving protonation of metal(0) complexes can lead to low yields. In these dicarboxylate species generally one carboxylate functions as a monodentate ligand, the other as a chelate resulting in 18-electron configurations at the metal. Few stable carbonyl oxides are known for the group 6 metals. Various synthetic strategies to these molecules have been employed including CO_2 splitting (Equation (15)) and direct aerial oxidation (Equation (16)). In the latter case the protection of the bulky pyrazolylborate ligand presumably helps prevent oxidative liberation of the final CO. Oxo species (among a number of other products) are occasionally also produced by reaction of homoleptic group 6 alkoxides.[211] Most of this chemistry is beyond the scope of this section; further information may be found in a review by Buhro and Chisholm.[181]

4.2.5.2 Structure and properties

The properties of oxygen containing group 6 complexes vary widely, some are only observed as transient intermediates in flash photolysis or matrix isolation experiments while others are prepared by refluxing at high temperatures in air. Generalizations cannot be made. Examples of structurally charaterized complexes may be found in Table 6. The structural diversity in carbonyl alkoxide complexes deserves special comment. Depending on the stereoselectonic properties of both the ligand and metal, dimeric (**20**), trimeric (**21**) and cubane (**22**) structures are formed. Unfortunately it is not possible to predict the exact structure for a given alkoxide or hydroxide. For example, recent studies indicate that Hieber's complex originally formulated as $[Mo_2(\mu\text{-}OH)_3(CO)_6]^{3-}$ has the cubane structure (**22**) (M = Mo). One final class of alkoxide complex, exemplified by the asymmetric structures (**23**)[212]

Table 6 Selected group 6 carbonyl complexes with oxygen ligands.

Ligand	*Complex*	*Comments*	*M–O$_{av}$* (nm)	*Ref.*
H_2O	$[Cr(CO)_5(OH_2)]$	flash photolysis impurity		183,184
H_2O	$[W(CO)_3(OH_2)(PPr^i_3)_2]$		0.2320	185
PrOH	$[Cr(CO)_5(PrOH)]$	reactive, H- and O-bound		186
DMF	*fac*-$[Mo(CO)_3(DMF)_3]$		0.2266	187
acetone	*fac*-$[W(CO)_3(O{=}CMe_2)(dppe)]$			188
CO_2	*trans*-$[W(CO)(\eta^2\text{-}CO_2)(dppe)_2]$	poor x ray		99
CO_2	$[W(CO)_5(CO_2)]$	gas phase		189
(14)	$[W(CO)_5(O{=}PPh_2NPPh_3)]$		0.2220	190
(15)	*cis*-$[M(CO)_4\{CH_2[P(O)Ph_2]_2\}]$	M = Mo, W		191
(16)	Na[M(CO)$_3$(**16**)]	M = Cr, Mo, W		192
OH	$M_4[Mo_4(\mu_3\text{-}OH)_4(CO)_{12}]$	M = Cs, $[NEt_4]$	0.2252	193,194
OH	$[NEt_4]_4[W_4(\mu_3\text{-}OH)_4(CO)_{12}]$		0.2243	193
OAr	$[NEt_4][W(OAr)(CO)_5]$	Ar = 2,6-$C_6H_3Ph_2$	0.2168	195
OPh	$[NEt_4][W(OPh)(CO)_4(PMePh_2)]$		0.2191	195
OCH_2CF_3	$[PPN]_2[W_2(\mu\text{-}OCH_2CF_3)_2(CO)_8]$	rare dimer	0.219	196
OEt	$[NMe_4]_3[W_3(OEt)_3(CO)_9]$	one μ_2-OEt, two μ_3	0.213/0.223[a]	197
OMe	$[NEt_4]_4[Cr_4(\mu_3\text{-}OMe)_4(CO)_{12}]$		0.2121	198
OMe	$[MEt_4]_2[MoBi(\mu_3\text{-}OMe)_3(CO)_{12}]$	μ_3-Bi capping ligand	0.2218	199
$O_2C_6H_4$	$[NEt_4]_2[W(1,2\text{-}O_2C_6H_4)(CO)_4]$	hydrogen bonding effects	0.2175	200
O_2CCF_3	$[Mo(O_2CCF_3)_2(CO)_2(PMe_3)_3]$	both η^1-O_2CCF_3	0.2185	201
O_2CH	$[Mo(O_2CH)_2(CO)_2(PEt_3)_2]$	one η^1-O_2CH, one η^2	0.2148[b]	80
O_2CH	$[W(O_2CH)(CO)_5]^-$	Na(cryptand-221) cation	0.2216	202
OAc	$[PPN][W(OAc)(CO)_5]$	also Cr, Mo	0.2207	203,204
OAc	$[PPN](W(OAc)(CO)_4(PEt_3)]$		0.2169	204
acac	$[WCl(acac)(CO)_2(PEt_3)_2]$	also F_6 acac and $O_2C_7H_5$		80
C_2O_4	$[NEt_4]_2[W(C_2O_4)(CO)_4]$		0.2196	205
C_2O_4	$[PPN]_2[W_2(\mu\text{-}C_2O_4)(CO)_8]$		0.2254	205
CO_3	$[Mo(CO_3)(CO)(PMe_3)_4]$			206
CO_3	$[NEt_4]_2[W(CO_3)(CO)_4]$		0.220	207
O	$[Cr(=O)_2(CO)_2]$	matrix isolation species		208
O	*trans*-$[M(=O)_2(CO)_4]$	M = Mo, W; in matrix		208
O	$[WCl_2(=O)(CO)(PMePh_2)_2]$		0.1689	209
O	$[WI(=O)(CO)\{HB(3,5\text{-}Me_2pz)_3\}]$		0.180	210

[a] First figure is for μ_2-ligand, second is for μ_3. [b] For η^1-ligand, average Mo–O for chelate is 0.225 nm.

and **(24)**,[213] are those which feature carbonyls coordinated to only one of the metals. Perhaps the most unusual structure to have been reported in recent years is **(25)**, which features oxide, μ-oxide, hydroxide ligands and a μ_4-carbonate.[206]

4.2.5.3 Reactions

A large number of recent studies have concentrated on the behaviour of alkoxide complexes. The reactivity of $[W(OPh)(CO)_5]^-$ is exemplary. This complex expels CO leading to the formation of the cubane $[W_4(\mu_3\text{-}OPh)_4(CO)_{12}]^{4-}$, a reaction which may be reversed by addition of pressurized CO. Reaction of CO_2 with $[W(OPh)(CO)_5]^-$ leads to the formation of $[W\{OC(O)OPh\}(CO)_5]^-$, while the reaction of water with this complex leads to the formation of $[W(CO_3)(CO)_4]^{2-}$. Addition of excess phenoxide to $[W(OPh)(CO)_5]^-$ yields the dimer complex $[W_2(\mu\text{-}OPh)_3(CO)_6]^{3-}$.[214]

A wide range of chemistry is also known for group 6 complexes of the Cp anion analogue (**16**), commonly referred to as Kläui's ligand. The complexes [M(CO)$_3$(**16**)]$^-$ are versatile starting materials easily modified to [MX(CO)$_3$(**16**)] (X = H, halogen), [M(CO)$_2$(NO)(**16**)] or [M$_2$(CO)$_4$(**16**)$_2$]; this chemistry has been reviewed.[192] The methyl complexes [MMe(CO)$_3$(**16**)] exist in equilibrium with dihapto acyls [M(η^2-COMe)(CO)$_2$(**16**)], reactivity that contrasts with the equivalent Cp and Cp* systems where only alkyl complexes are formed and pyrazolylborate borate complexes which result in the isolation of η^2-acyls.[215]

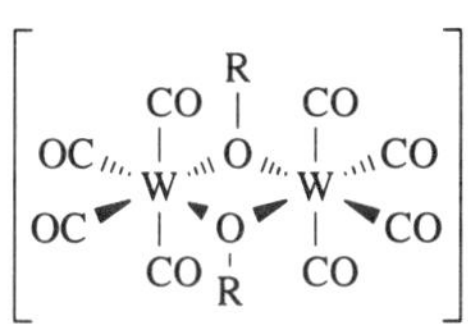

R = CH_2CF_3

(20)

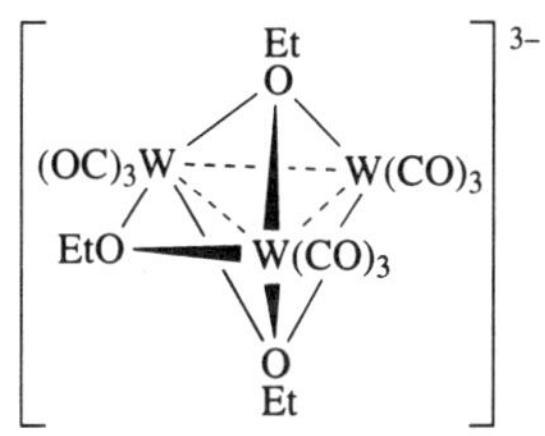

(21)

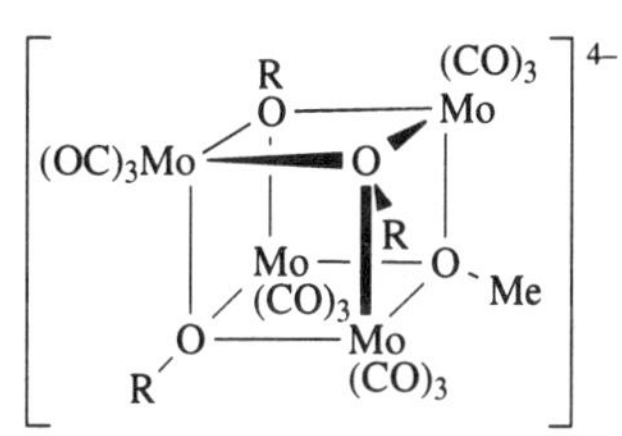

R = H, alkyl
M = Cr, Mo, W

(22)

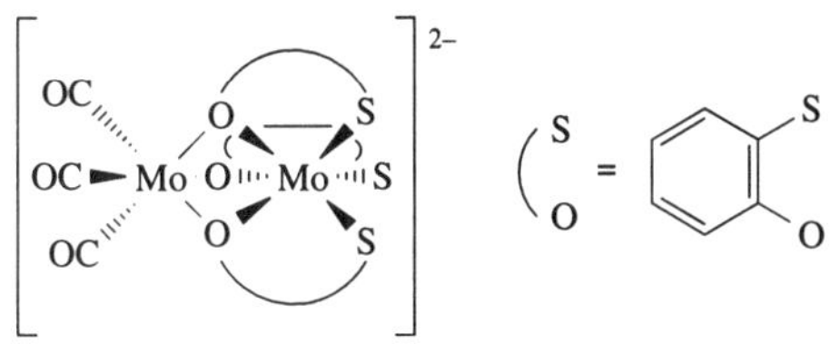

(23)

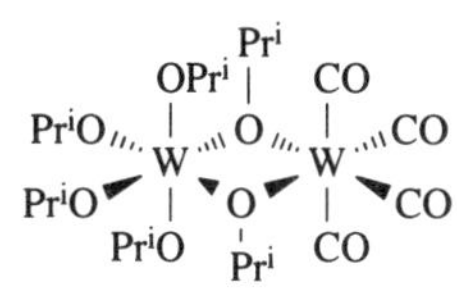

(24)

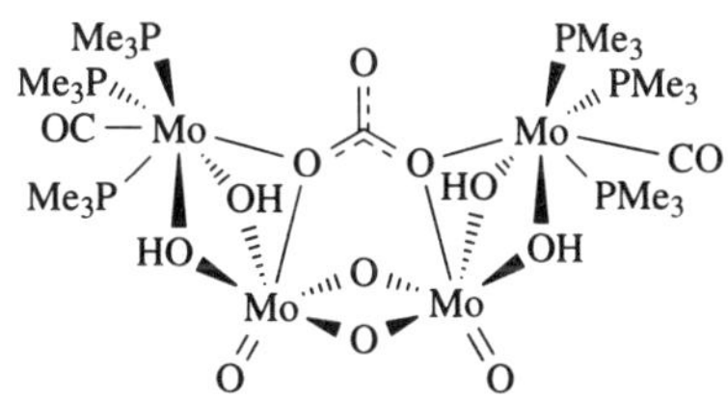

(25)

4.3 CARBONYL COMPLEXES WITH π-ACCEPTOR LIGANDS

4.3.1 Carbonyl Complexes with π-Acceptor Nitrogen Ligands

This section includes complexes of heterocyclic amine, including tris(pyrazolylborate) derivatives; dinitrogen; nitrene (imido); and isoelectronic nitrosyl and organodiazonium (RN_2^+) ligands. The degree of ligand π-acceptor character in group 6 complexes of these species varies widely. The ordering shown is often accepted with heterocyclic amines being regarded as very limited π-acceptors, their actual ability depending on the ligand structure. For convenience all heterocyclic species are considered here. For similar reasons of convenience, complexes containing linear imido functions (M≡NR), although best formulated as π-donor ligands, are considered here to allow comparison with their bent nitrene (M=NR) π-acceptor counterparts.

4.3.1.1 Synthesis

(i) Diimines and heterocyclic ligands

Complexes of neutral heterocyclic amines form the largest body of species in this section. Representative examples selected on criteria of unusual ligand structure, preparative route, or useful crystallographic data, appear in Table 7. Indication of exact synthetic conditions is excluded as essentially all members of this class may be obtained by simple thermal substitutions of $[M(CO)_6]$ or of labile derivatives. Nevertheless, even such simple reactions can show interesting features. For example, while thermolysis of $PhCH(3,5\text{-}Me_2pz)_2$ with $[Mo(CO)_6]$ leads to the expected tetracarbonyl (**26**), prolonged heating gives the alkene complex (**27**) featuring a labile Mo–C interaction (~0.280 nm) (Equation (17)).[217] Reaction with heterocycle radical anions is the most convenient route to the paramagnetic members of the class $[M(CO)_n(\text{heterocycle})]^-$, and these are best characterized by ESR spectroscopy. Alternatively these compounds may be accessed using $[M_2(CO)_{10}]^{2-}$, the two preparations of (**28**) (Equation (18)) being typical.[217] Thermodynamic studies of the system $[M(CO)_4(bipy)]$ and

$[bipy]^{\cdot-}$ indicate that the radical anions are often better ligands than their neutral analogues as evidenced by significant forward equilibrium constants in this particular case.[218] The ability of bipy and phen ligands to carry appreciable negative charge also allows the ready preparation of $[M(CO)_2(N–N)_2]$ (N–N = bipy, phen) (Equation (19)) in contrast with most amine complexes where trisubstitution is the highest degree of CO replacement encountered.[219] Several routes to mixed species are known. Use of *fac*-$[M(CO)_3(NCR)_3]$ as a starting material can lead to the isolation of partially substituted nitrile species, as in the synthesis of (**29**) (Equation (20)).[220] A further range of mixed ligand systems is available if heterocycle bears a pendant anionic function. The preparations in Equation (21) are representative of such approaches.[221]

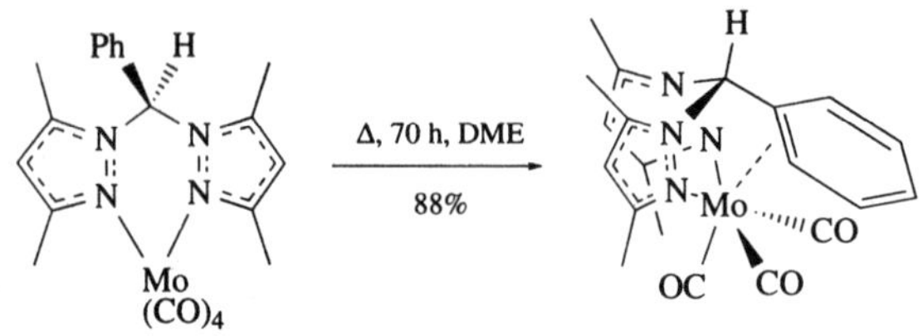

(**26**) (**27**)

(17)

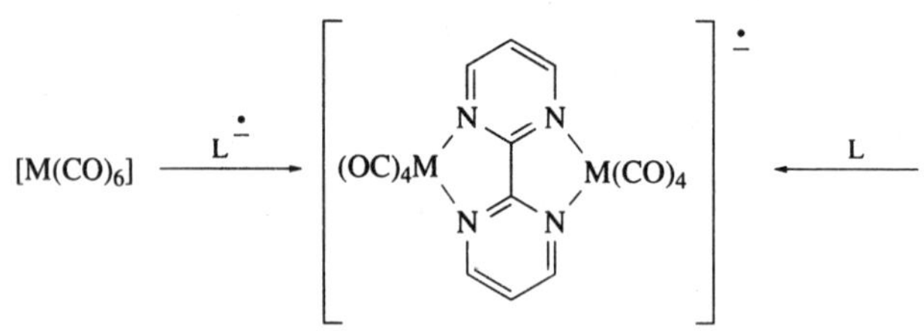

(**28**)

(18)

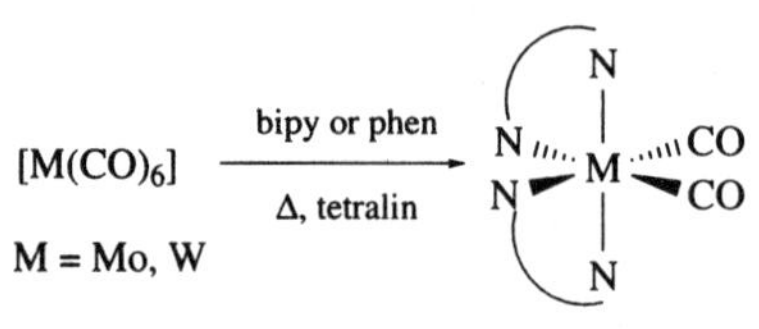

(19)

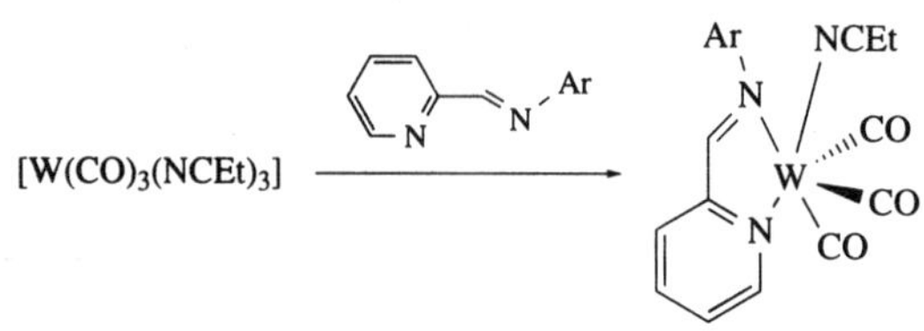

(**29**)

(20)

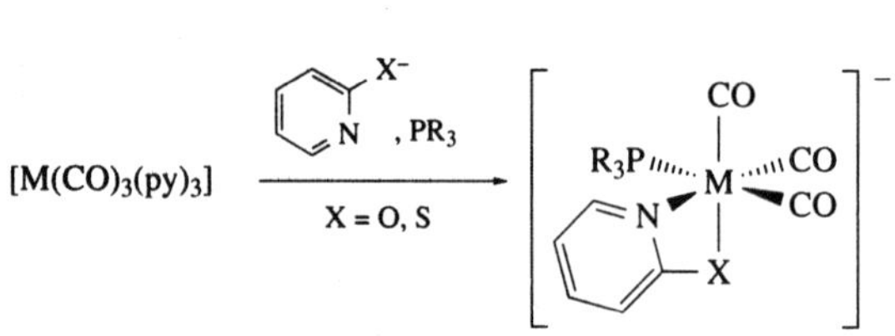

(21)

Complexes containing pyrazolyl derived ligands form another large family of heterocyclic donors. those bearing the tris(pyrazolylborate) ligands $[HBpz_3]^-$ and $[HB(3,5\text{-}Me_2pz)_3]^-$ being particularly important. Throughout this chapter Tp^R is used to represent an arbitrary tris(pyrazolylborate) ligand, Tp^1 for $[HBpz_3]^-$, and Tp^2 for $[HB(3,5\text{-}Me_2pz)_3]^-$. As these and related ligands are the subjects of a recent and comprehensive review,[243] only general comments are made here. The chromium chemistry remains the least developed, although the versatile starting materials $[M(CO)_3Tp^R]^-$ are easily prepared for all the group 6 metals.[201,244–7] The 17-electron $[M(CO)_3Tp^R]$ radicals are conveniently obtained via oxidation of the precursor anions with $[FeCp_2]PF_6$ at –40 °C, the preparation of the chromium complex is typical.[247]

Table 7 Representative group 6 complexes of heterocyclic amines.

Metal, M	*(CO)$_n$*	*M:L*[a]	*Ligand*[b]	*M–N or Comments* (nm)	*Ref.*
Monodentate and bridging donors					
Cr	5	1:1	py	0.2148, 0.2165	222
Cr	4	1:2	py	0.2158, 0.2171	222
Cr	5	2:1	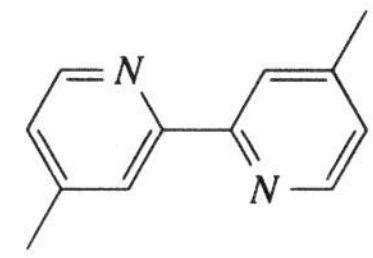	ring closes to chelate	223
Cr, Mo, W	5	(1 or 2):1	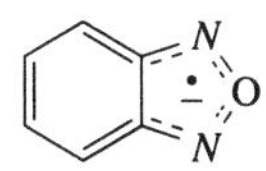	from [M(CO)$_6$] and ligand anion	224
Mo	4, 5	1:(1 or 2)	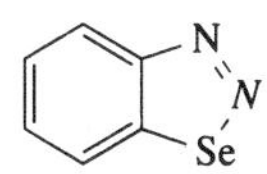	also M:L = 2:3 complex	224
Mo	3	2:3	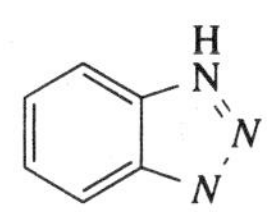	0.225	225
Mo	4	2:2	pyCH$_2$CH$_2$py		226
W	5	1:1	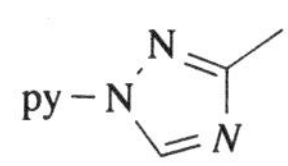	0.2273	227
W	5	2:1	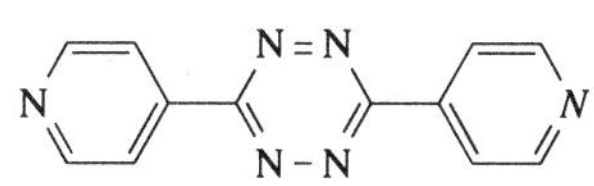	reduces to radical anion	228
Chelate and polydentate donors					
Cr	4	1:1	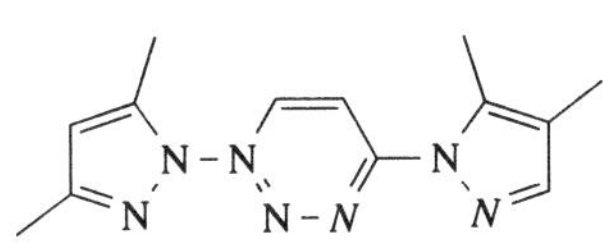	0.2082, 0.2083	229
Cr	4	1:1	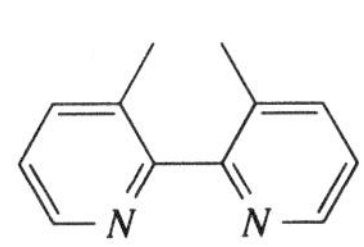	0.2123, 0.2115	230
Mo	2	1:2	bipy	0.2143–0.2239	219
Mo	4	1:1	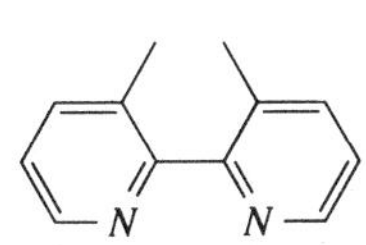	0.2238, 0.2272	230
Mo	4	1:1	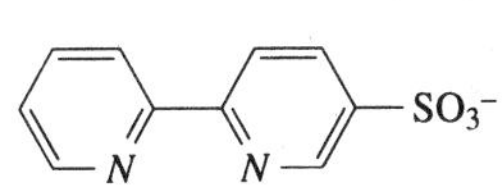	0.2223, 0.2228	231
Mo	4	1:1	pyNHpy	0.2269, 0.2274	232
Mo	4	(1 or 2):1	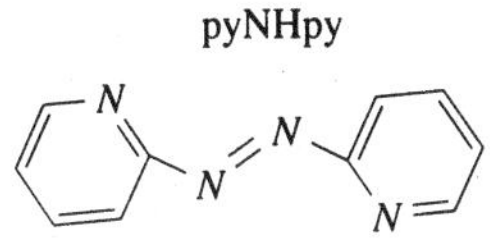	0.2190, 0.2197	233, 234

Table 7 (continued)

Metal, M	$(CO)_n$	*M:L*[a]	*Ligand*[b]	*M–N or Comments* (nm)	*Ref.*
Mo	4	(1 or 2):1	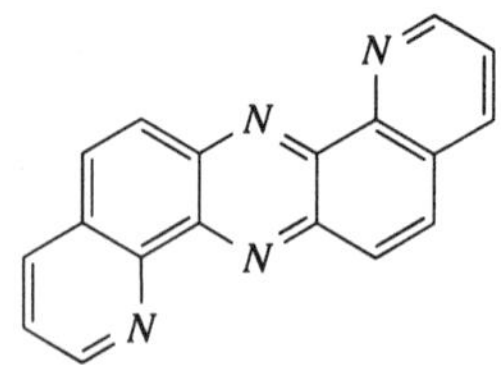		235
Mo	4	1:1	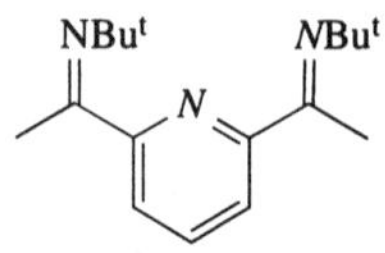	0.227 0.2289	236
Mo	4	1:1	$Pr^iN{=}CHCH({=}NPr^i)$	0.2264, 0.2266	237
Mo	4	1:1	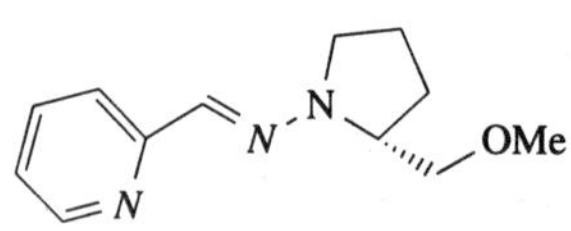	0.2308, 0.2235	238
Mo	4	1:1	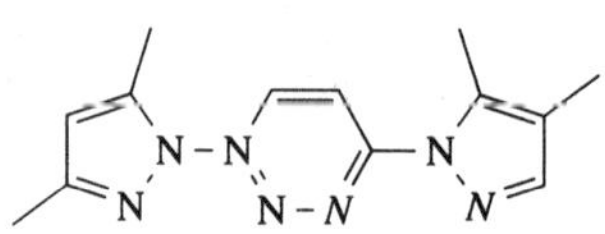	0.2322, 0.2336	229
Mo	4	2:1	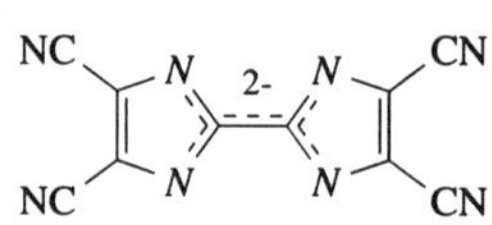	0.2316	239
Mo	4	(1 or 2):1	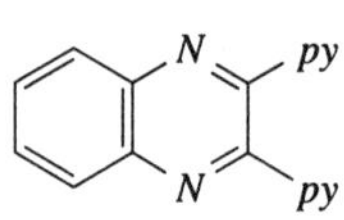		240
W	4	1:1	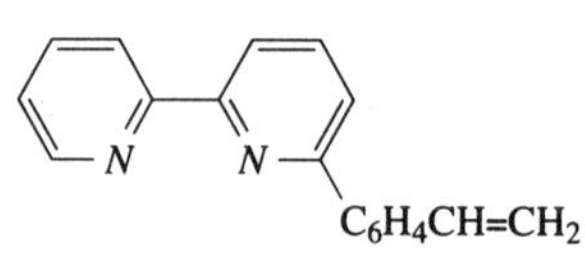	0.2290, 0.2230	241
W	4	1:1	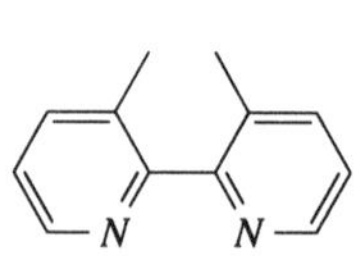	0.2272, 0.2238	230
Cr, Mo, W	3	1:1	$HC(py)_3$		242

[a] Ratio of coordinated metal fragments to ligands. [b] Coordinated nitrogens are shown in italics.

(ii) Nitrene and imido complexes

In this section the term 'nitrene' is used to distinguish complexes with bent M=NR ligands from linear M≡NR species, which are referred to here as 'imido' complexes. This nomenclature is not standardized and is used simply for convenience, both terms appear interchangeably in the chemical literature. Representative preparations of nitrene and imido complexes are shown in Table 8; for comparison two related species are also included. In addition to these rare isolated carbonyl complexes metal nitrenes have been implicated as possible intermediates in decomposition reactions of 8-azidoisoquinoline with $[Mo(CO)_3(NCMe)_2(PPh_3)]$ and $[Mo(CO)_4(dppm)]$.[248]

Table 8 Preparation of group 6 carbonyl complexes containing NR and related ligands.

Complex	*Synthesis*	*Comments*	*Ref.*
Nitrenes and imido donors			
$[W(CO)_5(=NPh)]$	$[W(CO)_5\{=CPh(OMe)\}] + PhN{=}NPh \xrightarrow{h\nu}$	transient prepared via a zwitterionic intermediate	249
$[W(CO)_5(=NNMe_2)]$	$[W(CO)_5\{=CPh(4\text{-Tol})\}] + PhN{=}NNMe_2 \xrightarrow{h\nu}$	limited solution stability	250
$[W(CO)_3(=NNMe_2)(L_2)]$	$[W(CO)_5(=NNMe_2)] + L_2 \longrightarrow$	L_2 = 2 PMe_3, dmpe, dppe; x ray	251
$[W(\equiv NR)(CO)_2Tp^2]PF_6$[a]	$[W(NHR)(CO)_2Tp^2] + [Ph_3C]PF_6 \longrightarrow$	R = Ph, Bu^t; x rays on both	252
$[W(\equiv NH)(CO)_2Tp^2]PF_6$[a]	$[W(NH_2)(CO)_2Tp^2] + [Ph_3C]PF_6 \longrightarrow$	deprotonates to nitride	253
Related donors			
$[W(CO)_5\{N(O)Bu^t\}]$	$[W(CO)_5(C_5H_{10}NH)] + O{=}NBu^t \longrightarrow$	x ray	254
$[WCl(\equiv NN{=}CMe_2)(CO)(L)_3]^+$	$[WCl_2(\equiv NN{=}CMe_2)(L)_3] + ZnCl_2 \longrightarrow$	L = PMe_2Ph; $[ZnCl_3(THF)]^-$ salt	255

[a] $Tp^2 = HB(3,5\text{-}Me_2pz)_3$.

(iii) Dinitrogen, nitrosyl and diazoligands

The poor σ-donor and π-acceptor qualities of dinitrogen result in most of their group 6 carbonyl complexes being rather labile. Photochemical procedures at low temperature or in the gas phase are the preferred protocols for the preparation of $[M(CO)_{6-n}(N_2)_n]$ (n = 1, 2),[256] although $[Cr(CO)_5(N_2)]$ can be observed in cyclohexane at room temperature.[257] The complex $[W(N_2)(CO)_3(PCy_3)_2]$ is also straightforward to prepare, allowing the binding enthalpy for the N_2 ligand to be determined.[140]

A large number of new nitrosyl complexes have appeared since the early 1980s; a general review in this area is available.[258] All the currently known group 6 thionitrosyl complexes are summarized in a short note.[259] General nitrosyl preparations involve addition of NO or NO^+ to group 6 precursors; a number of selected recent preparations appear in Table 9. The most commonly used reagents remain the commerically available NOX (X = Cl, BF_4, PF_6) salts. Of the many alternatives aqueous HNO_2, and even HNO_3, prove remarkably effective for the synthesis of $[M(CO)_2(NO)(N_3\text{-Macrocycle})]^+$ complexes provided the reaction conditions are controlled (see Section 4.2.4.3). The rare thionitrosyl complexes $[M(CO)_2(NS)Tp^2]$ (M = Mo, W) are best prepared by $N_3S_3Cl_3$ addition to $[M(CO)_3Tp^2]^-$, a protocol which interestingly fails for cyclopentadienyl analogues of these compounds. Recent standard preparations of alkyl and aryl diazonium species are listed in a recent review.[144] Of greater interest is the reaction of the α-lithio diazo compound $Li[Me_3SiC{=}N_2]$ which reacts with $[M(CO)_5(PPh_3)]$ (M = Cr, W) to an intermediate anion that is trapped with Me_3SiCl leading to *cis*-$[M(CO)_4\{N_2C(SiMe_3)_2\}(PPh_3)]$. The crystal structure of the tungsten complex indicates extensive bond delocalization in this class of compound.[260]

Table 9 Representative group 6 carbonyl complexes containing nitrosyl ligands.

Complex	*Synthesis*	*M–NO$_{av}$* (nm)	*Comments*	*Ref.*
$[CrH(CO)_2(NO)(PR_3)_2]$	$[Cr(CO)_2(NO)(\eta^6\text{-}C_6HMe_5)]^+ \xrightarrow[PR_3]{LiBH_4}$		R = Me, Et, OMe, OPh, Ph	127
$[Cr_2(CO)_8(NO)_2]$	$[Cr(CO)_4(NO)]^- + [CrBr(CO)_4(NO)] \longrightarrow$	0.1694	also $[PPN][Cr_2(CO)_9(NO)]$	261
$[PPN][Cr(CO)_4(NO)]$	$[Cr(CO)_6] + PPN[NO_2] \longrightarrow$			262
$[M(CO)_3(NO)(dppe)]PF_6$	$[M(CO)_4(dppe)] + NOPF_6 \longrightarrow$		M = Cr, Mo, W	263
$[MoCl(CO)_2(NO)(PR_3)_2]$	$[Mo(Cl\text{-}AlCl_3)(CO)_4(NO)] + 3\ PR_3 \longrightarrow$		hydride precursor	97
$[Mo(CO)_2(NO)(\mathbf{10})]$	$[Mo(CO)_3(\mathbf{10})] + HNO_2/NaPF_6 \longrightarrow$	0.1874	N_2S-Macrocycle	264
$[M(CO)_2(NO)(\mathbf{11})]X$	$[M(CO)_3(\mathbf{11})] + HNO_3$ or $NOPF_6 \longrightarrow$		M = Mo, W; X = BF_4, PF_6	36,153
$[W(F\text{-}SbF_5)(CO)_3(NO)(PCy_3)]$	$[W(CO)_5(PCy_3)] + NOSbF_6 \longrightarrow$	0.1788	x rays on four derivatives	265
$[WH(CO)_2(NO)(PPh_3)_2]$	$[W(CO)_3(NO)(PPh_3)_2]PF_6 + LiBH_4 \longrightarrow$	0.1842	x rays on two derivatives	98
$[W_2(\mu\text{-}H)(CO)_7(NO)(\mu\text{-}dppm)]$	$[W_2(CO)_8(\mu\text{-}dppm)]^{2-} + NOBF_4 \longrightarrow$	0.179/ 0.186	2 independent molecules	116
$[W(OMe)(CO)_2(NO)\{P(OPr^i)_3\}]$	$[W(CO)_3(NO)\{P(OPr^i)_3\}]PF_6 + MeO^- \longrightarrow$		versatile precursor	266

4.3.1.2 Structure and properties

(i) Diimines and heterocyclic ligands

Crystallographically characterized examples of heterocyclic amine complexes are included in Table 7. The structure of $[Mo(CO)_2(bipy)_2]$ merits further consideration. Although two of the Mo–N distances are within the expected range two are considerably shorter (0.2143, 0.2155 nm) as are the Mo–CO distances (0.192.2, 0.193.5 nm).[219] Such behaviour is indicative of π-back donation to the metal carbonyls and to a degree to the bipy ligands. Consistent with this charge transfer picture, $[M(CO)_2(N\text{–}N)_2]$ (N–N = bipy, phen) complexes are intensely coloured.

The photophysics of $[M(CO)_5(py)]$, $[M(CO)_4(bipy)]$, $[M(CO)_4(phen)]$ and their substituted derivatives have been extensively investigated in recent years.[267] All of these complexes show both ligand field (LF) and metal-to-ligand charge transfer (MLCT) states, the latter being of lower energy if the heterocyclic ligand bears electronegative substituents. In $[W(CO)_5(4\text{-CNpy})]$ emission from both LF and MLCF states proceeds at the same rate, suggesting rapid communication between the states.[268] Emissions from the MLCT state are commonly detected by luminescence studies upon selective irradiation of the MLCT band although some ligand dissociation always competes.[269,270] Even low-intensity irradiation of the LF state promotes M–L cleavage, for example, in the absence of added ligands $[M(CO)_5(Rpy)]$ disproportionates to $[M(CO)_6]$ and $[M(CO)_4(Rpy)_2]$.[271] Significant extensions (>0.010 nm) of the M–N and *trans* M–CO in the lowest excited state of $[M(CO)_5(py)]$ are predicted by theoretical calculations.[272] Results on $[M(CO)_4(phen)]$ suggest that although the MLCT and LF states effectively communicate LF states promote dissociative processes (CO loss in this case),[273] while MLCT excited states lead to less efficient associative substitution processes (Scheme 6) although such arguments are somewhat contentious.[274]

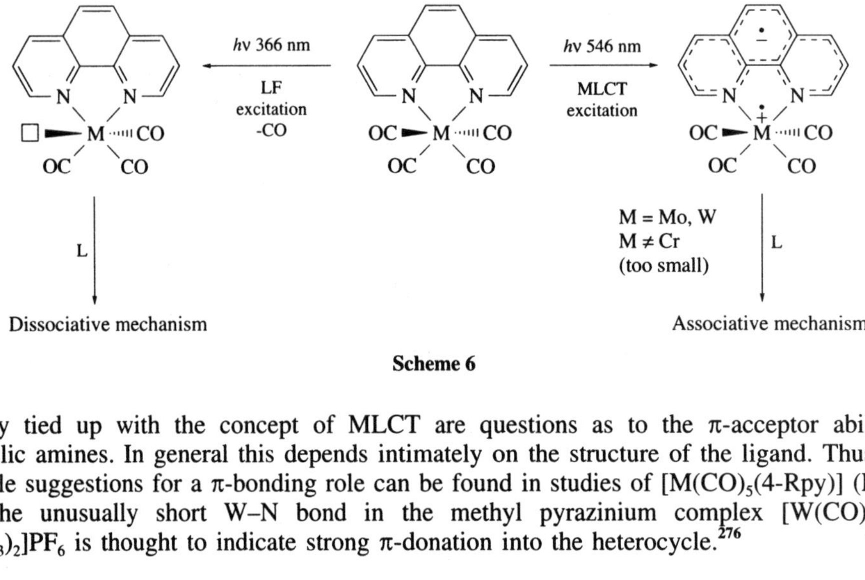

Scheme 6

Closely tied up with the concept of MLCT are questions as to the π-acceptor ability of the heterocyclic amines. In general this depends intimately on the structure of the ligand. Thus, although only subtle suggestions for a π-bonding role can be found in studies of $[M(CO)_5(4\text{-Rpy})]$ (R = H, Me, Bu^t),[275] the unusually short W–N bond in the methyl pyrazinium complex $[W(CO)_3(NC_4H_4N\text{-}Me)(PCy_3)_2]PF_6$ is thought to indicate strong π-donation into the heterocycle.[276]

(ii) Nitrene and imido complexes

Imido and nitrene ligands can be regarded as the two valence tautomers (**30**) and (**31**) (Figure 1) where formal charges have been assigned to the ligands. While such procedures are a matter of semantics it is helpful to consider the ligands this way. Thus, imido ligands are associated with electron deficient metal centres, while those in nitrene species will be rather electron rich. As nitrene ligands possess a nitrogen-centered lone pair, conversion to imido structures is prevented only in complexes where this would lead to metal configurations exceeding 18 valence electrons. Alternatively, the π-acceptor character of the nitrogen may be retained by removing the lone pair, as in the isoelectronic organonitroso complexes (**32**). The bonding in (**31**) and (**32**) is closely related to that in metal carbene complexes (Figure 1). Although the full molecular orbital picture of these species is more complicated, the LUMO in each of these cases is approximated to a good degree by an empty *p* orbital on nitrogen. The differences in the bonding modes are readily apparent in crystallographic studies; annotated core structures of $[W(\equiv NBu^t)(CO)_2Tp^2]PF_6$ (**33**) and $[W(CO)_3(=NNMe_2)(dppe)]$ (**34**) are shown in Figure 1.

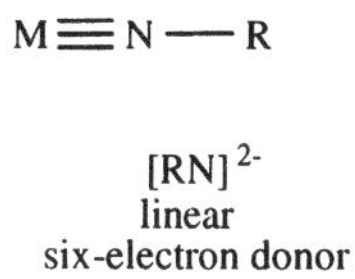

(30)

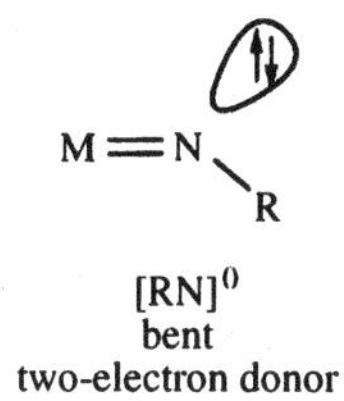

(31)

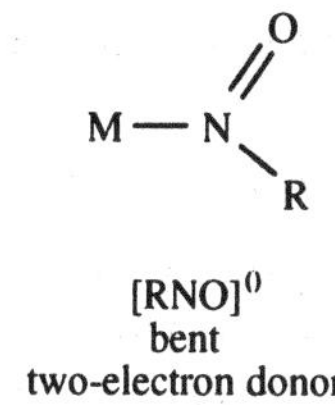

(32)

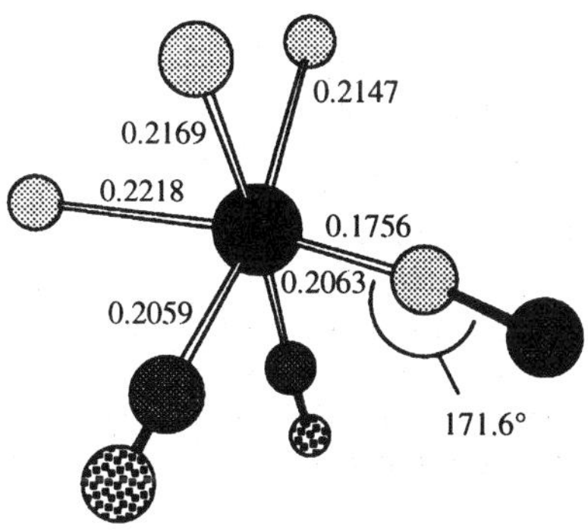

(33)

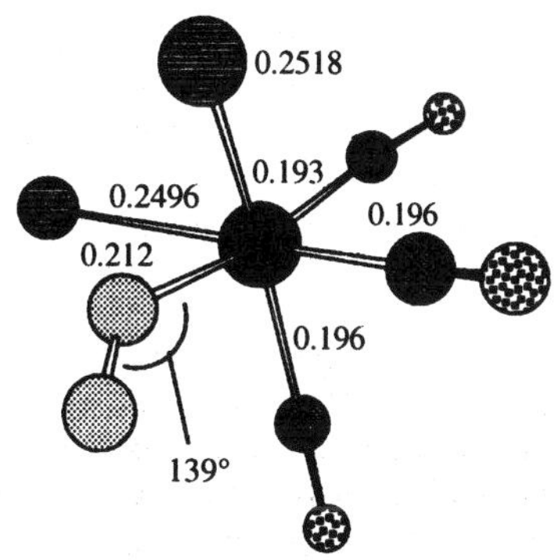

(34)

Figure 1 Bonding in imido and nitrene complexes and the core structures of $[W(\equiv NPh)(CO)_2Tp^2]PF_6$ (**33**) and $[W(CO)_3(=NNMe_2)(dppe)]$ (**34**) ($Tp^2 = [HB(3,5\text{-}Me_2pz)_3]$).

(iii) Dinitrogen, nitrosyl and diazoligands

Generalizations of the spectroscopic and structural properties of both linear and bent nitrosyl complexes can be found in a review by Mingos,[258] little further comment is required here. Some recent structural characterizations may also be found in Table 9.

4.3.1.3 Reactions

(i) Diimines and heterocyclic ligands

Electron-rich $[M(CO)_2(N\text{–}N)_2]$ complexes show diverse reactivity.[277,278] The behaviour of $[Mo(CO)_2(bipy)_2]$ is typical, and this is shown in Scheme 7. Oxidation to green paramagnetic *trans*-$[Mo(CO)_2(bipy)_2]BF_4$ is unusual in that it proceeds via the brown diamagnetic intermediate $[Mo_2(CO)_4(bipy)_4][BF_4]_2$. Reduction of this species with $NaC_{10}H_8$ leads to regeneration of the starting material, addition of a further electron, probably to a ligand-based LUMO, is possible leading to a radical anion. In addition to the chemistry of Scheme 7, the complex $[Mo(CO)_2(NCMe)(bipy)_2][BF_4]_2$ is versatile intermediate in its own right. Reactions with PEt_3, PPh_3, and $P(OMe)_3$ or dppe lead to $[Mo(PEt_3)_2(bipy)_2][BF_4]_2$, $[Mo(CO)_2(PPh_3)(bipy)_2][BF_4]_2$, or $[Mo(CO)(L_2)(bipy)_2][BF_4]_2$ [$L_2 = 2P(OMe)_3$, dppe], respectively.[279]

Recent reactivity of tris(pyrazolylborate) complexes are summarized in Scheme 8, in which Tp^R represents an arbitrary pyrazolylborate ligand, Tp^1 is $[HBpz_3]^-$, and Tp^2 is $[HB(3,5\text{-}Me_2pz)_3]^-$. Further reactions can be found in a recent review by Trofimenko.[243]

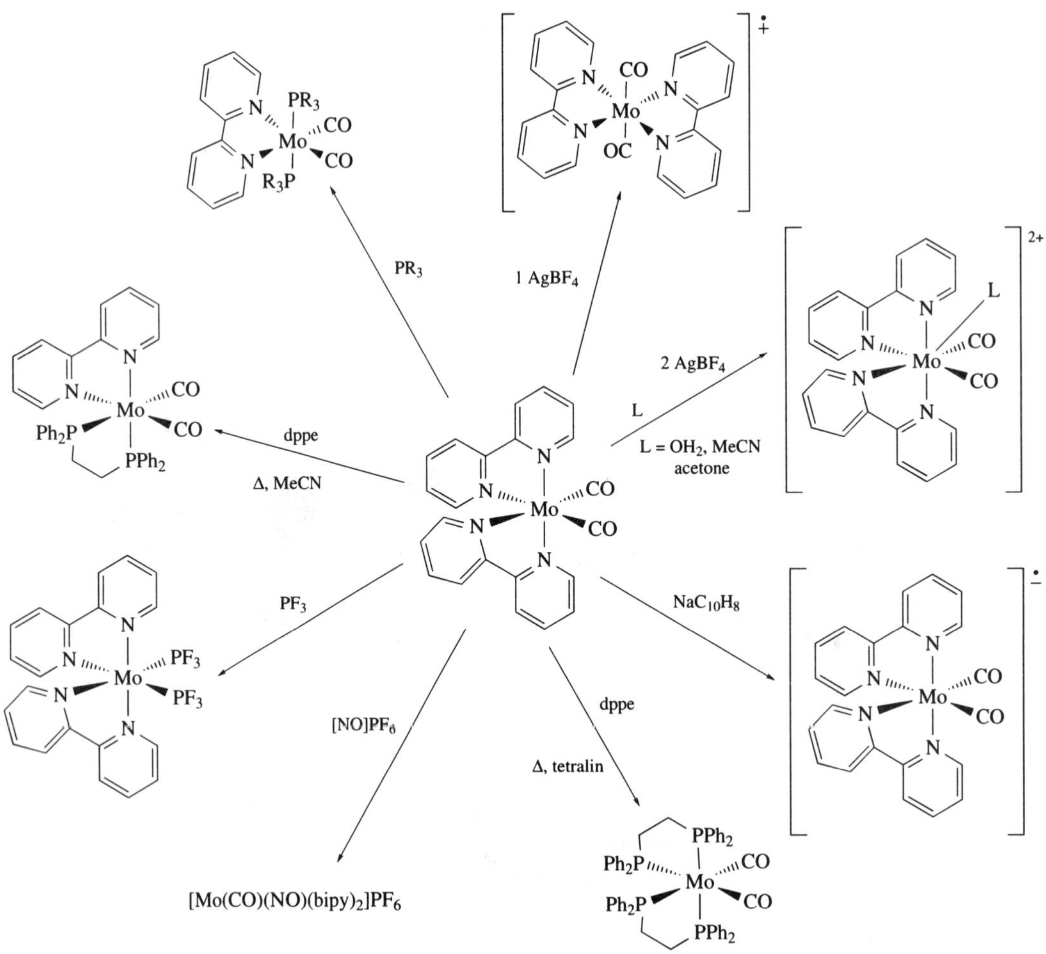

Scheme 7

(ii) Nitrene and imido complexes

Nitrene complexes are highly reactive. The presence of a high-energy HOMO–LUMO pair based on a nitrogen lone pair and an empty *p* orbital, respectively, allows these complexes to react both with electrophiles and nucleophiles. The nucleophilic reaction of phosphines to yield phosphinimines $[M\{NR(PR_3)\}(L)_n]$ is often used as an indication of the presence of a coordinated nitrene even when the complex itself is not isolable. Such reactivity is greatly attenuated by the presence of an α-heteroatom due to increased electron density in the LUMO. For example, $[W(CO)_5(=NNMe_2)]$ fails to react with PPh_3 at room temperature.[251] Nitrene complexes also react with electrophilic carbonyl compounds, in these cases imines are produced. Triply bonded imido ligand are normally rather less reactive. Treatment of $[W(\equiv NPh)(CO)_2Tp^2]^+$ with $LiBH_4$ causes initial formation of the formyl $[W(CHO)(\equiv NPh)(CO)Tp^2]$ which undergoes carbon to nitrogen hydrogen migration resulting in the amido complex $[W(NHPh)(CO)_2Tp^2]$.[288] In the special case of the terminal NH ligand in $[W(\equiv NH)(CO)_2Tp^2]PF_6$ deprotonation is effected by NEt_3, KH or $LiNPr^i_2$ yielding the relatively stable nitride complex $[W(\equiv N)(CO)_2Tp^2]$.[253]

(iii) Dinitrogen, nitrosyl and diazoligands

The lability of most dinitrogen complexes is such that most of their reactions involve simple displacement reactions as in the transformation of $[M(CO)_5(N_2)]$ complexes with a wide range of ligands to fashion $[M(CO)_5(L)]$ species. More interesting reactivity is shown by *trans*-$[W(CO)(N_2)(dppe)_2]$, loss of N_2 in the absence of added ligands leads to a mixture of two compounds believed to be the hydride

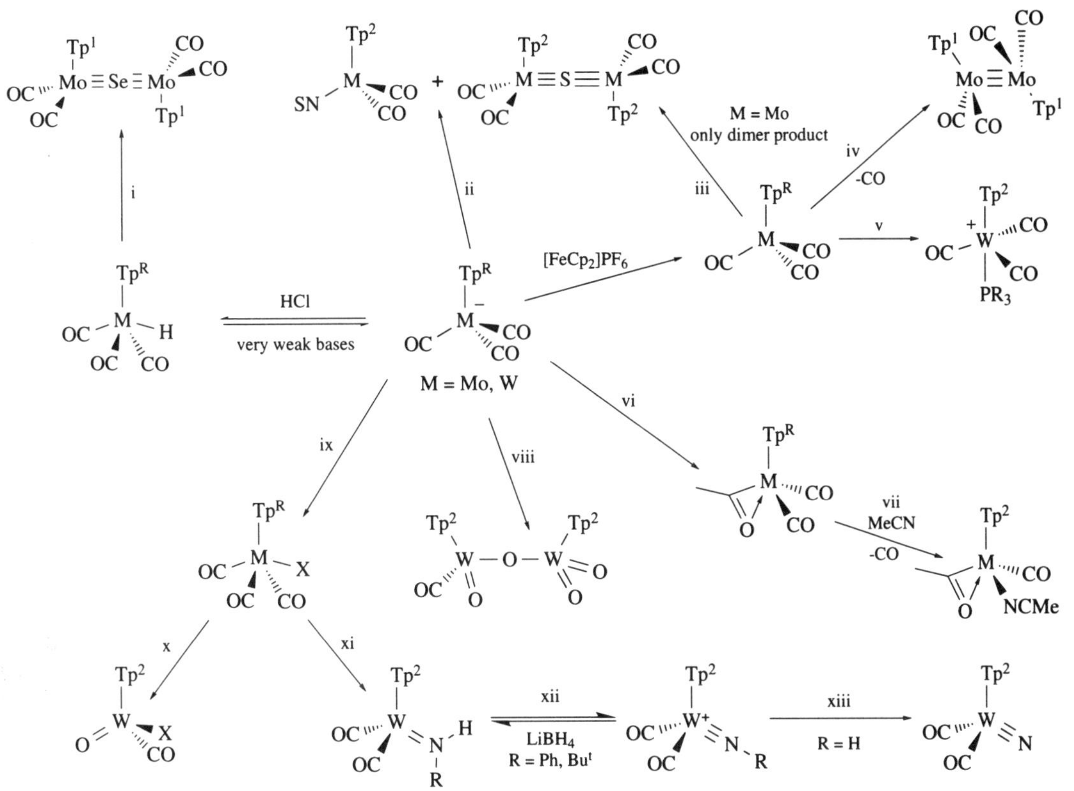

i, black Se;[280] ii, $N_3S_3Cl_3$;[281] iii, S_8;[282] iv, heat in MeCN;[283] v, $[FeCp_2]PF_6$ then PR_3;[284] vi, MeI or $[Me_3O]$;[283] vii, photolysis 0 °C;[285] viii, O_2-H_2O-acetone;[286] ix, X_2 (X = Br, I);[287] x, O_2–MeCN;[210] xi, RNH_2;[252] xii, $[Ph_3C]PF_6$ or I_2;[252] xiii, $LiNPr^i_2$ or KH or NEt_3[253]

Scheme 8

$[\overline{WH(CO)(dppe)\{C_6H_4P}(Ph)CH_2CH_2PPh_2)\}]$ and agostic $[W(CO)(dppe)_2]$. The compound *trans*-$[W(CO)(N_2)(dppe)_2]$ also reacts with CO_2 leading to the acetate $[WH(\eta^2\text{-}OCO_2Me)(CO)(dppe)_2]$ in the presence of MeOH, and the carbamates $[WH(O_2CNR_2)(CO)(\eta^1\text{-dppe})(dppe)]$ in the presence of secondary amines.

The reactions of nitrosyl complexes dominate this class. Reactions of nitrosyl hydrides are considered in Section 4.2.2. Cationic nitrosyl complexes are versatile starting points for many chemical transformations. Such complexes are particularly susceptible to attack by nucleophiles with liberation of CO. The behaviour of $[W(CO)_3(NO)\{P(OPr^i)_3\}_2]PF_6$ is exemplary (Scheme 9).[289]

Differences in the behaviour of chromium nitrosyls compared to their molybdenum and tungsten analogues are rather common. However, differences in the reactivity of analogous molybdenum and tungsten complexes, although far rarer, are known. For example, while treatment of $[Mo(CO)_2(NO)(\mathbf{11})]^+$ (R = Me) with Br_2 leads to the dibromide $[MoBr_2(NO)(\mathbf{11})]^+$ the equivalent tungsten complex scavenges trace amounts of the EtOH solvent stabilizer on addition of Br_2 to $CHCl_3$ solutions leading to $[WBr(OEt)(\mathbf{11})]^+$.[153] Oxidations of metal(0) nitrosyl complexes may be effected either electrochemical or with excess NOX (X = BF_4, PF_6) in the case of molybdenum and tungsten complexes, a good example is the behaviour of the complexes $[WI(CO)(NO)(\eta^1\text{-L–L})(L\text{–}L)]$ which initially lead to the species $[WI(CO)(NO)(\eta^1\text{-L–L})(L\text{–}L)]^+$ and on further oxidation to $[WI(NO)(L\text{–}L)_2]^{2+}$.[290] Such redox chemistry is not observed for chromium, the smaller metal radius resulting in the complexes being unable to simultaneously support the higher oxidation state and coordination number.

4.3.2 Carbonyl Complexes with Monodentate Phosphorus Ligands

Throughout this section the symbol Y is used to represent both carbon and heteroatom substituents attached to phosphorus, the symbol R being retained for alkyl and aryl functions. A number of useful reviews are available to supplement this section.[291–3]

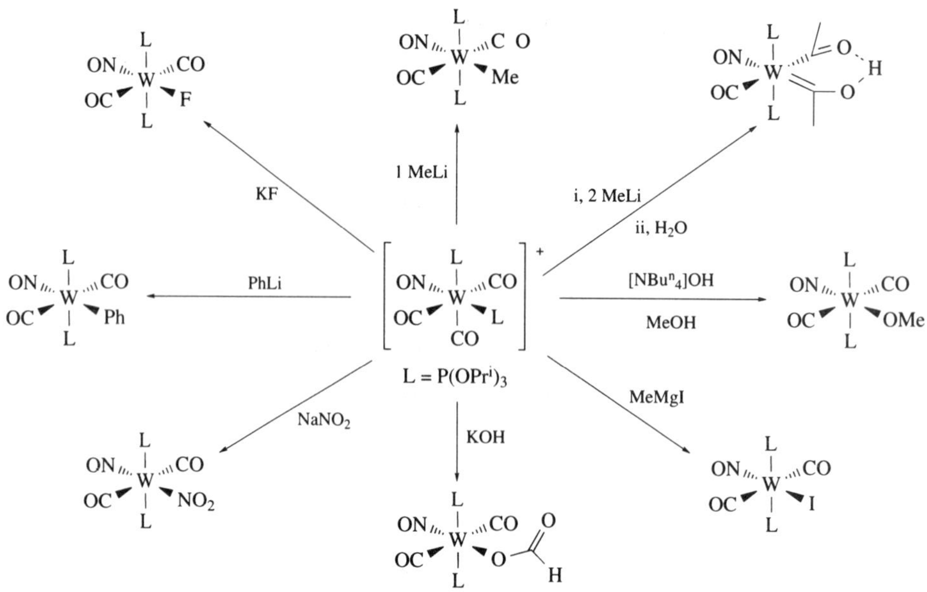

Scheme 9

4.3.2.1 Synthesis

(i) PY_3 species

Selected preparations of $[M(CO)_{6-n}(PR_3)_n]$ phosphine complexes and complexes of organophosphorus ligand with at least one P–Y linkage (Y = C, O, S, N, P, halogen or main-group metal) appear in Table 10. The examples are chosen on criteria of unusual reactivity, structure, PY_3 species, or examples that have expanded the crystallographic database since 1981. Some ligand structures pertinent to Table 10 are shown (structures (**35**)–(**46**). Exemplary synthetic approaches to these molecules, routes (a)–(g), are discussed below; the relevant references may be found within Table 10.

(a) Direct thermal, photochemical or chemically promoted substitution of $[M(CO)_6]$ or existing phosphine derivatives.
(b) Ligand substitution of $[M(CO)_5(L)]$, $[M(CO)_4(L)_2]$ or $[M(CO)_3(L)_3]$ (L = labile or solvent molecule) which can in addition be employed in stepwise reactions for the synthesis of mixed derivatives (Equations (22) and (23)). The cycloheptatriene complex $[M(CO)_3(chpt)]$ and $[M(CO)_3(NCR)_3]$ are generally the most effective choices for the preparation of *fac*-trisubstituted species, provided mild conditions are employed. Yellow crystalline $[Cr(CO)_5(cyclo\text{-}C_8H_{14})]$ also has recommendations for its use; this convenient $[Cr(CO)_5]$ source may be stored for prolonged periods.[340]
(c) Modification of halophosphine complexes (Equation (24)) has been exploited extensively by Gray as an alternative preparation for phosphites, thiophosphites and phosphorus amines. Such approaches are complimentary to carbonyl displacement strategies.
(d) Effective nonsubstitution routes to halophosphines involve cleavage of P–N and P–O bonds, typically with HCl or BF_3 (Equations (25) and (26)).
(e) Alkylation of P=O, P–OH, or PNHR (R = H, alkyl) functions with electrophilic reagents is related to routes (c) and (d) but in a reversed sense and is an effective route to pendant systems (Equation (27)).
(f) Metallospecies are obtained from the interaction of labile precursors with metallophosphines, of which the preparation of (**47**) where the labile amine retained is a special example (Equation (28)).
(g) Metallation of P–H bonds in the presence of a base is an alternative route to metallophosphines. Two points are of note here. First, reagents such as Bu^nLi and KH are simultaneously bases and metallating agents, although other species require preparation by transmetallation or the use of metal salts in the presence of a base as shown for just one case (Equation (29)). Second, such species can alternatively be viewed as heterometallic species held together by bidentate phosphido ligands. In this chapter nontransition metal functionalized phosphines are regarded as PR_3

Table 10 Representative carbonyl complexes containing monodentate PY_3 ligands.

Complex	*Synthesis*[a]	*Comments*	*M–P*$_{av}$ (nm)	*Ref.*
Phosphines				
$[Cr(CO)_5(\mathbf{35})]$	(b)	also W	0.2368	294
$[Cr(CO)_{6-n}(\mathbf{36})_n]$	(a)	n = 1–3; also Mo, W		295
$[Cr(CO)_5(\mathbf{37})]$	(a)	also Mo, W	0.2364	296
$[Cr(CO)_5(\mathbf{38})]$	(b)	also Mo, W		297
$[Cr(CO)_5\{P(C{\equiv}CPh)_3\}]$	(a)		0.2322	298
$[Mo(CO)_5\{P(n\text{-Tol})_3\}]$	(a), (b)	n = 2,3,4	0.2562	299
$[Mo(CO)_5\{PPh_n(C_5HMe_4)_{3-n}\}]$	(b)	n = 1,2	0.2255	300
$[W(CO)_5\{PPh_2(C_{14}H_9)\}]$	(a)	anthracene-based	0.2543	301
$[W(CO)_5(\mathbf{39})]$	(b)		0.2487	302
$[W(CO)_5\{P(3\text{-}C_6H_4SO_3)_3\}]^{3-}$	(b)	Na(cryptand-221) cations	0.2554	303
cis-$[Cr(CO)_4(\mathbf{37})_2]$	(a)	also Mo, W	0.2358	296
cis-$[Mo(CO)_4(PBu^n{}_3)_2]$	(b)	also PMe_3, PEt_3	0.2552	304
cis-$[Mo(CO)_4(PMePh_2)_2]$	(b)	also PPh_3, $PMePh_2$	0.2555	305
mer-$[Cr(CO)_3(\mathbf{37})_3]$	(a)	also Mo, W	0.2332	296
Phosphites and thiophosphites				
$[Cr(CO)_{6-n}\{\overline{RPSCH_2CH_2S}\}_n]$	(b)	R = Me, CF_3; n = 1,2		306
$[Mo(CO)_5\{PMe(OEt)(OH)\}]$	(c)	(*S*)-antipode 98% *ee*		307
$[Mo(CO)_5\{PPh_2(O\text{–}HNEt_3)\}]$	(c)	also Cr, W	0.2550	308
$[Mo(CO)_5\{P(XR)Ph_2\}]$	(c)	X = O, S; R = alkyl		309,310
cis-$[W(CO)_4\{P(CF_3)(SMe)_2\}_2]$	(b)	also Cr, Mo		311
Phosphaamines and related species				
$[Cr(CO)_5(\mathbf{40})]$	(c)		0.2353	312
$[Cr(CO)_5(\mathbf{41})]$	(a), (b)	phosphazene-based	0.2387	313
$[Cr(CO)_5(\mathbf{42})]$	(b)		0.2545	314
$[Mo(CO)_5\{P(NSO)Ph_2\}]$	(c)	also Cr	0.2482	315
$[Mo(CO)_5(\mathbf{43})]$	(a)	cyclen-based	0.2533	316
$[Mo(CO)_3\{P(NHPr^i)_3\}_3]$	(c)		0.2545	317
$[Mo(CO)_3(Me\overline{PNHC_6H_4N}H)_3]$	(b)		0.2470	318
$[Mo(CO)_5\{P[P(O)(OEt)_2]_nPh_{3-n}\}]$	(a)	n = 1–3	0.2524	319
Mixed and pendant systems				
$[Cr(CO)_5(\mathbf{44})]$	(e)	also Mo		320
trans-$[Cr(CO)_4(PPh_3)\{P(OPh)_3\}]$	(b)	also PBu_3, $P(OMe)_3$	0.2395[b]	23
cis-$[Mo(CO)_4(PPhMe_2)(PPh_3)]$	(b)	Mo–PMe_2Ph 253.8 pm	0.2562[b]	321
cis-$[W(CO)_4(PMe_3)(PMe_2Ph)]$	(b)	many other examples		322
mer-$[Cr(CO)_3(\eta^1\text{-dppm})(dppm)]$	(a)		0.2375[c]	323
$[Mo(CO)_5(\eta^1\text{-dppf})]$	(b)	also Cr, W	0.2563	324
$[W(CO)_5(\eta^1\text{-}Ph_2PCH_2PHPh_2)]^+$	(a)	also Cr, Mo		325
$[W(CO)_5\{\eta^1\text{-}PH(NPr^i{}_2)(PCl[NPr^i{}_2]_2)\}]$	(c)		0.2549	326
Halophosphines				
$[Cr(CO)_5\{PX(CF_3)_2\}]$	(b)	X = F, Cl, Br, I		327
$[Cr(CO)_5\{PHCl(NPr^i{}_2)\}]$	(d)	also Mo, W		328
$[W(CO)_5\{\overline{PCH_2CH_2}(Cl)\}]$	(d)		0.2455	329
$[W(CO)_5(PH_nX_{3-n})]$	(d)	n = 1,2; X = F, Cl		330
$[Mo(CO)_4(PFPh_2)\{P(NH_2)Ph_2\}]$	(d)	Mo–$P(NH_2)Ph_2$ 247.1 pm	0.2414	331
$[Mo(CO)_4\{PPh(Cl)(NHPr^i)\}_2]$	(b)	*rac*-diastereomer	0.2459	332
$[Mo(CO)_3(L)(dppe)]$	(a)	L = PF_3, PF_2NMe_2		333
fac-$[Mo(CO)_3(bipy)\{PF(OR)_2\}]$	(d)	R = Me, Et, Pr^i		334
Metallophosphines				
$[Cr(CO)_5\{P(Al[NMe_3]R_2)Ph_2\}]$	(f)	R = CH_2SiMe_3	0.2482	335
$[Cr(CO)_5(PKPh_2)]$	(g)	also Mo, W		336
$[Hg\{\mu\text{-}PHCyCr(CO)_5\}_3]^-$	(g)			337
$[Hg\{\mu\text{-}PPh_2Cr(CO)_5\}_2]$	(g)	also Mo, W		338
cis-$[Mo(CO)_4\{PLiR_2\}_2]$	(g)	R = H, alkyl, aryl		339

[a] For synthetic methods see text. [b] M–PPh_3 distance. [c] M–$P_{monodentate}$ distance.

derivatives, while transition metal heterometallics are considered to have bridging phosphido ligands. Complexes with metal–metal bonds have been excluded, these being reviewed elsewhere.[341]

$$[CrCl(CO)_5]^- \xrightarrow[-CO]{PR_3} cis\text{-}[CrCl(CO)_4(PR_3)]^- \xrightarrow[-NEt_4Cl]{EPh_3} trans\text{-}[Cr(CO)_4(PR_3)(EPh_3)] \quad (22)$$

R = Bu^n, OMe, OPh E = P, As

(35)

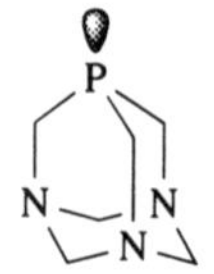

(36)

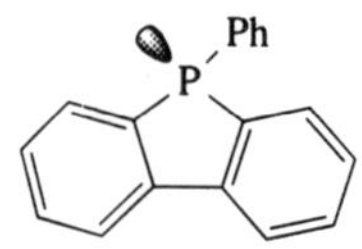

(37)

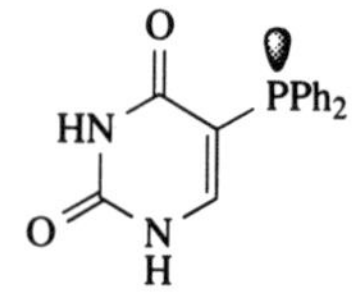

(38)

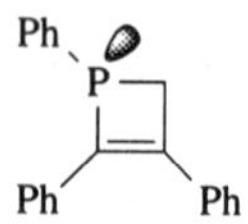

(39)

(40)

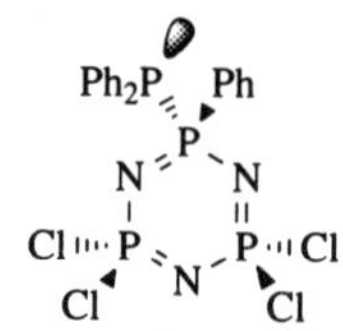

(41)

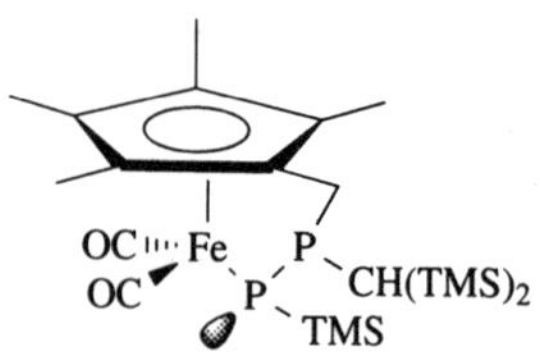

(42)

(43)

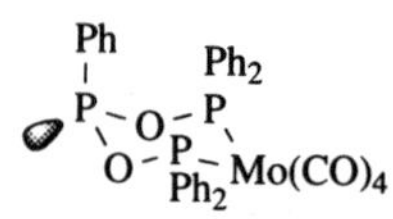

(44)

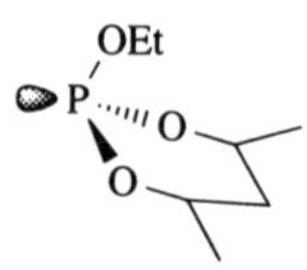

(45)

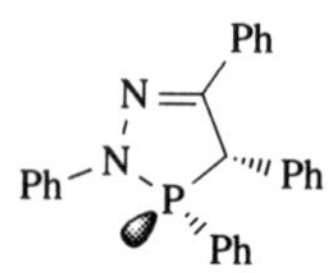

(46)

$$[W(CO)_4(py)_2] \xrightarrow[\text{-py}]{PR_3} cis\text{-}[W(CO)_4(py)(PR_3)] \xrightarrow[\text{-py}]{L} cis\text{-}[W(CO)_4(PR_3)(L)] \quad (23)$$

R = alkyl, aryl; L = PMe_2Ph, $PEtPh_2$, $PTol_3$

$$[Mo(CO)_5(PClPh_2)] \xrightarrow[\text{-base HCl}]{\text{RXH, base}} [Mo(CO)_5\{P(XR)Ph_2\}] \quad (24)$$

X = O, S, NH
R = alkyl, aryl

$$[M(CO)_5\{PH(NPr^i)_2\}] \xrightarrow{HCl} [M(CO)_5\{PHCl(NPr^i)\}] \quad (25)$$

$$fac\text{-}[Mo(CO)_3\{P(OR)_3\}(bipy)] \xrightarrow[\text{-}BF_2(OR)]{BF_3\bullet OEt_2} fac\text{-}[Mo(CO)_3\{PF(OR)_2\}(bipy)] \quad (26)$$

R = Me, Et, Pr^i

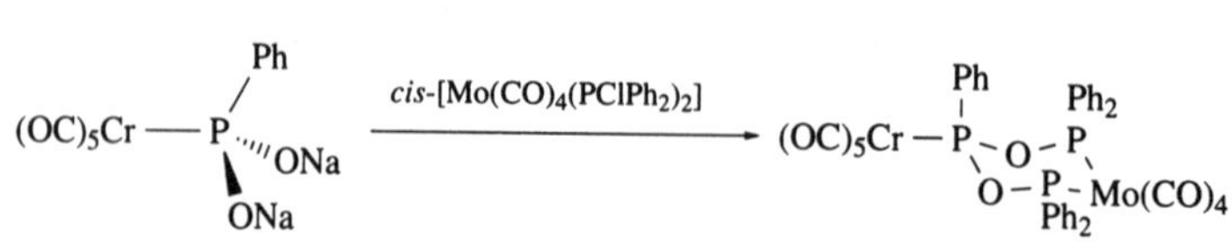

(27)

$[Cr(CO)_5(NMe_3)]$ $\xrightarrow{Ph_2PAl(CH_2TMS)_2}$ 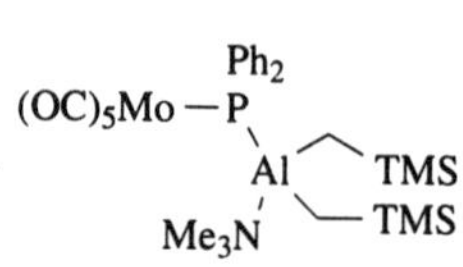

(47)

(28)

$$[M(CO)_5(PHPh_2)] \xrightarrow[\text{phen}\ NHEt_2]{AgO_3SCF_3} [M(CO)_5(\mu\text{-}PPh_2)Ag(phen)] \quad (29)$$

(ii) PY_2 species

Commonly present in η^1-bound phosphorins (also named as phosphinines or phosphabenzene derivatives) (**48**), phosphaalkenes (**49**) and phosphoranylidenes (**50**), the fragment (–P=) is also occasionally encountered in more exotic species (**51**) featuring heteroatom inclusion. For example, boron-based $[Cr(CO)_5\{(Et_3C)P{=}BNR_2\}]$ complexes have recently been isolated.[342] Typical routes to these compounds are as follows.

(a) Displacement reactions from $[M(CO)_5(THF)]$ under mild conditions, as shown for just one example (Equation (30)). This route is only applicable to phosphoraalkenes complexes where the free ligand is isolable (generally sterically stabilized versions).

(b) Phosphorus versions of the Wittig–Horner–Emmons reaction are highly effective routes to phosphaalkene complexes in cases where the free ligand is unavailable (Equation (31)). The intermediate anion (**52**) in these reactions is thought to be stabilized via $[M(CO)_5]$ coordination.

(c) Preparation of phosphoranylidenes by interaction of the phosphranorbonadiene species (**53**) with basic phosphines (Equation (32)). These compounds act as phospha–Wittig reagents in their own right and may be used in type (b) preparations (Equation (33)).

These and other examples are referenced in Table 11.

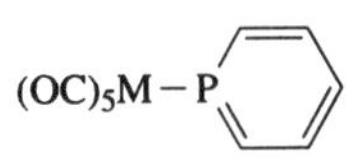

(48)

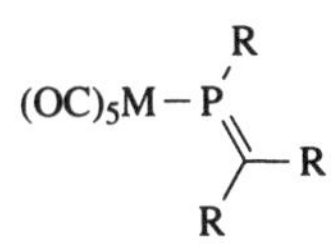

(49)

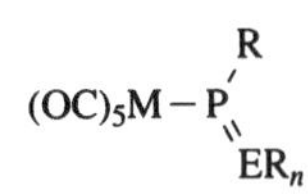

(**50**) $ER_n = PR_3$
(**51**) $ER_n = BNR_2$

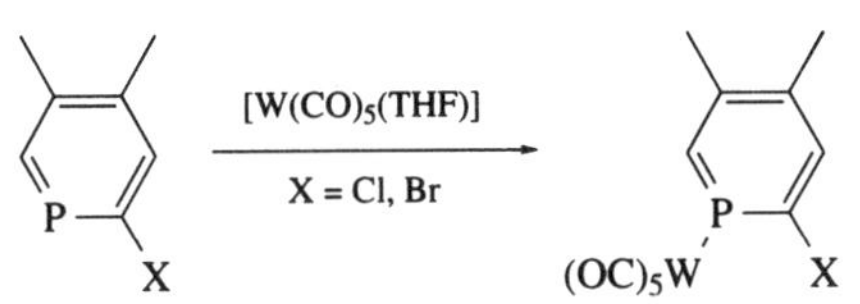

(30)

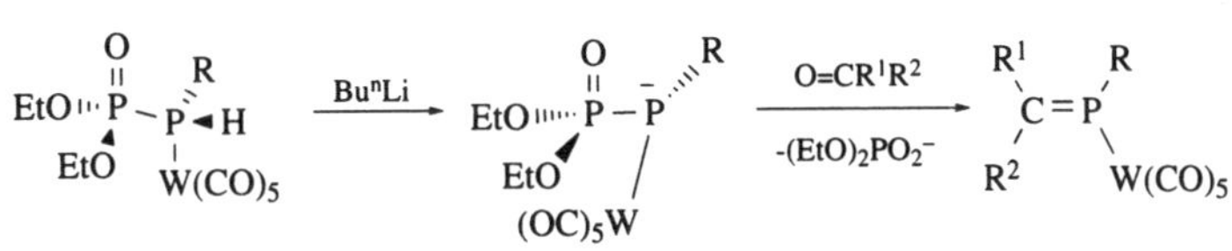

(**52**)

(31)

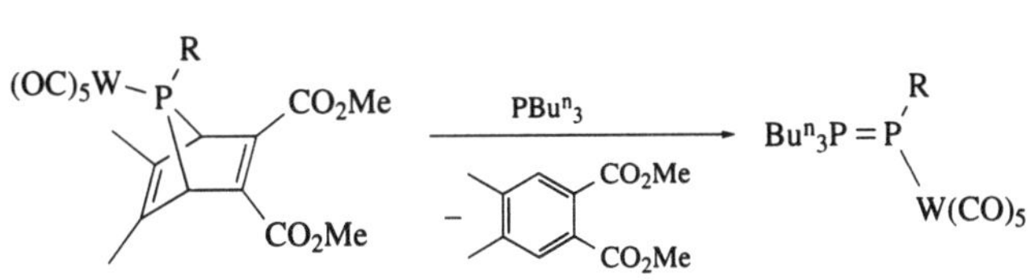

(**53**)

(32)

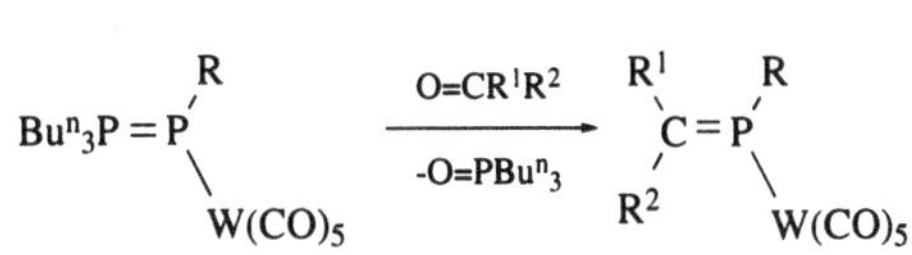

(33)

Table 11 Representative phosphorin, phosphaalkene and phosphoranylidene complexes.

Complex	*Synthesis*[a]	*M–P* (nm)	*P=C* (nm)	*M–P=C$_a$* (°)	*Ref.*
$[Cr(CO)_5(MesP{=}CPh_2)]$	(a)	0.2356	0.1679	130.8	343
$[Cr(CO)_5\{(F_3C)P{=}CF(NMe_2)\}]$	(a)	0.2454	0.1805	108.3	344
$[Cr(CO)_5\{[(Me_3Si)_2CH]P{=}CH(Fp^*)\}]$[b]	(a)	0.2399	0.1643	135.2	345
$[Mo(CO)_5(\eta^1\text{-}PC_5H_5)]$	(a)	0.2464	0.1706[c]		346
$[W(CO)_5(\eta^1\text{-}2\text{-}PC_5H_4Cl)]$	(a)	0.2457	0.1727[c]		347
$[W(CO)_5\{(PhCH{=}CH)P{=}CH(Pr^i)\}]$	(b)	0.2472	0.1799	125.1	348
$[W(CO)_5\{(CH_2)_nP{=}CMe\}]$ $n = 3,4$	(b) + (c)				349
$[W(CO)_5\{EtO_2C)P{=}PEt_3\}]$	(c)	0.2603	0.2156[d]	109.7[d]	350

[a] For synthetic methods see text. [b] $Fp^* = Fe(CO)_2(\eta\text{-}C_5Me_5)$. [c] Average. [d] Distance and angle are P–P and M–P–P, respectively.

(iii) PY species

Isolated terminal phosphinidene complexes are limited to the crystallographically characterized tungsten complex $[WCl_2(\equiv PAr)(CO)(PMePh_2)_2]$ ($Ar = 2,4,6\text{-}C_6H_2Bu^t_3$). Reaction of *trans*-$[WCl_2(PMePh_2)_4]$ with a sterically hindered phosphaketene species (ArPCO), which acts as a simultaneous source of both phosphinidene and carbonyl ligands, leads to the deep green phosphinidene. Details of the structure of this complex are discussed in Section 4.3.2.2, however, it is clear that this species owes its existence to extreme steric protection of the highly reactive W≡P bond. Due to this lack of stable starting materials a number of routes have been developed which allow for the production of transient phosphinidene complexes which are employed in trapping-type reactions. Either thermal activation of (**53**) or $[W(CO)_5(\overline{PRCH_2CH_2})]$ can be employed as routes to transient $[W(\equiv PR)(CO)_5]$ complexes. Alternatively, electrochemical methods may be used, as in the reduction of $[Cr(CO)_5(PBr_2R)]$.[351] The thermal reactions have been reviewed and are therefore not covered further here.[352,353]

4.3.2.2 Structure and properties

Numerous spectroscopic and theoretical techniques have been applied to group 6 phosphine and related complexes since the early 1980s, but only recent developments are highlighted here. Examples of kinetic and reaction mechanism studies are available in Section 4.1.1.

(i) PY_3 species

Steric factors are often the dominant effect leading to M–P bond extension in $[M(CO)_{6-n}(PY_3)_n]$ complexes. Molecular mechanics studies of $[Cr(CO)_5(PR_3)]$ show that a steady increase in Cr–P bond length results from increasing the cone angle of the phosphine (θ) from 118° (PMe_3) to 183° (PBu^t_3). The bond energy term is dominant in these calculations, other attractive and repulsive Van der Waals terms approximately cancelling one another out.[354] It is instructive to concentrate on the structures and reactivity of *cis*-$[Mo(CO)_4(PR_3)_2]$ complexes as significant amounts of information are available for comparisons of steric and electronic effects in these species. In general, larger cone angle phosphines and related PY_3 species result in increased Mo–P bond lengths, P–Mo–P angles, and increased tendency towards intramolecular rearrangement to the *trans* isomer.[355] Examples of PY_3 species in order of increasing Mo–P distance are shown in Table 12 together with their cone angles (θ), some phosphine properties and estimates of the Mo–P bond strength from calorimetric studies of the reaction of *fac*-$[Mo(CO)_3(\eta\text{-}C_6H_5Me)]$ with PR_3 where available.[356] The structures of two of the ligands are shown in Structures (**45**) and (**46**). Although the cone angle correlation is not totally dominant a clear trend is present for most of the group. Related investigations of $[W(CO)_4(PR_3)_2]$ have centred on both kinetic[357] and thermodynamic[322] aspects of *cis–trans* isomerism vs. phosphine cone angle correlations with similar conclusions being reached.

In addition to steric factors, the electronic properties of tertiary phosphine species are still an area of substantial debate, the relative worth of both the σ- and π-components to the bonding scheme being much discussed. The pK_a of the conjugate acid (HPY_3^+) of the phosphorus ligand provides a useful

Table 12 Structural and physical properties in *cis*-$[Mo(CO)_4(PR_3)_2]$ complexes.

Phosphine	θ (°)	$pK_a(HPY_3^+)$	*E(Mo-P)* (kJ mol^{-1})	*Mo–P$_{av.}$* (nm)	*P–Mo–P* (°)
$PPhCl(NHPr^i)$	~102[a]			0.2461	90.5
(**45**)	~101[a]			0.2464	90.0
(**46**)	~134[a]			0.2506	90.0
PMe_3	118	8.65	161	0.2522	97.54
PMe_2Ph	122	6.5	156	0.2529	94.78
PEt_3	132	8.69	161	0.2544	100.27
$PPh_3 + PMe_2Ph$[b]	134[b]	4.6[b]	147, 156	0.2550[b]	97.73
$PBu^n{}_3$	132	8.43	157	0.2552	99.29
$PMePh_2$	136	4.57	147	0.2555	92.52
$PPh_2(CH_2NHTol\text{-}4)$	~136[a]			0.2557	94.2
PPh_3	145	2.73	147	0.2577	104.62
PCy_3	170	9.70		0.2654	105.1

[a] Cone angle estimated from values and methods cited in Rahman *et al.*[358] and references within. [b] Averaged values for the complex *cis*-$[Mo(CO)_4(PPh_3)(PMe_2Ph)]$ and its parent phosphines.

measure of σ-basicity, resulting in the traditional ordering $P(alkyl)_3 > P(aryl)_3$. However, it is argued that as these figures are obtained in solution, they really reflect the solvation ability of a given phosphonium salt rather than phosphine σ-basicity and that protons are rather poor surrogates for softer more electronically diffuse low oxidation state transition metal centres. An alternative measure of the phosphorus σ-base character is derived from gas phase photoelectron spectra of $[Mo(CO)_5(PMe_nPh_{3-n})]$ (n = 0–3) complexes and shows an equally clear but opposite σ-basicity ordering: $P(aryl)_3 > P(alkyl)_3$.[359] Recent attempts to resolve these dilemmas shed light on the relative worth of both the σ- and π-components involved in $[M(CO)_{6-n}(PY_3)_n]$ bonding. By assuming that the π-bonding in $[W(CO)_3(NO)(PMe_3)(PY_3)]^+$ is dominated by the π-acceptor qualities of the nitrosyl ligand correlation of the trimethylphosphine ^{31}P NMR chemical shift and J_{PW} coupling constant can be used as a means of measuring σ-effects at the tungsten centre from the ligand (PY_3). The results support the traditional σ-basicity ordering.[359] Similarly, correlation of the reduction potentials and carbonyl stretching frequencies of $[Fe(COMe)(CO)(PR_3)Cp]$ leads to the conclusion that both PMe_3 and PPh_3 are both pure σ-donors.[358] These results are at odds with optical polarizability studies of $[Cr(CO)_5(PMe_3)]$ which indicate a measurable π-component to the bonding[360] and additional photoelectron studies of $[Mo(CO)_5(L)]$ (L = PMe_3, η^1-dmpm, η^1-dmpe) which show a small but significant π-back-donation (about one-quarter the effectiveness of CO).[361] Great care must be taken in the design of experiments comparing σ- with π-bonding. As can be seen, contradictions easily arise from the inability of the spectroscopic probe to measure one variable (σ- or π-bonding) with the total exclusion of the other. Traditional Cotton–Kraihanzel force constant analyses often fall victim to such problems. Alternatively the inability of experiments to deal simultaneously with combined σ/π-bonding effects also can lead to erroneous results. A recent successful 'exclusion' approach appears to be the semiquantitative relationship between ^{17}O quadrapole coupling constants and CO-π* orbital populations, which when interrogated for $[W(CO)_5(L)]$ (L = CO, NMe_3, PMe_3, $P(OMe)_3$] allow identification of PMe_3 as a slight π-acceptor.[362] A elegant 'combination' approach recently utilized by Alyea is to plot the ^{95}Mo NMR chemical shift for *fac*-$[Mo(CO)_3(L)_3]$ complexes against the summation of the Kabachnik σ^{ph} parameters (Hammett-type values derived from the ionisation of phosphoric acids) for the phosphine substituents to yield a smooth curve showing the variation in σ- and π-bonding for 15 complexes.[363] One surprising conclusion of this work is that PCl_3 is predicted to be both a weak σ-donor and π-acceptor.

Although both steric and electronic factors in isolation strongly affect the bonding in $[M(CO)_{6-n}(L)_n]$, in practice the two factors interact strongly giving rise to deviations in experiments designed to measure one facet (steric or electronic) in isolation. For example, Alyea's approach is not applicable to *fac*-$[Mo(CO)_3(PPh_3)_3]$ in which the phosphine has a cone angle above the 'steric threshold' of the experiment.

Applications of metal spectroscopy to these species are becoming routine,[364] ^{53}Cr NMR being the most limited due to the sensitivity problems associated with this nucleus. Complexes of the type $[Mo(CO)_{6-n}(PY_3)_n]$ (n = 1–3; Y = alkyl, aryl, OR) show ^{95}Mo NMR signals in the range δ_{Mo} −800 to −1900. The molybdenum–phosphorus coupling constants range from 110 to 250 Hz, the lower end of the range being associated with pentacarbonyl complexes or phosphorus ligands bearing heteroatoms. The equivalent tungsten(0) complexes show ^{183}W NMR spectra which cover a much smaller range (δ_W −2600 to −3500). This nucleus is less useful as the lower sensitivity and slow relaxation result in long accumulation times. Heteronuclear ^{183}W–^{31}P couplings are typically 1.75 times larger than the equivalent ^{95}Mo–^{31}P interactions.

(ii) PY_2 and PY species

As judged by their M–P and *trans*-M–CO structural data the bonding in phosphorin and phosphaalkene complexes seems to involve a greater degree of π-back donation character than in tertiary phosphines; those in phosphoranylidenes seem minimized. Given the caveats above, further investigations are needed to clarify the bonding scheme in these molecules. The free heterocyclic phosphorus ligand shows no basic properties; a pK_a of −10 has been calculated for $HPC_5H_5^+$ over that of 5.25 for $HNC_5H_5^+$. Selected structural data for these species appears in Table 11. Asymmetrical phosphaalkene ligands $[RP{=}CR^1(R^2)]$ can exist as a mixture of (*E/Z*) isomers giving the possibility of isomeric species upon complexation. In practice, normally only a single isomer is obtained if R^1 is a hydrogen atom; isomer identification is easily accomplished by ^{31}P NMR spectroscopy, the (*E*) species showing a large J_{PH} coupling of around 25 Hz.[345] The structure of phosphinidene complex $[WCl_2(\equiv PAr)(CO)(PMePh_2)_2]$ ($Ar = 2,4,6\text{-}C_6H_2Bu^t{}_2$) is shown in Figure 2. The short W–P distance (0.2169 nm) and W–P–C angle (168.2°) are consistent with the presence of a strong π-donor. The conclusion is reinforced by the observation of a low-frequency CO stretch (1903 cm^{-1}) for the complex indicating that the metal is electron rich. The effective π-donation from the phosphinidene leads to a significant shortening (~0.004 nm) of the W–Cl *trans* to it over the *cis* W–Cl bond.

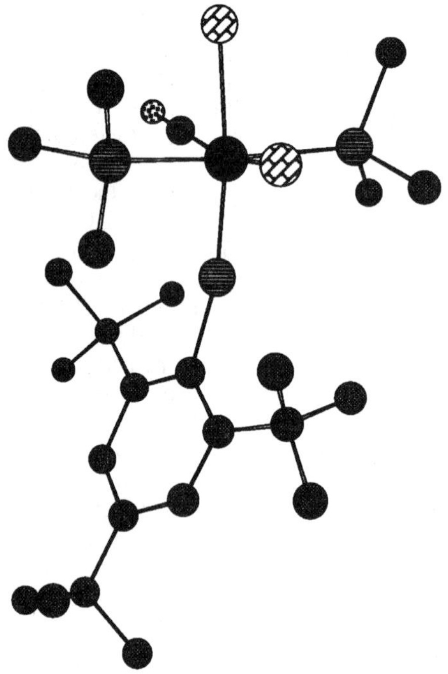

Figure 2 Structure of $[WCl_2(\equiv PAr)(CO)(PMePh_2)_2]$ ($Ar=2,4,6\text{-}C_6H_2Bu^t{}_3$). Only the *ipso* carbons of the $PMePh_2$ ligands are shown.

4.3.2.3 Reactions

(i) PY_3 species

Recent mechanistic results relating to group 6 PY_3 complexes are summarized in Section 4.1.1. Some new classes of recently developed reactions are worthy of independent mention. Electrochemical oxidation of $[Cr(CO)_3(\eta^1\text{-dppm})(dppm)]$ leads to the expected Cr^I species, but use of $AgClO_4$ leads instead to a species containing silver coordinated to the pendant diphosphines.[365] In comparison, oxidation of the equivalent tungsten complex generates $[W(CO)_3(dppm)_2]^{2+}$.[266] Functionalized phosphines often lead to unusual reactivity. For example, the metallophosphine complex $[Cr(CO)_5(PKPh_2)]$ activates THF, and in the presence of $AlBrR_2$ unusual bridged species result.[336] Additionally, interesting reactivity is shown by $[Mo(CO)_5(\mathbf{43})]$, the free NH may be reacted both with electrophiles and bases.[316,367] Interesting ligands are also synthesised in the reaction of $[Mo(CO)_4(PCl_2Ph)_2]$ with $Na_3[Mo(CO)_4(PPhO_2)_2]$.[368] In an impressive body of work Mathey and co-workers have demonstrated several classes of ligand centred reactions, which are summarized in Scheme 10.[329,353]

(ii) PY_2 and PY species

Recent representative reactions of phosphorin complexes are summarized in Scheme 11, reactions at phosphorus being somewhat more common than modifications of the heterocyclic ring due to the localized electron density at phosphorus.

The reactions of the related phosphaalkene chemistry have been reviewed.[353] These species quantitatively add activated X–H bonds across the P=C fragment and this reaction may be used to quantify particularly reactive complexes (Equation (34)). Electrocyclic additions (Equation (35)) are also well represented in addition to epoxidation (Equation (36)) and hydrogenation (Equation (37)) of the P=C bond. In an interesting reaction, the complex $[Cr(CO)_5\{(F_3C)P{=}CF(NMe_2)\}]$ decomposes slowly in solution to (**54**) (Equation (38)). The bridging mode results in lengthening of the Cr–P and P=C bonds to averages of 254.8 pm and 185.9 pm, respectively.[371]

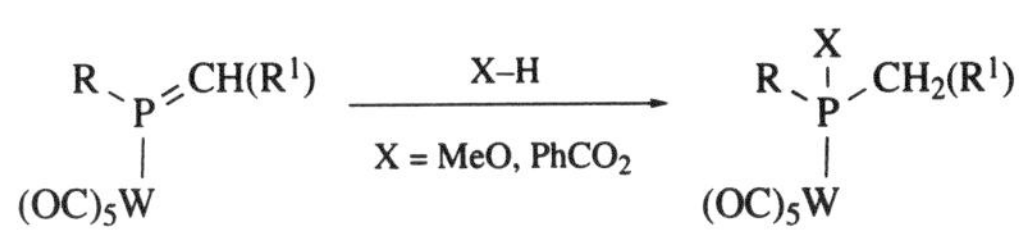

(34)

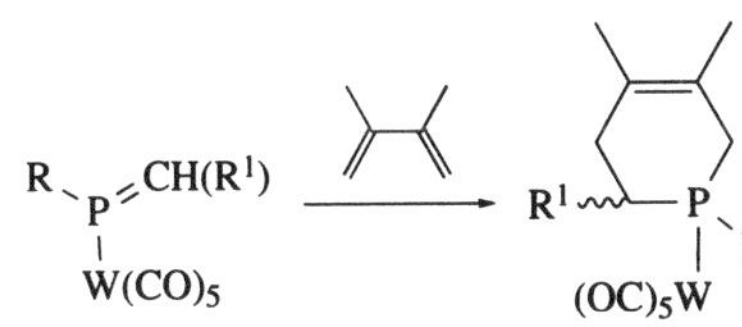

(35)

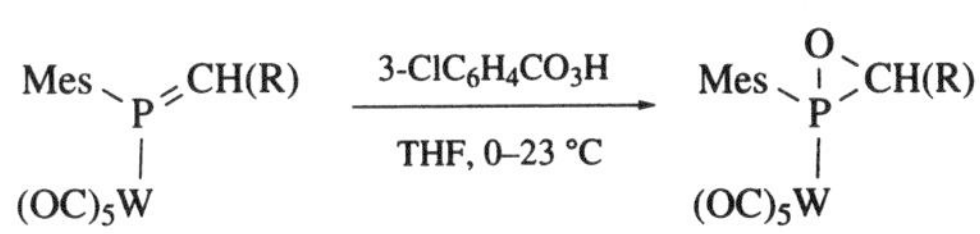

(36)

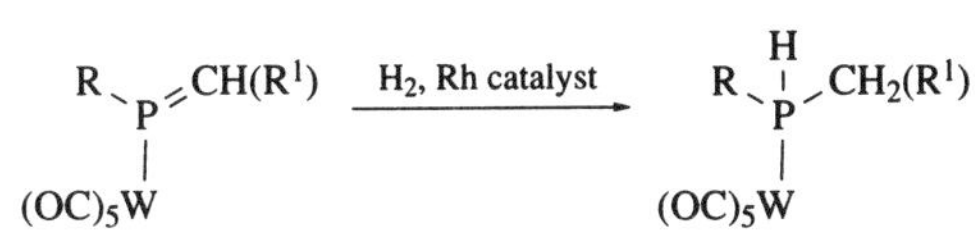

(37)

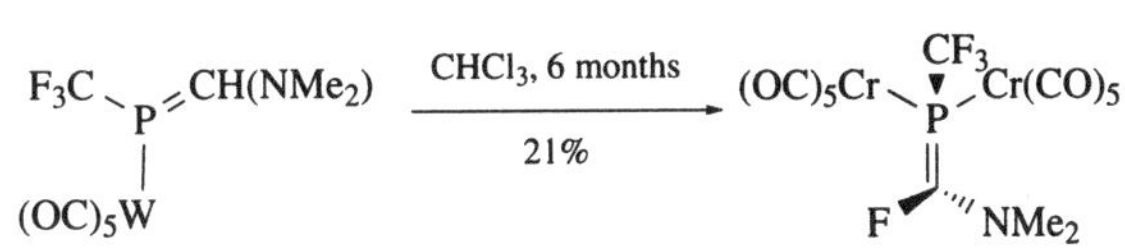

(38)

(**54**)

Transient $[W(\equiv PR)(CO)_5]$ undergoes numerous trapping reactions and these protocols are reviewed;[352] key and recent results[372–5] are summarized in Scheme 12.

4.3.3 Carbonyl Complexes with Polydentate Phosphorus Ligands

In addition to polyphosphine complexes bridging phosphinidene (PR) and phosphene (RP=PR), unusual phosphorus-rich ligands are included in this section. Although the phosphenes can act a monodentate ligands (η^1-RP=PR), these are included here for the sake of completeness. For similar reasons the arsenic and antimony analogues of phosphinidene and phosphene ligands are presented here rather than in Section 4.3.4

A number of useful general reviews are available including: comprehensive polyphosphine[376] and pyridylphosphine[377] ligand listings; a detailed overview of the literature to 1985 of R_2P–X–PR_2 complexes where X is an inorganic bridge such as O, NR, S, Se, Te, BR, SnR_2, O_2SiR_2 and other species;[378] reviews covering phosphinidenes, phosphenes and related species;[292,293] and a review listing a number of $[M^1(CO)_4(\mu\text{-}PR_2)M^2(L)_n]$, $[Cp_2M^2(C,P\text{-}\mu\text{-}CH_2PR_2)M^1(CO)_4]$, and $[M^1(CO)_4\{\mu\text{-}R_2P(CH_2)_nC_5H_4\}_2M^2X_2]$ complexes (M^1 = group 6 metal; M^{12} = nongroup 6 metal).[341]

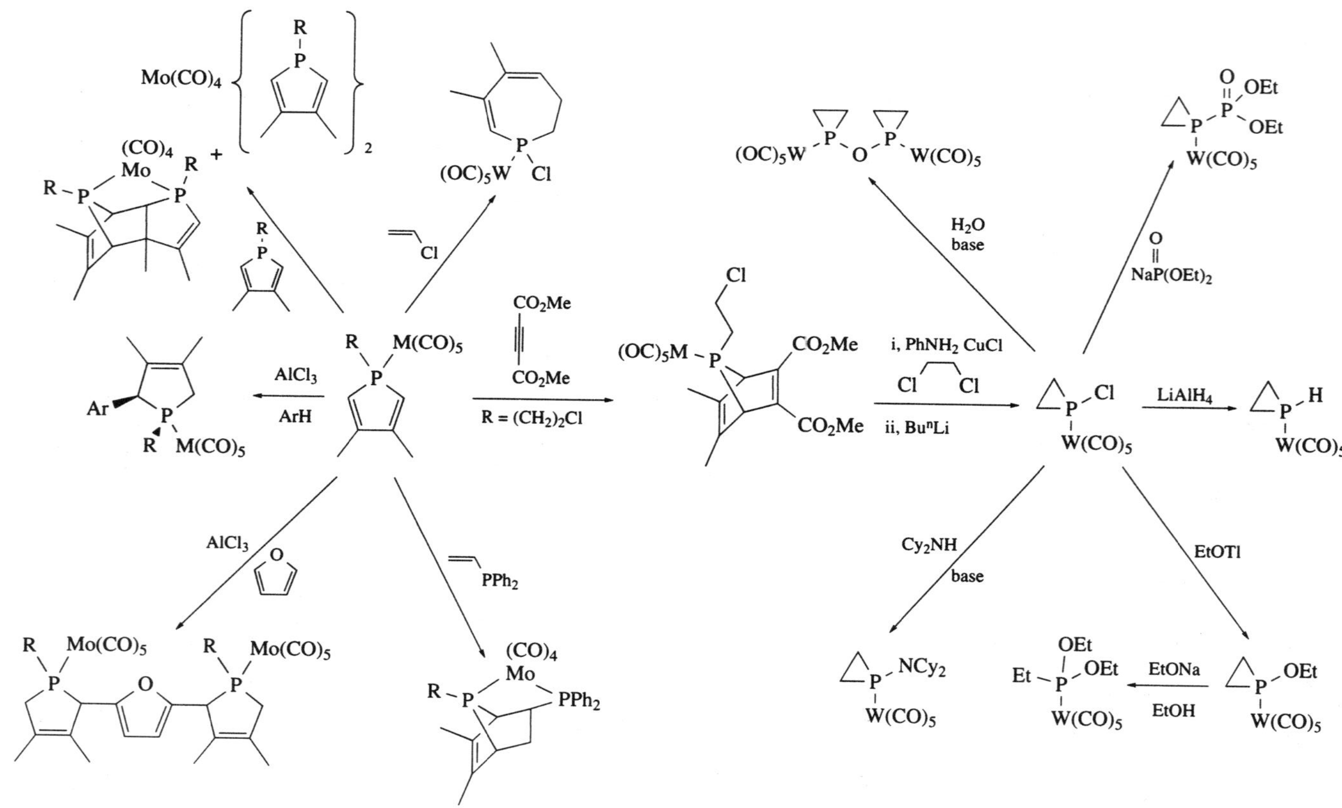

Mo(CO)4
R
P
2
(CO)4
Mo
R
+
R – P
P
(OC)5W
Cl
Cl
R
P
CO2Me
CO2Me
R = (CH2)2Cl
R
M(CO)5
AlCl3
ArH
Ar
R
M(CO)5
(OC)5M – P
CO2Me
CO2Me
i, PhNH2 CuCl
Cl
Cl
ii, BunLi
(OC)5W
P
O
P
W(CO)5
H2O
base
O
OEt
OEt
W(CO)5
O
NaP(OEt)2
Cl
W(CO)5
LiAlH4
H
W(CO)5
AlCl3
PPh2
R
Mo(CO)5
O
R
Mo(CO)5
(CO)4
Mo
R – P
PPh2
Cy2NH
base
NCy2
W(CO)5
EtOTl
OEt
Et – P
OEt
W(CO)5
EtONa
EtOH
OEt
W(CO)5

Scheme 10

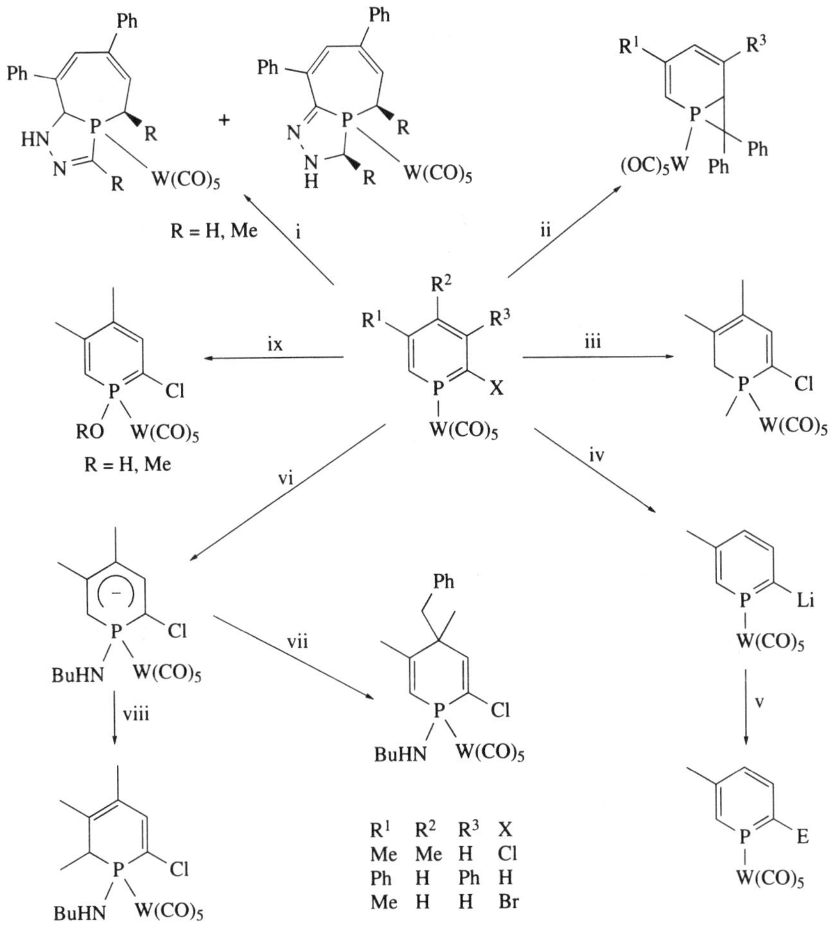

i, N_2CHR;[369] ii, N_2CPh_2;[369] iii, $LiNPr^i_2$ then MeI;[347] iv, X = Br, PhLi;[370]
v, I_2, $PClPh_2$, $ClCO_2Et$, or TMS-Cl;[370] vi, BuNHLi −78 °C;[347] vii, $PhCH_2Br$;[347] viii, MeI[347]

Scheme 11

4.3.3.1 Synthesis

(i) Multidentate phosphines and related species

Higher denticity ligands containing at least one phosphorus atom are the topic of this section. Known coordination modes are summarized in Structures (**55**)–(**69**), where X represents a generalized two-electron nonphosphorus donor and M a coordinatively saturated $[M(CO)_n]$ (n = 2–5) fragment. Where only a single phosphorus centre is present the ligand is counted as bidentate by virtue of bridging two metal centres as a phosphide (**57**) or being part of a heteroatom chelate as in (**64**), (**65**) or (**67**).

Representative examples of polydentate phosphine and related complexes appear in Table 13. The synthetic procedures closely parallel the preparation of monodentate complexes (Section 4.3.2.1) and are therefore not discussed in detail here. Briefly, typical preparations involve the following.

(a) Substitution reactions of $[M(CO)_6]$.
(b) Milder displacements from the labile precursors $[M(CO)_5(THF)]$, $[M(CO)_4(nbd)]$, $[M(CO)_4(C_5H_{10}NH)_2]$, $[M(CO)_3(NCMe)_3]$, $[M(CO)_3(chpt)]$ and related species.
(c) Halophosphine solvolysis within the metal coordination sphere.
(d) Modification of phosphites and phosphorus amines to halophosphines or halophosphine exchange.
(e) Alkylation of group 6 coordinated $PR_n(ONa)_{3-n}$ ligands with electrophiles leading to a new ligands with increased denticity.
(f) Interaction with metallophosphines.
(g) Modification reactions of P–H bonds via metallation or other reactions.

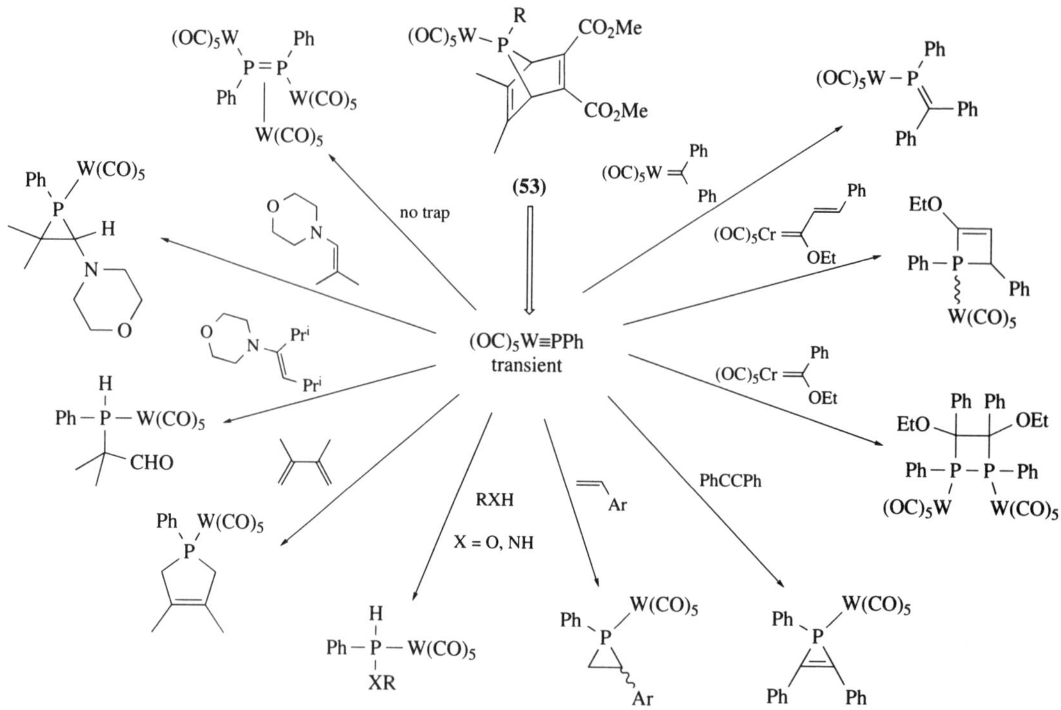

Scheme 12

(55)
$n = 1–3$

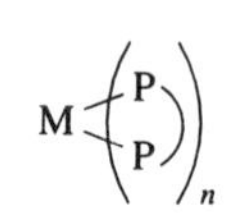

(56)
$n = 1–2$

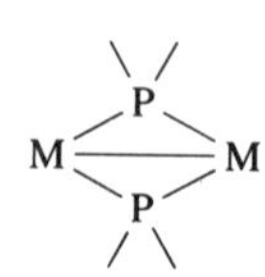

(57)

(58)

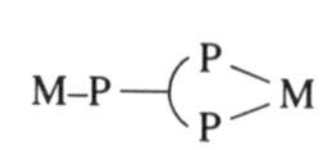

(59)

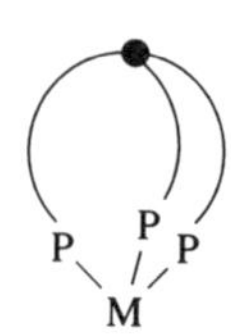

(60)

Some further points are of note. In addition to the more readily available diphosphines these procedures may be employed with unusual ligands, such as those based on metal complexes. Two recent examples are the use of the species $[W(S_2CNEt_2)_2(CO)(Ph_2PCCPPh_2)]$[379] and ligands based on dimolybdenum cores.[380] Diphosphines which are geometrically constrained against chelation normally result in the bridged structures (**55**), however, more unusual species can be isolated as in the preparation of $[W_3(CO)_{14}\{\mu_2\text{-}(Et_2N)_2PCCP(NEt_2)_2\}_2]$.[381]

Numerous nonsystematic routes to diphosphine and higher denticity systems exist. For example, electrocyclic rearrangement of the bis(phosphine) complex (**70**) leads to the Diels–Alder product (**71**) (Equation (39)). Note both valance-isomerization of the ligand and *cis* → *trans* isomerization of the complex are required.[419] Reaction of $[W(CO)_6]$ with $Ph_2PCH_2CH(R)CN$ (R = H, Me) in the presence of $NaBH_4$ leads to the phosphorus–nitrogen chelate (**72**) after prolonged reflux.[420] At shorter reaction times. the partial reduction product (**73**) can be isolated, the ethoxy group being derived from the solvent. Ligand modification is also a common feature of the preparation of phosphido complexes; base or thermal activation of P–H bonds are common routes to these compounds. A typical example is presented in Equation (40).[421] Ligand modification reactions prove to be of most use in nontraditional phosphorus macrocycle preparations. For example, the tris(phosphine) complex (**74**) cyclizes in the presence of the radical promoter AIBN to yield (**75**) as the major product (Equation (41)).[422] Some other examples of complex molecules that are derived form simple precursors are the products obtained from refluxing

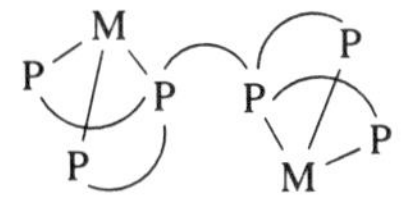
(61)

(62)

(63)

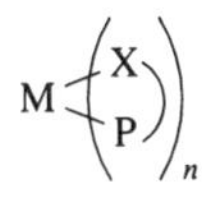
(64)
$n = 1–2$

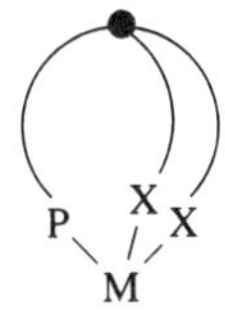
(65)

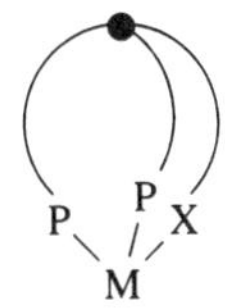
(66)

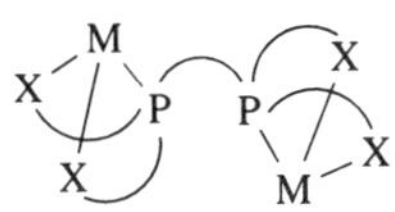
(67)

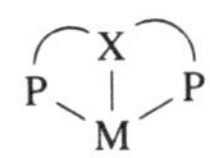
(68)

(69)

$(Pr^i_2N)_2P(O)OH$ and the hexacarbonyls together (Scheme 13).[423] The exact phosphorus–oxygen cage compound isolated depends both on the reaction conditions and metal.

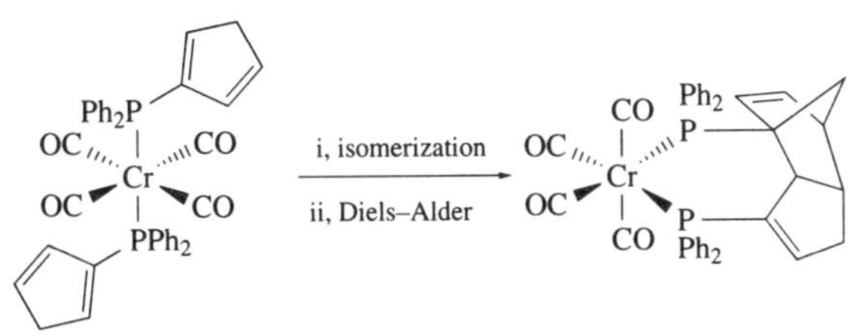

(70) **(71)** (39)

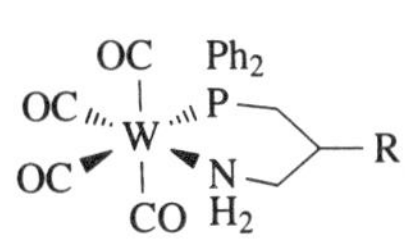
(72)

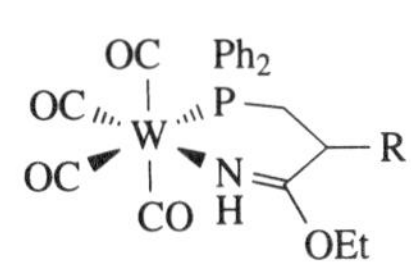
(73)

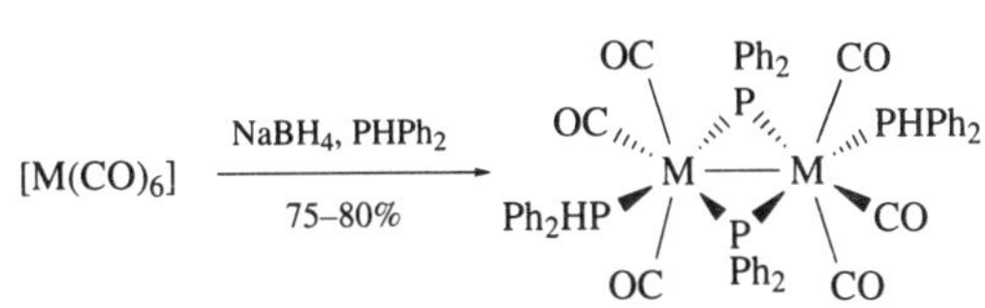

(40)

M = Mo, W

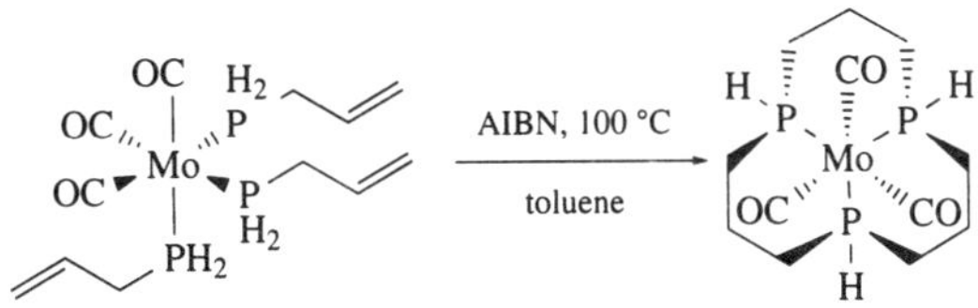

(74) **(75)** (41)

Table 13 Representative group 6 carbonyl complexes containing polydentate phosphorus ligands.

Ligand	*Metal*	*Synthesis*[a]	*Structure*	*Comments*	*M–P$_{av}$* (nm)	*Ref.*
Diphosphine donors						
dppf	Cr	(b)	**(55)** *n* = 1	also x ray on Mo analogue	0.2389	382
dppp	Mo	(b)	**(55)** *n* = 1	also dppe; dpmp gives η^1-species	0.2532	383, 384
$Ph_2PC\equiv C–C\equiv CPPh_2$	Mo, W	(b)	**(55)** *n* = 1	forms adducts with $[Co_2(CO)_8]$		385
(*E*)-$Ph_2PCH_2CH_2CH=CH$-$CH_2CH_2PPh_2$	Mo	(a)	**(55)** *n* = 2	ligand dehydrogenation	0.2469	386
(*E*)-$Ph_2CH=CHPPh_2$	Mo	(a)	**(55)** *n* = 2	also **(55)** *n* = 3 structure	0.2532	387
Me_2PPMe_2	Cr, Mo, W	(b)	**(55)** *n* = 2	prepared from pendant phosphine		388
$Ph_2PCH(Ph)CH(Ph)PPh_2$	Cr, Mo, W	(a)	**(56)** *n* = 1	*meso-* and *rac*-forms	0.2501	389
$Ph_2P(CH_2CH_2O)_3CH_2CH_2PPh_2$	Mo	(b)	**(56)** *n* = 1	crown-ether hybrid	0.2553	390
dppp	Mo	(b)	**(56)** *n* = 2	also dppm, dppe	0.2522	391
$Fe\{\eta-C_4HPMe_2(PPh_2)\}_2$	Mo	(a),(b)	**(56)** *n* = 1	phosphole-based		392
Other diphosphorus donors						
Ph_2POPPh_2	Cr, Mo, W	(g)	**(55)** *n* = 1	x ray on Cr; unusual reaction	0.2358	393
$Ph_2POP(Ph)(O)PPh_2$	Cr, Mo, W	(d)	**(56)** *n* = 1	x ray on Mo	0.2491	320
$PhP(OPPh_2)_2$	Mo	(c) +(d)	**(57)**			320
$MeN\{P(OR)_2\}_2$	Cr, Mo, W	(b)	**(58)**	R = Me, Pr^i, $-CH_2CH_2$-	b	394
$MeN(PF_2)_2$	Mo	(b)	**(55)** *n* = 3	bridging carbonyl	0.2371	395
$F_2PC\equiv CPF_2$	W	(b)	**(55)** *n* = 1	also other halophosphines		396
$PhN\{P(OPh)_2\}_2$	Mo, W	(b)	**(56)** *n* = 1	x ray on Mo	0.2434	397
$Pr^iN(PPh_2)_2$	Mo, W	(b)	**(56)** *n* = 1	x ray on W	0.2493	397
$Pr^iN\{PPh(NPr^i)_2\}_2$	Mo	(b)	**(56)** *n* = 1	*meso*-diastereoisomer	0.2498	332
$HP(PPh_2)_2$	Cr, W	(c)	**(56)** *n* = 1	x ray on Cr	0.2396	398
$R_2PN=S=NPR_2$	Cr, Mo	(c)	**(56)** *n* = 1	R = Me, Ph; fluxional		315
$Pr^i_2NP(N=S=N)_2PNPr^i_2$	Cr	(c)	**(55)** *n* = 1	also Bu^t derivative	0.2359	399
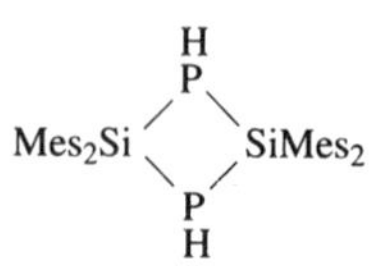	Cr, W	(b)	**(55)** *n* = 1			400
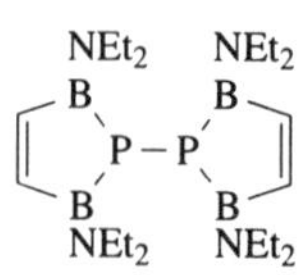	Cr	(b)	**(55)** *n* = 1		0.2520	401
Bidentate mixed donors						
$Ph_2PC_6H_4CH=N(CH_2)_2N=CH$-$C_6H_4PPh_2$	Cr	(b)	**(64)** *n* = 1	PN-bound; bridged	0.2372	402
$Me_2POC_5H_4N$	Mo	(b)	**(64)** *n* = 1	PN-bound; unusual reactivity		403
$(C_6H_4O_2)PNMe(CH_2CH_2NMe_2)$	Cr, Mo	(b)	**(64)** *n* = 1	PN-bound; x ray on Cr	0.2269	404
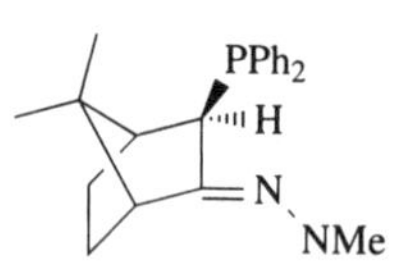	Cr, Mo	(a)	**(64)** *n* = 1	PN-bound; x ray on Mo	0.2514	405
$Ph_2PCH_2CMe(CH_2OMe)$-(CH_2SPh)	Cr, W	(a)	**(64)** *n* = 1	PS-bound; x rays on both	0.2385/0.2510	406
$Ph_2PC(S)NR_2$	Cr, Mo, W	(b)	**(64)** *n* = 1	PS-bound; R = H, Me		407
Polydentate phosphorus donors						
$MeSi\{P(OR)_2\}_3$	Mo	(b)	**(60)**	R = Me, Et; x ray on Et	0.2441	408
$[PCySiH(Mes)]_3$	Mo	(b)	**(62)**		0.2616	409
$(Et_2PCH_2CH_2)_2PCH_2$-$(CH_2CH_2PEt_2)_2$	Cr, Mo, W	(a)	**(61)**	Cr–P distance shown	0.2367	410

Table 13 (continued)

Ligand	*Metal*	*Synthesis*[a]	*Structure*	*Comments*	*M–P$_{av}$* (nm)	*Ref.*
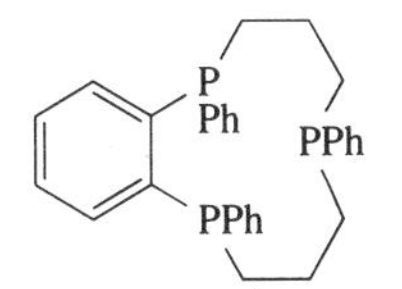	W	(a)	**(62)**		0.2469	411
P{N(Ph)PF$_2$}$_3$	Mo	(b)	**(60)**	heterobimetallic precursor		412
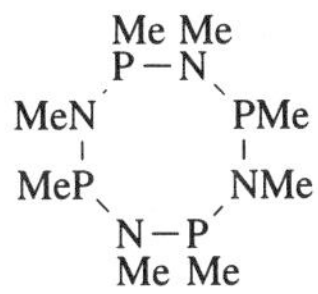	Mo	(a)	**(63)**	PPP-bound; uncoordinated phosphorus is reactive	0.2502	412
Polydentate mixed donors						
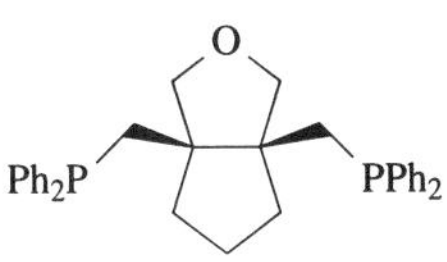	Mo, W	(b)	**(66)**	PPO-bound; x ray on W	0.2554	414
P{2,4,6-C$_6$H$_2$(OMe)$_3$}$_3$	Mo	(b)	**(65)**	POO-bound	0.2476	415
H$_2$C[P{2-C$_6$H$_4$(SMe)}$_2$]$_2$	W	(b)	**(67)**	PSS-bound	0.2420	416
Ph$_2$PCH$_2$CBu^t(NN)CBu^t-(CH$_2$PPh$_2$)	Cr, Mo,W	(b)	**(68)**	PNP-bound; x ray on Mo	0.2492	417
2,6-C$_6$H$_3$N(NHPPh$_2$)$_2$	Cr, Mo, W	(b)	**(68)**	PNP-bound; x ray on Mo	0.2389	418
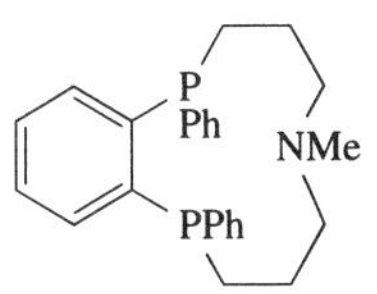	Mo	(a)	**(69)**	PPN-bound; x ray on Mo	0.2453	411

[a] For syntheic methods see text. [b] Structures of two compounds reported.

(ii) Phosphinidenes, phosphenes and related species

Reaction of $[M_2(CO)_{10}]^{2-}$ with $RECl_2$ (E = P, As, Sb) leads to structurally rewarding chemistry. These compounds are best dealt with as a complete set, hence coverage of the arsenic and antimony reactions are also considered here. The major product classes prepared from these reactions are shown in Scheme 14, with specific examples appearing in Table 14. Additionally, a number of alternative routes have appeared of which one of the most important is halophosphine reduction, the reaction of [Cr(CO)$_5${(–)-menthylPBr$_2$}] with magnesium being an early but important example.[424] Arsenes of type **(79)** are rather difficult to access from halophosphine reductions, whether coordinated or otherwise. In these cases an effective route is dehydrogenation of [(CO)$_5$M{μ-AsR(H)AsR(H)}M(CO)$_5$] species in the presence of catalytic amounts of platinum black.[425]

(iii) Complexes with phosphorus rich ligands

Representative preparations of complexes with ligands containing three or more P–P-bound phosphorus atoms appear in Table 15. Broadly speaking these products arise from ligand interaction with labile [M(CO)$_5$(THF)], [M(CO)$_4$(nbd)] or [M(CO)$_3$(chpt)]. The complex [{W(CO)$_4$}{W(CO)$_5$}$_4$(μ-η^1-η^4-P$_4$)], independently reported by two groups in 1991 remains the only example of an η^4-P$_4$ phosphorus unit bound to a group 6 metal centre. Fritz and co-workers have developed alternative routes to phosphorus-rich compounds via the reaction of phosphorus cage compounds with labile chromium carbonyl species. A wide range of products may be accessed. The behaviour of P$_7$R$_3$ clusters with [Cr(CO)$_5$(THF)] and [Cr(CO)$_4$(nbd)] is typical.[435]

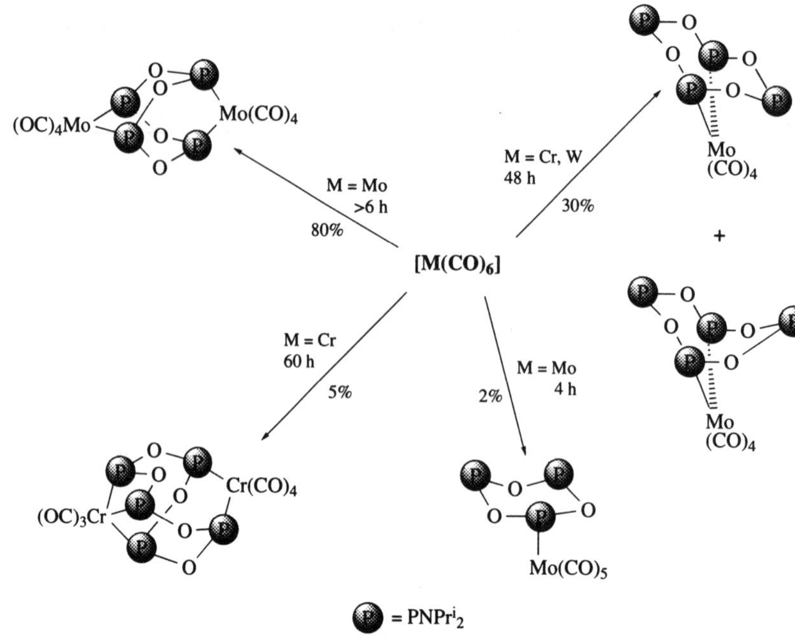

Scheme 13

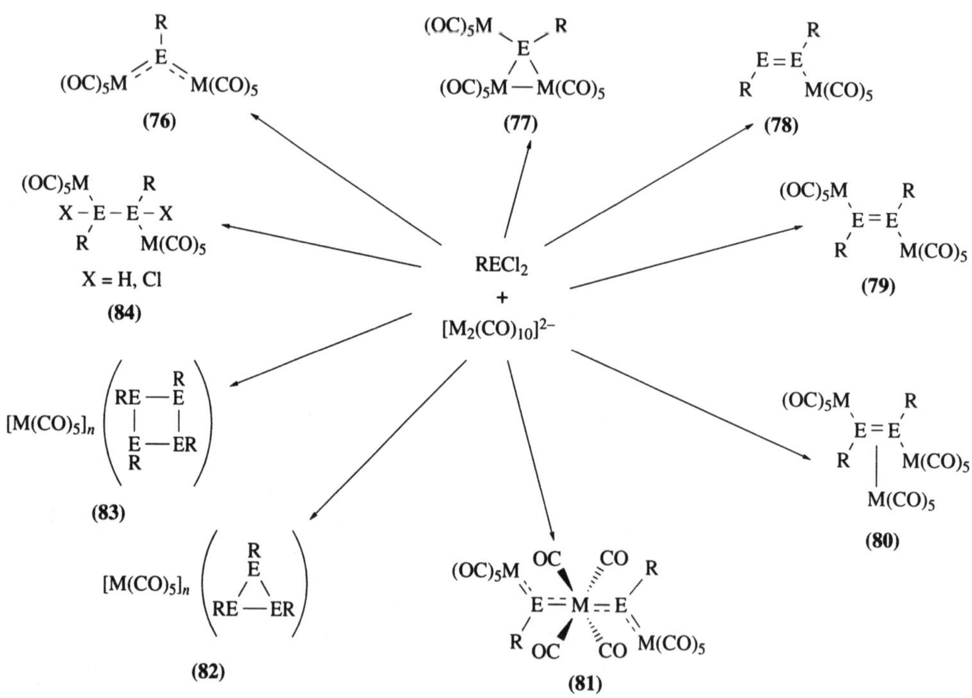

Scheme 14

4.3.3.2 Structure and properties

(i) Multidentate phosphines and related species

The subtle interplay of σ/π-bonding and steric effects in diphosphines is identical in many respects to that observed in monodentate ligand complexes if account is taken of the tendency of multidentate

Table 14 Representative products from the reaction of $RECl_2$ with $[M_2(CO)_{10}]^{2-}$ (E = P, As, Sb).[a]

M	*R*	*E*	*Products* (**76**)	(**77**)	(**78**)	(**79**)	(**80**)	*Other*	*Ref.*
Cr	Mes	P	●			●		(**81**)(**82**)	426,427
Cr	Bu^t	P	●						428
Cr	$(TMS)_2N$	P	●		○			(**81**)	426
Cr	$(TMS)_2CH$	P			●	○			426,427
Cr	$(TMS)CH_2$	P			●			(**83**)	426
Mo	Mes	P				●			427
Mo	$(TMS)_2CH$	P			●	●			427
W	pip[b]	P	●		●				429
W	Ph	P					●		430
Cr	$(TMS)_2CH$	As	○						431
Cr	Bu^t	As	●						432
Mo	$(TMS)_2CH$	As			●				431
W	Ph	As		●			●		431
Cr	Bu^t	Sb	○				●		433
W	Me	Sb		○					431
W	Bu^t	Sb		●			○		433
W[c]	$(TMS)_2CH$	Sb	●						434

[a] ○ = Prepared; ● = structurally characterized. [b] pip = 2,6-tetramethylpiperidyl. [c] Prepared from $Na_2[W(CO)_5]$.

Table 15 Preparations of group 6 complexes containing ligands with P–P bonds.[a]

Complex	*Synthesis*	*Core structure*	*M–P* (nm)	*Ref.*
$[Cr_2(CO)_{10}\{R_2PP(R)PR_2\}]$	$[Cr(CO)_5(THF)]$ + $R_2PP(R)PR_2$	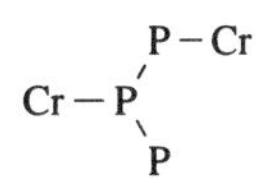		436
$[Cr(CO)_4\{\eta^2\text{-}[P(CO)_2FeCp^*]_3\}]$	$[Cr(CO)_4(nbd)]$ + $[\{FeP(CO)_2Cp^*\}_3]$	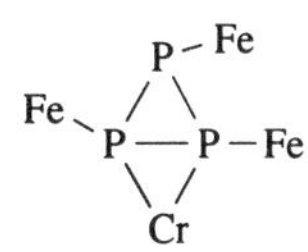	0.2445, 0.2476	437
$[Li(DME)_2][Cr_4(CO)_{20}(RP_4R)]$	$[Cr(CO)_5(PR_3)]$ i, BuLi; ii, $BrCH_2CH_2Br$	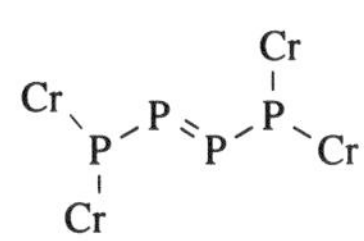	0.2219, 0.2511	438
$[W_5(CO)_{24}(P_4)]$	$[W(CO)_5(THF)]$ + P_4	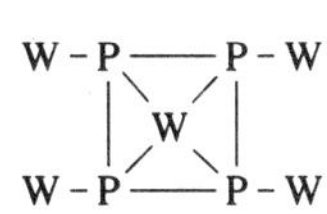	0.2604 apical 0.2445 basal	439, 440
$[Cr_3(CO)_{15}(P_7R_3)]$	$[Cr(CO)_5(THF)]$ + P_7R_3	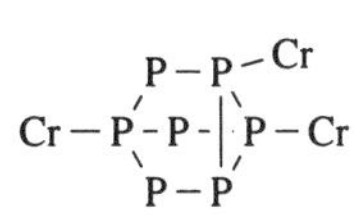		441
$[Cr(CO)_5\{P_7(PBu^t_2)_3\}]$	$[Cr(CO)_4(THF)]$ + $P_7(PBu^t_2)_3$	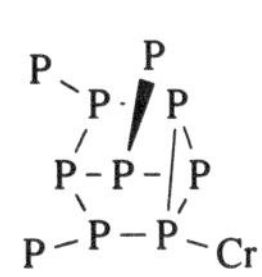	0.2392	442
$[Cr_2(CO)_6\{\mu\text{-}P_4(SiMe_2)_3\}_3]$	$[Cr(CO)_3(chpt)]$ + $P_4(SiMe_2)_3$	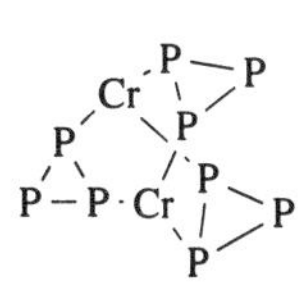	0.2365 0.2382 0.2354	443

[a] Throughout this table R = TMS; Cp* = η-C_5Me_5.

ligands to enforce *cis*-chelate geometries at the metal. Overall the outcome is reported in the bond length data for these species. For comparison the structural data for a number of *cis*-[M(CO)$_4$(P–Y)] (Y = P, N, S) complexes are given in Table 16. Carbonyl 'a' is defined as being *trans* to phosphorus, 'b' *trans* to the Y-donor, the two carbonyls *cis* to the P–M–Y plane being CO$_c$. The ligand bite is defined as the angle P–M–Y found in the crystal structure.

Table 16 Structural parameters for *cis*-[M(CO)$_4$(PY)] complexes (Y = P, N, S).

Ligand	*Metal*	*Chelate*	*M–P* (nm)	*M–Y* (nm)	*M–CO$_a$* (nm)	*M–CO$_b$* (nm)	*Av M–CO$_c$* (nm)	*Bite* (°)	*Ref.*
$Bu^t_2PN{=}S{=}NPBu^t_2$	Cr	PP	0.2478	0.24704	0.1843	0.1845	0.1883	92.0	444
$Ph_2PN{=}S{=}NPPh_2$	Cr	PP	0.2341	0.2349	0.1861	0.1853	0.1853	90.0	445
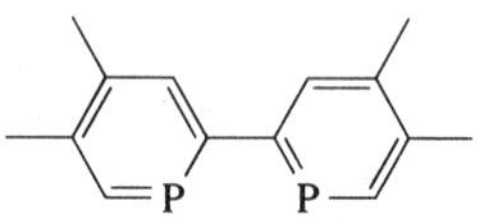	Cr	PP	0.22952	0.22952	0.1865	0.1865	0.1894	77.15	446
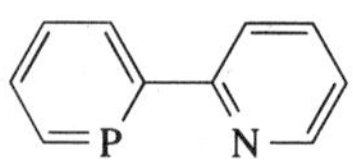	Cr	PN	0.2280	0.2139	0.1843	0.1825	0.1909	76.7	447
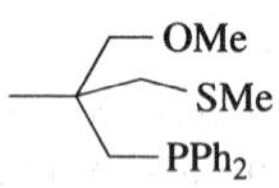	Cr	PS	0.2358	0.2420	0.1820	0.1806	0.1895	91.85	406
dppe	Mo	PP	0.2500	0.2495	0.1974	0.1999	0.2042	80.2	448
meso-$Ph_2PCH(Ph)CH(Ph)PPh_2$	Mo	PP	0.2486	0.2497	0.2002	0.1983	0.2034	79.8	389
rac-$Ph_2PCH(Ph)CH(Ph)PPh_2$	Mo	PP	0.2514	0.2509	0.2035	0.1986	0.2003	80.5	389
$ZrCl_2(\eta\text{-}C_5H_4PPh_2)$	Mo	PP	0.2547	0.2538	0.1982	0.1977	0.2031	94.0	449
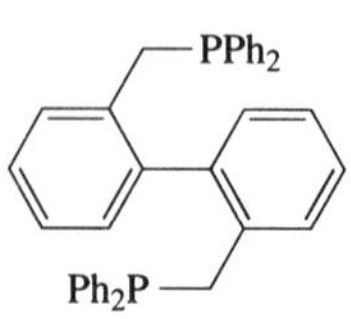	Mo	PP	0.2561	0.2570	0.1972	0.1992	0.2033	103.5	450
$Ph_2PCH_2CH_2NEt_2$	Mo	PN	0.2510	0.2436	0.2002	0.1916	0.2000	79.0	451
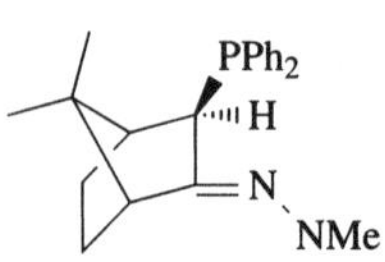	Mo	PN	0.2541	0.2331	0.1951	0.1924	0.2029	77.4	452
2,6-$C_5H_3N(NHPPh_2)\{NHP(O)Ph_2\}$	Mo	PN	0.2445	0.2352	0.1989	0.1903	0.1980	74.5	453
$Ph_2PCH_2CH_2SMe$	Mo	PS	0.2542	0.2560	0.1992	0.1967	0.2029	80.7	454
$Ph_2PCH_2CH_2S^+Me_2$	Mo	PS	0.2516	0.2425	0.2008	0.1988	0.2019	80.5	454
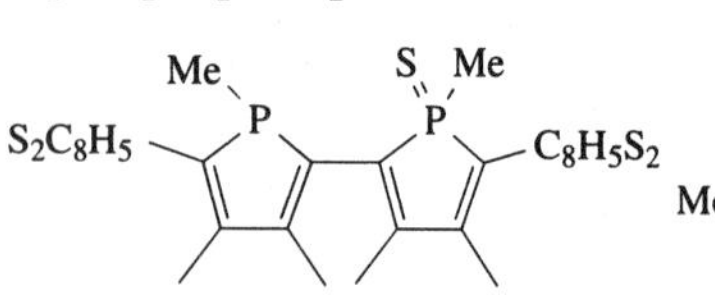	Mo	PS	0.2525	0.2532	0.1953	0.1909	0.2020	84.5	455
$(Ph_2P)_2C{=}CH(CH{=}CHPh)$	W	PP	0.2513	0.2509	0.2021	0.1980	0.2029	68.1	456
$Ph_2PN(Pr^i)PPh_2$	W	PP	0.2497	0.2489	0.1985	0.1994	0.2020	64.8	397
$Ph_2POC(Ph){=}CHPPh_2$	W	PP	0.2508	0.2470	0.1990	0.1997	0.2042	83.4	457
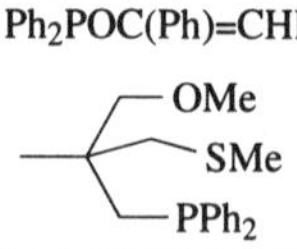	W	PS	0.2510	0.2546	0.1940	0.1880	0.1950	89.7	406

Although exceptions arise, a number of trends may be noted. In diphosphine ligands with similar bite angles increasing the steric bulk of the four substituents at phosphorus leads to longer M–P contacts in an identical manner of monodentate phosphorus species. Generally, increasing the length of the bridge (X) in $R_2P–X–PR_2$ species leads to larger bite angles. The four R substituents are pushed closer to the complex resulting in steric interaction and hence longer M–P bonds. Short M–CO_a and M–CO_b distances compared with M–CO_c reflect greater π-back donation, however, differences between the two *trans* carbonyl distances cannot be used to accurately assess the relative π-acceptor ability of P and Y.

The smaller atomic radius of chromium sometimes results in dramatic changes of coordination mode, especially for bulky ligands. For example, the triphosphine *cyclo*-[$P_3(SiHR)_3$] (R = Mes) reacts with [$Mo(CO)_3(NCMe)_3$] to yield the expected *fac*-[$Mo(CO)_3${*P,P,P*–$P_3(SiHR)_3$}], while reaction with [$Cr(CO)_3(NCMe)_3$] yields agostic [$Cr(CO)_3${*P,P,H*–$P_3(SiHR)_3$}]. Use of the less hindered ligand (R = Ph) leads to typical triphosphorus coordination for both metals as poor interligand clashes are alleviated.[409]

(ii) Phosphinidene, phosphene, and related species

The metal centres in the complexes (**76**) are widely separated (typically > 400 pm) with large M–E–M angles (typically ~130°) and M–E lengths ranging from 0.225–0.230 nm for Cr–P bonds to ~0.270 nm for W–Sb interactions. The heteroatom is planar, sp^2 hybridized, and behaves as a Lewis acid readily coordinating two electron donors (Scheme 15). The three-centre-4π-bonding in the complexes (**76**) is reminiscent of other allylic species. In these phosphinidene species the small HOMO–LUMO separation leads to paramagnetic shifted ^{31}P NMR signals at extremely high frequencies, normally around +1000 ppm. This bonding results in highly coloured species. Complexes of type (**76**) are formal valence tautomers of the Lewis base complex (**85**) (Scheme 15). In general such complexes are not isolable as the species (**85**) are easily trapped by adventitious [$M(CO)_5$] fragments resulting in the complexes (**77**) (Scheme 15). Alternatively, (**76**) and (**85**) may combine to the phosphene-type complex (**80**). The presence of a metal–metal bond μ_3-phosphinidene species (**77**) is witnessed by a shortening of the M–M vector and a greatly reduced M–E–M angle (typically ~70°).

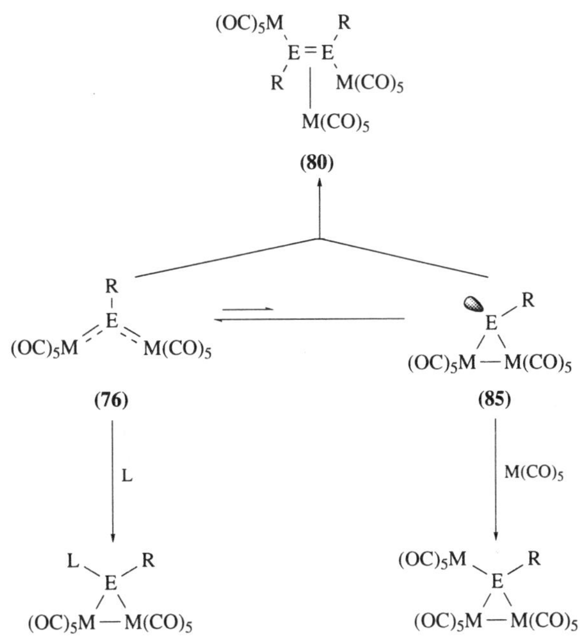

Scheme 15

Phosphene complexes show P–P distances of 0.200–0.205 nm when one or two terminal [$M(CO)_5$] fragments are coordinated to the phosphene, however, binding of a third pentacarbonyl unit dramatically increases the interelement distance indicative of P_2-antibonding orbital population by back donation from the metal in complexes of type (**80**). Such effects are entirely analogous to Chatt–Dewar bonding in metal–alkene complexes. Phosphene complexes are normally isolated with (*E*) configurations, [$Mo_2(CO)_{10}$(μ-MesP=PMes)] being a notable exception. Photochemically promoted (*E*) → (*Z*) interconversions are known.[458]

(iii) Complexes with phosphorus rich ligands

The core structures of representative examples may be found in Table 15. Bonding in [{W(CO)$_4$}{W(CO)$_5$}$_4$(μ-η^1-η^4-P$_4$)] deserves special comment. The P$_4$ ligand is a net 12-electron donor whose ^{31}P NMR signal resonates close to that of elemental phosphorus. Fenske–Hall studies of hypothetical [W(CO)$_4$(η^4-P$_4$)] show a HOMO comprised of slightly antibonding combinations of 3*p*-phosphorus orbitals with a doubly degenerate LUMO 6.1 eV higher showing W(5*d*π)–COπ* bonding character. The P$_4$ ligand is best rationalized as a planar antiaromatic unit..

4.3.3.3 Reactions

(i) Multidentate phosphines and related species

The chemistry of group 6 coordinated polydentate phosphines has changed little since the early 1980s; only a few recent highlights are added here. Photolysis of hexadentate [Cr$_2$(CO)$_6$(**61**)] in halogenated solvents leads to facile formation of [Cr$_2$Cl$_4$(**61'**)] where modification of the ligand framework takes place. The analogous molybdenum and tungsten react more slowly to give the carbonyl halides [M$_2$Cl$_4$(CO)$_4$(**61**)]. The different reaction pathways are indicative of the inaccessibility of seven coordinate chromium(II) species.[459] Ligand modification is a recurrent theme in many papers that have appeared since 1981; a few selected examples are included here. Reaction of the phosphite [Mo(CO)$_4$(Me$_2$POpy)] with allyl alkoxides enables the preparation of phosphino–alkene chelates [Mo(CO)$_4$(Me$_2$POCH$_2$CH=CH$_2$)].[403] The alkyne complex [W$_2$(CO)$_{10}$(μ-Ph$_2$PC≡C–C≡CPPh$_2$)] reacts with [Co$_2$(CO)$_8$] to yield an unusual adduct via coordination of one triple bond to a 'Co$_2$(CO)$_6$' fragment.[385] One of the most successful applications of ligand modification is Shaw's systematic elaboration of *cis*-[M(CO)$_4$(dppm)] to a host of species (Scheme 16).[457,460,461]

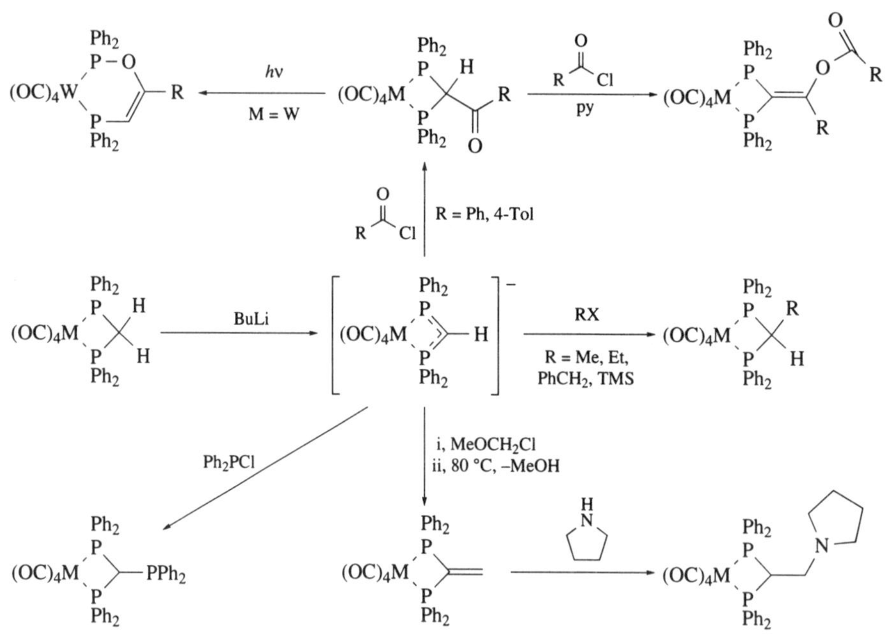

Scheme 16

(ii) Phosphinidene, phosphene and related species

Phosphinidene complexes are good cluster precursors. While gentle heating of (**76**) (M = Cr; E = P) leads to a phosphene complex of type (**80**), thermolysis at higher temperatures leads to the cluster [Cr$_3$(CO)$_{10}$(PBu^t)(Bu^{t}P=PBu^t)].[462] Alternatively, the halophosphinidene complexes (**76**) (M = Cr; E = P; R = Br) react with Na[W(CO)$_3$Cp] to yield trimetallic complexes [Cr$_2$W(μ_3-P)(CO)$_{12}$Cp].[463] Rather wider reactivity is shown by phosphene complexes (and their arsenic or antimony homologues). These undergo numerous addition reactions of the E=E bond; the behaviour of $[Cr_2(CO)_{10}(PhP{=}PPh)]$ is exemplary (Scheme 17).[464]

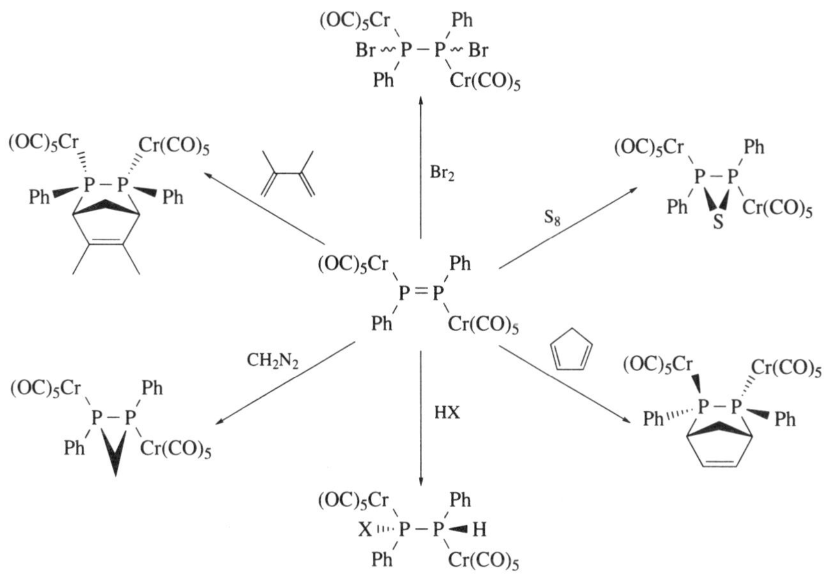

Scheme 17

(iii) Complexes with phosphorus rich ligands

The behaviour of these complexes is dominated by the dissociation of $[M(CO)_5]$ fragments.

4.3.4 Complexes with Arsenic, Antimony and Bismuth Ligands

This section describes the synthesis, properties and reactions of ER_3 (E = As, Sb, or Bi), their polydentate equivalents, and more exotic arsenic, antimony and bismuth species. Arsinidenes, stibinidenes, arsenes and stibenes are considered with their phosphorus relatives (see Section 4.3.3). Some reviews of relevance include complexes with unusual E ligands[293] and some material of relevance in a review of complexes with antimony and bismuth ligands.[465]

4.3.4.1 Synthesis

The preparation of simple tertiary arsine, stibine and bismuthine complexes and their polydentate relatives is changed little since 1981 and requires no further discussion. For the sake of completeness some selected recent examples of trisubstituted arsenic and antimony ligated complexes appear in Table 17.

In addition to arsinidenes, arsenes and their antimony analogues the reactions of $Na_2[M_2(CO)_{10}]$ with halo group 15 compounds can lead to more unusual, often unpredicted, products. For example, although $Na_2[Cr_2(CO)_{10}]$ and Bu^t_2AsCl combine to give an arsinidene complex, the reaction pathway is dramatically altered in the presence of [PPN]Cl leading to the chelate compound (**86**) in low yield (Equation (42)).[432] A number of other arsenic, antimony and bismuth species are also accessible by related chemistry. Reaction of $Na_2[W_2(CO)_{10}]$ with ECl_3 gives star-shaped (**87**)–(**89**) (Equation (43)).[472–4] These molecules formally contain the unknown six electron donors E≡E. The bismuth compound (**89**) is also isolated (5%) from the reaction of $Na_2[W(CO)_5]$ with $(TMS)_2CHBiCl_2$ and in this case small amounts (12%) of (**90**) are also isolated.[475] Such unpredictability is a common feature of bismuth chemistry where relatively weak Bi–C bonds often rupture under the reaction conditions. Another interesting bismuth compound has been synthesized by simple reflux of $[Mo(CO)_6]$ with sodium bismuthate resulting in the cluster complex shown in Equation (44).[199] Perhaps the most remarkable example of ligand architecture built solely of group 15 elements is the isolation of (**91**) from the reaction

Table 17 Preparations of arsine and stibine carbonyl complexes.

Complex	*Synthesis*	*Comments*	*Ref.*
cis/trans-$[Cr(CO)_4(PR_3)(AsPh_3)]$	$[NEt_4][CrCl(CO)_4(PR_3)] + AsPh_3$	R = Bu, OMe, OPh; *trans* major	23
$[Cr(CO)_5\{AsPh_2(NSO)\}]$	$[Cr(CO)_5(THF)] + Ph_2As(NSO)$	x ray	466
$[Cr(CO)_5\{AsR_2(NSO)_2\}]$	$[Cr(CO)_5(THF)] + RAs(NSO)_2$	R = Ph, Bu^t	466
$[Cr(CO)_5\{AsHPh(CH_2CMe{=}CMe_2)\}]$	$[Cr(CO)_5(AsH_2Ph)]$ + 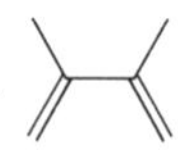		467
$[M(CO)_5\{AsMe_n(NMe_2)_{3-n}\}]$	$[NEt_4][MI(CO)_5] + AsMe_n(NMe_2)_{3-n}$	M = Cr, Mo, W; n = 1–3	468
$[M(CO)_5\{As(XMe_3)_3\}]$	$[M(CO)_6] + As(XMe_3)_3$	M = Cr, Mo, W; X = C, Sn	469
$[M(CO)_4\{Me_2AsGeMe_2(CH_2)_2AsMe_2\}]$	$[M(CO)_4(nbd)] + Me_2AsGeMe_2(CH_2)_2AsMe_2$	M = Cr, Mo, W	470
$[M_2(CO)_{10}(\mu\text{-}Me_2AsAsMe_2)]$	from $[M(CO)_5(\eta^1\text{-}AsMe_2AsMe_2)]$	M = Cr, Mo, W	388
fac-$[Cr(CO)_3(\eta^1\text{-L})(\eta^2\text{-L})]$	$[Cr(CO)_3(chpt)] + Ph_2AsCH_2CH_2AsPh_2$	L = $Ph_2AsCH_2CH_2AsPh_2$	365
fac-$[Mo(CO)_3\{SbPh_2(OR)\}]$	$[Mo(CO)_3(chpt)] + SbPh_2(OR)$	R = Et, Ph	471
fac-$[Mo(CO)_3\{SbPh_2(SR)\}]$	$[Mo(CO)_3(chpt)] + SbPh_2(SR)$	R = Ar, Et, Pr^i; x ray	471

of Rb_3As_7 with $[Cr(CO)_3(\eta\text{-}C_6H_3Me_6)]$ in the presence of cryptands (Equation (45)).[476] The crystallographically characterized anion shows the As_7 fragment to reside as a formal inorganic equivalent of norbornadiene.

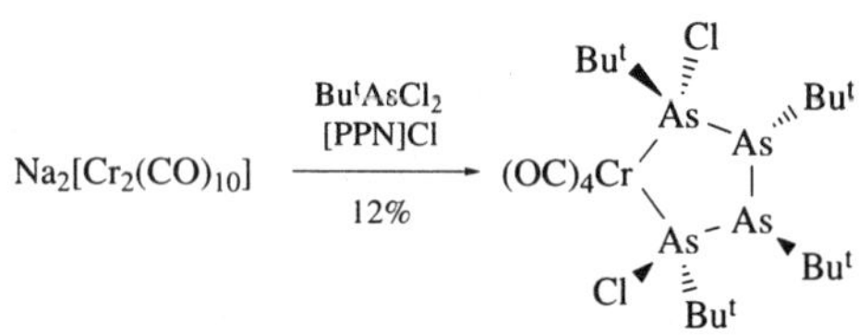

(86)

(42)

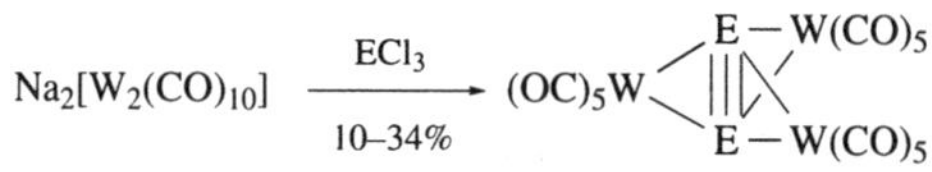

(43)

(87) E = As
(88) E = Sb
(89) E = Bi

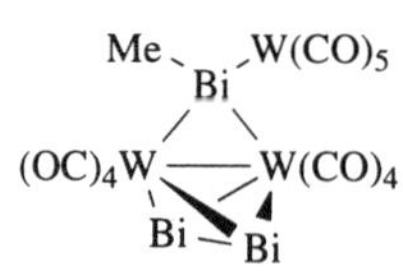

(90)

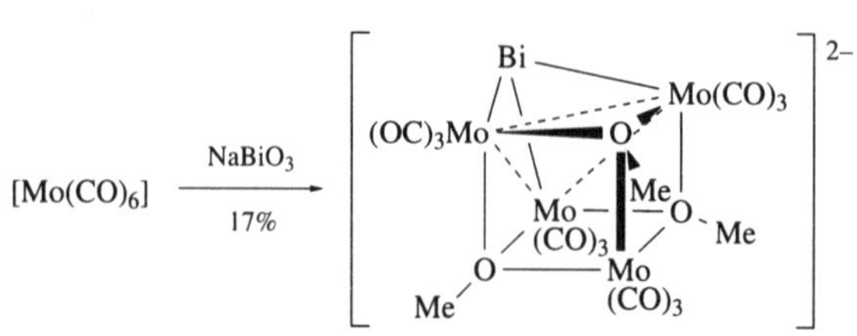

(44)

In a series of papers Rheingold has described the synthesis of molybdenum carbonyl complexes of cycloarsine ligands; this chemistry is reviewed.[477] The major species prepared in these high-temperature thermolysis reactions are summarized in Scheme 18.[478–80] Some related monodentate $[Cr(CO)_5\{cyclo\text{-}(AsPh)_n\}]$ (n = 4, 5) species have also been structurally characterized.[425]

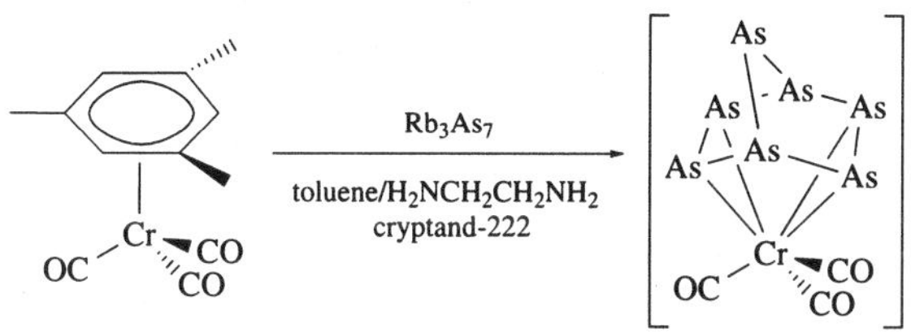

(45)

(91)

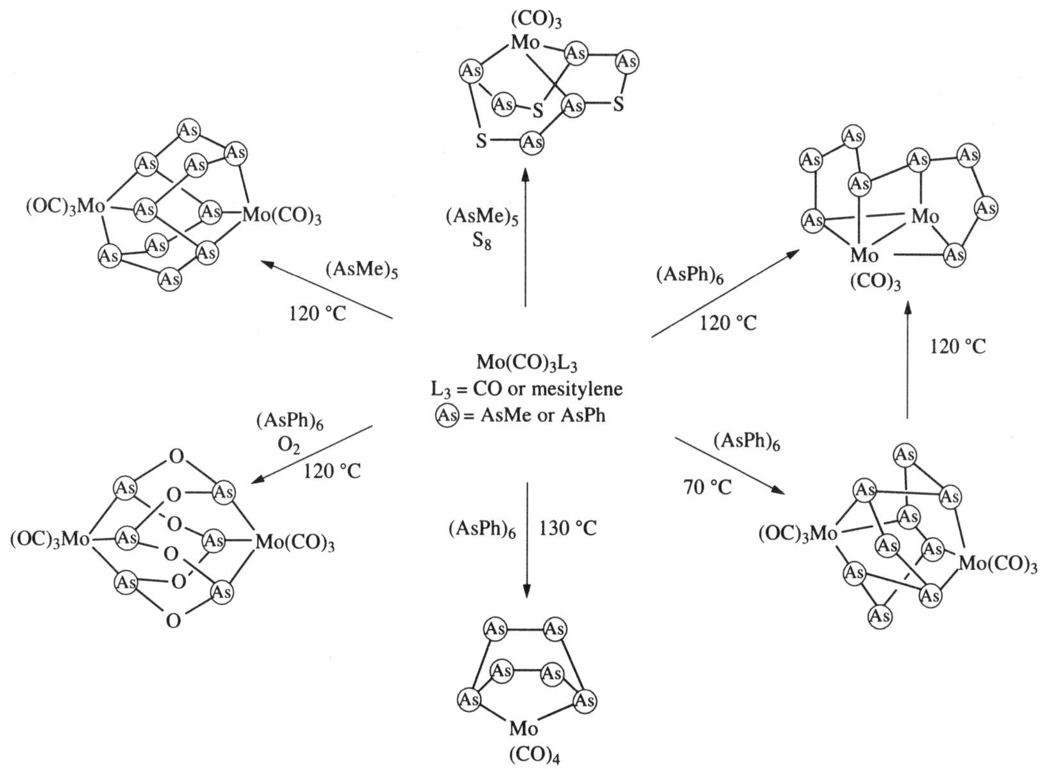

Scheme 18

4.3.4.2 Structure and properties

Representative examples of structurally characterized arsenic, antimony and bismuth complexes may be found in Table 17. An unusual coordination mode for $Sb(SBu^t)_3$ is found in the complex $[W(CO)_5(\mu\text{-}SBu^t)(SbSBu^t)_2]$.[481] The interelement distances in the complexes (**87**)–(**89**) are longer that would be expected for a triply bonded E≡E fragment but shorter than anticipated for just a single bond. The bonding in these complexes is best regarded as involving strong π-acceptor interaction of the E_2 unit with the metal centres, an idea supported by molecular orbital calculations.

The structure of the trianion (**91**) shows no contacts with the rubidium cryptand cations. Features of interest include the two-coordinate bridgehead arsenic atom and the Cr–As and Cr–C bond lengths. While the chromium–arsenic interactions are extended (av. Cr–As 0.267 nm) the carbonyls are tightly bound (av. Cr–C 0.174 nm) due to the extensive backbonding induced by the formal three minus charge. These effects are also reported in the IR spectrum of (**91**) where three low-frequency carbonyl stretches are present.

4.3.4.3 Reactions

Reactions of ER_3 complexes and polydentate ligands are reminiscent of their equivalent phosphines. Although ER_3 species are generally more labile than their phosphine analogues such ligands are capable of stabilizing higher oxidation states. For example, the iodide $[MI_2(CO)_3(PPh_2Cy)(SbPh_3)]$ (M = Mo, W) decomposes via phosphine loss to $[HPPh_2Cy][MI_3(CO)_3(SbPh_3)]$ rather than expulsion of the stibine.[482] The carbonyl halide $[WBr_2(CO)\{P(OMe)_3\}_2\{Me_2AsC(CF_3)=C(CF_3)AsMe_2\}]$ is readily prepared by bromine addition to the precursor carbonyl,[35] while addition of I_2 to (**87**) leads to $[W(CO)_4(\mu\text{-}I)(\mu\text{-}As_2)WI(CO)_3]$.[483]

4.3.5 Carbonyl Complexes with Sulfur, Selenium and Tellurium Ligands

Throughout this section, unless otherwise indicated, E refers to any member of the sulfur, selenium or tellurium triad. To avoid repetition, where ambiguity is not a problem, only the parental thioligand is referred to in the cursive text even if seleno and telluro examples are known. It is convenient to consider this chemistry in three areas: complexes with neutral, anionic and elemental ligands, respectively. Neutral ligands are taken to include E_nR_2 (n = 1, 2), thioketone and related $E=CR_2$ species, $E=PR_3$, carbon disulfide and its neutral adducts, and SO_2 complexes. Thiolates, thiocarboxylates, xanthates, thiocarbamates and their seleno and telluro analogues are typical anionic ligands. Naked sulfur, selenium and tellurium ligands are considered separately, as although these can all be thought of as sulfido (S^{2-}) analogues a number of these species are best regarded as containing cationic $(E_n)^{m+}$ ligands.

The chemistry of sulfur, selenium and tellurium ligands with group 6 metals has seen great growth since 1981. Is has therefore been necessary to be rather selective with the coverage here. In partial recompense for this situation a range of reviews are available to support this section including overviews of the coordination of sulfur,[484] selenium,[484,485] and tellurium.[465,485] More specific reviews are available on seleno and telluro ether ligands[486] and a summary of early results in SO_2 coordination chemistry.[487]

4.3.5.1 Synthesis

(i) Complexes with neutral ligands

Representative examples of thioether, disulfide, and their equivalent selenium and tellurium complexes can be found in Table 18, which also contains selected complexes with neutral $E=CR_2$ ligands along with complexes of CS_2, R_3PCS_2, and SO_2. Only a few CS_2 complexes are included as most of these feature metal–carbon bonds and are therefore outside the scope of this chapter. Some ligands of relevance to these species are shown in Structures (**92**)–(**101**). Complexes of $E=PR_3$ are not considered in depth here as two useful review articles are available.[78,179] It is worth mentioning in passing that although tellurium examples are rare the complexes $[M(CO)_5(Te=PBu^t_3)]$ (M = Cr, Mo, W) are easily synthesized via photolysis reactions.[488]

The usual general routes to neutral group 16 ligand carbonyl complexes may be employed:

(a) Heat or light induced substitution from $[M(CO)_6]$ by typical dissociative mechanisms. A more interesting example is the preparation of $[Cr(CO)_5\{Te(4\text{-}C_5H_4OMe)_2\}]$ via attack of the telluro oxide on coordinated CO (Equation (46)).
(b) Displacement from labile precursors including $[M(CO)_5(THF)]$, $[M(CO)_4(C_5H_{10}NH)_2]$, $[M(CO)_3(NCMe)_3]$, and related molecules.
(c) The combination of $[NEt_4][MBr(CO)_4(L)]$ (L = CO, PR_3), the sulfur ligand, and an electrophilic promoter, typically Et_3OBF_4, is particularly effective for the preparation of pentacarbonyl species.
(d) Modification of carbene complexes. This strategy is commonly employed in the synthesis of thioketone and related carbonyls; the preparations of (**102**) and (**103**) are exemplary (Equations (47) and (48)).
(e) Additions to coordinated $E=CR_2$ fragments lead to seleno and telluro ethers, for example Equation (49), but sulfur analogues rarely participate in these reactions.
(f) Carbonylation approaches, although uncommon, have also been employed. For high oxidation state precursors a reducing agent is often required as in the preparation of (**104**) (Equation (50)).

Table 18 Selected group 6 carbonyl complexes of neutral sulfur, selenium and tellurium ligands.

Complex	*Synthesis*[a]	*Comments*	$M–E_{av.}$ (nm)	*Ref.*
Thioether and related donors				
$[Cr(CO)_5(Bu^tSH)]$	(g)	also RSH; R = Et, Pr^i, Ph	0.2439	489
$[Cr(CO)_5(Bu^tEEBu^t)]$	(c)	also Mo, W; E = S, Se		490
$[Cr_2(CO)_8(\mu\text{-}SEt_2)_2]$	(b)	four other complexes	0.2422	491
$[Cr(CO)_5(\textit{Se}\text{–}PhSeSPh)]$	(b)	from carbyne, low yield	0.2531	492
$[Cr(CO)_5(SeEt_2)]$	(g)			493
$[Cr(CO)_4(SeEt_2)_2]$	(b)			491
$[Cr(CO)_5\{Te(4\text{-}C_6H_4OMe)_2\}]$	(a)	unusual preparation	0.2684	14
$[Cr_2(CO)_{10}(\mu\text{-}TeMe_2)]$	(b)	also S, Se compounds		494
$[Cr(CO)_4(Bu^tSCH{=}CHSBu^t)]$	(a)		0.2407	8
$[Cr(CO)_4(\mathbf{92})]$	(b)	also Mo; *S*-bound	0.2339	495
$[Cr(CO)_4\{\eta^2\text{-}C_6(SMe)_6\}]$	(a), (b)	also Mo, W	0.2378	496
$[Mo_2(CO)_8(\mu\text{-}TePh_2)_2]$	(a)	low yield	0.2744	497
$[Mo(CO)_4(\mathbf{93})]$	(b)	S-bound	0.2556	498
fac-$[Mo(CO)_3(\mathbf{94})]$	(b)	S-bound	0.2561	499
fac-$[Mo(CO)_3(\mathbf{95})]$	(b)	S-bound; *n* = 2	0.2536	500
fac-$[Mo(CO)_3(\mathbf{95})]$	(b)	S-bound; *n* = 1	0.2520	501
trans-$[Mo(CO)_2(\mathbf{96})]$	(f)	S-bound	0.2436	502
$[W(CO)_5\{MeS(CH_2S)_nMe\}]$	(c)	*n* = 2,3; also Cr		503
$[W(CO)_5(SCH_2CHPh)]$	(e)	from S=CHPh + N_2CH_2		504
$[W(CO)_5(PhSXMe_3)]$	(b)	X = C, Si, Ge, Sn		505
$[W(CO)_3\{SCH_2CH_2CH{=}CH)L_2]$	(b)	L_2 = dppe; also dmpe	0.2573	506
$[W(CO)_5(Bu^tSCH_2SBu^t)]$	(b)		0.2571	507
$[W(CO)_5(\mathbf{97})]$	(b)	*S*-bound	0.2533	508
$[W(CO)_5(\mathbf{98})]$	(b)	also Cr; S-bound	0.2525	509
$[W(CO)_5\{CH_2(SeMe)_2\}]$	(c)	also Cr		510
$[W(CO)_4(PPh_3)(\mathbf{99})]$	(b)	Se-bound	0.2681	511
$[W(CO)_5(\mathbf{100})]$	(e)	Se-bound	0.2674	512
$[W(CO)_5(\mathbf{100})]$	(e)	Te-bound		513
$[W(CO)_4\{(MeSeC_5H_4)_2Fe\}]$	(b)	also Cr, Mo	0.2683	514
Thioketone and related donors				
$[Cr(CO)_5\{S{=}C(NHR_2)_2\}]$	(b)	R = H, 4-Tol, Bu^t; also Mo		515,516
$[Cr(CO)_5\{S{=}C(NMe)(CH_2)_2NMe\}]$	(b)		0.2510	517
$[Cr(CO)_5(S{=}NNMe_2)]$	(b)		0.2386	518
$[Cr(CO)_5\{S{=}CR(Fc)\}]$	(c)	also Mo, W	0.2412	519
$[Cr(CO)_5(S{=}COCMe_2CMe_2O)]$	(a)	also Mo, W	0.2438	520
$[Cr(CO)_5\{S{=}CH(4\text{-}C_6H_4OMe)\}$	(d)	also W		521
$[Cr(CO)_5(Se{=}CPh_2)]$	(d)	also W		522
$[Cr(CO)_5\{S{=}CPh(Fc)\}]$	(b)	also Mo, W		523
$[W(CO)_5\{S{=}C(NMe)(CH_2)_2NMe\}]$	(b)		0.2617	517
$[W(CO)_5\{S{=}CHC(Ph){=}CPh(OEt)\}]$	(d)	by-product formed	0.2508	524
$[W(CO)_5\{S{=}C(NH_2)NHN{=}CMe_2\}]$	(a)		0.2568	525
$[W(CO)_5\{S{=}C(SR)C(PPh_3)CN\}]$	(b)	R = Me, Et, Pr^i, Cy		526
$[W(CO)_5(\mathbf{101})]$	(d)	S-bound	0.2528	527
$[W(CO)_5(Se{=}CHAr)]$	(d)	η^1–η^2 interconversions	0.2635	528
$[W(CO)_5(Te{=}CPh_2)]$	(d)			529
SO_2 complexes				
$[Cr(CO)_5(\eta^1\text{-}SO_2)]$	(b)		0.2192	530
cis-$[M(CO)_4(\eta^1\text{-}SO_2)\{P(OMe)_3\}]$	(c)	M = Cr, Mo, W		531
fac-$[M(CO)_3(\eta^2\text{-}SO_2)(dppe)]$	(b)	M = Cr, Mo, W		531
mer-$[M(CO)_3(\eta^1\text{-}SO_2)(dmpe)]$	(b)	M = Cr, Mo, W		532,533
mer-$[Mo(CO)_2(\eta^2\text{-}SO_2)(PMe_3)_3]$	(b)	two other isomers	0.2470	534
$[Mo(CO)_3(\eta^1\text{-}SO_2)(PPr^i_3)_2]$	(b)	also W	0.2285	535
CS_2 complexes				
$[W(CO)_3(\eta^2\text{-}CS_2)(dppe)]$	(b)	three other derivatives		536
$[W(CO)_3(\eta^2\text{-}CS_2)(PMe_3)_2]$	(b)	unstable		536
R_3PCS_2 Adducts				
$[Cr(CO)_5(\eta^1\text{-}S_2CPEt_3)]$	(b)	also Mo, W	0.2389	537
$[Mo_2(CO)_4(PEt_3)_2(\eta^3\text{-}\mu\text{-}S_2CPEt_3)_2]$	(a), (b)			537
$[MoSnBuCl_3(CO)_2(PCy_3)(\eta^3\text{-}\mu\text{-}L)]$	(b)	L = S_2CPCy_3	0.2538	538
$[Mo(CO)_2(PMe_3)_2(S_2CPMe_3)]$	(d)			539

[a] For synthetic methods see text.

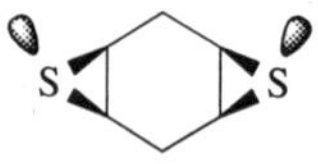

(92)

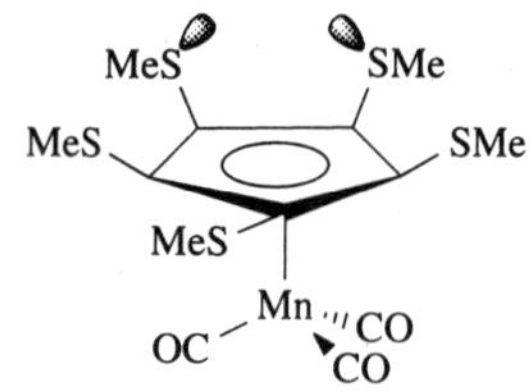

(93)

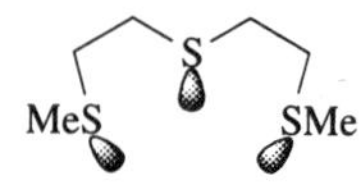

(94)

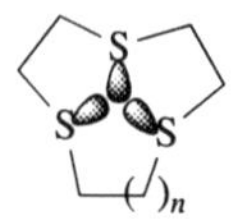

$n = 1,2$

(95)

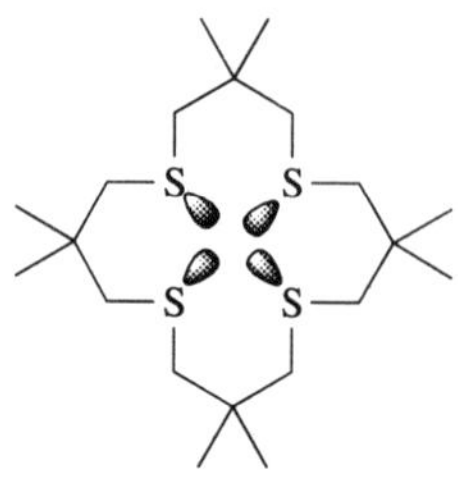

(96)

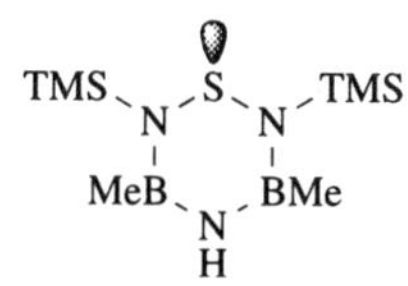

(97)

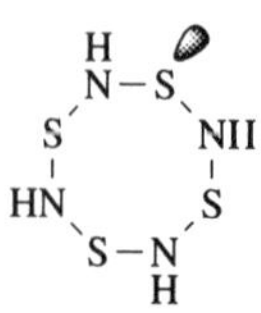

(98)

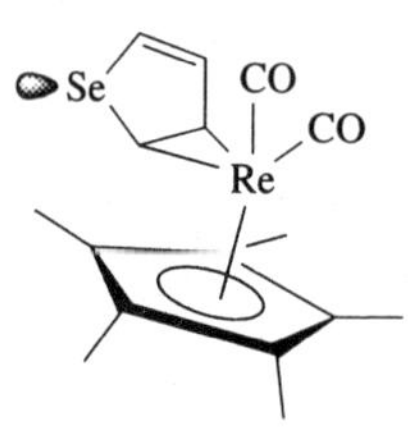

(99)

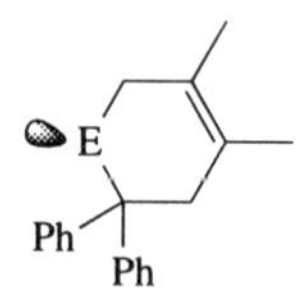

E = Se, Te

(100)

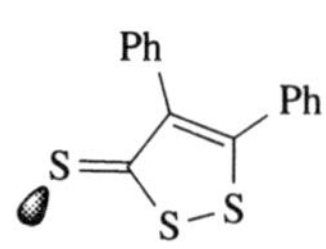

(101)

(g) Some thioether and related complexes may be prepared by protonation or alkylation of anionic complexes $[M(ER)(CO)_5]^-$; this is the best route to the coordinated thiol complexes $[M(CO)_5(RSH)]$.

The references for the examples cited above may be found in Table 18.

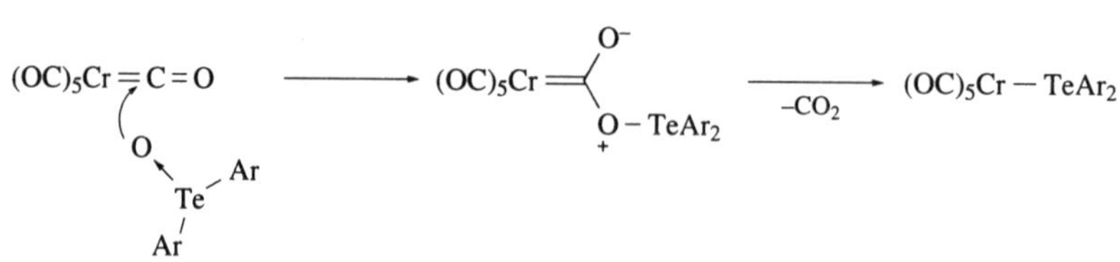

Ar = 4-C_6H_4OMe

(46)

(47)

(102)

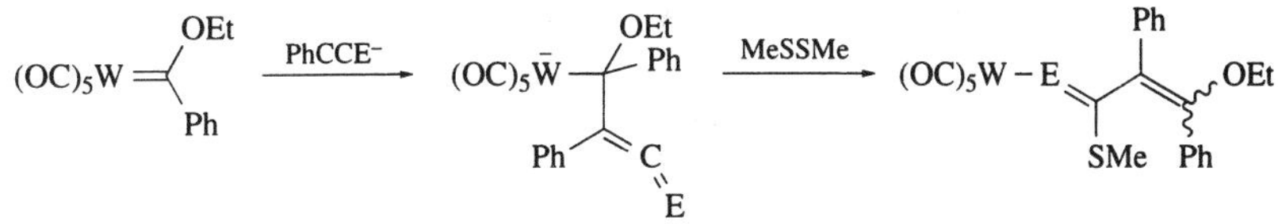

E = S, Se (103) (48)

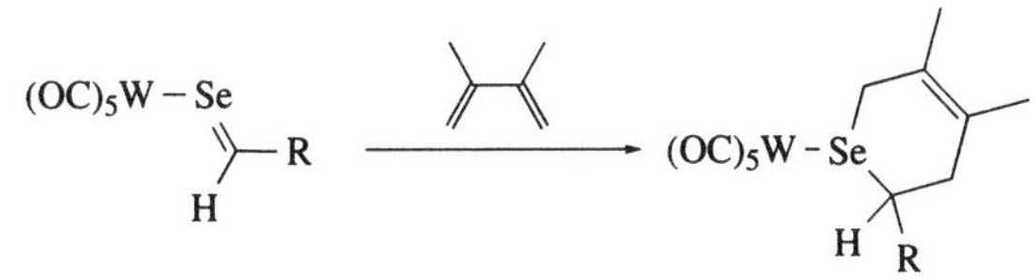

(49)

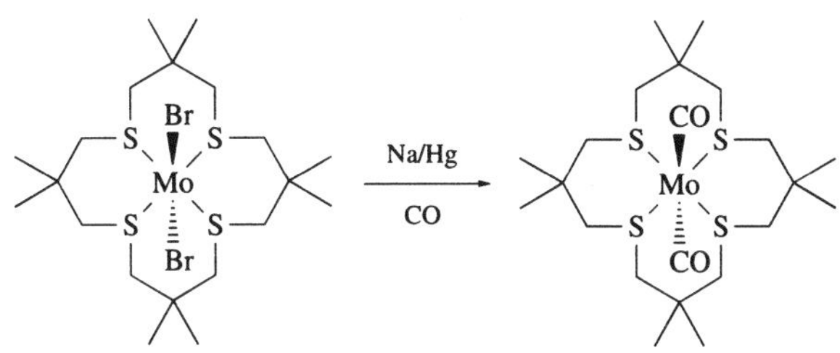

(104) (50)

(ii) Complexes with anionic ligands

Some selected preparations of thiolate and related complexes are collected in Table 19 along with examples of thiocarboxylic acid ligands and related derivatives. As the preparative procedures employed mirror those used to prepare neutral complexes, the same protocol code letters have been used here. A number of additional strategies (h)–(j) may be identified, some examples of which are shown in Equations (51)–(55) together with a few additional supporting examples, relevant references may be found in Table 19.

$$[M_2(\mu\text{-SR})_2(CO)_8]^{2-} \xrightarrow[\text{MeCN}]{I_2} [M_2(\mu\text{-SR})_2(CO)_6(NCMe)_2] \quad (51)$$

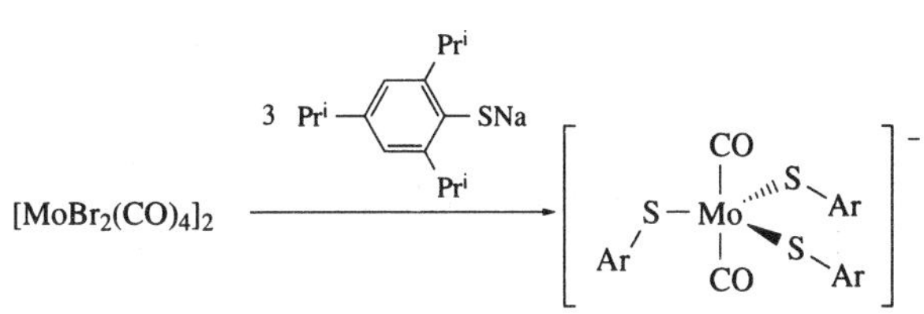

(52)

$$[MCl_2(CO)_2(PMe_3)_2] \xrightarrow{2\ ROCS_2K} [M(S_2COR)_2(CO)_n(PMe_3)_2] \quad (53)$$

M = Mo; n = 1
M = W; n = 2

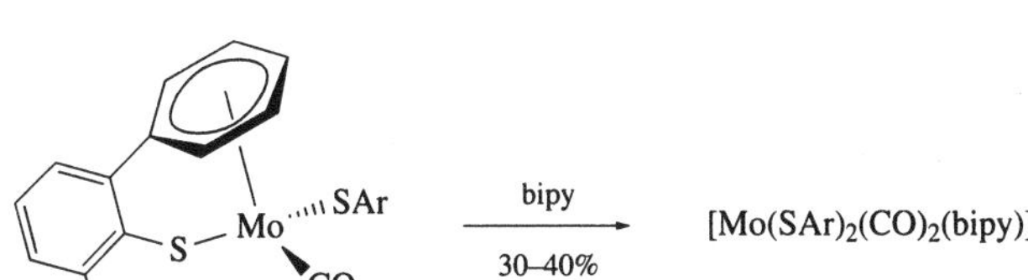

Ar = 2,6-$C_6H_4Ph_2$

(105) (106) (54)

(h) Metal–metal dimers can be accessed by oxidation of $[M(\mu\text{-SR})_2(CO)_8]^{2-}$ species with I_2. Oxidations carried out in acetonitrile lead to the partial CO substitution products (Equation (51)).

Table 19 Group 6 carbonyl complexes of anionic sulfur ligands.

Complex	*Synthesis*[a]	*Comments*	*M–S*$_{av}$ (nm)	*Ref.*
M^0 thiolate complexes				
$[Na(cryptand\text{-}221)][Cr(SH)(CO)_5]$	(a)	also Mo, W	0.2473	543
$[NEt_4][Cr(SR)(CO)_5]$	(a), (b)	also Mo, W; R = H, Ph		544
$[NEt_4][Cr(SR)(CO)_4(PMe_3)]$	(b)	also Mo, W; R = H, Ph		544
$[NEt_4][Cr(SR)(CO)_3\{P(OMe)_3\}_2]$	(b)	also Mo, W; R = H, Ph		544
$[NEt_4]_2[Cr(1,2\text{-}S_2C_6H_4)(CO)_3]$	(a)	16 e^- species	0.2295	545
$[NEt_4]_2[Mo(SR)_2(CO)_4]$	(b)	R = Ph, CH_2Ph		546
$[NEt_4]_2[Mo_2(\mu\text{-}SR)_2(CO)_8]$	(a)	also W; R = Bu^t, CH_2Ph		547,548
$[NEt_4]_2[Mo_2(\mu\text{-}SPh)_2(CO)_8]$	(a)		0.2608	549
$[Na(18\text{-}crown\text{-}6)][W(SH)(CO)_5]$	(a)		0.2567	550
$[NEt_4][W(SPh)(CO)_5]$	(b)		0.2591	544
$[Na(18\text{-}crown\text{-}6)][W_2(\mu\text{-}SH)(CO)_{10}]$	(a)		0.2592	550
M^I thiolate complexes				
$[Cr_2(\mu\text{-}SePh)_2(CO)_8]$	(i)	low yield	0.2469[b]	551
$[Mo_2(\mu\text{-}SR)_2(CO)_6(NCMe)_2]$	(h)	also W; R = Bu^t, CH_2Ph		547
$[Mo_2(\mu\text{-}SR)_2(CO)_6(NCMe)_2]$	(h)	R = CH_2CO_2Et	0.2461	552
$[Mo_2(\mu\text{-}SR)_2(CO)_8]$	(h)	R = CH_2CO_2Et	0.2449	553
$[Mo_2(\mu\text{-}SR)_2(CO)_6(NCMe)_2]$	(h)	R = Ph	0.2473	554
$[W_2(\mu\text{-}SMe)_2(CO)_8]$	(i)	also trimetallic by-product	0.247	555
$M^{(b)}$ thiolate complexes				
$[Mo(2,4,6\text{-}SC_6H_2Pr^i_3)_3(CO)_2]^-$	(f), (j)	one CO labile	0.2330	73
$[Mo(SAr)_2(CO)_2(bipy)]$	(b)	Ar = $2,6\text{-}C_6H_4Ph_2$	0.2351	75
$[Mo(SAr)_2(CO)_2(PMe_2Ph)]$	(f)	Ar = $2,4,6,\text{-}C_6H_2Pr^i_3$	0.2430	556
$[W(SAr)_2(CO)_2(PMe_2Ph)]$	(f)	Ar = $2,4,6,\text{-}C_6H_2Pr^i_3$	0.2428	556
$[W(2\text{-}Spy)_2(CO)_2(PMe_2Ph)]$	(i)		0.2529	557
Thioacetate and related complexes				
$[PPN][Cr\{SC(O)H\}(CO)_5]$	(k)		0.2447	558
$[Mo(S_2COR)_2(CO)(PMe_3)_2]$	(j)	R = Me, Et, Pr^i, Bu^t		559
$[NEt_4][Mo(S_2CNEt_2)(CO)_4]$	(a)		0.2593	560
$[PPh_4][W(S_2COEt)(CO)_5]$	(b)	also Mo and derivatives	0.2571	561
$[PPN][W(S_2CMe)(CO)_4]$	(k)	hν promoted	0.2561	562
$[W(CO)_2(S_2CNEt_2)Tp^2]$[c]	(i)		0.2566	563
$[W(S_2COPr^i)_2(CO)_2(PMe_3)_2]$	(j)	η^1 and η^2 ligands	0.2573	559
$[NEt_4][W\{SC(O)OAr\}(CO)_5]$	(k)	Ar = $2,6\text{-}C_6H_3Ph_2$	0.2582	195
$[NEt_4][W(S_2CNEt_2)(CO)_4]$	(a)		0.2614	564
$[NEt_4]_2[W(\eta^2\text{-}S_2WS_2)(CO)_4]$	(j)	also Mo	0.2535	564

[a] For synthetic methods see text. [b] av. Cr–Se distance. [c] $Tp^2 = HB(3,5\text{-}Me_2pz)_3$.

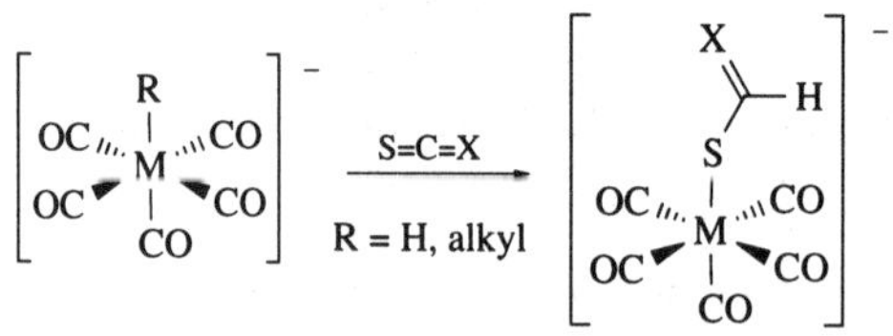

(55)

(i) Oxidative addition of RSSR or RSH leads to diverse products; complete decarbonylation, low yields and unexpected products are common features of this chemistry. Such reactions may be promoted thermally or photochemically as in the preparation of $[W_2(\mu\text{-}Ph)_2(CO)_8]$ and $[W_3(\mu\text{-}SPh)_4(CO)_{10}]$ from $[W(CO)_6]$ and PhSSPh.

(j) One effective route for the preparation of metal(II) complexes is direct metathesis of carbonyl halide precursors with thiolates or thiocarboxylic acid and related salt derivatives, shown for just two examples (Equations (52) and (53)). Use of hindered thiolates can result in unusual products, for example, (**105**). The aryl interaction can be displaced by bipy addition, leading to crystallographically characterized (**106**) (Equation (54)). Sellmann and co-workers have extensively explored the coordination chemistry of the mixed thiol-thioether (**107**). Simple displacement reactions lead to (**108**), (**109**) and (**110**).[540–2]

(k) Carbon disulfide and COS insertions are into alkyl and hydride bonds (Equation (55)) are effective for the preparation of thiocarboxylates and thioformates.

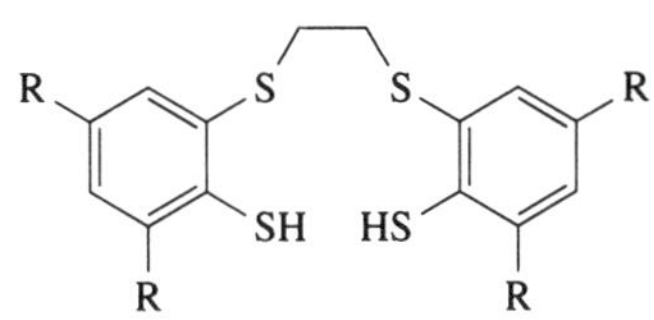

(107)

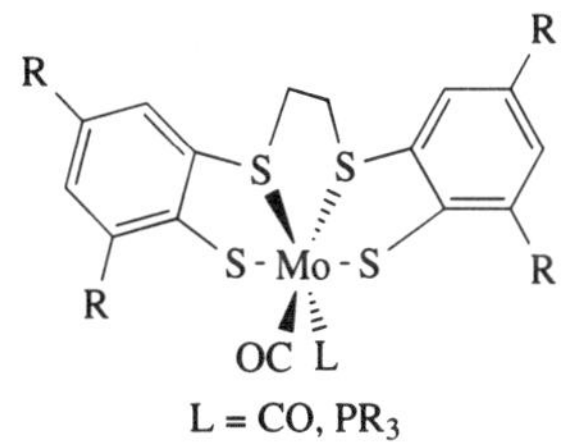

(108)

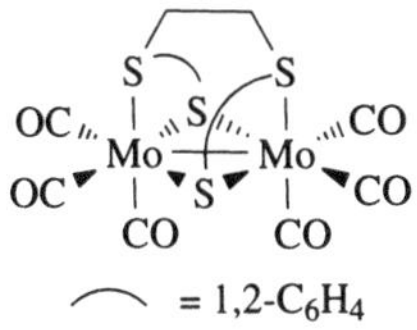

(109)

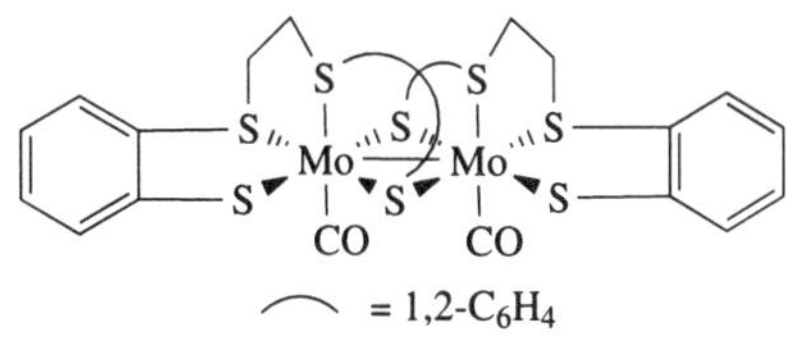

(110)

(iii) Complexes with elemental sulfur, selenium and tellurium ligands

The preparations of $[M(CO)_n(E)_n]$ complexes formally containg sufido, seleno and telluro ligands are rather nonsystematic; representative examples are given in Table 20. The absence of a convenient ^1H NMR spectroscopic probe requires that these species must normally be characterized by crystallographic studies. Solubility is an important criteron for successful reactions, for example, the cationic complexes can only be prepared in liquid SO_2.

4.3.5.2 Structure and properties

Crystallographically characterized examples of thioether, thioketone, thiolate, thiocarboxylate and sulfide complexes together with their equivalent selenium and tellurium partners can be identified in Tables 18 and 19 by the presence of an M–E bond length.

(i) Neutral complexes

Complexes of ER_2, REER and $RE(CH_2)ER$ have been extensively studied by the group of Abel. Both one- and two-dimensional NMR spectroscopic methods may be used to assay activation barriers for pyramidal inversion at the coordinated heteroatom (Equation (56)); 1,2-elemental shifts (Equation (57)) and 1,3-elemental shifts (Equation (58)). Early results and factors that affect the magnitude of the interconversion barriers are reviewed.[575] Recent results include studies of the behaviour of coordinated $TMS\text{-}CH_2EECH_2\text{-}TMS$,[490] $CH_n(SMe)_{4-n}$,[510] and $R^1SCH_2SR^2$ ligands.[576] Thio- and selenoketone complexes show M–E=C angles in the range ca. 104–124° indicative of sp^2 hybridization at the heteroatom. In solution these complexes show labile behaviour (Scheme 19), low temperatures and nonpolar solvents favour the η^2-structure, while conditions that favour $E = CR_2$ loss lead to isolation of the bridging compounds **(111)**.[521,528]

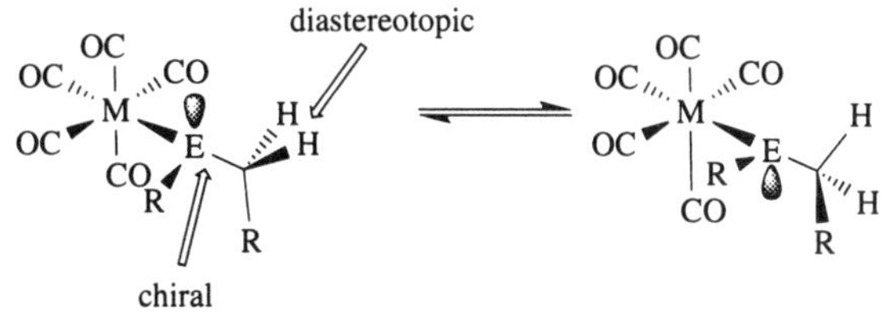

(56)

For 1,2-dithio ligands complications in the bonding scheme can arise as both dithione [S=C(R)–C(R)=S] and dithiolene [S–C(R)–C(R)–S] forms are recognized depending on where the internal redox equilibrium lies. It is suggested that the two structures can be differentiated on the basis of CO stretching frequencies or crystallographic studies. For example, on these criteria

Table 20 Preparation of sulfido and telluro group 6 carbonyls.

Complex	*Synthesis*	*Core structure*	*Ref.*
$[PPN]_2[Cr_3(\mu_3\text{-}S)(\mu\text{-}CO)_3(CO)_9]$	$Na[PPN][Cr_2(\mu\text{-}S)(CO)_{10}]$ + EtOH + [PPN]Cl	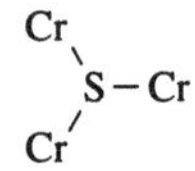	565
$[PPN]_2[Cr_4(\mu_4\text{-}S)(\mu_2\text{-}CO)_3(CO)_{14}]$	$Na[PPN][Cr_2(\mu\text{-}S)(CO)_{10}]$ + $[Cr(CO)_5(THF)]$ + [PPN]Cl	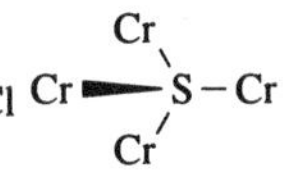	566
$[FeW(Se_2)(CO)_8][SbF_6]_2$	$[W(CO)_6]$ + $[Fe_2(CO)_9]$ + $Se_4[Sb_2F_{11}]_2$	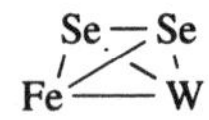	567
$[W_2(Se_2)_2(CO)_{10}][AsF_6]_2$ or $[SbF_6]_2$	$[W(CO)_6]$ + $Se_4[AsF_6]_2$ or $Se_4[Sb_2F_{11}]_2$	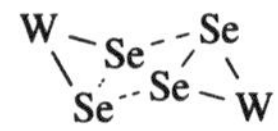	568, 569
$[M(Te_3)(CO)_4][SbF_6]_2$; M = Mo, W	$[M(CO)_6]$ + $Te_4[Sb_9F_{35}]$ or $Te_4[SbF_6]_2$	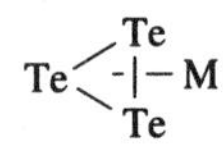	567, 570
$[W(CO)_5\{\eta^2\text{-}(CO)_5WTe_2W(CO)_5\}]$	$[W_3Sn(CO)_{15}]$ + H_2Te	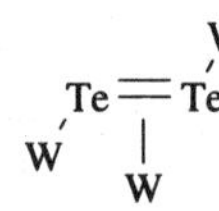	571
$[PPh_4]_2[M(Te_4)(CO)_4]$; M = Cr, W	$[M(CO)_6]$ + K_2Te_4	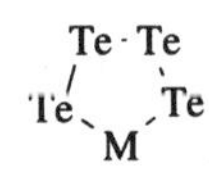	572
$K_2[M_4Te_2(CO)_{18}]$; M = Cr, W	3 $[M(CO)_6]$ + K_2Te_2	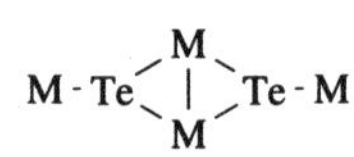	573
$K_2[\{Cr(CO)_5\}_4(Te_2)]$	3 $[Cr(CO)_6]$ + 2 K_2Te_2	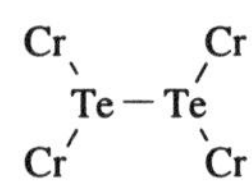	573
$K_2[\{Cr(CO)_5\}_4(Te_3)]$	3 $[Cr(CO)_6]$ + K_2Te_3	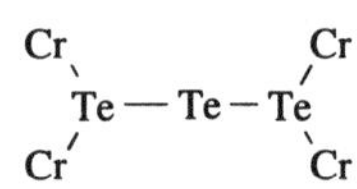	573
$[PPh_4]_2[\{W(CO)_3\}_6(Te_2)_4]$	$[W(CO)_6]$ + $[PPh_4]Te_4$	see Figure 3	574

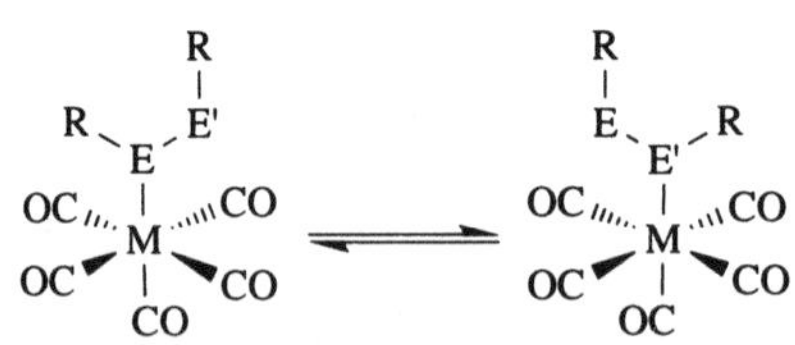

(57)

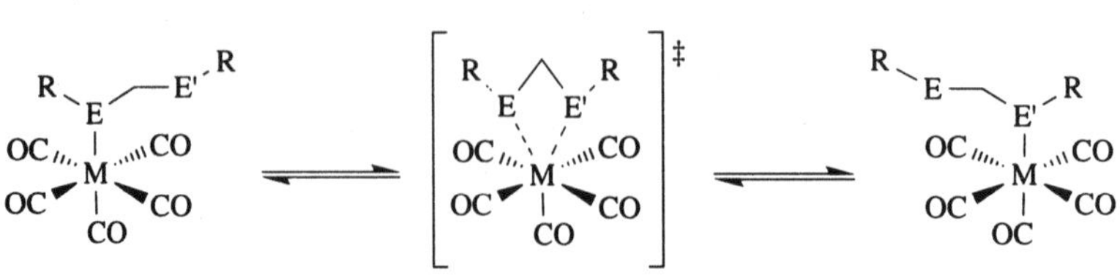

(58)

$[Mo(CO)_3(PR_3)\{S_2C_2(NR_2)_2\}]$ complexes are assigned dithione structures, while $[Mo(CO)_2(PR_3)_2\text{-}\{S_2C_2(NR_2)_2\}]$ species exist in a dithiolene form.[577]

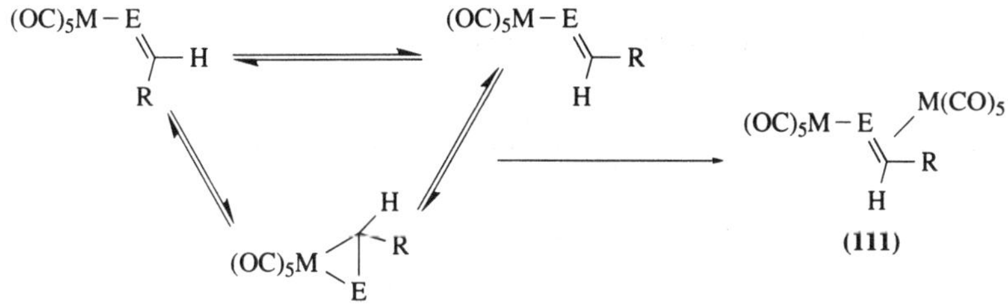

Scheme 19

(ii) Complexes with anionic ligands

Thiolates show a distinct tendency to agglomerate via bridging modes unless the thiolate is sufficiently sterically bulked to mitigate this pathway. Under such circumstances unusual reactivity can be observed in mononuclear species. A good example is the behaviour of (**108**) (R = Bu^t). Thermolysis leads to loss of CO and ethene leading to $[Mo(S–S)_2(CO)(PPh_3)]$ which shows a trigonal prismatic geometry in the solid state but converts to a higher symmetry isomer in solution. The high carbonyl stretching frequency, deep colour, and short Mo–S distances suggest that this species is best regarded as a genuine 14-electron thiolate complex.[578]

Complexes of the type $[M(CO)_3(S_2CNR_2)_2]$ (M = Mo, W) are also capable of undergoing decarbonylation to isolable unsaturated $[M(CO)_2(S_2CNR_2)_2]$ compounds. The greater reluctance of the tungsten complexes to lose CO is well documented, however, it must also be appreciated that substitution of the dithiocarbamate for another anionic S–S ligand can have a profound effect on the reaction. For example, while species with S–S=S_2COMe, $S_2P(OR)_2$ and $S_2PPr^i_2$ ligands decarbonylate, the complex $[Mo(CO)_3(S_2PPh_2)_2]$ does not participate in this reaction.[68] Isolated $[M(CO)_2(S–S)_2]$ complexes readily add two-electron ligands to recover their 18-electron configurations; studies on $[Mo(CO)_2(S_2CNEt_2)_2]$ suggest these are frontier orbital controlled processes, the ligand approaching a LUMO based mainly on a $d_{x^2-y^2}$ orbital.[579]

(iii) Complexes with elemental sulfur, selenium and tellurium ligands

The structure of $[PPh_4]_2[\{W(CO)_3\}_6(Te_2)_4]$ is worth special mention being one of the largest group 16 ligated tungsten complexes yet isolated. Two views of the anion are shown in Figure 3. The complex is unusual in that it shows two types of bonding for the Te_2 fragments, however, the W–W (0.2083 nm), η^4-Te_2 (Te–Te, 0.285, 0.2852, 0.2852 nm), and η^6-Te_2 (Te–Te, 0.2855 nm) distances are all indicative of the presence of single bonds.

4.3.5.3 Reactions

(i) Neutral complexes

Thioether and related complexes often show simple displacement reactivity as do $E=CR_2$ complexes, however, the latter also show more interesting behaviour associated mainly with additions to the E=C bond; typical reactions of thio-, seleno- and telluroketone complexes are summarized in Scheme 20.

Coordinated SO_2, CS_2 and S_2CPR_3 show a wide range of reactivity. When ligated only through sulfur, SO_2 is a far better π-acceptor than when bound in an η^2-manner. This is reflected in the substitution chemistry, η^2-SO_2 complexes are by far the most susceptible to ligand displacement. Reaction of η^2-SO_2 with electrophiles commonly leads to alkylation at oxygen. Much of the reactivity of carbon disulfide and its adducts involves metal complexes with metal–carbon bonds and therefore largely outside the scope of this chapter. In general, reactions with both nucleophiles (Equation (59)) and electrophiles (Equation (60)) are known, although unusual products can arise when either the metal complex or a given reagent has a high affinity for sulfur.[586,587] Complexes of R_3PCS_2 adducts also react with electrophiles at both carbon and sulfur, the behaviour of $[Mo(CO)_2(PMe_3)_2(S_2CPMe_3)]$ is typical (Equation (61)).[588]

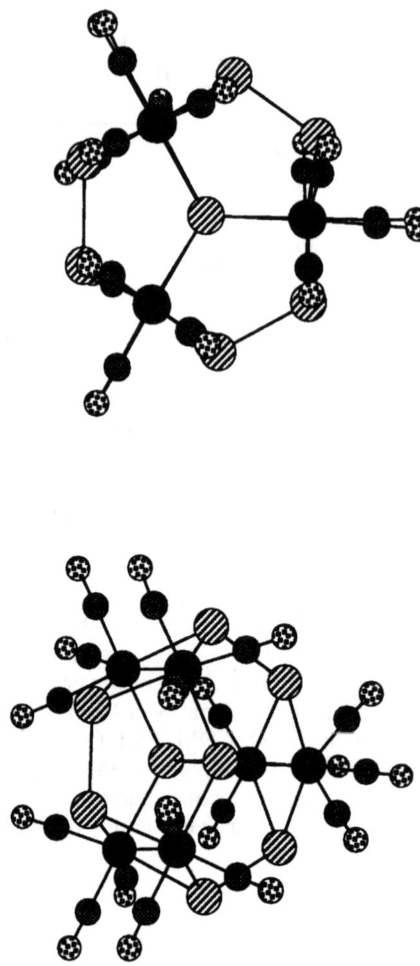

Figure 3 Two views of the anion in $[PPh_4]_2[\{W(CO)_3\}_6(Te_2)_4]$.

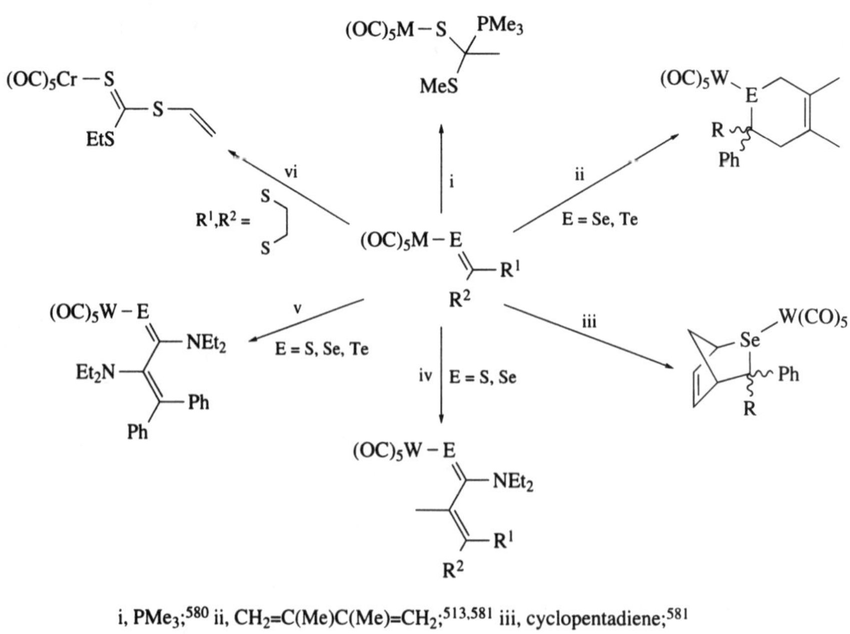

i, PMe_3;[580] ii, $CH_2{=}C(Me)C(Me){=}CH_2$;[513,581] iii, cyclopentadiene;[581]
iv, $MeC{\equiv}CNEt_2$;[582] v, $Et_2NC{\equiv}CNRt_2$;[583,584] vi, $LiNPr^i_2$ and $[Et_3O]BF_4$[585]

Scheme 20

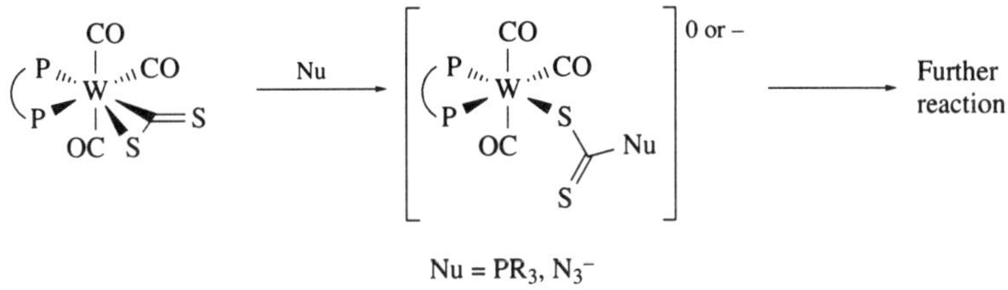

(59)

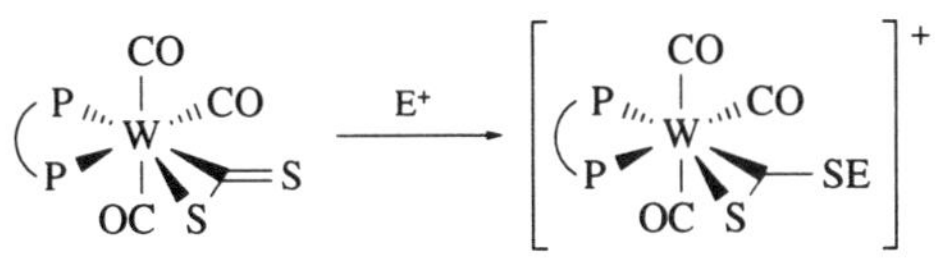

(60)

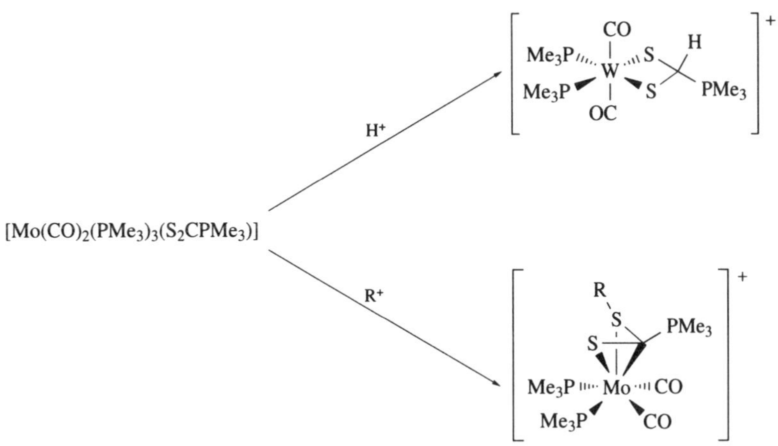

(61)

(ii) Complexes with anionic ligands

A full description of the reactivity of $[W(SH)(CO)_5]^-$, partially described in 1981, is now available. This study includes the behaviour of $[W_2(\mu\text{-}SH)(CO)_{10}]^-$ and its deprotonation product $[W(\mu\text{-}S)(CO)_{10}]^{2-}$.[589] It is interesting to compare their reactivity with the redox behaviour shown by $[Cr_2(\mu\text{-}SR)(CO)_{10}]^-$ (R = H, alkyl) complexes. Deprotonation of $[Cr_2(\mu\text{-}SH)(CO)_{10}]^-$ leads to $[Cr(\mu\text{-}S)(CO)_{10}]^{2-}$ which reacts with '$Cr(CO)_5$' to yield μ_3- and μ_4-sulfido cluster species (see Table 20). Exposure of crystalline $[PPN][Cr_2(\mu\text{-}SH)(CO)_{10}]$ to gamma rays leads to selective H$^\bullet$ loss. This compound is not isolable as it is only stable in the crystal matrix.[590] In some related chemistry, oxidation of $[Cr_2(\mu\text{-}SR)(CO)_{10}]^-$ (R = Et, Pr^i, Bu^t) with $[FcH]PF_6$ leads to the species $[Cr_2(\mu\text{-}SR)(CO)_{10}]$ which are stable at −25 °C.[591] In the absence of protic agents (which promote M–SR cleavage[544]) thiolates are effective spectator ligands. For example, when three bulky aryl groups are present in the complexes $[Mo(SAr)_3(CO)_2]^-$ (Ar = 2,6-$C_6H_3Pr^i$) manipulation of the carbonyls to other complexes including $[Mo(SAr)_3(CO)(L)]^-$ (L = PR_3, CNR) and $[Mo(SAr)_3(NCMe)(N_2Ph)]$ is facile.[73]

The complexes $[NEt_4][M(CO)_4(S\text{–}S)]$, $[M(CO)_3(S\text{–}S)_2]$ and $[M(CO)_2(S\text{–}S)_2]$ (S–S = thiocarboxylates, dithiocarbamates, xanthates, and related ligands) readily undergo substitution reactions with expulsion of carbon monoxide, coordinatively unsaturated species being the most reactive. Many reactions are present in this class, an unusual example is the preparation of the fluoro species $[NEt_4][MoF(CO)_2(S_2CNEt_2)_2]$, by reaction of $[NEt_4]F$ with $[Mo(CO)_2(S_2CNEt_2)_2]$.[31] An unusual trimetallic complex $[Fe_2Mo(CO)_2(S_2CNEt_2)(\mu_3\text{-}S)_2]^-$ is also known. This is prepared by the reaction of $[NEt_4][Mo(CO)_4(S_2CNEt_2)_2]$ with *in situ* $[Fe_2(CO)_6(\mu\text{-}S)_2]^{2-}$.[592]

(iii) Complexes with elemental sulfur, selenium and tellurium ligands

Few reactivity studies of these molecules have been reported.

4.4 REFERENCES

1. S. W. Kirtley, in 'COMC-I', vol. 3, p. 783.
2. S. W. Kirtley, in 'COMC-I', vol. 3, p. 1079.
3. S. W. Kirtley, in 'COMC-I', vol. 3, p. 1255.
4. J. A. S. Howell and P. M. Burkinshaw, *Chem. Rev.*, 1983, **83**, 557.
5. D. J. Darensbourg, E. V. Atnip, and J. H. Reibenspies, *Inorg. Chem.*, 1992, **31**, 4475.
6. K. J. Asali, H. H. Awad, J. F. Kimbrough, B. C. Lang, J. M. Watts and G. R. Dobson, *Organometallics*, 1991, **10**, 1822.
7. H. H. Awad *et al.*, *Inorg. Chem.*, 1989, **28**, 1654.
8. D. E. Halverson, G. M. Reisner, G. R. Dobson, I. Bernal and T. L. Mulcahy, *Inorg. Chem.*, 1982, **21**, 4285.
9. K. J. Asali, G. J. van Zyl and G. R. Dobson, *ibid.*, 1988, **27**, 3314.
10. G. R. Dobson and J. E. Cortés, *ibid.*, 1988, **27**, 3308.
11. G. R. Dobson and M. D. Spradling, *ibid.*, 1990, **29**, 880.
12. K. J. Asali *et al.*, *J. Am. Chem. Soc.*, 1987, **109**, 5386.
13. D. J. Darensbourg, D. J. Zalewski, C. Plepys and C. Campana, *Inorg. Chem.*, 1987, **26**, 3727.
14. J.-K. Shen, Y.-C. Gao, Q.-Z. Shi, A. L. Rheingold and F. Basolo, *ibid.*, 1991, **30**, 1868.
15. Y.-L. Shi, Y.-C. Gao, Q.-Z. Shi, D. L. Kershner and F. Basolo, *Organometallics*, 1987, **6**, 1528.
16. Y.-C. Gao, Q.-Z. Shi, D. L. Kershner and F. Basolo, *Inorg. Chem.*, 1988, **27**, 188.
17. Y. Sulfab, F. Basolo and A. L. Rheingold, *Organometallics*, 1989, **8**, 2139.
18. H.-T. Macholdt and H. Elias, *Inorg. Chem.*, 1984, **23**, 4315.
19. L. Chan and A. J. Lees, *Inorg. Chim. Acta*, 1986, **113**, L3.
20. M. R. Mason and J. G. Verkade, *Organometallics*, 1992, **11**, 1514.
21. K. B. Reddy, R. Hoffmann, G. Konya, R. van Eldik and E. M. Eyring, *ibid.*, 1992, **11**, 2319.
22. S. Oishi, M. Watanabe and T. Muraishi, *Chem. Lett.*, 1993, 713.
23. M. J. Wovkulich and J. D. Atwood, *Inorg. Synth.*, 1985, **23**, 37.
24. H. Kunkely and A. Vogler, *Organometallics*, 1992, **11**, 3172.
25. E. O. Fischer and J. K. R. Wanner, *J. Organomet. Chem.*, 1983, **252**, 175.
26. J. Powell, A. Lough and F. Wang, *Organometallics*, 1992, **11**, 2289.
27. R. N. Bagchi, A. M. Bond, R. Colton, D. L. Luscombe and J. E. Moir, *J. Am. Chem. Soc.*, 1986, **108**, 3352.
28. I. R. Beattie, P. J. Jones and N. A. Young, *ibid.*, 1992, **114**, 6146.
29. S. N. Misra and K. Venkatasubramanian, *Bull. Chem. Soc. Jpn.*, 1988, **61**, 1182.
30. E. Horn, F. L. Wimmer and M. R. Snow, *Inorg. Synth.*, 1989, **26**, 81.
31. S. J. N. Burgmayer and J. L. Templeton, *Inorg. Chem.*, 1985, **24**, 2224.
32. N. M. Doherty and N. W. Hoffman, *Chem. Rev.*, 1991, **91**, 553.
33. P. K. Baker, S. G. Fraser and E. M. Keys, *J. Organomet. Chem.*, 1986, **309**, 319.
34. F. A. Cotton and M. Matusz, *Polyhedron*, 1987, **6**, 261.
35. L. Mihichuk, M. Pizzey, B. Robertson and R. Barton, *Can. J. Chem.*, 1986, **64**, 991.
36. P. Chaudhuri, K. Wieghardt, Y.-H. Tsai and C. Krüger, *Inorg. Chem.*, 1984, **23**, 427.
37. M. Iglesias, A. Llorente, C. Del-Pino and A. Santos, *J. Organomet. Chem.*, 1984, **263**, 193.
38. K.-B. Shiu, K.-S. Liou, S.-L. Wang and S.-C. Wei, *Organometallics*, 1990, **9**, 669.
39. F. J. Arnaiz, G. Garcia, V. Riera, Y. Dromzée and Y. Jeannin, *J. Chem. Soc., Dalton Trans.*, 1987, 819.
40. G. L. Hillhouse, G. V. Goeden and B. L. Haymore, *Inorg. Chem.*, 1982, **21**, 2064.
41. M. G. B. Drew, P. K. Baker, E. M. Armstrong and S. G. Fraser, *Polyhedron*, 1988, **7**, 245.
42. R. S. Herrick, C. H. Peters and R. R. Duff, Jr., *Inorg. Chem.*, 1988, **27**, 2214.
43. R. S. Herrick, M. S. George, R. R. Duff, Jr., F. H. D'Aulnois, R. M. Jarret and J. L. Hubbard, *ibid.*, 1991, **30**, 3711.
44. B. J. Brisdon and A. G. W. Hodson, *Inorg. Chim. Acta*, 1987, **128**, 51.
45. P. K. Baker and S. G. Fraser, *J. Organomet. Chem.*, 1986, **299**, C23.
46. P. K. Baker, M. van Kampen and D. ap Kendrick, *ibid.*, 1991, **421**, 241.
47. F. J. Arnaiz, G. Garcia, V. Riera, C. Foces-Foces, F. H. Cano and M. Martinez-Ripoll, *ibid.*, 1987, **332**, 299.
48. F. A. Cotton and R. Poli, *Inorg. Chem.*, 1986, **25**, 3703.
49. F. C. Bradley, E. H. Wong, E. J. Gabe, F. L. Lee and Y. Lepage, *Polyhedron*, 1987, **6**, 1103.
50. L. K. Fong, J. R. Fox, B. M. Foxman and N. J. Cooper, *Inorg. Chem.*, 1986, **25**, 1880.
51. P. Umland and H. Vahrenkamp, *Chem. Ber.*, 1982, **115**, 3565.
52. S. W. Lee and W. C. Trogler, *Organometallics*, 1990, **9**, 1470.
53. F. A. Cotton and J. H. Meadows, *Inorg. Chem.*, 1984, **23**, 4688.
54. F. A. Cotton, L. R. Falvello and R. Poli, *Polyhedron*, 1987, **6**, 1135.
55. M. S. Balakrishna, S. S. Krishnamurthy and H. Manohar, *Organometallics*, 1991, **10**, 2522.
56. M. R. Churchill and H. J. Wasserman, *Inorg. Chem.*, 1982, **21**, 3913.
57. C. E. Osterberg, M. A. King, A. M. Arif and T. G. Richmond, *Angew. Chem., Int. Ed. Engl.*, 1990, **29**, 888.
58. R. Reina, O. Rossell, M. Seco and A. Perales, *J. Organomet. Chem.*, 1991, **415**, 101.
59. M. Melnik and P. Sharrock, *Coord. Chem. Rev.*, 1985, **65**, 49.
60. R. Colton and J. Kevekordes, *Aust. J. Chem.*, 1982, **35**, 895.
61. F. A. Cotton, L. R. Falvello and J. H. Meadows, *Inorg. Chem.*, 1985, **24**, 514.
62. F. Calderazzo, R. Poli and P. F. Zanazzi, *Gazz. Chim. Ital.*, 1988, **118**, 583.
63. A. B. van Oort, P. H. M. Budzelaar, J. H. G. Frijns and A. G. Orpen, *J. Organomet. Chem.*, 1990, **396**, 33.
64. D. M. Dawson, R. A. Henderson, A. Hills and D. L. Hughes, *J. Chem. Soc., Dalton Trans.*, 1992, 973.
65. P. Rosendorfer, C. Robl and W. Beck, *Gazz. Chim. Ital.*, 1993, **123**, 145.
66. W. Buchner and W. A. Schenk, *Inorg. Chem.*, 1984, **23**, 132.
67. P. K. Baker and D. ap Kendrick, *J. Organomet. Chem.*, 1991, **411**, 215.
68. C. Y. Chou and E. A. Maatta, *Inorg. Chem.*, 1984, **23**, 2912.
69. M. C. Durrant, D. L. Hughes, R. L. Richards, P. K. Baker and S. D. Harris, *J. Chem. Soc., Dalton Trans.*, 1992, 3399.
70. J. A. Broomhead, J. Budge and W. Grumley, *Inorg. Synth.*, 1990, **28**, 145.
71. F. A. Cotton and R. Poli, *Inorg. Chem.*, 1987, **26**, 1514.

72. H. P. M. M. Ambrosius, J. Willemse, J. A. Cras, W. P. Bosman and J. H. Noordik, *ibid.*, 1984, **23**, 2672.
73. P. J. Blower, J. R. Dilworth, J. Hutchinson, T. Nicholson and J. A. Zubieta, *J. Chem. Soc., Dalton Trans.*, 1985, 2639.
74. F. A. Cotton and R. Poli, *Inorg. Chem.*, 1987, **26**, 3228.
75. P. T. Bishop, J. R. Dilworth, T. Nicholson and J. Zubieta, *J. Chem. Soc., Dalton Trans.*, 1991, 385.
76. F. A. Cotton and R. Poli, *J. Am. Chem. Soc.*, 1988, **110**, 830.
77. P. B. Winston, S. J. N. Burgmayer and J. L. Templeton, *Organometallics*, 1983, **2**, 167.
78. A. Ford, E. Sinn and S. Woodward, *Polyhedron*, 1994, **13**, 635.
79. V. Riera, F. J. Arnaiz and G. G. Herbosa, *J. Organomet. Chem.*, 1986, **315**, 51.
80. D. C. Brower, P. B. Winston, T. L. Tonker and J. L. Templeton, *Inorg. Chem.*, 1986, **25**, 2883.
81. M. Y. Darensbourg and C. E. Ash, *Adv. Organomet. Chem.*, 1987, **27**, 1.
82. D. J. Darensbourg and R. A. Kudaroski, *ibid.*, 1983, **22**, 129.
83. J. E. Ellis, *ibid.*, 1990, **31**, 1.
84. M. Y. Darensbourg, R. Bau, M. W. Marks, R. R. Burch, Jr., J. C. Deaton and S. Slater, *J. Am. Chem. Soc.*, 1982, **104**, 6961.
85. M. Y. Darensbourg and S. Slater, *Inorg. Synth.*, 1983, **22**, 181.
86. D. J. Darensbourg and A. Rokicki, *Organometallics*, 1982, **1**, 1685.
87. M. D. Grillone, *Inorg. Synth.*, 1985, **23**, 27.
88. J. Rozière, P. Teulon and M. D. Grillone, *Inorg. Chem.*, 1983, **22**, 557.
89. D. H. Gibson, F. U. Ahmed and K. R. Phillips, *J. Organomet. Chem.*, 1981, **218**, 325.
90. D. H. Gibson, F. U. Ahmed and K. R. Phillips, *Organometallics*, 1982, **1**, 679.
91. R. C. Hynes, K. F. Preston, J. J. Springs and A. J. Williams, *J. Chem. Soc., Dalton Trans.*, 1990, 3655.
92. D. W. Hart, R. Bau and T. F. Koetzle, *Organometallics*, 1985, **4**, 1590.
93. C.-Y. Wei *et al.*, *Inorg. Chem.*, 1982, **21**, 2556.
94. J. T. Lin and J. E. Ellis, *J. Am. Chem. Soc.*, 1983, **105**, 6252.
95. H. Piana and U. Schubert, *J. Organomet. Chem.*, 1991, **411**, 303.
96. A. A. H. van der Zeijden, T. Bürgi and H. Berke, *Inorg. Chim. Acta*, 1992, **201**, 131.
97. M. R. Smith, T.-Y. Cheng and G. L. Hillhouse, *Inorg. Chem.*, 1992, **31**, 1535.
98. A. A. H. van der Zeijden *et al.*, *Helv. Chim. Acta*, 1991, **74**, 1194.
99. T. Ishida, T. Hayashi, Y. Mizobe and M. Hidai, *Inorg. Chem.*, 1992, **31**, 4481.
100. L. S. Van Der Sluys, K. A. Kubat-Martin, G. J. Kubas and K. G. Caulton, *ibid.*, 1991, **30**, 306.
101. A. J. M. Caffyn *et al.*, *Organometallics*, 1991, **10**, 2842.
102. T. Ishida, Y. Mizobe and M. Hidai, *Chem. Lett.*, 1989, 2077.
103. T. Ishida, Y. Mizobe, T. Tanase and M. Hidai, *J. Organomet. Chem.*, 1991, **409**, 355.
104. J. E. Salt, G. Wilkinson, M. Motevalli and M. B. Hursthouse, *J. Chem. Soc., Dalton Trans.*, 1986, 1141.
105. J. R. Sowa, Jr., J. B. Bonanno, V. Zanotti and R. J. Angelici, *Inorg. Chem.*, 1992, **31**, 1370.
106. P. Schreiber, K. Wieghardt, U. Flörke and H.-J. Haupt, *Z. Naturforsch., B*, 1987, **42**, 1391.
107. J. R. Sowa, V. Zanotti and R. J. Angelici, *Inorg. Chem.*, 1993, **32**, 848.
108. J. R. Sowa, V. Zanotti and R. J. Angelici, *ibid.*, 1991, **30**, 4108.
109. B. Jeżowska-Trzebiatowska and B. Nissen-Sobocińska, *J. Organomet. Chem.*, 1989, **369**, 69.
110. R. H. Crabtree and D. G. Hamilton, *Adv. Organomet. Chem.*, 1988, **28**, 299.
111. S. C. Kao, M. Y. Darensbourg and W. Schenk, *Organometallics*, 1984, **3**, 871.
112. G. Bombieri, G. Bruno, M. D. Grillone and G. Polizzotti, *J. Organomet. Chem.*, 1984, **273**, 69.
113. M. Green, A. G. Orpen, I. D. Salter and F. G. A. Stone, *J. Chem. Soc., Dalton Trans.*, 1984, 2497.
114. M. Shimoi, K. Katoh and H. Ogino, *J. Chem. Soc., Chem. Commun.*, 1990, 811.
115. M. Y. Darensbourg, R. El Mehdawi, T. J. Delord, F. R. Fronczek and S. F. Watkins, *J. Am. Chem. Soc.*, 1984, **106**, 2583.
116. J. T. Lin *et al.*, *Organometallics*, 1991, **10**, 2377.
117. H. Nagorski, M. J. Mirbach and M. F. Mirbach, *J. Organomet. Chem.*, 1985, **297**, 171.
118. T. Fuchikami, Y. Ubukata and Y. Tanaka, *Tetrahedron Lett.*, 1991, **32**, 1199.
119. D. E. Linn, R. B. King and A. D. King, *J. Mol. Cat.*, 1993, **80**, 165.
120. E. Keinan and D. Perez, *J. Org. Chem.*, 1987, **52**, 2576.
121. D. H. Gibson and Y. S. El-Omrani, *Organometallics*, 1985, **4**, 1473.
122. S. C. Kao, C. T. Spillett, C. Ash, R. Lusk, Y. K. Park and M. Y. Darensbourg, *ibid.*, 1985, **4**, 83.
123. S. G. Slater, R. Lusk, B. F. Schmann and M. Y. Darensbourg, *ibid.*, 1982, **1**, 1662.
124. J. T. Lin, S. Y. Wang, P. S. Huang, Y. M. Hsiao, Y. S. Wen and S. K. Yeh, *J. Organomet. Chem.*, 1990, **388**, 151.
125. J. T. Lin *et al.*, *Inorg. Chem.*, 1992, **31**, 4444.
126. A. A. H. van der Zeijden, H. W. Bosch and H. Berke, *Organometallics*, 1992, **11**, 2051.
127. A. A. H. van der Zeijden, D. Veghini and H. Berke, *Inorg. Chem.*, 1992, **31**, 5106.
128. A. A. H. van der Zeijden, H. W. Bosch and H. Berke, *Organometallics*, 1992, **11**, 563.
129. G. J. Kubas, R. R. Ryan, B. I. Swanson, P. J. Vergamini and H. J. Wasserman, *J. Am. Chem. Soc.*, 1984, **106**, 451.
130. P. G. Jessop and R. M. Morris, *Coord. Chem. Rev.*, 1992, **122**, 155.
131. J. Eckert, G. J. Kubas and R. P. White, *Inorg. Chem.*, 1992, **31**, 1550.
132. L. S. Van Der Sluys, M. M. Miller, G. J. Kubas and K. G. Caulton, *J. Am. Chem. Soc.*, 1991, **113**, 2513.
133. R. K. Upmacis, M. Poliakoff and J. J. Turner, *ibid.*, 1986, **108**, 3645.
134. S. A. Jackson, R. K. Upmacis, M. Poliakoff, J. J. Turner, J. K. Burdett and F. W. Greves, *J. Chem. Soc., Chem. Commun.*, 1987, 678.
135. S. M. Howdle, M. A. Healy and M. Poliakoff, *J. Am. Chem. Soc.*, 1990, **112**, 4804.
136. S. A. Jackson, P. M. Hodges, M. Poliakoff, J. J. Turner and F. W. Grevels, *ibid.*, 1990, **112**, 1221.
137. G. J. Kubas *et al.*, *ibid.*, 1993, **115**, 569.
138. G. R. K. Khalsa, G. J. Kubas, C. J. Unkefer, L. S. Van Der Sluys and K. A. Kubat-Martin, *ibid.*, 1990, **112**, 3855.
139. J. Eckert and G. J. Kubas, *J. Phys. Chem.*, 1993, **97**, 2378.
140. A. A. Gonzalez *et al.*, *Organometallics*, 1988, **7**, 2429.
141. G. J. Kubas, C. J. Unkefer, B. I. Swanson and E. Fukushima, *J. Am. Chem. Soc.*, 1986, **108**, 7000.
142. A. Syamal and M. R. Maurya, *Coord. Chem. Rev.*, 1989, **95**, 183.
143. H. Kisch and P. Holzmeier, *Adv. Organomet. Chem.*, 1992, **34**, 67.

144. D. Sutton, *Chem. Rev.*, 1993, **93**, 995.
145. B. Denise *et al. J. Organomet. Chem.*, 1990, **386**, 51.
146. T. Roth and W. Kaim, *Inorg. Chem.*, 1992, **31**, 1930.
147. K. Y. Hui, P. C. Chan and T. C. W. Mak, *Inorg. Chim. Acta*, 1984, **84**, 25.
148. J. D. Korp, I. Bernal, J. L. Mills and H. T. Weaver, *ibid.*, 1983, **75**, 173.
149. J.-J. Yaouanc, N. Le Bris, J. C. Clément, H. Handel and H. Des Abbayes, *J. Chem. Soc., Chem. Commun.*, 1993, 696.
150. S. P. Van Kouwenberg, E. H. Wong, G. R. Weisman, E. J. Gabe, F. L. Lee and P. Jackson, *Polyhedron*, 1989, **8**, 2333.
151. P. Hoffmann, A. Steinhoff and R. Mattes, *Z. Naturforsch., B*, 1987, **42**, 867.
152. G. Backes-Dahmann, W. Herrmann, K. Wieghardt and J. Weiss, *Inorg. Chem.*, 1985, **24**, 485.
153. G. Backes-Dahmann and K. Wieghardt, *ibid.*, 1985, **24**, 4044.
154. T. Beissel, B. S. P. C. D. Vedova, K. Wieghardt and R. Boese, *ibid.*, 1990, **29**, 1736.
155. R. W. Hay, I. Fraser and G. Ferguson, *J. Chem. Soc., Dalton Trans.*, 1989, 2183.
156. V. Zanotti, V. Rutar and R. J. Angelici, *J. Organomet. Chem.*, 1991, **414**, 177.
157. A. Solladié-Cavallo, J. Suffert and A. De Cian, *ibid.*, 1982, **236**, 83.
158. R. Aumann, H. Heinen and C. Krüger, *Angew. Chem., Int. Ed. Engl.*, 1984, **23**, 223.
159. H. G. Raubenheimer, G. J. Kruger, F. Scott and R. Otte, *Organometallics*, 1987, **6**, 1789.
160. P. Leoni, E. Grilli, M. Pasquali and M. Tomassini, *J. Chem. Soc., Dalton Trans.*, 1985, 2561.
161. P. Leoni, E. Grilli, M. Pasquali and M. Tomassini, *ibid.*, 1986, 879.
162. M. N. Ackermann *et al. Organometallics*, 1986, **5**, 966.
163. B. Bovio, *J. Organomet. Chem.*, 1983, **244**, 257.
164. M. R. Smith, III, R. L. Keys, G. L. Hillhouse and A. L. Rheingold, *J. Am. Chem. Soc.*, 1989, **111**, 8312.
165. U. Kunze, H. Jawad, W. Conzelmann and J. Strähle, *Z. Anorg. Allg. Chem.*, 1984, **516**, 93.
166. H. Fischer, F. Seitz and G. Müller, *J. Organomet. Chem.*, 1987, **320**, 185.
167. G. J. Kubas and L. S. Van Der Sluys, *Inorg. Synth.*, 1990, **28**, 29.
168. D. Sellmann, W. Weber, G. Liehr and H. P. Beck, *J. Organomet. Chem.*, 1984, **269**, 155.
169. A. Blagg, A. T. Hutton and B. L. Shaw, *Polyhedron*, 1987, **6**, 95.
170. F. Beck, W. P. Fehlhammer and W. Beck, *J. Organomet. Chem.*, 1985, **279**, C22.
171. M.-G. Choi and T. L. Brown, *Inorg. Chem.*, 1993, **32**, 1548.
172. T. J. Burkey, *Polyhedron*, 1989, **8**, 2681.
173. P. Chaudhuri and K. Wieghardt, *Prog. Inorg. Chem.*, 1987, **35**, 329.
174. G. Backes-Dahmann and K. Wieghardt, *Inorg. Chem.*, 1985, **24**, 4049.
175. K. Wieghardt *et al.*, *ibid.*, 1985, **24**, 3151.
176. R. Harrison *et al.*, *J. Chem. Soc., Chem. Commun.*, 1992, 1374.
177. D. Sellmann and J. Müller, *J. Organomet. Chem.*, 1984, **277**, 379.
178. T. S. Lobana, *Prog. Inorg. Chem.*, 1989, **37**, 495.
179. T. S. Lobana, in 'Chemistry of Organophosphorus Compounds', ed. F. R. Hartley, Wiley, New York, 1992, vol. 2, p. 409.
180. M. H. Chisholm, *Polyhedron*, 1983, **2**, 681.
181. W. E. Buhro and M. H. Chisholm, *Adv. Organomet. Chem.*, 1987, **27**, 311.
182. F. Bottomley and L. Sutin, *Adv. Organomet. Chem.*, 1988, **28**, 339.
183. M. Poliakoff and E. Weitz, *ibid.*, 1986, **25**, 277.
184. S. P. Church, F.-W. Grevels, H. Hermann and K. Schaffner, *Inorg. Chem.*, 1985, **24**, 418.
185. G. J. Kubas, C. J. Burns, G. R. K. Khalsa, L. S. Van Der Sluys, G. Kiss and C. D. Hoff, *Organometallics*, 1992, **11**, 3390.
186. X. L. Xie and J. D. Simon, *J. Am. Chem. Soc.*, 1990, **112**, 1130.
187. M. Pasquali, P. Leoni, P. Sabatino and D. Braga, *Gazz. Chim. Ital.*, 1992, **122**, 275.
188. K. R. Birdwhistell, *Inorg. Synth*, 1992, **29**, 141.
189. Y. F. Zheng, W. H. Wang, J. G. Lin, Y. B. She and K.-J. Fu, *Chem. Phys. Lett.*, 1993, **202**, 148.
190. D. J. Darensbourg, M. Pala, D. Simmons and A. L. Rheingold, *Inorg. Chem.*, 1986, **25**, 3537.
191. P. Planinić and H. Meider, *Polyhedron*, 1990, **9**, 1099.
192. W. Kläui, *Angew. Chem., Int. Ed. Engl.*, 1990, **29**, 627.
193. J. T. Lin, S. K. Yeh, G.-H. Lee and Y. Wang, *J. Organomet. Chem.*, 1989, **361**, 89.
194. A. Beyerholm, M. Brorson, M. Minelli and L. K. Skov, *Inorg. Chem.*, 1992, **31**, 3672.
195. D. J. Darensbourg, B. L. Mueller, C. J. Bischoff, S. S. Chojnacki and J. H. Reibenspies, *ibid.*, 1991, **30**, 2418.
196. D. J. Darensbourg, B. L. Mueller, J. H. Reibenspies and C. J. Bischoff, *ibid.*, 1990, **29**, 1789.
197. J. E. Ellis and G. L. Rochfort, *Organometallics*, 1982, **1**, 682.
198. T. J. McNeese, T. E. Mueller, D. A. Wierda, D. J. Darensbourg and T. J. Delord, *Inorg. Chem.*, 1985, **24**, 3465.
199. M. Shieh, F.-D. Mia, S.-M. Peng and G.-H. Lee, *ibid.*, 1993, **32**, 2785.
200. D. J. Darensbourg, K. K. Klausmeyer, B. L. Mueller and J. H. Reibenspies, *Angew. Chem., Int. Ed. Engl.*, 1992, **31**, 1503.
201. A. L. Beauchamp, F. Belanger-Gariepy and S. Arabi, *Inorg. Chem.*, 1985, **24**, 1860.
202. D. J. Darensbourg and M. Pala, *J. Am. Chem. Soc.*, 1985, **107**, 5687.
203. D. J. Darensbourg, S. A. Morse, C. Ovalles and H. C. Pickner *Inorg. Synth.*, 1990, **27**, 295.
204. F. A. Cotton, D. J. Darensbourg, B. W. S. Kolthammer and R. Kudaroski, *Inorg. Chem.*, 1982, **21**, 1656.
205. D. J. Darensbourg, J. A. Chojnacki and J. H. Reibenspies, *ibid.*, 1992, **31**, 3428.
206. R. Alvarez, J. L. Atwood, E. Carmona, P. J. Pérez, M. L. Poveda and R. D. Rogers, *ibid.*, 1991, **30**, 1493.
207. D. J. Darensbourg, K. M. Sanchez and A. L. Rheingold, *J. Am. Chem. Soc.*, 1987, **109**, 290.
208. M. J. Almond, J. A. Crayston, A. J. Downs, M. Poliakoff and J. J. Turner, *Inorg. Chem.*, 1986, **25**, 19.
209. J. C. Bryan, S. J. Geib, A. L. Rheingold and J. M. Mayer, *J. Am. Chem. Soc.*, 1987, **109**, 2826.
210. S. G. Feng, L. Luan, P. White, M. S. Brookhart, J. L. Templeton and C. G. Young, *Inorg. Chem.*, 1991, **30**, 2582.
211. M. H. Chisholm, C. E. Hammond, V. J. Johnston, W. E. Streib and J. C. Huffman, *J. Am. Chem. Soc.*, 1992, **114**, 7056.
212. B. Zhuang, L. Huang, L. He, P. Yu, A. Lan and J. Lu, *Inorg. Chim. Acta*, 1991, **179**, 179.
213. F. A. Cotton and W. Schwotzer, *J. Am. Chem. Soc.*, 1983, **105**, 5639.
214. D. J. Darensbourg, K. M. Sanchez, J. H. Reibenspies and A. L. Rheingold, *ibid.*, 1989, **111**, 7094.
215. W. Kläui and H. Hamers, *Z. Anorg. Allg. Chem.*, 1991, **595**, 151.
216. K.-B. Shiu, C.-C. Chou, S.-L. Wang and S.-C. Wei, *Organometallics*, 1990, **9**, 286.

217. W. Kaim, *Inorg. Chem.*, 1984, **23**, 3365.
218. A. E. Alegría, O. Lozada, H. Rivera and J. Sánchez, *J. Organomet. Chem.*, 1985, **281**, 229.
219. M. H. Chisholm, J. A. Connor, J. C. Huffman, E. M. Kober and C. Overton, *Inorg. Chem.*, 1984, **23**, 2298.
220. B. P. Buffin and T. G. Richmond, *Polyhedron*, 1990, **9**, 2887.
221. A. E. Sánchez-Peláez and M. F. Perpiñán, *J. Organomet. Chem.*, 1991, **405**, 101.
222. W. Ries, I. Bernal, M. Quast and T. A. Albright, *Inorg. Chim. Acta*, 1984, **83**, 5.
223. B. S. Creaven, C. Long, R. A. Howie, G. P. McQuillan and J. Low, *ibid.*, 1989, **157**, 151.
224. K. H. Pannell, A. J. Mayr, R. Hoggard, J. S. McKennis and J. C. Dawson, *Chem. Ber.*, 1983, **116**, 230.
225. G. G. Aleksandrov, V. N. Babin, A. P. Kharchevnikov, Y. T. Struchkov and N. S. Kochetkova, *J. Organomet. Chem.*, 1984, **266**, 109.
226. J. A. Connor, E. J. James, and C. Overton, *Polyhedron*, 1985, **4**, 69.
227. C. O'Flaherty, C. Long, J. G. Vos and R. A. Howie, *Inorg. Chim. Acta*, 1991, **180**, 195.
228. W. Kaim and S. Kohlmann, *Inorg. Chem.*, 1990, **29**, 1898.
229. M. P. Gamasa, J. Gimeno, E. Lastra, S. García-Granda and D. Moreiras, *J. Chem. Soc., Dalton Trans.*, 1991, 1557.
230. P. N. W. Baxter, J. A. Connor, J. D. Wallis, D. C. Povey and A. K. Powell, *Polyhedron*, 1992, **11**, 1771.
231. W. A. Herrmann, W. R. Thiel, J. G. Kuchler, J. Behm and E. Herdtweck, *Chem. Ber.*, 1990, **123**, 1963.
232. R. A. Howie, G. Izquierdo and G. P. McQuillan, *Inorg. Chim. Acta*, 1983, **72**, 165.
233. W. Kaim, S. Kohlmann, J. Jordanov and D. Fenske, *Z. Anorg. Allg. Chem.*, 1991, **599**, 217.
234. W. Kaim, S. Ernst and S. Kohlmann, *Polyhedron*, 1986, **5**, 445.
235. R. R. Ruminski, C. DeGroff and S. J. Smith, *Inorg. Chem.*, 1992, **31**, 3325.
236. S. Lu and J. Selbin, *Inorg. Chim. Acta*, 1987, **134**, 229.
237. H. tom Dieck, T. Mack, K. Peters and H. G. v. Schnering, *Z. Naturforsch., Teil B*, 1983, **38**, 568.
238. J. Ehlers and H. tom Dieck, *Z. Anorg. Allg. Chem.*, 1988, **560**, 80.
239. P. G. Rasmussen and J. C. Bayón, *Inorg. Chim. Acta*, 1984, **81**, L15.
240. J. Granifo, *Polyhedron*, 1993, **12**, 27.
241. C. Long, J. G. Vos, R. A. Howie and J. M. Kelly, *J. Organomet. Chem.*, 1984, **272**, 385.
242. J. W. Faller and Y. N. Ma, *J. Am. Chem. Soc.*, 1991, **113**, 1579.
243. S. Trofimenko, *Chem. Rev.*, 1993, **93**, 943.
244. C. P. Marabella and J. H. Enemark, *J. Organomet. Chem.*, 1982, **226**, 57.
245. T. J. Desmond, F. J. Lalor, G. Ferguson and M. Parvez, *ibid.*, 1984, **277**, 91.
246. E. C. Onyiriuka and A. Storr, *Can. J. Chem.*, 1987, **65**, 2464.
247. J. H. MacNeil, W. C. Watkins, M. C. Baird and K. F. Preston, *Organometallics*, 1992, **11**, 2761.
248. A. Saravanamuthu, D. M. Ho, M. E. Kerr, C. Fitzgerald, M. R. M. Bruce and A. E. Bruce, *Inorg. Chem.*, 1993, **32**, 2202.
249. B. A. Arndtsen, H. F. Sleiman, A. K. Chang and L. McElwee-White, *J. Am. Chem. Soc.*, 1991, **113**, 4871.
250. H. F. Sleiman, B. A. Arndtsen and L. McElwee-White, *Organometallics*, 1991, **10**, 541.
251. B. A. Arndtsen, T. K. Schoch and L. McElwee-White, *J. Am. Chem. Soc.*, 1992, **114**, 7041.
252. L. Luan, P. S. White, M. Brookhart and J. L. Templeton, *ibid.*, 1990, **112**, 8190.
253. P. J. Pérez, L. Luan, P. S. White, M. Brookhart and J. L. Templeton, *ibid.*, 1992, **114**, 7928.
254. R. S. Pilato, C. McGettigan, G. L. Geoffroy, A. L. Rheingold and S. J. Geib, *Organometallics*, 1990, **9**, 312.
255. T. Aoshima, T. Tanase, Y. Mizobe, Y. Yamamoto and M. Hidai, *J. Chem. Soc., Chem. Commun.*, 1992, 586.
256. Y.-I. Ishikawa, P. A. Hackett and D. M. Rayner, *J. Phys. Chem.*, 1989, **93**, 652.
257. S. P. Church, F.-W. Grevels, H. Hermann and K. Schaffner, *Inorg. Chem.*, 1984, **23**, 3830.
258. D. M. P. Mingos and D. J. Sherman, *Adv. Inorg. Chem.*, 1989, **34**, 293.
259. M. Minelli, J. L. Hubbard, D. L. Lichtenberger and J. H. Enemark, *Inorg. Chem.*, 1984, **23**, 2721.
260. M. J. Menu, G. Crocco, M. Dartiguenave, Y. Dartiguenave and G. Bertrand, *J. Chem. Soc., Chem. Commun.*, 1988, 1598.
261. A. P. Masters, M. Parvez and T. S. Sorensen, *Can. J. Chem.*, 1991, **69**, 2136.
262. A. P. Masters, M. Parvez, T. S. Sorensen and F. Sun, *ibid.*, 1993, **71**, 230.
263. A. M. Bond, R. Colton, J. E. Kevekordes and P. Panagiotidou, *Inorg. Chem.*, 1987, **26**, 1430.
264. P. Hoffmann and R. Mattes, *ibid.*, 1989, **28**, 2092.
265. R. V. Honeychuck and W. H. Hersh, *ibid.*, 1989, **28**, 2869.
266. P. Kundel and H. Berke, *J. Organomet. Chem.*, 1988, **339**, 103.
267. D. J. Stufkens, *Coord. Chem. Rev.*, 1990, **104**, 39.
268. P. Glyn, F. P. A. Johnson, M. W. George, A. J. Lees and J. J. Turner, *Inorg. Chem.*, 1991, **30**, 3543.
269. A. J. Lees, *Chem. Rev.*, 1987, **87**, 711.
270. K. A. Rawlins, A. J. Lees and A. W. Adamson, *Inorg. Chem.*, 1990, **29**, 3866.
271. N. Leventis and P. J. Wagner, *J. Am. Chem. Soc.*, 1985, **107**, 5807.
272. L. Tutt and J. I. Zink, *ibid.*, 1986, **108**, 5830.
273. E. Lindsay, A. Vlček, Jr., and C. H. Langford, *Inorg. Chem.*, 1993, **32**, 2269.
274. J. Víchová, F. Hartl, and A. Vlček, Jr., *J. Am. Chem. Soc.*, 1992, **114**, 10903.
275. M. J. Aroney, R. M. Clarkson, T. W. Hambley and R. K. Pierens, *J. Organomet. Chem.*, 1992, **426**, 331.
276. W. Bruns, H.-D. Hausen, W. Kaim and A. Schulz, *ibid.*, 1993, **444**, 121.
277. J. A. Connor, E. J. James, C. Overton and N. Elmurr, *J. Chem. Soc., Dalton Trans.*, 1984, 255.
278. J. A. Connor and C. Overton, *ibid.*, 1982, 2397.
279. J. A. Connor, E. J. James, C. Overton, J. M. A. Walshe and R. A. Head, *ibid.*, 1986, 511.
280. S. Lincoln, S.-L. Soong, S. A. Koch, M. Sato and J. H. Enemark, *Inorg. Chem.*, 1985, **24**, 1355.
281. D. L. Lichtenberger and J. L. Hubbard, *ibid.*, 1984, **23**, 2718.
282. K.-B. Shiu and L.-Y. Lee, *J. Organomet. Chem.*, 1988, **348**, 357.
283. M. D. Curtis, K.-B. Shiu, W. M. Butler and J. C. Huffman, *J. Am. Chem. Soc.*, 1986, **108**, 3335.
284. C. C. Philipp, P. S. White and J. L. Templeton, *Inorg. Chem.*, 1992, **31**, 3825.
285. C. A. Rusik, M. A. Collins, A. S. Gamble, T. L. Tonker and J. L. Templeton, *J. Am. Chem. Soc.*, 1989, **111**, 2550.
286. C. G. Young, R. W. Gable and M. F. Mackay, *Inorg. Chem.*, 1990, **29**, 1777.
287. M. D. Curtis and K.-B. Shiu, *ibid.*, 1985, **24**, 1213.
288. L. Luan, M. Brookhart and J. L. Templeton, *Organometallics*, 1992, **11**, 1433.

289. P. Kundel and H. Berke, *Z. Naturforsch., Teil B*, 1987, **42**, 993.
290. A. M. Bond, R. Colton and P. Panagiotidou, *Organometallics*, 1988, **7**, 1774.
291. F. R. Hartley (ed.), 'Chemistry of Organophosphorus Compounds', Wiley, New York, 1990, vol. 1.
292. M. Regitz and O. J. Scherer, 'Multiple Bonds and Low Coordination in Phosphorus Chemistry', G. Thieme Verlag, Stuttgart, 1990.
293. O. J. Scherer, *Angew. Chem., Int. Ed. Engl.*, 1985, **24**, 924.
294. A. Zschunke *et al.*, *Z. Anorg. Allg. Chem.*, 1992, **610**, 127.
295. E. C. Alyea, K. J. Fisher, S. Foo and B. Philip, *Polyhedron*, 1993, **12**, 489.
296. S. Affandi, J. H. Nelson, N. W. Alcock, O. W. Howarth, E. C. Alyea and G. M. Sheldrick, *Organometallics*, 1988, **7**, 1724.
297. J. Ellermann and A. A. M. Demuth, *J. Organomet. Chem.*, 1986, **309**, 307.
298. A. Hengefeld, J. Kopf and D. Rehder, *Organometallics*, 1983, **2**, 114.
299. E. C. Alyea, G. Ferguson and A. Somogyvari, *ibid.*, 1983, **2**, 668.
300. W.-K. Wong, F. L. Chow, H. Chen, B. W. Au-Yeung, R. J. Wang and T. C. W. Mak, *Polyhedron*, 1990, **9**, 2901.
301. J. Wesemann, P. G. Jones, D. Schomburg, L. Heuer and R. Schmutzler, *Chem. Ber.*, 1992, **125**, 2187.
302. K. M. Doxsee, E. M. Hanawalt, G. S. Shen, T. J. R. Weakley, H. Hope and C. B. Knobler, *Inorg. Chem.*, 1991, **30**, 3381.
303. D. J. Darensbourg, C. J. Bischoff and J. H. Reibenspies, *Inorg. Chem.*, 1991, **30**, 1144.
304. F. A. Cotton, D. J. Darensbourg, S. Klein and B. W. S. Kolthammer, *ibid.*, 1982, **21**, 2661.
305. F. A. Cotton, D. J. Darensbourg, S. Klein and B. W. S. Kolthammer, *ibid.*, 1982, **21**, 294.
306. J. Grobe, D. Le Van and J. Szameitat, *J. Organomet. Chem.*, 1985, **289**, 341.
307. H. S. Aaron and D. I. Rossman, *J. Org. Chem.*, 1987, **52**, 2939.
308. C. Zeiher, W. Hiller and I.-P. Lorenz, *Chem. Ber.*, 1985, **118**, 3127.
309. G. M. Gray, *Inorg. Chim. Acta*, 1984, **81**, 157.
310. G. M. Gray and R. J. Gray, *Organometallics*, 1983, **2**, 1026.
311. J. Grobe, D. Le Van and W. Meyring, *J. Organomet. Chem.*, 1986, **307**, 327.
312. A. Gieren, C. Ruiz-Pérez, T. Hübner, M. Herberhold, K. Schamel and K. Guldner, *ibid.*, 1989, **366**, 105.
313. H. R. Allcock, I. Manners, M. N. Mang and M. Parvez, *Inorg. Chem.*, 1990, **29**, 522.
314. L. Weber, R. Kirchoff and R. K. Boese, *J. Chem. Soc., Chem. Commun.*, 1992, 1182.
315. T. Chivers, C. Lensink and J. F. Richardson, *Organometallics*, 1987, **6**, 1904.
316. D. V. Khasnis, M. Lattman and U. Siriwardane, *ibid.*, 1991, **10**, 1326.
317. A. Tarassoli *et al.*, *Inorg. Chem.*, 1986, **25**, 3541.
318. E. G. Bent, R. Schaeffer, R. C. Haltiwanger and A. D. Norman, *ibid.*, 1990, **29**, 2608.
319. M. M. Turnbull, E. H. Wong, E. J. Gabe and F. L. Lee, *ibid.*, 1986, **25**, 3189.
320. E. H. Wong, F. C. Bradley and E. J. Gabe, *J. Organomet. Chem.*, 1983, **244**, 235.
321. F. A. Cotton, D. J. Darensbourg, S. Klein and B. W. S. Kolthammer, *Inorg. Chem.*, 1982, **21**, 1651.
322. M. L. Boyles, D. V. Brown, D. A. Drake, C. K. Hostetler, C. K. Maves and J. A. Mosbo, *ibid.*, 1985, **24**, 3126.
323. R. Colton, B. F. Hoskins and K. McGregor, *Aust. J. Chem.*, 1987, **40**, 1471.
324. L.-T. Phang, S. C. F. Au-Yeung, T. S. A. Hor, S. B. Khoo, Z.-Y. Zhou and T. C. W. Mak, *J. Chem. Soc., Dalton Trans.*, 1993, 165.
325. M. K. Rottink and R. J. Angelici, *Inorg. Chem.*, 1993, **32**, 2421.
326. H. Westermann *et al.*, *Organometallics*, 1989, **8**, 244.
327. J. Grobe, D. Le Van and W. Meyring, *Z. Anorg. Allg. Chem.*, 1990, **586**, 149.
328. R. B. King and W.-K. Fu, *J. Organomet. Chem.*, 1984, **272**, C33.
329. B. Deschamps, L. Ricard and F. Mathey, *Polyhedron*, 1989, **8**, 2671.
330. F. Mercier, F. Mathey, C. Afiong-Akpan and J. F. Nixon, *J. Organomet. Chem.*, 1988, **348**, 361.
331. F. C. Bradley, E. H. Wong, E. J. Gabe and F. L. Lee, *Inorg. Chim. Acta*, 1986, **120**, L21.
332. T. G. Hill, R. C. Haltiwanger, T. R. Prout and A. D. Norman, *Inorg. Chem.*, 1989, **28**, 3461.
333. G. Andolfatto, R. Granger and L. K. Peterson, *Inorg. Chim. Acta*, 1985, **99**, L1.
334. H. Nakazawa, M. Ohta and H. Yoneda, *Inorg. Chem.*, 1988, **27**, 973.
335. C. Tessier-Youngs, C. Bueno, O. T. Beachley, Jr. and M. R. Churchill, *ibid.*, 1983, **22**, 1054.
336. C. Tessier-Youngs, W. J. Youngs, O. T. Beachley, Jr. and M. R. Churchill, *Organometallics*, 1983, **2**, 1128.
337. D. Obendorf, P. Peringer, N. Müller and H. Falk, *J. Organomet. Chem.*, 1987, **326**, 375.
338. P. Peringer, *Polyhedron*, 1982, **1**, 819.
339. G. Hasselkuss, S. Hietkamp and O. Stelzer, *Z. Anorg. Allg. Chem.*, 1986, **534**, 50.
340. F.-W. Grevels and V. Skibbe, *J. Chem. Soc., Chem. Commun.*, 1984, 681.
341. D. W. Stephan, *Coord. Chem. Rev.*, 1989, **95**, 41.
342. G. Linti, H. Nöth, K. Polborn and R. T. Paine, *Angew. Chem., Int. Ed. Engl.*, 1990, **29**, 682.
343. T. C. Kiebach, R. Lourens, F. Bickelhaupt, C. H. Stam and A. Van Herk, *J. Organomet. Chem.*, 1981, **210**, 211.
344. J. Grobe, D. Le Van, J. Nientiedt, B. Krebs and M. Dartmann, *Chem. Ber.*, 1988, **121**, 655.
345. L. Weber, E. Lücke, A. Müller and H. Bögge, *Z. Anorg. Allg. Chem.*, 1990, **583**, 91.
346. A. J. Ashe, III, W. Butler, J. C. Colburn and S. Abu-Orabi, *J. Organomet. Chem.*, 1985, **282**, 233.
347. P. Le Floch, L. Ricard and F. Mathey, *Polyhedron*, 1990, **9**, 991.
348. A. Marinetti, S. Bauer, L. Ricard and F. Mathey, *J. Chem. Soc., Dalton Trans.*, 1991, 597.
349. A. Marinetti, P. Le Floch and F. Mathey, *Organometallics*, 1991, **10**, 1190.
350. P. Le Floch, A. Marinetti, L. Ricard and F. Mathey, *J. Am. Chem. Soc.*, 1990, **112**, 2407.
351. H. P. Fritz and T. Schöttle, *J. Organomet. Chem.*, 1984, **265**, 175.
352. F. Mathey, *Angew. Chem., Int. Ed. Engl.*, 1987, **26**, 275.
353. F. Mathey, A. Marinetti and F. Mercier, *Synlett*, 1992, 363.
354. K. J. Lee and T. L. Brown, *Inorg. Chem.*, 1992, **31**, 289.
355. F. A. Cotton, D. J. Darensbourg, S. Klein and B. W. S. Kolthammer, *Inorg. Chem.*, 1982, **21**, 2661.
356. S. P. Nolan and C. D. Hoff, *J. Organomet. Chem.*, 1985, **290**, 365.
357. D. J. Darensbourg and C. J. Bischoff, *Inorg. Chem.*, 1993, **32**, 47.
358. M. M. Rahman, H.-K. Lui, K. Eriks, A. Prock and W. P. Giering, *Organometallics*, 1989, **8**, 1.
359. G. M. Bancroft, L. Dignard-Bailey and R. J. Puddephatt, *Inorg. Chem.*, 1986, **25**, 3675.

360. M. J. Aroney, M. K. Cooper, R. K. Pierens and S. J. Pratten, *J. Organomet. Chem.*, 1986, **309**, 293.
361. D. L. Lichtenberger and M. E. Jatcko, *Inorg. Chem.*, 1992, **31**, 451.
362. S. P. Wang, M. G. Richmond and M. Schwartz, *J. Am. Chem. Soc.*, 1992, **114**, 7595.
363. E. C. Alyea and S. Song, *Inorg. Chem.*, 1992, **31**, 4909.
364. M. Minelli, J. H. Enemark, R. T. C. Brownlee, M. J. O'Connor and A. G. Wedd, *Coord. Chem. Rev.*, 1985, **68**, 169.
365. R. N. Bagchi, A. M. Bond, R. Colton, I. Creece, K. McGregor and T. Whyte, *Organometallics*, 1991, **10**, 2611.
366. A. M. Bond, R. Colton and R. S. Hutton, *Inorg. Chim. Acta*, 1992, **200**, 671.
367. D. V. Khasnis, J. M. Burton, H. M. Zhang and M. Lattman, *Organometallics*, 1992, **11**, 3745.
368. E. H. Wong *et al.*, *J. Am. Chem. Soc.*, 1988, **110**, 8422.
369. G. Märkl, H. J. Beckh, M. L. Ziegler and B. Nuber, *Angew. Chem., Int. Ed. Engl.*, 1987, **26**, 1134.
370. P. Le Floch, D. Carmichael and F. Mathey, *Organometallics*, 1991, **10**, 2432.
371. J. Grobe, D. Le Van, B. Krebs, R. Frohlich and A. Schiemann, *J. Organomet. Chem.*, 1990, **389**, C29.
372. J.-T. Hung and K. Lammertsma, *Organometallics*, 1992, **11**, 4365.
373. A. Marinetti, C. Charrier, F. Mathey and J. Fischer, *ibid.*, 1985, **4**, 2134.
374. N. H. T. Huy, L. Ricard and F. Mathey, *ibid.*, 1988, **7**, 1791.
375. A. Martinetti and F. Mathey, *ibid.*, 1984, **3**, 1492.
376. F. A. Cotton and B. Hong, *Prog. Inorg. Chem.*, 1992, **40**, 179.
377. G. R. Newkome, *Chem. Rev.*, 1993, **93**, 2067.
378. I. Haiduc and I. Silaghi-Dumitrescu, *Coord. Chem. Rev.*, 1986, **74**, 127.
379. A. K. Powell and M. J. Went, *J. Chem. Soc., Dalton Trans.*, 1992, 439.
380. K. Mashima, H. Nakano, T. Mori, H. Takaya and A. Nakamura, *Chem. Lett.*, 1992, 185.
381. E. Fluck, P. Kuhm, A. Müller and H. Bögge, *Z. Anorg. Allg. Chem.*, 1988, **567**, 13.
382. T. S. A. Hor, L.-T. Phang, L. K. Liu and Y. S. Wen, *J. Organomet. Chem.*, 1990, **397**, 29.
383. T. S. A. Hor, *ibid.*, 1987, **319**, 213.
384. E. C. Alyea, G. Ferguson, K. J. Fisher, R. A. Gossage and M. C. Jennings, *Polyhedron*, 1990, **9**, 2393.
385. C. J. Adams, M. I. Bruce, E. Horn and E. R. T. Tiekink, *J. Chem. Soc., Dalton Trans.*, 1992, 1157.
386. C.-H. Ueng and G.-Y. Hwang, *ibid.*, 1991, 1643.
387. G. B. Jacobsen, B. L. Shaw and M. Thornton-Pett, *J. Organomet. Chem.*, 1987, **323**, 313.
388. A. Trenkle and H. Vahrenkamp, *Chem. Ber.*, 1981, **114**, 1343.
389. J. L. Bookham, W. McFarlane, M. Thornton-Pett and S. Jones, *J. Chem. Soc., Dalton Trans.*, 1990, 3621.
390. A. Varshney and G. M. Gray, *Inorg. Chem.*, 1991, **30**, 1748.
391. T.-J. Chow, C.-Y. Wang, S.-C. Sheu and S.-M. Peng, *J. Organomet. Chem.*, 1986, **311**, 339.
392. B. Deschamps and F. Mathey, *Organometallics*, 1992, **11**, 1411.
393. C. Zeiher, J. Mohyla, I.-P. Lorenz and W. Hiller, *J. Organomet. Chem.*, 1985, **286**, 159.
394. J. T. Mague and M. P. Johnson, *Organometallics*, 1990, **9**, 1254.
395. M. G. Newton, R. B. King, T.-W. Lee, L. Norskov-Lauritzen and V. Kumar, *J. Chem. Soc., Chem. Commun.*, 1982, 201.
396. E. Fluck, P. Kuhm, A. Müller and H. Bögge, *Z. Anorg. Allg. Chem.*, 1988, **567**, 48.
397. M. S. Balakrishna, T. K. Prakasha, S. S. Krishnamurthy, U. Siriwardane and N. S. Hosmane, *J. Organomet. Chem.*, 1990, **390**, 203.
398. M. Scheer, S. Gremler, E. Herrmann and P. G. Jones, *ibid.*, 1991, **414**, 337.
399. T. Chivers, K. S. Dhathathreyan, C. Lensink, A. Meetsma, J. C. van de Grampel and J. L. de Boer, *Inorg. Chem.*, 1989, **28**, 4150.
400. M. Driess, H. Pritzkow and M. Reisgys, *Chem. Ber.*, 1991, **124**, 1931.
401. M. Driess, P. Frankhauser, H. Pritzkow and W. Siebert, *ibid.*, 1991, **124**, 1497.
402. W.-K. Wong, J.-X. Gao, Z.-Y. Zhou and T. C. W. Mak, *Polyhedron*, 1992, **11**, 2965.
403. R. T. DePue, D. B. Collum, J. W. Ziller and M. R. Churchill, *J. Am. Chem. Soc.*, 1985, **107**, 2131.
404. T. Kaukorat, A. Fischer, P. G. Jones and R. Schmutzler, *Chem. Ber.*, 1992, **125**, 301.
405. S. D. Perera, B. L. Shaw and M. Thornton-Pett, *J. Chem. Soc., Dalton Trans.*, 1991, 1183.
406. S.-T. Liu, C.-L. Tsao, M.-C. Cheng and S.-M. Peng, *Polyhedron*, 1990, **9**, 2579.
407. U. Kunze, H. Jawad and E. Boziaris, *J. Organomet. Chem.*, 1986, **312**, 67.
408. N. Greene, H. Taylor, T. P. Kee and M. Thornton-Pett, *J. Chem. Soc., Dalton Trans.*, 1993, 821.
409. M. Driess, M. Reisgys and H. Pritzkow, *Angew. Chem., Int. Ed. Engl.*, 1992, **31**, 1510.
410. S. E. Saum, F. R. Askham, F. R. Fronczek and G. G. Stanley, *Organometallics*, 1988, **7**, 1409.
411. E. P. Kyba, R. E. Davis, S.-T. Liu, K. A. Hassett and S. B. Larson, *Inorg. Chem.*, 1985, **24**, 4629.
412. R. B. King and M. Shimura, *J. Organomet. Chem.*, 1983, **256**, 71.
413. K. D. Gallicano, N. L. Paddock, S. J. Rettig and J. Trotter, *Can. J. Chem.*, 1982, **60**, 2415.
414. M. R. Mason, Y. Su, R. A. Jacobson and J. G. Verkade, *Organometallics*, 1991, **10**, 2335.
415. K. R. Dunbar, S. C. Haefner and D. J. Burzynski, *ibid.*, 1990, **9**, 1347.
416. M. F. M. Al-Dulaymmi, P. B. Hitchcock and R. L. Richards, *Polyhedron*, 1991, **10**, 1549.
417. S. D. Perera, B. L. Shaw and M. Thornton-Pett, *J. Chem. Soc., Dalton Trans.*, 1992, 1469.
418. W. Schirmer, U. Flörke and H.-J. Haupt, *Z. Anorg. Allg. Chem.*, 1987, **545**, 83.
419. W.-K. Wong and F.-W. Lee, *Polyhedron*, 1992, **11**, 863.
420. S. Akhavan, C.-S. Kim and B. N. Storhoff, *Inorg. Chim. Acta*, 1986, **115**, 173.
421. R. L. Keiter, E. A. Keiter, K. N. Mittelberg, J. S. Martin, V. M. Meyers and J.-G. Wang, *Organometallics*, 1989, **8**, 1399.
422. B. N. Diel, P. F. Brandt, R. C. Haltiwanger, M. L. J. Hackney and A. D. Norman, *Inorg. Chem.*, 1989, **28**, 2811.
423. E. H. Wong, X. Sun, E. J. Gabe, F. L. Lee and J.-P. Charland, *Organometallics*, 1991, **10**, 3010.
424. A.-M. Hinke, A. Hinke and W. Kuchen, *Z. Naturforsch., Teil B*, 1986, **41**, 629.
425. I. Jibril, L.-R. Frank, L. Zsolnai, K. Evertz and G. Huttner, *J. Organomet. Chem.*, 1990, **393**, 213.
426. R. A. Bartlett *et al.*, *J. Am. Chem. Soc.*, 1987, **109**, 5693.
427. H. Lang, O. Orama and G. Huttner, *J. Organomet. Chem.*, 1985, **291**, 293.
428. G. Huttner, J. Borm and L. Zsolnai, *ibid.*, 1984, **263**, C33.
429. J. Borm, G. Huttner, L. Zsolnai, K. Evertz and H. Berke, *ibid.*, 1987, **327**, 223.
430. G. Huttner, J. Borm and L. Zsolnai, *ibid.*, 1986, **304**, 309.

431. H. Lang *et al.*, *Z. Naturforsch., Teil B*, 1986, **41**, 191.
432. R. A. Jones and B. R. Whittlesey, *Organometallics*, 1984, **3**, 469.
433. U. Weber, G. Huttner, O. Scheidsteger and L. Zsolnai, *J. Organomet. Chem.*, 1985, **289**, 357.
434. A. M. Arif, A. H. Cowley, N. C. Norman and M. Pakulski, *Inorg. Chem.*, 1986, **25**, 4836.
435. G. Fritz, H.-W. Schneider, W. Hönle and H. G. v. Schnering, *Z. Anorg. Allg. Chem.*, 1990, **584**, 21.
436. G. Fritz and H. Bauer, *ibid.*, 1993, **619**, 22.
437. L. Weber, U. Sonnenberg, H.-G. Stammler and B. Neumann, *ibid.*, 1991, **605**, 87.
438. G. Fritz, E. Layher, H. Krautscheid, B. Mayer, E. Matern, W. Hönle and H. G. v. Schnering, *ibid.*, 1992, **611**, 56.
439. M. E. Barr, S. K. Smith, B. Spencer and L. F. Dahl, *Organometallics*, 1991, **10**, 3983.
440. M. Scheer, E. Herrmann, J. Sieler and M. Oehme, *Angew. Chem., Int. Ed. Engl.*, 1991, **30**, 969.
441. G. Fritz, B. Mayer and E. Matern, *Z. Anorg. Allg. Chem.*, 1992, **608**, 7.
442. G. Fritz, E. Layher, W. Hönle and H. G. v. Schnering, *ibid.*, 1991, **595**, 67.
443. G. Fritz, R. Uhlmann, K. D. Hoppe, W. Hönle and H. G. v. Schnering, *ibid.*, 1982, **491**, 83.
444. M. Herberhold, W. Bühlmeyer, A. Gieren, T. Hübner and J. Wu, *J. Organomet. Chem.*, 1987, **321**, 51.
445. T. Chivers, C. Lensink and J. F. Richardson, *J. Organomet. Chem.*, 1987, **325**, 169.
446. P. Le Floch, D. Carmichael, L. Ricard and F. Mathey, *J. Am. Chem. Soc.*, 1991, **113**, 667.
447. A. Brèque, C. C. Santini, F. Mathey, J. Fischer and A. Mitschler, *Inorg. Chem.*, 1984, **23**, 3463.
448. I. Bernal, G. M. Reisner, G. R. Dobson and C. B. Dobson, *Inorg. Chim. Acta*, 1986, **121**, 199.
449. W. Tikkanen, Y. Fujita and J. L. Petersen, *Organometallics*, 1986, **5**, 888.
450. W. A. Herrmann, C. W. Kohlpaintner, E. Herdtweck and P. Kiprof, *Inorg. Chem.*, 1991, **30**, 4271.
451. G. R. Dobson, I. Bernal, G. M. Reisner, C. B. Dobson and S. E. Mansour, *J. Am. Chem. Soc.*, 1985, **107**, 525.
452. S. D. Perera, B. L. Shaw and M. Thornton-Pett, *J. Organomet. Chem.*, 1992, **428**, 59.
453. W. Schirmer, U. Flörke and H.-J. Haupt, *Z. Anorg. Allg. Chem.*, 1989, **574**, 239.
454. R. D. Adams, C. Blankenship, B. E. Segmüller and M. Shiralian, *J. Am. Chem. Soc.*, 1983, **105**, 4319.
455. M.-O. Bevierre, F. Mercier, L. Ricard and F. Mathey, *Angew. Chem., Int. Ed. Engl.*, 1990, **29**, 655.
456. F. S. M. Hassan, B. L. Shaw and M. Thornton-Pett, *J. Chem. Soc., Dalton Trans.*, 1988, 89.
457. S. Al-Jibori, M. Hall, A. T. Hutton and B. L. Shaw, *ibid.*, 1984, 863.
458. M. Yoshifuji *et al.*, *Angew. Chem., Int. Ed. Engl.*, 1985, **24**, 211.
459. S. E. Saum, F. R. Fronczek, S. A. Laneman and G. G. Stanley, *Inorg. Chem.*, 1989, **28**, 1878.
460. S. Al-Jibori and B. L. Shaw, *Inorg. Chim. Acta*, 1983, **74**, 235.
461. G. R. Cooper, D. M. McEwan and B. L. Shaw, *ibid.*, 1983, **76**, L165.
462. J. Borm, G. Huttner and L. Zsolnai, *Angew. Chem., Int. Ed. Engl.*, 1985, **24**, 1069.
463. G. Huttner, U. Weber, B. Sigwarth, O. Scheidsteger, H. Lang and L. Zsolnai, *J. Organomet. Chem.*, 1985, **282**, 331.
464. J. Borm, G. Huttner, O. Orama and L. Zsolnai, *ibid.*, 1985, **282**, 53.
465. N. A. Compton, R. J. Errington and N. C. Norman, *Adv. Organomet. Chem.*, 1990, **31**, 91.
466. T. Chivers, K. S. Dhathathreyan, C. Lensink and J. F. Richardson, *Inorg. Chem.*, 1988, **27**, 1570.
467. L.-R. Frank, I. Jibril, L. Zsolnai and G. Huttner, *J. Organomet. Chem.*, 1987, **336**, 337.
468. F. Kober and M. Kerber, *Z. Anorg. Allg. Chem.*, 1983, **507**, 119.
469. H. Schumann, U. Frank, W.-W. Du Mont and F. Marschner, *J. Organomet. Chem.*, 1981, **222**, 217.
470. K. Betka and J. Grobe, *ibid.*, 1981, **210**, 193.
471. M. Wieber, H. Höhl and C. Burschka, *Z. Anorg. Allg. Chem.*, 1990, **583**, 113.
472. B. Sigwarth, L. Zsolnai, H. Berke and G. Huttner, *J. Organomet. Chem.*, 1982, **226**, C5.
473. G. Huttner, U. Weber, B. Sigwarth and O. Scheidsteger, *Angew. Chem., Int. Ed. Engl.*, 1982, **21**, 215.
474. G. Huttner, U. Weber and L. Zsolnai, *Z. Naturforsch., Teil B*, 1982, **37**, 707.
475. A. M. Arif, A. H. Cowley, N. C. Norman and M. Pakulski, *Inorg. Chem.*, 1986, **25**, 4836.
476. B. W. Eichhorn, R. C. Haushalter and J. C. Huffman, *Angew. Chem., Int. Ed. Engl.*, 1989, **28**, 1032.
477. A.-J. DiMaio and A. L. Rheingold, *Chem. Rev.*, 1990, **90**, 169.
478. A. L. Rheingold, M. E. Fountain and A.-J. DiMaio, *J. Am. Chem. Soc.*, 1987, **109**, 141.
479. A. L. Rheingold and M. E. Fountain, *Organometallics*, 1986, **5**, 2410.
480. A.-J. DiMaio and A. L. Rheingold, *Inorg. Chem.*, 1990, **29**, 798.
481. A. M. Arif, D. J. Chandler and R. A. Jones, *J. Coord. Chem.*, 1987, **16**, 213.
482. P. K. Baker, S. G. Fraser and M. G. B. Drew, *J. Chem. Soc., Dalton Trans.*, 1988, 2729.
483. G. Huttner, B. Sigwarth, O. Scheidsteger, L. Zolnai and O. Orama, *Organometallics*, 1985, **4**, 326.
484. L. Linford and H. G. Raubenheimer, *Adv. Organomet. Chem.*, 1991, **32**, 1.
485. L. C. Roof and J. W. Kolis, *Chem. Rev.*, 1993, **93**, 1037.
486. E. G. Hope and W. Levason, *Coord. Chem. Rev.*, 1993, **122**, 109.
487. R. R. Ryan, G. J. Kubas, D. C. Moody and P. G. Eller, *Struct. Bond.*, 1981, **46**, 47.
488. N. Kuhn, H. Schumann and G. Wolmershäuser, *J. Chem. Soc., Chem. Commun.*, 1985, 1595.
489. M. Y. Darensbourg *et al.*, *Inorg. Chem.*, 1990, **29**, 2721.
490. E. W. Abel, S. K. Bhargava, P. K. Mittal, K. G. Orrell and V. Šik, *J. Chem. Soc., Dalton Trans.*, 1985, 1561.
491. G. Bremer, P. Klüfers, and T. Kruck, *Chem. Ber.*, 1985, **118**, 4224.
492. D. Neugebauer and U. Schubert, *J. Organomet. Chem.*, 1983, **256**, 43.
493. H. Hausmann, M. Höfler, T. Kruck and H. W. Zimmerman, *Chem. Ber.*, 1981, **114**, 975.
494. G. Bremer, R. Boese, M. Keddo and T. Kruck, *Z. Naturforsch., Teil B*, 1986, **41**, 981.
495. E. W. Abel *et al.*, *Polyhedron*, 1989, **8**, 887.
496. M. E. Peach and C. Burschka, *Can. J. Chem.*, 1982, **60**, 2029.
497. T. Vogt and J. Strähle, *Z. Naturforsch., Teil B*, 1985, **40**, 1599.
498. K. Sünkel, A. Blum, K. Polborn and E. Lippmann, *Chem. Ber.*, 1990, **123**, 1227.
499. M. T. Ashby, J. H. Enemark, D. L. Lichtenberger and R. B. Ortega, *Inorg. Chem.*, 1986, **25**, 3154.
500. G. J. Grant, J. P. Carpenter, W. N. Setzer and D. G. VanDerveer, *ibid.*, 1989, **28**, 4128.
501. M. T. Ashby and D. L. Lichtenberger, *ibid.*, 1985, **24**, 636.
502. T. Yoshida, T. Adachi, T. Ueda, M. Watanabe, M. Kaminaka and T. Higuchi, *Angew. Chem., Int. Ed. Engl.*, 1987, **26**, 1171.
503. E. W. Abel, M. A. Z. Chowdhury, K. G. Orrell and V. Šik, *J. Organomet. Chem.*, 1984, **262**, 293.

504. H. Fischer, K. Treier and U. Gerbing, *ibid.*, 1992, **433**, 127.
505. C. R. Lucas, *Can. J. Chem.*, 1983, **61**, 1096.
506. G. N. Glavee, L. M. Daniels and R. J. Angelici, *Inorg. Chem.*, 1989, **28**, 1751.
507. E. W. Abel, K. G. Orrell, H. Rahoo, V. Šik, M. A. Mazid and M. B. Hursthouse, *J. Organomet. Chem.*, 1992, **437**, 191.
508. C. Habben, A. Meller, M. Noltemeyer and G. M. Sheldrick, *Z. Naturforsch., Teil B*, 1986, **41**, 799.
509. R. E. Schmidt and W. Massa, *ibid.*, 1984, **39**, 213.
510. E. W. Abel, T. E. MacKenzie, K. G. Orrell, and V. Šik, *Polyhedron*, 1987, **6**, 1785.
511. M.-G. Choi and R. J. Angelici, *J. Am. Chem. Soc.*, 1990, **112**, 7811.
512. H. Fischer, U. Gerbing, and J. Riede, *J. Organomet. Chem.*, 1989, **364**, 155.
513. H. Fischer and U. Gerbing, *ibid.*, 1986, **299**, C7.
514. E. W. Abel *et al.*, *ibid.*, 1989, **367**, 275.
515. J. A. Costamagna and J. Granifo, *Inorg. Synth.*, 1985, **23**, 1.
516. A. O. Baghlaf, M. Ishaq and A. S. Daifuliah, *Polyhedron*, 1984, **3**, 235.
517. T. C. W. Mak, K. S. Jasim and C. Chieh, *Inorg. Chim. Acta*, 1985, **99**, 31.
518. H. W. Roesky, R. Emmert, W. Isenberg, M. Schmidt and G. M. Sheldrick, *J. Chem. Soc., Dalton Trans.*, 1983, 183.
519. J. C. Barnes, W. Bell, C. Glidewell and R. A. Howie, *J. Organomet. Chem.*, 1990, **385**, 369.
520. J. Daub, J. Kappler, K. H. Jogun, J. J. Stezowski and H. Binder, *ibid.*, 1982, **240**, 239.
521. H. Fischer and S. Zeuner, *Z. Naturforsch., Teil B*, 1985, **40**, 954.
522. H. Fischer and S. Zeuner, *ibid.*, 1983, **38**, 1365.
523. P. V. Petrovskii, V. N. Setkina, S. P. Dolgova and D. N. Kursanov, *J. Organomet. Chem.*, 1981, **221**, 177.
524. H. G. Raubenheimer *et al.*, *J. Chem. Soc., Dalton Trans.*, 1989, 1565.
525. J. Valdés-Martínez, A. Sierra-Romero, C. Alvarez-Toledano, R. A. Toscano and H. García-Tapia, *J. Organomet. Chem.*, 1988, **352**, 321.
526. U. Kunze, R. Merkel and M. Moll, *ibid.*, 1983, **248**, 205.
527. H. G. Raubenheimer, G. J. Kruger, C. F. Marais, R. Otte and J. T. Z. Hattingh, *Organometallics*, 1988, **7**, 1853.
528. H. Fischer, S. Zeuner, U. Gerbing, J. Riede and C. J. Kreiter, *J. Organomet. Chem.*, 1989, **377**, 105.
529. H. Fischer and S. Zeuner, *ibid.*, 1983, **252**, C63.
530. Von Ch. Burschka, F.-E. Baumann and W. A. Schenk, *Z. Anorg. Allg. Chem.*, 1983, **502**, 191.
531. W. A. Schenk and F.-E. Baumann, *Chem. Ber.*, 1982, **115**, 2615.
532. W. A. Schenk and F.-E. Baumann, *J. Organomet. Chem.*, 1983, **256**, 261.
533. W. A. Schenk and F.-E. Baumann, *ibid.*, 1984, **260**, C6.
534. F.-E. Baumann, C. Burschka and W. A. Schenk, *Z. Naturforsch., Teil B*, 1986, **41**, 1211.
535. G. J. Kubas, G. D. Jarvinen and R. R. Ryan, *J. Am. Chem. Soc.*, 1983, **105**, 1883.
536. W. A. Schenk, D. Kuemmerle and C. Burschka, *J. Organomet. Chem.*, 1988, **349**, 183.
537. C. Bianchini, C. A. Ghilardi, A. Meli, S. Midollini and A. Orlandini, *Organometallics*, 1982, **1**, 778.
538. D. Miguel, J. A. Pérez-Martinez, V. Riera and S. García-Granda, *Angew. Chem., Int. Ed. Engl.*, 1992, **31**, 76.
539. E. Carmona, A. Galindo, A. Monge, M. A. Muñoz, M. L. Poveda and C. Ruiz, *Inorg. Chem.*, 1990, **29**, 5074.
540. D. Sellmann *et al.*, *J. Organomet. Chem.*, 1987, **323**, 323.
541. D. Sellmann, J. Keller, M. Moll, C. F. Campana and M. Haase, *Inorg. Chim. Acta*, 1988, **141**, 243.
542. D. Sellmann, R. Weiss and F. Knoch, *ibid.*, 1990, **175**, 65.
543. D. J. Darensbourg, A. Rokicki and R. Kudaroski, *Organometallics*, 1982, **1**, 1161.
544. D. J. Darensbourg, K. M. Sanchez and J. Reibenspies, *Inorg. Chem.*, 1988, **27**, 3636.
545. D. Sellmann, W. Ludwig, G. Huttner and L. Zsolnai, *J. Organomet. Chem.*, 1985, **294**, 199.
546. L. D. Rosenhein, W. E. Newton and J. W. McDonald, *ibid.*, 1985, **288**, C17.
547. D. A. Smith, B. Zhuang, W. E. Newton, J. W. McDonald and F. A. Schultz, *Inorg. Chem.*, 1987, **26**, 2524.
548. M. G. Hill, L. D. Rosenhein, K. R. Mann, X. H. Mu and F. A. Schultz, *ibid.*, 1992, **31**, 4108.
549. B. Zhuang, L. G. Huang, L. He, Y. Yang and J. Lu, *Inorg. Chim. Acta*, 1989, **157**, 85.
550. M. K. Cooper, P. A. Duckworth, K. Henrick and M. McPartlin, *J. Chem. Soc., Dalton Trans.*, 1981, 2357.
551. W. Röll, E. O. Fischer, D. Neugebauer and U. Schubert, *Z. Naturforsch., Teil B*, 1982, **37**, 1274.
552. B. Zhuang, L. Huang, Y. Yang and J. Lu, *Inorg. Chim. Acta*, 1986, **116**, L41.
553. B. Zhuang, L. Huang, L. He, Y. Yang and J. Lu, *ibid.*, 1987, **127**, L7.
554. I. L. Eremenko *et al.*, *J. Organomet. Chem.*, 1989, **365**, 297.
555. A. Winter, O. Scheidsteger and G. Huttner, *Z. Naturforsch., Teil B*, 1983, **38**, 1525.
556. T. E. Burrow *et al.*, *J. Chem. Soc., Dalton Trans.*, 1991, 2519.
557. A. J. Deeming, M. Karim and N. I. Powell, *ibid.*, 1990, 2321.
558. D. J. Darensbourg and A. Rokicki, *J. Am. Chem. Soc.*, 1982, **104**, 349.
559. E. Carmona, L. Contreras, L. J. Sánchez, E. Gútierrez-Puebla and A. Monge, *J. Chem. Soc., Dalton Trans.*, 1989, 2003.
560. B. Zhuang, L. Huang, L. He, Y. Yang and J. Lu, *Inorg. Chim. Acta*, 1988, **145**, 225.
561. A. E. Sánchez-Peláez, M. F. Perpiñán, E. Gútierrez-Puebla, A. Monge and C. Ruiz-Valero, *J. Organomet. Chem.*, 1990, **384**, 79.
562. D. J. Darensbourg, H. P. Wiegreffe and J. H. Reibenspies, *Organometallics*, 1991, **10**, 6.
563. C. G. Young, M. A. Bruck, P. A. Wexler, M. D. Carducci and J. H. Enemark, *Inorg. Chem.*, 1992, **31**, 587.
564. B. Zhuang, P. Yu, L. Huang, L. He and J. Lu, *Inorg. Chim. Acta*, 1990, **177**, 239.
565. D. J. Darensbourg, D. J. Zalewski, K. M. Sanchez and T. Delord, *Inorg. Chem.*, 1988, **27**, 821.
566. M. Hoefler, K.-F. Tebbe, H. Veit and N. E. Weiler, *J. Am. Chem. Soc.*, 1983, **105**, 6338.
567. A. Seigneurin, T. Makin, D. J. Jones and J. Rozière, *J. Chem. Soc., Dalton Trans.*, 1987, 2111.
568. M. J. Collins, R. J. Gillespie, J. W. Kolis and J. F. Sawyer, *Inorg. Chem.*, 1986, **25**, 2057.
569. C. Belin, T. Makani and J. Rozière, *J. Chem. Soc., Chem. Commun.*, 1985, 118.
570. R. Faggiani, R. J. Gillespie, C. Campana and J. W. Kolis, *J. Chem. Soc., Chem. Commun.*, 1987, 485.
571. O. Scheidsteger, G. Huttner, K. Dehnicke and J. Pebler, *Angew. Chem., Int. Ed. Engl.*, 1985, **24**, 428.
572. W. A. Flomer, S. C. O'Neal, J. W. Kolis, D. Jeter and A. W. Cordes, *Inorg. Chem.*, 1988, **27**, 969.
573. J. W. Kolis, *Coord. Chem. Rev.*, 1990, **105**, 195.
574. L. C. Roof, W. T. Pennington and J. W. Kolis, *J. Am. Chem. Soc.*, 1990, **112**, 8172.

575. E. W. Abel, S. K. Bhargava and K. G. Orrell, *Prog. Inorg. Chem.*, 1984, **32**, 1.
576. E. W. Abel, K. G. Orrell, H. Rahoo and V. Šik, *J. Organomet. Chem.*, 1992, **441**, 441.
577. K. R. Barnard, A. G. Wedd and E. R. T. Tiekink, *Inorg. Chem.*, 1990, **29**, 891.
578. D. Sellmann, F. Grasser, F. Knoch and M. Moll, *Inorg. Chim. Acta*, 1992, **195**, 25.
579. J. L. Templeton and S. J. N. Burgmayer, *Organometallics*, 1982, **1**, 1007.
580. W. A. Schenk, D. Rüb and C. Burschka, *J. Organomet. Chem.*, 1987, **328**, 305.
581. H. Fischer, U. Gerbing, K. Treier and J. Hofmann, *Chem. Ber.*, 1990, **123**, 725.
582. H. Fischer *et al.*, *Chem. Ber.*, 1988, **121**, 2095.
583. H. Fischer and I. Pashalidis, *J. Organomet. Chem.*, 1988, **348**, C1.
584. H. Fischer, J. Hofmann, U. Gerbing and A. Tiriliomis, *ibid.*, 1988, **358**, 229.
585. G. J. Kruger, L. Linford and H. G. Raubenheimer, *J. Chem. Soc., Dalton Trans.*, 1984, 2337.
586. W. A. Schenk, T. Schwietzke and H. Müller, *J. Organomet. Chem.*, 1982, **232**, C41.
587. W. A. Schenk, D. Kuemmerle and T. Schwietzke, *ibid.*, 1988, **349**, 163.
588. A. Galindo *et al.*, *J. Chem. Soc., Dalton Trans.*, 1992, 2307.
589. R. J. Angelici and R. G. W. Gingerich, *Organometallics*, 1983, **2**, 89.
590. R. C. Hynes, K. F. Preston, J. J. Springs and A. J. Williams, *ibid.*, 1991, **10**, 180.
591. J. Springs *et al.*, *J. Am. Chem. Soc.*, 1990, **112**, 5789.
592. B. Zhuang, P. Chen, L. Huang and J. Lu, *Polyhedron*, 1992, **11**, 127.

5

Organometallic Compounds of Chromium, Molybdenum and Tungsten without Carbonyl Ligands

MARK J. WINTER
University of Sheffield, UK
and
SIMON WOODWARD
University of Hull, UK

5.1 INTRODUCTION

The first part of this chapter covers the organometallic chemistry of group 6 metal compounds bearing no carbonyl ligands for which the carbon ligands are principally σ-only donors. The latter part of this chapter covers π-acceptor ligands and M–M complexes. By the nature of the chemistry, it is necessary to make frequent cross-references to other sections. A review covering several of the earlier sections in this chapter, but with special reference to σ-aryl complexes, is available.[1]

5.2 NEUTRAL HOMOLEPTIC COMPLEXES

The number of MR_4 complexes continues to expand, while the MR_6 systems remain restricted to trigonal prismatic $[WMe_6]$[2] and $[\overline{W(2\text{-}CH_2C_6H_4CH_2)_3}]$ (**1**).[3] Apparently no stable tri- or pentaorgano homoleptic complexes have appeared since 1981. Complexes such as $[\overline{Cr\{(CH_2)_3NMe_2\}_3}]$ and related species are treated as internal solvato complexes and these are considered in Section 5.3.

5.2.1 Synthesis

Recent representative preparations of homoleptic group 6 organometallic complexes are presented in Table 1. These syntheses typically use transmetallation reactions. Finding optimal conditions for the preparations of a given MR_n complex is often an empirical process. For example, the best syntheses of $[M(2\text{-Tol})_4]$ (M = Cr, Mo, W) employ aryllithium reagents in the case of chromium and molybdenum, but for tungsten, Grignard reagents are preferred. Even under optimum conditions low yields (<25%) are normally attained; these are attributable to inefficient disproportionation or oxidation reactions of $[MR_4]^-$ intermediates. The importance of identifying correct reaction protocols is further exemplified by the failure of $MoCl_5$ or $[Mo(acac)_2O_2]$ to yield $[Mo(1\text{-norbornyl})_4]$ when treated with the organolithium reagents,[4] compared with the successful preparation which uses $[MoCl_3(THF)_3]$ (Table 1).

Table 1 Syntheses of neutral homoleptic group 6 organocomplexes.

Synthesis	*Complex*	*Yield* (%)	*Colour*	μ_{eff} (μ_B)	*Comments*	*Ref.*
A. MR_2 Complexes						
WCl_6 + Li(1-norbornyl)	$[W(1\text{-norbornyl})_2]_n$	66	brown	0.64	poorly characterized	4
B. MR_4 Complexes						
$[CrCl_3(THF)_3]$ + CyMgBr	$[CrCy_4]$	22	dark red	2.77	x-ray	5
$[CrCl_3(THF)_3]$ + $Li(CPh{=}CMe_2)$	$[Cr(CPh{=}CMe_2)_4]$	2	green	2.81	x-ray	6
$[CrCl_3(THF)_3]$ + Li(2-Tol)	$[Cr(2\text{-Tol})_4]$	27	purple	2.7		7
$[MoCl_3(THF)_3]$ + Li(1-norbornyl)	$[Mo(1\text{-norbornyl})_4]$	25	red	2.62	partial x-ray	8
$[MoCl_4(THF)_2]$ + Li(2-Tol)	$[Mo(2\text{-Tol})_4]$	13	purple	2.9	x-ray	9
$[W(OPh)_6]$ + (2-Tol)MgBr	$[W(2\text{-Tol})_4]$	13	blue	2.9		7
$[W(OPh)_6]$ + $(2,5\text{-}C_6H_3Me_2)MgBr$	$[W(C_6H_3Me_2\text{-}2,5)_4]$	4	blue	2.9		7
C. MR_6 Complexes						
$[WCl_4(O)]$ + $[1,2\text{-}C_6H_4(CH_2MgCl)_2]$	$[\overline{W(2\text{-}CH_2C_6H_4CH_2})_3]$	35	orange	diamagnetic	x-ray	3

5.2.2 Structure and Properties

It is normal to employ steric protection of the metal centre in homoleptic organo complexes to provide stability against M–C bond homolysis and protonation reactions. In general, alkyl complexes are more reactive than their aryl counterparts. All the group 6 aryl complexes are stable at room temperature. The air and moisture sensitivity of these complexes varies in the order chromium ≫ molybdenum, tungsten. The bond enthalpy in $[WMe_6]$ is ~160 kJ mol^{-1}.[10]

Photoelectron spectroscopy suggests that $[MR_4]$ complexes exist in the gas phase with a distorted D_{2d} 0geometry by contraction of two R–M–R angles to ~96° rather than the expected tetrahedral array. The $4d^2$ electrons of $[Mo(1\text{-norbornyl})_4]$ reside in a pair of discrete molecular orbitals inconsistent with a degenerate $e^2(T_d)$ configuration.[11] Similar conclusions are drawn from electron diffraction studies of $[Cr(CH_2Bu^t)_4]$ and supported by *ab initio* molecular calculations on $[CrMe_4]$.[12] Distortions towards D_{2d} symmetry may also be inferred from the visible spectra of $[CrR_4]$ (R = 1-norbornyl, CH_2Bu^t). The strongest absorptions (~480 nm) of these complexes are assigned as $^3B_1 \rightarrow {}^3E$ and $^3B_1 \rightarrow A_2$ under D_{2d}.[13] These states are derived from the symmetry-forbidden $^3A_2(e^2) \rightarrow {}^3T_1(e^1t_2^1)$ transition under strict T_d symmetry.[14] In the solid state, however, there is little evidence for MR_4 distortion. The range of C–M–C angles is 105–116° but known structures all show the metal to be essentially tetrahedral rather than displaying clear D_{2d} distortion.

Naively one might expect $[MR_6]$ complexes to be octahedral, but this is not the case. Electron diffraction studies of $[WMe_6]$ indicate a trigonal prismatic structure (either D_{3h} or C_{3v}) with an average W–Me distance of 0.2146(3) nm.[2] This geometry is also present in the crystal structure of (**1**) (Figure 1). Such distortions accord with MO calculations on d^0 ML_6 systems with small strongly ligated hydride or alkyl functions, for which D_{3h} and C_{3v} geometries have the lowest energy (Figure 1).[15]

5.2.3 Reactions

The small atomic radii associated with M^{IV} in the presence of large organo ligands renders tetraorgano complexes unexpectedly unreactive (amines, alkynes and CO can fail to react). The known reactivity of $[MR_4]$ complexes is shown in Scheme 1. The oxidation of $[Mo(Mes)_4]$ to $[Mo(Mes)_4]^{+\cdot}$ is not general; so far nothing has proved isolable for the oxidation of other $[MR_4]$ complexes. The basic reactivity of $[WMe_6]$ has also expanded a little since 1980.[16,17]

5.3 SOLVATO, HALOGENO AND PSEUDOHALOGEN COMPLEXES

This section covers mono-, di- and triorgano complexes that also feature at least one other noncarbon ligand that is either a σ- or a moderate π-donor. However, aquo and phosphine complexes are regarded as special cases and have their own sections (Sections 5.6 and 5.7, respectively).

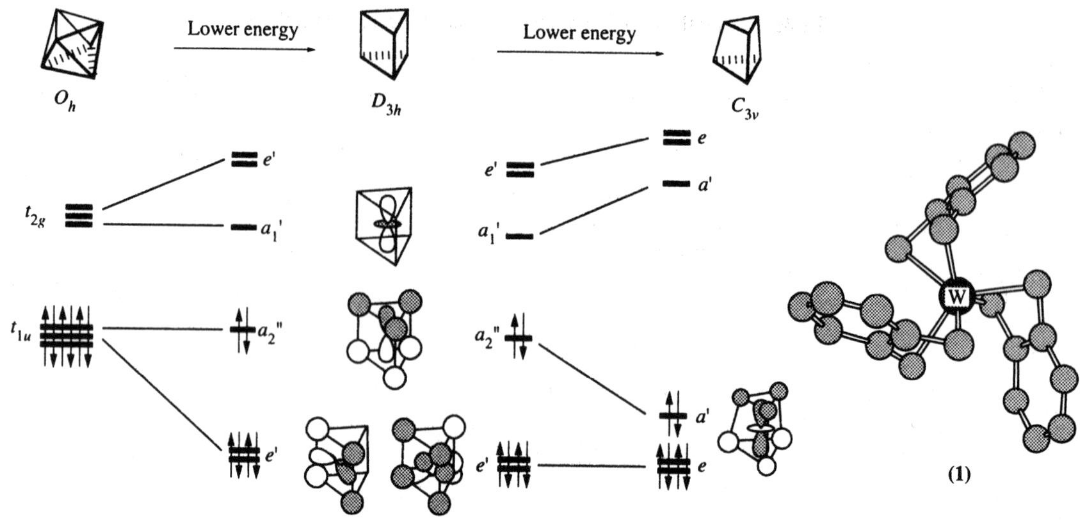

Figure 1 Molecular orbital diagram for d^0 ML_6 systems with O_h, D_{3h} and C_{3v} geometries, and the crystal structure of $[W(2\text{-}CH_2C_6H_4CH_2)_3]$ (**1**) (reproduced in part by permission of the American Chemical Society from *Inorg. Chem.* 1989, **28**, 1612).

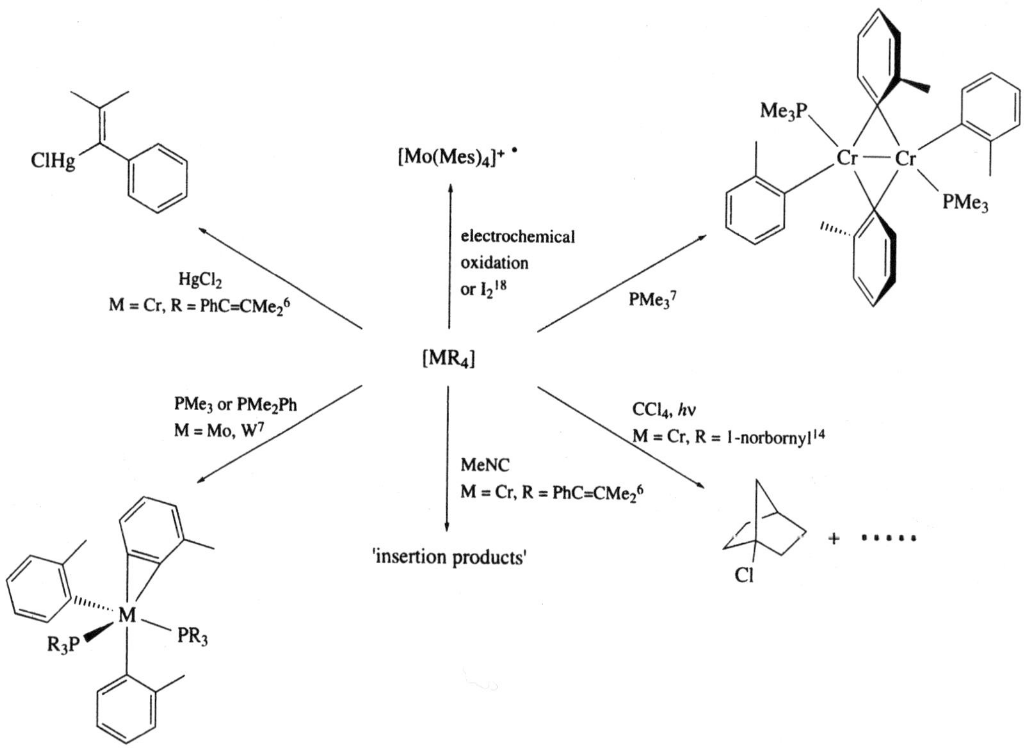

Scheme 1

5.3.1 Synthesis

Typical preparations of ligated organo group 6 complexes are shown in Table 2. Transmetallation reactions using organolithium, Grignard and organoaluminum reagents are commonly employed, but unusual reactivity is sometimes encountered. For example, the preparation of the unique μ_4-oxide (**2**) (Equation (1)) is rather dependent on the reaction stoichiometry.[19] Other approaches include preparation from pentaaquochromium complexes (Equation (2)) and C–H activation reactions (Equation (3)).

Table 2 Syntheses of substituted group 6 organo complexes.

Synthesis	*Complex*	*Comments*	*Ref.*
A. Monoorgano species			
$[CrCl_3(THF)_3] + MR$	$[CrCl_2(R)(THF)_n]$	M = Li, MgX, AlR_2; R = Me, Et, Pr^n, Bu^n, Bu^s, n-C_8H_{17}, CH_2Ph. Not isolated	20
$[Cr(acac)_3] + AlPh_3$	$[Cr(Ph)(acac)_2(THF)]$	orange (63%), $\mu_{eff} = 3.88\ \mu_B$	21
$[Cr(acac)_3]$ + $Li[Fe(C_5H_3CH_2NMe_2\text{-}2)(Cp)]$	$[Cr(acac)_2\{(2\text{-}Me_2NCH_2C_5H_3)Fe(Cp)\}]$	green (40%), $\mu_{eff} = 3.73\ \mu_B$	22
$[Cr(R)(acac)_2(OH_2)]$ + py	$[Cr(R)(acac)_2(py)]$	R = $CHCl_2$ (78%), CH_2Cl (73%), x-ray	23–5
$[CrCl(TTP)]^a + RMgX$	$[Cr(R)(TTP)]$	R = Ph, 4-$C_6H_4Bu^t$, CH_2TMS (30–50%), red, $\mu_{eff} = 3.3–4.0\ \mu_B$	26
$[Mo(TPP)(CO)_2]^b$ + chlorinated impurities	*trans*-$[MoCl(Ph)(TPP)]$	low yield, x-ray	27
$[Mo(NMe_2)_4]$ + ArOH	$[Mo(OAr)_2\{2,6\text{-}OC_6H_3Ph(C_6H_4)\}(L)]$	L = $NHMe_2$ (75%), py (70%), Ar = 2,6-$C_6H_3Ph_2$, red, diamagnetic, x-ray	28
B. Diorgano species			
$[CrCl_2(THF)_2] + Li[2\text{-}C_6H_4CH_2NMe_2]$	$[Cr(2\text{-}C_6H_4CH_2NMe_2)_2]$	yellow (67%), $\mu_{eff} = 4.69\ \mu_B$	29
$[Cr(2\text{-}C_6H_4CH_2NMe_2)_2]$ + py	$[Cr(2\text{-}C_6H_4CH_2NMe_2)_2(py)]$	red (63%), $\mu_{eff} = 2.57\ \mu_B$, x-ray	29
$[CrCl_2(THF)_2]$ + 2 MesLi	$[Cr(Mes)_2(THF)_2]$	violet (20–25%), $\mu_{eff} = 4.69\ \mu_B$, x-ray	30
$[Cr(Mes)_2(THF)_2]$ + bipy	$[Cr(Mes)_2(bipy)]$	blue in C_6H_6, $\mu_{eff} = 4.86\ \mu_B$, x-ray	30
$MoCl_5$ + 2 MeLi	$[MoCl_3(Me)_2(THF)_n]$	poorly characterized	31
C. Triorgano species			
$CrCl_3$ + PhMgBr	*fac*-$[Cr(Ph)_3(THF)_3]$	improved yields, x-ray	32
$[CrCl_3(THF)_3] + Li[CPh{=}CMe_2]$ + L	$[Cr(CPh{=}CMe_2)_3(L)]$	blue (L = THF, low yield), green (L = TMEDA, 28%)	6
$[Cr_2(OAc)_4]$ + 4 $Li[2\text{-}C_6H_4CH_2NMe_2]$	$[Cr(2\text{-}C_6H_4CH_2NMe_2)_3]$	low yield, x-ray structure	29
$[Cr_4(\mu\text{-}2\text{-}C_6H_4NMe_2)_4]$	$[Cr(2\text{-}C_6H_4NMe_2)_3]$	orange (32%), $\mu_{eff} = 3.73\ \mu_B$, x-ray	29
$[CrCl_3(THF)_3] + Me_2N(CH_2)_3Li$	$[Cr(CH_2CH_2CH_2NMe_2)_3]$	red (76%), $\mu_{eff} = 3.66\ \mu_B$	33

[a] TTP = dianion of *meso*-tetra-4-tolylporphyrin. [b] TPP = dianion of *meso*-tetraphenylporphyrin.

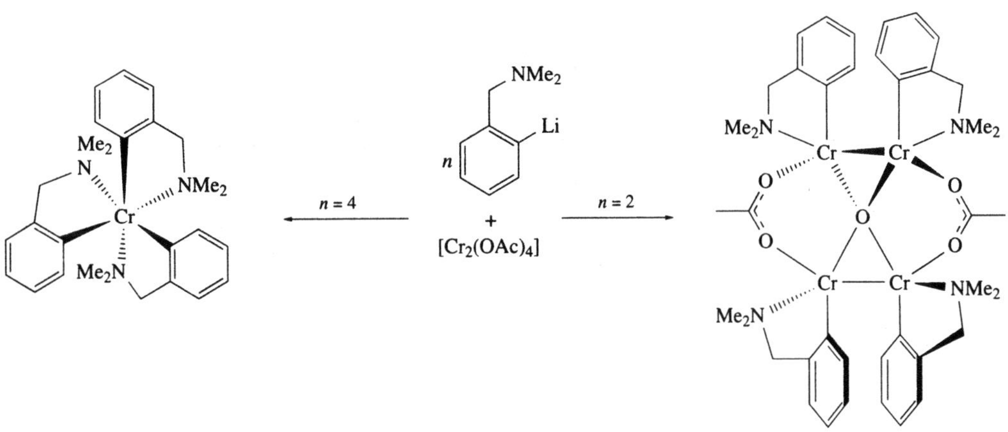

(2)

(1)

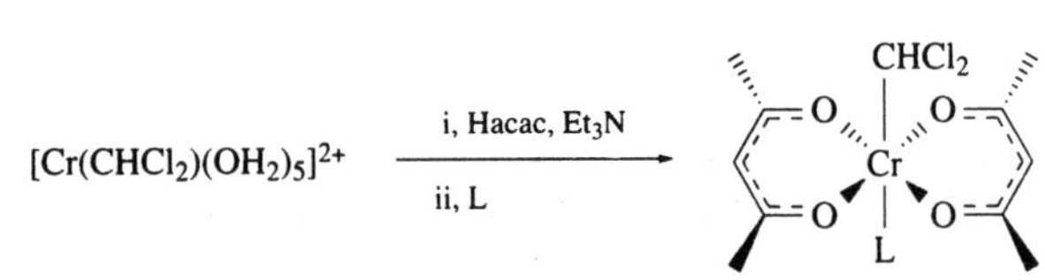

L = MeOH, py

(2)

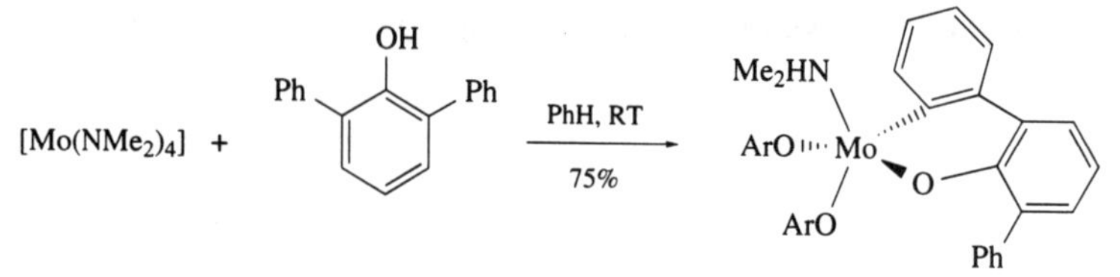

(3)

5.3.2 Structure and Properties

This class of compounds covers a rather diverse range of ligands with the metals in either oxidation state +2 or +3, with chromium complexes in the majority. As might be expected there is a broad spectrum of stability and structure.

A number of the arylchromium complexes have been structurally characterized and these are shown in Table 3. The range of Cr–C bonds is 0.204–0.213 nm and they are normally considerably shorter than other ligand contacts. One possible origin for this observation is *d–p* bonding in either a π-donor or π-acceptor sense. Monomeric alkylchromium(II) complexes show distorted square planar structures. These distortions are best described by the C–Cr–C vs. L–Cr–L interplanar angle, θ_{planes}, which is 0° in ideal square planar complexes and expands to 90° for tetrahedral species. For complexes (**3**) and (**4**) (Table 3) these angles are 27.6(6) and 31.5(3)°, respectively. In triorgano systems, $[CrR_3L_3]$, known structures all show exclusively the *fac* isomer. Apparently no structurally characterized *mer* example is known.

5.3.3 Reactions

Insertion reactions of $[Cr(Mes)_3(THF)]$ with Bu^{t}NC or Bu^{t}NCO lead to the expected products (Equation (4)).[34] The products have not been structurally characterized but mononuclear structures such as (**5**) are probable. Reaction of haloalkylchromium complexes with sources of $[Cp^*]^-$ leads to the cyclopentadienyl dimers (**6**) (Equation (5)).[35,36] The species $[CrCl_2(R)(THF)_n]$ is a versatile organic reagent for addition reactions to carbonyl compounds (Equation (6)).[20] These reagents are also generated by oxidative additions to $CrCl_2$ and this chemistry has been reviewed.[37] The postulated organochromium reagents in these transformations, such as (**7**), are normally neither isolated nor characterized. Molybdenum reagents are also of use in organic synthesis, but in this case for alkenation reactions (Equation (7)), for which the intermediate (**8**) can be detected by ^{1}H and ^{13}C NMR spectroscopy.[31,38,39]

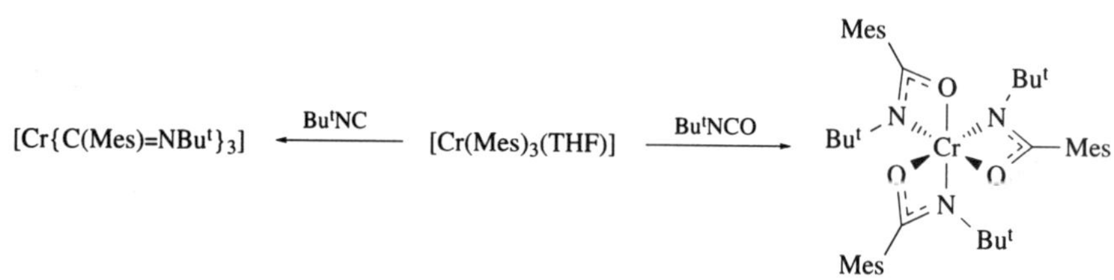

(4)

(**5**)

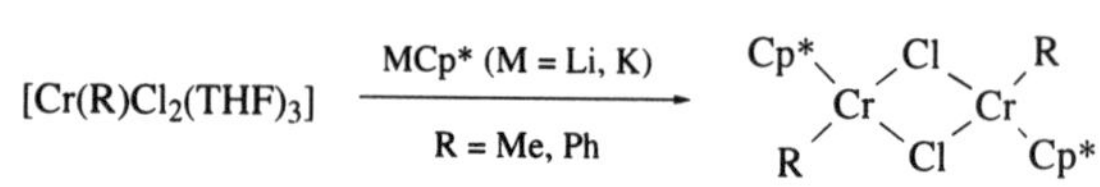

(5)

(**6**)

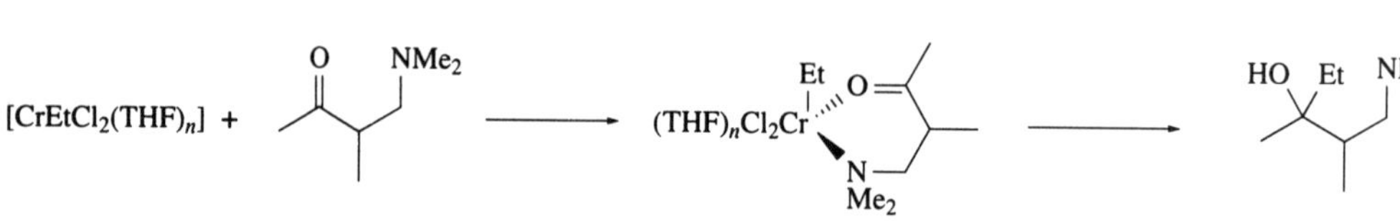

(6)

(**7**)

Table 3 Crystallographically characterized organochromium solvates.

Complex	*M–C* (nm)	*Other bond distances* (nm)		*Selected bond angles* (°)		*Ref.*
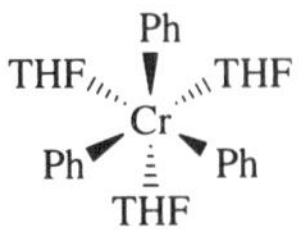	0.206(10)	Cr–O	0.2225(10)	C–Cr–C	93.9(4)	32
				C–Cr–O	91.2(6)	
				O–Cr–O	83.2(3)	
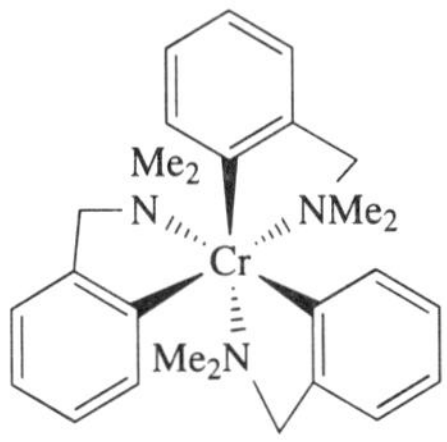	0.2078(5)	Cr–N	0.2355(4)	C–Cr–C	94.0(2)	19
				C–Cr–N_{cis}	77.2(2)	
					92.9(2)	
				N–Cr–N	98.7(2)	
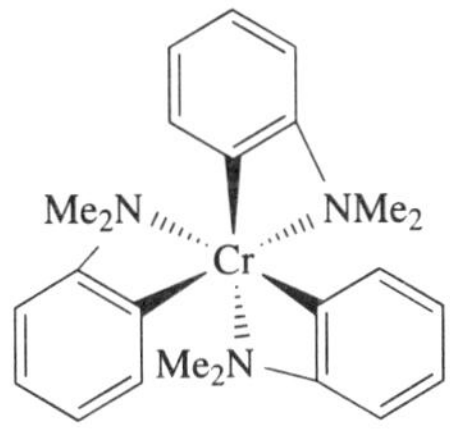	0.2037(12)	Cr–N	0.2286(10)	C–Cr–C	99.3(5)	29
				C–Cr–N_{cis}	85.0(5)	
					97.9(5)	
				N–Cr–N	100.2(4)	
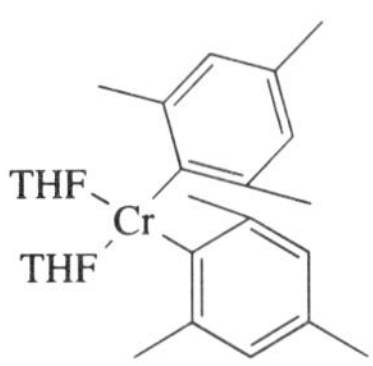 **(3)**	0.2083(11)	Cr–O	0.2111(8)	C–Cr–O	91.3(4)	30
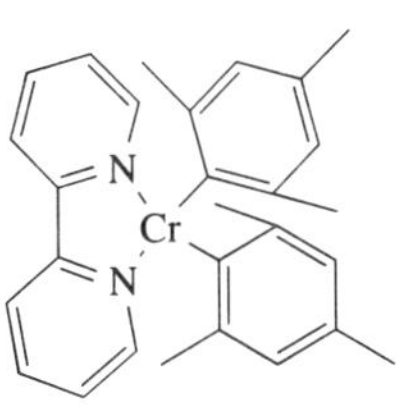 **(4)**	0.2099(5)	Cr–N	0.2137(4)	C–Cr–O	103.5(7)	30
		Cr–N	0.2137(4)	C–Cr–N_{cis}	96.2(2)	
	0.2130(4)	Cr–N	0.2131(4)		155.0(2)	
				N–Cr–N	76.2(2)	

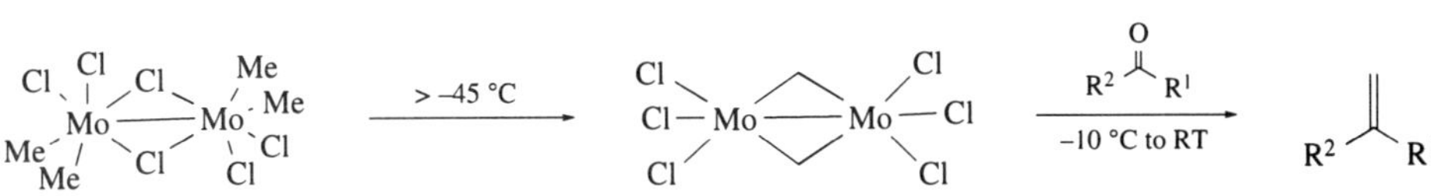

(8) (7)

5.4 ORGANOMETALLATE COMPLEXES

5.4.1 Synthesis

Organochromium species form the majority of recent examples of organometallate complexes. Representative syntheses appear in Table 4. It is general to employ methods where a metal halide salt is treated with excess organolithium species in an ethereal solvent to prepare these species.

Table 4 Syntheses of group 6 organometallate complexes.

Synthesis	*Complex*	*Colour*	*Yield (%)*	*Comments*	*Ref.*
$[CrCl_2(THF)_2]$ + MeLi + TMEDA	$[Li(TMEDA)]_2[CrMe_4]$	yellow	57	$\mu_{eff} = 4.98\ \mu_B$, x-ray, rearranges to $[Cr_2Me_8]^{4-}$ in THF	40
$[CrCl_2(THF)_2]$ + MesLi	$[Li(THF)]_2[Cr(Mes)_4]$	orange	60	$\mu_{eff} = 4.69\ \mu_B$, x-ray	30
$[CrCl_2(THF)_2]$ + PhLi	$[Li(THF)_2]_2[CrPh_4]$	yellow	67	$\mu_{eff} = 4.73\ \mu_B$, x-ray	30
$CrCl_2$ + PhLi	$[Li(OEt_2)]_3[CrPh_6]$	orange	40	x-ray	41
$MoCl_5$ + Li[4-C_6H_4Me]	$[Li(OEt_2)]_3[Mo(4\text{-Tol})_6]$	yellow	38	$\mu_{eff} = 3.51\ \mu_B$	42
WBr_5 + Li[4-C_6H_4Me]	$[Li(OEt_2)]_4[W(4\text{-Tol})_4(\eta^2\text{-}C_6H_3Me)_2]$	dark violet	6	diamagnetic	42

5.4.2 Structure and Properties

The structure and behaviour of $[Li(TMEDA)]_2[CrMe_4]$ (**9**) (Scheme 2) is quite remarkable.[40] The magnetic moment of (**9**) at 4.98 μ_B is typical for such a square planar chromium(II) complex. The bridging methyl groups show slightly distorted sp^3 geometries at the hinges of the double butterfly structure. While the Cr–Me (0.2204(5) nm, 0.2213(4) nm) and Li–Me (0.217.6(7) nm) distances indicate clear three-centre bonding interactions, Li–Me–Cr, the chromium–lithium distance of 0.2589(6) nm is suggestive of the presence of an additional weak Cr–Li interaction. However, the presence of short Li–H contacts (0.1992 pm, 0.2064 pm) in the bridge allows for an alternative bonding scheme featuring agostic interactions of two methyl hydrogens with lithium.

The structure of (**9**) is modified upon exposure to THF. The result is (**10**), while heating in toluene results in decomposition, possibly via (**11**) (Scheme 2). This latter molecule is available by treatment of $[CrCl_2(THF)_2]$ with MeLi and TMEDA. The incredibly facile cleavage of the formal quadruple bond in (**10**) indicates that the stability of the dimer arises from its lithium bridges rather than any strong Cr–Cr interaction. The reaction is completely reversible; treatment of (**10**) with TMEDA in Et_2O leads to (**9**). The molybdenum analogue of (**10**) fails to react, indicating that although the quadruple bond of (**10**) is innately weak, the equivalent interaction in the dimolybdenum species is much stronger.

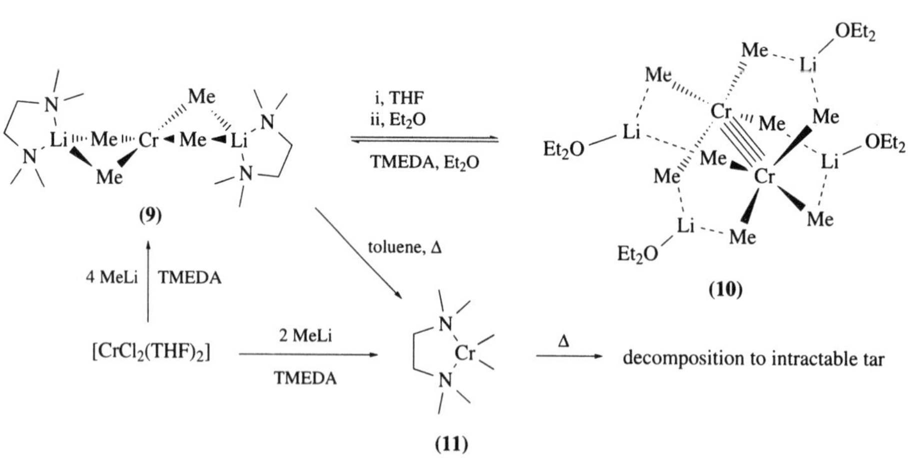

Scheme 2

The reaction of WBr_5 with excess 4-tolyllithium leads to a dark violet complex in low yield. Its ^1H NMR spectrum is in accord with structure (**12**), if rotation of the two axial tolyne units (R) is assumed to be rapid. The molecule is fluxional above −30 °C, which can be ascribed to dodecahedral versus square antiprism geometry interconversion.[42] However, an alternative fluxional process involving pairwise aryl-face exchange of lithium cations could also explain the dynamic ^1H NMR spectrum.

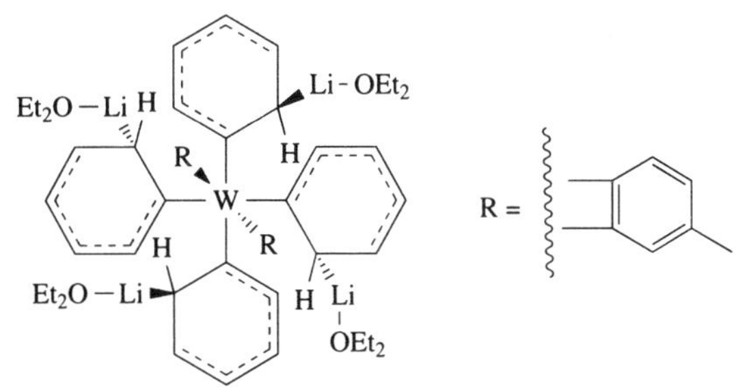

(12)

5.4.3 Reactions

In the presence of added donor ligands, organometallate complexes show a tendency to lose one or more equivalents of LiR leading to the formation of solvated species. For example, treatment of either $[Li(THF)]_2[Cr(Mes)_4]$ or $[Li(THF)_2]_2[Cr(Ph)_4]$ with bipy leads to isolation of $[Cr(Ar)_2(bipy)]$ (Ar = Mes, Ph).[30]

5.5 OXO AND IMIDO COMPLEXES

Perhaps the fastest growth in the chemistry of noncarbonyl group 6 organometallics since 1980 has been of complexes containing strongly π-donating oxo (=O) and imido (≡NR) functions. Although often represented in this chapter as having double and triple bonds, respectively, these selections have been chosen for convenience alone and do not indicate that imido species are more strongly bonded than oxo species; the bond order in both normally approaches that of a triple bond. Due to the large quantity of literature a representative selection has been made, and the coverage here is intended as comprehensive in scope if not in examples. Some of the complexes in this section are either precursors to alkylidene species or are derived themselves from alkylidene or alkylidyne complexes. In these cases cross-reference to Sections 5.11 and 5.12 may be necessary.

5.5.1 Synthesis

5.5.1.1 Oxo complexes

Syntheses of representative organometallic oxo compounds appear in Table 5, and further examples may be found in a review.[43] Most rely on metathesis reactions followed by recrystallization and commonly crystallographic analysis. Both aryl and alkyl lithium or magnesium reagents may be employed but these reactions are normally limited to organo ligands lacking β-hydrogens. Notable exceptions are the crystallographically characterized ethyl complexes (**13**) and (**14**). The stability of (**13**) is largely due to the steric protection of the hindered pyrazoylborate ligand. In the absence of x-ray data structural assignments of paramagnetic species can be problematic. For example, reaction of (**15**) with $[Mo(acac)_2(=O)_2]$ or $[WCl_4(=O)]$ leads to species tentatively formulated as (**16**) and (**17**), respectively. However, the observation of magnetic moments and the presence of a single IR Mo=O stretch for (**16**) militate somewhat against these structures.

The strongly reducing properties of organolithium and Grignard reagents can lead to products of lower oxidation state being isolated, even when M^{VI} starting materials are employed, for example, in the preparation of $[Mg(THF)_4][M(Me)_4(=O)]_2$ (Table 5) (M = Mo, W). The sensitivity of chromium(VI) oxo complexes to reduction renders transmetallation approaches to chromium species problematic. For example, while MesMgBr reacts cleanly with $[MoCl_2(=O)_2]$ to yield $[Mo(Mes)_2(=O)_2]$, use of $[CrCl_2(=O)_2]$ leads to chromium(V) and chromium(III) products.[66]

Table 5 Preparation of organo group 6 oxide complexes.

Synthesis	*Complex*	*Comments*	*Ref.*
A. Preparation by transmetallation			
$[MoCl_3(=O)(THF)_2] + MeLi$	$[MoCl(Me)_2(=O)(THF)_n]$	unstable above −70 °C	38,39
$[MoBr_2(=O)_2(bipy)] + RMgX$	$[MoBr(R)(=O)_2(bipy)]$	R = Me, Et, Pr^n, Pr^i, Bu^t, CH_2Bu^t	44
$[MoBr_2(=O)_2(bipy)] + RMgX$	$[Mo(R)_2(=O)_2(bipy)]$	R = Me, Et, Pr^n, Pr^i, Bu^n, Cy, CH_2Ph, CH_2Bu^t, CH_2TMS, x-rays	45–9
$[MBr_2(=O)_2(bipy)] + ArMgX$	$[MAr_2(=O)_2(bipy)]$	M = Mo, W; Ar = Ph, 4-Tol, 4-C_6H_4OMe; x-ray (M = Mo)	50,51
$[MCl_4(=O)] + MeLi$	$[Li(THF)_2][M(Me)_4(=O)]$	M = Mo, W; orange-red (24–26%)	52
$[MCl_4(=O)] + MeMgCl$	$[Mg(THF)_4][M(Me)_4(=O)]_2$	M = Mo, W; orange-red (22–28%)	52
$[WCl_4(=O)] + Mg[1,2\text{-}(CH_2)_2C_6H_4]_2$	$[Mg(THF)_4][W(2\text{-}CH_2C_6H_4CH_2)_2(=O)]_2$	orange (35%), x-ray	3
$[WCl_4(=O)] + C_8H_{12}(CH_2MgCl)_2$	$[Mg(THF)_4][W(2\text{-}CH_2C_{12}H_8CH_2)_2(=O)]_2$	x-ray	53
$[WBr_4(=O)(bipy)] + MeMgBr$	$[W(Me)_4(=O)(bipy)]$	x-ray, forms $[WMe_2(=O)_2(bipy)]$	54
$[WCl_4(=O)] + Mg(CH_2Bu^t)_2$	$[WCl(CH_2Bu^t)_3(=O)]$		55
$[WCl_4(=O)] + Mg(CH_2Bu^t)_2$	$[W(CH_2Bu^t)_3(OCH_2Bu^t)(=O)]$	initially misassigned as $[W(CH_2Bu^t)_4(=O)]$	55,56
$[WCl_4(=O)] + LiNEt_2 + Mg(CH_2Bu^t)_2$	$[WCl(CH_2Bu^t)_2(NEt_2)(=O)]$	colourless (30%)	57
$[WCl(CH_2Bu^t)_2(NEt_2)(=O)] + LiCH_2Bu^t$	$[W(CH_2Bu^t)_3(NEt_2)(=O)]$	orange (85%), x-ray	57
$[WCl_4(=O)] + [FcH + BuLi]$	$[W(X)(Fc)_3(=O)]$	X = OFc, OBu, Cl; red, x-ray (X = OFc)	58,59
$[MCl(=O)_2\{HB(3,5\text{-}Me_2pz)_3\}] + RMgX$	$[M(R)(=O)_2\{HB(3,5\text{-}Me_2pz)_3\}]$	M = Mo, W; R = Me, Et, Bu^t, Ph, CH_2Ph, CMe_2Ph, $CMe{=}CH_2$; x rays	60,61
B. Preparation by alternative routes			
$[Mo(CH_2Bu^t)Br(=O)_2(bipy)] + H_2O$	$[Mo(CH_2Bu^t)(=O)_3]^-$	observed in solution	44
$[W(OBu^t)_3(\equiv CR)] + NEt_4OH$	$[NEt_4][W(CH_2R)(=O)_3]$	R = Bu^t, TMS; colourless (60–90%)	44,62,63
$[W(=O)\{HB(Me_2pz)_3\}(=CHPh)] + H_2O$	$[W(CH_2Ph)(=O)_2\{HB(3,5\text{-}Me_2pz)_3\}]$	faster hydrolysis with OH^-	64
$[W_2(CH_2Bu^t)_6(=O)_2(\mu\text{-}O)] + TMS\text{-}Cl$	$[WCl(CH_2Bu^t)_3(=O)]$	colourless (90%)	62–4
$[W(OAr)_3(\equiv CBu^t)] + O{=}CR^1R^2$	$[W(\eta^1\text{-}CBu^t{=}CR^1R^2)(OAr)_2(=O)]^a$	R^1 = H, R^2 = H, Ph, OEt, NMe_2	65
$[W(CEtCEtCEt)(OAr)_3] + O{=}CR^1R^2$	$[W(\eta^1\text{-}CEt{=}CR^1R^2)(OAr)_2(=O)]^a$	R^1, R^2 = Me; R^1 = H, R^2 = Ph, OEt, NMe_2	65

[a] Ar = 2,6-$C_6H_4Pr^i_2$.

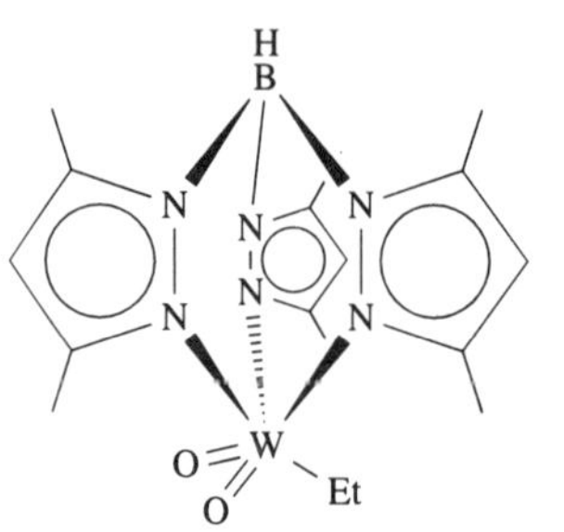

(13)

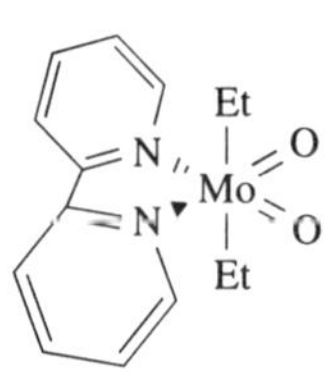

(14)

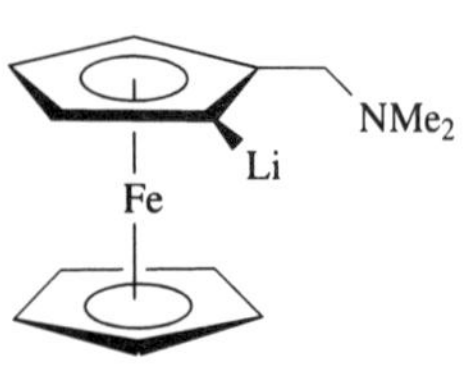

(15)

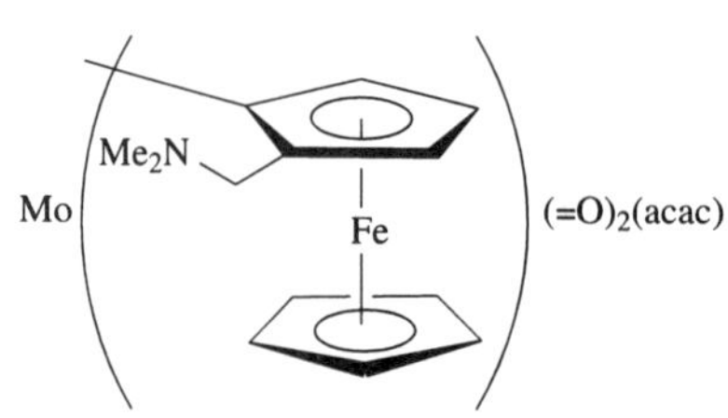

(16)

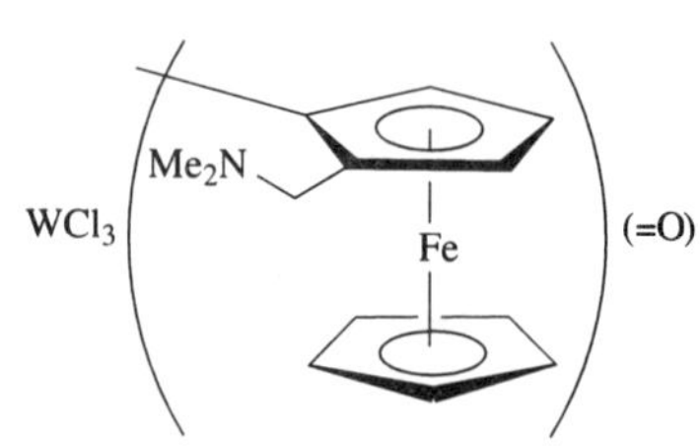

(17)

Alternative preparative routes to organometallic oxo species include protonation of alkyne complexes (Equation (8)) leading to (**18**);[67] the reaction of $[MoCl_2(PMe_3)_3]$ with dithioalkylcarbonate anions which yields the unusual η^3-complexes (**19**) (Equation (9));[68] or the hydrolysis of alkylidyne complex (**20**) to fashion (**21**) (Equation (10)).[63]

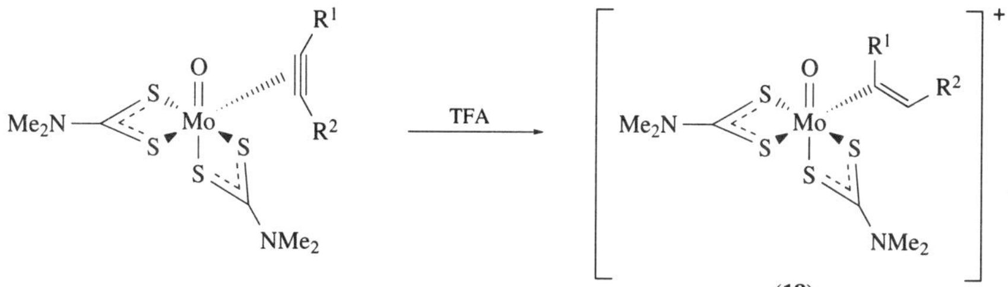

(18) (8)

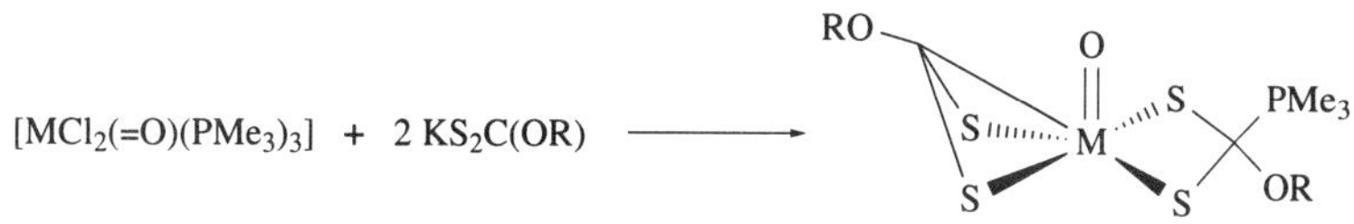

(19) (9)

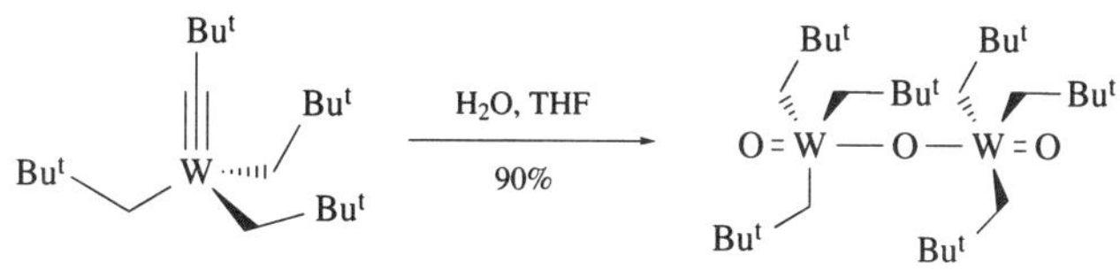

(20) (21) (10)

5.5.1.2 Imido complexes

Typical preparations of organoimido complexes appear in Table 6. For halide metathesis reactions, the choice of transmetallating reagent is important. For example, while reaction of MesMgBr with $[Cr(O\text{-}TMS)_2(\equiv NBu^t)_2]$ leads to the expected mononuclear complex, use of the less bulky phenyl group (from $ZnPh_2$) results only in $PhNHBu^t$ and biphenyl. As some *t*-butyl imido complexes are low melting point solids, in these cases high yield by-product-free preparations are required as few separation techniques are available to purify the crude materials. Occasionally, unexpected products can be isolated, as in the preparation of iminoacyl (**22**) (Equation (11)).[69] Although not strictly within the bounds of this section, the synthesis of the nitrido complex (**23**) is also of interest (Equation (12)).[70] Few examples of such reactivity are known. Preparation of alkyl imido complexes from $[MoCl_2(\equiv NR)(DME)]$ (R = Bu^t, $C_6H_4Me_2$-2,6, $C_6H_4Pr^i{}_2$-2,6, $C_6H_4Bu^t$-2) are particularly convenient as the starting materials are easily formulated in one-pot syntheses from inexpensive $(NH_4)_2[Mo_2O_7]$.[71] As they are prepared by alkene addition to alkylidene complexes, and are therefore clearly related to the material in Section 5.11, the metallocyclobutane species (**24**)–(**34**) have been collected together to aid cross-referencing (Table 7). Related chemistry is demonstrated by the preparation of (**35**) from diphenyl ketene (Equation (13)).[71] Stepwise treatment of $[WCl(Me)_3(\equiv NPh)]$ with MeLi and then hydrazine leads to novel dinuclear (**36**) (Equation (14)).

$$[CrCl_2(\equiv NBu^t)_2(CNBu^t)] \xrightarrow[Et_2O]{MeLi} [CrCl(\equiv NBu^t)(\eta^2\text{-}Bu^tN{=}CMe)] \quad (11)$$

(**22**)

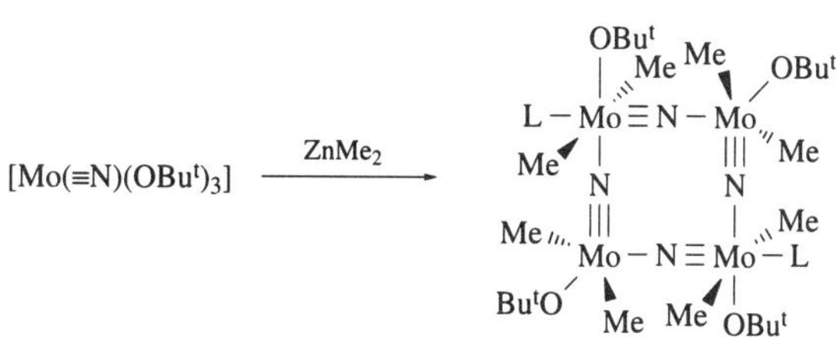

(**23**) L = 4-$Bu^tC_5H_5N$ (12)

Reaction of $[CrBr_2(py)(\equiv NBu^t)_2]$ with Cp_2Mg leads to $[CrBr(\equiv NBu^t)_2(Cp)]$, which is formally a 20-electron species. In these cases one of the imido ligands is expected to become appreciably bent to avoid unfavourable electronic configurations. The remaining halide may be alkylated to yield dark red

Table 6 Preparation of organo group 6 imido complexes.

Synthesis	*Complex*	*Comments*	*Ref.*
$[CrBr_2(\equiv NBu^t)_2(py)]$ + MCH_2R	$[Cr(CH_2R)_2(\equiv NBu^t)_2]$	R = Bu^t, TMS, CMe_2Ph, red oils (67–100%), M = MgCl, $ZnCH_2Bu^t$	72
$[CrBr_2(\equiv NBu^t)_2(py)]$ + $[Li(TMEDA)]_2[1,2\text{-}C_6H_4(CH\text{-}TMS)_2]$	$[Cr\{1,2\text{-}CH(TMS)C_6H_4CH(TMS)(\equiv NBu^t)_2]$	red (64%), x-ray	72
$[M(\equiv NBu^t)_2(O\text{-}TMS)_2]$ + ArMgBr	$[M(Ar)_2(\equiv NBu^t)_2]$	M = Cr, Mo; Ar = Mes, 2-6-$Me_2C_6H_3$; x-ray	66, 74
$[MoCl_2(\equiv NBu^t)(=O)(MeCN)]_3$ + MgR_2	$[MoCl(R)_3(\equiv NBu^t)]$	R = CH_2Bu^t; colourless (35%)	75
$[Mo(\equiv NBu^t)(OCH_2Bu^t)_2(=O)]_n$ + MgR_2	$[Mo(R)_3(\equiv NBu^t)(OR)]$	R = CH_2Bu^t, brown (75%)	75
$[MoCl_2(\equiv NAr)_2(DME)]$ + RCH_2MgCl	$[Mo(CH_2R)_2(\equiv NAr)_2]$	R = Bu^t, CMe_2Ph; Ar = 2,6-$Pr^i_2C_6H_3$; orange (70–80%)	76
$[Mo(R)_2(\equiv NBu^t)(=CHBu^t)_2]$ + H_2O	$[\{Mo(R)_3(\equiv NBu^t)\}_2MoO_4]$	R = CH_2Bu^t; colourless, x-ray	77
$[Mo(\equiv NBu^t)_2(O\text{-}TMS)_2]$ + $ZnMe_2$	$[Mo_2(\equiv NBu^t)_2(\mu\text{-}NBu^t)_2(Me)_4]$	violet (52%), x-ray	78
$[W(\equiv NBu^t)_2(OBu^t)_2]$ + $ZnMe_2$	$[W_2(\equiv NBu^t)_2(\mu\text{-}NBu^t)_2(Me)_4]$	red (60%)	78
$[W(\equiv NBu^t)_2(OSiMe_2Ph)]$ + ArMgBr	$[W(Ar)_2(\equiv NBu^t)_2]$	Ar = Mes; 2,6-$Me_2C_6H_3$; yellow-orange	66
$[WCl_2(\equiv NAr)(OBu^t)(THF)]$ + RMgCl	$[W(CH_2Bu^t)_2(\equiv NAr)(OBu^t)_2]$	R = CH_2Bu^t, orange (88%)	79
$[WCl_4(\equiv NR)]$ + $Mg(CH_2Bu^t)_2$	$[WCl(CH_2Bu^t)_3(\equiv NR)]$	R = Me, Pr^i, 2,6-$Pr^i_2C_6H_3$; colourless	80
$[WCl_2(\equiv NAr)_2(DME)]$ + RCH_2MgCl	$[W(CH_2R)_2(\equiv NAr)_2]$	R = Bu^t, CMe_2Ph; Ar = 2,6-$Pr^i_2C_6H_3$; yellow (78–90%)	79
$[WCl_4(\equiv NPh)]$ + $MgMe_2$ + PMe_3	$[W(Me)_2(\equiv NPh)(\mu\text{-}O)(PMe_3)]_3$	low yield, x-ray	81
$[WCl_2(\equiv NAr)(DME)(=CHBu^t)]$ + LiOR	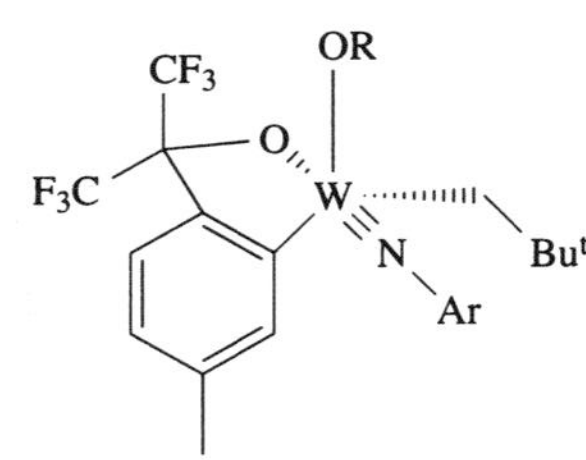	R = $C(CF_3)_2$(4-Tol); Ar = 2,6-$Pr^i_2C_6H_3$, x-ray	79

Table 7 Preparation of group 6 imido metallacyclo complexes.[a]

Synthesis	*Complex*	*Yield* (%)	*Comments*	*Ref.*
$[Mo(\equiv NAr)(OBu^t)_2(=CHCMe_2Ph)]$ + 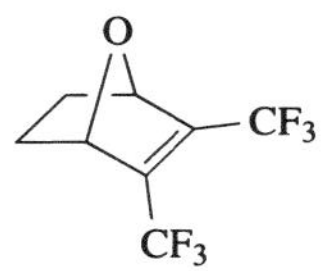	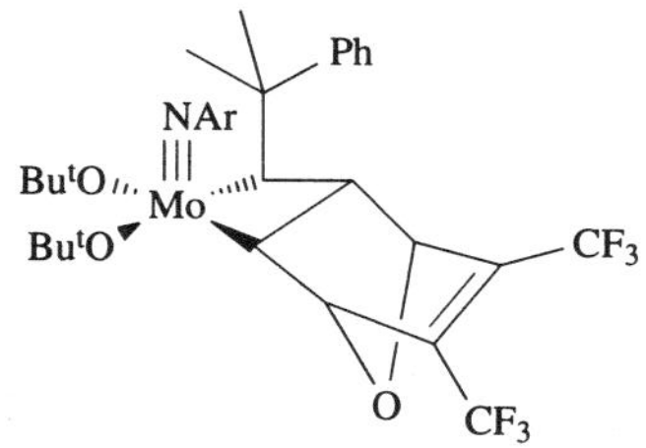 **(24)**	75	Red, SP, x-ray	82
$[Mo(\equiv NAr)(ORf_2)_2(=CHBu^t)]$ + excess $H_2C{=}CH_2$	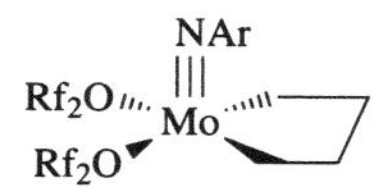 **(25)**	~100	Not isolable	83
$[M(\equiv NAr)(OAr)_2(=CHBu^t)]$ + $H_2C{=}CH_2$	 **(26)** M = Mo **(27)** M = W	~100	Not isolable, SP but in equilibrium with small amounts of TBP form	76, 84
$[W(\equiv NAr)(OR_f^{\,1})_2(=CHBu^t)]$ + $H_2C{=}CH(Bu^t)$	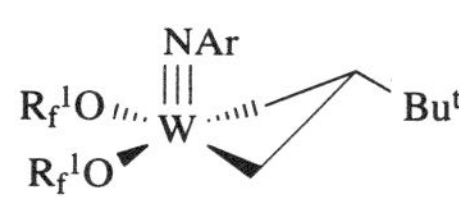 **(28)**	~100	SP, x-ray	84, 85

Table 7 (continued)

Synthesis	Complex	Yield (%)	Comments	Ref.
$[W(\equiv NAr)(OBu^t)_2(=CHBu^t)]$ + 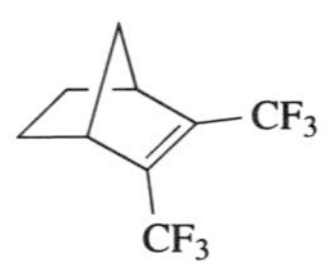	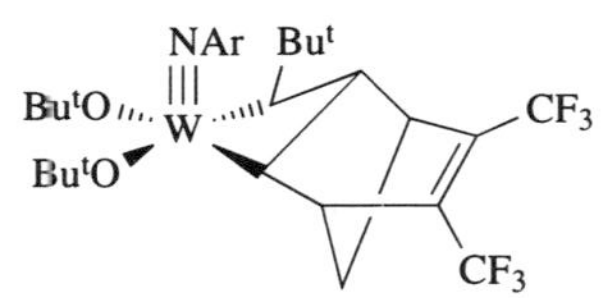 **(29)**	~100	SP, x-ray	86
$[W(\equiv NAr)(OR_f^1)_2(=CHBu^t)]$ + $H_2C=CH(CO_2Me)$	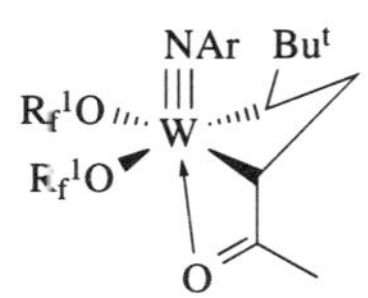 **(30)**	81	Yellow, SP, x-ray	87
$[W(\equiv NAr)(OR_f^1)_2(=CHBu^t)]$ + $H_2C=CH_2$	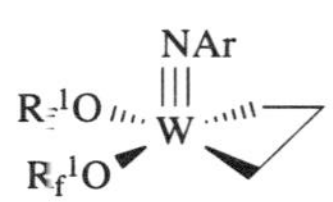 **(31)**	~100	Not isolable, SP, in equilibrium with TBP	84

$[W(\equiv NAr)(OR_f^3)_2(=CHBu^t)]$ + $H_2C{=}CH_2$	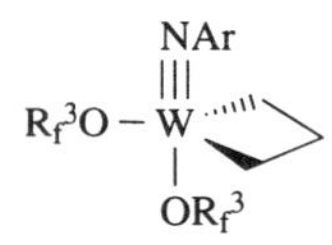 **(32)**	77	Pale yellow, TBP, x-ray	88
$[W(\equiv NAr)(OR_f^2)_2(=CHBu^t)$ + $H_2C{=}CH(TMS)$	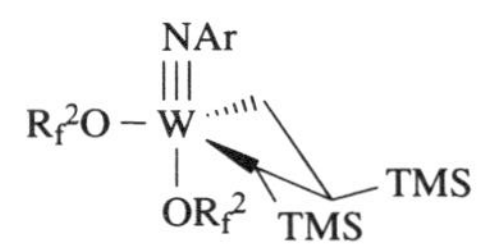 **(33)**	~100	Yellow, TBP, x-ray	88
$[Mo(\equiv NAr)(OBu^t)_2(=CHBu^t)]$ + C_6F_5CHO	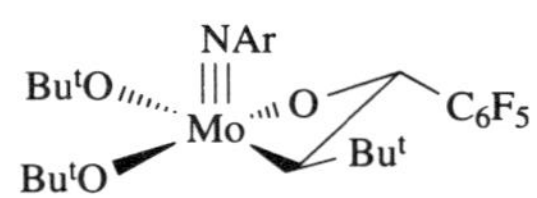 **(34)**	~46	Red, SP, x-ray	89

[a] Ar = 2,6-$Pr^i_2C_6H_3$; $R_f^1 = CMe_2(CF_3)$; $R_f^2 = CMe(CF_3)_2$; $R_f^3 = C(CF_3)_2(CF_2CF_2CF_3)$; SP = square pyramidal; TBP = trigonal bipyramidal.

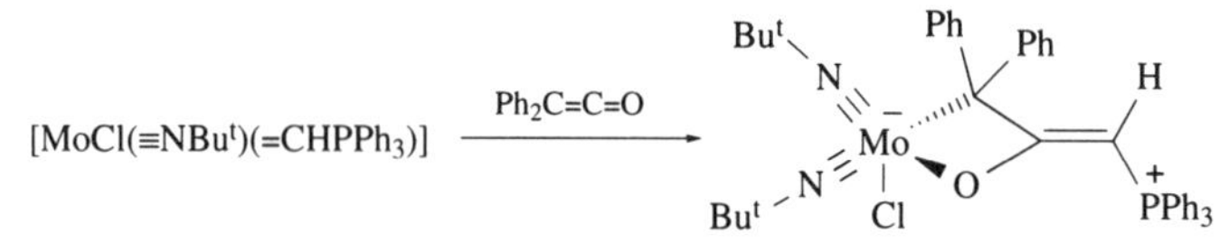

(35)

(13)

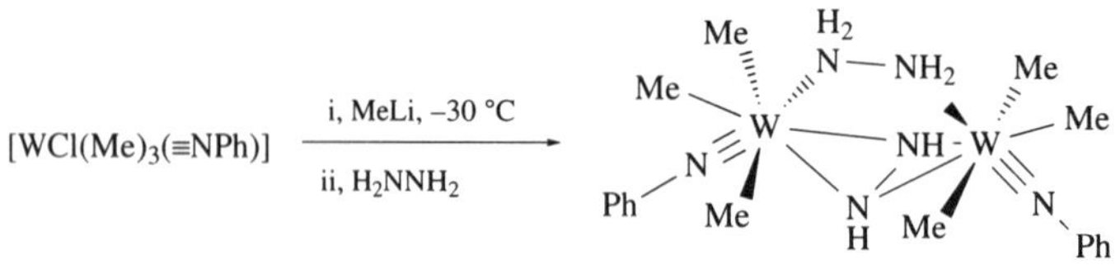

(36)

(14)

$[Cr(R)(\equiv NBu^t)_2(Cp)]$ complexes.[72] Analogous chemistry allows for the preparation of $[MBr(\equiv NBu^t)_2(Cp^*)]$ and $[M(Me)(\equiv NBu^t)_2(Cp^*)]$ (M = Mo, W). The yellow–orange methyl complexes show two $\nu_{M\equiv N}$ IR stretches around 1230 cm^{-1} suggesting that one imido ligand is also appreciably bent in these complexes to avoid 20-electron counts at the metal.[73]

5.5.2 Structure and Properties

5.5.2.1 Oxo complexes

Crystallographically characterized organo oxo complexes are indicated in Table 5. Additionally, the structure of $[Mo(Mes)_2(=O)_2]$ is known.[90] The structures of $[WX(CH_2Bu^t)_3(=O)]$ (X = Cl, OCH_2Bu^t) are based on trigonal bypyramidal (TBP) geometries, while those of $[WCl(CH_2Bu^t)_2(NEt_2)(=O)]$ and $[W(CH_2Bu^t)_3(NEt_2)(=O)]$ are square pyramidal (SP) as shown explicitly in the latter case by crystallography. The change in coordination geometry is mediated by the need of the two strong π-donors (=O and $-NEt_2$) to avoid competition for the same d_{xz}/d_{xy}-based MOs leading to the observed O–W–N angle of 118.8°. In this geometry the amido is free to interact with a metal d_{xy}-based orbital approximately orthogonal to the oxide. These complexes are also fluxional in solution, and variable-temperature NMR studies suggest two exchange processes: amido rotation and coordination geometry interconversion.

5.5.2.2 Imido complexes

The M–N–C_α angles of imido ligands in high oxidation state group 6 metal complexes lie in the range 155–170° or more; representative geometries are depicted in Table 8. Such angles are indicative of π-interactions from the imido nitrogen. At the upper limit of such a bonding scheme the imido may be regarded as a linear NR^{2-} six-electron donor.[91] Most group 6 complexes closely approach rather than attain this state. Although attempts have been made to correlate the degree of π-interaction in NBu^t imido ligands with ^{13}C NMR chemical shift differences between the quaternary and methyl signals in the *t*-butyl group these have not generally proved useful.[72,92] Detailed understanding of the metal–imido interaction is best acquired via theoretical calculations. For example, even totally linear M≡NR fragments may act as less than six-electron donors.[93]

Distortions in the structures of $[M(R^1)_2(\equiv NR^2)_2]$ complexes were originally attributed to agostic interactions[66] but later theoretical calculations indicate that steric factors account for these distortions as such contacts are apparently disfavoured in these systems.[95] No spectroscopic evidence for M–H–C interactions is apparent for these complexes.

A number of imido metallocycle complexes show TBP–SP interconversions. In an elegant series of papers Schrock and co-workers determined the structures of a number of these complexes together with the kinetic barriers to interconversions.[82,84,86] The relative structures of these metallocycles are easily determined by ^{13}C NMR spectroscopy. For TBP metallocycles the C_β carbons resonate at >95 ppm, while those due to C_α appear close to 0 ppm. For SP metallocycles the two resonances are far closer

Table 8 Crystallographically characterized organo group 6 imido complexes.[a]

Complex	*M≡N* (nm)	*M–N–C*$_\alpha$ (°)	*Other bond distances* (nm)		*Ref.*
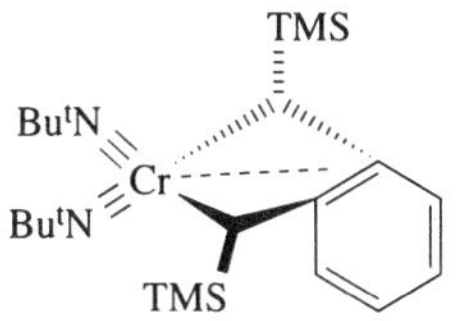	0.1636(4) 0.1651(3)	171.2(3) 155.8(3)	Cr–C π-contact	0.2078(4) 0.2076(3) ~0.237	72
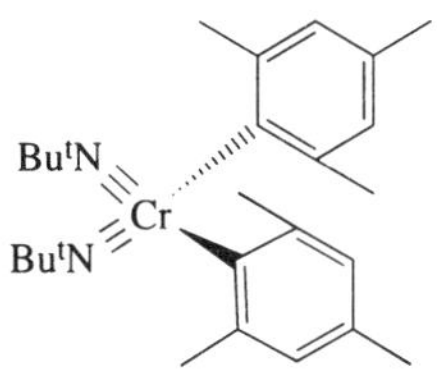	0.1628(5) 0.1623(5)	159.6(3) 159.7(3)	Cr–C	0.2032(6) 0.2036(6)	66
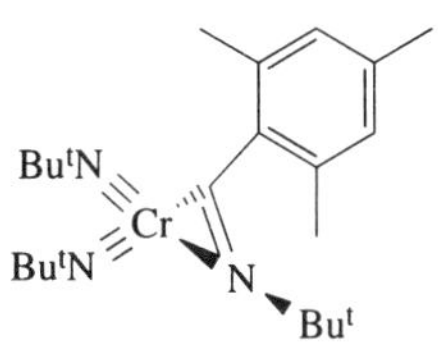	0.1628(6) 0.1630(6)	162.8(4) 161.5(4)	Cr–C Cr–N_{imine}	0.1982(7) 0.2057(6)	66
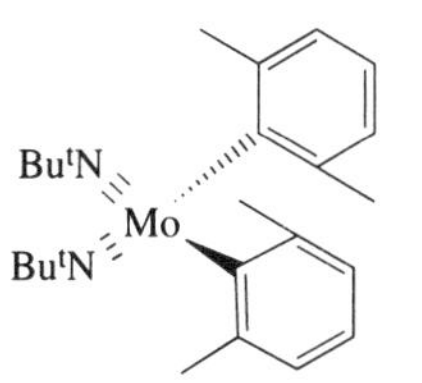	0.1733 0.1708(7)	161.1(4) 160.1(5)	Mo–C	0.2161(7) 0.2162(7)	66
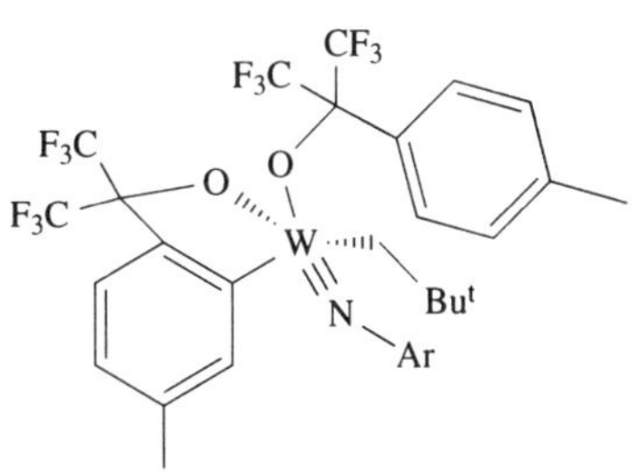	0.1746(5)	165.7(4)	W–C_{Ar} W–C_{CH_2} W–O_{ring} W–O	0.2123(6) 0.2195(7) 0.2012(4) 0.1866(4)	79
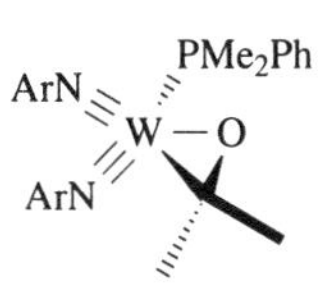	0.1785(6) 0.1776(7)	168.8(6) 155.8(5)	W–C W–O W–P	0.2118(9) 0.2025(5) 0.2509(2)	94

[a] Ar = 2,6-$Pr^i_2C_6H_3$.

together. The SP metallocycles show significantly negative entropies of activation ($\Delta S^\ddagger$ = −13 to −78 J K mol^{-1}) for reactions involving rearrangement to TBP structure or alkene loss, implying significant reorganization as the transition state is attained, and this is shown Scheme 3.

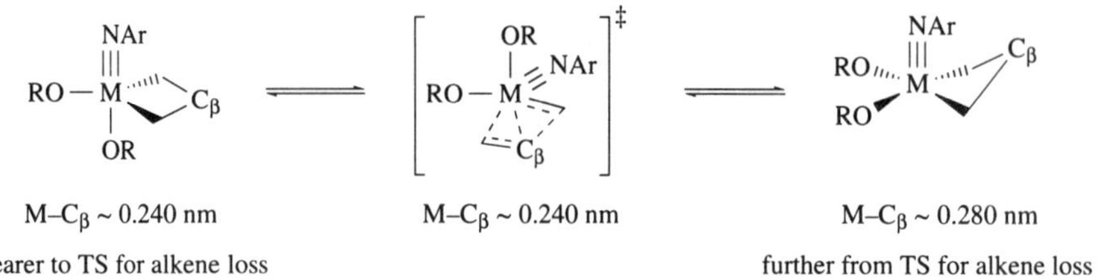

Scheme 3

5.5.3 Reactions

5.5.3.1 Oxo complexes

Diorgano complexes with α-hydrogens normally eliminate alkanes with the formation of alkylidenes ($[M(CH_2R)_2] \rightarrow [M{=}CHR] + RH$), and most of these reactions are considered in Section 5.11. However, the rates of these reactions are highly dependent on the ancillary ligands present in the complex or the presence of added Lewis acids. In special cases (Equation (15)) the intermediate adduct may be isolated and crystallographically characterized.[96–8] Related adducts (**36**) and (**37**) are postulated to be involved as *in situ* methylenation reagents in reactions with organic carbonyl compounds.[99] Monoorgano oxo complexes, being unable to generate alkylidenes, are surprisingly rather underinvestigated, but some typical reactivity for these species is shown in Equation (16).[100]

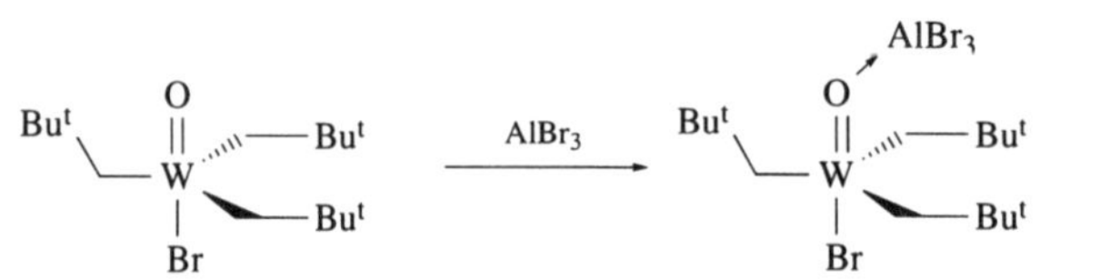

(15)

(**36**) (**37**)

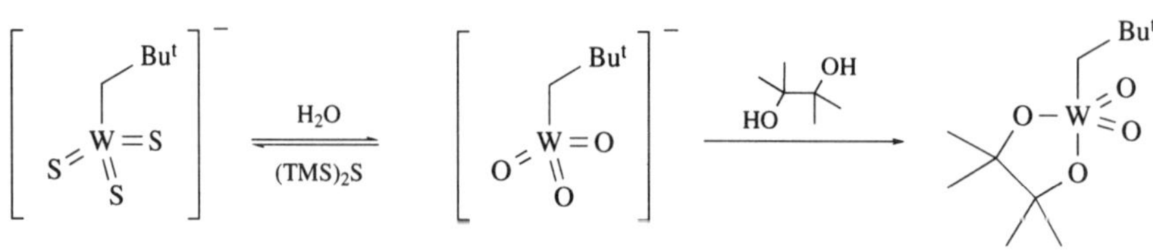

(16)

The reactivity of $[M(R)_2({=}O)_2(bipy)]$ and related species (M = Mo, W) (Table 5) is controlled by the presence of β-hydrogens on the organo substituents. In their presence, β-elimination processes dominate leading to alkene decomposition products, while in their absence metal–carbon bond homolysis dominates. The majority of the mechanistic studies on these compounds have concentrated on the organic products produced; few metal-containing by-products have proved isolable. Group 6 aryl complexes, unable to participate in either α- or β-hydrogen eliminations, show rather different reactivity. The behaviour of $[Mo(Mes)_2({=}O)_2]$ (Scheme 4) is exemplary and has been extensively studied by Laï.

5.5.3.2 Imido complexes

Like their oxo counterparts, imido species possessing α-hydrogens are valuable alkylidene precursors. For example, treatment of $[M(CH_2R)_2({\equiv}NAr)_2]$ (M = Mo, W; R = Bu^t, CMe_2Ph; Ar = $C_6H_3Pr^i_2$-2,6) with triflic acid in dimethoxyethane leads to the versatile alkylidene complex $[M(NAr)(OTf)_2(DME)(CHR)]$ (Scheme 5).[76,79] Alkylidene complexes are also generated via alkene loss from metallacyclobutane complexes in the presence of added ligands (Scheme 5). Further discussion of alkylidene chemistry is available in Section 5.11.

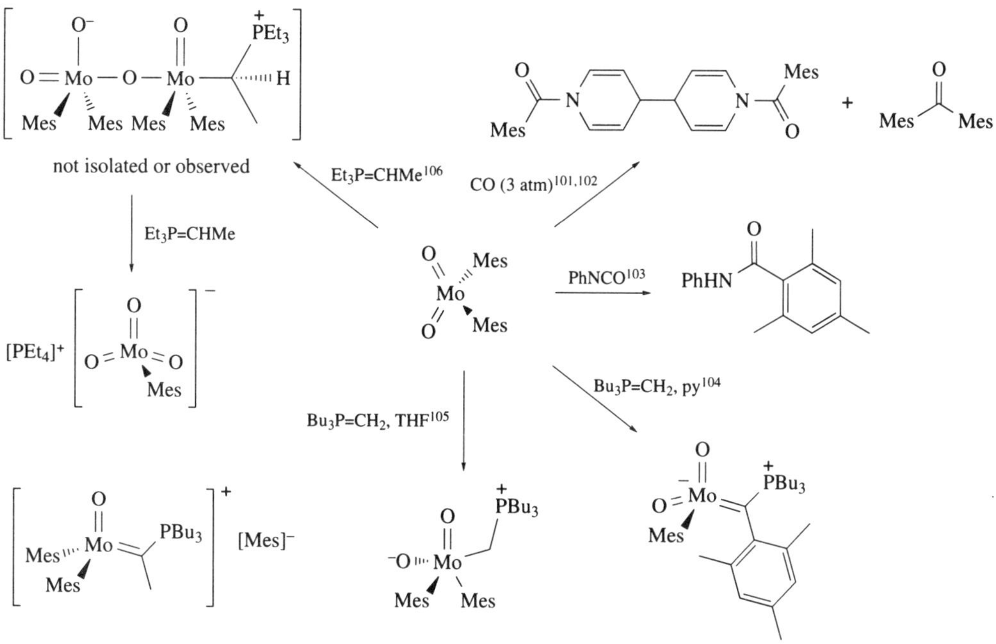

Scheme 4

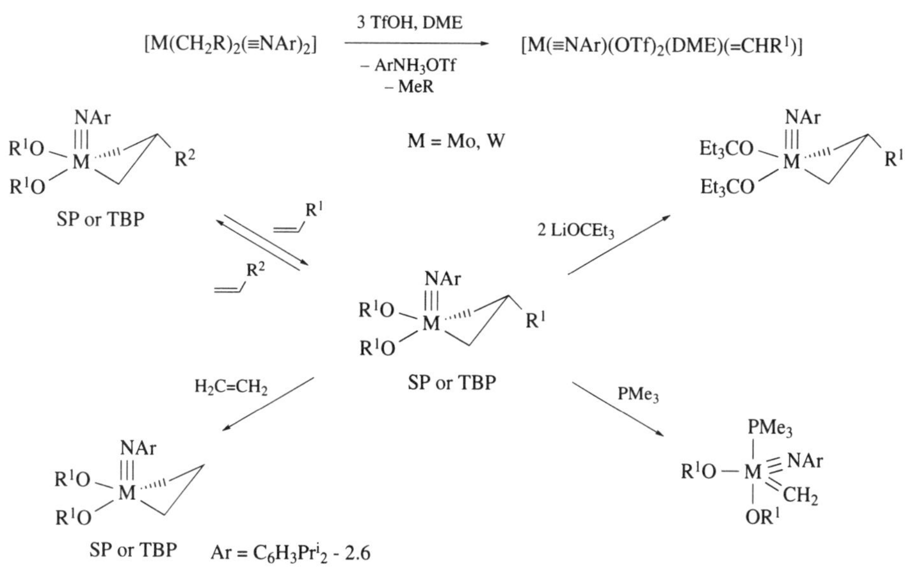

Ar = $C_6H_3Pr^i{}_2$ - 2.6
R^1 = Bu^t, $CMe_2(CF_3)$, $CMe(CF_3)_2$, $C(CF_3)_2(CF_2CF_2CF_3)$

Scheme 5

Nonalkylidene imido chemistry is dominated by the complexes $[M(R^1)_2(\equiv NR^2)_2]$ and $[MX(R^1)_3(\equiv NR^2)]$ (X = halide, pseudohalide). Representative chemistry, in an abridged form, for $[M(Mes)_2(\equiv NBu^t)_2]$ and $[WCl(CH_2Bu^t)(\equiv NR)]$ is presented in Schemes 6[66,104] and 7,[80] respectively.

5.6 AQUO COMPLEXES

Three recent reviews of $[Cr(R)(OH_2)_5]^+$ chemistry are available.[107–9]

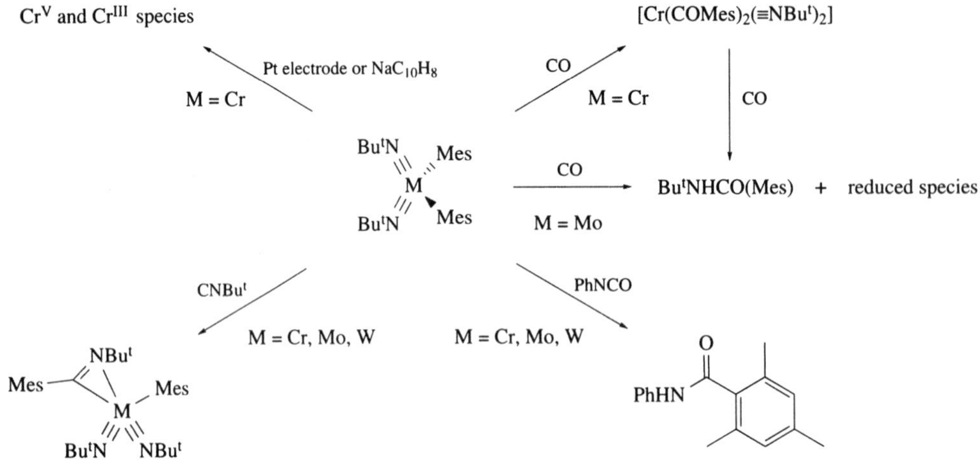

Scheme 6

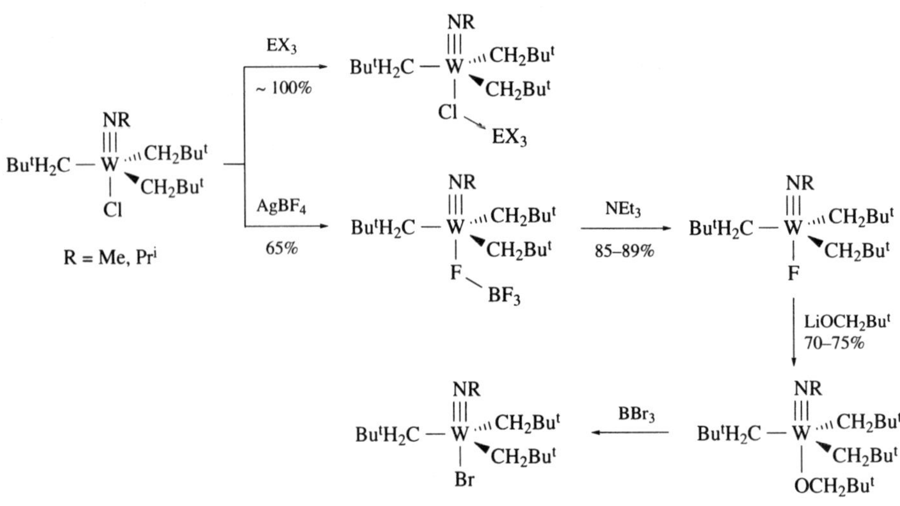

Scheme 7

5.6.1 Synthesis

Although synthetic procedures leading to $[Cr(R)(OH_2)_5]^+$ complexes are summarized in Table 9, some additional points are of note. Alkylations of $[Cr(OH_2)_6]^{2+}$ are characterized by large bimolecular rate constants (10^7–10^8 mol^{-1} dm^3 s^{-1}), but only activated organic halides, typically $CHCl_3$, CH_2Cl_2 or $ArCH_2Br$, react. Use of some functionalized halides can result in complex reactivity and polymer production.[110,111] The combination of $[Cr(OH_2)_5]^{2+}$, RH and H_2O_2 (modified Fenton conditions) is the most general route to pentaaquoorgano complexes proceeding via stabilized radicals generated by abstraction by $OH^•$. Equimolar quantities of $[Cr(R)(OH_2)_5]^+$ and $[Cr(OH)(OH_2)_5]^+$ result, together with smaller amounts of $[Cr(OH_2)_6]^{3+}$, $[(H_2O)_4Cr(\mu\text{-}OH)_2Cr(OH_2)_4]^{4+}$ and $[(H_2O)_5Cr{-}O{-}Cr(OH_2)_5]^{4+}$.[108] These reactions are normally carried out with an activated alkane present in excess to minimize the contribution to the $[Cr(OH)(OH_2)_5]^{2+}$ manifold by limiting the $OH^•$ concentration. The rate and volume of activation are rather insensitive to the nature of R, which is consistent with an I_d rather than an associative pathway. One possible origin for this observation is the tendency of $[Cr(OH_2)_6]^{2+}$ to show Jahn–Teller distortions giving rise to transition states with significantly lengthened Cr–OH_2 bonds.[112,113]

Whatever the preparative route, normal separative techniques employ the use of ion-exchange chromatography on strong acid resin columns. Alternatively $[Cr(R)(OH_2)_5]^{2+}$ complexes may be used *in situ* after quantification by UV–visual spectroscopy.

Table 9 Preparation of aquochromium complexes.

Synthesis	*Complex*	*Comments*	*Ref.*
A. Preparation by alkylation of Cr^{II}			
$[Cr(OH_2)_6]^{2+} + ICH_2CN$	$[Cr(CH_2CN)(OH_2)_5]^{2+}$	yield 25%, $[Cr(OH_2)_5(NCMe)]^{3+}$ (75%)	114
$[Cr(OH_2)_6]^{2+} + BrCH_2SO_3Na$	$[Cr(CH_2SO_3)(OH_2)_5]^{+}$	yield ~20%	115
$[CrCl_2(OH_2)_4]$ + *n*-$ClCH_2C_5H_4N$ + en	$[Cr(n\text{-}CH_2C_5H_4N)(OH_2)(en)_2]^{2+}$	$n = 2, 3$; impure pink–orange salt	116
$[CrCl_2(OH_2)_4]$ + *n*-$ClCH_2C_5H_4N$ + dap	$[Cr(n\text{-}CH_2C_5H_4N)(OH_2)(dap)_2]^{2+}$	$n = 2, 3$; dap = $H_2N(CH_2)_3NH_2$; orange	116
$[Cr(OH_2)_2(L)]^{2+}$ + 4-$BrCH_2C_6H_4Br$	$[Cr(4\text{-}CH_2C_6H_4Br)(OH_2)(L)]^{2+}$	L-[15]aneN_4 = (**39**), x-ray on perchlorate	117
B. Preparation using R_3CH *and* H_2O_2			
$[Cr(OH_2)_6]^{2+} + H_2O_2 + CH_2Cl_2$	$[Cr(CHCl_2)(OH_2)_5]^{2+}$	dichromium by-products	118
$[Cr(OH_2)_6]^{2+} + H_2O_2 + RH$	$[Cr(R)(OH_2)_5]^{2+}$	R = CH_2OH, CMe_2OH, and others	119
C. Miscellaneous preparations			
$CrCl_3 + CHCl_3$ + Zn/Hg + HCl	$[Cr(CHCl_2)(OH_2)_5]^{2+}$	no $[CrX(OH_2)_5]^{2+}$ by-product	23
$[Cr(CHCl_2)(OH_2)_5]^{2+}$ + Zn/Hg + $CrCl_3$	$[Cr(CH_2Cl)(OH_2)_5]^{2+}$	quantitative conversion	23
$[Cr(OH_2)_6]^{2+}$ + O=CH_2 + γ-radiation	$[Cr(CHO)(OH_2)_5]^{2+}$	via $[Cr\{CH(OH)_2\}(OH_2)_5]^{2+}$	120

5.6.2 Structure and Properties

No $[Cr(R)(OH_2)_5]^{2+}$ complex is structurally characterized. The lack of structural data is attributed to the extensive hydrogen bonding of these complexes to solvents in which they are soluble. Partial substitution of the aquo ligands reduces these effects and this has lead to crystalline *trans*-$[Cr(CH_2Cl)(acac)_2(OH_2)]$[25] and $[Cr(CH_2C_6H_4Br)(OH_2)([15]aneN_4)][ClO_4]_2$ (**38**)[117] ([15]aneN_4 = 1,4,8,12-tetraazacyclopentadecane (**39**)). Both of these aquo species show similar long Cr–OH_2 contacts (0.2134(6) nm and 0.213(2) nm respectively), some 7% longer than average Cr^{III}–OH_2 bond lengths, which are typically in the range 0.195–0.202 nm. Such values are indicative of a large *trans* influence from the organometallic ligand, which also manifests itself in a greater reactivity for the *trans* site as detected by kinetic and reactivity studies.

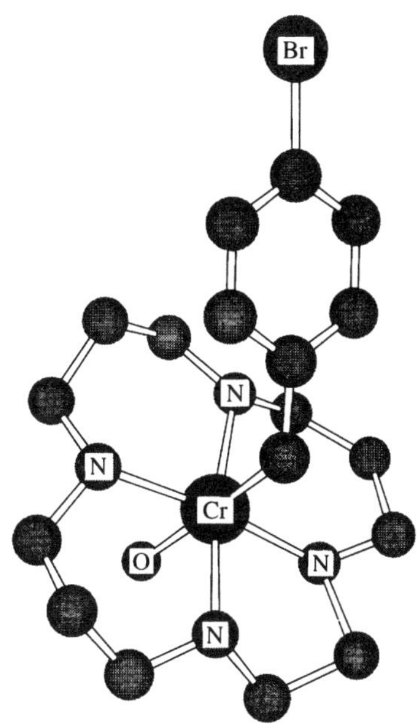

(38)

The UV–visible spectra of pentaquoorganochromium complexes are highly diagnostic and an effective method of routine characterization (NMR spectroscopic techniques are less useful due to the paramagnetic contact shifts these molecules experience). Typically three absorption bands are present: a weak *d–d* transition at 510–560 nm, a stronger *d–d* transition with some charge-transfer character at 380–410 nm, and a strong LMCT (ligand–metal charge transfer) band at 250–310 nm. For some substituents (including CH_2OH, CHMeOH, CMe_2OH) an additional band can be observed in the region 210–310 nm which may arise from the transition of a heteroatom π-electron to the chromium centre.[119]

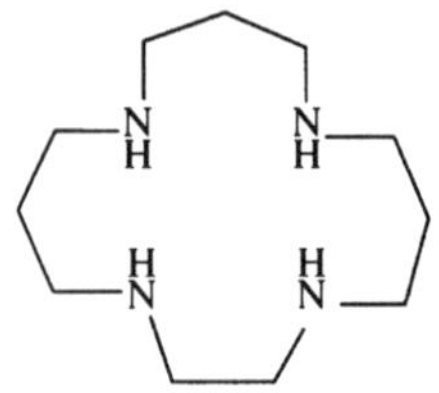

(39)

5.6.3 Reactions

Although the Cr–C bond in $[Cr(R)(OH_2)_5]^{2+}$ complexes is thermodynamically rather robust (for example, the bond enthalpy of the benzyl derivative is ~125 kJ mol^{-1}), three decomposition routes have been identified (Scheme 8). Bond homolysis is the reverse of many pentaaquo complex formation reactions and may therefore be suppressed by adding $[Cr(OH_2)_6]^{2+}$ or alternatively promoted by photolysis;[118] CH_2Ar and CHR_2 groups are particularly susceptible to homolysis. The rate of heterolysis (protonolysis) reactions to produce the alkane is dependent on a number of factors including pH and the nature of the substituent. Low acetate concentrations promote the hydrolysis of $[Cr(CH_2OH)(OH_2)_5]^{2+}$. This effect is presumed to arise from acetate *trans*-influence lengthening the Cr–C bond after substitution of $(H_2O)_{trans}$ by acetate. Similar observations are made for *trans*-$[Cr(CH_2OH)(OH_2)(L)]^{2+}$ (L = (**39**)). Kinetic studies of $[Cr(CMe_2OH)(OH_2)_5]^{2+}$ hydrolysis in the absence of acetate indicate that Cr–OH and $CHMe_2OH$ formation occur at comparable rates.[121] A number of transition states are consistent with these data (Scheme 8); (**40**) is thought to be dominant under acid conditions and (**41**) in neutral solution. Complexes with β-OH or halogen substituents are extremely sensitive to β-elimination.[122] Under appropriate conditions η^2-alkene intermediates may be detected.[123]

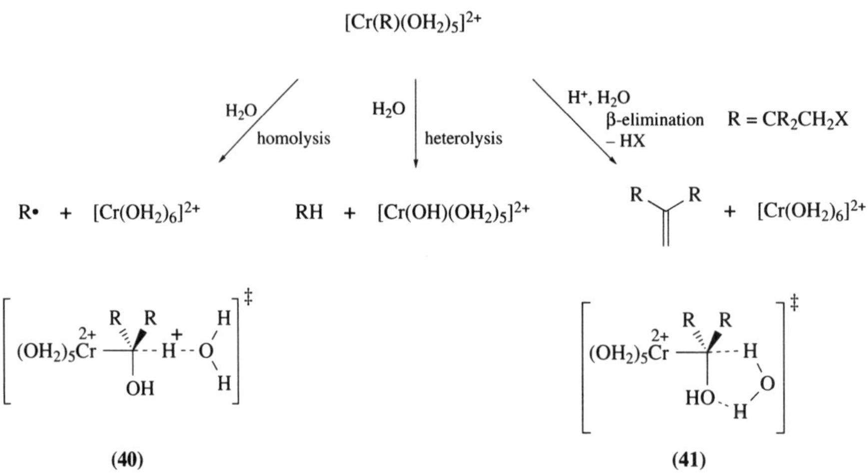

Scheme 8

Other reactions and behaviour of $[Cr(R)(OH_2)_5]^{2+}$ are summarized in Scheme 9. Although substitution of the water ligand *trans* to the organo function gives a common kinetic product, subsequent rearrangements to isomeric mixtures are normally observed in the reactions of $[Cr(R)(OH_2)_5]^{2+}$ complexes. Macrocyclic $[Cr(R)(OH_2)(L)]^{2+}$ complexes (L = (**39**)) show similar reactivity[124,125] and additionally undergo reduction in the presence of $EtS^{\bullet}$ radicals.[126]

5.7 PHOSPHINE COMPLEXES

5.7.1 Synthesis

Exemplary synthetic routes to organo group 6 phosphine complexes appear in Table 10. Additional examples of strategies employing ligand exchange or phosphine C–H oxidative additions are shown in Equations (17),[131] (18)[132] and (19).[133]

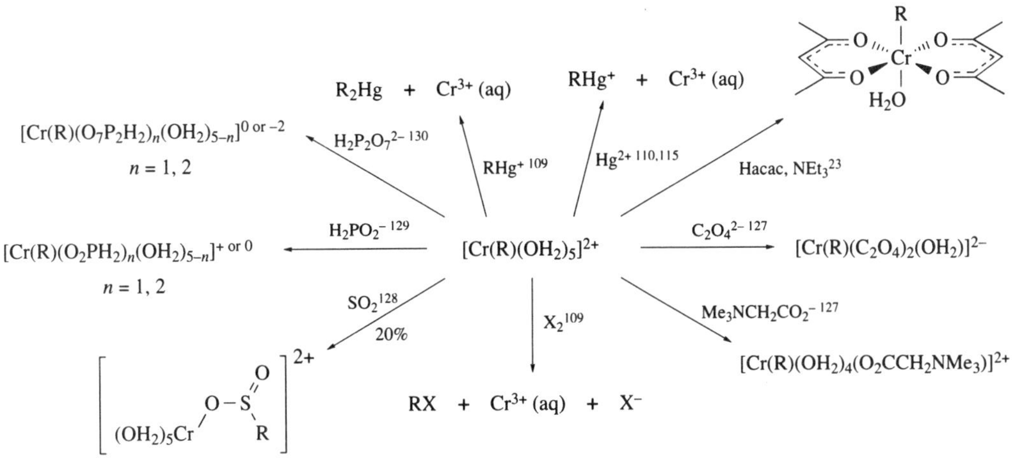

Scheme 9

Table 10 Preparation of organo group 6 phosphine complexes.

Synthesis	*Complex*	*Comments*	*Ref.*
A. Preparation by transmetallation			
$[Cr_2(OAc)_4]$ + 2-TolMgBr + PMe_3	$[Cr(2\text{-Tol})(PMe_3)_2]$	orange (7%), $\mu_{eff} = 3.7\ \mu_B$	7
$[CrCl_2(PMe_3)_n]$ + $MgMes_2$	$[Cr(Mes)_2(PMe_3)_2]$	orange (50%), x-ray, $\mu_{eff} = 4.9\ \mu_B$	134
$[CrCl_2(dippe)]$ + MgR_2	$[Cr(R)_2(dippe)]$[a]	R = Mes, $CH_2(TMS)$, CH_2Bu^t, x-rays, $\mu_{eff} = 4.7\ \mu_B$	134
$[CrCl_2(dmpe)_2]$ + MeLi	*trans*-$[CrMe_2(dmpe)_2]$	orange (85%), x-ray, $\mu_{eff} = 2.8\ \mu_B$	135
$[CrCl_3\{Bu^tC(CH_2PMe_2)_3\}]$ + RLi	$[Cr(R)_3\{Bu^tC(CH_2PMe_2)_3\}]$	red, R = Me, Bu, x-ray, $\mu_{eff} = 3.5\ \mu_B$	136
$[CrCl_3\{MeC(CH_2PMe_2)_3\}]$ + BuLi	$[CrCl(Bu)_2\{MeC(CH_2PMe_2)_3\}]$	x-ray	137
B. Preparation by C–H activation			
$[MoH_4(dppe)_2]$ + $H_2C{=}CMe(CO_2R)$	$[MoH\{CH{=}CMe(CO_2R)\}(dppe)_2]$	red (14–57%), R = Et, Pr^i, Bu^n, Cy	138
WCl_6 + Na–K + PMe_3	$[WH(\eta^2\text{-}CH_2PMe_2)(PMe_3)_4]$	yellow (40%), x-ray	139,140
$[WH(\eta^2\text{-}CH_2PMe_2)(PMe_3)_4]$ + MeOH	$[WH_2(\eta^2\text{-}OCH_2)(PMe_3)_4]$	yellow (80%)	141
$[WCl_4(OC_6H_3Ph_2\text{-}2,6)_2]$ + Na–Hg + L	$[W(O\text{-}2,6\text{-}PhC_6H_3C_6H_4)_2(L)_2]$	orange (50–65%), L = $PMePh_2$, $PMePh_2$, PMe_3, 0.5 dppe, x-ray (L = $PMePh_2$)	142
C. Preparation by alkyne protonation			
$[MoH_4(dppe)_2]$ + $HC{\equiv}CCO_2R$ + HBr	$[MoBr(CH{=}CHR)(dppe)_2]$	R = Me, Et	143
$[MoH_4(dppe)_2]$ + $HC{\equiv}CCO_2R$ + HBF_4	$[MoH_2\{CH{=}CH(CO_2R)\}(dppe)_2]^+$	R = Me, Et	143
D. Preparation by O=CR$_2$ coordination			
$[WCl_2(PMePh_2)_4]$ + $O{=}CR^1R^2$	$[WCl_2(\eta^2\text{-}OCR^1R^2)_2(PMePh_2)_2]$	R^1R^2 = H, Me; H, Bu^t; Me, Me; $-(CH_2)_4-$; $-(CH_2)_5-$; x-ray (diacetone complex)	144,145
trans-$[W(S)_2(PMe_3)_4]$ + O=CHR	$[W(S)_2(\eta^2\text{-}OCHR)(PMe_3)_2]$	R = H, Me, Ph, 4-Tol, C_6H_4OMe-4, x-ray (R = Ph)	146
$[W({\equiv}NAr)(PMe_2Ph)_2]$ + $O{=}CR^1R^2$	$[W(NAr)(\eta^2\text{-}OCR^1R^2)(PMe_2Ph)]$	Ar = $C_6H_3Pr^i_2$-2,6, R^1R^2 = H, Et; Me, Me; x-ray	94

[a] dippe = 1,2-bis(di-*i*-propylphosphino)ethane.

$$[CrMeCl_2(THF)_3] \xrightarrow[\text{THF, 61\%}]{Pr^i_2P\frown PPr^i_2} [CrCl_2(Me)(dippe)] \qquad (17)$$

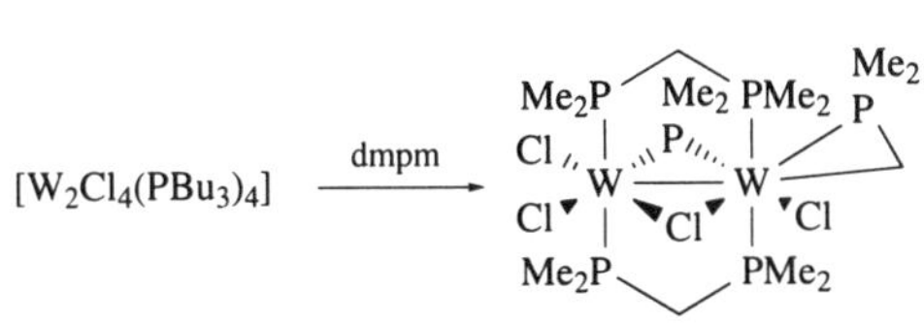

(18)

A number of unusual complexes are known. For example, the reaction of $[MoCl(NO)(PMe_3)_4]$ with CS_2 leads to the thioallyl compound $[MoCl(NO)(PMe_3)_2\{\eta^3\text{-}SC(PMe_3)S\}]$ which may be derivatized with NaS_2CNMe_2 to $[Mo(S_2CNMe_2)(NO)(PMe_3)\{\eta^3\text{-}SC(PMe_3)S\}]$.[68] Acetonitrile normally coordinates

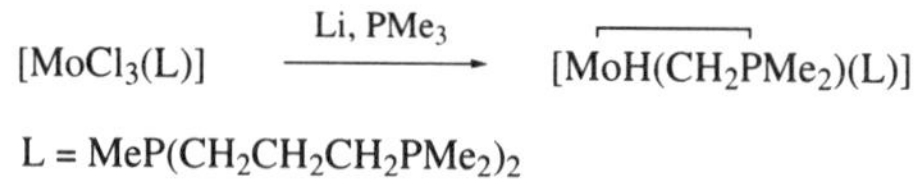

(19)

through nitrogen; however, crystallographically characterized η^2-MeCN complexes are known for both molybdenum[147] and tungsten[148] phosphine complexes. Heating *trans*-[$W(N_2)_2(dppe)_2$] in DMF at reflux for short periods leads to the related acyl [$WH(\eta^2$-$CONMe_2)(dppe)_2$].[149,150] Prolonging the reflux leads only to the isolation of [$W(DMF)(dppe)(CO)_2$].

5.7.2 Structure and Properties

Crystallographic investigations of the complexes [$Cr(R)_2(dippe)$] ($R = CH_2Bu^t$, CH_2TMS) reveal distorted square planar geometries.[134] The interplanar angle, θ_{planes}, for these complexes (28.4° and 15.8°) is clearly closer to the idealized square planar geometry than a tetrahedral structure. Despite their formal 12-electron counts, neither of the complexes shows any signs of agostic interactions in either the solid state or solution. The related structure of [$Cr(Mes)_2(PMe_3)_2$] shows essentially a perfect square planar geometry.

Reactive [$WH(\eta^2$-$CH_2PMe_2)(PMe_3)_4$] (**43**), and to a lesser extent its molybdenum analogue (**42**), are important precursors to many other species. Both these species exist in equilibrium with [$M(PMe_3)_6$] (M = Mo, W) (Equation (20)). For molybdenum only traces (a few percent) of (**42**) are present in C_6D_6 solutions of [$Mo(PMe_3)_6$].[152] In the case of tungsten only (**43**) is observed. Homoleptic [$W(PMe_3)_6$] is rather fragile even in neat PMe_3, and rapid formation of (**43**) and PMe_3 occurs ($K_{eq} = 17.8$ mol dm^{-3} at 30 °C).[140] The yellow crystalline tungsten hydride (**43**) shows pairwise exchange of the four PMe_3 ligands in the variable-temperature ^{31}P NMR spectrum while the couplings to the η^2-CH_2PMe_2 unit are statistically averaged. These results are consistent with the exchange process shown in Equation (21), where the arrows denote movement.[139]

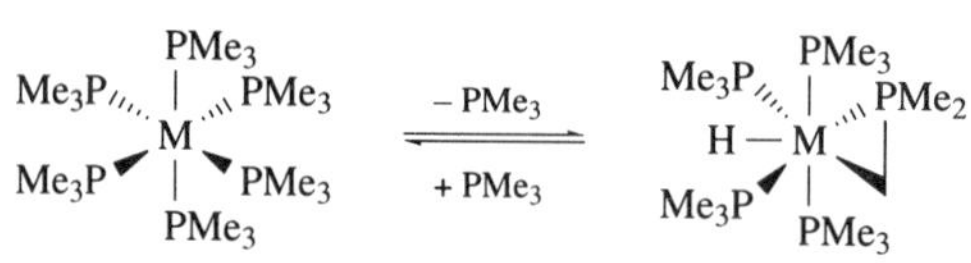

(20)

(**42**) M = Mo
(**43**) M = W

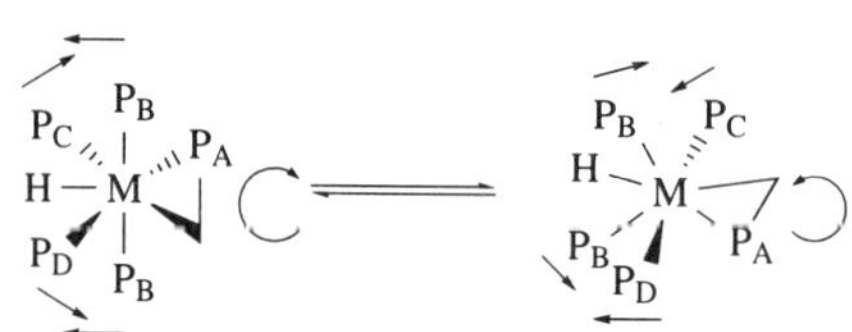

(21)

Equilibria are also common features of aldehyde or ketone coordination reactions, for example Equation (22). For asymmetric carbonyl compounds, O=CRH or O=CR^{1}R^2, mixtures of stereoisomeric complexes are formed. Crystallographic studies of these aldehyde and ketone complexes reveal the presence of C–O bonds considerably longer (>0.137 nm) than that found in acetone (0.121 nm), indicating that the complexes are best regarded as metallocycles rather than π-complexes.

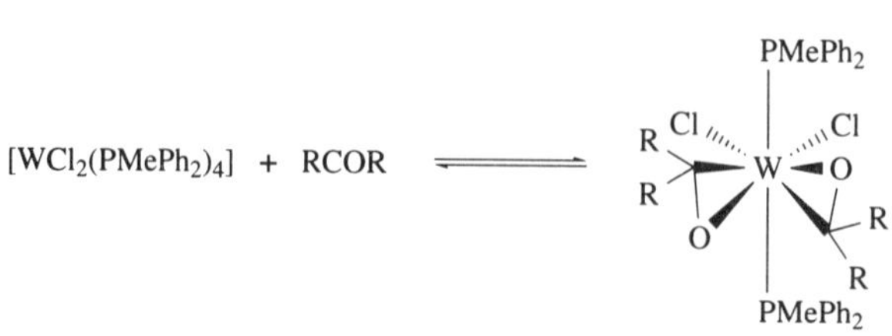

(22)

5.7.3 Reactions

An impressive literature on the reactivity of $[WH(\eta^2\text{-}CH_2PMe_2)(PMe_3)_4]$ (**43**) has built up in recent years and this is summarized in Scheme 10. The related oxometallocycle $[WH_2(\eta^2\text{-}OCH_2)(PMe_3)_4]$ (**44**), generated by MeOH addition to (**43**), is also a valuable starting material leading to the products shown in Scheme 11.[141,151] Other ketone complexes, $[WCl_2(\eta^2\text{-}OCR_2)_2(PMePh_2)_2]$, how a distinct tendency to lose one molecule of carbonyl compound leading eventually to the alkylidenes $[WCl_2(=O)(PMePh_2)_2(=CR_2)]$.[146] If these reactions are carried out in the presence of ethene then the reaction pathway can be modified leading to $H_2C=CH(CHR_2)$ and $[WCl_2(=O)(PMePh_2)_2(C_2H_4)]$. Labelling studies of this reaction indicate that the reaction proceeds via ethene insertion into W–C bond followed by β-elimination and rearrangement.

5.8 CYANIDE COMPLEXES

This section on cyanide complexes is conveniently subdivided into discussion of homoleptic complexes and others.

5.8.1 Synthesis

The reaction of $K_3[W_2Cl_9]$ with excess KCN in aqueous solution in the absence of air but under UV irradiation affords the heptacoordinate $K_5[W(CN)_7]\cdot 2H_2O$. The metal ion is pentagonal bipyramidal and isomorphous with the corresponding molybdenum ion.[156,157]

Cyanonitrosyl complexes such as $(Ph_4P)[Mo(CN)_5(NO)]\cdot 2H_2O$, $[Cr(CN)_4(NO)]^{2-}$, $[Cr(CN)_3(NO)(OH_2)]^-$, or $[Mo(CN)_3(NO)(L_2)]\cdot H_2O$ (L_2 = bipy, phen) may be formed directly from the ions $[MO_4]^{2-}$ (M = Cr or Mo) in aqueous–alkaline and aerobic media. At lower pH, species such as $(Ph_4P)_2[Mo(CN)_4(NO)_2]$ and $[Mo(CN)_2(NO)_2(phen)]\cdot 2H_2O$ are formed.[158,159]

Complexes of the type $[CrX_2(amine)_4]$ are quite common. Cyanide ligands may be introduced onto the metal by substitution of a ligand such as DMSO. Thus, reaction of *cis*- and *trans*-$[Cr(DMSO)_2(NH_3)_4][ClO_4]$ with excess cyanide gives a mixture of *cis*- and *trans*-$[Cr(CN)_2(NH_3)_4][ClO_4]$. The *trans*-isomer is relatively insoluble.[160] In a related process, treatment of *trans*-$[CrCl(DMSO)(NH_3)_4][ClO_4]_2$ with excess cyanide affords *trans*-$[CrCl(CN)(NH_3)_4][ClO_4]$.[161] The reaction of $[Cr(en)_3][Cl]_3\cdot 3.5H_2O$ with KCN in warm Me_2SO results in the bis(cyano) complex *cis*-$[Cr(CN)_2(en)_2][Cl]\cdot H_2O$ after workup. This material may be resolved through fractional crystallization of the mixture of diastereomers produced upon treatment with $(+)_{589}$-ammonium-3-bromo-*d*-camphor-8-sulfonate. After regeneration of the chloride form, the optical rotation, $[\alpha]^{20}_{589}$, is +39.7°.[162]

Salts of the $[Mo(CN)_8]^{3-}$ ion undergo cation exchange. Thus addition of protonated bipy to aqueous solutions of $K_3[Mo(CN)_8]$ results in $(bipyH)_3[Mo(CN)_8]\cdot 4H_2O$.[163]

5.8.2 Structure and Properties

The complex $Cs_2K[Cr(CN)_6]$ has been studied both by high precision x-ray diffraction[164] and by polarized neutron diffraction methods.[165] The former study shows that the $[Cr(CN)_6]^{3-}$ ion is substantially octahedral. To a first approximation the complex ion consists of a $Cr(t_{2g}{}^3e_g{}^0)$ atom surrounded by six free ion-like cyanide ions. The small differences between this description and the observed structure are a consequence of σ-donation (0.17(8) e^- per cyanide) from cyanide to chromium, a lesser back-bonding interaction (0.09(6) e^-) from chromium to cyanide π-orbitals, and a transfer of electron density (0.3(1) e^-) from cyanide π-orbitals to σ-orbitals. Analysis of the polarized neutron diffraction experiment suggests that the spin density is satisfactorily described by a model in which the $[Cr(CN)_6]^{3-}$ is octahedral with 3*d* and 4*p* orbitals on chromium and $(sp)_\sigma$ and p_π on the carbon and nitrogen atoms. Under this model the chromium is $t_{2g}{}^{2.30(3)}e_g{}^{0.12(4)}4p^{0.83(7)}$ with the 3*d* orbitals expanded radially by 8%.

It is well known that the structures of the ion $[Mo(CN)_8]^{4-}$ in the solid state are dependent on the nature of the countercation and crystal packing forces. However, variable-temperature NMR linewidth studies (^{95}Mo and ^{14}N) on the salt $K_4[Mo(CN)_8]\cdot 2H_2O$ suggest that, in aqueous solution, the ion $[Mo(CN)_8]^{4-}$ is dodecahedral.[166] A related study on the species $(Bu^n{}_4N)_4[W(CN)_8]$ shows that it exists in three different phases in the temperature range 90–353 K.[167]

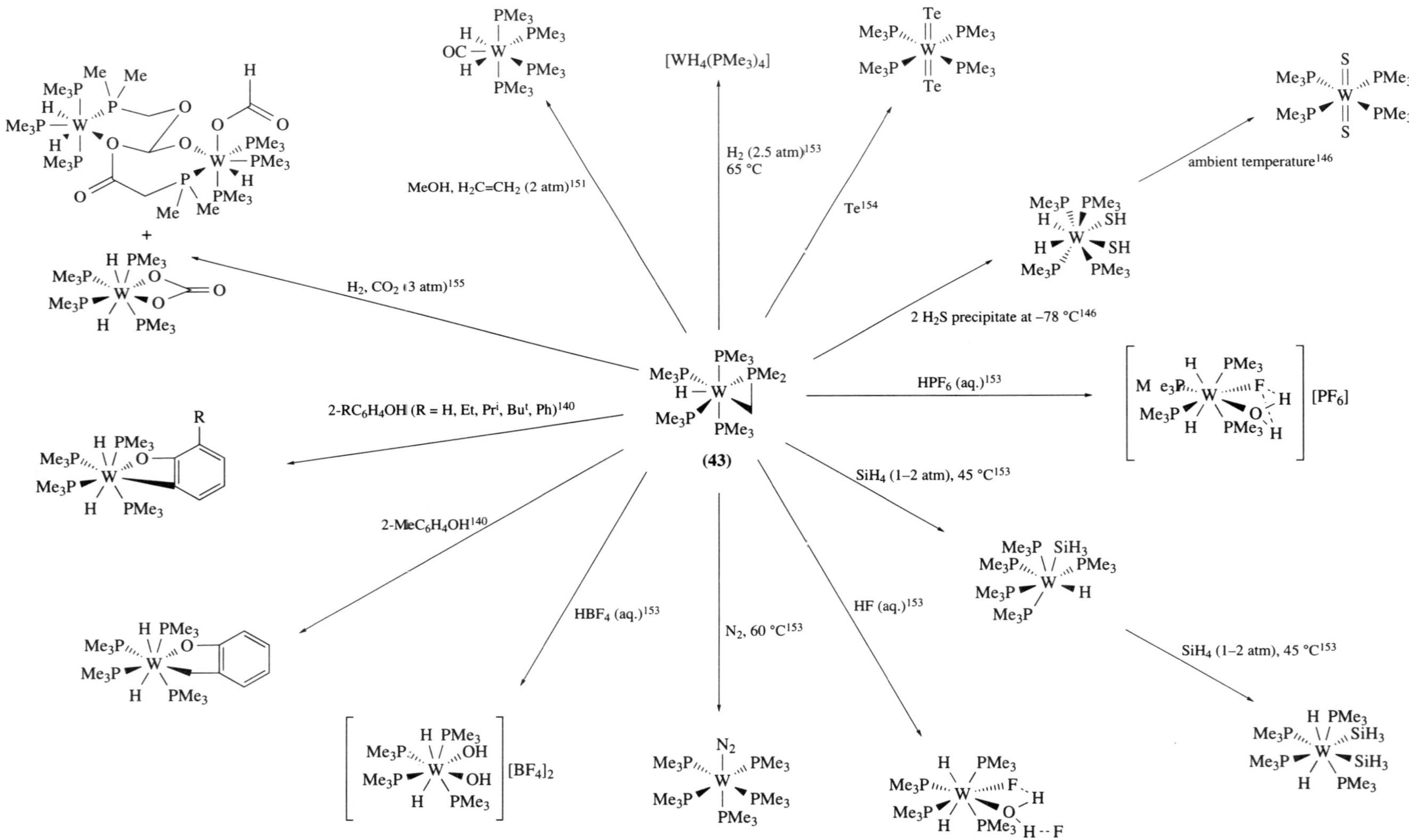

(43)
MeOH, $H_2C{=}CH_2$ (2 atm)[151]
$[WH_4(PMe_3)_4]$
H_2 (2.5 atm)[153]
65 °C
Te[154]
ambient temperature[146]
H_2, CO_2 (3 atm)[155]
2 H_2S precipitate at −78 °C[146]
HPF_6 (aq.)[153]
$[PF_6]$
2-RC_6H_4OH (R = H, Et, Pr^i, Bu^t, Ph)[140]
SiH_4 (1–2 atm), 45 °C[153]
2-MeC_6H_4OH[140]
HBF_4 (aq.)[153]
N_2, 60 °C[153]
HF (aq.)[153]
SiH_4 (1–2 atm), 45 °C[153]
$[BF_4]_2$

Scheme 10

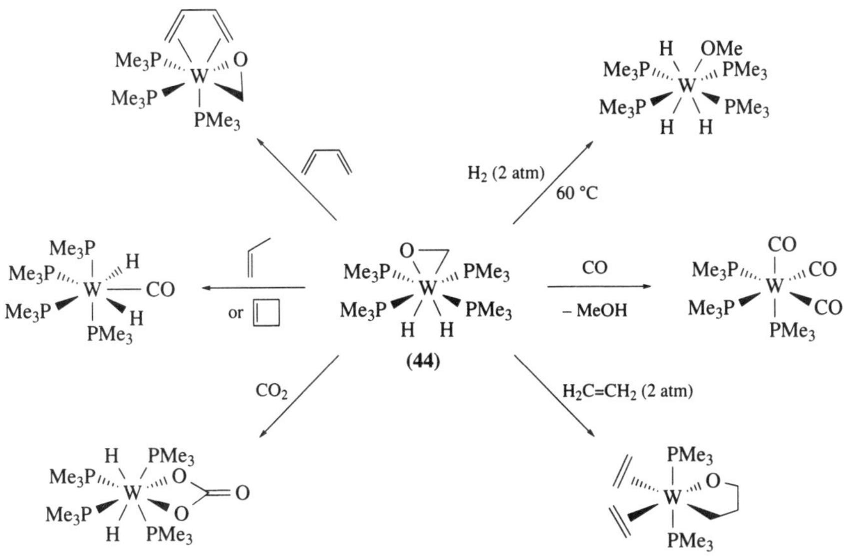

Scheme 11

5.8.3 Reactions

The molybdenum complex $[Mo(CN)_8]^{4-}$ is oxidized by $S_2O_8^{2-}$ in reactions which form $[Mo(CN)_8]^{3-}$. These reactions show a pronounced metal ion catalytic effect which varies in the order $Li^+ < Na^+ < K^+ < Rb^+ < Cs^+$. The alkali metal cation probably constitutes part of the transition state and it appears that larger alkali metal cations are more effective in forming a bridge for electron transfer.[168] Electron transfer between hydrazine and $[M(CN)_8]^{3-}$ (M = Mo or W) proceeds through an outer-sphere mechanism in reactions that are pH independent. The rate data for these reactions are consistent with the Marcus theory.[169]

Products formed from the ion $[W(CN)_8]^{3-}$ upon photolysis depend on the conditions employed. Irradiation in the charge-transfer-to-metal band proceeds largely through a photoredox process which affords $[W(CN)_8]^{4-}$. However, about 10% of the reaction involves formation of a transient $[W(CN)_7]^{2-}$. This either solvates to form $[W(CN)_7(OH_2)]^{2-}$ or in the presence of high concentrations of X^- (X = Cl, Br, N_3, SCN) or O_2 affords, respectively, $[WX(CN)_7]^{3-}$ or $[W(CN)_7(O_2)]^{2-}$.[170,171] High-intensity irradiation of an aqueous solution of $[W(CN)_8]^{3-}$ leads to tungsten(VI) complexes not formed under more normal irradiation intensities. The ion $[W(CN)_6(=O)]^{2-}$ is formed in aqueous solution and crystallographically characterized as $[PPh_4]_2[W(CN)_6(=O)]$. In a $MeOH/H_2O$ mixture (1:1) the result is the ion $[W(CN)_5(OMe)(=O)]^{2-}$, produced upon standing after photolysis and crystallographically characterized as $[PPh_4]_2[W(CN)_5(OMe)(=O)]$. Both complexes are approximately pentagonal bipyramidal with all cyano ligands equatorial.[172] Substitution reactions take place following photoexcitation of aqueous $K_4[M(CN)_8]$. The quantum yields of the photoreactions in the cases of phen and bipy increase with metal complex ion concentration but are not particularly dependent upon pH or ligand concentration.[173] If additional alkali is present, products such as $[W(CN)_7(OH)]^{4-}$, $[W(CN)_4(=O)_2]^{4-}$, $[W(CN)_4(OH)(=O)]^{3-}$, and $[W(CN)_4(OH)_2]^{2-}$ are formed.[174]

At 400 °C, H_2 reacts with $K_4[W(CN)_8]\cdot 2H_2O$ to give the hydride $K_4[WH(CN)_7]\cdot 2H_2O$. The main feature of the ^{13}C NMR spectrum is a doublet ($\delta = 154$, $J(^{13}C{-}^1H)$ 11.1 Hz) consistent with a monohydride. Further, IR and Raman studies show bands at 1830 cm^{-1} and 834 cm^{-1}, which shift to 1310 cm^{-1} and 615 cm^{-1} upon deuteration.[156] The corresponding molybdenum complex $K_4[MoH(CN)_7]\cdot 2H_2O$ is formed in similar fashion but is made in a purer state by addition of acids such as $MeCO_2H$ or HCl to $K_5[Mo(CN)_7]$.[175]

Photolysis of *trans*-$[Cr(CN)_2(NH_3)_4]^+$ in acidic solution results in NH_3 aquation. The quantum yield is 0.24 in a process which is wavelength-independent throughout the ligand field region. The corresponding dark reaction results in exclusive CN^- loss.[176] Aquation proceeds in two successive steps and each stage may be described by an acid-independent path and an acid-dependent path.[177] The photochemistry of *trans*-$[Cr(CN)_2(cyclam)][ClO_4]$ is quite different. There is no discernible photochemical activity but there is an intense and long-lived (335 μs) $^2E_g \rightarrow {}^4B_{1g}$ phosphorescence. Deuteration of the cyclam protons results in an increase in both intensity and lifetime of the emission.

This solution behaviour corresponds to the behaviour frequently observed of chromium(III) complexes at liquid nitrogen temperatures and in the solid phase.[178] Ligand field irradiation of *trans*-$[CrCl(CN)(NH_3)_4]^+$ in aqueous acidic solution results in release of mainly cyanide but some ammonia. The stereochemistry of the main product is largely *cis*. This appears to be a consequence of an excited-state *trans* effect.[161]

5.9 ISOCYANIDE COMPLEXES

5.9.1 Synthesis

Six- and seven-coordinate homoleptic isocyanide complexes are known. The neutral compounds are all six-coordinate M^0 compounds, whereas charged M^{II} complexes are known for both coordination numbers. Neutral octahedral $[M(CNR)_6]$ complexes are well known but a few new syntheses have appeared (Table 11). For instance, reduction of various metal halide derivatives MX (M = Cr, Mo, W) with sodium amalgam in the presence of the isonitriles CNR (R = Bu^t, $C_6H_3Me_2$-2,6, $C_6H_2Me_3$-2,4,6, C_6H_2Me-6-Bu^t_2-2,4, $C_6H_2Bu^t_3$-2,4,6) affords $[M(CNR)_6]$ in good yield.[179] A number of chromium complexes $[Cr(CNAr)_6]$ are formed by the electrochemical reduction of the corresponding dication.[180] Addition of isocyanide to the M≡M bonded complexes $[M_2(OR)_6]$ followed by workup with MeOH and KPF_6 results in the salts $[M(CNAlk)_7]^{2+}$ or $[M(CNPh)_6]$.[181]

Table 11 Formation of some isocyanide complexes.

Synthesis	*Complex*	*Comments*	*Ref.*
$[MX_nL_x]$ + Na/Hg in THF	$[M(CNR)_6]$	Cr, Mo, W; R = Bu^t, Ar	179
$[M_2(OAc)_4]$ + $CNBu^t$/Na/Hg	$[M(CNBu^t)_6]$	M = Cr, Mo; x-ray (M = Mo)	182
$[W_2(mhp)_4]$ + $CNBu^t$/Na/Hg	$[W(CNBu^t)_6)]$	mhp = anion of 2-hydroxy-6-methylpyridine; R = Bu^t, Cy; cleavage of M≡M bond	183
$[W_2(dmhp)_4]$ + CNPh/MeOH	$[W(CNPh)_6]$	dmhp = anion of 2,4-dimethyl-6-hydroxypyrimidine; also works with $[W_2(mhp)_4]$	184
$CrCl_3{\cdot}6H_2O$ + i, Zn dust/THF; ii, Bu^tDiNC; iii, Na/Hg	$[Cr(Bu^tDiNC)_3]$	Bu^tDiNC = 1,2-bis(4-*t*-butyl-2-isocyanophenoxy)ethane	185
$[WCl_4(PMe_3)_2]$ + CNMe/Na/Hg	$[W(PMe_3)_2(CNMe)_4]$	x-ray	186
$[Cr(Bu^tDiNC)_3]$ + nAgPF$_6$/Me$_2$C=O	$[Cr(Bu^tDiNC)_3][PF_6]_n$	n = 1 or 2	185
Cr^{II}/EtOH + CNR, KPF_6	$[Cr(CNR)_6][PF_6]_2$	R = Bu^t, Pr^i, Cy	187,188
$[Cr(CNAr)_6]$ + $SbCl_5/CH_2Cl_2$	$[Cr(CNAr)_6][SbCl_6]_3$	deep purple colour; Ar = Ph, $C_6H_3Me_2$-2,6, $C_6H_3Pr_2$-2,6	189
$CrCl_3{\cdot}6H_2O$ + Zn/Hg, CNPh, KPF_6 in EtOH	$[CrCl(CNPh)_5][PF_6]$	decomposes to $[Cr(CNPh)_6][PF_6]$	190
$[CrCl(CNPh)_5][PF]_6$ + PR_3 in CH_2Cl_2	$[CrCl(CNPh)_4(PR_3)][PF_6]$	R_3 = Bu^n_3, Pr^n_3, Et_3, Et_2Ph	190
$[Mo(SBu^t)_4]$ + $CNBu^t$	$[Mo(SBu^t)_2(CNBu^t)_4]$	d^4; distorted octahedral	191,192
$[Mo(SBu^t)_4]$ + $[CuBr(CNBu^t)_3]$ in Me$_2$C=O	$[(Bu^tNC)_4Mo(\mu\text{-}SBu^t)_2CuBr]$	two isomers	193
$[W_2(mhp)_4]$ + KPF_6/CNR/Me$_2$C=O	$[W(CNR)_7][PF_6]_2$	R = Bu^t, Cy; cleavage of M≡M bond	194
$[WI_2(CO)_4]$ + CNR/KPF_6/Me$_2$C=O	$[W(CNR)_7][PF_6]_2$	R = Bu^t, Cy	194
$[M(CNPh)_6]$ + $AgNO_3$/CNPh/KPF_6	$[M(CNPh)_7][PF_6]_2$	M = Mo or W	184
$[Mo_2(OAc)_4]$ + $CNBu^t$	$[Mo(OAc)_2(CNBu^t)_5]$	cleavage of M≡M bond	183,195
$[Mo_2(O_2CCF_3)_4]$ or $K_4[Mo_2Cl_8]$ + $CNBu^t$	$[MoX(CNBu^t)_6][X]$	X = Cl or O_2CCF_3; cleavage of M≡M bond	195
$[Mo(CNMe)_3(CO)_3]$ + i, I_2/toluene; ii, CNMe; iii, $h\nu$/MeOH; iv, $NaBF_4$	$[MoI(CNMe)_6][BF_4]$		196
$[M(CO)_6]$ + i, CNR; ii, I_2; iii, $h\nu$/EtOH	$[MI(CNR)_6][I]$	M = Mo; R = Bu^t, Cy; M = W, R = Bu^t	196
$[Mo(CO)_6]$ + i, CNR; ii, $PhICl_2$	$[MoCl(CNR)_6][Cl]$		196
$[MCl_2(PMe_3)_4]$ + $CNBu^t$	$[MCl_2(CNBu^t)_3][PMe_3]_2$	M = Mo or W	197
$[M(L_2)(CO)_4]$ + i, $PhICl_2$/MeOH; ii, CNR; iii, KPF_6	$[M(CNR)_5(L_2)][PF_6]_2$	M = Mo or W; L_2 = bipy, Me$_2$bipy, phen; R = Pr^i, Bu^t, Cy, Bz	198,199
$[MoH(I)(\eta\text{-}Cp)_2]$ + CNMe/THF	$[MoH(CNMe)(\eta\text{-}Cp)_2][I]$	Related reactions for $[MX_2(\eta\text{-}Cp)_2]$ (M = Mo or W)	200

Polymer-encapsulated $[M(CNR)_6]$ (M = Cr, Mo, W) complexes are formed in the reaction of the metal vapour with the appropriate isocyano-functionalized polysiloxane.[201] Another route to polymeric materials involves the reaction of group 6 metal aryldiisocyanides with the appropriate dimetal complex $[W_2$(2,4-dimethyl-6-hydroxypyrimidine)$_4]$ or $[M_2(OAc)_4]$ (M = Cr, Mo). These materials are hygroscopic and are typically formulated as $[M(1,4\text{-diisocyanobenzene})_x{\cdot}yH_2O]_n$ ($x = 3.15 \pm 0.05$, $y = 2.52$).[202]

A number of improved routes for the synthesis of $[M(CNR)_7]^{2+}$ complexes appeared in the literature. The first of these involves the substitution of the remaining carbonyl groups of $[M(CNR)_3(CO)_3]$ (M = Mo, R = Cy, Bu^t; M = W, R = Bu^t) with excess CNR in the presence of $PhICl_2$. The salts are normally isolated with $[PF_6]^-$ as counterion.[196] The chromium compounds (R = Cy, Bu^t) are conveniently made by reacting the salts $[Cr(CNR)_6][PF_6]_2$ with neat isonitrile.[187]

Other routes to isocyanide complexes involve substitution reactions, for instance, the successive replacement of PMe_3 ligands in $[M(PMe_3)_4(C_2H_4)_2]$ (Equation (23)) (M = Mo or W).[203] or of N_2 ligands in *trans*-$[Mo(N_2)_2(Me_8[16]aneS_4)]$ by CNPh, a structure notable for its bent CNPh ligands.[204]

$$[M(PMe_3)_4(C_2H_4)_2] \xrightarrow{\text{CNMe}} [M(PMe_3)_3(CNMe)(C_2H_4)_2] \xrightarrow{\text{CNMe}} [M(PMe_3)_2(CNMe)_2(C_2H_4)_2] \quad (23)$$

Addition of CNR (R = xylyl, Bu^t) to $[Mo_2(OBu^t)_6]$ or $[Mo_2(OBu^t)_6(py)_2]$ results in formation of the adducts $[Mo_2(OBu^t)_6(CNR)]$ or $[Mo_2(OBu^t)_6(py)(CNR)]$. In both cases there are bent μ-CNAr ligands with the C–N–C planes aligned along the M–M axis.[205]

5.9.2 Structure and Properties

In general, the complexes $[M(CNR)_6]^+$, $[M(CNR)_6]^{2+}$ and $[M(CNR)_6]^{3+}$ are yellow, orange or red. A number of crystal structures of such complexes are known. The coordination of the molybdenum in $[Mo(CNC_6H_3Me_2\text{-}2,6)_6]$ is octahedral with deviations of about 3° from an ideal angle of 90°.[179] The crystal structures of the three ions $[Cr(CNPh)_6]^{n+}$ (n = 1–3) completes a series of structures $[Cr(CNPh)_6]^{n+}$ (n = 0–3) and reveals interesting trends for bond lenghts.[189,206] The data show that increasing the oxidation state causes a systematic increase in the Cr–C and N–C_{ring} bond lengths but a decrease in the C≡N bond lengths. This is attributable to a decrease in metal $d\pi \rightarrow$ ligand π^* donation (backbonding) with increase in oxidation state. However, as the oxidation state increases, the Cr–C bond lengthens, the reverse of the trend normally seen for M–L σ-bonds in coordination compounds. The crystal structure of the complex $[Mo(CNBu^t)_6]$ shows considerable bending at the nitrogen atoms, indicative of charge delocalization onto the ligands.[182]

Crystallographic studies on the homoleptic seven-coordinate compounds $[M(CNBu^t)_7][PF_6]_2$ (M = Cr[207] and W[208]) and $[Mo(CNCy)_7][PF_6]_2$[208] give further information on this coordination number. The structure of the chromium compound is closer to a 4:3 (C_s) piano stool structure than other idealized geometries, while the molybdenum compound displays a distorted capped octahedral structure (C_{3v}). The tungsten cation exhibits a capped trigonal prismatic structure. In practice, there is little difference in energies between these various geometries and it is highly likely that crystal-packing forces are important in determining which coordination geometry is displayed.[208]

Cyclic voltammetry studies on a series of complexes $[Cr(CNAr)_6]^{n+}$ (n = 0, 1, 2, 3) show that the inductive effects from *para*-substituents on $E^{\circ\prime}$ for the various couples give linear correlations with the Hammett σ_p parameter. *ortho*-Substituents have a pronounced effect on $E^{\circ\prime}_{3+/2+}$ and the substitutional lability of the $[Cr(CNR)_6]^{3+}$ complexes. This appears to be a consequence of steric interactions among the *ortho*-substituents.[209] Electrochemical studies on various seven-coordinate complexes $[M(CNR)_7]^{2+}$ and $[MX(CNR)_5]^{2+}$ (M = Mo or W) indicate that most exhibit quasi-reversible one-electron reversible oxidation processes to form seven-coordinate M^{III} cations and also show irreversible reduction. The half-wave potentials increase in the order $[MX(CNR)_7]^+$ (X = halide) < $[M(CN)(CNR)_7]^+$ < $[M(CNR)_7]^+$ < $[M(SnCl_3)(CNR)_7]^+$, a result consistent with ligand electron donor–acceptor properties. There is a correlation between the ease of oxidation of these complexes and the tendency for the isocyanide ligands to undergo reductive coupling, although there is no evidence for electrochemical reductive coupling.[210]

The d^4 complex *cis*-$[Mo(SBu^t)_4(CNBu^t)_4]$ is distorted octahedral, but the reason for this seems not to be a consequence of a Jahn–Teller distortion, although the d^4 structure is a contributory factor.[192] Instead, the distortion stems from a repulsive interaction between two sulfur *p*-orbitals.

5.9.3 Reactions

The complexes $[Cr(CNC_6H_4Pr^i_2\text{-}2,6)_6]$ (structurally characterized[211]) and $[Cr(CNC_6H_4Pr^i_2\text{-}2,6)_6][BF_4]$ undergo electron self-exchange, as do $[Cr(CNC_6H_4Pr^i_2\text{-}2,6)_6][BF_4]$ and $[Cr(CNC_6H_4Pr^i_2\text{-}2,6)_6][BF_4]_2$. The Cr^0–Cr^I rates are higher by a factor of ten than the corresponding Cr^I–Cr^{II} rates in CH_2Cl_2 and acetone. The activation parameters correspond with predictions from the Marcus collision model,[212,213] as do the results of kinetic studies on the outer-sphere electron transfer between $[Cr(CNC_6H_4Pr^i_2\text{-}2,6)_6][BF_4]_2$ and cobalt(II) clathrochelates.[214]

Isocyanide complexes undergo substitution reactions. The seven-coordinate species often lose a ligand to form six-coordinate complexes. Thus, the reaction of either $[Cr(CNR^1)_n][PF_6]_2$ ($n = 6$ or 7; $R^1 = Bu^t$ or Cy) with PR^2_3 ($R^2 = Et$, Pr^n, Bu^n) afford the paramagnetic complexes *trans*-$[Cr(CNR^1)_4(PR^2_3)_2][PF_6]_2$, while the reaction with dppe affords *cis*-$[Cr(CNR)_4(dppe)][PF_6]_2$. This last complex reacts with neat CNR to revert to a seven-coordinate complex $[Cr(CNR)_5(dppe)][PF_6]_2$.[187] The seven-coordinate species $[Mo(CNCy)_7][PF_6]_2$ reacts with bidentate ligands L_2 such as bipy or phen to form complexes $[Mo(CNCy)_5(L_2)][PF_6]_2$.[215] The crystal structure of the related $[Mo(CNCy)_5(dppm)][PF_6]_2$ complex shows that the molybdenum is in a distorted pentagonal bipyramidal configuration with two isocyanide ligands axial.[216] The corresponding reaction of the Bu^t analogue $[Mo(CNBu^t)_7][PF_6]_2$ is more complex and results in a cyano-bridged species $[(L_2)(Bu^tNC)_4Mo(\mu\text{-CN})Mo(CNBu^t)_4(L_2)][PF_6]_3$. The solid-state structure of this species shows that the geometry of each molybdenum is monocapped trigonal prismatic. The bipyridyl complex $[Mo(CNCy)_5(bipy)][PF_6]_2$ reacts with Bu^tCl to form $[MoCl(CNBu^t)_4(bipy)][PF_6]$, a compound also formed through cleavage of the Mo≡Mo bond in $[MoCl_4(bipy)_2]$.[215] An analysis of the substitution reaction of X^- (X = Cl, Br, I) with $[Mo(CNBu^t)_7]^{2+}$ by conductance studies in acetonitrile shows that the reaction is reversible. The mechanism appears to involve dissociative processes of both ion-paired and non ion-paired species. The corresponding tungsten reaction does not proceed at an observable rate in either direction.[217]

Warming a mixture of $[Mo(CNBu^t)_7]^{2+}$ and zinc in ethanol results in a complex process involving dealkylation of one isocyanide ligand and coupling of two others to form ligated $Bu^tHNC{\equiv}CNHBu^t$ (Equation (24)).[218] The crystal structure shows the molybdenum coordination geometry is close to pentagonal bipyramidal. A related coupling reaction occurs in the reaction of $[MoCl(CNBu^t)_4(bipy)]^+$ with zinc in wet THF[219] but in this case there is no dealkylation step (Equation (25)).

$$[Mo(CNBu^t)_7]^{2+} + Zn/EtOH \longrightarrow [Mo(CN)(CNBu^t)_4(\eta^2\text{-}Bu^tHNC{\equiv}CNHBu^t)]^+ \quad (24)$$

$$[MoCl(CNBu^t)_4(bipy)]^{2+} + Zn/THF/H_2O \longrightarrow [Mo(Cl)(CNBu^t)_4(\eta^2\text{-}Bu^tHNC{\equiv}CNHBu^t)]^+ \quad (25)$$

The chemistry of *cis*-$[Mo(SBu^t)_4(CNBu^t)_4]$ is developing. It reacts with $[CuBr(CNBu^t)_3]$ to form two isomers of the bridged complex $[(Bu^tNC)_4Mo(\mu\text{-}SBu^t)_2CuBr]$[193] and undergoes substitution of isocyanide ligands in its reaction with RC≡CR (R = H or Ph) which forms $[Mo(SBu^t)_4(CNBu^t)_2(\eta^2\text{-}RCCR)]$. These complexes are approximately trigonal bipyramidal with the alkyne C≡C axis lying parallel to the main molecular axis.[220,221]

Oxidation of hexacoordinated complex $[Mo(CNPh)_6]$ with I_2 affords the seven-coordinate cation $[MoI(CNPh)_6]^+$.[184] In related reactions, oxidation of polymeric species such as $[W(4,4'\text{-diisocyanobiphenyl})_x{\cdot}yH_2O]_n$ ($x = 3.45 \pm 0.40$, $y = 2.80$) with iodine results in the oxidized polymers $[\{WI(4,4'\text{-diisocyanobiphenyl})_x\}I{\cdot}yH_2O]_n$ ($x = 2.90 \pm 0.32$, $y = 2.82$). Partial oxidation under similar conditions gives mixed valence $W^{0,II}$ polymers in which the $W^0{:}W^{II}$ ratio varies.

A number of isocyanide complexes oxidize in their reactions with nitrosyl sources. Thus, reaction of $[M(CNPh)_6]$ (M = Mo or W) with $NOPF_6$ affords $[M(NO)(CNPh)_5][PF_6]$,[184] while NO reacts with $[Cr(CNR)_6][PF_6]_2$ ($R = Bu^t$, Pr^i or Cy) to form $[Cr(NO)(CNR)_5][PF_6]$.[188] The chromium compounds undergo substitution reactions with monodentate phosphines PR_3 to form $[Cr(NO)(CNR)_4(PR_3)][PF_6]$, and with chelating phosphines L_2 such as dppm or dppe to form $[Cr(NO)(CNR)_3(L_2)][PF_6]$.[222,223] The *fac* isomers isomerize slowly to the corresponding *mer* isomers in a process which may be monitored by ESR spectroscopy.

Isocyanide ligands are prone to alkylation or protonation processes. This can form carbyne complexes. This is explored in more detail in Section 5.12. Addition of CF_3CO_2H to $[Mo(CNBu^t)_6]$ in THF results in isocyanide coupling to form the alkyne complex $[Mo(O_2CCF_3)(CNBu^t)_4(\eta^2\text{-}Bu^tHNC{\equiv}CNHBu^t)]^+$. It is likely that an intermediate aminocarbyne is involved in this process. The corresponding reaction with excess CF_3CO_2H in cold toluene results in the doubly protonated alkylidene complex $[Mo(CNBu^t)_5\{=C(H)NHBu^t\}]^{2+}$.[182] Related compounds $[MX(CNBu^t)_4(\eta^2\text{-}Bu^tHNC{\equiv}CNHBu^t)]^+$ ($M = Mo$, $R = Bu^t$, X = Cl, Br, I, CN; M = Mo, R = Cy, X = I; M = W, $R = Bu^t$, X = I) are formed through the reductive coupling (zinc dust, H_2O, THF) of isocyanides in the complexes $[MX(CNBu^t)_6]^+$.[224] The solid-state structures of several of these complexes are known and all possess capped trigonal prismatic coordination geometries.[225]

Occasionally reagents insert into the Mo–Cl bond of $[MoCl(CNBu^t)_6]^+$ and a good example is provided by $SnCl_2$. This reacts with $[MoCl(CNBu^t)_6]^+$ to form $[Mo(SnCl_3)(CNBu^t)_6]^+$, structurally characterized as the $[(Ph_3B)_2CN]^-$ salt. The coordination geometry of the molybdenum is a distorted capped octahedron with the $SnCl_3$ group as capping ligand.[226] This compound does not undergo a reductive coupling on treatment with zinc, but instead the cation $[Mo(CNBu^t)_7]^{2+}$ is isolable.

Calculations suggest that the β-protonation of isocyanide in *trans*-$[Mo(CNMe)_2(dppe)_2]$ is predominantly charge controlled.[227] Alkylation of the carbyne complex $[W(CNR)_2(\equiv CNEt_2)(\eta\text{-}Cp^*)]$ (R = Et, Bu) with Et_3OBF_4 results in the cationic bis(carbyne) species $[W(CNR)_2(\equiv CNEt_2)(\equiv CNEtR)(\eta\text{-}Cp^*)][BF_4]$.[228] Iodine reacts with $[WI(CNR)_3(\equiv CNEt_2)]$ (R = Me, Bu^t) to form the seven-coordinate $[WI_3(CNR)_3(\equiv CNEt_2)]$. This reacts with CNR to form $[WI_2(CNR)_4(\equiv CNEt_2)][I]$.[229]

5.10 ALKYNIDE COMPLEXES

Alkynide complexes which do not bear carbonyl ligands are comparatively rare for the group 6 elements. The chemistry of group 6 alkynide complexes containing carbonyl ligands is discussed in Chapter 4, this volume. The most common precursors for alkynide complexes of group 6 elements are alkynides, vinylidene complexes and alkynes.

Deprotonation of the vinylidene complex $[Mo(dppe)(=C=CHR)(\eta\text{-}C_7H_7)][PF_6]$ (R = Ph, Bu^t) affords the η^1-alkynyl $[Mo(C\equiv CR)(dppe)(\eta\text{-}C_7H_7)]$.[230,231] Protonation of the alkynide $[Mo(C\equiv CR)(dppe)(\eta\text{-}C_7H_7)]$[230,231] results in re-formation of $[Mo(dppe)(=C=CHR)(\eta\text{-}C_7H_7)]^+$. Both alkynide complexes $[Mo(C\equiv CR)(dppe)(\eta\text{-}C_7H_7)]$ undergo reversible one-electron oxidations to form the radical cations $[Mo(C\equiv CR)(dppe)(\eta\text{-}C_7H_7)][BF_4]$.

The reactions of alkynes RC≡CH (R = Ph, CO_2Me, CO_2Et) with *trans*-$[M(N_2)_2(dppe)_2]$ under photolytic conditions gives a mixture of the alkynyl complexes *trans*-$[M(C\equiv CR)_2(dppe)_2]$ and $[MH_2(C\equiv CR)_2(dppe)_2]$. The M–C≡CR units are essentially linear in these systems. The x-ray crystal structure of $[MoH_2(C\equiv CCO_2Me)_2(dppe)_2]$ shows that the coordination geometry of the molybdenum is> square antiprismatic. The corresponding tungsten species reacts with HBF_4 to give the fluoro complex $[WF(dppe)_2(\equiv CCH_2CO_2Me)]$ in high yield, whereas the corresponding reaction of $[WH_2(C\equiv CPh)_2(dppe)_2]$ affords the cationic alkylidene $[WF(=CHCH_2Ph)(dppe)_2][BF_4]$.[232,233]

The addition of four equivalents of LiCCR (R = Bu^t, TMS, Pr^i) to $[M_2Cl_4(PMe_3)_4]$ (M = Mo or W) affords the corresponding complexes $[M_2(C\equiv CR)_4(PMe_3)_4]$.[234] These compounds possess the same D_{2d} structures as other dimetal complexes with M≡M bonds. Addition of base to the bridging ethynyl complex $[(Cp)(\eta^2\text{-}PhC\equiv CPh)W(\eta^2\text{-}HC\equiv CM(L)_2(Cp))]$ (M = Fe, Ru; L = PMe_3, $P(OMe)_3$) results in deprotonation and formation of the bridged ethynediyl complex $[(Cp)(\eta^2\text{-}PhCCPh)(OC)W\text{–}C\equiv C\text{–}M(L)_2(Cp)]$. This reaction is reversed upon addition of $HBF_4{\cdot}OEt_2$.[235]

There is little other chemistry known of group 6 alkynyl complexes without carbonyl ligands, but prolonged stirring of the radical cation $[Mo(C\equiv CPh)(dppe)(\eta\text{-}C_7H_7)][PF_6]$ in THF results in dimerization to form the divinylidene complex $[(\eta^7\text{-}C_7H_7)(dppe)Mo\{=C=C(Ph)C(Ph)=C=\}Mo(dppe)(\eta\text{-}C_7H_7)]$.[236]

5.11 ALKYLIDENE COMPLEXES

Alkylidene complexes are here defined as $L_nM{=}CR^1R^2$ species in which R^1 and R^2 are H or carbon groups. The reader is referred to the discussion of Chapter 3, this volume which is concerned largely with carbonyl complexes of heteroatom-stabilized carbene complexes and to an excellent review on alkylidene complexes and metathesis reactions.[237] Alkylidenes have been mentioned in earlier sections of this chapter.

5.11.1 Synthesis

A number of syntheses of alkylidene complexes are gathered in Table 12. Syntheses of such complexes are many and varied in approach.

Many alkylidene complexes of the type $[WX_n(OR^1)_{4-n}(CHR^2)]$ (X = halide, R^1 = alkyl, R^2 = various) and related species such as $[WCl(CH_2R^1)(OC_6H_3R^2{}_2\text{-}2,6)(OR^3{}_2)(CHR^4)]$ are known. Their syntheses are summarized in Equations (26)–(29).[98,238,239] Timing is important in these reactions since metathesis of the alkene can interfere.

$$[WCl_2(OCH_2Bu^t)_2(=O)] + Mg(CH_2Bu^t)_2{\bullet}C_4H_8O_2 \longrightarrow [W(CH_2Bu^t)_2(OCH_2Bu^t)_2(=O)] \qquad (26)$$

Table 12 Syntheses and structural data for some alkylidene complexes.

Synthesis	*Complex*	*M=C* (nm)	*Comments*	*Ref.*
$[WCl_2(NC_6H_3Pr^i_2\text{-}2,6)(DME)(=CHPh)] + LiOCMe(CF_3)_2$	$[W(NC_6H_3Pr^i_2\text{-}2,6)\{OCMe(CF_3)_2\}_2(=CHPh)]$	0.1859(22)		88
$[WCl_2(NC_6H_3Me_2\text{-}2,6)\{OCMe(CF_3)_2\}_2] + Ph_3P{=}CH\{C_6H_4(OMe)\text{-}2\}$, Na/Hg	$[W{=}(CHC_6H_4OMe\text{-}2)(NC_6H_3Me_2\text{-}2,6)\{OCMe(CF_3)_2\}_2\text{-}(THF)]$	0.1943(5)	intramolecular coordination of methoxy group	252
$[Mo(NC_6H_3Pr^i_2\text{-}2,6)\{OCMe(CF_3)_2\}_2(=CHBu^t)] + PMe_3$	$[Mo(NC_6H_3Pr^i_2\text{-}2,6)\{OCMe(CF_3)_2\}_2(PMe_3)(=CHBu^t)]$	0.1878(9)	*syn*, distorted trigonal bipyramidal	247
$[W(OBu^t)_4(NPh)] + [TaCl_3(PMe_3)_2(=CHBu^t)]$	$[WCl_2(NPh)(PMe_3)_2(=CHBu^t)]$			243
$[WX_2(OCH_2Bu^t)_2(=CHBu^t)] + LiOCH_2Bu^t$	$[WX(OCH_2Bu^t)_3(=CHBu^t)]$		fluxional, X = Cl, Br, I	98
$[WCl_4(OAr)_2] + MgNp_2/OEt_2$	$[WCl\{OC_6H_3Ph(C_6H_4)\}(OC_6H_3Ph_2)(OEt_2)(=CHBu^t)]$		cyclometallated	239
$[W(NC_6H_3Pr^i_2\text{-}2,6)(OBu^t)(THF)(=CHBu^t)] + MgCl(CH_2Bu^t)$	$[W(NC_6H_3Pr^i_2\text{-}2,6)(OBu^t)_2(=CHBu^t)]$	0.1892(7), 0.1863(8)	pseudo tetrahedral, two molecules in unit cell	79
$[Mo(CH_2Bu^t)_2(NBu^t)_2]$ + i, $HOCH(CF_3)_2$; ii, MeCN; iii, vacuum	$[Mo(NBu^t)\{OCH(CF_3)_2\}_2(=CHBu^t)]$			254
$[Mo(CH_2R^2)_2(NR^1)_2] + TfOH$	$[Mo(NR^1)(OTf)_2(DME)(=CHR^2)]$		R^1 = aryl	213
$[WCl(CH_2TMS)_3(NPh)] + LiC_6H_4CH_2NMe_2\text{-}2$	$[W\{C_6H_4CH_2NMe_2\}(CH_2TMS)(NPh)(=CHTMS)]$	0.1872(15)	carbene axial	253
$[WBr_2(OCH_2Bu^t)_2(=CHBu^t)]$ + *exo*-methylenecyclopentane/CH_2Cl_2	$[WBr_2(OCH_2Bu^t)_2\{=C(CH_2)_3CH_2\}]_2$	0.1890(5)	weak bromide bridging groups, monomeric in solution	256
$[WBr_2(OCH_2Bu^t)_2\{=C(CH_2)_3CH_2\}]_2 + GaBr_3/CH_2Cl_2$	$[WBr(OCH_2Bu^t)_2(\eta^2\text{-}GaBr_4)\{=C(CH_2)_3CH_2\}]$	0.1896(9)		256
$[WCl_2(PMe_3)_2(=CHPh)(CO)] + PhC{\equiv}CPh$ + 1,8-bis-(NMe_2)-naphthalene	$[WCl_2(PMe_3)_2(=CHPh)(\eta^2\text{-}PhCCPh)]$	0.197(2)	carbyne intermediate	257
$[Mo(NO)_2(\mu\text{-}OPr^i)_2] + AlCl_2Et$	$[(AlCl_2)_2(\mu\text{-}OPr^i)_2Mo(NO)_2(=CHMe)]$		via $[(EtAlCl_2)_2(\mu\text{-}OPr^i)_2Mo(NO)_2]$	258,259
$[Mo(AlCl_3)_2(NO)_2(\mu\text{-}OPr^i)_2] + SnEt_4$	$[(Et_3Sn{\cdot}AlCl_3)_2(\mu\text{-}OPr^i)_2Mo(NO)_2(=CHMe)]_n$		polymeric	260
$[Mo(OAc)_2(NO)_2] + AlCl_2Et$	$[(AlCl_2)_2(OAc)_2Mo(NO)_2(=CHMe)]_n$		polymeric	261
$[WCl_2(PMePh_2)_4] + O{=}C(CH_2)_3CH_2$	$[WCl_2(=O)(PMePh_2)_2\{=C(CH_2)_3CH_2\}]$	0.1980(12)	W–O 0.1708(8) nm	144
$[WCl_2(=O)(PEt_3)_2(=CHBu^t)] + [PdCl_2(NCPh)_2]$	$[WCl_2(=O)(PEt_3)(=CHBu^t)]$	0.1882(14)		241
$[TaCl_3(PMe_3)_2(=CHBu^t)] + [W(=O)(OBu^t)_4]$	$[WCl_2(=O)(PMe_3)(=CHBu^t)]$	0.1986(21)	phosphines mutually *trans*, alkylidene and oxo mutually *cis*	240,262
$[Mo(Mes)_2(=O)_2] + Bu^n{}_3PCH_2$	$[Mo(Mes)(=O)_2\{=C(PBu^n)_3Mes\}]$	0.195(1)		104
$[WI(PMe_3)_4(\equiv CH)] + CF_3SO_3H$	$[WI(PMe_3)_4(=CH_2)][SO_3CF_3]$			263
$[W(Me)_4(\eta\text{-}Cp^*)]^+ + NEt_3$	$[W(Me)_3(=CH_2)(\eta\text{-}Cp^*)]$		methylene is distorted (T-shaped)	264
$[W(CH_2TMS)_2(NO)(\eta\text{-}Cp)] + \Delta$	$[W(CH_2TMS)(=O)\{=CH(TMS)\}(\eta\text{-}Cp)]$		slow at 20 °C	265
$[Mo(CH_2Bu^t)_2(NO)(\eta\text{-}Cp)] + \Delta$	$[(\eta\text{-}Cp)(ON)Mo(\mu\text{-}CHBu^t)_2Mo(NO)(\eta\text{-}Cp)]$	0.2186(3), 0.2192(3)	bridging alkylidene	266
$[Mo(CH_2Bu^t)_2(NO)(\eta\text{-}Cp)] + PMePh_2$	$[Mo(NO)(PMePh_2)(=CHBu^t)(\eta\text{-}Cp)]$	0.1950(3)	nearly linear nitrosyl	266
$[W(NC_6H_3Pr^i_2\text{-}2,6\}(OTf)_2\{=CH(CMe_2Ph)\}] + KTp^*$	$[W(NC_6H_3Pr^i_2\text{-}2,6)(Tp^*)\{CH(CMe_2Ph)\}][OTf]$	0.1964(9)	air-stable	246
$[WCl(CH_2CMe_2R)_3(NAr)] + KTp$	$[W(CH_2CMe_2R)(NAr)(Tp)\{=CH(CMe_2R)\}]$			267
$[(Cp^*)(Me)Cr(\mu\text{-}Me)_2Cr(Me)(Cp^*)] + C_6D_6$	$[(Cp^*)Cr(\mu\text{-}Me)_2(\mu\text{-}CH_2)Cr(Cp^*)]$		bridging methylene, Cr···Cr 0.2394(1) nm	268,269

$$[W(CH_2Bu^t)_2(OCH_2Bu^t)_2(=O)] + AlX_3 \longrightarrow [WX_2(OCH_2Bu^t)_2(=CHBu^t)] \quad (27)$$

$$[WX_2(OCH_2Bu^t)_2(=CHBu^t)] + H_2C{=}CR^1R^2 \longrightarrow [WX_2(OCH_2Bu^t)_2(=CR^1R^2)] \quad (28)$$

$$[WCl_4(OC_6H_3Ph_2\text{-}2,6)_2] + MgNp_2\bullet dioxane \longrightarrow [WCl(CH_2Bu^t)(OC_6H_3Ph_2\text{-}2,6)(OEt_2)(=CHBu^t)] \quad (29)$$

Oxo alkylidene complexes are comparatively unusual. Complexes such as $[WCl_2(=O)(PEt_3)_2(=CHBu^t)]$ are available through exchange between $[W(OCBu^t)_4(=O)]$ and $[TaCl_3(PEt_3)_2(=CHBu^t)]$.[240,241] One oxo complex with a bridging alkylidene is $[(Bu^tCH_2)_3W(\mu\text{-}OCH_2Bu^t)_2(\mu\text{-}CHR)W(OCH_2Bu^t)(=O)(py)]$ produced through cleavage of the aldehyde CO bond in the reaction of $[W_2(OCH_2Bu^t)_6(py)_2]$ with RCHO.[242]

The imido ligand is a better π-bonding ligand than the oxo ligand. One way to prepare such alkylidene complexes is by transfer of neopentylidene from tantalum to tungsten (Equation (30)).[243] Alkylidene ligands are also known to transfer between tungsten and molybdenum (Equation (31)).[244]

$$[W(NPh)(OBu^t)_4] + [TaCl_3(PR_3)_2(=CHBu^t)] \longrightarrow 0.5\,[TaCl(OBu^t)_4] + [WCl_2(NPh)(PR_3)_2(=CHBu^t)] \quad (30)$$

$$[Mo(OBu^t)_2(NR^1)_2] + [W(NR^1)(OBu^t)_2(=CHR^2)] \longrightarrow [W(NR^1)_2(OBu^t)_2] + [Mo(NR^1)(OBu^t)_2(=CHR^2)] \quad (31)$$

It is also possible to make related phosphine-free complexes (Equations (32) and (33)), making use of the fact that neopentylidyne complexes are precursors to neopentylidene complexes through a hydrogen migration process.[76,245] This reaction also proceeds at –40 °C for complexes in which DME is used in place of the phosphine ligands.[88] An alternative synthesis starts from WCl_6, which is first used to make $[WCl_2(NAr)(OBu^t)_2(THF)]$. Successive reactions with $MgCl(CH_2Bu^t)$ and PCl_5/DME results in the final product (Equation (34)). These complexes react with alkoxide reagents LiOR or KOR ($OR = OBu^t$, $OCMe_2CF_3$, $OCMe(CF_3)_2$, $OC(CF_3)_2(CF_2CF_2CF_3)$) by replacement of the chloride ligands and sometimes loss of DME to form $[W(NAr)(OR)_2(=CHBu^t)]$ (Equation (35)). Alternatively, the use of KTp* gives (**45**) (Equation (36)).[246]

$$[MCl_2(NHPh)(DME)(\equiv CBu^t)] \xrightarrow{NEt_3\ \text{catalyst}} [MCl_2(NAr)(DME)(=CHBu^t)] \quad (32)$$

$$[MCl_2(NHAr)(DME)(\equiv CBu^t)] \xrightarrow{NEt_3\ \text{catalyst}} [MCl_2(NAr)(DME)(=CHBu^t)] \quad (33)$$

$$[WCl_2(NAr)(OBu^t)_2(THF)] \xrightarrow{MgCl(CH_2Bu^t)} [W(NAr)(CH_2Bu^t)_2(OBu^t)_2] \xrightarrow{PCl_5,\ DME} [WCl_2(NAr)(DME)(=CHBu^t) \quad (34)$$

$$[WCl_2(NHAr)(DME)(=CHBu^t)] + LiOR \xrightarrow{x\,=\,0,\,1} [W(OR)(NHAr)(DME)_x(=CHBu^t)] \quad (35)$$

$$[W(NAr)(OTf)_2(DME)\{=CH(CMe_2Ph)\}] \xrightarrow{KTp^*} \quad (36)$$

$Ar = 2,6\text{-}Pr^i_2C_6H_3$

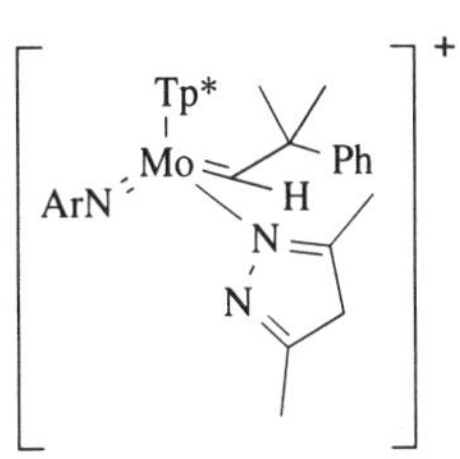

(45)

Addition of phosphines or quinuclidine (L) to complexes of the type $[M(NAr)(OR)_2(=CHBu^t)]$ (M = Mo or W) results in adduct formation to give five-coordinate fluxional complexes $[M(NAr)(OR)_2(L)(=CHBu^t)]$.[247,248] The related complexes $[Mo(CH_2Bu^t)_2(NBu^t)(=CHBu^t)]$ are available through the reaction of $LiCH_2Bu^t$ with the tris(alkyl)derivative $[MoCl(CH_2Bu^t)_3(NBu^t)]$.[75]

Metallacyclopentatrienes can be classed as special examples of alkylidenes. A few syntheses are known and the syntheses of complexes (**46**)[249] and (**47**)[250,251] are based on alkyne precursors (Equations (37) and (38)). Some other syntheses for alkylidene complexes are tabulated in Table 12.

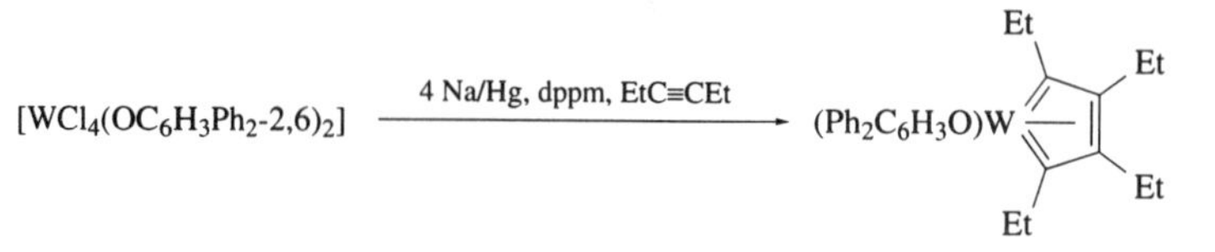

(46) (37)

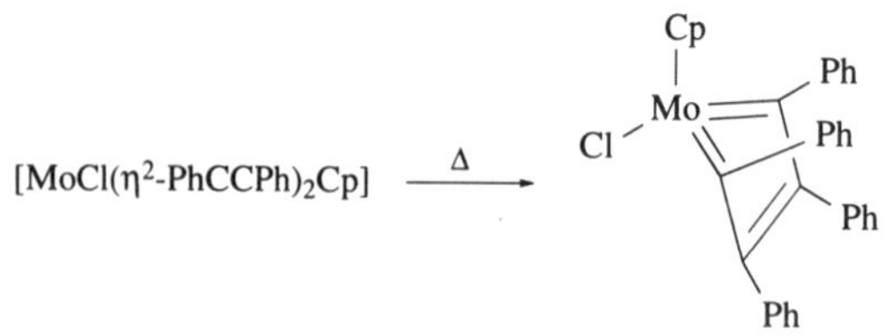

(47) (38)

5.11.2 Structure and Properties

Carbenes tend to be classified as either Fischer or Schrock type carbenes. The former (Chapter 3, this volume) possess substituents such as OR, NR_2 or Ar, the latter normally H or alkyl substituents. Schrock-type carbenes are often referred to as alkylidenes and for formal electron counting purposes, are treated as $(=CR^1R^2)^{2-}$ ligands. This leads to a description C⇉M in which the metal is in its highest oxidation state and often with a d^0 configuration. *Ab initio* calculations suggest several important resonance contributors and that contributors to the ground-state wave-function with a negative charge on carbon (nucleophilic carbon) are the most important. One of the most important resonance contributors is best described as a dative carbon-to-metal σ-bond together with a covalent metal–carbon π-bond (represented as C⇌M).[270]

Many calculations have been performed on alkylidene complexes or models for them. *Ab initio* calculations suggest that the $Cr(CH_2)^+$ fragment with $^6A_1$ symmetry possesses a single σ-donor bond between a singlet CH_2 and a high-spin d^5 Cr^+ ion but no backbond. This contrasts with the corresponding ruthenium system for which there is backbonding.[271]

A theoretical conformational study on trigonal bipyramidal $[WCl_4(=CH_2)]$ in its $^1A_1$ state suggests that the most stable configuration is with the methylene fragment in an equatorial position with all the CH_2 atoms in the equatorial plane. However, there is a low-lying triplet $^3B_2$ state whose energy is very similar.[272,273]

The x-ray structure of $[WBr_2(OCH_2Bu^t)_2\{=\overline{C(CH_2)_3C}H_2\}]$ consists of two monomeric units weakly linked through halide bridges, whereas the $GaBr_3$ adduct is monomeric. The dimeric solution is not preserved in solution and it is the monomer which produces the chain-carrying species involved in alkene metathesis. The $GaBr_3$ adduct is more efficient.[256] Studies by NMR spectroscopy of many other related species show that low-temperature limiting spectra are consistent with trigonal bipyramidal arrangements with the alkylidene equatorial.[98]

Crystallographic studies on imido complexes of the type $[M(NAr)(OR^1)_2(=CHR^2)]$ show them to be nominally tetrahedral. The compounds can exist as *syn*- or *anti*-isomers, and interconvert on NMR timescales, with the R^2 alkylidene substituent pointing towards or away from the ≡NAr group.[79,88,274] The complexes $[Mo(NAr)(OBu^t)_2(=CHBu^t)]$ are not oxidized at potentials up to 1.0 V and undergo a reversible one-electron reduction at −2.16 V at a platinum electrode.[275] Phosphines may reversibly bind to form five-coordinate species.[247,248] The DME adducts are octahedral.[276]

5.11.3 Reactions

Many alkylidene complexes catalyse the metathesis of alkenes but the reader is referred elsewhere for a full discussion of alkene metathesis (see Chapter 12.2, Volume 12). Discussion is limited here to isolable complexes, many of which are relevant to metathesis.

Alkylidene complexes of the type $[W(NAr)(OR)_2(=CHBu^t)]$ catalytically metathesize alkenes. When the R group is reasonably electron withdrawing, metathesis of terminal alkenes is rapid, whereas when R is a reasonably electron donating group, activity is restricted to strained cyclic alkenes such as norbornenes and norbornadiene.[237] Metallacyclobutanes are important in these processes and under controlled conditions may be isolated. Thus, reaction of ethene with $[W(NC_6H_3Pr^i{}_2\text{-}2,6)\text{-}$

$\{OCMe_2(CF_3)\}_2(=CHBu^t)]$ at –40 °C affords the tungstabutane (**48**) (Scheme 12). Spectroscopic studies suggest this to be a square pyramidal species. Warming in the presence of ethene leads to the unsubstituted species (**49**), a fluxional species whose NMR spectra are consistent with the presence of both trigonal bipyramidal and square-based pyramidal species.[84] This compound is not isolable pure since workup affords a mixture of (**31**) and (**28**). The reaction of ethene with $[W(NC_6H_3Pr^i_2\text{-}2,6)\{OCMe_2(CF_3)\}_2(=CHBu^t)]$ at –25 °C leads directly to good yields of the crystallographically characterized complex (**28**).[84] These data are consistent with Scheme 12, which also shows the results of using ^{13}C-labelled ethene, $*C_2H_4$.

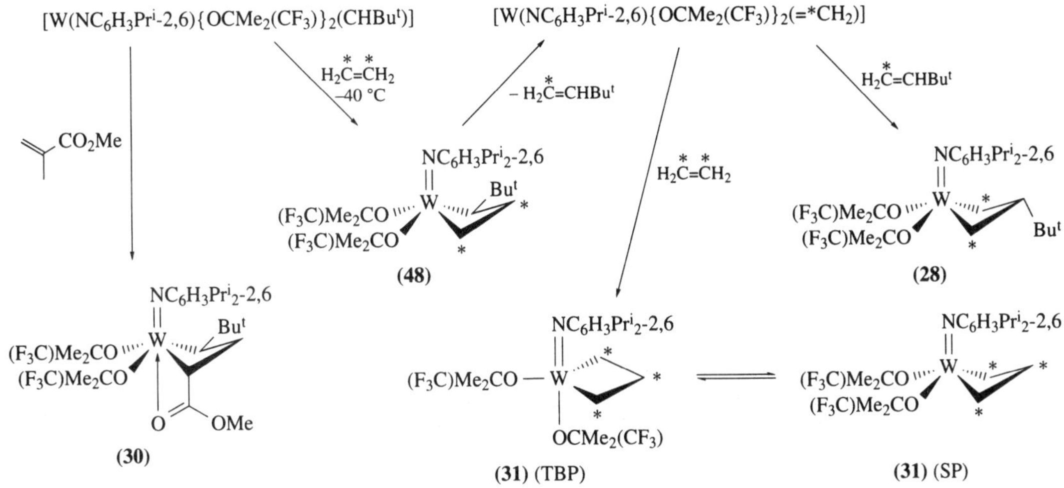

Scheme 12

The corresponding reaction of $CH_2=CH(CO_2Me)$ with $[W(NC_6H_3Pr^i_2\text{-}2,6)\{OCMe_2(CF_3)\}_2(=CHBu^t)]$ affords the crystallographically characterized species (**51**) (Scheme 12).[87] The carbonyl oxygen atom coordinates weakly to the metal. Similar results are found with $CH_2=CH\{C(=O)NMe_2\}$ and in this case the corresponding molybdenum species are also formed. In related reactions, treatment of $[W(NC_6H_3Pr^i_2\text{-}2,6)\{OCMe(CF_3)_2\}_2(=CHBu^t)]$ with $CH_2=CH(TMS)$ affords the metallacycle $[\overline{W\{CH_2CH(TMS)CH(TMS)\}}(NC_6H_3Pr^i_2\text{-}2,6)\{OCMe(CF_3)_2\}_2]$, (**33**), a compound which loses $CH_2=CH(TMS)$ very readily to form $[W(NC_6H_3Pr^i_2\text{-}2,6)\{OCMe(CF_3)_2\}_2\{=CH(TMS)\}]$.[88] See Table 7 for other examples of metallacyclobutanes made by this route.

Use of a "heteroene" such as C_6F_5CHO can give oxametallacyclobutanes. Thus, reaction of C_6F_5CHO with $[M(NC_6H_3Pr^i_2\text{-}2,6)(OBu^t)_2(=CHBu^t)]$ (M = Mo or W) affords the metallacycles $[\overline{M\{CHBu^tCH(C_6F_5)O\}}(NC_6H_3Pr^i_2\text{-}2,6)(OBu^t)_2]$ (**34**). The molybdenum complex is crystallographically characterized and reveals a square-based pyramidal geometry.[89] There are two isomers with the Bu^t and C_6F_5 ring substituents either *cis* or *trans* to each other (*trans* predominant).

5.12 ALKYLIDYNE COMPLEXES

The chemistry of complexes possessing a formal M≡C bond (carbyne or alkylidyne complexes) has increased dramatically since the mid-1980s and the literature up to 1987 is well reviewed.[276] The chemistry of carbyne complexes possessing carbonyl ligands is discussed in Chapter 4, this volume. Other reviews pertinent to this section have appeared.[277]

5.12.1 Synthesis

The syntheses and some notes regarding a number of carbyne complexes are gathered in Table 13. The majority of "high oxidation state" complexes (i.e., d^0 configuration) possess the formula $[MX_3(\equiv CR)]$ (M = Mo or W; X = alkyl, halide, alkoxide, carboxylate, thiolate, amido). Many adducts are also known.

There are comparatively few alkyl complexes of the type $[M(R^1)_3(\equiv CR^2)]$ (R^1 = alkyl). The tungsten complex $[W(Np)_3(\equiv CBu^t)]$ (50–70%) (Np = CH_2Bu^t (neopentyl)) is made conveniently through the

Table 13 Examples of carbyne complexes and adducts.

Synthesis	*Complex*	*M≡C* (nm)	*Comments*	*Ref.*
A. Examples of d^0 complexes $[MX_3(\equiv CR)]$				
$[Mo_2(OBu^t)_6] + PhC\equiv CH$	$[Mo(OBu^t)_3(\equiv CPh)]$		$\delta_C(C_6D_6)$ 276.7	278
$[W_2(OBu^t)_6] + MeC\equiv CMe$	$[W(OBu^t)_3(\equiv CMe)]_2$	0.1759(6)		279,280
$[W_2(OBu^t)_6] + Me_2NC\equiv N$	$[W(OBu^t)_3(\equiv CNMe_2)]_2$	0.177(2), 0.175(2)?		281
$[W_2(OBu^t)_6] + PhC\equiv CPh$	$[W(OBu^t)_3(\equiv CPh)]$	0.1758(5)	$\delta_C(C_6D_6)$ 257.0	282–4
$[MoCl_3(DME)(\equiv CBu^t)] + LiOBu^t$	$[Mo(OBu^t)_3(\equiv CBu^t)]$		$\delta_C(C_6D_6)$ 296.1	285
$[WCl_3(DME)(\equiv CBu^t)] + LiOC_6H_3Pr^i_2\text{-}2,6$	$[W(OC_6H_3Pr^i_2\text{-}2,6)_3(\equiv CBu^t)]$		$\delta_C(C_6D_6)$ 295.2	286
$[W_2(CH_2Bu^t)_2(OPr^i)_4] + EtC\equiv CEt$	$[W(CH_2Bu^t)(OPr^i)_2(\equiv CEt)]_2$	0.176(2)	$\delta_C(C_6D_5CD_3)$ 278.3	287
$[W_2(OBu^t)_6] + N\equiv C(CH_2)_nC\equiv N$	$[(Bu^tO)_3W\equiv C(CH_2)_nC\equiv W(OBu^t)_3]$		$n = 3$–6	288
$[W_2(OR)_6] + EtC\equiv C\text{–}C\equiv CEt$	$[(RO)_3W\equiv C\text{–}C\equiv W(OR)_3]$	0.179(1), 0.1819(16)	C–C, 0.138(2) nm; R = Bu^t, CMe_2CF_3, CMe_2Et	284,289–91
$[W(OBu^t)_3(\equiv CEt)] + [Ru(C\equiv CMe)(CO)_2(Cp)]$	$[(Bu^tO)_3W\equiv C\text{–}Ru(CO)_2(Cp)]$	0.175(2)	$\delta_C(C_6D_6)$ 237.3	292
$[WCl_3(PMe_3)_2(\equiv CBu^t)] + LiOMe$	$[W(OMe)_3(NMe_2H)(\equiv CBu^t)]_2$		$\delta_C(C_6D_6)$ 286	293
$[Mo_2(OBu^t)_6] + PhC\equiv CH$/quinuclidine	$[Mo(OBu^t)_3(\text{quinuclidine})(\equiv CH)]$		$\delta_C(C_6D_6)$ 267.2	278
$MoCl_5$ or $WCl_6 + LiCH_2TMS$	$[M(CH_2TMS)_3(\equiv C(TMS))]$		M = Mo, $\delta_C(C_6D_5CD_3)$ 363.9; M = W, $\delta_{cC}(C_6D_5CD_3)$ 344.6	294
$[WCl_3(OMe)_3] + 6MgClNp$	$[M(Np)_3(\equiv CBu^t)]$		M = Mo, $\delta_C(C_6D_6)$ 323.8; M = W, $\delta_C(C_6D_6)$ 316	285,293
$[W(OBu^t)_3(\equiv CPh)] + LiCH_2Bu^t/BF_3$/py	$[W(Np)_3(py)(\equiv CPh)]$	0.174(3)	$\delta_C(C_6D_6)$ 283.4	295
$[WCl_3(PMe_3)_3(\equiv CBu^t)] + C_2Cl_6$	$[WCl_3(PMe_3)(\equiv CBu^t)]$		$\delta_C(CDCl_3)$ 345	293
$[WCl_3(PMe_3)_3(\equiv CBu^t)] + C_2Cl_6$	$[WCl_3(PMe_3)_2(\equiv CBu^t)]$		$\delta_C(CDCl_3)$ 357	293
$[WCl_2(=O)(PEt_3)(=CHBu^t)] + PPr_3Cl_2$	$[WCl_3(PMe_3)_3(\equiv CBu^t)]$	0.1793(6)	$\delta_C(CDCl_3)$ 401	293,296
$[W(Np)_3(\equiv CBu^t)] + 3HCl$/DME	$[WCl_3(DME)(\equiv CBu^t)]$		$\delta_C(C_6D_6)$ 335	285,293
$[MoCl_3(DME)(\equiv CBu^t)] + KOCMe_2CF_3$	$[Mo\{OCMe_2(CF_3)\}_3(\equiv CBu^t)]$		$\delta_C(C_6D_6)$ 309.7	285
$[MoCl_3(DME)(\equiv CBu^t)] + KOCMe(CF_3)_2$	$[Mo\{OCMe(CF_3)_2\}_3(DME)(\equiv CBu^t)]$		$\delta_C(C_6D_6)$ 318.8	285
$[W(Np)_3(\equiv CBu^t)] + 3HCl/NEt_4Cl$	$[NEt_4][WCl_4(\equiv CBu^t)]$		$\delta_C(CD_2Cl_2)$ 337	285,293

[NEt$_4$][WCl$_4$(≡CBu^t)] + PMe$_3$	[NEt$_4$][WCl$_4$(PEt$_3$)(≡CBu^t)]		δ_C(CDCl$_3$) 335	293
[MCl$_3$(≡CBu^t)(DME)] + LiSC$_6$H$_2$Me$_3$-2,4,6)	[M(SC$_6$H$_2$Me$_3$-2,4,6)$_3$(≡CBu^t)]$_2$		M = Mo, δ_C(CD$_2$Cl$_2$) 336.0; M = W, δ_C(CD$_2$Cl$_2$) 324.3	297
[MCl$_3$(DME)(≡CBu^t)] + LiSC$_6$H$_2$Pr$^i_3$-2,4,6)	[M(SC$_6$H$_2$Pr$^i_3$-2,4,6)$_3$(≡CBu^t)]		M = Mo, δ_C(CD$_2$Cl$_2$) 341.5; M = W, δ_C(CD$_2$Cl$_2$) 329.4	297
[MCl$_3$(DME)(≡CBu^t)] + HSC$_6$H$_2$Me$_3$-2,4,6/NEt$_3$	[NEt$_3$H][MCl(TMT)$_3$(≡CBu^t)]		M = Mo, δ_C(CD$_2$Cl$_2$) 316.3; M = W, δ_C(CD$_2$Cl$_2$) 304.0	297
[M(CH$_2$Bu^t)$_3$(≡CBu^t)] + RCO$_2$H	[M(O$_2$CR)$_3$(≡CBu^t)]		M = Mo, W; R = Me, Pr^i, Bu^t, CF$_3$	298
[WCl$_4$(≡CBu^t)]$^-$ + PHPh$_2$/PEt$_3$/NEt$_3$	[WCl$_2$(PHPh$_2$)(PEt$_3$)$_2$(≡CBu^t)]	0.1808(6)	δ_C(C$_6$D$_6$) 289	245
[W(CH$_2$Bu^t)(dmpe)(=CHBu^t)(≡CBu^t)] + dmpe	[WH(dmpe)$_2$(≡CBu^t)]		δ_C(C$_6$D$_6$) 281.6	299
[WH(dmpe)$_2$(≡CBu^t)] + excess HCl/THF	[WH(Cl)(dmpe)$_2$(≡CBu^t)][Cl]		δ_C(CDCl$_3$) 285.5	299
[MoCl$_2$NH(C$_6$H$_3$Pr$^i_2$-2,6)}(≡CBu^t)] + LiOC$_6$H$_3$Pr$^i_2$-2,6	[Mo{NH(C$_6$H$_3$Pr$^i_2$-2,6)}(OC$_6$H$_3$Pr$^i_2$-2,6)$_2$(≡CBu^t)]		δ_C(CD$_2$Cl$_2$) 317.0	76
[MoCl$_3$(DME)(≡CBu^t)] + (TMS)NHC$_6$H$_3$Pr$^i_2$-2,6	[MoCl$_2${NH(C$_6$H$_3$Pr$^i_2$-2,6)}(DME)(≡CBu^t)]		δ_C(CD$_2$Cl$_2$, 213 K) 320.4	76
B. Examples of d^1 complexes [MX(L)$_4$(≡CR)]				
[MoH$_3$(C≡CBu^t)(dppe)$_2$] + [OEt$_2$H][BF$_4$]	*trans*-[MoF(dppe)$_2$(≡CCH$_2$Bu^t)][BF$_4$]	0.1821(7)	paramagnetic, Et$_2$O solvate	300
C. Examples of d^2 complexes [MX(L)$_4$(≡CR)]				
[WCl$_2$(PMe$_3$)$_4$] + i, AlMe$_3$/toluene; ii, TMEDA	[WCl(PMe$_3$)$_4$(≡CH)]	0.184	δ_C 250	301,302
[WCl$_2$(PMe$_3$)$_4$] + CH$_2$=CH{Si(OMe)$_3$}	[WCl(PMe$_3$)$_4$(≡CMe)]	0.176(4)	δ_C(C$_6$D$_6$) 253.3	303
[WCl(PMe$_3$)$_4$(≡CBu^t)] + TMS-I	[WI(PMe$_3$)$_4$(≡CH)]		δ_C(C$_6$D$_6$) 248.7	263
[WCl(PMe$_3$)$_4$(≡CBu^t)] + (TMS)OSO$_2$CF$_3$	[W(CF$_3$SO$_3$)(PMe$_3$)$_4$(≡CH)]		δ_C(C$_6$D$_6$) 264.9	263
[W(OSO$_2$CF$_3$)(PMe$_3$)$_4$(≡CBu^t)] + NaBH$_4$	[W(BH$_4$)(PMe$_3$)$_4$(≡CH)]		δ_C(C$_6$D$_6$) 259.5	263
[W(CNBu^t)$_6$] + [Et$_3$O][BF$_4$]	[W(CNBu^t)$_5$(≡CNEtBu^t)][BF$_4$]		δ_C(CD$_2$Cl$_2$) 244.4	304
[WBr$_2$(CNBu^t)$_3$(η^2-PhCCNHBu^t)] + LiPh	[WBr(CNBu^t)$_5$(≡CPh)]	0.1764(6)	δ_C(CD$_2$Cl$_2$) 260.5	305
[WCl(PMe$_3$)$_4$(≡CMe)] + MeC≡CMe	[WCl(PMe$_3$)$_2$(η^2-MeC≡CMe)(≡CMe)]		δ_C321.2	306
[WCl(PMe$_3$)$_4$(≡CH)] + 2AlMe$_3$	[WCl(PMe$_3$)$_3$(≡CH·AlMe$_{2-x}$Cl$_{1+x}$)]	0.1807(6)	$x \approx 0.18$	301,302
[WCl(AlMe$_3$)(PMe$_3$)$_3$(≡CH)] + C$_2$H$_4$	[WMe(PMe$_3$)$_3$(η^2-C$_2$H$_4$)(≡CAl$_2$Me$_4$Cl)]	0.1813(5)		307
D. Examples of [M(L)$_2$(≡CR)(η-Cp)] type complexes				
[NEt$_4$][WCl$_4$(≡CBu^t)] + i, (TMS)NEt$_2$; ii, TlCp; iii, HCl	[WCl$_2$(≡CBu^t)(η-Cp)]		δ_C(C$_6$D$_6$) 328.3	308
[MoCl{P(OMe)$_3$}$_4$(≡CPh)] + NaCp	[Mo{P(OMe)$_3$}$_2$(≡CPh)(η-Cp)]		δ_C(C$_6$D$_6$) 286.3 (t, J_{CP} 26.5 Hz)	309
[W(CO)$_2${η^2-C(SMe)$_2$-SMe}{HB(pz)$_3$}] in ClCH$_2$CH$_2$Cl + Δ	[W(SMe)$_2${HB(pz)$_3$}(≡CSMe)]	0.1788(7)	δ_C 268.3	310
[Mo{σ-(*E*)-CH=CHBu^t}{P(OMe)$_3$}$_2$(η-Cp)] + Δ	[Mo{P(OMe)$_3$)$_2$}$_2$(≡CCH$_2$Bu^t)(η-Cp)]	0.1796(8)	δ_C(C$_6$D$_6$) 299.8 (t, J_{CP} 27.0 Hz)	311
[Mo{σ-(*E*)-CH=CHBu^t}{P(OMe)$_3$}$_2$(η-C$_9$H$_7$)] + Δ	[Mo{P(OMe)$_3$}$_2$(≡CCH$_2$Bu^t)(η-C$_9$H$_7$)]		δ_C(C$_6$D$_6$) 302.29 (t, J_{CP} 31.5 Hz)	311
[MoCl{P(OMe)$_3$}$_4$(≡CPh)] + NaCp	[Mo{P(OMe)$_3$}$_2$(≡CPh)(η-Cp)]		δ_C(C$_6$D$_6$)286.3	309
[WI$_2$(≡CBu^t)(η-C$_5$Me$_4$Bu^t)] + NEt$_3$/N$_2$H$_4$	[{WI(≡CBu^t)(η-C$_5$Me$_4$Bu^t)}$_2$(μ-N$_2$H$_2$)]	0.1769(8)		312
[Mo{P(OMe)$_3$)$_2$}$_2$(≡CCH$_2$Bu^t)(η-Cp)] + HBF$_4$	[MoH{P(OMe)$_3$)$_2$}$_2$(≡CCH$_2$Bu^t)(η-Cp)][BF$_4$]	0.1798(2)	δ_C(CDCl$_3$) 346.4	313,314

addition of $[WCl_3(OMe)_3]$ to a solution of MgCl(Np) in ether.[293] The mechanism of formation of the alkylidyne ligand is not clear but involves the formal α-abstraction of two neopentyl hydrogen atoms. The corresponding reaction of $[MoCl_2(=O)_2]$ with MgCl(Np) proceeds to form $[Mo(Np)_3(\equiv CBu^t)]$ in 35% yield.[285] The addition of $MoCl_5$ to five equivalents of $LiCH_2TMS$ in hexane/ether at −78 °C or of WCl_6 to six equivalents of $LiCH_2TMS$ in hexane/ether at −78 °C results in the corresponding distillable derivatives $[M(CH_2TMS)_3(\equiv C\text{-}TMS)]$ (M = Mo or W) in relatively low yield.[294] Finally, the reaction between $[W(OBu^t)_3(\equiv CBu^t)]$ and MgClBz results in an exchange reaction to form $[W(Bz)_3(\equiv CBu^t)]$ in 60% yield.[295]

These alkyl complexes are precursors to halo complexes. Treatment of $[M(Np)_3(\equiv CBu^t)]$ (M = Mo[285] or W[293]) with dry HCl/NEt_4Cl affords the anionic complex $[NEt_4][MCl_4(\equiv CBu^t)]$. In the presence of DME, the product is the octahedral species $[MCl_3(DME)(\equiv CBu^t)]$. The analogous bromide is available through reaction with HBr, while both the bromide and iodide $[MX_3(DME)(\equiv CBu^t)]$ are formed in the reaction of X-TMS (X = Br, I) with $[M(Np)_3(\equiv CBu^t)]$ in the presence of DME.[96] The DME exchanges rapidly on the NMR timescale, and is readily replaced by donor ligands.

Alkoxide and aryloxide complexes of the type $[M(OR^1)_3(\equiv CR^2)]$ (R = alkyl or aryl) are important because of their relevance to alkyne metathesis. Many are known. The most common precursor is $[MCl_3(DME)(\equiv CR^1)]$ (M = Mo or W) and their reaction with $LiOR^2$ or KOR^2 leads to $[M(OR^2)_3(\equiv CR^1)]$ complexes. In order to isolate the products, it is necessary for the OR^2 group to be large (OBu^t, $OC_6H_3Pr^i_2$-2,6, etc.). Fluorinated alkoxides form a particularly interesting group of compounds within this structural type. Examples include $[M(OR_F)_3(DME)(\equiv CBu^t)]$ (M = Mo or W; $OR_F = OCH(CF_3)_2$, $OCMe_2(CF_3)$, $OCMe(CF_3)_2$).[315]

Another good way to make alkylidyne complexes is through a metathesis type process involving M≡M complexes. Thus, reaction of disubstituted alkynes RC≡CR (R = Me, Et, Pr^n) with $[W_2(OBu^t)_6]$ affords the alkylidynes $[W(OBu^t)_3(\equiv CR)]$.[280,284] Asymmetrically substituted alkynes seem to react more readily. In many cases the products require the presence of a donor ligand such as pyridine or quinuclidene to allow isolation. Diynes such as EtC≡C–C≡CEt react to form dark red $[(Bu^tO)_3W\equiv C\text{–}C\equiv W(OBu^t)_3]$.[284]

A similar reaction occurs for cyanides; thus reaction of $[W_2(OBu^t)_6]$ with $Me_2NC\equiv CN$ results in the formation of $[W(OBu^t)_3(\equiv CNMe_2)]$ together with the nitride $[W(\equiv N)(OBu^t)_3]_x$,[281] while reaction with $N\equiv C(CH_2)_nC\equiv N$ (n = 3–6) gives $[(Bu^tO)_3W\equiv C(CH_2)_nC\equiv W(OBu^t)_3]$.[288] Notably, the related nitride $[W(\equiv N)(OC_6H_3Pr^i_2\text{-}2,6)_3]_x$ is formed in the reaction of $[W(OC_6H_3Pr^i_2\text{-}2,6)_3(\equiv CBu^t)]$ with MeCN.[316]

Thiolate derivatives are less common, but are available through the reaction of, for instance, $[MCl_3(DME)(\equiv CBu^t)]$ (M = Mo or W) with LiSAr (Ar = $C_6H_3Pr^i_2$-2,6, Mes)[297] or $LiSBu^t$.[293]

Irradiation of $[W(Me)_6(PMe_3)]$ affords a high yield of the crystallographically characterized carbyne *trans*-$[W(Me)(PMe_3)_4(\equiv CMe)]$.[17] The related parent carbyne complex *trans*$[WCl(PMe_3)_4(\equiv CH)]$ is formed in the reaction of $[WCl_2(PMe_3)_4]$ with $AlMe_3$/toluene followed by TMEDA.[302] Related compounds, but of chelating phosphine ligands, such as $[W(C\equiv CR)(dmpe)_2(\equiv CH)]$ are formed in the reactions of $[W(C\equiv CR)(dmpe)_2(\equiv CH)]$ (R = H, TMS, Ph, etc.) with an alkynyllithium LiC≡CR,[317] while the electrochemical reduction of the cyano complex *trans*-$[MoCl(CN)(dppe)_2]$ in the presence of phenol affords *trans*-$[MoCl(dppe)_2(\equiv CNH_2)]$.[318] Treatment of the alkynide complex $[WH_2(C\equiv CCO_2Me)_2\text{-}(dppe)_2]$ with HBF_4 results in formation of the fluorotungsten carbyne $[WF(\equiv CCO_2Me)(dppe)_2]$.[319]

There are many octahedral carbyne complexes containing both carbonyl and isocyanide ligands (see Section 5.9) but just a few derived from the homoleptic species $[M(CNR)_6]$ (M = Mo or W). As an example, treatment of $[M(CNR)_6]$ (R = Et, Bu^t) with Et_3O^+ affords good yields of the cation $[W(CNR)_5(\equiv CNREt)]^+$ isolated as the $[BF_4]^-$ salt.[304]

5.12.2 Structure and Properties

The chief characteristic and diagnostic feature of carbyne complexes is their ^{13}C NMR shift, some examples of which are given in Table 13.

The yellow alkyl complexes $[M(CH_2R)_3(\equiv CSiR_3)]$ are generally very soluble in hydrocarbon solvents, and sensitive to both air and moisture, but stable when stored under an inert atmosphere. They may be prepared in multigram quantities. They and their alkoxy analogues are pseudotetrahedral when not coordinated by further ligands such as DME. A number of compounds have been characterized by x-ray crystallography (Table 13). The complex $[(Bu^tO)_3W\equiv C\text{–}C\equiv W(OBu^t)_3]$ is notable. The two $W(OBu^t)_3$ groups adopt a staggered conformation and calculations suggest that the C_2 fragment is both a strong π-donor and a strong π-acceptor.[291] Carbyne complexes are the target of numerous other calculations. For instance, extended Hückel methods suggest that in bis(carbyne) species (exemplified by $[WH_4(\equiv CH)_2]^{4-}$) there is a nonbonding $2p$ combination for both *cis* and *trans* isomers. The carbyne carbons should be either very electrophilic or very nucleophilic according to whether or not that orbital is empty or filled. However an oxo species such as $[WH_4(=O)(\equiv CH)]^{3-}$ exhibits $d\pi$ conflicts resulting in charge buildup on the oxygen atom.[320]

5.12.3 Reactions

Hydrogenation of the alkylidyne ligand in $[WCl_3(\equiv CBu^t)(PMe_3)_2]$ results in formation of the alkylidene $[WH(Cl)_3(PMe_3)_2(=CHBu^t)]$ (Section 5.11).[321] Some of the chloride ligands in derivatives of $[MCl_3(DME)(\equiv CR)]$ may be replaced (Equations (39)[308] and (40)[76]). In the presence of $H_2O/NEt_3/PEt_3$, the anion of $[NEt_4][WCl_4(\equiv CBu^t)]$ is converted into the oxo derivative $[WCl_2(=O)(PEt_3)_2(=CHBu^t)]$.[245]

$$KTp^* + [WCl_3(DME)(\equiv CBu^t)] \longrightarrow [WCl_2(Tp^*)(\equiv CBu^t)] \quad (39)$$

$$[MoCl_3(DME)(\equiv CBu^t)] + TMS\text{-}NHAr \longrightarrow [MoCl_2(NHAr)(DME)(\equiv CBu^t)] \quad (40)$$

The alkylidyne ligand of $[W(OBu^t)_3(\equiv CR)]$ (R = Bu^t or TMS) is converted into an alkyl in its reaction with Et_4NOH which affords colourless $[NEt_4][W(CH_2R)(=O)_3]$.[100] Addition of CO to the alkylidyne $[W(OBu^t)_3(\equiv CNMe_2)]$ results in CO and carbyne coupling to form the dimeric species $[(Bu^tO)_3W(\mu\text{-}OCCNMe_2)_2W(OBu^t)_3]$. This reaction is very sensitive to functionality; thus the corresponding reaction of $[W(OPr^i)_3(\equiv CNMe_2)]$ results in alkylidyne–alkylidyne coupling to give an alkyne complex (Equation (41)).[322]

$$[W(OPr^i)_3(\equiv CNMe_2)] + CO \longrightarrow [(OC)_2(Pr^iO)W(\mu\text{-}OPr^i)_2W(OPr^i)_2\{\eta^2\text{-}C_2(NMe_2)_2\}] \quad (41)$$

The polymethylene-bridged alkylidyne (**49**) (Equation (42)) (n = 4 or 5) reacts with CO in a metal–metal bond forming process which results in complex (**50**). This reaction does not proceed for n = 3 or 6, apparently a consequence of the steric requirements of the ring closure reaction.

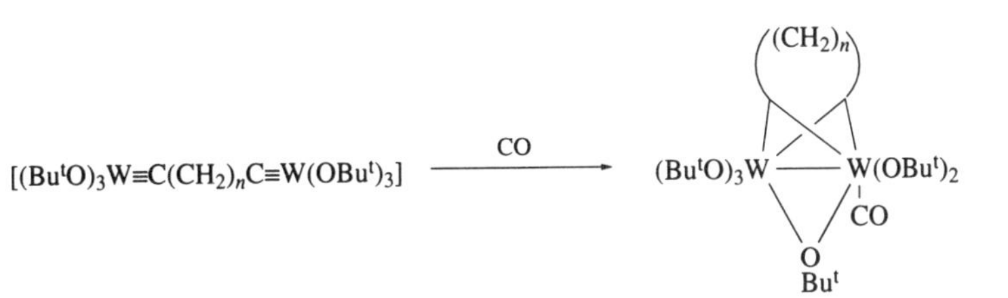

(42)

(49) **(50)**

The reaction of $[W(OR)_3(\equiv CBu^t)]$ (R = $C_6H_3Pr^i_2$-2,6) with two equivalents of EtC≡CEt results in the metallacyclobutadiene complex (**51**) (Equation (43)). It is felt that formation of (**51**) proceeds initially to form complex (**52**); this loses EtC≡CBu^t relatively quickly to form $[W(OR)_3(\equiv CEt)]$ (**53**), which in turn reacts rapidly with a second equivalent of EtC≡CEt to form the final product (**51**). An x-ray crystal structure reveals that the coordination geometry of the tungsten is a somewhat distorted trigonal bipyramid with the metallacyclic ring in the equatorial position. This metallacycle metathesizes alkynes, whereas a related species $[W(OC_6H_3Me_2\text{-}2,6)_3(C_3Et_3)]$ contains aryloxide ligands that are insufficiently bulky to yield a metathesis catalyst.

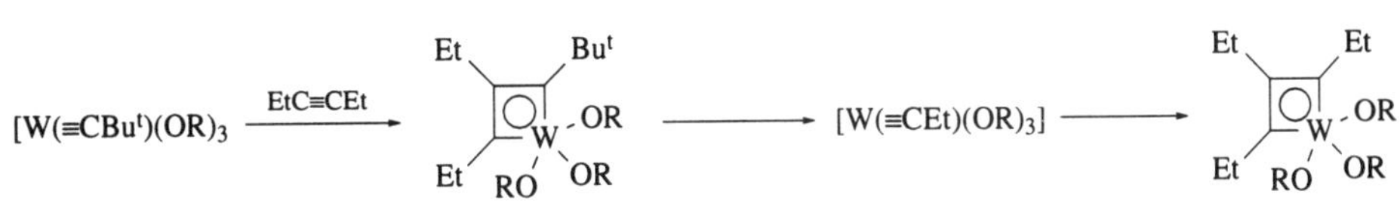

(43)

R = 2,6-Pr^{i}C$_6$H$_3$ **(52)** **(53)** **(51)**

Many other alkylidyne complexes $[MX_3(\equiv CR)]$ react with internal alkynes to form metallacyclobutadiene complexes, but not all the resulting products are alkyne metathesis catalysts.[323–8] Terminal alkynes also react to form metallacyclobutadiene complexes.[329,330] The requirements for alkyne metathesis are a crowded metallacyclobutadiene unable to react further with alkyne but whose steric and electronic properties render it liable to lose an alkyne from the metallacycle to form an alkylidyne complex. The metal centre also needs to be electrophilic. Metallacyclobutadienes $[MCl_3(C_3H_3)]$ (M = Mo or W) are the subject of calculations.[331,332] Calculations on the bonding and energetics of $[MoCl_3(C_3H_3)]$ metallacyclobutadiene and metallatetrahedrane structures place the two structures within 84 kJ mol^{-1} of each other.[333] Nonlocal density functional theory calculations have also been used to study alkyne metathesis catalysed by molybdenum alkylidyne complexes. These calculations suggest that formation of the metallacyclobutadiene is allowed formally by symmetry and has a calculated energy barrier of just 10 kJ mol^{-1}.[334] Other products (Equations (44)–(46)) may also ultimately be formed in reactions of alkylidyne complexes with alkynes. These include deprotiometallacyclobutadienes,[329,335] cyclopentadienyl[298,336–41] complexes and η^3-cyclopropenyl derivatives.[298,337]

$$[M\{OCH(CF_3)_2\}_3{}^{i}(DME)(\equiv CBu^t)] + Bu^tC\equiv CH/NEt_3/DME \longrightarrow [M\{OCH(CF_3)_2\}_2(DME)(C_3Bu^tPh)] \quad (44)$$

$$[WCl_3(DME)(\equiv CBu^t)] + MeC\equiv CMe \longrightarrow [WCl_4(\eta^5\text{-}C_5Me_4Bu^t)] + [WCl_2(\eta^2\text{-}MeC\equiv CMe)(\eta^5\text{-}C_5Me_4Bu^t)] \quad (45)$$

$$[M(OAc)_3(\equiv CBu^t)] + RC\equiv CR \longrightarrow [M(OAc)_3(\eta^3\text{-}C_3Bu^tR_2)] \quad (46)$$

The alkylidyne ligand in $[WX(PMe_3)_4(\equiv CH)]$ (X = Cl, I or OTf) protonates with triflic acid to form unusual alkylidene structures described as distorted since one of the alkylidene hydrogen atoms is distorted so that it lies over the octahedral WCP_2 face.[342,343] These compounds are fluxional by a process in which the two hydrogen environments exchange, and one is crystallographically characterized.[344] The protonation of $[WCl(dmpe)_2(\equiv CBu^t)]$ is a little different in that the result is the pentagonal bipyramidal cationic species $[WH(Cl)(dmpe)_2(\equiv CBu^t)]^+$. It is likely that related species are not formed for the PMe_3 complexes because, for steric reasons, a pentagonal bipyramidal species in not accessible.[299]

The reactions of $[WCl(PMe_3)_4(\equiv CMe)]$ with alkynes are interesting. Thus, the thermal replacement of phosphines by PhC≡CPh results in the alkyne derivative $[WCl(PMe_3)_2(\equiv CMe)(\eta^2\text{-}PhCCPh)]$. Further warming results in rearrangement to $[WCl(PMe_3)_2(\equiv CPh)(\eta^2\text{-}PhCCMe)]$. The same compound reacts with CO to form $[\overline{W\{OC(PMe_3)=CMe}C(O)\}Cl(PMe_3)_2(CO)_2]$.[303]

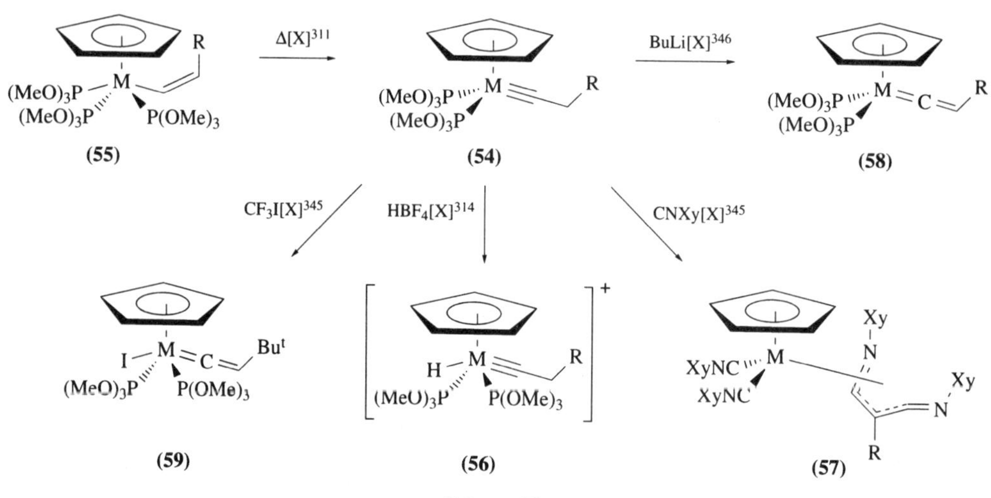

Scheme 13

There is an extensive chemistry of carbyne complexes $[ML_2(\equiv CR)(\eta\text{-}Cp)]$ systems. The chemistry for which one or more of the ligands L is carbonyl is addressed in Chapter 4, this volume. Complex (**54**) (M = Mo or W) is available through thermolysis of the σ-vinyl (**55**).[311] It has an extensive chemistry, some of which is depicted in Scheme 13. Protonation of (**54**) (M = Mo) with HBF_4 affords the crystallographically characterized cationic carbyne hydride (**56**). However, the nature of the acid is important since use of CF_3CO_2H causes loss of the carbyne fragment to form $[Mo\{\sigma\text{-}OC(=O)CF_3\}\{\eta^2\text{-}HOC(=O)CF_3\}\{P(OMe)_3\}_2(\eta\text{-}Cp)]$. Other reagents such as PPh_2H or TCNE also result in carbyne loss.[314] The phosphine ligands are substituted consecutively in reactions with CO; in the molybdenum case, the products are $[Mo\{P(OMe)_3\}(C=O)(\equiv CCH_2R)(\eta\text{-}Cp)]$ and then $[Mo(CO)_2(\equiv CCH_2R)(\eta\text{-}Cp)]$. The latter has been characterized crystallographically.[345] Reactions of (**54**) with isonitriles such as CNXy are complex and result in the fluxional 1,3-diimino-substituted η^3-allyl complex (**57**). NMR studies suggest that dissociative loss of $P(OMe)_3$ is not the rate-determining step in these reactions. Addition of CF_3I affords the vinylidene complex (**59**).[345] Deprotonation of (**54**) (M = Mo) using LiBu results in the vinylidene anion (**58**). This reacts with a number of electrophiles such as MeI, TMS-Cl, Bu^tCOCl, and others, generally through attack at the β-carbon atom of the vinylidene and formation of substituted carbynes $[Mo\{P(OMe)_3\}_2(\equiv CHR^1R^2)(\eta\text{-}Cp)]$.[346]

The carbyne $[Mo\{P(OMe)_3\}_2(\equiv CPh)(\eta\text{-}Cp)]$ does not undergo photochemical phosphite for phosphine exchange. However, photolysis of $[Mo\{P(OMe)_3\}_2(\equiv CPh)(\eta\text{-}Cp)]$ in the presence of PMe_3 in $CHCl_3$ results in formation of the cation $[MoCl(PMe_3)_2(\equiv CPh)(\eta\text{-}Cp)]^+$ via the 17-electron intermediate $[Mo\{P(OMe)_3\}_2(\equiv CPh)(\eta\text{-}Cp)]\cdot$ which undergoes the ligand exchange.[309]

5.13 COMPLEXES WITH HOMONUCLEAR METAL–METAL BONDS

Perhaps the greatest interest within this area involves the chemistry of M_2 complexes in which the M–M bond orders are formally triple or quadruple. A significant number of good reviews address this chemistry and so it is not appropriate to attempt full coverage here.[347–57] In the context of this chapter, most of the compounds of interest have M≡M bonds. Examples are given in Table 14.

Table 14 Structures of some new M≡M bonded complexes.

Synthesis	*Complex*	*M≡M*	*Comments*	*Ref.*
$[Mo_2(OPr^i)_6] + 3MgBz_2$	$[Mo_2Bz_6]$	0.2175(1)	S_6 symmetry	386
$[Mo_2Br_2(CH_2TMS)_4] + LiOC_6H_3Me_2\text{-}2,6$	$[Mo_2(CH_2TMS)_4(OC_6H_3Me_2\text{-}2,6)_2]$	0.2218(2)		387
$[Mo_2Br_2(CH_2TMS)_4] + LiOBu^t$	$[Mo_2(CH_2TMS)_4(OBu^t)_2]$	0.2209(2)		388
$[W_2(NMe_2)_4(Bu^i)_2] + Pr^iOH$	$[W_2(OPr^i)_4(Bu^i)_2]$	0.2309(1)	unstable over −20 °C	389
$[W_2(NMe_2)_4(Tol)_2] + Pr^iOH$	$[W_2(OPr^i)_4(Tol)_2(NHMe_2)]$	0.2317(2)		389
$[W_2(Bz)_2(NMe_2)_4] + (EtCO)_2O$	$[W_2(Bz)_2(O_2CEt)_4]$	0.2186(1)		390
$[Mo_2Cl_2(NMe_2)_4] + LiBz$	$[Mo_2(Bz)_2(NMe_2)_4]$	0.2200(1)		391
$[Mo_2(Bz)_2(OPr^i)_4] + PMe_3$	$[(Me_3P)(Pr^iO)(Bz)_2Mo{\equiv}Mo(Bz)(OPr^i)_3]$	0.2235(7)		392
$[Mo_2(Bz)_2(OPr^i)_4] + dmpm$	$[(\eta^1\text{-dmpm})(Pr^iO)(Bz)_2Mo{\equiv}Mo(Bz)(OPr^i)_3]$	0.2254(11)		393
$[Mo_2Br_2(CH_2TMS)_4]$ + excess PMe_3	$[Mo_2Br_2(PMe_3)_4({=}CH\text{-}TMS)_2]$	0.2276(1)		394
$[Mo(OAc)_2]_2 + PMe_3/MgMe_2$	$[Mo(Me)(PMe_3)_2]_2$	0.214 89(4)		395
$[Mo(OAc)_2]_2 + PMe_2Ph/MgMe_2$	$[Mo(Me)(PMe_2Ph)_2]_2$	0.2164(1)		395
$[Mo_2(Et)_2(NMe_2)_4$ + TolNNNHTol	$[Mo_2(Et)_2(NMe_2)_2(TolN_3Tol)]$	0.2171(1)		396
$[W_2(Bz)_2(NMe_2)_4]$ + excess $(EtCO)_2O$	$[W_2(Bz)_2(O_2CEt)_4]$	0.218 59(12)		359
$[W_2(Np)_2(NMe_2)_4]$ + excess $(EtCO)_2O$	$[W_2(Np)_2(O_2CEt)_4]$	0.219 11(7)		397
$[W_2(Np)_2(NMe_2)_4]$ + excess HCO_2H	$[W_2(Np)_2(O_2CH)_4]$	0.2187(23)		359
$[W_2(Bz)_2(NMe_2)_4]$ + excess $(PhCO)_2O$	$[W_2(Np)_2(O_2CPh)_4]$	0.220 26(11)		359
$[W_2(Np)_2(OAc)_4] + 2CF_3CO_2H$	$[W_2(Np)_2(OAc)_2(O_2CCF_3)_2]$	0.219 35(13)		359
$[Mo_2(Np)_2(NMe_2)_4]$ + excess Ac_2O	$[Mo_2(Np)_2(OAc)_4]$	0.213 02(6)		359

5.13.1 Synthesis

Most compounds of interest here with M≡M bonds have supporting alkoxide or amido ligands. Important amido complexes such as $[M_2(CH_2TMS)_2(NMe_2)_4]$ and $[Mo_2(CH_2TMS)_4(NMe_2)_2]$ are conveniently made from the reactions of the corresponding halide and an organolithium reagent (Equation (47)). The reaction is tolerant to R (R = alkyl, allyl, η^5-cyclopentadienyl, indenyl, benzyl, aryl) and organomagnesium reagents MgClR also work. The substitution pattern (1,1 or 1,2) is dependent on the precise reaction conditions employed. Alkoxides are available from these amido complexes through alcoholysis reactions (Equation (48)).

$$[M_2Cl_2(NMe_2)_4] + LiR^1 \longrightarrow [M_2(R)_2(NMe_2)_4] \qquad (47)$$

$$[M_2(R^1)_2(NMe_2)_4] + R^2OH \longrightarrow [M_2(R^1)_2(OR^2)_4] \qquad (48)$$

In the case of the reaction between $1,4\text{-}Li_2C_4Ph_4$ and $[W_2Cl_2(NMe_2)_4]$, however, the result is a dinuclear complex $[W_2(\mu\text{-}CPh)(\mu\text{-}C_3Ph_3)(NMe_2)_4]$ rather than the anticipated 1,2-ditungstabenzyne, although that species may be an intermediate in the formation of the final product.[358]

Carboxylate structures $[W_2(R^1)_2(O_2CR^2)_4]$ are available through one of four reaction types (Equations (49)–(52)). It is largely a requirement that the R^1 group lacks β-hydrogens.[359] In such cases, β-elimination generally takes place to form carboxylates $[W_2(O_2CR^2)_4]$.

$$[W_2(R^1)_2(NMe_2)_4] + 4\,R^2CO_2COR^2 \longrightarrow [W_2(R^1)_2(O_2CR^2)_4] + 4\,R^2CONMe_2 \quad (49)$$

$$[W_2(R^1)_2(OR^2)_4] + 4\,R^3CO_2H \longrightarrow [W_2(R^1)_2(O_2CR^3)_4] + 4\,R^2OH \quad (50)$$

$$[W_2(R^1)_2(O_2CR^2)_4] + 4\,R^3CO_2H \longrightarrow [W_2(R^1)_2(O_2CR^3)_4] + 4\,R^2CO_2H \quad (51)$$

$$[W_2R^1_6] + 4\,R^2CO_2H \longrightarrow [W_2(R^1)_2(O_2CR^2)_4] + 4\,R^1H \quad (52)$$

While there are very many systems containing quadruple bonds, there are relatively few new compounds with direct M–C bonds. A common source of these is from organolithium reagents (Equations (53)[29] and (54)[360]). Their bond lengths lie within the normal ranges for M≣M bonds: $[Cr_2\{C_6H_4NMe_2\text{-}2\}_4]$, 0.1887(1) nm and $[M_2\{CH_2C_6H_4NMe_2\text{-}2\}_2(OAc)_2]$, M = Cr, 0.1870(1) nm, M = Mo, 0.2065(1) nm. The complex $[Mo_2Me_4(PEt_3)_4]$ is formed in the reaction of $[Mo_2(O_2CBu^t)_4]$ with MgClMe and PEt_3.[361] NMR studies of this and related systems show that the phosphines undergo stepwise exchange with other phosphines in toluene solution by a dissociative process.[362]

$$CrCl_2(THF)_2 + [C_6H_4NMe_2\text{-}2]Li \longrightarrow [Cr_2\{C_6H_4NMe_2\text{-}2\}_4] \quad (53)$$

$$[M_2(OAc)_4] + Li[CH_2C_6H_4NMe_2\text{-}2] \longrightarrow [M_2\{CH_2C_6H_4NMe_2\text{-}2\}(OAc)_2] \quad (54)$$

In Section 5.11.1 it was pointed out that M≣M bonds are cleaved through their reactions with isocyanides. Thus, excess CNR reacts with $[Mo_2X_4(\mu\text{-dppm})_2]$ (X = Cl, Br, I) to form, eventually, $[M(CNR)_7]^{2+}$. However, a single equivalent of CNR (R = Pr^i or Bu^t) in the presence of $TlPF_6$ allows the isolation of the compound $[Mo_2X_3(dppm)_2(CNR)][PF_6]$ in which the Mo≣Mo bond is maintained.[363]

Alkynes may be cleaved by dimetal complexes to give terminal alkylidyne species. Sometimes, bridging alkylidynes are also accessible from alkynes (Equation (55)).[364]

$$[W_2(OBu^t)_6] + PhC{\equiv}CPh \longrightarrow [W_2(OBu^t)_4(\mu\text{-}CPh)_2] \quad (55)$$

Related complexes are also formed through the exchange of OR groups for CH_2TMS groups (Equations (56) and (57)) in existing dimetallacyclobutadiene species.[365,366] In the case of ethanol, a W_4 species is produced (Equation (58)), effectively $[W_2(OEt)_7(\mu\text{-}C(TMS))]_2$ with the two halves held together by μ-OEt groups.[367] The reaction of $[W_2(CH_2TMS)_4(\mu\text{-}C(TMS))_2]$ with X_2 (Equation (59)) (X = Cl, Br, I) is interesting, leading to a paramagnetic species in which the tungsten atoms are in a trigonal bipyramidal environment, joined along a common axial–equatorial edge by the μ-C(TMS) groups.[368] The structure of compounds containing bridging (μ-CR) groups is the subject of theoretical analysis.[369–71]

$$[W_2(CH_2TMS)_4(\mu\text{-}C(TMS))_2] + Bu^tOH \longrightarrow [W_2(CH_2TMS)_2(OBu^t)_2(\mu\text{-}C(TMS))_2] \quad (56)$$

$$[W_2(CH_2TMS)_4(\mu\text{-}C(TMS))_2] + \text{excess } Pr^iOH \longrightarrow [W_2(OPr^i)_4(\mu\text{-}C(TMS))_2] \quad (57)$$

$$[W_2(CH_2(TMS))_4(\mu\text{-}C(TMS))_2] + \text{excess EtOH} \longrightarrow [W_4(OEt)_{14}(\mu\text{-}C(TMS))_2] \quad (58)$$

$$[W_2(CH_2TMS)_4(\mu\text{-}C(TMS))_2] + X_2 \longrightarrow [W_2X_2(CH_2(TMS))_4(\mu\text{-}C(TMS))_2] \quad (59)$$

The bridging CMe complex $[W_2(OBu^t)(OCH_2Bu^t)_6(\mu\text{-}CMe)]$ is formed as a minor product in the reaction of $[W_2(OCH_2Bu^t)_6]$ with ethene but is better made by the route of Equation (60).[372–4]

$$[W_2(OCH_2Bu^t)_4(C_2H_4)_2] + Bu^tOH \longrightarrow [W_2(OBu^t)(OCH_2Bu^t)_6(\mu\text{-}CMe)] \quad (60)$$

In a related system, the reaction between $[W_2(OPr^i)_6(py)_2]$ and MeC≡CMe affords the cluster $[W_3(OPr^i)_9(\mu_3\text{-}CMe)]$ in good yield. This molecule has approximately C_{3v} symmetry and reacts with $[PdCl_2(NCPh)_2]$ to form palladium metal and $[W_3Cl_2(OPr^i)_9(\mu_3\text{-}CMe)]$, an unusual molecule in that the W_3 triangle is asymmetric.[367,375] Other M_3 complexes in which the M_3 triangle is bridged by μ_3-alkylidyne groups are formed in the "primeval" reactions[376] between $[M(CO)_6]$ or $[W(CO)_4(piperidine)_2]$ and mixtures of glacial acetic acid and acetic anhydride. Examples include $[Mo_3(OAc)_6(OH_2)_3(\mu_3\text{-}O)(\mu_3\text{-}CMe)][BF_4]\cdot 9H_2O$,[376] $[Mo_3(OAc)_6(OH_2)_3(\mu_3\text{-}CMe)_2]^{n+}$ (n = 1 or 2),[377] $[W_3(OAc)_6(OH_2)_3(\mu_3\text{-}O)(\mu_3\text{-}CMe)]^{2+}$ [378] and $[Mo_3(OH_2)_3(\mu_2\text{-}Br)_3(\mu_3\text{-}CMe)(\eta^2\text{-}\mu_2\text{-}OAc)_3][ClO_4]\cdot 4H_2O$.[379] The precise conditions employed and the workup are important in these reactions.

Other pertinent clusters include a few carbido complexes. The first tungsten carbido cluster, $[W_4(\mu_4\text{-}C)(OPr^i)_{12}(NMe)]$, was isolated as a by-product in the reaction between $[W_2(NMe_2)_6]$ and Pr^iOH.[380] The carbon monoxide bond is cleaved very easily (even at −30 °C) in its reaction with $[W_4(OCH_2\text{-c-Pe})_{12}]$ (Pe = pentyl) which affords, among other products, $[W_4(\mu_4\text{-}C)(OCH_2\text{-c-Pe})_{14}]$.[381] The core geometry of both these compounds is very similar.

5.13.2 Structure and Properties

Systems containing metal–metal multiple bonds continue to attract considerable theoretical interest. Hartree–Fock–Slater calculations on M_2X_6 systems (M = Cr or Mo; L = H, Me, Cl, OH, NH_2) give the staggered conformation as most stable for all systems. The inclusion of relativistic contributions improves the results.[382] The NMR spectra of M≡M $[Mo_2(R)_4(NMe_2)_4]$ derivatives show that there is restricted rotation about the triple bond, and that the barriers to rotation may be ascribed to steric factors.

Calculations (X_α-SW) on complexes of the type $[W_2(Me)_2(O_2CH)_4]$ suggest that while the molecule is formally a d^3–d^3 system (triple bond), an alternative description involves the interaction of a neutral $W_2(O_2CH)_4$ core with two alkyl radicals. The major W–C interaction is through donation from methyl into M–M σ* and empty *s*–*s* σ orbitals, preserving the formal bond order as four.[383] Calculations[384] have been performed on systems containing quadruple bonds such as $[M_2Me_8]^{4-}$ and they have also been modelled by XANES methods.[385]

Structures of some new M≡M bonded complexes are shown in Table 14.

5.13.3 Reactions

There is an extensive chemistry of organometallic M≡M complexes with alkynes.[397,398] The complex $[Mo_2Br_2(CH_2)_4]$ is an important precursor molecule for many compounds and typical reactions of it and the consequent product are depicted in Scheme 14.

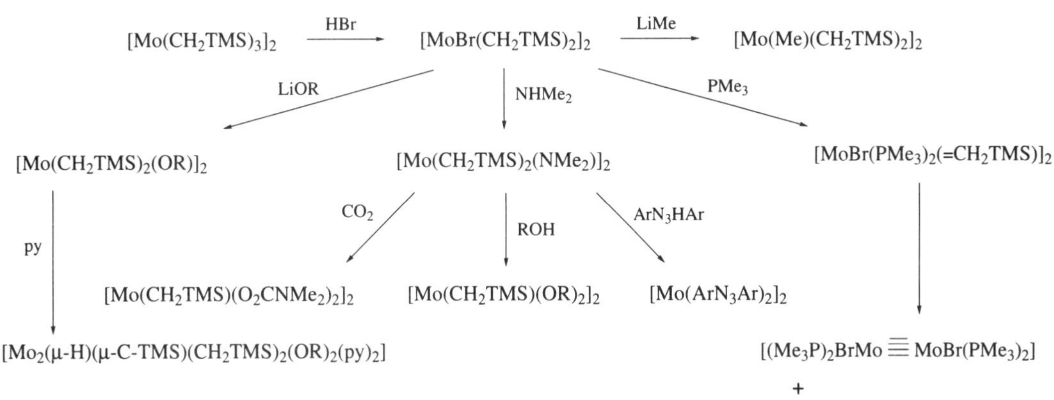

Scheme 14

The octamethyl complex $[Li(L)]_4[Cr_2Me_8]$ (L = THF, OEt_2), but not its molybdenum analogue, is reversibly cleaved in the presence of TMEDA to form an unusual monomeric, high-spin compound $[(TMEDA)Li(\mu\text{-}Me)_2Cr(\mu\text{-}Me)_2Li(TMEDA)]$ (**9**).[40] However, its reaction with two equivalents tetramethyldibenzotetraaza[14]annulene ($tmtaaH_2$) affords methane and the Cr≡Cr system $[Cr(tmtaa)]_2$.[400]

5.14 REFERENCES

1. S. U. Koschmieder and G. Wilkinson, *Polyhedron*, 1991, **10**, 135.
2. A. Haaland, A. Hammel, K. Rypdal and H. V. Volden, *J. Am. Chem. Soc.*, 1990, **112**, 4547.
3. M. F. Lappert, C. L. Raston, G. L. Rowbottom, B. W. Skelton and A. H. White, *J. Chem. Soc., Dalton Trans.*, 1984, 883.
4. K. Jacob and K. H. Thiele, *Z. Anorg. Allg. Chem.*, 1984, **508**, 50.
5. P. Stavropoulos, P. D. Savage, R. P. Tooze, G. Wilkinson, B. Hussain, M. Motevalli and M. B. Hursthouse, *J. Chem. Soc., Dalton Trans.*, 1987, 557.
6. C. J. Cardin *et al.*, *J. Chem. Soc., Dalton Trans.*, 1983, 671.

7. S. U. Koschmieder *et al.*, *J. Chem. Soc., Dalton Trans.*, 1990, 3427.
8. R. M. Kolodziej, R. R. Schrock and W. M. Davis, *Inorg. Chem.*, 1988, **27**, 3253.
9. J. Arnold, G. Wilkinson, B. Hussain and M. B. Hursthouse, *J. Chem. Soc., Dalton Trans.*, 1989, 2149.
10. J. Halpern, *Acc. Chem. Res.*, 1982, **15**, 238.
11. C. Cauletti, R. Zanoni and W. Seidel, *Z. Anorg. Allg. Chem.*, 1983, **496**, 143.
12. A. Haaland, K. Rypdal, H. V. Volden and R. A. Andersen, *Acta Chem. Scand.*, 1991, **45**, 955.
13. K. L. Brandenburg and H. B. Abrahamson, *J. Organomet. Chem.*, 1985, **293**, 371.
14. H. B. Abrahamson, K. L. Brandenburg, B. Lucero, M. E. Martin and E. Dennis, *Organometallics*, 1984, **3**, 1379.
15. S. K. Kang, T. A. Albright and O. Eisenstein, *Inorg. Chem.*, 1989, **28**, 1611.
16. K. W. Chiu, R. A. Jones, G. Wilkinson, A. M. R. Galas and M. B. Hursthouse, *J. Chem. Soc., Dalton Trans.*, 1981, 2088.
17. K. W. Chiu, R. A. Jones, G. Wilkinson, A. M. R. Galas, M. B. Hursthouse and K. M. A. Malik, *J. Chem. Soc., Dalton Trans.*, 1981, 1204.
18. V. R. Kirmse and W. Seidel, *Z. Anorg. Allg. Chem.*, 1982, **490**, 19.
19. F. A. Cotton and G. N. Mott, *Organometallics*, 1982, **1**, 38.
20. T. Kauffmann, C. Beirich, A. Hamsen, T. Möller, C. Philipp and D. Wingbermühle, *Chem. Ber.*, 1992, **125**, 157.
21. T. Ito, T. Ono, K. Maruyama and A. Yamamoto, *Bull. Chem. Soc. Jpn.*, 1982, **55**, 2212.
22. K. Jacob, W. Kretschmer, K. H. Thiele, I. Pavlik, A. Lycka and J. Holecek, *Z. Anorg. Allg. Chem.*, 1991, **606**, 133.
23. H. Ogino, M. Shoji, Y. Abe, M. Shimura and M. Shimoi, *Inorg. Chem.*, 1987, **26**, 2542.
24. M. Shoji, M. Shimura, H. Ogino and T. Ito, *Chem. Lett.*, 1986, 995.
25. Y. Abe and H. Ogino, *Bull. Chem. Soc. Jpn.*, 1989, **62**, 56.
26. P. J. Toscano, H. Brand and P. T. DiMauro, *Organometallics*, 1993, **12**, 30.
27. J. Colin and B. Chevrier, *Organometallics*, 1985, **4**, 1090.
28. J. L. Kerschner, J. S. Yu, P. E. Fanwick, I. P. Rothwell and J. C. Huffman, *Organometallics*, 1989, **8**, 1414.
29. J. J. H. Edema, S. Gambarotta, A. Meetsma and A. L. Spek, *Organometallics*, 1992, **11**, 2452.
30. J. J. H. Edema, S. Gambarotta, F. van Bolhuis, W. J. J. Smeets, A. L. Spek and M. Y. Chiang, *J. Organomet. Chem.*, 1990, **389**, 47.
31. T. Kauffmann, P. Fiegenbaum, M. Papenberg, R. Weischollek and J. Sander, *Chem. Ber.*, 1992, **125**, 143.
32. S. I. Khan and R. Bau, *Organometallics*, 1983, **2**, 1896.
33. V. E. Langguth, N. van Thu, A. Shakoor, K. Jacob and K.-H. Thiele, *Z. Anorg. Allg. Chem.*, 1983, **505**, 127.
34. S. U. Koschmieder, G. Wilkinson, B. Hussain-Bates and M. B. Hursthouse, *J. Chem. Soc., Dalton Trans.*, 1992, 19
35. D. S. Richeson, J. F. Mitchell and K. H. Theopold, *Organometallics*, 1989, **8**, 2570.
36. W. A. Herrmann, W. R. Thiel and E. Herdtweck, *J. Organomet. Chem.*, 1988, **353**, 323.
37. P. Cintas, *Synthesis*, 1992, 248.
38. T. Kauffmann, P. Fiegenbaum, M. Papenberg, R. Wieschollek and D. Wingbermühle, *Chem. Ber.*, 1993, **126**, 79.
39. T. Kauffmann, J. Baune, P. Fiegenbaum, U. Hansmermann, C. Neiteler, M. Papenberg and R. Wieschollek, *Chem. Ber.*, 1993, **126**, 89.
40. S. Hao, S. Gambarotta and C. Bensimon, *J. Am. Chem. Soc.*, 1992, **114**, 3556.
41. M. M. Olmstead, P. P. Power and S. C. Shoner, *Organometallics*, 1988, **7**, 1380.
42. V. B. Sarry and P. Velling, *Z. Anorg. Allg. Chem.*, 1983, **500**, 199.
43. F. Bottomley and L. Sutin, *Adv. Organomet. Chem.*, 1988, **28**, 339.
44. G. N. Schrauzer, L. A. Hughes and N. Strampach, *Z. Naturforsch., Teil B*, 1982, **37**, 380.
45. G. N. Schrauzer, L. A. Hughes, N. Strampach, P. R. Robinson and E. O. Schlemper, *Organometallics*, 1982, **1**, 44.
46. G. N. Schrauzer, L. A. Hughes, N. Strampach, F. Ross, D. Ross and E. O. Schlemper, *Organometallics*, 1983, **2**, 481.
47. G. N. Schrauzer, L. A. Hughes, E. O. Schlemper, F. Ross and D. Ross, *Organometallics*, 1983, **2**, 1163.
48. G. N. Schrauzer *et al.*, *Organometallics*, 1986, **5**, 2452.
49. F. Djafri, R. Laï, M. Pierrot and J. Regnier, *Acta Crystallogr., Sect. C*, 1991, **47**, 1374.
50. C. Zhang, X. Zhang, N. H. Liu, G. N. Schrauzer and E. O. Schlemper, *Organometallics*, 1990, **9**, 1307.
51. G. N. Schrauzer, X. Zhang, N. H. Liu and E. O. Schlemper, *Organometallics*, 1988, **7**, 279.
52. P. Stavropoulos, G. Wilkinson, M. Motevalli and M. B. Hursthouse, *Polyhedron*, 1987, **6**, 1081.
53. L. M. Engelhardt, R. I. Papasergio, C. L. Raston, G. Salem and A. H. White, *J. Chem. Soc., Dalton Trans.*, 1986, 789
54. C. Zhang, E. O. Schlemper and G. N. Schrauzer, *Organometallics*, 1990, **9**, 1016.
55. J. R. M. Kress, M. J. M. Russell, M. G. Wesolek and J. A. Osborn, *J. Chem. Soc., Chem. Commun.*, 1980, 431.
56. J. Kress, M. Wesolek, J. P. Le Ny and J. A. Osborn, *J. Chem. Soc., Chem. Commun.*, 1981, 1039.
57. J. P. Le Ny, M.-T. Youinou and J. A. Osborn, *Organometallics*, 1992, **11**, 2413.
58. M. Herberhold, H. Kniesel, L. Haumaier and U. Thewalt, *J. Organomet. Chem.*, 1986, **301**, 355.
59. M. Herberhold and H. Kniesel, *J. Organomet. Chem.*, 1989, **371**, 205.
60. A. A. Eagle, C. G. Young and E. R. T. Tiekink, *Organometallics*, 1992, **11**, 2934.
61. J. Sundermeyer, J. Putterlik and H. Pritzkow, *Chem. Ber.*, 1993, **126**, 289.
62. I. Feinstein-Jaffe, S. F. Pedersen and R. R. Schrock, *J. Am. Chem. Soc.*, 1983, **105**, 7176.
63. I. Feinstein-Jaffe, D. Gibson, S. J. Lippard, R. R. Schrock and A. Spool, *J. Am. Chem. Soc.*, 1984, **106**, 6305.
64. L. L. Blosch, A. S. Gamble and J. M. Boncella, *J. Mol. Catal.*, 1992, **76**, 229.
65. J. H. Freudenberger and R. R. Schrock, *Organometallics*, 1986, **5**, 398.
66. A. C. Sullivan, G. Wilkinson, M. Motevalli and M. B. Hursthouse, *J. Chem. Soc., Dalton Trans.*, 1988, 53.
67. G. J.-J. Chen, J. W. McDonald and W. E. Newton, *Organometallics*, 1985, **4**, 422.
68. E. Carmona, E. Gutiérrez-Puebla, A. Monge, P. J. Pérez and L. J. Sánchez, *Inorg. Chem.*, 1989, **28**, 2120.
69. A. A. Danopoulos, W.-H. Leung, G. Wilkinson, B. Hussain-Bates and M. B. Hursthouse, *Polyhedron*, 1990, **9**, 2625.
70. W. A. Herrmann, S. Bogdanovic, J. Behm and M. Denk, *J. Organomet. Chem.*, 1992, **430**, C33.
71. J. Sundermeyer, K. Weber and H. Pritzkow, *Angew. Chem., Int. Ed. Engl.*, 1993, **32**, 731.
72. N. Meijboon, C. J. Schaverien and A. G. Orpen, *Organometallics*, 1990, **9**, 774.
73. J. Sundermeyer, U. Radius and C. Burschka, *Chem. Ber.*, 1992, **125**, 2379.
74. M. B. Hursthouse, M. Montevalli, A. C. Sullivan and G. Wilkinson, *J. Chem. Soc., Chem. Commun.*, 1986, 1398.
75. D. Ehrenfeld, J. Kress, B. D. Moore, J. A. Osborn and G. Schoettel, *J. Chem. Soc., Chem. Commun.*, 1987, 129.
76. R. R. Schrock, J. S. Murdzek, G. C. Bazan, J. Robbins, M. DiMare and M. O'Regan, *J. Am. Chem. Soc.*, 1990, **112**, 3875.

77. G. Schoettel, J. Kress, J. Fischer and J. A. Osborn, *J. Chem. Soc., Chem. Commun.*, 1988, 914.
78. W. A. Nugent and R. L. Harlow, *J. Am. Chem. Soc.*, 1980, **102**, 1759.
79. R. R. Schrock *et al.*, *Organometallics*, 1990, **9**, 2262.
80. J. P. Le Ny and J. A. Osborn, *Organometallics*, 1991, **10**, 1546.
81. D. C. Bradley, M. B. Hursthouse, K. M. A. Malik and A. J. Nielson, *J. Chem. Soc., Chem. Commun.*, 1981, 103.
82. G. C. Bazan, J. H. Oskam, H.-N. Cho, L. Y. Park and R. R. Schrock, *J. Am. Chem. Soc.*, 1991, **113**, 6899.
83. J. Robbins, G. C. Bazan, J. S. Murdzek, M. B. O'Regan and R. R. Schrock, *Organometallics*, 1991, **10**, 2902.
84. J. Feldman, W. M. Davis, J. K. Thomas and R. R. Schrock, *Organometallics*, 1990, **9**, 2535.
85. J. Feldman, W. M. Davis and R. R. Schrock, *Organometallics*, 1989, **8**, 2266.
86. G. C. Bazan *et al.*, *J. Am. Chem. Soc.*, 1990, **112**, 8378.
87. J. Feldman, J. S. Murdzek, W. M. Davis and R. R. Schrock, *Organometallics*, 1989, **8**, 2260.
88. R. R. Schrock, R. T. DePue, J. Feldman, C. J. Schaverien, J. C. Dewan and A. H. Liu, *J. Am. Chem. Soc.*, 1988, **110**, 1423.
89. G. C. Bazan, R. R. Schrock and M. B. O'Regan, *Organometallics*, 1991, **10**, 1062.
90. F. Djafri, R. Laï and M. Pierrot, *Acta Crystallogr., Sect. C*, 1989, **45**, 585.
91. W. A. Nugent and J. M. Mayer, 'Metal–Ligand Multiple Bonds', Wiley, New York, 1988.
92. W. A. Nugent, R. J. McKinney, R. V. Kasowski and F. A. Van-Catledge, *Inorg. Chim. Acta*, 1982, **65**, L91.
93. J. T. Anhaus, T. P. Kee, M. H. Schofield and R. R. Schrock, *J. Am. Chem. Soc.*, 1990, **112**, 1642.
94. D. S. Williams, M. H. Schofield, J. T. Anhaus and R. R. Schrock, *J. Am. Chem. Soc.*, 1990, **112**, 6728.
95. O. Eisenstein and Y. Jean, *J. Am. Chem. Soc.*, 1985, **107**, 1177.
96. J. Fischer, J. Kress, J. A. Osborn, L. Ricard and M. Wesolek, *Polyhedron*, 1987, **6**, 1839.
97. J. Kress, M. Wesolek and J. A. Osborn, *J. Am. Chem. Soc., Chem. Commun.*, 1982, 514.
98. J. Kress and J. A. Osborn, *J. Am. Chem. Soc.*, 1987, **109**, 3953.
99. T. Kauffmann, M. Enk, W. Kaschube, E. Toliopoulos and D. Wingbermühle, *Angew. Chem., Int. Ed. Engl.*, 1986, **25**, 910.
100. I. Feinstein-Jaffe, J. C. Dewan and R. R. Schrock, *Organometallics*, 1985, **4**, 1189.
101. R. Laï, S. Le Bot and R. Faure, *J. Organomet. Chem.*, 1987, **336**, C5.
102. R. Laï, S. Le Bot and F. Djafri, *J. Organomet. Chem.*, 1991, **410**, 335.
103. R. Laï, S. Mabille, A. Croux and S. Le Bot, *Polyhedron*, 1991, **10**, 463.
104. H. Arzoumanian, A. Baldy, R. Laï, J. Metzger, M.-L. N. Peh and M. Pierrot, *J. Chem. Soc., Chem. Commun.*, 1985, 1151.
105. R. Laï, S. L. Bot, A. Baldy, M. Pierrot and H. Arzoumanian, *J. Chem. Soc., Chem. Commun.*, 1986, 1208.
106. H. Arzoumanian, R. Laï, S. Le Bot, M. Pierrot and F. Ridouane, *Inorg. Chim. Acta*, 1991, **182**, 173.
107. J. H. Espenson, *Adv. Inorg. Bioinorg. Mech.*, 1982, **1**, 1.
108. J. H. Espenson, *Prog. Inorg. Chem.*, 1983, **30**, 189.
109. J. H. Espenson, *Acc. Chem. Res.*, 1992, **25**, 222.
110. P. Kita and R. B. Jordan, *Inorg. Chem.*, 1986, **25**, 4791.
111. P. Kita and R. B. Jordan, *Inorg. Chem.*, 1985, **24**, 2701.
112. R. van Eldik, in 'Perspectives in Coordination Chemistry', eds. A. F. Williams, C. Fioriani and A. E. Merbach, Verlag Helvetica Chimica Acta, Basel, 1992, pp. 55–65.
113. R. van Eldik, W. Gaede, H. Cohen and D. Meyerstein, *Inorg. Chem.*, 1992, **31**, 3695.
114. W. C. Kupferschmidt and R. B. Jordan, *J. Am. Chem. Soc.*, 1984, **106**, 991.
115. M. J. Sisley and R. B. Jordan, *Inorg. Chem.*, 1988, **27**, 1963.
116. K. Crouse and L.-Y. Goh, *Inorg. Chim. Acta*, 1982, **60**, 205.
117. S. Shi, J. H. Espenson and A. Bakac, *J. Am. Chem. Soc.*, 1990, **112**, 1841.
118. A. Bakac and J. H. Espenson, *Inorg. Chem.*, 1983, **22**, 779.
119. A. Sauer, H. Cohen and D. Meyerstein, *Inorg. Chim. Acta*, 1989, **155**, 101.
120. H. Cohen, D. Meyerstein, A. J. Shusterman and M. Weiss, *J. Chem. Soc., Chem. Commun.*, 1985, 424.
121. M. J. Sisley, W. Rindermann, R. van Eldik and T. W. Swaddle, *J. Am. Chem. Soc.*, 1984, **106**, 7432.
122. D. A. Ryan and J. H. Espenson, *Inorg. Chem.*, 1982, **21**, 527.
123. H. Cohen, A. Feldman, R. Ish-Shalom and D. Meyerstein, *J. Am. Chem. Soc.*, 1991, **113**, 5292.
124. S. Shi, J. H. Espenson, D. Meyerstein and D. Meisel, *Inorg. Chem.*, 1991, **30**, 4468.
125. S. Shi, J. H. Espenson and A. Bakac, *Inorg. Chem.*, 1990, **29**, 4318.
126. P. Huston, J. H. Espenson and A. Bakac, *J. Am. Chem. Soc.*, 1992, **114**, 9510.
127. M. J. Sisley and R. B. Jordan, *Inorg. Chem.*, 1987, **26**, 2833.
128. C. A. Simmons, J. H. Espenson and A. Bakac, *Inorg. Chem.*, 1987, **26**, 4093.
129. M. J. Sisley and R. B. Jordan, *Inorg. Chem.*, 1987, **26**, 273.
130. M. J. Sisley and R. B. Jordan, *Inorg. Chem.*, 1988, **27**, 4483.
131. A. R. Hermes and G. S. Girolami, *Inorg. Chem.*, 1990, **29**, 313.
132. F. A. Cotton, J. A. M. Canich, R. L. Luck and K. Vidyasagar, *Organometallics*, 1991, **10**, 352.
133. L. Dahlenburg and B. Pietsch, *J. Organomet. Chem.*, 1989, **378**, 199.
134. A. R. Hermes, R. J. Morris and G. S. Girolami, *Organometallics*, 1988, **7**, 2372.
135. G. S. Girolami, G. Wilkinson, A. M. R. Galas, M. Thornton-Pett and M. B. Hursthouse, *J. Chem. Soc., Dalton Trans.*, 1985, 1339.
136. T. G. Gardner and G. S. Girolami, *J. Chem. Soc., Chem. Commun.*, 1987, 1758.
137. E. G. Thaler, K. Folting, J. C. Huffman and K. G. Caulton, *J. Organomet. Chem.*, 1989, **376**, 343.
138. T. Ito, H. Tosaka, S. Yoshida, K. Mita and A. Yamamoto, *Organometallics*, 1986, **5**, 735.
139. V. C. Gibson *et al.*, *J. Chem. Soc., Dalton Trans.*, 1985, 2025.
140. D. Rabinovich, R. Zelman and G. Parkin, *J. Am. Chem. Soc.*, 1992, **114**, 4611.
141. M. L. H. Green, G. Parkin, K. J. Moynihan and K. Prout, *J. Chem. Soc., Chem. Commun.*, 1984, 1540.
142. J. L. Kerschner, P. E. Fanwick, I. P. Rothwell and J. C. Huffman, *Organometallics*, 1989, **8**, 1431.
143. A. Hills, D. L. Hughes, N. Kashef, R. L. Richards, M. A. N. D. A. Lemos and A. J. L. Pombeiro, *J. Organomet. Chem.*, 1988, **350**, C4.
144. J. C. Bryan and M. C. Mayer, *J. Am. Chem. Soc.*, 1987, **109**, 7213.
145. J. C. Bryan and J. M. Mayer, *J. Am. Chem. Soc.*, 1990, **112**, 2298.
146. D. Rabinovich and G. Parkin, *J. Am. Chem. Soc.*, 1991, **113**, 5904.

147. S. J. Anderson, F. J. Wells, G. Wilkinson, B. Hussain and M. B. Hursthouse, *Polyhedron*, 1988, **7**, 2615.
148. J. Barrera, M. Sabat and W. D. Harman, *J. Am. Chem. Soc.*, 1991, **113**, 8178.
149. T. Ishida, Y. Mizobe, T. Tanase and M. Hidai, *J. Organomet. Chem.*, 1991, **409**, 355.
150. T. Ishida, Y. Mizobe, T. Tanase and M. Hidai, *Chem. Lett.*, 1988, 441.
151. M. L. H. Green and G. Parkin, *J. Chem. Soc., Chem. Commun.*, 1986, 90.
152. J. C. Bryan, J. B. Arterburn, G. K. Cook and J. M. Mayer, *Organometallics*, 1992, **11**, 3965.
153. M. L. H. Green, G. Parkin, M. Chen and K. Prout, *J. Chem. Soc., Dalton Trans.*, 1986, 2227.
154. D. Rabinovich and G. Parkin, *J. Am. Chem. Soc.*, 1991, **113**, 9421.
155. M. L. H. Green, G. Parkin, D. O'Hare, L.-L. Wong and A. E. Derome, *J. Organomet. Chem.*, 1986, **317**, 61.
156. A.-M. Soares, P. M. Kiernan, D. J. Cole-Hamilton and W. P. Griffith, *J. Chem. Soc., Chem. Commun.*, 1981, 84.
157. A.-M. Soares and W. P. Griffith, *J. Chem. Soc., Dalton Trans.*, 1981, 1886.
158. R. G. Bhattacharyya, G. P. Bhattacharjee and N. Ghosh, *Polyhedron*, 1983, **2**, 543.
159. R. G. Bhattacharyya, G. P. Bhattacharjee and A. M. Saha, *Polyhedron*, 1985, **4**, 583.
160. P. Riccieri and E. Zinato, *Inorg. Chem.*, 1981, **20**, 3722.
161. P. Riccieri, E. Zinato and A. Damiani, *Inorg. Chem.*, 1988, **27**, 3755.
162. Y. Sakabe and Y. Matsumoto, *Inorg. Chem.*, 1985, **24**, 2119.
163. J. Szklarzewicz, A. Samotus and A. Kanas, *Polyhedron*, 1986, **5**, 1733.
164. B. N. Figgis and P. A. Reynolds, *J. Chem. Soc., Dalton Trans.*, 1987, 1747.
165. B. N. Figgis, J. B. Forsyth and P. A. Reynolds, *Inorg. Chem.*, 1987, **26**, 101.
166. R. T. C. Brownlee, B. P. Shehan and A. G. Wedd, *Inorg. Chem.*, 1987, **26**, 2022.
167. A. Samotus, A. Kanas, S. Hodorowicz, J. Czerwonka, J. Scieski and C. Paluszkiewicz, *Polyhedron*, 1982, **1**, 779.
168. C. R. Dennis, J. G. Leipoldt, S. S. Basson and A. J. Van Wyk, *Inorg. Chem.*, 1986, **25**, 1268.
169. C. R. Dennis, A. J. Van Wyk, S. S. Basson and J. G. Leipoldt, *Inorg. Chem.*, 1987, **26**, 270.
170. T. J. Kemp, M. A. Shand and D. Rehorek, *J. Chem. Soc., Dalton Trans.*, 1988, 285.
171. V. Gáspár and M. T. Beck, *Polyhedron*, 1987, **6**, 269.
172. B. Sieklucka, N. W. Alcock, T. J. Kemp and D. J. Stufkens, *J. Chem. Soc., Dalton Trans.*, 1990, 2331.
173. S. I. Ali and Z. Murtaza, *Polyhedron*, 1985, **4**, 821.
174. A. E. Nya and H. Mohan, *Polyhedron*, 1984, **3**, 743.
175. M. J. Mockford and W. P. Griffith, *J. Chem. Soc., Dalton Trans.*, 1985, 717.
176. E. Zinato, P. Riccieri and M. Prelati, *Inorg. Chem.*, 1981, **20**, 1432.
177. P. Riccieri and E. Zinato, *Inorg. Chem.*, 1985, **24**, 441.
178. N. A. P. Kane-Maguire, W. S. Crippen and P. K. Miller, *Inorg. Chem.*, 1983, **22**, 696.
179. Y. Yamamoto and H. Yamazaki, *J. Organomet. Chem.*, 1985, **282**, 191.
180. J. P. Bullock and K. R. Mann, *Inorg. Chem.*, 1989, **28**, 4006.
181. M. H. Chisholm *et al.*, *Inorg. Chem.*, 1984, **23**, 1037.
182. E. M. Carnahan and S. J. Lippard, *J. Chem. Soc., Dalton Trans.*, 1991, 699.
183. K. W. Chiu, C. G. Howard, G. Wilkinson, A. M. R. Galas and M. B. Hursthouse, *Polyhedron*, 1982, **1**, 803.
184. D. D. Klendworth, W. W. Welters and R. A. Walton, *Organometallics*, 1982, **1**, 336.
185. D. T. Plummer and R. J. Angelici, *Inorg. Chem.*, 1983, **22**, 4063.
186. K. W. Chiu, D. Lyons, G. Wilkinson, M. Thornton-Pett and M. B. Hursthouse, *Polyhedron*, 1983, **2**, 803.
187. W. S. Mialki, D. E. Wigley, T. E. Wood and R. A. Walton, *Inorg. Chem.*, 1982, **21**, 480.
188. D. E. Wigley and R. A. Walton, *Organometallics*, 1982, **1**, 1322.
189. D. A. Bohling and K. R. Mann, *Inorg. Chem.*, 1983, **22**, 1561.
190. F. R. Lemke, D. E. Wigley and R. A. Walton, *J. Organomet. Chem.*, 1983, **248**, 321.
191. M. Kamata, T. Yoshida, S. Otsuka, K. Hirotsu and T. Higuchi, *J. Am. Chem. Soc.*, 1981, **103**, 3572.
192. M. Kamata *et al.*, *J. Am. Chem. Soc.*, 1981, **103**, 5772.
193. N. C. Payne, N. Okura and S. Otsuka, *J. Am. Chem. Soc.*, 1983, **105**, 245.
194. W. S. Mialki, R. E. Wild and R. A. Walton, *Inorg. Chem.*, 1981, **20**, 1380.
195. G. S. Girolami and R. A. Andersen, *Inorg. Chem.*, 1981, **20**, 2040.
196. C. M. Giandomenico, L. H. Hanau and S. J. Lippard, *Organometallics*, 1982, **1**, 142.
197. M. J. Fernandez-Trujillo, M. Jimenez-Tenorio, M. C. Puerta and D. L. Hughes, *J. Organomet. Chem.*, 1989, **377**, 241.
198. A. Bell and R. A. Walton, *J. Organomet. Chem.*, 1984, **263**, 359.
199. M. Minelli, J. H. Enemark, A. Bell and R. A. Walton, *J. Organomet. Chem.*, 1985, **284**, 25.
200. M. J. Calhorda, A. R. Dias, M. T. Duarte, A. M. Martins, P. M. Matias and C. C. Romaõ, *J. Organomet. Chem.*, 1992, **440**, 119.
201. C. G. Francis, D. Klein and P. D. Morand, *J. Chem. Soc., Chem. Commun.*, 1985, 1142.
202. I. Feinstein-Jaffe and S. E. Maisuls, *J. Organomet. Chem.*, 1988, **350**, 57.
203. E. Carmona, A. Galindo, J. M. Marin, E. Gutierrez, A. Monge and C. Ruiz, *Polyhedron*, 1988, **7**, 1831.
204. T. Adachi, N. Sasaki, T. Ueda, M. Kaminaka and T. Yoshida, *J. Chem. Soc., Chem. Commun.*, 1989, 1320.
205. M. H. Chisholm, D. L. Clark, D. Ho and J. C. Huffman, *Organometallics*, 1987, **6**, 1532.
206. D. A. Bohling and K. R. Mann, *Inorg. Chem.*, 1984, **23**, 1426.
207. J. C. Dewan, W. S. Mialki, R. A. Walton and S. J. Lippard, *J. Am. Chem. Soc.*, 1982, **104**, 133.
208. D. J. Szalda, J. C. Dewan and S. J. Lippard, *Inorg. Chem.*, 1981, **20**, 3851.
209. D. A. Bohling, J. F. Evans and K. R. Mann, *Inorg. Chem.*, 1982, **21**, 3546.
210. C. Caravana, C. M. Giandomenico and S. J. Lippard, *Inorg. Chem.*, 1982, **21**, 1860.
211. K. A. Anderson, B. Scott, S. Wherland and R. D. Willett, *Acta Crystallogr., Sect. C*, 1991, **47**, 2337.
212. K. A. Anderson and S. Wherland, *Inorg. Chem.*, 1989, **28**, 601.
213. K. A. Anderson and S. Wherland, *Inorg. Chem.*, 1990, **29**, 3822.
214. K. A. Anderson and S. Wherland, *Inorg. Chem.*, 1991, **30**, 624.
215. A. Bell, S. J. Lippard, M. Roberts and R. A. Walton, *Organometallics*, 1983, **2**, 1562.
216. P. W. R. Corfield, J. C. Dewan and S. J. Lippard, *Inorg. Chem.*, 1983, **22**, 3424.
217. A. F. Lindmark, *Inorg. Chem.*, 1992, **31**, 3507.
218. J. C. Dewan, C. M. Giandomenico and S. J. Lippard, *Inorg. Chem.*, 1981, **20**, 4069.

219. S. Warner and S. J. Lippard, *Organometallics*, 1986, **5**, 1716.
220. M. Kamata *et al.*, *Organometallics*, 1982, **1**, 227.
221. M. Kamata *et al.*, *Inorg. Chem.*, 1983, **22**, 2416.
222. D. E. Wigley and R. A. Walton, *Inorg. Chem.*, 1983, **22**, 3138.
223. W. R. Robinson, D. E. Wigley and R. A. Walton, *Inorg. Chem.*, 1985, **24**, 918.
224. C. M. Giandomenico, C. T. Lam and S. J. Lippard, *J. Am. Chem. Soc.*, 1982, **104**, 1263.
225. P. W. R. Corfield, L. M. Baltusis and S. J. Lippard, *Inorg. Chem.*, 1981, **20**, 922.
226. C. M. Giandomenico, J. C. Dewan and S. J. Lippard, *J. Am. Chem. Soc.*, 1981, **103**, 1407.
227. E. G. Bakalbassis, C. A. Tsipis and A. J. L. Pombeiro, *J. Organomet. Chem.*, 1991, **408**, 181.
228. A. C. Filippou, W. Grünleitner, C. Völkl and P. Kiprof, *Angew. Chem., Int. Ed. Engl.*, 1991, **30**, 1167.
229. A. C. Filippou and E. O. Fischer, *J. Organomet. Chem.*, 1990, **382**, 143.
230. J. S. Adams, M. Cunningham and M. W. Whiteley, *J. Organomet. Chem.*, 1985, **293**, C13.
231. J. S. Adams, C. Bitcon, J. R. Brown, D. Collison, M. Cunningham and M. W. Whiteley, *J. Chem. Soc., Dalton Trans.*, 1987, 3049.
232. N. A. Buang, D. L. Hughes, N. Kashef, R. L. Richards and A. J. L. Pombeiro, *J. Organomet. Chem.*, 1987, **323**, C47.
233. A. Hills, D. L. Hughes, N. Kashef, R. L. Richards, M. A. N. D. A. Lemos and A. J. L. Pombeiro, *J. Organomet. Chem.*, 1988, **350**, C4.
234. T. C. Stoner, R. F. Dallinger and M. D. Hopkins, *J. Am. Chem. Soc.*, 1990, **112**, 5651.
235. K. G. Frank and J. P. Selegue, *J. Am. Chem. Soc.*, 1990, **112**, 6414.
236. R. L. Beddoes, C. Bitcon, A. Ricalton and M. W. Whiteley, *J. Organomet. Chem.*, 1989, **367**, C21.
237. R. R. Schrock, *Acc. Chem. Res.*, 1990, **23**, 158.
238. F. Quignard, M. Leconte and J.-M. Basset, *J. Chem. Soc., Chem. Commun.*, 1985, 1816.
239. J.-L. Couturier, C. Paillet, M. Leconte, J.-M. Basset and K. Weiss, *Angew. Chem., Int. Ed. Engl.*, 1992, **31**, 628.
240. J. H. Wengrovius and R. R. Schrock, *Organometallics*, 1982, **1**, 148.
241. M. R. Churchill, J. R. Missert and W. J. Youngs, *Inorg. Chem.*, 1981, **20**, 3388.
242. M. H. Chisholm, J. C. Huffman, E. A. Lucas, A. Sousa and W. E. Streib, *J. Am. Chem. Soc.*, 1992, **114**, 2710.
243. S. F. Pedersen and R. R. Schrock, *J. Am. Chem. Soc.*, 1982, **104**, 7483.
244. M. Jolly, J. P. Mitchell and V. C. Gibson, *J. Chem. Soc., Dalton Trans*, 1992, 1331.
245. S. M. Rocklage, R. R. Schrock, M. R. Churchill and H. J. Wasserman, *Organometallics*, 1982, **1**, 1332.
246. L. L. Blosch, A. S. Gamble, K. Abboud and J. M. Boncella, *Organometallics*, 1992, **11**, 2342.
247. R. R. Schrock, W. E. Crowe, G. C. Bazan, M. DiMare, M. B. O'Regan and M. H. Schofield, *Organometallics*, 1991, **10**, 1832.
248. Z. Wu, D. R. Wheeler and R. H. Grubbs, *J. Am. Chem. Soc.*, 1992, **114**, 146.
249. J. L. Kerschner, P. E. Fanwick and I. P. Rothwell, *J. Am. Chem. Soc.*, 1988, **110**, 8235.
250. W. Hirpo and M. D. Curtis, *J. Am. Chem. Soc.*, 1988, **110**, 5218.
251. M. D. Curtis, J. Real, W. Hirpo and W. M. Butler, *Organometallics*, 1990, **9**, 66.
252. L. K. Johnson, S. C. Virgil, R. H. Grubbs and J. W. Ziller, *J. Am. Chem. Soc.*, 1990, **112**, 5384.
253. P. A. van der Schaaf, D. M. Grove, W. J. J. Smeets, A. L. Spek and G. van Koten, *Organometallics*, 1993, **12**, 3955.
254. G. Schoettel, J. Kress and J. A. Osborn, *J. Chem. Soc., Chem. Commun.*, 1989, 1062.
255. J. H. Oskam *et al.*, *J. Organomet. Chem.*, 1993, **459**, 185.
256. M. T. Youinou, J. Kress, J. Fischer, A. Aguero and J. A. Osborn, *J. Am. Chem. Soc.*, 1988, **110**, 1488.
257. A. Mayr, K. S. Lee, M. A. Kjelsberg and D. Van Engen, *J. Am. Chem. Soc.*, 1986, **108**, 6079.
258. A. Keller, *J. Organomet. Chem.*, 1990, **385**, 285.
259. A. Keller, *J. Organomet. Chem.*, 1991, **407**, 237.
260. A. Keller, *J. Organomet. Chem.*, 1991, **415**, 97.
261. A. Keller, *J. Organomet. Chem.*, 1992, **436**, 199.
262. M. R. Churchill and A. L. Rheingold, *Inorg. Chem.*, 1982, **21**, 1357.
263. S. J. Holmes, R. R. Schrock, M. R. Churchill and H. J. Wasserman, *Organometallics*, 1984, **3**, 476.
264. A. H. Liu, R. C. Murray, J. C. Dewan, B. D. Santarsiero and R. R. Schrock, *J. Am. Chem. Soc.*, 1987, **109**, 4282.
265. P. Legzdins, S. J. Rettig and L. Sánchez, *Organometallics*, 1985, **4**, 1470.
266. P. Legzdins, S. J. Rettig, J. E. Veltheer, R. J. Batchelor and F. W. B. Einstein, *Organometallics*, 1993, **12**, 3575.
267. A. S. Gamble and J. M. Boncella, *Organometallics*, 1993, **12**, 2814.
268. S. K. Noh, S. C. Sendlinger, C. Janiak and K. H. Theopold, *J. Am. Chem. Soc.*, 1989, **111**, 9127.
269. S. K. Noh, R. A. Heintz, C. Janiak, S. C. Sendlinger and K. H. Theopold, *Angew. Chem., Int. Ed. Engl.*, 1990, **29**, 775.
270. T. R. Cundari and M. S. Gordon, *J. Am. Chem. Soc.*, 1991, **113**, 5231.
271. E. A. Carter and W. A. Goddard, *J. Am. Chem. Soc.*, 1986, **108**, 4746.
272. M. Sodupe, J. M. Lluch, A. Oliva and J. Bertrán, *Organometallics*, 1989, **8**, 1837.
273. M. Sodupe, J. M. Lluch, A. Oliva and J. Bertrán, *Theochem. J. Mol. Struct.*, 1991, **83**, 37.
274. J. H. Oskam and R. R. Schrock, *J. Am. Chem. Soc.*, 1992, **114**, 7588.
275. D. Albagli, G. Bazan, M. S. Wrighton and R. R. Schrock, *J. Am. Chem. Soc.*, 1992, **114**, 4150.
276. H. Fischer, P. Hoffman, F. R. Kreissl, R. R. Schrock, U. Schubert and K. Weiss, 'Carbyne Complexes', VCH, Weinheim, Germany, 1988, p. 235.
277. R. R. Schrock, *Acc. Chem. Res.*, 1986, **19**, 342.
278. H. Strutz and R. R. Schrock, *Organometallics*, 1984, **3**, 1600.
279. M. H. Chisholm, D. M. Hoffman and J. C. Huffman, *Inorg. Chem.*, 1983, **22**, 2903.
280. R. R. Schrock, M. L. Listemann and L. G. Sturgeoff, *J. Am. Chem. Soc.*, 1982, **104**, 4291.
281. M. H. Chisholm, J. C. Huffman and N. S. Marchant, *J. Am. Chem. Soc.*, 1983, **105**, 6162.
282. F. A. Cotton, W. Schwotzer and E. S. Shamshoum, *Organometallics*, 1984, **3**, 1770.
283. F. A. Cotton, W. Schwotzer and E. S. Shamshoum, *J. Organomet. Chem.*, 1985, **296**, 55.
284. M. L. Listemann and R. R. Schrock, *Organometallics*, 1985, **4**, 74.
285. L. G. McCullough, R. R. Schrock, J. C. Dewan and J. C. Murdzek, *J. Am. Chem. Soc.*, 1985, **107**, 5987.
286. M. R. Churchill, J. W. Ziller, J. H. Freudenberger and R. R. Schrock, *Organometallics*, 1984, **3**, 1554.
287. M. H. Chisholm, B. W. Eichhorn, K. Folting and J. C. Huffman, *Organometallics*, 1989, **8**, 49.

288. M. H. Chisholm, K. Folting, J. C. Huffman and E. A. Lucas, *Organometallics*, 1991, **10**, 535.
289. T. M. Gilbert and R. D. Rogers, *J. Organomet. Chem.*, 1991, **421**, C1.
290. T. M. Gilbert and R. D. Rogers, *Acta Crystallogr., Sect. C*, 1993, **49**, 677.
291. K. G. Caulton, R. H. Cayton, M. H. Chisholm, J. C. Huffman, E. B. Lobkovsky and Z. Xue, *Organometallics*, 1992, **11**, 321.
292. S. L. Latesky and J. P. Selegue, *J. Am. Chem. Soc.*, 1987, **109**, 4731.
293. R. R. Schrock, D. N. Clark, J. Sancho, J. H. Wengrovius, S. M. Rocklage and S. F. Pedersen, *Organometallics*, 1982, **1**, 1645.
294. R. A. Andersen, M. H. Chisholm, J. F. Gibson, W. W. Reichert, I. P. Rothwell and G. Wilkinson, *Inorg. Chem.*, 1981, **20**, 3934.
295. M. H. Chisholm, J. C. Huffman and J. A. Klang, *Polyhedron*, 1990, **9**, 1271.
296. M. R. Churchill and Y.-J. Li, *J. Organomet. Chem.*, 1985, **282**, 239.
297. J. S. Murdzek, L. Blum and R. R. Schrock, *Organometallics*, 1988, **7**, 436.
298. R. R. Schrock, J. S. Murdzek, J. H. Freudenberger, M. R. Churchill and J. W. Ziller, *Organometallics*, 1986, **5**, 25.
299. S. J. Holmes, D. N. Clark, H. W. Turner and R. R. Schrock, *J. Am. Chem. Soc.*, 1982, **104**, 6322.
300. M. A. N. D. A. Lemos, A. J. L. Pombeiro, D. L. Hughes and R. L. Richards, *J. Organomet. Chem.*, 1992, **434**, C6.
301. M. R. Churchill, A. L. Rheingold and H. J. Wasserman, *Inorg. Chem.*, 1981, **20**, 3392.
302. P. R. Sharp, S. J. Holmes, R. R. Schrock, M. R. Churchill and H. J. Wasserman, *J. Am. Chem. Soc.*, 1981, **103**, 965.
303. L. M. Atagi, S. C. Critchlow and J. M. Mayer, *J. Am. Chem. Soc.*, 1992, **114**, 1483.
304. A. C. Filippou and W. Grünleitner, *J. Organomet. Chem.*, 1990, **393**, C10.
305. A. C. Filippou, C. Völkl, W. Grünleitner and P. Kiprof, *Angew. Chem., Int. Ed. Engl.*, 1990, **29**, 207.
306. L. M. Atagi, S. C. Critchlow and J. M. Mayer, *J. Am. Chem. Soc.*, 1992, **114**, 9223.
307. M. R. Churchill and H. J. Wasserman, *Inorg. Chem.*, 1981, **20**, 4119.
308. M. R. Churchill, J. W. Ziller, L. McCullough, S. F. Pedersen and R. R. Schrock, *Organometallics*, 1983, **2**, 1046.
309. C. J. Leep, K. B. Kingsbury and L. McElwee-White, *J. Am. Chem. Soc.*, 1988, **110**, 7535.
310. R. A. Doyle and R. J. Angelici, *J. Am. Chem. Soc.*, 1990, **112**, 194.
311. S. R. Allen, R. G. Beevor, M. Green, A. G. Orpen, K. E. Paddick and I. D. Williams, *J. Chem. Soc., Dalton Trans.*, 1987, 591.
312. M. R. Churchill, Y.-J. Li, L. Blum and R. R. Schrock, *Organometallics*, 1984, **3**, 109.
313. M. Green, A. G. Orpen and I. D. Williams, *J. Chem. Soc., Chem. Commun.*, 1982, 493.
314. M. Bottrill, M. Green, A. G. Orpen, D. R. Saunders and I. D. Williams, *J. Chem. Soc., Dalton Trans.*, 1989, 511.
315. J. H. Freudenberger, R. R. Schrock, M. R. Churchill, A. L. Rheingold and J. W. Ziller, *Organometallics*, 1984, **3**, 1563.
316. J. H. Freudenberger and R. R. Schrock, *Organometallics*, 1986, **5**, 398.
317. J. Manna, S. J. Geib and M. D. Hopkins, *J. Am. Chem. Soc.*, 1992, **114**, 9199.
318. A. Hills, D. L. Hughes, C. J. Macdonald, M. Y. Mohammed and C. J. Pickett, *J. Chem. Soc., Dalton Trans.*, 1991, 121.
319. A. Hills, D. L. Hughes, N. Kashef, M. A. N. D. A. Lemos, A. J. L. Pombeiro and R. L. Richards, *J. Chem. Soc., Dalton Trans.*, 1992, 1775.
320. D. C. Brower, J. L. Templeton and D. M. P. Mingos, *J. Am. Chem. Soc.*, 1987, **109**, 5203.
321. J. H. Wengrovius, R. R. Schrock, M. R. Churchill and H. J. Wasserman, *J. Am. Chem. Soc.*, 1982, **104**, 1739.
322. M. H. Chisholm, J. C. Huffman and N. S. Marchant, *J. Chem. Soc., Chem. Commun.*, 1986, 717.
323. P. G. Baraldi, A. Barco, S. Benetti, S. Manfredini and D. Simoni, *Synthesis*, 1987, 276.
324. R. Aumann and P. Hinterding, *Chem. Ber.*, 1993, **126**, 421.
325. S. Barber, J. Booth, D. R. Pyke, R. Reid and R. J. D. Tilley, *J. Catal.*, 1982, **77**, 180.
326. L. A. Barnes, M. Rosi and C. W. Bauschlicher, *J. Chem. Phys.*, 1991, **94**, 2031.
327. L. A. Barnes, B. Liu and R. Lindh, *J. Chem. Phys.*, 1993, **98**, 3978.
328. A. Avey, S. C. Tenhaeff, T. J. R. Weakley and D. R. Tyler, *Organometallics*, 1991, **10**, 3607.
329. P. A. Barnard, I.-W. Sun and C. L. Hussey, *Inorg. Chem.*, 1990, **29**, 3670.
330. A. Baretta, K. S. Chong, F. G. N. Cloke, A. Feigenbaum and M. L. H. Green, *J. Chem. Soc., Dalton Trans.*, 1983, 861.
331. E. V. Anslyn, M. J. Brusich and W. A. Goddard, *Organometallics*, 1988, **7**, 98.
332. B. E. Bursten, *J. Am. Chem. Soc.*, 1983, **105**, 121.
333. J. C. Barnes, W. Bell, C. Glidewell and R. A. Howie, *J. Organomet. Chem.*, 1990, **385**, 369.
334. T. Woo, E. Folga and T. Ziegler, *Organometallics*, 1993, **12**, 1289.
335. A. G. M. Barrett, J. Mortier, M. Sabat and M. A. Sturgess, *Organometallics*, 1988, **7**, 2553.
336. S. F. Pedersen, R. R. Schrock, M. R. Churchill and H. J. Wasserman, *J. Am. Chem. Soc.*, 1982, **104**, 6808.
337. R. R. Schrock, S. F. Pedersen, M. R. Churchill and J. W. Ziller, *Organometallics*, 1984, **3**, 1574.
338. M. R. Churchill, J. W. Ziller, S. F. Pedersen and R. R. Schrock, *J. Chem. Soc., Chem. Commun.*, 1984, 485.
339. M. R. Churchill and H. J. Wassermann, *Organometallics*, 1983, **2**, 755.
340. J. F. Buzinkai and R. R. Schrock, *Organometallics*, 1987, **6**, 1447.
341. S. A. MacLaughlin, R. C. Murray, J. C. Dewan and R. R. Schrock, *Organometallics*, 1985, **4**, 796.
342. R. Aumann, B. Trentmann, C. Krüger and F. Lutz, *Chem. Ber.*, 1991, **124**, 2595.
343. A. J. Ashe, J. W. Kampf, C. M. Kausch, H. Konishi, M. O. Kristen and J. Kroker, *Abstr. Papers Am. Chem. Soc.*, 1990, **200**, 236.
344. A. J. Schultz, J. M. Williams, R. R. Schrock and S. J. Holmes, *Acta Crystallogr., Sect. C*, 1984, **40**, 590.
345. P. K. Baker *et al.*, *J. Chem. Soc., Dalton Trans.*, 1989, 1321.
346. R. G. Beevor, M. J. Freeman, M. Green, C. E. Morton and A. G. Orpen, *J. Chem. Soc., Dalton Trans.*, 1991, 3021.
347. M. H. Chisholm, *Angew. Chem., Int. Ed. Engl.*, 1986, **25**, 21.
348. M. H. Chisholm, *Pure Appl. Chem.*, 1989, **61**, 1707.
349. M. H. Chisholm, *J. Organomet. Chem.*, 1990, **400**, 235.
350. M. H. Chisholm, *Acc. Chem. Res.*, 1990, **23**, 419.
351. M. H. Chisholm, *Pure Appl. Chem.*, 1991, **63**, 665.
352. F. A. Cotton and R. A. Walton, 'Multiple Bonds Between Metal Atoms', Wiley, New York, 1982.
353. F. A. Cotton and R. A. Walton, 'Metal–Metal Multiple Bonds in Dinuclear Clusters: Structure and Bonding', Berlin, 1985.
354. W. E. Buhro and M. H. Chisholm, *Adv. Organomet. Chem.*, 1987, **27**, 311.

355. M. H. Chisholm, *Polyhedron*, 1983, **2**, 681.
356. F. A. Cotton, *Polyhedron*, 1986, **5**, 3.
357. M. H. Chisholm, *Polyhedron*, 1986, **5**, 25.
358. M. H. Chisholm, R.-M. Jansen and J. C. Huffman, *Organometallics*, 1992, **11**, 2305.
359. M. H. Chisholm *et al.*, *J. Am. Chem. Soc.*, 1987, **109**, 6796.
360. F. A. Cotton and G. N. Mott, *Organometallics*, 1982, **1**, 302.
361. G. S. Girolami, V. V. Mainz and R. A. Andersen, *J. Am. Chem. Soc.*, 1981, **103**, 3953.
362. G. S. Girolami, V. V. Mainz and R. A. Andersen, *J. Am. Chem. Soc.*, 1982, **104**, 2041.
363. W. S. Harwood, J.-S. Qi and R. A. Walton, *Polyhedron*, 1986, **5**, 15.
364. F. A. Cotton, W. Schwotzer and E. S. Shamshoum, *Organometallics*, 1983, **2**, 1167.
365. M. H. Chisholm, J. A. Heppert, E. M. Kober and D. L. Lichtenberger, *Organometallics*, 1987, **6**, 1065.
366. M. H. Chisholm, J. C. Huffman and J. A. Heppert, *J. Am. Chem. Soc.*, 1985, **107**, 5116.
367. M. H. Chisholm, K. Folting, J. A. Heppert, D. M. Hoffman and J. C. Huffman, *J. Am. Chem. Soc.*, 1985, **107**, 1234.
368. M. H. Chisholm, J. A. Heppert, J. C. Huffman and P. Thornton, *J. Chem. Soc., Chem. Commun.*, 1985, 1466.
369. E. D. Jemmis and B. V. Prasad, *Organometallics*, 1992, **11**, 2528.
370. E. D. Jemmis and B. V. Prasad, *Polyhedron*, 1988, **7**, 871.
371. E. D. Jemmis and B. V. Prasad, *J. Am. Chem. Soc.*, 1987, **109**, 2560.
372. M. H. Chisholm, J. C. Huffman and M. J. Hampden-Smith, *J. Am. Chem. Soc.*, 1989, **111**, 5284.
373. R. H. Cayton, S. T. Chacon, M. H. Chisholm and J. C. Huffman, *Angew. Chem., Int. Ed. Engl.*, 1990, **29**, 1026.
374. S. T. Chacon, M. H. Chisholm, C. M. Cook, M. J. Hampden-Smith and W. E. Streib, *Angew. Chem., Int. Ed. Engl.*, 1992, **31**, 462.
375. M. H. Chisholm, K. Folting, J. C. Huffman and J. A. Klang, *Organometallics*, 1988, **7**, 1033.
376. A. Bino, F. A. Cotton, Z. Dori and B. W. S. Kolthammer, *J. Am. Chem. Soc.*, 1981, **103**, 5779.
377. M. Ardon *et al.*, *Inorg. Chem.*, 1981, **20**, 4083.
378. F. A. Cotton *et al.*, *Inorg. Chem.*, 1985, **24**, 4381.
379. A. Birnbaum *et al.*, *J. Am. Chem. Soc.*, 1985, **107**, 2405.
380. M. H. Chisholm, D. L. Clark, J. C. Huffman and C. A. Smith, *Organometallics*, 1987, **6**, 1280.
381. M. H. Chisholm, C. E. Hammond, V. J. Johnston, W. E. Streib and J. C. Huffman, *J. Am. Chem. Soc.*, 1992, **114**, 7056.
382. T. Ziegler, *J. Am. Chem. Soc.*, 1983, **105**, 7543.
383. M. D. Braydich, B. E. Bursten, M. H. Chisholm and D. L. Clark, *J. Am. Chem. Soc.*, 1985, **107**, 4459.
384. R. Beckhaus, S. Flatau, S. Trojanov and P. Hofmann, *Chem. Ber.*, 1992, **125**, 291.
385. J. Evans and J. F. W. Mosselmans, *J. Am. Chem. Soc.*, 1991, **113**, 3737.
386. S. M. Beshouri, I. P. Rothwell, K. Folting, J. C. Huffman and W. E. Streib, *Polyhedron*, 1986, **5**, 1191.
387. T. W. Coffindaffer, I. P. Rothwell and J. C. Huffman, *J. Chem. Soc., Chem. Commun.*, 1983, 1249.
388. K. A. Azam, P. C. Das, M. K. Hasan and S. E. Kabir, *Indian J. Chem., Sect. A*, 1989, **28**, 906.
389. M. H. Chisholm, B. W. Eichhorn, K. Folting, J. C. Huffman and R. J. Tatz, *Organometallics*, 1986, **5**, 1599.
390. B. Bachmann, G. Baum, J. Heck, W. Massa and B. Ziegler, *Z. Naturforsch., Teil A*, 1990, **45**, 221.
391. M. J. Chetcuti, M. H. Chisholm, K. Folting, J. C. Huffman and J. Janos, *J. Am. Chem. Soc.*, 1982, **104**, 4684.
392. I. T. Badejo, H. Choi and J. L. Fry, *Tetrahedron*, 1988, **29**, 4787.
393. M. H. Chisholm, K. Folting, J. C. Huffman, K. S. Kramer and R. J. Tatz, *Organometallics*, 1992, **11**, 4029.
394. K. J. Ahmed, M. H. Chisholm and J. C. Huffman, *Organometallics*, 1985, **4**, 1168.
395. F. A. Cotton and K. J. Wiesinger, *Inorg. Chem.*, 1990, **29**, 2594.
396. M. J. Chetcuti, M. H. Chisholm, K. Folting, D. A. Haitko and J. C. Huffman, *J. Am. Chem. Soc.*, 1982, **104**, 2138.
397. F. Baert, J. Lamiot, J. Lebibi and J. Brocard, *Acta Crystallogr., Sect. C*, 1988, **44**, 1389.
398. M. H. Chisholm, K. Folting, J. C. Huffman and I. P. Rothwell, *Organometallics*, 1982, **1**, 251.
399. B. Bachmann, F. Hahn, J. Heck and M. Wünsch, *Organometallics*, 1989, **8**, 2523.
400. J. J. H. Edema, S. Gambarotta, P. van der Sluis, W. J. J. Smeets and A. L. Spek, *Inorg. Chem.*, 1989, **28**, 3782.

6

π-Complexes of Chromium, Molybdenum and Tungsten, Excluding those of Cyclopentadienyls and Arenes

MARK W. WHITELEY
University of Manchester, UK

6.1 INTRODUCTION

The material in this chapter provides coverage of the literature from 1981 to 1992; the literature for 1993 is also included but with a cut-off point at the end of the June issue(s) of any journal. As far as possible, the classification of material follows that of Chapter 27.2 in *COMC-I*, but extensive developments in the chemistry of alkyne complexes necessitate some changes. The coverage of bimetallic complexes is confined to examples which contain only chromium triad metals.

6.2 η^2-COMPLEXES

6.2.1 Simple Alkene Complexes

Simple alkene–carbonyl complexes [$M(CO)_{6-n}(\eta^2$-alkene$)_n$] ($n = 1$ or 2) are relatively unstable, especially for M = Cr, but are accessible via photochemical syntheses. Photolysis of [$Cr(CO)_6$] in ethene-doped liquid xenon yields unstable [$Cr(CO)_5(\eta^2-C_2H_4)$] and [$Cr(CO)_4(\eta^2-C_2H_4)_2$] as a *cis–trans* mixture;[1] alternatively, extended photolysis of [$Cr(CO)_6$] in ethene-saturated pentane at −50 °C proceeds via [$Cr(CO)_5(\eta^2-C_2H_4)$] to give *trans*-[$Cr(CO)_4(\eta^2-C_2H_4)_2$] which,[2] in distinction to the previous report,[1] was found to be an isolable species in common with the molybdenum and tungsten analogues. The poor stability of *cis*-[$Cr(CO)_4(\eta^2-C_2H_4)_2$] has been corroborated via a gas-phase synthesis[3] involving a pulsed UV eximer laser beam; interestingly no *trans*-[$Cr(CO)_4(\eta^2-C_2H_4)_2$] was observed in this latter synthesis and it is probable that CO dissociation from [$Cr(CO)_5(\eta^2-C_2H_4)$] occurs preferentially *cis* to ethene leading to the formation of *cis*-[$Cr(CO)_4(\eta^2-C_2H_4)_2$], which undergoes a subsequent, photochemically induced *cis–trans* isomerization.[4] Infrared and Raman spectroscopic studies on *trans*-[$M(CO)_4(\eta^2-C_2H_4)_2$] (M = Mo or W) in hexane or liquid xenon suggest a D_{2d} symmetry with the ethene ligands in a *trans*-orthogonal orientation;[5] the stability of this conformation has been rationalized in terms of the efficiency of metal to alkene (π*) back donation.[4] Photolysis of *trans*-[$M(CO)_4(\eta^2-C_2H_4)_2$] (M = W) in liquid xenon, doped with H_2, affords the nonclassical dihydrogen complexes *mer*-[$W(CO)_3(\eta^2\ H_2)(\eta^2-C_2H_4)_2$] and *cis*-[$W(CO)_4(\eta^2-H_2)(\eta^2-C_2H_4)$]; analogous, but less stable, complexes are probably formed where M = Cr or Mo.[6] The photochemical generation and reactivity of the mixed diene–alkene complexes [$M(CO)_3(\eta^2$-alkene$)(\eta^4$-nbd)] (M = Cr, Mo or W) is discussed in Section 6.4.2.

Analogous complexes of the cycloalkene (*cis* or *trans*) cyclooctene appear to possess enhanced stability. Room-temperature photolysis of [$Cr(CO)_6$] with *cis*-cyclooctene (*cis*-coe) affords isolable [$Cr(CO)_5(\eta^2$-*cis*-coe)] and correspondingly, with *trans*-coe, [$Cr(CO)_5(\eta^2$-*trans*-coe)] followed by *trans*-[$Cr(CO)_4(\eta^2$-*trans*-coe$)_2$].[7] Time-resolved photoacoustic calorimetry studies reveal that the chromium–alkene bond strength in [$Cr(CO)_5(\eta^2$-*trans*-coe)] exceeds that of the corresponding *cis* complex by ca. 20 kJ mol^{-1}, an effect probably associated with ring strain.[8] The complexes [$M(CO)_5(\eta^2$-*trans*-coe)] (M = Cr or W[9]) have been employed in the attempted photochemical generation of η^2-dihydrogen species [$M(CO)_4(\eta^2-H_2)(\eta^2$-coe)],[10] and the carbene–alkene complex *trans*-[$W(CPh_2)(\eta^2$-*trans*-coe$)(CO)_4$] has been obtained by low-temperature photolysis of [$W(CPh_2)(CO)_5$] with *trans*-coe.[9] Low-temperature UV irradiation of [$W(CO)_5(\eta^2$-alkene)] (alkene = ethene, propene or pent-1-ene) reveals two competitive primary photoprocesses, CO and alkene loss. Carbonyl loss is favoured by short wavelength radiation and where alkene = C_2H_4.[11] The synthesis and subsequent reactions of the unstable η^2-cyclopropene complex (**1**) are shown in Scheme 1.[12]

Interest in the tetracyanoethene complexes [$M(CO)_5(\eta^2$-TCNE)] (M = Cr or W) has centred on metal–ligand charge-transfer excitation[13,14] and redox properties. The synthesis of [$M(CO)_5(\eta^2$-TCNE)] from [$M(CO)_5$(THF)] and TCNE probably proceeds via an electron transfer chain process[15] and

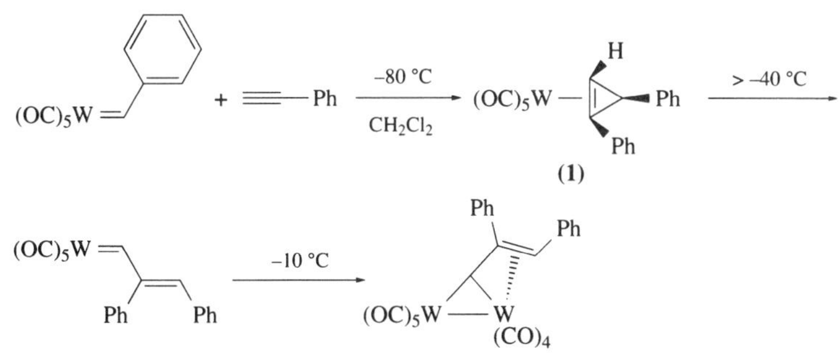

Scheme 1

[$M(CO)_5(\eta^2$-TCNE)] exhibits a rich electrochemistry.[16,17] Electron-transfer-induced isomerization {η^2-(C=C) to η^1-(N)} of the bonding mode of the TCNE ligand occurs on reduction of [$W(CO)_5(\eta^2$-TCNE)] to the corresponding mono- and dianionic species.[17]

A limited chemistry of monoalkene–carbonylphosphine complexes [$M(CO)_{5-n}(PR_3)_n(\eta^2$-alkene)] (M = Cr, Mo or W; PR_3 = phosphorus-donor ligand; n = 1 or 2) has been developed. The monophosphine complexes *cis*-[$M(CO)_4(PR_3)(\eta^2$-alkene)] (M = Cr, Mo or W; R = Et, Ph, Bu^t, etc.; alkene = dimethylmaleate, dimethylfumarate, etc.) have been obtained[18] via acetonitrile displacement from [$M(CO)_4(NCMe)(PR_3)$] and undergo facile alkene substitution,[19] but the majority of examples are complexes of tungsten with electron-poor alkenes which are synthesized via photolysis of [$W(CO)_4(PR_3)_2$] and the appropriate alkene. Thus, syntheses of [$W(CO)_3(PR_3)_2(\eta^2$-alkene)] (alkene = dimethylmalonate, fumarate, etc.; $PR_3 = PMe_3$[20], $P(OMe)_3$ or $P(OEt)_3$;[21] $(PR_3)_2$ = dmpe[22] or dppe[23]) have been reported, including chromium and molybdenum derivatives of the dppe complexes. Analogous complexes of haloalkenes [$W(CO)_3(PR_3)_2\{\eta^2$-*trans*-$(CO_2Me)CH{=}CX(CO_2Me)\}$] ($PR_3 = PMe_3$, $(PR_3)_2$ = dmpe; X = Br or Cl) are unstable with respect to loss of CO, so resulting in the formation of vinyl products [$WX(CO)_2(PR_3)_2\{\eta$-$C(CO_2Me){=}CH(CO_2Me)\}$].[24] The complexes [$W(CO)_3(PR_3)_2(\eta^2$-alkene)] have stimulated studies on the dynamics of alkene rotation ($PR_3 = PMe_3$,[20] $P(OMe)_3$ or $P(OEt)_3$;[25] $(PR_3)_2$ = dmpe[22]) and, in the case of [$W(CO)_3\{P(OMe)_3\}\{P(OEt)_3\}(\eta^2$-alkene)] (alkene = ethene, methylmaleate or methylfumarate), $\Delta G^{\ddagger}$ for alkene rotation lies in the range 28–59 kJ mol^{-1}.

Some chemistry of bis(alkene)–phosphine complexes *trans*-[$M(PR_3)_4(\eta^2$-alkene$)_2$] (M = Mo or W) has been developed. The trimethylphosphine complexes *trans*-[$M(PMe_3)_4(\eta^2$-$C_2H_4)_2$] (M = Mo[26] or W[27]) are accessible via reaction of *cis*-[$M(N_2)_2(PMe_3)_4$] with ethene, but the tungsten derivative is more effectively synthesized[27] by sodium/mercury reduction of [$WCl_4(PMe_3)_3$] in ethene-saturated THF; alternative syntheses involve reaction of ethene with [$Mo(PMe_3)_6$][28] and [$WH(\eta^2$-$CH_2PMe_2)(PMe_3)_4$].[29] The complexes [$M(PMe_3)_4(\eta^2$-$C_2H_4)_2$] exhibit an extensive reaction chemistry due to the electron-rich metal centre and the availability of a vacant coordination site by facile displacement of at least one PMe_3 ligand. Thus, reaction with CO affords *mer,trans*-[$M(CO)(PMe_3)_3(\eta^2$-$C_2H_4)_2$] (M = Mo[26] or W[27]) and *trans,trans,trans*-[$Mo(CO)_2(PMe_3)_2(\eta^2$-$C_2H_4)_2$];[27] analogous substitution products are formed with isocyanides.[30] More remarkably, the compounds [$M(PMe_3)_4(\eta^2$-$C_2H_4)_2$] (M = Mo or W) react with CO_2 to give (**2**) in which the bridging acrylate ligands originate from CO_2-ethene coupling.[31] A theoretical study on this reaction has considered the importance of rotation about the M–CO_2 bond in the initiation of the coupling process.[32] Head-to-tail dimerization of CS_2 is also promoted by [$M(PMe_3)_4(\eta^2$-$C_2H_4)_2$], leading to formation of (**3**).[33]

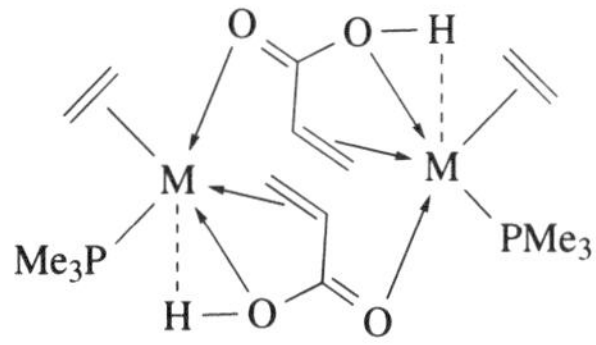

(**2**)

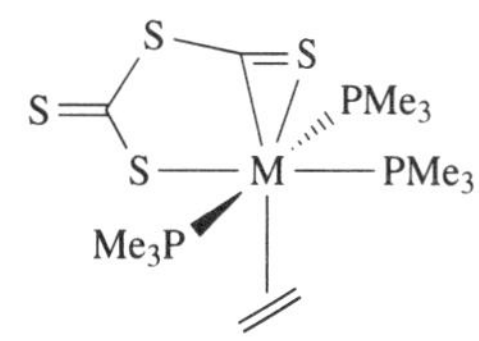

(**3**)

The chromium complex *trans*-[$Cr(dmpe)_2(\eta^2$-$C_2H_4)_2$] has been synthesized via reaction of *trans*-[$Cr(N_2)_2(dmpe)_2$] with ethene,[34] and a series of complexes *trans*-[$Mo(\eta^2$-dimethylfumarate$)_2(CO)_2$(N–N)] (N–N = bipy, en, phen, etc.) is accessible from [$Mo(CO)_3$(NCMe)(N–N)].[35] Recent interest in *trans*-[$M(P\text{–}P)_2(\eta^2$-$C_2H_4)_2$] (M = Mo or W; P–P = dppe or *cis*-$Ph_2PCH{=}CHPPh_2$ {dppen}) has centred on the protonation products [$MH(P\text{–}P)_2(\eta^2$-$C_2H_4)_2]^+$. Thus, the fluxional processes operative in

$[MoH(dppen)_2(\eta^2\text{-}C_2H_4)_2]^+$ have been reinvestigated and a novel mechanism proposed to account for hydrogen scrambling between hydride and ethene ligands; effectively 'in place' rotation of the methyl group of a β-agostic ethyl intermediate permits interchange of ethene and hydride hydrogens without recourse to a 16-electron σ-ethyl species.[36] Stopped flow spectrophotometric studies reveal that HX (X = Br or Cl) protonation of *trans*-$[W(dppe)_2(\eta^2\text{-}C_2H_4)_2]$ proceeds via two pathways: slow direct protonation at tungsten or alternatively rapid protonation at C_2H_4 and subsequent slow intramolecular β-H migration from the resulting ethyl ligand to tungsten;[37] investigations on the corresponding protonation of the molybdenum analogue[38] are complicated by ethene dissociation and the reversibility of β-H migration.

A limited number of high-oxidation-state alkene complexes, generally with halide ligands, have been reported. The complexes $[WCl_2(PR_3)_4]$ ($PR_3 = PMe_3$, PMe_2Ph or $PMePh_2$) react with ethene to give *cis,trans*-$[WCl_2(PR_3)_2(\eta^2\text{-}C_2H_4)_2]$[39] and reaction of $[WCl_2(O)(PMePh_2)_3]$ with $CH_2{=}CHR$ (R = H, Me or $CH{=}CH_2$) yields (**4**)[40] in which the *cis* arrangement of oxo and alkene ligands facilitates significant tungsten–alkene π* interaction. Analogous imido and sulfido complexes $[WCl_2(E)(PR_3)_2(\eta^2\text{-alkene})]$ ($PR_3 = PMe_3$ or $PMePh_2$; alkene = ethene, propene, etc.; E = NPh[41,42] or S[42]) have also been reported. High-oxidation-state, mixed alkene–carbene or alkene–carbyne complexes have been implicated in catalytic processes and the isolation and reactivity of such complexes is therefore of considerable interest. The tungsten(VI) carbene complex (**5**) ($R = CH_2Bu^t$) is an active catalyst for the ring-opening metathesis polymerization (ROMP) of cycloheptene; the reaction intermediate (**6**) has been identified spectroscopically, so providing unprecedented confirmation for the involvement of a carbene–alkene intermediate in ROMP reactions.[43] The carbyne–alkene species $[WCl_2(CO)(py)_2(\eta^2\text{-alkene})(\equiv CPh)]$ (alkene = maleic anhydride or fumaronitrile) and related anionic complexes have also been reported.[44]

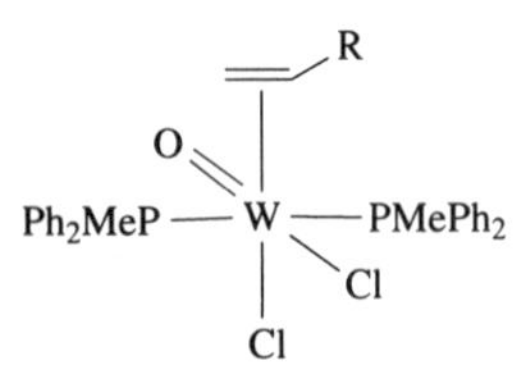

(**4**)

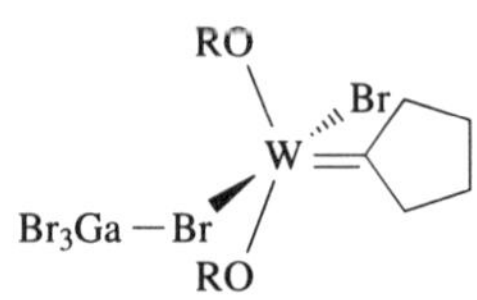

(**5**)

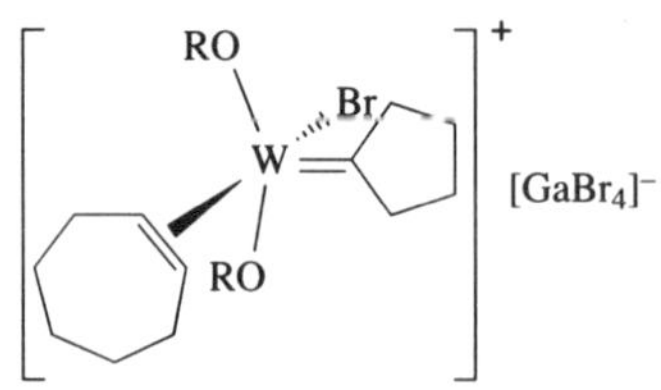

(**6**)

New syntheses of half-sandwich cyclopentadienyl–alkene complexes have been developed. The tetrafluoroborate complex $[Mo(CO)_3(FBF_3)Cp]$ is precursor to $[Mo(CO)_3(\eta^2\text{-butene})Cp][BF_4]$ via facile displacement of BF_4^- from the metal coordination sphere[45] and treatment of $[Mo(CO)_2(\eta^3\text{-}C_3H_5)Cp]$ with HBF_4 affords $[Mo(CO)_2(FBF_3)(\eta^2\text{-}CH_2{=}CHMe)Cp]$ which is a source of the 14-electron fragment $Mo(CO)_2Cp^+$ via displacement of both BF_4^- and propene ligands.[46] Chelate phosphine derivatives $[Mo(CO)(P\text{–}P)(\eta^2\text{-}RHC{=}CH_2)Cp]^+$ [P–P = dmpe, dppe or $(MeO)_2PCH_2CH_2P(OMe)_2$; R = H, Ph, CO_2Me, etc.] have been obtained by photolysis of $[Mo(CO)_2(P\text{–}P)Cp]^+$ in acetone followed by treatment with alkene; the $(MeO)_2PCH_2CH_2P(OMe)_2$ derivatives are relatively resistant to alkene dissociation and exhibit stereoselective binding of $RHC{=}CH_2$.[47,48] The chelating $\eta^{5:1}\text{-}C_5H_4C_2H_4$ ligand of $[M(CO)_3(\eta^{5:1}\text{-}C_5H_4C_2H_4)]$ (M = Mo or W) leads to distortion of the basal coordination sites so creating an enlarged site *trans* to the M–C σ-bond; this facilitates the photolytic generation of complexes of sterically demanding alkenes $[M(CO)_2(\eta^2\text{-alkene})(\eta^{5:1}\text{-}C_5H_4C_2H_4)]$ and of the bis(ethene) species $[W(CO)(\eta^2\text{-}C_2H_4)_2(\eta^{5:1}\text{-}C_5H_4C_2H_4)]$.[49] Photolytic techniques are also employed for the synthesis of $[Mo(CO)(NO)(\eta^2\text{-alkene})(\eta\text{-}C_5Me_4R)]$ (R = Me or Et) from $[Mo(CO)_2(NO)(\eta\text{-}C_5Me_4R)]$ and alkenes.[50] The imido complexes $[MoCl(\equiv NBu^t)(\eta^2\text{-}C_2H_4)(\eta\text{-}C_5H_4R)]$ (R = H or Me) are accessible through sodium/mercury reduction of $[MoCl_2(\equiv NBu^t)(\eta\text{-}C_5H_4R)]$ under an ethene atmosphere.[51] A series of investigations has centred on the photochemically induced conversion of $[MEt(CO)_3Cp]$ to the alkene-hydride $[MH(CO)_2(\eta^2\text{-}C_2H_4)Cp]$ and subsequently to $[MH(CO)_3Cp]$ (M = Mo or W).[52–4] It is probable that the transformation of $[MEt(CO)_3Cp]$ to *cis*-$[MH(CO)_2(\eta^2\text{-}C_2H_4)Cp]$ proceeds via an agostic ethyl intermediate and that a *cis* to *trans* isomerization of $[MH(CO)_2(\eta^2\text{-}C_2H_4)Cp]$ precedes ethene elimination.

6.2.2 Alkene Chelate Complexes

Most reported complexes of this type contain either phosphinoalkene or carbenealkene chelate ligands. Typically, reaction of $[W(CO)_4(pip)_2]$ (pip = piperidine) with ligands L ($L = PPh_2(CH_2)_nCH{=}CH_2$ (n = 1–4)[55] or *o*-$Ph_2P\text{–}C_6H_4\text{–}CH_2CH{=}CH_2$[56]) proceeds via $[W(CO)_4(pip)L]$, in which L is bonded through phosphorus, to give chelated products such as (**7**). The application of a phosphinoalkene chelate ligand in the synthesis of a very rare example of a stable, mixed alkene–alkyne complex is described in Section 6.8.2.3. Formation of the phosphinoalkene chelates (**8**) via reaction of lithium alkoxides $LiOCR^1R^2CH{=}CH_2$ (R^1, R^2 = H, Me, Ph, etc.) with (**9**) proceeds stereoselectively where the alkoxide incorporates an asymmetric centre.[57]

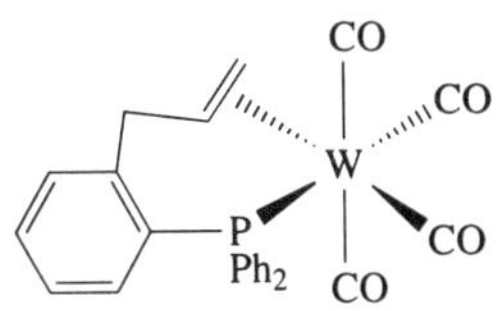

(7)

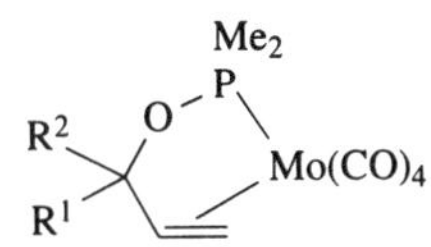

(8)

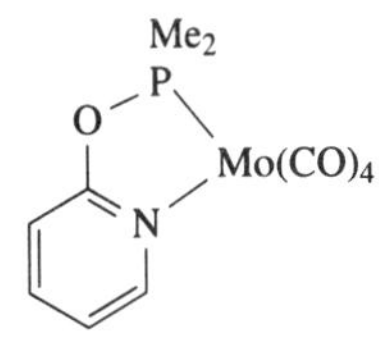

(9)

Recent examples of chelated carbene–alkene complexes are (**10**),[58] (**11**) and (**12**),[59] and (**13**),[60] all of which are generated by thermolysis or photolysis of carbene precursors bearing an unsaturated group on the carbene substituent. Crystallographic studies reveal a mutually perpendicular orientation of carbene and alkene functionalities with the exception of the almost coplanar arrangement of unstable (**12**). The additional stability conferred by chelation has also been exploited in the synthesis of (**14**), the first example of an isolable molybdenum–cyclopropene complex, which was generated via reaction of $[Mo(CO)_3(\eta\text{-}Cp^*)]^+$ with $NaO_2C\overline{C(H)C(Ph){=}C}Ph$.[61]

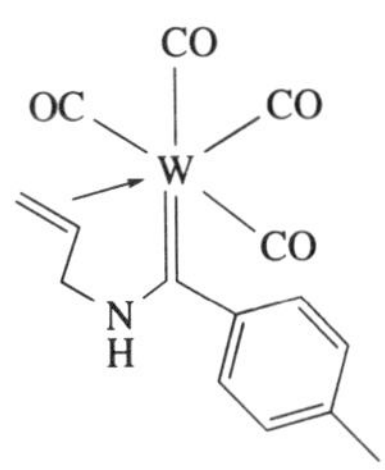

(10)

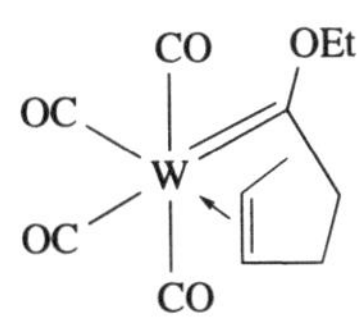

(11)

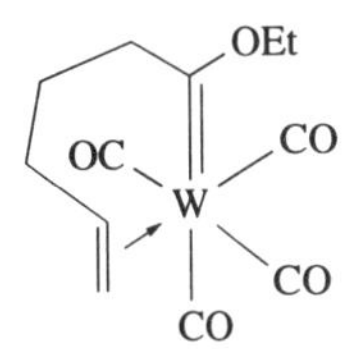

(12)

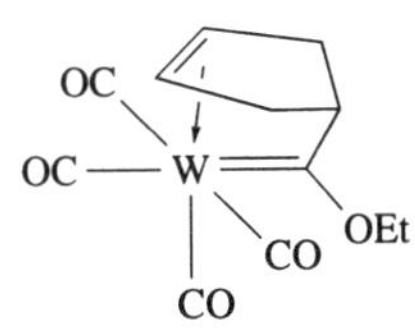

(13)

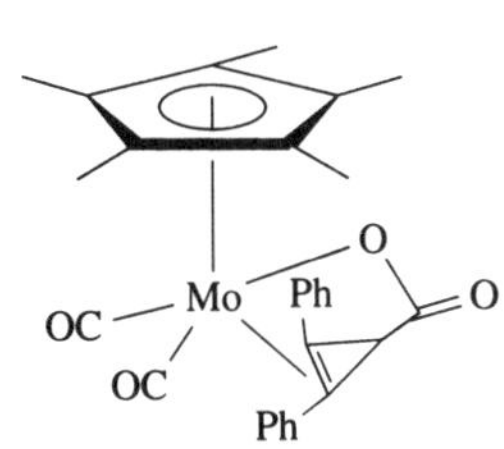

(14)

6.3 η^3-COMPLEXES

6.3.1 Homoleptic Allyls and Allyl Halides

Investigations on the electronic structure of $[Cr(\eta^3\text{-}C_3H_5)_3]$ by combined *ab initio* MO calculations and helium(I)/helium(II) photoelectron spectroscopic studies[62] conclude that the lowest energy features in the photoelectron spectra originate primarily from ionization from metal localized levels. Treatment of $[Cr(\eta^3\text{-}C_3H_5)_3]$ with monodentate phosphorus-donor ligands, PR_3 (R = Me, Ph, OMe, etc.) affords intractable adducts $[Cr(PR_3)(\eta^3\text{-}C_3H_5)_3]$ but reaction with dmpe affords[63] the allyl coupled product $[Cr(dmpe)(\eta^3\text{-}C_3H_5)(\eta^4\text{-}1,3\text{-hexadiene})]$. Thermolabile bis(allyl) complexes $[Cr(L)_2(\eta^3\text{-allyl})_2]$ (L = PMe_3, PMe_2Ph, etc.; L_2 = dmpe or $Pr^i_2PCH_2CH_2PPr^i_2$; allyl = C_3H_5 or 2-Me-C_3H_4) have been synthesized[64] via treatment of $[CrCl_2(L)_2]$ with propenyl Grignard reagents, but attempts to prepare $[Cr(PMe_3)_2(\eta^3\text{-}1\text{-Me-}C_3H_4)_2]$ result in the isolation of allyl-coupled products; allyl–alkyne coupling occurs in the reaction of MeC≡CMe with $[Cr(PMe_3)_2(\eta^3\text{-}C_3H_5)_2]$ to give (**15**). Similarly, 17-electron $[Cr(\eta^3\text{-}C_3H_5)_2Cp]$ reacts with L (L = CO, PMe_3 or $P(OMe)_3$) to give η^2,η^2-hexadiene products (**16**) and with butadiene to form (**17**);[65] $[Cr(\eta^3\text{-}C_3H_5)_2Cp]$ is very thermolabile and, above −10 °C, decomposes to binuclear $[\{Cr(\eta^3\text{-}C_3H_5)Cp\}_2]$.

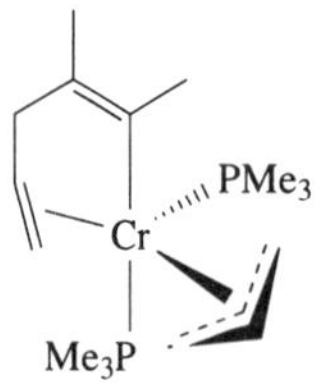

(15)

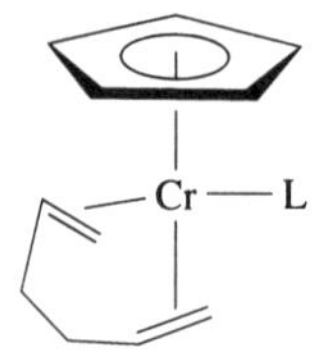

(16)

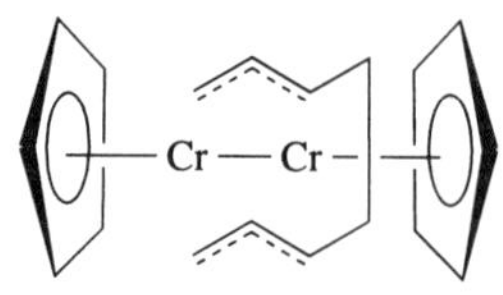

(17)

Magnetization transfer difference spectroscopy (MTDS) (^{1}H NMR) has been used to probe *syn–anti* and *syn–syn* (*anti–anti*) exchange in $[M(\eta^3\text{-}C_3H_5)_4]$ (M = Mo or W)[66] and ^{183}W and ^{95}Mo NMR studies on $[M(\eta^3\text{-}C_3H_5)_4]$ (M = Mo or W) have determined metal shift parameters δ(M) as −3186 (M = W) and −1405 (M = Mo).[67] The original, poorly documented, synthesis of $[W(\eta^3\text{-allyl})_4]$ has been superseded by a high-yield route involving reaction of $[W(OPh)_6]$ with MgBr(allyl)[68] and an improved synthesis of $[Mo(\eta^3\text{-}C_3H_5)_4]$ starting from $MoCl_4{\cdot}2THF$ and MgCl(allyl) has also been described.[69] A significant development in the chemistry of $[M(\eta^3\text{-}C_3H_5)_4]$ (M = Mo or W) is the reaction with HCl to give halide-bridged dimers $[\{M(\mu\text{-}Cl)(\eta^3\text{-}C_3H_5)_3\}_2]$ (M = Mo[69] or W[68]) and subsequent reactions involving bridge cleavage. Thus, $[\{Mo(\mu\text{-}Cl)(\eta^3\text{-}C_3H_5)_3\}_2]$ is a precursor to $[MoCl(L)(\eta^3\text{-}C_3H_5)_3]$ (L = py, MeCN or PMe_3), $[Mo(L)_2(\eta^3\text{-}C_3H_5)_3]^+$ (L = NCMe, L_2 = en or dmpe), $[Mo(OAc)(\eta^3\text{-}C_3H_5)_3]$[69] and 17-electron $[Mo(\eta^3\text{-}C_3H_5)_2Cp]$.[70] Treatment of $[MoCl(L)(\eta^3\text{-}C_3H_5)_3]$ (L = PMe_3 or $P(OMe)_3$) with Grignard reagents MgRX (R = Me, Et or allyl) yields $[Mo(R)(L)(\eta^3\text{-}C_3H_5)_3]$ but $[\{Mo(\mu\text{-}Cl)(\eta^3\text{-}C_3H_5)_3\}_2]$ reacts with MgRCl (R = Et, Pr^i, Bu^t, etc.) to give unstable products $[Mo(CR^1R^2CH_2\text{-}\mu\text{-}H)(\eta^3\text{-}C_3H_5)_3]$ (R^1, R^2 = H or Me) which incorporate an agostic C–H → Mo bond.[71] The analogous chemistry of the tungsten–alkyl complexes has also been developed,[68,72] but treatment of $[\{W(\mu\text{-}Cl)(\eta^3\text{-}C_3H_5)_3\}_2]$ with NaCp yields[73] $[W(\eta^1\text{-}C_3H_5)(\eta^3\text{-}C_3H_5)_2Cp]$ in contrast to the 17-electron product of the corresponding reaction in molybdenum chemistry.[70]

In principle, the allyl-bridged dimers $[M_2(\eta^3\text{-allyl})_4]$ may exhibit nine isomeric forms with respect to configurations of terminal and bridging allyl ligands but, in solution, $[M_2(\eta^3\text{-}C_3H_5)_4]$ (M = Cr or Mo) exists as an isomeric mixture of (**18**) and (**19**) in the ratio (**18**):(**19**) 60:40 (M = Cr) and 90:10 (M = Mo);[66] by contrast the methallyl derivative $[Mo_2(\eta^3\text{-2-Me-}C_3H_4)_4]$ exists as a 12:88 ratio of the 2-methyl derivatives of (**20**):(**19**).[74] Proton NMR MTDS studies on $[M_2(\eta^3\text{-}C_3H_5)_4]$ reveal *syn–syn* (*anti–anti*) exchange of the protons of the bridging allyl groups (M = Cr) and additionally isomer interconversion, (**18**) with (**19**), (M = Cr, 30 °C; M = Mo, 80 °C) via exchange of *syn* (*anti*) protons of the terminal allyl groups of (**18**) with *syn* (*anti*) protons of the terminal allyl groups of (**19**).[66] In the case of $[Cr_2(\eta^3\text{-}C_3H_5)_4]$ these exchange processes have been corroborated by two-dimensional dynamic ^{13}C NMR spectroscopy.[75] The synthesis and crystallographic characterization[74] of a further isomer of $[Mo_2(\eta^3\text{-}C_3H_5)_4]$, violet (**21**), has also been reported, so permitting a direct structural comparison with the green isomer (**18**) (M = Mo). The reactivity of $[Mo_2(\eta^3\text{-}C_3H_5)_4]$ is isomer dependent and the relatively slow reactions of (**21**) have been attributed to more strongly bonded and sterically shielded terminal allyl ligands. Reaction of (**18**) (M = Mo) with β-diketonates results in substitution of terminal allyl ligands to give products probably of structure (**22**) (R^1, R^2 = Me or CF_3).[76]

(18)

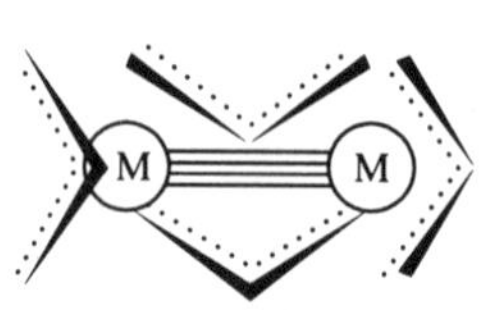

(19)

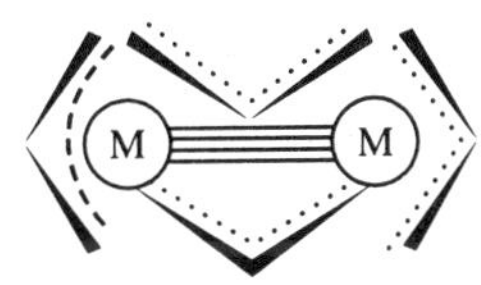

(20)

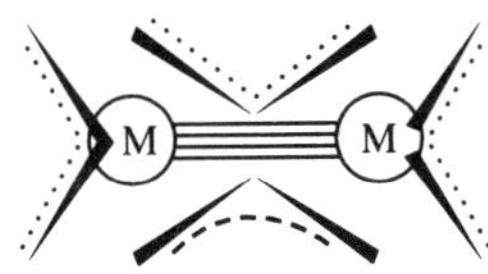

(21)

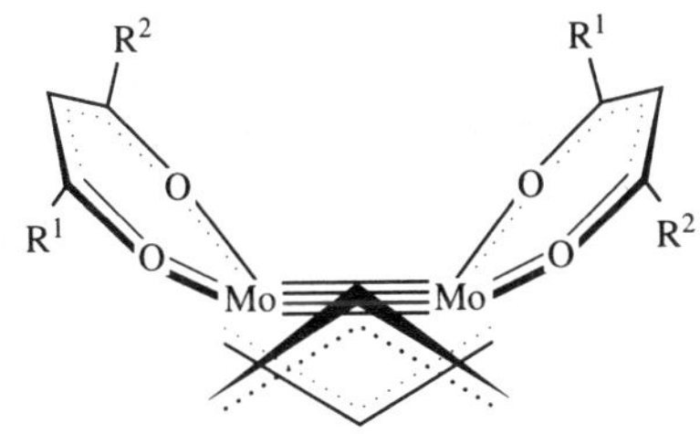

(22)

6.3.2 [MX(CO)$_2$(L)$_2$(η^3-allyl)] and Related Compounds

The chemistry of a wide range of complexes [MX(CO)$_2$(L)$_2$(η^3-allyl)] (M = Mo or W; X = halide or pseudohalide; L/L$_2$ = mono-/bidentate nitrogen- or phosphorus-donor ligand) is well established. Advances in this area have continued along a broad front of synthetic, spectroscopic and structural studies.

Most complexes [MX(CO)$_2$(L)$_2$(η^3-allyl)] are derived via substitution of NCMe in [MX(CO)$_2$(NCMe)$_2$(η^3-allyl)], and this route has been employed in new syntheses of [MoX-(CO)$_2$(R$^1{}_2$PCH$_2$CH$_2$PR$^2{}_2$)(η^3-C$_3$H$_5$)] (X = Br or Cl; R^1 = Ph, R^2 = C$_6$H$_4$F or C$_6$H$_4$CF$_3$;[77] R^1 = R^2 = C$_6$H$_4$Me[78]) and the semicarbazone[79] (E = O) and thiosemicarbazone[80] (E = S) complexes [MoX(CO)$_2${E=C(NH$_2$)NHN=CR$_2$}(η^3-allyl)] (X = halide; R = alkyl or aryl). The chloromethylallyl species [MoCl(CO)$_2$(NCMe)$_2${η^3-(2-CH$_2$Cl)C$_3$H$_4$}] may be obtained by reaction of [Mo(CO)$_6$] with CH$_2$=C(CH$_2$Cl)$_2$ in NCMe.[81] Initial investigations on allyl complexes of alkylxanthate and *N*-alkyldithiocarbamate ligands employed [MBr(CO)$_2$(L)$_2$(η^3-allyl)] (M = Mo, L = NCMe or py;[82] M = Mo or W, L$_2$ = bipy[83]) as starting materials but the reaction products are dependent upon L. Thus, reaction of [MBr(CO)$_2$(L)$_2$(η^3-allyl)] with S–S (S–S = alkylxanthate or *N*-alkyldithiocarbamate) affords [Mo(CO)$_2$(η^1-S–S)(η^2-S–S)(η^3-allyl)]$^-$ (L = NCMe), [Mo(CO)$_2$(py)(η^2-S–S)(η^3-allyl)] (L = py) and [M(CO)$_2$(bipy)(η^1-S–S)(η^3-allyl)] (L$_2$ = bipy). Recently, it has been shown that reaction of [MoBr(CO)$_2$(NCMe)$_2$(η^3-allyl)] with dithiocarbamates is also solvent dependent and, in NCMe, [Mo(CO)$_2$(η^2-S–S)(η^3-allyl)] (S–S = S$_2$CNEt$_2$ or S$_2$CNC$_4$H$_8$) is formed. Reaction of [Mo(CO)$_2$(η^2-S–S)(η^3-allyl)] with bidentate phosphines, or alternatively [MoBr(CO)$_2$(P–P)(η^3-allyl)] with dithiocarbamates, affords [Mo(CO)(η^2-S–S)(P–P)(η^3-allyl)] (P–P = dppe or dppm) in which one phosphorus of the chelating P–P ligand is located *trans* to the allyl ligand.[84] Reactions of [Mo(CO)$_2$(py)(η^2-S–S)(η^3-allyl)] with nitrogen- and phosphorus-donor ligands result in substitution of py and the generation of a range of further xanthate and dithiocarbamate allyl complexes.[85] The carboxylate complexes [Mo(O$_2$CR^2)(CO)$_2$(bipy)(η^3-C$_3$H$_4$R^1)] (R^1 = H or Me; R^2 = CF$_3$, C$_2$F$_5$, etc.) may be obtained by sequential treatment of [MoCl(CO)$_2$(bipy)(η^3-C$_3$H$_4$R)] with Ag[BF$_4$] then NaO$_2$CR^2; in the case of [Mo(O$_2$CCF$_3$)(CO)$_2$(bipy)(η^3-C$_3$H$_5$)], variable-temperature NMR studies reveal two coexistent isomeric forms.[86] Further studies on the ionization of [MoCl(CO)$_2$(NCMe)$_2$(η^3-allyl)] in methanol or deoxygenated water[87] establish the respective formation of [Mo(CO)$_2$(HOMe)$_3$(η^3-allyl)]$^+$ and

$[Mo(CO)_2(H_2O)_x(\eta^3\text{-allyl})]^+$ and cyclic voltammetric studies have been employed to monitor ionization to $[Mo(CO)_2(NCMe)_3(\eta^3\text{-allyl})][Mo_2Cl_3(CO)_4(\eta^3\text{-allyl})_2]$ in $CH_2Cl_2/[Bu^n{}_4N][PF_6]$.[88]

Cyclic voltammetric studies on a series of complexes $[MX(CO)_2(L\text{–}L)(\eta^3\text{-2-R–}C_3H_4)]$ (M = Mo or W; X = Br, Cl or O_2CCF_3; L–L = bipy, phen, $Bu^tN{=}CH{-}CH{=}NBu^t$, dppm, dppe, etc.; R = H or Me) reveal a reversible +1/0 redox couple in the potential range +0.47 V to +0.69 V (vs. a saturated calomel electrode).[88,89] The 17-electron radicals $[MX(CO)_2(L\text{–}L)(\eta^3\text{-2-R–}C_3H_4)]^+$ have been studied by ESR[88] and IR[89] spectroscopy and their stability found to be ligand dependent; phosphine chelate complexes give rise to unstable radicals which (M = Mo) decompose to $[Mo(CO)_3(P\text{–}P)(\eta^3\text{-allyl})]^+$ but radicals derived from complexes of nitrogen-donor chelate ligands are isolable via electrochemical generation in the cases X = Cl, L–L = bipy or phen; X = O_2CCF_3, L–L = bipy. X-ray photoelectron spectroscopic investigations on a series of complexes $[WX(CO)_2(L\text{–}L)(\eta^3\text{-allyl})]$ reveal a minimal dependence of tungsten 4*f* binding energies upon the identity of X and L–L.[90] X-ray crystallographic studies on $[Mo(NCS)(CO)_2(phen)\{\eta^3\text{-(1-}syn\text{-Ph)}C_3H_4\}]$[91] (**23**), $[\{Mo(CO)_2(bipy)(\eta^3\text{-}C_3H_5)\}_2(\mu\text{-Cl})]^+$[92] and the diazabutadiene complexes $[MoBr(CO)_2(RN{=}CHCH{=}NR)(\eta^3\text{-}C_3H_5)]$ (R = Bu^t[93] or Cy[94]) uncover the expected ligand arrangement in which the halide/pseudohalide ligand is located *trans* to the allyl ligand; an additional feature of interest in (**23**) is the asymmetry in molybdenum–allyl(C) distances [Mo–C-1, 0.242 9(4) nm; Mo–C-3, 0.231 1(5) nm]. By contrast, an x-ray structural study on (**24**) reveals a pyrazole ligand *trans* to the phenyl allyl ligand in preference to bromine.[95] The conformational preference of the allyl ligand in $[MX(CO)_2(L)_2(\eta^3\text{-allyl})]$ (in which the open face of the allyl ligand points towards the carbonyls, as in (**23**)) and the regioselectivity of nucleophilic addition at the terminal allyl carbon have been rationalized by a MO analysis;[96] the regioselectivity of nucleophilic addition is frontier orbital rather than charge controlled.

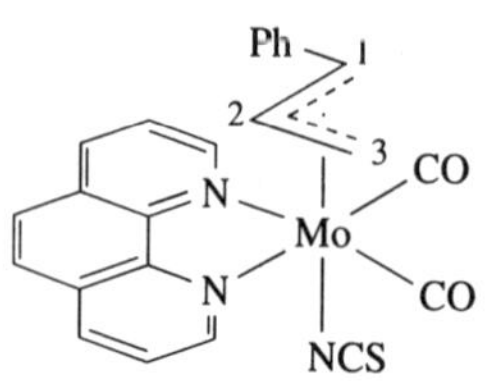

(**23**)

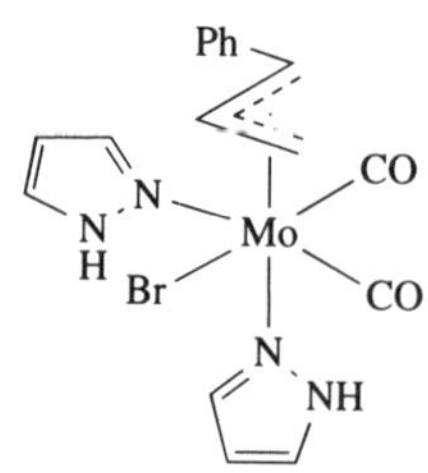

(**24**)

6.3.3 Allyl Complexes Containing Pyrazole-derived Bi- and Multidentate Ligands

Allyl complexes of the bis(pyrazolyl)methane ligand (**25**) (R^1 = H or Me; R^2 = H or Ph) are accessible via allyl bromination of $[Mo(CO)_4(\mathbf{25})]$ or alternatively reaction of (**25**) with $[MoBr(CO)_2(NCMe)_2(\eta^3\text{-allyl})]$;[97,98] in contrast to (**24**),[95] $[MoBr(CO)_2\{CHPh(3,5\text{-}Me_2\text{-}N_2C_3H)\}_2(\eta^3\text{-}C_3H_5)]$ adopts a structure in which bromine is *trans* to the allyl ligand.[97] Related syntheses of $[M(CO)_2(\mathbf{26})(\eta^3\text{-}C_3H_5)]^+$ (M = Cr, Mo or W; R = H or Me) have been reported; two structural types are observed in which the tridentate (**26**) may have a symmetric or asymmetric orientation with respect to a plane through the $Mo(CO)_2(\eta^3\text{-}C_3H_5)$ fragment.[99,100] Some complementary studies involving the tridentate, bis(2-pyridylmethyl)amine and the tetradentate, tris(2-pyridylmethyl)amine ligands have also been reported.[101] Reaction of tris(3,5-Me_2-1-pyrazolyl)phosphine oxide with $[MoX(CO)_2(NCMe)_2\{\eta^3\text{-(2-R)}C_3H_4\}]$ (X = Br or Cl; R = H, Me or Ph) unexpectedly yields (**27**), probably via partial hydrolysis of the pyrazolyl ligand,[102] and the benzotriazolyl complexes (**28**) (M = Mo or W) have been synthesized[103] from $K[M(CO)_3\{HB(C_6H_4N_3)_3\}]$ and allyl bromide. A series of hydrido-tris(pyrazolyl)-borate complexes (**29**) (R = H or Me; η^3-L = crotyl, cyclohexenyl, cycloheptenyl or cyclooctenyl) has been obtained.[104] Treatment of (**29**) (η^3-L = cyclohexenyl or cycloheptenyl) with $[Ph_3C]^+$ yields η^4-dienes which react with base or reducing agents to regenerate (**29**); where diene = cycloheptadiene, an η^3-cycloheptadienyl product is also isolated.[105]

6.3.4 $[M(CO)_2(\eta^3\text{-allyl})Cp]$ and Related Complexes

High-yield synthetic routes to $[Mo(CO)_2(\eta^3\text{-allyl})Cp]$ are well established but some modifications of interest have been reported. Trimethylamine-*N*-oxide promotes conversion of $[Mo(CO)_3(\eta^1\text{-allyl})Cp]$ to $[Mo(CO)_2(\eta^3\text{-allyl})Cp]$,[106] and the cyclopropylmethyl complexes (**30**) (M = Mo or W) undergo ring-

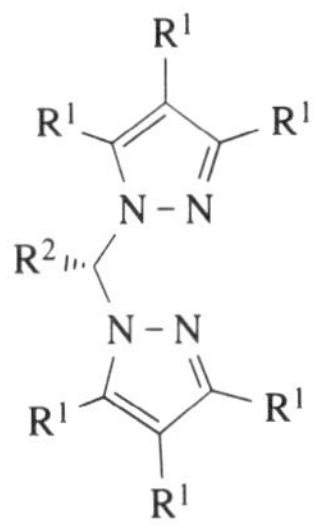

(25)

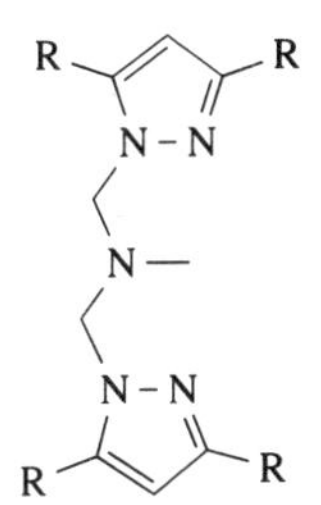

(26)

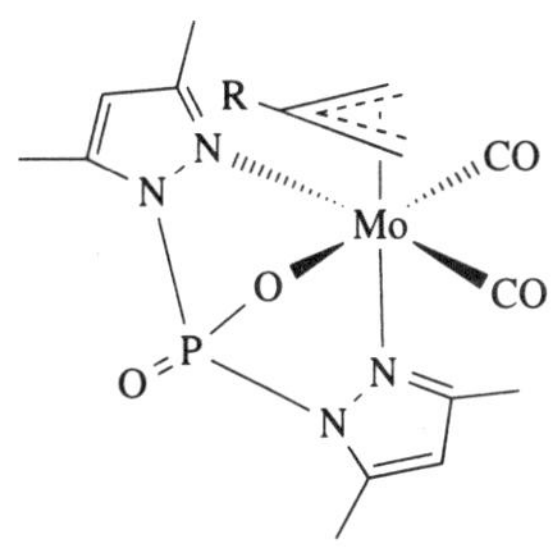

(27)

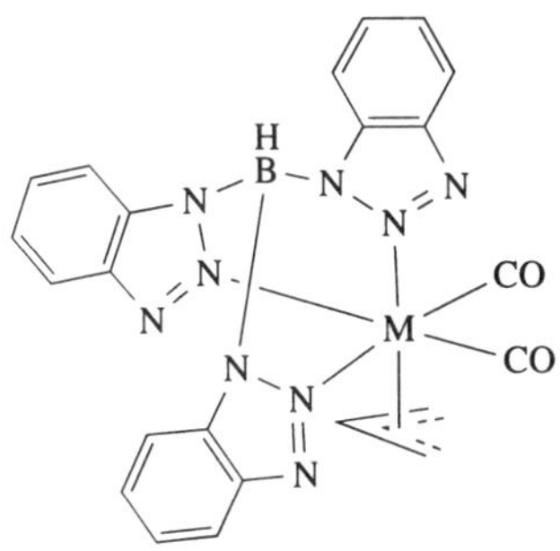

(28)

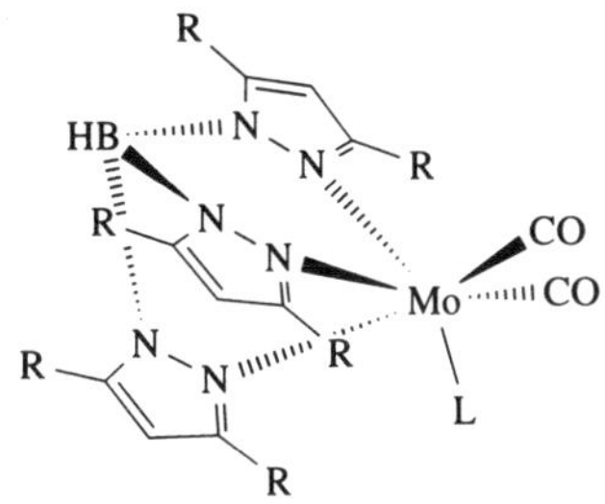

(29)

opening rearrangement to yield[107] [M(CO)$_2${η^3-(1-Me)C$_3$H$_4$}Cp]. Photolysis of [WH(CO)$_3$(η-C$_5$R$^1_5$)] (R^1 = H or Me) in the presence of alkynes MeC≡CR^2 (R^2 = H or Me) affords[108] [W(CO)$_2${η^3-(1-R^2)C$_3$H$_4$}(η-C$_5$R$^1_5$)], probably via hydrogen migration from tungsten to alkyne in the proposed intermediate [WH(CO)$_2$(η^2-R^{2}C≡CMe)(η-C$_5$R$^1_5$)]. The classical synthesis of [Mo(CO)$_2$(η^3-C$_3$H$_5$)Cp] via reaction of [MoX(CO)$_2$(NCMe)$_2$(η^3-C$_3$H$_5$)] with LiCp also generates, as a minor side product, ring-substituted [Mo(CO)$_2$(η^3-C$_3$H$_5$){η-C$_5$H$_4$(COMe)}]. Deuterium labelling studies reveal that the acetyl substituent originates from a coordinated NCMe ligand.[109] Investigations on factors which control the preference for *exo* vs. *endo* conformers, that is, (**31**) vs. (**32**) have continued. The x-ray crystal structure of [Mo(CO)$_2${η^3-(2-Me)C$_3$H$_4$}Cp] reveals an *endo* conformation of the allyl ligand in accord with the preferred solution conformation and the steric requirements of the 2-Me substituent; unusually, all the molybdenum–allyl carbon distances are equal.[110] Substitution at the cyclopentadienyl ring appears to enhance selectivity between conformers; thus, solution IR studies on [Mo(CO)$_2$(η^3-C$_3$H$_4$R){η^5-C$_5$H$_4$(COMe)}] reveal an almost exclusive *exo* conformation for (R = H or 1-Me) and a corresponding selectivity for the *endo* conformation where (R = 2-Me).[109] The solution *exo*:*endo* conformer ratio in [Mo(CO)$_2$(η^3-allyl)Cp] and [Mo(CO)(NO)(η^3-allyl)Cp]$^+$ can also be probed by ^{95}Mo NMR studies; *exo* isomers exhibit ^{95}Mo shifts to high field of corresponding *endo* species.[111] The molybdenum(IV) complexes [MoBr$_2$(CO)(η^3-2-R-C$_3$H$_4$)Cp] (R = Br, Cl or Me) are accessible[112] via bromine oxidation of [Mo(CO)$_2$(η^3-2-R-C$_3$H$_4$)Cp]. Alternatively, low-temperature photolysis of [MCl(CO)$_3$Cp] (M = Mo or W) in allyl chloride affords [MCl(CO)$_2$(η^2-C$_3$H$_5$Cl)Cp], which undergo M dependent thermolysis to give [MoCl$_2$(CO)(η^3-C$_3$H$_5$)Cp] and [WCl(CO)$_2$(η^3-C$_3$H$_5$)Cp][Cl].[113] A related reaction occurs on treatment of the anions [M(EPh$_3$)(CO)(NO)Cp]$^-$ (M = Mo or W; E = Si, Ge or Sn) with allyl halides leading to sequential formation of [M(η^1-C$_3$H$_5$)(EPh$_3$)(CO)(NO)Cp], [M(CO)(NO){η^2-C$_3$H$_5$(EPh$_3$)}Cp] and [M(EPh$_3$)(NO)(η^3-C$_3$H$_5$)Cp].[114]

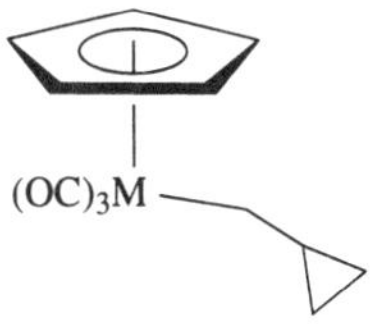

(30)

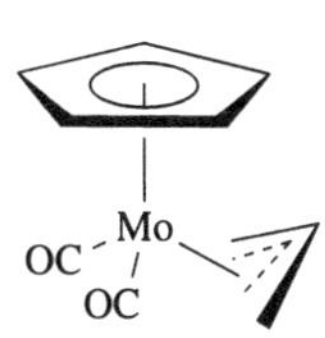

exo
(31)

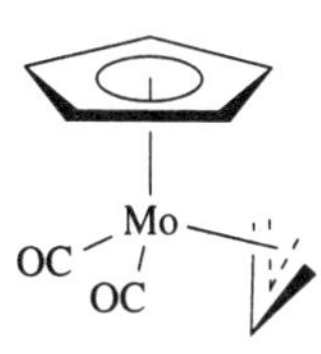

endo
(32)

Most advances in the chemistry of [Mo(CO)$_2$(η^3-cycloallyl)(η^5-L)] involve investigations on nucleophilic addition to [Mo(CO)$_2$(η^4-diene)(η^5-L)]$^+$ (diene = cyclohexadiene or cycloheptadiene;

L = Cp or indenyl). A seminal study in this area developed the reaction sequence shown in Scheme 2[115] and this prompted further work since difunctionalization proceeds both regio- and stereospecifically to give the *cis*-4,6-Me_2-cyclohexenyl product (**33**). A parallel reaction sequence occurs starting from the cycloheptadiene complex $[Mo(CO)_2(\eta^4\text{-}C_7H_{10})Cp]^+$ but the regiospecificity of difunctionalization is less well controlled and a product mixture, (**34**) with (**35**), is formed unless R^1, R^2 are sterically demanding, in which case (**34**) is favoured.[116] This relatively poor product selectivity has been discussed in terms of the conformational flexibility of the cycloheptenyl ring.[116,117] In the case of indenyl complexes, precursor diene species are conveniently prepared via direct reaction of $[Mo(CO)_2(NCMe)_2(\eta^5\text{-indenyl})]^+$ with the requisite diene and this provides syntheses of substituted cyclohexadiene complexes $[Mo(CO)_2(\eta^4\text{-}C_6H_7R)(\eta^5\text{-indenyl})]^+$ (R = Me, OMe or CO_2Me)[118] and $[Mo(CO)_2(\eta^4\text{-1,4-Me}_2\text{-cyclohexa-1,3-diene})(\eta^5\text{-indenyl})]^+$ [119] which undergo regioselective reactions with nucleophiles. An alternative, regio- and stereoselective synthesis of functionalized cyclohexenyl complexes proceeds via reaction of electrophiles with the carbanion $[Mo(CO)_2(1\text{-}3\text{-}\eta^3\text{-4-R-cyclohexenyl})Cp]^-$ (R = H or Me)[119] which is generated from $[Mo(CO)_2(1\text{-}3\text{-}\eta^3\text{-4-R,6-CN-cyclohexenyl})Cp]$ and Bu^nLi. The decomplexation of molybdenum-coordinated, functionalized allyl ligands represents a potential limitation to the synthetic applications of this chemistry but $[Mo(CO)_2(\eta^3\text{-allyl})Cp]$ (allyl = cyclohexenyl[120] or cycloheptenyl[116]) react with I_2 to yield iodocycloalkenes or lactones depending on ring functionality.

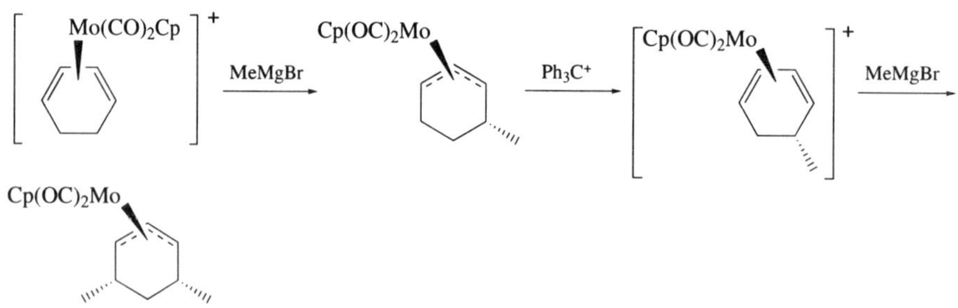

(**33**)

Scheme 2

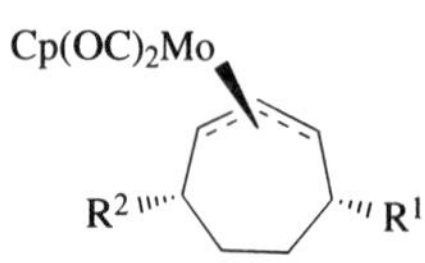

(**34**)

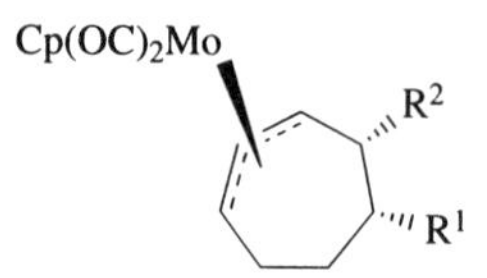

(**35**)

The η^3-5-oxo-cyclohexenyl complex (**36**) has been obtained by the reaction sequence in Scheme 3; treatment of (**36**) with strong bases yields enolate anions which react regio- and stereospecifically with electrophiles to give products such as (**37**) in which the electrophile is added α to the oxo group and on the opposite face of the C_6 ring to the coordinated molybdenum.[117,121] Analogous chemistry of the 4-oxo-cyclohexenyl complex $[Mo(CO)_2(\eta^3\text{-}C_6H_7O)(\eta^5\text{-indenyl})]$[122] and the 5-oxo-cycloheptenyl complex $[Mo(CO)_2(\eta^3\text{-}C_7H_9O)Cp]$[117] has also been developed. Treatment of $[Mo(CO)_2(NCMe)_2(\eta^5\text{-indenyl})]^+$ with 1-dimethylamino-cyclohexa-1,3-diene affords[123] the related iminium substituted allyl complex (**38**) which is deprotonated to the enamine (**39**). The stereospecific reaction of (**38**) with $K[BHBu^s_3]$ gives $[Mo(CO)_2(\eta^3\text{-4-NMe}_2\text{-}C_6H_9)(\eta^5\text{-indenyl})]$, derivatives of which are also accessible via reaction of $[Mo(CO)_2(\eta^4\text{-cyclohexadiene})(\eta^5\text{-indenyl})]^+$ with $[ZrX(NRPh)Cp_2]$ (X = Me or NRPh; R = H or Me).[124] A recent extension to this area heralds the possibility of enantiospecific syntheses; thus, the 2-H pyran complex (**40**) and its enantiomer exhibit enantioselectivity in reactions with nucleophiles.[125]

The development of the corresponding chemistry of acyclic allyl ligands has been impeded by poor regio- and stereocontrol of nucleophilic additions. An early study on (**41**) reported that sterically demanding nucleophiles exhibit poor selectivity between C-1 and C-4 and attempted generation of the carbanion $[Mo(CO)_2(1\text{-}3\text{-}\eta^3\text{-1-Me-}C_4H_5)Cp]^-$ via reaction of (**42**) with Bu^nLi (analogous to the successful strategy employed for cyclohexenyl complexes[119]) resulted in rearrangement.[126] One approach to address the problem of poor selectivity may be the generation of η^4-*s-trans*-1,3-diene complexes for reaction with nucleophiles; the synthesis of η^4-*s-trans*-1,3-diene complexes is described in Section 6.4.3. Typical complexes such as (**43**) react with nucleophiles R^- at both α and δ carbons although, in some examples, good regio- and stereocontrol is observed and a number of applications in organic transformations have been reported.[127–30]

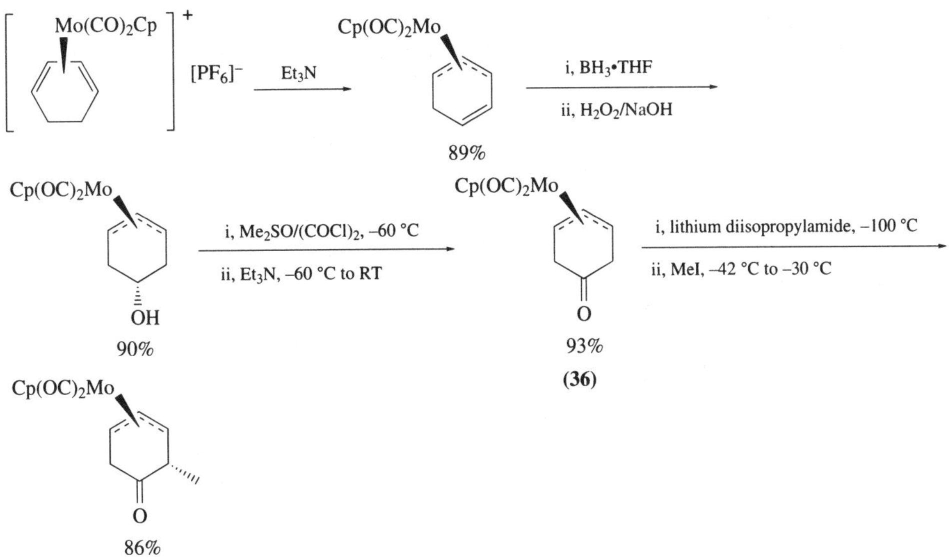

(37)

Scheme 3

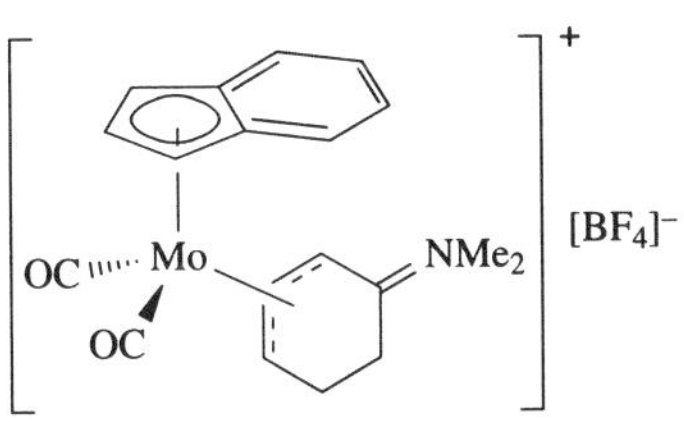

(38)

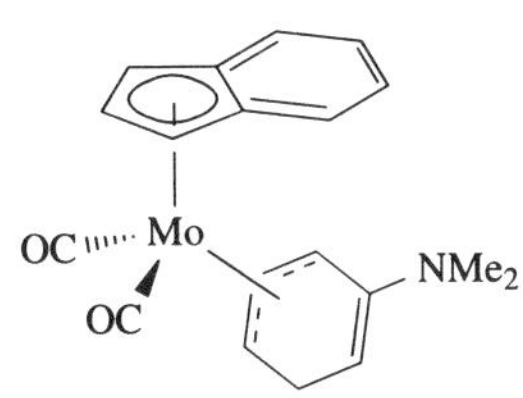

(39)

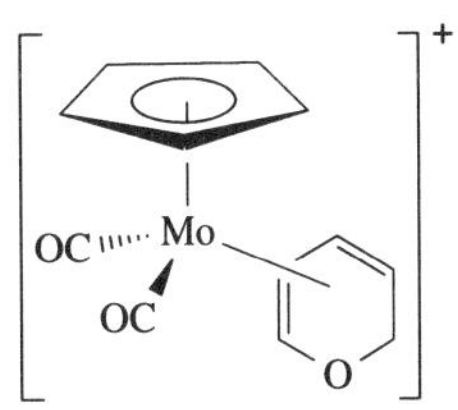

(40)

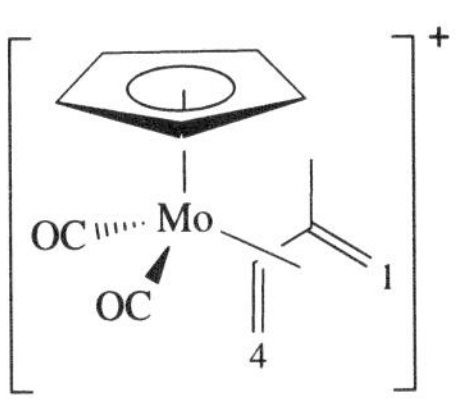

(41)

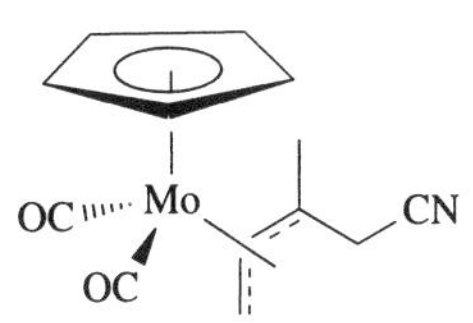

(42)

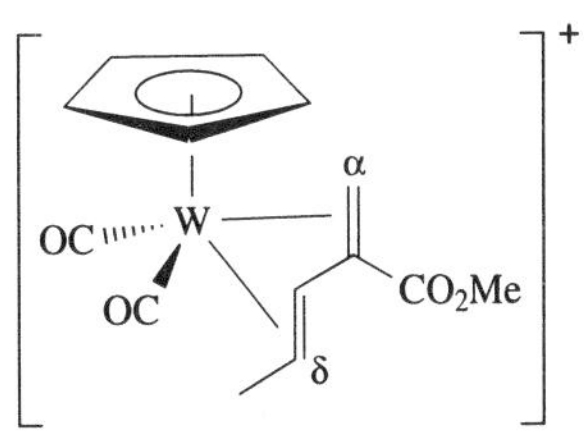

(43)

New syntheses of η^3-benzyl complexes have been developed. The anionic η^2-acetaldehyde complex (**44**) reacts with $PhCH_2Br$ to give (**45**) (M = Mo; R = H)[131] and protonation of the carbene $[M\{=C(OEt)Ph\}(CO)_2Cp]^-$ (M = Mo or W) affords (**45**) (M = Mo or W; R = OEt) via a hydride to carbene migration process.[132] The carbyne $[W(CO)_2(\equiv CC_6H_4Me\text{-}4)Cp]$ is a precursor to the η^3-benzyl (**46**) (R = Ph, Pr^i, etc.) through reaction with ZnR_2.[133] Proton NMR studies on $[Mo(CO)_2(1,2,7\text{-}\eta^3\text{-}2\text{-Me-benzyl})Cp]$ and $[Mo(CO)_2(1,2,7\text{-}\eta^3\text{-}3\text{-Me-benzyl})Cp]$ suggest[134] that, depending on the position of the methyl substituent, the major operative fluxional process is either allyl rotation or rotation in an η^1-benzyl intermediate.

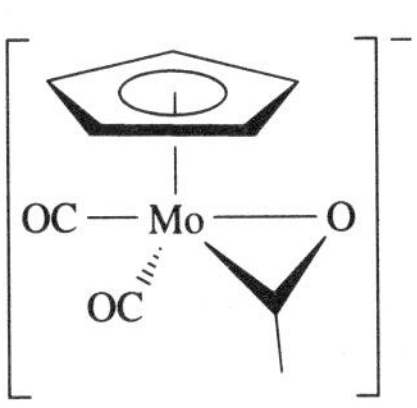

(44)

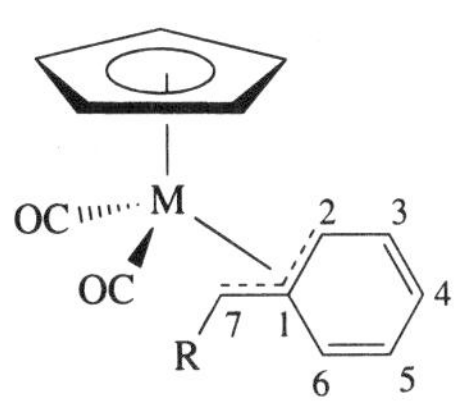

(45)

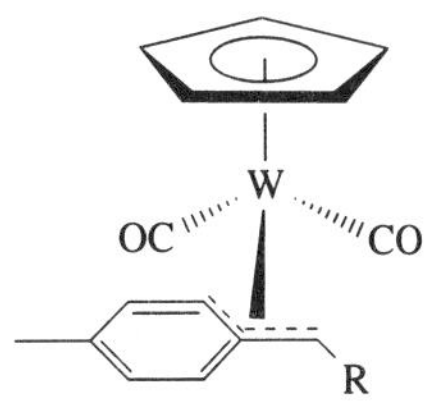

(46)

6.3.5 [Mo(CO)(NO)(η^3-allyl)Cp]$^+$ Cations

Recent investigations on [Mo(CO)(NO)(η^3-allyl)Cp]$^+$ complexes have concentrated on the selectivity of nucleophilic addition to the η^3-allyl ligand (allyl = C_3H_5 or 2-methylallyl,[135,136] 1,3-dimethylallyl,[136–8] cyclooctenyl[136,139,140] and η^3-1',2,3-(methylene-2γ-butyrolactone)-yl-3[141]) to yield η^2-alkene products. The following factors are fundamental to the development of this area. First, [Mo(CO)(NO)(η^3-allyl)Cp]$^+$ exhibits *exo–endo* conformational isomerism; in general, the *exo* form (**47**) is thermodynamically favoured over the *endo* form (**48**) and interconversion is relatively slow, but is catalysed[136] by anionic nucleophiles such as I^-. In most cases the kinetics of nucleophilic addition to [Mo(CO)(NO)(η^3-allyl)Cp]$^+$ are rapid by comparison with *exo–endo* interconversion but problems can arise when anionic nucleophiles are added to pure *endo* forms. Nucleophilic addition to the allyl ligand occurs on the opposite face to the coordinating molybdenum and, most importantly, is directed *cis* to the NO ligand as shown in Scheme 4. The basis of the directing effect of the NO ligand probably originates from the differing bonding properties of NO and CO leading to charge asymmetry and an asymmetry in molybdenum–allyl carbon bond distances.

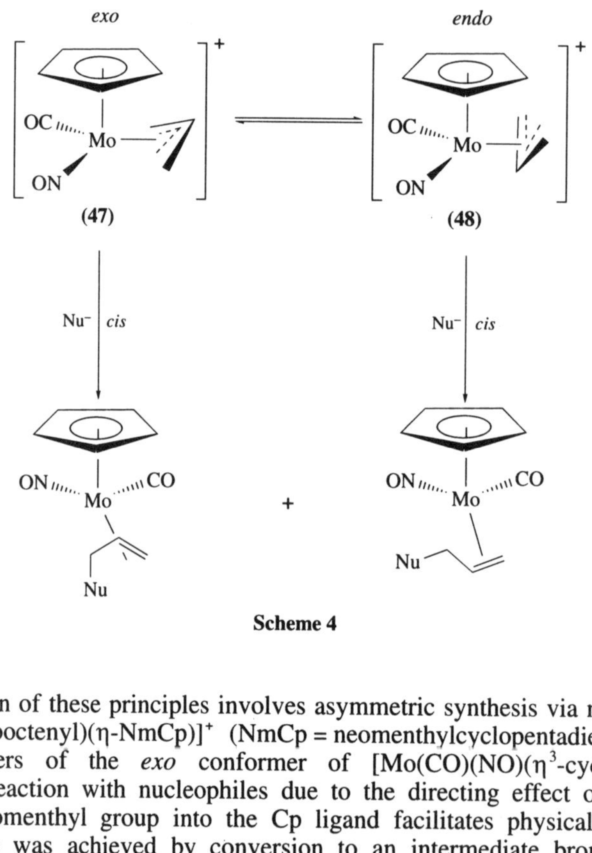

Scheme 4

A notable application of these principles involves asymmetric synthesis via nucleophilic addition to [Mo(CO)(NO)(η^3-cyclooctenyl)(η-NmCp)]$^+$ (NmCp = neomenthylcyclopentadienyl).[140] As shown in Scheme 5, enantiomers of the *exo* conformer of [Mo(CO)(NO)(η^3-cyclooctenyl)Cp]$^+$ retain enantiospecificity on reaction with nucleophiles due to the directing effect of the NO ligand. The incorporation of a neomenthyl group into the Cp ligand facilitates physical separation of (*R*)/(*S*) diastereomers and this was achieved by conversion to an intermediate bromide derivative which proceeds with retention of configuration.[142] Subsequent nucleophilic addition to the isolated (+)-(*R*)-[Mo(CO)(NO)(η^3-C_8H_{13})(η-NmCp)]$^+$ cation provides optically pure, allylically substituted cycloalkenes. A further development is the observation that the *cis* directing effect of the NO ligand can surmount substituent effects of the allyl ligand; thus, appropriate isomers can be compelled to undergo nucleophilic addition at a substituted terminal allyl carbon leading to the generation of a chiral centre (Scheme 6).[143] The feasibility of stereospecific synthesis via addition to the allyl ligand in [MoX(NO)(η^3-allyl)Cp] (X = halide)[144,145] and [Mo(OH_2)(NO)(η^3-allyl)Cp]$^{+}$[146] has also been explored.

6.3.6 Other η^3-Allyl Complexes

The allyl–hydride complex [MoH(dppe)$_2$(η^3-C_3H_5)] has been synthesized via sodium/mercury reduction of [$MoCl_4$(THF)$_3$] in the presence of dppe and propene and is a precursor to [MoX(dppe)$_2$(η^3-C_3H_5)] (X = Cl, Me, etc.).[147] Kinetics studies suggest that the formation of [MoH_2(Cl)$_2$(dppe)$_2$] and HC≡CMe via HCl protonation of [MoH(dppe)$_2$(η^3-C_3H_5)] proceeds[148] with loss of H_2 from the propene

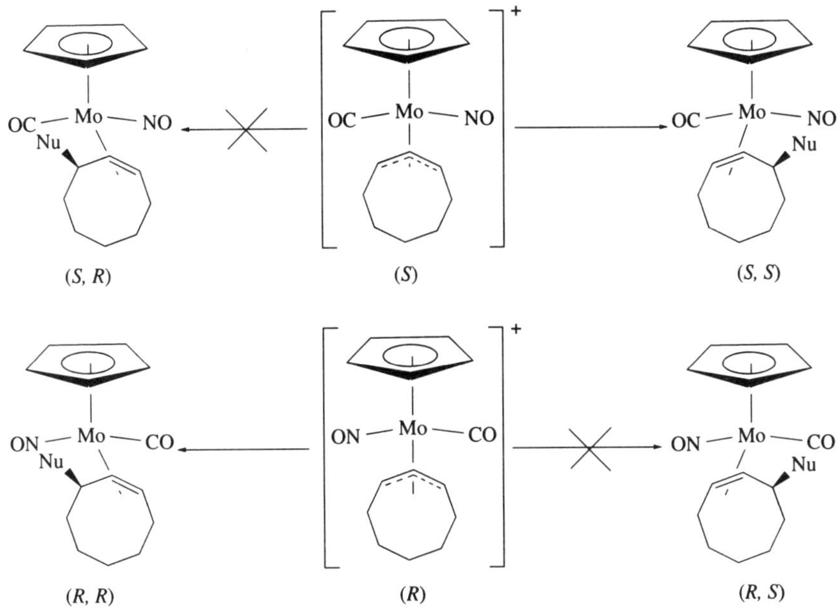

Scheme 5

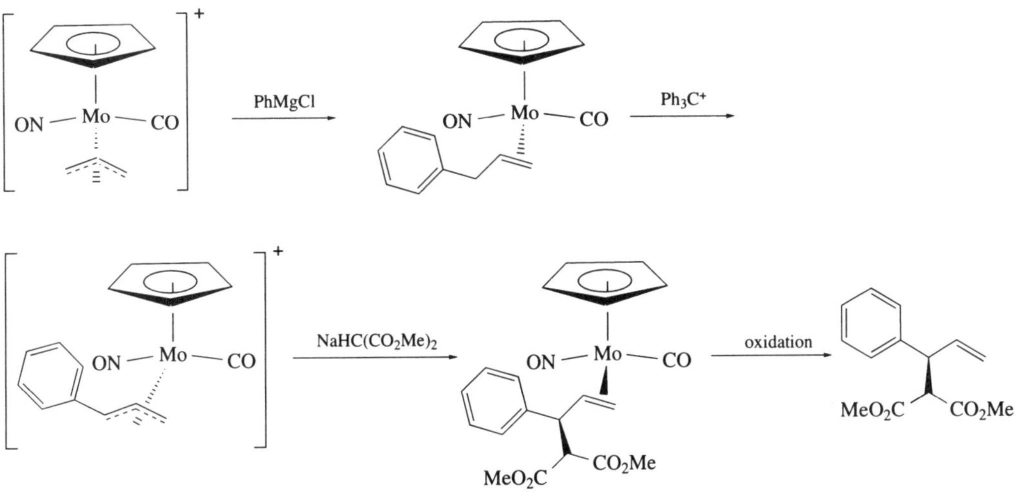

Scheme 6

intermediate $[MoH_2(dppe)_2(\eta^2\text{-}MeHC{=}CH_2)]^{2+}$. Hydride addition to $[Cr(CO)_3\{P(OMe)_3\}(\eta^4\text{-buta-1,3-diene})]$ proceeds as shown in Scheme 7; crystallographic studies establish that the initial product is the (*Z*)-2-buten-1-yl (**49**) which, at ambient temperature, isomerizes to the (*E*) form (**50**).[149]

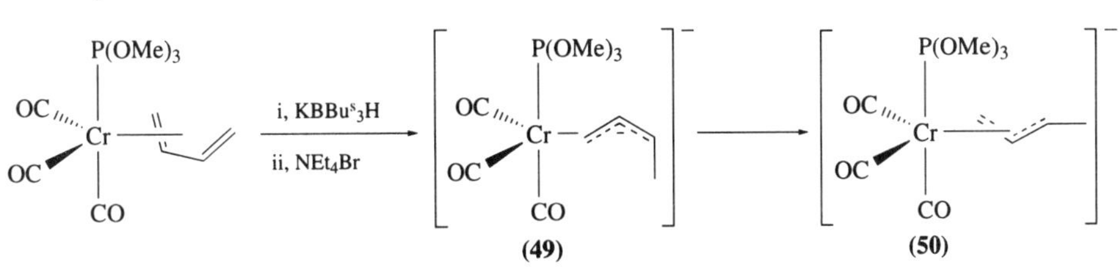

Scheme 7

Reaction of (**51**) with 3,3-dimethylcyclopropene affords[150] (**52**) via alkyne displacement and phenyl migration from sulfur to carbon of the ring-opened cyclopropene. The first examples of oxo-allyl compounds (**53**) ($L = PMe_3$ or $PMePh_2$) have been obtained[151] by reaction of $[WCl_2L_4]$ with $CH_2{=}CHCH(O\text{-}TMS)$ and the bis(arene) complex $[W(\eta^6\text{-}C_6H_5Me)_2]$ provides access to the chemistry of half-sandwich tungsten–arene species including crystallographically characterized $[W(NCMe)_2(\eta^3\text{-}C_3H_5)(\eta^6\text{-}C_6H_5Me)]^+$[152] and $[W(dmpe)(\eta^3\text{-}C_3H_5)(\eta^6\text{-}C_6H_5Me)]^+$.[153]

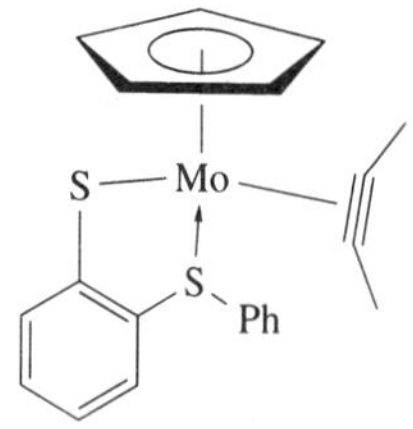

(51)

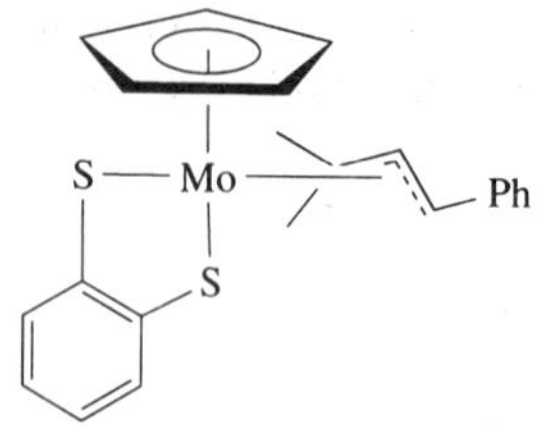

(52)

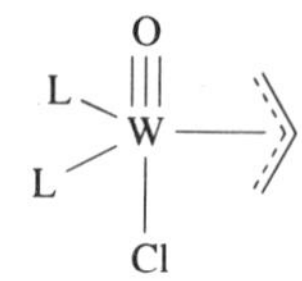

(53)

6.3.7 η^3-Butadienyl, η^3-Cyclopropenyl and Other η^3-Ligands

Several distinct routes to the synthesis of η^3-butadienyl complexes have been reported. The η^2-butatrienyl complex (**54**) reacts with nucleophiles R^- (R = OH, OMe or NEt_2) to give[154] η^3-butadienyl products (**55**), and related complexes (**56**) (R' = OMe[155] or NHMe[156]) are obtained by treatment of $[MoCl(CO)_3(bipy)]^-$ with $ClCH_2C{\equiv}CCH_2Cl$ in the presence of MeOH or NH_2Me. A similar reaction between $[MoCl(CO)_3(bipy)]^-$ and $HC{\equiv}CCH_2Cl$ in methanol affords the η^3-oxacyclobutenyl (**57**).[157] Fluoride anion-induced desilylation of the silyl–butadiene complex $[Mo(CO)_2\{\eta^4\text{-}(2\text{-}SiEt_3)C_4H_5\}(\eta\text{-}C_5Me_5)]^+$ has been found to give high yields of $[Mo(CO)_2(\eta^3\text{-}CH_2{\cdots}CH{\cdots}C{=}CH_2)(\eta\text{-}C_5Me_5)]$ which incorporates the parent, unsubstituted butadienyl ligand[158] but an alternative projected synthesis, starting from $[Mo(CO)_3(\eta\text{-}C_5Me_5)]^-$ and $CH_2{=}C{=}(Me)CH_2Cl$, generates the η^3-cyclobutenone (**58**). Addition of activated alkenes $R_2C{=}C(CN)_2$ (R = CN[159] or CF_3[160]) to the alkynyl ligand of $[W(C{\equiv}CPh)(CO)_3Cp]$ proceeds via ring opening to give (**59**). Crystallographic studies on η^3-butadienyl complexes[155,156,158] reveal a transoid butadienyl ligand with the uncoordinated C-3=C-4 bond length comparable to that in free 1,3-butadiene and theoretical[156] and structural[158] studies suggest a perpendicular orientation of substituents on C-4 with respect to the η^3-allyl plane; typical Mo–C-3 bond lengths are unusually short for a terminal allyl carbon suggesting some σ,η^2-character to the bonding. However, the η^3-butadienyl ligand of (**59**) (R = CN)[159] has also been described as a metallacyclobutane. Theoretical[156,158] and synthetic[154,158] studies demonstrate that nucleophiles add at C-1 and electrophiles at C-4 of the η^3-butadienyl ligand. Complexes of the η^4(5e)-butadienyl ligand are described in Section 6.4.4.

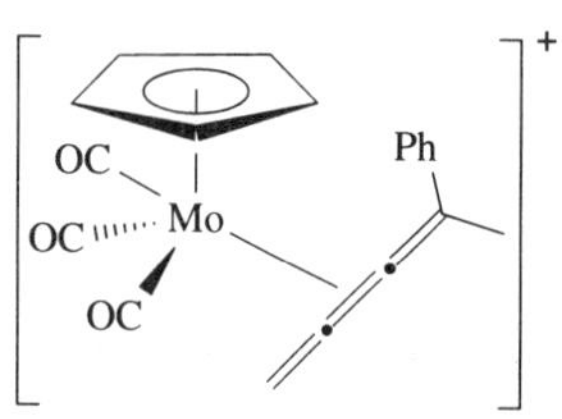

(54)

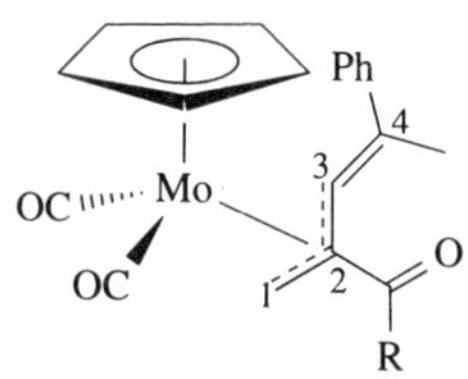

(55)

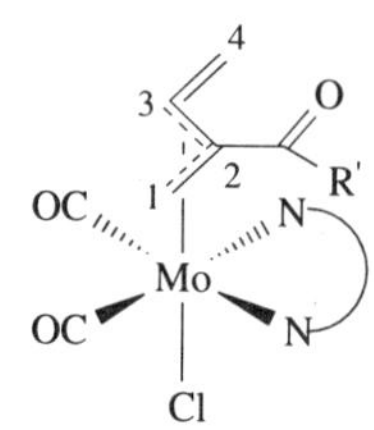

(56)

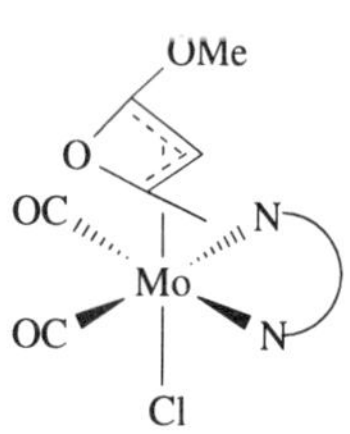

(57)

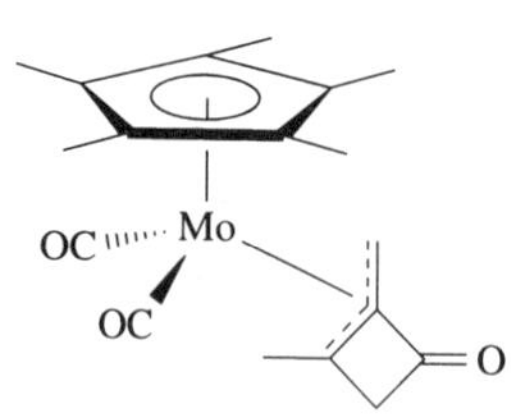

(58)

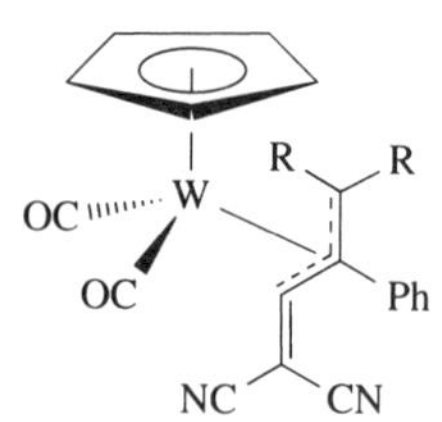

(59)

The η^3-prop-2-ynyl complexes (**60**) (n = 3, 5 or 6) have been obtained by photolysis of $[Mo(CO)_3\text{-}(\eta^6\text{-}C_6Me_nH_{6-n})]$ in the presence of $HC{\equiv}CCH_2OH/HBF_4{\cdot}Et_2O$; reaction with H_2O yields 2-hydroxyallyl products via specific nucleophilic addition to the central propynyl carbon.[161] The synthesis of the prop-2-ynyl complex (**61**)[162] involves dehydrohalogenation of $[MoBr(\eta^2\text{-}MeC{\equiv}CR)_2Cp]$ (R = Me or Ph) with $Li[N(TMS)_2]$. Deprotonation of the alkyne–vinylidene complex (**62**), in the presence of CO, yields the η^3-enynyl (**63**) via alkynyl–vinylidene coupling;[163] in both (**60**) and (**63**) the carbon chain of the η^3-ligand exhibits some localization of the C≡C bond as evidenced by asymmetry in the bond distances C-1–C-2 and C-2–C-3. The η^3-propenoyl complexes (**64**)[164] and $[Mo(CO)(PPh_3)\text{-}\{\eta^3\text{-}CH_2{\cdots}CH{\cdots}C(O)\}\text{-}Cp]$[165] are obtained, respectively, from photolysis of the vinyl complex $[W\{\eta^1\text{-}$

$C(CF_3)$=$C(CF_3)H\}(CO)_3Cp]$ and the acid-catalysed rearrangement of *trans*-[Mo{C-HMe(OMe)}$(CO)_2(PPh_3)Cp]$; both reactions probably involve a vinyl to carbonyl migration.

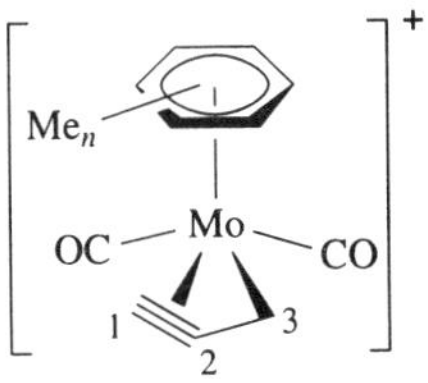

(60)

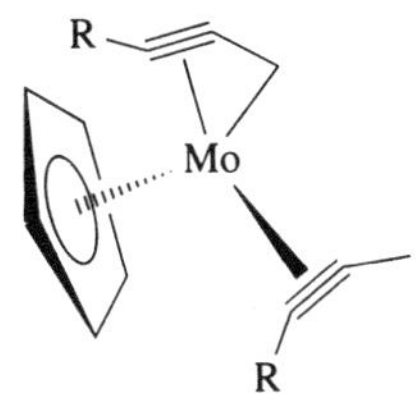

(61)

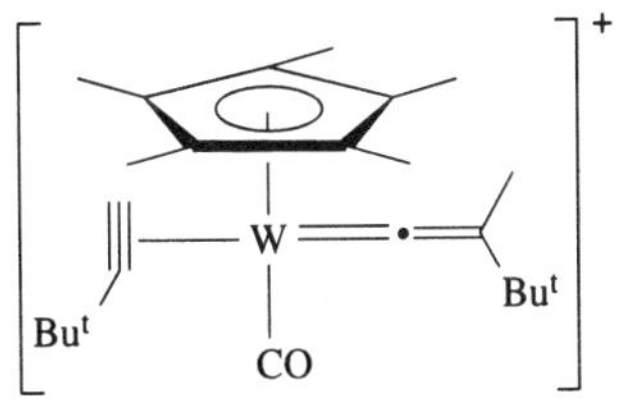

(62)

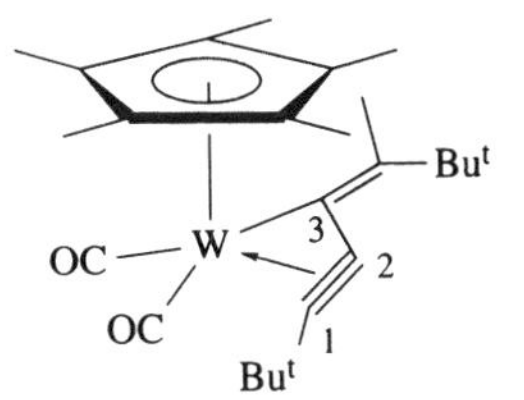

(63)

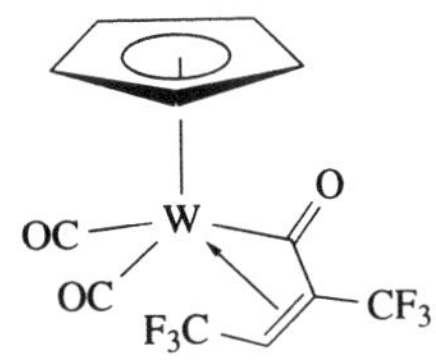

(64)

Complexes (**65**) (M = Mo, R = Ph or Bu^t; M = W, R = Bu^t) of the η^3-cyclopropenyl ligand are conveniently prepared via sequential treatment of $[M(CO)_3(NCMe)_3]$ with cyclopropenium cations $[C_3Ph_2R][PF_6]$ followed by TlCp.[166] A competing process involving CO insertion into the C_3 ring generates η^3-oxocyclobutenyls (**66**); thus, reaction of (**65**) (M = Mo, R = Ph) with PMe_2Ph yields (**66**) (M = Mo; L = PMe_2Ph; R = H).[166] Alternatively, high yields of Cp* derivatives of (**66**) (M = Cr, Mo or W; L = CO; R^2 = Me) are accessible via reaction of $[C_3Ph_3][PF_6]$ with $[M(CO)_3Cp^*]^-$.[167] The formation of η^3-oxocyclobutenyl products appears to be halide ion assisted[166] and reaction of $[Mo(CO)_4(bipy)]$ with $[C_3Ph_3]Br$ affords mixtures of cyclopropenyl and oxocyclobutenyl products.[168] In specific cases tungstenacyclobutadienes may be converted to η^3-cyclopropenyl products; thus, reaction of $[\overline{W\{C(Bu^t)C(Bu^t)C}(Me)\}(Cl)_2Cp]$ with PMe_3 affords (**67**),[169] and $[WCl_3(TMEDA)(\eta^3\text{-}C_3Me_2Bu^t)]$ has been similarly obtained[170] while $[W(O_2CMe)_3(\eta^3\text{-}C_3Et_2Bu^t)]$ is formed directly from the carbyne $[W(O_2CMe)_3(\equiv CBu^t)]$ and EtC≡CEt.[171] A related alkyne–carbyne coupling process may be involved in the formation of $[Mo\{P(OMe)_3\}_2(\eta^3\text{-}C_3Ph_2Me)Cp]$ via treatment of $[Mo\{P(OMe)_3\}_2(\eta^2\text{-}PhC\equiv CPh)Cp]^+$ with CH_2=CHMgBr; the reaction is thought to proceed sequentially through the vinyl species $[Mo(\eta^1\text{-}CH{=}CH_2)\{P(OMe)_3\}(\eta^2\text{-}PhC\equiv CPh)Cp]$ which undergoes α-H elimination to yield the carbyne intermediate $[Mo\{P(OMe)_3\}(\eta^2\text{-}PhC\equiv CPh)(\equiv CMe)Cp]$.[172] Crystallographic studies on the high-oxidation state η^3-cyclopropenyl complexes of tungsten reveal relatively short tungsten–cyclopropenyl carbon distances and a tungstenatetrahedrane description of bonding (as illustrated for (**67**)) may be appropriate in these cases.

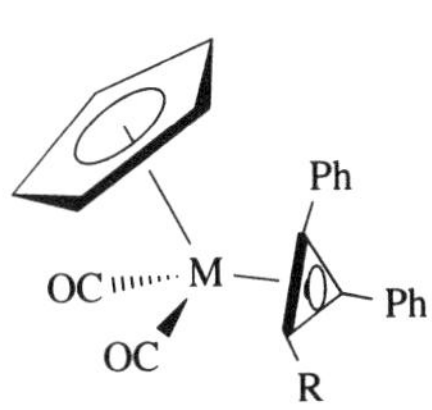

(65)

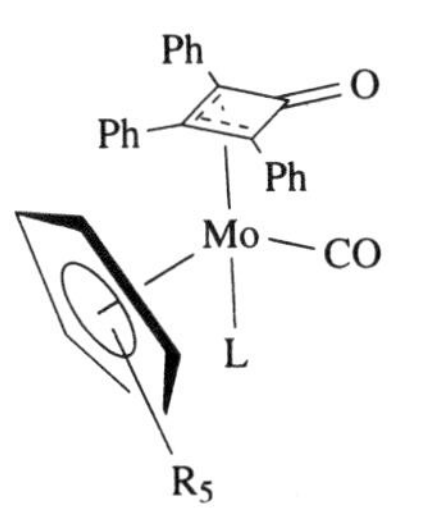

(66)

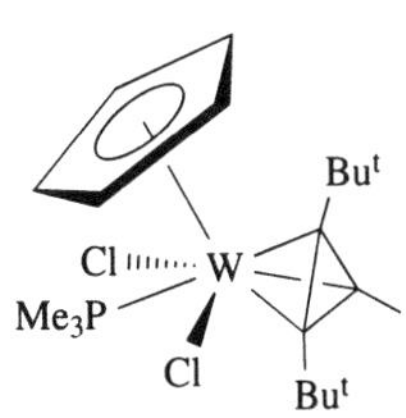

(67)

Complexes of the η^3-1-azaallyl ligand were first synthesized in low yield via reaction of $[Mo_2(CO)_4Cp_2]$ with 2-substituted azirines[173] but the route shown in Scheme 8 has found more general application.[174–6] Crystallographic studies[173–5] reveal a common *endo* arrangement of the azaallyl ligand with C–N azaallyl distances in the range 0.132–0.138 nm. Other examples of trihapto-bonding include complexes of acyclic and cyclic dienyl and of cycloheptatrienyl ligands but these are more conveniently discussed in Sections 6.5.1, 6.5.2 and 6.7.2 since the η^3-bonding mode may be observed as an intimate part of a hapticity interconversion process.

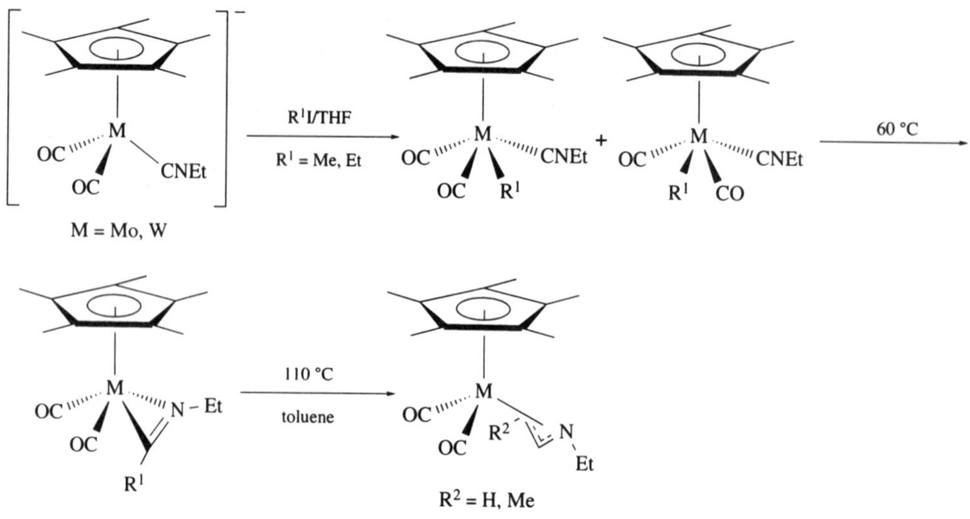

Scheme 8

6.4 η[4]-COMPLEXES

6.4.1 Homoleptic Diene Complexes

A new synthesis of [Mo(η[4]-butadiene)$_3$], not requiring metal-atom vapour techniques, has been developed. Thus, treatment of $MoCl_4\cdot 2THF$ with liquid C_4H_6 in the presence of activated magnesium affords [Mo(η[4]-C_4H_6)$_3$] in 30% yield[177] and related syntheses of [M(η[4]-diene)$_3$] (M = Mo or W; diene = isoprene or 2,3-dimethylbutadiene) have been reported.[178] The *o*-xylidene complex (**68**) is obtained from [WCl_4(O)] and *o*-$C_6H_4(CH_2MgCl)_2$[179] and new syntheses of η[4]-1-oxa-1,3-diene complexes such as (**69**) (M = Mo or W) have been described.[180] Recent structural studies on these trigonal prismatic complexes have been motivated by the details of metal–diene bonding. The methyl-substituted derivative [Mo(η[4]-2,3-Me_2-C_4H_4)$_3$] exhibits[178,181] the expected long-short-long C–C distances along the butadiene backbone, but the corresponding pattern in the unsubstituted complex [Mo(η[4]-C_4H_6)$_3$] has been reported[182] as short-long-short, in common with free butadiene. Proton NMR studies on [W(η[4]-C_4H_6)$_3$] suggest that the fluxional process leading to interconversion of terminal diene protons proceeds via a metallacyclopentene intermediate[183] and ^{183}W and ^{95}Mo chemical shifts have been determined for a series of complexes [M(η[4]-diene)$_3$].[67] The reaction chemistry of [M(η[4]-C_4H_6)$_3$] (M = Mo or W)[177] is dominated by ligand coupling; thus, [Mo(η[4]-C_4H_6)$_3$] reacts with CO to give [$Mo(CO)_6$] and 1,5,9-cyclododecatriene, probably via the intermediate (**70**).

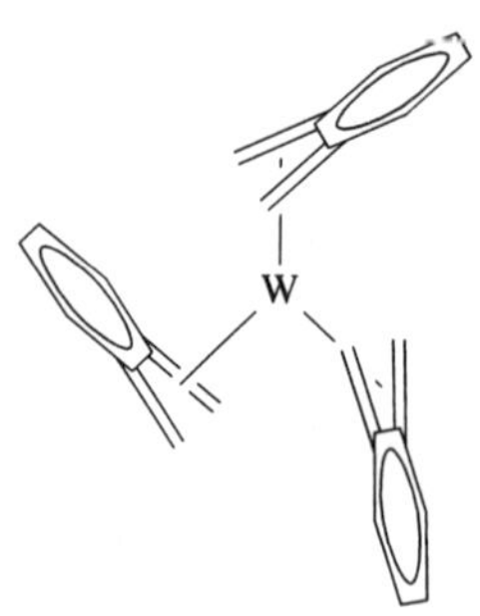

(68)

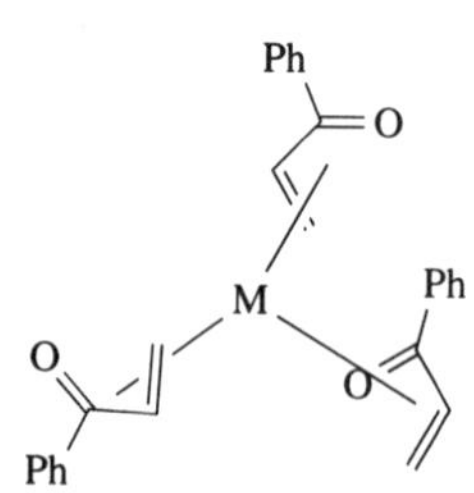

(69)

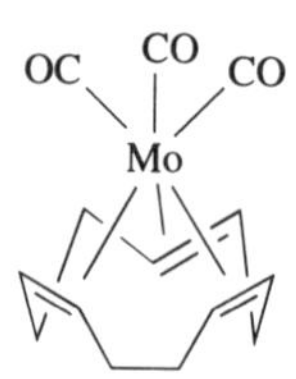

(70)

6.4.2 Diene Carbonyl and Related Complexes

A series of acyclic, conjugated diene complexes [$M(CO)_{4-n}(L)_n$(η[4]-diene)] (M = Cr, Mo or W; L = CO or phosphorus-donor ligand; n = 0–2; diene = buta-1,3-diene, 2-Me-buta-1,3-diene, penta-1,3-diene, (*E,E*)-hexa-2,4-diene, etc.) have been prepared by photolysis of [$M(CO)_{6-n}(L)_n$] with the appropriate diene in solution at −20 °C. The literature in this area has been reviewed[184] and therefore only a list of syntheses is given here. Complex types reported comprise [Cr(CO)$_4$(η^4-diene)],[185] [W(CO)$_4$(η^4-1,4-Ph$_2$-buta-1,3-diene)],[186] [Cr(CO)$_3$(L)(η^4-diene)] (L = PMe$_3$ or P(OMe)$_3$),[187] [W(CO)$_3${P(OMe)$_3$}(η^4-diene)],[188] [Cr(CO)$_2$(L)$_2$(η^4-diene)] (L = PMe$_3$ or P(OMe)$_3$),[189] [Cr(CO)$_2$(PMe$_3$){P(OMe)$_3$}(η^4-diene)],[190] [M(CO)$_2${P(OMe)$_3$}$_2$(η^4-diene)] (M = Mo or W)[191] and [M(CO)$_2$(dmpe)(η^4-diene)] (M = Cr[192] or W[193]). The thermolabile, bis(diene) complexes [M(CO)$_2$(η^4-2-Me-buta-1,3-diene)$_2$] (M = Mo or W) have also been reported.[194] Much interest has centred on the structure and fluxional properties of [M(CO)$_{4-n}$(L)$_n$(η^4-diene)]; in most cases solution NMR studies reveal preferred ligand arrangements as shown for **(71)** (n = 1) and **(72)** (n = 2) and these are supported by crystallographic studies on [Cr(CO)$_3${P(OMe)$_3$}(η^4-buta-1,3-diene)][149] and [Cr(CO)$_2$(PMe$_3$)$_2$(η^4-diene)] (diene = penta-1,3-diene or hexa-2,4-diene).[189]

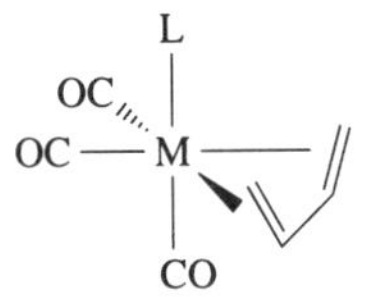

(71)

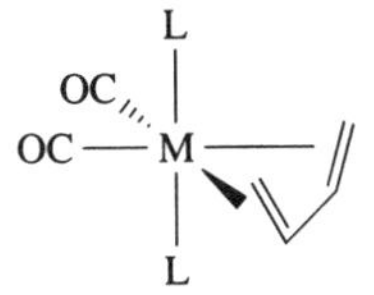

(72)

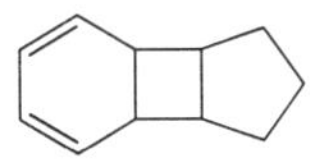

(73)

The majority of complexes [M(CO)$_4$(η^4-cyclodiene)] (M = Cr, Mo or W) incorporate nonconjugated diene ligands such as cod and nbd. However, the substitution of CO for a phosphorus-donor ligand appears to strongly influence bonding preferences. Low-temperature photolysis of [W(CO)$_6$] and cot sequentially affords[195] [W(CO)$_4$(η^4-cot)] and [W(CO)$_3$(η^6-cot)] but photolysis of [W(CO)$_5${P(OMe)$_3$}] with cot yields exclusively [W(CO)$_3${P(OMe)$_3$}(η^4-cot)] as a *fac–mer* mixture.[196] The cycloocta-1,5-diene complexes [M(CO)$_3${P(OMe)$_3$}(η^4-cod)] (M = Cr[197] or W[198]) have been synthesized via photolysis of [M(CO)$_5${P(OMe)$_3$}] and cod; the chromium derivative is isolated as the *mer* isomer but the tungsten analogue is formed as an interconvertible *fac–mer* mixture. Both complexes exhibit exchange of cod for acyclic, conjugated 1,3-dienes suggesting that the binding preference of the M(CO)$_3${P(OMe)$_3$} fragment is the inverse of M(CO)$_4$. These observations may explain the successful syntheses of the conjugated cyclodiene complexes [W(CO)$_{4-n}${P(OMe)$_3$}$_n$(η^4-cyclohexa-1,3-diene)] (n = 1,[188] n = 2[191]). However, a full range of derivatives [M(CO)$_{4-n}${P(OMe)$_3$}$_n${η^4-**(73)**}] (M = Cr, Mo or W; n = 0,[199] 1 or 2[200]) and [M(CO)$_2${η^4-**(73)**}$_2$] (M = Mo or W)[199] has been obtained. The 16-electron complex [WBr$_2$(CO)$_2$(η^4-nbd)] is obtained by treatment of [WBr$_2$(CO)$_4$] with nbd.[201] The diene–carbene complexes [M{=C(OEt)Ar}(CO)$_3$(η^4-nbd)] (M = Mo,[202] Cr or W[203]) are generated by sequential treatment of [M(CO)$_4$(η^4-nbd)] with LiAr (Ar = Ph, C$_6$H$_4$Me-4, C$_6$H$_4$OMe-4, etc.) then Et$_3$O[BF$_4$]; an analogous reaction sequence starting from [W(CO)$_4$(η^4-cod)] yields[203] [W{=C(OEt)Ar}(CO)$_3$(η^4-cod)].

Combined photochemical/matrix isolation investigations on [M(CO)$_4$(η^4-diene)] have elucidated the mechanism of photocatalytic hydrogenation by these complexes. In inert matrices, the predominant photoreaction of [M(CO)$_4$(η^4-diene)] (M = Cr, diene = buta-1,3-diene etc.;[204] M = Cr, Mo or W, diene = nbd[205]) is loss of CO to give *fac*-[M(CO)$_3$(η^4-diene)]; some complications are observed where diene = nbd but, in general, diene displacement only occurs in CO-doped matrices. Photolysis of [M(CO)$_4$(η^4-nbd)] in the presence of nbd and H$_2$ affords a product mixture of norbornene (NBN) and nortricyclene (NTC) which are generated as shown in Scheme 9. Important aspects of the mechanism are the intermediacy of established nonclassical dihydrogen complexes and the specific derivation of NBN and NTC from *mer*- and *fac*-[M(CO)$_3$(η^2-H$_2$)(η^4-nbd)], respectively.[6] Although significant quantities of *mer*-[M(CO)$_3$(η^4-nbd)] are not formed in the primary photoprocess, *fac–mer* isomerization can proceed via [M(CO)$_3$(η^2-nbd)(η^4-nbd)] which undergoes facile loss of the η^2-nbd ligand.[206] The tungsten complexes [W(CO)$_3$(η^2-alkene)(η^4-nbd)] (alkene = nbd,[206] C$_2$H$_4$ or cyclooctene[207]) have been isolated and the preference for a *mer* geometry verified crystallographically.[207] The unstable chromium derivative *mer*-[Cr(CO)$_3$(η^2-C$_2$H$_4$)(η^4-nbd)] has also been isolated and shown to catalyse norbornadiene hydrogenation without photochemical initiation, but a product mixture of NBN and NTC implies rapid *mer–fac* interconversion of the active [Cr(CO)$_3$(η^4-nbd)] species formed by alkene dissociation.[208] A further advance in the understanding of organic transformations based on [Mo(CO)$_4$(η^4-nbd)] involves the stereospecific dimerization of norbornadiene as shown in Scheme 10.[209] The intermediate bis(diene) complex binds both nbd ligands in the *endo* mode resulting in a highly specific *endo-cis-endo* coupling on thermolysis. This process is also operative for some 7-substituted norbornadienes but substituents attached at C-2/C-3 or C-5/C-6 appear to block ligand dimerization[210,211] in isolable [Mo(CO)$_2$(η^4-norbornadiene)$_2$] derivatives. The coupling process is also unsuccessful for [W(CO)$_2$(η^4-nbd)$_2$][209] and with other norbornadiene related ligands.[212]

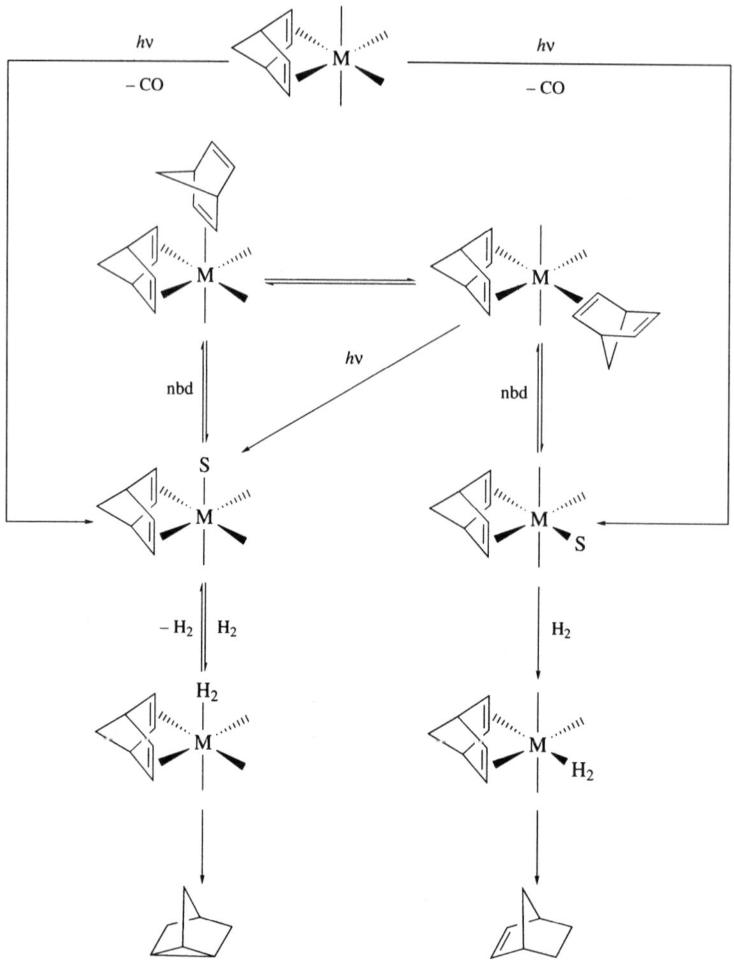

S = solvent NTC NBN

Scheme 9

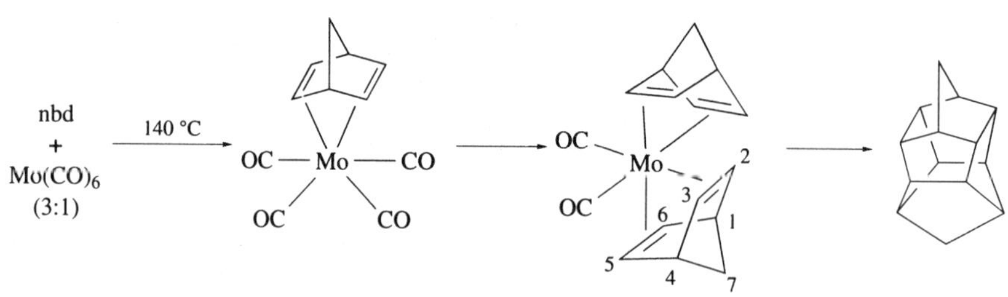

Scheme 10

Dienes such as H_2C=CH–CH=CHMe or cyclohepta-1,3-diene, with methyl or CH_2 groups adjacent to the terminal diene carbon, react photochemically with $[Cr(CO)_5(L)]$ (L = CO, PMe_3 or $P(OMe)_3$) to give $[Cr(CO)_2(L)(\eta^{4:CH}$-diene)] products[184,213,214] in which electron deficiency at chromium is reduced by a three-centre, two-electron C–H–Cr bond. For example, 2,4-Me_2-penta-1,3-diene reacts with $[Cr(CO)_5\{P(OMe)_3\}]$ to give (**74**);[213] crystallographic studies reveal that the diene occupies facial coordination sites with a notably short C-5–Cr distance (0.240 nm). It is probable that '$[CrH(PF_3)_3(C_8H_{11})]$', obtained by cocondensation of 1,5-cod and PF_3 with chromium atoms,[215] is also best formulated with an agostic C–H–Cr bond. A similar phenomenon is observed for (**75**),[216] which, in the solid state, is best described as $[Cr(CO)_2\{P(OMe)_3\}(\eta^{4:(2)}-C_{11}H_{10})]$ indicating that, in addition tc bonding η^4 to the C_7 ring, chromium is also weakly bonded to the arene ring but, in solution, interconversion with (**76**) is observed.

Several noncarbonyl-containing phosphine–diene complexes exist. The species $[M(PMe_3)_2(\eta^4$-buta-1,3-diene$)_2]$ (M = Mo or W) have been obtained via reaction of $[Mo(PMe_3)_6]$ with butadiene[28], ethene

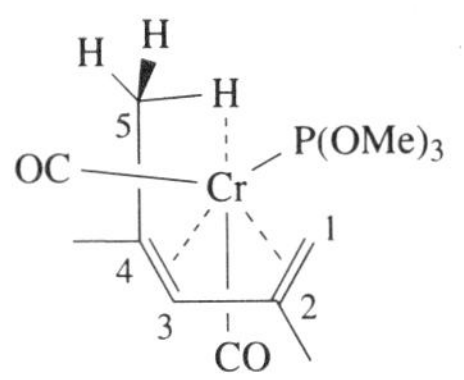

(74)

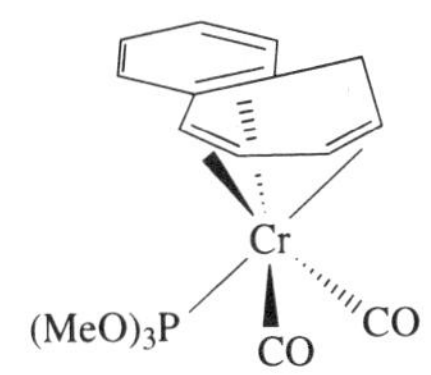

(75)

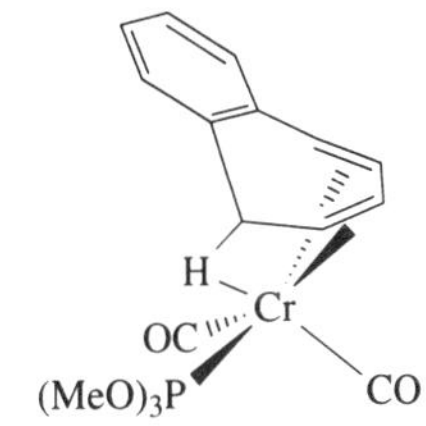

(76)

with $[WH(\eta^2\text{-}CH_2PMe_2)(PMe_3)_4]$[29] or $[WH_6(PMe_3)_3]$ with butadiene.[217] Hydrido–phosphine–diene complexes $[WH_2(PMe_3)_3(\eta^4\text{-diene})]$ (diene = 2,4-Me_2-buta-1,3-diene,[29] cyclopentadiene or cyclohexadiene[217,218]) have also been reported. Treatment of *trans*-$[Cr(N_2)_2(dmpe)_2]$ with buta-1,3-diene affords $[Cr(dmpe)_2(\eta^4\text{-}C_4H_6)]$.[34] The unstable complexes $[Cr(PMe_3)(\eta^4\text{-diene})_2]$ (diene = acyclic diene), $[CrCl(PMe_3)_2(\eta^4\text{-butadiene})]$ and $[\{Cr(\mu\text{-}Cl)(PEt_3)(\eta^4\text{-}2,3\text{-}Me_2\text{-}C_4H_4)\}_2]$ are accessible via reaction of $CrCl_2THF$ or $CrCl_3(THF)_3$ with appropriate dienes and phosphines in the presence of activated magnesium.[219]

6.4.3 Diene Derivatives of Cyclopentadienyl Complexes

The synthesis of cationic complexes $[Mo(CO)_2(\eta^4\text{-cyclodiene})(\eta^5\text{-L})]^+$ (L = Cp, C_5Me_5 or indenyl) and subsequent stereo- and regiospecific reactions with nucleophiles to give η^3-allyl products are discussed in Section 6.3.4. A significant advance in the chemistry of corresponding acyclic diene complexes is the synthesis of η^4-*s*-*trans*-1,3-diene species. As shown in Scheme 11, two synthetic routes are available involving low-temperature, CF_3SO_3H protonation of *syn*-η^3-4-oxo-allyl or *syn*-η^3-pentadienyl complexes.[127,128,220] The essence of the method is that the stereochemistry of the precursor ligand is retained on protonation and thus, correspondingly, precursors with an *anti* configuration at the η^3-ligand afford exclusively η^4-*s*-*cis*-1,3-diene products. The *s*-*trans*-diene complexes depicted in Scheme 11 are thermodynamically unstable with respect to the corresponding *s*-*cis* forms and, in solution above −30 °C, irreversible isomerization occurs, thus impeding structural characterization. However, spectroscopic data are consistent with an authentic π-bonded *trans*-diene ligand as opposed to the alternative metallacyclopentene formulation. A series of *trans*-diene complexes $[Mo(NO)(\eta^4\text{-}trans\text{-diene})Cp]$ (diene = buta-1,3-diene, 2,3-Me_2-buta-1,3-diene, 2,5-Me_2-hexa-2,4-diene, etc.) are accessible by sodium/mercury reduction of $[\{MoI_2(NO)Cp\}_2]$ in the presence of the appropriate diene. Where diene = 2,3-Me_2-buta-1,3-diene, an intermediate *cis*-diene derivative has been isolated but, remarkably, the *trans*-diene is the thermodynamically favoured product. The x-ray crystal structure of $[Mo(NO)(\eta^4\text{-}trans\text{-}2,5\text{-}Me_2\text{-hexa-}2,4\text{-diene})Cp]$ reveals almost equal C–C bond lengths in the η^4-coordinated unit, an observation attributed to extensive back donation into the π^*_3 MO of the diene; theoretical studies suggest that this interaction is most effective with a twisted *trans*-diene ligand.[221] Reactions of $[Mo(NO)(\eta^4\text{-}trans\text{-diene})Cp]$ with phosphines,[221] alkynes[222] and acetone[223] have been investigated.

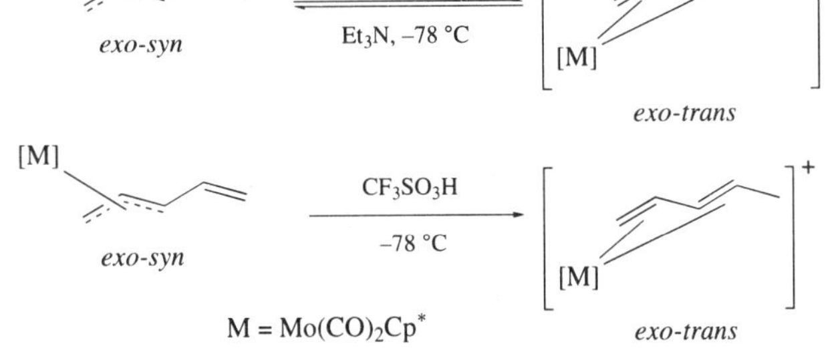

Scheme 11

Photolysis of $[MMe(CO)_3(\eta^5\text{-L})]$ (M = Mo, L = Cp*;[224] M = W, L = Cp[225]) in the presence of conjugated dienes (buta-1,3-diene, 2,3-Me_2-buta-1,3-diene, cyclohexa-1,3-diene, etc.) proceeds with alkyl migration to generate acetyl-η^4-diene products $[M(COMe)(CO)(\eta^4\text{-diene})(\eta^5\text{-L})]$. The corresponding reactions of $[MoMe(CO)_3Cp]$ are less straightforward with the exception of the reaction with cyclopentadiene which affords a product mixture of [Mo(R)(CO)(η^4-C_5H_6)Cp] (R = Me or COMe).[226] An interesting extension to this area involves the photochemical reactions of dienes with [$M(CO)_3$($\eta^{5:1}$-$C_5H_4C_2H_4$)] (M = Mo[227,228] or W[228,229]). In these examples, the stepwise formation of [$M(CO)_2$(η^2-diene)($\eta^{5:1}$-$C_5H_4C_2H_4$)] followed by [M(CO)(η^4-diene)($\eta^{5:1}$-$C_5H_4C_2H_4$)] can be observed but, in selected cases, diene insertion into the M–C σ-bond competes with CO loss from [$M(CO)_2$(η^2-diene)($\eta^{5:1}$-$C_5H_4C_2H_4$)].

Paramagnetic [MoX_2(η^4-diene)Cp] (X = Cl, Br or I; diene = buta-1,3-diene, isoprene or penta-1,3-diene) have been obtained by treatment of [MoX(η^2-$F_3CC{\equiv}CCF_3$)$_2$Cp] with dienes; selected derivatives undergo reversible one-electron reduction and structural studies on [$MoCl_2$(η^4-buta-1,3-diene)Cp] reveal a *cis-endo* conformation of the diene ligand.[230] The synthesis of related, diamagnetic [MoBr{$P(OMe)_3$}(η^4-diene)Cp] is detailed in Scheme 12.

6.4.4 η^4(5e)-Butadienyl, η^4-Cyclobutadiene, η^4-Trimethylenemethane and other η^4-Ligands

Complexes of the η^4(5e)-butadienyl ligand are generated by alkyne coupling in bis(alkyne) species. Protonation of [MoX(η^2-MeC≡CMe)$_2$(η^5-L)] (X = Br or Cl; L = Cp or indenyl) yields the η^2-vinyl-alkyne species (**77**) which, as shown in Scheme 12 reacts with phosphorus-donor ligands to give the η^4(5e)-butadienyl complex (**78**); η^4-vinylallene, (**79**), and η^4-diene products are formed reversibly from (**78**).[231,232] Early work on η^4(5e)-butadienyl complexes demonstrated that HBF_4 protonation of [W(S_2CNEt_2)$_2$(η^2-PhC≡CPh)$_2$] affords (**80**).[233] The structures of (**80**) and (**78**) represent the two extremes of bonding description for the η^4(5e)-butadienyl ligand; crystallographic studies reveal that in (**80**) there is a marked asymmetry in the tungsten–C-2 and tungsten–C-4 distances[233] but a corresponding asymmetry is not observed for (**78**) (X = Br, PR_3 = PMe_3, L = indenyl;[231] X = Br, PR_3 = $P(OMe)_3$, L = Cp[232]). Low-temperature treatment of [Mo(SC_6F_5)(η^2-$F_3CC{\equiv}CCF_3$)$_2$Cp] with $R^1C{\equiv}CR^2$ (R^1, R^2 = H, Me, Ph, CO_2Me) initially affords the unstable, *prone* (*endo*) η^4-butadienyl (**81**), crystallographically characterized for R^1 = Me, R^2 = CO_2Me, in which the carbene carbon C-1 is orientated *trans* to the Cp ligand;[234] subsequent conversion to *supine* (*exo*) butadienyl products such as (**82**), in which C-1 is orientated *trans* to $F_3CC{\equiv}CCF_3$, occurs on increase of temperature.

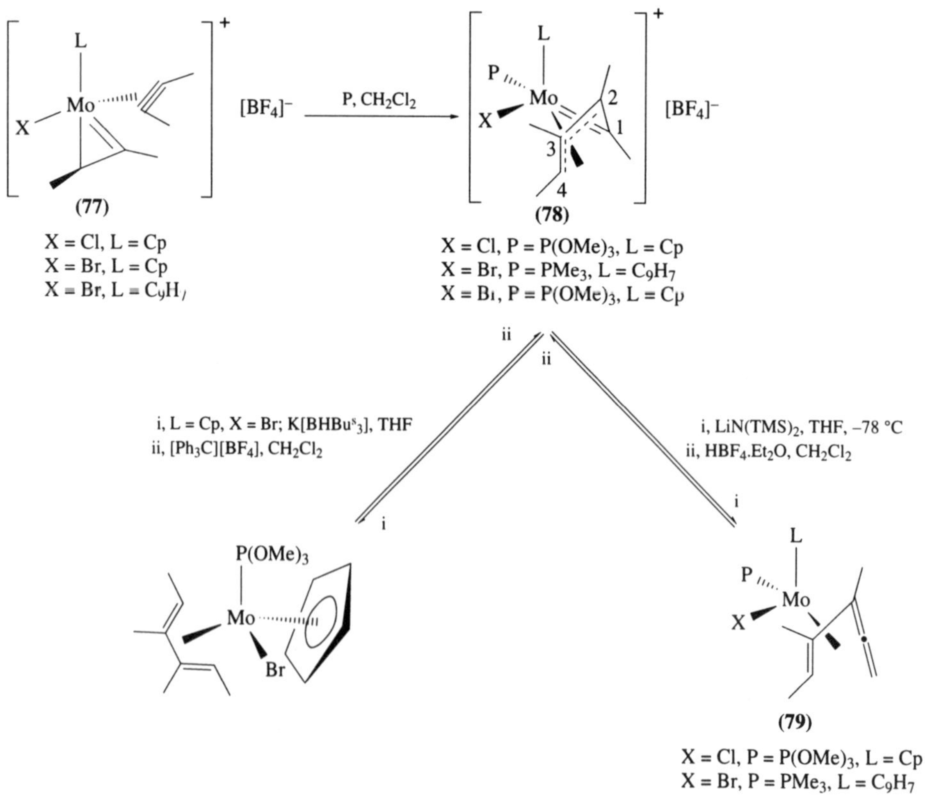

Scheme 12

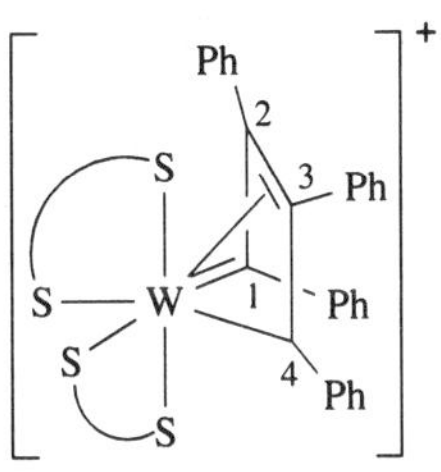

(80)

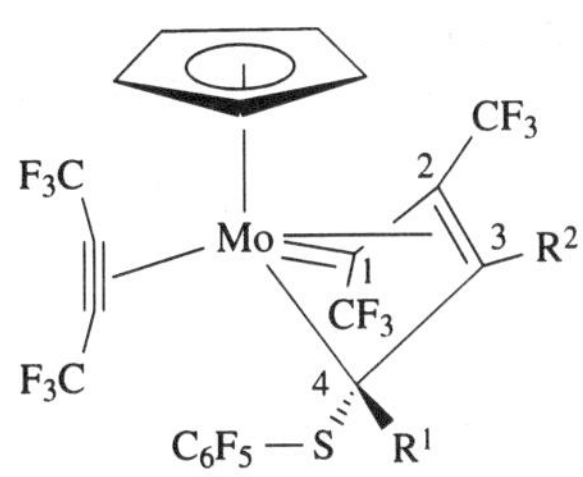

(81)

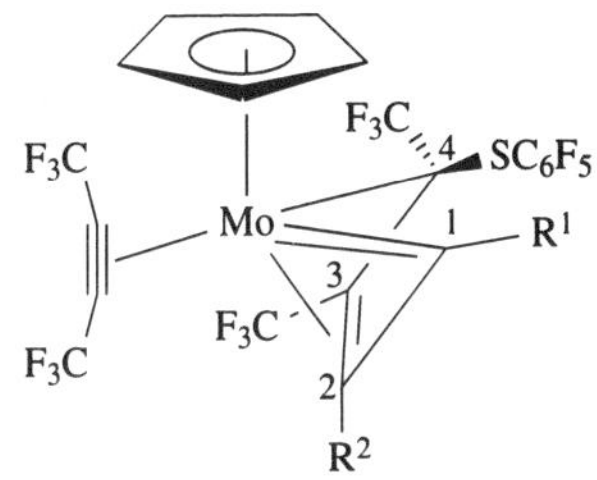

(82)

A common route to cyclobutadiene complexes involves metal-assisted alkyne dimerization and accordingly vacuum thermolysis of $[W(L)(\eta^2\text{-PhC}\equiv\text{CPh})_3]$ (L = CO or NCMe) with excess PhC≡CPh affords the tetraphenylcyclobutadiene products $[W(L)(\eta^2\text{-PhC}\equiv\text{CPh})_2(\eta^4\text{-C}_4\text{Ph}_4)]$.[235] Similarly, $[Cr(CO)_2(\eta^2\text{-PhC}\equiv\text{CPh})(\eta^4\text{-C}_4\text{Ph}_4)]$ has been prepared via reaction of PhC≡CPh with $[Cr(CO)_2(\eta^2\text{-TMS-C}\equiv\text{C-TMS})_2]$,[236] and the tetrakis(trifluoromethyl)cyclobutadiene complexes $[MoI(CO)\{\eta^4\text{-C}_4(CF_3)_4\}Cp]$,[237] $[Mo(S_2CNR_2)\{\eta^4\text{-C}_4(CF_3)_4\}Cp]$ (R = Me or Et)[238] and $[Mo(SC_6F_5)(PMe_2Ph)\{\eta^4\text{-C}_4(CF_3)_4\}Cp]$[239] are formed from $[MoI(\eta^2\text{-F}_3\text{CC}\equiv\text{CCF}_3)_2Cp]$, (**83**) and (**84**) (L = PMe_2Ph), respectively. An alternative construction of a cyclobutadiene ligand, starting from an oxocyclobutenyl complex, is illustrated in Scheme 13.[240]

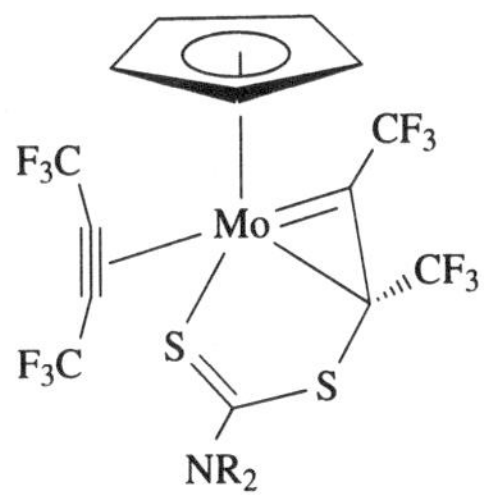

(83)

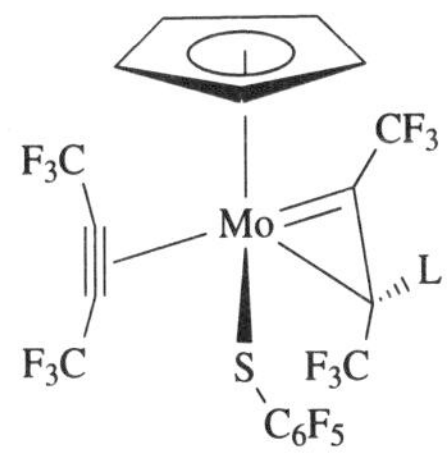

(84)

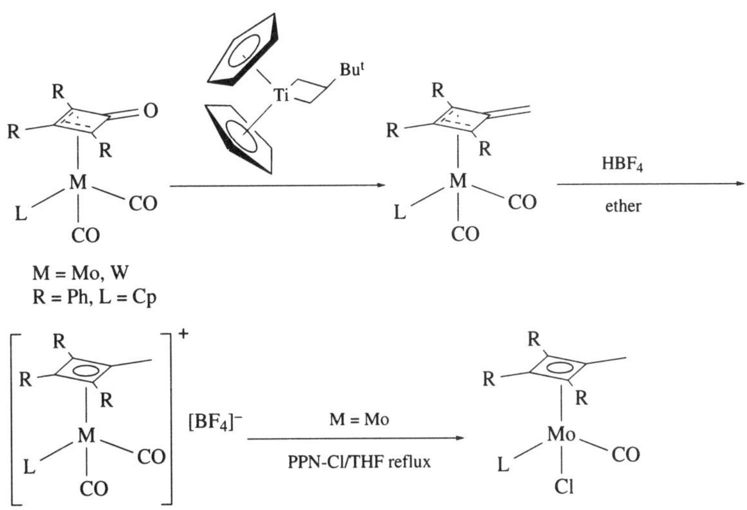

Scheme 13

Several approaches to trimethylenemethane complexes have been developed. Treatment of $[Mo(NCMe)_2(CO)_2Cp]^+$ or $[Mo_2(CO)_6(\eta\text{-Cp*}_2]/Ag[BF_4]$ with methylenecyclopropane or 2,2-dimethylenecyclopropane proceeds with cyclopropane ring opening to give (**85**) (M = Mo; X = H, $R^1 = R^2$ = H or Me; X = Me, $R^1 = R^2$ = H);[241] alternatively, protonation of the *anti*-η^3-1-Me-2-(hydroxymethyl)allyl (**86**)[242] or related complexes with more elaborate 2-hydroxy substituents[243] affords respectively (**85**) (M = Mo, X = H, R^1 = H, R^2 = Me) or complexes with more highly substituted trimethylenemethane ligands. The tungsten derivatives (**85**) (M = W, X = H, $R^1 = R^2$ = Me or Ph) have also been reported.[244] Some reactions of the cationic complexes (**85**) have been investigated; nucleophiles add at the R^1, R^2-substituted terminus of the trimethylenemethane ligand generating

η^3-allyl products[241,244] and deprotonation of (**85**) (M = Mo, X = H, R^1 = H, R^2 = Me) with $LiN(TMS)_2$ yields the 2-vinylallyl complex $[Mo(CO)_2\{\eta^3\text{-}2\text{-}(CH{=}CH_2)C_3H_4\}Cp]$.[242] A novel approach to the trimethylenemethane ligand involves reaction of carbene complexes $[M\{=C(OEt)Ph\}(CO)_5]$ (M = Cr, Mo or W) with substituted allenes $H_2C{=}C{=}CHX$ (X = Ph[245], CH_2OH,[246] etc.) to give isomeric mixtures of (**87**) ($R^1 \neq R^2$ = Ph or OEt); thus exceptionally, the trimethylenemethane ligand is derived from a combination of C_1 and C_3 fragments.

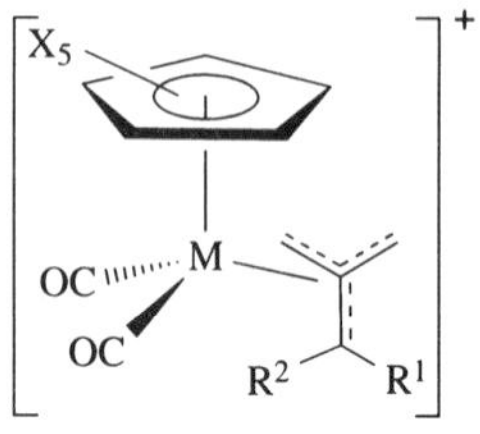

(**85**)

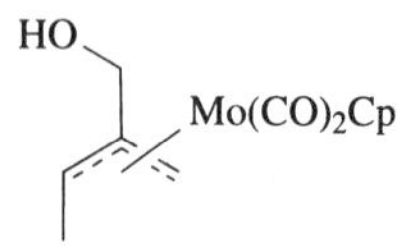

(**86**)

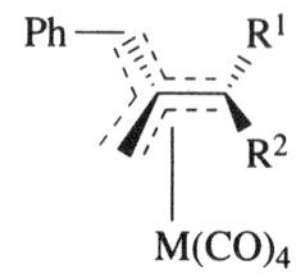

(**87**)

Treatment of $[Mo(CO)_6]$ with tetraphenylcyclopentadienone yields, depending on reaction stoichiometry, either (**88**) or binuclear (**89**);[247] (**88**) is a precursor to $[Mo(CO)_2(L)_2(\eta^4\text{-}C_4Ph_4CO)]$ ($L = MeO_2CC{\equiv}CCO_2Me$; L_2 = dppm or dppe). Tetrakis(trifluoromethyl)-cyclopentadienone complexes $[MoX(CO)\{\eta^4\text{-}C_4(CF_3)_4CO\}Cp]$ (X = Br or I) result from reaction of $[MoX(CO)_3Cp]$ with $F_3CC{\equiv}CCF_3$ in a sealed tube.[237] Addition of 4 -bromo-2-cyclopentenone to $[Mo(NCMe)_3(CO)_3]$ affords (**90**) which is a precursor to the unsubstituted cyclopentadienone complex $[Mo(CO)_2(\eta^4\text{-}C_4H_4CO)Cp]^+$ via sequential treatment with LiCp then $[Ph_3C][PF_6]$; $[Mo(CO)_2(\eta^4\text{-}C_4H_4CO)Cp]^+$ undergoes regiospecific addition of nucleophiles α to the cyclopentadienone CO group.[248]

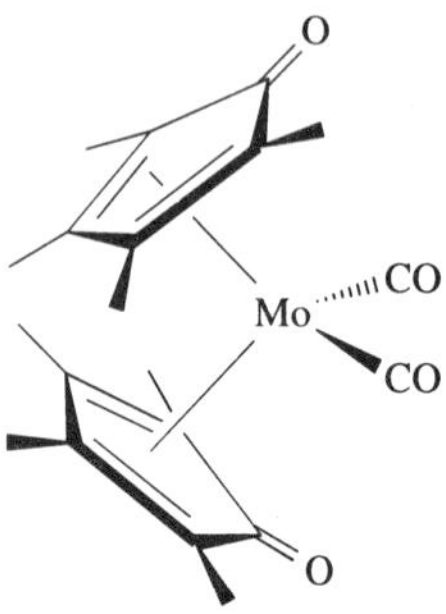

(**88**)

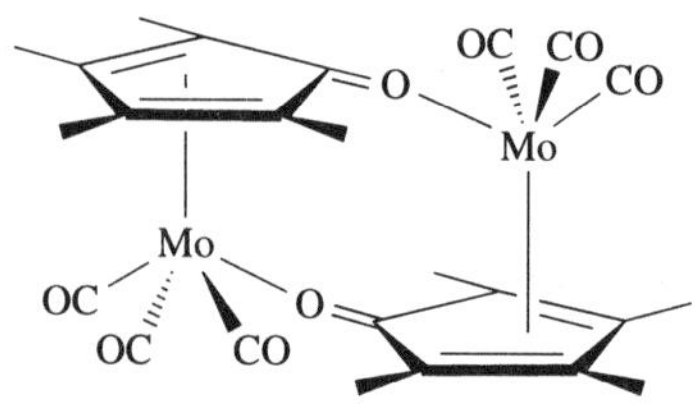

(**89**)

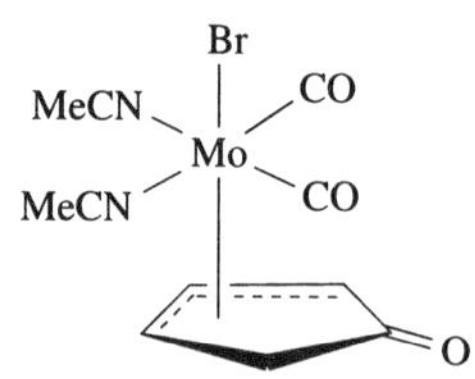

(**90**)

6.5 η^5-COMPLEXES

6.5.1 Acyclic Pentadienyl Complexes

The pentadienyl ligand C_5H_7 and its substituted derivatives possess several properties conducive to an interesting chemistry. The 'open' dienyl system facilitates hapticity interconversions between η^1-, η^3- and η^5- bonding modes and the acyclic character of the ligand imparts conformational flexibility. A general review of acyclic pentadienyl complexes is available.[249]

The anionic half-sandwich complex $[K(diglyme)][Mo(CO)_3(\eta^5\text{-}2,4\text{-}Me_2\text{-}pdyl)]$ (pdyl = pentadienyl) has been prepared. Treatment with I_2 affords $[MoI(CO)_3(\eta^5\text{-}2,4\text{-}Me_2\text{-}pdyl)]$ but reaction with MeI yields (**91**),[250] effectively via trialkylation of CO, and reaction with ICH_2CH_2I affords[251] the metallabenzene (**92**). Complex (**93**), which contains an unusual example of an η^5-'S'-pentadienyl ligand, has been synthesized from $[WBr(CO)_4({\equiv}CMe)]$ and 2-ethenyl-2'-ethynyl-1,1'-biphenyl.[252] Half-sandwich derivatives of chromium, $[CrCl(PR_3)_2(\eta^5\text{-}2,4\text{-}Me_2\text{-}pdyl)]$ ($PR_3 = PMe_3$, $(PR_3)_2$ = dmpe) have been obtained from $K[2,4\text{-}Me_2\text{-}pdyl]$ and $[CrCl_2(PMe_3)_2]$ or $[CrCl_2(dmpe)_2]$; subsequent halide substitution with $MgCl(C_3H_5)$[253] or MeI[254] has been reported.

The pseudooctahedral complexes $[MoX(CO)_2(NCMe)_2(\eta^3\text{-}C_5H_7)]$ (X = Br or Cl) have been synthesized via oxidative addition of 1-halopenta-2,4-dienes to $[Mo(CO)_3(NCMe)_3]$;[255] an analogous reaction with 1-halohexa-2,4-dienes affords corresponding η^3-hexadienyl complexes.[256] The substitution chemistry of $[MoX(CO)_2(NCMe)_2(\eta^3\text{-dienyl})]$ (dienyl = pentadienyl or hexadienyl) resembles that of

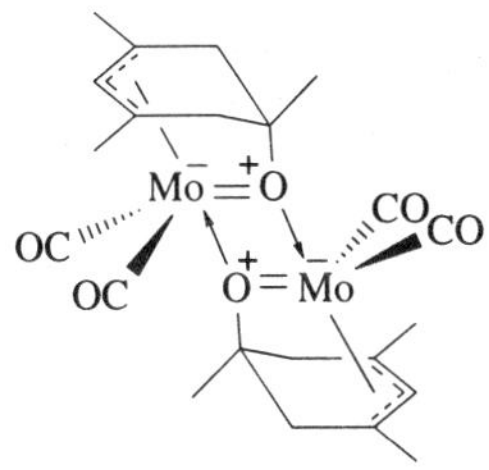

(91)

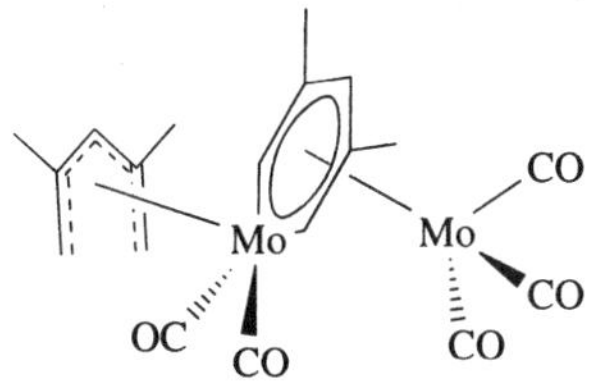

(92)

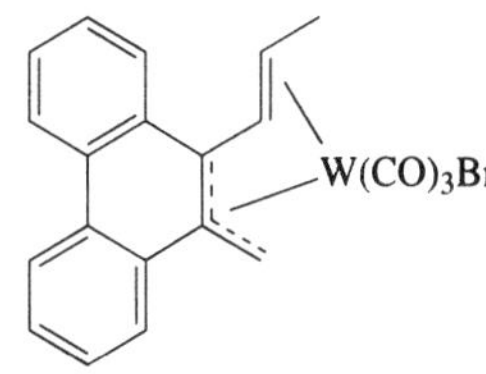

(93)

η^3-allyl analogues and, accordingly, reaction with chelate phosphines[255,256] or amines[256,257] yields $[MoX(CO)_2(L–L)(\eta^3\text{-dienyl})]$ (L–L = dmpe, dppe, bipy or phen). Most examples of $[MoX(CO)_2(L–L)(\eta^3\text{-}C_5H_7)]$ adopt a *syn* (or 'W' shaped) η^3-pentadienyl ligand as in (**94**) but exceptionally, crystallographic studies on (**95**) reveal an *anti* (or 'S') shaped ligand.[255] In CH_2Cl_2, $Ag[BF_4]$ promoted halide abstraction from $[MoBr(CO)_2(P–P)(\eta^3\text{-}C_5H_7)]$ (P–P = dmpe or dppe) affords[258] $[Mo(CO)_2(P–P)(\eta^5\text{-}C_5H_7)]^+$ which incorporate a 'U' shaped η^5-pentadienyl ligand. Reversible coordination of NCMe yields $[Mo(CO)_2(NCMe)(P–P)(\eta^3\text{-}C_5H_7)]^+$ in an illustration of the ease of hapticity interconversions in these complexes; the analogous chemistry of tungsten has also been reported.[259] Reaction of $[MoCl(CO)_2(NCMe)_2(\eta^3\text{-}C_5H_7)]$ with Li(L) (L = Cp or indenyl) generates $[Mo(CO)_2(\eta^3\text{-}syn\text{-}C_5H_7)(\eta^5\text{-}L)]$,[255,260] the Cp derivative of which has also been obtained via photolysis of $[Mo(\eta^1\text{-}C_5H_7)(CO)_3Cp]$.[261] The application of $[M(CO)_2(\eta^3\text{-}syn\text{-pdyl})Cp]$ (M = Mo or W) in the synthesis of *s-trans*-η^4-diene complexes has been discussed in Section 6.4.3.

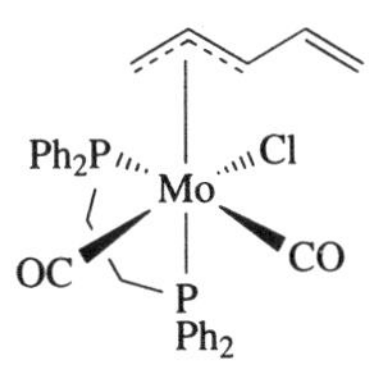

(94)

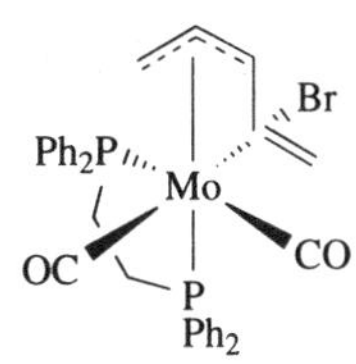

(95)

Crystallographic studies on the 'open' chromocene sandwich complex (**96**) reveal[254] a staggered ligand arrangement with an average Cr–C distance (0.216 3(3) nm) almost identical to that in $[CrCp_2]$. A derivative of (**96**), in which the pentadienyl ligands are linked by a CH_2CH_2 bridge, has also been reported and shown to exhibit enhanced reactivity towards ligand addition at chromium.[262] The 18-electron PEt_3 adduct, $[Mo(PEt_3)(\eta^5\text{-}2,4\text{-}Me_2\text{-pdyl})_2]$, in which one pentadienyl ligand adopts an η^5-'S' conformation, has been obtained[263] but attempts to synthesize an analogous chromium complex resulted[264] in the formation of dimeric (**97**) probably via coupling of η^5-'S'-pentadienyl ligands in the mononuclear intermediate. The formation of 'half-open' chromocene derivatives $[Cr(\eta^5\text{-pdyl})\ (\eta^5\text{-L})]$ (pdyl = C_5H_7, 2,4-Me_2-Cp, etc.; L = Cp or Cp*) from a reaction mixture of '$CrCl_2$', K[pdyl] and NaCp or KCp*, is surprisingly selective. An x-ray structural determination on $[Cr(\eta^5\text{-}C_5H_7)Cp^*]$ defines a 'U' shaped pentadienyl ligand with the average Cr–C(pdyl) distance of 0.211 7(3) nm shorter than the corresponding Cr–C(Cp*) distance (0.219 4(3) nm). In contrast with $[CrCp_2]$ and (**96**), $[Cr(\eta^5\text{-pdyl})Cp]$ readily forms stable, 18-electron monoadducts $[Cr(L)(\eta^5\text{-pdyl})Cp]$ (L = CO, isocyanide, etc.) with an η^5-'S' pentadienyl ligand.[265] Analogous molybdenum complexes have been obtained via extended photolysis of $[Mo(CO)_2(\eta^3\text{-}syn\text{-}C_5H_7)Cp]$[261] or hydride addition to (**98**).[266] One-electron oxidation of $[Cr(CO)(\eta^5\text{-}2,4\text{-}Me_2\text{-pdyl})Cp]$ affords the corresponding 17-electron radical cation in which the 'U' conformation of the pentadienyl ligand is regenerated; an unexpected retardation of CO substitution rate resulting from one-electron oxidation has been attributed to the operation of a sterically impeded associative mechanism in the cationic species.[267]

6.5.2 Cyclohexadienyl and Cycloheptadienyl Complexes

Most cyclohexadienyl complexes of chromium, molybdenum or tungsten are obtained from arene precursors. Addition of R^- to $[M(CO)_3(\eta^6\text{-arene})]$ (M = Cr, R = $\overline{C(Me)S(CH_2)_3S}$;[268] M = W, R = H[269]) affords unstable, anionic tricarbonyl-η^5-cyclohexadienyl products which undergo subsequent electrophilic addition at the metal centre; the chromium complexes have been employed in the metal-assisted synthesis of *trans*-disubstituted cyclohexadienes. An unusual extension to this chemistry is

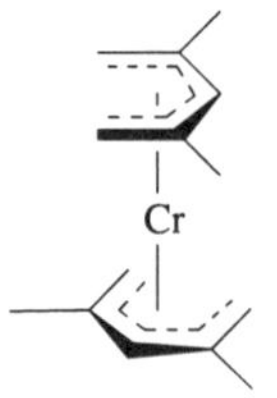

(96)

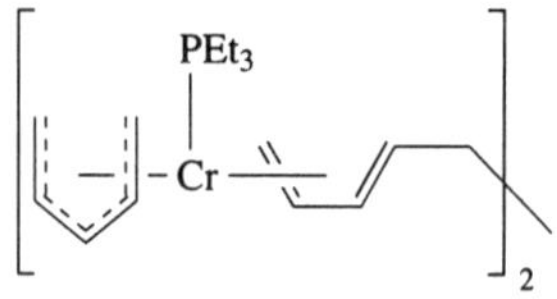

(97)

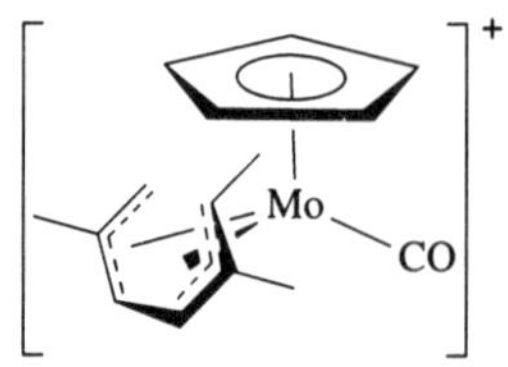

(98)

provided by η^5-1-Me-thiabenzene-1-oxide complexes.[270,271] Hydride addition to [W(dmpe)(η^3-allyl)(η^6-toluene)]$^+$ is directed to the C_6 ring providing a synthesis of neutral cyclohexadienyl complexes.[272] An alternative entry to cyclohexadienyl chemistry involves reaction of the 6,6-Me_2-cyclohexadienyl anion, (dmCh)$^-$, with [Mo(CO)$_3$(diglyme)] to give isolable [K(diglyme)$_n$][Mo(CO)$_3$(η^5-dmCh)], which exhibits an extensive reaction chemistry leading to syntheses of [MoI(CO)$_3$(η^5-dmCh)], [Mo$_2$(CO)$_4$(η^5-dmCh)$_2$] (Mo≡Mo) and [Mo(CO)$_2$(NO)(η^5-dmCh)].[273]

Most cycloheptadienyl complexes of the group 6 transition metals are derived from cycloheptatriene (chpt). The half-sandwich [Mo(CO)$_4$(η^5-C_7H_9)]$^+$ is obtained by HBF_4 protonation of the cycloheptatriene precursor [Mo(CO)$_3$(η^6-C_7H_8)] under a CO atmosphere and reacts with NCMe to yield the trihapto-bonded cycloheptadienyl species [Mo(CO)$_2$(NCMe)$_3$(η^3-C_7H_9)]$^+$, which is precursor to a range of substituted products [Mo(CO)$_2$(L)$_2$(η^5-C_7H_9)]$^+$ (L = PPh_3, CNBu^t; L_2 = dppm, bipy or diene). Reversible coordination of NCMe to [Mo(CO)$_2$(bipy)(η^5-C_7H_9)]$^+$ affords [Mo(CO)$_2$(NCMe)(bipy)(η^3-C_7H_9)]$^+$, thus exemplifying the potential for $\eta^3 \rightleftharpoons \eta^5$ hapticity interconversions akin to acyclic dienyl ligands.[274] Reaction of [Mo(CO)$_2$(NCMe)$_3$(η^3-C_7H_9)]$^+$ with NaCp affords [Mo(CO)$_2$(η^3-C_7H_9)Cp], which is also accessible via deprotonation of the cycloheptadiene complex [Mo(CO)$_2$(η^4-C_7H_{10})Cp]$^+$;[117] analogous species [M(CO)$_2$(η^3-C_7H_9)(η-C_5H_4Me)] (M = Mo[275] or W[276]) are formed on photolysis of [M$_2$(CO)$_6$(η-C_5H_4Me)$_2$] with chpt. The sandwich complexes [M(η^5-C_7H_9)(η^7-C_7H_7)] (C_7H_7 = cycloheptatrienyl; M = Mo[277] or W[278]) are formed by respective sodium/mercury reduction of $MoCl_5$ or WCl_6 in the presence of chpt; the molybdenum derivative is also accessible via intramolecular hydrogen migration in the bis(cycloheptatriene) complex [Mo(η^6-C_7H_8)$_2$].[279] Both [M(η^5-C_7H_9)(η^7-C_7H_7)] (M = Mo or W) undergo displacement of the cycloheptadienyl ligand providing syntheses of a range of half-sandwich cycloheptatrienyl complexes[277,278] and the molybdenum derivative has been reversibly oxidized to the corresponding radical cation.[279]

6.6 η^6-COMPLEXES

6.6.1 Complexes of Cycloheptatriene and Related Ligands

Basic syntheses of cycloheptatriene complexes of the group 6 transition metals are well developed but a few modifications of interest have been reported. The reaction of 1,3,5,7-Me_4-chpt with [Mo(CO)$_6$] proceeds as shown in Scheme 14; the isomeric product mixture arises from a series of 1,5-shifts of the 7-*endo* hydrogen of (**99**);[280] there is some evidence that such 1,5-shifts are metal-assisted, since matrix isolation studies on [Mo(CO)$_3$(η^6-C_7H_8)], cocondensed with matrix gas at 12 K, advocate formation of (**100**).[281] Reaction of chpt with [W(CO)$_3$(NCR)$_3$] (R = Et or Pr^n) provides an improved synthesis[282] of [W(CO)$_3$(η^6-C_7H_8)] and sodium/mercury amalgam promoted reductive coupling of the cycloheptatrienyl ring in [Mo(CO)$_3$(η^7-C_7H_7)]$^+$ proceeds stereospecifically to yield (**101**) (M = Mo) in which the hydrogens attached to each bridgehead carbon are arranged *endo* with respect to the Mo(CO)$_3$ group.[283] In contrast to an early report, cocondensation of molybdenum atoms with chpt affords both the mixed sandwich [Mo(η^5-C_7H_9)(η^7-C_7H_7)] and the novel bis(cycloheptatriene) sandwich [Mo(η^6-C_7H_8)$_2$]. One-electron oxidation of [Mo(η^6-C_7H_8)$_2$]$^{n+}$ (n = 0) generates the corresponding 17-electron radical (n = 1) and both species isomerize via intramolecular hydrogen migration to [Mo(η^5-C_7H_9)(η^7-C_7H_7)]$^{n+}$; unexpectedly, however, isomerization of the neutral species (n = 0) proceeds more rapidly than for the radical (n = 1).[279] Other syntheses of note are the formation of carbenes [M{=C(OEt)Ar}(CO)$_2$(η^6-C_7H_8)] (M = Cr, Ar = $C_6H_4CF_3$-4; M = Mo, Ar = Ph, $C_6H_4CF_3$-4, etc.)[202] and the generation of (**102**), which contains an exceptional example of cycloheptatriene acting as a μ-η^2:η^2 ligand.[284] Entry into the chemistry of half-sandwich cycloheptadienyl complexes via protonation of [Mo(CO)$_3$(η^6-C_7H_8)] in the presence of CO has been discussed in Section 6.5.2 and the synthesis of a range of ring-substituted cycloheptatriene complexes [M(CO)$_3${η^6-(7-*exo*-R)C_7H_7}] via addition of nucleophiles R$^-$ to [M(CO)$_3$(η^7-C_7H_7)]$^+$ is described in Section 6.7.1.

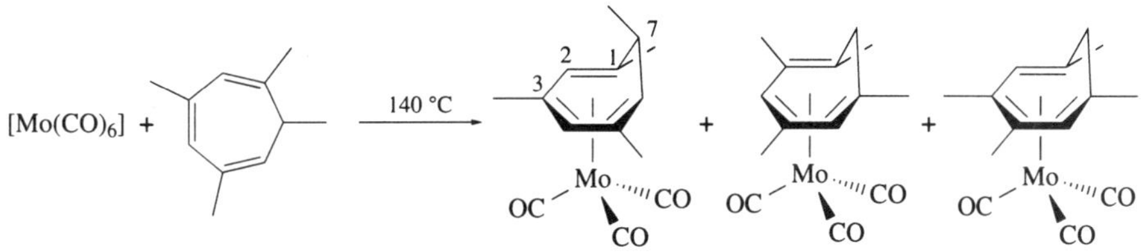

(99)

Scheme 14

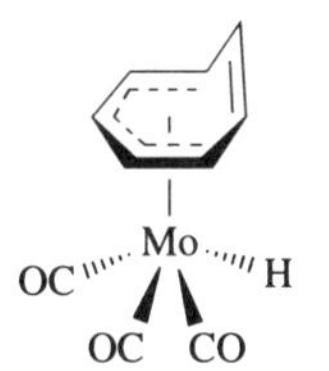

(100)

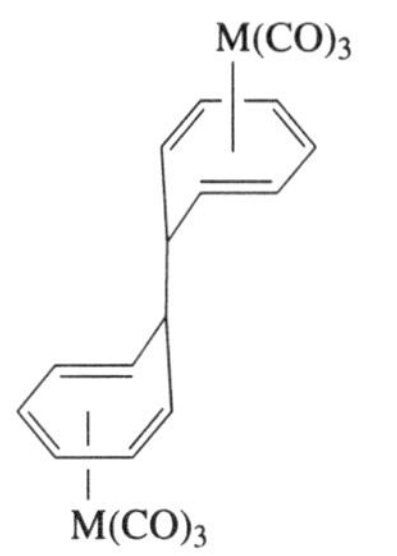

(101)

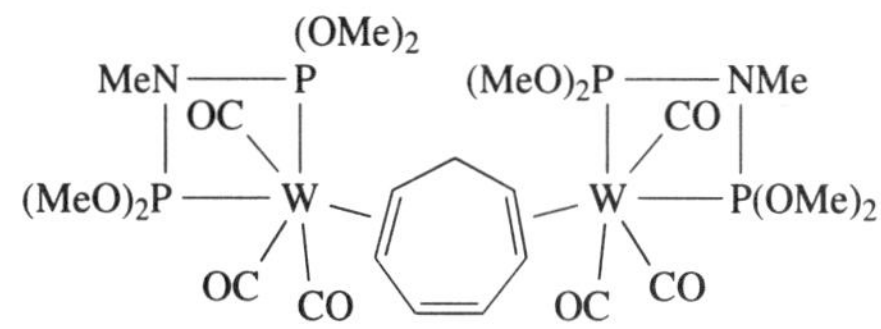

(102)

Recent physical studies on $[M(CO)_3(\eta^6\text{-}C_7H_8)]$ comprise matrix isolation IR,[285,286] NMR and thermochemical investigations. In the ring-substituted complexes $[Cr(CO)_3\{\eta^6\text{-}(7\text{-}exo\text{-}R)C_7H_7\}]$ (R = CN or Bu^t) the barrier to rotation of the $Cr(CO)_3$ group with respect to the cycloheptatriene ring, as determined from ^{13}C NMR spectra, is R-dependent; electron-withdrawing substituents appear to facilitate rotation via stabilization of a norcaradiene type transition state.[287] The ^{95}Mo chemical shift of $[Mo(CO)_3(\eta^6\text{-}C_7H_8)]$ has been reported as δ −1675;[288] by contrast the corresponding shift in $[Mo(\eta^6\text{-}C_7H_8)_2]$ is δ +358 and this significant deshielding correlates with a decrease in complex stability.[289] Thermochemical studies concerned with the strength of the M–cycloheptatriene bond of $[M(CO)_3(\eta^6\text{-}C_7H_8)]$ continue. There is general agreement from enthalpies of reaction with I_2[290] (M = Cr) and pyridine[291] (M = Mo), as measured by solution calorimetry, that, in contention with earlier reports, the stability of $[M(CO)_3(\eta^6\text{-}L)]$ lies in the order (L = benzene < toluene < chpt) and accordingly chpt can substitute toluene in $[Mo(CO)_3(\eta^6\text{-toluene})]$.[291] However, in agreement with earlier investigations, solution calorimetry studies on ligand substitution reactions suggest that metal to cycloheptatriene bond strengths in $[M(CO)_3(\eta^6\text{-}C_7H_8)]$ increase in the order Cr < Mo < W.[292]

A significant development in the chemistry of $[Cr(CO)_3(\eta^6\text{-}C_7H_8)]$ involves photochemically-induced, metal-assisted [6 + 2]- and [6 + 4]-cycloadditions at the cycloheptatriene ring and the basic principles of this work hold promise of application in a range of organic transformations.[293] Photolysis of $[Cr(CO)_3(\eta^6\text{-}C_7H_8)]$ with alkynes RC≡CR affords the [6 + 2] adducts (**103**) (R = TMS,[294] Ph or C_6H_4Me-4[295]) from which the organic cycloadducts can be released. Reports on the photochemical reactions of $[Cr(CO)_3(\eta^6\text{-}C_7H_8)]$ and its 8,8-Me_2-heptafulvene congener (**104**) with conjugated dienes are more involved and a detailed review of some of this work is available.[184] At 263 K, photolysis of $[Cr(CO)_3(\eta^6\text{-}C_7H_8)]$ in the presence of acyclic dienes (buta-1,3-diene and its methyl-substituted derivatives) affords [6 + 4] adducts (**105**) via C–C bond formation at C-1 and C-6 of coordinated chpt.[296] An analogous [6 + 4] adduct is formed between (**104**) and buta-1,3-diene but sterically bulky dienes such as 2,3-Me_2-buta-1,3-diene yield complexes (**106**) which are converted to the normal [6 + 4] adduct on reaction with CO which induces formation of a second C–C bond.[297] Analogous [6 + 4] adducts, albeit with some decomplexation of $Cr(CO)_3$, are generated via photochemical reaction of buta-1,3-diene with ligand-bridged binuclear complexes (**101**) (M = Cr)[298] and its heptafulvalene congener (**107**);[299] the addition proceeds stepwise but bis(adducts) are formed preferentially. However, reaction of (**101**) (M = Cr) with 2,3-Me_2-buta-1,3-diene yields (**108**) as the major product.[298] Matrix isolation/IR studies on photolysis products derived from $[M(CO)_3(\eta^6\text{-}C_7H_8)]$ (M = Cr[285,300] or Mo[286]) provide some insight into the mechanism of the formal [6 + 4]-addition with acyclic dienes. In inert matrices, at 12 K, the primary photoprocess for $[M(CO)_3(\eta^6\text{-}C_7H_8)]$ involves CO dissociation with reversible formation of $[M(CO)_2(\eta^6\text{-}C_7H_8)]$; in CO doped matrices conversion to $[M(CO)_6]$ proceeds via $[M(CO)_4(\eta^4\text{-}C_7H_8)]$. Subsequent studies involving acyclic dienes support the mechanism depicted in Scheme 15[300] but [6 + 4]-additions can also be initiated thermally and the requirement for CO dissociation remains in contention.[293]

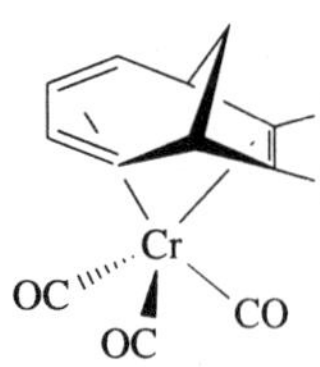

(103)

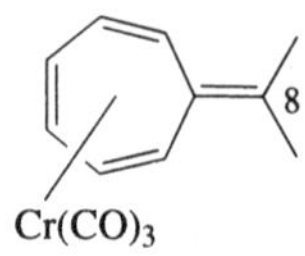

(104)

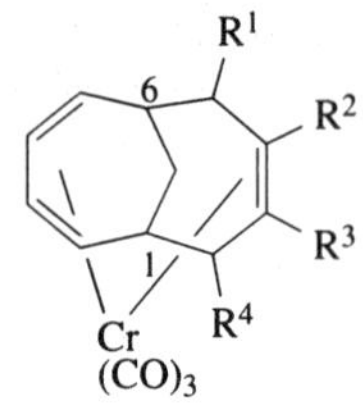

(105)

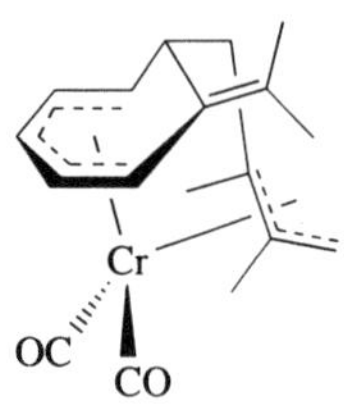

(106)

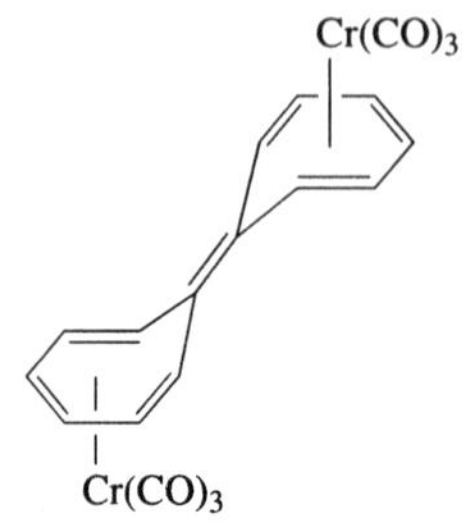

(107)

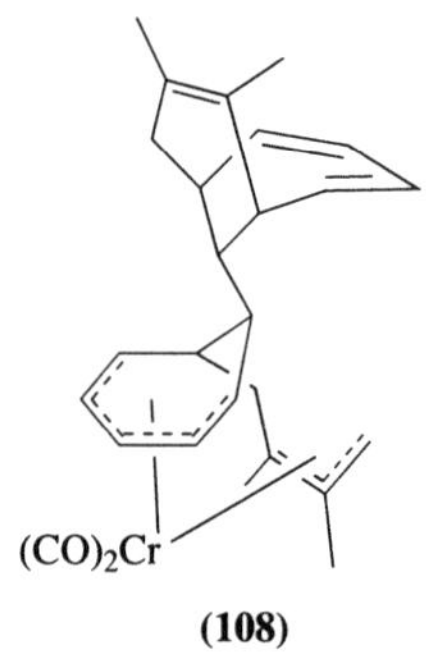

(108)

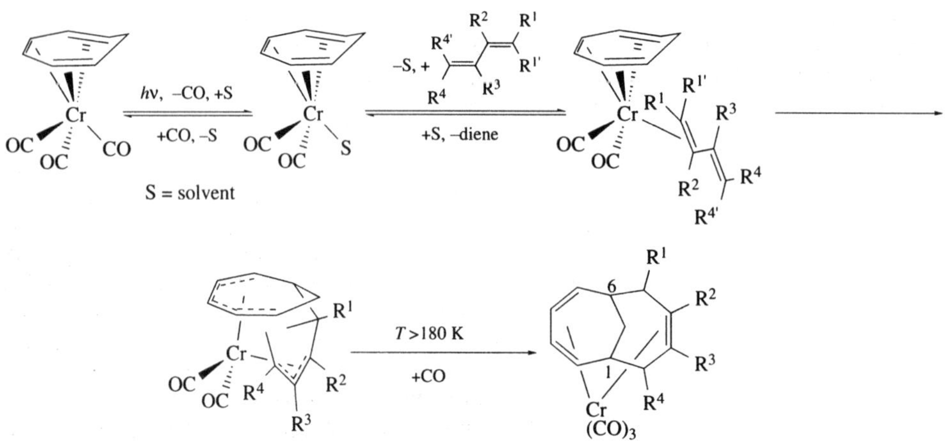

(105)

Scheme 15

The corresponding reactions of $[Cr(CO)_3(\eta^6\text{-}C_7H_8)]$ with cyclic dienes are less straightforward; although a normal [6 + 4]-addition is observed with (**73**),[301] reaction with cyclohexa-1,3-diene proceeds with 1,5-H shifts from C_6- to C_7-derived fragments to give (**109**).[302] Photolysis of $[Cr(CO)_3(\eta^6\text{-}C_7H_8)]$ and cyclopentadiene results in displacement of chpt but with substituted fulvenes, 6,6-R^1,R^2-C_6H_4, products of type (**110**) are formed via C–C bond formation between C-1 of chpt and C-6 of the fulvene;[303] identical products are obtained in the inverse reaction of $[Cr(CO)_3(\eta^6\text{-}6,6\text{-}Me_2\text{-fulvene})]$ with 7-substituted cycloheptatrienes[304] and reaction of (**104**) with 6,6-substituted fulvenes also yields related species.[305]

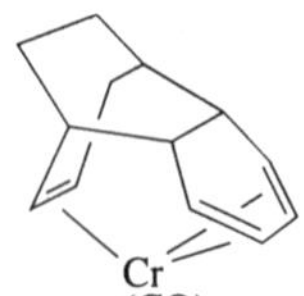

(109)

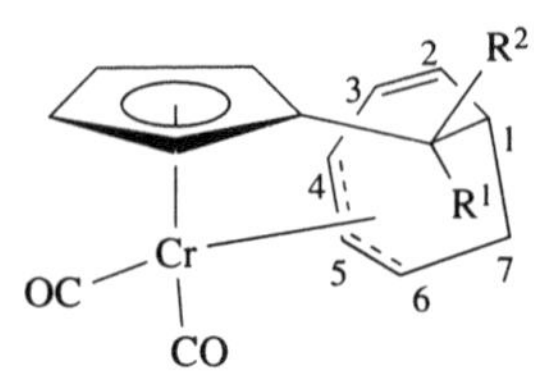

(110)

Corresponding diene additions to $[M(CO)_3(\eta^6\text{-}C_7H_8)]$ (M = Mo or W) are either not observed,[296] or proceed in poor yield,[293] but $M(CO)_3$ groups can be complexed to the preformed ligands of [6 + 4]-

addition at a $Cr(CO)_3$ centre. For example, decomplexation of the organic ligand in (**109**) and subsequent reaction with [$Mo(CO)_3$(diglyme)] affords the molybdenum analogue of (**109**) as a mixture with the rearranged formal [6 + 4] adduct (**111**).[302] The tropone complexes (**112**) (M = Cr, R = H, Me, etc.) also exhibit [6 + 4]-additions with acyclic, conjugated dienes.[293] The ketonic base strength of tropone is significantly increased via complexation as in (**112**) (M = Cr, Mo or W);[306,307] this effect probably originates from metal-assisted stabilization of a hydroxycycloheptatrienyl product (**113**) (R^2 = H), for which the structure of the methylated derivative (**113**) (M = Cr, R^1 = H, R^2 = Me) has been established crystallographically.[308]

(**111**)

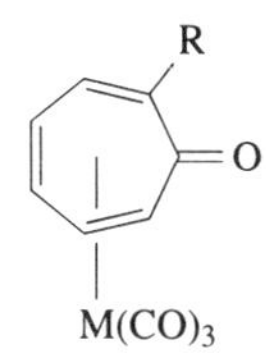

(**112**)

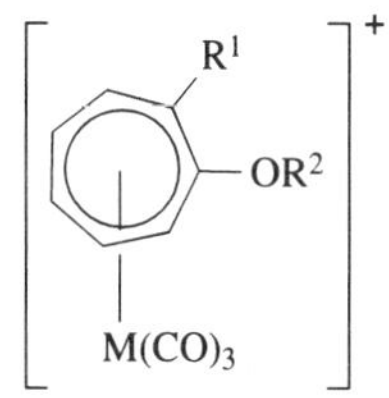

(**113**)

6.6.2 Complexes of Cyclooctatetraene and Other η^6-Ligands

The cyclooctatetraene complexes [$M(CO)_3(\eta^6\text{-}C_8H_8)$] (M = Cr, Mo or W) are relatively labile towards displacement of the cot ring but some reactivity of coordinated cot in the chromium derivative has been explored. Friedel–Crafts acylation of [$Cr(CO)_3(\eta^6\text{-}C_8H_8)$] gives the substituted arene complex [$Cr(CO)_3(\eta^6\text{-}C_6H_5R)$] (R = CH=CH(COMe)),[309] probably via ring contraction of the ring-substituted intermediate [$Cr(CO)_3\{\eta^6\text{-}C_8H_7(COMe)\}$] and photochemical reaction of [$Cr(CO)_3(\eta^6\text{-}C_8H_8)$] with 6,6-$R^1$,$R^2$-substituted fulvenes affords ligand-coupled products (**114**).[310] Investigations on the mechanism of metal migration in [$M(CO)_3(\eta^6\text{-}C_8H_8)$] continue. There is general agreement from two-dimensional NMR exchange spectroscopy (2D-EXSY)[311] and theoretical studies on potential energy surfaces[312] that, at least for the chromium derivative, migration occurs via 1,2- and 1,3-metal shifts with the 1,3-shift as the preferred process. However, the nature of the transition states remains under debate; a novel proposal[311] implicates the bicyclo[4.2.0]octatriene (**115**) as a common intermediate for 1,2- and 1,3-shifts. Limiting low-temperature ^{1}H NMR studies on the ring-substituted derivatives [$Cr(CO)_3(\eta^6\text{-}C_8H_7R)$] (R = Me or Ph) conclude that the preferred substituent position is at a carbon of the uncoordinated C=C bond.[309] The benzocyclooctatetraene complex (**116**) exhibits a thermally initiated, intramolecular haptotropic shift of the $Cr(CO)_3$ group from the C_8 to the C_6 ring.[313] Treatment of [$\{CrCl(\mu\text{-}Cl)(TMEDA)\}_2$] with K_2cot followed by $M'C_5R_5$ (M' = Na, R = H; M' = Li, R = Me) affords the paramagnetic sandwich complexes [$Cr(\eta\text{-}C_5R_5)(\eta^6\text{-}cot)$].[314] However, where R = H, the major product is the 'fly over' bridge species (**117**) which has been thermally isomerized to (**118**).[314,315] Reaction of [$Cr(\eta\text{-}C_5R_5)(\eta^6\text{-}cot)$] with [$M(CO)_3(NCEt)_3$] yields paramagnetic, 31-valence-electron products (**119**) (R = H, M = Cr or W; R = Me, M = Cr); the heterobimetallic derivative exhibits an extensive redox chemistry.[316]

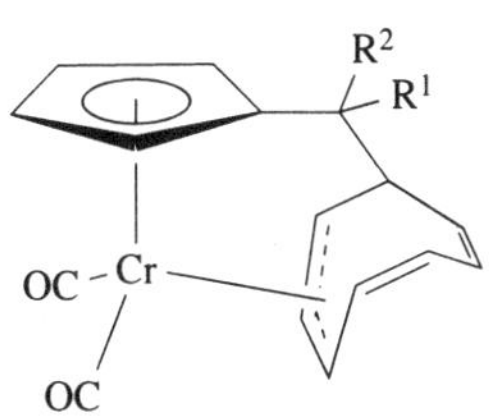

(**114**)

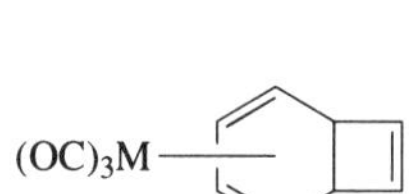

(**115**)

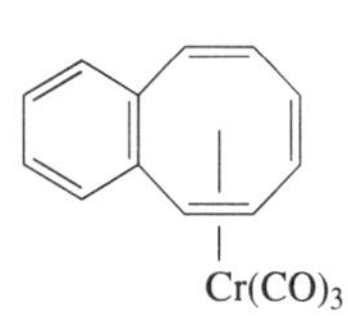

(**116**)

Further studies on bi- and polycyclic η^6-bonded ligands have been reported. X-ray crystallographic studies on the bicyclo[6.1.0]nona-2,4,6-triene complex (**120**) reveal that the cyclopropane ring is orientated *syn* with respect to the metal and moreover the 9-Br derivative has structure (**121**). The mechanism of thermal rearrangement of (**120**) to (**122**) has been reconsidered and it now seems probable that the function of the $Mo(CO)_3$ group is the selective inhibition of rearrangement pathways available to the uncomplexed hydrocarbon.[317] The coordination of $M(CO)_3$ to exocyclic C=C bonds has also stimulated attention. The η^6-3,4-dimethylenecyclobutene complex (**123**) exhibits significant bending of the exocyclic C=C bonds towards chromium[318] and the 9-methylenebicyclo[4.2.1]nona-2,4,7-triene

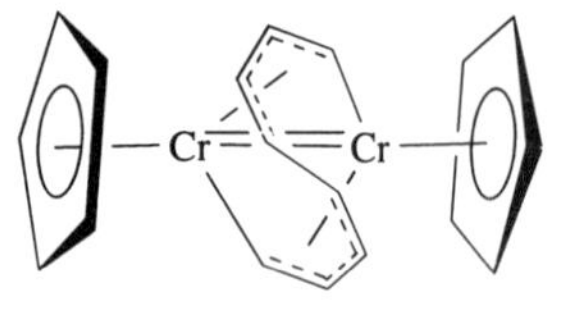

(117)

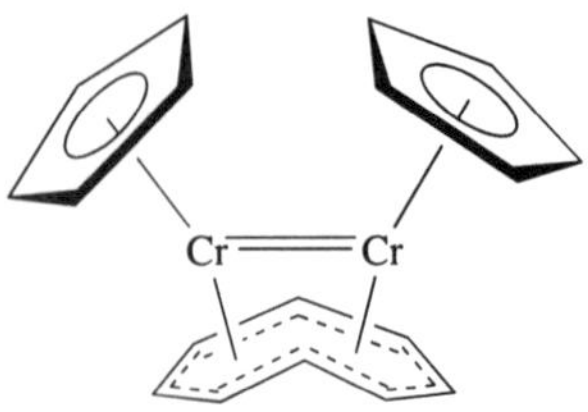

(118)

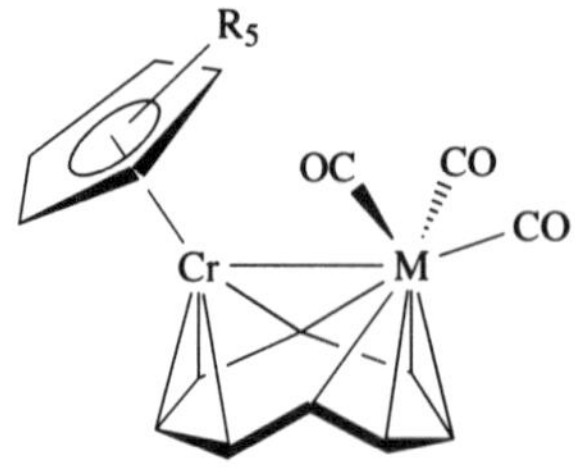

(119)

ligand of (**124**) exhibits preferential coordination of the exocyclic C=C to chromium.[319] A similar result is obtained on coordination of $Cr(CO)_3$ to (**125**) but η^6-coordination of $Mo(CO)_3$ is less selective.[320]

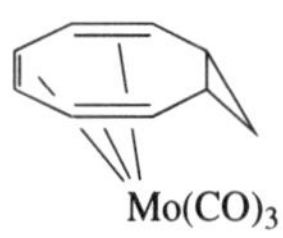

(120)

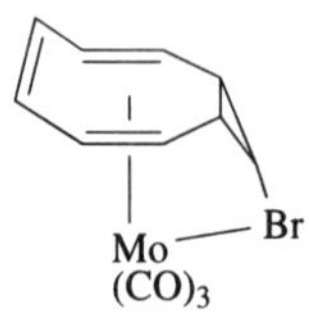

(121)

(122)

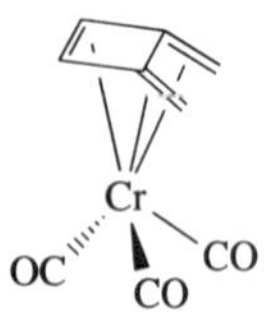

(123)

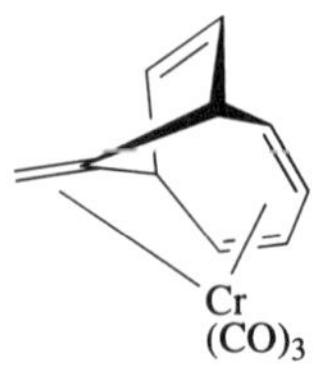

(124)

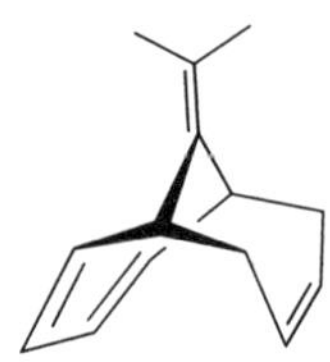

(125)

6.7 η^7-COMPLEXES

6.7.1 $[M(CO)_3(\eta^7\text{-Cycloheptatrienyl})]^+$ Complexes

Syntheses of $[M(CO)_3(\eta^7\text{-}C_7H_7)]^+$ (M = Cr, Mo or W) via $[Ph_3C]^+$ initiated, H^- abstraction from cycloheptatriene precursors $[M(CO)_3(\eta^6\text{-}C_7H_8)]$ are well established but a recent development is the formation of ring-substituted derivatives (**126**) (M = Mo[280] or W[321]). A potential problem in syntheses of ring-substituted cycloheptatrienyl complexes is the inaccessibility of the 7-*endo*-H in species of type (**99**) towards reaction with $[Ph_3C]^+$ but this has been overcome by thermally induced rearrangement as shown in Scheme 14 or alternatively via a novel direct addition of the substituted cycloheptatrienyl ligand to $[W(CO)_3(NCEt)_3]$.[321] Some general chemistry of (**126**) (M = Mo) involving ligand substitution at the molybdenum centre has been described.[280] Other derivatives of interest are the η^7-1-phenylborepin complex (**127**) in which the Mo–B distance (0.252 3(3) nm) is slightly longer than the typical Mo–C(ring) distance (0.233–0.242 nm)[322] and the anionic benzocycloheptatrienyl species (**128**), which is an exceptional example of a group 6 transition metal η^5 bonded to a cycloheptatrienyl type ring.[323] Cycloheptatrienyl ring rotation in solid-state samples of $[M(CO)_3(\eta^7\text{-}C_7H_7)][BF_4]$ (M = Cr or Mo) has been probed by NMR and vibrational spectroscopy; the activation energy of ~13 kJ mol^{-1} is almost metal independent.[324] Electrochemical investigations on the mechanism of reduction of $[M(CO)_3(\eta^7\text{-}C_7H_7)]^+$ (M = Cr, Mo or W) to ligand-bridged, binuclear (**101**) advocate a coupling process between $[M(CO)_3(\eta\text{-}C_7H_7)]^-$ and $[M(CO)_3(\eta^7\text{-}C_7H_7)]^+$ which is impeded by ring substituents.[325]

Much interest has centred on the kinetics and mechanism of nucleophilic addition of the methoxide anion to $[M(CO)_3(\eta^7\text{-}C_7H_7)]^+$; the overall process is probably best described by Scheme 16 which illustrates three potential sites of addition. In the cases M = Mo or W, there is good spectroscopic evidence for the low-temperature generation of the alkoxymetal adduct (**129**); subsequent warming affords the unstable carbomethoxy species (**130**) and the thermodynamically favoured ring-adduct $[M(CO)_3\{\eta^7\text{-}(7\text{-}exo\text{-}OMe)C_7H_7\}]$ (**131**).[326] Attempts have been made to rationalize the observed pattern of addition via MO calculations[327] but it is probable that (**129**) and (**130**) are not required intermediates in the formation of (**131**) in view of the *exo* stereochemistry of ring addition and the results of a stopped

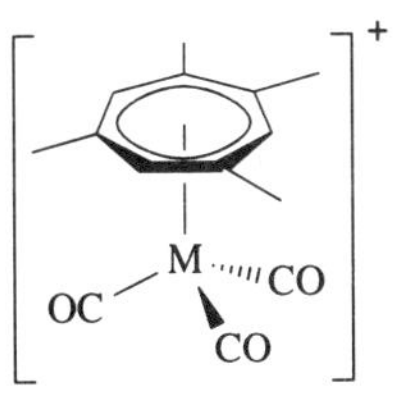

(126)

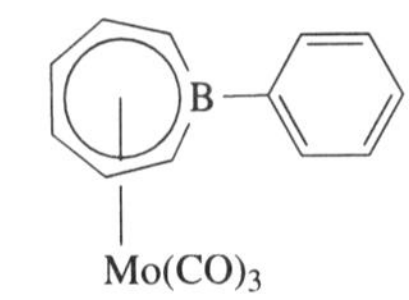

(127)

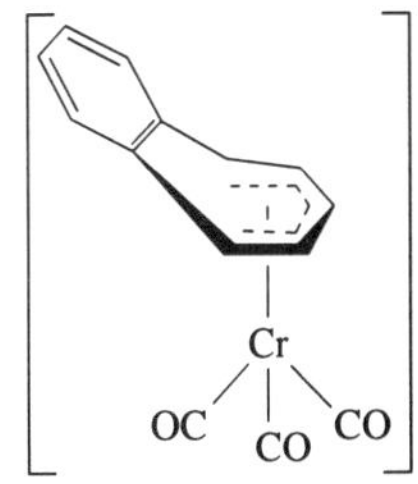

(128)

flow kinetics study[328] which are consistent with Scheme 16. There is good evidence that the kinetics of addition of methoxide to $[M(CO)_3(\eta^7\text{-}C_7H_7)]^+$ to give (**131**) are metal dependent although a consistent trend is elusive.[328,329] However, it is clear that metal complexation significantly reduces the susceptibility of $[C_7H_7]^+$ to nucleophilic addition; in MeOH, $[Cr(CO)_3(\eta^7\text{-}C_7H_7)]^+$ is approximately 2×10^3 times less reactive than free, uncoordinated $[C_7H_7]^+$ towards addition of MeOH.[330] Other recent kinetics studies on reactions of $[M(CO)_3(\eta^7\text{-}C_7H_7)]^+$ with nucleophiles comprise addition of PPh_3[331] or primary aromatic amines, NH_2R (R = Ph, C_6H_4Me-4, etc.)[332] to $[W(CO)_3(\eta^7\text{-}C_7H_7)]^+$; the former reaction results in C_7H_7 displacement, probably via sequential PPh_3 addition at tungsten, while the latter affords neutral ring adducts $[W(CO)_3\{\eta^6\text{-}(7\text{-}exo\text{-}NHR)C_7H_7\}]$.

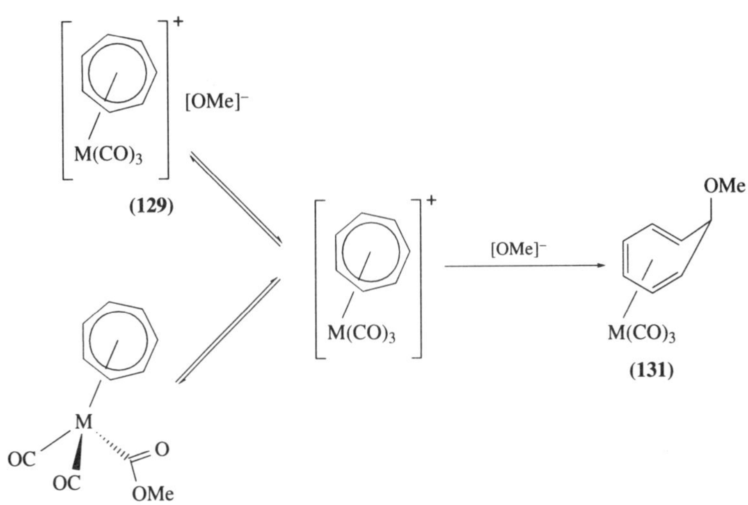

Scheme 16

An extensive series of ligand-bridged bimetallic complexes (**132**) [$M^2 = Re(CO)_5$;[333] $M^2 = Ru(CO)_2Cp^*$[334]], (**133**) (M^1 = Cr, Mo or W);[335] (**134**) (M = Cr or Mo),[336] and (**135**) (M^1 = Cr or Mo, M^2 = Cr or W)[337] have been synthesized by nucleophilic addition of appropriate metalloanions to $[M^1(CO)_3(\eta^7\text{-}C_7H_7)]^+$. However, treatment of $[M^1(CO)_3(\eta^7\text{-}C_7H_7)]^+$ with $[Mn(CO)_5]^-$ or $[W(CO)_3Cp]^-$ yields, respectively, $[Mn_2(CO)_{10}]$ or $[W_2(CO)_6Cp_2]$ together with (**101**).[338] Reaction of $[Cr(CO)_3(\eta^7\text{-}C_7H_7)]^+$ with $Ph_2P(CH_2)_3PPh_2/[Fe(CO)_2(THF)Cp]^+$ affords (**136**), which exhibits a novel bridging mode of a bidentate phosphine ligand.[339]

6.7.2 $[MX(L)_2(\eta^7\text{-}C_7H_7)]$ (X = Halide, L = CO or Phosphorus-donor Ligand) and Related Complexes

The chemistry of this area is, almost without exception, restricted to molybdenum and tungsten. The ^{95}Mo NMR chemical shift of $[MoI(CO)_2(\eta^7\text{-}C_7H_7)]$ has been reported[288] as δ −1348 and fast scan rate cyclic voltammetric studies on $[MoX(CO)_2(\eta^7\text{-}C_7H_7)]$ reveal a one-electron oxidation with $E°$ in the range +0.43–0.46 V with respect to the $FeCp_2/[FeCp_2]^+$ couple.[340] The complexes $[MoR(CO)_2(\eta^7\text{-}C_7H_7)]$ (R = η^1-C ligand) are generally highly unstable but, in addition to the archetypal $[Mo(C_6F_5)(CO)_2(\eta^7\text{-}C_7H_7)]$, $[MR(CO)_2(\eta^7\text{-}C_7H_7)]$ (M = Mo, R = Me or Ph;[341] M = Mo or W, R = C≡CPh[342]) are now known. The unstable complexes $[MoR(CO)_2(\eta^7\text{-}C_7H_7)]$ (R = Me or Ph) undergo facile SO_2 insertion into the

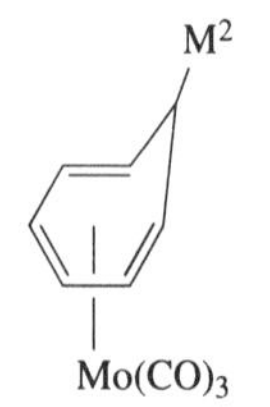

(132)

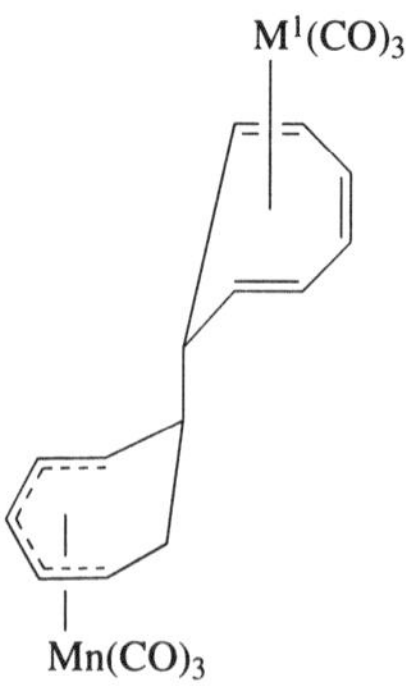

(133)

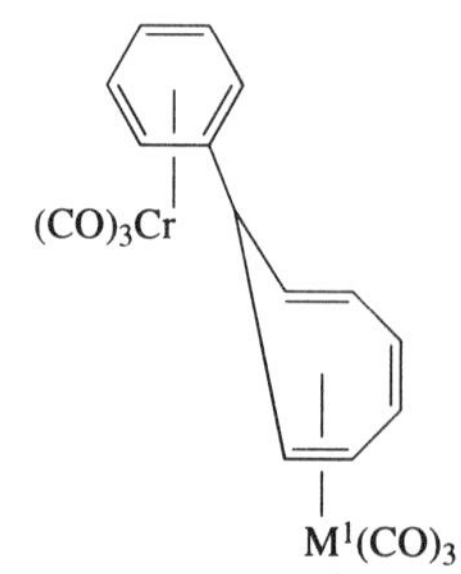

(134)

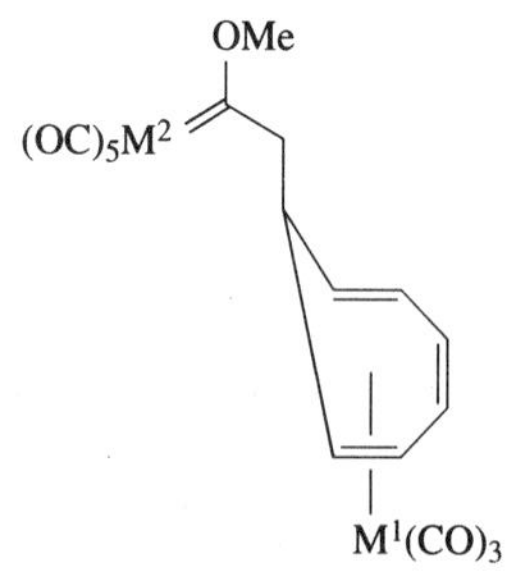

(135)

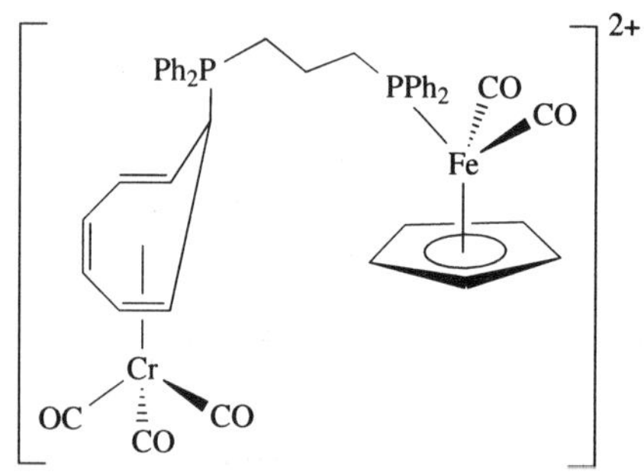

(136)

Mo–C bond[341] and investigations on the redox chemistry of $[Mo(C_6F_5)(CO)_2(\eta^7\text{-}C_7H_7)]$[343] conclude that CO insertion into the Mo–C bond can be oxidatively induced. A novel development is the formation of 17-electron dihalides $[MX_2L(\eta^7\text{-}C_7H_7)]$ (M = Mo,[277] X = I, L = THF,[344] NCMe, PPh_3 or PMe_3; M = Mo, X = Br, L = NCMe; M = W,[278] X = I, L = NCMe or PMe_3, X = Br, L = PMe_3) which are accessible via halogen oxidation of $[M(\eta^5\text{-}C_7H_9)(\eta^7\text{-}C_7H_7)]$ in NCMe or THF; the phosphine derivatives are generated by subsequent NCMe displacement.

Thermal reaction of $[MX(CO)_2(\eta^7\text{-}C_7H_7)]$ with monodentate phosphine ligands PR_3 affords carbonyl substituted $[MX(CO)(PR_3)(\eta^7\text{-}C_7H_7)]$,[321,345] but reaction with chelate phosphine and some related ligands proceeds at room temperature to give 'ring-slipped' $\eta^3\text{-}C_7H_7$ products; thus, a range of complexes $[MX(CO)_2(L\text{–}L)(\eta^3\text{-}C_7H_7)]$ (**137**) (M = Mo or W, X = I or Cl, L–L = en, dppm or dppe) has been synthesized.[346,347] Spectroscopic studies are consistent with the ligand arrangement of (**137**) in which one end of the chelate ligand is located *trans* to the $\eta^3\text{-}C_7H_7$ ring in accord with the analogous complexes $[MX(CO)_2(L^1\text{–}L^1)(\eta^3\text{-}L^2)]$ (L^2 = allyl or pentadienyl[255]). In common with the η^3-allyl analogues, (**137**) (L–L = dppe) undergoes a trigonal twist process (which results in the interconversion of inequivalent phosphorus environments) and, in addition, exhibits a low-energy 1,2-shift of the metal group around the $\eta^3\text{-}C_7H_7$ ring; TCNE adds to the uncoordinated diene unit of the $\eta^3\text{-}C_7H_7$ ring in the derivatives (M = W, L–L = dppm or dppe). Structural studies on the related $\eta^3\text{-}C_7H_7$ complexes (**138**) (X = NH_2[348] or SEt[349]) have been reported. Treatment of $[MoI(CO)_2(\eta^7\text{-}C_7H_7)]$ with $Ag[BF_4]$ in NCMe affords an equilibrium mixture of $[Mo(NCMe)(CO)_2(\eta^7\text{-}C_7H_7)]^+$ and $[Mo(NCMe)_3(CO)_2(\eta^3\text{-}C_7H_7)]^+$ which, when dissolved in CH_2Cl_2, yields $[Mo(NCMe)(CO)_2(\eta^7\text{-}C_7H_7)]^+$ as the exclusive product; the corresponding chemistry of tungsten differs in that, in NCMe, formation of $[W(NCMe)_3(CO)_2(\eta^3\text{-}C_7H_7)]^+$ is complete.[350] The complexes $[M(NCMe)(CO)_2(\eta^7\text{-}C_7H_7)]^+$ (M = Mo or W) are useful precursors to a range of substituted products via facile displacement of NCMe; notably reaction with $CNBu^t$ affords $[M(CNBu^t)(CO)_2(\eta^7\text{-}C_7H_7)]^+$, which undergo associative substitution with $CNBu^t$ via the isolated intermediates $[M(CNBu^t)_3(CO)_2(\eta^3\text{-}C_7H_7)]^+$ to form $[M(CNBu^t)_2(CO)(\eta^7\text{-}C_7H_7)]^+$.[351] Reaction of $[MoBr(CO)_2(\eta^7\text{-}C_7H_7)]$ with sodium amides yields amide-bridged, mixed valence products (**139**) (R = H, $C_6H_4NO_2$-4, etc.).[352]

Few reports discuss the chemistry of monosubstituted complexes $[MX(CO)(PR_3)(\eta^7\text{-}C_7H_7)]$. Cyclic voltammetric studies on $[WI(CO)(PPh_3)(\eta^7\text{-}C_7H_7)]$ in CH_2Cl_2 disclose a reversible one-electron oxidation process ($E°$ = +0.49 V vs. saturated calomel electrode) and the chemically generated radical has been observed by IR and ESR spectroscopy.[345] The unusual hydride species $[WH(CO)(L)(\eta^7\text{-}C_7H_3R_4)]$ (R = H, L = $P(OMe)_3$, PPh_3 or $P(OPr^i)_3$; R = Me, L = $P(OPr^i)_3$ or CO) are formed via reaction of $[WX(CO)(L)(\eta^7\text{-}C_7H_3R_4)]$ with sodium dihydridobis(2-methoxyethoxy)aluminum; typical ^1H NMR chemical shifts of the hydride hydrogen lie in the range δ –3.5 to –4.8.[321]

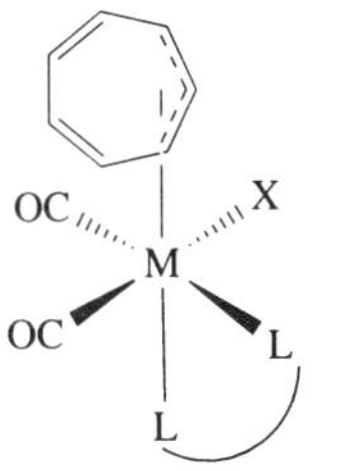

(137)

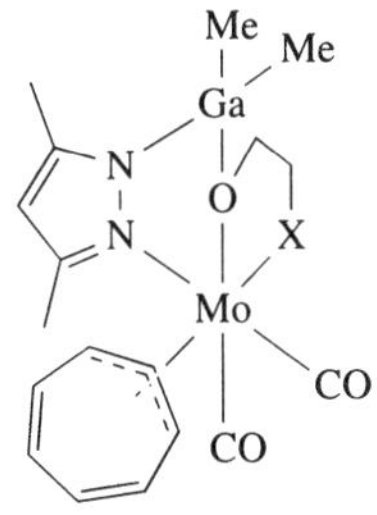

(138)

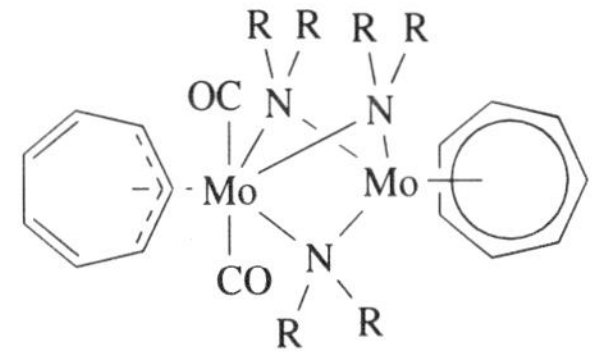

(139)

The chemistry of disubstituted complexes $[MX(PR_3)_2(\eta^7\text{-}C_7H_7)]$ is more extensive but appears to be restricted to chelating ligands or sterically undemanding monodentate phosphorus-donor ligands. Reports of halide derivatives comprise $[MoX(PR_3)_2(\eta^7\text{-}C_7H_7)]$ ($PR_3 = P(OMe)_3$, X = I or Cl;[345] $PR_3 = PMe_3$, X = I[277]) and $[MX(P\text{–}P)(\eta^7\text{-}C_7H_7)]$ (M = Mo, P–P = dppe, X = halide;[353] M = Mo, P–P = dmpe, X = I;[277] M = W, P–P = dmpe, X = I;[278] M = W, P–P = dppe, X = Cl[345]); most examples are accessible via the sandwich complexes $[Mo(\eta^6\text{-toluene})(\eta^7\text{-}C_7H_7)]^+$ or $[M(\eta^5\text{-}C_7H_9)(\eta^7\text{-}C_7H_7)]$. Halide substitution in $[MoX(dppe)(\eta^7\text{-}C_7H_7)]$ provides syntheses of $[MoR(dppe)(\eta^7\text{-}C_7H_7)]$ (R = H or Me).[353] The alkynyl complexes $[Mo(C{\equiv}CR)(dppe)(\eta^7\text{-}C_7H_7)]$ (R = Ph or Bu^t) have been obtained via deprotonation of vinylidene precursors $[Mo({=}C{=}CHR)(dppe)(\eta^7\text{-}C_7H_7)]^+$.[342] The formation of vinylidene complexes from reaction of $[Mo(sol)(dppe)(\eta^7\text{-}C_7H_7)]^+$ (sol = acetone or NCMe) with terminal alkynes HC≡CR[342] is consistent with alkyne–vinylidene isomerization promoted by a metal d^6 centre, but the assignment of a formal positive charge to the C_7H_7 ligand in these complexes is not without contention; other studies[353] assert that the covalent bond formed between the $\eta^7\text{-}C_7H_7$ ligand and a transition metal requires three electrons from the metal. The redox chemistry of $[MX(L)_2(\eta^7\text{-}C_7H_7)]$ (L = phosphorus-donor ligand) is a fruitful area due to a facile one-electron oxidation process leading to the generation of isolable 17-electron radicals $[MX(L)_2(\eta^7\text{-}C_7H_7)]^+$ (M = Mo, L = $P(OMe)_3$, X = I or Cl;[345] M = Mo, L_2 = dppe, X = I, Cl,[353] C≡CPh or C≡CBu^t;[342] M = W, L_2 = dppe, X = Cl[345]). The alkynyl radical $[Mo(C{\equiv}CPh)(dppe)(\eta^7\text{-}C_7H_7)]^+$ has additional importance in that a slow, ligand-centred dimerization proceeds to yield divinylidene-bridged $[\{Mo(dppe)(\eta^7\text{-}C_7H_7)\}_2\{\mu\text{-}C{=}C(Ph)\text{-}C(Ph){=}C\}]^{2+}$.[354] This reaction provides a unique example of the redox-induced, ligand-centred dimerization of an organometallic complex in which the intermediate radical is isolable.[354,355]

6.7.3 Sandwich Complexes Containing the Cycloheptatrienyl Ligand

Investigations on the mixed sandwich $[CrCp(\eta^7\text{-}C_7H_7)]$ continue with recent studies on cyclic voltammetric[356] and vapour-phase electronic absorption[357] properties. The synthesis and reactions of the related cycloheptadienyl complexes $[M(\eta^5\text{-}C_7H_9)(\eta^7\text{-}C_7H_7)]$ (M = Mo or W) have been discussed in Sections 6.5.2 and 6.7.2; they also serve as efficient precursors to $[M(\eta\text{-}C_5H_4R)(\eta^7\text{-}C_7H_7)]$ (M = Mo or W, R = H or Me). Complex (**140**), in effect a bis(adduct) of $[W(\eta^5\text{-}C_7H_9)(\eta^7\text{-}C_7H_7)]$, has been obtained as a mixture with $[W(PMe_3)(\eta^3\text{-cycloheptenyl})(\eta^7\text{-}C_7H_7)]$ from reaction of $[WH_6(PMe_3)_3]$ with chpt.[358] Molybdenum-95 NMR studies on $[Mo(\eta^5\text{-}C_7H_9)(\eta^7\text{-}C_7H_7)]$ and $[Mo(\eta^6\text{-arene})(\eta^7\text{-}C_7H_7)]^+$ (arene = benzene or toluene) reveal ^{95}Mo chemical shifts in the range δ –450 to –490;[289] these values suggest a deshielded molybdenum centre and are consistent with the experimentally observed, relatively facile displacement of arene or C_7H_9 ligands. The carborane sandwich (**141**) has been prepared[359] via treatment of $CrCl_3$ with a $K_2C_8H_8/Na[Et_2C_2B_4H_5]$ mixture.

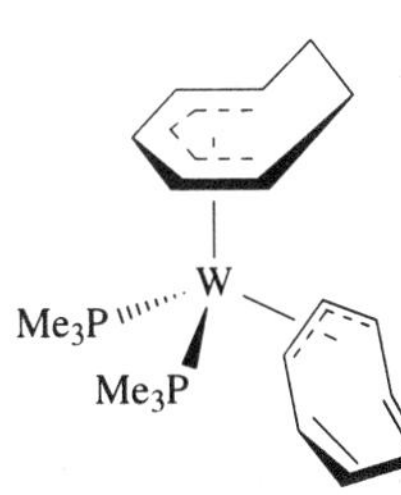

(140)

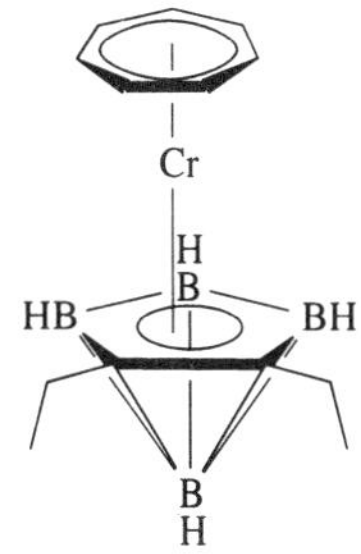

(141)

6.7.4 Metal–Metal Bonded, Bimetallic Cycloheptatrienyl Complexes

The availability of $[M(NCMe)(CO)_2(\eta^7\text{-}C_7H_7)]^+$ (M = Mo or W) has stimulated new syntheses of metal–metal bonded derivatives. Treatment of $[M^1(NCMe)(CO)_2(\eta^7\text{-}C_7H_7)]^+$ with $[M^2(CO)_3Cp]^-$ (M^2 = Mo or W) affords[360] $[M^1(CO)_2(\eta^7\text{-}C_7H_7)M^2(CO)_3Cp]$; crystallographic studies on the derivative $M^1 = M^2$ = Mo reveal an unsupported Mo–Mo bond of length 0.316 0(1) nm.[361] Complexes (**142**) (M = Mo or W, $R^1 = C_6H_4Me\text{-}4$, R^2 = H; M = W, $R^1 = C_6H_4Me\text{-}4$ or C≡CBu^t, R^2 = Me) have been synthesized via reaction of $[Mo(NCMe)(CO)_2(\eta^7\text{-}C_7H_7)]^+$ with anionic carbynes $[M(CO)_2(\equiv CR^1)(\eta^5\text{-}C_2B_9H_9R^2{}_2)]^-$.[362] Reaction of (**142**) (M = W, R^2 = Me) with alkynes MeC≡CMe or EtC≡CEt results in CO substitution at the tungsten centre but, with further MeC≡CMe, the derivative ($R^1 = C_6H_4Me\text{-}4$) yields (**143**) in which the Mo–W separation is 0.292 2(1) nm;[363] other alkyne-bridged species are generated via treatment of (**142**) with $PHPh_2$[364] or diazomethane.[365] There is a fairly extensive chemistry of group 6 bridged complexes (**144**) (M = Mo or W, L = CO or phosphorus-donor ligand, E = S, Se or Te, R = $C_6H_4Me\text{-}4$, Bu^t, etc.) and recent reports comprise structural studies on the derivatives (M = Mo, L = CO, E = Se, R = Ph),[366] (M = Mo, L = $P(OMe)_3$, E = S, R = Bu^t),[367] and (M = W, L = CO, E = S, R = $C_6H_4Me\text{-}4$);[368] the structure of trinuclear (**145**) has also been reported.[368] A recent development is the synthesis of thiolato-bridged $[\{Mo(\eta^7\text{-}C_7H_3R^1{}_4)\}_2(\mu\text{-}SR^2)_3]^+$ (Mo–Mo) (R^1 = H or Me, R^2 = Et, Ph, etc.) via reaction of $[Mo(\eta^6\text{-toluene})(\eta^7\text{-}C_7H_3R^1{}_4)]^+$ with thiols R^2SH; cyclic voltammetric investigations reveal two reversible one-electron reduction processes, probably associated with Mo–Mo bond fission.[369]

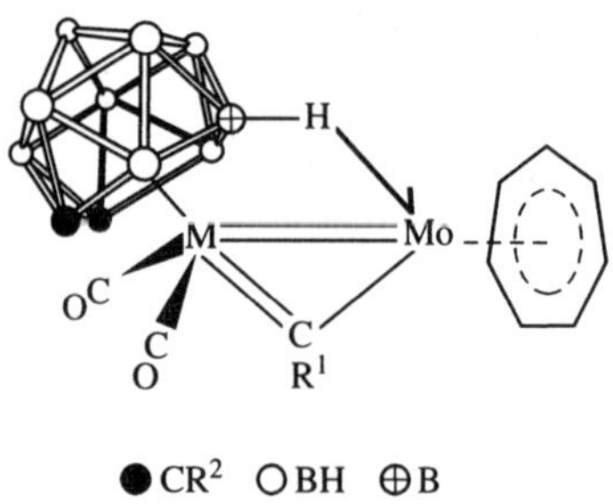

(142)

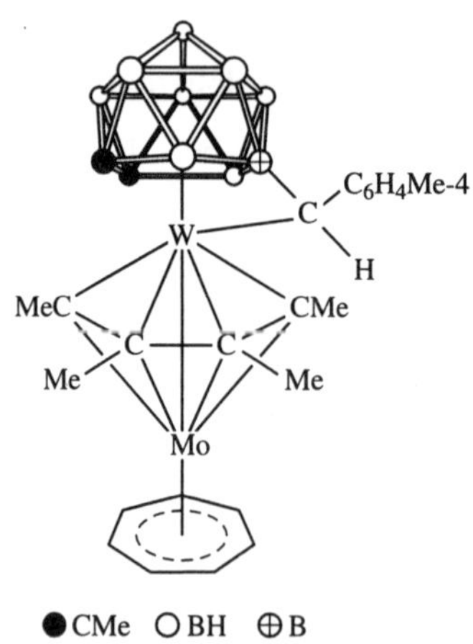

(143)

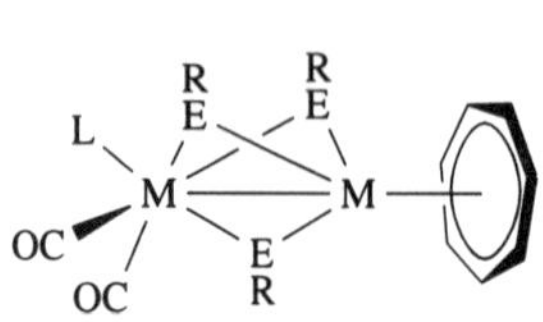

(144)

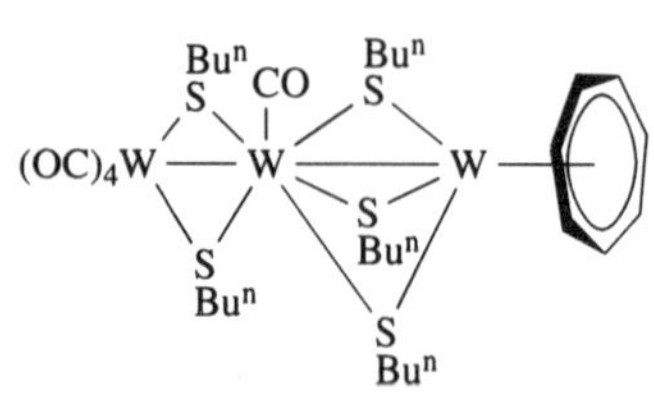

(145)

6.8 MONOMETALLIC ALKYNE COMPLEXES

A general concept, fundamental to this section, is the ability of alkynes to function as formal four-electron ligands to a monometallic centre via $\pi_{\parallel}$ and $\pi_{\perp}$ interactions. This concept is more rigorously justified by MO studies[370,371] and a comprehensive account of the properties of four-electron alkyne ligands in M^{II} (d^4) complexes is also available.[372]

6.8.1 Alkyne Complexes of M^0 (d^6) Metal Centres

Simple M^0 alkyne complexes in which one, two or three alkyne ligands are attached at the metal centre are known. The chemistry of monometallic chromium–alkyne complexes is dominated by derivatives of chromium(0).

Unstable [Cr(CO)$_5$(η^2-HC≡CCO$_2$Me)] is formed as one product of photolysis of [Cr(CO)$_6$] and HC≡CCO$_2$Me;[373] the thermally sensitive, mixed alkyne–carbene complex [W{C(OMe)Ph}(CO)$_4$(η^2-PhC≡CPh)] has also been obtained photochemically from [W{C(OMe)Ph}(CO)$_5$] and PhC≡CPh.[374] More stable species [Cr(CO)$_2${P(OMe)$_3$}(η^2-RC≡CPh)] (R = H or Ph) are accessible via treatment of [Cr(CO)$_2${P(OMe)$_3$}(η^4-cod)] with RC≡CPh.[375] The complexes *fac*-[W(CO)$_3$(dppe)(η^2-HC≡CR)] (R = H, Ph or Bu^n) have been obtained from reaction of HC≡CR with *fac*-[W(CO)$_3$(dppe)(acetone)] but rapid isomerization to vinylidenes, *mer*-[W(=C=CHR)(CO)$_3$(dppe)] proceeds to relieve unfavourable two-centre, four-electron repulsion between the filled alkyne $\pi_{\perp}$ orbital and a filled metal $d\pi$ orbital.[376]

Most bis(alkyne)Cr^0 complexes have the formulation [Cr(L)$_2$(η^2-RC≡CR)$_2$] (L = CO, R = TMS;[377] L$_2$ = dmpe, R = Ph;[34] L = PMe$_2$Ph, R = Me; L = PMe$_3$, R = Ph[219]); structural studies reveal a staggered alkyne arrangement which is probably promoted by an interaction between alkyne $\pi_{\perp}$ orbitals and vacant metal orbitals.[378] With a few exceptions such as *trans*-[Mo(η^2-HC≡CPh)$_2$(dppe)$_2$],[379] most bis(alkyne) derivatives of molybdenum(0) and tungsten(0) have the formulation [M(CO)$_2$(L–L)(η^2-R^{1}C≡CR^2)$_2$] (M = Mo, L–L = bipy or phen, R^1 = R^2 = CO$_2$Me;[380] M = W, L–L = bipy, phen, etc., R^1 = CO$_2$Me, CO$_2$Et, R^2 = H, CO$_2$Me;[381] M = W, L–L = dppe, R^1 = R^2 = CO$_2$Me[376]). The common structural features of these complexes are *cis*-disposed CO ligands with mutually *trans*, orthogonal alkynes. The electron-withdrawing alkyne substituents play a crucial role in enhancing stability via the promotion of metal $d\pi$ to alkyne $\pi^*_{\parallel}$ backbonding and in the delocalization of $\pi_{\perp}$ alkyne electrons away from $d\pi$ metal electron density; effectively, therefore, the alkynes function as two-electron donor alkene analogues. Reaction of HC≡CBu^t with the thioaldehyde complexes [W(CO)$_5$(η^2-X=CHPh)] (X = S or Se) yields [W(CO)(η^2-X=CHPh)(η^2-HC≡CBu^t)$_2$], which are structural analogues of [W(CO)(η^2-alkyne)$_3$].[382]

The formation of tris(alkyne) complexes [W(CO)(η^2-alkyne)$_3$] by reaction of alkynes with [W(CO)$_3$(NCMe)$_3$] is well established. Recent extensions of this basic reaction comprise syntheses of [W(CO)(η^2-R^{1}C≡CR^2)$_3$] (R^1 = R^2 = Pr^i, C$_6$H$_4$Me-4[383] or TMS;[384] R^1 = H, R^2 = Ph[384]). Of particular note is the formation of [W(CO)(η^2-TMS-C≡C-TMS)$_3$]; the successful generation of this complex of a sterically demanding alkyne contrasts with the result for chromium(0)[377] and NMR studies on barriers to alkyne rotation in [W(CO)(η^2-RC≡CR)$_3$] suggest that steric constraints are relatively small.[384] However, the C_{3v} structure of [W(CO)(η^2-RC≡CR)$_3$] is crucial for bonding interactions; MO studies suggest that a structure in which one alkyne is rotated by 90° is destabilized by an unfavourable interaction with a filled $\pi_{\perp}$ orbital.[378] Carbonyl substitution in [W(CO)(η^2-RC≡CR)$_3$] can be effected by direct reaction with phosphorus-donor ligands[384] or alternatively Me$_3$NO promoted generation of [W(NCMe)(η^2-PhC≡CPh)$_3$] may precede reaction with phosphorus-donor ligands;[385] [W(PMe$_3$)(η^2-HC≡CPh)$_3$] has been obtained by sodium/mercury reduction of [WCl$_4$(PMe$_3$)$_2$] in the presence of HC≡CPh.[386] Alkyne coupling in reactions of [W(CO)(R^{1}C≡CR^2)$_3$] is also observed; the derivative (R^1 = R^2 = Ph) reacts with excess PhC≡CPh to give[235] the cyclobutadiene product [W(CO)(η^2-PhC≡CPh)$_2$(η^4-C$_4$Ph$_4$)] and thermolysis of a range of derivatives (R^1, R^2 = Me, Et or Ph) in the presence of the carbyne [W(CO)$_2$(≡CR^3)Cp] (R^3 = Me or C$_6$H$_4$Me-4) proceeds with alkyne–carbyne coupling as detailed in Section 6.9.2.[387] Lithium naphthalenide reduction of [W(CO)(PhC≡CPh)$_3$] gives the tentatively formulated, homoleptic alkyne complex [W(η^2-PhC≡CPh)$_3$]$^{2-}$ which reacts with SnPh$_3$Cl/[NEt$_4$]Br to give isolable [NEt$_4$][W(SnPh$_3$)(η^2-PhC≡CPh)$_3$].[388]

The alkyne complexes [Cr(CO)$_2$(η^2-R^{1}C≡CR^1)(η^6-C$_6$R$^2_6$)] (R^1 = H, Pr^i, etc.; R^2 = H, Me, etc.) are accessible photochemically from [Cr(CO)$_3$(η^6-C$_6$R$^2_6$)] and the appropriate alkyne but extended photolysis affords bimetallic products.[389] Exceptionally, [Cr(CO)$_3$(η^6-C$_6$Me$_6$)] reacts with TMS-C≡C-TMS to yield the vinylidene [Cr{C=C(TMS)$_2$}(CO)$_2$(η^6-C$_6$Me$_6$)] but subsequent treatment with [FeCp$_2$]$^+$ generates 17-electron [Cr(CO)$_2$(η^2-TMS-C≡C-TMS)(η^6-C$_6$Me$_6$)]$^+$ in an unprecedented example of redox-induced vinylidene–alkyne isomerization.[390] The cyclobutadiene complex [Cr(CO)$_2$(η^2-PhC≡CPh)(η^4-C$_4$Ph$_4$)] undergoes two successive, reversible one-electron reductions and the stability of the resulting mono- and dianions has been attributed to stepwise conversion of the alkyne from a four-electron to a two-electron ligand.[236]

6.8.2 Alkyne Complexes of M^{II} (d^4) Metal Centres

6.8.2.1 *[MX₂(CO)(L)₂(η²-RC≡CR)], [M(L)(S–S)₂(η²-RC≡CR)] and related simple monoalkyne complexes*

Several routes to $[MX_2(CO)_{3-n}(L)_n(\eta^2\text{-RC}\equiv\text{CR})]$ (X = halide, n = 1 or 2) have been developed. In the initial report, reaction of $[WI_2(CO)_4(L)]$ (L = PMe_3, $CNBu^t$ or $AsMe_3$) with RC≡CH (R = H or Ph) was shown to give $[WI_2(CO)_2(L)(\eta^2\text{-RC}\equiv\text{CH})]$[391] and subsequently a series of complexes $[MX_2(CO)(L)_2(\eta^2\text{-R}^1\text{C}\equiv\text{CR}^2)]$ (M = Mo or W; X = Br or Cl; L = PEt_3, PPh_3, L_2 = dppe; R^1, R^2 = H, Me, Et, Ph, etc.) was prepared via treatment of $[MX_2(CO)_n(L)_2]$ (n = 2 or 3) with free alkyne.[392] Alternative syntheses involve cleavage of halide-bridged binuclear precursors or alkyne substitution in bis(alkyne) species. Thus, $[\{WBr_2(CO)_4\}_n]$ reacts with alkynes to give $[\{WBr(CO)(\eta^2\text{-RC}\equiv\text{CR})_2\}_2(\mu\text{-Br})_2]$, which subsequently yields $[WBr_2(CO)(L)_2(\eta^2\text{-RC}\equiv\text{CR})]$ (L = $P(OMe)_3$, PPh_3 or $CNBu^t$)[393] or $[W(L\text{–}L)_2(CO)(\eta^2\text{-RC}\equiv\text{CR})]$ (L–L = S_2CNMe_2, S_2PMe_2, etc.).[394] Similarly, a series of phosphite and phosphine substituted complexes $[WI_2(CO)(L)_2(\eta^2\text{-RC}\equiv\text{CR})]$ (R = Me or Ph; L = $P(OMe)_3$, etc.;[395] L = PMe_3, etc., L_2 = dppm, dppe, etc.[396]) has been synthesized by reaction of $2L/L_2$ with iodide-bridged $[\{WI(CO)(NCMe)(\eta^2\text{-RC}\equiv\text{CR})\}_2(\mu\text{-I})_2]$[397] or by displacement of NCMe and one alkyne from the bis(alkyne) precursor $[WI_2(CO)(NCMe)(\eta^2\text{-RC}\equiv\text{CR})_2]$. Assuming a distorted octahedral arrangement of ligands, several isomeric structures may be proposed for the complexes $[MX_2(CO)_{3-n}(L)_n(\eta^2\text{-RC}\equiv\text{CR})]$; structurally characterized examples comprise (**146**),[392] (**147**)[395] and (**148**).[396] The common structural features are *cis*-halide ligands and a *cis*-coplanar arrangement of alkyne and CO groups. With monodentate ligands L, a *trans*-L arrangement, as in (**146**), is usually preferred, but several studies[391,393,395] report *cis–trans* isomerism. The carbonyl-free complexes *cis-trans*-$[WCl_2(PMe_3)_2(L)(\eta^2\text{-PhC}\equiv\text{CPh})]$ (L = PMe_3;[398] C_2H_4 or propene[399]) have been obtained by sodium/mercury reduction of $[WCl_3(PMe_3)_2(\eta^2\text{-PhC}\equiv\text{CPh})]$ in the presence of L, and the Mo–Ge complex $[MoCl(GeCl_3)(CO)(NCMe)(PPh_3)(\eta^2\text{-MeC}\equiv\text{CMe})]$ has also been described.[400]

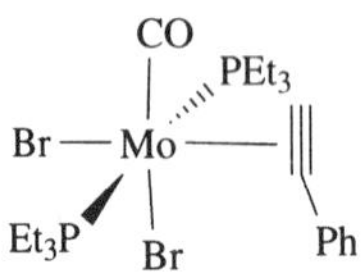

(**146**)

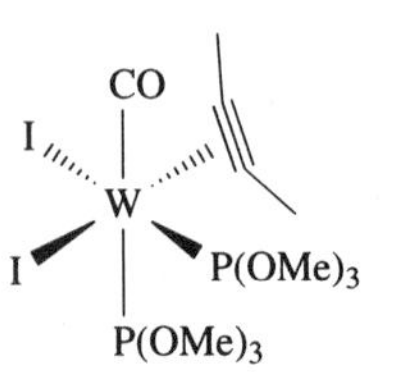

(**147**)

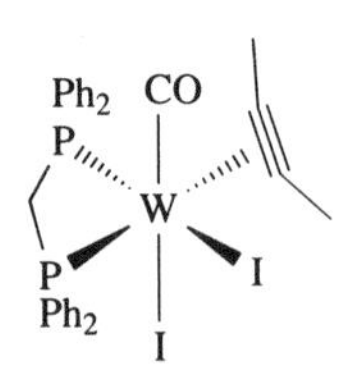

(**148**)

Some ligand substitution chemistry of $[WI_2(CO)(L)_2(\eta^2\text{-RC}\equiv\text{CR})]$ has been developed. Treatment of $[WI_2(CO)(L)_2(\eta^2\text{-RC}\equiv\text{CR})]$ with $Ag[BF_4]$ in NCMe affords $[WI(NCMe)(CO)(L)_2(\eta^2\text{-RC}\equiv\text{CR})]^+$ (R = Me or Ph; L = PEt_3 or PMe_2Ph;[401] L_2 = dppm, dppe, etc.[402]) and, in the case of $[WI(NCMe)(CO)(dppm)(\eta^2\text{-MeC}\equiv\text{CMe})]^+$, subsequent NCMe substitution with ligands L (phosphites,[402] phosphines,[403,404] CO or Bu^tCN[405] and $SC(NR_2)_2$[406]) yields $[WI(L)(CO)(dppm)(\eta^2\text{-MeC}\equiv\text{CMe})]^+$. The structurally characterized,[407] dithiocarbamate complex (**149**) has been prepared via treatment of $[WI(NCMe)(CO)(dppm)(\eta^2\text{-MeC}\equiv\text{CMe})]^+$ with $NaS_2CNC_4H_8$,[408] but a more general synthetic route involves reaction of (**150**) with 2L or L_2 to give $[W(CO)(S_2CNC_4H_8)(L)_2(\eta^2\text{-MeC}\equiv\text{CMe})]^+$ (L = phosphine[409] or imidazole;[410] L_2 = dppm, dppe, bipy, etc.[407]). Deprotonation of $[W(CO)(S_2CNR_2)(dppe)(\eta^2\text{-MeOC}\equiv\text{CCH}_2\text{Ph})]^+$ (R = Me, Pr^i, etc.) yields η^2-allenyl products (**151**), which add electrophiles E (E = H or Me) at C_γ to give $[W(CO)(S_2CNR_2)(dppe)(\eta^2\text{-OMeC}\equiv\text{CCHEPh})]^+$.[411] An extensive data set is available for NMR-derived barriers to alkyne rotation in $[MX_2(CO)(L)_2(\eta^2\text{-RC}\equiv\text{CR})]$ and related derivatives. Typical values lie in the range 40–65 kJ mol^{-1} and exhibit some correlation with ligand electronic and steric properties.

The formation of dithiocarbamate complexes $[Mo(CO)(S_2CNEt_2)_2(\eta^2\text{-RC}\equiv\text{CR})]$ via reaction of $[Mo(CO)_n(S_2CNEt_2)_2]$ (n = 2 or 3) with alkynes is well established. Advances in this area comprise syntheses of $[M(CO)(S_2CNR_2)_2(\eta^2\text{-cyclooctyne})]$ (M = Mo or W; R = Me or Et)[412] and the unusual mixed alkene–alkyne complexes $[M(S_2CNR_2)_2(\eta^2\text{-alkene})(\eta^2\text{-alkyne})]$ (M = Mo or W; R = Me or Et; alkene = maleic anhydride, TCNE or *trans*-dicyanoethene). Structurally characterized (**152**) reveals the characteristic *cis*, parallel arrangement of alkene and alkyne ligands in these complexes; selected

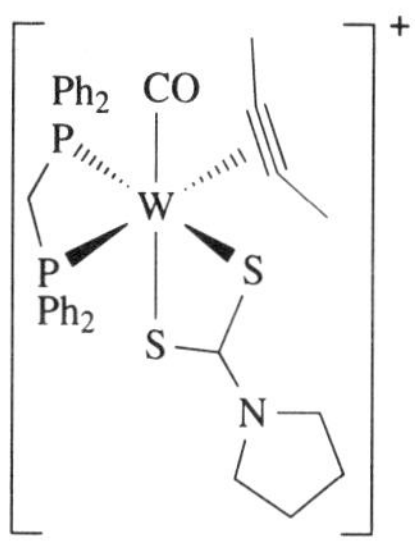

(149)

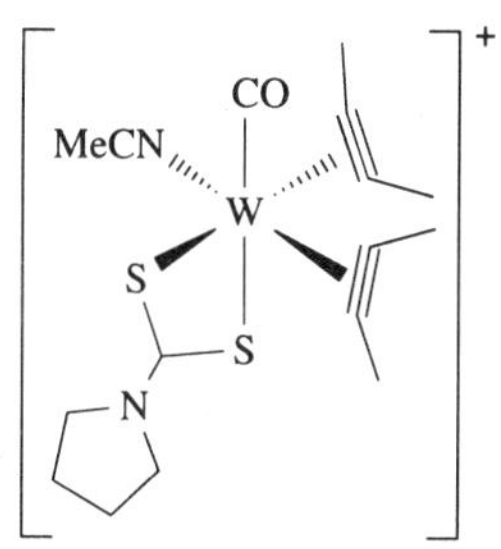

(150)

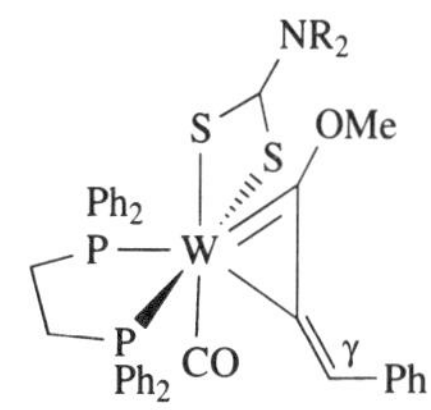

(151)

examples undergo nucleophilic addition at the alkyne ligand.[413] Ligand substitution in $[Mo(CO)(S_2CNMe_2)_2(\eta^2\text{-alkyne})]$ may proceed via a rapid associative process resulting in displacement of bound alkyne or alternatively, slow dissociative loss of CO in the presence of excess alkyne provides a synthesis of $[Mo(S_2CNMe_2)_2(\eta^2\text{-alkyne})_2]$.[414] Cyclic voltammetric studies on a series of complexes $[M(CO)(S_2CNEt_2)_2(\eta^2\text{-alkyne})]$ (M = Mo or W) reveal a reversible reduction process with $E°$ dependent upon alkyne substituents; this observation is consistent with an MO description in which the LUMO energy is dependent upon alkyne $\pi_\perp$ donation.[415] Reports on related alkyne complexes with sulfur-donor ligands comprise syntheses of $[Mo(CO)(S_2PPh_2)_2(\eta^2\text{-PhC≡CH})]$[416] and the bis(thiolato) complexes $[Mo(SBu^t)_2(CNBu^t)_2(\eta^2\text{-R}^1C≡CR^2)]$ (R^1, R^2 = H or Ph).[417] Further examples of simple monoalkyne M^{II} species comprise *trans*-$[MoF(\eta^2\text{-PhC≡CH})(dppe)_2]^+$[418] and some mixed alkyne–carbene/carbyne complexes. Deprotonation of $[WCl_2(=CHPh)(CO)(PMe_3)_2]$ in the presence of PhC≡CPh proceeds via the intermediate carbyne $[WCl(CO)(PMe_3)_2(≡CPh)(\eta^2\text{-PhC≡CPh})]$ to give (**153**) in which the tungsten–alkyne π bonds are directed orthogonally to the W=C bond of the carbene ligand[419] in contrast to the parallel alkene–alkyne arrangement of (**152**). Treatment of $[WCl(PMe_3)_4(≡CMe)]$ with RC≡CR (R = Me or Ph) yields unstable, mixed carbyne–alkyne complexes $[WCl(PMe_3)_2(≡CMe)(\eta^2\text{-RC≡CR})]$, which exhibit stoichiometric alkyne metathesis,[420] but reaction of PhC≡CH with $[WCl(CO)(PMe_3)_3(≡CPh)]$ generates products such as (**154**), which result from formal coupling of CO, carbyne and alkyne groups.[421]

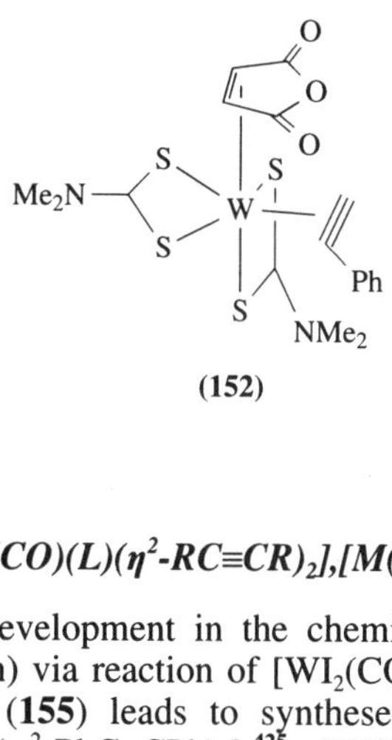

(152)

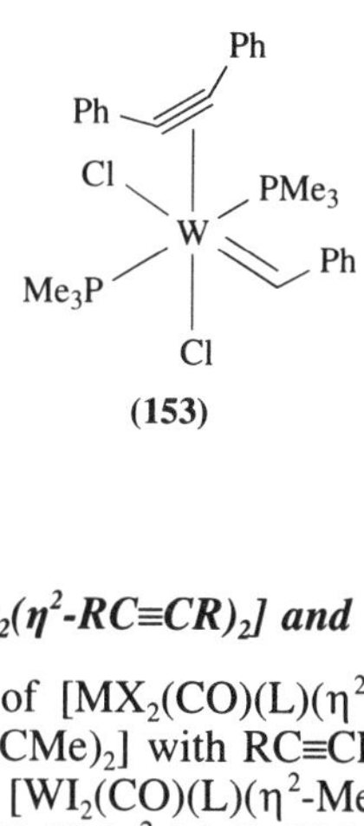

(153)

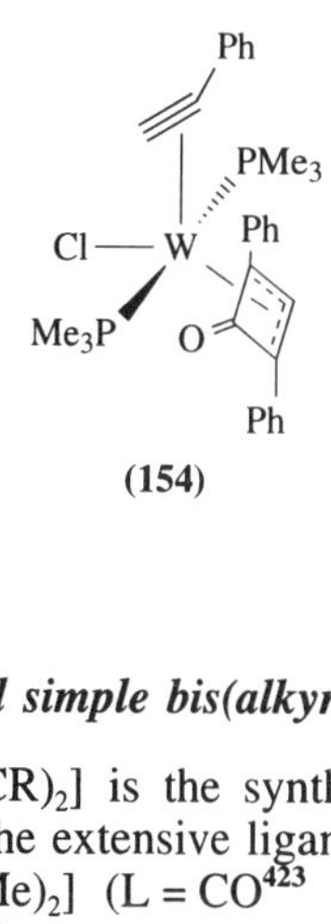

(154)

6.8.2.2 *$[MX_2(CO)(L)(\eta^2\text{-RC≡CR})_2]$, $[M(S–S)_2(\eta^2\text{-RC≡CR})_2]$ and related simple bis(alkyne) complexes*

A major development in the chemistry of $[MX_2(CO)(L)(\eta^2\text{-RC≡CR})_2]$ is the synthesis of (**155**) (R = Me or Ph) via reaction of $[WI_2(CO)_3(NCMe)_2]$ with RC≡CR.[422] The extensive ligand substitution chemistry of (**155**) leads to syntheses of $[WI_2(CO)(L)(\eta^2\text{-MeC≡CMe})_2]$ (L = CO[423] or $CNBu^t$[424]); $[WI_2(NCMe)_2(\eta^2\text{-PhC≡CPh})_2]$;[425] $[WI(CO)(N–N)(\eta^2\text{-MeC≡CMe})_2]^+$ (N–N = phen, bipy etc.;[426] or diazabutadiene[427]) and $[WI(CO)(L–L)(\eta^2\text{-RC≡CR})_2]$ (R = Me, L–L = acac;[428] R = Me or Ph, L–L = dithiocarbamate[429]). The structures of (**155**),[422] (**156**)[426] and (**157**)[429] have been established crystallographically. In NCMe in the presence of $Ag[BF_4]$[407] or $Na[BPh_4]$,[430] (**157**) undergoes iodide substitution to yield (**150**); ligand substitution reactions of both (**155**) and (**150**), leading to monoalkyne products, have been referred to in Section 6.8.2.1. The bis(alkyne) complexes $[WBr(L–L)(CO)(\eta^2\text{-MeC≡CMe})_2]$ (L–L = S_2CNMe_2, S_2PMe_2 or acac)[431] and $[W(2\text{-}SC_5H_4N)_2(\eta^2\text{-PhC≡CPh})_2]$[394] are accessible via bridge cleavage reactions of $[\{WBr(CO)(\eta^2\text{-RC≡CR})\}_2(\mu\text{-Br})_2]$ (R = Me or Ph). Reaction of $[\{WBr_2(CO)_4\}_2]$ with $F_3CC≡CCF_3$ yields $[WBr_2(CO)(\eta^2\text{-}F_3CC≡CCF_3)_2]$ which, exceptionally, is probably monomeric rather than halide bridged; subsequent reaction with $P(OR)_3$ (R = Me or Et) is thought to proceed as shown in Scheme 17.[432] Mixed, bis(alkyne) complexes, *cis-trans*-$[WCl_2(PMe_3)_2(\eta^2\text{-PhC≡CPh})(\eta^2\text{-R}^1C≡CR^2)]$ (R^1, R^2 = H or Ph) have been synthesized via reaction of $[WCl_2(PMe_3)_3(\eta^2\text{-PhC≡CPh})]$ with $R^1C≡CR^2$.[399]

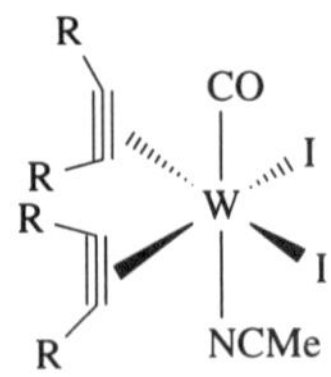

(155)

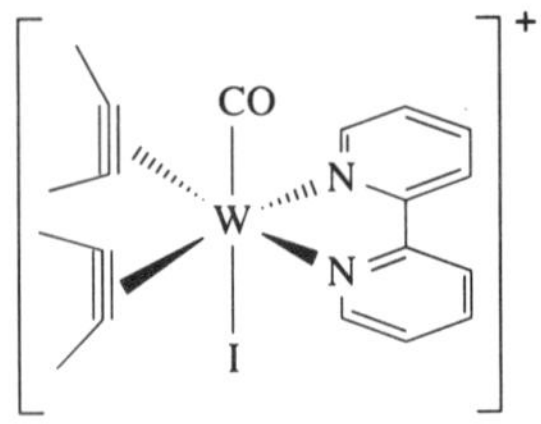

(156)

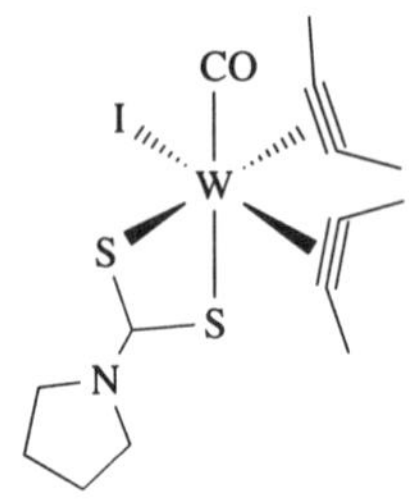

(157)

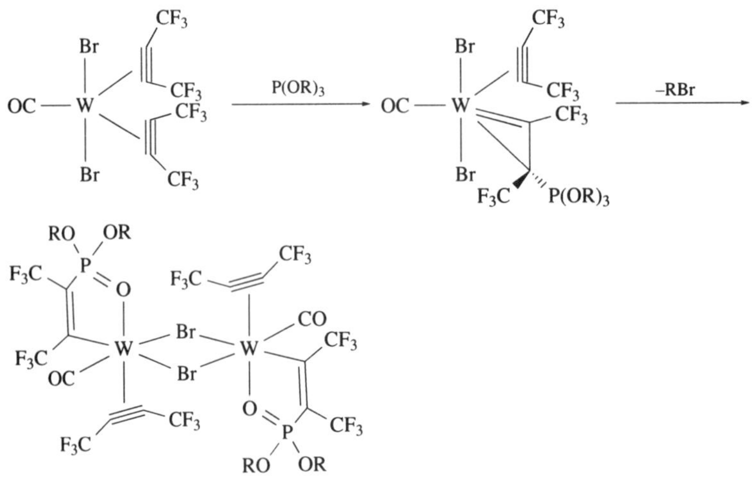

Scheme 17

The synthesis of a series of complexes $[Mo(S_2CNR_2)_2(\eta^2\text{-alkyne})_2]$ (R = Me or Et) by reaction of $[Mo(CO)_2(S_2CNR_2)_2]$ with alkynes is described in a seminal paper by Templeton and co-workers.[433] The formation of $[Mo(S_2CNR_2)_2(\eta^2\text{-alkyne})_2]$ (R = Me or Et) proceeds via the monoalkyne intermediate $[Mo(CO)(S_2CNR_2)_2(\eta^2\text{-alkyne})]$ for which dissociative CO substitution is slow.[414] By contrast, the rapid generation of (**158**) from $[Mo(CO)_3(S_2CNC_4H_4)_2]$ and MeC≡CMe may be attributed to increased lability of CO in the intermediate $[Mo(CO)(S_2CNC_4H_4)_2(\eta^2\text{-MeC≡CMe})]$ due to the relatively poor donor properties of the pyrrole-*N*-carbodithioate ligand.[434] Some reaction chemistry of $[M(S_2CNR_2)_2(\eta^2\text{-alkyne})_2]$ has been described. Thermolysis of $[Mo(S_2CNR^1{}_2)_2(\eta^2\text{-}R^2C≡CR^3)_2]$ (R^1 = Me or Et; R^2, R^3 = H, Me, Ph, etc.) in the presence of PEt_3 or $P(OMe)_3$ yields binuclear products such as (**159**); the thiocarboxamido and bridging sulfur ligands originate from oxidative addition of one S_2CNR_2 ligand across the dimetal centre.[435] Alkyne coupling in $[M(S_2CNEt_2)_2(\eta^2\text{-PhC≡CPh})_2]$ (M = Mo or W) is promoted by HBF_4 protonation and, for M − W, the (5e)-butadienyl product (**80**) has been isolated;[233] the formation of (**160**) (M = Mo or W; L = CO or η^2-cyclooctyne; R = Me or Et) via reaction of $[M(CO)_2(L)(S_2CNR_2)_2]$ (L = CO or PPh_3) with excess cyclooctyne presumably also involves a metal-based alkyne coupling reaction.[436] Thermolysis of the pseudotetrahedral complex $[W(OC_6H_3Ph_2\text{-}2,6)_2(\eta^2\text{-EtC≡CEt})_2]$ promotes alkyne coupling to give a tungstacyclopentatriene product.[437]

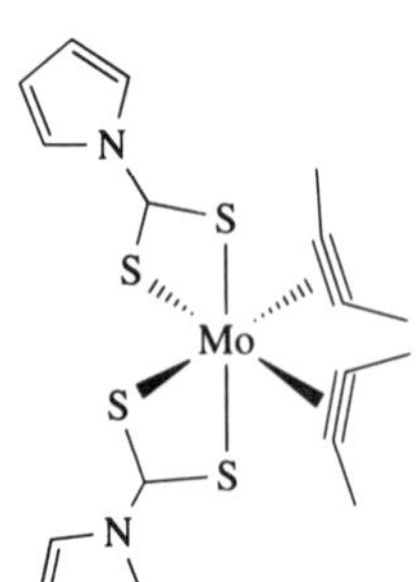

(158)

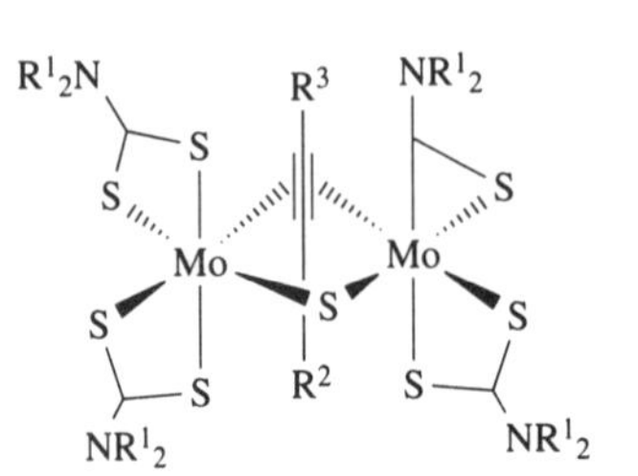

(159)

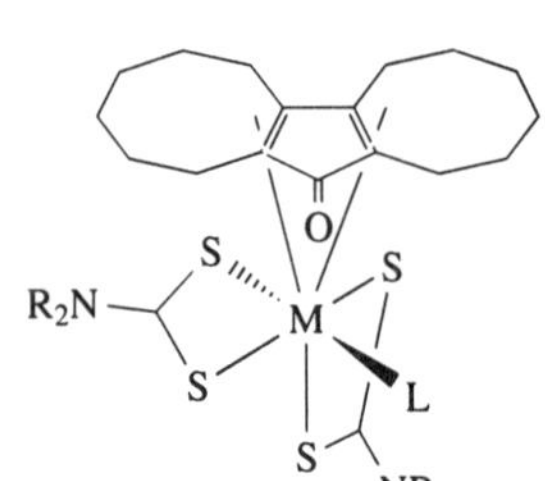

(160)

6.8.2.3 Monoalkyne complexes containing cyclopentadienyl and related ligands

The major development in this area involves the synthesis and reactions of cationic monoalkyne complexes $[Mo(CO)(L)(\eta^2\text{-alkyne})(\eta^5\text{-dienyl})]^+$ ($L = PEt_3$, PPh_3, PCy_3; dienyl = Cp or indenyl) and $[M(L)_2(\eta^2\text{-alkyne})(\eta^5\text{-dienyl})]^+$ (dienyl = Cp or indenyl; M = Mo, $L = P(OMe)_3$, PMe_3, PEt_3 etc., L_2 = dppe, etc.; M = W, $L = P(OMe)_3$). The complexes $[Mo(CO)(L)(\eta^2\text{-alkyne})(\eta^5\text{-dienyl})]^+$ are obtained via reaction of $[Mo(CO)(\eta^2\text{-alkyne})_2(\eta^5\text{-dienyl})]^+$ with L, while reaction of $[Mo(CO)(NCMe)(\eta^2\text{-alkyne})(\eta^5\text{-dienyl})]^+$ with excess L provides a synthesis of $[Mo(L)_2(\eta^2\text{-alkyne})(\eta^5\text{-dienyl})]^+$.[438] The tungsten derivatives $[W\{P(OMe)_3\}_2(\eta^2\text{-HC}{\equiv}\text{CBu}^t)(\eta^5\text{-dienyl})]^+$ have been obtained from a reaction sequence commencing with HBF_4 protonation of $[WMe(CO)_3(\eta^5\text{-dienyl})]$.[439] Crystallographic studies on $[Mo(L)(PR_3)(\eta^2\text{-MeC}{\equiv}\text{CMe})(\eta^5\text{-indenyl})]^+$ (L = CO, R = Et; $L = PR_3 = PMe_3$) reveal that the alkyne ligand is oriented approximately coplanar with Mo–CO or Mo–PMe_3 bonds with typical molybdenum–alkyne carbon bond lengths in the range 0.200–0.206 nm, consistent with a four-electron alkyne ligand. Substitution reactions of $[Mo\{P(OMe)_3\}_2(\eta^2\text{-MeC}{\equiv}\text{CMe})Cp]^+$ with sulfur-donor ligands lead to the formation of $[Mo(SR)\{P(OMe)_3\}(\eta^2\text{-MeC}{\equiv}\text{CMe})Cp]$ (R = Me, $C_6H_4NO_2$-4, etc.) and $[Mo(S_2CNMe_2)(\eta^2\text{-MeC}{\equiv}\text{CMe})Cp]$.[440] Two very unusual examples of mixed alkene–alkyne complexes have been prepared from $[Mo(CO)(\eta^2\text{-MeC}{\equiv}\text{CMe})_2Cp]^+$. Thus, treatment with diphenylphosphinostyrene yields (**161**) which undergoes alkene–alkyne coupling in refluxing NCMe to form (**162**).[441] With MgRX or MgX_2 (X = halide), $[Mo(CO)(\eta^2\text{-MeC}{\equiv}\text{CMe})_2Cp]^+$ reacts to give $[MoX(CO)(\eta^2\text{-MeC}{\equiv}\text{CMe})Cp]$; subsequent treatment with $Ag[BF_4]/C_2H_4$ yields unstable $[Mo(CO)(\eta^2\text{-}C_2H_4)(\eta^2\text{-MeC}{\equiv}\text{CMe})Cp]^+$ but analogous reactions with dienes such as isoprene proceed with alkyne–diene coupling to give η^3,η^3-hexatrienyl products (**98**).[266] Deprotonation of $[Mo\{P(OMe)_3\}_2(\eta^2\text{-PhC}{\equiv}\text{CCHR}^1\text{R}^2)Cp]^+$ (R^1, R^2 = H or Ph) with KH/Bu^tOH gives η^2-allenyl products (**163**).[442]

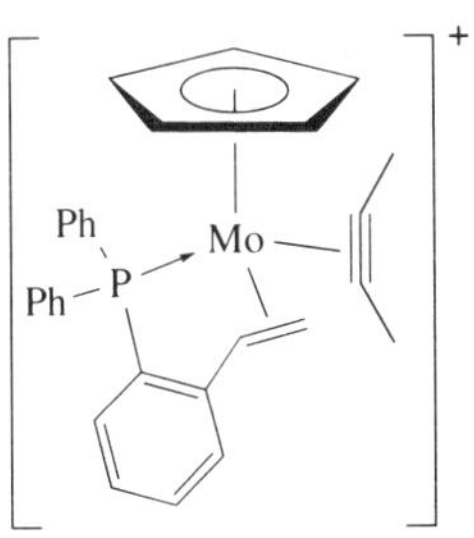

(**161**)

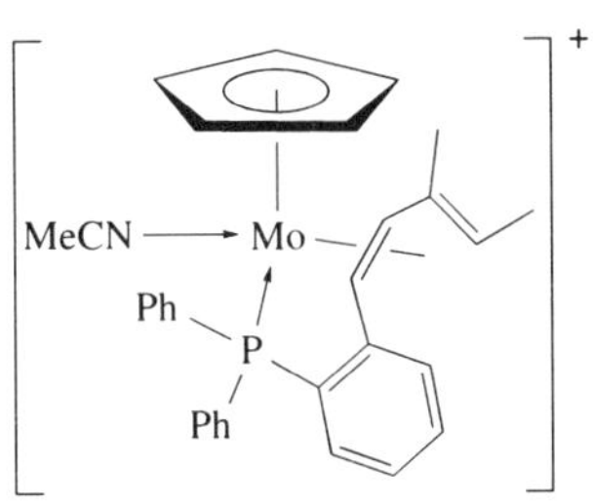

(**162**)

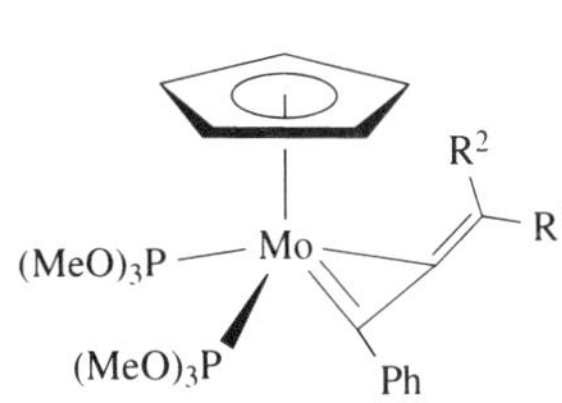

(**163**)

The complexes $[Mo\{P(OMe)_3\}_2(\eta^2\text{-alkyne})(\eta^5\text{-dienyl})]^+$ (dienyl = Cp or indenyl) have been employed in a series of investigations which elegantly illustrate the reactions of nucleophilic reagents with coordinated alkyne ligands. The details of this work have been reviewed[443] but the essential reaction types are summarized in Scheme 18.[439,444–6] The precise outcome of each reaction is critically dependent upon the alkyne substituents R^1, R^2 and upon the nucleophile employed but six basic product types are observed. These comprise η^1-vinyl,[444,439] η^2-vinyl,[445,446,439] η^3-allyl,[444,445] carbyne,[439,446] bromovinylidene[446] and alkyl[172] complexes. Formal alkyne–methyl and alkyne–vinyl coupling processes resulting in the respective formation of η^4-cyclopentadiene and η^3-cyclopropenyl ligands probably proceed via α-H elimination from Me or $CH{=}CH_2$ ligands.[172] The formation of carbyne complexes from η^1-vinyl,[439] η^2-vinyl[439] and bromovinylidene[446] intermediates is a further major aspect of this work.

The chemistry of $[Mo\{P(OMe)_3\}_2(\eta^2\text{-alkyne})Cp]^+$ is complemented by a recent development involving complexes of the hydrido-tris(3,5-Me_2-pyrazolyl)borate ligand, Tp'. Thus, $[WI(CO)(\eta^2\text{-R}^1\text{C}{\equiv}\text{CR}^2)Tp']$ and $[W(CO)_2(\eta^2\text{-R}^1\text{C}{\equiv}\text{CR}^2)Tp']^+$ have been obtained via respective reaction of $[WI(CO)_3Tp']$ with $R^1C{\equiv}CR^2$ or $R^1C{\equiv}CR^2/Ag[BF_4]$ (R^1 = Ph, R^2 = H or Me; $R^1 = R^2$ = Me) and subsequent reactions of $[W(CO)_2(\eta^2\text{-R}^1\text{C}{\equiv}\text{CR}^2)Tp']^+$ yield $[W(CO)(L)(\eta^2\text{-R}^1\text{C}{\equiv}\text{CR}^2)Tp']^+$ (L = NCMe or $P(OMe)_3$).[447] The reactions of $[W(CO)(L)(\eta^2\text{-R}^1\text{C}{\equiv}\text{CR}^2)Tp']^+$ (L = CO or NCMe) with nucleophiles are summarized in Scheme 19.[448–53] The selectivity between η^3-allyl and η^2-vinyl products, formed by reaction of $[W(CO)_2(\eta^2\text{-R}^1\text{C}{\equiv}\text{CR}^2)Tp']^+$ with $Li[BHEt_3]$, etc., is controlled by the alkyne substituents; η^3-allyls result where R^1 or R^2 = Me.[448] Protonation of selected η^2-vinyl complexes affords unusual agostic carbenes,[449,450] and the acyls $[W\{\eta^1\text{-C(O)R}^3\}(CO)(\eta^2\text{-PhC}{\equiv}\text{CMe})Tp']$ (R^3 = Me or Ph) are precursors to alkoxycarbene–alkyne complexes.[451] The outcome of H^- addition to $[W(CO)(NCMe)(\eta^2\text{-MeC}{\equiv}\text{CPh})Tp']^+$ is reagent dependent. Substitution of NCMe by hydrogen occurs on reaction with $Na[BH_4]$,[452] but $Li[BHEt_3]$ delivers H^- to the NCMe ligand and subsequent transformations with HBF_4 and $Li[BHEt_3]$ effect an overall, stepwise reduction of NCMe to ethylamine.[453] Other developments in

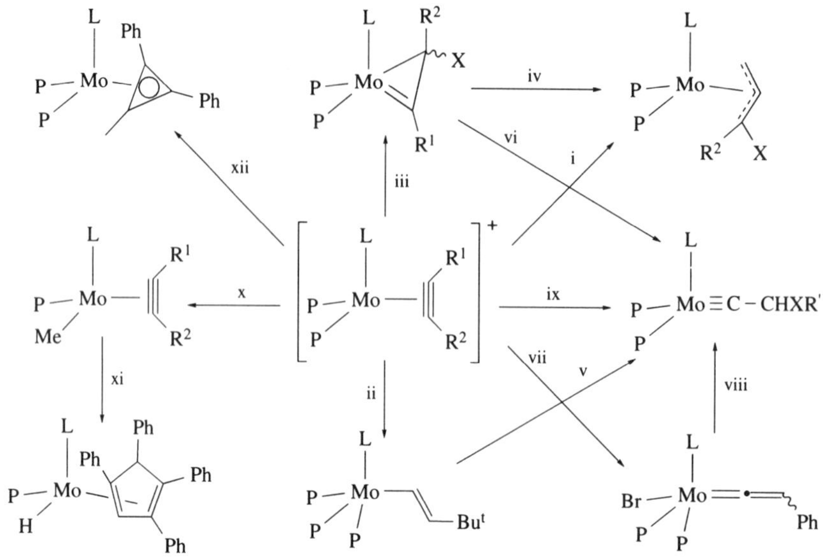

P = $P(OMe)_3$. i, $K[BHBu^s_3]$, L = Cp, indenyl, R^1 = Me, R^2 = Me, Et, Pr^i, X = H; ii, $K[BHBu^s_3]$ + excess $P(OMe)_3$, L = Cp, indenyl, R^1 = H, R^2 = Bu^t; iii, $K[BHBu^s_3]$, L = Cp, R^1 = R^2 = Ph, X = H; L = indenyl, R^1 = TMS, R^2 = H, X = H, $LiCuX_2$; L = Cp, R^1 = Br, Me, R^2 = X = Ph; L = Cp, R^1 = Pr^i, Bu^t, R^2 = H, X = Ph, C_6H_4Me-4; iv, RT, 3 d L = Cp, R^1 = Me, R^2 = X = Ph; v, heat, 100 °C, L = Cp, indenyl, R^1 = Bu^t, X = H; vi, heat, 25 °C, L = indenyl, R^1 = TMS, R^2 = X = H; vii, $K[BHBu^s_3]$, L = Cp, R^1 = Ph, R^2 = Br; viii, $K[BHBu^s_3]$, L = Cp, R^1 = Ph, X = H; ix, $K[BHBu^s_3]$, L = Cp, R^1 = Bu^t, R^2 = Br, X = H; x, MeMgI, L = Cp, R^1 = R^2 = Ph; xi, PhC≡CPh, benzene reflux, L = Cp, R^1 = R^2 = Ph; xii, CH_2=CHMgBr, L = Cp, R^1 = R^2 = Ph

Scheme 18

this area comprise a route to enantiomerically pure [W(CO)(NCMe)(η^2-MeC≡CPh)Tp']$^+$,[454] and an illustration of the diastereoselective elaboration of a coordinated alkyne substituent;[455] this latter reaction proceeds via Bu^nLi-initiated deprotonation of the alkyne Me substituent of [WI(CO)(η^2-MeC≡CPh)Tp'] to give an anionic η^2-allenyl intermediate followed by subsequent electrophilic addition.

Further additions to the chemistry of monocyclopentadienyl–monoalkyne complexes comprise synthetic and spectroscopic studies on [M(SR^1)(CO)(η^2-$R^2C≡CR^2$)Cp] (M = Mo or W; R^1 = C_6F_5, R^2 = CF_3;[456] R^1 = Me, Et, etc., R^2 = Ph[457]); the substitution of CO by PEt_3 in [W(SC_6F_5)(CO)(η^2-$CF_3C≡CCF_3$)Cp] proceeds via an isolable η^2-vinyl intermediate. Photolysis of [WR$(CO)_3$Cp] (R = Me, Et, etc.) in the presence of HC≡CH affords[458] the acyl product [W(COR)(CO)(η^2-HC≡CH)Cp], which is a precursor to metallacyclic,[458] carbene[459] and nitrosyl[460] complexes. New syntheses of metallocene derivatives [M(η^2-alkyne)Cp_2] (M = Mo or W) have also been reported. In the presence of $F_3CC≡CCF_3$, sodium/mercury reduction of [$MoCl_2$(η-C_5H_4Me)$_2$] yields [Mo(η^2-$CF_3C≡CCF_3$)(η-C_5H_4Me)$_2$], in which molybdenum–alkyne carbon bond lengths of 0.213 nm are indicative of the alkyne functioning as a two-electron donor.[461] More general, convenient syntheses of [Mo(η^2-alkyne)Cp_2] comprise reaction of [Mo$(H)_2Cp_2$] with either methyl acrylate in the presence of alkynes[462] or with PhC≡CPh to give [Mo(η^2-(Z)-PhHC=CHPh)Cp_2], in which the stilbene ligand is readily substituted by alkynes;[463] an analogous application of this latter synthesis provides a route to [W(η^2-PhC≡CPh)Cp_2].[464] Protonation of [Mo(η^2-alkyne)Cp_2] yields [MoH(η^2-alkyne)Cp_2]$^+$, isolable where alkyne = TMS-C≡C-TMS.[462]

6.8.2.4 Bis(alkyne) complexes containing cyclopentadienyl and related ligands

A range of synthetic routes to [MX(η^2-alkyne)$_2$Cp] (X = halide, thiolato, phenoxide, etc.) has been developed. Reaction of [Mo(NCMe)(η^2-MeC≡CMe)$_2$(η^5-L)]$^+$ (L = Cp or indenyl) with LiX (X = Br or Cl) proceeds with substitution of NCMe by X[231] and sodium/mercury reduction of [WX_4Cp*] (X = Cl or OPh) in the presence of two equivalents of $R^1C≡CR^2$ (R^1, R^2 = H, Me or TMS) yields [WX(η^2-$R^1C≡CR^2$)$_2$Cp*].[465] Complexes of the electron-withdrawing $F_3CC≡CCF_3$ ligand [MX(η^2-$F_3CC≡CCF_3$)$_2$Cp] (M = Mo or W; X = halide) are accessible via reaction of [MX$(CO)_3$Cp] with $F_3CC≡CCF_3$ in a sealed tube, but where M = Mo, X = Br or I, cyclopentadienone and cyclobutadiene (X = I) products result from alkyne–CO coupling unless the reaction mixture is purged of CO.[237]

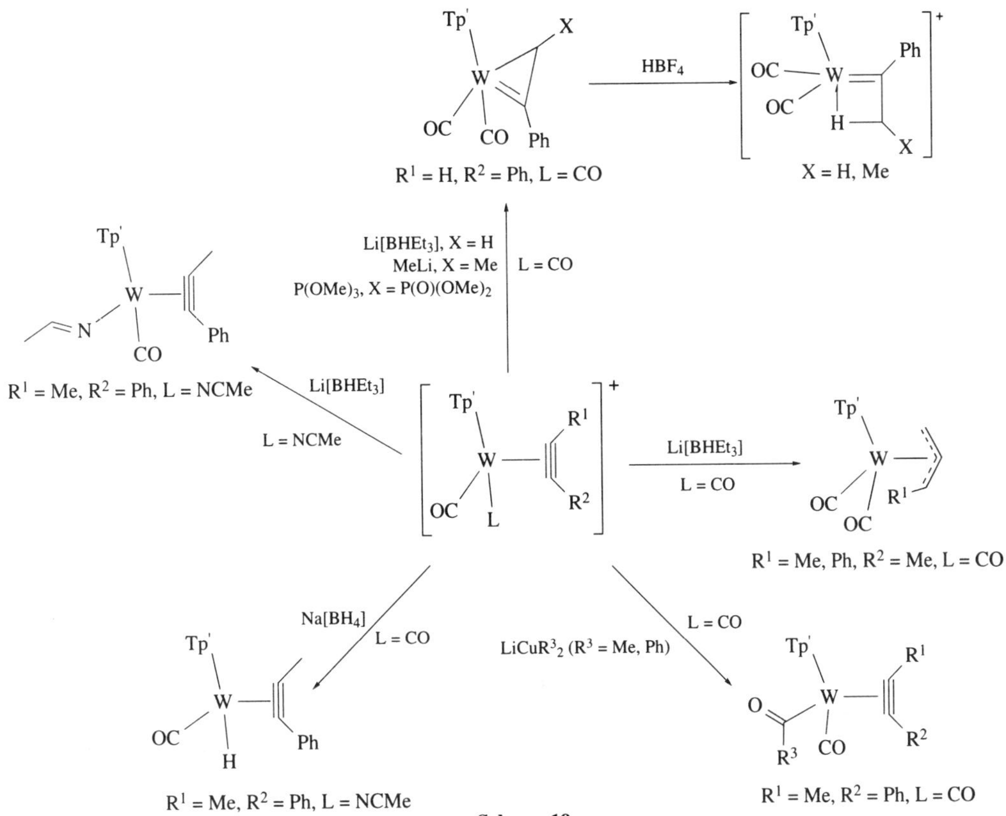

Scheme 19

The further reactions of $[MX(\eta^2\text{-}RC{\equiv}CR)_2Cp]$ are strongly dependent upon the identity of the alkyne substituents. Where R = Me, activation towards reactions with nucleophiles, as summarized in Scheme 12 (Section 6.4.4), requires initial protonation to give the η^2-vinyl–alkyne complex (**77**);[231] the dehydrohalogenation of $[MoBr(\eta^2\text{-}MeC{\equiv}CR^1)_2Cp]$ (R^1 = Me or Ph) to give[162] prop-2-ynyl complexes (**61**) has been discussed in Section 6.3.7. By contrast, complexes of the $F_3CC{\equiv}CCF_3$ ligand undergo addition at the alkyne without prior activation. The halide substitution reactions of $[MCl(\eta^2\text{-}F_3CC{\equiv}CCF_3)_2Cp]$ (M = Mo or W) with a series of anionic ligands L (L = thiolato[466–8] or $MeCO_2^-$, $Me_2NCS_2^-$, etc.[238]) yield two product types dependent upon the identity of L; thus, either the simple, substituted species $[M(L)(\eta^2\text{-}F_3CC{\equiv}CCF_3)_2Cp]$ or η^2-vinyl complexes are formed. For example, reaction of $[WCl(\eta^2\text{-}F_3CC{\equiv}CCF_3)_2Cp]$ with arylthiolato reagents, TlSR, yields $[W(SR)(\eta^2\text{-}F_3CC{\equiv}CCF_3)_2Cp]$ (R = Ph, C_6H_4Me-4 or C_6F_5) but alkylthiolato reagents yield (**164**) (R = Me, Et, Pr^i, Bu^t) together with a second isomer where R = Me or Et.[466] Addition of neutral ligands PEt_3, $CNBu^t$ etc. to $[MX(\eta^2\text{-}F_3C{\equiv}CCF_3)_2Cp]$ (M = Mo or W; X = Cl, C_6H_4Me-4 etc.) or (**164**) also yields η^2-vinyl products;[466] in the case of the reaction of $[MoCl(\eta^2\text{-}F_3CC{\equiv}CCF_3)_2Cp]$ with PEt_3, two isomeric forms, (**165**) and (**166**), have been isolated and crystallographically characterized. Isomer (**165**) is the kinetic product which converts to the thermodynamic product (**166**) via a proposed η^1-vinylphosphonium intermediate.[469] Reactions of these η^2-vinyl complexes leading to metallacyclic,[239,470,471] cyclobutadiene[239] and ethenethiolato[472] products have been described. A summary of the reaction sequence deduced for the coupling reactions of $[M(SR^1)(\eta^2\text{-}F_3CC{\equiv}CCF_3)_2Cp]$ (M = Mo or W; R^1 = alkyl) with alkynes $R^2C{\equiv}CR^3$ (R^2, R^3 = H, Me, Ph, etc.) is given in Scheme 20.[473] The relative stabilities of possible products exhibit a marked dependence upon the identity of R^1, R^2 and R^3.

Some chemistry of cationic $[M(L)(\eta^2\text{-alkyne})_2(\eta^5\text{-dienyl})]^+$ (M = Mo or W; L = CO or phosphorus-donor ligand; dienyl = Cp, C_5Me_5 or indenyl) has also been developed. The established route to $[Mo(CO)(\eta^2\text{-alkyne})_2Cp]^+$ involves $Ag[BF_4]$-induced cleavage of the Mo–Mo bond of $[Mo_2(CO)_6Cp_2]$ in the presence of alkyne but alternative syntheses, which provide access to an extended range of complexes, comprise treatment of $[M(CO)_3Cp^*]^+$ with $R^1C{\equiv}CR^2$ (M = Mo, $R^1 = R^2$ = Ph;[474] M = W, R^1 = H, R^2 = Bu^t[163]) or reaction of $[Mo(L)(CO)_2(FBF_3)Cp]$ (L = CO or phosphorus-donor ligand) with $R^1C{\equiv}CR^2$ although this latter reaction exhibits poor product selectivity.[475] The acetonitrile complexes $[Mo(NCMe)(\eta^2\text{-}R^1C{\equiv}CR^2)_2(\eta^5\text{-dienyl})]^+$ (R^1, R^2 = H, Me, Ph, etc.; dienyl = Cp or indenyl) are accessible[438] via acetonitrile reflux of $[Mo(CO)(\eta^2\text{-}R^1C{\equiv}CR^2)_2(\eta^5\text{-dienyl})]^+$ in the presence of $R^1C{\equiv}CR^2$.

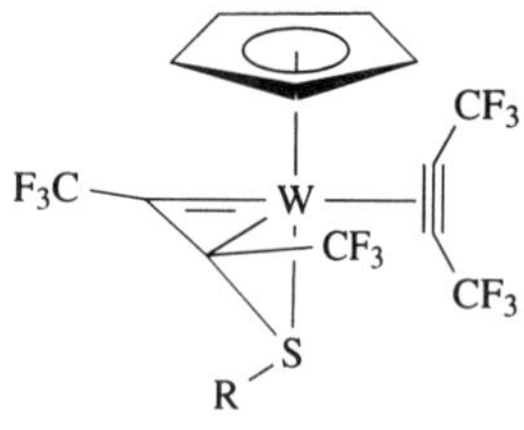

(164)

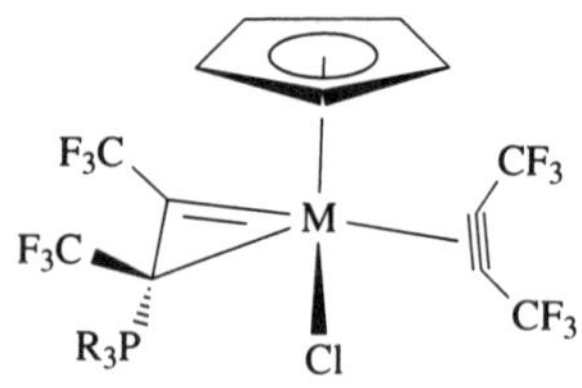

(165)

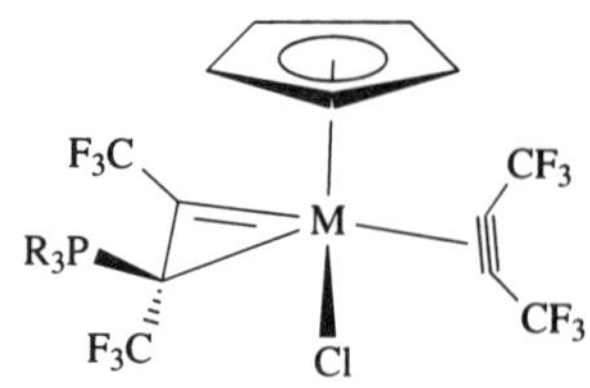

(166)

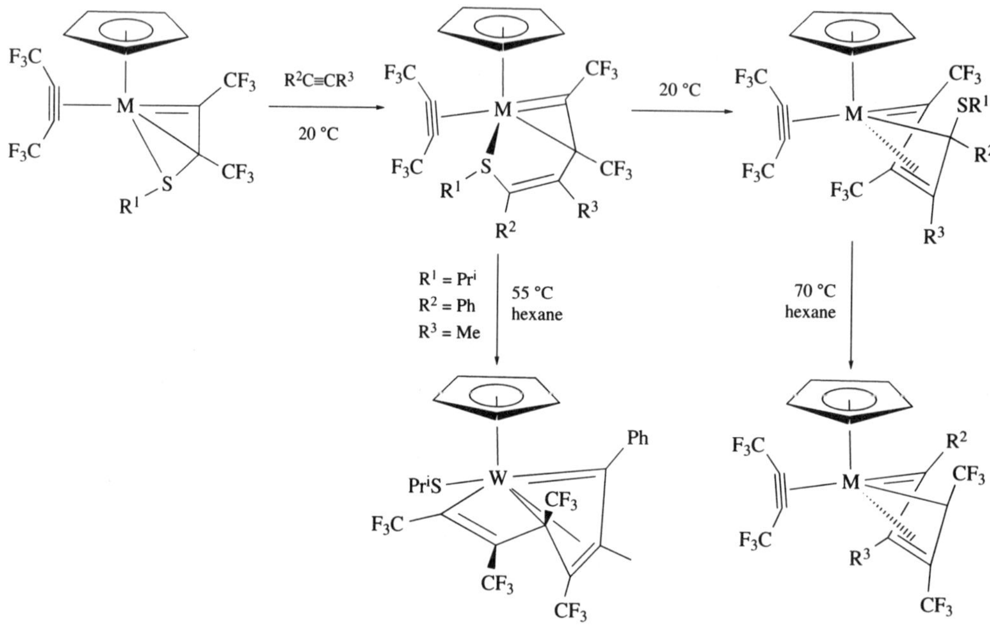

Scheme 20

Recent physical investigations on $[M(L)(\eta^2\text{-alkyne})_2(\eta^5\text{-dienyl})]^+$ comprise crystallographic studies on $[Mo(L)(\eta^2\text{-MeC≡CMe})_2Cp]^+$ (L = CO or NCMe) which reveal that the alkyne C≡C axis is oriented parallel to the Mo–L bond direction,[476] cyclic voltammetric investigations on $[Mo(CO)(\eta^2\text{-PhC≡CPh})_2Cp^*]^+$, which uncover two sequential, quasireversible, one-electron reduction processes,[474] and the application of two-dimensional ^1H NMR exchange techniques to probe independent alkyne rotations in $[Mo(CO)(\eta^2\text{-HC≡CR})_2Cp]^+$ (R = Me, Ph, Bu^t or CO_2Me).[477]

In addition to their application in the synthesis of monoalkyne complexes,[438] $[M(L)(\eta^2\text{-alkyne})_2(\eta^5\text{-dienyl})]^+$ exhibit a rich chemistry of ligand-coupling processes. Treatment of $[Mo(CO)(\eta^2\text{-MeC≡CMe})_2Cp]^+$ with $K[BHBu^s_3]$ or $LiMe_2Cu$ in the presence of CO yields (**167**) (R = H or Me).[478] The complexes $[Mo(NCMe)(\eta^2\text{-}R^1C≡CR^2)_2(\eta^5\text{-L})]^+$ (L = Cp or indenyl) exhibit metal-promoted alkyne–polyene coupling. Reaction with cot is proposed to proceed via the 16-electron complexes (**168**) (R^1, R^2 = Me, Et or Ph) which either eliminate bicyclo[4.2.2]deca-2,4,7,9-trienes (formal [6 + 2] adducts of cot with alkyne) or, where $R^1 = R^2$ = Me or Et, form (**169**) (R^3 = H or Me) via metal-assisted hydrogen migration from the R^1, R^2 substituents of (**168**); subsequent reactions of (**169**) with $[Ph_3C]^+$, $Ag[BF_4]$ and H^+ have been described.[479] The one-electron reduction of $[Mo(NCMe)(\eta^2\text{-MeC≡CMe})_2(\eta^5\text{-dienyl})]^+$ (dienyl = Cp or indenyl) leading to binuclear products with alkyne-derived C_6 and C_8 ligands will be described in Section 6.9.2. Deprotonation of $[W(CO)(\eta^2\text{-HC≡CBu}^t)_2Cp^*]^+$ yields the mixed alkynyl–alkyne product $[W(C≡CBu^t)(CO)(\eta^2\text{-HC≡CBu}^t)Cp^*]$; subsequent sequential methylation and deprotonation lead to formation of (**63**)[163] as discussed in Section 6.3.7.

Bis(alkyne) complexes containing the hydro-tris(3,5-dimethylpyrazolyl)borato ligand (Tp'), $[W(CO)(\eta^2\text{-alkyne})_2Tp']^+$ may be prepared via reaction of $[WI(CO)(\eta^2\text{-alkyne})Tp']$ with $Ag[BF_4]$/alkyne or $[W(CO)_2(\eta^2\text{-alkyne})Tp']^+$ with alkyne. Deprotonation of $[W(CO)(\eta^2\text{-HC≡CPh})_2Tp']^+$ yields $[W(C≡CPh)(CO)(\eta^2\text{-HC≡CPh})Tp']$, whereas reaction with $LiBEt_3H$ affords the η^1-vinyl $[W\{C(Ph){=}CH_2\}(CO)(\eta^2\text{-HC≡CPh})Tp']$, which is a precursor to the mixed alkyne–carbene $[W\{{=}C(Me)Ph\}(CO)(\eta^2\text{-HC≡CPh})Tp']^+$.[450]

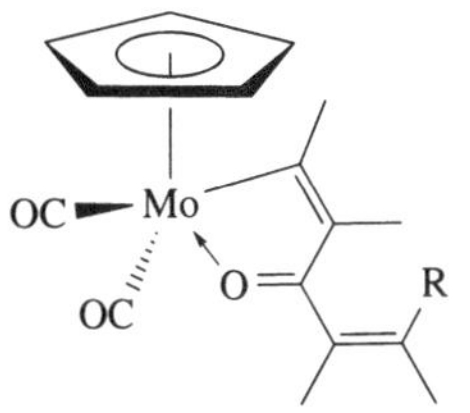

(167)

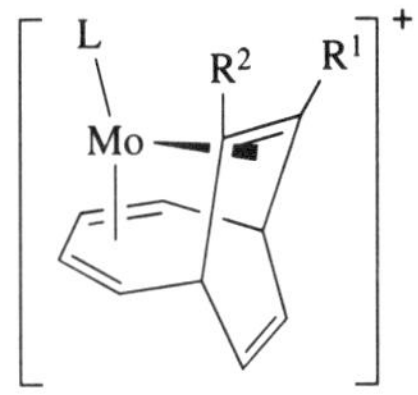

(168)

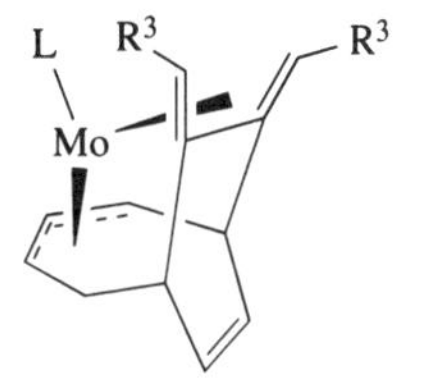

(169)

6.8.3 Alkyne Complexes of M^{IV} (d^2) and Other High Oxidation State Metal Centres

A range of bimetallic, chloride bridged complexes $[\{WCl_3(\eta^2\text{-alkyne})\}_2(\mu\text{-Cl})_2]$ (alkyne = PhC≡CPh,[480] PhC≡CH, TMS-C≡CH,[481] TMS-C≡C-TMS,[482] HC≡CH, DC≡CD,[483] ClC≡CCl,[484] ClC≡C-TMS, PhC≡C–C≡C-TMS, etc.[485]) has been prepared by reaction of WCl_6 with the appropriate alkyne in CCl_4 or CH_2Cl_2, sometimes in the presence of C_2Cl_4. Alternatively, treatment of WCl_4 with RC≡CR (R = Me or Et) in diethylether yields monomeric $[WCl_4(\eta^2\text{-RC≡CR})(OEt_2)]$, which is converted to $[\{WCl_3(\eta^2\text{-RC≡CR})\}_2(\mu\text{-Cl})_2]$ by removal of Et_2O *in vacuo*.[486] Some molybdenum analogues have also been described. Reaction of $MoCl_5$ with PhC≡CPh in CH_2Cl_2, or ClC≡C-TMS in CCl_4, gives $[\{MoCl_3(\eta^2\text{-R}^1\text{C≡CR}^2\}_2(\mu\text{-Cl})_2]$ ($R^1 = R^2$ = Ph;[480] R^1 = Cl, R^2 = TMS[485]); alternatively, $MoCl_5$ reacts with PhC≡CPh in the presence of $POCl_3$ to give $[MoCl_4(POCl_3)(\eta^2\text{-PhC≡CPh})]$, which readily dissociates $POCl_3$ to form $[\{MoCl_3(\eta^2\text{-PhC≡CPh})\}_2(\mu\text{-Cl})_2]$.[487] Reaction of $MoCl_4$ with RC≡CR (R = Me, Et or Ph) in Et_2O yields thermally sensitive, monomeric $[MoCl_4(\eta^2\text{-RC≡CR})(OEt_2)]$, which are precursors to $[Mo(L)_4(\eta^2\text{-RC≡CR})]$ ($L = OC_6H_3Pr^i{}_2$-2,6 or $SC_6H_2Pr^i{}_3$-2,4,6).[488] Related tungsten complexes $[WCl_2(OC_6H_3R_2\text{-}2,6)_2(\eta^2\text{-alkyne})]$ (R = Pr^i or Ph) are accessible via sodium/mercury reduction of $[WCl_4(OC_6H_3R_2\text{-}2,6)_2]$ in the presence of alkyne.[437]

Some reaction chemistry of $[\{WCl_3(\eta^2\text{-alkyne})\}_2(\mu\text{-Cl})_2]$ has been developed. Treatment with Cl^- affords anionic $[WCl_5(\eta^2\text{-alkyne})]^-$ [481,486] which react with ZnR_2 (R = Me or CH_2TMS) to give $[WCl(R)_3(\eta^2\text{-alkyne})]$.[486] Reduction of $[\{WCl_3(\eta^2\text{-PhC≡CPh})\}_2(\mu\text{-Cl})_2]$ in the presence of phosphines PR_3 ($PR_3 = PMe_3$, PMe_2Ph or $PMePh_2$) yields paramagnetic *mer-trans*-$[WCl_3(PR_3)_2(\eta^2\text{-PhC≡CPh})]$,[398] which may be further reduced in the presence of L (L = PR_3,[398] ethene or propene[399]) to *cis-trans*-$[WCl_2(PR_3)_2(L)(\eta^2\text{-PhC≡CPh})]$; XPS data suggest that complexes of formulation $[WCl_5(\eta^2\text{-alkyne})]^-$, $[WCl_3(PR_3)_2(\eta^2\text{-alkyne})]$ and $[WCl_2(PR_3)_3(\eta^2\text{-alkyne})]$ may be best described as possessing respective d^0, d^1 and d^2 configurations with the alkyne ligand acting as a dianion.[398]

A series of complexes $[M(E)(\eta^2\text{-RC≡CR})(S_2CNEt_2)_2]$ (M = W, E = O, R = H;[489] M = W, E = S, R = Ph;[490] M = Mo, E = NC_6H_4Me-4, R = CO_2Me[491]) and *cis-trans*-$[WCl_2(PMe_3)_2(NR^2)(\eta^2\text{-R}^1\text{C≡CR}^1)]$ (R^2 = Ph or $CHMe_2$, R^1 = H or Ph)[492] has been reported. The common structural feature of these species is a *cis* arrangement of alkyne and π-donor E ligands with the alkyne C≡C axis arranged perpendicular to the M–E vector. As such, the alkyne and E ligands compete as π-donors to the metal centre and structural data for M to alkyne carbon distances suggest that in $[WCl_2(PMe_3)_2(NPh)(\eta^2\text{-PhC≡CPh})]$ the alkyne functions essentially as a two-electron donor, while in $[W(S)(S_2CNEt_2)_2(\eta^2\text{-PhC≡CPh})]$ the alkyne competes more effectively with sulfur in acting as a π-donor to the metal centre. Treatment of *cis-trans*-$[WCl_2(PMe_3)_2(NPh)(\eta^2\text{-PhC≡CPh})]$ with LiC≡CPh affords *trans-cis*-$[W(C≡CPh)_2(PMe_3)_2(NPh)(\eta^2\text{-PhC≡CPh})]$; the stability of this complex with respect to alkyne–alkynyl coupling has been attributed to the bonding properties of the *cis* imido–alkyne structural unit.[493] A few high-oxidation-state cyclopentadienyl alkyne complexes are known. The unusual, paramagnetic tungsten(III) species $[WCl_2(\eta^2\text{-MeC≡CMe})(\eta\text{-}C_5Me_4Bu^t)]$, in which the alkyne functions as a four-electron donor, results from reaction of $[WCl_3(DME)(≡CBu^t)]$ with excess MeC≡CMe[494] and air oxidation of $[M(C_6F_5)(CO)(\eta^2\text{-}F_3CC≡CCF_3)Cp]$ (M = Mo or W) affords the M^{IV} oxo complexes $[M(C_6F_5)(O)(\eta^2\text{-}F_3CC≡CCF_3)Cp]$.[456]

6.8.4 Alkyne Complexes Derived from Intramolecular Coupling of C_1 Ligands

Intramolecular coupling processes between carbyne, CO and isocyanide ligands result in coordinated alkoxyalkyne (RC≡COR), diaminoalkyne ($R_2NC≡CNR_2$) and aminoalkyne ($R_2NC≡CR$) products. A detailed, general review of these ligand coupling processes is available.[495]

A general scheme for carbyne–carbonyl coupling leading to alkoxyalkynes is given in Scheme 21. Most examples involve nucleophilic-assisted formation of an isolable η^2-ketenyl intermediate which is subsequently alkylated at the ketenyl oxygen (Scheme 21 (i)). Thus, $[W(CO)_2(\equiv CR^1)Cp]$ (R^1 = Me, C_6H_4Me-4, etc.) reacts with one equivalent of PMe_3 to give the η^2-ketenyl complex $[W(CO)(PMe_3)\{\eta^2\text{-}C(O)CR^1\}Cp]$ which, on treatment with $[R^2_3O][BF_4]$ (R^2 = Me or Et), yields $[W(CO)(PMe_3)(\eta^2\text{-}R^1C{\equiv}COR^2)Cp]^+$.[496] Related syntheses of $[W(CO)(PR_3)(\eta^2\text{-}HOC{\equiv}CC_6H_4Me\text{-}4)Cp]^+$ ($PR_3 = PMe_3$, PPr^i_3, etc.),[497] $[W(S_2CNMe_2)(CO)(dppe)(\eta^2\text{-}ROC{\equiv}CCH_2Ph)]^+$ (R = H or Me),[498] $[W(CO)(N\text{–}N)(\eta^2\text{-}RC{\equiv}COMe)]$ (R = Me or Ph, N–N = pyrrole-2-carboxyaldehyde methyliminato),[499] and $[M(S_2CNR^3{}_2)_2(CO)(\eta^2\text{-}R^1C{\equiv}COCOR^2)]$ (M = Mo or W, R^3 = Me or Et, R^1 = Ph or $C_6H_4Bu^t$-4, R^2 = Me, Ph, etc.)[500] have been described. Similarly, nucleophile-induced carbonyl–thiocarbyne coupling provides syntheses of $[W(CO)(PR_3)(\eta^2\text{-}MeOC{\equiv}CSMe)\{HE(pyrazolyl)_3\}]^{n+}$ (R = Et, E = B, n = 1;[501] R = Me, E = C, n = 2[502]). The formation of $[WCl(CO)(PMe_3)_3(\eta^2\text{-}HC{\equiv}COAlCl_3)]$ via reaction of $[WCl(PMe_3)_4(\equiv CH)]$ with CO in the presence of $AlCl_3$[503] is probably a unique example of electrophile-induced carbyne-CO coupling (Scheme 21 (ii)) although the photoinduced formation of *cis-trans*-$[WCl_2(PMe_2R)_2(CO)(\eta^2\text{-}PhC{\equiv}COE)]$ (R = Me or Ph, E = H, C(O)Bu^t, $SiBu^tPh$) from $[WCl(CO)_2(PMe_2R)_2(\equiv CPh)]$ in the presence of ECl[504] may be electrophile assisted (Scheme 21 (iii)).

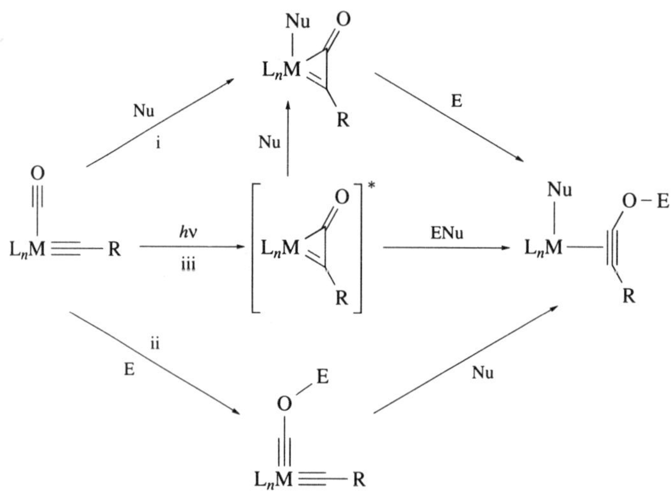

Scheme 21

The formation of diaminoalkyne ligands via coupling of two isocyanide units was first reported in the synthesis of $[MX(CNR)_4(\eta^2\text{-}RHNC{\equiv}CNHR)]^+$ (M = Mo, X = halide, R = Bu^t or Cy; M = W, X = I, R = Bu^t) from reaction of $[MX(CNR)_6]^+$ with zinc in wet THF[505,506] and subsequently extended to analogous reactions of $[MoCl(CNBu^t)_4(bipy)]^+$, $[Mo(CNR)_5(bipy)]^{2+}$ (R = Me, Et, Bu^t, etc.)[507] and $[Mo(CNBu^t)_7]^{2+}$.[508] There is now good evidence that isocyanide coupling in $[MX(CNR)_6]^+$ proceeds as shown in Scheme 22 (L = CNR, n = 4). The two-electron reduction of $[MX(CNR)_6]^+$ yields[509,510] isolable $[M(CNR)_6]$ and subsequent reaction of $[M(CNR)_6]$ (M = Mo or W, R = Et or Bu^t) with Et_3O^+ results in ethylation at an isocyanide nitrogen to give the aminocarbyne $[M(CNR)_5(\equiv CNEtR)]^+$, which is converted to $[MI(CNR)_4(\eta^2\text{-}RHNC{\equiv}CNEtR)]^+$ via reaction with HI.[510] Further examples of aminocarbyne–isocyanide coupling are provided by the complexes $[WI(CO)_2(CNBu^t)_2(\equiv CNEt_2)]$[511] and $[M(CNEt)_5(\equiv CNEt_2)]^+$ (M = Mo or W);[512] in the latter case reaction of $[M(CNEt)_5(\equiv CNEt_2)]^+$ with Et_3O^+ produces a spectroscopically observed species, tentatively formulated as the bis(aminocarbyne) $[M(CNEt)_4(\equiv CNEt_2)_2]^{2+}$, which reacts with NEt_4Br to yield $[MBr(CNEt)_4(\eta^2\text{-}Et_2NC{\equiv}CNEt_2)]^+$, consistent with the final two steps in Scheme 22.

In addition to aminocarbyne–isocyanide coupling, carbyne–isocyanide coupling, leading to coordinated aminoalkynes, $R_2NC{\equiv}CR$, has been demonstrated by the formation of $[WBr_2(CNBu^t)_3(\eta^2\text{-}PhC{\equiv}CNHBu^t)]$ from reaction of $[WBr(CNBu^t)_4(\equiv CPh)]$ with HBr in CH_2Cl_2.[513] Presumably this latter reaction proceeds via an aminocarbyne–carbyne intermediate but, in $[WCl(CO)(PMe_3)_2(CNBu^t)(\equiv CPh)]$, the site of protonation ($CNBu^t$ vs. CPh) is a competitive process.[514,515] Two remarkable examples of alkyne formation via direct carbyne–carbyne coupling have been established. As shown in Scheme 23, the first depends on the oxallyl bromide initiated, *in situ* conversion of an acyl to a carbyne ligand,[516] while the bis(aminocarbyne) complex $[W(CNEt)(\equiv CNEt_2)_2Cp^*]^+$ reacts with CNEt at room temperature to yield $[W(CNEt)_2(\eta^2\text{-}Et_2NC{\equiv}CNEt_2)Cp^*]^+$.[517]

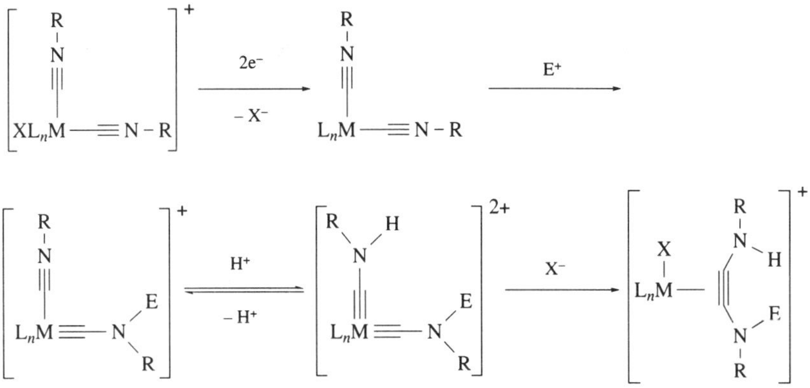

Scheme 22

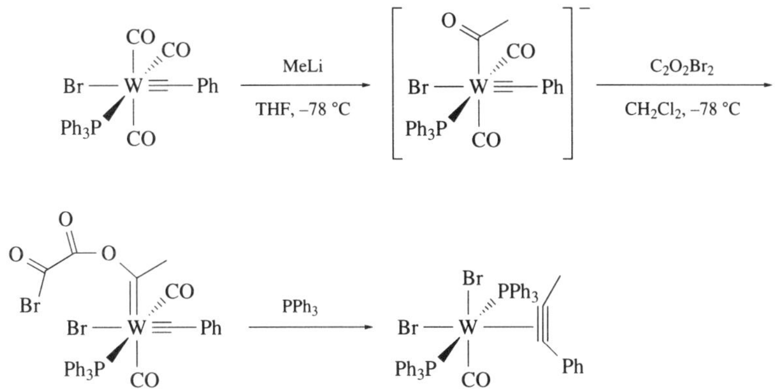

Scheme 23

6.9 METAL–METAL-BONDED BINUCLEAR COMPLEXES CONTAINING π-BONDED LIGANDS

6.9.1 Complexes Derived from $[M_2X_6]$ (M≡M) (X = Alkoxide, Alkyl, Halide, etc.)

An extensive chemistry has developed involving addition of unsaturated molecules to the metal–metal triple bond of hexa-alkoxide d^3-d^3 dimers $[M_2(OR)_6]$ (M = Mo or W, R = Pr^i, Bu^t, CH_2Bu^t, etc.) and related complexes. This work is set apart from the analogous chemistry of $[M_2(CO)_4Cp_2]$ by the π-donating capability of alkoxide ligands and by the extreme sensitivity of the course of reaction to substituent effects, both in the alkoxide ligand and the reactant alkyne, allene, etc. A series of reviews[518] include relevant aspects of this chemistry and a specific review is available on reactions of alkynes with $[W_2(OR)_6]$.[519]

Reaction of $[M_2(OBu^t)_6]$ with alkynes $R^1C{\equiv}CR^2$ in nonpolar solvents generally results in metathesis-like scission of C≡C and M≡M bonds. Thus, treatment of $[Mo_2(OBu^t)_6]$ with PhC≡CH yields $[Mo(OBu^t)_3({\equiv}CPh)]$ although, with less sterically demanding HC≡CH, unstable $[Mo_2(OBu^t)_6(\mu\text{-}HC_2H)]$ may be formed.[520] Reaction of $[W(OBu^t)_6]$ with PhC≡CPh yields (**170**) ($R^1 = OBu^t$, R^2 = Ph), together with the alkyne scission product, carbyne-bridged $[W_2(OBu^t)_4(\mu\text{-}CPh)_2]$.[521] A more extensive range of alkyne bridged products is accessible via reaction of $R^2C{\equiv}CR^2$ with $[M_2(OR^1)_6]$ (M = Mo, $R^1 = Pr^i$ or CH_2Bu^t;[522] M = W, $R^1 = Pr^i$, Bu^t;[523] M = W, $R^1 = CH_2Bu^t$;[523,524] M = W, R^1 = O-TBDMS[525]) in the presence of pyridine. Three product types, (**171**), (**172**) and (**173**), each with a perpendicular μ-alkyne and bridging alkoxide ligands, are observed dependent upon M, R^1 and the alkyne substituent R^2. All molybdenum complexes adopt structure (**171**), while sterically demanding R^1 or R^2 substituents in tungsten complexes favour structure (**173**) in which back-donation to alkyne π* orbitals (and therefore the alkyne C≡C bond length) is at a maximum.[526] The combined steric requirements of R^1 and R^2 may promote alkyne scission and there is good evidence for a solution equilibrium between $[W_2(OR^1)_6(\mu\text{-}R^2C_2R^2)(py)_n]$ and the carbyne $[W(OR^1)_3({\equiv}CR^2)]$ ($R^1 = Bu^t$, R^2 = H;[523] $R^1 = CH_2Bu^t$, R^2 = Me, Et or Ph[524]). Additional confirmation of a carbyne ⇌ μ-alkyne equilibrium is provided by the generation of (**174**) from reaction of $[W(OBu^t)_3({\equiv}CMe)]$ with CO[527] and this synthetic strategy has been extended to the formation of cycloalkyne-bridged bimetallics;[528] in both these cases the μ-alkyne species is probably stabilized by the π-acceptor CO ligand. However, by contrast, reaction of $[W(OPr^i)_3(py)_2(\equiv CNMe_2)]$ with CO leads to formation of (**175**).[527]

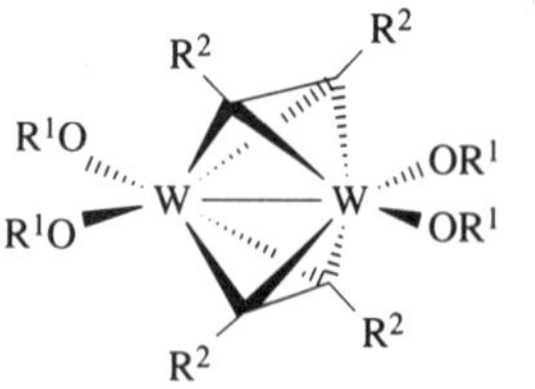

(**170**)

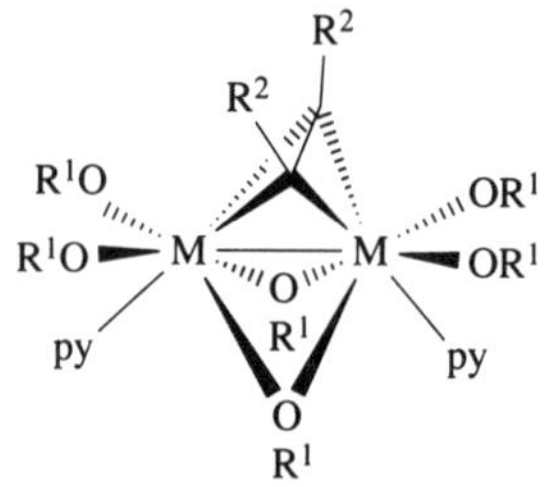

(**171**)

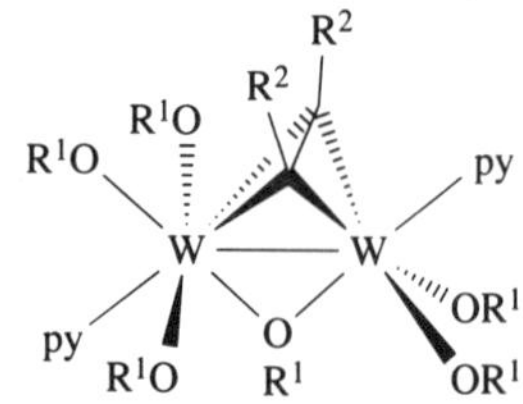

(**172**)

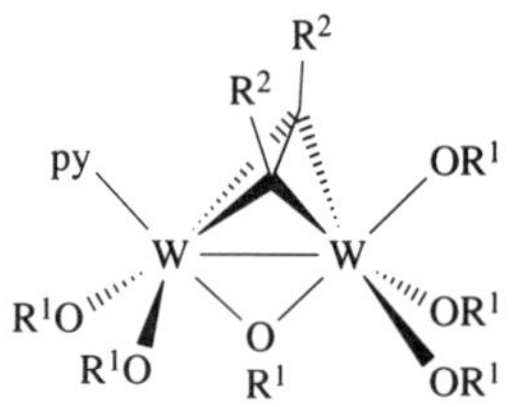

(**173**)

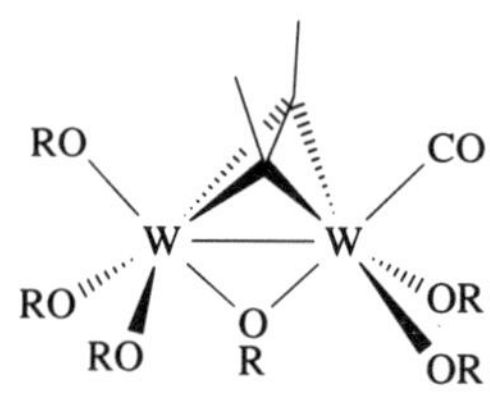

R = Bu^t

(**174**)

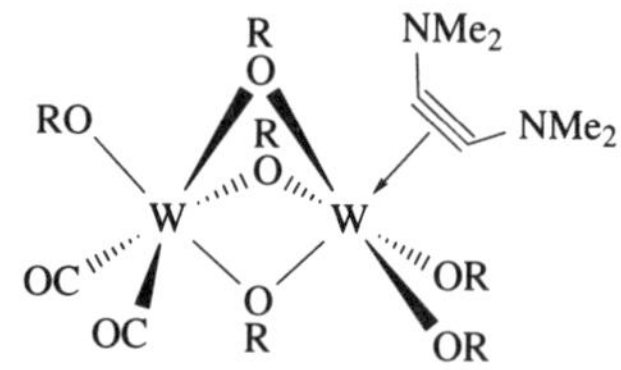

R = Pr^i

(**175**)

Alkyne-bridged complexes of types (**171**)–(**173**) undergo C–C coupling reactions at the μ-alkyne ligand with nitriles[529] and alkynes.[522,530] Thus, reaction of (**171**) (M = Mo, $R^1 = CH_2Bu^t$, R^2 = H) with HC≡CH affords (**176**) (L = py, R^2 = H) and the analogous tungsten complex is an isolable intermediate in the formation of (**177**) from (**172**) ($R^1 = CH_2Bu^t$, R^2 = H) and HC≡CH; other tungsten derivatives of structure type (**177**) have been isolated but the corresponding molybdenum complexes have not been observed, probably due to the enhanced reactivity of the molybdenum system towards alkyne cyclotrimerization. An alternative approach to (**176**) ($L = NCNEt_2$, $R^1 = CH_2Bu^t$, R^2 = Me) involves reaction of $[Mo_2(OCH_2Bu^t)_6(\mu\text{-}\eta^1,\eta^2\text{-}NCNEt_2)]$ with two equivalents of MeC≡CMe but a subsequent insertion process results in the formation of (**178**).[531]

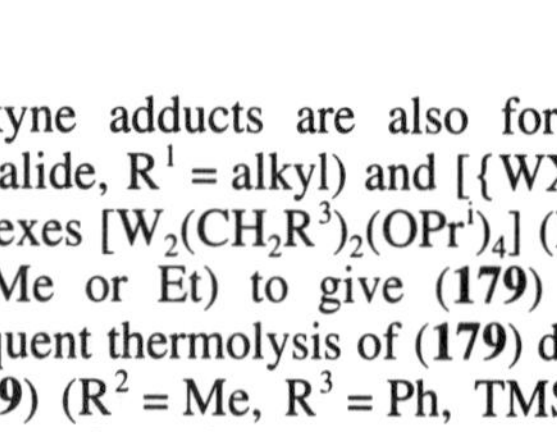

(**176**)

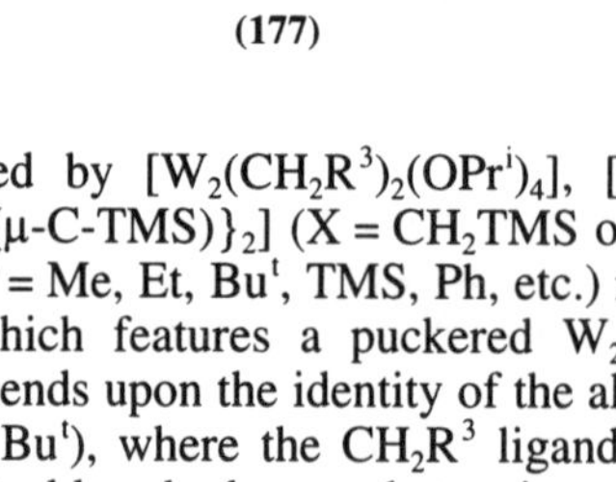

(**177**)

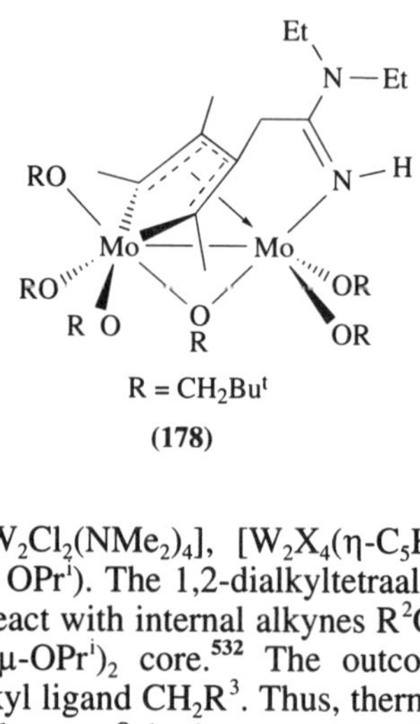

R = CH_2Bu^t

(**178**)

Alkyne adducts are also formed by $[W_2(CH_2R^3)_2(OPr^i)_4]$, $[W_2Cl_2(NMe_2)_4]$, $[W_2X_4(\eta\text{-}C_5H_4R^1)_2]$ (X = halide, R^1 = alkyl) and $[\{WX_2(\mu\text{-C-TMS})\}_2]$ ($X = CH_2TMS$ or OPr^i). The 1,2-dialkyltetraalkoxide complexes $[W_2(CH_2R^3)_2(OPr^i)_4]$ (R^3 = Me, Et, Bu^t, TMS, Ph, etc.) react with internal alkynes $R^2C\equiv CR^2$ (R^2 = Me or Et) to give (**179**) which features a puckered $W_2(\mu\text{-}OPr^i)_2$ core.[532] The outcome of subsequent thermolysis of (**179**) depends upon the identity of the alkyl ligand CH_2R^3. Thus, thermolysis of (**179**) (R^2 = Me, R^3 = Ph, TMS, Bu^t), where the CH_2R^3 ligand has no β hydrogen, proceeds with coupling of terminal alkynes and a double α-hydrogen abstraction to give (**180**) (X = H or CH_2R^3)[533] but, by contrast, complexes (**179**) (R^2 = Me, R^3 = Me or Et; R^2 = Me or Et, R^3 = Pr^i) are converted to (**170**) (R^1 = Pr^i, R^2 = Me or Et) via β-elimination although, where R^3 = Me or Et, a competitive α-elimination process also operates.[534] Unstable alkyne bridged complexes $[\{WCl(py)(NMe_2)_2\}_2(\mu\text{-}NMe_2)_2(\mu\text{-}R^1C_2R^2)]$ (R^1 = H, R^2 = H, Me or Ph; $R^1 = R^2$ = Me) are generated in the reaction of $[W_2Cl_2(NMe_2)_4]$ with $R^1C\equiv CR^2$ in the presence of pyridine but, with the exception of the derivative $R^1 = R^2$ = H, these undergo a thermally initiated, alkyne substituent dependent, ligand redistribution process to give (**181**)

(L = py, R^2 = Ph) or (**182**).[535] An unusual structural feature of (**182**) is a twisting of the alkyne C–C axis by 35° away from a perpendicular orientation with respect to the W–W bond and this observation has been rationalized in terms of a second-order Jahn–Teller effect.[536] However, the related complexes (**181**) (L = py, R^2 = Ph;[535] L = PMe_3 or PMe_2Ph, R^2 = H[537]) and (**183**) (R^1 = H or Me)[538] do not exhibit any significant, corresponding twisting of the alkyne C–C axis. Complex (**181**) (L = PMe_3 or PMe_2Ph, R^2 = H) slowly rearranges to the μ-η^1,η^2-vinyl species (**184**) via hydrogen atom transfer from NMe_2 to coordinated HC≡CH.[537] Reaction of [W_2X_4(η-$C_5H_4R^1$)$_2$] (W≡W) (X = Br or Cl; R^1 = Me or Pr^i) with $R^2C{\equiv}CR^2$ (R^2 = Me, Et, Ph, etc.) yields [W_2X_4(μ-$R^2C_2R^2$)(η-$C_5H_4R^1$)$_2$]; the derivative (X = Br, R^1 = Pr^i, R^2 = Ph) exhibits a deviation of 26° of the alkyne C–C axis away from orthogonality with respect to the W–W vector. With excess MeC≡CMe, [W_2X_4(η-$C_5H_4R^1$)$_2$] forms (**185**) (X = Cl, R^1 = Me or Pr^i; X = Br, R^1 = Pr^i) which feature a symmetrically bonded C_4 'fly-over' bridge;[539,540] this highly unusual structure, of which [Mo_2Cl_2(μ-C_4Ar_4)Cp_2] (Ar = Ph or C_6H_4Me-4) is the only other example,[541] contrasts with the normal σ^2:η^4-metallacyclopentadiene arrangement of (**176**). The carbyne bridged complexes [{WX_2(μ-C-TMS)}$_2$] (W–W) (X = CH_2TMS or OPr^i) react with alkynes $R^1C{\equiv}CR^2$ to give dimetallaallyl bridged products (**186**) (X = CH_2TMS, R^1, R^2 = H, Me or Ph; X = OPr^i, R^1 = R^2 = H) via alkyne insertion into the tungsten to bridging carbyne bond; in the case where X = CH_2TMS, R^1 = Ph, R^2 = Me, an intermediate, terminal η^2-alkyne complex has been identified.[542,543]

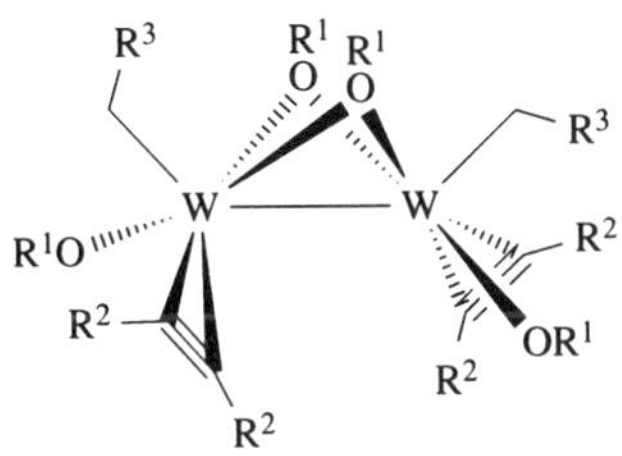

R^1 = Pr^i

(179)

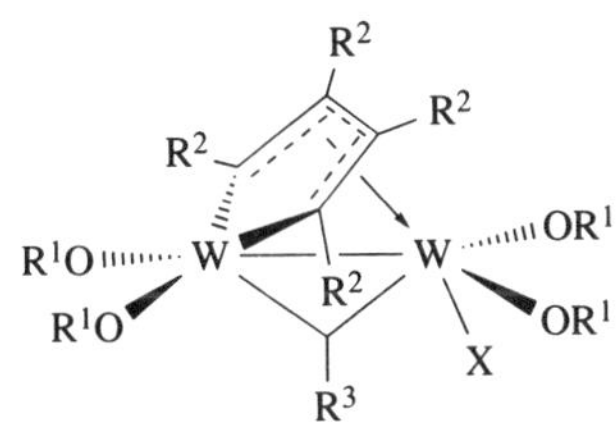

R^1 = Pr^i

(180)

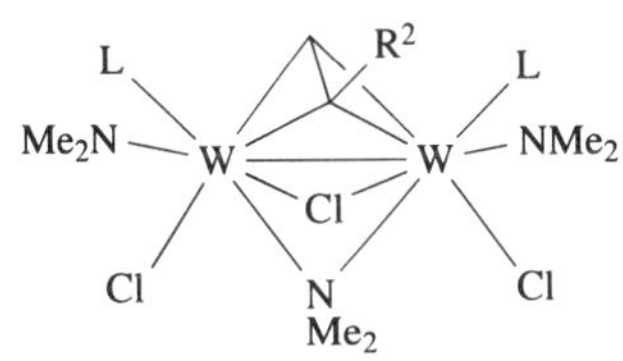

(181)

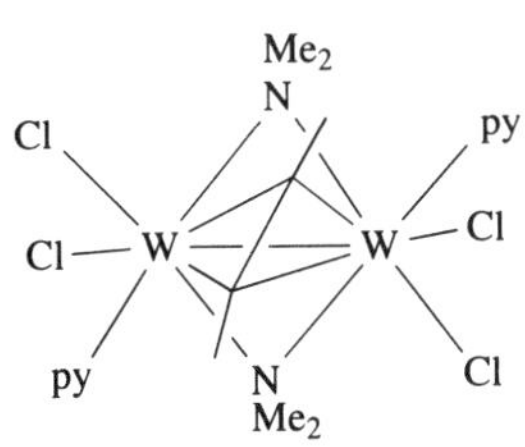

(182)

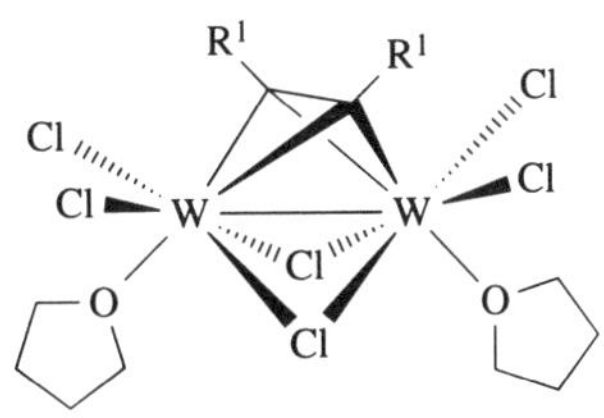

(183)

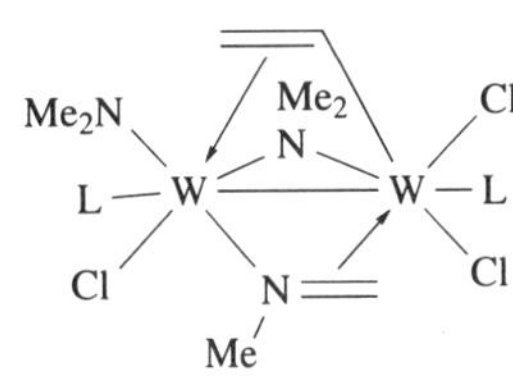

(184)

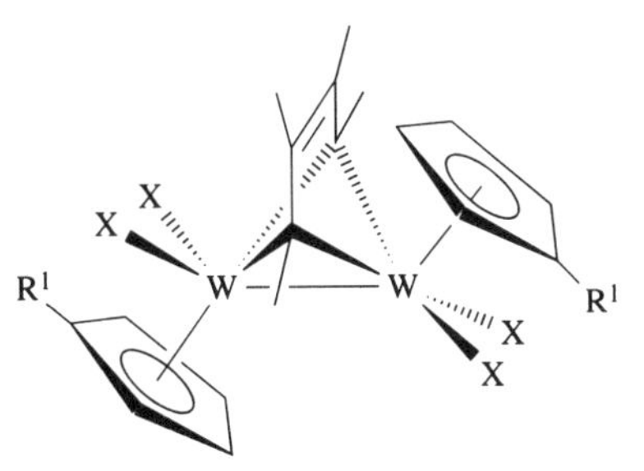

(185)

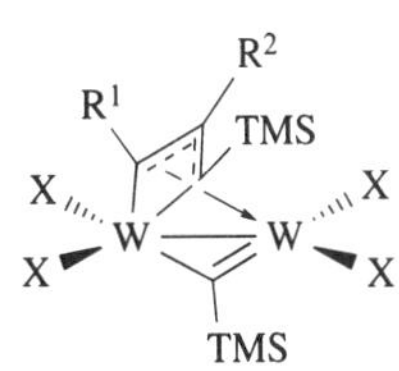

(186)

Reactions of other unsaturated hydrocarbons with [$W_2(OR)_6$], [{WX_2(μ-C-TMS)}$_2$] and [$W_2Cl_2(NMe_2)_4$] have also been explored. At 0 °C, ethene reacts reversibly with [$W_2(OCH_2Bu^t)_6$] to give unstable (**187**) (R = CH_2Bu^t), which is of note for the four bridging OCH_2Bu^t ligands and the orthogonal arrangement of terminal, η^2-C_2H_4 ligands;[544] above 0 °C a C–C bond activation/ethene coupling process proceeds as shown in Scheme 24 and this has been shown to operate for a range of alkoxide ligands.[545] Butadiene reacts reversibly with [$W_2(OCH_2Bu^t)_6(py)_2$] to give (**188**) (R = CH_2Bu^t).[546] Treatment of [$W_2(OBu^t)_6$] with allene affords [$W_2(OBu^t)_6(C_3H_4)$] which, on the basis of theoretical arguments, was originally considered to possess a μ-allene ligand bonded perpendicular to the W–W axis[547] but subsequent crystallographic studies revealed the μ-parallel structure of (**189**) (R = Bu^t); all other hexaalkoxide complexes [$W_2(OR)_6$] (R = Pr^i, Cy, CH_2Bu^t, etc.) react with allene in a 1:2 ratio to give (**190**).[548] Allene and its substituted derivatives insert into the tungsten–carbyne bond of [{$W(X^1)(X^2)$(μ-C-TMS)}$_2$] ($X^1 = X^2 = CH_2TMS$ or OPr^i; $X^1 = CH_2TMS$, $X^2 = OBu^t$) to give (**191**).[549] Some reactions of [$W_2Cl_2(NMe_2)_4$] with carbanions have been reported: with MgBr(allyl), allyl bridged (**192**) is formed,[550] although this structure is not accessible with substituted allyls; with the dilithiotetraphenylbutadiene reagent 1,4-$Li_2C_4Ph_4$, (**193**) is generated, probably via a dimetallabenzyne intermediate.[551]

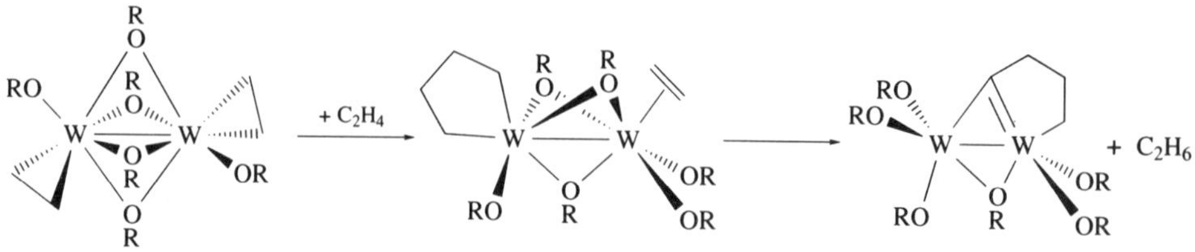

(187)

Scheme 24

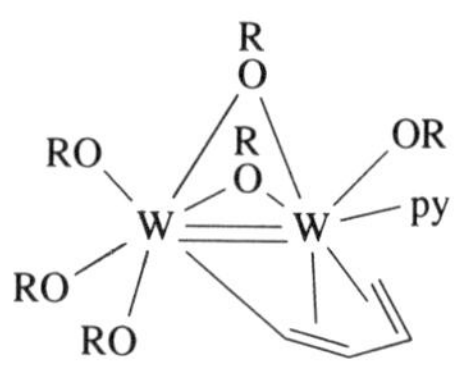

(188)

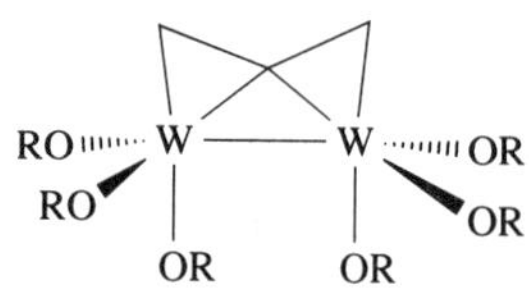

(189)

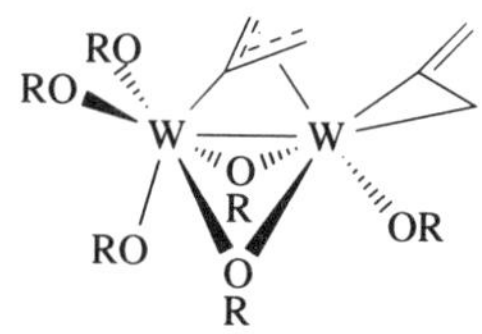

(190)

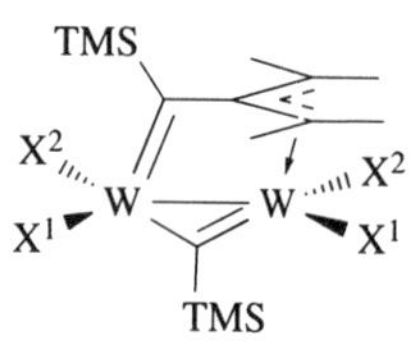

(191)

(192)

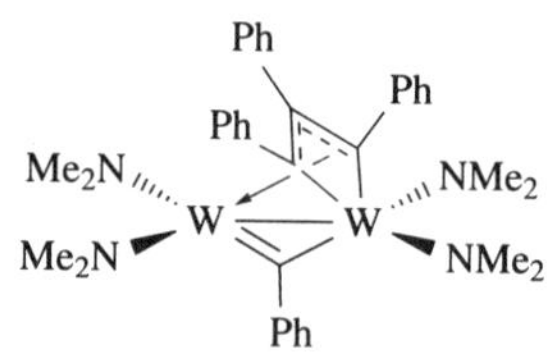

(193)

6.9.2 Complexes Derived from [$M_2(CO)_4L_2$] (M≡M) (L = Cp, Indenyl, etc.)

The synthesis of alkyne bridged complexes (**194**) (M = Mo or W, R^1, R^2 = alkyl, aryl, etc.; M = Cr, $R^1 = R^2$ = Ph only[552]) via thermal or photochemical reaction of [$M_2(CO)_nCp_2$] (n = 4 or 6) with alkynes $R^1C{\equiv}CR^2$, is well established. The salient structural features of (**194**) are the perpendicular μ-alkyne and a semibridging CO ligand; a notable exception is [$Mo_2(CO)_4$(μ TMS C_2TMS)Cp_2] in which all CO ligands are terminal, probably due to the sterically bulky alkyne.[553] Some new aspects of synthesis of complexes (**194**) have been reported. The metallocyclobutenone complex (**195**) ($R^1 = CO_2Me$) has been isolated as an intermediate in the photochemical generation of (**194**) (M = W, $R^1 = R^2 = CO_2Me$) from [$W_2(CO)_6Cp_2$] and $MeO_2CC{\equiv}CCO_2Me$[554] and thermolysis of the alkynyl complex [Mo(C≡CPh)$(CO)_3$Cp] yields (**194**) (M = Mo, R^1 = Ph, R^2 = C≡CPh), probably via an alkynyl bridged binuclear intermediate.[555] The carbyne complexes [$W(CO)_2(\equiv CR^3)$Cp] (R^3 = Me or C_6H_4Me-4) provide interesting alternative syntheses of some tungsten derivatives of (**194**). Thus, HBF_4 protonation of [$W(CO)_2(\equiv CR^3)$Cp] affords [$W_2(CO)_4$(μ-H)(μ-$R^3C_2R^3$)Cp_2]$^+$ which is reversibly deprotonated to [$W_2(CO)_4$(μ-$R^3C_2R^3$)Cp_2][497] or reaction of [$W(CO)_2(\equiv CR^3)$Cp] (R^3 = Me or C_6H_4Me-4) with [W(CO)(η^2-$R^1C{\equiv}CR^1$)$_3$] (R^1 = Et or Ph) yields (**196**) (not R^1 = Ph, $R^3 = C_6H_4$Me-4), in which the substituted cyclopentadienyl ligand is derived from alkyne–carbyne coupling.[387] Some related syntheses of mixed metal (W–Mo) species have also been described.[556,557]

Recent physical investigations on (**194**) and its derivatives comprise thermochemical, electrochemical and NMR studies. Heats of reaction for the addition of RC≡CH (R = H or Ph) to [$Mo_2(CO)_4Cp_2$] lie in the range 130–140 kJ mol^{-1}.[558] Cyclic voltammetric investigations on [$Mo_2(CO)_3(L^1)$(μ-$R^1C_2R^2$)(η^5-L^2)$_2$] (L^1 = CO or $CNBu^t$; R^1, R^2 = Me, CO_2Me, etc.; L^2 = Cp, Cp* or indenyl) reveal two sequential one-electron oxidation processes; spectroscopic studies on the isolable radical [$Mo_2(CO)_3(CNBu^t)$(μ-$MeO_2CC_2CO_2Me$)Cp^*_2]$^+$ suggest that the structure of (**194**), including the semibridging CO, is retained.[559] The dynamic behaviour of complexes (**194**) has received further attention; remarkably

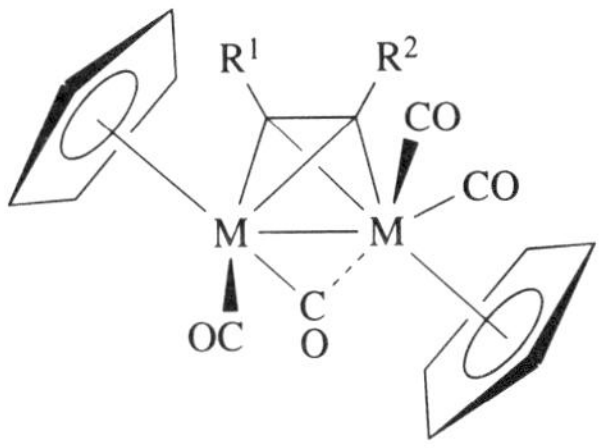

(194)

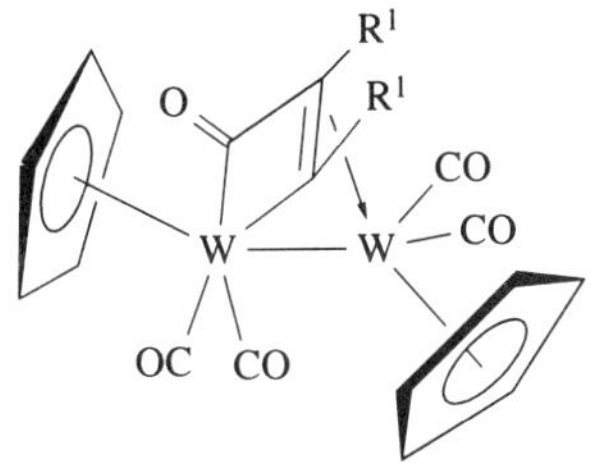

(195)

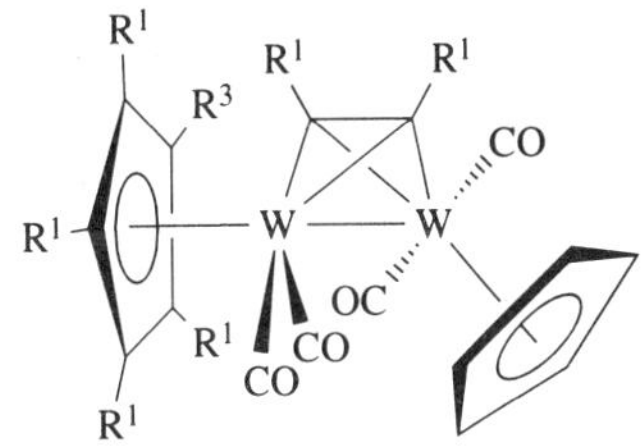

(196)

$[W_2(CO)_4(\mu\text{-}HC_2Ph)Cp^*_2]$ exhibits a single CO resonance in the ^{13}C NMR spectrum, even at −100 °C, whereas in the molybdenum analogue, four discrete CO resonances are resolved.[560]

An extensive reaction chemistry of (**194**) and related Cp* and indenyl derivatives has been developed. High-temperature reflux of $[M_2(CO)_4(\mu\text{-}RC_2R)Cp_2]$ with RC≡CR yields (**197**) (Scheme 25) (M = Cr, R = Ph;[552] M = Mo, R = Et[561]), in which a M≡M core is linked by a C_4 'fly over' bridge derived from coupling of two alkynes; complexes (**197**) are also obtained directly from reaction of $[M_2(CO)_4Cp_2]$ with RC≡CR.[552] As shown in Scheme 25, further alkyne linking, resulting in C_6 and C_8 bridged species, has been observed with electron-withdrawing alkynes RC≡CR (R = CO_2Me or Ph) at the dimolybdenum centre of $[Mo_2(CO)_4(\mu\text{-}RC_2R)Cp_2]$.[552,553] Four product types have been characterized and some selectivity is observed dependent upon the identity of alkyne substituents (R = combinations of H, Ph or CO_2Me). Complexes (**198**) and (**199**) are related by CO addition across the Mo≡Mo bond of (**198**) (CO addition across the Mo≡Mo bond in (**197**) is an established process[561]) but the relationship between the two forms of the C_8 bridged complexes (**200**) and (**201**), both of which have been crystallographically characterized,[562] is less clear. There is good evidence that the formation of a C_8 chain at a dimolybdenum centre proceeds via sequential alkyne linking[552,563] but the function of C_6 'fly over' bridge complexes (**198**) or (**199**) as direct precursors to (**201**), in which both ends of the C_8 chain are bonded to a single molybdenum, has been questioned.[564] An alternative synthesis of (**201**) (R = Me) involves the one-electron reduction[565] of $[Mo(NCMe)(\eta^2\text{-}MeC{\equiv}Me)_2Cp]^+$ with $Na[Fe(CO)_2Cp]$[566] and a related product is formed[567] in the analogous reaction of $[Mo(NCMe)(\eta^2\text{-}HC{\equiv}CBu^t)_2Cp]^+$. However, sodium/mercury amalgam reduction of the indenyl complex $[Mo(NCMe)(\eta^2\text{-}MeC{\equiv}CMe)_2(\eta^5\text{-}C_9H_7)]^+$ results in the formation of a C_6 bridged product analogous to (**198**),[564] which is also the final product in a sequence of thermally initiated alkyne linkage reactions starting from $[Mo_2(CO)_4(\eta^5\text{-indenyl})_2]$ and MeC≡CMe.[563] Some reaction chemistry of the C_8 bridged complex (**201**) (R = Me) has been developed. Protonation occurs at a terminal carbon of the C_8 chain,[566] while reaction with Ph_3C^+ or excess CF_3SO_3H gives (**202**), probably via an intermediate radical monocation.[568] A few examples of alkyne linking at a ditungsten centre have also been described,[554,569] and reduction of $[CrCl_2(\eta\text{-}C_5Me_5)]$ with activated magnesium in the presence of MeC≡CMe affords C_4 and C_6 bridged complexes;[570] a chromium analogue of C_8 bridged (**200**) has been referred to in Section 6.6.2.

Examples of alkyne–diene and alkyne–phosphorus coupling at the dimetal centre of (**194**) are also known. Although HC≡CH is a poor dienophile, thermal reaction of (**194**) (M = Mo, $R^1 = R^2$ = H) with cyclic dienes (cot, chpt, chd or 1,3-cod) results in a formal Diels–Alder 1,4-addition of HC≡CH to the diene, thus with cot, (**203**) is formed.[571] No corresponding products are observed with acyclic dienes but photolysis of (**194**) (M = Mo or W; $R^1 = R^2$ = H) in the presence of 1,3-butadiene affords a range of complexes such as (**204**) (M = Mo or W), which are typified by a C_6 bridging ligand derived from linking of HC≡CH with 1,3-butadiene.[569] The reaction of PPh_2Cl with (**194**) (M = Mo, R^1 = H, R^2 = H, Me or Ph;[572] M = W, $R^1 = R^2 = CO_2Me$[573]) results in P–Cl bond cleavage and formation of three general product types, alkyne-bridged (**205**), and phosphido-alkyne linked (**206**) and (**207**) together with other products in selected examples. Where M = Mo, R^1 = H, R^2 = Ph, (**205**) is a proven precursor to (**207**) and, where M = W, $R^1 = R^2 = CO_2Me$, thermolysis of (**206**) has been shown to give (**205**). Treatment of (**194**) (M = Mo, $R^1 = R^2$ = Me) with P_2Ph_4 does not result in phosphido alkyne linking[574]; the main product is the terminal alkyne complex $[Mo(\eta^2\text{-}MeC{\equiv}CMe)Cp(\mu\text{-}PPh_2)_2Mo(CO)Cp]$ together with minor products (**208**)[575] and (**209**) (R^3 = Ph, L = O).[576] However, protonation of $[Mo(\eta^2\text{-}MeC{\equiv}CMe)Cp(\mu\text{-}PPh_2)_2Mo(CO)Cp]$ yields (**210**).[574] An alternative strategy for the formation of phosphido coupled products involving introduction of the phosphido bridge prior to reaction with alkynes[577] or dienes[578,579] is not successful; treatment of $[\{Mo(CO)_2Cp\}_2(\mu\text{-}PR^3_2)(\mu\text{-}H)]$ with MeC≡CMe affords vinyl-bridged (**209**) (R^3 = Me, L = CO or O).[577]

Alkyne-centred protonation or hydride abstraction in (**194**) has also been explored. Protonation of (**194**) (M = Mo, R^1 = H, R^2 = Ph) with HBF_4 followed by treatment with H_2O yields vinyl-bridged (**211**) (z = 1, L = OH_2).[580] A series of η^2,η^3-propargyl bridged complexes (**212**) (M = Mo; L = C_5H_4Me,

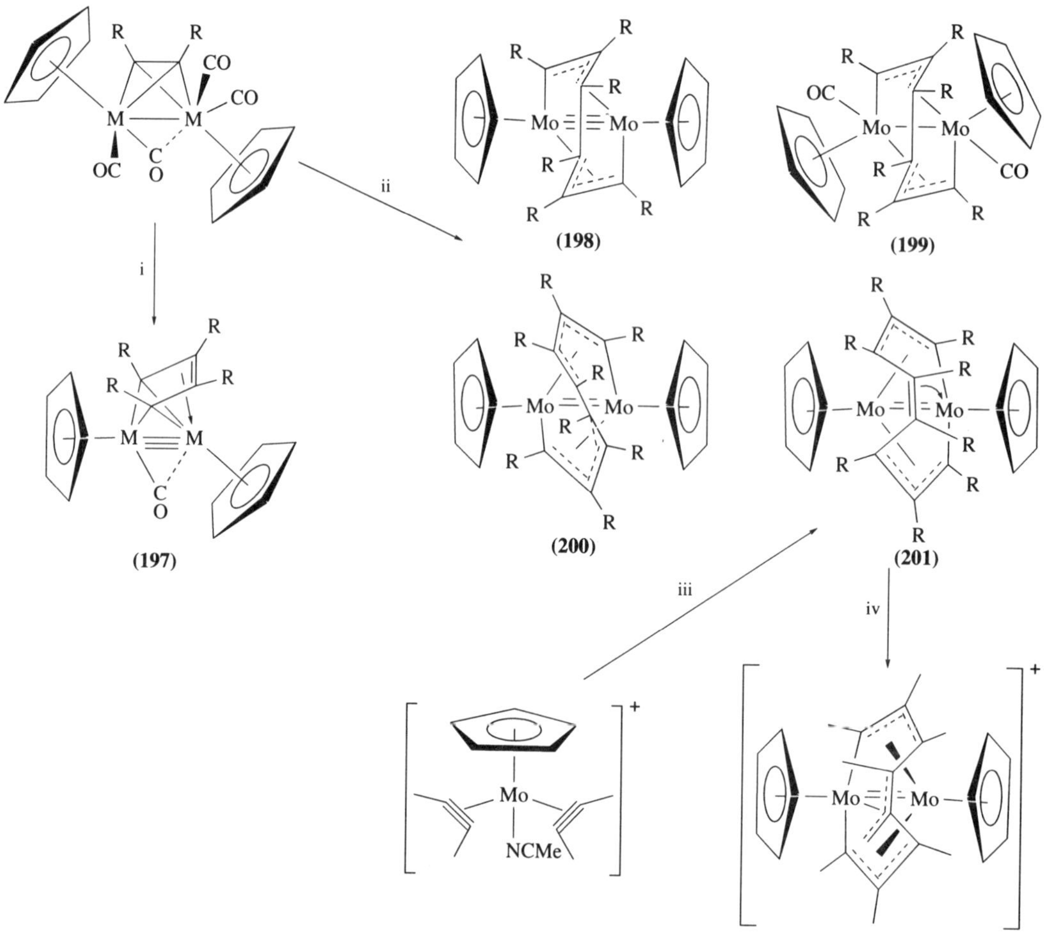

i, RC≡CR, M = Cr, R = Ph, heptane reflux, M = Mo, R = Et, mesitylene reflux; ii, RC≡CR, R = CO_2Me, M = Mo, octane reflux; iii, $Na[Fe(CO)_2Cp]$, THF; iv, R = Me, $[Ph_3C]^+$ or CF_3SO_3H in CH_2Cl_2

Scheme 25

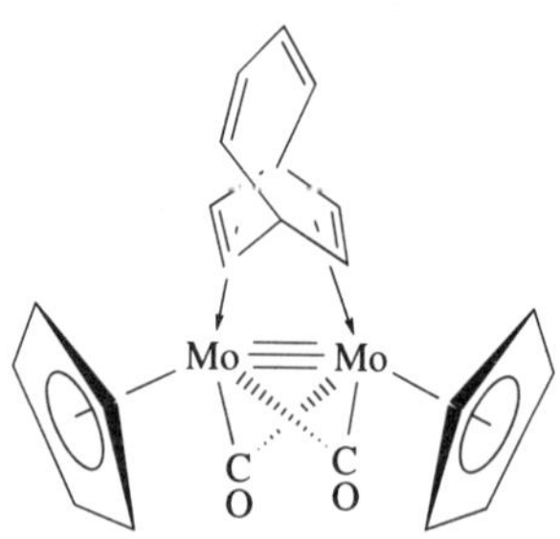

(203)

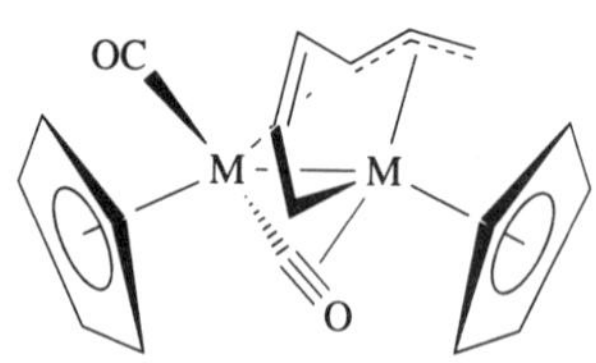

(204)

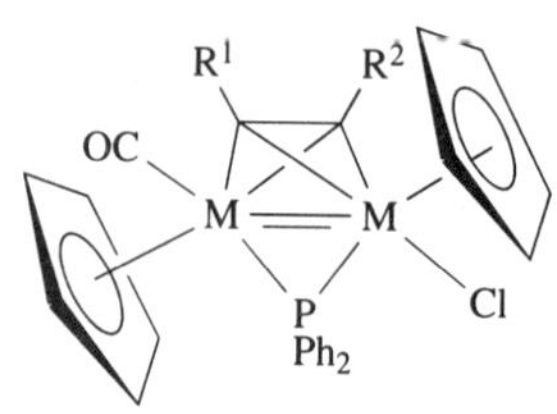

(205)

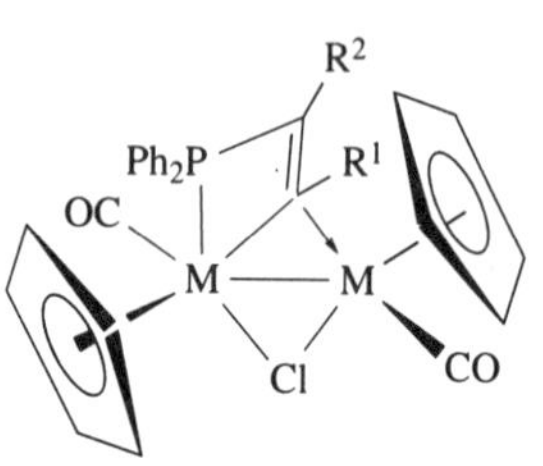

(206)

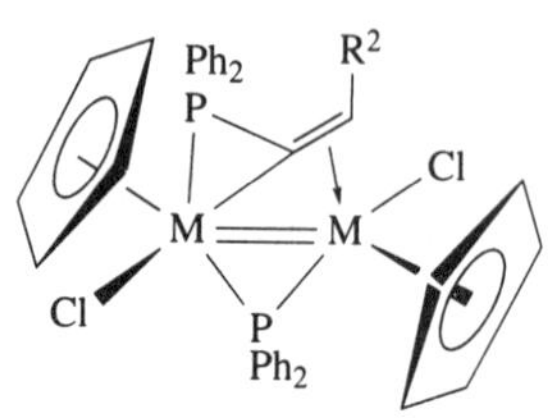

(207)

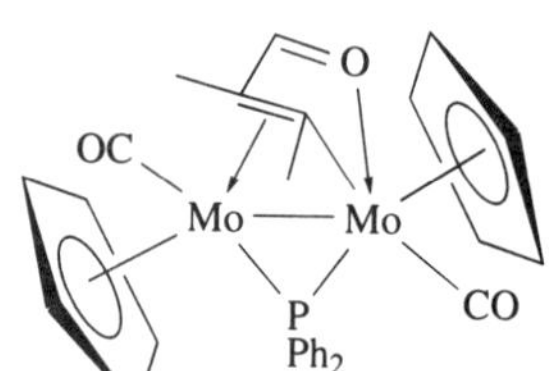

(208)

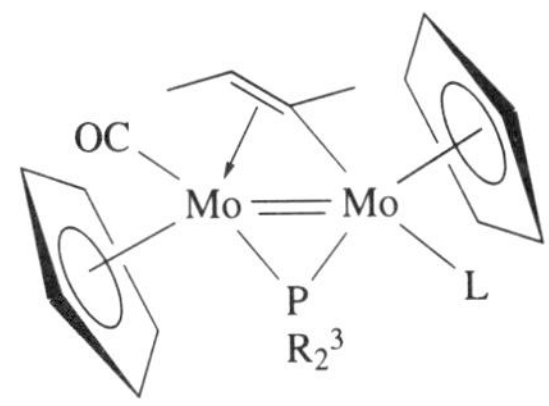

(209)

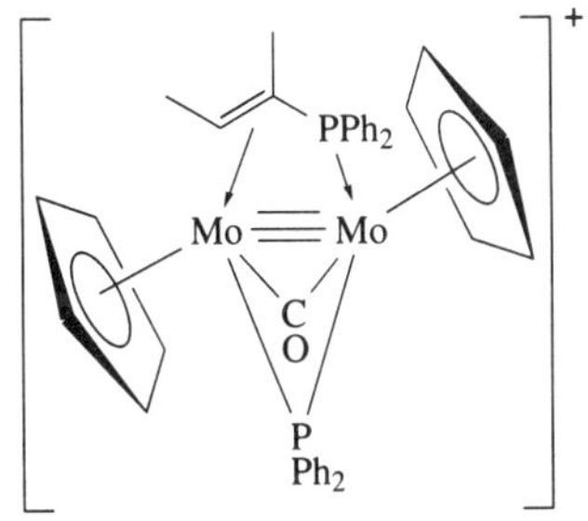

(210)

$R^1 = R^2 = H$;[581] L = Cp, $R^1 = H$, $R^2 = Me$;[582] L = Cp, $R^1 = R^2 = H$ or Me[583]) has been obtained via protonation of alkoxyalkyne[581] or vinylalkyne[582] derivatives of (**194**) or alternatively via hydride abstraction from $[Mo_2(CO)_4(\eta^2,\eta^2\text{-allene})Cp_2]$ or (**194**) (M = Mo, $R^1 = H$, $R^2 = Me$).[583] The dynamic behaviour of propargyl bridged complexes (**212**) has been examined;[581,584] a range of processes operate of which rotation about a C–C bond in the propargyl ligand has the highest activation energy. Deprotonation of (**194**) ($R^1 = H$, $R^2 = CH_2OMe$[585] or Ph[586]) results in the formation of anionic, alkynyl bridged (**213**) ($R^2 = CH_2OMe$ or Ph); derivatives of (**213**) (R^2 = Ph or Me) have also been synthesized via treatment of $[Mo_2(CO)_4Cp_2]$ with $LiC{\equiv}CR^2$ and isolated as PPN salts.[586]

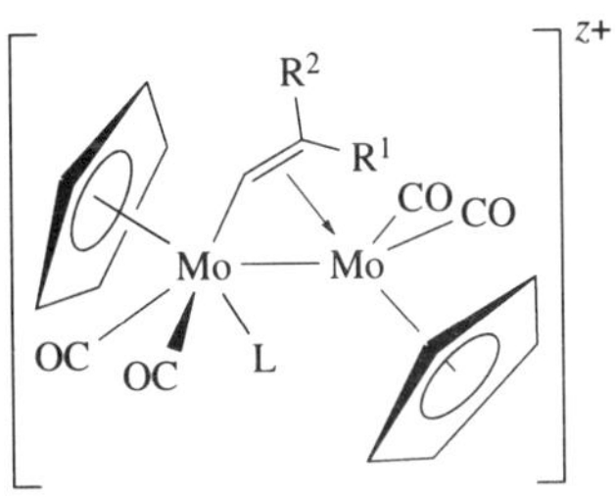

(211)

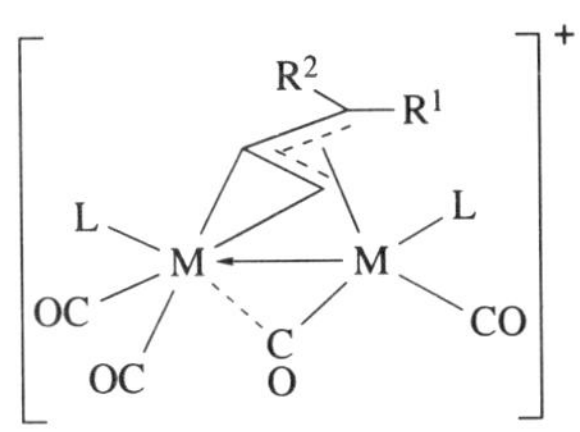

(212)

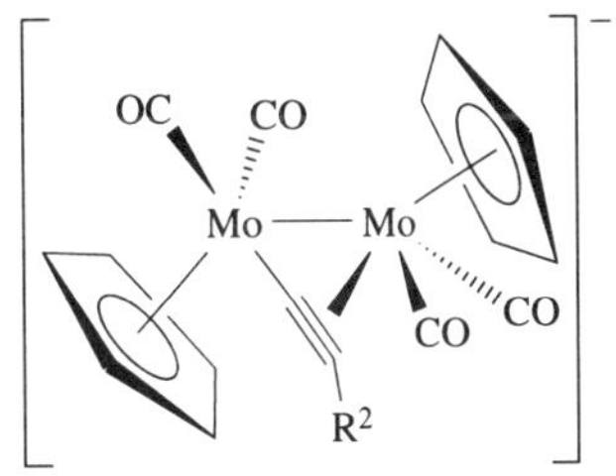

(213)

The synthesis and reaction chemistry of μ-η^1,η^2-(4e)-vinylidene complexes (**214**), isomeric with alkyne bridged (**194**), has been developed. The unsubstituted derivatives (**214**) ($R^1 = R^2 = H$, L = Cp*,[587] L = Cp[588]) have been obtained by photolysis of $[Mo_2(CO)_4L_2]$ in the presence of $HC{\equiv}CH$, while substituted derivatives (L = Cp; R^2 = Ph, R^1 = H, Me or $(CH_2)_4OMe$; $R^1 = R^2 = Me$) are accessible via protonation or alkylation of (**213**) (R^2 = Ph or Me).[586] Derivatives of (**214**) for which R^1 or $R^2 = H$ are relatively unstable with respect to rearrangement of the μ-vinylidene ligand to a μ-alkyne ligand[586,587] result in conversion to (**194**). In general, CF_3CO_2H protonation occurs at C_α to yield[587,589] vinyl bridged (**211**) ($z = 0$, $L = CO_2CF_3$), thus parallelling the protonation of isomeric alkyne bridged (**194**); however, HBF_4 protonation of (**214**) (L = Cp, $R^1 = R^2 = Me$) yields (**215**), which exhibits a novel, asymmetrically bridged allyl ligand.[589] Nucleophilic attack of L (L = $CNBu^t$, PMe_3 or $P(OMe)_3$) is also directed at C_α of (**214**) ($R^1 = R^2 = H$ or Me) to generate vinyl bridged products $[\{Mo(CO)_2Cp\}_2\{\mu\text{-}\eta^1,\eta^2\text{-}C(L){=}CR^1R^2\}]$.[590] The μ-η^1,η^2(4e) 'side on' bonded allenylidene complexes (**216**) (M = Mo or W) are accessible via treatment of $[M_2(CO)_4Cp_2]$ with $LiC{\equiv}C(Me)C{=}CH_2$ and subsequent chromatography of the reaction mixture on alumina;[586] protonation of (**216**) provides a novel, alternative synthesis of η^2,η^3-propargyl bridged complexes (**212**) (M = Mo or W, $R^1 = R^2 = Me$).[591]

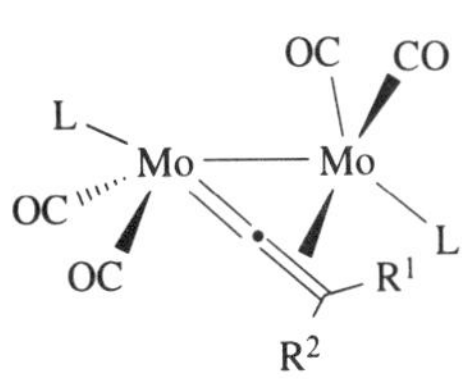

(214)

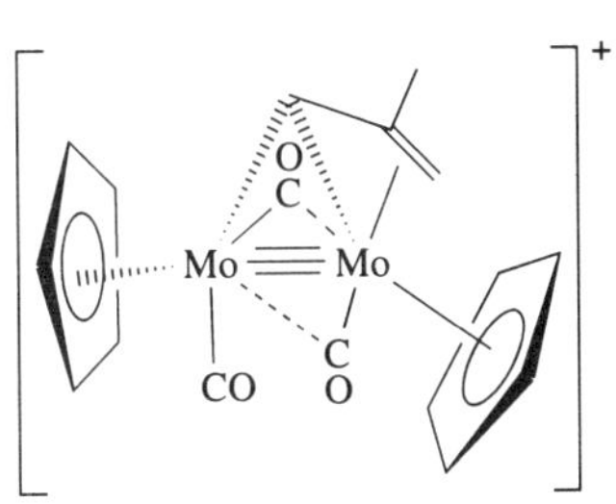

(215)

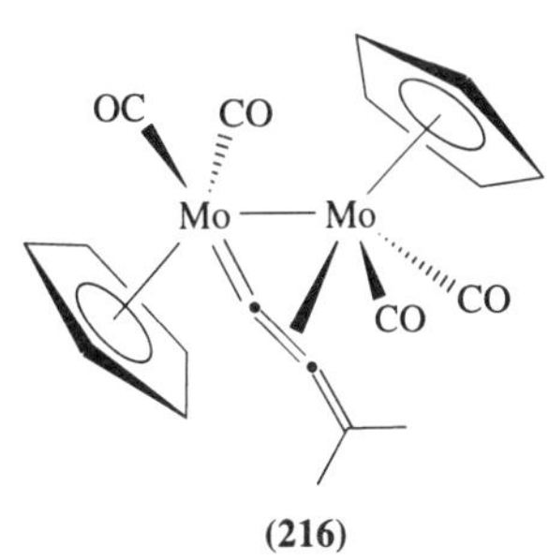

(216)

The reactions of fulvalene bridged complexes $[M_2(CO)_6(\eta^5,\eta^5\text{-}C_{10}H_8)]$ (M = Mo or W) with alkynes provide some interesting contrasts with the chemistry of cyclopentadienyl analogues. Photolysis of $[Mo_2(CO)_6(\eta^5,\eta^5\text{-}C_{10}H_8)]$ in the presence of PhC≡CPh yields (**217**) (M = Mo, $n = 0$, $R^1 = R^2$ = Ph, L = CO), which is notable for the absence of a semibridging carbonyl ligand, but the corresponding reaction with the tungsten derivative affords the terminal alkyne product $[W_2(CO)_3(\eta^2\text{-PhC≡CPh})(\eta^5,\eta^5\text{-}C_{10}H_8)]$. Both these monoalkyne adducts react with further PhC≡CPh to give (**217**) (M = Mo or W, $n = 0$, $R^1 = R^2$ = Ph, L = PhC≡CPh) but alkyne linkage is not observed.[592] However, complexes of type (**217**) (M = Mo, $n = 1$, $R^1 = R^2 = Pr^n$ or Ph, L = CO) possess increased flexibility in the bridging fulvalene ligand and alkyne linkage can proceed, albeit with the formation of unusual products such as (**218**) (R = Pr^n).[593] Protonation of (**217**) (M = Mo, $n = 0$, R^1 = Me, $R^2 = CH_2OMe$, L = CO) affords a propargyl bridged complex comparable to the cyclopentadienyl analogues (**212**) except for the absence of a semibridging CO ligand in the fulvalene bridged derivative.[594]

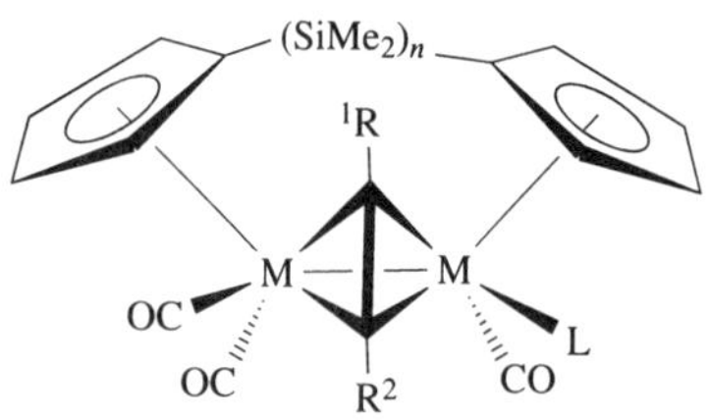

(217)

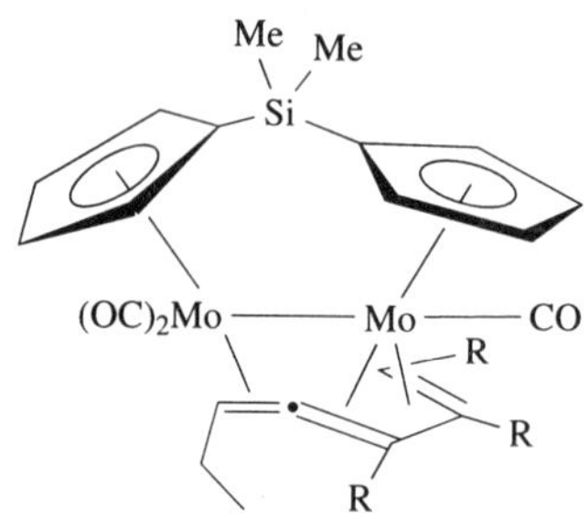

(218)

Although the organic chemistry of the dimetal centre in $[M_2(CO)_4Cp_2]$ is dominated by alkyne derived ligands, some chemistry which stems from reaction with, for example, cycloalkenes and dienes, has been developed. Treatment of $[Mo_2(CO)_4Cp_2]$ with 3,3-Me_2-cyclopropene affords the μ-allylidene complex (**219**) which, on thermolysis, gives a product mixture including isoprene bridged (**220**).[595] The outcome of the reactions of (**219**) with alkynes $R^1C{\equiv}CR^2$ is substituent dependent but, with $HC{\equiv}CBu^t$, the product is (**221**), which exhibits a novel bonding mode for a bridging 1,3-diene ligand.[596] Photolysis of $[M_2(CO)_6(\eta\text{-}C_5H_4Me)_2]$ (M = Mo[597] or W[276]) with acyclic dienes (diene = buta-1,3-diene, 2,3-Me_2-buta-1,3-diene, etc.) in general affords complexes of type (**222**), which are notable for the (4e)-$\eta^{1:2}$ bridging carbonyl ligand. By contrast, 1,3-cycloheptadiene reacts photochemically with $[Mo_2(CO)_6(\eta\text{-}C_5H_4Me)_2]$ to give (**223**).[275] Thermal reaction of $[Mo_2(CO)_6Cp_2]$ with cod or cot yields, respectively, (**224**)[598] and (**225**);[599] these two reactions are related by the formation of a bridging cycloalkyne and probably involve metal-assisted hydrogen migration. The allene bridged complex $[Mo_2(CO)_4(\mu\text{-}CH_2CCH_2)Cp_2]$ is much less reactive towards alkynes than alkyne bridged (**194**); allene–alkyne coupling is only observed with $MeO_2CC{\equiv}CCO_2Me$ to give (**226**) (R = CO_2Me), together with a related compound resulting from additional linking with CO.[600]

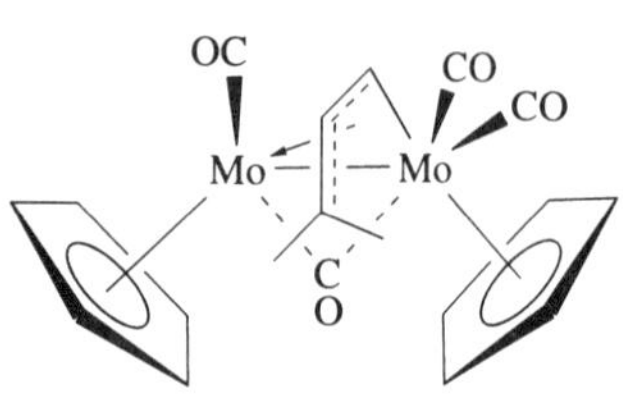

(219)

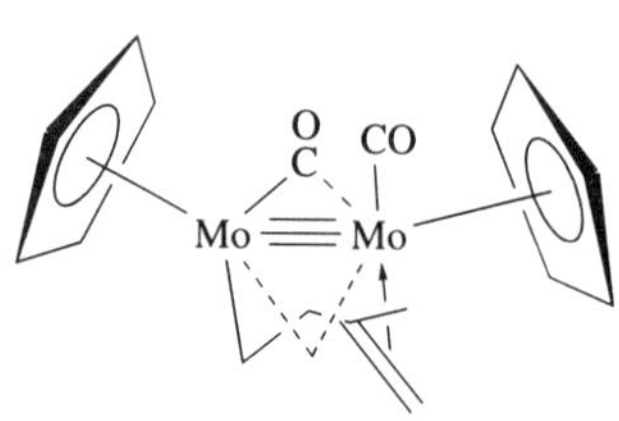

(220)

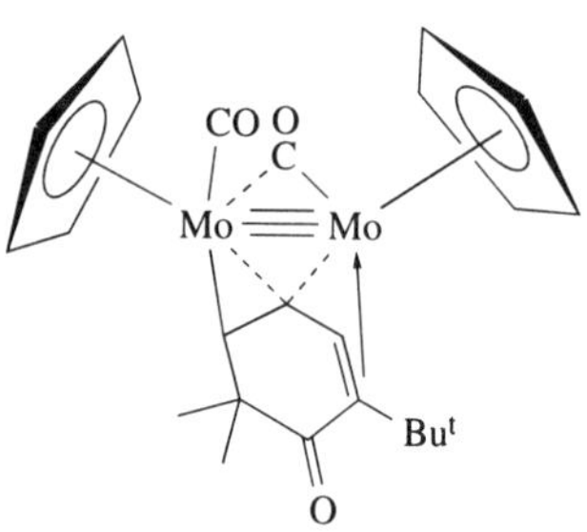

(221)

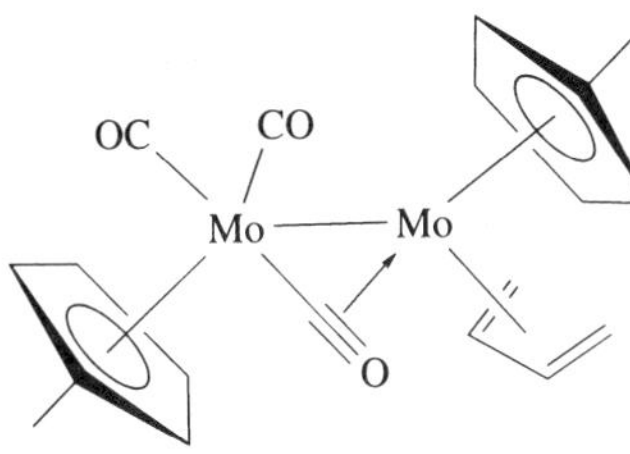

(222)

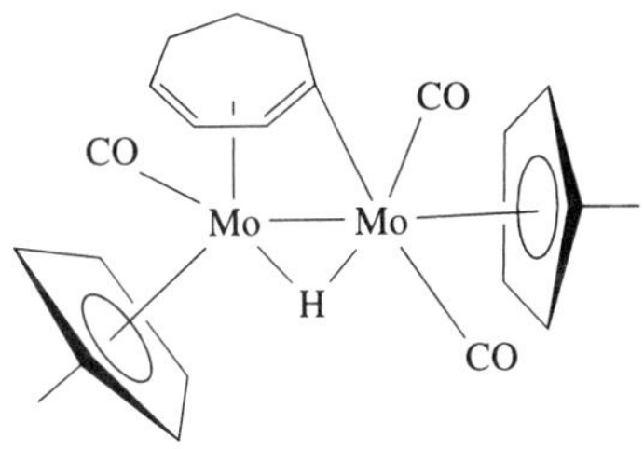

(223)

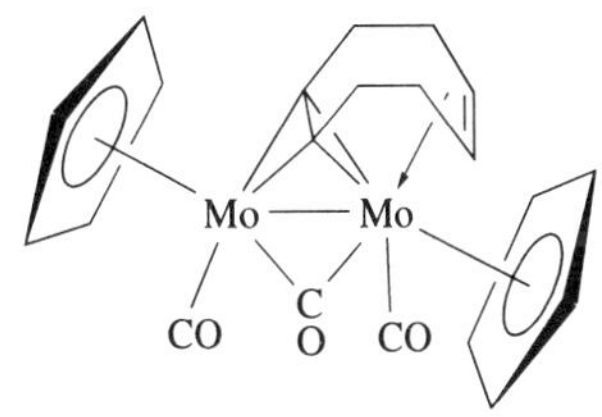

(224)

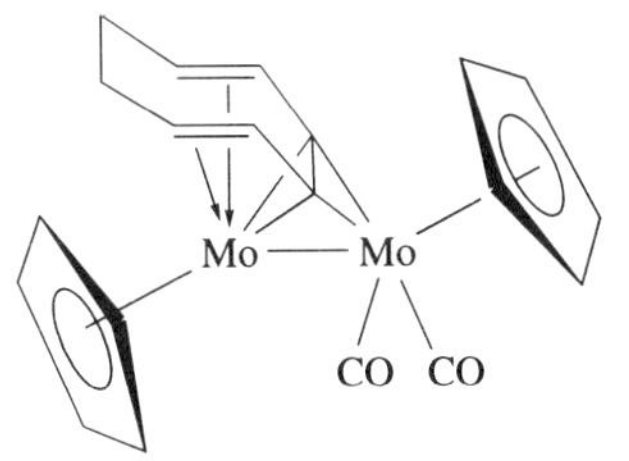

(225)

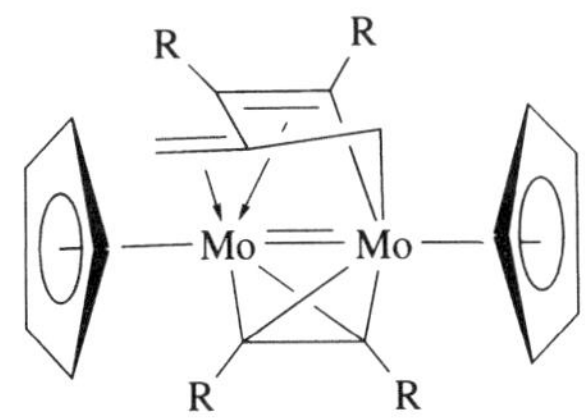

(226)

6.10 REFERENCES

1. M. F. Gregory, S. A. Jackson, M. Poliakoff and J. J. Turner, *J. Chem. Soc., Chem. Commun.*, 1986, 1175.
2. F.-W. Grevels, J. Jacke and S. Özkar, *J. Am. Chem. Soc.*, 1987, **109**, 7536.
3. B. H. Weiller and E. R. Grant, *J. Am. Chem. Soc.*, 1987, **109**, 1252.
4. F.-W. Grevels, J. Jacke, W. E. Klotzbücher, S. Özkar and V. Skibbe, *Pure Appl. Chem.*, 1988, **60**, 1017.
5. G. Davidson and C. L. Davies, *Inorg. Chim. Acta*, 1989, **165**, 231.
6. S. A. Jackson, P. M. Hodges, M. Poliakoff, J. J. Turner and F.-W. Grevels, *J. Am. Chem. Soc.*, 1990, **112**, 1221.
7. F.-W. Grevels and V. Skibbe, *J. Chem. Soc., Chem. Commun.*, 1984, 681.
8. J. K. Klassen and G. K. Yang, *Organometallics*, 1990, **9**, 874.
9. K. Angermund, F.-W. Grevels, C. Krüger and V. Skibbe, *Angew. Chem., Int. Ed. Engl.*, 1984, **23**, 904.
10. S. A. Jackson, R. K. Upmacis, M. Poliakoff, J. J. Turner, J. K. Burdett and F.-W. Grevels, *J. Chem. Soc., Chem. Commun.*, 1987, 678.
11. K. R. Pope and M. S. Wrighton, *Inorg. Chem.*, 1985, **24**, 2792.
12. H. Fischer, J. Hofmann and E. Mauz, *Angew. Chem., Int. Ed. Engl.*, 1991, **30**, 998.
13. D. J. Stufkens, T. L. Snoeck, W. Kaim, T. Roth and B. Olbrich-Deussner, *J. Organomet. Chem.*, 1991, **409**, 189.
14. W. Kaim, B. Olbrich-Deussner and T. Roth, *Organometallics*, 1991, **10**, 410.
15. B. Olbrich-Deussner, R. Gross and W. Kaim, *J. Organomet. Chem.*, 1989, **366**, 155.
16. B. Olbrich-Deussner, W. Kaim and R. Gross-Lannert, *Inorg. Chem.*, 1989, **28**, 3113.
17. T. Roth and W. Kaim, *Inorg. Chem.*, 1992, **31**, 1930.
18. W. A. Schenk and H. Müller, *Z. Anorg. Allg. Chem.*, 1981, **478**, 205.
19. W. A. Schenk and H. Müller, *Inorg. Chem.*, 1981, **20**, 6.
20. U. Koemm and C. G. Kreiter, *J. Organomet. Chem.*, 1982, **240**, 27.
21. H. Berke and C. Sontag, *Z. Naturforsch., Teil B*, 1985, **40**, 794.
22. C. G. Kreiter and U. Koemm, *Z. Naturforsch., Teil B*, 1983, **38**, 943.
23. W. A. Schenk and H. Müller, *Chem. Ber.*, 1982, **115**, 3618.
24. C. G. Kreiter and U. Koemm, *J. Organomet. Chem.*, 1982, **238**, C67.
25. H. Berke, G. Huttner, C. Sontag and L. Zsolnai, *Z. Naturforsch., Teil B.*, 1985, **40**, 799.
26. (a) E. Carmona, J. M. Marin, M. L. Poveda, J. L. Atwood, R. D. Rogers and G. Wilkinson, *Angew. Chem., Int. Ed. Engl.*, 1982, **21**, 441; (b) E. Carmona, J. M. Marin, M. L. Poveda, J. L. Atwood and R. D. Rogers, *J. Am. Chem. Soc.*, 1983, **105**, 3014.
27. E. Carmona, A. Galindo, M. L. Poveda and R. D. Rogers, *Inorg. Chem.*, 1985, **24**, 4033.
28. M. Brookhart *et al.*, *J. Chem. Soc., Dalton Trans.*, 1985, 423.
29. M. L. H. Green, G. Parkin, C. Mingqin and K. Prout, *J. Chem. Soc., Chem. Commun.*, 1984, 1400. M. L. H. Green, G. Parkin, M. Chen and K. Prout, *J. Chem. Soc., Dalton Trans.*, 1986, 2227.
30. E. Carmona, A. Galindo, J. M. Marin, E. Gutierrez, A. Monge and C. Ruiz, *Polyhedron*, 1988, **7**, 1831.
31. R. Alvarez *et al.*, *J. Am. Chem. Soc.*, 1985, **107**, 5529.
32. V. Branchadell and A. Dedieu, *Inorg. Chem.*, 1987, **26**, 3966.
33. E. Carmona, A. Galindo, A. Monge, M. A. Muñoz, M. L. Poveda and C. Ruiz, *Inorg. Chem.*, 1990, **29**, 5074.
34. J. E. Salt, G. S. Girolami, G. Wilkinson, M. Motevalli, M. Thornton-Pett and M. B. Hursthouse, *J. Chem. Soc., Dalton Trans.*, 1985, 685.
35. C.-H. Lai, C.-H. Cheng, W.-C. Chou and S.-L. Wang, *Organometallics*, 1993, **12**, 1105.
36. (a) M. L. H. Green and L.-L. Wong, *J. Chem. Soc., Chem. Commun.*, 1988, 677; (b) A. E. Derome, M. L. H. Green and L.-L. Wong, *New J. Chem.*, 1989, **13**, 747.
37. R. A. Henderson and K. E. Oglieve, *J. Chem. Soc., Chem. Commun.*, 1991, 584.

38. K. E. Oglieve and R. A. Henderson, *J. Chem. Soc., Dalton Trans.*, 1991, 3295.
39. P. R. Sharp, *Organometallics*, 1984, **3**, 1217.
40. F.-M. Su, C. Cooper, S. J. Geib, A. L. Rheingold and J. M. Mayer, *J. Am. Chem. Soc.*, 1986, **108**, 3545.
41. G. R. Clark, A. J. Nielson, C. E. F. Rickard and D. C. Ware, *J. Chem. Soc., Dalton Trans.*, 1990, 1173.
42. F.-M. Su, J. C. Bryan, S. Jang and J. M. Mayer, *Polyhedron*, 1989, **8**, 1261.
43. J. Kress and J. A. Osborn, *Angew. Chem., Int. Ed. Engl.*, 1992, **31**, 1585.
44. A. Mayr, A. M. Dorries, A. L. Rheingold and S. J. Geib, *Organometallics*, 1990, **9**, 964.
45. K. Sünkel, H. Ernst and W. Beck, *Z. Naturforsch., Teil B*, 1981, **36**, 474.
46. J. Markham, K. Menard and A. Cutler, *Inorg. Chem.*, 1985, **24**, 1581.
47. S. E. Kegley, D. T. Bergstrom, L. S. Crocker, E. P. Weiss, W. G. Berndt and A. L. Rheingold, *Organometallics*, 1991, **10**, 567.
48. S. E. Kegley, K. A. Walter, D. T. Bergstrom, D. K. MacFarland, B. G. Young and A. L. Rheingold, *Organometallics*, 1993, **12**, 2339.
49. C. G. Kreiter, M. Wenz and W. Michels, *Z. Naturforsch., Teil B*, 1989, **44**, 1247.
50. K. H. Dötz, C. Lyon and J. Rott, *J. Organomet. Chem.*, 1988, **345**, 117.
51. M. L. H. Green, P. C. Konidaris, P. Mountford and S. J. Simpson, *J. Chem. Soc., Chem. Commun.*, 1992, 256.
52. H. G. Alt and M. E. Eichner, *Angew. Chem., Int. Ed. Engl.*, 1982, **21**, 78.
53. K. A. Mahmoud, A. J. Rest, H. G. Alt, M. E. Eichner and B. M. Jansen, *J. Chem. Soc., Dalton Trans.*, 1984, 175.
54. F. P. A. Johnson, C. M. Gordon, P. M. Hodges, M. Poliakoff and J. J. Turner, *J. Chem. Soc., Dalton Trans.*, 1991, 833.
55. (a) P. H. Wermer, C. B. Dobson and G. R. Dobson, *J. Organomet. Chem.*, 1986, **311**, C47; (b) I.-H. Wang, P. H. Wermer, C. B. Dobson and G. R. Dobson, *Inorg. Chim. Acta*, 1991, **183**, 31.
56. I.-H. Wang and G. R. Dobson, *J. Organomet. Chem.*, 1988, **356**, 77.
57. R. T. DePue, D. B. Collum, J. W. Ziller and M. Rowen Churchill, *J. Am. Chem. Soc.*, 1985, **107**, 2131.
58. C. P. Casey, A. J. Shusterman, N. W. Vollendorf and K. J. Haller, *J. Am. Chem. Soc.*, 1982, **104**, 2417.
59. (a) C. A. Toledano, J. Levisalles, M. Rudler, H. Rudler, J.-C. Daran and Y. Jeannin, *J. Organomet. Chem.*, 1982, **228**, C7; (b) C. A. Toledano, H. Rudler, J.-C. Daran and Y. Jeannin, *J. Chem. Soc., Chem. Commun.*, 1984, 574.
60. C. A. Toledano, A. Parlier, H. Rudler, J.-C. Daran and Y. Jeannin, *J. Chem. Soc., Chem. Commun.*, 1984, 576.
61. R. P. Hughes, J. W. Reisch and A. L. Rheingold, *Organometallics*, 1985, **4**, 241.
62. M. Berry, C. D. Garner, I. H. Hillier and A. A. MacDowell, *Inorg. Chem.*, 1981, **20**, 1962.
63. F. Iczek, P. W. Jolly and C. Krüger, *J. Organomet. Chem.*, 1990, **382**, C11.
64. P. Betz, P. W. Jolly, C. Krüger and U. Zakrzewski, *Organometallics*, 1991, **10**, 3520.
65. K. Angermund, A. Döhring, P. W. Jolly, C. Krüger and C. C. Romão, *Organometallics*, 1986, **5**, 1268.
66. R. Benn, A. Rufińska and G. Schroth, *J. Organomet. Chem.*, 1981, **217**, 91.
67. R. Benn, C. Brevard, A. Rufińska and G. Schroth, *Organometallics*, 1987, **6**, 938.
68. R. Benn *et al.*, *Polyhedron*, 1990, **9**, 11.
69. R. Benn *et al.*, *Polyhedron*, 1986, **5**, 461.
70. P. W. Jolly, C. Krüger, C. C. Romão and M. J. Romão, *Organometallics*, 1984, **3**, 936.
71. R. Benn, S. Holle, P. W. Jolly, R. Mynott and C. C. Romão, *Angew. Chem., Int. Ed. Engl.*, 1986, **25**, 555.
72. R. Benn, T. H. Brock, P. W. Jolly, A. Rufińska and G. Schroth, *Polyhedron*, 1990, **9**, 23.
73. M. C. Azevedo, T. H. Brock, P. W. Jolly, A. Rufińska and G. Schroth, *Polyhedron*, 1991, **10**, 459.
74. R. J. Blau, M. S. Goetz, R. R. Howe, C. J. Smith, R.-J. Tsay and U. Siriwardane, *Organometallics*, 1991, **10**, 3259.
75. R. Benn, *Angew. Chem., Int. Ed. Engl.*, 1982, **21**, 626.
76. R. J. Blau, M. S. Goetz and R.-J. Tsay, *Polyhedron*, 1991, **10**, 605.
77. P. N. Kapoor, D. D. Pathak, G. Gaur and P. C. Mercykutty, *J. Organomet. Chem.*, 1987, **322**, 71.
78. P. N. Kapoor and P. C. Mercykutty, *J. Organomet. Chem.*, 1988, **339**, 97.
79. M. J. M. Campbell, E. C. Morrison, V. Rogers and P. K. Baker, *Polyhedron*, 1988, **7**, 1719.
80. M. J. M. Campbell, E. Morrison, V. Rogers and P. K. Baker, *Transition Met. Chem.*, 1986, **11**, 381.
81. P. K. Baker, *Inorg. Chim. Acta*, 1986, **118**, L3.
82. M. F. Perpiñán and A. Santos, *J. Organomet. Chem.*, 1981, **218**, 185.
83. M. F. Perpiñán and A. Santos, *J. Organomet. Chem.*, 1981, **221**, 63.
84. K.-B. Shiu, K.-H. Yih, S.-L. Wang and F.-L. Liao, *J. Organomet. Chem.*, 1991, **420**, 359.
85. M. F. Perpiñán, L. Ballester and A. Santos, *J. Organomet. Chem.*, 1983, **241**, 215.
86. B. J. Brisdon and A. Day, *J. Organomet. Chem.*, 1981, **221**, 279.
87. B. J. Brisdon, M. Cartwright and A. G. Hodson, *J. Organomet. Chem.*, 1984, **277**, 85.
88. B. J. Brisdon, K. A. Conner and R. A. Walton, *Organometallics*, 1983, **2**, 1159.
89. B. J. Brisdon, S. K. Enger, M. J. Weaver and R. A. Walton, *Inorg. Chem.*, 1987, **26**, 3340.
90. B. J. Brisdon, G. F. Griffin, J. Pierce and R. A. Walton, *J. Organomet. Chem.*, 1981, **219**, 53.
91. N. W. Murrall and A. J. Welch, *J. Organomet. Chem.*, 1986, **301**, 109.
92. M. D. Curtis and N. A. Fotinos, *J. Organomet. Chem.*, 1984, **272**, 43.
93. A. J. Graham, D. Akrigg and B. Sheldrick, *Acta Crystallogr., Sect. C*, 1983, **39**, 192.
94. A. J. Graham, D. Akrigg and B. Sheldrick, *Acta Crystallogr., Sect. C*, 1985, **41**, 995.
95. F. A. Cotton and R. L. Luck, *Acta Crystallogr., Sect. C*, 1990, **46**, 138.
96. M. D. Curtis and O. Eisenstein, *Organometallics*, 1984, **3**, 887.
97. K.-B. Shiu, C.-J. Chang, S.-L. Wang and F.-L. Liao, *J. Organomet. Chem.*, 1991, **407**, 225.
98. V. S. Joshi, A. Sarkar and P. R. Rajamohanan, *J. Organomet. Chem.*, 1991, **409**, 341.
99. K.-B. Shiu *et al.*, *Organometallics*, 1989, **8**, 1219.
100. K.-B. Shiu, C.-J. Chang, Y. Wang and M.-C. Cheng, *J. Organomet. Chem.*, 1991, **406**, 363.
101. B. J. Brisdon, M. Cartwright, A. G. W. Hodson, M. F. Mahon and K. C. Molloy, *J. Organomet. Chem.*, 1992, **435**, 319.
102. V. S. Joshi, V. K. Kale, K. M. Sathe, A. Sarkar, S. S. Tavale and C. G. Suresh, *Organometallics*, 1991, **10**, 2898.
103. K.-B. Shiu, F.-M. Shen, S.-L. Wang and S.-C. Wei, *J. Organomet. Chem.*, 1989, **372**, 251.
104. J. Ipaktschi, A. Hartmann and R. Boese, *J. Organomet. Chem.*, 1992, **434**, 303.
105. J. Ipaktschi and A. Hartmann, *J. Organomet. Chem.*, 1992, **431**, 303.
106. T.-Y. Luh and C. S. Wong, *J. Organomet. Chem.*, 1985, **287**, 231.

107. K. H. Pannell, R. N. Kapoor, M. Wells, T. Giasolli and L. Párkányi, *Organometallics*, 1987, **6**, 663.
108. H. G. Alt, H. E. Engelhardt, B. Wrackmeyer and R. D. Rogers, *J. Organomet. Chem.*, 1989, **379**, 289.
109. W. E. VanArsdale and J. K. Kochi, *J. Organomet. Chem.*, 1986, **317**, 215.
110. N. W. Murrall and A. J. Welch, *Acta Crystallogr., Sect. C*, 1984, **40**, 401.
111. J. W. Faller and B. C. Whitmore, *Organometallics*, 1986, **5**, 752.
112. J. W. Faller and Y. Ma, *Organometallics*, 1986, **5**, 1949.
113. R. H. Hill, A. Becalska and N. Chiem, *Organometallics*, 1991, **10**, 2104.
114. F. Carré, E. Colomer, R. J. P. Corriu and A. Vioux, *Organometallics*, 1984, **3**, 970.
115. J. W. Faller, H. H. Murray, D. L. White and K. H. Chao, *Organometallics*, 1983, **2**, 400.
116. (a) A. J. Pearson and M. N. I. Khan, *Tetrahedron Lett.*, 1985, **26**, 1407; (b) A. J. Pearson and M. N. I. Khan, *J. Org. Chem.*, 1985, **50**, 5276.
117. A. J. Pearson, S. Mallik, R. Mortezaei, M. W. D. Perry, R. J. Shively, Jr. and W. J. Youngs, *J. Am. Chem. Soc.*, 1990, **112**, 8034.
118. M. Green, S. Greenfield and M. Kersting, *J. Chem. Soc., Chem. Commun.*, 1985, 18.
119. (a) A. J. Pearson and V. D. Khetani, *J. Org. Chem.*, 1988, **53**, 3395; (b) A. J. Pearson and V. D. Khetani, *J. Chem. Soc., Chem. Commun.*, 1986, 1772; (c) A. J. Pearson and V. D. Khetani, *J. Am. Chem. Soc.*, 1989, **111**, 6778.
120. (a) A. J. Pearson and M. N. I. Khan, *J. Am. Chem. Soc.*, 1984, **106**, 1872; (b) A. J. Pearson and M. N. I. Khan, *Tetrahedron Lett.*, 1984, **25**, 3507; (c) A. J. Pearson, M. N. I. Khan, J. C. Clardy and H. Cun-heng, *J. Am. Chem. Soc.*, 1985, **107**, 2748.
121. A. J. Pearson and M. W. D. Perry, *J. Chem. Soc., Chem. Commun.*, 1989, 389.
122. M. Green, S. Greenfield, M. J. Grimshire, M. Kersting, A. G. Orpen and R. A. Rodrigues, *J. Chem. Soc., Chem. Commun.*, 1987, 97.
123. S. A. Benyunes and M. Green, *J. Organomet. Chem.*, 1990, **390**, C32.
124. J. Feldman and J. C. Calabrese, *J. Chem. Soc., Chem. Commun.*, 1991, 134.
125. S. Hansson, J. F. Miller and L. S. Liebeskind, *J. Am. Chem. Soc.*, 1990, **112**, 9660.
126. A. J. Pearson, M. S. Holden and R. D. Simpson, *Tetrahedron Lett.*, 1986, **27**, 4121.
127. (a) W.-J. Vong, S.-M. Peng and R.-S. Liu, *Organometallics*, 1990, **9**, 2187; (b) W.-J. Vong, S.-H. Lin, R.-S. Liu, G.-H. Lee and S.-M. Peng, *J. Chem. Soc., Chem. Commun.*, 1990, 1285; (c) W.-J. Vong, S.-M. Peng, S.-H. Lin, W.-J. Lin and R.-S. Liu, *J. Am. Chem. Soc.*, 1991, **113**, 573.
128. M.-H. Cheng, Y.-H. Ho, G.-H. Lee, S.-M. Peng and R.-S. Liu, *J. Chem. Soc., Chem. Commun.*, 1991, 697.
129. S.-H. Lin, Y.-J. Yang and R.-S. Liu, *J. Chem. Soc., Chem. Commun.*, 1991, 1004.
130. M.-H. Cheng *et al.*, *J. Chem. Soc., Chem. Commun.*, 1992, 45.
131. (a) J. T. Gauntlett, B. F. Taylor and M. J. Winter, *J. Chem. Soc., Chem. Commun.*, 1984, 420; (b) J. T. Gauntlett, B. F. Taylor and M. J. Winter, *J. Chem. Soc., Dalton Trans.*, 1985, 1815.
132. H. Adams, N. A. Bailey, M. J. Winter and S. Woodward, *J. Organomet. Chem.*, 1991, **418**, C39.
133. J. C. Jeffery, A. L. Ratermann and F. G. A. Stone, *J. Organomet. Chem.*, 1985, **289**, 367.
134. B. E. Mann and S. D. Shaw, *J. Organomet. Chem.*, 1987, **326**, C13.
135. W. E. VanArsdale, R. E. K. Winter and J. K. Kochi, *J. Organomet. Chem.*, 1985, **296**, 31.
136. W. E. VanArsdale, R. E. K. Winter and J. K. Kochi, *Organometallics*, 1986, **5**, 645.
137. J. W. Faller and K.-H. Chao, *J. Am. Chem. Soc.*, 1983, **105**, 3893.
138. J. W. Faller and C. Lambert, *Tetrahedron*, 1985, **41**, 5755.
139. J. W. Faller, K. H. Chao and H. H. Murray, *Organometallics*, 1984, **3**, 1231.
140. J. W. Faller and K.-H. Chao, *Organometallics*, 1984, **3**, 927.
141. C. Benamou and J. Benaïm, *J. Organomet. Chem.*, 1985, **280**, 377.
142. J. W. Faller, Y. Shvo, K. Chao and H. H. Murray, *J. Organomet. Chem.*, 1982, **226**, 251.
143. J. W. Faller, C. Lambert and M. R. Mazzieri, *J. Organomet. Chem.*, 1990, **383**, 161.
144. J. W. Faller and D. L. Linebarrier, *J. Am. Chem. Soc.*, 1989, **111**, 1937.
145. J. W. Faller, J. Thu Nguyen, W. Ellis and M. R. Mazzieri, *Organometallics*, 1993, **12**, 1434.
146. J. W. Faller and D. L. Linebarrier, *Organometallics*, 1990, **9**, 3182.
147. A. R. Barron, S. J. Anderson, K. McGregor and G. Wilkinson, *Polyhedron*, 1989, **8**, 2599.
148. K. E. Oglieve and R. A. Henderson, *J. Chem. Soc., Chem. Commun.*, 1993, 474.
149. (a) D. J. Wink, N.-F. Wang and J. P. Springer, *Organometallics*, 1989, **8**, 259; (b) N.-F. Wang, D. J. Wink and J. C. Dewan, *ibid.*, 1990, **9**, 335.
150. M. Green, N. C. Norman and A. G. Orpen, *J. Organomet. Chem.*, 1981, **221**, C11.
151. L. M. Atagi, S. C. Critchlow and J. M. Mayer, *J. Am. Chem. Soc.*, 1992, **114**, 1483.
152. K. Prout, A. Gourdon, C. Couldwell, B. Meunier, F. M. Miao and J. Woolcock, *Acta Crystallogr., Sect. B*, 1982, **38**, 456.
153. A. Gourdon and K. Prout, *Acta Crystallogr., Sect. C*, 1983, **39**, 865.
154. F. Giuliéri and J. Benaïm, *New J. Chem.*, 1985, **9**, 335.
155. M. G. B. Drew, B. J. Brisdon, D. W. Brown and C. R. Willis, *J. Chem. Soc., Chem. Commun.*, 1986, 1510.
156. (a) B. J. Brisdon, A. G. W. Hodson, M. F. Mahon and K. C. Molloy, *J. Organomet. Chem.*, 1988, **344**, C8; (b) B. J. Brisdon, R. J. Deeth, A. G. W. Hodson, C. M. Kemp, M. F. Mahon and K. C. Molloy, *Organometallics*, 1991, **10**, 1107.
157. B. J. Brisdon, D. W. Brown, C. R. Willis and M. G. B. Drew, *J. Chem. Soc., Dalton Trans.*, 1986, 2405.
158. (a) S. A. Benyunes, M. Green, M. McPartlin and C. B. M. Nation, *J. Chem. Soc., Chem. Commun.*, 1989, 1887; (b) S. A. Benyunes, R. J. Deeth, A. Fries, M. Green, M. McPartlin and C. B. M. Nation, *J. Chem. Soc., Dalton Trans.*, 1992, 3453.
159. M. I. Bruce, T. W. Hambley, M. R. Snow and A. G. Swincer, *Organometallics*, 1985, **4**, 494.
160. M. I. Bruce, M. J. Liddell, M. R. Snow and E. R. T. Tiekink, *Organometallics*, 1988, **7**, 343.
161. V. V. Krivykh, E. S. Taits, P. V. Petrovskii, Y. T. Struchkov and A. I. Yanovskii, *Mendeleev Commun.*, 1991, 103.
162. C. Carfagna, M. Green, M. F. Mahon, S. Rumble and C. M. Woolhouse, *J. Chem. Soc., Chem. Commun.*, 1993, 879.
163. A. K. McMullen, J. P. Selegue and J.-G. Wang, *Organometallics*, 1991, **10**, 3421.
164. F. Y. Pétillon *et al.*, *J. Organomet. Chem.*, 1984, **255**, 231.
165. H. Adams, N. A. Bailey, J. T. Gauntlett, I. M. Harkin, M. J. Winter and S. Woodward, *J. Chem. Soc., Dalton Trans.*, 1991, 1117.
166. R. P. Hughes, J. W. Reisch and A. L. Rheingold, *Organometallics*, 1985, **4**, 1754.
167. R. P. Hughes, W. Kläui, J. W. Reisch and A. Müller, *Organometallics*, 1985, **4**, 1761.

168. M. G. B. Drew, B. J. Brisdon and A. Day, *J. Chem. Soc., Dalton Trans.*, 1981, 1310.
169. M. Rowen Churchill, J. C. Fettinger, L. G. McCullough and R. R. Schrock, *J. Am. Chem. Soc.*, 1984, **106**, 3356.
170. (a) M. Rowen Churchill, J. W. Ziller, S. F. Pedersen and R. R. Schrock, *J. Chem. Soc., Chem. Commun.*, 1984, 485; (b) R. R. Schrock, S. F. Pedersen, M. Rowen Churchill and J. W. Ziller, *Organometallics*, 1984, **3**, 1574.
171. R. R. Schrock, J. S. Murdzek, J. H. Freudenberger, M. Rowen Churchill and J. W. Ziller, *Organometallics*, 1986, **5**, 25.
172. F. J. Feher, M. Green and A. G. Orpen, *J. Chem. Soc., Chem. Commun.*, 1986, 291.
173. M. Green, R. J. Mercer, C. E. Morton and A. G. Orpen, *Angew. Chem., Int. Ed. Engl.*, 1985, **24**, 422.
174. E. Carmona, P. J. Daff, A. Monge, P. Palma, M. L. Poveda and C. Ruiz, *J. Chem. Soc., Chem. Commun.*, 1991, 1503.
175. A. C. Filippou, W. Grünleitner, C. Völkl and P. Kiprof, *J. Organomet. Chem.*, 1991, **413**, 181.
176. A. C. Filippou, W. Grünleitner and P. Kiprof, *J. Organomet. Chem.*, 1991, **410**, 175.
177. W. Gausing and G. Wilke, *Angew. Chem., Int. Ed. Engl.*, 1981, **20**, 186.
178. B. Bogdanović, H. Bönnemann, R. Goddard, A. Startsev and J. M. Wallis, *J. Organomet. Chem.*, 1986, **299**, 347.
179. M. F. Lappert, C. L. Raston, B. W. Skelton and A. H. White, *J. Chem. Soc., Chem. Commun.*, 1981, 485.
180. T. Schmidt and S. Neis, *J. Organomet. Chem.*, 1992, **430**, C5.
181. S. S. Yun, S. K. Kang, I.-H. Suh, Y. D. Choi and I. S. Chang, *Organometallics*, 1991, **10**, 2509.
182. J. C. Green, M. R. Kelly, P. D. Grebenik, C. E. Briant, N. A. McEvoy and D. M. P. Mingos, *J. Organomet. Chem.*, 1982, **228**, 239.
183. R. Benn and G. Schroth, *J. Organomet. Chem.*, 1982, **228**, 71.
184. C. G. Kreiter, *Adv. Organomet. Chem.*, 1986, **26**, 297.
185. M. Kotzian, C. G. Kreiter and S. Özkar, *J. Organomet. Chem.*, 1982, **229**, 29.
186. S. Özkar and N. B. Peynircioğlu, *Inorg. Chim. Acta*, 1986, **119**, 127.
187. M. Kotzian, C. G. Kreiter, G. Michael and S. Özkar, *Chem. Ber.*, 1983, **116**, 3637.
188. S. Özkar and C. G. Kreiter, *J. Organomet. Chem.*, 1986, **303**, 367.
189. C. G. Kreiter, M. Kotzian, U. Schubert, R. Bau and M. A. Bruck, *Z. Naturforsch., Teil B*, 1984, **39**, 1553.
190. G. Michael and C. G. Kreiter, *Z. Naturforsch., Teil B*, 1984, **39**, 1738.
191. S. Özkar and C. G. Kreiter, *J. Organomet. Chem.*, 1983, **256**, 57.
192. C. G. Kreiter and M. Kotzian, *J. Organomet. Chem.*, 1985, **289**, 295.
193. S. Özkar, C. Kayran and C. G. Kreiter, *J. Organomet. Chem.*, 1992, **434**, 79.
194. C. G. Kreiter and S. Özkar, *Z. Naturforsch., Teil B*, 1983, **38**, 1424.
195. S. Özkar and C. G. Kreiter, *Z. Anorg. Allg. Chem.*, 1983, **502**, 215.
196. S. Özkar, A. Erdoğan and C. G. Kreiter, *Inorg. Chim. Acta*, 1985, **98**, 147.
197. D. J. Wink, N.-F. Wang and B. T. Creagan, *Organometallics*, 1989, **8**, 561.
198. D. J. Wink and C. K. Oh, *Organometallics*, 1990, **9**, 2403.
199. S. Özkar, N. K. Tunali and C. G. Kreiter, *J. Organomet. Chem.*, 1992, **428**, 345.
200. C. G. Kreiter, M. Kotzian, S. Özkar and I. Abu-Mour, *J. Organomet. Chem.*, 1992, **431**, 159.
201. F. A. Cotton and J. H. Meadows, *Inorg. Chem.*, 1984, **23**, 4688.
202. C. Jiabi *et al.*, *J. Organomet. Chem.*, 1987, **329**, 69.
203. J. Chen and B. Wang, *J. Organomet. Chem.*, 1992, **440**, 67.
204. W. Gerhartz, F.-W. Grevels, W. E. Klotzbücher, E. A. Koerner von Gustorf and R. N. Perutz, *Z. Naturforsch., Teil B*, 1985, **40**, 518.
205. F.-W. Grevels, J. Jacke, W. E. Klotzbücher, K. Schaffner, R. H. Hooker and A. J. Rest, *J. Organomet. Chem.*, 1990, **382**, 201.
206. P. M. Hodges, S. A. Jackson, J. Jacke, M. Poliakoff, J. J. Turner and F.-W. Grevels, *J. Am. Chem. Soc.*, 1990, **112**, 1234.
207. F.-W. Grevels, J. Jacke, P. Betz, C. Krüger and Y.-H. Tsay, *Organometallics*, 1989, **8**, 293.
208. D. Chmielewski, F.-W. Grevels, J. Jacke and K. Schaffner, *Angew. Chem., Int. Ed. Engl.*, 1991, **30**, 1343.
209. (a) T. J. Chow, M.-Y. Wu and L.-K. Liu, *J. Organomet. Chem.*, 1985, **281**, C33; (b) T. J. Chow, L.-K. Liu and Y.-S. Chao, *J. Chem. Soc., Chem. Commun.*, 1985, 700; (c) T. J. Chow and Y.-S. Chao, *J. Organomet. Chem.*, 1985, **296**, C23; (d) T. S. Chow, Y.-S. Chao and L.-K. Liu, *J. Am. Chem. Soc.*, 1987, **109**, 797.
210. T. J. Chow, T.-H. Lin, S.-M. Peng and M.-C. Cheng, *J. Organomet. Chem.*, 1986, **316**, C29.
211. T. J. Chow and C.-C. Cheng, *J. Organomet. Chem.*, 1989, **368**, 323.
212. T. J. Chow and M.-F. Ding, *J. Organomet. Chem.*, 1987, **329**, 217.
213. G. Michael, J. Kaub and C. G. Kreiter, *Angew. Chem., Int. Ed. Engl.*, 1985, **24**, 502.
214. G. Michael, J. Kaub and C. G. Kreiter, *Chem. Ber.*, 1985, **118**, 3944.
215. J. R. Blackborow *et al.*, *J. Chem. Soc., Dalton Trans.*, 1981, 661.
216. C. G. Kreiter, G. Michael and J. Kaub, *J. Organomet. Chem.*, 1988, **355**, 149.
217. M. L. H. Green and G. Parkin, *J. Chem. Soc., Dalton Trans.*, 1987, 1611.
218. M. L. H. Green, D. O'Hare and G. Parkin, *J. Chem. Soc., Chem. Commun.*, 1985, 356.
219. P. W. Jolly and U. Zakrzewski, *Polyhedron*, 1991, **10**, 1427.
220. (a) S. A. Benyunes, M. Green and M. J. Grimshire, *Organometallics*, 1989, **8**, 2268; (b) S. A. Benyunes, A. Binelli, M. Green and M. J. Grimshire, *J. Chem. Soc., Dalton Trans.*, 1991, 895.
221. (a) A. D. Hunter, P. Legzdins, C. R. Nurse, F. W. B. Einstein and A. C. Willis, *J. Am. Chem. Soc.*, 1985, **107**, 1791; (b) A. D. Hunter, P. Legzdins, F. W. B. Einstein, A. C. Willis, B. E. Bursten and M. C. Gatter, *ibid.*, 1986, **108**, 3843; (c) N. J. Christensen, A. D. Hunter and P. Legzdins, *Organometallics*, 1989, **8**, 930.
222. N. J. Christensen, P. Legzdins, F. W. B. Einstein and R. H. Jones, *Organometallics*, 1991, **10**, 3070.
223. N. J. Christensen, P. Legzdins, J. Trotter and V. C. Yee, *Organometallics*, 1991, **10**, 4021.
224. C. G. Kreiter, G. Wendt and W. S. Sheldrick, *J. Organomet. Chem.*, 1987, **333**, 47.
225. C. G. Kreiter, K. Nist and J. Kögler, *Z. Naturforsch., Teil B*, 1986, **41**, 599.
226. C. G. Kreiter, J. Kögler and K. Nist, *J. Organomet. Chem.*, 1986, **310**, 35.
227. C. G. Kreiter, W. Michels and M. Wenz, *Chem. Ber.*, 1986, **119**, 1994.
228. C. G. Kreiter, M. Wenz and P. Bell, *J. Organomet. Chem.*, 1990, **394**, 195.
229. C. G. Kreiter, M. Wenz and P. Bell, *J. Organomet. Chem.*, 1990, **387**, 175.
230. (a) J. L. Davidson, K. Davidson and W. E. Lindsell, *J. Chem. Soc., Chem. Commun.*, 1983, 452; (b) J. L. Davidson, K. Davidson, W. E. Lindsell, N. W. Murall and A. J. Welch, *J. Chem. Soc., Dalton Trans.*, 1986, 1677.

231. G. C. Conole, M. Green, M. McPartlin, C. Reeve and C. M. Woolhouse, *J. Chem. Soc., Chem. Commun.*, 1988, 1310.
232. M. Green, M. F. Mahon, K. C. Molloy, C. B. M. Nation and C. M. Woolhouse, *J. Chem. Soc., Chem. Commun.*, 1991, 1587.
233. J. R. Morrow, T. L. Tonker and J. L. Templeton, *J. Am. Chem. Soc.*, 1985, **107**, 5004.
234. (a) N. M. Agh-Atabay, J. L. Davidson and K. W. Muir, *J. Chem. Soc., Chem. Commun.*, 1990, 1399; (b) N. M. Agh-Atabay and J. L. Davidson, *J. Chem. Soc., Dalton Trans.*, 1992, 3531.
235. W.-Y. Yeh and L.-K. Liu, *J. Am. Chem. Soc.*, 1992, **114**, 2267.
236. D. J. Wink, J. R. Fox and N. J. Cooper, *J. Am. Chem. Soc.*, 1985, **107**, 5012.
237. J. L. Davidson, *J. Organomet. Chem.*, 1991, **419**, 137.
238. J. L. Davidson, *J. Chem. Soc., Dalton Trans.*, 1987, 2715.
239. (a) W. A. W. A. Bakar, L. Carlton, J. L. Davidson, L. Manojlović-Muir and K. W. Muir, *J. Organomet. Chem.*, 1988, **352**, C54; (b) N. M. Agh-Atabay, L. J. Canoira, L. Carlton and J. L. Davidson, *J. Chem. Soc., Dalton Trans.*, 1991, 1175.
240. R. P. Hughes, J. W. Reisch and A. L. Rheingold, *Organometallics*, 1984, **3**, 1761.
241. S. R. Allen *et al.*, *J. Chem. Soc., Dalton Trans.*, 1984, 1157.
242. (a) G.-M. Yang, G.-H. Lee, S.-M. Peng and R.-S. Liu, *Organometallics*, 1991, **10**, 2531; (b) G.-M. Yang, G.-M. Su and R.-S. Liu, *ibid.*, 1992, **11**, 3444.
243. G.-M. Su, G.-H. Lee, S.-M. Peng and R.-S. Liu, *J. Chem. Soc., Chem. Commun.*, 1992, 215.
244. I.-Y. Wu, M.-C. Cheng, Y.-C. Lin and Y. Wang, *Organometallics*, 1993, **12**, 1686.
245. R. Aumann and J. Uphoff, *Angew. Chem., Int. Ed. Engl.*, 1987, **26**, 357.
246. R. Aumann and B. Trentmann, *Chem. Ber.*, 1989, **122**, 1977.
247. (a) R. L. Beddoes, E. S. Cook and M. J. Morris, *Polyhedron*, 1989, **8**, 1810; (b) H. Adams *et al.*, *J. Chem. Soc., Dalton Trans.*, 1993, 91.
248. L. S. Liebeskind and A. Bombrun, *J. Am. Chem. Soc.*, 1991, **113**, 8736.
249. R. D. Ernst, *Chem. Rev.*, 1988, **88**, 1255.
250. M. S. Kralik, J. P. Hutchinson and R. D. Ernst, *J. Am. Chem. Soc.*, 1985, **107**, 8296.
251. M. S. Kralik, A. L. Rheingold and R. D. Ernst, *Organometallics*, 1987, **6**, 2612.
252. T. M. Sivavec, T. J. Katz, M. Y. Chiang and G. X.-Q. Yang, *Organometallics*, 1989, **8**, 1620.
253. P. W. Jolly, C. Krüger and U. Zakrzewski, *J. Organomet. Chem.*, 1991, **412**, 371.
254. T. D. Newbound *et al.*, *Organometallics*, 1987, **6**, 2432.
255. G.-H. Lee, S.-M. Peng, F.-C. Liu, D. Mu and R.-S. Liu, *Organometallics*, 1989, **8**, 402.
256. M. A. Paz-Sandoval, P. Juárez Saavedra, G. Durán Pomposo, P. Joseph-Nathan and P. Powell, *J. Organomet. Chem.*, 1990, **387**, 265.
257. S.-F. Lush, S.-H. Wang, G.-H. Lee, S.-M. Peng, S.-L. Wang and R.-S. Liu, *Organometallics*, 1990, **9**, 1862.
258. G.-H. Lee, S.-M. Peng, I.-C. Tsung, D. Mu and R.-S. Liu, *Organometallics*, 1989, **8**, 2248.
259. S. L. Wu, C.-Y. Cheng, S.-L. Wang and R.-S. Liu, *Inorg. Chem.*, 1991, **30**, 311.
260. G.-H. Lee, S.-M. Peng, F.-C. Liu and R.-S. Liu, *J. Organomet. Chem.*, 1989, **377**, 123.
261. (a) G.-H. Lee, S.-M. Peng, T.-W. Lee and R.-S. Liu, *Organometallics*, 1986, **5**, 2378; (b) T. W. Lee and R.-S. Liu, *ibid.*, 1988, **7**, 878.
262. W. Weng, A. M. Arif and R. D. Ernst, *Organometallics*, 1993, **12**, 1537.
263. (a) L. Stahl, J. P. Hutchinson, D. R. Wilson and R. D. Ernst, *J. Am. Chem. Soc.*, 1985, **107**, 5016; (b) T. E. Waldman, L. Stahl, D. R. Wilson, A. M. Arif, J. P. Hutchinson and R. D. Ernst, *Organometallics*, 1993, **12**, 1543.
264. T. D. Newbound, A. M. Arif, D. R. Wilson, A. L. Rheingold and R. D. Ernst, *J. Organomet. Chem.*, 1992, **435**, 73.
265. J. W. Freeman, N. C. Hallinan, A. M. Arif, R. W. Gedridge, R. D. Ernst and F. Basolo, *J. Am. Chem. Soc.*, 1991, **113**, 6509.
266. (a) M. Green, K. R. Nagle, C. M. Woolhouse and D. J. Williams, *J. Chem. Soc., Chem. Commun.*, 1987, 1793; (b) C. Carfagna, M. Green, K. R. Nagle, D. J. Williams and C. M. Woolhouse, *J. Chem. Soc., Dalton Trans.*, 1993, 1761.
267. J. K. Shen *et al.*, *Organometallics*, 1992, **11**, 3215.
268. E. P. Kündig, A. F. Cunningham, Jr., P. Paglia, D. P. Simmons and G. Bernardinelli, *Helv. Chim. Acta*, 1990, **73**, 386.
269. C. G. Kreiter and T. Hellmann, *J. Organomet. Chem.*, 1991, **405**, C6.
270. L. Weber, *Z. Naturforsch., Teil B*, 1985, **40**, 373.
271. L. Weber and R. Boese, *Chem. Ber.*, 1985, **118**, 1545.
272. M. L. H. Green *et al.*, *J. Chem. Soc., Dalton Trans.*, 1992, 2063.
273. P. T. DiMauro, P. T. Wolczanski, L. Párkányi and H. Hurrell Petach, *Organometallics*, 1990, **9**, 1097.
274. (a) J. R. Hinchliffe and M. W. Whiteley, *J. Organomet. Chem.*, 1991, **402**, C50; (b) R. L. Beddoes, J. R. Hinchliffe and M. W. Whiteley, *J. Chem. Soc., Dalton Trans.*, 1993, 501.
275. C. G. Kreiter, G. Wendt and J. Kaub, *J. Organomet. Chem.*, 1988, **352**, 307.
276. C. G. Kreiter, G. Wendt and J. Kaub, *Chem. Ber.*, 1989, **122**, 215.
277. M. L. H. Green, D. K. P. Ng and R. C. Tovey, *J. Chem. Soc., Chem. Commun.*, 1992, 918.
278. M. L. H. Green and D. K. P. Ng, *J. Chem. Soc., Chem. Commun.*, 1992, 1116.
279. (a) M. L. H. Green and P. A. Newman, *J. Chem. Soc., Chem. Commun.*, 1984, 816; (b) M. L. H. Green, P. A. Newman and J. A. Bandy, *J. Chem. Soc., Dalton Trans.*, 1989, 331.
280. M. L. H. Green and D. K. P. Ng, *J. Chem. Soc., Dalton Trans.*, 1993, 17.
281. R. H. Hooker and A. J. Rest, *J. Organomet. Chem.*, 1982, **234**, C23.
282. G. J. Kubas, *Inorg. Chem.*, 1983, **22**, 692.
283. H. Adams, N. A. Bailey, D. G. Willett and M. J. Winter, *J. Organomet. Chem.*, 1987, **333**, 61.
284. J. T. Mague and M. P. Johnson, *Organometallics*, 1991, **10**, 349.
285. P. E. Bloyce, R. H. Hooker, A. J. Rest, T. E. Bitterwolf, N. J. Fitzpatrick and J. E. Shade, *J. Chem. Soc., Dalton Trans.*, 1990, 833.
286. R. H. Hooker and A. J. Rest, *J. Chem. Soc., Dalton Trans.*, 1982, 2029.
287. S. D. Reynolds and T. A. Albright, *Organometallics*, 1985, **4**, 980.
288. S. Dysart, I. Georgii and B. E. Mann, *J. Organomet. Chem.*, 1981, **213**, C10.
289. J. C. Green, R. A. Grieves and J. Mason, *J. Chem. Soc., Dalton Trans.*, 1986, 1313.
290. C. D. Hoff, *J. Organomet. Chem.*, 1983, **246**, C53.
291. C. D. Hoff, *J. Organomet. Chem.*, 1985, **282**, 201.

292. S. L. Mukerjee, R. F. Lang, T. Ju, G. Kiss, C. D. Hoff and S. P. Nolan, *Inorg. Chem.*, 1992, **31**, 4885.
293. J. H. Rigby *et al.*, *J. Am. Chem. Soc.*, 1993, **115**, 1382.
294. I. Fischler, F.-W. Grevels, J. Leitich and S. Özkar, *Chem. Ber.*, 1991, **124**, 2857.
295. K. Chaffee, J. B. Sheridan and A. Aistars, *Organometallics*, 1992, **11**, 18.
296. C. G. Kreiter and H. Kurz, *Chem. Ber.*, 1983, **116**, 1494.
297. (a) E. Michels and C. G. Kreiter, *J. Organomet. Chem.*, 1983, **252**, C1; (b) E. Michels, W. S. Sheldrick and C. G. Kreiter, *Chem. Ber.*, 1985, **118**, 964.
298. C. G. Kreiter, E. Michels and J. Kaub, *J. Organomet. Chem.*, 1986, **312**, 221.
299. C. G. Kreiter and E. Michels, *J. Organomet. Chem.*, 1986, **312**, 59.
300. T. van Houwelingen, D. J. Stufkens and A. Oskam, *Organometallics*, 1992, **11**, 1146.
301. S. Özkar and C. G. Kreiter, *J. Organomet. Chem.*, 1985, **293**, 229.
302. C. G. Kreiter, E. Michels and H. Kurz, *J. Organomet. Chem.*, 1982, **232**, 249.
303. C. G. Kreiter and H. Kurz, *J. Organomet. Chem.*, 1981, **214**, 339.
304. C. G. Kreiter and H. Kurz, *Z. Naturforsch., Teil B*, 1982, **37**, 1322.
305. C. G. Kreiter and E. Michels, *Chem. Ber.*, 1984, **117**, 344.
306. K. Lal, N. T. Leckey and W. E. Watts, *J. Organomet. Chem.*, 1983, **254**, 193.
307. K. Lal, N. T. Leckey and W. E. Watts, *J. Organomet. Chem.*, 1983, **258**, 205.
308. U. Behrens, J. Kopf, K. Lal and W. E. Watts, *J. Organomet. Chem.*, 1984, **276**, 193.
309. P. Hackett, B. F. G. Johnson and J. Lewis, *J. Chem. Soc., Dalton Trans.*, 1982, 603.
310. C. G. Kreiter and H. Kurz, *Z. Naturforsch., Teil B*, 1983, **38**, 841.
311. E. W. Abel, K. G. Orrell, K. B. Qureshi, V. Šik and D. Stephenson, *J. Organomet. Chem.*, 1988, **353**, 337.
312. M. S. Lawless and D. S. Marynick, *J. Am. Chem. Soc.*, 1991, **113**, 7513.
313. P. Berno, A. Ceccon, F. Daprà, A. Gambaro and A. Venzo, *J. Chem. Soc., Chem. Commun.*, 1986, 1518.
314. J. Heck and G. Rist, *J. Organomet. Chem.*, 1988, **342**, 45.
315. C. Elschenbroich, J. Heck, W. Massa and R. Schmidt, *Angew. Chem., Int. Ed. Engl.*, 1983, **22**, 330.
316. P. M. J. A. Hermans *et al.*, *Chem. Ber.*, 1993, **126**, 553.
317. F. J. Liotta, Jr., G. Van Duyne and B. K. Carpenter, *Organometallics*, 1987, **6**, 1010.
318. D. Rau and U. Behrens, *Angew. Chem., Int. Ed. Engl.*, 1991, **30**, 870.
319. G. B. Jameson and A. Salzer, *Organometallics*, 1982, **1**, 689.
320. C. G. Kreiter, E. Michels and J. Kaub, *Z. Naturforsch., Teil B*, 1986, **41**, 722.
321. E. Schwed, H.-U. Hund, H. W. Bosch and H. Berke, *Helv. Chim. Acta*, 1991, **74**, 189.
322. A. J. Ashe III, J. W. Kampf, Y. Nakadaira and J. M. Pace, *Angew. Chem., Int. Ed. Engl.*, 1992, **31**, 1255.
323. A. Ceccon, A. Gambaro and A. Venzo, *J. Organomet. Chem.*, 1985, **281**, 221.
324. P. D. Harvey, I. S. Butler and D. F. R. Gilson, *Inorg. Chem.*, 1987, **26**, 32.
325. A. M. Romanin, A. Ceccon and A. Venzo, *J. Electroanal. Chem.*, 1981, **130**, 245.
326. (a) D. A. Brown, N. J. Fitzpatrick, W. K. Glass and T. H. Taylor, *J. Organomet. Chem.*, 1984, **275**, C9; (b) D. A. Brown, N. J. Fitzpatrick, W. K. Glass and T. H. Taylor, *Organometallics*, 1986, **5**, 158.
327. (a) D. A. Brown, N. J. Fitzpatrick and M. A. McGinn, *J. Organomet. Chem.*, 1984, **275**, C5; (b) D. A. Brown, N. J. Fitzpatrick, M. A. McGinn and T. H. Taylor, *Organometallics*, 1986, **5**, 152.
328. P. Powell, M. Stephens and K. H. Yassin, *J. Organomet. Chem.*, 1986, **301**, 313.
329. C. A. Bunton, K. Lal and W. E. Watts, *J. Organomet. Chem.*, 1983, **247**, C14.
330. C. A. Bunton, M. M. Mhala, J. R. Moffatt and W. E. Watts, *J. Organomet. Chem.*, 1983, **253**, C33.
331. T. I. Odiaka, *Inorg. Chim. Acta*, 1985, **103**, 9.
332. T. I. Odiaka and L. A. P. Kane-Maguire, *J. Organomet. Chem.*, 1985, **284**, 35.
333. H.-J. Müller, U. Nagel, M. Steimann, K. Polborn and W. Beck, *Chem. Ber.*, 1989, **122**, 1387.
334. R. L. Beddoes, E. S. Davies, M. Helliwell and M. W. Whiteley, *J. Organomet. Chem.*, 1991, **421**, 285.
335. M. Wieser, K. Sünkel, C. Robl and W. Beck, *Chem. Ber.*, 1992, **125**, 1369.
336. J. Breimair, M. Wieser and W. Beck, *J. Organomet. Chem.*, 1992, **441**, 429.
337. J. Breimair, T. Weidmann, B. Wagner and W. Beck, *Chem. Ber.*, 1991, **124**, 2431.
338. B. Olgemöller and W. Beck, *Chem. Ber.*, 1981, **114**, 867.
339. D. A. Brown *et al.*, *J. Chem. Soc., Chem. Commun.*, 1992, 701.
340. Y. Y. Lau, W. W. Huckabee and S. L. Gipson, *Inorg. Chim. Acta*, 1990, **172**, 41.
341. (a) W. Dell and M. L. Ziegler, *Angew. Chem., Int. Ed. Engl.*, 1981, **20**, 471; (b) W. H. Dell and M. L. Ziegler, *Z. Naturforsch., Teil B*, 1982, **37**, 1.
342. (a) J. S. Adams, M. Cunningham and M. W. Whiteley, *J. Organomet. Chem.*, 1985, **293**, C13; (b) J. S. Adams, C. Bitcon, J. R. Brown, D. Collison, M. Cunningham and M. W. Whiteley, *J. Chem. Soc., Dalton Trans.*, 1987, 3049.
343. S. L. Gipson and K. Kneten, *Inorg. Chim. Acta*, 1989, **157**, 143.
344. A. Gourdon and K. Prout, *Acta Crystallogr., Sect. B*, 1982, **38**, 1596.
345. C. Bitcon, R. Breeze, P. F. Miller and M. W. Whiteley, *J. Organomet. Chem.*, 1989, **364**, 181.
346. R. Breeze, S. Endud and M. W. Whiteley, *J. Organomet. Chem.*, 1986, **302**, 371.
347. R. A. Brown, S. Endud, J. Friend, J. M. Hill and M. W. Whiteley, *J. Organomet. Chem.*, 1988, **339**, 283.
348. K. S. Chong, S. J. Rettig, A. Storr and J. Trotter, *Can. J. Chem.*, 1981, **59**, 1665.
349. S. J. Rettig, A. Storr and J. Trotter, *Can. J. Chem.*, 1981, **59**, 2391.
350. (a) R. Breeze, A. Ricalton and M. W. Whiteley, *J. Organomet. Chem.*, 1987, **327**, C29; (b) R. Breeze, M. S. Plant, A. Ricalton, D. J. Sutton and M. W. Whiteley, *ibid.*, 1988, **356**, 343.
351. J. R. Hinchliffe, A. Ricalton and M. W. Whiteley, *Polyhedron*, 1991, **10**, 267.
352. W. Dell and M. L. Ziegler, *Z. Naturforsch., Teil B*, 1982, **37**, 7.
353. M. L. H. Green and R. B. A. Pardy, *Polyhedron*, 1985, **4**, 1035.
354. R. L. Beddoes, C. Bitcon, A. Ricalton and M. W. Whiteley, *J. Organomet. Chem.*, 1989, **367**, C21.
355. R. L. Beddoes, C. Bitcon and M. W. Whiteley, *J. Organomet. Chem.*, 1991, **402**, 85.
356. C. Elschenbroich, E. Bilger and B. Metz, *Organometallics*, 1991, **10**, 2823.
357. S. Y. Ketov, *J. Organomet. Chem.*, 1992, **429**, C38.
358. M. L. H. Green, D. K. Siriwardene, D. O'Hare and L.-L. Wong, *J. Chem. Soc., Dalton Trans.*, 1988, 2851.

359. R. G. Swisher, E. Sinn and R. N. Grimes, *Organometallics*, 1984, **3**, 599.
360. A. Ricalton and M. W. Whiteley, *J. Organomet. Chem.*, 1989, **361**, 101.
361. R. L. Beddoes, A. Ricalton and M. W. Whiteley, *Acta Crystallogr., Sect. C*, 1988, **44**, 2025.
362. S. J. Dossett, I. J. Hart and F. G. A. Stone, *J. Chem. Soc., Dalton Trans.*, 1990, 3481.
363. S. J. Dossett, I. J. Hart, M. U. Pilotti and F. G. A. Stone, *J. Chem. Soc., Dalton Trans.*, 1991, 511.
364. S. A. Brew, S. J. Dossett, J. C. Jeffery and F. G. A. Stone, *J. Chem. Soc., Dalton Trans.*, 1990, 3709.
365. S. J. Dossett, M. U. Pilotti and F. G. A. Stone, *Polyhedron*, 1990, **9**, 2953.
366. A. Rettenmeier, K. Weidenhammer and M. L. Ziegler, *Z. Anorg. Allg. Chem.*, 1981, **473**, 91.
367. I. B. Benson, S. A. R. Knox, P. J. Naish and A. J. Welch, *J. Chem. Soc., Dalton Trans.*, 1981, 2235.
368. M. Hohmann, L. Krauth-Siegel, K. Weidenhammer, W. Schulze and M. L. Ziegler, *Z. Anorg. Allg. Chem.*, 1981, **481**, 95.
369. M. L. H. Green and D. K. P. Ng, *J. Chem. Soc., Dalton Trans.*, 1993, 11.
370. J. L. Templeton, P. B. Winston and B. C. Ward, *J. Am. Chem. Soc.*, 1981, **103**, 7713.
371. K. Tatsumi, R. Hoffmann and J. L. Templeton, *Inorg. Chem.*, 1982, **21**, 466.
372. J. L. Templeton, *Adv. Organomet. Chem.*, 1989, **29**, 1.
373. H. Berke, P. Härter, G. Huttner and L. Zsolnai, *Z. Naturforsch., Teil B*, 1981, **36**, 929.
374. H. C. Foley, L. M. Strubinger, T. S. Targos and G. L. Geoffroy, *J. Am. Chem. Soc.*, 1983, **105**, 3064.
375. D. J. Wink and B. T. Creagan, *J. Am. Chem. Soc.*, 1990, **112**, 8585.
376. K. R. Birdwhistell, T. L. Tonker and J. L. Templeton, *J. Am. Chem. Soc.*, 1987, **109**, 1401.
377. K. H. Dötz and J. Mühlemeier, *Angew. Chem., Int. Ed. Engl.*, 1982, **21**, 929.
378. D. J. Wink and B. T. Creagan, *Organometallics*, 1990, **9**, 328.
379. A. Hills, D. L. Hughes, N. Kashef, M. A. N. D. A. Lemos, A. J. L. Pombeiro and R. L. Richards, *J. Chem. Soc., Dalton Trans.*, 1992, 1775.
380. J. S. Lain, C. H. Cheng, C. Y. Cheng and S. L. Wang, *J. Organomet. Chem.*, 1990, **390**, 333.
381. T.-Y. Hsiao, P.-L. Kuo, C.-H. Lai, C.-H. Cheng, C.-Y. Cheng and S.-L. Wang, *Organometallics*, 1993, **12**, 1094.
382. H. Fischer, U. Gerbing, G. Müller and H. G. Alt, *Chem. Ber.*, 1987, **120**, 1905.
383. J. Okuda and K. H. Zimmermann, *J. Organomet. Chem.*, 1990, **384**, C21.
384. D. J. Wink and N. J. Cooper, *Organometallics*, 1991, **10**, 494.
385. W.-Y. Yeh, C.-S. Ting and C.-F. Chih, *J. Organomet. Chem.*, 1991, **427**, 257.
386. K. W. Chiu, D. Lyons, G. Wilkinson, M. Thornton-Pett and M. B. Hursthouse, *Polyhedron*, 1983, **2**, 803.
387. G. A. Carriedo, J. A. K. Howard, D. B. Lewis, G. E. Lewis and F. G. A. Stone, *J. Chem. Soc., Dalton Trans.*, 1985, 905.
388. J. M. Maher, J. R. Fox, B. M. Foxman and N. J. Cooper, *J. Am. Chem. Soc.*, 1984, **106**, 2347.
389. H. G. Alt, H. E. Engelhardt, R. D. Rogers and S. T. Abu-Orabi, *J. Organomet. Chem.*, 1989, **378**, 33.
390. N. G. Connelly, A. G. Orpen, A. L. Rieger, P. H. Rieger, C. J. Scott and G. M. Rosair, *J. Chem. Soc., Chem. Commun.*, 1992, 1293.
391. P. Umland and H. Vahrenkamp, *Chem. Ber.*, 1982, **115**, 3580.
392. (a) P. B. Winston, S. J. Nieter Burgmayer and J. L. Templeton, *Organometallics*, 1983, **2**, 167; (b) P. B. Winston, S. J. N. Burgmayer, T. L. Tonker and J. L. Templeton, *ibid.*, 1986, **5**, 1707.
393. (a) J. L. Davidson and G. Vasapollo, *Polyhedron*, 1983, **2**, 305. (b) J. L. Davidson and G. Vasapollo, *J. Chem. Soc., Dalton Trans.*, 1985, 2239.
394. L. Carlton and J. L. Davidson, *J. Chem. Soc., Dalton Trans.*, 1988, 2071.
395. (a) P. K. Baker and E. M. Keys, *Inorg. Chim. Acta*, 1986, **116**, L49; (b) P. K. Baker, E. M. Armstrong and M. G. B. Drew, *Inorg. Chem.*, 1989, **28**, 2406.
396. E. M. Armstrong, P. K. Baker, M. E. Harman and M. B. Hursthouse, *J. Chem. Soc., Dalton Trans.*, 1989, 295.
397. P. K. Baker and E. M. Keys, *Polyhedron*, 1986, **5**, 1233.
398. A. J. Nielson *et al.*, *Polyhedron*, 1992, **11**, 1419.
399. G. R. Clark, A. J. Nielson, A. D. Rae and C. E. F. Rickard, *J. Chem. Soc., Chem. Commun.*, 1992, 1069.
400. (a) P. K. Baker and D. ap Kendrick, *Polyhedron*, 1991, **10**, 2519; (b) P. K. Baker and D. ap Kendrick, *J. Chem. Soc., Dalton Trans.*, 1993, 1039.
401. P. K. Baker, E. M. Armstrong and K. R. Flower, *Polyhedron*, 1988, **7**, 2769.
402. P. K. Baker, K. R. Flower, M. G. B. Drew and G. Forsyth, *J. Chem. Soc., Dalton Trans.*, 1989, 1903.
403. P. K. Baker and K. R. Flower, *Polyhedron*, 1990, **9**, 2233.
404. P. K. Baker and K. R. Flower, *Polyhedron*, 1990, **9**, 2507.
405. P. K. Baker and K. R. Flower, *Inorg. Chim. Acta*, 1989, **165**, 241.
406. P. K. Baker, K. R. Flower, P. A. Bates and M. B. Hursthouse, *J. Organomet. Chem.*, 1989, **372**, 263.
407. P. K. Baker, K. R. Flower, M. E. Harman and M. B. Hursthouse, *J. Chem. Soc., Dalton Trans.*, 1990, 3169.
408. P. K. Baker and K. R. Flower, *J. Organomet. Chem.*, 1990, **397**, 59.
409. P. K. Baker and K. R. Flower, *J. Organomet. Chem.*, 1991, **405**, 299.
410. P. K. Baker and K. R. Flower, *Polyhedron*, 1991, **10**, 711.
411. A. S. Gamble, K. R. Birdwhistell and J. L. Templeton, *J. Am. Chem. Soc.*, 1990, **112**, 1818.
412. M. A. Bennett and I. W. Boyd, *J. Organomet. Chem.*, 1985, **290**, 165.
413. J. R. Morrow, T. L. Tonker and J. L. Templeton, *J. Am. Chem. Soc.*, 1985, **107**, 6956.
414. R. S. Herrick, D. M. Leazer and J. L. Templeton, *Organometallics*, 1983, **2**, 834.
415. J. L. Templeton, R. S. Herrick and J. R. Morrow, *Organometallics*, 1984, **3**, 535.
416. C. Y. Chou and E. A. Maatta, *Inorg. Chem.*, 1984, **23**, 2912.
417. (a) M. Kamata *et al.*, *Organometallics*, 1982, **1**, 227; (b) M. Kamata *et al. Inorg. Chem.*, 1983, **22**, 2416.
418. R. Ellis, R. A. Henderson, A. Hills and D. L. Hughes, *J. Organomet. Chem.*, 1987, **333**, C6.
419. A. Mayr, K. S. Lee, M. A. Kjelsberg and D. Van Engen, *J. Am. Chem. Soc.*, 1986, **108**, 6079.
420. L. M. Atagi, S. C. Critchlow and J. M. Mayer, *J. Am. Chem. Soc.*, 1992, **114**, 9223.
421. A. Mayr, K. S. Lee and B. Kahr, *Angew. Chem., Int. Ed. Engl.*, 1988, **27**, 1730.
422. E. M. Armstrong, P. K. Baker and M. G. B. Drew, *Organometallics*, 1988, **7**, 319.
423. E. M. Armstrong, P. K. Baker and M. G. B. Drew, *J. Organomet. Chem.*, 1987, **336**, 377.
424. E. M. Armstrong, P. K. Baker, T. Callow, K. R. Flower, P. D. Jackson and L. M. Severs, *J. Organomet. Chem.*, 1992, **434**, 321.

425. P. K. Baker, G. A. Cartwright, P. D. Jackson, K. R. Flower, N. Galeotti and L. M. Severs, *Polyhedron*, 1992, **11**, 1043.
426. P. K. Baker, E. M. Armstrong and M. G. B. Drew, *Inorg. Chem.*, 1988, **27**, 2287.
427. P. K. Baker and E. M. Armstrong, *Polyhedron*, 1988, **7**, 63.
428. E. M. Armstrong and P. K. Baker, *Inorg. Chim. Acta*, 1988, **143**, 13.
429. E. M. Armstrong, P. K. Baker, K. R. Flower and M. G. B. Drew, *J. Chem. Soc., Dalton Trans.*, 1990, 2535.
430. (a) P. K. Baker, M. G. B. Drew and K. R. Flower, *J. Organomet. Chem.*, 1990, **391**, C12; (b) P. K. Baker, K. R. Flower and M. G. B. Drew, *Organometallics*, 1993, **12**, 276.
431. J. L. Davidson and G. Vasapollo, *J. Chem. Soc., Dalton Trans.*, 1988, 2855.
432. J. L. Davidson, G. Vasapollo, J. C. Millar and K. W. Muir, *J. Chem. Soc., Dalton Trans.*, 1987, 2165.
433. R. S. Herrick and J. L. Templeton, *Organometallics*, 1982, **1**, 842.
434. R. S. Herrick, S. J. Nieter Burgmayer and J. L. Templeton, *Inorg. Chem.*, 1983, **22**, 3275.
435. R. S. Herrick, S. J. Nieter-Burgmayer and J. L. Templeton, *J. Am. Chem. Soc.*, 1983, **105**, 2599.
436. M. A. Bennett, I. W. Boyd, G. B. Robertson and W. A. Wickramasinghe, *J. Organomet. Chem.*, 1985, **290**, 181.
437. C. E. Kriley, J. L. Kerschner, P. E. Fanwick and I. P. Rothwell, *Organometallics*, 1993, **12**, 2051.
438. S. R. Allen *et al.*, *J. Chem. Soc., Dalton Trans.*, 1981, 873.
439. (a) S. R. Allen, M. Green, A. G. Orpen and I. D. Williams, *J. Chem. Soc., Chem. Commun.*, 1982, 826; (b) S. R. Allen, R. G. Beevor, M. Green, A. G. Orpen, K. E. Paddick and I. D. Williams, *J. Chem. Soc., Dalton Trans.*, 1987, 591.
440. S. R. Allen *et al.*, *J. Chem. Soc., Dalton Trans.*, 1984, 2747.
441. S. R. Allen, M. Green, G. Moran, A. G. Orpen and G. E. Taylor, *J. Chem. Soc., Dalton Trans.*, 1984, 441.
442. F. J. Feher, M. Green and R. A. Rodrigues, *J. Chem. Soc., Chem. Commun.*, 1987, 1206.
443. M. Green, *J. Organomet. Chem.*, 1986, **300**, 93.
444. S. R. Allen *et al.*, *J. Chem. Soc., Dalton Trans.*, 1983, 927.
445. (a) M. Green, N. C. Norman and A. G. Orpen, *J. Am. Chem. Soc.*, 1981, **103**, 1267; (b) S. R. Allen, R. G. Beevor, M. Green, N. C. Norman, A. G. Orpen and I. D. Williams, *J. Chem. Soc., Dalton Trans.*, 1985, 435.
446. (a) R. G. Beevor, M. Green, A. G. Orpen and I. D. Williams, *J. Chem. Soc., Chem. Commun.*, 1983, 673; (b) R. G. Beevor, M. Green, A. G. Orpen and I. D. Williams, *J. Chem. Soc., Dalton Trans.*, 1987, 1319.
447. S. G. Feng, C. C. Philipp, A. S. Gamble, P. S. White and J. L. Templeton, *Organometallics*, 1991, **10**, 3504.
448. S. G. Feng and J. L. Templeton, *Organometallics*, 1992, **11**, 2168.
449. S. G. Feng, P. S. White and J. L. Templeton, *J. Am. Chem. Soc.*, 1990, **112**, 8192.
450. S. G. Feng, P. S. White and J. L. Templeton, *J. Am. Chem. Soc.*, 1992, **114**, 2951.
451. S. G. Feng, P. S. White and J. L. Templeton, *Organometallics*, 1993, **12**, 2131.
452. A. J. M. Caffyn *et al.*, *Organometallics*, 1991, **10**, 2842.
453. S. G. Feng and J. L. Templeton, *J. Am. Chem. Soc.*, 1989, **111**, 6477.
454. J. L. Caldarelli, P. S. White and J. L. Templeton, *J. Am. Chem. Soc.*, 1992, **114**, 10097.
455. M. A. Collins, S. G. Feng, P. A. White and J. L. Templeton, *J. Am. Chem. Soc.*, 1992, **114**, 3771.
456. J. L. Davidson, *J. Chem. Soc., Dalton Trans.*, 1986, 2423.
457. J. L. Davidson and F. Sence, *J. Organomet. Chem.*, 1991, **409**, 219.
458. (a) H. G. Alt, M. E. Eichner and B. M. Jansen, *Angew. Chem., Int. Ed. Engl.*, 1982, **21**, 861; (b) H. G. Alt, *J. Organomet. Chem.*, 1985, **288**, 149.
459. H. G. Alt, *J. Organomet. Chem.*, 1983, **256**, C12.
460. H. G. Alt and H. I. Hayen, *Angew. Chem., Int. Ed. Engl.*, 1983, **22**, 1008.
461. J. L. Petersen and J. W. Egan, Jr., *Inorg. Chem.*, 1981, **20**, 2883.
462. G. E. Herberich, U. Englert and W. Fassbender, *J. Organomet. Chem.*, 1991, **420**, 303.
463. G. E. Herberich and J. Okuda, *Chem. Ber.*, 1984, **117**, 3112.
464. G. E. Herberich, W. Barlage and K. Linn, *J. Organomet. Chem.*, 1991, **414**, 193.
465. M. B. O'Regan, M. G. Vale, J. F. Payack and R. R. Schrock, *Inorg. Chem.*, 1992, **31**, 1112.
466. (a) L. Carlton, J. L. Davidson, J. C. Millar and K. W. Muir, *J. Chem. Soc., Chem. Commun.*, 1984, 11; (b) J. L. Davidson, G. Vasapollo, L. Manojlović-Muir and K. W. Muir *ibid.*, 1982, 1025; (c) L. Carlton and J. L. Davidson, *J. Chem. Soc., Dalton Trans.*, 1987, 895.
467. L. Carlton, W. A. W. Abu Bakar and J. L. Davidson, *J. Organomet. Chem.*, 1990, **394**, 177.
468. (a) J. L. Davidson, I. E. P. Murray, P. N. Preston, M. V. Russo, L. Manojlović-Muir and K. W. Muir, *J. Chem. Soc., Chem. Commun.*, 1981, 1059; (b) J. L. Davidson, I. E. P. Murray, P. N. Preston and M. V. Russo, *J. Chem. Soc., Dalton Trans.*, 1983, 1783.
469. J. L. Davidson, W. F. Wilson, L. Manojlović-Muir and K. W. Muir, *J. Organomet. Chem.*, 1983, **254**, C6.
470. L. J. Canoira, J. L. Davidson, G. Douglas and K. W. Muir, *J. Organomet. Chem.*, 1989, **362**, 135.
471. (a) J. L. Davidson, W. F. Wilson and K. W. Muir, *J. Chem. Soc., Chem. Commun.*, 1985, 460; (b) J. L. Davidson and W. F. Wilson, *J. Chem. Soc., Dalton Trans.*, 1988, 27.
472. (a) N. M. Agh-Atabay, J. L. Davidson, G. Douglas and K. W. Muir, *J. Chem. Soc., Chem. Commun.*, 1987, 1526; (b) N. M. Agh-Atabay and J. L. Davidson, *J. Chem. Soc., Dalton Trans.*, 1989, 1027.
473. L. Carlton, J. L. Davidson, P. Ewing, L. Manojlović-Muir and K. W. Muir, *J. Chem. Soc., Chem. Commun.*, 1985, 1474. N. M. Agh-Atabay, J. L. Davidson, G. Douglas and K. W. Muir, *ibid.*, 1989, 549. L. Carlton, N. M. Agh-Atabay and J. L. Davidson, *J. Organomet. Chem.*, 1991, **413**, 205.
474. P. Leoni, F. Marchetti, M. Pasquali and P. Zanello, *J. Chem. Soc., Dalton Trans.*, 1988, 635.
475. K. Sünkel, U. Nagel and W. Beck, *J. Organomet. Chem.*, 1981, **222**, 251.
476. K. A. Mead, H. Morgan and P. Woodward, *J. Chem. Soc., Dalton Trans.*, 1983, 271.
477. A. M. Kook, P. N. Nickias, J. P. Selegue and S. L. Smith, *Organometallics*, 1984, **3**, 499.
478. S. R. Allen, M. Green, N. C. Norman, K. E. Paddick and A. G. Orpen, *J. Chem. Soc., Dalton Trans.*, 1983, 1625.
479. (a) D. G. Bourner *et al.*, *J. Chem. Soc., Chem. Commun.*, 1985, 1409; (b) L. Brammer, *et al.*, *J. Chem. Soc., Dalton Trans.* 1993, 1747.
480. E. Hey, F. Weller and K. Dehnicke, *Z. Anorg. Allg. Chem.*, 1984, **514**, 25.
481. M. Kersting, K. Dehnicke and D. Fenske, *J. Organomet. Chem.*, 1988, **346**, 201.
482. E. Hey, F. Weller and K. Dehnicke, *Z. Anorg. Allg. Chem.*, 1984, **514**, 18.
483. M. Kersting, A. El-Kholi, U. Müller and K. Dehnicke, *Chem. Ber.*, 1989, **122**, 279.

484. K. Stahl, F. Weller and K. Dehnicke, *Z. Anorg. Allg. Chem.*, 1986, **533**, 73.
485. M. Kersting, K. Dehnicke and D. Fenske, *J. Organomet. Chem.*, 1986, **309**, 125.
486. K. H. Theopold, S. J. Holmes and R. R. Schrock, *Angew. Chem., Int. Ed. Engl.*, 1983, **22**, 1010.
487. I. Pauls, K. Dehnicke and D. Fenske, *Chem. Ber.*, 1989, **122**, 481.
488. E. C. Walborsky, D. E. Wigley, E. Roland, J. C. Dewan and R. R. Schrock, *Inorg. Chem.*, 1987, **26**, 1615.
489. J. L. Templeton, B. C. Ward, G. J.-J. Chen, J. W. McDonald and W. E. Newton, *Inorg. Chem.*, 1981, **20**, 1248.
490. J. R. Morrow, T. L. Tonker and J. L. Templeton, *Organometallics*, 1985, **4**, 745.
491. D. D. Devore, E. A. Maatta and F. Takusagawa, *Inorg. Chim. Acta*, 1986, **112**, 87.
492. (a) G. R. Clark, A. J. Nielson, C. E. F. Rickard and D. C. Ware, *J. Chem. Soc., Chem. Commun.*, 1989, 343; (b) A. J. Nielson and D. C. Ware, *Polyhedron*, 1990, **9**, 603.
493. G. R. Clark, A. J. Nielson and C. E. F. Rickard, *J. Chem. Soc., Chem. Commun.*, 1989, 1157.
494. M. Rowen Churchill and H. J. Wasserman, *Organometallics*, 1983, **2**, 755.
495. A. Mayr and C. M. Bastos, *Prog. Inorg. Chem.*, 1992, **40**, 1.
496. (a) F. R. Kreissl, W. Sieber and M. Wolfgruber, *Angew. Chem., Int. Ed. Engl.*, 1983, **22**, 493; (b) F. R. Kreissl, W. J. Sieber and M. Wolfgruber, *Z. Naturforsch., Teil B*, 1983, **38**, 1419.
497. (a) J. C. Jeffery, J. C. V. Laurie, I. Moore and F. G. A. Stone, *J. Organomet. Chem.*, 1983, **258**, C37; (b) J. A. K. Howard, J. C. Jeffery, J. C. V. Laurie, I. Moore, F. G. A. Stone and A. Stringer, *Inorg. Chim. Acta*, 1985, **100**, 23.
498. K. R. Birdwhistell, T. L. Tonker and J. L. Templeton, *J. Am. Chem. Soc.*, 1985, **107**, 4474.
499. A. Mayr, G. A. McDermott, A. M. Dorries and D. Van Engen, *Organometallics*, 1987, **6**, 1503.
500. K. A. Belsky, M. F. Asaro, S. Y. Chen and A. Mayr, *Organometallics*, 1992, **11**, 1926.
501. H. P. Kim, S. Kim, R. A. Jacobson and R. J. Angelici, *Organometallics*, 1986, **5**, 2481.
502. R. A. Doyle and R. J. Angelici, *Organometallics*, 1989, **8**, 2207.
503. (a) M. Rowen Churchill, H. J. Wasserman, S. J. Holmes and R. R. Schrock, *Organometallics*, 1982, **1**, 766; (b) M. Rowen Churchill and H. J. Wasserman, *Inorg. Chem.*, 1983, **22**, 41.
504. A. Mayr, C. M. Bastos, R. T. Chang, J. X. Haberman, K. S. Robinson and D. A. Belle-Oudry, *Angew. Chem., Int. Ed. Engl.*, 1992, **31**, 747.
505. C. M. Giandomenico, C. T. Lam and S. J. Lippard, *J. Am. Chem. Soc.*, 1982, **104**, 1263.
506. P. W. R. Corfield, L. M. Baltusis and S. J. Lippard, *Inorg. Chem.*, 1981, **20**, 922.
507. S. Warner and S. J. Lippard, *Organometallics*, 1986, **5**, 1716.
508. J. C. Dewan, C. M. Giandomenico and S. J. Lippard, *Inorg. Chem.*, 1981, **20**, 4069.
509. E. M. Carnahan and S. J. Lippard, *J. Chem. Soc., Dalton Trans.*, 1991, 699.
510. (a) A. C. Filippou and W. Grünleitner, *J. Organomet. Chem.*, 1990, **393**, C10; (b) A. C. Filippou and W. Grünleitner, *Z. Naturforsch., Teil B*, 1991, **46**, 216.
511. A. C. Filippou, *Polyhedron*, 1990, **9**, 727.
512. A. C. Filippou, C. Völkl, W. Grünleitner and P. Kiprof, *J. Organomet. Chem.*, 1992, **434**, 201.
513. A. C. Filippou and W. Grünleitner, *Z. Naturforsch., Teil B*, 1989, **44**, 1023.
514. A. Mayr and C. M. Bastos, *J. Am. Chem. Soc.*, 1990, **112**, 7797.
515. C. M. Bastos, N. Daubenspeck and A. Mayr, *Angew. Chem., Int. Ed. Engl.*, 1993, **32**, 743.
516. (a) G. A. McDermott and A. Mayr, *J. Am. Chem. Soc.*, 1987, **109**, 580; (b) A. Mayr, C. M. Bastos, N. Daubenspeck and G. A. McDermott, *Chem. Ber.*, 1992, **125**, 1583.
517. A. C. Filippou, W. Grünleitner, C. Völkl and P. Kiprof, *Angew. Chem., Int. Ed. Engl.*, 1991, **30**, 1167.
518. (a) W. E. Buhro and M. H. Chisholm, *Adv. Organomet. Chem.*, 1987, **27**, 311; (b) M. H. Chisholm, *New J. Chem.*, 1987, **11**, 459; (c) M. H. Chisholm, *Pure Appl. Chem.*, 1989, **61**, 1707; (d) M. H. Chisholm, *J. Organomet. Chem.*, 1990, **400**, 235.
519. M. H. Chisholm *et al.*, *Polyhedron*, 1987, **6**, 783.
520. H. Strutz and R. R. Schrock, *Organometallics*, 1984, **3**, 1600.
521. F. A. Cotton, W. Schwotzer and E. S. Shamshoum, *Organometallics*, 1983, **2**, 1167.
522. (a) M. H. Chisholm, J. C. Huffman and I. P. Rothwell, *J. Am. Chem. Soc.*, 1981, **103**, 4245; (b) M. H. Chisholm, K. Folting, J. C. Huffman and I. P. Rothwell, *ibid.*, 1982, **104**, 4389.
523. (a) M. H. Chisholm, K. Folting, D. M. Hoffman, J. C. Huffman and J. Leonelli, *J. Chem. Soc., Chem. Commun.*, 1983, 589; (b) M. H. Chisholm, K. Folting, D. M. Hoffman and J. C. Huffman, *J. Am. Chem. Soc.*, 1984, **106**, 6794.
524. M. H. Chisholm, B. K. Conroy, K. Folting, D. M. Hoffman and J. C. Huffman, *Organometallics*, 1986, **5**, 2457.
525. M. H. Chisholm, C. M. Cook, J. C. Huffman and W. E. Streib, *J. Chem. Soc., Dalton Trans.*, 1991, 929.
526. M. H. Chisholm, B. K. Conroy, D. L. Clark and J. C. Huffman, *Polyhedron*, 1988, **7**, 903.
527. (a) M. H. Chisholm, B. K. Conroy, J. C. Huffman and N. S. Marchant, *Angew. Chem., Int. Ed. Engl.*, 1986, **25**, 446; (b) M. H. Chisholm, J. C. Huffman and N. S. Marchant, *J. Chem. Soc., Chem. Commun.*, 1986, 717; (c) M. H. Chisholm, D. Ho, J. C. Huffman and N. S. Marchant, *Organometallics*, 1989, **8**, 1626.
528. M. H. Chisholm, K. Folting, J. C. Huffman and E. A. Lucas, *Organometallics*, 1991, **10**, 535.
529. M. H. Chisholm, D. M. Hoffman and J. C. Huffman, *J. Am. Chem. Soc.*, 1984, **106**, 6815.
530. M. H. Chisholm, D. M. Hoffman and J. C. Huffman, *J. Am. Chem. Soc.*, 1984, **106**, 6806.
531. (a) M. H. Chisholm, K. Folting, J. C. Huffman and N. S. Marchant, *Organometallics*, 1986, **5**, 602; (b) M. H. Chisholm, J. C. Huffman and N. S. Marchant, *Polyhedron*, 1988, **7**, 919.
532. (a) M. H. Chisholm, B. W. Eichhorn and J. C. Huffman, *J. Chem. Soc., Chem. Commun.*, 1985, 861; (b) M. H. Chisholm, B. W. Eichhorn, K. Folting and J. C. Huffman, *Organometallics*, 1989, **8**, 49.
533. M. H. Chisholm, B. W. Eichhorn and J. C. Huffman, *Organometallics*, 1989, **8**, 67.
534. (a) M. H. Chisholm, B. W. Eichhorn and J. C. Huffman, *Organometallics*, 1987, **6**, 2264; (b) M. H. Chisholm, B. W. Eichhorn and J. C. Huffman, *ibid.*, 1989, **8**, 80.
535. (a) K. J. Ahmed, M. H. Chisholm and J. C. Huffman, *Organometallics*, 1985, **4**, 1312; (b) K. J. Ahmed, M. H. Chisholm, K. Folting and J. C. Huffman, *ibid.*, 1986, **5**, 2171.
536. M. J. Calhorda and R. Hoffmann, *Organometallics*, 1986, **5**, 2181.
537. (a) K. J. Ahmed, M. H. Chisholm, K. Folting and J. C. Huffman, *J. Chem. Soc., Chem. Commun.*, 1985, 152; (b) K. J. Ahmed, M. H. Chisholm, K. Folting and J. C. Huffman, *J. Am. Chem. Soc.*, 1986, **108**, 989.

538. (a) S. G. Bott, D. L. Clark, M. L. H. Green and P. Mountford, *J. Chem. Soc., Chem. Commun.*, 1989, 418; (b) S. G. Bott, D. L. Clark, M. L. H. Green and P. Mountford, *J. Chem. Soc., Dalton Trans.*, 1991, 471.
539. (a) M. L. H. Green and P. Mountford, *Organometallics*, 1990, **9**, 886; (b) Q. Feng, M. L. H. Green and P. Mountford, *J. Chem. Soc., Dalton Trans.*, 1992, 2171.
540. M. L. H. Green and P. Mountford, *Chem. Soc. Rev.*, 1992, 29.
541. W. Hirpo and M. D. Curtis, *J. Am. Chem. Soc.*, 1988, **110**, 5218.
542. (a) M. H. Chisholm, J. A. Heppert and J. C. Huffman, *J. Am. Chem. Soc.*, 1984, **106**, 1151; (b) M. H. Chisholm, J. C. Huffman and J. A. Heppert, *ibid.*, 1985, **107**, 5116.
543. M. H. Chisholm and C. D. Ontiveros, *Polyhedron*, 1988, **7**, 1015.
544. (a) R. H. Cayton, S. T. Chacon, M. H. Chisholm and J. C. Huffman, *Angew. Chem., Int. Ed. Engl.*, 1990, **29**, 1026; (b) S. T. Chacon, M. H. Chisholm, O. Eisenstein and J. C. Huffman, *J. Am. Chem. Soc.*, 1992, **114**, 8497.
545. (a) M. H. Chisholm and M. Hampden-Smith, *Angew. Chem., Int. Ed. Engl.*, 1987, **26**, 903; (b) M. H. Chisholm and M. Hampden-Smith, *J. Am. Chem. Soc.*, 1987, **109**, 5871; (c) M. H. Chisholm, J. C. Huffman and M. J. Hampden-Smith, *ibid.*, 1989, **111**, 5284.
546. M. H. Chisholm, J. C. Huffman, E. A. Lucas and E. B. Lubkovsky, *Organometallics*, 1991, **10**, 3424.
547. R. H. Cayton, M. H. Chisholm and M. J. Hampden-Smith, *J. Am. Chem. Soc.*, 1988, **110**, 4438.
548. (a) R. H. Cayton *et al. Angew. Chem., Int. Ed. Engl.*, 1989, **28**, 1523; (b) S. T. Chacon, M. H. Chisholm, K. Folting, J. C. Huffman and M. J. Hampden-Smith, *Organometallics*, 1991, **10**, 3722.
549. M. H. Chisholm, K. Folting, J. A. Heppert and W. E. Streib, *J. Chem. Soc., Chem. Commun.*, 1985, 1755.
550. M. H. Chisholm, M. J. Hampden-Smith, J. C. Huffman and K. G. Moodley, *J. Am. Chem. Soc.*, 1988, **110**, 4070.
551. M. H. Chisholm, R.-M. Jansen and J. C. Huffman, *Organometallics*, 1992, **11**, 2305.
552. S. A. R. Knox, R. F. D. Stansfield, F. G. A. Stone, M. J. Winter and P. Woodward, *J. Chem. Soc., Dalton Trans.*, 1982, 173.
553. J. A. Beck, S. A. R. Knox, R. F. D. Stansfield, F. G. A. Stone, M. J. Winter and P. Woodward, *J. Chem. Soc., Dalton Trans.*, 1982, 195.
554. S. R. Finnimore, S. A. R. Knox and G. E. Taylor, *J. Chem. Soc., Dalton Trans.*, 1982, 1783.
555. N. A. Ustynyuk, V. N. Vinogradova, V. N. Korneva, D. N. Kravtsov, V. G. Andrianov and Yu. T. Struchkov, *J. Organomet. Chem.*, 1984, **277**, 285.
556. J. C. Jeffery, J. C. V. Laurie and F. G. A. Stone, *Polyhedron*, 1985, **4**, 1135.
557. E. Delgado, M. E. Garcia, J. C. Jeffery, P. Sherwood and F. G. A. Stone, *J. Chem. Soc., Dalton Trans.*, 1988, 207.
558. S. P. Nolan, R. López de la Vega and C. D. Hoff, *Inorg. Chem.*, 1986, **25**, 4446.
559. C. Coates, N. G. Connelly and M. Carmen Crespo, *J. Chem. Soc., Dalton Trans.*, 1988, 2509.
560. P. Bougeard, S. Peng, M. Mlekuz and M. J. McGlinchey, *J. Organomet. Chem.*, 1985, **296**, 383.
561. S. Slater and E. L. Muetterties, *Inorg. Chem.*, 1981, **20**, 946.
562. A. M. Boileau, A. G. Orpen, R. F. D. Stansfield and P. Woodward, *J. Chem. Soc., Dalton Trans.*, 1982, 187.
563. M. Green, P. A. Kale and R. J. Mercer, *J. Chem. Soc., Chem. Commun.*, 1987, 375.
564. L. Brammer, M. Green, A. G. Orpen, K. E. Paddick and D. R. Saunders, *J. Chem. Soc., Dalton Trans.*, 1986, 657.
565. D. Pufahl, W. E. Geiger and N. G. Connelly, *Organometallics*, 1989, **8**, 412.
566. (a) M. Green, N. C. Norman and A. G. Orpen, *J. Am. Chem. Soc.*, 1981, **103**, 1269; (b) M. Green, N. C. Norman, A. G. Orpen and C. J. Schaverien, *J. Chem. Soc., Dalton Trans.*, 1984, 2455.
567. M. Green, N. K. Jetha, R. J. Mercer, N. C. Norman and A. G. Orpen, *J. Chem. Soc., Dalton Trans.*, 1988, 1843.
568. (a) S. G. Bott *et al. J. Chem. Soc., Chem. Commun.*, 1983, 378; (b) S. G. Bott *et al.*, *J. Chem. Soc., Dalton Trans.*, 1990, 1957.
569. C. G. Kreiter, A. Würtz and P. Bell, *Chem. Ber.*, 1992, **125**, 377.
570. G. Wilke, H. Benn, R. Goddard, C. Krüger and B. Pfeil, *Inorg. Chim. Acta*, 1992, **198–200**, 741.
571. S. A. R. Knox, R. F. D. Stansfield, F. G. A. Stone, M. J. Winter and P. Woodward, *J. Chem. Soc., Dalton Trans.*, 1982, 167.
572. G. Conole, K. A. Hill, M. McPartlin, M. J. Mays and M. J. Morris, *J. Chem. Soc., Chem. Commun.*, 1989, 688.
573. M. J. Mays, M. J. Morris, P. F. Reinisch, M. McPartlin and H. R. Powell, *Polyhedron*, 1989, **8**, 1807.
574. G. Conole, M. McPartlin, M. J. Mays and M. J. Morris, *J. Chem. Soc., Dalton Trans.*, 1990, 2359.
575. G. R. Doel, N. D. Feasey, S. A. R. Knox, A. G. Orpen and J. Webster, *J. Chem. Soc., Chem. Commun.*, 1986, 542.
576. K. Endrich, R. Korswagen, T. Zahn and M. L. Ziegler, *Angew. Chem., Int. Ed. Engl.*, 1982, **21**, 919.
577. G. Conole, K. Henrick, M. McPartlin, A. D. Horton and M. J. Mays, *New J. Chem.*, 1988, **12**, 559.
578. A. D. Horton and M. J. Mays, *J. Chem. Soc., Dalton Trans.*, 1990, 155.
579. G. Conole, K. Henrick, M. McPartlin, A. D. Horton, M. J. Mays and E. Sappa, *J. Chem. Soc., Dalton Trans.*, 1990, 2367.
580. R. F. Gerlach, D. N. Duffy and M. D. Curtis, *Organometallics*, 1983, **2**, 1172.
581. A. Meyer, D. J. McCabe and M. D. Curtis, *Organometallics*, 1987, **6**, 1491.
582. N. Le Berre-Cosquer, R. Kergoat and P. L'Haridon, *Organometallics*, 1992, **11**, 721.
583. I. V. Barinov, O. A. Reutov, A. V. Polyakov, A. I. Yanovsky, Y. T. Struchkov and V. I. Sokolov, *J. Organomet. Chem.*, 1991, **418**, C24.
584. M. V. Galakhov, V. I. Bakhmutov, I. V. Barinov and O. A. Reutov, *J. Organomet. Chem.*, 1991, **421**, 65.
585. M. D. Curtis, A. Meyer and W. M. Butler, *Organometallics*, 1992, **11**, 4343.
586. (a) R. J. Mercer, M. Green and A. G. Orpen, *J. Chem. Soc., Chem. Commun.*, 1986, 567; (b) S. F. T. Froom, M. Green, R. J. Mercer, K. R. Nagle, A. G. Orpen and S. Schwiegk, *ibid.*, 1986, 1666; (c) S. F. T. Froom, M. Green, R. J. Mercer, K. R. Nagle, A. G. Orpen and R. A. Rodrigues, *J. Chem. Soc., Dalton Trans.*, 1991, 3171.
587. N. M. Doherty, C. Elschenbroich, H.-J. Kneuper and S. A. R. Knox, *J. Chem. Soc., Chem. Commun.*, 1985, 170.
588. U. Kern, C. G. Kreiter, S. Müller-Becker and W. Frank, *J. Organomet. Chem.*, 1993, **444**, C31.
589. G. C. Conole, S. F. T. Froom, M. Green and M. McPartlin, *J. Chem. Soc., Chem. Commun.*, 1989, 92.
590. M. Bamber, S. F. T. Froom, M. Green, M. Schulz and H. Werner, *J. Organomet. Chem.*, 1992, **434**, C19.
591. S. F. T. Froom, M. Green, K. R. Nagle and D. J. Williams, *J. Chem. Soc., Chem. Commun.*, 1987, 1305.
592. J. S. Drage, M. Tilset, P. C. Vollhardt and T. W. Weidman, *Organometallics*, 1984, **3**, 812. J. S. Drage and K. P. C. Vollhardt, *ibid.*, 1986, **5**, 280.
593. W. Abriel, G. Baum, J. Heck and K.-A. Kriebisch, *Chem. Ber.*, 1990, **123**, 1767.

594. H. El Amouri, J. Vaissermann, Y. Besace, K. P. C. Vollhardt and G. E. Ball, *Organometallics*, 1993, **12**, 605.
595. (a) M. Green, A. G. Orpen, C. J. Schaverien and I. D. Williams, *J. Chem. Soc., Chem. Commun.*, 1983, 583; (b) W. E. Carroll, M. Green, A. G. Orpen, C. J. Schaverien, I. D. Williams and A. J. Welch, *J. Chem. Soc., Dalton Trans.*, 1986, 1021.
596. (a) M. Green, A. G. Orpen, C. J. Schaverien and I. D. Williams, *J. Chem. Soc., Chem. Commun.*, 1983, 181; (b) M. Green, R. J. Mercer, A. G. Orpen, C. J. Schaverien and I. D. Williams, *J. Chem. Soc., Dalton Trans.*, 1986, 1971.
597. C. G. Kreiter, G. Wendt and J. Kaub, *J. Organomet. Chem.*, 1988, **346**, 35.
598. M. Griffiths, S. A. R. Knox, R. F. D. Stansfield, F. G. A. Stone, M. J. Winter and P. Woodward, *J. Chem. Soc., Dalton Trans.*, 1982, 159.
599. R. Goddard, S. A. R. Knox, R. F. D. Stansfield, F. G. A. Stone, M. J. Winter and P. Woodward, *J. Chem. Soc., Dalton Trans.*, 1982, 147.
600. N. D. Feasey, S. A. R. Knox, A. G. Orpen and M. J. Winter, *New J. Chem.*, 1988, **12**, 581.

7
Cyclopentadienyl Complexes of Chromium, Molybdenum and Tungsten

MICHAEL J. MORRIS
University of Sheffield, UK

7.1 INTRODUCTION

This chapter covers chromium, molybdenum and tungsten complexes with cyclopentadienyl and substituted cyclopentadienyl ligands, including indenyl species, and corresponds to Sections 26.2.5, 27.2.5, and 28.2.5 of *COMC-I*. Literature coverage extends from 1981 to the time of writing, with a final cut-off point at the end of 1993. The chemistry of chromium, molybdenum and tungsten Cp complexes has continued to thrive during this decade; if any trend can be discerned, it is perhaps the increased emphasis on higher oxidation state species with halide, oxo, imido and sulfido ligands. Annual reviews of chromium, molybdenum and tungsten organometallic chemistry have appeared throughout the decade; moreover, a new volume of the *Gmelin Handbook of Inorganic and Organometallic Chemistry* dealing with Cp molybdenum complexes has appeared and contains detailed information about many of the complexes discussed in this chapter.[1] Throughout the chapter the abbreviations Cp, Cp', and Cp* refer to η-C_5H_5, η-C_5H_4Me, and η-C_5Me_5, respectively. It should be emphasized that many complexes have been prepared with Cp and Cp* ligands; in general, these are only explicitly referred to where there are significant differences in structure or reactivity between the two. It should also be recognized that in many cases a parallel chemistry of complexes containing the tris-pyrazolylborate ligand exists; for developments in this field the reader is referred to a recent review.[2]

7.2 CYCLOPENTADIENYL METAL CARBONYL AND RELATED COMPOUNDS

7.2.1 [M$_2$(CO)$_n$Cp$_2$] (n = 4–6)

7.2.1.1 [M$_2$(CO)$_6$Cp$_2$]

(i) Synthesis

There are two general methods for the synthesis of [$M_2(CO)_6Cp_2$] (**1**). Reaction of the metal hexacarbonyl with lithium or sodium cyclopentadienide in diglyme or THF followed by oxidation with iron(III) salts (method A) gives excellent yields for all three metals (60–85% for Cr, 90–98% for Mo and W).[3] The direct thermolysis of the cyclopentadiene with [$M(CO)_6$] or more commonly [$M(CO)_3(NCMe)_3$] (method B) can also give good yields under controlled conditions.[4] For [$Cr_2(CO)_6Cp_2$] and its Cp* analogue, dehydrogenation of the intermediate [$CrH(CO)_3Cp$] with isoprene gives the best yield,[5,6] and a related procedure for [$Mo_2(CO)_6Cp^*_2$] with cyclohexadiene also exists.[7] [$W_2(CO)_6Cp_2$] has also been prepared by thermolysis of [$W(CO)_3(CH_2Ph)Cp$], with [$W_2(CO)_6(\mu$-fulvalene)] appearing as a by-product.[8]

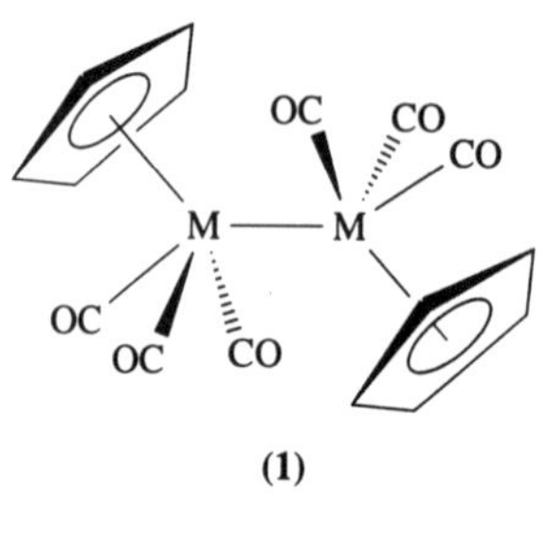

(**1**)

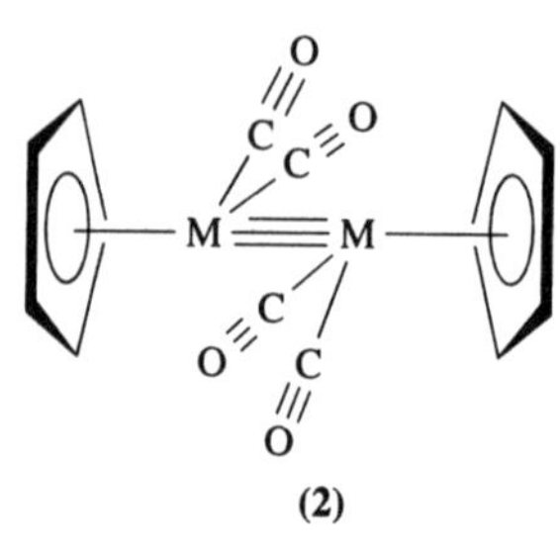

(**2**)

A number of new substituted derivatives, listed in Table 1, have been prepared by these methods or slight modifications thereof.

Table 1 New $[M_2(CO)_6Cp_2]$ derivatives.

Metal	*Cp derivative*	*Method*	*Ref.*
Cr	C_5H_4COMe, $C_5H_4CO_2Me$	A	9
Cr, Mo, W	C_5H_4CHO	A	10
Cr, Mo, W	$C_5H_4CH_2C_5H_4$	A, B	11,12
Cr, Mo, W	$C_5H_4SiMe_2C_5H_4$	A, B	13,14
Cr, Mo, W	C_5H_4–C_5H_4 (fulvalene)	B	8,15
Cr, W	1,1',3,3'–$C_5H_2Bu^t_2$–$C_5H_2Bu^t_2$	B	16
Mo	$C_5H_4COCH_2OH$, $C_5H_4(CH_2)_nOH$, $n = 1$–3	A, B	17,18
Mo	$C_5H_4CH_2CH_2NH_3^+$	A	19
Mo	$C_5Me_4CHEtPh$	A	20
Mo	$C_5H_4Bu^t$, 1,3-$C_5H_3Bu^t_2$	B	21
Mo	$C_5H_4SiMe_2SiMe_2C_5H_4$	A	22
Mo	C_5Ph_5	A	23
W	C_5H_4COOH	A	24

In an interesting development, Tyler and co-workers have prepared oligomeric urethanes containing $Mo_2(CO)_6$ units in the backbone using functionalized derivatives from Table 1; like their dimeric precursors, these undergo decarbonylation to $Mo_2(CO)_4$ groups and photochemical metal–metal bond scission.[18,25,26]

(ii) Structure and characterization

The crystal structures of several of the compounds in Table 1 have been determined; in general they are similar to those of the unsubstituted analogues, with an *anti* disposition of the Cp ligands and similar bond lengths (Table 2). In all structures of the *anti* conformer the four carbonyl ligands situated *cis* to the M–M bond are slightly nonlinear (typical M–C–O angle around 170°), as if weakly semibridging. The x-ray structure of the *gauche* conformer of $[Mo_2(CO)_6Cp_2]$ has also been determined; the metal–metal bond length of 0.32239(11) nm is slightly longer than that in the more stable *anti* conformer reported previously.[27] As shown in Table 2, however, a shortening was observed for the *gauche* conformer of $[W_2(CO)_6Cp^*_2]$.[28] The metal–metal bond lengths in $[M_2(CO)_6Cp^*_2]$ follow a similar trend to those in $[M_2(CO)_6Cp_2]$, with the Cr–Cr bond being particularly long and weak. When the two Cp ligands are linked by CH_2 or $SiMe_2$ spacers, the resulting ligand is still flexible enough to allow a *gauche* conformation to be adopted; a slight decrease in M–M bond length is observed in these cases. Alternatively, in the highly strained fulvalene complexes, which are forced into a *syn* conformation, the metal–metal bond is unusually long.

Table 2 Structural studies of $[M_2(CO)_6Cp_2]$ complexes.

Metal	*Cp derivative*	*Conformation*	*M–M length* (nm)	*Ref.*
Cr	Cp*	*anti*	0.33107(7)	29
			0.3310(1)	30
Mo	Cp*	*anti*	0.3278(14)	31
			0.3281(1)	32
Mo	$C_5H_4CH_2CH_2OH$	*anti*	0.3211(1)	33
Mo	$C_5H_4CH_2CH_2CH_2OH$	*anti*	0.3213(8)	17
Mo	$C_5H_4CHPh_2$	*anti*	0.3227(1)	34
Mo	C_9H_7	*anti*	0.3251(1)	35
Mo	$C_5H_4C_5H_4$ (fulvalene)	*syn*	0.3371(1)	36
W	Cp*	*anti*	0.3288(1)	37
W	Cp*	*gauche*	0.3273(1)	28
W	C_5H_4COOH	*anti*	0.3215(3)	24
W	$C_5H_4CO_2Me$	*anti*	0.3216(1)	38
W	$C_5H_4C_5H_4$ (fulvalene)	*syn*	0.3347(1)	8
W	$C_5H_4CH_2C_5H_4$	*gauche* (48°)	0.3166(1)	12
W	2-$MeCOC_5H_4CH_2C_5H_4$	*gauche* (47°)	0.3175(1)	12
W	$C_5H_4SiMe_2C_5H_4$	*gauche*	0.3196(1)	14
W	$C_5H_4CMe_2C_{13}H_9$	*anti*	0.32257(6)	39

The related compounds [MM'(CO)$_5$Cp(η-C$_7$H$_7$)] (M,M' = Mo, W) have been prepared, of which the Mo$_2$ complex has been structurally characterized [Mo–Mo 0.3160(1) nm].[40–2]

Carbon-13 NMR studies of the dimers show that the δ(CO) values decrease down the triad, and shift downfield with increasing methyl substitution of the Cp ligands.[43] They also confirm that simple rotation about the metal–metal bond is responsible for the isomerization of [W$_2$(CO)$_6$Cp$_2$] in solution.[44] The solid state ^{1}H and ^{13}C NMR spectra of [W$_2$(CO)$_6$Cp$_2$] confirm that the *anti* rotamer is present; the solid state ^{13}C NMR spectrum of [Mo$_2$(CO)$_6$Cp$_2$] shows three equal intensity CO environments due to lower symmetry in the crystal.[45,46] The barrier to Cp ring rotation in the molybdenum complex calculated from its ^{1}H spin lattice relaxation time is 13.9 kJ mol^{-1}.[47]

The thermochemistry of [M$_2$(CO)$_6$Cp$_2$] and related compounds has been thoroughly studied by Hoff and co-workers, with the results summarized in a review article.[48] Thermochemical cycles show that hydrogenation of the dimers to [M(H)(CO)$_3$Cp] is a slightly exothermic process for chromium and tungsten, but slightly endothermic for molybdenum; this is a consequence of increasing M–H bond strength but also the weak Cr–Cr bond in the chromium dimer.[49] Hydrogenation of the Cp* analogues is more favourable. The heats of reaction of [Mo$_2$(CO)$_4$L$_2$] (L = Cp, Cp*, indenyl) with CO and alkynes have also been measured.[50] The estimated bond strengths of the Mo–X bond in various [Mo(X)(CO)$_3$Cp] complexes were derived (Table 3), and values for the single and triple Mo–Mo bonds in [Mo$_2$(CO)$_n$Cp$_2$] (n = 6, 4) were found to be 136 and 284 kJ mol^{-1} respectively.

Table 3 Estimated Mo–X bond strengths in [MoX(CO)$_3$Cp].

X	*Bond strength* (kJ mol^{-1})	*X*	*Bond strength* (kJ mol^{-1})
H	276	Me	196
Cl	303	Et	155
Br	253	σ-CH$_2$CH=CH$_2$	146
I	217	CH$_2$Ph	134

Irradiation of [M$_2$(CO)$_6$Cp$_2$] in the near-UV–visible region can cause both cleavage to 17-electron M(CO)$_3$Cp radicals or carbonyl loss to give [M$_2$(CO)$_5$Cp$_2$]. Some wavelength dependence is observed, with longer wavelengths producing the former and shorter wavelengths the latter.[51,52] By photolysing in an inert gas matrix or a PVC film at 12 K, [M$_2$(CO)$_5$Cp$_2$] was shown to contain two terminal CO ligands, two semibridging CO ligands, and one four-electron-donor bridging CO, whereas in the analogous complex with the linked Cp ligand C$_5$H$_4$CH$_2$C$_5$H$_4$, the pentacarbonyl species has three terminal and two bridging CO ligands.[53,54] Flash photolysis studies confirmed these results and also detected the *gauche* rotamer of [Mo$_2$(CO)$_6$Cp$_2$].[55,56] The pentacarbonyl species is the intermediate in the known decarbonylation of [M$_2$(CO)$_6$Cp$_2$] to [M$_2$(CO)$_4$Cp$_2$], which was originally thought to proceed by a radical mechanism.

7.2.1.2 [M$_2$(CO)$_4$Cp$_2$]

(i) Synthesis

The reactive tetracarbonyl dimers (**2**), which formally require a M≡M bond to achieve 18-electron configuration, have an extensive chemistry which is covered in two reviews.[57,58] The dimolybdenum complex has been most studied as it is the easiest to prepare in relatively large quantity; the chromium analogue is much more air-sensitive while the tungsten complex requires longer or more forcing conditions for its synthesis. The most convenient preparation of [M$_2$(CO)$_4$Cp] consists of heating a solution of [M$_2$(CO)$_6$Cp$_2$] in a high-boiling solvent (usually toluene, xylene or diglyme) with a purge of inert gas to sweep away the liberated carbon monoxide; the duration of reaction depends on the temperature and purging efficiency.[4] In many cases the products are then used *in situ* for further reactions, though they are sufficiently stable to be purified by column chromatography or recrystallization if required. Sometimes [M$_2$(CO)$_4$Cp$_2$] can be obtained directly from [M(CO)$_6$]; thus reaction of indene with [Mo(CO)$_6$] in diglyme/octane at 165 °C gives the tetracarbonyl dimer and not, as previously thought, the hexacarbonyl,[59] and in the case of C$_5$H$_4$Bu^t or 1,3-C$_5$H$_3$Bu$^t_2$ the desired product can be selected by controlling the temperature.[21] Similarly the chromium dimer [Cr$_2$(CO)$_4$Cp*$_2$] can be obtained directly by heating C$_5$Me$_5$H with [Cr(CO)$_3$(NCMe)$_3$] in toluene.[60] Cothermolysis or photolysis of $[Mo_2(CO)_6Cp_2]$ with $[W_2(CO)_6Cp_2]$ gives the mixed metal species $[MoW(CO)_4Cp_2]$.[61,62] Compounds in which the Cp ligands are linked by CH_2 or $SiMe_2$ bridges, the azulene complexes $[M_2(CO)_6(\mu\text{-azulene})]$ and the fulvalene complexes $[M_2(CO)_6(\mu\text{-}C_5H_4C_5H_4)]$, cannot be decarbonylated to the corresponding tetracarbonyls.

(ii) Structure

The structures of $[M_2(CO)_4Cp_2]$ and their Cp* analogues were all previously known, with the exception of $[Mo_2(CO)_4Cp^*_2]$; this has since been determined, along with several others (Table 4). As noted previously,[58] the CpMMCp axis is only linear in the case of $[Mo_2(CO)_4Cp_2]$, with all other examples showing angles of 153–168°, and there are two sets of CO ligands differing in their degree of semibridging, as shown by the M'–M–CO angle; all of the M–C–O angles remain virtually linear. Molecular orbital calculations, backed up by UV photoelectron spectra, suggest that donation of electron density from the metal into the π* orbitals of the semibridging carbonyls is maximized when the CpMMCp axis is linear.[63]

Table 4 Structural studies of $[M_2(CO)_4Cp_2]$ complexes.

Complex	*M–M' distance* (nm)	*M'–M–Cp angle* (°)	*M–C–O angle* (°)^a	*M'–M–CO angles* (°)^a	*Ref.*
$[Cr_2(CO)_4(\eta\text{-}1,2,4\text{-}C_5H_2Me_3)_2]$	0.2267(1)	158.2	173.6(5), 173.4(5)	78.5(2), 75.9(2)	64
$[Mo_2(CO)_4(\eta\text{-}C_9H_7)_2]$	0.2500(1)	not given	171, 174	71, 75	59
$[Mo_2(CO)_4Cp^*_2]$	0.2488(3)	168	154, 169	57, 70	65
$[Mo_2(CO)_4(\eta\text{-}1,3\text{-}C_5H_3Bu^t_2)_2]$	0.2533(1)	160.9	168.9(4), 174.6(4)	68.3(1), 81.0(1)	21
$[W_2(CO)_4(\eta\text{-}C_5H_4CO_2Et)_2]$	0.25064(6)	159.1(2)	172.2(6), 174.9(6)	71.3(2), 81.9(2)	66

^{a} For noncentrosymmetric molecules the average values for each set of two CO ligands is given.

A spin–lattice relaxation and molecular orbital study of $[Cr_2(CO)_4Cp^*_2]$ revealed two molecular motions in the solid state: rotation of the methyl groups and reorientation of the Cp* ligands about the Cr–Cp* axis.[67]

(iii) Reactions

The reactions of $[M_2(CO)_4Cp_2]$ almost invariably involve addition of two two-electron ligands or one four-electron ligand across the M≡M bond (Scheme 1). Carbonylation to form $[M_2(CO)_6Cp_2]$ occurs readily in most cases under 1 atm of CO at room temperature, and crossover experiments with mixtures of $[Mo_2(CO)_4Cp_2]$ and its Cp' analogue, as well as the mixed ligand dimer $[Mo_2(CO)_6CpCp']$, have shown that Mo–Mo bond cleavage is not involved. This implies that radical formation does not occur during the decarbonylation reaction of the hexacarbonyl dimers to form the tetracarbonyls.[68]

Addition of two phosphine ligands also occurs readily to give $[M(CO)_2(L)Cp]_2$ (if one equivalent of phosphine is used $[M_2(CO)_5(L)Cp_2]$ is produced by a carbonyl redistribution process). The main synthetic utility of the substituted dimers lies in cleavage with sodium amalgam to form $[M(CO)_2(PR_3)Cp]^-$ or photolysis to form radical species.[69] New examples prepared include $[Mo(CO)_2(L)Cp]_2$ (L = PPh_2(2-pyridyl),[70] PPh_2H[71]) and $[M(CO)_2(L)Cp]_2$ (L = PH_3, PH_2Me, and $PHMe_2$ for Mo and W).[72] The latter compounds readily undergo oxidative addition of a P–H bond and phosphine elimination to give $[M_2(CO)_4(\mu\text{-H})(\mu\text{-PR}^1R^2)Cp_2]$ (R^1, R^2 = H, Me). Diphosphines such as dppm or tedip $[(EtO)_2PCH_2P(OEt)_2]$, and $MeN(PF_2)_2$ add to $[Mo_2(CO)_4Cp_2]$ to give $[Mo_2(CO)_4(\mu\text{-L-L})Cp_2]$.[73–5] Simple addition also occurs for phosphines of the type R_2PCH_2PHR and $HPRCH_2PHR$ (R = various alkyl); there is no subsequent P–H bond addition.[76] The triphosphine $HC(PMe_2)_3$ also coordinates in a bidentate μ, η^2-mode, leaving one pendant arm. A substantial chemistry, shown in Scheme 2, has been developed from $[Mo_2(CO)_4(\mu\text{-dppm})Cp_2]$.[75,77–9] and, more recently, the tungsten analogue has been explored too; it can be oxidized to a stable dication $[W_2(CO)_2(\mu\text{-CO})_2(\mu\text{-dppm})Cp_2]^{2+}$, and undergoes C–H activation of one of the Cp rings on photolysis.[80,81]

Uniquely, addition of the cyclic phosphine $Ph\overline{PH(OCH_2CH_2)_2N}$ (phoran) to $[Mo_2(CO)_4Cp_2]$ produces mainly the substitution product $[Mo_2(CO)_3(L)Cp_2]$ in which the phosphine is bound in its open-chain $PhP(OCH_2CH_2)_2NH$ form.[82] The Cp* analogue reacts similarly with phoran, and also with $P(OMe)_3$. On treatment of $[Mo_2(CO)_3(L)Cp_2]$ with $P(OMe)_3$ the phoran ligand is substituted, whereas addition occurs

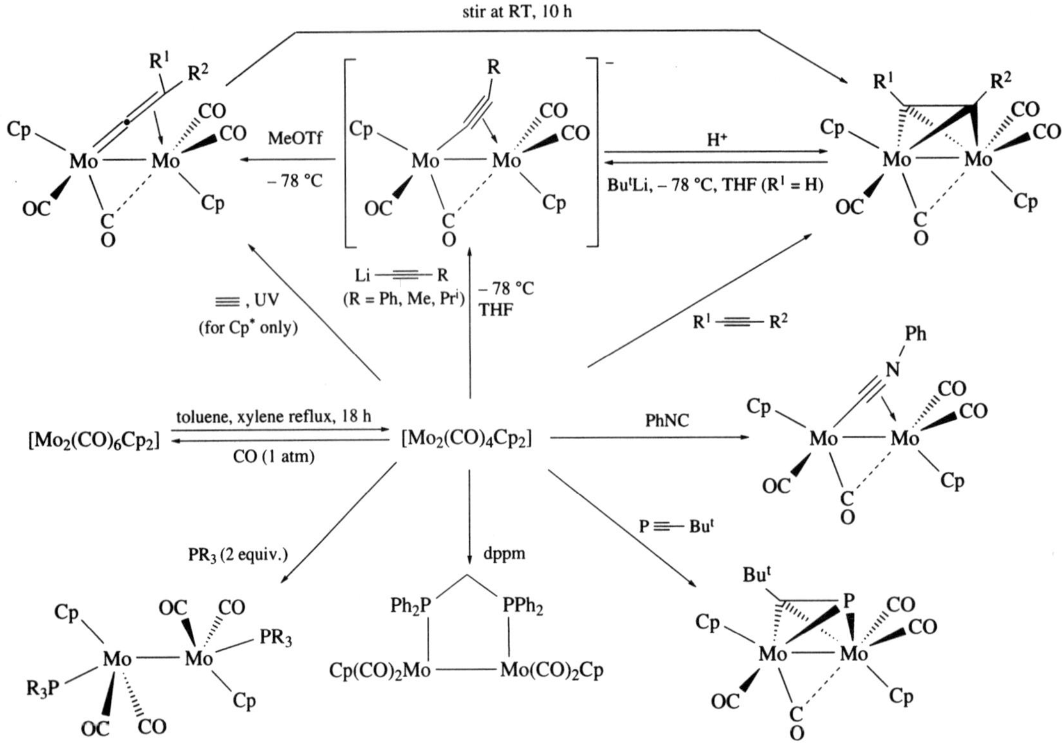

Scheme 1

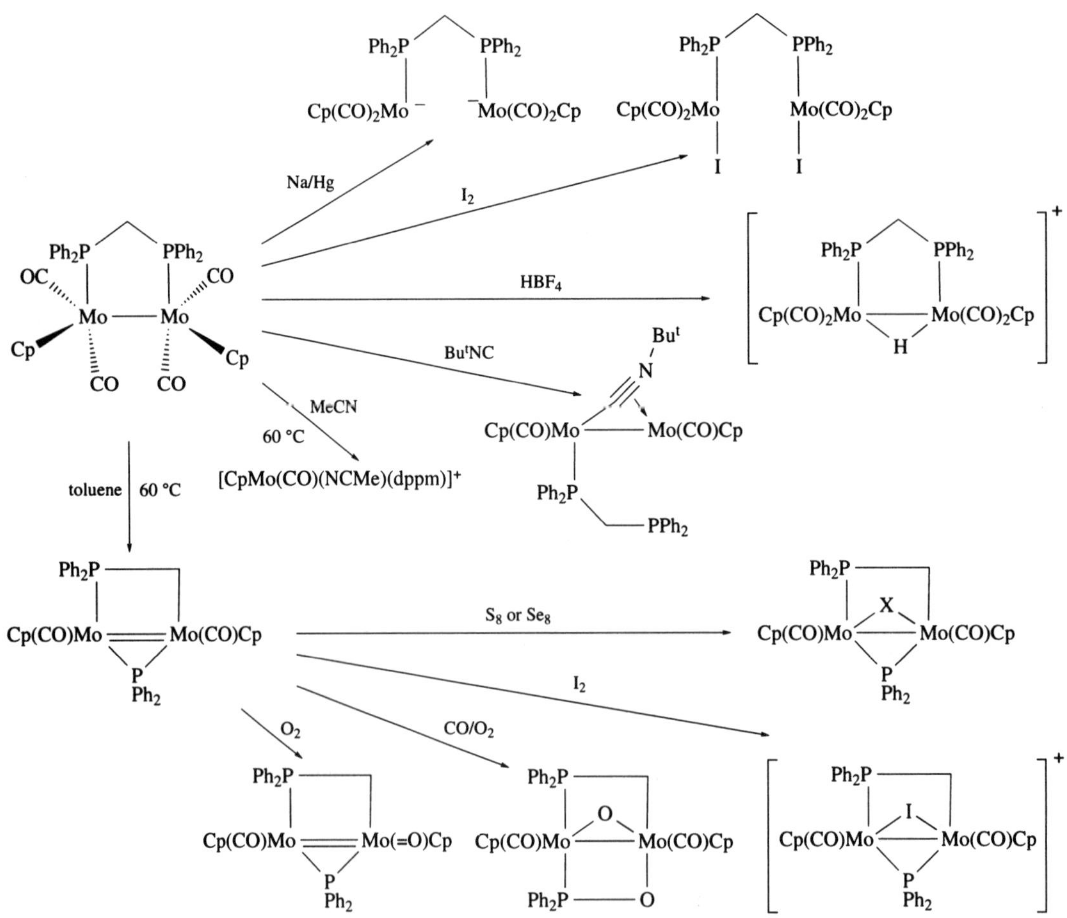

Scheme 2

with CO or isonitrile.[83] The structures of both $[Mo_2(CO)_3(L)Cp_2]$ complexes have been determined and unlike $[Mo_2(CO)_4Cp_2]$ have a nonlinear CpMoMoCp unit; the Mo–Mo distances are 0.2506(1) nm and 0.2502(1) nm for L = phoran and $P(OMe)_3$ respectively.

Isonitriles RNC (R = Me, Ph, CF_3, etc.) react with $[M_2(CO)_4Cp_2]$ (M = Mo, W) and their Cp* analogues to give initially $[M_2(CO)_4(\mu\text{-}\sigma, \eta^2\text{-CNR})Cp_2]$ in which the ligand acts as a four-electron donor through both the carbon and the C≡N bond.[84–7] Several structure determinations show a lengthened C≡N bond for these compounds, and their IR spectra show low values for ν(CN) in the range 1650–1740 cm^{-1}. Prolonged reaction with an excess of RNC gives $[M_2(CO)_4(CNR)_2Cp_2]$.

The μ-σ, η^2 bonding mode adopted by the isonitrile ligand above is also available to a number of related species, notably acetylide and vinylidene. Reaction of $[Mo_2(CO)_4Cp_2]$ with LiC≡CR (R = Me, Ph) (or deprotonation of the μ-alkyne complex $[Mo_2(CO)_4(\mu\text{-HCCR})Cp_2]$) gives the anionic acetylide-bridged species $Li[Mo_2(CO)_4(\mu\text{-C≡CR})Cp_2]$, which can be alkylated to the side-bound vinylidenes $[Mo_2(CO)_4(\mu\text{-C=C=CR}^1R^2)Cp_2]$.[88,89] Photochemical reaction of $[Mo_2(CO)_4Cp*_2]$ with HC≡CH affords the side-bound vinylidene directly, which slowly isomerizes to the μ-alkyne complex.[90]

The formation of the alkyne complexes $[M_2(CO)_4(\mu\text{-R}^1CCR^2)Cp_2]$ by addition of $R^1C≡CR^2$ to $[M_2(CO)_4Cp_2]$ has been well documented and is covered in Chapter 6, this volume. However, one notable new example is the phospha–alkyne complex $[Mo_2(CO)_4(\mu\text{-PCBu}^t)Cp_2]$.[91,92] The phosphorus atom can subsequently be coordinated to a variety of other metal fragments.[93,94]

Species such as thioesters, dithioesters, thiolactones, thioketones, Lawesson's reagent and ketimines all react with $[M_2(CO)_4Cp_2]$ (M = Mo, W) to produce related compounds in which the reagent acts as a bridging four-electron ligand (Scheme 3).[95–100] Another consistent feature is the presence of a semibridging carbonyl ligand. The bonding can be regarded as π-donation from a double bond (C=S, C=N, or P=S) to one metal, with lone pair donation from the heteroatom to the other metal. In some cases two of these ligands are incorporated into a $Mo_2(CO)_2Cp_2$ framework. With certain reagents, however, such as tetraalkyl thiuram disulfides, fragmentation of the dimolybdenum structure occurs.[86,101,102]

The reactions of $[M_2(CO)_4Cp_2]$ and their Cp* analogues with diazoalkanes have been extensively studied by the groups of Herrmann and Curtis; as shown in Scheme 4 (alkyl diazoalkanes) and Scheme 5 (aryl diazo compounds), a wide variety of products can be isolated depending on both the metal complex employed and the substituents on the diazoalkane, with little apparent correlation.[103–16] Only in a few cases is the simple incorporation of a carbene ligand and elimination of N_2 observed; more often the nitrogen is retained in the product. The reader is referred to the several reviews of the area for a detailed discussion of each structural type.[57,58,117] Little further chemistry of any of the resulting products has been reported subsequently.

Compounds related to those in Schemes 4 and 5 can be made by treating $[M_2(CO)_4Cp_2]$ with nitrogen-containing species such as azides,[110,118] diethyl azodicarboxylate,[119] diphenyl carbodiimide,[120–2] selenadiazoles[123] and azirines[124,125] (Scheme 6).

When $[Mo_2(CO)_4Cp_2]$ is treated with very electrophilic metal centres, complexes with isocarbonyl bridges are formed, for example $[Ti(CO)_2Cp_2]$ in THF gives $[Cp_2Ti(THF)(\mu\text{-OC})Mo(CO)_2Cp]$, essentially an adduct in which the Lewis acid centre is coordinated to the carbonyl oxygen. The Ti–O–C link is bent, but the Mo–C–O axis is linear with a short Mo–C bond (0.1874(7) nm) indicating some contribution of a carbyne structure, Mo≡C–O–Ti.[126] The reaction is solvent dependent: in benzene $[Cp_2Ti(\mu\text{-OC})Mo(CO)(\mu\text{-CO})Cp]_2$ (**3**), containing a 12-membered ring held together by similar linkages, is formed.[127] Similar ring structures are found in $[L_2Ce(\mu\text{-OC})W(CO)(\mu\text{-CO})Cp]_2$ (L = $\eta\text{-C}_5H_3\text{-1,3-(TMS)}_2$) and in $[Zr\{(\mu\text{-OC})Mo(CO)_2Cp\}(NEt_2)_2(NHEt_2)(\mu\text{-OC})Mo(CO)(\mu\text{-CO})Cp]_2$, which contains both ring and terminal isocarbonyl–Mo units.[128,129] Isocarbonyl linkages are also present in $[Cp_{4-n}U\{(\mu\text{-OC})Mo(CO)_2Cp\}_n]$ (n = 1, 2).[130]

The complex $[Cp_2Ti(THF)(\mu\text{-OC})Mo(CO)_2Cp]$ is also formed when $[Mo_2(CO)_6Cp_2]$ is treated with $[Cp_2Ti(\mu\text{-PEt}_2)]_2$ in THF; in C_6H_6, the phosphido-bridged $[CpMo(CO)(\mu\text{-PEt}_2)(\mu\text{-}\eta^1, \eta^2\text{-CO})TiCp_2]$ is produced instead.[131–3] In contrast, reaction of $[Mo_2(CO)_4Cp_2]$ with $[Ti(CH_2Bu^t)_2Cp*_2]$ affords $[Mo_2(CO)_2(\mu\text{-COTiCp*}_2)_2Cp_2]$ (**4**) in which the titanium is coordinated to bridging carbonyls with a linear Ti–O–C axis.[134]

7.2.1.3 The {M(CO)₃Cp} radical and related species

Investigations by Baird and Tyler have revealed an extensive chemistry of radical species derived from $[M_2(CO)_6Cp_2]$, which forms the subject of several reviews.[135–42] For this section the convention that radical species are enclosed in curly brackets is adopted.

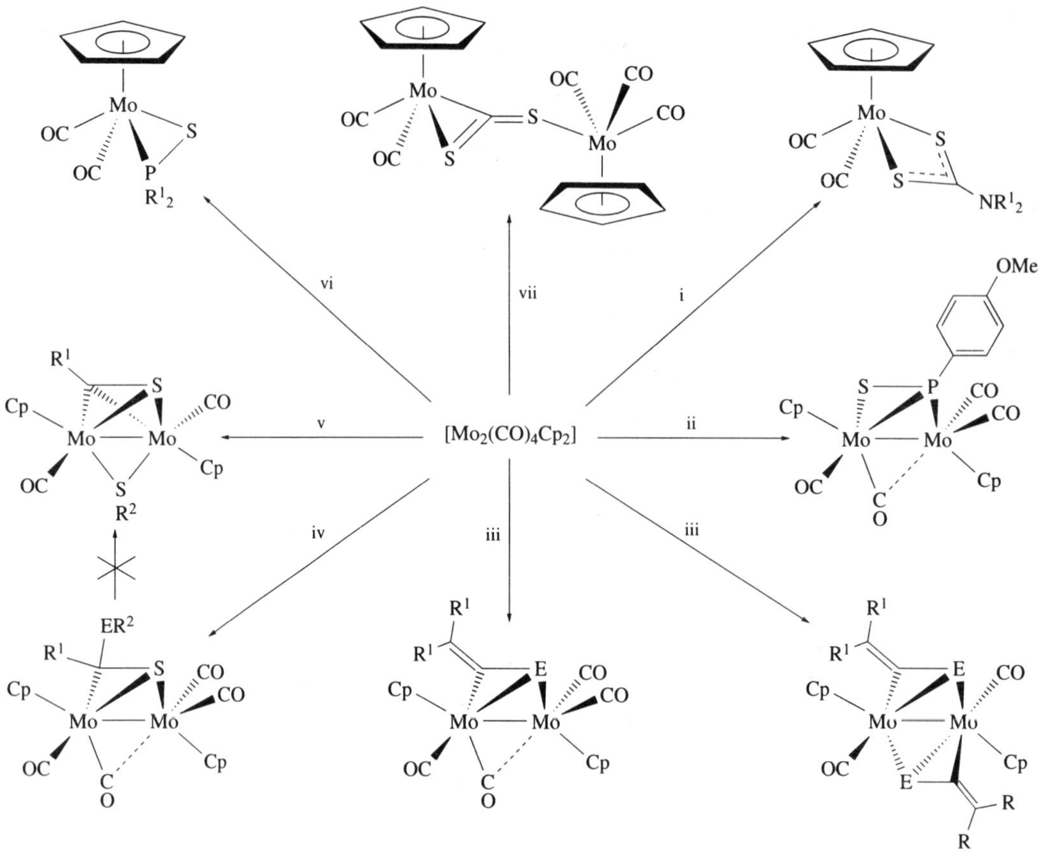

i, $R^1{}_2NC(S)C(S)NR^1{}_2$ (R^1 = Me, Et, Pr^i); ii, $(S_2P\text{-}p\text{-}C_6H_4OMe)_2$ (Lawesson's reagent), 70 °C; iii, $Ph_2C{=}C{=}NTol$ or $S{=}C{=}CCMe_2(CH_2)_3CMe_2$; iv, $R^1C({=}S)SR^2$ or $R^1C({=}S)OR^2$, RT (R^1 = aryl, R^2 = Me, Et); v, $R^1C({=}S)SR^2$, refluxing toluene; vi, $R^1{}_2P({=}S)P({=}S)R^1{}_2$, refluxing toluene ($R^1$ = Me, Et); vii, CS_2

Scheme 3

As explained previously, the long Cr–Cr bond length in $[Cr_2(CO)_6Cp_2]$ leads to appreciable dissociation in solution to $\{Cr(CO)_3Cp\}$ radicals, whereas the molybdenum and tungsten analogues require photolysis. The chromium dimer is more dissociated than previously thought (approximately 10% at room temperature for a 0.01 M solution)[143] and its high reactivity really reflects that of the radical monomer (Scheme 7).[5,144–6] Various thermodynamic studies indicate a bond strength of 46–63 kJ mol^{-1} for the Cr–Cr bond,[29,147] though it should be noted that NMR data do not appear to give reliable results in this system due to the multiplicity of states in the radical species. The kinetics of dissociation and recombination have also been measured, and a kinetic study of the reaction with alkyl halides suggested that it occurs by a simple halide abstraction mechanism.[148,149] The radical species can be stabilized by employing bulky Cp ligands or by phosphine substitution (which occurs by an associative mechanism); thus $[Cr_2(CO)_6Cp^*{}_2]$, which in the solid state has a Cr–Cr bond even longer than its Cp analogue, is approximately 90% dissociated in solution,[6,29,150] and $\{Cr(CO)_3(\eta\text{-}C_5Ph_5)\}$ exists solely as the monomer even in the solid state.[151] Similarly $\{Cr(CO)_2(PPh_3)Cp\}$, which was previously assumed to be dimeric, has been shown to be a stable radical; the crystal structures of this compound and of $\{Cr(CO)_2(PMe_3)Cp^*\}$ consist of three-legged piano stools with closed-in OC–Cr–CO angles of 80.9° and 79.9° respectively, showing that the singly occupied orbital lies in this region, consistent with theoretical predictions.[144,152] The single crystal ESR spectra of $\{Cr(CO)_2LCp\}$ (L = CO, PPh_3) and $\{Cr(CO)_3Cp^*\}$ doped into their isostructural diamagnetic manganese analogues also show $^2A''$ ground states consistent with the unpaired electron residing in an orbital of largely d_{xy} character.[153,154] As might be expected, electrochemical reduction of all these radicals occurs readily to form $[Cr(CO)_2(L)Cp]^-$.[155]

With the exception of $[Mo_2(CO)_4(\eta\text{-}C_5Ph_5)_2]$, which is in equilibrium with its radical monomer,[23] compounds of the type $[M_2(CO)_{6-n}(L)_nCp_2]$ (M = Mo, W; n = 1, 2) exist essentially as the dimers both in the solid state and in solution. By radical trapping experiments it can be shown that even for $[Mo_2(CO)_6Cp'_2]$ there is a small degree of dissociation, though the equilibrium constant for homolysis

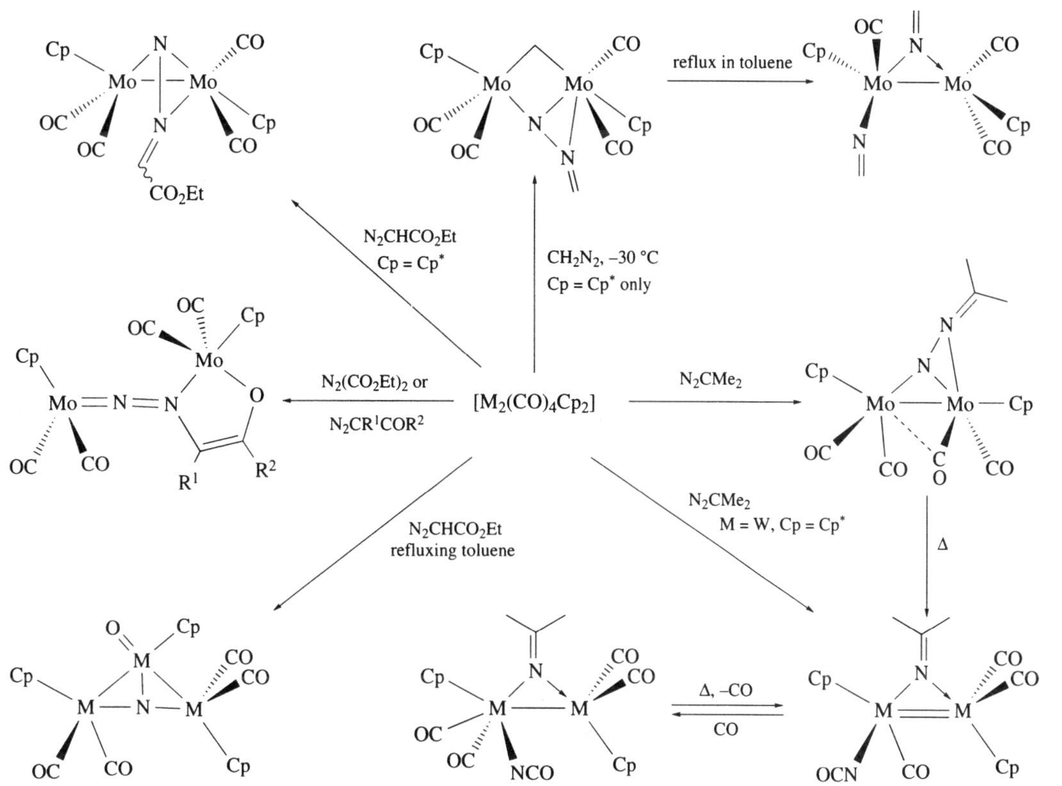

Scheme 4

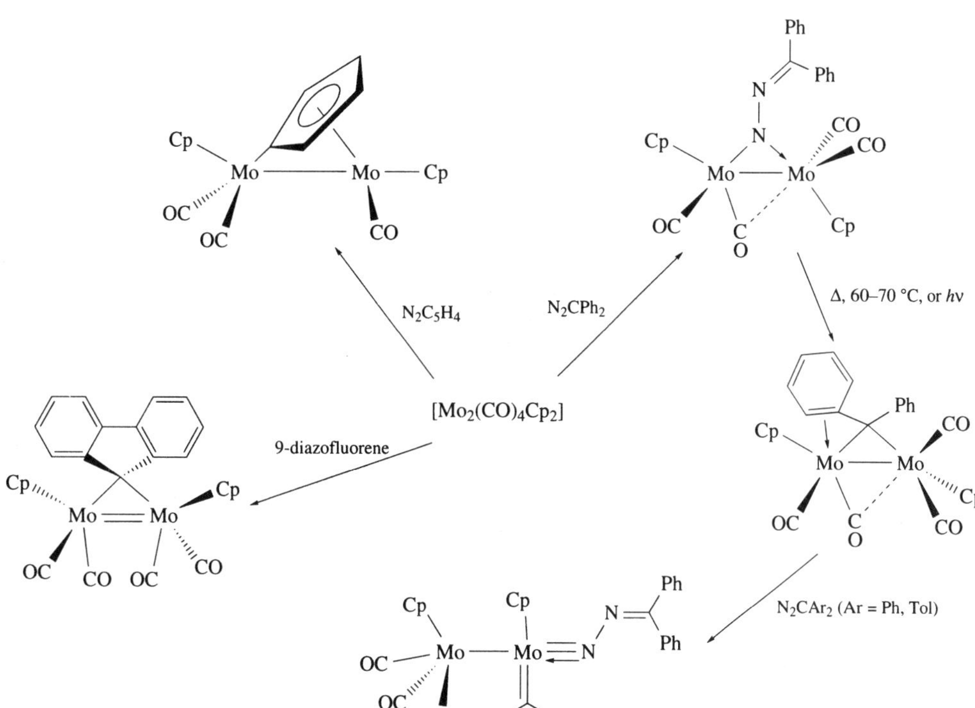

Scheme 5

is calculated as 1×10^{-17}.[156] The radical species can, however, be generated conveniently either by photolysis of the appropriate dimer (some reactions which had previously been considered to involve thermal dissociation of the dimers have been found to be photochemically activated by fluorescent lights) or by hydrogen abstraction from $[MH(CO)_2(L)Cp]$ using trityl radical. When so produced they

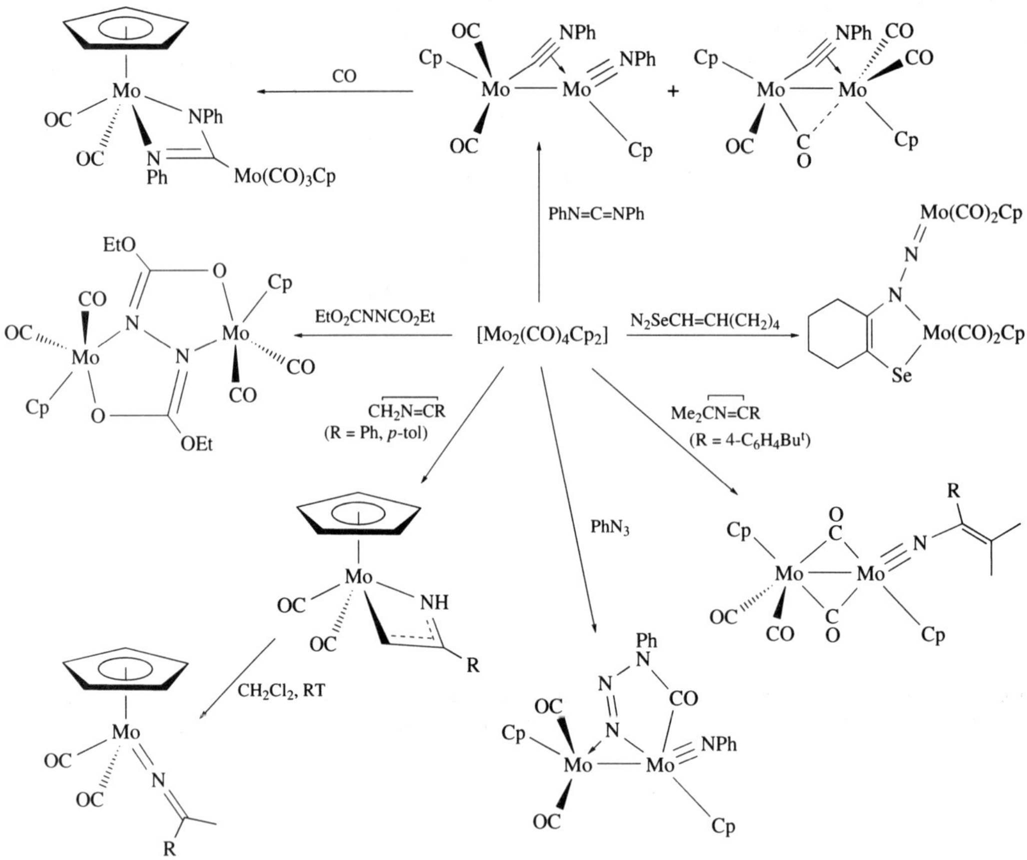

Scheme 6

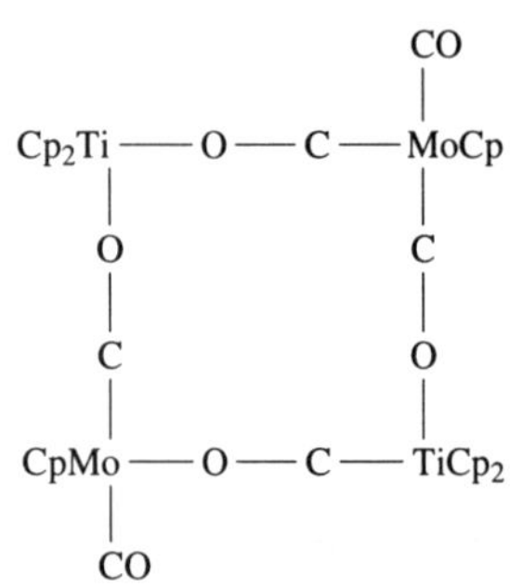

(3)

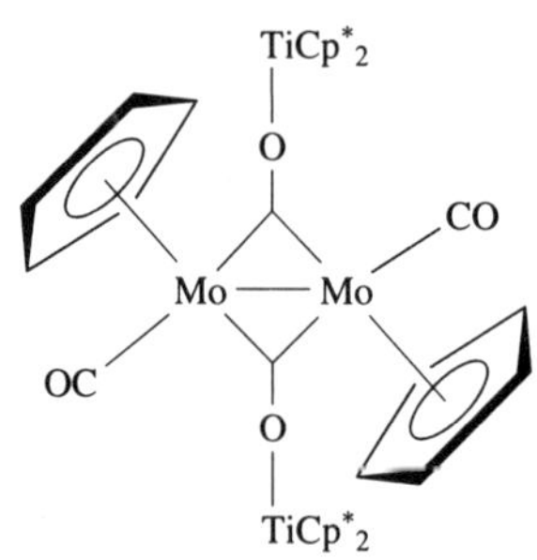

(4)

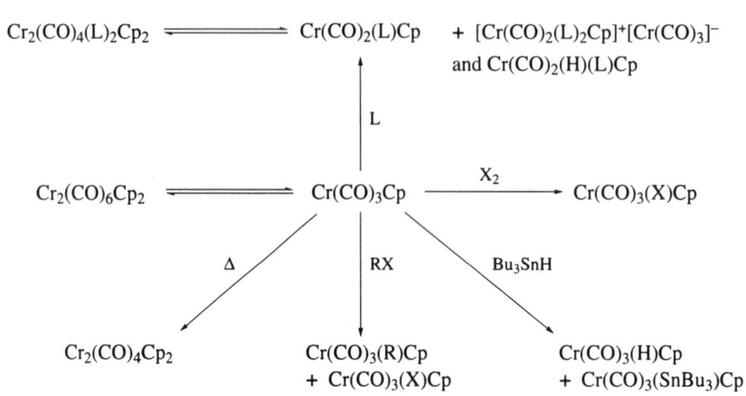

Scheme 7

undergo typical radical reactions (e.g., dimerization, hydrogen abstraction, halide abstraction) in much the same way as the chromium species described above, and are also substitutionally labile.[69,157,158]

Tyler and co-workers have studied the complex disproportionation reactions that occur when $[Mo_2(CO)_6Cp_2]$ (or its more soluble Cp' analogue) is irradiated in the presence of ligands such as phosphines to form $[Mo(CO)_2(L)Cp]^+$ and $[Mo(CO)_3Cp]^-$, accompanied in some cases by the substitution products $[Mo_2(CO)_{6-n}(L)_nCp_2]$ (n = 1, 2).[159,160] A radical chain pathway operates in which the key step is the reaction of the 17-electron $\{Mo(CO)_3Cp\}$ with L to form the 19-electron species $\{Mo(CO)_3(L)Cp\}$ (Equations (1)–(5)).[161,162] A similar scheme was invoked for the related photochemical reaction with halide ions (giving $[MoX(CO)_3Cp]$ and $[Mo(CO)_3Cp]^-$) in which either a donor solvent or a halide ion was acting as L;[163] for the latter case $\{Mo(CO)_3Cp\}$ and its halide adduct were shown to differ only slightly in energy, hence the 19-electron intermediate is thermodynamically feasible.[164]

$$[Mo_2(CO)_6Cp_2] \longrightarrow 2\{Mo(CO)_3Cp\} \quad (1)$$

$$\{Mo(CO)_3Cp\} + L \longrightarrow \{Mo(CO)_3(L)Cp\} \quad (2)$$

$$\{Mo(CO)_3(L)Cp\} + L \longrightarrow \{Mo(CO)_2(L)_2Cp\} + CO \quad (3)$$

$$\{Mo(CO)_2(L)_2Cp\} + [Mo_2(CO)_6Cp_2] \longrightarrow [Mo(CO)_2(L)_2Cp]^+ + [Mo_2(CO)_6Cp_2]^- \quad (4)$$

$$[Mo_2(CO)_6Cp_2]^- \longrightarrow [Mo(CO)_3Cp]^- + \{Mo(CO)_3Cp\} \quad (5)$$

One puzzling feature was the apparent dependence on the ligand and on the wavelength of light used: for example, PPh_3 caused disproportionation at 290 nm but not at λ >366 nm.[159] As an alternative to Equations (3) and (4), the monosubstituted $\{Mo(CO)_3(L)Cp\}$ may act as the reducing agent, and then undergo additional photoinduced substitution (Equations (6) and (7)).

The apparent nonreaction of PPh_3 and other ligands has been explained in terms of a rapid back reaction between the ionic products of Equations (5) and (6), $[Mo(CO)_3(PPh_3)Cp]^+$ and $[Mo(CO)_3Cp]^-$, to reform $[Mo_2(CO)_6Cp_2]$ and PPh_3 (in CH_2Cl_2 this back-reaction is sufficiently slow to observe the ionic species). Evidently $[Mo(CO)_3(PPh_3)Cp]^+$ does not absorb at long wavelengths and thus does not undergo reaction (7), but does do so at lower wavelength.[165] In the complex $[Mo_2(CO)_6(\eta\text{-}C_5H_4CH_2CH_2NH_3^+)_2][PF_6]_2$, photochemical disproportionation occurs in THF and the tentacle substituent inhibits the back reaction.[166]

$$\{Mo(CO)_3(L)Cp\} + [Mo_2(CO)_6Cp_2] \longrightarrow [Mo(CO)_3(L)Cp]^+ + [Mo_2(CO)_6Cp_2]^- \quad (6)$$

$$[Mo(CO)_3(L)Cp]^+ + L \longrightarrow [Mo(CO)_2(L)_2Cp]^+ + CO \quad (7)$$

In order for reactions (4) or (6) to occur, $\{Mo(CO)_2(L)_2Cp\}$ or $\{Mo(CO)_3(L)Cp\}$ must be good reducing agents, and indeed it was shown that irradiated mixtures of $[Mo_2(CO)_6Cp_2]$ and PMe_2Ph were capable of effecting electron-transfer-catalysed substitution of $[Ru_3(CO)_{12}]$ and of reducing ferricyanide to ferrocyanide.[167] Moreover, irradiation of $[Mo_2(CO)_6(\eta\text{-}C_5Ph_4H)_2]$ with bis(diphenylphospino)maleic anhydride allowed the isolation and structural characterization of $\{Mo(CO)_2L_2(\eta\text{-}C_5Ph_4H)\}$.[168] Due to the unsymmetrical positioning of the C_5Ph_4H ring, the two phosphorus atoms are inequivalent in the solid state and also in solution at room temperature by ESR. At 185 °C they become equivalent, allowing calculation of the barrier to ring rotation.[169] In the symmetrical C_5Ph_5 analogue, which is in equilibrium with the 17-electron $\{Mo(CO)_2(\eta^1\text{-L-L})(\eta\text{-}C_5Ph_5)\}$ the phosphorus atoms are equivalent at all temperatures.[23] However, the spectroscopic data also indicate that these are not true 19-electron complexes, but are more accurately described as '$18 + \delta$' complexes in which the extra electron is localized largely on the phosphine ligand. Consequently they do not display the reducing ability of the supposed intermediates, though they are substitutionally labile.

Stable radical species are also formed on photolysis of $[M_2(CO)_6Cp_2]$ (M = Mo, W) in the presence of quinones. With *p*-benzoquinones the products are monodentate $\{Mo(CO)_3(quinone)Cp\}$;[170,171] these can be detected by low-temperature ESR for *o*-quinones but rapidly rearrange to bidentate $\{Mo(CO)_2(quinone)Cp\}$.[172–5] Similar products can be obtained from diketones and related species.[176,177] Photolysis of $[Hg\{Cr(CO)_3Cp\}_2]$ or $[Mo_2(CO)_6Cp_2]$ followed by oxygenation gives radical species which may contain peroxo ligands.[178] Sonication of $[Mo_2(CO)_6Cp'_2]$ causes homolysis of the Mo–Cp' bond instead of the Mo–Mo bond.[179]

7.2.2 The $[M(CO)_3Cp]^-$ Anion and Related Compounds

The remainder of Section 7.2 covers chemistry involving the cyclopentadienyl metal tricarbonyl unit. Most reactions in this area stem from one of the basic starting materials shown in Scheme 8, which also shows the interconversions between them.

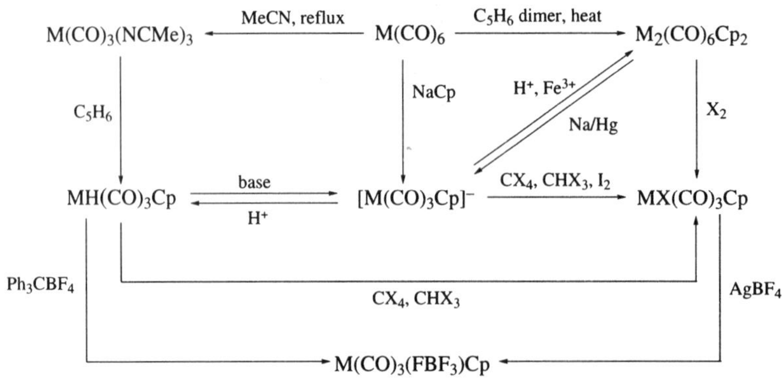

Scheme 8

The anions $[M(CO)_3Cp]^-$ are conveniently prepared either by reaction of the alkali metal cyclopentadienide with $[M(CO)_6]$ or $[M(CO)_3(NCMe)_3]$ or by reduction of the dimers $[M_2(CO)_6Cp_2]$. New reducing agents used for the latter include $[BH_4]^-$ for $[Mo(CO)_3Cp]^-$,[180] and sonochemically activated potassium for the Cp* analogue.[181] The anions can be isolated as NEt_4 or PPh_4 salts[182] or as conveniently stable solvates $Na[M(CO)_3Cp]{\cdot}2DME$ which as solids can even be briefly handled in air.[183] In general exposure to air results in oxidation to $[M_2(CO)_6Cp_2]$, but with large counterions such as PPN^+, $[Mo(CO)_3Cp]^-$ is rapidly oxidized by air in solution to $[Mo(CO)_2(\eta^2\text{-}CO_3)Cp]^-$ and then to $[Mo_2O_7]^{2-}$.[184,185]

A large number of substituted Cp derivatives have been prepared, often by the same methods that are used for the parent compounds. However, several additional routes to substituted derivatives exist, for example nucleophilic attack on the fulvene species $[M(CO)_3(\eta\text{-}C_5H_4{=}CR_2)]$ and the haptotropic rearrangement of η^6-complexes of indene, fluorene and related species on deprotonation. The latter is dealt with in Chapter 8, this volume. As an example of the former, attack of $KPPh_2$ or NaC_5H_5 on the $Cr(CO)_3$ complex of 6,6-dimethylfulvene produces the anionic derivatives $[Cr(CO)_3(\eta^5\text{-}C_5H_4CMe_2PPh_2)]^-$ and $[Cr(CO)_3(\eta^5\text{-}C_5H_4CMe_2C_5H_5)]^-$, respectively.[186] Treatment of $[Cr(CO)_3(\eta^4\text{-}C_5H_4CHNMe_2)]$ with NaCp gives $[Cr(CO)_3(\eta^5\text{-}C_5H_4CH{=}C_5H_4)]^-$, to which a second $Cr(CO)_3$ fragment can be coordinated.

The anions are not usually isolated but instead are converted into the $[MMe(CO)_3Cp]$, $[MX(CO)_3Cp]$ or $[M(CO)_2(NO)Cp]$ derivatives for characterization (Table 5).

In a series of studies Atwood and co-workers have examined the outer sphere electron transfer reduction of coordination complexes such as $[Fe(o\text{-phen})_3]^{3+}$ and $[CrCl_3(THF)_3]$ by various carbonylate anions, including $[Mo(CO)_3Cp]^-$; the organometallic product is $[Mo_2(CO)_6Cp_2]$. The reaction shown in Equation (8), where M and M' are fragments such as $M(CO)_3Cp$ (M = Cr, Mo, W) or $Mn(CO)_5$, enables derivation of two-electron reduction potentials for $M_2 \rightarrow 2M^-$ which correctly predict the direction of reaction. The values for Cr, Mo, and W are –0.70, –0.79, and –1.3 V respectively, though the authors stress that the last of these is probably inaccurate.[220–3]

$$2M^- + M'_2 \longrightarrow M_2 + 2M'^- \qquad (8)$$

7.2.3 The $[M(CO)_4Cp]^+$ Cation and Related Compounds

The major advance in this area has been the synthesis and use of complexes containing a loosely coordinated anion.[224] Protonolysis of $[MMe(CO)_2(L)Cp]$ (M = Mo, W; L = CO, phosphines) with strong acids (e.g., HBF_4, HOTf, HPF_6) gives the complexes $[M(A)(CO)_3Cp]$ (A = FBF_3, OTf, FPF_5, etc.).[225–7] Alternative syntheses involve hydride abstraction from $[MH(CO)_3Cp]$ with $[Ph_3C][BF_4]$, halide abstraction from $[M(Cl)(CO)_3Cp]$ with $AgBF_4$ or $AlCl_3$,[228] or cleavage of $[Hg\{Mo(CO)_2(PPh_3)Cp\}_2]$ by $[FeCp_2][BF_4]$.[229] The coordination mode of the anion can be proven in the FBF_3 complex (M = Mo, L = PMe_3) where the bridging fluorine can be distinguished as a doublet of quartets in the low-

Table 5 Substituted Cp derivatives of $[M(CO)_3Cp]^-$.

Cp ligand	*M*	*Derivative made*	*Ref.*
$C_5H_4CMe_2PPh_2$	Cr	isolated as salt	186
$C_5H_4CMe_2C_5H_5$			
C_5H_4CHO	Cr, Mo, W	I, NO	10, 187, 188
$C_5H_4CO_2R$ (R = Me, Et)	Cr, Mo, W	NO, halide	188, 189
C_5H_4COMe			
C_5HPh_4	Cr, Mo, W	NO	190, 191
C_5Ph_5	Cr, Mo, W	Me	192
$C_5H_4–C_6H_6Mn(CO)_3$	Cr, Mo, W	Me, NO	193
$C_9H_6CH_2CH_2C_9H_6$	Cr, Mo, W	Me	194
$C_5H_4CMe_2C_{13}H_9$	Cr, Mo, W	H, Me	195
$C_5H_4CH{=}C_5H_4$	Cr, Mo	Me	186, 196
$C_5H_{5-n}Et_n$ (n = 3–5)	Mo	Me	197
$C_5H_{5-n}Pr^i{}_n$ (n = 1–4)	Mo	Me	198
$C_5H_4PPh_2$, $C_5H_4PTol_2$	Mo	I, NO	199–202
$C_5H_4CH{=}C_9H_6$	Mo	Me, NO	203
C_5Me_4R (R = Me, Et, 4-butenyl)	Mo	NO	204
C_5Me_4-3-pinanyl	Mo	Me	205
$C_5H_4N{=}NPh$	Mo	Me	206
1,1',3,3'-$C_{10}H_4Bu^t{}_4$	Mo	Me	207
$C_5H_4SiMe_2SiMe_2C_5H_4$	Mo	Me, halide	22
$C_5H_4CH_2CH_2Fe(CO)_2Cp$	Mo	Me	208
$C_5H_4CMe_2C_9H_7$	Mo	Me	39, 209
$C_5H_4CMe_2C_{13}H_9$			
$C_5H_4(CH_2)_3C_{13}H_9$			
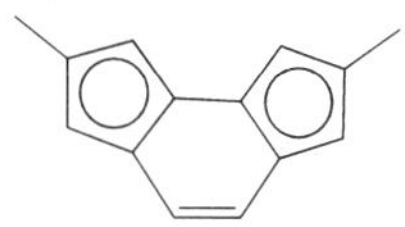	Mo	Me	210
$C_5H_4CH{=}CH_2$, $C_5H_4CMe{=}CH_2$	Mo, W	NO	188, 211, 212
$C_5H_4CMe_2$-2-pyridyl	Mo, W	Me	213, 214
1,3-$C_5H_3\{CMe(CH_2)_4CH_2\}_2$			
$C_5H_4SiMe_2C_5H_4$	Mo, W	H, Me, halide, NO	14, 215, 216
$C_5H_4SiMe_2OPh$	Mo, W	Me	217
1,4-$C_6H_4(OSiMe_2C_5H_4)_2$			
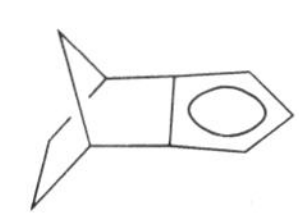	Cr, Mo, W	NO	218
$C_5H_4SiMe_2Ph$	W	Me	219
1,4-$C_6H_4(SiMe_2C_5H_4)_2$			
1,2,3-$C_6H_3(SiMe_2C_5H_4)_3$			

temperature ^{19}F NMR spectrum whereas at ambient temperature the four fluorine atoms are equivalent.[230,231] These compounds are extremely useful for further synthesis, as the loosely bound anion is readily displaced by a variety of ligands L' including not only CO, phosphines and isonitriles but also relatively weak ligands such as H_2O, Et_2O, EtOH, RCHO, H_2S and RSH to give $[M(CO)_2(L)(L')Cp][A]$.[232–8] Alkynes produce $[M(CO)(RC{\equiv}CR)_2Cp][BF_4]$ and $[M(CO)(L)(RC{\equiv}CR)Cp][BF_4]$,[239,240] and reaction with $[MoR(CO)_3Cp]$ (R = Me, Et, CH_2Ph) forms $[Mo_2(CO)_4(\mu\text{-}COR)Cp_2][BF_4]$.[241–3] Heterometallic μ-acetyl complexes can also be prepared from $[FeMe(CO)_2Cp]$ or $[Fe(COMe)(CO)(L)Cp]$ (L = CO, PPh_3).[244,245]

Nucleophilic attack on $[M(CO)_3(L)Cp]^+$ by H^- occurs at the CO ligand to give neutral formyls $[M(CO)_2(CHO)(L)Cp]$; these tend to be unstable, but the Cp* analogues are more robust and can be converted into carbene complexes.[246–9] Nucleophilic attack at CO also occurs by OH^- and OMe^-, giving metallocarboxylates.[250–2] Azide attacks at both CO and the metal, giving a mixture of *cis*-$[Mo(NCO)(CO)_2(PPh_3)Cp^*]$ and *cis*-$[Mo(N_3)(CO)_2(PPh_3)Cp^*]$.[253] Electrochemical reduction of $[Mo(CO)_3(L)Cp^*]^+$ is a one-electron process leading to either $[Mo_2(CO)_6Cp^*_2]$ (L = CO, CNR) or $[Mo_2(CO)_4(L)_2Cp^*_2]$ when L = $P(OMe)_3$.[31]

The useful intermediates $[Mo(CO)_2(NCMe)_2Cp][BF_4]$ (and its Cp* and η-C_9H_7 analogues), formed by cleavage of $[Mo_2(CO)_6Cp_2]$ with $AgBF_4$ in MeCN, or by refluxing $[Mo(CO)_3(NCMe)Cp^*][BF_4]$ in MeCN[254] have been used extensively by Green and co-workers as $Mo(CO)_2Cp^+$ equivalents in reactions with dienes, metal alkylidynes, etc.; for example, reaction with $[Li][Mo_2(CO)_4(\mu\text{-}C{\equiv}CR)Cp_2]$ affords $[Mo_3(\mu_3\text{-}C{\equiv}CR)(CO)_6Cp_3]$.[255] Another useful $Mo(CO)_2Cp^+$ synthon is the propene complex $[Mo(FBF_3)(CO)_2(\eta^2\text{-}CH_2{=}CHMe)Cp]$, prepared by protonation of $[Mo(CO)_2(\eta^3\text{-}C_3H_5)Cp]$ with HBF_4.[256]

Oxidative cleavage of $[Mo_2(CO)_nCp_2]$ (n = 4, 6) can also be achieved with $[FeCp_2][BF_4]$ in the presence of L (L = phosphines, $AsPh_3$, SMe_2, etc.) to give $[Mo(CO)_2L_2Cp]^+$ in over 90% yield.[257–60] In some cases the disproportionation reaction of $[Mo_2(CO)_6Cp_2]$ can be used to good effect, for example PMe_3 gives good yields of $[Mo(CO)_2(PMe_3)_2Cp][Mo(CO)_3Cp]$ without any added oxidant.[261] Reaction of $[Mo_2(CO)_6Cp_2]$ with RNC followed by crystallization in the presence of KPF_6 produces $[Mo(CO)(CNR)_3Cp][PF_6]$ if R = alkyl or $[Mo(CNR)_4Cp][PF_6]$ if R = aryl.[262,263] A different route to the complexes with R = Bu^t starting from $[MI_3(CO)(CNBu^t)Cp]$ has been described.[264]

7.2.4 $[MH(CO)_3Cp]$ and Related Compounds

The protonation of $[M(CO)_3Cp]^-$, usually with acetic acid, produces $[MH(CO)_3Cp]$; this method works well for both Cp and Cp* derivatives.[265] For pure samples, the reaction of $[Mo(CO)_3(\eta^6\text{-}C_7H_8)]$ with C_5Me_5H has also been used.[266] A combined x-ray and neutron diffraction study of $[MoH(CO)_3Cp^*]$ at 163 K has been published.[267]

Molecular orbital calculations indicate that four-legged piano-stool complexes, for example $[MH(CO)_2(L)Cp]$, should function as proton donors whereas three-legged ones, for example $[MH(NO)_2Cp]$, will be hydride donors.[268,269] The pK_a values of $[MH(CO)_3Cp]$, their Cp* analogues, and $[MH(CO)_2(PPh_3)Cp]$ have been measured. Electrochemical oxidation of these species gives very acidic radical cations whose pK_a values are in the range −10 to +5; preparative scale oxidation of $[WH(CO)_2(PMe_3)Cp]$ in MeCN forms a 1:1 mixture of $[W(H)_2(CO)_2(PMe_3)Cp]^+$ and $[W(CO)_2(PMe_3)(MeCN)Cp]^+$, the latter in a 95:5 *cis*:*trans* ratio, whereas hydride abstraction with $[Ph_3C]^+$ produces a 10:90 ratio of the same isomers.[270] A similar situation pertains for $[MoH(CO)_2(PPh_3)Cp]$.[271]

The hydride complexes are useful intermediates: dissolution in halogenated solvents forms $[MX(CO)_3Cp]$, deprotonation affords $[M(CO)_3Cp]^-$, and hydride abstraction with $[Ph_3C][BF_4]$ provides $[M(CO)_3(FBF_3)Cp]$ (see Scheme 8). The reaction of $[Mo(CO)_2(L)(FBF_3)Cp]$ (L = CO, PPh_3) with further hydride complex produces $[\{Mo(CO)_2(L)Cp\}_2(\mu\text{-}H)][BF_4]$.[272]

The reaction of $[MH(CO)_3Cp]$ (M = Cr, Mo, W for Cp and Cp*) with phosphines occurs both thermally and photochemically to give *cis*- and *trans*-$[MH(CO)_2(L)Cp]$ and then $[MH(CO)(L)_2Cp]$, which exists as a single isomer.[273–5] The disubstituted complex can also be made by $LiAlH_4$ reduction of $[Mo(CO)_3(L)Cp^*]^+$ in the presence of L, which occurs by an electron transfer mechanism.[276] The *cis* and *trans* isomers of the monosubstituted product interconvert rapidly at room temperature but the fluxionality can be frozen out at −50 °C, when two hydride peaks become visible in the ^{1}H NMR with J(HP) of ~65 Hz and 25 Hz, respectively; a detailed NMR study has determined the signs of the coupling constants.[277] Similar spectra are observed for other complexes of this type. The isonitrile species $[MH(CO)_2(CNR)Cp]$ (M = Mo, W; R = Me, Bu^t for Cp, Cp*) have also been prepared.[278]

The photolysis of $[MH(CO)_3Cp]$ (M = Mo, W) in a CH_4 matrix doped with C_2H_4 at 12 K affords both the CO loss product $[MH(CO)_2Cp]$ and *cis*- and *trans*-$[MH(CO)_2(\eta\text{-}C_2H_4)Cp]$, which on further irradiation forms the insertion product $[M(Et)(CO)_2Cp]$.[279] In an argon matrix saturated with H_2, the dihydrogen adduct *cis*- and *trans*-$[MH(H_2)(CO)_2Cp]$ is formed.[280] On a preparative scale the major products of photolysis of $[MH(CO)_3Cp]$ are $[M_2(CO)_6Cp_2]$ and $[W_2(\mu\text{-}H)_2(CO)_4Cp_2]$ (**5**).[281] This dinuclear hydride undergoes insertion of MeNC into one of the W–H bonds to give (**6**) and reacts with halogens to give $[W_2(CO)_4(\mu\text{-}X)_2Cp_2]$.[282,283]

A small number of polyhydride species of molybdenum and tungsten have been prepared. For molybdenum they are limited to $[MoH_3(dmpe)Cp]$ (and its PMe_3 and PMe_2Ph analogues), $[MoH_3(PMe_3)_2Cp^*]$ and $[MoH_5(PMe_3)Cp^*]$. The Cp trihydride is prepared by reduction of the corresponding trichloride, and catalyses the H/D exchange reaction of hydrocarbons, ethers, and other species with C_6D_6 solvent; it can also be protonated to $[MoH_4(dmpe)Cp]^+$.[284,285] The Cp* trihydride is made by reduction of $[MoCl(PMe_3)_3Cp^*]$ and does not react with H_2 to give the pentahydride.[286] The pentahydride is made by reduction of $[MoCl_2(PMe_3)Cp^*]$ and is a classical polyhydride having a pentagonal bipyramidal structure with five equivalent hydrides, in accord with theoretical predictions; it does not react with PMe_3 to give the trihydride.[287,288] In contrast photolysis of $[W(H)(PMe_3)_3Cp]$ under

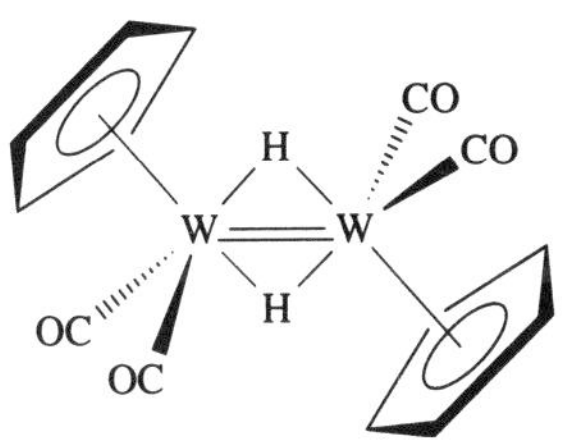

(5)

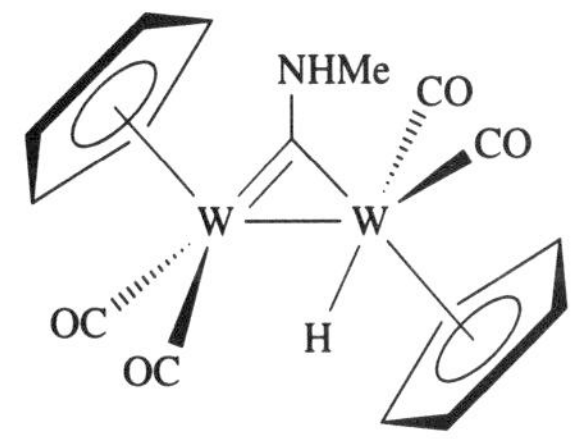

(6)

H_2 affords the trihydride $[WH_3(PMe_3)_2Cp]$ and the pentahydride $[WH_5(PMe_3)Cp]$ sequentially; they can also be made by condensation of tungsten atoms with PMe_3 and cyclopentene or cyclopentane, respectively.[289,290] The pentahydride can be deprotonated with KH.[291] Compounds in which the pentahydride unit is linked to a rhodium fragment through a $C_5Me_4CH_2CH_2C_5Me_4$ ligand are also known.[292] Reduction of $[WCl_4(PMe_3)Cp^*]$ gives the analogous $[WH_5(PMe_3)Cp^*]$.[293] Reduction of $[WCl_4Cp^*]$ with $LiAlH_4$ however produces $[WH_4Cp^*]_2$ which reacts with CO to give $[WH(CO)_2Cp^*]_2$ and with PMe_3 to give $[W_2H_6(PMe_3)Cp^*_2]$; it can also be protonated with HBF_4. Hydrogenolysis of $[WMe_4Cp^*]$ produces a small amount of the dimeric tetrahydride, but the major product is $[WH_3Cp^*]_3$.[293,294]

7.2.5 [MX(CO)₃Cp] (X = Halide) and Related Compounds

The tricarbonyl halides can be made by cleavage of $[M_2(CO)_6Cp_2]$ with halogens or by exposing $[MH(CO)_3Cp]$ to CX_4 or other halogenating agents; a one-pot procedure for $[W(Cl)(CO)_3Cp]$ has been devised.[295] The first method is most convenient for the iodides, but over-oxidation can occur with Cl_2 or Br_2, for which the second route is preferred. The structures of $[MCl(CO)_3Cp]$ (M = Mo, W) and of $[MoBr(CO)_3Cp]$ have both been determined, confirming the expected four-legged piano-stool arrangement.[296,297] The halides undergo rapid self exchange with $[M(CO)_3Cp]^-$ anions and $\{M(CO)_3Cp\}$ radicals, the rate of which can be measured by NMR line-broadening.[298,299] Exposure of the solids to γ-rays produced radical anions whose ESR spectra suggest the electron is taken up into a M–X σ^* orbital.[300]

The tricarbonyl halides undergo substitution reactions with phosphines and diphosphines to produce $[M(X)(CO)_2(L)Cp]$ and $[M(X)(CO)(L_2)Cp]$.[274,301,302] The substitution can be intramolecular, as in $[Mo(I)(CO)_3(\eta\text{-}C_5H_4CH_2CH_2NMe_2)]$.[303] The indenyl tungsten complexes react much faster than either their indenyl molybdenum or Cp tungsten analogues, by a mixed associative/dissociative mechanism.[304] The $[M(X)(CO)_{3-n}(L)_nCp]$ complexes generally exist as either exclusively *cis* or a mixture of *cis* and *trans* isomers (unusually the diphosphine $(C_6F_5)_2PCH_2CH_2P(C_6F_5)_2$ is able to span the *trans* positions in $[Mo(Cl)(CO)(L_2)Cp]$[305]), and they display a reversible electrochemical oxidation with potentials depending on the number and donor ability of L.[306,307] Photolysis of *cis*- and *trans*-$[MX(CO)_2(L)Cp]$ in a matrix at 12 K gives two different isomers of $[MX(CO)(L)Cp]$, which each revert exclusively to their original isomers on addition of CO.[308]

The substitution reactions of $[M(I)(CO)_3Cp]$ (M = Mo, W) with RNC in the presence of Me_3NO or catalysed by $[Fe_2(CO)_4Cp_2]$ give $[M(I)(CO)_2(CNR)Cp]$, whereas with a $[M_2(CO)_6Cp_2]$ catalyst multiple substitution occurs, giving ultimately $[M(CNR)_4Cp][I]$.[309,310] As shown in Scheme 9, a number of RNC-substituted complexes with R = Et, Bu^t have subsequently been prepared en route to aminocarbynes.[264,311–19]

Oxidation of $[MX(CO)_3Cp]$ with halogens, or photochemically with CCl_4, produces $[MX_3(CO)_2Cp]$; the geometry is pseudooctahedral with *fac*-halides.[320,321] Allyl halides undergo oxidative addition to $[Mo(Cl)(CO)_3Cp]$ to form $[Mo(Cl)(X)(CO)(\eta^3\text{-}C_3H_5)Cp]$, whereas the tungsten analogue oxidizes to $[W(Cl)_3(CO)_2Cp]$; the same processes can be observed on photolysis in an allyl halide matrix.[322,323]

7.2.6 Metallo Derivatives of [M(CO)₃Cp]

The $[M(CO)_3Cp]^-$ anions have been extensively used to construct heterometallic dimers and clusters by reaction with main group and transition metal halides; the latter are outside the scope of this chapter but are covered in Volume 10. In the case of the former the structural types observed have been well established for many years.

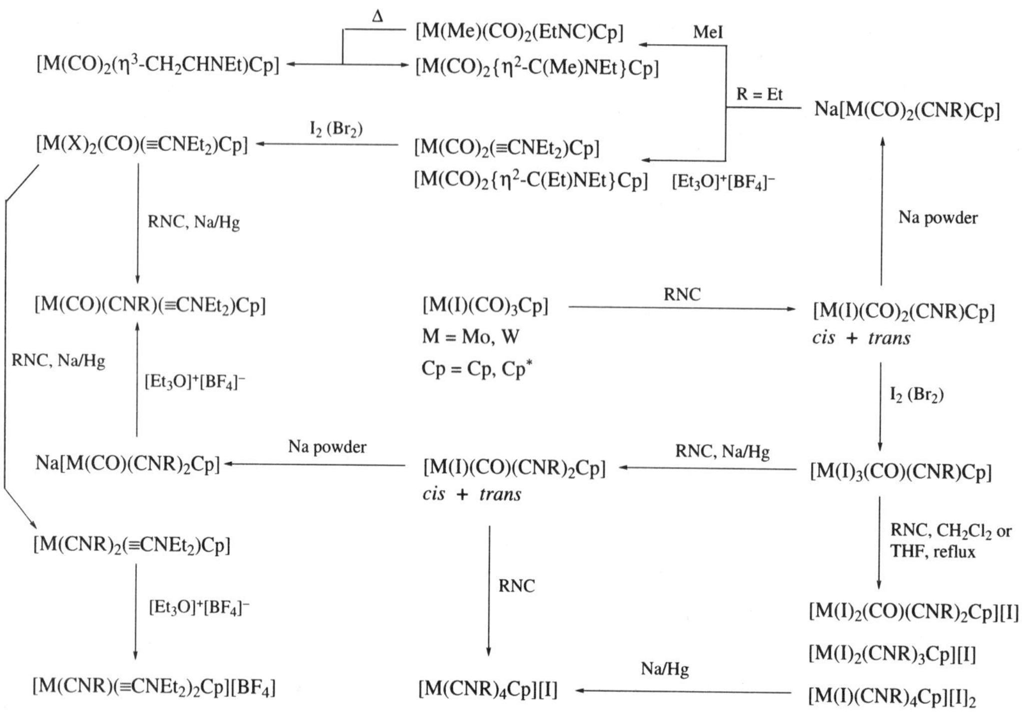

Scheme 9

7.2.6.1 [M(CO)$_3$Cp] complexes of group 11 elements

The compounds [CpM(CO)$_3$CuL$_2$] (**7**) (M = Cr, Mo, W; L = PPh$_3$ or L$_2$ = tmeda) have been synthesized; the x-ray structures of the Mo, tmeda and two forms of the W, PPh$_3$ complexes all show the presence of two semibridging carbonyl ligands.[324,325] According to a theoretical analysis, the Cu centre is donating electron density from a high energy $d\pi$ orbital to the unoccupied π^* level of the CO.[326,327] The reaction of [Mo(CO)$_3$Cp]$^-$ with [Cu$_2$(OAc)$_4$py$_2$] produces [CpMo(CO)$_2$(μ-OAc)$_2$Cu(py)] (**8**), which has one semibridging CO as well as two bridging acetate ligands.[328,329]

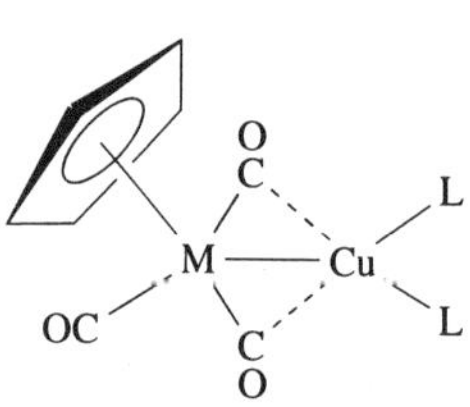

(7)

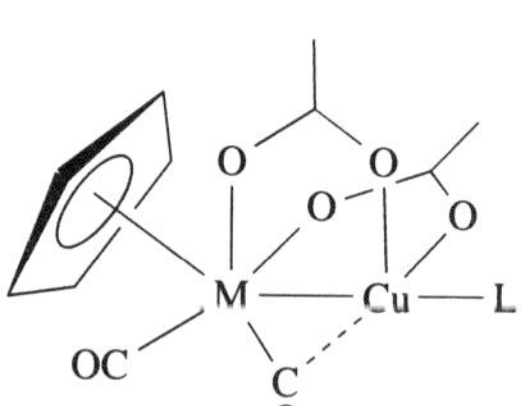

(8)

The preparation of [(η-C$_5$H$_4$CHO)MAu(PPh$_3$)] has been described and the chromium complex structurally characterized.[10] The reaction of [AuX$_2$]$^-$ or [Au(C$_6$F$_5$)(tht)] with [M(CO)$_3$Cp]$^-$ produces the aurate complexes [Au{M(CO)$_3$Cp}$_2$]$^-$; similarly the gold(III) derivative [Au(C$_6$F$_5$)$_3$(tht)] forms [Au(C$_6$F$_5$)$_3${M(CO)$_3$Cp}]$^-$.[330,331]

7.2.6.2 [M(CO)$_3$Cp] complexes of group 12 elements

Reaction of ZnCl$_2$ with [W(CO)$_3$Cp]$^-$ affords [Zn{W(CO)$_3$Cp}$_2$], while the molybdenum analogue has been made by treatment of [Mo$_2$(CO)$_6$Cp$_2$] with ZnEt$_2$;[332,333] it reacts with MeOH to give tetrameric [Zn(OMe){Mo(CO)$_3$Cp}]$_4$ and [MoH(CO)$_3$Cp].[334] The addition reaction of zinc, cadmium and mercury complexes [E{ML$_n$}$_2$] with further carbonylmetalates [M'L$_n$]$^-$ to give the 'ate' complexes [E{ML$_n$}$_2${M'L$_n$}]$^-$ is an equilibrium process with E, M, M', temperature and solvent all determining the position.[335]

Many new complexes of the type $[Hg\{M(CO)_2(L)Cp\}_2]$ and $[XHg\{M(CO)_2(L)Cp\}]$ (X = halide, SCN, CN, OAc, S_2CNEt_2, SBu^t, aryl, $M'L_n$, etc.; L = various phosphines) have been synthesized, usually through the reaction of $[M(CO)_2LCp]^-$ with HgX_2 or redistribution of $[Hg\{M(CO)_2(L)Cp\}_2]$ with HgX_2.[9,336–48] Comprehensive multinuclear NMR investigations show that the ^{199}Hg shift and $^2J(HgP)$ both show a linear correlation with phosphine cone angle.[349–53] Treatment of $[Hg\{M(CO)_2(L)Cp\}_2]$ with L' and $Hg(BF_4)_2$ affords $[\{M(CO)_2(L)Cp\}HgL'][BF_4]$ whereas a similar reaction with L' in the presence of $[FeCp_2][BF_4]$ produces $[Mo(CO)_2(L)(L')Cp][BF_4]$ by oxidative cleavage and elimination of mercury; electrochemistry shows this to be a two-step process.[229,354,355]

A number of clusters bridged by $Hg\{M(CO)_3Cp\}$ fragments have been prepared by reaction of $[M(CO)_3Cp]^-$ with HgX-bridged clusters, for example $[Ru_3(CO)_9(\mu\text{-}C{\equiv}CBu^t)(\mu_3\text{-}Hg\{Mo(CO)_3Cp\})]$[356,357] or by the reaction of $[XHg\{M(CO)_3Cp\}]$ with cluster anions, for example $[Fe_2(CO)_8(\mu\text{-}Hg\{M(CO)_3Cp\})]^-$.[358]

7.2.6.3 *[M(CO)₃Cp] complexes of group 13 elements*

Only two new gallium complexes have been synthesized recently: insertion of transient 'GaI' into $[Mo(I)(CO)_3Cp]$ gave $[Mo(GaI_2)(CO)_3Cp]$ as its Et_2O adduct,[359] while reaction of $GaCl_3$ with $[Mo(CO)_3Cp]^-$ gave $[Ga\{Mo(CO)_3Cp\}_3]$.[360]

The indium complexes $[InCl_n\{M(CO)_3Cp\}_{3-n}]$ (M = Cr, Mo; $n = 0, 1$) and the 'ate' complex $[InCl_2\{M(CO)_3Cp\}_2]^-$ have been synthesized by treating $InCl_3$ with the appropriate carbonylmetalate. Calculations show that the M–In bonds are weak and IR spectra show some dissociation back to $[M(CO)_3Cp]^-$ in solution.[361,362] An alternative route involves reaction of $[Mo_2(CO)_6Cp_2]$ with InCl or indium metal;[333,363] with InR_3 (R = Me, Bu^t, etc.) the complexes $[Mo(InR_2)(CO)_3Cp]$ are formed by a radical substitution mechanism. Two oxo-capped clusters, $[Mo_3(CO)_4(\mu\text{-}Cl)(\mu_3\text{-}O)Cp_3]$ and $[Mo_3(CO)_3(\mu\text{-}CO)_3(\mu_3\text{-}O)Cp_3][Mo(InCl_3)(CO)_3Cp]$ were isolated from a similar reaction with $InCl_3$.[364]

A number of complexes have been prepared in which $M(CO)_3Cp$ units are joined to indium or thallium porphyrin complexes; in some cases they undergo reversible electrochemical reduction or oxidation without metal–metal bond cleavage, and can be fabricated into Langmuir–Blodgett films.[365–72] A review of this area has been published.[373]

7.2.6.4 *[M(CO)₃Cp] complexes of group 14 elements*

Compounds of the type $[M(CO)_3(SiR_3)Cp]$ are well known. Those containing Si–H bonds such as $[M(CO)_2L(SiRH_2)Cp]$ (M = Cr, Mo, W; L = CO, PMe_3; R = Me, Ph) have attracted some interest as they can be used to make μ_3-SiR capped clusters through reaction with $[Co_2(CO)_8]$.[374,375] A selection of similar Cp*W complexes have been prepared including $[W(CO)_2(PMe_3)(SiH_3)Cp^*]$ by $LiAlH_4$ reduction of the analogous $W\text{–}SiHCl_2$ species.[376]

Reaction of GeI_2 with $[Mo(CO)_3Cp]^-$ gives the metal-substituted germylene $[Mo(GeI)(CO)_3Cp]$ which can be trapped by cycloaddition with 2,3-dimethylbutadiene.[377] Deprotonation of $[M(ER_3)(CO)_3Cp]$ (M = Mo, W; E = Ge, Sn, Pb; R = Me, Ph) is followed by migration of ER_3 onto the Cp ligand to give the ring-substituted anions $[M(CO)_3(\eta\text{-}C_5H_4ER_3)]^-$.[378]

The complexes $[W(SnR_nCl_{3-n})(CO)_3Cp]$ (R = Me, Ph) in combination with Bu^iAlCl_2 and O_2 act as alkene metathesis catalysts with high conversion and good selectivity; the halide $[W(Cl)(CO)_3Cp]$ is probably produced *in situ*.[379–82] The reaction of $[WH(CO)_3Cp]$ with $Sn(C_5H_5)_2$ produces $[\{W(CO)_3Cp\}_3SnCl]$ rather than the metallated stannylene $[Sn\{W(CO)_3Cp\}_2]_n$ previously claimed,[383,384] but the tin(II) complexes $[\{M(CO)_3Cp\}Sn\{(PR_2)_2CX\}]$ (M = Mo, R = Me, X = TMS; M = W, R = Ph, X = PPh_2) have been made.[385] A number of complexes involving different metal fragments attached to tin atoms have been synthesized by reaction of molybdenum and tungsten carbonylmetalate anions with appropriate tin halides,[386–8] including one containing a linear OsOsOsSnM (M = Mo, W) chain of five metal atoms.[389] Oxidation or reduction of $[\{Mo(CO)_3Cp\}_nSnPh_{4-n}]$ (n = 1, 2) proceeds in two-electron steps with rupture of the Mo–Sn bonds.[390]

The complexes $[M(PbR_3)(CO)_3Cp]$ (M = Cr, Mo, W; R = Me, Et, Ph) undergo either plumbylene elimination to form $[M(R)(CO)_3Cp]$ or redistribution to $[\{M(CO)_3Cp\}_2PbR_2]$ on irradiation.[391,392] The characterization of some of these by ^{207}Pb and ^{95}Mo NMR has also been reported.[393,394] The plumbylene complex $[Pb\{Mo(CO)_3Cp\}_2(THF)]$ can be made by treatment of $[MoH(CO)_3Cp]$ with $Pb\{N(TMS)_2\}_2$.[395]

7.2.7 Organic Derivatives of $[M(CO)_3Cp]$

Simple alkyl complexes $[MR(CO)_3Cp]$ are generally made by nucleophilic attack of $[M(CO)_3Cp]^-$ on the appropriate alkyl halide; usually the iodide is preferable (see Table 5 for examples involving substituted Cp ligands). As might be expected this occurs with inversion of configuration at the carbon atom: thus, *erythro*-PhCHDCHDOTs produces *threo*-[W(CHDCHDPh)$(CO)_3$Cp] when treated with the tungsten anion.[396] In a study of metal carbonylates, Atwood has shown that $[MoR(CO)_3Cp]$ (R = Me, Et, CH_2Ph) will rapidly transfer its alkyl group to the stronger nucleophile $[Fe(CO)_2Cp]^-$, but self-exchange with $[Mo(CO)_3Cp]^-$ is very slow.[397]

Some recent examples prepared are given in Table 6. Notable among these are the C-bound enolate complexes $[M(CH_2COR)(CO)_3Cp]$, which are best made by treatment of the metal carbonylates with α-chloroesters or α-chloroketones. These undergo a range of reactions without disruption of the M-CH_2COR unit. Loss of CO gives η^3-oxoallyl complexes, which are not isolable but which react with added ligands to form substituted enolate complexes.

Table 6 Alkyl derivatives of $[M(CO)_3Cp]$.

Metal	*Alkyl ligand*	*Ref.*
Cr	Et	398
Mo, W	CH_2Cl	399,400
Mo, W	CH_2COR (R = Me, Ph, OEt, NEt_2)	401–3
Mo, W	$CH_2CHCH_2CH_2$, $CH_2CMeCH_2CH_2$ (cyclopropylmethyl)	404,405
Mo, W	CH_2CH_2OMe, $CH_2CH_2SO_2OMe$	406
Mo, W	$CH_2C_6H_4NC$	407
Mo, W	$CH_2NCOC_6H_4CO$ (phthalimidomethyl)	408
Mo, W	$CH_2CHCH_2OCH_2CHCH_2M(CO)_3Cp$	409
Mo, W	CF_3	410–12
Mo	CH_2CO_2TMS, CH_2COOH	413
Mo	CH_2O_2CMe	414
W	$CH_2Si_2Me_5$	415
W	$CH_2SiMe_2GeMe_3$, CH_2GeMe_2TMS	416
W	CH_2SR, $CH_2S(O)R$, $CH_2S(O)_2R$ (R = Ph, Me)	417,418
W	CH_2COFc, $CH_2COC_5H_4Mn(CO)_3$	419

Spectroscopic studies of these compounds include the complete assignment of the IR spectrum of $[MMe(CO)_3Cp]$ including the frequencies and bond strengths in the methyl group, and a ^{95}Mo NMR study of $[Mo(CH_2C_6H_4X)(CO)_3Cp]$, which showed that the chemical shift correlated with the substituent constant of X.[420–2] Thermochemical studies of the iodination and insertion reactions of $[MoR(CO)_3Cp]$ (R = Me, Et, CH_2Ph, allyl) have allowed an estimate of the Mo–R bond energies (see Table 3).[48] Calculation of the bond dissociation energies from kinetic data for the thermolysis of [M(CHMePh)$(CO)_3$Cp] gave values of 104.5 and 121 kJ mol^{-1} for molybdenum and tungsten respectively.[423]

Chemical or electrochemical oxidation of $[WMe(CO)_3Cp]$ in MeCN occurs by a two-electron process allied with ligand addition and substitution to give $[WMe(CO)_3(NCMe)Cp]^{2+}$ and $[WMe(CO)_2(NCMe)_2Cp]^{2+}$, the ratio of these depending on the conditions used; while the former does not undergo migratory insertion, the latter does to yield ultimately $[W(\eta^2\text{-COMe})(NCMe)_3Cp]^{2+}$, and it also undergoes further sequential substitution on photolysis to give $[WMe(NCMe)_4Cp]^{2+}$.[424,425]

Functionalization of the Cp ring in $[WMe(CO)_3Cp]$ can be achieved by deprotonation with Bu^{s}Li. The lithiated species can then be derivatized with, for example, $MeSiCl_3$ to give $[MeSi\{\eta\text{-}C_5H_4WMe(CO)_3\}_3]$,[426] with PPh_2Cl to give $[WMe(CO)_3(\eta\text{-}C_5H_4PPh_2)]$,[427] or by addition of iodine to give $[WMe(CO)_3(\eta\text{-}C_5H_4I)]$. The last of these is a useful starting point for further elaboration as it undergoes palladium-catalysed Stille cross-coupling with aryl-, vinyl- or alkynylstannanes, ultimately allowing the construction of dimetallic species linked by unsaturated spacer groups between the Cp ligands, for example Scheme 10.[428–30] In the presence of a palladium catalyst, $[WMe(CO)_3(\eta\text{-}C_5H_4I)]$ also isomerizes to $[W(I)(CO)_3(\eta\text{-}C_5H_4Me)]$.[431] The ring opening of spiro [2.4] hepta-4,6-dienes with $[Mo(CO)_6]$ gives complexes $[Mo(CO)_3(\eta\text{-}C_5H_4CH_2CH_2)]$ in which the alkyl group is a substituent on the Cp ring.[432]

A number of complexes in which a $M(CO)_3Cp$ (M = Mo, W) unit is linked to another metal–ligand fragment by an alkyl chain, that is, $[Cp(CO)_3M(CH_2)_mM'L_n]$, have been prepared, usually by reaction of $[M(CO)_3Cp]^-$ with the appropriate α, ω-dihalide or $[M'(CH_2)_mX]$; for complexes with $m = 2$ the anion is reacted with the ethylene complexes $[M'(CO)_3(\eta\text{-}C_2H_4)Cp]^+$.[433] Apart from the homonuclear species (m = 2, 5, 10) and the related $[Cp(CO)_3MoCH_2C_6H_4CH_2Mo(CO)_3Cp]$, heterodinuclear species in which

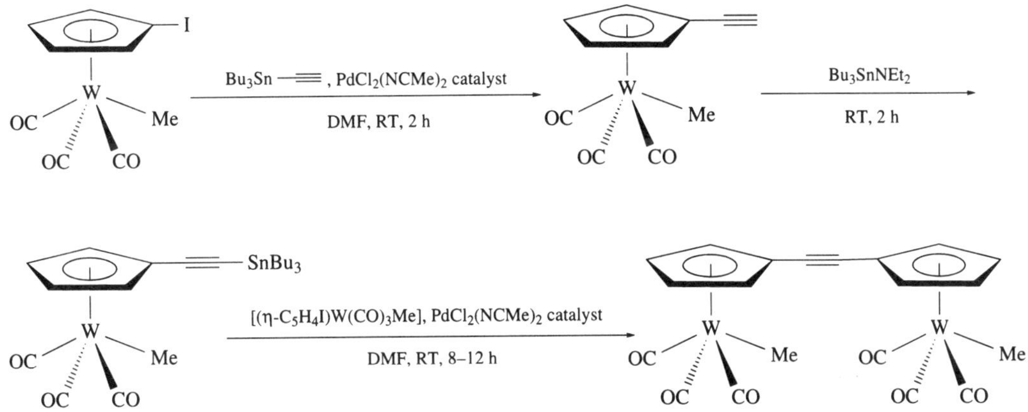

Scheme 10

m = generally 3–6 and M' can be $Fe(CO)_2Cp$, $Ru(CO)_2Cp$, $Re(CO)_5$, etc., have also been prepared. These compounds undergo insertion reactions of the M–C bond usually in the same way as their mononuclear counterparts, for example with PPh_3 or SO_2. In some cases reactions can be selective for the most reactive metal, for example FeMo complexes react with PPh_3 at Mo whereas FeW ones react at Fe.[434–41]

Attempts to prepare shorter chain homonuclear complexes with $m = 3$ by reaction of $[M(CO)_3Cp]^-$ with 1,3-diiodopropane lead instead to the cyclic carbene complexes $[M(I)(CO)_2(=\overline{COCH_2CH_2CH_2})Cp]$ (Scheme 11). The haloalkyl compounds $[M(CH_2CH_2CH_2I)(CO)_3Cp]$ are intermediates in this transformation.[436,442] Using an excess of metal anion, or treating isolated $[M(CH_2CH_2CH_2Br)(CO)_3Cp]$ with further anion, gives the unusual dinuclear terminal carbene complexes $[M_2(CO)_5(=\overline{COCH_2CH_2CH_2})Cp_2]$.[443] In the same way, reaction of $[M(CH_2CH_2CH_2Br)(CO)_3Cp]$ with PPh_2H affords the cationic carbene $[Mo(CO)_2(PPh_2H)(=\overline{COCH_2CH_2CH_2})Cp]^+$.[444] If the substituted anion $[M(CO)_2(CNMe)Cp]^-$ is employed with $I(CH_2)_3I$, the isonitrile ligand is incorporated into the ring to give the aminocarbenes $[M(I)(CO)_2(=\overline{CNMeCH_2CH_2CH_2})Cp]$.[445,446] The chemistry of these complexes is described in Chapter 3, this volume.

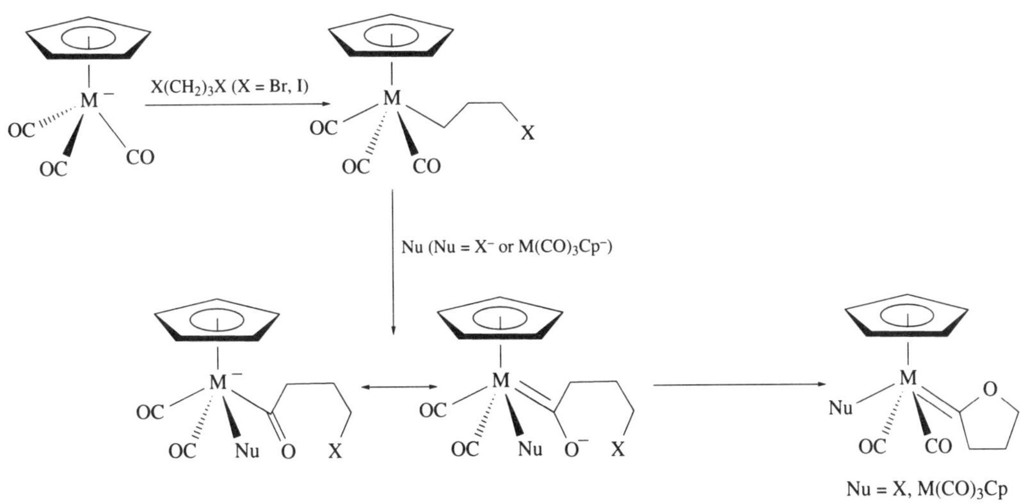

Scheme 11

The reaction of $[W(CO)_3Cp]^-$ with CH_2I_2 in MeCN produces the μ-ketene complex $[W_2(CO)_4(\mu\text{-}CO)(\mu\text{-}COCH_2)Cp_2]$ by way of the intermediate iodomethyl species $[W(CH_2I)(CO)_3Cp]$. Warming the ketene complex above −5 °C causes conversion into the dimetalla-alkyne $[Cp(CO)_3WC{\equiv}CW(CO)_3Cp]$ (Scheme 12). The phosphite-substituted anion $[W(CO)_2\{P(OMe)_3\}Cp]^-$ reacts with CH_2Br_2 to give the μ-methylene complex $[\{W(CO)_2(L)Cp\}_2(\mu\text{-}CH_2)]$ [L = $P(OMe)_3$], which on warming gives an analogous substituted μ-ketene complex, but does not proceed to the alkyne.[447]

Substituted alkyl complexes $[MoR(CO)_2(L)Cp]$ can be prepared as mixtures of *cis* and *trans* isomers by photolysis of $[MR(CO)_3Cp]$ in the presence of phosphines; in the case of diphosphines, dinuclear species linked through the phosphine are obtained.[448] This method can be rendered ineffective by

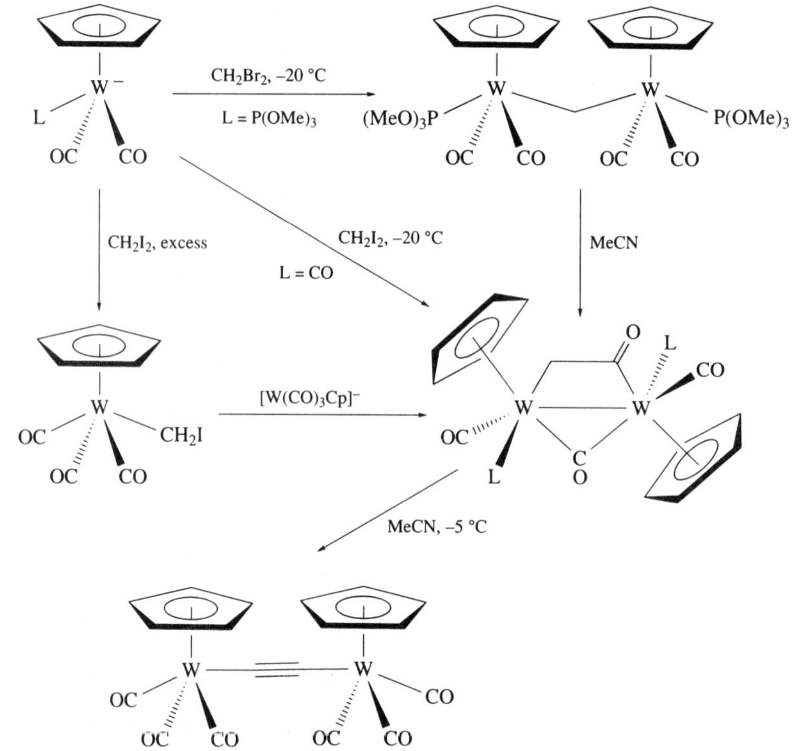

Scheme 12

competing photochemical loss of phosphine, and an alternative route consists of alkylation of the substituted anions $[M(CO)_2(L)Cp]^-$, made in turn by sodium amalgam cleavage of $[M(CO)_2(L)Cp]_2$.[71] While this works well for phosphines, when L is RNC, the product is dependent on the alkylating agent and conditions used. Thus, in the case of $[W(CO)_2(CNEt)Cp^*]^-$, alkylation with $[Et_3O]BF_4$ occurs at the isonitrile ligand to give the carbyne complex $[W(CO)_2(\equiv CNEt_2)Cp^*]$, whereas alkylation with MeI occurs at the metal to give *cis*- and *trans*-$[WMe(CO)_2(CNEt)Cp^*]$. However, on heating the latter, insertion of the isonitrile into the metal alkyl bond occurs to give, sequentially, η^2-iminoacyl and η^3-aza-allyl complexes. Each of these will in turn undergo further insertion of RNC to give novel metallacycles (see Scheme 9).[314–16,449,450] An alternative route to the η^3-aza-allyl complexes consists of treating $[Mo(CH_2COR)(CO)_3Cp]$ (R = H, Me) with Pr^iNH_2 to give the corresponding imine $[Mo(CH_2CR{=}NPr^i)(CO)_3Cp]$ followed by decarbonylation.[451]

Lindner and co-workers have prepared a series of chelate complexes of the type $[\overline{M(CO)_2\{PPh_2(C}H_2)_n\}Cp]$ (M = Mo, W, n = 2–4), in which the alkyl group and phosphine ligand are constrained into a *cis* geometry, by reductive cyclization of $[M(Cl)(CO)_2\{PPh_2(CH_2)_nCl\}Cp]$ with sodium; $[\overline{M(CO)_2\{PPh_2O(C}H_2)_n\}Cp]$ (n = 2, 3) can be prepared in the same way.[452–5]

The photochemistry of $[MR(CO)_3Cp]$ has been extensively studied by a variety of IR-based techniques both in solution and at low temperature in frozen gas matrices or PVC films. For molybdenum and tungsten, the primary photoprocess in all cases at low temperature is carbonyl loss to give $[MR(CO)_2Cp]$, which in a nitrogen matrix exists as its N_2 adduct.[456] In the case of R = Me in a PVC film, the products are $[M(Cl)(CO)_3Cp]$, $[MH(CO)_3Cp]$ and ultimately $[M_2(CO)_6Cp_2]$.[457] If the alkyl group possesses β-hydrogens (R = Et has been widely studied) two forms of this intermediate exist, one of which has an agostic interaction of one ethyl β-H with the metal. This intermediate then undergoes β-elimination to give $[MH(CO)_2(\eta\text{-}C_2H_4)Cp]$, which exists as *cis* and *trans* isomers at low temperature for tungsten but only as the *trans* isomer for molybdenum. This complex can also be made by photolysis of $[MH(CO)_3Cp]$ in the presence of ethylene, and is sufficiently stable to isolate and compare with related known compounds such as $[WMe(CO)_2(\eta\text{-}C_2H_4)Cp]$.[458,459] Prolonged photolysis again leads to $[MH(CO)_3Cp]$, and $[M_2(CO)_6Cp_2]$.[279,457,460–6] Similar chemistry also occurs when the tungsten alkyl complex is attached to a polymer support.[467]

Photolysis of $[W(CH_2SiMe_2H)(CO)_3Cp]$ in hydrocarbons at 77 K or 196 K also proceeds by CO loss and β-elimination; in this case the product is the silene complex $[WH(CO)_2(\eta\text{-}CH_2SiMe_2)Cp]$.[468]

Solution photolysis of $[Mo(CH_2Ph)(CO)_3Cp]$ in pentane gave $[Mo_2(CO)_6Cp_2]$ and toluene, whereas the tungsten analogue also produced $[W(CO)_2(\eta^3\text{-}CH_2Ph)Cp]$ and bibenzyl.[469] Although this appears to involve M–C bond homolysis, photolysis of *threo*-$[W(CHDCHDPh)(CO)_3Cp]$ did not cause any racemization, implying the absence of such homolysis.[396] Photolysis of $[W(CH_2CH{=}CH_2)(CO)_3Cp]$ also proceeds initially by CO dissociation followed by conversion to the η^3-allyl complex.[470]

Evidence has been described that photolysis of $[CrMe(CO)_3Cp]$ in a low temperature matrix initially gives $[CrMe(CO)_2Cp]$, which then undergoes α-elimination to form *trans*-$[CrH(CO)_2({=}CH_2)Cp]$. On warming in the presence of ligands, this is reversible and $[CrMe(CO)_2(L)Cp]$ is produced. Solution photolysis affords only $[Cr_2(CO)_6Cp_2]$, which is thought to arise through reaction of $[CrMe(CO)_2Cp]$ with further starting complex rather than simple Cr–Me homolysis.[471] An α-elimination process was also invoked in the case of $[Mo(CF_3)(CO)_3Cp]$.[472]

The photolysis of *trans*-$[MoMe(CO)_2(L)Cp]$ (L = $P(OEt)_3$, PPh_3, PBu_3) in a 1,2-epoxyethylbenzene glass at 12 K leads to CO loss; on warming in the presence of CO the *trans* isomer is reformed. Similarly the *cis* isomer gives an intermediate which returns to the *cis* isomer when L = $P(OEt)_3$; in the other cases phosphine loss, rather than CO loss, is the primary photochemical process.[473]

Acyl complexes of the type $[M(COR)(CO)_3Cp]$ can be prepared by treatment of $[M(CO)_3Cp]^-$ with acyl chlorides.[474] An unusual recent example is the ketoacyl complex $[Mo(COCOPh)(CO)_3Cp]$, which undergoes ready decarbonylation to $[Mo(COPh)(CO)_3Cp]$.[475] The reaction of $[Mo(CO)_3Cp]^-$ with $CF_2(COCl)_2$ gives $[CF_2\{COMo(CO)_3Cp\}_2]$, which can be decarbonylated to the μ-ketene complex $[Cp(CO)_3MoCF_2COMo(CO)_3Cp]$.[476] This procedure has also been used to make heterometallic μ-ketene complexes with R = $CH_2Fe(CO)_2Cp$ from $[Fe(CH_2COCl)(CO)_2Cp]$, and the heterometallic complex $[Mo\{COC_5H_4Mn(CO)_3\}(CO)_3Cp]$.[477–9] With succinoyl or glutaroyl chloride, cyclization occurs to give $[M\{\overline{C{=}CH(CH_2)_nCOO}\}(CO)_3Cp]$ (M = Mo, W; n = 2, 3).[480] A study of the photolysis of $[Mo(COCF_3)(CO)_3Cp]$ in inert gas matrices showed formation of $[Mo(CF_3)(CO)_3Cp]$ via observable $[Mo(COCF_3)(CO)_2Cp]$.[472]

The reaction of the alkyl complexes $[MR(CO)_3Cp]$ with phosphines in THF or MeCN at room temperature gives acyl complexes $[M(COR)(CO)_2(L)Cp]$. A large variety of phosphines has been used, including metallophosphines: the reaction of $[MoMe(CO)_3Cp]$ with $[Fe(PPh_2)(CO)_2Cp]$ produces the bridging acetyl complex $[Cp(CO)_2Mo(\mu\text{-}PPh_2)(\mu\text{-}COMe)FeCp]$.[481] In cases where R has a low migratory aptitude (e.g., CF_3, aryl), and under photochemical conditions when R = alkyl, the substituted compounds $[MR(CO)_2(L)Cp]$ are formed instead. Other exceptions include the chloromethyl derivatives $[M(CH_2Cl)(CO)_3Cp]$ which react with PPh_3 to give either $[M(Cl)(CO)_2(PPh_3)\ Cp]$ (for M = Mo) or the ylid species $[M(CH_2PPh_3)(CO)_3Cp]^+$ (for M = W).[399,400] The latter reaction can be encouraged for other nucleophiles by adding $TlBF_4$.[482]

The mechanism of the migratory insertion has been studied in detail for the reaction of phosphines with $[Mo(CH_2Ph)(CO)_3Cp]$; similar conclusions apply to reactions with isonitriles.[483] In polar solvents it proceeds as shown in Equations (9) and (10) via a solvent-stabilized acyl intermediate which subsequently reacts with phosphine:

$$[Mo(CH_2Ph)(CO)_3Cp] + S \longrightarrow [Mo(COCH_2Ph)(S)(CO)_2Cp] \quad (9)$$

$$[Mo(COCH_2Ph)(S)(CO)_2Cp] + L \longrightarrow [Mo(COCH_2Ph)(CO)_2(L)Cp] + S \quad (10)$$

The reaction is very susceptible to steric influence on the benzyl group, the Cp ligand, the phosphine and the solvent. X-ray structures of $[MoR(CO)_3Cp]$ where R = CH_2Ph, CH_2-*o*-Tol, and CH_2Mes show a lengthening of the Mo–C bond, which correlates with increasing rate of reaction.[484,485] The rate is also enhanced by electron-donating substituents in the benzyl group.[486] The rate of the second step correlates inversely with the cone angle of the phosphine ligand but is not influenced by its electronic properties.[487] Changing the Cp ligand to C_5H_4Me or Cp* also increases the rate of the dissociative step for both steric and electronic reasons, but the second step is retarded, so much so that the Cp* complex only reacts with phosphines with small cone angles.[488] The proportion of *cis*-acyl product is also increased. Finally, using 2-methyl THF or 2,5-dimethyl THF as solvent also slows the first stage, an observation also noted for $[MoMe(CO)_3Cp]$.[489,490] In nonpolar solvents the phosphine acts as a direct nucleophile; again the rate correlates inversely with cone angle.

Unlike the corresponding iron acyls, complexes such as $[Mo(COMe)(CO)_2(PPh_3)Cp]$ cannot usefully be deprotonated with Bu^nLi to enolate anions. Evidently loss of ketene occurs with formation of $[Mo(CO)_2(PPh_3)Cp]^-$.[491] Treatment of the acyl complex with ylids also gives the same anion as its acylphosphonium salt.[492] Deprotonation of $[Mo(COR^1)(CO)_2(PPh_2H)Cp]$ at low temperature occurs at the phosphine, producing $[Mo(COR^1)(CO)_2(PPh_2R^2)Cp]$ on alkylation with R^2I (R^1, R^2 = Me, Et); if the same reaction is carried out at room temperature, migration of the acyl group to phosphorus occurs to yield $[Mo(R^2)(CO)_2(PPh_2COR^1)Cp]$.[71]

The acyl complex $[Mo(COMe)(CO)_2(PPh_3)Cp]$ undergoes hydrosilylation with Ph_2SiH_2 in the presence of Wilkinson's catalyst to give the α-siloxyalkyl species $[Mo(CHMeOSiPh_2H)(CO)_2(PPh_3)Cp]$.[493,494] Alkylation of the acyl complexes with Et_3O^+ or similar reagents occurs at the acyl carbonyl oxygen to give cationic carbenes.[495]

The reaction of $[MoMe(CO)_3Cp]$ with $LiBHEt_3$ in THF at room temperature produces the η^2-acetaldehyde complex $[Mo(CO)_2(\eta^2\text{-}MeCHO)Cp]^-$. Contrary to an earlier report,[496] low temperature IR and NMR monitoring reveals the sequential formation of intermediate formyl and hydrido acyl species *trans*-$[MoMe(CHO)(CO)_2Cp]^-$ and *trans*-$[MoH(COMe)(CO)_2Cp]^-$.[497,498] The tungsten analogue reacts similarly except that the reaction does not proceed beyond the hydrido–acyl complex, which can be protonated or alkylated to give carbene complexes.[499] Isolable formyl species $[Mo(CHO)(CO)_2(L)Cp]$ ($L = P(OPh)_3$, etc.) can be made by addition of hydride to $[Mo(CO)_3(L)Cp]^+$; they can be converted into alkoxymethyl ($-CH_2OMe$) or halomethyl complexes and carbenes.[249,500,501]

The photochemical reaction of $[WMe(CO)_3Cp]$ with acetylene in pentane gives three complexes, the yields of which are dependent on the reaction temperature. At −30 °C the major product is $[W(COMe)(CO)(\eta\text{-}C_2H_2)Cp]$, whereas at 15 °C $[WMe(CO)(\eta\text{-}C_2H_2)Cp]$ and the alkenylketone complex $[\overline{W(CH{=}CHCO}Me)(CO)_2Cp]$ are formed; they can be interconverted as shown in Scheme 13, which also shows some of their further reactions.[502–9]

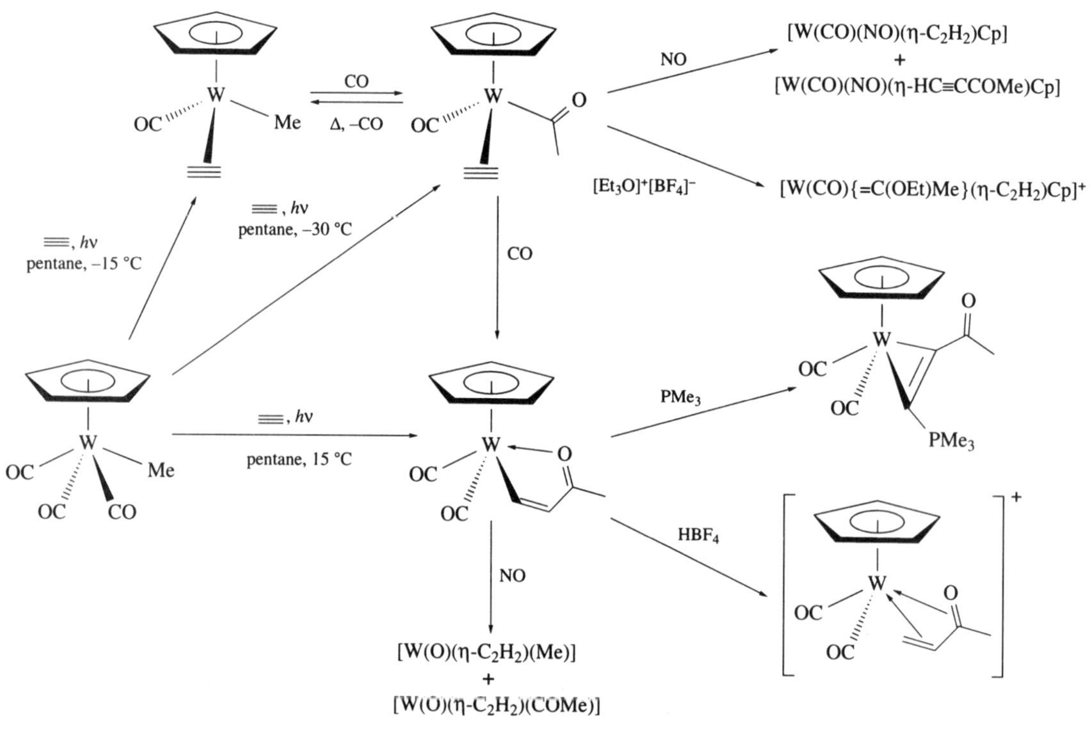

Scheme 13

Similar complexes can be made for other alkynes;[510,511] their chemistry has been reviewed.[512] In general the products from $[MR(CO)_3Cp]$ (M = Cr, Mo, W) with alkynes $R^1C{\equiv}CR^2$ are the alkenylketone compounds $[\overline{M(CR^1{=}CR^2CO}R)(CO)_2Cp]$ (**9a**).[513,514] With unsymmetrical alkynes such as PhC≡CH, two regioisomers are possible; their relative ratio depends on M as well as the R groups.[515] The α-carbon of the metallacycle shows a degree of carbene-like character (low field ^{13}C resonance, short M–C bond) due to the contribution of the resonance structure (**9b**). Insertion of alkynes into the enolate complexes $[M(CH_2COR)(CO)_3Cp]$ produces related compounds.[401]

Several other routes to alkenylketone complexes have been described. The reaction of $[MR(CO)_3Cp]$ with the ylid $Me_3P{=}CH_2$ gives $[PMe_4][\overline{M(CO{=}CHCO}R)(CO)(PMe_3)Cp]$, which can be silylated at the acylate oxygen with TMS-Cl to give an alkenylketone complex.[516] Alkenylketone complexes can also be prepared from the cyclopropylcarbynes $[Mo({\equiv}C\overline{CHCH_2CH_2})(CO)(L)Cp]$ ($L = P(OMe)_3$) by deprotonation and acylation to give $[Mo\{{\equiv}C\overline{C(COR)CH_2CH_2}\}(CO)(L)Cp]$ followed by ring opening by addition of Cl^-. This sequence produces chloroethyl-substituted complexes $[\overline{Mo\{CH{=}C(CH_2CH_2Cl)CO}R\}(CO)(L)Cp]$.[517] Sequential addition of Me_2CuLi and $KBHBu^s_3$ to $[Mo(CO)_2(\eta\text{-}C_2Me_2)Cp]^+$ gives the alkenylketone complex $[\overline{Mo(CMe{=}CMeCO}CMe{=}CHMe)(CO)_2Cp]$.[518]

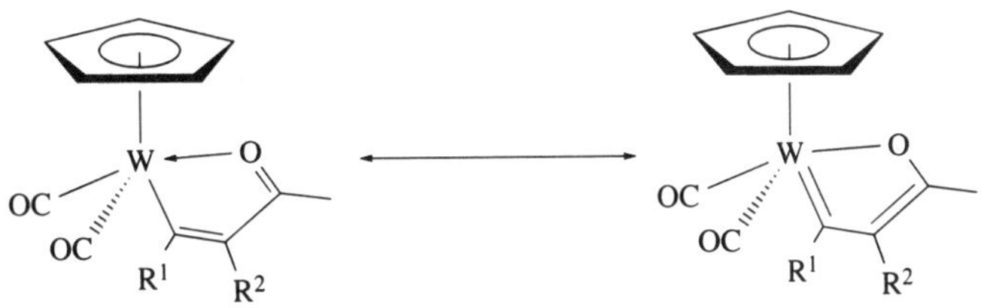

(9a) **(9b)**

The reaction of [WH(CO)$_3$Cp] with alkynes can also give alkenylketone species, as well as η^3-allyl complexes.[519] However, if the alkyne is F$_3$CC≡CCF$_3$ insertion occurs to give the interconvertible σ-vinyl and η^3-acryloyl species [W(CR=CHR)(CO)$_3$Cp] and [W(CO)$_2$(η^3-COCR=CHR)Cp] (R = CF$_3$) (Scheme 14); the latter undergoes substitution reactions with phosphines with retention of the acryloyl moiety intact.[520] Acryloyl complexes also result with the alkyne HC≡CNEt$_2$, and the unsubstituted compound [Mo(CO)(PPh$_3$)(η^3-COCH=CH$_2$)Cp] has been prepared by a different route.[521,522]

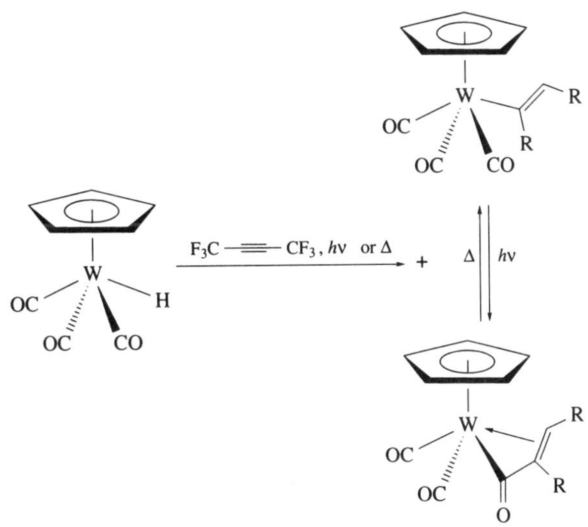

Scheme 14

Reaction of [MMe(CO)$_3$Cp] (M = Mo, W) with [W(≡CTol)(CO)$_2$Cp] affords [MW(CO)$_4$(μ-CTolCOMe)Cp$_2$] through coupling of the carbyne with the acetyl ligand.[523]

The reactions of [MMe(CO)$_3$Cp] with dienes have been thoroughly explored by Kreiter;[524] they are dealt with in detail in Chapter 6, Section 4.3.

It is well known that treatment of [M(CO)$_3$Cp]$^-$ (M = Mo, W) with allyl halides produces initially σ-allyl tricarbonyl complexes, which can then be decarbonylated to the corresponding η^3-allyls by photolysis; an alternative procedure uses Me$_3$NO to effect this conversion.[525] With M = Cr, electron transfer occurs and [Cr$_2$(CO)$_6$Cp$_2$] is formed.[526] If the reaction of [Mo(CO)$_3$Cp]$^-$ with allyl bromide is carried out in the presence of PPh$_3$, the insertion product [Mo(COCH$_2$CH=CH$_2$)(CO)$_2$(PPh$_3$)Cp] is obtained; this insertion is reversible, as heating causes loss of PPh$_3$ and again gives [Mo(CO)$_2$(η^3-CH$_2$CHCH$_2$)Cp]. Propargyl halides did not undergo a similar insertion.[527] In an interesting related reaction of the carbonyl anion with epibromohydrin, BrCH$_2\overline{\text{CHCH}_2\text{O}}$, in the presence of PPh$_3$, the 2-furyl complex [Mo($\overline{\text{C=CHCH=CHO}}$)(CO)$_2$(PPh$_3$)Cp] was formed, whereas in the absence of PPh$_3$ the product was [Mo(CO)$_2$(σ,η^2-COOCH$_2$CH=CH$_2$)Cp].[528]

In the same way as for the allyl complexes, σ-pentadienyl and hexadienyl species [Mo(CH$_2$CH=CHCH=CHR)(CO)$_3$Cp] (R = H, Me) have been made and converted to η^3 and η^5-complexes.[529] They also undergo cycloaddition reactions with TCNE.[530,531] The reaction of hexadienyl iron tricarbonyl cation with [Mo(CO)$_3$Cp]$^-$ gives complex (**10**), in which the hexadienyl ligand is σ-bound to Mo and η^4-bound to Fe.[532] Attempts to make the related cyclohexadienyl complex failed for molybdenum due to interference by electron transfer processes, but its tungsten analogue has been prepared.[533]

The propargyl complexes [M(CH$_2$C≡CR)(CO)$_3$Cp] are readily prepared from propargyl halides and the carbonylmetalate anions. The complex [W(CH$_2$C≡CH)(CO)$_3$Cp] readily isomerizes to the allenyl species [W(CH=C=CH$_2$)(CO)$_3$Cp] on dissolution in benzene or heating in MeCN.[534] The two isomers

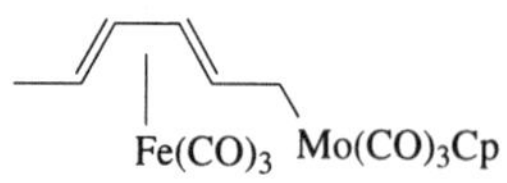

(10)

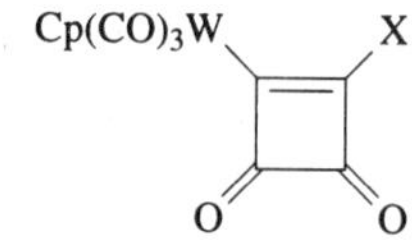

(11a) X = Cl
(11b) X = W(CO)$_3$Cp

show different reactivity towards nucleophiles, undergoing attack at the β- and α-carbons, respectively (Scheme 15).[535,536] The propargyl complexes also take part in [3 + 2] cycloaddition reactions with TCNE, with *N*-sulfinylsulfonamides, with RCHO in the presence of BF_3 and with sources of S_2O,[537–40] and react with metal carbonyls to give clusters.[541–4]

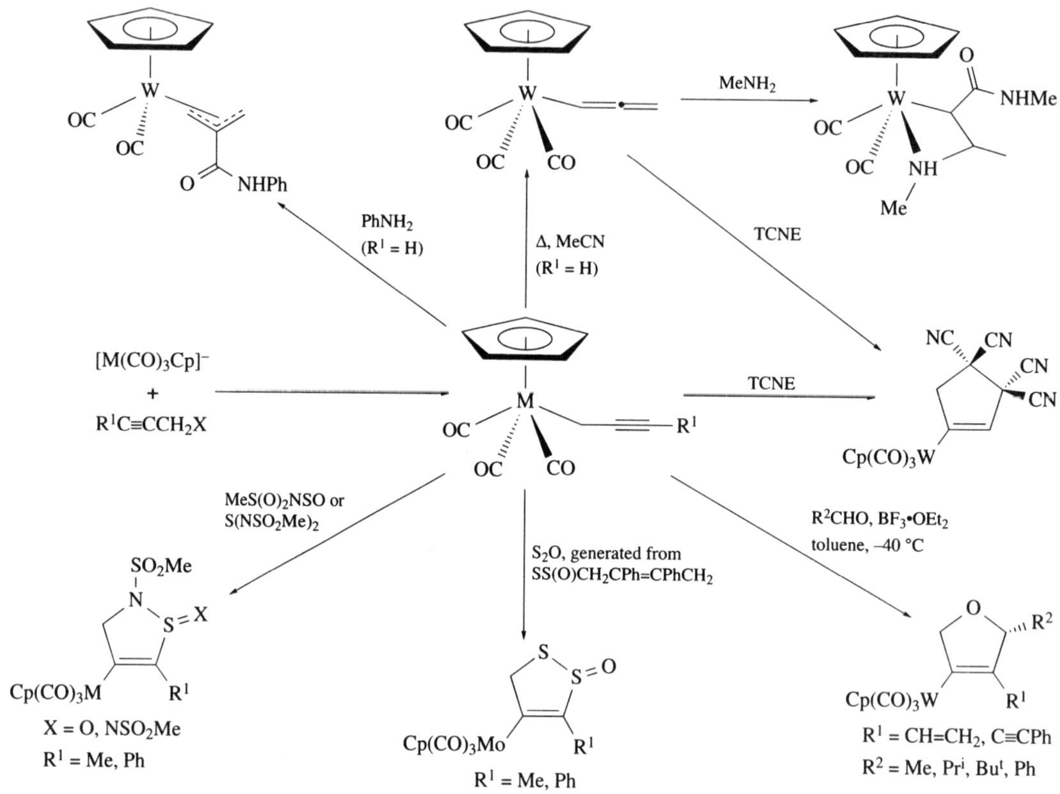

Scheme 15

Several routes to σ-vinyl (alkenyl) complexes have been described above. Direct synthesis from vinyl halides and $[M(CO)_3Cp]^-$ is usually unsuccessful, as the anion is not a strong enough nucleophile for the unreactive halides. Two successful examples involve dichlorocyclobutenedione, which gives mono- and disubstituted complexes **(11a)** and **(11b)**, and perfluroro-1-methylcyclopentene; however, an addition–elimination mechanism is thought to operate for the latter.[545,546] Addition of H^- to $[MoL_2(\eta$-$HC{\equiv}CBu^t)Cp]^+$ [L = $P(OMe)_3$] in the presence of L affords $[Mo(CH{=}CHBu^t)L_3Cp]$, which on thermolysis rearranges to the carbyne $[Mo({\equiv}CCH_2Bu^t)L_2Cp]$ (see Chapter 5, this volume).[547,548]

Similar restrictions of halide reactivity apply to the synthesis of aryl complexes $[M(aryl)(CO)_3Cp]$; often these are prepared by decarbonylation of the corresponding acyls. However, C_5F_5N does react with $[W(CO)_3Cp]^-$ to give a quantitative yield of the monosubstituted pyridyl, and also reacts with perfluorotriazines. In these reactions the degree of ion pairing with the counterion is important, with contact ion pairs being more reactive.[549,550]

Treatment of the well-known carbyne complex $[W({\equiv}CTol)(CO)_2Cp]$ with PMe_3 produces the η^2-ketenyl species $[W(CO)(PMe_3)(\eta^2$-$CTol{=}C{=}O)Cp]$, which can be alkylated with Et_3O^+ to the alkyne complex $[W(CO)(PMe_3)(\eta$-$TolC{\equiv}COEt)Cp]^+$.[551] Addition of further ligands such as alkynes, CN^- or CO to the η^2-ketenyls forms the corresponding η^1-ketenyls $[W(\eta^1$-$CTol{=}C{=}O)(CO)(PMe_3)(L)Cp]$ (Scheme 16).[552–5] Related η^1-ketenyls $[W(\eta^1$-$CTol{=}C{=}O)(CO)(L_2)Cp]$ are formed by treating the carbyne complex with bidentate phosphines.[556,557] The reaction of the dicarbonyl $[W(\eta^1$-CTol-

$=C=O)(CO)_2(PMe_3)Cp]$ with further CO produces CO_2 and the acetylides $[W(C{\equiv}C\,Tol)(CO)_2(PMe_3)Cp]$, while H_2O, MeOH or Me_2NH add to the ketenyl ligand to give $[W(CHTolCONu)(CO)_2(PMe_3)Cp]$ (Nu = OH, OMe, NMe_2).[558–60]

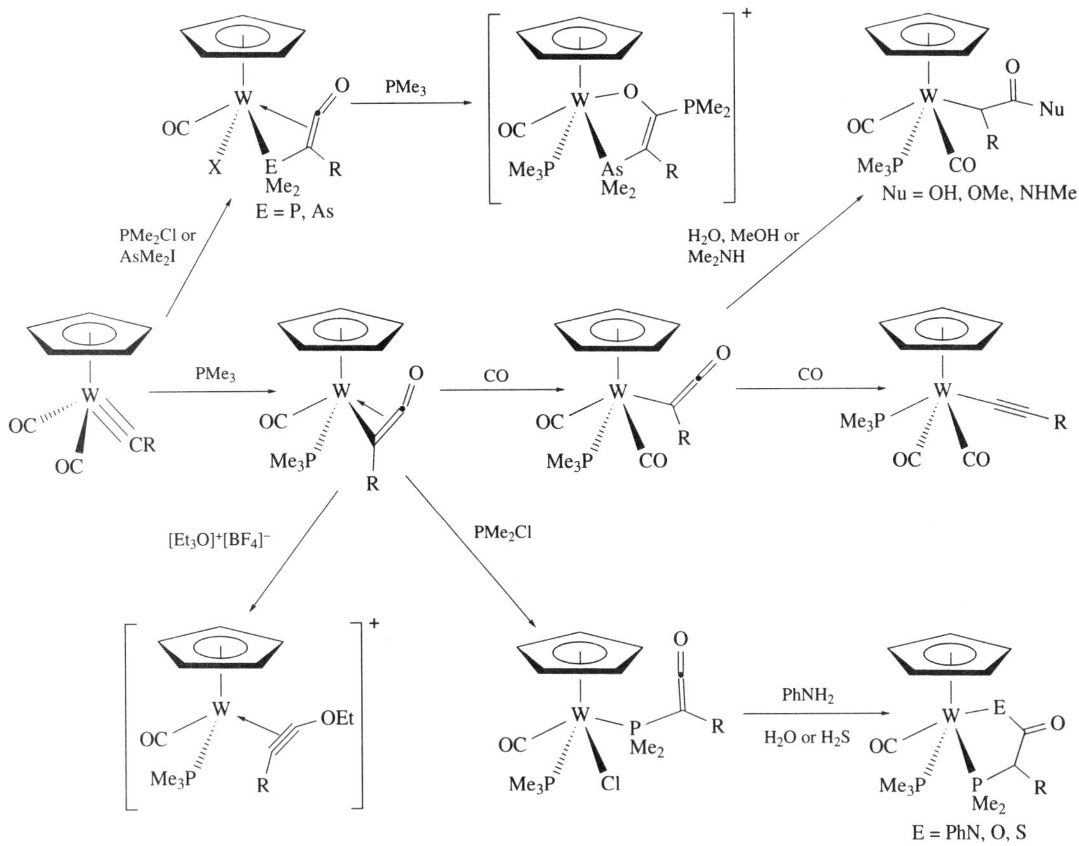

Scheme 16

The reaction of the carbyne complex with PMe_2Cl yields $[W(Cl)(CO)(\eta^2\text{-}PMe_2CTol{=}C{=}O)Cp]$, in which the ketenyl ligand has migrated to phosphorus; $AsMe_2I$ gives an analogous product.[561] A similar oxidative addition occurs on treating $[W(CO)(PMe_3)(\eta^2\text{-}CMe{=}C{=}O)Cp]$ with PMe_2Cl to give $[W(Cl)(CO)(PMe_3)(PMe_2CMe{=}C{=}O)Cp]$.[562,563] An isolable cationic intermediate is involved and if the nucleophilic Cl^- counterion is exchanged so that it cannot recoordinate, the carbene $[\overline{W(CO)(PMe_3)(\eta^2\text{-}{=}CMePPh_2)Cp}]^+$ is formed as its PF_6 salt.[564] Addition of PMe_3 to $[W(I)(CO)(PMe_3)(AsMe_2CTol{=}C{=}O)Cp]$ produces the metallacycle $[\overline{W(CO)(PMe_3)\{AsMe_2CTol{=}C(PMe_3)O}\}Cp]^+$.[565–7]

The acetylide (alkynyl) complexes $[M(C{\equiv}CR)(CO)_3Cp]$ are readily prepared by reaction of $[M(Cl)(CO)_3Cp]$ with $RC{\equiv}CH$ in NEt_3 in the presence of a CuI catalyst.[568,569] Alternatively the lithium acetylide can be used: $[Mo(C{\equiv}CH)(CO)_3Cp]$ was prepared by treatment of the halide complex with $LiC{\equiv}CH$, though with the tungsten analogue disproportionation of the acetylide produced $[Cp(CO)_3WC{\equiv}CW(CO)_3Cp]$ instead.[570] The butadiynyl complex $[Cp(CO)_3MC{\equiv}CC{\equiv}CFe(CO)_2Cp]$ can also be prepared from $[Fe(C{\equiv}CC{\equiv}CH)(CO)_2Cp]$ in this way.[571] A further method consists of the reaction of $[M(CO)_3Cp]^-$ with $ClC{\equiv}CCl$; $[M(C{\equiv}CCl)(CO)_3Cp]$ is formed for M = molybdenum or tungsten, but the major product for chromium is $[Cp(CO)_3CrC{\equiv}CCr(CO)_3Cp]$.[572] Another route to $[Cp(CO)_3WC{\equiv}CW(CO)_3Cp]$ has been described above.[447] Deprotonation of cationic terminal alkyne complexes such as $[MoL_2(\eta\text{-}HC{\equiv}CBu^t)Cp]^+$ (L = $P(OMe)_3$, PPh_2Me)) or $[W(CO)(\eta\text{-}Bu^tC{\equiv}CH)_2Cp]^+$ also gives neutral acetylide complexes.[240,573] Conversely protonation of $[M(C{\equiv}CPh)(CO)_3Cp]$ with HBF_4 in the presence of alkynes produces the bis(alkyne) species $[M(CO)(\eta\text{-}PhC{\equiv}CH)(\eta\text{-alkyne})Cp]^+$.[574] If no additional alkyne is present, a second molecule of the acetylide complex becomes coordinated.[575]

Thermolysis of $[Mo(C{\equiv}CPh)(CO)_3Cp]$ causes coupling of the acetylide ligands to give the alkyne complex $[Mo_2(CO)_4(\mu\text{-}PhC{\equiv}CC{\equiv}CPh)Cp_2]$.[576] Coupling of the acetylides in $[Cr(C{\equiv}CPh)(CO)_3Cp]$ can also be induced by reduction with sodium–potassium alloy, which on work up yields the linked carbyne complex $[Cp(CO)_2Cr\{{\equiv}CC(Ph){=}C(Ph)C{\equiv}\}Cr(CO)_2Cp]$.[577] Other reactions, shown in Scheme 17, include cycloadditions with nitrile oxides and TCNE.[578–80] More recently, larger clusters have been formed by treatment of the acetylide complexes with a variety of metal carbonyls.[581–5]

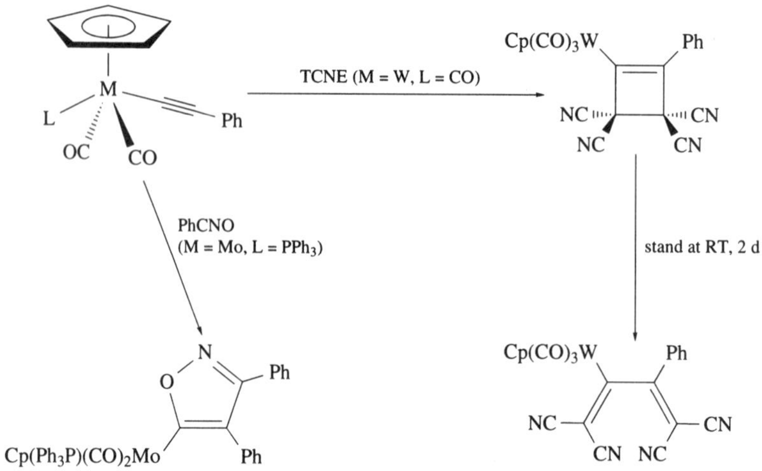

Scheme 17

7.3 CYCLOPENTADIENYL METAL NITROSYL COMPLEXES

Developments in this area over the 1980s have been largely due to Legzdins and co-workers, who have also produced a general review of organometallic nitrosyl chemistry.[586]

7.3.1 [M(CO)$_2$(NO)Cp] and Related Compounds

Treatment of the [M(CO)$_3$Cp]$^-$ anions with Diazald (p-MeC$_6$H$_4$SO$_2$N(NO)Me) in THF is the best route to [M(CO)$_2$(NO)Cp], with yields of 60%, 93% and 84% attainable for chromium, molybdenum and tungsten, respectively.[587] Other reactions which produce [Mo(CO)$_2$(NO)Cp] include irradiation of [Mo$_2$(CO)$_6$Cp$_2$] with nitrite or with [Co(NO)$_2$X]$_2$.[588,589] The photoelectron spectra and ^{14}N NMR spectra of [Cr(CO)$_2$(NO)Cp] and its thionitrosyl analogue have been recorded.[590–2] Chemical or electrochemical reduction of the chromium and molybdenum complexes gives a radical anion in which the electron is largely localized on the nitrosyl ligand.[593,594]

Carbonyl loss is the primary photoprocess on photolysis of [Cr(CO)$_2$(NO)Cp] either in a matrix or in solution,[595] though for the tungsten analogue formation of an isocyanate species by intramolecular attack of CO on NO can also be observed.[596] In coordinating solvents, useful solvent-stabilized intermediates can be prepared photochemically and then used in subsequent reactions, for example treatment of [Cr(CO)(NO)(THF)Cp] with N$_2$CPh$_2$ gave the carbene complex [Cr(CO)(NO)(=CPh$_2$)Cp].[597] Photochemical CO substitution was also employed in the reaction of [Cr(CO)$_2$(NO)Cp] with [W(CO)$_2$(≡CTol)Cp] and in the preparation of [Mo(CO)(NO)(η-alkene)Cp*].[598,599] Phosphine substitution in [W(CO)$_2$(NO)Cp] with PMe$_3$ occurs rapidly at room temperature by an associative mechanism; in neat PMe$_3$ the ring-slipped complex [W(CO)$_2$-(NO)(PMe$_3$)$_2$(η^1-C$_5$H$_5$)] can be isolated.[600]

The CO ligands in [M(CO)$_2$(NO)Cp] are also susceptible to nucleophilic attack, for example by ylids or by organolithium reagents.[601,602] Alkylation in the latter case gives Fischer carbenes such as [Mo(CO)(NO){=C(OEt)Me}Cp].[603] Reaction of [M(CO)$_2$(NO)Cp] (M = Mo, W) with LiER$^1_3$ (E = Ge, Sn) followed by alkyl halides R^{2}X gives chiral, configurationally stable square based pyramidal species [M(R^2)(CO)(NO)(ER$^1_3$)Cp]; if the ER$^1_3$ group is optically active GeMePh(1-naphthyl), the diastereomers can be separated by crystallization.[604–6]

A range of derivatives with substituents on the Cp ring have been prepared; some are listed in Table 5 and others have been made by subsequent modification of C$_5$H$_4$COOH or other functionalized ligands.[607–9] Studies on the carbonium ions [Cr(CO)$_2$(NO)(η-C$_5$H$_4$CR^{1}R^2)]$^+$ suggest that the so-called cynichrodene unit is an electron releaser comparable to Mn(CO)$_3$Cp.[610,611] Complexes bearing vinyl groups on the Cp ligand have been homo- and copolymerized.[612,613] Treatment of [Cr(CO)$_2$(NO)(η-C$_5$H$_4$COOH)] with PCl$_5$ gives the corresponding acyl chloride;[607] Friedel–Crafts acylation of ferrocene with this reagent followed by modification of the products has given rise to a wide range of linked species.[614–25]

7.3.2 [MX(NO)$_2$Cp] (X = Halide) and Related Compounds

Treatment of [Cr(CO)$_2$(NO)Cp] with halogens affords [CrX(NO)$_2$Cp]; in the case of iodine the paramagnetic intermediate [Cr(I)(CO)(NO)Cp]$_2$ can be isolated, and reacts with phosphines to form 17-electron [Cr(I)(L)(NO)Cp].[626] The conversion of [Cr(CO)$_2$(NO)Cp] to [Cr(Cl)(NO)$_2$Cp] can also be effected by HCl and isoamyl nitrite.[627] Moderate yields of [W(Cl)(NO)$_2$Cp] can be obtained by treating [WCl$_2$(NO)$_2$L$_2$] (L = THF, OEt$_2$, NCMe) with Bu$_3$SnC$_5$H$_5$.[628] The conventional route to [M(Cl)(NO)$_2$Cp] (M = Mo, W) involves the reaction of [M(CO)$_2$(NO)Cp] with NOCl; in some cases NO$^+$ followed by a source of Cl$^-$ is preferable.[191]

Photoelectron spectra of [M(Cl)(NO)$_2$Cp] show that backbonding to the NO groups increases down the group.[629,630] Chemical or electrochemical reduction of [MX(NO)$_2$Cp] (X = Cl, H, alkyl) gives isolable radical anions in which the extra electron resides on the NO ligands.[631,632] Reduction of [M(NO)$_2$(PPh$_3$)Cp]$^+$ affords the stable radical {W(NO)$_2$(PPh$_3$)Cp}; the electron probably still resides on the NO ligands, even though they are both virtually linear.[632,633]

Replacement of the halide ligand in [MX(NO)$_2$Cp] is readily effected by treatment with anionic nucleophiles. New examples synthesized include compounds with X = CF$_3$ (M = Cr, Mo),[634] ferrocenyl (M = W),[635] and various thiolate ligands.[636,637] Reaction of [Cr(Cl)(NO)$_2$Cp*] with NaOPr^i, however, unexpectedly gave [Cr(NO)(OPr^i)$_2$Cp*], which is rendered relatively unreactive by virtue of $p\pi$ donation from the alkoxide ligands into an empty Cr *d*-orbital; a similar situation operates in [Cr(NO)(SPh)$_2$Cp].[638,639]

Treatment of [Cr(Cl)(NO)$_2$Cp] with diazomethane in the presence of copper powder affords the insertion product [Cr(CH$_2$Cl)(NO)$_2$Cp]; on halide abstraction with Ag$^+$ the methylene group undergoes intramolecular migration to the Cp ring.[640] Several electrophiles insert into the Cr–C bond of [Cr(Me)(NO)$_2$Cp]; in the case of NO the initial product, the formaldoxime complex [Cr(NO)$_2${N(=CH$_2$)OH}][PF$_6$], reacts with further methyl complex to form [{Cr(NO)$_2$Cp}$_2${μ-N(=CH$_2$)O}][PF$_6$] (Scheme 18).[641,642]

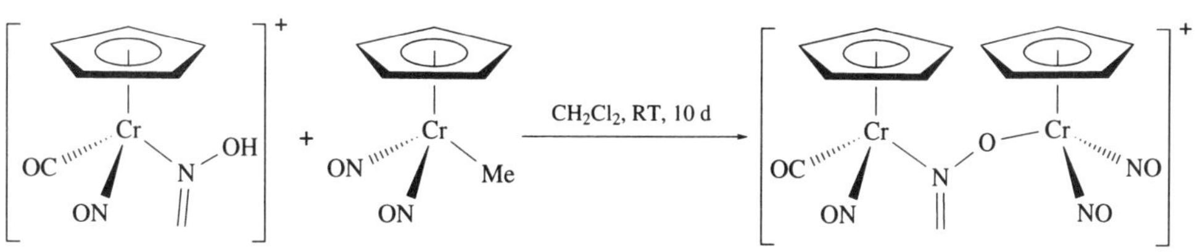

Scheme 18

Abstraction of halide from [MX(NO)$_2$Cp] (M = Cr, Mo, W) with AgBF$_4$ gives a loosely bound species [M(BF$_4$)(NO)$_2$Cp], analogous to the Beck complex [M(CO)$_3$(FBF$_3$)Cp], which behaves as the electrophilic M(NO)$_2$Cp$^+$ fragment in reactions, for example addition of phosphines gives [M(NO)$_2$(L)Cp][BF$_4$].[643,644] A similar reaction with AgAsF$_6$ in liquid SO$_2$ gave [Cr(NO)$_2$(SO$_2$)Cp][AsF$_6$]; replacement of the SO$_2$ ligand by NSF followed by a second halide abstraction yields the thionitrosyl complex [Cr(NO)$_2$(NS)Cp][AsF$_6$]$_2$.[645,646] Hydride abstraction from [MH(NO)$_2$Cp] gives the bridging hydride complex [{M(NO)$_2$Cp}$_2$(μ-H)]$^+$.[647]

The electronic structure of [Cr(NO)(μ-NO)Cp]$_2$ has been analysed in relation to isoelectronic iron and manganese carbonyl complexes.[648] It undergoes a two-electron oxidation to [Cr(NO)$_2$(NCMe)Cp]$^+$ and also a reversible one-electron reduction to a radical anion; chemical reduction gives only low yields of known [Cr$_2$(NO)$_2$(μ-NO)(μ-NH$_2$)Cp$_2$] and [Cr(NO)(μ-NH$_2$)Cp]$_2$, which can also be prepared by catalytic hydrogenation.[649,650]

7.3.3 [MX$_2$(NO)Cp]$_2$ (X = Halide) and Related Compounds

Treatment of [M(CO)$_2$(NO)Cp] (M = Mo, W) with halogens affords [MX$_2$(NO)Cp]$_n$; a more convenient new procedure for the chlorides utilizes PCl$_5$.[651–3] In the solid state most of these are halide-bridged dimers (though [WI$_2$(NO)Cp] exists as the 16-electron monomer[654]); in solution they behave as the solvated monomers. An extensive chemistry of these systems, shown in Scheme 19, has been developed.[653,655–71]

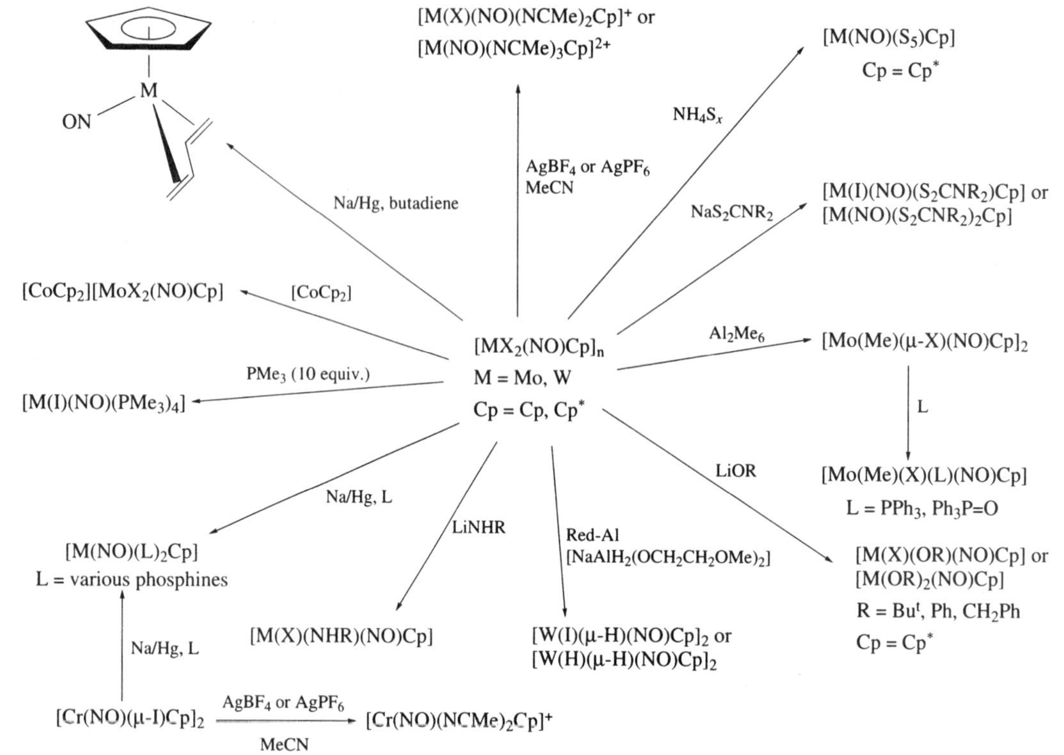

Scheme 19

Notable among these reactions are the formation of the stable *trans*-diene complexes [M(NO)(η^4-diene)Cp], and the synthesis of the 16-electron diorganyl species [MR_2(NO)Cp]. Calculations show that the LUMO of the latter is metal-centred and nonbonding, and consequently they undergo two reversible one-electron reductions and act as Lewis acids. The degree of Lewis acidity is reflected in the stability and reactivity of the complexes: the Cp*W dialkyl complexes are the most stable, whereas the tungsten diaryls are more reactive and the CpMo diaryls have yet to be isolated. Complexes in which R = CH_2Ph contain the unusual η^2-coordination of the benzyl group. The chemistry of these reactive compounds has been reviewed;[672] some important examples are shown in Scheme 20. A feature of these reactions is the facile insertion of a variety of molecules (CO, CS_2, CO_2, PhNCO, S_8) into one or both of the M–carbon bonds.[673–91]

7.4 CYCLOPENTADIENYL METAL COMPOUNDS WITH GROUP 15 ELEMENTS

7.4.1 Cyclopentadienyl Metal Amido and Imido Complexes

There appear to be only three examples of cyclopentadienyl chromium imido complexes. The reaction of azobenzene with [$CrCp_2$] affords [$Cr_3(\mu$-NPh$)_3(\mu_3$-NPh)Cp_3], which oxidizes on chromatogaphy, while the same reagent with [$Cr_2(\mu$-OBu$^t)_2Cp_2$] produces [Cr_2(OBu$^t)_2(\mu$-NPh$)_2Cp_2$]. Oxidation of this product gives [$Cr_2(O)_2(\mu$-NPh$)_2Cp_2$].[692]

The reaction of [Mo(NBu$^t)_2Cl_2$] with LiCp (or LiCp*) forms [Mo(Cl)(NBu$^t)_2$Cp] (**12**); the tungsten analogues with NBu^t or N(2,6-C_6H_3Pr$^i_2$) ligands can be made similarly.[693–5] Protonolysis of one imido group occurs with HCl to give [MCl_3(NBu^t)Cp], while alkylation with RLi affords [MR(NBu$^t)_2$Cp]. Conversion into a range of oxo species, including [NBu_4][MO_3Cp*] and [M_2O_4Cp*$_2$], is also possible.[694–6] Further routes into tungsten imido chemistry are provided by [W(NPh)R_3Cl] (R = CH_2Bu^t) which can be converted sequentially into [W(R)(NPh)(=CHBu^t)Cp], [WR_2(Cl)(NPh)Cp], and [W(Cl)(NPh)(=CHBu^t)Cp],[697] and by the oxo–imido exchange reaction of [W(CH_2-TMS)(O)$_2$Cp*] with PhNCO to give [W(CH_2-TMS)(NPh)$_2$Cp*]; in the latter case, further incorporation of PhNCO occurs to give a metallacycle.[698] The reaction of [WMe_4Cp*]$^+$ with $PhNH_2$ also leads to imido complexes (see below).

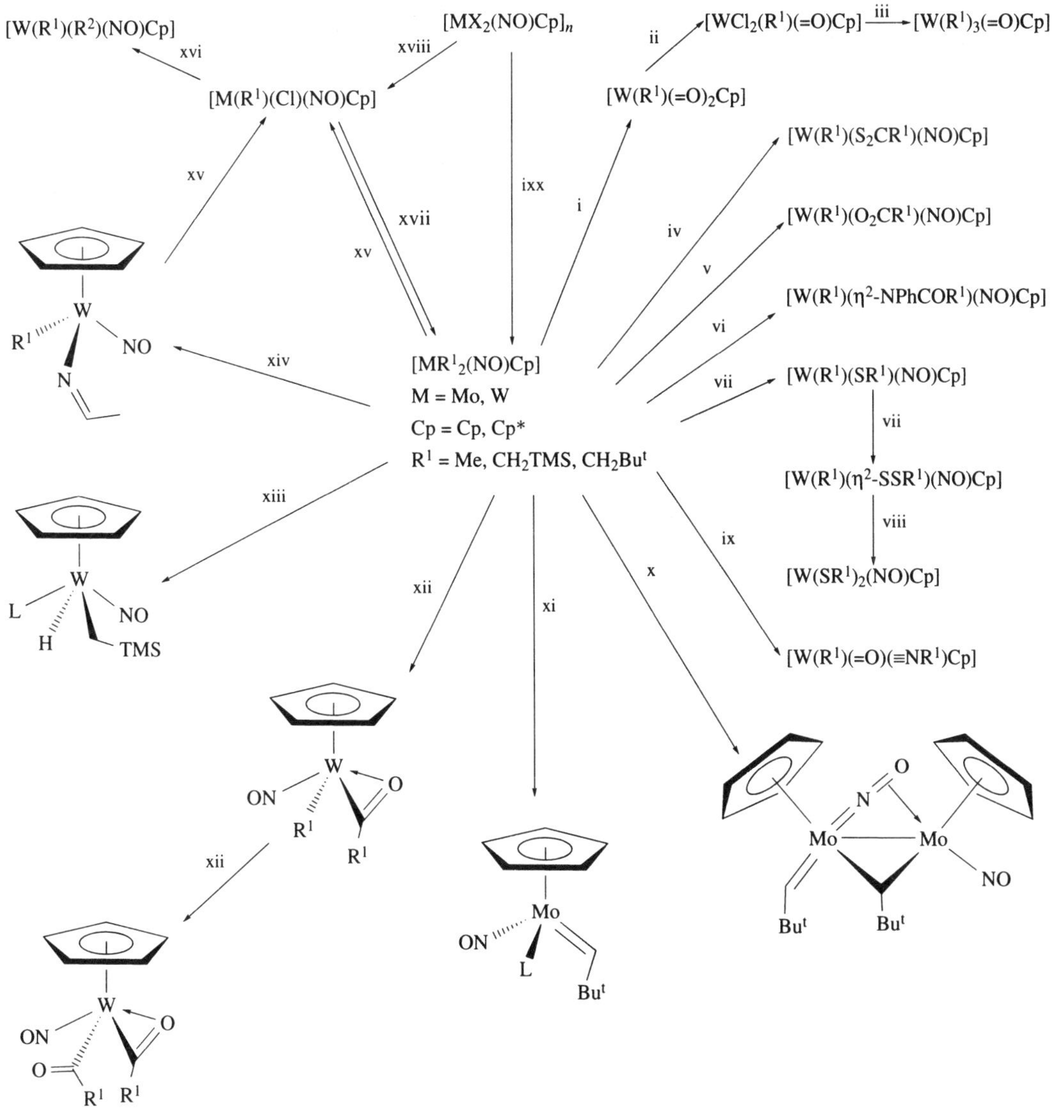

i, O_2; ii, PCl_5, TMS-Cl or HCl; iii, R^1MgBr; iv, CS_2; v, CO_2; vi, PhNCO; vii, S_8; viii, Δ, ix, H_2O vapour (R^1 = *o*-Tol); x, stir at RT, $-CMe_4$; xi, stir at RT with L (L = phosphines); xii, CO; xiii, H_2, L (R^1 = CH_2TMS, L = phosphines); xiv, H_2, MeCN; xv, HCl; xvi, $R^2{}_2Mg{\bullet}dioxan$; xvii, R^1MgCl or $R^1{}_2Mg{\bullet}dioxan$; xviii, $R^1{}_2Mg{\bullet}dioxan$ (0.5 equiv.); ixx, $R^1{}_2Mg{\bullet}dioxan$ (1 equiv.)

Scheme 20

Green and co-workers have recently reported the synthesis of [$MCl_2(NR)Cp$] (R = Mo, W; R = Bu^t, Pr, Ph) and from it a number of derivatives of the type [M(Cl)(NR)(L)Cp], where L = η-C_2H_4 or η-C_2Me_2.[699] Photolysis of the ethylene complex with NaCp produces [M(NBu^t)Cp_2], which, if the imido ligand is a four-electron donor, is nominally a 20-electron complex. A similar reaction sequence with NaC_9H_7 gave the indenyl complex [Mo(NBu^t)(η-$C_5H_4Pr^i$)(η^3-C_9H_7)] (**13**) in which the indenyl ligand is coordinated in a ring-slipped η^3 mode due to the strong π-donation of the imido group.[700]

Starting from [WMe_4Cp^*]$^+$, Schrock and co-workers have synthesized a range of complexes with amine, amido, imido and nitrido ligands (Scheme 21).[701] The WMe_3Cp^* unit in these complexes is remarkably inert to protonation and deprotonation. Similarly a number of species with hydrazine, hydrazide, diazene and dinitrogen ligands have been made, most of which can be reduced to ammonia (Scheme 22).[702–8] Molybdenum analogues of some of these species also exist, but they tend to be less stable than the corresponding tungsten derivatives. A rare example of a heteronuclear dinitrogen complex, [$Cp^*WMe_3(\mu$-$N_2)MoMe_3Cp^*$], has also been prepared.[709]

The most convenient route to dinuclear molybdenum imido complexes consists of the reaction of [$Mo_2(CO)_nCp_2$] (n = 4 or 6) with $ArNO_2$ or ArNO, which produces predominantly [$Mo_2(O)_2(\mu$-O)(μ-

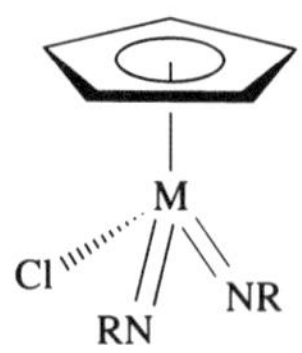

(12)

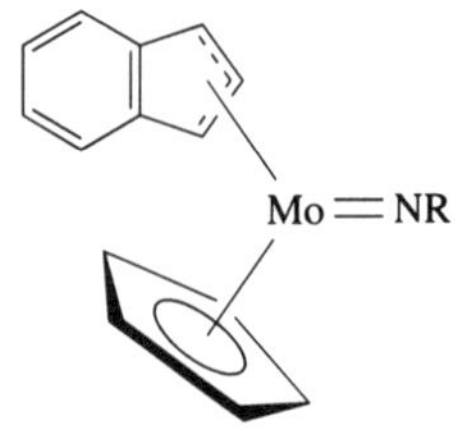

(13)

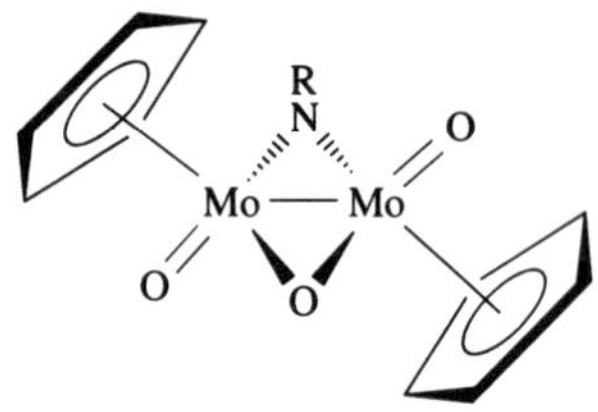

(14)

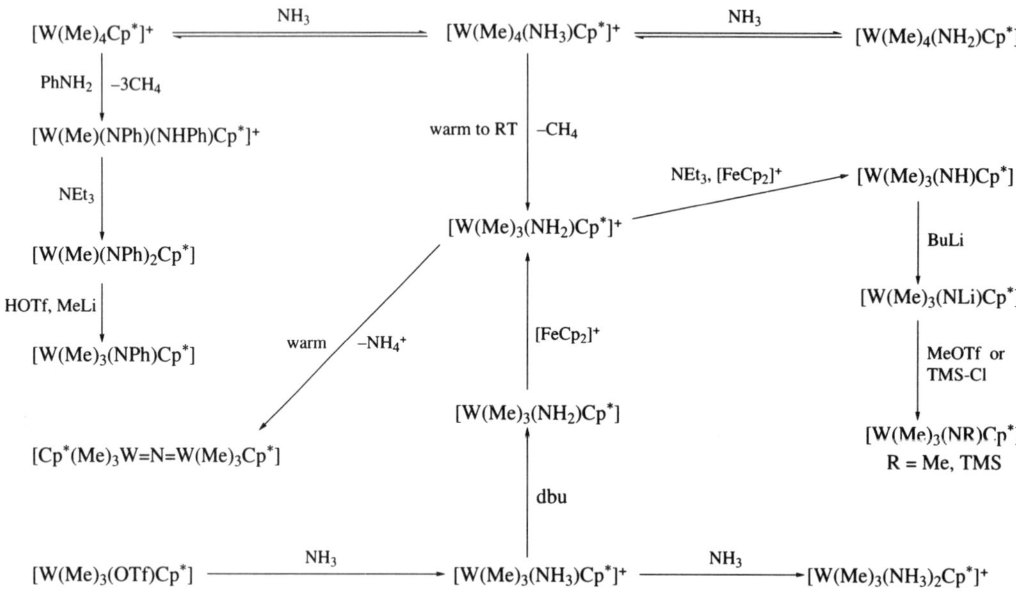

Scheme 21

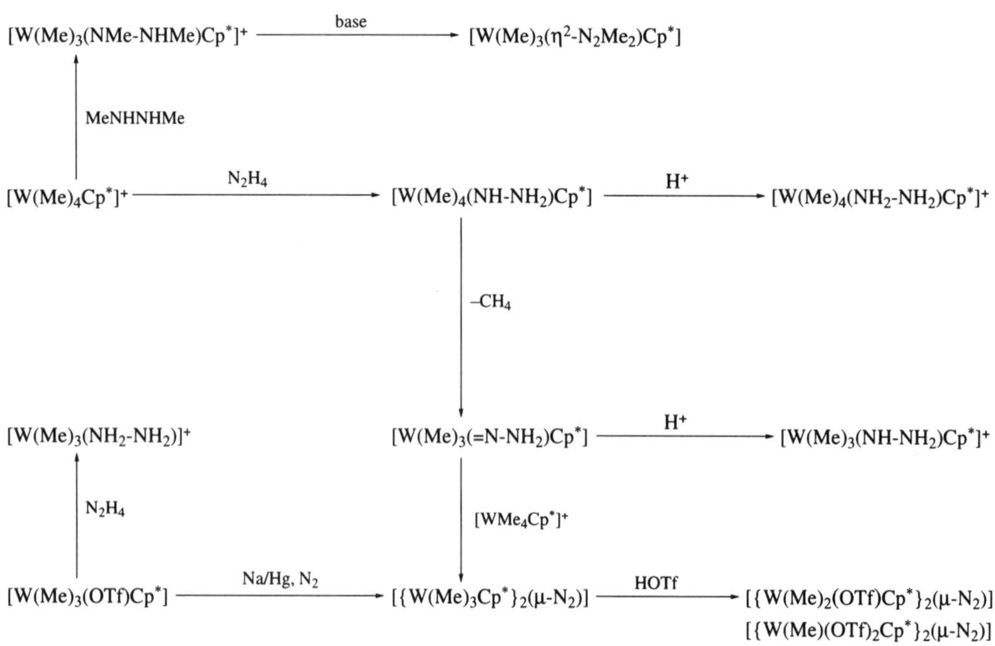

Scheme 22

NAr)Cp$_2$] (**14**).[710–12] The oxo ligands in this, and in [Mo$_2$(O)$_2$(μ-O)$_2$Cp$_2$], can be replaced with imido groups by heating with ArNCO and eliminating CO_2.[713–17] The terminal imido ligands in [Mo$_2$(NAr)$_2$(μ-NAr)$_2$Cp$_2$] exchange with Ar'NCO, but the bridging ones do not.[718] Hydrolysis of the terminal imido groups allows the preparation of other mixed oxo–imido species [Mo$_2$(O)$_2$(μ-NAr)$_2$Cp$_2$] and

$[Mo_2(O)(NAr)(\mu\text{-}NAr)_2Cp_2]$; replacement of the oxo group by sulfide is also possible, giving $[Mo_2(O)(S)(\mu\text{-}O)(\mu\text{-}NAr)Cp_2]$ and $[Mo_2(S)(NAr)(\mu\text{-}NAr)_2Cp_2]$.[713,716] $[Mo_2(O)_2(\mu\text{-}NPh)_2Cp_2]$ can also be made by oxidation of $[Mo_2(NPh)_2(\mu\text{-}PhNNPh)Cp_2]$, itself produced by UV irradiation of $[Mo_2(CO)_6Cp_2]$ with azobenzene.[719] Unlike $[Mo_2(O)_4Cp_2]$, which has a folded structure, $[Mo_2(NPh)_2(\mu\text{-}NPh)_2Cp_2]$ and the mixed ligand species have a planar core; this is thought to be due to electronic factors, and electrochemistry shows that the order of effectiveness of π-donation is NR > S > O in this system.[720]

The reaction of $[Mo(CO)_3Cp]^-$ with $[ArN_2]^+$ produces the diazonium complexes $[Mo(CO)_2(NNAr)Cp]$; with Ar = p-C_6H_4F, the minor product $[MoCl(NNAr)_2Cp]$ was also isolated.[721] The reactions of $[Mo(CO)_2(NNAr)Cp]$ (Ar = p-Tol mainly) are shown in Scheme 23.[722–6] The related complex $[W(CO)_2(NNMe)Cp]$ has also been studied; protonation with HOTf occurs at the NMe group, and a $Cr(CO)_5$ unit can be attached at the same site.[727,728]

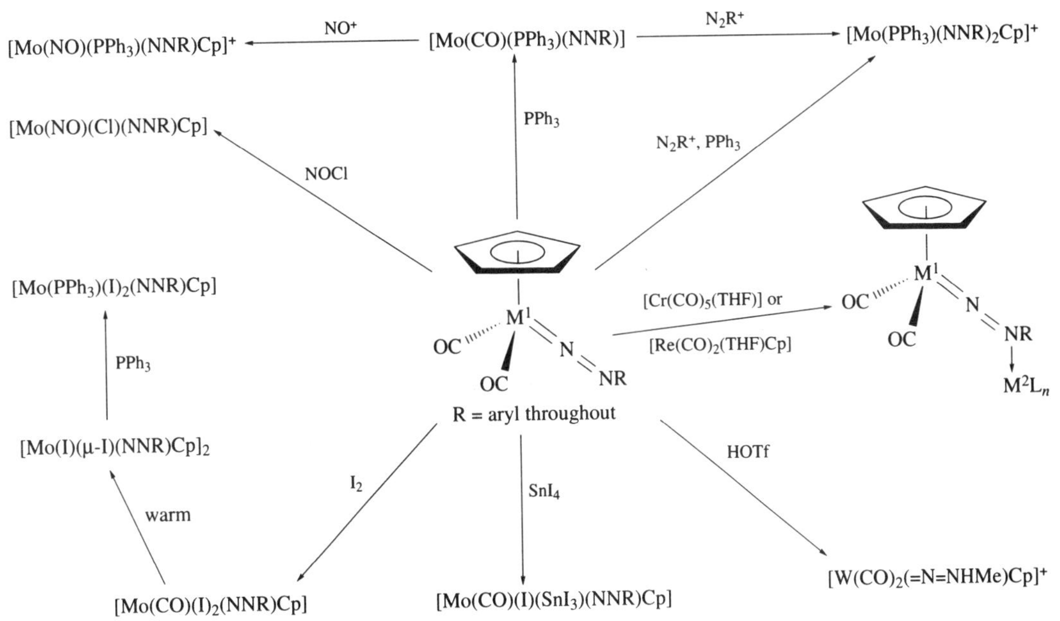

Scheme 23

7.4.2 Cyclopentadienyl Metal Complexes with N–N Chelate Ligands

Brunner, Bernal and co-workers have extensively studied the synthesis and conformational preferences of complexes of the types $[Mo(CO)_2(NN)Cp]^+$ and $[Mo(CO)_2(NN)Cp]$, where NN is a neutral or uninegative chelate, often a Schiff's base. For example, the complexes $[Mo(CO)_2(R^*N{=}CR\text{-}2\text{-}pyridyl)Cp]^+$ (R = H, Me, Ph, etc.; R* = a chiral group such as CHMePh) exist as diastereomers, which in some cases can be separated by fractional crystallization.[729–34] Reduction of these cations causes a change from N,N coordination to C,N in $[Mo(CO)_2\{R^*NHCH(2\text{-}pyridyl)\}Cp]$,[735,736] and reaction with PhMgBr unexpectedly gave the η^2-benzaldehyde complex $[Mo(CO)(\eta^2\text{-}PhCHO)\{MeNHCPh(2\text{-}pyridyl)\}Cp]$.[737] The conformational preferences of the amidine complexes $[Mo(CO)_2(R^*NCRNR^*)Cp]$ (R* is a chiral group, R = Ph or Me) have also received much attention; the diastereomers are configurationally stable in the solid state but undergo exchange at 70 °C.[738–42] In $[Mo(CO)_2(RNNNNNR)Cp]$ (R = p-Tol) the pentaazenido ligand acts as a 1,3-N,N chelate.[743] A number of related chelate complexes have been prepared; they are listed in Table 7.

7.4.3 Cyclopentadienyl Metal Complexes with Phosphorus, Arsenic, Antimony and Bismuth Atoms

A number of compounds with highly novel structures have been isolated as a result of intense activity in this area.[762–5] Two general synthetic approaches have been employed: thermolysis of the $[M_2(CO)_nCp_2]$ (n = 4 or 6) dimers with elemental pnictogens in high-boiling solvents, and nucleophilic reaction of metal carbonylate anions with the nonmetal halides. The thermolysis approach is

Table 7 Chelate complexes of the type [$M(CO)_2$(L–L)Cp].

Complex	*Chelate mode*	*Ref.*
[$Mo(CO)_2(PhN{=}NC_6H_4)Cp^*$]	*N, C*	744
[$Mo(CO)_2$(MeN=CPhNMeCPhR)Cp] (R = H, Ph)	*N, C*	745,746
[Mo(Cl)(CO)(MeN=CPhNMeCPh)Cp]	*N, C*	747
[$Mo(CO)_2(R^1R^2NCR^3R^4COO)Cp$] ($R^1$–$R^4$ = H, Me, etc); some also made for W	*N, O*	748
[$Mo(CO)_2(R^1N{=}CR^2COO)Cp$] ($R^1$, R^2 = H, alkyl, aryl, etc.)	*N, O*	749
[$Mo(CO)_2$(8-quinolinethiolate)Cp]	*N, S*	750
[$Mo(CO)_2(R^*NCRS)Cp$] (R = Me, R* = CHMePh, CHEtPh)	*N, S*	751,752
[$Mo(CO)_2\{Me_2NC(=NR)S\}Cp$] (R = Me, Ph)	*N, S*	753
[$Mo(CO)_2(RNC\{P(=O)Ph_2\}S)Cp$]	*N, S*	754
[$Mo(CO)_2\{Ph_2PC(=NR)S\}Cp$] (R = Me, Ph, CHMePh)	*P, S*	753,755,756
[$Mo(CO)_2\{PhP(OCH_2CH_2)_2N\}Cp$]	*P, O*	757
[$Mo(CO)_2\{RP(OCH_2CH_2)_2N\}Cp$] (R = Ph, CH=$CH_2$, etc.)	*P, N*	758,759
[$Mo(CO)_2\{PN(CH_2CH_2NCH_2CH_2)_2N\}Cp$]	*P, N*	760
[$Mo(CO)_2\{PN(CH_2)_2N(CH_2)_3N(CH_2)_2N(CH_2)_3\}Cp$]	*P, N*	761

documented in Table 8. As a result of the extreme reaction conditions, a mixture of products is usually obtained with the yields depending on the duration and conditions of reaction; with a few exceptions they are generally low.

Compounds of the type [$M(CO)_2(\eta\text{-}E_3)Cp$] (**15**), [$M_2(CO)_4(\mu\text{-}E_2)Cp_2$] (**16**) and [$M_3(CO)_6(\mu_3\text{-}E)Cp_3$] (**17**) can be regarded as tetrahedranes in which one, two or three vertices of the E_4 tetrahedron have been replaced by $M(CO)_2Cp$ units.

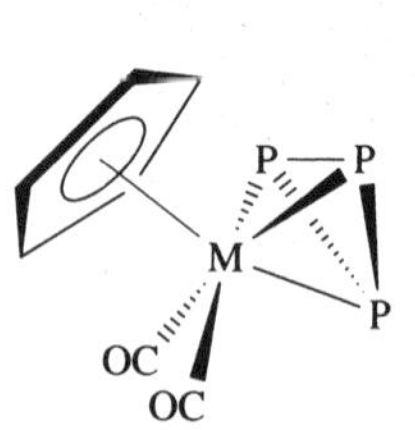

(15)

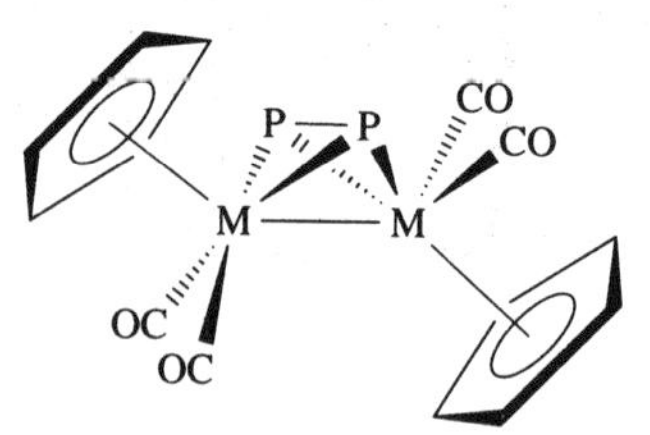

(16)

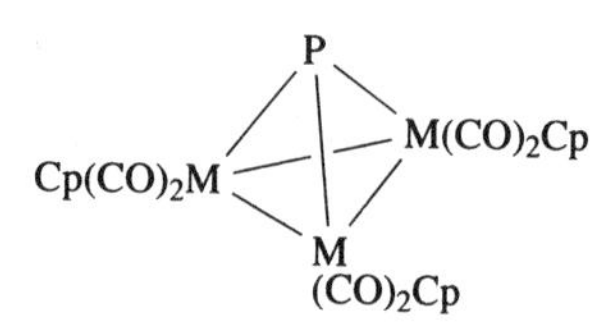

(17)

The η-E_3 complexes are obtained usually in low yields by thermolysis, but they can also be formed by the reaction of [$M(CO)_3(NCMe)_3$] with $(PC_5Me_5)_3$,[794] and it has been found that treatment of [$Cr(CO)_5(PCl_3)$] with K[$Mo(CO)_3Cp$] unexpectedly produces [$Mo(CO)_2(\eta\text{-}P_3)Cp$] in good yield.[795] The P_3 ring can be spectroscopically identified by its ^{31}P NMR shift of around –350 ppm. The structure of [$Mo(CO)_2(\eta\text{-}P_3)Cp$] is shown in Figure 1.[796] Only a limited amount of chemistry has yet been reported, but photolysis of [$Mo(CO)_2(\eta\text{-}E_3)Cp^*$] produced [$Cp^*Mo(\mu,\eta^6\text{-}E_6)MoCp^*$],[788,794] while [$Cr(CO)_2(\eta\text{-}As_3)Cp^*$] gave [$Cp^*Cr(\mu,\eta^5\text{-}As_5)CrCp^*$].[774] All three E atoms retain a lone pair of electrons and each can coordinate to a $Cr(CO)_5$ or $Mn(CO)_2Cp$ fragment.[797]

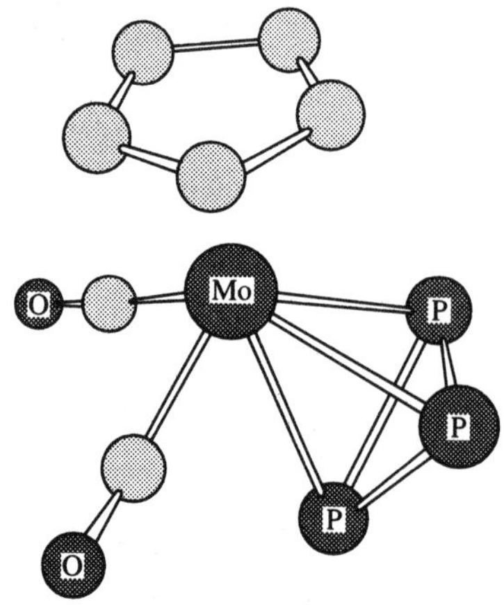

Figure 1 X-ray crystal structure of [$Mo(CO)_2(\eta\text{-}P_3)Cp$].

Table 8 Thermolytic routes to complexes with naked group 15 elements.

Starting complex	*Reagent*	*Solvent, conditions*	*Products*	*Ref.*
$[Cr_2(CO)_6Cp_2]$	P_4 (1 equiv.)	Toluene, 80–90 °C, 3–3.5 h	$[Cr_2(CO)_4(\mu\text{-}P_2)Cp_2]$, $[Cr(CO)_3(\eta\text{-}P_3)Cp]$	766,767
$[Cr_2(CO)_6Cp_2]$	P_4 (1.5–2 equiv.)	Toluene, 80–90 °C, 4–5 h	As above, + $[Cr_2(CO)_4Cp_2]$, $[CpCr(\mu,\eta^5\text{-}P_5)CrCp]$ and $[\{Cr(CO)_2Cp\}_5P_{10}]$	768–70
$[Cr_2(CO)_6Cp^*_2]$ or Cp' analogue	P_4	Xylene, 140 °C, 2h	$[Cr(CO)_2(\eta\text{-}P_3)Cp^*]$, $[Cp^*Cr(\mu,\eta^5\text{-}P_5)CrCp^*]$	771
$[Cr_2(CO)_6Cp_2]$	As_4	Toluene, 110 °C, 1–16 h	$[Cr_2(CO)_4(\mu\text{-}As_2)Cp_2]$, $[Cr(CO)_3(\eta\text{-}As_3)Cp]$, + $[Cr_2(CO)_4Cp_2]$; longer times also gave $[CpCr(\mu,\eta^5\text{-}As_5)CrCp]$	772
$[Cr_2(CO)_4Cp_2]$ or Cp' analogue	As_4	Xylene, 150 °C, 2.5–3.5 h	$[CpCr(\mu,\eta^5\text{-}As_5)CrCp]$	773
$[Cr_2(CO)_4(\eta\text{-}C_5Me_4R)_2]$ R = Me, Et	As_4	Xylene, 150 °C, 2.5 h	$[Cr(CO)_2(\eta\text{-}As_3)(\eta\text{-}C_5Me_4R)]$, $[(\eta\text{-}C_5Me_4R)Cr(\mu,\eta^5\text{-}As_5)Cr(\eta\text{-}C_5Me_4R)]$	774
$[Cr_2(CO)_6Cp_2]$	P_4S_3	1 equiv., RT, 13 d	$[Cr_4(CO)_9(P_4S_3)Cp_4]$	775
$[Mo_2(CO)_4Cp_2]$	P_4	Toluene, 110 °C, 8 h	$[Mo(CO)_2(\eta\text{-}P_3)Cp]$, $[Mo_2(CO)_4(\mu\text{-}P_2)Cp_2]$	776,777
$[Mo_2(CO)_4Cp^*_2]$	P_4	Xylene, 140 °C, 5 h	$[Mo(CO)_2(\eta\text{-}P_3)Cp^*]$, $[Mo_2(CO)_4(\mu\text{-}P_2)Cp^*_2]$, $[Mo_2(CO)_2(\mu\text{-}P_2)_2Cp^*_2]$, $[Cp^*Mo(\mu,\eta^6\text{-}P_6)MoCp^*]$	778
$[M_2(CO)_6Cp_2]$ (M = Mo, W)	As_4	Xylene, 140 °C, 24 h	$[M_2(CO)_4(\mu\text{-}As_2)Cp_2]$, $[M_3(CO)_6(\mu_3\text{-}As)Cp_3]$	779,780
$[Mo_2(CO)_6Cp_2]$ or $[WH(CO)_3Cp]$	$(AsPh)_6$	Toluene, 150–180 °C, sealed tube, 24 h	$[M_2(CO)_4(\mu\text{-}As_2)Cp_2]$	781
$[Mo_2(CO)_6Cp_2]$	$(AsMe)_5$	Toluene, 140 °C, sealed tube, 4 d Toluene, 190 °C, sealed tube, 2 d	$[Mo_2(CO)_4(\mu,\eta^2\text{-}As_5Me_5)Cp_2]$, $[Mo_4(\mu_3\text{-}As)_4Cp_4]$ $[CpMo(\mu,\eta^5\text{-}As_5)MoCp]$	782,783
$[Mo_2(CO)_6Cp'_2]$	$(AsMe)_5$	Toluene, 130 °C, sealed tube, 48 h	$[Mo_2(CO)_2(\mu\text{-}As_2)_2Cp'_2]$	784
$[Mo_2(CO)_6(\eta,\eta\text{-}C_5H_4CH_2C_5H_4)]$	$(AsMe)_5$	Toluene, 130 °C, sealed tube, no time given	$[Mo_2(CO)_4(\mu,\eta^2\text{-}As_4Me_4)(\eta,\eta\text{-}C_5H_4CH_2C_5H_4)]$, $[Mo_2(CO)_4(\mu\text{-}As_2)(\eta,\eta\text{-}C_5H_4CH_2C_5H_4)]$	765
$[Mo_2(CO)_6Cp_2]$	$(MeAsS)_n$ (n = 3, 4)	Toluene, 125 °C, sealed tube, 48 h	$[Mo_2(CO)_4(\mu\text{-}As_2)Cp_2]$, $[Mo_2(CO)_2(\mu\text{-}As_2)_2Cp'_2]$, and $[Mo_2(\mu,\eta^2\text{-}As_3)(\mu\text{-}AsS)Cp_2]$	785
$[Mo_2(CO)_6Cp_2]$	As_4	Toluene/H_2O, 110 °C, 24 h	$[Mo_3(\mu\text{-}O)_3(\mu_3\text{-}As)Cp_3]$	786
$[Mo_2(CO)_6Cp_2]$	As_4 and As_4S_4	Toluene/H_2O, 110 °C, 48 h	$[Mo_3(\mu\text{-}O)_2(\mu\text{-}S)(\mu_3\text{-}As)Cp_3]$	786
$[Mo_2(CO)_4(\eta\text{-}C_5Me_4R)_2]$ (R = Me, Et)	As_4	Xylene, 140 °C, 5–30 h	$[Mo(CO)_2(\eta\text{-}As_3)(\eta\text{-}C_5Me_4R)]$, $[Mo_2(CO)_4(\mu\text{-}As_2)(\eta\text{-}C_5Me_4R)_2]$, $[Mo_2(CO)_2(\mu\text{-}As_2)_2(\eta\text{-}C_5Me_4R)_2]$, $[\eta\text{-}C_5Me_4R)Mo(\mu,\eta^6\text{-}As_6)Mo(\eta\text{-}C_5Me_4R)]$	787,788
$[Mo_2(CO)_4Cp^*_2]$	P_4S_3 or As_4S_4	Toluene, 110 °C, 5 h	$[Mo(CO)_2(\eta\text{-}E_3)Cp^*]$, $[Mo_2(CO)_4(\mu\text{-}E_2)Cp^*_2]$, $[Mo_2(\mu\text{-}S_2)(\mu\text{-}E_2S)Cp^*_2]$, $[Mo_2(\mu\text{-}PS)(\mu\text{-}P_3)Cp^*_2]$	789,790
$[Mo_2(CO)_4(\eta\text{-}C_5Me_4Et)_2]$	As_4S_4	Toluene, 110 °C, 24 h	$[Mo(CO)_3(\eta\text{-}As_3)(\eta\text{-}C_5Me_4Et)]$, $[Mo_2(CO)_4(\mu\text{-}As_2)(\eta\text{-}C_5Me_4Et)_2]$, $[Mo_2(S)_{4-n}(O)_n(\eta\text{-}C_5Me_4Et)_2]$ (n = 0–2)	791
$[Mo_2(CO)_6Cp_2]$	Sb	Toluene, 180 °C, sealed tube, 5 d	$[Mo_2(CO)_4(\mu\text{-}Sb_2)Cp_2]$	792
$[W_2(CO)_nCp^*_2]$ (n = 4 or 6)	P_4	Xylene, 140 °C, 16 h.	$[Cp^*W(\mu,\eta^6\text{-}P_6)WCp^*]$	793

In $[M_2(CO)_4(\mu\text{-}E_2)Cp_2]$ the E_2 unit lies perpendicular to the M–M bond and acts as a four-electron donor; these compounds are thus related to the alkyne complexes $[M_2(CO)_4(\mu\text{-}C_2R_2)Cp_2]$ and the phospha-alkyne complex $[M_2(CO)_4(\mu\text{-}PCBu^t)Cp_2]$. Several structure determinations (e.g., Figure 2) show a metal–metal single bond and a short E–E bond (typically 0.208 nm for P_2, 0.232 nm for As_2, 0.268 nm for Sb_2). Each E atom again retains a lone pair, and mono- and bis-M'(CO)$_5$ adducts (M' = Cr, Mo, W) can be prepared by reaction with the appropriate number of equivalents of $[M'(CO)_5(THF)]$. Abstraction of $Cr(CO)_5$ units from $[As\{Cr(CO)_5\}_2\{M(CO)_3Cp\}]$ (M = Mo, W) also gives the $[M_2(CO)_4(\mu\text{-}As_2)Cp_2]$ complexes or their mono- or bis-$Cr(CO)_5$ adducts depending on the stoichiometry.[798] Structure determinations show that the M_2E_2 tetrahedron is hardly perturbed by attachment of the $M'(CO)_5$ units.[799–802] The coordination of the E_2 fragments in $[M_2(CO)_2(\mu\text{-}E_2)_2Cp_2]$ (**18**) is also of the above type with similar E–E bond lengths, and they are therefore analogues of the unknown (but postulated as intermediates) bis(alkyne) species $[M_2(CO)_2(\mu\text{-}C_2R_2)_2Cp_2]$. As above, attachment of $Cr(CO)_5$ groups does not affect the structure of the $M_2(E_2)_2$ portion of the molecule.[787] Again, little chemistry has been studied, but treating $[Mo_2(CO)_2(\mu\text{-}P_2)_2Cp^*_2]$ with P_4 does afford $[Cp^*Mo(\mu,\eta^6\text{-}P_6)MoCp^*]$ in a reaction reminiscent of alkyne cyclotrimerization.

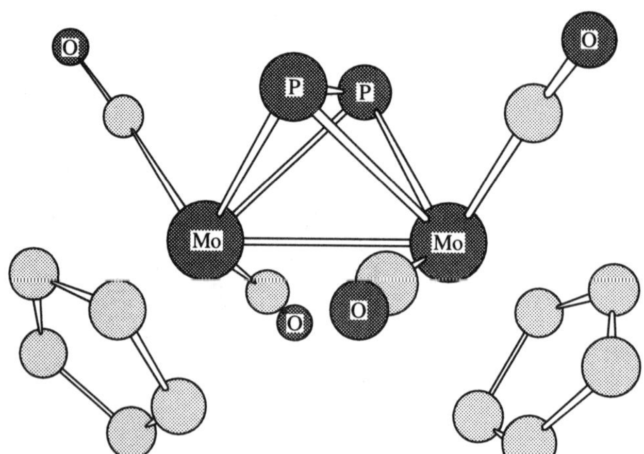

Figure 2 X-ray crystal structure of $[Mo_2(CO)_4(\mu\text{-}P_2)Cp_2]$.

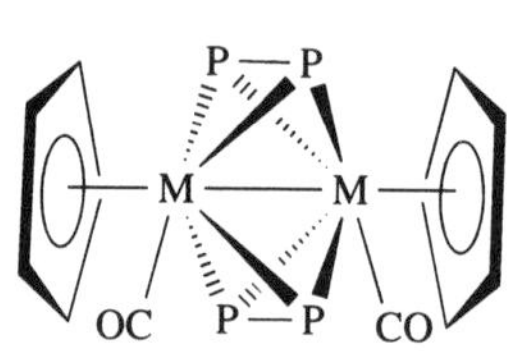

(18)

The clusters $[M_3(CO)_6(\mu_3\text{-}E)Cp_3]$ only occur for M = Mo, W and E = As. The As atom can be capped by ligation to $M(CO)_5$ (M = Cr, Mo, W) or $Mn(CO)_2Cp$ fragments.[803] As they are readily available in reasonable yield they have also been used to construct larger clusters through reactions with species such as iron carbonyls,[780,804,805] $[Rh(CO)_2Cp]$,[806] $[Ni_2(CO)_2Cp_2]$ and $[Pt(\eta\text{-}C_2H_4)(PPh_3)_2]$.[807]

Undoubtedly the most interesting products in these reactions are those containing bridging E_5 or E_6 units, the phosphorus or arsenic analogues of cyclopentadienyl or benzene. The structures of $[CpMo(\mu\text{-}\eta^5\text{-}As_5)MoCp]$ and $[Cp^*W(\mu,\eta^6\text{-}P_6)WCp^*]$ are shown in Figures 3 and 4. The As_5 ring in $[CpMo(\mu,\eta^5\text{-}As_5)MoCp]$ is distorted, with two very long, one medium and two short As–As distances. This is attributed to a second-order Jahn–Teller effect caused by the smaller range of orbital energies associated with the As_5 ring compared to the P_5 ring;[808,809] the structure of $[CpCr(\mu,\eta^5\text{-}P_5)CrCp]$ is, in contrast, much more regular, but so are those of $[Cp'Cr(\mu,\eta^5\text{-}As_5)CrCp']$ and its C_5Me_4Et analogue, which have been determined since.

The formation of complexes containing naked E atoms by attack of $[M(CO)_3Cp]^-$ on ECl_3 has been the subject of renewed interest.[810] Compounds of the type $[E\{M(CO)_3Cp\}_3]$ can be made in this way for molybdenum and tungsten where E = P,[811] Bi[812] and Sb;[813] the chromium compounds are also known for bismuth, which has received most attention.[814] The mixed species $[BiCl_{3-n}\{M(CO)_3Cp\}_n]$ (n = 1, 2) can also be prepared either by adjusting the stoichiometry of the reaction, or by redistribution of the BiM_3 complex with appropriate amounts of $BiCl_3$. They undergo further addition of halide to form the bismuth

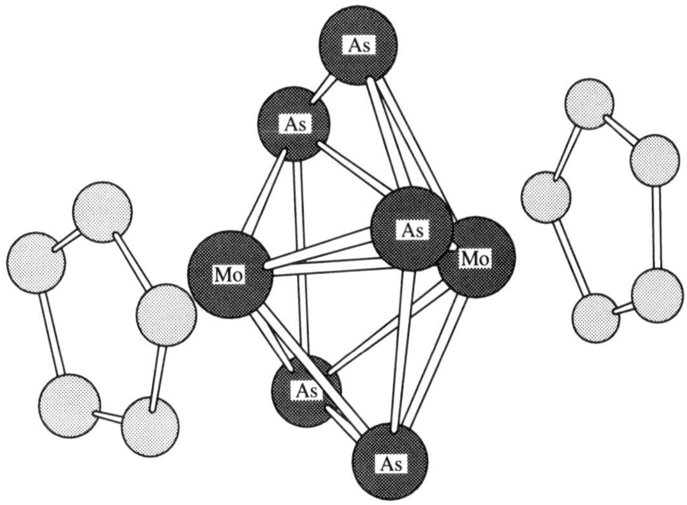

Figure 3 X-ray crystal structure of [CpMo(μ,η^5-As_5)MoCp].

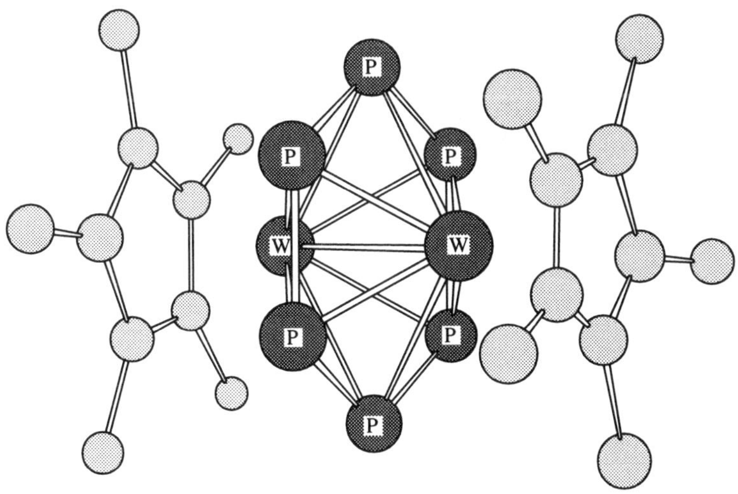

Figure 4 X-ray crystal structure of [Cp*W(μ,η^6-P_6)WCp*].

'ate' complexes [$BiCl_{4-n}${$Mo(CO)_3Cp$}$_n$]$^-$, which can be crystallized as their PPN or PPh_3CH_2Ph salts.[815] The nitrate compounds [$Bi(NO_3)${$M(CO)_3Cp$}$_2$] (M = Cr, Mo, W) have also been made.[816] Yields are generally good, though even here the outcome is not always predictable, for example the isolation of the cluster [$Mo_5(CO)_{14}(\mu_4$-$Sb)_2Cp_4$] from the reaction of $SbCl_3$ with [$Mo(CO)_3Cp$]$^-$.[817]

On photolysis or thermolysis, the complex [BiCl{$Mo(CO)_3Cp$}$_2$] is converted into the dibismuth compound [$Mo_2(CO)_4(\mu$-$Bi_2)Cp_2$], analogous to the μ-P_2 and μ-As_2 species described above,[814,818] while treatment with $S(TMS)_2$ gives a compound with a bridging BiS unit.[819] The hydride donor $LiBHEt_3$ unexpectedly reacts by ethyl transfer to give [BiEt{$Mo(CO)_3Cp$}$_2$].[820]

Related to the reactions above are those in which [$M(CO)_3Cp$]$^-$ reacts with a halophosphinidene, -arsinidene or -stibinidene complex [XE{$M'L_n$}$_2$] to produce compounds with trigonal planar E atoms; such processes are known for M = Mo, W and $M'L_n$ = $Cr(CO)_5$, $W(CO)_5$ or $Mn(CO)_2Cp$.[821,822] The Jutzi Cp* transfer process has also been employed in a similar way: reaction of [{$W(CO)_5$}$_2P(C_5Me_5)$] with [$Cr(CO)_3(NCMe)_3$] affords [$Cp^*Cr(CO)(\mu_3$-P){$W(CO)_5$}$_2$].[823]

7.4.4 Cyclopentadienyl Metal Complexes with Phosphinidene, Phosphido and Related Ligands

The reaction of K[$M(CO)_3Cp$] with $RPCl_2$ (R = Mes*) gives the μ-phosphinidene complexes [$M_2(CO)_4(\mu$-PR)Cp_2] (**19**). A similar complex can be prepared for R = tetramethylpiperidyl, accompanied by [$Mo(CO)_2$(=PRCl)Cp] as co-product. The phosphinidene group in the dinuclear complexes is planar, with a characteristic low field ^{31}P NMR shift (e.g., δ 687 ppm for M = Mo); there is some degree of M–P multiple bonding but only a M–M single bond.[824–6]

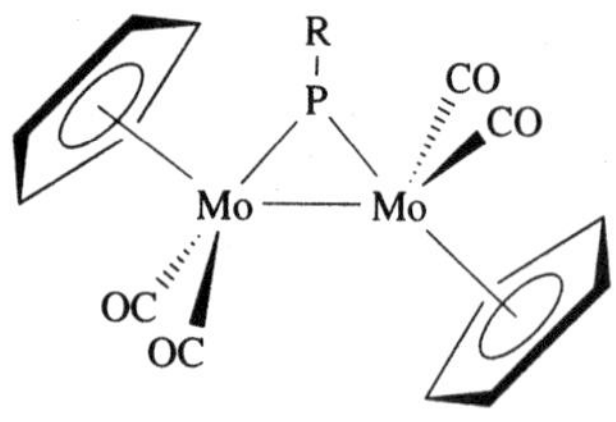

(19)

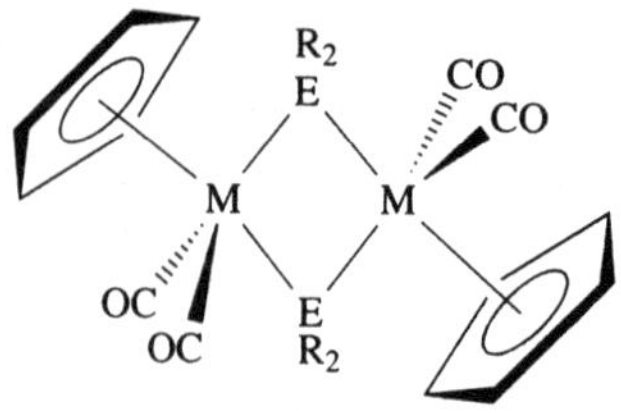

(20)

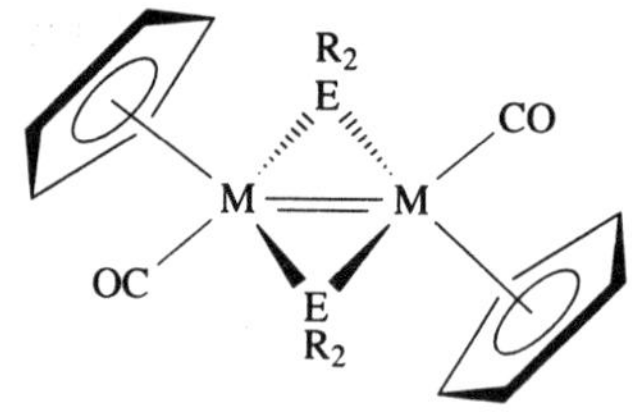

(21)

Complexes of the type $[M(ER_2)(CO)_3Cp]$ (E = P, As) are usually prepared by reaction of $[M(CO)_3Cp]^-$ with R_2ECl, though reaction of E_2R_4 with $[MH(CO)_3Cp]$ or with $[M_2(CO)_6Cp_2]$ (photochemically induced for molybdenum and tungsten) has also been used.[62,827] A wide variety of substituents R can be tolerated (see Table 9), with the stability of the complex depending on the bulk both of the R groups and the Cp ligand. As described below, most are prone to either decarbonylation or dimerization. The metal is most often molybdenum or tungsten, with very few examples known for chromium.

In these compounds the pyramidal ER_2 fragment is σ-bound to the metal and retains its lone pair, rendering it nucleophilic. The nucleophilicity can be enhanced by substitution of a CO ligand for PMe_3. Consequently the E atom can be quaternized, oxidized, undergo cyclization reactions with electron-deficient alkynes, and participate in bonding to Lewis acids (e.g., boranes) or other suitable metal fragments, forming phosphido-bridged complexes (Scheme 24). The complex $[Mo(AsMe_2)(CO)_3Cp]$ has been used as a metal exchange reagent in cluster chemistry; first the complex is attached to the cluster in a substitution reaction, and then thermolysis or photolysis results in either metal exchange or cluster expansion.[828,829]

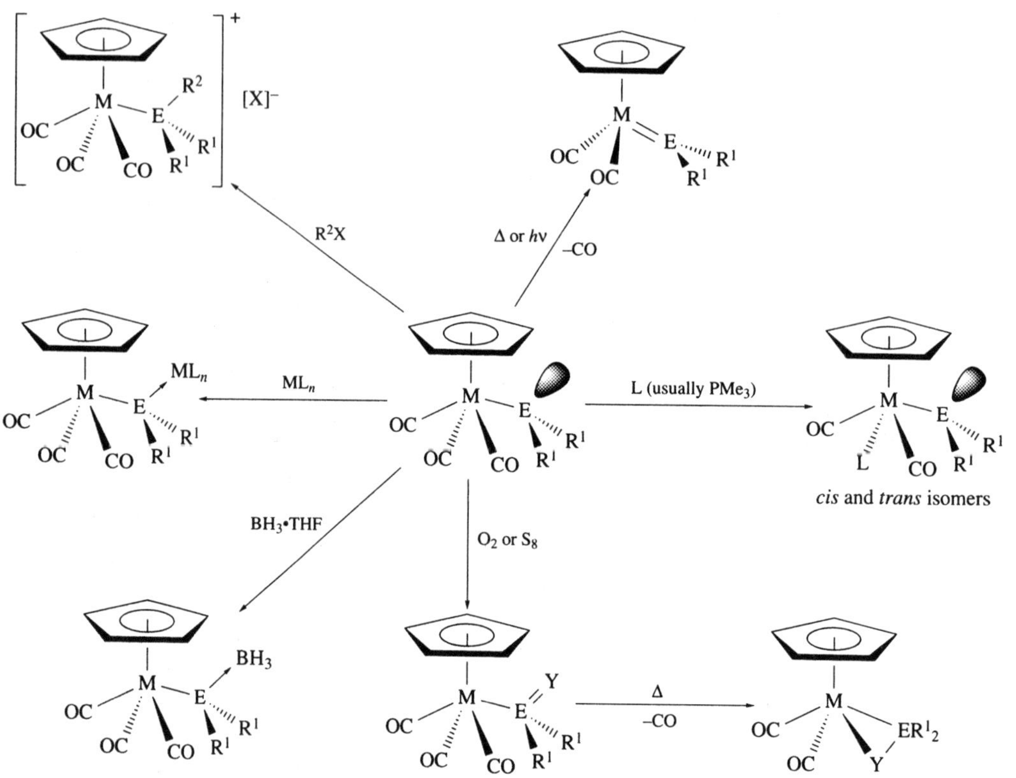

Scheme 24

Examples of $[M(ER^1R^2)(CO)_3Cp]$ complexes which have been prepared and their reported reactions are listed in Table 9.

Reduction of $[MoBr(CO)_2(PPh_2H)Cp]$ with sodium or deprotonation of $[MH(CO)_2(PPh_2H)Cp]$ affords the phosphido dianion $[Mo(CO)_2(PPh_2)Cp]^{2-}$ which can be used to make a number of metallacycles by reaction with dihalides (Scheme 25).[860–3]

Table 9 Complexes of the type $[M(CO)_3(ER^1R^2)Cp]$.

E	R^1	R^2	*Reactions*[a]	*Ref.*
P	Cl	Cl	O, B	830
P	Cl	Me, Ph, Bu^t, C_5Me_5		831
P	Ph	Ph	O, B, M, A, S, Q, D, Di	832–40
P	Bu^t or SBu^t	SBu^t	D	841
P	CH=CHCH=CH (phosphole)		O, Q, A	842
P	$=C(TMS)_2$		O, Q, D	843
$P\{W(CO)_5\}$	$CH=CH_2$	Ph	O	844, 845
P, As	CF_3 or CN	CF_3 or CN	O, M, Q, D	827, 846
P, As	=PMes*			847
As	Me	Me	O, M, Q, A, S	215, 848–51
As	Bu^t	Bu^t	M, S, D	849, 850, 852–4
As	Pr^i	Pr^i	Di	855
As	Me	Bu^t	Di	855
As	SCH_2CH_2S		D	856
Sb	Cl	Cl	M	857
Sb	Cl	Bu^t	Q, S	858
Sb, Bi	$NBu^tCH_2NBu^t$			859

[a] Key to reactions: Q: quaternization; O: oxidation or addition of sulfur; B: adduct formation with boranes; M: adduct formation with ML_n; A: cycloaddition with alkynes; S: substitution of CO ligand; D: decarbonylation; Di: dimerization.

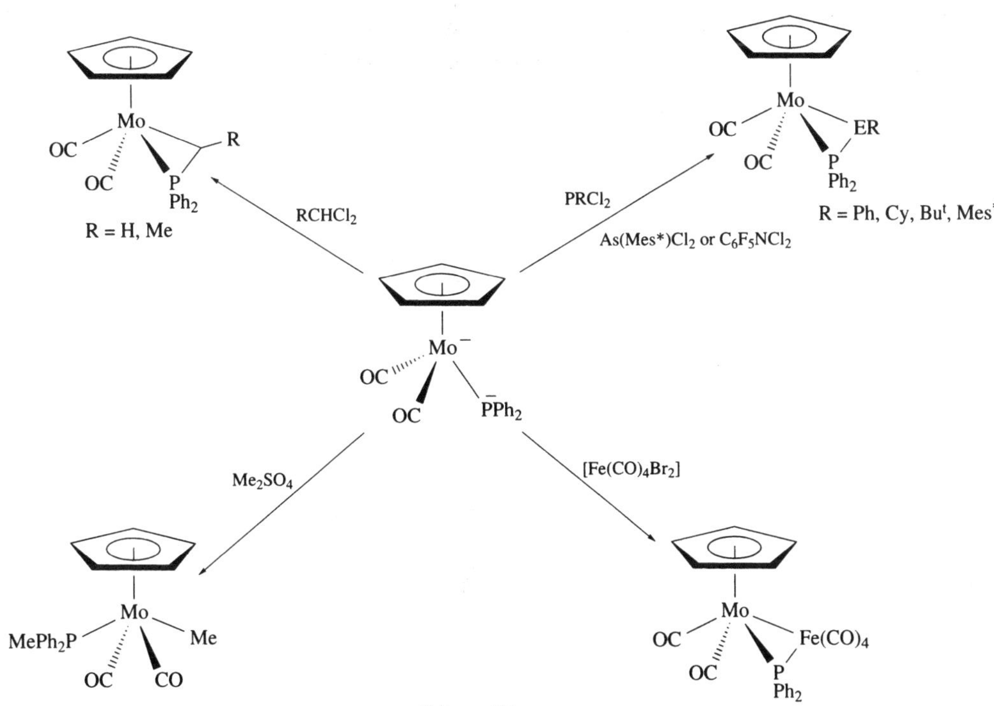

Scheme 25

If the R groups (or the Cp ligand) are sufficiently bulky, reaction of $[M(CO)_3Cp]^-$ with ER_2Cl affords the phosphenium (arsenium) complexes $[M(=ER_2)(CO)_2Cp]$; in some cases the isolable tricarbonyl complex can be induced to decarbonylate by heating or photolysis.[864] An alternative synthesis involves the dehydrohalogenation of $[M(Cl)(CO)_2(ER_2H)Cp]$, easily prepared from $[M(Cl)(CO)_3Cp]$ and ER_2H. The M atom in $[M(=ER_2)(CO)_2Cp]$ is always molybdenum or tungsten; no examples exist for chromium. As shown in Table 10, a range of substituents, including organometallic ones, can again be tolerated provided they have sufficient bulk.

In these complexes the lone pair of E is donated to the metal. As a result the E atom adopts a planar geometry with a short M–E bond, and as shown by the reactions in Scheme 26, it now undergoes nucleophilic attack.[883,885,866]

Complexes of the type $[M(ER_2)(CO)_3Cp]$ which do not undergo thermal decarbonylation to $[M(=ER_2)(CO)_2Cp]$ instead dimerize to either $[M_2(CO)_4(\mu\text{-}ER_2)_2Cp_2]$ or $[M_2(CO)_2(\mu\text{-}ER_2)_2Cp_2]$ depending on the steric requirements of R.[855] Like the analogous SR-bridged species, the tetracarbonyl

Table 10 Complexes of the type $[M(CO)_2(ER^1R^2)Cp]$ (M = Mo, W).

E	*R*[1]	*R*[2]	*Ref.*
P	NMe_2, $N(CH_2Ph)_2$	NMe_2, $N(CH_2Ph)_2$	865, 866
P	OCH_2CH_2NR (R = Me, Et, Bu^t, Ph)		867
P	$OCMe_2CMe_2O$		868
P	Ph	$N(TMS)_2$	869
P	Mes* or OMes*	$CH_2C{\equiv}CH$, $C{\equiv}CPh$, CH=CHPh, $CH{=}CH_2$, CH=PMes*	870–7
P	Cl	$CH_2(TMS)$, tmp	878
P	$=C(TMS)_2$		843, 864, 879–81
P	Ph, *o*-Tol, Bu^t, Pr^i, Mes	Ph, *o*-Tol, Bu^t, Pr^i, Mes	836, 882–6
P	Mes	$Fe(CO)_2Cp$, $M(CO)_3Cp$ (M = Mo, W)	887–90
As	Bu^t	Bu^t	852,884

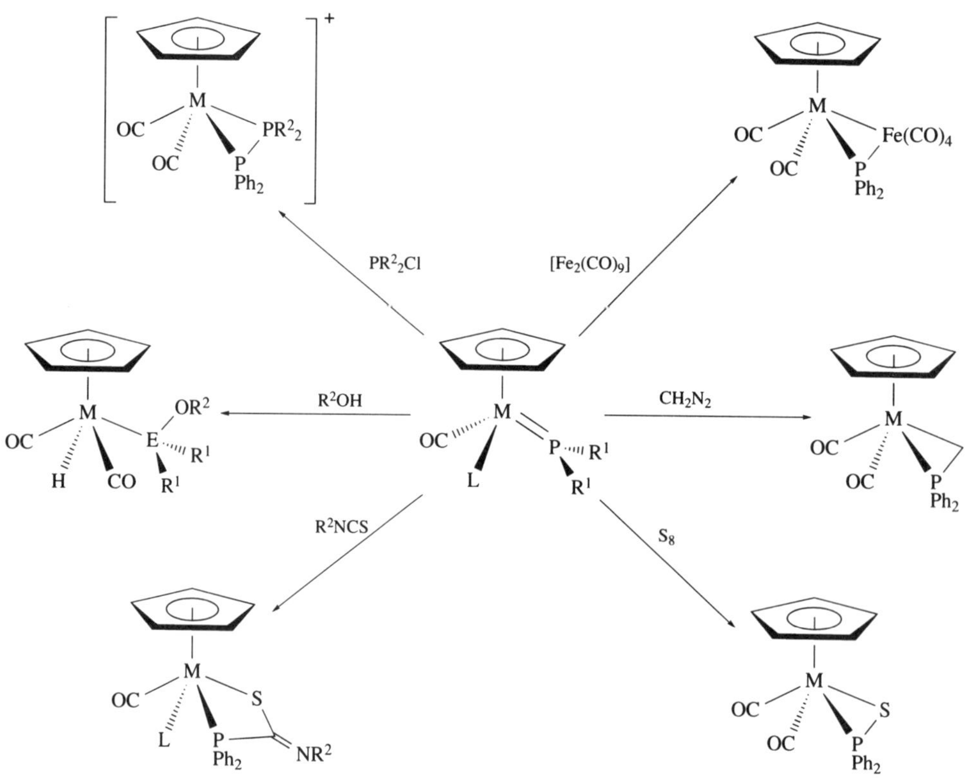

Scheme 26

complexes (**20**) contain a folded M_2E_2 core and no M–M bond, whereas the dicarbonyls (**21**) contain a planar core and a formal metal–metal double bond.

The reaction of $[M_2(CO)_6(\mu\text{-}C_5H_4SiMe_2C_5H_4)]$ (M = Mo, W) with P_2Me_4 produces $[M_2(CO)_4(\mu\text{-}PMe_2)_2(\mu\text{-}C_5H_4SiMe_2C_5H_4)]$, which undergoes further decarbonylation to $[M_2(CO)_3(\mu\text{-}PMe_2)_2(\mu\text{-}C_5H_4SiMe_2C_5H_4)]$ with a single metal–metal bond, but does not proceed to a dicarbonyl species.[891] The complex $[Mo_2(CO)_2(\mu\text{-}PPh_2)_2Cp_2]$ can be prepared by heating $[Mo_2(CO)_6Cp_2]$ with P_2Ph_4 or by thermolysis of $[Mo(PPh_2)(CO)_3Cp]$;[833] it was previously thought to be trimeric $[Mo_3(CO)_3(\mu\text{-}PPh_2)_3Cp_3]$. Aerial oxidation forms the oxo complex $[Mo_2(CO)(O)(\mu\text{-}PPh_2)_2Cp_2]$, and the alkyne derivative $[Mo_2(CO)(\eta\text{-}C_2Me_2)(\mu\text{-}PPh_2)_2Cp_2]$ can also be prepared.[892] X-ray structures of all three and of related species[893,894] show that the Mo_2P_2 core is always planar, though with variations in Mo–Mo distance and Mo–P–Mo bond angles. Photochemical decarbonylation of $[Mo_2(CO)_2(\mu\text{-}PPh_2)_2Cp_2]$ gives $[Mo_2(\mu\text{-}CO)(\mu\text{-}PPh_2)_2Cp_2]$ (**22**) which has a folded core and a short Mo–Mo bond [0.2515(2) nm].[833] The related $[Mo_2(\mu\text{-}CO)(\mu\text{-}RPC_6H_4PR)Cp_2]$ can be prepared by thermal reaction of $[Mo_2(CO)_6Cp_2]$ with *o*-$C_6H_4(PRH)_2$ (R = Ph, Bu^t) and has been used to construct a number of heterometallic clusters by reaction with, for example, $[Fe_2(CO)_9]$ or $[Co_2(CO)_8]$.[895–7]

Co-products in some of the above reactions are the hydrido-bridged species $[M_2(\mu\text{-}H)(\mu\text{-}ER_2)(CO)_4Cp_2]$ (**23**) (M = Mo, W), which are better made by reaction of $[M_2(CO)_nCp_2]$ (*n* = 4 or 6) with

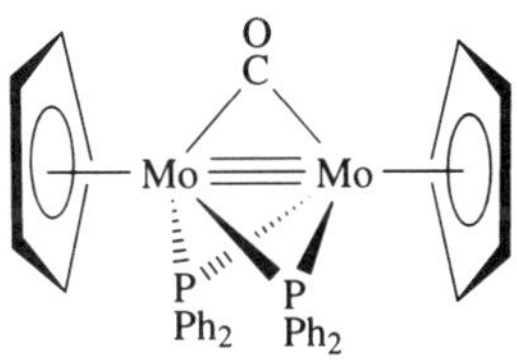

(22)

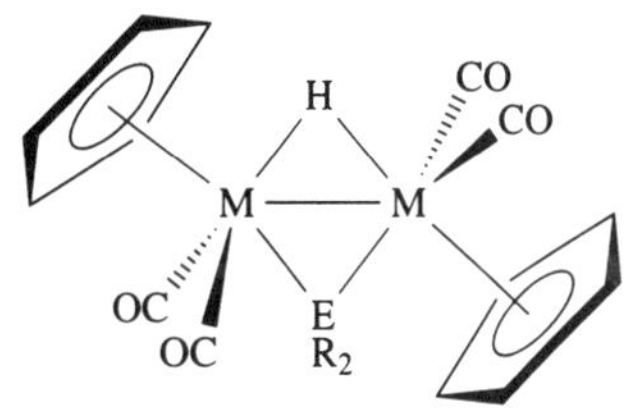

(23)

ER_2H; examples with a number of substituents are known.[72,898–903] The reactions of $[Mo_2(\mu\text{-}H)(\mu\text{-}PR_2)(CO)_4Cp_2]$ (R = Me, Ph) with alkynes, dienes and allenes, which generally proceed by insertion into the bridging hydride, have been explored.[904–6]

7.5 CYCLOPENTADIENYL METAL COMPOUNDS WITH GROUP 16 ELEMENTS

7.5.1 Cyclopentadienyl Metal Oxides

Cyclopentadienyl metal oxides have received increased attention during the 1980s.[907,908] A large number exist for molybdenum and tungsten in particular, where the M=O unit is especially stable. Many of them can be made simply by exposing lower oxidation state complexes to atmospheric oxygen.

Exposure of chromocene to a variety of oxidizing agents (O_2, N_2O, Me_3NO) affords low to moderate yields of the cubane $[Cr_4(\mu_3\text{-}O)_4Cp_4]$; variable amounts of $[Cr_4(\mu_3\text{-}O)_3(\mu_3\text{-}\sigma,\sigma,\eta^2\text{-}C_5H_4)Cp_4]$ are formed as a by-product.[909–12] The cubane displays a distorted structure with D_2 symmetry and is antiferromagnetic, which was explained by MO calculations.[913] The Cp' and Cp* analogues can be made similarly, have D_{2d} and T_d symmetry, respectively, and are also antiferromagnetic,[914–16] necessitating a revision of the molecular orbital picture.[912] It appears that the Cr_4O_4 core is largely unaffected by the identity of the Cp ligands and that the different solid-state symmetries are due to crystal packing. More recent photoelectron spectra, electrochemical, theoretical and structural studies of these and a number of other cubanes have shown that the HOMO is a metal-based nonbonding orbital, allowing oxidation to cation and dication with little structural change.[917] Mixed ligand $Cr_4O_nS_{4-n}$ (n = 1, 2) and $Cr_4O_2Se_2$ cubanes also exist.[918–20] The reaction of $[Cr_2(CO)_4Cp^*_2]$ with H_2O in toluene produces the unusual adamantane-like $[Cr_4(\mu\text{-}OH)_6Cp^*_4][Cr(CO)_3Cp^*]_2$.[921]

Air oxidation of $[Cr(CO)_2(NO)Cp^*]$ produces $[Cr_2(O)_2(\mu\text{-}O)_2Cp^*_2]$, which has a centrosymmetric *trans* structure with short Cr–Cr [250.5(1) pm] and Cr=O [159.4(3) pm] bonds.[922] The preparations of the oxo halides $[Cr(O)Br_2Cp]$ (from $[CrCp_2]$ and Br_2) and $[Cr(O)Br_2Cp^*]$ (from $[Cr_2(CO)_4Cp^*_2]$ via $[Cr_2Br_4Cp^*_2]$) have been described.[923]

Reaction of $[MCl_4Cp^*]_2$ with Bu^tNH_2 and water in air gives $[Bu^tNH_3][MO_3Cp^*]$ (64% for molybdenum, 80% for tungsten). The tungsten anion shows nucleophilic character (attributed to the fact that each oxo ligand acts as only a two-electron donor) as shown in Scheme 27.[924,925]

A large variety of oxo-halides of molybdenum and tungsten have been prepared; most are derived from the M^{VI} complexes $[MX(O)_2Cp]$ (**24**) or the M^V species $[MX_2(O)Cp]$ (**25**).

The $[MX(O)_2Cp]$ compounds and their Cp* analogues are most conveniently prepared by air oxidation of $[M_2(CO)_4Cp_2]$ or related species in halogenated solvents.[926,927] Cleavage of the $[\{M(O)_2Cp^*\}_2(\mu\text{-}O)]$ co-product (**26**) with PCl_5 also gives $[M(Cl)(O)_2Cp^*]$. Irradiation of $[Mo(Cl)(CO)_3Cp^*]$ with O_2 produced $[Mo(Cl)(O)_2Cp^*]$, which acted as a catalyst for the epoxidation of alkenes with alkyl hydroperoxides, but the related peroxide $[Mo(Cl)(O)(\eta^2\text{-}O_2)Cp^*]$ was inactive.[928] Irradiation of $[Mo_2(CO)_4Cp^*_2]$ in $CHCl_3$ in the presence of air gave two new dinuclear oxohalides, $[Mo_2Cl_2(\mu\text{-}Cl)(\mu\text{-}X)(\mu\text{-}O)Cp^*_2]$ where X = Cl or HCO_3.[929] A series of alkoxides $[W(OBu^t)(O)_2L]$, where L is a substituted Cp ligand formed by insertion of dialkylacetylenes into a tungstenacyclobutadiene ring, have also been made.[930,931]

Alkylation of $[W(Cl)(O)_2Cp^*]$ with RMgCl gives $[WR(O)_2Cp^*]$ (R = Me, CH_2TMS), which can also be made by exposure of $[WR(NO)_2Cp^*]$ to air.[674,932] Treatment of these with H_2S gives the mixed species $[WR(O)(S)Cp^*]$, whereas with B_2S_3 in CS_2 both oxo ligands are replaced to yield $[WR(S)_2Cp^*]$ and $[WR(S)(\eta^2\text{-}S_2)Cp^*]$; the latter can be easily converted into the former by sulfur abstraction with PPh_3 or CN^-.[933,934] Replacement of the oxo ligand by sulfido in this system causes a large change in the ^{183}W NMR chemical shift.[935] Surprisingly the related peroxo complexes $[MR(O)(\eta^2\text{-}O_2)Cp^*]$ can be made by treating $[MR(O)_2Cp^*]$ with H_2O_2; the M–R bond is unaffected.[936] The trihalides $[MCl_3(O)Cp^*]$ can also be prepared by reaction of $[M(Cl)(O)_2Cp^*]$ with HCl, and can be alkylated to $[M(R)_3(O)Cp^*]$ in the same way.[937]

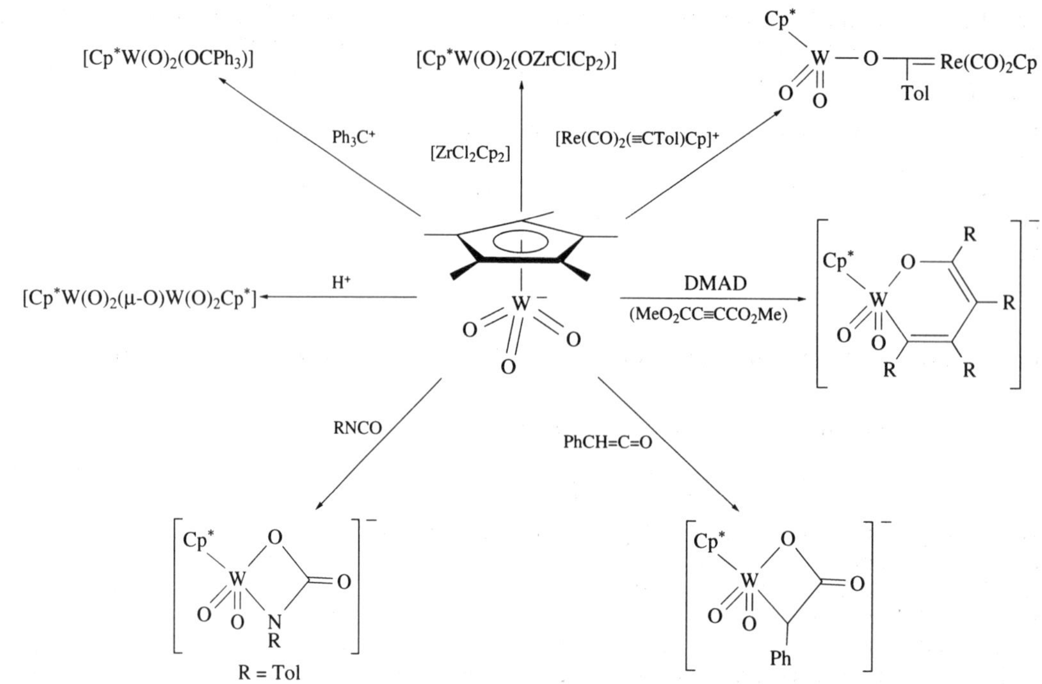

Scheme 27

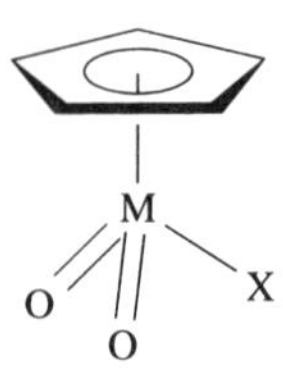

(24)

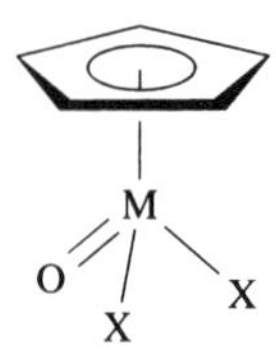

(25)

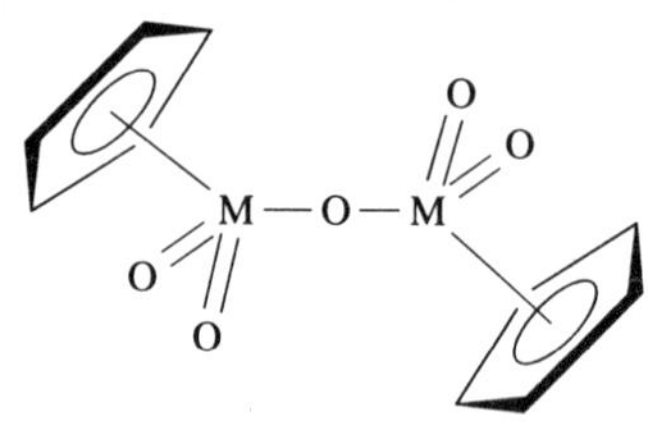

(26)

Improved preparations of $[MoCl_2(O)Cp]$ in up to 40% yield from $[Mo(O)Cl_3]_n$ and $Bu_3SnC_5H_5$ have been reported.[938,939] Reduction with zinc powder in THF affords $[(MoClCp)_4(\mu\text{-}O)_6][ZnCl(THF)]_2$, which can be reoxidized to $[\{Mo(Cl)(O)Cp\}_2(\mu\text{-}O)]$. The Cp* analogue of the latter can also be made.[940]

Oxidation of $[Mo_2(CO)_4Cp^*_2]$ with the peroxo complex $[Ni(CNBu^t)_2(O_2)]$ produces *cis*-$[Mo_2(O)_2(\mu\text{-}O)_2Cp^*_2]$, which like its Cp analogue has a slightly puckered structure;[941] a *trans* isomer has also been isolated serendipitously as a cocrystallate.[671] The mixed ligand species $[Mo(O)_2(\mu\text{-}S)_2L_2]$ and $[Mo(O)(S)(\mu\text{-}S)_2L_2]$ (L = Cp', Cp*) are also known; they both exist as *cis* and *trans* isomers, and can be prepared in a variety of ways, for example by air oxidation of $[Mo_2(\mu\text{-}S)_2(\mu\text{-}SH)_2Cp'_2]$ and $[Mo_2(S)_2(\mu\text{-}S)_2Cp^*_2]$ or reaction of $[Mo_2(CO)_4Cp^*_2]$ with propene sulfide.[791,942–5] The selenium analogues have also been made, as have *trans*-$[W_2(O)_2(\mu\text{-}Se)_2Cp_2]$ and *trans*-$[W_2(O)_2(\mu\text{-}Te)_2Cp^*_2]$.[945,946] All have the same basic structure, depicted as (**27**).

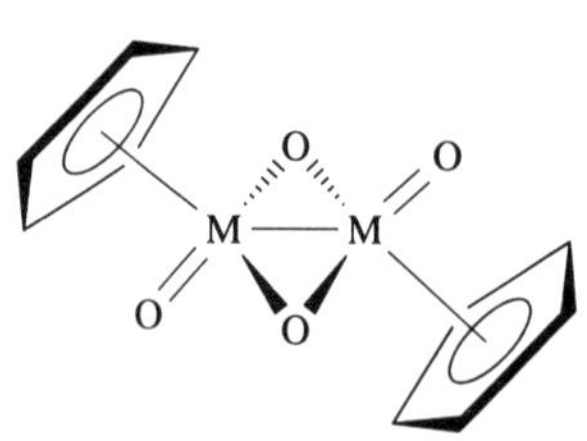

(27)

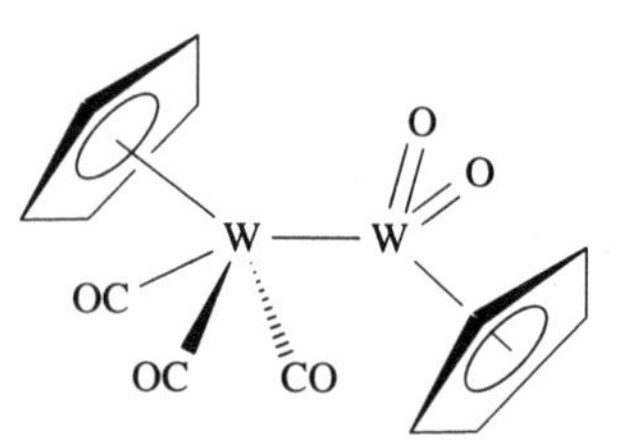

(28)

Stirring a chloroform solution of $[M_2(CO)_4Cp^*_2]$ in air affords $[\{M(O)_2Cp^*\}_2(\mu\text{-}O)]$ and $[M(Cl)(O)_2Cp^*]$; after treatment of the crude product with wet acetone to convert the latter to the former, yields of ~70% can be realized.[926] The Cp* complexes are easier to prepare and separate than their Cp analogues. Longer reaction times lead to the cluster $[C_5Me_5O][Mo_6O_{18}Cp^*]$, in which the cation is a pyrylium salt formed by insertion of an oxygen atom into a Cp* ring.[929] Several other routes are known to $[\{Mo(O)_2Cp^*\}_2(\mu\text{-}O)]$, including oxidation of $[Mo(CO)_2(NO)Cp^*]$, and three different polymorphs have been structurally characterized. All show an essentially linear MoOMo axis.[251,922,927,940,947]

Heating $[M_2(CO)_4Cp^*_2]$ with the arsaoxane $(AsMeO)_4$ gave two unusual oxo clusters, $[Mo_8O_{16}Cp^*_6]$ and $[W_6O_{17}Cp^*_2]$.[948]

Air oxidation of $[W_2(CO)_6Cp^*_2]$ unexpectedly gave $[Cp^*(CO)_3W\text{–}W(O)_2Cp^*]$ (**28**), containing a direct $W^I\text{–}W^V$ bond;[949] it can be further oxidized to $[\{W(O)_2Cp^*\}_2(\mu\text{-}O)]$ or cleaved with halogens to $[WX_3(CO)_2Cp^*]$ and $[WX(O)_2Cp^*]$.[950]

7.5.2 Cyclopentadienyl Metal Compounds with Sulfur, Selenium or Tellurium Ligands

7.5.2.1 Chromium complexes with sulfur, selenium or tellurium ligands

The reaction of $[Cr_2(CO)_6Cp_2]$ with elemental sulfur gives a number of products depending on the stoichiometry and conditions employed, including $[Cr_2(CO)_4(\mu\text{-}S)Cp_2]$, $[Cr_2(CO)_5(\mu\text{-}S_2)Cp_2]$, and $[Cr_2(CO)_4(\mu\text{-}S_2)Cp_2]$; their interrelationships are shown in Scheme 28.[951–4] Thermolysis of all these compounds leads ultimately to the cubane cluster $[Cr_4(\mu_3\text{-}S)_4Cp_4]$,[955] accompanied by $[Cr_4(\mu_3\text{-}CO)_4(\mu_3\text{-}S)_2Cp_4]$. The x-ray structure of $[Cr_2(CO)_4(\mu\text{-}S)Cp_2]$ shows an almost linear Cr–S–Cr axis (angle 174.4°) with short Cr–S bonds indicative of multiple bonding. Calculations show little or no barrier to rotation about this bond, however, in accord with the observed fluxionality in solution.[956] The linked Cp complex $[Cr_2(CO)_6(\eta,\eta\text{-}C_5H_4CH_2C_5H_4)]$ reacts in an analogous manner to those above.[957] Access to the analogous Cp* chemistry can be gained through reaction of $[Cr(CO)_3Cp^*]^-$ with trithiazyl chloride $(N_3S_3Cl_3)$.[958,959] The reaction of $[Cr(CO)_3Cp]^-$ with S_2Cl_2 is claimed to give $[\{Cr(CO)_3Cp\}_2(\mu\text{-}S)]$.[960]

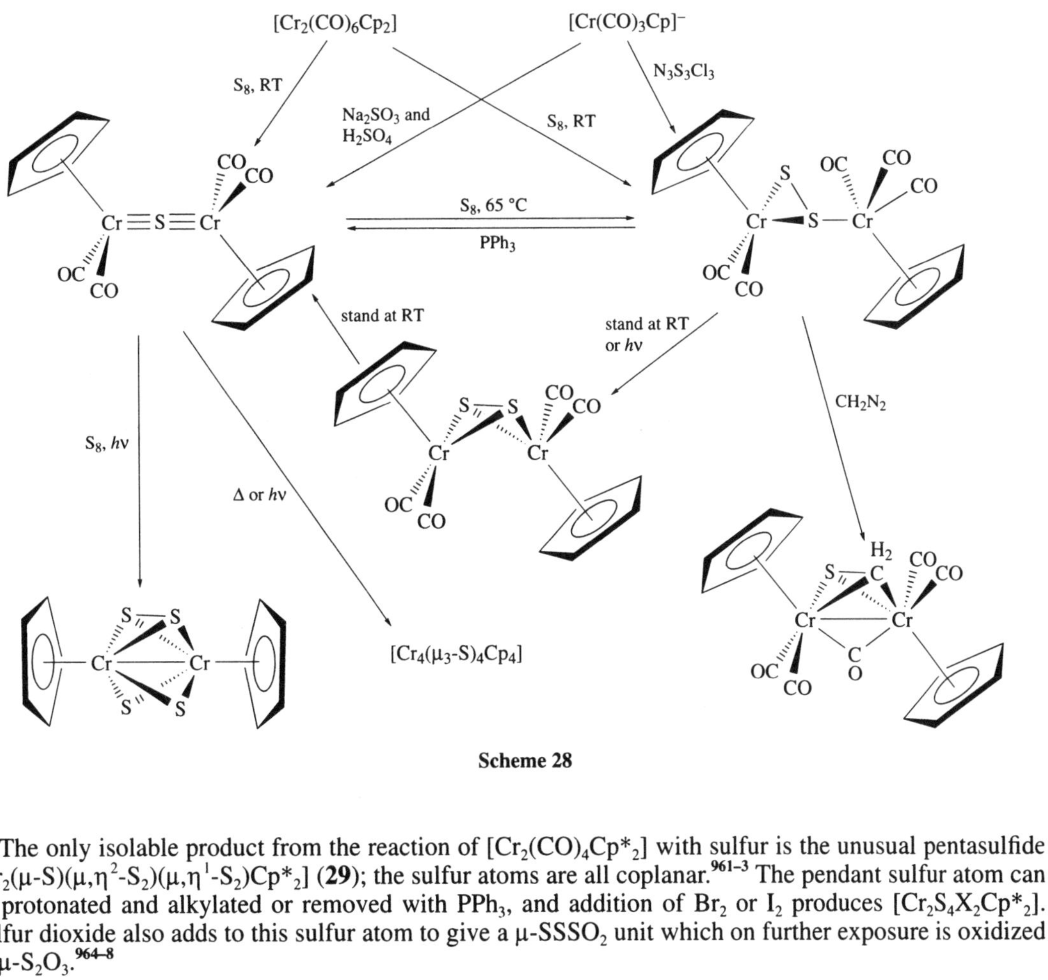

Scheme 28

The only isolable product from the reaction of $[Cr_2(CO)_4Cp^*_2]$ with sulfur is the unusual pentasulfide $[Cr_2(\mu\text{-}S)(\mu,\eta^2\text{-}S_2)(\mu,\eta^1\text{-}S_2)Cp^*_2]$ (**29**); the sulfur atoms are all coplanar.[961–3] The pendant sulfur atom can be protonated and alkylated or removed with PPh_3, and addition of Br_2 or I_2 produces $[Cr_2S_4X_2Cp^*_2]$. Sulfur dioxide also adds to this sulfur atom to give a $\mu\text{-}SSSO_2$ unit which on further exposure is oxidized to $\mu\text{-}S_2O_3$.[964–8]

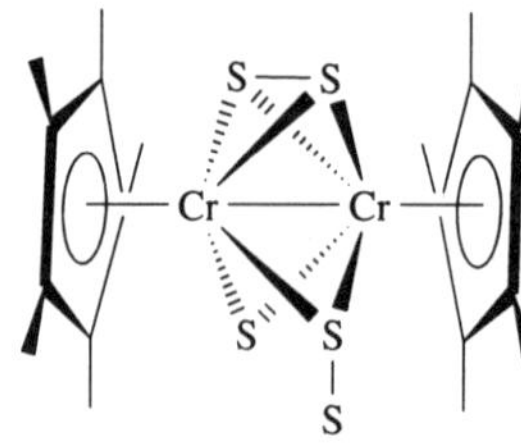

(29)

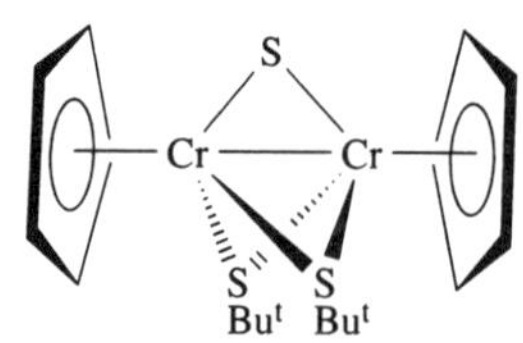

(30)

Stoichiometric reaction of $[Cr_2(CO)_6Cp_2]$ with one or two equivalents of elemental selenium gives the analogous species $[Cr_2(CO)_4(\mu\text{-}Se)Cp_2]$, $[Cr_2(CO)_4(\mu\text{-}Se_2)Cp_2]$ and unstable $[Cr_2(CO)_5(\mu\text{-}Se_2)Cp_2]$.[969,970] Other routes to the μ-Se complex and its Cp* analogue include reaction of $[Cr(CO)_3Cp]^-$ with acidified sodium selenite[958,971] or with Se_2Cl_2[972] and elimination of H_2Se from $[Cr(CO)_3(SeH)Cp]$.[973,974] Interconversions and reactions of these species are shown in Scheme 29.

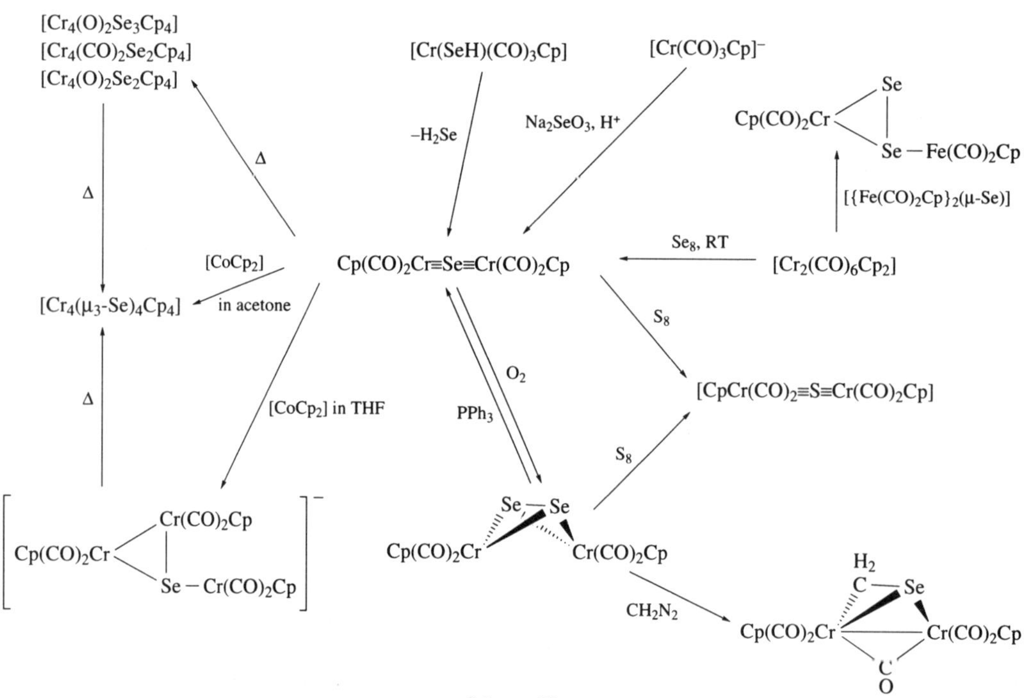

Scheme 29

Similar routes to tellurium analogues have been explored, but as yet $[Cr_2(CO)_4(\mu\text{-}Te)Cp_2]$ is unknown and all reactions lead instead to $[Cr_2(CO)_6(\mu\text{-}Te)Cp_2]$ with a bent Cr–Te–Cr framework;[974,975] the tellurium atom can be protonated or methylated.[976] Its ^{125}Te Mössbauer spectrum has also been recorded.[977]

Reaction of $[Cr_2(CO)_6Cp_2]$ with S_2Ph_2 produces $[Cr_2(CO)_4(\mu\text{-}SPh)_2Cp_2]$, $[Cr_2(CO)_2(\mu\text{-}SPh)_2Cp_2]$ and $[Cr_2(\mu\text{-}S)(\mu\text{-}SPh)_2Cp_2]$ in varying yield depending on the temperature; on continued heating all three degrade to the cubane $[Cr_4(\mu_3\text{-}S)_4Cp_4]$.[978] A similar picture emerges for Se_2Ph_2 except that $[Cr(CO)_3(SePh)Cp]$ can be isolated as an intermediate en route to the other species.[979,980]

Treatment of chromocene with Bu^{t}SH yields $[Cr_2(\mu\text{-}S)(\mu\text{-}SBu^t)_2Cp_2]$ (**30**);[914] this compound has been used by Pasynskii and co-workers to prepare a large number of heterometallic clusters.[981,982] On exposure to NO, $[Cr_2(NO)_2(\mu\text{-}SBu^t)(\mu\text{-}SSBu^t)Cp_2]$ is formed, as well as $[Cr(NO)_2(NO_2)Cp]$.[983] The related alkoxide species $[Cr_2(\mu\text{-}OBu^t)_2Cp_2]$ reacts with sulfur to give $[Cr_2(\mu\text{-}S_4)(\mu\text{-}OBu^t)_2Cp_2]$, whereas selenium and tellurium produce $[Cr_2(\mu\text{-}E)(\mu\text{-}OBu^t)_2Cp_2]$.[984]

7.5.2.2 *Molybdenum complexes with sulfur, selenium or tellurium ligands*

The polychalcogenide anions $[Mo(CO)_2(E_2)Cp]$ and $[Mo(E_4)_2Cp]$ (E = S, Se) have been prepared by reaction of $[Mo_2(CO)_6Cp_2]$ with 'K_2S_3', $[Se_6]^{2-}$, or sodium and selenium. The related complex $[Mo(S_2CO)(S_4)Cp^*]$ was formed from $[MoMe(CO)_3Cp^*]$ in a similar way, presumably via the disulfide species.[985–8]

The chemistry of dinuclear cyclopentadienyl molybdenum sulfur complexes has undergone a rapid expansion since the early 1980s, largely due to the groups of Rakowski Dubois,[989] Curtis[57] and Wachter.[990] The impetus for much of this research has been the possibility that dinuclear complexes might model the processes occurring on heterogeneous molybdenum sulfide catalysts during the hydrodesulfurization of crude oil.

Whereas the reaction of $[Mo_2(CO)_nCp_2]$ (n = 4, 6) with sulfur gives an insoluble polymeric material, the Cp* analogue gives three isomers of $[Mo_2S_4Cp^*_2]$ (structural types (**31a–c**)).[963,991] Treatment of $[MoH(CO)_3Cp^*]$ with sulfur also gives isomer **c**, as well as $[Mo_2(\eta^2\text{-}S_2)_2(\mu\text{-}\eta^2\text{-}S_2)_2(\mu,\eta^1\text{-}S_2)Cp^*_2]$.[992] The isomers are interconvertible either by further heating (**a** → **b**) or by photolysis (**c** → **b** → **a** → **c** so that an equilibrium is established);[993] theoretical calculations on this rearrangement show that a terminal sulfur to molybdenum charge transfer excitation is involved.[994–6] The reaction of $[Mo_2(CO)_4Cp^*_2]$ with selenium gives only the type **a** complex $[Mo_2(\mu\text{-}Se)_2(\mu\text{-}Se_2)Cp^*_2]$.

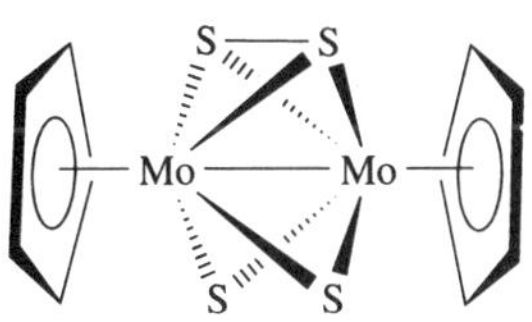

(31a)

$[Mo_2(\mu\text{-}S)_2(\mu\text{-}S_2)Cp^*_2]$

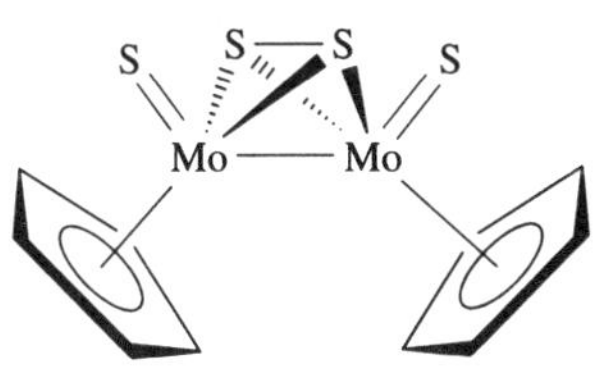

(31b)

cis-$[Mo_2S_2(\mu\text{-}S_2)Cp^*_2]$

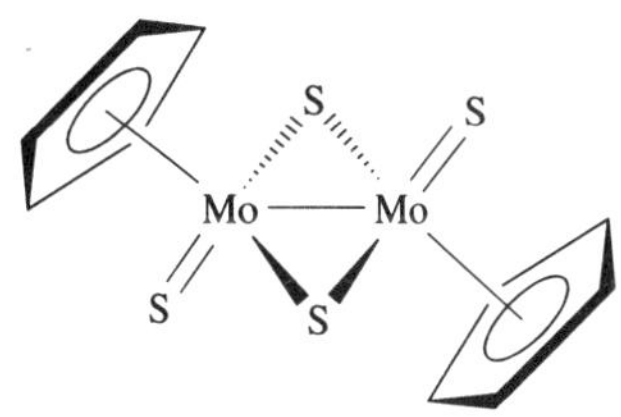

(31c)

trans-$[Mo_2S_2(\mu\text{-}S)_2Cp^*_2]$

On sequential treatment with MeI and MeLi, isomer (**31a**) forms $[Mo_2(\mu\text{-}S)_2(\mu\text{-}SMe)_2Cp^*_2]$.[997] The S–S bond can also be broken by reaction with I_2 to give $[Mo_2S_4I_2Cp^*_2]$ of unknown structure, which forms $[Mo_2(\mu\text{-}S)_2(\mu\text{-}\eta^1\text{-}SSR)_2Cp^*_2]$ on reaction with NaSR (R = alkyl, Ph).[998] Isomer (**31a**) disproportionates SO_2 to form sulfur and $[Mo_2(\mu\text{-}S_2)(\mu\text{-}S)(\mu\text{-}SSO_3)Cp^*_2]$,[999] and the same thiosulfate adduct is formed by air oxidation of (**31a**).[1000]

The discovery that the insoluble polymeric $[Mo_2S_xCp_2]_n$ mentioned above could be converted into $[Mo_2(\mu\text{-}S)_2(\mu\text{-}SH)_2Cp_2]$ by reaction with either H_2[1001,1002] or $LiBHEt_3$[1003] allowed further development of the Cp system; the more soluble Cp' complexes are often used for convenience. The Cp* analogue can be made by hydrogenation of isomer (**31a**).[964] As shown in Scheme 30, a wide variety of stoichiometric reactions involving S–H and C–S bonds have been discovered, as well as the catalytic reduction of SO_2 to sulfur and H_2O.[944,1004–18] Thiol exchange and deprotonation/alkylation reactions provide useful routes to the mixed sulfido-thiolate species $[Mo_2(\mu\text{-}S)_2(\mu\text{-}SR)_2Cp_2]$, including the methanedithiolate complex $[Mo_2(\mu\text{-}S)_2(\mu\text{-}S_2CH_2)Cp_2]$, which also displays interesting chemistry, such as the homogeneous hydrogenolysis of CS_2. The bis(dithiolate) species $[Mo(\mu\text{-}SCH_2CHRS)_2Cp_2]$ (R = H, Me) can also be prepared by the reactions of ethylene sulfide or propylene sulfide with $[Mo_2(CO)_6Cp_2]$ or $[Mo(H)(CO)_3Cp]$;[1002,1019] similarly, bridged dithiolene species can be prepared from 3,4-dimercapto-toluene or by ring opening of the dithiete $S_2C_2(CF_3)_2$.[1020,1021] In more recent work, a substantial chemistry of cationic species formed by protonation or alkylation of $[Mo_2(\mu\text{-}S)_2(\mu\text{-}S_2CH_2)Cp_2]$, shown in Scheme 31, has been developed, including the ring opening of THF.[1022–31] The deuteration of thiophene is catalysed by $[Mo_2(\mu\text{-}S)(\mu\text{-}SH)(\mu\text{-}S_2CH_2)Cp'_2][OTf]$, probably by an acid exchange pathway.[1032]

The complexes $[Mo_2(\mu\text{-}S)_2(\mu\text{-}SR)_2Cp_2]$ generally exist as two isomers, usually attributed to the *syn* or *anti* disposition of the R groups on opposing sulfur atoms. No structure exists for R = H, but crystallographic confirmation has been obtained for both possibilities in *anti*-$[Mo_2(\mu\text{-}S)_2(\mu\text{-}SMe)_2L_2]$ (L = Cp' or Cp*) and *syn*-$[Mo_2(\mu\text{-}S)_2(\mu\text{-}SBu^t)_2(\eta\text{-}C_5H_4Pr^i)_2]$.[997,1001,1033] For R = H it is also possible to distinguish the two isomers in the ^{95}Mo NMR spectrum; a number of Mo/S/SR species have been studied and show a 2000 ppm wide range of chemical shifts which depend on oxidation state and also on R. The shielding pattern can be correlated with electronic spectra, redox potentials and reactivity by a MO scheme.[1034] The IR and ESR spectra of a number of mixed valence Mo^{III}/Mo^{IV} dimers have also been interpreted.[1035]

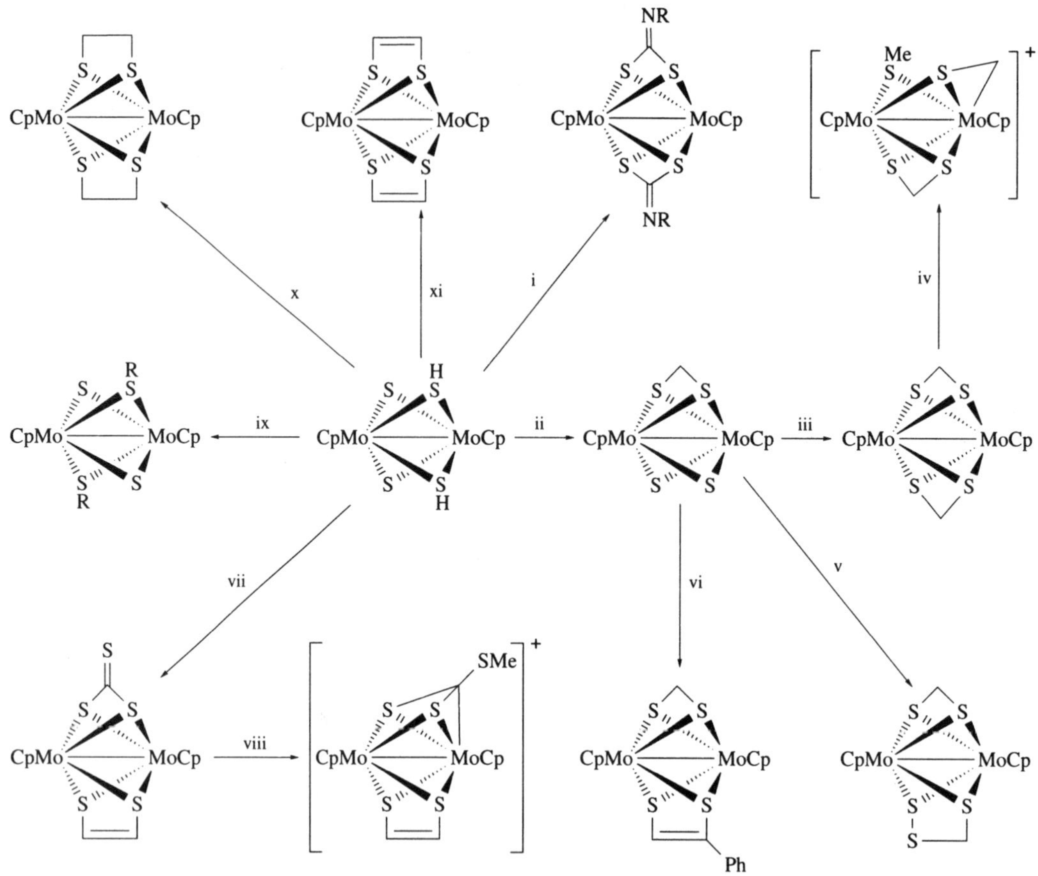

i, $PhCH_2NC$; ii, CH_2X_2, base; iii, CH_2Cl_2, H_2 (1–2 atm, 50 °C); iv, MeOTf; v, CS_2, H_2; vi, PhC≡CH; vii, HC≡CH, $Cl_2C{=}S$; viii, MeI; ix, RSH; x, C_2H_4; xi, HC≡CH

Scheme 30

Compounds of the type $[Mo_2S_4Cp_2]$ and $[Mo_2(\mu\text{-}S)_2(\mu\text{-}SR)_2Cp_2]$ (R = H, alkyl) are also useful precursors to a variety of mixed metal clusters, some of which display desulfurization activity when supported on alumina.[1036] Related clusters can be obtained by reaction of $[Mo_2(CO)_nCp_2]$ with species such as $[Fe_2(\mu\text{-}S)_2(CO)_6]$. The reaction of $[Mo_2S_4Cp^*_2]$ or $[Cr_2S_5Cp^*_2]$ with $[M_2(CO)_4Cp_2]$ (M = Cr, Mo) constitutes a rational synthesis of the mixed cubanes $[M_2M'_2(\mu_3\text{-}S)_4Cp_2Cp^*_2]$.[965,1037] Similarly treating $[Mo_2(\mu\text{-}SCH_2CHMeS)_2Cp_2]$ with $[Mo_2(CO)_4Cp'_2]$ affords $[Mo_4(\mu_3\text{-}S)_4Cp_2Cp'_2]$ by displacement of propene.[1038] The $[M_4(\mu_3\text{-}S)_4Cp_4]$ cubanes are readily oxidized to mono- and dications; there is little structural change on oxidation, a consequence of the HOMO being a metal-centred nonbonding orbital.[917,1039] Consequently a series of charge transfer salts of $[M_4(\mu_3\text{-}E)_4Cp_4]$ (M = Cr, Mo; E = O, S, Se) with anions such as TCNE, TCNQ or other metal clusters have been prepared.[1040–2]

7.5.2.3 Molybdenum complexes with thiolate and related ligands

The mononuclear complexes $[Mo(SR)(CO)_3Cp]$ are known, but they dimerize rapidly with loss of CO to form $[Mo_2(CO)_4(\mu\text{-}SR)_2Cp_2]$ (**32**),[1043] which can also be made by heating $[Mo_2(CO)_6Cp_2]$ or $[MoH(CO)_3Cp]$ with RSSR.[827,1044,1045] These compounds are analogous to the phosphido-bridged complexes (**20**): the Mo_2S_2 core is folded with the Cp ligands in a *trans* orientation; the two R groups can be *syn* or *anti*, and there is easy exchange between the two forms. Two-electron oxidation, however, causes isomerization to a *cis* structure with a Mo–Mo bond.[1046] This isomerization has been rationalized by molecular orbital calculations.[1047] Both the neutral complex and the dication can be activated electrochemically towards substitution by MeCN or Bu^tNC via a common radical cation intermediate.[1048] The substituted dications can then be reduced electrochemically back to the neutral complexes.[1049] Selenium and tellurium complexes $[Mo_2(CO)_4(\mu\text{-}ER)_2Cp_2]$ have also been prepared and undergo a similar oxidative *trans* → *cis* isomerization.[1044,1050–3]

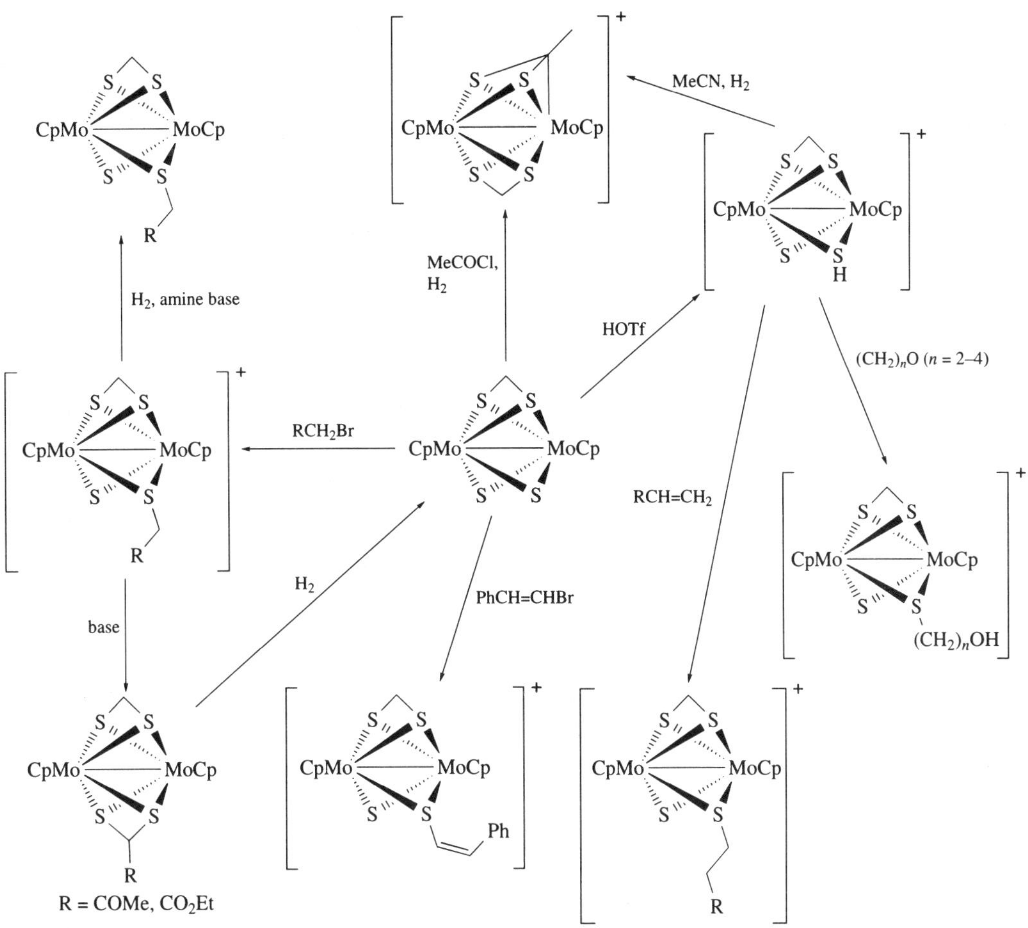

Scheme 31

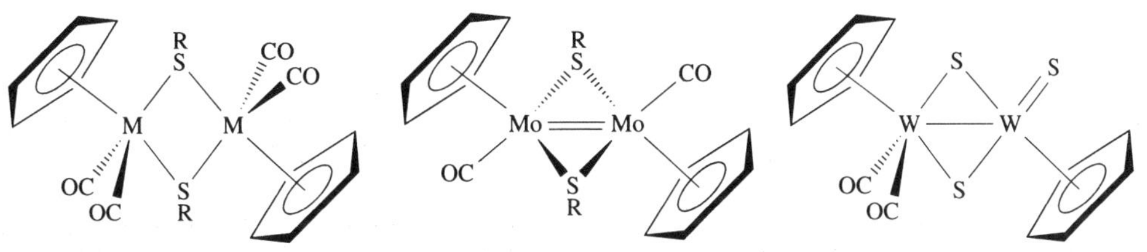

(32) **(33)** **(34)**

Reaction of $[Mo_2(CO)_4(\mu\text{-SMe})_2Cp_2]$ with Me_2S_2 in the presence of allyl halides gives $[Mo_2(CO)_2(\mu\text{-SMe})_3Cp_2]^+$.[1054] Reduction of this complex or of $[Mo_2(CO)_2X_2(\mu\text{-SMe})_2Cp_2]^+$ (X = Cl, Br) gives a common intermediate, probably *cis*-$[Mo_2(CO)_2(\mu\text{-SMe})_2Cp_2]$, which rapidly undergoes a CO-catalysed isomerization to the isolable *trans* isomer via the adduct $[Mo_2(CO)_3(\mu\text{-SMe})_2Cp_2]$.[1055,1056] Extended heating of $[Mo_2(CO)_4(\mu\text{-SR})_2Cp_2]$ also results in conversion into $[Mo_2(CO)_2(\mu\text{-SR})_2Cp_2]$ (**33**). These complexes have a formal Mo=Mo double bond with a planar Mo_2S_2 unit, and are thus analogous to the phosphido-bridged species (**21**).[1057] Alkylation with Me_3O^+ occurs stepwise to give $[Mo_2(CO)_2(\mu\text{-SMe}_2)_2Cp_2]^{2+}$.[1058] On reaction with sulfur, $[Mo_2(\mu\text{-S})_2(\mu\text{-SMe})_2Cp_2]$ is formed.

The reaction of $[Mo(Cl)(CO)_3Cp]$ with TlSR ($R = C_6F_5$) unexpectedly produced $[CpMo(CO)_2(\mu\text{-SR})_2Tl]$ and $[CpMo(\mu\text{-SR})_4Tl]$ in which the thallium atom is coordinated to the thiolate ligands; the analogous tungsten complex can be made from $[WBr_3(CO)_2Cp]$ with $[W(SR)_3(CO)_2Cp]$ as by-product.[1059,1060] There is a solvent-dependent dissociation of the Tl^+ cation, and related examples with potassium, caesium or rubidium as the counter-ion can also be prepared.[1061]

On heating, complexes of the type $[CpM(CO)_3(\mu\text{-SMe})M'(CO)_5]$ (M, M' = Cr, Mo, W) undergo decarbonylation and M–M' bond formation to give $[CpM(CO)_2(\mu\text{-SMe})M'(CO)_5]$; adding CO or PMe_3 reverses the process.[1062,1063] The electrochemistry of these species has also been studied.[1064,1065]

The reaction of $[Mo_2(CO)_6Cp_2]$ with $NHEt_2$ and CS_2 in refluxing THF produces $[Mo(CO)_2(S_2CNEt_2)Cp]$ in over 90% yield.[1066] Treatment of $[M_2(CO)_nCp_2]$ (M = Mo, W; n = 4 or 6) with RNCS or thiuram disulfides also gives the same dithiocarbamate complexes.[86,101,1067] Monodentate $[W(\eta^1\text{-}SCSNMe_2)(CO)_3Cp]$ was isolated as an intermediate;[1068] it forms an adduct with $[W(CO)_5(THF)]$ but transfers the dithiocarbamate ligand to iron or manganese carbonyls.[1069] Related complexes with S_2CPh and $S_2P(OPr^i)_2$ ligands can be made by the action of $Zn(S_3CPh)$ or $P(OPr^i)_2(S)(SH)$ on $[Mo_2(CO)_6Cp_2]$.[1020,1070] Compounds of the type $[M(CO)_2(E_2CR)Cp]$ can also be made by insertion of sulfur (or selenium) into the carbyne complexes $[M(CO)_2(\equiv CR)Cp]$[1071,1072] and by reaction of RCS_2^- with $[M(CO)_3(L)Cp]^+$ (L = solvent).[238]

7.5.2.4 *Tungsten complexes with sulfur and/or thiolate ligands*

Tungsten–sulfur chemistry is less well documented than that of molybdenum. Reaction of $[W_2(CO)_4Cp^*_2]$ with sulfur gives the unsymmetrical complex $[W_2(\mu\text{-}S)_2(S)(CO)_2Cp^*_2]$ (**34**), which on further reaction produces two isomers of $[W_2S_4Cp^*_2]$, analogous to (**31a**) and (**31c**) above.[991,1073] Like its molybdenum analogue, $[W_2(S)_2(\mu\text{-}S)_2Cp_2]$ reacts with acetylene to give the μ-dithiolene complex $[W_2(\mu\text{-}SCH{=}CHS)_2Cp_2]$ but the chelating dithiolene complexes $[W_2(S)(\mu\text{-}S)_2(SCH{=}CHS)Cp_2]$ and $[W_2(\mu\text{-}S)_2(SCH{=}CHS)_2Cp_2]$ are also formed; the analogous SCH_2CH_2S species are available from the reaction of $[WH(CO)_3Cp]$ with ethylene sulfide.[1074]

The reaction of $[WH(CO)_3Cp]$ with $S(NMe_2)_2$ produces $[\{W(CO)_3Cp\}_2(\mu\text{-}S)]$, which reacts with CH_2N_2 to give the thioformaldehyde complex $[W_2(CO)_4(\mu\text{-}SCH_2)Cp_2]$.[1075] Prolonged treatment of $[WH(CO)_3Cp]$ with SO_2 also gives $[\{W(CO)_3Cp\}_2(\mu\text{-}S)]$, together with $[\{W(CO)_2Cp\}_2(\mu\text{-}SSO_2)_2]$; under milder conditions $[W(SO_2H)(CO)_3Cp]$ and $[\{W(CO)_3Cp\}_2(\mu\text{-}S_2O_4)]$ can also be isolated.[1076,1077] A similar insertion reaction of $R^1N{=}S{=}O$ and $R^1N{=}S{=}NR^2$ (R^1 = aryl, R^2 = SO_2aryl) has been reported.[1078] Low yields of the complexes $[W_2({=}E)_2(\mu\text{-}S)_2Cp_2]$ (E = N-TMS, S or O) were formed in the reaction of $[W_2(CO)_6(Cp_2]$ or $[W(Cl)(CO)_3Cp]$ with $S({=}N\text{-}TMS)_2$.[1079] The synthesis of $[\{W(CO)_3Cp\}_2(\mu\text{-}S_2)]$ has also been reported[960] and a compound containing an unsupported WSZr unit, $[CpW(CO)_2(PMe_3)(\mu\text{-}S)ZrClCp_2]$, has also been prepared by reacting $[W(CO)_2(PMe_3)Cp]^-$ with Cp_2ZrCl_2 and propene sulfide.[1080]

Reaction of $[WH(CO)_3Cp]$ with sulfur affords $[W(SH)(CO)_3Cp]$; the selenium analogue can also be made.[973,974] Elaboration of the sulfur ligand can be achieved to give $[W(SSR)(CO)_3Cp]$, $[W(SSSR)(CO)_3Cp]$, or $[W\{SS(O)R\}(CO)_3Cp]$.[1081,1082] Oxidation of $[W(SR)(CO)_3Cp]$ (R = aryl) gives the corresponding $[W\{S(O)_2R\}(CO)_3Cp]$.[1083]

The $[W(SR)(CO)_3Cp]$ complexes are less prone to dimerization than their molybdenum analogues and treatment of $[W_2(CO)_6Cp_2]$ or $[W(CO)_3Cp]^-$ with R_2S_2 or (phthalimido)SR produces them as isolable complexes.[1043,1084,1085] They react with PPh_3 to form $[W(SR)(CO)_2(PPh)_3Cp]$,[1086] which can also be made by deprotonation of $[WH(CO)_2(PPh)_3Cp]$ followed by reaction with (phthalimido)SR.[1087] Treatment of $[W(SMe)(CO)_3Cp]$ or $[W_2(CO)_6Cp_2]$ with $[Me_2SSMe][BF_4]$ produces $[W(CO)_3(S_2Me_2)Cp]^+$ which undergoes a rapid exchange process with free thiolate.[1088–90]

The cycloaddition of $[W(SR^1)(CO)_3Cp]$ (R^1 = Me, Et) with electron deficient alkynes ($R^2C{\equiv}CR^2$ where R^2 = CF_3, CO_2Me, CN) yields $[W(CO)_2(SR^1CR^2{=}CR^2CO)Cp]$, which subsequently undergoes a series of isomerizations of the organic ligand (see Chapter 6, Section 8.2).[1091,1092]

Conversion of $[W(SR)(CO)_3Cp]$ into first $[W_2(CO)_4(\mu\text{-}SR)_2Cp_2]$ and then $[W_2(CO)_2(\mu\text{-}SR)_2Cp_2]$ can be carried out photochemically or thermally, with the CO lability depending on R.[1086] For example, warming $[W(SPr^i)(CO)_2(PPh_3)Cp]$ in THF gives the tetracarbonyl dimer, and extended refluxing gives the dicarbonyl.[1093] As in the molybdenum analogue, the structures show folded and planar W_2S_2 cores, respectively; unlike the molybdenum analogue, however, the Cp ligands in the tetracarbonyl dimer are in a *cis* disposition.

7.6 CYCLOPENTADIENYL METAL HALIDES

The halides $[CrX(\mu\text{-}X)Cp]_2$ (X = Cl, Br I) and their Cp* analogues can be prepared from $[CrCl_3(THF)_3]$ by treatment with LiC_5H_5, from $[CrCp_2]$ by reaction with PX_3 or by the action of X_2 on $[Cr_2(CO)_nCp^*_2]$ (n = 4 or 6).[923,1094–6] The $[CrCl(\mu\text{-}Cl)Cp]_2$ complex has been shown to be a halide-bridged dimer in the solid state,[1097] but the monomeric derivatives $[CrCl_2(L)Cp]$ (L = THF, MeCN) have also been prepared.[1094,1098,1099] Cleavage of the dimer with phosphines in the presence of MeMgI or MeLi affords the alkyls $[CrMe_2(L)Cp]$ (L = PMe_3, PEt_3), which are paramagnetic but display interpretable shifted ^1H NMR spectra.[1100] The reaction of $CrCl_3$, LiCp, THF and dioxan produced [Li(THF)(dioxan)][$CrCl_3Cp$] which could be alkylated to $[CrR_3Cp]^-$ (R = Me, Ph).[1101] Reaction of $CrCl_2$ with LiCp*, or reduction of [CrCl(μ-Cl)Cp*]$_2$ affords the chromium(II) species [Cr(μ-Cl)Cp*]$_2$, which can be alkylated to [Cr(μ-R)Cp*]$_2$ (R = Me, Bu, CH_2TMS); hydrogenolysis of this forms the unusual cubane [Cr(μ_3-H)Cp*]$_4$.[1102]

The chemistry of cyclopentadienyl chromium(III) alkyls and halides has been studied by Theopold and co-workers.[1103] The starting materials are dimeric [Cr(R)(μ-Cl)Cp]$_2$ (or the Cp* analogues), which can be made by reaction of [Cr(R)Cl_2(THF)$_3$] with TlCp* or by the action of KCp* and MeLi on [$CrCl_3$(THF)$_3$].[1104–6] The dimers feature unusually short Cr–Cr distances and are also paramagnetic; this has been explained by a change from an edge-sharing bioctahedral structure to a face-sharing one as the Cr–Cr distance decreases.[1107] The tendency to dimerization decreases with increasing bulk of the Cp ligand and R, so that [Cr(CH_2TMS)(Cl)Cp*] is almost completely monomeric.[1108] Typical reactions of these compounds are shown in Scheme 32.[1109–16]

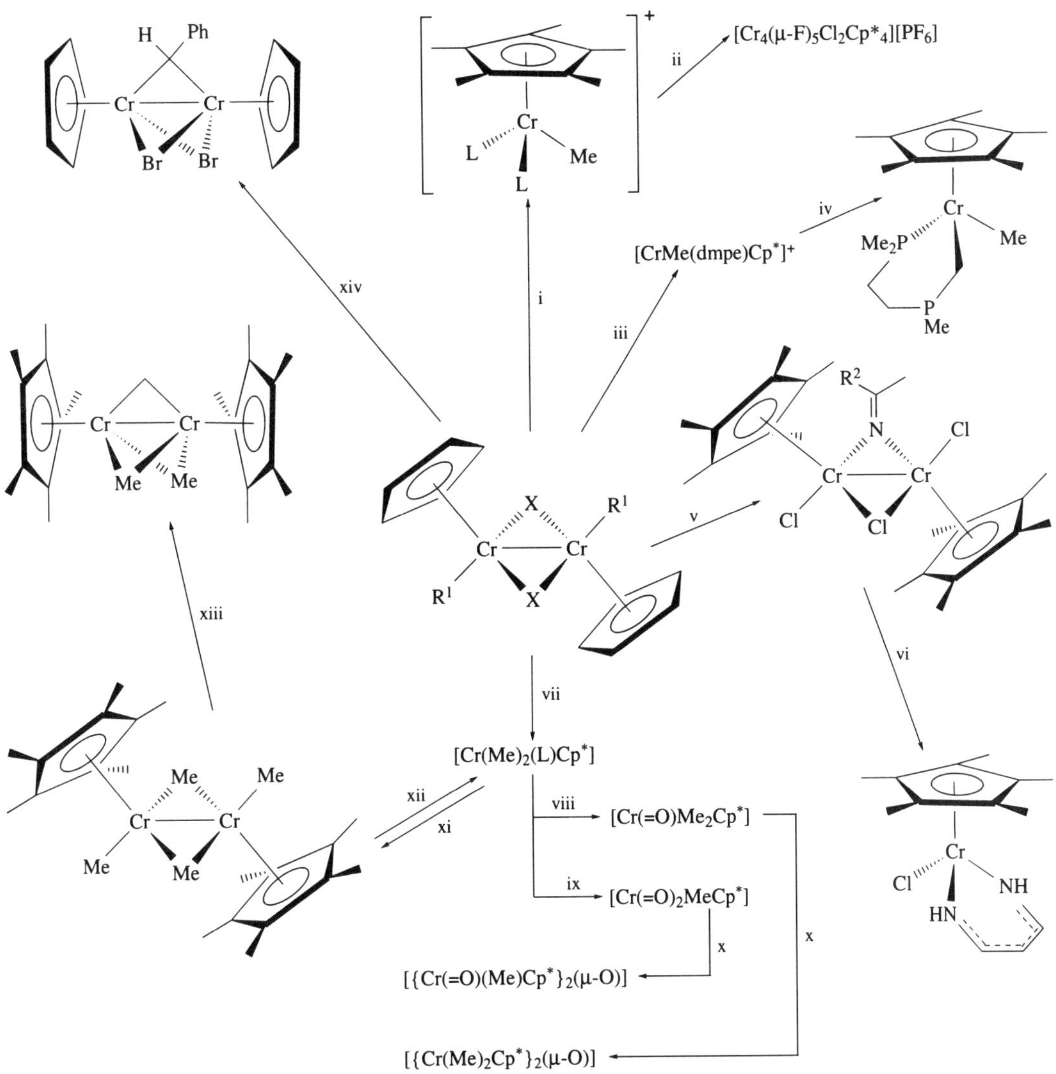

i, L (2 equiv.), $TlPF_6$ (L = py, THF, PMe_3); ii, stir (R = Me, L = THF); iii, dmpe, $TlPF_6$; iv, LiN(TMS)$_2$; v, R^2CN (R^2 = Me, Ph); vi, MeCN (excess); vii, MeLi, L = THF or py; viii, Me_3NO; ix, O_2; x, PPh_3 or slow decomposition; xi, vacuum; xii, THF; xiii, Δ; xiv, Δ, R^1 = CH_2Ph, X = Br

Scheme 32

Higher valent molybdenum and tungsten cyclopentadienyl halide chemistry has undergone a rapid development since the early 1980s, particularly due to efficient syntheses of [MCl_4Cp] and [$M_2Cl_4Cp_2$]. Further reduction of these compounds in the presence of phosphines allows access to other halide complexes of M^{II}, M^{III} and M^{IV}. Two reviews cover the following material in greater detail.[1117,1118]

Oxidation of [$MMe(CO)_3Cp^*$] with PCl_5 affords [MCl_4Cp^*] (90% for molybdenum, 80% for tungsten).[1119] Similar treatment of [$M_2(CO)_6Cp^*_2$] in MeCN produces the adduct [$MCl_4(NCMe)Cp^*$].[1099] The reactions of [MCl_4Cp^*] are shown in Scheme 33.[293,1120–2]

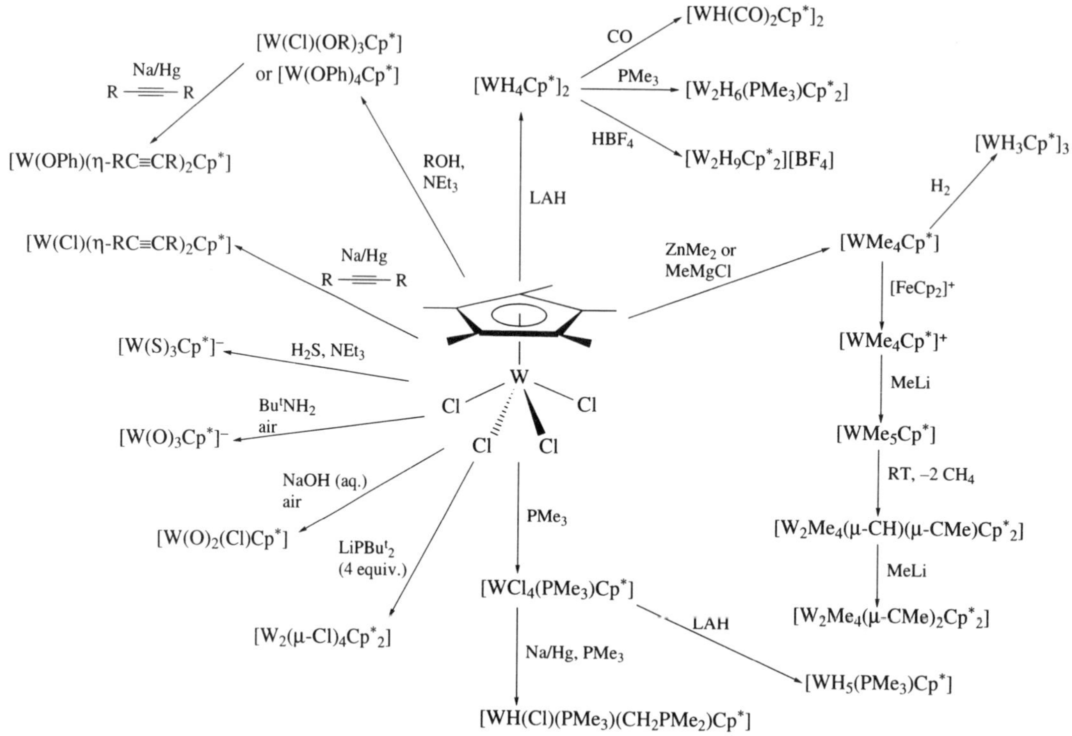

Scheme 33

The reaction of [$Mo(\eta\text{-}C_6H_6)_2$], available from metal vapour synthesis, with 6,6-dimethylfulvene followed by HCl, produces [$Mo_2(\mu\text{-}Cl)_4(\eta\text{-}C_5H_4Pr^i)_2$] (**35**);[285] more efficiently the same complex results from sodium/mercury reduction of [$MoCl_4(\eta\text{-}C_5H_4Pr^i)$].[1123] The structure has four bridging chloride ligands with a Mo–Mo bond length of 260.7(1) pm. It can be cleaved by phosphines to form [$MoCl_2(L)_2(\eta\text{-}C_5H_4Pr^i)$] ($L = PMe_3$; L_2 = dmpe, etc.) which can then be reduced to [$MoH_3(L)_2(\eta\text{-}C_5H_4Pr^i)$].[285] The dimer also reacts with LiSH to give the cubane [$Mo_4(\mu_3\text{-}S)_4(\eta\text{-}C_5H_4Pr^i)_4$].[1039] The crystal structure of [$Mo_2(\mu\text{-}Cl)_4(\eta\text{-}C_5Me_4Et)_2$], produced by reducing [$MoCl_4(NCMe)(\eta\text{-}C_5Me_4Et)$] with lithium phosphides, has also been determined; if $LiPPh_2$ is employed the related complex [$Mo_2(\mu\text{-}Cl)_3(\mu\text{-}PPh_2)(\eta\text{-}C_5Me_4Et)_2$] is formed.[1124] The cationic iodide dimer [$Mo_2(\mu\text{-}I)_4Cp^*_2$][$I_3$] has been isolated from attempted crystallization of [MoI_3Cp^*].[1125]

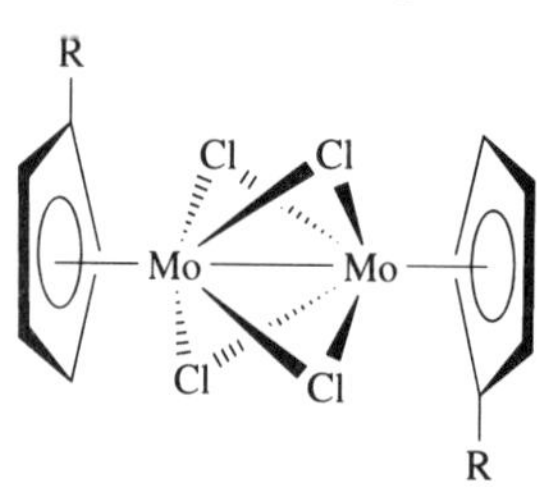

(35)

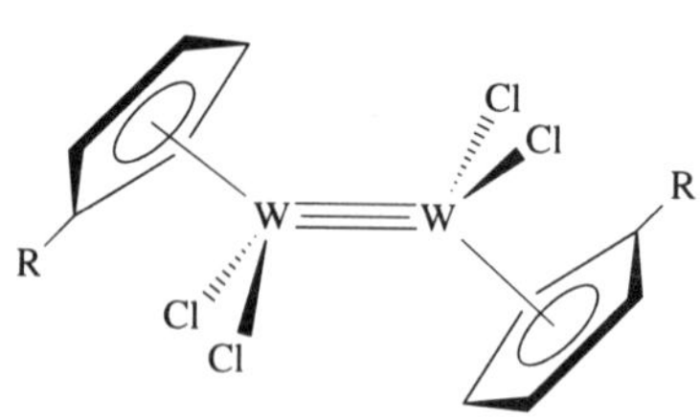

(36)

In contrast to the above, reduction of [$WCl_4(\eta\text{-}C_5H_4R)$] (R = Me, Pr^i) produces [$W_2Cl_4(\eta\text{-}C_5H_4R)_2$] (**36**), which has no bridging chlorides but instead an unsupported W≡W bond of length 0.23678(8) nm (R = Pr^i).[1123] Extended Hückel MO calculations and photoelectron spectra explain the difference in structure between the molybdenum and tungsten dimers in terms of their electronic configurations; the bonding in the former is predominantly ligand based with a Mo_2 configuration of $\sigma^2\delta^{*2}\delta^2$, while in the latter the bonding is metal–metal based with a $\sigma^2\pi^4$ configuration.[1126] The delicate balance between the two forms is illustrated by [$W_2(\mu\text{-}Cl)_4Cp^*_2$], which adopts the bridged structure.[1127] The reactions of [$W_2Cl_4(\eta\text{-}C_5H_4R)_2$] are shown in Scheme 34; particularly noteworthy are the novel bonding modes adopted by C_4R_4 and nitrile ligands.[1128–32]

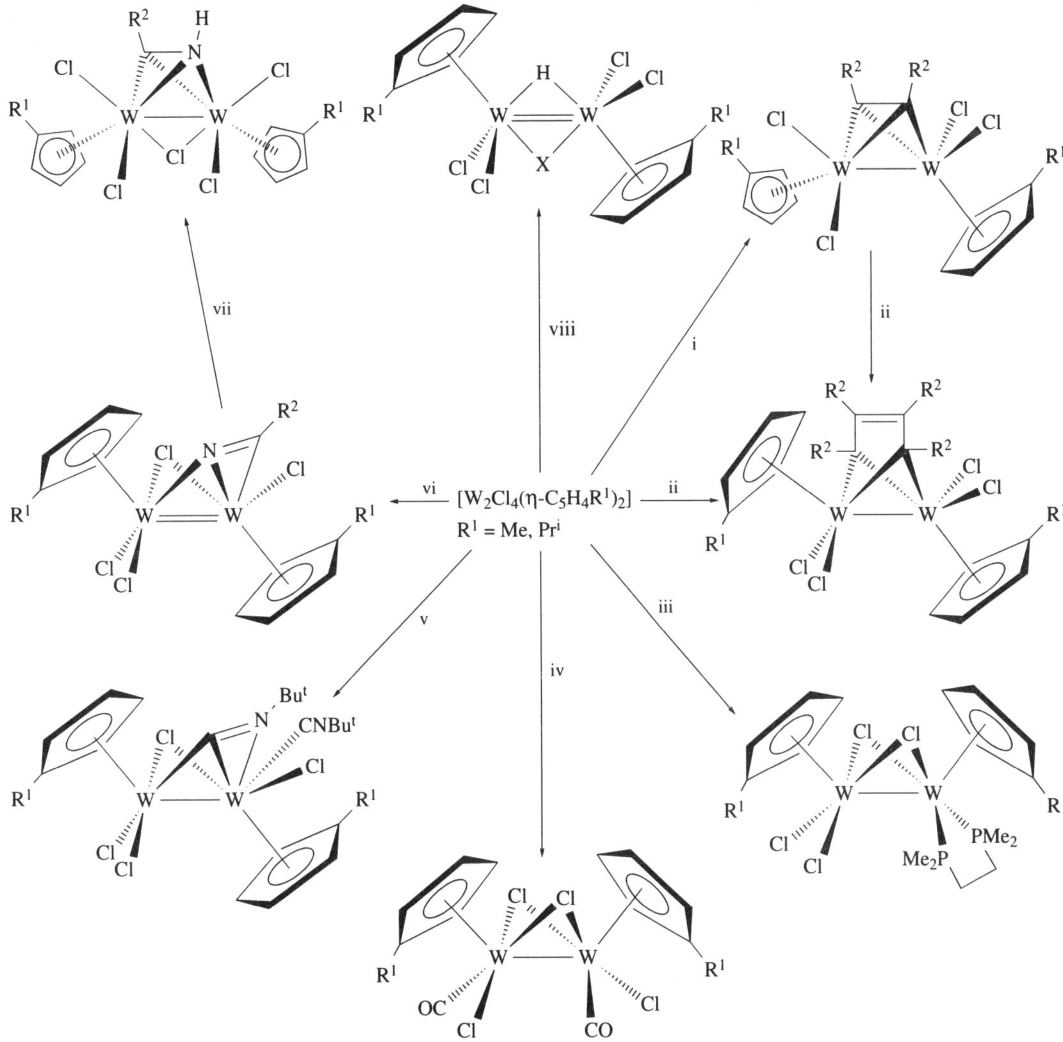

i, $R^2C{\equiv}CR^2$ (R^2 = Me, Et, TMS, Ph); ii, MeC≡CMe, excess; iii, dmpe; iv, CO; v, Bu^tNC; vi, R^2CN (R^2 = Me, Et, Ph); vii, HCl; viii, HX (X = H, Cl, SR^3, PPh_2)

Scheme 34

Reduction of the unsubstituted Cp complex [$MoCl_4Cp$] with zinc produces a material, presumably oligomeric, formulated as [$MoCl_2Cp$].[1133] It reacts with [PPN][Cl] to produce [PPN][$Mo_2(\mu\text{-}Cl)Cl_4Cp_2$], and with phosphines to give [$MoCl_2L_2Cp$] (L = PMe_3, PPh_3; L_2 = dppe), which can also be prepared from [$MoCl_3L_3$] and TlCp.[1134] These are low-spin, paramagnetic 17-electron complexes; ESR analysis of the Cp* analogues shows that the lone electron resides mainly in the Mo d_{z^2} orbital.[1135] Unusually they cannot easily be reduced to 18-electron species, but are readily oxidized to 16-electron [$MoCl_2(L)_2Cp$]$^+$, which has two unpaired electrons;[1136,1137] indeed, the substitution of [$MoI_2(PMe_3)_2Cp$] with Cl^- is catalysed by one-electron oxidants.[1138] This exchange can be followed by ESR spectroscopy, with [$Mo(I)(Cl)(PMe_3)_2Cp$] observed as intermediate.[1139,1140] The crystal structures of the oxidized species [$MoI_2(PMe_3)_2Cp$][I] and [$MoI_2(PMe_3)_2Cp^*$][BF_4] have been determined.[1141]

Reaction of [$MoCl_2Cp$] with $PhICl_2$ affords [$MoCl_3Cp$], which can also be made by reduction of [$MoCl_4Cp$] or its conproportionation with [$MoCl_2Cp$].[1142] Perhaps most usefully, simple thermal decarbonylation of [$MoX_3(CO)_2Cp$] (X = Cl, Br I) in refluxing CH_2Cl_2 provides [MoX_3Cp] and also allows access to the Cp* analogues.[1143,1144] Addition of ligands produces [$MoCl_3(L)_2Cp$] (L_2 = dmpe, dppe; L = PMe_3) or [$MoCl_3(L)Cp$] for larger phosphines; the Cp* analogues form only the mono-adducts.[1145] An alternative route to [$MoCl_3(dmpe)Cp$] from [$Mo(dmpe)Cp_2$][PF_6]$_2$ and its reduction to [$MoH_3(dmpe)Cp$] have also been described.[284] Addition of dmpe to [$Mo_2(\mu\text{-}Cl)Cl_4Cp_2$]$^-$ also gave [$MoCl_3(dmpe)Cp$],[1146] and [$MoX_2(Cl)(dppe)Cp$] (X = Cl, Br) can be prepared by X_2 oxidation of [$MoCl(CO)(dppe)Cp$].[302] In warm THF, [$MoCl_3Cp$] will abstract oxygen from the solvent to form paramagnetic $[Mo_3Cl_3(\mu\text{-}Cl)_4(\mu_3\text{-}O)Cp_3]$ in which two unpaired electrons are localized on the unique unsaturated molybdenum atom.[1147]

Reduction of $[MoCl_2Cp]$ or $[MoCl_4Cp^*]$ with sodium amalgam and PMe_3 affords $[MoCl(PMe_3)_3Cp]$ and its Cp* analogue.[286,1148] Addition of L causes displacement of one (L = N_2, C_2H_4) or two (L = 1,3-butadiene) PMe_3 ligands. The chloride complex can be methylated to $[Mo(Me)(PMe_3)_3Cp]$, which can also be made from $[MoCl_2(PMe_3)_2Cp]$ and MeLi via $[Mo(Me)_2(PMe_3)_2Cp]$.[1149] It can also be reduced to $[MoH(PMe_3)_3Cp]$; this, and the tungsten analogue, had previously been made by reaction of C_5H_6 with $[Mo(PMe_3)_6]$, $[WH_2(PMe_3)_4]$ or $[WH(PMe_3)_4(CH_2PMe_2)]$.[289,1150,1151]

7.7 BIS(CYCLOPENTADIENYL) METAL COMPOUNDS

7.7.1 [MCp₂] and Related Compounds

A number of new chromocenes have been synthesized by reaction of the appropriate alkali metal cyclopentadienide with $CrCl_2$ or $[Cr_2(\mu\text{-}OAc)_4]$ (Table 11). In some cases a mixture of cyclopentadienides was used to prepare mixed ligand complexes.

Table 11 New chromocenes.

Complex	*Yield (%)*	*Ref.*
$[Cr(\eta\text{-}C_5H_4EMe_3)_2]$ (E = Si, Ge, Sn)	31, 66, 54	1152
$[Cr(\eta\text{-}C_5Me_4Et)Cp]$	38	1153
$[Cr(\eta\text{-}C_5Me_4H)_2]$	69	1154
$[CrCp^*_2]$	64	1155
$[Cr(\eta\text{-}C_5Me_4Et)_2]$	26	1156
$[Cr(\eta\text{-}C_5HPh_4)_2]$	47	1157
$[Cr(\eta^5\text{-}C_9Me_7)_2]$	70	1158
$[Cr_2(\mu, \eta^5, \eta^5\text{-}C_{10}H_8)_2]$ (fulvalene)	38	1159
$[CpCr\{\mu, \eta^5, \eta^5\text{-}C_5H_3(SiMe_2)_2C_5H_3\}CrCp]$	43	1160
$[Cr\{(\mu,\eta^5,\eta^5\text{-}C_5H_3(SiMe_2)_2C_5H_3)FeCp\}_2]$	47	1161
$[Cr\{(\mu,\eta^5,\eta^5\text{-}C_5H_3(SiMe_2)_2C_5H_3)NiCp\}_2]$	42	1162
$[Cr(\eta\text{-}C_2Bu^t_2P_3)_2]$	not stated	1163

The ^1H and ^{13}C NMR spectra of these chromocenes are broad and paramagnetically shifted; however, it has been discovered that the ^2D NMR spectra, measurable at natural abundance, are over 30 times sharper.[1164,1165] The first example of a ^{29}Si paramagnetic shift has been observed in the C_5H_4TMS complex.[1166]

Chromocene itself has been investigated by a number of spectroscopic techniques. Analysis of the absolute intensities of its IR spectrum shows that the charge separation in the molecule is not large.[1167] The ionization energy of $[CrCp_2]$ can be found from the Rydberg transitions in its gas phase absorption spectrum,[1168] while electron transmission spectroscopy allows determination of the electron affinity; negative states with sizeable energy splittings are formed by capture of an electron into the e_{1g} orbital.[1169,1170]

Chromocene forms adducts with species such as N_2, CO, O_2 and H_2 which can be detected by low-temperature IR measurements.[1171,1172] It reacts with protic compounds such as diols, carboxylic acids, and acetylacetone by Cp displacement to give initially Cr^{II} complexes, for example $[Cr_2(\mu\text{-}O_2CCPh_3)_4]$, and ultimately the Cr^{III} species, for example $[Cr(acac)_3]$.[1173–9] Ethylenediamine also displaces both rings to give $[Cr(en)_3]^{2+}$.[1180] In some cases only one ring is removed, for example, reaction with Bu^tSH forms $[Cr_2(\mu\text{-}S)(\mu\text{-}SBu^t)_2Cp_2]$.[914] The reactions of various metallocenes, including chromocene, towards CdR_2 and HgR_2 has been reviewed.[1181]

At 0 °C in THF, chromocene reacts with CO to give $[Cr(CO)Cp_2]$, which readily loses CO on oxidation. The bent *ansa*-metallocene $[Cr(\eta^5,\eta^5\text{-}C_5Me_4CMe_2CMe_2C_5Me_4)]$, however, forms a more stable CO adduct, which can be oxidized reversibly to a radical cation and irreversibly to a dication.[1182–4] This stronger CO binding is a consequence of the distorted ground state geometry.[1185] Under CO pressure, the dicarbonyl adduct $[Cr(CO)_2Cp_2]$ is formed which, like its tungsten analogue, contains one η^3-bonded Cp ring, as shown by the x-ray structure of the Cp* analogue.[1186,1187] Geometry optimization shows that this ring is not planar, but there is a rapid interchange between the ring bonding modes involving an intermediate in which both rings are planar.[1188,1189]

A number of half-open chromocenes of the type [Cr(η^5-pentadienyl)Cp] have been prepared; these too form 18-electron adducts with CO, PF_3, RNC or PMe_2Ph, which can then be oxidized to 17-electron cations.[1190–3]

The elusive molybdenocene and tungstenocene have been trapped in inert argon matrices at low temperature on photolysis of [MH_2Cp_2] or [M(CO)Cp_2]; [W(η-C_2H_4)Cp_2] and [WH(Me)Cp_2] also give [WCp_2]. The intensity of the magnetic circular dichroism spectra of both metallocenes is temperature dependent, indicating the paramagnetic $^3E_{2g}$ ground state which had been previously predicted. The molybdenum complex is subject to a Jahn–Teller distortion but this is quenched by spin–orbit coupling in the tungsten compound.[1194–6]

Improved syntheses of some of the dimeric molybdenocenes previously reported have been developed. Reaction of [$Mo_2(OAc)_4$] with NaCp in the presence of PPh_3 provides a 55% yield of the new dimer [Cp_2Mo(H)(μ-σ,η^5-C_5H_4)MoCp] (**37**) which can then be converted into the isomeric *cis*- and *trans*-[{Mo(H)Cp}$_2$(μ-σ,η^5-C_5H_4)$_2$] and [{Mo(H)Cp}(μ,η^5,η^5-$C_{10}H_8$)] as described earlier. Photolysis of the latter product produces the former one, and then [Mo_2(μ,σ,η^5-C_5H_4)$_2Cp_2$] (**38**).[1197,1198] The structure of a methyl-substituted example of the last of these has been confirmed crystallographically.[1199] The initial product can also be treated with HX or NaOH, giving useful yields of [MoX_2Cp_2] and [MoH_2Cp_2]. Reaction of [$Mo_2(OAc)_4$] with NaCp under a CO atmosphere produces [$Mo_2(CO)_3$(μ-σ,η^5-C_5H_4)Cp_2].[1197]

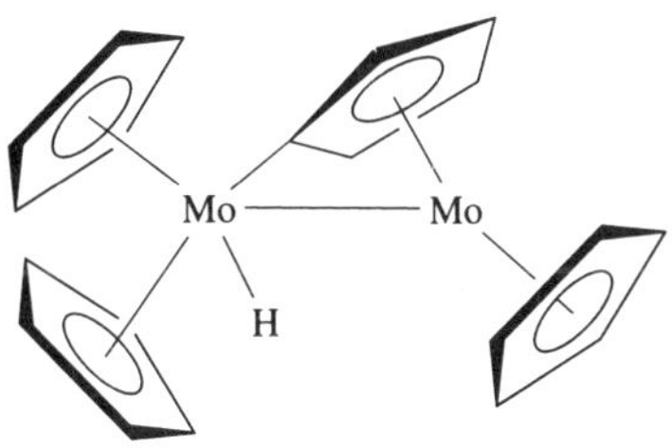

(37)

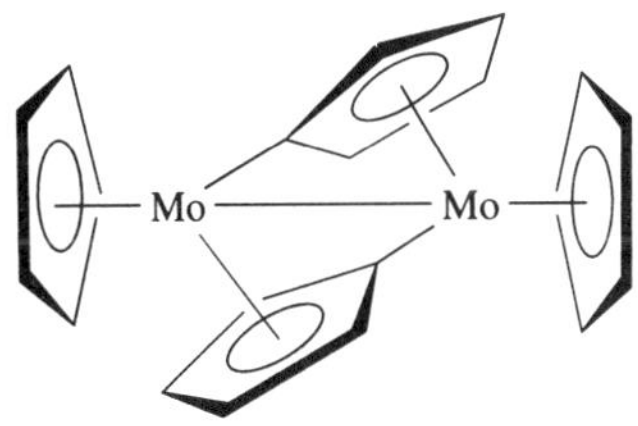

(38)

7.7.2 [MH_2Cp_2] and Related Compounds

An improved procedure for the synthesis of [MoH_2Cp_2] involves treatment of $MoCl_5$ with NaCp.DME and $NaBH_4$ in a THF–hexane mixture, with yields of around 40% attainable in 5–6 days.[1200] The C_5H_4Me analogue can be made in the same way.[1201] More recently, an improved preparation of [WCl_4(DME)] has allowed the synthesis of [WH_2Cp_2] in 42% yield by reaction with LiCp and $LiAlH_4$.[1202] The [$MH_2Cp^*_2$] analogues are more difficult to prepare, but can be made by cocondensation of molybdenum or tungsten atoms with C_5Me_5H; on photolysis they give the 'tucked-in' species [MH(η-$C_5Me_4CH_2$)Cp*] and [M{η-$C_5Me_4CH_2$}$_2$], and the same oxidative addition process can be observed in their mass spectra.[1203,1204] Reduction of [$WCl_2Cp^*_2$] has also been used to prepare [$WH_2Cp^*_2$].[1205] The unsymmetrical [MoH_2Cp(η-C_5Ph_4H)] has been made by cyclization of two C_2Ph_2 ligands.[1206] The IR and solution Raman spectra of [MH_2Cp_2] and its protonation product have been analysed in detail, allowing assignment of the M–H and M–Cp regions.[1207]

The thermochemistry of many of the following metallocene complexes have been studied by the group of Dias; the data are collected and analysed in a review article.[1208] By a combination of calorimetry and theoretical arguments, it is possible to derive values for bond enthalpies (which do not take into account reorganization of the MCp_2 fragment) and mean bond dissociation enthalpies of the various M–L groups, and to predict the thermochemistry of new compounds with reasonable accuracy. In many cases the ΔH_f° values of the crystalline complexes [ML_2Cp_2] (L = halide, O_2CR, etc.) show a linear correlation with the ΔH_f° of HL.[1209] Since the publication of the above review, the values of the M–H bond dissociation enthalpy in [MH_2Cp_2] have been re-evaluated as 256 ± 8 and 311 ± 4 kJ mol^{-1} for molybdenum and tungsten, respectively.[1210]

Some reactions of [MH_2Cp_2] are displayed in Scheme 35. Due to the presence of the lone pair of electrons in an orbital situated between the two hydride ligands, the complexes are able to act as Lewis bases. Protonation of [$WH_2Cp^*_2$] is proposed to proceed by attack at the two W–H bonds rather than at the metal, however, as deuterium labelling shows that the H^+ becomes attached first in the central position and subsequently slowly isomerizes to an external site. An intermediate which involves the H_3^+ ligand is one possibility.[1211] The room temperature NMR spectrum of [WH_3Cp_2]$^+$ shows an AB_2 pattern for the hydrides with a coupling of 8.5 Hz, whereas the molybdenum analogue collapses to an AB_2 pattern only at low temperature with an anomalously large J(HH) of at least 450 Hz due to exchange coupling.[1212] The trihydride cation can reduce ketones, and reacts with allyl alcohols to give cyclic $[Cp_2Mo(CH_2CH_2CMe_2OH)]^+$.[1213–16] Little new chemistry involving the deprotonation of $[MoH_2Cp_2]$ has been performed; the x-ray structure of the K^+/18-crown-6 salt has been determined, and reaction of the dihydride with $MgMe_2$ and TMEDA gives $[Mg\{MoHCp_2\}_2]$.[1217,1218] Formation of $[Mo(PMe_3)Cp_2]$ from $[MoH_2Cp_2]$ and $Me_3P{=}CH_2$ probably proceeds via the analogous phosphonium salt $[PMe_4][MoHCp_2]$.[1219]

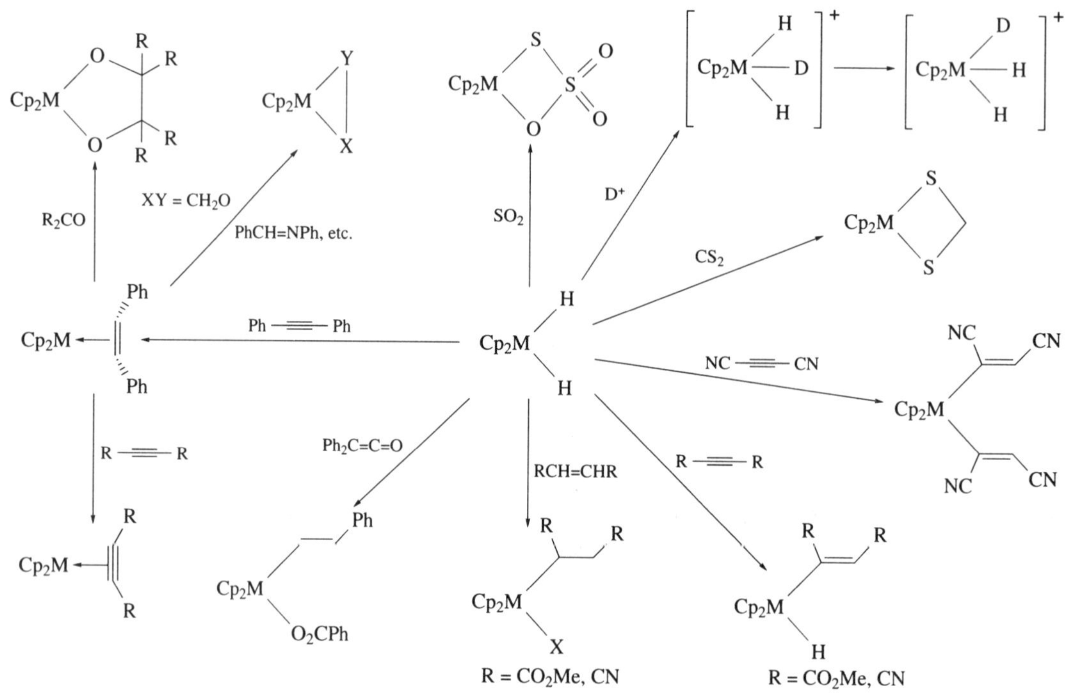

Scheme 35

Numerous further examples of the insertion of activated alkynes into the M–H bonds of $[MH_2Cp_2]$ have been explored; $C_2(CN)_2$ inserts into both M–H bonds, but in most cases $[M(H)(CR{=}CHR)Cp_2]$ is formed.[1220–4] Protonation of the latter with HX affords the alkyls $[MX(CHRCH_2R)Cp_2]$ (X = Cl, OTf, SPh, SH).[1225,1226] Diphenylacetylene gives the useful [M(η-*Z*-PhCH=CHPh)Cp_2], from which the labile *cis*-stilbene ligand can be displaced by a variety of other ligands, including heteroallenes, formaldehyde, nonenolizable ketones and alkynes.[1227–32] The resulting alkyne species [M(η-RC≡CR)Cp_2] can also be made by treating the hydrides with methyl acrylate in the presence of the alkyne.[1233–5] Activated alkenes $RCH{=}CH_2$ (R = CN, CO_2Me) also insert into one M–H bond of $[MH_2Cp_2]$ via intermediate donor–acceptor complexes;[1236] other substrates that undergo insertion include CS_2 and $Ph_2C{=}C{=}O$.[1237,1238] The η^2-thiosulfate species formed from SO_2 decomposes on addition of HX to give $[M({=}S)XCp_2]^+$, SO_2 and H_2O.[1239]

The reaction of $[MoH_2Cp_2]$ with $BH_3{\cdot}THF$ affords initially $[Mo(H)(B_2H_5)Cp_2]$, which can also be made from $[MoCl_2Cp_2]$ and $B_3H_8^-$; on prolonged reaction, however, incorporation of one Cp ring into the borane framework occurs to give the two unusual complexes (**39**) and (**40**).[1240,1241]

As Lewis bases, the $[MH_2Cp_2]$ compounds are about as strong as Et_2O and form adducts with a number of metal fragments, such as CuCl, $[Cu(NCMe)_4]^{2+}$, Ag^+ and $[PtR(acetone)(PEt_3)_2]^+$.[1242–4] The adducts with CuCl were thought to be simple hydride-bridged species, but later work formulates them as salts of the type $[\{Mo(H)Cp_2\}_2H][CuCl_2]$.[1245] Whereas $[RhH_2(acetone)_2(PPh_3)_2]^+$ forms a simple adduct with μ-H ligands, the analogous iridium complex causes activation of one Cp ring to provide $[CpM(\mu\text{-}H)_2(\mu\text{-}\sigma,\eta^5\text{-}C_5H_4)IrH(PPh_3)_2]^+$.[1246,1247] Transfer of a hydride to the acyl complex $[Zr(X)(COR)Cp_2]$ (X = Me, Cl) occurs to yield $[Cp_2MCHROZr(X)Cp_2]$.[1248] Reactions with $[Fe_2(CO)_9]$ or $[Co_2(CO)_8]$ however give only the $[HFe_3(CO)_{10}]^-$ or $[Co(CO)_4]^-$ salts of $[MH(CO)Cp_2]^+$.[1249,1250]

Exposure of $[WH_2Cp_2]$ to CO gives $[W(CO)Cp_2]$, which is resistant to further carbonylation except in the presence of oxidants, when it forms the ring-slipped compound $[W(CO)_2(\eta^3\text{-}Cp)Cp]$.[1251] The carbonyl complex undergoes [2 + 2] cycloaddition reactions with a number of heteroallenes to give metallacycles of the type $[W\{C({=}X)NRC({=}Y)\}Cp_2]$ (X, Y = O, S, NR), and reacts with $[ZrH_2Cp^*_2]$ to produce $[Cp_2W{=}CHOZrHCp^*_2]$.[1252–4]

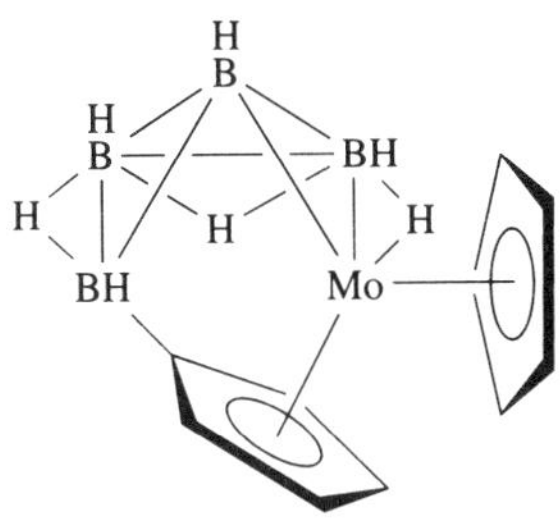

(39)

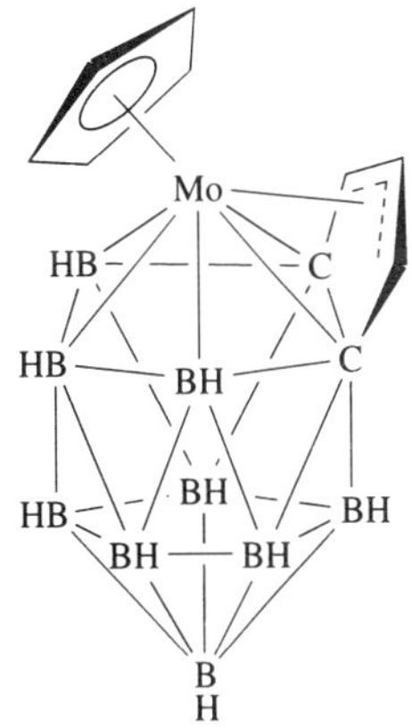

(40)

Irradiation of $[MoH_2Cp_2]$ under a CO_2 atmosphere produces initially the side-bound η^2-CO_2 complex $[Mo(\eta^2\text{-}CO_2)Cp_2]$, which gradually changes to a mixture of $[Mo(CO)Cp_2]$ and the carbonate complex $[Mo(\eta^2\text{-}CO_3)Cp_2]$, the identity of which was confirmed by independent synthesis from $[MoCl_2Cp_2]$ and Na_2CO_3.[1255] The CO_2 disproportionation reaction only occurs photochemically. It was found that exposure of $[Mo(CO)Cp_2]$ to O_2 also affords small amounts of the CO_2 and carbonate complexes as well as $[Mo(=O)Cp_2]$.[1256] The CO_2 complex reacts with electrophiles (HCl, TMS-Cl) to give $[Mo(Cl)(CO)Cp_2]^+$; with RCOCl $[Mo(O_2CR)(CO)Cp_2]^+$ is formed.[1257] It also reacts with $[HCo(CO)_4]$ to give $[Cp_2Mo(\mu\text{-}H)(\mu\text{-}CO)Co(CO)_3]$ with an unsymmetrical bridging hydride, via $[Cp_2Mo(H)(CO)][Co(CO)_4]$.[1258] The tungsten carbonate $[W(\eta^2\text{-}CO_3)Cp_2]$ can be prepared by treating $[WH(Ph)Cp_2]$ with CO_2 in wet acetone.[1259]

The reaction of $[MX_2Cp_2]^+$ (X = halide) with $NaBH_4$ in THF containing 3% water allows the synthesis of $[MH(X)Cp_2]$ in reasonable yields for both metals.[1260,1261] Stirring $[WH_2Cp_2]$ with CCl_4, $CHBr_3$ or MeI can also give good yields of $[WH(X)Cp_2]$ as the second halogenation takes place more slowly.[1262] A convenient route to $[MoH(L)Cp_2]^+$ and $[MoH(L_2)Cp_2]^+$ consists of treating $[MoH(I)Cp_2]$ with L or L_2 in the presence of $TlBF_4$; the reaction is slower for the analogous chlorine or bromine species, and does not occur for tungsten. With L_2 = dppe, hydride transfer onto one Cp ring occurs to produce $[Mo(dppe)(\eta^4\text{-}C_5H_6)Cp]^+$.[1263]

7.7.3 $[MX_2Cp_2]$ (X = Halide) and Related Compounds

The dihalides $[MX_2Cp_2]$ can be prepared by stirring the corresponding dihydrides in $CHCl_3$ or other appropriate halogenating agent; the reverse transformation can be achieved by reduction with $LiAlH_4$. Reduction of $[WCl_4(PMe_3)Cp^*]$ with magnesium and reaction with LiCp* provides a convenient route to $[WCl_2Cp^*_2]$, from which a range of other derivatives can be prepared.[1205] The barrier to ring rotation in solid $[MoCl_2Cp_2]$ has been measured by spin–lattice relaxation methods.[1264] The *in vitro* antitumour activity of various metallocene dihalides, including $[MoCl_2Cp_2]$, has been investigated; this activity does not appear to be simply due to the release of free C_5H_6 by decomposition.[1265,1266]

Electrochemical oxidation of $[MX_2Cp_2]$ in MeCN is quasireversible, forming $[MX(NCMe)Cp_2]^+$; these useful adducts can also be isolated on a preparative scale.[1261,1267] Complex formation between $[WX_2Cp_2]$ and electron acceptors such as TCNE and TCNQ has also been demonstrated.[1268] In the presence of very strong oxidants such as AsF_5 or SbF_5, salts of the dicationic M^{VI} species $[MCl_2Cp_2]^{2+}$ can be isolated.[1269] Treatment of $[MCl_2Cp_2]$ with WF_6 in liquid SO_2, however, affords the M^V complex $[MCl_2Cp_2]^+$ as either the $[WF_6]^-$ or $[W_4F_{18}]^{2-}$ salts.[1270,1271]

An extremely useful route to chiral metallocene complexes has been discovered by Cooper and co-workers. In the presence of $MgCl_2$, certain Grignard reagents RMgBr (R = allyl, $CH_2CMe{=}CH_2$, CH_2Ph) add to $[WCl_2Cp_2]$ at the Cp ring to produce $[W(H)(Cl)(\eta\text{-}C_5H_4R)Cp]$; after reconversion to the dihalide the process can be repeated to give $[W(H)Cl(\eta\text{-}C_5H_4R)_2]$.[1272,1273] In the same way the substituted complex $[WH_2\{\eta\text{-}C_5H_4Si(TMS)_3\}_2]$ can be made by treating $[WCl_2Cp_2]$ with $Li[Si(TMS)_3]$.[1274] Similarly, a range of nucleophiles such as OBu^{t-}, SPr^{i-}, and PPh_2^- as well as RMgBr add to the Cp ring of $[W(Br)(SMe_2)Cp_2]^+$ to give initially a η^4-C_5H_5R intermediate which loses SMe_2 to produce $[W(H)(Br)(\eta\text{-}C_5H_4R)Cp]$.[1275] The diastereomeric product can be used to show that substitution of the MeCN ligand in $[W(Me)(NCMe)(\eta\text{-}C_5H_4R)Cp]^+$ occurs with retention of configuration.[1276]

A number of species of the type $[MX(L)Cp_2]^+$ and $[ML_2Cp_2]^{2+}$ have been synthesized either by treating the L = MeCN complexes mentioned above with L (L = CO, PPh_3, PMe_3) or simply by the reaction of the dihalides with ligands such as RNC.[1261,1277,1278] Both $[MX_2Cp_2]$ and $[MX_2Cp_2]^+$ probably react photochemically by dissociation of a chloride radical followed by either charge transfer or substitution.[1279] Kinetics show that the MeCN substitution reaction is dissociative under both thermal and photochemical conditions.[1280] Two groups have reported their failure to reproduce the published preparation of the M^{VI} oxo-halides $[M(O)(Cl)_2Cp_2]$ or the indenyl analogues from $[M(O)Cl_4]$, and attempts to prepare them by halogenation of $[Mo(=O)Cp_2]$ also failed.[1281,1282] Alternatively, numerous new derivatives have been described, with the same preparation cited.[1283,1284]

7.7.4 $[MCp_2]$ Derivatives of Group 14 Elements

The isonitrile complexes $[MoX(CNMe)Cp_2]^+$ (X = halide, H, Me) and $[Mo(CNMe)_2Cp_2]^{2+}$ can be prepared by alkylation of the appropriate cyanide complex precursors.[1285] Sodium amalgam reduction of the halides $[MoX(CNR)Cp_2]^+$ produces $[Mo(CNR)Cp_2]$, in which the RNC ligand is probably end-bound; it undergoes alkylation at the metal with MeI, but heating causes methyl migration to the RNC ligand to give an η^2-iminoacyl species.[1277] The side-bound acetonitrile complexes $[M(MeCN)Cp_2]$ can be made by reduction of $[MoCl_2Cp_2]$ or decomposition of $[W(H)(Me)Cp_2]$ in MeCN; protonation in MeCN affords $[Mo(NCMe)(NH{=}CHMe)Cp_2]^{2+}$.[1286–8]

Alkylation of $[MX_2Cp_2]$ with LiR or AlR_3 produces the dialkyls $[MR_2Cp_2]$. Metallacyclobutanes $[\overline{M(CH_2CMe_2CH_2})Cp_2]$, metallacyclobutabenzenes $[\overline{M(CH_2C_6H_4})Cp_2]$, and metallasilacyclobutanes $[\overline{M(CH_2SiMe_2CH_2})Cp_2]$ can all be made by employing the appropriate di-Grignard reagents.[1289–93] Convenient syntheses of metallocene dialkyls have also been developed by Green and co-workers. Cocondensation of molybdenum or tungsten atoms with spiro[2.4]hepta-4,6-diene produces $[\overline{M(\eta^5,\sigma\text{-}C_5H_4CH_2CH_2})_2]$ which can be treated with halogens X_2 to give $[MX_2(\eta\text{-}C_5H_4CH_2CH_2X)_2]$.[1294–6] Similarly spiro[4.4]nona-1,3-diene yields $[\overline{M(\eta^5,\sigma\text{-}C_5H_4CH_2CH_2CH_2CH_2})_2]$[1297] and C_5Me_6 produces $[MoMe_2Cp^*_2]$.[1298]

Electrochemical studies show that $[MR_2Cp_2]$ can be readily oxidized to the monocations, with a more negative potential for R = Me than for higher alkyls (–0.27 V compared to around –0.21 V).[1299–301] With AsF_5, $[MoMe_2Cp_2]$ oxidizes to the Mo^{VI} dication.[1302] Exposure of $[MMe_2Cp_2]$ to ^{60}Co γ-rays produces radical anions, but also $[M(Me)(CH_2)Cp_2]$ radicals according to the ESR spectra.[1303]

Studies on the elimination reactions of $[W(H)(Me)Cp_2]$ show that scrambling of the hydride and methyl protons occurs intramolecularly, and in concentrated solutions intermolecularly too. The most plausible explanation for this and the observed inverse isotope effect is an intermediate η^2-CH_4 σ-bond complex.[1304] Reductive elimination of RH from $[W(H)(Me)Cp^*_2]$ at 100 °C producing $[\overline{W(H)(\sigma,\eta^5\text{-}C_5Me_4CH_2})Cp^*]$, also proceeds via a similar intermediate.[1305]

Elegant mechanistic studies, particularly by Cooper *et al.*, have uncovered several more examples of α-elimination reactions in cationic tungstenocene dialkyls.[1306] Treatment of $[WMe_2Cp_2]^+$ with trityl radical (in the form of its dimer) affords $[W(H)(\eta\text{-}C_2H_4)Cp_2]^+$ by an intramolecular process involving hydrogen atom abstraction, insertion of the methylene group into the W–Me bond, and β-elimination to form the alkene, as shown in Scheme 36.[1307] The same mechanism operates from $[WMe_2Cp_2]$ and Ph_3C^+ as an initial electron transfer step occurs.[1308] The intermediate $W{=}CH_2$ complex can be trapped as $[W(Me)(CH_2PMe_2Ph)Cp_2]^+$ by addition of the appropriate phosphine. With $[W(Me)(Et)Cp_2]$, hydride abstraction occurs only from the methyl group, as $[W(Et)(CH_2PMe_2Ph)Cp_2]^+$ is formed on addition of phosphine. In the case of $[W(Me)(Ph)Cp_2]^+$, where β-elimination is not possible for the last step, the MeCN solvent coordinates to give $[W(CH_2Ph)(NCMe)Cp_2]^+$;[1309] in CH_2Cl_2, $[W(Cl)(CH_2Ph)Cp_2]^+$ is formed instead.[1310] To check whether the β-elimination step might in fact be another α-elimination followed by a 1,2-H shift, the stereochemistry of the propene complex $[WH(\eta^2\text{-}C_3H_6)Cp_2]^+$ formed from *n*-propyl and *i*-propyl intermediates was examined; the former gave exclusively *endo*, the latter exclusively *exo* as expected for a true β-elimination (see Scheme 36).[1311]

Reaction of $[W(H)(\eta\text{-}C_2H_4)Cp_2]^+$ with I_2 produces $[W(I)(=CHMe)Cp_2]^+$, which can be deprotonated to $[W(I)(CH{=}CH_2)Cp_2]$. The proposed mechanism involves firstly insertion to give $[W(Et)Cp_2]^+$, followed by α-elimination, hydrogen atom abstraction and reaction with iodine radical. This shows that α-elimination can occur even when β-elimination also appears a viable possibility.[1312]

Further examples of the reactions of $[W(Me)(\eta\text{-}C_2H_4)Cp_2]^+$ with phosphines, namely PMe_3 and $PMePh_2$, have been documented. The compounds $[W(Me)(CH_2CH_2PMe_3)Cp_2]^+$ and $[W(H)(CH_2PMe_3)Cp_2]^+$ are formed sequentially, and the latter is in thermal equilibrium with $[W(Me)(PMe_3)Cp_2]^+$.[1313,1314] The kinetics of the isomerization of $[W(H)(CH_2PMe_2Ph)Cp_2]^+$ to

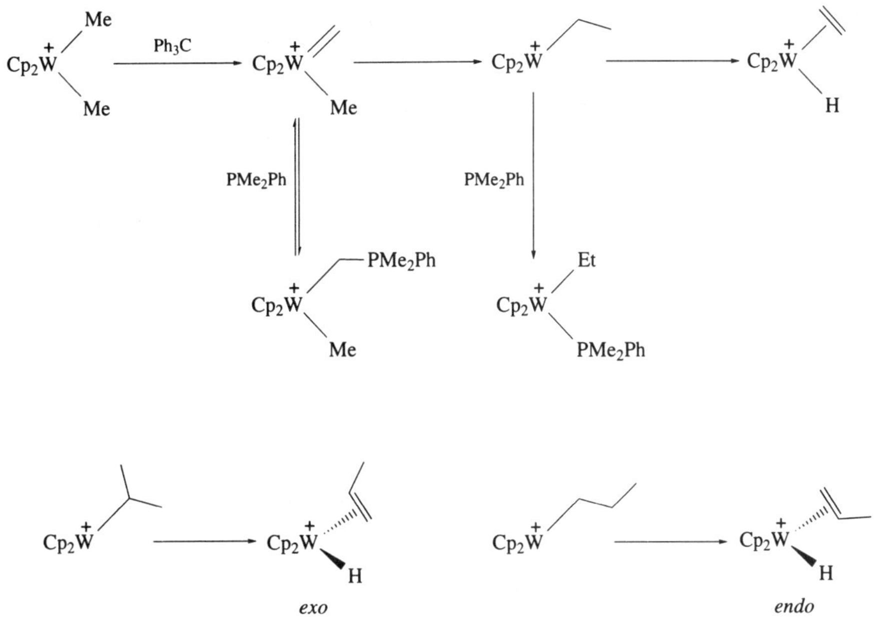

Scheme 36

$[W(Me)(PMe_2Ph)Cp_2]^+$ are consistent with an intermediate involving an agostic methyl group, or a carbene hydride in fast equilibrium with a W–Me unit, but not with simple phosphine dissociation.[1315]

Reductive cyclization of $[W(Cl)(CH_2SiMe_2Cl)Cp_2]$ with magnesium affords the silene complex $[W(CH_2SiMe_2)Cp_2]$;[1316] the disilene complexes $[M(SiMe_2SiMe_2)Cp_2]$ can be prepared similarly.[1317,1318] The reactivity of these species is shown in Scheme 37.

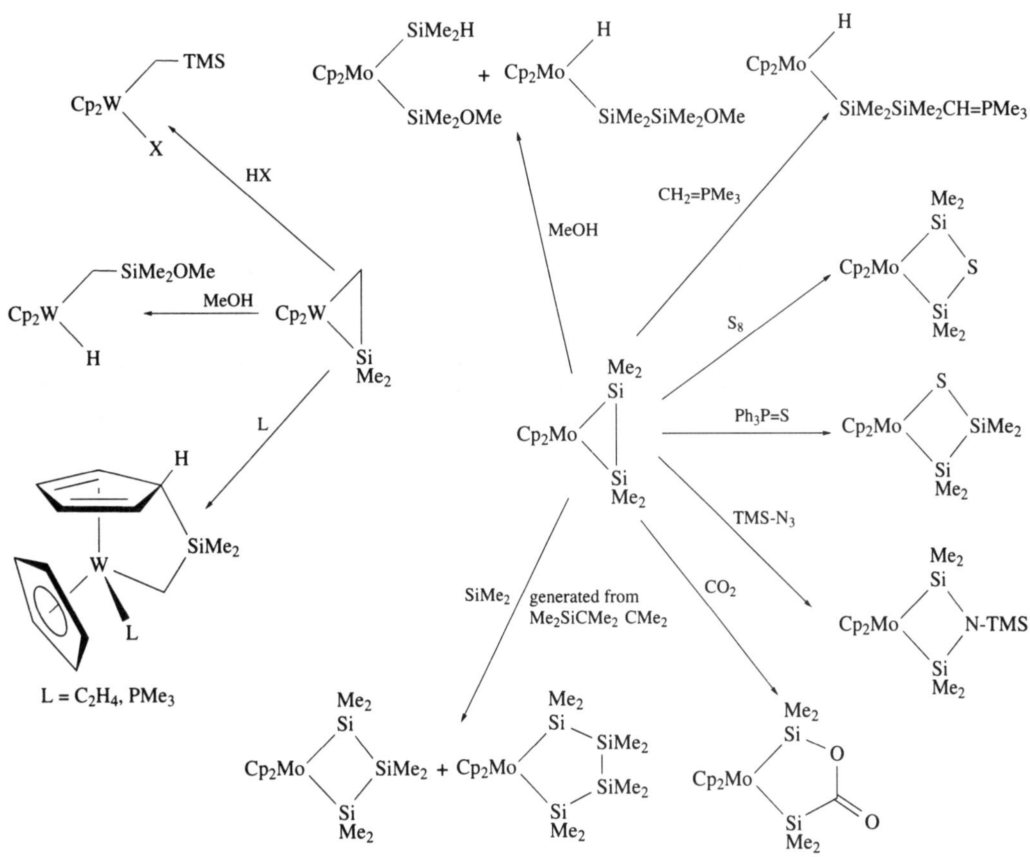

Scheme 37

Interaction of $[MoH_2Cp_2]$ with tin halides affords compounds of the type $[Mo(H)(SnR_3)Cp_2]$ and $[Mo(SnR_3)_2Cp_2]$ with direct Mo–Sn bonds; several of these have been crystallographically characterized.[1319–22] The complex $[Mo(H)(SnPh_3)Cp_2]$ can be deprotonated and functionalized at the metal, whereas the analogous $[Mo(H)(SiMePh_2)Cp_2]$ undergoes migration of the silyl group to the Cp ring instead.[1323] The dihydride also reacts with $PbMe_3(OAc)$ to form $[\{MoHCp_2\}_2Pb(OAc)_2]$.[1324]

7.7.5 $[MCp_2]$ Derivatives of Group 15 Elements

A variety of complexes containing a *N,N*-chelate ligand bound to the MCp_2 unit have been made, including the amidinate species $[Mo(R^1NCR^2NR^1)Cp_2]^+$ (R^1 = aryl, R^2 = H, Me, Ph) and the analogous triazenido complexes $[Mo(RNNNR)Cp_2]^+$.[1325–7] Unusual reactions are observed between $[MoCl_2Cp_2]$ and nucleotides, for example with 9-methyladenine the N(1), N(6) chelate is the kinetic product and the N(7), N(6) chelate the thermodynamic one.[1328,1329] A variety of mixed species $[MX(py)Cp_2]^+$ (X = H, halide) have also been made for various 3- and 4-aminopyridines; chelate compounds are obtained with the 2-isomers.[1330,1331] Crystal structures of isostructural $[M(phen)Cp_2]^{2+}$ have been determined and the absence of backbonding to the low-lying orbital on the phen ligand confirmed by EHMO calculations.[1332]

Little new chemistry of $[Mo(NO)(I)Cp_2]$ has been reported, but its oxidation with $AgBF_4$ in MeCN gives the adduct $[Mo(NO)(NCMe)Cp_2]^+$.[1333] The synthesis of the mixed complex [Mo(NO)(X)CpCp*] has also been achieved.[671]

The first terminal phosphinidene complexes, $[M(=PMes^*)Cp_2]$ (**41**), have been prepared from $[M(H)(Li)Cp_2]_4$ and Mes^*PCl_2; they display extremely low-field ^{31}P NMR shifts of 799.5 and 666.1 ppm for molybdenum and tungsten, respectively.[1334] The related $[W(=PNHMes^*)Cp^*_2]$ was prepared by an analogous route.[1335] An alternative approach, treating $[WCl_2Cp'_2]$ with LiP(Mes*)TMS gave instead 1,2- and 1,3-isomers of $[WH(\sigma,\eta^5\text{-}PMes^*C_5H_3Me)Cp]$ (**42**).[1336] The reaction of $[MH_2Cp_2]$ with PPh_2Cl affords $[MH(PPh_2H)Cp_2]^+$, which undergoes alkyne insertion into the M–H bond, but also deprotonation to the pyramidal phosphido complex $[MH(PPh_2)Cp_2]$; addition of $[Mn(CO)_2(THF)Cp]$ gives heterodinuclear species.[1337,1338] The chelating phosphido species $[\overline{M(PhPC_6H_4PPh)}Cp_2]$ have also been made.[1339]

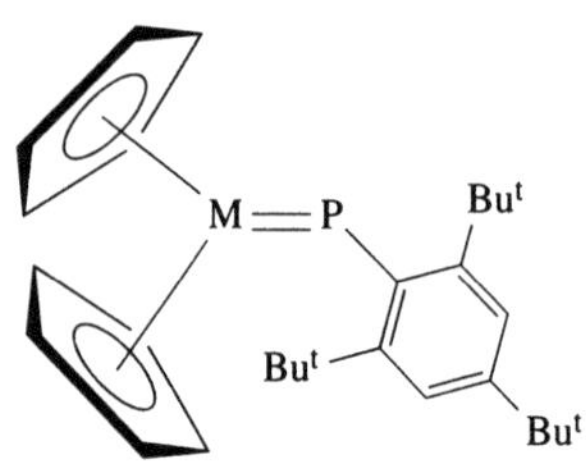

(41)

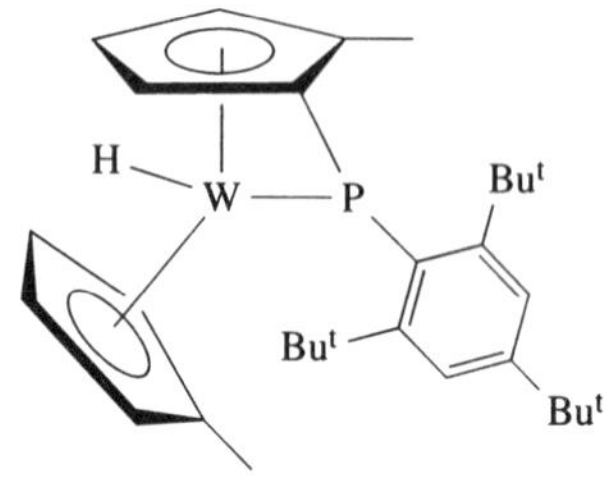

(42)

The monohydride [Mo(H)(OTs)Cp_2], formed by protonation of the dihydride with TsOH in ethanol at 80 °C, reacts with PPh_3 to give $[MoH(PPh_3)Cp_2]^+$, deprotonation of which provides a useful route to $[Mo(PPh_3)Cp_2]$.[1340] This compound undergoes oxidative additions of HCl, MeI, etc. to give $[M(R)(PPh_3)Cp_2]^+$.[1341] The phosphine in $[Mo(PEt_3)Cp_2]$ is labile; it can, for example, be displaced by ketenes.[1342] The reaction of $[MoI_3(THF)(dppe)]$ with indenyl sodium produces $[Mo(dppe)(\eta\text{-}C_9H_7)_2]$, which contains one η^3- and one η^5-indenyl ring; rapid coordination mode exchange between the two occurs even at low temperature.[1343]

7.7.6 $[MCp_2]$ Derivatives of Group 16 Elements

Treatment of $[MCl_2Cp_2]$ with aqueous KOH affords the oxo complexes $[M(=O)Cp_2]$.[1200,1344] Geoffroy has shown that the oxo ligand in $[M(=O)Cp_2]$ undergoes attack by electrophiles and [2 + 2] cycloadditions, including with the CO groups of electrophilic metal complexes to afford heterobimetallic μ-CO_2 species (Scheme 38).[1345] The analogous $[W(=O)Cp^*_2]$ also shows unusual reactivity; the oxo ligand exchanges extremely rapidly with $H_2{}^{18}O$ and can be reduced with H_2 to give $[W(H)_2Cp^*_2]$. Its reaction with oxygen is particularly novel as insertion of an oxo group into the W–Cp* bond occurs to give $[W(=O)_2(OC_5Me_5)Cp^*]$ (**43**); with H_2O_2 only ring slippage occurs to produce

$[W(=O)_2(\eta^1\text{-}C_5Me_5)Cp^*]$.[1205,1346,1347] The reactivity of the oxo ligand is attributed to the fact that, barring ring slippage of the Cp rings, it can only donate a maximum of two electrons to the metal. The weakness of the M=O bond is reflected in both its IR frequency and length, for example 0.172(2) nm for $[Mo(=O)Cp'_2]$.[1344] Irradiation of $[Mo(=O)Cp_2]$ forms molybdenocene (which dimerizes) and O_2, which reacts with more oxo complex to yield $[Cp_2Mo(\mu\text{-}O)_2MoO_2]_2$, which can be better made from $[MoO_4]^{2-}$ and $[MoCl_2Cp_2]$.[1348,1349]

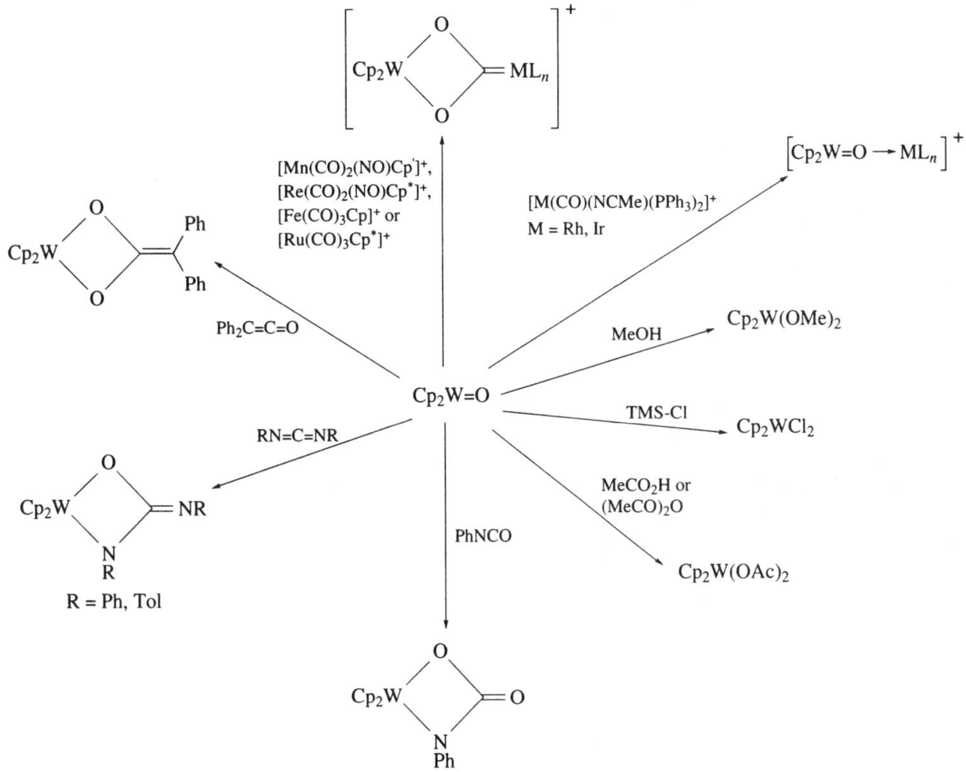

Scheme 38

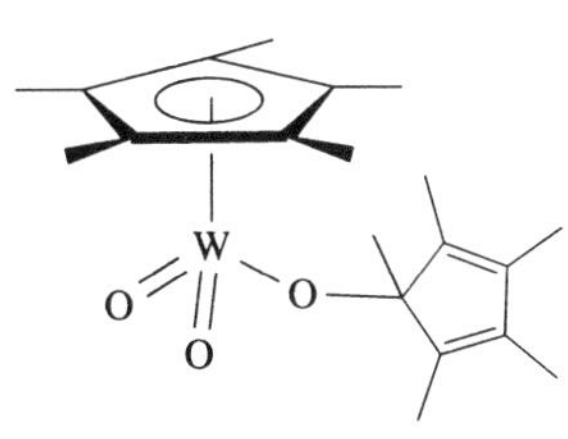

(43)

Routes to the carboxylate complexes $[M(O_2CR)_2Cp_2]$ and $[MH(O_2CR)Cp_2]$ have been described.[344] In the dark, $[WH_2Cp_2]$ reacts with RCO_2H by protonation, but exposure to light and one equivalent of O_2 as hydride acceptor gives good yields of the bis(carboxylates).[1350] The mixed hydrido-carboxylates are prepared from $[WH(Ph)Cp_2]$, which also reacts with ROH to give the alkoxides $[WH(OR)Cp_2]$.[1259,1351] The heats of formation of the carboxylates and of complexes with chelating diol ligands have been determined by calorimetry; the latter show some degree of ring strain, reflected in lower M–O bond strengths.[1352,1353]

Two dicyclopentadienyl molybdenum sulfides have been reported, $[Mo(S_2)Cp_2]$ and $[Mo(S_4)Cp_2]$; the former can be converted to the latter photochemically or by treatment with S_8, while PPh_3 effects the reverse transformation.[1354–6] The S_4 complex undergoes alkyne insertion to form tri- and dithiolene species.[1357] Recently, indirect evidence for the existence of $[Mo(=S)Cp_2]$ has been obtained by treating $[Mo(S_2)Cp_2]$ with PMe_3 in the presence of TolN=C=NTol; $S=PMe_3$ is formed and the carbodiimide undergoes cyclization to give $[\overline{Mo\{SC(=NTol)N}Tol\}Cp_2]$. Attempts to isolate the monosulfide were unsuccessful, but it could be detected in solution.[1358] Only the S_4 complex is known for tungsten.

Numerous complexes of the type $[M(SR)_2Cp_2]$ have been prepared, ranging from the simplest (R = H)[1359] to species where the SR groups form part of a crown ether or thiacrown ligand; attempts to influence the electrochemistry of the molybdenum moiety by complexation of alkali metals were, however, only partially successful.[1360,1361] The Mo–SR bond enthalpies in $[Mo(SR)_2Cp_2]$ increase as the length of the alkyl chain R increases, but there is also a steric effect when R = Bu^t.[1362–4]

The dithiolate complexes generally display three redox waves in MeCN, the first corresponding to the Mo^{IV}/Mo^{V} couple, the second to a second oxidation and also to $[Mo(SR)(NCMe)Cp_2]^+$, and the third to $[Cp_2Mo(\mu\text{-}SR)_2MoCp_2]^{2+}$, which could be isolated.[1267] Oxidation of $[Mo(SR)_2Cp_2]$ with AsF_5 in liquid SO_2 gives the Mo^{VI} species $[Mo(SR)_2Cp_2][AsF_6]_2$, which can also be made by reacting $[MoCl_2Cp_2]^{2+}$ with LiSR.[1365]

Unusual reactions were observed between $[Mo(\eta^2\text{-}CS_2)Cp_2]$ and activated alkynes to produce the metallacycle $[\overline{Mo\{C(=S)SCR=C}R\}Cp_2]$ and its *S, C*-bound isomer. With HC≡CCN incorporation of two alkyne molecules occurs and the metallacycle becomes linked to one Cp ligand.[1366]

The dithiolate complexes are capable of acting as chelate ligands to other metal fragments; a large number of such complexes are known with metals such as rhodium, nickel, platinum, palladium and copper.[344,1367,1368] However, the outcome is not always straightforward; reaction of $[W(SH)_2Cp_2]$ with $[Pd(NCMe)_2(dppe)]^{2+}$ produced $[W(S_2CNHMe)Cp_2]^+$ by hydride transfer to the MeCN.[1369]

A number of mixed ligand species of the type $[M(SR)(L)Cp_2]^+$ have been prepared by oxidation of the dithiolate complexes in the presence of L, including some with L = NH_3.[1370,1371] Reports of selenium complexes are limited to $[\overline{M(SeC_6H_4}Se)Cp_2]$ and $[W(Cl)(SePh)Cp_2]$; the mass spectra of the latter and of some related dithiolenes have been analysed in detail.[1372,1373]

7.8 OTHER CYCLOPENTADIENYL METAL COMPLEXES

A comparative study of the electrochemistry of $[M(\eta\text{-}C_6H_6)_2]$ and $[M(\eta\text{-}C_7H_7)Cp]$ (M = Cr, V) has shown that the oxidation potentials correlate with ionization energies from photoelectron spectra.[1374] A well-defined Rydberg structure has been observed in the gas-phase absorption spectrum of the latter chromium complex.[1375] One-pot syntheses of $[M(\eta\text{-}C_7H_7)(\eta\text{-}C_7H_9)]$ from $MoCl_5$ or WCl_6 have been developed, which in turn lead to improved access to $[M(\eta\text{-}C_7H_7)Cp]$ via the dihalides $[MI_2(NCMe)(\eta\text{-}C_7H_7)]$.[1376] The naphthalene complex $[Cr(\eta^6\text{-}C_{10}H_8)Cp]$ has been prepared from chromocene and $Li[C_{10}H_8]$; subsequent reaction with acetylene produces $[Cr_2(\mu\text{-}C_8H_8)Cp_2]$.[1377]

A one-pot synthesis of the flyover species $[Cr_2(\mu\text{-}C_8H_8)Cp_2]$ in 46% yield from $[CrCl_2(tmeda)]$, Na and $K_2[C_8H_8]$ has been devised; subsequent thermolysis gives the ring-closed cot complex $[Cr_2(\mu\text{-}C_8H_8)Cp_2]$ in 55% yield.[1378,1379] Related compounds in which the Cp rings are linked by CH_2, $SiMe_2$ or $GeMe_2$ spacers have also been prepared, but the linking group has little effect on either the overall geometry or the redox properties; electrochemically all undergo one reduction and two oxidations, all reversible, and chemical oxidation with $[FeCp_2][BF_4]$ produces stable monocations.[1380,1381] Heteronuclear derivatives can also be made by treating $[Cr(\eta^6\text{-}C_8H_8)Cp]$ with $Fe_2(CO)_9$, $[Fe(\eta\text{-}C_6H_6)Cp]$, $[Co(\eta\text{-}C_2H_4)_2Cp]$ or $[M(CO)_3(NCMe)_3]$ (M = Cr, W),[1382,1383] and recently the related flyover species $[Cr_2(\mu\text{-}C_nMe_n)Cp_2]$ (n = 4 or 6) have been synthesized by oligomerization of C_2Me_2 with $[CrCl_2Cp^*]$ and active magnesium.[1384]

The reaction of $CrCl_2$, NaCp and $B_8H_8C_4Et_4{}^{2-}$ gives the paramagnetic 13-vertex *nido*-cluster $[Cr(\eta\text{-}B_8H_8C_4Et_4)Cp]$, with the 12-vertex $[Cr(\eta\text{-}B_7H_7C_4Et_4)Cp]$ isolable as a by-product if air is admitted.[1385] The reaction of $[CrCp_2]$ with 5,6-$C_2B_8H_{12}$ affords firstly $[Cr(\eta\text{-}C_2B_8H_{11})Cp]$ and then $[(CrCp)_2(C_2B_8H_{10})]$.[1386]

7.9 REFERENCES

1. H. Schumann, 'Mo: Organomolybdenum Compounds; Gmelin Handbook of Inorganic and Organometallic Chemistry', 8th edn. Springer Verlag, Berlin, 1991, vol. 7.
2. S. Trofimenko, *Chem. Rev.*, 1993, **93**, 943.
3. A. R. Manning, P. Hackett, R. Birdwhistell and P. Soye, *Inorg. Synth.*, 1990, **28**, 148.
4. M. D. Curtis and M. S. Hay, *Inorg. Synth.*, 1990, **28**, 150.
5. N. A. Cooley, P. T. F. MacConnachie and M. C. Baird, *Polyhedron*, 1988, **7**, 1965.
6. T. J. Jaeger and M. C. Baird, *Organometallics*, 1988, **7**, 2074.
7. S. P. Nolan and C. D. Hoff, in 'Organometallic Syntheses', eds. R. B. King and J. J. Eisch, Elsevier, Amsterdam, 1988, Vol. 4, p. 67.
8. H. B. Abrahamson and M. J. Heeg, *Inorg. Chem.*, 1984, **23**, 2281.
9. R. M. Medina, J. R. Masaguer, M. Morán and J. Losada, *Inorg. Chim. Acta*, 1988, **146**, 115.
10. F. Edelmann, S. Töfke and U. Behrens, *J. Organomet. Chem.*, 1986, **309**, 87.

11. H. Brunner, R. Grassl, J. Wachter, B. Nuber and M. L. Ziegler, *J. Organomet. Chem.*, 1992, **431**, 143.
12. T. E. Bitterwolf and A. L. Rheingold, *Organometallics*, 1991, **10**, 3856.
13. J. Heck, K.-A. Kriebisch and H. Mellinghoff, *Chem. Ber.*, 1988, **121**, 1753.
14. W. Abriel and J. Heck, *J. Organomet. Chem.*, 1986, **302**, 363.
15. K. P. C. Vollhardt and T. W. Weidman, *Organometallics*, 1984, **3**, 82.
16. P. Jutzi and J. Schnittger, *Chem. Ber.*, 1989, **122**, 629.
17. T. S. Coolbaugh, R. J. Coots, B. D. Santarsiero and R. H. Grubbs, *Inorg. Chim. Acta*, 1985, **98**, 99.
18. S. C. Tenhaeff and D. R. Tyler, *Organometallics*, 1991, **10**, 473.
19. A. Avey and D. R. Tyler, *Organometallics*, 1992, **11**, 3856.
20. A. Dormond, A. El Bouadili and C. Moise, *Tetrahedron Lett.*, 1983, **24**, 3087.
21. M. Scheer, K. Schuster, K. Schenzel, E. Herrmann and P. G. Jones, *Z. Anorg. Allg. Chem.*, 1991, **600**, 109.
22. X. Zhou, R. Jin, H. Wang and X. Yao, *Huaxue Xuebao*, 1992, **50**, 930.
23. M. Fei, S. K. Sur and D. R. Tyler, *Organometallics*, 1991, **10**, 419.
24. A. Avey, S. C. Tenhaeff, T. J. R. Weakley and D. R. Tyler, *Organometallics*, 1991, **10**, 3607.
25. S. C. Tenhaeff and D. R. Tyler, *Organometallics*, 1991, **10**, 1116.
26. S. C. Tenhaeff and D. R. Tyler, *Organometallics*, 1992, **11**, 1466.
27. R. O. Gould, J. Barker and M. Kilner, *Acta Crystallogr., Sect. C*, 1988, **44**, 461.
28. I. R. Lyatifov, T. K. Gasanov, T. K. Kurbanov, A. N. Shnulin and V. K. Shil'nikov, *Koord. Khim.*, 1990, **16**, 1343 (*Chem. Abstr.*, 1991, **114**, 15 336k).
29. W. C. Watkins *et al.*, *J. Am. Chem. Soc.*, 1992, **114**, 907.
30. L. Y. Goh, T. W. Hambley, D. J. Darensbourg and J. Reibenspies, *J. Organomet. Chem.*, 1990, **381**, 349.
31. P. Leoni, F. Marchetti, M. Pasquali and P. Zanello, *J. Chem. Soc., Dalton Trans.*, 1988, 635.
32. W. Clegg, N. A. Compton, R. J. Errington and N. C. Norman, *Acta Crystallogr., Sect. C*, 1988, **44**, 568.
33. S. C. Tenhaeff, D. R. Tyler and T. J. R. Weakley, *Acta Crystallogr., Sect. C*, 1991, **47**, 303.
34. R. Drews and U. Behrens, *Chem. Ber.*, 1985, **118**, 888.
35. W. S. Striejewske, R. F. See, M. R. Churchill and J. D. Atwood, *Organometallics*, 1993, **12**, 4413.
36. J. S. Drage and K. P. C. Vollhardt, *Organometallics*, 1986, **5**, 280.
37. A. L. Rheingold and J. R. Harper, *Acta Crystallogr., Sect. C*, 1991, **47**, 184.
38. T. J. R. Weakley, A. A. Avey, Jr. and D. R. Tyler, *Acta Crystallogr., Sect. C*, 1992, **48**, 154.
39. H. G. Alt, J. S. Han and R. D. Rogers, *J. Organomet. Chem.*, 1993, **454**, 165.
40. R. Breeze, A. Ricalton and M. W. Whiteley, *J. Organomet. Chem.*, 1987, **327**, C29.
41. A. Ricalton and M. W. Whiteley, *J. Organomet. Chem.*, 1989, **361**, 101.
42. R. L. Beddoes, A. Ricalton and M. W. Whiteley, *Acta Crystallogr., Sect. C*, 1988, **44**, 2025.
43. I. R. Lyatifov, T. K. Gasanov, P. V. Petrovskii and A. I. Lutsenko, *J. Organomet. Chem.*, 1989, **361**, 181.
44. W. E. Lindsell and P. J. Tomb, *J. Organomet. Chem.*, 1989, **378**, 245.
45. S. Aime, L. Cordero, R. Gobetto and G. Szalontai, *Spectrochim. Acta*, 1993, **49A**, 1307.
46. H. C. Dorn, B. E. Hanson and E. Motell, *J. Organomet. Chem.*, 1982, **224**, 181.
47. D. F. R. Gilson and G. Gomez, *J. Organomet. Chem.*, 1982, **240**, 41.
48. S. P. Nolan, R. Lopez de la Vega, S. L. Mukerjee, A. A. Gonzalez, K. Zhang and C. D. Hoff, *Polyhedron*, 1988, **7**, 1491.
49. J. T. Landrum and C. D. Hoff, *J. Organomet. Chem.*, 1985, **282**, 215.
50. S. P. Nolan, R. Lopez de la Vega and C. D. Hoff, *Inorg. Chem.*, 1986, **25**, 4446.
51. S. L. Scott, J. H. Espenson and W.-J. Chen, *Organometallics*, 1993, **12**, 4077.
52. S. L. Scott, J. H. Espenson and Z. Zhu, *J. Am. Chem. Soc.*, 1993, **115**, 1789.
53. M. L. Baker, P. E. Bloyce, A. K. Campen, A. J. Rest and T. E. Bitterwolf, *J. Chem. Soc., Dalton Trans.*, 1990, 2825.
54. R. H. Hooker and A. J. Rest, *J. Chem. Soc., Dalton Trans.*, 1990, 1221.
55. J. Knorr and T. L. Brown, *J. Am. Chem. Soc.*, 1993, **115**, 4087.
56. B. A. Van Vlierberge and H. B. Abrahamson, *J. Photochem. Photobiol. A*, 1990, **52**, 69.
57. M. D. Curtis, *Polyhedron*, 1987, **6**, 759.
58. M. J. Winter, *Adv. Organomet. Chem.*, 1989, **29**, 101.
59. M. A. Greaney, J. S. Merola and T. R. Halbert, *Organometallics*, 1985, **4**, 2057.
60. S. A. R. Knox, R. F. D. Stansfield, F. G. A. Stone, M. J. Winter and P. Woodward, *J. Chem. Soc., Dalton Trans.*, 1982, 173.
61. M. D. Curtis, N. A. Fotinos, L. Messerle and A. P. Sattelberger, *Inorg. Chem.*, 1983, **22**, 1559.
62. F. Calderazzo, A. Juris, R. Poli and F. Ungari, *Inorg. Chem.*, 1991, **30**, 1274.
63. B. J. Morris-Sherwood, C. B. Powell and M. B. Hall, *J. Am. Chem. Soc.*, 1984, **106**, 5079.
64. T. K. Gasanov, I. R. Lyatifov and A. N. Shnulin, *J. Organomet. Chem.*, 1989, **361**, 173.
65. J.-S. Huang and L. F. Dahl, *J. Organomet. Chem.*, 1983, **243**, 57.
66. L.-C. Song *et al.*, *J. Organomet. Chem.*, 1993, **453**, 105.
67. S. Aime *et al.*, *Inorg. Chem.*, 1992, **31**, 3054.
68. N. N. Turaki and J. M. Huggins, *Organometallics*, 1985, **4**, 1766.
69. P. R. Drake and M. C. Baird, *J. Organomet. Chem.*, 1989, **363**, 131.
70. C. G. Arena *et al.*, *J. Chem. Soc., Dalton Trans.*, 1992, 1847.
71. P. Blenkiron, M. H. Lavender and M. J. Morris, *J. Organomet. Chem.*, 1992, **426**, C28.
72. E. A. V. Ebsworth, A. P. McIntosh and M. Schröder, *J. Organomet. Chem.*, 1986, **312**, C41.
73. J. T. Mague and Z. Lin, *Organometallics*, 1992, **11**, 4139.
74. K. A. Azam, A. J. Deeming, M. S. B. Felix, P. A. Bates and M. B. Hursthouse, *Polyhedron*, 1988, **7**, 1793.
75. V. Riera, M. A. Ruiz and F. Villafañe, *Organometallics*, 1992, **11**, 2854.
76. D. J. Brauer, G. Hasselkuss and O. Stelzer, *J. Organomet. Chem.*, 1987, **321**, 339.
77. V. Riera, M. A. Ruiz, F. Villafañe, C. Bois and Y. Jeannin, *J. Organomet. Chem.*, 1990, **382**, 407.
78. V. Riera, M. A. Ruiz, F. Villafañe, C. Bois and Y. Jeannin, *J. Organomet. Chem.*, 1989, **375**, C23.
79. V. Riera, M. A. Ruiz, F. Villafañe, C. Bois and Y. Jeannin, *Organometallics*, 1993, **12**, 124.
80. M. A. Alvarez, M. E. Garcia, V. Riera, M. A. Ruiz, C. Bois and Y. Jeannin, *J. Am. Chem. Soc.*, 1993, **115**, 3786.
81. M. A. Alvarez, M. E. Garcia, V. Riera, M. A. Ruiz, C. Bois, and Y. Jeannin, *Angew. Chem., Int. Ed. Engl.*, 1993, **32**, 1156.

82. J. Wachter, A. Mitschler and J. G. Riess, *J. Am. Chem. Soc.*, 1981, **103**, 2121.
83. J. Wachter, J. G. Riess and A. Mitschler, *Organometallics*, 1984, **3**, 714.
84. H. Adams *et al.*, *J. Chem. Soc., Dalton Trans.*, 1987, 341.
85. R. D. Adams, D. A. Katahira and L.-W. Yang, *Organometallics*, 1982, **1**, 231.
86. H. Brunner, H. Buchner, J. Wachter, I. Bernal and W. H. Ries, *J. Organomet. Chem.*, 1983, **244**, 247.
87. I. Bernal, H. Brunner and J. Wachter, *J. Organomet. Chem.*, 1984, **277**, 395.
88. S. F. T. Froom, M. Green, R. J. Mercer, K. R. Nagle, A. G. Orpen and R. A. Rodrigues, *J. Chem. Soc., Dalton Trans.*, 1991, 3171.
89. S. F. T. Froom, M. Green, R. J. Mercer, K. R. Nagle, A. G. Orpen and S. Schwiegk, *J. Chem. Soc., Chem. Commun.*, 1986, 1666.
90. N. M. Doherty, C. Elschenbroich, H.-J. Kneuper and S. A. R. Knox, *J. Chem. Soc., Chem. Commun.*, 1985, 170.
91. J. T. C. R. Burckett-St. Laurent, P. B. Hitchcock, H. W. Kroto, M. F. Meidine and J. F. Nixon, *J. Organomet. Chem.*, 1982, **238**, C82.
92. G. Becker *et al.*, *Angew. Chem., Int. Ed. Engl.*, 1983, **22**, 413.
93. M. F. Meidine, C. J. Meir, S. Morton and J. F. Nixon, *J. Organomet. Chem.*, 1985, **297**, 255.
94. P. B. Hitchcock, M. F. Meidine and J. F. Nixon, *J. Organomet. Chem.*, 1987, **333**, 337.
95. H. Alper, F. W. B. Einstein, R. Nagai, J.-F. Petrignani and A. C. Willis, *Organometallics*, 1983, **2**, 1291.
96. H. Alper, F. W. B. Einstein, F. W. Hartstock and A. C. Willis, *J. Am. Chem. Soc.*, 1985, **107**, 173.
97. W. Ziegler, D. Wörmsbächer and U. Behrens, *J. Organomet. Chem.*, 1992, **431**, C11.
98. H. Alper, F. W. B. Einstein, J.-F. Petrignani and A. C. Willis, *Organometallics*, 1983, **2**, 1422.
99. T. Sielisch, W. Ziegler and U. Behrens, *J. Organomet. Chem.*, 1989, **359**, 315.
100. R. Hussong, H. Heydt, G. Maas and M. Regitz, *Chem. Ber.*, 1987, **120**, 1263.
101. H. Alper, F. W. B. Einstein, F. W. Hartstock and R. H. Jones, *Organometallics*, 1987, **6**, 829.
102. Y. Shi *et al.*, *J. Organomet. Chem.*, 1993, **455**, 115.
103. W. A. Herrmann and L. K. Bell, *J. Organomet. Chem.*, 1982, **239**, C4.
104. L. K. Bell, W. A. Herrmann, G. W. Kriechbaum, H. Pfisterer and M. L. Ziegler, *J. Organomet. Chem.*, 1982, **240**, 381.
105. W. A. Herrmann, L. K. Bell, M. L. Ziegler, H. Pfisterer and C. Pahl, *J. Organomet. Chem.*, 1983, **247**, 39.
106. W. A. Herrmann and G. Ihl, *J. Organomet. Chem.*, 1983, **251**, C1.
107. L. K. Bell, W. A. Herrmann, M. L. Ziegler and H. Pfisterer, *Organometallics*, 1982, **1**, 1673.
108. M. D. Curtis, L. Messerle, J. J. D'Errico, W. M. Butler and M. S. Hay, *Organometallics*, 1986, **5**, 2283.
109. M. D. Curtis and L. Messerle, *Organometallics*, 1987, **6**, 1713.
110. J. J. D'Errico, L. Messerle and M. D. Curtis, *Inorg. Chem.*, 1983, **22**, 849.
111. L. Messerle and M. D. Curtis, *J. Am. Chem. Soc.*, 1980, **102**, 7789.
112. L. Messerle and M. D. Curtis, *J. Am. Chem. Soc.*, 1982, **104**, 889.
113. J. J. D'Errico and M. D. Curtis, *J. Am. Chem. Soc.*, 1983, **105**, 4479.
114. M. D. Curtis, L. Messerle, J. J. D'Errico, H. E. Solis, I. D. Barcelo and W. M. Butler, *J. Am. Chem. Soc.*, 1987, **109**, 3603.
115. N. D. Feasey, S. A. R. Knox and A. G. Orpen, *J. Chem. Soc., Chem. Commun.*, 1982, 75.
116. N. D. Feasey and S. A. R. Knox, *J. Chem. Soc., Chem. Commun.*, 1982, 1062.
117. M. D. Curtis, L. Messerle, N. A. Fotinos and R. F. Gerlach, in 'Reactivity of Metal–Metal Bonds', ed. M. H. Chisholm, American Chemical Society, Washington D.C., 1981, p. 221.
118. W. A. Herrmann, G. W. Kriechbaum, R. Dammel, H. Bock, M. L. Ziegler and H. Pfisterer, *J. Organomet. Chem.*, 1983, **254**, 219.
119. M. D. Curtis, J. J. D'Errico and W. M. Butler, *Organometallics*, 1987, **6**, 2151.
120. H. Brunner, B. Hoffmann and J. Wachter, *J. Organomet. Chem.*, 1983, **252**, C35.
121. I. Bernal, M. Draux, H. Brunner, B. Hoffmann and J. Wachter, *Organometallics*, 1986, **5**, 655.
122. H. Brunner, W. Meier, J. Wachter, I. Bernal and E. Raabe, *J. Organomet. Chem.*, 1989, **362**, 95.
123. A. J. Mayr, B. Carrasco-Flores, L. Párkányi and K. H. Pannell, *J. Am. Chem. Soc.*, 1992, **114**, 5467.
124. M. D. Curtis, M. S. Hay, W. M. Butler, J. Kampf, A. L. Rheingold and B. S. Haggerty, *Organometallics*, 1992, **11**, 2884.
125. M. Green, R. J. Mercer, C. E. Morton and A. G. Orpen, *Angew. Chem., Int. Ed. Engl.*, 1985, **24**, 422.
126. J. S. Merola, R. A. Gentile, G. B. Ansell, M. A. Modrick and S. Zentz, *Organometallics*, 1982, **1**, 1731.
127. J. S. Merola, K. S. Campo, R. A. Gentile, M. A. Modrick and S. Zentz, *Organometallics*, 1984, **3**, 334.
128. P. N. Hazin, J. C. Huffman and J. W. Bruno, *J. Chem. Soc., Chem. Commun.*, 1988, 1473.
129. W. J. Sartain and J. P. Selegue, *Organometallics*, 1984, **3**, 1922.
130. A. Dormond and C. Moise, *Polyhedron*, 1985, **4**, 595.
131. D. G. Dick, Z. Hou and D. W. Stephan, *Organometallics*, 1992, **11**, 2378.
132. D. G. Dick and D. W. Stephan, *Organometallics*, 1990, **9**, 1910.
133. P. Y. Zheng, T. T. Nadasdi and D. W. Stephan, *Organometallics*, 1989, **8**, 1393.
134. E. J. M. de Boer, J. de With and A. G. Orpen, *J. Chem. Soc., Chem. Commun.*, 1985, 1666.
135. M. C. Baird, *Chem. Rev.*, 1988, **88**, 1217.
136. M. C. Baird, in 'Organometallic Radical Processes', ed. W. C. Trogler, Elsevier, Amsterdam, 1990, p. 49.
137. A. E. Stiegman and D. R. Tyler, *Acc. Chem. Res.*, 1984, **17**, 61.
138. A. E. Stiegman and D. R. Tyler, *Coord. Chem. Rev.*, 1985, **63**, 217.
139. D. R. Tyler, in 'Organometallic Radical Processes', ed. W. C. Trogler, Elsevier, Amsterdam, 1990, p. 338.
140. D. R. Tyler, *Prog. Inorg. Chem.*, 1988, **36**, 125.
141. D. R. Tyler and F. Mao, *Coord. Chem. Rev.*, 1990, **97**, 119.
142. D. R. Tyler, *Acc. Chem. Res.*, 1991, **24**, 325.
143. S. J. McLain, *J. Am. Chem. Soc.*, 1988, **110**, 643.
144. N. A. Cooley, K. A. Watson, S. Fortier and M. C. Baird, *Organometallics*, 1986, **5**, 2563.
145. W. C. Watkins, K. Hensel, S. Fortier, D. H. Macartney, M. C. Baird and S. J. McLain, *Organometallics*, 1992, **11**, 2418.
146. C. A. MacConnachie, J. M. Nelson and M. C. Baird, *Organometallics*, 1992, **11**, 2521.
147. L. Y. Goh and Y. Y. Lim, *J. Organomet. Chem.*, 1991, **402**, 209.
148. Q. Yao, A. Bakac and J. H. Espenson, *Organometallics*, 1993, **12**, 2010.
149. T. A. Huber, D. H. Macartney and M. C. Baird, *Organometallics*, 1993, **12**, 4715.

150. L. Y. Goh, S. K. Khoo and Y. Y. Lim, *J. Organomet. Chem.*, 1990, **399**, 115.
151. R. J. Hoobler *et al.*, *Organometallics*, 1993, **12**, 116.
152. S. Fortier *et al.*, *J. Am. Chem. Soc.*, 1991, **113**, 542.
153. P. J. Krusic, S. J. McLain, J. R. Morton, K. F. Preston and Y. Le Page, *J. Magn. Reson.*, 1987, **74**, 72.
154. N. A. Cooley, M. C. Baird, J. R. Morton, K. F. Preston and Y. Le Page, *J. Magn. Reson.*, 1988, **76**, 325.
155. K. A. E. O'Callaghan, S. J. Brown, J. A. Page, M. C. Baird, T. C. Richards and W. E. Geiger, *Organometallics*, 1991, **10**, 3119.
156. S. C. Tenhaeff *et al.*, *Organometallics*, 1993, **12**, 5000.
157. R. W. Wegman and T. L. Brown, *Organometallics*, 1982, **1**, 47.
158. N. N. Turaki and J. M. Huggins, *Organometallics*, 1986, **5**, 1703.
159. A. E. Stiegman and D. R. Tyler, *J. Am. Chem. Soc.*, 1982, **104**, 2944.
160. A. E. Stiegman and D. R. Tyler, *J. Photochem.*, 1984, **24**, 311.
161. A. S. Goldman and D. R. Tyler, *J. Am. Chem. Soc.*, 1984, **106**, 4066.
162. A. E. Stiegman, M. Stieglitz and D. R. Tyler, *J. Am. Chem. Soc.*, 1983, **105**, 6032.
163. A. E. Stiegman and D. R. Tyler, *J. Am. Chem. Soc.*, 1985, **107**, 967.
164. C. E. Philbin, C. A. Granatir and D. R. Tyler, *Inorg. Chem.*, 1986, **25**, 4806.
165. C. E. Philbin, A. S. Goldman and D. R. Tyler, *Inorg. Chem.*, 1986, **25**, 4434.
166. A. Avey, T. J. R. Weakley and D. R. Tyler, *J. Am. Chem. Soc.*, 1993, **115**, 7706.
167. A. E. Stiegman, A. S. Goldman, D. B. Leslie and D. R. Tyler, *J. Chem. Soc., Chem. Commun.*, 1984, 632.
168. F. Mao, C. E. Philbin, T. J. R. Weakley and D. R. Tyler, *Organometallics*, 1990, **9**, 1510.
169. F. Mao, S. K. Sur and D. R. Tyler, *J. Am. Chem. Soc.*, 1989, **111**, 7627.
170. M. Hanaya and M. Iwaizumi, *Organometallics*, 1989, **8**, 672.
171. M. Hanaya and M. Iwaizumi, *Chem. Lett.*, 1989, 1381.
172. M. Hanaya and M. Iwaizumi, *J. Organomet. Chem.*, 1991, **417**, 407.
173. M. Hanaya and M. Iwaizumi, *J. Organomet. Chem.*, 1992, **435**, 337.
174. M. Hanaya and M. Iwaizumi, *Bull. Chem. Soc. Jpn.*, 1990, **63**, 275.
175. K. Sarbasov, S. P. Solodovnikov, B. L. Tumanskii, N. N. Bubnov and A. I. Prokof'ev, *Izv. Akad. Nauk SSSR, Ser. Khim.*, 1982, 1509 (*Chem. Abstr.*, 1982, **97**, 198 351j).
176. M. Hanaya, S. Tero-Kubota and M. Iwaizumi, *Organometallics*, 1988, **7**, 1500.
177. I. V. Karsanov, Y. P. Ivakhnenko, V. S. Khandkarova, A. I. Prokof'ev, A. Z. Rubezhof and M. I. Kabachnik, *J. Organomet. Chem.*, 1989, **379**, 1.
178. L. Carlton, W. E. Lindsell and P. N. Preston, *J. Chem. Soc., Dalton Trans.*, 1982, 1483.
179. D. Rehorek, S. Di Martino, S. Sostero, O. Traverso and T. J. Kemp, *Inorg. Chim. Acta*, 1990, **178**, 1.
180. D. H. Gibson, F. U. Ahmed and K. R. Phillips, *J. Organomet. Chem.*, 1981, **218**, 325.
181. C. Roger, M.-J. Tudoret, V. Guerchais and C. Lapinte, *J. Organomet. Chem.*, 1989, **365**, 347.
182. U. Behrens and F. Edelmann, *J. Organomet. Chem.*, 1984, **263**, 179.
183. P. Braunstein, R. Bender and J.-M. Jud, *Inorg. Synth.*, 1989, **26**, 341.
184. M. D. Curtis and K. R. Han, *Inorg. Chem.*, 1985, **24**, 378.
185. P. Braunstein, C. de Méric de Bellefon, M. Lanfranchi and A. Tiripicchio, *Organometallics*, 1984, **3**, 1772.
186. F. Edelmann, P. Behrens, S. Behrens and U. Behrens, *J. Organomet. Chem.*, 1986, **310**, 333.
187. F. Edelmann and U. Behrens, *Chem. Ber.*, 1984, **117**, 3463.
188. D. W. Macomber and M. D. Rausch, *Organometallics*, 1983, **2**, 1523.
189. L. Song, Q. Dong and Q. Hu, *Huaxue Xuebao*, 1991, **49**, 1129.
190. O. I. Trifonova, Y. F. Oprunenko and Y. A. Ustynyuk, *Metalloorg. Khim.*, 1990, **3**, 465.
191. P. Legzdins, R. Reina, M. J. Shaw, R. J. Batchelor and F. W. B. Einstein, *Organometallics*, 1993, **12**, 1029.
192. D. W. Slocum *et al.*, *Polym. Mater. Sci. Eng.*, 1983, **49**, 353.
193. T.-M. Chung and Y. K. Chung, *Organometallics*, 1992, **11**, 2822.
194. H. G. Alt, S. J. Palackal and R. D. Rogers, *J. Organomet. Chem.*, 1990, **388**, 105.
195. H. G. Alt, H. E. Maisel, J. S. Han, B. Wrackmeyer and A. Razavi, *J. Organomet. Chem.*, 1990, **399**, 131.
196. P. Härter and P. Kiprof, *J. Organomet. Chem.*, 1990, **397**, C1.
197. D. Stein and H. Sitzmann, *J. Organomet. Chem.*, 1991, **402**, 249.
198. H. Sitzmann, *J. Organomet. Chem.*, 1988, **354**, 203.
199. C. P. Casey, R. M. Bullock, W. C. Fultz and A. L. Rheingold, *Organometallics*, 1982, **1**, 1591.
200. C. P. Casey, R. M. Bullock and F. Nief, *J. Am. Chem. Soc.*, 1983, **105**, 7574.
201. T. J. Duckworth, M. J. Mays, G. Conole and M. McPartlin, *J. Organomet. Chem.*, 1992, **439**, 327.
202. B. Brumas, D. de Caro, F. Dahan, D. de Montauzon and R. Poilblanc, *Organometallics*, 1993, **12**, 1503.
203. P. Härter, J. Behm and K. J. Burkert, *J. Organomet. Chem.*, 1992, **438**, 297.
204. K. H. Dötz and J. Rott, *J. Organomet. Chem.*, 1988, **338**, C11.
205. R. Lai, L. Bousquet and A. Heumann, *J. Organomet. Chem.*, 1993, **444**, 115.
206. G. R. Knox, P. L. Pauson and D. Willison, *J. Organomet. Chem.*, 1993, **450**, 177.
207. P. Jutzi, J. Schnittger, B. Neumann and H.-G. Stammler, *J. Organomet. Chem.*, 1991, **410**, C13.
208. P. Eilbracht and W. Totzauer, *Chem. Ber.*, 1982, **115**, 1669.
209. H. G. Alt, J. S. Han and R. D. Rogers, *J. Organomet. Chem.*, 1993, **445**, 115.
210. W. L. Bell *et al.*, *Organometallics*, 1988, **7**, 691.
211. D. W. Macomber, W. C. Spink and M. D. Rausch, *J. Organomet. Chem.*, 1983, **250**, 311.
212. D. W. Macomber, W. P. Hart, M. D. Rausch, R. D. Priester and C. U. Pittman, Jr., *J. Am. Chem. Soc.*, 1982, **104**, 884.
213. T. J. Clark, T. A. Nile, D. McPhail and A. T. McPhail, *Polyhedron*, 1989, **8**, 1804.
214. T. J. Clark, C. M. Killian, S. Luthra and T. A. Nile, *J. Organomet. Chem.*, 1993, **462**, 247.
215. R. Baumann and W. Malisch, *J. Organomet. Chem.*, 1986, **303**, C33.
216. P. Gomez-Sal, E. de Jésus, A. I. Pérez and P. Royo, *Organometallics*, 1993, **12**, 4633.
217. H. Plenio, *Chem. Ber.*, 1991, **124**, 2185.
218. L. A. Paquette, S. J. Hathaway, P. F. T. Schirch and J. C. Gallucci, *Organometallics*, 1986, **5**, 500.
219. H. Plenio, *J. Organomet. Chem.*, 1992, **435**, 21.

220. M. S. Corraine *et al.*, *Organometallics*, 1992, **11**, 35.
221. C. K. Lai, M. S. Corraine and J. D. Atwood, *Organometallics*, 1992, **11**, 582.
222. M. S. Corraine and J. D. Atwood, *Organometallics*, 1991, **10**, 2315.
223. M. S. Corraine and J. D. Atwood, *Organometallics*, 1991, **10**, 2647.
224. W. Beck and K. Sünkel, *Chem. Rev.*, 1988, **88**, 1405.
225. K.-H. Griessmann, A. Stasunik, W. Angerer and W. Malisch, *J. Organomet. Chem.*, 1986, **303**, C29.
226. M. Appel, K. Schloter, J. Heidrich and W. Beck, *J. Organomet. Chem.*, 1987, **322**, 77.
227. W. Beck, K. Schloter, K. Sünkel and G. Urban, *Inorg. Synth.*, 1989, **26**, 96.
228. J. A. Clark and M. Kilner, *J. Organomet. Chem.*, 1988, **347**, 107.
229. C. C. Lin, S. Hathcox and S. L. Gipson, *Inorg. Chim. Acta*, 1990, **170**, 151.
230. K. Sünkel, G. Urban and W. Beck, *J. Organomet. Chem.*, 1983, **252**, 187.
231. M. Appel and W. Beck, *J. Organomet. Chem.*, 1987, **319**, C1.
232. P. Leoni, E. Aquilini, M. Pasquali, F. Marchetti and M. Sabat, *J. Chem. Soc., Dalton Trans.*, 1988, 329.
233. K. Sünkel, G. Urban and W. Beck, *J. Organomet. Chem.*, 1985, **290**, 231.
234. G. Urban, K. Sünkel and W. Beck, *J. Organomet. Chem.*, 1985, **290**, 329.
235. M. Appel, W. Sacher and W. Beck, *J. Organomet. Chem.*, 1987, **322**, 351.
236. M. Appel, W. Sacher and W. Beck, *J. Organomet. Chem.*, 1987, **333**, 237.
237. F. Götzfried, R. Grenz, G. Urban and W. Beck, *Chem. Ber.*, 1985, **118**, 4179.
238. M. Herberhold, J. Ott and L. Haumaier, *Chem. Ber.*, 1985, **118**, 3143.
239. K. Sünkel, U. Nagel and W. Beck, *J. Organomet. Chem.*, 1981, **222**, 251.
240. A. K. McMullen, J. P. Selegue and J.-G. Wang, *Organometallics*, 1991, **10**, 3421.
241. K. Sünkel, K. Schloter, W. Beck, K. Ackermann and U. Schubert, *J. Organomet. Chem.*, 1983, **241**, 333.
242. K. Sünkel, U. Nagel and W. Beck, *J. Organomet. Chem.*, 1983, **251**, 227.
243. J. Markham, W. Tolman, K. Menard and A. Cutler, *J. Organomet. Chem.*, 1985, **294**, 45.
244. S. J. LaCroce and A. R. Cutler, *J. Am. Chem. Soc.*, 1982, **104**, 2312.
245. T. C. Forschner and A. R. Cutler, *Inorg. Synth.*, 1989, **26**, 231.
246. P. Leoni, A. Landi and M. Pasquali, *J. Organomet. Chem.*, 1987, **321**, 365.
247. A. Asdar, C. Lapinte and L. Toupet, *Organometallics*, 1989, **8**, 2708.
248. W. Tam, G.-Y. Lin and J. A. Gladysz, *Organometallics*, 1982, **1**, 525.
249. D. H. Gibson, K. Owens, S. K. Mandal, W. E. Sattich and J. O. Franco, *Organometallics*, 1989, **8**, 498.
250. D. H. Gibson, K. Owens and T.-S. Ong, *J. Am. Chem. Soc.*, 1984, **106**, 1125.
251. P. Leoni, M. Pasquali, L. Salsini, C. di Bugno, D. Braga and P. Sabatino, *J. Chem. Soc., Dalton Trans.*, 1989, 155.
252. T. S. Coolbaugh, B. D. Santarsiero and R. H. Grubbs, *J. Am. Chem. Soc.*, 1984, **106**, 6310.
253. P. Leoni, M. Pasquali, D. Braga and P. Sabatino, *J. Chem. Soc., Dalton Trans.*, 1989, 959.
254. S. A. Benyunes, A. Binelli, M. Green and M. J. Grimshire, *J. Chem. Soc., Dalton Trans.*, 1991, 895.
255. A. W. Al-Saadoon, M. Green, R. J. Mercer and A. G. Orpen, *J. Organomet. Chem.*, 1990, **384**, C12.
256. J. Markham, K. Menard and A. Cutler, *Inorg. Chem.*, 1985, **24**, 1581.
257. H. Schumann, *J. Organomet. Chem.*, 1987, **323**, 193.
258. H. Schumann, *J. Organomet. Chem.*, 1986, **304**, 341.
259. N. Kuhn, H. Schumann and E. Zaunder, *J. Organomet. Chem.*, 1988, **354**, 161.
260. H. Schumann, J. H. Enemark, M. J. Labarre, M. Bruck and P. Wexler, *Polyhedron*, 1991, **10**, 665.
261. U. Schubert, K. Ackermann, R. Janta, S. Voran and W. Malisch, *Chem. Ber.*, 1982, **115**, 2003.
262. G. McNally and A. R. Manning, *Inorg. Chim. Acta*, 1984, **85**, L7.
263. G. Lawless, G. McNally, A. R. Manning, D. Cunningham and P. McArdle, *Polyhedron*, 1986, **5**, 1741.
264. A. C. Filippou and W. Grünleitner, *J. Organomet. Chem.*, 1989, **378**, 387.
265. A. Asdar, M.-J. Tudoret and C. Lapinte, *J. Organomet. Chem.*, 1988, **349**, 353.
266. P. Leoni, E. Grilli, M. Pasquali and M. Tomassini, *J. Chem. Soc., Dalton Trans.*, 1986, 1041.
267. L. Brammer, D. Zhao, R. M. Bullock and R. K. McMullen, *Inorg. Chem.*, 1993, **32**, 4819.
268. B. E. Bursten and M. G. Gatter, *J. Am. Chem. Soc.*, 1984, **106**, 2554.
269. B. E. Bursten and M. G. Gatter, *Organometallics*, 1984, **3**, 941.
270. O. B. Ryan, M. Tilset and V. D. Parker, *J. Am. Chem. Soc.*, 1990, **112**, 2618.
271. K.-T. Smith and M. Tilset, *J. Organomet. Chem.*, 1992, **431**, 55.
272. J. Markham and A. R. Cutler, *Organometallics*, 1984, **3**, 736.
273. H. G. Alt, H. E. Engelhardt, W. Kläui and A. Müller, *J. Organomet. Chem.*, 1987, **331**, 317.
274. J. Febvay, F. Casabianca and J. G. Riess, *Inorg. Chem.*, 1985, **24**, 3235.
275. J. Febvay, F. Casabianca and J. G. Riess, *J. Am. Chem. Soc.*, 1984, **106**, 7985.
276. M.-J. Tudoret, M.-L. Robo and C. Lapinte, *Organometallics*, 1992, **11**, 1419.
277. B. Wrackmeyer, H. G. Alt and H. E. Maisel, *J. Organomet. Chem.*, 1990, **399**, 125.
278. H. G. Alt, H. E. Engelhardt, T. Frister and R. D. Rogers, *J. Organomet. Chem.*, 1989, **366**, 297.
279. K. A. Mahmoud, A. J. Rest and H. G. Alt, *J. Chem. Soc., Dalton Trans.*, 1984, 187.
280. R. L. Sweany, *J. Am. Chem. Soc.*, 1986, **108**, 6986.
281. H. G. Alt, K. A. Mahmoud and A. J. Rest, *Angew. Chem., Int. Ed. Engl.*, 1983, **22**, 544.
282. H. G. Alt and T. Frister, *J. Organomet. Chem.*, 1985, **293**, C7.
283. H. G. Alt, T. Frister, E. E. Trapl and H. E. Engelhardt, *J. Organomet. Chem.*, 1989, **362**, 125.
284. G. S. B. Adams and M. L. H. Green, *J. Chem. Soc., Dalton Trans.*, 1981, 353.
285. P. D. Grebenik, M. L. H. Green, A. Izquierdo, V. S. B. Mtetwa and K. Prout, *J. Chem. Soc., Dalton Trans.*, 1987, 9.
286. F. Abugideiri, M. A. Kelland, R. Poli and A. L. Rheingold, *Organometallics*, 1992, **11**, 1303.
287. F. Abugideiri, M. A. Kelland and R. Poli, *Organometallics*, 1993, **12**, 2388.
288. Z. Lin and M. B. Hall, *Organometallics*, 1993, **12**, 4046.
289. M. L. H. Green and G. Parkin, *J. Chem. Soc., Dalton Trans.*, 1987, 1611.
290. M. L. H. Green and G. Parkin, *J. Chem. Soc., Chem. Commun.*, 1984, 1467.
291. A. Berry, M. L. H. Green, J. A. Bandy and K. Prout, *J. Chem. Soc., Dalton Trans.*, 1991, 2185.
292. J. F. Buzinkai and R. R. Schrock, *Inorg. Chem.*, 1989, **28**, 2837.

293. J. Okuda, R. C. Murray, J. C. Dewan and R. R. Schrock, *Organometallics*, 1986, **5**, 1681.
294. S. J. Holmes and R. R. Schrock, *Organometallics*, 1983, **2**, 1463.
295. N. W. Hoffman, *Inorg. Chim. Acta*, 1984, **88**, 59.
296. C. Bueno and M. R. Churchill, *Inorg. Chem.*, 1981, **20**, 2197.
297. T. J. Boyle, F. Takusagawa and J. A. Heppert, *Acta Crystallogr., Sect. C*, 1990, **46**, 892.
298. C. L. Schwarz, R. M. Bullock and C. Creutz, *J. Am. Chem. Soc.*, 1991, **113**, 1225.
299. J.-S. Song, R. M. Bullock and C. Creutz, *J. Am. Chem. Soc.*, 1991, **113**, 9862.
300. M. C. R. Symons, S. W. Bratt and J. L. Wyatt, *J. Chem. Soc., Dalton Trans.*, 1983, 1377.
301. N. J. Coville, M. S. Loonat and L. Carlton, *S. Afr. J. Chem.*, 1992, **45**, 63.
302. K. Stärker and M. D. Curtis, *Inorg. Chem.*, 1985, **24**, 3006.
303. T.-F. Wang, T.-Y. Lee, J.-W. Chou and C.-W. Ong, *J. Organomet. Chem.*, 1992, **423**, 31.
304. N. N. Turaki, J. M. Huggins and L. Lebioda, *Inorg. Chem.*, 1988, **27**, 424.
305. M. F. Ernst and D. M. Roddick, *Inorg. Chem.*, 1990, **29**, 3627.
306. Y. Y. Lau and S. L. Gipson, *Inorg. Chim. Acta*, 1989, **157**, 147.
307. Y. Y. Lau, W. W. Huckabee and S. L. Gipson, *Inorg. Chim. Acta*, 1990, **172**, 41.
308. W. Xia and R. H. Hill, *Polyhedron*, 1992, **11**, 1319.
309. N. J. Coville, M. O. Albers and E. Singleton, *J. Chem. Soc., Dalton Trans.*, 1983, 947.
310. N. J. Coville, *J. Organomet. Chem.*, 1981, **218**, 337.
311. A. C. Filippou, W. Grünleitner and E. Herdtweck, *J. Organomet. Chem.*, 1989, **373**, 325.
312. A. C. Filippou, E. O. Fischer and W. Grünleitner, *J. Organomet. Chem.*, 1990, **386**, 333.
313. A. C. Filippou and W. Grünleitner, *J. Organomet. Chem.*, 1991, **407**, 61.
314. A. C. Filippou, W. Grünleitner and P. Kiprof, *J. Organomet. Chem.*, 1991, **410**, 175.
315. A. C. Filippou, W. Grünleitner, E. O. Fischer, W. Imhof and G. Huttner, *J. Organomet. Chem.*, 1991, **413**, 165.
316. A. C. Filippou, W. Grünleitner, C. Völkl and P. Kiprof, *J. Organomet. Chem.*, 1991, **413**, 181.
317. A. C. Filippou, W. Grünleitner and E. O. Fischer, *J. Organomet. Chem.*, 1992, **428**, C37.
318. A. C. Filippou, W. Grünleitner, C. Völkl and P. Kiprof, *Angew. Chem., Int. Ed. Engl.*, 1991, **30**, 1167.
319. A. C. Filippou and W. Grünleitner, *Z. Naturforsch., Teil B*, 1989, **44**, 666.
320. A. S. Goldman and D. R. Tyler, *Organometallics*, 1984, **3**, 449.
321. B. S. Erler, J. C. Dewan, S. J. Lippard and D. R. Tyler, *Inorg. Chem.*, 1981, **20**, 2719.
322. J. L. Davidson and G. Vasapollo, *J. Organomet. Chem.*, 1985, **291**, 43.
323. R. H. Hill, A. Becalska and N. Chiem, *Organometallics*, 1991, **10**, 2104.
324. G. Doyle, K. A. Eriksen and D. van Engen, *Organometallics*, 1985, **4**, 2201.
325. L. Carlton, W. E. Lindsell, K. J. McCullough and P. N. Preston, *J. Chem. Soc., Chem. Commun.*, 1983, 216.
326. A. L. Sargent and M. B. Hall, *J. Am. Chem. Soc.*, 1989, **111**, 1563.
327. A. L. Sargent and M. B. Hall, *Polyhedron*, 1990, **9**, 1799.
328. H. Werner, J. Roll, K. Linse and M. L. Ziegler, *Angew. Chem., Int. Ed. Engl.*, 1983, **22**, 982.
329. H. Werner, J. Roll, R. Zolk, P. Thometzek, K. Linse and M. L. Ziegler, *Chem. Ber.*, 1987, **120**, 1553.
330. P. Braunstein, U. Schubert and M. Burgard, *Inorg. Chem.*, 1984, **23**, 4057.
331. R. Uson, A. Laguna, M. Laguna, P. G. Jones and G. M. Sheldrick, *J. Chem. Soc., Dalton Trans.*, 1981, 366.
332. G. A. Artamkina, A. Y. Milchenko, N. A. Bumagin, I. P. Beletskaya and O. A. Reutov, *Metalloorg. Khim.*, 1988, **1**, 34 (*Chem. Abstr.*, 1989, **110**, 213 016m).
333. D. L. Thorn, *J. Organomet. Chem.*, 1991, **405**, 161.
334. J. M. Burlitch, S. E. Hayes and G. E. Whitwell, II, *Organometallics*, 1982, **1**, 1074.
335. R. B. Petersen, J. M. Ragosta, G. E. Whitwell, II and J. M. Burlitch, *Inorg. Chem.*, 1983, **22**, 3407.
336. O. Rossell, M. Seco and P. Braunstein, *J. Organomet. Chem.*, 1984, **273**, 233.
337. M. Cano, R. Criado, E. Gutierrez-Puebla, A. Monge and M. P. Pardo, *J. Organomet. Chem.*, 1985, **292**, 375.
338. S. Alvarez, M. Ferrer, R. Reina, O. Rossell, M. Seco and X. Solans, *J. Organomet. Chem.*, 1989, **377**, 291.
339. J. Granifo and M. E. Vargas, *J. Organomet. Chem.*, 1991, **408**, 357.
340. L.-C. Song, H. Yang, Q. Dong and Q.-M. Hu, *J. Organomet. Chem.*, 1991, **414**, 137.
341. P. Braunstein, O. Rossell, M. Seco, I. Torra, X. Solans and C. Miravitlles, *Organometallics*, 1986, **5**, 1113.
342. G. Salem, B. W. Skelton, A. H. White and S. B. Wild, *J. Chem. Soc., Dalton Trans.*, 1983, 2117.
343. M. Cano and J. A. Campo, *Polyhedron*, 1991, **10**, 133.
344. J. Roll and H. Werner, *Inorg. Chim. Acta*, 1988, **147**, 93.
345. R. M. Medina and J. R. Masaguer, *An. Quim., Ser. B*, 1988, **84**, 223 (*Chem. Abstr.*, 1989, **111**, 174 300y).
346. R. M. Medina and J. R. Masaguer, *An. Quim.*, 1990, **86**, 73 (*Chem. Abstr.*, 1991, **114**, 185 688u).
347. L. Song, Q. Dong and Q. Hu, *Huaxue Xuebao*, 1992, **50**, 193.
348. M. M. Kubicki, P. Richard and M. Jannin, *Acta Crystallogr., Sect. C*, 1991, **47**, 2362.
349. M. M. Kubicki, J. Y. Le Gall, R. Pichon, J. Y. Salaün, M. Cano and J. A. Campo, *J. Organomet. Chem.*, 1988, **348**, 349.
350. M. M. Kubicki, R. Kergoat, J. Y. De Vanssey, J. Y. Le Gall and J. E. Guerchais, *Polyhedron*, 1984, **3**, 365.
351. J. D. Cotton and E. A. Miles, *Inorg. Chim. Acta*, 1990, **173**, 129.
352. M. Cano, J. A. Campo, J. Y. Le Gall, R. Pichon, J. Y. Salaün and M. M. Kubicki, *Inorg. Chim. Acta*, 1992, **193**, 207.
353. M. M. Kubicki, R. Kergoat, J.-Y. Le Gall, J. E. Guerchais, J. Douglade and R. Mercier, *Aust. J. Chem.*, 1982, **35**, 1543.
354. D. F. Mullica, S. L. Gipson, L. C. Snell, E. L. Sappenfield and D. H. Leschnitzer, *Polyhedron*, 1992, **11**, 2265.
355. C. C. Lin and S. L. Gipson, *Inorg. Chem.*, 1992, **31**, 167.
356. S. Ermer *et al.*, *Inorg. Chem.*, 1983, **22**, 1339.
357. E. Rosenberg and B. Novak, *Inorg. Synth.*, 1989, **26**, 328.
358. R. Reina, O. Rossell and M. Seco, *J. Organomet. Chem.*, 1990, **398**, 285.
359. M. L. H. Green, P. Mountford, G. J. Smout and S. R. Speed, *Polyhedron*, 1990, **9**, 2763.
360. R. M. Campbell, L. M. Clarkson, W. Clegg, D. C. R. Hockless, N. L. Pickett and N. C. Norman, *Chem. Ber.*, 1992, **125**, 55.
361. L. M. Clarkson, W. Clegg, D. C. R. Hockless, N. C. Norman and T. B. Marder, *J. Chem. Soc., Dalton Trans.*, 1991, 2229.
362. L. M. Clarkson *et al.*, *J. Chem. Soc., Dalton Trans.*, 1991, 2241.
363. L. M. Clarkson, N. C. Norman and L. J. Farrugia, *J. Organomet. Chem.*, 1990, **390**, C10.

364. B. Nuber, W. Schatz and M. L. Ziegler, *Z. Naturforsch., Teil B*, 1990, **45**, 508.
365. S. Takagi, Y. Kato, H. Furuta, S. Onaka and T. K. Miyamoto, *J. Organomet. Chem.*, 1992, **429**, 287.
366. P. Cocolios, D. Chang, O. Vittori, R. Guilard, C. Moise and K. M. Kadish, *J. Am. Chem. Soc.*, 1984, **106**, 5724.
367. C. Lecomte, A. Habbou, P. Mitaine, P. Richard and R. Guilard, *Acta Crystallogr., Sect. C*, 1989, **45**, 1226.
368. R. Guilard *et al.*, *Inorg. Chem.*, 1987, **26**, 2647.
369. S. Y. Luk, F. R. Mayers, and J. O. Williams, *J. Chem. Soc., Chem. Commun.*, 1987, 215.
370. R. Guilard, A. Tabard, A. Zrineh and M. Ferhat, *J. Organomet. Chem.*, 1990, **389**, 315.
371. P. Richard, A. Zrineh, R. Guilard, A. Habbou and C. Lecomte, *Acta Crystallogr., Sect. C*, 1989, **45**, 1224.
372. R. Guilard *et al.*, *Inorg. Chem.*, 1988, **27**, 697.
373. R. Guilard, C. Lecomte and K. M. Kadish, *Structure and Bonding*, 1987, **64**, 205.
374. P. Gusbeth and H. Vahrenkamp, *Chem. Ber.*, 1985, **118**, 1143.
375. W. Malisch, H.-U. Wekel, I. Grob and F. H. Köhler, *Z. Naturforsch., Teil B*, 1982, **37**, 601.
376. S. Schmitzer *et al.*, *Inorg. Chem.*, 1993, **32**, 303.
377. D. Lei and M. J. Hampden-Smith, *J. Chem. Soc., Chem. Commun.*, 1989, 1211.
378. J. Cervantes, S. P. Vincenti, R. N. Kapoor and K. H. Pannell, *Organometallics*, 1989, **8**, 744.
379. L. M. Vandereyse, L. Verdonck, A. R. Bossuyt, F. M. G. van Wijnsberghe and G. P. van der Kelen, *Bull. Soc. Chim. Belg.*, 1988, **97**, 317.
380. L. M. Vandereyse, T. Haemers, A. R. Bossuyt, L. Verdonck and G. P. van der Kelen, *Bull. Soc. Chim. Belg.*, 1988, **97**, 723.
381. L. M. Vandereyse, L. Verdonck, A. R. Bossuyt, A. Praet and G. P. van der Kelen, *J. Mol. Catal.*, 1991, **65**, 269.
382. L. Verdonck *et al.*, *J. Mol. Catal.*, 1992, **76**, 319.
383. P. G. Harrison, *J. Organomet. Chem.*, 1981, **212**, 183.
384. T. S. Dory, J. J. Zuckerman, C. D. Hoff and J. W. Connolly, *J. Chem. Soc., Chem. Commun.*, 1981, 521.
385. H. H. Karsch and A. Appelt, *J. Organomet. Chem.*, 1986, **312**, C6.
386. F. H. Carré, R. J. P. Corriu and B. J. L. Henner, *J. Organomet. Chem.*, 1982, **228**, 139.
387. J. R. Chipperfield, S. Clark, D. E. Webster and H. Yusof, *J. Organomet. Chem.*, 1991, **421**, 205.
388. S. Onaka and Y. Kato, *Bull. Chem. Soc. Jpn.*, 1988, **61**, 304.
389. D. B. Firfiray, A. Irving and J. R. Moss, *J. Chem. Soc., Chem. Commun.*, 1990, 377.
390. J. P. Bullock, M. C. Palazzotto and K. R. Mann, *Inorg. Chem.*, 1990, **29**, 4413.
391. K. H. Pannell and R. N. Kapoor, *J. Organomet. Chem.*, 1981, **214**, 47.
392. K. H. Pannell and R. N. Kapoor, *J. Organomet. Chem.*, 1984, **269**, 59.
393. K. H. Pannell, J. M. Rozell, S. Cortez and R. Kapoor, *Organometallics*, 1990, **9**, 1322.
394. M. M. Kubicki, J.-Y. Le Gall, R. Kergoat and L. C. Gomes de Lima, *Can. J. Chem.*, 1987, **65**, 1292.
395. P. B. Hitchcock, M. F. Lappert and M. J. Michalczyk, *J. Chem. Soc., Dalton Trans.*, 1987, 2635.
396. S.-C. H. Su and A. Wojcicki, *Organometallics*, 1983, **2**, 1296.
397. P. Wang and J. D. Atwood, *J. Am. Chem. Soc.*, 1992, **114**, 6424.
398. R. M. Medina and J. R. Masaguer, *J. Organomet. Chem.*, 1986, **299**, 341.
399. S. Pelling, C. Botha and J. R. Moss, *J. Chem. Soc., Dalton Trans.*, 1983, 1495.
400. J. R. Moss, M. L. Niven and P. M. Stretch, *Inorg. Chim. Acta*, 1986, **119**, 177.
401. E. R. Burkhardt, J. J. Doney, R. G. Bergman and C. H. Heathcock, *J. Am. Chem. Soc.*, 1987, **109**, 2022.
402. E. R. Burkhardt, J. J. Doney, J. G. Stack, C. H. Heathcock and R. G. Bergman, *J. Mol. Catal.*, 1987, **41**, 41.
403. E. N. Jacobsen, M. K. Trost and R. G. Bergman, *J. Am. Chem. Soc.*, 1986, **108**, 8092.
404. K. H. Pannell, R. N. Kapoor, M. Wells, T. Giasolli and L. Párkányi, *Organometallics*, 1987, **6**, 663.
405. R. Poli, G. Wilkinson, M. Motevalli and M. B. Hursthouse, *J. Chem. Soc., Dalton Trans.*, 1985, 931.
406. W. A. Schenk and J. Pfeffermann, *J. Organomet. Chem.*, 1992, **440**, 341.
407. G. Facchin, P. Uguagliati and R. A. Michelin, *Organometallics*, 1984, **3**, 1818.
408. W. Petri and W. Beck, *Chem. Ber.*, 1984, **117**, 3265.
409. E. Lindner, M. Pabel, R. Fawzi and M. Steimann, *J. Organomet. Chem.*, 1992, **441**, 63.
410. D. Naumann and H. G. Varbelow, *J. Fluorine Chem.*, 1988, **41**, 415.
411. T. G. Richmond, A. M. Crespi and D. F. Shriver, *Organometallics*, 1984, **3**, 314.
412. J. D. Koola and D. M. Roddick, *Organometallics*, 1991, **10**, 591.
413. M. Akita, N. Kakinuma and Y. Moro-oka, *J. Organomet. Chem.*, 1988, **348**, 91.
414. S. E. Himmel, G. B. Young, D. C. M. Fung and C. Hollinshead, *Polyhedron*, 1985, **4**, 349.
415. S. Sharma, R. N. Kapoor, F. Cervantes-Lee and K. H. Pannell, *Polyhedron*, 1991, **10**, 1177.
416. S. Sharma and K. H. Pannell, *Organometallics*, 1993, **12**, 3979.
417. H.-J. Kneuper, C. Zimmermann, K. Harms and G. Boche, *Chem. Ber.*, 1989, **122**, 1043.
418. W. A. Schenk, J. Frisch, W. Adam and F. Prechtl, *Inorg. Chem.*, 1992, **31**, 3329.
419. W. B. Rybakov, A. I. Tursina, L. A. Aslanov, S. A. Yeremin, H. Schrauber and L. Kutschabsky, *Z. Anorg. Allg. Chem.*, 1982, **487**, 217.
420. D. C. McKean, G. P. McQuillan, A. R. Morrisson and I. Torto, *J. Chem. Soc., Dalton Trans.*, 1985, 1207.
421. R. T. C. Brownlee, A. F. Masters, M. J. O'Connor, A. G. Wedd, H. A. Kimlin and J. D. Cotton, *Org. Magn. Reson.*, 1982, **20**, 73.
422. R. T. C. Brownlee, M. J. O'Connor, B. P. Shehan and A. G. Wedd, *J. Magn. Reson.*, 1985, **61**, 516.
423. C. Mancuso and J. Halpern, *J. Organomet. Chem.*, 1992, **428**, C8.
424. V. Skagestad and M. Tilset, *Organometallics*, 1991, **10**, 2110.
425. V. Skagestad and M. Tilset, *Organometallics*, 1992, **11**, 3293.
426. M. E. Wright and V. W. Day, *J. Organomet. Chem.*, 1987, **329**, 43.
427. J. K. Stille, C. Smith, O. P. Anderson and M. M. Miller, *Organometallics*, 1989, **8**, 1040.
428. C. Lo Sterzo and J. K. Stille, *Organometallics*, 1990, **9**, 687.
429. E. C. Brehm, J. K. Stille and A. I. Meyers, *Organometallics*, 1992, **11**, 938.
430. C. Lo Sterzo, *Organometallics*, 1990, **9**, 3185.
431. C. Lo Sterzo, *J. Organomet. Chem.*, 1991, **408**, 253.
432. P. Eilbracht, W. Fassmann and W. Diehl, *Chem. Ber.*, 1985, **118**, 2314.

433. B. Olgemöller and W. Beck, *Chem. Ber.*, 1981, **114**, 867.
434. J. R. Moss and L. G. Scott, *Coord. Chem. Rev.*, 1984, **60**, 171.
435. E. Lindner, M. Pabel, R. Fawzi, H. A. Mayer and K. Wurst, *J. Organomet. Chem.*, 1992, **435**, 109.
436. N. A. Bailey, D. A. Dunn, C. N. Foxcroft, G. R. Harrison, M. J. Winter and S. Woodward, *J. Chem. Soc., Dalton Trans.*, 1988, 1449.
437. N. A. Bailey, P. A. Chell, A. Mukhopadhyay, H. E. Tabbron and M. J. Winter, *J. Chem. Soc., Chem. Commun.*, 1982, 215.
438. H. B. Friedrich, J. R. Moss and B. K. Williamson, *J. Organomet. Chem.*, 1990, **394**, 313.
439. J. R. Moss and L. G. Scott, *J. Organomet. Chem.*, 1989, **363**, 351.
440. H. B. Friedrich and J. R. Moss, *J. Chem. Soc., Dalton Trans.*, 1993, 2863.
441. K. Isobe, S. Tero-Kubota, H. Tanaka, Y. Nakamura, S. Okeya and K. Saito, *J. Organomet. Chem.*, 1987, **329**, C21.
442. N. A. Bailey *et al.*, *J. Chem. Soc., Dalton Trans.*, 1983, 2397.
443. H. Adams, N. A. Bailey and M. J. Winter, *J. Chem. Soc., Dalton Trans.*, 1984, 273.
444. J. Powell, E. Fuchs and J. F. Sawyer, *Organometallics*, 1990, **9**, 1722.
445. H. Adams, N. A. Bailey, V. A. Osborn and M. J. Winter, *J. Chem. Soc., Dalton Trans.*, 1986, 2127.
446. V. A. Osborn and M. J. Winter, *Polyhedron*, 1986, **5**, 435.
447. M.-C. Chen *et al.*, *Organometallics*, 1991, **10**, 378.
448. G. Schmidt, H. Pritzkow and H. P. Latscha, *Z. Naturforsch., Teil B*, 1987, **42**, 1115.
449. A. C. Filippou, C. Völkl and P. Kiprof, *J. Organomet. Chem.*, 1991, **415**, 375.
450. E. Carmona, P. J. Daff, A. Monge, P. Palma, M. L. Poveda and C. Ruiz, *J. Chem. Soc., Chem. Commun.*, 1991, 1503.
451. G.-M. Yang, G.-H. Lee, S.-M. Peng and R. S. Liu, *Organometallics*, 1991, **10**, 1305.
452. E. Lindner, G. Funk and F. Bouachir, *Chem. Ber.*, 1981, **114**, 2653.
453. E. Lindner, R. Fawzi, H. A. Mayer, K. Eichele and K. Pohmer, *Inorg. Chem.*, 1991, **30**, 1102.
454. E. Lindner and A. Brösamle, *Z. Naturforsch., Teil B*, 1984, **39**, 535.
455. E. Lindner, R. Fawzi, H. A. Mayer, K. Eichele and K. Pohmer, *J. Organomet. Chem.*, 1990, **386**, 63.
456. K. A. Mahmoud, R. Narayanaswamy and A. J. Rest, *J. Chem. Soc., Dalton Trans.*, 1981, 2199.
457. R. H. Hooker and A. J. Rest, *J. Chem. Soc., Dalton Trans.*, 1984, 761.
458. H. G. Alt and M. E. Eichner, *Angew. Chem., Int. Ed. Engl.*, 1982, **21**, 78.
459. C. G. Kreiter, K. Nist and H. G. Alt, *Chem. Ber.*, 1981, **114**, 1845.
460. K. A. Mahmoud, A. J. Rest, H. G. Alt, M. E. Eichner and B. M. Jansen, *J. Chem. Soc., Dalton Trans.*, 1984, 175.
461. F. P. A. Johnson, C. M. Gordon, P. M. Hodges, M. Poliakoff and J. J. Turner, *J. Chem. Soc., Dalton Trans.*, 1991, 833.
462. R. J. Kazlauskas and M. S. Wrighton, *J. Am. Chem. Soc.*, 1982, **104**, 6005.
463. G. K. Yang, K. S. Peters and V. Vaida, *J. Am. Chem. Soc.*, 1986, **108**, 2511.
464. H. G. Alt and M. E. Eichner, *J. Organomet. Chem.*, 1981, **212**, 397.
465. H. G. Alt and M. E. Eichner, *Angew. Chem., Int. Ed. Engl.*, 1981, **21**, 78.
466. R. B. Hitam, R. H. Hooker, K. A. Mahmoud, R. Narayanaswamy and A. J. Rest, *J. Organomet. Chem.*, 1981, **222**, C9.
467. B. Klein, R. J. Kazlauskas and M. S. Wrighton, *Organometallics*, 1982, **1**, 1338.
468. C. Lewis and M. S. Wrighton, *J. Am. Chem. Soc.*, 1983, **105**, 7768.
469. T. E. Gismondi and M. D. Rausch, *J. Organomet. Chem.*, 1985, **284**, 59.
470. R. B. Hitam, K. A. Mahmoud and A. J. Rest, *J. Organomet. Chem.*, 1985, **291**, 321.
471. R. H. Hooker, K. A. Mahmoud, A. J. Rest and H. G. Alt, *J. Organomet. Chem.*, 1991, **419**, 101.
472. A. K. Campen, K. A. Mahmoud, A. J. Rest and P. A. Willis, *J. Chem. Soc., Dalton Trans.*, 1990, 2817.
473. R. H. Hill and J. D. Debad, *Polyhedron*, 1991, **10**, 1705.
474. S. Mapolie and J. R. Moss, *Polyhedron*, 1991, **10**, 717.
475. Y. S. Yeh, J. T. Chen, G.-H. Lee and Y. Wang, *J. Chin. Chem. Soc. (Taipei)*, 1992, **39**, 271.
476. W. Schulze, H. Hartl and K. Seppelt, *J. Organomet. Chem.*, 1987, **319**, 77.
477. M. Akita, A. Kondoh, T. Kawahara, T. Takagi and Y. Moro-oka, *Organometallics*, 1988, **7**, 366.
478. M. Akita, H. Andoh, O. Mitani, T. Oku and Y. Moro-oka, *J. Mol. Catal.*, 1989, **56**, 107.
479. L. I. Leont'eva and E. G. Perevalova, *Izv. Akad. Nauk SSSR, Ser. Khim.*, 1984, 2352 (*Chem. Abstr.*, 1985, **102**, 132 197d).
480. A. Wong, R. L. Morgan, II, J. M. Golder, G. E. Quimbita and R. V. Pawlick, *Organometallics*, 1989, **8**, 844.
481. T. S. Targos, G. L. Geoffroy and A. L. Rheingold, *J. Organomet. Chem.*, 1986, **299**, 223.
482. E. K. Barefield, P. McCarten and M. C. Hillhouse, *Organometallics*, 1985, **4**, 1682.
483. J. D. Cotton and P. R. Dunstan, *Inorg. Chim. Acta*, 1984, **88**, 223.
484. J. D. Cotton, H. A. Kimlin and R. D. Markwell, *J. Organomet. Chem.*, 1982, **232**, C75.
485. J. D. Cotton, C. H. L. Kennard, R. D. Markwell, G. Smith and A. H. White, *J. Organomet. Chem.*, 1986, **309**, 117.
486. J. D. Cotton, G. T. Crisp and V. A. Daly, *Inorg. Chim. Acta*, 1981, **47**, 165.
487. J. D. Cotton and R. D. Markwell, *Inorg. Chim. Acta*, 1982, **63**, 13.
488. J. D. Cotton and H. A. Kimlin, *J. Organomet. Chem.*, 1985, **294**, 213.
489. J. D. Cotton, M. M. Kroes, R. D. Markwell and E. A. Miles, *J. Organomet. Chem.*, 1990, **388**, 133.
490. M. J. Wax and R. G. Bergman, *J. Am. Chem. Soc.*, 1981, **103**, 7028.
491. S. Abbot *et al.*, *J. Organomet. Chem.*, 1985, **289**, C13.
492. W. Malisch, H. Blau and F. J. Haaf, *Chem. Ber.*, 1981, **114**, 2956.
493. M. Akita, O. Mitani and Y. Moro-oka, *J. Chem. Soc., Chem. Commun.*, 1989, 527.
494. M. Akita, O. Mitani, M. Sayama and Y. Moro-oka, *Organometallics*, 1991, **10**, 1394.
495. G. Grötsch and W. Malisch, *J. Organomet. Chem.*, 1983, **258**, 297.
496. J. C. Selover, M. Marsi, D. W. Parker and J. A. Gladysz, *J. Organomet. Chem.*, 1981, **206**, 317.
497. J. T. Gauntlett, B. F. Taylor and M. J. Winter, *J. Chem. Soc., Dalton Trans.*, 1985, 1815.
498. J. T. Gauntlett and M. J. Winter, *Polyhedron*, 1986, **5**, 451.
499. J. T. Gauntlett, B. E. Mann, M. J. Winter and S. Woodward, *J. Chem. Soc., Dalton Trans.*, 1991, 1427.
500. D. H. Gibson, S. K. Mandal, K. Owens, W. E. Sattich and J. O. Franco, *Organometallics*, 1989, **8**, 1114.
501. D. H. Gibson, K. Owens, S. K. Mandal, W. E. Sattich and J. O. Franco, *Organometallics*, 1991, **10**, 1203.
502. H. G. Alt, M. E. Eichner and B. M. Jansen, *Angew. Chem., Int. Ed. Engl.*, 1982, **21**, 861.
503. H. G. Alt in 'Organometallic Syntheses', eds. R. B. King and J. J. Eisch, Elsevier, Amsterdam, 1986, vol. 3, p. 122.
504. H. G. Alt and H. I. Hayen, *J. Organomet. Chem.*, 1986, **316**, 301.

505. H. G. Alt and U. Thewalt, *J. Organomet. Chem.*, 1984, **268**, 235.
506. H. G. Alt, *J. Organomet. Chem.*, 1983, **256**, C12.
507. H. G. Alt, G. S. Herrmann and U. Thewalt, *J. Organomet. Chem.*, 1987, **327**, 237.
508. H. G. Alt and H. I. Hayen, *J. Organomet. Chem.*, 1986, **316**, 105.
509. H. I. Hayen, H. G. Alt and R. D. Rogers, *J. Organomet. Chem.*, 1987, **323**, 339.
510. H. G. Alt, *J. Organomet. Chem.*, 1985, **288**, 149.
511. H. G. Alt, J. S. Han and H. E. Maisel, *J. Organomet. Chem.*, 1991, **409**, 197.
512. H. G. Alt, *J. Organomet. Chem.*, 1990, **383**, 125.
513. H. G. Alt and H. I. Hayen, *J. Organomet. Chem.*, 1986, **315**, 337.
514. H. G. Alt, H. E. Engelhardt, U. Thewalt and J. Riede, *J. Organomet. Chem.*, 1985, **288**, 165.
515. H. G. Alt, G. S. Herrmann, H. E. Engelhardt and R. D. Rogers, *J. Organomet. Chem.*, 1987, **331**, 329.
516. G. Grötsch, W. Malisch and H. Blau, *J. Organomet. Chem.*, 1983, **252**, C19.
517. J. D. Carter, T. K. Schoch and L. McElwee-White, *Organometallics*, 1992, **11**, 3571.
518. S. R. Allen, M. Green, N. C. Norman, K. E. Paddick and A. G. Orpen, *J. Chem. Soc., Dalton Trans.*, 1983, 1625.
519. H. G. Alt, H. E. Engelhardt, B. Wrackmeyer and R. D. Rogers, *J. Organomet. Chem.*, 1989, **379**, 289.
520. F. Y. Pétillon *et al.*, *J. Organomet. Chem.*, 1984, **255**, 231.
521. H. Brix and W. Beck, *J. Organomet. Chem.*, 1982, **234**, 151.
522. H. Adams, N. A. Bailey, J. T. Gauntlett, I. M. Harkin, M. J. Winter and S. Woodward, *J. Chem. Soc., Dalton Trans.*, 1991, 1117.
523. I. J. Hart, J. C. Jeffery, R. M. Lowry and F. G. A. Stone, *Angew. Chem., Int. Ed. Engl.*, 1988, **27**, 1703.
524. C. G. Kreiter, *Adv. Organomet. Chem.*, 1986, **26**, 297.
525. T.-Y. Luh and C. S. Wong, *J. Organomet. Chem.*, 1985, **287**, 231.
526. A. D. Hunter and P. Legzdins, *Organometallics*, 1983, **2**, 525.
527. I. Michelini-Rodriguez, A. L. Romero, R. N. Kapoor, F. Cervantes-Lee and K. H. Pannell, *Organometallics*, 1993, **12**, 1221.
528. K. H. Pannell, R. Cea-Olivares, R. A. Toscano and R. N. Kapoor, *Organometallics*, 1987, **6**, 1821.
529. T.-W. Lee and R.-S. Liu, *Organometallics*, 1988, **7**, 878.
530. G.-H. Lee, S.-M. Peng, S.-F. Lush and R.-S. Liu, *J. Chem. Soc., Chem. Commun.*, 1988, 1513.
531. G.-H. Lee, S.-M. Peng, G.-M. Yang, S.-F. Lush and R.-S. Liu, *Organometallics*, 1989, **8**, 1106.
532. R. E. Lehmann and J. K. Kochi, *Organometallics*, 1991, **10**, 190.
533. B. Niemer, J. Breimair, B. Wagner, K. Polborn and W. Beck, *Chem. Ber.*, 1991, **124**, 2227.
534. R.-S. Keng and Y.-C. Lin, *Organometallics*, 1990, **9**, 289.
535. T.-W. Tseng *et al.*, *Organometallics*, 1991, **10**, 43.
536. I.-Y. Wu, T.-W. Tseng, Y.-C. Lin, M.-C. Cheng and Y. Wang, *Organometallics*, 1993, **12**, 478.
537. T. W. Leung, G. G. Christoph, J. Gallucci and A. Wojcicki, *Organometallics*, 1986, **5**, 846.
538. R.-S. Keng and Y. C. Lin, *Organometallics*, 1990, **9**, 289.
539. M.-H. Cheng, G.-M. Yang, J.-F. Chow, G.-H. Lee, S.-M. Peng and R.-S. Liu, *J. Chem. Soc., Chem. Commun.*, 1992, 934; correction: 1162.
540. M. E. Raseta, R. K. Mishra, S. A. Cawood, M. E. Welker and A. L. Rheingold, *Organometallics*, 1991, **10**, 2936.
541. G. H. Young, A. Wojcicki, M. Calligaris, G. Nardin and N. Bresciani-Pahor, *J. Am. Chem. Soc.*, 1989, **111**, 6890.
542. M.-H. Cheng, G.-H. Lee, S.-M. Peng and R.-S. Liu, *Organometallics*, 1991, **10**, 3600.
543. G. H. Young, M. V. Raphael, A. Wojcicki, M. Calligaris, G. Nardin and N. Bresciani-Pahor, *Organometallics*, 1991, **10**, 1934.
544. T. M. Wido, G. H. Young, A. Wojcicki, M. Calligaris and G. Nardin, *Organometallics*, 1988, **7**, 452.
545. W. Beck, M. J. Schweiger and G. Müller, *Chem. Ber.*, 1987, **120**, 889.
546. G. A. Artamkina, A. Y. Milchenko, I. P. Beletskaya and O. A. Reutov, *Metalloorg. Khim.*, 1988, **1**, 908 (*Chem. Abstr.*, 1990, **112**, 7 659m).
547. S. R. Allen *et al.*, *J. Chem. Soc., Dalton Trans.*, 1983, 927.
548. S. R. Allen, R. G. Beevor, M. Green, A. G. Orpen, K. E. Paddick and I. D. Williams, *J. Chem. Soc., Dalton Trans.*, 1987, 591.
549. G. A. Artamkina, A. Y. Mil'chenko, I. P. Beletskaya and O. A. Reutov, *J. Organomet. Chem.*, 1986, **311**, 199.
550. G. A. Artamkina, A. Y. Mil'chenko, I. P. Beletskaya and O. A. Reutov, *J. Organomet. Chem.*, 1987, **321**, 371.
551. F. R. Kreissl, W. Sieber and M. Wolfgruber, *Angew. Chem., Int. Ed. Engl.*, 1983, **22**, 493.
552. W. J. Sieber *et al.*, *J. Organomet. Chem.*, 1988, **340**, 341.
553. F. R. Kreissl, G. Reber and G. Müller, *Angew. Chem., Int. Ed. Engl.*, 1986, **25**, 643.
554. F. R. Kreissl, W. J. Sieber and H. G. Alt, *Chem. Ber.*, 1984, **117**, 2527.
555. F. R. Kreissl and M. Wolfgruber, *J. Organomet. Chem.*, 1988, **355**, 267.
556. J. C. Jeffery, D. W. I. Sams and K. D. V. Weerasuria, *J. Organomet. Chem.*, 1991, **418**, 395.
557. J. B. Sheridan, D. B. Pourreau, G. L. Geoffroy and A. L. Rheingold, *Organometallics*, 1988, **7**, 289.
558. K. Eberl, W. Uedelhoven, M. Wolfgruber and F. R. Kreissl, *Chem. Ber.*, 1982, **115**, 504.
559. K. Eberl, M. Wolfgruber, W. Sieber and F. R. Kreissl, *J. Organomet. Chem.*, 1982, **236**, 171.
560. W. J. Sieber, M. Wolfgruber, F. R. Kreissl and O. Orama, *J. Organomet. Chem.*, 1984, **270**, C41.
561. M. Wolfgruber and F. R. Kreissl, *J. Organomet. Chem.*, 1988, **349**, C4.
562. F. R. Kreissl, M. Wolfgruber, W. Sieber and K. Ackermann, *J. Organomet. Chem.*, 1983, **252**, C39.
563. F. R. Kreissl, M. Wolfgruber, W. Sieber and H. G. Alt, *Angew. Chem., Int. Ed. Engl.*, 1983, **22**, 149.
564. F. R. Kreissl, M. Wolfgruber and W. J. Sieber, *J. Organomet. Chem.*, 1984, **270**, C4.
565. F. R. Kreissl, M. Wolfgruber and U. Thewalt, *J. Organomet. Chem.*, 1986, **317**, 175.
566. M. Wolfgruber and F. R. Kreissl, *J. Organomet. Chem.*, 1983, **258**, C9.
567. F. R. Kreissl, M. Wolfgruber and W. J. Sieber, *Organometallics*, 1983, **2**, 1266.
568. M. I. Bruce, M. G. Humphrey, J. G. Matisons, S. K. Roy and A. G. Swincer, *Aust. J. Chem.*, 1984, **37**, 1955.
569. D. Villemin and E. Schigeko, *J. Organomet. Chem.*, 1988, **346**, C24.
570. N. A. Ustynyuk, V. N. Vinogradova, D. N. Kravtsov, Y. F. Oprunenko and V. A. Piven, *Metalloorg. Khim.*, 1988, **1**, 884.
571. A. Wong, P. C. W. Kang, C. D. Tagge and D. R. Leon, *Organometallics*, 1990, **9**, 1992.

572. N. A. Ustynyuk, V. N. Vinogradova and D. N. Kravtsov, *Metalloorg. Khim.*, 1988, **1**, 85 (*Chem. Abstr.*, 1989, **110**, 231 804w).
573. P. N. Nicklas, J. P. Selegue and B. A. Young, *Organometallics*, 1988, **7**, 2248.
574. N. E. Kolobova, *et al.*, *Koord. Khim.*, 1982, **8**, 1655.
575. N. E. Kolobova, V. V. Skripkin, T. V. Rozantseva, Y. T. Struchkov, G. G. Aleksandrov and V. G. Andrianov, *J. Organomet. Chem.*, 1981, **218**, 351.
576. N. A. Ustynyuk, V. N. Vinogradova, V. N. Korneva, D. N. Kravtsov, V. G. Andrianov and Y. T. Struchkov, *J. Organomet. Chem.*, 1984, **277**, 285.
577. N. A. Ustynyuk, V. N. Vinogradova, V. G. Andrianov and Y. T. Struchkov, *J. Organomet. Chem.*, 1984, **268**, 73.
578. V. N. Kalinin, T. V. Rozantseva, P. V. Petrovskii, A. S. Batsanov and Y. T. Struchkov, *J. Organomet. Chem.*, 1989, **372**, 287.
579. M. I. Bruce, T. W. Hambley, M. R. Snow and A. G. Swincer, *Organometallics*, 1985, **4**, 494.
580. M. I. Bruce, M. J. Liddell, M. R. Snow and E. R. T. Tiekink, *Organometallics*, 1988, **7**, 343.
581. G. H. Young and A. Wojcicki, *J. Organomet. Chem.*, 1990, **390**, 351.
582. D. K. Hwang, Y. Chi, S.-M. Peng and G.-H. Lee, *Organometallics*, 1990, **9**, 2709.
583. Y. Chi, C.-H. Wu, S.-M. Peng and G.-H. Lee, *Organometallics*, 1990, **9**, 2305.
584. Y. Chi, R.-C. Lin, C.-C. Chen, S.-M. Peng and G.-H. Lee, *J. Organomet. Chem.*, 1992, **439**, 347.
585. Y. Chi, D.-K. Hwang, S.-F. Chen and L.-K. Liu, *J. Chem. Soc., Chem. Commun.*, 1989, 1540.
586. G. B. Richter-Addo and P. Legzdins, *Chem. Rev.*, 1988, **88**, 991.
587. T. T. Chin, J. K. Hoyano, P. Legzdins and J. T. Malito, *Inorg. Synth.*, 1990, **28**, 196.
588. K. Isobe, S. Kimura and Y. Nakamura, *J. Organomet. Chem.*, 1987, **331**, 221.
589. A. Brauel and D. Rehder, *Z. Naturforsch., Teil B*, 1987, **42**, 605.
590. D. L. Lichtenberger and J. L. Hubbard, *Inorg. Chem.*, 1985, **24**, 3835.
591. H. W. Chen, W. L. Jolly, S.-F. Xiang and P. Legzdins, *Inorg. Chem.*, 1981, **20**, 1133.
592. M. Minelli, J. L. Hubbard, D. L. Lichtenberger and J. H. Enemark, *Inorg. Chem.*, 1984, **23**, 2721.
593. W. E. Geiger, P. H. Rieger, B. Tulyathan and M. D. Rausch, *J. Am. Chem. Soc.*, 1984, **106**, 7000.
594. S. P. Solodovnikov, B. L. Tumanskii, N. N. Bubnov and M. I. Kabachnik, *Izv. Akad. Nauk SSSR, Ser. Khim.*, 1986, 2147.
595. M. Herberhold, W. Kremnitz, H. Trampisch, R. B. Hitam, A. J. Rest and D. J. Taylor, *J. Chem. Soc., Dalton Trans.*, 1982, 1261.
596. R. B. Hitam, A. J. Rest, M. Herberhold and W. Kremnitz, *J. Chem. Soc., Chem. Commun.*, 1984, 471.
597. W. A. Herrmann, J. L. Hubbard, I. Bernal, J. D. Korp, B. L. Haymore and G. L. Hillhouse, *Inorg. Chem.*, 1984, **23**, 2978.
598. J. C. Jeffery, J. C. V. Laurie, I. Moore, H. Razay and F. G. A. Stone, *J. Chem. Soc., Dalton Trans.*, 1984, 1563.
599. K. H. Dötz, C. Lyon and J. Rott, *J. Organomet. Chem.*, 1988, **345**, 117.
600. C. P. Casey, W. D. Jones and S. G. Harsy, *J. Organomet. Chem.*, 1981, **206**, C38.
601. W. Malisch, H. Blau, P. Weickert and K. H. Griessmann, *Z. Naturforsch., Teil B*, 1983, **38**, 711.
602. P. V. Bonnesen, A. T. Baker and W. H. Hersh, *J. Am. Chem. Soc.*, 1986, **108**, 8304.
603. F. R. Kreissl and N. Ullrich, *J. Organomet. Chem.*, 1992, **440**, 335.
604. G. Cerveau, E. Colomer, R. J. P. Corriu and A. Vioux, *J. Organomet. Chem.*, 1987, **321**, 327.
605. E. Colomer, R. J. P. Corriu and A. Vioux, *Bull. Soc. Chim. Fr.*, 1984, **II**, 247.
606. F. Carré, E. Colomer, R. J. P. Corriu and A. Vioux, *Organometallics*, 1984, **3**, 970.
607. M. D. Rausch and Y.-P. Wang, *J. Organomet. Chem.*, 1991, **413**, 111.
608. Y.-P. Wang, Y. H. Yang and J. H. Liaw, *Donghai Xuebao*, 1991, **32**, 495 (*Chem. Abstr.*, 1991, **116**, 41 674e).
609. R. D. Rogers, R. Shakir, J. L. Atwood, D. W. Macomber, Y. P. Wang and M. D. Rausch, *J. Crystallogr. Spectrosc. Res.*, 1988, **18**, 767 (*Chem. Abstr.*, 1989, **111**, 214 626t).
610. K. England and P. C. Reeves, *J. Organomet. Chem.*, 1987, **319**, 355.
611. M. D. Rausch, D. J. Kowalski and E. A. Mintz, *J. Organomet. Chem.*, 1988, **342**, 201.
612. C. U. Pittman, Jr. and M. D. Rausch, *Pure Appl. Chem.*, 1986, **58**, 617.
613. K. Gonsalves, L. Zhan-Ru, R. W. Lenz and M. D. Rausch, *J. Polym. Chem., Polym. Chem. Ed.*, 1985, **23**, 1707.
614. Y.-P. Wang, T.-S. Lin, R.-S. Shyu, J.-M. Hwu, Y. Wang and M.-C. Cheng, *J. Organomet. Chem.*, 1989, **371**, 57.
615. Y.-P. Wang, J.-M. Hwu and S.-L. Wang, *J. Organomet. Chem.*, 1989, **371**, 71.
616. Y. P. Wang and J.-M. Hwu, *J. Organomet. Chem.*, 1990, **385**, 61.
617. Y.-P. Wang, R.-S. Shyu, J.-M. Hwu, Y. Wang and M.-C. Cheng, *J. Organomet. Chem.*, 1990, **386**, 79.
618. Y.-P. Wang, J.-M. Hwu, S.-L. Wang and Y.-J. Wu, *J. Organomet. Chem.*, 1991, **414**, 33.
619. M. D. Rausch and Y. P. Wang, *Organometallics*, 1991, **10**, 1438.
620. Y. P. Wang, T. S. Lin, R.-S. Shyu and J. M. Hwu, *Tung-hai Hsueh Pao*, 1989, **30**, 367 (*Chem. Abstr.*, 1990, **113**, 212 257r).
621. Y.-P. Wang, R.-S. Shyu and J.-M. Hwu, *Tung-hai Hsueh Pao*, 1989, **30**, 377 (*Chem. Abstr.*, 1990, **113**, 231 611t).
622. Y.-P. Wang and J.-M. Hwu, *Tung-hai Hsueh Pao*, 1990, **31**, 481 (*Chem. Abstr.*, 1991, **114**, 143 630p).
623. Y.-P. Wang, J.-M. Hwu and S.-L. Wang, *J. Organomet. Chem.*, 1990, **390**, 179.
624. Y.-P. Wang, J.-M. Hwu and S.-L. Wang, *Acta Crystallogr., Sect. C*, 1990, **46**, 2305.
625. Y.-P. Wang and J.-M. Hwu, *Tung-hai Hsueh Pao*, 1990, **31**, 493 (*Chem. Abstr.*, 1991, **114**, 102 318z).
626. P. Legzdins and C. R. Nurse, *Inorg. Chem.*, 1985, **24**, 327.
627. Y.-P. Wang, Y.-H. Yang, S.-L. Wang and F.-L. Liao, *J. Organomet. Chem.*, 1991, **419**, 325.
628. A. D. Hunter and P. Legzdins, *Inorg. Chem.*, 1984, **23**, 4198.
629. H. W. Chen, W. L. Jolly, S.-F. Xiang and P. Legzdins, *Inorg. Chem.*, 1981, **20**, 1779.
630. B. J. Morris-Sherwood, B. W. S. Kolthammer and M. B. Hall, *Inorg. Chem.*, 1981, **20**, 2771.
631. P. Legzdins, B. Wassink, F. W. B. Einstein and R. H. Jones, *Organometallics*, 1988, **7**, 477.
632. P. Legzdins and B. Wassink, *Organometallics*, 1988, **7**, 482.
633. Y. S. Yu, R. A. Jacobson and R. J. Angelici, *Inorg. Chem.*, 1982, **21**, 3106.
634. D. C. Loizou, J. Castillo, A. R. Oki, N. S. Hosmane and J. A. Morrison, *Organometallics*, 1992, **11**, 4189.
635. M. Herberhold, H. Kneisel, L. Haumaier, A. Gieren and C. Ruiz-Peréz, *Z. Naturforsch., Teil B*, 1986, **41**, 1431.
636. J. Rott, E. Guggolz, A. Rettenmeier and M. L. Ziegler, *Z. Naturforsch., Teil B*, 1982, **37**, 13.
637. M. Herberhold and J. Ott, *Z. Naturforsch., Teil B*, 1988, **43**, 682.
638. J. L. Hubbard and W. K. McVicar, *Inorg. Chem.*, 1992, **31**, 910.

639. M. T. Ashby and J. H. Enemark, *J. Am. Chem. Soc.*, 1986, **108**, 730.
640. J. L. Hubbard and W. K. McVicar, *Organometallics*, 1990, **9**, 2683.
641. P. Legzdins, G. B. Richter-Addo, B. Wassink, F. W. B. Einstein, R. H. Jones and A. C. Willis, *J. Am. Chem. Soc.*, 1989, **111**, 2097.
642. P. Legzdins, G. B. Richter-Addo, F. W. B. Einstein and R. H. Jones, *Organometallics*, 1987, **6**, 1807.
643. P. Legzdins, G. B. Richter-Addo, F. W. B. Einstein and R. H. Jones, *Organometallics*, 1990, **9**, 431.
644. P. Legzdins, W. S. McNeil, E. G. Vessey, R. J. Batchelor and F. W. B. Einstein, *Organometallics*, 1992, **11**, 2718.
645. G. Hartmann, R. Froböse, R. Mews and G. M. Sheldrick, *Z. Naturforsch., Teil B*, 1982, **37**, 1234.
646. K. Beran, G. Steinke and R. Mews, *Chem. Ber.*, 1989, **122**, 1613.
647. B. W. Hames and P. Legzdins, *Organometallics*, 1982, **1**, 116.
648. B. Bursten, R. H. Cayton and M. G. Gatter, *Organometallics*, 1988, **7**, 1342.
649. P. Legzdins and B. Wassink, *Organometallics*, 1984, **3**, 1811.
650. J. Müller, I. Sonn and M. Strampfer, *J. Organomet. Chem.*, 1992, **427**, C15.
651. P. Gomez-Sal, E. de Jésus, W. Michiels, P. Royo, A. Vázquez de Miguel and S. Martinez-Carrera, *J. Chem. Soc., Dalton Trans.*, 1990, 2445.
652. N. H. Dryden, P. Legzdins, R. J. Batchelor and F. W. B. Einstein, *Organometallics*, 1991, **10**, 2077.
653. A. D. Hunter, P. Legzdins, J. T. Martin and L. Sánchez, in 'Organometallic Syntheses' eds. R. B. King and J. J. Eisch, Elsevier, Amsterdam, 1986, vol. 3, p. 58.
654. N. H. Dryden, P. Legzdins, F. W. B. Einstein and R. H. Jones, *Can. J. Chem.*, 1988, **66**, 2100.
655. T. T. Chin, P. Legzdins, J. Trotter and V. C. Yee, *Organometallics*, 1992, **11**, 913.
656. W. de Oliveira, J.-L. Migot, M. B. Gomes de Lima, J. Sala-Pala, J. E. Guerchais and J. Y. Le Gall, *J. Organomet. Chem.*, 1985, **284**, 313.
657. B. Alegre, E. de Jésus, A. Vázquez de Miguel, P. Royo, A. M. Manotti Lanfredi and A. Tiripicchio, *J. Chem. Soc., Dalton Trans.*, 1988, 819.
658. M. Herberhold, G.-X. Jin, W. Kremnitz, A. L. Rheingold and B. S. Haggerty, *Z. Naturforsch., Teil B*, 1991, **46**, 500.
659. P. Legzdins, J. T. Martin and J. C. Oxley, *Organometallics*, 1985, **4**, 1263.
660. P. Legzdins, J. T. Martin, F. W. B. Einstein and A. C. Willis, *J. Am. Chem. Soc.*, 1986, **108**, 7971.
661. B. E. Bursten and R. H. Cayton, *Organometallics*, 1988, **7**, 1349.
662. F. G. Herring, P. Legzdins and G. B. Richter-Addo, *Organometallics*, 1989, **8**, 1485.
663. A. D. Hunter and P. Legzdins, *Organometallics*, 1986, **5**, 1001.
664. A. D. Hunter, P. Legzdins, F. W. B. Einstein, A. C. Willis, B. E. Bursten and M. G. Gatter, *J. Am. Chem. Soc.*, 1986, **108**, 3843.
665. N. J. Christensen, P. Legzdins, F. W. B. Einstein and R. H. Jones, *Organometallics*, 1991, **10**, 3070.
666. N. J. Christensen, A. D. Hunter and P. Legzdins, *Organometallics*, 1989, **8**, 930.
667. A. D. Hunter, P. Legzdins, C. R. Nurse, F. W. B. Einstein and A. C. Willis, *J. Am. Chem. Soc.*, 1985, **107**, 1791.
668. N. J. Christensen, A. D. Hunter, P. Legzdins and L. Sánchez, *Inorg. Chem.*, 1987, **26**, 3344.
669. P. Legzdins, S. J. Rettig and K. J. Ross, *Organometallics*, 1993, **12**, 2103.
670. P. Legzdins, P. J. Lundmark and S. J. Rettig, *Organometallics*, 1993, **12**, 3545.
671. E. de Jésus, A. Vázquez de Miguel, P. Royo, A. M. Manotti Lanfredi and A. Tiripicchio, *J. Chem. Soc., Dalton Trans.*, 1990, 2779.
672. P. Legzdins and J. Veltheer, *Acc. Chem. Res.*, 1993, **26**, 41.
673. J. D. Debad, P. Legzdins, R. J. Batchelor and F. W. B. Einstein, *Organometallics*, 1993, **12**, 2094.
674. P. Legzdins, S. J. Rettig and L. Sánchez, *Organometallics*, 1985, **4**, 1470.
675. P. Legzdins, E. C. Phillips and L. Sánchez, *Organometallics*, 1989, **8**, 940.
676. P. Legzdins, J. T. Martin, F. W. B. Einstein and R. H. Jones, *Organometallics*, 1987, **6**, 1826.
677. S. V. Evans, P. Legzdins, S. J. Rettig, L. Sánchez and J. Trotter, *Organometallics*, 1987, **6**, 7.
678. N. H. Dryden, P. Legzdins, E. C. Phillips, J. Trotter and V. C. Yee, *Organometallics*, 1990, **9**, 882.
679. J. D. Debad, P. Legzdins, R. J. Batchelor and F. W. B. Einstein, *Organometallics*, 1992, **11**, 6.
680. N. H. Dryden, P. Legzdins, P. J. Lundmark, A. Riesen and F. W. B. Einstein, *Organometallics*, 1993, **12**, 2085.
681. P. Legzdins, S. J. Rettig and J. E. Veltheer, *J. Am. Chem. Soc.*, 1992, **114**, 6922.
682. J. D. Debad, P. Legzdins, S. J. Rettig and J. E. Veltheer, *Organometallics*, 1993, **12**, 2714.
683. P. Legzdins, S. J. Rettig and L. Sánchez, *Organometallics*, 1988, **7**, 2394.
684. P. Legzdins, R. H. Jones, E. C. Phillips, V. C. Yee, J. Trotter and F. W. B. Einstein, *Organometallics*, 1991, **10**, 986.
685. N. H. Dryden, P. Legzdins, J. Trotter and V. C. Yee, *Organometallics*, 1991, **10**, 2857.
686. N. H. Dryden, P. Legzdins, S. J. Rettig and J. E. Veltheer, *Organometallics*, 1992, **11**, 2583.
687. P. Legzdins, P. J. Lundmark, E. C. Phillips, S. J. Rettig and J. E. Veltheer, *Organometallics*, 1992, **11**, 2991.
688. P. Legzdins, S. J. Rettig, K. J. Ross and J. E. Veltheer, *J. Am. Chem. Soc.*, 1991, **113**, 4361.
689. P. Legzdins, S. J. Rettig, J. E. Veltheer, R. J. Batchelor and F. W. B. Einstein, *Organometallics*, 1993, **12**, 3575.
690. E. B. Brouwer, P. Legzdins, S. J. Rettig and K. J. Ross, *Organometallics*, 1993, **12**, 4234.
691. N. Brunet, J. D. Debad, P. Legzdins, J. Trotter, J. E. Veltheer and V. C. Yee, *Organometallics*, 1993, **12**, 4572.
692. I. L. Eremenko *et al.*, *J. Organomet. Chem.*, 1991, **411**, 193.
693. J. Sundermeyer, *Chem. Ber.*, 1991, **124**, 1977.
694. J. Sundermeyer, U. Radius and C. Burschka, *Chem. Ber.*, 1992, **125**, 2379.
695. S. R. Huber, T. C. Baldwin and D. E. Wigley, *Organometallics*, 1993, **12**, 91.
696. U. Radius and J. Sundermeyer, *Chem. Ber.*, 1992, **125**, 2183.
697. S. F. Pedersen and R. R. Schrock, *J. Am. Chem. Soc.*, 1982, **104**, 7483.
698. P. Legzdins, E. C. Phillips, S. J. Rettig, J. Trotter, J. E. Veltheer and V. C. Yee, *Organometallics*, 1992, **11**, 3104.
699. M. L. H. Green, P. C. Konidaris, P. Mountford and S. J. Simpson, *J. Chem. Soc., Chem. Commun.*, 1992, 256.
700. J. C. Green, M. L. H. Green, J. T. James, P. C. Konidaris, G. H. Maunder and P. Mountford, *J. Chem. Soc., Chem. Commun.*, 1992, 1361.
701. T. E. Glassman, M. G. Vale and R. R. Schrock, *Organometallics*, 1991, **10**, 4046.
702. R. C. Murray and R. R. Schrock, *J. Am. Chem. Soc.*, 1985, **107**, 4557.
703. M. B. O'Regan, A. H. Liu, W. C. Finch, R. R. Schrock and W. M. Davis, *J. Am. Chem. Soc.*, 1990, **112**, 4331.

704. M. R. Churchill and Y.-J. Li, *J. Organomet. Chem.*, 1986, **301**, 49.
705. R. R. Schrock, T. E. Glassman and M. G. Vale, *J. Am. Chem. Soc.*, 1991, **113**, 725.
706. T. E. Glassman, M. G. Vale and R. R. Schrock, *Inorg. Chem.*, 1992, **31**, 1985.
707. R. R. Schrock, A. H. Liu, M. B. O'Regan, W. C. Finch and J. F. Payack, *Inorg. Chem.*, 1988, **27**, 3574.
708. T. E. Glassman, M. G. Vale and R. R. Schrock, *J. Am. Chem. Soc.*, 1992, **114**, 8098.
709. R. R. Schrock, R. M. Kolodziej, A. H. Liu, W. M. Davis and M. G. Vale, *J. Am. Chem. Soc.*, 1990, **112**, 4338.
710. H. Alper, J. F. Petrignani, F. W. B. Einstein and A. C. Willis, *J. Am. Chem. Soc.*, 1983, **105**, 1701.
711. J.-F. Petrignani and H. Alper, *Inorg. Chim. Acta*, 1983, **77**, L243.
712. J. Fletcher, G. Hogarth and D. A. Tocher, *J. Organomet. Chem.*, 1991, **403**, 345.
713. J. Fletcher, G. Hogarth and D. A. Tocher, *J. Organomet. Chem.*, 1991, **405**, 207.
714. M. L. H. Green, G. Hogarth and G. C. Saunders, *J. Organomet. Chem.*, 1991, **421**, 233.
715. M. L. H. Green and K. J. Moynihan, *Polyhedron*, 1986, **5**, 921.
716. M. L. H. Green, G. Hogarth, P. C. Konidaris and P. Mountford, *J. Organomet. Chem.*, 1990, **394**, C9.
717. M. L. H. Green, G. Hogarth, P. C. Konidaris and P. Mountford, *J. Chem. Soc., Dalton Trans.*, 1990, 3781.
718. G. Hogarth and P. C. Konidaris, *J. Organomet. Chem.*, 1990, **399**, 149.
719. E. A. Vasyutinskaya, I. L. Eremenko, A. A. Paynskii, S. E. Nefedov, A. I. Yanovskii and Y. T. Struchkov, *Zh. Neorg. Khim.*, 1991, **36**, 1707 (*Chem. Abstr.*, 1991, **115**, 232 431n).
720. G. Hogarth, P. C. Konidaris and G. C. Saunders, *J. Organomet. Chem.*, 1991, **406**, 153.
721. F. J. Lalor, D. Condon, G. Ferguson and M. A. Khan, *Inorg. Chem.*, 1981, **20**, 2178.
722. F. J. Lalor, M. E. Deane and D. M. Condon, *J. Organomet. Chem.*, 1983, **259**, 305.
723. F. J. Lalor and D. M. Condon, *J. Organomet. Chem.*, 1985, **284**, 327.
724. M. E. Deane, F. J. Lalor, G. Ferguson, B. L. Ruhl and M. Parvez, *J. Organomet. Chem.*, 1990, **381**, 213.
725. G. Ferguson, B. L. Ruhl, M. Parvez, F. J. Lalor and M. E. Deane, *J. Organomet. Chem.*, 1990, **381**, 357.
726. C. F. Barrientos-Penna, F. W. B. Einstein, T. Jones and D. Sutton, *Inorg. Chem.*, 1983, **22**, 2614.
727. W. A. Herrmann, G. W. Kriechbaum and I. Steffl, *Z. Anorg. Allg. Chem.*, 1984, **508**, 100.
728. G. L. Hillhouse, B. L. Haymore, S. A. Bistram and W. A. Herrmann, *Inorg. Chem.*, 1983, **22**, 314.
729. I. Bernal, W. Ries, H. Brunner and D. K. Rastogi, *J. Organomet. Chem.*, 1985, **290**, 353.
730. H. Brunner and H. Fisch, *J. Organomet. Chem.*, 1987, **335**, 15.
731. D. K. Rastogi and S. Rastogi, *Polyhedron*, 1982, **1**, 233.
732. J. D. Korp and I. Bernal, *Polyhedron*, 1982, **1**, 617.
733. H. Brunner *et al.*, *Chim. Acta*, 1985, **96**, L5.
734. M. Minelli, T. W. Rockway, J. H. Enemark, H. Brunner and M. Muschiol, *J. Organomet. Chem.*, 1981, **217**, C34.
735. H. Brunner, J. Wachter, J. Schmidbauer, G. M. Sheldrick and P. G. Jones, *Organometallics*, 1986, **5**, 2212.
736. G. M. Reisner and I. Bernal, *Inorg. Chim. Acta*, 1981, **51**, 201.
737. M. W. Creswick and I. Bernal, *Inorg. Chim. Acta*, 1983, **71**, 41.
738. H. Brunner, G. Agrifoglio, R. Benn and A. Rufinska, *J. Organomet. Chem.*, 1981, **217**, 365.
739. M. W. Creswick and I. Bernal, *Inorg. Chim. Acta*, 1983, **74**, 241.
740. M. Draux and I. Bernal, *Inorg. Chim. Acta*, 1986, **114**, 75.
741. H. Brunner and B. Schönhammer, *Z. Naturforsch., Teil B*, 1983, **38**, 852.
742. B. Gaylani, M. Kilner, C. I. French, A. J. Pick and S. C. Wallwork, *Acta Crystallogr., Sect. C*, 1991, **47**, 257.
743. J. Beck, *Z. Naturforsch., Teil B*, 1988, **43**, 1029.
744. H. Kisch and D. Garn, *J. Organomet. Chem.*, 1991, **409**, 347.
745. I. Bernal and G. M. Reisner, *Inorg. Chim. Acta*, 1983, **71**, 65.
746. H. Brunner, J. Wachter, I. Bernal, G. M. Reisner and R. Benn, *J. Organomet. Chem.*, 1983, **243**, 179.
747. H. Brunner, W. Meyer and J. Wachter, *J. Organomet. Chem.*, 1983, **243**, 437.
748. W. Petri *et al.*, *Chem. Ber.*, 1982, **115**, 846.
749. J. Meder, W. Petri and W. Beck, *Chem. Ber.*, 1984, **117**, 827.
750. Y. Shi, S. Lu, H. Guo and N. Hu, *Chin. Chem. Lett.*, 1992, **3**, 665.
751. M. Creswick and I. Bernal, *Inorg. Chim. Acta*, 1982, **57**, 171.
752. G. M. Reisner and I. Bernal, *J. Organomet. Chem.*, 1981, **220**, 55.
753. H. P. M. M. Ambrosius, A. H. I. M. van der Linden and J. J. Steggerda, *J. Organomet. Chem.*, 1981, **204**, 211.
754. H. P. M. M. Ambrosius, W. P. Bosman and J. A. Cras, *J. Organomet. Chem.*, 1981, **215**, 201.
755. U. Kunze, H. Jawad and R. Burghardt, *Z. Naturforsch., Teil B*, 1986, **41**, 1142.
756. U. Kunze and R. Burghardt, *Z. Naturforsch., Teil B*, 1987, **42**, 860.
757. J. Wachter, B. F. Mentzen and J. G. Riess, *Angew. Chem., Int. Ed. Engl.*, 1981, **20**, 284.
758. F. Jeanneaux, A. Grand and J. G. Riess, *J. Am. Chem. Soc.*, 1981, **103**, 4272.
759. P. Vierling and J. G. Riess, *Organometallics*, 1986, **5**, 2543.
760. M. Lattman, S. K. Chopra, A. H. Cowley and A. M. Arif, *Organometallics*, 1986, **5**, 677.
761. J.-M. Dupart, A. Grand, S. Pace and J. G. Riess, *J. Am. Chem. Soc.*, 1982, **104**, 2316.
762. O. J. Scherer, *Angew. Chem., Int. Ed. Engl.*, 1985, **24**, 924.
763. O. J. Scherer, *Angew. Chem., Int. Ed. Engl.*, 1990, **29**, 1104.
764. M. Di Vaira, P. Stoppioni and M. Peruzzini, *Polyhedron*, 1987, **6**, 351.
765. A. J. DiMaio and A. L. Rheingold, *Chem. Rev.*, 1990, **90**, 169.
766. L. Y. Goh, C. K. Chu, R. C. S. Wong and T. W. Hambley, *J. Chem. Soc., Dalton Trans.*, 1989, 1951.
767. L. Y. Goh and R. C. S. Wong, *Inorg. Synth.*, 1992, **29**, 247.
768. L. Y. Goh, R. C. S. Wong, C. K. Chu and T. W. Hambley, *J. Chem. Soc., Dalton Trans.*, 1990, 977.
769. L. Y. Goh, R. C. S. Wong and E. Sinn, *J. Chem. Soc., Chem. Commun.*, 1990, 1484.
770. L. Y. Goh, R. C. S. Wong and E. Sinn, *Organometallics*, 1993, **12**, 888.
771. O. J. Scherer, J. Schwalb, G. Wolmershäuser, W. Kaim and R. Gross, *Angew. Chem., Int. Ed. Engl.*, 1986, **25**, 363.
772. L. Y. Goh, R. C. S. Wong, W.-H. Yip and T. C. W. Mak, *Organometallics*, 1991, **10**, 875.
773. O. J. Scherer, W. Wiedemann and G. Wolmershäuser, *J. Organomet. Chem.*, 1989, **361**, C11.
774. O. J. Scherer, W. Wiedemann and G. Wolmershäuser, *Chem. Ber.*, 1990, **123**, 3.
775. L. Y. Goh, W. Chen and R. C. S. Wong, *Angew. Chem., Int. Ed. Engl.*, 1993, **32**, 1728.

776. O. J. Scherer, H. Sitzmann and G. Wolmershäuser, *J. Organomet. Chem.*, 1984, **268**, C9.
777. O. J. Scherer, J. Schwalb and H. Sitzmann, *Inorg. Synth.*, 1990, **27**, 224.
778. O. J. Scherer, H. Sitzmann and G. Wolmershäuser, *Angew. Chem., Int. Ed. Engl.*, 1984, **24**, 351.
779. K. Blechschmitt, H. Pfisterer, T. Zahn and M. L. Ziegler, *Angew. Chem., Int. Ed. Engl.*, 1985, **24**, 66.
780. M. L. Ziegler, K. Blechschmitt, B. Nuber and T. Zahn, *Chem. Ber.*, 1988, **121**, 159.
781. P. J. Sullivan and A. L. Rheingold, *Organometallics*, 1982, **1**, 1547.
782. A. L. Rheingold, M. J. Foley and P. J. Sullivan, *J. Am. Chem. Soc.*, 1982, **104**, 4727.
783. A. L. Rheingold and M. R. Churchill, *J. Organomet. Chem.*, 1983, **243**, 165.
784. A. J. DiMaio and A. L. Rheingold, *J. Chem. Soc., Chem. Commun.*, 1987, 404.
785. A. J. DiMaio and A. L. Rheingold, *Inorg. Chem.*, 1990, **29**, 798.
786. H.-P. Neumann and M. L. Ziegler, *J. Chem. Soc., Chem. Commun.*, 1988, 498.
787. O. J. Scherer, H. Sitzmann and G. Wolmershäuser, *J. Organomet. Chem.*, 1986, **309**, 77.
788. O. J. Scherer, H. Sitzmann and G. Wolmershäuser, *Angew. Chem., Int. Ed. Engl.*, 1989, **28**, 212.
789. I. Bernal, H. Brunner, W. Meier, H. Pfisterer, J. Wachter and M. L. Ziegler, *Angew. Chem., Int. Ed. Engl.*, 1984, **23**, 438.
790. H. Brunner, U. Klement, W. Meier, J. Wachter, O. Serhadle and M. L. Ziegler, *J. Organomet. Chem.*, 1987, **335**, 339.
791. M. Gorzellik, B. Nuber and M. L. Ziegler, *J. Organomet. Chem.*, 1992, **429**, 153.
792. J. R. Harper and A. L. Rheingold, *J. Organomet. Chem.*, 1990, **390**, C36.
793. O. J. Scherer, J. Schwalb, H. Swarowsky, G. Wolmershäuser, W. Kaim and R. Gross, *Chem. Ber.*, 1988, **121**, 443.
794. P. Jutzi and R. Kroos, *Chem. Ber.*, 1988, **121**, 1399.
795. M. Scheer, G. Friedrich and K. Schuster, *Angew. Chem., Int. Ed. Engl.*, 1993, **32**, 593.
796. O. J. Scherer, H. Sitzmann and G. Wolmershäuser, *Acta Crystallogr., Sect. C*, 1985, **41**, 1761.
797. O. J. Scherer, J. Schwalb and G. Wolmershäuser, *New J. Chem.*, 1989, **13**, 399.
798. G. Huttner, B. Sigwarth, O. Scheidsteger, L. Zsolnai and O. Orama, *Organometallics*, 1985, **4**, 326.
799. L. Y. Goh, R. C. S. Wong and T. C. W. Mak, *J. Organomet. Chem.*, 1989, **364**, 363.
800. L. Y. Goh, R. C. S. Wong and T. C. W. Mak, *J. Organomet. Chem.*, 1989, **373**, 71.
801. O. J. Scherer, H. Sitzmann and G. Wolmershäuser, *Angew. Chem, Int. Ed. Engl.*, 1984, **23**, 968.
802. L. Y. Goh, R. C. S. Wong, R. Wang and T. C. W. Mak, *Jiegou Huaxue*, 1991, **10**, 89.
803. K. Blechschmitt, T. Zahn and M. L. Ziegler, *Angew. Chem., Int. Ed. Engl.*, 1985, **24**, 702.
804. M. Gorzellik, B. Nuber and M. L. Ziegler, *J. Organomet. Chem.*, 1992, **436**, 207.
805. C. Caballero, D. Lehne, B. Nuber and M. L. Ziegler, *Chem. Ber.*, 1991, **124**, 1327.
806. H.-P. Neumann and M. L. Ziegler, *J. Organomet. Chem.*, 1989, **377**, 255.
807. M. Gorzellik, B. Nuber and M. L. Ziegler, *J. Organomet. Chem.*, 1992, **429**, 173.
808. W. Tremel, R. Hoffmann and M. Kertesz, *J. Am. Chem. Soc.*, 1989, **111**, 2030.
809. E. D. Jemmis and A. C. Reddy, *Organometallics*, 1988, **7**, 1561.
810. N. A. Compton, R. J. Errington and N. C. Norman, *Adv. Organomet. Chem.*, 1990, **31**, 91.
811. V. Grossbruchhaus and D. Rehder, *Inorg. Chim. Acta*, 1988, **141**, 9.
812. W. Clegg, N. A. Compton, R. J. Errington, N. C. Norman, A. J. Tucker and M. J. Winter, *J. Chem. Soc., Dalton Trans.*, 1988, 2941; correction: 1989, 1887.
813. A. M. Barr, M. D. Kerlogue, N. C. Norman, P. M. Webster and L. J. Farrugia, *Polyhedron*, 1989, **8**, 2495.
814. W. Clegg, N. A. Compton, R. J. Errington, G. A. Fisher, N. C. Norman and T. B. Marder, *J. Chem. Soc., Dalton Trans.*, 1991, 2887.
815. W. Clegg *et al.*, *J. Chem. Soc., Dalton Trans.*, 1992, 3515.
816. R. J. Errington, L. J. Farrugia, G. A. Fisher, A. Niklaus and N. C. Norman, *J. Chem. Soc., Dalton Trans.*, 1993, 1201.
817. M. Gorzellik, B. Nuber and M. L. Ziegler, *J. Organomet. Chem.*, 1992, **431**, 171.
818. W. Clegg, N. A. Compton, R. J. Errington and N. C. Norman, *Polyhedron*, 1988, **7**, 2239.
819. K. Merzweiler and L. Brands, *Z. Naturforsch., Teil B*, 1992, **47**, 978.
820. W. Clegg, N. A. Compton, R. J. Errington, G. A. Fisher, N. C. Norman and N. Wishart, *J. Organomet. Chem.*, 1990, **399**, C21.
821. G. Huttner, U. Weber, B. Sigwarth, O. Scheidsteger, H. Lang and L. Zsolnai, *J. Organomet. Chem.*, 1985, **282**, 331.
822. G. Huttner, B. Sigwarth, J. von Seyerl and L. Zsolnai, *Chem. Ber.*, 1982, **115**, 2035.
823. P. Jutzi and R. Kroos, *J. Organomet. Chem.*, 1990, **390**, 317.
824. A. M. Arif, A. H. Cowley and M. Pakulski, *J. Chem. Soc., Chem. Commun.*, 1985, 1707.
825. A. M. Arif, A. H. Cowley, N. C. Norman, A. G. Orpen and M. Pakulski, *Organometallics*, 1988, **7**, 309.
826. A. M. Arif, A. H. Cowley, M. Pakulski and G. J. Thomas, *Polyhedron*, 1986, **5**, 1651.
827. J. Grobe and R. Haubold, *Z. Anorg. Allg. Chem.*, 1985, **522**, 159.
828. H. Beurich and H. Vahrenkamp, *Chem. Ber.*, 1982, **115**, 2385.
829. F. Richter and H. Vahrenkamp, *Chem. Ber.*, 1982, **115**, 3224.
830. R. Maisch, M. Barth and W. Malisch, *J. Organomet. Chem.*, 1984, **260**, C35.
831. R. Schemm, R. Maisch, M. Luksza and W. Malisch, *Phosphorus and Sulfur*, 1987, **30**, 788.
832. R. Maisch, E. Ott, W. Buchner, W. Malisch, I. J. Colquhoun and W. McFarlane, *J. Organomet. Chem.*, 1985, **286**, C31.
833. T. Adatia, M. McPartlin, M. J. Mays, M. J. Morris and P. R. Raithby, *J. Chem. Soc., Dalton Trans.*, 1989, 1555.
834. W. Malisch *et al.*, *Phosphorus and Sulfur*, 1983, **18**, 299.
835. W. Malisch, K. Jörg, E. Gross, M. Schmeusser and A. Meyer, *Phosphorus and Sulfur*, 1986, **26**, 25.
836. M. Schmeusser, R. Maisch, P. Vierling and W. Malisch, *Phosphorus and Sulfur*, 1987, **30**, 790.
837. P. Braunstein, E. de Jésus, A. Dedieu, M. Lanfranchi and A. Tiripicchio, *Inorg. Chem.*, 1992, **31**, 399.
838. W. Malisch, R. Maisch, I. J. Colquhoun and W. McFarlane, *J. Organomet. Chem.*, 1981, **220**, C1.
839. J. Grobe and R. Haubold, *Z. Anorg. Allg. Chem.*, 1985, **526**, 145.
840. S.-G. Shyu, J.-Y. Hsu and Y.-S. Wen, *J. Organomet. Chem.*, 1993, **453**, 97.
841. S. Amann and W. Malisch, *Phosphorus, Sulfur, Silicon & Relat. Elem.*, 1990, **51/52**, 346.
842. F. Mercier, L. Ricard and F. Mathey, *Organometallics*, 1993, **12**, 98.
843. D. Gudat *et al.*, *J. Chem. Soc., Chem. Commun.*, 1985, 1687.
844. C. Hugel-Le Goff, F. Mercier, L. Ricard and F. Mathey, *J. Organomet. Chem.*, 1989, **363**, 325.
845. F. Mercier, L. Ricard and F. Mathey, *Polyhedron*, 1990, **9**, 2859.

846. J. Grobe and R. Haubold, *Z. Anorg. Allg. Chem.*, 1986, **534**, 121.
847. P. Jutzi and U. Meyer, *Chem. Ber.*, 1988, **121**, 559.
848. R. Janta, M. Luksza, W. Malisch, D. Kempf and G. Künzel, *J. Organomet. Chem.*, 1984, **266**, C22.
849. M. Luksza, U. Wanka and W. Malisch, *J. Organomet. Chem.*, 1984, **266**, 237.
850. K. A. Mahmoud, A. J. Rest, M. Luksza, K. Jörg and W. Malisch, *Organometallics*, 1984, **3**, 501.
851. L. Carlton, J. L. Davidson and M. Shiralian, *J. Chem. Soc., Dalton Trans.*, 1986, 1577.
852. M. Luksza, S. Himmel and W. Malisch, *Angew. Chem., Int. Ed. Engl.*, 1983, **22**, 416.
853. W. Malisch, M. Luksza and W. S. Sheldrick, *Z. Naturforsch., Teil B*, 1982, **36**, 1580.
854. M. Luksza and W. Malisch, *Z. Naturforsch., Teil B*, 1984, **39**, 1123.
855. M. Luksza, K. Jörg and W. Malisch, *Inorg. Chim. Acta*, 1984, **85**, L49.
856. E. W. Abel, S. R. Allen, P. A. Bates, M. B. Hursthouse and B. Khandelwal, *J. Chem. Soc., Chem. Commun.*, 1987, 983.
857. A. Lombard, G. Huttner and L. Zsolnai, *J. Organomet. Chem.*, 1988, **352**, 295.
858. R. Schemm and W. Malisch, *J. Organomet. Chem.*, 1985, **288**, C9.
859. A.-M. Caminade, M. Veith, V. Huch and W. Malisch, *Organometallics*, 1990, **9**, 1798.
860. E. Lindner, E. U. Küster, W. Hiller and R. Fawzi, *Chem. Ber.*, 1984, **117**, 127.
861. E. Lindner, M. Stängle, W. Hiller and R. Fawzi, *Chem. Ber.*, 1988, **121**, 1421.
862. E. Lindner, M. Stängle, W. Hiller and R. Fawzi, *Chem. Ber.*, 1989, **122**, 823.
863. E. Lindner and M. Heckmann, *Chem. Ber.*, 1991, **124**, 1715.
864. A. M. Arif, A. H. Cowley, C. M. Nunn, S. Quashie, N. C. Norman and A. G. Orpen, *Organometallics*, 1989, **8**, 1878.
865. L. D. Hutchins, H.-U. Reisacher, G. L. Wood, E. N. Duesler and R. T. Paine, *J. Organomet. Chem.*, 1987, **335**, 229.
866. S. Amann, K. Jörg and W. Malisch, *Phosphorus and Sulfur*, 1987, **30**, 791.
867. L. D. Hutchins, E. N. Duesler and R. T. Paine, *Organometallics*, 1984, **3**, 399.
868. E. Gross, K. Jörg, K. Fiederling, A. Göttlein, W. Malisch and R. Boese, *Angew. Chem., Int. Ed. Engl.*, 1984, **23**, 738.
869. W. F. McNamara, E. N. Duesler, R. T. Paine, J. V. Ortiz, P. Kölle and H. Nöth, *Organometallics*, 1986, **5**, 380.
870. H. Lang, M. Leise and L. Zsolnai, *J. Organomet. Chem.*, 1990, **386**, 349.
871. H. Lang, M. Leise and L. Zsolnai, *J. Organomet. Chem.*, 1990, **389**, 325.
872. H. Lang, M. Leise and L. Zsolnai, *Polyhedron*, 1992, **11**, 1281.
873. H. Lang, M. Leise and W. Imhof, *Z. Naturforsch., Teil B*, 1991, **46**, 1650.
874. H. Lang, M. Leise, L. Zsolnai and M. Fritz, *J. Organomet. Chem.*, 1990, **395**, C30.
875. F. Mercier and F. Mathey, *Organometallics*, 1990, **9**, 863.
876. H. H. Karsch, H.-U. Reisacher, B. Huber, G. Müller, K. Jörg and W. Malisch, *New J. Chem.*, 1989, **13**, 319.
877. R. El-Ouatib, D. Ballivet-Tkatchenko, G. Etemad-Moghadam and M. Koenig, *J. Organomet. Chem.*, 1993, **453**, 77.
878. A. H. Cowley, D. M. Giolando, C. M. Nunn, M. Pakulski, D. Westmoreland and N. C. Norman, *J. Chem. Soc., Dalton Trans.*, 1988, 2127.
879. A. M. Arif, A. H. Cowley and S. Quashie, *J. Chem. Soc., Chem. Commun.*, 1986, 1437.
880. C. J. Carrano, A. H. Cowley, C. M. Nunn, M. Pakulski and S. Quashie, *J. Chem. Soc., Chem. Commun.*, 1988, 170.
881. A. H. Cowley, N. C. Norman and S. Quashie, *J. Am. Chem. Soc.*, 1984, **106**, 5007.
882. U. Hofmockel and W. Malisch, *Phosphorus and Sulfur*, 1987, **30**, 789.
883. K. Jörg, W. Malisch, W. Reich, A. Meyer and U. Schubert, *Angew. Chem., Int. Ed. Engl.*, 1986, **25**, 92.
884. W. Malisch, K. Hindahl and R. Schemm, *Chem. Ber.*, 1992, **125**, 2027.
885. H. Pfister and W. Malisch, *J. Organomet. Chem.*, 1992, **439**, C11.
886. A. Fried, W. Malisch, M. Schmeusser and U. Weis, *Phosphorus, Sulfur, Silicon & Relat. Elem.*, 1992, **65**, 75.
887. U.-A. Hirth and W. Malisch, *J. Organomet. Chem.*, 1992, **439**, C16.
888. U.-A. Hirth, W. Malisch and H. Käb, *J. Organomet. Chem.*, 1992, **439**, C20.
889. W. Malisch *et al.*, *Angew. Chem., Int. Ed. Engl.*, 1992, **31**, 738.
890. W. Malisch, M. Märkl, S. Amann, U. Hirth and M. Schmeusser, *Phosphorus, Sulfur and Silicon*, 1990, **49/50**, 441.
891. W. Abriel, G. Baum, H. Burdorf and J. Heck, *Z. Naturforsch., Teil B*, 1991, **46**, 841.
892. G. Conole, M. McPartlin, M. J. Mays and M. J. Morris, *J. Chem. Soc., Dalton Trans.*, 1990, 2359.
893. W.-K. Wong, F. L. Chow, H. Chen, R.-J. Wang and T. C. W. Mak, *Polyhedron*, 1990, **9**, 2469.
894. W.-K. Wong, F. L. Chow, H. Chen, B. W. Au-Yeung, R.-J. Wang and T. C. W. Mak, *Polyhedron*, 1990, **9**, 2901.
895. E. P. Kyba, J. D. Mather, K. L. Hassett, J. S. McKennis and R. E. Davis, *J. Am. Chem. Soc.*, 1984, **106**, 5371.
896. E. P. Kyba, M. C. Kerby, R. P. Kashyap, K. L. Hassett and R. E. Davis, *J. Organomet. Chem.*, 1988, **346**, C19.
897. E. P. Kyba, M. C. Kerby, R. P. Kashyap, J. A. Mountzouris and R. E. Davis, *J. Am. Chem. Soc.*, 1990, **112**, 905.
898. H. Brunner and H. Peter, *J. Organomet. Chem.*, 1990, **393**, 401.
899. K. Henrick, M. McPartlin, A. D. Horton and M. J. Mays, *J. Chem. Soc., Dalton Trans.*, 1988, 1083.
900. S. Woodward and M. D. Curtis, *J. Organomet. Chem.*, 1992, **439**, 319.
901. H. Hartung *et al.*, *Polyhedron*, 1992, **11**, 1563.
902. R. A. Jones, S. T. Schwab, A. L. Stuart, B. R. Whittlesey and T. C. Wright, *Polyhedron*, 1985, **4**, 1689.
903. R. A. Jones, S. T. Schwab and A. L. Stuart, *Inorg. Synth.*, 1989, **25**, 167.
904. G. Conole, K. Henrick, M. McPartlin, A. D. Horton and M. J. Mays, *New J. Chem.*, 1988, **12**, 559.
905. A. D. Horton and M. J. Mays, *J. Chem. Soc., Dalton Trans.*, 1990, 155.
906. G. Conole, K. Henrick, M. McPartlin, A. D. Horton, M. J. Mays and E. Sappa, *J. Chem. Soc., Dalton Trans.*, 1990, 2367.
907. F. Bottomley and L. Sutin, *Adv. Organomet. Chem.*, 1988, **28**, 339.
908. F. Bottomley, *Polyhedron*, 1992, **11**, 1707.
909. F. Bottomley, D. E. Paez and P. S. White, *J. Am. Chem. Soc.*, 1981, **103**, 5581.
910. F. Bottomley, D. E. Paez and P. S. White, *J. Am. Chem. Soc.*, 1982, **104**, 5651.
911. I. L. Eremenko *et al.*, *Izv. Akad. Nauk SSSR, Ser. Khim.*, 1983, 1445 (*Chem. Abstr.*, 1983, **99**, 175 972q).
912. F. Bottomley *et al.*, *Organometallics*, 1990, **9**, 2443.
913. F. Bottomley and F. Grein, *Inorg. Chem.*, 1982, **21**, 4170.
914. A. A. Pasynskii *et al.*, *J. Organomet. Chem.*, 1983, **248**, 309.
915. I. L. Eremenko *et al.*, *J. Organomet. Chem.*, 1989, **368**, 185.
916. F. Bottomley, J. Chen, S. M. MacIntosh and R. C. Thompson, *Organometallics*, 1991, **10**, 906.
917. C. E. Davies *et al.*, *Inorg. Chem.*, 1992, **31**, 3779.

918. L. Y. Goh and T. C. W. Mak, *J. Organomet. Chem.*, 1989, **363**, 77.
919. I. L. Eremenko, A. A. Pasynskii, A. S. Katugin, B. Orazsakhatov, V. E. Shklover and Y. T. Struchkov, *Metalloorg. Khim.*, 1988, **1**, 166.
920. A. A. Pasynskii *et al.*, *J. Organomet. Chem.*, 1988, **344**, 195.
921. D. Wörmsbächer, K. M. Nicholas and A. L. Rheingold, *J. Chem. Soc., Chem. Commun.*, 1985, 721.
922. M. Herberhold, W. Kremnitz, A. Razavi, H. Schöllhorn and U. Thewalt, *Angew. Chem., Int. Ed. Engl.*, 1985, **24**, 601.
923. D. B. Morse, T. B. Rauchfuss and S. R. Wilson, *J. Am. Chem. Soc.*, 1988, **110**, 8234.
924. M. S. Rau, C. M. Kretz, L. A. Mercando, G. L. Geoffroy and A. L. Rheingold, *J. Am. Chem. Soc.*, 1991, **113**, 7420.
925. C. E. Housmekerides *et al.*, *Inorg. Chem.*, 1992, **31**, 4453.
926. J. W. Faller and Y. Ma, *J. Organomet. Chem.*, 1988, **340**, 59.
927. P. Gomez-Sal, E. de Jésus, P. Royo, A. Vázquez de Miguel, S. Martinez-Carrera and S. Garcia-Blanco, *J. Organomet. Chem.*, 1988, **353**, 191.
928. M. K. Trost and R. G. Bergman, *Organometallics*, 1991, **10**, 1172.
929. F. Bottomley and J. Chen, *Organometallics*, 1992, **11**, 3404.
930. R. R. Schrock, S. F. Pedersen, M. R. Churchill and J. W. Ziller, *Organometallics*, 1984, **3**, 1574.
931. S. A. MacLaughlin, R. C. Murray, J. C. Dewan and R. R. Schrock, *Organometallics*, 1985, **4**, 796.
932. J. W. Faller and Y. Ma, *Organometallics*, 1988, **7**, 559.
933. J. W. Faller and Y. Ma, *Organometallics*, 1989, **8**, 609.
934. J. W. Faller, R. R. Kucharczyk and Y. Ma, *Inorg. Chem.*, 1990, **29**, 1662.
935. Y. Ma, P. Demou and J. W. Faller, *Inorg. Chem.*, 1991, **30**, 62.
936. P. Legzdins, E. C. Phillips, S. J. Rettig, L. Sánchez, J. Trotter and V. C. Yee, *Organometallics*, 1988, **7**, 1877.
937. J. W. Faller and Y. Ma, *J. Organomet. Chem.*, 1989, **368**, 45.
938. M. J. Bunker and M. L. H. Green, *J. Chem. Soc., Dalton Trans.*, 1981, 847.
939. F. Bottomley, E. C. Ferris and P. S. White, *Organometallics*, 1990, **9**, 1166.
940. K. Umakoshi and K. Isobe, *J. Organomet. Chem.*, 1990, **395**, 47.
941. H. Arzoumanian, A. Baldy, M. Pierrot and J. F. Petrignani, *J. Organomet. Chem.*, 1985, **294**, 327.
942. H. Brunner, H. Kauermann, J. Pfauntsch, J. Wachter, J. H. Enemark and C. G. Young, *J. Organomet. Chem.*, 1987, **331**, 45.
943. Z. Zhu, X. You, C. F. Campana, M. F. Fredrich and L. F. Dahl, *Jiegou Huaxue*, 1987, **4**, 295.
944. L. D. Tanner, R. C. Haltiwanger and M. Rakowski DuBois, *Inorg. Chem.*, 1988, **27**, 1741.
945. M. Gorzellik, H. Bock, L. Gang, B. Nuber and M. L. Ziegler, *J. Organomet. Chem.*, 1991, **412**, 95.
946. H. Brunner, J. Wachter and H. Wintergerst, *J. Organomet. Chem.*, 1982, **235**, 77.
947. A. L. Rheingold and J. R. Harper, *J. Organomet. Chem.*, 1991, **403**, 335.
948. J. R. Harper and A. L. Rheingold, *J. Am. Chem. Soc.*, 1990, **112**, 4037.
949. H. G. Alt, H. I. Hayen and R. D. Rogers, *J. Chem. Soc., Chem. Commun.*, 1987, 1795.
950. H. G. Alt, H. E. Engelhardt, H. I. Hayen and R. D. Rogers, *J. Organomet. Chem.*, 1989, **366**, 287.
951. L. Y. Goh, M. Y. Darensbourg and Y. M. Hsiao, *Inorg. Synth.*, 1992, **29**, 251.
952. W. Chen, L. Y. Goh and T. C. W. Mak, *Organometallics*, 1986, **5**, 1997.
953. L. Y. Goh, T. W. Hambley and G. B. Robertson, *Organometallics*, 1987, **6**, 1051.
954. L. Y. Goh, T. W. Hambley and G. B. Robertson, *J. Chem. Soc., Chem. Commun.*, 1983, 1458.
955. C. H. Wei, L. Y. Goh, R. F. Bryan and E. Sinn, *Acta Crystallogr., Sect. C*, 1986, **42**, 796.
956. N. M. Kostic and R. F. Fenske, *J. Organomet. Chem.*, 1982, **233**, 337.
957. H. Brunner, R. Grassl and J. Wachter, *J. Organomet. Chem.*, 1992, **431**, 151.
958. W. A. Herrmann, J. Rohrmann and A. Schäfer, *J. Organomet. Chem.*, 1984, **265**, C1.
959. W. A. Herrmann, J. Rohrmann, H. Nöth, C. K. Nanila, I. Bernal and M. Draux, *J. Organomet. Chem.*, 1985, **284**, 189.
960. M. A. El-Hinnawi and A. K. El-Quaseer, *J. Organomet. Chem.*, 1985, **296**, 393.
961. J. Wachter, H. Brunner and W. Meier, *Inorg. Synth.*, 1990, **27**, 69.
962. H. Brunner, J. Wachter, E. Guggolz and M. L. Ziegler, *J. Am. Chem. Soc.*, 1982, **104**, 1765.
963. H. Brunner, R. Grassl, W. Meier, J. Wachter, B. Nuber and M. L. Ziegler, *J. Organomet. Chem.*, 1992, **434**, 63.
964. H. Brunner, H. Kauermann, W. Meier and J. Wachter, *J. Organomet. Chem.*, 1984, **263**, 183.
965. H. Brunner, H. Kauermann and J. Wachter, *J. Organomet. Chem.*, 1984, **265**, 189.
966. H. Brunner, J. Pfauntsch, J. Wachter, B. Nuber and M. L. Ziegler, *J. Organomet. Chem.*, 1989, **359**, 179.
967. H. Brunner, H. Kauermann and J. Wachter, *Angew. Chem., Int. Ed. Engl.*, 1983, **22**, 549.
968. A. Toupadakis, G. J. Kubas and C. J. Burns, *Inorg. Chem.*, 1992, **31**, 3810.
969. W. Chen, L. Y. Goh and E. Sinn, *Organometallics*, 1988, **7**, 2020.
970. L. Y. Goh, C. Wei and E. Sinn, *J. Chem. Soc., Chem. Commun.*, 1985, 462.
971. N. Albrecht, P. Hübener, U. Behrens and E. Weiss, *Chem. Ber.*, 1985, **118**, 4059.
972. W. A. Herrmann and J. Rohrmann, *Chem. Ber.*, 1986, **119**, 1437.
973. R. A. Fischer and W. A. Herrmann, *J. Organomet. Chem.*, 1987, **330**, 377.
974. R. A. Fischer, H.-J. Kneuper and W. A. Herrmann, *J. Organomet. Chem.*, 1987, **330**, 365.
975. W. A. Herrmann, J. Rohrmann, M. L. Ziegler and T. Zahn, *J. Organomet. Chem.*, 1984, **273**, 221.
976. W. A. Herrmann, J. Rohrmann and C. Hecht, *J. Organomet. Chem.*, 1985, **290**, 53.
977. C. H. W. Jones and R. D. Sharma, *Organometallics*, 1986, **5**, 1194.
978. L. Y. Goh, M. S. Tay, T. C. W. Mak and R.-J. Wang, *Organometallics*, 1992, **11**, 1711.
979. L. Y. Goh, Y. Y. Lim, M. S. Tay, T. C. W. Mak and Z.-Y. Zhou, *J. Chem. Soc., Dalton Trans.*, 1992, 1239.
980. L. Y. Goh, M. S. Tay, Y. Y. Lim, W. Chen, Z.-Y. Zhou and T. C. W. Mak, *J. Organomet. Chem.*, 1992, **441**, 51.
981. A. A. Pasynskii *et al.*, *Koord. Khim.*, 1984, **10**, 634 (*Chem. Abstr.*, 1984, **101**, 90 998n).
982. A. A. Pasynskii and I. L. Eremenko, *Russ. Chem. Rev.*, 1989, **58**, 181.
983. I. L. Eremenko, A. A. Pasynskii, V. T. Kalinnikov, Y. T. Struchkov and G. G. Aleksandrov, *Inorg. Chim. Acta*, 1981, **52**, 107.
984. S. E. Nefedov *et al.*, *J. Organomet. Chem.*, 1990, **384**, 295.
985. S. L. Gales, W. T. Pennington and J. W. Kolis, *J. Organomet. Chem.*, 1991, **419**, C10.
986. S. C. O'Neal, W. T. Pennington and J. W. Kolis, *Organometallics*, 1989, **8**, 2281.

987. J. Adel, F. Weller and K. Dehnicke, *J. Organomet. Chem.*, 1988, **347**, 343.
988. R. M. Herath Banda, J. Cusick, M. L. Scudder, D. C. Craig and I. G. Dance, *Polyhedron*, 1989, **8**, 1999.
989. M. Rakowski DuBois, *Chem. Rev.*, 1989, **89**, 1.
990. J. Wachter, *Angew. Chem., Int. Ed. Engl.*, 1989, **28**, 1613.
991. H. Brunner, W. Meier, J. Wachter, E. Guggolz, T. Zahn and M. L. Ziegler, *Organometallics*, 1982, **1**, 1107.
992. M. Rakowski DuBois, D. L. DuBois, M. C. van der Veer and R. C. Haltiwanger, *Inorg. Chem.*, 1981, **20**, 3064.
993. A. E. Bruce and D. R. Tyler, *Inorg. Chem.*, 1984, **23**, 3433.
994. W. Tremel, R. Hoffmann and E. D. Jemmis, *Inorg. Chem.*, 1989, **28**, 1213.
995. B. E. Bursten and R. H. Cayton, *Inorg. Chem.*, 1989, **28**, 2846.
996. M. R. M. Bruce, A. E. Bruce and D. R. Tyler, *Polyhedron*, 1985, **4**, 2073.
997. H. Brunner *et al.*, *J. Organomet. Chem.*, 1986, **309**, 313.
998. H. Brunner, R. Grassl, J. Wachter, B. Nuber and M. L. Ziegler, *J. Organomet. Chem.*, 1992, **427**, 57.
999. G. J. Kubas, R. R. Ryan and K. A. Kubat-Martin, *J. Am. Chem. Soc.*, 1989, **111**, 7823.
1000. H. Brunner, U. Klement, J. Pfauntsch and J. Wachter, *Angew. Chem., Int. Ed. Engl.*, 1987, **26**, 230.
1001. M. Rakowski DuBois, M. C. VanDerveer, D. L. DuBois, R. C. Haltiwanger and W. K. Miller, *J. Am. Chem. Soc.*, 1980, **102**, 7456.
1002. B. Cowans and M. Rakowski DuBois, in 'Organometallic Syntheses', eds. R. B. King and J. J. Eisch, Elsevier, Amsterdam, 1986, Vol. 3, p. 262.
1003. M. D. Curtis and P. D. Williams, *Inorg. Chem.*, 1983, **22**, 2661.
1004. G. J. Kubas and R. R. Ryan, *J. Am. Chem. Soc.*, 1985, **107**, 6138.
1005. G. J. Kubas and R. R. Ryan, *Polyhedron*, 1986, **5**, 473.
1006. D. L. DuBois, W. K. Miller and M. Rakowski DuBois, *J. Am. Chem. Soc.*, 1981, **103**, 3429.
1007. M. McKenna, L. L. Wright, D. J. Miller, L. Tanner, R. C. Haltiwanger and M. Rakowski DuBois, *J. Am. Chem. Soc.*, 1983, **105**, 5329.
1008. L. L. Wright, R. C. Haltiwanger, J. Noordik and M. Rakowski DuBois, *J. Am. Chem. Soc.*, 1987, **109**, 282.
1009. B. V. Prasad, C. S. Reddy and E. D. Jemmis, *Inorg. Chim. Acta*, 1989, **162**, 281.
1010. M. Rakowski DuBois, *J. Am. Chem. Soc.*, 1983, **105**, 3710.
1011. M. Maciejewski Farmer, R. C. Haltiwanger, F. Kvietok and M. Rakowski DuBois, *Organometallics*, 1991, **10**, 4066.
1012. P. Bernatis, R. C. Haltiwanger and M. Rakowski DuBois, *Organometallics*, 1992, **11**, 2435.
1013. L. D. Tanner, R. C. Haltiwanger, J. Noordik and M. Rakowski DuBois, *Inorg. Chem.*, 1988, **27**, 1736.
1014. D. E. Coons, J. C. V. Laurie, R. C. Haltiwanger and M. Rakowski DuBois, *J. Am. Chem. Soc.*, 1987, **109**, 282.
1015. J. C. V. Laurie, L. Duncan, R. C. Haltiwanger, R. T. Weberg and M. Rakowski DuBois, *J. Am. Chem. Soc.*, 1986, **108**, 6234.
1016. C. J. Casewit, D. E. Coons, L. L. Wright, W. K. Miller and M. Rakowski DuBois, *Organometallics*, 1986, **5**, 951.
1017. C. J. Casewit and M. Rakowski DuBois, *J. Am. Chem. Soc.*, 1986, **108**, 5482.
1018. L. L. Lopez *et al.*, *Organometallics*, 1993, **12**, 4764.
1019. W. Danzer, W. P. Fehlhammer, A. T. Liu, G. Thiel and W. Beck, *Chem. Ber.*, 1982, **115**, 1682.
1020. W. K. Miller, R. C. Haltiwanger, M. C. van Derveer and M. Rakowski DuBois, *Inorg. Chem.*, 1983, **22**, 2973.
1021. K. Roesselet *et al.*, *Organometallics*, 1987, **6**, 480.
1022. R. T. Weberg, J. C. V. Laurie and M. Rakowski DuBois, *J. Coord. Chem.*, 1988, **19**, 39 (*Chem. Abstr.*, 1990, **112**, 77 434h).
1023. R. T. Weberg, R. C. Haltiwanger, J. C. V. Laurie and M. Rakowski DuBois, *J. Am. Chem. Soc.*, 1986, **108**, 6242.
1024. J. Birnbaum, J. C. V. Laurie and M. Rakowski DuBois, *Organometallics*, 1990, **9**, 156.
1025. J. Birnbaum, G. Godziela, M. Maciejewski, T. L. Tonker, R. C. Haltiwanger and M. Rakowski DuBois, *Organometallics*, 1990, **9**, 394.
1026. G. Godziela, T. Tonker and M. Rakowski DuBois, *Organometallics*, 1989, **8**, 2220.
1027. P. Bernatis, J. C. V. Laurie and M. Rakowski DuBois, *Organometallics*, 1990, **9**, 1607.
1028. L. L. Lopez, P. Bernatis, J. Birnbaum, R. C. Haltiwanger and M. Rakowski DuBois, *Organometallics*, 1992, **11**, 2424.
1029. C. J. Casewit, R. C. Haltiwanger, J. Noordik and M. Rakowski DuBois, *Organometallics*, 1985, **4**, 326.
1030. D. E. Coons, R. C. Haltiwanger and M. Rakowski DuBois, *Organometallics*, 1987, **6**, 2417.
1031. J. Birnbaum, R. C. Haltiwanger, P. Bernatis, C. Teachout, K. Parker and M. Rakowski DuBois, *Organometallics*, 1991, **10**, 1779.
1032. L. Lopez, G. Godziela and M. Rakowski DuBois, *Organometallics*, 1991, **10**, 2660.
1033. M. Chen, J. A. Bandy and K. Prout, *Jiegou Huaxue*, 1986, **5**, 18.
1034. C. J. Casewit, M. Rakowski DuBois, R. A. Grieves and J. Mason, *Inorg. Chem.*, 1987, **26**, 1889.
1035. C. J. Casewit and M. Rakowski DuBois, *Inorg. Chem.*, 1986, **25**, 74.
1036. M. D. Curtis, *Appl. Organomet. Chem.*, 1992, **6**, 429.
1037. P. D. Williams and M. D. Curtis, *Inorg. Chem.*, 1986, **25**, 4562.
1038. P. D. Williams, M. D. Curtis, D. N. Duffy and W. M. Butler, *Organometallics*, 1983, **2**, 165.
1039. J. A. Bandy, C. E. Davies, J. C. Green, M. L. H. Green, K. Prout and D. P. S. Rodgers, *J. Chem. Soc., Chem. Commun.*, 1983, 1395.
1040. M. L. H. Green, J. Qin and D. O'Hare, *J. Organomet. Chem.*, 1988, **358**, 375.
1041. P. Baird *et al. J. Chem. Soc., Dalton Trans.*, 1991, 2377.
1042. I. L. Eremenko *et al.*, *Metalloorg. Khim.*, 1988, **1**, 172.
1043. P. M. Treichel and P. C. Nakagaki, *Organometallics*, 1986, **5**, 711.
1044. P. Jaitner and W. Wohlgenannt, *Inorg. Chim. Acta*, 1985, **101**, L43.
1045. F. Y. Pétillon, J. L. Le Quèré, J. Roué, J. E. Guerchais and D. W. A. Sharp, *J. Organomet. Chem.*, 1981, **204**, 207.
1046. J. Courtot-Coupez, M. Guéguen, J. E. Guerchais, F. Y. Pétillon, J. Talarmin and R. Mercier, *J. Organomet. Chem.*, 1986, **312**, 81.
1047. M. El Khalifa, F. Y. Pétillon, J.-Y. Saillard and J. Talarmin, *Inorg. Chem.*, 1989, **28**, 3849.
1048. M. El Khalifa, M. Guéguen, R. Mercier, F. Y. Pétillon, J.-Y. Saillard and J. Talarmin, *Organometallics*, 1989, **8**, 140.
1049. M. Guéguen, F. Y. Pétillon and J. Talarmin, *Organometallics*, 1989, **8**, 148.
1050. P. Jaitner, *J. Organomet. Chem.*, 1982, **233**, 333.

1051. P. Jaitner and W. Winder, *Inorg. Chim. Acta*, 1987, **128**, L17.
1052. M. Guéguen, F. Y. Pétillon, J. Talarmin, P. Jaitner and W. Winder, *Inorg. Chim. Acta*, 1988, **144**, 5.
1053. P. Jaitner, W. Wohlgenannt, A. Gieren, H. Betz and T. Hübner, *J. Organomet. Chem.*, 1985, **297**, 281.
1054. M. B. Gomes de Lima, J. E. Guerchais, R. Mercier and F. Y. Pétillon, *Organometallics*, 1986, **5**, 1952.
1055. F. Gloaguen, C. Le Floc'h, F. Y. Pétillon, J. Talarmin, M. El Khalifa and J.-Y. Saillard, *Organometallics*, 1991, **10**, 2004.
1056. M. El Khalifa, J.-Y. Saillard, F. Gloaguen, C. Le Floc'h, F. Y. Pétillon and J. Talarmin, *New J. Chem.*, 1992, **16**, 847.
1057. I. B. Benson, S. D. Killops, S. A. R. Knox and A. J. Welch, *J. Chem. Soc., Chem. Commun.*, 1980, 1137.
1058. N. Kuhn, E. Zauder, R. Boese and D. Bläser, *J. Chem. Soc., Dalton Trans.*, 1988, 2171.
1059. W. A. W. A. Bakar, J. L. Davidson, W. E. Lindsell, K. J. McCullough and K. W. Muir, *J. Chem. Soc., Dalton Trans.*, 1989, 991.
1060. W. A. W. A. Bakar, J. L. Davidson, W. E. Lindsell and K. J. McCullough, *J. Chem. Soc., Dalton Trans.*, 1990, 61.
1061. A. S. F. Boyd, J. L. Davidson, C. H. McIntosh, P. C. Leverd, W. E. Lindsell and N. J. Simpson, *J. Chem. Soc., Dalton Trans.*, 1992, 2531.
1062. J. E. Guerchais, J.'L. Le Quèré, F. Y. Pétillon, L. Manojlovic-Muir, K. W. Muir and D. W. A. Sharp, *J. Chem. Soc., Dalton Trans.*, 1982, 283.
1063. J. L. Le Quèré, F. Y. Pétillon, J. E. Guerchais, L. Manojlovic-Muir, K. W. Muir and D. W. A. Sharp, *J. Organomet. Chem.*, 1983, **249**, 127.
1064. J. Courtot-Coupez, J. E. Guerchais, F. Y. Pétillon and J. Talarmin, *J. Chem. Soc., Dalton Trans.*, 1986, 1917.
1065. J. Courtot-Coupez, M. Guéguen, J. E. Guerchais, F. Y. Pétillon and J. Talarmin, *J. Chem. Soc., Dalton Trans.*, 1986, 1923.
1066. E. Román, D. Catheline and D. Astruc, *J. Organomet. Chem.*, 1982, **236**, 229.
1067. H. B. Abrahamson and M. L. Freeman, *Organometallics*, 1983, **2**, 679.
1068. H. B. Abrahamson, M. L. Freeman, M. B. Hossain and D. van der Helm, *Inorg. Chem.*, 1984, **23**, 2286.
1069. P. Mathur and B. H. S. Thimmappa, *Inorg. Chim. Acta*, 1988, **148**, 119.
1070. S. Woodward, U. Riaz, M. D. Curtis, B. S. Haggerty and A. L. Rheingold, *Organometallics*, 1990, **9**, 2703.
1071. F. R. Kreissl and N. Ullrich, *J. Organomet. Chem.*, 1989, **361**, C30.
1072. D. S. Gill *et al.*, *J. Chem. Soc., Dalton Trans.*, 1984, 1343.
1073. K. Endrich, E. Guggolz, O. Serhadle, M. L. Ziegler and R. P. Korswagen, *J. Organomet. Chem.*, 1988, **349**, 323.
1074. O. A. Rajan, M. McKenna, J. Noordik, R. C. Haltiwanger and M. Rakowski DuBois, *Organometallics*, 1984, **3**, 831.
1075. M. Herberhold, W. Jellen and H. H. Murray, *J. Organomet. Chem.*, 1984, **270**, 65.
1076. G. J. Kubas, H. J. Wasserman and R. R. Ryan, *Organometallics*, 1985, **4**, 2012.
1077. G. J. Kubas, H. J. Wasserman and R. R. Ryan, *Organometallics*, 1985, **4**, 419.
1078. W. P. Ehrenreich, M. Herberhold and A. F. Hill, *J. Organomet. Chem.*, 1989, **371**, 303.
1079. M. Herberhold, W. Jellen and M. L. Ziegler, *Inorg. Chim. Acta*, 1986, **118**, 15.
1080. J. A. Kovacs and R. G. Bergman, *J. Am. Chem. Soc.*, 1989, **111**, 1131.
1081. A. Shaver, J. Hartgerink, R. D. Lal, P. Bird and N. Ansari, *Organometallics*, 1983, **2**, 938.
1082. A. Shaver and J. Hartgerink, *Can. J. Chem.*, 1987, **65**, 1190.
1083. D. J. Weinmann and H. B. Abrahamson, *Inorg. Chem.*, 1987, **26**, 3034.
1084. P. M. Treichel, P. C. Nakagaki and K. J. Haller, *J. Organomet. Chem.*, 1987, **327**, 327.
1085. K. L. Brandenburg, M. J. Heeg and H. B. Abrahamson, *Inorg. Chem.*, 1987, **26**, 1064.
1086. D. J. Weinmann and H. B. Abrahamson, *Inorg. Chem.*, 1987, **26**, 2133.
1087. A. Shaver, B. S. Lum, P. Bird and K. Arnold, *Inorg. Chem.*, 1989, **28**, 1900.
1088. P. M. Treichel and E. K. Rublein, *J. Organomet. Chem.*, 1989, **359**, 195.
1089. P. M. Triechel, M. S. Schmidt, P. C. Nakagaki and E. K. Rublein, *J. Organomet. Chem.*, 1986, **311**, 193.
1090. P. M. Treichel and E. K. Rublein, *J. Organomet. Chem.*, 1992, **423**, 391.
1091. J. L. Davidson, M. Shiralian, L. Manojlovic-Muir and K. W. Muir, *J. Chem. Soc., Dalton Trans.*, 1984, 2167.
1092. M. M. Kubicki, R. Kergoat, H. Scordia, L. C. Gomes de Lima, J. E. Guerchais and P. L'Haridon, *J. Organomet. Chem.*, 1988, **340**, 41.
1093. A. Shaver, B. S. Lum, P. Bird, E. Livingstone and M. Schweitzer, *Inorg. Chem.*, 1990, **29**, 1832.
1094. F. H. Köhler, *et al.*, *J. Organomet. Chem.*, 1989, **365**, C15.
1095. D. B. Morse, T. B. Rauchfuss and S. R. Wilson, *J. Am. Chem. Soc.*, 1990, **112**, 1860.
1096. M. Morán, *Transition Met. Chem.*, 1981, **6**, 173.
1097. F. H. Köhler, R. de Cao, K. Ackermann and J. Sedlmair, *Z. Naturforsch., Teil B*, 1983, **38**, 1406.
1098. L. Y. Goh and T. W. Hambley, *J. Organomet. Chem.*, 1990, **395**, 269.
1099. M. Scheer *et al.*, *Z. Anorg. Allg. Chem.*, 1990, **591**, 221.
1100. A. Grohmann, F. H. Köhler, G. Müller and H. Zeh, *Chem. Ber.*, 1989, **122**, 897.
1101. P. Arndt, E. Kurras, and J. Otto, *Z. Chem.*, 1983, **23**, 443.
1102. R. A. Heintz, B. S. Haggerty, H. Wan, A. L. Rheingold and K. H. Theopold, *Angew. Chem., Int. Ed. Engl.*, 1992, **31**, 1077.
1103. K. H. Theopold, *Acc. Chem. Res.*, 1990, **23**, 263.
1104. W. A. Herrmann, W. R. Thiel and E. Herdtweck, *J. Organomet. Chem.*, 1988, **353**, 323.
1105. D. S. Richeson, S.-W. Hsu, N. H. Fredd, G. Van Duyne and K. H. Theopold, *J. Am. Chem. Soc.*, 1986, **108**, 8273.
1106. D. S. Richeson, J. F. Mitchell and K. H. Theopold, *Organometallics*, 1989, **8**, 2570.
1107. C. Janiak, J. Silvestre and K. H. Theopold, *Chem. Ber.*, 1993, **126**, 631.
1108. J. A. Barrera and D. E. Wilcox, *Inorg. Chem.*, 1992, **31**, 1745.
1109. S. K. Noh, R. A. Heintz and K. H. Theopold, *Organometallics*, 1989, **8**, 2071.
1110. F. H. Köhler, C. Krüger and H. Zeh, *J. Organomet. Chem.*, 1990, **386**, C13.
1111. B. J. Thomas, J. F. Mitchell, K. H. Theopold and J. A. Leary, *J. Organomet. Chem.*, 1988, **348**, 333.
1112. B. J. Thomas and K. H. Theopold, *J. Am. Chem. Soc.*, 1988, **110**, 5902.
1113. B. J. Thomas, S. K. Noh, G. K. Schulte, S. C. Sendlinger and K. H. Theopold, *J. Am. Chem. Soc.*, 1991, **113**, 893.
1114. S. K. Noh, R. A. Heintz, C. Janiak, S. C. Sendlinger and K. H. Theopold, *Angew. Chem., Int. Ed. Engl.*, 1990, **29**, 775.
1115. S. K. Noh, S. C. Sendlinger, C. Janiak and K. H. Theopold, *J. Am. Chem. Soc.*, 1989, **111**, 9127.
1116. S. K. Noh, R. A. Heintz, B. S. Haggerty, A. L. Rheingold and K. H. Theopold, *J. Am. Chem. Soc.*, 1992, **114**, 1892.
1117. R. Poli, *Chem. Rev.*, 1991, **91**, 509.
1118. M. L. H. Green and P. Mountford, *Chem. Soc. Rev.*, 1992, **21**, 29.

1119. R. C. Murray, L. Blum, A. H. Liu and R. R. Schrock, *Organometallics*, 1985, **4**, 953.
1120. M. B. O'Regan, M. G. Vale, J. F. Payack and R. R. Schrock, *Inorg. Chem.*, 1992, **31**, 1112.
1121. R. T. Baker, J. C. Calabrese, R. L. Harlow and I. D. Williams, *Organometallics*, 1993, **12**, 830.
1122. A. H. Liu, R. C. Murray, J. C. Dewan, B. D. Santarsiero and R. R. Schrock, *J. Am. Chem. Soc.*, 1987, **109**, 4282.
1123. M. L. H. Green, J. D. Hubert and P. Mountford, *J. Chem. Soc., Dalton Trans.*, 1990, 3793.
1124. K. Fromm and E. Hey-Hawkins, *Z. Anorg. Allg. Chem.*, 1993, **619**, 261.
1125. R. Poli, J. C. Gordon, J. U. Desai and A. L. Rheingold, *J. Chem. Soc., Chem. Commun.*, 1991, 1518.
1126. J. C. Green, M. L. H. Green, P. Mountford and M. J. Parkington, *J. Chem. Soc., Dalton Trans.*, 1990, 3407.
1127. C. J. Harlan, R. A. Jones, S. U. Koschmeider and C. M. Nunn, *Polyhedron*, 1990, **9**, 669.
1128. Q. Feng, M. L. H. Green and P. Mountford, *J. Chem. Soc., Dalton Trans.*, 1992, 2171.
1129. M. L. H. Green and P. Mountford, *Organometallics*, 1990, **9**, 886.
1130. Q. Feng, M. Ferrer, M. L. H. Green, P. Mountford and V. S. B. Mtetwa, *J. Chem. Soc., Dalton Trans.*, 1992, 1205.
1131. Q. Feng, M. Ferrer, M. L. H. Green, P. C. McGowan, P. Mountford and V. S. B. Mtetwa, *J. Chem. Soc., Chem. Commun.*, 1991, 552.
1132. Q. Feng, M. Ferrer, M. L. H. Green, P. Mountford, V. S. B. Mtetwa and K. Prout, *J. Chem. Soc., Dalton Trans.*, 1991, 1397.
1133. R. Poli and A. L. Rheingold, *J. Chem. Soc., Chem. Commun.*, 1990, 552.
1134. R. Poli, B. E. Owens, S. T. Krueger and A. L. Rheingold, *Polyhedron*, 1992, **11**, 2301.
1135. R. T. Baker, J. R. Morton, K. F. Preston, A. J. Williams and Y. Le Page, *Inorg. Chem.*, 1991, **30**, 113.
1136. S. T. Krueger, R. Poli, A. L. Rheingold and D. L. Staley, *Inorg. Chem.*, 1989, **28**, 4599.
1137. S. T. Krueger, B. E. Owens and R. Poli, *Inorg. Chem.*, 1990, **29**, 2001.
1138. R. Poli, B. E. Owens and R. G. Linck, *J. Am. Chem. Soc.*, 1992, **114**, 1302.
1139. R. Poli, B. E. Owens and R. G. Linck, *Inorg. Chem.*, 1992, **31**, 662.
1140. R. G. Linck, B. E. Owens, R. Poli and A. L. Rheingold, *Gazz. Chim. Ital.*, 1991, **121**, 163.
1141. R. Poli, A. L. Rheingold and B. E. Owens-Waltermire, *Inorg. Chim. Acta*, 1993, **203**, 223.
1142. R. Poli and M. A. Kelland, *J. Organomet. Chem.*, 1991, **419**, 127.
1143. J. C. Gordon, V. T. Lee and R. Poli, *Inorg. Chem.*, 1993, **32**, 4460.
1144. R. Poli, J. C. Gordon, J. U. Desai and A. L. Rheingold, *J. Chem. Soc., Chem. Commun.*, 1991, 1518.
1145. F. Abugideiri, J. C. Gordon, R. Poli, B. E. Owens-Waltermire and A. L. Rheingold, *Organometallics*, 1993, **12**, 1575.
1146. B. E. Owens and R. Poli, *Inorg. Chim. Acta*, 1991, **179**, 229.
1147. A. A. Cole, J. C. Gordon, M. A. Kelland, R. Poli and A. L. Rheingold, *Organometallics*, 1992, **11**, 1754.
1148. F. Abugideiri, M. A. Kelland and R. Poli, *Organometallics*, 1992, **11**, 1311.
1149. R. Poli, S. T. Krueger, F. Abugideiri, B. S. Haggerty and A. L. Rheingold, *Organometallics*, 1991, **10**, 3041.
1150. V. C. Gibson *et al.*, *J. Chem. Soc., Dalton Trans.*, 1985, 2025.
1151. M. Brookhart *et al.*, *J. Chem. Soc., Dalton Trans.*, 1985, 423.
1152. F. H. Köhler and W. A. Geike, *J. Organomet. Chem.*, 1987, **328**, 35.
1153. F. H. Köhler, R. de Cao and G. Manlik, *Inorg. Chim. Acta*, 1984, **91**, L1.
1154. F. H. Köhler and K. H. Doll, *Z. Naturforsch., Teil B*, 1982, **37**, 144.
1155. J. L. Robbins, N. Edelstein, B. Spencer and J. C. Smart, *J. Am. Chem. Soc.*, 1982, **104**, 1882.
1156. F. H. Köhler, K. H. Doll and W. Prössdorf, *J. Organomet. Chem.*, 1982, **224**, 341.
1157. M. P. Castellani, S. J. Geib, A. L. Rheingold and W. C. Trogler, *Organometallics*, 1987, **6**, 1703.
1158. D. O'Hare, V. J. Murphy and N. Kaltsoyannis, *J. Chem. Soc., Dalton Trans.*, 1993, 383.
1159. F. H. Köhler, K. H. Doll, W. Prössdorf and J. Müller, *Angew. Chem., Int. Ed. Engl.*, 1982, **21**, 151.
1160. H. Atzkern, J. Hiermeier, B. Kanellakopulos, F. H. Köhler, G. Müller and O. Steigelmann, *J. Chem. Soc., Chem. Commun.*, 1991, 997.
1161. M. Fritz *et al.*, *Chem. Ber.*, 1991, **124**, 1531.
1162. P. Bergerat *et al.*, *Angew. Chem., Int. Ed. Engl.*, 1992, **31**, 1258.
1163. R. Bartsch, P. B. Hitchcock and J. F. Nixon, *J. Organomet. Chem.*, 1988, **356**, C1.
1164. N. Hebendanz, F. H. Köhler, F. Scherbaum and B. Schlesinger, *Magn. Reson. Chem.*, 1989, **27**, 798.
1165. F. H. Köhler and W. Geike, *J. Magn. Reson.*, 1983, **53**, 297.
1166. J. Blümel, P. Hofmann and F. H. Köhler, *Magn. Reson. Chem.*, 1993, **31**, 2.
1167. V. T. Aleksanyan and I. I. Greenwald, *J. Mol. Struct.*, 1982, **90**, 35.
1168. S. Y. Ketkov and G. A. Domrachev, *Inorg. Chim. Acta*, 1990, **178**, 233.
1169. J. C. Giordan, J. H. Moore, J. A. Tossell and J. Weber, *J. Am. Chem. Soc.*, 1983, **105**, 3431.
1170. A. Modelli, A. Foffani, M. Guerra, D. Jones and G. Distefano, *Chem. Phys. Lett.*, 1983, **99**, 58.
1171. B. V. Lokshin and I. I. Greenwald, *Metalloorg. Khim.*, 1989, **2**, 118.
1172. I. I. Grinval'd, B. V. Lokshin, O. A. Usacheva, I. A. Abronin, L. M. Golubinskaya and V. I. Bregadze, *Metalloorg. Khim.*, 1990, **3**, 166.
1173. F. A. Cotton, X. Feng, P. A. Kibala and M. Matusz, *J. Am. Chem. Soc.*, 1988, **110**, 2807.
1174. F. A. Cotton and W. Wang, *Polyhedron*, 1985, **4**, 1735.
1175. L. Benes, J. Kalousova and J. Votinsky, *J. Organomet. Chem.*, 1985, **290**, 147.
1176. I. L. Eremenko, A. A. Pasynskii, O. G. Volkov, O. G. Ellert and V. T. Kalinnikov, *J. Organomet. Chem.*, 1981, **222**, 235.
1177. L. Benes, J. Votinsky, J. Kalousova, M. Nadvornik and J. Klikorka, *Coll. Czech. Chem. Commun.*, 1982, **47**, 3381.
1178. J. Kalousova, L. Benes, J. Votinsky and M. Nadvornik, *Coll. Czech. Chem. Commun.*, 1985, **50**, 828.
1179. J. Votinsky, J. Kalousova, L. Benes and M. Nadvornik, *Z. Chem.*, 1981, **21**, 445.
1180. J. Kalousova, L. Benes and J. Votinsky, *Coll. Czech. Chem. Commun.*, 1986, **51**, 314.
1181. V. P. Maryin, *J. Organomet. Chem.*, 1992, **441**, 241.
1182. E. U. van Raaij, S. Mönkeberg, H. Kiesele and H.-H. Brintzinger, *J. Organomet. Chem.*, 1988, **356**, 307.
1183. H. Schwemlein, L. Zsolnai, G. Huttner and H. H. Brintzinger, *J. Organomet. Chem.*, 1983, **256**, 285.
1184. S. Gutmann, P. Burger, M.-H. Prosenc and H.-H. Brintzinger, *J. Organomet. Chem.*, 1990, **397**, 21.
1185. K. M. Simpson, M. F. Rettig and R. M. Wing, *Organometallics*, 1992, **11**, 4363.
1186. E. U. van Raaij and H.-H. Brintzinger, *J. Organomet. Chem.*, 1988, **356**, 315.
1187. E. U. van Raaij, H.-H. Brintzinger, L. Zsolnai and G. Huttner, *Z. Anorg. Allg. Chem.*, 1989, **577**, 217.

1188. M. S. Lawless and D. S. Marynick, *Organometallics*, 1991, **10**, 543.
1189. J. M. Millar, R. V. Kastrup, S. Harris and I. T. Horváth, *Angew. Chem., Int. Ed. Engl.*, 1990, **29**, 194.
1190. J. W. Freeman, N. C. Hallinan, A. M. Arif, R. W. Gedridge, R. D. Ernst and F. Basolo, *J. Am. Chem. Soc.*, 1991, **113**, 6509.
1191. N. C. Hallinan, G. Morelli and F. Basolo, *J. Am. Chem. Soc.*, 1988, **110**, 6585.
1192. J. K. Shen *et al.*, *Organometallics*, 1992, **11**, 3215.
1193. S. C. Bovino, G. W. Coates, J. P. Banovetz, R. M. Waymouth, D. A. Straus and J. W. Ziller, *Inorg. Chim. Acta*, 1993, **203**, 179.
1194. R. N. Perutz and J. C. Scaiano, *J. Chem. Soc., Chem. Commun.*, 1984, 457.
1195. J. Chetwynd-Talbot, P. Grebenik and R. N. Perutz, *Inorg. Chem.*, 1982, **21**, 3647.
1196. P. A. Cox, P. Grebenik, R. N. Perutz, M. D. Robinson, R. Grinter and D. R. Stern, *Inorg. Chem.*, 1983, **22**, 3614.
1197. J. Bashkin, M. L. H. Green, M. L. Poveda and K. Prout, *J. Chem. Soc., Dalton Trans.*, 1982, 2485.
1198. M. del C. Berral, M. L. H. Green and R. Jimenez, *J. Chem. Soc., Dalton Trans.*, 1982, 2495.
1199. J. W. Egan, Jr. and J. L. Petersen, *Organometallics*, 1986, **5**, 906.
1200. N. D. Silavwe, M. P. Castellani and D. R. Tyler, *Inorg. Synth.*, 1992, **29**, 204.
1201. T. Ito and T. Yoden, *Bull. Chem. Soc. Jpn.*, 1993, **66**, 2365.
1202. C. Persson and C. Andersson, *Organometallics*, 1993, **12**, 2370.
1203. F. G. N. Cloke, J. P. Day, J. C. Green, C. P. Morley and A. C. Swain, *J. Chem. Soc., Dalton Trans.*, 1991, 789.
1204. F. G. N. Cloke, J. P. Day, A. M. Greenway, K. R. Seddon, A. A. Shimran and A. C. Swain, *J. Organomet. Chem.*, 1989, **372**, 231.
1205. G. Parkin and J. E. Bercaw, *Polyhedron*, 1988, **7**, 2053.
1206. F. J. Feher, M. Green and A. G. Orpen, *J. Chem. Soc., Chem. Commun.*, 1986, 291.
1207. R. B. Girling, P. Grebenik and R. N. Perutz, *Inorg. Chem.*, 1986, **25**, 31.
1208. A. R. Dias and J. A. Martinho Simões, *Polyhedron*, 1988, **7**, 1531.
1209. A. R. Dias, J. A. Martinho Simões, C. Teixeira, C. Airoldi and A. P. Chagas, *J. Organomet. Chem.*, 1989, **361**, 319.
1210. A. R. Dias, H. P. Diogo, M. E. M. da Piedade, J. A. Martinho Simões, A. S. Carson and E. H. Jamea, *J. Organomet. Chem.*, 1990, **391**, 361.
1211. G. Parkin and J. E. Bercaw, *J. Chem. Soc., Chem. Commun.*, 1989, 255.
1212. D. M. Heinekey, *J. Am. Chem. Soc.*, 1991, **113**, 6074.
1213. T. Ito, M. Koga, S. Kurishima, M. Natori, N. Sekizuka and K. Yoshioka, *J. Chem. Soc., Chem. Commun.*, 1990, 988.
1214. T. Igarashi and T. Ito, *Chem. Lett.*, 1985, 1699.
1215. T. Ito, T. Igarashi and F. Suzuki, *J. Organomet. Chem.*, 1987, **320**, C16.
1216. T. Ito, K. Haji, F. Suzuki and T. Igarashi, *Organometallics*, 1993, **12**, 2325.
1217. J. A. Bandy, A. Berry, M. L. H. Green, R. N. Perutz, K. Prout and J.-N. Verpeaux, *J. Chem. Soc., Chem. Commun.*, 1984, 729.
1218. O. N. D. Mackey and C. P. Morley, *Polyhedron*, 1992, **11**, 389.
1219. K. Fiederling, I. Grob and W. Malisch, *J. Organomet. Chem.*, 1983, **255**, 299.
1220. G. E. Herberich and W. Barlage, *Organometallics*, 1987, **6**, 1924.
1221. G. E. Herberich, B. Hessner and J. Okuda, *J. Organomet. Chem.*, 1983, **254**, 317.
1222. M. Cariou, M. M. Kubicki, R. Kergoat, J. E. Guerchais and P. L'Haridon, *J. Organomet. Chem.*, 1986, **304**, 127.
1223. L. Gomes de Lima, M. Cariou, H. Scordia, R. Kergoat, M. M. Kubicki and J. E. Guerchais, *J. Organomet. Chem.*, 1985, **290**, 321.
1224. H. Scordia, R. Kergoat, M. M. Kubicki and J. E. Guerchais, *J. Organomet. Chem.*, 1983, **249**, 371.
1225. M. Cariou, M. M. Kubicki, R. Kergoat, L. C. Gomes de Lima, H. Scordia and J. E. Guerchais, *Inorg. Chim. Acta*, 1985, **104**, 185.
1226. M. M. Kubicki *et al.*, *Inorg. Chim. Acta*, 1985, **104**, 191.
1227. J. Okuda and G. E. Herberich, *Organometallics*, 1987, **6**, 2331.
1228. G. E. Herberich and J. Okuda, *Angew. Chem., Int. Ed. Engl.*, 1985, **24**, 402.
1229. S. Gambarotta, C. Floriani, A. Chiesi-Villa and C. Guastini, *J. Am. Chem. Soc.*, 1985, **107**, 2985.
1230. J. Okuda, G. E. Herberich, E. Raabe and I. Bernal, *J. Organomet. Chem.*, 1988, **353**, 65.
1231. J. L. Petersen and J. W. Egan Jr., *Inorg. Chem.*, 1981, **20**, 2883.
1232. A. De Cian, J. Colin, M. Schappacher, L. Ricard and R. Weiss, *J. Am. Chem. Soc.*, 1981, **103**, 1850.
1233. G. E. Herberich, W. Barlage and K. Linn, *J. Organomet. Chem.*, 1991, **414**, 193.
1234. G. E. Herberich and J. Okuda, *Chem. Ber.*, 1984, **117**, 3112.
1235. G. E. Herberich, U. Englert and W. Fassbender, *J. Organomet. Chem.*, 1991, **420**, 303.
1236. J. J. Ko, T. M. Bockman and J. K. Kochi, *Organometallics*, 1990, **9**, 1833.
1237. J. Okuda, *Z. Naturforsch., Teil B*, 1988, **45**, 753.
1238. G. E. Herberich and K. Linn, *J. Organomet. Chem.*, 1991, **418**, 409.
1239. G. J. Kubas and R. R. Ryan, *Inorg. Chem.*, 1984, **23**, 3181.
1240. P. D. Grebenik *et al.*, *J. Chem. Soc., Chem. Commun.*, 1988, 799.
1241. P. D. Grebenik, M. L. H. Green, M. A. Kelland, J. B. Leach and P. Mountford, *J. Chem. Soc., Chem. Commun.*, 1989, 1397.
1242. L. F. Rhodes, J. C. Huffman and K. G. Caulton, *Inorg. Chim. Acta*, 1992, **198–200**, 639.
1243. A. Albinati, R. Naegeli, A. Togni and L. M. Venanzi, *J. Organomet. Chem.*, 1987, **330**, 85.
1244. G. L. Soloveichik and B. M. Bulychev, *Russian Chem. Rev.*, 1983, **52**, 43.
1245. V. M. Ishchenko, B. M. Bulychev, G. L. Soloveichik, V. K. Bel'sky and O. G. Ellert, *Polyhedron*, 1984, **3**, 771.
1246. O. W. Howarth, C. H. McAteer, P. Moore, G. E. Morris and N. W. Alcock, *J. Chem. Soc., Dalton Trans.*, 1982, 541.
1247. A. Albinati, A. Togni and L. M. Venanzi, *Organometallics*, 1986, **5**, 1785.
1248. J. A. Marsella, J. C. Huffman, K. Folting and K. G. Caulton, *Inorg. Chim. Acta*, 1985, **96**, 161.
1249. Y. Mugnier and C. Moise, *J. Organomet. Chem.*, 1983, **248**, C33.
1250. A. S. Antsyshkina *et al.*, *Koord. Khim.*, 1985, **11**, 82.
1251. S. Mönkeberg, E. van Raaij, H. Kiesele and H.-H. Brintzinger, *J. Organomet. Chem.*, 1989, **365**, 285.
1252. P. Jernakoff and N. J. Cooper, *J. Am. Chem. Soc.*, 1987, **109**, 2173.

1253. P. Jernakoff and N. J. Cooper, *J. Am. Chem. Soc.*, 1989, **111**, 7424.
1254. P. T. Wolczanski, R. S. Threlkel and B. D. Santarsiero, *Acta Crystallogr., Sect. C*, 1983, **39C**, 1330.
1255. K. A. Belmore, R. A. Vanderpool, J.-C. Tsai, M. A. Khan and K. M. Nicholas, *J. Am. Chem. Soc.*, 1988, **110**, 2004.
1256. P.-F. Fu, M. A. Khan and K. M. Nicholas, *J. Am. Chem. Soc.*, 1992, **114**, 6579.
1257. J.-C. Tsai, M. Khan and K. M. Nicholas, *Organometallics*, 1989, **8**, 2967.
1258. J.-C. Tsai, R. A. Wheeler, M. A. Khan and K. M. Nicholas, *Organometallics*, 1991, **10**, 1344.
1259. T. Ito, S. Sugimoto, T. Ohki, T. Nakano and K. Osakada, *J. Organomet. Chem.*, 1992, **428**, 69.
1260. A. R. Dias and C. C. Romão, *J. Organomet. Chem.*, 1992, **233**, 223.
1261. M. J. Calhorda *et al.*, *J. Organomet. Chem.*, 1987, **320**, 63.
1262. M. A. Bennett and H. P. Schwemlein, *Aust. J. Chem.*, 1989, **42**, 587.
1263. C. G. De Azevedo, A. R. Dias, A. M. Martins and C. C. Romão, *J. Organomet. Chem.*, 1989, **368**, 57.
1264. I. S. Butler, P. J. Fitzpatrick, D. F. R. Gilson, G. Gomez and A. Shaver, *Mol. Cryst. Liq. Cryst.*, 1981, **71**, 213.
1265. P. Köpf-Maier and H. Köpf, *J. Organomet. Chem.*, 1988, **342**, 167.
1266. P. Köpf-Maier and H. Köpf, *Chem. Rev.*, 1987, **87**, 1137.
1267. J. C. Kotz, W. Vining, W. Coco, R. Rosen, A. R. Dias and M. H. Garcia, *Organometallics*, 1983, **2**, 68.
1268. A. A. Tumanov *et al.*, *J. Organomet. Chem.*, 1984, **263**, 213.
1269. P. Gowik, T. Klapötke and P. White, *Chem. Ber.*, 1989, **122**, 1649.
1270. T. S. Cameron, T. M. Klapötke, A. Schulz and J. Valkonen, *J. Chem. Soc., Dalton Trans.*, 1993, 659.
1271. T. M. Klapötke, A. Schulz, T. S. Cameron and P. K. Bakshi, *J. Organomet. Chem.*, 1993, **463**, 115.
1272. T. C. Forschner and N. J. Cooper, *J. Am. Chem. Soc.*, 1989, **111**, 7420.
1273. T. C. Forschner, J. A. Corella, II and N. J. Cooper, *Organometallics*, 1990, **9**, 2478.
1274. U. Schubert and A. Schenkel, *Chem. Ber.*, 1988, **121**, 939.
1275. J. P. McNally, D. Glueck and N. J. Cooper, *J. Am. Chem. Soc.*, 1988, **110**, 4838.
1276. J. P. McNally and N. J. Cooper, *J. Am. Chem. Soc.*, 1989, **111**, 4500.
1277. A. M. Martins, M. J. Calhorda, C. C. Romão, C. Völkl, P. Kiprof and A. C. Filippou, *J. Organomet. Chem.*, 1992, **423**, 367.
1278. M. J. Calhorda, A. R. Dias, M. T. Duarte, A. M. Martins, P. M. Matias and C. C. Romão, *J. Organomet. Chem.*, 1992, **440**, 119.
1279. M. J. Calhorda, S. M. B. Costa, A. R. Dias and F. J. S. Pina, *New J. Chem.*, 1984, **8**, 619.
1280. S. M. B. Costa, A. R. Dias and F. J. S. Pina, *J. Organomet. Chem.*, 1981, **217**, 357.
1281. R. H. Crabtree and G. G. Hlatky, *Polyhedron*, 1985, **4**, 521.
1282. P. Diversi, G. Ingrosso, A. Lucherini and M. Landucci, *Polyhedron*, 1987, **6**, 281.
1283. S. K. Sengupta and S. Kumar, *Synth. React. Inorg. Met.-Org. Chem.*, 1982, **12**, 297.
1284. O. P. Pandey, S. K. Sengupta and S. C. Tripathi, *J. Indian Chem. Soc.*, 1985, **62**, 413.
1285. M. J. Calhorda, A. R. Dias, A. M. Martins and C. C. Romão, *Polyhedron*, 1989, **8**, 1802.
1286. T. C. Wright, G. Wilkinson, M. Motevalli and M. B. Hursthouse, *J. Chem. Soc., Dalton Trans.*, 1986, 2017.
1287. B. S. McGilligan, T. C. Wright, G. Wilkinson, M. Motevalli and M. B. Hursthouse, *J. Chem. Soc., Dalton Trans.*, 1988, 1737.
1288. R. M. Bullock, C. E. L. Headford, S. E. Kegley and J. R. Norton, *J. Am. Chem. Soc.*, 1985, **107**, 727.
1289. H. J. R. de Boer, B. J. J. van de Heisteeg, G. Schat, O. S. Akkerman and F. Bickelhaupt, *J. Organomet. Chem.*, 1988, **346**, 197.
1290. P. Diversi, G. Ingrosso, A. Lucherini, W. Porzio and M. Zocchi, *J. Chem. Soc., Dalton Trans.*, 1983, 967.
1291. W. R. Tikkanen, J. Z. Liu, J. W. Egan Jr. and J. L. Petersen, *Organometallics*, 1984, **3**, 825.
1292. E. Ciliberto, S. Di Bella, A. Gulino, I. Fragalá, J. L. Petersen and T. J. Marks, *Organometallics*, 1992, **11**, 1727.
1293. W. R. Tikkanen, J. W. Egan Jr. and J. L. Petersen, *Organometallics*, 1984, **3**, 1646.
1294. K. S. Chong and M. L. H. Green, *Organometallics*, 1982, **1**, 1586.
1295. A. Baretta, K. S. Chong, F. G. N. Cloke, A. Feigenbaum and M. L. H. Green, *J. Chem. Soc., Dalton Trans.*, 1983, 861.
1296. A. Gourdon and K. Prout, *Acta Crystallogr., Sect. B*, 1981, **37**, 1982.
1297. M. L. H. Green and D. O'Hare, *J. Chem. Soc., Dalton Trans.*, 1985, 1585.
1298. J. C. Green, M. L. H. Green and C. P. Morley, *J. Organomet. Chem.*, 1982, **233**, C4.
1299. A. R. Dias, M. H. Garcia, A. M. Martins, C. I. Pinheiro, C. C. Romão and L. F. Veiros, *J. Organomet. Chem.*, 1987, **327**, C59.
1300. M. J. Calhorda *et al.*, *Organometallics*, 1991, **10**, 483.
1301. M. F. Asaro, S. R. Cooper and N. J. Cooper, *J. Am. Chem. Soc.*, 1986, **108**, 5187.
1302. P. Gowik and T. Klapötke, *J. Organomet. Chem.*, 1989, **375**, C20.
1303. M. M. Maguire *et al.*, *J. Organomet. Chem.*, 1984, **269**, 183.
1304. R. M. Bullock, C. E. L. Headford, K. M. Hennessy, S. E. Kegley and J. R. Norton, *J. Am. Chem. Soc.*, 1989, **111**, 3897.
1305. G. Parkin and J. E. Bercaw, *Organometallics*, 1989, **8**, 1172.
1306. J. C. Hayes, P. Jernakoff, G. A. Miller and N. J. Cooper, *Pure Appl. Chem.*, 1984, **56**, 25.
1307. J. C. Hayes, G. D. N. Pearson and N. J. Cooper, *J. Am. Chem. Soc.*, 1981, **103**, 4648.
1308. J. C. Hayes and N. J. Cooper, *J. Am. Chem. Soc.*, 1982, **104**, 5570.
1309. P. Jernakoff and N. J. Cooper, *J. Am. Chem. Soc.*, 1984, **106**, 3026.
1310. P. Jernakoff and N. J. Cooper, *Organometallics*, 1986, **5**, 747.
1311. J. P. McNally and N. J. Cooper, *Organometallics*, 1988, **7**, 1704.
1312. G. A. Miller and N. J. Cooper, *J. Am. Chem. Soc.*, 1985, **107**, 709.
1313. M. Canestrari and M. L. H. Green, *J. Chem. Soc., Dalton Trans.*, 1982, 1789.
1314. M. L. H. Green, *Pure Appl. Chem.*, 1984, **56**, 47.
1315. J. C. Green, M. L. H. Green and C. P. Morley, *Organometallics*, 1985, **4**, 1302.
1316. T. S. Koloski, P. J. Carroll and D. H. Berry, *J. Am. Chem. Soc.*, 1990, **112**, 6405.
1317. D. H. Berry, J. H. Chey, H. S. Zipin and P. J. Carroll, *J. Am. Chem. Soc.*, 1990, **112**, 452.
1318. D. H. Berry, J. Chey, H. S. Zipin and P. J. Carroll, *Polyhedron*, 1991, **10**, 1189.
1319. V. K. Bel'skii, A. N. Protskii, B. M. Bulychev and G. L. Soloveichik, *J. Organomet. Chem.*, 1985, **280**, 45.
1320. A. I. Gusev, N. I. Kirillova, A. N. Protsky, B. M. Bulychev and G. L. Soloveichik, *Polyhedron*, 1984, **3**, 765.

1321. A. N. Protsky, B. M. Bulychev and G. L. Soloveichik, *Inorg. Chim. Acta*, 1983, **71**, 35.
1322. V. K. Bel'skii, A. N. Protskii, B. M. Bulychev and G. L. Soloveichik, *Dokl. Akad. Nauk SSSR, Ser. Khim.*, 1986, **287**, 857.
1323. S. Seebald, B. Mayer and U. Schubert, *J. Organomet. Chem.*, 1993, **462**, 225.
1324. M. M. Kubicki, R. Kergoat, J. E. Guerchais and P. L'Haridon, *J. Chem. Soc., Dalton Trans.*, 1984, 1791.
1325. A. R. Dias and M. A. Queirós, *J. Organomet. Chem.*, 1990, **390**, 193.
1326. M. A. A. F. de C. T. Carrondo and V. Félix, *Acta Crystallogr., Sect. C*, 1991, **47**, 2451.
1327. M. A. M. Queirós, J. E. J. Simão and A. R. Dias, *J. Organomet. Chem.*, 1987, **329**, 85.
1328. L. Y. Kuo, M. G. Kanatzidis and T. J. Marks, *J. Am. Chem. Soc.*, 1987, **109**, 7207.
1329. L. Y. Kuo, M. G. Kanatzidis, M. Sabat, A. L. Tipton and T. J. Marks, *J. Am. Chem. Soc.*, 1991, **113**, 9027.
1330. R. G. Costa, A. R. Dias and J. C. Kotz, *J. Organomet. Chem.*, 1988, **354**, 341.
1331. M. J. Calhorda, M. A. A. F. de C. T. Carrondo, R. Gomes da Costa, A. R. Dias, M. T. L. S. Duarte and M. B. Hursthouse, *J. Organomet. Chem.*, 1987, **320**, 53.
1332. M. J. Calhorda *et al.*, *J. Organomet. Chem.*, 1992, **426**, 195.
1333. P. Legzdins and C. R. Nurse, *Inorg. Chem.*, 1982, **21**, 3110.
1334. P. B. Hitchcock, M. F. Lappert and W.-P. Leung, *J. Chem. Soc., Chem. Commun.*, 1987, 1282.
1335. E. Niecke, J. Hein and M. Nieger, *Organometallics*, 1989, **8**, 2290.
1336. A. M. Arif, A. H. Cowley, C. M. Nunn and M. Pakulski, *J. Chem. Soc., Chem. Commun.*, 1987, 994.
1337. M. M. Kubicki, R. Kergoat, M. Cariou, J. E. Guerchais and P. L'Haridon, *J. Organomet. Chem.*, 1987, **322**, 357.
1338. C. Barre, M. M. Kubicki, J.-C. Leblanc and C. Moise, *Inorg. Chem.*, 1990, **29**, 5244.
1339. H. Köpf and V. Richtering, *Z. Naturforsch., Teil B*, 1988, **43**, 1653.
1340. T. Ito, T. Tokunaga, M. Minato and T. Nakamura, *Chem. Lett.*, 1991, 1893.
1341. C. G. Azevedo, M. J. Calhorda, M. A. A. F. de C. T. Carrondo, A. R. Dias, V. Félix and C. C. Romão, *J. Organomet. Chem.*, 1990, **391**, 345.
1342. J. M. Galante, J. W. Bruno, P. N. Hazin, K. Folting and J. C. Huffman, *Organometallics*, 1988, **7**, 1066.
1343. R. Poli, S. P. Mattamana and L. R. Falvello, *Gazz. Chim. Ital.*, 1992, **122**, 315.
1344. N. D. Silavwe, M. Y. Chiang and D. R. Tyler, *Inorg. Chem.*, 1985, **24**, 4219.
1345. R. S. Pilato, C. E. Housmekerides, P. Jernakoff, D. Rubin, G. L. Geoffroy and A. L. Rheingold, *Organometallics*, 1990, **9**, 2333.
1346. G. Parkin and J. E. Bercaw, *J. Am. Chem. Soc.*, 1989, **111**, 391.
1347. G. Parkin, R. E. Marsh, W. P. Schaefer and J. E. Bercaw, *Inorg. Chem.*, 1988, **27**, 3262.
1348. N. D. Silavwe, M. R. M. Bruce, C. E. Philbin and D. R. Tyler, *Inorg. Chem.*, 1988, **27**, 4669.
1349. G. J. S. Adam and M. L. H. Green, *J. Organomet. Chem.*, 1981, **208**, 299.
1350. T. Ito and T. Nakano, *J. Chem. Soc., Dalton Trans.*, 1987, 1857.
1351. T. Ito, T. Ohki and T. Nakano, *Synth. React. Inorg. Met.-Org. Chem.*, 1986, **16**, 169.
1352. J. C. G. Calado, A. R. Dias, M. S. Salema and J. A. Martinho Simões, *J. Chem. Soc., Dalton Trans.*, 1981, 1174.
1353. M. A. V. Ribeiro da Silva, M. de D. M. C. Ribeiro da Silva and A. R. Dias, *J. Organomet. Chem.*, 1988, **345**, 105.
1354. A. E. Bruce, M. R. M. Bruce, A. Sclafani and D. R. Tyler, *Organometallics*, 1984, **3**, 1610.
1355. A. Shaver, J. M. McCall and G. Marmolejo, *Inorg. Synth.*, 1990, **27**, 59.
1356. I. S. Butler, P. D. Harvey, J. M. McCall and A. Shaver, *J. Raman. Spectrosc.*, 1986, **17**, 221.
1357. R. S. Pilato *et al.*, *J. Am. Chem. Soc.*, 1991, **113**, 9372.
1358. R. S. Pilato, K. A. Eriksen, E. I. Stiefel and A. L. Rheingold, *Inorg. Chem.*, 1993, **32**, 3799.
1359. A. Shaver, G. Marmolejo and J. M. McCall, *Inorg. Synth.*, 1990, **27**, 65.
1360. E. Fu *et al.*, *J. Organomet. Chem.*, 1988, **355**, 205.
1361. M. L. H. Green, W. B. Heuer and G. C. Saunders, *J. Chem. Soc., Dalton Trans.*, 1990, 3789.
1362. A. R. Dias, J. A. Martinho Simões and C. Teixeira, *J. Chem. Soc., Dalton Trans.*, 1981, 1178.
1363. M. J. Calhorda, A. R. Dias, J. A. Martinho Simões and C. Teixeira, *J. Chem. Soc., Dalton Trans.*, 1984, 2659.
1364. M. J. Calhorda, M. A. A. F. de C. T. Carrondo, A. R. Dias, A. M. T. S. Domingos, J. A. Martinho Simões and C. Teixeira, *Organometallics*, 1986, **5**, 660.
1365. P. K. Gowik and T. Klapötke, *Inorg. Chim. Acta*, 1990, **169**, 1.
1366. F. Conan *et al.*, *Organometallics*, 1989, **8**, 1929.
1367. D. O'Hare, M. L. H. Green and F. G. N. Cloke, *J. Organomet. Chem.*, 1985, **282**, 225.
1368. H. Werner, B. Ulrich, U. Schubert, P. Hofmann and B. Zimmer-Gasser, *J. Organomet. Chem.*, 1985, **297**, 27.
1369. T. B. Rauchfuss and C. J. Ruffing, *Organometallics*, 1982, **1**, 400.
1370. M. A. A. F. de C. T. Carrondo *et al.*, *J. Organomet. Chem.*, 1990, **395**, 279.
1371. M. J. Calhorda, M. A. A. F. de C. T. Carrondo, M. H. Garcia and M. B. Hursthouse, *J. Organomet. Chem.*, 1988, **342**, 209.
1372. H. Köpf and T. Klapötke, *J. Organomet. Chem.*, 1986, **310**, 303.
1373. T. Klapötke and H. Köpf, *Z. Anorg. Allg. Chem.*, 1988, **558**, 217.
1374. C. Elschenbroich, E. Bilger and B. Metz, *Organometallics*, 1991, **10**, 2823.
1375. S. Y. Ketkov, *J. Organomet. Chem.*, 1992, **429**, C38.
1376. M. L. H. Green, D. K. P. Ng and R. C. Tovey, *J. Chem. Soc., Chem. Commun.*, 1992, 918.
1377. K. Jonas, *Pure Appl. Chem.*, 1990, **62**, 1169.
1378. C. Elschenbroich, J. Heck, W. Massa and R. Schmidt, *Angew. Chem., Int. Ed. Engl.*, 1983, **23**, 330.
1379. J. Heck and G. Rist, *J. Organomet. Chem.*, 1988, **342**, 45.
1380. B. Bachmann, F. Hahn, J. Heck and M. Wünsch, *Organometallics*, 1989, **8**, 2523.
1381. B. Bachmann, G. Baum, J. Heck, W. Massa and B. Ziegler, *Z. Naturforsch., Teil B*, 1990, **45**, 221.
1382. J. Heck *et al.*, *Z. Anorg. Allg. Chem.*, 1992, **611**, 35.
1383. P. M. J. A. Hermans *et al.*, *Chem. Ber.*, 1993, **126**, 553.
1384. G. Wilke, H. Benn, R. Goddard, C. Krüger and B. Pfeil, *Inorg. Chim. Acta*, 1992, **198–200**, 741.
1385. R. B. Maynard, Z.-T. Wang, E. Sinn and R. N. Grimes, *Inorg. Chem.*, 1983, **22**, 873.
1386. G. K. Barker, N. R. Godfrey, M. Green, H. E. Parge, F. G. A. Stone and A. J. Welch, *J. Chem. Soc., Chem. Commun.*, 1983, 277.

8

Arene and Heteroarene Complexes of Chromium, Molybdenum and Tungsten

MICHAEL J. MORRIS
University of Sheffield, UK

8.1 INTRODUCTION

The vast majority of arene complexes of chromium, molybdenum and tungsten fall into two general structural types: the bis(arene) complexes [M(η^6-arene)$_2$] (**1**) and the tricarbonyl derivatives [M(CO)$_3$(η^6-arene)] (**2**). In both cases those of chromium greatly outnumber those of molybdenum and tungsten. However, there also exist a number of molybdenum and tungsten complexes of the type [ML$_3$(η-arene)] where L = a phosphine, which have no chromium counterparts and are covered in Section 8.4.

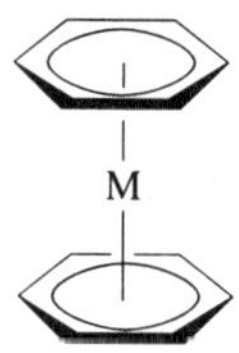

(1)

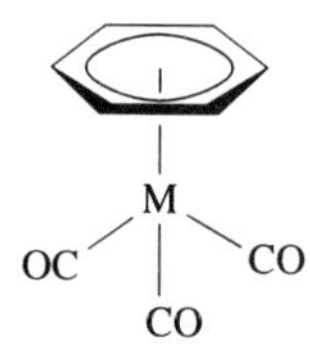

(2)

The general structural, stereochemical and electronic aspects of arene complexes have been reviewed.[1] Developments in chromium, molybdenum and tungsten arene chemistry during the 1980s have been covered by the annual review of metal π-complexes by Rockett and Marr.[2–12]

8.2 BIS(ARENE) COMPLEXES

8.2.1 Bis(arene) Chromium(0) Complexes

8.2.1.1 Synthesis

Of the five synthetic routes discussed in the first edition of this work, only the co-condensation of metal atoms with arenes has continued to receive detailed attention, though one recent paper describes a modification of the Fischer–Hafner synthesis: reduction of $CrCl_3$ with aluminum powder and $AlCl_3$ in boiling toluene followed by treatment with THF produced a quantitative yield of [Cr(η-PhMe)$_2$]. This procedure also worked, though only in 40% yield, for the analogous molybdenum complex.[13]

The condensation of metal atoms onto cold liquid arenes or arene solutions is the most versatile method for the synthesis of bis(arene) complexes.[14] The method can also be used without solvent for solid arenes: chromium sandwiches with durene, C_6Me_5H and C_6Me_6 have been prepared this way, in an improved yield of 10%.[15] The path of complex formation in the co-condensate can be followed spectroscopically, but the monoarene species [Cr(η-arene)] cannot be detected.[16] The structure of the matrix surface plays a part in complex formation as against metal particle aggregation, with rough surfaces favouring the former.[17,18] An unusual approach to bis(arene) complexes has been described by Timms; co-condensation of potassium atoms into solutions of metal halides and arenes in THF allows the synthesis of [M(η^6-$C_{10}H_8$)$_2$] in about 40% yield from $CrCl_3$ or $MoCl_5$.[19] Low or moderate yields of bis(arene) complexes can also be made by condensing chromium atoms with cyclohexenes, showing that C–H activation occurs and demonstrating the affinity of chromium for arene ligands.

Some examples of chromium complexes prepared by the metal atom route are listed in Table 1. In addition to those listed, condensation of metal atoms into liquid polymers, usually siloxanes, bearing phenyl or naphthyl groups result in polymer-bound [M(arene)$_2$] species.[20–3]

As the lower half of Table 1 shows, co-condensation provides a convenient route to complexes of polycyclic arenes and those with two arenes linked by spacer groups, including cyclophanes. The

Table 1 Homoleptic bis(arene) chromium complexes prepared by metal atom co-condensation.

Arene	*Notes*	*Ref.*
Ph-TMS, 1,4-$C_6H_4(TMS)_2$, 1,3,5-$C_6H_3(TMS)_3$	a	24
$PhSiMe_2Cl$, 1,4-$ClSiMe_2C_6H_4SiMe_2Cl$		25
$SiPh_4$		26
9,9-dimethyl-9-silafluorene		27,28
o-$C_6H_4(SiMe_2)_2C_6H_4$		28,29
PhCN	b	30
PhC≡CH, PhC≡CPh	c	31–3
1,3,5-$C_6H_3Bu^t_3$	d	34
$PhN(TMS)_2$, $PhNMe_2$	e	35
m-FC_6H_4R (R = Me, OMe, Cl)	f	36
2-Cl-1,4-$Me_2C_6H_3$, 4-Cl-1,2-$Me_2C_6H_3$, 2,4-Cl_2-1-MeC_6H_3, 2,6-Cl_2-1-MeC_6H_3		37
p-$MeC_6H_4Bu^t$	g	38
2,4,8,10-tetramethyl biphenyl, bimesityl	g	39
biphenylene		40
phenanthrene		41
anthracene		42
cyclobutabenzene [1,2:4,5]-dicyclobutabenzene	d	43
[10]-paracyclophane		44

[a] Undergo ready desilylation to $[Cr(\eta\text{-}C_6H_6)_2]$. [b] Accompanied by trimerization to 2,4,6-triphenyltriazine. [c] Products contain cyclotrimerized alkynes, i.e., C_6Ph_6 and the 1,2,4- and 1,3,5-isomers of $C_6H_3Ph_3$. [d] For Cr, Mo and W. [e] Desilylation gives $[Cr(\eta\text{-}PhNH_2)_2]$. [f] Isomers observed in ^{19}F NMR spectrum due to planar chirality. [g] Hindered rotation observed by variable-temperature NMR.

simplest of the polycyclic species is the naphthalene complex $[Cr(\eta^6\text{-}C_{10}H_8)_2]$ (**3**). In the solid state the two rings unexpectedly adopt a *syn* structure with C_{2v} symmetry.[45] This complex has received much attention due to the lability of one naphthalene ligand, which enables the synthesis of mixed ligand $[Cr(\eta\text{-arene})(\eta^6\text{-naphthalene})]$ complexes by arene exchange reactions. Comparative MO calculations suggest the reason for this lability is a lower charge on the chromium atom and weaker bonding than in $[Cr(\eta\text{-}C_6H_6)_2]$.[46,47] Exchange occurs thermally at 150 °C, or at 80 °C in the presence of THF (presumably via an η^4-intermediate); a variety of arenes can be added, including C_6H_6, PhMe, xylene, mesitylene and C_6Me_6.[48,49] In some cases, for example Ph_2SiMe_2 and 1,4-dimethylnaphthalene, exchange of both $C_{10}H_8$ rings occurs. Coproducts of the exchange reactions are hydrogenated sandwiches with tetralin ligands, and also slipped triple-decker sandwiches $[(\eta\text{-arene})Cr(\mu,\eta^6,\eta^6\text{-}C_{10}H_8)Cr(\eta\text{-arene})]$ (**4**). These are proposed to form via an $[(\eta\text{-arene})Cr(\mu,\eta^6,\eta^6\text{-}C_{10}H_8)Cr(\eta^6\text{-}C_{10}H_8)]$ intermediate which then undergoes exchange of the remaining $C_{10}H_8$ ligand, rather than by combination of two $[Cr(\eta\text{-arene})(\eta^6\text{-naphthalene})]$ units, since the naphthalene ligands in the latter do not exchange easily. Support for this idea came from the realization that the insoluble material formed in the reaction contains polymeric $[Cr(\mu,\eta^6,\eta^6\text{-}C_{10}H_8)]_n$, which incorporates this structure.[50] Earlier theoretical work had analysed the band-gap structure of such a polymer, and predicted that it would conduct when oxidized.[51] Electrochemistry of all these naphthalene complexes includes oxidation to mono- and dications as well as reduction of the $C_{10}H_8$ ligands, but there is no mixed valence interaction between metal sites.[52]

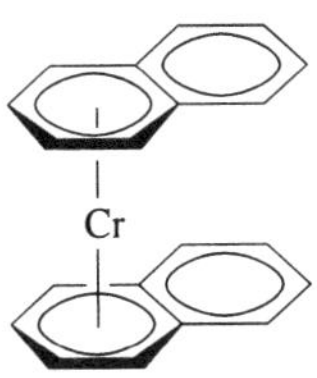

(**3**)

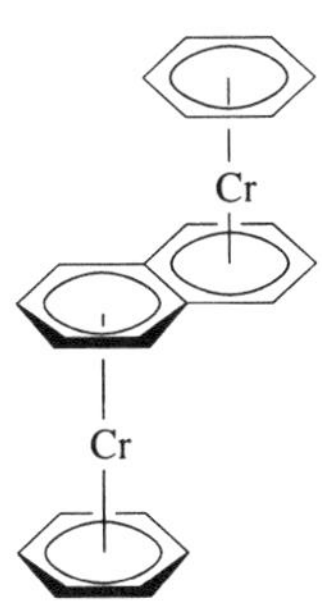

(**4**)

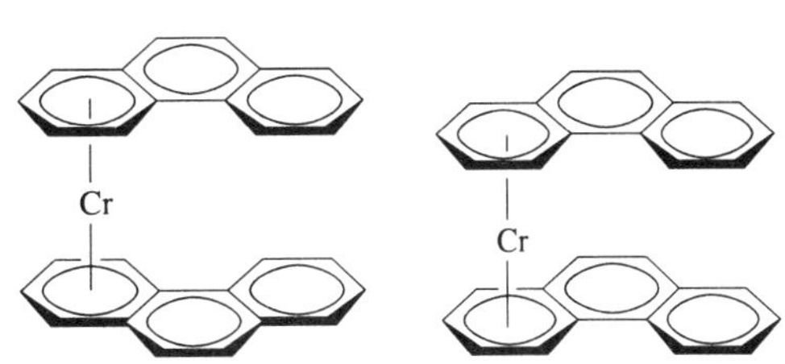

(**5**)

Of the other polycyclic species, only anthracene undergoes a similar facile displacement reaction; like the naphthalene complex, it also undergoes ready hydrogenation to produce $[Cr(\eta^6\text{-}9,10\text{-}$

dihydroanthracene)$_2$].[42] The related [Cr(η^6-phenanthrene)$_2$] (**5**) exists as two diastereomeric forms, *rac* and *meso*, in a ratio of 2:1.[41]

The ring current in [Cr(η-arene)$_2$] species can be investigated by examining the complexation shift ($\Delta\delta$) of protons in specific environments in polycyclic complexes. For example, in the cyclobutabenzene complex (**6**), the *exo* and *endo* CH_2 protons exhibit complexation shifts of opposite sign, explained by a movement of the centre of anisotropy towards the chromium atom on complexation.[43] The methyl protons in complex (**7**) lie above the arene ring and display a complexation shift consistent with a reduction in ring current of 25% compared to the free arene.[53] The well-defined disposition of the twenty CH_2 protons in the bis(arene) complex of [10]-paracyclophane have also been used to study this quenching, and complex (**8**) allows both the radial and angular dependence of the complexation shifts to be analysed.[44,54]

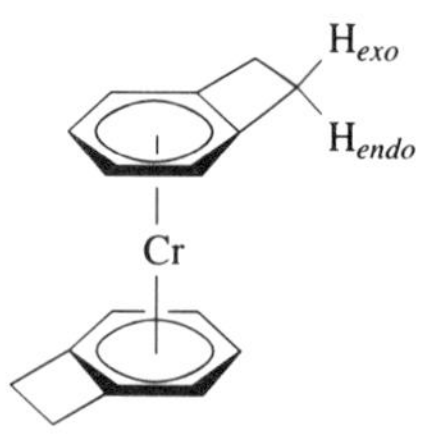

(**6**)

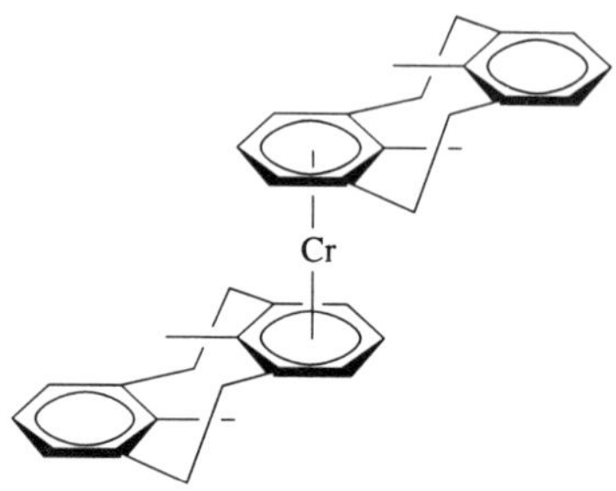

(**7**)

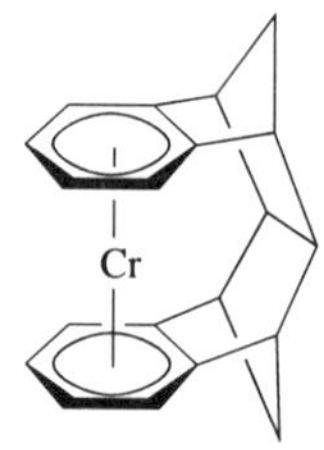

(**8**)

The synthesis of complexes with arenes linked by spacer groups, [Cr(η^6-PhXPh)$_2$] (X = bond, CH_2, CH_2CH_2, CH=CH) can also be achieved.[55] Attempts to coordinate $Cr(CO)_3$ units to the uncomplexed rings of [Cr(η^6-PhPh)$_2$] were successful with [$Cr(CO)_3(NCMe)_3$], whereas [$Cr(CO)_3(NH_3)_3$] produced only arene transfer products.[56]

Co-condensation of two different arenes with chromium atoms affords the mixed ligand species [Cr(η-arene)(η-arene')] as mixtures with the two homoleptic compounds. Most of these involve monosubstituted arenes (Table 2). The mixture has to be separated either by chromatography or by oxidation to the more robust Cr^I cations followed by TLC. The rates of reaction of various arenes with molybdenum atoms at −170 °C in competition with PhF span only a small range; consequently the reaction is not very selective and a virtually statistical product distribution is obtained.[57] However, in competition reactions of chromium atoms between $PhCF_3$ and *o*-xylene, the bis(xylene) complex was completely absent; lower than expected amounts of homoleptic products were also obtained with other methylbenzenes.[58]

Table 2 Heteroleptic bis(arene) chromium complexes prepared by metal atom co-condensation.

Arene	*Second arene*	*Ref.*
C_6H_6	$Ph(CH_2)_nCO_2Me$, $Ph(CH_2)_nCO_2Et$ (n = 0–2), *o*-, *m*-, *p*-$C_6H_4(CO_2Me)_2$; $PhNMe_2$	59
C_6H_6	*o*-$MeC_6H_4CO_2R$ (R = Me, menthyl), tetralone	60
C_6H_6	PhCl, ClC_6H_4X, $MeO_2CC_6H_4X$ (X = *m*- or *p*-Me, Cl; *m*-OMe)	61
C_6H_6	FC_6H_4X (X = *m*- or *p*-Me, OMe, F; *p*-CN, *m*-Cl)	62
C_6H_6	biphenylene	40
C_6H_6	*o*-$C_6H_4(SiMe_2)_2C_6H_4$	29
C_6H_6	pyridine	63
PhF	C_6H_6, PhEt, $PhNMe_2$, PhOMe, PhCl, $PhCF_3$	64
$PhCF_3$	$C_6H_{6-n}Me_n$ (n = 1–3)	58,65,66
PhCN	C_6H_6, PhMe, PhF, PhCl, $PhCF_3$, PhCOMe	67–9
$PhCO_2H$	C_6H_6, PhMe, PhCl, $PhCF_3$, $PhCO_2Me$	70

The crystal structure of the mixed biphenyl complex [(η-C_6H_6)Cr(μ,η^6,η^6-PhPh)Cr(η-C_6H_6)] shows an antiperiplanar configuration, whereas the ESR spectrum of the cation indicates a dihedral angle of only 60° in solution.[71]

Co-condensation of further chromium atoms into a solution of [Cr(η-PhMe)$_2$] produces [Cr_2(η-PhMe)$_2$], which can be detected by UV–visible spectra, by an apparent insertion reaction. If vanadium atoms are added instead, [CrV(η-PhMe)$_2$] is formed.[72,73] Condensation of chromium atoms with mesitylene or 1,3,5-tri-*t*-butylbenzene at high metal–ligand ratios gives the triple-decker [(η-

arene)Cr(μ,η^6,η^6-arene)Cr(η-arene)] (**9**) sandwich as well as the usual [Cr(η-arene)$_2$]; in the solid state the rings are parallel and eclipsed.[34,74,75] Extended Hückel calculations show that four benzene σ-electrons can be donated to the metal, that is, the central ring can act as an overall 10-electron donor.[76] Co-condensation of chromium atoms with arene and P(OMe)$_3$ affords good yields of the [CrL$_3$(η-arene)] complexes.[77]

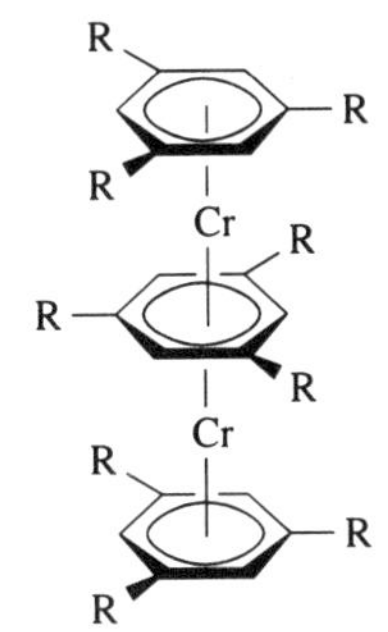

(9)

8.2.1.2 Structure

The solid-state structures of many [Cr(η-arene)$_2$] complexes have been confirmed by x-ray crystallography; all essentially show that the rings are planar, parallel and in an eclipsed conformation. The reorientational motions in solid [Cr(η-C$_6$H$_6$)$_2$] have been studied by potential energy calculations based on the published crystal structures; the barrier for libration is estimated as 8.3 kJ mol^{-1}, which can be compared with the values of 12.6 kJ mol^{-1} and 15.5 kJ mol^{-1} respectively for arene rotation in the neutral complex and [Cr(η-C$_6$H$_6$)$_2$][I] obtained from inelastic neutron scattering measurements.[78]

Several new molecular orbital studies of bis(arene) complexes have appeared (see Figure 1 for a molecular orbital diagram of [Cr(η-C$_6$H$_6$)$_2$]). Two of these use the SCF Xα method to compare the electronic structures of [Cr(η-C$_6$H$_6$)$_2$] and [Cr(η^6-C$_{10}$H$_8$)$_2$] (the latter in C_{2h} and C_{2v} geometries). The results for the former differ slightly from those previously reported in that a ligand π^* orbital ($3e_{2u}$) is designated the LUMO instead of the largely metal-based $4e_{1g}$ orbital, in keeping with the ESR spectrum of the radical anion.[47,79–81] This, however, conflicts with the results obtained from electron transmission spectroscopy (backed up with MS Xα calculations) and negative ion mass spectra that characterize the LUMO as the e_{1g} orbital.[82,83] Bonding occurs by overlap of the filled benzene $1e_{1g}$ π-orbitals with the chromium and d_{yz}, but unusually there is also delocalization of electrons from the filled chromium $3e_{2g}$ orbital into the benzene π^* system. A total geometry optimization calculation on [Cr(η-C$_6$H$_6$)$_2$] has been carried out and shows good agreement with the experimental Cr–arene distances.[84]

The He$^{\rm I}$ and He$^{\rm II}$ photoelectron spectra of [M(arene)$_2$] have been discussed in several articles.[85–7] The lowest energy ionization occurs from the a_{1g} orbital, and its sharpness and intensity is consistent with the view that it corresponds essentially to the metal d_{z^2} orbital, which is practically nonbonding. The second band corresponds to ionization from the e_{2g} orbital, and for tungsten is split due to spin orbit coupling. The third zone corresponds to ionization from the e_{1g} and e_{1u} orbitals. The first and second ionizations depend predominantly on the electronegativity of the arene ligands, that is, transmission of electronic effects to the chromium atom is primarily inductive in nature.[88] In the case of the mixed species [Cr(η-PhCN)(η-PhR)] these can be correlated to substituent constants by an additive scheme; derivation of the molecular energy level diagram by the MNDO (modified neglect of differential overlap) method was also described.[89] A study of the spatial distribution of the orbitals by Penning ionization electron spectroscopy showed the a_{1g} zone enhanced compared to other orbitals, which implies that it is more exposed.[90]

Extensive studies of the UV–visible absorption spectra of [Cr(η-arene)$_2$] complexes in the gas phase have identified four band systems, of which the shortest wavelength group correspond to transitions from the a_{1g} HOMO to a vacant Rydberg orbital; these form a Rydberg series, allowing calculation of the ionization energies, which show good agreement with the values obtained from photoelectron spectra. The next band is due to the allowed e_{2g} to e_{2u} transition and corresponds to the strong absorption observed in solution around 31 000 cm^{-1}. The other two band systems were also assigned.[91–103] Wittmann *et al.* reassigned the strong solution absorbance to a Rydberg series on the basis of its solvent dependence, particularly in EtOH;[104] however, it was later pointed out that the spectrum given was identical to that of the [Cr(η-C$_6$H$_6$)$_2$]$^+$ cation, and in anaerobic conditions the spectrum was unchanged

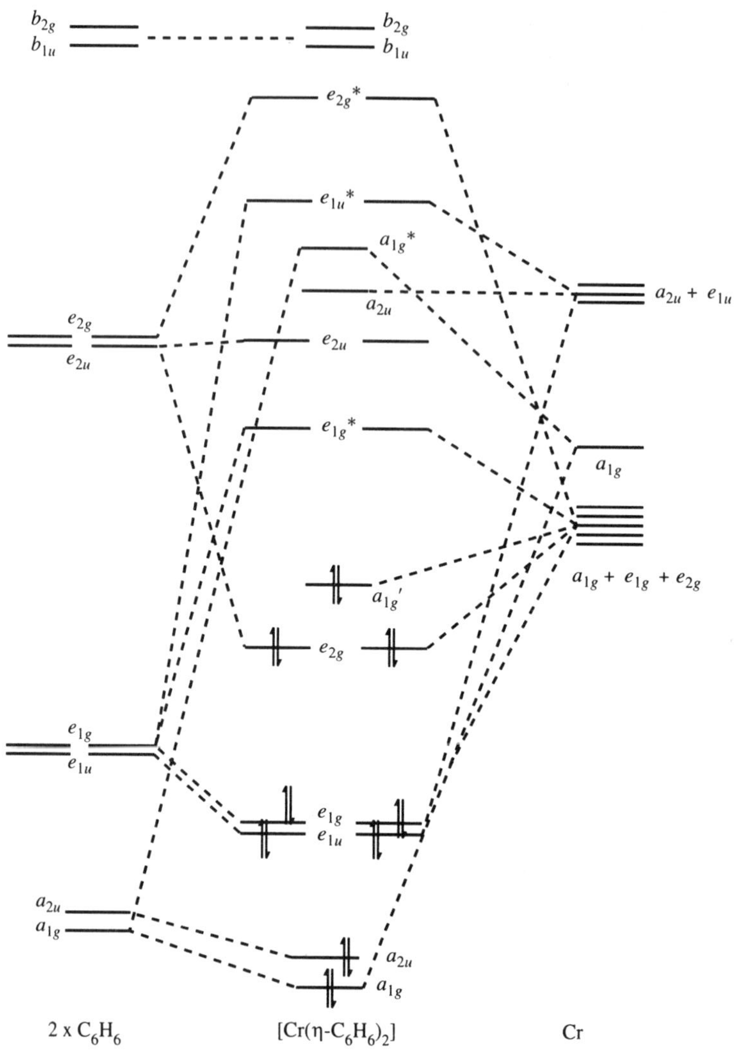

Figure 1 Molecular orbital diagram for [Cr(η-C_6H_6)$_2$]. The ligand σ-orbitals are not included. The z-axis is the C_6 rotational axis (after Muetterties *et al.* 1982[1]).

from that in nonpolar solvents. The Rydberg transition can also be observed in the gas phase laser fluorescence emission spectrum.[105]

The effective charge on the metal in [M(η-C_6H_6)$_2$] has been calculated by integration of the absolute IR intensities; the values are somewhat higher than for metallocenes. Values of 167 kJ mol^{-1} and 247 (±4) kJ mol^{-1} for the M–arene bond energies were also obtained for chromium and molybdenum, respectively.[106,107] This correlates well with a microcalorimetry study, which found values of 247 ± 6 kJ mol^{-1} and 304 ± 6 kJ mol^{-1} for the molybdenum and tungsten complexes.[108]

The electrochemistry of [Cr(η-arene)$_2$] species has been widely studied.[109] All undergo reversible oxidation to the corresponding [Cr(η-arene)$_2$]$^+$; the oxidation potentials correlate with substituent constants and the ionization potentials determined by PES.[110–12] A comparative study of [Cr(η-C_6H_6)$_2$] and [Cr(η-C_7H_7)Cp] showed that the symmetrical species was easier to oxidize.[113] The complex [Cr(η-C_6Me_6)$_2$] can be oxidized to a stable dication in two one-electron steps; other complexes form dications at low temperature.[114,115] There have been proposals that the oxidation of [Cr(η^6-PhPh)$_2$] should be adopted as an electrochemical standard in nonaqueous solvents.[116,117] The redox potentials of the Cr^0/Cr^I couple in polycyclic arene complexes is almost independent of the ligand, supporting the view that the chromium $3d_{z^2}$ is the HOMO.[118]

In contrast to the failure of previous attempts, reduction of [Cr(η-C_6H_6)$_2$] in DME by a potassium mirror in the presence of dicyclohexyl-18-crown-6 produced the radical anion; the ESR spectrum of this species indicates a lowering of symmetry, thought to be due to a folding of the arene ligand as in (**10**) (cf. reduction of [Cr(CO)$_3$(η-C_6H_6)]). The spectrum is consistent with the electron occupying the e_{2u} level, which has its degeneracy lifted by a Jahn–Teller distortion.[80] One notable feature of the ESR spectra is that for the d^5 cation, $g_\perp \sim 2.00 > g_\perp$, whereas for the d^7 anion $g_\perp \sim 2.00 > g_\perp$, reflecting the different nature of the HOMO. This is also observed in the anions of silylbenzene complexes such as [Cr(Ph-TMS)$_2$] and in [Cr(η^6-phenanthrene)$_2$], which are electrochemically accessible.[24,41,81,119] The biphenyl complex [Cr(η^6-PhPh)$_2$] reduces in two one-electron steps, of which the first is reversible.[120] The reduction of a number of complexes [Cr(η-PhR)$_2$] where R is electrochemically active (e.g., R = CHO, COMe, CH=CHCO$_2$Et, CH=CHCOPh, etc.) has been investigated in DMSO using rotating disc and rotating ring disc techniques and compared with that of the free arene; the change in reduction potential on complexation correlates with the oxidation potential of the complexes.[121,122] With R = CN, the complexes differ from the free arene in that a two-electron reduction ECE mechanism involving elimination of the cyano group becomes operative.[123] Reduction of the mixed species [Cr(η-C$_6$H$_6$)(η-PhCOR)] (R = aryl, 4-pyridyl) produces a radical anion with the electron localized on the substituted ring; the situation is more complex in the corresponding [Cr(η-PhCOR)$_2$] where fast or slow electron transfer between the rings can be observed.[119]

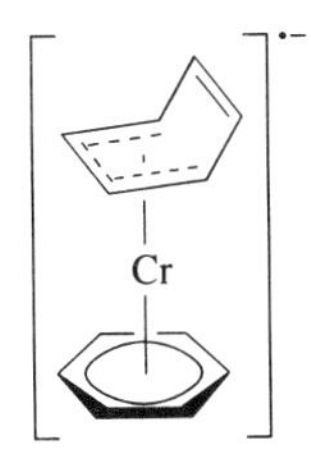

(10)

8.2.1.3 Reactions

The reactions of bis(arene) complexes are still underexplored compared to the corresponding arene tricarbonyl species, due no doubt to the difficulties in their synthesis and handling compared to the latter.

Generally coordination of an arene to a neutral Cr(arene) fragment results in an increase in electron density on the ring, whereas a Cr(arene)$^+$ fragment acts as a strong electron-withdrawing group. Moreover, the effect of substituents on the first arene ring is transmitted in amplified form to the second. Thus, the rate of nucleophilic substitution of chloride by OMe$^-$ in [Cr(η-PhCl)(η-PhR)] and the corresponding cations increases in the order R = NMe$_2$ < OPr^i < OBu < OMe < H < Cl < CO$_2$Me < CF$_3$; the rate can be enhanced by up to 10^{15} in [Cr(η-PhCl)$_2$]$^+$ compared to the free arene (the high degree of electron withdrawal in this cation can also be demonstrated by ^{35}Cl NQR spectroscopy).[124] This is synthetically useful as replacement of one chloride then deactivates the other, allowing the reaction to be carried out stepwise with two different nucleophiles.[61,125–31] A further measure of the effect is provided by the pK_a values of [Cr(η-PhCO$_2$H)(η-PhR)]$^+$, which follow the same trend and can be correlated with substituent constants.[125,132,133] A Cr(η-C$_6$F$_6$) fragment is about as electron-withdrawing as a NO$_2$ group, according to an ion cyclotron resonance study of the acidity of the benzylic protons in toluene and xylene complexed to this fragment.[134] An unusual decarboxylation reaction of [Cr(η-C$_6$H$_6$)(η-C$_6$F$_5$CO$_2$Et)] occurred on attempted base hydrolysis to the C$_6$F$_5$H complex.[135]

Lithiation of [Cr(η-C$_6$H$_6$)$_2$] with Bu^{n}Li and TMEDA gives predominantly the dimetallated species, which reacts readily with electrophiles as shown in Scheme 1. Some of the resulting complexes with η^6-PMe$_2$Ph and PhSMe arenes can themselves be used as ligands for other transition metal fragments.[26,119,136–41] Similarly the complex [Cr{η-*p*-C$_6$H$_4$(PPh$_2$)$_2$}$_2$], which can be made from [Cr(η-*p*-C$_6$H$_4$Cl$_2$)$_2$] and LiPPh$_2$, can be employed as a ligand to two Ni(CO)$_2$ fragments.[142] The lithiation of other [Cr(η-arene)$_2$] complexes does not appear to have been explored.

Dibenzene chromium catalyses the oligomerization of perfluoropropene to dimers and trimers. Defluorinated trimers are also formed, apparently via Cr–H species formed by C–H activation of the arene ligand.[143–6] The polymerization of RC≡CCN (R = perfluoroalkyl) has also been carried out.[147] While there appear to be no reports of catalysis by [Cr(arene)$_2$]$^+$, reduction of [Cr(η-C$_6$H$_6$)$_2$] with potassium sand in DME produces a species, presumably the radical anion described above, which acts as a dehydrogenation catalyst.[148]

Thermolysis of [Cr(η^6-PhCOPh)$_2$] (or related arylketone complexes) produces Ph$_2$C=CPh$_2$ (together with some Ph$_2$CH$_2$ and Ph$_2$CO) by intermolecular coupling that may involve deoxygenation of the ketones to carbenelike fragments by Cr^0.[149] Continued studies of the production of chromium films by decomposition of [Cr(η-arene)$_2$] have been reviewed;[150,151] theoretical studies indicate that steric factors are important in alkyl-substituted derivatives.[152] The formation of films of Cr$_2$O$_3$ by decomposition of arene complexes with carboxylic acid derivatives and quinones has also been investigated.[153,154]

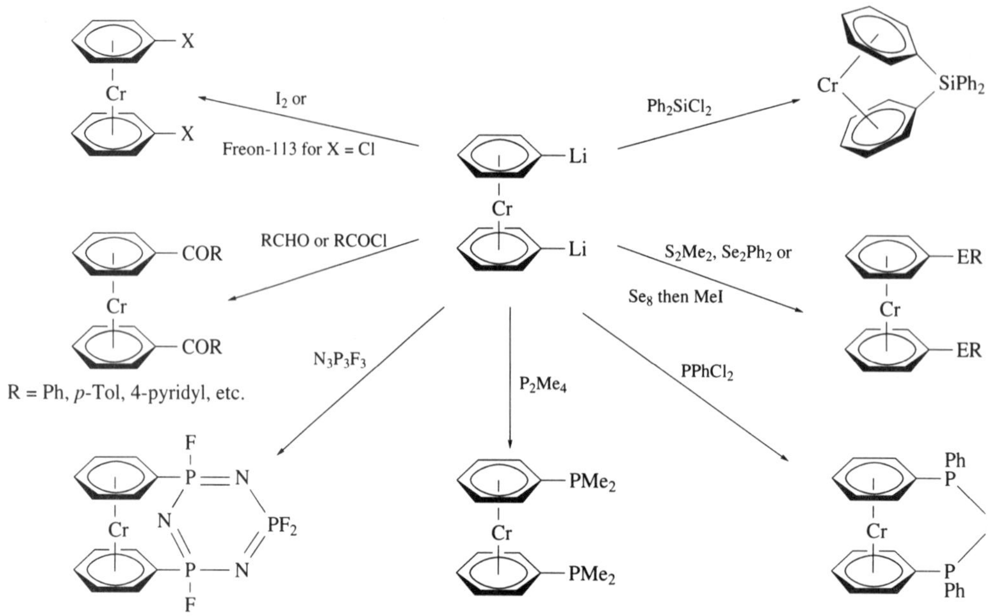

Scheme 1

8.2.1.4 Bis(arene) chromium(I) complexes

On a preparative scale, many oxidizing agents can be used to convert [Cr(η-arene)$_2$] to the corresponding cations, including [FeCp$_2$]$^+$ or I_2; simple exposure to air is most common.[155] The neutral bis(arene) complexes are strong reducing agents and will also convert metal carbonyls such as [Mn$_2$(CO)$_{10}$] and [Co$_2$(CO)$_8$] and arenes such as 9-cyanofluorene to the corresponding anions.[156–60] Reactions of [Cr(η-C$_6$H$_6$)$_2$] with HgX$_2$ (X = halide, Ge(CF$_3$)$_3$) give the bis(arene) cation salts of [Hg$_2$X$_6$]$^{2-}$ or [Hg{Ge(CF$_3$)$_3$}$_4$]$^{2-}$.[161–4] Similarly with SO_2 the salt of the unusual S_4O_{10} dianion was produced.[165]

The search for superconducting 'organic metals' has prompted renewed interest in charge transfer complexes of [Cr(arene)$_2$] with electron acceptors such as TCNE and TCNQ.[166] Two structural types have been revealed by x-ray crystallography: the first contains alternating chains of donors and acceptors (DADA) while the other has dimeric acceptors in the chains (DDA$_2$DD). The compound [Cr(mesitylene)$_2$][TCNQ] displays both forms; it also displays anomalous magnetic behaviour which does not obey the Curie–Weiss law but can be described by an expression including a term for the paramagnetic cations and one for a thermally populated triplet state in the dimeric anions.[167–9] The magnetic properties of [Mo(η-C$_6$H$_6$)$_2$]$^+$ with [FeBr$_4$]$^-$ as the counterion have also been studied; at low temperature ordering of the anions occurs.[170] Several papers have reported the intercalation of [Cr(η-C$_6$H$_6$)$_2$] into the layered structures of M^{II}PS$_3$ (M^{II} = Zn, Mn, Ni, etc).[171,172] Intercalation into FePS$_3$ occurs normally at 130 °C but at 150 °C decomposition occurs to give a semiconducting material.[173]

The self-exchange redox reaction between [Cr(η-arene)$_2$] and the corresponding chromium(I) cations has been studied by ESR line-broadening. The kinetics are consistent with Marcus theory for an outer-sphere reactive collision model, with discrepancies due to steric and nonadiabatic effects.[112,174] In pyridine, [Cr(η-C$_6$H$_6$)$_2$][PF$_6$] undergoes disproportionation to [Cr(py)$_4$(PF$_6$)$_2$], [Cr(η-C$_6$H$_6$)$_2$] and [CrF$_2$(py)$_4$]$^+$.[175]

The electronic spectra of [Cr(η-arene)$_2$]$^+$ species show a band in the 340–360 nm region assigned to ligand to metal ($e_{1u} \rightarrow a_{1g}$) charge transfer. This is rather insensitive to substituents, but shifts to lower frequency in highly alkylated rings.[176,177] The ESR parameters of the cations are also relatively unaffected by arene substitution.

8.2.2 Bis(arene) Molybdenum and Tungsten Complexes

The modification of the metal vapour synthesis apparatus by incorporation of an electron beam furnace instead of resistive heating facilitates the evaporation of refractory metals such as molybdenum and tungsten.[178,179] Even so, the number of examples prepared is relatively small compared to those of chromium.

Apart from the Fischer–Hafner synthesis of [Mo(η-PhMe)$_2$] mentioned earlier, compounds of the type [Mo(η-PhR)$_2$] can be synthesized by direct metal atom co-condensation methods (e.g., R = H, Cl, SiMe$_2$H), by lithiation of [Mo(η-C$_6$H$_6$)$_2$] and reaction with RX (R = D, Me, TMS), or by nucleophilic attack on [Mo(η-PhCl)$_2$] (for R = CH=CH$_2$, *p*-C$_6$H$_4$CH=CH$_2$, CH$_2$CH=CH$_2$).[180] Recently Timms has reported the metal vapour synthesis of the first bis(arene) complexes of C$_6$F$_6$, together with the crystal structure of [W(η-C$_6$F$_6$)$_2$]; mixed molybdenum and tungsten sandwiches with C$_6$F$_6$ and C$_6$H$_6$ or mesitylene were also prepared.[181]

X-ray structures of [M(η-PhMe)$_2$] have been determined for molybdenum and tungsten, allowing a direct comparison.[182,183] In both cases the rings are planar and parallel, and the M–arene centre distances are effectively the same in both cases (0.178–0.180 nm). There is no evidence for bond alternation in the arene as claimed previously for Cr–arene complexes.

The bis(naphthalene) complex [Mo(η^6-C$_{10}$H$_8$)$_2$] can be prepared by co-condensation or by reduction of MoCl$_5$ with activated magnesium in the presence of naphthalene. As with the chromium analogue the C$_{10}$H$_8$ rings are labile; PF$_3$, PMe$_3$ or P(OMe)$_3$ displace one to produce [MoL$_3$(η^6-C$_{10}$H$_8$)], whereas CO displaces both.[184]

The molybdenum and tungsten bis(arene) species are very electron rich, and undergo both oxidation and protonation readily. The structures of two protonated complexes, [WH(η-PhMe)$_2$]$^+$ and [WH(η-PhF)$_2$]$^+$, have been determined and show angles of 156–157° between the arene planes.[182]

The other reactions of molybdenum and tungsten bis(arene) complexes generally involve the displacement of one ring (Scheme 2). These include reaction with the fulvenes C$_5$H$_4$=CR$_2$ (R = Me, Ph) to produce [M(η^5,η^1-C$_5$H$_4$CR$_2$)(η-arene)]; subsequent treatment with HCl comprised the original route to [Mo$_2$Cl$_4$(η-C$_5$H$_4$CHR$_2$)$_2$].[185,186] The ^{95}Mo and ^{13}C NMR spectra of a number of these arene–fulvene species, together with some bis(arene) complexes, have been interpreted using a ligand field model.[187] Azulene also displaces one ring to give [Mo(η-C$_6$H$_6$)(η^6-azulene)], in which the azulene is bonded in a novel η^4,η^2 mode.[188] Addition of [M(CO)$_3$(NCMe)$_3$] (M = Cr, Mo, W) then produces dinuclear [(η-C$_6$H$_6$)Mo(μ-azulene)M(CO)$_3$].[189]

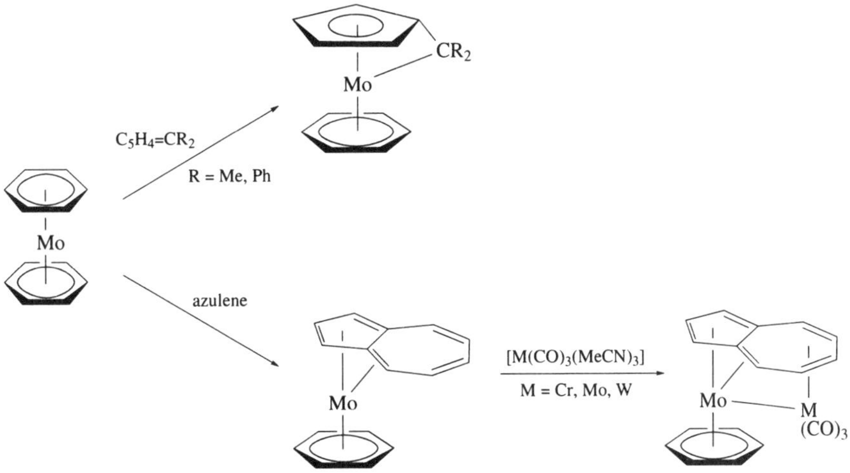

Scheme 2

8.2.3 Cyclopentadienyl Metal Arene Sandwich Complexes

The gas phase electronic spectrum of [CpCr(η-C$_6$H$_6$)] has been recorded and some transitions assigned to a Rydberg series.[190] The reaction of [CrCp$_2$] with LiC$_{10}$H$_8$ to produce [Cr(η^6-C$_{10}$H$_8$)Cp] has been mentioned in a review.[191] The analogous [Cr(η-C$_6$Me$_6$)Cp*] has been made by the cyclotrimerization of 2-butyne with [Cr(Me)(μ-Cl)Cp*]$_2$; the 17-electron sandwich exhibits reversible oxidation and reduction in THF.[192]

8.3 ARENE METAL TRICARBONYL COMPLEXES

8.3.1 Introduction

The chemistry of [$Cr(CO)_3$(η-arene)] complexes has continued to attract much research effort; however, since the first edition of this work only one general review of bonding, conformation and reactions has appeared, in 1985.[193] The main practical attractions of arene chromium tricarbonyl complexes lie in their ease of synthesis, ease of handling due to their relative stability to air and water, and diamagnetism (allowing NMR studies). From the organic chemist's point of view, the fact that the reactivity of arenes is greatly modified (and usually greatly enhanced) by coordination, and the product arenes can be easily decomplexed by exposure of solutions to air and sunlight or other oxidizing agents such as iodine (or, a more recent innovation, dimethyldioxirane[194]) is also important. Consequently the use of such complexes in organic synthesis has expanded tremendously and this aspect has been the subject of numerous review articles, and is covered in greater detail in other volumes of this work.[195–204] The chemistry of the analogous molybdenum and tungsten species is much less extensive, probably due to the longer reaction times required for their synthesis and, perhaps, the greater cost of the parent hexacarbonyls.

8.3.2 Synthesis of Arene Metal Tricarbonyl Complexes

The methods of synthesis of [$M(CO)_3$(η-arene)] have in general changed little since the early 1980s. There are two main approaches: direct complexation of the arene to carbonyl fragments, and assembly of the arene at the metal centre using the Dötz carbene annelation sequence.

8.3.2.1 Thermal reaction from [$M(CO)_6$]

The method of Pauson and Mahaffy, which involves refluxing [$Cr(CO)_6$] with the arene in a mixture of di-*n*-butyl ether and THF (approximately 10:1 by volume), is now established as the best way to prepare [$Cr(CO)_3$(η-arene)] complexes routinely.[205] No special apparatus is required, reaction time is not unreasonably long, and it can be applied to an enormous range of arenes with reasonable to excellent yield. Usually a slight excess of [$Cr(CO)_6$] is used as it can be easily sublimed away from the product. Table 3 shows a representative sample of simple arene complexes subsequently prepared by Mahaffy and co-workers using this method.

A variety of other solvent systems have been explored for the complexation reaction, including the arene itself, Bu_2O, 1,2-dimethoxyethane,[221] THF/diglyme[223] and heptane/diglyme.[224] For derivatives of the amino acid phenylalanine, a mixture of water and THF (80:20) proved successful.[225] Using a specially designed apparatus, Toma and co-workers have carried out a systematic study of complexation reactions in decalin; good yields can be obtained for many arenes, but those with electron withdrawing substituents (e.g., PhCHO, $PhCO_2H$, $PhNO_2$) and, anomalously, benzene itself, gave low yields.[226,227] Although the reaction time is reduced to about 5–6 h, there is then the problem of removing the high-boiling solvent. A further study of toluene complexation showed that addition of 1–3 equivalents of butyl acetate to the decalin catalysed the reaction effectively; longer reaction times were needed in neat BuOAc.[228,229] Acetic acid was also a useful catalyst for PhCOR or $PhCO_2R$.[230] Good yields of [$M(CO)_3$(η-C_6H_6)] (M = Cr, Mo) have also been obtained by refluxing [$M(CO)_6$] in benzene in the presence of pyridine; for the tungsten analogue, NH_4BF_4 proved a better initiator.[24,231,232]

8.3.2.2 Thermal reaction from [$M(CO)_3L_3$] species

Replacement of three carbonyls of [$M(CO)_6$] by more labile ligands enables subsequent thermal reactions with arenes to be carried out at lower temperatures. Particular attention has been given to those complexes where L = MeCN, py or NH_3. For molybdenum and tungsten, the MeCN complex is the most common, though the tricarbonyl complex of $MeN(CH_2CH_2NMe_2)_2$ has also been used.[233] The complexes with MeCN or py are readily prepared simply by extended reflux of [$M(CO)_6$] in the appropriate solvent (as expected, the time required for complete reaction increases from chromium to molybdenum and particularly tungsten, for which EtCN is often used instead). An efficient synthesis of the tris(ammine) complex [$Cr(CO)_3(NH_3)_3$] consists of treatment of [$Cr(CO)_6$] with ethanolic KOH, then NH_4OH; yields

Table 3 Arene tricarbonyl chromium complexes of simple arenes prepared by the Pauson–Mahaffy procedure.

Table 3(a) Complexes of monosubstituted arenes, PhR.

R	*Ref.*
$NHPr^i$, NHCy, $NHBu^s$	203,204
Cy	206,207
CO_2Et	208

Table 3(b) Complexes of disubstituted arenes, $R^1C_6H_4R^2$.

R^1	R^2	*Isomers*	*Ref.*
NH_2	F, Et, Pr^i, OMe, OEt, CO_2Me	*o, m, p*	206,209–14
NH_2	CN, Bu^s, Bu^t CO_2Bu^i	*p*	212,214,215
F	Me, OMe	*o, m, p*	209
F	F, CO_2Et	*o*	215
TMS	Me, OMe, Cl	*o, m, p*	216
Cy	Cy	*p*	217

Table 3(c) Complexes of trisubstituted arenes, $R^1C_6H_3R^2R^3$.

R^1	R^2	R^3	*Isomers*	*Ref.*
NH_2, OMe, COMe	Me	Me	all six	218,219
NH_2	F	F	all six	210
Me	OMe	OMe	all six	220
F	Me	Me	2,6; 3,4	206
NH_2	Me, Et, Pr^i	Et	2,6	211,213,218
NH_2	Me, Et, Pr^i	Pr^i	2,6	212,213
NMe_2	NMe_2	NMe_2	3,5	221

Table 3(d) Complexes of tetrasubstituted arenes, $R^1C_6H_2R^2R^3R^4$.

R^1	R^2	R^3	R^4	*Isomers*	*Ref.*
NH_2, NMe_2, NHEt, F, CO_2Et, COMe, TMS	Me	Me	Me	2,4,6	208,214–16, 218,219,222
COMe	Pr^i	Pr^i	Pr^i	2,4,6	219
COMe	OMe	OMe	OMe	2,3,4; 3,4,5; 2,4,6	219
Me	OMe	OMe	OMe	3,4,5; 2,4,6	220
NH_2	Bu^t	Bu^t	Bu^t	2,4,6	215

Table 3(e) Complexes of pentasubstituted arenes, $R^1C_6HR^2_4$.

R^1	R^2	*Isomers*	*Ref.*
NH_2, COMe	Me	2,3,4,5	213,219

of over 90% on a 10 g scale are attainable. The product was then used to complex indenes and dihydronaphthalenes in refluxing dioxane in 65–85% yield.[234]

The advantages of the milder conditions required by these reagents have been exploited in two ways. Firstly, thermally sensitive arenes, for example those prone to polymerization or which undergo thermal ring opening of cyclopropyl or cyclobutyl rings, can be successfully complexed. Improved yields of some arene complexes, for example that of 1,4-diphenylbutadiene, can thus be obtained.[235] The first example of a $Cr(CO)_3$ cycloproparene complex, (**11**) (R = TMS) was made by treating $[Cr(CO)_3(NCMe)_3]$ with the arene. Desilylation then affords the parent compound (R = H), which could not be prepared directly as insertion of CO into the three-membered ring occurred instead. The complex with R = H can be deprotonated and alkylated at the CH_2 group.[236–8] The complexation of various cyclobutabenzenes can also be achieved without ring opening from $[Cr(CO)_3(NH_3)_3]$.[239,240] Complexation of phenylcyclopropane can be achieved with $[Cr(CO)_3(NH_3)_3]$ in refluxing DME, though the yields are only moderate; 3-methyl-3-phenylcyclopropene underwent ring opening or reduction to 1-methyl-1-phenylcyclopropane in a similar reaction.[241]

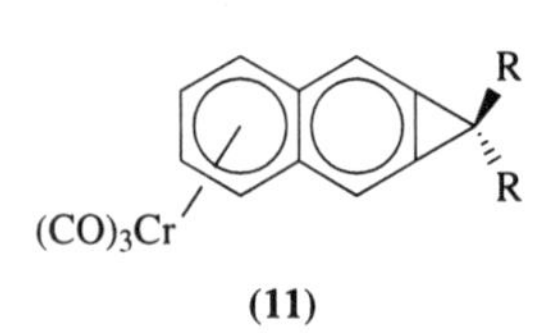

(11)

The ring opening reactions of cyclobutabenzene derivatives can, however, be desirable; thus, treatment of the bicyclo[5.2.0] compound (**12**) with [$M(CO)_3(NCMe)_3$] (M = Cr, Mo, W) affords complexes of the isomeric [4.2.1] bicycle (**13**).[242] Moreover, thermal ring opening of the all cisoid isomer of the trimer of benzocyclobutene (**14**) gave the 2,2,2-tribenz[12]annulene (**15**), which was isolated as its $Cr(CO)_3$ complex, though in this case the actual ring-opening was not metal-mediated.[243] An unusual rearrangement reaction of hexaethenylidene cyclohexane occurs with [$Cr(CO)_3(NCMe)_3$] involving firstly partial and then complete rearrangement into (**16**), the complex of 1,3,5-triethyl-2,4,6-trivinylbenzene.[244]

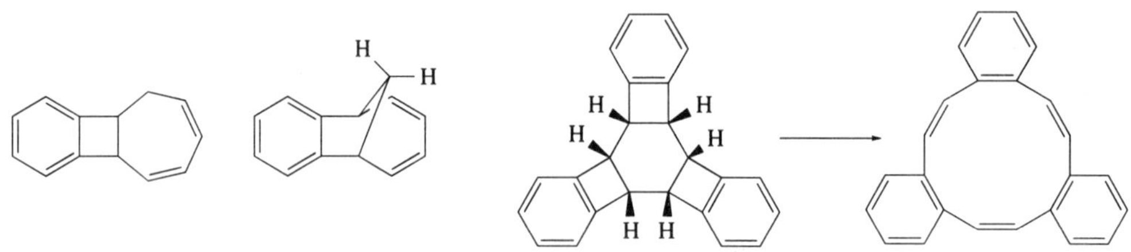

(12) **(13)** **(14)** **(15)**

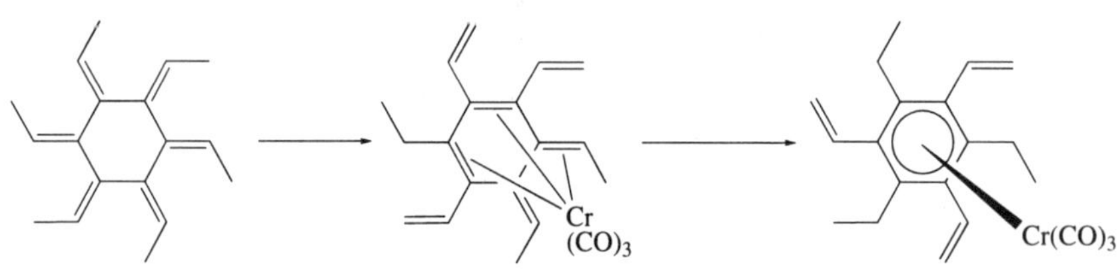

(16)

The second advantage of using [$Cr(CO)_3L_3$] as complexing agent is that in favourable cases a different product distribution is obtained under milder conditions, that is, the kinetic as opposed to the thermodynamic isomer is formed. This, however, has been exploited particularly in arene exchange reactions from naphthalene chromium tricarbonyl, which occur under milder conditions still.

8.3.2.3 Arene exchange reactions

Treating chromium tricarbonyl arene complexes with different arenes leads in some cases to arene exchange reactions; the feasibility and rate of such exchanges depends greatly on the identities of the arenes involved. As explained in the first edition of this work, kinetic studies of the arene exchange reaction in hydrocarbon solvents reveals both arene-dependent and arene-independent terms in the rate law. Generally the rate in hydrocarbon solvents is given by Equation (1).

$$\text{Rate} = k_1[\text{Cr(CO)}_3(\eta\text{-arene})] + k_2[\text{Cr(CO)}_3(\eta\text{-arene})][\text{arene}'] + k_3[\text{Cr(CO)}_3(\eta\text{-arene})]^2 \quad (1)$$

For labile arenes such as naphthalene, the third term is absent, whereas for more strongly bound ligands such as C_6H_6, it is the first term which disappears.[245]

Extensive studies by Traylor and co-workers demonstrated that the second-order term in the rate equation is not a true second-order reaction, but is due to general catalysis by all the available [$Cr(CO)_3(\eta$-arene)] complexes.[246] Thus, replacement of *p*-xylene in [$Cr(CO)_3(\eta$-$MeC_6H_4Me)$] by benzene is catalysed by [$Cr(CO)_3(\eta$-$C_6Me_6)$] without any exchange of the less labile C_6Me_6.[247] The formation of a dinuclear species (**17**) bonded through an isocarbonyl linkage was proposed to explain these results. However, CO exchange between ^{13}CO-enriched [$Cr(CO)_3(\eta$-$C_6H_6)$] and [$Cr(CO)_3(\eta$-$C_6D_6)$] during an exchange reaction with mesitylene was shown to be 10 times faster than arene exchange.[248] This cannot be explained by an isocarbonyl structure, and an intermediate with two μ-CO ligands, (**18**), was proposed instead. The mechanism proposed by Pauson, which involves CO transfer and formation of the ring-slipped intermediate $[Cr(CO)_4(\eta^4\text{-arene})]$ and unsaturated $[Cr(CO)_2(\eta^6\text{-arene})]$ can be disproved by labelling studies, which show that arene exchange can occur without concomitant CO exchange.[246]

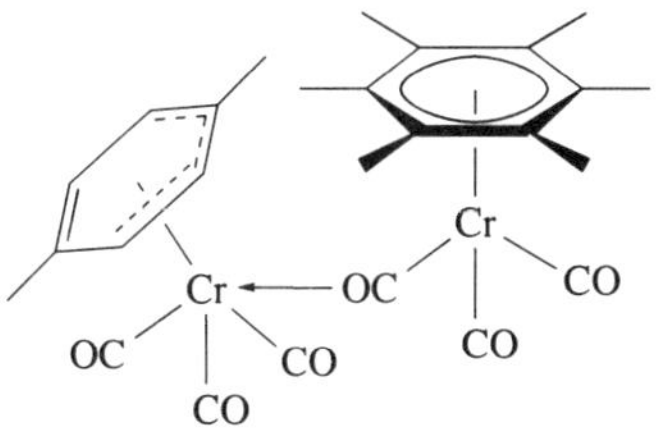

(17)

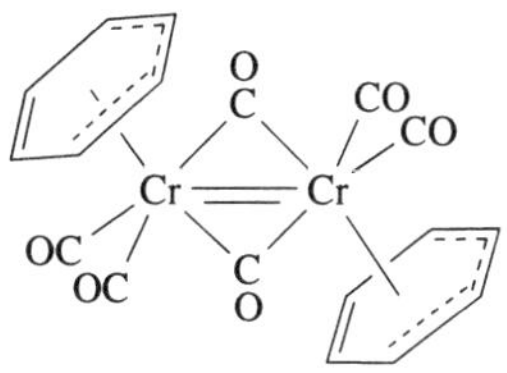

(18)

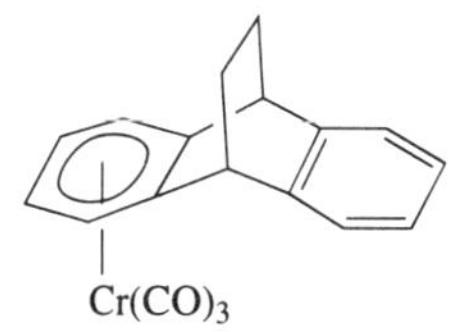

(19)

The arene exchange is also catalysed by added donors such as cyclohexanone. It can also be catalysed by iodine at room temperature, presumably by oxidation to a substitutionally labile 17-electron species.[249] Neighbouring group participation by donor groups is also important. Both intra- and intermolecular effects due to the alkene functionality were observed in the exchange reaction of $[Cr(CO)_3(\eta\text{-}E\text{-PhCH=CHMe})]$ with C_6H_6, and exchange of *cis-* or *trans*-stilbene is several hundred times faster than that of diphenylethane.[250] The internal exchange of the arenes in $[Cr(CO)_3(\eta^6\text{-PhCH}_2\text{Ph})]$ proceeds by a similar assisted pathway, through an intermediate in which both arenes are η^2-bonded to the metal.[251,252] For complexes of $Ph(CH_2)_nPh$, the effectiveness of the neighbouring group effect varies according to the ring size of the transition state in a conventional organic manner; thus, the relative rates for internal exchange are 64:0.62:3.1: < 0.1 for $n = 1\text{–}4$.[253] This interaction can also be observed in the mass spectra of these compounds.[254] When the arene used is one that allows the η^4,η^2 and η^2,η^2 bonding situations to be facilitated, such as (**19**), internal exchange is faster, but surprisingly so is external exchange with C_6H_6; this raises the possibility of an intermediate in which the original arene is η^2,η^2-bound, with the incoming arene also η^2 bound.[251] X-ray structural studies of both the *endo* and *exo* isomers of this complex showed that while the *exo* isomer displayed a distorted structure, which could explain rapid arene exchange reactions, the *endo* isomer was not distorted, implying that its rapid exchange is due to π-participation.[255]

All this evidence indicates that exchange is initiated by ring slippage of the arene, with the remaining coordination sites being occupied by any available donor molecules. For the labile arenes, this can occur by competing associative and dissociative ring slippage pathways. The degree of slippage involved is a matter of some debate; certainly in the case of naphthalene, molecular orbital calculations indicate that an intermediate η^2 species is most accessible.[245]

8.3.2.4 Photochemically from $[M(CO)_6]$

UV irradiation of a mixture of $[Cr(CO)_6]$ and arene in THF, or prior irradiation of the carbonyl in THF followed by addition of the arene, does produce $[Cr(CO)_3(\eta\text{-arene})]$; however, the yields are less than 50% and are generally lower than with thermal methods.[256]

8.3.2.5 From carbene complexes

The reactions of chromium carbene complexes with alkynes to give naphthol derivatives (Dötz annelation) have been extensively studied and refined over the past decade. They form the subject of numerous review articles and a full account is given in Chapter 3, of this volume.[257,258] Although the products are η-arene complexes, in many cases the desired organic products are deliberately decomplexed from the metal (by exposure to air and sunlight, or iodine) before isolation; this section deals only with cases where the arene complex has been isolated intact.

In the basic reaction, a carbene complex of the type $[Cr\{=C(OR)Ph\}(CO)_5]$ reacts with an alkyne $R^1C{\equiv}CR^2$ as shown in Equation (2).[259–61] A large variety of substituents can be tolerated both on the carbene and the alkyne, and some degree of stereoselectivity is usually observed in the positioning of R^1 and R^2 in the product. If the alkyne used is $HC{\equiv}C(CH_2)_4CN$, the annelation can be followed by nucleophilic substitution at the arene nucleus by generation of a carbanion from the cyanoalkyl group.[262]

A dissociative mechanism involving loss of one CO has been proposed for the annelation.[263] If the aryl group of the carbene complex contains a donor substituent in the *o*-position, for example [Cr{=C(OMe)C_6H_4-2-OMe}$(CO)_5$], this can coordinate to the metal on heating to form an intermediate $Cr(CO)_4$-chelate species before addition of the alkyne.[264,265] Indeed, heating [Cr{=C(OMe)(C_6H_4-2-Ph)}$(CO)_5$] causes coordination of the aryl ring to give (**20**).[266]

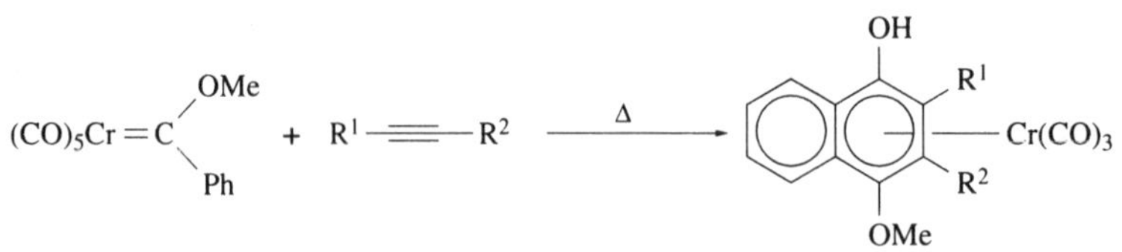

(2)

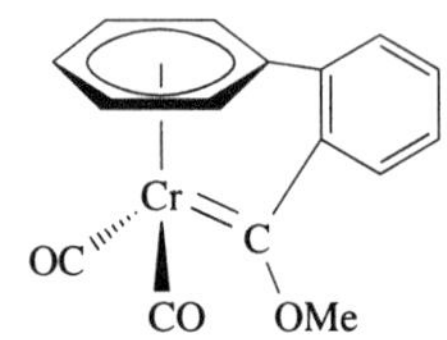

(20)

Related to the basic Dötz annelation is the reaction of vinylcarbene complexes [Cr{=C(OMe)CR^3=CHR^4}$(CO)_5$] with alkynes to produce complexes of 1,4-hydroquinone monoethers (Equation (3)).[267–9] The Diels–Alder reaction of carbenes such as [Cr{=C(OMe)C≡C-TMS}$(CO)_5$] with dienes and then alkynes also proceeds ultimately to the benzannelated complex.[270] An interesting variation involves the reaction of allenes (e.g., PhCH=C=CH_2) with [M{=C(OMe)Ph}$(CO)_5$] (M = Cr, Mo, W) which initially forms trimethylenemethane complexes and ultimately dimethylenecyclopentane derivatives (Scheme 3).[271]

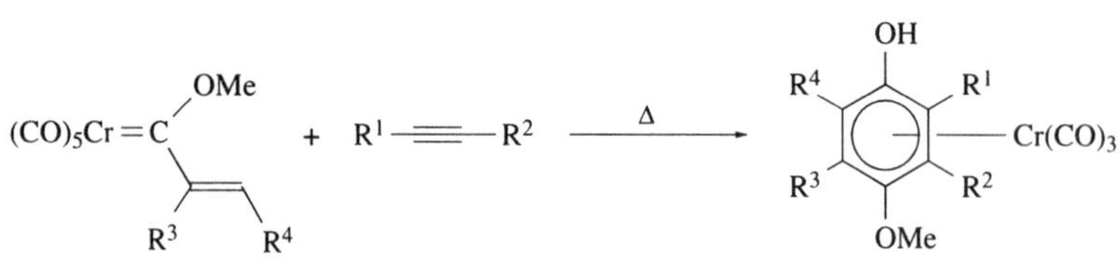

(3)

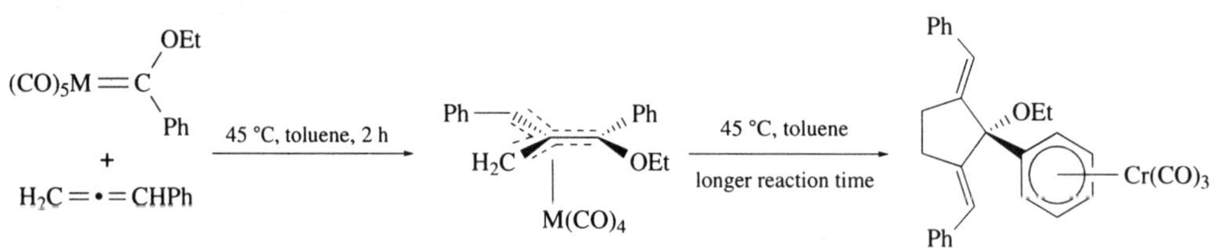

Scheme 3

The reactions of aminocarbene species with alkynes are more complex, and involve incorporation of CO and migration of groups from nitrogen to carbon. The products include nitrogen ylides, pyrollidines and indanone derivatives; a typical selection is shown in Scheme 4.[272–83] By incorporation of the alkyne function into the starting carbene, fused rings are obtained.

Thermolysis of the cyclic bis(carbene) [Cr{=C(OEt)CHPhCHPhC(OEt)=}$(CO)_4$] produces the mono- and bis-$Cr(CO)_3$ complexes of PhCH=C(OEt)C(OEt)=CHPh as well as the free diene.[284] Thermolysis of [Cr{=C(OEt)$SiPh_3$}$(CO)_5$] produces free $Ph_3SiCOEt$ as well as its tricarbonyl chromium complex.[285]

8.3.2.6 Miscellaneous methods

Friedel–Crafts acylation of [$Cr(CO)_3(\eta^6$-$C_8H_8)$] unexpectedly produced a 50% yield of [$Cr(CO)_3(\eta^6$-*E*-PhCH=CHCOMe)] by ring contraction.[286] There are also several examples of the preparation of arene complexes by the cyclization of alkynes, including the ring closure of $C_2(SiMe_2OSiMe_2C≡CH)_2$ by the action of [$Mo(CO)_6$].[287,288]

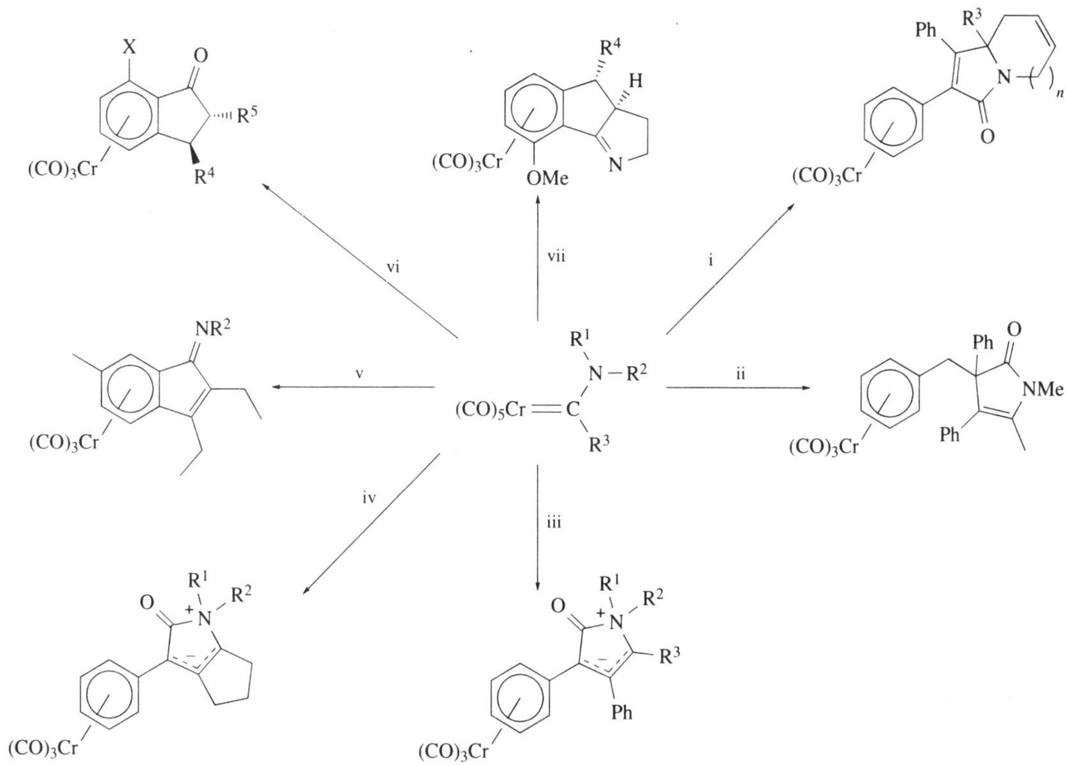

i, C_2Ph_2, benzene reflux, 12 h, $R^1R^2 = (CH_2)_nCH{=}CHCH_2$, R^3 = Me, Ph; ii, C_2Ph_2, benzene reflux, 12 h, $R^1 = R^3$ = Me, R^2 = Bz; iii, C_2Ph_2, cyclohexane reflux, $R^1 = R^2$ = Me or $R^1R^2 = (CH_2)_5$, R^3 = H, Me, Ph; iv, Δ, $R^1 = R^2$ = Me or $R^1R^2 = (CH_2)_4$, $R^3 = (CH_2)_3C{\equiv}CPh$; v, C_2Et_2, toluene, 70 °C, R^1 = H, R^2 = o-$C_6H_4NMe_2$, R^3 = p-Tol; vi, $R^4C{\equiv}CR^5$, Δ, R^1 = H, $R^2 = CH_2CH{=}CH_2$, R^3 = o-C_6H_4X (X = H, OMe), $R^4 = R^5$ = Me, Et; vii, Δ, R^1 = H, $R^2 = CH_2CH_2C{\equiv}CR^4$ where R^4 = Me, Et, TMS, R^3 = o-C_6H_4OMe

Scheme 4

8.3.3 Range of Arene Complexes Prepared

A vast number of arenes have been successfully attached to the $Cr(CO)_3$ fragment, most of them by simple thermal complexation (Sections 8.3.2.1 to 8.3.2.3), which is successful in the majority of cases. However, arenes with substituents which are electron-withdrawing and π-accepting (e.g., NO_2) are either too electron-poor or too oxidizing to form complexes; 1,3,5-$C_6H_3(CO_2Me)_3$ is perhaps the most electron-poor arene to be successfully used, though it has been claimed that the reaction of $[Cr(CO)_3(NCMe)_3]$ with trinitrotoluene gives the $Cr(CO)_3$ and $[Cr(\eta\text{-arene})_2]$ complexes of TNT.[289] Some other arenes, notably PhCHO and $PhCO_2H$, do not undergo direct complexation readily and must be protected as acetals or esters before reaction. Phenol can be complexed directly, but complexes of 1,2- or 1,4-dihydroxyarenes, prepared by complexation of their silyl ethers and deprotection, were too sensitive to isolate.[290,291]

The attempted complexation of benzylic halides with $[Cr(CO)_3(NH_3)_3]$ led instead to reasonable yields of 1,2-diphenylethane coupling products, or, if the chromium reagent was in excess, their $Cr(CO)_3$ complexes.[292]

Apart from the simple arenes described in Table 3, several families of complexes have been prepared which deserve further comment.

8.3.3.1 Biologically relevant arenes

A number of papers deal with $Cr(CO)_3$ complexes of molecules of biological interest. On the simplest level these include complexes of protected aromatic amino acids such as tryptophan and phenylalanine.[293,294] Methods have also been developed for the attachment of a $Cr(CO)_3$ marker to drugs such as amphetamines by transfer from esters of *N*-hydroxysuccinimide.[295,296] Several other derivatives themselves possess biological activity, for example complexes of Hantzsch esters (4-aryl-1,4-dihydropyridines) act as calcium antagonists.[297]

However, most attention has focused on the steroid framework, particularly those derived from methyl podocarpate (**21**)[298–300] and estradiol (**22**).[301,302] Both of these exist as α and β-isomers depending on which face of the arene is complexed (the α isomer is illustrated in both cases). The isomers can be separated by crystallization or HPLC, and distinguished by their high-field two-dimensional NMR spectra, Raman spectra and electrochemical properties.[303–6] Further modification of the complexed ring can be achieved by nucleophilic attack or lithiation.[307–10]

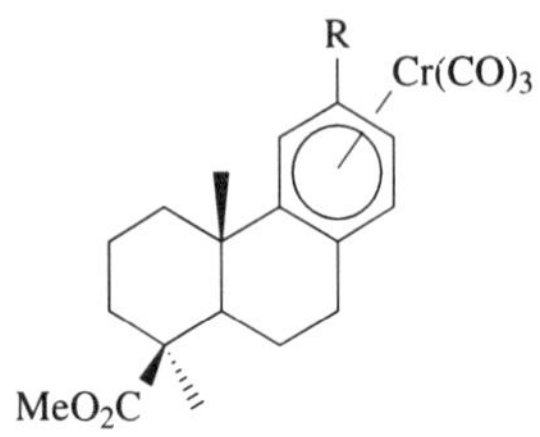

(21)

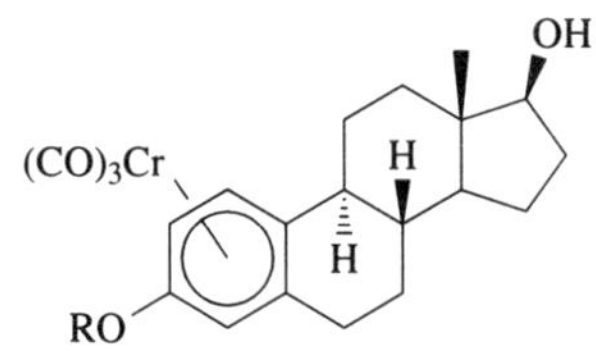

(22)

The strong IR absorption of the $Cr(CO)_3$ group, coupled with the advent of Fourier transform IR spectroscopy, enables compounds such as (**22**) (R = $CH_2CH_2CH_2OH$) to be used as biological assays for the estrogen receptor.[311] Moreover, it can be readily radiolabelled with deuterium or tritium.[312] Chromium tricarbonyl complexes of hexestrol (**23**) (R = H, Bz) can also serve the same purpose and show reasonable affinity for the receptor.[313]

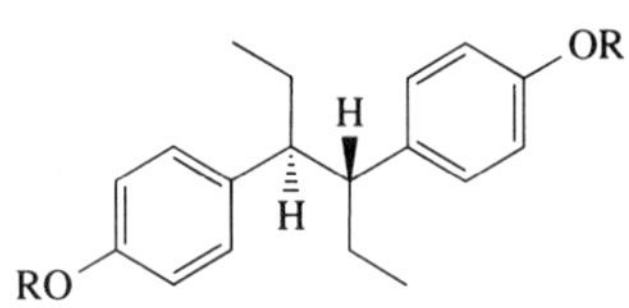

(23)

8.3.3.2 Polycyclic hydrocarbons

A large number of $Cr(CO)_3$ complexes of polycyclic hydrocarbons have been prepared. Direct complexation of polycyclic arenes with $[Cr(CO)_6]$ sometimes gives low yields; as alternatives the tris(pyridine) or tris(ammine) complexes in conjunction with a Lewis acid have been used, and arene exchange from $[Cr(CO)_3(\eta^6\text{-}C_{10}H_8)]$ has also proven useful.

The site of attachment of the $Cr(CO)_3$ fragment to polycyclic aromatics is usually one of the terminal rings; this is conventionally rationalized as the one for which the highest number of Kekulé resonance structures can be drawn. However, a recent study of the complexation of 21 substituted polycyclic aromatics to $[Cr(CO)_6]$ under thermodynamic conditions has shown that the site of attachment depends more precisely on the π-electron density of the rings. Thus, in hydrocarbons such as 1,2,3,4-tetrahydrophenanthrene, complexation occurs exclusively to the more substituted central ring to give (**24**), whereas an electron-withdrawing substituent placed in the hydrogenated ring directs complexation to the other end of the molecule. Moreover, 9-methyl- and 9,10-dimethylanthracene were found to undergo complexation at the central, substituted electron-rich ring.[314]

In the case of bent diarenes, which have concave and convex faces, complexation usually occurs to the concave face. Thus, for *cis*-9,10-dihydro-9,10-dimethylanthracene and related complexes such as (**19**), the concave isomer, that is, that with the $Cr(CO)_3$ *anti* to the alkyl groups, is favoured; however, it is not possible to attach two metals to this face and for the bis-$Cr(CO)_3$ complex the *cis, syn, syn* and *trans* isomers exist.[315,316] A similar situation pertains for the bowl-like triarene 10-methyltribenzoquinacene (**25**).[317] However, with 9,10-dihydro-9,10-methanoanthracene and *N*-methylanthracene-9,10-imine, formation of the *syn* isomer occurs apparently exclusively.[318,319] In the case of the 10,11-dibenzocycloheptenes *o,o*-$\overline{C_6H_4XC_6H_4CH_2C}H_2$, complexation occurs on the convex face if X = CH_2 or CO, but to the concave face if X = CHOH; this is attributed to the directing ability of the OH group.[320]

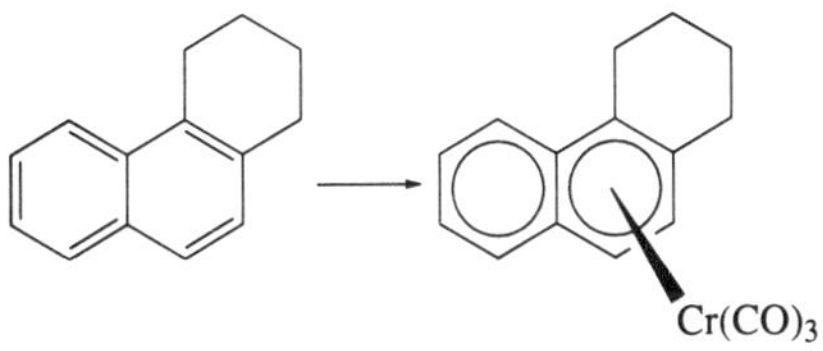

(24)

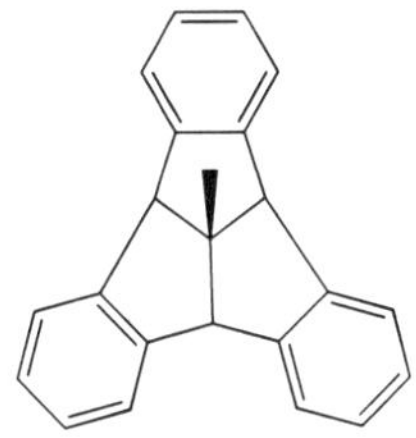

(25)

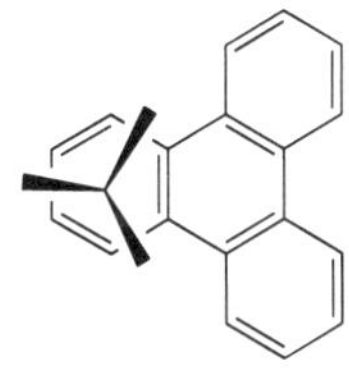

(26)

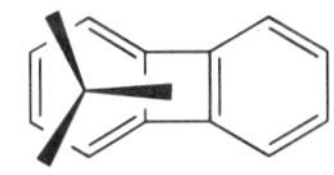

(27)

The x-ray structures of complexes of polycyclic aromatics such as naphthalene and biphenylene show that the structure is somewhat distorted in that the $Cr(CO)_3$ unit is not situated directly beneath the centre of the ring; the Cr–C bonds to carbon atoms at the ring junction are slightly longer than those to the periphery. This slight distortion towards an η^4-bonded structure is related to the ability of these complexes to undergo facile arene exchange reactions and haptotropic shifts between rings.[321] The distortions of the aromatic ligand in the $M(CO)_3$ complexes of octamethylnaphthalene are the same as those observed in the free arene.[322]

The conformation of the $Cr(CO)_3$ fragment in complexes of polycyclic aromatics has been studied theoretically. This was prompted by the finding that whereas complexes of naphthalene, anthracene, phenanthrene, triphenylene and most other aromatics have a conformation in which the unique CO ligand points away from the uncomplexed rings,[323] for example (**26**), that of biphenylene (**27**) has the CO below the ring. This can be rationalized by an electron-counting approach; if, after complexation of one ring as an η^6-arene, the remaining ring has $4n$ π-electrons, the conformation with the CO pointing away from that ring will be favoured; if the uncomplexed ring has $4n + 2$ π-electrons, the CO will lie underneath that ring.[324] The bis-$Cr(CO)_3$ complex of phenanthrene has one tripod in each conformation, but this may well be due to crystal packing forces.[325] The complex of decahydrotriphenylene shows a completely staggered arrangement with none of the COs lying under the saturated rings.[326]

8.3.3.3 Cyclophanes

Numerous tricarbonyl chromium complexes of cyclophanes have been prepared, including both mono- and bis-$Cr(CO)_3$ derivatives; the metal always coordinates to the external face of the arene.

The magnitude of the ^{13}C-NMR complexation shifts in cyclophane complexes has been related to the distance between the chromium atom and the arene carbons.[327] In paracyclophane complexes the arene ring is forced into a slight boatlike conformation by the constraints of the spacer groups; the NMR complexation shifts of these carbons can be correlated with the degree of bending back.[328,329] However, the solvent dependence of complexation shifts means that their use as a measure of the quenching of ring current must be treated with caution.[330]

The increased basicity of these bent arene units can be shown by the selective complexation of the 1,2:9,10-dibenzo[2.2] paracyclophane-1,9-dienes (**28**) at the *para*-substituted rings.[331] With the related cyclophanes (**29**), complexation is directed to the nonsubstituted electron-rich ring.[332] The coordinated ring can then be functionalized by lithiation followed by quenching with electrophiles, but attack can also occur in the spacer groups. The x-ray structures of some of these complexes shows that as π-electron density is removed, the decks of the phane move closer together.[333]

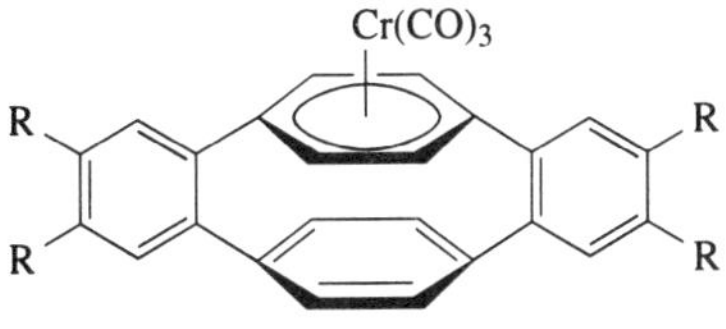

(28)

R = Ph, $C_6H_4Bu^t$

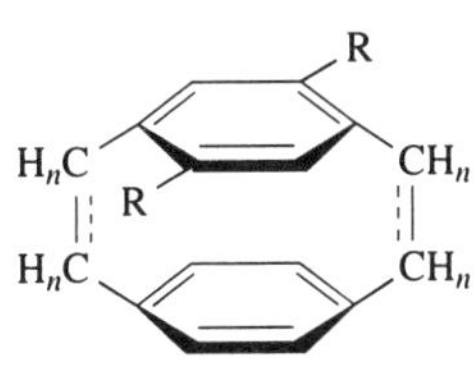

(29)

R = OH, OMe, OBu

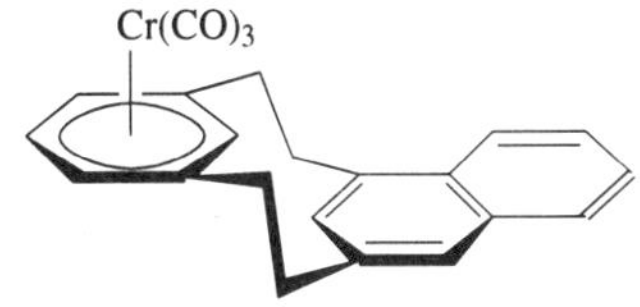

(30)

The regioselective complexation of [2.2] metacyclophanes such as (**30**) and (**31**) has been studied; given the choice between benzene and naphthalene rings, complexation occurs at the benzene, whereas tetrahydronaphthalene proves more favourable. Introduction of dithiane substituents on the ethylene bridges in (**32**) causes complexation at the ring furthest from the substituents for steric reasons.[334–6] Again, in these cases lithiation can be used to functionalize the coordinated ring.

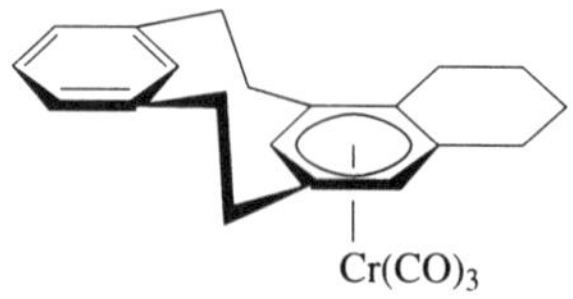

(**31**)

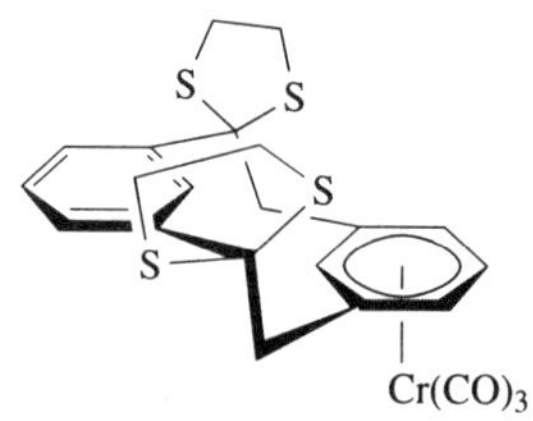

(**32**)

Mitchell has examined the complexation of *syn*-2,11-dithia[3.3]metacyclophane (**33**), which retains its *syn* form in the complex. On treatment with $[CH(OMe)_2][BF_4]$ followed by $KOBu^t$ it undergoes a Stevens rearrangement to produce the $Cr(CO)_3$ complex of the SMe-substituted *syn* [2.2]-metacyclophane (**34**), which cannot be made in any other way; it rapidly isomerizes to the *trans* isomer (**35**).[337–40] By a similar approach it was possible to prepare the first complexes of a metacyclophane diene. Complexation to a $Cr(CO)_3$ fragment can also be used to separate chiral [3.3]-metacyclophanes into diastereomers.[341]

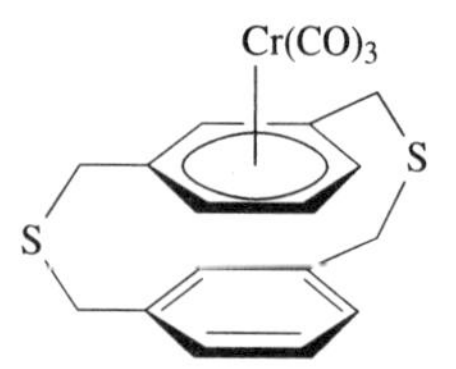

(**33**)

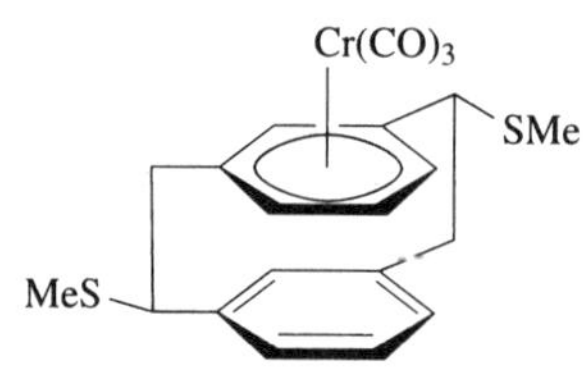

(**34**)

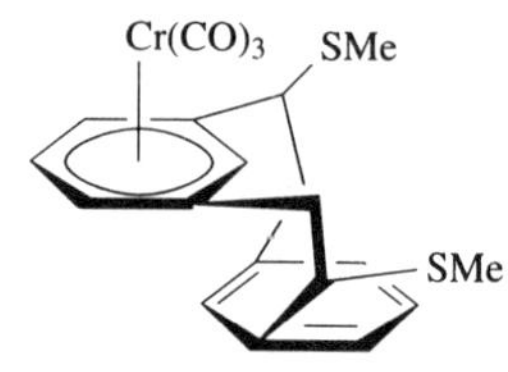

(**35**)

The first example of a $Cr(CO)_3$ complex of a heteroarene cyclophane is provided by the [2.2] pyrroloparacyclophane (**36**), which forms both η^6- and η^5-isomers on complexation.[342]

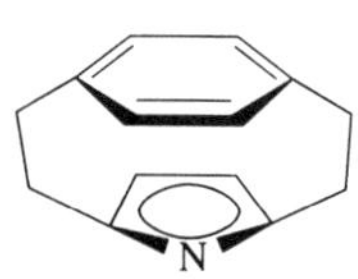

(**36**)

The attachment of $Cr(CO)_3$ fragments to compounds of interest in supramolecular chemistry is typified by complexes of α- and β-perbenzoyl cyclodextrin and three conformers of a calixarene.[343,344]

8.3.3.4 Complexes of main group aryl compounds

The ability to complex virtually any arene to a $Cr(CO)_3$ unit has led to the preparation of many complexes of main group aryl compounds, particularly those of group 14. With groups 15 and 16, η^6-coordination of the arene ring is in competition with coordination through the donor main group atom; the latter usually predominates, for example thermal reaction of $[Cr(CO)_6]$ with PPh_3 gives compounds of the type $[Cr(CO)_{6-n}(PPh_3)_n]$.

Although no examples of complexed arylboranes have been reported recently, the mono- and bis-$Cr(CO)_3$ complexes of the carborane *nido*-2,3-$B_4H_6C_2(CH_2Ph)_2$ have been prepared.[345] The sandwich complex of this carborane with $Fe(\eta^6\text{-}C_8H_{10})$ can also be complexed.[346] Complexation of the diboradiazanaphthalene below to $M(CO)_3$ fragments occurs exclusively at the ring shown in (**37**).[347] A number of diazaboroles have also been complexed to $Cr(CO)_3$.[348]

A wide range of arylsilanes have been complexed to $M(CO)_3$ units, usually by direct reaction with $[M(CO)_6]$ but also from $[M(CO)_3(NCMe)_3]$ (Table 4). It should be noted that $[Cr(CO)_3(\eta\text{-Ph-TMS})]$ undergoes ready desilylation with KF/H_2O under conditions where the free silane is unreactive, which allows the use of TMS as a protecting group for H.

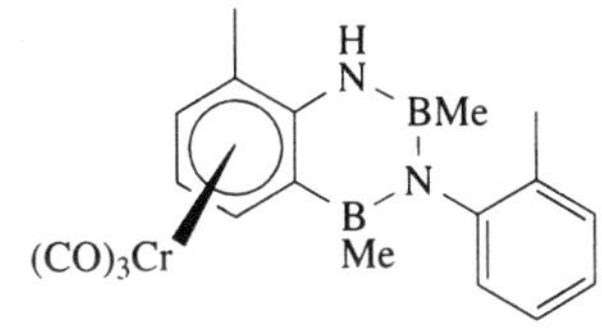

(37)

Table 4 Tricarbonyl metal complexes of arylsilanes and related compounds.

Arene in [M(CO)$_3$(η-arene)]	*Metal*	*Ref.*
$C_6(SiMe_2H)_6$	Cr, Mo, W	349
$PhSi(OMe)_3$, $PhSi(OEt)_3$, $PhSiMe(OMe)_2$, $PhSiMe_2H$, $PhSiMe_2(CH{=}CH_2)$, $PhSiMe(NMe_2)_2$ $PhCH_2CH_2Si(OMe)_3$, 1,4-$C_6H_4(SiMe_2H)_2$	Cr	350–2
Ph_3SiH, Ph_2SiMeH, $PhSiMe_2H$	Cr	353
Ph_3SiOH	Cr	354
$(PhSiMeO)_n$ (n = 3, 4)	Cr	355–7
$PhSi(OCH_2CH_2)_3N$ (phenylsilatrane)	Cr, W	358, 359
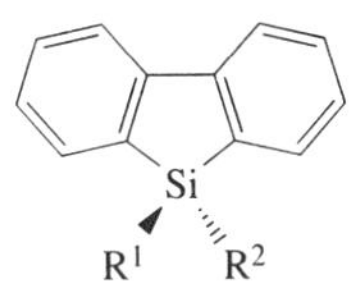 R^1 = Me, R^2 = Me, TMS, SSiMe($C_{12}H_{10}$)	Cr, W	360–7
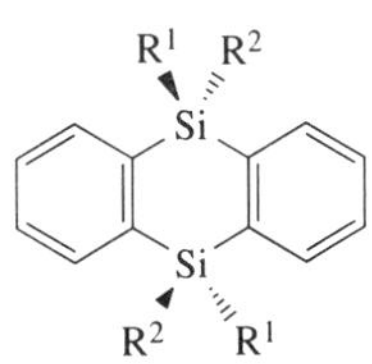 $R^1 = R^2$ = Me, OEt; R^1 = Me, R^2 = OPr^i	Cr, Mo, W	364, 368–72
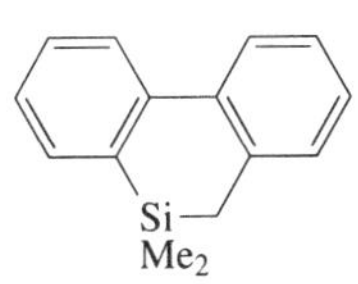	Cr	362, 373–6
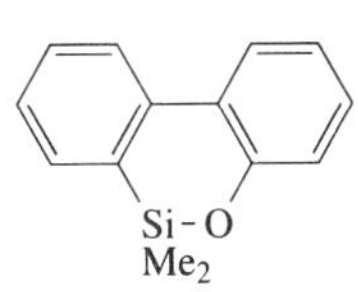	Cr, W	367, 377, 378
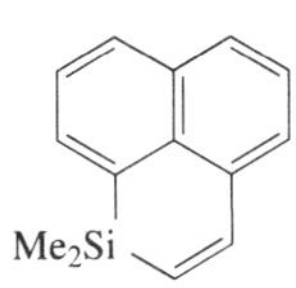	Cr	379–81
Ph_2CH-TMS	Cr	382

A comprehensive study of the complexation of Me_2EPh_2 (E = C, Si, Ge, Sn, Pb) to $Cr(CO)_3$ has been undertaken, using $[Cr(CO)_6]$ in diglyme/THF or $[Cr(CO)_3(NH_3)_3]$ in dioxane as the metal source. Formation of mono- and bis-complexes was observed.[223,383,384]

The reaction of Schiff's bases of the type ArCH=NCH$_2$CH$_2$N=CHAr with $[M(CO)_6]$ gives chelate complexes of the type $[M(CO)_4(L)]$ if M = Mo or W, but η^6-arene complexes if M = Cr, demonstrating the affinity of the $Cr(CO)_3$ unit for the arene ring.[385] A number of other aryl Schiff's bases have been complexed to $Cr(CO)_3$.[386] Attempted complexation of 2-aroylpyrroles led instead to their coordination as N,O chelate ligands. However, 3-aroylpyrroles behave normally as η^6 ligands; the IR and NMR spectra of a variety of substituted ones suggest that there is little electronic communication between the two rings.[387–9]

Complexation of the diphosphene ArP=PAr (Ar = 2,4,6-$C_6H_2Bu^t_3$) occurs through the arene rings rather than phosphorus, presumably for steric reasons. Both mono- and bis-$Cr(CO)_3$ species are formed; comparison of their structures shows that the chromium fragments do have some effect on the arene conformations.[390,391] Cyclotriphosphazenes with aryl substituents can also be complexed to $Cr(CO)_3$.[392]

The lower donor propensity of antimony and bismuth means that $Cr(CO)_3$ complexes of $SbPh_3$ and $BiPh_3$ can be prepared directly from $[Cr(CO)_3(NH_3)_3]$.[393,394]

8.3.3.5 Complexes containing other metal atoms

A number of tricarbonyl arene complexes have been synthesized in which the arene ligand is itself part of another metal complex. In some of these the second metal (M^2) is directly σ-bonded to the arene ring, in others it is separated by alkyl chains or other spacer groups and can be σ- or π-bound. These are listed in Tables 5–7, with their methods of preparation. The main synthetic routes involve (A) nucleophilic attack of metal carbonylates on preformed $[M^1(CO)_3(\eta\text{-arene})]$ complexes; (B) attachment of intact complexes to the $M^1(CO)_3$ fragment; (C) reaction of lithiated arene complexes with metal halides; and (D) from the mercury complex $[Hg\{PhCr(CO)_3\}_2]$. Other methods (E) include oxidative addition of haloarene complexes to palladium centres, and reaction of $[Cp_2ZrMe_2]$ or $[Cp_2Zr(NMe_2)_2]$ with protic arene complexes with elimination of CH_4.

Table 5 Complexes with M^2 σ-bonded directly to the aryl ring.

Arene in $[Cr(CO)_3(\eta\text{-arene})]$	*Synthetic method*	*Ref.*
$PhFe(CO)_4^-$, $PhCr(CO)_5^-$, $PhW(CO)_5^-$	A	395,396
PhFp, 1,4-XC_6H_4Fp (X = Me, OMe, F, Cl, CF_3) 1,3-XC_6H_4Fp (X = Me, OMe, Cl, CF_3)	A,B	397–400
1,3-FpC_6H_4Fp, 1,4-FpC_6H_4Fp, 1,3,5-$C_6H_3Fp_3$	B	398,401,402
1,4-PhC_6H_4Fp, 4,4'-$FpC_6H_4C_6H_4Fp$, 2,6-$FpC_{10}H_6Fp$	B	403
$CpFe(CO)C_6H_4$-2-COC_6H_4-2-C_6H_4F-2	C	404
$XC_6H_4TiClCp_2$ (X = H, *m*- or *p*-Me, *o*-F) $(XC_6H_4)_2TiCp_2$ (X = H, *o*-F)	C	405
$PhMn(CO)_5$	C	406
$PhW(X)Cp_2$ (X = H, Cl, Br, I)	B	407
$PhPd(Cl)L_2$ (L = PPh_3, PPh_2Et, $PPhEt_2$)	E	408
$XC_6H_4Au(PPh_3)$ (X = H, *o*-F)	C	405
Ph_2Hg	B	409,410
$LnPh_2\cdot n$THF (Ln = Sm, Eu, Yb; *n* = 1 or 2)	D	411,412

Two further examples exist of η-arene complexes being used as ligands in coordination complexes; in one case preformed compounds of the type $[M(salen)_2]$ (M = Co, Cr, Ni, Cu, Zn) were complexed to $Cr(CO)_3$ units, and in the other the chromium complex of Schiff's bases derived from pyrazole aldehydes were then used to prepare complexes of zinc(II), cadmium(II), nickel(II) and mercury(II).[442–6]

The fact that $[Cr(CO)_3(\eta\text{-PhCl})]$ undergoes enhanced oxidative addition to palladium centres has been used to modify the arene ligand by palladium-catalysed cross-coupling with organotin reagents, for example $Bu_3Sn(CH{=}CH_2)$;[447–9] in $[Cr(CO)_3(o\text{-}C_6H_4Cl_2)]$ only one chloride is replaced, and a reasonable degree of asymmetric induction can be achieved by employing a chiral phosphine with the palladium source.[450] Chloroarene complexes also undergo carbonylation to esters and amides, and double carbonylation to oxoamides, when treated with $[PdCl_2(PPh_3)_2]$, PPh_3, CO, NEt_3, and MeOH or $NHEt_2$.[408,451,452] Alkoxycarbonylation can also be achieved with alkyl formates and alkali metal alkoxides, though in this case varying amounts of by-products are formed depending on the bulkiness of the alkyl group and the reagent ratio.[453,454] Conversion of $[Cr(CO)_3(\eta\text{-PhCl})]$ to $[Cr(CO)_3(\eta\text{-PhC}{\equiv}CR)]$ by treatment with terminal alkynes, CuI, a palladium catalyst, and NEt_3 has also been reported.[455,456] An unusual palladium-catalysed Arbusov reaction between haloarene complexes and $P(OMe)_3$ to give aryl dimethyl phosphonates has been reported, but yields are modest due to competing arene displacement by the phosphite.[457]

Table 6 Complexes with M^2 σ-bonded to the aryl ring through a spacer group.

Arene in $[Cr(CO)_3(\eta\text{-arene})]$	*Synthetic method*	*Ref.*
$o\text{-}MeOC_6H_4C{\equiv}M(CO)_2(O_2CCF_3)_2L_2$ (L_2 = bipy, TMEDA)	C	413
$o\text{-}MeOC_6H_4C{\equiv}M(CO)_2Cp$ (M = Mo, W), $o\text{-}MeOC_6H_4C{\equiv}Mo(CO)_2(HBpz_3)$		
$PhCCo_3(CO)_9$	D	414
$p\text{-}MeC_6H_4CReCo_2(CO)_{10}$	E	415
$PhPRu_3(CO)_9(\mu_3\text{-}C_6H_4)$, $Bu^tPRu_3(CO)_9(\mu_3\text{-}C_6H_4)$	E	416,417
$o\text{-}XC_6H_4COFe(CO)_4^-$ (X = H, F, Cl, OMe)	C	396
$o\text{-}XC_6H_4COMn(CO)_5$ (X = H, F)	A	406
1,3-$FpCOC_6H_4COFp$, 1,4-$FpCOC_6H_4COFp$, 1,3,5-$C_6H_3(COFP)_3$	B	401
$p\text{-}XC_6H_4CH_2Co(CO)_4$ (X = H, Me)	A	418,419
$BzCo(dmgH)_2(py)$	A	420
$Me_5C_6CH_2COFe(CO)_4^-$	C	421
$BzOM^2(X)Cp_2$ (M^2 = Ti, X = Cl, Br; M^2 = Zr, X = Bz), $(PhOCH_2)_2ZrCp_2$	E	422
$PhCOOTi(X)Cp_2$ (X = Cl, Br; Me), $(PhCOO)_2M^2Cp_2$ (M^2 = Ti, Zr)	E	423
$(PhCOCHCOMe)Zr(O_2CMe)_3$, $(PhCOCHCOMe)_2Ti(OBu^t)_2$	B	424
$(PhCOCHCOMe)_2Al(OBu^t)$		
$PhOW(CO)_5^-$	E	425
$(PhO)_2ZrCp_2$	E	290
$Ph(CH_2)_nFp$ (n = 2, 3)	A	426
$\{Ph(CH_2)_n\}_2Hg$ (n = 1, 3, 4)	B	427,428
1,4-$C_6H_4\{SiMe_2Co(CO)_4\}_2$	E	352
$(3,5\text{-}Me_2C_6H_3CH_2CN)Os_3(CO)_{11}$	E	429
1,4-$[-Ph_2PC_6H_4PPh_2Rh(CO)(Cl)-]_{n>}$	E	430
$BzPPh(CH_2)_3PR_2\{PtMe_2\}$	E	431

Table 7 Complexes with M^2 π-bound to the aryl fragment through a spacer chain.

Arene in $[Cr(CO)_3(\eta\text{-arene})]$	*Synthetic method*	*Ref.*
$o\text{-}XC_6H_4CH{=}CHCOPh\{Fe(CO)_3\}$ (X = Me, OMe)	E	432,433
exo-PhC_5H_5CoCp	B	434
$PhC_5H_4Mn(CO)_3$	B	435
$XC_6H_4COC_5H_4Re(CO)_3$ (X = H, *o*- or *p*-Me, *o*-F, *p*-Cl)	B	436
$PhCHOHC_5H_4M^2Cp$ (M^2 = Fe, Ru, Os)	B	437
$PhCOC_5H_4FeCp$, $PhC_7H_7M^2(CO)_3$ (M^2 = Cr, Mo, W) + various others	C	438,439
$\{PhCHC(CO_2Me){=}C(CO_2Me)CH_2\}Fe(CO)_4$	E	440
$\{o\text{-}MeC_6H_4CH{=}CHCPh{=}O\}Fe(CO)_3$	E	441

8.3.4 Characterization of Arene Metal Tricarbonyl Complexes

Arene tricarbonyl chromium complexes are yellow or red solids that are usually sufficiently air stable for handling in the solid state and that decompose only relatively slowly when exposed to air in solution. Their air- and thermal stability enables them to be purified by techniques such as column chromatography, recrystallization or sublimation. Separation by HPLC has also been reported, and it is even possible to employ gas chromatography on very inert columns.[458–61]

8.3.4.1 Solid state and electronic structure

Literally hundreds of tricarbonyl chromium arene complexes have been crystallographically characterized. For simple monocyclic arenes, each displays the classical piano-stool structure in which the metal lies directly beneath the ring centroid at a distance between 0.170 nm and 0.174 nm. The major points of interest are the planarity of the arene ring, the bond lengths in that ring and the conformation of the $Cr(CO)_3$ tripod in relation to the ring carbons. The last of these is dealt with in a later section.

Although the arene ring is essentially planar, there are slight deviations towards a boat conformation. A recent analysis has proposed that these are due to the π-donating or π-accepting properties of the substituents; π-donor substituents (e.g., NMe_2, OMe) and their associated *ipso*-carbons are bent away from the metal atom by up to 6°, attributed to a resonance contribution from the zwitterionic form (**38b**); conversely π-acceptor substituents (CO_2Me, COMe) are displaced slightly towards the metal, though this effect is not so large. For *p*-disubstituted complexes a boatlike conformation results. There is little or no dependence on the steric bulk of the substituents except in extreme cases, for example $CHBu^t_2$.[462] Arenes with the $CpFe(CO)_2$ substituent show a substantial deviation from planarity in their $Cr(CO)_3$ complexes. In the case of cyclophane complexes, the *ipso*-carbons are bent away from the metal for steric reasons, and the charge density (and ^{13}C complexation shift) of these carbons can be related to the degree of bending; however, this distortion also occurs in the free ligands.[328,329]

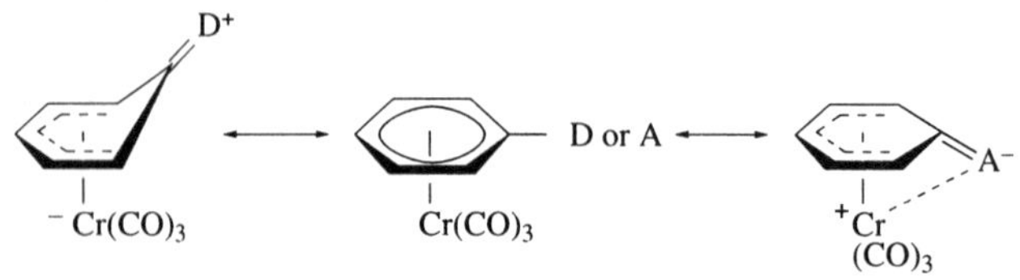

(38b) **(38a)** **(38c)**

The question of bond length alternation in the arene rings of [$Cr(CO)_3$(η-arene)] complexes has remained controversial. A low-temperature x-ray structure of [$Cr(CO)_3$(η-C_6Me_6)], which adopts a staggered conformation, showed an average bond length alternation of 0.0021 nm, with the C–C bonds situated above the CO ligands being lengthened.[463] A theoretical analysis showed that this is due to electronic factors, namely interruption of conjugation in the π-ligand, and indeed suggested it should be a general phenomenon for polyene complexes. It also showed that the degree of alternation depends more on the metal atom than on the arene, in keeping with the larger alternation seen in [$Mo(CO)_3$(η-C_6Me_6)].[464]

The molecular orbital diagram of [$M(CO)_3$(η-arene)] complexes is shown in Figure 2. Comparative UV-photoelectron spectroscopy studies and Fenske–Hall MO calculations have been carried out on [$M(CO)_3$(η-arene)] where M = Cr, Mo, W and arene = C_6H_6, mesitylene, C_6Me_6. The spectra show two bands of interest, the first being due to ionizations from the closely spaced a_1 HOMO and *e* orbitals and the second to those from the 1*e*; the first band is split for tungsten due to spin–orbit coupling.[465] This increasing spin–orbit contribution is also responsible for changes in the quantum yields of photophysical processes on descending the group.[466] The x-ray photoelectron spectra of substituted arene complexes show little dependence on the substituent, which may suggest a disruption of conjugation between the arene and the substituent.[467,468] The application of UV/PES and related techniques to this area has been reviewed.[469]

A theoretical study of the electron density distribution in [$Cr(CO)_3$(η-C_6H_6)] disagreed with the results previously obtained by x-ray diffraction, particularly in the region close to the metal atom;[470] however, this prompted an experimental redetermination which this time was in agreement with the theory.[471] Total geometry optimization calculations have been carried out for [$Cr(CO)_3$(η-C_6H_6)]; the chromium–arene and Cr–CO distances are in good agreement with those observed experimentally.[84]

As mentioned in the previous edition of this work, [$Cr(CO)_3$(η-arene)] complexes have dipole moments in the direction arene → chromium, in keeping with the electron-withdrawing effect of the $Cr(CO)_3$ unit. The M–arene binding in a number of compounds has been studied by examination of their IR spectra, dipole moments and polarizabilities.[472–5] The polarizability of the complex is greatly enhanced along the chromium–arene axis, but lowered perpendicular to it. A method of determining the net charge on arene rings based on the ^{19}F NMR spectra of fluorinated analogues led to a value of 0.26 for [$Cr(CO)_3$(η-C_6H_6)], and rather less for [$Cr(CO)_2(PPh_3)$(η-C_6H_6)].[476]

The relationship between hyperpolarizability and molecular structure has also been studied theoretically; the nonlinear optical responses of a range of [$Cr(CO)_3$(η-arene)] complexes were calculated and found to show good agreement with the experimental values.[477]

8.3.4.2 Spectroscopy

The IR spectra of [$Cr(CO)_3$(η-arene)] complexes usually show two CO absorptions, A_1 and E, corresponding to C_{3v} symmetry. Occasionally, depending on the arene concerned, the E band is split. As might be expected, the frequencies of these bands (and the associated Cotton–Kraihanzel force constants) depend on the electron richness of the arene. An extensive study has shown that this depends on the π-donor or π-acceptor properties of the arene substituents as well as their inductive effects.[462,478,479] By studying a series of related compounds, Taft substituent constants for the phenyl chromium tricarbonyl group can be calculated, and show an increased electron-withdrawing inductive

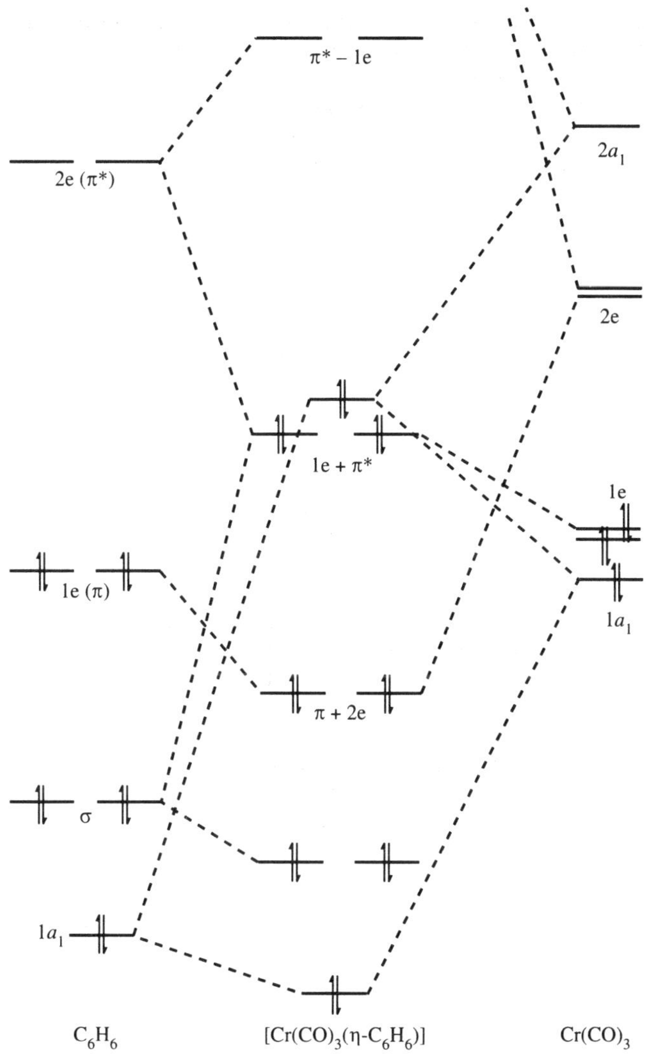

Figure 2 Orbital interaction diagram for $[Cr(CO)_3(\eta\text{-}C_6H_6)]$ in the valence orbital region; the z axis is the C_3 rotation axis passing through the ring centroid (after Muetteries *et al.* 1982[1], Albright and Carpenter 1980[907]).

effect and a decreased resonance effect compared to the uncomplexed arene; substitution of a CO by a phosphine reverses these changes to some extent.[480,481]

Full-range IR spectra and Fourier transform Raman spectra have also been recorded and assigned for $[M(CO)_3(\eta\text{-arene})]$ (arene = C_6H_6, mesitylene, etc.; M = Cr, Mo, W),[482–5] and also for the related thio- and selenocarbonyl species $[Cr(CO)_2(CX)(\eta\text{-}C_6H_6)]$ (X = S, Se).[486] At high pressure the thiocarbonyl complex shows increased backbonding to the CO ligands in the IR spectrum.[487] The photoacoustic spectrum of $[Cr(CO)_3(\eta\text{-}C_6H_6)]$ in the near-IR region has been recorded.[488,489]

The electronic spectra in solution of a wide range of polysubstituted $[Cr(CO)_3(\eta\text{-arene})]$ complexes have been analysed and the correlation between the extinction coefficient of the longest wavelength absorption, which occurs at around 320 nm and is attributed to a MLCT (metal to ligand charge transfer) process, and the sum of the Hammett σ_p^+ parameters extended.[490,491] This band is sensitive to the formation of a charge-transfer adduct with $[Eu(fod)_3]$, changing position by up to 1000 cm^{-1}, and also indicates weak chemisorptive interactions between the arene complexes and silica.[492,493] The gas phase electronic spectra of $[Cr(CO)_3(\eta\text{-PhR})]$ (R = H, Cl, Me) have also been analysed in terms of at least five accessible electronic states, only one of which has an energy that is dependent on arene substitution.[494]

As discussed in *COMC-I*, coordination of an arene to the $Cr(CO)_3$ fragment causes an upfield shift of the ^1H and ^{13}C resonances in the NMR spectra; though there are number of possible causes (quenching of ring current, the electron-withdrawing effect of the $Cr(CO)_3$ group, magnetic anisotropy of the Cr–arene bond, etc.) it is difficult to assess the importance of each one. Analysis of the complexation shifts suggests that electron withdrawal by the metal is through a largely inductive mechanism. A study of the complexes of substituted phenylalkynes, however, produced evidence of some interaction with the alkyne π-system.[495]

The ^1H and ^{13}C NMR spectra of the complexes of polycyclic aromatics can be used to determine the increase or decrease in aromatic or antiaromatic character caused by complexation. For example, the mono- and bis-$Cr(CO)_3$ complexes of biphenylene demonstrate a marked decrease in paratropic character.[496] Complex (**39**) is a derivative of 14-annulene; its resonance energy can be calculated from the anisotropic effect of the $Cr(CO)_3$ unit on the methyl resonances and appears to be larger than that of benzene.[497]

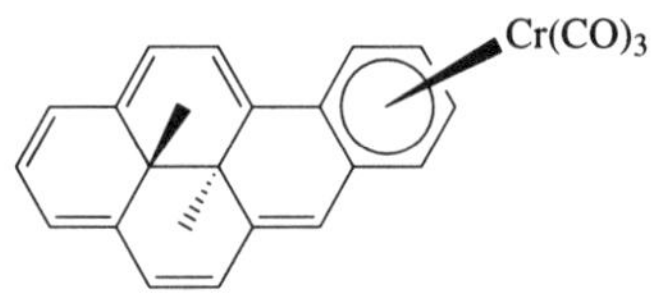

(39)

Spin lattice relaxation times (T_1) for a number of half-sandwich complexes, including $[Cr(CO)_3$-(η-mesitylene)], have been calculated.[498] An intensive ^1H, ^{13}C, ^{17}O and ^{95}Mo NMR study, including T_1 measurements, of $[Mo(CO)_3$(η-arene)] complexes with methylated arenes has appeared; correlations were observed with the number of methyl groups and also with the CO force constants.[499,500] The quadrupole coupling tensor of ^{95}Mo in $[Mo(CO)_3$(η-mesitylene)] has also been examined.[501]

The negative ion EI (electron impact) mass spectra of a selection of $[Cr(CO)_3$(η-arene)] complexes show the molecular ion for certain arenes (e.g., PhCOR, $PhCH_2COR$, PhCl and PhF) but otherwise the highest peak corresponds to $[M-CO]^-$ and the other carbonyl loss fragments.[502–5] Complete fragmentation patterns have been produced for a number of arene complexes.[506–8] Enhanced molecular ions can be observed in the negative chemical ionization mass spectra with CH_4.[509] The positive ion CI (chemical ionization) spectra with NH_3 or isobutane exhibit $[M + NH_4]^+$ and $[M + H]^+$, respectively.[510] The ready loss of $Cr(CO)_3$ units was seen in the fast atom bombardment (FAB) mass spectra of compounds such as $[Cr\{\mu,\eta^6,\eta^6\text{-PhPhCr(CO)}_3\}_2]$ and $[Fe(\eta\text{-}C_5H_4Ph)_n\{\mu,\eta^5,\eta^6\text{-}C_5H_4PhCr(CO)_3\}_{2-n}]$ (n = 0–2), though the molecular ions could be observed.[511]

The volatility of $[Cr(CO)_3$(η-arene)] complexes has allowed their gas phase reactions with a number of molecules to be studied by mass spectrometric techniques. The reaction of $[Cr(CO)_3(\eta\text{-}C_6H_6)]$ with hydride ion in a flowing afterglow apparatus produces $HCr(CO)_3^-$ by a process that involves addition to the arene, as complete isotope scrambling can occur before arene loss.[512] Ion–molecule reactions with propene, C_6H_6 or toluene result mainly in CO displacement by the reagent gas.[513,514] Reactions with ammonia under a variety of conditions lead to proton addition and arene displacement.[515,516] Addition reactions of the benzyne complex $Cr\text{–}C_6H_4^+$, generated from the $Cr(CO)_3$ complex of 1,2-dioxocyclobutabenzene, have also been studied by FT-ICR spectrometry.[517]

Laser multiphoton dissociation/ionization (MPD/MPI) spectroscopy is a technique which examines the distribution of excited states of chromium atoms produced from $[Cr(CO)_3$(η-arene)] complexes; this depends on the vibrational properties of the ligand.[518–20] Similar results were obtained for the analogous molybdenum complexes.[521] Complete dissociation of the arene complex has also been studied by helium plasma emission spectroscopy.[522]

8.3.4.3 Electrochemistry

Compounds of the type $[Cr(CO)_3$(η-arene)] can be oxidized in a one-electron process to the corresponding cations. The ease of oxidation and stability of the cations increases with increasing methyl substitution, and $[Cr(CO)_3(\eta\text{-mesitylene})]^+$ can be observed spectroelectrochemically in a special cell.[523] In general, however, the cations are particularly susceptible to attack by nucleophiles, with stability only attainable in CH_2Cl_2 by the use of low temperatures, with $[NBu_4][PF_6]$ as supporting electrolyte, or in the presence of activated alumina.[524] A recent paper presents data from 63 complexes obtained in this way.[479] Electrochemical reversibility can also be achieved in CF_3CO_2H solvent; the oxidation potentials correlate with those of the free arenes.[525,526] For $[Me_2Sn\{\eta^6\text{-ArCr(CO)}_3\}_2]$ the stability of the cation depends on the steric bulk of the aryl group; the chromium centres are non-interacting.[383,527] Attaching a $CpFe(CO)_2$ unit to the arene ring also results in an increase in cation stability.[399,400] Compounds of the type $[Cr(CO)_2L(\mu\text{-}\eta^6,\eta^6\text{-PhPh})Cr(CO)_2L]$ (L = CO, PPh_3) show two successive one-electron oxidations; the monocation is a localized system, whereas $[Cr(CO)_2(\mu\text{-}\eta^6,\eta^6\text{-}$

$PhPh)(\mu\text{-dppm})Cr(CO)_2]^+$ is delocalized and oxidation is accompanied by a change in the biphenyl torsion angle to 3.7° from 50.8° in the neutral species (**40**).[528,529]

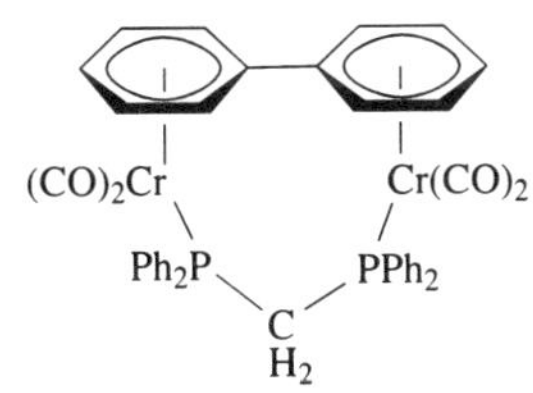

(40)

In nucleophilic solvents (or with even weakly nucleophilic counterions—it should be noted that in much of this work reproducible data can only be obtained in the absence of a supporting electrolyte due to electrode coating) the cations decompose. Thus, in DMF $[Cr(CO)_3(\eta\text{-arene})]$ compounds undergo a net three-electron oxidation to free arene, chromium(III) and CO.[530] In MeCN a two-electron ECE mechanism operates: initial oxidation is followed by an associative nucleophilic attack, decomposition and a second oxidation of the resulting chromium(I) to chromium(II).[531–5] Recent work suggests that the chromium(I) species is $[Cr(CO)_3(NCMe)_3]^+$, detected by voltammetry.[536] In the presence of phosphines, CO substitution in the initial cation can be achieved; however, this does not in most cases form a useful preparative procedure.[531,537,538] The substitution reaction of $[Cr(CO)_2\{P(OMe)_3\}(\eta\text{-}C_6Me_6]$ with TCNE operates by an electron transfer mechanism with substitution occurring in the labile 17-electron cation.[539–41]

The electrochemistry of tungsten tricarbonyl arene complexes differs somewhat from that of their chromium analogues in that complete ligand dissociation does not occur. Initially it was reported that $[W(CO)_3(\eta\text{-}C_6Me_6)]$ underwent only a one-electron oxidation in MeCN followed by coordination of solvent or other available nucleophile to give an observable $[W(CO)_3L(\eta\text{-}C_6Me_6)]^+$ intermediate.[542] Later studies on $[W(CO)_3(1,3,5\text{-}C_6H_3Me_3)]$ using spectroelectrochemistry and digital simulation techniques demonstrated that these 19-electron species undergo spontaneous oxidation to $[W(CO)_3L(\eta\text{-arene})]^{2+}$, resulting in an overall two-electron process.[531,543]

The chemical oxidation of $[Mo(CO)_3(\eta\text{-arene})]$ complexes with iodine has been shown to be more complex than previously assumed; although the cation $[MoI(CO)_3(\eta\text{-arene})]^+$ is always formed, the counterion is variable, and examples of I^-, I_5^-, $[MoI_2(I_3)(CO)_4]^-$, $[Mo_2I_5(CO)_6]^-$ and $[MoI_3(CO)_4]^-$ salts have all been crystallographically characterized.[544,545] The reaction of the $[Mo_2I_5(CO)_6]^-$ salt with CO under pressure gives good yields of $[Mo_2I_4(CO)_8]$, which can disproportionate to $[MoI_3]$ and $[Mo_2I_2(CO)_8]$.[546] Several of these species act as catalysts for the carbonylation of arene tricarbonyl molybdenum complexes to $[Mo(CO)_6]$.[547] The synthesis of $[MoI_3(THF)_3]$ by iodination of $[Mo(CO)_3(\eta\text{-}PhMe)]$ in THF has also been reported.[548] Oxidation of $[Mo(CO)_3(\eta\text{-}C_6Et_6)]$ with the stronger oxidizing agent Cl_2 produces the molybdenum(II)–molybdenum(V) species $[Mo(CO)_3(Cl)(\eta\text{-}C_6Et_6)][MoCl_6]$.[549]

Electrochemical reduction of $[Cr(CO)_3(\eta\text{-arene})]$ complexes usually occurs by an irreversible two-electron process. However, $[Cr(CO)_3(\eta^6\text{-}C_{10}H_8)]$ undergoes a two-electron reduction to a stable dianion by an ECE process in which the chemical step is rearrangement of the naphthalene to an η^4 bonding mode.[550] Subsequently it was discovered that chemical reduction of $[Cr(CO)_3(\eta\text{-}C_6H_6)]$ with potassium naphthalenide or lithium sand produces $[Cr(CO)_3(\eta^4\text{-}C_6H_6)]^{2-}$, which reacts with electrophiles to give anionic cyclohexadienyl species (Scheme 5); H^+ attacks from the *endo* face via the metal, whereas BzCl attacks directly at the ring. Arene exchange is also facilitated in the dianion.[551–4] Reduction of the substituted complex $[Cr(CO)_2(py)(\eta\text{-}C_6H_6)]$ produces the dicarbonyl dianion $[Cr(CO)_2(\eta\text{-}C_6H_6)]^{2-}$, which reacts with electrophiles such as Ph_3SnCl at the metal, forming $[Cr(CO)_2(SnPh_3)_2(\eta\text{-}C_6H_6)]$.[555]

Rieke has shown that reduction of bis-$Cr(CO)_3$ complexes of conjugated arenes such as biphenyl gives stable dianions in which each arene is bonded in a cyclohexadienyl η^5-manner to the chromium. Addition of electrophiles produces substituted biphenyls or cyclohexadienes (Scheme 6).[556–62] Similarly reduction of the bis-$Cr(CO)_3$ complex of 1,4-diphenylbutadiene followed by treatment with electrophiles can be used to make 1,4-disubstituted 1,4-diphenylbut-2-enes.[563,564]

Electrochemical reduction of $Cr(CO)_3$ complexes of ketones (e.g., Ph_2CO, fluorenone) in the presence of electrophiles induces attack on the carbonyl group, producing alcohols, ethers or esters.[565–8] This reduction can also be carried out chemically with potassium in DME, and the ESR spectra of the resulting radicals recorded.[569] Reduction of $[Cr(CO)_3(\eta^6\text{-PhCH=CHPh})]$ causes dimerization of the stilbene ligand.[570]

Low-temperature IR studies reveal that tricarbonyl chromium arene complexes form weak adducts with simple species such as CH_4 and N_2.[571,572] Some Lewis acids (e.g., $AlBr_3$ and $GaCl_3$) form adducts

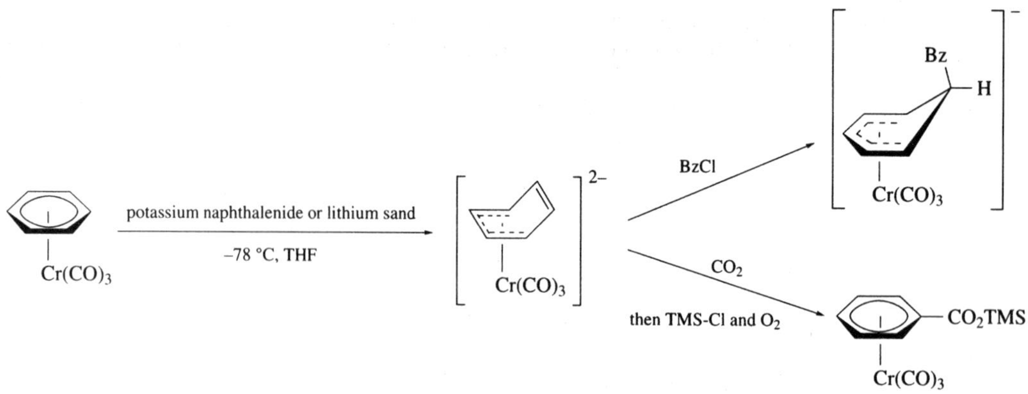

Scheme 5

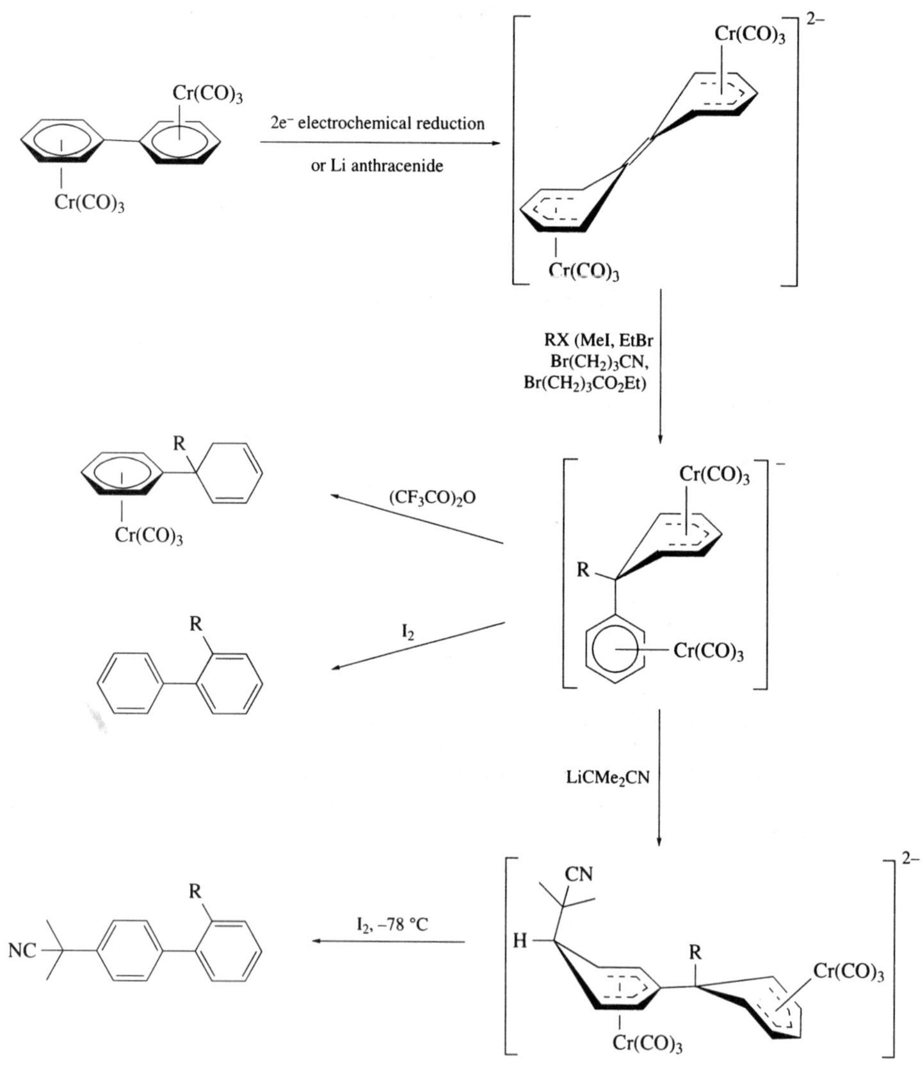

Scheme 6

in which they are attached to the metal, whereas others (e.g., $CaCp^*_2$) are associated with the CO ligands.[573–6] Charge transfer complexes with diaryl alkynes, in which the arene complex acts as electron acceptor, also exist, with the strength of interaction reflected in the activity in alkyne polymerization.[577,578] There is evidence from NMR studies that tricarbonyl chromium arene complexes can form charge-transfer adducts with other arenes such as C_6D_6.[579,580]

Tricarbonyl chromium arene complexes undergo selective 1:1 inclusion into β- or γ-cyclodextrins depending on their size; the former will accept complexes of C_6H_6, PhMe and *o*-xylene whereas bulkier arenes complex better with the latter.[581–3] Host–guest inclusion compounds of chromium arene complexes with thiourea and tris-*o*-thymotide have also been made in a search for nonlinear optical materials.[584] Encapsulation of [$M(CO)_3$(η-arene)] into zeolites has also been achieved; alternatively encapsulated [$M(CO)_6$] can be treated with arenes *in situ* to give the same product, as shown by diffuse reflectance FTIR studies.[585,586] In some zeolites, and also on alumina, loss of the arene occurs to give grafted $M(CO)_3$ units.[587]

8.3.4.4 Thermochemistry

Thermochemical data and EHMO (extended Hückel molecular orbital) calculations have been used to compare the Cr–CO and Cr–C_6H_6 bond enthalpies in [$Cr(CO)_6$], [$Cr(CO)_3$(η-C_6H_6)] and [Cr(η-$C_6H_6)_2$]; both quantities increase in the order given, in keeping with the bond length variations determined crystallographically.[588] Further microcalorimetry measurements on the sublimation, decomposition and halogenation of $Cr(CO)_3$ complexes have been made; they confirm that the order of bond strengths is mesitylene > cycloheptatriene > PhMe > C_6H_6.[589,590] A thermochemical study of the reactions of [$Mo(CO)_3$(η-PhMe)] with a range of phosphines concluded that the stability of the *fac*-[$Mo(CO)_3(PR_3)_3$] products depend on both steric and electronic factors of the phosphine ligands.[591] Such substitution reactions get more thermodyamically favourable from chromium to molybdenum to tungsten.[592]

8.3.5 Fluxionality of Arene Metal Tricarbonyl Complexes

The reorientational motion of the arene ring in solid [$Cr(CO)_3$(η-C_6H_6)] has been investigated by a number of techniques. Quasielastic neutron scattering (QENS) studies indicate that this occurs in jumps with an activation energy of 16 kJ mol^{-1};[484,593] this can be compared with values of 27.5 kJ mol^{-1} from inelastic neutron scattering, 19.5 kJ mol^{-1} from Raman spectra, 17.6 kJ mol^{-1} from NMR measurements and 19.2 kJ mol^{-1} from potential energy calculations.[594–6] An absorption attributed to arene rotation was also observed in the mechanical spectrum.[597] The molybdenum analogue has a very similar structure with a slightly higher potential energy barrier of 24.7 kJ mol^{-1} to reorientational motion.[598]

A variable-temperature CP-MAS (cross polarization magic angle spinning) ^{13}C NMR study of [$M(CO)_3$(η-PhMe)] suggested that it is the three CO ligands that rotate while the arene remains static; rotational barriers of 65 kJ mol^{-1} and 71 kJ mol^{-1} for chromium and molybdenum respectively were calculated.[599] However, potential energy calculations suggest that this is prevented by crystal packing, and that large amplitude librational motions of the arene and the tripod are actually what occur.[596] It is interesting to note that the tripod is eclipsed in the chromium complex but staggered in the molybdenum.[600] The solid-state dynamic behaviour of the $Cr(CO)_3$ complexes of C_6Me_6, mesitylene and durene has recently been studied by variable temperature spin lattice relaxation time measurements and potential energy calculations. For C_6Me_6 the lowest energy process is rotation of the methyl groups, with reorientation of the arene occurring at higher energy, while for the other two arenes reorientation is forbidden by crystal packing.[601,602]

The conformation of the $Cr(CO)_3$ tripod in the solid state is controlled electronically and depends on the arene substituents. Unsubstituted arene complexes adopt the staggered conformation (**41a**), but monosubstituted ones are almost always eclipsed. If the substituent is electron donating, the *syn*-eclipsed isomer (**41b**) is favoured, whereas if it is electron-withdrawing the *anti*-eclipsed isomer (**41c**) is preferred. This can be extended to compounds such as [$M(CO)_3$(η-mesitylene)] (M = Cr, W) and even [$Mo(CO)_3$(η-C_6H_3-1,3,5-$Bu^t{}_3$)] in which all the substituents are eclipsed.[603–5] Extended Hückel calculations on the preferred conformations in the gas phase produce largely comparable results, showing that crystal packing effects are not usually important.[606]

(41a)

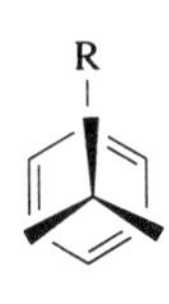

(41b)

(41c)

The conformation of the $Cr(CO)_3$ tripod in solution has been extensively studied, since it exerts a strong influence on the site of nucleophilic or electrophilic attack. In solution the conformers interconvert readily, but the preferred conformation is usually the same as in the solid state. In favourable cases the population of the conformers can be deduced from the ^1H (or ^{13}C) NMR spectrum; however, the calculation is not always straightforward.[607] It is well known that coordination of an arene to the $Cr(CO)_3$ fragment results in shielding of the arene protons which causes them to resonate at higher field. However, protons that lie beneath eclipsing CO ligands are deshielded relative to those that are staggered. Unfortunately, the difference in the chemical shifts of eclipsed and staggered protons depends not only on the conformer populations but also on the different substituent effects at the *ortho* and *meta* positions. Studies by Solladié-Cavallo have led to the conclusion that substituent effects for alkyl and OMe substituents are negligible, allowing the calculation that the deshielding effect experienced by an eclipsed proton is between 0.84 ppm and 0.88 ppm, depending on the solvent. Using this value in conjunction with estimated substituent effects it is possible to calculate the proportion of the *syn*-eclipsed conformer for various [$Cr(CO)_3$(η-PhR)] species (Table 8).[193]

Table 8 Proportion of the *syn*-eclipsed conformer in [$Cr(CO)_3$(η-PhR)].

R	CHO	CO_2Me	Me	Pr	Bu	Bu^t	OMe	NH_2
%	30	37	64	61	62	33	80	90

An elegant use of this principle was the use of 500 MHz two-dimensional NMR to distinguish the α and β isomers of the complex of methyl *O*-methylpodocarpate, which adopt different conformations due to steric interference by the proximal methyl group in the α-isomer.[300]

Similar calculations can be carried out for disubstituted arenes, which if possible adopt conformations that satisfy the preferences of both substituents (e.g., in *m*-xylene it is possible to obtain a *syn*-eclipsed arrangement with respect to both methyls); if there is a conflict of interests between substituents there may be no conformational preference. Steric effects can also be important.[193]

A more recent attempt to deduce the solution conformation of substituted arene complexes by NMR uses the observation of ^{13}C–^1H or ^{13}C–^{19}F coupling between the CO ligands and the ring substituents. Although these couplings are small (about 0.6 Hz and 2–5 Hz, respectively) they can in certain cases be correlated with the electron donating properties of the arene and the tripod conformation.[608–11]

The fluxionality of chromium tricarbonyl arene complexes in solution has been the subject of much attention and some controversy, as a recent review points out.[612] Several types of fluxionality have been observed, including hindered rotation of substituents on the arene ring, of phosphine ligands on the metal atom, and of the $Cr(CO)_3$ tripod in relation to the arene.

In dealing with the rotation of the chromium tripod in [$Cr(CO)_3$(η-arene)], it is important to realize that temperature-dependent NMR spectra can be obtained by kinetic or thermodynamic restriction of rotation. The former case arises when a large energy barrier exists for the interconversion of two equivalent conformers, for example 120° rotation of an eclipsed-*anti* conformer to its symmetry-related isomer. Thermodynamic restriction involves an equilibrium between two different conformers such as eclipsed-*syn* and eclipsed-*anti*, when the equilibrium populations of the two isomers are not equal, and thus change with temperature.

For most arenes, rotation of the $Cr(CO)_3$ tripod in solution has a very low activation energy and can be regarded as essentially unhindered. In order to slow the rotation of the $Cr(CO)_3$ unit by steric means it is therefore necessary to use extremely bulky arenes. Many complexes, even those of sterically demanding arenes such as 1,3,5-$C_6H_3Ph_3$, show no sign of slowed tripod rotation, even at very low temperatures.[613]

In the complex [$Cr(CO)_3$(η-C_6Ph_6)] the chromium is bonded to one of the outer rings, which is orthogonal to the central ring and exhibits slowed rotation with $\Delta G^{\ddagger}$ of 51 kJ mol^{-1} relative to the other rings; this, however, is less than the barrier in free hexa-*o*-tolyl benzene, for instance.[614] Similarly the barriers to rotation in a number of bis-2,6-diarylmethanes, -ethanes and ketones and their $Cr(CO)_3$ complexes were found to show little difference between free and complexed arenes in most cases.[615] The fluxional process in [$Cr(CO)_2(PPh_3)$(η-C_6H_3OMe-2,6-Bu^t_2)] is assigned as restricted rotation about the OMe bond, especially as a similar process occurs in the complex of C_6H_2OMe-2,6-Bu^t_2-4-CH_2OPPh_2, which acts as a σ,π-chelate thus ruling out arene rotation.[616] In some cases the barrier to alkyl rotation may be less in the $Cr(CO)_3$ complex than in the free arene, for example for C_6Me_4-1,2-$(CH_2Bu^t)_2$; this is due to an increase in the ground state energy caused by steric interaction between the alkyl groups and the CO ligands.[617] Barriers to amide rotation in complexes of the type [$Cr(CO)_3$(η-$PhCONR_2$)] have also been measured.[618]

Complexes of biphenyls, diarylmethanes and diarylketones which have bulky groups in the 2 and/or 6 positions of each ring often display restricted rotation both in the free state and in their mono- and bis-$Cr(CO)_3$ complexes. In some cases, for example the mono-$Cr(CO)_3$ complex of $(2,4\text{-}Me_2C_6H_2\text{-}6\text{-}Bu^t)_2CO$, stable torsional isomers exist due to this phenomenon.[615,619] Attachment of the metal fragments, however, seems to make little difference overall to the dynamics of the ligand. The crystal structures of the mono- and bis-$Cr(CO)_3$ complexes of benzophenone itself have been determined, and show torsion angles of 58° and 56° respectively between the two rings.[620] The chiral complexes of bridged biphenyls such as 9,10-dihydrophenanthrene and diphenic acid anhydride display a similar phenomenon; the $Cr(CO)_3$ unit can be attached to either the *exo* or *endo* face, and the inversion barrier for the twisting of the two aryl rings can be studied.[621–5]

Much attention has, however, focused on complexes of hexaethylbenzene and related arenes. In $[Cr(CO)_3(\eta\text{-}C_6Et_6)]$ (**42**) the arene ligand adopts a conformation in the solid state in which alternate methyls of the ethyl are pointing towards (proximal) or away from (distal) the chromium atom, with the CO ligands eclipsed beneath the distal methyls (i.e., C_{3v} symmetry). In solution at room temperature the ethyl groups rotate ($\Delta G^{\ddagger} = 48$ kJ mol^{-1}), which renders all six equivalent.[626] The related $[Cr(CO)_3(\eta\text{-}C_6Pr_6)]$ also has an alternating arrangement, with $\Delta G^{\ddagger}$ of 49.6 kJ mol^{-1} for propyl rotation, in good agreement with a theoretical prediction of 57.7 kJ mol^{-1}.[627] Protonation of $[Cr(CO)_3(\eta\text{-}C_6Et_6)]$ with strong acids in liquid SO_2 occurred to give a cation that displayed threefold symmetry, and was fluxional even at −100 °C.[628] An NMR study of $[Cr(CO)_3(B_3N_3Et_6)]$ showed that two conformations of the borazine ligand were present, with one and two proximal methyls respectively.[629]

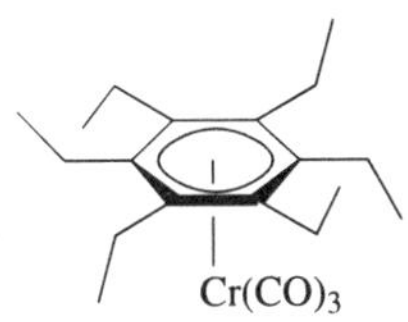

(42)

In order to prove that tripod rotation is slowed at low temperature it is necessary to break the threefold symmetry by substitution of a CO for another ligand, or use an arene of the type C_6Et_5X. Both approaches have been useful. McGlinchey *et al.* thus prepared $[Cr(CO)_2(CS)(\eta\text{-}C_6Et_6)]$, which has a very similar solid-state structure with the CO and CS ligands again lying beneath the distal ethyls (C_s symmetry). In solution at low temperature the ^{13}C NMR spectrum shows a 1:2:2:1 pattern for each of the three types of arene carbon. This would be evidence of slowed rotation, assuming that the conformation of the arene remains the same in solution as in the solid state.[630] In support of this, the solid state ^{13}C NMR spectrum is indeed very similar to that in solution.[631] However, as pointed out by Hunter and Mislow, if the arene conformation does change, the spectra could still be explicable without invoking slowed rotation.[632] Indeed, replacing one CO by a series of phosphines gives a number of crystallographically characterized $[Cr(CO)_2L(\eta\text{-}C_6Et_6)]$ complexes with different arene conformations. With L = PPh_3 an all-distal isomer is observed, for PEt_3 a 1:1 mixture of this and an isomer with one proximal and five distal ethyls was found, and with PMe_3 a 1,3-proximal 2,4,5,6-distal arrangement was present.[626,633,634] Alternatively,in $[\{Cr(CO)_2(\eta\text{-}C_6Et_6)\}_2(\mu\text{-}N_2)]$ the ethyls again alternate.[635] Lineshape analysis of the low temperature ^{13}C NMR spectra of these and related complexes showed the presence of several different arene conformations, and again could be explained without slowed tripodal rotation.[636] In $[Cr(CO)_2(PPh_3)(\eta\text{-}C_6R_6)]$, restricted rotation about the Cr–P bond was observed for R = Me, Et and Pr with $\Delta G^{\ddagger}$ of around 38 kJ mol^{-1}.[637] Recently, slowed tripod rotation has been observed in $[Cr(CO)_2(PPh_3)(\eta\text{-}p\text{-}C_6H_4Bu^t_2)]$ and a number of related complexes with $P(o\text{-}Tol)_3$ ligands.[638]

A number of related molybdenum complexes have also been studied. The structure of $[Mo(CO)_3(\eta\text{-}C_6Et_6)]$ has the alternating proximal–distal arrangement,[626] but $[Mo(CO)_3(Cl)(\eta\text{-}C_6Et_6)][MoCl_6]$ has a unique 1,4-proximal-2,3,5,6-distal conformation.[549] More recently it has been demonstrated that $[Mo(CO)_2(PPh_3)\{C_6\text{-}1,3,5\text{-}Et_3\text{-}2,4,6\text{-}(CH_2TMS)_3\}]$ has a conformation in which two of the ethyl groups are proximal, and the other, as well as the CH_2TMS groups, are distal; it exhibits three fluxional processes (ethyl rotation, phosphine rotation, and tripod rotation), all of which can be slowed at low temperature.[639] The corresponding complex of $C_6\text{-}1,3,5\text{-}Et_3\text{-}2,4,6\text{-}(CH_2Bu^t)_3$ has also been studied, and two structures of $[Mo(CO)_2(\eta^2\text{-cyclooctene})\{\eta\text{-}C_6\text{-}1,3,5\text{-}Et_3\text{-}2,4,6\text{-}(CH_2TMS)_3\}]$ have been described, which differ only in the orientation of the alkene ligand.[640,641] Restricted rotation about the Mo–P bond has also been detected in the more highly methylated complexes of the type $[Mo(CO)_2(PPh_3)(C_6H_nMe_{6-n})]$; if two adjacent positions on the arene are unsubstituted rotation remains rapid.[642]

The cationic derivatives $[Cr(CO)_2(NO)(\eta\text{-}C_6Et_6)][BF_4]$ and $[Cr(CO)(NO)(CS)(\eta\text{-}C_6Et_6)][BF_4]$ both show an alternating arrangement of ethyls in the solid state. The former shows a similar solution behaviour to the thiocarbonyl complex $[Cr(CO)_2(CS)(\eta\text{-}C_6Et_6)]$. In the latter, however, all six ethyls are rendered inequivalent at low temperature, with a chemical shift pattern that still corresponds to a 1,3,5-proximal, 2,4,6-distal conformation. This is best explained by two fluxional processes: tripod rotation with $\Delta G^{\ddagger}$ of 39.7 kJ mol^{-1}, and ethyl rotation with $\Delta G^{\ddagger}$ of 48 kJ mol^{-1}.[643] There appears to be no evidence that the two motions are meshed in a cogwheel arrangement.

In an attempt to model the C_6Et_6 situation, the $Cr(CO)_3$ complexes of the trimer of bicyclo[2.2.1]hept-2-yne were prepared for both the all-*syn* and *syn-anti-syn* isomer (**43, 44**). The all-*syn* isomer, in which the methylene bridges are proximal to the chromium atom, shows no evidence of slowed rotation, whereas the *syn-anti-syn* isomer, in which one of the two-carbon bridges is distal, exhibits slowed rotation at low temperature with $\Delta G^{\ddagger}$ of 27 kJ mol^{-1}.[644,645]

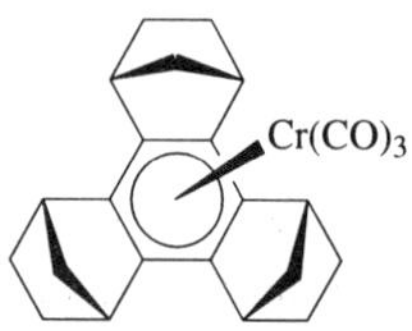

(43)

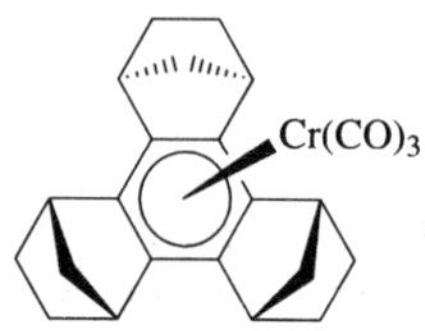

(44)

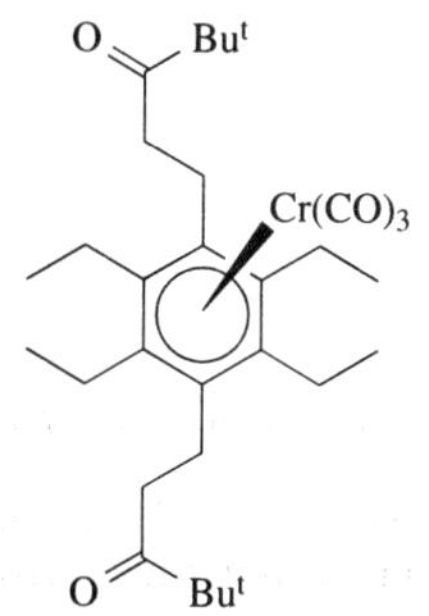

(45)

As mentioned above, the second way to break the C_{3v} symmetry in $[Cr(CO)_3(\eta\text{-}C_6Et_6)]$ is to use an arene ligand such as C_6Et_5COMe, in which case the highest possible symmetry of the complex is C_s. In $[Cr(CO)_3(\eta\text{-}C_6Et_5COMe)]$, the acetyl group and the ethyls in the 3 and 5 positions are proximal, while those in the 2, 4 and 6 positions are distal; the carbonyl ligands are again located beneath the distal ethyls. In the solid state and at low temperature in solution the CO ligands appear as two signals in a 2:1 ratio in the ^{13}C NMR spectrum, thus proving slowed tripod rotation ($\Delta G^{\ddagger}$ = about 38 kJ mol^{-1}) unequivocally.[646] It is possible to assign the proximal and distal resonances in the spectrum and then apply the results to the analogous C_6Et_6 species.[643]

More recently Siegel has prepared the $Cr(CO)_3$ complex (**45**) from 1,4-bis(4,4-dimethyl-3-oxopentyl)-2,3,5,6-tetraethyl benzene. This ingeniously designed ligand contains separate probes for alkyl rotation and tripod rotation. Cooling a solution to −6 °C slows the alkyl rotation, while at −90 °C the tripod rotation also stops and a 2:1 pattern of CO ligands is observed. The activation energies for the two processes are 48 kJ mol^{-1} and 40 kJ mol^{-1}, respectively.[647] The $Mo(CO)_3$ complex of the same ligand behaves similarly, ethyl rotation stopping at −80 °C and tripod rotation at −140 °C, with activation energies of 48 kJ mol^{-1} and 28 kJ mol^{-1}.[648] In a more extensive study of similar ligands with OMe, CH_2OMe and neohexyl groups in the 1 and 4 positions, it was found that the barriers to alkyl rotation do not change much on complexation; this was attributed to a cogwheel-like complementarity of rotation between the ethyl groups and the $Cr(CO)_3$ tripod.[649]

The barrier to tripod rotation in $Cr(CO)_3$ complexes can be electronic rather than steric. In accord with theoretical calculations that predict slowed rotation if the double bonds in the arene can be localized, the complex of the bent terphenylene illustrated as (**46**) has a rotational barrier of 39 ± 2 kJ mol^{-1}, and is assigned a cyclohexatriene type structure.[650] The complex of the ligand (**47**), termed starphenylene, also shows this phenomenon with an activation energy of 46.4 ± 2.1 kJ mol^{-1}. When R = TMS it also exhibits an irreversible migration of the $Cr(CO)_3$ unit from the inner to the outer ring on heating.[651] Extended Hückel calculations on cyclobutabenzene complexes show that the difference in bond populations between long and short bonds in the arene ring correlates well with the rotational barrier and provides a measure of the ability of the arene to bind to the metal during rotation.[652]

The second example of an electronic barrier to rotation arises in benzylic carbonium cations of the type shown in (**48**). At low temperatures the CO peaks of the tripod occur as a 2:1 pattern in the ^{13}C NMR spectrum, indicative of slowed rotation, due again to the degree of charge localization in the arene. Calculations show that if R^1 = OMe, the charge localization should be intensified whereas if R^2 = OMe it should be lessened, and this indeed proves to be the case experimentally.[653]

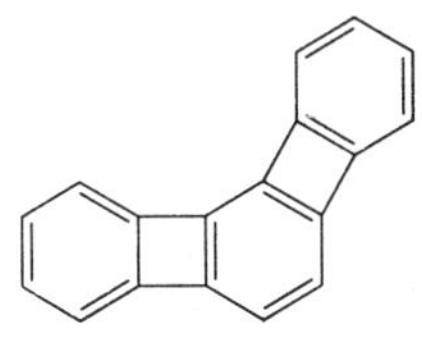

(46)

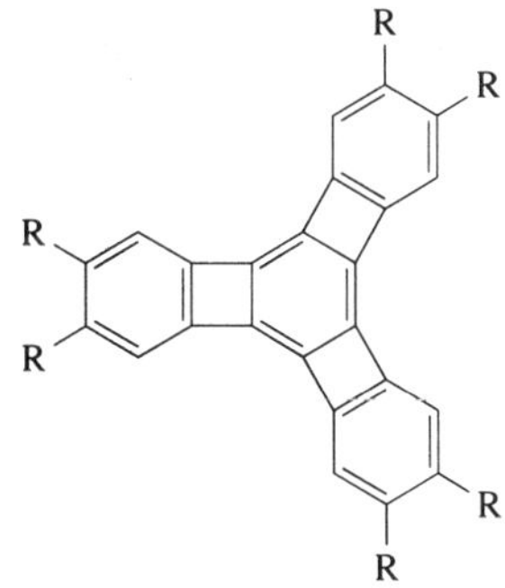

(47)

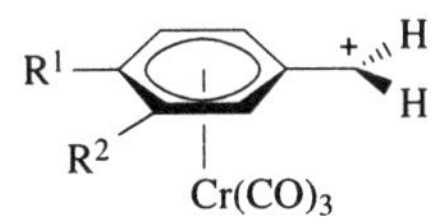

(48)

8.3.6 Haptotropic Rearrangements of Arene Metal Tricarbonyl Complexes

Complexes which contain a $Cr(CO)_3$ fragment bonded to fused aromatic rings sometimes undergo migration of the metal from one ring to the other, termed a haptotropic shift. The most carefully investigated systems are anionic complexes, exemplified by those derived from indene and fluorene, which undergo migration at or below room temperature, and neutral naphthalene complexes, which require heating to induce migration.

The reaction of indene with $[Cr(CO)_3L_3]$ ($L = NH_3$, MeCN) produces $[Cr(CO)_3(\eta^6\text{-}C_9H_8)]$ in which the chromium is bonded to the arene ring. The analogous molybdenum complex can be prepared from $[Mo(CO)_3(py)_3]$, but the tungsten species undergoes oxidative addition to give $[WH(CO)_3(\eta^5\text{-}C_9H_7)]$.[234,654,655]

Deprotonation of the η^6-indene complex $[Cr(CO)_3(\eta^6\text{-}C_9H_8)]$ with KH at –30 °C produces the stable η^6 anion (Scheme 7). On warming, however, the chromium migrates to the five-membered ring by an intramolecular pathway involving an η^3 intermediate.[656] Alkylation with MeI or BzBr initially occurs at the metal to afford $[Cr(R)(CO)_3(\eta^5\text{-}C_9H_7)]$, but this rearranges to $[Cr(CO)_3(\eta^6\text{-}C_9H_7\text{-}1\text{-}R)]$ in which the R group has been stereospecifically transferred to the *endo* position and the $Cr(CO)_3$ unit has migrated back to the arene ring.[657] Initial formation of the η^6-product was also observed with the bis(indenyl)ethane species $[Cr(CO)_3(\eta^5,\eta^5\text{-}C_9H_6CH_2CH_2C_9H_6)Cr(CO)_3]^{2-}$.[658]

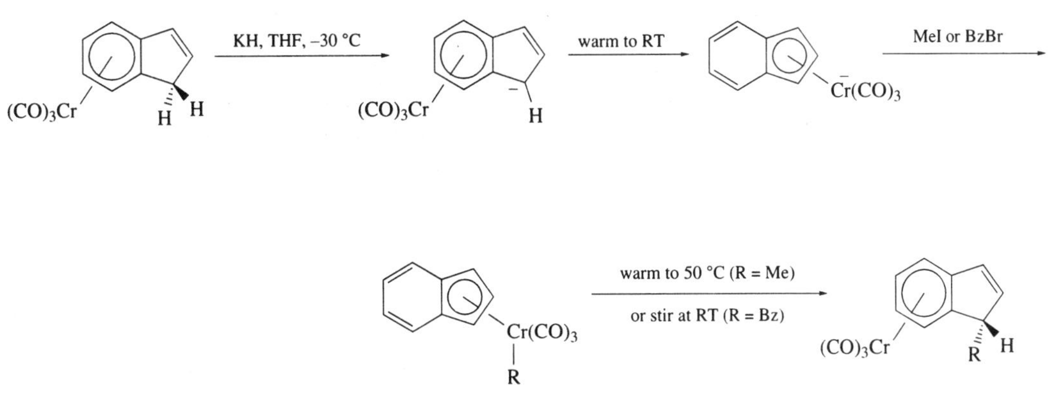

Scheme 7

There is evidence that in the η^6 anion, ion pairing occurs in which the counterion is situated above the five-membered ring; this can be disrupted by steric hindrance in the complexes derived from 3-phenylindene and 1,3-diphenylindene.[656] In passing it should be noted that the rearrangement of 1-phenylindene to 3-phenylindene occurs at room temperature in the presence of $[Cr(CO)_3(NCMe)_3]$, and attempts to coordinate the 1-isomer led only to complexes of the 3-isomer.[659]

Interestingly both the η^6 and η^5 isomers of $[Cr(CO)_3(\eta\text{-}C_9H_7)]^-$ react with $[RhCl(cod)]_2$ to produce $[Cr(CO)_3(\mu,\eta^6,\eta^3\text{-}C_9H_7)Rh(cod)]$, which showed an increased efficiency in catalysis of alkyne cyclotrimerization, and faster displacement of the cod ligands by CO, attributed to a possible η^4-Cr, η^5-Rh isomer; in fact the x-ray structure shows the rhodium to be within bonding range of all five carbons, though the rhodium–indenyl bond is somewhat weakened by coordination of the chromium.[660–2] In these compounds the rhodium and chromium are bonded to opposite faces of the indenyl ligand, but if $[RhCl(nbd)]_2$ is used as the rhodium source, the product is $[Cr(CO)_3(\mu,\eta^6,\eta^3\text{-}C_9H_7)Rh(nbd)]$, in which both metals are on the same face; the indenyl ligand is somewhat bent due to the steric interactions between them.[663]

A similar $\eta^6 \rightleftharpoons \eta^5$ haptotropic rearrangement occurs on deprotonation (or reduction) of the fluorene complex $[Cr(CO)_3(\eta^6\text{-}C_{11}H_{10})]$ (Scheme 8).[664–7] The rate of deprotonation is approximately 10 times that of free fluorene, demonstrating the increased kinetic acidity of the benzyl protons on coordination. Crossover experiments confirmed that dissociation of the chromium fragment was not occurring.[668] Alkylation was shown to occur from the *exo* face for the η^6 anion, but from the *endo* face (via the metal) for the η^5; moreover deprotonation of the *exo* isomer, followed by reprotonation, gave the *endo*-substituted species, that is, H^+ attacks from the *exo* face.[669] Again ion pairing effects are observable, with the cation situated above the five-membered ring in the η^6 anion and associated with a CO in the η^5 anion.[664,670] Exposure of the η^6 anion to air causes oxidation of the ligand to give the $Cr(CO)_3$ complex of fluorenone.[671]

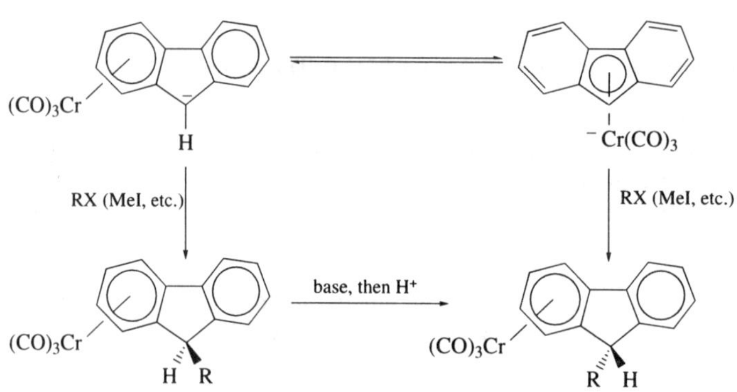

Scheme 8

As might be expected, substituted complexes $[Cr(CO)_2(L)(\eta^6\text{-}C_{11}H_{10})]$ ($L = PPh_3$, PBu_3) show a stabilization of the η^6 anion as the electron-accepting capability of the chromium atom decreases.[672] The bis-$Cr(CO)_3$ complex of fluorene does not undergo a haptotropic shift on deprotonation, but NMR analysis suggests the chromium atoms both move slightly towards the five-membered ring.[673] Extending the ligand π-system also disfavours the rearrangement; thus, the indenyl complex rearranges rapidly below room temperature, the fluorenyl complex reaches an approximately 1:1 equilibrium of η^6 and η^5 anions, and the related fluoradenyl complex (**49**) produces a stable η^6 anion that does not rearrange at all.[670]

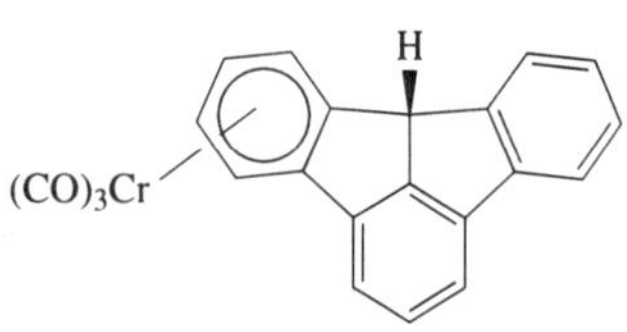

(49)

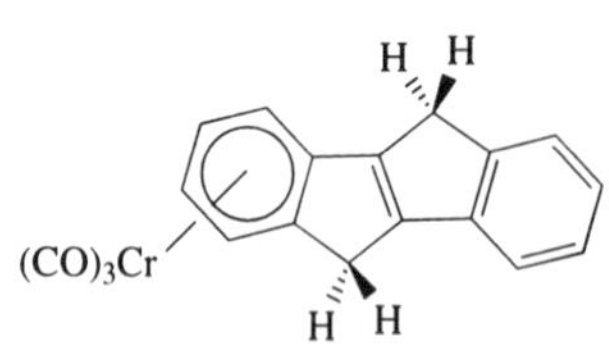

(50)

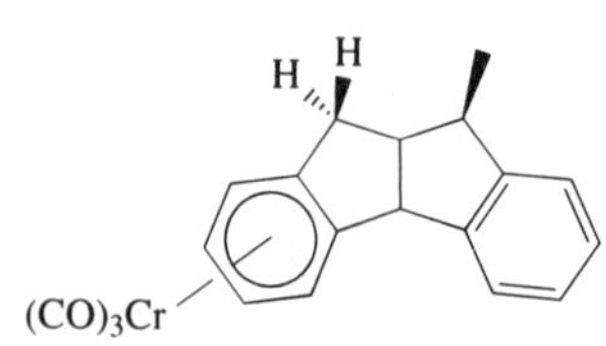

(51)

A range of structurally related complexes also undergo haptotropic rearrangements. Complex (**50**) rearranges irreversibly to the η^5 complex on deprotonation; with excess base the η^5-dianion is formed.[674–7] Complex (**51**) undergoes not only a $\eta^6 \rightarrow \eta^5$ shift, but also a $\eta^5 \rightarrow \eta^5$ one subsequently.[675] Migration from an arene ring to a seven-membered one is observed in the benzocycloheptatrienyl complex (**52**).[678] The two rings do not have to be fused; thus, deprotonation of (**53**) (made by nucleophilic attack of $C_5H_5^-$ on $[Cr(CO)_3(\eta\text{-}PhF)]$) still resulted in chromium migration across the σ-bond to give the η^5-anion.[679,680] This also occurred for the open-chain 3-phenylpentadienyl species (**54**), and for tetraphenylcyclopentadiene, which coordinates to the $Cr(CO)_3$ fragment as an η^6 complex and rearranges to the η^5 anion on deprotonation.[681]

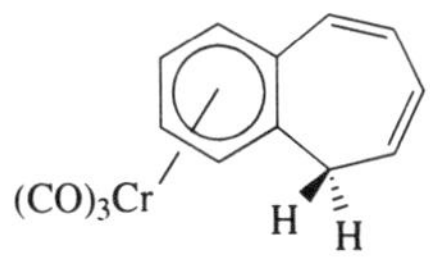

(52)

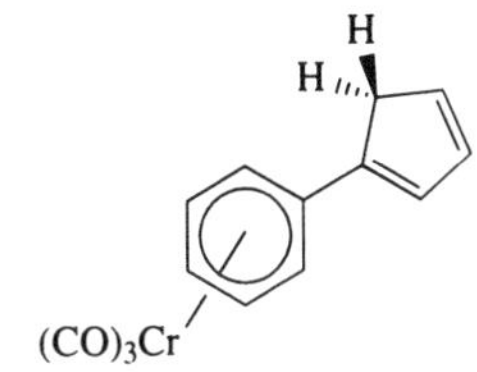

(53)

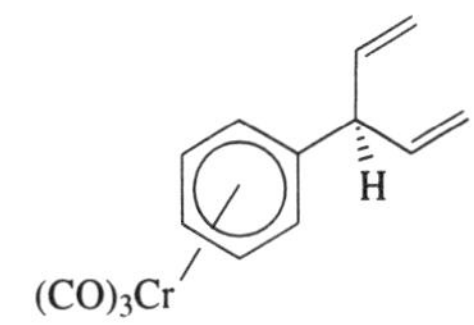

(54)

In an elegant experiment, substituents were introduced into the naphthalene complex $[Cr(CO)_3(\eta\text{-}C_{10}H_8)]$ by lithiation, which occurs only in the coordinated ring, followed by quenching with electrophiles, for example D^+. On heating, redistribution of the chromium unit between the labelled and unlabelled rings (**55, 56**) occurs by a sliding process.[682] A similar result was obtained with other substituents, introduced by lithium-halide exchange of coordinated 1- or 2-bromonaphthalene.[683] The two isomers of the $Cr(CO)_3$ complex of 2-methyl naphthalene could be separated by gas chromatography, and the rearrangement of each one studied separately; the activation energies for both isomers was the same within experimental error.[460] A surprising rate difference was observed for 2,6 and 2,7-dimethoxynaphthalene, attributed to the differing effect on the transition state of the substituents; again no exchange with the C_6D_6 solvent was observed, implying an intramolecular process rather than complete arene dissociation.[684]

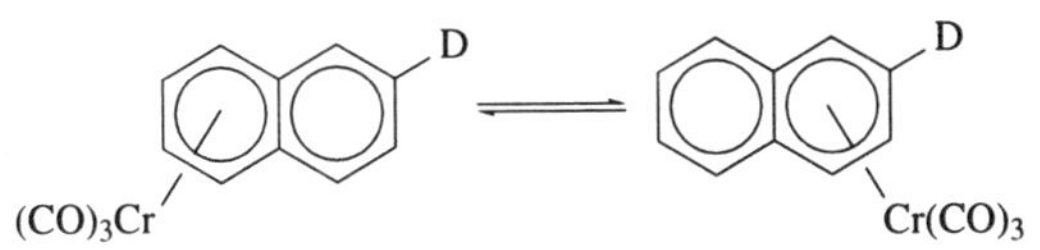

(55) (56)

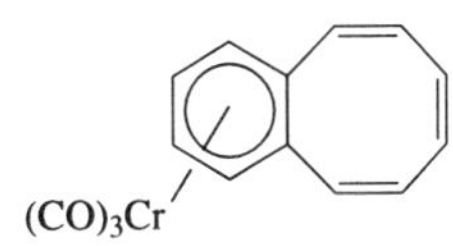

(57)

A theoretical analysis of these haptotropic processes has provided a simple electron-counting prediction of the magnitude of the activation energy for migration compared to dissociative intermolecular transfer. It also demonstrates that for the $[Cr(CO)_3(\eta\text{-}C_{10}H_8)]$ complex, a least-motion pathway in which the $Cr(CO)_3$ crosses the central naphthalene bond is unfavourable compared to a more circuitous route around the periphery. Alternatively, formation of an η^2-complex (and thence to arene dissociation in the presence of coordinating solvents) is also a low energy process, explaining the facile arene exchange in this complex.[321]

Again a range of compounds undergo related migrations. Two isomers of the $Cr(CO)_3$ complex of benzocyclooctatetraene (**57**), in which the chromium is bonded to either the six- or eight-membered ring, have been characterized; on heating the latter converts into the former.[685] The stereochemistry of the 1-methylacenaphthene complexes (**58**) is preserved on migration of the chromium, again implying a sliding process rather than complete dissociation.[686,687] The thermally induced migration of the $Cr(CO)_3$ unit between the rings of the silicon heterocycles (**59**) has also been shown to be a first-order intramolecular process.[374,376,688] Migration of chromium across the σ-bond of biphenyls also occurs in some cases; combined with preferential coordination of the electron-rich ring of substituted biphenyls to $Cr(CO)_3$, this provides a way of making a range of derivatives with substituents on the coordinated or uncoordinated rings.[689,690] In the $Cr(CO)_3$ complex of 9-phenylanthracene (**60**), interconversion occurs between a yellow form in which the metal is bonded to the pendant ring and a purple one which is bonded through a fused ring.[691]

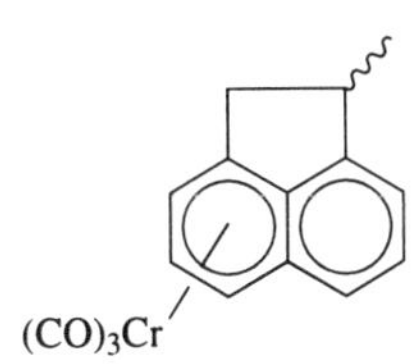

(58)

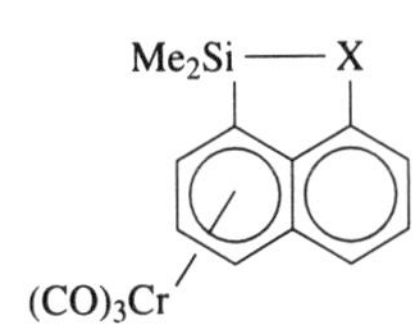

X = CH_2, O
(59)

While the deprotonation of complexes of indene or fluorene produces an aromatic system, the case of 5H-dibenzo[a,d]cycloheptene also deserves mention. Deprotonation of its $Cr(CO)_3$ complex (**61**) is some 800 times faster than the free arene, compared to a rate enhancement of around 88 for the analogous saturated dibenzocycloheptane. This is interpreted as a quenching of the antiaromaticity by

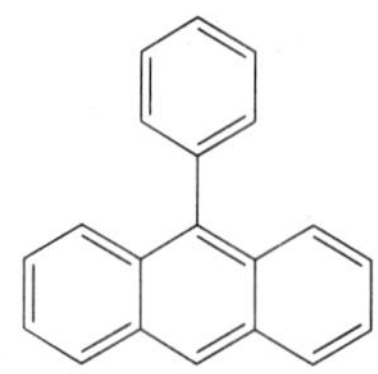

(60)

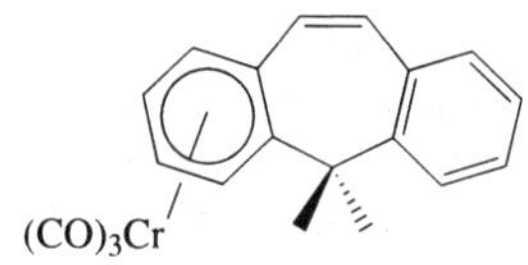

(61)

coordination to the metal in the former case.[667,692] Hydride abstraction from the same complex with Ph_3C^+ affords the corresponding cation, in which the aromaticity is also quenched.[693]

8.3.7 Chirality in Arene Metal Tricarbonyl Complexes

Chromium tricarbonyl arene complexes with different substituents in the *o*- or *m*-positions possess planar chirality; if the arene is also chiral, diastereomers result. In favourable cases, notably benzyl alcohols, diastereoselective complexation of the $Cr(CO)_3$ unit occurs to one face of the arene (Equation (4)), due to chelation control through a preliminary interaction between the metal and the oxygen.[694,695] Better selectivity can be achieved by complexation under kinetic conditions with $[Cr(CO)_3(\eta^6\text{-}C_{10}H_8)]$ than $[Cr(CO)_6]$, where extended refluxing gives an equilibrium mixture of diastereomers.

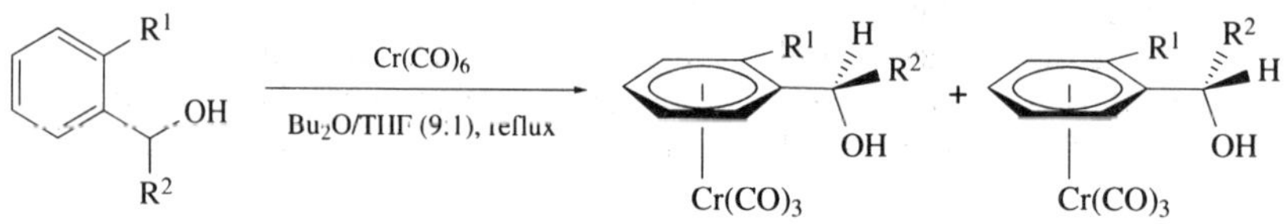

(4)

(*R*)-isomer (major) (*S*)-isomer (minor)

R^1 = Me, OMe, OPr^i
R^2 = Me, Pr^i

Directed complimentary diastereoselectivity of *m*-substituted benzyl alcohols can be achieved by temporary introduction of a TMS group in either the 2- or 6-position (Scheme 9).[696] The 2-substituted complexes sometimes also give the cyclic siloxanes (**62**), which can be reconverted to the desired products with NBu_4F. In the case of the tetralol derivatives (**63**), *syn*:*anti* ratios of >200:1 and 10:1 are obtained for R = H and Me respectively; however, steric effects are dominant in R = TBDMS.[697] Oxidation with acetic anhydride in DMSO gives optically pure tetralone complexes.[698]

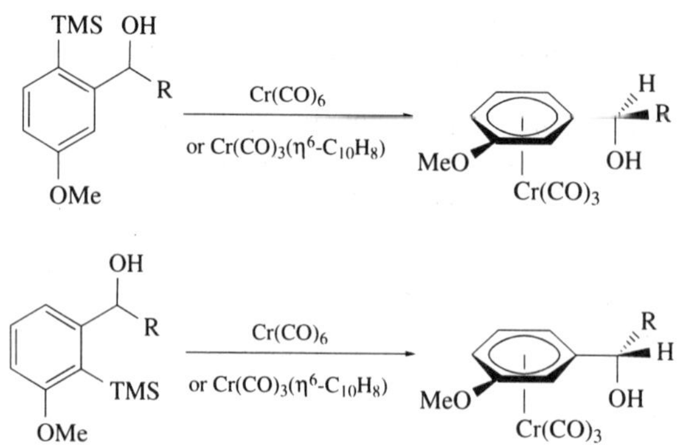

Scheme 9

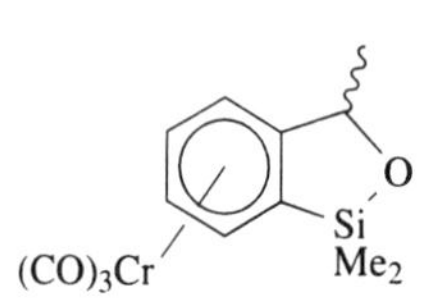

(62)

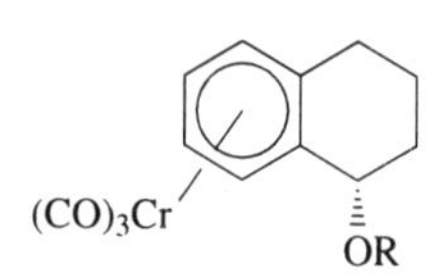

(63)

The hydroxyl group does not have to be in the benzylic position for efficient diastereoselective complexation to occur. Thus, under kinetic conditions with $[Cr(CO)_3(\eta^6\text{-}C_{10}H_8)]$, complexation of the 3-hydroxyketones (protected as the ethylene acetal; the unprotected arenes do not give any coordinated product) occurs to give the diastereomer shown in 80–90% yield with a stereoselectivity of over 90:10 (Equation (5)). The steric effect of the acetal is important in this reaction: changing the acetal lowered the diastereoselection, but the related *o*-$MeOC_6H_4CMe_2CH_2CH(OH)Et$ also gave a good result.[699,700] Similar ketones with the hydroxyl group in the 2-position also complexed to give predominantly one diastereomer, but no selectivity was observed when the hydroxyl group was moved to the 4-position.[200]

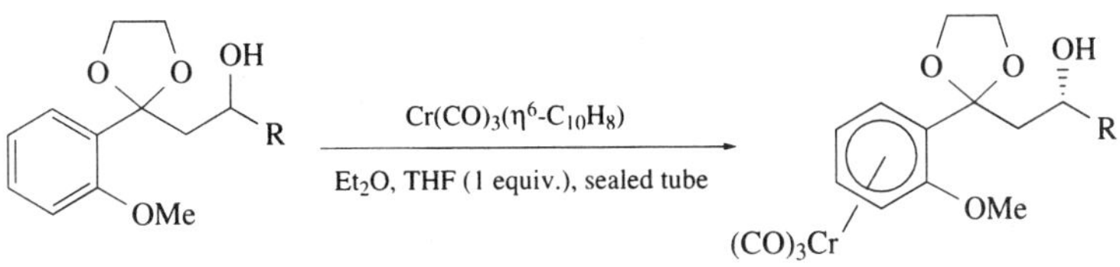

(5)

Several methods have been developed for the resolution of enantiomeric complexes, with attention concentrated on those of *o*-substituted benzaldehydes, that is, $[Cr(CO)_3(\eta\text{-}1,2\text{-}RC_6H_4CHO)]$, where R = Me, OMe, TMS, $SiPr^i_3$. The enantiomeric purity of such complexes can be determined by ^1H NMR in the presence of the chiral shift reagent $Eu(hfc)_3$.[701] The absolute configuration of these enantiomers can be deduced by application of an extended form of the Cahn–Ingold–Prelog rules in which each chromium–arene carbon bond is regarded as single. The enantiomer of $[Cr(CO)_3(\eta\text{-}o\text{-}MeC_6H_4CHO)]$ shown in Figure 3 is therefore designated as 1*S* (or 2*R*).[202] The absolute configuration of chiral $Cr(CO)_3$ complexes must usually be determined unambiguously by x-ray diffraction and chemical correlation as techniques such as optical rotatory dispersion and circular dichroism cannot be generally used. However, a method for assignment of *meta*-substituted complexes on the basis of their Cotton effects has been described.

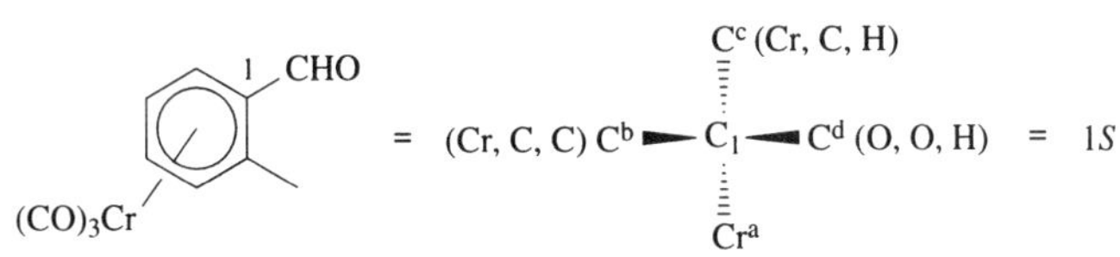

Figure 3 Stereochemical assignment of $[Cr(CO)_3(\eta\text{-}o\text{-}MeC_6H_4CHO)]$ (after Solladié-Cavallo 1989[202]).

The resolution method developed by Davies involves formation of a Schiff's base with L-valinol. Chromatography on alumina achieves kinetic resolution due to the differing hydrolysis rates of the diastereomers.[702–4] Related reactions use the hydrazone formed from $H_2NNHCOCONHCHMePh$ or the aminal from (*R*,*R*)-1,2-bis(*N*-methylamino)cyclohexane.[705–7] The aminals can also be selectively complexed under either thermodynamic conditions from $[Cr(CO)_6]$ or kinetic ones from $[Cr(CO)_3(\eta^6\text{-}C_{10}H_8)]$.

Resolution of chromium tricarbonyl arene complexes by enzymatic methods is finding increasing application. The substituted benzaldehyde complexes can be enantioselectively reduced by baker's yeast to the corresponding benzyl alcohols; kinetic resolution is effected by one enantiomer reacting preferentially while the other remains unchanged (Scheme 10); dehydrogenase enzymes can be used instead of yeast.[708–13] The oximes of the aldehyde complexes can also be reduced enzymatically.[714] Complexes of aryl ketones undergo reduction enantioselectively in some cases, for example $[Cr(CO)_3(\eta\text{-}PhCOMe)]$ produces (*S*)-$[Cr(CO)_3(\eta\text{-}PhCHOHMe)]$ in 96% yield with 99% *ee*, while others, for example indanone chromium tricarbonyl, again undergo kinetic resolution.[715] Complexes of *o*-substituted benzyl alcohols can themselves be resolved by enantioselective esterification with vinyl acetate in the presence of a lipase enzyme.[716–18] The hydrolysis of $[Cr(CO)_3\{\eta\text{-}o\text{-}C_6H_4(CO_2R)_2\}]$ (R = Me, Et) with pig liver esterase gives the monoester in good yield and high *ee*.[719]

A further approach to resolution is CO substitution by a chiral ligand; the $Cr(CO)_3$ complexes of *o*- or *m*-$MeC_6H_4CO_2Me$ have been resolved in this way using (+)-neomenthylPPh_2.[720]

8.3.8 Reactions of Arene Metal Tricarbonyl Complexes

Coordination of an arene ligand to a chromium tricarbonyl fragment modifies its reactivity in a variety of ways, summarized in Figure 4. As a result of the electron-withdrawing effect of the $Cr(CO)_3$

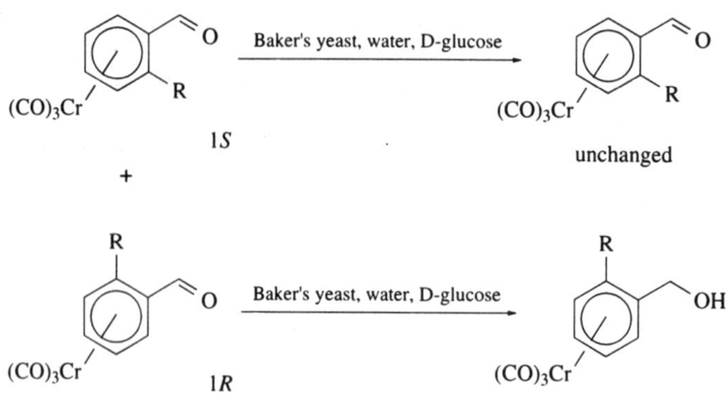

Scheme 10

group acidity of both the aromatic hydrogens and those in the benzylic position is increased, allowing deprotonation reactions to occur, and nucleophilic attack on the ring is facilitated. Alternatively, the arene is deactivated towards electrophilic attack, though the $Cr(CO)_3$ unit is also able to stabilize carbonium ions in the benzylic position. Finally, one face of the arene ligand is sterically hindered by the presence of the metal fragment.

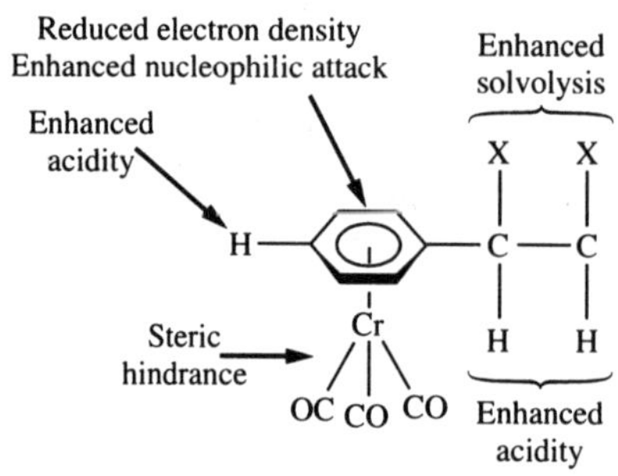

Figure 4 Effects of coordination to a $Cr(CO)_3$ fragment on arene reactivity.

8.3.8.1 Ligand replacement

Treatment of [$M(CO)_3$(η-arene)] complexes with ligands such as phosphines under thermal conditions generally leads to displacement of the arene; consequently the arene complexes can be employed as useful starting materials for *fac*-[$M(CO)_3L_3$], particularly for reactive or thermally sensitive phosphines, for example $PH_2CH_2CH{=}CH_2$.[721–5] The special lability of the naphthalene complex [$Cr(CO)_3(\eta^6\text{-}C_{10}H_8)$] is demonstrated by its reaction with PCy_3, which occurs to give [$Cr(CO)_3(PCy_3)_2$], whereas cycloheptatriene is not displaced from [$Cr(CO)_3$(chpt)] by this phosphine; indeed reaction of the phosphine complex with the polyene produces [$Cr(CO)_3(\eta^6$-chpt)].[726] Refluxing [$Mo(CO)_3$(η-arene)] with polyfunctional amines and aldehydes produces $Mo(CO)_3$ complexes of their Schiff base condensation products, including macrocycles.[727,728] Displacement of the arene ligand and the carbonyls by α,β-unsaturated ketones gives [Mo(ketone)$_3$] species.[729,730]

One successful thermal CO substitution reaction is that of [$Cr(CO)_3$(η-arene)] with isonitriles (RNC; R = Bu^t or 2,6-xylyl), which can be carried out in refluxing heptane in the presence of a catalyst system consisting of [$Fe_2(CO)_4Cp_2$] and palladium oxide, giving [$Cr(CO)_2$(CNR)(η-arene)] in >50% yield. Only small amounts of arene displacement products are observed in this case.[731]

Under photochemical conditions, a carbonyl ligand can be cleanly removed from [$Cr(CO)_3$(η-arene)] complexes. The resulting dicarbonyl species has been detected in a number of matrix isolation, laser flash photolysis and time-resolved IR studies.[732–5] Shorter wavelength (266 nm) photolysis in the gas phase also produces a monocarbonyl species.[736] Not surprisingly the dicarbonyl species forms loosely stabilized adducts with coordinating solvents, including an η^2-benzene complex, and also with molecules such as N_2, H_2, cyclooctene and styrene, and the x-ray structure of a sufficiently stable one of these, [$\{(\eta\text{-}C_6Et_6)Cr(CO)_2\}_2(\mu\text{-}N_2)$], has been determined.[635,737–41] Rate constants for the back-reaction of the dicarbonyl complexes with CO have been determined in a number of cases; they are highly dependent on the alkane solvent and also on the substitution pattern of the arene.[735] In several of these studies the formation of a dinuclear secondary product was observed; it has been proposed that this ultimately yields [$Cr(CO)_6$] and free arene on decomposition.[742]

Photochemical CO substitution in the presence of other ligands leads to [$Cr(CO)_2L(\eta$-arene)] complexes; some recent examples are listed in Table 9. Sometimes labile [$Cr(CO)_2(THF)(\eta$-arene)] is prepared separately before addition of L. The acyl isonitrile complexes are made by reaction of [$Cr(CO)_3(\eta$-arene)] with KCN followed by RCOCl; although NEt_4N_3 reacted to produce an anionic intermediate, it could not then be successfully alkylated.[743] The η^1-TCNE complex is prepared by treatment of [$Cr(CO)_2\{P(OMe)_3\}(\eta-C_6Me_6)$] with TCNE; it is the phosphite ligand that is substututed by a self-induced homogeneous electron transfer mechanism.[541] Attempts to make [$Cr(CO)_2(NO)(\eta-C_6H_6)]^+$ by reaction of the tricarbonyl benzene complex with NO^+ were unsuccessful and led to arene dissociation; even so, useful intermediates are produced that can be converted to a range of chromium nitrosyl complexes.[744] Alternatively, the same procedure successfully produced [$Cr(CO)_2(NO)(\eta-C_6Et_6)]^+$. Bitterwolf and co-workers have prepared a number of compounds of the type [$Cr(CO)_2(\mu-\eta^6,\eta^6$-PhPh)($\mu$-L–L)$Cr(CO)_2$] where L–L is a bidentate ligand such as P_2Me_4, dppm or dppe (**40**). Addition of strong acids to these results in release of H_2, presumably by protonation of both centres followed by reductive elimination.[528,761,762]

Table 9 Ligand substitution reactions of chromium tricarbonyl arene complexes.

Arene	*L*	*Ref.*
C_6H_6	PPh_2Me, η^1-dppe, η^1-dppm	745–7
C_6H_6	$Cr(CO)_3(\eta^6-PhPPh_2)$	748
C_6H_6	$Mn(CO)_3(\eta^5-C_5H_4PPh_2)$, $Mn(CO)_3(\eta^5-C_5H_4CH_2PPh_2)$	749
PhCOMe	PPh_3, PBu_3, $P(OMe)_3$, $P(OPh)_3$	750
$PhCO_2Me$	PF_3	751
o- or *m*-$C_6H_4Me(CO_2Me)$	(+)-neomenthylPPh_2[a]	329
1,3,5-$C_6H_3Me_3$	$CpV(CO)_2\{\eta^2-Ph_2P(CH_2)_2PPh(CH_2)_2PPh_2\}$	752
$PhSi(OMe)_3$, $PhSi(OEt)_3$ and other arylsilanes	$P(OMe)_3$	350
C_6Et_6	PEt_3, PPh_3	633
C_6Et_6	η^1-TCNE	539–41
C_6H_6	$Mn(CO)_3(\eta-C_4H_4N)$	753
C_6H_6 + others	CNCOR (R = Ph, NMe_2, OEt)	754,755
$PhCO_2Me$	$CN(CH_2)_3Si(OEt)_3$[b]	756
1,3,5-$C_6H_3Me_3$	$CNCF_3$	757
C_6H_6, PhMe, $PhCO_2Me$	SPh_2, $S(Bz)_2$, dithiane	758,759
1,3,5-$C_6H_3Me_3$	$H_2C{=}C(CN)(SBu^t)$[c]	760

[a] Allows resolution of the diastereomers. [b] Can be anchored to a silica support through the $Si(OEt)_3$ functionality. [c] Produces S-bound, N-bound and η^2-bound isomers.

The problem of arene displacement in substitution reactions of naphthalene chromium tricarbonyl can be circumvented by using the 1,4-epoxytetrahydronaphthalene complex (**64**); after photochemical CO substitution the epoxy group is removed with HBF_4 to give [$Cr(CO)_2L(\eta^6-C_{10}H_8)$].[763,764]

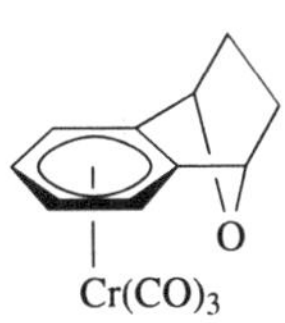

(64)

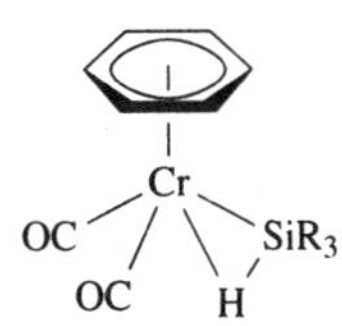

(65)

Photolysis of [$Cr(CO)_3(\eta$-arene)] complexes in the presence of hydrosilanes leads to incomplete oxidative addition and formation of [$Cr(CO)_2(HSiR_3)(\eta$-arene)] (**65**), in which the SiH bond is coordinated to the metal.[765] With $HSiCl_3$ further photolysis causes elimination of H_2 and formation of [$Cr(CO)_2(SiCl_3)_2(\eta$-arene)].[766] The enthalpies of chromium–alkane and chromium–silane bonding have been estimated as 37–54 kJ mol^{-1} and 88–117 kJ mol^{-1} respectively from a study of the reaction of [$Cr(CO)_3(\eta-C_6H_6)$] with Et_3SiH in heptane.[767] NMR and x-ray studies of [$Cr(CO)_2(HSiPh_2H)(\eta-C_6Me_6)$] have established the presence of a three-centre, two-electron bond, for example the *J*(SiH) values for the terminal and agostic hydrogens are 197 Hz and 71 Hz, respectively.[768,769] These silyl complexes act as effective catalysts for the formation of silicon–heteroatom bonds by coupling of hydrosilanes with amines, alcohols, etc.[770]

The analogous stannyl species [$Cr(CO)_2(HSnPh_3)(\eta$-arene)], which contain an η^2-Sn–H interaction, are prepared by photolysis of the tricarbonyl arene complexes with Ph_3SnH. They can be deprotonated with KH to give [$Cr(CO)_2(SnPh_3)(\eta$-arene)]$^-$, which can be derivatized by reaction with Me_3SnCl or

Me_2SnCl_2; reaction with halides such as the iminium salt [Me_2N=CHCl][Cl] affords carbene complexes such as [Cr(CO)$_2$(=CHNMe$_2$)(η-arene)].[771–3]

The unstable compounds [XSb{Cr(CO)$_2$(η-arene)}$_2$] are made by photochemical reaction of the tricarbonyl arene complexes with SbX_3 (X = halide). They are open stibinidene complexes with no Cr–Cr bond.[774] Halide abstraction with $NaBPh_4$ produces the naked antimony complexes [(η-arene)(CO)$_2$Cr=Sb=Cr(CO)$_2$(η-arene)]$^+$, which can be isolated for bulky methylated arenes or trapped with ligands such as bipy or phen which coordinate to the antimony.[775,776]

In an unusual reaction, α-chlorovinylsilanes TMS-C(Cl)=CR$_2$ (R = Me, Ph) undergo photochemical oxidative addition to [Cr(CO)$_3$(η-arene)] followed by elimination of TMS-Cl to produce the vinylidene complexes [(η-arene)Cr(CO)$_2$(=C=CR$_2$)], or, with an excess of silane, the η^2-butatrienes [(η-arene)Cr(CO)$_2$(η^2-R$_2$C=C=C=CR$_2$)].[777]

The photochemical reaction between [Cr(CO)$_3$(η-arene)] (arene = C_6H_6, 1,3,5-$C_6H_3Me_3$, C_6Me_6) with acetylene yields [Cr(CO)$_2$(η-C_2H_2)(η-arene)]; the mesitylene complex reacts with phosphines either by substitution of the alkyne or addition to it; Me_2NH adds to give the aminocarbene [Cr(CO)$_2${=C(Me)NMe$_2$}(η-arene)].[778,779] Other alkynes also give η^2-complexes, the further reactions of which depend on their substituents.[780] They undergo one-electron oxidation to give stable cations in which the unpaired electron is highly delocalized onto the alkyne. Irradiation of [Cr(CO)$_3$(η-C_6Me_6)] with TMS-C≡C-TMS affords the vinylidene complex [Cr(CO)$_2${=C=C(TMS)$_2$}(η-C_6Me_6)]; however this undergoes reversible isomerization to the alkyne cation on oxidation.[781]

A number of chelate complexes have been prepared in which the donor group is attached to the arene nucleus by an alkyl chain, generally 1–3 CH_2 units. Some examples are given in Table 10. In most cases these are prepared by attachment of the donor groups to a preformed arene complex, followed by photochemical CO loss leading to chelate formation.

Table 10 Arene chromium complexes with intramolecular coordination.

Arene in [Cr(CO)$_2$(η-arene-L)]	*Ref.*
$R_2PCH_2PR_2$ (R = *o*-Tol)	782
o, m, and *p*-$FC_6H_4CH_2P(OPh)_2$	783
$Ph(CH_2)_nCPh$=NPh (*n* = 2, 3)	784
3,5-$Me_2C_6H_3(CH_2)_2OPR_2$ (R = F, OEt, OPh)	537
3,5-$Me_2C_6H_3(CH_2)_3PR_2$ (R = H, TMS, OEt, OPh)	785
1,3-C_6H_4{$(CH_2)_nOPF_2$}$_2$, 1,3,5-C_6H_3{$(CH_2)_nOPF_2$}$_3$	786
3,5-$Me_2C_6H_3CH_2OCH_2$C≡CH, 3,5-$Me_2C_6H_3CH_2OCH$=C=CH_2	787
3,5-$Me_2C_6H_3CH_2OCH_2CH$=CH_2	788

Protonation of [Cr(CO)$_2$(η^6,η^2-PhCH$_2$OCH$_2$CH=CH$_2$)] with HBF_4 causes cleavage of the ether linkage to give [Cr(CO)$_2$(η^3-allyl)(η^6-PhCH$_2$OH)][BF$_4$].[789,790] Cationic allyls of this type can also be made by irradiation of [Cr(CO)$_3$(η-arene)] with allyl alcohol or substituted dienes in the presence of HBF_4.[790–3] They exist as mixtures of *endo* and *exo* conformers with the former predominant for the analogous molybdenum and tungsten complexes (which can be prepared in a similar way) and can be used as catalysts for the carboxylation of alkenes to esters.[794,795] Irradiation of [Cr(CO)$_3$(η-C_6H_6)] with HC≡CCH$_2$OH in HBF_4 produces the 2-hydroxyallyl species [Cr(CO)$_2$(η^3-CH$_2$COHCH$_2$)(η-C_6H_6)]$^+$, but if PPh_3 is also present the allenylphosphonium complex [Cr(CO)$_2$(η^2-H$_2$C=C=CHPPh$_3$)(η-C_6H_6)]$^+$ is produced instead.[796] With the analogous [Mo(CO)$_3$(η-arene)], [(η-arene)Mo(CO)$_2$(σ,η^2-CH$_2$C≡CH)]$^+$ can be isolated as an intermediate in the same transformations.[797,798]

Treatment of [Cr(CO)$_2$(η^2-cyclooctene)(η-arene)] with CS_2 or CSe_2 produces [Cr(CO)$_2$(CX)(η-arene)] where X = S or Se.[799] A number of papers deal with a detailed comparison of their properties with the corresponding tricarbonyl, including IR and Raman spectra,[486,487,800,801] mass spectra,[802] solid state ^{13}C NMR spectra[803] and electronic spectra.[804] These data suggest that the order of π-acceptor ability is CSe > CS > CO. The mass spectrum of the selenocarbonyl derivative contains no peak for CSe^+ in keeping with the known instability of CSe compared to CS and CO. Displacement of the arene by other ligands provides a useful route to [Cr(CO)$_2$(CX)L$_3$] (L = CO, phosphines, phosphites, etc.); kinetics indicate a ring-slippage mechanism, with the CS and CSe complexes more reactive due to the weaker chromium–arene bond.[805–8] The [Cr(CO)$_2$(CS)(η-arene)] species undergoes reaction with [Cr(CO)$_5$(THF)] to give [(η-arene)Cr(CO)$_2$(μ-CS)Cr(CO)$_5$], which contains a nonlinear μ-CS ligand, and with activated alkynes to form the dithiocarbenes [(η-arene)Cr(CO)$_2$(=$\overline{\text{CSCR=CRS}}$)].[809,810]

Warming or irradiation of [Cr(CO)$_2$(NCMe)(η-arene)] produces moderate yields of black dinuclear [Cr$_2$(CO)$_2$(μ-CO)(η-arene)$_2$] (**66**); the x-ray structure of the C_6H_6 complex is consistent with the presence of a Cr≡Cr bond (av. Cr–Cr 0.2221 pm).[811] Similarly photolysis of [Cr(CO)$_3$(η-arene)] complexes with a number of suitable metal fragments leads to homo- or heterometallic clusters (Table 11).

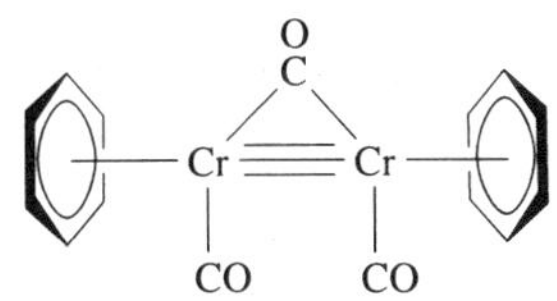

(66)

Table 11 Synthesis of clusters from chromium tricarbonyl arene complexes.

Arene in [Cr(CO)$_3$(η-arene)]	*Reagent*	*Product*	*Ref.*
C_6H_6	[Cr$_2$(μ-S)(μ-SBu^t)$_2$Cp$_2$]	[Cr{(μ$_3$-S)$_2$Cr$_2$(μ-SBu^t)Cp$_2$}$_2$]	812
C_6Me_6	[Cp*Mn(CO)$_3$]	[Cp*Mn(μ-CO)$_3$Cr(η-C$_6$Me$_6$)]	813
C_6H_6, C_6Me_6, 1,3,5-$C_6H_3Me_3$	[Cp*Rh(CO)$_2$] or [(η-C$_9$H$_7$)Rh(CO)$_2$]	[Cp*Rh(CO)(μ-CO)$_2$Cr(CO)(η-arene)] or η-C$_9$H$_7$ analogue	814–816
C_6H_6	[CpRh(=C=CH$_2$)(PPr$^i_3$)]	[CpRh(PPr$^i_3$)(μ-CO)(μ-C=CH$_2$)Cr(CO)(η-C$_6$H$_6$)]	817
C_6Me_6	[CpW(≡CTol)(CO)$_2$]	[CpW(CO)$_2$(μ-CTol)Cr(CO)$_2$(η-C$_6$Me$_6$)]	818

8.3.8.2 Arene lithiation

The enhanced kinetic acidity of the aromatic protons in chromium tricarbonyl arene complexes allows the ring to be deprotonated with strong bases (e.g., organolithium reagents) and functionalized by quenching with electrophiles. Measurements show that the acidity of the arene protons is increased by 6–9 pK_a units on coordination.[819,820] The usual base employed is *n*-butyl lithium in THF at low temperature, sometimes in the presence of TMEDA. Under these conditions deprotonation occurs at the ring rather than at any benzylic protons, even though the latter is thermodynamically more favourable. Despite its easy nucleophilic substitution, [Cr(CO)$_3$(η-PhF)] can be readily deprotonated whereas [Cr(CO)$_3$(η-PhI)] undergoes halide–lithium exchange. The decomposition of [Cr(CO)$_3$(*o*-ClC$_6$H$_4$Li)] was found to be first order with a half life of 136 min at 0 °C; the intermediacy of a benzyne complex was suggested, but trapping experiments were largely unsuccessful.[821] The dilithiated species [Cr(CO)$_3$(1,4-LiC$_6$H$_4$Li)] and its 1,3-isomer can be prepared by Li–SnBu$_3$ exchange from the appropriate Bu$_3$SnC$_6$H$_4$SnBu$_3$ complex.[822]

The lithiation of complexes of substituted arenes has received much attention. In practice this is only synthetically useful where the lithiation is directed to a particular position in relation to the substituents, especially if this is different to that observed in the free arene. For simple alkylbenzenes, lithiation occurs mainly at the *meta* position, though with poor selectivity.[823] Heteroatom substituents, notably F, Cl and OMe, undergo direct lithiation to the *ortho* position either by their inductive effect or by stabilization of the lithiated species with a chelating interaction;[823–5] the order of *ortho*-directing ability F > CONHR > NHCOR > CH$_2$NR$_2$ > OMe > CH$_2$OMe was established in one study.[826] Thus, in complexes of 2-, 3- or 4-FC$_6$H$_4$OMe, lithiation occurs *ortho* to fluorine.[827] Subsequent nucleophilic displacement of the fluoride by amines or thiolates allows access to a number of highly substituted arenes.[828] Lithiation of the complex of 3-MeOC$_6$H$_4$CH$_2$OH occurs at the 4- and 2-positions in a ratio of 77:23 compared to 1:9 for the uncomplexed arene; a similar selectivity change is observed for 7-methoxytetralol.[829–31]

At low temperature [Cr(CO)$_3$(η-*o*-FC$_6$H$_4$Li)] was found to be in equilibrium with a dinuclear form in which attack on a CO ligand of a second molecule had occurred. This could be trapped by addition of ClCO$_2$Me which caused rearrangement to the benzophenone complex [{Cr(CO)$_3$}$_2${(η-*o*-FC$_6$H$_4$)$_2$CO}] (Scheme 11).[832] An example of attack of a lithiated arene complex on free CO has also been reported.[833]

If the heteratom substituent is very bulky, lithiation is directed to the *meta* position. Complexes of the arenes PhNMe(TBDMS) and PhO-TBDMS undergo functionalization at the *meta* position on treatment with Bu^{n}Li and electrophiles such as PhCHO, MeI, Me$_2$NCHO, PhCN and S$_2$Me$_2$ with a selectivity of over 85% (>98% in some cases).[834–8]

The explanation advanced for the regioselectivity of deprotonation is that attack occurs at a carbon atom that is eclipsed by a CO ligand, with which the lithium reagent forms a preliminary adduct.

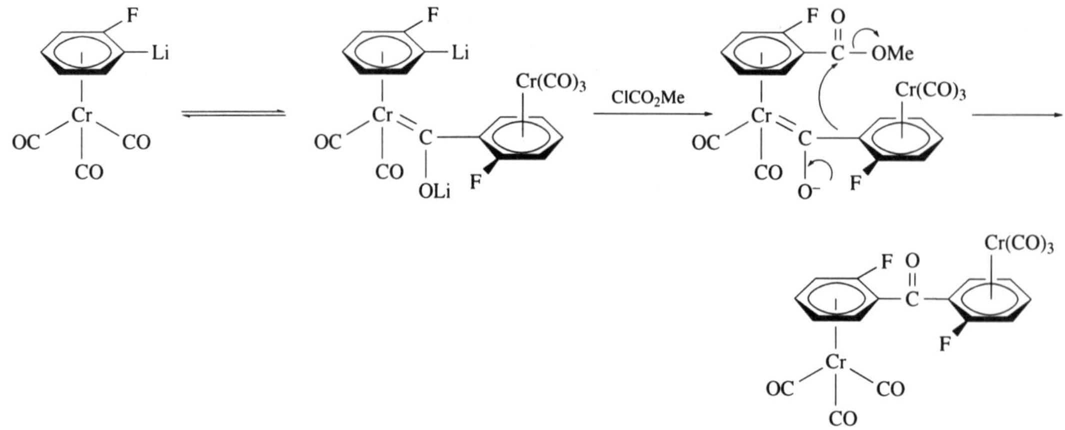

Scheme 11

Introduction of a substituent directs lithiation to the *ortho* position, but if this is sterically hindered attack reverts to the *meta* position; since most substituted arene complexes adopt the *syn*, eclipsed conformation (**41b**) this is also an eclipsed site.

The lithiation of [$Cr(CO)_3$(η-PhCHMeNMe$_2$)] with Bu^{t}Li is not only regioselective for the *ortho* position, but also diastereoselective. As shown in Scheme 12, the reaction is chelate controlled, with the transition state minimizing the interaction between the methyl group and the metal.[839–41] If the added electrophile is PPh$_2$Cl, the resulting chiral phosphine can be used as a P,N-bidentate ligand in palladium-catalysed asymmetric cross-coupling.[842] A similar system (**67**) has been designed using chiral ketals derived from tartaric acid. In this case the lithium is chelated by the ketal oxygen and one of the X groups, and in only one of these positions is it close to the ring proton.[843,844] A further example of this principle is provided by the complex of (−)-(1*S*, 2*S*)-N,O-dimethylephedrine.[845] Attempts to carry out a similar lithiation with complexes of PhCH(Me)OMe were complicated by benzylic deprotonation and by nucleophilic addition which, after quenching with H$^+$, gave the methylene–cyclohexadiene complex (**68**).[841,846]

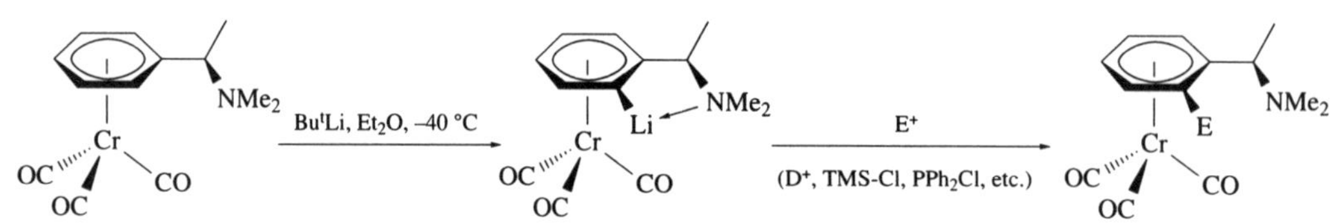

Scheme 12

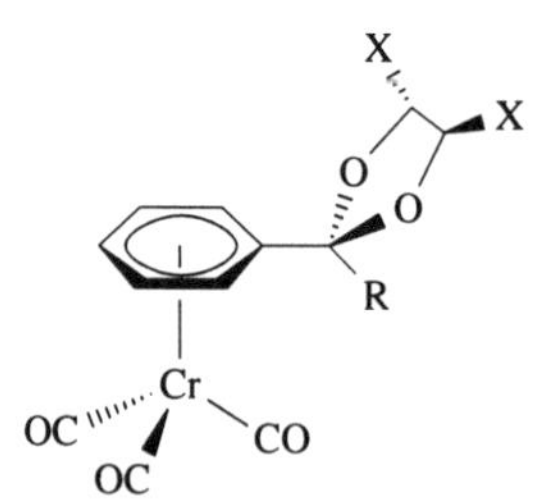

X = CH$_2$OMe, CMe$_2$OMe, CH$_2$NMe$_2$
(67)

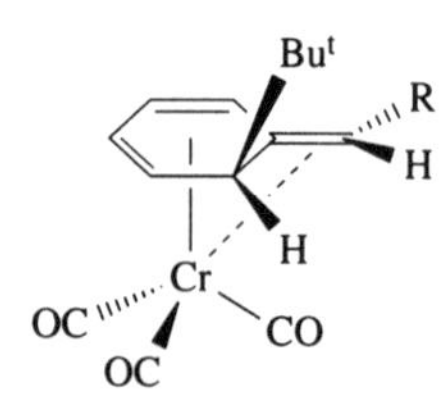

(68)

The lithiation of [$Cr(CO)_3$(η^6-$C_{10}H_8$)] with LiTMP is regioselective at the sterically more accessible 2-position, whereas Bu^{n}Li deprotonates at both the 1- and 2-positions in a 3:7 ratio. Both these reactions are kinetically controlled and addition of Pr$^i_2$NH establishes an equilibrium between the two isomers; some degree of product selectivity is achievable by changing the reaction conditions.[682,847–50] Reports of the lithiation of other polycyclic arenes are less widespread, but complexes of acenaphthene, anthracene, fluoranthrene and pyrene all undergo sterically controlled regioselective lithiation.[850,851] Lithiation of complexes of cyclobutabenzene and 1,4-benzodioxin occur at positions *ortho* to the ring junction, but that of indane is less selective.[239,852,853]

The regiochemistry of lithiation of the indole complex (**69**) depends on the identity of the protecting group R. If R = Me, lithiation occurs first at the 2-position; if this is blocked or if R = CH_2OMe, lithiation is directed to the 7-position. Finally, if R = $SiPr^i_3$ the reaction moves to the 4-position, allowing the synthesis of a variety of substituted derivatives.[854–7] Lithiation of complexed benzothiophene also occurs at the 2- and 7-positions.[858]

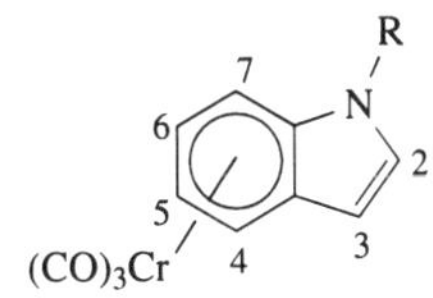

(69)

Apart from simple derivation with electrophiles, the lithiated species can be converted to cuprate reagents with CuBr·SMe_2, which allows arylation and acylation, and also undergo palladium-induced cross-coupling with vinyl or aryl halides.[859] Widdowson has used the regioselective *o*-lithiation of the fluorobenzene complex to prepare cyclic species by combining it with intramolecular nucleophilic substitution (Scheme 13).[827,860,861]

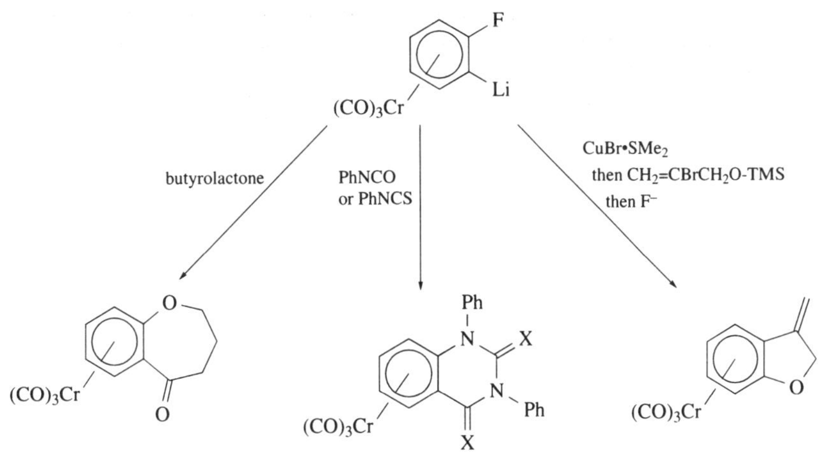

Scheme 13

8.3.8.3 Nucleophilic attack

Several reviews of the reactions of $[Cr(CO)_3(\eta\text{-arene})]$ complexes with nucleophiles have appeared during the 1980s, the latest being by Semmelhack (see also Volume 12 of this work).[196,862–4]

It has long been known that addition of anionic nucleophiles to tricarbonyl chromium arene complexes produces cyclohexadienyl anions. Recently, the parent unsubstituted cyclohexadienyl has been prepared by addition of $LiBHEt_3$ to $[Cr(CO)_3(\eta\text{-}C_6H_6)]$ and trapped with Ph_3SnCl, which adds at the metal.[865] Virtually all the reactions in this section involve chromium complexes, but the addition of $LiBHEt_3$ or MeLi to $[W(CO)_3(\eta\text{-arene})]$ complexes has also been shown to give cyclohexadienyl anions which can be methylated at the metal to give $[WMe(CO)_3(\eta^5\text{-}C_6H_6R)]$ (R = H, Me).[866] As might be expected, replacement of one CO ligand by a phosphine deactivates the ring to attack, but does not stop it completely.[745,867]

The fate of the cyclohexadienyl anions depends on the arene and on the subsequent procedure employed (Scheme 14). If a good leaving group (usually F or Cl) is available at the *ipso* position, elimination of X^- occurs, effectively giving substitution of Nu for X in an S_NAr-type process. Alternatively, if X = H, the intermediate can be oxidized to give the arene where Nu has replaced hydrogen, or trapped by electrophiles to give *trans*-cyclohexadienes. Both of these procedures result in loss of the tricarbonyl chromium moiety.

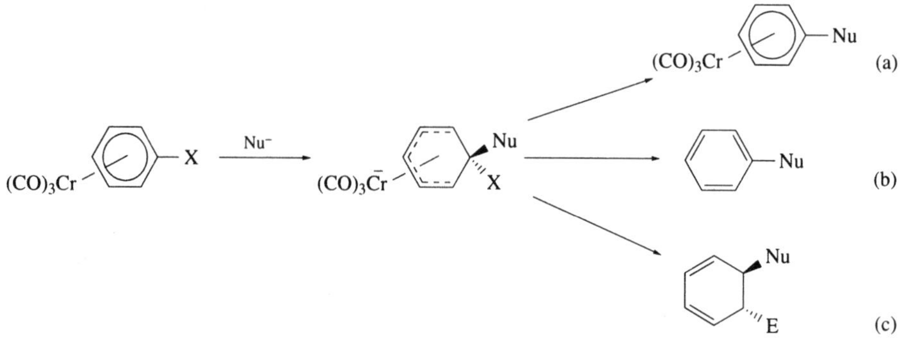

Scheme 14

(i) Nucleophilic substitution of halobenzenes and related species

The complexation of haloarenes to the $Cr(CO)_3$ group has long been known to activate them towards *ipso* nucleophilic substitution, allowing the halide to be replaced by a wide range of functional groups; some examples involving carbanions, and oxygen, nitrogen and sulfur-based nucleophiles are given in Table 12 (see Table 5 for examples involving metal-based nucleophiles). In general fluoride is more easily displaced than chloride. The reaction can also be intramolecular, for example cyclization of $[Cr(CO)_3\{\eta\text{-}o\text{-}FC_6H_4(CH_2)_3OH\}]$ to chroman.[868] Starting from $[Cr(CO)_3(\eta\text{-PhF})]$, cyclic compounds can be made by sequential lithiation at the *ortho* position, followed by treatment with PhNCS or butyrolactone (see Scheme 13).[860,861]

Table 12 Nucleophilic substitution of haloarene chromium tricarbonyl complexes.

Arene	*Nucleophile*	*Remarks*	*Ref.*
PhF	Cp^-, $C_9H_7^-$, fluorenyl, pentadienyl		679
PhF, PhCl	$^-CH(CO_2Et)_2$, $^-CH(CN)Ph$, $^-CH(CO_2Et)CN$	Phase transfer conditions	872
PhF	$LiC(N{=}CHC_6H_4Cl)(Me)(CO_2Me)$	Produces α-substituted aryl amino acids after hydrolysis	873,874
MeC_6H_4F	$LiC(R^1)(R^2)(N{=}CPh_2)$ (R^1 = CN, CO_2Me; R^2 = H, Me)	Produces α-substituted aryl imino esters or nitriles	875,876
PhCl, *o*- and *p*-$C_6H_4F_2$	*o*-$LiCB_{10}H_{10}CMe$	In *p*-$C_6H_4F_2$, both fluorines are replaced	877,878
PhF, PhCl	NaOMe, NaOEt, NaOPr	Phase transfer conditions	879
o-, *m*-, *p*-$C_6H_4Cl_2$	MeOH or Pr^iOH, KOH	Phase transfer conditions. With *m*-isomer, both chlorines are replaced; with *o*- and *p*-, only one	880
RC_6H_4F, RC_6H_4Cl	NaOMe	Rate of substitution is R = *p*-OMe > *o*-Me > *p*-Me > *m*-Me > H	881,882
p-$C_6H_4Cl_2$	NaOPh, then NaOMe	Stepwise replacement of chlorines observed	883
o-$C_6H_4Cl_2$	Na or K salts of *o*-$C_6H_4(OH)_2$, *o*-$HOC_6H_4NH_2$, *o*-HOC_6H_4SH		884
o-$C_6H_4Cl_2$	$NaX(CH_2)_2O(CH_2)_2XNa$ (X = O, S)	Gives bis-$Cr(CO)_3$ complex of 2 + 2 dibenzo crown ether macrocycles, e.g., (**70**)	885
o-, *m*-, *p*-MeC_6H_4F *m*-$MeOC_6H_4C$	$Me_2C{=}NOH$, KOH	Phase transfer conditions. Can be cyclized to benzofuran derivatives after decomplexation	886,887
PhCl	$HONHCO_2Bu^t$	Produces *O*-aryl hydroxylamines after decomplexation and hydrolysis of ester	888
o-RC_6H_4F (R = H, Me, $CH(OEt)_2$)	$NaNHCOCF_3$, $NaNHP(O)(OBz)_2$	Produces substituted aniline complexes after deprotection	889
PhF, PhCl	NaSR (R = Me, Bu, Pr^i, Bu^t)	Can be done under phase transfer conditions.	828,890, 891
o-, *m*-, *p*-$C_6H_4Cl_2$	NaSR	Occurs stepwise under phase transfer conditions	892

Only a few cases of *ipso* substitution involving leaving groups other than halide are known, and all involve phenoxides or alkoxides. One example is the substitution of phenoxide in $[Cr(CO)_3(\eta\text{-PhOPh})]$

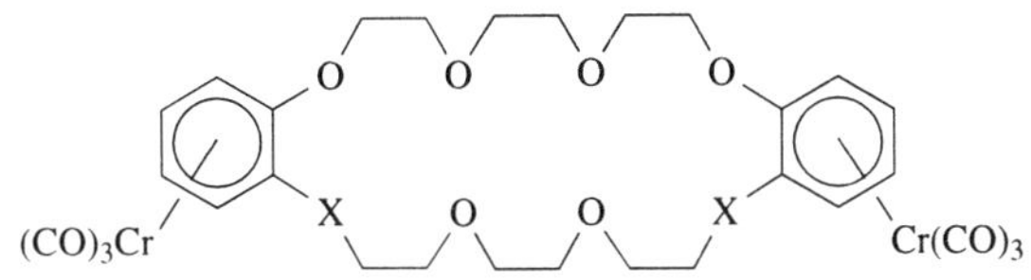

(70)

on reaction with 2-methyl-1,3-dithianyl anion at room temperature;[869] more recently, examples of replacement of OMe by $LiCH_2CN$, $LiBHEt_3$ or LAH have come to light.[309,870,871] Ring opening of the $Cr(CO)_3$ complex of dibenzofuran to $[Cr(CO)_3(\eta^6\text{-}PhC_6H_4\text{-}2\text{-}OH)]$ occurs on treatment with $LiBHEt_3$; the intermediate cyclohexadienyl anion can be trapped with Ph_3SnCl, but initial addition of H^- is reversible, allowing eventual *ipso*-substitution (Scheme 15).[865] Attempts to induce loss of hydride from the cyclohexadienyl intermediates with Ph_3C^+ have only recently been successful with complexes of imines or hydrazones, leading to *ortho*-substituted species.[893]

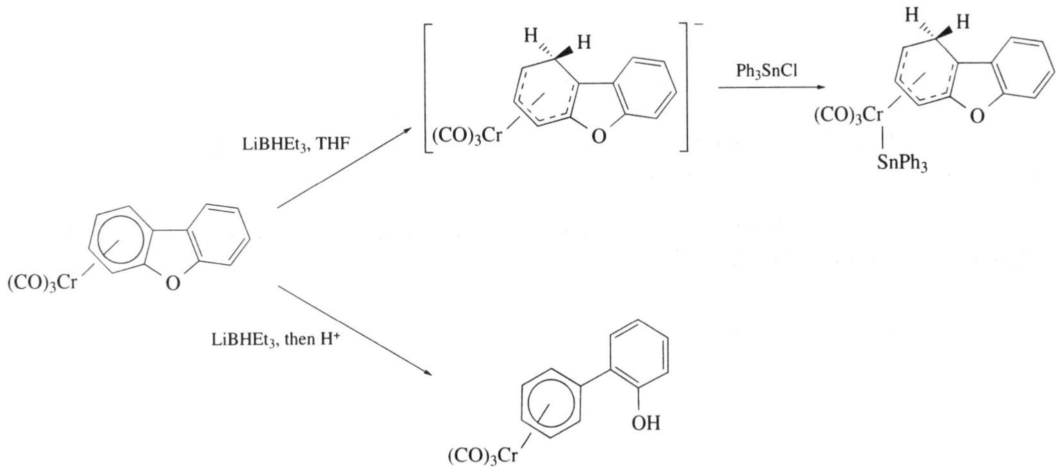

Scheme 15

The reaction of the halobenzene complexes with carbanions is more complex than Table 12 would indicate. Initial attack of the nucleophile is favoured at the *ortho* and *meta* positions (see below), and the subsequent course of reaction depends on whether the initial addition is reversible, which in turn depends on the reactivity of the carbanion. Stabilized carbanions, for example $Na[CH(CO_2Et)_2]$, add reversibly; eventually addition occurs at the *ipso* position and rapid loss of halide ensues. More reactive carbanions, such as $LiCMe_2CN$, add rapidly to the *ortho* and *meta* positions, but the resulting cyclohexadienyl anions rearrange slowly to the *ipso*-substituted ones, and then lose halide. The most reactive carbanions, for example 2-methyl-1,3-dithianyl, $LiCH_2CN$, $LiCH_2CO_2Bu^t$, add irreversibly to the *ortho* or *meta* positions; consequently they produce products in which the nucleophile has replaced hydrogen. There is thus an upper limit on the reactivity of the carbanion, above which *ipso* substitution does not occur because the initial step is no longer reversible.

However, loss of halogen from the cyclohexadienyl intermediates can be induced even in these cases by protonation, hydrogen migration to the *ipso* position, and loss of HX. The result is substitution in which the carbanion occupies a site which is *ortho* to the original halide (called cine substitution) or *meta* or *para* (both termed tele substitution, referring to the fact that the nucleophile is more than one carbon atom removed from the halide). The regiochemical course of the reaction can be conveniently followed by using D^+ or by introducing substituents into the ring. An example of cine substitution with D^+ as protonating agent is shown in Scheme 16.

The regioselectivity of the substitution depends on the halide; fluorine directs strongly *meta*, whereas chlorine directs weakly *ortho* or *meta*. Thus, reaction of 2,6-dimethylhalobenzene complexes with dithianyl carbanions produces tele-*meta* substitution when X = F, and tele-*para* when X = Cl.[894–6] With *o*-, *m*-, or *p*-chlorotoluene, cine and tele-*meta* substitution are observed.[897–9] Treatment of $[Cr(CO)_3(\eta\text{-}o\text{-}XC_6H_4TMS)]$ (X = F, Cl) with $LiBHEt_3$ provides an unusual example of tele-*meta* substitution of halide by hydride.[900]

The elimination of phenol from complexes of diphenyl ether also occurs by tele-*meta* substitution. By labelling the *meta* position with deuterium, it can be shown that the mechanism involves 1,3-H

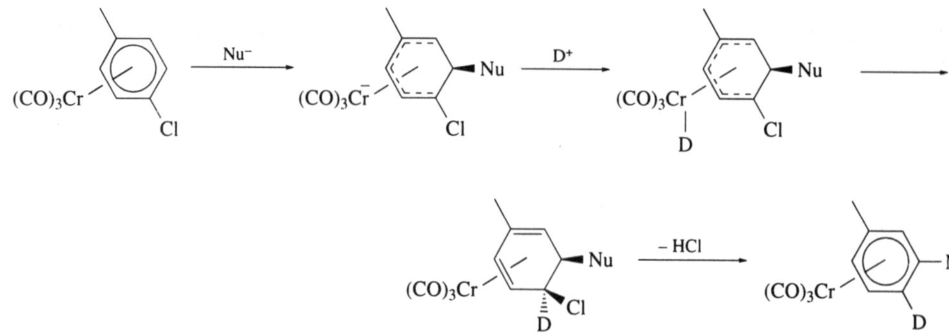

Scheme 16

migration (Scheme 17).[869,901,902] More recently the elimination of $HNMe_2$ has been achieved by cine and tele-*meta* substitution of $[Cr(CO)_3(\eta\text{-}PhNMe_2)]$ with $LiBHEt_3$.[903]

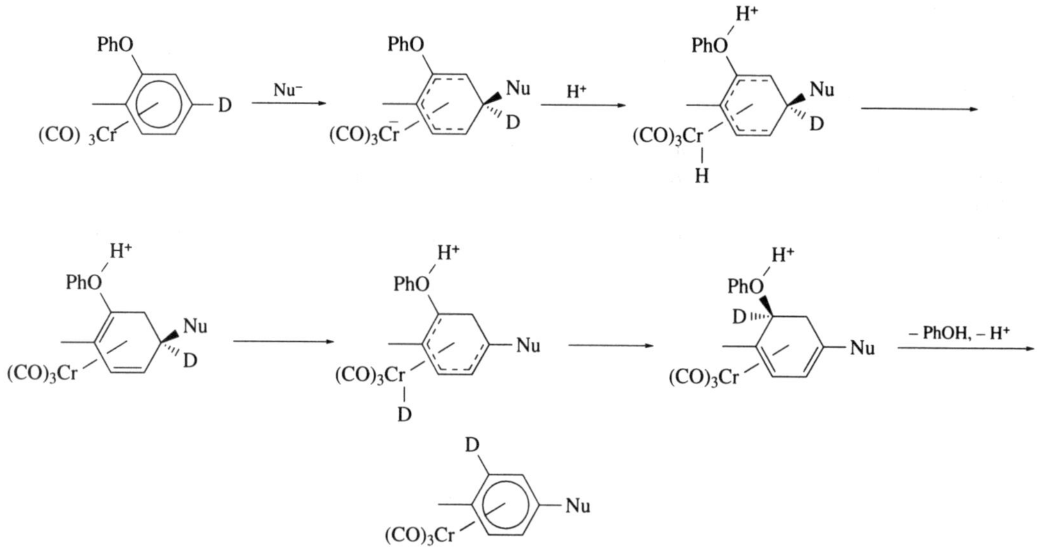

Scheme 17

An unusual example of cine substitution is provided by the lithiation of $[Cr(CO)_3(\eta\text{-}o\text{-}MeOC_6H_4OMe)]$ with BuLi and quenching with $Pr^i{}_3SiCl$, which in addition to substitution at the 3 and 6 positions also produced dinuclear complex (**71**).[824,904]

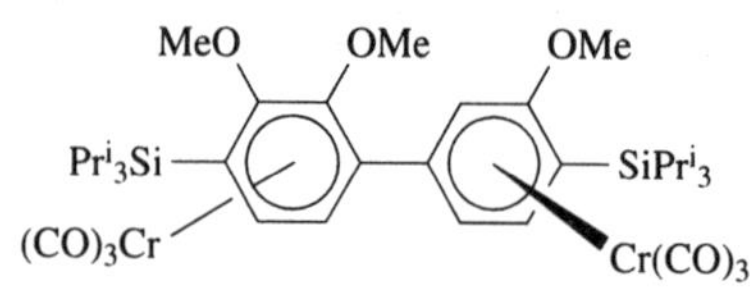

(71)

(ii) Nucleophilic addition followed by oxidation

Oxidation of the anionic cyclohexadienyl intermediates produces free arenes in which formal substitution of the incoming nucleophile for hydrogen has occurred (pathway (b) in Scheme 14). The oxidation process is usually carried out with iodine, and the chromium is irretrievably lost as chromium(III). Only reactive carbanions, such as 1,3-dithianyl, $LiCH_2CN$, $LiCMe_2CN$ and $LiCH_2CO_2Bu^t$ are usually effective as nucleophiles in this reaction; with simple organolithium reagents such as MeLi or BuLi, competing ring lithiation or benzylic deprotonation can occur instead of nucleophilic attack. Only one example of successful substitution by a heteroatom nucleophile has been reported: reaction of $[Cr(CO)_3(\eta\text{-}C_6H_6)]$ with LiNHCOPh to give PhNHCOPh.[905]

In terms of regioselectivity of attack, resonance donor substituents such as NMe_2 or OMe direct attack to the *meta* position, sometimes exclusively. Methyl or chloro substituents also tend to favour the *meta* position, though in these cases a greater proportion of *ortho* attack can occur. The electron-withdrawing CF_3 group directs attack to the *para* position. An extensive study of alkylbenzene complexes by Semmelhack showed that increasing the size of the alkyl substituent to Bu^t or $CHBu^t_2$ increases the proportion of *para* attack, and that the electronic properties of the incoming nucleophile are also important: more reactive anions favour the *meta* position.[906,907] The biphenyl complex $[Cr(CO)_3(\eta^6\text{-PhPh})]$ undergoes attack mainly at the *ortho* position, except with tertiary carbanions which prefer the *para*.[908]

The question of whether the regioselectivity of nucleophilic attack is under charge control or frontier orbital control has been extensively discussed, and the true answer appears to involve a balance of the two. An additional factor in the charge distribution of the arene ring is the conformation of the $Cr(CO)_3$ tripod, and in many cases the preferential position of attack has been explained on this basis. In monosubstituted complexes, the eclipsed, *anti* conformation (**41c**) is favoured by electron-withdrawing groups (and by large alkyl substituents), while the eclipsed, *syn* conformation (**41b**) is preferred for donor substituents. Attack will occur at δ^+ carbon atoms that are eclipsed by COs as predicted by EHMO calculations, that is, *para* and *meta* respectively for the two conformations.[907]

In many cases the regioselectivity predicted on a conformational basis is the same as that predicted by a frontier orbital picture or by a charge control approach. For example, assuming that the LUMO of the arene ligand resembles that of the free arene (an assumption which again is backed up by EHMO calculations[906,909]) attack on PhMe or PhCl is predicted to be at the *ortho* and *meta* sites. With donor substituents such as OMe, charge control of the reaction would favour the *meta* position as the *o*- and *p*-sites are deactivated, and this is also predicted on the basis of tripod conformation. It can thus be argued that in the former case orbital control is predominating whereas in the latter charge control is operative.

In disubstituted benzenes with identical substituents such as veratrole or *o*-xylene, attack occurs at an adjacent position.[910] This is also true for bicyclic rings such as cyclobutabenzene, indane and tetralin.[911–13] By introduction of substituents at the benzylic position (especially in an *exo* orientation), for example 1,1-dimethylindane, the attack can be relocated to the *meta* position (Equation (6)).[913–16] Similarly, Uemura has used the positioning of an *exo*-Pr^i substituent in the complex of 1-*exo*-isopropyl-5-methoxy tetralin to direct attack of a dithianyl anion to the *meta* position in a synthesis of (±)-dihydroxyserrulatic acid.[917,918] The effect of the substituent on the $Cr(CO)_3$ unit was demonstrated by crystallography, with the *exo* isomer adopting an eclipsed structure whereas the *endo* isomer is forced into a staggered arrangement.

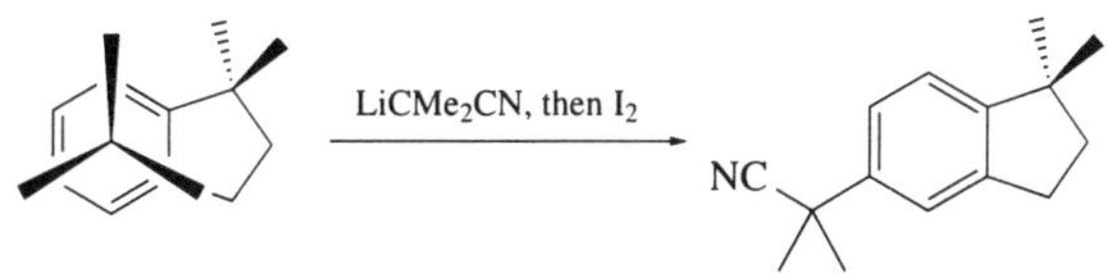

(6)

In the case of $[Cr(CO)_3(\eta\text{-}o\text{-}MeC_6H_4OMe)]$, attack occurs *meta* to the donor OMe substituent, which evidently is the dominating influence on the position of attack. This fact has been used by Semmelhack in a synthesis of deoxyfrenolicin.[919]

The situation with *o*-$MeC_6H_4NMe_2$ in compound (**72**) is more complex, as different products are formed depending on whether kinetic or thermodynamic control of the reaction is operative. The distribution obtained thus depends on the reversibility of initial anion addition, as well as the duration and temperature of the reaction. With $LiCMe_2CN$, reaction at −78 °C for 2.5 h gives almost equal amounts of attack at C-3, C-4 and C-5, whereas if the mixture is allowed to equilibrate for 24 h at higher temperature, the major product is the 1,2,4-isomer. Addition at C-3 is favoured for $LiCH_2CN$ and 2-lithio-2-methyl-1,3-dithianyl, though in these cases there was no evidence for reversibility as the product distribution remained the same under all conditions.[920]

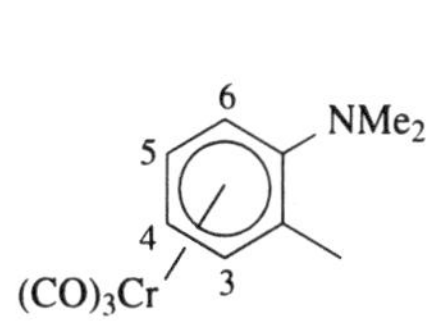

(72)

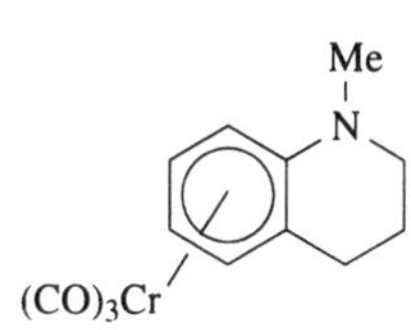

(73)

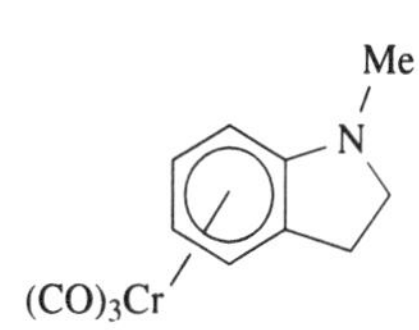

(74)

In the case of the complex of *N*-methyltetrahydroquinoline (**73**), a similar situation is observed with $LiCMe_2CN$ in THF at −78 °C; under kinetic conditions addition at C-5 is favoured, but rapid equilibration occurs and under thermodynamic conditions addition at this position can be made almost exclusive. The rate of equilibration can be slowed substantially by the addition of hexamethylphosphoramide (HMPA). Addition at C-3 is favoured for the less bulky $LiCH_2CN$.[920,921] In the five-membered ring analogue *N*-methylindoline (**74**) attack is favoured at C-3 in all cases. The different substitution patterns observed for *o*-$MeC_6H_4NMe_2$ and its cyclic analogues can be related to the conformation of the $Cr(CO)_3$ tripod, which in turn is due to the different orientations of the nitrogen lone pair.[920]

Extensive studies have also been made of the $Cr(CO)_3$ complex of indole, (**69**). Substitution occurs mainly at C-4 or C-7 depending on the anion type, for example $LiCMe_2CN$ attacks C-4, whereas the 1,3-dithianyl anion attacks C-7. However, by placing bulky substituents at C-3 or on the indole nitrogen, this selectivity can be reversed.[922,923] A recent synthesis of teleocidin uses the *N*-substituent to change the position of attack from C-7 to C-6; in this case C-4 is already blocked by an NMe_2 group.[924]

The addition of nucleophiles to the naphthalene complex $[Cr(CO)_3(\eta^6\text{-}C_{10}H_8)]$ (**75**; R = H) also gives different products depending on reaction conditions. Under kinetic control (short reaction time, −65 °C, in THF containing HMPA) addition of $LiCMe_2CN$ occurs at C-α and C-β in virtually equal amounts; at 0 °C with no HMPA the C-α product is formed almost exclusively. Firm evidence for dissociation and readdition of the nitrile-stabilized carbanion was obtained by crossover experiments, whereas the more reactive dithianyl anions again showed no evidence of reversibility.[863,925,926] A similar range of reactivity was observed for 1,4-dimethoxynaphthalene (R = OMe). Thus, nitrile-stabilized anions and $LiCH_2CO_2Bu^t$ undergo rapid equilibration to give the α-substituted product almost exclusively, whereas dithianyl anions and phenyllithium show no evidence of equilibration. The site of attack does differ however, with 2-methyl-1,3-dithianyl preferring the β-site, PhLi the α.[927] The attack of dithianyl anions on 1-methoxynaphthalene is not so selective, but has been used in a synthesis of a section of aklavinone.[928] Acenaphthalene undergoes carbanion attack predominantly at the 3-position.[851]

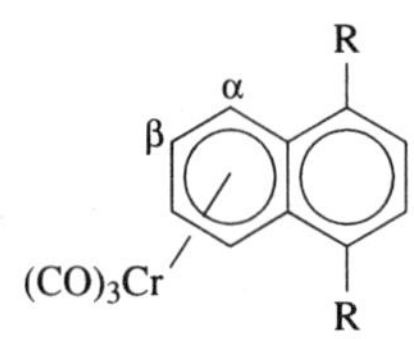

(75)

(iii) Nucleophilic addition followed by protonation or acetylation

Protonation of the anionic cyclohexadienyl complexes produces substituted cyclohexadienes (pathway (c) in Scheme 14), which can subsequently be oxidized to arenes with DDQ (2,3-dichloro-5,6-dicyanobenzoquinone). Although the $Cr(CO)_3$ unit is lost, it can be recovered as $[Cr(CO)_6]$ if the reaction is carried out under a CO atmosphere.[929] Recently it has been shown that addition of $LiBHEt_3$ or PhLi to $[Cr(PF_3)_3(\eta\text{-}C_6H_6)]$ followed by protonation produces cyclohexadiene complexes such as (**76**) in which there is an agostic interaction of one CH bond with the metal, and which are fluxional by 1,2-H exchange and 1,5-H migration.[930]

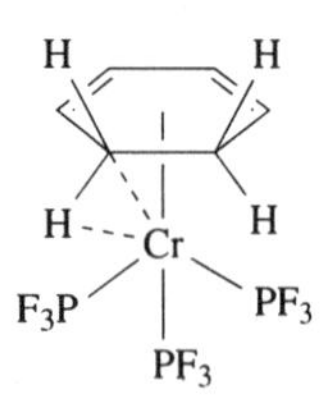

(76)

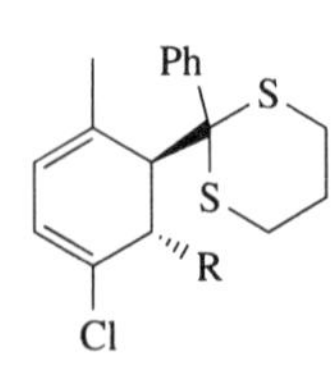

(77)

As shown above, in the case of halobenzene complexes, protonation can result in cine or tele substitution. However, the addition of 2-phenyl-1,3-dithianyl anion to $[Cr(CO)_3(\eta\text{-}p\text{-}MeC_6H_4Cl)]$ and protonation with CF_3CO_2H produces (**77**) as a single isomer.[931]

If the addition of the nucleophile is reversible, quenching with electrophiles other than H^+ is usually not successful as reaction occurs preferentially with the free nucleophile rather than the cyclohexadienyl complex. However, in cases where nucleophile addition is essentially irreversible (e.g., with dithianyl anions, PhLi or Bu^{t}Li) or if equilibration can be slowed by addition of HMPA, Kündig has developed a procedure for addition of MeI which produces acetyl-substituted cyclohexadienes (Equation (7)).[932,933] These arise from addition of MeI at the metal, followed by CO insertion and migration to the cyclohexadienyl ligand; in keeping with this mechanism the acetyl group is located on the *endo* face. Since the nucleophile initially approaches from the *exo* face, *trans*-substituted cyclohexadienes result. Higher yields can be obtained by addition of extra ligands (CO or PPh_3) to drive the insertion step forward.

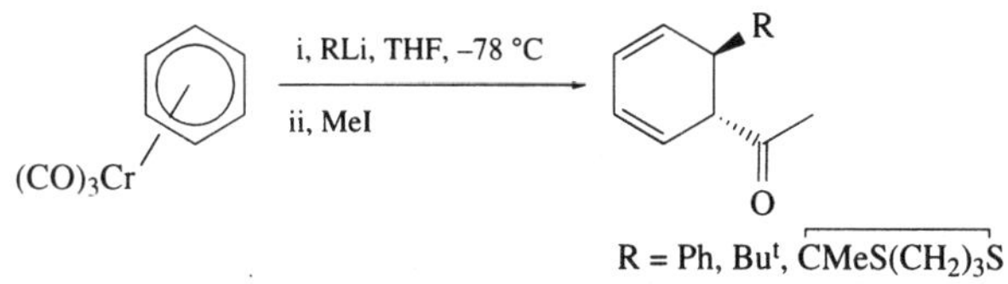

(7)

Kündig has since extended this method to phenyloxazoline complexes. Whereas CO insertion still occurs with MeI, groups with lower migratory aptitude such as allyl or benzyl alkylate directly, an effect attributed to the electron-withdrawing ability of the oxazoline unit. Moreover, since the starting materials can be prepared enantiomerically pure and attack can be induced selectively at one of the *ortho* positions by the lithium–nitrogen interaction, which involves less steric hindrance between R^2 and R^1 when R^1 is *syn* to the $Cr(CO)_3$ fragment, the reaction can produce one diastereomer of the resulting cyclohexadiene (Equation (8)).[934–6] Oxidation of the intermediate anions with I_2 as described above similarly produces *ortho*-substituted phenyloxazolidines with excellent selectivity. It is again noticeable that even simple organolithium reagents can be used without competition from ring lithiation.[937]

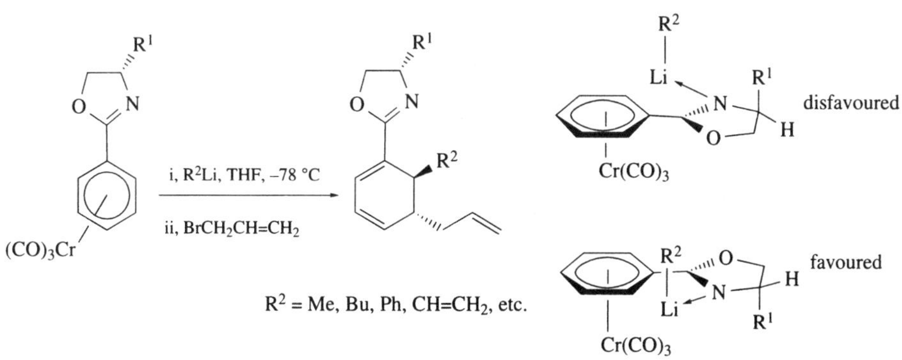

(8)

An unusual example of nucleophilic addition is provided by reaction of [$Cr(CO)_3$(η-BzOH)] with two equivalents of Bu^{t}Li. This produces an intermediate methylenecyclohexadiene complex, which undergoes acid-catalysed isomerization to the substituted arene complex (Scheme 18).[846]

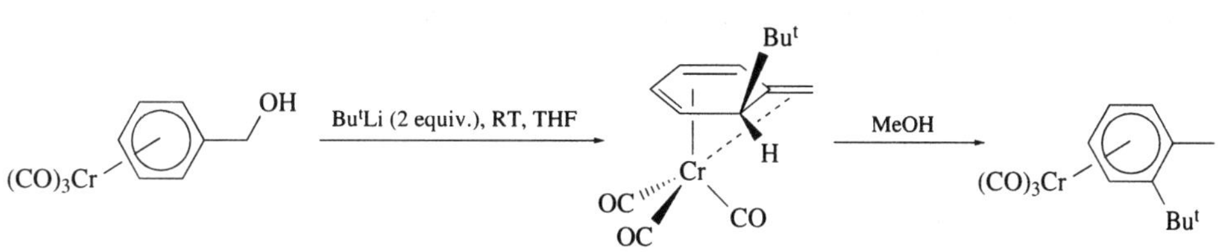

Scheme 18

8.3.8.4 Electrophilic attack

Chromium tricarbonyl arene complexes do undergo electrophilic attack, for example Friedel–Crafts acetylation, but less readily than the free arene due to the electron-withdrawing effect of the $Cr(CO)_3$ group. For example, in [$Cr(CO)_3$(η^6-$PhCH_2Ph$)], acylation occurred to almost equal extents on the complexed and uncomplexed rings.[938] Intramolecular reactions are more successful; for example, cyclization of [$Cr(CO)_3$(η-$PhCHRCH_2COOH$)] in strong acid gives the corresponding indanone complex, with some degree of selectivity for the *exo* diastereomer.[939] The regioselectivity of electrophilic attack, like that of nucleophilic attack, is controlled by the conformation of the tripod; whereas the latter occurs at the carbons which are eclipsed by CO ligands in the preferred conformation, the former is predicted to occur at those which are staggered.[907] Experimental verification is provided by the complex of 1,1-dimethylindane, which has a highly preferred conformation; nucleophilic attack occurs at C-5 and electrophilic attack at C-4 and C-6.[914,915]

8.3.8.5 Deprotonation at the benzylic position

The increased acidity of benzylic protons in chromium tricarbonyl arene complexes enables this position to be functionalized by deprotonation and reaction with electrophiles. The generation, stabilization and use of the benzylic carbanions has recently been reviewed.[197,940]

The increase in benzylic acidity caused by complexation to the electron-withdrawing $Cr(CO)_3$ unit has been studied in a number of cases. For benzene, toluene and xylenes, the increase in kinetic acidity on complexation was estimated to be 10^7–10^9 for the aryl protons and 5×10^4 to 5×10^5 for benzylic ones.[820] For Ph_2CH_2, Ph_3CH and fluorene, the ratios of the rate constants for deprotonation of complexed and uncomplexed arene by a KH–THF–crown ether system were estimated as 460, 240 and 10, respectively.[667] A study on diphenylmethanes bearing either *p*-NO_2 substituents or $Cr(CO)_3$ groups showed that the kinetic effect of the latter was greater, but the thermodynamic acidities of the nitroarenes were higher, showing the greater resonance stabilization of the anion afforded by the NO_2 group.[941] Alternatively, a study of the bis-$Cr(CO)_3$ complex of PhCH(CN)Ph ascribed the two groups similar resonance properties.[942] It should be noted that for steric reasons, the kinetic effects of attaching more than one chromium unit to diphenylmethanes is not cumulative. A further study of diphenylmethanes concluded that the $Cr(CO)_3$ unit was better than a *p*-TMS substituent in stabilizing the carbanion.[943] The increased acidity of the benzyl protons of a coordinated ring can be judged by the fact that $[Cr(CO)_3(\eta^6\text{-BzBz})]$ can be selectively deuterated to form $[Cr(CO)_3(\eta^6\text{-PhCD}_2\text{Bz})]$.[253] Moreover, the related complex derived from $CH_2(p\text{-}C_6H_4Me)_2$ undergoes deuteriation at the methylene group and the methyl of the coordinated ring, but not that of the uncoordinated ring.

Examination of the NMR spectra of the benzylic anion (**78**) derived from $[Cr(CO)_3(\eta^6\text{-PhBz})]$ and related complexes showed that the *ortho*-protons and carbons were inequivalent. This implies restricted rotation about the *ipso* to α-carbon bond, and an η^5-cyclohexadienylidene structure (**78b**) was suggested as the dominant resonance form for the anion.[667]

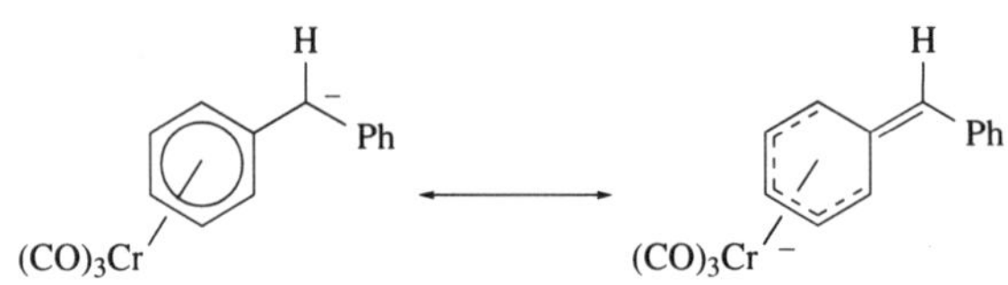

(78a) **(78b)**

A number of bases have been used for benzylic deprotonation; choice of base and conditions can be crucial in order to avoid either deprotonation of the aryl protons or nucleophilic attack at the ring.[940] The most widely applied system is $KOBu^t$ in DMSO, which is selective and compatible with many functional groups (except $OSiR_3$) but others employed include $KOBu^t$ in THF, KH in THF,[667] NaH in DMF, Bu^nLi or Bu^tLi in THF or ether, KOH in DME,[944] and amide bases such as LDA, $NaN(TMS)_2$, or $NaNH_2$,[945] with the addition of crown ethers or TMEDA in some cases. Certain substrates undergo alkylation at the benzylic position on electrochemical reduction, such as $BzCO_2Me$.[946,947]

Reactions are usually carried out under thermodynamic control at ambient temperature. In a seminal study, Semmelhack showed that treatment of $[Cr(CO)_3(\eta\text{-PhMe})]$ with BuLi in THF at –60 °C followed by CO_2/H^+ gave a mixture of *m*- and *p*-$[Cr(CO)_3(\eta\text{-MeC}_6\text{H}_4\text{CO}_2\text{H})]$, whereas at 0 °C $[Cr(CO)_3(\eta\text{-BzCO}_2\text{H})]$ was formed exclusively, showing that the benzylic carbanion is the thermodynamically favoured one.[948] Formation of the carbanion is accompanied by a colour change to red, which returns to yellow on quenching with electrophiles.

It has been noted that π-electron donor substituents such as NMe_2 or OMe deactivate benzylic protons to attack when in the *para* position, whereas with π-electron withdrawing groups (e.g., CO_2Bu^t) attack occurs preferentially at the *para* position. This appears to be solely due to the electronic effects of these groups rather than their influence on the conformation of the $Cr(CO)_3$ tripod.[949–52] Thus, in $[Cr(CO)_3(\eta\text{-}p\text{-XC}_6\text{H}_4\text{Me})]$, reactions proceed more readily if $X = CO_2Bu^t$ than if X = OMe or NMe_2; although the latter do still undergo deprotonation, yields are lower. Substituents have little overall influence when in the *meta* position; thus, $[Cr(CO)_3(\eta\text{-}m\text{-}XC_6H_4Me)]$ undergoes reaction quite readily even if X = OMe.

A useful illustration of the effect of substituents is provided by the reactions of 4-substituted *o*-xylenes (Scheme 19). If substituent X is OMe, NMe_2 or Cl, the 1-position is deactivated and reaction occurs only at the 2-position, but if X = CO_2Bu^t or TMS, it occurs exclusively at the 1-position. If X is alkyl (e.g., Bu^t) it exerts no influence and a mixture of substituted products is obtained.

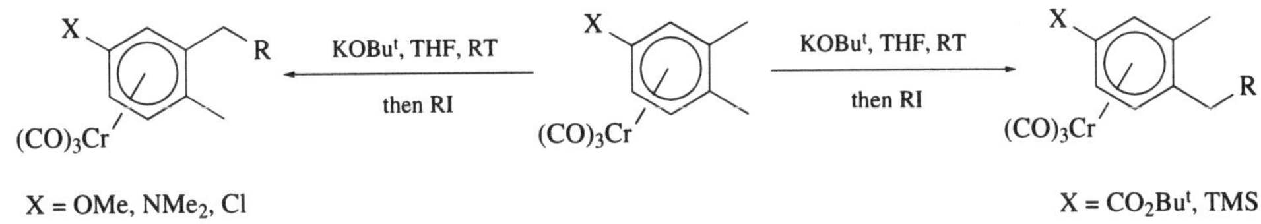

Scheme 19

Deprotonation of the benzylic position and subsequent attack of electrophiles generally takes place from the *exo* face due to the steric protection offered by the $Cr(CO)_3$ unit. An example of this is shown in Scheme 20: the phthalan complex can be alkylated at both benzylic positions to afford the *cis* (*exo, exo*) isomer only.[953] Similarly, functionalization of *N*-methyl-1,2,3,4-tetrahydroisoquinoline also occurs initially in the 4-*exo* position and then in the 1-*exo* site. Complexes of the related heterocycles *N*-methyltetrahydroquinoline, chroman and isochroman also undergo selective *exo*-benzylic functionalization.[954–7] Confirmation of the *exo* attack is provided by the two isomeric methoxytetralin complexes; the *syn* isomer is readily deprotonated whereas the *anti* is inert.[958] The substituent effects described above can be extended to complexes of indane and tetralin to achieve the selective *exo*-functionalization of one benzylic position, for example the one *meta* to an OMe substituent (Equation (9)).[949,959] The chromium tricarbonyl complex of indanone undergoes deprotonation in the five-membered ring, but adjacent to the CO group rather than at the benzylic position.[960]

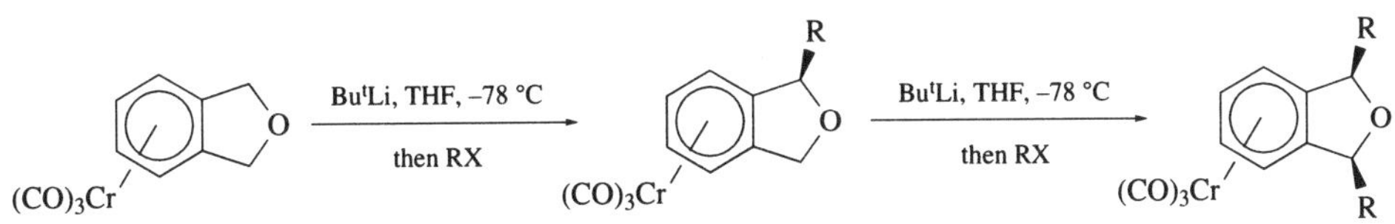

Scheme 20

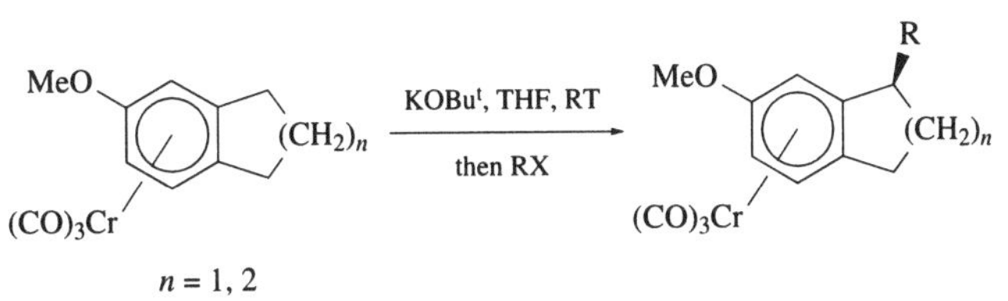

(9)

Although attack has been generally thought to occur exclusively from the *exo* face, very recently it has been reported that treatment of the $Cr(CO)_3$ complex of tetralin with $KOBu^t$ and d^6-DMSO does in fact result in exchange of both the *syn* and *anti* protons. However, the importance of steric factors is shown by the suppression of *syn* exchange if a methyl substituent is placed in the *exo* position.[961]

The benzylic alkylation reaction has been used to construct 6,7-benzomorphans from protected tetralones,[962,963] and to modify a number of substrates of biological interest, such as estradiols, canadine and codeine.[964–6]

The electronic influence of a substituent in the *ortho* position is the same as in the *para* position, but there is an additional stereochemical consequence because the benzylic protons become diastereotopic. and in favourable cases selective removal of one of these is possible, leading to excellent stereoselectivity. Thus, deprotonation of $[Cr(CO)_3(\eta\text{-}o\text{-}XC_6H_4Et)]$ where X = OMe, OPr^i or NEt_2 with $KOBu^t$ occurs stereoselectively and on treatment with HCHO one diastereomer of the product is formed exclusively (Equation (10)). Stereoselection is also observed in $[Cr(CO)_3(\eta\text{-}o\text{-}MeOC_6H_4CH_2OMe)]$, $[Cr(CO)_3\{\eta\text{-}o\text{-}C_6H_4(CH_2OMe)_2\}]$, and related compounds.[958,967,968] The origin of the effect in these cases is not chelation but steric interaction between the two *ortho* substituents, which leads to selective removal of the *exo*-hydrogen, giving the least hindered anion.

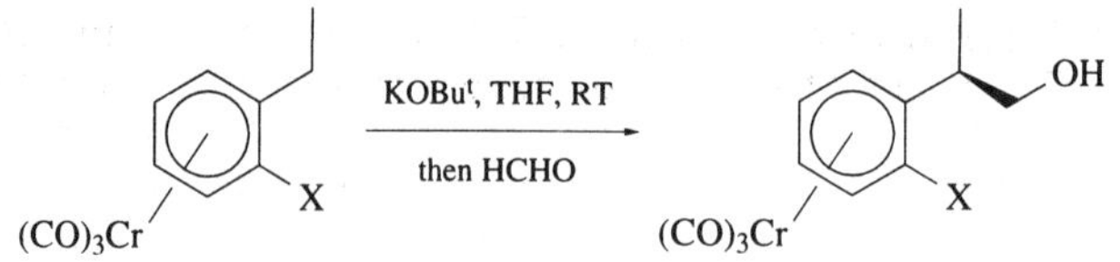

(10)

The presence of a heteroatom substituent on the benzyl carbon can direct attack of BuLi to that site; thus, complexes of benzyl alkyl ethers can be functionalized at the benzylic position rather than the ring. Moreover, the Wittig rearrangement (migration of the alkyl group from oxygen to carbon) is suppressed.[958] Consequently, lithiation of [Cr(CO)$_3$(η-BzOMe)] with Bu^{n}Li at −78 °C followed by quenching with electrophiles gives good yields of [Cr(CO)$_3$(η-PhCHROMe)] (R = Me, TMS, etc.) Intramolecular examples also exist, such as cyclization of [Cr(CO)(η-BzOCH$_2$CH$_2$CH$_2$Cl)] into a complex of 2-phenyl THF.[958] The presence of an optically active substituent on the heteratom can help to enhance the stereoselectivity of the reaction.[969]

The [2,3] Wittig rearrangement of benzyl-(*E*)-crotyl ether does still occur in its Cr(CO)$_3$ complex (Equation (11)); however, the major product is the *syn* isomer, as opposed to the *anti* for the free arene.[970] Complete stereoselectivity is observed in *ortho*-substituted complexes of allyl and crotyl ethers.[971,972]

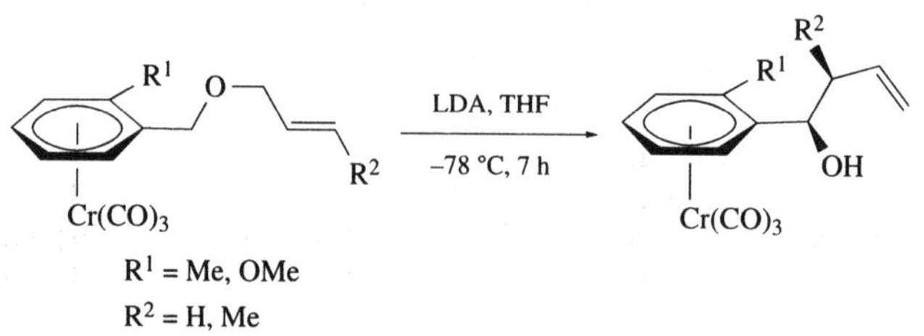

(11)

In favourable cases the presence of a chelating heteroatom on the β-carbon can induce completely stereoselective benzylic functionalization. One such case is the Cr(CO)$_3$ complex of (+)-*N,N*-dimethylamphetamine, which undergoes selective removal of the pro-(*R*) hydrogen on treatment with BuLi (Equation (12)); as shown in the accompanying Newman projections, an unfavourable steric interaction arises between the arene ring and the methyl group if the other hydrogen is to react (left).[973]

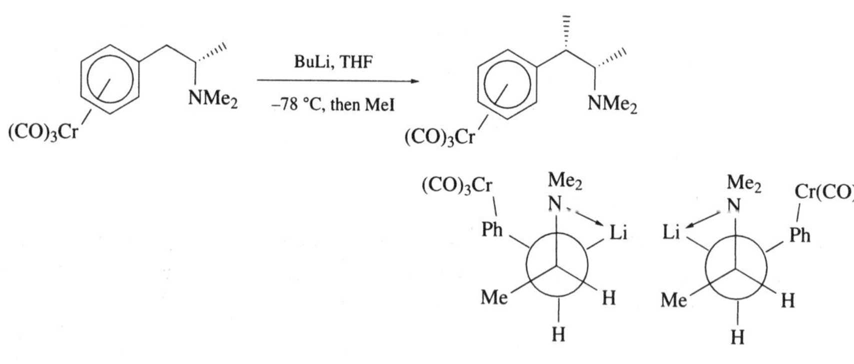

(12)

The carbanions produced by benzylic deprotonation readily undergo the expected reactions with electrophiles, for example deuteration and alkylation, without loss of the chromium fragment. More elaborate arene complexes can thus be built up, for example the synthesis of a complex of pentaphenylethane by deprotonation of [Cr(CO)$_3$(η^6-PhCHPh$_2$)] followed by alkylation with Ph$_2$CHBr.[974] Occasionally multiple substitution can occur at the benzylic centre, for example deprotonation and methylation of [Cr(CO)$_3$(η-PhMe)] gives a small amount of the complex of [Cr(CO)$_3$(η-PhPr^i)] as well as [Cr(CO)$_3$(η-PhEt)]. However, the presence of a methyl group at the benzyl site deactivates it towards further reaction, so that many substrates undergo a single alkylation; others, such as allylbenzene, readily undergo two.[975]

The most widely studied reactions of the benzylic carbanions are with carbonyl compounds. With the KOBu^t–DMSO system, aldehydes react with alkylbenzene complexes to give mixtures of complexed alcohols and alkenes (Scheme 21).[949,950,952,976,977] From [Cr(CO)$_3$(η-PhEt)] and HCHO, the major product is [Cr(CO)$_3${η-PhCMe(CH$_2$OH)$_2$}], whereas from [Cr(CO)$_3$(η-PhMe)] the major product is [Cr(CO)$_3$-

{η-PhC(CH_2OH)=CH_2}].[977] If the solvent is changed to THF, a third type of product is also formed, the ketone arising from *in situ* Woodward–Oppenauer oxidation (transfer of a hydride to excess aldehyde present). This is favoured by the presence of an alkyl substituent at the benzylic position, or electron-releasing substituents on the benzene ring, or electron-withdrawing substituents on the aldehyde, notably PhCHO.[978,979] The addition of the benzylic anions to ketones is limited to nonenolizable ones owing to the high basicity of the potassium salts.

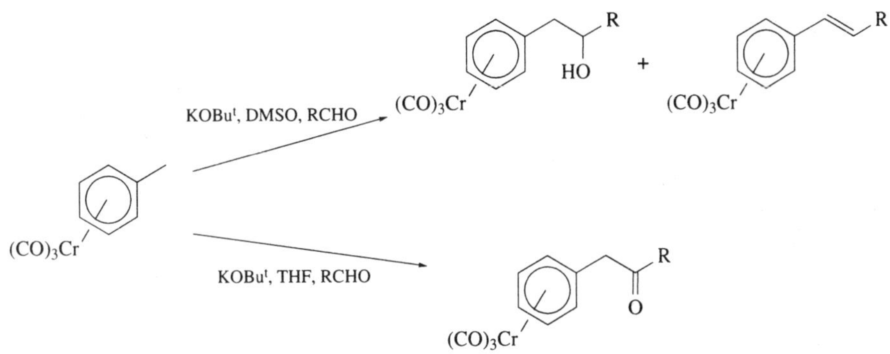

Scheme 21

Reaction of the benzylic anions with diethyl oxalate in either DMSO or THF produces derivatives of ethyl phenylpyruvate which, on protonation, give complexes of stable enols in some cases or the corresponding ketone in others. Treatment with TBDMS-Cl affords the corresponding silyl enol ether (Equation (13)).[980–2] Recently it has been shown that the benzylic anions react with Bu^{t}ONO to afford oximes.[983]

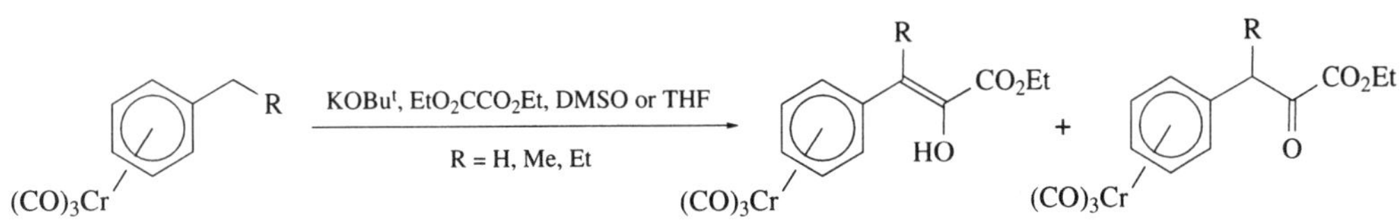

(13)

The case of [Cr$(CO)_3$(η-BzCH=CH_2)] has received detailed attention. The allylbenzene ligand can be deprotonated by a number of reagents, with the subsequent regioselectivity depending on the counterion; with Na^+, K^+ or NBu_4^+, the reaction is under thermodynamic control and addition occurs at the γ-carbon, whereas with Li^+ some degree of α-alkylation is also observed.[975,984–6]

8.3.8.6 Formation of benzylic carbonium ions

Despite the electron-withdrawing nature of the Cr$(CO)_3$ group, it is also able to act as a strong electron donor in order to stabilize benzylic carbonium ions.[987] These are usually made by protonation of the corresponding carbinol complexes with strong acids such as HBF_4, or by treatment of benzylic acetals with Lewis acids. After generation *in situ* by either of these routes, the carbonium ions can be trapped by the addition of nucleophiles, which as expected occurs from the *exo* face, for example the reaction of benzylic acetate with silyl enol ethers and $ZnCl_2$ occurs to give the *exo*-substituted products in 100% optical purity after decomplexation (Equation (14)).[988,989]

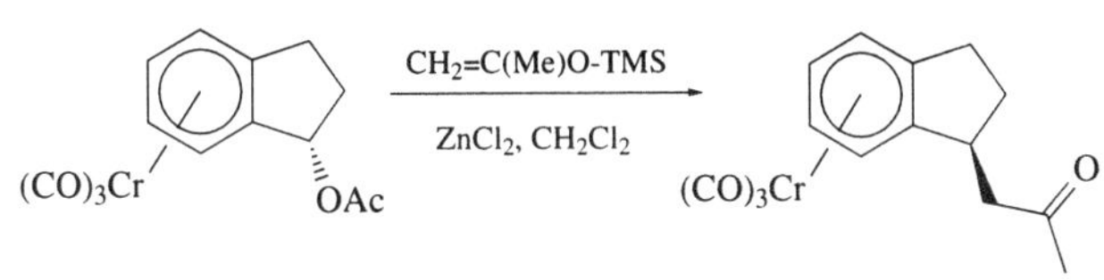

(14)

In this way, as shown in Scheme 22, the two diastereomers of [Cr$(CO)_3$(η-PhCHMeEt)] can be prepared from appropriate precursors with complete stereoselectivity. The intermediate carbonium ion does not have free rotation around the carbon–arene bond; elimination of acetyl and attack of the nucleophile both occur from the *exo* face.

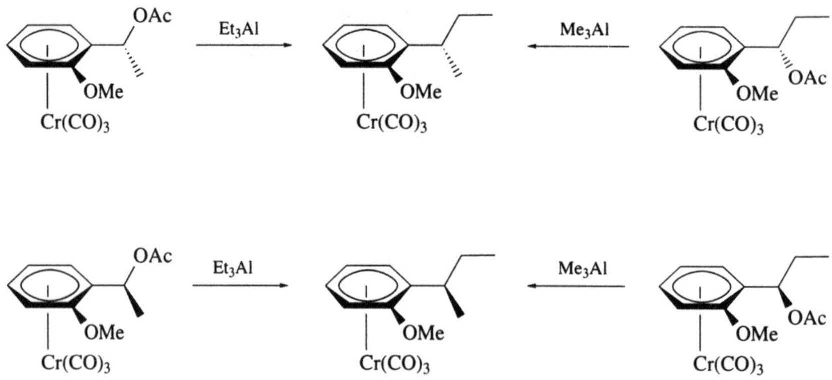

Scheme 22

Uemura and co-workers have explored the reactions of *endo*-tetralol derivatives with alkylating agents such as $AlMe_3$ and $ZnEt_2$; in each case the nucleophile adds in the *exo* position, though sometimes the addition of an additional Lewis acid, usually $TiCl_4$, was also required (Equation (15)).[990] Allyl- or propargylsilanes also act as nucleophiles, with BF_3 being the best added acid in this case.[991] When combined with the stereospecific *exo* attack of nucleophiles on the $Cr(CO)_3$ complex of tetralone, this allows production of both isomers of the disubstituted benzylic product depending on the order of introduction of the nucleophiles (Scheme 23). The same methodology can be applied to 3-hydroxyketone complexes, which can be made by diastereoselective complexation of the corresponding acetals (Scheme 24).

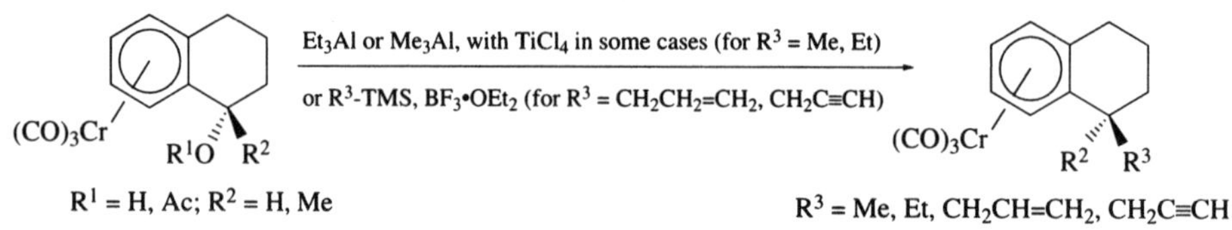

(15)

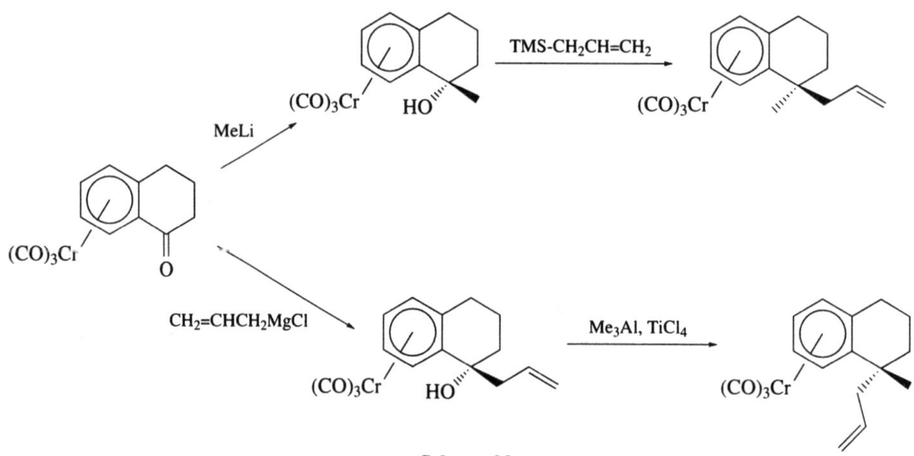

Scheme 23

The reaction of the benzylic carbonium ions with the Grignard reagent BuMgBr occurs by two competing pathways: simple nucleophilic attack and electron transfer leading to dimerization of the resulting radical and formation of biphenyl complexes.[992]

The stabilization of the carbonium ions allows Friedel–Crafts acylation of double bonds that are conjugated to an arene which bears a $Cr(CO)_3$ unit, for example complexes of indene, dihydronaphthalene and styrene.[993]

The chromium-stabilized carbonium ions can be used in intramolecular Friedel–Crafts attack on another arene ring. Thus, protonation of the $Cr(CO)_3$ complex of *N*-(3,4-dimethoxybenzyl)ephedrine (**79**) causes cyclization to give the complex of *cis*-2,3-dimethyl-4-phenyl-6,7-dimethoxy-1,2,3,4-tetrahydroisoquinoline (**80**); cyclization also occurs in the free arene but forms the *trans* isomer of the

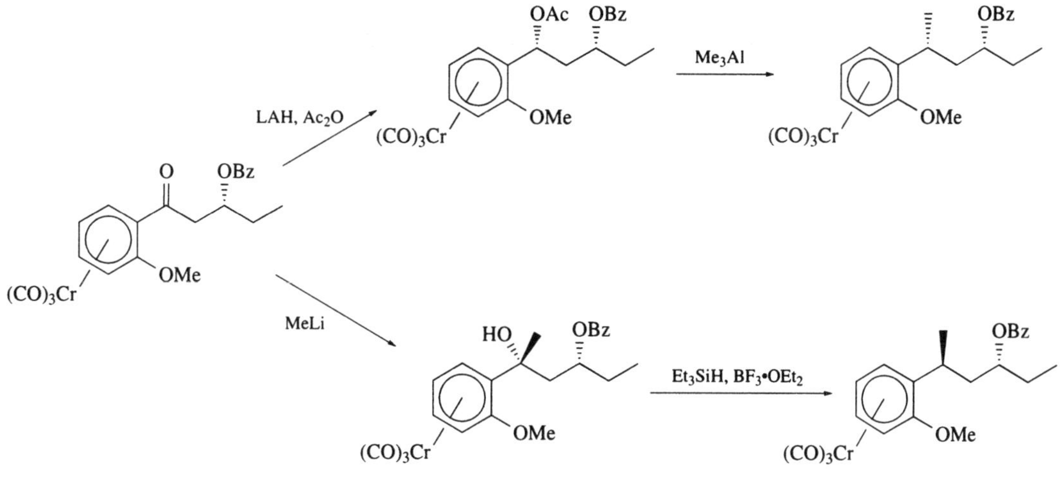

Scheme 24

product instead.[994] A similar cyclization of the complex of (R)-*N*-(3,4-dimethoxyphenethyl) halostachine has been reported.[995,996]

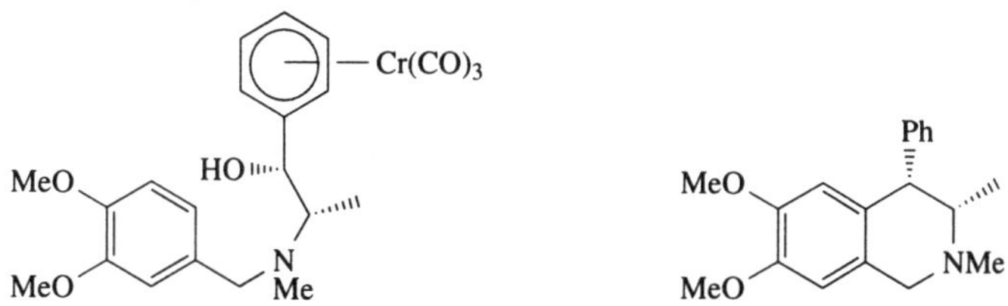

(79) **(80)**

Acetals derived from complexed benzaldehydes react with $TiCl_4$ and $AlMe_3$ by replacement of one of the OR groups by methyl. If the acetal derived from (R,R)-butane-2,3-diol is used, the pro-R oxygen is replaced selectively; subsequent Ritter reaction releases the diol and forms (R)-*N*-acetyl-1-phenylethylamine (Scheme 25).[997] A similar reaction of *o*-substituted aldehyde complexes with $TiCl_4$ and allylic alcohols leads to tetrahydropyran derivatives.[998,999]

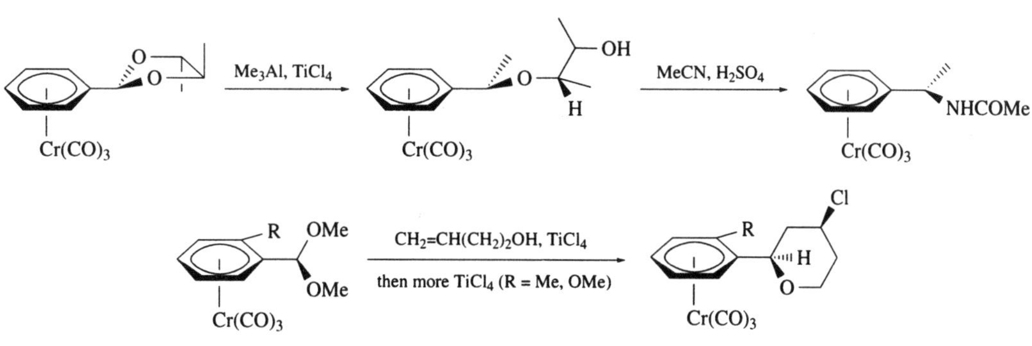

Scheme 25

The effect of the $Cr(CO)_3$ fragment has led to its employment in organic chemistry to stabilize carbonium ions for further study; for example, the structure of the complex of 2-cumyladamantan-2-ol has been determined, and the corresponding carbonium ion produced.[1000]

8.3.8.7 Reactions at arene substituents

As described in Section 8.3.6, chiral *o*-substituted benzaldehyde complexes can be resolved by a number of methods. They can then be used in a variety of further reactions, such as addition of alkyl Grignards, perfluoroalkyllithiums, enolates and carbanions, for further elaboration of the arene ligand, with stereoselectivities ranging from moderate to 100%. Solladié-Cavallo has proposed a method of determining which of the resulting diastereomers will predominate (Scheme 26).[202] This assumes that the carbonyl group will be coplanar with the arene ring and that its preferred conformation can be predicted on the basis of effects such as H-bonding, steric repulsion and dipolar interactions, and that reagents will approach from the *exo* face. In the case of the *o*-substituted benzaldehydes, this usually means that the carbonyl group is pointing away from the substituent. Using this approach it is expected that *meta*-substituted complexes will show only low stereoselectivities, which is indeed the case.

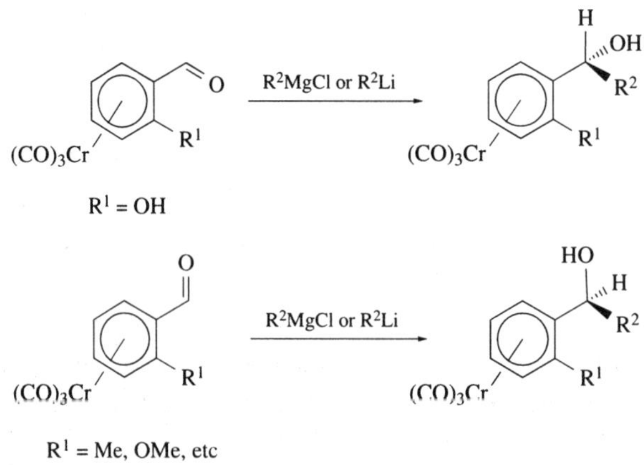

Scheme 26

A variety of organometallic reagents can be used in addition reactions. By way of illustration (Schemes 27 and 28), addition of the carbanion of nitromethane occurs with 95% diastereomeric excess at $-40\,°C$, and the product can be reduced with LAH and decomplexed in good yield. The stereoselectivity is slightly lower with the carbanion of $EtNO_2$, but purification gives analogues of ephedrine in good optical purity.[1001] Grignard reagents add to the $Cr(CO)_3$ complex of anisaldehyde to give enantiomerically pure (*o*-methoxyphenyl)ethanol or -propanol.[702] In the case of $[Cr(CO)_3(\eta$-*o*-TMS-$C_6H_4CHO)]$ this occurs with complementary diastereoselectivity depending on whether Lewis acids are present, thus allowing synthesis of (+) and (−)-PhMeCHOH.[1002] Addition of organozincs from 2-bromoesters or 2-bromonitriles (Reformatsky reagents) and the ZnI_2-catalysed addition of TMS-CN also occur with high stereoselectivity.[1003–5] The organolithium $LiCH_2NMeCH{=}NBu^t$ can be used to prepare analogues of halostachin in this way.[1006] Addition of perfluroalkyllithium reagents is also stereoselective, and leads to optically pure α-perfluoralkyl benzyl alcohols.[1007–10] Use of perfluoroalkyl zinc reagents is faster, but the optical yields are lower.[706,1011] The complexes resulting from these additions can themselves be used as recoverable chiral auxiliaries in Prelog-type syntheses of chiral carboxylic acids and the reduction of α-ketoesters.[1010,1012] Stereoselective addition of alkynyl lithium reagents and ethynyl magnesium bromide has also been reported recently.[1013]

Other reagents that also react stereoselectively include silyl enol ethers,[1014–16] titanium enolates,[1017,1018] allylboranes,[1019] methyl acrylate and acrylonitrile,[1020] and the bulky isonitriles RCH_2NC ($R = CO_2Et$ or Ts), which give *trans*-oxazolines in which the large R group points away from the chromium.[1021–3] Reaction of the aldehydes with MeNHOH gives the complexed nitrone, which reacts stereoselectively with alkenes to form isoxazolidines.[1024,1025] The Darzens condensation of $[Cr(CO)_3(\eta$-*o*-$MeC_6H_4CHO)]$ with RCH_2Cl to give epoxides occurs in 40–80% *ee*, or better when R = COPh.[1026]

Base-induced condensation of complexed benzaldehydes with PhCOMe or related species produces chalcone complexes such as $[Cr(CO)_3(\eta^6\text{-PhCH=CHCOPh})]$, which can also be made using arsonium ylids.[1027–30] Enantiomers of the *ortho*-substituted species can be separated by chiral chromatography and show high stereoselectivity in Michael additions with dimethyl malonate.[1031]

Reaction of the benzylamine complex shown in Scheme 29 with alkyllithiums occurs stereospecifically to provide the *anti* isomer with >90% stereoselection. The corresponding *syn* complex can be produced by *ortho*-lithiation of $[Cr(CO)_3(\eta\text{-PhCHMeNMe}_2)]$ followed by treatment with RCHO; decomplexation of the resulting ligand and stereoselective recomplexation with $[Cr(CO)_3(\eta^6\text{-}C_{10}H_8)]$ produces the isomer with the $Cr(CO)_3$ unit attached to the opposite face.[1032] By acting as chiral ligands

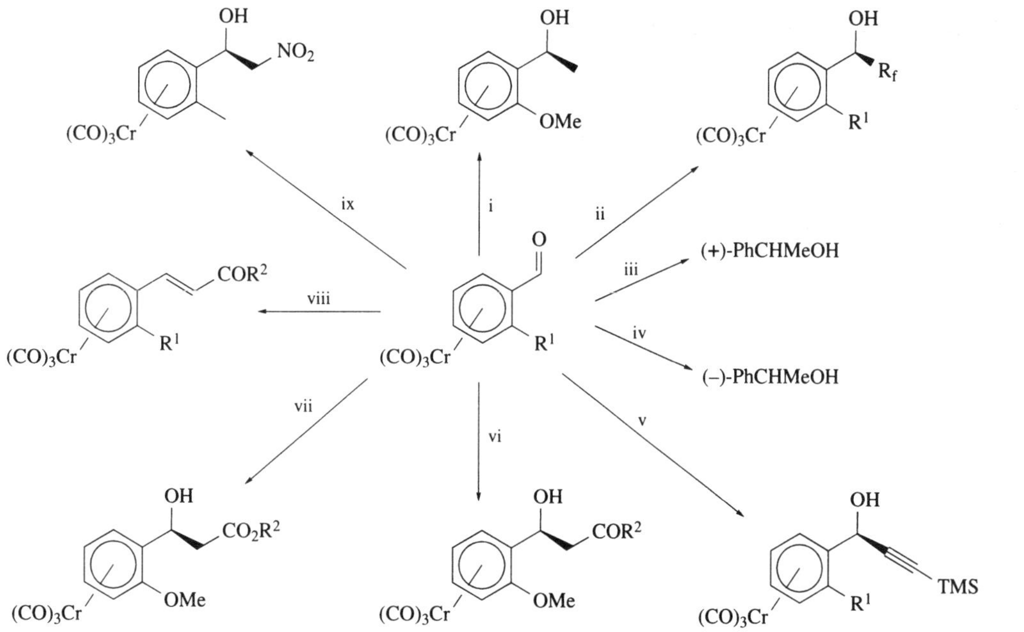

i, MeMgBr (R^1 = OMe); ii, LiR_f, Et_2O, –125 °C to 0 °C, or R_fI and Zn, ultrasound ($R_f = C_2F_5$, *i*-C_3F_7; R^1 = Me, OMe); iii, R^1 = TMS or $SiPr^i_3$: MeLi, THF, –78 °C, then decomplexation and desilylation with NBu_4F; iv, R^1 = TMS or $SiPr^i_3$: MeLi, $MgBr_2{\bullet}OEt_2$, Et_2O, or MeMgI, then decomplexation and desilylation with NBu_4F; v, LiC≡C-TMS (R^1 = Me, OMe); vi, R^2COMe, base (R^2 = Ph, Bu^t; R^1 = Me, OMe); vii, Zn, $BrCH_2CO_2R^2$, THF, Δ (R^2 = Me, Bu^t; R^1 = OMe); viii, R^2COMe, NaOH, EtOH (R^2 = Ph) or $[Ph_3AsCH_2COR^2]^+$, K_2CO_3, THF/H_2O (R^2 = Me, Ph, aryl);ix, $NaCH_2NO_2$, EtOH, –40 °C (R^1 = Me)

Scheme 27

to zinc, these complexes catalyse the addition of $ZnEt_2$ to benzaldehyde with a high degree of enantioselective induction.[1033] A related catalyst system (**81**) with similar effectiveness has been made by the selective complexation of norephedrine.[1034,1035]

As shown in Scheme 30, additions to complexed ketones can also be stereospecific. For example, *o*-alkoxyacetophenone complexes undergo stereospecific aldol reactions with boron enolates (generated from Bu_2BOTf, RCHO and Pr^i_2NEt) in good yields.[1036,1037] Dehydration of the products with TsOH in refluxing benzene produces complexed phenyl vinyl ketones. These in turn undergo diastereoselective 1,4-addition of organocuprates; when the *o*-substituent is OMe the isomer produced depends on the reagent used, for example, Gilman reagents such as Bu_2CuLi attack one face, whereas $BuCu{\cdot}BF_3$ attacks the other. X-ray studies show that the benzylic carbonyl oxygen is *anti* to the *o*-OMe group, whereas for R = Me (where both reagents attack the same face) it is situated *syn* to the methyl.[1038,1039]

Examples of 1,5-stereocontrol have also been produced by the same group. Treatment of the acetal complex in Scheme 30, obtained by selective complexation, with $NaCH(CO_2Me)_2$ and palladium catalyst followed by Ph_3CBF_4 proceeded regio- and stereospecifically to give a product that could subsequently be functionalized stereospecifically at the benzylic position to *syn* and *anti* isomers by the carbonium ion methodology discussed in the previous section.[700]

The stereoselection of the addition of 2-butenylmagnesium chloride to acetophenone to give (**82**) can be enhanced by coordination to $Cr(CO)_3$. If the reaction is carried out in the presence of $AlEt_3$, an 89:11 ratio of *anti* (illustrated) vs. *syn* isomers is obtained. The selection is explained by the intermediacy of a cyclic transition state of the aluminum 'ate' complex; interestingly, in the presence of BEt_3 instead, the *syn* isomer is the major product.[1040]

Ketone complexes $[Cr(CO)_3(\eta\text{-}PhCOR)]$ (R = alkyl) undergo reductive coupling to stilbene complexes $[Cr(CO)_3(\eta^6,\eta^6\text{-}PhCR{=}CRPh)Cr(CO)_3]$ on treatment with $TiCl_3$ and lithium; typically a Z:E ratio of 3:1 is obtained.[1041] Hypervalent iodine oxidation of ketone complexes with $PhI(OAc)_2$/-KOH/MeOH converts them to α-hydroxymethyl dimethylacetals, for example $[Cr(CO)_3(\eta\text{-}PhCOMe)]$ gives $[Cr(CO)_3\{\eta\text{-}PhC(OMe)_2CH_2OH\}]$.[1042] In the case of tetralone, the reaction is stereospecific with the OH placed on the *endo* face.[1043]

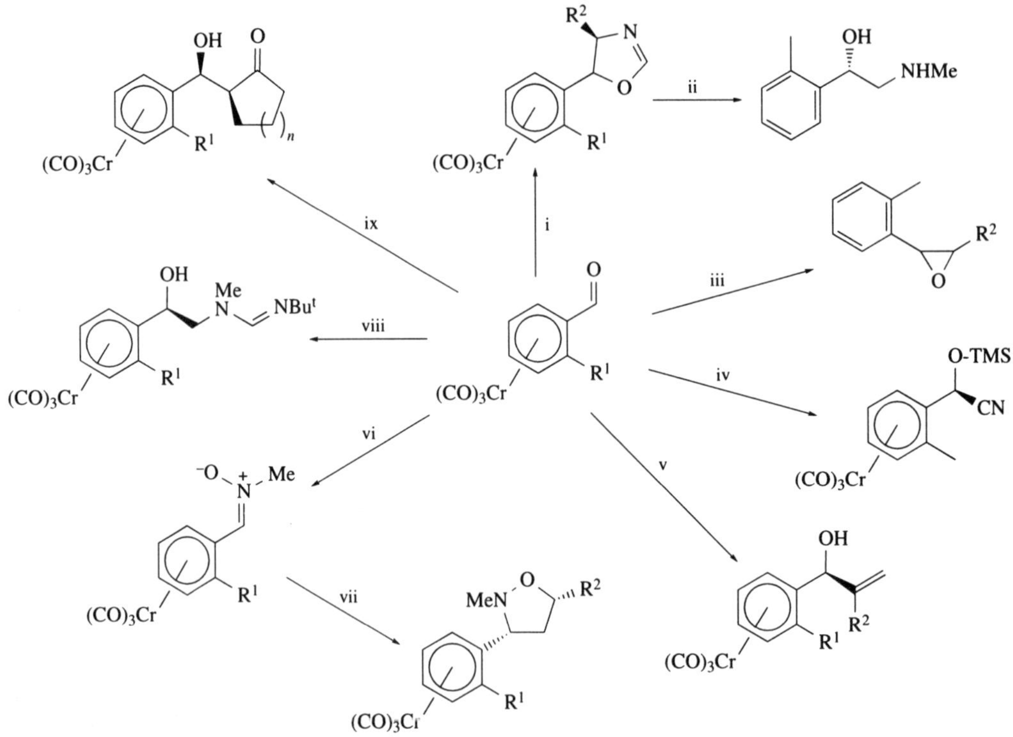

i, R^2CH_2NC, base (R^2 = Ts, CO_2Et); ii, decomplex then LAH; iii, $ClCH_2R^2$, KOH or $KOBu^t$, $(C_8H_{17})_4NBr$, then decomplex with H_2O_2 (R^2 = CN, CO_2Bu^t, COPh, R^1 = Me); iv, ZnI_2, TMS-CN, CH_2Cl_2 (R^1 = Me); v, $R^2CH{=}CH_2$, dabco (R^2 = CO_2Me, CN, R^1 = Me, OMe); vi, MeNHOH, $NaHCO_3$, CH_2Cl_2 reflux (R^1 = H, TMS); vii, $R^2CH{=}CH_2$ (R^2 = Ph, OEt, OAc, TMS); viii, $LiCH_2NMeCH{=}NBu^t$ (R^1 = Me, OMe, F); ix, $CH{=}C(O\text{-}TMS)CH_2(CH_2)_nCH_2$, $BF_3{\bullet}OEt_2$, CH_2Cl_2, –78 °C (n = 1–3, R^1 = Me, OMe, TMS)

Scheme 28

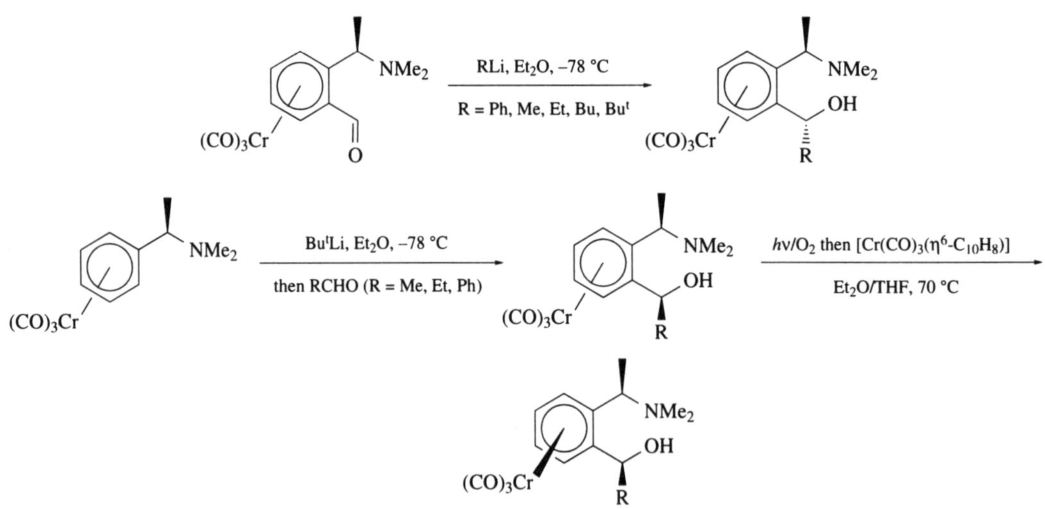

Scheme 29

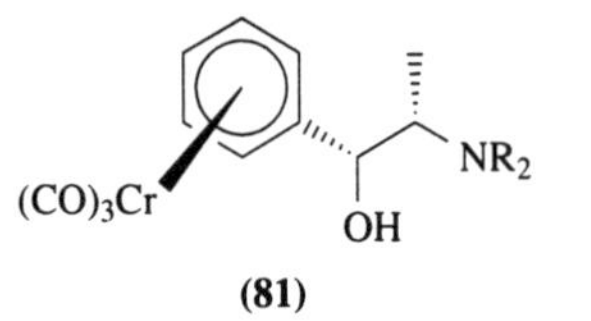

(81)

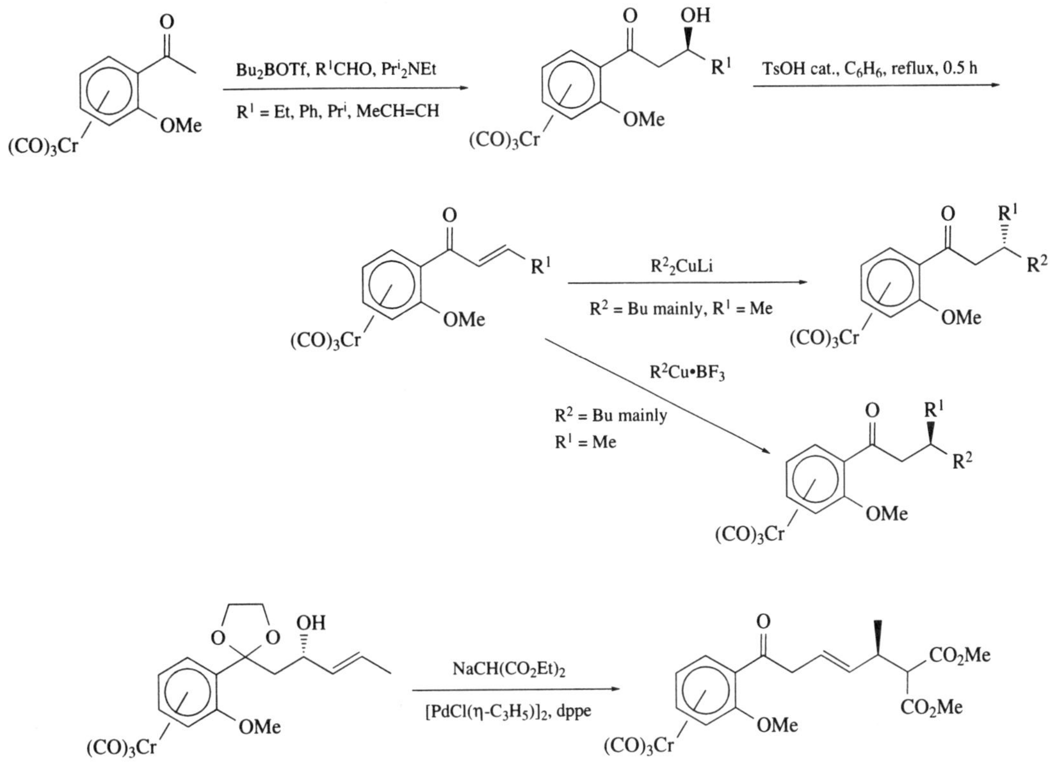

Scheme 30

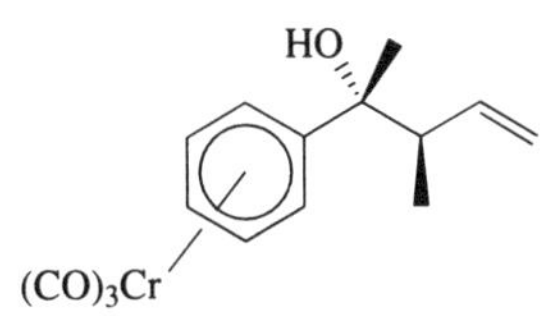

(82)

Chromium tricarbonyl complexes of cyclic ketones such as indanone and tetralone undergo nucleophilic attack at the carbonyl group from the *exo* face, for example reduction with LAH or $NaBH_4$ produces the *endo* alcohol.[1044–8] Addition of Grignard reagents and organolithiums also occurs from the *exo* face, and has been extensively employed by Uemura and co-workers in the synthesis of a range of natural products such as anthracyclinones and hydroxycalamenenes.[1049–52] Chromium tricarbonyl complexes of 2-arylidene tetralones undergo cyclopropanation of the exocyclic double bond with Me_3SO^+ exclusively from the *endo* face, and the conjugate addition of $MeNO_2$ to this double bond is also stereoselective.[1053,1054]

Complexation of ethylenedioxy cyclobutabenzene followed by deprotection allows access to complexes of 1-oxocyclobutabenzene, which can be reduced to the *endo*-1-hydroxy derivative with LAH in >99% *de*.[1055] A similar sequence produces the 1,2-dioxocomplex, which reacts with excess $LiCH{=}CH_2$ or $BrMgCH{=}CH_2$ by a double oxy-Cope rearrangement to give the benzocyclooctane-1,4-dione complex (Scheme 31). If only a small excess of vinyl reagent is used, *exo*-addition to one carbonyl group occurs.[1056,1057]

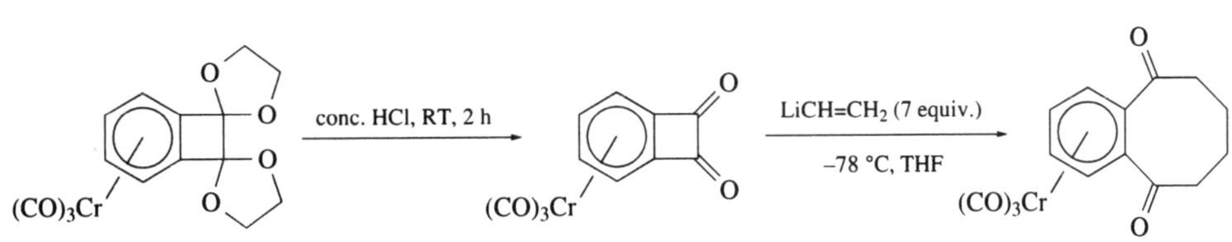

Scheme 31

Two groups have demonstrated that chromium tricarbonyl complexes of cyclobutabenzenes with 1-OEt or 1-OLi substituents undergo distal ring opening to give transient *o*-quinodimethane species that can be trapped by cycloaddition with alkenes as *trans*-substituted tetrahydronaphthalenes (Scheme 32). In the case of the 1-ethoxy derivative this occurs during thermal isomerization of the *endo* to the *exo* isomer.[1055,1058,1059] In contrast, ring opening of the proximal bond occurs on treatment of the 1-hexenyl-1-hydroxy complex with butyl lithium (Scheme 33), and has also been observed on reaction of the 1-oxo species with primary amines, giving [Cr(CO)$_3$(η-PhCH$_2$CONHR)].[1060]

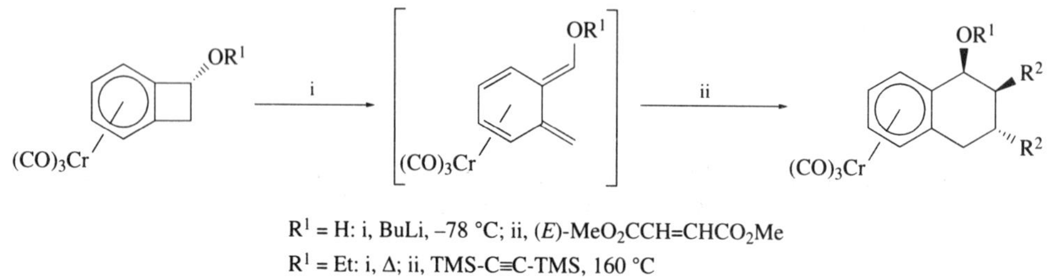

Scheme 32

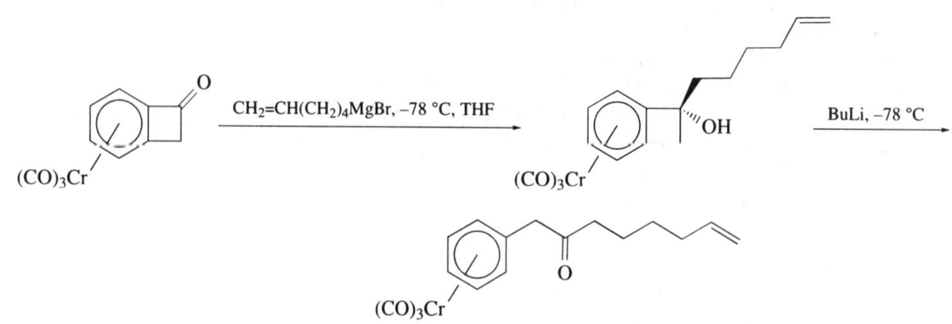

Scheme 33

Addition of benzylamine to the complexed benzaldehydes described above gives the corresponding imines [Cr(CO)$_3$(η-*o*-MeC$_6$H$_4$CH=NBz)], which undergo stereospecific alkylation at the α-carbon; hydrolysis and decomplexation yields optically active benzylamines.[1061] Attack of ester enolate carbanions on the related imines [Cr(CO)$_3$(η-*o*-MeC$_6$H$_4$C=NPh)] provides an enantioselective synthesis of β-lactams.[1062] Addition of ArCNO (Ar = the bulky C$_6$-2,4-Cl$_2$Me$_3$) to chiral styrene complexes [Cr(CO)$_3$(η^6-*o*-RC$_6$H$_4$CH=CH$_2$)] (R = Me, OMe, Cl) occurs stereospecifically to produce derivatives of dihydroisoxazoles.[1063]

The imine complexes can also be derived from [Cr(CO)$_3${*o*-MeC$_6$H$_4$NH$_2$}] and RCHO (R = Ph, Bu^t), and undergo highly diastereoselective addition of nucleophiles such as LiMe or NaBH$_4$.[1064] Starting from the same complex, Ziessel and co-workers have prepared molecular tweezer ligands derived from pyridine or 1,10-phenanthroline such as 2,6-NC$_5$H$_3${CH$_2$NHC$_6$H$_4$MeCr(CO)$_3$}$_2$, 2,6-NC$_5$H$_3${CH=NC$_6$H$_4$MeCr(CO)$_3$}$_2$, and 2,9-N$_2$C$_{12}$H$_6${CH$_2$NHC$_6$H$_4$MeCr(CO)$_3$}$_2$ (**83**) which can be complexed to rhodium or copper fragments.[1065,1066] In a related assembly, Schiff's base condensation of [Cr(CO)$_3${*p*-C$_6$H$_4$(CHO)$_2$}] with diamines such as 1,3-C$_6$H$_4$(CH$_2$NH$_2$)$_2$ or 1,6-diaminohexane produced macrocycles such as (**84**) in a one-pot procedure. From N(CH$_2$NH$_2$)$_3$ a bicyclic three-dimensional cage macrocycle was formed which could encapsulate a copper atom.[1067]

8.3.8.8 Catalysis by chromium arene complexes

Several types of reaction are catalysed by chromium arene complexes, and these have recently been reviewed.[1068] Their major use has been in hydrogenation, particularly the partial reduction of conjugated dienes or alkynes to mono-enes.[1069–73] Usually the more active complexes of arenes such as naphthalene and phenanthrene are employed, and as in arene exchange, ketones act as accelerators, in keeping with the idea that the reaction proceeds through a similar ring-slipped intermediate.[1074] The system is also capable of diene isomerization.[1070,1075–7]

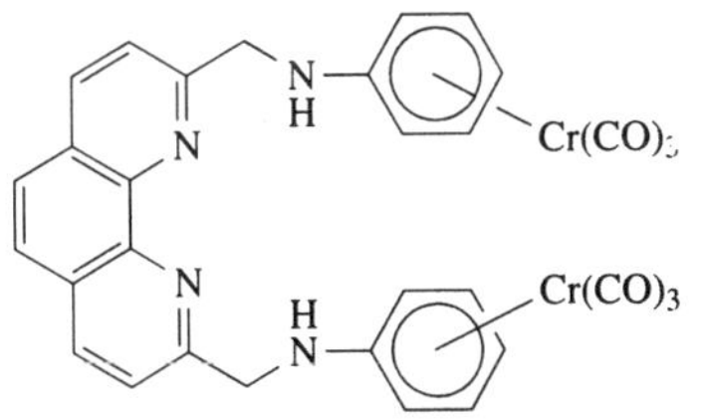

(83)

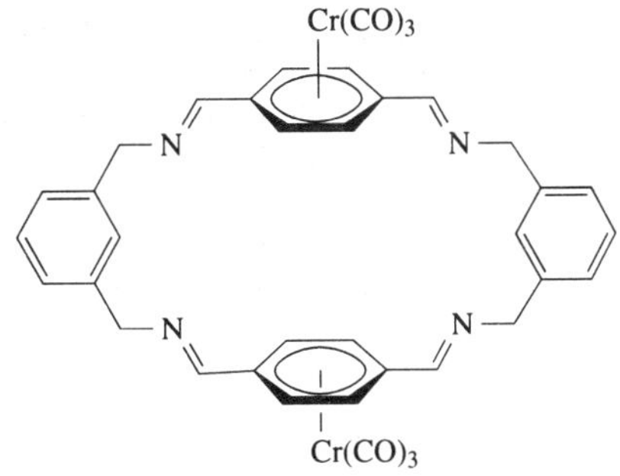

(84)

Other reactions catalysed by $[Cr(CO)_3(\eta^6\text{-}C_{10}H_8)]$ include cycloaddition of alkenes to azepins,[1078] the Kharasch addition of CCl_4 to alkenes,[1079] and the polymerization of phenylacetylene.[1080]

8.4 OTHER MOLYBDENUM AND TUNGSTEN ARENE COMPLEXES

Routes to a number of half-sandwich molybdenum and tungsten arene complexes with phosphine ligands have been developed; most of these start from the bis(arene) complexes, which are readily available by the metal atom co-condensation route. For molybdenum, one of the arene rings can be displaced, for example in reaction with $(C_2F_5)_2PCH_2CH_2P(C_2F_5)_2$ under N_2, which gives $[\{Mo(L_2)(\eta\text{-}C_6H_6)\}_2(\mu\text{-}N_2)]$ or $[Mo(L_2)(N_2)(\eta\text{-PhMe})]$ depending on the arene. The N_2 ligand can then be replaced by other donors such as CO, H_2, NCMe or py.[1081] One naphthalene can be displaced from $[Mo(\eta^6\text{-}C_{10}H_8)_2]$ by PMe_3, PF_3 or $P(OMe)_3$, but CO displaces both.[184]

The chemistry developed by Green and co-workers is summarized in Scheme 34 and leads to the cationic ethylene–hydride complex $[MoH(\eta\text{-}C_2H_4)(L_2)(\eta\text{-arene})]^+$. This compound is fluxional, the dynamic process involving alkene rotation and insertion to give an Mo–Et group.[1082,1083] This insertion process is also implicated in its reaction with PMe_3, which gives $[MoH(L_2)(PMe_3)(\eta\text{-PhEt})]^+$ when arene = C_6H_6; the ethyl group is transferred to the ring to form an *endo*-C_6H_6Et cyclohexadienyl ligand, and a proton migrates back to the molybdenum. This mechanism is supported by the fact that $[MoH\{o\text{-}C_6H_4(PMe_2)_2\}(\eta\text{-}C_2H_4)(\eta\text{-}C_6H_6)]^+$ reacts with further phosphine to give $[Mo\{o\text{-}C_6H_4(PMe_2)_2\}_2(\eta^5\text{-}endo\text{-}C_6H_6Et)]^+$.[1083] The tungsten chemistry differs only slightly from that of molybdenum; $[W(\eta\text{-PhMe})_2]$ does not react cleanly with allyl chloride, but does do so with allyl acetate; subsequent treatment with $EtAlCl_2$ gives a reasonable overall yield of $[W(\mu\text{-Cl})(\eta\text{-}C_3H_5)(\eta\text{-PhMe})]_2$, which can be converted to other cationic allyl species.[1084]

The arene rings in $[W(\eta\text{-arene})_2]$ are less labile than for molybdenum; displacement of one arene from $[W(\eta\text{-PhMe})_2]$ with PMe_3 occurs only at 200 °C to produce $[W(PMe_3)_3(\eta\text{-PhMe})]$, which can also be made by co-condensation methods.[1085] If done in the presence of a potassium mirror, however, displacement occurs at room temperature, presumably due to weakening of the tungsten–arene bond caused by reductive occupation of antibonding orbitals, to give a 50% yield of $[WH_2(PMe_3)_2(\eta\text{-PhMe})]$. This compound also results in 30% yield from reaction of $[WH_4(PMe_3)_4]$ with cyclohexadiene, or Na/K alloy reduction of $[WCl_4(PMe_3)_4]$ in the presence of cyclohexadiene.[1086,1087] Like the bis(arene) complexes, $[W(PMe_3)_3(\eta\text{-}C_6H_6)]$ is appreciably basic and can be reversibly protonated to a monocation and then to *mer*-$[W(H)_2(PMe_3)_3(\eta\text{-}C_6H_6)]^{2+}$.

A number of compounds of the general type $[MoL_3(\eta^6\text{-ArPR}_2)]$ (L = phosphine) that contain an arylphosphine ligand bonded in an η^6 manner are known. Perhaps the most accessible is $[Mo(PPh_2Me)_3(\eta^6\text{-PhPMePh})]$, which can be prepared in 75% yield by reduction of $MoCl_5$ with magnesium under argon in the presence of PPh_2Me.[1088] The $PPhMe_2$ analogue can be prepared similarly in 67% yield.[1089] The related complexes $[Mo(triphos)\{\eta^6\text{-}o\text{-TolP(o-Tol)}_2\}]$, $[Mo\{PhP(CH_2CH_2PPh_2)_2\}(\eta^6\text{-PhPPh}_2)]$, $[Mo(pp_3)(\eta\text{-arene})]$ ($pp_3 = MeP(CH_2CH_2CH_2PMe_2)_2$, arene = C_6H_6, PhMe) and $[Mo(dppe)(PPhR_2)(\eta^6\text{-PhPR}_2)]$ (R = Ph, Et) can be prepared by reduction of precursors such as $[MoCl_3L_3]$ and $[MoCl_4(dppe)]$ in the presence of the appropriate phosphine.[1090–4] The same route provides $[W(PPhEt_2)_2(N_2)(\eta^6\text{-PhPEt}_2)]$.[1095]

The chemistry of these half-sandwich complexes is summarized in Scheme 35.[1096–8] One of the phosphines in $[Mo(PPh_2Me)_3(\eta^6\text{-PhPMePh})]$ can be replaced by a variety of ligands such as PMe_3, $P(OMe)_3$, Bu^{t}NC, CO, MeCN, N_2 (hence the obligatory use of an argon atmosphere for the synthesis) and H_2, which yields the dihydride $[Mo(H)_2(PPh_2Me)_2(\eta^6\text{-PhPMePh})]$. These reactions are reversible, but can be driven forward by addition of a Lewis acid to absorb the liberated phosphine.[1099,1100] Reaction with $[WCl_4(PPh_3)_2]$ provides a useful route to mixed metal $[MoWCl_4(PPh_2Me)_4]$ species.[1101,1102]

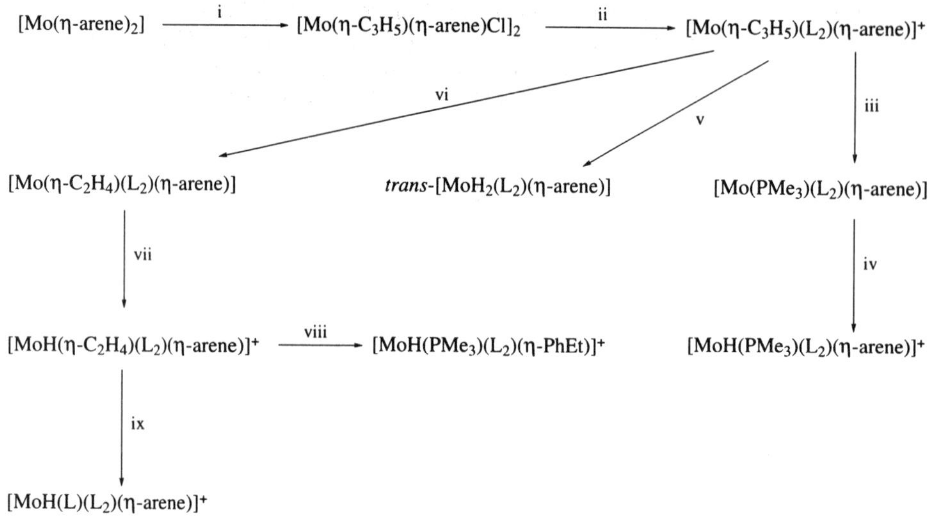

i, CH_2=$CHCH_2Cl$; ii, L_2 (L_2 = dmpe, dppm, *o*-$C_6H_4(EMe_2)_2$ where E = P or As); iii, Na/Hg, PMe_3 (L_2 = dmpe); iv, dil. HCl, NH_4PF_6; v, Na/Hg, H_2 (L_2 = dmpe); vi, Na/Hg, C_2H_4 (L_2 = dmpe, *o*-$C_6H_4(EMe_2)_2$ where E = P or As); vii, HBF_4•OEt_2, NH_4PF_6; viii, PMe_3 (arene = C_6H_6); ix, L (L = $P(OMe)_3$, CO, 2, 6-xylylNC)

Scheme 34

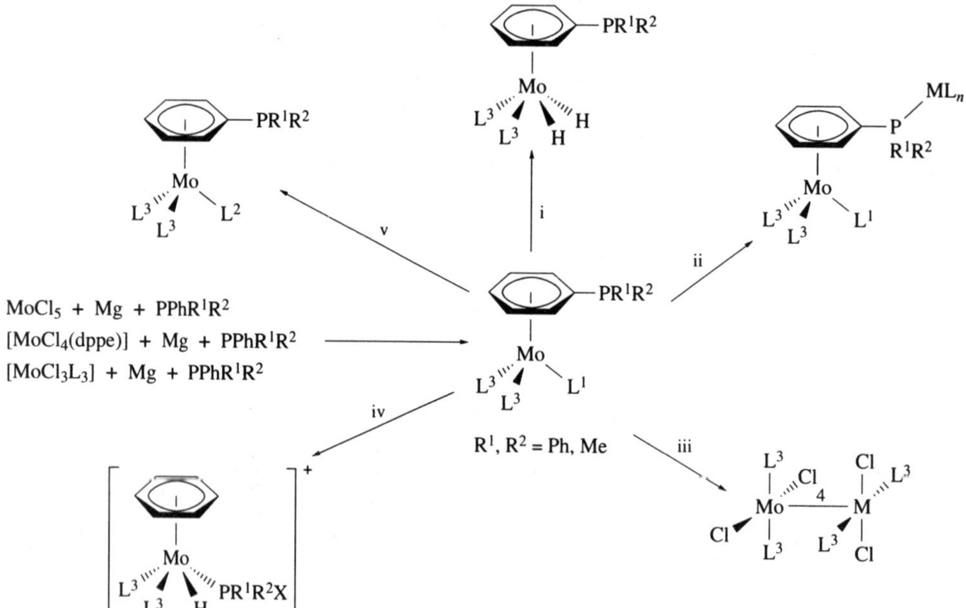

i, H_2 (L^3 = PPh_2Me); ii, $[RhCl(cod)]_2$ ($L^3{}_2$ = dppe, L^1 = CO, PMe_2Ph; L^3 = PPh_2Me, L^1 = $P(OMe)_3$, $CNBu^t$) or $[M(CO)_6]$, *hν* (M = Cr, Mo, W; $L^3{}_2$ = dppe, L^1 = CO, PMe_3); iii, $[MoCl_4(THF)_2]$ or $[WCl_4(PPh_3)_2]$ (L^3 = PPh_2Me, PMe_2Ph); iv, HBF_4 (aq.) (X =OH) or HBF_4•OEt_2 (X = F) (R^1 = R^2 =Ph, $L^3{}_2$ = dppm, dppe, dppp); v, L^2 (L^2 = N_2, $P(OMe)_3$, CO, PMe_3, $CNBu^t$

Scheme 35

Dinuclear species with μ,σ,η^6-arylphosphine ligands have also been prepared. Reduction of $[Mo(acac)_3]$ with $AlMe_3$ and PPh_3 under nitrogen produces $[Mo(N_2)(PPh_3)(\mu,\sigma,\eta^6\text{-}PhPPh_2)]_2$, from which the N_2 ligands can be displaced by isonitriles.[1103]

The reaction of $[MoBr_2(CO)_4]$ with NaSR (R = 2,6-$C_6H_3Ph_2$) produces 60–70% yields of $[Mo(SR)(\sigma,\eta^6\text{-}SR)(CO)]$ (**85**), in which one thiolate ligand acts as a σ,η^6 chelate.[1104] The arene coordination is relatively labile and can be displaced by ligands such as CO, bipy or dppe.[1105,1106] Related complexes with alkoxide ligands have been made by Rothwell; $[MH_4(PPh_2Me)_4]$ (M = Mo, W) reacts with ROH (same R) in a melt at 150 °C to give $[MH(PPh_2Me)_2(\sigma,\eta^6\text{-OR})]$, and reduction of $[WCl_4(OR)_2]$ in the presence of PPh_2Me or dppm gives $[W(OR)(\sigma,\eta^6\text{-OR})(L)]$; the η^1-dppm complex reacts with 3-hexyne to give $[W(OR)_2(C_4Et_4)]$ containing a tungstacyclopentatriene ligand.[1105,1107–11]

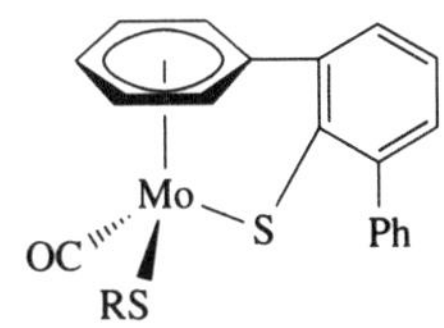

(85)

8.5 HETEROARENE COMPLEXES

8.5.1 Bis(heteroarene) Complexes

Co-condensation of chromium atoms with $2,6\text{-NC}_5H_3R_2$ (R = Me, TMS) affords the appropriate bis(η^6-pyridine) complex, which in the case of R = TMS can be desilylated to unsubstituted $[Cr(\eta^6\text{-py})_2]$. These complexes are very similar in behaviour to $[Cr(\eta\text{-C}_6H_6)_2]$, for example they can be readily oxidized to the corresponding chromium(I) cations.[63,1112,1113] In the x-ray structure of $[Cr(\eta^6\text{-2,6-NC}_5H_3Me_2)_2]^+$ the rings are eclipsed (synperiplanar). Condensation of a mixture of benzene and pyridine produces $[Cr(\eta^6\text{-py})(\eta\text{-C}_6H_6)]$. An improved synthesis of $[Mo(\eta^6\text{-2,6-NC}_5H_3Me_2)_2]$ has been reported, together with the displacement of one lutidine by PMe_3 or allyl chloride in a similar manner to other bis(arene) molybdenum complexes.[1114]

Co-condensation routes to bis(arene) complexes of the phosphabenzene $2,4,6\text{-PC}_5H_2Bu^t_3$ and unsubstituted arsabenzene have also been described; in the former the Bu^t groups stop arene rotation occurring, and a 3:1 distribution of isomers in which the rings are synclinal or antiperiplanar is obtained; the deviation from a statistical 2:1 ratio is attributed to the electronic perturbation of the phosphorus atoms.[1115,1116]

8.5.2 Heteroarene Metal Tricarbonyl Complexes

To achieve η^6-coordination of pyridine to the $Cr(CO)_3$ fragment rather than coordination through nitrogen, protection of the nitrogen atom by substitution at the 2 and 6 positions is necessary. Thus, refluxing $[Cr(CO)_6]$ with pyridine produces $[Cr(CO)_3(py)_3]$ whereas collidine (2,4,6-trimethylpyridine), $2,6\text{-NC}_5H_3Bu^t_2$ and $2,4,6\text{-NC}_5H_2Bu^t_3$ produce $[Cr(CO)_3(\eta^6\text{-pyridine})]$.[1117,1118] The x-ray structure of the $2,6\text{-NC}_5H_3Bu^t_2$ derivative shows that the chromium tripod eclipses the 2 and 6 positions with the CO ligands situated between the methyl groups of the Bu^t substituent.[1119] It has been reported that 4-vinylpyridine also gives an η^6-complex with $[M(CO)_3(NCMe)_3]$.[1120] The pyridine ligands are more labile than naphthalene and exchange readily with other arenes.[245] Although the nitrogen atom in these species is sterically hindered, it still forms adducts with Lewis acids.[1118,1121] Photolysis of the pyridine complexes with PPh_3 leads to $[Cr(CO)_2(PPh_3)(\eta^6\text{-py})]$.[1122]

Complexation of 2,6-bis(trimethylsilyl)pyridine followed by desilylation with F^- gives the unsubstituted $[Cr(CO)_3(\eta^6\text{-py})]$.[1123] As shown in Scheme 36, it undergoes both deprotonation/alkylation and stereoselective alkyllithium attack at the 2-position; the resulting 2-methylpyridine complex also undergoes stereoselective deprotonation/alkylation at the methyl group.[1124–6] The first η^6-pyridine complex of molybdenum, $[Mo(PPh_2Me)_3(\eta^6\text{-py})]$, has been made by dinitrogen loss from *trans, mer*-$[Mo(N_2)_2(py)(PPh_2Me)_3]$ *in vacuo*.[1127]

Complexation of phosphabenzenes $PC_6H_2\text{-2,4,6-R}_3$ (R = Ph, Bu^t) to $[M(CO)_3(NCMe)_3]$ affords η^6-complexes. These undergo stereoselective *exo*-addition of nucleophiles at phosphorus, followed by *endo*-addition of electrophiles, to give λ^5-phosphorin complexes, which can also be made by complexation of the preformed ligands.[1128–31] Chromium complexes of phosphabenzene derivatives can also be made by Dötz annelation of carbene complexes with $P{\equiv}CBu^t$, which reacts in a similar manner to ordinary alkynes; however, there is competing oxaphosphole formation; thus, $[Cr\{=C(OEt)R\}(CO)_5]$ produces (**86**) when R = 1-naphthyl, but (**87**) when R = $o\text{-C}_6H_4OMe$.[1132,1133] Similarly annelation of $[Cr(CO)_5\{=C(OMe)CMe=CHMe\}]$ with the phosphaalkyne gives (**88**).[1134] A unique cyclotrimerization

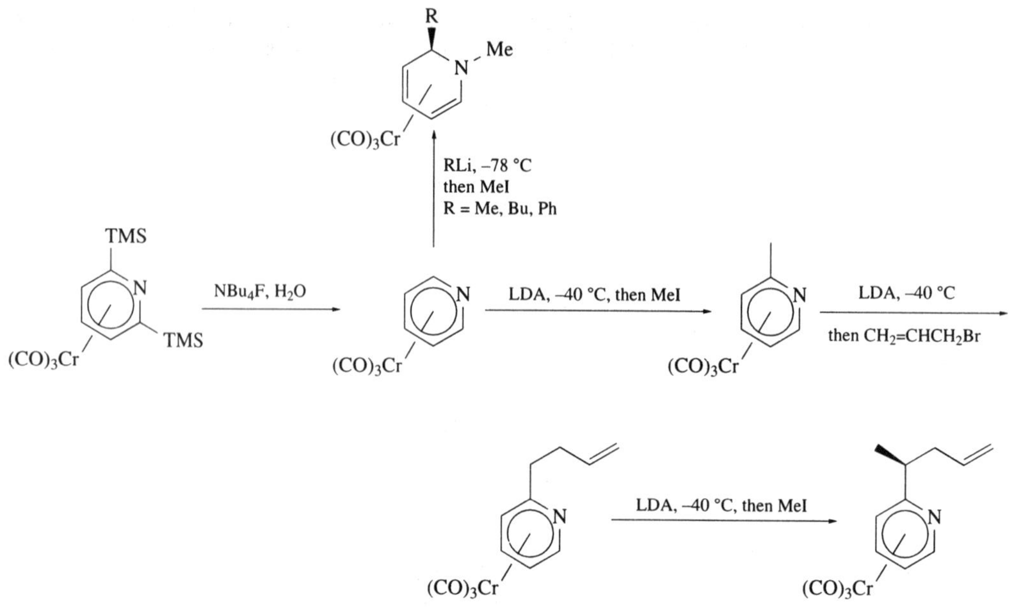

Scheme 36

of P≡CBu^t occurred on reaction with [Mo(CO)$_3$(cht)] to give triphosphabenzene complex (**89**); unfortunately no x-ray structure could be obtained.[1135]

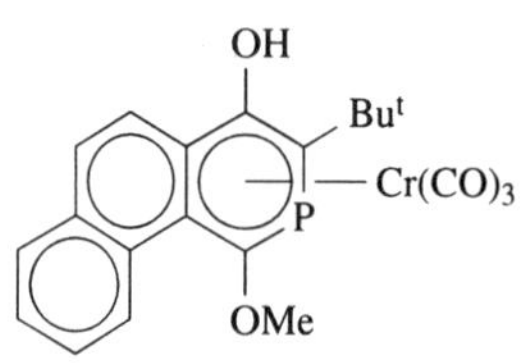

(86)

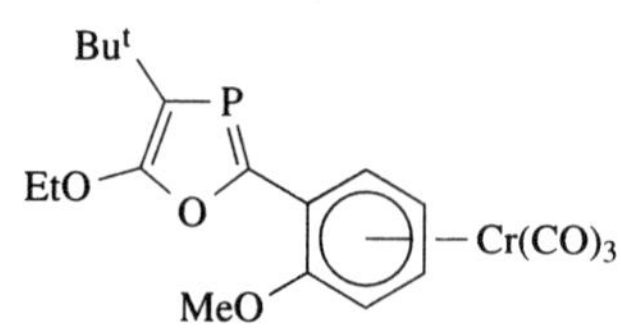

(87)

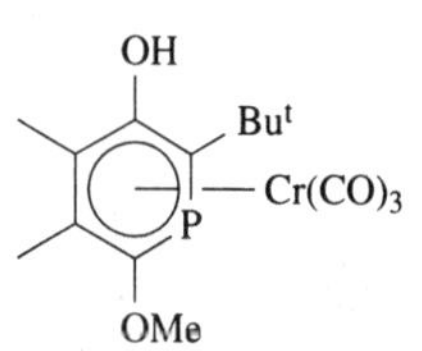

(88)

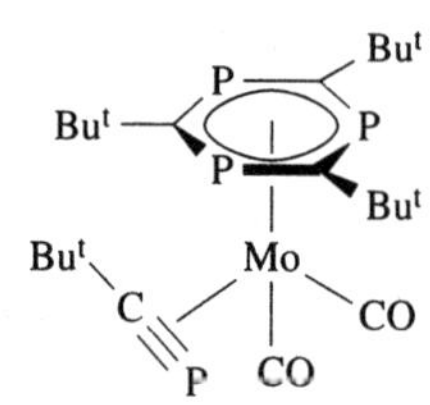

(89)

Arsabenzene complexes [M(CO)$_3$(η^6-AsC$_5$H$_4$-4-R)] (R = Bu^t, Ph, Cy) can be made by coordination of the ligand to [M(CO)$_3$(NCMe)$_3$].[1136]

Only two examples of Cr(CO)$_3$ pyrrole complexes appear to have been prepared; a complex of ligand (**36**) was produced along with its η^6 isomer on reaction of the pyrrolophane with [Cr(CO)$_3$(NCMe)$_3$], and more recently the Cr(CO)$_3$ complex of the better donor pentamethylpyrrole has been prepared.[342,1137] A number of η^5-thiophene complexes have been made, including some of bithiophenes.[1138] Few reactions have been reported, but metallation with BuLi occurs at the 2-position.[1139,1140] Migration of the silyl group around the ring unexpectedly occurred on attempted deprotonation of [Cr(CO)$_3$(η^5-SC$_4$H$_2$-2-Me-5-TMS)].[1141] Theoretical calculations on the binding of η^1- and η^5-thiophene ligands to molybdenum carbonyl fragments suggest that the η^5 form should be accessible, and could be susceptible to nucleophilic attack at the 2 and 5 positions.[1142] Benzothiophene coordinates to Cr(CO)$_3$ preferentially through the six-membered ring, and undergoes subsequent lithiation in the five-membered ring at C-2, though evidence for a second deprotonation in the arene ring, at C-7, was also obtained.[858,1143]

The reaction of the pyridine adducts of borabenzene or 2-boranaphthalene with [M(CO)$_3$(NCMe)$_3$] produces η^6-borabenzene complexes; the metal is coordinated exclusively to the heterocyclic ring in the latter case.[1144] The anionic species [Cr(CO)$_3$(η^6-MeBC$_5$H$_5$)]$^-$ can also be made.[1145] Calculations comparing borabenzene complexes to analogous Cp and arene compounds indicate similarities to both.[1146] The iridabenzene [$Ir\overline{(CHCHCHCHCH)}(PEt_3)_3$] has been complexed to a $Mo(CO)_3$ fragment.[1147]

8.6 REFERENCES

1. E. L. Muetterties, J. R. Bleeke, E. J. Wucherer and T. A. Albright, *Chem. Rev.*, 1982, **82**, 499.
2. G. Marr and B. W. Rockett, *J. Organomet. Chem.*, 1981, **207**, 343.
3. B. W. Rockett and G. Marr, *J. Organomet. Chem.*, 1982, **230**, 261.
4. G. Marr and B. W. Rockett, *J. Organomet. Chem.*, 1983, **242**, 1.
5. B. W. Rockett and G. Marr, *J. Organomet. Chem.*, 1983, **257**, 275.
6. G. Marr and B. W. Rockett, *J. Organomet. Chem.*, 1984, **278**, 331.
7. B. W. Rockett and G. Marr, *J. Organomet. Chem.*, 1986, **305**, 199.
8. G. Marr and B. W. Rockett, *J. Organomet. Chem. Libr.*, 1987, **19**, 403.
9. B. W. Rockett and G. Marr, *J. Organomet. Chem.*, 1988, **343**, 553.
10. G. Marr and B. W. Rockett, *J. Organomet. Chem.*, 1988, **357**, 319.
11. B. W. Rockett and G. Marr, *J. Organomet. Chem.*, 1990, **392**, 161.
12. G. Marr and B. W. Rockett, *J. Organomet. Chem.*, 1991, **416**, 399.
13. F. Calderazzo *et al.*, *Gazz. Chim. Ital.*, 1993, **123**, 53.
14. G. A. Ozin, M. P. Andrews, L. F. Nazar, H. X. Hüber and C. G. Francis, *Coord. Chem. Rev.*, 1983, **48**, 203.
15. R. J. Markle, T. M. Pettijohn and J. J. Lagowski, *Organometallics*, 1985, **4**, 1529.
16. M. Dalibart, J. Derouault, H. Labarrere and M. Tranquille, *J. Mol. Struct.*, 1988, **174**, 17.
17. G. A. Domrachev, B. S. Kaverin, V. L. Karnatsevich, Y. A. Shevelev and B. A. Nesterov, *Zh. Fiz. Khim.*, 1986, **60**, 3108.
18. V. L. Karnatsevich, B. S. Kaverin, L. N. Zakharov and G. A. Domrachev, *Zh. Fiz. Khim.*, 1989, **63**, 1282.
19. P. N. Hawker and P. L. Timms, *J. Chem. Soc., Dalton Trans.*, 1983, 1123.
20. C. G. Francis, P. D. Morand and N. J. Spare, *Organometallics*, 1985, **4**, 1958.
21. G. A. Ozin, M. P. Andrews and R. West, *Inorg. Chem.*, 1986, **25**, 580.
22. M. P. Andrews and G. A. Ozin, *Inorg. Chem.*, 1986, **25**, 2587.
23. G. A. Ozin and M. P. Andrews, *Angew. Chem., Int. Ed. Engl.*, 1982, **21**, 550.
24. C. Elschenbroich and J. Koch, *J. Organomet. Chem.*, 1982, **229**, 139.
25. C. Elschenbroich, J. Hurley, W. Massa and G. Baum, *Angew. Chem., Int. Ed. Engl.*, 1988, **27**, 684.
26. C. Elschenbroich, J. Hurley, B. Metz, W. Massa and G. Baum, *Organometallics*, 1990, **9**, 889.
27. E. A. Chernyshev *et al.*, *Metalloorg. Khim.*, 1991, **4**, 700.
28. G. K.-I. Magomedov *et al.*, *Metalloorg. Khim.*, 1992, **5**, 151.
29. E. A. Chernyshev *et al.*, *Metalloorg. Khim.*, 1988, **1**, 1407.
30. A. N. Nesmeyanov *et al.*, *J. Organomet. Chem.*, 1981, **217**, 195.
31. V. A. Sergeev, L. P. Yur'eva, N. N. Zaitseva, A. Y. Vasil'kov, A. S. Kogan and D. N. Kravtsov, *Izv. Akad. Nauk SSSR, Ser. Khim.*, 1986, 2307.
32. L. P. Yur'eva, N. N. Zaitseva, N. V. Zakurin, A. Y. Vasil'kov and N. I. Vasyukova, *J. Organomet. Chem.*, 1983, **247**, 287.
33. T. T.-T. Li and C. H. Brubaker, Jr., *Inorg. Chim. Acta*, 1982, **65**, L113.
34. F. G. N. Cloke, K. A. E. Courtney, A. A. Sameh and A. C. Swain, *Polyhedron*, 1989, **8**, 1641.
35. C. Elschenbroich, S. Hoppe and B. Metz, *Chem. Ber.*, 1993, **126**, 399.
36. L. P. Yur'eva, I. A. Uralets, A. S. Peregudov, D. N. Kravtsov, A. Y. Vasil'kov and V. A. Sergeev, *Metalloorg. Khim.*, 1989, **2**, 1067.
37. P. Lumme, P. Von Bagh, J. Kahima and H. Karrus, *Inorg. Chim. Acta*, 1983, **71**, 209.
38. U. Zenneck, C. Elschenbroich and R. Möckel, *J. Organomet. Chem.*, 1981, **219**, 177.
39. C. Elschenbroich, J. Schneider and H. Burdorf, *J. Organomet. Chem.*, 1990, **391**, 195.
40. C. Elschenbroich, J. Schneider, W. Massa, G. Baum and H. Mellinghoff, *J. Organomet. Chem.*, 1988, **355**, 163.
41. C. Elschenbroich, E. Bilger and R. Möckel, *Z. Naturforsch., Teil B*, 1983, **38B**, 1357.
42. C. Elschenbroich, R. Möckel and E. Bilger, *Z. Naturforsch., Teil B*, 1984, **39B**, 375.
43. C. Elschenbroich, J. Koch, J. Schneider, B. Spangenberg and P. Schiess, *J. Organomet. Chem.*, 1986, **317**, 41.
44. C. Elschenbroich, B. Spangenberg and H. Mellinghoff, *Chem. Ber.*, 1984, **117**, 3165.
45. C. Elschenbroich, R. Möckel, W. Massa, M. Birkhahn and U. Zenneck, *Chem. Ber.*, 1982, **115**, 334.
46. J. Weber, E. P. Kündig, A. Goursot and E. Penigault, *Can. J. Chem.*, 1985, **63**, 1734.
47. J. H. Osborne, W. C. Trogler, P. D. Morand and C. G. Francis, *Organometallics*, 1987, **6**, 94.
48. E. P. Kündig, C. Perret, S. Spichiger and G. Bernardinelli, *J. Organomet. Chem.*, 1985, **286**, 183.
49. B. F. Bush, V. M. Lynch and J. J. Lagowski, *Organometallics*, 1987, **6**, 1267.
50. B. F. Bush and J. J. Lagowski, *Organometallics*, 1988, **7**, 1945.
51. J. K. Burdett and E. Canadell, *Organometallics*, 1985, **4**, 805.
52. B. F. Bush and J. J. Lagowski, *J. Organomet. Chem.*, 1990, **386**, 37.
53. C. Elschenbroich, J. Schneider and H. Mellinghoff, *J. Organomet. Chem.*, 1987, **333**, 37.
54. C. Elschenbroich, J. Schneider, H. Prinzbach and W. D. Fessner, *Organometallics*, 1986, **5**, 2091.
55. T. M. Pettijohn and J. J. Lagowski, *J. Organomet. Chem.*, 1988, **356**, 67.
56. L. P. Yur'eva, N. N. Zaitseva and D. N. Kravtsov, *Metalloorg. Khim.*, 1989, **2**, 1328.
57. B. E. Wilburn and P. S. Skell, *J. Am. Chem. Soc.*, 1982, **104**, 6989.
58. H. Ssekaalo, J. J. Lagowski and C. M. Seymour, *Inorg. Chim. Acta*, 1990, **173**, 237.
59. H. Brunner and H. Koch, *Chem. Ber.*, 1982, **115**, 65.
60. H. Brunner, J. Ernst, T. Wischert and U. Thewalt, *J. Organomet. Chem.*, 1987, **328**, 331.
61. V. V. Litvak, P. P. Kun, I. I. Oleynik, O. V. Volkov and V. D. Shteingarts, *J. Organomet. Chem.*, 1986, **310**, 189.
62. L. P. Yur'eva, I. A. Uralets, A. S. Peregudov, D. N. Kravtsov, A. Y. Vasil'kov and V. A. Sergeev, *Metalloorg. Khim.*, 1990, **3**, 290.

63. C. Elschenbroich *et al.*, *Chem. Ber.*, 1988, **121**, 1983.
64. V. A. Sergeev *et al.*, *Izv. Akad. Nauk SSSR, Ser. Khim.*, 1987, 2576.
65. H. Ssekaalo, J. J. Lagowski and C. M. Seymour, *Inorg. Chim. Acta*, 1991, **183**, 91.
66. S. B. Larson, C. M. Seymour and J. J. Lagowski, *Acta Crystallogr., Sect. C*, 1987, **43C**, 1626.
67. A. Y. Vasil'kov *et al.*, *Izv. Akad. Nauk SSSR, Ser. Khim.*, 1982, 2144.
68. V. A. Sergeev, A. S. Kogan, I. A. Uralets, A. Y. Vasil'kov, N. V. Zakurin and L. P. Yur'eva, *Izv. Akad. Nauk SSSR, Ser. Khim.*, 1984, 420.
69. V. A. Sergeev *et al.*, *Izv. Akad. Nauk SSSR, Ser. Khim.*, 1986, 237.
70. V. V. Litvak, P. P. Kun and V. D. Shteingarts, *Zh. Org. Khim.*, 1982, **18**, 2153.
71. C. Elschenbroich, J. Heck, W. Massa and M. Birkhahn, *Chem. Ber.*, 1990, **123**, 2321.
72. G. A. Ozin and M. P. Andrews, *Angew. Chem., Int. Ed. Engl.*, 1982, **21**, 212.
73. M. P. Andrews and G. A. Ozin, *J. Phys. Chem.*, 1986, **90**, 1245.
74. W. M. Lamanna, *J. Am. Chem. Soc.*, 1986, **108**, 2096.
75. W. M. Lamanna, W. B. Gleason and D. Britton, *Organometallics*, 1987, **6**, 1583.
76. E. D. Jemmis and A. C. Reddy, *Organometallics*, 1988, **7**, 1561.
77. S. D. Ittel, F. A. Van-Catledge and C. A. Tolman, *Inorg. Chem.*, 1985, **24**, 62.
78. J. Howard, K. Robson and T. C. Waddington, *J. Chem. Soc., Dalton Trans.*, 1982, 977.
79. J. Weber, E. P. Kündig, A. Goursot and E. Penigault, *Can. J. Chem.*, 1985, **63**, 1734.
80. C. Elschenbroich, E. Bilger, J. Koch and J. Weber, *J. Am. Chem. Soc.*, 1984, **106**, 4297.
81. S. P. Solodovnikov, L. P. Yur'eva and N. N. Zaitseva, *Izv. Akad. Nauk SSSR, Ser. Khim.*, 1982, 2392.
82. P. D. Burrow, A. Modelli, M. Guerra and K. D. Jordan, *Chem. Phys. Lett.*, 1985, **118**, 328.
83. N. L. Asfandiarov *et al.*, *Metalloorg. Khim.*, 1991, **4**, 316.
84. R. L. Williamson and M. B. Hall, *Int. J. Quantum Chem., Quantum Chem. Symp.*, 1987, **21**, 503.
85. J. C. Green, *Struct. and Bonding (Berlin)*, 1981, **43**, 37.
86. F. G. N. Cloke, A. N. Dix, J. C. Green, R. N. Perutz and E. A. Seddon, *Organometallics*, 1983, **2**, 1150.
87. M. C. Böhm, *J. Chem. Phys.*, 1983, **78**, 7044.
88. D. E. Cabelli, A. H. Cowley and J. J. Lagowski, *Inorg. Chim. Acta*, 1982, **57**, 195.
89. Y. V. Chizhov *et al.*, *J. Organomet. Chem.*, 1989, **361**, 79.
90. S. Masuda, M. Aoyama and Y. Harada, *J. Am. Chem. Soc.*, 1990, **112**, 6445.
91. G. A. Domrachev, S. Y. Ketkov and G. A. Razuvaev, *J. Organomet. Chem.*, 1987, **328**, 341.
92. G. A. Domrachev, S. Y. Ketkov and G. A. Razuvaev, *Zh. Obshch. Khim.*, 1987, **57**, 967.
93. S. Y. Ketkov, G. A. Domrachev, G. A. Razuvaev, L. P. Yur'eva and I. A. Uralets, *Metalloorg. Khim.*, 1988, **1**, 906.
94. S. Y. Ketkov and G. A. Domrachev, *Metalloorg. Khim.*, 1988, **1**, 1190.
95. S. Y. Ketkov and G. A. Domrachev, *Opt. Spectrosk.*, 1989, **67**, 475.
96. S. Y. Ketkov, G. A. Domrachev and G. A. Razuvaev, *Metalloorg. Khim.*, 1988, **1**, 40.
97. S. Y. Ketkov, G. A. Domrachev and G. A. Razuvaev, *Dokl. Akad. Nauk SSSR*, 1987, **292**, 890.
98. G. A. Domrachev, S. Y. Ketkov and G. A. Razuvaev, *Zh. Fiz. Khim.*, 1987, **61**, 1682.
99. S. Y. Ketkov, G. A. Domrachev and G. A. Razuvaev, *Opt. Spectrosk.*, 1987, **63**, 284.
100. S. Y. Ketkov, G. A. Domrachev and I. G. Andreev, *Metalloorg. Khim.*, 1989, **2**, 453.
101. S. Y. Ketkov, G. A. Domrachev and G. A. Razuvaev, *J. Mol. Struct.*, 1989, **195**, 175.
102. S. Y. Ketkov and G. A. Domrachev, *Metalloorg. Khim.*, 1989, **2**, 919.
103. G. A. Domrachev *et al.*, *Metalloorg. Khim.*, 1990, **3**, 827.
104. G. T. W. Wittmann, G. N. Krynauw, S. Lotz and W. Ludwig, *J. Organomet. Chem.*, 1985, **293**, C33.
105. S. Y. Ketkov and G. A. Domrachev, *Metalloorg. Khim.*, 1990, **3**, 957.
106. I. I. Grinval'd, B. V. Lokshin and N. K. Rudnevskii, *Izv. Akad. Nauk SSSR, Ser. Khim.*, 1986, 1047.
107. I. I. Greenwald and B. V. Lokshin, *J. Mol. Struct.*, 1990, **218**, 255.
108. J. A. Connor, N. I. El-Saied, J. A. Martinho-Simões and H. A. Skinner, *J. Organomet. Chem.*, 1981, **212**, 405.
109. L. P. Yur'eva *et al.*, *Metalloorg. Khim.*, 1990, **3**, 783.
110. L. P. Yur'eva *et al.*, *J. Organomet. Chem.*, 1987, **336**, 371.
111. L. P. Yur'eva *et al.*, *J. Organomet. Chem.*, 1981, **219**, 43.
112. T. T. T. Li and C. H. Brubaker, Jr., *J. Organomet. Chem.*, 1981, **216**, 223.
113. C. Elschenbroich, E. Bilger and B. Metz, *Organometallics*, 1991, **10**, 2823.
114. R. J. Markle and J. J. Lagowski, *Organometallics*, 1986, **5**, 595.
115. A. B. Gavrilov, S. V. Kukharenko and V. V. Strelets, *Metalloorg. Khim.*, 1990, **3**, 406.
116. S. I. Bailey, W. P. Leung and I. M. Ritchie, *Electrochim. Acta*, 1985, **30**, 861.
117. G. Gritzner and J. Kuta, *Pure Appl. Chem.*, 1982, **54**, 1527.
118. N. Ito, T. Saji, K. Suga and S. Aoyagui, *J. Organomet. Chem.*, 1982, **229**, 43.
119. C. Elschenbroich, E. Bilger, J. Heck, F. Stohler and J. Heinzer, *Chem. Ber.*, 1984, **117**, 23.
120. H. S. Hao, Y. C. Fann and R. N. Kapoor, *Bull. Electrochem.*, 1989, **5**, 382.
121. L. N. Nekrasov, L. P. Yur'eva and S. M. Peregudova, *J. Organomet. Chem.*, 1982, **238**, 185.
122. L. P. Yur'eva *et al.*, *J. Organomet. Chem.*, 1982, **238**, 197.
123. L. N. Nekrasov, S. M. Peregudova, L. P. Yur'eva, D. N. Kravtsov, I. A. Uralets and N. N. Zaitseva, *J. Organomet. Chem.*, 1989, **365**, 269.
124. G. K. Semin, E. V. Bryukhova, D. N. Kravtsov, L. P. Yur'eva, N. N. Zaitseva and I. A. Uralets, *Izv. Akad. Nauk SSSR, Ser. Khim.*, 1986, 2801.
125. V. V. Litvak, P. P. Kun and V. D. Shteingarts, *J. Organomet. Chem.*, 1988, **348**, 219.
126. V. V. Litvak, P. P. Kun and V. D. Shteingarts, *Zh. Org. Khim.*, 1982, **18**, 1966.
127. V. V. Litvak, P. P. Kun and V. D. Shteingarts, *Zh. Org. Khim.*, 1983, **19**, 1784.
128. P. P. Kun, V. V. Litvak and V. D. Shteingarts, *Zh. Org. Khim.*, 1984, **20**, 768.
129. P. P. Kun, V. V. Litvak, I. I. Oleinik, O. V. Volkov and V. D. Shteingarts, *Zh. Org. Khim.*, 1987, **23**, 555.
130. I. I. Oleinik, P. P. Kun, V. V. Litvak and V. D. Shteingarts, *Zh. Org. Khim.*, 1987, **23**, 2580.
131. R. L. Harris, *J. Organomet. Chem.*, 1986, **299**, 105.
132. P. P. Kun, V. V. Litvak and V. D. Shteingarts, *Zh. Org. Khim.*, 1984, **20**, 759.

133. P. P. Kun, V. V. Litvak and V. D. Shteingarts, *Zh. Org. Khim.*, 1987, **23**, 547.
134. S. D. Kahn, W. J. Hehre, J. E. Bartmess and G. Caldwell, *Organometallics*, 1984, **3**, 1740.
135. M. J. McGlinchey and H. Nguyen, *Can. J. Chem.*, 1986, **64**, 1170.
136. G. H. Riding, M. Parvez and H. R. Allcock, *Organometallics*, 1986, **5**, 2153.
137. C. Elschenbroich, G. Heikenfeld, M. Wünsch, W. Massa and G. Baum, *Angew. Chem., Int. Ed. Engl.*, 1988, **27**, 414.
138. C. Elschenbroich, J. Sebbach and B. Metz, *Helv. Chim. Acta*, 1991, **74**, 1718.
139. L. P. Yur'eva, N. N. Zaitseva and D. N. Kravtsov, *Metalloorg. Khim.*, 1989, **2**, 411.
140. H. Burdorf and C. Elschenbroich, *Z. Naturforsch., Teil B*, 1981, **36B**, 94.
141. C. Elschenbroich, H. Burdorf, H. Burdorf, D. Mahrwald and B. Metz, *Z. Naturforsch., Teil B*, 1992, **47B**, 1157.
142. C. Elschenbroich, J. Sebbach, B. Metz and G. Heikenfeld, *J. Organomet. Chem.*, 1992, **426**, 173.
143. Y. Huang, J. Li, J. Zhou, Z. Zhu and G. Hou, *J. Organomet. Chem.*, 1981, **205**, 185.
144. Y. Huang, J. Li, J. Zhou, Q. Wang and M. Gui, *J. Organomet. Chem.*, 1981, **218**, 169.
145. Y. Huang, J. Zhou, Z. Zhu and J. Li, *J. Fluorine Chem.*, 1986, **30**, 455.
146. Y.-Z. Huang and J. Q. Zhou, *J. Organomet. Chem.*, 1988, **348**, 235.
147. Y. Huang, J. Li, Q. Zhou and J. Zhou, *J. Polym. Sci., Polym. Chem. Ed.*, 1985, **23**, 1853.
148. G. Fochi, *Organometallics*, 1988, **7**, 2255.
149. C. Elschenbroich, J. Heck and F. Stohler, *Organometallics*, 1982, **1**, 1399.
150. G. A. Domrachev, O. N. Zakharov and Y. A. Shevelev, *Primen. Metalloorg. Soedin. Poluch. Neorg. Pokrytii. Mater.*, 1986, 5.
151. G. A. Domrachev, *Primen. Metalloorg. Soedin. Poluch. Neorg. Pokrytii. Mater.*, 1986, 18.
152. G. A. Domrachev, L. N. Zakharov, Y. A. Shevelev and G. A. Razuvaev, *Zh. Strukt. Khim.*, 1986, **27**, 14.
153. M. Nandi, J. A. Conklin, L. Salvati Jr. and A. Sen, *Chem. Mater.*, 1990, **2**, 772.
154. R. R. Burch, *Chem. Mater.*, 1990, **2**, 633.
155. H. Schumann, *J. Organomet. Chem.*, 1986, **304**, 341.
156. H. Kunkely and A. Vogler, *J. Organomet. Chem.*, 1989, **372**, C29.
157. D. N. Kravtsov, N. N. Zaitseva and L. P. Yur'eva, *Metalloorg. Khim.*, 1989, **2**, 415.
158. A. S. Batsanov, Y. T. Struchkov, N. N. Zaitseva, L. P. Yur'eva and D. N. Kravtsov, *Metalloorg. Khim.*, 1989, **2**, 586.
159. D. N. Kravtsov, N. N. Zaitseva and L. P. Yur'eva, *Metalloorg. Khim.*, 1988, **1**, 654.
160. D. N. Kravtsov, E. S. Shubina, L. N. Saitkulova, L. M. Epshtein, N. N. Zaitseva and L. P. Yur'eva, *Metalloorg. Khim.*, 1988, **1**, 866.
161. S. Y. Kochev, E. A. Chernyshev, V. E. Zavodnik, V. K. Belskii, N. S. Magomedova and G. K. Magomedov, *Metalloorg. Khim.*, 1992, **5**, 1149.
162. N. L. Ermolaev, M. N. Bochkarev and G. A. Razuvaev, *Zh. Obshch. Khim.*, 1987, **57**, 2729.
163. L. N. Zakharov, A. I. Yanovskii, Y. T. Struchkov, L. V. Pankratov and M. N. Bochkarev, *Metalloorg. Khim.*, 1988, **1**, 1231.
164. G. A. Razuvaev, M. N. Bochkarev and L. V. Pankratov, *J. Organomet. Chem.*, 1983, **250**, 135.
165. C. Elschenbroich, R. Gondrum and W. Massa, *Angew. Chem., Int. Ed. Engl.*, 1985, **24**, 967.
166. M. L. H. Green, J. Qin and D. O'Hare, *J. Organomet. Chem.*, 1988, **358**, 375.
167. D. O'Hare, M. D. Ward and J. S. Miller, *Chem. Mater.*, 1990, **2**, 758.
168. J. S. Miller, D. M. O'Hare, A. Chakraborty and A. J. Epstein, *J. Am. Chem. Soc.*, 1989, **111**, 7853.
169. D. O'Hare and J. S. Miller, *Mol. Cryst. Liq. Cryst.*, 1989, **176**, 381.
170. D. O'Hare, M. Kurmoo, R. Lewis and H. Powell, *J. Chem. Soc., Dalton Trans.*, 1992, 1351.
171. C. Sourisseau, Y. Mathey and C. Poinsignon, *Chem. Phys.*, 1982, **71**, 257.
172. C. Sourisseau, J. P. Forgerit and Y. Mathey, *J. Phys. Chem. Solids*, 1983, **44**, 119.
173. R. Clement *et al.*, *J. Chem. Soc., Chem. Commun.*, 1984, 1354.
174. T. T. T. Li, M. J. Weaver and C. H. Brubaker, Jr., *J. Am. Chem. Soc.*, 1982, **104**, 2381.
175. G. Fochi, J. Strähle and F. Gingl, *Inorg. Chem.*, 1991, **30**, 4669.
176. T. T. T. Li, W. Kung, D. L. Ward, B. McCulloch and C. H. Brubaker Jr., *Organometallics*, 1982, **1**, 1229.
177. G. A. Domrachev, S. Y. Ketkov and Y. A. Shevelev, *Zh. Obshch. Khim.*, 1986, **56**, 1116.
178. F. G. N. Cloke and M. L. H. Green, *J. Chem. Soc., Dalton Trans.*, 1981, 1938.
179. G. A. Ozin, M. P. Andrews, C. G. Francis, H. X. Hüber and K. Molnar, *Inorg. Chem.*, 1990, **29**, 1068.
180. M. L. H. Green, I. Treurnicht, J. A. Bandy, A. Gourdon and K. Prout, *J. Organomet. Chem.*, 1986, **306**, 145.
181. J. J. Barker, A. G. Orpen, A. J. Seeley and P. L. Timms, *J. Chem. Soc., Dalton Trans.*, 1993, 3097.
182. K. Prout, A. Gourdon, C. Couldwell, B. Meunier, F. M. Miao and J. Woolcock, *Acta Crystallogr., Sect. B*, 1982, **38B**, 456.
183. F. M. Miao, *Cryst. Struct. Commun.*, 1982, **11**, 153.
184. N. P. D. Thi, S. Spichiger, P. Paglia, G. Bernardinelli, E. P. Kündig and P. L. Timms, *Helv. Chim. Acta*, 1992, **75**, 2593.
185. J. A. Bandy *et al.*, *J. Chem. Soc., Dalton Trans.*, 1985, 2037.
186. P. D. Grebenik, M. L. H. Green, A. Izquierdo, V. S. B. Mtetwa and K. Prout, *J. Chem. Soc., Dalton Trans.*, 1987, 9.
187. J. C. Green, R. A. Grieves and J. Mason, *J. Chem. Soc., Dalton Trans.*, 1986, 1313.
188. S. Töfke and U. Behrens, *Angew. Chem., Int. Ed. Engl.*, 1987, **26**, 147.
189. S. Töfke and U. Behrens, *J. Organomet. Chem.*, 1988, **338**, 29.
190. S. Y. Ketkov, *Metalloorg. Khim.*, 1992, **5**, 1337.
191. K. Jonas, *Pure Appl. Chem.*, 1990, **62**, 1169.
192. D. S. Richeson, J. F. Mitchell and K. H. Theopold, *Organometallics*, 1989, **8**, 2570.
193. A. Solladié-Cavallo, *Polyhedron*, 1985, **4**, 901.
194. A. M. Lluch, F. Sánchez-Baeza, F. Camps and J. Messeguer, *Tetrahedron Lett.*, 1991, **32**, 5629.
195. J. P. Collman, L. S. Hegedus, J. R. Norton and R. G. Finke, 'Principles and Applications of Organotransition Metal Chemistry', University Science Books, Mill Valley, CA, 1987.
196. M. F. Semmelhack, in 'Comprehensive Organic Synthesis', eds. B. M. Trost and I. Fleming, Pergamon, Oxford, 1991, vol. 4, p. 517.
197. S. G. Davies, S. J. Coote and C. L. Goodfellow, in 'Advances in Metal-Organic Chemistry', ed. L. S. Liebeskind, JAI Press, Greenwich, CT, 1991, vol. 2, p. 1.
198. S. G. Davies, 'Organotransition Metal Chemistry: Applications to Organic Synthesis', Pergamon, Oxford, 1982.
199. V. N. Kalinin, *Russian Chem. Rev.*, 1987, **56**, 682.

200. M. Uemura, in 'Advances in Metal-Organic Chemistry', ed. L. S. Liebeskind, JAI Press, Greenwich, CT, 1991, vol. 2, p. 195.
201. S. Maiorana, C. Baldoli and E. Licandro, *Catal. Met. Complexes*, 1991, **12**, 261.
202. A. Solladié-Cavallo, in 'Advances in Metal-Organic Chemistry'; ed L. S. Liebeskind, JAI Press, Greenwich, CT, 1989, vol. 1, p. 99.
203. S. G. Davies, *J. Organomet. Chem.*, 1990, **400**, 223.
204. U. S. Gill, R. M. Moriarty, Y. Y. Ku and I. R. Butler, *J. Organomet. Chem.*, 1991, **417**, 313.
205. C. A. L. Mahaffy and P. Pauson, *Inorg. Synth.*, 1990, **28**, 137.
206. C. A. L. Mahaffy and J. Hamilton, *Synth. React. Inorg. Met.-Org. Chem.*, 1986, **16**, 137.
207. E. L. Eliel, M. Manoharan, S. G. Levine and A. Ng, *J. Org. Chem.* , 1985, **50**, 4978.
208. C. A. L. Mahaffy and J. Hamilton, *Synth. React. Inorg. Met.-Org. Chem.*, 1987, **17**, 43.
209. C. A. L. Mahaffy, *J. Organomet. Chem.*, 1984, **262**, 33.
210. C. A. L. Mahaffy, *Synth. React. Inorg. Met.-Org. Chem.*, 1985, **15**, 155.
211. C. A. L. Mahaffy, *Synth. React. Inorg. Met.-Org. Chem.*, 1984, **14**, 895.
212. C. A. L. Mahaffy, *Synth. React. Inorg. Met.-Org. Chem.*, 1985, **15**, 301.
213. J. B. Hamilton, L. Harvey, C. A. L. Mahaffy, D. T. Radzykewycz and J. Rawlings, *Synth. React. Inorg. Met.-Org. Chem.*, 1993, **23**, 629.
214. C. A. L. Mahaffy, *Synth. React. Inorg. Met.-Org. Chem.*, 1985, **15**, 945.
215. C. A. L. Mahaffy and J. Hamilton, *Synth. React. Inorg. Met.-Org. Chem.*, 1987, **17**, 849.
216. F. Effenberger and K. Schöllkopf, *Chem. Ber.*, 1985, **118**, 4356.
217. C. A. L. Mahaffy and J. Rawlings, *Synth. React. Inorg. Met.-Org. Chem.*, 1985, **15**, 1097.
218. C. A. L. Mahaffy, *Synth. React. Inorg. Met.-Org. Chem.*, 1984, **14**, 679.
219. B. M. Armstrong *et al.*, *Inorg. Chim. Acta*, 1990, **170**, 37.
220. J. Hamilton and C. A. L. Mahaffy, *Synth. React. Inorg. Met.-Org. Chem.*, 1988, **18**, 243.
221. K. Schöllkopf, J. J. Stezowski and F. Effenberger, *Organometallics*, 1985, **4**, 922.
222. B. Armstrong *et al.*, *Inorg. Chim. Acta*, 1992, **191**, 189.
223. S. B. Nagelburg, C. E. Reinhold, B. R. Willeford, M. P. Bigwood, K. C. Molloy and J. J. Zuckerman, *Organometallics*, 1982, **1**, 851.
224. V. S. Kaganovich and M. I. Rybinskaya, *J. Organomet. Chem.*, 1991, **407**, 215.
225. C. Segheraert, J.-C. Brunet and A. Tartar, *J. Chem. Soc., Chem. Commun.*, 1982, 1417.
226. M. Hudecek and S. Toma, *J. Organomet. Chem.*, 1990, **393**, 115.
227. P. Hrnciar, M. Hudecek, G. K. I. Magomedov and S. Toma, *Coll. Czech. Chem. Commun.*, 1991, **56**, 1477.
228. M. Hudecek and S. Toma, *J. Organomet. Chem.*, 1991, **406**, 147.
229. M. Hudecek, V. Gajda and S. Toma, *J. Organomet. Chem.*, 1991, **413**, 155.
230. P. Hrnciar and S. Toma, *J. Organomet. Chem.*, 1991, **413**, 161.
231. G. K. I. Magomedov, *J. Organomet. Chem.*, 1990, **385**, 113.
232. G. K.-I. Magomedov, A. S. Frenkel and E. I. Blokhina, *Zh. Obsch. Khim.*, 1985, **55**, 1323.
233. V. Zanotti, V. Rutar and R. J. Angelici, *J. Organomet. Chem.*, 1991, **414**, 177.
234. J. Vebrel, R. Mercier and J. Belleney, *J. Organomet. Chem.*, 1982, **235**, 197.
235. C. Kayran and S. Ozkar, *Z. Naturforsch., Teil B*, 1992, **47B**, 1051.
236. P. Müller, G. Bernardinelli and Y. Jacquier, *Helv. Chim. Acta*, 1988, **71**, 1328.
237. P. Müller, G. Bernardinelli, Y. Jacquier and A. Ricca, *Helv. Chim. Acta*, 1989, **72**, 1618.
238. P. Müller, G. Bernardinelli and Y. Jacquier, *Helv. Chim. Acta*, 1992, **75**, 1995.
239. H. G. Wey and H. Butenschön, *J. Organomet. Chem.*, 1988, **350**, C8.
240. H. G. Wey, P. Betz and H. Butenschön, *Chem. Ber.*, 1991, **124**, 465.
241. Y. A. Ustynyuk, O. I. Trifonova and Y. F. Oprunenko, *Metalloorg. Khim.*, 1992, **5**, 654.
242. P. Berno *et al.*, *J. Organomet. Chem.*, 1986, **301**, 161.
243. M. Iyoda, Y. Kuwatani, T. Yamauchi and M. Oda, *J. Chem. Soc., Chem. Commun.*, 1988, 65.
244. M. Yalpani, R. Benn, R. Goddard and G. Wilke, *J. Organomet. Chem.*, 1982, **240**, 49.
245. J. A. S. Howell, N. F. Ashford, D. T. Dixon, J. C. Kola, T. A. Albright and S. K. Kang, *Organometallics*, 1991, **10**, 1852.
246. T. G. Traylor, K. J. Stewart and M. J. Goldberg, *J. Am. Chem. Soc.*, 1984, **106**, 4445.
247. T. G. Traylor and K. Stewart, *Organometallics*, 1984, **3**, 325.
248. T. G. Traylor, K. J. Stewart and M. J. Goldberg, *Organometallics*, 1986, **5**, 2062.
249. J. J. Harrison, *J. Am. Chem. Soc.*, 1984, **106**, 1487.
250. T. G. Traylor and K. J. Stewart, *J. Am. Chem. Soc.*, 1986, **108**, 6977.
251. T. G. Traylor and M. J. Goldberg, *Organometallics*, 1987, **6**, 2413.
252. T. G. Traylor and M. J. Goldberg, *Organometallics*, 1987, **6**, 2531.
253. T. G. Traylor and M. J. Goldberg, *J. Am. Chem. Soc.*, 1987, **109**, 3968.
254. V. A. Piven and Y. S. Nekrasov, *Izv. Akad. Nauk SSSR, Ser. Khim.*, 1987, 1649.
255. T. G. Traylor, M. J. Goldberg, A. R. Mikisztal and C. E. Strouse, *Organometallics*, 1989, **8**, 1101.
256. G. B. M. Kostermans, M. Bobeldijk, P. J. Kwakman, W. H. de Wolf and F. Bickelhaupt, *J. Organomet. Chem.*, 1989, **363**, 291.
257. K. S. Chan *et al.*, *J. Organomet. Chem.*, 1987, **334**, 9.
258. K. H. Dötz, M. Popall and G. Müller, *J. Organomet. Chem.*, 1987, **334**, 57.
259. K. H. Dötz, J. Mühlemeier, U. Schubert and O. Orama, *J. Organomet. Chem.*, 1983, **247**, 187.
260. K. H. Dötz, I. Pruskil and J. Mühlemeier, *Chem. Ber.*, 1982, **115**, 1278.
261. N. H. Tran Huy and P. Lefloch, *J. Organomet. Chem.*, 1988, **344**, 303.
262. S. Chamberlin and W. D. Wulff, *J. Am. Chem. Soc.*, 1992, **114**, 10 667.
263. H. Fischer, J. Mühlemeier, R. Märkl and K. H. Dötz, *Chem. Ber.*, 1982, **115**, 1355.
264. K. H. Dötz and M. Popall, *Angew. Chem., Int. Ed. Engl.*, 1987, **26**, 1158.
265. K. H. Dötz and M. Popall, *Chem. Ber.*, 1988, **121**, 665.
266. C. A. Merlic, D. Xu and S. I. Khan, *Organometallics*, 1992, **11**, 412.
267. K. H. Dötz and M. Kuhn, *J. Organomet. Chem.*, 1983, **252**, C78.

268. K. H. Dötz and W. Sturm, *J. Organomet. Chem.*, 1985, **285**, 205.
269. K. H. Dötz and W. Kuhn, *Angew. Chem., Int. Ed. Engl.*, 1983, **22**, 732; *Angew. Chem. Suppl.*, 1983, 1045.
270. W. D. Wulff and D. C. Yang, *J. Am. Chem. Soc.*, 1984, **106**, 7565.
271. R. Aumann and J. Uphoff, *Angew. Chem., Int. Ed. Engl.*, 1987, **26**, 357.
272. H. Rudler *et al.*, *J. Organomet. Chem.*, 1988, **358**, 245.
273. B. Denise, R. Goumont, A. Parlier, H. Rudler, J. C. Daran and J. Vaissermann, *J. Organomet. Chem.*, 1989, **377**, 89.
274. E. Chelain, A. Parlier, H. Rudler, J. C. Daran and J. Vaissermann, *J. Organomet. Chem.*, 1991, **419**, C5.
275. A. Parlier, H. Rudler, R. Yefsah, J. C. Daran and C. Knobler, *J. Chem. Soc., Chem. Commun.*, 1988, 635.
276. C. Alvarez, A. Parlier, H. Rudler, R. Yefsah, J. C. Daran and C. Knobler, *Organometallics*, 1989, **8**, 2253.
277. K. H. Dötz, H. G. Erben and K. Harms, *J. Chem. Soc., Chem. Commun.*, 1989, 692.
278. B. Denise, R. Goumont, A. Parlier, H. Rudler, J. C. Daran and J. Vaissermann, *J. Chem. Soc., Chem. Commun.*, 1990, 1238.
279. H. Rudler, A. Parlier, R. Goumont, J. C. Daran and J. Vaissermann, *J. Chem. Soc., Chem. Commun.*, 1991, 1075.
280. K. H. Dötz, T. Schäfer and K. Harms, *Angew. Chem., Int. Ed. Engl.*, 1990, **29**, 176.
281. E. Chelain *et al.*, *J. Am. Chem. Soc.*, 1992, **114**, 8088.
282. K. H. Dötz, A. Rau and K. Harms, *J. Organomet. Chem.*, 1992, **439**, 263.
283. E. Chelain, A. Parlier, M. Audouin, H. Rudler, J. C. Daran and J. Vaissermann, *J. Am. Chem. Soc.*, 1993, **115**, 10 568.
284. H. Tran Huy, P. Lefloch, J. M. Louis and M. Fetizon, *J. Organomet. Chem.*, 1986, **311**, 79.
285. U. Schubert and H. Hörnig, *J. Organomet. Chem.*, 1984, **273**, C11.
286. P. Hackett, B. F. G. Johnson and J. Lewis, *J. Chem. Soc., Dalton Trans.*, 1982, 603.
287. R. L. Beddoes, E. S. Cook and M. J. Morris, *Polyhedron*, 1989, **8**, 1810.
288. H. Sakurai, Y. Nakadaira, A. Hosomi, Y. Eriyama, K. Hirama and C. Kabuto, *J. Am. Chem. Soc.*, 1984, **106**, 8315.
289. K. A. Azam, S. E. Kabir, A. B. Kazi, A. H. Molla and S. S. Ullah, *J. Bangladesh Acad. Sci.*, 1987, **11**, 239.
290. J. A. Heppert, T. J. Boyle and F. Takusagawa, *Organometallics*, 1989, **8**, 461.
291. H. Schumann, A. M. Arif and T. G. Richmond, *Polyhedron*, 1990, **9**, 1677.
292. H. G. Wey and H. Butenschön, *Chem. Ber.*, 1990, **123**, 93.
293. C. Sergheraert and A. Tartar, *J. Organomet. Chem.*, 1982, **240**, 163.
294. A. J. Pearson and J. G. Park, *J. Org. Chem.*, 1992, **57**, 1744.
295. I. Lavastre, J. Besancon, P. Brossier and C. Moise, *Appl. Organomet. Chem.*, 1990, **4**, 9.
296. I. Lavastre, J. Besancon, P. Brossier and C. Moise, *Appl. Organomet. Chem.*, 1991, **5**, 143.
297. T. L. Hubler, S. B. Meikrantz, T. E. Bitterwolf, N. R. Natale, D. J. Triggle and Y.-K. Kwon, *J. Med. Chem.*, 1992, **35**, 1165.
298. R. C. Cambie, G. R. Clark, A. C. Gourdie, P. S. Rutledge and P. D. Woodgate, *J. Organomet. Chem.*, 1985, **297**, 177.
299. R. C. Cambie, G. R. Clark, S. R. Gallagher, P. S. Rutledge, M. J. Stone and P. D. Woodgate, *J. Organomet. Chem.*, 1988, **342**, 315.
300. B. Mailvaganam, R. E. Perrier, B. G. Sayer, B. E. McCarry, R. A. Bell and M. J. McGlinchey, *J. Organomet. Chem.*, 1988, **354**, 325.
301. G. Jaouen, A. Vessières, S. Top, A. A. Ismail and I. S. Butler, *J. Am. Chem. Soc.*, 1985, **107**, 4778.
302. G. Jaouen and A. Vessières, *Pure Appl. Chem.*, 1985, **57**, 1865.
303. S. Top *et al.*, *Organometallics*, 1985, **4**, 2143.
304. A. M. Bond, G. Jaouen, T. F. Mann, E. Mocellin and S. Top, *Organometallics*, 1989, **8**, 2382.
305. A. M. Bond, E. Mocellin and C. B. Pascual, *Organometallics*, 1987, **6**, 385.
306. I. S. Butler *et al.*, *J. Raman Spectrosc.*, 1990, **21**, 355.
307. R. C. Cambie, A. D. Erson, A. C. Gourdie, P. S. Rutledge and P. D. Woodgate, *J. Organomet. Chem.*, 1988, **348**, 317.
308. R. C. Cambie, P. I. Higgs, P. S. Rutledge and P. D. Woodgate, *J. Organomet. Chem.*, 1990, **384**, C6.
309. H. Künzer and M. Thiel, *Tetrahedron Lett.*, 1988, **29**, 1135.
310. H. Künzer and M. Thiel, *Tetrahedron Lett.*, 1988, **29**, 3223.
311. A. Vessières, S. Top, A. A. Ismail, I. S. Butler, M. Louer and G. Jaouen, *Biochemistry*, 1988, **27**, 6659.
312. S. Top, A. Vessières and G. Jaouen, *J. Labelled Compound Radiopharm.*, 1987, **24**, 1257.
313. G. Jaouen, S. Top, A. Vessières, B. G. Sayer, C. S. Frampton and M. J. McGlinchey, *Organometallics*, 1992, **11**, 4061.
314. Z. Y. Own, S. M. Wang, J. F. Chung, D. W. Miller and P. P. Fu, *Inorg. Chem.*, 1993, **32**, 152.
315. A. J. Whitton, O. Kumberger, G. Müller and H. Schmidbaur, *Chem. Ber.*, 1990, **123**, 1931.
316. T. E. Bitterwolf, R. Herzog and P. D. Rockswold, *J. Organomet. Chem.*, 1987, **320**, 197.
317. A. Ceccon *et al.*, *J. Chem. Soc., Perkin Trans. 2*, 1991, 233.
318. J. W. Merkert, W. E. Geiger, M. N. Paddon-Row, A. M. Oliver and A. L. Rheingold, *Organometallics*, 1992, **11**, 4109.
319. C. H. Sun, G. Z. Yang and J. C. Wang, *Bull. Inst. Chem. Acad. Sin.*, 1990, **37**, 27.
320. W. Weissensteiner, O. Hofer and U. G. Wagner, *J. Org. Chem.*, 1988, **53**, 3988.
321. T. A. Albright, P. Hofmann, R. Hoffmann, C. P. Lillya and P. A. Dobosh, *J. Am. Chem. Soc.*, 1983, **105**, 3396.
322. J. W. Hull, Jr. and W. L. Gladfelter, *Organometallics*, 1982, **1**, 264.
323. R. A. Gancarz, J. F. Blount and K. Mislow, *Organometallics*, 1985, **4**, 2028.
324. R. D. Rogers, J. L. Atwood, T. A. Albright, W. A. Lee and M. D. Rausch, *Organometallics*, 1984, **3**, 263.
325. D. J. Peitz, R. T. Palmer, L. J. Radonovich and N. F. Woolsey, *Organometallics*, 1993, **12**, 4580.
326. L. H. P. Gommans, L. Main, B. K. Nicholson, W. T. Robinson and A. D. Rae, *J. Organomet. Chem.*, 1988, **346**, 385.
327. N. Mori, M. Takamori and T. Takemura, *J. Chem. Soc., Dalton Trans.*, 1985, 1065.
328. M. Takamori and N. Mori, *J. Organomet. Chem.*, 1986, **301**, 321.
329. I. Ono, S. Mita, S. Kondo and N. Mori, *J. Organomet. Chem.*, 1989, **367**, 81.
330. L. W. Jenneskens, W. H. de Wolf and F. Bickelhaupt, *J. Organomet. Chem.*, 1990, **390**, 171.
331. O. Reiser and A. de Meijere, *J. Organomet. Chem.*, 1991, **413**, 137.
332. M. Stöbbe, O. Reiser, T. Thiemann, R. G. Daniels and A. de Meijere, *Tetrahedron Lett.*, 1986, **27**, 2353.
333. A. de Meijere *et al.*, *Acta Chim. Scand.*, 1988, **A42**, 611.
334. F. Vögtle, J. Schulz and M. Nieger, *Chem. Ber.*, 1991, **124**, 1415.
335. J. Schulz, M. Nieger and F. Vögtle, *Chem. Ber.*, 1991, **124**, 2797.
336. J. Schulz, S. Bartram, M. Nieger and F. Vögtle, *Chem. Ber.*, 1992, **125**, 2553.

337. R. H. Mitchell, T. K. Vinod and G. W. Bushnell, *J. Am. Chem. Soc.*, 1985, **107**, 3340.
338. R. H. Mitchell *et al.*, *Pure Appl. Chem.*, 1986, **58**, 15.
339. R. H. Mitchell, T. K. Vinod, G. J. Bodwell and G. W. Bushnell, *J. Org. Chem.*, 1989, **54**, 5871.
340. R. H. Mitchell, T. K. Vinod and G. W. Bushnell, *J. Am. Chem. Soc.*, 1990, **112**, 3487.
341. K. Sako, T. Shinmyozu, H. Takemura, M. Suenaga and T. Inazu, *J. Org. Chem.*, 1992, **57**, 6536.
342. M. J. Zaworotko, R. J. Stamps, M. T. Ledet, H. Zhang and J. L. Atwood, *Organometallics*, 1985, **4**, 1697.
343. V. I. Sokolov and V. L. Bondareva, *Metalloorg. Khim.*, 1988, **1**, 1195.
344. H. Iki, T. Kikuchi and S. Shinkai, *J. Chem. Soc., Perkin Trans.* 1, 1992, 669.
345. T. Whelan, J. T. Spencer, M. R. Pourian and R. N. Grimes, *Inorg. Chem.*, 1987, **26**, 3116.
346. J. T. Spencer and R. N. Grimes, *Organometallics*, 1987, **6**, 328.
347. S. Allaoud, A. Karim, A. Mortreux, F. Petit and B. Frange, *J. Organomet. Chem.*, 1989, **379**, 89.
348. J. Schulze and G. Schmid, *Chem. Ber.*, 1981, **114**, 495.
349. B. Kahr, J. M. Chance and K. Mislow, *Mol. Cryst. Liq. Cryst. Sci. Technol., Sect. A*, 1992, **210**, 195.
350. M. Morán, I. Cuadrado, M. C. Pascual, C. M. Casado and J. Losada, *Organometallics*, 1992, **11**, 1210.
351. M. Morán, C. M. Casado, I. Cuadrado and J. Losada, *Inorg. Chim. Acta*, 1991, **185**, 33.
352. A. M. Mance, N. D. Miro, C. H. van Dyke and N. Viswanathan, *Inorg. Chem.*, 1981, **20**, 635.
353. G. K.-I. Magomedov, A. S. Frenkel, O. V. Shkol'nik, E. V. Bulycheva and V. D. Sheludyakov, *Metalloorg. Khim.*, 1988, **1**, 938.
354. K. L. Malisza *et al.*, *Organometallics*, 1993, **12**, 2462.
355. J. F. Harrod, A. Shaver and A. Tucka, *Organometallics*, 1985, **4**, 2166.
356. L. M. Antipin, V. V. Fedoseevskii, V. N. Mochalov, A. S. Frenkel, G. K. Magomedov and M. F. Shostakovskii, *Izv. Akad. Nauk SSSR, Ser. Khim.*, 1983, 642.
357. V. S. Kaganovich, M. I. Rybinskaya, A. A. Zhdanov, M. M. Levitskii and B. D. Lavrukin, *Izv. Akad. Nauk SSSR, Ser. Khim.*, 1987, 2060.
358. T.-M. Chung, Y.-A. Lee, Y. K. Chung and I. N. Jung, *Organometallics*, 1990, **9**, 1976.
359. J.-S. Lee, Y. K. Chung, D. Whang and K. Kim, *J. Organomet. Chem.*, 1993, **445**, 49.
360. M. Ishikawa and T. Tabohashi, *J. Organomet. Chem.*, 1984, **271**, C4.
361. B. Becker, R. J. P. Corriu, B. J. L. Henner, W. Wojnowski, K. Peters and H. G. von Schnering, *J. Organomet. Chem.*, 1986, **312**, 305.
362. O. B. Afanasova *et al.*, *Dokl. Akad. Nauk SSSR*, 1984, **279**, 904.
363. N. I. Kirillova, A. I. Gusev and O. B. Afanasova, *Koord. Khim.*, 1988, **14**, 989.
364. E. A. Chernyshev, O. B. Afanasova, A. S. Frenkel, Y. E. Zubarev and G. K.-I. Magomedov, *Zh. Obshch. Khim.*, 1987, **57**, 2531.
365. N. I. Kirillova, A. I. Gusev and O. B. Afanasova, *Metalloorg. Khim.*, 1988, **1**, 1278.
366. O. B. Afanasova, Y. E. Zubarev, N. I. Kirillova, T. M. Sokolova, A. I. Gusev and E. A. Chernyshev, *Metalloorg. Khim.*, 1988, **1**, 1057.
367. N. I. Kirillova *et al.*, *Metalloorg. Khim.*, 1990, **3**, 335.
368. V. A. Sharapov *et al.*, *Organometallics*, 1984, **3**, 1755.
369. O. B. Afanasova *et al.*, *Zh. Obshch. Khim.*, 1982, **52**, 2137.
370. N. I. Kirillova, A. I. Gusev, O. B. Afanasova and E. A. Chernyshev, *Metalloorg. Khim.*, 1988, **1**, 1172.
371. O. B. Afanasova *et al.*, *Zh. Obshch. Khim.*, 1988, **58**, 1679.
372. V. M. Nosova, A. V. Kisin, N. V. Alekseev, O. B. Afanasova, N. G. Komalenkova and I. A. Shashkov, *Metalloorg. Khim.*, 1989, **2**, 1168.
373. N. I. Vasyukova, Y. S. Nekrasov, O. B. Afanasova and E. A. Chernyshev, *Izv. Akad. Nauk SSSR, Ser. Khim.*, 1984, 1795.
374. E. A. Chernyshev *et al.*, *Izv. Akad. Nauk SSSR, Ser. Khim.*, 1986, 2324.
375. O. B. Afanasova, Y. E. Zubarev, T. M. Sokolova and E. A. Chernyshev, *Zh. Obshch. Khim.*, 1987, **57**, 1909.
376. Y. F. Oprunenko, O. K. Sokolikova, S. G. Malyugina, Y. A. Ustynyuk, N. A. Ustynyuk and O. B. Afanasova, *Metalloorg. Khim.*, 1990, **3**, 1396.
377. N. I. Kirillova *et al.*, *J. Organomet. Chem.*, 1986, **306**, 55.
378. O. B. Afanasova *et al.*, *Zh. Obshch. Khim.*, 1984, **54**, 1924.
379. N. I. Kirillova, A. I. Gusev, O. B. Afanasova and O. N. Kalinina, *Metalloorg. Khim.*, 1991, **4**, 1040.
380. O. B. Afanasova, O. N. Kalinina, Y. F. Oprunenko, T. M. Sokolova, Y. A. Ustynyuk and E. A. Chernyshev, *Metalloorg. Khim.*, 1990, **3**, 399.
381. N. I. Kirillova, A. I. Gusev, O. B. Afanasova and O. N. Kalinina, *Metalloorg. Khim.*, 1990, **3**, 1310.
382. N. H. Buttrus, C. Eaborn, M. E. Molla, P. B. Hitchcock and J. D. Smith, *J. Organomet. Chem.*, 1988, **344**, 313.
383. R. D. Rieke, I. Tucker, S. N. Milligan, D. R. Wright, B. R. Willeford, L. J. Radonovich and M. W. Eyring, *Organometallics*, 1982, **1**, 938.
384. U. Schubert, B. R. Willeford and J. J. Zuckerman, *J. Organomet. Chem.*, 1981, **215**, 367.
385. S. C. Srivastava, A. K. Shrimal and P. Tiwari, *Polyhedron*, 1992, **11**, 1181.
386. D. Y. Movshovich, S. B. Bulgarevich, M. G. Gruntfest, S. E. Filippov, A. I. Uraev and A. D. Garnovskii, *Zh. Obshch. Khim.*, 1991, **61**, 2579.
387. N. J. Gogan, I. S. Jayasinghe and P. K. Sayal, *J. Organomet. Chem.*, 1986, **317**, 55.
388. N. J. Gogan, I. S. Jayasinghe and P. K. Sayal, *J. Organomet. Chem.*, 1987, **336**, 137.
389. N. J. Gogan, R. McDonald, H. J. Anderson and C. E. Loader, *Can. J. Chem.*, 1989, **67**, 433.
390. M. Yoshifuji and N. Inamoto, *Tetrahedron Lett.*, 1983, **24**, 4855.
391. M. Yoshifuji, N. Inamoto, K. Hirotsu and T. Higuchi, *J. Chem. Soc., Chem. Commun.*, 1985, 1109; correction 1986, p. 640.
392. H. R. Allcock, A. A. Dembek, J. L. Bennett, I. Manners and M. Parvez, *Organometallics*, 1991, **10**, 1865.
393. A. A. Alad'in, A. N. Artemov and G. A. Razuvaev, *Izv. Akad. Nauk SSSR, Ser. Khim.*, 1981, 1874.
394. L. N. Zakharov, Y. T. Struchkov, A. A. Alad'in and A. N. Artemov, *Koord. Khim.*, 1981, **7**, 1705.
395. J. A. Heppert, M. E. Thomas-Miller, D. M. Scherubel, F. Takusagawa, M. A. Morgenstern and M. R. Shaker, *Organometallics*, 1989, **8**, 1199.
396. J. A. Heppert, M. E. Thomas-Miller, P. N. Swepston and M. W. Extine, *J. Chem. Soc., Chem. Commun.*, 1988, 280.

397. J. A. Heppert, M. A. Morgenstern, D. M. Scherubel, F. Takusagawa and M. R. Shaker, *Organometallics*, 1988, **7**, 1715.
398. J. Li, A. D. Hunter, R. McDonald, B. D. Santarsiero, S. G. Bott and J. L. Atwood, *Organometallics*, 1992, **11**, 3050.
399. G. B. Richter-Addo and A. D. Hunter, *Inorg. Chem.*, 1989, **28**, 4063.
400. G. B. Richter-Addo, A. D. Hunter and N. Wichrowska, *Can. J. Chem.*, 1990, **68**, 41.
401. A. D. Hunter, *Organometallics*, 1989, **8**, 1118.
402. A. D. Hunter and J. L. McLernon, *Organometallics*, 1989, **8**, 2679.
403. A. D. Hunter, D. Ristic-Petrovic and J. L. McLernon, *Organometallics*, 1992, **11**, 864.
404. I. R. Butler, U. Gill, Y. LePage, W. E. Lindsell and P. N. Preston, *Organometallics*, 1990, **9**, 1964.
405. P. H. van Rooyen, M. Schindehutte and S. Lotz, *Organometallics*, 1992, **11**, 1104.
406. S. Lotz, M. Schindehutte and P. H. van Rooyen, *Organometallics*, 1992, **11**, 629.
407. A. S. Smirnov, T. G. Kasatkina and A. N. Artemov, *Metalloorg. Khim.*, 1988, **1**, 1169.
408. V. Dufaud, J. Thivolle-Cazat, J.-M. Basset, R. Mathieu, J. Jaud and J. Waissermann, *Organometallics*, 1991, **10**, 4005.
409. L. G. Kuzmina and Y. T. Struchkov, *Koord. Khim.*, 1982, **8**, 1542.
410. G. A. Razuvaev, N. I. Sirotkin, A. N. Artemov and A. S. Smirnov, *Izv. Akad. Nauk SSSR, Ser. Khim.*, 1981, 2149.
411. G. Z. Suleimanov, V. N. Khandozhko, P. V. Petrovskii, R. Y. Mekhdiev, N. E. Kolobova and I. P. Beletskaya, *J. Chem. Soc., Chem. Commun.*, 1985, 596.
412. G. Z. Suleimanov *et al.*, *Izv. Akad. Nauk SSSR, Ser. Khim.*, 1985, 1376.
413. J. R. Fernandez and F. G. A. Stone, *J. Chem. Soc., Dalton Trans.*, 1988, 3035.
414. S. Sun, Y. Yao, H. Zhu, Q. Meng and X. You, *Transition Met. Chem. (London)*, 1992, **17**, 216.
415. M. R. Bradford, J. C. Jeffery, K. M. Johnson and S. E. Phillips, *J. Organomet. Chem.*, 1989, **362**, 205.
416. W. R. Cullen, S. J. Rettig and H. Zhang, *Organometallics*, 1991, **10**, 2965.
417. W. R. Cullen, S. J. Rettig and H. Zhang, *Organometallics*, 1992, **11**, 1000.
418. V. Galamb and G. Pályi, *J. Chem. Soc., Chem. Commun.*, 1982, 487.
419. V. Galamb, G. Pályi, F. Ungváry, L. Markó, R. Boese and G. Schmid, *J. Am. Chem. Soc.*, 1986, **108**, 3344.
420. M. R. Domingo, A. Irving, Y.-H. Liao, J. R. Moss and A. Nash, *J. Organomet. Chem.*, 1993, **443**, 233.
421. J.-J. Yaouanc, J.-C. Clement and H. des Abbayes, *J. Chem. Soc., Chem. Commun.*, 1988, 1379.
422. H.-M. Gau, C.-T. Chen and C.-C. Schei, *J. Organomet. Chem.*, 1992, **424**, 307.
423. H.-M. Gau, C.-T. Chen, T.-T. Jong and M.-Y. Chien, *J. Organomet. Chem.*, 1993, **448**, 99.
424. U. Schubert, H. Buhler and B. Hirle, *Chem. Ber.*, 1992, **125**, 999.
425. D. J. Darensbourg, B. L. Mueller and J. H. Reibenspiess, *J. Organomet. Chem.*, 1993, **451**, 83.
426. V. S. Kaganovich and M. I. Rybinskaya, *Izv. Akad. Nauk SSSR, Ser. Khim.*, 1985, 1644.
427. V. S. Kaganovich and M. I. Rybinskaya, *Izv. Akad. Nauk SSSR, Ser. Khim.*, 1984, 1601.
428. G. A. Razuvaev, A. N. Artemov, T. G. Kasatkina and E. N. Poselenova, *Izv. Akad. Nauk SSSR, Ser. Khim.*, 1982, 1131.
429. M. I. Rybinskaya, V. S. Kaganovich, Y. T. Struchkov and A. I. Yanovsky, *Metalloorg. Khim.*, 1993, **6**, 218.
430. M. E. Wright, L. Lawson, R. T. Baker and D. C. Roe, *Polyhedron*, 1992, **11**, 323.
431. M. G. McDougall, J. C. Gallucci and D. W. Meek, *Inorg. Chim. Acta*, 1991, **188**, 123.
432. M. A. Mahmoudi, J. C. Lamiot, F. Baert, L. A. Maciejewski and J. S. Brocard, *J. Chem. Soc., Chem. Commun.*, 1990, 1051.
433. J. Lamiot, F. Baert, J. Brocard, L. Maciejewski and M. Mahmoudi, *Acta Crystallogr., Sect. C*, 1991, **C47**, 1833.
434. V. P. Mar'in, P. V. Petrovskii, L. I. Vyshinskaya, M. A. Doronicheva and A. A. Alad'in, *Metalloorg. Khim.*, 1991, **4**, 676.
435. T. E. Bitterwolf, *Polym. Mater. Sci. Eng.*, 1983, **49**, 368.
436. N. Gogan and S. I. de Silva, *Organometallics*, 1990, **9**, 1970.
437. E. S. Shubina *et al.*, *J. Organomet. Chem.*, 1991, **401**, 155.
438. J. Breimair, M. Wieser and W. Beck, *J. Organomet. Chem.*, 1992, **441**, 429.
439. M. Wieser, K. Karaghiosoff and W. Beck, *Chem. Ber.*, 1993, **126**, 1081.
440. S. S. Ullah, I. Ahmed, S. E. Kabir, M. M. Karim and A. H. Molla, *Ind. J. Chem. Soc., Section A*, 1985, **24A**, 1052.
441. J. Lamiot, F. Baert, J. Brocard, L. Maciejewski and M. Mahmoudi, *Acta Crystallogr., Sect. C*, 1991, **C47**, 1833.
442. Z. U. Siddiqui and N. J. Gogan, *Inorg. Chim. Acta*, 1986, **120**, 119.
443. A. I. Uraev *et al.*, *J. Organomet. Chem.*, 1989, **368**, 303.
444. A. I. Uraev, A. L. Nivorozhkin, A. S. Frenkel, G. K.-I. Magomedov and A. D. Garnovskii, *Zh. Obshch. Khim.*, 1988, **58**, 1175.
445. A. I. Uraev, A. L. Nivorozhkin, A. S. Frenkel, L. E. Konstantinovskii, G. K.-I. Magomedov and A. D. Garnovskii, *Metalloorg. Khim.*, 1988, **1**, 934.
446. A. S. Antsyshkina *et al.*, *Koord. Khim.*, 1990, **16**, 1492.
447. W. J. Scott, *J. Chem. Soc., Chem. Commun.*, 1987, 1755.
448. M. E. Wright, *Macromolecules*, 1989, **22**, 3256.
449. J. M. Clough, I. S. Mann and D. A. Widdowson, *Tetrahedron Lett.*, 1987, **28**, 2645.
450. M. Uemura, H. Nishimura and T. Hayashi, *Tetrahedron Lett.*, 1993, **34**, 107.
451. R. Mutin, C. Lucas, J. Thivolle-Cazat, V. Dufaud, F. Dany and J. M. Basset, *J. Chem. Soc., Chem. Commun.*, 1988, 896.
452. F. Dany, R. Mutin, C. Lucas, V. Dufaud, J. Thivolle-Cazat and J. M. Basset, *J. Mol. Catal.*, 1989, **51**, L15.
453. J.-F. Carpentier, Y. Castanet, J. Brocard, A. Mortreux and F. Petit, *Tetrahedron Lett.*, 1991, **32**, 4705.
454. J.-F. Carpentier, Y. Castanet, J. Brocard, A. Mortreux and F. Petit, *Tetrahedron Lett.*, 1992, **33**, 2001.
455. D. Villemin and E. Schigeko, *J. Organomet. Chem.*, 1985, **293**, C10.
456. M. E. Wright, *J. Organomet. Chem.*, 1989, **376**, 353.
457. R. Chauvin, *J. Organomet. Chem.*, 1990, **387**, C1.
458. A. C. Chang, H. Abdel-Aziz, N. Melchor, Q. Wu, K. H. Pannell and D. W. Armstrong, *J. Chromatogr.*, 1986, **347**, 51
459. E. Patsalides, S. J. Pratten and K. Robards, *J. Organomet. Chem.*, 1989, **364**, 169.
460. P. J. Marriot and Y.-H. Lai, *Inorg. Chem.*, 1986, **25**, 3680.
461. T. R. Crompton, 'Gas Chromatography of Organometallic Compounds', Wiley, New York, 1982.
462. A. D. Hunter, L. Shilliday, W. S. Furey and M. J. Zaworotko, *Organometallics*, 1992, **11**, 1550.
463. B. P. Byers and M. B. Hall, *Inorg. Chem.*, 1987, **26**, 2186.
464. J. W. Chinn Jr. and M. B. Hall, *J. Am. Chem. Soc.*, 1983, **105**, 4930.
465. B. P. Byers and M. B. Hall, *Organometallics*, 1987, **6**, 2319.

466. E. A. Domogatskaya, A. U. Dzhalmukhambetov, V. N. Tremblover and B. M. Yavorskii, *Khim. Fiz.*, 1987, **6**, 1592.
467. G. C. Allen, I. S. Butler and C. Kirby, *Inorg. Chim. Acta*, 1987, **134**, 289.
468. Y. V. Chizhov *et al.*, *Izv. Akad. Nauk SSSR, Ser. Khim.*, 1985, 1425.
469. J. C. Maire, *J. Organomet. Chem.*, 1985, **281**, 45.
470. R. A. Kok and M. B. Hall, *J. Am. Chem. Soc.*, 1985, **107**, 2599.
471. Y. Wang, K. Angermund, R. Goddard and C. Krüger, *J. Am. Chem. Soc.*, 1987, **109**, 587.
472. M. J. Aroney, M. K. Cooper, R. K. Pierens, S. J. Pratten and S. W. Filipczuk, *J. Organomet. Chem.*, 1985, **295**, 333.
473. M. J. Aroney, R. M. Clarkson, R. J. Klepetko, A. F. Masters and R. K. Pierens, *J. Organomet. Chem.*, 1990, **393**, 371.
474. M. J. Aroney, M. K. Cooper, P. A. Englert and R. K. Pierens, *J. Mol. Struct.*, 1981, **77**, 99.
475. M. J. Aroney, M. K. Cooper, R. K. Pierens, S. J. Pratten and S. W. Filipczuk, *J. Organomet. Chem.*, 1986, **307**, 191.
476. N. A. Ogorodnikova and D. N. Kravtsov, *Polyhedron*, 1990, **9**, 1.
477. D. R. Kanis, M. A. Ratner and T. J. Marks, *J. Am. Chem. Soc.*, 1992, **114**, 10 338.
478. J. Rawlings and C. A. L. Mahaffy, *Spectrosc. Lett.*, 1988, **21**, 597.
479. A. D. Hunter, V. Mozol and S. D. Tsai, *Organometallics*, 1992, **11**, 2251.
480. T. E. Bitterwolf, *Polyhedron*, 1988, **7**, 409.
481. T. E. Bitterwolf, *Polyhedron*, 1988, **7**, 1377.
482. R. S. Armstrong, M. J. Aroney, C. M. Barnes and K. W. Nugent, *Appl. Organomet. Chem.*, 1990, **4**, 569.
483. I. S. Butler, P. D. Harvey and G. C. Allen, *J. Raman. Spectrosc.*, 1987, **18**, 1.
484. K. Chhor, C. Sourisseau and G. Lucazeau, *J. Mol. Struct.*, 1982, **82**, 485.
485. S. M. Barnett, F. Dicaire and A. A. Ismail, *Can. J. Chem.*, 1990, **68**, 1196.
486. A. M. English, K. R. Plowman and I. S. Butler, *Inorg. Chem.*, 1982, **21**, 338.
487. Y. Huang, I. S. Butler and D. F. R. Gilson, *Inorg. Chem.*, 1992, **31**, 303.
488. L. N. Lewis, *J. Organomet. Chem.*, 1982, **234**, 355.
489. I. S. Butler, Z. H. Xu, R. S. Werbowyj and F. St. Germain, *Appl. Spectrosc.*, 1987, **41**, 149.
490. C. A. L. Mahaffy and J. Rawlings, *Spectrosc. Lett.*, 1984, **17**, 765.
491. J. Rawlings and C. A. L. Mahaffy, *Spectrosc. Lett.*, 1986, **19**, 517.
492. R. Schreiner and A. B. Ellis, *J. Am. Chem. Soc.*, 1982, **104**, 3374.
493. M. A. Zawadzki and A. B. Ellis, *Organometallics*, 1984, **3**, 192.
494. D. Rooney, J. Chaiken and D. Driscoll, *Inorg. Chem.*, 1987, **26**, 3939.
495. J. Szewczyk and A. Gryff-Keller, *J. Organomet. Chem.*, 1992, **424**, 41.
496. A. Ceccon, A. Gambaro, A. M. Romanin and A. Venzo, *J. Organomet. Chem.*, 1982, **239**, 345.
497. R. H. Mitchell, P. Zhou, S. Venugopalanan and T. W. Dingle, *J. Am. Chem. Soc.*, 1990, **112**, 7812.
498. H. Schumann, *J. Organomet. Chem.*, 1991, **421**, C7.
499. A. F. Masters, R. T. C. Brownlee, M. J. O'Connor and A. G. Wedd, *Inorg. Chem.*, 1981, **20**, 4183.
500. R. T. C. Brownlee, M. J. O'Connor, B. P. Shehan and A. G. Wedd, *Aust. J. Chem.*, 1986, **39**, 931.
501. L. Huis, P. J. W. Pouwels and C. MacLean, *Mol. Phys.*, 1990, **70**, 879.
502. P. M. Lausarot, L. Operti, G. A. Vaglio and M. Valle, *Inorg. Chim. Acta*, 1982, **60**, 185.
503. G. A. Vaglio, P. Volpe and L. Operti, *Org. Mass Spectrom.*, 1982, **17**, 617.
504. Y. S. Nekrasov *et al.*, *J. Organomet. Chem.*, 1981, **212**, 373.
505. V. I. Khvostenko, Y. S. Nekrasov, I. I. Furley, N. I. Vasyukova and G. A. Tolstikov, *J. Organomet. Chem.*, 1981, **212**, 369.
506. J. Müller and F. Lüdemann, *J. Organomet. Chem.*, 1981, **204**, 361.
507. J. Müller, G. Krebs, F. Lüdemann and E. Baumgartner, *J. Organomet. Chem.*, 1981, **218**, 61.
508. N. I. Vasyukova, Y. S. Nekrasov, Y. N. Sukharev, G. K. Magomedov and A. S. Frenkel, *Izv. Akad. Nauk SSSR, Ser. Khim.*, 1985, 1548.
509. J. Tirouflet, J. Besancon, B. Gautheron, F. Gomez and D. Fraisse, *J. Organomet. Chem.*, 1982, **234**, 143.
510. O. Gambino, P. Michelin-Lausarot, G. A. Vaglio, M. Valle, P. Volpe and L. Operti, *Transition Metal Chem. (Weinheim)*, 1982, **7**, 330.
511. M. I. Bruce and N. N. Zaitseva, *Aust. J. Chem.*, 1992, **45**, 1095.
512. K. R. Lane and R. R. Squires, *J. Am. Chem. Soc.*, 1985, **107**, 6403.
513. L. Operti, G. A. Vaglio, J. R. Gord and B. S. Freiser, *Organometallics*, 1991, **10**, 104.
514. G. Cetini *et al.*, *Inorg. Chim. Acta*, 1985, **104**, 69.
515. V. A. Piven, Y. S. Nekrasov, V. I. Losilkina and D. V. Zagorevskii, *Izv. Akad. Nauk SSSR, Ser. Khim.*, 1987, 664.
516. V. A. Piven and Y. S. Nekrasov, *Metalloorg. Khim.*, 1988, **1**, 115.
517. D. Wittneben, H.-F. Grützmacher, H. Butenschön and H. G. Wey, *Organometallics*, 1992, **11**, 3111.
518. B. Samoriski, J. M. Hossenlopp, D. Rooney and J. Chaiken, *J. Chem. Phys.*, 1986, **85**, 3326.
519. J. M. Hossenlopp, B. Samoriski, D. Rooney and J. Chaiken, *J. Chem. Phys.*, 1986, **85**, 3331.
520. G. W. Tyndall, C. E. Larson and R. L. Jackson, *J. Phys. Chem.*, 1989, **93**, 5508.
521. B. Samoriski and J. Chaiken, *J. Chem. Phys.*, 1989, **90**, 4079.
522. A. M. Ob'edkov, B. V. Zhuk and G. A. Domrachev, *Metalloorg. Khim.*, 1988, **1**, 592.
523. J. P. Bullock, D. C. Boyd and K. R. Mann, *Inorg. Chem.*, 1987, **26**, 3084.
524. N. J. Stone, D. A. Sweigart and A. M. Bond, *Organometallics*, 1986, **5**, 2553.
525. S. N. Milligan, I. Tucker and R. D. Rieke, *Inorg. Chem.*, 1983, **22**, 987.
526. J. O. Howell, J. M. Goncalves, C. Amatore, L. Klasinc, R. M. Wightman and J. K. Kochi, *J. Am. Chem. Soc.*, 1984, **106**, 3968.
527. R. D. Rieke, S. N. Milligan, I. Tucker, K. A. Dowler and B. R. Willeford, *J. Organomet. Chem.*, 1981, **218**, C25.
528. N. Van Order Jr., W. E. Geiger, T. E. Bitterwolf and A. L. Rheingold, *J. Am. Chem. Soc.*, 1987, **109**, 5680.
529. W. E. Geiger, N. Van Order Jr., D. T. Pierce, T. E. Bitterwolf, A. L. Rheingold and N. D. Chasteen, *Organometallics*, 1991, **10**, 2403.
530. C. Degrand, A. Radecki-Sudre and J. Besancon, *Organometallics*, 1982, **1**, 1311.
531. C. G. Zoski *et al.*, *J. Am. Chem. Soc.*, 1988, **110**, 2109.
532. C. Degrand and A. Radecki-Sudre, *J. Organomet. Chem.*, 1984, **268**, 63.
533. T. Ikeshoji and V. D. Parker, *Acta Chem. Scand., Ser. B*, 1983, **37**, 715.
534. T. Ikeshoji and V. D. Parker, *Acta Chem. Scand., Ser. B*, 1984, **38**, 175.

535. T. Ikeshoji and V. D. Parker, *Bull. Chem. Soc. Jpn.*, 1984, **57**, 1112.
536. F. Rourke, R. Gash and J. A. Crayston, *J. Organomet. Chem.*, 1992, **423**, 223.
537. M. G. Peterleitner, M. V. Tolstaya, V. V. Krivykh, L. I. Denisovich and M. I. Rybinskaya, *J. Organomet. Chem.*, 1983, **254**, 313.
538. M. G. Peterleitner, M. V. Tolstaya, V. V. Krivykh, L. I. Denisovich, M. I. Rybinskaya and D. N. Kravtsov, *Izv. Akad. Nauk SSSR, Ser. Khim.*, 1986, 662.
539. B. Olbrich-Deussner, R. Grob and W. Kaim, *J. Organomet. Chem.*, 1989, **366**, 155.
540. B. Olbrich-Deussner, W. Kaim and R. Gross-Lannert, *Inorg. Chem.*, 1989, **28**, 3113.
541. B. Schwederski, W. Kaim, B. Olbrich-Deussner and T. Roth, *J. Organomet. Chem.*, 1992, **440**, 145.
542. K. M. Doxsee, R. H. Grubbs and F. C. Anson, *J. Am. Chem. Soc.*, 1984, **106**, 7819.
543. Y. Zhang, D. K. Gosser, P. H. Rieger and D. A. Sweigart, *J. Am. Chem. Soc.*, 1991, **113**, 4062.
544. A. Barbati, F. Calderazzo, R. Poli and P. F. Zanazzi, *J. Chem. Soc., Dalton Trans.*, 1986, 2569.
545. F. Calderazzo, R. Poli and P. F. Zanazzi, *Gazz. Chim. Ital.*, 1988, **118**, 595.
546. F. Calderazzo, R. Poli and P. F. Zanazzi, *Gazz. Chim. Ital*, 1988, **118**, 583.
547. A. Barbati, F. Calderazzo and R. Poli, *Gazz. Chim. Ital.*, 1988, **118**, 589.
548. F. A. Cotton and R. Poli, *Inorg. Chem.*, 1987, **26**, 1514.
549. W. A. Herrmann, W. R. Thiel and E. Herdtweck, *Polyhedron*, 1988, **7**, 2027.
550. R. D. Rieke, W. P. Henry and J. S. Arney, *Inorg. Chem.*, 1987, **26**, 420.
551. V. S. Leong and N. J. Cooper, *J. Am. Chem. Soc.*, 1988, **110**, 2644.
552. H. G. Wey and H. Butenschön, *Angew. Chem., Int. Ed. Engl.*, 1990, **29**, 1444.
553. V. S. Leong and N. J. Cooper, *Organometallics*, 1988, **7**, 2058.
554. J. A. Corella, II and N. J. Cooper, *J. Am. Chem. Soc.*, 1990, **112**, 2832.
555. V. S. Leong and N. J. Cooper, *Organometallics*, 1987, **6**, 2000.
556. S. N. Milligan and R. D. Rieke, *Organometallics*, 1983, **2**, 171.
557. R. D. Rieke, S. N. Milligan and L. D. Schulte, *Organometallics*, 1987, **6**, 699.
558. L. D. Schulte and R. D. Rieke, *J. Org. Chem.*, 1987, **52**, 4827.
559. L. D. Schulte and R. D. Rieke, *Tetrahedron Lett.*, 1988, **29**, 5483.
560. R. D. Rieke, K. P. Daruwala, L. D. Schulte and S. M. Pankas, *Organometallics*, 1992, **11**, 284.
561. S. S. Yang, B. T. Dawson and R. D. Rieke, *Tetrahedron Lett.*, 1991, **32**, 3341.
562. R. D. Rieke, L. D. Schulte, B. T. Dawson and S. S. Yang, *J. Am. Chem. Soc.*, 1990, **112**, 8388.
563. R. D. Rieke, K. P. Daruwala and M. W. Forkner, *J. Org. Chem.*, 1989, **54**, 21.
564. R. D. Rieke, K. P. Daruwala and M. W. Forkner, *Organometallics*, 1991, **10**, 2946; correction p. 3946.
565. C. Degrand, F. Gasquez and P.-L. Compagnon, *J. Organomet. Chem.*, 1985, **280**, 87.
566. C. Degrand, M. Bikrani and B. Gautheron, *J. Organomet. Chem.*, 1986, **299**, 111.
567. B. Gautheron, C. Degrand and M. Bikrani, *J. Organomet. Chem.*, 1987, **329**, 349.
568. M. G. Peterleitner, L. I. Denisovich and D. N. Kravtsov, *Izv. Akad. Nauk SSSR, Ser. Khim.*, 1988, 442.
569. N. J. Gogan, I. L. Dickinson, J. Doull and J. R. Patterson, *J. Organomet. Chem.*, 1981, **212**, 71.
570. Y. L. Chiu, A. E. G. Sant'ana and J. H. P. Utley, *Tetrahedron Lett.*, 1987, **28**, 1349.
571. I. I. Grinval'd, B. V. Lokshin, N. K. Rudnevskii, V. P. Mar'in, A. A. Alad'in and G. A. Razuvaev, *Dokl. Akad. Nauk SSSR*, 1988, **298**, 366.
572. I. I. Grinval'd, B. V. Lokshin, O. A. Usacheva, I. A. Abronin, L. M. Golubinskaya and V. I. Bregadze, *Metalloorg. Khim.*, 1990, **3**, 166.
573. A. G. Ginzburg, V. N. Setkina and D. N. Kursanov, *Izv. Akad. Nauk SSSR, Ser. Khim.*, 1983, 1155.
574. E. N. Gur'yanova, E. S. Isaeva and A. G. Ginzburg, *Koord. Khim.*, 1987, **13**, 23.
575. A. G. Ginzburg, V. N. Setkina and D. N. Kursanov, *Izv. Akad. Nauk SSSR, Ser. Khim.*, 1985, 447.
576. S. C. Sockwell, P. S. Tanner and T. P. Hanusa, *Organometallics*, 1992, **11**, 2634.
577. A. A. Tumanov, P. G. Sennikov, A. N. Egorochkin, M. V. Panicheva, G. V. Belysheva and N. I. Sirotkin, *Izv. Akad. Nauk SSSR, Ser. Khim.*, 1983, 1634.
578. P. G. Sennikov, M. V. Panicheva, A. N. Egorochkin, N. I. Sirotkin and Y. Y. Zhiltsova, *J. Organomet. Chem.*, 1985, **280**, 81.
579. W. J. Bland, R. Davis and J. L. A. Durrant, *J. Organomet. Chem.*, 1982, **234**, C20.
580. H. Kobayashi, K. Kobayashi and Y. Kaizu, *Inorg. Chem.*, 1981, **20**, 4135.
581. A. Harada, K. Saeki and S. Takahashi, *Chem. Lett.*, 1985, 1157.
582. A. Harada, K. Saeki and S. Takahashi, *Organometallics*, 1989, **8**, 730.
583. V. I. Sokolov, V. L. Bondareva and I. F. Golovaneva, *Metalloorg. Khim.*, 1989, **2**, 1252.
584. W. Tam, D. F. Eaton, J. C. Calabrese, I. D. Williams, Y. Wang and A. G. Anderson, *Chem. Mater.*, 1989, **1**, 128.
585. S. Ozkar, G. A. Ozin, K. Moller and T. Bein, *J. Am. Chem. Soc.*, 1990, **112**, 9575.
586. H. Boulet, C. Bremard, C. Depecker and P. Legrand, *J. Chem. Soc., Chem. Commun.*, 1991, 1411.
587. G. W. Wagner and B. E. Hanson, *Organometallics*, 1987, **6**, 2494.
588. M. J. Calhorda, C. A. Frazão and J. A. Martinho Simões, *J. Organomet. Chem.*, 1984, **262**, 305.
589. C. D. Hoff, *J. Organomet. Chem.*, 1983, **246**, C53.
590. G. Al-Takhin, J. A. Connor, H. A. Skinner and M. T. Zafanari-Moattar, *J. Organomet. Chem.*, 1984, **260**, 189.
591. S. P. Nolan and C. D. Hoff, *J. Organomet. Chem.*, 1985, **290**, 365.
592. S. L. Mukerjee, R. F. Lang, T. Ju, G. Kiss, C. D. Hoff and S. P. Nolan, *Inorg. Chem.*, 1992, **31**, 4885.
593. G. Lucazeau, K. Chhor, C. Sourisseau and A. J. Dianoux, *Chem. Phys.*, 1983, **76**, 307.
594. J. Howard, K. Robson and T. C. Waddington, *J. Chem. Soc., Dalton Trans.*, 1982, 967.
595. D. F. R. Gilson, G. Gomez, I. S. Butler and P. J. Fitzpatrick, *Can. J. Chem.*, 1983, **61**, 737.
596. D. Braga and F. Grepioni, *Polyhedron*, 1990, **9**, 53.
597. A. Shaver, A. Eisenberg, K. Yamada, A. J. F. Clark and S. Farrokyzad, *Inorg. Chem.*, 1983, **22**, 4154.
598. H.-B. Burgi, A. Raselli, D. Braga and F. Grepioni, *Acta Crystallogr.*, 1992, **48**, 428.
599. G. W. Wagner and B. E. Hanson, *Inorg. Chem.*, 1987, **26**, 2019.
600. D. Braga and F. Grepioni, *J. Chem. Soc., Dalton Trans.*, 1990, 3143.
601. D. Braga and F. Grepioni, *Organometallics*, 1991, **10**, 2563.

602. S. Aime, D. Braga, R. Gobetto, F. Grepioni and A. Orlandi, *Inorg. Chem.*, 1991, **30**, 951.
603. F. Calderazzo, R. Poli, A. Barbati and P. F. Zanazzi, *J. Chem. Soc., Dalton Trans.*, 1984, 1059.
604. M. J. Zaworotko, J. L. Atwood and R. D. Priester, *J. Coord. Chem.*, 1990, **22**, 209.
605. N. I. Kirillova *et al.*, *Metalloorg. Khim.*, 1992, **5**, 1107.
606. J. Y. Saillard, R. Lissillour and D. Grandjean, *J. Organomet. Chem.*, 1981, **210**, 365.
607. J. Brocard, A. Laconi and D. Couturier, *Org. Magn. Reson.*, 1984, **22**, 369.
608. P. Szczecinski and A. Gryff-Keller, *Magn. Reson. Chem.*, 1988, **26**, 990.
609. P. Szczecinski, *J. Organomet. Chem.*, 1992, **423**, 23.
610. P. Szczecinski and K. Wisniewski, *J. Organomet. Chem.*, 1992, **423**, C13.
611. P. Szczecinski and J. Zachara, *J. Organomet. Chem.*, 1993, **447**, 241.
612. M. J. McGlinchey, *Adv. Organomet. Chem.*, 1992, **34**, 285.
613. B. Mailvaganam, B. E. McCarry, B. G. Sayer, R. E. Perrier, R. Faggiani and M. J. McGlinchey, *J. Organomet. Chem.*, 1987, **335**, 213.
614. B. Mailvaganam, B. G. Sayer and M. J. McGlinchey, *J. Organomet. Chem.*, 1990, **395**, 177.
615. W. Weissensteiner, J. Scharf and K. Schlögl, *J. Org. Chem.*, 1987, **52**, 1210.
616. E. M. Campi, B. M. K. Gatehouse, W. R. Jackson, I. D. Rae and M. G. Wong, *Can. J. Chem.*, 1984, **62**, 2566.
617. D. J. Iverson and K. Mislow, *Organometallics*, 1982, **1**, 3.
618. A. Gryff-Keller, P. Szczecinski and H. Koziel, *J. Organomet. Chem.*, 1989, **372**, 225.
619. J. Scharf, K. Schlögl and W. Weissensteiner, *Monatsh. Chem.*, 1986, **117**, 883.
620. X. Yang, J. Huang and J. Huang, *Jiegou Huaxue*, 1985, **4**, 52, 60; *Chem. Abstr.* 1985, **103**, 224 930m; 1986, **105**, 172 647u.
621. H. Kalchhauser, K. Schlögl, W. Weissensteiner and A. Werner, *J. Chem. Soc., Perkin Trans. 1*, 1983, 1723.
622. K. Schlögl, A. Werner and M. Widhalm, *J. Chem. Soc., Perkin Trans. 1*, 1983, 1731.
623. W. Weissensteiner and A. Werner, *Monatsh. Chem.*, 1984, **115**, 223.
624. M. Eyer, K. Schlögl and M. Widhalm, *Monatsh. Chem.*, 1984, **115**, 1429.
625. E. Halwax and H. Völlenkle, *Monatsh. Chem.*, 1983, **114**, 687.
626. D. J. Iverson, G. Hunter, J. F. Blount, J. R. Damewood, Jr. and K. Mislow, *J. Am. Chem. Soc.*, 1981, **103**, 6073.
627. G. Hunter, T. J. R. Weakley and W. Weissensteiner, *J. Chem. Soc., Perkin Trans. 2*, 1987, 1633.
628. B. Mailvaganam, B. G. Sayer and M. J. McGlinchey, *Organometallics*, 1990, **9**, 1089.
629. G. Hunter, W. S. Wadsworth Jr. and K. Mislow, *Organometallics*, 1982, **1**, 968.
630. M. J. McGlinchey *et al.*, *J. Chem. Soc., Chem. Commun.*, 1983, 634.
631. M. J. McGlinchey, P. Bougeard, B. G. Sayer, R. Hofer and C. J. L. Lock, *J. Chem. Soc., Chem. Commun.*, 1984, 789.
632. G. Hunter and K. Mislow, *J. Chem. Soc., Chem. Commun.*, 1984, 172.
633. G. Hunter, J. F. Blount, J. R. Damewood, Jr., D. J. Iverson and K. Mislow, *Organometallics*, 1982, **1**, 448.
634. J. F. Blount, G. Hunter and K. Mislow, *J. Chem. Soc., Chem. Commun.*, 1984, 170.
635. S. Denholm, G. Hunter and T. J. R. Weakley, *J. Chem. Soc., Dalton Trans.*, 1987, 2789.
636. G. Hunter, T. J. R. Weakley, K. Mislow and M. G. Wong, *J. Chem. Soc., Dalton Trans.*, 1986, 577.
637. G. Hunter, T. J. R. Weakley and W. Weissensteiner, *J. Chem. Soc., Dalton Trans.*, 1987, 1545.
638. J. A. S. Howell, M. G. Palin, P. McArdle, D. Cunningham, Z. Goldschmidt, H. E. Gottlieb and D. Hezroni-Langerman, *Organometallics*, 1993, **12**, 1694.
639. J. A. Chudek, G. Hunter, R. L. MacKay, G. Färber and W. Weissensteiner, *J. Organomet. Chem.*, 1989, **377**, C69.
640. J. A. Chudek, G. Hunter, R. L. MacKay, P. Kremminger and W. Weissensteiner, *J. Chem. Soc., Dalton Trans.*, 1991, 3337.
641. J. C. Barnes, G. Hunter, W. Keller, J. D. Paton and W. Weissensteiner, *Monatsh. Chem.*, 1992, **123**, 443.
642. J. A. Chudek, G. Hunter, R. L. MacKay, P. Kremminger, K. Schlögl and W. Weissensteiner, *J. Chem. Soc., Dalton Trans.*, 1990, 2001.
643. B. Mailvaganam, C. S. Frampton, S. Top, B. G. Sayer and M. J. McGlinchey, *J. Am. Chem. Soc.*, 1991, **113**, 1177.
644. P. Kremminger, W. Weissensteiner, C. Kratky, G. Hunter and R. L. MacKay, *Monatsh. Chem.*, 1989, **120**, 1175.
645. G. Hunter, R. L. MacKay, P. Kremminger and W. Weissensteiner, *J. Chem. Soc., Dalton Trans.*, 1991, 3349.
646. P. A. Downton, B. Mailvaganam, C. S. Frampton, B. G. Sayer and M. J. McGlinchey, *J. Am. Chem. Soc.*, 1991, **112**, 27.
647. K. V. Kilway and J. S. Siegel, *J. Am. Chem. Soc.*, 1991, **113**, 2332.
648. K. V. Kilway and J. S. Siegel, *Organometallics*, 1992, **11**, 1426.
649. K. V. Kilway and J. S. Siegel, *J. Am. Chem. Soc.*, 1992, **114**, 255.
650. M. Nambu and J. S. Siegel, *J. Am. Chem. Soc.*, 1988, **110**, 3675.
651. M. Nambu, D. L. Mohler, K. Hardcastle, K. K. Baldridge and J. S. Siegel, *J. Am. Chem. Soc.*, 1993, **115**, 6138.
652. M.-J. Liang, S.-Y. Chu, M.-C. Ou and M.-S. Tsai, *J. Phys. Chem.*, 1992, **96**, 9213.
653. P. A. Downton, B. G. Sayer and M. J. McGlinchey, *Organometallics*, 1992, **11**, 3281.
654. G. J. Kubas, G. Kiss and C. D. Hoff, *Organometallics*, 1991, **10**, 2870.
655. L. N. Novikova, N. A. Ustynyuk, V. E. Zvorykin, L. S. Dneprovskaya and Y. A. Ustynyuk, *J. Organomet. Chem.*, 1985, **292**, 237.
656. A. Ceccon, A. Gambaro, F. Gottardi, S. Santi and A. Venzo, *J. Organomet. Chem.*, 1991, **412**, 85.
657. A. N. Nesmeyanov *et al.*, *J. Organomet. Chem.*, 1982, **226**, 239.
658. H. G. Alt, S. J. Palackal and R. D. Rogers, *J. Organomet. Chem.*, 1990, **388**, 105.
659. P. Berno, A. Ceccon, A. Gambaro and A. Venzo, *Tetrahedron Lett.*, 1988, **29**, 3489.
660. A. Ceccon, A. Gambaro, S. Santi, G. Valle and A. Venzo, *J. Chem. Soc., Chem. Commun.*, 1989, 51.
661. A. Ceccon, A. Gambaro, S. Santi and A. Venzo, *J. Mol. Catal.*, 1991, **69**, L1.
662. A. Ceccon, C. J. Elsevier, J. M. Ernsting, A. Gambaro, S. Santi and A. Venzo, *Inorg. Chim. Acta*, 1993, **204**, 15.
663. C. Bonifaci *et al.*, *Organometallics*, 1993, **12**, 4211.
664. Y. F. Oprunenko, Y. N. Luzikov, Y. A. Ustynyuk and N. A. Ustynyuk, *J. Organomet. Chem.*, 1982, **231**, 137.
665. B. Gautheron, C. Degrand and M. Bikrani, *J. Organomet. Chem.*, 1986, **312**, 207.
666. A. Ceccon, A. Gambaro, G. Agostini and A. Venzo, *J. Organomet. Chem.*, 1981, **217**, 79.
667. A. Ceccon, A. Gambaro and A. Venzo, *J. Organomet. Chem.*, 1984, **275**, 209.
668. N. A. Ustynyuk, L. N. Novikova, Y. F. Oprunenko, S. G. Malyugina and Y. A. Ustynyuk, *J. Organomet. Chem.*, 1984, **277**, 75.

669. N. A. Ustynyuk *et al.*, *J. Organomet. Chem.*, 1985, **294**, 31.
670. N. A. Ustynyuk, Y. F. Oprunenko, S. G. Malyugina, O. I. Trifonova and Y. A. Ustynyuk, *J. Organomet. Chem.*, 1984, **270**, 185.
671. C. Degrand, B. Gautheron, M. Bikrani, F. Gasquez and P. L. Compagnon, *J. Organomet. Chem.*, 1984, **273**, 319.
672. A. Ceccon, A. Gambaro, A. Venzo, V. Lucchini, T. E. Bitterwolf and J. Shade, *J. Organomet. Chem.*, 1987, **327**, 55.
673. A. Ceccon, A. Gambaro, A. Venzo, V. Lucchini, T. E. Bitterwolf and J. Shade, *J. Organomet. Chem.*, 1988, **349**, 315.
674. O. I. Trifonova, R. A. Galiullin, Y. A. Ustynyuk, N. A. Ustynyuk, P. V. Petrovskii and D. N. Kravtsov, *J. Organomet. Chem.*, 1987, **328**, 321.
675. O. I. Trifonova, Y. F. Oprunenko, Y. A. Ustynyuk and N. A. Ustynyuk, *Metalloorg. Khim.*, 1990, **3**, 686.
676. O. I. Trifonova, R. A. Galliulin, Y. A. Ustynyuk, N. A. Ustynyuk, P. V. Petrovskii and D. N. Kravtsov, *Izv. Akad. Nauk SSSR, Ser. Khim.*, 1988, 1383.
677. A. Ceccon *et al.*, *J. Chem. Soc., Perkin Trans. 2*, 1992, 1111.
678. A. Ceccon, A. Gambaro and A. Venzo, *J. Organomet. Chem.*, 1985, **281**, 221.
679. A. Ceccon, A. Gambaro, F. Gottardi, F. Manoli and A. Venzo, *J. Organomet. Chem.*, 1989, **363**, 91.
680. A. Ceccon, A. Gambaro, F. Gottardi, S. Santi, A. Venzo and V. Lucchini, *J. Organomet. Chem.*, 1989, **379**, 67.
681. O. I. Trifonova, Y. F. Oprunenko and Y. A. Ustynyuk, *Metalloorg. Khim.*, 1990, **3**, 465.
682. Y. F. Oprunenko, S. G. Malyugina, Y. A. Ustynyuk, N. A. Ustynyuk and D. N. Kravtsov, *J. Organomet. Chem.*, 1988, **338**, 357.
683. Y. F. Oprunenko, S. G. Malyugina, Y. A. Ustynyuk and N. A. Ustynyuk, *Izv. Akad. Nauk SSSR, Ser. Khim.*, 1985, 2156.
684. R. U. Kirss and P. M. Treichel, *J. Am. Chem. Soc.*, 1986, **108**, 853.
685. P. Berno, A. Ceccon, A. Gambaro, A. Venzo, P. Ganis and G. Valle, *J. Chem. Soc., Perkin Trans. 2*, 1987, 935.
686. Y. F. Oprunenko *et al.*, *Izv. Akad. Nauk SSSR, Ser. Khim.*, 1988, 438.
687. Y. F. Oprunenko *et al.*, *Metalloorg. Khim.*, 1988, **1**, 384.
688. Y. F. Oprunenko *et al.*, *Metalloorg. Khim.*, 1988, **1**, 815.
689. Y. F. Oprunenko, I. A. Shaposhnikova and Y. A. Ustynyuk, *Metalloorg. Khim.*, 1991, **4**, 1377.
690. Y. F. Oprunenko, I. A. Shaposhnikova, Y. A. Ustynyuk, N. A. Ustynyuk and D. N. Kravtsov, *Izv. Akad. Nauk SSSR, Ser. Khim.*, 1987, 703.
691. S. D. Cunningham, K. Öfele and B. R. Willeford, *J. Am. Chem. Soc.*, 1983, **105**, 3724; correction p. 7474.
692. A. Ceccon, A. Gambaro, L. Pizzato, A. Romanin and A. Venzo, *J. Chem. Soc., Chem. Commun.*, 1982, 907.
693. A. Ceccon, A. Gambaro, A. Romanin and A. Venzo, *Angew. Chem., Int. Ed. Engl.*, 1983, **22**, 559; *Angew. Chem. Suppl.*, 1983, 734.
694. J. Brocard, J. Lebibi, L. Pelinski and M. Mahmoudi, *Tetrahedron Lett.*, 1986, **27**, 6325.
695. J. Brocard, L. Pelinski, J. Lebibi, M. Mahmoudi and L. Maciejewski, *Tetrahedron*, 1989, **45**, 709.
696. M. Uemura, T. Kobayashi, K. Isobe, T. Minami and Y. Hayashi, *J. Org. Chem.*, 1986, **51**, 2859.
697. S. G. Davies and C. L. Goodfellow, *J. Organomet. Chem.*, 1988, **340**, 195.
698. H.-G. Schmalz, B. Millies, J. W. Bats and G. Dürner, *Angew. Chem., Int. Ed. Engl.*, 1992, **31**, 631.
699. M. Uemura, T. Minami and Y. Hayashi, *J. Am. Chem. Soc.*, 1987, **109**, 5277.
700. M. Uemura, T. Minami, K. Hirotsu and Y. Hayashi, *J. Org. Chem.*, 1989, **54**, 469.
701. A. Solladié-Cavallo and J. Suffert, *Magn. Reson. Chem.*, 1985, **23**, 739.
702. S. G. Davies and C. L. Goodfellow, *J. Chem. Soc., Perkin Trans. 1*, 1989, 192.
703. S. G. Davies and C. L. Goodfellow, *J. Chem. Soc., Perkin Trans. 1*, 1990, 393.
704. L. A. Bromley, S. G. Davies and C. L. Goodfellow, *Tetrahedron Asymmetry*, 1991, **2**, 139.
705. A. Solladié-Cavallo, G. Solladié and E. Tsamo, *Inorg. Synth.*, 1985, **23**, 85.
706. A. Solladié-Cavallo, D. Farkhani, A. C. Dreyfuss and F. Sanch, *Bull. Soc. Chim. Fr.*, 1986, 906.
707. A. Alexakis *et al.*, *J. Am. Chem. Soc.*, 1992, **114**, 8288.
708. S. Top, G. Jaouen, J. Gillois, C. Baldoli and S. Maiorana, *J. Chem. Soc., Chem. Commun.*, 1988, 1284.
709. S. Top, G. Jaouen, C. Baldoli, P. Del Buttero and S. Maiorana, *J. Organomet. Chem.*, 1991, **413**, 125.
710. Y. Yamazaki and K. Hosono, *Biotechnol. Lett.*, 1989, **11**, 679.
711. Y. Yamazaki and K. Hosono, *Agric. Biol. Chem.*, 1988, **52**, 3239.
712. Y. Yamazaki and K. Hosono, *Tetrahedron Lett.*, 1989, **30**, 5313.
713. Y. Yamazaki, M. Uebayasi, J. Someya and K. Hosono, *Agric. Biol. Chem.*, 1990, **54**, 1781.
714. C. Baldoli, S. Maiorana, G. Carrea and S. Riva, *Tetrahedron Asymmetry*, 1993, **4**, 767.
715. J. Gillois, D. Buisson, R. Azerad and G. Jaouen, *J. Chem. Soc., Chem. Commun.*, 1988, 1224.
716. K. Nakamura, K. Ishihara, A. Ohno, M. Uemura, H. Nishimura and Y. Hayashi, *Tetrahedron Lett.*, 1990, **31**, 3603.
717. Y. Yamazaki and K. Hosono, *Tetrahedron Lett.*, 1990, **31**, 3895.
718. Y. Yamazaki, N. Morohashi and K. Hosono, *Biotechnol. Lett.*, 1991, **13**, 81.
719. B. Malezieux *et al.*, *Tetrahedron Asymmetry*, 1992, **3**, 375.
720. G. Jaouen, J. A. S. Howell, M.-C. Tirvengadum, P. McArdle and D. Cunningham, *J. Organomet. Chem.*, 1989, **370**, 51.
721. I. S. Butler and H. L. Uhm, *Comments Inorg. Chem.*, 1988, **7**, 1.
722. B. N. Diel, R. C. Haltiwanger and A. D. Norman, *J. Am. Chem. Soc.*, 1982, **104**, 4700.
723. B. N. Diel, P. F. Brandt, R. C. Haltiwanger, M. L. J. Hackney and A. D. Norman, *Inorg. Chem.*, 1989, **28**, 2811.
724. A. L. Rheingold, M. E. Fountain and A. J. DiMaio, *J. Am. Chem. Soc.*, 1987, **109**, 141.
725. J. Febvay, D. Grec and J. G. Riess, *Inorg. Chem.*, 1981, **20**, 4285.
726. A. A. Gonzalez, S. L. Mukerjee, S.-J. Chou, Z. Kai and C. D. Hoff, *J. Am. Chem. Soc.*, 1988, **110**, 4419.
727. P. Leoni, E. Grilli, M. Pasquali and M. Tomassini, *J. Chem. Soc., Dalton Trans.*, 1985, 2561.
728. P. Leoni, E. Grilli, M. Pasquali and M. Tomassini, *J. Chem. Soc., Dalton Trans.*, 1986, 879.
729. T. Schmidt and S. Neis, *J. Organomet. Chem.*, 1992, **430**, C5.
730. T. Schmidt, C. Krüger and P. Betz, *J. Organomet. Chem.*, 1991, **402**, 97.
731. G. W. Harris, M. O. Albers, J. C. A. Boeyens and N. J. Coville, *Organometallics*, 1983, **2**, 609.
732. T. E. Bitterwolf, K. A. Lott, A. J. Rest and J. Mascetti, *J. Organomet. Chem.*, 1991, **419**, 113.
733. C. H. Bamford and K. G. Al-Lamee, *J. Chem. Soc., Faraday Trans. 1*, 1984, **80**, 2175.
734. P. E. Bloyce, R. H. Hooker, A. J. Rest, T. E. Bitterwolf, N. J. Fitzpatrick and J. E. Shade, *J. Chem. Soc., Dalton Trans.*, 1990, 833.

735. B. S. Creaven *et al.*, *Organometallics*, 1993, **12**, 3127.
736. W. Wang, P. Jin, Y. Liu, Y. She and K.-J. Fu, *J. Phys. Chem.*, 1992, **96**, 1278.
737. S. Oishi, *J. Organomet. Chem.*, 1987, **335**, 207.
738. Y. Zheng, W. Wang, J. Lin, Y. She and K. Fu, *Chin. Phys. Lett.*, 1992, **9**, 329.
739. J. D. Black, M. J. Boylan and P. S. Braterman, *J. Chem. Soc., Dalton Trans.*, 1981, 673.
740. S. M. Howdle, M. A. Healy and M. Poliakoff, *J. Am. Chem. Soc.*, 1990, **112**, 4804.
741. C. H. Bamford and K. G. Al-Lamee, *J. Chem. Soc., Faraday Trans, 1*, 1984, **80**, 2187.
742. E. A. Domogatskaya *et al.*, *J. Organomet. Chem.*, 1983, **248**, 161.
743. D. Sellmann and A. Holzmeier, *Z. Naturforsch., Teil B*, 1990, **45B**, 635.
744. M. Herberhold and L. Haumaier, *Chem. Ber.*, 1982, **115**, 1399.
745. M. F. Semmelhack, W. Seufert and L. Keller, *J. Organomet. Chem.*, 1982, **226**, 183.
746. M. N. Estekhina, N. K. Baranetskaya and V. N. Setkina, *Izv. Akad. Nauk SSSR, Ser. Khim.*, 1986, 1178.
747. V. I. Losilkina, P. V. Petrovskii, N. K. Baranetskaya, V. N. Setkina and D. N. Kursanov, *Izv. Akad. Nauk SSSR, Ser. Khim.*, 1983, 2825.
748. V. I. Losilkina, M. N. Estekhina, N. K. Baranetskaya and V. N. Setkina, *J. Organomet. Chem.*, 1986, **299**, 187.
749. M. N. Estekhina, V. I. Losilkina, N. M. Loim, N. K. Baranetskaya and V. N. Setkina, *Izv. Akad. Nauk SSSR, Ser. Khim.*, 1986, 2371.
750. B. Boyer, G. Lamaty, B. Caro and G. Jaouen, *J. Organomet. Chem.*, 1981, **216**, C51.
751. J. Y. Saillard, D. Grandjean, A. Le Beuze and G. Simonneaux, *J. Organomet. Chem.*, 1981, **204**, 197.
752. A. Kececi, D. Rehder, W. Roose and R. Talay, *Chem. Ber.*, 1982, **115**, 3257.
753. V. I. Losilkina, N. I. Pyshnograeva, N. K. Baranetskaya and V. N. Setkina, *Izv. Akad. Nauk SSSR, Ser. Khim.*, 1985, 2780.
754. P. Le Maux, G. Simonneaux and G. Jaouen, *J. Organomet. Chem.*, 1981, **217**, 61.
755. P. Caillet and P. Le Maux, *J. Organomet. Chem.*, 1983, **243**, 51.
756. C. Andersson, R. Larsson and B. Yom-Tov, *Inorg. Chim. Acta*, 1983, **76**, L185.
757. D. Lentz and R. Marschall, *Z. Anorg. Allg. Chem.*, 1991, **593**, 181.
758. S. Lotz, M. Schindehutte, M. M. van Dyk, J. L. M. Dillen and P. H. van Rooyen, *J. Organomet. Chem.*, 1985, **295**, 51.
759. P. H. van Rooyen, J. L. M. Dillen, S. Lotz and M. Schindehutte, *J. Organomet. Chem.*, 1984, **273**, 61.
760. J. Herbig, J. Köhler, B. Nuber, H. G. Stammler and M. L. Ziegler, *J. Organomet. Chem.*, 1993, **444**, 107.
761. T. E. Bitterwolf, *J. Organomet. Chem.*, 1983, **252**, 305.
762. T. E. Bitterwolf and K. S. Raghuveer, *Inorg. Chim. Acta*, 1990, **172**, 59.
763. P. C. Nirchio and D. J. Wink, *Organometallics*, 1991, **10**, 336.
764. P. C. Nirchio and D. J. Wink, *Organometallics*, 1991, **10**, 2499.
765. R. H. Hill and M. S. Wrighton, *Organometallics*, 1987, **6**, 632.
766. G. N. Glavee, B. R. Jagirdar, J. J. Schneider, K. J. Klabunde, L. J. Radonovich and K. Dodd, *Organometallics*, 1992, **11**, 1043.
767. T. J. Burkey, *J. Am. Chem. Soc.*, 1990, **112**, 8329.
768. U. Schubert, J. Müller and H. G. Alt, *Organometallics*, 1987, **6**, 469.
769. E. Matarasso-Tchiroukine and G. Jaouen, *Can. J. Chem.*, 1988, **66**, 2157.
770. E. Matarasso-Tchiroukine, *J. Chem. Soc., Chem. Commun.*, 1990, 681.
771. H. Piana, U. Kirchgassner and U. Schubert, *Chem. Ber.*, 1991, **124**, 743.
772. U. Schubert and J. Schubert, *J. Organomet. Chem.*, 1992, **434**, 169.
773. J. Schubert, S. Mock and U. Schubert, *Chem. Ber.*, 1993, **126**, 657.
774. F. Bringewski, G. Huttner, W. Imhof and L. Zsolnai, *J. Organomet. Chem.*, 1992, **439**, 33.
775. F. Bringewski, G. Huttner and L. Zsolnai, *J. Organomet. Chem.*, 1993, **447**, 233.
776. F. Bringewski, G. Huttner and W. Imhof, *J. Organomet. Chem.*, 1993, **448**, C3.
777. U. Schubert and J. Grönen, *Chem. Ber.*, 1989, **122**, 1237.
778. H. G. Alt, H. E. Engelhardt and A. C. Filippou, *J. Organomet. Chem.*, 1988, **355**, 139.
779. R. D. Rogers, H. G. Alt and H. E. Engelhardt, *J. Organomet. Chem.*, 1989, **366**, 305.
780. H. G. Alt, H. E. Engelhardt, R. D. Rogers and S. T. Abu-Orabi, *J. Organomet. Chem.*, 1989, **378**, 33.
781. N. G. Connelly, A. G. Orpen, A. L. Rieger, P. H. Rieger, C. J. Scott and G. M. Rosair, *J. Chem. Soc., Chem. Commun.*, 1992, 1293.
782. G. R. Clark and P. W. Clark, *J. Organomet. Chem.*, 1983, **255**, 205.
783. P. Szczecinski, *J. Organomet. Chem.*, 1990, **393**, 111.
784. A. Solladié-Cavallo, J. Suffert and A. de Cian, *J. Organomet. Chem.*, 1982, **236**, 83.
785. V. V. Krivykh, M. V. Tolstaya and M. I. Rybinskaya, *Izv. Akad. Nauk SSSR, Ser. Khim.*, 1986, 2356.
786. V. V. Krivykh, M. V. Tolstaya and M. I. Rybinskaya, *Izv. Akad. Nauk SSSR, Ser. Khim.*, 1986, 2353.
787. M. I. Rybinskaya, V. V. Krivykh, E. S. Gusev and E. S. Il'minskaya, *Izv. Akad. Nauk SSSR, Ser. Khim.*, 1981, 2839.
788. V. V. Krivykh, E. S. Il'minskaya and M. I. Rybinskaya, *Izv. Akad. Nauk SSSR, Ser. Khim.*, 1983, 1911.
789. V. V. Krivykh, *Metalloorg. Khim.*, 1992, **5**, 213.
790. A. N. Nesmeyanov, V. V. Krivykh, E. S. Il'minskaya and M. I. Rybinskaya, *J. Organomet. Chem.*, 1981, **209**, 309.
791. V. V. Krivykh, O. V. Gusev, P. V. Petrovskii and M. I. Rybinskaya, *Izv. Akad. Nauk SSSR, Ser. Khim.*, 1983, 2635.
792. M. I. Rybinskaya, V. S. Kaganovich, V. V. Krivykh, A. R. Kudinov and O. V. Gusev, *Metalloorg. Khim.*, 1990, **3**, 876.
793. M. I. Rybinskaya, *J. Organomet. Chem.*, 1990, **383**, 113.
794. V. V. Krivykh, O. V. Gusev, P. V. Petrovskii and M. I. Rybinskaya, *J. Organomet. Chem.*, 1989, **366**, 129.
795. G. K. Magomedov, L. V. Morozova, S. A. Sigachev, V. V. Krivykh, E. S. Taits and M. I. Rybinskaya, *Zh. Obshch. Khim.* 1986, **56**, 1119.
796. V. V. Krivykh, E. S. Taits and M. I. Rybinskaya, *Izv. Akad. Nauk SSSR, Ser. Khim.*, 1987, 2866.
797. V. V. Krivykh, E. S. Taits, P. V. Petrovskii and M. I. Rybinskaya, *Metalloorg. Khim.*, 1989, **2**, 939.
798. V. V. Krivykh, E. S. Taits, P. V. Petrovskii, Y. T. Struchkov and A. I. Yanovskii, *Mendeleev Commun.*, 1991, 103.
799. I. S. Butler, A. M. English and K. R. Plowman, *Inorg. Synth.*, 1982, **21**, 1.
800. I. M. Baibich, A. M. English and I. S. Butler, *Organometallics*, 1984, **3**, 1786.
801. D. Albers, N. J. Coville, T. A. Ford, M. V. M. Hall and D. C. Levendis, *Spectrochim. Acta*, 1989, **45A**, 249.
802. D. Cozak and I. S. Butler, *Can. J. Spectrosc.*, 1982, **27**, 141.

803. Y. Huang, D. F. R. Gilson, I. S. Butler and F. G. Morin, *Inorg. Chem.*, 1992, **31**, 322.
804. I. M. Baibich and I. S. Butler, *Inorg. Chim. Acta*, 1984, **89**, 73.
805. I. S. Butler and A. A. Ismail, *Inorg. Chem.*, 1986, **25**, 3910.
806. A. M. English *et al.*, *J. Organomet. Chem.*, 1981, **205**, 177.
807. P. H. Bird, A. A. Ismail and I. S. Butler, *Inorg. Chem.*, 1985, **24**, 2911.
808. A. A. Ismail and I. S. Butler, *J. Organomet. Chem.*, 1988, **346**, 185.
809. S. Lotz, R. R. Pille and P. H. van Rooyen, *Inorg. Chem.*, 1986, **25**, 3053.
810. C. C. Frazier, N. D. Magnussen, L. N. Osuji and K. O. Parker, *Organometallics*, 1982, **1**, 903.
811. P. Klüfers, L. Knoll, C. Reiners and K. Reiss, *Chem. Ber.*, 1985, **118**, 1825.
812. A. A. Pasynskii, I. L. Eremenko, B. Orazsakhatov, G. S. Gasanov, V. E. Shklover and Y. T. Struchkov, *J. Organomet. Chem.*, 1984, **269**, 147.
813. W. A. Herrmann, C. E. Barnes, R. Serrano and B. Koumbouris, *J. Organomet. Chem.*, 1983, **256**, C30.
814. R. D. Barr, M. Green, K. Marsden, F. G. A. Stone and P. Woodward, *J. Chem. Soc., Dalton Trans.*, 1983, 507.
815. R. D. Barr, T. B. Marder, A. G. Orpen and I. D. Williams, *J. Chem. Soc., Chem. Commun.*, 1984, 112.
816. R. D. Barr, M. Green, T. B. Marder and F. G. A. Stone, *J. Chem. Soc., Dalton Trans.*, 1984, 1261.
817. H. Werner, F. J. Garcia Alonso, H. Otto, K. Peters and H. G. von Schnering, *J. Organomet. Chem.*, 1985, **289**, C5.
818. J. A. Abad *et al.*, *J. Chem. Soc., Dalton Trans.*, 1983, 2075.
819. R. R. Fraser and T. S. Mansour, *J. Organomet. Chem.*, 1986, **310**, C60.
820. M. V. Antonova, S. K. Moiseev, V. N. Kalinin and I. O. Shapiro, *Izv. Akad. Nauk SSSR, Ser. Khim.*, 1989, 2458.
821. M. F. Semmelhack and C. Ullenius, *J. Organomet. Chem.*, 1982, **235**, C10.
822. M. E. Wright, *Organometallics*, 1989, **8**, 407.
823. R. J. Card and W. S. Trahanovsky, *J. Org. Chem.*, 1980, **45**, 2560.
824. F. Rose-Munch, E. Rose and A. Semra, *J. Organomet. Chem.*, 1989, **377**, C9.
825. J. C. Gill, B. A. Marples and J. R. Traynor, *Tetrahedron Lett.*, 1987, **28**, 2643.
826. J. P. Gilday, J. T. Negri and D. A. Widdowson, *Tetrahedron*, 1989, **45**, 4605.
827. J. P. Gilday and D. A. Widdowson, *J. Chem. Soc., Chem. Commun.*, 1986, 1235.
828. J. P. Gilday and D. A. Widdowson, *Tetrahedron Lett.*, 1986, **27**, 5525.
829. M. Uemura *et al.*, *J. Org. Chem.*, 1983, **48**, 2349.
830. M. Uemura, K. Take, K. Isobe, T. Minami and Y. Hayashi, *Tetrahedron*, 1985, **41**, 5771.
831. M. Uemura, K. Take and Y. Hayashi, *J. Chem. Soc., Chem. Commun.*, 1983, 858.
832. L. M. Sandilands *et al.*, *J. Organomet. Chem.*, 1982, **224**, 267.
833. J.-C. Boutonnet, F. Rose-Munch, E. Rose and G. Precigoux, *J. Organomet. Chem.*, 1985, **284**, C25.
834. N. F. Masters and D. A. Widdowson, *J. Chem. Soc., Chem. Commun.*, 1983, 955.
835. M. Fukui, T. Ikeda and T. Oishi, *Tetrahedron Lett.*, 1982, **23**, 1605.
836. M. Fukui, T. Ikeda and T. Oishi, *Chem. Pharm. Bull.*, 1983, **31**, 466.
837. M. Fukui, Y. Yamada, A. Asakura and T. Oishi, *Heterocycles*, 1981, **15**, 415.
838. J. Levisalles *et al.*, *J. Organomet. Chem.*, 1987, **328**, 109.
839. J. Blagg, S. G. Davies, C. L. Goodfellow and K. H. Sutton, *J. Chem. Soc., Perkin Trans. 1*, 1987, 1805.
840. J. A. Heppert, M. E. Thomas-Miller, M. L. Milligan, D. Vander Velde and J. Aubé, *Organometallics*, 1988, **7**, 2581.
841. J. A. Heppert, J. Aubé, M. E. Thomas-Miller, M. L. Milligan and F. Takusagawa, *Organometallics*, 1990, **9**, 727.
842. M. Uemura, R. Miyake, H. Nishimura, Y. Matsumoto and T. Hayashi, *Tetrahedron Asymmetry*, 1992, **3**, 213.
843. J. Aubé, J. A. Heppert, M. L. Milligan, M. J. Smith and P. Zenk, *J. Org. Chem.*, 1992, **57**, 3563.
844. Y. Kondo, J. R. Green and J. Ho, *J. Org. Chem.*, 1991, **56**, 7199.
845. S. J. Coote, S. G. Davies, C. L. Goodfellow, K. H. Sutton, D. Middlemiss and A. Naylor, *Tetrahedron Asymmetry*, 1990, **1**, 817.
846. J. Blagg, S. G. Davies, C. L. Goodfellow and K. H. Sutton, *J. Chem. Soc., Perkin Trans. 1*, 1990, 1133.
847. E. P. Kündig, C. Grivet and S. Spichiger, *J. Organomet. Chem.*, 1987, **332**, C13.
848. P. M. Treichel and R. U. Kirss, *Organometallics*, 1987, **6**, 249.
849. E. P. Kündig, V. Desobry, C. Grivet, B. Rudolph and S. Spichiger, *Organometallics*, 1987, **6**, 1173.
850. J. A. Morley and N. F. Woolsey, *J. Org. Chem.*, 1992, **57**, 6487.
851. R. U. Kirss, P. M. Treichel and K. J. Haller, *Organometallics*, 1987, **6**, 242.
852. E. P. Kündig, C. Perret and B. Rudolph, *Helv. Chim. Acta*, 1990, **73**, 1970.
853. T. V. Lee, A. J. Leigh and C. B. Chapleo, *Tetrahedron Lett.*, 1989, **30**, 5519.
854. N. F. Masters, N. Mathews, G. Nechvatal and D. A. Widdowson, *Tetrahedron*, 1989, **45**, 5955.
855. G. Nechvatal, D. A. Widdowson and D. J. Williams, *J. Chem. Soc., Chem. Commun.*, 1981, 1260.
856. P. J. Beswick, C. S. Greenwood, T. J. Mowlem, G. Nechvatal and D. A. Widdowson, *Tetrahedron*, 1988, **44**, 7325.
857. G. Nechvatal and D. A. Widdowson, *J. Chem. Soc., Chem. Commun.*, 1982, 467.
858. J. L. Davidson, H. Patel and P. N. Preston, *J. Organomet. Chem.*, 1987, **336**, C44.
859. P. J. Beswick, S. J. Leach, N. F. Masters and D. A. Widdowson, *J. Chem. Soc., Chem. Commun.*, 1984, 46.
860. M. Ghavshou and D. A. Widdowson, *J. Chem. Soc., Perkin Trans. 1*, 1983, 3065.
861. P. J. Beswick and D. A. Widdowson, *Synthesis*, 1985, 492.
862. M. F. Semmelhack *et al.*, *Tetrahedron*, 1981, **37**, 3957.
863. E. P. Kündig, *Pure Appl. Chem.*, 1985, **57**, 1855.
864. L. A. P. Kane-Maguire, E. D. Honig and D. A. Sweigart, *Chem. Rev.*, 1984, **84**, 525.
865. J.-P. Djukic, F. Rose-Munch, E. Rose and Y. Dromzee, *J. Am. Chem. Soc.*, 1993, **115**, 6434.
866. C. G. Kreiter and T. Hellmann, *J. Organomet. Chem.*, 1991, **405**, C6.
867. G. Bernardinelli, A. Cunningham, C. Dupre, E. P. Kündig, D. Stussi and J. Weber, *Chimia*, 1992, **46**, 126.
868. R. P. Houghton, M. Voyle and R. Price, *J. Organomet. Chem.*, 1983, **259**, 183.
869. J.-C. Boutonnet, F. Rose-Munch and E. Rose, *Tetrahedron Lett.*, 1985, **26**, 3989.
870. F. Rose-Munch, J. P. Djukic and E. Rose, *Tetrahedron Lett.*, 1990, **31**, 2589.
871. M. Persson, U. Hacksell and I. Csoregh, *J. Chem. Soc., Perkin Trans. 1*, 1991, 1453.
872. C. Baldoli, P. Del Buttero, E. Licandro and S. Maiorana, *Gazz. Chim. Ital.*, 1988, **118**, 409.
873. M. Chaari, A. Jenhi, J.-P. Lavergne and P. Viallefont, *J. Organomet. Chem.*, 1991, **401**, C10.

874. M. Chaari, J.-P. Lavergne and P. Viallefont, *Synth. Commun.*, 1989, **19**, 1211.
875. F. Rose-Munch, K. Aniss and E. Rose, *J. Organomet. Chem.*, 1990, **385**, C1.
876. F. Rose-Munch, K. Aniss, E. Rose and J. Vaissermann, *J. Organomet. Chem.*, 1991, **415**, 223.
877. L. I. Zakharkin and G. G. Zhigareva, *Zh. Obshch. Khim.*, 1983, **53**, 953.
878. T. J. Henly, C. B. Knobler and M. F. Hawthorne, *Organometallics*, 1992, **11**, 2313.
879. B. R. Cho and S. W. Kim, *Bull. Korean Chem. Soc.*, 1984, **5**, 170.
880. A. Alemagna, C. Baldoli, P. Del Buttero, E. Licandro and S. Maiorana, *Gazz. Chim. Ital.*, 1985, **115**, 555.
881. V. V. Litvak, P. P. Kun and V. D. Shteingarts, *Zh. Org. Khim.*, 1984, **20**, 753.
882. I. I. Oleinik, P. P. Kun, V. V. Litvak and V. D. Shteingarts, *Zh. Org. Khim.*, 1988, **24**, 929.
883. F. Hossner and M. Voyle, *J. Organomet. Chem.*, 1988, **347**, 365.
884. V. V. Litvak, L. I. Goryunov and V. D. Shteingarts, *Zh. Org. Khim.*, 1986, **22**, 157.
885. C. Baldoli, P. Del Buttero, S. Maiorana and A. Papagni, *J. Chem. Soc., Chem. Commun.*, 1985, 1181.
886. A. Alemagna, C. Baldoli, P. Del Buttero, E. Licandro and S. Maiorana, *J. Chem. Soc., Chem. Commun.*, 1985, 417.
887. A. Alemagna, C. Baldoli, P. Del Buttero, E. Licandro and S. Maiorana, *Synthesis*, 1987, 192.
888. C. Baldoli, P. Del Buttero, E. Licandro and S. Maiorana, *Synthesis*, 1988, 344.
889. C. Baldoli, P. Del Buttero and S. Maiorana, *Tetrahedron Lett.*, 1992, **33**, 4049.
890. M. J. Dickens, J. P. Gilday, T. J. Mowlem and D. A. Widdowson, *Tetrahedron*, 1991, **47**, 8621.
891. A. Alemagna, P. Del Buttero, C. Gorini, D. Landini, E. Licandro and S. Maiorana, *J. Org. Chem.*, 1983, **48**, 605.
892. A. Alemagna, P. Cremonesi, P. Del Buttero, E. Licandro and S. Maiorana, *J. Org. Chem.*, 1983, **48**, 3114.
893. E. P. Kündig, R. Liu and A. Ripa, *Helv. Chim. Acta*, 1992, **75**, 2657.
894. F. Rose-Munch, E. Rose, A. Semra and M. Filoche, *J. Organomet. Chem.*, 1989, **363**, 123.
895. F. Rose-Munch, E. Rose, A. Semra, L. Mignon, J. Garcia-Oricain and C. Knobler, *J. Organomet. Chem.*, 1989, **363**, 297.
896. F. Rose-Munch, E. Rose and A. Semra, *J. Chem. Soc., Chem. Commun.*, 1987, 942.
897. F. Rose-Munch, E. Rose, A. Semra and C. Bois, *J. Organomet. Chem.*, 1989, **363**, 103.
898. F. Rose-Munch, E. Rose and A. Semra, *J. Chem. Soc., Chem. Commun.*, 1986, 1551.
899. R. Khourzom, F. Rose-Munch and E. Rose, *Tetrahedron Lett.*, 1990, **31**, 2011.
900. J.-P. Djukic, P. Geysermans, F. Rose-Munch and E. Rose, *Tetrahedron Lett.*, 1991, **32**, 6703.
901. F. Rose-Munch, E. Rose and A. Semra, *J. Chem. Soc., Chem. Commun.*, 1986, 1108.
902. J.-C. Boutonnet, F. Rose-Munch, E. Rose and A. Semra, *Bull. Soc. Chim. Fr.*, 1987, 640.
903. J.-P. Djukic, F. Rose-Munch and E. Rose, *J. Chem. Soc., Chem. Commun.*, 1991, 1634.
904. F. Rose-Munch, O. Bellot, L. Mignon, A. Semra, F. Robert and Y. Jeannin, *J. Organomet. Chem.*, 1991, **402**, 1.
905. L. Keller, K. Times-Marshall, S. Behar and K. Richards, *Tetrahedron Lett.*, 1989, **30**, 3373.
906. M. F. Semmelhack, J. L. Garcia, D. Cortes, R. Farina, R. Hong and B. K. Carpenter, *Organometallics*, 1983, **2**, 467.
907. T. A. Albright and B. K. Carpenter, *Inorg. Chem.*, 1980, **19**, 3092; see also T. A. Albright, *Acc. Chem. Res.*, 1982, **15**, 149.
908. R. D. Rieke and K. P. Daruwala, *J. Org. Chem.*, 1988, **53**, 2073.
909. P. Y. Morgantini, P. Flükiger, J. Weber and E. P. Kündig, *Stud. Phys. Theor. Chem.*, 1990, **71**, 313.
910. J.-C. Boutonnet, J. Levisalles, F. Rose-Munch, E. Rose, G. Precigoux and F. Leroy, *J. Organomet. Chem.*, 1985, **290**, 153.
911. E. P. Kündig, C. Grivet, E. Wenger, G. Bernardinelli and A. F. Williams, *Helv. Chim. Acta*, 1991, **74**, 2009.
912. C. Guimon, G. Pfister-Guillouzo and E. Rose, *J. Organomet. Chem.*, 1982, **224**, 125.
913. B. Ohlsson, C. Ullenius, S. Jagner, C. Grivet, E. Wenger and E. P. Kündig, *J. Organomet. Chem.*, 1989, **365**, 243.
914. W. R. Jackson, I. D. Rae, M. G. Wong, M. F. Semmelhack and J. N. Garcia, *J. Chem. Soc., Chem. Commun.*, 1982, 1359.
915. W. R. Jackson, I. D. Rae and M. G. Wong, *Aust. J. Chem.*, 1986, **39**, 303.
916. J.-C. Boutonnet, L. Mordenti, E. Rose, O. Le Martret and G. Precigoux, *J. Organomet. Chem.*, 1981, **221**, 147.
917. M. Uemura, T. Minami, Y. Shinoda, H. Nishimura, M. Shiro and Y. Hayashi, *J. Organomet. Chem.*, 1991, **406**, 371.
918. M. Uemura, H. Nishimura, T. Minami and Y. Hayashi, *J. Am. Chem. Soc.*, 1991, **113**, 5402.
919. M. F. Semmelhack and A. Zask, *J. Am. Chem. Soc.*, 1983, **105**, 2034.
920. B. Ohlsson and C. Ullenius, *J. Organomet. Chem.*, 1988, **350**, 35.
921. B. Ohlsson and C. Ullenius, *J. Organomet. Chem.*, 1984, **267**, C34.
922. M. F. Semmelhack, W. Wulff and J. L. Garcia, *J. Organomet. Chem.*, 1982, **240**, C5.
923. J.-C. Boutonnet, J. Levisalles, E. Rose, G. Precigoux, C. Courseille and N. Platzer, *J. Organomet. Chem.*, 1983, **255**, 317.
924. M. F. Semmelhack and H. Rhee, *Tetrahedron Lett.*, 1993, **34**, 1395.
925. V. Desobry and E. P. Kündig, *Helv. Chim. Acta*, 1981, **64**, 1288.
926. E. P. Kündig, V. Desobry, D. P. Simmons and E. Wenger, *J. Am. Chem. Soc.*, 1989, **111**, 1804.
927. E. P. Kündig, V. Desobry and D. P. Simmons, *J. Am. Chem. Soc.*, 1983, **105**, 6962.
928. E. P. Kündig, M. Inage and G. Bernardinelli, *Organometallics*, 1991, **10**, 2921.
929. J.-C. Boutonnet, J. Levisalles, J.-M. Normant and E. Rose, *J. Organomet. Chem.*, 1983, **255**, C21.
930. E. P. Kündig, D. Amurrio, G. Bernardinelli and R. Chowdhury, *Organometallics*, 1993, **12**, 4275.
931. F. Rose-Munch, E. Rose, A. Semra, Y. Jeannin and F. Robert, *J. Organomet. Chem.*, 1988, **353**, 53.
932. E. P. Kündig and D. P. Simmons, *J. Chem. Soc., Chem. Commun.*, 1983, 1320.
933. E. P. Kündig, A. F. Cunningham, Jr., P. Paglia, D. P. Simmons and G. Bernardinelli, *Helv. Chim. Acta*, 1990, **73**, 386.
934. E. P. Kündig, G. Bernardinelli, R. Liu and A. Ripa, *J. Am. Chem. Soc.*, 1991, **113**, 9676.
935. E. P. Kündig, D. Amurrio, R. Liu and A. Ripa, *Synlett*, 1991, 657.
936. E. P. Kündig, A. Ripa and G. Bernardinelli, *Angew. Chem., Int. Ed. Engl.*, 1992, **31**, 1071.
937. E. P. Kündig, A. Ripa, R. Liu, D. Amurrio and G. Bernardinelli, *Organometallics*, 1993, **12**, 3724.
938. J. L. von Rosenberg and A. R. Pinder, *J. Chem. Soc., Perkin Trans. 1*, 1987, 747.
939. B. Caro and G. Jaouen, *J. Organomet. Chem.*, 1982, **228**, 87.
940. M.-C. Sénéchal-Tocquer, D. Senechal, J.-Y. Le Bihan, D. Gentric and B. Caro, *Bull. Soc. Chim. Fr.*, 1992, **129**, 121.
941. F. Terrier, P. G. Farrell, J. Lelievre, S. Top and G. Jaouen, *Organometallics*, 1985, **4**, 1291.
942. C. F. Bernasconi, R. D. Bunnell and F. Terrier, *J. Am. Chem. Soc.*, 1988, **110**, 6514.
943. S. Top, G. Jaouen, B. G. Sayer and M. J. McGlinchey, *J. Am. Chem. Soc.*, 1983, **105**, 6426.
944. G. A. Artamkina, A. A. Grinfel'd and I. P. Beletskaya, *Tetrahedron Lett.*, 1984, **25**, 4989.
945. V. N. Kalinin, N. I. Udalov and A. V. Usatov, *Izv. Akad. Nauk SSSR, Ser. Khim.*, 1987, 1672.

946. A. A. Vasil'ev and V. A. Petrosyan, *Izv. Akad. Nauk SSSR, Ser. Khim.*, 1991, 2160.
947. A. A. Vasil'ev, V. I. Tatarinova and V. A. Petrosyan, *Mendeleev Commun.*, 1993, 27.
948. M. F. Semmelhack *et al.*, *Tetrahedron*, 1982, **37**, 3957.
949. J. Brocard, L. Pelinski and J. Lebibi, *J. Organomet. Chem.*, 1986, **309**, 299.
950. J. Brocard, A. Laconi, D. Couturier, S. Top and G. Jaouen, *J. Chem. Soc., Chem. Commun.*, 1984, 475.
951. G. Jaouen, S. Top, A. Laconi, D. Couturier and J. Brocard, *J. Am. Chem. Soc.*, 1984, **106**, 2207.
952. J. Brocard and J. Lebibi, *J. Organomet. Chem.*, 1987, **320**, 295.
953. S. J. Coote, S. G. Davies, D. Middlemiss and A. Naylor, *J. Organomet. Chem.*, 1989, **379**, 81.
954. J. Blagg, S. J. Coote, S. G. Davies and B. E. Mobbs, *J. Chem. Soc., Perkin Trans. 1*, 1986, 2257.
955. J. Blagg, S. J. Coote, S. G. Davies, D. Middlemiss and A. Naylor, *J. Chem. Soc., Perkin Trans. 1*, 1987, 689.
956. S. J. Coote, S. G. Davies and K. H. Sutton, *J. Chem. Soc., Perkin Trans. 1*, 1988, 1481.
957. V. N. Kalinin, I. A. Cherepanov and S. K. Moiseev, *Metalloorg. Khim.*, 1991, **4**, 177.
958. J. Blagg, S. G. Davies, N. J. Holman, C. A. Laughton and B. E. Mobbs, *J. Chem. Soc., Perkin Trans. 1*, 1986, 1581.
959. M. Persson and U. Hacksell, *J. Chem. Soc., Perkin Trans. 1*, 1992, 131.
960. J.-M. Lamarche and B. Laude, *Bull. Soc. Chim. Fr.*, 1982, II-97.
961. J. Brocard, L. Pelinski, S. Goetgheluck and L. Maciejewski, *J. Organomet. Chem.*, 1993, **456**, C24.
962. V. N. Kalinin, S. K. Moiseev, V. I. Bakhmutov and I. A. Cherepanov, *J. Organomet. Chem.*, 1990, **383**, 85.
963. V. N. Kalinin, S. K. Moiseev, V. I. Bakhmutov and N. I. Udalov, *Rev. Roum. Chem.*, 1989, **34**, 503.
964. S. Top, A. Vessières, J.-P. Abjean and G. Jaouen, *J. Chem. Soc., Chem. Commun.*, 1984, 428.
965. J. Blagg and S. G. Davies, *J. Chem. Soc., Chem. Commun.*, 1986, 492.
966. N. Mathews and M. Sainsbury, *J. Chem. Res. Synop.*, 1988, 22.
967. S. G. Davies and C. L. Goodfellow, *J. Organomet. Chem.*, 1989, **370**, C5.
968. J. Lebibi, L. Pelinski, L. Maciejewski and J. Brocard, *Tetrahedron*, 1990, **46**, 6011.
969. J. Albert and S. G. Davies, *Tetrahedron Lett.*, 1989, **30**, 5945.
970. M. Uemura, H. Nishimura and Y. Hayashi, *J. Organomet. Chem.*, 1989, **376**, C3.
971. J. Brocard, M. Mahmoudi, L. Pelinski and L. Maciejewski, *Tetrahedron Lett.*, 1989, **30**, 2549.
972. M. Mahmoudi, L. Pelinski, L. Maciejewski and J. Brocard, *J. Organomet. Chem.*, 1991, **405**, 93.
973. J. Blagg and S. G. Davies, *Tetrahedron*, 1987, **43**, 4463.
974. B. Misra, H. E. Zieger and J. Bordner, *J. Organomet. Chem.*, 1988, **348**, 79.
975. N. Mathews and D. A. Widdowson, *Synlett*, 1990, 467.
976. J. Brocard, J. Lebibi and D. Couturier, *J. Chem. Soc., Chem. Commun.*, 1981, 1264.
977. J. Lebibi, J. Brocard and D. Couturier, *Bull. Soc. Chim. Fr.*, 1982, **II**, 357.
978. M.-C. Sénéchal-Tocquer, D. Senechal, J.-Y. Le Bihan, D. Gentric and B. Caro, *J. Organomet. Chem.*, 1985, **291**, C5.
979. M.-C. Sénéchal-Tocquer, D. Senechal, J.-Y. Le Bihan, D. Gentric and B. Caro, *J. Organomet. Chem.*, 1987, **321**, 353.
980. J.-F. Halet, J.-Y. Saillard, B. Caro, J.-Y. Le Bihan, S. Top and G. Jaouen, *J. Organomet. Chem.*, 1984, **267**, C37.
981. B. Caro, J.-Y. Le Bihan, J.-P. Guillot, S. Top and G. Jaouen, *J. Chem. Soc., Chem. Commun.*, 1984, 602.
982. J.-Y. Le Bihan *et al.*, *Tetrahedron*, 1988, **44**, 3565.
983. M.-C. Sénéchal-Tocquer *et al.*, *J. Organomet. Chem.*, 1992, **433**, 261.
984. M.-C. Sénéchal-Tocquer, D. Senechal, J.-Y. Le Bihan, D. Gentric and B. Caro, *J. Organomet. Chem.*, 1991, **420**, 185.
985. D. Gentric, J.-Y. Le Bihan, M.-C. Sénéchal-Tocquer, D. Senechal and B. Caro, *Tetrahedron Lett.*, 1986, **27**, 3849.
986. M.-C. Sénéchal-Tocquer *et al.*, *J. Organomet. Chem.*, 1993, **458**, 105.
987. S. G. Davies and T. J. Donohoe, *Synlett*, 1993, 323.
988. M. T. Reetz and M. Sauerwald, *Tetrahedron Lett.*, 1983, **24**, 2837.
989. M. T. Reetz and M. Sauerwald, *J. Organomet. Chem.*, 1990, **382**, 121.
990. M. Uemura, K. Isobe and Y. Hayashi, *Tetrahedron Lett.*, 1985, **26**, 767.
991. M. Uemura, T. Kobayashi and Y. Hayashi, *Synthesis*, 1986, 386.
992. S. Top and G. Jaouen, *J. Organomet. Chem.*, 1987, **336**, 143.
993. M.-C. Sénéchal-Tocquer, J. Y. Le Bihan, D. Gentric, D. Senechal and B. Caro, *J. Organomet. Chem.*, 1988, **356**, C5.
994. S. J. Coote and S. G. Davies, *J. Chem. Soc., Chem. Commun.*, 1988, 648.
995. S. J. Coote, S. G. Davies, D. Middlemiss and A. Naylor, *Tetrahedron Lett.*, 1989, **30**, 3581.
996. S. J. Coote, S. G. Davies, D. Middlemiss and A. Naylor, *Tetrahedron Asymmetry*, 1990, **1**, 33.
997. S. G. Davies, R. F. Newton and J. M. J. Williams, *Tetrahedron Lett.*, 1989, **30**, 2967.
998. S. G. Davies, T. J. Donohoe and M. A. Lister, *Tetrahedron Asymmetry*, 1991, **2**, 1085.
999. S. G. Davies, T. J. Donohoe and M. A. Lister, *Tetrahedron Asymmetry*, 1991, **2**, 1089.
1000. I. T. Badejo, H. Choi, C. M. Hockensmith, R. Karaman, A. A. Pinkerton and J. L. Fry, *J. Org. Chem.*, 1991, **56**, 4688.
1001. A. Solladié-Cavallo, G. Lapitajs, P. Buchert, A. Klein, S. Colonna and A. Manfredi, *J. Organomet. Chem.*, 1987, **330**, 357.
1002. S. G. Davies and C. L. Goodfellow, *Synlett*, 1989, 59.
1003. J. Brocard, L. Pelinski and J. Lebibi, *J. Organomet. Chem.*, 1987, **336**, C47.
1004. A. Solladié-Cavallo, A.-C. Dreyfus, F. Sanch and A. Klein, *Chem. Lett.*, 1987, 1583.
1005. T. E. Bitterwolf and X. Dai, *J. Organomet. Chem.*, 1992, **440**, 103.
1006. A. Solladié-Cavallo and M. Bencheqroun, *J. Organomet. Chem.*, 1991, **406**, C15.
1007. A. Solladié-Cavallo and S. Quazzotti, *Synthesis*, 1991, 177.
1008. A. Solladié-Cavallo and J. Suffert, *Synthesis*, 1985, 659.
1009. A. Solladié-Cavallo, J. Suffert and D. Farkhani, *Ann. Chim. (Paris)*, 1984, **9**, 683.
1010. A. Solladié-Cavallo and J. Suffert, *Tetrahedron Lett.*, 1984, **25**, 1897.
1011. A. Solladié-Cavallo, D. Farkhani, S. Fritz, T. Lazrak and J. Suffert, *Tetrahedron Lett.*, 1984, **25**, 4117.
1012. A. Solladié-Cavallo and J. Suffert, *Tetrahedron Lett.*, 1985, **26**, 429.
1013. C. Baldoli, P. Del Buttero, E. Licandro, S. Maiorana, A. Papagni and M. Torchio, *Tetrahedron Lett.*, 1993, **34**, 7943.
1014. C. Mukai, W. J. Cho, I. J. Kim, M. Kido and M. Hanaoka, *Tetrahedron*, 1991, **47**, 3007.
1015. C. Mukai, W. J. Cho and M. Hanaoka, *Tetrahedron Lett.*, 1989, **30**, 7435.
1016. C. Mukai, M. Miyakawa, A. Mihira and M. Hanaoka, *J. Org. Chem.*, 1992, **57**, 2034.
1017. C. Mukai, I. J. Kim and M. Hanaoka, *Tetrahedron Asymmetry*, 1992, **3**, 1007.
1018. C. Mukai, I. J. Kim and M. Hanaoka, *Tetrahedron Lett.*, 1993, **34**, 6081.

1019. W. R. Roush and J. C. Park, *J. Org. Chem.*, 1990, **55**, 1143.
1020. E. P. Kündig, L. H. Xu, P. Romanens and G. Bernardinelli, *Tetrahedron Lett.*, 1993, **34**, 7049.
1021. S. Colonna, A. Manfredi, A. Solladié-Cavallo and S. Quazzotti, *Tetrahedron Lett.*, 1990, **31**, 6185.
1022. A. Solladié-Cavallo, S. Quazzotti, S. Colonna and A. Manfredi, *Tetrahedron Lett.*, 1989, **30**, 2933.
1023. A. Solladié-Cavallo, S. Quazzotti, S. Colonna, A. Manfredi, J. Fischer and A. DeCian, *Tetrahedron Asymmetry*, 1992, **3**, 287.
1024. C. Mukai, W. J. Cho, I. J. Kim and M. Hanaoka, *Tetrahedron Lett.*, 1990, **31**, 6893.
1025. C. Mukai, I. J. Kim, W. J. Cho, M. Kido and M. Hanaoka, *J. Chem. Soc., Perkin Trans. 1*, 1993, 2495.
1026. C. Baldoli, P. Del Buttero, E. Licandro, S. Maiorana and A. Papagni, *J. Chem. Soc., Chem. Commun.*, 1987, 762.
1027. J. Federic, S. Toma and B. Gautheron, *J. Organomet. Chem.*, 1988, **338**, 211.
1028. J. Huang and Y.-Z. Huang, *J. Organomet. Chem.*, 1991, **414**, 49.
1029. J. Huang, H.-X. Jiang and Y.-Z. Huang, *J. Organomet. Chem.*, 1991, **419**, 337.
1030. J. Besancon, A. Radecki-Sudre, J. Szymoniak and J. Padiou, *J. Organomet. Chem.*, 1992, **429**, 335.
1031. V. Gajda, S. Toma and M. Widhalm, *Monatsh. Chem.*, 1989, **120**, 147.
1032. M. Uemura, R. Miyake, M. Shiro and Y. Hayashi, *Tetrahedron Lett.*, 1991, **32**, 4569.
1033. M. Uemura, R. Miyake and Y. Hayashi, *J. Chem. Soc., Chem. Commun.*, 1991, 1696.
1034. S. B. Heaton and G. B. Jones, *Tetrahedron Lett.*, 1992, **33**, 1693.
1035. G. B. Jones and S. B. Heaton, *Tetrahedron Asymmetry*, 1993, **4**, 261.
1036. M. Uemura, T. Minami, M. Shiro and Y. Hayashi, *J. Org. Chem.*, 1992, **57**, 5590.
1037. M. Uemura, T. Minami and Y. Hayashi, *Tetrahedron Lett.*, 1989, **30**, 6383.
1038. M. Uemura, H. Oda, T. Minami, M. Shiro and Y. Hayashi, *Organometallics*, 1992, **11**, 3705.
1039. M. Uemura, H. Oda, T. Minami and Y. Hayashi, *Tetrahedron Lett.*, 1991, **32**, 4565.
1040. M. Uemura, T. Minami, K. Isobe, T. Kobayashi and Y. Hayashi, *Tetrahedron Lett.*, 1986, **27**, 967.
1041. J. Besancon, J. Szymoniak and C. Moise, *J. Organomet. Chem.*, 1992, **426**, 325.
1042. R. M. Moriarty, S. G. Engerer, O. Prakash, I. Prakash, U. S. Gill and W. A. Freeman, *J. Chem. Soc., Chem. Commun.*, 1985, 1715.
1043. R. M. Moriarty, S. C. Engerer, O. Prakash, I. Prakash, U. S. Gill and W. A. Freeman, *J. Org. Chem.*, 1987, **52**, 153.
1044. B. Caro and G. Jaouen, *Nouv. J. Chim.*, 1981, **5**, 457.
1045. B. Caro and G. Jaouen, *J. Organomet. Chem.*, 1981, **220**, 309.
1046. B. Caro and G. Jaouen, *Tetrahedron*, 1983, **39**, 83.
1047. B. Caro and G. Jaouen, *Tetrahedron*, 1983, **39**, 93.
1048. P. J. Dickens, A. M. Z. Slawin, D. A. Widdowson and D. J. Williams, *Tetrahedron Lett.*, 1988, **29**, 103.
1049. M. Uemura, T. Minami and Y. Hayashi, *J. Chem. Soc., Chem. Commun.*, 1984, 1193.
1050. M. Uemura, K. Isobe and Y. Hayashi, *Chem. Lett.*, 1985, 91.
1051. M. Uemura, K. Isobe, K. Take and Y. Hayashi, *J. Org. Chem.*, 1983, **48**, 3855.
1052. M. Uemura, H. Nishimura and Y. Hayashi, *Tetrahedron Lett.*, 1990, **31**, 2319.
1053. S. Ganesh, K. M. Sathe, M. Nandi, P. Chakrabarti and A. Sarkar, *J. Chem. Soc., Chem. Commun.*, 1993, 224.
1054. S. Ganesh and S. Sarkar, *Tetrahedron Lett.*, 1991, **32**, 1085.
1055. H. G. Wey and H. Butenschön, *Angew. Chem., Int. Ed. Engl.*, 1991, **30**, 880.
1056. H. G. Wey, P. Betz, I. Topalovic and H. Butenschön, *J. Organomet. Chem.*, 1991, **411**, 369.
1057. M. Brands, R. Goddard, H. G. Wey and H. Butenschön, *Angew. Chem., Int. Ed. Engl.*, 1993, **32**, 267.
1058. E. P. Kündig, G. Bernardinelli, J. Leresche and P. Romanens, *Angew. Chem., Int. Ed. Engl.*, 1990, **29**, 407.
1059. E. P. Kündig, G. Bernardinelli and J. Leresche, *J. Chem. Soc., Chem. Commun.*, 1991, 1713.
1060. M. Brands, H. G. Wey and H. Butenschön, *J. Chem. Soc., Chem. Commun.*, 1991, 1541.
1061. A. Solladié-Cavallo and D. Farkhani, *Tetrahedron Lett.*, 1986, **27**, 1131.
1062. C. Baldoli and P. Del Buttero, *J. Chem. Soc., Chem. Commun.*, 1991, 982.
1063. C. Baldoli, P. Del Buttero, S. Maiorana, G. Zecchi and M. Moret, *Tetrahedron Lett.*, 1993, **34**, 2529.
1064. P. Bloem, D. M. David, L. A. P. Kane-Maguire, S. G. Pyne, B. W. Skelton and A. H. White, *J. Organomet. Chem.*, 1991, **407**, C19.
1065. J. Suffert and R. Ziessel, *J. Organomet. Chem.*, 1989, **359**, C45.
1066. C. Diaz, J.-P. Gisselbrecht, M. Gross, J. Suffert and R. Ziessel, *J. Organomet. Chem.*, 1991, **401**, C54.
1067. M.-T. Youinou, J. Suffert and R. Ziessel, *Angew. Chem., Int. Ed. Engl.*, 1992, **31**, 775.
1068. M. Sodeoka and M. Shibasaki, *Synthesis*, 1993, 643.
1069. M. Sodeoka and M. Shibasaki, *J. Org. Chem.*, 1985, **50**, 1147.
1070. M. Sodeoka, S. Satoh and M. Shibasaki, *J. Am. Chem. Soc.*, 1984, **110**, 4823.
1071. M. Shibasaki, M. Sodeoka and Y. Ogawa, *J. Org. Chem.*, 1984, **49**, 4096.
1072. M. Sodeoka, Y. Ogawa, Y. Kirio and M. Shibasaki, *Chem. Pharm. Bull.*, 1991, **39**, 309.
1073. M. Sodeoka, T. Iimori and M. Shibasaki, *Chem. Pharm. Bull.*, 1991, **39**, 323.
1074. P. Le Maux, R. Dabard and G. Simonneaux, *J. Organomet. Chem.*, 1983, **241**, 37.
1075. J. R. Tucker and D. P. Riley, *J. Organomet. Chem.*, 1985, **279**, 49.
1076. M. Sodeoka, H. Yamada and M. Shibasaki, *J. Am. Chem. Soc.*, 1990, **112**, 4906.
1077. H. Yamada, M. Sodeoka and M. Shibasaki, *J. Org. Chem.*, 1991, **56**, 4569.
1078. J. H. Rigby, K. M. Short, H. S. Ateeq and J. A. Henshilwood, *J. Org. Chem.*, 1992, **57**, 5290.
1079. W. J. Bland, R. Davis and J. L. A. Durrant, *J. Organomet. Chem.*, 1985, **280**, 95.
1080. S. Ganesamoorthy and G. Sundararajan, *J. Mol. Catal.*, 1992, **76**, 335.
1081. M. F. Ernst and D. M. Roddick, *Organometallics*, 1990, **9**, 1586.
1082. M. Canestrari, M. L. H. Green and A. Izquierdo, *J. Chem. Soc., Dalton Trans.*, 1984, 2795.
1083. H. E. Bunting, M. L. H. Green and P. A. Newman, *J. Chem. Soc., Dalton Trans.*, 1988, 557.
1084. P. R. Brown, F. G. N. Cloke, M. L. H. Green and N. J. Hazel, *J. Chem. Soc., Dalton Trans.*, 1983, 1075.
1085. M. L. H. Green, D. O'Hare and G. Parkin, *J. Chem. Soc., Chem. Commun.*, 1985, 356.
1086. M. L. H. Green and G. Parkin, *J. Chem. Soc., Chem. Commun.*, 1985, 771.
1087. M. L. H. Green *et al.*, *J. Chem. Soc., Dalton Trans.*, 1992, 2063.
1088. R. L. Luck, R. H. Morris and J. F. Sawyer, *Organometallics*, 1984, **3**, 247.

1089. F. A. Cotton, R. L. Luck and R. H. Morris, *Organometallics*, 1989, **8**, 1282.
1090. M. C. Davies and T. A. George, *J. Organomet. Chem.*, 1982, **224**, C25.
1091. J. J. Frizzell, R. L. Luck, R. H. Morris and S. H. Peng, *J. Organomet. Chem.*, 1985, **284**, 243.
1092. L. Dahlenburg and B. Pietsch, *J. Organomet. Chem.*, 1989, **378**, 199.
1093. R. H. Morris, J. F. Sawyer and A. Sella, *Acta Crystallogr., Sect. C*, 1988, **44C**, 23.
1094. J. Talarmin, T. I. Al-Salih, C. J. Pickett, G. E. Bossard, T. A. George and C. M. Duff-Spence, *J. Chem. Soc., Dalton Trans.*, 1992, 2263.
1095. S. N. Anderson, R. L. Richards and D. L. Hughes, *J. Chem. Soc., Dalton Trans.*, 1986, 245.
1096. R. Luck and R. H. Morris, *J. Organomet. Chem.*, 1983, **255**, 221.
1097. R. L. Luck and R. H. Morris, *J. Organomet. Chem.*, 1988, **347**, 349.
1098. R. H. Morris, J. F. Sawyer, C. T. Schweitzer and A. Sella, *Organometallics*, 1989, **8**, 2099.
1099. R. Luck and R. H. Morris, *Inorg. Chem.*, 1984, **23**, 1489.
1100. R. Luck and R. H. Morris, *Inorg. Chem.*, 1984, **23**, 1788.
1101. R. L. Luck and R. H. Morris, *J. Am. Chem. Soc.*, 1984, **106**, 7978.
1102. R. L. Luck, R. H. Morris and J. F. Sawyer, *Inorg. Chem.*, 1987, **26**, 2422.
1103. R. L. Luck, R. H. Morris and J. F. Sawyer, *Organometallics*, 1984, **3**, 1009.
1104. P. T. Bishop, J. R. Dilworth and J. A. Zubieta, *J. Chem. Soc., Chem. Commun.*, 1985, 257.
1105. J. L. Kerschner, P. E. Fanwick, I. P. Rothwell and J. C. Huffman, *Organometallics*, 1989, **8**, 1431.
1106. P. T. Bishop, J. R. Dilworth, T. Nicholson and J. Zubieta, *J. Chem. Soc., Dalton Trans.*, 1991, 385.
1107. J. L. Kerschner, P. E. Fanwick and I. P. Rothwell, *J. Am. Chem. Soc.*, 1987, **109**, 5840.
1108. J. L. Kerschner, I. P. Rothwell, J. C. Huffman and W. E. Streib, *Organometallics*, 1988, **7**, 1871.
1109. J. L. Kerschner, P. E. Fanwick and I. P. Rothwell, *J. Am. Chem. Soc.*, 1988, **110**, 8235.
1110. J. L. Kerschner, E. M. Torres, P. E. Fanwick, I. P. Rothwell and J. C. Huffman, *Organometallics*, 1989, **8**, 1424.
1111. C. E. Kriley, J. L. Kerschner, P. E. Fanwick and I. P. Rothwell, *Organometallics*, 1993, **12**, 2051.
1112. E. J. Wucherer and E. L. Muetterties, *Organometallics*, 1987, **6**, 1691.
1113. E. J. Wucherer and E. L. Muetterties, *Organometallics*, 1987, **6**, 1696.
1114. W. D. McGhee *et al.*, *J. Organomet. Chem.*, 1993, **459**, 125.
1115. C. Elschenbroich, F. Bär, E. Bilger, D. Mahrwald, M. Nowotny and B. Metz, *Organometallics*, 1993, **12**, 3373.
1116. C. Elschenbroich, J. Kroker, W. Massa, M. Wünsch and A. J. Ashe, III, *Angew. Chem., Int. Ed. Engl.*, 1986, **25**, 571.
1117. K. Dimroth, R. Thamm and H. Kaletsch, *Z. Naturforsch., Teil B*, 1984, **39B**, 207.
1118. A. I. Uraev, O. A. Osipov, A. P. Sadimenko, V. K. Belskii, N. S. Magomedova and A. D. Garnovskii, *Zh. Obshch. Khim.*, 1987, **57**, 2153.
1119. R. E. Schmidt and W. Massa, *Z. Naturforsch., Teil B*, 1984, **39**, 213.
1120. A. K. M. M. Haque, S. E. Kabir, A. B. Kazi, K. Kundu, M. W. Miah and S. S. Ullah, *Indian J. Chem., Sect. A*, 1990, **29A**, 25.
1121. A. I. Uraev, A. D. Garnovskii and A. P. Sadimenko, *Metalloorg. Khim.*, 1992, **5**, 934.
1122. H. W. Choi and M. S. Sollberger, *J. Organomet. Chem.*, 1983, **243**, C39.
1123. S. G. Davies and M. R. Shipton, *J. Chem. Soc., Perkin Trans. 1*, 1991, 501.
1124. S. G. Davies and M. R. Shipton, *J. Chem. Soc., Chem. Commun.*, 1990, 1780.
1125. S. G. Davies and M. R. Shipton, *J. Chem. Soc., Perkin Trans. 1*, 1991, 757.
1126. S. G. Davies, A. J. Edwards and M. R. Shipton, *J. Chem. Soc., Perkin Trans. 1*, 1991, 1009.
1127. R. H. Morris and J. M. Ressner, *J. Chem. Soc., Chem. Commun.*, 1983, 909.
1128. K. Dimroth and H. Kaletsch, *J. Organomet. Chem.*, 1983, **247**, 271.
1129. K. Dimroth, M. Lückoff, and H. Kaletsch, *Phosphorus and Sulfur*, 1981, **10**, 285.
1130. K. Dimroth, S. Berger and H. Kaletsch, *Phosphorus and Sulfur*, 1981, **10**, 295.
1131. T. Debaerdemaeker, *Cryst. Struct. Commun.*, 1982, **11**, 1209.
1132. K. H. Dötz, A. Tiriliomis and K. Harms, *J. Chem. Soc., Chem. Commun.*, 1989, 788.
1133. K. H. Dötz, A. Tiriliomis, K. Harms, M. Regitz and U. Annen, *Angew. Chem., Int. Ed. Engl.*, 1988, **27**, 713.
1134. K. H. Dötz, A. Tiriliomis and K. Harms, *Tetrahedron*, 1993, **49**, 5577.
1135. A. R. Barron and A. H. Cowley, *Angew. Chem., Int. Ed. Engl.*, 1987, **26**, 907.
1136. G. Märkl, H. Baier, R. Liebl and K. K. Mayer, *J. Organomet. Chem.*, 1981, **217**, 333.
1137. N. Kuhn, J. Kreutzberg, E.-M. Lampe, D. Bläser and R. Boese, *J. Organomet. Chem.*, 1993, **458**, 125.
1138. W. A. Herrmann and D. Andrejewski, *Chem. Ber.*, 1986, **119**, 878.
1139. M. N. Nefedova, V. N. Setkina and D. N. Kursanov, *J. Organomet. Chem.*, 1983, **244**, C21.
1140. M. N. Nefedova, Y. I. Lyakhovetskii and Y. S. Nekrasov, *Izv. Akad. Nauk SSSR, Ser. Khim.*, 1987, 1679.
1141. M. S. Loft, D. A. Widdowson and T. J. Mowlem, *Synlett*, 1992, 135.
1142. F. Ruette, N. Valencia and R. Sanchez-Delgado, *J. Am. Chem. Soc.*, 1989, **111**, 40.
1143. P. Kumar Das, A. Mukhopadhyay, S. Ray, S. Bhattacharya, S. Datta and A. De, *Acta Crystallogr., Sect. C*, 1992, **48**, 266.
1144. R. Boese, N. Finke, T. Keil, P. Paetzold and G. Schmid, *Z. Naturforsch., Teil B*, 1985, **40**, 1327.
1145. G. E. Herberich and D. Söhnen, *J. Organomet. Chem.*, 1983, **254**, 143.
1146. D. W. Clack and K. D. Warren, *J. Organomet. Chem.*, 1981, **208**, 183.
1147. J. R. Bleeke, Y.-F. Xie, L. Bass and M. Y. Chiang, *J. Am. Chem. Soc.*, 1991, **113**, 4703.

Author Index

This Author Index comprises an alphabetical listing of the names of the authors cited in the text and the references listed at the end of each chapter in this volume.

Each entry consists of the author's name, followed by a list of numbers, for example

Templeton, J. L., 366, 385[233] (350, 366), 387[370] (363)

For each name, the page numbers for the citation in the reference list are given, followed by the reference number in superscript and the page number(s) in parentheses of where that reference is cited in the text. Where a name is referred to in text only, the page number of the citation appears with no superscript number. References cited both in the text and in the tables are included.

Although much effort has gone into eliminating inaccuracies resulting from the use of different combinations of initials by the same author, the use by some journals of only one initial, and different spellings of the same name as a result of the transliteration processes, the accuracy of some entries may have been affected by these factors.

Subject Index

JOHN NEWTON
David John (Services), Slough, UK

This Subject Index contains individual entries to the text pages of Volume 5. The index covers general types of organometallic compound, specific organometallic compounds, general and specific organic compounds where their synthesis or use involves organometallic compounds, types of reaction (insertion, oxidative addition, etc.), spectroscopic techniques (NMR, IR, etc.), and topics involving organometallic compounds.

Because authors may have approached similar topics from different viewpoints, index entries to those topics may not always appear under the same headings. Both synonyms and alternatives should therefore be considered to obtain all the entries on a particular topic. Commonly used synonyms include alkyne/acetylene, compound/complex, preparation/synthesis, etc. Entries where the oxidative state of a metal has been specified occur after all the entries for the unspecified oxidation state, and the same or similar compounds may occur under both types of heading. Thus $Cr(C_6H_6)_2$ occurs under Chromium, bis(η-benzene) and again under Chromium(0), bis(η-benzene). Similar ligands may also occur in different entries. Thus a carbene–metal complex may occur under Carbene complexes, Carbene ligands, or Carbenes, as well as under the specific metal. Individual organometallic compounds may also be listed in the Cumulative Formula Index in Volume 14.
